OPERATIONS RESEARCH

"OPERATIONS RESEARCH, a branch of Applied Mathematics, has witnessed a remarkable phenomenal growth during the last two decades. Its usefulness in the context of our present-day socio-economic problems has come to be widely recognised. The inevitable result has been a scarcity of up-to-date text material for university students sufficient to permit a comprehensive study of its basic methodology and techniques.

This book intends to provide an introduction to the subject — its theory, techniques and their applications. It encompasses the needs of the degree students of Mathematics, Statistics and Engineering. The text will be found equally useful for the post-graduate students of Economics, Commerce and Management. It may also serve as a good reference for professionals".

— from PREFACE to the First Edition

उपदेशलवं शास्त्रं कुरुते धीमतो यतः।
तत्तु प्राप्यैव विस्तारं स्वयमेवोपगच्छति।।

जले तेलं खले गुह्यं पात्रे दामं मनागपि।
प्राज्ञे शास्त्रं स्वयं याति विस्तारं वस्तुशक्तिनः।।

"A little instruction and guidance in Science is sufficient for the intelligent student, for this alone will help him develop his knowledge of his own accord. Science instilled into the intelligent mind has sufficient vitality in it to grow and expand by its own force even as a drop of oil on a sheet of water, a piece of secret confined to a villain, or a little act of charity of the deserving person".

— Bhaskara

Introduction to Management Science

Operations Research

[*A textbook for the students of* Science & Technology, *Economics, Accountancy, and Management of all Universities, Professional Bodies and Civil Services*]

KANTI SWARUP
Formerly Professor of Operations Research
Indian Institute of Public Administration, New Delhi

MAN MOHAN
Formerly Associate Professor, Ramjas College
University of Delhi, Delhi

P K GUPTA
Associate Prof. (Rtd.), J V Jain College, Saharanpur
Formerly visiting faculty member
Department of Mathematics
Cochin University of Science and Technology, Kochi

With the assistance of
PRIYANSHU GUPTA
B.Tech. (IIT – BHU), PGDM (IIM, Bangalore),
Ph.D. (IIM, Calcutta)
Asst. Professor, Faculty of Business Sustainability
Indian Institute of Management, Lucknow

Twentieth Thoroughly Revised Edition

Dedicated to the Service of Education
SULTAN
• 1950 •

Sultan Chand & Sons
Educational Publishers
New Delhi

SULTAN CHAND & SONS®
Educational Publishers
23, Daryaganj, New Delhi-110 002
Phones : 011-23281876, 23266105, 41625022 *(Showroom & Shop)*
011-23247051, 40234454 *(Office)*
E-mail : sultanchand74@yahoo.com; info@sultanchandandsons.com
Fax : 011-23266357; Website: www.sultanchandandsons.com

ISBN : 978-93-5161-183-7 (TC-525)

***Price :* ₹ 795.00**

First Edition: 1977
Ninteenth Edition: 2017; Reprint: 2018, 2019, 2020, 2021
Twentieth Revised Edition: 2022; Reprint: 2023, 2024, 2025

EVERY GENUINE COPY OF THIS BOOK HAS A HOLOGRAM

In our endeavour to protect you against counterfeit/fake books, we have pasted a copper hologram over the cover of this book. The hologram displays the full visual image, unique 3D multi-level, multi-colour effects of our logo from different angles when tilted or properly illuminated under a single light source, such as 3D depth effect, kinetic effect, pearl effect, gradient effect, trailing effect, emboss effect, glitter effect, randomly sparking tiny dots, micro text, laser numbering, etc.

A fake hologram does not display all these effects.

Always ask the bookseller to put his stamp on the first page of this book.

Author's Acknowledgement: The writing of a Textbook always involves creation of a huge debt towards innumerable author's and publications. We owe our gratitude to all of them. We acknowledge our indebtedness in extensive footnotes throughout the book. If, for any reason, any acknowledgement has been left out we beg to be excused. We assure to carry out correction in the subsequent edition, as and when it is known.

Printer: Taj Press, Noida (UP)

Preface

Over the last seven decades of its existence, the field of Operations Research (OR) has grown by leaps and bounds. As scientific decision-making permeates every sphere of human economic activity, the use of various tools, techniques, and methods of operations research has become increasingly important. Whether someone is an engineer, a manager, an economist, or a public administrator, it is now expected that one will be required to make sound data-driven decisions within the contextual constraints for effective problem-solving. Therefore, it is unsurprising that OR has become ever specialized and intermeshed with several other academic areas. As a result, the field of Operations Research has witnessed the development of advanced algorithms and software applications targeted at solving a multitude of problems spanning various disciplinary boundaries. However, the decision-maker is often required to possess a sound conceptual understanding to ensure that operations research techniques are suitably employed within a specific context for informed and optimal problem-solving.

This book has been designed as an introductory text to the world of operations research. From its humble beginnings, this book has emerged as one of the most popular textbooks on Operations Research in India. Over the last 45 years of this book's existence, it has been revised numerous times to incorporate contemporary developments in the discipline and customized to student's and reader's ever-changing needs. If one were to compare the first edition to the latest edition, it would appear to be a completely different book altogether. However, across these last four decades, one thing has remained the same: the commitment to introduce a new reader to the world of operations research in the most simple straightforward approach without compromising with the rigour demanded by this discipline. Only the reader's enthusiastic response and feedback have motivated us to undertake the periodic revisions and adapt to the latest requirements.

It gives me great pleasure and satisfaction to present the thoroughly revised ***Twentieth Edition*** of this book to our esteemed readers. In this edition of the book, entire text has been overhauled to adapt it to the requirements of today's business/management executives. This edition now comes with a fresh look. Before, we introduce this book and talk about the contents of the latest edition, it is perhaps important to look at the journey that this book has undertaken over the last four decades.

A brief context of the Twentieth Edition

The idea of this book came out of the author's experience in teaching operations research, since the late 1960s to undergraduate and post-graduate students, as well as to several practitioners of public administration. We perceived two key challenges in teaching operations research for a predominantly Indian audience. The first was to introduce the domain of operations research in a simple and lucid manner, while enabling the readers to develop a strong foundation on the quantitative tools and techniques required by the discipline. The second was to communicate to the students the exciting advances in the dynamic and rapidly changing field of operations research. We felt that the existing books and pedagogical material did not adequately meet this challenge.

At the time of writing the first edition, the author's goal was to provide a standard text on Operations Research for the students of Engineering, Economics, and Mathematical Sciences. The new scientific discipline of operations research had emerged during World War II and had thrown up a wide range of techniques and methods. It was hoped that these theories, models, and techniques could be employed for the rapid development of the nascent Indian industry, thereby contributing to our country's fast-paced economic growth. At the same time, it was hoped that our defence sector could also improve its effectiveness by using the principles and methods of Operations Research. Following the publication of this text, it remained unclear as to how effectively these goals were met. However, it soon became apparent that this book's demand was far greater than what we had initially anticipated. The first edition of the book's enthusiastic response had been most gratifying, and hence, the book was revised, updated, and reprinted several times thereafter. We also tried to keep abreast with the rapid changes in the discipline.

After nearly thirty years of the first edition of the book, a significant revision of the text was felt necessary as the subject of Operations Research had crept into the field of management, which itself was becoming increasingly popular and important at the turn of the century. Management science requires the conduct of a wide range of functions, including planning, organizing, directing, and controlling the activities of an enterprise or a firm scientifically and efficiently. Evidently, all these functions can be carried out more effectively by using Operations Research techniques and methods. Therefore, the entire text of this book was re-organized and re-written, keeping in view the requirements of business executives. Thus, an improved and updated edition of the book was brought out as the Fourteenth Edition in 2008. This edition sought to provide a comprehensive exposition of Operations Research applications, while avoiding the complexity of mathematical tools as far as possible.

In current times, the context of management is increasingly becoming complex and fast-changing owing to rapid advances in technology and globalization. Management practitioners are required to adapt to the ever-changing realities that such a dynamic environment presents. Consequently, the decision-making process has become increasingly challenging, owing to a high level of variability and uncertainty associated with the twenty-first century industry. The present decade has also seen the emergence of new-age businesses like the platform-based business models (sometimes also called the information economy) like food delivery companies (Zomato, Swiggy, etc.), cab-sharing companies (Ola, Uber, etc.) or online e-commerce companies (Amazon, Flipkart, etc.). One thing is shared across all these new-age business models: the central role of operations management. All these businesses are critically dependent on various aspects of operations planning and optimization to explore efficiencies in distribution, logistics, inventory planning, along with quick turnaround times and a wide assortment to cater to an extremely demanding consumer. Thus, Operations Research techniques are fundamental armor and an irreplaceable toolkit for every business/ management executive to undertake better informed, timely, and effective decision-making. Therefore, the current circumstances have necessitated a significant overhaul of the book's contents to adapt it to the changing times and cater to the requirements of the twenty-first century economy. The present, thoroughly revised Twentieth Edition attempts to respond by incorporating the latest developments towards the application of Operations Research at appropriate places in the text.

At this moment, I must also inform the readers of my misfortune of having lost the company of my co-authors and fellow travelers in this journey to improve this book continuously. Professor Kanti Swarup had sadly left this world in December, 2009. He was one of the pioneers of the academic discipline of Operations Research in India, having been a faculty member in the Department of Operational Research, Faculty of Mathematical Sciences at University of Delhi. During his illustrious career, he also taught Operations Research to the students of the University of Queensland, Australia, and students of Indian Institute of Management, Ahmedabad. At the time of retirement, he was associated with the Indian Institute of Public Administration, New Delhi, where he trained practitioners—the public administration professionals and civil servants—in the domain of

Quantitative Techniques and Operations Research. He had been the editor in-chief of *OPSERCH* (an official journal of OR Society of India) for five years. Also, he served as the President of the Operational Research Society of India and was actively engaged in research in mathematical programming. His demise has been an irreparable loss both to me personally as well as to the esteemed scholars in the discipline. My other co-author and dear friend Dr. Man Mohan also died in February, 2015. During his illustrious career following a master's degree in Statistics and a Doctorate in Information Theory from the University of Delhi, he researched extensively on several topics related to Statistics and Quantitative Techniques. He was also the managing editor of *Ganita Bharati*, the official journal of the Indian Society for History of Mathematics (ISHMA).

The loss of my co-authors has undoubtedly left a void, but also given me a fresh inspiration to continue working on this book in their memory. With renewed vigour, I have now undertaken the responsibility of revising the book. During revision of the book, I received valuable assistance from Dr. Priyanshu Gupta, Assistant Professor at Indian Institute of Management, Lucknow.

What's new in the Twentieth Edition

For this twentieth edition, the book has been improved in several ways. Most obviously, it has been restructured, re-written, and updated by incorporating the latest developments in various areas of management. Some of the significant changes in the reorganization of the contents of the book are the following:

- The present book now comprises of thirty-three chapters and four appendices. Appendix B is completely devoted to the answers to unsolved problems that appear in the text.
- Chapter 1 highlights the nature, scope, and limitations of Operations Research. "A new sub-field of Operations Research, namely Community Based Operations Research" is introduced in this chapter. This Chapter also provides a broad classification of various OR models
- Chapter 2 concentrates on the formulation of practical problems (based on actual real-life situations) that require the application of linear programming models. Chapters 3 to 9 deal with the development of solution methods of various linear programming models; presented in an easy to understand manner.
- The vital concepts of Integer Programming and Goal Programming are discussed in Chapters 11 and 12, respectively.
- The concept of network in scheduling problems is introduced in Chapter 13, wherein Scheduling techniques using PERT/CPM methods are discussed. Chapter 14 on Resource Analysis in Project Management now includes a detailed discussion on Precedence Planning, Updating, and Resource Allocation. As project planning and execution have become the dominant mode of industrial and entrepreneurial activity in recent times, such topics have gained increasing prominence.
- A special topic on Network Routing Problems is introduced in Chapter 16. The Chapter includes Network Flow Problems, Minimal Spanning Tree Problems, Shortest Path Problems, and Maximal Flow Problems. This Chapter also provides insights into Big Networks. Such type of problems are becoming increasingly important in recent times with the advent of platform-based business models.
- The concept of Operations Scheduling is introduced in Chapter 10. The objectives and methods for Scheduling, Single Processor Scheduling, and FlowShop Scheduling models are discussed in this Chapter.
- Decision-making rules under conflict, under uncertainty, and under risk are discussed in Chapter 19. As management practitioners today face an increasingly uncertain and risky environment, this Chapter has been strengthened.

- Chapter 20 presents an introduction to Queuing Theory — both Poisson and non-Poisson Queuing systems are discussed.
- Replacement Decisions are discussed in Chapter 21. The Chapter also includes problems related to recruitment and promotion, besides equipment renewal problems. Reliability and Systems Failure rate are also discussed in this chapter.
- Chapter on Inventory Management problems has been re-written afresh in the light of this topic's crucial importance for new-age businesses. Deterministic Inventory Control problems — both with and without price-breaks — are covered in detail. The probability-based Inventory Management problems are extensively discussed in Chapter 23. Systems of Inventory Control, One-period Inventory Control problem and application of Inventory Control to Dynamic Programming and Queuing Theory are some of the newly added features of this chapter.
- Chapter 32 on Statistical Quality Control is a fresh addition to this book. Operations Research (OR) is often described as a combination of several disciplines. For the survival of any firm/ industry, it is essential to maintain the quality of its output and processes; therefore, this addition bridges a critical gap in the text.
- A novel feature of this book is the inclusion of case problems in Operations Research, for which an entire Chapter 33 has been dedicated. These cases have been derived from real-life examples of critical decision-making situations that management professionals have encountered recently. This Chapter offers challenging decision-making problems to the would be Operations Research consultants.

In addition to the changes mentioned above, old stereotype problems have now been replaced with new ones. Several of these have been taken from the latest examination question papers of various universities and institutes of management and technology. Each section is followed by solved examples based on the concept introduced in that particular section of the chapter. After that, a set of unsolved problems is provided to enable the reader to grasp theory discussed in the section and prepare him for the next section of the Chapter. The set of unsolved problems are, thus, an integral part of the text. A set of multiple-choice questions based on the concepts discussed in the chapter, along with their answers, are also given at the end of the same chapter. It is followed by a well-graded review exercise that presents questions based on the theory discussed in the entire chapter.

To enhance further reading, carefully selected references have been provided in Appendix D. Necessary statistical tables are included in Appendix C. The prerequisite — Linear Algebra tools are provided in Appendix A. With these appendices; the effort is to make the book self-contained.

Acknowledgements

While writing and revising the book, I have benefitted immensely by referring to several books on Operations Research and Quantitative Techniques. I express my gratitude to all the authors, publications, and research papers whose work I have used freely while preparing this text. I have also benefited tremendously from the input of many reviewers, as well as gaining from the critical comments by the readers of earlier editions of the book. I would like to thank each of those who gave up their valuable time to help me in improving the text.

I express my gratitude to various practitioners from the corporate sector and academicians engaged in teaching. These include Dr. Priyanshu Gupta, Assistant Professor in the Faculty of Businss Sustainability, Indian Institute of Management, Lucknow; Dr. Rajendra Kumar, Associate Professor and Head, Department of Mathematics, JV Jain College, Saharanpur (CCS University); Dr. Akash Krishnan, Assistant Professor at Masters Union School of Business, Gurugram; Dr. Rakesh Gupta,

Senior Technical Director, National Informatic Centre, New Delhi; Mr. Manish Gupta, Senior Director at Affirmed Networks (a Microsoft company); Mr. Rajesh Meena, Director, Supply Head at Blackbuck (Zinka Logistics) and formerly Director, Ola Commute Solutions; Mrs. Prachi Gupta, co-founder of Eye-play Sports Technology, Bengaluru. Finally, I am also thankful to Ms Aparna for compiling the index of the book. She is having almost 14 years of work experience comprising of Philanthropic grant-making, Management consulting & corporate strategy and is now Co-CEO at Karta Initiative, based in UK and India.

It is earnestly hoped that the present version of the book will be found more useful and of greater utility to the readers as well as various business executives and consultants. Further suggestions for the improvement of the book will be much appreciated and duly acknowledged.

P K GUPTA

January, 2022

Contents in Brief

	Figures	Solved Examples	Unsolved Problems	Multiple Choise Questions	Review Exerrcises	Page
1. Operations Research — An Overview	—	—	—	12	20	1
2. Linear Programming	1	21	25	12	20	15
3. Solving Linear Programming Problems	12	30	60	20	25	49
4. Theory of Simplex Method	12	30	60	20	25	109
5. Duality in Linear Programming	—	14	36	12	16	133
6. Post-optimal Analysis	—	12	18	10	6	163
7. Transportation Problem	3	18	—	10	18	191
8. Assignment Problem	9	10	40	12	14	243
9. Other Methods of Solving Linear Programming Problems	1	8	26	12	12	275
10. Operations Scheduling	14	15	40	10	18	303
11. Integer Programming	3	13	24	6	10	339
12. Goal Programming	3	8	20	6	12	363
13. Scheduling Techniques	41	16	52	15	25	387
14. Resource Analysis in Project Management	23	5	15	7	8	439
15. Dynamic Programming	2	11	26	8	14	457
16. Network Routing Problems	82	10	24	10	24	489
17. Probability	12	2	—	10	14	531
18. Markov Analysis	1	9	18	12	12	551
19. Decision Theory	4	22	68	18	30	575
20. Queuing Theory	5	41	50	14	20	631
21. Replacement Decisions	2	15	35	10	12	687
22. Inventory Management — Deterministic	12	30	66	15	30	715
23. Inventory Management — Probabilistic	10	18	27	10	12	777
24. Simulation	2	18	25	12	18	805
25. Non-linear Programming and Optimization	—	16	29	8	10	841
26. Solving Non-linear Programming Problems	3	6	30	6	8	869
27. Geometric Programming	—	5	11	6	3	895
28. Information Theory	3	6	8	6	8	909
29. Investment Analysis and Break-even Analysis	5	23	36	8	10	927
30. Business Forecasting	6	10	23	10	5	961
31. Learning Curve Theory	1	14	6	8	12	981
32. Statistical Quality Control	9	6	12	—	12	999
33. Some Case Problems	—	—	—	—	—	1017
Appendices:						
Appendix A. Prerequisite — Linear Algebra	7	—	16	10	—	1049
Appendix B. Answers to Problems	77	—	—	—	—	1069
Appendix C. Statistical Tables	1	—	—	—	—	1139
Appendix D. Select References	—	—	—	—	—	1143
Index	—	—	—	—	—	1145
TOTAL	**366**	**462**	**926**	**345**	**483**	

Contents

		Pages
Chapter 1. Operations Research — An Overview		**1 – 13**
1 : 1. Development and Growth of Operations Research (OR)	...	1
1 : 2. Community Operations Research	... ...	2
1 : 3. Nature and Characteristic Features of OR	... ...	3
1 : 4. What is Decision Optimization	... ...	4
1 : 5. Modelling in Operations Research	... ...	4
1 : 6. General Solution Methods for OR Models	... ...	5
1 : 7. Methodology of Operations Research	... ...	6
1 : 8. Operations Research Approach to Problem Solving	... ...	8
1 : 9. Operations Research and Business/Industry	... ...	9
1 : 10. Operations Research and Decision-making	... .	10
1 : 11. Limitations of Operations Research	... ...	11
Test Your Understanding	... ...	11
Review Exercise	... ...	13
Chapter 2. Linear Programming		**15 – 48**
2 : 1. Linear Programming Problem (LPP)	... ...	15
2 : 2. Assumptions of Linear Programming Problem	... ...	16
2 : 3. Three Stages of Linear Programming Problem	... ...	16
2 : 4. Limitations of Linear Programming	... ...	17
2 : 5. Formulating a Problem as Linear Programming Model	...	17
2 : 6. Illustrative Examples of Linear Programming Model Formulation.	...	17
2 : 7. General Linear Programming Problem	... ...	41
2 : 8. Canonical and Standard Forms of LPP	... ...	42
2 : 9. Terminology for the Solution of LPP	... ...	45
Test Your Understanding	... ...	46
Review Exercise	... ...	48
Chapter 3. Solving Linear Programming Problems		**49 – 107**
3 : 1. Graphical Solution Method	... ...	49
3 : 2. Insights into the Simplex Method	... ...	66
3 : 3. The Computational Procedure	... ...	67
3 : 4. Simplex Algorithm (Maximization Case)	... ...	68
3 : 5. Use of Artificial Variables	... ...	74
Two-Phase Method	... ...	75
Big-M Method (Method of Penalties)	... ...	76
3 : 6. Solution of Simultaneous Linear Equations by Simplex Method	...	102
Test Your Understanding	... ...	104
Review Exercise	... ...	106
Chapter 4. Theory of Simplex Method		**109 – 132**
4 : 1. Foundations of the Simplex Method	... ...	109
4 : 2. Terminology of Simplex Method	... ...	110

4 : 3. Fundamental Theorem of Linear Programming 116
4 : 4. Improving of Basic Feasible Solution (BFS) 116
4 : 5. Unbounded Solution 118
4 : 6. Condition of Optimality 119
4 : 7. Revised Simplex Method 120
4 : 8. Simplex Method *Versus* Revised Simplex Method 130
Test Your Understanding 130
Review Exercise 131

Chapter 5. Duality in Linear Programming 133 – 161
5 : 1. Writing of Dual Linear Programming Problem 133
5 : 2. General Primal-Dual Pair 134
5 : 3. Formulating a Dual Problem 135
5 : 4. Properties of Dual Problem 139
5 : 5. Duality and Simplex Method 139
5 : 6. Summary of Primal-Dual Relationship 147
5 : 7. Economic Interpretation of Duality 148
5 : 8. Dual Simplex Method 155
Test Your Understanding 159
Review Exercise 161

Chapter 6. Post-Optimal Analysis 163 – 190
6 : 1. Sensitivity Analysis 163
6 : 2. Changes in the Cost/Profit Coefficients c_j 164
6 : 3. Changes in the Right-hand Side of Constraints b_i (Availability of Resources) 167
6 : 4. Changes in the Activity Coefficients a_{ij}'s 171
6 : 5. Structural Changes 177
6 : 6. Managerial Problems and Post-Optimal Analysis 181
Test Your Understanding 189
Review Exercise 190

Chapter 7. Transportation Problem 191 – 242
7 : 1. General Structure of a Transportation Problem 192
7 : 2. Existence of Solution in Transportation Problem 192
7 : 3. Duality in Transportation Problem 194
7 : 4. Economic Interpretation of u_i's and v_j's 194
7 : 5. Degeneracy in Transportation Problem 194
7 : 6. Standard Transportation Table 195
7 : 7. Some Definitions 195
7 : 8. Solution of a Transportation Problem 196
7 : 9. Methods for Finding Initial Basic Feasible Solution 197
7 : 10. Optimality Test 207
Stepping Stone Method 208
Modified Distribution Method (or MODI Method) 210
7 : 11. Post-Optimal Analysis 232
7 : 12. Time-Minimization Transportation Problem 235
7 : 13. Transhipment Problem 237
Test Your Understanding 241
Review Exercise 242

Chapter 8. Assignment Problem 243 – 274
8 : 1. Model Formulation of an Assignment Problem 243
8 : 2. Assumptions in Assignment Problem 244
8 : 3. Methods of Solving an Assignment Problem 244
8 : 4. Special Cases in Assignment Problems 255
8 : 5. A Typical Assignment Problem 257
8 : 6. Dual of the Assignment Problem 265
8 : 7. Branch and Bound Method 267
Test Your Understanding 272
Review Exercise 274

Chapter 9. Other Methods of Solving Linear Programming Problems 275 – 301
9 : 1. Bounded Variables LPP 275
9 : 2. Parametric Linear Programming 281
9 : 3. Decomposition Method 287
9 : 4. An Interior-Point Algorithm 288
9 : 5. Linear Fractional Programming 290
9 : 6. Shortest Cyclic Route Model (Traveling Salesman Problem) ... 295
Test Your Understanding 299
Review Exercise 301

Chapter 10. Operations Scheduling 303 – 337
10 : 1. Problem of Sequencing 303
10 : 2. Basic Terminology and Assumptions 303
10 : 3. Gantt Chart 304
10 : 4. Criteria and Objective for Scheduling 306
10 : 5. Methods of Scheduling 306
10 : 6. Single Processor Scheduling 307
10 : 7. Flow Shop Scheduling (Sequencing) 312
10 : 8. Processing of Two Jobs through '*m*' Machines 328
10 : 9. Personnel Scheduling 332
10 : 10. Routing Problems and Sequencing 334
10 : 11. Problems of Complex Sequencing 334
Test Your Understanding 335
Review Exercise 337

Chapter 11. Integer Programming 339 – 362
11 : 1. Integer Linear Programming Problem 339
11 : 2. Formulation of Integer Linear Programming Problem ... 339
11 : 3. Methods of Solving an Integer Programming Problem... ... 347
11 : 4. Branch and Bound Method 347
11 : 5. Cutting Plane Method 353
Test Your Understanding 362
Review Exercise 362

Chapter 12. Goal Programming 363 – 386
12 : 1. Why Goal Programming? 363
12 : 2. Categorization of Goal Programming Problem 364
12 : 3. Formulation of Linear Goal Programming Problem 365
12 : 4. Graphical Goal Attainment Method 371
12 : 5. Simplex Method for Solving Goal Programming Problem ... 375

12 : 6. Limitations of Goal Programming Problems 385
Test Your Understanding 385
Review Exercise 386

Chapter 13. Scheduling Techniques **387 – 438**
13 : 1. Why Networks? 387
13 : 2. Basic Components of Network 388
13 : 3. Logical Sequencing 389
13 : 4. Rules of Network Construction 390
13 : 5. Network Scheduling 396
13 : 6. Critical Path Analysis 396
13 : 7. Critical Path Calculations 398
13 : 8. Procedure for Determining Critical Path (CPM) 401
13 : 9. Programme Evaluation and Review Technique (PERT) ... 408
13 : 10. Procedure for the Development of *PERT* 409
13 : 11. Time-Cost Relationship in *PERT* 424
13 : 12. Distinction Between *PERT* and *CPM* 434
13 : 13. Linear Programming Formulation and Network 435
Test Your Understanding 435
Review Exercise 437

Chapter 14. Resource Analysis in Project Management **439 – 456**
14 : 1. Project Planning and Project Scheduling 439
14 : 2. Updating of the Project 439
14 : 3. Resource Allocation and Scheduling 443
14 : 4. Management Operation System Technique (MOST) 452
14 : 5. Graphical Evaluation and Review Technique (GERT) 453
14 : 6. Precedence Planning 454
14 : 7. Line of Balance (LOB) 454
Test Your Understanding 455
Review Exercise 456

Chapter 15. Dynamic Programming **457 – 488**
15 : 1. Characteristics of Dynamic Programming Problems 457
15 : 2. Shortest Route Problem 457
15 : 3. Formulation of Dynamic Programming Problem 460
15 : 4. Solving Dynamic Programming Problems 461
15 : 5. Optimal Sub-division Problem 479
15 : 6. Solving LLP by Dynamic Programming 484
Test Your Understanding 487
Review Exercise 488

Chapter 16. Network Routing Problems **489 – 529**
16 : 1. Origin of the Network Flow Problems 489
16 : 2. Network Flow Problems 491
16 : 3. Minimal Spanning Tree Problem 494
16 : 4. Shortest Route Problem 500
16 : 5. Applications of Shortest Route Problem 508
16 : 6. Maximal-Flow Problem 511
16 : 7. Maximum-Flow Minimum-Cut Method 516
16 : 8. Minimum-Cost Flow Problem 518

16 : 9.	Transportation Problem as a Minimum - Cost Flow Problem	...	519
16 : 10.	More Network Flow Problems	... ...	521
16 : 11.	Insights into Big Networks	... ...	525
16 : 12.	Concluding Remarks	... ...	527
	Test Your Understanding	... ...	527
	Review Exercise	... ...	529
Chapter 17.	**Probability**		**531 – 550**
17 : 1.	Uncertainty and Probability	... ...	531
17 : 2.	Sample Space and Probability	... ...	532
17 : 3.	Algebra of Events	... ...	533
17 : 4.	Conditional Probability	... ...	535
17 : 5.	Random Variables	... ...	536
17 : 6.	Expectation of a Random Variable	... ...	539
17 : 7.	Central Tendency and Dispersion	... ...	540
17 : 8.	Some Probability Distributions	... ...	541
	Test Your Understanding	... ...	549
	Review Exercise	... ...	550
Chapter 18.	**Markov Analysis**		**551 – 574**
18 : 1.	Markov Process	... ...	551
18 : 2.	State and Transition Probabilities	... ...	551
18 : 3.	Characteristics of a Markov Process	... ...	552
18 : 4.	State Transition Matrix	... ...	552
18 : 5.	Transition Diagram	... ...	553
18 : 6.	Construction of a State - Transition Matrix	... ...	553
18 : 7.	n-Step Transition Probabilities	... ...	555
18 : 8.	Steady State (Equilibrium) Conditions	... ...	556
18 : 9.	Markov Analysis Algorithm	... ...	556
18 : 10.	Special Cases in Markov Analysis	... ...	567
	Test Your Understanding	... ...	572
	Review Exercise	... ...	573
Chapter 19.	**Decision Theory**		**575 – 630**
19 : 1.	Elements of Decision-Making Problem	... ...	575
19 : 2.	Pay-off Table and Opportunity Loss Table	... ...	575
19 : 3.	Decision-making Process	... ...	578
19 : 4.	Decision-making under Certainty	... ...	578
19 : 5.	Decision-making under Uncertainty	... ...	578
19 : 6.	Decision-making under Risk	... ...	585
19 : 7.	Decision Tree	... ...	594
19 : 8.	Decision-making with Utilities	... ...	602
19 : 9.	Decision-making under Conflict (Competitive Game)	...	603
19 : 10.	Two-Person Zero-Sum Game	... ...	604
19 : 11.	Solution of Two-Person Zero-Sum Game	... ...	605
19 : 12.	Arithmetic Method for $n \times n$ Game	... ...	622
19 : 13.	Iterative Method of Approximation Solution of Games	...	624
19 : 14.	Competitors Against Passivity	... ...	627
	Test Your Understanding	... ...	627
	Review Exercise	... ...	629

Chapter 20. Queuing Theory **631 – 686**
20 : 1. Queuing System 631
20 : 2. Deterministic Queuing System 634
20 : 3. Operating Characteristics of Queuing System 634
20 : 4. Probability Distributions in Queuing System 635
20 : 5. Transient-State and Steady-State of the System 636
20 : 6. Classification of Queuing Models 636
20 : 7. Probabilistic Queuing Models 636
20 : 8. Poisson-Exponential Single Server and Infinite Population, *i.e.*, $(M/M/1) : (\infty/FCFS)$ Model 637
20 : 9. Poisson-Exponential Single Server and Finite Population, *i.e.*, $(M/M/1) : (N/FCFS)$ Model 659
20 : 10. Poisson-Exponential Multiple Server and Infinite Population, *i.e.*, $(M/M/S) : (\infty/FCFS)$ Model 664
20 : 11. Poisson-Exponential Multiple Server and Finite Population, *i.e.*, $(M/M/S) : (N/FCFS)$ Model 671
20 : 12. Poisson-Exponential Multiple Server with Maximum Limit on Arrivals, *i.e.*, $(M/M/R) : (K/GD, K > R)$ Model 673
20 : 13. Non-Poisson Queuing Systems 677
20 : 14. Poisson-Erlangian Service, Single Server and Infinite Population, *i.e.*, $(M/E_k/1) : (\infty/FCFS)$ Model 678
20 : 15. Poisson Arrivals, Arbitrary Service Times, and Single Server, *i.e.*, $(M/G/1) : (\infty/GD)$ Model 681
20 : 16. Cost Models in Queuing 683
Test Your Understanding 684
Review Exercise 685

Chapter 21. Replacement Decisions **687 – 714**
21 : 1. Situations which make the Replacement of Items Necessary ... 687
21 : 2. Failure Mechanism of Items 687
21 : 3. Assumptions and Objectives of Replacement Decisions... ... 688
21 : 4. Types of Replacement Problems 688
21 : 5. Replacement Policy for Equipment that Deteriorate Gradually ... 688
21 : 6. Replacement Policy for Equipment that Breaks-Down/Fail Suddenly 697
21 : 7. Replacement Policy of Staffing/Man-Power 706
21 : 8. Equipment Renewal Policy 708
21 : 9. Reliability and System Failure Rates 709
Test Your Understanding 712
Review Exercise 714

Chapter 22. Inventory Management — Deterministic **715 – 775**
22 : 1. Types of Inventory 715
22 : 2. Reasons for Carrying Inventories 716
22 : 3. The Inventory Decisions 717
22 : 4. Objectives of Scientific Inventory Control 717
22 : 5. Costs Associated with Inventories 718
22 : 6. Factors Affecting Inventory Control 718
22 : 7. An Inventory Control Problem 719
22 : 8. The Concept of *EOQ* 720
22 : 9. Economic Order Quantity (*EOQ*) Model 720

22 : 10. *EOQ* and Quantity Discount 729
22 : 11. *EOQ* and Price Breaks 734
22 : 12. *EOQ* Model with Replenishment 740
22 : 13. *EOQ* Model and Planned Shortages 744
22 : 14. Multi-Item *EOQ* Model 753
22 : 15. Dynamic Order Quantity 761
22 : 16. *EOQ* Model with Uncertain Demand 762
22 : 17. Selective Inventory Control Techniques 766
Test Your Understanding 772
Review Exercise 774

Chapter 23. Inventory Management — Probabilistic **777 – 804**
23 : 1. Systems of Inventory Control 777
Fixed Order Quantity System (*Q*-System) 777
Periodic Review System (*P*-System) 780
23 : 2. One Period Problem 786
23 : 3. One Period Problem without Set-up Cost 792
23 : 4. One Period Problem with Set-up Cost 799
23 : 5. Inventory Control and Dynamic Programming 800
23 : 6. Inventory Control and Queuing Theory 801
Test Your Understanding 802
Review Exercise 804

Chapter 24. Simulation **805 – 840**
24 : 1. What is Simulation? 805
24 : 2. Reasons for Using Simulation 805
24 : 3. Steps in Simulation Process 806
24 : 4. Generation of Random Numbers 806
24 : 5. Event-Type Simulation 807
24 : 6. Monte-Carlo Simulation 809
24 : 7. Advantages of Simulation 837
24 : 8. Limitations of Simulation 837
Test Your Understanding 838
Review Exercise 839

Chapter 25. Non-Linear Programming and Optimization **841 – 868**
25 : 1. Formulating a Non-Linear Programming Problem (*NLPP*) ... 841
25 : 2. General Non-Linear Programming Problem 844
25 : 3. Types of Non-Linear Programming Problems 845
25 : 4. One-variable Unconstrained Optimization 846
25 : 5. Multi-variable Unconstrained Optimization 848
25 : 6. Constrained Optimization with Equality Constraints 850
25 : 7. Constrained Optimization with Inequality Constraints ... 858
25 : 8. Saddle Point Problems 864
25 : 9. Saddle Points and *NLPP* 865
Test Your Understanding 866
Review Exercise 867

Chapter 26. Solving Non-linear Programming Problems **869 – 894**
26 : 1. Graphical Solution of Non-linear Programming Problem ... 869
26 : 2. Kuhn-Tucker Conditions with Non-negative Constraints ... 873

26 : 3. Quadratic Programming Problem 874
26 : 4. Wolfe's Modified Simplex Method 876
26 : 5. Beale's Method 880
26 : 6. Separable Convex Programming 885
26 : 7. Separable Programming Algorithm 888
26 : 8. Stochastic Programming 892
Test Your Understanding 893
Review Exercise 894

Chapter 27. Geometric Programming **895 – 908**
27 : 1. Geometric-Arithmetic Mean Inequality 895
27 : 2. Unconstrained Geometric Programming Problem (GPP) 896
27 : 3. Constrained Geometric Programming Problem 901
27 : 4. Complementary Geometric Programming 904
27 : 5. Complementary Geometric Programming Algorithm 905
Test Your Understanding 907
Review Exercise 908

Chapter 28. Information Theory **909 – 926**
28 : 1. Communication System 909
28 : 2. A Measure of Information 910
28 : 3. Entropy — The Expected Information 911
28 : 4. Entropy as a Measure of Uncertainty 913
28 : 5. Joint and Conditional Entropies 914
28 : 6. Mutual Information 917
28 : 7. Encoding 920
28 : 8. Shannon-Fano Encoding Procedure 922
Test Your Understanding 925
Review Exercise 926

Chapter 29. Investment Analysis and Break-even Analysis **927 – 960**
29 : 1. Time Value of Money 927
29 : 2. Annuities and Sinking Fund 929
29 : 3. Methods of Investment Analysis 932
29 : 4. Investment under Uncertainty (Risky Investments) 939
29 : 5. Decision Trees and Investment Analysis 949
29 : 6. Simulation Approach to Risky Investments 951
29 : 7. Break-even Analysis 952
Test Your Understanding 959
Review Exercise 960

Chapter 30. Business Forecasting **961 – 980**
30 : 1. Forecasting Methods 961
30 : 2. Components of Time Series 962
30 : 3. Smoothing Methods in Forecasting 962
30 : 4. Trend Projection in Forecasting 968
30 : 5. Seasonal Components in Forecasting 971
30 : 6. Casual Forecasting Method 975
30 : 7. Qualitative Methods of Forecasting 979
Test Your Understanding 979
Review Exercise 980

Chapter 31. Learning Curve Theory **981 – 998**
31 : 1. Graphical and Tabular Presentation of Learning Curve ... 981
31 : 2. Learning Curve Equation 982
31 : 3. Specific Learning Curves 983
31 : 4. Regression and Learning Curve 984
31 : 5. Learning Curve Table (When Production is not Exactly Doubled) ... 985
31 : 6. Uses of Learning Curve 996
31 : 7. Costs Affected by Learning Curve 996
31 : 8. Learning Curve Theory in Manufacturing Organization ... 996
31 : 9. Advantages of Learning Curve Theory 996
31 : 10. Limitations of Learning Curve Theory 997
Test Your Understanding 997
Review Exercise 998

Chapter 32. Statistical Quality Control **999 – 1016**
32 : 1. Meaning of Statistical Quality Control (*SQC*) 999
32 : 2. Uses of Quality Control 999
32 : 3. Variation in Quality 999
32 : 4. Techniques of *SQC* 1000
32 : 5. Control Charts 1000
32 : 6. Control Charts for Variables 1001
32 : 7. Control Charts for Attributes 1006
32 : 8. Inspection 1012
32 : 9. Acceptance Sampling 1012
32 : 10. Single Sampling Plan 1013
32 : 11. Operating Characteristics Curve (*OC*-Curve) 1014
32 : 12. Double Sampling Plan 1015
Review Exercise 1016

Chapter 33. Some Case Problems **1017 – 1048**
33 : 1. Case Problems in Linear Programming 1017
33 : 2. Case Problems in Transportation 1021
33 : 3. Case Problem in Integer Programming 1023
33 : 4. Case Problems in Goal Programming 1024
33 : 5. Case Problems in Network Scheduling through *PERT/CPM* ... 1025
33 : 6. Case Problems in Decision Theory 1031
33 : 7. Case Problems in Queuing Theory 1036
33 : 8. Case Problems in Simulation 1037
33 : 9. Case Problem in Replacement Decision 1041
33 : 10. Case Problem in Dynamic Programming 1042
33 : 11. Case Problems in Inventory Control 1043
33 : 12. Case Problem in Investment Analysis 1045
33 : 13. Case Problem in Management Science Application 1046

Appendices: **1049 – 1144**
A. Prerequisite — Linear Algebra 1049
B. Answers to Problems 1069
C. Statistical Tables 1139
D. Select References 1143
Index **1145 – 1156**

Business Statistics

S.P. Gupta • M.P. Gupta

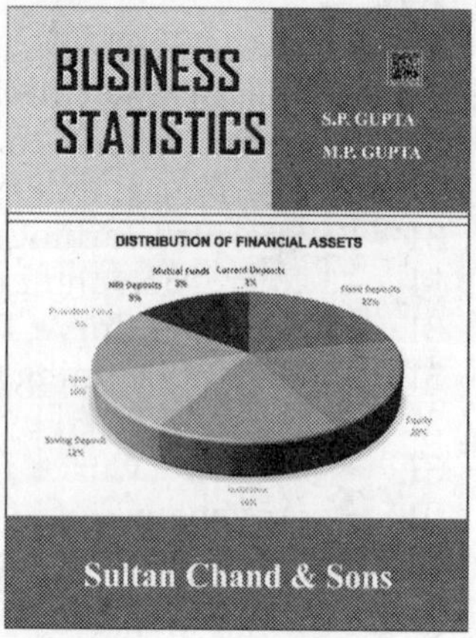

About the Book

The book is essentially non-mathematical in character and an attempt has been made to illustrate the application of statistical techniques with the help of business data. Various types of study material are given at the end of each chapter to aid the students in applying the principles discussed in the text. The object is to develop the faculty of thinking amongst the students and to develop the skill of performing the calculations needed for various methods of analysis.

Salient Features

- The entire text of this edition has been thoroughly revised. In particular more attention has been given to chapters like Tests of Hypothesis, Estimation, Chi-Square Test, Statistical Decision Decision Theory.
- A unique feature of the book is that most chapters end with an interesting write up in the form of Ethical Consideration.
- Every effort has been made to reduce to the minimum the printing or calculation mistakes.

Contents

- Business Statistics – What & Why
- Collection of Data
- Presentation of Data
- Measures of Central Tendency
- Measures of Variations
- Skewness, Moments and Kurtosis
- Correlation analysis
- Regression Analysis
- Index Numbers: Concepts and Applications
- Business Forecasting and Time Series Analysis
- Probability
- Probability Distributions
- Sampling and Sampling Distributions
- Estimation of Parameters
- Test of Hypothesis
- Small Sampling Theory
- Chi-Square Test
- Analysis of Variance
- Statistical Quality Control
- Partial and Multiple Correlation and Regression
- Statistical Decision Theory

Appendix: Statistical Tables.

Edition: 19th, 2017; Pages: xxiv + 800; Size: 185 mm × 240 mm; ISBN: 978-93-5161-106-6 (TC 518)

A Premier Guide to Data Analysis

Applications with STATA and R

Sajal Jana • Jhumur Sengupta

About the Book

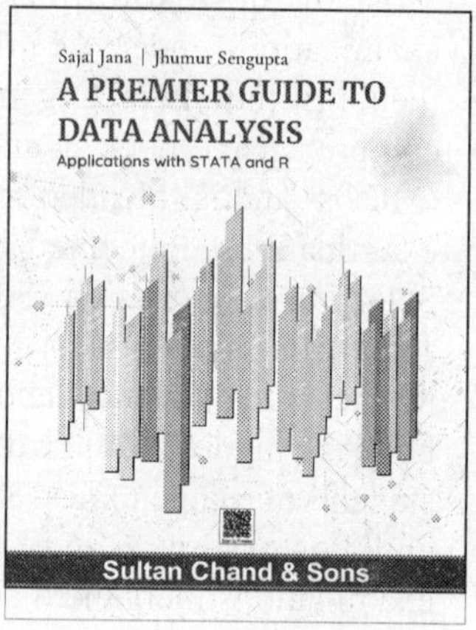

The book is written to provide a strong foundation of data analysis techniques based on statistical and econometric models using STATA and R. The objective is to explain the concepts and their applications with practical illustrations. It covers topics such as data representation, statistical techniques,and regression analysis including non-linear, time series, and panel data models. All the chapters include real-life illustrations and use real-world data sets to provide examples of how to explore data, build models, find results, and evaluate using codes in STATA and R. The book attempts to provide an easier learning experience to the readers. The practical approach would enable readers to develop the required skills to perform data analysis using STATA and R.

Contents

- Data Management and Data Source
- Basic Data Analysis
- Getting Started with MS-Excel
- An Introduction to STATA
- An Introduction to R
- Distribution Functions
- Sampling Techniques and Survey Design
- Index Number
- Hypothesis Testing and Statistical Inference
- Linear Correlation and Regression
- Time Series, Panel and Non-Linear Regression Model
- Analysis of National Accounts Statistics in RStudio
- Analysis of National Sample Survey Data Using STATA
- Reserve Bank of India Data Analysis in STATA
- Analysis of Census Data Using RStudio.

Edition: 1st, 2024; Pages: xxviii + 372; Size: 185 mm × 240 mm; ISBN: 978-93-91820-89-3 (TC 1281)

Operations Research

(Quantitative Techniques for Management)

V.K. Kapoor

The thoroughly revised edition of this authoritative, and best selling text-book thoroughout India and abroad, while retaining its analytical and modern approach now includes extensively revised and rewritten material, updated information and various distinct pedagogical features. During the revision process, no pains have been spared to make the book useful for all categories of students.

About the Book

The revised edition of this book is intended to serve as a comprehensive text for students of undergraduate, postgraduate in commerce and management courses, i.e., MBA/PGDBM, M.Com., CA. It will also meet the requirement of students of MA/M.Sc. (Statistics, Applied Operations Research), MCA, B.Tech. who need both theoretical and practical knowledge of Operations Research. The book will also meet the requirements of academicians as well as the research scholars & teachers, and also practitioners in the field of management, etc. The book will also be found useful by the students preparing for various competitive examinations.

Salient Features

- The book is oriented towards applied approach.
- The book maintains user friendly simple language, lucid style, cogent presentation, clarity of exposition, self-contained & practical approach involving hundreds of business applications which provides both motivation and physical insight to the readers.
- No special mathematical knowledge is assumed and the techniques are developed as situation generates. Each chapter begins with an introduction highlighting the development, application needs and prospects. In a nut-shell mathematical proofs are carefully avoided.
- One of the novel features of the text-book is that all major steps in the computational procedure of important techniques are presented in a step-by-step algorithm format.
- To enhance the understating of the subject by the students of the various disciplines, a conceptual rather than mathematical approach is adopted. In fact, throughout the text, each concept is followed by numerous illustrations from different functional areas of management.
- The book is student friendly containing plenty of problems from real-life situations, which are included in each chapter to enable the reader to gain further insight and clarity of the application of various O.R. techniques.
- The revised edition of the book reveals a refreshing new touch to every topic discussed with a score of valuable points which the students is unlikely to find in any other competitive title and which will give him the confident feeling of knowing something extra about the subject.

Edition: 9th; *Pages:* *xvi* + 632; *Size:* 220 × 280 mm; ISBN: 978-93-5161-011-3 (TC-531)

Problems in Operations Research

(Quantitative Approaches to Decision Making)

P.K. Gupta • Man Mohan

About the Book

The book includes all basic features of earlier editions and most of the topics of Management Science. The chapter on Business Forecasting and Learning Curve Theory has been included in the revised edition of the book.

Salient Features

- Book covers the latest syllabus in Operations Research Quantitative Techniques/ Management Science of most of the universities, Civil Services examination and other Professional examinations.
- Each chapter begins with a brief synopsis of the techniques discussed therein.
- Each chapter consists of sufficiently large number of solved problems mostly taken from recent examination papers.
- At the end of each chapter, unsolved problems followed by their answers are presented for self practice.

Edition: 14th; *Pages:* *viii* + 936; *Size:* 160 × 240 mm; ISBN: 978-93-5161-007-6 (TC-495)

Quantitative Techniques and Operations Research

P.K. Gupta • S.P. Gupta

About the Book

This book on Quantitative Techniques and Operations Research provides a conceptual understanding of basic quantitative techniques/methods used in solving managerial problems. It discusses comprehensively the essential topics of quantitative decision-making by using illustrations and examples. With revised and updated contents, this edition of the book will be useful to the students pursuing the undergraduate programme in Management and Commerce.

Salient Features

- The chapters on Linear Programming Problems are rewritten and have been designed to take the reader through gradual increase in complexity. Chapter 2 of part II is completely dealt with the formulation of linear programming models from business and economics.
- Given its importance in the current business environment, chapter on Operations Scheduling has been dealt with at length.

Edition: 4th, 2022; *Pages:* xvi + 600; *Size:* 185 × 240 mm; ISBN: 978-93-5161-182-0 (TC-1017)

Business Statistics and Operations Research

P.K. Gupta • S.P. Gupta • Man Mohan

About the Book

The text book on "Business Statistics and Operations Research" provides a conceptual understanding of basic techniques of Statistics and Operations Research for solving managerial and business problems. The book discusses comprehensively the essential topics of quantitative decision-making by using various illustrations and examples. The book covers complete syllabus of B.Com. & BBA prescribed by most Indian Universities.

Salient Features

- Book is divided into two parts: Part I gives basic concepts of Statistics and Part II provides various Operations Research Techniques.
- There are detailed self contained chapters covering the syllabus completely.
- The language used in the text is simple and subject-matter has been presented in a lucid style.
- Special care has been taken to develop the concepts in an easy to understand manner.
- Chapter on Network Analysis is discussed in detail due to its importance in solving complex business problems.

Edition: 6th, 2022; *Pages:* xvi + 768; *Size:* 185 × 240 mm; ISBN: 978-93-91820-35-0 (TC-508)

CHAPTER 1

Operations Research — An Overview

"Ideas illuminate, while interest only befog"

The activity "Operations Research" (abbreviated as OR) has become increasingly important in the face of fast moving technology and increasing complexities in business and industry. This chapter starts with the historical development of Operations Research and provides an overall view of the subject. It covers some general ideas on the subject, thus, providing a perspective.

1:1. DEVELOPMENT AND GROWTH OF OPERATIONS RESEARCH (OR)

The term, *Operational Research*, was first coined in 1940 by McClosky and Trefthen in a small town, Bowdsey, of the United Kingdom. This new science came into existence as a result of research on military operations during World War II. During the war there were strategic and tactical problems which were greatly complicated, to expect adequate solutions from individuals or specialists was unrealistic. Therefore, the military management called on scientists from various disciplines and organized them into teams to assist in solving strategic and tactical military projects. By their joint efforts, experience and deliberations, they suggested certain approaches that showed remarkable progress. This new approach to systematic and scientific study of the operations of the system was called the ***Operations Research*** or ***Operational Research*** (abbreviated as OR).

Following the end of World War II, OR was adopted by the industry and some of its techniques that had been applied to the complex problems of war were successfully transferred and assimilated for use in the industrial sector.

The dramatic development and refinement of the various techniques of OR and the advent of digital computer are the two prime factors that have contributed to the growth and application of OR in the post-war period. In the 1950's, OR achieved recognition as a subject for study in the universities. Since then, the subject has gained more and more importance for the students of management, public administration, behavioural sciences, engineering, and economics.

In recent years, Operations Research has had an increasingly great impact on the management of organizations. Both the number and variety of its applications continue to grow rapidly, and no slowdown is in sight. The subject is also being used widely in other types of organizations, including business and industry. Besides industry, the financial institutions, governmental agencies, hospitals and so on are rapidly increasing the use of Operations Research Techniques.

For the rapid growth of Operations Research, at least two factors can be identified. These are: First, the substantial progress was made early in improving the techniques of OR. For example, George Dantzig developed ***simplex method*** for solving linear programming problems in 1947. Many other techniques of OR, such as game theory, queuing theory, dynamic programming, inventory models, network optimization models were also well developed before the end of 1950's.

Second factor that gave great impetus to the growth of OR was the onslaught of the ***computer revolution***. The development of electronic digital computers, with their ability to perform arithmetic calculations millions of time faster than a human being can, was a tremendous boon to OR. A further boost came in 1980's with the development of increasingly powerful personal computers accompanied by good software packages for solving OR problems. This brought the use of OR within the easy reach of much larger number of people. The progress further accelerated with the invent of laptops. Today, literally millions of individuals have already access to OR software.

In India, OR came into existence in 1949 with the opening of an Operational Research unit at the Regional Research Laboratory, Hyderabad for using the techniques/methods of this new science in planning and organizing research. At the same time, another OR group was set up at the then Defence Science Laboratory which devoted itself to the problems of stores, purchase and planning. In 1953, an OR unit was established at the Indian Statistical Institute, Kolkata with a view to apply OR techniques to National Planning under the able guidance of Prof. Mahalonobis. Formal recognition to this field of science was given with the formation of Operational Research Society of India (ORSI) in 1957. The society became a member of International Federation of OR societies in 1959 and organized its first conference at all-India level during the same year. Later on, in 1964, the ORSI started the publication of its quarterly journal, "OPSEARCH" to highlight the activities of OR in India.

Towards the formal education in India, University of Delhi started a full-time Master's Degree programme in OR in 1963. A graduate programme in OR was introduced by the ORSI in 1973. Also, Diploma course in Operations Research & Computer Applications was started by Cochin University of Science & Technology at Kochi in 1979. Short-term training programmes are being run by a number of professional institutions such as Indian Institute of Public Administration, National Productivity Council, etc. Today, almost all the universities and institutions in India are offering the OR techniques/methods in their carricula in some form or the other.

1:2. COMMUNITY OPERATIONS RESEARCH

Community based Operations Research (CBOR) is a new sub-discipline within Operations Research and Management Science, and also at the time an emerging field of Public Policy. The term CBOR was first coined in late 1980's in the UK, when public and private sector Operations Research was on decline. The new discipline of Management Science gained importance with the appearance of books on community based operations research by Midgley[1] in 2004 and subsequently by Michael P. Johnson[2] in 2011.

The emerging sub-field of Public Sector Operations Research focusses on community issues such as poverty, food scarcity, homelessness, etc. It addresses these public sector issues in which outcome measure to be optimized, is not a direct representation. Here, problems from our daily lives have to be identified and the solutions have to be implemented in a manner to simultaneously optimize all these variables of economic efficiency, social equality and administrative work. As such, it represents a specific domain within public sector operations research.

The important implications that distinguish community based operations research from other areas of public sector research are as follows:

1. Our focus should be on human stakeholders that implies a fundamental interest in human *vs.* physical resources. This might be represented by decision models for natural resources management, energy policy and infrastructure design.
2. Out focus should be on disadvantaged, under-served or vulnerable population groups, who may have distinct social and political preferences and are less able to use political

---

1 *Community Operational Research* by Midgley, Ochoa-Avias & Alejandro (2004), Springer.

2 *Community Based Operations Research* by Michael P. Jhonson (2011), Springer.

influence to get Public Policy in line with their own needs, and may suffer stigma that may impede access to resources.

3. Our focus should be on 'place'. This implies that models should account for community characteristics like socio-economic status. Salient decision models may vary over place in systematic ways. For example, a public library location model does not distinguish between demands originating from neighbourhoods that differ in socio-economic status.

Like other areas of public sector operations research, the outcome measures of CBOR are not directly related to profit maximization, but rather focus on welfare maximization more broadly.

1:3. NATURE AND CHARACTERISTIC FEATURES OF OPERATIONS RESEARCH

After tracing the process of establishment and growth of Operations Research, we can consider it as a source to other new sciences. Literally, the word 'operation' may be defined as some action that we apply to some problems or hypotheses and the word 'research' is an organized process of seeking out facts about the same. OR has been variously described as the "science of use", "scientific approach to decision-making problems", "application of scientific principles to business management providing a quantitative basis for complex decisions", and so on. A classical definition of OR is given by Churchman, C.W., Ackoff, R.L. and Arnoff, E.L., *viz.*,

"Operations Research is the application of scientific methods, techniques and tools to problems involving the operations of system so as to provide those in control of operations with optimum solution to the problems".

Operations Research Society of America (ORSA) has described this field of science as under:

"Operations Research is an experimental and applied science devoted to observing, understanding and predicting the behaviour of purposeful man-machine systems".

The significant features of OR are given as under:

1. **Decision-Making.** Primarily, OR is addressed to managerial decision-making or problem-solving. A major premise of OR is that decision-making, irrespective of the situation involved, can be considered as a general systematic process. It consists of the following four steps:
   (*i*) Define the problem and establish a criterion to be used.
   (*ii*) Select the alternative courses of action for consideration.
   (*iii*) Determine the model to be used and the values of the parameters involved.
   (*iv*) Solve the model to choose the best (optimal) alternative.
2. **Scientific Approach.** Operations Research employs ***scientific methods*** for the purpose of solving problems. It is a formalized process of reasoning.
3. **Objective.** Operations Research attempts to locate the best or optimal solution to the problem under consideration. For this purpose, it is necessary that a measure of effectiveness is defined which is based on the goals of the organization. This measure is then used as the basis to compare the alternative courses of action.
4. **Interdisciplinary Team Approach.** Operations Research is interdisciplinary in nature and requires a team approach to a solution of the problem. Managerial problems have economic, physical, psychological, biological, sociological and engineering aspects. This requires a blend of people with expertise in the areas of mathematics, statistics, engineering, economics, management, computer science, and so on.
5. **Wholistic Approach.** While arriving at a decision, an OR team examines the relative importance of all conflicting and multiple objectives and validity of the claims of various departments of the organization from the perspective of the whole organization.

1:4. WHAT IS DECISION OPTIMIZATION?

Decision optimization is exactly what it sounds like . . . optimizing decision by using mathematical and constraint programming. It follows the operations research process. For example, you can utilize decision optimization while creating staff schedule. For this, there are several constraints that need to be considered, *e.g.*, availability of employees, time off, skill set, need, overtime, budget. Using decision optimization technology, you will be able to create the schedule that meets all these constraints at the click of a button.

1:5. MODELLING IN OPERATIONS RESEARCH

A ***model*** in OR is a simplified representation of an operation or a process in which only the basic aspects or the most important features of a typical problem under investigation are considered. Models, in general, cannot represent every aspect of reality because of innumerable and changing characteristics of real life problems to be represented. However, they may be classified as under:

1. **Models by Degree of Abstraction.** These models are based on the past data/information of the problems under consideration and can be categorized into: (*a*) language models, and (*b*) case studies.

   A book may be regarded as an example of a language model.

2. **Models of Function.** These models consist of: (*a*) Descriptive models, (*b*) Predictive models, and (*c*) Normative models.

   (*a*) ***Descriptive Models.*** A descriptive model simply describes some aspects of a situation based on observation, survey, questionnaire results, or other available data. The result of an opinion poll represents a descriptive model.

   (*b*) ***Predictive Models.*** Such models can answer 'if and what' type of questions, *i.e.*, they make predictions regarding certain events. For example, based on survey results, television networks attempt to explain and predict the election outcome before all the votes are actually counted.

   (*c*) ***Normative Models.*** Finally, when a predictive model has been repeatedly successful it can be used to prescribe a source of action. Linear programming is a normative or prescriptive model, because it prescribes what the managers ought to do.

3. **Models by Structure.** These models are represented by: (*a*) Iconic models, (*b*) Analogue models, and (*c*) Symbolic models.

   (*a*) ***Iconic*** or ***Physical models*** are pictorial representation of real systems and have the appearance of the real thing. Examples of such models are: city maps, houses blueprint's, globe, and so on. An iconic model is said to be 'scaled-down' or 'scaled-up' according as the dimensions of the model are smaller or greater than those of the real item. For instance, in biology, the structure of a cell may be illustrated by an enlarged (scaled-up) iconic model for teaching purposes.

   Iconic models are easy to observe, build and describe, but are difficult to manipulate and not very useful for the purposes of prediction. Commonly, these models represent a ***static*** event.

   (*b*) ***Analogue models*** are more abstract than the iconic ones for there is no 'look-alike' correspondence between these models and real life items. They are built by utilizing one set of properties to represent another set of properties. For instance, a network of pipes through which water is running could be used as a parallel for understanding the distribution of electric currents. Graphs and maps in various colours are analogue models, in which different colours correspond to different characteristics. A flow process chart is an analogue model which represents the order of occurrence of various events to make a product.

(*c*) ***Mathematical*** or ***Symbolic models*** are most abstract in nature. They employ a set of mathematical symbols to represent the components (and relationships between them) of the real system. These models are most general and precise. However, it is not always possible to depict a real system in mathematical formulation, sometimes it is easier to use mathematical symbols for describing the relationships among variables of the system to describe its properties or behaviour. Sometimes an analogue model may express the pattern of its relationship in a better way.

4. **Models by Nature of the Environment.** These models can be classified into: (*a*) Deterministic models, and (*b*) Probabilistic models.

   In ***deterministic models,*** all the parameters and functional relationships are assumed to be known with certainty when the decision is to be made. Linear Programming and Break-even models are the examples of deterministic models.

   On the other hand, models in which at least one parameter or decision variable is a random variable are called ***probabilistic*** or ***stochastic models***. These models reflect, to some extent, the complexity of the real world and the uncertainty surrounding it.

5. **Models by the Extent of Generality.** These models can be categorized into: (*a*) Specific models, and (*b*) General models.

   When a model presents a system at some specific time, it is known as a ***specific model***. In these models if the time factor is not considered, then they are termed as ***static models,*** and ***dynamic models*** otherwise. An inventory problem of determining economic order quantity for the next period, assuming that the demand in planning period would remain same as that of today, is an example of a static model. Dynamic Programming may be considered as an example of dynamic model.

6. **Simulation and Heuristic Models** fall under the category of general models. These models are mainly used to explore alternative strategies (courses of action) which have been overlooked previously. These models do not yield any optimum solution to the problem, but give a solution to a problem depending on assumptions based on the past experience.

Avantages of Models in Operations Research

Models in Operations Research are used as an aid for analyzing complex problems. Generally advantages of a model are:

(*i*) It provides some logical and systematic approach to the problem.
(*ii*) It indicates the limitations and scope of an activity.
(*iii*) It helps in incorporating useful tools that eliminate duplication of methods applied to solve any specific problem.
(*iv*) It helps in finding avenues for new research and improvement in a system.
(*v*) It provides economic description and explanation of the operations of the system they represent.

1:6. GENERAL SOLUTION METHODS FOR OR MODELS

There are, in general, three methods for deriving the solution to an OR model: ***analytical methods***, ***numerical methods*** and ***Monte Carlo methods***. The first one is deductive by nature, while the second one is inductive. Monte Carlo method is a statistical method that rely on randomness to obtain numerical results.

Analytical or Deductive Methods

In these methods classical optimization techniques such as Calculus, Finite Differences, etc. are used for solving an OR model. The kind of mathematics required depends upon the nature of the model. For example, in the inventory model, in order to calculate the ***Economic Order Quantity*** (EOQ), the analytical method requires the first derivative of the mathematical expression

$$TC = (D/Q)\, C_S + (Q/2)\, C_1,$$

equated to zero, as the first step towards identifying the optimum value of *EOQ, viz.,* $\sqrt{2DC_1 / C_S}$. This is because of the concept of maxima and minima for optimality.

Here, TC = total annual cost, D = annual demand, Q = size of an order, C_S = set up cost per production run, and C_1 = holding cost for a unit in inventory.

Numerical Methods

Numerical methods are concerned with the iterative or trial and error procedure through the use of numerical computation at each step. These numerical methods are used when some analytical methods fail to derive the solution.

The numerical method starts with a trial (initial) solution and continued with a set of rules for improving it towards optimality. The trial solution is then replaced by the improved one and the process is repeated until either no further improvement is possible or the cost of further computations cannot be justified. Thus, the numerical methods are hit and trial methods that end at a certain step after which no further improvement can be made.

Monte Carlo Methods

These involve the use of probability and sampling concepts. The various steps associated with a Monte Carlo method are as follows:

(*i*) For appropriate model of the system, make sample observations and determine the probability distribution for the variables of interest.

(*ii*) Convert the probability distribution to cumulative distribution.

(*iii*) Select the sequence of random numbers with the help of random tables.

(*iv*) Determine the sequence of values of variables of interest with the sequence of random numbers obtained in the above step.

(*v*) Fit an appropriate standard mathematical function to the values obtained in *Step (iv).*

The Monte Carlo method is essentially a ***simulation technique*** in which statistical distribution functions are created by generating a series of random numbers.

1:7. METHODOLOGY OF OPERATIONS RESEARCH

The systematic methodology developed for an OR study deals with problems involving conflicting multiple objectives, policies and alternatives. OR is a scientific methodology which is applied to the study of the operations of large complex organizations or activities with a view to assessing the overall implications of various alternative courses of action, thus providing an improved basis for managerial decisions.

The OR approach to problem solving consists of the following **eight** steps:

1. Formulating the problem
2. Obtaining the input data
3. Constructing a mathematical model
4. Deriving the solution from the model

5. Testing the model
6. Validation of the model
7. Establishing control over the solution
8. Implementation of the final results.

1. ***Formulating the Problem.*** The first step in OR is to develop a clear and concise statement of the problem, *i.e.*, to identify the problem correctly. Once the problem is rightly identified, we can say that fifty per cent of the problem is solved. After identification, formulation of the problem is to be carried out. For this it is necessary to study comprehensively the components of the problem, *viz.*, (*a*) the decision-maker, (*b*) the objective, (*c*) the alternative courses of action, and (*d*) the environment. After identifying the components of the problem, the relationship that exists among the components of the problem should be analysed. Also, there must be complete agreement on the above points between persons initiating the OR study and the persons performing the study.

2. ***Obtaining the Input Data.*** Once the mathematical model of the problem has been formulated, the next step is to obtain the data to be used in the model as input. Since, the quality of data determines the quality of output, the importance of obtaining accurate and complete data cannot be over-emphasized.

   Obtaining correct and relevant data may, however, be a difficult exercise when relatively large problems are involved. A number of sources, such as: company reports and documents, interviews with the company personnel and so on may be used for collecting data.

3. ***Constructing a Mathematical Model.*** The next step is to build a suitable mathematical model. Model construction consists of hypothesizing relationships between variables subject to, and not subject to control by decision-maker.

   A mathematical model should include mainly the following three basic sets of elements:

   (*a*) Decision variables and Parameters,

   (*b*) Constraints or Restrictions, and

   (*c*) Objective function.

   A generalized mathematical model might take the form: $E = f(x_i, y_i)$, where f represents a system of mathematical relationships between the measures of effectiveness of the objective sought (E) and the variables, both controllable (x_i) and uncontrollable (y_i).

4. ***Deriving the Solution from the Model.*** Having collected the input data, the next step is to determine the values of decision variables that optimize the given objective function. This deals with the mathematical calculations for obtaining the solution to the model.

   It may be noted that the solutions can be categorized as being: (*a*) feasible or infeasible, (*b*) optimum or non-optimum, and (*c*) unique or multiple.

   (*a*) A solution that satisfies all the constraints of the problem including the condition of non-negativity is known as a ***feasible solution***, while the infeasible solution is one that does not satisfy the condition of non-negativity of variables.

   (*b*) An ***optimum solution*** to the problem is one of the feasible solutions which also optimizes the objective function. The feasible solutions other than the optimum solution are called non-optimum solutions.

   (*c*) If there exists only one optimum solution to be problem, it is said to be ***unique optimum solution.*** *On the other other hand, if* ***two or more optimum solutions*** *to the problem exist, then* ***multiple optimum solutions*** *are said to exist*. Thus, sensitivity analysis is an essential part of this step of the methodology and must not be overlooked.

5. ***Testing the Model.*** Since, a model is only a partial and simplified representation of reality, the results are to be tested against the real world experience in order to establish the model's credibility. Several simplifying assumptions are made while initiating the model-building. These assumptions can be relaxed one by one to see the reaction of the model results to such relaxation. Similarly, inclusion of some inappropriate variables in the model, and exclusion of some appropriate variables from the model, just for purposes of abstraction and simplification, may require review so that their impact on the model *vis-à-vis* reality is analysed from a proper perspective. Model distortion through simplification, intentional or otherwise, has to be properly assessed with the help of previous experience, judgement, actual test data and similar devices.
6. ***Validation of the Model.*** Validating a model requires of determining as if the model can reliably predict the actual system's performance. It also involves testing the structural assumptions of the model to ascertain their validity. A common approach for testing the validity of a model is to compare its performance with past data available for the actual system.

   If during validation, the solution cannot be implemented, one needs to find if some of the original constraints were incorrect and need to be modified. In such a case, one must return to the problem formulation step and carefully make the appropriate modification to represent more realistic situation.

   A model must be applicable for a reasonable time period and should be updated from time to time, taking into consideration the past, present and future aspects of the problem.
7. ***Establishing Control over the Solution.*** After testing the model and its solution, the next step of the study is to establish control over the solution, by proper feedback of the information on variables which deviated significantly. As soon as one or more of the controlled variables change significantly, the solution goes out of control. In such a situation the model may accordingly be modified.
8. ***Implementation of the Final Results.*** Finally, the tested results of the model are implemented to work. This would basically involve a careful explanation of the solution to be adopted and its relationship with the operating realities. This stage of OR investigation is executed primarily through the co-operation of both the OR experts and those who are responsible for managing and operating the system.

1:8. OPERATIONS RESEARCH APPROACH TO PROBLEM SOLVING

The most important aspect of solving an OR problem is the use of scientific method and the formulation of decision models. The operations research approach to problem solving is based on the following **three** phases:

1. ***Judgement Phase.*** This phase consists of: (*i*) determination of the operation, (*ii*) identification of the real-life problem, (*iii*) establishment of the objectives and values related to the operation, (*iv*) determination of the suitable measures of effectiveness, (*v*) formulation of the problems relative to the objectives, (*vi*) building of an appropriate model of the problem and abstracting the essential information, so that a solution to the decision-maker's goals can be attained.
2. ***Research Phase.*** This phase utilizes: (*i*) operations and data collection for a better understanding of the problems, (*ii*) formulation of hypothesis and model, (*iii*) observation and experimentation to test the hypothesis on the basis of additional data, (*iv*) analysis of the available information and verification of the hypothesis using pre-established measures

of effectiveness, and predictions of various results from the hypothesis, (*v*) generalization of the various results and consideration of alternative methods.

3. ***Action Phase.*** This phase consists of making recommendations for the decision process by those who first posed the problem for consideration or by anyone in a position to make a decision, influencing the operation in which the problem occurred.

1:9. OPERATIONS RESEARCH AND BUSINESS / INDUSTRY

The importance of Operations Research in business/industry may be summarized as follows:

(*i*) Operations Research models provide a structural, logical and systematic approach for solving a decision-making problem.

(*ii*) Operations Research models help in changing the decision of a problem with the application of improved information.

(*iii*) Operations Research models can qualify and quantify various measurable quantities related to a given problem.

(*iv*) Operations Research models help in the physical and economic understanding of a problem and in the improvement of the system.

Some of the areas of management where techniques of Operations Research have been successfully applied are listed below:

1. ***Marketing and Sales.*** (*i*) Advertising and media planning, (*ii*) Product selection, timing and competitive strategies, (*iii*) Recruitment of salesman, (*iv*) Forecasting and decision trends, (*v*) Pricing and competitive decisions, (*vi*) Market research decisions.
2. ***Finance and Accounting.*** (*i*) Cash flow planning, (*ii*) Credit policy analysis, (*iii*) Investment policy for maximum return, (*iv*) Dividend policies, (*v*) Portfolio analysis, (*vi*) Profit planning.
3. ***Personnel.*** (*i*) Selection of personnel, their age and skills, (*ii*) Recruitment policies and assignment of jobs, (*iii*) Manpower planning, wage administration, (*iv*) Scheduling of training programmes.
4. ***Construction.*** (*i*) Allocation of resources to projects, (*ii*) Determination of proper workforce, (*iii*) Employment of workforce, (*iv*) Project scheduling, monitoring and control.
5. ***Facilities Planning.*** (*i*) Factory size location decision, (*ii*) Location and size of warehouse, distribution centres and retail outlets, (*iii*) Transportation, loading and unloading, (*iv*) Logistics system design, layout.
6. ***Manufacturing.*** (*i*) Employment, training, layoffs, quality control, (*ii*) Aggregate production planning, assembly line, blending, purchasing, (*iii*) Production scheduling, production smoothing, (*iv*) Inventory control.
7. ***Purchasing and Procurement.*** (*i*) Bidding policies, (*ii*) Vendor analysis, replacement policies, (*iii*) Optimal buying and reordering, (*iv*) Materials transfer.
8. ***Maintenance and Project Scheduling.*** (*i*) Preventive maintenance and maintenance policies, (*ii*) Project scheduling and allocation of resources, (*iii*) Maintenance, crew size and scheduling, (*iv*) Project management and strategic planning.
9. ***Production.*** (*i*) Scheduling and sequencing the production run by proper allocation of machines, (*ii*) calculating the optimum product mix, (*iii*) Selection, location and design of the sites for the production plant.
10. ***Research and Development.*** (*i*) Control of research and development projects, (*ii*) Product introduction planning, (*iii*) Organizational design and control, (*iv*) Decision support system and MIS.

Besides the above-mentioned applications of Operations Research in the context of modern management, its use has now extended to a wide range of problems, such as the problems of communication and information, socio-economic fields and national planning.

1:10. OPERATIONS RESEARCH AND DECISION-MAKING

Every industrial enterprise or business house works on a properly established management information system. Such a system usually assumes a three-tier position at the level of: (*i*) strategic planning, (*ii*) managerial control, and (*iii*) operational control. These levels are identified with the top management, middle management and supervisory management respectively. Many prefer to call these three-tier activities as policy-making, decision-making and activity oriented. For a competent management, decision-making is a major task.

Decision-making is not the headache of only the management, rather all of us make decisions. We daily decide about many minor and major issues. The essential characteristics of all decisions are: (*i*) objective; (*ii*) alternatives at the disposal; and (*iii*) influencing factors.

Once these characteristics are known, one can think of improving the characteristics so as to improve upon the decision itself.

Let us consider a situation where a decision concerns spending summer vacations at a hill resort. The problem may be to decide the mode of conveyance from amongst the three alternatives: train, bus and a taxi.

At the *first* level of decision-making, bus is chosen as the mode of conveyance just by intuition. At the *second* level of decision-making, the three modes are compared and it is decided qualitatively that the bus will be preferred since, it is less time consuming compared to the train and cheaper as compared to taxi. At the *third* level of decision-making, the three alternatives are compared and it is suggested that the bus will be chosen, as it will be taking only half the time taken by train and shall be 40% less constlier than the taxi.

Although outcome of all these decisions is the same, one can easily judge the quality of each decision. We may brand the first decision as 'bad' since it is highly emotional, while we may call the second decision as 'good' since, it is scientific, though qualitative. The third decision is doubtlessly the best as it is scientific as well as quantitative.

It is this ***scientific quantification*** used in OR that helps management to make better decisions.

Operations Research may be regarded as a tool employed to increase the effectiveness of managerial decisions as an objective supplement to the subjective feelings of the decision-maker.

For instance, in distribution or allocation areas, OR may suggest the best locations for agencies, warehouses as well as the most economical kind of transportation; in marketing areas, it may aid in indicating the most profitable type, use and size of advertising campaigns, in regard to available financial limits. OR may suggest alternative courses of action when a problem is analysed and solution is attempted. However, the study of complex problems by OR techniques becomes useful only when a choice between two or more courses of action is possible.

Advantages of Operations Research

OR may be regarded as a tool that enables the decision-maker to be objective in choosing an alternative from among many that he can conceive of. Following are the salient advantages of an operations research study approach in decision-making:

1. ***Better Decisions.*** OR models frequently yield actions that do improve on intuitive decision-making. A situation may be so complex that the human mind can never hope to assimilate all the significant factors without the aid of OR guided computer analysis.

2. ***Better Coordination.*** Sometimes, operations research has been instrumental in bringing order out of chaos. For instance, an OR-oriented planning model becomes a vehicle for coordinating marketing decisions within the limitations imposed on manufacturing capabilities.
3. ***Better Control.*** The management of large organizations recognize that it is extremely costly to provide continuous executive supervision over routine decisions. An OR approach, thereby regained new freedom to the executive to devote his attention to more pressing matters. The most frequently adopted application in this category deal with production scheduling and inventory replenishment.
4. ***Better Systems.*** Often, an OR study is initiated to analyse a particular decision-problem, such as whether to open a new warehouse. Afterwards, the approach is further developed into a system to be employed repeatedly. Thus, the cost of undertaking the first application may produce benefits.

1:11. LIMITATIONS OF OPERATIONS RESEARCH

Although the use of OR techniques to improve decision-making has become almost universal today, yet there are certain limitations. These limitations result from lack of awareness on the part of managers about their roles, while others result due to avoidance of the behavioural and organizational issues which are part of every successful application. Some of the limitations of OR techniques are given below:

1. There are certain problems which a decision-maker may have to solve only once. Constructing a complex OR model for solving such problems is often too expensive when compared with the cost of other less sophisticated approaches available to solve them.
2. Sometimes OR specialists become too much enamoured with the model they have built and forget the fact that their model does not represent the "real world problem" in which decisions have to be made.
3. Sometimes the basic data are subject to frequent changes. In such cases, modification of OR models is a costly affair.
4. Many OR models are so complex that they cannot be solved without the use of computer. Also, the solutions obtained from these models are difficult to explain to managers and hence fail to gain their support and confidence.
5. Magnitude of computation involved, lack of consideration for non-quantifiable factors and psychological issues involved in implementation are some of the other limitations of OR techniques.

TEST YOUR UNDERSTANDING

Choose the correct alternative:

1. Operations Research came into existence
   (*a*) in the year 1940 (*b*) in the business context
   (*c*) during World War I (*d*) during World War II
2. Operations Research achieved recognition as a subject of academic study in the year
   (*a*) 1949 (*b*) 1950
   (*c*) 1953 (*d*) 1957

3. Operations Research is
   (*a*) applied decision theory
   (*b*) a scientific approach to problems solving for executive management
   (*c*) the science of use
   (*d*) all of the above
4. Operations Research approach is
   (*a*) intuitive
   (*b*) objective
   (*c*) multi-disciplinary
   (*d*) all of the above
5. A model in Operations Research is
   (*a*) an approximation
   (*b*) an idealization
   (*c*) an essence of reality
   (*d*) all of the above
6. A physical model is an example of
   (*a*) iconic model
   (*b*) analogue model
   (*c*) verbal model
   (*d*) symbolic model
7. Decision variables in an OR model are
   (*a*) controllable
   (*b*) uncontrollable
   (*c*) parameters
   (*d*) constants
8. The scientific method in Operations Research consists of
   (*a*) judgement phase
   (*b*) research phase
   (*c*) action phase
   (*d*) all of the above
9. Operations Research techniques are not applicable in the following situation:
   (*a*) Resources available are unlimited
   (*b*) Objective can be defined for maximization or minimization
   (*c*) Sufficient input data is available for formulating the problem
   (*d*) Scientific methods, techniques and tools may be applied
10. An optimization model in OR
    (*a*) provides decision within its limited context
    (*b*) mathematically provides the best decision
    (*c*) helps in evaluating various alternatives constantly
    (*d*) all of the above
11. Which of the following is not correct?
    (*a*) A model should be simple and coherent
    (*b*) A model should not take much time in its construction for any problem
    (*c*) A model representing the typical budget of business accounts is called 'account model'
    (*d*) A model which has the probability of measuring observations is called 'qualitative model'
12. Which of the following is not correct?
    (*a*) Models are only an attempt in understanding operations and should never be considered as absolute in any sense
    (*b*) Iconic models are pictorial representation of real system
    (*c*) Heuristic model are particular cases of iconic models
    (*d*) Analogue models are most abstract than the iconic ones

ANSWERS

1. (*d*)	**2.** (*b*)	**3.** (*d*)	**4.** (*c*)	**5.** (*d*)	**6.** (*a*)
7. (*a*)	**8.** (*d*)	**9.** (*a*)	**10.** (*d*)	**11.** (*d*)	**12.** (*c*)

REVIEW EXERCISE

1. What is Operations Research? What are the characteristics and limitations of Operations Research techniques? *[B.Tech. (Mech.), Himachal Univ. 2015]*
2. Define Operations Research. Briefly discuss techniques and tools of Operations Research. *[M.Com., Bangalore Univ. 2017; BCA, Madras Univ. 2010]*
3. Discuss the scope and limitations of Operations Research. *[MBA, Panjab Tech. Univ. 2015]*
4. Mention different phases of Operations Research study? *[BE (Nov.), Visvesvaraya Tech. Univ. 2020]*
5. Explain different type of models used in Operations Research. *[MBA, Madras Univ. 2012]*
6. What is an Operations Research model? Give the main advantages of an Operations Research model. *[MBA, Panjab Univ. 2015]*
7. "Model building is the essence of Operations Research". Discuss.
8. Discuss various classification scheme of models in Operations Research study. *[B.Tech., Himachal Univ. 2015; M.Tech., Panjab Tech. Univ. 2012]*
9. Define a scientific model. Discuss in detail three types of models with special emphasis on the important logical properties and relationships the three types bear to each other and to modelled entities. *[MBA, Panjab Tech. Univ. 2011]*
10. Distinguish between: (*a*) Analogue and Iconic models, (*b*) Deterministic and Stochastic models. *[B.Tech. (Mech.), Himachal Univ. 2013]*
11. Give any four comprehensive definitions of Operations Research. *[MBA, Ranchi Univ. 2011]*
12. List and explain the steps in conducting an Operations Research study.
13. Describe briefly different phases of Operations Research. *[BE (Jan.), Visvesvaraya Tech. Univ. 2020]*
14. Write short note on methodology of Operations Research.
15. Write short-note on the following:
    (*a*) Importance of Operations Research in project management.
    (*b*) Scope of Operations Research in public utilities and in industry. *[B.Tech. (Mech.), Himachal Univ. 2010]*
    (*c*) Scope of Operations Research in business. *[MBA, Madras Univ. 2012]*
    (*d*) Scope of Operations Research in financial management.
16. What are the limitations of using results from a mathematical model to make decision about operations? *[MCA, Hyderabad Univ. 2012]*
17. Write short note on the role of Operations Research in:
    (*a*) marketing management, (*b*) production planning, and (*c*) production management.
18. "Operations Research is more than just mathematics". Justify the statement with an example. *[B.Com., Guru Nanak Dev Univ. 2014]*
19. Explain the role and importance of operations research in managerial decisions. *[MBA, Bangalore Univ. 2018]*
20. Discuss the usefulness of Operations Research in decision-making process and the role of computer in this field.

CHAPTER 2

Linear Programming

"Resources are scarce and property of the society and their abuse is a social evil"

Linear Programming is one of the most versatile, popular and widely used operations research technique where planning is an important or rather essential aspect of any business or industrial activity. It means a particular plan of action is to be selected from amongst several alternatives that may be available. Of course, in any such activity, there is always an ***objective*** to be achieved, *i.e.*, the purpose of that activity. For example, a production manager of an industrial concern has 'profit' as his objective. He has made available to himself limited resources in men, money, and materials. The problem of the production manager is to use these resources in such a way as to have the maximum profit, that may be possible. A farmer, a businessman, a transporter, etc., each have an objective and limited resources of one kind or the other. In order to achieve the objective, a linear programming model is to be formulated that provides an efficient method for determining an ***optimum (or optimal) decision***. An optimum strategy (or optimum plan) is chosen from a large number of possible decisions. The optimum decision is one that meets a specified objective of management subject to various restrictions (or ***constraints***).

Programming is just another word of 'planning'. But it means something more, *i.e.*, planning in a scientific way. The ***objective*** (goal) must be identified as a function of its variables (the quantities of the relevant factors) and all the limitations on resources must be stated as equations and/or inequalities involving those variables.

The word ***linear*** indicates that the objective function is linear in its variables, and that the equations and/or inequalities that represent limitations on resources must also be linear.

2:1. LINEAR PROGRAMMING PROBLEM (LPP)

A Linear Programming Problem (LPP) consists of ***three components***, namely the decision variables (activities), the objective (goal), and the constraints (restrictions).

1. The ***decision variables*** refer to the activities that are competing one another for sharing the resources available. These variables are usually inter-related in terms of utilization of resources and need simultaneous solutions. All the decision variables are considered as continuous, controllable and non-negative.
2. A linear programming problem must have an ***objective*** which should be clearly identifiable and measurable in quantitative terms. It could be of profit (sales) maximization, cost (time) minimization, and so on. The relationship among the variables representing objective must be linear.
3. There are always certain limitations (or ***constraints***) on the use of resources, such as labour, space, raw material, money, etc. that limit the degree to which an objective can be achieved. These are expressed as linear inequalities or equalities in terms of decision variables.

2:2. ASSUMPTIONS OF LINEAR PROGRAMMING PROBLEM

A linear programming problem (or linear programming model) is based on the following five assumptions:

1. **Certainty.** In all linear programming problems it is assumed that all the parameters; such as availability of resources, profit (or cost) contribution of a unit of decision variable and consumption of resources by a unit decision variable must be known and fixed. In other words, all the coefficients in the objective function as well as in the constraints are completely known with certainty and do not change during the period of study.
2. **Proportionality.** A basic assumption of linear programming problem is that the relationship is directly proportional, *i.e.*, if resource availability increases by some percentage, then the output shall also increase by the same percentage. For example, if one unit of decision variable consumes 'x' hours of capacity, then five units shall consume '$5x$' hours of capacity.
3. **Divisibility (or Continuity).** This implies that the decision variables are not restricted to integer values, and as such their fractional values are permitted in the solution. For example, it is possible to produce 3.25 quintals of wheat or rice, 5.62 kilolitres of milk, and so on. But it is not possible to produce 3.4 refrigerators, 7.4 computers, etc. Such variables are not divisible and hence are to be assigned integer values. When it becomes necessary to have integer values, the integer programming problem is considered to attain the desired values.
4. **Additivity.** The value of the objective function for the given values of decision variables and the total sum of resources used, must be equal to the sum of the contributions (profit or cost) earned from each decision variable and the sum of the resources used by each decision variable respectively. For example, the total profit earned by the sale of two products A and B must be equal to the sum of the profits earned separately from products A and B. Similarly, the amount of a resource consumed by products A and B must be equal to the sum of resources used for products A and B individually.
5. **Finite Choices.** A linear programming problem also assumes that a finite (limited) number of choices (alternatives) are available to the decision-maker and that the decision variables are inter-related and non-negative. The non-negativity shows that linear programming problem deals with real-life situations as it is not possible to produce/use negative quantities.

2:3. THREE STAGES OF LINEAR PROGRAMMING PROBLEM

Each linear programming problem involves three stages, *viz.*: (1) identification, (2) formulation, and (3) solution.

1. **Identification** involves identification of the available alternatives of the problem, establishing linear relationship between variables and specification of the constraints (*e.g.*, space, material, money, etc.).
2. **Formulation** of the problem involves construction of a mathematical model from the given data. It requires identification of the decision variables, specifying the objective and setting up linear equalities/inequalities for the constraints.
3. **Solution** of the problem involves that particular set of values of decision variables which lead to achieve the objective — maximum profit, least cost, etc.

2:4. LIMITATIONS OF LINEAR PROGRAMMING

Though having a wide field of applications, linear programming has the following limitations:

1. In linear programming, uncertainty is not allowed, *i.e.*, linear programming methods are applicable only when values for cost or profit, constaints (or restrictions), etc. are known, but in real life such factors may be unknown.
2. According to the linear programming problem, the solution variables can have any values, whereas sometimes it happens that some of the variables may have only integral values.
3. There must be alternative courses of action. For example, a given product may be processed by two different machines and problem may be to determine how much of the product to be allocated to which machine.
4. Another necessary requirement is that decision variables should be non-negative. The non-negativity condition shows that linear programming deals with real-life situations for which negative quantities are generally illogical.
5. Parameters appearing in linear programming problem are assumed to be constant but in real-life situations, they are frequently neither known nor constant.
6. As stated earlier, the resources must be in limited supply. For example, if a firm starts producing greater number of a particular product, it must make smaller number of other products as the total production capacity is limited.

2:5. FORMULATING A PROBLEM AS LINEAR PROGRAMMING MODEL

Formulation of a linear programming problem means constructing a mathematical model from the given data. This may be done in the following seven steps:

Step* 1.** Study the situation to find the ***key decision to be made.

Step* 2.** Designate the various activities in terms of variable (known as ***decision variables).

Step* 3.** Express the ***feasible alternatives in terms of decision variables.

Step* 4.** Identify the ***contribution coefficients associated with each decision variable.

Step* 5.** Identify the ***objective and express it s a linear function of decision variables involving their unit cost (or profit) contributions.

Step* 6.** Express in words the influencing factors or ***restrictions (constraints). Put these restrictions in the form of linear equations/inequalities involving the decision variables.

Step* 7.** Express the objective function, constraints, and non-negative restrictions (identified in *steps 1* to 6) in ***linear programming format.

2:6. ILLUSTRATIVE EXAMPLES OF LP MODEL FORMULATION

Example 2.1 (*Product Mix Problem*). *A firm produces high quality glass products, including windows and glass doors. It has three plants. Aluminium frames and hardware are made in Plant 1, Wooden frames are made in Plant 2, and Plant 3 produces the glass and assemble the products.*

Because of declining earnings, top management of the firm decided to discontinue the production of unprofitable products and at the same time launch two new products having large sales potential. These new products are:

Product A: 7-foot glass door with aluminium framing,

Product B: 4 × 6 foot double-hung wood-framed window.

Each unit of product A requires 2 hours in Plant 1 and 5 hours of in Plant 3, while product B requires 4 hours in Plant 2 as well as in Plant 3. Every week, the firm has an availability of 30 hours in plant 1, 76 hours in plant 2 and 90 hours in plant 3. One unit of product A yields ₹ 300 and one unit of product B sold gives ₹ 500 as profit.

Formulate the above problem as a linear programming problem to determine what the production rates should be for the two products in order to maximize their total profit, subject to the restrictions imposed by the limited production capacities available in three plants.

Formulation of Linear Programming Model

We formulate the given problem as an LPP in the following steps:

Step 1. The key decision to be made is to determine the number of units of products *A* and *B* to be produced per week.

Step 2. Let x_1 designate the number of units of product *A* and x_2 the number of units of product *B* which the firm decides to produce per week.

Step 3. Feasible alternatives are the sets of values of x_1 and x_2, such that $x_1 \geq 0$ and $x_2 \geq 0$; since it is not possible to produce negative quantities.

Step 4. Each unit of product *A* yields ₹ 300 as profit, therefore, total profit from product *A* will be ₹ $300x_1$. Similarly, total profit from product *B* will be ₹ $500x_2$.

Step 5. The objective is to maximize the total profit from the production of two products *A* and *B*. Therefore, the objective function is to maximize

$$z = 300x_1 + 500x_2$$

Step 6. Each unit of product *A* produced uses 2 hours of production time per week in Plant 1, whereas only 30 hours are available. This restriction is expressed mathematically by the inequality $2x_1 \leq 30$. Similarly, Plant 2 imposes the restriction that $4x_2 \leq 76$. The number of hours of production time used per week in Plant 3 by products *A* and *B* would be $5x_1 + 4x_2$. The mathematical statement of the Plant 3 restriction, therefore, is: $5x_1 + 4x_2 \leq 90$.

Step 7. The given problem, therefore, can be written as an linear programming model (or linear programming problem) as follows:

Determine the values of x_1 and x_2 so as to

Maximize $z = 300x_1 + 500x_2$ *(Objective function)*

subject to the constraints:

$2x_1 \leq 30$ *(Time restriction in Plant 1)*

$4x_2 \leq 76$ *(Time restriction in Plant 2)*

$5x_1 + 4x_2 \leq 90$ *(Time restriction in Plant 3)*

$x_1 \geq 0$ and $x_2 \geq 0$ *(Non-negative restrictions)*

Example 2.2 (*Product Mix Problem*). *Production of a certain chemical mixture should contain at least 80 mg chloride, 28 mg nitrate and 36 mg of sulphate per gram. The company can use two substances and a base (assume this is costless). Substance X contains 8 mg chloride, 4 mg nitrate and 6 mg sulphate per gram. Substance Y contains 10 mg chloride, 2 mg nitrate and 2 mg sulphate per gram. Both substances cost ₹ 20 per gram. It is required to produce the mixture using substances X and Y so that the cost is minimized. Formulate the problem as an LP problem.* ***[MBA, GGSIP Univ. 2008]***

Formulation of Linear Programming Model

The information given in the above problem may be summarized as below:

Substance	*Contents (in mg)*			*Cost per unit*
	Chloride	*Nitrate*	*Sulphate*	*(₹)*
X	8	4	6	20
Y	10	2	2	20
Minimum requirement	80	28	36	

We formulate the problem as an LPP in the following seven steps:

Step 1. The key decision to be made is to determine the number of units of substances X and Y, respectively, so that the cost is minimum.

Step 2. Let us designate the number of units of substances X and Y to be produced by x_1 and x_2, respectively.

Step 3. Since, it is not possible to produce negative quantities, feasible alternatives are sets of values of x_1 and x_2 satisfying $x_1 \geq 0$ and $x_2 \geq 0$.

Step 4. Since each of the two substances X and Y cost ₹ 20 per gram, total cost of substance X is $20x_1$ and that of substance Y is $20x_2$.

Step 5. Since both the substances cost ₹ 20 per gram each, the objective is to minimize cost and, therefore, the objective function is

$$z = 20x_1 + 20x_2$$

Step 6. The total amount of chloride supplied from substances X and Y in chemical mixture is $8x_1 + 10x_2$, and this is required to be at least as large as the minimum amount of chloride required, namely 80 kg. Therefore, the constraint on chloride is: $8x_1 + 10x_2 \geq 80$. Similarly, the constraints on nitrate and sulphate respectively are: $4x_1 + 2x_2 \geq 28$ and $6x_1 + 2x_2 \geq 36$

Step 7. The linear programming problem to produce the mixture of substances X and Y can thus be put in the following linear programming model:
Determine the value of x_1 and x_2 so as to

$$\text{Minimize } z = 20x_1 + 20x_2 \quad \textit{(Objective function)}$$

subject to the constraints:

$$8x_1 + 10x_2 \geq 80 \quad \textit{(Restriction on Chloride)}$$
$$4x_1 + 2x_2 \geq 28 \quad \textit{(Restriction on Nitrate)}$$
$$6x_1 + 2x_2 \geq 36 \quad \textit{(Restriction on Sulphate)}$$
$$x_1 \geq 0 \text{ and } x_2 \geq 0 \quad \textit{(Non-negative restriction)}$$

Example 2.3 (*Product Mix Problem*). *Consider the following problem faced by a production planner in a soft drink plant. He has two bottling machines A and B. A is designed for 8-ounce bottles and B for 16-ounce bottles. However, each can be used on both types with some loss of efficiency. The following data is available:*

Machine	*8-ounce bottles*	*16-ounce botles*
A	*100/minute*	*40/minute*
B	*60/minute*	*75/minute*

Each machine can be run 8-hours per day, 5 days per week. Profit on a 8-ounce bottle is 25 paise and on a 16-ounce bottle is 35 paise. Weekly production of the drink cannot exceed 3,00,000 ounces and the market can absorb 25,000 8-ounce bottles and 7,000 16-ounce bottles per week. The planner wishes to maximize his product subject, of course, to all the production and marketing restrictions. Formulate this as a linear programming problem. **[M.Sc. (Math.), CCS Univ. 1998]**

Formulation of Linear Programming Model

The data of the problem is summarized as follows:

Resource/Constraint	*Production*		*Availability*
	8-ounce bottle	*16-ounce bottle*	
Machine *A* time	100/minute	40/minute	8 × 5 × 60 = 2400 minutes
Machine *B* time	60/minute	75/minute	8 × 5 × 60 = 2400 minutes
Production	1	1	3,00,000 ounces/week
Marketing	1	—	25,000 units/week
	—	1	7,000 units/week
Profit/unit (₹)	0.25	0.35	

Step 1. The key decision to be made is to determine the number of bottles (*8-ounce* and *16-ounce*) to be produced per week. Let x_1 and x_2 be number of *8-ounce* and *16-ounce* bottles, respectively, produced per week.

Step 2. Feasible alternatives are the sets of values of $x_1 \geq 0$ and $x_2 \geq 0$.

Step 3. Constraints are on the availability of machine time and production.

(*i*) *Machine-time constraints.* An *8-ounce* bottle takes 1/100 minutes on machine *A* and 1/60 minutes on machine *B*, while a *16-ounce* bottle takes 1/40 minutes on machine *A* and 1/75 minutes on machine *B*. Since both the machines can run 8 hours per day for 5 days per week, the time available on both the machines is 2,400 minutes per week individually. Thus, the machine time constraints are:

$$\frac{x_1}{100} + \frac{x_2}{40} \leq 2,400 \qquad (\textit{Machine A})$$

$$\frac{x_1}{60} + \frac{x_2}{75} \leq 2,400 \qquad (\textit{Machine B})$$

(*ii*) *Production constraints.* It is given that the weekly production of the drink should not exceed 3,00,000 ounces and the market can absorb only up to 25,000 (*8-ounce* botles) and 7,000 (*16-ounce* bottles) per week. Therefore, the two production constraints are:

$$8x_1 + 16x_2 \leq 3,00,000 \qquad (\textit{Production})$$

$$x_1 \leq 25,000 \text{ and } x_2 \leq 7,000 \qquad (\textit{Market})$$

Step 4. The objective is to maximize the total profit, *viz.*, $0.25x_1 + 0.35x_2$.

The linear programming problem, therefore, can be put in the following linear programming model:

$$\text{Maximize } z = 0.25x_1 + 0.35x_2 \text{ subject to the constraints:}$$

$$4x_1 + 10x_2 \leq 9,60,000$$

$$5x_1 + 4x_2 \leq 7,20,000.$$

$$8x_1 + 16x_2 \leq 3,00,000$$

$$x_1 \leq 25,000 \text{ and } x_2 \leq 7,000$$

$$x_1 \geq 0 \text{ and } x_2 \geq 0$$

Example 2.4 (*Product Allocation Problem*). *An electronics company produces three types of parts for automatic washing machine. It purchases casting of the parts from a local foundry and then finishes the part of drilling, shaping and polishing machines.*

The selling prices of part A, B and C, respectively are ₹8, ₹10 and ₹14. All parts made can be sold. Castings for parts A, B and C costs ₹5, ₹6 and ₹10, respectively.

The shop possesses only one of each type of machine. Costs per hour to run each of the three machines are ₹20 for drilling, ₹30 for shaping and ₹30 for polishing. The capacities (per hour) for each part on each machine are shown in the following table:

Machine	*Capacity per hour*		
	Part A	*Part B*	*Part C*
Drilling	25	40	25
Shaping	25	20	20
Polishing	40	30	40

The management of the shop wants to know how many parts of each type it should produce per hour in order to maximize profit for an hour's run. Formulate this problem as an Linear Programming model so as to maximize total profit to the company. **[MBA (Nov.), Delhi Univ. 2007]**

Formulation of Linear Programming Model

Step 1. The key decision is to produce the three type of parts for the automatic washing machine.

Step 2. Decision variables: Let x_1, x_2 and x_3 be number of type *A*, *B* and *C* parts to be produced per hour, respectively.

Step 3. Feasible alternatives: $x_1 \geq 0$, $x_2 \geq 0$ and $x_3 \geq 0$.

Step 4. Profit must allow not only for the cost of the casting but for the cost of drilling, shaping, and polishing also. Since, 25 type *A* parts per hour can be run on the drilling machine at a cost of ₹ 20, then ₹ 20 × 1/25 = ₹ 0.80 is the drilling cost per type *A* part. Similar reasoning for shaping and polishing gives

$$\text{Profit per type } A \text{ part} = (8-5)-\left(\frac{20}{25}+\frac{30}{25}+\frac{30}{40}\right)=3-2.75=0.25$$

$$\text{Profit per type } B \text{ part} = (10-6)-\left(\frac{20}{40}+\frac{30}{20}+\frac{30}{30}\right)=4-3.00=1.00$$

$$\text{Profit per type } C \text{ part} = (14-10)-\left(\frac{20}{25}+\frac{30}{20}+\frac{30}{40}\right)=4-3.05=0.95$$

The objective is to maximize the total profit from sales, *viz.*, $0.25x_1 + x_2 + 0.95x_3$

Step 5. The constraints are on the time. All the three parts are to be processed on each of the three machines. On the drilling-machine, one type *A* part consumes (1/25)th of the available hour, a type *B* part consumes (1/40)th, and type *C* part consumes (1/25)th of an hour. Thus, the drilling-machine constraint is

$$\frac{x_1}{25}+\frac{x_2}{40}+\frac{x_3}{25} \leq 1, \quad \textit{i.e.,} \quad 0.04x_1 + 0.025x_2 + 0.04x_3 \leq 1 \qquad (\textit{Drilling-machine})$$

Similarly, the other two constrains are:

$$\frac{1}{25}x_1+\frac{1}{20}x_2+\frac{1}{20}x_3 \leq 1, \quad \textit{i.e.,} \quad 0.04x_1 + 0.05x_2 + 0.05x_3 \leq 1 \qquad (\textit{Shaping-machine})$$

and $$\frac{1}{40}x_1+\frac{1}{30}x_2+\frac{1}{40}x_3 \leq 1, \quad \textit{i.e.,} \quad 0.025x_1 + 0.033x_2 + 0.025x_3 \leq 1 \qquad (\textit{Polishing-machine})$$

The linear programming problem, therefore, can be put in the following linear programming format:

Maximize $z = 0.25x_1 + x_2 + 0.95x_3$ subject to the constraints:

$$0.04x_1 + 0.025x_2 + 0.04x_3 \leq 1,$$
$$0.04x_1 + 0.05x_2 + 0.05x_3 \leq 1,$$
$$0.025x_1 + 0.033x_2 + 0.025x_3 \leq 1,$$

and $$x_1 \geq 0, \quad x_2 \geq 0, \quad x_3 \geq 0.$$

Example 2.5 (*Blending Problem*). *The manager of an oil refinery must decide on the optimum mix of two possible blending processes of which the input and output production runs are as follows:*

Process	*Input*		*Output*	
	Crude A	*Crude B*	*Gasoline X*	*Gasoline Y*
1	6	4	6	9
2	5	6	5	5

The maximum amounts available of crudes A and B are 250 units and 200 units respectively. Market demand shows that at least 150 units of gasoline X and 130 units of gasoline Y must be produced. The profits per production run from process 1 and process 2 are ₹4 and ₹5, respectively. Formulate the problem for maximizing the profit.

Formulation of Linear Programming Model

***Step* 1.** The key decision is to determine the number of units of gasoline produced from process *1* and process 2.

***Step* 2.** Decision variables: Let x_1, x_2 be number of units of gasoline produced from process *1* and 2, respectively.

***Step* 3.** Feasible alternatives: $x_1 \geq 0$ and $x_2 \geq 0$.

***Step* 4.** The objective is to maximize the total profit from the production of gasoline, *viz.*, $4x_1 + 5x_2$.

***Step* 5.** The constraints are on the availability of crude oil and demand of crude oil, *viz.*,

$$6x_1 + 5x_2 \leq 250 \quad \text{(Availability of crude A)}$$
$$4x_1 + 6x_2 \leq 200 \quad \text{(Availability of crude B)}$$
$$6x_1 + 5x_2 \geq 150 \quad \text{(Demand of gasoline X)}$$
and
$$9x_1 + 5x_2 \geq 130 \quad \text{(Demand of gasoline Y)}$$

The required linear programming problem, therefore, is

$$\text{Maximize } z = 4x_1 + 5x_2 \text{ subject to the constraints:}$$
$$6x_1 + 5x_2 \leq 250, \quad 4x_1 + 6x_2 \leq 200,$$
$$6x_1 + 5x_2 \geq 150, \quad 9x_1 + 5x_2 \geq 130,$$
$$x_1 \geq 0 \text{ and } x_2 \geq 0.$$

Example 2.6 (*Agriculturist Problem*). *An agriculturist has a farm with 126 acres. He produces Radish, Muttar and Potato. Whatever he produces is fully sold in the market. He gets ₹5 for Radish per kg, ₹4 for Muttar per kg and ₹5 for Potato per kg. The average yield is 1,500 kg of Radish per acre, 1,800 kg of Muttar per acre and 1,200 kg of Potato per acre. To produce each 100 kg of Radish and Muttar, and to produce 80 kg of Potato, a sum of ₹12.50 has to be used for manure. Labour required for each acre to raise the crop is 6 man-days for Radish and Potato each and 5 man-days for Muttar. A total of 500 man-days of labour at a rate of ₹40 per man-day are available. Formulate this as a Linear Programming model to maximize the Agriculturist's total profit.* ***[MBA, Visvesvarayya Univ. 2011]***

Formulation of Linear Programming Model

***Step* 1.** The key decision is to determine the number of acres allotted for cultivating Radish, Muttar and Potato, respectively.

***Step* 2.** Let x_1, x_2 and x_3 be number of acres of three respective crops.

***Step* 3.** Feasible alternatives are the sets of values of x_1, x_2 and x_3 satisfying $x_1 \geq 0$, $x_2 \geq 0$ and $x_3 \geq 0$.

***Step* 4.** For objective function, the working table (in ₹) is as under:

	Radish	*Muttar*	*Potato*
Selling price	5 × 1,500 = 7,500	4 × 1,800 = 7,200	5 × 1,200 = 6,000
Manure cost	12.5 × 15 = 187.50	12.50 × 18 = 225	12.50 × 120/8 = 187.50
Labour cost	40 × 6 = 240	40 × 5 = 200	40 × 6 = 240
Profit	7,500 – 427.50 = 7,072.50	7,200 – 425 = 6,775	6,000 – 427.50 = 5,572.50

Using the information given in the above table, the objective function is to maximize

$$z = 7{,}072.5x_1 + 6{,}775x_2 + 5{,}572.5x_3$$

Step 5. The constraints on the availability of land and man-days respectively are:

$$x_1 + x_2 + x_3 \le 125 \quad \text{and} \quad 6x_1 + 5x_2 + 6x_3 \le 500$$

The required linear programming problem, therefore, is

$$\text{Maximize } z = 7{,}072.5x_1 + 6{,}775x_2 + 5{,}572.5x_3 \text{ subject to the constraints:}$$

$$x_1 + x_2 + x_3 \le 125,$$

$$6x_1 + 5x_2 + 6x_3 \le 500,$$

$$x_1 \ge 0, \quad x_2 \ge 0 \quad \text{and} \quad x_3 \ge 0$$

Example 2.7 (*Blending Problem*). *Three grades of coal A, B and C contain ash and phosphorus as impurities. In a particular industrial process a fuel obtained by blending the above grades containing not more than 25% ash and 0.03% phosphorus is required. The maximum demand of the fuel is 100 tons. Percentage impurities and costs of the various grades of coal are shown below. Assuming that there is an unlimited supply of each grade of coal and there is no loss in blending, formulate the blending problem to minimize the cost.*

Coal grade	*% ash*	*% phosphorus*	*Cost per ton (in ₹)*
A	30	0.02	240
B	20	0.04	300
C	35	0.03	280

[MBA, Pune Univ. 2009]

Formulation of Linear Programming Model

Decision variables: Let x_1 = tons of grade *A* coal,

x_2 = tons of grade *B* coal,

and x_3 = tons of grade C coal.

Objective function: Minimize $z = 240x_1 + 300x_2 + 280x_3$

Constraints: $0.3x_1 + 0.2x_2 + 0.35x_3 \le 0.25\,(x_1 + x_2 + x_3)$

or $x_1 - x_2 + 2x_3 \le 0$ *(Ash)*

$$\frac{0.02}{100}x_1 + \frac{0.04}{100}x_2 + \frac{0.03}{100}x_3 \le \frac{0.03}{100}(x_1 + x_2 + x_3)$$

or $-x_1 + x_2 \le 0$ *(Phosphorus)*

$x_1 + x_2 + x_3 \le 100$ *(Demand of fuel)*

$x_1 \ge 0, \quad x_2 \ge 0 \quad \text{and} \quad x_3 \ge 0$ *(Non-negative restrictions)*

Example 2.8 (*Advertising/Media Selection Problem*). *The marketing department of Everest company has collected informaιon on the problem of advertising for its products. This relates to the advertising media available, the number of families expected to be reached with each alternative, cost per advertisement, the maximum availability of each medium and the expected exposure of each one (Measured as the relative value of one advertisement in each of the media). The information is given as under:*

Advertising media	*No. of families to cover*	*Cost/ad. (₹)*	*Maximum availability (No. of times)*	*Expected exposure (Units)*
TV (30 sec.)	3,000	8,000	8	80
Radio (15 sec.)	7,000	3,000	30	20
Sunday Edition (1/4 page)	5,000	4,000	4	50
Magazine (1 page)	2,000	3,000	2	60

Other information and requirements:

(*a*) *The advertising budget is ₹70,000.* (*b*) *At least 40,000 families should be covered.*

(*c*) *At least 2 insertions be given in Sunday Edition of Daily, but not more than 4 advertisements should be given on the TV.*

Formulate this as a linear programming problem. The company's objective is to maximize the expected exposure.

Formulation of Linear Programming Model

The key decision to be made is to determine the number of advertisements to be brought in Television, Radio, Sunday edition and Magazine.

Decision variables. Let x_1, x_2, x_3 and x_4 represent the number of advertisements in TV, Radio, Sunday Edition and Magazine, respectively.

Using the given information, the appropriate linear programming problem is:

$$\text{Maximize (total expected exposure)} \quad z = 80x_1 + 20x_2 + 50x_3 + 60x_3$$

subject to the constraints:

$$3{,}000x_1 + 7{,}000x_2 + 5{,}000x_3 + 2{,}000x_4 \geq 40{,}000 \qquad \textit{(Families)}$$

$$8{,}000x_1 + 3{,}000x_2 + 4{,}000x_3 + 3{,}000x_4 \geq 70{,}000 \qquad \textit{(Budget)}$$

$$x_1 \leq 8, \quad x_2 \leq 30, \quad x_3 \leq 4, \quad x_4 \leq 2 \qquad \textit{(Availability)}$$

$$x_3 \geq 2 \quad \text{and} \quad x_1 \leq 4 \qquad \textit{(No. of advt.)}$$

Also, $$x_1 \geq 0, \quad x_2 \geq 0, x_3 \geq 0 \quad \text{and} \quad x_4 \geq 0 \qquad \textit{(Non-negative restrictions)}$$

Example 2.9 (*Crop and Livestock Problem*). *A co-operative farm owns 100 acres of land and has ₹25,000 in funds available for investment. The farm members can produce a total of 3,500 man-hours worth of labour during the months September–May and 4,000 man-hours during June–August. If any of these man-hours are not needed, some members of the farm will use them to work on a neighbouring farm for ₹2/hour during September–May and ₹3/hour during June–August. Cash income can be obtained from the three main crops and two types of livestock: dairy cows and laying hens. No investment funds are needed for the crops. However, each cow will require an investment outlay of ₹3,200 and each hen will require ₹15.*

Moreover, each cow will require 1.5 acres of land, 100 man-hours of work during September–May and another 50 man-hours during the summer. Each cow will produce a net annual cash income of ₹3,500 for the farm. The corresponding figures for each hen are: no acreage, 0.6 man-hours during September–May. 0.4 man-hours during June–August, and an annual net cash income of ₹200. The chicken house can accommodate a maximum of 4,000 hens and the size of the cattle-shed limits the numbers to a maximum of 32 cows.

Estimated man-hours and income per acre planted in each of the three crops are:

	Paddy	*Bajra*	*Jowar*
Man-hours:			
September–May	40	20	25
June–August	50	35	40
Net annual cash income (₹)	1,200	800	850

The co-operative farm wishes to determine how much acreage should be planted of each of the crops and how many cows and hens should be kept to maximize its net cash income.

Formulate this problem as an LPP to maximize net annual cash income.

Formulation of Linear Programming Model

The data of the problem is summarized below:

Constraints	*Cows*	*Hens*	*Crop*			*Extra hours*		*Total availability*
			Paddy	*Bajra*	*Jowar*	*Sept.–May*	*June–Aug.*	
Man-hours:								
Sept.–May	100	0.6	40	20	25	1	—	3,500
June–Aug.	50	0.4	50	35	40	—	1	4,000
Land	1.5	—	1	1	1	—	—	100
Cows	1	—	—	—	—	—	—	32
Hens	—	1	—	—	—	—	—	4,000
Net annual cash income (₹)	3,500	200	1,200	800	850	2	3	

Decision variables: Let x_1 and x_2 = number of cows and hens respectively.

x_3, x_4 and x_5 = acreage planting of paddy, bajra and jowar, respectively.

x_6 and x_7 = extra man-hours utilized during Sept.–May and May–June, respectively.

Objective function: Maximization of net annual cash, *viz.*,

$$z = 3{,}500x_1 + 200x_2 + 1{,}200x_3 + 800x_4 + 850x_5 + 2x_6 + 3x_7$$

Constraints:

(*i*) $$\left.\begin{aligned} 100x_1 + 0.6x_2 + 40x_3 + 20x_4 + 25x_5 + x_6 = 3{,}500 \\ 50x_1 + 0.4x_2 + 50x_3 + 35x_4 + 40x_5 + x_7 = 4{,}000 \end{aligned}\right\} \quad \text{(Man-hours)}$$

(*ii*) $$1.5x_1 + x_3 + x_4 + x_5 \le 100 \quad \text{(Land availability)}$$

(*iii*) $$x_1 \le 32 \quad \text{and} \quad x_2 \le 4{,}000 \quad \text{(Livestock constraints)}$$

$$x_1 \ge 0,\; x_2 \ge 0,\; x_3 \ge 0,\; x_4 \ge 0,\; x_5 \ge 0,\; x_6 \ge 0 \text{ and } x_7 \ge 0 \quad \text{(Non-negative restrictions)}$$

Example 2.10 (*Trim Loss Problem*). *Rolls of paper having a fixed length and width of 180 cm are being manufactured by a paper mill. The rolls have to be cut to satisfy the following demand:*

Width :	*80 cm*	*45 cm*	*27 cm*
No. of rolls :	*200*	*120*	*130*

Obtain the linear programming formulation of the problem to determine the cutting pattern, so that the demand is satisfied and wastage of paper is minimum.

Formulation of Linear Programming Model

Here, the key decision is to determine how the paper rolls be cut to the required width so that trim loss (wastage) is minimum. Various alternatives for the number of rolls are given below:

Feasible patterns of cutting	*No. of rolls cut*	*Wastage per roll*	*Rolls obtained from each roll of width*		
			80 cm	*45 cm*	*27 cm*
80 + 80	x_1	20	2	—	—
80 + 45 + 45	x_2	10	1	2	—
80 + 45 + 27 + 27	x_3	1	1	1	2
80 + 27 + 27 + 27	x_4	19	1	—	3
45 + 45 + 45 + 45	x_5	0	—	4	—
45 + 45 + 45 + 27	x_6	18	—	3	1
45 + 45 + 27 + 27 + 27	x_7	9	—	2	3
45 + 27 + 27 + 27 + 27 + 27	x_8	0	—	1	5
27 + 27 + 27 + 27 + 27 + 27	x_9	18	—	—	6

Decision variables: Let x_j ($j = 1, 2, ..., 9$) represent the number of rolls each cutting alternative is to be used.

Objective function: The objective is to minimize the wastage per roll, *i.e.*,

$$\text{Minimize } z = 20x_1 + 10x_2 + x_3 + 19x_4 + 18x_6 + 9x_7 + 18x_9$$

Constraints: Constraints on the availability of rolls and the requirement of desired width of rolls are:

$$2x_1 + x_2 + x_3 + x_4 = 200 \quad \text{(80 cm rolls)}$$
$$2x_2 + x_3 + 4x_5 + 3x_6 + 2x_7 + x_8 = 120 \quad \text{(45 cm rolls)}$$
$$2x_3 + 3x_4 + x_6 + 3x_7 + 5x_8 + 6x_9 = 130 \quad \text{(27 cm rolls)}$$
$$x_j \geq 0; \; j = 1, 2, 3, ..., 9 \quad \text{(Non-negativity)}$$

Example 2.11 (*Transportation Problem*). *A manufacturer of wooden boxes has 1,800 boxes of superior quality at one factory and another 1,400 boxes at its second factory. The manufacturer has orders for this product from three different retailers in quantities of 1,200, 1,100 and 900 boxes, respectively. The unit shipping costs (in rupees per box) from the factories to retailers are given as follows:*

	Retailer 1	*Retailer 2*	*Retailer 3*
Factory 1	15	14	12
Factory 2	12	13	11

Determine a minimum cost shipping schedule for satisfying all demands from current inventory. Formulate the problem as a linear programming model.

Formulation of Linear Programming Model

Decision variables: Let x_{ij} be the number of boxes shipped from factory i ($i = 1, 2$) to retailer j ($j = 1, 2, 3$), *i.e.*, let

x_{11} = Number of boxes shipped from Factory *1* to Retailer *1*

We let the first subscript represent the origin (factory) and the second subscript the destination (retailer). Thus, in general, x_{ij} refers to the number of boxes shipped from origin i to destination j. Thus,

x_{12} = Number of boxes shipped from Factory *1* to Retailer 2.
x_{13} = Number of boxes shipped from Factory *1* to Retailer 3.
x_{21} = Number of boxes shipped from Factory 2 to Retailer *1*.
x_{22} = Number of boxes shipped from Factory 2 to Retailer 2.
x_{23} = Number of boxes shipped from Factory 2 to Retailer 3.

Objective function: The objective is to minimize the total shipping cost from factories to retailers, *i.e.*, to minimize the linear expression

$$z = (15x_{11} + 14x_{12} + 12x_{13}) + (12x_{21} + 13x_{22} + 11x_{23})$$

Constraints: The constraints for shipping boxes from factories to retailers are:

(*i*) Supply constraints:
$$x_{11} + x_{12} + x_{13} \leq 1{,}800$$
$$x_{21} + x_{22} + x_{23} \leq 1{,}400$$

(*ii*) Demand constraints:
$$x_{11} + x_{21} = 1{,}200$$
$$x_{12} + x_{22} = 1{,}100$$
$$x_{13} + x_{23} = 900$$

(*iii*) Non-negative restrictions are: $x_{ij} \geq 0 \quad (i = 1, 2 \text{ and } j = 1, 2, 3)$

Example 2.12 (*Transportation Problem*). *ABC manufacturing company wishes to develop a monthly production schedule for the next three months. Depending upon the sales commitments, the company can either keep the production constant, allowing fluctuation in inventory or inventories can be maintained at a constant level, with fluctuating production. Fluctuating production makes overtime work necessary, the cost of which is estimated to be double the normal production cost of ₹12 per unit. Fluctuating inventories result in an inventory carrying cost of ₹2 per unit/month. If the company fails to fulfil its sales commitment, it incurs a shortage cost of ₹4 per unit/month. The production capacities for the next three months are shown below:*

Month	*Production capacity*		*Sales*
	Regular	*Overtime*	
1	50	30	60
2	50	0	120
3	60	30	40

Formulate this problem as an LP model so as to minimize the total cost. **[MBA, Delhi Univ. 2008]**

Formulation of Linear Programming Model

We are given the following information:

Normal production cost: ₹ 12 per unit, Overtime cost: ₹ 24 per unit.
Inventory carrying cost: ₹ 2 per unit per month, Shortage cost: ₹ 4 per unit per month.

Let there be five sources of supply — three regular and two overtime (because second month overtime production is zero) production capacities. The demand for the three months will be the sales during these months.

It is assumed that all supplies against order have to be fulfilled and can be made in the subsequent month, if not possible during the current month of order with additional cost equivalent to shortage cost, *i.e.*, in month 2. The cumulative production of first two months in regular and overtime is of 130 units, while the orders are for 180 units. The balance can be supplied during month 3 at an additional production cost of ₹ 4.

The total cost (normal cost + inventory carrying cost + shortage cost, if any) is given below:

Month	*Cost (₹)*			*Overtime cost (₹)*		*Sales*
	Month 1	*Month 2*	*Month 3*	*Month 1*	*Month 3*	
Month 1	12	16	20	24	32	60
Month 2	14	12	16	26	28	120
Month 3	16	14	12	28	24	40
Supply (units)	50	50	60	30	30	220

Decision variables: Let x_{ij} represents the number of units to be shipped from source of supply i ($i = 1, 2, 3$) to destination j ($j = 1, 2, 3, 4, 5$).

Objective function: The objective is to minimize the total cost involved, *i.e.*, to minimize the expression

$$z\,\{(12x_{11} + 16x_{12} + 20x_{13} + 24x_{14} + 32x_{15}) + (14x_{21} + 12x_{22} + 16x_{23} + 26x_{24} + 28x_{25}) + (16x_{31} + 14x_{32} + 12x_{33} + 28x_{34} + 24x_{35})\}$$

Constraints: The constraints on demand and supply are:

(*i*) $x_{11} + x_{12} + x_{13} + x_{14} + x_{15} = 60,$

$x_{21} + x_{22} + x_{23} + x_{24} + x_{25} = 120,$

and $x_{31} + x_{32} + x_{33} + x_{34} + x_{35} = 40.$

(*ii*) $x_{11} + x_{21} + x_{31} = 50,\ \ x_{12} + x_{22} + x_{32} = 50,$

$x_{13} + x_{23} + x_{33} = 60,\ \ x_{14} + x_{24} + x_{34} = 30,$

$x_{15} + x_{25} + x_{35} = 30,$

(*iii*) $x_{ij} \geq 0$ for all i and j (*Non-negative restrictions*)

Example 2.13 (*Manpower Scheduling Problem*). *A city hospital has the following minimal daily requirements for nurses:*

Period	*Clock time (24 hour day)*	*Minimal number of nurses required*
1	6 A.M. — 10 A.M.	2
2	10 A.M. — 2 P.M.	7
3	2 P.M. — 6 P.M.	15
4	6 P.M. — 10 P.M.	8
5	10 P.M. — 2 A.M.	20
6	2 A.M. — 6 A.M.	6

Nurses report to the hospital at the beginning of each period and work for 8 consecutive hours. The hospital wants to determine the minimal number of nurses to be employed so that there will be sufficient number of nurses available for each period. Formulate this problem as a linear programming model by setting up appropriate constraints and objective function. ***[MBA, Delhi Univ. 2002; B.Com. (Hons.), Delhi Univ. 2006]***

Formulation of Linear Programming Model

The *key decision* to be made is to determine the number of nurses to be employed in the city hospital.

Decision variables: Let x_j $(j = 1, 2, 3, 4, 5, 6)$ be the number of nurses reporting to work on shift j. Also, since negative values of x_j are meaningless, we must have $x_j \geq 0$ for all j.

Objective function: If the number of nurses reporting to work on shifts 1, 2, 3, ..., 6 is known, then the manpower schedule can be decided. Thus, the objective is to minimize the total number of nurses (manpower for all the six periods together). As such, the objective is to minimize the linear function: $z = x_1 + x_2 + x_3 + x_4 + x_5 + x_6$.

Constraints: In any 8-hour period, the number of nurses reporting at the beginning of the period and the number of nurses continuing from the earlier period should be at least equal to the minimum requirement.

Thus, we have the following constraints:

$$x_1 + x_2 \geq 7, \quad x_2 + x_3 \geq 15, \quad x_3 + x_4 \geq 8,$$
$$x_4 + x_5 \geq 20, \quad x_5 + x_6 \geq 6, \quad x_6 + x_1 \geq 2.$$

∴ The given *linear programming problem* is:

Minimize $z = x_1 + x_2 + x_3 + x_4 + x_5 + x_6$ subject to the constraints:

$$x_1 + x_2 \geq 7, \quad x_2 + x_3 \geq 15, \quad x_3 + x_4 \geq 8,$$
$$x_4 + x_5 \geq 20, \quad x_5 + x_6 \geq 6, \quad x_6 + x_1 \geq 2;$$
$$x_1 \geq 0, x_2 \geq 0, x_3 \geq 0, x_4 \geq 0, x_5 \geq 0, x_6 \geq 0.$$

Example 2.14 (*Cargo Loading Problem*). *A ship has three cargo holds, forward, aft and centre; the capacity limits are:*

Forward	*2,000 tonnes*	*100,000 m^3*
Centre	*3,000 tonnes*	*135,000 m^3*
Aft	*1,500 tonnes*	*30,000 m^3*

The following cargoes are offered; the ship owners may accept all or any part of each commodity:

Commodity	*Amount (tonnes)*	*Value per tonne (m^3)*	*Profit per tonne (₹)*
A	6,000	60	60
B	4,000	50	80
C	2,000	25	50

In order to preserve the trim of the ship, the weight in each hold must be proportional to the capacity in tonnes. The objective is to maximize the profit. Formulate the linear programming model for this problem.

Formulation of Linear Programming Model

Let x_{1A}, x_{2A}, x_{3A} be the weights (in tonne) of the commodity *A* to be accommodated in *forward, centre* and *aft* portions respectively. Similarly let x_{1B}, x_{2B}, x_{3B} and $x_{1C}, x_{2C}\ x_{3C}$ be the corresponding weights (in tonne) of *B* and C. Then the *objective* is to maximize the linear expression:

$$z = 60(x_{1A} + x_{2A} + x_{3A}) + 80(x_{1B} + x_{2B} + x_{3B}) + 50(x_{1C} + x_{2C} + x_{3C})$$

Various constraints are:

$$\left.\begin{array}{r} x_{1A} + x_{2A} + x_{3A} \le 6000 \\ x_{1B} + x_{2B} + x_{3B} \le 4000 \\ x_{1C} + x_{2C} + x_{3C} \le 2000 \end{array}\right\} \quad \text{(Commodity cargo)}$$

$$\left.\begin{array}{r} x_{1A} + x_{1B} + x_{1C} \le 2000 \\ x_{2A} + x_{2B} + x_{2C} \le 3000 \\ x_{3A} + x_{3B} + x_{3C} \le 1500 \end{array}\right\} \quad \text{(Weight capacity)}$$

$$\left.\begin{array}{r} 60x_{1A} + 50x_{1B} + 25x_{1C} \le 1,00,000 \\ 60x_{2A} + 50x_{2B} + 25x_{2C} \le 1,35,000 \\ 60x_{3A} + 50x_{3B} + 25x_{3C} \le \ \ 30,000 \end{array}\right\} \quad \text{(Volume capacity)}$$

$$x_{iA} \ge 0,\ x_{iB} \ge 0 \text{ and } x_{iC} \ge 0 \text{ for } i = 1, 2 \text{ and } 3. \quad \text{(Non-negative restrictions)}$$

Example 2.15 (*Personnel Problem*). *A company engaged in producing tinned food delicacies has 300 trained employees on rolls each of whom can produce one can of food in a week. Due to the developing taste of the public for this kind of food, the company plans to add the existing labour force by employing 150 people, in a phased manner, over the next five weeks. The newcomers would have to undergo a two-week training programme before being put to work. The training is to be given by employees from amongst the existing ones and it is known that one employee can train three trainees. Assume that there would be no production coming forth from the trainers and the trainees during training period as the training is off-the-job. However, the trainees would be remunerated at the rate of ₹ 300 per week, the same rates as is for the trainers.*

The company has booked the following number of cans to supply during the next five weeks.

Week :	1	2	3	4	5
No. of Cans :	280	298	305	360	400

Assume that the production in any week would not be more than the number of cans ordered for so that every delivery of the food delicacy would be 'fresh'.

Using the information, draft a LPP to develop a training schedule that will minimize the labour costs over the five week period. **[MBA, Delhi Univ. 2005]**

Formulation of Linear Programming Model

Information given in the problem is summarized as given below:

Week:	1	2	3	4	5
Cans required	280	298	305	360	400
No. of weeks for which newcomer would be employed*	5	4	3	2	1

Other information:

(*a*) Each trainee to be trained for 2 weeks.
(*b*) One employee required to train 3 trainees.
(*c*) Every trained worker producing 1 cm per week, but no production from trainers and trainees during training.
(*d*) Number of people to be employed = 150.
(*e*) The production in any week not to exceed the cans required.
(*f*) Number of weeks for which new comers would be employed: 5, 4, 3, 2, 1.

Let x_1, x_2, x_3, x_4 and x_5 be the number of trainees appointed in the beginning of week 1, 2, 3, 4 and 5 respectively. Then, the above problem is written as an LPP as follows:

Minimize $z = 5x_1 + 4x_2 + 3x_3 + 2x_4 + x_5$ subject to the constraints:

$$\left.\begin{aligned}300 - \frac{1}{3}x_1 &\geq 280\\ 300 - \frac{1}{3}x_1 - \frac{1}{3}x_2 &\geq 298\\ 300 + x_1 - \frac{1}{3}x_2 - \frac{1}{3}x_3 &\geq 305\\ 300 + x_1 + x_2 - \frac{1}{3}x_3 - \frac{1}{3}x_4 &\geq 360\\ 300 + x_1 + x_2 + x_3 - \frac{1}{3}x_4 - \frac{1}{3}x_5 &\geq 400\end{aligned}\right\} \quad \text{(Capacity constraints)}$$

$$x_1 + x_2 + x_3 + x_4 + x_5 = 150 \quad \text{(New employment condition)}$$

$$x_1, x_2, x_3, x_4, x_5 \geq 0 \quad \text{(Non-negativity restrictions)}$$

Notes

1. All the 300 employees capable of producing 300 cans a week remain employed. The production capacity in different weeks (as shown on LHS of the constraints) is affected (negatively) by the fact that some of them are away for training and is affected (positively) by the fact that newcomers join the team of experienced workers.
2. Inequalities have been used in the constraints because some workers might remain idle in some week(s).
3. The training schedule as obtained by solving this problem would involve minimum cost in respect of the trainees. The cost would be obtained by multiplying the objective function by 300, because each person would get a salary of ₹ 300 per week.

Example 2.16 (*Investment Problem*). *Suppose an investor, on the basis of his experience and his broker's advice, considers various investment alternatives to which he assigns corresponding expectations of yield (income or growth), risk (safety), and average term (liquidity) values as shown in the table on ext page.*

* Workers employed at the beginning of the first week would get salary for all the 5 weeks under consideration, those employed at the beginning of the second week would get salary for 4 weeks, and so on.

Investment alternatives	*Expected annual yields (in ₹)*	*Proportion (%)*of wrong estimate*	*Average term (in years)*
Saving deposits x_1	5	5	1
Income stock x_2	8	10	3
Bonds x_3	4	3	15
Growth stock x_4	15	20	5
Speculative venture x_5	40	40	10
Cash on hand x_6	0	0	0

Let x_1 *be defined as the proportion of the portfolio invested in savings deposits:* x_2, x_3, x_4, x_5 *and* x_6 *be defined in a corresponding manner. Further, it is assumed that the following restrictions will satisfy his investment requirements:*

(*i*) *The investment decision is to be limited to the six investment alternatives,* x_1 *through* x_6, *as listed above.*

(*ii*) *The safety requirement is such that the average risk of the portfolio cannot exceed 15.*

(*iii*) *The average length of the investment should not exceed eight years.*

The investor's problem is to maximize his annual yield by satisfying the above mentioned requirements. Formulate the above problem in a linear programming format.

Formulation of Linear Programming Model

Since x_1, x_2, x_3, x_4, x_5 and x_6 denote the six different investment alternatives, the objective is to maximize the annual yield of the investor (in rupees) given by the linear expression:

$$z = 5x_1 + 8x_2 + 4x_3 + 15x_4 + 40x_5 + 0.0x_6$$

The constraints of the problem are:

(*i*) *Investment Decision:* $x_1 + x_2 + x_3 + x_4 + x_5 + x_6 = 1$

(*ii*) *Average Risk of the Portfolio:* $5x_1 + 10x_2 + 3x_3 + 20x_4 + 40x_5 \leq 15$

(*iii*) *Average Length of the Investment:* $x_1 + 3x_2 + 15x_3 + 5x_4 + 10x_5 \leq 8$

Investment alternatives are assumed to be non-negative, *i.e.*, $x_j \geq 0$ ($j = 1, 2, 3, 4, 5, 6$).

Example 2.17 (*Network Problem*). *Given below is a network in which the figures written against the arrows indicate the cost of travelling of a person (in rupees) from the preceding to the following node.*

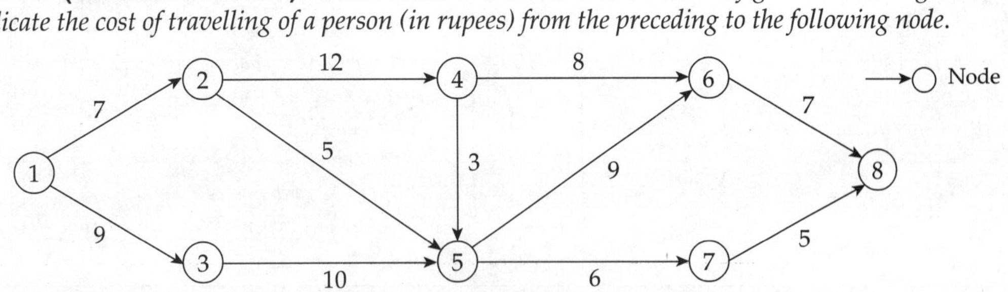

Formulate the linear programming model to find the least cost route from node 1 to node 8.

Formulation of Linear Programming Model

Let the person travel from node *1* to node *8* along a succession of links. Let x_{ij} represent the link joining node *i* to node *j*. If the person uses link x_{ij}, then x_{ij} will be given a value *'1'* and if not then a value zero '0'. The person reaches a node, say *j via* one of the links, say x_{ij} and leaves it by another

* Proportion of wrong estimate implies proportion of actual yield being significantly lower than expected yield.

link, say x_{jk}. So, if a node j has several arrows, such as ij, hj, pj pointing towards it and several arrows, such as jk, jl, jm pointing away from it, then the equation for node j can be written as:

$$x_{hj} + x_{ij} + x_{pj} - x_{jk} - x_{jl} - x_{jm} = 0.$$

This is because, for each of the links x_{ij} and x_{jk}, the value is 1 (the person uses this link) as for each of the links x_{hj}, x_{pj}, x_{jl} and x_{jm} the value is zero (the person does not use these links).

For the network given above, we have the following equations for

Node 2: $x_{12} - x_{24} - x_{25} = 0$

Node 3: $x_{13} - x_{35} = 0$

Node 4: $x_{24} - x_{45} - x_{46} = 0$

Node 5: $x_{25} + x_{45} + x_{35} - x_{56} - x_{57} = 0$

Node 6: $x_{46} + x_{56} - x_{68} = 0$

Node 7: $x_{57} - x_{78} = 0$

Further, since the person has to start from node 1 along one of the links starting from it, we have

$$-x_{12} - x_{13} = -1, \text{ i.e., } x_{12} + x_{13} = 1.$$

Similarly, as the person has to reach node 8 along one of the links, we get

$$x_{68} + x_{78} = 1.$$

Given above are the eight constraints that must be satisfied. The objective is to minimize the total cost of travelling from node 1 to node 8. This is given by

$$\text{Minimize } z = 7x_{12} + 9x_{13} + 12x_{24} + 5x_{25} + 10x_{35} + 8x_{46} + 3x_{45} + 9x_{56} + 6x_{57} + 7x_{68} + 5x_{78}$$

This is the required linear programming model for the network problem wherein every variable has a value 1 or 0.

Example 2.18 (*Funding Problem*). *The most recent audited summarized Balance Sheet of Express Financial Services is given below:*

Balance Sheet *as on March 31, 2019*

Liabilities	*(in lakh ₹)*	*Assets*	*(in lakh ₹)*
Equity Share Capital	65	*Fixed Assets:*	
Reserves and Surplus	110	*Assets on Lease (original cost: ₹ 550 lakh)*	375
Term Loan from IFCI	80	*Other Fixed Assets*	50
Public Deposits	150	*Investments (on wholly owned subsidiaries)*	20
Bank Borrowings	147	*Current Assets:*	
Other Current Liabilities	50	*Stock on Hire*	80
		Receivables	30
		Other Current Assets	35
		Miscellaneous expenditure (not written off)	12
	602		602

The company intends to enhance its investment in the lease portfolio by another ₹1,000 lakh. For this purpose, it would like to raise a mix of debt and equity in such a way that the overall cost of raising additional funds is minimized. The following constraints apply to the way the funds can be mobilized:

1. *Total debt divided by net owner funds, cannot exceed 10.*
2. *Amount borrowed from financial institutions cannot exceed 25% of the net worth.*
3. *Maximum amount of bank borrowings cannot exceed three times the net owned funds.*
4. *The company would like to keep the total public deposit limited to 40% of the total debt.*

The post-tax costs of the different sources of finance are as follows:

Equity	:	25%
Term Loans	:	8.5%
Public Deposits	:	7%
Bank Borrowings	:	10%

Formulate the funding problem as an LP model.

Note. Total Debt = Term Loans from Financial Institutions + Public Deposits + Bank Borrowings
Net Worth = Equity Share Capital + Reserves and Surplus
Net owned Funds = Net Worth – Miscllaneous Expenditures

Formulation of Linear Programming Model

Let x_1, x_2, x_3 and x_4 be the quantity of funds raised by way of equity, term loans, public deposits and bank borrowings respectively. Then, the objective is to minimize the post-tax costs, *i.e.*,

$$\text{Minimize } z = 0.25x_1 + 0.085x_2 + 0.07x_3 + 0.1x_4.$$

Various constraints are:

(*i*) $\dfrac{x_2 + x_3 + x_4 + 80 + 150 + 147}{(65 + 110 + x_1) - 12} \le 10;$ *i.e.*, $-10x_1 + x_2 + x_3 + x_4 \le 1{,}253$

(*ii*) $(80 + x_2) \le 0.25(65 + 110 + x_1);$ *i.e.*, $x_1 - 4x_2 \ge 145$

(*iii*) $(147 + x_4) \le 3(x_1 + 65 + 110 - 12);$ *i.e.*, $-3x_1 + x_4 \le 342$

(*iv*) $(150 + x_3) \le 0.4(x_2 + x_3 + x_4 + 80 + 150 + 147);$ *i.e.*, $-4x_2 + 6x_3 - 4x_4 \le 8$

(*v*) $x_1 + x_2 + x_3 + x_4 = 1{,}000$ and $x_j \ge 0\ (j = 1, 2, 3, 4)$

Example 2.19. *A computer company produces three types of models, which are first required to be machined and then assembled. The time (in hours) for these operations for each model is given below:*

Model	*Machine time*	*Assembly time*
P III	20	5
P II	15	4
P IV	12	3

The total available machine time and assembly time are 1,000 hours and 1,500 hours respectively. The selling price and other variable costs for three models are as follows:

	P III	*P II*	*P IV*
Selling Price (₹)	3,000	5,000	15,000
Labour, Material and other Variable Costs (₹)	2,000	4,000	8,000

The company has taken a loan of ₹50,000 from a Nationalized Bank, which is required to be repaid on 1.4.2020. In addition, the company has borrowed ₹1,00,000 from XYZ Cooperative Bank. However, this bank has given its consent to renew the loan.

The Balance Sheet of the company as on 31.3.2020 is as follows:

Liabilities	₹	*Assets*	₹
Equity Share Capital	1,00,000	*Land*	80,000
Capital Reserve	20,000	*Building*	50,000
Profit and Loss Account	30,000	*Plant and Machinery*	1,00,000
Long-term Loan	2,00,000	*Furniture, etc.*	20,000
Loan from XYZ Cooperative Bank	1,00,000	*Vehicles*	40,000
Loan from Nationalized Bank	50,000	*Cash*	2,10,000
	₹ 5,00,000		₹ 5,00,000

The company is required to pay a sum of ₹ 15,000 towards the salary. Interest on long-term loan is to be paid every month @ 18% per annum. Interest on loan from XYZ Cooperative and Nationalized Banks may be taken as ₹ 1,500 per month. The company has already promised to deliver three P III, two P II and five Celeron type of computers to M/s ABC Ltd. next month. The level of operation in the company is subject to the availability of cash next month.

The company manager is willing to know that how many units of each model must be manufactured next month, so as to maximize the profit.

Formulate a linear programming problem for the above.

Formulation of Linear Programming Model

Let x_1, x_2 and x_3 designate the number of *P III*, *P II* and *P IV* computers respectively to be manufactured by the company. The objective function is to maximize the profit, *viz.*,

$$\begin{aligned} z &= (3{,}000 - 2{,}000)x_1 + (5{,}000 - 4{,}000)x_2 + (15{,}000 - 8{,}000)x_3 \\ &\quad - (\text{Interest on loan from bank} + \text{Interest on long term loans} + \text{Salary to staff}) \\ &= 1{,}000x_1 + 1{,}000x_2 + 7{,}000x_3 - (₹\,15{,}000 + ₹\,3{,}000 + ₹\,1{,}500) \\ &= 1{,}000x_1 + 1{,}000x_2 + 7{,}000x_3 - ₹\,19{,}500 \end{aligned}$$

Constraints are:

(*i*) $20x_1 + 15x_2 + 12x_3 \leq 1{,}000$ (*Machine time constraint*)

(*ii*) $5x_1 + 4x_2 + 3x_3 \leq 1{,}500$ (*Assembly time constraint*)

(*iii*) The cash required for the manufacture of various models of computer should not exceed the cash availability for the next month, *i.e.*,

Cash availability = Cash balance – Loan of repayment to nationalized bank
– Interest on loan from Cooperative and nationalized bank
– Interest on loan term loans – Salary to staff
= ₹ 2,10,000 – ₹ (50,000 + 1,500 + 3,000 + 15,000)
= ₹ 2,10,000 – ₹ 69,500 = ₹ 1,40,500

∴ $2{,}000x_1 + 4{,}000x_2 + 8{,}000x_3 \leq 1{,}40{,}500$ (*Cash constraint*)

(*iv*) $x_1 \geq 3$, $x_2 \geq 2$ and $x_3 \geq 5$ (*Delivery constraint*)

(*v*) x_1, x_2 and x_3 can take only positive integral values.

PROBLEMS

2.01. The XYZ Company Ltd. manufactures two products *A* and *B*, and are processed on the machine. It takes 20 minutes to process one unit of product *A* and 15 minutes for each unit of product *B*. Machine operates for a maximum of 80 hours in a week. Product *A* requires 3 kg and product *B* requires 2 kg of the raw material per unit, the supply of which is 1,200 kg per week. Market constraint on product *B* is known to be 1500 units every week. If product *A* costs ₹ 10 per unit and can be sold at ₹ 15, while product *B* costs ₹ 15 per unit and can be sold in the market at a unit price of ₹ 22, the problem is to find out the number of units of *A* and *B* that should be produced per week in order to maximize the profit. Formulate the problem as a linear programming model. ***[M.Com., Karnataka Univ. 2013]***

2.02. A firm manufactures headache pills in two sizes *A* and *B*. Size *A* contains 2 grains of aspirin, 5 grains of bicarbonate and 1 grain of codeine. Size *B* contains 1 grain of aspirin, 8 grains of bicarbonate and 6 grains of codeine. It is found by users that it requires at least 12 grains of aspirin, 74 grains of bicarbonate and 24 grains of codeine for providing immediate effect. It is required to determine the least number of pills a patient should take to get immediate relief. Formulate the problem as a standard LPP. ***[MBA, GGSIP Univ. 2011]***

2.03. Old hens can be bought for ₹ 20 each and young ones cost ₹ 50 each. The old hens lay 3 eggs per week and the young ones, 5 eggs per week, each being worth ₹ 3. A hen costs ₹ 10 per week to feed. If I have only ₹ 800 to spend for hens, how many of each kind should I buy to give a profit of more than ₹ 60 per week, assuming that I cannot house more than 20 hens? Write a linear programming model of the problem. ***[BE (Jan.), Visvesvaraya Tech. Univ. 2020]***

2.04. An advertising company wishes to plan an advertising campaign in three different media — television, radio and magazine. The purpose of the advertising is to reach as many potential customers as possible. Results of a market study are given below (in ₹):

	Television		*Radio*	*Magazine*
	Prime time	*Normal time*		
Cost of an advertising unit	40,000	75,000	30,000	15,000
Number of potential customers reached per unit	4,00,000	9,00,000	5,00,000	2,00,000
Number of women customers reached per unit	3,00,000	4,00,000	2,00,000	1,00,000

The company does not want to spend more than ₹ 8 lakh on advertising. It is further required that

(*i*) at least 2 million exposures take place among women;
(*ii*) advertising on television be limited to ₹ 5,00,000;
(*iii*) at least 3 advertising units be bought in prime time and two units during normal time;
(*iv*) the number of advertising units on radio and magazine should each be between 5 and 10.

2.05. Formulate the following linear programming problem. Do not solve.

A publishing house publishes three weekly magazines — *Daily Life, Agriculture Today* and *Surf's Up*. Publication of one issue of each of the magazines require the following amounts of production time and paper:

Magazine	*Production (hour)*	*Paper (kg)*
Daily Life	0.01	0.2
Agriculture Today	0.03	0.5
Surf's Up	0.02	0.3

Each week the publisher has available 120 hours of production time and 3,000 kg of paper. Total circulation for all three magazines must exceed, 5,000 issues per week, if the company is to keep its advertisers. The selling price per issue is ₹ 22.50 for *Daily Life*, ₹ 40.00 for *Agriculture Today*, and ₹ 15.00 for *Surf's Up*. Based on past sales, the publisher knows that the maximum weekly demand for *Daily Life* is 3,000 issues; for *Agriculture Today*, 2,000 issues; and for *Surf's Up*, 6,000 issues. The production manager wants to know the number of issues of each magazine to produce weekly in order to maximize total sales revenue.

[MBA (Nov.), Delhi Univ. 2008]

2.06. The ABC Company manufacture three lines of washing machines: Manual, Semi-automatic and Automatic, which sells for ₹ 5,000, ₹ 10,000 and ₹ 25,000 respectively. The marketing department estimates that at least 2,500 Manual machines, 4,000 Semi-automatic machines, and 1,500 Automatic machines needs to be produced each week.

Producing each washing machine requires a certain amount of time to (1) manufacture the body parts; (2) assemble the parts (some components are purchased from outside sources and can be ignored in the production scheduling decision); and (3) inspect, test, and package the final product. The following table shows the requirements (in hours) for each line of washing machine. There are 4,000 hours of manufacturing time, 5,000 hours of assembly time, and 1,500 hours of time to inspect, test, and package per week.

Washing machine	*Manufacture*	*Assemble*	*Test*
Manual	0.2	0.4	0.1
Semi-Automatic	0.3	0.4	0.2
Automatic	0.6	0.1	0.3

ABC Company has decided that, in addition to the minimum production levels of each type of washing machine, Automatic machines should make up at least 25% of the total number of machines produced each week.

Formulate (do not solve) this problem as a linear programming problem.

[MBA (PT), Delhi Univ. 2015]

2.07. A public limited company is planning its capital structure that will consist of equity capital, 15% debentures and term loan. Debentures are to be repaid on face value, interest rate is payable half yearly and annualized cost of issue of debenture is 0.5%. Interest on term-loan is 18% p.a. to be paid annually, while the cost of equity is estimated at 20%. It is decided not to have outsiders funds more than 2 times of equity funds; also the amount of term-loan must be at least 50% of the debenture amount. Formulate a suitable linear programming model so as to minimize average cost of capital of the company.

2.08. In a chemical industry, two products *A* and *B* are made involving two operations. The production of *B* also results in a by-product C. The product *A* can be sold at a profit of ₹ 3 per unit and *B* at a profit of ₹ 8 per unit. The by-product C has a profit of ₹ 2 per unit. Forecasts show that up to 5 units of C can be sold. The company gets 3 units of C for each unit of *B* produced. The manufacturing times are 3 hours per unit for *A* on each of the operation one and two, 4 hours and 5 hours per unit for *B* on operation one and two respectively. Because the product C results from producing *B*, no time is used in producing C. The available times are 18 hours and 21 hours of operation one and two respectively. The company desires to know as to how much *A* and *B* should be produced keeping C in mind to make the highest profit. Formulate LP model for this problem.

2.09. Dr. Shilpa Soni, the head administrator at XYZ Hospital must determine a schedule for nurses to make sure that there are enough nurses on duty throughout the day. During the day, the demand for nurses varies. Dr. Shilpa has broken the day into 12 two-hour periods. The slowest time of the day encompasses the three periods from 12:00 mid-night to 6:00 A.M., which beginning at mid-night require a minimum of 30, 20 and 40 nurses, respectively.

The demand for nurses steadily increases during the next four day time periods. Beginning with the 6:00 A.M. to 8:00 A.M. period, a minimum of 50, 60, 80 and 80 nurses are required for these four periods, respectively. After 2:00 P.M. the demand for nurses decreases during the afternoon and evening hours. For the five two-hour periods beginning at 2:00 P.M. and ending at midnight 70, 70, 60, 50 and 50 nurses are required, respectively. A nurse reports for duty at the beginning of one of the two hour periods and works eight consecutive hours (which is required in the nurses contract). Dr. Shilpa Soni wants to determine a nursing schedule that will meet the hospital's minimum requirements throughout the day, while using the minimum number of nurses.

[MBA (Nov.), Delhi Univ. 2008]

2.10. A media specialist plans to allocate advertising expenditure in three media whose unit costs of a message are ₹ 7,500, ₹ 6,250 and ₹ 5,000 respectively. The total advertising budget available for the year is ₹ 2,50,000. The first medium is monthly magazine and it is desired to advertise not more than once in one issue. At least five advertisements should appear in the second medium and the number of advertisements in the third medium should strictly lie between 6 and 10. The effective audience for unit advertisement in the three media is given below:

Medium :	*I*	*II*	*III*
Expected affective audience :	1,00,000	80,000	50,000

Formulate a linear programming problem to find the optimum allocation of advertisements in three media that would maximize the total effective audience.

[PG Dip. in Glob. Bus. Oper., Delhi Univ. 2011]

2.11. Evening shift resident doctors in a government hospital work for five consecutive days and have two consecutive days off. Their five days work can start on any day of the week and the schedule rotates indefinitely. The hospital requires the following minimum number of doctors working:

Sun.	*Mon.*	*Tues.*	*Wed.*	*Thurs.*	*Fri.*	*Sat.*
35	55	60	50	60	50	45

Not more than 40 doctors can start their five working days on the same day. Formulate the general linear programming model to minimize the number of doctors employed by the hospital.

2.12. A city hospital has the following minimal daily requirements for doctors:

Period	*Clock time (24-hour day)*	*Minimal number of doctors required*
1	6 A.M. — 10 A.M.	72
2	10 A.M. — 2 P.M.	77
3	2 P.M. — 6 P.M.	85
4	6 P.M. — 10 P.M.	68
5	10 P.M. — 2 A.M.	25
6	2 A.M. — 6 A.M.	23

Doctors report to the hospital at the beginning of each period and work for 8 consecutive hours. The hospital wants to determine the minimum number of doctors to be employed so that there will be sufficient number of doctors available for each period. Formulate this as a Linear Programming Problem by setting up appropriate constraints and objective function. Do not solve the LPP. ***[MBA, Lucknow Univ. 2008]***

2.13. A 24-hour BPO company has the following minimal requirements for IT personnel and software engineers:

Time of day	*Minimum number of IT personnel and software engineers required*
Midnight — 4 A.M.	7
4 A.M. — 8 A.M.	20
8 A.M. — Noon	14
Noon — 4 P.M.	20
4 P.M. — 8 P.M.	10
8 P.M. — Midnight	5

Shift 1 follows immediately after shift 6. The BPO employee works eight consecutive hours, starting at the beginning of one of the six periods. The company wants to determine how many IT experts should work each shift in order to minimize the total number of persons employed, while still satisfying the staffing requirements.

Formulate the problem as a linear programming problem.

[PG Dip. in Glob. Bus. Oper., Delhi Univ. 2010]

2.14. A company manufactures three grades of paints — *Venus, Diana* and *Aurora*. The plant operates on a three-shift basis and the following data are available from the production records:

Requirement of resource	*Grade*			*Availability (capacity per month)*
	Venus	*Diana*	*Aurora*	
Special additive (kgs per litre)	0.30	0.15	0.75	600 tonnes
Milling (kilolitres per machine shift)	2.00	3.00	5.00	100 machine shifts
Packing (kilolitres per shift)	12.00	12.00	12.00	80 shifts

There are no limitations on other resources. The particulars of sale forecasts and estimated contribution to overheads and profits are given below:

	Venus	*Diana*	*Aurora*
Maximum possible sales per month (in kilolitres)	100	400	600
Contribution (₹ per kilolitre)	4,000	3,500	2,000

Due to commitments already made, a minimum of 200 kilolitres per month of *Aurora* has to be necessarily supplied during the next year.

Just as the company was able to finalize the monthly production programme for the next 12 months, an offer was received from a nearby competitor for hiring 40 machine shifts per month of milling capacity for grinding *Diana* paint, that can be spared for at least a year. However, due to additional handling and the profit margin of the competitor involved by using this facility, the contribution from *Diana* will get reduced by ₹ 1 per litre. Formulate (do not solve) the linear programming model for determining the monthly production programme to maximize contribution. *[MBA (Nov.), Delhi Univ. 2005]*

2.15. Two alloys *A* and *B* are made from four different metals, *I, II, III* and *IV* according to the following specifications:

A: at most 80% of *I*; at most 30% of *II*; at least 50% of *III*

B: between 40% & 60% of *II*; at least 30% of *III*; at most 70% of *IV*

The four metals are extracted from three different ores whose constituents percentage of these metals, maximum available quantity and cost per tonne are as follows:

Ore	*Maximum quantity (tonnes)*	*Constituent %*					*Price (₹/tonne)*
		I	*II*	*III*	*IV*	*Others*	
1	1,000	20	10	30	30	10	30
2	2,000	10	20	30	30	10	40
3	3,000	5	5	70	20	0	50

Assuming the selling prices of alloys *A* and *B* are ₹ 200 and ₹ 300 per tonne, respectively. Formulate the above as a linear programming problem selecting appropriate objective function and constraints. *[MBA, Pune Univ. 1998]*

2.16. A company manufacturing laminated sheets is considering to adopt an optimum advertising strategy to advertise its product in South India for which it has an advertising budget of ₹ 5 lakh for the coming year. The company has decided to advertise in three magazines, say M_1,

M_2 and M_3. Magazines M_1 and M_3 are monthly and M_2 is fortnightly. The company is targeting the audience having age 20–40 years, monthly income above ₹ 10,000, and education above SSC. These three characteristics have been assigned weights as 30 per cent, 50 per cent and 20 per cent respectively. The following table gives, for each of the three magazines, the percentage of readers having the above characteristics:

Characteristics	*Percentage of readers of magazine*		
	M_1	M_2	M_3
Age: 20–40 years	70	60	90
Monthly income: > ₹ 10,000	50	40	75
Education: above SSC	80	70	80

The advertisement may be in colour, or black and white. The efficiency index for colour advertisement may be taken to be 0.3 and that for black and white 0.2. The cost per insertion of colour, black and white advertisement and readership for the three magazines are given below:

Magazine	*Cost (in ₹) per insertion*		*Readership (in lakh)*
	Colour	*Black and white*	
M_1	18,000	12,000	4
M_2	16,000	10,000	3
M_3	19,000	15,000	2

To create the desired impact on the audience, it has been felt that during the coming year, at least five insertions are necessary in each of M_1 and M_3, and at least four in M_2. No issue will have more than one insertion.

Formulate the above information as a linear programming problem to find the optimum advertising strategy for the coming year in order to maximize the expected effective exposure. (Do not solve the LPP). ***[MIB, Delhi Univ. 2001]***

2.17. Five projects are being evaluated over a 3-year planning horizon. The following table gives the expected returns for each project and the associated yearly expenditure:

Project	*Expenditures (million ₹/year)*			*Returns (million ₹/year)*
	1	2	3	
1	5	1	8	20
2	4	7	10	40
3	3	9	2	20
4	7	4	1	15
5	8	6	10	30
Available funds	25	25	25	

Project *1* and *2* are mutually exclusive in nature. Which project should be selected over the 3-year planning horizon? Formulate the above problem as an LP model.

[MBA (PT), Delhi Univ. 2015]

2.18. A firm places an order for a particular product at the beginning of each month and the product is received at the end of the month. The firm sells the same product during the month from the stock and it can sell any quantity.

The prices at which the firm buys and sells vary every month. The following table shows the projected buying and selling prices for the next four months:

Month	Selling price (₹) (during the month)	Purchase price (₹) (beginning of the month)
April	—	75
May	90	75
June	60	60
July	75	—

The firm has no stock on hand as on April 1, and does not wish to have any stock at the end of July. The firm has a warehouse of limited size, which can hold a maximum of 150 units of the product.

Formulate the problem as an LP model to determine the number of units to buy and sell each month so as to maximize the profit from its operations. *[MBA (Nov.), Delhi Univ. 2009]*

2.19. A mutual fund has cash resources of ₹ 200 million for investment in a diversified portfolio. Table given below shows the opportunities available, their estimated annual yields, risk factors and term period details. Formulate a linear programming model to find the optimal portfolio that will maximize the return.

Investment type	Annual yield (percentage)	Risk factor	Time period (years)
Bank deposit	9.5	0.02	6
Treasury notes	8.5	0.01	4
Corporate deposit	12.0	0.08	3
Blue-chip stock	15.0	0.25	5
Speculative stock	32.5	0.45	3
Real estate	35.0	0.40	10

Restriction on the fund investment are:

(*a*) All the funds available may be invested.

(*b*) Weighted average period of at least 5 years as planning horizon.

(*c*) Weighted average risk factor not to exceed 0.20.

(*d*) Investment in real estate and speculative stocks to be not more than 25% of the money invested in total.

2.20. A Mutual Fund company has ₹ 20 lakh available for investment in Government Bonds, Blue Chip stocks, Speculative stocks and Short-term bank deposits. The annual expected return and risk factor are given below:

Type of investment	Annual expected return (%)	Risk factor (0.100)
Government Bonds	14	12
Blue Chip stocks	19	24
Speculative stocks	23	48
Short-term deposits	12	6

Mutual fund is required to keep at least ₹ 2 lakh in short-term deposits and not to exceed an average risk factor of 42. Speculative stocks must be at most 20 per cent of the total amount invested. How should mutual fund invest the funds so as to maximize its total expected annual return? Formulate this as a Linear Programming Problem. Do not solve it.

[B.Tech. (Mech.), Himachal Univ. 2006]

2:7. GENERAL LINEAR PROGRAMMING PROBLEM

We now consider the linear programming problem in general context, that is, when the number of resources are 'm' and the number of activities are 'n'. The problem is of determining the allocation of resources choosing the levels of the activities in such a way that the best possible value of the overall measure of performance is achieved.

Given below are certain symbols which are commonly used to denote various components of a linear programming problem:

x_j = level of activity j ($j = 1, 2, ..., n$), *i.e.*, decision variables.

b_i = amount of resource i ($i = 1, 2, ..., m$) available for allocation to activity j ($j = 1, 2, ..., n$).

a_{ij} = amount of resource i consumed by each unit of activity j, *i.e.*, the constants a_{ij} are the technological (or substitution or structural) coefficients.

z = value of overall measure of performance.

c_j = per unit contribution (profit or cost) of the activity j ($j = 1, 2, ..., n$) to the overall measure of performance.

Using above symbols, the general linear programming problem is written as follows:

Select the value of decision variables $x_1, x_2, ..., x_n$ so as to maximize or minimize the linear expression:

(*a*) $$z = c_1x_1 + c_2x_2 + ... + c_nx_n \text{ subject to the restrictions:}$$

(*b*) $$\begin{aligned} a_{11}x_1 + a_{12}x_2 + ... + a_{1n}x_n &\ \{\le \text{ or } \ge \text{ or } =\}\ b_1 \\ a_{21}x_1 + a_{22}x_2 + ... + a_{2n}x_n &\ \{\le \text{ or } \ge \text{ or } =\}\ b_2 \\ \vdots \qquad \vdots \qquad \cdots \qquad \vdots & \\ a_{m1}x_1 + a_{m2}x_2 + ... + a_{mn}x_n &\ \{\le \text{ or } \ge \text{ or } =\}\ b_m \end{aligned}$$

and let (*c*) $$x_1 \ge 0,\ x_2 \ge 0, ..., x_n \ge 0.$$

The problem of determining the values of decision variables x_j ($j = 1, 2, ..., n$) which makes z a maximum (or minimum) and satisfies (*b*) and (*c*) is called the general linear programming problem, denoted by GLPP.

The linear function: $z = c_1x_1 + c_2x_2 + ... + c_nx_n$ which is to be maximized (or minimized) is called the ***objective function***. The restrictions (*b*) are referred to as ***constraints***. The first m constraints (those with a ***function*** of all the variables, $a_{i2}x_1 + a_{i2}x_2 + ... + a_{in}x_n$, on the left-hand side) are sometimes called ***functional constraints*** (or structural constraints). The set of inequations (*c*) are usually known as ***non-negative restrictions*** (or non-negativity constraints).

Remarks

1. By using symbol Σ for summation, the general linear programming problem may be put in the following compact form:

$$\text{Maximize or Minimize } z = \sum_{j=1}^{n} c_j x_j \text{ subject to linear constraints:}$$

$$\sum_{j=1}^{n} a_{ij} x_j \ \{\le \text{ or } \ge \text{ or } =\} b_i,\quad i = 1, 2, ..., m$$

and $$x_j \ge 0,\ j = 1, 2, ..., n.$$

2. If any of the constraint of a linear programming problem is expressed by equality sign, then it can be replaced by two weak inequalities. For example, $a_{11}x_1 + a_{12}x_2 + ... + a_{1n}x_n = b_1$ is equivalent to two simultaneous linear inequalities (constraints), *viz.*,

$$a_{11}x_1 + a_{12}x_2 + ... + a_{1n}x_n \le b_1, \text{ and } a_{11}x_1 + a_{12}x_2 + ... + a_{1n}x_n \ge b_1.$$

3. When some decision ***variables are unrestricted in sign*** (positive, negative or zero), they can always be expressed as the difference of two non-negative variables, *e.g.*, if x_2 is unrestricted in sign, then we define two non-negative variables x_2' and x_2'', such that $x_2 = x_2' - x_2''$, where $x_2' \geq 0$ and $x_2'' \geq 0$.
4. Any minimization linear programming problem can be converted into maximization problem by simply changing the sign of c_j's in the objective function. Thus, Minimize $z = \sum_{j=1}^{n} c_j x_j$ is equivalent to Maximize $(-z) = \sum_{j=1}^{n} (-c_j) x_j$; because the smaller value of z is the larger value of $(-z)$.

Slack and Surplus Variables

Definition 1 (*Slack Variable*). Let the constraint of a general LPP be

$$a_{i1}x_1 + a_{i2}x_2 + \ldots + a_{in}x_n \leq b_i, \quad i = 1, 2, \ldots, m.$$

Then, the non-negative variables, s_i, that satisfy

$$a_{i1}x_1 + a_{i2}x_2 + \ldots + a_{in}x_n + s_i \leq b_i, \quad i = 1, 2, \ldots, m.$$

are called **slack variables.**

The slack variables can be thought of as imaginary products, each requiring for its production of one unit of capacity from only one of the resources of zero units of capacity from the other, and each yielding a profit of zero. Thus, a slack variable corresponds to the amount of unused capacity for the constraint to which it is added.

In economic terms, slack variable may be interpreted as:

Slack = Requirement – Production

Definition 2 (*Surplus Variable*). Let the constraint of a general LPP be

$$a_{i1}x_1 + a_{12}x_2 + \ldots + a_{in}x_n \geq b_i, \quad i = 1, 2, \ldots, m$$

Then, the non-negative variables, s_i, that satisfy

$$a_{i1}x_1 + a_{i2}x_2 + \ldots + a_{in}x_n - s_i = b_i, \quad i = 1, 2, \ldots, m$$

are called ***surplus*** (or negative slack) variables.

In economic terms, surplus variables may be interpreted as:

Surplus = Production – Requirement

Note. It may be noted that the contribution (cost or profit) associated with slack and surplus variables are assumed to be zero.

2:8. CANONICAL AND STANDARD FORMS OF LPP

The general formulation of linear programming problem discussed in the previous section can always be put in either ***canonical form*** or ***standard form*** by using some elementary transformations.

The Canonical Form

The general linear programming problem in the form

Maximize $z = c_1x_1 + c_1x_2 + \ldots + c_nx_n$ subject to the constraints:

$$a_{i1}x_1 + a_{i2}x_2 + \ldots + a_{in}x_n \leq b_i, \quad i = 1, 2, \ldots, m$$

and $\quad x_1 \geq 0, x_2 \geq 0, \ldots, x_n \geq 0$

is called the ***canonical*** form of LPP.

The characteristics of this form are:

(*i*) Objective function is always of the maximization type.
(*ii*) All the constraints are inequalities of '≤' type, except for the non-negative restrictions.
(*iii*) All the decision variables are non-negative.

The Standard Form

The general linear programming problem is the form

$$\text{Maximize or Minimize } z = c_1x_1 + c_2x_2 + \dots + c_nx_n \text{ subject to the constraints:}$$
$$a_{i1}x_1 + a_{i2}x_2 + \dots + a_{in}x_n = b_i, \quad i = 1, 2, \dots, m$$
$$x_1, x_2, \dots, x_n \geq 0$$

is known as in its ***standard form***. The characteristics of this form are:

(*i*) Objective function may be of maximization or minimization.
(*ii*) All the constraints are expressed in the form of equations, except for non-negative restrictions.
(*iii*) The right-hand side of each constraint equation must be non-negative.
(*iv*) All the decision variables are non-negative.

Note. The inequality constraint can be changed into equation by introducing a non-negative variable on the left-hand side of such constraint. It is to be added (slack variable), if the constraint is of '≤' type and is to be subtracted (surplus or negative slack variable), if the constraint is of '≥' type.

In tabular form, the general linear programming problem can be expressed as follows:

Resource	*Resource usage per unit of activity*					*Amount of resources available*
	1	*2*	*3*	...	*n*	
1	a_{11}	a_{12}	a_{13}	...	a_{1n}	b_1
2	a_{21}	a_{22}	a_{23}	...	a_{2n}	b_2
3	a_{31}	a_{32}	a_{33}	...	a_{3n}	b_3
⋮	⋮	⋮	⋮	⋮	⋮	⋮
m	a_{m1}	a_{m2}	a_{m3}	...	a_{mn}	b_m
Contribution of z per unit of activity	c_1	c_2	c_3	...	c_n	

Remarks

1. In the above table, the values of c_j, b_i and a_{ij} ($i = 1, 2, \dots, m$, and $j = 1, 2, \dots, n$) are the input constants for the linear programming model. The c_j, b_i and a_{ij} are also referred to as the ***parameters*** of the model.

   In matrix notations, the standard form of general linear programming problem can be expressed as:

   Maximize or Minimize $z = \mathbf{cx}$ subject to the constraints: (*objective function*)

   $\mathbf{Ax} = \mathbf{b}$, (*constraints*)

   and $\mathbf{x} \geq \mathbf{0}$ (*non-negative restrictions*)

   where $\mathbf{x} = [x_1, x_2, \dots, x_n]$, $\mathbf{c} = (c_1, c_2, \dots, c_n)$, $\mathbf{b} = [b_1, b_2, \dots, b_m]$,

   and $\mathbf{A} = (a_{ij})$; $i = 1, 2, \dots, m$ and $j = 1, 2, \dots, n$.

2. The linear programming problem in the form

   $$\text{Maximize } z = \mathbf{cx} \text{ subject to the constraints:}$$
   $$\mathbf{Ax} \leq \mathbf{b} \text{ and } \mathbf{x} \geq \mathbf{0}$$

   is known as ***canonical form*** of linear programming problem.

Example 2.20. *Rewrite in standard form (maximization) the following linear programming problem:*

$$\text{Minimize } z = 2x_1 + x_2 + 4x_3 \text{ subject to the constraints:}$$
$$-2x_1 + 4x_2 \leq 4,$$
$$x_1 + 2x_2 + x_3 \geq 5,$$
$$2x_1 + 3x_3 \leq 2,$$
$$x_1 \geq 0,\ x_2 \geq 0 \text{ and } x_3 \text{ unrestricted in sign.}$$

Standard Form of LPP. As constraints are in the form of inequalities, we introduce slack variables s_1 (≥ 0) and s_3 (≥ 0) in the first and third inequality constraints respectively, while a surplus (negative slack) variable s_2 (≥ 0) in the second inequality constraint.

Further, since x_3 is unrestricted in sign, we write $x_3 = x_3' - x_3''$, where $x_3' \geq 0$ and $x_3'' \geq 0$. Thus, the constraints of the problem become:

$$-2x_1 + 4x_2 + s_1 = 4$$
$$x_1 + 2x_2 + (x_3' - x_3'') - s_2 = 5$$
$$2x_1 + 3(x_3' - x_3'') + s_3 = 2$$

The non-negative restrictions are:

$$x_1 \geq 0,\ x_2 \geq 0,\ x_3' \geq 0,\ x_3'' \geq 0,\ s_1 \geq 0,\ s_2 \geq 0 \text{ and } s_3 \geq 0.$$

The objective function of the problem is converted into that of maximization by multiplying it by (–1). Also, we assign a cost zero to each slack/surplus variable. Thus, the revised objective function is:

$$\text{Maximize } z^* = -2x_1 - x_2 - 4(x_3' - x_3'') + 0 \cdot s_1 + 0 \cdot s_2 + 0 \cdot s_3$$

where $z^* = -z$.

Thus, the given LPP in standard form (maximization) is:

$$\text{Maximize } z^* = -2x_1 - x_2 - 4x_3' + 4x_3'' + 0 \cdot s_1 + 0 \cdot s_2 + 0 \cdot s_3 \text{ subject to the constraints:}$$
$$-2x_1 + 4x_2 + s_1 = 4,\ x_1 + 2x_2 + x_3' - x_3'' - s_2 = 5,$$
$$2x_1 + 3x_3' - 3x_3'' + s_3 = 2,$$
$$x_j \geq 0;\ (j = 1, 2),\ x_3' \geq 0 \text{ and } x_3'' \geq 0;\ s_1 \geq 0, s_2 \geq 0 \text{ and } s_3 \geq 0.$$

Example 2.21. *Reduce the following LPP into its standard form:*

Maximize $z = 3x_1 + 4x_2 + 6x_3$ *subject to the constraints:*

$$2x_1 + x_2 + 2x_3 \geq 6,$$
$$3x_1 + 2x_2 = 8$$
$$7x_1 - 3x_2 + 5x_3 \geq 9;$$
$$x_1 \geq 0, x_2 \geq 0 \textit{ and } x_3 \textit{ unrestricted in sign.}$$

Solution. Since, first and third constraints are inequalities with $\geq$ sign, we introduce surplus variables $s_1 \geq 0$ and $s_2 \geq 0$ in the respective inequalities.

Further, since x_3 is unrestricted in sign, we write $x_3 = x_3' - x_3''$, where $x_3' \geq 0$ and $x_3'' \geq 0$.

Thus, constraints of the given LPP are:

$$2x_1 + x_2 + 2(x_3' - x_3'') - s_1 = 6,$$
$$3x_1 + 2x_2 = 8,$$
$$7x_1 - 3x_2 + 5(x_3' - x_3'') - s_2 = 9,$$

The objective function will be:

$$\text{Maximize } z = 3x_1 + 4x_2 + 6(x_3' - x_3'') + 0 \cdot s_1 + 0 \cdot s_2$$

Thus, the given LPP in its standard form is:

$$\text{Maximize } z = 3x_1 + 4x_2 + 6x_3' - 6x_3'' + 0 \cdot s_1 + 0 \cdot s_2 \text{ subject to the constraints:}$$
$$2x_1 + x_2 + 2x_3' - 2x_3'' - s_1 = 6,\ 3x_1 + 2x_2 = 8$$
$$7x_1 - 3x_2 + 5x_3' - 5x_3'' - s_2 = 9$$
$$x_1 \geq 0,\ x_2 \geq 0,\ x_3' \geq 0,\ x_3'' \geq 0,\ s_1 \geq 0 \text{ and } s_2 \geq 0.$$

PROBLEMS

2.21 Reduce the following LPP to its standard form:

$$\text{Maximize } z = x_1 - 3x_2 \text{ subject to the constraints:}$$
$$-x_1 + 2x_2 \le 15,\ x_1 + 3x_2 = 10$$
$$x_1 \text{ and } x_2 \text{ unrestricted in sign.}$$

2.22. Write down the following LPP in standard form:

$$\text{Maximize } z = 3x_1 + 2x_2 + 5x_3 \text{ subject to the constraints:}$$
$$2x_1 - 3x_2 \le 3,\ x_1 + 2x_2 + 3x_3 \ge 5,\ \ 3x_1 + 2x_3 \le 2$$
$$x_1 \ge 0,\ x_2 \ge 0 \text{ and } x_3 \ge 0.$$

2.23. Transform the following LPP into the form where all the constraints are of equaliity type:

$$\text{Maximize } z = 2x_1 + x_2 - 6x_3 - 4x_4 \text{ subject to the constraints:}$$
$$3x_1 + x_4 \le 25,\ \ x_1 + x_2 + x_3 + x_4 = 20$$
$$4x_1 + 6x_3 \ge 5,\ \ 2 \le x_1 + 3x_3 + 2x_4 \le 30;\ \ x_j \ge 0.\ \ (j = 1, 2, 3, 4).$$

2.24. Consider the following linear programming problem:

$$\text{Minimize } z = -3x_1 + x_2 + x_3 \text{ subject to the constraints:}$$
$$x_1 - 2x_2 + x_3 \le 11$$
$$-4x_1 + x_2 + 2x_3 \ge 3,\ \ 2x_1 - x_3 = -1$$
$$x_1 \ge 0,\ x_2 \ge 0, \text{ and } x_3 \text{ unrestricted in sign.}$$

Cast it in standard form, adding variables and/or constraints, if necessary. How many decision variables and constraints are present in this problem?

2.25. Transform the following LP problem into the form where all constraints are of equality type:

$$\text{Maximize } z = 2x_1 + x_2 - 5x_3 + 3x_4 \text{ subject to the constraints:}$$
$$3x_1 + 2x_2 \le 15,\ 4x_1 + 5x_2 \ge 20,\ x_1 + x_2 - x_3 + 2x_4 = 10$$
$$2 \le 2x_1 + 4x_2 - x_3 \le 30,\ x_i \ge 0;\ \ i = 1, 2, 3, 4$$

Identify the variables introduced.

2:9. TERMINOLOGY FOR THE SOLUTION OF LPP

When there is a term ***solution***, we usually mean the final answer to a problem. But in linear programming any specification of values for decision variables: $x_1, x_2, ..., x_n$ is called a *solution*, regardless of whether it is a desirable or even an allowable choice. Different type of solutions are then identified by using an appropriate adjective.

1. **Feasible Solution.** A solution for which all the constraints (including non-negative restrictions) of a LPP are satisfied is called ***feasible solution.***
2. **Infeasible Solution.** A solution for which at least one of the constraints of an LPP is violated is known as ***infeasible solution***, *i.e.*, at least one of the decision variables does not satisfy the condition of non-negative.
3. **Optimum (or Optimal) Solution.** A feasible solution that has the *most favourable value* of the objective function is called the ***optimum (or optimal)*** solution to the LPP.

   The *most favourable value* is the *largest value*, if the objective function is to be maximized, whereas it is the *smallest value*, if the objective function is to be minimized.
4. **Basic Solution.** Given a system of m simultaneous linear equations in n decision variables $(m < n)$, a solution obtained by setting $(n - m)$ variables equal to zero and solving the m equations in remaining m variables is called a ***basic solution*** and the corresponding variables

are known as ***basic variables.*** Remaining $(n - m)$ variables having zero value are known as ***non-basic variables.***

5. **Basic Feasible Solution.** A feasible solution to an LPP which is also the basic solution is called the ***basic feasible solution*** (BFS). That is, all basic variables assume non-negative values. If the value of at least one basic variable is zero, then the basic feasible solution is called ***degenerate*** basic feasible solution, otherwise ***non-degenerate*** basic feasible solution.
   In matrix notations, a system of m simultaneous linear equations in n variables $(m < n)$ can be written as: $\mathbf{Ax} = \mathbf{b}$, where

$$\mathbf{A} = \begin{bmatrix} a_{11} & a_{12} & \cdots & a_{1n} \\ a_{21} & a_{22} & \cdots & a_{2n} \\ \vdots & \vdots & \cdots & \vdots \\ a_{m1} & a_{m2} & \cdots & a_{mn} \end{bmatrix}, \quad \mathbf{x} = \begin{bmatrix} x_1 \\ x_2 \\ \vdots \\ x_n \end{bmatrix}, \quad \text{and} \quad \mathbf{b} = \begin{bmatrix} b_1 \\ b_2 \\ \vdots \\ b_m \end{bmatrix}.$$

   Let $\mathbf{B}$ be any $m \times m$ submatrix formed by m linearly independent columns of $m \times n$ matrix $\mathbf{A}$. Then, a solution obtained by setting $(n - m)$ variables not associated with the columns of $\mathbf{B}$ to zero and then solving the system of equations is called the basic solution. The m variables which may be all different from zero are called ***basic variables.*** The $m \times m$ submatrix must be non-singular and is known as ***basis matrix*** with the columns of $\mathbf{B}$ as ***basis vectors.***

   Remarks

   1. If $\mathbf{B}$ is basis matrix, where $\mathbf{B}$ is sub-matrix of $\mathbf{A}$, then the basic solution to the system of equations $\mathbf{Ax} = \mathbf{b}$ is given by
   $$\mathbf{x_B} = \mathbf{B}^{-1}\,\mathbf{b},$$
   since, $\mathbf{B}$ is non-singular, *i.e.*, $\mathbf{B}^{-1}$ exists.
   2. If $\mathbf{x_B} \geq \mathbf{0}$, then basic solution is also feasible.

6. **Optimum Basic Feasible Solution.** A basic feasible solution that optimizes (maximizes or minimizes) the objective function value of the given LPP is called an ***optimum basic feasible solution.***
7. **Unbounded Solution.** A solution that can increase or decrease infinitely the value of the objective function of a LPP is called an ***unbounded solution.***
8. **Associated Cost Vector.** Let $\mathbf{x_B}$ be a basic feasible solution to the LPP:
   $$\text{Maximize } z = \mathbf{cx} \text{ subject to the constraints:}$$
   $$\mathbf{Ax} = \mathbf{b} \text{ and } \mathbf{x} \geq \mathbf{0}$$
   Then the cost vector $\mathbf{c_B} = (c_{B_1}\ c_{B_2}\ \cdots\ c_{B_m})$, where c_{Bi} are the components of $\mathbf{c} = (c_1\ c_2\ \cdots\ c_n)$ associated with the basic variables is called the cost vector associated with the basic feasible solution $\mathbf{x}_B$.
   Value of the objective function for the basic feasible solution will be given by $\mathbf{c}_B\mathbf{x}_B\ (= z)$.
9. **Improved Basic Feasible Solution.** Let $\mathbf{x}_B$ and $\hat{\mathbf{x}}_B$ be two basic feasible solutions to the standard LPP. Then $\hat{\mathbf{x}}_B$ is said to be an improved basic feasible solution, if $\hat{\mathbf{c}}_B\,\hat{\mathbf{x}}_B \geq \mathbf{c}_B\,\mathbf{x}_B$, where $\hat{\mathbf{c}}_B$ is constituted of cost components corresponding to $\hat{\mathbf{x}}_B$.

TEST YOUR UNDERSTANDING

1. Linear programming problem (LPP) must have an
   (*a*) Objective (goal) that we aim to maximize or minimize
   (*b*) Constraints (restrictions) that we need to specify
   (*c*) Decision variables (activities) that we need to determine
   (*d*) All of the above

2. Which of the following is not associated with an LPP?
   (*a*) Proportionality (*b*) Uncertainty
   (*c*) Additivity (*d*) Divisibility
3. A constraint in an LPP restricts
   (*a*) value of objective function (*b*) value of a decision variable
   (*c*) use of available resource (*d*) uncertainty of optimum value
4. A constraint in an LPP is expressed as
   (*a*) an equation with '=' sign (*b*) inequality with '≥' sign
   (*c*) inequality with '≤' sign (*d*) any or all of the above
5. Non-negativity condition is an important component of LPP, because
   (*a*) variables are inter-related in terms of limited resources
   (*b*) value of variables make sense and correspond to real world problems
   (*c*) value of variables should remain under the control of decision-maker
   (*d*) none of the above
6. Minimization of objective function in LPP means
   (*a*) least value chosen among the allowable decisions
   (*b*) greatest value chosen among the allowable decisions
   (*c*) both (*a*) and (*b*) (*d*) none of the above
7. Which of the following is not correct about LPP?
   (*a*) all constraints must be linear relationships (*b*) Objective function must be linear
   (*c*) All the constraints and decision variables must be of either '≤' or '≥' type
   (*d*) All decision variables must be non-negative
8. The general linear programming problem is in standard form, if
   (*a*) the constraints are strict equations
   (*b*) the constraints are inequalities of '≤' type
   (*c*) the constraints are inequalities of '≥' type
   (*d*) the decision variables are unrestricted in sign
9. Which of the following statements is wrong?
   (*a*) slack variables are used to convert the inequalities of the type '≤' into equations
   (*b*) surplus variables are used to convert the inequalities of the type '≥' into equations
   (*c*) all LPP with all its constraints are of the type '≥' is said to be in *standard form*
   (*d*) an LPP with all its constraints are of the type '≤' is said to be in *canonical form*
10. A feasible solution to an LPP
    (*a*) must satisfy all of the problem's constraints simultaneously
    (*b*) must be a corner point of the feasible region
    (*c*) need not satisfy all of the constraints, only some of them
    (*d*) must optimize the value of the objective function
11. Give a system of m simultaneous linear equations in n unknowns ($m < n$), the number of basic variables will be
    (*a*) m (*b*) n
    (*c*) $n - m$ (*d*) $n + m$
12. Which of the following is not correct?
    (*a*) The general LPP with m constraints in n unknowns ($m < n$) give $\binom{n}{m}$ basic solutions.

(*b*) A basic solution that satisfies all the constraints and non-negative restrictions is called a basic feasible solution.
(*c*) A basic solution in which at least one basic variable vanishes is zero is called a non-degenerate basic solution.
(*d*) The basic feasible solution which maximizes or minimizes the objective function is called an optimum basic feasible solution.

ANSWERS

1. (*d*)	**2.** (*b*)	**3.** (*c*)	**4.** (*d*)	**5.** (*b*)	**6.** (*a*)
7. (*c*)	**8.** (*a*)	**9.** (*c*)	**10.** (*a*)	**11.** (*a*)	**12.** (*c*)

REVIEW EXERCISE

1. (*a*) What is the term 'Linear Programming'? ***[M.Com., Bangalore Univ. 2018]***
(*b*) What is linear programming? What are its limitations? Discuss. ***[MBA (PT.), Delhi Univ. 2008]***
2. Discuss the importance of linear programming using one application to a business problem. ***[MBA, Delhi Univ. 2015]***
3. What are the components of linear programming model? Explain them in brief. ***[MBA, Madras Univ. 2004]***
4. Major limitations of linear programming are: assumptions of 'additivity' and 'single objective'. Elaborate by giving appropriate examples. ***[MBA, Delhi Univ. 2008]***
5. Give the general form of a basic model of linear programming.
6. State any two applications of linear programming. ***[M.Com., Bharthidasan Univ. 2007]***
7. Explain how the linear programming technique can be helpful in decision-making, bringing out limitations, if any. ***[MBA, Bangalore Univ. 2018]***
8. Describe the major applications of linear programming in business and industry, pointing out limitations, if any. ***[MBA, Delhi Univ. 2002]***
9. Explain the major characteristics of a linear programming problem.
10. What are the major assumptions and limitations of linear programming? ***[MBA, Madras Univ. 2004]***
11. "Linear Programming has no real application". Do you agree? Discuss. ***[MBA, Delhi Univ. 2009]***
12. "Linear Programming is useful management science technique, but it has some limitations". Discuss. ***[MBA (Nov.), Delhi Univ. 2009]***
13. Write down the standard form of an LPP. ***[MBA, Annamalai Univ. 2010]***
14. Define slack and surplus variables in LPP.
15. State the general linear programming problem in (*a*) Standard form, (*b*) Canonical form.
16. Define the following terms:
(*a*) Basic variable, (*b*) Basic solution, (*c*) Basic feasible solution, (*d*) Degenerate solution.
17. Differentiate between basic and non-basic variables. ***[MBA (Dec.), Panjab Tech. Univ. 2011]***
18. (*a*) What is meant by a feasible solution of an LPP?
(*b*) What is meant by non-degenerate Basic Feasible Solution? ***[M.Com., Bangalore Univ. 2017]***
19. Explain the phenomenon of 'infeasibility' in an LPP. ***[M.Com., Delhi Univ. 2000]***
20. "Linear programming is one of the most frequently and successfully applied Operations Research technique to managerial decisions". Elucidate the statement with some examples. ***[MBA, Delhi Univ. 2008]***

CHAPTER 3

Solving Linear Programming Problems

"Graph is the best tool to visualize any concept"

Once a given decision-making problem is formulated in a linear programming model, the next step is to solve the problem to get an optimum (or optimal) solution. An optimum as well as feasible solution to a linear programming problem is obtained by choosing one set of values from several possible values of decision variables that satisfy the given constraints of the problem simultaneously and also provides an optimum (maximum or minimum) value of the objective function.

For linear programming problems involving not more than two variables, it is possible to display the entire set of feasible solutions graphically by plotting linear constraints on a graph paper in order to locate the best (optimum) solution. The technique used to identify the optimum solution is known as ***graphical*** (or ***geometrical***) ***method***. It, however, becomes too complex to be used in case of more than two variables. But, it provides an understanding for solving linear programming problems, involving more than two variables, algebraically.

The ***Simplex Method*** is an algebraic iterative procedure which either solves an LPP in a finite number of steps or gives an indication that there is an unbounded solution to the given LPP. The simplex method admits one of the edges (extreme point) lying on the feasible region (solution space) and tests whether an improvement in the objective function is possible by moving to an adjacent edge of the feasible region. If an improvement is possible, a new edge is tested to see whether further improvement can be made. When this leads to a solution where no further improvement is possible by moving to an adjacent edge, it is concluded that an optimum solution is reached.

3:1. GRAPHICAL SOLUTION METHOD

For the solution of a linear programming problem by graphical method, major steps are as follows:

Step 1. Identify the problem, *i.e.*, decision variables, objective function and restrictions (constraints).

Step 2. Set up the linear programming model formulation of the problem.

Step 3. Graph the linear inequalities and decide the feasible region. This is being done in the following four steps:

(*i*) Ignore the inequality sign temporarily and draw the straight line on the graph paper.

(*ii*) Consider a pair of numbers (x_1, x_2) and test whether it satisfies the inequality. If it satisfies, then shade the region, otherwise go to next step.

(*iii*) Consider some other pair of numbers (x_1, x_2) on the other side of straight line. If it satisfies the inequality then shade the region.

(*iv*) Do *Steps* (*i*), (*ii*) and (*iii*) for all the inequalities.

The common region of all the inequalities if it lies in the first quadrant is termed as ***feasible region*** or ***solution space*** (or ***polygonal convex set***).

Step 4. The feasible region obtained in *Step 3* may be bounded or unbounded. Compute the coordinates of all the corner points of the feasible region, *i.e.*, corner-point feasible (or CPF) solutions.

Step 5. Find out the value of the objective function at each CPF solution (commonly referred to as ***extreme point*** or ***vertex***).

Step 6. Select that CPF solution which optimizes (maximizes or minimizes) the value of objective function. This gives the optimum (optimal) basic feasible solution.

Remarks

1. The method described above is known as ***Search Approach Method***.
2. The collection of all feasible solutions to an LPP constitutes a convex set whose extreme points correspond to the basic feasible solution.
3. If the optimum solution occurs at more than one extreme point, the value of the objective function will be same for all convex combinations of these extreme points.
4. Another method known as ***Iso-profit*** or ***Iso-cost approach***, involves the following steps:
   (*i*) First ***four steps*** are same as in the Search Approach Method. In the ***fifth step***, we choose a convenient profit (or cost) and draw iso-profit (or iso-cost) line so that it falls within the feasible region.
   (*ii*) Move this Iso-profit (or Iso-cost) line parallel to itself farther (or closer) from (to) the origin.
   (*iii*) Identify the optimum solution, *i.e.*, the co-ordinate of that point on the feasible region which is touched by the highest possible iso-profit line (or lower possible iso-cost line).
   (*iv*) Compute the ***optimum feasible solution***.

Example 3.1 (*Maximization Case*). *A company makes two kinds of leather belts. Belt A is belt of high quality and Belt B is of lower quality. The respective profits are ₹ 4.00 and ₹ 3.00 per belt. Each belt of type A requires twice as much time as a belt of type B, and if all belts were of type B, the company could make 1000 belts per day. The supply of leather is sufficient for only 800 belts per day (Both A and B combined). Belt A requires a fancy buckle and only 400 buckles per day are available. There are only 700 buckles a day available for Belt B. Determine the optimal product mix.*

[MBA, Madras Univ. 2006; MBA (Nov.), Delhi Univ. 2009; BBA, GGSIP Univ. 2014]

Solution.

Step 1. The appropriate mathematical formulation of the given linear programming problem is:

Maximize $z = 4x_1 + 3x_2$ subject to the constraints:

$$2x_1 + x_2 \le 1{,}000 \qquad \text{(Time constraint)}$$

$$x_1 + x_2 \le 800 \qquad \text{(Availability of leather)}$$

$$x_1 \le 400 \text{ and } x_2 \le 700 \qquad \text{(Availability of buckles)}$$

$$x_1 \ge 0 \text{ and } x_2 \ge 0, \qquad \text{(Non-negativity)}$$

where x_1 = number of belts of type *A*, and x_2 = number of belts of type *B*.

Step 2. Next, we construct the graph by considering the cartesian rectangular axis OX_1X_2 in the plane. As each point has the coordinates of type (x_1, x_2); any point satisfying the conditions $x_1 \ge 0$ and $x_2 \ge 0$ lies in the first quadrant.

Now, the inequalities are graphed considering them as equations, *e.g.*, the first constraint $2x_1 + x_2 \le 1{,}000$ will be graphed as $2x_1 + x_2 = 1{,}000$. The equation is re-written as:

$$\frac{x_1}{500} + \frac{x_2}{1{,}000} = 1.$$

This equation indicates that when it is plotted on the graph, it cuts an x_1-intercept of 500 and x_2-intercept of 1,000. These two points are then connected by a straight line which is shown in *Fig. 3.1(a)* as line *AB*. Any point (representing a combination of x_1 and x_2) that falls on this line or in the area below it, is acceptable in so far as this constraint is concerned. The region *OAB* formed by two axes and the line representing the equation $2x_1 + x_2 = 1{,}000$ is the region containing acceptable values of x_1 and x_2 in respect of this constraint.

Similarly, the constraint $x_1 + x_2 \leq 800$ can be plotted. The line *CD* in *Fig. 3.1(b)* represents the equation $x_1 + x_2 = 800$. The region *OCD,* formed by the two axes and this line represents the area in which any point would satisfy this constraint of leather availability. Further, the constraints $x_1 \leq 400$ and $x_2 \leq 700$ are also plotted on the graph which represents the area between the two axes and the lines $x_1 = 400$ and $x_2 = 700$ as shown in *Fig. 3.1(b)*.

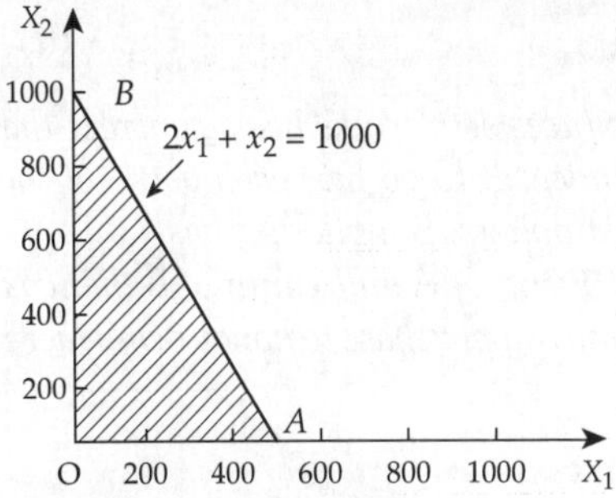

Fig. 3.1(*a*)

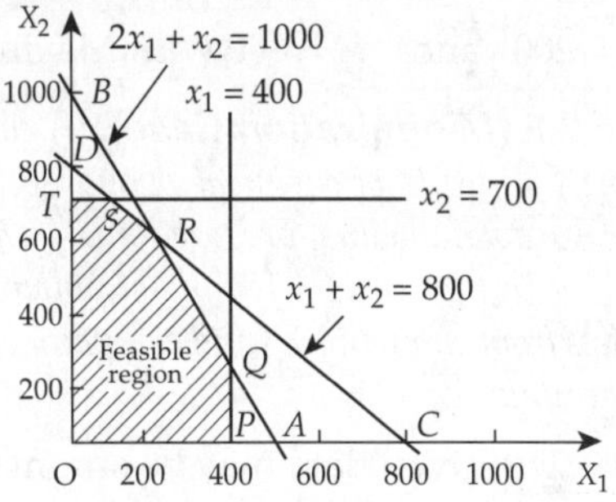

Fig. 3.1(*b*)

Now, all the constraints have been graphed. The area bounded by all these constraints, called feasible ***region*** or ***solution space,*** is as shown in *Fig. 3.1(b)* by the shaded area *OPQRST*.

Step 3. The optimum value of objective function occurs at one of the extreme (corner) points of the feasible region. The coordinates of the extreme points or *CPF* solutions are at

$$0 = (0, 0),\ P = (400, 0),\ Q = (400, 200),\ R\ (200, 600),\ S\ (100, 700), \text{ and } T = (0, 700)$$

Step 4. We now compute the *z*-value corresponding to the *CPF* solutions

Extreme point	*CPF solution*	$z = 4x_1 + 3x_2$
0	(0, 0)	0
P	(400, 0)	1600
Q	(400, 200)	2200
R	(200, 600)	**2600** ⟵ Maximum
S	(100, 700)	2500
T	(0, 700)	2100

Step 5. The optimum solution is that *CPF* solution for which the objective function has the largest value. Thus, the optimum solution occurs at the point *R, i.e.,* $x_1 = 200$ and $x_2 = 600$ with the objective function value of ₹ 2,600.

Hence, to maximize profit, the company should produce 200 belts of type *A* and 600 belts of type *B* per day.

Alternative Method (Iso-profit Approach)

The feasible region (solution space) obtained in *step* 2 is as shown in *Fig. 3.1(c)* by the shaded area *OPQRST*.

Let the profit to the company (arbitrary) is ₹ 1200. The objective function then becomes:

$$4x_1 + 3x_2 = 1200$$

We draw this equation as a straight line in the feasible region shown in *Fig. 3.1(c)*. This line is known as ***Iso-profit line.***

It may be noted that the iso-profit function (objective value function) is a straight line on which every point has the same total profit.

Now, we move the iso-profit line parallel to itself farther from the origin. We observe that one of the iso-profit line touches only point *R* before leaving the feasible region. This iso-profit line is termed as highest possible iso-profit line and point *R* gives the extreme point of the solution space.

Hence, the optimum feasible solution is:

$x_1 = 200$ and $x_2 = 600$ with Maximum $z =$ ₹ 2600

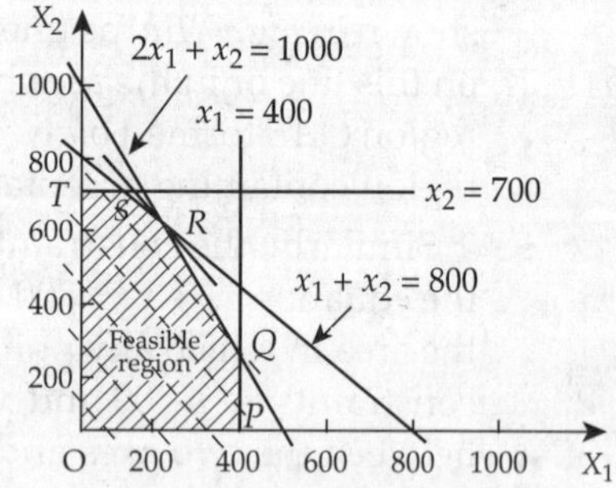

Fig. 3.1(*c*)

Example 3.2 (*Minimization Case*). *A company that produces soft drinks has a contract that requires a minimum of 80 units of chemical A and 60 units of the chemical B to go into each bottle of the drink. The chemicals are available in a prepared mix from two different suppliers. Supplier X_1 has a mix of 4 units of A and 2 units of B that costs ₹ 10, and supplier X_2 has a mix of 1 unit of A and 1 unit of B that costs ₹ 4. How many mixes from company X_1 and company X_2 should the company purchase to honour contract requirement and yet minimize cost?*

Solution. The given data may be summarized as follows:

Supplier	*Chemical A (units)*	*Chemical B (units)*	*Cost per mix (₹)*
X_1	4	2	10
X_2	1	1	4
Units required	80	60	

Let x_1 be the number of mixes to be purchased from supplier X_1 and x_2 be those from supplier X_2. Then, since the objective is to minimize the total cost, the given problem has the following linear programming formulation:

Minimize $z = 10x_1 + 4x_2$ subject to the constraints:

$$4x_1 + x_2 \geq 80 \quad \textit{(Chemical A)}$$

$$2x_1 + x_2 \geq 60 \quad \textit{(Chemical B)}$$

$$x_1 \geq 0 \text{ and } x_2 \geq 0. \quad \textit{(Non-negativity)}$$

Step 1. We first graph the constraints. Consider an OX_1X_2 plane. Clearly, $\{4x_1 + x_2 \geq 80\}$ and $\{2x_1 + x_2 \geq 60\}$ represent the regions 'on the right' of the lines $4x_1 + x_2 = 80$ and $2x_1 + x_2 = 60$, respectively.

Step 2. The solution space is the intersection of all the regions represented by the given constraints in the first quadrant. This is shaded in *Fig.* 3.2 which appears to be unbounded.

Now, we try to determine the additional hidden conditions in the problem for which the solution space (feasible region) becomes bounded.

Each bottle of the drink contains 80 units of chemical *A* and 60 units of chemical *B*. If the drinks are made from the mixes of supplier X_1 only, each bottle would require 20 mixes to meet the chemical *A* requirement and 30 mixes to

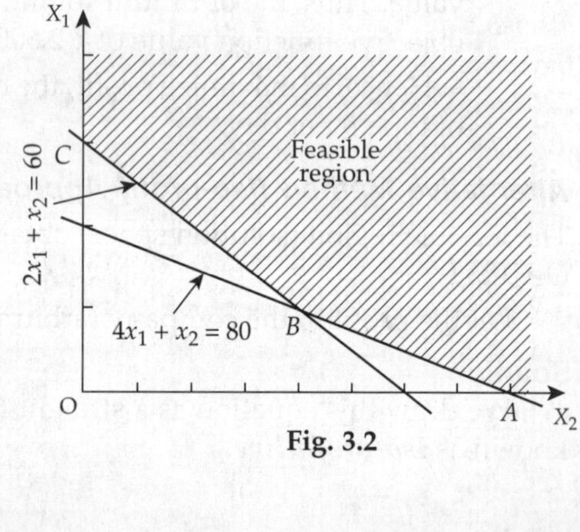

Fig. 3.2

meet the chemical B requirement. Therefore, 30 mixes would meet all minimum requirements. Any lesser amount of mixes would not meet the minimum requirement for chemical B.

Similarly, if the drinks are made from the mixes of supplier X_2 only, each bottle would require 80 mixes to meet both the chemical requirements.

Since, the company that manufactures drinks wants to make a blending of the mixes purchased from both the supplier X_1 and X_2, it is possible that the company would need less than the maximum amount of both supplies to meet the guaranteed minimum.

Thus, we add the two additional hidden conditions $x_1 \le 30, x_2 \le 80$. With these additional restrictions, the solution space becomes bounded as shown shaded in *Fig. 3.3*.

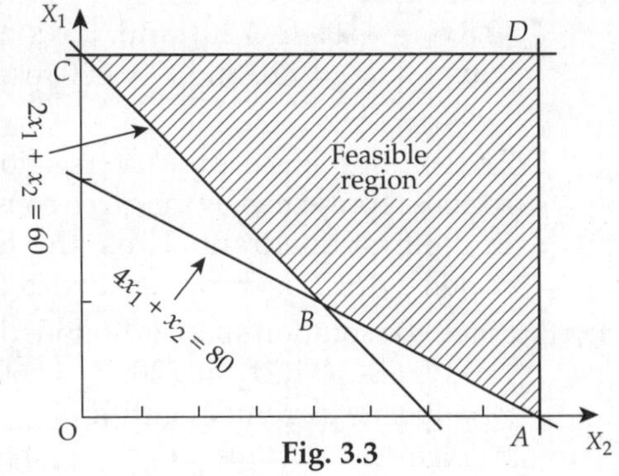

Fig. 3.3

***Step* 3.** As the feasible region *ABCD* is bounded by the lines $4x_1 + x_2 = 80$, $2x_1 + x_2 = 60$, $x_1 = 60$ and $x_2 = 80$, its extreme points are:

$$A = (0, 80), \quad B = (10, 40), \quad C = (30, 0) \quad \text{and} \quad D = (30, 80).$$

***Step* 4.** Values of the objective function at the extreme points (CPF solutions) are:

Extreme point	*CPF solutions*	*Cost*	
A	(0, 80)	320	
B	(10, 40)	260	⟵ Minimum
C	(30, 0)	300	
D	(30, 80)	620	

***Step* 5.** Since, objective function is least for CPF solution (10, 40), the optimal solution is:

$$x_1 = 10 \quad \text{and} \quad x_2 = 40 \text{ with minimum } z = ₹\,260.$$

Thus, to honour the contract and yet to minimize cost, the company should purchase 10 mixes from X_1 and 40 mixes from X_2.

Remark. It may be observed that it was not essential to identify ***hidden*** constraints to make the feasible region bounded. In fact unbounded solution space does not pose any problem in the case of ***Minimization problems***.

Example 3.3 (*Alternative Optimal Solution*). *A company manufactures two models of voltage stabilizers: Ordinary and Auto-cut. All components of the stabilizers are purchased from outside sources, whereas only assembly and testing is carried out at the company's own works. The assembly and testing time required for the two models are 0.8 hour each for Ordinary and 1.20 hours each for Auto-cut. Manufacturing capacity of 720 hours at present is available per week.*

The market for the two models has been surveyed which suggests maximum weekly sale of 600 units of ordinary and 400 units of Auto-cut. Profit per unit for Ordinary and Auto-cut models has been estimated at ₹ 100 and ₹ 150 respectively.

(i) Find the optimal product mix using graphical method.

(ii) Is there an alternative optimal solution? If yes, specify it.

Solution. Let x_1 be the number of Ordinary stabilizers and x_2 be the number of Auto-cut stabilizers to be manufactured. Then, the linear programming formulation of the problem is as follows:

Maximize $z = 100x_1 + 150x_2$ subject to the constraints:

$$0.8x_1 + 1.2x_2 \le 720 \quad \text{(Manufacturing capacity)}$$
$$x_1 \le 600 \quad \text{(Ordinary models)}$$
$$x_2 \le 400 \quad \text{(Auto-cut models)}$$
$$x_1 \ge 0 \text{ and } x_2 \ge 0 \quad \text{(Non-negativity)}$$

Step 1. To graph the constraints of the LPP, consider the OX_1X_2 plane. In this plane, $\{0.8x_1 + 1.2x_2 \le 720\}$ and $\{x_1 \le 600\}$ represents the regions 'on the left' of the lines $0.8x_1 + 1.2x_2 = 720$ and $x_1 = 600$ respectively. Also $\{x_2 \le 400\}$ represents the region below the line $x_2 = 400$.

Step 2. The solution space is the intersection of all regions represented by the given constraints in the first quadrant. Thus, the feasible region *OABCD* is shown shaded in *Fig. 3.4*.

Step 3. Since, the solution space is bounded by the lines $0.8x_1 + 1.2x_2 = 720$, $x_1 = 600$ and $x_2 = 400$, its extreme points are:

$O = (0, 0)$, $A = (600, 0)$, $B = (600, 200)$, $C = (300, 400)$ and $D = (0, 400)$.

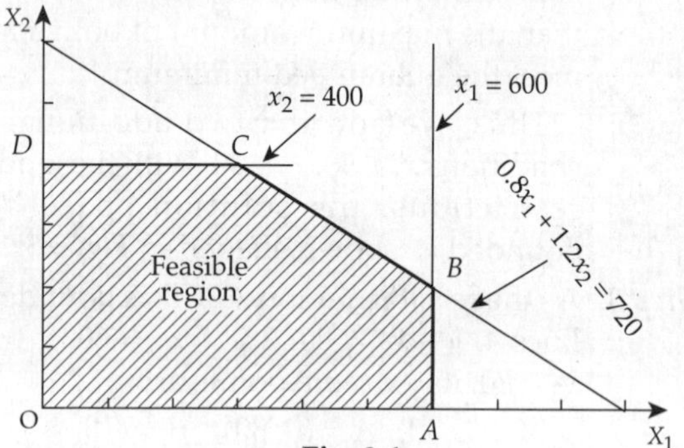

Fig. 3.4

Step 4. Values of the objective function at the extreme points of CPF solutions are:

Extreme point	*CPF solutions*	*Profit: $(100x_1 + 150x_2)$*	
O	(0, 0)	0	
A	(600, 0)	60,000	
B	(600, 200)	90,000 ⟵	Maximum
C	(300, 400)	90,000 ⟵	
D	(0, 400)	60,000	

Step 5. The objective function attains its maximum at CPF solutions $B = (600, 200)$ and $C = (300, 400)$. This suggests the presence of more than one optimum solution and we say that *B* and *C* are two alternative optimum solutions.

Hence, the company should manufacture either 600 Ordinary and 200 Auto-cut stabilizers or 300 Ordinary and 400 Auto-cut stabilizers to achieve a maximum profit of ₹ 90,000 per week.

Note. It may be observed that any point on the line segment joining extreme points B and C also gives a maximum *z*-value of ₹ 90,000. Thus, in the case there is an ***infinite*** number of optimal solutions. In fact there are two ***corner point*** feasible solutions and an infinite number of ***non-corner point*** feasible solutions.

Example 3.4 (*Unbounded Solution*). *Find the maximum value of*

$z = 3x_1 + 2x_2$ subject to the constraints:

$x_1 - x_2 \le 1$, $x_1 + x_2 \ge 3$; and $x_1 \ge 0$, $x_2 \ge 0$.

Solution.

Step 1. To graph the constraints of the LPP, consider the OX_1X_2 plane. Clearly, $\{x_1 - x_2 \le 1\}$ represents the region 'on the left' of the line $x_1 - x_2 = 1$ and $\{x_1 + x_2 \ge 3\}$ represents the region 'on the right' of the line $x_1 + x_2 = 3$.

Step 2. The solution space (feasible region) is the intersection of all the constraint regions in the first quadrant. This is shown shaded in the *Fig. 3.5*.

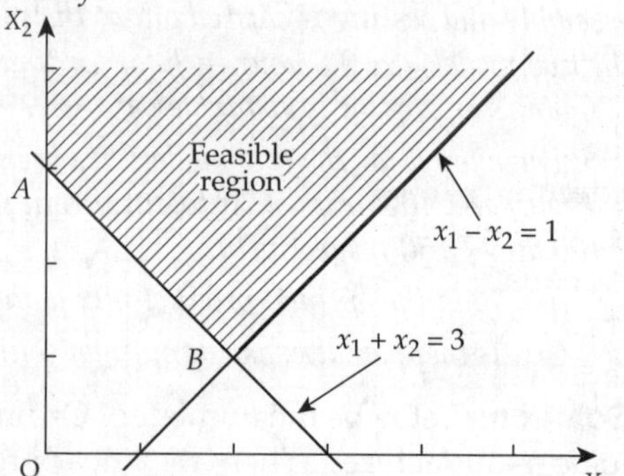

Fig. 3.5

***Step* 3.** The feasible region is unbounded to the right of the line joining extreme points *A* and *B*.

***Step* 4.** Value of the objective function *z* at the extreme points (2, 1) and (0, 3) are 8 and 6 respectively.

But, there exist points in the solution space for which the value of the objective function is more than 8. For instance, at the point (4, 5), the value for *z* is 22. It can be further increased by considering still higher points. Thus, we conclude that the maximum value of *z* occurs at infinity. We conclude that the given problem has an ***unbounded solution.***

Example 3.5 (*No Solution*). *Find the minimum value of*

$$z = -3x_1 + 4x_2 \text{ subject to the constraints:}$$

$$x_1 + x_2 \le 4,\ 2x_1 + 3x_2 \ge 18; \text{ and } x_1 \ge 0, x_2 \ge 0.$$

[B.Com. (Nov.), Madras Univ. 2007]

Solution.

***Step* 1.** Consider the OX_1X_2 plane. Here, $\{x_1 + x_2 \le 4\}$ represents the region 'on the left' of the line $x_1 + x_2 = 4$ and $\{2x_1 + 3x_2 \ge 18\}$ represents the region 'on the right' of the line $2x_1 + 3x_2 = 18$.

***Step* 2.** The feasible regions for the two inequalities $x_1 + x_2 \le 4$ and $2x_1 + 3x_2 \ge 18$ are depicted in *Fig. 3.6*. The two shaded regions indicate two half-spaces which do not intersect and thus, have no point in common.

Since, there is no point (x_1, x_2) which can lie in both the regions simultaneously, we conclude that there is no feasible solution to the given LPP.

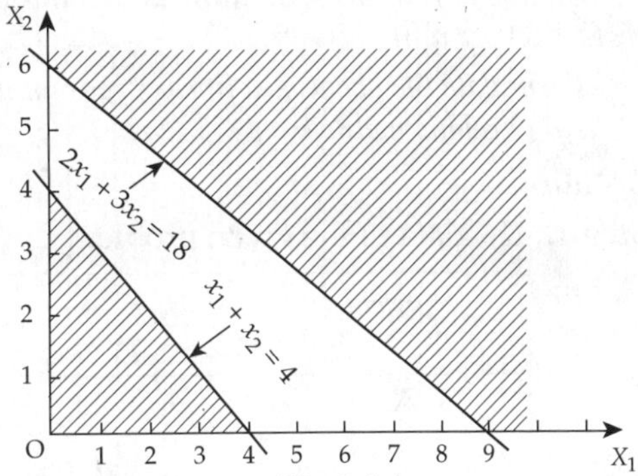

Fig. 3.6

REVIEW ILLUSTRATIONS

Example 3.6. *Solve the following LPP by graphical method:*

$$\text{Maximize } z = 3x_1 + 5x_2 \text{ subject to the constraints:}$$

$$3x_1 + 2x_2 \le 18,\ x_1 \le 4,\ x_2 \le 6 \text{ and } x_1 \ge 0, x_2 \ge 0$$ **[BBA, GGSIP Univ. 2008]**

Solution. We first draw the straight lines corresponding to equations:

$$3x_1 + 2x_2 = 18,\ x_1 = 4 \text{ and } x_2 = 6$$

Then, we shade the common region *OABCD* satisfying all the given inequalities:

$$3x_1 + 2x_2 \le 18,\ x_1 \le 4,\ x_2 \le 6$$

and $$x_1 \ge 0, x_2 \ge 0$$

By inspection of the graph of the feasible region, we find the coordinates of the three extreme points (vertices) *O*, *A* and *D* as (0, 0), (4, 0) and (0, 6). For the corner point *B*, we solve the equations $x_1 = 4$ and $3x_1 + 2x_2 = 18$ which gives (4, 3). For corner point *C*, we solve the equations $x_2 = 6$ and $3x_1 + 2x_2 = 18$ which gives (2, 6). Thus, the CPF solutions are given on the next page and we search for the optimum CPF solution.

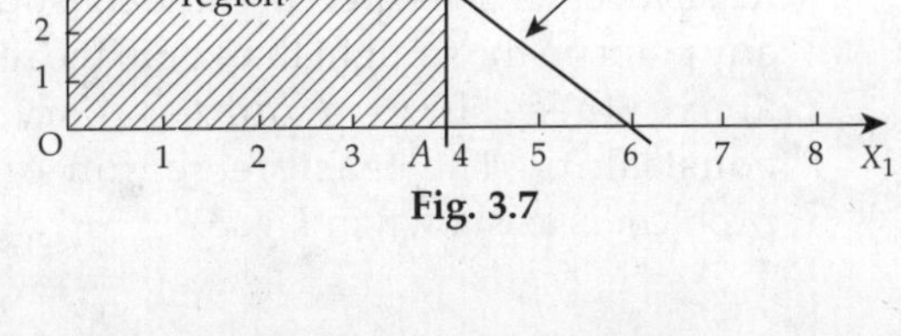

Fig. 3.7

Extreme point	*CPF solution*	$z\,(= 3x_1 + 5x_2)$
O	(0, 0)	$3 \times 0 + 5 \times 0 = 0$
A	(4, 0)	$3 \times 4 + 5 \times 0 = 12$
B	(4, 3)	$3 \times 4 + 5 \times 3 = 27$
C	(2, 6)	$3 \times 2 + 5 \times 6 = \boxed{36} \longleftarrow$ Maximum
D	(0, 6)	$3 \times 0 + 5 \times 6 = 30$

Since, 36 is the largest in the z-column; the optimum (optimal) solution is:

$$x_1 = 2 \text{ and } x_2 = 6 \text{ with maximum } z = 36.$$

Example 3.7. *A farm is engaged in breeding pigs. The pigs are fed on various products grown on the farm. In view of the need to ensure certain nutrient constituents (call them X, Y and Z), it is necessary to buy two additional products, say, A and B. One unit of product A contains 36 units of X, 3 units of Y and 20 units of Z. One unit of product B contains 6 units of X, 12 units of Y and 10 units of Z. The minimum requirement of X, Y and Z is 108 units, 36 units and 100 units respectively. Product A costs ₹20 per unit and product B costs ₹40 per unit.*

Formulate the above as a linear programming problem to minimize the total cost, and solve the problem by using graphic method. ***[C.A. Final (May) 2002]***

Solution.

Step 1. The data of the given problem can be summarized as follows:

Nutrient constituents	*Nutrient content in product*		*Minimum amount of nutrient*
	A	*B*	
X	36	06	108
Y	03	12	36
Z	20	10	100
Cost of product	₹ 20	₹ 40	

Making use of above information, the appropriate linear programming formulation of the model is:

$$\text{Minimize } z = 20x_1 + 40x_2 \text{ subject to the constraints:}$$

$$36x_1 + 6x_2 \geq 108,\ \ 3x_1 + 12x_2 \geq 36,\ \ 20x_1 + 10x_2 \geq 100, \text{ and } x_1, x_2 \geq 0$$

where x_1 = number of units of product A, and x_2 = number of units of product B.

Step 2. Consider now a set of cartesian rectangular axis OX_1X_2 in the plane. As each point has the coordinates of the type (x_1, x_2), any point satisfying the conditions $x_1 \geq 0$ and $x_2 \geq 0$ lies in the first quadrant only.

The constraints of the given problem are plotted as described earlier by treating them as equations:

$$36x_1 + 6x_2 = 108,\ \ 3x_1 + 12x_2 = 36$$

and $$20x_1 + 10x_2 = 100$$

or $$\frac{x_1}{3} + \frac{x_2}{18} = 1,\ \ \frac{x_1}{12} + \frac{x_2}{3} = 1 \text{ and } \frac{x_1}{5} + \frac{x_2}{10} = 1$$

The area beyond these lines represents the feasible region in respect of these constraints; any point on the straight lines or in the region above these lines would satisfy the constraints. The feasible region of the problem is as shown in *Fig. 3.8*.

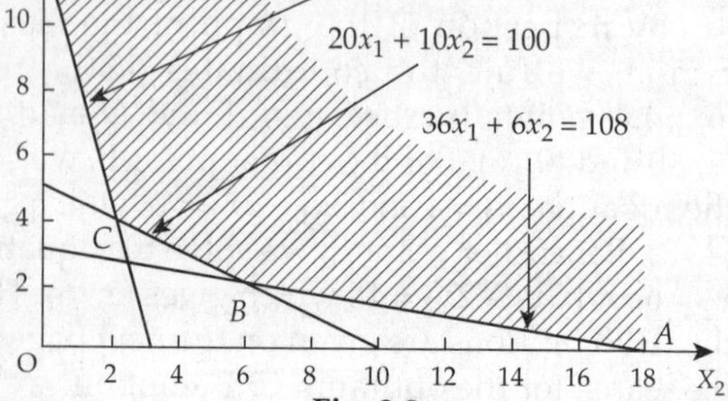

Fig. 3.8

Step 3. The coordinates of the extreme points or CPF solutions are at

$$A = (0, 18),\ B = (2, 6),\ C = (4, 2) \text{ and } D = (12, 0).$$

Step 4. The value of objective function at each of the extreme points can be evaluated as follows:

Extreme point	*CPF solution*	$z = 20x_1 + 40x_2$
A	(0, 18)	720
B	(2, 6)	280
C	(4, 2)	160 ⟵ Minimum
D	(12, 0)	240

Step 5. The optimum solution is that CPF solution for which the objective function has the least value. Thus, the optimum solution occurs at the extreme point C.

Hence, the optimum solution is to purchase 4 units of product *A* and 2 units of product *B* in order to maintain a minimum cost of ₹ 160.

Example 3.8. *A company has two grades of inspectors 1 and 2 who are to be assigned to a quality inspection work. It is required that at least 1,800 pieces are inspected per 8-hour day. Grade 1 inspectors can check pieces at the rate of 25 per hour with an accuracy of 98%. Grade 2 inspectors can check at the rate of 15 pieces per hour with an accuracy of 95%. The wage rate for grade 1 inspector is ₹ 40 per hour, while that of grade 2 is ₹ 30 per hour. Each time an error is caused by the inspector cost to the company is ₹ 20. The company has eight grade 1 and ten grade 2 inspectors. The company wants to determine the optimal assignment of inspectors to minimize total inspection cost. Formulate it as LPP and solve using graphical method.*

[MBA, Annamalai Univ. 2002; MBA, Delhi Univ. 2004]

Solution.

Step 1. The data of the given problem can be summarized as follows:

	Grade 1 inspector	*Grade 2 inspector*
Number of inspectors	8	10
Rate of checking per hour	25 pieces	15 pieces
Inaccuracy in checking	1 – 0.98 = 0.02	1 – 0.95 = 0.05
Cost of inaccuracy in checking	₹ 20	₹ 20
Wage rate per hour	₹ 40	₹ 30

Let x_1 and x_2 designate the number of Grade *1* and Grade 2 inspectors, respectively.

Hourly cost of each Grade *1* and Grade 2 inspectors are given by:

Grade *1* inspector: ₹ (40 + 20 × 0.02 × 25) = ₹ 50.

Grade 2 inspecto: ₹ (30 + 20 × 0.05 × 15) = ₹ 45.

Using the above information, an appropriate linear programming problem is:

Minimize $z = 8 \times 50x_1 + 8 \times 45x_2 = 400x_1 + 360x_2$ subject to the constraints:

$$8 \times 25x_1 + 8 \times 15x_2 \geq 1{,}800 \text{ or } 5x_1 + 3x_2 \geq 45 \quad \textit{(Number of pieces)}$$

$$x_1 \leq 8,\ x_2 \leq 10 \quad \textit{(Number of inspectors)}$$

$$x_1 \geq 0 \text{ and } x_2 \geq 0. \quad \textit{(Non-negative restriction)}$$

Step 2. Consider a set of cartesian rectangular axis OX_1X_2 in the plane. As each point has the coordinates of the type (x_1, x_2), any point satisfying the conditions $x_1 \geq 0$ and $x_2 \geq 0$ lies in the first quadrant only.

The first constraint $5x_1 + 3x_2 \geq 45$ can be graphed by plotting the straight line $\frac{x_1}{9} + \frac{x_2}{15} = 1$.

This gives an x_1-intercept of 9 and an x_2-intercept of 15. The area away from the origin, *i.e.*, above this line represents the feasible area of this constraint. Similarly, the other two constraints are depicted by plotting the straight lines corresponding to the equations $x_1 = 8$ and $x_2 = 10$. Here, the area below these lines (towards the origin) gives the feasible area. The common area *ABC* which satisfies all the three constraints is the solution space.

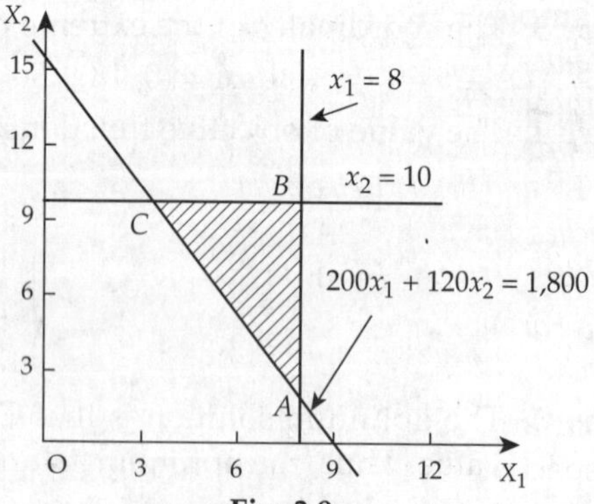

Fig. 3.9

***Step* 3.** The CPF solutions are: $A = (8, 5/3)$, $B = (8, 10)$, and $C = (3, 10)$.

***Step* 4.** The *z*-value corresponding to the feasible region (solution space) is:

Extreme point	*CPF solutions*	$z = 400x_1 + 360x_2$	
A	(8, 5/3)	3,800	← Minimum
B	(8, 10)	6,800	
C	(3, 10)	4,800	

Hence, the optimum solution is that 8 grade *1* inspectors and 5/3 (= 2) approx. grade *2* inspectors should be assigned to have ₹ 3,800 as the total minimum inspection cost.

Example 3.9. *Use the graphical method to solve the following LPP:*

$$\text{Maximize } z = 2x_1 + 3x_2 \text{ subject to the constraints:}$$

$$x_1 + x_2 \le 30, \quad x_1 - x_2 \ge 0, \quad x_2 \ge 3,$$

$$0 \le x_1 \le 20 \text{ and } 0 \le x_2 \le 12.$$

Solution. To graph the given inequalities, we first treat them as equations $x_1 + x_2 = 30$, $x_1 - x_2 = 0$, $x_2 = 3$, $x_1 = 20$ and $x_2 = 12$ and plot each of these equations as straight lines. We use the inequality condition of each constraint to plot the corresponding feasible region. The common region *ABCDE* satisfying all the inequalities (feasible region) is shown in shaded *Fig. 3.10*.

The coordinates of the extreme points of the feasible region (CPF solutions) are:

$A = (3, 3)$, $B = (20, 3)$, $C = (20, 10)$, $D = (18, 12)$, and $E = (12, 12)$.

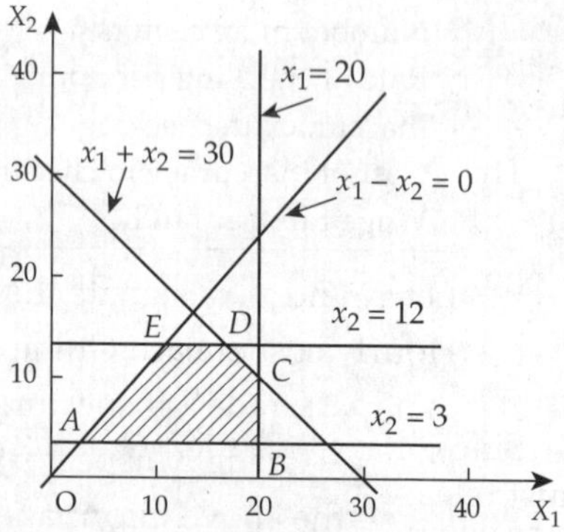

Fig. 3.10

The *z*-values corresponding to extreme points are:

Extreme point	*CPF solutions*	$z = 2x_1 + 3x_2$	
A	(3, 3)	15	
B	(20, 3)	49	
C	(20, 10)	70	
D	(18, 12)	72	← Maximum
E	(12, 12)	60	

The maximum value of *z* occurs at CPF solution (18, 12). Hence, the optimum solution is:

$$x_1 = 18, x_2 = 12 \text{ and maximum } z = 72.$$

Example 3.10. *The advertising agency wishes to reach two types of audiences — customers with annual incomes greater than ₹ 40,000 (target audience A) and customers with annual incomes of less than ₹ 40,000 (target audience B). the total advertising budget is ₹ 2,00,000. One programme of TV advertising costs ₹ 50,000; one programme of radio advertising costs ₹ 20,000. For contract reasons, at least 3 programmes ought to be on TV and the number of radio programmes must be limited to 5. Surveys indicate that a single TV programme reaches 7,50,000 customers in target audience A and 1,50,000 in target audience B. One radio programme reaches 40,000 in target audience A and 2,60,000 in target audience B.*

Formulate this as a linear programming problem and determine the media mix to maximize the total reach. **[MBA (Nov.), Delhi Univ. 2010]**

Solution. Let x_1 and x_2 represent the number of programmes to be released on TV and radio, respectively. The formulation of the linear programming model is:

$$\text{Maximize } z = \text{TV programme viewers} + \text{Radio programme viewers}$$
$$= (7{,}50{,}000 + 1{,}50{,}000)\, x_1 + (40{,}000 + 2{,}60{,}000)\, x_2$$
$$= 9{,}00{,}000x_1 + 3{,}00{,}000x_2$$

subject to the constraints:

$$50{,}000x_1 + 20{,}000x_2 \le 2{,}00{,}000 \quad \textit{(Budget constraint)}$$
$$x_1 \ge 3 \text{ and } x_2 \le 5 \quad \textit{(Programme constraints)}$$
$$x_1 \ge 0 \text{ and } x_2 \ge 0 \quad \textit{(Non-negative restriction)}$$

For solving this LPP graphically, we first consider the inequalities as equations and draw the straight lines of these equations. Now, consider the inequalities and shade the feasible region satisfying all the inequalities, *viz.*, the region *ABC* consisting of the extreme points *A*, *B* and *C*.

The coordinates of the extreme points of feasible region (or CPF solutions) are:

$A = (3, 0)$, $B\ (4, 0)$ and $C = (3, 5/2)$.

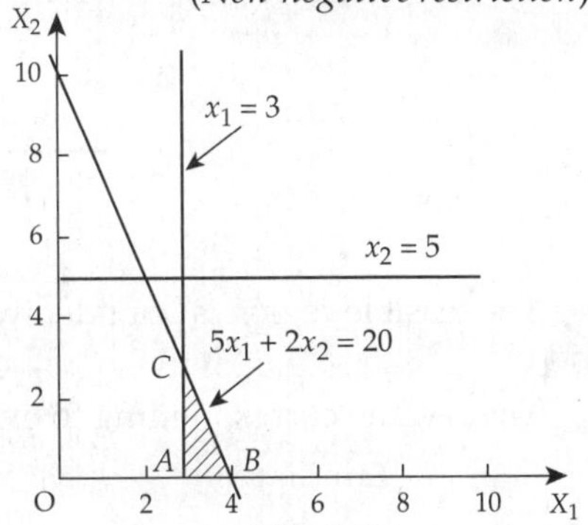

Fig. 3.11

The z-value corresponding to extreme points are:

Extreme point	*CPF solutions*	$z = 9{,}00{,}000x_1 + 3{,}00{,}000x_2$
A	(3, 0)	27,00,000
B	(4, 0)	36,00,000 ⟵ Maximum
C	(3, 5/2)	34,50,000

Since, the maximum value of *z* occurs at the extreme point *B*, *i.e.*, CPF solution (4, 0); the agency must release 4 programmes on TV and no programme on radio to achieve the maximum reach of 36,00,000 audiences.

Example 3.11. *A firm plans to purchase at least 200 quintals of scrap containing high quality metal X and low quality metal Y. It decides that the scrap to be purchased must contain at least 100 quintals of metal X and not more than 35 quintals of metal Y. The firm can purchase the scrap from two suppliers A and B in unlimited quantities. The percentage of X and Y metals in terms of weight in the scrap supplied by A and B is given below:*

Metals	*Supplier A*	*Supplier B*
X	25%	75%
Y	10%	20%

The price of A's scrap is ₹ 200 per quintal and that of B's scrap is ₹ 400 per quintal. The firm wants to determine the quantities that it should buy from the two suppliers so that the total cost is minimized.

[MBA, Delhi Univ. 2001]

Solution. Let x_1 and x_2 represent the quantity (in quintals) of scrap to be purchased from suppliers *A* and *B* respectively. Then, the formulation of linear programming model is:

Minimize $z = 200x_1 + 400x_2$ subject to the constraints:

$$x_1 + x_2 \geq 200 \quad \textit{(Constraint on purchase)}$$

$$(1/4)x_1 + (3/4)x_2 \geq 100 \text{ or } x_1 + 3x_2 \geq 400 \quad \textit{(Constrain on metal X)}$$

and

$$(1/10)x_1 + (1/5)x_2 \leq 35 \text{ or } x_1 + 2x_2 \leq 350 \quad \textit{(Constraint on metal Y)}$$

$$x_1 \geq 0 \text{ and } x_2 \geq 0. \quad \textit{(Non-negative restrictions)}$$

For solving this LPP graphically, plot on graph paper each of the three constraints by first treating them as linear equations and then consider the linear inequalities. The feasible region satisfying all the inequalities is given in *Fig. 3.12.*

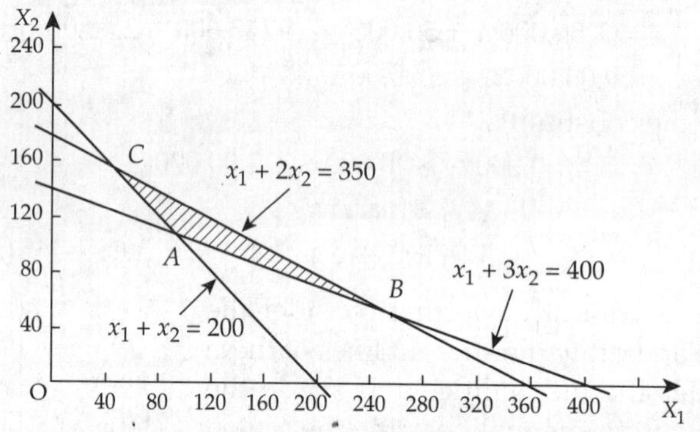

Fig. 3.12

The feasible region is bounded with its extreme points *A*, *B* and *C*, *i.e.*, CPF solutions are at *A*, *B* and *C*.

The *z*-value corresponding to extreme points (CPF solutions) is:

Extreme point	*CPF solution*	$z = 200x_1 + 400x_2$	
A	(100, 100)	60,000	← Minimum
B	(250, 50)	70,000	
C	(50, 150)	70,000	

Since, minimum value of *z* occurs at the CPF solution (100, 100), the firm should buy 100 quintals of scrap each from suppliers *A* and *B*.

Example 3.12. *Show that the following LPP has no feasible solution:*

Maximize $z = 4x_1 + 3x_2$ *subject to the constraints:*

$$x_1 + 2x_2 \geq 80, \quad x_1 + x_2 \leq 50,$$

$$3x_1 + 2x_2 \geq 140, \text{ and } x_1 \geq 0, x_2 \geq 0.$$

Solution. Do it yourself.

Example 3.13. *Show that the following LPP has an unbounded solution:*

Maximize $z = 5x_1 + 7x_2$ *subject to the constraints:*

$$x_1 - x_2 \leq -1, \quad -x_1 + x_2 \leq 0,$$

$$x_1 \geq 0 \text{ and } x_2 \geq 0.$$

Solution. Do it yourself.

Example 3.14. *Show graphically that the maximum or minimum values of the objective function for the following LPP are same:*

Maximize (or minimize) $z = 5x_1 + 3x_2$ *subject to the constraints:*

$x_1 + x_2 \leq 6,\ 2x_1 + 3x_2 \geq 3,$

$0 \leq x_1 \geq 3,$ *and* $0 \leq x_2 \geq 3.$ ***[MBA (DPL), Madurai Univ. 2009]***

Solution. Do it yourself.

PROBLEMS

Use graphical method to solve the following linear programming problems:

3.01. Maximize $z = 2x_1 + 3x_2$ subject to the constraints:

$x_1 + 2x_2 \leq 10,\ 2x_1 + x_2 \leq 14;$

$x_1 \geq 0$ and $x_2 \geq 0.$ ***[BE (Mech.), Kurukshetra Univ. 2012]***

3.02. Maximize $z = 5x_1 + 3x_2$ subject to the constraints:

$x_1 + x_2 \leq 6,\ 2x_1 + 3x_2 \geq 6,$

$0 \leq x_1 \leq 4$ and $0 \leq x_2 \leq 3.$ ***[MBA (DPL), Madurai Univ. 2009]***

3.03. Minimum $z = -x_1 + 2x_2;$ subject to the constraints:

$-x_1 + 3x_2 \leq 10,\ x_1 + x_2 \leq 6,\ x_1 - x_2 \leq 2,$

$x_1 \geq 0$ and $x_2 \geq 0.$ ***[BE, Gujarat Tech. Univ. 2012]***

3.04. Maximize $z = 10x_1 + 15x_2$ subject to the constraints:

$2x_1 + x_2 \leq 26,\ 2x_1 + 4x_2 \leq 56,$

$x_1 - x_2 \geq -5,\ x_1 \geq 0$ and $x_2 \geq 0.$ ***[MBA, Visvesvaraya Tech. Univ. 2011]***

3.05. Maximize $z = 60x_1 + 90x_2$ subject to the constraints:

$x_1 - 2x_2 \leq 40,\ 2x_1 + 3x_2 \leq 90,\ x_1 - x_2 \geq 10$

$x_1 \geq 0$ and $x_2 \geq 0.$ ***[MBA, Annamalai Univ. 2007]***

3.06. Maximum $z = 3x_1 + 5x_2$ subject to the constraints:

$-3x_1 + 4x_2 \leq 12,\ 2x_1 - x_2 \geq -2$

$2x_1 + 3x_2 \geq 12,\ 4 \geq x_1 \geq 0,\ 2 \geq x_2 \geq 0.$

3.07. Maximum $z = x_1 - 2x_2$ subject to the constraints:

$-x_1 + x_2 \leq 1,\ 6x_1 + 4x_2 \geq 24,$

$0 \leq x_1 \leq 5$ and $2 \leq x_2 \leq 4.$ ***[MCA, Panjab Tech. Univ. 2010]***

3.08. Minimum $z = x_1 + x_2$ subject to the constraints:

$5x_1 + 10x_2 \leq 50,\ x_1 + x_2 \geq 1,$

$x_1 \geq 0$ and $0 \leq x_2 \leq 4.$ ***[MBA, Gujarat Tech. Univ. 2010]***

3.09. Maximum $z = 3x_1 + 2x_2$ subject to the constraints:

(*a*) $-2x_1 + 3x_2 \leq 9,\ 3x_1 - 2x_2 \leq -20,$

$x_1 \geq 0$ and $x_2 \geq 0.$ ***[B.Tech. (Pred.), Panjab Tech. Univ. 2012]***

(*b*) $-2x_1 + 3x_2 \leq 9,\ x_1 - 5x_2 \geq -20,$

$x_1 \geq 0$ and $x_2 \geq 0.$ ***[MBA (Nov.), Panjab Tech. Univ. 2009]***

3.10. Consider the following LPP:

Maximize $z = 3x_1 + 7x_2$ subject to the constraints:

$4x_1 + 5x_2 \leq 20,\ 2x_1 + x_2 \leq 6,\ 2x_1 \geq 7,\ 2x_2 \leq 7;\ x_1 \geq 0$ and $x_2 \geq 0.$

Demonstrate whether it is possible to get a solution to the above LPP.

[MBA (HCA), Delhi Univ. 2015]

3.11. Two products A and B are to be manufactured. One unit of product A requires 2.4 minutes of punch press time and 5 minutes of assembly time. The profit for product A is ₹ 0.60 per unit. One unit of product B requires 3 minutes of punch press time and 2.5 minuts of welding time. The profit for product B is ₹ 0.70 per unit. The capacity of the punch press deptt. available for these products is 1,200 minutes/week. The welding deptt. has an ideal capacity of 600 minutes/week and assembly deptt. has 1,500 minutes/week.

(*i*) Formulate the problem as linear programming problem.

(*ii*) Determine the quantities of products A and B so that total profit is maximized.

[B.Tech. (Mech.), Panjab Tech. Univ. (Nov.) 2011]

3.12. A consumer durable manufacturer has potential market for two products P_1 and P_2. In the production shop he finds that two of his machines M_1 and M_2 which are necessary for the products P_1 and P_2 are having limited capacity. The following table gives the machine hours required and machine hours available:

Machine-hour required per unit

Product	M_1	M_2
P_1	3	2
P_2	4	1
Available machine-hours	36	16

The costing section has arrived at the contribution of P_1 as ₹ 30 per piece and of P_2 as ₹ 50 per piece. The sales division has indicated that they can sell any amount of P_2 but only 10 units of P_1. Find out the best manufacturing strategy.

3.13. Swasthya Hospitals imports mechanical heart valves. Different heart operations require valves of different sizes. Swasthya purchases valves from two different suppliers. The cost and size mix of the valves purchased are given in the following table. It can be observed from the table that for every order of 100 valves Supplier A sends 40 each of large and medium valves, and 20 small valves. All valves have the same per unit cost. Supplier B also offers a similar cost structure.

Supplier	*Cost per valve*	*Percentage*		
		Large	*Medium*	*Small*
A	₹ 5.0	40	40	20
B	₹ 4.0	20	20	60

Each month Swasthya places one order with each supplier. At least 100 large valves, 200 medium and 200 small valves must be purchased each month. Because of limited availability, at most 400 valves per month can be purchased from each supplier. Find the optimal number of valves to be purchased every month from Suppliers A and B. ***[MBA (PT), Delhi Univ. 2015]***

3.14. G J Breweries Ltd. has two bottling plants, one located at G and the other at J. Each plant produces three drinks, whisky, beer and fruit juices named A, B and C respectively. The number of bottles produced per day are as follows:

	Plant at	
	G	J
Whisky, A	1,500	1,500
Beer, B	3,000	1,000
Fruit juices, C	2,000	5,000

A market survey indicates that during the month of April, there will be a demand of 20,000 bottles of whisky, 40,000 bottles of beer and 44,000 bottles of fruit juices. The operating costs per day for plants at G and J are 600 and 400 monetary units. For how many days each plant be run in April so as to minimize the production cost, while still meeting the market demand?

Formulate this problem as an LPP, and solve it using graphical method.

[MBA, Pune Univ. 2009]

3.15. A factory has decided to diversify its activities. The data collected for the sales and production departments are summarized below:

Potential demand exists for two products A and B. Market can absorb any quantity of A, whereas the share of B for this organization is expected to be not more than 400 units per month.

For every three units of B produced, there is one unit of a by-product which sells at ₹ 3 per unit and only 100 units of this by-product can be sold per month. Contribution per unit of products A and B is expected to be ₹ 6 and ₹ 8, respectively. These products require three different processes and the time required per unit of product is given in the table below:

Process	*Product A*	*Product B*	*Available hours*
I	2	3	900
II	1	2	600
III	2	2	1,200

Find the product mix to optimize the contribution. *[B.Tech. (Mech.), Himachal Univ. 2010]*

3.16. A manufacturer employs three inputs: man-hours, machine-hours and cloth material to manufacture two types of dresses. Type A dress fetches him a profit of ₹ 160 per piece, while that of type B fetches ₹ 180 per piece. The manufacturer has enough man-hours to manufacture 50 pieces of type A or 20 pieces of type B dresses per day, while the machine-hours he possesses suffice only for 36 pieces of type A or for 24 pieces for type B dresses. Cloth material available per day is limited but sufficient enough for 30 pieces of either type of dresses. Formulate the linear programming model and solve it graphically. *[M.Com., Delhi Univ. 2005]*

$$\left[\textbf{Hint}: \text{Constraints are } \frac{x_1}{50} + \frac{x_2}{20} \le 1,\ \frac{x_1}{36} + \frac{x_2}{24} \le 1,\ \frac{x_1}{30} + \frac{x_2}{30} \le 1 \right]$$

3.17. The standard weight of a special purpose brick is 5 kg and it contains two basic ingredients B_1 and B_2. B_1 costs ₹ 5 per kg and B_2 costs ₹ 8 per kg. Strength consideration dictate that the brick contains not more than 4 kg of B_1 and minimum of 2 kg of B_2. If demand for the product is likely to be related to the price of the brick, find out graphically minimum cost of the brick satisfying the above conditions. *[MBA, Panjab Tech. Univ. 2011]*

3.18. A small scale manufacturer has production facilities for producing two different products. Each of the products requires three different operations: grinding, assembling and testing. Product A requires 15, 20 and 10 minutes to grinding, assembling and testing respectively, whereas product B requires 7.5, 40 and 45 minutes for grinding, assembling and testing. The production run calls for at least 7.5 hours of grinding time, at least 20 hours of assembling time, and at least 15 hours of testing time. If product A costs ₹ 60 and product B costs ₹ 90 to manufacture, determine the number of units of each product the firm should produce in order to minimize the cost of operations. *[MBA, IGNOU 2008]*

3.19. An oil refinery can blend three grades of crude oil to produce quality R and quality S petrol. Two possible blending processes are available. For each production run the older process uses 5 units of crude A, 7 units of crude B and 2 units of crude C to produce 9 units of R and 7 units of S. The newer process uses 3 units of crude A, 9 units of crude B and 4 units of crude C to produce 5 units of R and 9 units of S petrol. Because of prior contract commitments, the refinery must produce at least 500 units of R and at least 300 units of S for the next month. It has available 1,500 units of crude A, 1,900 units of crude B and 1,000 units of crude C. For each unit of R the refinery receive ₹ 60, while for each unit of S it receives ₹ 90. Write down the linear programming formulation of the problem so as to maximize the revenue and solve it by graphical method.

3.20. The manager of an oil refinery must decide on the optimal mix of two possible blending processes of which the inputs and outputs per production run are as follows:

Process	*Input (units)*		*Output (units)*	
	Grade A	*Grade B*	*Gasoline X*	*Gasoline Y*
1	5	3	5	8
2	4	5	4	4

The maximum amounts available of crude A and B are 200 units and 150 units respectively. Market requirements show that at least 100 units of gasoline X and 80 units of gasoline Y must be produced. The profit per production run for process *1* and process 2 are ₹ 300 and ₹ 400, respectively. Solve the LPP by the graphical method.

[B.Tech. (Mech.), Himachal Univ. 2009; MBA, Rajasthan Univ. 2008]

3.21. A scrap metal dealer has received an order from a customer for at least 2,000 kilograms of scrap metal. The customer requires that at least 1,000 kilograms of the shipment of metal must be of high quality copper that can be melted down and used to produce copper tubings. Furthermore, the customer will not accept delivery of the order if it contains more than 175 kilograms of metal that he deems unfit for commercial use, *i.e.*, metal that contains an excessive amount of impurities and cannot be melted down and refined profitably.

The dealer can purchase scrap metal from two different supplies in unlimited quantities with the following percentages (by weight) of high quality copper and unfit scrap:

Scrap metal	*Supplier A*	*Supplier B*
Copper	25%	75%
Unfit scrap	5%	10%

The costs per kilogram of scrap metal purchase from supplier A and supplier B are ₹ 1 and ₹ 4 respectively. The problem is to determine (using graphical method) the optimum quantities of scrap metal for the dealer to purchase from each of the two suppliers.

[MBA, Allahabad Univ. 2009]

3.22. A manufacturer produces two different models, X and Y, of the same product. The raw materials r_1 and r_2 are required for production. At least 18 kg of r_1 and 12 kg of r_2 must be used daily. Also, at most 34 hours of labour are to be utilized. 2 kg of r_1 are needed for each model X and 1 kg of r_1 for each model Y. For each model of X and Y, 1 kg of r_2 is required. It takes 3 hours to manufacture a model X and 2 hours to manufacture a model Y. The profit is ₹ 50 for each model X and ₹ 30 for each model Y. How many units of each model should be produced to maximize the profit. Use graphical model. ***[MBA, Gujarat Univ. 2008]***

3.23. A farmer has a supply of chemical fertilizer of Type *I* which contains 10% nitrogen and 6% phosphoric acid and Type *II* fertilizer which contains 5% nitrogen and 10% phosphoric acid. After testing the soil conditions of a field, it is found that at least 14 kg of nitrogen and 14 kg of phosphoric acid is required for a good crop. The fertilizer Type *I* costs ₹ 2 per kg and the Type *II* costs ₹ 3 per kg. How many kilograms of each fertilizer should be used to meet the requirement and the cost be minimum? (use Graphic method). ***[MBA, Jiwaji Univ. 2010]***

3.24. The ABC Company has been a producer of picture tubes for television sets and certain printed circuits for radios. The company has just expanded into full scale production and marketing of *AM* and *AM-FM* radios. It has built a new plant that can operate 48 hours per week. Production of an *AM* radio in the new plant will require 2 hours and production of an *AM-FM* radio will require 3 hours. Each *AM* radio will contribute ₹ 40 to profits, while an *AM-FM* radio will contribute ₹ 80 to profits. The marketing departments, after extensive research, has determined that a maximum of 15 *AM* radios and 10 *AM-FM* radios can be sold each week.

(*i*) Formulate a linear programming model to determine the optimum production mix of *AM-FM* radios that will maximize profits.

(*ii*) Solve the above problem using graphic method. ***[MBA, Delhi Univ. 2008]***

3.25. A firm makes two products *X* and *Y* and has a total production capacity of 9 tonnes per day, *X* and *Y* requiring the same production capacity. The firm has a permanent contract to supply at least 2 tonnes of *X* and at least 3 tonnes of *Y* per day to another company. Each tonne of *X* requires 20 machine hours production time and each tonne of *Y* requires 50 machine hours of production time. The daily maximum possible number of machine hours is 360. All the firm's output can be sold, and the product made is ₹ 80 per tonne of *X* and ₹ 120 per tonne of *Y*. It is required to determine the production schedule for maximum profit and to calculate this profit. ***[MBA, Delhi Univ. 2009]***

3.26. An aeroplane can carry a maximum of 200 passengers. A profit of ₹ 400 is made on each first class ticket and a profit of ₹ 300 is made on each economy class ticket. The airline reserves at least 20 tickets for first class seats. However, at least 4 times as many passengers prefer to travel by economy class as to the first class. Determine how many tickets of each type must be sold in order to maximize the total profit for the airline.

Formulate a linear programming model to determine the optimal mix that will maximize profit. Solve it by graphic method. ***[MBA (Nov.), MD Univ. 2010]***

3.27. A firm assembles and sells two different types of outboard motors *A* and *B* using four resources. The production process can be described as follows:

	Resources	*Capacity per month*
I	Motor unit shop	400 type *A* or 250 type *B* units or any linear combination of the two
II	Type *A* gear and drive shop resource	175 type *A* units
III	Type *B* gear and drive shop resource	225 type *B* units
IV	Final assembly resource	200 type *A* units or 350 type *B* units or any combination of the two

Type *A* units bring in a profit of ₹ 90 each and Type *B* units ₹ 60 each.

Formulate the above as an LPP to maximize profit and solve the same by graphic method. ***[B.Com. (Hons.), Delhi Univ. 2009]***

3.28. A company is manufacturing two products X and Y, which are processed through three departments A, B and C. The cost data for the two products are given below:

Cost data	*Product X (₹)*	*Product Y (₹)*
Selling price per unit	200	240
Cost per unit:		
(*i*) Material	45	50
(*ii*) Departmental overheads:		
Department A	16	10
Department B	22.50	13.50
Department C	10	30
(*iii*) General variable overheads	33	42

Fixed overheads are ₹ 2,750 per annum. Table below shows the available man-hours and rate per hour in departments A, B and C.

Department	*Available man-hours*	*Rate per hour (₹)*
A	800	2.00
B	600	2.25
C	720	2.50

Find the product mix to maximize the net profit.

[B.Tech. (Mech.) (Nov.), Himachal Univ. 2010]

[***Hint:*** Divide the departmental overheads by the rate per hour to find the processing time for the two products in three departments:

Department:	A	B	C
Product X:	8	10	4
Product Y:	5	6	12

Formulation of linear programming model:

$$\text{Maximize } z = \left(200 - 45 - 33 - \frac{16}{2} - \frac{22.50}{2.25} - \frac{10}{2.50}\right)x_1 + \left(240 - 50 - 42 - \frac{10}{2} - \frac{13.50}{2.25} - \frac{30}{2.50}\right)x_2$$
$$= 100x_1 + 125x_2$$

Constraints are: $8x_1 + 5x_2 \le 800,\ \ 10x_1 + 6x_2 \le 600,\ \ 4x_1 + 12x_2 \le 720,\ \ x_1 \ge 0, x_2 \ge 0$]

3:2. INSIGHTS INTO THE SIMPLEX METHOD

While solving an LPP graphically, we noticed the following important characteristics:

(*a*) Whenever feasible solutions existed, the region of feasible solutions was ***convex*** and bounded by ***lines*** **or** ***planes*** (more than two variable case). For each such convex region, there were ***corners*** (or vertices) on the boundary and ***edges*** joining these corners.

(*b*) For each value of z, the objective function could be represented by a line or plane and whenever the maximum or minimum value of z was finite, the optimal (or optimum) solution occurred at some corner (or vertex) of the convex region of feasible solutions. If the optimal solution was not unique, there were points (on an edge) that were optimal, but in every event at least one vertex was optimal. In the case of unbounded optimal solution; however, no corner point was optimal.

Interestingly, these observations, derived from simple graphical examples, hold true for the general LPP also, if we think of a geometrical representation in n dimensional space. But, it becomes too complex to solve the linear programming problems involving more than two variables.

The ***Simplex Method*** provides an efficient technique which can be applied for solving linear programming problems of any magnitude — involving two or more decision variables. The basic concept underlying the simplex method is:

"Since, feasible region is created by a set of straight lines representing the constraints and also because the objective function line is also a straight line, the point of maximum (or minimum) must lie at one of the corner points."

The evaluation of corner points always starts at the point of origin (initial basic feasible solution), which is one of the corner point of the feasible solution space. This solution is then tested for optimality, *i.e., it tests whether an improvement in the objective function is possible by moving to adjacent corner point of the feasible solution space*. If the improvement is possible, then at a new corner point, the solution is again tested for optimality. This iterative search for a better corner point is repeated until an optimal solution, if it exists, is determined.

3:3. THE COMPUTATIONAL PROCEDURE

The two fundamental conditions on which simplex method is based are:

(*i*) ***Condition of feasibility.*** It assumes that if the initial (starting) solution is basic feasible, then during computation only basic feasible solutions will be obtained.

(*ii*) ***Condition of optimality.*** It guarantees that only better solutions will be encountered.

The computational procedure of simplex method requires the construction of the simplex table (or tableau). The initial simplex table is constructed by writing out the coefficients of the variables in the objective function and constraints of LPP in a systematic tabular form.

Simplex Tableau

Consider the objective function as one of the constraint equation and transfer all the decision variables on left-hand side so that the right-hand side becomes zero.

To set up the tableau, we first list horizontally all the variables contained in the LPP. Next, the coefficients in the constraint equations are written listing vertically the coefficients under their respective variables. It may be noted that each of the slack variables, if any, appear only in one constraint equation. Therefore, the coefficient of each of the slack variable is taken to be zero in all the equations except the one in which it appears. After writing the coefficients of variables, the constraint values are mentioned on the right-hand side against the rows.

The last column of this tableau gives the current basic feasible solution, while the last row represents the coefficient of variables in the objective function known as net-evaluation row numbers. First column gives the coefficient of those variables of product-mix (or basic variables) whose values are given in the last column of the tableau.

Given below is a specimen of the ***simplex tableau***, which conveniently displays the basic feasible solution $\mathbf{x}_B$ of the LPP: Maximize $z = \mathbf{c}\,\mathbf{x}$ subject to the constraints: $\mathbf{A}\,\mathbf{x} = \mathbf{b}$ and $\mathbf{x} \geq \mathbf{0}$, where, $\mathbf{A} = (a_{ij})$ is an $m \times n$ matrix, $\mathbf{x} = [x_j]$ is $n \times 1$, $\mathbf{c} = (c_j)$ is $1 \times n$ and $\mathbf{b} = [b_i]$ is $m \times 1$ matrices, while $\mathbf{B}$ is an $m \times m$ non-singular matrix formed by m columns of $\mathbf{A}$, where $m < n$.

Coeff. of basic variable	*Basic variables, x_{Bi}*	x_1	x_2	x_3	...	x_n	z	*Solution*
c_{B_1}	x_{B_1}	a_{11}	a_{12}	a_{13}	...	a_{1n}	0	b_1
c_{B_2}	x_{B_2}	a_{21}	a_{22}	a_{23}	...	a_{2n}	0	b_2
$\vdots$	$\vdots$	$\vdots$	$\vdots$	$\vdots$	$\vdots$	$\vdots$	$\vdots$	$\vdots$
c_{B_m}	x_{B_m}	a_{m1}	a_{m2}	a_{m3}	...	a_{mn}	0	b_m
	Net evaluation: $(z_j - c_j)$ or z-equation row	$-c_1$	$-c_2$	$-c_3$	...	$-c_n$	1	0

Remarks

1. Basic feasible solution is given by $\mathbf{B}\,\mathbf{x}_B = \mathbf{b}$, where $\mathbf{x}_B \in$ variables corresponding to columns of $\mathbf{B} \subset \mathbf{A}$, *i.e.*, $\mathbf{x}_B = \mathbf{B}^{-1}\,\mathbf{b}$, as $\mathbf{B}$ is non-singular matrix ($|\mathbf{B}| \neq 0$).
2. Numbers, a_{ij} in the columns under each variable are called *substitution rates (or exchange coefficients)* because these represent the rate at which resource i ($i = 1, 2, \ldots, m$) is consumed by each unit of activity j ($j = 1, 2, \ldots, n$).
3. Net evaluations are given by $(z_j - c_j)$, where z_j denotes the *outgoing total profit/cost* and c_j *represents the incoming unit profit/cost* in respect of variable x_j. The values of z_j are computed by multiplying the coefficient of basic variables with the coefficient of variables x_j ($j = 1, 2, \ldots, n$) in (a_{ij}), $i = 1, 2, \ldots, m$ and $j = 1, 2, \ldots, n$. These are given in the last row of the tableau which is known as *z*-equation row.
4. The value of objective function *z* is given in the last column in the row of net evaluations. This is obtained by multiplying the coefficients of basic variables with the corresponding current solution given in the last column of the tableau.

Note. To simplify the computations, we usually consider the starting basis matrix $\mathbf{B}$ as an identity matrix $\mathbf{I}$, *i.e.*, $\mathbf{B} = \mathbf{I}$. Therefore, $\mathbf{x}_B = \mathbf{B}^{-1}\,\mathbf{b} = \mathbf{I}^{-1}\,\mathbf{b} = \mathbf{I}\mathbf{b} = \mathbf{b}$.

3:4. SIMPLEX ALGORITHM (MAXIMIZATION CASE)

The computational procedure for determining an optimum/optimal solution (if it exists) to an LPP, when objective function is of maximization and all the constraints are of '≤' type is obtained in the following eight steps:

Step 1. **Formulation of LPP.** Identify the objective function and constraints of the given problem. Express the objective function, constraints and non-negative restrictions in the format of linear programming problem.

Step 2. **Standardize the LPP.** Introduce slack variables on the left-hand side of the constraints to convert inequalities into equations. Assign zero coefficient to the slack variables in the objective function. The reformulated LPP is now in its standard form.

Step 3. **Determine the Initial Basic Feasible Solution.** The initial basic feasible solution is obtained by assigning zero value to all the decision variables just to initiate the solution procedure from the origin (as shown in graphical method).

Step 4. Set up the initial simplex tableau as discussed in *Section 3:3*.

Step 5. **Test the Solution for Optimality.** Examine the net-evaluation $(z_j - c_j)$ row of the simplex tableau obtained in *step 4*. If all $(z_j - c_j)$ are non-negative, *i.e.*, ≥ 0; the current solution is optimum (optimal). If at least one $(z_j - c_j)$ is negative, the current solution can be further improved by removing one of the basic variable and replacing it by some other non-basic variable. Thus, following three cases may arise:

(*i*) If all $(z_j - c_j) \geq 0$, current basic feasible solution is optimum.

(*ii*) If at least one $(z_j - c_j) < 0$ and all the elements in the column corresponding to this net evaluation are negative, there is an unbounded solution.

(*iii*) If at least one $(z_j - c_j) < 0$ and also at least one element in the column corresponding to this net evaluation is positive, go to *next step*.

Step 6. **Locate Scope for Improvement.** Choose the largest negative net evaluation, *i.e.*, $\max_j.\{(z_j - c_j) < 0\}$. Let it be $(z_k - c_k)$. The column corresponding to this net-evaluation, *i.e.*, k^{th} column becomes the ***key column***. Now, determine the non-negative ratios by dividing each element in the last column by the corresponding positive element in the *key* column and, then select the smallest of these, *i.e.*, $\min_i.\left\{\frac{x_{Bi}}{a_{ik}}, a_{ik} > 0\right\}$. The row corresponding to this smallest ratio determines the ***key row***. The non-basic variable corresponding to key column becomes the basic variable and the current basic variable corresponding to key row becomes non-basic variable in the next iteration.

The element at the intersection of key column and key row is called ***key number*** (or ***pivotal element***). This number is always a non-zero positive element.

Step 7. **Derive the Improved BF Solution.** Use key number to arrive at the improved BF solution by systematically transforming the simplex tableau with the aid of **Gauss-Jordan *complete elimination procedure***.

(*i*) If key number is 1, then the row in the revised simplex tableau remains the same.

(*ii*) If key number is other than 1, then divide each element in the key row (including the element of last column, *i.e.*, current BF solution) by the key number. New row of the simplex tableau so derived is called the ***replacement row***.

(*iii*) For each row other than the key row, the new values for each of the remaining rows (including z-equation row) are computed by using the following:

$$\begin{pmatrix}\text{New row}\\ \text{element}\end{pmatrix} = \begin{pmatrix}\text{Old row}\\ \text{element}\end{pmatrix} - \begin{pmatrix}\text{Row element in}\\ \text{the key column}\end{pmatrix} \times \begin{pmatrix}\text{Corresponding}\\ \text{replacement row value}\end{pmatrix}$$

Step 8. *Go to Step 5* and repeat the computational procedure untill all the net evaluations are non-negative, *i.e.*, $(z_j - c_j) \geq 0$, *i.e.*, an ***optimum solution exists***.

Remarks

1. If the objective function is that of minimization, then the LPP can be solved either by Big-*M* method or Two-phase method to be discussed in the *next section*, or by converting minimization problem into maximization by writing Minimum $z = -$ Maximum $(-z)$. For example, if Minimum $z = c_1x_1 + c_2x_2$, then Maximum $z^* = -c_1x_1 - c_2x_2$, where $z^* = (-z)$.
2. If any of the decision variable of the LPP is unrestricted in sign, express it as the difference of two non-negative variables. For example, if x_r is unrestricted, then write $x_r = x_r' - x_r''$, where $x_r' \geq 0$ and $x_r'' \geq 0$.
3. When there is a tie among non-basic variables to enter the basis (choosing the key column), selection for key column can be made arbitrary. However, the number of iterations to arrive at the optimum solution can be minimized by adopting the following rules:
   (*i*) If there is a tie between a decision variable and a slack (or surplus) variable, then select the decision variable to enter the basis.
   (*ii*) If there is a tie among two decision variables or among two slack/surplus variables, then selection can be made arbitrarily.
4. When there is a tie among basic variables to leave the basis (choosing the key row), it becomes the case of degeneracy. It generally occurs, when the minimum of the ratios does not become unique. In such a case, selection can be made arbitrarily. However, when there is a tie between a slack variable and artificial variable to leave the basis, preference should be given to the artificial variable to leave the basis.

Economic Interpretation of the Optimum Simplex Tableau

We have the following interesting interpretations for various terms associated with the optimum simplex tableau:

(*a*) Values of the decision variables appearing in the second (product-mix) column, *i.e.*, basic variables are given in the last (solution) column. All variables that do not appear in the second column are assumed to have zero value. In the case of slack variables, if some of them appear in the solution, it means there is some unutilized capacity/resource, otherwise there is no unused capacity/resource.

(*b*) Numbers under slack variables in the net evaluation row are known as ***shadow prices*** or ***accounting prices***. Thus, a negative number in the net evaluation row indicates an algebraic increase in the objective function, if one unit of the variable at the head of that column, were introduced in the basis (*solution*).

(*c*) The number in the net evaluation row, under the solution column gives the value of the objective function. This number indicates the improvement in the objective function.

(*d*) Remaining numbers in the coefficient matrix represent the marginal rate of exchange between the variables in the row and column.

Example 3.15 (*Maximization Problem*). *A factory engaged in the manufacture of pistons, rings and valves for which the profit per unit are ₹10, ₹6 and ₹4, respectively, wants to decide the most profitable mix. It takes one hour of preparatory work, ten hours of machining, two hours of packing and allied formalities for a piston. Corresponding time requirement of rings and valves are 1, 4 and 2; and 1, 5 and 6 hours respectively. The total number of hours available for preparatory work, machining, packing and allied formalities are 100, 600 and 300 respectively. Determine the most profitable mix, assuming that what is produced can be sold.*

Solution. The data of the given problem can be summarized in the following table:

Product	*Time required (in hrs.)*			*Profit*
	Preparatory work	*Machining*	*Packing*	
Pistons	1	10	2	₹ 10.00
Rings	1	4	2	₹ 6.00
Valves	1	5	6	₹ 4.00
Time available (hrs.)	100	600	300	

Step 1. Let x_1, x_2 and x_3 be the required quantities of pistons, rings and valves that need to be produced. Then, since the objective is to maximize the total profit, given problem has the following linear programming formulation:

Maximize $z = 10x_1 + 6x_2 + 4x_3$ subject to the constraints:

$$x_1 + x_2 + x_3 \leq 100 \quad \textit{(Preparatory work)}$$
$$10x_1 + 4x_2 + 5x_3 \leq 600 \quad \textit{(Machining)}$$
$$2x_1 + 2x_2 + 6x_3 \leq 300 \quad \textit{(Packing)}$$
$$x_1, x_2, x_3 \geq 0$$

Step 2. Introduce slack variables $s_1 \geq 0$, $s_2 \geq 0$ and $s_3 \geq 0$ in the constraint inequalities to convert these into equations. Also, since, slack variables represent unused resources, their contribution in the objective function is zero. The formulated LPP, therefore, is in its ***standard form***, *viz.*,

Maximize $z = 10x_1 + 6x_2 + 4x_3 + 0 \cdot s_1 + 0 \cdot s_2 + 0 \cdot s_3$ subject to the constraints:

$$x_1 + x_2 + x_3 + s_1 = 100$$
$$10x_1 + 4x_2 + 5x_3 + s_2 = 600$$
$$2x_1 + 2x_2 + 6x_3 + s_3 = 300$$
$$x_1, x_2, x_3, s_1, s_2, s_3 \geq 0$$

Step 3. The ***initial basic feasible solution*** is obtained by setting $x_1 = x_2 = x_3 = 0$. Thus, we have

$$s_1 = 100, \quad s_2 = 600 \quad \text{and} \quad s_3 = 300$$

Step 4. Write the objective function of the LPP obtained in *step 2* as one of the constraint equation. Transpose all terms of this equation to the left side and then consider it as the ***last constraint equation***. Thus, we have:

$$x_1 + x_2 + x_3 + s_1 = 100$$
$$10x_1 + 4x_2 + 5x_3 + s_2 = 600$$
$$2x_1 + 2x_2 + 6x_3 + s_3 = 300$$

and $\quad -10x_1 - 6x_2 - 4x_3 + 0 \cdot s_1 + 0 \cdot s_2 + 0 \cdot s_3 + z = 0$

Making use of the above system of equations, the ***initial simplex tableau*** is:

Coeff. of basic variable	*Basic variable* (x_{Bi})	x_1	x_2	x_3	s_1	s_2	s_3	z	*Solution*	*Ratio* $\frac{x_{Bi}}{a_{i1}}, a_{i1} > 0$
0	s_1	1	1	1	1	0	0	0	100	100/1 = 100
0	$\leftarrow s_2$	(10)	4	5	0	1	0	0	600	600/10 = 60
0	s_3	2	2	6	0	0	1	0	300	300/2 = 150
	$(z_j - c_j)$ or z-equation row	−10 ↑	−6	−4	0	0	0	1	$z = 0$	

Step 5. Since, first three net evaluation numbers are negative, current solution can be improved by removing one of the basic variables.

Step 6.

(*a*) The largest negative number in the net-evaluation row is (−10) under column x_1. This suggests that the column corresponding to this variable x_1 becomes ***key column*** and the non-basic variable x_1 becomes basic at the next iteration. In other words, x_1 (Pistons) produced first in order to increase total profit at the fastest rate, since, each unit of x_1 added into the solution mix will contribute ₹ 10 to the over all profit.

(*b*) Divide each element of the ***last column*** by the corresponding positive element of the ***Key-column*** (column under x_1) and choose the minimum of these, *i.e.*,

$$\text{Minimum} \left\{ \frac{100}{1}, \frac{600}{10}, \frac{300}{2} \right\} \text{ or } 60.$$

Since, minimum ratio is corresponding to second row, this row becomes ***key row*** and the basic variable s_2 leaves the basis at the next iteration, *i.e.*, becomes non-basic.

(*c*) The element at the intersection of first column (***key column***) and second row (***key row***) becomes the ***key number***, *i.e.*, number (10) becomes ***key number***.

Step 7.

(*a*) Divide each element of the key row (second row) by the key number (10). The replacement row would be:

10/10   4/10   5/10   0   1/10   0   0   60

(*b*) The new values (elements) of the ***first row*** under column

$$x_1 = 1 - 1 \times 1 = 0;\ x_2 = 1 - 1 \times \frac{4}{10} = 6/10;\ x_3 = 1 - 1 \times \frac{5}{10} = 5/10;$$

$$s_1 = 1 - 1 \times 0 = 1;\ s_2 = 0 - 1 \times \frac{1}{10} = -1/10;\ s_3 = 0 - 1 \times 0 = 0;$$

Solution = 100 − 1 × 60 = 40.

Similarly, new values of ***third row*** under column

$$x_1 = 2 - 2 \times 1 = 0;\ x_2 = 2 - 2 \times \frac{4}{10} = 6/5;\ x_3 = 6 - 2 \times \frac{5}{10} = 5;$$

$$s_1 = 0 - 2 \times 0 = 0;\ s_2 = 0 - 2 \times \frac{1}{10} = -1/5;\ s_3 = 1 - 2 \times 0 = 1;$$

Solution = 300 − 2 × 60 = 180.

New values of the *index row* under column

$$x_1 = -10-(-10)\times 1 = 0;\ x_2 = -6-(-10)\times\frac{4}{10} = -2;\ x_3 = -4-(-10)\times\frac{5}{10} = 1;$$

$$s_1 = 0-(-10)\times 0 = 0;\quad s_2 = 0-(-10)\times\frac{1}{10} = 1;\quad s_3 = 0-(-10)\times 0 = 0;$$

Solution = 0 – (– 10) × 60 = 600.

Making use of the above informations, the revised improved solution is shown in the simplex tableau given below:

Coeff. of basic variable	*Basic variable* (x_{Bi})	x_1	x_2	x_3	s_1	s_2	s_3	z	*Solution*	*Ratio* $\frac{x_{Bi}}{y_{i2}}, y_{i2} > 0$
0	← s_1	0	(6/10)	5/10	1	–1/10	0	0	40	$\frac{40}{6/10}$
10	x_1	1	4/10	5/10	0	1/10	0	0	60	$\frac{60}{4/10}$
0	s_3	0	6/5	5	0	–1/5	1	0	180	$\frac{180}{6/5}$
	$(z_j - c_j)$	0	–2 ↑	1	0	1	0	1	z = 600	

Step 8. The $(z_j - c_j)$ of the revised simplex tableau indicates a negative number under x_2-column. This implies that x_2-column becomes key column. Key row is then determined by choosing the minimum of the ratios obtained by dividing the elements of current solution by the corresponding positive elements of the key column. The key number, thus, is 6/10 which is at the intersection of key row and key column. Therefore, the non-basic variable x_2 enters the basis and the basic variable s_1 leaves the basis at the next iteration.

Use, now ***Gauss-Jordan Complete*** elimination procedure to convert the key number into unity and all other elements of the key column as zero. The revised improved solution, therefore, is shown in the simplex tableau given below:

Coeff. of basic variable	*Basic variable* (x_{Bi})	x_1	x_2	x_3	s_1	s_2	s_3	z	*Solution*
6	x_2	0	1	5/6	5/3	–1/6	0	0	200/3
10	x_1	1	0	1/6	–2/3	1/6	0	0	100/3
0	s_3	0	0	4	–2	0	1	0	100
	$(z_j - c_j)$	0	0	8/3	10/3	2/3	0	1	z = 2200/3

The net evaluation numbers of this tableau gives all its elements as non-negative, thereby sign of no further improvement in the profit. Thus, it follows that an optimum solution has been arrived at, *viz.*, $x_1 = 100/3$, $x_2 = 200/3$, $x_3 = 0$ and maximum $z = 2200/3$ = ₹ 733.33.

Step 9 The solution obtained above can be interpreted as follows:

The maximum profit of ₹ 733.33 is obtained when factory manufactures 100/3 (or 33.3) Pistons, 200/3 (66.6) Rings and no unit of Valves. The solution also indicates that 100 hours of Packing are still remained unutilized as shown by $s_3 = 100$.

Remark. *The coefficients of non-basic variables s_1 and s_2 in the net evaluation row are the '**shadow prices**'. They represent the decrease in the optimum value of the objective function resulting from a unit increase in a non-basic variable.*

Example 3.16 (Maximization Problem). *The XYZ Company manufactures two products A and B. These are processed on the same machine. Product A takes 10 minutes per item and product B takes 2 minutes per item on machine. Machine can run for a maximum of 35 hours in a week. Product A requires 1 kg and product B requires 0.5 kg of the raw material per item, the supply of which is 600 kg per week. Not more than 800 items of product B are required per week. If product A costs ₹5 per item and can be sold for ₹10; and product B costs ₹6 per item and can be sold for ₹8 per item, determine by simplex method how many items per week should be produced of A and B in order to maximize the profit.* **[BBA, GGSIP Univ. 2006]**

Solution. The information given in the product-mix problem may be summarized as under:

Product	*Processing time (minutes)*	*Raw material (kg)*	*Items per week*	*Profit (₹)*
A	10	1	—	10 – 5 (= 5)
B	2	0.5	1	8 – 6 (= 2)
Required	35 × 60 = 2,100	600	800	

Step 1. Let the company manufacture x_1 items of product A and x_2 items of product B. Then, since the objective is to maximize the total profit, given problem has the following linear programming formulation:

$$\text{Maximize } z = 5x_1 + 2x_2 \text{ subject to the constraints:}$$

$$10x_1 + 2x_2 \le 2100 \quad (Time)$$

$$x_1 + 0.5x_2 \le 600 \quad (Raw\ material)$$

$$x_2 \le 800 \quad (Manufacturing\ condition)$$

and $x_1, x_2 \ge 0.$

Step 2. Introducing slack variables $s_1 \ge 0$, $s_2 \ge 0$ and $s_3 \ge 0$, the reformulated LPP is:

$$\text{Maximize } z = 5x_1 + 2x_2 + 0 \cdot s_1 + 0 \cdot s_2 + 0 \cdot s_3 \text{ subject to the constraints :}$$

$$10x_1 + 2x_2 + s_1 = 2100$$

$$x_1 + 0.5x_2 + s_2 = 600$$

$$x_2 + s_3 = 800$$

$$x_1 \ge 0,\ x_2 \ge 0,\ s_1 \ge 0,\ s_2 \ge 0 \text{ and } s_3 \ge 0.$$

Step 3. From the above system of equations, the initial basic feasible solution is:

$$s_1 = 2100,\ s_2 = 600 \text{ and } s_3 = 800.$$

Initial Iteration. x_1 enters and s_1 leaves.

Step 4. Making use of the above information, the ***initial simplex tableau*** is:

Coeff. of basic variable	*Basic variable* (x_{Bi})	x_1	x_2	s_1	s_2	s_3	z	Solution	Ratio $\frac{x_{Bi}}{a_{i1}}, a_{i1} > 0$
0	$\leftarrow s_1$	(10)	2	1	0	0	0	2100	2100/10
0	s_2	1	0.5	0	1	0	0	600	600/1
0	s_3	0	1	0	0	1	0	800	—
	$(z_j - c_j)$	–5 ↑	–2	0	0	0	1	$z = 0$	

Step 5. Since, the most negative net evaluation number lies under x_1 column, the current non-basic variable x_1 will become basic at the next iteration. Dividing each element of the ***current solution*** column (last column) by the corresponding positive element in the ***key column***, we get

$$\frac{2,100}{10} = 210; \text{ and } \frac{600}{1} = 600.$$

The smallest ratio (= 210) identifies that 10 in the first row becomes the ***key number***.

First Iteration. x_2 enters and s_3 leaves.

Step 6. Using key number to arrive at an improved revised basic feasible solution by systematically transforming the simplex tableau with the aid of ***Gauss-Jordan*** complete elimination procedure, we get the next iterative tableau:

Coeff. of basic variable	*Basic variable* (x_{Bi})	x_1	x_2	s_1	s_2	s_3	z	*Solution*	*Ratio* $\frac{x_{Bi}}{a_{i2}}, a_{i2} > 0$
5	x_1	1	1/5	1/10	0	0	0	210	$\frac{210}{1/5}$
0	s_2	0	3/10	−1/10	1	0	0	390	$\frac{390}{3/10}$
0	$\leftarrow s_3$	0	(1)	0	0	1	0	800	$\frac{800}{1}$
	$(z_j - c_j)$	0	−1 ↑	1/2	0	0	1	$z = 1,050$	

Step 7. The last row of the above tableau indicates that there is only one negative net evaluation number which lies under non-basic variable x_2. So, the current non-basic variable x_2 will become basic at the next iteration. Also, by dividing the positive elements in the column under x_2 into the extreme right element of the same row, we get

$$\frac{210}{1/5}, \frac{390}{3/10} \text{ and } \frac{800}{1}, \text{ i.e., } 1{,}050,\ 1{,}300 \text{ and } 800.$$

The smallest ratio (= 800) identifies that element 1 in the third row becomes ***key number.***

Step 8. Using the key number to arrive at an improved revised basic feasible solution, the next iterative tableau is:

Coeff. of basic variable	*Basic variable* (x_{Bi})	x_1	x_2	s_1	s_2	s_3	z	*Solution*
5	x_1	1	0	1/10	0	−1/5	0	50
0	s_2	0	0	−1/10	1	−3/10	0	150
2	x_2	0	1	0	0	1	0	800
	$(z_j - c_j)$	0	0	1/2	0	1	1	$z = 1,850$

The above simplex tableau yields a new basic feasible solution with an increased value of z. Moreover, no further improvement in the value of z is possible, since, all the numbers in the net evaluation row are non-negative. Hence, the desired optimum solution is:

$$x_1 = 50,\ x_2 = 800 \text{ and } s_3 = 0 \text{ with max. } z = 1{,}850.$$

In other words, manufacturer must produce 50 items of product *A* and 800 items of product *B* in order to achieve a maximum profit of ₹ 1,850.

3:5. USE OF ARTIFICIAL VARIABLES

We have seen that in the computational procedure of the simplex method, it is most convenient to have slack variables as the starting (initial) basic variables. Thus, if the original constraint is an equation or is of the type (≥) we may no longer have a ***ready*** starting basic feasible solution.

In order to obtain an initial basic feasible solution, we first put the given LPP into its standard form and then a non-negative variable is added to the left side of each of equation that lacks the much needed starting basic variables. The so-added variable is called an ***artificial variable*** and plays the same role as a slack variable in providing the initial basic feasible solution. However, since, such artificial variables have no physical meaning from the standpoint of the original problem, the method will be valid only if we are able to force these variables to be out or at zero level when the optimum solution is attained. In other words, to get back to the original problem, artificial variables must be driven to zero in the final solution; otherwise the resulting solution may be infeasible.

Two methods are generally employed for the solution of linear programming problems for removing artificial variables from the solution, *viz.*,

1. Two-Phase Method; and
2. Big-*M* Method or Method of Penalties.

1. Two-Phase Method

In the first phase of this method, the sum of the artificial variables is minimized subject to the given constraints (known as auxiliary LPP) to get a basic feasible solution to the original LPP. Second phase then optimizes the original objective function starting with the basic feasible solution obtained at the end of *Phase 1*.

The iterative procedure of the algorithm may be summarize as below:

***Step* 1.** Write the given LPP into its standard form and check whether there exists a starting basic feasible solution.

(*a*) If there is a ready starting basic feasible solution, go to *Phase* 2.

(*b*) If there does not exist a ready starting basic feasible solution, go on to the next step.

PHASE 1

***Step* 2.** Add the artificial variable to the left side of the each equation that lacks the needed starting basic variables. Construct an auxiliary objective function aimed at minimizing the sum of all artificial variables.

Thus, the new objective function is to

$$\text{Minimize } z^* = A_1 + A_2 + \ldots + A_n$$

where, A_i ($i = 1, 2, \ldots, m$) are the non-negative artificial variables.

***Step* 3.** Apply simplex algorithm to the specially constructed LPP. The following three cases may arise at the last iteration:

(*a*) minimum $z^* > 0$ and at least one artificial variable is present in the basis with positive value. In such a case, the original LPP does not possess any feasible solution.

(*b*) minimum $z^* = 0$ and at least one artificial variable is present in the basis at zero value. In such a case, the original LPP possess the feasible solution. In order to get basic feasible solution, we may proceed directly to *Phase 2* or else eliminate the artificial basic variable and then proceed to *Phase* 2.

(*c*) minimum $z^* = 0$ and no artificial variable is present in the basis. In such a case, a basic feasible solution to the original LPP has been found. Go to *Phase* 2.

PHASE 2

***Step* 4.** Consider the optimum basic feasible solution of *Phase 1* as a starting basic feasible solution for the original LPP. Assign actual coefficient to the variables in the objective function and a value zero to the artificial variables that appear at zero value in the final simplex tableau of *Phase 1*.

Apply usual simplex algorithm to the modified simplex tableau to get the optimum solution of the original problem.

Note. Artificial variables that do not appear in the basic feasible solution may be removed from the simplex tableau. An artificial variable is never considered for re-entry into the basis.

2. Big-M Method (*Method of Penalties*)

The Big-M method is an alternative method of solving a linear programming problem involving artificial variables. In this method we assign a very high penalty (say M) to the artificial variables in the objective function.

The computational procedure for determining an optimum solution to an LPP when objective function is of minimization is obtained in the following seven steps:

Step 1. ***Formulation of LPP.*** Identify the objective function and constraints of the given problem. Express the objective function, constraints and non-negative restrictions in the format of linear programming.

Step 2. ***Standardize the LPP.*** Introduce slack and/or surplus (negative slack) variables on the left-hand side of the constraints to convert inequalities into equations. Assign a zero coefficient to slack and/or surplus variables in the objective function. Also, add an ***artificial*** variable to each of the constraint equations and assign high penalty cost '**M**' to artificial variables in the objective function.

Step 3. ***Determine the Initial Solution.*** The initial feasible solution is obtained by assigning zero value to all the decision as well as slack and/or surplus variables just to initiate the solution procedure.

Step 4. ***Set Up the Initial Tableau.*** Eliminate the ***artificial variables*** in the objective function of the standardized linear programming problem obtained in *step* 2 with the help of constraint equations. Then, develop the initial simplex tableau for this reformulated LPP in the same way as discussed in maximization case.

Step 5. ***Test the Solution for Optimality.*** Examine the net evaluation (index) row numbers of the simplex tableau obtained in *step* 4. If all the $(z_j - c_j)$ are '≤ 0', current basic feasible solution is optimum. Otherwise, select the most positive $(z_j - c_j)$. The column corresponding to this number determines the ***key column***. Non-basic variable corresponding to key column becomes basic in the next iteration.

Step 6. ***Derive the Improved BFS.*** Determine the ***key row*** and ***key number*** in the same way as discussed in the maximization case. Thereafter, derive the revised tableau for improved solution.

Step 7. ***Go to step*** 5 and repeat the computational procedure until an ***optimum solution*** is obtained or there is an indication of an ***unbound solution***.

Remarks

1. At any iteration of the simplex method, any one of the following cases may arise:
   (*a*) There is at least one artificial variable in the basic feasible solution at ***zero level***, *i.e.*, the value of this artificial variable in the solution is zero. Also, the coefficient of M in each number of z-equation row is ***non-positive***, *i.e.*, ≤ 0. Then, the solution of the problem under consideration is said to degenerate.
   (*b*) If at least one artificial variable is present in the basic feasible solution at ***positive level*** but not at zero level, and in the z-equation row, the coefficient of M in each number of the z-equation row is ***non-positive***, then the problem under consideration does not possess optimum basic feasible solution.
2. It is to be noted that an artificial variable, once driven out at any iteration, can never re-enter, because of the ***huge penalty cost (M)*** associated with it in the objective function. Advantage can be taken of this fact by not computing its column in iteration subsequent to the one from which it was driven out.
3. Method of solving LPP when a high penalty cost 'M' has been assigned to the artificial variable, is known as ***method of Penalties*** or ***Big 'M' method***.

Example 3.17. *Use simplex method to solve the following LPP:*

$$\text{Minimize } z = 3x_1 + 2.5x_2 \text{ subject to the constraints:}$$
$$2x_1 + 4x_2 \geq 40,$$
$$3x_1 + 2x_2 \geq 50,$$
$$x_1 \geq 0 \text{ and } x_2 \geq 0.$$

Solution. For writing the given LPP into standard form, we introduce surplus variables $s_1 \geq 0$ and $s_2 \geq 0$ in the constraints and assign zero as the coefficient to these variables in the objective function. Thus, we get

$$\text{Minimize } z = 3x_1 + 2.5x_2 + 0 \cdot s_1 + 0 \cdot s_2 \text{ subject to the constraints:}$$
$$2x_1 + 4x_2 - s_1 = 40$$
$$3x_1 + 2x_2 - s_2 = 50$$
$$x_1 \geq 0,\ x_2 \geq 0,\ s_1 \geq 0 \text{ and } s_2 \geq 0.$$

We observe that there is no ready basic feasible solution to the LPP. So, we introduce two new variables $A_1 \geq 0$ and $A_2 \geq 0$ to give us the starting basic feasible solution. These extraneous variables (known as artificial variables) provide us the starting basic feasible solution. However, since, such artificial variables have no physical meaning, we assign a very high penalty (say M) to these variables in the objective function of the given LPP.

For removing artificial variables from the optimum solution of the given LPP, there are two methods known as: ***Two-Phase method*** and ***Big-M method***. We shall discuss both of these here.

1. Two-Phase Method. *Phase 1* consider the objective function as minimization of the sum of artificial variables only, *i.e.*,

$$\text{Minimize } z^* = A_1 + A_2, \text{ considering } x_1 = x_2 = 0.$$

This is known as ***auxiliary objective function***.

By considering ***auxiliary objective function*** as one of the constraints, we have following system of equations:

$$2x_1 + 4x_2 - s_1 + A_1 = 40,$$
$$3x_1 + 2x_2 - s_2 + A_2 = 50,$$

and
$$z^* - A_1 - A_2 = 0.$$

We observe that the starting (initial) basic feasible solution is: $A_1 = 40$ and $A_2 = 50$.

Now, eliminating A_1 and A_2 in the z-equation with the help of constraint equations, we get

$$z^* - (40 - 2x_1 - 4x_2 + s_1) - (50 - 3x_1 - 2x_2 + s_2) = 0$$

i.e.,
$$z^* + 5x_1 + 6x_2 - s_1 - s_2 = 90.$$

The initial (starting) simplex tableau, therefore, is:

Initial Iteration. x_2 enters and A_1 leaves.

Coeff. of basic variable	*Basic variable* (x_{Bi})	x_1	x_2	s_1	s_2	A_1	A_2	z^*	*Solution*	*Ratio* $\frac{x_{Bi}}{a_{i2}}, a_{i2} > 0$
1	$\leftarrow A_1$	2	(4)	–1	0	1	0	0	40	40/4 = 10
1	A_2	3	2	0	–1	0	1	0	50	50/2 = 25
	$(z_j - c_j)$	5	6 ↑	–1	–1	0	0	1	$z^* = 90$	

Since, it is a minimization problem, all the net-evaluations, *i.e.*, $(z_j - c_j)$ must be non-positive. Here, most positive $(z_j - c_j)$ is given by $(z_2 - c_2)$, *viz.*, 6. This implies that x_2 enters the basis and column under x_2 becomes key column. Further, since, minimum of ratio is in the first row (*i.e.*, A_1-row), first row becomes the key row and thereby current basic variable A_1 leaves the basis and the element at the intersection of key column and key row becomes key number. Thus, a_{12} (= 4) becomes the key number.

Using, now, key number to arrive at a new basic feasible solution by systematically transforming the simplex tableau with the aid of Gauss-Jordan elimination procedure, we arrive at the next simplex tableau as follows:

First Iteration. x_1 enters and A_2 leaves.

Coeff. of basic variable	*Basic variable* (x_{Bi})	x_1	x_2	s_1	s_2	A_2	z^*	Solution	*Ratio* $\frac{x_{Bi}}{a_{i1}}, a_{i1} > 0$
0	x_2	1/2	1	–1/4	0	0	0	10	10/½
1	$\leftarrow A_2$	(2)	0	1/2	–1	1	0	30	30/2
	$(z_j - c_j)$	2 ↑	0	1/2	–1	0	1	$z = 30$	

In the above tableau, we have omitted all the entries in the column under A_1, because the artificial variable A_1 has left the basis and we shall not like it to re-enter into the basis again in the subsequent iterations.

Now, since, the most positive coefficient in the last row lies under x_1, first column becomes the key column. Also, minimum of the ratio is in second row, therefore, A_2 leaves the basis, *i.e.*, element 2 (= a_{21}) is the key number.

Final Iteration. Optimum solution.

Coeff. of basic variable	*Basic variable* (x_{Bi})	x_1	x_2	s_1	s_2	z^*	*Solution*
0	x_2	0	1	–3/8	1/4	0	5/2
0	x_1	1	0	1/4	–1/2	0	15
	$(z_j - c_j)$	0	0	0	0	1	$z = 0$

Since, minimum $z = 0$ and no artificial variable is in the basis, we move on to ***Phase 2.***

Phase 2. In *phase* 2, we assign the actual coefficient of the decision variables in the objective function and consider the final simplex tableau of phase 1 as the initial simplex tableau of phase 2. Thus, we have

Coeff. of basic variable	*Basic variable* (x_{Bi})	x_1	x_2	s_1	s_2	z	*Solution*
2.5	x_2	0	1	–3/8	1/4	0	5/2
3	x_1	1	0	1/4	–1/2	0	15
	$(z_j - c_j)$	–3	–2.5	0	0	1	$z = 0$

Since x_1 and x_2 are basic variables, we derive $z_1 - c_1 = 0$ and $z_2 - c_2 = 0$.

For this, we multiply all the elements of first row by 2.5 and all the elements of second row by 3, and then add to the last row. We get the following revised simplex tableau:

Coeff. of basic variable	Basic variable (x_{Bi})	x_1	x_2	s_1	s_2	z	Solution
2.5	x_2	0	1	–3/8	1/4	0	5/2 = 2.5
3	x_1	1	0	1/4	–1/2	0	15
	$(z_j - c_j)$	0	0	–1/2	–1/4	1	$z = 51.25$

Here, since, all the $(z_j - c_j)$ are non-positive, an optimum solution to the given LPP is attained, *viz.*, $x_1 = 15$ and $x_2 = 2.5$, with minimum $z = 51.25$.

Alternative Method. Another method of removing artificial variables from the original linear programming problem is ***Big-M method*** (or method of penalties).

In this method, we consider the objective function as:

$$\text{Minimize } z = 3x_1 + 2.5x_2 + 0 \cdot s_1 + 0 \cdot s_2 + MA_1 + MA_2$$

We, now, eliminate A_1 and A_2 in this z-equation by using the constraint equations. Thus, we have

$$z - 3x_1 - 2.5x_2 - M(40 - 2x_1 - 4x_2 + s_1) - M(50 - 3x_1 - 2x_2 + s_2) = 0$$

i.e.,
$$z - (3 - 5M)\, x_1 - (2.5 - 6M)\, x_2 - Ms_1 - Ms_2 = 90M$$

The initial basic feasible solution, therefore, is given in the starting simplex tableau.

Initial Iteration. x_2 enters and A_1 leaves.

Coeff. of basic variable	Basic variable (x_{Bi})	x_1	x_2	s_1	s_2	A_1	A_2	z	Solution	Ratio $\frac{x_{Bi}}{a_{i2}}, a_{i2} > 0$
M	$\leftarrow A_1$	2	(4)	–1	0	1	0	0	40	40/4
M	A_2	3	2	0	–1	0	1	0	50	50/2
	$(z_j - c_j)$	–3 + 5M	–2.5 + 6M ↑	–M	–M	0	0	1	$z = 90M$	

Observe that this is a minimization problem, and therefore, the non-basic variable that enters the basis must have most positive coefficient of M in the net evaluation row, *i.e.*, most positive $(z_j - c_j)$.

In the above tableau, since, the most positive value in the last row lies under x_2, the current non-basic variable x_2 will become basic in the next iteration. Further, dividing the positive elements in the column under x_2 into the extreme right element of the same row, we get

$$40/4 = 10; \quad \text{and} \quad 50/2 = 25$$

The smallest ratio, 10, identifies 4 as the key number. In other words, x_2 will become basic variable in place of A_1 in the next iteration.

Using, now, key number to arrive at a new basic feasible solution by systematically transforming the simplex tableau with the aid of ***Gauss-Jordan*** complete elimination procedure, we get the next simplex tableau as follows:

First Iteration. x_1 enters and A_2 leaves.

Coeff. of basic variable	Basic variable (x_{Bi})	x_1	x_2	s_1	s_2	A_2	z	Solution
2.5	x_2	0.5	1	– 0.25	0	0	0	10
M	$\leftarrow A_2$	(2)	0	0.50	–1	1	0	30
	$(z_j - c_j)$	–1.75 + 2M ↑	0	– 0.62 + 0.5M	0 – M	0	1	$z = 25 + 30M$

In the above tableau, we have omitted all the entries in the column under A_1 because the artificial variable A_1 has left the basis and we shall not like it to re-enter it into basis again in the subsequent iterations.

Now, since the most positive coefficient in the last row lies under x_1, our next iteration would be as follows:

Final Iteration. Optimum solution.

Coeff. of basic variable	Basic variable (x_{Bi})	x_1	x_2	s_1	s_2	z	Solution
2.5	x_2	0	1	−0.38	0.25	0	2.50
3	x_1	1	0	0.25	−0.50	0	15.00
	$(z_j - c_j)$	0	0	−0.20	−0.87	1	$z = 51.25$

In the last row of the above tableau, since, all the entries under non-basic variables s_1 and s_2 are non-positive, the optimum solution is reached, *viz.*,

$$x_1 = 15 \text{ and } x_2 = 2.50 \text{ with minimum } z = 51.25.$$

Example 3.18 (*Infeasible Solution*). *Use simplex method to solve the following LPP:*

$$\text{Maximize } z = 3x_1 + 2x_2 \text{ subject to the constraints:}$$
$$2x_1 + x_2 \le 2,$$
$$3x_1 + 4x_2 \ge 12,$$
$$x_1, x_2 \ge 0.$$

Solution. Introducing slack variable $s_1 \ge 0$, surplus variable $s_2 \ge 0$ and an artificial variable $A_1 \ge 0$, the problem can be reformulated as follows:

$$\text{Maximize } z = 3x_1 + 2x_2 + 0 \cdot s_1 + 0 \cdot s_2 - MA_1 \text{ subject to the constraints:}$$
$$2x_1 + x_2 + s_1 = 2$$
$$3x_1 + 4x_2 - s_2 + A_1 = 12$$
$$x_1, x_2, s_1, s_2, A_1 \ge 0.$$

An obvious starting basic feasible solution is: $s_1 = 2$ and $A_1 = 12$ which can be displayed in the form of simplex tableau as follows:

Initial Iteration. x_2 enters and s_1 leaves.

Coeff. of basic variable	Basic variable (x_{Bi})	x_1	x_2	s_1	s_2	A_1	z	Solution	Ratio $\frac{x_{Bi}}{a_{i2}}, a_{i2} > 0$
0	← s_1	2	(1)	1	0	0	0	2	2/1
−M	A_1	3	4	0	−1	1	0	12	12/4
	z-equation row	−3 − 3M	−2 − 4M ↑	0	0 + M	0	1	z = −12M	

Final Iteration. Infeasible solution.

Coeff. of basic variable	Basic variable (x_{Bi})	x_1	x_2	s_1	s_2	A_1	z	Solution
2	x_2	2	1	1	0	0	0	2
−M	A_1	−5	0	−4	−1	1	0	4
	z-equation row	1 + 5M	0	2 + 4M	0 + M	0	1	z = 4 − 4M

Since, it is maximization problem and all entries under non-basic variables in the last row are non-negative, an ***optimum solution*** is reached. But an artificial variable A_1 happens to be basic variable at positive level at the final iteration.

Hence, the given LPP does not possess any optimum basic feasible solution.

Note. Because of the maximization problem, we have assigned a very large ***negative coefficient*** (–M) to the artificial variable.

REVIEW ILLUSTRATIONS

Example 3.19. *The ABC manufacturing company can make two products P_1 and P_2. Each of the products requires time on a cutting machine and a finishing machine. Relevant data are:*

	Product	
	P_1	P_2
Cutting hrs. (per unit)	2	1
Finishing hrs. (per unit)	3	3
Profit (per unit)	₹ 6	₹ 4
Maximum sales (units per week)	—	200

The number of cutting hours available per week is 390 and number of finishing hours available per week is 810. How much should be produced of each product in order to achieve maximum profit for the company?

Solution. Model formulation of the linear programming problem is:

$$\text{Maximize } z = 6x_1 + 4x_2 \text{ subject to the constraints:}$$

$$2x_1 + x_2 \le 390$$
$$3x_1 + 3x_2 \le 810$$
$$x_2 \le 200$$
$$x_1, x_2 \ge 0$$

where x_1 = number of units of product P_1, and x_2 = number of units of product P_2.

By introducing slack variables $s_1 \ge 0, s_2 \ge 0$, and $s_3 \ge 0$, the linear programming problem becomes that of finding a solution of the system

$$2x_1 + x_2 + s_1 = 390$$
$$3x_1 + 3x_2 + s_2 = 810$$
$$x_2 + s_3 = 200$$

with $\quad x_1 \ge 0, x_2 \ge 0, s_1 \ge 0, s_2 \ge 0 \text{ and } s_3 \ge 0$

and makes $\quad z - 6x_1 - 4x_2 + 0 \cdot s_1 + 0 \cdot s_2 + 0 \cdot s_3 = 0$

as large as possible.

From the above system of equations, a starting basic feasible solution is

$$s_1 = 390, \quad s_2 = 810 \quad \text{and} \quad s_3 = 200.$$

i.e., s_1, s_2 and s_3 are basic variables and x_1, x_2 are non-basic variables.

Making use of the above information, the starting simplex tableau is obtained as follows:

Coeff. of basic variable	*Basic variable* (x_{Bi})	x_1	x_2	s_1	s_2	s_3	z	*Solution*	*Ratio* $\frac{x_{Bi}}{a_{i1}}, a_{i1} > 0$
0	$\leftarrow s_1$	(2)	1	1	0	0	0	390	390/2
0	s_2	3	3	0	1	0	0	810	810/3
0	s_3	0	1	0	0	1	0	200	
	Net evaluation row	– 6 ↑	– 4	0	0	0	1	z = 0	

The last row of the above tableau indicates that the highest negative value lies under x_1 and hence, x_1 enters the basis. Further, since, the smallest of ratios among 390/2 and 810/3 is in the first row (s_1 row), s_1 leaves the basis and the element 2 becomes key number. Converting the pivot element (key number) into unity and all other elements of entering column to zero, the next iteration is given below:

First Iteration. x_2 enters and s_2 leaves.

Coeff. of basic variable	*Basic variable* (x_{Bi})	x_1	x_2	s_1	s_2	s_3	z	*Solution*	*Ratio* $\frac{x_{Bi}}{a_{i2}}, a_{i2} > 0$
6	x_1	1	1/2	1/2	0	0	0	195	$\frac{195}{1/2}$
0	$\leftarrow s_2$	0	(3/2)	–3/2	1	0	0	225	$\frac{225}{3/2}$
0	s_3	0	1	0	0	1	0	200	200/1
	Net evaluation row	0	–1 ↑	3	0	0	1	z = 1,170	

Since, only negative entry in the last row is under x_2, non-basic variable x_2 enters the basis. Further, the smallest of the ratios of positive element in the column under x_2 into the extreme right element of the same row is $\frac{225}{3/2}$. This indicates that the element '3/2' becomes pivot element and current basic variable s_2 becomes non-basic in the next iteration.

Final Iteration. Optimum Solution.

Coeff. of basic variable	*Basic variable* (x_{Bi})	x_1	x_2	s_1	s_2	s_3	z	*Solution*
6	x_1	1	0	1	–1/3	0	0	120
4	x_2	0	1	–1	2/3	0	0	150
0	s_3	0	0	1	–2/3	1	0	50
	Net evaluation row	0	0	2	2/3	0	1	z = 1,320

The optimum solution is that *ABC* company should manufacture 120 units of P_1 and 150 units of P_2 in order to get the maximum profit of ₹ 1,320.

Example 3.20. *Given the simplex tableau for a maximization problem of linear programming:*

c_j	*Product mix*	x_1	x_2	s_1	s_2	*Quantity*
4	x_2	1	1	1	0	6
0	s_2	1	–1	0	1	2
	c_j	3	4	0	0	

(*i*) *Test whether the above table gives optimum solution. If not, improve it and find the optimum solution.*

(*ii*) *If s_1 and s_2 are slack variables at machine I and machine II, what is the maximum price you will like to pay for 1 hour of each machine?* **[B.Com. (Hons.), Delhi Univ. 2006]**

Solution. Using the information given in the simplex tableau, we compute the net evaluation row numbers under respective variable (both basic as well as non-basic) as follows:

$$\text{Net Evaluation under } x_1 \equiv (z_1 - c_1) = 4 \times 1 + 0 \times 1 - 3 = 1$$
$$\text{Net Evaluation under } x_2 \equiv (z_2 - c_2) = 4 \times 1 + 0 - 4 = 0$$

Net Evaluation under $s_1 \equiv (z_3 - c_3) = 4 \times 1 - 0 = 4$

and Net Evaluation under $s_2 \equiv (z_4 - c_4) = 0$

(*i*) Since, all $(z_j - c_j) \geq 0$, the above tableau gives an optimum basic feasible solution, *viz.*, $x_1 = 0$, $x_2 = 6$ and maximum $z = 4 \times 6 + 0 = 24$.

(*ii*) The slack variable s_1 corresponds to the first constraint, and it has $(z_3 - c_3) = 4$. Thus, the shadow price for machine *I* is 4. This means that we could increase the objective function by 4, if we had an additional unit of this machine; hence it is the maximum price one should pay for one hour of this machine. Similarly, for the second machine, the shadow price is given by $(z_4 - c_4) = 0$. This implies that an additional unit of the second machine won't help, since, this machine has not been fully utilized.

Example 3.21. *An animal feed company must produce 200 kg of a mixture consisting of ingredients X_1 and X_2. The ingredient X_1 costs ₹ 3 per kg and X_2 costs ₹ 5 per kg. No more than 80 kg of X_1 can be used and at least 60 kg of X_2 must be used. Find the minimum cost mixture.*

Solution. The model formulation of the linear programming problem is to

$$\text{Minimize } z = 3x_1 + 5x_2 \text{ subject to the constraints:}$$

$$x_1 + x_2 = 200$$

$$0 \leq x_1 \leq 80$$

and

$$0 \leq x_2 \geq 60$$

where, x_1 = number of kg of ingredient X_1, and x_2 = number of kg of ingredient X_2.

Introducing slack variable $s_2 \geq 0$ and surplus variable $s_2 \geq 0$; the problem can be reformulated as:

$$\text{Minimize } z = 3x_1 + 5x_2 + 0 \cdot s_1 + 0 \cdot s_2 \text{ subject to the constraints:}$$

$$x_1 + x_2 = 200$$

$$x_1 + s_1 = 80$$

$$x_2 - s_2 = 60$$

$$x_1, x_2, s_1, s_2 \geq 0$$

Now, using the constraint equation $x_2 = s_2 + 60$, we eliminate variable x_2 from the rest of the linear equations. The linear programming problem then becomes of finding the solution of the system

$$x_1 + s_2 = 200 - 60, \text{ and } x_1 + s_1 = 80;$$

where,

$$x_1 \geq 0, s_1 \geq 0, s_2 \geq 0$$

and makes

$$z = 3x_1 + 5(200 - x_1) + 0 \cdot s_1 + 0 \cdot s_2$$

at least possible.

Making use of the above system of equations, the iterative simplex tableau are:

Initial Iteration. x_1 enters and s_1 leaves.

Coeff. of basic variable	*Basic variable* (x_{Bi})	x_1	s_2	s_1	z	*Solution*
0	s_2	1	1	0	0	140
0	$\leftarrow s_1$	(1)	0	1	0	80
	Net evaluation row	2 ↑	0	0	1	z = 1,000

Final Iteration. Optimum solution.

Coeff. of basic variable	Basic variable (x_{Bi})	x_1	s_2	s_1	z	Solution
0	s_2	0	1	–1	0	60
–2	x_1	1	0	1	0	80
	Net evaluation row	0	0	–2	1	$z = 840$

The optimum solution is $x_1 = 80$, $s_2 = 60$ with minimum $z = 840$. But $x_2 = s_2 + 60$, therefore, $x_2 = 60 + 60 (= 120)$. Hence, the animal feed company should produce 80 kg of a mixture consisting of ingredient X_1 and 120 kg of a mixture consisting of ingredient X_2. The minimum cost, according to this allocation, is ₹ 840.

Example 3.22 (*Degeneracy*).

Maximize $z = 5x_1 + 2x_2 + 10x_3$ *subject to the constraints:*

$$x_1 - x_3 \le 10$$
$$x_2 - x_3 \ge 10$$
$$x_1 + x_2 + x_3 \le 10$$
$$x_1, x_2, x_3 \ge 0$$

Solution. Introducing slack variables $s_1 \ge 0$, $s_3 \ge 0$, surplus variable $s_2 \ge 0$ and an artificial variable $A_1 \ge 0$, the reformulated LPP is:

Maximize $z = 5x_1 + 2x_2 + 10x_3 + 0 \cdot s_1 + 0 \cdot s_2 + 0 \cdot s_3 - M(10 - x_2 + x_3 + s_2)$

subject to the constraints:

$$x_1 - x_3 + s_1 = 10$$
$$x_2 - x_3 - s_2 + A_1 = 10$$
$$x_1 + x_2 + x_3 + s_3 = 10$$
$$x_1, x_2, x_3, s_1, s_2, s_3, A_1 \ge 0.$$

Obviously, the initial basic feasible solution is:

$$s_1 = 10, A_1 = 10 \text{ and } s_3 = 10.$$

Using the above information, the iterative simplex tableau are:

Initial Iteration. x_2 enters and A_1 leaves.

Coeff. of basic variable	Basic variable (x_{Bi})	x_1	x_2	x_3	s_1	s_2	s_3	A_1	z	Solution
0	s_1	1	0	–1	1	0	0	0	0	10
–M	← A_1	0	(1)	–1	0	–1	0	1	0	10
0	s_3	1	1	1	0	0	1	0	0	10
	Net evaluation row	–5	–2 – M ↑	–10 + M	0	M	0	0	1	$z = -10M$

First Iteration. x_3 enters and s_3 leaves.

Coeff. of basic variable	Basic variable (x_{Bi})	x_1	x_2	x_3	s_1	s_2	s_3	z	Solution
0	s_1	1	0	–1	1	0	0	0	10
2	x_2	0	1	–1	0	–1	0	0	10
0	← s_3	1	0	(2)	0	1	1	0	0
	Net evaluation row	–5	0	–12 ↑	0	–2	0	1	20

Final Iteration. Optimum solution.

Coeff. of basic variable	*Basic variable* (x_{Bi})	x_1	x_2	x_3	s_1	s_2	s_3	z	*Solution*
0	s_1	3/2	0	0	1	1/2	1/2	0	10
2	x_2	1/2	1	0	0	–1/2	1/2	0	10
10	x_3	1/2	0	1	0	1/2	1/2	0	0
	Net evaluation row	1	0	0	0	4	6	1	$z = 20$

Since all the coefficients in net evaluation row are ≥ 0, optimum solution is attained, *viz.*,

$$x_1 = 0, x_2 = 10, x_3 = 0 \quad \text{and} \quad \text{maximum } z = 20.$$

Note. We observe from the final tableau that one of the basic variables has zero as its value, *i.e.*, $x_3 = 0$. Hence, the optimum basic feasible solution obtained above is said to be degenerate.

Example 3.23 (*Alternative Optimum Solution*). *A company manufactures two products A and B. Both products are processed on two machine centres,* M_1 *and* M_2*. Data is given below:*

Products	*Hours required per unit* M_1	M_2	*Profit*
A	1	3	30
B	2	2	20
Capacity per week	80	120	

Solve the problem using simplex method to determine number of units of products A and B to be produced per week to maximize profit. Is there an alternate optimal solution? If yes, find it.

Solution. Let x_1 and x_2 denote the quantity of products A and B to be manufactured. Then, the linear programming formulation of the problem is:

$$\text{Maximize} \quad z = 30x_1 + 20x_2 \quad \text{subject to the constraints:}$$

$$x_1 + 2x_2 \le 80$$

$$3x_1 + 2x_2 \le 120$$

$$x_1, x_2 \ge 0.$$

Introducing slack variables $s_1 \ge 0$ and $s_2 \ge 0$, an initial basic feasible solution is:

$$s_1 = 80 \quad \text{and} \quad s_2 = 120.$$

Using the above information, iterative simplex tableau are:

Initial Iteration. x_1 enters and s_2 leaves.

Coeff. of basic variable	*Basic variable* (x_{Bi})	x_1	x_2	s_1	s_2	z	*Solution*	*Ratio* $\frac{x_{Bi}}{a_{i1}}, a_{i1} > 0$
0	s_1	1	2	1	0	0	80	80/1
0	← s_2	(3)	2	0	1	0	120	120/3
	Net evaluation row	–30 ↑	–20	0	0	1	$z = 0$	

Final Iteration. Optimum solution.

Coeff. of basic variable	*Basic variable* (x_{Bi})	x_1	x_2	s_1	s_2	z	*Solution*
0	s_1	0	4/3	1	–1/3	0	40
30	x_1	1	2/3	0	1/3	0	40
	Net evaluation row	0	0	0	10	1	$z = 1,200$

Since, all the numbers in the net evaluation row are non-negative, optimum solution is reached, *viz.*,

$$x_1 = 40, x_2 = 0 \text{ and maximum } z = 1{,}200.$$

Note. In the final simplex tableau, we observe that the number in the net evaluation row under non-basic variable x_2 is zero. This is an indication of the existence of an alternative optimum solution. Thus, if x_2 enters the basis, then s_1 will leave the basis, since, min. $\left\{\frac{40}{4/3}, \frac{40}{2/3}\right\} = 30$ which corresponds to first row, *i.e.*, s_1-row. Thus, the revised simplex tableau giving an alternative optimum solution will be:

Coeff. of basic variable	*Basic variable* (x_{Bi})	x_1	x_2	s_1	s_2	z	*Solution*
20	x_2	0	1	3/4	−1/4	0	30
30	x_1	1	0	−1/2	1/2	0	20
	Net evaluation row	0	0	0	10	1	$z = 1{,}200$

Since, all the numbers in the net evaluation row are non-negative, the optimum solution (Alternative Optimum Solution) is:

$$x_1 = 20 \text{ and } x_2 = 30 \text{ with maximum } z = 1{,}200.$$

Example 3.24 (*Minimization Case*). *Use simplex method to*

$$\textit{Minimize } z = x_2 - 3x_3 + 2x_5 \textit{ subject to the constraints:}$$

$$3x_2 - x_3 + 2x_5 \le 7, \quad -2x_2 + 4x_3 \le 12,$$

$$-4x_2 + 3x_3 + 8x_5 \le 10, \quad x_2 \ge 0, x_3 \ge 0 \textit{ and } x_5 \ge 0.$$

[MBA, Panjab Tech. Univ. 2014; B.Tech. (Mech.), HP Univ. 2015]

Solution. Introducing slack variables $s_1 \ge 0$, $s_2 \ge 0$ and $s_3 \ge 0$ in the respective inequalities, the standard form of the given LPP is:

$$\text{Minimum } z = x_2 - 3x_3 + 2x_5 + 0 \cdot s_1 + 0 \cdot s_2 + 0 \cdot s_3 \text{ subject to the constraints:}$$

$$3x_2 - x_3 + 2x_5 + s_1 = 7,$$

$$-2x_2 + 4x_3 + s_2 = 12,$$

$$-4x_2 + 3x_3 + 8x_5 + s_3 = 10;$$

$$x_j \ge 0 \text{ and } s_j \ge 0 \ (j = 1, 2, 3).$$

The starting basic feasible solution is obtained by setting $x_1 = x_2 = x_3 = 0$. Thus, we have

$$s_1 = 7, \ s_2 = 12 \text{ and } s_3 = 10.$$

Now, considering the objective function as an equation, the initial basic feasible solution is displayed in the following simplex tableau:

Initial Iteration. x_3 enters and s_2 leaves.

Coeff. of basic variable	*Basic variable* (x_{Bi})	x_2	x_3	x_5	s_1	s_2	s_3	z	*Solution*	*Ratio* $\frac{x_{Bi}}{a_{i2}}, a_{i2} > 0$
0	s_1	3	−1	2	1	0	0	0	7	—
0	← s_2	−2	(4)	0	0	1	0	0	12	12/4
0	s_3	−4	3	8	0	0	1	0	10	10/3
	Net evaluation row or $(z_j - c_j)$	−1	3 ↑	−2	0	0	0	1	$z = 0$	

Since, it is a minimization problem, all the coefficients in the last row should be non-positive for optimum solution. Here, the net evaluation number under x_3-column is only positive, therefore, non-basic variable x_3 enters basis in the next iteration.

Further, since, minimum of ratio corresponds to second row, current basic variable s_2 leaves the basis and the element a_{22} becomes the key number.

Converting the key number to '1' and all other elements of entering column to zero, the next iteration is:

First Iteration. x_2 enters and s_1 leaves.

Coeff. of basic variable	*Basic variable* (x_{Bi})	x_2	x_3	x_5	s_1	s_2	s_3	z	*Solution*	*Ratio* $\frac{x_{Bi}}{a_{i1}}, a_{i1} > 0$
0	$\leftarrow s_1$	(5/2)	0	2	1	1/4	0	0	10	$\frac{10}{5/2}$
–3	x_3	–1/2	1	0	0	1/4	0	0	3	—
0	s_3	–5/2	0	8	0	–3/4	1	0	1	—
	$(z_j - c_j)$	1/2 ↑	0	–2	0	–3/4	0	1	$z = -9$	

Observe that $(z_1 - c_1)$ is positive and thus, the current basic feasible solution is not optimum. The column under x_2 enters the basis. Further, since, only $a_{11} > 0$ both $a_{12} < 0$ and $a_{13} < 0$; current basic variable s_1 leaves the basis. This gives $a_{11}\left(=\frac{5}{2}\right)$ as the leading (key) element. Using e-row operations, we convert the leading element into unity and all other enteries in its column to zero. The improved basic feasible solution is obtained as shown in the next simplex tableau.

Final Iteration. Optimum solution.

Coeff. of basic variable	*Basic variable* (x_{Bi})	x_2	x_3	x_5	s_1	s_2	s_3	z	*Solution*
1	x_2	1	0	4/5	2/5	1/10	0	0	4
–3	x_3	0	1	2/5	1/5	3/10	10	0	5
0	s_3	0	0	10	2/5	–1/2	1	0	11
	$(z_j - c_j)$	0	0	–12/5	–1/5	–8/10	0	1	$z = -11$

Since, all $(z_j - c_j) \leq 0$, an optimal basic feasible solution has been attained. Thus, the optimum solution to the given LPP is

$$\text{Minimum} \quad z = -11 \text{ with } x_2 = 4, x_3 = 5 \text{ and } x_5 = 0.$$

Example 3.25 (*Unbounded Solution*). *Find the maximum value of* $z = 107x_1 + x_2 + 2x_3$ *subject to the constraints:*

$$14x_1 + x_2 - 6x_3 + 3x_4 = 7, \quad 16x_1 + x_2 - 6x_3 \leq 5, \quad 3x_1 - x_2 - x_3 \leq 0;$$
$$x_1, x_2, x_3, x_4 \geq 0.$$

[BE (Dec.), Karnataka Univ. 2011]

Solution. By introducing slack variables $s_1 \geq 0$ and $s_2 \geq 0$ respectively, in second and third constraints, the set of constraints of the given linear programming problem is converted into the system of equations:

$$14x_1 + x_2 - 6x_3 + 3x_4 = 7$$
$$16x_1 + x_2 - 6x_3 + s_1 = 5$$
$$3x_1 - x_2 - x_3 + s_2 = 0$$
$$z - 107x_1 - x_2 - 2x_3 + 0 \cdot x_4 + 0 \cdot s_1 + 0 \cdot s_2 = 0.$$

Dividing the first equation by 3, the initial basic feasible solution is:

$$x_4 = 7/3,\ s_1 = 5 \text{ and } s_2 = 0.$$

Using, now, simplex method, the iterative simplex tableau are:

Initial Iteration. x_1 enters and s_2 leaves.

Coeff. of basic variable	*Basic variable* (x_{Bi})	x_1	x_2	x_3	x_4	s_1	s_2	z	*Solution*
0	x_4	14/3	1/3	−2	1	0	0	0	7/3
0	s_1	16	1	−6	0	1	0	0	5
0	← s_2	(3)	−1	−1	0	0	1	0	0
	$(z_j - c_j)$	−107 ↑	−1	−2	0	0	0	1	$z = 0$

It is apparent from the tableau that there are three $(z_j - c_j)$ which are negative. We choose the most negative of these, *viz.*, −107. The corresponding column under x_1, therefore, enters the basis. Further, since $\left\{\frac{x_{Bi}}{a_{i1}}, a_{i1} > 0\right\}$ is $\frac{0}{3}$ which occurs for the element a_{31} (= 3), s_2 leaves the basis and the element a_{31} becomes the leading element. Reducing the leading element into unity and all other entries of this column to zero, we get the next tableau.

Final Iteration. Unbounded solution.

Coeff. of basic variable	*Basic variable* (x_{Bi})	x_1	x_2	x_3	x_4	s_1	s_2	z	*Solution*
0	x_4	0	17/9	−4/9	1	0	−14/9	0	7/3
0	s_1	0	19/3	−2/3	0	1	−16/3	0	5
107	x_1	1	−1/3	−1/3	0	0	1/3	0	0
	$(z_j - c_j)$	0	−110/3	−113/3	0	0	107/3	1	$z = 0$

It is apparent from the tableau that the net evaluations $(z_2 - c_2)$ and $(z_3 - c_3)$ both are negative, $(z_3 - c_3)$ being the most negative one. The column under x_3 corresponding to $(z_3 - c_3)$, therefore, becomes key column. But, x_3 will not enter the basis, since a_{i3} $(i = 1, 2, 3)$ are non-positive. Instead, this is an indication that there is an unbounded solution to the given LPP.

Example 3.26. *Use two-phase simplex method to*

$$\text{Maximize } z = 5x_1 - 4x_2 + 3x_3 \text{ subject to the constraints:}$$
$$2x_1 + x_2 - 6x_3 = 20,\ 6x_1 + 5x_2 + 10x_3 \le 76$$
$$8x_1 - 3x_2 + 6x_3 \le 50,\ x_1, x_2, x_3 \ge 0.$$

[B.Tech. (Mech.) (Nov.), HP Univ. 2007]

Solution. Introducing slack variables $s_1 \ge 0$ and $s_2 \ge 0$, the LPP in its standard form is:

$$\text{Maximize } z = 5x_1 - 4x_2 + 3x_3 \text{ subject to the constraints:}$$
$$2x_1 + x_2 - 6x_3 = 20,\ 6x_1 + 5x_2 + 10x_3 + s_1 = 76$$
$$8x_1 - 3x_2 + 6x_3 + s_2 = 50,\ x_1, x_2, x_3, s_1, s_2 \ge 0.$$

From the above system of equations, we observe that there is no ready basic feasible solution. So, we introduce an artificial variable A_1 (≥ 0) in the first constraint equation. Thus, we have

$$2x_1 + x_2 - 6x_3 + A_1 = 20,$$
$$6x_1 + 5x_2 + 10x_3 + s_1 = 76,$$
$$8x_1 - 3x_2 + 6x_3 + s_2 = 50.$$

This system of equations gives us an initial basic feasible solution as:

$$A_1 = 20, s_1 = 76 \text{ and } s_2 = 50.$$

Phase 1. The objective function of the auxiliary LPP is $z^* = -A_1$.

The iterative simplex tableau for the auxiliary LPP are:

Initial Iteration. x_1 enters and s_2 leaves.

Coeff. of basic variable	*Basic variable* (x_{Bi})	x_1	x_2	x_3	s_1	s_2	A_1	z	*Solution*
–1	A_1	2	1	–6	0	0	1	0	20
0	s_1	6	5	10	1	0	0	0	76
0	$\leftarrow s_2$	(8)	–3	6	0	1	0	0	50
	$(z_j - c_j)$	$-2\uparrow$	–1	6	0	0	0	1	$z = -20$

Since, $(z_1 - c_1)$ is most negative, the column under x_1 enters the basis. Further, since $\min.\left\{\frac{x_{Bi}}{a_{i1}}, a_{i1} > 0\right\} = \frac{50}{8}$; s_2 leaves the basis. The element a_{31} (= 8) becomes the leading element.

First Iteration. x_2 enters and A_1 leaves.

Coeff. of basic variable	*Basic variable* (x_{Bi})	x_1	x_2	x_3	s_1	s_2	A_1	*Solution*
–1	$\leftarrow A_1$	0	(7/4)	–15/2	0	–1/4	1	15/2
0	s_1	0	29/4	11/2	1	–3/4	0	77/2
0	x_1	1	–3/8	3/4	0	1/8	0	25/4
	$(z_j - c_j)$	0	$-7/4\uparrow$	15/2	0	1/4	0	$z = -15/2$

Here, $(z_2 - c_2)$ is the only negative. This indicates that x_2 enters the basis. Also, $\min.\left\{\frac{x_{Bi}}{a_{i2}}, a_{i2} > 0\right\} = \frac{15/2}{7/4}$ suggest that A_1 must leave the basis, thereby a_{12} (= 7/4) becomes the leading element.

Final Iteration. Optimum solution.

Coeff. of basic variable	*Basic variable* (x_{Bi})	x_1	x_2	x_3	s_1	s_2	z	*Solution*
0	x_2	0	1	–30/7	0	–1/7	0	30/7
0	s_1	0	0	256/7	1	2/7	0	52/7
0	x_4	1	0	–6/7	0	1/14	0	55/7
	$(z_j - c_j)$	0	0	0	0	0	1	$z = 0$

Since, all $(z_j - c_j) \geq 0$ an optimum solution to the auxiliary LPP has been reached. Furthermore, it is apparent from the tableau that no artificial variable appears in the basis.

Phase 2. Now, we consider the actual costs associated with the original variables. So, the objective function is

$$z = 5x_1 - 4x_2 + 3x_3 + 0 \cdot s_1 + 0 \cdot s_2$$

The iterative simplex tableau for this phase is:

Coeff. of basic variable	Basic variable (x_{Bi})	x_1	x_2	x_3	s_1	s_2	z	Solution
–4	x_2	0	1	–30/7	0	–1/7	0	30/7
0	s_1	0	0	256/7	1	2/7	0	52/7
5	x_1	1	0	–6/7	0	1/14	0	55/7
	$(z_j - c_j)$	0	0	69/7	0	13/4	1	$z = 155/7$

Since, all $z_j - c_j \geq 0$ an optimum basic feasible solution has been reached. Hence, an optimum basic feasible solution to the given LPP is

$$x_1 = 55/7, x_2 = 30/7, x_3 = 0; \text{ and maximum } z = 155/7.$$

Example 3.27. *Use penalty (or Big M) method to*

$$\textit{Maximize } z = 6x_1 + 4x_2 \textit{ subject to the constraints:}$$
$$2x_1 + 3x_2 \leq 30, \; 3x_1 + 2x_2 \leq 24, \; x_1 + x_2 \geq 3,$$
$$x_1 \geq 0 \textit{ and } x_2 \geq 0$$

Is the solution unique? If not, give two different solutions.

Solution. Introducing slack variables $s_1 \geq 0$, $s_2 \geq 0$ and surplus variable $s_3 \geq 0$ in the constraint inequalities, the standard form of LPP is:

$$\text{Maximize } z = 6x_1 + 4x_2 + 0\cdot s_1 + 0\cdot s_2 + 0\cdot s_3 \text{ subject to the constraints:}$$
$$2x_1 + 3x_2 + s_1 = 30, \; 3x_1 + 2x_2 + s_2 = 24, \; x_1 + x_2 - s_3 = 3.$$
$$x_1, x_2, s_1, s_2, s_3 \geq 0.$$

Clearly, we do not have initial (starting) basic feasible solution. So, we introduce an artificial variable A_1 (≥ 0) in the third constraint. Then, an obvious initial basic feasible solution is

$$s_1 = 30, \; s_2 = 24 \text{ and } A_1 = 3.$$

The iterative simplex tableau are:

Initial Iteration. x_1 enters and A_1 leaves.

Coeff. of basic variable	Basic variable (x_{Bi})	x_1	x_2	s_1	s_2	s_3	A_1	z	Solution
0	s_1	2	3	1	0	0	0	0	30
0	s_2	3	2	0	1	0	0	0	24
–M	$\leftarrow A_1$	(1)	1	0	0	–1	1	0	3
	$(z_j - c_j)$	$-M-6\uparrow$	$-M-4$	0	0	M	0	1	$z = -3M$

In the above tableau $(z_1 - c_1)$ and $(z_2 - c_2)$ are negative. Among these two $(z_1 - c_1)$ gives the most negative value (since, M is very large), therefor, x_1 enters the basis. Further, $\min_i \left\{ \frac{x_{Bi}}{a_{i1}}, a_{i1} > 0 \right\} = \frac{3}{1}$.
This indicates A_1 leaves the basis and a_{31} becomes the leading element. Since, A_1 is an artificial variable, we drop it from the objective function for subsequent iterations.

First Iteration.

Coeff. of basic variable	Basic variable (x_{Bi})	x_1	x_2	s_1	s_2	s_3	z	Solution
0	s_1	0	1	1	0	2	0	24
0	$\leftarrow s_2$	0	–1	0	1	(3)	0	15
6	x_1	1	1	0	0	–1	0	3
	$(z_j - c_j)$	0	2	0	0	–6	1	$z = 18$

Since, $(z_5 - c_5) < 0$, s_3 enters the basis. Further, since least ratio is in second row, s_2 leaves the basis and a_{25} becomes the leading element.

Final Iteration. Optimum solution.

Coeff. of basic variable	*Basic variable* (x_{Bi})	x_1	x_2	s_1	s_2	s_3	z	*Solution*
0	s_1	0	5/3	1	–2/3	0	0	14
0	s_3	0	–1/3	0	1/3	1	0	5
6	x_1	1	2/3	0	1/3	0	0	8
	$(z_j - c_j)$	0	0	0	2	0	1	$z = 48$

Since all $(z_j - c_j) \geq 0$, an optimum solution has been reached. Thus, an optimum basic feasible solution to the given LPP is

$$x_1 = 8 \text{ and } x_2 = 0 \text{ with max. } z = 48.$$

Alternative Solution. It is evident from the net evaluations of the optimum tableau that the net evaluation corresponding to non-basic variable x_2 is zero. This is an indication that the current solution is not unique and an alternative solution exists. Thus, we can bring non-basic variable x_2 into the basis in place of s_1 or s_2. The resulting new basic feasible solution will also be an optimum one.

Therefore, by introducing x_2 into the basis in place of s_1 the new optimum simplex tableau is

Coeff. of basic variable	*Basic variable* (x_{Bi})	x_1	x_2	s_1	s_2	s_3	z	*Solution*
4	x_2	0	1	3/5	–2/5	0	0	42/5
0	s_3	0	0	1/5	1/5	1	0	39/5
6	x_1	1	0	–2/5	3/5	0	0	12/5
	$(z_j - c_j)$	0	0	0	2	0	1	$z = 48$

We observe from this tableau that another basic feasible solution results, but the optimum value of z remains the same. Hence, the two basic (optimum) feasible solutions to the given LPP are:

$$x_1 = [8,\ 0] \text{ and } x_2 = [12/5,\ 42/5] \text{ with maximum } z = 48$$

Example 3.28. *An advertising agency wishes to reach two types of audiences: Customers with annual income greater than ₹60,000 (category A) and customers with annual income less than ₹60,000 (category B). The total advertising budget is ₹2,00,000. One programme of TV advertising costs ₹50,000; one programme of radio advertising costs ₹20,000. For contract reasons, at least 3 programmes ought to be of TV, and the number of radio programmes must be limited to five. Surveys indicate that a single TV programme reaches 4,50,000 customers in category A and 50,000 customers in category B. One radio programme reaches 20,000 customers of category A and 80,000 customers of category B. Determine the media mix to maximize the total reach. Is there any alternative optimal solution to this problem (Use Simplex Method).*

[MIB, Delhi Univ. 2011]

Formulation of LP Model

Let the number of insertions in TV and radio be x_1 and x_2, respectively. Then, using the given information, an appropriate linear programming model formulation of the given problem is as follows:

$$\text{Maximize } z = (4{,}50{,}000 + 50{,}000)\, x_1 + (20{,}000 + 80{,}000)\, x_2$$
$$= 5{,}00{,}000x_1 + 1{,}00{,}000x_2 \text{ subject to the constraints:}$$
$$50{,}000x_1 + 20{,}000x_2 \le 2{,}00{,}000 \qquad \textit{(Constraint on cost)}$$
$$x_1 \ge 3 \text{ and } x_2 \le 5 \qquad \textit{(Constraint on programmes)}$$

where, $x_1 \ge 0,\ x_2 \ge 0.$ *(Non-negative constraint)*

Solution by Simplex Method. Introducing slack variables $s_1 \ge 0, s_3 \ge 0$, surplus variable $s_2 \ge 0$ and an artificial variable $A_1 \ge 0$, standard form of the LPP is:

$$\text{Maximize } z = 5x_1 + x_2 \text{ subject to the constraints:} \qquad \textit{(Considering 1,00,000 as one unit)}$$
$$5x_1 + 2x_2 + s_1 = 20$$
$$x_1 - s_2 + A_1 = 3$$
$$x_2 + s_3 = 5$$
$$x_1, x_2, s_1, s_2, s_3, A_1 \ge 0.$$

An obvious starting basic feasible solution is:

$$s_1 = 20, A_1 = 3 \text{ and } s_3 = 5$$

The iterative simplex tableau are:

Initial Iteration. x_1 enters and A_1 leaves.

Coeff. of basic variable	*Basic variable* (x_{Bi})	x_1	x_2	s_1	s_2	s_3	A_1	z	*Solution*
0	s_1	5	2	1	0	0	0	0	20
$-M$	$\leftarrow A_1$	(1)	0	0	-1	0	1	0	3
0	s_3	0	1	0	0	1	0	0	5
	$(z_j - c_j)$	$-M-5\uparrow$	-1	0	M	0	0	1	$-3M$

Since, $(z_1 - c_1)$ is most negative, x_1 enters the basis. Further, since $\min.\left\{\dfrac{x_{Bi}}{a_{i1}},\ a_{i1} > 0\right\} = 3$, A_1 leaves the current basis.

First Iteration. s_2 enters and s_1 leaves.

Coeff. of basic variable	*Basic variable* (x_{Bi})	x_1	x_2	s_1	s_2	s_3	z	*Solution*	*Ratio* $\dfrac{x_{Bi}}{a_{i4}}, a_{i4} > 0$
0	$\leftarrow s_1$	0	2	1	(5)	0	0	5	5/5
5	x_1	1	0	0	-1	0	0	3	—
0	s_3	0	1	0	0	1	0	5	—
	$(z_j - c_j)$	0	-1	0	$-5\uparrow$	0	1	$z = 15$	

Final Iteration. Optimum solution.

Coeff. of basic variable	*Basic variable* (x_{Bi})	x_1	x_2	s_1	s_2	s_3	z	*Solution*
0	s_2	0	2/5	1/5	1	0	0	1
5	x_1	1	2/5	1/5	0	0	0	4
0	s_3	0	1	0	0	1	0	5
	$(z_j - c_j)$	0	1	1	0	0	1	$z = 20$

From the last simplex tableau, we observe that all the numbers in net evaluation row are non-negative, *i.e.*, all $(z_j - c_j) \geq 0$, an optimum solution is reached. Thus, $x_1 = 4$ and $x_2 = 0$. Hence, the total, number of insertions in TV should be 4, while there should not be any insertion in radio. Maximum audience will be 20,00,000.

Example 3.29. *Air Force is experimenting with three types of bombs P, Q and R in which three kinds of explosive, viz., A, B and C will be used. Taking the various factors into consideration, it has been decided to use at most 600 kg of explosive A, at least 480 kg of explosive B and exactly 540 kg of explosive C. Bomb P requires 3 kg, 2 kg and 2 kg of A, B and C respectively. Bomb Q requires 1 kg, 4 kg and 3 kg of A, B and C respectively. Bomb R requires 6 kg, 2 kg and 3 kg of A, B and C respectively. Now, bomb P gives the equivalent of a 2 tonne explosion, bomb Q will give a 3 tonne explosion and bomb R will give a 4 tonne explosion. Under what production schedule can Air Force make the biggest target? Solve by Two-Phase Simplex Method.*

[B.Tech. (Mech.), Panjab Tech. Univ. 2011; Dipl. in Glob. Bus. Oper., Delhi Univ. 2011]

Linear Programming Model Formulation

Let the Air Force experiment with x_1 units of bomb *P*, x_2 units of bomb *Q* and x_3 units of bomb *R*. Then, using the given information, an appropriate linear programming formulation of the given problem is:

$$\text{Maximize } z = 2x_1 + 3x_2 + 4x_3 \text{ subject to the constraints:}$$

$$3x_1 + x_2 + 6x_3 \leq 600 \quad \textit{(Explosive A constraint)}$$

$$2x_1 + 4x_2 + 2x_3 \geq 480 \quad \textit{(Explosive B constraint)}$$

$$2x_1 + 3x_2 + 3x_3 = 540 \quad \textit{(Explosive C constraint)}$$

and $\quad x_1 \geq 0, \; x_2 \geq 0, \text{ and } x_3 \geq 0.$

Solution by Two-Phase Simplex Method. Introducing slack variable $s_1 \geq 0$, surplus variable $s_2 \geq 0$, artificial variables $A_1 \geq 0$ and $A_2 \geq 0$, standard form of the LPP is:

$$\text{Maximize } z = 2x_1 + 3x_2 + 4x_3 + 0 \cdot s_1 + 0 \cdot s_2 - MA_1 - MA_2 \text{ subject to the constraints:}$$

$$3x_1 + x_2 + 6x_3 + s_1 = 600$$

$$2x_1 + 4x_2 + 2x_3 - s_2 + A_1 = 480$$

$$2x_1 + 3x_2 + 3x_3 + A_2 = 540$$

$$x_1, x_2, x_3, s_1, s_2, A_1, A_2 \geq 0.$$

An obvious initial (starting) basic feasible solution is: $s_1 = 600$, $A_1 = 480$ and $A_2 = 540$.

Phase 1. The objective function of the auxiliary LPP is to maximize $z^* = -A_1 - A_2$, since the value of all the decision variables, *i.e.*, x_1, x_2 and x_3 are considered as zero.

The iterative simplex tableau for the auxiliary LPP are as follows:

Initial Iteration. x_2 enters and A_1 leaves.

Coeff. of basic variable	*Basic variable* (x_{Bi})	x_1	x_2	x_3	s_1	s_2	A_1	A_2	z	*Solution*	*Ratio* $\frac{x_{Bi}}{a_{i2}}, a_{i2} > 0$
0	s_1	3	1	6	1	0	0	0	0	600	600/1
–1	$\leftarrow A_1$	2	(4)	2	0	–1	1	0	0	480	480/4
–1	A_2	2	3	3	0	0	0	1	0	540	540/3
	$(z_j - c_j)$	–4	–7↑	–5	0	1	0	0	1	z = –1,020	

First Iteration. x_3 enters and s_1 leaves.

Coeff. of basic variable	Basic variable (x_{Bi})	x_1	x_2	x_3	s_1	s_2	A_2	z	Solution	Ratio $\frac{x_{Bi}}{a_{i3}}, a_{i3} > 0$
0	$\leftarrow s_1$	5/2	0	(11/2)	1	1/4	0	0	480	480 × 2/11
0	x_2	1/2	1	1/2	0	–1/4	0	0	120	120 × 2/1
–1	A_2	1/2	0	3/2	0	3/4	1	0	180	180 × 2/3
	$(z_j - c_j)$	–1/2	0	–3/2 ↑	0	–3/4	0	1	z = –180	

Second Iteration. s_2 enters and A_2 leaves.

Coeff. of basic variable	Basic variable (x_{Bi})	x_1	x_2	x_3	s_1	s_2	A_2	z	Solution	Ratio $\frac{x_{Bi}}{a_{i5}}, a_{i5} > 0$
0	x_3	5/11	0	1	2/11	1/22	0	0	87.27	87.2 × 22/1
0	x_2	3/11	1	0	–1/11	–3/11	0	0	76.36	—
–1	$\leftarrow A_2$	–2/11	0	0	–3/11	(15/22)	1	0	49.00	49 × 22/15
	$(z_j - c_j)$	2/11	0	0	3/11	–15/22 ↑	0	1	z = –49	

Final Iteration. Optimum solution.

Coeff. of basic variable	Basic variable (x_{Bi})	x_1	x_2	x_3	s_1	s_2	z	Solution
0	x_3	7/15	0	1	1/5	0	0	84.00
0	x_2	1/5	1	0	–1/5	0	0	56.76
0	s_2	–4/15	0	0	–2/5	1	0	71.86
	$(z_j - c_j)$	0	0	0	0	0	1	z = 0

Since, all $(z_j - c_j)$ are non-negative, an optimum solution to the auxiliary LPP has been reached. It is apparent from the optimum simplex tableau that no artificial variable appears in the basis. So, we move onto *Phase 2*.

Phase 2. Now, we restore actual coefficients of the original decision variables in the objective function of given LPP. Thus, we consider

$$z = 2x_1 + 3x_2 + 4x_3 + 0 \cdot s_1 + 0 \cdot s_2.$$

The initial (starting) simplex tableau for this phase is:

Coeff. of basic variable	Basic variable (x_{Bi})	x_1	x_2	x_3	s_1	s_2	z	Solution
4	x_3	7/15	0	1	1/5	0	0	84.00
3	x_2	1/5	1	0	–1/5	0	0	56.76
0	s_2	–4/15	0	0	–2/5	1	0	71.86
	$(z_j - c_j)$	–2	–3	–4	0	0	1	z = 506.28

Here, since, x_2, x_3 and s_2 are basic variables, coefficients under column x_2, x_3 and s_2 must be zero. For this, we multiply each element of first row of the simplex tableau by '4' and second row elements by '3' and then add to the corresponding elements of the z-equation row of the tableau. Thus, we get coefficients under non-basic variables x_1 and s_1 as follows:

$$4\times\frac{7}{15}+3\times\frac{1}{5}+(-2),\ i.e.,\ \frac{7}{15}\ \text{and}\ 4\times\frac{1}{5}+3\times\left(-\frac{1}{5}\right)+0,\ i.e.,\ \frac{1}{5},\ \text{respectivley.}$$

Thus, we have

$$(z_1 - c_1) = 7/15, (z_2 - c_2) = 4 \times 0 + 3 \times 1 - 3 = 0, (z_3 - c_3) = 4 \times 1 + 3 \times 0 - 4 = 0,$$
$$(z_4 - c_4) = 1/5 \text{ and } (z_5 - c_5) = 0.$$

The new value of z under solution column is: $4 \times 84 + 3 \times 56.76 + 0$, *i.e.*, 506.28.

Hence, an optimum basic feasible solution is:

$$x_1 = 0, x_2 = 56.76 \text{ and } x_3 = 84, \text{ with maximum } z = 506.28.$$

Thus, Air Force should use 56.76 (or approximately 57) units of Bomb Q and 84 units of Bomb R only to achieve the maximum target of 506 targets.

PROBLEMS

3.29. Use simplex method to solve the following LPP:

(*a*) Maximize $z = 20x_1 + 30x_2$ subject to the constraints:
$$3x_1 + 3x_2 \le 36,\ 5x_1 + 2x_2 \le 50$$
$$2x_1 + 6x_2 \le 60,\ x_1 \ge 0 \text{ and } x_2 \ge 0.$$

(*b*) Maximize $z = 4x_1 + 10x_2$ subject to the constraints:
$$2x_1 + x_2 \le 50,\ 2x_1 + 5x_2 \le 100,$$
$$2x_1 + 3x_2 \le 90;\ x_1 \ge 0 \text{ and } x_2 \ge 0.$$
[BE, Panjab Univ. 2004]

3.30. Use simplex method to:

(*a*) Maximize $z = 5x_1 + 3x_2$ subject to the constraints:
$$2x_1 + x_2 \le 1,\ x_1 + 4x_2 \ge 60;$$
$$x_1 \ge 0 \text{ and } x_2 \ge 0.$$
[B.Tech. (Comp. Sc.), Panjab Tech. Univ. 2010]

(*b*) Maximize $z = x_1 + 2x_2$ subject to the constraints:
$$-x_1 + 2x_2 \le 8,\ x_1 + 2x_2 \le 12,$$
$$x_1 - 2x_2 \le 3;\ x_1, x_2 \ge 0.$$
[B.Tech. (Mech.) (Nov.), HP Univ. 2010]

3.31. Solve the following LPP by simplex method:

Maximize $z = 5x_1 + 10x_2 + 8x_3$ subject to the constraints:
$$3x_1 + 5x_2 + 2x_3 \le 60,\ 4x_1 + 4x_2 + 4x_3 \le 72,$$
$$2x_1 + 4x_2 + 5x_3 \le 100;\ x_1, x_2, x_3 \ge 0.$$

3.32. Solve the following LPP by simplex method:

Maximize $z = 5x_1 + 3x_2 + 7x_3$ subject to the constraints:
$$x_1 + x_2 + 2x_3 \le 22,\ 3x_1 + 2x_2 + x_3 \le 26$$
$$x_1 + x_2 + x_3 \le 18,\ x_1, x_2, x_3 \ge 0.$$
[MBA, Madras Univ. 2006]

3.33. Solve the following LPP:

Maximize $z = 2x_1 - x_2 + x_3$ subject to the constraints:
$$3x_1 + x_2 + x_3 \le 60,\ x_1 - x_2 + 2x_3 \le 10$$
$$x_1 + x_2 - x_3 \le 20,\ x_1, x_2, x_3 \ge 0.$$
[M.Com., Kerala Univ. 1994]

3.34. Use simplex method to solve the following LPP:

(*a*) Maximize $z = 30x_1 + 40x_2$ subject to the constraints:
$$60x_1 + 120x_2 \le 12{,}000,\ 8x_1 + 5x_2 \le 600,$$
$$3x_1 + 4x_2 \le 500,\ x_1 \ge 0 \text{ and } x_2 \ge 0.$$
[MBA, Osmania Univ. 2010]

(*b*) Maximize $z = 100x_1 + 60x_2 + 40x_3$ subject to the constraints:

$$x_1 + x_2 + x_3 \le 100,\quad 10x_1 + 4x_2 + 5x_3 \le 600$$
$$2x_1 + 2x_2 + 6x_3 \le 300;\quad x_1, x_2, x_3 \ge 0.$$

[MBA, UP Tech. Univ. 2006]

3.35. Solve the following LPP:

Maximize $z = 22x_1 + 30x_2 + 25x_3$ subject to the constraints:

$$2x_1 + 2x_2 \le 100,\quad 2x_1 + x_2 + x_3 \le 100$$
$$x_1 + 2x_2 + 2x_3 \le 100;\quad x_1, x_2, x_3 \ge 0.$$

[M.Com., Delhi Univ. 2008]

3.36. Use simplex method to solve the following LPP:

Maximize $z = 2x_1 + 4x_2 + x_3 + x_4$ subject to the constraints:

$$x_1 + 3x_2 + x_4 \le 4,\quad 2x_1 + x_2 \le 3$$
$$x_2 + 4x_3 + x_4 \le 3,\quad x_1, x_2, x_3, x_4 \ge 0.$$

[MBA, Madras Univ. 2010]

3.37. Show that the following LPP:

Maximize $z = 4x_1 + x_2 + 3x_3 + 5x_4$ subject to the constraints:

$$4x_1 - 6x_2 - 5x_3 - 4x_4 \ge -20,\quad 3x_1 - 2x_2 + 4x_3 + x_4 \le 10$$
$$8x_1 - 3x_2 + 3x_3 + 2x_4 \le 20,\quad x_1, x_2, x_3, x_4 \ge 0$$

have an unbounded solution. ***[MCA, Purvanchal Univ. 1996]***

3.38. Use two-phase simplex method to

(*a*) Maximize $z = 10x_1 + 20x_2$ subject to the constraints:

$$2x_1 + x_2 = 1,\quad x_1 + 2x_2 = 5,$$
$$x_1 \ge 0, x_2 \ge 0.$$

[M.Com., Madurai Univ. 2002]

(*b*) Minimize $z = 2x_1 + 4x_2$ subject to the constraints:

$$2x_1 + x_2 \ge 14,\ x_1 + 3x_2 \ge 18,\ x_1 + x_2 \ge 12;$$
$$x_1 \ge 0, x_2 \ge 0.$$

[MBA (Dec.), Panjab Tech. Univ. 2010]

(*c*) Minimize $z = 4x_1 + x_2$ subject to the constraints:

$$3x_1 + x_2 = 3,\quad 4x_1 + 3x_2 \ge 6,$$
$$x_1 + 2x_2 \le 3;\quad x_1, x_2 \ge 0.$$

[B.Tech. (Chem.), Panjab Tech. Univ. 2012]

3.39. Use penalty method to solve the following LPP:

Maximize $z = x_1 + 2x_2 + 3x_3 - x_4$ subject to the constraints:

$$x_1 + 2x_2 + 3x_3 = 15,\quad 2x_1 + x_2 + 5x_3 = 20$$
$$x_1 + 2x_2 + x_3 + x_4 = 10,\quad x_1, x_2, x_3, x_4 \ge 0.$$

[BE (Nov.), Visvesvaraya Univ. 2020]

3.40. Solve the following linear programming problem:

Maximize $z = 5x_1 - 2x_2 + 3x_3$ subject to the constraints:

$$2x_1 + 2x_2 - x_3 \ge 2,\quad 3x_1 - 4x_2 \le 3,$$
$$x_2 + 3x_3 \le 5;\quad x_1, x_2, x_3 \ge 0.$$

[MBA (Nov.), Annamalai Univ. 2010]

3.41. Given the simplex tableau, write the LPP from which it would have been derived:

Basis	x_1	x_2	s_1	s_2	A_1	A_2	*Solution*
s_1	1	1	1	0	0	0	8
A_1	6	4	0	–1	1	0	12
A_2	5	8	0	0	0	1	20
$(c_j - z_j)$	$11M + 30$	$12M + 20$	0	$-M$	0	0	$-32M$

Find the optimum solution. ***[B.Com. (Hons.), Delhi Univ. 2010]***

3.42. A chemical company must produce exactly 1,000 kg of a special mixture of chemical *A* and chemical *B* for a customer. Chemical *A* costs ₹ 50 per kg and chemical *B* costs ₹ 60 per kg. No more than 300 kg of chemical *A* can be used and at least 150 kg of chemical *B* must be used.

(*a*) Construct a linear programming formulation of the problem.

(*b*) Using simplex method, determine the least cost blend of the two ingredients.

3.43. An automobile manufacturer makes automobiles and trucks in a factory that is divided into two shops. Shop *A*, which performs the basic assembly operation must work for 5 man-days on each truck, but only 2 man-days on each automobile. Shop *B*, which performs finishing operation must work for 3 man-days for each automobile or truck that it produces. Because of men and machine limitations, shop *A* has 180 man-days per week available, while shop *B* has 135 man-days per week. If the manufacturer makes profit of ₹ 300 on each truck and ₹ 200 on each automobile, how many of each should he produce to maximize his profit?

3.44. Product *A* yields a profit of ₹ 25 per unit and product *B* yields a profit of ₹ 40 per unit. To manufacture the products — leather, wood and glue are required the amount of which is shown below:

Resources required for one unit

Product	*Leather (in kg)*	*Wood (in sq. metres)*	*Glue (in litres)*
A	0.50	4	0.2
B	0.25	7	0.2

Available resources include 2,200 kg of leather; 28,000 sq metres of wood and 1,400 litres of glue. Answer the following:

(*i*) State the objective function and constraints in mathematical form.

(*ii*) Find the optimum solution.

(*iii*) Which of the resources are fully consumed? How much of each resource remains unused?

(*iv*) What are the shadow prices of resources? ***[C.S. Final (June) 1997]***

3.45. A manufacturer of leather belts makes three types of belts *A*, *B* and *C* which are processed on three machines M_1, M_2 and M_3. Belt *A* requires 2 hours on machine M_1 and 3 hours on machine M_3. Belt *B* requires 3 hours on machine M_1, 2 hours on machine M_2 and 2 hours on machine M_3; Belt *C* requires 5 hours on machine M_2 and 4 hours on machine M_3. There are 8 hours of time per day available on machine M_1, 10 hours of time per day available on machine M_2 and 15 hours of time per day available on machine M_3. The profit gained from Belt *A* is ₹ 3.00 per unit, from Belt *B* is ₹ 5.00 per unit, and from Belt *C* is ₹ 4.00 per unit. What should be the daily production of each type of belts so that the profit is maximum?

[MBA, UP Tech. Univ. 2005; B.E. (Mech.), Nagpur Univ. 2010]

3.46. A cabinet manufacturer produces wood cabinets for TV sets, stereosystems and radios, each of which must be assembled, decorated and crated. Each TV cabinet requires 3 hours to assemble, 5 hours to decorate and '1/10' hour to crate and returns a profit of ₹ 10. Each stereo cabinet requires 10 hours to assemble, 8 hours to decorate and '3/5' hour to crate and returns a profit of ₹ 25. Each radio cabinet requires 1 hour to assemble, 1 hour to decorate and '1/10' hour to crate and returns a profit of ₹ 3. The manufacturer has maximum of 30,000 hours, 40,000 hours and 120 hours available for assembling, decorating and crating respectively.

(*i*) Formulate the above problem as a linear programming problem.

(*ii*) Use simplex method to find how many units of each product should be manufactured to achieve maximum profit.

(*iii*) Does the problem have multiple solutions? ***[MBA (Nov.), Delhi Univ. 1998]***

3.47. Ashok chemicals company manufacture two chemicals *A* and *B* which are sold to the manufacturers of soaps and detergents. On the basis of the next month's demand, the management has decided that the total production for chemicals *A* and *B* should be at least 350 kilogram. Moreover, a major customer's order for 125 kg of product *A* must also be supplied. Product *A* requires 2 hours of processing time per kg and product *B* requires 1 hour of processing time per kg. For the coming month, 600 hours of processing time is available.

The company wants to meet the above requirements at minimum total production cost. The production costs are ₹ 2 per kg for product *A* and ₹ 3 per kg for product *B*.

Ashok Chemicals Company wants to determine its optimal product mix and the total minimum cost relevant there to.

(*i*) Formulate the above as an LPP.
(*ii*) Solve the problem using simplex method.
(*iii*) Does the problem have multiple optimal solution? Why? ***[M.Com., Delhi Univ. 1998]***

3.48. A factory manufactures three products which are processed through three different production stages. The time required to manufacture one unit of each of the three products and the daily capacity of the stages are given in the following table:

Stage	*Time per unit (minutes)*			*Stage capacity (in minutes)*
	Product 1	*Product 2*	*Product 3*	
1	1	2	1	430
2	3	—	2	460
3	1	4	—	420
Profit per unit	₹ 3	₹ 2	₹ 5	

(*i*) Set up the data in simplex tableau.
(*ii*) Find the tableau for optimum solution.
(*iii*) State from the optimum simplex tableau — maximum profit, production pattern and surplus capacity at any stage.
(*iv*) What is the meaning of shadow price? Where is it shown in this table? Explain it in respect of resource of stages having shadow price.
(*v*) How many units of other resources will be required so as to completely utilize the surplus resource? ***[MBA, Osmania Univ. 1997]***

[***Hint:*** Let x_1, x_2 and x_3 denote the number of units of products 1, 2 and 3, respectively. Then, the linear programming model of the problem is:

$$\text{Maximize } z = 3x_1 + 2x_2 + 5x_3 \text{ subject to the constraints:}$$
$$x_1 + 2x_2 + x_3 \leq 430,\ 3x_1 + 2x_3 \leq 460$$
$$x_1 + 4x_2 \leq 420;\ x_1, x_2, x_3 \geq 0.$$

Introducing slack variables $s_1 \geq 0$, $s_2 \geq 0$ and $s_3 \geq 0$; initial basic feasible solution is:

$$s_1 = 430, s_2 = 460 \text{ and } s_3 = 420.$$

Starting simplex tableau is as under:

Coeff. of basic variable	*Basic variable* (x_{Bi})	x_1	x_2	x_3	s_1	s_2	s_3	z	*Solution*	*Ratio* $\frac{x_{Bi}}{a_{i3}}, a_{i3} > 0$
0	s_1	1	2	1	1	0	0	0	430	430/1 = 430
0	$\leftarrow s_2$	3	0	(2)	0	1	0	0	460	460/2 = 230
0	s_3	1	4	0	0	0	1	0	420	—
	$(z_j - c_j)$	–3	–2	–5 ↑	0	0	0	1	z = 0	

Applying simplex method, the optimum solution is obtained in the following simplex tableau:

Coeff. of basic variable	*Basic variable* (x_{Bi})	x_1	x_2	x_3	s_1	s_2	s_3	z	*Solution*
2	x_2	–1/4	1	0	1/2	–1/4	0	0	100
5	x_3	3/2	0	1	0	1/2	0	0	230
0	s_3	1	0	0	–2	1	1	0	20
	$(z_j - c_j)$	4	0	0	1	2	0	1	$z = 1{,}350$

Since, slack variable s_3 appears in the optimum solution, resource has not been fully utilized. Shadow price for the first resource is the net evaluation number 1, *i.e.*, $(z_j - c_j)$ value under s_1. This implies that we could increase the objective function by 1. Similarly, corresponding to slack variable s_2, the shadow price for the second resource is 2, *i.e.*, we can increase the objective function by 2. Third resource corresponding to slack variable s_3 is fully utilized, since net evaluation number under s_3 is zero.]

3.49. A manufacturer produces three products X, Y and Z daily. The three products are each processed through three production operations with time constraints and then stored. The problem has been formulated as under:

Maximize $z = 32x_1 + 35x_2 + 45x_3$ subject to the constraints: (*Profit in ₹*)

$2x_1 + 3x_2 + 2x_3 \le 120$ (*Operation 1*)

$4x_1 + 3x_2 + x_3 \le 160$ (*Operation 2*)

$3x_1 + 2x_2 + 4x_3 \le 100$ (*Operation 3*)

$x_1 + x_2 + x_3 \le 40$ (*Storage*)

$x_1, x_2, x_3 \ge 0.$

The final optimal simplex tableau for this problem is:

Coeff. of basic variable	*Basic variable* (x_{Bi})	x_1	x_2	x_3	s_1	s_2	s_3	s_4	z	*Solution*
0	s_1	–1/2	0	0	1	0	1/2	–4	0	10
0	s_2	2	0	0	0	1	1	–5	0	60
45	x_3	1/2	0	1	0	0	1/2	–1	0	10
35	x_2	1/2	1	0	0	0	–1/2	2	0	30
	$(z_j - c_j)$	8	0	0	0	0	5	25	1	$z = 1{,}500$

on the basis of the above information, answer the following questions:

(*i*) Is the above solution feasible?
(*ii*) Is the above solution optimum? If yes, what it is?
(*iii*) Is the above solution unbounded?
(*iv*) Is the above solution degenerate?
(*v*) Does it have multiple optimum solutions?
(*vi*) Why does the company find product X not profitable for the production?
(*vii*) If at all it is decided to produce product X, then by what minimum amount its price should be increased? ***[Dip. in Glob. Bus. Oper., Delhi Univ. 2011]***

3.50. A mining company produces 100 tons of red ore and 80 tons of black ore each week. These can be treated in different ways to produce three different alloys — Soft, Hard and Strong. To produce 1 ton of Soft alloy requires 5 tons of red ore and 3 tons of black ore. For Hard alloy the requirements are 3 tons of red ore and 5 tons of black ore, whereas for Strong alloy they are 5 tons of red ore and 5 tons of black ore. The profit per ton from selling the alloys are ₹ 250, ₹ 300 and ₹ 400 for Soft, Hard and Strong alloys respectively. Formulate the problem of deciding how much of each alloy to make each week as a LP problem and use Simplex Method to find the optimal solution. *[MBA (Dec.), Panjab Tech. Univ. 2011]*

3.51. A metal products company produces waste cans, filing cabinets, file boxes for correspondence, and lunch boxes. Its inputs are sheet metal of two different thickness, called *A* and *B*, and manual labour. Input-output relationships for the company are shown in the table given below:

	Waste cans	*Filing cabinets*	*Correspondence boxes*	*Lunch boxes*
Sheet metal — *A*	6	0	2	3
Sheet metal — *B*	0	10	0	0
Manual labour	4	8	2	3

The sales revenue per unit of waste cans, filing cabinets, correspondence boxes, and lunch boxes are ₹ 20, ₹ 400, ₹ 90 and ₹ 20 respectively. There are 225 units of sheet metal *A* available in the company's inventory, 300 of sheet metal *B*, and 190 units of manual labour. What is the company's optimal sales revenue? *[MBA, Delhi Univ. 2000]*

3.52. Food *A* contains 20 units of vitamin *X* and 40 units of vitamin *Y* per gram. Food *B* contains 30 units each of vitamin *X* and *Y*. The daily minimum human requirements of vitamins *X* and *Y* are 900 units and 1,200 units respectively. How many grams of each type of food should be consumed so as to minimize the cost, if food *A* costs 60 paise per gram and food *B* costs 80 paise per gram.

3.53. A diet for a sick person must contain at least 4,000 units of vitamins, 50 units of minerals and 1,400 units of calories. Two foods *A* and *B* are available at a cost of ₹ 4 and ₹ 3 per unit respectively. If one unit of *A* contains 200 units of vitamins, 1 units of mineral and 40 units of calories, and one unit of food *B* contains 100 units of vitamins, 2 units of minerals and 40 units of calories, find by simplex method, what combination of food be used to have least cost. *[M.Com., Calicut Univ. 2009]*

3.54. A Pharmaceutical company has 100 kg of *A*, 180 kg of *B* and 120 kg of *C* available per month. They can use these materials to make three basic pharmaceutical products namely *5-10-5*, *5-5-10* and *20-5-10*; where the numbers in each case represent the percentage by weight of *A*, *B* and *C* respectively in each of the products. The costs of these raw materials are given below:

Ingredients	*Cost per kg (₹)*
A	80
B	20
C	50
Inert ingredient	20

Selling prices of these products are ₹ 40.5, ₹ 43 and ₹ 45 per kg respectively. There is a capacity restriction of the company for the product *5-10-5*, so as they cannot produce more than 30 kg per month. Determine how much of each product should be produced to maximize profit. *[MBA, Delhi Univ. 2008]*

[*Hint:* Let the product manufactured be P_1, P_2 and P_3. Then, the given information can be displayed as under:

Product	Ingredients			Inert
	A	B	C	
P_1	5%	10%	5%	80%
P_2	5%	5%	10%	80%
P_3	20%	5%	10%	65%
Cost per kg	₹ 80	₹ 20	₹ 50	₹ 20

Cost of P_1 $\quad 5\% \times 80 + 10\% \times 20 + 5\% \times 50 + 80\% \times 20 = 4 + 2 + 2.50 + 16 =$ ₹ 24.50 per kg

Cost of P_2 $\quad 5\% \times 80 + 5\% \times 20 + 10\% \times 50 + 80\% \times 20 = 4 + 1 + 5 + 16 =$ ₹ 26 per kg

Cost of P_3 $\quad 20\% \times 80 + 5\% \times 20 + 10\% \times 50 + 65\% \times 20 = 16 + 1 + 5 + 13 =$ ₹ 35 per kg

Margin of P_1: ₹ 40.50 – ₹ 24.50 = ₹ 16, *Margin of* P_2: ₹ 43 – ₹ 26 = ₹ 17, and

Margin of P_3: ₹ 45 – ₹ 35 = ₹ 10.

The given LPP can, thus, be put in the following form:

$$\text{Maximize } z = 16x_1 + 17x_2 + 10x_3 \text{ subject to the constraints:}$$

$$0.05x_1 + 0.05x_2 + 0.20x_3 \le 100 \quad \text{or} \quad x_1 + x_2 + 4x_3 \le 2{,}000$$

$$0.10x_1 + 0.05x_2 + 0.05x_3 \le 180 \quad \text{or} \quad 2x_1 + x_2 + x_3 \le 3{,}600$$

$$0.05x_1 + 0.10x_2 + 0.10x_3 \le 120 \quad \text{or} \quad x_1 + 2x_2 + 2x_3 \le 2{,}400$$

$$0 \le x_1 \le 30,\ x_2 \ge 0 \text{ and } x_3 \ge 0,$$

where, x_1, x_2 and x_3 denote the quantity in kgs of the three products, respectively.]

3.55. A marketing manager wishes to allocate his annual advertising budget of ₹ 20,000 in two media vehicles *A* and *B*. The unit cost of a message in media *A* is ₹ 1,000 and that of *B* is ₹ 1,500. Media *A* is a monthly magazine and not more than one insertion is desired in one issue. At least 5 messages should appear in media *B*. The expected effective audience for a unit message in media *A* is 40,000 and for media *B* is 55,000.

(*i*) Develop a linear programming model of the above problem.

(*ii*) Solve the LPP for maximizing the total effective audience.

3.56. A farmer has 1,000 acres of land on which he can grow corn, wheat or soyabeans. Each acre of corn costs ₹ 100 for preparation, requires 7 man-days of work and yields a profit of ₹ 30. An acre of wheat costs ₹ 120 to prepare, requires 10 man-days of work and yields a profit of ₹ 40. An acre of soyabeans costs ₹ 70 to prepare, requires 8 man-days of work and yields a profit of ₹ 20. If farmer has ₹ 1,00,000 for preparation and can count on 8,000 man-days of work, how many acres should be allocated to each crop to maximize profits?

[MBA, Jammu Univ. 2006]

[*Hint:* The appropriate model formulation of LPP is:

$$\text{Maximize } z = 30x_1 + 40x_2 + 20x_3 \text{ subject to the constraints:}$$

$$100x_1 + 120x_2 + 70x_3 \le 1{,}00{,}000$$

$$7x_1 + 10x_2 + 8x_3 \le 8{,}000$$

$$x_1 + x_2 + x_3 \le 1{,}000 \text{ and } x_1, x_2, x_3 \ge 0;$$

where, x_1, x_2 and x_3 designate the acreage of corn, wheat and soyabeans, respectively.]

3.57. A firm wants to reach two types of customers: Households having ₹ 5 lakh or more of family income and those having less than ₹ 5 lakh income. The first group purchases twice as much as the second group. One unit of TV advertisement reaches 2,000 families of first group and

8,000 families of second group and costs ₹ 20,000, while advertisements in a magazine costs ₹ 12,000 and reaches 6,000 and 3,000 families respectively. No more than 12 advertisements can be given in magazine and at least 6 TV advertisements have to be given. The advertisement budget is ₹ 1,80,000. Assume that every contracted family purchases the company product. The firm wants to maximize its sales.

Write the above problem as LP model and solve by simplex method.

[M.Com., Delhi Univ. 2006]

[**Hint:** The appropriate model formulation of LPP is:

Maximize $z = (2 \times 2{,}000 + 8{,}000)\, x_1 + (2 \times 6{,}000 + 3{,}000)\, x_2$ subject to the constraints:

$$20{,}000x_1 + 12{,}000x_2 \le 1{,}80{,}000;$$
$$0 \le x_1 \ge 6 \text{ and } 0 \le x_2 \le 12;$$

where, x_1 and x_2 denote the number of TV and magazine advertisements.]

3.58. An advertising company wants to cover two types of audiences: Type *A*, consisting of customers with annual income greater than ₹ 1,50,000 and Type *B*, consisting of customers with annual income less than ₹ 1,50,000. The total advertising budget of the company is ₹ 2,10,000. One programme of TV advertising costs ₹ 50,000 and one programme of radio advertising costs ₹ 20,000. As per the terms of the contract, at least three programmes must be on TV and the number of radio programmes must be limited to five. On the basis of some surveys, it is found that a single TV programme reaches 4,50,000 customers of Type *A* and 50,000 customers of Type *B*, while one radio programme reaches 80,000 customers of Type *A* and 2,20,000 customers of Type *B*. Use simplex method and determine the optimal media mix to maximize the total reach. What is the total reach in each type of audience?

[M.Com., Delhi Univ. 2008; MBA (Nov.), Delhi Univ. 2010]

3:6. SOLUTION OF SIMULTANEOUS LINEAR EQUATIONS BY SIMPLEX METHOD

Let us consider the following system of *m* simultaneous linear equations in *n*-unknowns:

$$a_{i1}x_1 + a_{i2}x_2 + \dots + a_{in}x_n = b_i$$

where, x_j $(j = 1, 2, \dots, n)$ are unrestricted in sign.

Since, there is no objective function and decision variables are unrestriced in sign, we consider a dummy objective function having zero cost to all the decision variables and convert all the unrestricted variable into non-negative variables by writing $x_j = x'_j - x''_j$, where $x'_j \ge 0$ and $x''_j \ge 0$.

The system of equations, therefore, becomes a LPP and can be solved by simplex method.

We observe that there is no ready basic feasible solution. So, we introduce artificial variables A_i $(i = 1, 2, \dots, m)$ in the respective equation. Objective function, then, can be considered as of minimizing $z = A_1 + A_2 + \dots + A_m$.

The starting basic feasible solution is:

$$A_1 = b_1, A_2 = b_2, \dots, A_m = b_m.$$

Solve, now, the auxiliary linear programming problem by usual simplex method. The optimum solution to this LPP gives the solution of the given system of equations.

Example 3.30. *Use simplex method to solve the following system of linear equations:*

$$x_1 + x_2 = 1 \text{ and } 2x_1 + x_2 = 3.$$

Solution. Since, we don't know the nature of variables x_1 and x_2, let us consider these being unrestricted in sign. So, we let

$$x_1 = x'_1 - x''_1 \text{ and } x_2 = x'_2 - x''_2,$$

where $x'_1, x''_1, x'_2, x''_2 \ge 0$.

We observe that there is no ready basic feasible solution, so we introduce artificial variable $A_1 \geq 0$ and $A_2 \geq 0$ in the respective equation and also write a dummy objective function as of minimizing $z = A_1 + A_2$.

Obviously, the starting basic feasible solution is: $A_1 = 1$ and $A_2 = 3$. Considering the objective function as one of the constraint equation and eliminating A_1 and A_2 in this equation, we get

$$z-[1-(x_1'-x_1'')-(x_2'-x_2'')]-[3-2(x_1'-x_1'')-(x_2'-x_2'')]=0$$

i.e.,
$$3x_1'-3x_1''+2x_2'-2x_2''=4$$

Making use of above information, the iterative simplex tableau are:

Initial Iteration. x_1' enters and A_1 leaves.

Coeff. of basic variable	*Basic variable* (x_{Bi})	x_1'	x_1''	x_2'	x_2''	A_1	A_2	z	*Solution*	*Ratio* $\frac{x_{Bi}}{a_{i1}}, a_{i1} > 0$
1	$\leftarrow A_1$	(1)	–1	1	–1	1	0	0	1	1/1
1	A_2	2	–2	1	–1	0	1	0	3	3/2
	$(z_j - c_j)$	3 ↑	–3	2	–2	0	0	1	4	

Since, $(z_1 - c_1)$ is the most positive, non-basic variable x_1' enters the basis, *i.e.*, column under x_1' becomes key column.

The minimum ratio now corresponds to A_1-row, the current basic variable A_1 leaves the basis. Element at the intersection of first row and first column becomes the key number.

First Iteration. x_2'' enters and A_2 leaves.

Coeff. of basic variable	*Basic variable* (x_{Bi})	x_1'	x_1''	x_2'	x_2''	A_2	z	*Solution*	*Ratio* $\frac{x_{Bi}}{a_{i4}}, a_{i4} > 0$
0	x_1'	1	–1	1	–1	0	0	1	—
1	$\leftarrow A_2$	0	0	–1	(1)	1	0	1	1/1
	$(z_j - c_j)$	0	0	–1	1 ↑	1	1	1	

Here, $(z_4 - c_4)$ is the only positive, so column under x_2'' becomes the key column and A_2-row becomes the key row.

Final Iteration. Optimum solution.

Coeff. of basic variable	*Basic variable* (x_{Bi})	x_1'	x_1''	x_2'	x_2''	z	*Solution*
0	x_1'	1	–1	0	0	0	2
0	x_2''	0	0	–1	1	0	1
	$(z_j - c_j)$	0	0	0	0	1	0

Since, all $(z_j - c_j) \geq 0$, *i.e.*, net evaluation numbers in the last row are zero, the optimum solution to the auxiliary LPP is obtained. Hence, the optimum solution is:

$$x_1' = 2 \text{ and } x_2'' = 1$$

$\therefore$
$$x_1 = x_1' - x_1'' = 2 - 0 = 2 \text{ and } x_2 = x_2' - x_2'' = 0 - 1 = -1$$

i.e.,
$$x_1 = 2 \text{ and } x_2 = -1$$

PROBLEMS

Use simplex method to solve the following system of linear equations:

3.59. $2x_1 + x_2 = 1$ and $x_1 + 2x_2 = 5$.

3.60. $x_1 - x_3 + 4x_4 = 3$, $2x_1 - x_2 = 3$, and $3x_1 - 2x_2 - x_4 = 1$;
where $x_1 \geq 0, x_2 \geq 0, x_3 \geq 0$ and $x_4 \geq 0$.

TEST YOUR UNDERSTANDING

1. Graphical method is useful to solve the LPP, when there are
(*a*) only two decision variables
(*b*) more than two decision variables
(*c*) objective function is non-linear
(*d*) decision variables are not positive

2. The graphic approach to an LPP is useful because
(*a*) it provides general way to solve linear programming problems
(*b*) it does not provide unbounded solution
(*c*) it gives geometric insight into the given LPP and the meaning of optimality
(*d*) none of the above

3. A feasible solution to an LPP
(*a*) must be a corner point of the feasible region.
(*b*) must optimize the value of the objective function.
(*c*) must satisfy all the constraints of the LPP simultaneously.
(*d*) need not satisfy all of the constraints, only some of them.

4. An iso-profit line represents
(*a*) an infinite number of solutions all of which yield same profit
(*b*) an infinite number of optimum solutions
(*c*) an infinite number of solutions all of which yield the same cost
(*d*) a boundary line of the feasible region

5. If an iso-profit line yielding the optimum solution coincides with one of the constraint equation line, then
(*a*) the solution is unbounded
(*b*) the solution is infeasible
(*c*) the constraint which coincides is redundant
(*d*) none of the above

6. While solving a LPP graphically, the region bounded by the constraints is called
(*a*) feasible region
(*b*) infeasible region
(*c*) unbounded solution
(*d*) solution space

7. Alternative solution exists of an LPP, when
(*a*) one of the constraints is redundant
(*b*) two of the constraints are parallel
(*c*) objective function equation is parallel to one of the constraints
(*d*) all of the above

8. Given a system of m simultaneous linear equations in n unknowns ($m < n$), the number of basic variables will be
(*a*) m
(*b*) n
(*c*) $n - m$
(*d*) $n + m$

9. For maximization linear programming problem, the simplex method is terminated when all the net-evaluations are
(*a*) negative
(*b*) non-negative
(*c*) zero
(*d*) non-positive

10. Which of the following is not correct?

(*a*) Simplex method is an iterative process which involves the substitution of variables for obtaining successively better solutions.

(*b*) For solving an LPP by simplex method, it is essential that all variables involved are non-negative.

(*c*) Simplex method requires that all the constraints should be first converted into equations.

(*d*) Solution by simplex method requires that an LPP should have at least one non-negative value in the right hand side of the constraints.

11. A necessary and sufficient condition for a basic feasible solution to a minimization LPP to be an optimum is that (for all j):

(*a*) $(z_j - c_j) \geq 0$ (*b*) $(z_j - c_j) \leq 0$

(*c*) $(z_j - c_j) = 0$ (*d*) $(z_j - c_j) > 0$ or $(z_j - c_j) < 0$

12. At any iteration of the usual simplex method of maximization LPP, if there is at least one basic variable in the basis at zero level and all $(z_j - c_j) \geq 0$, the current solution is

(*a*) Infeasible (*b*) Unbounded

(*c*) Non-degenerate (*d*) Degenerate

13. The role of artificial variable in simplex method is

(*a*) to aid in finding initial basic feasible solution

(*b*) to start phases of simplex method

(*c*) to find shadow prices from the final simplex tableau

(*d*) none of the above

14. For maximization LPP, the objective function coefficient for an artificial variable is

(*a*) $+M$ (*b*) $-M$

(*c*) $+1$ (*d*) Zero

15. In a maximization LPP, if at least one artificial variable is in the basis, but not at zero level and the coefficient of M in each of the net evaluation $(z_j - c_j)$ is non-negative, then we have

(*a*) a feasible solution (*b*) no feasible solution

(*c*) an unbounded solution (*d*) an optimum solution

16. In final (optimum) simplex tableau, if $(z_j - c_j) = 0$ for at least one non-basic variable, then there will be

(*a*) an unbounded solution (*b*) infeasible solution

(*c*) alternative solution (*d*) cycling

17. If a negative value appears in the solution values column of the simplex method, then

(*a*) the basic solution is optimum (*b*) the basic solution is infeasible

(*c*) the basic solution is unbounded (*d*) all of the above

18. If an optimum solution is degenerate, then

(*a*) the solution is infeasible (*b*) there are alternative optimum solutions

(*c*) the solution is of no use to the decision maker

(*d*) none of the above

19. Which of the following is not correct?

(*a*) Degeneracy in an LPP may arise at the initial stage.

(*b*) A degenerate solution can never be optimum.

(*c*) Degeneracy may be a temporary phenomenon.

(*d*) The Big-M method and the Two-Phase method do not require the same number of iterations for solving an LPP.

20. When we solve a system of simultaneous linear equations by using *Two-Phase Simplex Method*, the values of decision variables may be
(*a*) positive (*b*) negative
(*c*) zero (*d*) positive and/or negative

ANSWERS

1. (*a*)	**2.** (*c*)	**3.** (*c*)	**4.** (*a*)	**5.** (*d*)
6. (*a*) & (*d*)	**7.** (*c*)	**8.** (*a*)	**9.** (*b*)	**10.** (*d*)
11. (*b*)	**12.** (*d*)	**13.** (*a*)	**14.** (*b*)	**15.** (*b*)
16. (*c*)	**17.** (*b*)	**18.** (*c*)	**19.** (*b*)	**20.** (*d*)

REVIEW EXERCISE

1. Explain the graphical method of solving a LPP involving two variables.
[MBA, Gujarat Tech. Univ. 2010; M.Com. (Nov.), Madurai Univ. 2002]

2. What is feasible region? Is it necessary that it should always be a convex set?
[M.Com., Madurai Univ. 2003]

3. Graphical solution is not possible for LPP with more than two constraints. True of False? Justify your answer. *[MBA (Nov.), Annamalai Univ. 2002]*

4. What is meant by the term feasible space? What determines the region?
[MBA (June), Panjab Tech. Univ. 2011]

5. Explain with an example of LPP which has no feasible solution. Use graphical method to explain. *[MBA, Dayal Bagh Edu. Institute 2014]*

6. What do you understand by convex region? Discuss one special case in the graphical solution of such a problem. *[MBA, Gujarat Tech. Univ. 2010]*

7. With the help of suitable sketches, define convex, non-convex, and infeasible region in relation to the graphical solution of linear programming problem. *[M.Com., Delhi Univ. 2008]*

8. Give outlines of simplex method in linear programming. *[BBA, GGSIP Univ. 2011]*

9. Explain simplex method of solving linear programming problem.
[B.E. (Mech.), Panjab Univ. 2012]

10. What is the criterion for testing optimality of solution to a linear programming problem in case of maximization as well as minimization? *[MIB, Delhi Univ. 2011]*

11. Write the difference in the simplex solution procedure for a maximization problem and a minimization problem of linear programming. *[M.Com., Delhi Univ. 1998]*

12. "To locate scope for improvement in the initial basic feasible solution, net evaluation row is introduced in the initial iteration." Where are these net evaluation numbers placed and how are they computed?

13. In solving a linear programming problem by simplex method, explain how you will move from a given basic feasible solution to another basic feasible solution with an improved value of the objective function. *[M.Com., Delhi Univ. 1997]*

14. Explain the use of artificial variables in linear programming.
[MBA, Panjab Tech. Univ. 2015; MBA, Mumbai Univ. 2012]

15. Explain the concept of degeneracy in simplex method. How is it resolved?
[B.Tech. (Mech.), HP Univ. 2012; B.E. Gujarat Tech. Univ. 2012]

16. What do you mean by two-phase method for solving a given LPP by simplex method? Why is it used?

17. Write the steps involved in solving an LPP by two-phase simplex method.
[M.Com., Kerala Univ. 2012]
18. Explain the penalty method or 'Big-M' method for solving a linear programming problem by simplex method.
19. How does the simplex method indicate that the problem has:
(*a*) unbounded solution, (*b*) no feasible solution, and (*c*) alternative optimum solution?
[PG Dipl. in Glob. Buis. Oper., Delhi Univ. 2011]
20. While solving a LPP by simplex method, what indicates the following:
(*a*) Unique optimum solution, (*b*) Presence of redundant constraints?
21. How will you identify the existence of the following in the optimum simplex tableau and what are the effects of these cases on the final answer:
(*a*) Alternative optimum solution, (*b*) Degeneracy,
(*c*) Infeasible solution, (*d*) Unbounded solution?
[PG Dipl. in Glob. Buis. Oper., Delhi Univ. 2010]
22. With reference to the solution of LPP by simplex method, when do you conclude that the given linear programming problem has multiple optimal solutions?
23. What is shadow price? How can we read the shadow price from a simplex tableau?
[MIB, Delhi Univ. 2011]
24. What is pseudo-optimal solution? *[B.Sc. (Math.), Madras Univ. 2012]*
25. What is the physical significance of shadow price of a resource? Explain with the help of an example. *[MBA (HCA), Delhi Univ. 2015]*

CHAPTER 4

Theory of Simplex Method

"Understanding is deeper than knowledge"

Basic mechanics of simplex method had been discussed in *Chapter 3*. Now, we shall go a little more deeply into this method (algorithm) by examining some of its underlying theory. For this, we first develop the general geometric and algebraic properties that form the foundation of the simplex method. Then, we shall describe the matrix form of the simplex method, which streamlines the procedure considerably for computer implementation.

4:1. FOUNDATIONS OF THE SIMPLEX METHOD

Recall that while solving a linear programming problem (LPP) by graphical (geometric) method, we graph the constraints of the problem to find the feasible region (or convex region). This feasible region is bounded by the constraint equations whose corners give the corner-point feasible (CPF) solutions. Clearly a CPF solution is a feasible solution that does not lie on any line segment connecting two other feasible solutions. In other words, a feasible solution that does lie on a line segment connecting two other feasible solutions is not a CPF solution. The CPF solutions play a key role in the computational procedure of simplex method.

For any linear programming problem that has feasible solutions and a bounded feasible region, the key properties of CPF solutions are given below:

Property 1.

(*a*) If there is exactly one (unique) optimum solution, it must be a CPF solution.

(*b*) If there are multiple optimum solutions and a bounded feasible region, then at least two CPF solutions must be adjacent.

Property 2. There are only a finite number of CPF solutions.

Property 3. If a CPF solution has no adjacent CPF solutions that are better (as measured by z), then there are no better CPF solutions anywhere. Therefore, such a CPF solution is guaranteed to be an optimum solution (*Property 1*), assuming that the problem possesses at least one optimum solution.

The simplex method consists of moving along an 'edge' of the convex region of feasible solutions from one CPF solution to an adjacent one. Of all the adjacent CPF solutions, the one yielding an ***improved value*** of the objective function over that of the preceding CPF solution, is chosen. At each CPF solution, the method tells us whether that CPF solution is optimum, and if not, the procedure of moving from one CPF solution to another is repeated. Since, the number of CPF solutions (vertices) are finite, simplex method leads to an ***optimum CPF solution*** (vertex) in a finite number of steps. If at any stage, the procedure leads us to a CPF solution (vertex) which has an edge leading to infinity and, if the objective function value can be further improved by moving along that edge, the simplex method tells us that there is an ***unbounded solution***.

4:2. TERMINOLOGY OF SIMPLEX METHOD

The general linear programming problem in standard form expressed in Section 2:8 of Chapter 2 is:

Maximize or Minimize $z = c_1x_1 + c_2x_2 + ... + c_nx_n$ subject to the constraints:

$$a_{i1}x_1 + a_{i2}x_2 + ... + a_{in}x_n = b_i, \quad (i = 1, 2, ..., m)$$

$$x_1, x_2, x_3, ..., x_n \geq 0$$

In matrix notations, the general LPP can be expressed as to

Maximize or Minimize $z = \mathbf{cx}$ subject to the constraints:

$$\mathbf{Ax} = \mathbf{b} \text{ and } \mathbf{x} \geq \mathbf{0},$$

where, $\mathbf{A} = \begin{bmatrix} a_{11} & a_{12} & \cdots & a_{1n} \\ a_{21} & a_{22} & \cdots & a_{2n} \\ \vdots & \vdots & \vdots & \vdots \\ a_{m1} & a_{m2} & \cdots & a_{mn} \end{bmatrix} = (\mathbf{a}_1 \ \ \mathbf{a}_2 \ \ \cdots \ \ \mathbf{a}_m), \ \mathbf{x} = \begin{bmatrix} x_1 \\ x_2 \\ \vdots \\ x_n \end{bmatrix}, \ \mathbf{b} = \begin{bmatrix} b_1 \\ b_2 \\ \vdots \\ b_m \end{bmatrix},$

and $\mathbf{c} = (c_1 \ \ c_2 \ \ \cdots \ \ c_n)$

1. **Basic Solution.** Given a system of m simultaneous linear equations in n ($n > m$) variables:

$$\mathbf{Ax} = \mathbf{b}, \ \mathbf{x} \in \mathbf{R}^n,$$

where $\mathbf{A}$ is an $m \times n$ matrix of rank 'm'. Let $\mathbf{B}$ be any $m \times m$ sub-matrix formed by m linearly independent columns of $\mathbf{A}$. By setting $(n - m)$ variables not associated with the columns of $\mathbf{B}$ equal to zero, the given system of equations reduces to $\mathbf{Bx}_B = \mathbf{b}$, where $\mathbf{B} = (\boldsymbol{\beta}_1 \ \ \boldsymbol{\beta}_2 \ \ ... \ \ \boldsymbol{\beta}_m)$ and $\boldsymbol{\beta}_i$ $(i = 1, 2, ..., m)$ are columns of $\mathbf{A}$.

By solving the system of equations $\mathbf{Bx}_B = \mathbf{b}$, we get $\mathbf{x}_B = \mathbf{B}^{-1}\mathbf{b}$ (since, $\mathbf{B}^{-1}\mathbf{B} = \mathbf{I}$). The solution $\mathbf{x}_B$ is called the ***basic solution*** to the given system of linear equations.

The m variables which may be all different from zero, are called ***basic variables*** and the remaining $(n - m)$ variables are known as ***non-basic variables***. The $m \times m$ non-singular sub-matrix $\mathbf{B}$ is called the ***basis matrix*** with the columns of $\mathbf{B}$ as ***basis vectors***.

If all the basic variables also satisfy the non-negative conditions, *i.e.*, $\mathbf{x} \geq 0$, then the basic solution is called ***basic feasible solution***.

2. **Degenerate Solution.** A basic feasible solution $\mathbf{x}_B = \mathbf{B}^{-1}\,\mathbf{b}$ is said to be ***degenerate***, if at least one component of $\mathbf{x}_B$ (basic variables) is zero. If all the components of $\mathbf{x}_B$ are positive, *i.e.*, $\mathbf{x}_B > \mathbf{0}$, then it is called a non-degenerate basic feasible solution.

Value of the Objective Function. Let c_{Bi} $(i = 1, 2, ..., m)$ are the components of $\mathbf{C}$ associated with the basic variables in the objective function coefficients corresponding to x_{Bi} $(i = 1, 2, ..., m)$. Then, the value of objective function for basic solution is: $z = \mathbf{c}_B\,\mathbf{x}_B = \mathbf{c}_B\,\mathbf{B}^{-1}\,\mathbf{b}$.

3. **Optimum Basic Feasible Solution.** A basic feasible solution $\mathbf{x}_B$ to an LPP is called an ***Optimum basic feasible solution***, if $z_0 = \mathbf{c}_B\,\mathbf{x}_B > z^*$ for maximization problem and $z_0 = \mathbf{c}_B\,\mathbf{x}_B < z^*$ for minimization problem, where z^* is the value of objective function for any feasible solution.

4. **Reduction of Feasible Solution to Basic Feasible Solution.** Consider a system of m simultaneous linear equations in n variables (unknowns) with $m < n$ as: $\mathbf{A}\,\mathbf{x} = \mathbf{b}$, where $\mathbf{A} = (\mathbf{a}_1 \ \ \mathbf{a}_2 \ \ ... \ \ \mathbf{a}_n)$ is an $m \times n$ matrix and $\mathbf{b}$ is an $m \times 1$ column vector (column matrix). Then, the given system of equation can expressed as:

$$\mathbf{a}_1x_1 + \mathbf{a}_2x_2 + ... + \mathbf{a}_nx_n = \mathbf{b}.$$

Suppose there exists a feasible solution to the above system of m equations. Then, this feasible solution can be reduced to basic feasible solution in the following four steps:

Step 1. Since, $m < n$, the m column vectors of $\mathbf{A}$ are linearly dependent and we write

$$\lambda_1 \mathbf{a}_1 + \lambda_2 \mathbf{a}_2 + ... + \lambda_m \mathbf{a}_m = 0,$$

where λ_j ($j = 1, 2, ..., m$) are scalars.

Step 2. For any index k, let $\lambda_k \neq 0$. Then

$$\mathbf{a}_k = -\sum_{j \neq k}^{m} \frac{\lambda_j}{\lambda_k} \mathbf{a}_j, \text{ for } j \neq k.$$

$$\therefore \quad \sum_{j \neq k} \mathbf{a}_j x_j - \sum_{j \neq k} \left(\frac{\lambda_j}{\lambda_k}\right) \mathbf{a}_j x_k = \mathbf{b} \quad \text{or} \quad \sum_{j \neq k} \left(x_j - \frac{\lambda_j}{\lambda_k} x_k\right) \mathbf{a}_j = \mathbf{b}.$$

Step 3. The solution with not more than $(m - 1)$ non-zero components is obtained in *step* 2. To ensure that these are positive, we must have

$$x_j - \left(\frac{\lambda_j}{\lambda_k}\right) x_k \geq 0, \text{ for } j \neq k.$$

This requires that either $\lambda_j = 0$ or

$$\frac{x_j}{\lambda_j} \geq \frac{x_k}{\lambda_k}, \text{ if } \lambda_j > 0 \text{ and } \frac{x_j}{\lambda_j} \leq \frac{x_k}{\lambda_k}, \text{ if } \lambda_j < 0.$$

Thus, if we select $\mathbf{a}_k$ such that

$$\frac{x_k}{\lambda_k} = \min_j \left\{\frac{x_j}{\lambda_j}, \lambda_j > 0\right\},$$

then we have a feasible solution with not more than $(m - 1)$ non-zero components.

Step 4. Procedure is repeated until all the m column vectors are linearly independent, *i.e.*, a basic feasible solution is attained.

5. Basic Feasible Solution (BFS) and Extreme Point. Consider the system of m simultaneous linear equations in n variables with $m < n$ as : $\mathbf{Ax} = \mathbf{b}, \mathbf{x} \in \mathbf{R}^n$.

Let $\mathbf{S}$ be the set of all feasible solutions and suppose that $\mathbf{x}$ is a basic feasible solution.

Then, $\mathbf{x} = \begin{bmatrix} \mathbf{x}_B \\ \mathbf{0} \end{bmatrix}$, where $\mathbf{x}_B$ is an $m \times 1$ column vector such that for a non-singular sub-matrix $\mathbf{B}$ of $\mathbf{A}$,

$$\mathbf{B}\mathbf{x}_B = \mathbf{b} \text{ or } \mathbf{x}_B = \mathbf{B}^{-1}\mathbf{b}.$$

If possible, let $\mathbf{x}$ be a point of S, such that there exist $\mathbf{x}_1, \mathbf{x}_2 \in S$ such that $\mathbf{x}_1 \neq \mathbf{x}_2$ and

$$\mathbf{x} = \lambda \mathbf{x}_1 + (1 - \lambda)\mathbf{x}_2 \qquad 0 < \lambda < 1.$$

Let

$$\mathbf{x}_1 = \begin{bmatrix} \mathbf{u}_1 \\ \mathbf{v}_1 \end{bmatrix} \text{ and } \mathbf{x}_2 = \begin{bmatrix} \mathbf{u}_2 \\ \mathbf{v}_2 \end{bmatrix}$$

where $\mathbf{u}_1, \mathbf{u}_2$ are $m \times 1$ vectors and $\mathbf{v}_1, \mathbf{v}_2$ are $(n - m) \times 1$ vectors. Then

$$\begin{bmatrix} \mathbf{x}_B \\ 0 \end{bmatrix} = \lambda \begin{bmatrix} \mathbf{u}_1 \\ \mathbf{v}_1 \end{bmatrix} + (1 - \lambda) \begin{bmatrix} \mathbf{u}_2 \\ \mathbf{v}_2 \end{bmatrix}$$

$$\therefore \quad \mathbf{x}_B = \lambda \mathbf{u}_1 + (1 - \lambda)\mathbf{u}_2$$

$$\text{and} \quad \mathbf{0} = \lambda \mathbf{v}_1 + (1 - \lambda)\mathbf{v}_2 \qquad 0 < \lambda < 1.$$

Since, $\mathbf{x}_1, \mathbf{x}_2$ are feasible solutions, therefore, $\mathbf{u}_1, \mathbf{u}_2, \mathbf{v}_1, \mathbf{v}_2 \geq \mathbf{0}$.

Now $\quad 0 < \lambda < 1$ and $\mathbf{0} = \lambda \mathbf{v}_1 + (1-\lambda)\mathbf{v}_2$

∴ We must have $\quad \mathbf{v}_1 = \mathbf{v}_2 = \mathbf{0}.$

Thus,
$$\mathbf{x}_1 = \begin{bmatrix} \mathbf{u}_1 \\ \mathbf{0} \end{bmatrix} \text{ and } \mathbf{x}_2 = \begin{bmatrix} \mathbf{u}_2 \\ \mathbf{0} \end{bmatrix}$$

Again, since, $\mathbf{x}_1, \mathbf{x}_2$ satisfy $\quad \mathbf{A}\mathbf{x}_1 = \mathbf{b}$ and $\mathbf{A}\mathbf{x}_2 = \mathbf{b}$, therefore,

$$\mathbf{B}\mathbf{u}_1 = \mathbf{b} \text{ and } \mathbf{B}\mathbf{u}_2 = \mathbf{b}.$$

Also, since $\mathbf{B}\mathbf{x}_B = \mathbf{b}$, and expression of $\mathbf{b}$ (as a linear combination of basis vectors) must be unique, therefore, $\mathbf{u}_1 = \mathbf{u}_2 = \mathbf{x}_B$. Hence

$$\mathbf{x} = \mathbf{x}_1 = \mathbf{x}_2.$$

This is a contradiction, for by assumption $\mathbf{x}_1 \neq \mathbf{x}_2$.

Hence, $\mathbf{x}$ is an extreme point of S.

Remark. *The number of extreme points of convex set of feasible solutions is finite.* It follows from the fact that since, a feasible solution can be reduced to basic feasible solution, there is only one extreme point for a given basic feasible solution. So, there is one-to-one correspondence between the extreme points and the basic feasible solutions. But, basic feasible solutions are finite in number, therefore, the extreme points of the convex set of feasible solutions are also finite in number.

Example 4.1. *Obtain all the basic solutions to the following system of linear equations:*

$$x_1 + 2x_2 + x_3 = 4$$
$$2x_1 + x_2 + 5x_3 = 5.$$

Solution. The given system of equations can be written in matrix form as $\mathbf{A}\mathbf{x} = \mathbf{b}$, where

$$\mathbf{A} = \begin{pmatrix} 1 & 2 & 1 \\ 2 & 1 & 5 \end{pmatrix}, \quad \mathbf{x} = \begin{bmatrix} x_1 \\ x_2 \\ x_3 \end{bmatrix} \text{ and } \mathbf{b} = \begin{bmatrix} 4 \\ 5 \end{bmatrix}.$$

Since, rank of $\mathbf{A}$ is 2, maximum number of linearly independent columns of $\mathbf{A}$ is 2. Thus, we can consider any of the following 2×2 sub-matrices as ***basis matrix*** $\mathbf{B}$:

$$\begin{pmatrix} 1 & 2 \\ 2 & 1 \end{pmatrix}, \begin{pmatrix} 1 & 1 \\ 2 & 5 \end{pmatrix} \text{ and } \begin{pmatrix} 2 & 1 \\ 1 & 5 \end{pmatrix}$$

The variables not associated with the columns of $\mathbf{B}$ are x_3, x_2 and x_1 respectively, in the three different cases.

Let us first consider $\mathbf{B} = \begin{pmatrix} 1 & 2 \\ 2 & 1 \end{pmatrix}$. A basic solution to the given system is now obtained by setting $x_3 = 0$, and solving the system

$$\begin{pmatrix} 1 & 2 \\ 2 & 1 \end{pmatrix} \begin{bmatrix} x_1 \\ x_2 \end{bmatrix} = \begin{bmatrix} 4 \\ 5 \end{bmatrix}.$$

Thus, a basic (non-basic) solution to the given system is:

(Basic) $x_1 = 2$ and $x_2 = 1$; *(Non-basic)* $x_3 = 0$

Similarly, the other two basic and non-basic solutions are:

(Basic) $x_1 = 5$ and $x_3 = -1$ *(Non-basic)* $x_2 = 0$

and (Basic) $x_2 = 5/3$ and $x_3 = 2/3$ *(Non-basic)* $x_1 = 0$

We observe that all the three basic solutions obtained above are non-degenerate.

Remark. Above example shows that one of the basic solution does not satisfy the condition of non-negative, *i.e.*, $x_1 = 5$ and $x_3 = -1$. So, it is not feasible solution. Hence, there are only two basic feasible solutions, *viz.*,

$$x_1 = 2 \text{ and } x_1 = 1, \quad x_2 = 5/3 \text{ and } x_3 = 2/3.$$

Example 4.2. *Show that the following system of linear equations have a degenerate solution:*

$$2x_1 + x_2 - x_3 = 2$$
$$3x_1 + 2x_2 + x_3 = 3.$$

Solution. The given system of equations can be written as $\mathbf{Ax} = \mathbf{b}$, where

$$\mathbf{A} = \begin{pmatrix} 2 & 1 & -1 \\ 3 & 2 & 1 \end{pmatrix}, \quad \mathbf{x} = \begin{bmatrix} x_1 \\ x_2 \\ x_3 \end{bmatrix} \text{ and } \mathbf{b} = \begin{bmatrix} 2 \\ 3 \end{bmatrix}.$$

Since, rank of **A** is 2, maximum number of linearly independent columns of **A** is 2. Thus, we can take any of the following 2 × 2 sub-matrices of **A**, as basis matrix **B**:

$$\begin{pmatrix} 2 & 1 \\ 3 & 2 \end{pmatrix}, \begin{pmatrix} 1 & -1 \\ 2 & 1 \end{pmatrix} \text{ and } \begin{pmatrix} 2 & -1 \\ 3 & 1 \end{pmatrix}.$$

The variables not associated with the columns of these submatrices are, respectively, x_3, x_1 and x_2.

Considering $\mathbf{B} = \begin{pmatrix} 2 & 1 \\ 3 & 2 \end{pmatrix}$, a basic solution to the given system is obtained by setting $x_3 = 0$ and solving the system

$$\begin{pmatrix} 2 & 1 \\ 3 & 2 \end{pmatrix} \begin{bmatrix} x_1 \\ x_2 \end{bmatrix} = \begin{bmatrix} 2 \\ 3 \end{bmatrix}.$$

Thus, basic solution to the given problem is

(Basic) $x_1 = 1$ and $x_2 = 0$ *(Non-basic)* $x_3 = 0$

Similarly, the other two solutions are:

(Basic) $x_2 = 5/3$ and $x_3 = -1/3$ *(Non-basic)* $x_1 = 0$

and (Basic) $x_1 = 1$ and $x_3 = 0$ *(Non-basic)* $x_2 = 0$

In each of the two basic solutions, at least one of the basic variables is zero. Hence, two of the basic solutions are degenerate solutions.

Remark. In the above ***example***, we observe that $x_2 = 5/3$ and $x_3 = -1/3$ is not feasible as $x_3 < 0$; while $x_1 = 1$ and $x_2 = 0$ as well as $x_1 = 1$ and $x_3 = 0$ both are basic feasible solutions. But, since, one of the basic variable in both the solutions have zero value, both are ***degenerate basic feasible solutions***.

Example 4.3. *Let $x_1 = 2$, $x_2 = 3$ and $x_3 = 1$ be a feasible solution to the following system of equations:*

$$2x_1 + x_2 + 4x_3 = 11,$$
$$3x_1 + x_2 + 5x_3 = 14.$$

Reduce the feasible solution to basic feasible solution. ***[BE (Jan.), Visvesvaraya Univ. 2020]***

Solution. In matrix notations, the equations are written as:

$$\mathbf{Ax} = \mathbf{b}, \text{ where } \mathbf{A} = (\mathbf{a}_1 \ \mathbf{a}_2 \ \mathbf{a}_3), \ \mathbf{x} = [x_1 \ x_2 \ x_3] \text{ and } \mathbf{b} = [11 \ 14]$$

So, we have $\mathbf{a}_1 x_1 + \mathbf{a}_2 x_2 + \mathbf{a}_3 x_3 = \mathbf{b}$

Given: $x_1 = 2$, $x_2 = 3$ and $x_3 = 1$, therefore, we get

$$2\mathbf{a}_1 + 3\mathbf{a}_2 + \mathbf{a}_3 = \mathbf{b}$$

Since, there are only 2 rows and 3 columns, *i.e.*, 2 < 3, there can be only two linearly independent column vectors of **A**. Assume that these three vectors are linearly dependent. Then, we express one of the column vector as a linear combination of remaining two column vectors. That is, for scalars λ_1 and λ_2; $\mathbf{a}_3 = \lambda_1\mathbf{a}_1 + \lambda_2\mathbf{a}_2$. Thus,

$$\begin{bmatrix} 4 \\ 5 \end{bmatrix} = \lambda_1 \begin{bmatrix} 2 \\ 3 \end{bmatrix} + \lambda_2 \begin{bmatrix} 1 \\ 1 \end{bmatrix},$$

where, λ_1 and λ_2 are scalars, *i.e.*, $4 = 2\lambda_1 + \lambda_2$ and $5 = 3\lambda_1 + \lambda_2$.

On solving, we get $\lambda_1 = 1$ and $\lambda_2 = 2$.

Substituting these values of λ_1 and λ_2 in the linear combination of vectors, we have:

$$1 \times \mathbf{a}_1 + 2 \times \mathbf{a}_2 - \mathbf{a}_3 = 0$$

Now $$\frac{x_3}{\lambda_3} = \text{Min.}\left\{\frac{x_1}{\lambda_1}, \frac{x_2}{\lambda_2}\right\} = \text{Min.}\left\{\frac{2}{1}, \frac{3}{2}\right\} = \frac{3}{2}.$$

Since, minimum of the ratios computed above corresponds to $\mathbf{a}_2$, it can be removed to obtain a new solution with not more than (3 – 1), *i.e.*, two variables having non-zero value. Thus, the value of new variables are:

$$x_1^* = x_1 - \frac{x_2}{\lambda_2}\lambda_1 = 2 - \frac{3}{2} \times 1 = \frac{1}{2},$$

$$x_2^* = x_2 - \frac{x_2}{\lambda_2}\lambda_2 = 3 - \frac{3}{2} \times 2 = 0,$$

and $$x_3^* = x_3 - \frac{x_2}{\lambda_2}\lambda_3 = 1 - \frac{3}{2} \times (-1) = \frac{5}{2}.$$

Thus, the basic feasible solution is:

$$x_1 = 1/2 \text{ and } x_3 = 5/2 \text{ with } x_2 = 0 \qquad \textit{(Non-basic)}$$

Note. Another basic feasible solution to the given system of equations can easily be obtained as:

$$x_1 = 3 \text{ and } x_2 = 5 \text{ with } x_3 = 0 \qquad \textit{(Non-basic)}$$

Example 4.4. *Show that the feasible solution: $x_1 = 1$, $x_2 = 0$ and $x_3 = 1$ to the system of equations*

$$x_1 + x_2 + x_3 = 1 \quad \text{and} \quad x_1 - x_2 + x_3 = 2$$

is not basic feasible solution.

Solution. In matrix notations, the system of equations are written as $\mathbf{Ax} = \mathbf{b}$, where $\mathbf{A} = (\mathbf{a}_1 \ \mathbf{a}_2 \ \mathbf{a}_3)$, $\mathbf{x} = [x_1 \ x_2 \ x_3]$, and $\mathbf{b} = [b_1 \ b_2 \ b_3]$

Therefore, we write: $\mathbf{a}_1x_1 + \mathbf{a}_2x_2 + \mathbf{a}_3x_3 = \mathbf{b}$.

Using the given values of x_1, x_2 and x_3, we have $\mathbf{a}_1 + \mathbf{a}_3 = \mathbf{b}$.

Since, rank (**A**) = 2, only two out of three column vectors $\mathbf{a}_1$, $\mathbf{a}_2$ and $\mathbf{a}_3$ are linearly independent. Assume that the three vectors are linearly dependent. Then by expressing one of them as a linear combination of the remaining two, we have

$$\mathbf{a}_3 = \lambda_1\mathbf{a}_1 + \lambda_2\mathbf{a}_2, \text{ where } \lambda_1 \text{ and } \lambda_2 \text{ are scalars.}$$

$$1 = \lambda_1 + \lambda_2 \text{ and } 1 = \lambda_1 - \lambda_2.$$

On solving these equations, we get: $\lambda_1 = 1$ and $\lambda_2 = 0$

$$\therefore \quad \mathbf{a}_3 = \lambda_1\mathbf{a}_1 + \lambda_2\mathbf{a}_2 = \mathbf{a}_1 + 0, \text{ i.e., } \mathbf{a}_1 - \mathbf{a}_3 = 0,$$

where, $\lambda_1 = 1$, $\lambda_2 = 0$ and $\lambda_3 = -1$.

To reduce the number of positive variables, we compute the ratio of x_3 and λ_3.

Now, $$\frac{x_3}{\lambda_3} = \text{Min.}\left\{\frac{x_1}{\lambda_1}, \frac{x_2}{\lambda_2}\right\} = \text{Min.}\left\{\frac{1}{1}\right\} = 1.$$

Since, $x_3/\lambda_3 = 1$ corresponds to $\mathbf{a}_1$, it should be removed to obtain a new solution with not more than two non-negative variables. The value of the new variables are:

$$x_1^* = x_1 - \frac{x_1}{\lambda_1} \times 1 = 1 - 1 \times 1 = 0$$

$$x_2^* = x_2 - \frac{x_1}{\lambda_1} \times 0 = 0 - 1 \times 0 = 0$$

$$x_3^* = x_3 - \frac{x_1}{\lambda_1} \times (-1) = 1 - 1 \times (-1) = 2$$

The basic feasible solution, therefore, is

$$x_2 = 0 \text{ and } x_3 = 2 \text{ with } x_1 = 0 \qquad \textit{(Non-basic)}$$

Another basic feasible solution can easily be obtained as

$$x_1 = 2, x_2 = 0 \text{ with } x_3 = 0 \qquad \textit{(Non-basic)}$$

We observe that both these basic feasible solutions are different from the given feasible solution, and hence, the given solution is not basic feasible solution.

PROBLEMS

4.01. Compute all the basic feasible solutions to the system of linear equations:

$$4x_1 + 2x_2 + x_3 = 7, \text{ and } -x_1 + 4x_2 + 2x_3 = 14.$$

4.02. Find all the basic feasible solutions for the system of equations:

(*i*) $2x_1 + 6x_2 + 2x_3 + x_4 = 3$, and $6x_1 + 4x_2 + 4x_3 + 6x_4 = 2$

(*ii*) $3x_1 + x_2 - x_3 = 8$, and $x_1 + x_2 + x_3 = 4$

4.03. Find all the basic feasible solutions of the following LPP without using simplex alogorithm:

Maximize $z = 2x_1 + 3x_2 + 4x_3 + 7x_4$ subject to the constraints:

$$2x_1 + 3x_2 + x_3 + 4x_4 = 8$$
$$x_1 - 2x_2 + 6x_3 - 7x_4 = -3$$
$$x_1 \geq 0, x_2 \geq 0, x_3 \geq 0 \text{ and } x_4 \geq 0$$

and choose that one which maximizes z.

4.04. Let $x_1 = 0, x_2 = 0, x_3 = 0, x_4 = 7, x_5 = 12$ and $x_6 = 10$ is an extreme point solution of the set of equations:

$$3x_1 - x_2 + 2x_3 + x_4 = 7$$
$$-2x_1 + 4x_2 + x_5 = 12$$
$$-4x_1 + 3x_2 + 8x_3 + x_6 = 10$$

Find two more extreme point solutions of the above problem.

4.05. The column vector [1, 1, 1] is a feasible solution to the system of equations:

$$x_1 + x_2 + 2x_3 = 4$$
$$2x_1 - x_2 + x_3 = 2$$

Reduce the given feasible solution to a basic feasible solution.

4.06. Consider the system of equations:

$$x_1 + 2x_2 + 4x_3 + x_4 = 7$$
$$2x_1 - x_2 + 3x_3 - 2x_4 = 4$$

Here, $x_1 = 1, x_2 = 1, x_3 = 1$ and $x_4 = 0$ is a feasible solution. Reduce this feasible solution to two different basic feasible solutions. *[M.Sc. (Math.), CCS Univ. 1998]*

4:3. FUNDAMENTAL THEOREM OF LINEAR PROGRAMMING

Consider the LPP to determine **x**, so as to

$$\text{Maximize } z = \mathbf{cx} \text{ subject to the constraints:}$$
$$\mathbf{Ax} = \mathbf{b} \text{ and } \mathbf{x} \geq \mathbf{0}$$

where, $\mathbf{c}, \mathbf{x} \in \mathbf{R}^n$.

The feasible region S of the LPP is given by

$$S = \{ \mathbf{x} \mid \mathbf{Ax} = \mathbf{b}, \mathbf{x} \geq \mathbf{0} \}$$

Since, S is a convex polyhedron, it is non-empty, closed and bounded. Therefore, the objective function $z = \mathbf{cx}$, $\mathbf{x} \in S$ is continuous on S, which is non-empty, closed and bounded and hence, z attains its maximum on S. This proves the existence of an optimum solution.

Now, since, S is a convex polyhedron, it has finite number of extreme points. Let these be $\mathbf{x}_1, \mathbf{x}_2, ..., \mathbf{x}_k \in S$. Clearly $S = < \mathbf{x}_1, \mathbf{x}_2, ..., \mathbf{x}_k >$. Therefore, any $\mathbf{x} \in S$ can be expressed as a convex combination of the extreme points, say

$$\mathbf{x} = \sum_j \alpha_j \mathbf{x}_j; \ \alpha_j \geq 0, \ \sum_j \alpha_j = 1 \text{ for all } j = 1, 2, ..., k$$

Let $z_0 = \max. \{ \mathbf{cx}_j ;\ j = 1, 2, ..., k \}$. Then for any $\mathbf{x} \in S$,

$$z = \mathbf{cx} = \mathbf{c}\left(\sum_j \alpha_j \mathbf{x}_j\right) = \sum_j \alpha_j (\mathbf{cx}_j) \leq \sum_j \alpha_j z_0 \text{ or } z_0 \sum_j \alpha_j$$

$\therefore$ $$z \leq z_0 \text{ for any } \mathbf{x} \in S$$

Thus, the maximum value of z is attained only at one of the extreme points of S. That is, at least one extreme point of S yields an optimum solution.

Hence, if the feasible region of an LPP is a convex polyhedron, then there exists an optimum solution to the LPP and at least one basic feasible solution must be optimum.

Remark. It is easily seen that if the optimum value of z is attained at more than one extreme point of S, then every convex combination of such extreme points also provides an optimum solution to the LPP. Thus, the optimum solution to an LPP is either ***unique*** or ***infinite*** in number.

4:4. IMPROVING OF BASIC FEASIBLE SOLUTION (BFS)

(*i*) **Replacing of current basis vector with a column vector not in the basis.** Suppose that we have a basic feasible solution to the following LPP:

$$\text{Maximize } z = \mathbf{cx} \text{ subject to the constraints:}$$
$$\mathbf{Ax} = \mathbf{b} \text{ and } \mathbf{x} \geq \mathbf{0},$$

where, **A** is an $m \times n$ matrix of rank 'm'. Since, **B** (sub-matrix of **A**) is a basis matrix, any column vector of **A** can be expressed as a linear combination of basis vectors. Therefore, if $\mathbf{B} = (\boldsymbol{\beta}_1\ \boldsymbol{\beta}_2\ \cdots\ \boldsymbol{\beta}_m)$, then

$$\mathbf{a}_j = y_{1j}\,\beta_1 + y_{2j}\,\beta_2 + ... + y_{mj}\,\beta_m = \sum_{i=1}^{m} y_{ij}\,\beta_i = \mathbf{y}_j\,\mathbf{B},$$

where, y_{ij} are scalars and $\mathbf{y}_j = (y_{1j}\ \ y_{2j}\ \ ...,\ \ y_{mj})$.

We know that a column vector $\mathbf{a}_j$ of $\mathbf{A}$ can replace any column vector of $\mathbf{B}$, say $\boldsymbol{\beta}_k$ for which $y_{kj} \neq 0$, and the new set of vectors will still form a basis matrix. So, select any vector $\mathbf{a}_j$ of $\mathbf{A}$ (not in $\mathbf{B}$) for which at least one $y_{kj} \neq 0$, and introduce this into the basis matrix $\mathbf{B}$ to replace the basis vector $\boldsymbol{\beta}_k$.

Therefore,
$$\mathbf{a}_j = \sum_{i \neq k} y_{ij}\, \boldsymbol{\beta}_i + y_{kj}\, \boldsymbol{\beta}_k$$

or
$$\boldsymbol{\beta}_k = \frac{1}{y_{kj}} \mathbf{a}_j - \sum_{i \neq k} \left(\frac{y_{ij}}{y_{kj}} \right) \boldsymbol{\beta}_i, \; y_{kj} \neq 0$$

The basic feasible solution can also be written as
$$\sum_{i = k} x_{Bi}\, \boldsymbol{\beta}_i = \mathbf{b} \quad \text{or} \quad \sum_{i \neq k} x_{Bi}\, \boldsymbol{\beta}_i + x_{Bk}\, \boldsymbol{\beta}_k = \mathbf{b}$$

Eliminating $\boldsymbol{\beta}_k$ from this equation, we get
$$\sum_{i \neq k} x_{Bi}\, \boldsymbol{\beta}_i + x_{Bk} \left[\frac{1}{y_{kj}} \mathbf{a}_j - \sum_{i \neq k} \left(\frac{y_{ij}}{y_{kj}} \right) \boldsymbol{\beta}_i \right] = \mathbf{b}$$

or
$$\sum_{i \neq k} \left(x_{Bi} - \frac{y_{ij}}{y_{kj}} x_{Bk} \right) \boldsymbol{\beta}_i + \frac{1}{y_{kj}} x_{Bk}\, \mathbf{a}_j = \mathbf{b}$$

For this solution to be basic feasible, we must have
$$x_{Bi} - \frac{y_{ij}}{y_{kj}} x_{Bk} \geq 0, \qquad \text{for } i \neq k \;\; (i = 1, 2, ..., m)$$

and
$$\frac{1}{y_{kj}} x_{Bk} \geq 0 \qquad \text{for } i = k$$

We observe that if $x_{Bk} \neq 0$, then we must have $y_{kj} > 0$. In case, when all $y_{ij} \leq 0$ $(i \neq k)$, the new solution will be feasible. However, if not all $y_{ij} \leq 0$, then we must choose a basis vector (column) to be removed from the basis. Otherwise, we may get an infeasible solution. If $y_{ij} > 0$, we shall select the column vector $\boldsymbol{\beta}_k$ to be replaced by a non-basis column vector $\mathbf{a}_j$ in such a way that
$$\frac{x_{Bi}}{y_{ij}} - \frac{x_{Bk}}{y_{kj}} \geq 0, \quad \text{for} \quad y_{ij} > 0 \text{ and } i \neq k$$

and
$$\frac{x_{Bi}}{y_{ij}} - \frac{x_{Bk}}{y_{kj}} \leq 0, \quad \text{for} \quad y_{ij} < 0 \text{ and } i \neq k$$

If we select the basis vector $\boldsymbol{\beta}_k$ $(y_{kj} \neq 0)$ in such a way that
$$\frac{x_{Bk}}{y_{kj}} = \text{min.} \left\{ \frac{x_{Bi}}{y_{ij}};\; y_{ij} > 0,\; i \neq k \right\}$$

then, the set of basic variables are non-negative, and hence the new basic solution is feasible.

Remark. If after replacement of basis vector $\boldsymbol{\beta}_k$ with a non-basis column vector $\mathbf{a}_j$, the new basis matrix is
$$\mathbf{B}^* = (\boldsymbol{\beta}_1^* \; \boldsymbol{\beta}_2^* \; ... \; \boldsymbol{\beta}_m^*),$$

where, $\boldsymbol{\beta}_i^* = \boldsymbol{\beta}_i$ $(i \neq k)$ and $\boldsymbol{\beta}_k = \mathbf{a}_j$. The corresponding new basic feasible solution is: $\mathbf{x}_B^* = \mathbf{B}^{*-1}\mathbf{b}$. The new basic variables would be:
$$x_{Bi}^* = x_{Bi} - x_{Bk} \frac{y_{ij}}{y_{kj}}, \; (i \neq k) \text{ and } x_{Bk}^* = \frac{x_{Bk}}{y_{kj}}, \; (i = k)$$

(*ii*) **Improved Basic Feasible Solution (BFS).** Suppose that we have a basic feasible solution $\mathbf{x}_B = \mathbf{B}^{-1}\mathbf{b}$ of the LPP discussed above. The value of the objective function for this solution is given by $z = \mathbf{c}_B\,\mathbf{x}_B = \mathbf{c}_B\,\mathbf{B}^{-1}\,\mathbf{b}$ or $\sum_{i=1}^{m} c_{Bi}\,x_{Bi}$

If a basis vector $\boldsymbol{\beta}_k$ is replaced by a column vector $\mathbf{a}_j$ of $\mathbf{A}$ (not in $\mathbf{B}$), then the new value of the objective function is:

$$z^* = \sum_{i=1}^{m} c_{Bi}^*\,x_{Bi}^* = \sum_{i \neq k} c_{Bi}^*\,x_{Bi}^* + c_{Bk}^*\,x_{Bk}^*,$$

where, $c_{Bi}^* = c_{Bi}\ (i \neq k)$ and $c_{Bk}^* = c_j$, since only the coefficient of x_j entering the basis changes the coefficient of x_{Bk}^*.

Using, now, the values of $x_{Bi}^*\ (i \neq k)$ and x_{Bk}^* obtained above in the value of z^*, we have

$$z^* = \sum_{i \neq k} c_{Bi}^* \left(x_{Bi} - x_{Bk}\frac{y_{ij}}{y_{kj}} \right) + c_{Bk}^*\frac{x_{Bk}}{y_{kj}}$$

$$= \sum_{i \neq k} c_{Bi} \left(x_{Bi} - x_{Bk}\frac{y_{ij}}{y_{kj}} \right) + c_j\frac{x_{Bk}}{y_{kj}}, \qquad \text{since, } c_{Bi}^* = c_{Bi}\ (i \neq k) \text{ and } c_{Bk}^* = c_j.$$

$$= \sum_{i=1}^{m} c_{Bi}\,x_{Bi} - \frac{x_{Bk}}{y_{kj}}\left(\sum_{i=1}^{m} c_{Bi}\,y_{ij} - c_j \right), \qquad \text{since, } x_{Bk} - x_{Bk}\frac{y_{kj}}{y_{kj}} = 0$$

$$= z - \frac{x_{Bk}}{y_{kj}}(z_j - c_j), \quad \text{where } z_j = \sum_{i=1}^{m} c_{Bi}\,y_{ij} = \mathbf{c}_B\,\mathbf{B}^{-1}\,\mathbf{a}_j = \mathbf{c}_B\,\mathbf{y}_j.$$

From this expression, we observe that if $(z_j - c_j) > 0$, an improved value of z^* is obtained because z has to be maximized. The value of $(z_j - c_j)$ may be viewed as the change in z^* due to per unit increase in x_j.

Remarks

1. If given basic feasible solution is non-degenerate, then $(x_{Bk}/y_{kj}) > 0$ and a definite increase in the value of z occurs, *i.e.*, $z^* > z$.
2. If given basic feasible solution is degenerate, then an increase in the value of z depends on x_{Bk}/y_{kj}. In any case, if $(z_j - c_j) < 0$, then z^* can never be less than z.
3. If $(z_j - c_j) = 0$ for at least one j for which $y_{ij} > 0$, $i = 1, 2, ..., m$; then another basic feasible solution is obtained (alterative optimum solution) which gives an unchanged value of the objective function.

4:5. UNBOUNDED SOLUTION

Let the linear programming problem

$$\text{Maximize } z = \mathbf{cx} \text{ subject to the constrains:}$$

$$\mathbf{Ax} = \mathbf{b} \text{ and } \mathbf{x} \geq \mathbf{0},$$

where, $\mathbf{A}$ is an $m \times n$ matrix of rank m have a basic feasible solution. Then

$$\mathbf{B}\,\mathbf{x}_B = \mathbf{b} \text{ with } \mathbf{x}_B \geq \mathbf{0}.$$

Suppose, we introduce a column vector $\mathbf{a}_j$ of $\mathbf{A}$ (not in $\mathbf{B}$) for which all $y_{ij} \leq 0$ $(i = 1, 2, ..., m)$ into the basis. Then, this column vector $\mathbf{a}_j$ must enter into the basis either at a negative level or at the zero level. Thus, the new basic solution will be infeasible unless $\mathbf{a}_j$ enters at zero level. We have

$$\mathbf{b} = \mathbf{B}\mathbf{x}_B = \sum_{i=1}^{m} x_{Bi}\,\boldsymbol{\beta}_i, \qquad \text{where } \mathbf{B} = (\boldsymbol{\beta}_1\ \boldsymbol{\beta}_2\ \cdots\ \boldsymbol{\beta}_m)$$

$$= \sum_{i=1}^{m} x_{Bi}\,\boldsymbol{\beta}_i + \lambda \mathbf{a}_j - \lambda \mathbf{a}_j, \qquad \text{where } \mathbf{a}_j \in \mathbf{A} \text{ (not in } \mathbf{B}) \text{ and } \lambda \text{ is any scalar}$$

$$= \sum_{i=1}^{m} x_{Bi}\,\boldsymbol{\beta}_i + \lambda \mathbf{a}_j - \lambda \sum_{i=1}^{m} y_{ij}\,\boldsymbol{\beta}_i, \qquad \text{since, } \mathbf{a}_j = \sum_{i=1}^{m} y_{ij}\,\boldsymbol{\beta}_i, \text{ and } y_{ij} \text{ are scalars}$$

$$= \sum_{i=1}^{m} (x_{Bi} - \lambda y_{ij})\,\boldsymbol{\beta}_i + \lambda \mathbf{a}_j.$$

If $\lambda > 0$, then $(x_{Bi} - \lambda\, y_{ij}) \geq 0$ since, $y_{ij} \leq 0$. This shows that there exists a feasible solution whose $(m + 1)$ components may be strictly positive. But, in general, it may not be a basic solution.

The value of the objective function for these $(m + 1)$ variables is given by

$$z^* = \sum_{i=1}^{m} c_{Bi}\,(x_{Bi} - \lambda y_{ij}) + \lambda c_j = \sum_{i=1}^{m} c_{Bi}\, x_{Bi} - \lambda\left(\sum_{i=1}^{m} c_{Bi}\, y_{ij} - c_j\right)$$

$$= z - \lambda (z_j - c_j)$$

But $\qquad (z_j - c_j) < 0 \ \text{ and } \ \lambda > 0$

$\therefore \qquad z^* \to +\infty \ \text{ as } \ \lambda \to +\infty$

Hence, there is no limit to the optimum value of z and, hence there exists an unbounded solution to the given LPP.

Remarks

1. If for some $\mathbf{a}_j$ for which $(z_j - c_j) > 0$ and $y_{ij} \leq 0$, $i = 1, 2, \ldots, m$, then there exists a feasible solution to the LPP, such that the $(m + 1)$ components of the solution may be strictly positive and

$$z^* \to -\infty \ \text{ as } \ \lambda \to +\infty$$

2. If $y_{ij} \leq 0$ for at least one j for which $(z_j - c_j) = 0$, then $z^* = z$.

4:6. CONDITION OF OPTIMALITY

Let the linear programming problem,

$$\text{Maximize } \ z = \mathbf{c}\,\mathbf{x} \ \text{ subject to the constraints:}$$

$$\mathbf{A}\mathbf{x} = \mathbf{b} \ \text{ and } \ \mathbf{x} \geq \mathbf{0},$$

where, $\mathbf{A}$ is an $m \times n$ matrix of rank m have a basic feasible solution. Then

$$\mathbf{B}\,\mathbf{x}_B = \mathbf{b} \ \text{ with } \ \mathbf{x}_B \geq \mathbf{0} \ \text{ and } \ z = \mathbf{c}_B\,\mathbf{x}_B.$$

Suppose that for all those j for which $\mathbf{a}_j$ of $\mathbf{A}$ (but not in $\mathbf{B}$), we have $(z_j - c_j) \geq 0$.

For all column vectors of $\mathbf{A}$ in $\mathbf{B}$, *i.e.*, when $\mathbf{a}_j \in \mathbf{B}$, we have

$$\mathbf{y}_j = \mathbf{B}^{-1}\,\boldsymbol{\beta}_i = \mathbf{e}_i \qquad \textit{(unit vector)}$$

provided $\mathbf{a}_j$ is in column i for $\mathbf{B}$.

In this case, $\qquad z_j - c_j = \mathbf{c}_B\,\mathbf{y}_j - c_j$

$$= \mathbf{c}_B\,\mathbf{e}_i - c_j = c_j - c_j = 0$$

Thus, $z_j - c_j = 0$ for all columns of $\mathbf{A}$ which are in $\mathbf{B}$.

Now, let **x** be a feasible solution to the given LPP. Then

$$\sum_{j=1}^{n} (z_j - c_j) x_j \geq 0, \qquad \text{(since, } x_j \geq 0\text{)}$$

or

$$\sum_{j=1}^{n} z_j x_j \geq \sum_{j=1}^{n} c_j x_j \quad \text{or} \quad \sum_{j=1}^{n} \mathbf{c}_B \mathbf{y}_j x_j \geq \sum_{j=1}^{n} c_j x_j \qquad \text{(since, } z_j = \mathbf{c}_B \mathbf{y}_j\text{)}$$

or

$$\sum_{i=1}^{m} c_{Bi} \sum_{j=1}^{n} y_{ij} x_j \geq \sum_{j=1}^{n} c_j x_j, \qquad \left(\text{since, } \mathbf{c}_B \mathbf{y}_j = \sum_{i=1}^{m} c_{Bi} y_{ij}\right)$$

for all j for which $\mathbf{a}_j \notin \mathbf{B}$.

Now

$$x_B = \mathbf{B}^{-1} (\mathbf{Ax}) = (\mathbf{B}^{-1} \mathbf{A})\, \mathbf{x} = \mathbf{Yx}, \qquad \text{(since, } \mathbf{Ax} = \mathbf{b}\text{)}$$

or

$$x_{Bi} = \sum_{j=1}^{n} y_{ij} x_j \quad i = 1, 2, ..., m$$

Therefore, the above inequality can be written as,

$$\sum_{i=1}^{m} c_{Bi} x_{Bi} \geq \sum_{j=1}^{n} c_j x_j \quad \text{or} \quad \mathbf{c}_B \mathbf{x}_B \geq \mathbf{cx} \quad \text{or} \quad z^* \geq z$$

where, z is the value of the objective function for the feasible solution **x**.

Hence, z^* is an optimum for that basic feasible solution for which $(z_j - c_j) > 0$ for all j such that $\mathbf{a}_j \notin \mathbf{B}$.

4:7. REVISED SIMPLEX METHOD

Revised simplex method is a modification of the simplex method in the sense that it is more economical on the computer, as it computes and stores only the relevant information needed currently for testing and/or updating the current solution. Moreover, in the revised simplex method, we need to compute the values of B^{-1}, x_B, $c_B B^{-1}$ and z only. All these values can be computed directly from their definition, provided B^{-1} is known.

Matrix Formulation of LPP for Revised Simplex Method

Consider the linear programming problem in its standard form as to

$$\text{Maximize } z = \mathbf{cx} \text{ subject to the constraints:}$$

$$\mathbf{Ax} = \mathbf{b} \text{ and } \mathbf{x} \geq 0,$$

where, **A** is an $m \times n$ matrix of rank m, **c** is $1 \times n$ row vector, **b** is $m \times 1$ column vector, and **x** is $n \times 1$ column vector.

If we consider objective function $z = \mathbf{cx}$ as one of the constraint equation, then the extended system of linear equations are

$$\mathbf{Ax} = \mathbf{b} \text{ and } -\mathbf{cx} + z = 0$$

Obviously, these are $(m + 1)$ equations in $(n + 1)$ variables [z being considered as $(n + 1)^{th}$ variable].

The given system of equation in matrix form is:

$$\begin{pmatrix} \mathbf{A} & \mathbf{0} \\ -\mathbf{c} & 1 \end{pmatrix} \begin{pmatrix} \mathbf{x} \\ z \end{pmatrix} = \begin{pmatrix} \mathbf{b} \\ 0 \end{pmatrix}$$

It may be noted that all column vectors, now, have $(m + 1)$ components instead of m components and the basis matrix $\mathbf{B}$ is also of order $(m + 1)$.

Let $\mathbf{B}$ be an initial basis sub-matrix of $\mathbf{A}$ and $\mathbf{x}_B = \mathbf{B}^{-1}\,\mathbf{b}$ be an initial basic feasible solution to the given LPP. Then, an initial basic feasible solution to the reformulated problem is given by $\hat{\mathbf{x}} = \hat{\mathbf{B}}^{-1}\,\hat{\mathbf{b}}$,

where, $\hat{\mathbf{x}}_B = \begin{bmatrix} \hat{\mathbf{x}}_B \\ z \end{bmatrix}$, $\hat{\mathbf{b}} = \begin{bmatrix} \mathbf{b} \\ 0 \end{bmatrix}$ and $\hat{\mathbf{B}} = \begin{pmatrix} \mathbf{B} & 0 \\ -\mathbf{c}_B & 1 \end{pmatrix}$.

Since, $\mathbf{B}$ is invertible, therefore, we can write

$$\hat{\mathbf{B}}^{-1} = \begin{pmatrix} \mathbf{B}^{-1} & 0 \\ \mathbf{c}_B\,\mathbf{B}^{-1} & 1 \end{pmatrix}$$

By assumption $\mathbf{B}^{-1}$ is known, $\mathbf{c}_B\,\mathbf{B}^{-1}$ is known and hence, all the elements of $\hat{\mathbf{B}}^{-1}$ are known. Further, it may be verified that $\hat{\mathbf{B}}\hat{\mathbf{B}}^{-1} = \mathbf{I}_{m+1}$.

Let us define a new $(m + 1) \times n$ matrix

$$\hat{\mathbf{Y}} = \hat{\mathbf{B}}^{-1}\hat{\mathbf{A}} \quad \text{where } \hat{\mathbf{A}} = \begin{bmatrix} \mathbf{A} \\ -\mathbf{c} \end{bmatrix}$$

$$\therefore \qquad \hat{\mathbf{y}}_j = \begin{pmatrix} \mathbf{B}^{-1} & 0 \\ \mathbf{c}_B\,\mathbf{B}^{-1} & 1 \end{pmatrix}\begin{bmatrix} \mathbf{A} \\ -\mathbf{c} \end{bmatrix} = \begin{bmatrix} \mathbf{B}^{-1}\mathbf{A} \\ \mathbf{c}_B\,\mathbf{B}^{-1}\,\mathbf{A} - \mathbf{c} \end{bmatrix}$$

or

$$\hat{\mathbf{Y}} = \begin{bmatrix} \mathbf{y}_j \\ z_j - c_j \end{bmatrix}, \quad j = 1, 2, ..., n$$

where, $\mathbf{y}_j = \mathbf{B}^{-1}\,\mathbf{a}_j$ and $z_j = \mathbf{c}_B\,\mathbf{B}^{-1}\,\mathbf{a}_j$.

Thus, we arrive at an interesting conclusion that the first m components of $\hat{\mathbf{y}}_j$ constitute the vector $\mathbf{y}_j$ and the $(m + 1)^{\text{th}}$ component is $(z_j - c_j)$; the first m components of $\hat{\mathbf{x}}_B$ constitute $\mathbf{x}_B$ and the $(m + 1)^{\text{th}}$ component is z (being treated as a variable).

The above discussion enables us to give the computational procedure for solving linear programming problem by ***Revised Simplex Method.***

Computational Procedure

Major steps for the computation of an optimum solution of any LPP by revised simplex method are summarized below:

Step 1. Introduce slack and surplus variables, if needed, and restate the given LPP in maximization standard form.

Step 2. Begin with an initial basis $\mathbf{B} = \mathbf{I}_m = \mathbf{B}^{-1}$ and form the auxiliary matrix $\hat{\mathbf{B}}$. Write down $\hat{\mathbf{B}}^{-1}$.

Evidently, $$\hat{\mathbf{B}}^{-1} = \begin{pmatrix} \mathbf{B}^{-1} & 0 \\ \mathbf{c}_B\,\mathbf{B}^{-1} & 1 \end{pmatrix} = \begin{pmatrix} \mathbf{I} & 0 \\ \mathbf{c}_B & 1 \end{pmatrix}, \quad \text{since } \mathbf{B} = \mathbf{I}_m = \mathbf{B}^{-1}$$

or $$\hat{\mathbf{B}}^{-1} = \begin{pmatrix} \mathbf{I} & 0 \\ 0 & 1 \end{pmatrix}, \quad \text{since } \mathbf{c}_{Bi} = 0 \text{ for } i = 1, 2, ..., m$$ *(Coefficients of slack variables)*

This shows that $\hat{\mathbf{B}}^{-1} = \mathbf{I}_{m+1}$, while $\mathbf{B}^{-1} = \mathbf{I}_m$.

$$\text{Compute } \hat{\mathbf{x}}_B = \hat{\mathbf{B}}^{-1}\,\hat{\mathbf{b}} = \begin{pmatrix} \mathbf{I} & 0 \\ 0 & 1 \end{pmatrix}\begin{bmatrix} \mathbf{b} \\ 0 \end{bmatrix} = \begin{bmatrix} \mathbf{b} \\ 0 \end{bmatrix}.$$

Step 3. Compute the net evaluations $(z_j - c_j)$, $j = 1, 2, ..., n$ by multiplying the successive columns of $\hat{\mathbf{A}}$ with the last row of $\hat{\mathbf{B}}^{-1}$, that is,

$$(z_j - c_j) = (\mathbf{c}_B\, \mathbf{B}^{-1} \;\; 1)\begin{pmatrix} \mathbf{A} \\ -\mathbf{c} \end{pmatrix}.$$

If all $(z_j - c_j)$ are non-negative, the current basic feasible solution is an optimum one.

If at least one $(z_j - c_j)$ is negative, determine the most negative of them, say $(z_k - c_k)$, the corresponding column vector $\mathbf{y}_k$ enters the basis. Go to *Step 4*.

If there is a tie for the most negative $(z_j - c_j)$, resolve the tie by any standard method. Go to *Step 4*.

Step 4. Compute $\hat{\mathbf{y}}_k = \hat{\mathbf{B}}_{curr}^{-1} \cdot \hat{\mathbf{a}}_k$. If all $y_{ik} \le 0$, there exists an unbounded optimum solution to the given linear programming problem.

If at least one $y_{ik} > 0$, consider the current $\hat{\mathbf{x}}_B$ (solution column) and determine the row vector that leaves the basis by choosing $\min.\left\{\dfrac{x_{Bi}}{y_{ik}},\ y_{ik} > 0\right\}$. Go to *Step 5*.

Step 5. Write down the results obtained in *Step 2* through *Step 4* in a tabular form known as revised simplex tableau.

Step 6. Convert the leading (key) element to unity and all other elements of the entering column to zero by suitable row operations and update the current basic feasible solution.

Step 7. Go to *Step 3* and repeat the procedure until an optimum basic feasible solution is obtained or there is an indication of an unbounded solution.

Example 4.5. *Use revised simplex method to solve the following LPP:*

Maximize $z = 3x_1 + 5x_2$ *subject to the constraints:*

$$3x_1 + 2x_2 \le 18;$$

$$0 \le x_1 \le 4,\ 0 \le x_2 \le 6, \text{ and } x_3 \ge 0.$$

Solution.

Step 1. Introducing slack variables $s_1 \ge 0$, $s_2 \ge 0$ and $s_3 \ge 0$, the given LPP can be restated in the standard form as:

Maximize $z = \mathbf{cx}$ subject to the constraints:

$$\mathbf{Ax} = \mathbf{b} \text{ and } \mathbf{x} \ge 0,$$

where, $\mathbf{A} = \begin{pmatrix} 1 & 0 & 1 & 0 & 0 \\ 0 & 1 & 0 & 1 & 0 \\ 3 & 2 & 0 & 0 & 1 \end{pmatrix}$, $\mathbf{b} = \begin{bmatrix} 4 \\ 6 \\ 18 \end{bmatrix}$ and $\mathbf{c} = (3\ 5\ 0\ 0\ 0)$.

Step 2. An initial basic feasible solution is:

$s_1 = 4$, $s_2 = 6$ and $s_3 = 18$ with $\mathbf{I}_3$ as the initial basis matrix.

$$\hat{\mathbf{A}} = \begin{pmatrix} \mathbf{A} \\ -\mathbf{c} \end{pmatrix} = \begin{pmatrix} 1 & 0 & 1 & 0 & 0 \\ 0 & 1 & 0 & 1 & 0 \\ 3 & 2 & 0 & 0 & 1 \\ \hline -3 & -5 & 0 & 0 & 0 \end{pmatrix},\quad \hat{\mathbf{B}}_{curr}^{-1} = \begin{pmatrix} \mathbf{B}^{-1} & \mathbf{0} \\ \mathbf{c}_B\,\mathbf{B}^{-1} & 1 \end{pmatrix} = \begin{pmatrix} 1 & 0 & 0 & 0 \\ 0 & 1 & 0 & 0 \\ 0 & 0 & 1 & 0 \\ \hline 0 & 0 & 0 & 1 \end{pmatrix}, \text{ and } \mathbf{c}_B = (0\ 0\ 0).$$

Initial Iteration.

Step 3. The net evaluations are given by

$$(z_j - c_j) = (\mathbf{c}_B\,\mathbf{B}^{-1} \;\; 1)\,\hat{\mathbf{A}} = (\mathbf{c}_B\,\mathbf{B}^{-1} \;\; 1)\left[\hat{\mathbf{a}}_1\ \hat{\mathbf{a}}_2\ \hat{\mathbf{a}}_3\ \hat{\mathbf{a}}_4\ \hat{\mathbf{a}}_5\right] = (-3\ -5\ 0\ 0\ 0).$$

Since, $(z_2 - c_2)$ is the most negative, x_2 enters the basis.

Step 4. Now,
$$\hat{\mathbf{y}}_2 = \hat{\mathbf{B}}_{curr}^{-1}\,\hat{\mathbf{a}}_2 = \begin{pmatrix} 1 & 0 & 0 & 0 \\ 0 & 1 & 0 & 0 \\ 0 & 0 & 1 & 0 \\ \hline 0 & 0 & 0 & 1 \end{pmatrix}\begin{pmatrix} 0 \\ 1 \\ 2 \\ \hline -5 \end{pmatrix} = \begin{pmatrix} 0 \\ 1 \\ 2 \\ \hline -5 \end{pmatrix},$$

and
$$\hat{\mathbf{x}}_B = \hat{\mathbf{B}}_{curr}^{-1}\,\hat{\mathbf{b}} = \begin{pmatrix} 1 & 0 & 0 & 0 \\ 0 & 1 & 0 & 0 \\ 0 & 0 & 1 & 0 \\ \hline 0 & 0 & 0 & 1 \end{pmatrix}\begin{pmatrix} 4 \\ 6 \\ 18 \\ \hline 0 \end{pmatrix} = \begin{pmatrix} 4 \\ 6 \\ 18 \\ \hline 0 \end{pmatrix},$$

or
$$\hat{\mathbf{x}}_B = [\underbrace{4 \;\; 6 \;\; 18}_{\mathbf{x}_B} \; \vdots \; \underbrace{0}_{z}] \qquad (s_1 \;\; s_2 \;\; s_3 \;\; \text{basic variables})$$

Basic variable to be removed from the basis is determined by using
$$\min_i . \left\{ \frac{x_{Bi}}{a_{i2}},\; a_{i2} > 0 \right\} = \min . \left\{ \frac{6}{1}, \frac{18}{2} \right\} = 6.$$

Since, this minimum ratio corresponds to basic variable s_2, second row leaves the basis. Thus, a_{22} becomes the leading element.

Step 5. The initial revised simplex tableau, therefore, is:

Basic variable	$\hat{\mathbf{B}}^{-1}$				$\hat{\mathbf{a}}_2$	*Solution*	*Ratio*
s_1	1	0	0	0	0	4	—
s_2	0	1	0	0	(1)	6	6/1
s_3	0	0	1	0	2	18	18/2
$(z_j - c_j)$	0	0	0	1	–5	0	

First Iteration.

Step 3. Converting the leading element to unity and all other elements of the entering column to zero, we get
$$\hat{\mathbf{B}}_{next}^{-1} = \begin{pmatrix} 1 & 0 & 0 & 0 \\ 0 & 1 & 0 & 0 \\ 0 & -2 & 1 & 0 \\ \hline 0 & 5 & 0 & 1 \end{pmatrix}$$

This directly gives the new current quantities as
$$\hat{\mathbf{B}}^{-1} = \begin{pmatrix} 1 & 0 & 0 \\ 0 & 1 & 0 \\ 0 & -2 & 1 \end{pmatrix} \text{ and } \mathbf{c}_B\,\hat{\mathbf{B}}_{curr}^{-1} = (0 \;\; 5 \;\; 0)$$

Using these current quantities, an improved solution is
$$\hat{\mathbf{x}}_B = \hat{\mathbf{B}}_{curr}^{-1}\,\hat{\mathbf{b}} = \hat{\mathbf{B}}_{curr}^{-1}\begin{bmatrix} \mathbf{b} \\ 0 \end{bmatrix} = \begin{pmatrix} 1 & 0 & 0 & 0 \\ 0 & 1 & 0 & 0 \\ 0 & -2 & 1 & 0 \\ \hline 0 & 5 & 0 & 1 \end{pmatrix}\begin{pmatrix} 4 \\ 6 \\ 18 \\ \hline 0 \end{pmatrix} = \begin{pmatrix} 4 \\ 6 \\ 6 \\ \hline 30 \end{pmatrix}$$

Therefore,
$$\hat{\mathbf{x}}_B = [\underbrace{4 \;\; 6 \;\; 6}_{\mathbf{x}_B} \; \vdots \; \underbrace{30}_{z}] \qquad (s_1, x_2, s_3 \;\; \text{basic varaibles})$$

Step 4. The net evaluations are given by

$$(z_j - c_j) = (\mathbf{c}_B \mathbf{B}^{-1} \; 1)[\hat{\mathbf{a}}_1 \; \hat{\mathbf{a}}_2 \; \hat{\mathbf{a}}_3 \; \hat{\mathbf{a}}_4 \; \hat{\mathbf{a}}_5]$$
$$= (-3 \; 0 \; 0 \; 5 \; 0).$$

Since, $(z_1 - c_1)$ is the only negative, x_1 enters the basis.

Step 5. Now,

$$\hat{\mathbf{y}}_1 = \hat{\mathbf{B}}_{curr}^{-1} \hat{\mathbf{a}}_1 = \begin{pmatrix} 1 & 0 & 0 & 0 \\ 0 & 1 & 0 & 0 \\ 0 & -2 & 1 & 0 \\ \hline 0 & 5 & 0 & 1 \end{pmatrix} \begin{pmatrix} 1 \\ 0 \\ 3 \\ \hline -3 \end{pmatrix} = \begin{pmatrix} 1 \\ 0 \\ 3 \\ \hline -3 \end{pmatrix}$$

and

$$\hat{\mathbf{x}}_B = \hat{\mathbf{B}}_{curr}^{-1} \hat{\mathbf{b}} = \begin{pmatrix} 1 & 0 & 0 & 0 \\ 0 & 1 & 0 & 0 \\ 0 & -2 & 1 & 0 \\ \hline 0 & 5 & 0 & 1 \end{pmatrix} \begin{pmatrix} 4 \\ 6 \\ 18 \\ \hline 0 \end{pmatrix} = \begin{pmatrix} 4 \\ 6 \\ 6 \\ \hline 30 \end{pmatrix}$$

or

$$\hat{\mathbf{x}}_B = [\underbrace{4 \; 6 \; 6}_{\mathbf{x}_B} \; : \; \underbrace{30}_{z}] \qquad (s_1, x_2, s_3 \text{ basic varaibles})$$

Basic variable to be removed from the basis is determined by using

$$\min_i . \left\{ \frac{x_{Bi}}{a_{i1}}, \; a_{i1} > 0 \right\} = \min . \left\{ \frac{4}{1}, \frac{6}{3} \right\} = 2.$$

Since, this minimum ratio corresponds to the basic variable s_3, $\mathbf{a}_1$ leaves the basis. The revised simplex tableau, therefore, is

Basic variable	$\hat{\mathbf{B}}^{-1}$				$\hat{\mathbf{a}}_1$	*Solution*	*Ratio*
s_1	1	0	0	0	1	4	4/1
x_2	0	1	0	0	0	6	—
s_3	0	–2	1	0	(3)	6	6/3
$(z_j - c_j)$	0	5	0	1	–3	30	

Final Iteration.

Step 3. Converting the leading element to unity and all other elements of the entering column to zero, we get

$$\hat{\mathbf{B}}_{next}^{-1} = \begin{pmatrix} 1 & 2/3 & -1/3 & 0 \\ 0 & 1 & 0 & 0 \\ 0 & -2/3 & 1/3 & 0 \\ \hline 0 & 3 & 1 & 1 \end{pmatrix}$$

$$\therefore \qquad \hat{\mathbf{x}}_B = \hat{\mathbf{B}}_{next}^{-1} \hat{\mathbf{b}} = \begin{pmatrix} 1 & 2/3 & -1/3 & 0 \\ 0 & 1 & 0 & 0 \\ 0 & -2/3 & 1/3 & 0 \\ \hline 0 & 3 & 1 & 1 \end{pmatrix} \begin{pmatrix} 4 \\ 6 \\ 18 \\ \hline 0 \end{pmatrix} = \begin{pmatrix} 4 \\ 6 \\ 2 \\ \hline 36 \end{pmatrix}$$

or

$$\hat{\mathbf{x}}_B = [\underbrace{2 \; 6 \; 2}_{\mathbf{x}_B} \; : \; \underbrace{36}_{z}] \qquad (s_1, x_2, x_1 \text{ basic varaibles})$$

Step 4. The net evaluations are given by

$$(z_j - c_j) = (0\ \ 3\ \ 1\ \ 1)\,\hat{\mathbf{A}} = (0\ \ 0\ \ 0\ \ 3\ \ 1).$$

Since, all $(z_j - c_j) \geq 0$, an optimum basic feasible solution is obtained which is given by

$$x_1 = 2,\ \ x_2 = 6,\ \text{ and }\ \text{maximum } z = 36.$$

Example 4.6. *Use revised simplex method to solve the LPP,*

$$\textit{Maximize}\ \ z = 3x_1 + 2x_2 + 5x_3\ \ \textit{subject to the constraints:}$$
$$x_1 + 2x_2 + x_3 \leq 430,\ \ 3x_1 + 2x_3 \leq 460,$$
$$x_1 + 4x_2 \leq 420, \textit{ and } x_1, x_2, x_3 \geq 0.$$

Solution.

Step 1. By introducing slack variables $s_1 \geq 0$, $s_2 \geq 0$ and $s_3 \geq 0$, given LPP can be re-written as

$$\text{Maximize}\ \ z = \mathbf{cx}\ \ \text{subject to the constraints:}$$
$$\mathbf{Ax} = \mathbf{b}\ \text{ and }\ \mathbf{x} \geq \mathbf{0}.$$

where
$$\mathbf{A} = \begin{pmatrix} 1 & 2 & 1 & 1 & 0 & 0 \\ 3 & 0 & 2 & 0 & 1 & 0 \\ 1 & 4 & 0 & 0 & 0 & 1 \end{pmatrix},\ \mathbf{b} = \begin{bmatrix} 430 \\ 460 \\ 420 \end{bmatrix} \text{ and } \mathbf{c} = (3\ \ 2\ \ 5\ \ 0\ \ 0\ \ 0).$$

Step 2. An obvious initial basic feasible solution is: $s_1 = 430$, $s_2 = 460$ and $s_3 = 420$ with $\mathbf{B} = \mathbf{I}_3$ as the initial basis matrix. Now

$$\hat{\mathbf{A}} = \begin{pmatrix} \mathbf{A} \\ -\mathbf{c} \end{pmatrix} = \left(\begin{array}{cccccc} 1 & 2 & 1 & 1 & 0 & 0 \\ 3 & 0 & 2 & 0 & 1 & 0 \\ 1 & 4 & 0 & 0 & 0 & 1 \\ \hline -3 & -2 & -5 & 0 & 0 & 0 \end{array}\right) \text{ and } \hat{\mathbf{B}}_{curr}^{-1} = \left(\begin{array}{cccc} 1 & 0 & 0 & 0 \\ 0 & 1 & 0 & 0 \\ 0 & 0 & 1 & 0 \\ \hline 0 & 0 & 0 & 1 \end{array}\right)$$

with $\quad \mathbf{c}_B = (0\ \ 0\ \ 0).$

Initial Iteration.

Step 3. The net evaluations for non-basic variables are:

$$(z_j - c_j) = (\mathbf{c}_B\,\mathbf{B}^{-1}\ \ 1)\,\hat{\mathbf{a}}_j\,; \qquad j = 1, 2, 3$$

$$= (0\ \ 0\ \ 0\ \ 1)\left(\begin{array}{ccc} 1 & 2 & 1 \\ 3 & 0 & 2 \\ 1 & 4 & 0 \\ \hline -3 & -2 & -5 \end{array}\right) = (-3\ \ -2\ \ -5)$$

Since, $(z_3 - c_3)$ is the most negative, x_3 enters the basis.

Step 4. Now,
$$\hat{\mathbf{y}}_3 = \hat{\mathbf{B}}_{curr}^{-1}\,\hat{\mathbf{a}}_3 = \left(\begin{array}{cccc} 1 & 0 & 0 & 0 \\ 0 & 1 & 0 & 0 \\ 0 & 0 & 1 & 0 \\ \hline 0 & 0 & 0 & 1 \end{array}\right)\left(\begin{array}{c} 1 \\ 2 \\ 0 \\ \hline -5 \end{array}\right) = \left(\begin{array}{c} 1 \\ 2 \\ 0 \\ \hline -5 \end{array}\right)$$

$$\hat{\mathbf{x}}_B = \hat{\mathbf{B}}_{curr}^{-1}\ \hat{\mathbf{b}} = \begin{pmatrix} 1 & 0 & 0 & 0 \\ 0 & 1 & 0 & 0 \\ 0 & 0 & 1 & 0 \\ \hline 0 & 0 & 0 & 1 \end{pmatrix}\begin{pmatrix} 430 \\ 460 \\ 420 \\ \hline 0 \end{pmatrix} = \begin{pmatrix} 430 \\ 460 \\ 420 \\ \hline 0 \end{pmatrix}$$

or $\quad \hat{\mathbf{x}}_B = [\underbrace{430\ \ 460\ \ 420}_{\mathbf{x}_B} \ \vdots\ \underbrace{0}_{z}] \qquad (s_1, s_2, s_3 \text{ basic varaibles})$

Basic variable to be removed from the basis is determined by using

$$\min_i . \left\{ \frac{x_{Bi}}{a_{i3}},\ a_{i3} > 0 \right\} = \min . \left\{ \frac{430}{1}, \frac{460}{2} \right\} = 230.$$

Since, this minimum ratio corresponds to basic variables s_2, s_2 leaves the basis. Thus, a_{23} becomes the leading element.

The initial revised simplex tableau, therefore, is:

Basic variable	$\hat{\mathbf{B}}^{-1}$				$\hat{\mathbf{a}}_3$	*Solution*	*Ratio*
s_1	1	0	0	0	1	430	430/1
x_3	0	1	0	0	(2)	460	460/2
s_3	0	0	1	0	0	420	—
$(z_j - c_j)$	0	0	0	1	–5	0	

First Iteration.

Step 2. Converting the leading element to unity and all other elements of the entering column to zero, we get

$$\hat{\mathbf{B}}_{next}^{-1} = \begin{pmatrix} 1 & -1/2 & 0 & 0 \\ 0 & 1/2 & 0 & 0 \\ 0 & 0 & 1 & 0 \\ \hline 0 & 5/2 & 0 & 1 \end{pmatrix}.$$

This directly gives the new current quantities updated basis inverse as

$$\hat{\mathbf{B}}^{-1} = \begin{pmatrix} 1 & -1/2 & 0 \\ 0 & 1/2 & 0 \\ 0 & 0 & 1 \end{pmatrix}, \quad \mathbf{c}_B\, \hat{\mathbf{B}}_{curr}^{-1} = (0\ \ 5/2\ \ 0).$$

Making use of these current quantities, an improved solution is:

$$\hat{\mathbf{x}}_B = \hat{\mathbf{B}}_{curr}^{-1}\ [\mathbf{b}\ \ 0] = \begin{pmatrix} 1 & -1/2 & 0 & 0 \\ 0 & 1/2 & 0 & 0 \\ 0 & 0 & 1 & 0 \\ 0 & 5/2 & 0 & 1 \end{pmatrix}\begin{pmatrix} 430 \\ 460 \\ 420 \\ 0 \end{pmatrix}$$

or $\quad = [\underbrace{200\ \ 230\ \ 420}_{\mathbf{x}_B} \ \vdots\ \underbrace{1150}_{z}] \qquad (s_1, x_3, s_3 \text{ basic variables})$

Step 3. The net evaluations for the non-basic variables are given by

$$(z_j - c_j) = (\mathbf{c}_B \mathbf{B}^{-1} \ \ 1)[\hat{\mathbf{a}}_1 \ \ \hat{\mathbf{a}}_2 \ \ \hat{\mathbf{a}}_5]$$

$$= (0 \ \ 5/2 \ \ 0 \ \ 1)\begin{bmatrix} 1 & 2 & 0 \\ 3 & 0 & 1 \\ 1 & 4 & 0 \\ -3 & -2 & 0 \end{bmatrix} = (9/2 \ \ -2 \ \ 5/2).$$

Since, $(z_2 - c_2)$ is the only negative, x_2 enters the basis.

Step 5. Now, $\hat{\mathbf{y}}_2 = \hat{\mathbf{B}}_{curr}^{-1}\, \hat{\mathbf{a}}_2 = [2 \ \ 0 \ \ 4 \ \vdots \ -2]$ (current = next of *previous step*)

Also, $\min_i .\left\{\dfrac{x_{Bi}}{a_{i2}},\ a_{i2} > 0\right\} = \min_i .\left\{\dfrac{200}{2},\ \dfrac{420}{4}\right\} = 100.$

As minimum ratio corresponds to basic variable s_1, s_1 leaves the basis. The revised simplex tableau is:

Basic variable	$\hat{\mathbf{B}}^{-1}$				$\hat{\mathbf{a}}_2$	*Solution*	*Ratio*
s_1	1	–1/2	0	0	(2)	200	200/2
x_3	0	1/2	0	0	0	230	—
s_3	0	0	1	0	4	420	420/4
$(z_j - c_j)$	0	5/2	0	1	–2	1150	

Final Iteration. Converting the leading element to unity and all other elements of the entering column to zero, we get

$$\hat{\mathbf{B}}_{next}^{-1} = \begin{pmatrix} 1/2 & -1/4 & 0 & 0 \\ 0 & 1/2 & 0 & 0 \\ -2 & 1 & 1 & 0 \\ \hline 1 & 2 & 0 & 1 \end{pmatrix}$$

The net evaluations for non-basic variables are given by

$$(z_j - c_j) = (1 \ \ 2 \ \ 0 \ \ 1)[\hat{\mathbf{a}}_1 \ \ \hat{\mathbf{a}}_4 \ \ \hat{\mathbf{a}}_5] = (4 \ \ 1 \ \ 2)$$

Since, all $(z_j - c_j) \geq 0$, an optimum basic feasible solution is attained:

$$\therefore \qquad \hat{\mathbf{x}}_B = \hat{\mathbf{B}}_{curr}^{-1}\, \hat{\mathbf{b}} = \begin{pmatrix} 1/2 & -1/4 & 0 & 0 \\ 0 & 1/2 & 0 & 0 \\ -2 & 1 & 1 & 0 \\ \hline 1 & 2 & 0 & 1 \end{pmatrix}\begin{pmatrix} 430 \\ 460 \\ 420 \\ \hline 0 \end{pmatrix}$$

$$= [\underbrace{100 \ \ 230 \ \ 20}_{\mathbf{x}_B} \ \vdots \ \underbrace{1350}_{z}] \qquad (x_2, x_3, s_3 \text{ basic variables})$$

Hence, $x_1 = 0$, $x_2 = 100$, $x_3 = 230$ and maximum $z = 1350$.

Example 4.7. *Use revised simplex method to solve the following LPP:*

Minimize $z = x_1 + x_2$ *subject to the constraints:*

$x_1 + 2x_2 \geq 7,\ \ 4x_1 + x_2 \geq 6,$ *and* $x_1, x_2 \geq 0.$

[MCA, Panjab Tech. Univ. 2010]

Solution. By introducing surplus variables $s_1 \geq 0$, $s_2 \geq 0$ and artificial variables $A_1 \geq 0$ and $A_2 \geq 0$, given LPP is written in standard form as under:

$$\text{Maximize} \quad z^* = -\mathbf{cx} \quad \text{subject to the constraints:}$$

$$\mathbf{Ax} = \mathbf{b} \quad \text{and} \quad \mathbf{x} \geq \mathbf{0}$$

where, $\mathbf{A} = \begin{pmatrix} 1 & 2 & -1 & 0 & 1 & 0 \\ 4 & 1 & 0 & -1 & 0 & 1 \end{pmatrix}$, $\mathbf{b} = \begin{bmatrix} 7 \\ 6 \end{bmatrix}$, $\mathbf{c} = (-1 \ \ -1 \ \ 0 \ \ 0 \ \ -M \ \ -M)$.

An initial basic feasible solution is: $A_1 = 7$ and $A_2 = 6$ with $\mathbf{I}_2$ as the initial basis.

Now,
$$\hat{\mathbf{A}} = \begin{pmatrix} \mathbf{A} \\ -\mathbf{c} \end{pmatrix} = \left[\begin{array}{cccccc} 1 & 2 & -1 & 0 & 1 & 0 \\ 4 & 1 & 0 & -1 & 0 & 1 \\ \hline 1 & 1 & 0 & 0 & M & M \end{array}\right]$$

and
$$\hat{\mathbf{B}}^{-1}_{curr} = \begin{pmatrix} \mathbf{B} & \mathbf{0} \\ \mathbf{c}_B \mathbf{B}^{-1} & 1 \end{pmatrix} = \left[\begin{array}{ccc} 1 & 0 & 0 \\ 0 & 1 & 0 \\ \hline -M & -M & 1 \end{array}\right]$$

with $\mathbf{c}_B = (-M, \ -M)$.

Initial Iteration. The net evaluations for non-basic variables are computed as:

$$(z_j - c_j) = (-M \ \ -M \ \ 1)\,[\hat{\mathbf{a}}_1 \ \ \hat{\mathbf{a}}_2 \ \ \hat{\mathbf{a}}_3 \ \ \hat{\mathbf{a}}_4] = (-5M+1 \ \ -3M+1 \ \ M \ \ M)$$

This shows that $\hat{\mathbf{a}}_1$ enters the basis, since $(z_1 - c_1)$ is the most negative.

Now, $\hat{\mathbf{y}}_1 = \hat{\mathbf{B}}^{-1}_{curr}\, \hat{\mathbf{a}}_1 = [1 \ \ 4 \ \vdots \ -5M+1]$

and $\hat{\mathbf{x}}_B = \hat{\mathbf{B}}^{-1}_{curr}\, [\mathbf{b} \ \ 0] = [7 \ \ 6 \ \vdots \ -13M]$ (A_1 and A_2 basic variables)

The initial revised simplex tableau, therefore, is

Basic variable		$\hat{\mathbf{B}}^{-1}$		$\hat{\mathbf{a}}_1$	*Solution*	*Ratio*
A_1	1	0	0	1	7	7/1
A_2	0	1	0	(4)	6	6/4
$(z_j - c_j)$	$-M$	$-M$	1	$-5M+1$	$-13M$	

First Iteration. Converting the leading element to unity and all other elements of the entering column to zero, the updated basis inverse is

$$\hat{\mathbf{B}}^{-1}_{next} = \left(\begin{array}{ccc} 1 & -1/4 & 0 \\ 0 & 1/4 & 0 \\ \hline -M & (M-1)/4 & 1 \end{array}\right), \text{ and therefore}$$

$$(z_j - c_j) = \left(-M \quad \frac{M-1}{4} \quad 1\right) [\hat{\mathbf{a}}_2 \ \ \hat{\mathbf{a}}_3 \ \ \hat{\mathbf{a}}_4] = \left(\frac{-7M+3}{4} \quad M \quad \frac{-M+1}{4}\right).$$

Since, $(z_2 - c_2)$ is most negative, $\hat{\mathbf{a}}_2$ enters the basis.

Therefore, the next revised simplex tableau is:

Basic variable		$\hat{\mathbf{B}}^{-1}$		$\hat{\mathbf{a}}_2$	*Solution*	*Ratio*
A_1	1	$-1/4$	0	(7/4)	11/2	$\frac{11/2}{7/4}$
x_1	0	$1/4$	0	1/4	3/2	$\frac{1/4}{3/2}$
$(z_j - c_j)$	$-M$	$\frac{M-1}{4}$	1	$\frac{-7M+3}{4}$	$\frac{-11M-3}{2}$	

Final Iteration. The updated basis inverse is

$$\hat{\mathbf{B}}_{next}^{-1} = \begin{pmatrix} 4/7 & -1/7 & 0 \\ -1/7 & 2/7 & 0 \\ -3/7 & -1/7 & 1 \end{pmatrix}$$

Therefore, $(z_j - c_j) = (-3/7 \quad -1/7 \quad 1)[\hat{\mathbf{a}}_3 \quad \hat{\mathbf{a}}_4] = (3/7 \quad 1/7)$

Since, all $(z_j - c_j) \geq 0$ an optimum basic feasible solution is attained.

$$\therefore \qquad \hat{\mathbf{x}}_B = \hat{\mathbf{B}}_{curr}^{-1}\,\hat{\mathbf{b}} = \begin{bmatrix} 4/7 & -1/7 & 0 \\ -1/7 & 2/7 & 0 \\ -3/7 & -1/7 & 1 \end{bmatrix} \begin{bmatrix} 7 \\ 6 \\ 0 \end{bmatrix} = \begin{bmatrix} 22/7 \\ 5/7 \\ -27/7 \end{bmatrix}$$

$$= \begin{bmatrix} \frac{22}{7} & \frac{5}{7} & \vdots & \frac{-27}{7} \end{bmatrix} \qquad (x_2 \text{ and } x_1 \text{ basic variables})$$

Hence, an optimum basic feasible solution is:

$$x_1 = \frac{5}{7} \text{ and } x_2 = \frac{22}{7} \text{ with minimum } z = -\left(-\frac{22}{7}\right) = \frac{22}{7}.$$

PROBLEMS

Use revised simplex method to solve the following linear programming problems:

4.07. Maximize $z = x_1 + 2x_2$ subject to the constraints:

$x_1 + x_2 \leq 3,\ x_1 + 2x_2 \leq 5,\ 3x_1 + x_2 \leq 6,$ and $x_1 \geq 0, x_2 \geq 0.$

4.08. Maximize $z = x_1 + x_2$ subject to the constraints:

$3x_1 + 2x_2 \leq 6,\ x_1 + 4x_2 \leq 4,$ and $x_1 \geq 0, x_2 \geq 0.$

4.09. Maximize $z = 2x_1 + x_2$ subject to the constraints:

$3x_1 + 4x_2 \leq 6,\ 6x_1 + x_2 \leq 3,$ and $x_1 \geq 0, x_2 \geq 0.$

4.10. Maximize $z = 7x_1 + 6x_1$ subject to the constraints:

$x_1 + 4x_2 \leq 1,\ 2x_1 + x_2 \leq 1,$ and $x_1 \geq 0, x_2 \geq 0.$

4.11. Maximize $z = 3x_1 + 2x_2$ subject to the constraints:

$x_1 + 2x_2 \leq 6,\ 2x_1 + x_2 \leq 8,\ -x_1 + x_2 \leq 1,\ x_2 \leq 2;$ and $x_1 \geq 0, x_2 \geq 0.$

4.12. Maximize $z = 6x_1 - 2x_2 + 3x_3$ subject to the constraints:

$2x_1 - x_2 + 2x_3 \leq 2,\ x_1 + 4x_3 \leq 4,$ and $x_1, x_2, x_3 \geq 0.$

[M.Sc. (Math.), Garhwal Univ. 2002]

4.13. Maximize $z = 2x_1 - 6x_2$ subject to the constraints:

$$x_1 - 3x_2 \le 6,\ \ 2x_1 + 4x_2 \ge 8,\ \ -x_1 + 3x_2 \le 6,$$
$$x_1 \ge 0 \text{ and } x_2 \ge 0.$$

4.14. Maximize $z = x_1 + 2x_2$ subject to the constraints:

$$3x_1 + 2x_2 \ge 6,\ \ x_1 + 6x_2 \ge 3,$$
$$x_1 \ge 0 \text{ and } x_2 \ge 0.$$

4.15. Minimize $z = 2x_1 + x_2$ subject to the constraints:

$$3x_1 + x_2 \ge 3,\ \ 4x_1 + 3x_2 \ge 6,\ \ x_1 + 2x_2 \ge 2$$
$$x_1 \ge 0 \text{ and } x_2 \ge 0.$$

4.16. Maximize $z = -4x_1 + x_2 + 2x_3$ subject to the constraints:

$$2x_1 - 3x_2 + 2x_3 \le 12,\ \ -5x_1 + 2x_2 + 3x_3 \ge 4,\ \ 3x_1 - 2x_3 = -1;$$
$$x_1, x_2, x_3 \ge 0.$$

[BE (IT) (Nov.), Panjab Univ. 2004]

4:8. SIMPLEX METHOD *VERSUS* REVISED SIMPLEX METHOD

Consider the general LPP as of maximizing $z = \mathbf{cx}$, subject to the constraints: $\mathbf{Ax} = \mathbf{b}$ and $\mathbf{x} \ge 0$, where $\mathbf{A}$ is an $m \times n$ matrix ($m < n$) and $\mathbf{x}$ as well as $\mathbf{b}$ are $m \times 1$ matrices. In solving the LPP by simplex method suppose that artificial variables are not needed. Then, we have to carry out the calculations of $(n + 1)$ columns (n columns corresponding to columns of $\mathbf{A}$ and one column corresponding to $\mathbf{x}_B$, the basic solution) at each iteration. At each iteration of simplex method, one non-basic variable is introduced into basis and one current basic variable is removed from the basis.

Thus, in total we compute for $n - m + 1$ columns. Furthermore, for each of these columns, we have to transform $m + 1$ elements (m corresponding to $\mathbf{y}_j$ and one corresponding to $z_j - c_j$). For moving from one iteration to another we also need to calculate minimum ratio $\mathbf{x}_{Bi}/\mathbf{a}_{ik}$. Hence, in all we have to perform multiplication $(m + 1)(n - m + 1)$ times and addition $m(n - m + 1)$ times.

In the revised simplex method, there are $m + 1$ rows and $m + 2$ columns. So, for moving from one iteration to another we have to make $(m + 1)^2$ multiplication operations to get an improved solution in addition to $m\,(n - m)$ operations for calculating $(z_j - c_j)$. The major difference between two methods of solution are the following:

(*i*) In revised simplex method we need to make $(m + 1) \times (m + 2)$ entries in each tableau while in simplex method there are $(m + 1) \times (m + 1)$ entries in each tableau.

(*ii*) In simplex method all $\mathbf{a}_j$ columns are updated at each iteration, whereas in the revised simplex method only the column of entering variable is updated.

(*iii*) If the number of variables is significantly larger than the number of constraints, then the computational efforts of the revised simplex method is smaller than that of the simplex method.

(*iv*) The inverse of the current basis matrix is obtained automatically.

One of the disadvantage of the revised simplex method is that while updating the tableau to move from one basic feasible solution to another, an additional tableau of original non-basic variable, not in the basis, is required. This may cause some computational errors.

TEST YOUR UNDERSTANDING

1. Given a basic solution $\mathbf{x}_B$ to a linear programming problem it will be feasible, if

(*a*) $\mathbf{x}_B \ge 0$ (*b*) $\mathbf{x}_B \le 0$

(*c*) $\mathbf{x}_B < 0$ (*d*) $\mathbf{x}_B = 0$

2. In an LPP, if all basic variables are strictly positive, then the solution is called
   (*a*) degenerate (*b*) non-degenerate
   (*c*) infeasible (*d*) unbounded
3. The number of CPF solutions (extreme points) of a convex set of feasible solutions are
   (*a*) finite (*b*) infinite
   (*c*) undefined (*d*) none of these
4. For minimization LPP, the simplex method gives an optimum solution, when
   (*a*) $(z_j - c_j) = 0$ (*b*) $(z_j - c_j) \geq 0$
   (*c*) $(z_j - c_j) \leq 0$ (*d*) $z_j \leq 0$
5. Revised simplex method
   (*a*) is the streamlined version of the simplex method.
   (*b*) is very convenient to do work on the desk.
   (*c*) does not permit the computation of large size linear programming problems.
   (*d*) is applicable to only those LPP's which involve artificial variables.
6. Revised simplex method
   (*a*) is more economical when LPP is solved manually, but is not very convenient when computer is used.
   (*b*) generates the inverse of the current basis matrix automatically.
   (*c*) is based exactly on the same steps as are used in the simplex tableau.
   (*d*) both (*b*) and (*c*).
7. In the revised simplex method
   (*a*) the crux of computations is rooted in the basis matrix and its inverse.
   (*b*) the number of arithmetic operations in the revised simplex method may be smaller than simplex method in an iteration.
   (*c*) the computations of large size linear programming problems are possible.
   (*d*) all of the above.

REVIEW EXERCISE

1. Establish the difference between (*i*) Feasible solution, (*ii*) Basic feasible solution, (*iii*) Degenerate basic feasible solution, and (*iv*) Optimum basic feasible solution. ***[M.Com., Delhi Univ. 1998]***
2. Show that if there is a feasible solution of the system: $\mathbf{Ax} = \mathbf{b}, \mathbf{x} \geq \mathbf{0}$; where $\mathbf{A}$ is an $m \times n$ matrix and $m < n$, then there is also a basic feasible solution.
3. Prove that every basic feasible solution of an LPP is an extreme point of the convex set of feasible solutions. ***[B.Sc. (Stat.), Delhi Univ. 2000]***
4. Show that a linear objective function takes its minimum at an extreme point of the convex set generated by the set of feasible solutions to the linear programming problem. If it assumes its minimum at more than one extreme point, then it takes on the same value for every convex combination of these particular points.
5. In solving a linear programming problem by simplex method, explain how you will move from a given basic feasible solution to another basic feasible solution with an improved value of the objective function.
6. What is the criterion for testing optimaltiy of solution to a linear programming problem in case of maximization as well as minimization? ***[MJB, Delhi Univ. 2011]***

7. Let there exist a basic feasible solution to a maximization LPP. If for at least one j, for which $a_{ij} \leq 0$ $(i = 1, 2, ..., m)$ and $(z_j - c_j) < 0$, then there does not exist any optimum solution to the given LPP.
8. Prove that the sufficient condition for a basic feasible solution of an LPP to be maximum is $(z_j - c_j) \geq 0$ for all j.
9. If there is an optimum basic feasible solution to an LPP and for some $\mathbf{a}_j$ not in the basis $(z_j - c_j) = 0$ and at least one $a_{ij} > 0$ for all i, then an alternative optimal solution exists.
10. Formulate a linear programming problem in the form of revised simplex tableau.
11. Explain revised simplex method and compare it with the simplex method.
12. Develop a computational algorithm for solving an LPP by revised simplex method.
13. Give a brief outline of the revised simplex method.
14. In what way do you consider the revised simplex method is superior to the usual simplex method?

CHAPTER 5

Duality in Linear Programming

"Man is simply to recognize that there are different approaches for solving problems"

For every linear programming problem of maximization (or minimization), there is a related unique problem of minimization (or maximization) based on the same data. The original (given) problem is then called as ***primal problem***, while the other is called its ***dual problem***. If we state the dual problem as the primal problem, then the other problem can be considered as the dual of this primal.

5:1. WRITING OF DUAL LINEAR PROGRAMMING PROBLEM

Following are the three simple rules for obtaining the dual from the primal:

Rule 1. Select as many variables as the number of constraints in the primal problem and restrict them to be non-negative.

Rule 2. Objective function of dual is obtained from the right-hand side constants of the primal constraints. It may be noted here, that the maximization changes to minimization and *vice versa* in the dual formulation.

Rule 3. Constraints to the dual would be as many in numbers as the number of variables in the primal. Thus, each constraint of the dual is derived from the coefficients of variables in the primal. The coefficient of the primal variables in the objective function constitute the RHS of the dual problem. It may be noted here that the direction of the inequalities also changes during dual formulation.

In short, the dual is obtained by interchanging the rows and columns of the constraint coefficients, interchanging the coefficients of the objective function and RHS constants of the constraints, reversing the inequalities and finally minimizing instead of maximizing and *vice versa*.

Illustration. Consider the following linear programming problem:

$$\text{Maximize } z = c_1x_1 + c_2x_2 \text{ subject to the constraints:}$$
$$a_{11}x_1 + a_{12}x_2 \le b_1$$
$$a_{21}x_1 + a_{22}x_2 \le b_2$$
$$x_1 \ge 0 \text{ and } x_2 \ge 0.$$

Dual Formulation. Considering the given linear programming problem as primal problem, we observe that there are two constraints in the primal. So, there will be two ***dual variables***. Let $y_1 \ge 0$ and $y_2 \ge 0$ be dual variables.

Using, right-hand side constants of the constraints of primal, new objective function will be:

$$\text{Minimize } z^* = b_1y_1 + b_2y_2.$$

Now, considering the coefficient of x_1 only in the primal problem, first constraint of the dual will be:

$$a_{11}y_1 + a_{21}y_2 \ge c_1.$$

Further, considering the coefficient of x_2 only in the primal problem, second constraint will be:

$$a_{12}y_1 + a_{22}y_2 \geq c_2.$$

Making use of the above information, complete dual formulation is as under:

$$\text{Minimize } z^* = b_1y_1 + b_2y_2 \text{ subject to the constraints:}$$

$$a_{11}y_1 + a_{21}y_2 \geq c_1,$$
$$a_{12}y_1 + a_{22}y_2 \geq c_2,$$
$$y_1 \geq 0 \text{ and } y_2 \geq 0.$$

Observing the primal and dual problems given above, we find that

(*a*) The primal problem is of maximization, while the dual is of minimization.

(*b*) The RHS constants b_1 and b_2 of the primal have become the coefficients of the dual variables y_1 and y_2 in the objective function of dual in that order; while the coefficients of the variables in the objective function of the primal have become RHS constants in the dual.

(*c*) The first column of the coefficients in the constraints of the primal has become the first row coefficients in the constraints of dual, and the second column of the coefficients in the primal, become the second row coefficients in the dual.

(*d*) The inequalities in the primal (maximization) are of the type '≤', while they are of '≥' type in the dual.

(*e*) The dual variables are non-negative.

In matrix notations, the primal-dual relationship is represented as follows:

Primal. Maximize $z = \mathbf{cx}$ subject to the constraints:

$$\mathbf{Ax} \leq \mathbf{b} \text{ and } \mathbf{x} \geq \mathbf{0},$$

where, $\mathbf{A} = \begin{bmatrix} a_{11} & a_{12} \\ a_{21} & a_{22} \end{bmatrix}$, $\mathbf{b} = \begin{bmatrix} b_1 \\ b_2 \end{bmatrix}$, $\mathbf{x} = \begin{bmatrix} x_1 \\ x_2 \end{bmatrix}$, and $\mathbf{c} = [c_1 \ \ c_2]$.

Dual. Minimize $z^* = \mathbf{b}^T \mathbf{y}$ subject to the constraints:

$$\mathbf{A}^T \mathbf{y} \geq \mathbf{c}^T \text{ and } \mathbf{y} \geq \mathbf{0},$$

where, $\mathbf{A}^T = \begin{bmatrix} a_{11} & a_{21} \\ a_{12} & a_{22} \end{bmatrix}$, $\mathbf{c}^T = \begin{bmatrix} c_1 \\ c_2 \end{bmatrix}$, $\mathbf{y} = \begin{bmatrix} y_1 \\ y_2 \end{bmatrix}$, and $\mathbf{b}^T = [b_1 \ \ b_2]$.

Remark. The primal-dual pair given above is said to be in ***symmetric form***, where primal problem is given in *canonical form*.

5:2. GENERAL PRIMAL – DUAL PAIR

Since, all the primal – dual computations are obtained directly from the simplex tableau, it is logical that we define the dual in a way that is consistent with the standard form of the primal. Given below are two important definitions of primal – dual pair based on the standard form of primal:

Definition 1. **Standard Primal Problem**

$$\text{Maximize } z = c_1x_1 + c_2x_2 + \ldots + c_nx_n \text{ subject to the constraints:}$$

$$a_{i1}x_1 + a_{i2}x_2 + \ldots + a_{in}x_n = b_i \qquad i = 1, 2, \ldots, m$$

and $$x_j \geq 0. \qquad j = 1, 2, \ldots, n$$

Dual Problem

$$\text{Minimize } z^* = b_1y_1 + b_2y_2 + \ldots + b_my_m \text{ subject to the constraints:}$$

$$a_{1j}y_1 + a_{2j}y_2 + \ldots + a_{mj}y_m \geq c_j \qquad j = 1, 2, \ldots, m$$

and y_i unrestricted in sign. $\qquad i = 1, 2, \ldots, m$

Note that x_j's are ***primal variables***, and y_i's the ***dual variables***, while other constants have their usual meanings.

Definition 2. **Standard Primal Problem**

$$\text{Minimize } z = c_1x_1 + c_2x_2 + \ldots + c_nx_n \text{ subject to the constraints:}$$
$$a_{i1}x_1 + a_{i2}x_2 + \ldots + a_{in}x_n = b_i \text{ and } x_j \geq 0,$$
$$\text{for } i = 1, 2, \ldots, m \text{ and } j = 1, 2, \ldots, n.$$

Dual Problem

$$\text{Maximize } z^* = b_1y_1 + b_2y_2 + \ldots + b_my_m \text{ subject to the constraints:}$$
$$a_{1j}y_1 + a_{2j}y_2 + \ldots + a_{mj}y_m \leq c_j, \quad j = 1, 2, \ldots, n$$
$$y_i \text{ unrestricted in sign.} \quad i = 1, 2, \ldots, m$$

The information regarding the primal-dual objective, type of constraints and the signs of the dual variables may be summarized in the following table:

Standard primal objective	*Dual*		
	Objective	*Constraints*	*Variables*
Maximization	Minimization	≥	Unrestricted
Minimization	Maximization	≤	Unrestricted

Remark. From the definitions of primal-dual pair one may easily observe the following:

(*a*) For every primal constraint, there is a dual variable.

(*b*) For every primal variable, there is a dual constraint.

(*c*) The coefficients of the dual variables in the constraints are the same as the coefficients of the primal variables except that they are transposed, *i.e.*, columns in primal coefficient become the rows in the dual coefficient.

(*d*) The number of dual constraints is exactly equal to the number of primal variables, whereas the number of dual variables is exactly equal to the number of primal constraints.

(*e*) The objective coefficients of the primal variables become the right-hand side constants of dual constraints, whereas the right-hand side constants of the dual constraints become the objective coefficients of the primal variables.

5:3. FORMULATING A DUAL PROBLEM

Various steps involved in the formulation of a primal-dual pair are:

Step 1. Write the given linear programming problem into its standard form. Consider it as the primal problem.

Step 2. Identify the variables to be used in the dual problem. The number of these variables equals the number of constraint equations in the primal.

Step 3. Write down the objective function of the dual, using the right-hand side constants of the primal constraints.

If the primal problem is of maximization type, the dual will be a minimization problem and *vice versa*.

Step 4. Making use of dual variables identified in *step 2*, write the constraints for the dual problem.

(*a*) If the primal is a maximization problem, the dual constraints must be all of '≥' type. If the primal is a minimization problem, then dual constraints must be all of '≤' type.

(*b*) The column coefficient of the primal constraints become the row coefficients of the dual constraints.

(c) The coefficients of the primal objective function becomes the right-hand side constants of the dual constraints.

(d) The dual variables are defined to be unrestricted in sign.

Step 5. using *steps 3* and *4*, write down the dual of the given LPP.

Note. The dual constraint corresponding to an artificial variable in the standard form of the primal is always *redundant*, hence, it is never necessary to consider the dual constraint associated with an artificial variable.

Remarks.

1. If given linear programming problem is in its canonical form, the primal-dual pair is said to be ***symmetric*** and dual variables are non-negative.
2. If given linear programming problem is in its standard form, the primal-dual pair is said to be ***unsymmetric*** and dual variables are unrestricted in sign.

Example 5.1. *Formulate the dual of the following LPP:*

$$\text{Maximize } z = 3x_1 - x_2 + x_3 \text{ subject to the constraints:}$$
$$4x_1 - x_2 \le 8, \quad 8x_1 + x_2 + 3x_3 \ge 12,$$
$$5x_1 - x_3 \le 13, \quad x_1, x_2, x_3 \ge 0.$$

Dual Formulation. Here, the objective function is of maximization with first and third inequalities being of '≤' type, while second inequality is of '≥' type. So, multiplying the second inequality by '–1' on both sides, we have $-8x_1 - x_2 - 3x_3 \le -12$. Following the rules of duality, the primal-dual pair is written as follows:

Primal

Maximize $z = 3x_1 - x_2 + x_3$
subject to the constraints:

$$4x_1 - x_2 \le 8$$
$$-8x_1 - x_2 - 3x_3 \le -12$$
$$5x_1 - x_3 \le 13$$
$$x_1, x_2, x_3 \ge 0.$$

Dual

Minimize $z^* = 8y_1 - 12y_2 + 13y_3$
subject to the constraints:

$$4y_1 - 8y_2 + 5y_3 \ge 3$$
$$-y_1 - y_2 \ge -1$$
$$-3y_2 - y_3 \ge 1$$
$$y_1, y_2, y_3 \ge 0.$$

In dual formulation, y_j $(j = 1, 2, 3)$ are the dual variables and second inequality can be written as $y_1 + y_2 \le 1$.

Example 5.2. *Write the dual of the following LPP:*

$$\text{Minimize } z = 3x_1 - 2x_2 + 4x_3 \text{ subject to the constraints:}$$
$$3x_1 + 5x_2 + 4x_3 \ge 7, \quad 6x_1 + x_2 + 3x_3 \ge 4, \quad 7x_1 - 2x_2 - x_3 \le 10,$$
$$x_1 - 2x_2 + 5x_3 \ge 3, \quad 4x_1 + 7x_2 - 2x_3 \ge 2; \quad x_1 \ge 0, x_2 \ge 0 \text{ and } x_3 \ge 0.$$

*[MBA (Dec.), **Visvesvaraya Univ. 2011**]*

Dual Formulation. Here, the objective function is of minimization with first, second, fourth and fifth constraint are in the right direction, *i.e.*, of '≥' type. But third inequality is not in the right direction. Therefore, multiplying both sides of this inequality by –1, this can be written as: $-7x_1 + 2x_2 + x_3 \ge -10$. Now, following the rules of duality, the primal-dual pair is written as follows:

Primal

Minimize $z = 3x_1 - 2x_2 + 4x_3$
subject to the constraints:

$$3x_1 + 5x_2 + 4x_3 \ge 7$$
$$6x_1 + x_2 + 3x_3 \ge 4$$
$$-7x_1 + 2x_2 + x_3 \ge -10$$
$$x_1 - 2x_2 + 5x_3 \ge 3$$
$$4x_1 + 7x_2 - 2x_3 \ge 2$$
$$x_1, x_2, x_3 \ge 0.$$

Dual

Maximize $z^* = 7y_1 + 4y_2 - 10y_3 + 3y_4 + 2y_5$
subject to the constraints:

$$3y_1 + 6y_2 - 7y_3 + y_4 + 4y_5 \le 3$$
$$5y_1 + y_2 + 2y_3 - 2y_4 + 7y_5 \le -2$$
$$4y_1 + 3y_2 + y_3 + 5y_4 - 2y_5 \le 4$$
$$y_j \ge 0 \; (j = 1, 2, 3, 4, 5).$$

Example 5.3. *Write the dual of the following LPP:*

$$\text{Maximize } z = x_1 + x_2 + x_3 \text{ subject to the constraints:}$$
$$x_1 - 3x_2 + 4x_3 = 5, \quad x_1 - 2x_2 \leq 3$$
$$2x_2 - x_3 \geq 4, \quad x_1, x_2, x_3 \geq 0.$$

[M.Sc. (Math.), CCS Univ. 2005]

Dual Formulation. Here, objective function is of maximization. This implies that all the inequalities must be of '≤' type. But first constraint is equality and third is of '≥' type. So, we multiply the third inequality on both sides by –1 and then write the given linear programming in standard form by introducing slack variables $s_1 \geq 0$ in second constraint and $s_2 \geq 0$ in third constraint. The primal-dual pair, therefore, is written as follows:

Standard Primal	***Dual***
Maximize $z = x_1 + x_2 + x_3$	Minimize $z^* = 5y_1 + 3y_2 - 4y_3$
subject to the constraints:	subject to the constraints:
$x_1 - 3x_2 + 4x_3 = 5$	$y_1 + y_2 \geq 1$
$x_1 - 2x_2 + s_1 = 3$	$-3y_1 - 2y_2 - 2y_3 \geq 1$
$-2x_2 + x_3 + s_2 = -4$	$4y_1 + y_3 \geq 1$
$x_1, x_2, x_3, s_1, s_2 \geq 0$	$y_2 \geq 0, y_3 \geq 0$
	y_1, y_2 and y_3 unrestricted in sign.

Since, dual variables y_2 and y_3 unrestricted in sign are dominated by the restricted variables $y_2 \geq 0$ and $y_3 \geq 0$; eliminating redundancy we have y_1 unrestricted and $y_2 \geq 0, y_3 \geq 0$. Thus, we write the dual formulation of the given LPP as under:

$$\text{Minimize } z^* = 5y_1 + 3y_2 - 4y_3 \text{ subject to the constraints:}$$
$$y_1 + y_2 \geq 1, \quad 3y_1 + 2y_2 + 2y_3 \leq -1,$$
$$4y_1 + y_3 \geq 1, \quad y_2 \geq 0, y_3 \geq 0 \text{ and } y_1 \text{ is unrestricted.}$$

Example 5.4. *Obtain the dual problem of the following LPP:*

$$\text{Minimize } z = x_1 - 3x_2 - 2x_3 \text{ subject to the constraints:}$$
$$3x_1 - x_2 + 2x_3 \leq 7, \quad 2x_1 - 4x_2 \geq 12$$
$$-4x_1 + 3x_2 + 8x_3 = 10; \quad x_1 \geq 0, x_2 \geq 0 \text{ and } x_3 \text{ is unrestricted.}$$

Dual Formulation. Here, objective function is of minimization and x_3 is unrestricted. So, we write $x_3 = x_3' - x_3''$, where $x_3' \geq 0$ and $x_3'' \geq 0$. Further, multiply first inequality by '–1'. Then, introducing surplus variables $s_1 \geq 0$ and $s_2 \geq 0$ in the first and second inequality constraint respectively, the primal-dual pair is written as follows:

Standard Primal	***Dual Problem***
Minimize $z = x_1 - 3x_3 - 2(x_3' - x_3'') + 0 \cdot s_1 + 0 \cdot s_2$	Maximize $z^* = -7y_1 + 12y_2 + 10y_3$
subject to the constraints:	subject to the constraints:
$-3x_1 + x_2 - 2(x_3' - x_3'') - s_1 = -7$	$-3y_1 + 2y_2 - 4y_3 \leq 1$
$2x_1 - 4x_2 - s_2 = 12$	$y_1 - 4y_2 + 3y_3 \leq -3$
$-4x_1 + 3x_2 + 8(x_3' - x_3'') = 10$	$\left.\begin{array}{l} -2y_1 + 8y_3 \leq -2 \\ 2y_1 - 8y_3 \leq 2 \end{array}\right\} 2y_1 - 8y_3 = 2$
$x_1, x_2, x_3', x_3'' \geq 0$ and $s_1, s_2 \geq 0$	
	$-y_1 \leq 0, \ -y_2 \leq 0; \ y_1, y_2$ and y_3 unrestricted.

Since, dual variables y_1 and y_2 unrestricted in sign are dominated by the restricted variables $y_1 \geq 0$ and $y_2 \geq 0$, eliminating redundancy, the complete dual formulation is given on next page.

$$\text{Maximize } z^* = -7y_1 + 12y_2 + 10y_3 \text{ subject to the constraints:}$$
$$-3y_1 + 2y_2 - 4y_3 \le 1, \quad -y_1 + 4y_2 - 3y_3 \ge 3,$$
$$2y_1 - 8y_3 = 2; \quad y_1 \ge 0, y_2 \ge 0 \text{ and } y_3 \text{ unrestricted.}$$

Example 5.5. *Write the dual of the following LPP:*

$$\textit{Maximize } z = 3x_1 + x_2 + 2x_3 - x_4 \textit{ subject to the constraints:}$$
$$2x_1 - x_2 + 3x_3 + x_4 = 1, \quad x_1 + x_2 - x_3 + x_4 = 3$$
$$x_1, x_2 \ge 0 \textit{ and } x_3, x_4 \textit{ unrestricted in sign.}$$

Dual Formulation. Here, x_3 and x_4 are unrestricted variables, therefore, third and fourth constraints in the dual shall be equalities.

If y_1 and y_2 are dual variables corresponding to the two primal constraints in the given order, the dual of the given primal LPP is:

$$\text{Minimize } z^* = y_1 + 3y_2 \text{ subject to the constraints:}$$
$$2y_1 + y_2 \ge 3, \quad -y_1 + y_2 \ge 1,$$
$$3y_1 - y_2 = 2, \quad y_1 + y_2 = -1,$$
$$y_1 \text{ and } y_2 \text{ are unrestricted in sign.}$$

PROBLEMS

5.01. Write the dual of the following LPP:

(*a*)
$$\text{Maximize } z = 2x_1 + 3x_2 \text{ subject to the constraints:}$$
$$x_1 \le 4, \quad x_2 \ge 1, \quad x_1 + 2x_2 \ge 2;$$
$$x_1 \ge 0 \text{ and } x_2 \ge 0.$$
[M.Com., Bangalore Univ. 2018]

(*b*)
$$\text{Maximize } z = x_1 - x_2 + 3x_3 \text{ subject to the constraints:}$$
$$x_1 + x_2 + x_3 \le 10, \quad 2x_1 - x_3 \le 2,$$
$$2x_1 - 2x_2 + 3x_3 \le 6; \quad x_1, x_2, x_3 \ge 0.$$
[MBA, Hyderabad Univ. 2012]

5.02. Obtain the dual of the following LPP:

$$\text{Minimize } z = 20x_1 + 40x_2 \text{ subject to the constraints:}$$
$$2x_1 + 20x_2 \ge 40, \quad 20x_1 + 3x_2 \ge 20,$$
$$4x_1 + 15x_2 \ge 30; \quad x_1 \ge 0 \text{ and } x_2 \ge 0.$$
[BCA, Madras Univ. 2010]

5.03. Write the dual of the following LPP:

$$\text{Minimize } z = x_1 + 2x_2 \text{ subject to the constraints:}$$
$$2x_1 + 4x_2 \le 160, \quad x_1 - x_2 = 30,$$
$$0 \le x_1 \ge 10 \text{ and } x_2 \ge 0.$$
[MIB, Delhi Univ. 2011]

5.04. Write the dual of the following LPP:

$$\text{Minimize } z = 80x_1 + 90x_2 + 100x_3 \text{ subject to the constraints:}$$
$$2x_1 + 5x_2 - 3x_3 \le 100, \quad 8x_1 + 9x_2 \ge 50,$$
$$5x_1 + 6x_2 + 10x_3 \le 150; \quad x_1, x_2, x_3 \ge 0.$$
[MBA, Mumbai Univ. 2010]

5.05. Write the dual problem for the following LPP:

$$\text{Maximize } z = 3x_1 + 5x_2 + 7x_3 \text{ subject to the constraints:}$$
$$x_1 + x_2 + 3x_3 \le 10, \quad 4x_1 - x_2 + 2x_3 \ge 15$$
$$x_1, x_2 \ge 0 \text{ and } x_3 \text{ is unrestricted.}$$
[PG Dip. in Glob. Bus. Oper., Delhi Univ. 2010]

5.06. *Formulate the dual of the LPP:*

$$\text{Minimize } z = 3x_1 - 2x_2 + x_3 \text{ subject to the constraints:}$$
$$2x_1 - 3x_2 + x_3 \le 5, \quad 4x_1 - 2x_2 \ge 9, \quad -8x_1 + 4x_2 + 3x_3 = 8;$$
$$x_1 \ge 0, x_2 \ge 0 \text{ and } x_3 \text{ is unrestricted.}$$
[B.E., Karnataka Univ. 2011]

5.07. Give the dual of the following LPP:

Minimize $z = 2x_1 + 3x_2 + 4x_3$ subject to the constraints:

$2x_1 + 3x_2 + 5x_3 \geq 2,\ \ 3x_1 + x_2 + 7x_3 = 3,\ \ x_1 + 4x_2 + 6x_3 \leq 5;$

$x_1, x_2 \geq 0$ and x_3 is unrestricted. *[BE, Karnataka Univ. 2010]*

5.08. Obtain the dual problem of the following LPP:

Maximize $z = 7x_1 + 2x_2 - 5x_3$ subject to the constraints:

$x_1 + 2x_2 + 3x_3 \leq 50,\ \ 2x_1 + 3x_2 - 4x_3 \geq 40,\ \ 3x_1 + 4x_2 + 5x_3 = 75;$

$x_1 \geq 0, x_2 \geq 0$ and x_3 is unrestricted. *[M.Com., Delhi Univ. 2009]*

5.09. Write the dual of the following LPP:

Minimize $z = x_1 + 2x_2 + 3x_3 - x_4$ subject to the constraints:

$x_1 + 2x_2 + 3x_3 \geq 15,\ \ 2x_1 + x_2 + 3x_3 = 20,\ \ x_1 + 2x_2 + x_3 + x_4 \leq 10;$

$x_j \geq 0\ (j = 1, 2, 3, 4)$ *[MBA, Annamalai Univ. 2010]*

5.10. A machine producing either product *A* or *B* can produce *A* by using 2 units of chemicals and one unit of a compound. It can produce *B* by using one unit of chemicals and 2 units of the compound. Only 800 units of chemicals and 1,000 units of compound are available. The profit per unit of product *A* and *B* are ₹ 30 and ₹ 20 respectively. Write the primal and dual of the problem. *[MBA (Oct.), Madras Univ. 2009]*

5:4. PROPERTIES OF DUAL PROBLEM

Since, there is a close relationship between primal and dual of a linear programming problem, the major properties of dual problem are:

1. The dual of the dual LPP is again primal LPP.
2. If either the primal or dual linear programming problem has a finite optimum solution, the other one also posses the same, and the optimum value of the objective function of the two linear programming problems are equal, *i.e.*, max. z = min. z^* (z corresponds to primal problem and z^* corresponds to dual problem).
3. If either the primal or the dual linear programming problem has an unbounded solution, the solution to the other linear programming problem is infeasible.
4. For a positive basic variable in the primal, the corresponding dual variable will be equal to zero. Alternatively, for a non-basic variable in the primal (which is zero), the corresponding dual variable will be basic and positive.

The standard results on duality are summarized as under:

Dual problem	*Primal problem*	
	Feasible solution	*No feasible solution*
Feasible Solution	Optimum	Dual Unbounded
No Feasible Solution	Primal Unbounded	Unbounded or Infeasible

5:5. DUALITY AND SIMPLEX METHOD

Since, the dual of a given primal problem is derived from and related to it, the optimum solution to the two problems shall also be related to each other.

If primal is a ***maximization*** problem, then following are the set of rules that govern the derivation of the optimum solution of the dual.

Rule 1. Corresponding net evaluations $(z_j - c_j)$ of the starting ***primal*** variables = Difference between the left and right sides of the dual constraints associated with dual starting variables.

Rule 2. Negative of the corresponding net evaluations $(z_j - c_j)$ of the starting ***dual*** variables is equal to difference between the left and right sides of the primal constraints associated with primal starting variables.

Rule 3. If the primal (dual) problem is unbounded, then the dual (primal) problem does not have any ***feasible solution***.

Note. In *Rule 2*, dual problem is to be solved by changing the objective function from minimization to maximization, *i.e.*, maximum z = – minimum (–z).

Example 5.6. *Given the linear programming problem:*

$$\text{Maximize } z = x_1 - 3x_2 + 2x_3 \text{ subject to the constraints:}$$
$$3x_1 - x_2 + 2x_3 \le 7, \quad -2x_1 + 4x_2 \le 12$$
$$-4x_1 + 2x_2 + 8x_3 \le 10; \quad x_1, x_2, x_3 \ge 0$$

Defining this LPP is primal, write its dual and derive the optimum value of the dual variables from the information contained in the optimum simplex tableau of the primal problem.

Solution. Introducing slack variables $s_1 \ge 0$, $s_2 \ge 0$ and $s_3 \ge 0$ in the respective constraints of the given LPP, we write the LPP in its standard form, *viz.*,

$$\text{Maximize } z = x_1 - 3x_2 + 2x_3 + 0 \cdot s_1 + 0 \cdot s_2 + 0 \cdot s_3 \text{ subject to the constraints:}$$
$$3x_1 - x_2 + 2x_3 + s_1 = 7$$
$$-2x_1 + 4x_2 + 0 \cdot x_3 + s_2 = 12$$
$$-4x_1 + 2x_2 + 8x_3 + s_3 = 10$$
$$x_1, x_2, s_1, s_2, s_3 \ge 10.$$

Now, considering this linear programming problem as primal, its dual formulation will be as under:

$$\text{Minimize } z^* = 7y_1 + 12y_2 + 10y_3 \text{ subject to the constraints:}$$
$$3y_1 - 2y_2 - 4y_3 \ge 1, \quad -y_1 + 4y_2 + 2y_3 \ge -3$$
$$2y_1 + 8y_3 \ge 2, \quad y_1 \ge 0, y_2 \ge 0 \text{ and } y_3 \ge 0$$
$$y_1, y_2, y_3 \text{ are unrestricted (redundant).}$$

It may be noted that y_1, y_2 and y_3 are the dual variables.

Solution of Primal. An obvious initial basic feasible solution of primal *LPP* is $s_1 = 7$, $s_2 = 12$ and $s_3 = 10$. Using simplex method, the iterative simplex tableau of the primal problem are:

Initial Iteration. x_3 enters and s_3 leaves.

Coeff. of basic variable	*Basic variable*	x_1	x_2	x_3	s_1	s_2	s_3	z	*Solution*	*Ratio*
0	s_1	3	–1	2	1	0	0	0	7	7/2
0	s_2	–2	4	0	0	1	0	0	12	—
0	← s_3	–4	2	(8)	0	0	1	0	10	10/8
	$(z_j - c_j)$	–1	3	–2 ↑	0	0	0	1	z = 0	

First Iteration. x_1 enters and s_1 leaves.

Coeff. of basic variable	*Basic variable*	x_1	x_2	x_3	s_1	s_2	s_3	z	*Solution*	*Ratio*
0	← s_1	(4)	–3/2	0	1	0	–1/4	0	9/2	9/8
0	s_2	–2	4	0	0	1	0	0	12	—
2	x_3	–1/2	1/4	1	0	0	1/8	0	5/4	—
	$(z_j - c_j)$	–2 ↑	7/2	0	0	0	1/4	1	z = 5/2	

Final Iteration. Optimum solution.

Coeff. of basic variable	*Basic variable*	x_1	x_2	x_3	s_1	s_2	s_3	z	*Solution*
1	x_1	1	–3/8	0	1/4	0	–1/16	0	9/8
0	s_2	0	13/4	0	1/2	1	–1/8	0	57/4
2	x_3	0	1/16	1	1/8	0	3/32	0	29/16
	$(z_j - c_j)$	0	11/4	0	1/2	0	1/8	1	$z = 19/4$

Since, all the net evaluations $(z_j - c_j)$ are non-negative, optimum solution is achieved, *viz.*,

$$x_1 = \frac{9}{8},\ x_2 = 0,\ x_3 = \frac{29}{16} \text{ and maximum } z = \frac{19}{4}.$$

For the optimum solution of the dual, we write

Starting primal variables	s_1	s_2	s_3
Corresponding $(z_j - c_j)$ in the optimum tableau	1/2	0	1/8
Difference between the left and right sides of the dual constraints associated with primal starting variables	$y_1 - 0$	$y_2 - 0$	$y_3 - 0$

$$\therefore \qquad \frac{1}{2} = y_1 - 0,\ 0 = y_2 - 0 \text{ and } \frac{1}{8} = y_3 - 0$$

This implies that $$y_1 = \frac{1}{2},\ y_2 = 0 \text{ and } y_3 = \frac{1}{8}$$

Further, $$\text{Minimum } z^* = 7 \times \frac{1}{2} + 12 \times 0 + 10 \times \frac{1}{8} = \frac{19}{4} = \text{Maximum } z$$

Example 5.7. *Write down the dual of the following LPP and solve it:*

Maximize $z = 4x_1 + 2x_2$ *subject to the constraints:*

$x_1 + x_2 \geq 3$ *and* $x_1 - x_2 \geq 2;$

$x_1 \geq 0$ *and* $x_2 \geq 0.$ **[MBA, Annamalai Univ. 2009]**

Solution. Since, objective is of maximization, its constraints should be of '≤' type. So, we multiply first as well as second constraint on both sides by '–1'. Therefore, we have

Primal. Maximize $z = 4x_1 + 2x_2$ subject to the constraints:

$$-x_1 - x_2 \leq -3 \text{ and } -x_1 + x_2 \leq -2;$$
$$x_1 \geq 0 \text{ and } x_2 \geq 0.$$

Dual of this primal problem is as follows:

Dual. Minimize $z^* = -3y_1 - 2y_2$ subject to the constraints:

$$-y_1 - y_2 \geq 4,\ -y_1 + y_2 \geq 2;$$
$$y_1 \geq 0 \text{ and } y_2 \geq 0.$$

For the solution of dual problem by simplex method, we introduce surplus variables $s_1 \geq 0$, $s_2 \geq 0$ and artificial variables $A_1 \geq 0$, $A_2 \geq 0$ in the respective constraints. Therefore, the initial basic feasible solution will be:

$$A_1 = 4 \text{ and } A_2 = 2.$$

The iterative simplex tableau are as follows:

Initial Iteration. y_2 enters and A_2 leaves.

Coeff. of basic variable	Basic variable	y_1	y_2	s_1	s_2	A_1	A_2	z^*	Solution	Ratio
M	A_1	-1	-1	-1	0	1	0	0	4	—
M	$\leftarrow A_2$	-1	(1)	0	-1	0	1	0	2	$2/1$
	$(z_j - c_j)$	$3 - 2M$	$2\uparrow$	$-M$	$-M$	0	0	1	$z^* = 6M$	

Final Iteration. Pseudo-optimum solution.

Coeff. of basic variable	Basic variable	y_1	y_2	s_1	s_2	A_1	A_2	z^*	Solution
M	A_1	-2	0	-1	-1	1	1	0	6
-2	y_2	-1	1	0	-1	0	1	0	2
	$(z_j - c_j)$	$5 - 2M$	0	$-M$	$2 - M$	0	-2	1	$z^* = 6M - 4$

From this tableau, we observe that the coefficient of M in each of the net evaluation given in last row is negative, *i.e.*, condition of optimality is satisfied. But, a non-zero artificial variable appears in the basis. Therefore, the dual problem does not possess any optimum basic feasible solution. In other words, we say that there exists ***pseudo-optimum solution***. Hence, we conclude that there exists an unbounded solution to the primal problem.

Example 5.8. *Use duality to solve the following LPP:*

$$\text{Maximize } z = 2x_1 + x_2 \text{ subject to the constraints:}$$
$$x_1 + 2x_2 \le 10,\; x_1 + x_2 \le 6,\; x_1 - x_2 \le 2,\; x_1 - 2x_2 \le 1;$$
$$x_1, x_2 \ge 0.$$

[B.Sc. (Stat.), Delhi Univ. 2006]

Solution. The dual problem of the given primal is:

$$\text{Minimize } z^* = 10y_1 + 6y_2 + 2y_3 + y_4 \text{ subject to the constraints:}$$
$$y_1 + y_2 + y_3 + y_4 \ge 2,\; 2y_1 + y_2 - y_3 - 2y_4 \ge 1;\; y_1, y_2, y_3, y_4 \ge 0.$$

Introducing surplus variables $s_1 \ge 0, s_2 \ge 0$ and artificial variables $A_1 \ge 0, A_2 \ge 0$, an initial basic feasible solution is: $A_1 = 2$ and $A_2 = 1$. (The primal constraints associated with s_1, s_2, A_1 and A_2 are: $-x_1 \le 0, -x_2 \le 0, x_1 \le M$ and $x_2 \le M$).

The iterative simplex tableau are as follows:

Initial Iteration. y_1 enters and A_2 leaves.

Coeff. of basic variable	Basic variable	y_1	y_2	y_3	y_4	s_1	s_2	A_1	A_2	z^*	Solution
M	A_1	1	1	1	1	-1	0	1	0	0	2
M	$\leftarrow A_2$	(2)	1	-1	-2	0	-1	0	1	0	1
	z^*-equation $(z_j - c_j)$	$3M\uparrow$ $- 10$	$2M$ $- 6$	-2	$-M-1$	$-M$	$-M$	0	0	1	$z^* = 3M$

First Iteration. y_3 enters and A_1 leaves.

Coeff. of basic variable	Basic variable	y_1	y_2	y_3	y_4	s_1	s_2	A_1	A_2	z^*	Solution
M	A_1	0	$1/2$	(3/2)	2	-1	$1/2$	1	$-1/2$	0	$3/2$
10	y_1	1	$1/2$	$-1/2$	-1	0	$-1/2$	0	$1/2$	0	$1/2$
	z^*-equation $(z_j - c_j)$	0	$M/2$ $- 1$	$3M/2$ $- 7$	$2M$ $- 11$	$-M$	$M/2$ $- 5$	0	$-3M/2$ $+ 5$	1	$z^* = 3M/2$ $+ 5$

Second Iteration. y_2 enters and y_1 leaves.

Coeff. of basic variable	*Basic variable*	y_1	y_2	y_3	y_4	s_1	s_2	A_1	A_2	z^*	*Solution*
2	y_3	0	1/3	1	4/3	–2/3	1/3	2/3	–1/3	0	1
10	y_1	1	(2/3)	0	–1/3	–1/3	–1/3	1/3	1/3	0	1
	z^*-equation $(z_j - c_j)$	0	4/3	0	–5/3	–14/3	–8/3	$-M + 14/3$	$-M + 8/3$	1	$z^* = 12$

Final Iteration. Optimum solution.

Coeff. of basic variable	*Basic variable*	y_1	y_2	y_3	y_4	s_1	s_2	A_1	A_2	z^*	*Solution*
2	y_3	–1/2	0	1	3/2	–1/2	1/2	1/2	–1/2	0	1/2
6	y_2	3/2	1	0	–1/2	–1/2	–1/2	1/2	1/2	0	3/2
	z^*-equation $(z_j - c_j)$	–2	0	0	–1	–4	–2	$4 - M$	$2 - M$	1	$z^* = 10$

From the last simplex tableau, an optimum basic feasible solution to the dual LPP is:

$$y_1 = 0,\ y_2 = 3/2 \text{ and } y_3 = 1/2 \text{ with min. } z^* = 10.$$

Also, the primal constraints associated with the dual variables A_1 and A_2 are $x_1 \le M$ and $x_2 \le M$. Thus, using duality rules, the optimum solution to the primal is obtained below:

Starting dual variables	A_1	A_2
Corresponding $(z_j - c_j)$	$4 - M$	$2 - M$
Left minus right sides of the primal constraint associated with starting dual variables	$x_1 - M$	$x_2 - M$

Making use of duality rules, we get

$$x_1 - M = -M + 4 \text{ and } x_2 - M = -M + 2, \text{ i.e., } x_1 = 4 \text{ and } x_2 = 2$$

Hence, Maximum z = Minimum $z^* = 10$.

Example 5.9. *Consider the LPP:*

$$\text{Maximize } z = 2x_1 + 4x_2 + 4x_3 - 3x_4 \text{ subject to the constraints:}$$

$$x_1 + x_2 + x_3 = 4,\quad x_1 + 4x_2 + x_4 = 8;\quad x_1, x_2, x_3, x_4 \ge 0.$$

By using x_3 and x_4 as the starting variables, the optimum table is given by

Basis	*Solution*	x_1	x_2	x_3	x_4
x_3	2	3/4	0	1	–1/4
x_2	2	1/4	1	0	1/4
z	16	2	0	0	3

Write the dual problem and find its solution from the optimum primal table.

[B.Sc. (Stat.), Delhi Univ. 2002]

Solution. Since, the primal problem is in standard form with non-negative variables, the dual problem is:

$$\text{Minimize } z^* = 4y_1 + 8y_2 \text{ subject to the constraints:}$$

$$y_1 + y_2 \ge 2,\quad y_1 + 4y_2 \ge 4,\quad y_1 \ge 4 \text{ and } y_2 \ge -3$$

$$y_1 \text{ and } y_2 \text{ unrestricted (redundant).}$$

Now, x_3 and x_4 are the starting variables in the primal, and their respective net evaluations are 0 and 3 in the optimum primal tableau. The dual constraints associated with x_3 and x_4 are $y_1 \geq 4$ and $y_2 \geq -3$. Therefore, we write this information as below:

Starting primal variables	x_3	x_4
Corresponding net evaluations	0	3
Left minus right sides of the dual constraint associated with primal variables	$y_1 - 4$	$y_2 - (-3)$

Making use of *Rule 1*, we get

$$y_1 - 4 = 0 \text{ and } y_2 - (-3) = 3, \text{ i.e., } y_1 = 4 \text{ and } y_2 = 0$$

The optimum value of the dual objective is Min. $z^* = 16 = \max. z$

Example 5.10. *Consider the following LP problem:*

$$\text{Maximize } z = 3x_1 + 2x_2 + 5x_3 \text{ subject to the constraints:}$$

$$x_1 + 2x_2 + x_3 \leq a_1, \quad 3x_1 + 2x_3 \leq a_2, \quad x_1 + 4x_2 \leq a_3,$$

where, a_1, a_2 and a_3 are constants. For specific values of a_1, a_2 and a_3 the optimal solution is:

Basic	x_1	x_2	x_3	s_1	s_2	s_3	z	*Solution*
z	4	0	0	c_1	c_2	0	1	1,350
x_2	b_1	1	0	1/2	–1/4	0	0	100
x_3	b_2	0	1	0	1/2	0	0	c_3
s_3	b_3	0	0	–2	1	1	0	20

where, b_i's and c_j's are constants. Determine:

(*i*) *The values of a_1, a_2 and a_3 that yield the given optimal solution.*

(*ii*) *The value of b_1, b_2, b_3 and c_1, c_2, c_3 in the optimal tableau.*

(*iii*) *The optimal dual solution.*

Solution. The optimal simplex tableau, indicates that slack variables s_1, s_2 and s_3 are introduced in the respective primal constraints. They happen to be the starting primal basic variables also. Thus, the optimal basis inverse is given by $\mathbf{B}^{-1} = \begin{bmatrix} \text{Column under } s_1 & \text{Column under } s_2 & \text{Column under } s_3 \end{bmatrix}$ from the optimum tableau.

(*i*) We have
$$\mathbf{B}^{-1}\,\mathbf{b} = \mathbf{x}_B$$

$$\Rightarrow \begin{pmatrix} \frac{1}{2} & \frac{-1}{4} & 0 \\ 0 & \frac{1}{2} & 0 \\ -2 & 1 & 1 \end{pmatrix} \begin{bmatrix} a_1 \\ a_2 \\ a_3 \end{bmatrix} = \begin{bmatrix} 100 \\ c_3 \\ 20 \end{bmatrix}$$

$$\Rightarrow \quad \frac{1}{2}a_1 - \frac{1}{4}a_2 = 100, \quad \frac{1}{2}a_2 = c_3, \quad -2a_1 + a_2 + a_3 = 20$$

Also, $\quad z = \mathbf{c}_B\,\mathbf{x}_B$. But $\mathbf{c}_B = (2 \;\; 5 \;\; 0)$ and $\mathbf{x}_B = [100 \;\; c_3 \;\; 20]$

$\therefore \quad 1{,}350 = 200 + 5c_3$. This gives $c_3 = 230$

Thus, we get $\quad a_1 = 430, \; a_2 = 460$ and $a_3 = 420$

(*ii*) The *z*-row gives:

$$4 = \mathbf{c}_B\mathbf{y}_1 - c_1 = (2 \;\; 5 \;\; 0)\,[b_1 \;\; b_2 \;\; b_3] - 3 = 2b_1 + 5b_2 - 3. \text{ This gives } 2b_1 + 5b_2 = 7$$

$$c_1 = \mathbf{c}_B\,\mathbf{y}_4 - c_4 = (2 \;\; 5 \;\; 0)\begin{bmatrix} \frac{1}{2} & 0 & -2 \end{bmatrix} - 0 = 1 - 0 = 1$$

$$c_2 = \mathbf{c}_B\,\mathbf{y}_5 - c_5 = (2 \;\; 5 \;\; 0)\begin{bmatrix} -\frac{1}{4} & \frac{1}{2} & 1 \end{bmatrix} - 0 = -\frac{1}{2} + \frac{5}{2} - 0 = 2$$

To obtain the values of b_1, b_2 and b_3, we perform iteration on the starting primal tableau:

Initial Iteration. Introduce x_3 and drop s_2.

Coeff. of basic variable	*Basic variable*	x_1	x_2	x_3	s_1	s_2	s_3	*Solution*
0	s_1	1	2	1	1	0	0	430
0	$\leftarrow s_2$	3	0	(2)	0	1	0	460
0	s_3	1	4	0	0	0	1	420
	$(z_j - c_j)$	–3	–2	–5 ↑	0	0	0	0

First Iteration. Introduce x_2 and drop s_1.

Coeff. of basic variable	*Basic variable*	x_1	x_2	x_3	s_1	s_2	s_3	*Solution*
0	$\leftarrow s_1$	–1/2	(2)	0	1	–1/2	0	200
5	x_3	3/2	0	1	0	1/2	0	230
0	s_3	1	4	0	0	0	1	420
	$(z_j - c_j)$	9/2	–2 ↑	0	0	5/2	0	1150

Second Iteration. Optimum solution.

Coeff. of basic variable	*Basic variable*	x_1	x_2	x_3	s_1	s_2	s_3	*Solution*
2	x_2	–1/4	1	0	1/2	–1/4	0	100
5	x_3	3/2	0	1	0	1/2	0	230
0	s_3	2	0	0	–2	1	1	20
	$(z_j - c_j)$	4	0	0	1	2	0	1350

Comparing it with the given optimal tableau, we get

$$b_1 = -1/4,\ b_2 = 3/2 \text{ and } b_3 = 2$$

(Note that the values of c_1, c_2 and c_3 are also readily available, *viz.*, $c_1 = 1$, $c_2 = 2$ and $c_3 = 230$.)

(*iii*) The dual problem is:

$$\text{Minimize } z^* = a_1y_1 + a_2y_2 + a_3y_3 \text{ subject to the constraints:}$$
$$y_1 + 3y_2 + y_3 \geq 3,\ 2y_1 + 4y_3 \geq 2,\ y_1 + 2y_2 \geq 5$$

The dual constraints associated with the starting primal variables s_1, s_2 and s_3 are $y_1 \geq 0$, $y_2 \geq 0$ and $y_3 \geq 0$.

Thus, we have the following information:

Starting primal variables	s_1	s_2	s_3
Left minus right sides of the associated dual constraint	$y_1 - 0$	$y_2 - 0$	$y_3 - 0$
Net evaluation numbers from primal optimal tableau	c_1	c_2	0

Using *Rule 1*, we get

$$c_1 = y_1 - 0 \Rightarrow y_1 = c_1 = 1$$
$$c_2 = y_2 - 0 \Rightarrow y_2 = c_2 = 2$$
$$0 = y_3 - 0 \Rightarrow y_3 = 0$$

The optimal dual objective is Minimum $z^* = 1{,}350$

PROBLEMS

5.11. Use duality to solve the following LPP:

(*a*) Minimize $z = x_1 - x_2$ subject to the constraints:

$$2x_1 - x_2 \geq 2, \quad -x_1 + x_2 \geq 1; \quad x_1 \geq 0 \text{ and } x_2 \geq 0$$

[MBA, Panjab Tech. Univ. 2011]

(*b*) Maximize $z = 3x_1 + x_2$ subject to the constraints:

$$x_1 - x_2 \leq 1, \quad x_1 + x_2 \leq 4, \quad x_1 - 3x_2 \leq 3; \quad x_1 \geq 0 \text{ and } x_2 \geq 0$$

[B.Tech. Hyderabad Univ. 2011]

5.12. Show that both the primal and its dual of the following LPP have same optimal value of z and the solution can be read from the primal solution:

Maximize $z = 2x_1 + x_2$ subject to the constraints:

$$x_1 + 5x_2 \leq 10, \quad x_1 + 3x_2 \leq 6, \quad 2x_1 + 2x_2 \leq 8,$$
$$x_1 \geq 0 \text{ and } x_2 \geq 0$$

[BE (Nov.), Karnataka Univ. 2012]

5.13. Using duality, solve the following LPP:

Maximize $z = 3x_1 + 2x_2$ subject to the constraints:

$$x_1 + x_2 \geq 1, \quad x_1 + x_2 \leq 7, \quad x_1 + 2x_2 \leq 10,$$
$$0 \leq x_2 \leq 3 \text{ and } x_1 \geq 0$$

[MBA, Madurai Univ. 2008]

5.14. Construct the dual of the following LPP and solve both the primal and its dual:

Minimize $z = 4x_1 + 2x_2 + 3x_3$ subject to the constraints:

$$2x_1 + 4x_3 \geq 5, \quad 2x_1 + 3x_2 + x_3 \geq 4,$$
$$x_1, x_2, x_3 \geq 0$$

[B.Tech. (Nov.), Hyderabad Univ. 2010]

5.15. Solve the following LPP by using its dual:

Maximize $z = 5x_1 - 2x_2 + 3x_3$ subject to the constraints:

$$2x_1 + 2x_2 - x_3 \geq 2, \quad 3x_1 - 4x_2 \leq 3, \quad x_2 + 3x_3 \leq 5;$$
$$x_1 \geq 0, x_2 \geq 0 \text{ and } x_3 \geq 0$$

[B.Tech. (Nov.), HP Univ. 2015]

5.16. Solve the following LPP by using its dual problem:

Minimize $z = 10x_1 + 15x_2 + 30x_3$ subject to the constraints:

$$x_1 + 3x_2 + x_3 \geq 90, \quad 2x_1 + 5x_2 + 3x_3 \geq 120, \quad x_1 + x_2 + x_3 \geq 60;$$
$$x_1, x_2, x_3 \geq 0$$

[BE (Mech.), Jammu Univ. 2004]

5.17. Write the dual corresponding to the following LPP:

$$x_1 + x_2 + 2x_3 \leq 120, \quad 3x_1 - 2x_2 - x_3 \geq 90,$$
$$2x_1 + 4x_2 + 2x_3 = 100, \quad 5x_1 + 8x_2 + 10x_3 = \text{maximum } z,$$
$$x_1, x_2, x_3 \geq 0$$

Use simplex method and obtain zeroth and first iteration of the dual.

[B.Tech., HP Univ. 2015; B.Tech. (Prod.), Panjab Tech. Univ. 2011]

5.18. Recast the maximization problem given below into its dual:

Maximize $z = 100x_1 + 400x_2 + 250x_3$ subject to the constraints:

$$2x_1 + 6x_2 + 3x_3 \leq 100, \quad x_1 + 5x_2 + 5x_3 \leq 80; \quad x_1, x_2, x_3 \geq 0$$

Solve the dual so obtained by simplex method. Can you read the solution of the primal from that of the dual? Find the solution of the primal problem. *[M.Com., Delhi Univ. 2008]*

5.19. Given below is the final simplex tableau of a LPP. Write the optimum solution of primal as well as of dual.

Basis	*Coefficients of* x_1	x_2	x_3	s_1	s_2	s_3	z	*Solution*
s_1	0	1/6	0	1	–1/6	–1/3	0	11
x_3	0	–4/3	1	0	1/3	–1/3	0	5
x_1	1	19/6	0	0	–1/6	2/3	0	20
$(z_j - c_j)$	0	–35	0	0	–10	–10	1	
c_j	40	25	50	0	0	0		

[MBA, Mumbai Univ. 2010]

5.20. Three food products are available at cost of ₹ 10, ₹ 36 and ₹ 24 per unit respectively. They contain 1000, 4000 and 2000 calories per unit respectively; and also 200, 900 and 500 protein units per unit of product respectively. It is required to find the minimum cost diet containing at least 20,000 calories and 3,000 units of protein. Formulate and solve the given problem as a LPP. Write the dual and use it to check the optimum solution of the given problem.

[MBA, Delhi Univ. 2009]

5:6. SUMMARY OF PRIMAL – DUAL RELATIONSHIP

Untill now we have discussed about the relationship among the primal linear programming problem and dual linear programming problem. We shall now summarize the key relationship between these primal and dual problems.

1. Dual of Dual is Primal. Consider the ***primal*** LPP in its ***canonical form***:

$$\text{Maximize } f(\mathbf{x}) = \mathbf{cx} \text{ subject to the constraints:}$$
$$\mathbf{Ax} \le \mathbf{b} \text{ and } \mathbf{x} \ge \mathbf{0}$$

Dual of this linear programming problem is:

$$\text{Minimize } g(\mathbf{y}) = \mathbf{b}^T\mathbf{y} \text{ subject to the constraints:}$$
$$\mathbf{A}^T\mathbf{y} \ge \mathbf{c}^T \text{ and } \mathbf{y} \ge \mathbf{0}$$

This LPP can also be written as:

$$\text{Maximize } g(-\mathbf{y}) = (-\mathbf{b}^T)\,\mathbf{y} \text{ subject to the constraints:}$$
$$(-\mathbf{A}^T)\,\mathbf{y} \le (-\mathbf{c}^T) \text{ and } \mathbf{y} \ge \mathbf{0}$$

Considering this LLP as primal, its dual is:

$$\text{Minimize } h(\mathbf{w}) = (-\mathbf{c}^T)^T\,\mathbf{w} \text{ subject to the constraints:}$$
$$(-\mathbf{A}^T)^T\,\mathbf{w} \ge (-\mathbf{b}^T)^T \text{ and } \mathbf{w} \ge \mathbf{0}$$

or
$$\text{Maximize } h(-\mathbf{w}) = \mathbf{cw} \text{ subject to the constraints:}$$
$$-\mathbf{Aw} \ge -\mathbf{b} \text{ or } \mathbf{Aw} \le \mathbf{b} \text{ and } \mathbf{w} \ge \mathbf{0}.$$

The LPP obtained above is identical to the given LPP. This property of duality is generally known as ***symmetry property***.

2. Weak Duality Property. If x is a feasible solution to the primal LPP and y is feasible solution to its dual, then

$$\mathbf{cx} \le \mathbf{b}^T\mathbf{y}$$

For $\quad \mathbf{Ax} \le \mathbf{b}, \mathbf{x} \ge \mathbf{0}$ and $\mathbf{A}^T\mathbf{y} \ge \mathbf{c}^T, \mathbf{y} \ge \mathbf{0},$

we have $\quad (\mathbf{c}^T)^T \le (\mathbf{A}^T\mathbf{y})^T$ or $\mathbf{c} \le \mathbf{y}^T\mathbf{A}$

This gives $\quad \mathbf{cx} \le \mathbf{y}^T\mathbf{Ax}\ (= \mathbf{y}^T\mathbf{b})$ or $\mathbf{cx} \le \mathbf{b}^T\mathbf{y},\quad$ since $\mathbf{y}^T\mathbf{b} = \mathbf{b}^T\mathbf{y}$ and $\mathbf{Ax} \le \mathbf{b}$.

3. Strong Duality Property. If $\mathbf{x}^0$ is an optimum solution to the primal LPP and $\mathbf{y}^0$ is an optimum solution to its dual LPP, then $\mathbf{cx}^0 = \mathbf{b}^T \mathbf{y}^0$.

Properties 2 and 3 imply that $\mathbf{cx} < \mathbf{b}^T\mathbf{y}$ for feasible solution if one or both of them are optimum for their respective linear programming problems, where equality holds when both are optimum.

Observe that the ***weak duality*** property gives the relationship between any pair of solutions for the primal and dual problems, where both solutions are ***feasible*** for their respective problems.

4. Complementary Solutions Property. While solving any LPP by simplex method, each iteration identifies a CPF solution $\mathbf{x}$ for the primal problem and a complementary solution $\mathbf{y}$ for the dual problem (found in the net evaluations row under the slack variables), where $\mathbf{cx} = \mathbf{b}^T\mathbf{y}$.

If $\mathbf{x}$ is not ***optimum*** for the primal problem, then $\mathbf{y}$ is ***not feasible*** for the dual problem.

The complementary solutions property also holds at the final iteration of the simplex method, where an optimum solution is found for the primal problem.

5. Complementary Optimum Solutions Property. The final iteration of the simplex method identifies an optimum solution $\mathbf{x}^0$ for the primal problem and a ***complementary optimum solution*** $\mathbf{y}^0$ for the dual problem (found in the net-evaluations row under slack variables). where $\mathbf{cx}^0 = \mathbf{b}^T\mathbf{y}$. The y_i^0 are the ***shadow prices*** for the primal problem.

6. Existence Property. If either primal or dual linear programming problem have an unbounded objective function value, then the other linear programming problem has no feasible solution.

Let the given primal problem have unbounded solution. Then, for any value of the objective function, say $+\infty$, there exists a feasible solution say $\mathbf{x}$ yielding this solution, *i.e.*, $\mathbf{cx} \to \infty$. If possible, let the dual problem have a feasible solution. Then from property 2, for each feasible solution $\mathbf{y}^0$ of the dual, there exists a feasible solution x^0 to the primal, such that $\mathbf{cx}^0 \le \mathbf{b}^T \mathbf{y}^0$. That is, $\mathbf{b}^T \mathbf{y}^0 \to \infty$. As $\mathbf{b}$ is constant and $\mathbf{y}^0$ has to satisfy the constraint $\mathbf{A}^T \mathbf{y}^0 \ge \mathbf{c}^T$, therefore, the objective function of the dual problem must be finite. This contradicts the result $\mathbf{b}^T \mathbf{y}^0 \to \infty$. This implies that the dual problem has no feasible solution.

Next property summarizes the primal-dual relationship under all the possibilities discussed in the above-mentioned six properties.

7. Duality Theorem. Following are the only possible relationships between the primal and dual problems:

(*a*) If one problem has ***feasible solution*** and a ***bounded*** objective function (*i.e.*, has an optimum solution), then so does the other problem, and therefore, both the weak as well as strong duality properties are applicable.

(*b*) If one problem has ***feasible solutions*** and ***unbounded*** objective function (*i.e.*, no ***optimum solution***), then the other problem has no feasible solution.

(*c*) If one problem has ***no feasible solution***, then the other problem has either no ***feasible solution*** or an ***unbounded*** objective function.

Remark. Relationship between the primal and dual as explained above is also known as ***fundamental theorem*** of duality.

5:7. ECONOMIC INTERPRETATION OF DUALITY

The primal – dual relationship discussed so far is not only related mathematically, they are also related in economic sense.

Illustration

Let the following table give the amounts of two vitamins v_1 and v_2 per unit, present in two different foods f_1 and f_2, respectively:

Vitamin	*Food* f_1	f_2	*Daily requirement*
v_1	6	8	100
v_2	7	12	120
Cost per unit	12	20	

The last column of the table represents number of units of the minimum daily requirement for the two vitamins, whereas last row represents the cost per unit for the two foods. The problem is to determine the minimum quantities of the two foods f_1 and f_2 so that minimum daily requirement of two vitamins is met and that, at the same time, cost of purchasing these quantities of f_1 and f_2 is minimum. To formulate the problem mathematically let x_j be the number of units of food f_j $(j = 1, 2)$ to be purchased. The above problem, therefore, is to determine two real numbers x_1 and x_2 so as to

$$\text{Minimize } z = 12x_1 + 20x_2 \text{ subject to the constraints:}$$
$$6x_1 + 8x_2 \geq 100, \quad 7x_1 + 12x_2 \geq 120; \text{ and } x_1, x_2 \geq 0$$

Now, let us consider a different problem, which is associated with the above problem. Suppose there is a wholesale dealer who sells two vitamins v_1 and v_2 along with some other commodities. The local shopkeepers purchase the vitamins from him and form two foods f_1 and f_2 (the details are same as given in the table). The dealer knows very well that the foods f_1 and f_2 have their market values because of their vitamin contents. The problem of the dealer is to fix the maximum per unit selling price for the two vitamins v_1 and v_2 in such a way that the resulting price of foods f_1 and f_2 do not exceed their existing market prices.

The selling prices are generally known as ***dual prices*** or ***shadow prices*** or ***imputed prices***, indicating the worth of the resources.

To formulate the problem mathematically, let the dealer decide to fix up the two prices as y_1 and y_2 per unit respectively. Then, the dealer's problem can be stated mathematically as to determine two real numbers y_1 and y_2 so as to

$$\text{Maximize } z^* = 100y_1 + 120y_2 \text{ subject to the constraints:}$$
$$6y_1 + 7y_2 \leq 12, \quad 8y_1 + 12y_2 \leq 20; \text{ and } y_1, y_2 \geq 0$$

In the context of general LPP (in standard form), economic interpretation of the primal problem may be defined as follows:

x_j : level of activity j $(j = 1, 2, ..., n)$
c_j : unit profit from activity j $(j = 1, 2, ..., n)$
b_i : amount of resource i $(i = 1, 2, ..., m)$
a_{ij} : amount of resource i by each unit of activity j
z : total profit from all the activities

Dual objective function is written as:

$$z = b_1y_1 + b_2y_2 + ... + b_my_m$$

Here, each b_iy_i $(i = 1, 2, ..., m)$ can be interpreted as the current ***contribution to profit*** by having b_i units of resource i available for the primal problem. Thus, dual variable y_i is interpreted as the contribution to profit per unit of resource i $(i = 1, 2, ..., m)$, when the current set of basic variables is used to obtain the optimum solution.

In other words, y_i values (or y_i^0 values in the optimum solution) are just the ***shadow prices***.

Hence, the importance of duality in linear programming problem is in terms of the information which the dual LPP provides about the value of the resources. The managerial significance of the study of duality may be given as follows:

(*a*) dual variables provide the decision-maker a basis for deciding how much to pay for additional units of resources.

(*b*) dual variables equal the marginal values of resources. Total marginal value of the resources equals the optimum value of the objective function.

(*c*) the maximum amount that should be paid for one additional unit of a resource is termed as ***shadow price*** or ***imputed price***.

Note. The value of ith dual variable is the rate at which the primal objective function value will increase as the resource i increases, assuming that all other information is unchanged.

Example 5.11. *One unit of product A contributes ₹7 and requires 3 units of raw material and 2 hours of labour. One unit of product B contributes ₹5 and requires 2 units of raw material and 3 hours of labour. Availability of raw material at present is of 240 units and there are 200 hours of labour.*

(i) Formulate it as a linear programming problem.

(ii) Write its dual and also its economic interpretation.

(iii) Solve the dual by simplex method and find the optimal product mix, the shadow prices of the raw material and labour.

(iv) Use shadow prices for checking the optimal solution of the given problem.

Solution. (*i*) Let x_1 denote the number of units of product A and x_2 denote the number of units of product B. Then, the appropriate mathematical formulation of LPP is:

$$\text{Maximize } z = 7x_1 + 5x_2 \text{ subject to the constraints:} \quad \textit{(Profit function)}$$
$$3x_1 + 2x_2 \le 240 \quad \textit{(Raw material constraint)}$$
$$2x_1 + 3x_2 \le 200 \quad \textit{(Labour hours constraint)}$$
$$x_1 \ge 0,\ x_2 \ge 0$$

(*ii*) Now, suppose the producer is approached by an individual who would like to rent the facilities for one week. The owner has its assets in the form of 240 units of raw material and 200 labour hours. If we let y_1 represent the rental rate per unit of raw material and y_2 the rental rate per labour hour, the owner would receive a total rental equal to $(240y_1 + 200y_2)$. We shall compute the minimum value of the rental so that the owner will know as to what minimum offer shall be economically acceptable to him. The constraints can be set up by keeping in mind that the alternative of renting must be at least as favourable as the other ones. The rental rates of the resources should be at least as attractive as producing of products A and B. We know that production of one unit of product A requires 3 units of raw material and 2 labour hours. Thus, the total rental for these amounts of resources should be greater than, or equal to, the profit obtainable from one unit of product A, *i.e.*, ₹ 7. Thus, we should have

$$3y_1 + 2y_2 \ge 7$$

In a similar way, the resources consumed in producing one unit of product B are 2 units of raw material and 3 labour hours. The total rental of these resources should be greater than, or equal to at least ₹ 5, the unit profit of product B. Thus, we should have

$$2y_1 + 3y_2 \ge 5$$

Also, the rentals can never be negative. This puts the condition that y_1 and y_2 both should be non-negative.

The dual problem, therefore, is

$$\text{Minimize } z^* = 240y_1 + 200y_2 \text{ subject to the constraints:}$$
$$3y_1 + 2y_2 \ge 7,\ 2y_1 + 3y_2 \ge 5,$$

and

$$y_1 \ge 0,\ y_2 \ge 0.$$

(*iii*) Using simplex method, an optimum simplex tableau of the dual LPP can easily be obtained as under:

Final Iteration. Optimum Solution.

Basic variable	y_1	y_2	s_1	s_2	A_1	A_2	z^*	*Solution*
y_1	1	0	–3/5	2/5	3/5	–2/5	0	11/5
y_2	0	1	2/5	–3/5	–2/5	3/5	0	1/5
$(z^* - c_j)$	0	0	64	24	$M - 64$	$M - 24$	1	–568

The optimum solution of the dual is:

$$y_1 = \frac{11}{5} \text{ and } y_2 = \frac{1}{5}$$

For the optimum solution of primal, we write

Starting dual variables	A_1	A_2
Corresponding $\{-(z_j - c_j)\}$	$-(M - 64)$	$-(M - 24)$
For starting primal variables	$x_1 - M$	$x_2 - M$

$$\therefore \quad x_1 - M = -(M - 64) \text{ and } x_2 - M = -(M - 24), \text{ i.e., } x_1 = 64 \text{ and } x_2 = 24.$$

$$\text{Maximum } z = -\text{ Minimum } z^* = 568$$

Hence, The optimum solution to the given LPP indicates that producing of 64 units of product *A* and 24 units of product *B* per week yield the maximum profit of ₹ 568. The optimum solution to the dual indicates that the rental rates y_1 and y_2 should be 11/5 (= 2.20) and 1/5 (= 0.20), respectively, so as to have the minimum rental of ₹ 568.

(*iv*) The individual rental rates y_1 and y_2 are called the ***shadow prices*** or ***imputed prices*** for raw material and labour hour, indicating the worth of resources. The economic nature of the shadow prices of raw material and labour hour in more detail is discussed below:

Of the two products *A* and *B*, we know that each unit of product *A* contributes ₹ 7 to the profit. The imputed price of raw material and labour being, ₹ 2.20 per unit and ₹ 0.20 per hour, respectively. The total imputed cost of resources used in making a unit of product would be:

3 units @ ₹ 2.20 per unit = ₹ 6.60
2 units @ ₹ 0.20 per unit = ₹ 0.40
Total ₹ 7.00

Thus, the total imputed cost of producing one unit of product *A* equals the unit profit obtainable from it.

Similarly, for each unit of product *B*, the total imputed cost of resources employed would be:

2 units @ ₹ 220 per unit = ₹ 4.40
3 units @ ₹ 0.20 per unit = ₹ 0.60
Total ₹ 5.00

This, obviously equals the profit per unit of product *B*.

Example 5.12. *A manufacturing company makes three products, each of which requires three operations as part of manufacturing process. The company can sell all the products it can manufacture but its production capacity is limited by the capacity of its operations centre. Additional data concerning the company is shown in the table as follows:*

Product	Manufacturing requirements (hours/unit)			Cost (₹)	Selling price (₹)
	Centre I	Centre II	Centre III		
A	1	3	2	11	15
B	3	4	1	12	20
C	2	2	2	10	16
Hours available	160	120	80		

What should the product mix be? Write dual of the given problem. Give its economic interpretation and use it for checking the optimal solution of the given problem. ***[M.Com., Delhi Univ. 2009]***

Solution. If x_1 units of product *A*, x_2 units of product *B*, and x_3 units of product C are produced, the LPP may be stated as follows:

$$\text{Maximize } z = 4x_1 + 8x_2 + 6x_3 \text{ subject to the constraints:}$$
$$x_1 + 3x_2 + 2x_3 \le 160, \quad 3x_1 + 4x_2 + 2x_3 \le 120,$$
$$2x_1 + x_2 + 2x_3 \le 80; \quad x_1 \ge 0, x_2 \ge 0 \text{ and } x_3 \ge 0$$

Using simplex method, the optimum solution is obtained in the following simplex tableau:

Final Iteration. Optimum solution.

Basic variable	x_1	x_2	x_3	s_1	s_2	s_3	Solution
s_1	–5/3	0	0	1	–2/3	–1/3	160/3
x_2	1/3	1	0	0	1/3	–1/3	40/3
x_3	5/6	0	1	0	–1/6	2/3	100/3
$(z_j - c_j)$	11/3	0	0	0	5/3	4/3	920/3

Optimum solution to the problem is to produce Product *A* = nil, Product *B* = 40/3 units and Product C = 100/3 units. Total profit = ₹ 920/3.

Dual: If y_1, y_2 and y_3 are dual variables, then the dual may be stated as:

$$\text{Minimize } z^* = 160y_1 + 120y_2 + 80y_3 \text{ subject to the constraints:}$$
$$y_1 + 3y_2 + 2y_3 \ge 4, \quad 3y_1 + 4y_2 + y_3 \ge 8,$$
$$2y_1 + 2y_2 + 2y_3 \ge 6 \text{ and } y_1, y_2, y_3 \ge 0.$$

For the optimum solution of the dual, we write

Starting primal variables	s_1	s_2	s_3
Corresponding $(z_j - c_j)$	0	5/3	4/3
For starting dual variables	$y_1 - 0$	$y_2 - 0$	$y_3 - 0$

$$\therefore \quad y_1 = 0, \; y_2 = \frac{5}{3} \text{ and } y_3 = \frac{4}{3}$$

and

$$\text{minimum } z^* = 160 \times 0 + 120 \times \frac{5}{3} + 80 \times \frac{4}{3} = \frac{920}{3}$$

Economic Interpretation

The dual variable represent the worth of resources of the organization. Variables y_1, y_2 and y_3 represent the marginal profitability of the capacity of centres 1, 2 and 3 respectively. If the company were to rent out its capacity, the minimum rental for each centre would be equal to its marginal profitability. Since, centre 1 has unused capacity, its marginal profitability is nil, while for centres 2 and 3, these are: ₹ 5/3 and ₹ 4/3 per hour respectively. With these shadow prices or imputed values, we can make the following observations:

For Product A. Decision reached is not to produce it. This is because it yields a profit of ₹ 4, while the value of resource it consumes is:

1 hour @ ₹ 0	= ₹ 1 × 0		= ₹ 0
3 hours @ ₹ 5/3	= ₹ 2 × 5/3		= ₹ 5
2 hours @ ₹ 4/3	= ₹ 2 × 4/3	= ₹ 8/3	= ₹ 2.67
		Total:	₹ 7.67

Since, the total value of resource required by a unit of product *A* is ₹ 7.67 while the profit associated is only ₹ 4, obviously it should not be produced. In case a unit of product *A* is produced, the manufacturer will incurr a loss of ₹ 7.67 – ₹ 4), *i.e.*, of ₹ 3.67.

For Products B and C, the value of resources required for their production are as follows:

Product B		*Product C*	
3 hours @ ₹ 0	= ₹ 0	2 hours @ ₹ 0	= ₹ 0
4 hours @ ₹ 5/3	= ₹ 20/3	2 hours @ ₹ 5/3	= ₹ 10/3
1 hours @ ₹ 4/3	= ₹ 4/3	2 hours @ ₹ 4/3	= ₹ 8/3
Total:	= ₹ 8.00	Total:	= ₹ 6.00

Profit per unit = ₹ 8.00 Profit per unit = ₹ 6.00

Clearly, for each of the product *B* and *C*, the opportunity cost is equal to zero.

PROBLEMS

5.21. A manufacturing company has three major departments for the manufacture of its two products *A* and *B*. The weekly capacities are given as follows:

Department	*Per unit time requirement (hours)*		*Hours available in a week*
	Model A	*Model B*	
I	2	3	42
II	7	7	70
III	7	5	70

The marginal profit per unit from Models *A* and *B* are ₹ 800 and ₹ 500 respectively. Assuming that the company can sell any quantity that it produces, determine the optimum outputs for both the models. (Use simplex method)

Write dual of the linear programming problem given above. What is the economic interpretation of the dual variables that a manager can use for making better decisions?

[*MBA (Nov.), Delhi Univ. 2008*]

5.22. A diet-conscious housewife wishes to ensure certain minimum intake of vitamins *A*, *B* and *C* for the family. The minimum daily needs of vitamins *A*, *B* and *C* for the family are 60, 40 and 32 units respectively. For supply of these vitamins, the housewife relies on two fresh foods F_1 and F_2. One unit of F_1 contains 14, 10 and 4 units of vitamins *A*, *B* and *C* respectively. One unit of F_2 contains 4, 8 and 16 units of the three vitamins, respectively. F_1 costs ₹ 3 per unit and F_2 costs ₹ 2 per unit. The problem is to find how many units of each food the housewife should buy every day to keep her food bill as low as possible.

(*a*) Formulate it as a linear programming problem to minimize the total cost.

(*b*) Write the dual of the problem as formulated in (*a*) and present the economic interpretation of the dual.

(*c*) Solve the dual by using simplex method and read therefrom the answer to the primal.

5.23. An electronics firm is undecided as to the most profitable mix for its products. The products, now, manufactured are transistors, resistors, and electron tubes with a profit (per 100 units) of ₹ 100, ₹ 60 and ₹ 40, respectively. To produce a shipment of transistors containing 100 units requires 1 hour of engineering service, 10 hours of direct labour, and 2 hours of administration service. To produce 100 resistors requires 1 hour, 4 hours and 2 hours of engineering, direct labour and administration time respectively. To produce one shipment of electron tubes (100 units) requires 1 hour of engineering, 5 hours of direct labour and 6 hours of administration service. There are 100 hours of engineering services available, 600 hours of direct labour and 300 hours of administration service. What is the most profitable product mix? Write the ***dual*** of the given problem and use it to check the optimum solution and give its economic interpretation. ***[MBA (Nov.), Delhi Univ. 2005]***

5.24. A company manufactures three models of cars. There is a backlog of orders with the company. Model *A* requires 60, 100 and 80 worker-days in three production processes *I*, *II* and *III* respectively. Model *B* requires 100, 240 and 100 worker-days, while Model *C* requires 200, 360 and 160 worker-days, respectively in the three production processes. The number of workers employed in the three production processes are 15, 30 and 15, respectively, and on an average, worker is on the job for 200 working-days in a year. The expected profit per car of the three models is ₹ 7,500, ₹ 15,000 and ₹ 30,000, respectively.

(*i*) Determine the optimal product-mix and total maximum profit obtainable.

(*ii*) What are the shadow prices of the resources? Determine the range over which each of these is valid.

(*iii*) Write the dual to the above LPP and obtain the optimal values of dual variables from the solution in (*i*).

(*iv*) If there is any other optimal product-mix possible, identify that too.

[MFC, Delhi Univ. 2004; M.Com., Delhi Univ. 2010]

5.25. In a product-mix problem x_1, x_2, x_3 and x_4 indicate the units of products *A*, *B*, *C* and *D*, respectively, and we have to:

Maximize $z = 4x_1 + 6x_2 + 3x_3 + x_4$ subject to the constraints:

$1.5x_1 + 2x_2 + 4x_3 + 3x_4 \le 550$ (*Constraint on Machine I, hours*)

$4x_1 + x_2 + 2x_3 + x_4 \le 700$ (*Constraint on Machine II, hours*)

$2x_1 + 3x_2 + x_3 + 2x_4 \le 200$ (*Constraint on Machine III, hours*)

$x_1, x_2, x_3, x_4 \ge 0$

(*a*) Formulate the dual of the above problem.

(*b*) Solve the dual. From the solution of the dual, read the solution of the primal problem.

(*c*) If machine *I* remains out of order for 24 hours, what will be the loss to the manufacturer?

(*d*) If profit contribution from product *A* increases from ₹ 4 to ₹ 5 per unit, what will be the effect on optimal product mix? ***[M.Com., Delhi Univ. 2007]***

5.26. Given below are the objective function, constraints and simplex tableau for a linear programming product mix problem:

Maximize $z = 2x_1 + 5x_2 + 8x_3 + 0 \cdot s_1 + 0 \cdot s_2 + 0 \cdot s_3$ subject to the constraints:

$6x_1 + 8x_2 + 4x_3 + s_1 \le 96$ (*hours/week, Department I*)

$2x_1 + x_2 + 2x_3 + s_2 \le 40$ (*hours/week, Department II*)

$5x_1 + 3x_2 + 2x_3 + s_3 \le 60$ (*hours/week, Department III*)

$x_1, x_2, x_3, s_1, s_2, s_3 \ge 0.$

c_j	Product mix	x_1	x_2	x_3	s_1	s_2	s_3	Quantity
5	x_2	1/3	1	0	1/6	–1/3	0	8/3
8	x_3	5/6	0	1	–1/12	2/3	0	56/3
0	s_3	7/3	0	0	–1/3	–1/3	1	44/3
	c_j	2	5	8	0	0	0	
	z_j	25/3	5	8	1/6	11/3	0	488/3
	$c_j - z_j$	–19/3	0	0	–1/6	–11/3	0	

Answer the following questions, giving reasons in brief:

(*i*) Is the above solution optimal? Explain.

(*ii*) Write the optimal product mix and profit contribution shown by the optimal solution.

(*iii*) If the company wishes to expand production capacity, which of the three departments should be given priority.

(*iv*) What are the shadow prices of three constraints?

(*v*) Write the dual of the given problem.

(*vi*) Obtain the optimal values of dual variables from the final simplex tableau of the primal problem. ***[PG Dip. in Glob. Bus. Oper., Delhi Univ. 2010]***

5:8. DUAL SIMPLEX METHOD

The computational procedure of simplex method is applicable only when the basic solution is feasible and we obtain an optimum basic feasible solution. But, there are situations when we have infeasible basic optimum solution, *i.e.*, solution to the given LPP is optimum and yet not feasible. In such cases, it is possible to find a starting basic solution, but not a feasible solution (that is dual feasible) and at the same time satisfies the condition of optimality, *i.e.*, $(z_j - c_j) \geq 0$ for a maximization problem and $(z_j - c_j) \leq 0$ for a minimization problem. In all such cases a variant of the simplex method called the ***dual simplex method*** would be used. It may be noted that in dual simplex method we always attempt to retain optimality, while bringing infeasible solution to feasible solution.

Computational Procedure

Given below are the various steps involved in solving a LPP by dual-simplex method:

Step 1. Write the given linear programming problem in its standard form and obtain a starting basic solution.

Step 2. (*a*) If the current basic solution is feasible, use simplex method to obtain an optimum solution.

(*b*) If the current basic solution is infeasible, *i.e.*, values of basic variables are ≤ 0, go to the next *step*.

Step 3. Check whether the solution is optimum.

(*a*) If the solution is not optimum, add an artificial constraint in such a way that the condition of optimality is satisfied.

(*b*) If the solution is ***optimum***, go to next *step*.

Step 4. Select the basic variable having the most negative value. This basic variable becomes the leaving variable and the row corresponding to it becomes the ***key row***.

Step 5. Determine the ratios of net evaluations to those corresponding elements of key row, which are negative. Then, select the maximum value of these ratios for maximization problems and minimum value of these ratios for minimization problems. The column corresponding to this value becomes the ***key column***. The element at the intersection of key row and key column becomes the ***leading element*** (or ***key number***).

The basic variable corresponding to key row will leave the basis, *i.e.*, becomes non-basic and the non-basic variable corresponding to key column will enter the basis (*i.e.*, becomes basic).

Step 6. Reduce the key number (leading element) into unity and all other elements of the ***key column*** to zero by elementary row operations.

Step 7. Go to *step* 2 and repeat the procedure until an optimum basic feasible solution is attained.

Example 5.13. *Use dual simplex method to solve the following LPP:*

$$\text{Minimize } z = 3x_1 + x_2 \text{ subject to the constraints:}$$
$$x_1 + x_2 \geq 1,\ \ 2x_1 + 3x_2 \geq 2;\ \ x_1 \geq 0 \text{ and } x_2 \geq 0.$$

[BE (Mech.), Panjab Univ. 2012]

Solution. To apply dual simplex method, convert the constraints of the linear programming problem into '≤' type by multiplying on both sides of the inequalities by '–1'. Introducing slack variables $s_1 \geq 0$ and $s_2 \geq 0$, the starting basic solution is $s_1 = -1$ and $s_2 = -2$. Clearly it is infeasible as both $s_1 < 0$ and $s_2 < 0$. To get optimum basic feasible solution, the iterative dual simplex tableau are:

Initial Iteration. Drop s_2 and enter x_2.

Coeff. of basic variable	*Basic variable*	x_1	x_2	s_1	s_2	z	*Solution*
0	s_1	–1	–1	1	0	0	–1
0	$\leftarrow s_2$	–2	(–3)	0	1	0	–2
	$(z_j - c_j)$	–3	–1 ↑	0	0	1	$z = 0$

Since, all $(z_j - c_j) \leq 0$, and both basic variables x_{B1} (= s_1) and x_{B2} (= s_2) are negative, an optimum ***basic infeasible*** solution has been attained, since, it is a minimization problem. In order to obtain a basic feasible optimum solution, we select a basic variable to leave the basis and a non-basic variable to enter the basis. For this since most negative x_{Bi} is x_{B2} (= s_2), basic variable s_2 leaves the basis. Further,

$$\text{min.}\left\{\frac{(z_j - c_j)}{a_{2j}},\ a_{2j} < 0\right\} = \text{min.}\left\{\frac{-3}{-2},\ \frac{-1}{-3}\right\} = \frac{1}{3}$$

This corresponds to column under x_2. Hence, non-basic variable x_2 enters the basis.

Now, convert the leading element to unity and all other elements of entering column to zero by row operations to get the next iteration.

First Iteration. Drop s_1 and enter s_2.

Coeff. of basic variable	*Basic variable*	x_1	x_2	s_1	s_2	z	*Solution*
0	$\leftarrow s_1$	–1/3	0	1	(–1/3)	0	–1/3
1	x_2	2/3	1	0	–1/3	0	2/3
	$(z_j - c_j)$	–7/3	0	0	–1/3 ↑	1	$z = 2/3$

From this simplex tableau, we have

$$\text{min.}\{x_{Bi},\ \text{for } x_{Bi} < 0\} = x_{B1}\ (= s_1),\ \text{and min.}\left\{\frac{(z_j - c_j)}{a_{1j}},\ a_{1j} < 0\right\} = \text{min.}\left\{\frac{-7/3}{-1/3},\ \frac{-1/3}{-1/3}\right\} = 1,$$

which corresponds to column under s_2. So, the non-basic variable s_2 enters the basis.

Final Iteration. Optimum solution.

Coeff. of basic variable	*Basic variable*	x_1	x_2	s_1	s_2	z	*Solution*
0	s_2	1	0	–3	1	0	1
1	x_2	1	1	–1	0	0	1
	$(z_j - c_j)$	–2	0	–1	0	1	$z = 1$

It is now apparent that all $(z_j - c_j) \le 0$ and also all $x_{Bi} \ge 0$, an optimum basic feasible solution has been reached. Hence, the optimum basic feasible solution is:

$$x_1 = 0 \text{ and } x_2 = 1 \text{ with minimum } z = 1$$

Example 5.14. *Use dual simplex method to solve the following LPP:*

$$\text{Maximize } z = -3x_1 - 2x_2 \text{ subject to the constraints:}$$
$$x_1 + x_2 \ge 1,\ x_1 + x_2 \le 7,\ x_1 + 2x_2 \ge 10,$$
$$0 \le x_2 \le 3 \text{ and } x_1 \ge 0.$$

[M.Tech., Panjab Tech. Univ. 2012]

Solution. To apply dual simplex method, convert all the constraints of the problem into '≤' type by multiplying on both sides of '≥' type inequalities by '–1'. Introducing, now, slack variables $s_1 \ge 0$, $s_2 \ge 0$, $s_3 \ge 0$ and $s_4 \ge 0$ in the constraint inequalities, the initial basic solution is:
$s_1 = -1, s_2 = 7, s_3 = -10$ and $s_4 = 3$.

Clearly, it is not feasible. Therefore, to get the optimum basic feasible solution, iterative dual simplex tableau are:

Initial Iteration. Drop s_3 and enter x_2.

Coeff. of basic variable	*Basic variable*	x_1	x_2	s_1	s_2	s_3	s_4	z	*Solution*
0	s_1	–1	–1	1	0	0	0	0	–1
0	s_2	1	1	0	1	0	0	0	7
0	$\leftarrow s_3$	–1	(–2)	0	0	1	0	0	–10
0	s_4	0	1	0	0	0	1	0	3
	$(z_j - c_j)$	3	2 ↑	0	0	0	0	1	$z = 0$

Since, all $(z_j - c_j) \ge 0$ and x_{B1} $(= s_1)$, x_{B3} $(= s_3)$ are two basic variables having negative value; we have an optimum ***basic infeasible*** solution. Note that it is a maximization LPP. In order to obtain a basic feasible optimum solution, we select the most negative basic variable to leave the basis. Here, since, x_{B3} $(= s_3)$ gives the most negative value, s_3 leaves the basis. Further, since,

$$\max.\left\{\frac{(z_j - c_j)}{a_{3j}},\ a_{3j} < 0\right\} = \max.\left\{\frac{3}{-1}, \frac{2}{-2}\right\} = -1,$$

the column under x_2 becomes the key column and the non-basic variable x_2 enters the basis.

We now convert the leading element to unity and all the elements of column under x_2 to zero by elementary row operations. Thus, we get the next iteration.

First Iteration. Drop s_4 and enter x_1.

Coeff. of basic variable	Basic variable	x_1	x_2	s_1	s_2	s_3	s_4	z	Solution
0	s_1	–1/2	0	1	0	–1/2	0	0	4
0	s_2	1/2	0	0	1	1/2	0	0	2
–2	x_2	1/2	1	0	0	–1/2	0	0	5
0	← s_4	(–1/2)	0	0	0	1/2	1	0	–2
	$(z_j - c_j)$	2 ↑	0	0	0	1	0	1	$z = -10$

From the above tableau, we observe that the optimum solution is still infeasible. Here, since, x_{B4} (= s_4) gives only the negative value; basic variable s_4 leaves the basis. Further, since, $\left\{\frac{(z_j - c_j)}{a_{4j}}, a_{4j} < 0\right\}$ = max. $\left\{\frac{2}{-1/2}\right\} = -4$; column under x_1 becomes the key column and non-basic variable x_1 enters the basis. The next iterative tableau is:

Final Iteration. Optimum basic feasible solution.

Coeff. of basic variable	Basic variable	x_1	x_2	s_1	s_2	s_3	s_4	z	Solution
0	s_1	0	0	1	0	–1	–1	0	6
0	s_2	0	0	0	1	1	1	0	0
–2	x_2	0	1	0	0	0	1	0	3
–3	x_1	1	0	0	0	–1	–2	0	4
	$(z_j - c_j)$	0	0	0	0	3	4	1	$z = -18$

Since, all $(z_j - c_j) \geq 0$ and both $x_1 \geq 0, x_2 \geq 0$; an optimum basic feasible solution is reached. Hence, the optimum basic feasible solution is:

$$x_1 = 4 \text{ and } x_2 = 3 \text{ with maximum } z = -18.$$

PROBLEMS

5.27. Use dual simplex method to solve the linear programming problem:

Minimize $z = x_1 + x_2$ subject to the constraints:

$$2x_1 + x_2 \geq 4, \quad x_1 + 7x_2 \geq 7; \quad x_1, x_2 \geq 0.$$

5.28. Use dual simplex method to solve the linear programming problem:

Minimize $z = x_1 + x_2$ subject to the constraints:

$$2x_1 + x_2 \geq 2, \quad -x_1 - x_2 \geq 1;$$

$$x_1 \geq 0, x_2 \geq 0.$$ ***[MBA (Nov.), Annamalai Univ. 2001]***

5.29. Use dual simplex method to solve the linear programming problem:

Maximize $z = -2x_1 - x_2$ subject to the constraints:

$$3x_1 + x_2 \geq 3, \quad 4x_1 + 3x_2 \geq 6, \quad x_1 + 2x_2 \geq 3;$$

$$x_1, x_2 \geq 0.$$ ***[BE (Jan.), Kurukshetra Univ. 2010]***

5.30. Use dual simplex method to solve the linear programming problem:

Minimize $z = 10x_1 + 6x_2 + 2x_3$ subject to the constraints:

$$-x_1 + x_2 + x_3 \geq 1, \quad 3x_1 + x_2 - x_3 \geq 2;$$

$$x_1, x_2, x_3 \geq 0.$$ ***[B.Tech. (Comp. Sc.), Panjab Tech. Univ. 2009]***

5.31. Use dual simplex method to solve the linear programming problem:

Minimize $z = 6x_1 + 7x_2 + 3x_3 + 5x_4$ subject to the constraints:

$$5x_1 + 6x_2 - 3x_3 + 4x_4 \geq 12, \quad x_2 + 5x_3 - 6x_4 \geq 10,$$
$$2x_1 + 5x_2 + x_3 + x_4 \geq 8; \quad x_1, x_2, x_3, x_4 \geq 0.$$

[BE (Mech.), Kurukshetra Univ. 2012]

5.32. Use dual simplex method to solve the LPP:

Minimize $z = 3x_1 + 2x_2 + x_3 + 4x_4$ subject to the constraints:

$$2x_1 + 4x_2 + 5x_3 + x_4 \geq 10, \quad 3x_1 - x_2 + 7x_3 - 2x_4 \geq 2,$$
$$5x_1 + 2x_2 + x_3 + 6x_4 \geq 15; \quad x_1, x_2, x_3, x_4 \geq 0.$$

[M.Tech. (Nov.), Panjab Tech. Univ. 2012; M.Sc. (Math.), Garhwal Univ. 2002]

5.33. Minimize $z = 2x_1 + 9x_2 + 24x_3 + 8x_4 + 5x_5$ subject to the constraints:

$$x_1 + x_2 + 2x_3 - x_5 \geq 1, \quad -2x_1 + x_3 + x_4 + x_5 \geq 2;$$
$$x_j \geq 0, \quad (j = 1, 2, 3, 4, 5).$$

5.34. Maximize $z = 2x_1 + x_2 + x_3$ subject to the constraints:

$$2x_1 + 3x_2 - 5x_3 = 4, \quad -x_1 + 9x_2 - x_3 \geq 3,$$
$$4x_1 + 6x_2 + 3x_3 \leq 8, \quad x_1, x_2, x_3 \geq 0.$$

[***Hint:*** Since, the initial solution is infeasible and also not optimum, add an artificial constraint $x_1 + x_2 + x_3 \leq M$ (where $M > 0$ is a large number). With the help of this additional constraint, eliminate x_1 from the objective function and constraints of the given problem.]

5.35. Use dual simplex method to solve the LPP:

Minimize $z = x_1 + 2x_2 + 3x_3$ subject to the constraints:

$$x_1 - x_2 + x_3 \geq 4, \quad x_1 + x_2 + 2x_3 \leq 8,$$
$$x_2 - x_3 \geq 2; \; x_1, x_2, x_3 \geq 0.$$

[M.Tech. and BE (Mech.), Panjab Tech. Univ. 2011]

5.36. Use dual simple method to solve the following LPP:

Minimize $z = 3x_1 + 5x_2 + 4x_3$ subject to the constraints:

$$-2x_1 - x_2 + 5x_3 \geq 2, \quad 3x_1 + 2x_2 + 4x_3 \geq 16;$$
$$x_1 \geq 0, x_2 \geq 0 \text{ and } x_3 \geq 0.$$

TEST YOUR UNDERSTANDING

1. Dual of the primal maximization LPP having m constraints and n non-negative variables should
   (*a*) be a minimization LPP
   (*b*) have n constraints and m non-negative variables
   (*c*) both (*a*) and (*b*)
   (*d*) none of the above
2. The right hand side constant of a constraint in a primal problem appears in the corresponding dual as
   (*a*) a coefficient in the objective function
   (*b*) a right hand side constant of a constraint in a dual problem
   (*c*) an input-output coefficient
   (*d*) none of the above

3. If dual has an unbounded solution, primal has
(*a*) an unbounded solution (*b*) an infeasible solution
(*c*) a feasible solution (*d*) none of the above

4. For any primal and its dual,
(*a*) optimum value of the objective function is same.
(*b*) both primal and dual cannot be feasible.
(*c*) primal will have an optimum solution, if and only if, dual does too.
(*d*) all the above.

5. If dual has unbounded solution,then primal has
(*a*) feasible solution (*b*) unbounded solution
(*c*) no feasible solution (*d*) none of the above

6. If primal LPP has a finite solution, then the dual LPP should have
(*a*) unbounded solution (*b*) infeasible solution
(*c*) finite solution (*d*) infinite solution

7. If value of primal slack variable in the optimum simplex tableau is zero, then
(*a*) the linear programming problem has multiple solutions
(*b*) dual variable value associated with that constraint is zero
(*c*) given primal problem has unique solution
(*d*) resource measured by that constraint is not fully utilized.

8. The value of dual variable
(*a*) represents worth of each additional unit of resource.
(*b*) can be obtained by examining the net evaluation row of primal optimum simplex tableau.
(*c*) can be obtained by examining that row which represents the coefficients of objective function.
(*d*) all of the above.

9. The shadow price of a '≥' constraint in a maximization LPP can be
(*a*) zero (*b*) positive
(*c*) negative (*d*) all of the above

10. Shadow price indicates as to how much one unit change in the resource value will change
(*a*) the value of the basic variable in the optimum solution
(*b*) optimum value of the objective function
(*c*) optimality range of an objective function
(*d*) none of the above

11. The shadow price for a resource is the
(*a*) increase in the objective function value by providing one additional unit of resource.
(*b*) price that is paid for the purchase of a resource.
(*c*) saving by eliminating one of the excess quantities of resource.
(*d*) none of the above.

12. Dual simplex method is applicable to those LPP's that start with
(*a*) an infeasible solution (*b*) an infeasible but optimum solution
(*c*) a feasible solution (*d*) a feasible and optimum solution

ANSWERS

1. (*c*)	**2.** (*a*)	**3.** (*b*)	**4.** (*a*)	**5.** (*c*)	**6.** (*c*)
7. (*d*)	**8.** (*a*)	**9.** (*d*)	**10.** (*b*)	**11.** (*a*)	**12.** (*b*)

REVIEW EXERCISE

1. Define dual. Give any two relationships between a primal and its dual. *[MIB, Delhi Univ. 2011]*
2. What is duality? What is the significance of dual variables in simplex method? *[MBA, Panjab Tech. Univ. 2014; B.Tech., Hyderabad Univ. 2011]*
3. What do you understand by duality? Write some important features of primal and dual problems. *[MBA (Dec.), Gujarat Univ. 2010]*
4. State the general rules for forming a dual LPP from its primal LPP.
5. What is duality in LPP? If dual of an LPP is solved, where will we get the values of decision variables of the primal LPP? *[B.Tech. (Comp. Sc.), Panjab Tech. Univ. 2009]*
6. State and prove the weak duality property. Using it, show that if the primal problem is unbounded, then the dual problem is infeasible. *[B.Sc. (Math.), Madras Univ. 2012]*
7. State the fundamental theorem of duality.
8. Explain the economic interpretation of duality. *[M.Tech., Panjab Tech. Univ. 2012]*
9. What is shadow price? How does the concept relate to the dual of an LPP? How does it relate to primal?
10. How the concept of duality can be useful in managerial decision-making? *[MBA, Panjab Univ. 2010]*
11. Give the applications of duality theory. *[MBA (Nov.), Annamalai Univ. 2009]*
12. What is shadow price? How can we read the shadow price from a simplex tableau? *[PG Dip. in Glob. Bus. Oper., Delhi Univ. 2010; MIB, Delhi Univ. 2011]*
13. What is the physical significance of shadow price of a resource? Explain with the help of an example. *[MBA (HCA), Delhi Univ. 2015]*
14. Explain dual simplex method briefly. *[MCA, Panjab Univ. 2001]*
15. Write the various steps involved in the dual simplex method. *[BE (IT) (Dec.), Panjab Univ. 2008]*
16. What is the essential difference between simplex method and dual simplex method. *[M.Tech. (Nov.), Panjab Tech. Univ. 2012]*

CHAPTER 6

Post-Optimal Analysis

"I prefer the dream of future than the history of past"

Following the formulation of linear programming problem and attainment of an optimum solution of it, two situations may arise which requires additional computations. These are:

1. While formulating the linear programming model (problem), it is assumed that the parameters such as market demand, resource consumption, resource availability, the relevant costs/profits are all known with certainty and do not change over time. But, in actual practice the market fluctuate, material and labour costs go up or down, production time change and equipment availability varies from time to time. Therefore, it is desirable to study how the current optimum solution changes when the parameters of the problem get changed. Such an analysis is known as ***post-optimal analysis***.
2. Sometimes after attaining the optimum solution one may discover that a wrong value of a cost/profit coefficient was used or a particular decision variable/constraint was omitted or at least one of the right-hand side constants used were wrong. In such cases it is advisable to solve the problem a new after replacing the correct values of the wrong ones in the data.

6:1. SENSITIVITY ANALYSIS

The study of the effect of ***discrete changes*** (variations) in parameters on the optimum solution of the given linear programming problem is called ***sensitivity analysis***. The ultimate goal of sensitivity analysis is to determine the range within which the linear programming model parameters can change without effecting the current optimum solution of the LPP. The changes (variations) in LPP which are usually studied by post-optimality analysis include:

1. Changes in objective function coefficients, c_j's, *i.e.*, profit (or cost) per unit associated with decision variables.
2. Changes in the b_i values (or right-hand side constants of the constraints), *i.e.*, availability of resources.
3. Changes in the coefficients of decision variables on the left-hand side of the constraints, a_{ij}'s, *i.e.*, consumption of resources per unit of decision variable x_j.
4. Structural changes due to addition and deletion of some decision variables.
5. Structural changes due to addition and deletion of some linear constraints.

Remark. It is possible to imagine a great variety of more complicated problems involving different combinations of structural and parametric changes.

After the above-mentioned changes have been effected in a LPP, we may have any of the following cases:

(*i*) Optimum solution remains unaltered, that is, the basic variables and their values are essentially unchanged.

(*ii*) Basic variables remain the same but their values undergo changes.

(*iii*) Basic solution gets changed entirely.

6:2. CHANGES IN THE COST/PROFIT COEFFICIENTS c_j

Changes in the coefficients of decision variables in the objective function may take place due to a change in cost (or profit) of either basic variable or non-basic variable.

Let $\mathbf{c}_B = (c_{B1} \quad c_{B2} \quad \dots \quad c_{Bm})$ be the optimum basic feasible solution to an LPP and suppose that the coefficient c_k of a variable x_k in the objective function is changed from c_k to $c_k + \Delta c_k$, where Δc_k represents the positive (or negative) amount of change in the value of c_k. We observe from the optimum simplex tableau of an LPP that the feasibility of the solution remains unaffected due to changes in the coefficients c_j of basic variables in the objective function, because $\mathbf{x}_B = \mathbf{B}^{-1}\mathbf{b}$ is independent of $\mathbf{c}$. However, any change in these coefficients (*i.e.*, c_j's) will affect the optimality of the solution. Therefore, such a change requires recomputing z_j values in $(z_j - c_j)$ row of the optimum simplex tableau, because z_k is computed as $\mathbf{c}_B \mathbf{B}^{-1}\mathbf{a}_k$. Thus, any change in c_j may definitely bring some change in the optimality condition, since, $(z_k - c_k)$ is equal to $(\mathbf{c}_B \mathbf{B}^{-1}\mathbf{a}_k - c_k)$.

Now, there are two possibilities: 1. c_k is not in $\mathbf{c}_B$, and 2. c_k is in $\mathbf{c}_B$.

Case 1. When c_k is not in $\mathbf{c}_B$, *i.e.*, c_k is not the coefficient of basic variable in the objective function, the net evaluation corresponding to non-basic variable x_k is given by

$$z_k - c_k = z_k - (c_k + \Delta c_k), \qquad \text{for all } c_k \notin \mathbf{c}_B.$$

Since, z_k is not affected for any change in c_k, the current solution $\mathbf{x}_B$ remains optimum for the new problem, if

$$z_k - (c_k + \Delta c_k) \geq 0 \quad \text{or} \quad \Delta c_k \leq z_k - c_k, \qquad \text{for all } c_k \notin \mathbf{c}_B.$$

Hence, for an LPP with objective function of maximization, value of c_k may be increased upto the value of z_k, and decreased to negative infinity, *i.e.*, $(-\infty)$ without affecting the optimum solution.

Further, since, $z = \mathbf{c}_B\mathbf{x}_B$ is independent of c_k, value of the objective function and optimum solution will remain unchanged.

Case 2. When c_k is in $\mathbf{c}_B$, *i.e.*, c_k is the coefficient of basic variable in the objective function, the net evaluation corresponding to basic variable x_k is given by

$$z_k^* - c_k = \mathbf{c}_B^* \mathbf{y}_k - c_k, \quad \text{where } \mathbf{c}_B^* = (c_{B1} \;\; c_{B2} \;\; \dots \;\; c_{Bk} + \Delta c_{Bk}), \text{ and } \mathbf{y}_k = \mathbf{B}^{-1}\mathbf{a}_k$$

$$\therefore \quad z_j^* - c_j = \sum_{\substack{i=1 \\ i \neq k}}^{m} c_{Bi}\, y_{ij} + (c_{Bk} + \Delta c_{Bk})\, y_{kj} - c_j$$

$$= \sum_{i=1}^{m} c_{Bi}\, y_{ij} - c_j + \Delta c_{Bk}\, y_{kj} = (z_j - c_j) + \Delta c_k\, y_{kj} \qquad (\because c_{Bk} = c_k)$$

For the current basic feasible solution to remain optimum, we must have $(z_j^* - c_j) \geq 0$.

That is, $\quad (z_j - c_j) + \Delta c_k\, y_{kj} \geq 0 \quad \text{or} \quad \Delta c_k\, y_{kj} \geq -(z_j - c_j)$

Thus, $\quad \Delta c_k \geq \dfrac{-(z_j - c_j)}{y_{kj}}, \text{ if } y_{kj} > 0 \quad \text{and} \quad \Delta c_k \leq \dfrac{-(z_j - c_j)}{y_{kj}}, \text{ if } y_{kj} < 0$

Hence, the value of Δc_k that satisfies the optimality criterion can be determined, when following inequality holds:

$$\text{Max.} \left\{ \frac{-(z_j - c_j)}{y_{kj} > 0} \right\} \leq \Delta c_k \leq \text{Min.} \left\{ \frac{-(z_j - c_j)}{y_{kj} < 0} \right\}$$

The revised value of the objective function becomes

$$z^* = \sum_{\substack{i=1 \\ i \neq k}}^{m} c_{Bi}\, x_{Bi} + (c_{Bk} + \Delta c_{Bk}) x_{Bk}$$

$$= \sum_{i=1}^{m} c_{Bi}\, x_{Bi} + \Delta c_{Bk} x_{Bk} = z + x_k\, \Delta c_k, \qquad \text{since } c_{Bk} = c_k \text{ and } x_{Bk} = x_k$$

Hence, if Δc_k satisfies the inequality given above, the optimum solution will remain unchanged, but value of z will be improved by an amount of $x_k \cdot \Delta c_k$.

Note. It may be noted that y_{ij} ($i = 1, 2, ..., m$ and $j = 1, 2, ..., n$) are constants and are equivalent to a_{ij} ($i = 1, 2, ..., m$ and $j = 1, 2, ..., n$) which are considered in *chapter 3*.

Example 6.1. *Given the linear programming problem:*

Maximize $z = 3x_1 + 5x_2$ *subject to the constraints:*

$$x_1 \le 4,\ \ x_2 \le 6,\ \ 3x_1 + 2x_2 \le 18;\ \ x_1, x_2 \ge 0.$$

Discuss the effect on the optimality of the solution, when the objective function is changed to $3x_1 + x_2$.

[B.Sc. (Stat.), Delhi Univ. 2006]

Solution. After introducing slack variables $s_1 \ge 0$, $s_2 \ge 0$ and $s_3 \ge 0$ in the constraints of the above LPP and then solving the LPP by simplex method, optimum simplex tableau is:

Basic variable	x_1	x_2	s_1	s_2	s_3	z	*Solution* (x_{Bi})
s_1	0	0	1	2/3	–1/3	0	2
x_2	0	1	0	1	0	0	6
x_1	1	0	0	–2/3	1/3	0	2
$(z_j - c_j)$	0	0	0	3	1	1	$z = 36$

Since, the objective function is changed to $3x_1 + x_2$; c_2 has been changed to 1 keeping c_1 fixed. So, we find the variation in c_2.

Variation in c_2. Since, $c_2 \in \mathbf{c}_B$, the range of Δc_2 is given by

$$\underset{y_{2j}>0}{\text{Max.}} \left\{\frac{-(z_j - c_j)}{y_{2j}}\right\} \le \Delta c_2 \le \underset{y_{2j}<0}{\text{Min.}} \left\{\frac{-(z_j - c_j)}{y_{2j}}\right\};\ \ i.e.,\ \text{Max.} \left\{\frac{-3}{1}\right\} \le \Delta c_2 < \infty \ \text{ or } \ -3 \le \Delta c_2 < \infty.$$

$$\therefore \qquad 5 - 3 \le c_2 < 5 + \infty \ \text{ or } \ 2 \le c_2 < \infty \ \text{ or } \ 2 \le 5 < \infty, \ \text{ since } c_2 = 5$$

This indicates that the solution will remain optimum so long as c_2 does not fall below the lower limit of the range, *i.e.*, $c_2 = 2$, otherwise the optimum solution will be different.

Example 6.2. *A company wants to produce three products A, B and C. The unit profits on these products are ₹4, ₹6 and ₹2 respectively. These products require two types of resources, manpower and raw material. The LP model formulated for determining the optimal product mix is as under:*

Maximize $z = 4x_1 + 6x_2 + 2x_3$ *subject to the constraints:*

$$x_1 + x_2 + x_3 \le 3 \qquad \textit{(Manpower constraints)}$$

$$x_1 + 4x_2 + 7x_3 \le 9 \qquad \textit{(Raw material constraints)}$$

$$x_1 \ge 0 \ \text{ and } \ x_2 \ge 0,$$

where, x_1, x_2 *and* x_3 *denote the number of units of product A, B and C respectively to be produced.*

(*a*) *Find the optimal product mix and the corresponding profit of the company.*

(*b*) *Find the range of profit contribution of products A and C in the objective function, such that the current optimal product mix remains unchanged.*

(*c*) *What shall be the new optimal product mix when profit per unit from product C is ₹ 12 and not ₹ 2.*

[MBA, Panjab Tech. Univ. 2003; BE (Mech.) (Dec.), MD Univ., 2006]

Solution. (*a*) Introducing slack variables $s_1 \geq 0$ and $s_2 \geq 0$, and then solving the LPP by simplex method, the optimum solution is displayed in the table below:

Basic variable	x_1	x_2	x_3	s_1	s_2	z	*Solution* (x_{Bi})
x_1	1	0	–1	4/3	–1/3	0	1
x_2	0	1	2	–1/3	1/3	0	2
$(z_j - c_j)$	0	0	6	10/3	2/3	1	$z = 16$

Optimum solution is $x_1 = 1$ and $x_2 = 2$; maximum $z = 16$; *i.e.*, to get the maximum profit of ₹ 16, the company should produce one unit of product *A* and two units of product *B*, while the company should not produce product *C*.

(*b*) (*i*) From the optimum simplex tableau, we observe that x_1 is a basic variable. This means that a further decrease in its profit contribution, c_1, will make it less profitable to produce and, therefore, the current optimal product mix will be affected. Also, an increase in the value of c_1 beyond a limit will make the product *A* much profitable and may force the decision-maker to decide to produce product *A* only. Thus, in either case the current optimal product mix will be affected and hence, we need to know both the lower as well as upper limit on the value of c_1 within which the optimum solution will not be affected.

Let c_1 changes to $c_1 + \Delta c_1$. Then, for the current basic feasible solution to remain optimum, we must have $z_j^* - c_j \geq 0$; *i.e.*, $(z_j - c_j) + \Delta c_1 y_{1j} \geq 0$.

The range of Δc_1, therefore, is given by

$$\max_{y_{1j}>0} \left\{ \frac{-(z_j - c_j)}{y_{1j}} \right\} \leq \Delta c_1 \leq \min_{y_{1j}<0} \left\{ \frac{-(z_j - c_j)}{y_{1j}} \right\}$$

i.e.,
$$\max. \left\{ \frac{-(10/3)}{4/3} \right\} \leq \Delta c_1 \leq \min. \left\{ \frac{-6}{-1}, \frac{-2/3}{-1/3} \right\} \quad \text{or} \quad -\frac{5}{2} \leq \Delta c_1 \leq 2$$

Thus, the range over which c_1 can vary maintaining the condition of optimality is given by

$$4 - \frac{5}{2} \leq c_1 \leq 4 + 2, \; i.e., \; \frac{3}{2} \leq c_1 \leq 6$$

(*ii*) Again, from the optimum simplex tableau, we observe that x_3 (decision variable corresponding to product C) is not a basic variable.

Now, let c_3 (coefficient of x_3 in the objective function) changes to $c_3 + \Delta c_3$. Then, the current solution will remain optimum only if $\Delta c_3 \leq (z_3 - c_3)$, *i.e.*, $\Delta c_3 \leq 6$

Further, since c_3 is changed to $c_3 + \Delta c_3$, *i.e.*, $\Delta c_3 = c_3 - 2$, after substituting this in the above inequality, we get $c_3 - 2 \leq 6$ or $c_3 \leq 8$

This implies that as long as the profit contribution per unit of product *C* is less than ₹ 8 (*i.e.*, change should not be more than ₹ 8), it is not profitable to produce this product and the current optimum solution will remain unchanged.

(*c*) In case the value of c_3 is ₹ 12 in place of ₹ 2, then $(z_3 - c_3)$ will be computed as:

$$\therefore \quad (z_3 - c_3) = [4 \;\; 6] \begin{bmatrix} -1 \\ 2 \end{bmatrix} - 12 = -4 (< 0)$$

This shows that the current optimum solution will not remain optimum and the current non-basic variable x_3 will become basic in the next iteration of simplex method.

The new optimum solution will be:

$$x_1 = 2, \; x_2 = 0 \; \text{and} \; x_3 = 1; \; \text{Maximum} \; z = 20;$$

i.e., the company should produce 2 units of product *A*, one unit of product *C* and no unit of product *B* to get the maximum profit of ₹ 20.

6:3. CHANGES IN THE RIGHT-HAND SIDE OF CONSTRAINTS b_i (AVAILABILITY OF RESOURCES)

Any change in the right hand constants of the constraints, *i.e.*, $\mathbf{b} = [b_1 \;\; b_2 \;\; \dots \;\; b_m]$ does not affect the optimality condition, since b_i values are not associated with the computations of $(z_j - c_j)$ or $\mathbf{c}_B \mathbf{B}^{-1} \mathbf{a}_j - c_j$.

However, it may affect the feasibility conditions and the optimum value, because $\mathbf{x}_B = \mathbf{B}^{-1}\mathbf{b}$ and optimum $z = \mathbf{c}_B \mathbf{x}_B = \mathbf{c}_B \mathbf{B}^{-1}\mathbf{b}$.

Thus, if resource b_k is changed to $b_k + \Delta b_k$, the new basic feasible solution is given by

$$\overset{*}{\mathbf{x}}_B = \mathbf{B}^{-1}\mathbf{b}^*, \text{ where } \mathbf{b}^* = [b_1, b_2, \dots, b_k + \Delta b_k \dots, b_m]$$

i.e.,

$$\begin{aligned}\overset{*}{\mathbf{x}}_B &= \mathbf{B}^{-1}\mathbf{b} + \mathbf{B}^{-1}[0 \;\; 0 \;\; \dots \;\; \Delta b_k \;\; \dots \;\; 0] \\ &= \mathbf{B}^{-1}\mathbf{b} + (\boldsymbol{\beta}_1 \;\; \boldsymbol{\beta}_1 \;\; \dots \;\; \boldsymbol{\beta}_m)[0 \; 0 \; 0 \; \dots \; \Delta b_k \; \dots \; 0], \text{ where } \mathbf{B}^{-1} = (\boldsymbol{\beta}_1 \;\; \boldsymbol{\beta}_1 \;\; \dots \;\; \boldsymbol{\beta}_m) \\ &= \mathbf{x}_B + \boldsymbol{\beta}_k \Delta b_k, \text{ where } \boldsymbol{\beta}_k \text{ is the } k^{\text{th}} \text{ column vector of } \mathbf{B}^{-1}.\end{aligned}$$

$$\therefore \quad \overset{*}{x}_{Bi} = x_{Bi} + \beta_{ik}\,\Delta b_k, \text{ where } \beta_{ik} \text{ is the } (i, k)^{\text{th}} \text{ element of } \mathbf{B}^{-1}.$$

In order to maintain the feasibility of the solution at each iteration, $\overset{*}{\mathbf{x}}_B$ must be non-negative.

Thus, we must have

$$x_{Bi} + \beta_{ik}\,\Delta b_k \geq 0; \qquad i = 1, 2, \dots, m$$

or

$$\Delta b_k \geq -\frac{x_{Bi}}{\beta_{ik}} \text{ for } \beta_{ik} > 0 \text{ and } \Delta b_k \leq -\frac{x_{Bi}}{\beta_{ik}}, \text{ for } \beta_{ik} < 0.$$

Hence, for the current solution to remain feasible, we must have

$$\text{Max.} \left\{ \frac{-x_{Bi}}{\beta_{ik} > 0} \right\} \leq \Delta b_k \leq \text{Min.} \left\{ \frac{-x_{Bi}}{\beta_{ik} < 0} \right\}.$$

Remarks

1. It is easy to observe that the new solution $\overset{*}{\mathbf{x}}_B$ still remains optimum. It is so, because for optimality all $(z_j - c_j) \geq 0 \Rightarrow (\mathbf{c}_B \mathbf{B}^{-1} \mathbf{a}_j - c_j) \geq 0$ which is independent of $\mathbf{b}$.
2. If some entries in the $\mathbf{x}_B$ column of the simplex tableau are negative, then the dual simplex method can be used to get an optimum solution to the new problem by maintaining feasibility.
3. A resource whose shadow price is bigger in comparison to others, should be increased first to ensure the best marginal increase in the objective function value.

Example 6.3. *Consider the LPP:*

$$\textit{Maximize} \;\; z = -x_1 + 2x_2 - x_3 \;\; \textit{subject to the constraints:}$$

$$3x_1 + x_2 - x_3 \leq 10, \quad -x_1 + 4x_2 + x_3 \geq 6$$

$$x_2 + x_3 \leq 4 \;\; \textit{and} \;\; x_j \geq 0; \; j = 1, 2, 3.$$

Determine the ranges for discrete changes in the components b_2 and b_3 of the requirement vector (RHS) constants so as to maintain the feasibility of the current optimum solution. **[M.Sc. (Math.), CCS Univ. 2005]**

Solution. Introduce slack variables $s_1 \geq 0$, $s_3 \geq 0$, surplus variable $s_2 \geq 0$ and an artificial variable $A_1 \geq 0$ in the constraints of the problem. Then, making use of simplex method, following optimum simplex tableau is obtained:

Basic variable	x_1	x_2	x_3	s_1	s_2	s_3	A_1	z	*Solution* (x_{Bi})
s_1	3	0	–2	1	0	–1	0	0	6
x_2	0	1	1	0	0	1	0	0	4
s_2	1	0	3	0	1	4	–1	0	10
$(z_j - c_j)$	1	0	3	0	0	2	M	1	$z = 8$

From the above simplex tableau, we observe that

$$\mathbf{x}_B = [6 \quad 4 \quad 10] \text{ and } \mathbf{B}^{-1} = [s_1 \quad A_1 \quad s_2] = \begin{pmatrix} 1 & 0 & -1 \\ 0 & 0 & 1 \\ 0 & -1 & 4 \end{pmatrix},$$

where columns of matrix $\mathbf{B}^{-1}$ correspond to columns of the starting basis matrix $\mathbf{B}$.

The individual effects of changes in b_2 and b_3 where $\mathbf{b} = [b_1 \quad b_2 \quad b_3]$ such that the optimality of the basic feasible solution is not violated, are given by

$$\text{Max.} \left\{ -\frac{x_{Bi}}{\beta_{ik} > 0} \right\} \leq \Delta b_k \leq \text{Min.} \left\{ -\frac{x_{Bi}}{\beta_{ik} < 0} \right\}.$$

i.e., $$\Delta b_2 \leq \frac{-10}{-1} \text{ and Max.} \left\{ -\frac{4}{1}, -\frac{10}{4} \right\} \leq \Delta b_3 \leq \frac{-6}{-1}$$

i.e., $$\Delta b_2 \leq 10 \text{ and } -5/2 \leq \Delta b_3 \leq 6.$$

Now, since $b_2 = 6$ and $b_3 = 4$, the required range of variation is

$$b_2 \leq 10 + 6 \text{ and } 4 - 5/2 \leq b_3 \leq 4 + 6$$

i.e., $$b_2 \leq 16 \text{ and } 3/2 \leq b_3 \leq 10.$$

Example 6.4. *Consider the linear programming problem given in **example 6.2**, i.e.,*

Maximize $z = 4x_1 + 6x_2 + 2x_3$ *subject to the constraints:*

$x_1 + x_2 + x_3 \leq 3$, $x_1 + 4x_2 + 7x_3 \leq 9$; $x_1 \geq 0$ *and* $x_2 \geq 0$.

(*a*) *Discuss the effect of discrete change in the availability of resources from [3 9] to [9 3].*

(*b*) *Which resource should be increased (or decreased) to get best marginal increase in the value of the objective function.*

Solution. The optimum solution to the above LPP is displayed in the following simplex table:

Basic variable	x_1	x_2	x_3	s_1	s_2	z	*Solution* (x_{Bi})
x_1	1	0	–1	4/3	–1/3	0	1
x_2	0	1	2	–1/3	1/3	0	2
$(z_j - c_j)$	0	0	6	10/3	2/3	1	$z = 16$

From the optimum simplex tableau, we observe that $\mathbf{x}_B = [1 \quad 2]$ and $\mathbf{B}^{-1} = [s_1 \quad s_2] = \begin{bmatrix} 4/3 & -1/3 \\ -1/3 & 1/3 \end{bmatrix}$,

since s_1 and s_2 were in the initial basis matrix $\mathbf{B}$.

(*a*) When the RHS constants are changed to [9 3], new basic solution is

$$\begin{bmatrix} x_1 \\ x_2 \end{bmatrix} = \begin{pmatrix} 4/3 & -1/3 \\ -1/3 & 1/3 \end{pmatrix} \begin{bmatrix} 9 \\ 3 \end{bmatrix} = \begin{bmatrix} 11 \\ -2 \end{bmatrix}, \; \textit{i.e.}, \; x_1 = 11 \text{ and } x_2 = -2$$

As the value of basic variable x_2 is negative, optimum solution obtained earlier becomes infeasible. To remove the infeasibility use dual simplex method. Accordingly, x_2 becomes a non-basic variable and s_1 becomes a basic variable in the next iteration, because

$$\text{Max.}\left\{\frac{-(z_j-c_j)}{y_{2j}},\ y_{2j}<0\right\}=\frac{-10/3}{-1/3}=10$$

The new (revised) optimum solution is given in the following simplex tableau:

Basic variable	x_1	x_2	x_3	s_1	s_2	z	*Solution* (x_{Bi})
x_1	1	4	7	0	1	0	3
s_1	0	–3	–6	1	–1	0	6
$(z_j - c_j)$	0	10	26	0	4	1	$z = 12$

As the slack variable s_1 is present at positive level in the optimal product mix, the sensitivity limits for the corresponding right hand side, *i.e.*, first constraint are determined as follows:

Lower limit = 9 – 6 = 3 and Upper limit = ∞

(*b*) From the optimum simplex tableau of the given LPP, we observe that $z_4 - c_4 = 10/3$ and $z_5 - c_5 = 2/3$. These values denote the shadow prices of resources 1 and 2 respectively. Increasing the amount of resources *1* and 2 will increase the value of objective function z by ₹ 10/3 and ₹ 2/3 respectively. For determining the amount of increase in the resources and at the same time maintaining the condition of feasibility in the optimum solution, we proceed as follows:

Let Δb_1 be an increase in the first resource, *i.e.*, b_1 becomes $b_1 + \Delta b_1$. Then

$$\mathbf{x}_B^* = \mathbf{B}^{-1}\mathbf{b}^* \quad \text{or} \quad \begin{bmatrix} x_1 \\ x_2 \end{bmatrix} = \begin{bmatrix} 4/3 & -1/3 \\ -1/3 & 1/3 \end{bmatrix}\begin{bmatrix} 3+\Delta b_1 \\ 9 \end{bmatrix}, \text{ since } \mathbf{b} = \begin{bmatrix} 3 \\ 9 \end{bmatrix}$$

For feasibility of the new basic solution,

$$\mathbf{x}_B^* \geq 0,\ i.e.,\ \begin{bmatrix} x_1 \\ x_2 \end{bmatrix} = \begin{bmatrix} 1+\frac{4}{3}\Delta b_1 \\ 2-\frac{1}{3}\Delta b_1 \end{bmatrix} \geq 0$$

or $$1+\frac{4}{3}\Delta b_1 \geq 0 \text{ and } 2-\frac{1}{3}\Delta b_1 \geq 0$$

or $$\Delta b_1 \geq -\frac{3}{4} \text{ and } \Delta b_1 \leq 6$$

These together give $$-\frac{3}{4} \leq \Delta b_1 \leq 6$$

$$\therefore \quad -\frac{3}{4} \leq \Delta b_1 \leq 6 \quad 3-\left(\frac{3}{4}\right) \leq b_1 \leq 3+6 \text{ or } \frac{9}{4} \leq b_1 \leq 9$$

Hence, value of first resource can be increased from 3 units to a maximum of 9 units. Similarly, second resource can be increased from 9 units to a maximum of 12 units.

PROBLEMS

6.01. Find the optimum solution to the LPP:

Maximize $z = 15x_1 + 45x_2$ subject to the constraints:

$x_1 + 16x_2 \leq 240,\ 5x_1 + 2x_2 \leq 162,\ x_2 \leq 50;\ x_1, x_2 \geq 0.$

If maximum $z = \Sigma\, c_j\, x_j$, $j = 1, 2$ and c_2 is kept fixed at 45, determine how much can c_1 be changed without affecting the optimum solution.

6.02. Discuss the effect on the optimum solution of discrete changes in the requirement vector for the following LPP:

$$\text{Maximize } z = 2x_1 + x_2 \quad \text{subject to the constraints:}$$
$$3x_1 + 5x_2 \le 15,\ 6x_1 + 2x_2 \le 24;\ x_1, x_2 \ge 0.$$

6.03. Consider the LPP:

$$\text{Maximize } z = 3x_1 + 5x_2 + 4x_3 \quad \text{subject to the constraints:}$$
$$2x_1 + 3x_2 \le 8,\ 2x_2 + 5x_3 \le 10,$$
$$3x_1 + 2x_2 + 4x_3 \le 15;\ x_1, x_2, x_3 \ge 0.$$

Find the ranges over which c_3, c_4 and b_2 can be changed maintaining the optimality of the current solution.

6.04. Consider the LPP:

$$\text{Maximize } z = 2x_1 + 4x_2 + 3x_3 \quad \text{subject to the constraints:}$$
$$x_1 + 4x_2 + 3x_3 \le 240,\ 2x_1 + x_2 + 5x_3 \le 300,\ x_1, x_2, x_3 \ge 0.$$

(*a*) Suppose the vector **b** = [240 300] is changed to **b** + Δ **b** = [300 400]. Find the effect of this change on the optimal solution.

(*b*) Suppose the cost vector **c** = (2 4 3) is changed to **c** + Δ **c** = (1 4 4). Find the effect of this change on the optimal solution.

6.05. Given the LPP:

$$\text{Minimize } x = 3x_1 + 6x_2 + x_3 \quad \text{subject to the constraints:}$$
$$x_1 + x_2 + x_3 \ge 6,\ x_1 + 5x_2 - x_3 \ge 4,$$
$$x_1 + 5x_2 + x_3 \ge 24;\ x_1, x_2, x_3 \ge 0.$$

(*a*) Solve the LPP.

(*b*) Discuss the effect of changing the requirement vector from [6 4 24] to [6 2 12] on the optimum solution.

6.06. Consider the LPP:

$$\text{Maximize } z = -x_2 + 3x_3 - 2x_5 \quad \text{subject to the constraints:}$$
$$3x_2 - x_3 + 2x_5 \le 7,\ -2x_2 + 4x_3 \le 12,$$
$$-4x_2 + 3x_3 + 8x_5 \le 10;\ x_2, x_3, x_5 \ge 0.$$

The optimum simplex tableau for this LPP is:

Basic variable	x_2	x_3	x_5	s_1	s_2	s_3	z	*Solution* (x_{Bi})
x_2	1	0	4/5	2/5	1/10	0	0	4
x_3	0	1	2/5	1/5	3/10	0	0	5
s_3	0	0	10	2/5	–1/2	1	0	11
$(z_j - c_j)$	0	0	12/5	1/5	4/5	0	1	$z = 11$

(*a*) Formulate the dual problem for this primal problem.

(*b*) What are optimal values of dual variables?

(*c*) How much can c_5 be decreased before s_2 goes into basis?

(*d*) How much can the requirement 7 in the first constraint be increased before the basis would change?

6.07. Consider the LPP:

$$\text{Maximize } z = 2x_1 + x_2 + 4x_3 - x_4 \quad \text{subject to the constraints:}$$
$$x_1 + 2x_2 + x_3 - 3x_4 \le 8,\ -x_2 + x_3 + 2x_4 \le 0,$$
$$2x_1 + 7x_2 - 5x_3 - 10x_4 \le 21;\ x_1, x_2, x_3, x_4 \ge 0.$$

The optimum solution to this linear programming problem is contained in the following simplex tableau:

Basic variable	x_1	x_2	x_3	x_4	s_1	s_2	s_3	z	*Solution* (x_{Bi})
x_1	1	0	3	1	1	2	0	0	8
x_2	0	1	–1	–2	0	–1	0	0	0
s_3	0	0	–4	2	–2	3	1	0	5
$(z_j - c_j)$	0	0	1	1	2	3	0	1	$z = 16$

For each of the discrete parameter changes listed below, make the necessary corrections in the optimum simplex tableau and solve the resulting problem:

(*a*) Change c_1 to 1 (*b*) Change **b** to [3, –2, 4]
(*c*) Change b_2 to 11 (*d*) Change **c** to [1, 2, 3, 4]

6.08. Consider the following LPP:

$$\text{Maximize } z = 3x_1 + 2x_2 \text{ subject to the constraints:}$$
$$4x_1 + 3x_2 \le 12 \text{ (Resource 1)},$$
$$4x_1 + x_2 \le 8 \text{ (Resource 2)},$$
$$4x_1 - x_2 \le 8 \text{ (Resource 3)},$$
$$\text{and } x_1 \ge 0,\ x_2 \ge 0.$$

(*a*) Find the optimum solution using simplex method.
(*b*) Which of the resources are scarce or abundant?
(*c*) Determine the optimum values of x_1 and x_2 when resource 1 is increased by two units and simultaneously resource 2 is decreased by one unit, while the rest of the problem remains unchanged.

6:4. CHANGES IN THE ACTIVITY COEFFICIENTS a_{ij}'s

After determining the optimum solution of an LPP, suppose that the coefficients of a variable in the coefficient matrix **A** of order $m \times n$ are changed, *i.e.*, there is a change in the elements of one of the columns of **A**, say $\mathbf{a}_k$. Now, if **B** is an $m \times m$ submatrix of **A**, such that its columns forms the initial basis vectors (columns) corresponding to basic variables; then, there are two cases, *viz.*

(*i*) Change in the coefficient of non-basic variable x_k, *i.e.*, $\mathbf{a}_k \notin \mathbf{B}$.
(*ii*) Change in the coefficient of basic variable x_k, *i.e.*, $\mathbf{a}_k \in \mathbf{B}$.

Case 1. When $\mathbf{a}_k \notin \mathbf{B}$, there is no effect on the feasibility of the current solution. However, optimality of the current solution will change. Current solution will remain optimum, only if $\overset{*}{z}_k - c_k \ge 0$, where

$$\overset{*}{z}_k = \mathbf{c}_B\, \mathbf{B}^{-1}\, \overset{*}{\mathbf{a}}_k = \mathbf{c}_B\, \mathbf{B}^{-1}\, \mathbf{a}_k + \mathbf{c}_B\, \boldsymbol{\beta}_k\, \Delta a_{rk},$$

where, $\boldsymbol{\beta}_k$ is the kth column of $\mathbf{B}^{-1}$ when a_{rk} is changed to $a_{rk} + \Delta a_{rk}$.

$$\therefore \qquad \overset{*}{z}_k - c_k = (z_k - c_k) + \mathbf{c}_B\, \boldsymbol{\beta}_k\, \Delta a_{rk}$$

The condition of optimality, thus, requires

$$(z_k - c_k) + \mathbf{c}_B\, \boldsymbol{\beta}_k\, \Delta a_{rk} \ge 0$$

Thus, $$\Delta a_{rk} \ge -\frac{(z_k - c_k)}{\mathbf{c}_B\, \boldsymbol{\beta}_k}, \text{ for } \mathbf{c}_B\, \boldsymbol{\beta}_k > 0,$$

and $$\Delta a_{rk} \le -\frac{(z_k - c_k)}{\mathbf{c}_B\, \boldsymbol{\beta}_k}, \text{ for } \mathbf{c}_B\, \boldsymbol{\beta}_k < 0.$$

Therefore, the required range of discrete change Δa_{rk} in the kth column and r^{th} constraint is given by

$$\max.\left\{\frac{-(z_k - c_k)}{\mathbf{c}_B \boldsymbol{\beta}_k},\ \mathbf{c}_B \boldsymbol{\beta}_k > 0\right\} \leq \Delta a_{rk} \leq \min.\left\{\frac{-(z_k - c_k)}{\mathbf{c}_B \boldsymbol{\beta}_k},\ \mathbf{c}_B \boldsymbol{\beta}_k < 0\right\}.$$

Case 2. When $\mathbf{a}_k \in \mathbf{B}$ is changed to $\mathbf{a}_k^*$, where $\mathbf{a}_k^* = \mathbf{a}_k + \Delta a_{rk}$ both feasibility as well as optimality conditions may be destroyed. In this case, we wish to specify a range for discrete changes Δa_{rk} in a_{rk} so as to maintain the feasibility as well as the optimality of the original solution of LPP.

(*a*) *Conditions for maintaining feasibility of the solution:*

Let $\mathbf{B} = (\mathbf{b}_1, \mathbf{b}_2, ..., \mathbf{b}_m)$, where $\mathbf{b}_j \in \mathbf{A}$; $j = 1, 2, ..., m$ and $\mathbf{B}^{-1} = (\boldsymbol{\beta}_1, \boldsymbol{\beta}_2, ..., \boldsymbol{\beta}_m)$, where $\boldsymbol{\beta}_j$ is the jth column vector of $\mathbf{B}^{-1}$.

Since $\mathbf{a}_k \in \mathbf{B}$, therefore, any change in a_{rk} will affect only the kth column of $\mathbf{B}$. Thus, after making the desired post-optimal change, basis matrix $\mathbf{B}$ becomes

$$\mathbf{B}^* = (\mathbf{b}_1, \mathbf{b}_2, ..., \mathbf{b}_{k-1}, \mathbf{b}_k^*, \mathbf{b}_{k+1}, ..., \mathbf{b}_m)$$

where, $\mathbf{b}_k^* = [b_{1k}, b_{2k}, ..., b_{r-1,k}, b_{rk} + \Delta a_{rk}, b_{r+1,k}, ..., b_{mk}]$ and can be expressed as a linear combination of basis vectors of $\mathbf{B}$. Therefore

$$\mathbf{b}_k^* = \lambda_1 \mathbf{b}_1 + \lambda_2 \mathbf{b}_2 + ... + \lambda_m \mathbf{b}_m = \mathbf{B}\boldsymbol{\lambda},$$

where $\lambda = \{\lambda_1, \lambda_2, ..., \lambda_m\}$.

$$\therefore \quad \lambda = \mathbf{B}^{-1}\mathbf{b}_k^* = \mathbf{B}^{-1}(\mathbf{b}_k + [\mathbf{O}_{r-1,k} \quad \Delta a_{rk} \quad \mathbf{O}_{m-r,k}])$$

$$= \mathbf{B}^{-1}\mathbf{b}_k + \boldsymbol{\beta}_k \Delta a_{rk} = \mathbf{e}_k + \boldsymbol{\beta}_k \Delta a_{rk},$$

since, $$\mathbf{B}^{-1}\mathbf{b}_k = \sum_{i=1}^{m} \boldsymbol{\beta}_i b_{ik} = \mathbf{e}_k$$

Now, for maintaining the feasibility of the new solution, we require

$$\mathbf{x}_B^* = \mathbf{B}^{*-1}\mathbf{b} \geq \mathbf{0},$$

where, $\mathbf{B}^{*-1}$ exists only if $\mathbf{B}^*$ is non-singular. So, we must have $\lambda_k \neq 0$.

We can write $\mathbf{x}_B^* = \mathbf{B}^{*-1}\mathbf{b} = \mathbf{E}\mathbf{B}^{-1}\mathbf{b}$, since $\mathbf{B}^{*-1} = \mathbf{E}\mathbf{B}^{-1}$ and

where
$$\mathbf{E} = \begin{pmatrix} 1 & 0 & \cdots & -\lambda_1/\lambda_k & \cdots & 1 \\ 0 & 1 & \cdots & -\lambda_2/\lambda_k & \cdots & 0 \\ \vdots & \vdots & \cdots & \vdots & \cdots & \vdots \\ 0 & 0 & \cdots & 1/\lambda_k & \cdots & 0 \\ \vdots & \vdots & \cdots & \vdots & \cdots & \vdots \\ 0 & 0 & \cdots & -\lambda_m/\lambda_k & \cdots & 0 \end{pmatrix}, \ \lambda_k \neq 0$$

Now, $\mathbf{b}_k^* = \lambda_1 \mathbf{b}_1 + \lambda_2 \mathbf{b}_2 + \ldots + \lambda_m \mathbf{b}_m$. This implies that

$$\mathbf{b}_k = -\frac{\lambda_1}{\lambda_k}\mathbf{b}_1 - \frac{\lambda_2}{\lambda_k}\mathbf{b}_2 - \ldots + \frac{1}{\lambda_k}\mathbf{b}_k^* - \ldots - \frac{\lambda_m}{\lambda_k}\mathbf{b}_m$$

This gives
$$x_{Bi}^* = \begin{cases} x_{Bi} - \dfrac{\lambda_i}{\lambda_k} x_{Bk} & i \neq k \\ (1/\lambda_k)\, x_{Bk} & i = k \end{cases}$$

But $\lambda_k = 1 + \beta_{kr}\,\Delta a_{rk}$ and $\lambda_i = \beta_{ir}\,\Delta a_{rk}$ for $i \neq k$,

$$\therefore \quad x^*_{Bi} = \begin{cases} x_{Bi} - \dfrac{\beta_{ir}\,\Delta a_{rk}}{1+\beta_{kr}\,\Delta a_{rk}}\, x_{Bk} & i \neq k \\[2ex] \dfrac{x_{Bk}}{1+\beta_{kr}\,\Delta a_{rk}} & i = k \end{cases}$$

For maintaining the feasibility of $\mathbf{x}^*_B$, we must have $\mathbf{x}^*_B \geq 0$

Therefore, $x_{Bi}(1+\beta_{kr}\,\Delta a_{rk}) - \beta_{ir}\,x_{Bk}\,\Delta a_{rk} \geq 0$ or $\Delta a_{rk}(\beta_{kr}\,x_{Bi} - \beta_{ir}\,x_{Bk}) \geq -x_{Bi}$

$$\text{or} \quad \begin{cases} \Delta a_{rk} \geq \dfrac{-x_{Bi}}{\beta_{kr}\,x_{Bi} - \beta_{ir}\,x_{Bk}} & \text{for } \beta_{kr}\,x_{Bi} - \beta_{ir}\,x_{Bk} > 0 \\[2ex] \Delta a_{rk} \leq \dfrac{-x_{Bi}}{\beta_{kr}\,x_{Bi} - \beta_{ir}\,x_{Bk}} & \text{for } \beta_{kr}\,x_{Bi} - \beta_{ir}\,x_{Bk} < 0 \end{cases}$$

Combining these two, we see that in order to maintain the feasibility, Δa_{rk} must not violate the double inequality.

$$\text{Hence,} \quad \text{Max.}\left\{\frac{-x_{Bi}}{Q_i},\; Q_i > 0\right\} \leq \Delta a_{rk} \leq \text{Min.}\left\{\frac{-x_{Bi}}{Q_i},\; Q_i < 0\right\},$$

where, $Q_i = \beta_{kr}\,x_{Bi} - \beta_{ir}\,x_{Bk}$

(*b*) *Conditions for maintaining optimality of the solution*:

For maintaining the optimality of the new solution, we have $z^*_j - c_j \geq 0$, for all $j = 1, 2, ..., n$.

$$\text{Now} \quad z^*_j - c_j = \mathbf{c}_B\,\mathbf{B}^{*-1}\mathbf{a}_j - c_j = \mathbf{c}_B \sum_j \mathbf{y}^*_j - c_j = \sum_{i \neq k}^{m} c_{Bi}\,y^*_{ij} - c_j$$

$$\text{where} \quad y^*_{ij} = \begin{cases} y_{ij} - \dfrac{\Delta a_{rk}\,\beta_{ir}}{1+\beta_{kr}\,\Delta a_{rk}}\, y_{kj}, & i \neq k \\[2ex] \dfrac{y_{kj}}{1+\beta_{kr}\,\Delta a_{rk}}, & i = k \end{cases}$$

$$\therefore \quad z^*_j - c_j = \sum_{i=1}^{m} c_{Bi}\left\{y_{ij} - \frac{\beta_{ir}\,\Delta a_{rk}}{1+\beta_{kr}\,\Delta a_{rk}}\, y_{kj}\right\} - c_j$$

$$= (z_j - c_j) - \left\{\frac{y_{kj}\,\Delta a_{rk}\,\mathbf{c}_B\boldsymbol{\beta}_r}{1+\beta_{kr}\,\Delta a_{rk}}\right\}, \quad \text{since } \mathbf{c}_B\,\boldsymbol{\beta}_r = \sum_i c_{Bi}\,\beta_{ir}$$

Thus, the condition of optimality requires

$$z^*_j - c_j \geq 0, \quad i.e., \quad (z_j - c_j) - \frac{y_{kj}\,\Delta a_{rk}\,\mathbf{c}_B\,\boldsymbol{\beta}_r}{1+\beta_{kr}\,\Delta a_{rk}} \geq 0$$

$$\text{or} \quad (z_j - c_j) + \Delta a_{rk}\{(z_j - c_j)\beta_{kr} - y_{kj}\,\mathbf{c}_B\,\boldsymbol{\beta}_r\} \geq 0$$

This implies that

$$\Delta a_{rk} \geq \frac{-(z_j - c_j)}{(z_j - c_j)\beta_{kr} - y_{kj}\,\mathbf{c}_B\,\boldsymbol{\beta}_r}, \quad (z_j - c_j)\beta_{kr} - y_{kj}\,\mathbf{c}_B\,\boldsymbol{\beta}_r > 0$$

$$\text{and} \quad \Delta a_{rk} \leq \frac{-(z_j - c_j)}{(z_j - c_j)\beta_{kr} - y_{kj}\,\mathbf{c}_B\,\boldsymbol{\beta}_r}, \quad (z_j - c_j)\beta_{kr} - y_{kj}\,\mathbf{c}_B\,\boldsymbol{\beta}_r < 0$$

Hence, in order to maintain the optimality of the new solution, Δa_{rk} must satisfy

$$\text{Max.}\left\{\frac{-(z_j-c_j)}{Q_j},\ Q_j>0\right\}\le \Delta a_{rk}\le \text{Min.}\left\{\frac{-(z_j-c_j)}{Q_j},\ Q_j<0\right\},$$

where, $Q_j = (z_j - c_j)\beta_{kr} - y_{kj}\, \mathbf{c}_B\, \boldsymbol{\beta}_r$.

Note. The sub-script r in $\mathbf{B}^{-1}$ stands for column whereas k stands for row.

Example 6.5. *Given the following LPP:*

$$\text{Maximize } z = 45x_1 + 100x_2 + 30x_3 + 50x_4 \text{ subject to the constraints:}$$
$$7x_1 + 10x_2 + 4x_3 + 9x_4 \ge 1200,$$
$$3x_1 + 40x_2 + x_3 + x_4 \ge 0;$$
$$x_1, x_2, x_3, x_4 \ge 0.$$

The optimum solution of this LPP is displayed in the simplex tableau given below:

Basic variable	x_1	x_2	x_3	x_4	s_1	s_2	z	*Solution* (x_{Bi})
x_3	5/3	0	1	7/3	4/15	–1/15	0	800/3
x_2	1/30	1	0	–1/30	–1/150	2/75	0	40/3
$(z_j - c_j)$	25/3	0	0	55/3	22/3	2/3	1	z = 28,000/3

Discuss the effect of discrete change in the activity coefficient of variable x_1, when [7 3] changes to [7 5] and also when [7 3] changes to [5 8].

Solution. Since x_1 is non-basic variable in the final simplex tableau, current solution remains feasible, but it may not be an optimum, unless $(z_1^* - c_1) \ge 0$, where $z_1^* = \mathbf{c}_B\,\mathbf{B}^{-1}\,\mathbf{a}_1^*$

When $\mathbf{a}_1^* = [7\ \ 5]$, we get

$$z_1^* = \mathbf{c}_B\,\mathbf{B}^{-1}\,\mathbf{a}_1^* = (30\ \ 100)\begin{bmatrix} 4/15 & -1/15 \\ -1/150 & 2/75 \end{bmatrix}[7\ \ 5]$$

$$= \begin{pmatrix} \frac{22}{3} & \frac{2}{3} \end{pmatrix}[7\ \ 5] = \frac{164}{3}$$

$$\therefore \qquad z_1^* - c_1 = \frac{164}{3} - 45 = \frac{29}{3} > 0$$

As $(z_1^* - c_1) > 0$, current solution remains optimum for the revised LPP also.

Further, when $\mathbf{a}_1^* = [5\ \ 8]$, we get

$$z_1^* = \mathbf{c}_B\,\mathbf{B}^{-1}\,\mathbf{a}_1^* = \begin{pmatrix} \frac{22}{3} & \frac{2}{3} \end{pmatrix}[5\ \ 8] = \frac{126}{3}$$

$$\therefore \qquad z_1^* - c_1 = \frac{126}{3} - 45 = -3 < 0$$

In this case, since, $(z_1^* - c_1) < 0$, current solution is not optimum and can be improved. For this, we compute the coefficients of variable x_1 in the optimum simplex tableau and then use regular simplex method to obtain optimum solution. Here,

$$\mathbf{B}^{-1}\,\mathbf{a}_1^* = \begin{bmatrix} 4/15 & -1/15 \\ -1/150 & 2/75 \end{bmatrix}\begin{bmatrix} 5 \\ 8 \end{bmatrix} = \begin{bmatrix} 4/5 \\ 27/150 \end{bmatrix} \text{ and } z_1^* = (30\ \ 100)\begin{bmatrix} 4/5 \\ 27/150 \end{bmatrix} = 42$$

Modified final simplex tableau, given above, therefore, is as follows:

Basic variable	x_1	x_2	x_3	x_4	s_1	s_2	z	Solution
x_3	4/5	0	1	7/3	4/15	–1/15	0	800/3
$\leftarrow x_2$	(27/150)	1	0	–1/30	–1/150	2/75	0	40/3
$(z_j - c_j)$	–3 ↑	0	0	55/3	22/3	2/3	1	28,000/3

Clearly, non-basic variable x_1 enters the basis and x_2 leaves the basis. Thus, the new simplex tableau will be:

Basic variable	x_1	x_2	x_3	x_4	s_1	s_2	z	Solution
x_3	0	–40/9	1	67/27	8/27	–5/27	0	5,600/27
x_1	1	50/9	0	–5/7	–1/27	4/27	0	2,000/27
$(z_j - c_j)$	0	50/3	0	145/9	65/9	10/9	1	86,000/9

Since, all $(z_j - c_j) \geq 0$, the solution obtained is optimum, *viz.*, $x_4 = 0, x_3 = 5{,}600/27, x_2 = 0$, and $x_1 = 2{,}000/27$ with maximum $z = 86{,}000/9$.

Example 6.6. *Given the linear programming problem:*

$$\text{Maximize } z = 3x_1 + 4x_2 + x_3 + 7x_4 \text{ subject to the constraints:}$$
$$8x_1 + 3x_2 + 4x_3 + x_4 \leq 7, \quad 2x_1 + 6x_2 + x_3 + 5x_4 \leq 3,$$
$$x_1 + 4x_2 + 5x_3 + 2x_4 \leq 8; \; x_1, x_2, x_3, x_4 \geq 0.$$

The following simplex tableau gives the optimum solution:

Basic variable	x_1	x_2	x_3	x_4	s_1	s_2	s_3	z	Solution
x_1	1	9/38	1/2	0	5/38	–1/38	0	0	16/19
x_4	0	21/19	0	1	–1/19	8/38	0	0	5/19
s_3	0	59/38	9/2	0	–1/38	–15/38	1	0	126/19
$(z_j - c_j)$	0	169/38	1/2	0	1/38	53/38	0	1	83/19

Discuss the effect of discrete changes in the activity coefficient a_{ij} of **A** *on the current optimum basic feasible solution to the given LPP.*

Solution. From the above simplex tableau, we have

$$\mathbf{B}^{-1} = \begin{pmatrix} \boldsymbol{\beta}_1 & \boldsymbol{\beta}_2 & \boldsymbol{\beta}_3 \\ 5/38 & -1/38 & 0 \\ -1/19 & 8/38 & 0 \\ -1/38 & -15/38 & 1 \end{pmatrix}, \text{ where } \mathbf{B} \text{ is the basis matrix for the given LPP.}$$

Thus, we have

$$\mathbf{c}_B \boldsymbol{\beta}_1 = 3(5/38) + 7(-1/19) + 0(-1/38) = 1/38$$
$$\mathbf{c}_B \boldsymbol{\beta}_2 = 3(-1/38) + 7(8/38) + 0(-15/38) = 53/38$$
$$\mathbf{c}_B \boldsymbol{\beta}_3 = 3(0) + 7(0) + 0(1) = 0$$

Let the element a_{rk} of **A** be changed to $a^*_{rk} = a_{rk} + \Delta a_{rk}$.

Case 1. a_{rk} is not in the basis **B**.

In the optimum simplex tableau; $\mathbf{y}_2, \mathbf{y}_3, \mathbf{y}_5$ and $\mathbf{y}_6$ are not in the basis, therefore, the ranges for discrete changes in the coefficient a_{ij} corresponding to these non-basis coloumns are given by the following inequalities:

$$\frac{-169/38}{1/38} \le \Delta a_{12} \quad \Rightarrow \quad \Delta a_{12} \ge -169,$$

$$\frac{-169/38}{53/38} \le \Delta a_{22} \quad \Rightarrow \quad \Delta a_{22} \ge -169/53,$$

$$\frac{-169/38}{0} \le \Delta a_{32} < \infty \quad \Rightarrow \quad -\infty \le \Delta a_{32} < \infty,$$

$$\frac{-1/2}{1/38} \le \Delta a_{13} \quad \Rightarrow \quad \Delta a_{13} \ge -19,$$

$$\frac{-1/2}{53/38} \le \Delta a_{23} \quad \Rightarrow \quad \Delta a_{23} \ge -19/53,$$

$$\frac{-1/2}{0} \le \Delta a_{33} < \infty \quad \Rightarrow \quad -\infty \le \Delta a_{33} < \infty,$$

$$\therefore \quad \begin{array}{lll} a_{12} \ge 6-169 & \text{or} & a_{12} \ge -163; \\ a_{22} \ge \dfrac{-169}{53}+6 & \text{or} & a_{22} \ge \dfrac{149}{53}; \text{ and } -\infty \le a_{32} < \infty, \\ a_{13} \ge 4-19 & \text{or} & a_{13} \ge -15; \\ a_{23} \ge 1-\dfrac{19}{53} \text{ or } a_{23} \ge \dfrac{34}{53}; & \text{and} & -\infty \le a_{33} < \infty \end{array}$$

Case 2. a_{rk} *is in the basis* **B**.

Since, $\mathbf{y}_1$, $\mathbf{y}_4$ and $\mathbf{y}_7$ are in the basis, therefore, any discrete change in a_{rk} belonging to any of these vectors may affect feasibility as well as optimality of the original optimum basic feasible solution $\mathbf{x}_B$.

Consider the discrete changes in a_{rk} belonging to $\mathbf{y}_4 = \boldsymbol{\beta}_2$, we have

$$\beta_{22}(z_2-c_2)-y_{22}\,\mathbf{c}_B\,\boldsymbol{\beta}_2 = \frac{8}{38}\times\frac{169}{38}-\frac{21}{19}\times\frac{53}{38} = -\frac{23}{38}$$

$$\beta_{22}(z_3-c_3)-y_{23}\,\mathbf{c}_B\,\boldsymbol{\beta}_2 = \frac{8}{38}\times\frac{1}{2}-0\times\frac{53}{38} = \frac{2}{19}$$

$$\beta_{22}(z_5-c_5)-y_{25}\,\mathbf{c}_B\,\boldsymbol{\beta}_2 = \frac{8}{38}\times\frac{1}{38}-\frac{-1}{19}\times\frac{53}{38} = \frac{3}{38}$$

and

$$\beta_{22}(z_6-c_6)-y_{26}\,\mathbf{c}_B\,\boldsymbol{\beta}_2 = \frac{8}{38}\times\frac{53}{38}-\frac{8}{38}\times\frac{53}{38} = 0$$

The range for the discrete change in the element a_{24} in order to maintain the optimality of the solution is, therefore, given by

$$\text{Max.}\left\{\frac{-1/2}{2/19}, \frac{-1/38}{3/38}\right\} \le \Delta a_{24} \le \text{Min.}\left\{\frac{-169/38}{-23/38}\right\}$$

i.e.,

$$-\frac{1}{3} \le \Delta a_{24} \le \frac{169}{23}$$

Further, since for the elements a_{24}, $r = 2$ and $k = 4$, therefore, we have

$$x_{B1}\,\beta_{22}-x_{B2}\,\beta_{31} = \frac{16}{19}\times\frac{8}{38}-\frac{5}{19}\times\frac{-1}{38} = \frac{7}{38}$$

$$x_{B3}\,\beta_{22}-x_{B2}\,\beta_{32} = \frac{126}{19}\times\frac{8}{38}-\frac{5}{19}\times\frac{-15}{38} = \frac{3}{2}$$

The range for the discrete change in the element a_{24} in order to maintain the feasibility of the solution is, therefore, given by

$$\text{Max.}\left\{\frac{-16/19}{7/38}, \frac{-126/19}{3/2}\right\} \le \Delta a_{24} \text{ or } \frac{-84}{19} \le \Delta a_{24}$$

$$\therefore \quad 5-\frac{84}{19} \le a_{24} < \infty \text{ or } \frac{11}{19} \le a_{24} < \infty \qquad (\text{since } a_{24} = 5)$$

Similarly, the ranges for changes in the elements a_{14}, a_{34}, etc., can also be obtained.

PROBLEMS

6.09. Given the following LPP:

$$\text{Maximize } z = 4x_1 + 3x_2 + 4x_3 + 6x_4 \text{ subject to the constraints:}$$
$$x_1 + 2x_2 + 2x_3 + 4x_4 \le 8, \; 2x_1 + 2x_3 + x_4 \le 6,$$
$$3x_1 + 3x_2 + x_3 + x_4 \le 8; \; x_1, x_2, x_3, x_4 \le 0.$$

(*a*) Determine the separate ranges for discrete changes in a_{12}, a_{22} and a_{23} consistent with the optimum solution of the given LPP.

(*b*) If a_{11} is changed to $a_{11} + \Delta a_{11}$, determine the limits for discrete change of Δa_{11} so as to maintain the optimality of the current optimum solution.

6.10. Given the following LPP:

$$\text{Maximize } z = 10x_1 + 3x_2 + 6x_3 + 5x_4 \text{ subject to the constraints:}$$
$$x_1 + 2x_2 + x_4 \le 6, \; 3x_1 + 2x_3 \le 5,$$
$$x_2 + 4x_3 + 5x_4 \le 3; \; x_1, x_2, x_3, x_4 \ge 0.$$

(*a*) Determine an optimum solution to the problem.

(*b*) If the element a_{11} is changed to $a_{11} + \Delta a_{11}$, determine the limits for discrete change of Δa_{11} so as to maintain the optimality of the current optimum solution.

(*c*) Determine the separate ranges for discrete changes in a_{13}, a_{23} and a_{33} consistent with the optimality of the solution obtained in (*a*).

6:5. STRUCTURAL CHANGES

This section deals with the post-optimal effects on an LPP, when

(*i*) a new activity or constraint is introduced.

(*ii*) one of the current activity or constraint is deleted.

Addition of a New Activity

The addition of a new activity in an LPP is equivalent to adding a new variable. Intuitively, the addition of new activity is desirable only if it is profitable — that is, if it improves the optimum value of the objective function. This condition can be checked by computing $(z_j - c_j)$, where $z_j = \mathbf{B}^{-1}\mathbf{a}_j$ for the new activity, where $\mathbf{B}^{-1}\mathbf{a}_j$ and c_j represent the resource usages and profit per unit of the new activity. If the computed $(z_j - c_j) \ge 0$, then the new activity is not desirable. Otherwise, the new activity is profitable and must be brought into the basis.

Example 6.7. *Given the following LPP:*

$$\textit{Maximize } z = 3x_1 + 5x_2 \textit{ subject to the constraints:}$$
$$x_1 \le 4, \; 3x_1 + 2x_2 \le 18 \textit{ and } x_1, x_2 \ge 0.$$

If a new variable x_3 is introduced, with $c_3 = 7$ and $\mathbf{a}_3 = [1, 2]$; discuss the effect of adding the new variable and obtain the revised solution, if any.

Solution. Introducing slack variables $s_1 \geq 0, s_2 \geq 0$ and then solving by simplex method; the optimum solution is displayed in the tableau given below:

Coeff. of basic variable	*Basic variable*	x_1	x_2	s_1	s_2	z	*Solution* (x_{Bi})
0	s_1	1	0	1	0	0	4
5	x_2	3/2	1	0	1/2	0	9
	$(z_j - c_j)$	9/2	0	0	5/2	1	$z = 45$

From the above simplex tableau, we observe that

$$x_B = [4, 9] \text{ and } \mathbf{B}^{-1} = \begin{pmatrix} 1 & 0 \\ 0 & 1/2 \end{pmatrix}$$

Since, a new variable x_3 is introduced, we compute

$$\mathbf{B}^{-1}\mathbf{a}_3 = \begin{pmatrix} 1 & 0 \\ 0 & 1/2 \end{pmatrix} [1, 2] = [1, 1]$$

and
$$(z_5 - c_5) = (0, 5)\,[1, 1] - 7 = -2$$

This indicates that the condition of optimality is violated and hence, it is profitable to bring the new activity (non-basic variable) into basis. Therefore, a new (improved) optimum solution can easily be obtained by entering the non-basic variable x_3 into basis, *viz.*,

Coeff. of basic variable	*Basic variable*	x_1	x_2	s_1	s_2	x_3	z	*Solution* (x_{Bi})
← 0	s_1	1	0	1	0	(1)	0	4
5	x_2	3/2	1	0	1/2	1	0	9
	$(z_j - c_j)$	9/2	0	0	5/2	−2 ↑	1	$z = 45$

Final Iteration. Optimum solution.

Coeff. of basic variable	*Basic variable*	x_1	x_2	s_1	s_2	x_3	z	*Solution* (x_{Bi})
7	x_3	1	0	1	0	1	0	4
5	x_2	1/2	1	−1	1/2	0	0	5
	$(z_j - c_j)$	13/2	0	2	5/2	0	1	$z = 53$

Optimum solution is

$$x_1 = 0, \; x_2 = 5, \text{ and } x_3 = 4 \text{ with maximum } z = 53$$

Remark. It is of interest to observe that if we have $c_3 = 3$ instead of 7, then we would have $z_5 - c_5 = 2 > 0$. Thus, the optimality of the given problem is not affected by post-optimal addition of variable x_3, *i.e.*, it would not have been profitable to bring x_3 into basis.

Addition of a New Constraint

Addition of a new constraint may or may not affect the feasibility of the current optimum solution. For this, it is sufficient to check whether new constraint is satisfied by the current optimum solution or not. If it is satisfied, the inclusion of the constraint has no effect on the current optimum solution, *i.e.*, it remains feasible as well as optimum. If, however, the constraint is not satisfied, the current optimum solution becomes infeasible. Dual simplex method is then used to find the new optimum solution. The concept is best explained through an example.

Example 6.8. *Consider the following table which presents an optimal solution to some linear programming problem:*

Coeff. of basic variable	Basic variable	x_1	x_2	x_3	x_4	x_5	s_1	s_2	s_3	z	Solution
	$c_j \rightarrow$	2	4	1	3	2	0	0	0		
2	x_1	1	0	0	–1	0	0.5	0.2	–1	0	3
4	x_2	0	1	0	2	1	–1	0	0.5	0	1
1	x_3	0	0	1	–1	–2	5	–0.3	2	0	7
	$(z_j - c_j)$	0	0	0	2	0	2	0.1	2	1	$z = 17$

If the additional constraint $2x_1 + 3x_2 - x_3 + 2x_4 + 4x_5 \leq 5$ were annexed to the system, would there be any change in the optimal solution? Justify your answer.

Solution. The optimum solution from the above table is,

$$x_1 = 3, \ x_2 = 1 \ \text{ and } \ x_3 = 7; \qquad \textit{(basic)}$$

and

$$x_4 = x_5 = x_6 = x_7 = x_8 = 0. \qquad \textit{(Non-basic)}$$

This also satisfies the new additional constraint,

$$2x_1 + 3x_2 - x_3 + 2x_4 + 4x_5 \leq 5$$

Thus, the optimum basic feasible solution of the original linear programming problem does not change, if we introduce the above constraint to it.

Hence, the additional constraint is redundant and the optimum solution of the given LPP is also optimum for the new LPP.

Remarks

1. If the right-hand side of the additional constraint is 1 instead 5, then the optimum solution does not satisfy the additional constraint.

   In this case the modified optimum simplex tableau is obtained by inserting the entries corresponding to the new additional constraint. Dual simplex method is then used to obtain the revised optimum solution. It may be observed that the optimum value of the objective function remains same, while the new optimum basic feasible solution will be

   $$x_1 = 3, \ x_2 = 0, \ x_3 = 9 \ \text{ and } \ x_5 = 1$$

2. If the new constraint added is an equation and an artificial variable appears in the basis of the new (revised) LPP then following cases may arise:
   - (*i*) If an artificial basis appears in the basis at negative value, a zero cost may be assigned to it. Then apply dual simplex method.
   - (*ii*) If the artificial variable appears in the basis at a positive value, then cost '– *M*' may be assigned to it and apply usual simplex method to obtain an optimum solution.

Deletion of a Variable

Deletion of a non-basic variable is a totally superfluous operation and does not affect the feasibility and/or optimality of the current optimum solution. However, if the variable to be deleted is a basic one, it may affect the optimality and a new optimum solution may have to be found out. For this, we assign a cost '–*M*' to the basic variable (to be deleted) and then apply simplex method, taking the modified current optimum simplex tableau as the starting simplex tableau.

Deletion of a Constraint

The constraint to be deleted may be either binding or unbinding on the optimum solution. The deletion of an unbinding constraint can only enlarge the feasible region, but will not affect the optimum solution. This can be easily verified graphically. Moreover, if the constraint under consideration has a slack or surplus variable of zero value in the basis matrix, it cannot be binding and hence will not affect the optimum solution.

The deletion of a binding constraint will, however, cause post-optimality problem. The simple way to proceed in this case is via the addition of one or two new variables. For example, the constraint

$$a_{i1}x_1 + a_{i2}x_2 + \dots + a_{in}x_n \le b_i$$

can be written as $a_{i1}x_1 + a_{i2}x_2 + \dots + a_{in}x_n + x_{n+1} - x_{n+2} = b_i$; $x_{n+1} \ge 0$ and $x_{n+2} \ge 0$

where x_{n+1} and x_{n+2} are slack and surplus variables respectively. The problem can then be solved by the procedure laid down earlier for addition of new variable.

Example 6.9. *Let the variable x_2 be deleted from the LPP of the* ***Example 6.8****. Obtain an optimum solution to the resulting LPP.*

Solution. Since the variable to be deleted is a basic variable, we assign a cost $-M$ to x_2 and consider the optimum simplex tableau as the initial simplex tableau for the new LPP. Thus, we have

Initial Iteration. Introduce x_4 and drop x_2.

		2	$-M$	1	3	2	0	0	0		
Coeff. of basic variable	*Basic variable*	x_1	x_2	x_3	x_4	x_5	s_1	s_2	s_3	z	*Solution*
2	x_1	1	0	0	-1	0	0.5	0.2	-1	0	3
$-M$	$\leftarrow x_2$	0	1	0	(2)	1	-1	0	0.5	0	1
1	x_3	0	0	1	-1	-2	5	-0.3	2	0	7
	$(z_j - c_j)$	0	0	0	$-2M - 6\uparrow$	$-M - 4$	$M + 6$	0.1	$-0.5M + 1$	1	$-M + 13$

First Iteration. Introduce x_5 and drop x_4.

Coeff. of basic variable	*Basic variable*	x_1	x_2	x_3	x_4	x_5	s_1	s_2	s_3	z	*Solution*
2	x_1	1	1/2	0	0	1/2	0	0.2	$-3/4$	0	7/2
3	$\leftarrow x_4$	0	1/2	0	1	(1/2)	$-1/2$	0	1/4	0	1/2
1	x_3	0	1/2	1	0	$-3/2$	9/2	-0.4	9/4	0	15/2
	$(z_j - c_j)$	0	$M + 3$	0	0	$-1\uparrow$	3	0.1	3/2	1	$z = 16$

Final Iteration. Optimum solution.

Coeff. of basic variable	*Basic variable*	x_1	x_2	x_3	x_4	x_5	s_1	s_2	s_3	z	*Solution*
2	x_1	1	0	0	-1	0	0.5	0.2	-1	0	3
2	x_5	0	1	0	2	1	-1	0	0.5	0	1
1	x_3	0	2	1	3	0	3	-0.3	3	0	9
	$(z_j - c_j)$	0	$M + 4$	0	2	0	2	0.1	2	1	$z = 17$

Optimum solution obtained above is:

$$x_1 = 3,\ x_2 = 0,\ x_3 = 9,\ x_4 = 0 \text{ and } x_5 = 1;\ \text{maximum } z = 17.$$

PROBLEMS

6.11. Let the optimum simplex tableau for a maximization problem (with one of the constraints of ≤ type) be:

Coeff. of basic variable	*Basic variable*	x_1	x_2	x_3	s_1	A	z	*Solution*
12	x_2	0	1	–1/5	2/5	–1/5	0	8/5
5	x_1	1	0	7/5	1/5	2/5	0	9/5
	$(z_j - c_j)$	0	0	3/5	29/5	$M - 2/5$	1	141/5

where, s_1 is slack and A an artificial variable. Let a new variable $x_4 \geq 0$ be introduced in the LPP with a cost 30 assigned to it in the objective function. Coefficient of x_4 in the two constraints are 5 and 7 respectively.

Discuss the effect of adding the new variable and obtain the revised solution, if any.

6.12. Consider the LPP:

$$\text{Maximize } z = 3x_1 + 4x_2 + x_3 + 7x_4 \text{ subject to the constraints:}$$
$$8x_1 + 3x_2 + 4x_3 + x_4 \leq 7, \quad 2x_1 + 6x_2 + x_3 + 5x_4 \leq 3,$$
$$x_1 + 4x_2 + 5x_3 + 2x_4 \leq 8, \text{ and } x_1, x_2, x_3, x_4 \geq 0.$$

(*a*) Discuss the effect of discrete changes in c_j on the optimality of the optimum basic feasible solution.

(*b*) Let the variable x_4 be deleted from the LPP of the linear programming problem. Obtain an optimum solution to the resulting LPP.

(*c*) Let a linear constraint $2x_1 + 3x_2 + x_3 + 5x_4 \leq 4$ be added to the constraints of the problem. Check whether there is any change in the optimum solution of the original LPP.

(*d*) Also, discuss the case when the upper limit of the above constraint is reduced to 2.

6:6. MANAGERIAL PROBLEMS AND POST-OPTIMAL ANALYSIS

While solving linear programming models of managerial problems, sometimes discrepancy in data is found after obtaining optimum solution. This section deals as to how to remove these discrepancies, while retaining the optimum solution of such problems. Given below are few examples to illustrate such discrepancies.

Example 6.10. *Three products A, B and C are manufactured by an enterprise for which the data is:*

		Product A	*Product B*	*Product C*
Material required (kg/unit)	:	5	4	3
Labour hours require per unit	:	3	2	4
Machine hours required per unit	:	2	3	1

If 2500 kg of material, 2100 labour hours and 1275 hours of machine time is available, find the optimum mix if the profit per unit is ₹32, ₹30 and ₹40 for products A, B and C respectively.

Discuss the effect on the optimum solution, if

(*a*) *machine hours are 1350 instead of 1275,*

(*b*) *labour hours are 2000 instead of 2100,*

(*c*) *it is essential to produce 10 units of product A.*

Solution. Let x_1, x_2 and x_3 are the number of units of product *A*, *B* and *C* to be produced, respectively. Then, using the given data, linear programming model formulation is:

$$\text{Maximize } z = 32x_1 + 30x_2 + 40x_3 \text{ subject to the constraints:}$$
$$5x_1 + 4x_2 + 3x_3 \le 2{,}500,\quad 3x_1 + 2x_2 + 4x_3 \le 2{,}100,$$
$$2x_1 + 3x_2 + x_3 \le 1{,}275;\; x_1 \ge 0, x_2 \ge 0 \text{ and } x_3 \ge 0.$$

Introducing slack variables $s_1 \ge 0, s_2 \ge 0, s_3 \ge 0$ and then solving the LPP by simplex method, the optimum solution of the LPP is displayed in the following simplex tableau:

Coeff. of basic variable	*Basic variable*	x_1	x_2	x_3	s_1	s_2	s_3	z	*Solution*
0	s_1	3/2	0	0	1	–1	–1/2	0	125
40	x_3	1/2	0	1	0	–1/5	3/10	0	375
30	x_2	1/2	1	0	0	2/5	–1/10	0	300
	$(z_j - c_j)$	3	0	0	0	4	9	1	z = 24,000

Clearly, the optimum solution is: $x_1 = 0$, $x_2 = 300$, $x_3 = 375$, $s_1 = 125$ and maximum $z = 24{,}000$.

From this solution we observe that $s_1 = 125$, while $s_2 = 0$ and $s_3 = 0$. This implies that machine hours and labour hours are fully utilized, while 125 kg of material still unutilized. Now, if

(*a*) one machine hour is increased, then the column under s_2 shows that one kg of material will remain unconsumed and product *B* will increase by 2/5 unit, while product *C* will reduce by 1/5 unit. Thus, if machine hours available are increased by 75 (= 1,350 – 1,275) hours, the product *B* will increase by (2/5) × 75, *i.e.*, 30 units and product *C* will decrease by 1/5 × 75, *i.e.*, 15 units. Thus, the optimum solution will be: x_2 = 300 + 30 (= 330) and x_3 = 375 – 15, *i.e.*, 360. For every additional machine hour used, there is a gain of ₹ 4 and hence profit will increase by ₹ 75 × 4 (= ₹ 300) to get a value of z = ₹ (0 × 125 + 40 × 360 + 30 × 330), *i.e.*, ₹ 24,300.

(*b*) one labour hour is decreased, then the column under s_3 shows that it would change the value of x_2 by (–1) × (–1/10), *i.e.*, 1/10 unit and x_3 by (–1) × (3/10), *i.e.*, (–3/10) units. Therefore, a reduction of 100 (2,100 – 2,000) labour hours will increase the value of x_2 by (1/10) × 100, *i.e.*, 10 units and decrease the value of x_3 by (3/10) × 100, *i.e.*, 30 units. The new optimum solution would be: $x_1 = 0$, $x_2 = 310$ units and $x_3 = 345$ units. Value of z will become ₹ 23,100.

(*c*) one unit of product *A* is produced, it would reduce the product of *B* and *C* each by 1/2 unit. Therefore, if x_1= 10 units, then x_2 = 300 – (1/2) × 10, *i.e.*, 295 units and x_3 = 375 – (1/2) × 10, *i.e.*, 370 units. The profit contribution of ₹ 3 by product *A* (net evaluation under x_1-column) will also reduces and hence it will reduce to ₹ 30.

Therefore, the value of z will be ₹ (32 × 10 + 30 × 295 + 40 × 370), *i.e.*, ₹ 23,970.

Example 6.11. *A firm engaged in manufacturing office furniture provides the following information regarding resource consumption, availability, and profit contribution:*

Resources	*Usage per unit*			*Daily availability*
	Tables	*Chairs*	*Bookcases*	
Timber (cu ft)	8	4	3	640
Assembly department (man-hours)	4	6	2	540
Finishing department (man-hours)	1	1	1	100
Profit contribution per unit (₹)	30	20	12	
Minimum production requirement	0	50	0	

The firm wants to determine its optimal product-mix.

(a) Formulate a linear programming problem with the help of above data.

(b) Solve the linear programming problem by simplex method and find the optimal product-mix, and total maximum profit contribution.

(c) Identify the shadow prices of resources.

(d) Examine the sensitivity of the optimal solution to the changes in profit contribution per unit for chairs, tables and bookcases.

(e) What other information can be obtained from the optimal solution of the LPP?

[M.Com., Delhi Univ. 2003]

Solution. (*a*) Let the daily output of tables, chairs, and bookcases be x_1, x_2 and x_3, respectively. With the given contribution rates, resource requirements, and resource availability, the problem can be expressed as an LPP as follows:

$$\text{Maximize } z = 30x_1 + 20x_2 + 12x_3 \quad \text{subject to the constraints:} \qquad \textit{(Total contribution)}$$

$$8x_1 + 4x_2 + 3x_3 \le 640 \qquad \textit{(Timber availability)}$$

$$4x_1 + 6x_2 + 2x_3 \le 540 \qquad \textit{(Labour: Assembly)}$$

$$x_1 + x_2 + x_3 \le 100 \qquad \textit{(Labour: Finishing)}$$

$$x_2 \ge 50 \qquad \textit{(Minimum production output)}$$

$$x_1, x_2, x_3 \ge 0 \qquad \textit{(Non-negativity condition)}$$

(*b*) Introducing necessary slack, surplus, and artificial variables, we have

$$\text{Maximize } z = 30x_1 + 20x_2 + 12x_3 + 0 \cdot s_1 + 0 \cdot s_2 + 0 \cdot s_3 + 0 \cdot s_4 - MA_1$$

subject to the constraints:

$$8x_1 + 4x_2 + 3x_3 + s_1 = 640,$$

$$4x_1 + 6x_2 + 2x_3 + s_2 = 540,$$

$$x_1 + x_2 + x_3 + s_3 = 100,$$

$$x_2 - s_4 + A_1 = 50;$$

$$x_1, x_2, x_3, s_1, s_2, s_3, s_4, A_1 \ge 0.$$

Now, using simplex method, an optimum simplex tableau is obtained as follows:

Coeff. of basic variable	*Basic variable*	x_1	x_2	x_3	s_1	s_2	s_3	s_4	A_1	*Solution*
0	s_1	0	0	–5	1	0	–8	–4	4	40
0	s_2	0	0	–2	0	1	–4	2	–2	40
30	x_1	1	0	1	0	0	1	1	–1	50
20	x_2	0	1	0	0	0	0	–1	1	50
	$(z_j - c_j)$	0	0	18	0	0	30	10	$M - 10$	2500

The optimum basic feasible solution is:

$$x_1 = 50, x_2 = 50 \text{ and } x_3 = 0; \text{ Maximum } z = 2{,}500.$$

Thus, in order to achieve the maximum profit of ₹ 2,500, the firm should produce 50 chairs and 50 tables only.

(*c*) Shadow prices of the resources are: Timber: Nil, Assembly Department = Nil, and Finishing Department = ₹ 30 per hour.

(*d*) Sensitivity of optimal solution to changes in profit contributions:

For variables x_1 and x_2, which are in the basis, we have

$(z_j - c_j)$	0	0	18	0	0	30	10	$(M - 10)$
Coeff. of basic variable								
x_1-row	1	0	1	0	0	1	1	–1
Quotient	0	—	18	—	—	30	10	$(-M + 10)$
	Least positive = 10			*and*			*Least negative* = $-M + 10$	
$(z_j - c_j)$	0	0	18	0	0	30	10	$M - 10$
Coeff. of basic variable								
x_2-row	0	1	0	0	0	0	–1	1
Quotient	—	0	—	—	—	—	(–10)	$(M - 10)$
	Least positive = $M - 10$			*and*			*Least negative* = –10	

Accordingly, the range of values for which current solution remains unchanged are given as under:

Item	*Variable*	*Present contribution*	*Minimum*	*Maximum*
Table	x_1	30	20	∞
Chair	x_2	20	– ∞	30
Bookcase	x_3	12	– ∞	30

Further, since x_3 is not in the basis, and it has a $(z_j - c_j)$ value equal to +18, it implies that to be included in the basis, its c_j value should be increased by greater than 18. Thus, the present solution would remain unchanged up to 12 + 18, *i.e.*, 30.

(*e*) The optimal solution calls for producing only tables and chairs, and not bookcases. Each bookcase produced would lead to a reduction of ₹ 18 of profit, as given by the $(z_j - c_j)$ value under x_3. This is because each unit of x_3 requires foregoing one unit of x_1, as evident from the final simplex tableau and accordingly, a gain of ₹ 12 is accompanied by a loss of ₹ 30.

Further, $(z_j - c_j)$ value under s_4 has the implication that production of one unit of x_2 in excess of the minimum requirement of 50 would lead to a loss of ₹ 10 per unit in the total profit. As indicated by the substitution ratios in the table, an increase of one unit of s_4 would entail foregoing one unit of x_1 (resulting in a loss of ₹ 30) and a gain of one unit of x_2 (giving a profit of ₹ 20).

Finally, we can obtain the optimum value of dual variables from the optimum solution. The given problem being a three variable and four constraint problem, the dual problem would have four variables and three constraints. The optimum values of the dual variables are: $y_1 = 0$, $y_2 = 0$, $y_3 = 30$ and $y_4 = -10$, assuming y_1, y_2, y_3 and y_4 being the dual variables. With these, the objective function value of the dual is:

$$640 \times 0 + 540 \times 0 + 100 \times 30 - 50 \times 10 = 2{,}500,$$

Observe that this is same as that of primal linear programming problem.

Example 6.12. *Cookwell, a manufacturer of large-size pressure cookers, produces and sells three models of cookers. While market demands pose no restriction, the capacity to produce is currently constrained by the limited supplies of special grade aluminium limited to 1,500 kg per week and machine processing time limited to 1,200 hours per week. In order to determine the optimal product-mix to maximize weekly contribution, a linear programme model as under was formulated:*

Maximize $z = 60x_1 + 40x_2 + 80x_3$ *subject to the constraints:*

$6x_1 + 3x_2 + 5x_3$ *less than equal to 1,500* *(Aluminium)*

$3x_1 + 4x_2 + 5x_3$ *less than equal to 1,200* *(Machine hours)*

x_1, x_2, x_3 *greater than equal to 0.*

Using simplex method, the following tableau was obtained:

Coeff. of basic variable	*Basic variable*	x_1	x_2	x_3	s_1	s_2	z	*Solution*
*	x_1	1	–1/3	0	1/3	–1/3	0	100
*	x_3	0	1	1	–1/5	2/5	0	180
	$(z_j - c_j)$	*	*	*	*	*	1	

(*a*) *Fill in all the numerical values in starred positions.*

(*b*) *Is the current solution optimal? If not, carry out the iterations until an optimal is reached.*

(*c*) *Analyze the sensitivity of optimal solution to the following changes, giving the new solution.*

[**Note.** Each condition given below is independent and the analysis alongwith justification should be shown separately for each condition.]

(*i*) *Due to a machine breakdown, machine hours available gets reduced to 1,050 hours.*

(*ii*) *An additional quantity of 50 kg of aluminium can be obtained.*

(*iii*) *The second model does not feature in the current optimal solution. What should be the minimum increase in unit contribution on this model for this to feature in the optimal solution?*

Solution. (*a*) Given simplex tableau is reproduced below. Missing values are also provided.

Coeff. of basic variable	*Basic variable*	x_1	x_2	x_3	s_1	s_2	z	*Solution*
60	x_1	1	–1/3	0	1/3	–1/3	0	100
80	x_3	0	1	1	–1/5	2/5	0	180
	$(z_j - c_j)$	0	20	0	4	12	1	

(*b*) The solution given in the tableau is optimal since, all $(z_j - c_j) \geq 0$. Thus, the solution requires no improvement. The optimal solution is: $x_1 = 100$, $x_2 = 0$ and $x_3 = 180$.

(*c*) The shadow prices of the resources, as obtained from the simplex tableau are:

Aluminium: ₹ 4 per kg and *Machine-hours:* ₹ 12 per hour.

The RHS ranging is done below:

x_{Bi}	For s_1		For s_2	
	a_{ij}	x_{Bi}/a_{ij}	a_{ij}	x_{Bi}/a_{ij}
100	1/3	300	–1/3	–300
180	–1/5	–900	2/5	450

Thus, shadow prices are valid over the following ranges:

Aluminium : (1,500 – 300) to 1,500 – (–900) or 1,200 to 2,400 kg

Machine-hours : (1,200 – 450) to 1,200 – (–300) or 750 to 1,500 hours

(*i*) Since, machine hours available are limited to 1,050 hours, and this availability is within the range obtained above, a reduction of 150 hours (1,200 – 1,050) would reduce the profit by 150 × 12 = ₹ 1,800.

(*ii*) An additional capacity in the form of 150 kg of aluminium would raise the availability to 1,650 kg, which is in the range established earlier. This would, therefore, increase the total profit by 150 × 4 = ₹ 600.

(*iii*) Since, $(z_2 - c_2)$ is equal to 20, it implies that currently a unit of x_2 produced would lead to a loss of ₹ 20. Thus, the minimum increase in unit contribution on this model is ₹ 20 for it to feature in the optimum solution.

PROBLEMS

6.13. (*a*) Mahendra is in jewellery business and makes rings and bracelets of silver and gold. Each ring takes 3 units of silver and 1 unit of gold and each bracelet takes 1 unit of silver and 2 units of gold. Mahendra has 9 units of silver and 8 units of gold. Each ring bring ₹ 40 as profit and each bracelet ₹ 50 as profit. His objective is to maximize profit. Develop a linear programming model and obtain the optimum solution and then answer the following:

(*i*) What happens if Mahendra has 1 unit less silver?

(*ii*) What happens if he has 1 unit more gold?

(*iii*) What happens if the contribution of rings increases by ₹ 10 per ring?

(*b*) JOBCO produces two products on two machines. A unit of product *1* requires 2 hours on machine *A* and one hour on machine *B*. For product *2*, a unit requires one hour on machine *A* and 3 hours on machine *B*. The revenue per unit of products *1* and *2* are ₹ 30 and ₹ 20 respectively. Total daily processing time available for each machine is 8 hours.

(*i*) Determine the number of units of product *1* and 2 that maximizes the revenue.

(*ii*) Determine the dual prices for machine *A* and *B*.

(*iii*) Suppose that the unit revenue of product 2 is changed to ₹ 40. Will the current optimum solution remain the same. If not, determine the new optimum solution.

[MBA (Jan.), Gujarat Technological Univ. 2011]

6.14. A manufacturer produces four products *A*, *B*, *C* and *D*, each of which is processed on three machine *X*, *Y* and *Z*. The time required to manufacture one unit of each of the four products and the capacity of each of the three machines are indicated in the following table:

Product	*Processing time (in hrs)*		
	Machine X	*Machine Y*	*Machine Z*
A	1.5	4	2
B	2	1	3
C	4	2	1
D	3	1	2
Capacity (hours)	550	700	200

The profit contribution per unit of the four products *A*, *B*, *C* and *D* is ₹ 4, ₹ 6, ₹ 3 and ₹ 1 respectively. The manufacturer wants to determine its optimal product-mix.

(*a*) Formulate the above as a linear programming problem.

(*b*) Solve it with the simplex method.

(*c*) Find out the optimal product-mix and the total maximum profit contribution.

(*d*) Which of the constraints are binding?

(*e*) Which machine(s) have excess capacity available? How much?

(*f*) If the profit contribution from product *Y* increases by ₹ 2 per unit, will the optimal product-mix change?

(*g*) If machine *X* is to be shutdown for 50 hours due a repairs, will the product-mix change?

(*h*) What are the shadow prices of the machine-hours on the three machines?

6.15. Cookwell Co. manufacture and sells three models of large-size pressure cookers for canteen use. While market demand poses no constraint, supplies of aluminium limited to 750 kg per week and availability of machine-time limit to 600 hours per week restrict the product-mix. The resource usages of the three models are given in table below:

	Models		
	M_1	M_2	M_3
Aluminium kg/unit	6	3	5
Machine-time hr/unit	3	4	5
Contribution ₹ /Unit	60	20	80

A linear programming model and the optimal solution obtained by simplex method are shown below:

$$\text{Maximize } z = 60x_1 + 20x_2 + 80x_3 \text{ subject to the constraints:}$$
$$6x_1 + 3x_2 + 5x_3 + s_1 = 750, \quad 3x_1 + 4x_2 + 5x_3 + s_2 = 600;$$
$$x_1, x_2, x_3, s_1, s_2 \geq 0.$$

Optimum Simplex Tableau:

Coeff. of basic variable	*Basic variable*	x_1	x_2	x_3	s_1	s_2	z	*Solution*
60	x_1	1	–1/3	0	1/3	–1/3	0	50
80	x_3	0	1	1	–1/5	2/5	0	90
	$(z_j - c_j)$	0	40	0	4	12	1	

Using the information given above, determine whether the current optimal solution would be sensitive to the following changes. Each condition given below is independent of the others, and therefore, the analysis for each should be stated separately.

[***Note.*** The conclusions stated must be supported by the details of the analysis done:

(*a*) An additional 150 kg of aluminium would become available.

(*b*) The machine time available will reduce from current level of 600 hours to 450 hours.

(*c*) Model M_2 at current contribution rate of ₹ 20 is not in the optimal solution. What would be the minimum contribution of model M_2 to feature in the optimal solution?

(*d*) Following a reduction in selling price of model M_3, its unit contribution decreases by ₹ 15.

(*e*) A new model has been developed requiring 3 kg of aluminium and 3 hours of machine-time per unit, with an estimated unit contribution of ₹ 40. Would it be worthwhile manufacturing this new model?] ***[M.Com., Delhi Univ. 2004]***

6.16. Consider the following linear programming problem to determine optimal product-mix of the products *A*, *B* and *C* which require processing in three departments of fabricating, finishing and packing:

$$\text{Maximize } z = 5x_1 + 10x_2 + 8x_3 \text{ subject to the constraints:} \quad (\textit{Total profit})$$
$$3x_1 + 5x_2 + 2x_3 \leq 60 \quad (\textit{Fabrication hours})$$
$$4x_1 + 4x_2 + 4x_3 \leq 72 \quad (\textit{Finishing hours})$$
$$2x_1 + 4x_2 + 5x_3 \leq 100 \quad (\textit{Packaging hours})$$
$$x_1, x_2, x_3 \geq 0.$$

The application of simplex algorithm to solve the LPP for optimal solution results in the following tableau:

Optimum Simplex Tableau:

Coeff. of basic variable	Basic variable	x_1	x_2	x_3	s_1	s_2	s_3	z	Solution
10	x_2	1/3	1	0	1/3	–1/6	0	0	8
8	x_3	2/3	0	1	–1/3	5/12	0	0	10
0	s_3	–8/3	0	0	1/3	–17/12	1	0	18
	$(z_j - c_j)$	11/3	0	0	2/3	5/3	0	1	

In the context of this linear programming problem, determine whether the current basis would change for each of the following independent changes:

(*a*) Profit for *A* increases by ₹ 2 per unit and profit for *B* increases by ₹ 3 per unit.
(*b*) New profit values are: ₹ 14 for product *B* and ₹ 6 for product *C*.
(*c*) Profit rate for each product decreases by rupees two.
(*d*) Fabrication hours increased by 5 and finishing hours decreased by 15.
(*e*) Availability becomes 80 hours in each department.
(*f*) Fabrication hours increased by 10 and finishing hours reduced by 10.

6.17. The following linear programming problem is used for analyzing the product-mix of Anupam Biscuit Company, which produces three types of biscuits. In the problem given below, x_j denotes the number of packets of biscuit type *j*. The LPP is:

$$\text{Maximize } z = 7x_1 + 5x_2 + 2x_3 \text{ subject to the constraints:}$$
$$3x_1 + 5x_2 + x_3 \leq 150,$$
$$5x_1 + 3x_2 + 2x_3 \leq 100,$$
$$x_1 + 2x_2 + x_3 \leq 160;$$
$$x_1 \geq 0, x_2 \geq 0 \text{ and } x_3 \geq 0.$$

(*a*) Determine the optimum solution to the given LPP.
(*b*) If profit of biscuit type 3 was increased to ₹ 2.70, what will be change in current optimum solution?
(*c*) What will be the profit, if RHS of the constraint 1 is increased to 160?
(*d*) If coefficient of x_1 in the objective function is reduced to 6.6. and coefficient of x_2 is reduced to 4.5 simultaneously, what will be value of x_3 in the optimum solution?
(*e*) What are the shadow prices in the optimum simplex tableau of the given LPP?

6.18. Himalyan Pharmaceutical Company can manufacture its latest product using any of the three processes. Process 1 costs ₹ 14,000 per batch, requires 3 tons of the major ingredient I_1 and one ton of the other ingredient I_2. It yields 2 tons of the product. Process 2 costs ₹ 30,000 per batch, requires 2 and 7 tons of the two ingredients respectively, and yields 5 tons of the product. Process 3 costs ₹ 11,000 per batch, requires 9 and 2 tons of the ingredients respectively and yields one ton of the product. The company wants to find the least costly way to produce at least 50 tons of the new product given that there are 75 tons of I_1 and 60 tons of I_2. It can be assumed that it is possible to run the processes a fractional number of times.

(*a*) Formulate the given decision problem as an LPP, and then determine the optimum solution.
(*b*) How much would it cost to produce 70 tons of the new pharmaceutical product?
(*c*) If an extra 15 tons of ingredient 1 is made available, how will this affect the optimum solution?

(*d*) Can the process 3 be made more attractive by reducing the amount of ingredient 1?
(*e*) Write the dual of the given decision problem and define the dual variables.

TEST YOUR UNDERSTANDING

1. Post-optimal analysis is a technique to:
   (*a*) determine how optimum solution to an LPP changes in response to problem inputs
   (*b*) allocate resources optimally
   (*c*) minimize cost of operations
   (*d*) spell out the relation between dual and its primal
2. To maintain optimality of current optimum solution for a change Δc_k in the coefficient c_k of non-basic variable x_k, we must have
   (*a*) $\Delta c_k = z_k - c_k$ (*b*) $\Delta c_k \leq z_k - c_k$
   (*c*) $\Delta c_k \geq z_k - c_k$ (*d*) $\Delta c_k = z_k$
3. While performing sensitivity analysis, the upper bound infinity on the value of RHS of a constraint means that
   (*a*) the constraint is redundant (*b*) there is slack in the constraint
   (*c*) the shadow price for the constraint is zero (*d*) none of the above.
4. When an additional variable is introduced in the LPP, the existing optimum solution can further be improved, if
   (*a*) $z_j - c_j \leq 0$ (*b*) $z_j - c_j \geq 0$
   (*c*) $z_j - c_j = 0$ (*d*) both (*a*) and (*b*).
5. If additional constraint is an equation and an artificial variable appears in the basis of the modified LPP, new optimum solution is obtained by
   (*a*) assigning zero as the cost coefficient to artificial variable, if it appears in the basis at negative value
   (*b*) assigning $(-M)$ as the cost coefficient to artificial variable, if it appears in the basis at positive value
   (*c*) both (*a*) and (*b*) (*d*) none of the above.
6. Addition of a new constraint in the existing constraints will ensure a
   (*a*) change in the coefficients 'a_{ij}' (*b*) change in the objective function coefficients 'c_j'
   (*c*) both (*a*) and (*b*) (*d*) either (*a*) or (*b*).
7. A non-basic variable should be brought into the new solution mix, provided its contribution rate c_j is given by
   (*a*) $c_j^* > c_j + (z_j - c_j)$ (*b*) $c_j^* < c_j + (z_j - c_j)$
   (*c*) $c_j^* = c_j + (z_j - c_j)$ (*d*) none of the above.
8. Any change in the right-hand side constants gives
   (*a*) improved solution (*b*) degenerate solution
   (*c*) infeasible solution (*d*) multiple solutions.
9. A change in the coefficient of non-basic variable in objective function may effect
   (*a*) $(z_j - c_j)$ values of non-basic variables (*b*) $(z_j - c_j)$ values of all basic variables
   (*c*) only $(z_j - c_j)$ value of that non-basic variable
   (*d*) the feasibility of solution.

10. To ensure best marginal increase in the objective function value, a resource may be increased whose shadow price is comparatively
    (*a*) smaller (*b*) larger
    (*c*) both (*a*) and (*b*) (*d*) neither (*a*) and (*b*).

ANSWERS

1. (*a*)	2. (*b*)	3. (*d*)	4. (*a*)	5. (*b*)
6. (*c*)	7. (*b*)	8. (*c*)	9. (*c*)	10. (*b*)

REVIEW EXERCISE

1. What is sensitivity analysis? Discuss its significance fully from managerial viewpoint. ***[MBA, Osmania Univ. 2000]***
2. Explain with suitable examples the basic philosophy behind sensitivity analysis. ***[MBA, Panjab Tech. Univ. 2014]***
3. Given an LPP: {Minimize $z = \mathbf{cx}$ subject to: $\mathbf{Ax} = \mathbf{b}$ and $\mathbf{x} \geq \mathbf{0}$}, discuss the effect of (*i*) variation in the requirement b_i, (*ii*) variation in the cost c_j and (*iii*) addition of a new activity on the current optimum solution. ***[B.Tech. (Comp. Sc.), Panjab Tech. Univ. 2010]***
4. Write a short note on shadow prices in simplex method. What is their role in sensitivity analysis? ***[B.Tech. (Mech.), Panjab Tech. Univ. 2006]***
5. Explain the basic concept of sensitivity analysis. What are the different factors affecting the given solution and how do we resolve them? Give a brief comment on each of them. ***[MBA, Gujarat Tech. Univ. 2010]***
6. Write a short note on post-optimal sensitivity analysis. ***[M.Sc. (Math.), CCS Univ. 2009]***

CHAPTER 7

Transportation Problem

"Commodities are transported while people travel"

The ***Transportation Problem*** is one of the sub-classes of *LPP* in which the objective is to ***transport*** various amounts of a single homogeneous commodity, that are initially stored at various ***origins*** to different ***destinations*** in such a way that the total transportation cost is minimum. Although, the name of the problem is derived from transport to which it was first applied, can also be used for machine allocation, plant location, product mix, and many others, so that the problem is not confined to transport or distribution only. A few industrial situations which establish the relevance and utility of this problem are given below:

Illustration 1 (*Cost Minimization Problem*). A dairy firm have three plants located throughout a state. Daily milk production at each plant is as follows:

Plant *A*	...	5 million litres
Plant *B*	...	2 million litres
Plant *C*	...	8 million litres

Each day the firm must fulfil the needs of its four distribution centres. Minimum requirement at each centre is as follows:

Distribution Centre 1	...	6 million litres
Distribution Centre 2	...	4 million litres
Distribution Centre 3	...	2 million litres
Distribution Centre 4	...	3 million litres

Cost of shipping one million litres of milk from each plant to each distribution centre is given in the following table (in hundreds of rupees):

Plant	*Distribution centres*			
	1	2	3	4
A	2	3	9	5
B	0	1	5	2
C	5	7	12	7

The dairy firm wishes to decide as to how much should be the shipment from which milk plant to which distribution centre, so that the cost of shipment may be ***minimum.***

Illustration 2 (*Profit Maximization Problem*). A company is manufacturing TV sets at two plants located at Hyderabad and Salem with a capacity of 300 units and 100 units per week respectively. The company supplies TV sets to its four showrooms situated at Ernakulam, Delhi, Bengaluru and Chennai which have a maximum demand of 85, 150, 150 and 55 units respectively. Due to the differences in raw material cost and transportation cost, the ***profit*** per unit (in rupee) differs which is shown in the following table:

City	*Ernakulam*	*Delhi*	*Bengaluru*	*Chennai*
Hyderabad	80	60	190	175
Salem	70	85	50	65

Plan the production programme so as to ***maximize the profit***. The company may have its production capacity at both plants partly or wholly unused.

Note. It may be noted that the places from where the commodity/goods are transported are called the ***sources*** or ***origins***; and the places where they are to be supplied are called ***destination***. In this terminology, the problem of the manufacturer or distributor is to decide as to how many units to be transported from different origins to different destinations so that the total transportation cost is minimum (or profit is maximum).

Remark. Transportation method (technique) helps in the (*i*) fulfillment of demand, (*ii*) to build a good logistic network, (*iii*) to reduce the distribution cost, and to centralize production operation. Also, it ensures effective supply chain.

7:1. GENERAL STRUCTURE OF A TRANSPORTATION PROBLEM

A typical transportation problem involving '*m*' origins and '*n*' destinations can be stated mathematically as follows:

Let a_i = quantity (units) of a product available at origin i.

b_j = quantity (units) of a product required at destination j.

c_{ij} = the cost of transporting one unit of a product from source (origin) i to destination j.

x_{ij} = the quantity (units) transported from origin i to destination j.

Assume that $a_1 + a_2 + ... + a_m = b_1 + b_2 + ... + b_n$. It means that the total quantity (units) available at origins is precisely equal to the total quantity (units) required at the destinations.

Using the above information, the problem can be written in the ***linear programming format***, *viz.*,

$$\text{Minimize (total cost) } z = \sum_{i=1}^{m} (c_{i1}x_{i1} + c_{i2}x_{i2} + ... + c_{in}x_{in}) \text{ subject to the constraints:}$$

$$x_{i1} + x_{i2} + ... + x_{in} = a_i \quad \text{for } i = 1, 2, ..., m$$

$$x_{1j} + x_{2j} + ... + x_{mj} = b_j \quad \text{for } j = 1, 2, ..., n$$

and
$$x_{ij} \geq 0 \text{ for all } i = 1, 2, ..., m \text{ and } j = 1, 2, ..., n.$$

Remarks

1. When the total supply equals total demand, the problem is called a ***balanced transportation problem***, otherwise an ***unbalanced transportation problem***.
2. The coefficients of x_{ij} ($i = 1, 2, ..., m$ and $j = 1, 2, ..., n$) in the constraints of any transportation problem are unity.
3. The objective function and the constraints being all linear.

7:2. EXISTENCE OF SOLUTION IN TRANSPORTATION PROBLEM

The general transportation problem always have a solution. In this section, we shall discuss the theorems on the existence of feasible solution, optimum solution and basic feasible solution.

***Theorem 7.1* (Existence of Feasible Solution).** *A necessary and sufficient condition for the existence of a feasible solution to the general transportation problem is that*

$$\textit{Total Supply} = \textit{Total Demand, i.e.,} \quad \sum_{i=1}^{m} a_i = \sum_{j=1}^{n} b_j = \lambda \quad \text{(say).}$$

Proof. *The condition is necessary.* Let there exist a feasible solution to the TP. Then, we have

$$\sum_{i=1}^{m}\sum_{j=1}^{n} x_{ij} = \sum_{i=1}^{m} a_i \quad \text{and} \quad \sum_{j=1}^{n}\sum_{i=1}^{m} x_{ij} = \sum_{j=1}^{n} b_j$$

yielding
$$\sum_{i=1}^{m} a_i = \sum_{j=1}^{n} b_j = \lambda \text{ (say)}$$

Sufficiency. We assert that there exists a feasible solution given by $x_{ij} = a_i b_j / \lambda$ for all i and j. Clearly, $x_{ij} \geq 0$, since, $a_i > 0$ and $b_j > 0$ for all i and j.

Also,
$$\sum_{j=1}^{n} x_{ij} = \sum_{j=1}^{n} (a_i b_j / \lambda) = \frac{a_i}{\lambda} \sum_{j=1}^{n} b_j = a_i, \qquad i = 1, 2, ..., m$$

and
$$\sum_{i=1}^{m} x_{ij} = \sum_{i=1}^{m} (a_i b_j / \lambda) = \frac{b_j}{\lambda} \sum_{i=1}^{m} a_i = b_j. \qquad j = 1, 2, ..., n$$

Thus, x_{ij} satisfies all the constraints of the TP and hence, is a feasible solution.

Corollary. (Existence of an Optimum Solution). *There always exists an optimum solution to a TP.*

Proof. Let $\sum_i a_i = \sum_j b_j$, so that a feasible solution x_{ij} exists. It follows from the constraints of the problem that each x_{ij} is bounded, *viz.*,

$$0 \leq x_{ij} \leq \min.(a_i, b_j)$$

Therefore, the feasible region of the problem is closed, bounded and non-empty and hence, there exists an optimum solution.

Theorem 7.2 (Basic Feasible Solution). *The number of basic (decision) variables of the general transportation problem at any stage of feasible solution must be m + n –1.*

Proof. Consider the $m + n$ constraints of the general transportation problem, *viz.*:

$$\sum_{i=1}^{m} x_{ij} = b_j \quad \text{and} \quad \sum_{j=1}^{n} x_{ij} = a_i \qquad (j = 1, 2, ..., n; i = 1, 2, ..., m)$$

Taking summation on both sides, these yield

$$\sum_{j=1}^{n-1}\sum_{i=1}^{m} x_{ij} = \sum_{j=1}^{n-1} b_j \quad \text{and} \quad \sum_{i=1}^{m}\sum_{j=1}^{n} x_{ij} = \sum_{i=1}^{m} a_i$$

Subtracting the two, we have

$$\sum_{j=1}^{n-1}\sum_{i=1}^{m} x_{ij} - \sum_{i=1}^{m}\sum_{j=1}^{n} x_{ij} = \sum_{j=1}^{n-1} b_j - \sum_{i=1}^{m} a_i$$

or
$$\sum_{i=1}^{m}\left(\sum_{j=1}^{n-1} x_{ij} - \sum_{j=1}^{n} x_{ij}\right) = \sum_{j=1}^{n-1} b_j - \sum_{j=1}^{n} b_j \qquad \left(\because \sum_{i=1}^{m} a_i = \sum_{j=1}^{n} b_j\right)$$

or
$$\sum_{i=1}^{m} x_{in} = b_n$$

which happens to be the last constraint of the problem.

This indicates that if the first $m + n - 1$ constraints are satisfied, then $\sum_i a_i = \sum_j b_j$ ensures that the $(m + n)^{\text{th}}$ constraint will be automatically satisfied.

Thus, out of $m + n$ equations, we have only $(m + n - 1)$ linearly independent equations. Therefore, a basic feasible solution will consist of at most $(m + n - 1)$ positive variables, the rest being zero. Further, a feasible solution involving exactly $(m + n - 1)$ positive variables is known as **non-degenerate** basic feasible solution. Otherwise, it is said to be **degenerate** basic feasible solution.

Remark. The allocated cells in the transportation table will be called **occupied** cells and empty cells will be called **non-occupied** *cells.*

7:3. DUALITY IN TRANSPORTATION PROBLEM

Since, a transportation problem is itself a linear programming problem, there always exist another linear programming problem based upon the same data and having the same solution. Let the original transportation problem is considered as **primal problem** and its associated one as its **dual problem**. Then, for the general transportation problem, the primal-dual pair is:

Primal Problem. Consider the general transportation problem as to

$$\text{Minimize } z = \sum_i \sum_j x_{ij}\, c_{ij} \text{ subject to the constraints:}$$

$$\sum_j x_{ij} = a_i\,,\ \sum_i x_{ij} = b_j\,;\ x_{ij} \ge 0 \qquad \text{for all } i = 1, 2, ..., m \text{ and } j = 1, 2, ..., n$$

Dual Problem. Since, the primal problem is in its standard form, dual problem will be as follows:

$$\text{Maximize } z^* = \sum_i u_i\, a_i + \sum_j v_j\, b_j \text{ subject to the constraints:}$$

$$u_i + v_j \le c_{ij}\,, \qquad \text{for all } i = 1, 2, ..., m \text{ and } j = 1, 2, ..., n$$

u_i and v_j unrestricted in sign for all i, j;

where, u_i $(i = 1, 2, ..., m)$ and v_j $(j = 1, 2, ..., n)$ are dual variables.

The dual solution of the transportation problem implicitly measures the comparative locational advantage of the sites of factories (source of capacities) and the delivered price of the product.

7:4. ECONOMIC INTERPRETATION OF u_i's AND v_j's

The dual variables u_i's and v_j's represent the shadow prices (value of the commodity) for the supply centres and demand centres. The value of each variable u_i measures the comparative implicit contribution (value or location advantages) of a unit of capacity at supply centre i, and therefore, may be termed **location rent**. Similarly, the value of each variable v_j measures the comparative implicit contribution of an additional unit of commodity transported to demand centre j and therefore, may be termed **market price**.

7:5. DEGENERACY IN TRANSPORTATION PROBLEM

A basic feasible solution for the general transportation problem must consist of $(m + n - 1)$ occupied cells. The basic solution will be called degenerate when the number of occupied cells is less than the required number of basic cells, *viz.*, $m + n - 1$. Degeneracy can occur in initial solution or it may arise in some subsequent iterations. We now discuss a procedure to deal with the problem of degeneracy.

Case 1. **Degeneracy at the Initial Solution.** To resolve degeneracy at the initial solution, a very small quantity ε (> 0) is allocated in an unoccupied cell so as to get $m + n - 1$ number of occupied cells. In a minimization transportation problem, it is better to allocate ε to unoccupied cell that have lowest transportation cost. In some cases, ε must be added in one of those unoccupied cells which make possible the determination of u_i and v_j uniquely.

The quantity ε is considered to be so small that if it is transferred to an occupied cell, it does not change the quantity of allocation. That is, $x_{ij} + \varepsilon = x_{ij} - \varepsilon = x_{ij}$ but $\varepsilon - \varepsilon = 0$. Also, ε does not affect the total transportation cost of the allocation. Hence, the quantity ε is used to evaluate unoccupied cells and once the purpose is over, ε must be removed form the scene.

Case 2. **Degeneracy at Subsequent Iterations.** To resolve degeneracy which occur during optimality test, the quantity ε may be allocated to one or more cells which have become unoccupied recently to have $m + n - 1$ number of occupied cells in the new solution. It may be removed once the purpose is over.

7:6. STANDARD TRANSPORTATION TABLE

The transportation problem is explicitly represented by the following transportation table:

	D_1	D_2	D_3	.....	D_j	.....	D_n	Supply (a_i)
O_1	x_{11} c_{11}	x_{12} c_{12}	x_{13} c_{13}	.....	x_{1j} c_{1j}	.....	x_{1n} c_{1n}	a_1
O_2	x_{21} c_{21}	x_{22} c_{22}	x_{23} c_{23}	.....	x_{2j} c_{2j}	.....	x_{2n} c_{2n}	a_2
⋮	⋮	⋮	⋮	⋮	⋮	⋮	⋮	⋮
O_m	x_{m1} c_{m1}	x_{m2} c_{m2}	x_{m3} c_{m3}	.....	x_{mj} c_{mj}	.....	x_{mn} c_{mn}	a_m
Demand (b_j)	b_1	b_2	b_3	.....	b_j	.....	b_n	

Table 7.1

This table can be thought of as an $m \times n$ matrix having mn square matrices. Each square matrix is known as a ***cell***. Thus, in all there are mn cells in the table. In each cell the unit transportation cost c_{ij} from i^{th} source (origin) to the jth destination is displayed in the ***lower right side*** of the $(i, j)^{th}$ cell. Any feasible solution is shown in the table by entering the value of x_{ij} in the ***upper left side*** of the $(i, j)^{th}$ cell. Various a_i's and b_j's are called ***rim requirements***.

7:7. SOME DEFINITIONS

1. A set of non-negative values x_{ij} $(i = 1, 2, ..., m; j = 1, 2, ..., n)$ that satisfies the constraints (rim conditions and also the non-negative restrictions) is called a ***feasible solution*** to the transportation problem. More precisely the necessary and sufficient condition for the existence of a feasible solution to the general transportation problem is that

$$a_1 + a_2 + ... + a_m = b_1 + b_2 + ... + b_n.$$

2. The constraints of the general transportation problem shows that each x_{ij} is bounded, *viz.*, $0 \le x_{ij} \le \min.\ (a_i, b_j)$. This implies that the feasible region of the transportation problem is closed, bounded and non-empty. Thus, there exists an ***optimum solution***.

3. The ***aggregate transportation cost*** of a general transportation problem is determined by multiplying various values of x_{ij} with corresponding c_{ij}'s and then adding them up all, *i.e.*, $\sum_{i=1}^{m} \sum_{j=1}^{n} c_{ij} x_{ij}$.
4. In a general transportation problem, the number of constraints are $m + n$, *i.e.*, the number of rows plus the number of columns. This problem can also be represented by an $m \times n$ transportation table having mn cells. The number of allocations (variables) in various cells required for forming a basis are $(m + n - 1)$, *i.e.*, the number of rows plus number of columns less one. This is so, because there are only $(m + n - 1)$ independent allocations (variables) in the solution basis. A feasible solution to a general transportation problem that contains no more than $(m + n - 1)$ non-negative allocations is called a ***basic feasible solution***.

   The number of ***basic variables*** representing transportation routes that are utilized equals $(m + n - 1)$ and all other variables be non-basic or zero representing the un-utilized routes. Further, the cells corresponding to basic variables are termed as ***occupied cells*** and the remaining cells are called ***unoccupied (or empty)*** cells.
5. A basic feasible solution to a general transportation problem is said to be ***non-degenerate***, if it contains exactly $(m + n - 1)$ non-negative allocations. If there are less than $(m + n - 1)$ non-negative allocations, then the basic feasible solution is ***degenerate*** one.

Note. The number of basic cells (variables) in an $m \times n$ transportation problem are $(m + n - 1)$, whereas the number of non-basic cells (variables) are $mn - (m + n - 1)$.

7:8. SOLUTION OF A TRANSPORTATION PROBLEM

As transportation problem can be put in the linear programming format, it can be solved by simplex method. But the solution is going to be very lengthy and a cumbersome process because of the involvement of a large number of decision as well as artificial variables. Hence, we modify the procedure of simplex method — popularly known as ***Modified Simplex Method***. This method yields results faster and with less computational effort. A significant point of difference between this method and the simplex method is regarding the determination of the initial basic feasible solution.

The solution to a transportation problem involve the following seven steps:

Step 1. Formulate the given problem as a linear programming problem.

Step 2. Set up the given LPP in the tabular form known as a ***transportation table.***

Step 3. Examine whether total supply equals total demand. If not, introduce dummy row/column having all its cost elements as zero and Supply/Demand as the (+ve) difference of Supply and Demand.

Step 4. Find an initial basic feasible solution that must satisfy all the supply and demand conditions.

Step 5. Examine the solution obtained in *Step 4* for optimally, *i.e.*, examine whether an improved transportation schedule with lower cost is possible.

Step 6. If the solution is not optimal, modify the shipping schedule by including those unoccupied cells whose inclusion may result in an improved solution.

Step 7. Examine the improved solution further for optimality and repeat the procedure until no further improvement is possible.

We shall now discuss various methods available for finding an initial basic feasible solution and then attaining an optimum solution.

7:9. METHODS FOR FINDING INITIAL BASIC FEASIBLE SOLUTION

The transportation problem has a solution, if and only if the problem is balanced, *i.e.*, total supply equals total demand. Therefore, before starting to find the initial basic feasible solution, check whether the given transportation problem is balanced. If not, one has to balance the transportation problem.

There are several methods available to obtain an initial basic feasible solution of a transportation problem. We discuss here only the following three:

1. North-West Corner Method (Rule),
2. Least Cost Method, and
3. Vogel's Approximation Method.

1. North-West Corner Method (*NWC Rule*)

It is simple and an efficient method to obtain an initial basic feasible solution. Various steps of the method are:

Step 1. The first allocation is made in the cell occupying the upper left-hand (***north-west***) corner of the transportation table. The maximum feasible amount is allocated there. That is, $x_{11} = \text{min.}\ (a_1, b_1)$. This value of x_{11} is entered at the upper left-hand corner of cell (1, 1) in the transportation table.

Step 2. If $b_1 > a_1$, move down ***vertically*** to the second row and make the second allocation of magnitude $x_{21} = \text{min.}\ (a_2, b_1 - x_{11})$ in the cell (2, 1).

If $b_1 < a_1$, move to the right ***horizontally*** to the second column and make the second allocation of magnitude $x_{12} = \text{min.}\ (a_1 - x_{11}, b_2)$ in the cell (1, 2).

If $b_1 = a_1$, there is a tie for second allocation. An arbitrary tie-breaking choice is made. Make the second allocation of magnitude $x_{12} = \text{min.}\ (a_1 - a_1, b_2)$, *i.e.*, 0 in the cell (1, 2) or $x_{21} = \text{min.}\ (a_2, b_1 - b_1)$, *i.e.*, 0 in the cell (2, 1).

Step 3. Start from the new north-west corner of the transportation table and go to *Step 1*.

Step 4. Continue the procedure until all the rim requirements are satisfied.

Example 7.1. *Determine an initial basic feasible solution to the following transportation problem using north-west corner rule:*

	D_1	D_2	D_3	D_4	*Availability*
O_1	6	4	1	5	14
O_2	8	9	2	7	16
O_3	4	3	6	2	5
Requirement	6	10	15	4	35

Here, O_i and D_j represent i^{th} origin and j^{th} destination respectively.

[M.Com., Bangalore Univ. 2017; MCA, Panjab Tech. Univ. 2010]

Solution. Since, there are three rows and four columns in the transportation problem, the transportation tableau will consist of 12 cells.

Following north-west corner method, the first allocation is made in the cell (1, 1), the magnitude being $x_{11} = \text{min.}\ (14, 6)$, *i.e.*, 6. This exhausts the requirement of first column. So, we move on to second column for the next allocation. The second allocation is made in the cell (1, 2) and the magnitude of allocation is given by $x_{12} = \text{min.}\ (14 - 6, 10)$, *i.e.*, 8. This exhausts the availability of the first row. So, we move to second row, and hence, next allocation will be made in the second row and second column. Therefore, third allocation is made in the cell (2, 2), the magnitude being

x_{22} = min. (16, 10 – 8), *i.e.*, 2. This exhausts the requirement of second column. So, we move on to third column and allocation is made in the cell at the second row and third column. Therefore, the magnitude of fourth allocation in the cell (2, 3) is given by x_{23} = min. (16 – 2, 15), *i.e.*, 14. The fifth allocation is made in the cell (3 3), the magnitude being x_{33} = min. (5, 15 – 14), *i.e.*, 1. This exhausts the availability of third row. So, we move on to fourth column for the sixth allocation. Therefore, the sixth and final allocation in the cell (3, 4) is given by x_{34} = min. (5 – 1, 4), *i.e.*, 4.

[6] 6	[8] 4	1	5
8	[2] 9	[14] 2	7
4	3	[1] 6	[4] 2

Table 7.2

Now, all the rim requirements have been satisfied and hence an initial basic feasible solution to the transportation problem has been obtained. The solution is displayed in the transportation *Table 7.2*.

The transportation cost according to above allocation is given by

$$z = x_{11}c_{11} + x_{12}c_{12} + x_{22}c_{22} + x_{23}c_{23} + x_{33}c_{33} + x_{34}c_{34}$$
$$= 6 \times 6 + 8 \times 4 + 2 \times 9 + 14 \times 2 + 1 \times 6 + 4 \times 2 = 128$$

Remarks

1. The cells which get allocation are basic cells, since, number of allocated cells are (3 + 4 – 1), *i.e.*, 6.
2. The initial basic feasible solution obtained by North-West corner method/rule may be far from optimum, because costs are completely ignored.

2. Least-Cost Method

The least-cost method is more economical than north-west corner rule, since, it gives a lower beginning cost. Various steps of this method can be summarized as under:

Step 1. Determine ***lowest* (*smallest*) cost** among all the rows of the transportation table.

Step 2. Identify that row which have lowest cost obtained in *Step 1*, and allocate the maximum feasible quantity in the cell corresponding to this smallest cost in this particular row. Then, eliminate (cross out) that row/column, where an allocation is made.

Step 3. Repeat *Step 1* and *2* for the reduced transportation table until all the rim requirements are satisfied. Whenever, the minimum cost is not unique, make an arbitrary choice among the minimum (least) costs.

Example 7.2. *Obtain an initial basic feasible solution to the following transportation problem, using least cost method:*

	D_1	D_2	D_3	D_4	*Supply*
O_1	1	2	3	4	6
O_2	4	3	2	0	8
O_3	0	2	2	1	10
Demand	4	6	8	6	—

Here, O_i and D_j denote the ith origin and j^{th} destination, respectively.

Solution. The transportation table of the given transportation problem has 12 cells. Following the least-cost method, we observe that the lowest of all cost elements is 0 for the cell (3, 1) and (2, 4). So, the first allocation can be made at either of the two cells. Let us choose cell (2, 4). As the corresponding supply and demand being 8 and 6 respectively, we allocate 6 to this cell. In other words x_{24} = min. (6, 8) = 6. Cross off the fourth column, since it exhausts the demand of 4th column. Of the remaining cost elements, the lowest cost element is corresponding to cell (3, 1). The second allocation, therefore, is made in this cell of the magnitude x_{31} = min. (10, 4) = 4. Cross off the first column, since, it exhausts the demand of first column and adjust the remaining availability. This yields *Table 7.3*. There is, again a tie for the third allocation. We choose arbitrarily the cell (1, 2) and allocate x_{12} = min. (6, 6) = 6 in this cell. Cross off either the second column or the first row. We choose to cross off the first row of the table. The next allocation of magnitude x_{32} (= 0) is made in cell (3, 2). Cross off the second column getting *Table 7.4*.

1	2	3	4	6
4	3	2	[6] 0	2
[4] 0	2	2	1	10
4	6	8		

Table 7.3

1	[6] 2	3	4	6
4	3	2	[6] 0	2
[4] 0	[0] 2	2	1	6
	6	8		

Table 7.4

We choose arbitrarily again, to make the next allocation in cell (2, 3) of magnitude x_{23} = min. (2, 8) = 2. Cross off the second row. This given *Table 7.5*. The last allocation of magnitude x_{33} = min. (6, 6) = 6 is made in the cell (3, 3).

Now, all the rim requirements have been satisfied and hence, an initial basic feasible solution has been determined. The solution to the given transportation problem is displayed in transportation *Table 7.6*.

1	[6] 2	3	4	
4	3	[2] 2	[6] 0	2
[4] 0	[0] 2	[6] 2	1	6
		8		

Table 7.5

1	[6] 2	3	4
4	3	[2] 2	[6] 0
[4] 0	[0] 2	[6] 2	1

Table 7.6

The transportation cost according to the above route is given by

$$z = 6 \times 2 + 2 \times 2 + 6 \times 0 + 4 \times 0 + 0 \times 2 + 6 \times 2 = 28$$

3. Vogel's Approximation Method

Vogel's approximation method yields an initial basic feasible solution which is very close to the optimum solution. The method is given as follows:

First consider each row of the cost matrix individually and find the difference between two least cost cells in it. Then, repeat the exercise for each of the column. Identify the row and column with the largest difference value. Now, consider the cell with the minimum cost in that column (or row, as the case may be) and allocate the maximum quantity (units) possible, looking to the demand and availability corresponding to that cell. In case of a tie in the largest cost differences, choose the cost difference corresponding to which the largest number of units may be allocated or corresponding to which the cell chosen has the minimum cost. The method can be summarized into following six major steps:

Step 1. For each row (column) of the transportation table, identify the smallest and next to smallest costs. Calculate difference between these two for each row and each column.

Step 2. Select the row or the column with the largest difference (penalty).

Step 3. Allocate the maximum feasible quantity in the cell with the minimum (least) cost in the selected row or column.

Step 4. Eliminate (cross out) that row or column, where an allocation is made.

Step 5. Recompute row and column differences for each row and each column of the reduced transportation table.

Step 6. Go to *Step* 2 and repeat the procedure until all the rim requirements are satisfied.

Remark. A row/column "difference" indicate the minimum unit penalty incurred by failing to make an allocation to the smallest cost cell in that row/column.

Example 7.3. *Obtain an initial basic feasible solution to the following transportation problem, using Vogel's approximation method:*

Warehouses	*Stores* I	II	III	IV	*Availability*
A	5	1	3	3	34
B	3	3	5	4	15
C	6	4	4	3	12
D	4	–1	4	2	19
Requirement	21	25	17	17	80

[M.Com., Mahatma Gandhi Univ. 2010]

Solution. The transportation table of the given transportation problem has 16 cells. The differences between the smallest and next-to-smallest costs in each row and each column are computed and displayed inside the parentheses against the respective rows and columns. The largest of these differences is (3) and is associated with the fourth row of the transportation table as shown in *Table 7.7*.

5	1	3	3	34(2)
3	3	5	4	15(0)
6	4	4	3	12(1)
4	[19] –1	4	2	19(3)
21(1)	25(2)	17(1)	17(1)	

Table 7.7

5	[6] 1	3	3	34(2)
3	3	5	4	15(0)
6	4	4	3	12(1)
21(2)	6(2)	17(1)	17(0)	

Table 7.8

The minimum cost in the fourth row is c_{42} (= –1). Accordingly, we allocate x_{42} = min. (19, 25) = 19 in the cell (4, 2). This exhausts the availability of warehouse *D*. Cross out the fourth row. The row and column differences are now computed for the resulting reduced transportation *Table 7.7*. The largest of these is (2) and is associated with the first row as well as the first and second columns. We select arbitrarily the first row whose minimum cost is c_{12} (= 1). Thus, the second allocation of magnitude $x_{12} = \min.\ (34,\ 6) = 6$ is made in the cell (1, 2). This exhausts the requirement of second column. Cross out the second column from the table. Continuing in this manner, the subsequent reduced transportation tables and the differences for the surviving rows and columns are shown in *Table 7.9* to *7.11*.

5	3	3	28(0)
15 (3)	5	4	15(1)
6	4	3	12(1)
21(2)	17(1)	17(0)	

Table 7.9

5	3	3	28(0)
6	4	12 (3)	12(1)
6(1)	17(1)	17(0)	

Table 7.10

6 (5)	17 (3)	5 (3)	28
6	17	5	

Table 7.11

Eventually, the basic feasible solution is obtain as shown in *Table 7.12* below:

6 (5)	6 (1)	17 (3)	5 (3)
15 (3)	3	5	4
6	4	4	12 (3)
4	19 (–1)	4	2

Table 7.12

The transportation cost according to above allocation is given by

$$\begin{aligned} z &= x_{11}c_{11} + x_{12}c_{12} + x_{13}c_{13} + x_{14}c_{14} + x_{21}c_{21} + x_{34}c_{34} + x_{42}c_{42} \\ &= 6 \times 5 + 6 \times 1 + 17 \times 3 + 5 \times 3 + 15 \times 3 + 12 \times 3 + 19 \times (-1) \\ &= 30 + 6 + 51 + 15 + 45 + 36 - 19 = 164 \end{aligned}$$

REVIEW ILLUSTRATIONS

Example 7.4. *Determine an initial basic feasible solution to the following transportation problem by (a) North-West Corner Method, and (b) Least-Cost Method:*

	D	E	F	G	*Available*
A	11	13	17	14	250
B	16	18	14	10	300
C	21	24	13	10	400
Required:	200	225	275	250	

Solution. Since, $\Sigma\, a_i = \Sigma\, b_j = 950$, there exists a feasible solution to the transportation problem. We obtain initial basic feasible solution as follows :

(a) North-West Corner Method. The transportation table of the given problem has 12 cells. Following north-west corner rule, the first allocation is made in the cell (1, 1), the magnitude being

x_{11} = min. (250, 200) = 200. The second allocation is made in the cell (1, 2) and the magnitude of the allocation is: x_{12} = min. (250 – 200, 225) = 50. The third allocation is made in the cell (2, 2), the magnitude is x_{22} = min. (300, 225 – 50) = 175. The magnitude of fourth allocation in the cell (2, 3) is: x_{23} = min. (300 – 175, 275) = 125. The fifth allocation is made in the cell (3, 3), the magnitude being x_{33} = min. (400, 275 – 125) = 150 and the sixth (last) allocation is made in the cell (3, 4) with magnitude x_{34} = min. (400 – 150 , 250) = 250. Hence, an initial basic feasible solution to the given transportation problem has been obtained and is displayed in *Table 7.13*.

(200) 11	(50) 13	17	14
16	(175) 18	(125) 14	10
21	24	(150) 13	(250) 10

Table 7.13

The transportation cost according to the above route is given by

$$z = 200 \times 11 + 50 \times 13 + 175 \times 18 + 125 \times 14 + 150 \times 13 + 250 \times 10 = 12{,}200$$

(b) ***Least-Cost Method.*** Following the least cost method, the first allocation is made in the cell (3, 4), the magnitude being x_{34} = min. (400, 250) = 250. This satisfies the requirement of G and thus, we cross out the fourth column from the table. This yields *Table 7.14*. The second allocation is made in the cell (1, 1) of magnitude x_{11} = min. (250, 200) = 200 which satisfies the requirement of D and, therefore, we cross out the first column. This yields *Table 7.15*.

11	13	17	14	250
16	18	14	10	300
21	24	13	(250) 10	150
200	225	275		

Table 7.14

(200) 11	13	17	14	50
16	18	14	10	300
21	24	13	(250) 10	150
	225	275		

Table 7.15

Now, there is a tie for third allocation. We choose arbitrarily the cell (1, 2) and allocate x_{12} = min. (50, 225) = 50. Cross out the first row. Fourth allocation is made in the cell (3, 3) with magnitude x_{33} = min. (150, 275) = 150 of the reduced *Table 7.15*. After crossing out the third row, we get *Table 7.16*.

Fifth allocation is made in the cell (2, 3) of magnitude x_{23} = min. (300, 125) = 125 and the sixth (last) allocation is made of magnitude x_{22} = min. (175, 175) = 175.

(200) 11	(50) 13	17	14	
16	18	14	10	300
21	24	(150) 13	(250) 10	
	175	125		

Table 7.16

(200) 11	(50) 13	17	14
16	(175) 18	(125) 14	10
21	24	(150) 13	(250) 10

Table 7.17

Thus, we get the initial basic feasible solution as displayed in *Table 7.17*. The transportation cost according to above route is given by

$$z = 200 \times 11 + 50 \times 13 + 175 \times 18 + 125 \times 14 + 150 \times 13 + 250 \times 10 = 12{,}200$$

Remark. We observe that the transportation cost in the above transportation problem using the North-West Corner method and Least Cost method is same, but in general it is not true.

Example 7.5. *Determine an initial basic feasible solution for the following transportation problem using Least Cost Method:*

Factory	*Distribution centres*				*Capacity (Units)*
	Bengaluru	*Mumbai*	*Delhi*	*Chennai*	
Ahmedabad	6	8	8	5	30
Ernakulam	5	11	9	7	40
Hyderabad	8	9	7	13	50
Demand (units/day)	35	28	32	25	120

Costs are expressed in terms of rupees per unit shipped. (For example, Ernakulam factory can ship one unit of the product to Delhi at ₹ 9.) The problem is to determine as to how much quantity should be shipped from the factory to various distribution centres.

Solution. The review of transportation costs row by row reveals that cells *AC* and *EB* have lowest cost of ₹ 5 per unit. Therefore, we allocate maximum feasible quantity of 25 units in cell *AC* and 35 units in cell *EB*. These two allocations satisfy the demand of column B and C both and hence are crossed out in the transportation table for the next allocation. This give *Table 7.18*.

	B	M	D	C	*Capacity*
A	6	8	8	[25] 5	5
E	[35] 5	11	9	7	5
H	8	9	7	13	50
Demand:		28	32		

Table 7.18

	B	M	D	C	*Capacity*
A	6	8	8	[25] 5	5
E	[35] 5	11	9	7	5
H	8	9	[32] 7	13	18
Demand:		28			

Table 7.19

In the reduced *Table 7.18*, the lowest cost (₹ 7) is in the cell *HD*. The next allocation of 32 units, therefore, is made in this cell. This allocation satisfies the demand of *D*. Crossing out of this column gives rise to a transportation table as shown in *Table 7.19*.

In the reduced transportation *Table 7.19*, since, there is only one column (*i.e.*, column *M*), we allocate the balance of 5 units to cell *AM* from *A*, 5 units to *EM* from *E* and the remaining 18 units to *HM* from *H*. This gives *Table 7.20*.

6	[5] 8	8	[25] 5
[35] 5	[5] 11	9	7
8	[18] 9	[32] 7	13

Table 7.20

The above routing of units meets all the rim requirements and entails 6 (= 3 + 4 – 1) shipments as there are 6 basic cells. It involves a total cost of

$$₹\,(5 \times 8 + 25 \times 5 + 35 \times 5 + 5 \times 11 + 18 \times 9 + 32 \times 7), \textit{ i.e.}, ₹\,781$$

Example 7.6. *Find the starting solution in the following transportation problem by (a) North-West Corner Method, (b) Least-Cost Method, and (c) Vogel's Approximation Method:*

	D_1	D_2	D_3	D_4	*Supply*
S_1	3	7	6	4	5
S_2	2	4	3	2	2
S_3	4	3	8	5	3
Demand	3	3	2	2	

Solution. (*a*) Following the ***North-West Corner Method***, the first allocation is made in the cell (S_1, D_1), the magnitude being x_{11} = min. (5, 3) = 3. This means that 3 units are to be transported from S_1 to D_1. This allocation meets the demand requirement at D_1. However, this allocation leaves 5 – 3 (= 2) units at S_1. Moving horizontally, now, allocate as much as possible in (S_1, D_2), *i.e.*, x_{12} = min. (2, 3) = 2. Proceeding on to second row, since demand requirement of D_1 has been met, nothing further, can be allocated to D_1. The requirement of D_2 is 3 – 2 (= 1) unit. Thus, we allocate x_{22} = min. (2, 1) = 1. Fourth allocation is made in the cell (S_2, D_3), the magnitude being x_{23} = min. (2 – 1, 2) = 1. Fifth allocation is made in the cell (S_3, D_3), the magnitude being x_{33} = min. (3, 2 – 1) = 1 and the sixth (last) allocation is made in the cell (S_3, D_4) of magnitude x_{34} = min. (3 – 1, 2) = 2. The number of positive allocations (occupied cells) is equal to $m + n - 1$, *i.e.*, 3 + 4 – 1 = 6. This ensures that the solution is basic feasible. Hence, an initial basic feasible solution to the transportation problem has been obtained. It is displayed in *Table 7.21*.

	D_1	D_2	D_3	D_4
S_1	$\boxed{3}$ 3	$\boxed{2}$ 7	6	4
S_2	2	$\boxed{1}$ 4	$\boxed{1}$ 3	2
S_3	4	3	$\boxed{1}$ 8	$\boxed{2}$ 5

Table 7.21

The transportation cost according to the above route is given by

$$z = 3 \times 3 + 2 \times 7 + 1 \times 4 + 1 \times 3 + 1 \times 8 + 2 \times 5 = 48.$$

(*b*) Following the ***Least-Cost Method***, the first allocation is made in the cell (S_2, D_1) being the cell having least cost in the cost matrix. Thus, the first allocation is of magnitude x_{21} = min. (2, 3) = 2. This satisfies the supply of S_2. We cross out the second row. Second allocation is then made in the cell (S_1, D_1) being the least element in first as well as third row of the reduced cost matrix of magnitude x_{11} = min. (5, 3 – 2) = 1. This satisfies the requirement of D_1 and so we cross out the first column. Third allocation is made in the cell (S_3, D_2), the magnitude being x_{32} = min. (3, 3) = 3. This satisfies the supply at S_3 as well as the requirement of D_2. So, we cross out the third row. The requirement at D_2 will be 3 – 3 = 0 or ε, where $\varepsilon \to 0$. Thus, we allocate ε in the cell (S_1, D_2). Finally, we allocate 2 in the cell (S_1, D_3) and 2 in the cell (S_1, D_4).

Since, the number of allocations is equal to six, we have obtained an initial basic feasible solution to the transportation problem as given in *Table 7.22*.

	D_1	D_2	D_3	D_4
S_1	$\boxed{1}$ 3	$\boxed{\varepsilon}$ 7	$\boxed{2}$ 6	$\boxed{2}$ 4
S_2	$\boxed{2}$ 2	4	3	2
S_3	4	$\boxed{3}$ 3	8	5

Table 7.22

The transportation cost according to above route is given by

$$z = 1 \times 3 + \varepsilon \times 7 + 2 \times 6 + 2 \times 4 + 2 \times 2 + 3 \times 3 = 36 + 7 \times \varepsilon = 36 \text{ as } \varepsilon \to 0.$$

(*c*) Following ***Vogel's Approximation Method,*** the differences between the two successive lowest costs for each row and each column are computed. These are written besides the corresponding rows and columns under the heading 'row difference' and 'column difference':

	D_1	D_2	D_3	D_4	Row difference			
S_1	[3] 3	7	6	[2] 4	(1)	(1)	(1)	(1)
S_2	2	4	[2] 3	[ε_1] 2	(0)	(0)	—	—
S_3	4	[3] 3	8	[ε_2] 5	(1)	(1)	(1)	(1)
Column difference	(1)	(1)	③	2				
	(1)	(1)	—	②				
	(1)	④	—	(1)				
	①	—	—	(1)				

Table 7.23

The largest row difference and column difference is 3 which corresponds to third column. Therefore, allocation of as much as possible to the lowest cost cell in the third column gives $x_{23} = 2$. This satisfies the supply at S_2 and the requirement of D_3. So, arbitrarily we cross off the third column and consider ε_1 ($\to 0$) as the small quantity to be supplied from S_2. The row and column differences are now recomputed for the reduced cost matrix, and we choose the largest of these, *i.e.*, 2 which corresponds to the fourth column. We, therefore, allocate $x_{24} = \varepsilon_1$ and second row is thus eliminated. Again, we compute the row and column differences for the reduced 2×3 cost matrix and choose the largest of these, *i.e.*, 4 corresponding to the second column. We thus, allocate $x_{32} = 3$ and eliminate second column while consider ε_2 ($\to 0$) being the negligible positive quantity to be supplied from S_3. Again, computing the column and row differences in the reduced 2×2 cost matrix, we assign $x_{11} = 3$ and thus, eliminate first column. This leaves now only fourth column, which can be filled by inspection and thus, we assign $x_{14} = 2$ and $x_{34} = \varepsilon_2$.

The transportation cost according to above allocation is given by

$$z = 3 \times 3 + 2 \times 4 + 2 \times 3 + \varepsilon_1 \times 2 + 3 \times 3 + \varepsilon_2 \times 5 = 32 + 2\varepsilon_1 + 5\varepsilon_2 = 32 \text{ as } \varepsilon_1 \to 0 \text{ and } \varepsilon_2 \to 0.$$

This cost is less than the cost associated with the earlier solution obtained by North-West Corner Method and Least-Cost Method.

Example 7.7. *Find the initial basic feasible solution to the following transportation problem using North-West Corner Method (rule):*

Source	Destination					Supply
	1	2	3	4	5	
S_1	15	14	14	11	16	50
S_2	13	12	14	17	14	60
S_3	18	17	19	22	M	50
S_4	M	0	M	0	0	50
Demand	30	20	70	30	60	

Solution. Since, total demand = total supply, there exists a feasible solution to the transportation problem. Also, because $m = 4$ and $n = 5$, number of basic cells (variables) will be $m + n - 1$, *i.e.*, $4 + 5 - 1$ or 8.

Following North-West Corner Method (rule), the first allocation is $x_{11} = 30$, which exhausts the demand in column 1. This first iteration leaves a supply of 20 remaining in S_1-row.

Second allocation is made in the cell (1, 2). Since the supply in S_1-row is, now, no larger than the demand of 20, all of it is allocated, *i.e.*, $x_{12} = 20$, and this row is eliminated from further consideration. For the third allocation, we move to S_2-row in second column. Now, since the remaining demand in second column is of 0 is less than the supply of 60 in S_2-row, we allocate $x_{22} = \varepsilon$ (where $\varepsilon \to 0$) and eliminate second column for further consideration. For the fourth allocation, we move now in the third column and S_2-row. Fourth allocation will be x_{23} = min. (60, 70), *i.e.*, 60 which exhausts S_2-row and leaves a demand of 10 in third column. So, we eliminate S_2-row and move to third row in third column.

Continuing in this manner, we eventually obtain the initial basic feasible solution (BFS) as shown in *Table 7.24*.

	D_1	D_2	D_3	D_4	D_5
S_1	[30] 15	[20] 14	14	11	16
S_2	13	[ε] 12	[60] 14	17	14
S_3	18	17	[10] 19	[30] 22	[10] M
S_4	M	0	M	0	[50] 0

Table 7.24

The transportation cost according to above route is given by

$$z = 30 \times 15 + 20 \times 14 + \varepsilon \times 12 + 60 \times 14 + 10 \times 19 + 30 \times 22 + 10 \times M + 50 \times 0$$
$$= 2420 + 10M + 12\varepsilon = 2420 + 10M\,, \text{ as } \varepsilon \to 0.$$

Note. It may be noted that $x_{11} = 30, x_{12} = 20, x_{22} = 0, x_{23} = 60, x_{33} = 10, x_{34} = 30, x_{35} = 10$ and $x_{45} = 50$ are the only basic variables and since one of the basic variable has zero value, there exists degenerate basic feasible solution. All other non-basic variables are set equal to zero.

PROBLEMS

7.01. Using north-west corner rule, develop an initial basic feasible solution to the following transportation problem:

Source	*Destination*			*Supply*
	D_1	D_2	D_3	
S_1	7	9	3	19
S_2	4	8	7	13
S_3	2	5	9	7
Demand	15	13	11	39

7.02. Find the initial basic feasible solution using Vogel's Approximation Method:

Origin	*Destination*				*Supply*
	1	2	3	4	
O_1	10	22	0	22	8
O2	15	20	12	8	13
O_3	20	12	10	15	11
Demand	5	11	8	8	

[MBA, Madras Univ. 2010]

7.03. Determine an initial basic feasible solution to the following transportation problem by Vogel's Approximation Method:

2	3	11	7	6
11	0	6	1	1
5	8	15	9	10
7	5	3	2	

[MBA, Madurai Univ. 2009]

7.04. Find the initial basic feasible solution to the following transportation problem using VAM, given the cost matrix:

	D_1	D_2	D_3	D_4	*Supply*
S_1	20	25	28	31	200
S_2	32	28	32	41	180
S_3	18	35	24	32	110
Demand:	150	40	180	170	

[MBA, Osmania Univ. 2001]

7.05. Obtain an initial basic feasible solution to the following transportation problem by (*i*) Least-Cost Method, and (*ii*) Vogel's Approximation Method:

Warehouse	*Destination*				*Availability*
	I	*II*	*III*	*IV*	
A	7	3	5	5	34
B	5	5	7	6	15
C	8	6	6	5	12
D	6	1	6	4	19
Demand	21	25	17	17	80

[M.Com., Mahatma Gandhi Univ. 2010]

7.06. A construction firm has currently undertaken three projects, one each in Noida, Greater Noida and Gurugram. Each project requires a specific supply of gravel which the firm sources from Dehradun, Aligarh and Kanpur. Shipping costs in rupees differ from location to location. The following table summarizes the problem, the manager of the construction firm faces:

From	*To*			*Available Supply (tons)*
	Noida	*Greater Noida*	*Gurugram*	
Dehradun	450	400	350	1,500
Aligarh	350	550	300	1,750
Kanpur	200	150	600	2,750
Demand (tons)	2,000	3,000	1,000	

Formulate the above problem with a view of minimizing the total shipping cost. Also, find out an initial basic feasible solution by applying any of the three methods.

[MBA, Delhi Univ. 2015]

7:10. OPTIMALITY TEST

Following the determination of an initial basic feasible solution to a transportation problem, the next question is to arrive at the optimum solution. There are two methods (algorithms) for testing the optimality of a solution to a transportation problem, *viz.*,

1. Stepping Stone Method.
2. Modified Distribution Method (or MODI Method).

1. Stepping Stone Method

Starting from the initial basic feasible solution of a transportation problem, following steps are involved in arriving at optimum solution to the problem:

Step 1. Check row-wise and select an unoccupied cell. Draw a closed path using only horizontal and vertical straight lines beginning and ending at the selected unoccupied cell and having its other corners at some occupied cells only. The cells at the corner points are called ***stepping stones*** on the closed path.

Step 2. Assign '+' and '–' sign alternatively on each corner point of the cell, starting with a '+' sign at the unoccupied cell to be evaluated.

Step 3. Compute the "net change in cost" along the closed path by adding the unit cost figures of each cell containing a '+' sign and then subtracting the unit costs of each cell containing the '–' sign.

Step 4. Repeat *Steps 1* to *3* for all unoccupied cells.

Step 5. Check the sign of each of the net changes. In case all the net changes are non-negative, an optimum solution is attained. Otherwise, it is possible to improve the current solution and decrease the total shipping cost.

Step 6. Select that unoccupied cell which have the highest negative net cost change and determine the maximum number of units that can be allocated to this cell. This is being done by assigning minimum number of units to the cells marked '–' in such a way that one of these cells becomes zero and the remaining cells does not become negative.

Add this number to the unoccupied cell and all other occupied cells marked '+' on the closed path and subtract the same number form those cells on the closed path which are marked '–'.

Step 7. Go to **Step 1** and repeat the procedure until an optimum solution is attained.

Note. It there is allocation with zero increase or decrease in the total cost, then the transportation problem has multiple solutions having the same total cost.

Example 7.8. *A company has factories at F_1, F_2 and F_3 which supply warehouses at W_1, W_2 and W_3. Weekly factory capacities are 200, 160 and 90 units respectively. Weekly warehouses requirements are 180, 120 and 150 units, respectively. Unit shipping costs (in rupees) are given as follows:*

Factory	*Warehouse*			*Supply*
	W_1	W_2	W_3	
F_1	16	20	12	200
F_2	14	8	18	160
F_3	26	24	16	90
Demand	180	120	150	450

Determine the optimum distribution for the company to minimize shipping costs.

[M.Com., Osmania Univ. 2010]

Solution. Following *Least-Cost Method*, an initial basic feasible solution of the transportation problem is displayed in *Table 7.25*.

	W_1	W_2	W_3
F_1	50 / 16	20	150 / 12
F_2	40 / 14	120 / 8	18
F_3	90 / 26	24	16

Table 7.25

The Stepping Stone Method starts with an evaluation of each of the unoccupied cells, *viz.*, (F_1, W_2), (F_2, W_3), (F_3, W_2), and (F_3, W_3) to examine whether it would improve the initial basic feasible solution by introducing any of these cells into the current solution basis. To demonstrate the application of this method let us evaluate the unoccupied cell (F_1, W_2) of *Table 7.25* and assign θ units to be shipped from factory F_1 to warehouse W_2 [Note that an addition of a unit to cell (F_1, W_2) adds ₹ 20 to the cost]. Since the requirements of the warehouse must be satisfied, we adjust the allocation of cell (F_1, W_1) by subtracting θ units from the initial assignment of 50 units which decreases the cost by ₹ 16 per unit. In order to make this adjustment and to satisfy the supply restriction at factories F_1 and F_2, we subtract θ units from cell (F_2, W_2) and then add θ units to cell (F_2, W_1). The closed path followed in making these adjustments is given in *Table 7.26*.

By following the closed path and analysing the unit shipment, we can compute the incremental cost evaluation of unoccupied cell (F_1, W_2) as follows:

Cell (i, j)	*Change in x_{ij}*	*Change in total cost*
(F_1, W2)	+1	+20 × 1
(F_1, W_1)	–1	–16 × 1
(F_2, W_1)	+1	+14 × 1
(F_2, W_2)	–1	–8 × 1
		+10 Net change

This shows that shift of one unit into unoccupied cell (F_1, W_2) increases the transportation cost by ₹ 10, and hence cell (F_1, W_2) is not included in the new solution. Now, we compute the net cost change for the unoccupied cell (F_2, W_3). As before, we allocate θ units to the unoccupied cell (F_2, W_3) and make necessary adjustments in the occupied cells (F_1, W_3), (F_1, W_1) and (F_2, W_1). The closed path associated with this assignment is shown in *Table 7.27*. The net change in cost for a unit shipment cost will be us under:

$$₹\,18 \times 1 - ₹\,12 \times 1 + ₹\,16 \times 1 - ₹\,14 \times 1, \textit{ i.e., } ₹\,8.$$

As this route is more expensive than our initial solution, we shall not plan any change based on this evaluation.

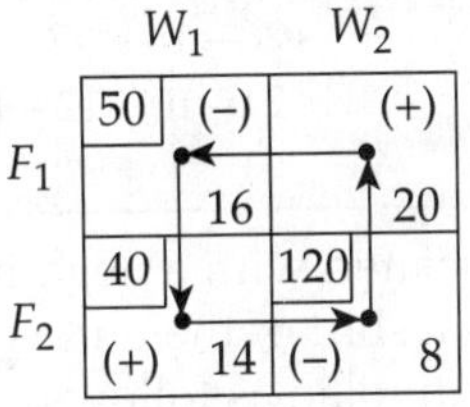

Table 7.26

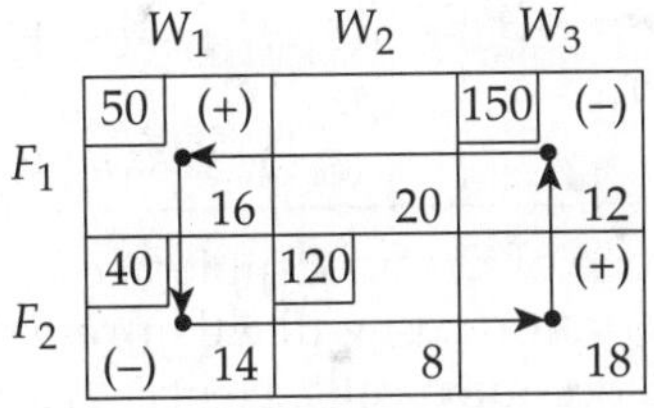

Table 7.27

We, now, evaluate the remaining unoccupied cells (F_3, W_2) and (F_3, W_3). The closed path associated with these cells are shown below:

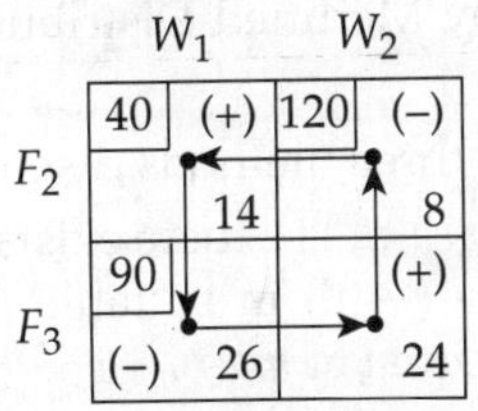

Table 7.28

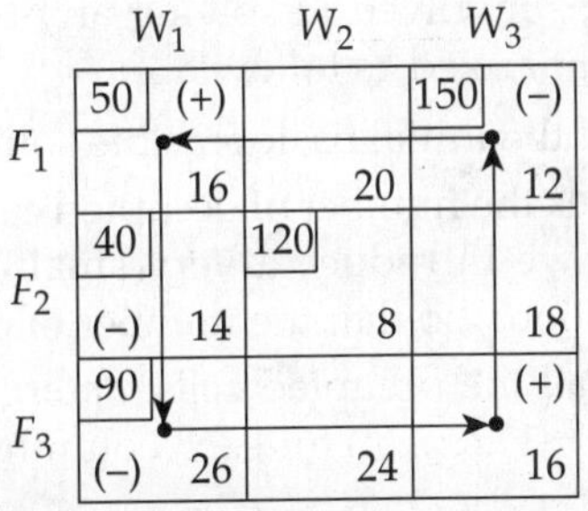

Table 7.29

The closed path and net changes in cost for each of the unoccupied cells is given below:

Unoccupied cell	*Closed path*	*Net change in cost (in ₹)*
(F_1, W_2)	$(F_1, W_2) \rightarrow (F_1, W_1) \rightarrow (F_2, W_1) \rightarrow (F_2, W_2)$	$+20 - 16 + 14 - 8 = 10$
(F_2, W_3)	$(F_2, W_3) \rightarrow (F_1, W_3) \rightarrow (F_1, W_1) \rightarrow (F_2, W_1)$	$+18 - 12 + 16 - 14 = 8$
(F_3, W_2)	$(F_3, W_2) \rightarrow (F_2, W_2) \rightarrow (F_2, W_1) \rightarrow (F_3, W_1)$	$+24 - 8 + 14 - 26 = 4$
(F_3, W_3)	$(F_3, W_3) \rightarrow (F_1, W_3) \rightarrow (F_1, W_1) \rightarrow (F_3, W_1)$	$+16 - 12 + 16 - 26 = -6$

Since, the unoccupied cell (F_3, W_3) is associated with negative value, this cell will be included in the new solution. This is because the shipping of one unit from factory F_3 to warehouse W_3 reduces the transportation cost by ₹ 6. So, we let θ units are shipped from F_3 to W_3. Make +θ and –θ adjustments among the occupied cells on the closed path which starts from (F_3, W_3) as shown in *Table 7.29*. Then, since $(150 - \theta) \geq 0$ and $(90 - \theta) \geq 0$, we choose $\theta = 90$ as θ is to be made as large as possible. The revised basic feasible solution, therefore, is displayed in *Table 7.30*.

	W_1	W_2	W_3
F_1	50 + 90 / 16	20	150 – 90 / 12
F_2	40 / 14	120 / 8	18
F_3	26	24	90 / 16

Table 7.30

Now, (F_1, W_2), (F_2, W_3), (F_3, W_1) and (F_3, W_2) are the unoccupied cells. So, examine each of these cells as to whether or not shipping of units from any one of them results in further savings in the transportation cost.

The closed path and an incremental cost change for each of the unoccupied cells are given below:

Unoccupied cell	*Closed path*	*Net change in cost (in ₹)*
(F_1, W_2)	$(F_1, W_2) \rightarrow (F_2, W_2) \rightarrow (F_2, W_1) \rightarrow (F_1, W_1)$	$+20 - 8 + 14 - 16 = 10$
(F_2, W_3)	$(F_2, W_3) \rightarrow (F_1, W_3) \rightarrow (F_1, W_1) \rightarrow (F_2, W_1)$	$+18 - 12 + 16 - 14 = 8$
(F_3, W_1)	$(F_3, W_1) \rightarrow (F_3, W_3) \rightarrow (F_1, W_3) \rightarrow (F_1, W_1)$	$+26 - 16 + 12 - 16 = 6$
(F_3, W_2)	$(F_3, W_2) \rightarrow (F_3, W_3) \rightarrow (F_1, W_3) \rightarrow (F_1, W_1) \rightarrow (F_2, W_1) \rightarrow (F_2, W_2)$	$+24 - 16 + 12 - 16 + 14 - 8 = 10$

Since, none of the unoccupied cells in the above table is associated with a negative value, the total transportation cost will either remain the same or increase if a shift of the unit is made to any one of the unoccupied cells. Hence, *Table 7.30* provides an optimum plan as follows:

F_1: $x_{11} = 140$, $x_{12} = 0$, $x_{13} = 60$; F_2: $x_{21} = 40$, $x_{22} = 120$, $x_{23} = 0$; F_3: $x_{31} = 0$, and $x_{32} = 0$, $x_{33} = 90$ with total transportation cost = ₹ 5,920.

2. Modified Distribution Method (or MODI Method)

Various steps involved in solving any transportation problem by Modified Distribution Method may be summarized as follows:

Step 1. Find the initial basic feasible solution by using any of the three methods discussed earlier.

Step 2. Check the number of occupied cells. If there are less than $(m + n - 1)$, there exists degeneracy and we introduce a very small positive assignment of ε $(\rightarrow 0)$ in suitable independent positions, so that the number of occupied cells is exactly equal to $m + n - 1$.

Step 3. For all the occupied cells, ***determine a set of numbers*** u_i $(i = 1, 2, ..., m)$ for each row and v_j $(j = 1, 2, ..., n)$ for each column by solving the system of equations:

$$u_i + v_j = c_{ij}$$

Because the number of unknown u_i and v_j are $m + n$, we can assign an arbitrary value (conveniently zero) to one of the unknowns without violating the equations.

Step 4. Compute the ***opportunity costs (net evaluations)*** for all the unoccupied cells by using the relationship:

$$d_{ij} = u_i + v_j - c_{ij}, \qquad \text{for } i = 1, 2, ..., m \text{ and } j = 1, 2, ..., n.$$

Step 5. Check the sign of each opportunity cost. If all the opportunity costs are less than or equal to zero, an optimum solution is reached. Otherwise go to ***next step***.

Step 6. Select the ***largest positive opportunity cost*** computed in *Step 4*. The unoccupied (non-basic) cell corresponding to this positive value becomes occupied (basic) in the next iteration.

Step 7. Determine a closed path for the current entering cell that starts and ends at this unoccupied cell. Right angle turns in this closed path are permitted only at the occupied cells and at the ***current entering (unoccupied) cell***.

Step 8. Assign alternate plus and minus signs at the corner points of occupied cells on the closed path starting with plus sign at the current entering cell (unoccupied cell).

Step 9. Determine the ***maximum number of units*** that should be shipped to the selected unoccupied cell. This is being done by choosing the least number of units that can be allocated to the occupied cells having minus sign. Add this quantity to all the cells on the closed path marked with (+) sign and subtract the same from the cells marked with (–) sign.

Step 10. Go to ***Step 3*** and repeat the procedure until an optimum solution is attained.

Remarks

1. The method described above is commonly known as MODI method.
2. For ***maximization transportation*** problems, replace each element of the profit matrix by its difference from the highest element of the profit matrix. Then, apply the steps of minimization transportation problem on the modified transportation problem.
3. If the optimum solution to a given transportation problem has ***zero opportunity cost*** for one or more unoccupied cells, then there are ***alternative optimum solutions*** to this transportation problem having the same total transportation cost (optimum).
4. At any stage, if number of occupied cells are less than $m + n - 1$, ***degeneracy exists***, *i.e.*, there is ***degenerate basic feasible solution***. In such a situation, we introduce a very small positive quantity ε $(\rightarrow 0)$ in suitable independent position, so that the number of occupied cells are exactly equal to $m + n - 1$. Allocation of ε should be in the lowest cost unoccupied cell.
5. There are situations, when aggregate supply is not equal to aggregate demand. Two possible situations, then, are:
   (*a*) The aggregate supply > aggregate demand.
   (*b*) The aggregate supply < aggregate demand.

The problems under the above two categories are called ***unbalanced transportation problems***.

In the ***first category***, the excess supply is assumed to go to inventory and costs nothing for shipping. A column is added to the transportation table which represents a ***dummy destination*** with a requirement equal to the amount of excess supply and transportation costs equal to zero. In the ***second category***, the excess of demand is achieved by adding an additional row to the transportation table. The additional row represents a ***dummy source (or capacity or origin)*** with capacity equal to the excess demand and each of the cells having zero unit costs.

Once the transportation problem is balanced, its solution proceeds in exactly the same manner as discussed above.

Example 7.9. *Solve the following transportation problem:*

From	*To* A	B	C	*Available*
I	50	30	220	1
II	90	45	170	3
III	250	200	50	4
Requirement	4	2	2	

[B.Com., Madras Univ., 2004]

Solution. Using ***north-west corner rule***, an initial basic feasible solution is obtained in *Table 7.31*.

	50 [1]	30	220	1
	90 [3]	45 [0]	170	3
	250	200 [2]	50 [2]	4
	4	2	2	

Table 7.31

Since, the number of occupied cells are 5 (*i.e.*, 3 + 3 – 1); the solution obtained in *Table 7.31* is basic feasible. Moreover, as one of the basic cell have zero value, there exists degeneracy. So we assign a small positive quantity $\varepsilon\ (\to 0)$ to the cell (2, 2) to overcome degeneracy.

Initial Iteration. Compute the unknown numbers, $u_i\ (i = 1, 2, 3)$ and $v_j\ (j = 1, 2, 3)$ using successively the equations $u_i + v_j = c_{ij}$ for all the occupied cells, starting with $u_1 = 0$. Thus, we have

$$u_1 + v_1 = c_{11} \quad\Rightarrow\quad 0 + v_1 = 50 \quad\Rightarrow\quad v_1 = 50$$
$$u_2 + v_1 = c_{21} \quad\Rightarrow\quad u_2 + 50 = 90 \quad\Rightarrow\quad u_2 = 40$$
$$u_2 + v_2 = c_{22} \quad\Rightarrow\quad 40 + v_2 = 45 \quad\Rightarrow\quad v_2 = 5$$
$$u_3 + v_2 = c_{32} \quad\Rightarrow\quad u_3 + 5 = 200 \quad\Rightarrow\quad u_3 = 195$$
$$u_3 + v_3 = c_{33} \quad\Rightarrow\quad 195 + v_3 = 50 \quad\Rightarrow\quad v_3 = -145$$

Making use of these values of $u_i\ (i = 1, 2, 3)$ and $v_j\ (j = 1, 2, 3)$; the opportunity cost for each of the unoccupied cells is computed as follows:

Unoccupied cell	*Opportunity cost* (d_{ij})
(1, 2)	$d_{12} = u_1 + v_2 - c_{12} = 0 + 5 - 30 = -25$
(1, 3)	$d_{13} = u_1 + v_3 - c_{13} = 0 + (-145) - 220 = -365$
(2, 3)	$d_{23} = u_2 + v_3 - c_{23} = 40 - 145 - 170 = -275$
(3, 1)	$d_{31} = u_3 + v_1 - c_{31} = 195 + 50 - 250 = -5$

It is more convenient to present a more compact form for computing the unknown u_i's and v_j's and then evaluate the opportunity cost for each of the unoccupied cells. For this the transportation table is augmented by ***an additional rightmost column*** giving the values of u_i's and an additional ***bottom row*** giving the values of v_j's. The revised transportation table is *Table 7.32*.

	A	B	C	u_i
I	50 [1]	30 [–25]	220 [–ve]	0
II	90 [3]	45 [ε]	170 [–275]	40
III	250 [–5]	200 [2]	50 [2]	195
v_j	50	5	–145	

Table 7.32

Since, all the opportunity costs in *Table 7.32* are negative, an optimum solution is reached. The optimum solution is given by

$$x_{11} = 1,\ x_{21} = 3,\ x_{22} = \varepsilon,\ x_{32} = 2 \text{ and } x_{33} = 2.$$

Total minimum cost according to this schedule is

$$z = 1 \times 50 + 3 \times 90 + \varepsilon \times 45 + 2 \times 200 + 2 \times 50$$
$$= 820 + 45\,\varepsilon = 820, \text{ as } \varepsilon \to 0.$$

Example 7.10. *Solve the following transportation problem:*

From	*To* A	B	C	*Supply*
I	6	9	4	14
II	4	9	8	12
III	1	2	6	5
Demand	6	10	15	

Solution. Using ***Vogel's Approximation Method***, an initial basic feasible solution to the transportation problem is obtained. See *Table 7.33*:

Since, the number of occupied cells are 3 + 3 – 1; the solution obtained in *Table 7.33* is non-degenerate basic feasible one. Using, now, MODI method, an optimum solution is obtained in the following iteration.

	A	B	C	*Row difference*
I	6	9	14 [4]	2 2 ⑤
II	6 [4]	5 [9]	1 [8]	4 ④ 1
III	1	5 [2]	6	1 — —
Column difference	3 2 —	⑦ 0 0	2 4 4	

Table 7.33

Initial Iteration. Compute the unknown numbers u_i $(i = 1, 2, 3)$ and v_j $(j = 1, 2, 3)$; using successively the equations $u_i + v_j = c_{ij}$ for all the occupied cells, starting with $u_2 = 0$. Thus, we have

$$u_1 + v_3 = c_{13} \Rightarrow u_1 + 8 = 4 \Rightarrow u_1 = -4$$
$$u_2 + v_1 = c_{21} \Rightarrow 0 + v_1 = 4 \Rightarrow v_1 = 4$$
$$u_2 + v_2 = c_{22} \Rightarrow 0 + v_2 = 9 \Rightarrow v_2 = 9$$
$$u_2 + v_3 = c_{23} \Rightarrow 0 + v_3 = 8 \Rightarrow v_3 = 8$$
$$u_3 + v_2 = c_{32} \Rightarrow u_3 + 9 = 2 \Rightarrow u_3 = -7$$

Using these values of u_i $(i = 1, 2, 3)$ and v_j $(j = 1, 2, 3)$; the opportunity cost for each of the unoccupied cells is computed as follows:

Unoccupied Cell	*Opportunity Cost* (d_{ij})
(1, 1)	$d_{11} = u_1 + v_1 - c_{11} = -4 + 4 - 6 = -6$
(1, 2)	$d_{12} = u_1 + v_2 - c_{12} = -4 + 9 - 9 = -4$
(3, 1)	$d_{31} = u_3 + v_1 - c_{31} = -7 + 4 - 1 = -4$
(3, 3)	$d_{33} = u_3 + v_3 - c_{33} = -7 + 8 - 6 = -5$

All these calculations are presented in more compact form in the transportation *Table 7.34*, where values of u_i are presented by an additional right most column and values of v_j are given at the additional bottom row. Opportunity costs are given in the lower left corner of each unoccupied cell.

	A	B	C	u_i
I	(–6) 6	(–4) 9	14 [4]	–4
II	6 [4]	5 [9]	1 [8]	0
III	(–4) 1	5 [2]	(–5) 6	–7
v_j	4	9	8	

Table 7.34

Since, all the opportunity costs in *Table 7.34* are negative, current solution is an optimum one. Hence, the optimum allocation is:

$$x_{13} = 14, x_{21} = 6, x_{22} = 5, x_{23} = 1 \text{ and } x_{32} = 5.$$

The transportation cost according to this optimum allocation is given by

$$z = 14 \times 4 + 6 \times 4 + 5 \times 9 + 1 \times 8 + 5 \times 2, \text{ i.e., } 143.$$

Example 7.11 ***(Cost Minimization).*** *ABC Limited has three production shops supplying a product to five warehouses. The cost of production varies from shop to shop, and cost of transportation from one shop to a warehouse also varies. Each shop has a specific production capacity, and each warehouse has certain amount of requirement. Cost of transportation are as given below:*

Shop	*Warehouse*					*Supply*
	I	*II*	*III*	*IV*	*V*	
A	6	4	4	7	5	100
B	5	6	7	4	8	125
C	3	4	6	3	4	175
Demand	60	80	85	105	70	400

The cost of manufacturing of the product at different production shops are:

Shop	:	A	B	C
Variable cost	:	14	16	15
Fixed cost	:	7,000	4,000	5,000

Find the optimum quantity to be supplied from each shop to different warehouses at minimum total cost.

[MBA, Delhi Univ. 2008; MBA (Nov.), Nagpur Univ. 1998]

Solution. In the transportation cost matrix, the data pertaining to fixed costs is of no use. We shall include only the transportation costs plus the variable costs as shown below:

Shop	*Warehouse*					*Supply*
	I	*II*	*III*	*IV*	*V*	
A	6 + 14 = 20	4 + 14 = 18	4 + 14 = 18	7 + 14 = 21	5 + 14 = 19	100
B	5 + 16 = 21	6 + 16 = 22	7 + 16 = 23	4 + 16 = 20	8 + 16 = 24	125
C	3 + 15 = 18	4 + 15 = 19	6 + 15 = 21	3 + 15 = 18	4 + 15 = 19	175
Demand	60	80	85	105	70	400

Using ***Vogel's Approximation Method,*** the following initial basic feasible solution to the transportation problem is obtained as in *Table 7.35*.

Now, using *MODI* method, the optimum solution as given in *Table 7.36* is obtained where $c_{ij} = u_i + v_j$ for basic cells and $d_{ij} = u_i + v_j - c_{ij}$ for non-basic cells.

20	[15] 18	[85] 18	21	19
21	[20] 22	23	[105] 20	24
[60] 18	[45] 19	21	18	[70] 19

Table 7.35

						u_i
	(−3) 20	[15] 18	[85] 18	(−5) 21	(−1) 19	0
	(0) 21	[20] 22	(−1) 23	[105] 20	(−2) 24	4
	[60] 18	[45] 19	(−2) 21	(−1) 18	[70] 19	1
v_j	17	18	18	16	18	

Table 7.36

Since, all the opportunity costs $d_{ij} \le 0$, an optimum solution is attained, *viz*, $x_{12} = 15$, $x_{13} = 85$, $x_{22} = 20$, $x_{24} = 105$, $x_{31} = 60$, $x_{32} = 45$ and $x_{35} = 70$. The transportation cost associated with this solution is:

$$15 \times 18 + 85 \times 18 + 20 \times 22 + 105 \times 20 + 60 \times 18 + 45 \times 19 + 70 \times 19, \textit{ i.e.}, 7605.$$

Therefore, Total minimum cost = Total Transportation Cost + Total Fixed Cost

= 7605 + 16,000 = 23,605

Hence, an optimum solution is to transport from shop *A*: 15 units to warehouse *II* and 85 units to warehouse *III*, from shop *B*: 20 units to warehouse *II* and 105 units to warehouse *IV*, from shop *C*: 60 units to warehouse *I*, 45 units to warehouse *II* and 70 units to warehouse *V*. Total minimum cost will be 23,605.

Example 7.12 *(Alternative Optimum Solution).* *Consider the problem given in* ***Example 7.11*** *and find an alternative optimum solution.*

Solution. We observe that the optimum solution of the transportation problem has opportunity cost of the unoccupied cell (2, 1) as zero, *i.e.*, $d_{12} = 0$. This is the indication for the existence of an ***alternative optimum solution***. So, we assign θ units to the unoccupied cell (2, 1), *i.e.*, $x_{21} = \theta$ and make + θ and – θ adjustments to the occupied (basic) cells to have a closed path that starts and ends at the cell (2, 1). Thus, we get the transportation *Table 7.37*.

In *Table 7.37*, since (20 – θ) and (60 – θ) should be non-negative, we choose θ = 20 the maximum number of units that should be shipped to satisfy the non-negative condition. Thus, $x_{22} = 20 - 20\ (= 0)$ and $x_{31} = 60 - 20\ (= 40)$. Therefore, in the next solution current occupied cell (2, 2) becomes unoccupied, *i.e.*, x_{22} will be non-basic variable.

The revised transportation *Table 7.38* is given below:

					u_i
(−3) 20	[15] 18	[85] 18	(−5) 21	(−1) 19	0
(0) (+) 21	[20] (−) 22	(−1) 23	[105] 20	(−2) 24	4
[60] (−) 18	[45] (+) 19	(−2) 21	(−1) 18	[70] 19	1
v_j: 17	18	18	16	18	

Table 7.37

					u_i
(−3) 20	[15] 18	[85] 18	(−5) 21	(−1) 19	0
[20] 21	(0) 22	(−1) 23	[105] 20	(−2) 24	4
[40] 18	[65] 19	(−2) 21	(−1) 18	[70] 19	1
v_j: 17	18	18	16	18	

Table 7.38

Here, since all the opportunity costs $d_{ij} \le 0$, an optimum solution is reached. Thus, the alternative optimum solution is:

$$x_{12} = 15,\ x_{13} = 85,\ x_{21} = 20,\ x_{24} = 105,\ x_{31} = 40,\ x_{32} = 65 \text{ and } x_{35} = 70.$$

The transportation cost associated with this solution is:

$$15 \times 18 + 85 \times 18 + 20 \times 21 + 105 \times 20 + 40 \times 18 + 65 \times 19 + 70 \times 19,\ i.e.,\ 7605;$$

which is same as was obtained in previous example. Hence, an alternative optimum solution is to transport from shop *A*: 15 units to *II*, and 85 units to *III*; from shop *B*: 20 units to *I*, and 105 units to *IV*; from shop *C*: 40 units to *I*, 65 units to *II* and 70 units to *V*.

Total Minimum Cost = 7605 + 16000 = 23,605

Example 7.13 *(Profit Maximization).* *A company has four factories situated in four different locations in the country and four sales agencies located in four other locations in the country. The cost of production (₹ per unit), sale price (₹ per unit), shipping cost (₹ per unit) in the cells of matrix, monthly capacities and monthly requirements are as follows:*

Factory	*Sales agency* 1	2	3	4	*Monthly capacity (units)*	*Cost of production*
A	7	5	6	4	10	10
B	3	5	4	2	15	15
C	4	6	4	5	20	16
D	8	7	6	5	15	15
Monthly Requirements (units)	8	12	18	22		
Sales Price	20	22	25	18		

Find the monthly production and distribution schedule which will maximize profit.

Solution. The profit matrix is derived by using the following equation:

Profit = Sales Price – Production Cost – Shipping Cost

For example, profit at (1, 1) = 20 – 10 – 7 = 3, profit at (2, 1) = 20 – 15 – 3 = 2,
profit at (1, 2) = 22 – 10 – 5 = 7, profit at (4, 1) = 20 – 15 – 8 = –3, etc.

The transportation problem (maximization), therefore, is as follows:

Factory	*Sales agency* 1	2	3	4	*Capacity (units)*
A	3	7	9	4	10
B	2	2	6	1	15
C	0	0	5	–3	20
D	–3	0	4	–2	15
Requirement (units)	8	12	18	22	60

In order to determine initial basic feasible solution, we convert the maximization problem into that of minimization. For this, subtract all the elements from the maximum element, *viz.*, '9'. Then, apply ***Vogel's Approximation Method*** for initial basic feasible solution. The following initial basic feasible solution to the modified transportation problem is obtained in *Table 7.39*.

					Capacity
	6	(10) 2	0	5	10
	7	7	3	(15) 9	15
	(2) 9	9	(18) 4	12	20
	(6) 12	(2) 9	5	(7) 11	15
Requirement	8	12	18	22	

Table 7.39

Now, using ***MODI*** method, the starting transportation table is displayed in *Table 7.40*, where u_i and v_j are computed using successively the equations $u_i + v_j = c_{ij}$ for all the occupied cells starting with $u_4 = 0$. Then, using these values of u_i and v_j the opportunity costs d_{ij} are computed for each of the unoccupied cell.

In *Table 7.40*, since most positive opportunity cost is for the unoccupied cell (2, 1), we assign θ units to this cell and make (+ θ) and (– θ) adjustments to occupied cells to have a closed path that starts and ends at the cell (2, 1). Thus, the unoccupied cell (2, 1) becomes occupied at the next iteration with $x_{21} = 6$ and the occupied cell (4, 1) becomes unoccupied. Thus, we have *Table 7.41*.

In *Table 7.41*, since none of the unoccupied cell is positive, we reached to an optimum solution, *viz.*,

From factory *A*: 10 units to Agency 2;
From factory *B*: 6 units to Agency 1 and 9 units to Agency 4;
From factory *C*: 2 units to Agency 1 and 18 units to Agency 3;
From factory *D*: 2 units to Agency 2 and 13 units to Agency 4.

					u_i
	(−1) 6	[10] 2	(0) 0	(−1) 5	−7
	(+) (2) 7	(−1) 7	(1) 3	[15] (−) 8	−3
	[2] 9	(−3) 9	[18] 4	(−4) 12	−3
	[6] (−) 12	[2] 9	(2) 5	[7] (+) 11	0
v_j	12	9	7	11	

Table 7.40

					u_i
	(−3) 6	[10] 2	(−2) 0	(−1) 5	−4
	[6] 7	(−1) 7	(−1) 3	[9] 8	0
	[2] 9	(−1) 9	[18] 4	(−2) 12	2
	(−2) 12	[2] 9	(0) 5	[13] 11	3
v_j	7	6	2	8	

Table 7.41

Total profit earned according to above shipment schedule will be as under:

$$10 \times ₹7 + 6 \times ₹2 + 9 \times ₹1 + 2 \times ₹0 + 18 \times ₹5 + 2 \times ₹0 + 13 \times ₹(-2), \textit{i.e.}, ₹155.$$

Remark. In the optimum transportation *Table 7.41*, since the opportunity cost $d_{43} = 0$, an alternative optimum solution exists. This can be obtained as under:

From factory *A*: 10 units to Agency 2;
From factory *B*: 15 units to Agency 4;
From factory *C*: 8 units to Agency 1 and 12 units to Agency 3;
From factory *D*: 2 units to Agency 2, 6 units to Agency 3 and 7 units to Agency 4.

Example 7.14 *(Unbalanced Transportation Problem).* *Solve the following transportation problem (cell entries give unit costs):*

Warehouse	*Market*				*Supply*
	I	*II*	*III*	*IV*	
W_1	3	4	5	6	6
W_2	2	4	4	3	10
Demand	4	6	8	6	

Solution. Since total demand exceeds total supply, we introduce a dummy warehouse having all the transportation costs equal to zero, and supply equal to (24 – 16), *i.e.*, 8. The following modified (balanced) transportation problem, therefore, is obtained is as under:

	I	*II*	*III*	*IV*	*Supply*
W_1	3	4	5	6	6
W_2	2	4	4	3	10
Dummy	0	0	0	0	8
Demand	4	6	8	6	24

Using, now, ***Vogel's Approximation Method***, an initial basic feasible solution is displayed in the transportation *Table 7.42*.

The number of occupied cells are 3 + 4 – 1, *i.e.*, 6. This indicates that the basic feasible solution obtained in *Table 7.42* is non-degenerate. So, using *MODI* method, an optimum solution is obtained in following iterations:

[4] 3	4	[2] 5	6
2	4	[4] 4	[6] 3
0	[6] 0	[2] 0	0

Table 7.42

Initial Iteration. Compute numbers u_i ($i = 1, 2, 3$) and v_j ($j = 1, 2, 3, 4$) for the occupied cells and then compute opportunity costs for the non-occupied cells. These are shown in *Table 7.43*, where we have set $u_1 = 0$.

Here, since $d_{12} > 0$, cell (1, 2) becomes occupied in the next iteration. Making usual (+) and (–) adjustments in the occupied cells forming a closed path around the cell (1, 2), we find that the current occupied cell (1, 3) becomes non-occupied as shown in *Table 7.44*.

					u_i
	4 3	(+) 1 4	2 (–) 5	–2 6	0
	0 2	0 4	4 4	6 3	–1
	–2 0	6 (–) 0	2 (+) 0	–1 0	–5
v_j	3	5	5	4	

Table 7.43

					u_i
	4 (–) 3	2 (+) 4	–1 5	–3 6	0
	(+) 1 2	0 4	4 (–) 4	6 3	0
	–1 0	4 (–) 0	4 (+) 0	–1 0	–4
v_j	3	4	4	3	

Table 7.44

First Iteration. Compute the numbers u_i ($i = 1, 2, 3$) and v_j ($j = 1, 2, 3, 4$) for the occupied cells and opportunity costs d_{ij} for non-occupied cells of the revised transportation table. Thus, we get *Table 7.44*.

Here, since $d_{21} > 0$, cell (2, 1) becomes occupied in the next iteration. Making usual (+) and (–) adjustments in the occupied cells forming a closed path around the cell (2, 1), we observe that current occupied cells (1, 1), (2, 3) and (3, 2) poses to become non-occupied. But only one cell is to become non-occupied. So, let the cell (1, 1) becomes non-occupied at the next iteration and assign the quantity $\varepsilon\ (\to 0)$ to the cells (2, 3) and (3, 2) to satisfy the condition of basic solution, *i.e.*, number of rows plus number of coloumns less one. Thus, the improved transportation table is as shown in *Table 7.45*.

					u_i
	–1 3	6 4	–1 5	–3 6	0
	4 2	0 4	ε 4	6 3	0
	–2 0	ε 0	8 0	–1 0	–4
v_j	2	4	4	3	

Table 7.45

Final Iteration. Compute numbers u_i ($i = 1, 2, 3$) and v_j ($j = 1, 2, 3, 4$) and then compute the opportunity costs d_{ij} for the non-occupied cells for the revised transportation *Table 7.45*.

Here, we observe that all the opportunity costs are negative or zero. Hence, an optimum solution is attained. Thus, the optimum degenerate solution is:

$$x_{12} = 6,\ x_{21} = 4,\ x_{23} = \varepsilon\ (\to 0),\ x_{32} = \varepsilon\ (\to 0),\ x_{33} = 8, \text{ and } x_{24} = 6,$$

i.e., from W_1, ship 6 units to *II*; from W_2, ship 4 units to *I* and 6 units to *IV*.

$$\text{Total minimum cost} = 6 \times 4 + 4 \times 2 + \varepsilon \times 4 + 6 \times 3 = 50 + 4\,\varepsilon = 50 \text{ as } \varepsilon \to 0$$

Example 7.15 ***(Unbalanced Transportation Problem).*** *ABC Enterprises is having three plants manufacturing dry-cells, located at different locations. Production cost differs from plant to plant. There are five sales offices of the company located in different regions of the country. The sales prices can differ from region to region. The shipping cost from each plant to each sales office and other data are given below:*

Demand	:	80	100	75	45	125
Sales Price	:	30	32	31	34	29
Production cost per unit (₹)	:	20	22	18		
Max. capacity in no. of units	:	150	200	125		
Plant number	:	1	2	3		

Shipping Costs (in ₹)

	Sales offices				
	A	*B*	*C*	*D*	*E*
Plant 1:	1	1	5	9	4
Plant 2:	9	7	8	3	6
Plant 3:	4	5	3	2	7

Find the production and distribution schedule most profitable to the company.

[M.Com. Poona Univ. 2009; M.I.B. Delhi Univ. 2008; M.Com. Delhi Univ. 2006]

Solution. In order to solve the transportation problem, we use the given information to derive the profit matrix. This is done as follows:

Profit = Sales Price – Production Cost – Shipping Cost

Thus, if we transport one unit of dry-cell from each of the three plants to each of the 5 sales offices, the following profit matrix is obtained:

$$\begin{array}{c} \\ 1 \\ 2 \\ 3 \end{array} \begin{array}{c} \begin{array}{ccccc} A & B & C & D & E \end{array} \\ \begin{bmatrix} 9 & 11 & 6 & 5 & 5 \\ -1 & 3 & 1 & 9 & 1 \\ 8 & 9 & 10 & 14 & 4 \end{bmatrix} \end{array}$$

Here, if we transport one unit from plant 2 to sales office *D*, the profit obtained is:

Profit = ₹ 34 – ₹ 22 – ₹ 3, *i.e.*, ₹ 9

Now, the above matrix is converted into a loss matrix by subtracting each element of the profit matrix from 14 (largest element). Thus, we have

$$\begin{array}{c} \\ 1 \\ 2 \\ 3 \end{array} \begin{array}{c} \begin{array}{ccccc} A & B & C & D & E \end{array} \\ \begin{bmatrix} 5 & 3 & 8 & 9 & 9 \\ 15 & 11 & 13 & 5 & 13 \\ 6 & 5 & 4 & 0 & 10 \end{bmatrix} \end{array}$$

Since, total demand (of 425 units) at five sales offices is less than the total supply (of 475 units) at the three plants, the problem is unbalanced. A dummy sales office with its demand of 50 units with zero transportation cost from each of the three plants to this sales office is added.

Using, now, ***Vogel's Approximation Method*** to this balanced transportation problem, the initial basic feasible solution is obtained and displayed in *Table 7.46*.

Since, there are 8 allocation, the solution is basic feasible and non-degenerate. Using, now, *MODI* method, we have the iterative transportation *Table 7.47*.

50 / 5	100 / 3	8	9	9	0
25 / 15	11	13	5	125 / 13	50 / 0
5 / 6	5	75 / 4	45 / 0	10	0

Table 7.46

						u_i
50 / 5	100 / 3	−5 / 8	−10 / 9	−6 / 9	−10 / 0	5
25 (−) / 15	2 / 11	0 / 13	4 (+) / 5	125 / 13	50 / 0	15
5 (+) / 6	−1 / 5	75 / 4	45 (−) / 0	−6 / 10	−9 / 0	6
v_j: 0	−2	−2	−6	−2	−15	

Table 7.47

In *Table 7.47*, the numbers u_i (i = 1, 2, 3) and v_j (j = 1, 2, 3, 4, 5, 6) have been computed for the occupied cells wherein we have set v_1 = 0. The opportunity costs d_{ij} are computed for the non-occupied cells.

Clearly, since d_{24} (= 4) is the largest positive opportunity cost, non-occupied cell (2, 4) becomes occupied in the next integration. Making of usual (+) and (–) adjustments in the occupied cells forming a closed path around the cell (2, 4), we find that cell (2, 1) becomes non-occupied in the revised transportation *Table 7.48*.

						u_i
50 / 5	100 / 3	(–5) 8	(–10) 9	(–2) 9	(–6) 0	–6
(–4) 15	(–2) 11	(–4) 13	25 / 5	125 / 13	50 / 0	0
30 / 6	(–1) 5	75 / 4	20 / 0	(–2) 10	(–5) 0	–5
v_j: 11	9	9	5	13	0	

Table 7.48

In *Table 7.48*, the numbers u_i and v_j have been computed for occupied cells wherein we have set u_2 = 0. The opportunity costs d_{ij} have been computed for the non-occupied cells.

Here, since all the opportunity costs are non-positive, an optimum solution is reached. The optimum allocations are:

$$x_{11} = 50,\ x_{12} = 100,\ x_{24} = 25,\ x_{25} = 125,\ x_{31} = 30,\ x_{33} = 75 \ \text{and}\ x_{34} = 20$$

$$\text{Maximum profit } = ₹\,(50 \times 9 + 100 \times 11 + 25 \times 9 + 125 \times 1 + 30 \times 8 + 75 \times 10 + 20 \times 14)$$

$$= ₹\,(450 + 1100 + 225 + 125 + 240 + 750 + 280) = ₹\,3{,}170$$

Example 7.16. *ABC Company wishes to develop a monthly production schedule for the next three months. Depending upon the sales commitments, the company can either keep the production constant, allowing fluctuations in inventory, or inventories can be maintained at a constant level, with fluctuating production. Fluctuating production necessitates in working overtime, the cost of which is estimated to be double the normal production cost of ₹ 12 per unit. Fluctuating inventories result in inventory carrying cost of ₹ 2 per unit. If the company fails to fulfil its sales commitment, it incurs a shortage cost of ₹ 4 per unit per month. The production capacities for the next three months are shown below:*

Month	*Production capacity*		*Sales (units)*
	Regular	*Overtime*	
1	50	30	60
2	50	0	120
3	60	50	40

Determine the optimal production schedule. **[*MBA, Delhi Univ. 2001*]**

Solution. The cost elements in each of the cell are determined as follows:

(*i*) If items are produced in a month for sales during the same month, then there will be no inventory carrying cost. Thus, total cost will be either the normal production cost or overtime production cost. Thus, cost elements for cells (R_1, 1), (R_2, 2) and (R_3, 3) are ₹ 12 each and for cells (O_1, 1), (O_3, 3) are ₹ 24 each.

(*ii*) If items are produced in a month for sales during subsequent month in addition to the production cost (normal or overtime), inventory carrying cost at the rate of ₹ 2 per month will be incurred.

Cell	*Production cost (₹)*	*Inventory carrying cost (₹)*	*Total cost (₹)*
$(R_1, 2)$	12	2	14
$(R_1, 3)$	12	4	16
$(O_1, 2)$	24	2	26
$(O_1, 3)$	24	4	28
$(R_2, 3)$	12	2	14

(*iii*) If the company fails to fulfil its sales commitment, it incurs a shortage cost of ₹ 4 per unit per month, in addition to the production costs (normal or overtime), carrying (storage) cost at the rate of ₹ 2 per month will be incurred.

Cell	*Production cost (₹)*	*Inventory carrying cost (₹)*	*Total cost (₹)*
$(R_2, 1)$	12	4	16
$(R_3, 2)$	12	4	16
$(R_3, 1)$	12	8	20
$(O_3, 2)$	24	4	28
$(O_3, 1)$	24	8	32

Using above information, the given problem can be presented in tabular form as follows:

Month	*Total cost (₹)*					*Demand (units)*
	R_1	O_1	R_2	R_3	O_3	
M_1	12	24	16	20	32	60
M_2	14	26	12	16	28	120
M_3	16	28	14	12	24	40
Supply (units)	50	30	50	60	50	

Since, total supply exceeds total demand, we add a dummy month with its capacity as 20 units having zero transportation costs from dummy month to R_1, O_1, R_2, R_3 and O_3. Using ***Vogel's Approximation Method***, the initial basic feasible solution is obtained as displayed in the *Table 7.49* given below:

	R_1	O_1	R_2	R_3	O_3	u_i
M_1	[50] 12	[10] 24	(−6) 16	(−6) 20	(−6) 32	−2
M_2	(0) 14	[20] 26	[50] 12	[20] 16	[30] 28	0
M_3	(−6) 16	(−6) 28	(−6) 14	[40] 12	(0) 24	−4
Dummy	(−14) 0	(−2) 0	(−16) 0	(−12) 0	[20] 0	−28
v_j	14	26	12	16	28	

Table 7.49

Since, there are 4 + 5 – 1 (= 8) allocations, the solution is non-degenerate. Using *MODI*'s algorithm, we observe that all opportunity costs in the unoccupied cells are non-positive. Therefore, the optimum solution is reached and the associated total minimum cost is ₹ 3,400. Optimum production schedule is:

	Production in		*No. of units*	*For use of*
	M_1	R_1	50	50 units in M_1
		O_1	30	10 units in M_1 and 20 units in M_2
Month	M_2	R_2	50	50 units in M_2
	M_3	R_3	60	20 units in M_2 and 40 units in M_3
		O_3	30	30 units in M_2

PROBLEMS

7.07 (*a*) Solve the following transportation problem using Vogel's Approx. Method:

Factory	*Distributor (cost of transportation (₹))*				*Supply*
	P	*Q*	*R*	*S*	
A	10	18	11	7	20
B	9	12	14	6	40
C	8	9	12	10	35
Demand	16	18	31	30	

[B.Tech. (Mech.), HP Univ. 2015; MBA, Osmania Univ. 2010]

(*b*) A company has three factories at Amethi, Baghpat and Gwalior; and four distribution centres at Prayagraj, Mumbai, Kolkata and Delhi. With identical cost of production at the three factories, only variable cost involved is transportation cost. The production at the factories is: 5,000 tonnes; 6,000 tonnes and 2,500 tonnes, respectively. The demand at four distribution centres is: 6,000 tonnes; 4,000 tonnes; 2,000 tonnes and 1,500 tonnes, respectively. The transportation costs per tonne from different factories to different centres are given below:

Factory	*Distribution centre*			
	Prayagraj	*Mumbai*	*Kolkata*	*Delhi*
Amethi	3	2	7	6
Baghpat	7	5	2	3
Gwalior	2	5	4	5

Suggest the optimum transportation schedule and find the minimum cost of transportation. ***[MBA, Panjab Tech. Univ. 2011; MBA (PT), Delhi Univ. 2008]***

7.08. A company has three cement factories *A*, *B* and *C*, and four area distributors *W*, *X*, *Y* and *Z*. With identical costs of production at three factories, the only variable cost involved is the transportation cost. The monthly production capacity (in tonne) of the three factories, monthly demand of the four distributors, and the transportation cost per tonne (in rupees) from different factories to different distribution centres, are given below:

Factory	*Distributor*				*Monthly availability*
	W	*X*	*Y*	*Z*	
A	20	25	50	10	45,000
B	45	50	15	40	50,000
C	22	10	45	35	55,000
Monthly demand	50,000	40,000	30,000	30,000	

Suggest an optimal transport schedule and find the minimum transportation cost.

[M.Com., Delhi Univ. 1999]

7.09. A wholesale company has three warehouses from which supplies are drawn for four retail customers. The company deals in a single product, the supplies of which at each warehouse are given below:

Warehouse	*Supply (units)*	*Customer*	*Demand (units)*
1	20	1	15
2	28	2	19
3	17	3	13
		4	18

Conveniently, total supply at the warehouses is equal to total demand from the customers. Following table gives the transportation costs per unit shipment from each warehouse to each customer:

Warehouse	*Customer*			
	1	*2*	*3*	*4*
1	3	6	8	5
2	6	1	2	5
3	7	8	3	9

Determine what supplies to despatch from each of the warehouse to each customer so as to minimize overall transportation cost. ***[M.Com., BHU 2010]***

7.10. A company is spending ₹ 1,000 on transportation of his product from three plants to four distribution centres. The supply and demand of its product with unit cost of transportation are given as below:

Plant	*Distribution centre*				*Capacity (tons)*
	D_1	D_2	D_3	D_4	
P_1	19	30	50	12	7
P_2	70	30	40	60	10
P_3	40	10	60	20	18
Demand	5	8	7	15	35

What can be the maximum saving by optimum scheduling?

[B.E. (Mech.), Nagpur Univ. 2011; B.E. (Mech.), Panjab Univ. 2012; BBA, GGSIP Univ. 2019]

7.11. A manufacturer has distribution centres at X, Y and Z. These centres have available 40, 20 and 40 units, respectively, of his product. His retail outlets at *A, B, C, D* and *E* requires 25, 10, 20, 30 and 15 units respectively. The transportation cost (in rupees) per unit between each centre and each outlet is given in the following table:

Distribution Centres	*Retail outlets*				
	A	B	C	D	E
X	55	30	40	50	50
Y	35	30	100	45	60
Z	40	60	95	35	30

Determine the optimal distribution plan. ***[MBA, Delhi Univ. (Nov.) 2003]***

7.12. The table given in this problem records transportation costs per unit of a product from origins O_1, O_2, O_3 and O_4 to destinations D_1, D_2, D_3, D_4 and D_5. The capacity of four origins are: 55, 45, 30 and 50, respectively, while the requirement of the five destinations are: 40, 20, 50, 30 and 40, respectively. Employing ***Vogel's Approximation Method,*** or otherwise, make the initial allocations to the origins to satisfy the requirements of the destinations and test the optimality of these allocations. Use *MODI* method for obtaining an optimal solution that would minimize the total cost of transportation.

Origin	*Destination*					*Capacity*
	D_1	D_2	D_3	D_4	D_5	
O_1	12	4	9	5	9	55
O_2	8	1	6	6	7	45
O_3	1	12	4	7	7	30
O_4	10	15	6	9	1	50
Requirement	40	20	50	30	40	

Find the total cost implied by the solution. Is the solution unique? If not, find an alternative solution as well. ***[M.Com., Delhi Univ. 2008]***

7.13. The following table shows all the necessary information on the available supply to each warehouse, the requirement of each market, and the unit transportation cost from each warehouse to each market:

Warehouse	*Market*				*Supply*
	I	*II*	*III*	*IV*	
A	5	2	4	3	22
B	4	8	1	6	15
C	4	6	7	5	8
Requirement	7	12	17	9	

The shipping clerk, has worked out the following schedule from experience: 12 units from *A* to *II*, 1 unit from *A* to *III*, 9 units from *A* to *IV*, 15 units from *B* to *III*, 7 units from *C* to *I* and one unit from *C* to *III*.

(*a*) Check and see if the clerk has the optimal schedule.

(*b*) Find the optimal schedule and minimum total shipping cost.

(*c*) If the clerk is approached by a Carrier of route C to *II* who offers to reduce his rate in the hope of getting some business, by how much must the rate be reduced before the clerk should consider giving him an order? ***[MBA, Panjab Tech. Univ. 2009; MBA (Nov.), Delhi Univ. 2008; MBA (Nov.,), Madras Univ. 2006]***

7.14. *Table 7.50* gives one of the possible solutions to a transportation problem involving three sources and four destinations:

Answer the following questions:

(*i*) Is the above solution degenerate?

(*ii*) Is the above solution optimal? Is it a unique solution? If yes, why?

(*iii*) What is the opportunity cost of transporting one unit from source S_1 to destination D_4?

(*iv*) How would the cost be affected, if it is decided to transport one unit from S_2 to D_2?

[MIB, Delhi Univ. 2011]

	D_1	D_2	D_3	D_4
S_1	10	200 (6)	100 (18)	23
S_2	150 (4)	9	50 (13)	10
S_3	7	13	50 (15)	350 (5)

Table 7.50

7.15. The following table gives the cost of transporting material from supply points A, B, C and D to demand points E, F, G, H and J:

Form	To E	F	G	H	J
A	8	10	12	17	15
B	15	13	18	11	9
C	14	20	6	10	13
D	13	19	7	6	12

The present allocation is as follows:

A to E — 90; A to F — 10; B to F — 150; C to F — 10; C to G — 50; C to J — 120; D to H — 210; and D to J — 70.

(*a*) Check if this allocation is optimum. If not, find an optimum schedule.

(*b*) If in the above problem the transportation cost from A to G is reduced to 10, what will be the new optimum schedule? ***[MBA, Annamalai Univ. 2002]***

7.16. Find the minimum cost distribution plan to satisfy demand for cement at three construction sites from available capacities at the three cement plants, given the following transportation costs (in rupees) per ton moved from plants to sites:

From	*Construction sites* 1	2	3	*Capacity (tons per month)*
P_1	300	360	425	600
P_2	390	340	310	300
P_3	255	295	275	1,000
Demand (tons per month)	400	500	800	

[MBA (Nov.), Annamalai Univ. 2013]

7.17. An Enterprise has three factories at locations A, B and C which supply three warehouses located at D, E and F. Monthly factory capacities are: 10, 80 and 15 units respectively. Monthly warehouse requirements are: 75, 20 and 50 units, respectively. Unit shipping costs (in rupees) are given below:

Factory	*Warehouse* D	E	F
A	5	1	7
B	6	4	6
C	3	2	5

The penalty costs for not satisfying demand at the warehouse D, E and F are ₹ 5, ₹ 3 and ₹ 2 per unit respectively. Determine the optimal distribution for the enterprise, using any of the known algorithms. ***[PG Dip. in Glob. Bus. Oper., Delhi Univ. 2010]***

7.18. A company has three factories at A, B and C which supply warehouses at D, E, F and G. Monthly capacities are 250, 300 and 400 units, respectively for regular production. If overtime production is utilized, factories A and B can produce 50 and 75 additional units with an incremental cost of ₹ 4 and ₹ 5, respectively. The current warehouse requirements are 200, 225, 275 and 300 units, respectively. Unit transportation costs (in rupees) from factories to warehouses are as follows:

From	To			
	D	E	F	G
A	11	13	17	14
B	16	18	14	10
C	21	24	13	10

Determine the optimum distribution for this company, to minimize cost.

[B.Tech., Panjab Tech. Univ. 2010; MBA, Anantpur Univ. 2012; MBA, Pondicherry Univ. 2007]

[**Hint:** Modify the transportation table which takes into account the overtime production and the corresponding production costs.]

7.19. A particular product is manufactured in factories *A*, *B* and *C* and is sold at centres 1, 2, 3, 4 and 5. The cost in rupees of product per unit and capacity in units is given. Also, are given the sale price per unit at the centres, demand at centres and the unit transportation cost in rupees.

Factory :	A	B	C
Production cost/unit (₹) :	20	22	18
Capacity (units) :	150	200	125

Shipping cost is given in the table given below:

1	1	5	9	4
9	7	8	3	6
4	5	3	2	7

Demand (units) :	80	100	75	45	125
Sale price (₹/unit) :	30	32	31	34	29

Find the production and distribution schedule which is most profitable to the company.

[M.Tech., Panjab Tech. Univ. 2012]

7.20. The ABC Tool Company has a sales force of 25 men who work out form three regional offices. The company produces four basic product lines of hand tools. Mr. Jain, the sales manager feels that 6 salesmen are needed to distribute product line 1, 10 salesmen to distribute product line 2, 4 salesmen for product line 3, and 5 salesmen for product line 4. The cost (in ₹) per day of assigning salesmen from each of the offices for selling each of the product lines are as follows:

	Product lines			
	1	2	3	4
Regional office *A*	20	21	16	18
Regional office *B*	17	28	14	16
Regional office *C*	29	23	19	20

At the present time, 10 salesmen are allocated to office *A*, 9 salesmen to office *B*, and 7 salesmen to office *C*. How many salesmen should be assigned from each office to sell at each of the product line in order to minimize the total cost? Identify alternative optimum solution, if any.

[M.Com., Delhi Univ. 2006; MBA, Delhi Univ. 2001]

7.21. There are five plants located in different cities of the country manufacturing a common product. The product is to be stored in five zonal warehouses for distribution to retailers. These warehouses are located at five other cities. The freight cost in rupees per unit for shipping the product from a plant to a warehouse are given as follows:

Plant	*Warehouse*					*Availability (units)*
	W_1	W_2	W_3	W_4	W_5	
P_1	14	10	9	7	7	800
P_2	12	8	5	5	7	200
P_3	9	7	8	6	6	600
P_4	4	7	9	9	15	100
P_5	8	9	9	10	16	100
Requirement (units)	300	500	300	400	600	

[M.Com. (CBCS), Karnataka Univ. 2013]

7.22. A company manufacturing air-coolers has two plants located at Bengaluru and Kolkata with a weekly capacity of 200 units and 100 units respectively. The company supplies air-coolers to its 4 showrooms situated at Hyderabad, Chennai, Madurai and Ernakulam which have a demand of 75, 100, 100 and 30 units respectively. Due to the difference in raw material cost and transportation cost, the profit per unit (in rupees) differ which is shown in the following table:

Plant	*Showrooms*			
	Hyderabad	*Chennai*	*Madurai*	*Ernakulam*
Bengaluru	90	90	100	110
Kolkata	50	70	130	85

Plan the production programme so as to maximize the profit. The company may have its production capacity at both plants partially or wholly unused.

[B.Tech., HP Univ. 2015; MBA, Panjab Tech. Univ. 2014; M.Com., Pune Univ. 2009]

7.23. Solve the following transportation problem for maximizing profit by using "VAM" for initial basic feasible solution, where entries are the profit per unit in rupees:

Warehouse	*Market*				*Available (units)*
	A	B	C	D	
X	12	18	6	25	200
Y	8	7	10	18	500
Z	14	3	11	20	300
Demand (units)	180	320	100	400	

[MBA, Visvesvaraya Univ. 2011]

7.24. A particular product is manufactured in factories *A*, *B*, *C* and *D* and is sold at centres 1, 2 and 3. The cost (in rupees) of product per unit and capacity (in kg.) per unit time of each plant is given below:

Factory	:	A	B	C	D
Cost per unit	:	₹ 12	₹ 15	₹ 11	₹ 13
Capacity (kg.) per unit	:	100	20	60	80

The sale price in rupees per unit and the demand in kg. per unit time are as follows:

Sales centre	:	1	2	3
Sale price per unit	:	₹ 15	₹ 14	₹ 16
Demand (kg.) per unit	:	120	140	60

Find an optimum sales distribution. ***[MBA, Bangalore Univ. 2007]***

[***Hint:*** Profit = Sales Price – Cost Price]

7.25. A departmental store wishes to purchase the following type of sarees:

Type of sarees :	*A*	*B*	*C*	*D*	*E*
Quantity :	150	100	75	250	200

Tenders are submitted by four different manufacturers who undertake to supply not more than the quantities mentioned below (all type of sarees combined):

Manufacturer :	*W*	*X*	*Y*	*Z*
Total quantity :	300	250	150	200

The store estimate that its profit per saree will vary with the manufacturer as shown below:

Manufacturer	*Saree*				
	A	*B*	*C*	*D*	*E*
W	275	350	425	225	150
X	300	325	450	175	100
Y	250	350	475	200	125
Z	325	275	400	250	175

Use transportation technique to determine how the orders should be placed? What is the maximum profit? ***[MBA, Pune Univ. 1999; MBA, Panjab Univ. 1996]***

7.26. A company has four factories F_1, F_2, F_3 and F_4 manufacturing the same product. Production and raw material costs differ from factory to factory, and are given in the first two rows of the table given below. The transportation cost from factories to sales depot S_1, S_2 and S_3 are also given. The last two columns of the table gives the sale price and total requirement at each depot. The production capacity of each factory is given in the last row:

Item		*Factories*				*Sales price per unit*	*Requirement*
		F_1	F_2	F_3	F_4		
Production cost per unit		15	18	14	13		
Raw material cost per unit		10	9	12	9		
Transportation cost per unit	S_1	3	9	5	4	34	80
	S_2	1	7	4	5	32	120
	S_3	5	8	3	6	31	150
Production capacity		10	150	50	100		

Determine the most profitable production and distribution schedule and the corresponding profit. Surplus be taken to yield zero profit. ***[MBA (PT), Delhi Univ. 2009]***

7.27. XYZ and Co. has provided the following data seeking your advice on optimum investment strategy:

Investment made at beginning of year	*Net return data (in paise) of selected investments*				*Amount available (lacs)*
	P	*Q*	*R*	*S*	
1	95	80	70	60	70
2	75	65	60	50	40
3	70	45	50	40	90
4	60	40	40	30	30
Maximum investment (lacs)	40	50	60	60	—

The following additional information is also provided:

- P, Q, R and S represent the selected investments.
- The company has decided to have four years investment plan.
- The policy of the company is that the amount invested in any year will remain so, until the end of fourth year.
- The values (paise) in the table represent net return on investment of one rupee till the end of the planning horizon (for example, a rupee invested in investment P at the beginning of year 1 will grow to ₹ 1.95 by the end of the fourth year, yielding a return of 95 paise).

Using the above information, determine the optimum investment strategy.

[B.Com. (Hons.), Delhi Univ. 2010]

7.28. A company has decided to initiate the production of some or all of the four new products at 3 branch plants with excess production capacity. The available capacities in the three plants are 40, 50 and 30 units per week, while the demand for the four products are 50, 30, 40 and 20 units respectively. Plant 2 cannot produce product 2. The cost per unit of production in the plants are given in the following table:

Plant	*Cost per unit of production of the products*			
	1	2	3	4
1	3	6	5	2
2	4	M	5	4
3	3	4	4	4

Formulate the above problem and find out the initial basic feasible solution.

[MBA (HCA), Delhi Univ. 2015]

7.29. XYZ Tobacco Company purchases tobacco and stores it in warehouses located in the following four cities:

Warehouse location	:	City A	City B	City C	City D
Capacity (tonnes)	:	90	50	80	60

The warehouses supply tobacco to cigarette companies in three cities that have the following demand:

Cigarette company	:	*Bharat*	*Janta*	*Red Lamp*
Demand (tonnes)	:	120	100	110

The railroad shipping costs per tonne (in hundred rupees) have been determined as follows:

Warehouse	*Company*		
	Bharat	*Janata*	*Red Lamp*
City A	7	10	5
City B	12	9	4
City C	7	3	11
City D	9	5	7

Because of railroad construction, shipments are temporarily prohibited from warehouse at City A to Bharat Cigarette Company.

(*i*) Find the optimum distribution for XYZ Tobacco Company.

(*ii*) Are there multiple optimum solutions? If there are alternative optimum solutions, identify them. *[MBA (Nov.), Delhi Univ. 2009; MBA, Panjab Univ. 1997]*

[**Hint:** Since shipments are prohibited from city A to Bharat, we consider the shipment cost as very large, say M from city A to Bharat.]

7.30. A company has factories at four different places 1, 2, 3 and 4 which produces 200, 175, 105 and 230 items respectively and supply these items to warehouses *A*, *B*, *C*, *D* and *E*. Monthly warehouse requirements are 110, 90, 120, 230 and 160, respectively. Unit shipping costs are given in the following table (in rupees):

		Warehouses				
		A	*B*	*C*	*D*	*E*
	1	13	—	31	8	20
Factories	2	14	9	17	26	10
	3	25	11	12	17	15
	4	10	21	13	—	17

Shipments from 1 to *B* and from 4 to *D* are not possible. Determine the optimum distribution plan to minimize shipping cost. ***[MBA, Madras Univ. 2008]***

[***Hint:*** Since, shipments from 1 to *B* and 4 to *D* are not possible, we consider very large cost to these two warehouses, say *M*.]

7.31. Solve the following transportation problem:

Factory	*Godowns* 1	2	3	4	5	6	*Stock available*
A	7	5	7	7	5	3	60
B	9	11	6	11	—	5	20
C	11	10	6	2	2	8	90
D	9	10	9	6	9	12	50
Demand	60	20	40	20	40	40	

It is not possible to transport any quantity from factory *B* to Godown 5.

[MBA, Annamalai Univ. 2009]

7.32. A manufacturer must produce a certain product in sufficient quantity to meet contracted sales in next four months. The production facilities available for this product are limited, but by different amounts in respective months. The unit cost of production also varies in each month. The product may be produced in one month and then held for sale in a later month, but at an estimated storage cost of ₹ 1 per unit per month. No storage cost is incurred for goods sold in the same month in which they are produced. There is no initial inventory and none is desired at the end of 4 months. Given the following information, show how much to produce in each of the four months in order to minimize total cost:

Month	*Contracted sales (in units)*	*Maximum production (in units)*	*Unit cost of production (in ₹)*
1	20	40	14
2	30	50	16
3	50	30	15
4	40	50	17

Formulate the problem as a transportation problem by constructing the appropriate cost and requirements table and solve the formulated transportation problem.

[**Hint:** Formulated transportation problem is as follows:

	To (Months)				Production
From (Months)	14	15	16	17	40
	—	16	17	18	50
	—	—	15	16	30
	—	—	—	17	50
Sales	20	30	50	40	

Here, the product produced in month 1 have a total cost of ₹ 14, ₹ 15, ₹ 16 and ₹ 17 of sold in the 1st, 2nd, 3rd and 4th month respectively. This forms the first row. Product produced in second month can not be sold in first month, second row consists of the total cost of ₹ 16, ₹ 17 and ₹ 18, if sold in 2nd, 3rd and 4th months respectively, etc.]

7.33. The demand and production costs vary from month to month in an industry. The following table contains budgeted information of a firm in this industry on the quantity demanded, the production cost per unit and the production capacity in each of the coming five months.

Month	*January*	*February*	*March*	*April*	*May*
Demand	200	250	150	80	120
Production cost	24	27	32	30	34
Capacity	250	225	250	200	225

It is known that the production in any month can meet demand in that month or can be held for the future. The holding cost is ₹ 5 per unit per month.

Using the above information:

(*i*) Formulate a transportation model to assist in the production planning, and

(*ii*) Determine what would the optimal plan be. ***[MBA (Nov.), Delhi Univ. 2003]***

[**Hint:** Formulated transportation problem is as follows:

	January	*February*	*March*	*April*	*May*	*Capacity*
January	24	29	34	39	44	250
February	—	27	32	37	42	225
March	—	—	32	37	42	250
April	—	—	—	30	35	200
May	—	—	—	—	35	225
Demand	200	250	150	80	120	

7.34. A manufacturer of jeans is interested in developing an advertising campaign that will reach four different age groups. Advertising campaigns can be conducted through TV, Newspapers and Magazines. The following table gives the estimated cost in rupees per exposure for each group according to the medium employed. In addition, maximum exposure level possible in each of the media, namely TV, Newspapers and Magazines are 40, 30 and 20 millions, respectively. Also, the minimum desired exposures within each age group, namely 18-25, 26-35, 36-50, 51 and older are 30, 25, 15 and 10 millions. The objective is to minimize the cost of obtaining the desired minimum exposure level in each age group.

Media	*Age groups*			
	18-25	*26-35*	*36-50*	*51 and older*
TV	12	7	10	10
Newspapers	10	9	12	10
Magazines	14	12	9	12

(*i*) Formulate the above as a transportation problem and find the optimal solution.

(*ii*) If the policy is to provide at least 4 million exposures through TV in the age group 18-25 and at least 8 million exposures through TV in the age group 26-35, then find the total minimum cost of the revised media exposure. ***[MBA, Delhi Univ. 2011]***

[**Hint:** (*ii*) Optimal solution obtained to the given transportation problem gives 25 million people in the age group 26-35 through TV. So, it entirely meets the requirements of at least 8 million such exposures. Now, to ensure 4 million exposures through TV in the age group 18-25, we allocate 4 million in the unoccupied cell (1, 1) and draw a closed path and make adjustments in the cells which lie on it. The occupied cells include (1, 5), (2, 5) and (2, 1).]

7:11. POST-OPTIMAL ANALYSIS

Similar to the study of post-optimal analysis of linear programming problems, in transportation problems, we investigate the effects on the solution of events like changes in the cost elements, increasing or decreasing the availability and demand on a given route, etc. Here, we discuss the following aspects of post-optimal analysis (sensitivity analysis) in respect of a transportation problem:

1. Change in the coefficient of a non-basic variable of objective function.
2. Change in the coefficient of a basic variable of objective function.
3. Change in the supply of particular source (or origin) and a particular destination.
4. Effect of a decision to ship certain maximum or minimum number of units on a given route.

We illustrate the above-mentioned changes in the context of ***Example 7.12***. The optimum solution is reproduced in *Table 7.51*.

						Supply	u_i
	(−3) 20	[15] 18	[85] 18	(−5) 21	(−1) 19	100	0
	[20] 21	(0) 22	(−1) 23	[105] 20	(−2) 24	125	4
	[40] 18	[65] 19	(−2) 21	(−1) 18	[70] 19	175	1
Demand	60	80	85	105	70		
v_j	17	18	18	16	18		

Table 7.51

1. Change in the Coefficient of a Non-basic Variable of the Objective Function. If there is a change in the cost element of an unoccupied cell (coefficient of a non-basic variable), it will have no effect on the u_i and v_j values, because these values are computed using only the occupied cells. So, the solution will remain unchanged as long as the opportunity costs remain non-positive. To ***illustrate***, let the cost element c_{23} is changed from 23 to $23 + k$, where k is some constant. The revised opportunity cost of this cell will be $4 + 18 - (23 + k)$, *i.e.*, $-1 - k$. The current solution remains unchanged only if $(-1 - k) \leq 0$. *i.e.*, $k \geq -1$ and $c_{23} \geq (23 - 1)$. Evidently an increase in the cost element c_{23} will in any case not affect the current solution and so will decline in it up to 1. In case the cost element were to decline below 22, then the solution will change.

2. Change in the Coefficient of a Basic Variable of the Objective Function. A change in the cost coefficient of a basic variable (*i.e.*, of an occupied cell) would cause a change in u_i and v_j values. This will in turn affect the opportunity costs of the unoccupied cells. However, so long as one or more of the revised opportunity costs do not turn to be positive, the solution would remain unchanged. To understand this, let us change c_{12} from 18 to 18 + k, where k is some constant.

By setting $u_1 = 0$ revised values of u_i and v_j are given by the following equations:

$$u_1 + v_2 = 18 + k,\ u_1 + v_3 = 18,\ u_2 + v_1 = 21,$$
$$u_2 + v_4 = 20,\ u_3 + v_1 = 18,\ u_3 + v_2 = 19 \text{ and } u_3 + v_5 = 19.$$

Solving these equations, we get: $u_1 = 0$, $u_2 = 4 - k$, $u_3 = 1 - k$; $v_1 = 17 + k$, $v_2 = 18 + k$, $v_3 = 18$, $v_4 = 16 + k$ and $v_5 = 18 + k$.

Now, using revised values of u_i and v_j, the opportunity costs of the unoccupied cells are given by

$$d_{11} = u_1 + v_1 - c_{11} = 0 + (17 + k) - 20 \quad = -3 + k \le 0 \text{ or } k \le 3$$
$$d_{14} = u_1 + v_4 - c_{14} = 0 + (16 + k) - 21 \quad = -5 + k \le 0 \text{ or } k \le 5$$
$$d_{15} = u_1 + v_5 - c_{15} = 0 + (18 + k) - 19 \quad = -1 + k \le 0 \text{ or } k \le 1$$
$$d_{22} = u_2 + v_2 - c_{22} = (4 - k) + (18 + k) - 22 = 0$$
$$d_{23} = u_2 + v_3 - c_{23} = (4 - k) + 18 - 23 \quad = -1 - k \le 0 \text{ or } k \ge -1$$
$$d_{25} = u_2 + v_5 - c_{25} = (4 - k) + (18 + k) - 24 = -2$$
$$d_{33} = u_3 + v_3 - c_{33} = (1 - k) + 18 - 21 \quad = -2 - k \le 0 \text{ or } k \ge -2$$
$$d_{34} = u_3 + v_4 - c_{34} = (1 - k) + (16 + k) - 18 = -1$$

We observe from the above opportunity costs that the least positive value for k is 1 and the least negative value is –2. Therefore, $-2 \le k \le 1$ will cause no change in the current solution. Thus, revised value of c_{12} may be 18 + k (= 18 – 2), *i.e.*, 16 and 18 + 1 (= 19). In other words, if the cost element 18 changes between 16 and 19, there will be no change in the current optimum solution.

3. Change in source of Supply by *k* units and destination of Demand by *k* units. Consider an equal change of 'k units' in the ith source of supply and j^{th} destination of demand, so that the transportation problem remains balanced. Now, u_i's and v_j's may be thought of as negative of shadow prices of the supply and demand constraints, then the new value of objective function is equal to old value of the objective function plus ($ku_i + kv_j$) and the current basic solution remains optimum. To illustrate, in our example, let there is an increase of 5 units at Shop *A* and the demand at Warehouse *II* also increases by 5 units. Then, the revised total cost will be 23,605 + (5 × 0 + 5 × 18), *i.e.*, 23,695.

We, now, discuss as to how the new values of decision variables x_{ij} would change.

(*a*) If x_{ij} is a basic variable, so that the route represented by source and destination where change takes place is in occupied cell, then the value of x_{ij} is revised to $x_{ij} + k$. To illustrate, if the supply at Shop *A* is increased by 5 units and the demand at Warehouse *III* by 5 units, then the revised quantity sent on this route will be of 90 units instead of 85 units. The revised total cost will be:

$$23{,}605 + 0 \times 5 + 18 \times 5, \textit{ i.e.}, 23{,}695$$

(*b*) If the route where change takes place is an unoccupied cell, then we draw a closed path starting with that cell. Once this is done, we select the cell on the closed path in the same row in which supply is increased by k units and add the same quantity of k units to the x_{ij} value in that cell. Then, we go round the closed path alternatively subtracting and adding k to the x_{ij} values in those cells.

To illustrate, suppose we add 7 units to the supply of Shop *B* and also the demand at Warehouse *II*. As the cell (2, 2) is unoccupied, we draw a closed path starting from this cell and turning points at the occupied cells, *viz.*, (2, 1), (3 1) and (3, 2). We notice that the occupied cell (2, 1) on the closed path is in the same row in which supply is increased and therefore, we add 7 units to the value of x_{21}. The value of x_{21}, then, become 20 + 7, *i.e.*, 27. Now, alternatively subtract 7 units from the value of x_{31}, and further add to x_{32}. The revised solution will be:

$$x_{12} = 15, x_{13} = 85, x_{21} = 27, x_{24} = 105, x_{31} = 33, x_{32} = 72 \text{ and } x_{35} = 70$$

This change has brought about an increase of 7 × (21 + 19 – 18) or 154 in the total cost which works out to be 23,759.

4. Setting Limits on the shipment of certain number of units on a particular route. The problem of assigning certain number of units on a particular route is illustrated in the context of ***Example 12***, whose optimum solution is given in *Table 7.51*. Suppose 40 units are shipped from Shop C to Warehouse *III*. By doing this, we find that 40 units are to be assigned to the unoccupied cell (3, 3). So, we draw a closed path beginning with this cell, namely (3, 3) → (1, 3) → (1, 2) → (3, 2) → (3, 3). After adjusting '+' and '–' values, new values of x_{ij} will be: $x_{33} = 40$, $x_{13} = 45$, $x_{12} = 55$ and $x_{32} = 25$. It will have the effect of increasing the cost by 80. Also, it may be noted that a maximum of 40 units can be dispatched on this route, *i.e.*, 40 units can be assigned to the unoccupied cell (3, 3) in the context of the present solution. However, if more than 40 units, say 80 units are planned to be sent on this route, the problem would need to be resolved after reducing the supply at Shop C and demand at Warehouse *III* by 80 units. To the total cost obtained from such a solution, we shall add the cost of sending 80 units on the route from Shop C to Warehouse *III*.

Now, suppose that it is desired to assign not more than 80 units to the occupied cell (1, 3). This can be done by reducing the supply at Shop *A* and demand at Warehouse *III* by 80 units each, and then re-solving the problem after replacing the cost element of the cell (1, 3) by *M*, *i.e.*, considering this route as prohibited.

PROBLEMS

7.35. The table given below shows all the necessary information on the available supply to each warehouse, the requirement of each market and the unit cost from each warehouse to each market:

Warehouse	*Market*				*Supply*
	I	*II*	*III*	*IV*	
A	5	2	4	3	22
B	4	8	1	6	15
C	4	6	7	5	8
Requirement	7	12	17	9	

Find the optimal solution and minimum total shipping cost.

(*a*) If supply from Warehouse *B* reduces to 12 units and simultaneously the requirement of Market *III* reduced to 14 units, find the optimal transportation schedule.

(*b*) If supply of Warehouse *B* is increased by 1 unit and simultaneously the demand of Market *III* is also increased by 1 unit, will the optimal solution change?

(*c*) If supply of Warehouse *B* increases by 3 units and demand of Market *IV* also increases by 3 units, will there be change in the optimal solution?

7.36. Consider the following transportation problem:

Factory	*Distribution centres*				*Capacity*
	D_1	D_2	D_3	D_4	
F_1	10	12	14	18	200
F_2	15	16	17	19	150
F_3	12	10	11	16	250
Demand	225	75	150	150	600

Find the optimal solution and on the basis of final table, answer the following with reasons:

(*i*) Is the optimal solution obtained a degenerate solution?

(*ii*) Does it have multiple solutions?

(*iii*) What would be the increase in the cost, if the management decides to transport one unit from F_3 to D_4?

(*iv*) If a transporter offers a discount of 20% on the route F_3 to D_4, when at least 20 units are transported, should this offer be accepted?

(*v*) If supply at F_1 is reduced by one unit and the demand of D_1 is also reduced by one unit, then what would be the optimal solution? Will it reduce the total cost? If yes, by what amount? *[MIB, Delhi Univ. 2011]*

7.37. A transportation problem involving three sources and four destination centres is presented in *Table 7.52*:

Answer the following questions:

(*i*) Is the solution optimal? Is it unique? Why?

(*ii*) Is the solution degenerate?

(*iii*) How would the cost be affected, if it is decided to transport one unit from S_2 to D_2?

(*iv*) The transporter is willing to provide a discount of 25% on the freight on the route S_1 to D_4, provided at least 50 units are sent on this route. Should the offer be accepted?

(*v*) If the output at source S_1 is reduced by 2 units and to compensate for the loss of output, the supply at source S_2 is increased by as many units; would it lead to a change in the cost? By how much?

	D_1	D_2	D_3	D_4	*Supply*
S_1	12	[250] 8	[150] 20	25	400
S_2	[200] 6	11	[100] 15	12	300
S_3	9	15	[100] 17	[400] 7	500
Demand	200	250	350	400	

Table 7.52

7:12. TIME-MINIMIZATION TRANSPORTATION PROBLEM

Time-minimization transportation problems are similar to the problems discussed so far, except that the unit transportation cost c_{ij} is replaced by the unit time t_{ij} required to transport the items from i^{th} origin to j^{th} destination.

An initial basic feasible solution (feasible plan) for this problem can be obtained by using ***Vogel's Approximation Method*** or any of the methods. Now, if T_j be the largest transportation time associated with the jth feasible plan, then we have to determine that plan which gives minimum (least) T_j.

A systematic procedure for the solution of such problems is as follows:

Step 1. Determine an initial basic feasible solution. Enter the solution in the upper left corners of the basic cells.

Step 2. Determine T_j for this basic feasible solution and cross out all the non-basic cells for which $t_{ij} \geq T_j$.

Step 3. Construct a loop for the basic cells corresponding to T_j in such a way that when the values at the corner cells are shifted around, value at this cell tends towards zero and no variable becomes zero. If no such closed path can be formed, the solution under test is optimal, otherwise move on to the next step.

Step 4. Go to *step* 2 and repeat the procedure until an optimum basic feasible solution is attained.

Example 7.17. If the elements of a matrix represent the unit transportation times solve the following transportation problem:

From	*To* D_1	D_2	D_3	D_4	*Available*
O_1	10	0	20	11	15
O_2	1	7	9	20	25
O_3	12	14	16	18	5
Required	12	8	15	10	45

Solution. Using *Vogel's Approximation Method* an initial basic feasible solution is given in *Table 7.53*.

The numbers written at the lower right corners of each cell represent the corresponding times. The time for this feasible plan is obtained in *Table 7.53:*

Since, $t_{12} = 0, t_{14} = 11, t_{21} = 1, t_{23} = 9, t_{33} = 16$ and $t_{34} = 18$.

$\therefore$ T_1 = max. {0, 11, 1, 9, 16, 18} = 18. Clearly, all the shipments for this plan are completed within the time $T_1 = 18$. So, we cross out the cell (1, 3) and (2, 4) as $t_{13} > T_1$ and $t_{24} > T_1$. Now, the closed path (loop) for the cell (3, 4) for which $t_{34} + T_1 = 18$ is shown in *Table 7.54*.

10	8 / 0	20	7 / 11
12 / 1	7	13 / 9	20
12	14	2 / 16	3 / 18

Table 7.53

It is evident from *Table 7.54* that only 3 units be shifted around. The next feasible plan with times is shown in *Table 7.55*.

10	8 (−) / 0	20	7 (+) / 11
12 / 1	7	13 / 9	20
12	(+) / 14	2 / 16	3 (−) / 18

Table 7.54

10	5 / 0	✗ 20	10 / 11
12 (−) / 1	7	13 (+) / 9	✗ 20
(+) / 12	3 / 14	2 (−) / 16	✗ 18

Table 7.55

Here, $$T_2 = \max. \{t_{12}, t_{14}, t_{21}, t_{23}, t_{32}, t_{33}\}$$
$$= \max. \{0, 11, 1, 9, 14, 16\} = 16$$

As all the shipments for the feasible solution are completed within time $T_2 = 16$, we cross out the cell (3, 4), since $t_{34} > T_2$. The closed loop starts from the cell (3, 3) as shown in *Table 7.55*.

Clearly, only 2 units can be shifted around. The next feasible plan, therefore, is as shown in *Table 7.56*.

10	5 / 0	✗ 20	10 / 11
10 (−) / 1	(+) / 7	15 / 9	✗ 20
2 (+) / 12	3 (−) / 14	✗ 16	✗ 18

Table 7.56

10	5 / 0	✗ 20	10 / 11
7 / 1	3 / 7	15 / 9	✗ 20
5 / 12	✗ 14	✗ 16	✗ 18

Table 7.57

Here, $T_3 = \text{max.} \{t_{12}, t_{14}, t_{21}, t_{23}, t_{31}, t_{32}\}$
$= \text{max.} \{0, 11, 1, 9, 12, 14\} = 14$

Since, $t_{32} = T_3 = 14$, the closed loop starts from the cell (3, 2) as shown in *Table 7.56*. It is clear that only 3 units are shifted around. The next feasible solution, therefore, is as shown in *Table 7.57*.

Here, $T_4 = \text{max.} \{t_{12}, t_{14}, t_{21}, t_{22}, t_{23}, t_{31}\}$
$= \text{max.} \{0, 11, 1, 7, 9, 12\} = 12$

Now, we cannot form any other closed loop without increasing the present minimum shipping time. Hence, the feasible plan at this stage is optimal.

Thus, all the shipments can be made within 12 time units.

7:13. TRANSHIPMENT PROBLEM

A transportation problem wherein available commodity that moves from each source to each destination has to pass through intermediate points is known as ***transhipment problem***. These intermediate points may be other sources and/or destinations, as well as additional transfer points. Therefore, the transhipment problem is a transportation problem, where shipment is possible from one source to another source or destination as well as from one destination to another source or destination.

So, in transhipment problems the distinction between source and destination is dropped and given problem is written as a modified transportation problem having number of sources and destinations as the sum of given sources and destinations. Following are the chief characteristics of the transhipment problems.

(*a*) Number of sources and destinations in the transportation problem are m and n respectively. In transhipment problem, however, we have $m + n$ sources and equal number of destinations.

(*b*) If S_i denotes the ith source and D_j denotes the jth destination then commodity can move along the route $S_i \to D_i \to D_j$, $S_i \to S_j \to D_i \to D_j$, $S_i \to D_i \to S_i \to D_j$, or in various other ways. Clearly, transportation cost from S_i to S_i is zero and the transportation costs from S_i to S_j and S_j to S_i do not have to be symmetrical, *i.e.*, in general $S_i \to S_j \neq S_j \to S_i$.

Solving Transhipment Problems

For solving a transhipment problem, following are the major steps:

Step 1. Construct a modified transportation table of order ($m \times n$), where there are m sources and n destinations.

Step 2. Supply at each source and demand at each destination are added with the total number of units to be transported from all sources to destinations.

Step 3. Find the initial basic feasible solution. This basic feasible solution contains $(2m + 2n - 1)$ basic variables, and if we omit the variables appearing in the $(m + n)$ diagonal cells, we are left with $(m + n - 1)$ basic variables.

Step 4. Solve the modified transportation problem by using *MODI method* for determining an optimal solution.

Step 5. In the final (optimum) solution, delete the entries which are present in the diagonal cells of the modified transportation table. The remaining entries gives the solution of the transhipment problem.

Example 7.18. *A firm having two sources S_1 and S_2 wishes to ship its products to two destinations D_1 and D_2. The number of units available at S_1 and S_2 are 5 and 25 respectively. The product demanded at D_1 and D_2 are 20 and 10 units respectively. The firm, instead of shipping directly from sources to destinations, decides to investigate the possibility of transhipment. The unit transportation cost (in rupees) are given in the following table. Find the optimal shipping schedule:*

		Source		Destination		
		S_1	S_2	D_1	D_2	Available
Source	S_1	0	2	3	4	5
	S_2	2	0	2	4	25
Destination	D_1	3	2	0	1	—
	D_2	4	4	1	0	—
	Demand	—	—	20	10	

Solution. The transportation problem can be extended, which includes four supply points and four demand points. Since, all the demand may concentrate at any one supply point or demand point, the fictitious supply and demand quantities may be assumed for each of these supply (source) points and demand (destination) points. These quantities may be regarded as *buffer stocks*. Each of these buffer stocks must, at least, be equal to total supply and demand in the original problem. For the given problem the buffer stock will be 30 units. The role of buffer stock in the transhipment problem is only to convert the problem into a standard form and once the optimum solution is obtained, necessary adjustments will be made to conform to the original problem.

Since, the buffer stock for the given transhipment problem is of 30 units, the extended (modified) transportation problem is portrayed in the following modified transportation *Table 7.58:*

Using ***Vogel's Approximation Method*** for an initial basic feasible solution and *MODI* method for an optimum solution to the extended transportation problem, we obtain the transportation *Table 7.59*.

	D_1	D_2	D_3	D_4	Supply + buffer stock
S_1	0	2	3	4	5 + 30
S_2	2	0	2	4	25 + 30
S_3	3	2	0	1	0 + 30
S_4	4	4	1	0	0 + 30
Demand + buffer stock	0 + 30	0 + 30	20 + 30	10 + 30	

Table 7.58

	S_1	S_2	D_1	D_2	u_i
S_1	30 / 0	(−1) / 2	(0) / 3	5 / 4	0
S_2	(−3) / 2	30 / 0	25 / 2	(−1) / 4	−1
D_1	(−6) / 3	(−4) / 2	25 / 0	5 / 1	−3
D_2	(−8) / 4	(−7) / 4	(−2) / 1	30 / 0	−4
v_j	0	1	3	4	

Table 7.59

From Table *7.59*, we observe that all the opportunity costs are non-positive for unoccupied cells, and hence solution obtained is optimum.

The allocation in the diagonal cells are ignored, since they do not have any physical meaning (*i.e.*, no transportation). The remaining allocations can be interpreted as follows:

(*a*) Transport 25 units from S_2 and D_1. It increases the capacity of S_2 to 55 units including 30 units as buffer stock.

(*b*) From S_1 transport (ship) 5 units to D_2, and
(*c*) Out of 25 units available at D_1, transport 5 units to D_2.

EXERCISE

7.38. Consider a firm having two factories. The firm is to ship its products from the factories to three retail stores. The number of units available at factories *X* and *Y* are 200 and 300 respectively, while those demanded at retail stores *A*, *B* and *C* are 100, 150 and 250 respectively. Rather than shipping directly from factories to retail stores, it is desired to investigate the possibility of transhipment. The transportation cost (in rupees) per unit is given below:

		Factory		*Retail store*		
		X	*Y*	*A*	*B*	*C*
Factory	*X*	0	8	7	8	9
	Y	6	0	5	4	3
	A	7	2	0	5	1
Retail store	*B*	1	5	1	0	4
	C	8	9	7	8	0

Find an optimal shipping schedule.

7.39. A firm owns facilities at six places. it has manufacturing plants at places *A*, *B* and *C* with daily production of 50, 40 and 60 units respectively. At places *D*, *E* and *F*, it has three warehouses with daily demand of 20, 95 and 35 units respectively. Per unit shipping cost is given in the following table:

		Warehouse		
		D	E	F
	A	6	4	1
Plant	B	3	8	7
	C	4	4	2

The following table gives additional per unit shipping costs data:

From Plant	*To plant*			*From warehouse*	*To warehouse*		
	A	B	C		D	E	F
A	0	3	2	D	0	2	5
B	3	0	4	E	2	0	1
C	2	4	0	F	5	1	0

(*i*) Formulate it as a transhipment problem.
(*ii*) Give the shipment schedule to minimize the transportation cost when transhipment is allowed. ***[MBA (Dec.), Visvesvaraya Univ. 2011; MBA (PT.), Delhi Univ. 2010]***

7.40. Using the information given below, determine the optimal solution under the assumption that inter-plant and inter-warehouse transfers are not allowed, and that the units may be transhipped.

Plant	:	P_1	P_2	
Capacity	:	240 units	160 units	
Warehouses	:	W_1	W_2	W_3
Requirements	:	80 units	120 units	200 units
Transportation costs	:	Exactly proportional to the distances (given in kilometres in the *Figure 7.1*)		

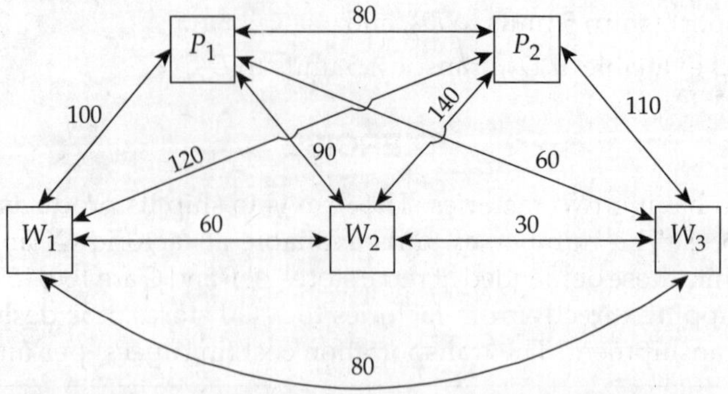

Fig. 7.1

7.41. There are two sources and two destinations in a transportation problem. Availabilities at sources are 4, 5 and requirements are 3, 6. The transportation costs are indicated by arrow in the network given below. Solve the problem.

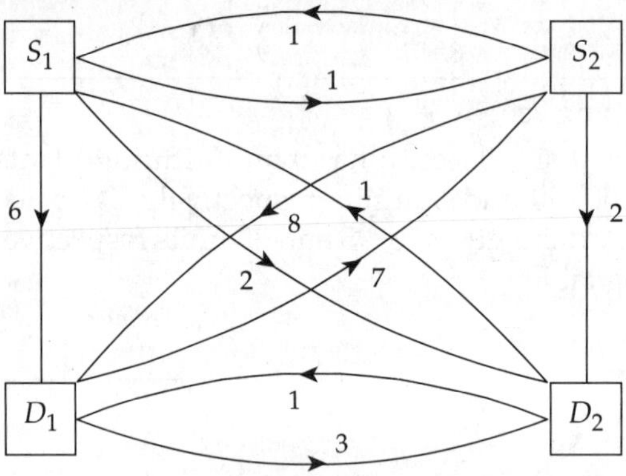

Fig. 7.2

7.42. Consider the oil pipeline network shown in the following figure. The different nodes represent the pumping and receiving stations. The distances in hundreds of kilometers between different stations are shown on the network. The transportation cost per gallon between two nodes is directly proportional to the length of the pipeline. Develop the associated transhipment model (Do not solve). Nodes *1* and *3* are source nodes, whereas there are demands for 90,000 and 20,000 gallons at nodes *2* and *4* respectively.

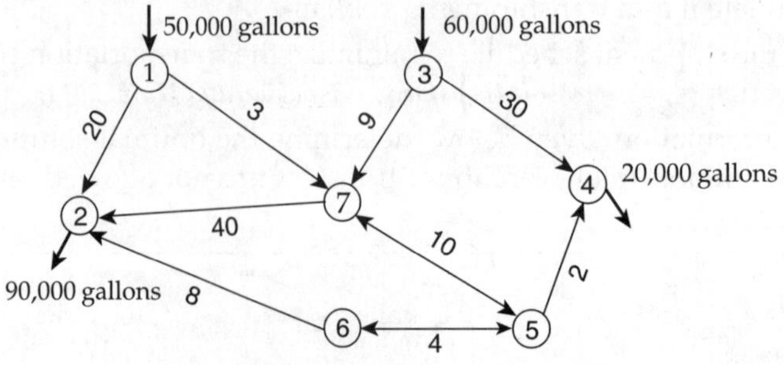

Fig. 7.3

[MBA, Delhi Univ. 2012]

TEST YOUR UNDERSTANDING

1. The transportation problem deals with the transportation of
   (*a*) a single product from several sources to a destination
   (*b*) a multi-product from several sources to several destinations
   (*c*) a single product form several sources to several destinations
   (*d*) a single product from a source to several destinations.
2. The solution to a transportation problem with m-sources and n-destinations is feasible, if the number of allocations are
   (*a*) $m + n - 1$ (*b*) $m + n + 1$
   (*c*) $m + n$ (*d*) $m \times n$
3. While solving a transportation problem, the occurrence of degeneracy means that
   (*a*) total supply equals total demand (*b*) solution so obtained is not feasible
   (*c*) few allocations become negative (*d*) none of the above.
4. In Vogel's Approximation method
   (*a*) cost difference indicate the penalties for not using the respective least cost routes
   (*b*) initial solution to transportation problem is not applicable, if some routes are prohibited
   (*c*) degeneracy never occurs (*d*) none of the above.
5. The initial solution of a transportation problem obtained by
   (*a*) North-West corner rule would invariably be optimum
   (*b*) least cost method does not provide the least cost solution to a transportaton problem
   (*c*) VAM would invariably be very near to optimum solution
   (*d*) MODI method is infeasible.
6. In transportation problem
   (*a*) a solution is said to degenerate, if the number of occupied cells is smaller than the number of rows plus the number of columns minus one
   (*b*) a degenerate solution may or may not be optimum
   (*c*) to remove degeneracy, an infinitesimally small quantity 'ε' is placed in each of the required number of independent cells
   (*d*) all of the above hold.
7. The dummy source or destination in a TP is introduced to
   (*a*) prevent solution to become degenerate
   (*b*) to satisfy rim conditions
   (*c*) ensure that total cost does not exceed a limit
   (*d*) solve the balanced transportation problem.
8. If we were to use opportunity cost value for non-basic cell to test optimality, it should be
   (*a*) most negative number (*b*) most positive number
   (*c*) equal to zero (*d*) any value.
9. In transportation problem
   (*a*) if all the cost elements, c_{ij}, are multiplied by a constant, the total cost of transportation in the optimum solution shall be multiplied by the same constant
   (*b*) if some $u_i + v_j - c_{ij}$ is equal to zero in the optimum solution, then the problem has multiple optimum solutions
   (*c*) if certain routes are prohibited, then their cost elements are replaced by M (an extremely large value)
   (*d*) all of the above.

10. For a transhipment problem, choose the statement which is not correct:
    (*a*) The problem allows for the shipment of goods from one source to another, and from one destination to another
    (*b*) There is no real distinction between sources and destinations
    (*c*) An '*m*' source, '*n*' destination transportation problem, when written as a transhipment problem would have $m + n$ sources and n destinations
    (*d*) A transhipment problem is not likely to involve a lower cost than a transportation problem, in a given situation.

ANSWERS

1. (*c*)	**2.** (*a*)	**3.** (*b*)	**4.** (*a*)	**5.** (*c*)
6. (*d*)	**7.** (*b*)	**8.** (*a*)	**9.** (*d*)	**10.** (*c*)

REVIEW EXERCISE

1. What is transportation problem? *[M.Com., Bharthidasan Univ. 2007]*
2. Explain transportation problem and show that it can be considered as an LLP.
3. Give the model formulation of a transportation problem.
4. State all the constraints in a transportation problem and how they are different from linear programming problem.
5. What is balanced transportation problem? What are its applications? *[MBA (Dec.), Panjab Tech. Univ. 2007]*
6. Explain North-West Corner method to solve transportation problem for an initial solution. *[MBA, Panjab Univ. 2006]*
7. Explain in brief, with examples (*i*) North-West Corner Rule, (*ii*) Least Cost Entry Method, and (*iii*) Vogel's Approximation Method.
8. What are the conditions to be satisfied for proceeding with optimality test in a transportation problem. *[MBA, Annamalai Univ. 2009]*
9. Give a brief outline of the procedure for solving a transportation problem.
10. How the problem of degeneracy arises in a transportation problem? Explain how does one overcome it? *[MBA (Dec.), Panjab Tech. Univ. 2010]*
11. What is stepping stone in transportation problem? *[MBA, Osmania Univ. 2001]*
12. What is meant by unbalanced transportation problem? *[MBA (Nov.), Annamalai Univ. 2002]*
13. Describe the method of solving unbalanced transportation problem.
14. How the transportation problem with a prohibited transportation route is solved?
15. What is a transhipment problem? Explain how a transhipment problem can be formulated and solved as a transportation problem? *[MBA, Annamalai Univ. 2010]*
16. Write a short note on Time-Minimization in transportation problem? *[MBA (Nov.), Delhi Univ. 2007]*
17. Explain the following in the context of TP:
    (*i*) Stepping Stone Method, (*ii*) Degenerate Transportation Problem, and (*iii*) Modified Distribution Method. *[MCA, IGNOU, 2003; MBA, Pondicherry Univ. 2006]*
18. What is the indicator of multiple optimal solution to a transportation problem in case of maximization? How do you find out the alternate optimal solution in such a case? *[MIB, Delhi Univ. 2011]*

CHAPTER 8

Assignment Problem

"OR produces an integrated solution for the good of entire organization and not just to effect local improvement"

Assignment problem is a special case of the transportation problem, wherein the objective is to ***assign*** a number of resources to the ***equal*** number of activities at a minimum cost (or maximum profit).

Assignment problem is completely degenerate form of a transportation problem. The units available at each origin (resource) and units demanded at each destination (activity) are all equal to one. That means exactly one occupied cell in each row and each column of the transportation table, *i.e.*, only n occupied (basic) cells in place of the required $n + n - 1$, *i.e.*, $2n - 1$.

8:1. MODEL FORMULATION OF AN ASSIGNMENT PROBLEM

Consider a problem of assignment of n resources (workers) to n activities (jobs) so as to minimize the overall cost or time in such a way that each resource can associate with one and only one activity. The cost (or effectiveness) matrix (c_{ij}) is given as under:

		Activity				
		A_1	A_2	...	A_n	*Available*
	R_1	c_{11}	c_{12}	...	c_{1n}	1
	R_2	c_{21}	c_{22}	...	c_{2n}	1
Resource	⋮	⋮	⋮	⋮	⋮	⋮
	R_n	c_{n1}	c_{n2}	...	c_{nn}	1
	Required	1	1	...	1	

This cost matrix is same as that of a transportation problem except that availability at each of the resources and the requirement at each of the destinations is unity (due to the fact that assignments are made on a one-to-one basis).

Let x_{ij} denote the assignment of ith resource to jth activity, such that

$$x_{ij} = \begin{cases} 1, & \text{if resource } i \text{ is assigned to activity } j \\ 0, & \text{otherwise} \end{cases}$$

Then, the model formulation of the assignment problem is:

$$\text{Minimize } z = \sum_{i=1}^{n} \sum_{j=1}^{n} c_{ij}\, x_{ij} \quad \text{subject to the constraints:}$$

$$\sum_{j=1}^{n} x_{ij} = 1 \qquad \text{for all } i \text{ (resource availability)}$$

$$\sum_{i=1}^{n} x_{ij} = 1 \qquad \text{for all } j \text{ (activity requirement)}$$

and $$x_{ij} = 0 \text{ or } 1; \qquad (i = 1, 2, ..., n;\ j = 1, 2, ..., n)$$

where, c_{ij} is the cost associated with assigning ith resource of jth activity.

Illustration. *Given below is an assignment problem, write it as a transportation problem:*

	A_1	A_2	A_3
R_1	1	2	3
R_2	4	5	1
R_3	2	1	4

Solution. Let x_{ij} denote the assignment of R_i $(i = 1, 2, 3)$ to A_j $(j = 1, 2, 3)$, such that

$$x_{ij} = \begin{cases} 1, & \text{if } R_i \text{ is assigned to } A_j \\ 0, & \text{otherwise} \end{cases}$$

Then, the transportation problem is:

Minimize $z = 1x_{11} + 2x_{12} + 3x_{13} + 4x_{21} + 5x_{22} + x_{23} + 2x_{31} + x_{32} + 4x_{33}$

subject to the constraints:

$$\left.\begin{aligned} x_{11} + x_{12} + x_{13} &= 1 \\ x_{21} + x_{22} + x_{23} &= 1 \\ x_{31} + x_{32} + x_{33} &= 1 \end{aligned}\right\}, \quad \left.\begin{aligned} x_{11} + x_{21} + x_{31} &= 1 \\ x_{12} + x_{22} + x_{32} &= 1 \\ x_{13} + x_{23} + x_{33} &= 1 \end{aligned}\right\},$$

$$x_{ij} = 0 \text{ or } 1 \text{ for } i = 1, 2, 3 \text{ and } j = 1, 2, 3.$$

Remark. If we add or subtract a constant to each element of any row (or column) of the cost matrix, then it would not change the optimum assignment and at the same time the value of objective function remains the same, *i.e.*, optimum value remains the same.

8:2. ASSUMPTIONS IN ASSIGNMENT PROBLEM

To fit the definition of assignment problem, the problem need to be formulated in a way that satisfies the following assumptions:

1. Number of resources (assignees) and the number of activities (tasks) must be equal.
2. Each of the resource is to be assigned exactly one activity.
3. Each resource is to be performed by exactly one activity.
4. The cost c_{ij} is associated with the resource i $(i = 1, 2, ..., n)$ performing the activity j $(j = 1, 2, ..., n)$.
5. The objective is to determine how all n resources should be utilized to have the minimum cost.

Any problem satisfying all the above mentioned assumptions can be solved by using the assignment methods specially designed for assignment problems, wherein the first three assumptions are fairly restrictive.

8:3. METHODS OF SOLVING AN ASSIGNMENT PROBLEM

An assignment problem can be solved using any of the following four methods:

1. Complete Enumeration Method. In this method, a list of all possible assignments among the given resources and activities is prepared. Then, an assignment involving the minimum cost/time or distance (or maximum profit) is selected. It represents the optimum solution. In case there are more than one assignment patterns involving the same least cost, then they all represent the optimum solution, *i.e.*, the problem has multiple optima.

In general, if there are n jobs and n workers, there are n ! possible assignments. Thus, the listing and evaluation of all the possible assignments is a simple matter when n is small. When n is large, this method is not very practical. For example, if there are 8 jobs and 8 workers, we have to evaluate a total of 8 ! or 40,320 assignments. The method, therefore, is not suitable for real world situations.

2. Transportation Method. Since, an assignment problem is a special case of the transportation problem, it can be solved by transportation methods discussed in the previous chapter. However, every basic feasible solution of a general assignment problem having a square payoff matrix of order n should have $m + n - 1 = n + n - 1 = 2n - 1$ assignments or basic cells. But due to the special structure of this problem, any basic feasible solution cannot have more than n assignments. Thus, the assignment problem is inherently degenerate. In order to remove degeneracy, $(n - 1)$ dummy allocations will be required to proceed with the transportation method. However, because of the large number of dummy allocations in the solution, the transportation method becomes computationally inefficient for solving an assignment problem.

3. Simplex Method. An assignment problem can be formulated as a transportation problem which, in turn, is itself a special case of an LPP. Accordingly, an assignment problem can be formulated as an LPP with integer valued variables and may be solved using simplex method. Here, the decision variables take only one of the two values: 1 or 0.

In general, let

$$x_{ij} = \begin{cases} 1, & \text{if } i\text{th person is assigned } j\text{th job} \\ 0, & \text{if } i\text{th person is not assigned } j\text{th job} \end{cases}$$

The mathematical formulation of the assignment problem as a 0 – 1 integer linear programming problem would be:

$$\text{Minimize } z = \sum_{i=1}^{n} \sum_{j=1}^{n} c_{ij}\, x_{ij} \text{ subject to the constraints:}$$

$$x_{i1} + x_{i2} + \ldots + x_{in} = 1; \qquad i = 1, 2, \ldots, n$$

$$x_{1j} + x_{2j} + \ldots + x_{nj} = 1; \qquad j = 1, 2, \ldots, n$$

$$x_{ij} = 0 \text{ or } 1 \text{ for all } i \text{ and } j.$$

As can be seen in the general mathematical formulation of the assignment problem, there are $n \times n$ decision variables and $n + n$ or $2n$ equalities/equations. In particular, for a problem involving 5 workers/jobs, there will be 25 decision variables and 10 equalities. That means a simplex table having 25 columns and 10 rows. It is difficult to solve manually and hence this approach to the solution is not considered.

4. Hungarian Assignment Method. An efficient method for solving an assignment problem is the ***Hungarian Assignment Method*** (also known as reduced matrix method), which is based on the concept of opportunity cost. Opportunity costs show the relative penalties associated with assignment of resource to an activity as opposed to making the best or least cost assignment. If we can reduce the cost matrix to the extent of having at least one zero in each row and each column, then it will be possible to make optimum assignments (opportunity costs are all zero).

The method of solving an assignment problem (minimization case) consists of the following steps:

Step 1. Determine the cost matrix from the given problem.

(*i*) If number of sources is equal to number of destinations, go to *Step 3*.

(*ii*) If number of sources is not equal to number of destinations, go to *Step 2*.

Step 2. Add a dummy source (or dummy destination), so that the cost matrix becomes a square matrix. The cost entries of dummy source/destination are always zero.

Step 3. Locate the smallest element in each row of the given cost matrix and then subtract the same from each element of that row.

Step 4. In the reduced matrix obtained in *Step 3*, locate the smallest element of each column, and then subtract the same from each element of that column. Each column and each row now have at least one zero.

Step 5. In the modified cost matrix obtained in *Step 4*, search for an optimum assignment as follows:
(*a*) Examine the rows successively from top to bottom until a row with a single zero is found. Make an assignment to this single zero by making a square (□) around it and cross off (×) all other zeros in its column. Continue in this manner until all the rows have been taken care of.
(*b*) Repeat the procedure for each column successively from left to right of the reduced matrix.
(*c*) If a row and/or column have two or more zeros and one cannot be chosen by inspection, then assign arbitrarily any one of these zeros and cross off all other zeros of that row/column.
(*d*) Repeat (*a*) through (*c*) above successively until the chain of assigning (□) or cross (×) ends.

Step 6. If the number of assignments (□) is equal to n (order of the cost matrix), an optimum solution is reached.

If the number of assignments is less than n (order of the matrix), go to next step.

Step 7. Draw minimum number of horizontal and/or vertical lines to cover all the zeros of the reduced cost matrix. This can be conveniently done by using a simple procedure given below:
(*a*) Mark (✓) rows that do not have any assigned zero.
(*b*) Mark (✓) columns that have zeros in the marked rows.
(*c*) Mark (✓) rows that have assigned zeros in the marked columns.
(*d*) Repeat (*b*) and (*c*) above until the chain of marking is completed.
(*e*) Draw lines through all the ***unmarked rows*** and ***marked columns***. This gives us the desired minimum number of lines.

Step 8. Develop the new revised cost matrix as follows:
(*a*) Find the smallest element of the reduced cost matrix not covered by any of the lines.
(*b*) Subtract this element from all the ***uncovered*** elements and add the same to all the elements lying at the ***intersection*** of horizontal and vertical lines.

Step 9. Go to *Step 6* and repeat the procedure until an ***optimum solution*** is attained.

Example 8.1. *A company wishes to assign 3 jobs to 3 machines in such a way that each job is assigned to some machine and no machine works on more than one job. The cost of assigning job i to machine j is given by the matrix below (ij^th entry):*

$$\text{Cost matrix: } \begin{bmatrix} 8 & 7 & 6 \\ 5 & 7 & 8 \\ 6 & 8 & 7 \end{bmatrix}$$

Draw the associated network. Formulate the given problem as an LPP and find the minimum cost of making the assignment. **[BBA, GGSIP Univ. 2011]**

Solution. (*a*) Network formulation of the given problem is:

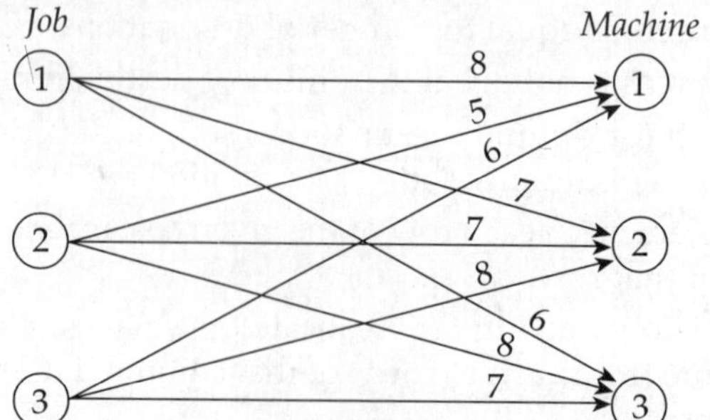

Fig. 8.1

(*b*) Linear programming formulation of the given problem is:

Minimize $z = (8x_{11} + 7x_{12} + 6x_{13}) + (5x_{21} + 7x_{22} + 8x_{23}) + (6x_{31} + 8x_{32} + 7x_{33})$

subject to the constraints:

$$x_{i1} + x_{i2} + x_{i3} = 1; \qquad i = 1, 2, 3$$
$$x_{1j} + x_{2j} + x_{3j} = 1; \qquad j = 1, 2, 3$$
$$x_{ij} = 0 \text{ or } 1, \text{ for all } i \text{ and } j.$$

(*c*) Reduce the cost matrix by subtracting smallest element of each row (column) from the corresponding row (column) elements. In the reduced cost matrix, make assignments in rows and columns having single zero and cross off all other zeros in rows and columns in which the assignment has been made. See *Table 8.1*. Now, draw the minimum number of lines to cover all the zeros. For this, we proceed as follows:

(*i*) Mark (✓) third row since it has no assignment.
(*ii*) Mark (✓) first column, since third row has a zero in this column.
(*iii*) Mark (✓) second row, since marked column has an assignment in the second row.
(*iv*) Since, no other row or column can be marked, draw straight lines through the unmarked row and marked column as shown in *Table 8.1*.

2	0	[0]	
[0]	1	3	✓
~~0~~	1	1	✓
✓			

Table 8.1

3	[0]	~~0~~
[0]	~~0~~	2
~~0~~	~~0~~	[0]

Table 8.2

Modify the reduced cost matrix (*Table 8.1*) by selecting the smallest element among all the uncovered elements. Subtract this element from all the uncovered elements including itself and add it to the element at (1, 1) which lies at the intersection of two lines. The modified cost matrix so obtained is shown in *Table 8.2*.

In *Table 8.2*, we observe that there is no row and column which has single zero. So, we make an assignment arbitrarily at (1, 2) and cross off all zeros of first row and second column. Now, we get a single zero in the second row and therefore, an assignment is made at (2, 1). Cross off all zeros in the first column. Finally, we make an assignment at (3, 3) being the single zero in the third row.

Clearly, the number of assignments in *Table 8.2* is equal to order of the cost matrix. Hence, an optimum assignment has been attained, *viz.*

Job 1 → Machine 2, Job 2 → Machine 1, and Job 3 → Machine 3.

Total minimum cost will be (7 + 5 + 7), i.e., 19.

Remark. Alternative Optimum assignment schedule:

Job 1 → Machine 3, Job 2 → Machine 1, Job 3 → Machine 2,

and Job 1 → Machine 3, Job 2 → Machine 2, Job 3 → Machine 1.

Example 8.2 (*Distance Minimization*). *A car hire company has one car at each of five depots a, b, c, d and e. A customer in each of the five towns A, B, C, D and E requires a car. The distance (in kms.) between the depots (origins) and the towns (destinations), where the customers are, is given in the following distance* *matrix:*

	a	*b*	*c*	*d*	*e*
A	*160*	*130*	*175*	*190*	*200*
B	*135*	*120*	*130*	*160*	*175*
C	*140*	*110*	*155*	*170*	*185*
D	*50*	*50*	*80*	*80*	*110*
E	*55*	*35*	*70*	*80*	*105*

How should the cars be assigned to customers so as to minimize the distance travelled?

[*MBA (PT.), Delhi Univ. 2008*]

Solution. Subtract the smallest element of each row from all the elements of that row and then subtract the smallest element of each column from all the elements of that column. In the reduced matrix, make assignments in rows and columns that have single zero.

Initial Iteration. Draw the minimum number of horizontal and vertical lines that cover all the zeros. See *Table 8.3.*

30	[0]	35	30	15	✓
15	~~0~~	[0]	10	~~0~~	
30	~~0~~	35	30	20	✓
[0]	~~0~~	20	~~0~~	5	
20	~~0~~	25	15	15	✓
	✓				

Table 8.3

15	~~0~~	20	15	[0]
15	15	[0]	10	~~0~~
15	[0]	20	15	5
[0]	15	20	~~0~~	5
5	~~0~~	10	[0]	~~0~~

Table 8.4

Final Iteration. Modify *Table 8.3* by subtracting the smallest element among those elements which are not covered by lines, *viz.*, 15 from all the elements not covered by lines. Also, add the same element at the intersection of horizontal and vertical lines. In the modified table, now, make assignments in rows and columns that have single zero. See *Table 8.4.*

From *Table 8.4,* we observe that the number of assignments equals the order of matrix. Hence, an optimum solution is attained, *viz.*, assign *A* to *e*, *B* to *c*, *C* to *b*, *D* to *a* and *E* to *d*.

Minimum total distance travelled will be of 570 kms.

Example 8.3 (*Cost Minimization*). *A private firm employs typists on hourly page rate basis for their daily work. Five typists are working in that firm and their charges and speeds are different. On the basis of some earlier understanding, only one job is given to one typist and the typist is paid for full hour even when he or she works for a fraction of an hour. Find the least cost allocation for the following data:*

Typist	*Rate per hour (₹)*	*Number of pages typed per hour*	*Job*	*No. of pages*
A	*5*	*12*	*P*	*199*
B	*6*	*14*	*Q*	*175*
C	*3*	*8*	*R*	*145*
D	*4*	*10*	*S*	*298*
E	*4*	*11*	*T*	*178*

[*BE, Karnataka Univ. 2011; M.Com., Delhi Univ. 2009*]

Solution. Making use of the information given above, the cost matrix is obtained as follows:

		Job P	Q	R	S	T
Typist	A	85	75	65	125	75
	B	90	78	66	132	78
	C	75	66	57	114	69
	D	80	72	60	120	72
	E	76	64	56	112	68

Here, element at (1, 1) is computed as (199/12 pages × ₹ 5), *i.e.*, 17 × 5 or 85, where fraction of hour is considered as complete hour. Similarly, element at (2, 2) is computed as (175/14) pages × ₹ 6, *i.e.*, 13 × 6 or 78.

Subtracting the smallest element of each row and each column from all the elements of that particular row/column, we get the reduced cost matrix. In the reduced cost matrix, make assignments in rows and columns that have single zeros.

Initial Iteration. Draw the minimum number of lines that cover all the zeros of the reduced cost matrix. See *Table 8.5*.

2	2	~~0~~	4	[0]	
6	4	[0]	10	2	✓
[0]	1	~~0~~	1	2	
2	4	~~0~~	4	2	✓
2	[0]	~~0~~	~~0~~	2	
		✓			

Table 8.5

2	2	2	4	[0]
4	2	[0]	8	~~0~~
[0]	1	2	1	2
~~0~~	2	~~0~~	2	~~0~~
2	[0]	2	~~0~~	2
✓		✓		✓

Table 8.6

First Iteration. Modify the reduced cost matrix (*Table 8.5*) by subtracting the least uncovered element '2' from all the elements not covered by the lines and adding the same at the intersection of horizontal and vertical lines. Now, draw the minimum number of lines to cover all the zeros of the modified cost matrix. See *Table 8.6*.

Final Iteration. Modify *Table 8.6* by subtracting the least uncovered element '1' from all the elements not covered by lines and adding the same at the intersection horizontal and vertical lines, and then make assignments. See *Table 8.7*.

	P	Q	R	S	T
A	2	1	2	3	[0]
B	4	1	[0]	7	~~0~~
C	~~0~~	~~0~~	2	[0]	2
D	[0]	1	~~0~~	1	~~0~~
E	3	[0]	3	~~0~~	3

Table 8.7

From *Table 8.7*, we observe that the optimum solution is to assign

A to *T*, *B* to *R*, *C* to *S*, *D* to *P* and *E* to *Q*.

Alternative optimum assignment schedule is to assign

A to *T*, *B* to *R*, *C* to *Q*, *D* to *P* and *E* to *S*.

Total minimum cost according to above assignment schedule is ₹ 399.

Example 8.4 (*Unbalanced Assignment Problem*). *A department head has four tasks to be performed and three subordinates, the subordinates differ in efficiency. The estimates of the time, each subordinate would take to perform task, is given below. How should he allocate the tasks one to each subordinate, so as to minimize the total man-hours?*

	Subordinate	1	2	3
Task	I	9	26	15
	II	13	27	6
	III	35	20	15
	IV	18	30	20

Solution. Here, we have three subordinates who have to perform four tasks. So, the given assignment problem is unbalanced and therefore, we add a dummy subordinate (column) with all its entries as zero. The resulting balanced problem is:

	Subordinate	1	2	3	Dummy
Task	I	9	26	15	0
	II	13	27	6	0
	III	35	20	15	0
	IV	18	30	20	0

Now, reduce the balanced time-matrix by subtracting the smallest element of each column from all the elements of that column. In the reduced matrix, make assignments in rows and columns having single zeros and cross off all other zeros of the rows and columns, where assignment have been made. We get the following assignment solution:

	1	2	3	Dummy
I	[0]	6	9	~~0~~
II	4	7	[0]	~~0~~
III	26	[0]	9	~~0~~
IV	9	10	14	[0]

The optimum assignment is: $I \to 1, II \to 3$ and $III \to 2$; while task *IV* should be assigned to a dummy subordinate, *i.e.*, it remains to be done. The minimum time is: 35 hours.

PROBLEMS

8.01. *PQR* is an accounting firm that has three new clients. Project leaders will be assigned to three clients. Based on the different backgrounds and experiences of the project leaders, the various leader — client assignments differ in terms of projected completion times. The possible assignments and the estimated completion times in days are given in the adjacent table:

		Clients 1	2	3
Project Leader	Gita	10	16	32
	Vijay	14	22	40
	Sita	22	24	34
	Ajay	14	18	36

(*i*) Formulate the problem as a linear program.

(*ii*) Find the optimal assignment using Hungarian method. **[*MBA, Delhi Univ. 2012*]**

8.02. Four teachers are capable of teaching anyone of the 4 different subjects. Class preparation time-hours for different topics varies from teacher to teacher and is given in the table below. Each teacher is assigned only one topic so as to minimize the total course preparation time for all the topics.

		Topics OR	CG	SM	PM
Teacher	A	10	24	30	15
	B	16	22	28	12
	C	12	20	32	10
	D	9	26	34	16

[*M.Com. (CBCS), Karnataka Univ. 2013*]

8.03. An automobile dealer wishes to put four repairmen to four different jobs. The repairmen have somewhat different kinds of skills and they exhibit different levels of efficiency from one job to another. The dealer has estimated the number of man-hours that would be required for each Job-Man combination. This is given in the following table:

Man	*Job*			
	A	B	C	D
1	5	3	2	8
2	7	9	2	6
3	6	4	5	7
4	5	7	7	8

Find the optimum assignment that will result in minimum man-hours needed.

[MBA, Lucknow Univ. 2008]

8.04. Solve the minimal assignment problem whose effectiveness cost matrix is given below:

(*a*)

	1	2	3	4
I	2	3	4	5
II	4	5	6	7
III	7	8	9	8
IV	3	5	8	4

[B.Tech. (Mech.), HP Univ. 2015]

(*b*)

	M_1	M_2	M_3	M_4
W_1	9	5	6	7
W_2	8	5	7	7
W_3	6	8	5	6
W_4	8	10	7	6

[M.Com., Bangalore Univ. 2018]

8.05. The average time (in minutes) to contact a potential new client by four sales-persons in four territories is given in the adjacent table. How each sales person should be assigned to the four territories in order to minimize the total time required to contact a potential client?

Sales person	*Sales territory* A	B	C	D
1	8	10	12	16
2	11	11	15	8
3	9	6	5	14
4	15	14	9	7

[MBA (HCA), Delhi Univ. 2015]

8.06. Solve the following assignment problem using Hungarian method. The matrix entries are processing times in hours:

Job	*Operator* O_1	O_2	O_3	O_4	O_5
J_1	20	22	35	22	18
J_2	4	26	24	24	7
J_3	23	14	17	19	19
J_4	17	15	16	18	15
J_5	16	19	21	19	25

[MBA, Madras Univ. 2006; BCA (Nov.), Madras Univ. 2010]

8.07. The *XYZ* company has 5 jobs *I, II, III, IV* and *V* to be done and 5 men *A, B, C, D* and *E* to do these jobs. The number of hours each man would take to accomplish each job is given by the adjacent table:

Work out the optimal assignment and the total minimum time taken. *[MBA, Lucknow Univ. 2009]*

	A	B	C	D	E
I	16	13	17	19	20
II	14	12	13	16	17
III	14	11	12	17	18
IV	5	5	8	8	11
V	5	3	8	8	10

8.08. A certain equipment needs five repairing jobs which have to be assigned to five mechanics. The estimated time (in hours) that each mechanic requires to complete the repair job is given in the following table:

Mechanics	Jobs:	I	II	III	IV	V
	A	2	9	2	7	1
	B	6	8	7	6	1
	C	4	6	5	3	1
	D	4	2	7	3	1
	E	5	3	9	5	1

Find the assignment of each mechanic to one job that will minimize the total time taken.

[BBA, GGSIP Univ. 2018; MBA (PT.), Delhi Univ. 2009]

8.09. Consider the problem of assigning five jobs to five persons. The assignment costs are given as follows:

Person	Jobs:	1	2	3	4	5
	A	8	4	2	6	1
	B	0	9	5	5	4
	C	3	8	9	2	6
	D	4	3	1	0	3
	E	9	5	8	9	5

Determine the optimum assignment schedule. *[MBA, Bangalore Univ. 2018]*

8.10. A company is producing a single product and is selling it through five agencies situated in different cities. All of a sudden, there is a demand for the product in another five cities not having any agency of the company. The company is faced with the problem of deciding on how to assign the existing agencies to despatch the product to needy cities in such a way that the total travelling distance is minimized. The distance between the surplus and deficit cities (in kilometres) is given in the following table:

Surplus city	Deficit city				
	A′	B′	C′	D′	E′
A	10	5	9	18	11
B	13	19	6	12	14
C	3	2	4	4	5
D	18	9	12	17	15
E	11	6	14	19	10

Determine the optimum assignment schedule. *[MBA, Madurai Univ. 2009]*

8.11. Solve the following assignment problem. Assign one machine to one worker so that time in hours is minimized.

	M_1	M_2	M_3	M_4	M_5
A	3	2	7	4	8
B	5	4	3	8	5
C	3	7	9	1	2
D	4	2	6	5	7
E	2	8	4	6	6

[B.Tech., Panjab Tech. Univ. 2012; MBA (Dec.), Gujarat Technological Univ. 2010]

8.12. There are five jobs to be assigned to five machines. Cost of completion of the jobs on respective machine are as given in the table given below:

		Machine				
		M_1	M_2	M_3	M_4	M_5
	J_1	65	40	90	80	90
	J_2	60	35	100	85	85
Job	J_3	60	38	105	90	95
	J_4	70	45	120	90	100
	J_5	65	40	105	87	90

Employ Hungarian method of optimal assignment of jobs to machines so as to minimize the total cost of assigning all the jobs. Is the solution obtained unique? If not, work out an alternative solution and calculate the total cost implied by it. ***[M.Com., Delhi Univ. 2008]***

8.13. A company has 4 machines on which to do 3 jobs. Each job can be assigned to one and only one machine. The cost of each job on each machine is given in the following table:

Job	*Machine*			
	A	B	C	D
1	18	24	28	32
2	8	13	17	19
3	10	15	19	22

Determine the optimum assignment. ***[MBA, Hyderabad Univ. 2011]***

8.14. A national truck-rental service has a surplus of one truck in each of the cities *1, 2, 3, 4, 5* and *6*; and a deficit of one truck in each of the cities *7, 8, 9, 10, 11* and *12*. The distances (in kilometres) between the cities with a surplus and the cities with a deficit are displayed below:

		7	*8*	*9*	*10*	*11*	*12*
	1	31	62	29	42	15	41
	2	12	19	39	55	71	40
From	3	17	29	50	41	22	22
	4	35	40	38	42	27	33
	5	19	30	29	16	20	23
	6	72	30	30	50	41	20

How should the trucks be dispersed so as to minimize the total distance travelled?

[MBA, Osmania Univ. 2009; MBA (Feb.), Panjab Tech. Univ. 2011]

8.15. *XYZ* company has 4 sales managers to be assigned to 5 different sales territories. The weekly wages estimated for each sales manager in different territories in thousands of rupees are shown in the following table:

Sales managers	*Sales territories*				
	I	*II*	*III*	*IV*	*V*
A	75	80	85	70	90
B	91	71	82	75	85
C	78	90	85	80	80
D	65	75	88	85	90

Suggest optimal assignment for the sales manager. Which sales territory will remain unassigned? ***[PG Dip. in Glob. Bus. Oper., Delhi Univ. 2011]***

8.16. A company has one surplus truck in each of the cities *A, B, C, D* and *E* and one deficit truck in each of the cities, *1, 2, 3, 4, 5* and *6*. The distance between the cities in kilometres is shown in the table given below:

	Truck →	1	2	3	4	5	6
	A	12	10	15	22	18	8
	B	10	18	25	15	16	12
City	C	11	10	3	8	5	9
	D	6	14	10	13	13	12
	E	8	12	11	7	13	10

Find the assignment of trucks from cities in surplus to cities in deficit so that the total distance covered by the vehicles is minimum. ***[MBA (Nov.), Annamalai Univ. 2008; B.Tech., Panjab Tech. Univ. 2012]***

8.17. The marketing director of a multi-unit company is faced with a problem of assigning 5 senior managers to six zones. From the past experience, he knows that the efficiency percentage judged by sales, operating cost, etc., depends on manager-zone combination. The efficiency of different managers is displayed as follows:

Senior Manager	Zone					
	I	II	III	IV	V	VI
A	73	91	87	82	78	80
B	81	85	69	76	74	85
C	75	72	83	84	78	91
D	93	96	86	91	83	82
E	90	91	79	89	69	76

Find out which zone will be managed by a junior manager due to non-availability of senior manager. ***[MBA, IGNOU 2008]***

8.18. Gurugram city municipal corporation has decided to carry out repairs on four main roads of the city. The State Government has agreed to make a special grant of ₹ 50 lakh towards the costs with a condition that the repairs must be done at the lowest cost and quickest time. If conditions warrant, then a supplementary token grant will also be considered favourably. The corporation has floated tenders and five contractors have sent in their bids. In order to expedite work, one road will be awarded to only one contractor.

Contractor	Cost of repairs (₹ in lakh)			
	Road R_1	Road R_2	Road R_3	Road R_4
C_1	9	14	19	15
C_2	7	17	20	19
C_3	9	18	21	18
C_4	10	12	18	19
C_5	10	15	21	16

(*i*) Find the best way of assigning the repair work to the contractors and the costs.

(*ii*) If it is necessary to seek supplementary grants, then what would be the amount sought?

(*iii*) Which of the five contractors will be unsuccessful in his bid?

[PG Dip. in Glob. Bus. Oper., Delhi Univ. 2010]

8.19. You are given information about cost of performing different jobs by different persons. The job-person marking '*X*' indicates that individual involved cannot perform particular job. Using this information state the optimal assignment of jobs. Also, find the cost of optimal assignment.

Person	*Job*				
	J_1	J_2	J_3	J_4	J_5
P_1	27	18	X	20	21
P_2	31	24	21	12	17
P_3	20	17	20	X	16
P_4	22	28	20	16	27

[MBA, Visvesvarayya Univ. 2011]

8:4. SPECIAL CASES IN ASSIGNMENT PROBLEMS

Maximization Case in Assignment Problem

In some cases, the pay off elements of the assignment problem may represent revenues or profits instead of costs so that the objective will be to maximize the total revenue or profit. The Hungarian method explained earlier can also be used for maximization case. The problem of maximization can be converted into that of minimization by selecting the largest element of the profit matrix and then subtracting from it all the elements of the profit matrix. This gives us an ***opportunity loss matrix***. We then proceed as usual and obtain the optimum assignment schedule. Maximum revenue/profit is then obtained by restoring the original values of those cells in which the assignments have been made.

Example 8.5 (*Maximization Problem*). *A manufacturing company has four zones A, B, C, D and four sales engineers P, Q, R, S respectively for assignment. Since, the zones are not equally rich in sales potential, it is estimated that a particular engineer operating in a particular zone will bring the following sales:*

Zone A: 4,20,000 Zone B: 3,36,000 Zone C: 2,94,000 Zone D: 4,62,000

The engineers are having different sales ability. Working under the same conditions their yearly sales are proportional to 14, 9, 11 and 8 respectively. The criteria of maximum expected total sales is to be met by assigning the best engineer to the richest zone, the next best to the second richest zone and so on.

Find the optimum assignment and the maximum sales. *[C.A. Final, May 1998]*

Solution.

Step 1 **(*Construct the Effectiveness Matrix*).** To avoid the fractional values of annual sales of each sales engineer in each zone; for convenience consider their yearly sales as 42 (*i.e.*, the sum of sales proportions), taking ₹ 1,000 as one unit. Now, divide the individual sales in each zone by 42 to obtain the required annual sales by each sales engineer. For example, if sales engineer *P* is assigned to zone *A*, the annual sales is 14 × 4,20,000/42, *i.e.*, 1,40,000. Thus, the proportional sales in the four zones are given in the following effective table (matrix):

		Zones				*Sales*
		A	*B*	*C*	*D*	*proportion*
	P	140	112	98	154	14
Sales	*Q*	90	72	63	99	9
engineer	*R*	110	88	77	121	11
	S	80	64	56	88	8

Step 2 **(*Converting Maximization Problem into Minimization Problem*).** The given maximization assignment problem can be converted into a minimization assignment problem by subtracting from the higher element (*i.e.*, 154), all the elements of the matrix given above. The resulting opportunity loss matrix is:

	A	B	C	D
P	14	42	56	0
Q	64	82	91	55
R	44	66	77	33
S	74	90	98	66

Step 3 ***Optimum Assignment.*** In the opportunity loss matrix subtract the smallest element of each row from all the elements of that row. Then, subtract the smallest element of each column from all the elements of that column. In the reduced matrix make assignments in rows and columns that have single zero as usual. Thus, we have:

Initial Iteration. Draw minimum number of lines to cover all the zeros of the reduced matrix. See *Table 8.8.*

First Iteration. Modify the reduced opportunity loss matrix by subtracting element '1' from all the elements not covered by the lines and adding the same at the intersection of two lines. Thus, we get *Table 8.9*.

	A	B	C	D	
P	6	18	24	[0]	✓
Q	1	3	4	~~0~~	✓
R	3	9	12	~~0~~	✓
S	[0]	~~0~~	~~0~~	~~0~~	
				✓	

Table 8.8

	A	B	C	D	
P	5	17	23	[0]	✓
Q	[0]	2	3	~~0~~	
R	2	8	11	~~0~~	✓
S	~~0~~	[0]	~~0~~	1	
				✓	

Table 8.9

Second Iteration. Modify further the reduced *Table 8.9* by subtracting element '2' from all the elements not covered by the lines and adding the same at the intersection of two lines. Thus, we get *Table 8.10.*

Final Iteration. Modify further the reduced *Table 8.10* by subtracting element '2' from all the elements not covered by the lines and adding the same at the intersection of two lines. Thus, we get *Table 8.11.*

	A	B	C	D	
P	3	15	21	[0]	✓
Q	[0]	2	3	2	✓
R	~~0~~	6	9	~~0~~	✓
S	~~0~~	[0]	~~0~~	3	
	✓			✓	

Table 8.10

	A	B	C	D
P	3	13	19	[0]
Q	~~0~~	[0]	1	2
R	[0]	4	7	~~0~~
S	2	~~0~~	[0]	5

Table 8.11

The optimum solution is obtained and the assignment is:

$$P \to D, \quad Q \to B, \quad R \to A \quad \text{and} \quad S \to C$$

The solution shows that the best salesman *P* is assigned to the richest zone *D* and the worst salesman *S* to the poorest zone *C*. The second best salesman *R* to the next richest zone *A*, and so on.

Maximum sales = ₹ (154 + 72 + 110 + 56) = ₹ 392 thousands.

Prohibited Assignments (*Restrictions on Assignment*)

Sometimes due to certain reasons, a particular resource (say a man or a machine) cannot be assigned to perform a particular activity (say a territory or a job). In such cases, the cost of performing that particular activity by a particular resource is considered to be very large (written as M or ∞) so as to prohibit the entry to this pair of resource-activity into the final (optimum) solution.

Example 8.6. *In the modification of a plant layout of a factory, four new machines M_1, M_2, M_3 and M_4 are to be installed in a machine shop. There are five vacant places A, B, C, D and E available. Because of limited space, machine M_2 cannot be placed at C and M_3 cannot be placed at A. The cost of placing of machine i at place j (in hundred rupees) is shown below:*

		Location				
		A	*B*	*C*	*D*	*E*
	M_1	9	11	15	10	11
Machine	M_2	12	9	—	10	9
	M_3	—	11	14	11	7
	M_4	14	8	12	7	8

Find the optimal assignment schedule. **[MBA (Nov.), Delhi Univ. 2009; MIB, Delhi Univ. 2011]**

Solution. Since, the number of machines ≠ number of vacant places, the cost matrix is unbalanced. So, we add a dummy machine (row) with all its elements as zero. Also, assign a very high cost, M, to places (3, 1) and (2, 3). The cost matrix so obtained is given in *Table 8.12*. Reduce this cost matrix by subtracting the smallest element of each row (column) from all the elements of that row (column). In the reduced matrix, make assignments in rows and columns that have single zero and cross off all other zeros of the row/column in which an assignment is made. Thus, we get *Table 8.13*.

	A	*B*	*C*	*D*	*E*
M_1	9	11	15	10	11
M_2	12	9	*M*	10	9
M_3	*M*	11	14	11	7
M_4	14	8	12	7	8
Dummy	0	0	0	0	0

Table 8.12

	A	*B*	*C*	*D*	*E*
M_1	[0]	2	6	1	2
M_2	3	[0]	*M*	1	~~0~~
M_3	*M*	4	7	4	[0]
M_4	7	1	5	[0]	1
Dummy	~~0~~	~~0~~	[0]	~~0~~	~~0~~

Table 8.13

Since, number of assignments in *Table 8.13* is equal to order of the matrix, an optimum solution is reached. The optimum solution is:

Assign M_1 at *A*, M_2 at *B*, M_3 at *E* and M_4 at *D*; at a minimum total cost of ₹ 3,200.

8:5. A TYPICAL ASSIGNMENT PROBLEM

Example 8.7. *XYZ Airline operating 7 days a week has given the following time-table. Crews must have a minimum layover of 5 hours between flights. Obtain the pairing of flights that minimizes layover time away from home. For any given pairing the crew will be based at the city having the smaller layover:*

Chennai–Mumbai			*Mumbai–Chennai*		
Flight No.	*Depart*	*Arrive*	*Flight No.*	*Depart*	*Arrive*
A_1	6 AM	8 AM	B_1	8 AM	10 AM
A_2	8 AM	10 AM	B_2	9 AM	11 AM
A_3	2 PM	4 PM	B_3	2 PM	4 PM
A_4	8 PM	10 PM	B_4	7 PM	9 PM

[M.Com., Delhi Univ. 2010]

Solution. It is assumed that a plane flying from Chennai to Mumbai must come back to Chennai at the immediate next opportunity. It is further assumed that each plane will make only one forward and one return trip, and thus there must be four planes for four forward and return flights. The problem is to determine the optimum pairing.

As the objective is to minimize the total layover time, let us at first instance determine the layover time both at Chennai and Mumbai for all possible pairings. For Flight No. A_1 and B_1 pair, the plane leaves Chennai at 6 AM and reaches Mumbai at 8 AM. If it has to avail of B_1 return flight, it can start at 8 AM next day because the minimum layover time between flights is 5 hours. Thus, the layover time at Mumbai will be 24 hours. Similarly, for the same pairing, the plane reaches Chennai at 10 AM and leaves for Mumbai at 6 AM next day making the layover of (24 – 4), *i.e.*, 20 hours at Chennai. Likewise, we determine layover time at Mumbai and Chennai for all possible pairings. These are shown in the following table:

Flight No.	*Crew based at Chennai*				*Crew based at Mumbai*			
	B_1	B_2	B_3	B_4	B_1	B_2	B_3	B_4
A_1	24	25	6	11	20	19	14	9
A_2	22	23	28	9	22	21	16	11
A_3	16	17	22	27	28	27	22	17
A_4	10	11	16	21	10	9	28	23

With the help of above information, we construct the table for minimum layover time between the pair of flights. The layover time marked '*' denotes that the crew is based at Mumbai. Thus, we get *Table 8.14:*

	B_1	B_2	B_3	B_3
A_1	20*	19*	6	9*
A_2	22	21*	16*	9
A_3	16	17	22	17*
A_4	10	9*	16	21

Table 8.14

Now, apply assignment method to find the optimum solution. For this, we subtract the smallest element of each row from all the elements of that row. Then, subtract the smallest element of each column from all the elements of that column. In this way, we have at least one zero in each row and each column. Thus, we have *Table 8.15*.

In *Table 8.15*, make assignments in rows and columns that have single zero and cross off all other zeros of that row/column. Thus, we get *Table 8.16*.

	B_1	B_2	B_3	B_4
A_1	14	13	0	3
A_2	13	12	7	0
A_3	0	1	6	1
A_4	1	0	7	12

Table 8.15

	B_1	B_2	B_3	B_4
A_1	14	13	[0]	3
A_2	13	12	7	[0]
A_3	[0]	1	6	1
A_4	1	[0]	7	12

Table 8.16

Since, the number of assignments in *Table 8.16* equals the order of matrix, an optimum solution is obtained with the following assignment schedule:

Flight No.	:	A_1 *to* B_3	A_2 *to* B_4	A_3 *to* B_1	A_4 *to* B_2
Base crew	:	Chennai	Chennai	Chennai	Mumbai
Layover time	:	6 hours	9 hours	16 hours	9 hours

Total layover time will be of 40 hours.

PROBLEMS

8.20. Suggest the optimal assignment schedule for the following assignment problem:

Sales (₹ in lakh)

Salesmen	*Markets*			
	I	*II*	*III*	*IV*
A	80	70	75	72
B	75	75	80	85
C	78	78	82	78

What will be the total maximum sales? ***[M.Com., Delhi Univ. 2001]***

8.21. Swasthya hospitals has three serious surgeons who have different levels of competence on four types of operations: *01, 02, 03* and *04*. The hospital receives a large number of patients who undergo these operations and many among them have to wait a long time due to unavailability of operation theaters. Swasthya wants to assign each surgeon to a particular operation type only, so that the utilization of operation theaters is maximized. The competence level is decided on time taken to complete an operation. The adjacent table gives the time taken in hours by each surgeon to conduct the operation. Find out best assignment for the only type of operation that will not get an in-hour surgeon, Swasthya will contract services of surgeon from nearby hospitals. ***[MBA (Oct.), Delhi Univ. 2010]***

		Dr. A	*Dr. B*	*Dr. C*
Operations	*01*	10	12	19
	02	5	10	7
	03	12	14	13
	04	8	15	11

8.22. A company has four sales representatives who are to be assigned to four different sales territories *I, II, III* and *IV*. The monthly sales increase estimated for each sales representative for different sales territories (in lakh rupees), are shown in the following table:

	Territory:	*I*	*II*	*III*	*IV*
Sales representative	A	200	150	170	220
	B	160	120	150	140
	C	190	195	190	200
	D	180	175	160	190

(*i*) Suggest optimum assignment and the total maximum sales increase per month.

(*ii*) If for certain reasons, sales representative *B* cannot be assigned to sales territory *III*, will the optimum assignment schedule be different? If so, find that schedule and the effect on total sales. ***[M.Com., Delhi Univ. 1997]***

8.23. Suggest optimal assignment for four salesmen *A, B, C* and *D* to sales territories *I, II, III, IV* and *V*, where the estimated sales (in lakh rupees) to be made by each of the salesmen in different sales territories are given below:

Territory:	*I*	*II*	*III*	*IV*	*V*
A	16	15	17	10	8
B	16	16	20	15	12
C	12	8	10	13	15
D	18	16	17	12	10

What will be total maximum sales? ***[MBA (Nov.), Delhi Univ. 2003]***

8.24. Five jobs are to be processed and five machines are available. Any machine can process any job with the resulting profit (in rupees) as follows:

Job	*Machine*				
	A	B	C	D	E
1	32	38	40	28	40
2	40	24	28	21	36
3	41	27	33	30	37
4	22	38	41	36	36
5	29	33	40	35	39

What is the maximum profit that may be expected, if an optimum assignment is made?

[MBA, UP Tech. Univ. 2006; M.Com., Delhi Univ. 2005]

8.25. The Captain of a cricket team has to allot five middle batting positions to five batsmen. The average runs scored by each batsman at these positions are as follows:

Batsman	*Batting positions*				
	I	*II*	*III*	*IV*	*V*
P	40	40	35	25	50
Q	42	30	16	25	27
R	50	48	40	60	50
S	20	19	20	18	25
T	58	60	59	55	53

(*i*) Find the assignment of batsman to positions, which would give the maximum number of runs. ***[BE (Dec.), Gujarat Tech. Univ. 2012; MBA (PT.), Delhi Univ. 2015]***

(*ii*) If another batsman '*U*' with the following average runs in batting positions:

Batting position :	*I*	*II*	*III*	*IV*	*V*
Average runs :	45	52	38	50	49

is added to the team, should he be included to play in the team? If so, who will be replaced by him? ***[MBA, PG Dip. in Glob. Bus. Oper., Delhi Univ. 2011]***

8.26. A company is faced with the problem of assigning 4 machines to 6 different jobs (one machine to one job only). The profits are estimated as follows:

Job	*Machine*			
	A	*B*	*C*	*D*
1	3	6	2	6
2	7	1	4	4
3	3	8	5	8
4	6	4	3	7
5	5	2	4	3
6	5	7	6	4

Solve the problem to maximize the total profits.

[MBA (Nov.), Annamalai Univ. 2010; M.Com., Madurai Univ., 2003]

8.27. A company has four zones *A*, *B*, *C*, *D* and four salesman *P*, *Q*, *R S* for assignment. The zones are not equally rich in their sale potentials. It is estimated that a typical salesman operating in each zone would bring in the following annual sales:

Zone A: 1,26,000; *Zone B*: 1,05,000; *Zone C*: 84,000; *Zone D*: 63,000

The four salesman are also considered to differ in ability. It is estimated that working under the same conditions, their yearly sales would be proportionately as follows:

Salesman *P*: 7; Salesman *Q*: 5; Salesman *R*: 5; Salesman *S*: 4.

If criterion is of maximum expected total sales, the intuitive answer is to assign the best salesman to the richest zone, the next best to the second richest zone and so on. Verify this by the method of assignment. *[M.Com., Kerala Univ. 2010; BFIA, Delhi Univ. 2008; MBA (Nov.), Delhi Univ. 2005]*

[*Hint:* Consider the yearly sales as 21 (*i.e.*, the sum of proportions 7 + 5 + 5 + 4), and take ₹ 1,000 as one unit. Then, the proportional sales in the four zones are:

$$\begin{bmatrix} \frac{7}{21}\times 126 & \frac{7}{21}\times 105 & \frac{7}{21}\times 84 & \frac{7}{21}\times 63 \\ \frac{5}{21}\times 126 & \frac{5}{21}\times 105 & \frac{5}{21}\times 84 & \frac{5}{21}\times 63 \\ \frac{5}{21}\times 126 & \frac{5}{21}\times 105 & \frac{5}{21}\times 84 & \frac{5}{21}\times 63 \\ \frac{4}{21}\times 126 & \frac{4}{21}\times 105 & \frac{4}{21}\times 84 & \frac{4}{21}\times 63 \end{bmatrix}.]$$

8.28. Parul Corporation has four plants each of which can manufacture any one of four products. Production costs differ from one plant to another as do sales revenue. Given the revenue and cost data below, obtain which product each plant should produce to maximize profit:

Plant	*Sales revenue (₹ '000s) Product*				*Production cost (₹ '000s) Product*			
	1	2	3	4	1	2	3	4
A	50	68	49	62	49	60	45	61
B	60	70	51	74	55	63	45	69
C	55	67	53	60	52	62	49	58
D	58	65	54	69	55	64	48	66

[B.Tech. (Comp. Sc.), Hyderabad Univ. 2011]

[*Hint:* Profit = Sales revenue – Production cost]

8.29. A small garment making unit has five tailors stitching five different types of garments. All the five tailors are capable of stitching all the five types of garments. The output per day per tailor and the profit (in ₹) for each type of garment are given below:

Tailors	*Garments*				
	1	2	3	4	5
A	7	9	4	8	6
B	4	9	5	7	8
C	8	5	2	9	8
D	6	5	8	10	10
E	7	8	10	9	9
Profit per garment (in ₹)	2	3	2	3	4

(*i*) Which type of garment should be assigned to which tailor in order to maximize profit, assuming that there are no other constraints?

(*ii*) If tailor *D* is absent for a specified period and no other substitute tailor is available, what should be the optimal assignment? *[B.Tech. (Mech.), HP Univ. 2012]*

8.30. A company is taking bids on four construction jobs. Three people have placed bids on the jobs. Their bids (in lakh of rupees) are given in the following table ('*' indicates that the person did not bid for the given job):

Person	*Job*			
	1	2	3	4
A	50	46	42	40
B	51	48	44	*
C	*	47	45	45

(*i*) Assuming each person can do only one job, determine the minimum cost assignment of persons to jobs.

(*ii*) Can the same method you used to solve (*i*) be applied, if the stated assumption is relaxed? Formulate the assignment problem when Persons *B* and *C* can do as many as two jobs, while *A* can do only one job. *[MBA (Nov.), Delhi Univ. 2012]*

8.31. An airline company has drawn up a new flight schedule involving five flights. To assist in allocating five pilots to the flights, it has asked them to state their preference scores by giving each flight a number out of 10. The higher the number, the greater is the preference. Some of these flights are not suitable to some pilots due to domestic reasons. These have been marked with X.

		Flights 1	2	3	4	5
	A	8	2	×	5	4
	B	10	9	2	8	4
Pilots	C	5	4	9	6	×
	D	3	6	2	8	7
	E	5	6	10	4	3

What should be the allocation of pilots to the flights in order to meet as many preferences as possible? *[MIB, Delhi Univ. 2011]*

8.32. Four new machines M_1, M_2, M_3 and M_4 are to be installed in a machine shop. There are five vacant places *A*, *B*, *C*, *D* and *E* available. Because of limited space, machine M_2 cannot be placed at *C* and M_3 cannot be place at *A*. The assignment cost of machine '*i*' to place at *j*th place (in rupees) is given in adjacent table. Find the optimal assignment schedule.

	A	B	C	D	E
M_1	4	6	10	5	6
M_2	7	4	—	5	4
M_3	—	6	9	6	2
M_4	9	3	7	2	3

[B.Tech. (Mech.) (Nov.), HP Univ. 2014; B.Tech. (Mech.), Karnataka Univ. 2012]

8.33. Four operators O_1, O_2, O_3 and O_4 are available to a manager who has to get four jobs J_1, J_2, J_3 and J_4 done by assigning one job to each operator. Given the times (in hours) needed by different operators for different jobs in the adjacent matrix:

(*i*) How should the manager assign the jobs to operators so that the total time needed for all the jobs is minimum.

	J_1	J_2	J_3	J_4
O_1	12	10	10	8
O_2	14	12	15	11
O_3	6	10	16	4
O_4	8	10	9	7

(*ii*) If J_2 is not assigned to O_2, what should be the assignment schedule and how much additional time will be required? ***[M.Com., Kerala Univ. 2012]***

8.34. Imagine yourself to be the Executive Director of a 5-Star Hotel which have four banquet halls that can be used for all functions including weddings. The halls are all of same size and the facilities in each hall differ. During a heavy marriage season, 4 parties approached you to reserve a hall for the marriage to be celebrated on the same day. These marriage parties were told that the first choice among these 4 halls would cost ₹ 10,000 for the day. They were also required to indicate the second, third and fourth preferences and the price that they would be willing to pay. Marriage parties *A* and *D* indicated that they won't be interested in Halls 3 and 4. Other particulars are given in the following table:

REVENUE/HALL

Marriage party	*Hall*			
	1	2	3	4
A	10,000	9,000	×	×
B	8,000	10,000	8,000	5,000
C	7,000	10,000	6,000	8,000
D	10,000	8,000	×	×

Here, × indicates that the party does not want that hall. Decide on an allocation that will maximize the revenue to your hotel. ***[PG Dip. in Glob. Bus. Oper., Delhi Univ. 2011]***

8.35. The secretary of a school is taking bids on the City's four school bus routes. Four companies have made the bids as detailed in the following table:

	Bids			
	Route 1	*Route 2*	*Route 3*	*Route 4*
Company 1	₹ 4,000	₹ 5,000	—	—
Company 2	—	₹ 4,000	—	₹ 4,000
Company 3	₹ 3,000	—	₹ 2,000	—
Company 4	—	—	₹ 4,000	₹ 5,000

Suppose each bidder can be assigned only one route. Use the assignment model to minimize the school's cost of running the four bus routes. ***[PG Dip. in Glob. Bus. Oper., Delhi Univ. 2010]***

8.36. Welldone Company has taken the third floor of a multi-storyed building for rent with a view to locate one of their zonal offices. There are five main rooms on this floor to be assigned to five managers. Each room has its own advantages and disadvantages. Some have windows, some are closer to washrooms or to the canteen or secretarial pool. The rooms are all of different sizes and shapes. Each of the five managers were asked to rank their room preferences amongst the rooms 301, 302, 303, 304 and 305. Their preferences were recorded as indicated below:

Manager				
M_1	M_2	M_3	M_4	M_5
302	302	303	302	301
303	304	301	305	302
304	305	304	304	304
—	301	305	303	—
—	—	302	—	—

Most of the managers did not list all the five rooms, since, they were not satisfied with some of these rooms and they left off these from the list. Assuming that their preferences can be quantified by numbers, find out as to which manager should be assigned to which room so that their total preference ranking is minimum. *[B.Tech. (Comp. Sc.) (Dec.), Hyderabad Univ. 2011]*

[***Hint:*** In terms of preference ranking, assignment problem is:

	M_1	M_2	M_3	M_4	M_5
301	—	4	2	—	1
302	1	1	5	1	2
303	2	—	1	4	—
304	3	2	3	3	3
305	—	3	4	2	—

8.37. A trip from Chandigarh to Delhi takes six hours by bus. A typical time table of the bus service in both directions is given below:

Departure from Chandigarh	*Route number*	*Arrival at Delhi*	*Arrival at Chandigarh*	*Route number*	*Departure from Delhi*
06:00	$\xrightarrow{a}$	12:00	11:30	$\xleftarrow{1}$	05:30
07:30	$\xrightarrow{b}$	13:30	15:00	$\xleftarrow{2}$	09:00
11:30	$\xrightarrow{c}$	17:30	21:00	$\xleftarrow{3}$	15:00
19:00	$\xrightarrow{d}$	01:00	00:30	$\xleftarrow{4}$	18:30
00:30	$\xrightarrow{e}$	06:30	06:00	$\xleftarrow{5}$	00:00

The cost of providing the service by the transport company depends upon the time spent by the bus crew (driver and conductor) away from their places in addition to service time. There are five crews. There is a constraint that every crew should be provided with more than 4 hours of rest before the return trip again and should not wait for more than 24 hours for the return trip. The company has residential facilities for the crew at Chandigarh as well as at Delhi. Find which crew be assigned which line of service or which service line be connected with which other line, so as to reduce the waiting time to be minimum.

[B.Tech. (Dec.), Nagpur Univ. 2005]

[***Hint:*** When crew is based at Chandigarh (*i.e.*, they start from and come back to Chandigarh with minimum halt at Delhi), waiting time for different services is obtained in *Table 8.17*. Similarly, when crew is based at Delhi, waiting time for different services is obtained in *Table 8.18*.

17.5	21	3	6.5	12
16	19.5	1.5	5	10.5
12	15.5	21.5	1	6.5
4.5	8	14	17.5	23
23	2.5	8.5	12	17.5

Table 8.17

18.5	15	9	5.5	0
20	16.5	10.5	7	1.5
0	20.5	14.5	11	5.5
7.5	4	22	18.5	13
13	9.5	3.5	0	18.5

Table 8.18

The minimum waiting time for different route connections by choosing minimum out of the two waiting times is given in *Table 8.19*, where waiting time is more than 4 hours (minimum desired rest).]

17.5	15	9	5.5	12
16	16.5	10.5	5	10.5
12	15.5	14.5	11	5.5
4.5	8	14	17.5	13
13	9.5	8.5	12	17.5

Table 8.19

8.38. Bright Airlines that operates seven days a week has a time-table as shown below. Crews must have a minimum layover of 5 hours between flights. Obtain the pairing of flights that minimizes layover time away from home. For any given pairing, the crew will be based at the city that results in the smaller layover:

Delhi–Jaipur			*Jaipur–Delhi*		
Flight No.	*Depart*	*Arrive*	*Flight No.*	*Depart*	*Arrive*
1	7:00 A.M.	8:00 A.M.	101	8:00 A.M.	9:15 A.M.
2	8:00 A.M.	9:00 A.M.	102	8.30 A.M.	9.45 A.M.
3	1:30 P.M.	2:30 P.M.	103	12:00 Noon	1:15 P.M.
4	6:30 P.M.	7:30 P.M.	104	5:30 P.M.	6:45 P.M.

For each pair, also, mention the town where the crew should be based.

[B.Tech., Nagpur Univ. 2003; M.Sc. (Math), CCS Univ. 2009]

8:6. DUAL OF THE ASSIGNMENT PROBLEM

While solving an assignment problem, we formulate it in the form of matrix (x_{ij}), such that

$$\sum_{i=1}^{n} x_{ij} = \sum_{j=1}^{n} x_{ij} = 1 \quad (\text{for } i, j = 1, 2, ..., n)$$

with $\quad$ minimum $z = \sum_{i=1}^{n} \sum_{j=1}^{n} c_{ij} x_{ij}, \quad$ where $c_{ij} \geq 0$ for i and $j = 1, 2, ..., n$.

The constraint equations are the conditions that jointly specify:

$$x_{ij} = \begin{cases} 1, & \text{if resource } i \text{ is assigned to activity } j. \\ 0, & \text{if resource } i \text{ is not assigned to activity } j. \end{cases}$$

Also, each row and each column of the assignment table has a single unity element, while all other elements are zero.

For the dual of an assignment problem, we proceed as in the case of a transportation problem and determine a set of u_i's and v_j's such that the following conditions are satisfied:

$$c_{ij} = u_i + v_j \text{ for } x_{ij} > 0 \quad \text{and} \quad c_{ij} \geq u_i + v_j \text{ for } x_{ij} = 0$$

or
$$\sum_i \sum_j u_i x_{ij} + \sum_i \sum_j v_j x_{ij} \leq \sum_i \sum_j c_{ij} x_{ij}$$

or
$$\sum_i u_i \left(\sum_j x_{ij}\right) + \sum_j v_j \left(\sum_i x_{ij}\right) \leq z$$

or
$$\sum_i u_i + \sum_j c_j \leq z, \quad \text{since } \sum_i x_{ij} = \sum_j x_{ij} = 1$$

If we denote the LHS of this expression by z^*, then the problem associated with minimizing $\sum_i \sum_j c_{ij} x_{ij}$ is identical to the following dual problem:

$$\text{Maximize } z^* = \sum_i u_i + \sum_j v_j \text{ subject to the constraints:}$$

$$u_i + v_j \leq c_{ij}, \qquad (i = 1, 2, ..., n;\ j = 1, 2, ..., n)$$

where, u_i and v_j unrestricted in sign.

Example 8.8. *Write the dual of the following assignment problem and obtain the optimal values of the dual variables:*

		Programmers		
		A	*B*	*C*
Programmes	*1*	*120*	*100*	*80*
	2	*80*	*90*	*110*
	3	*110*	*140*	*120*

[BBA, GGSIP Univ. 2019]

Solution. The given problem can be written as an transportation problem:

Programmes	*Programmers* A	B	C	*Supply*
1	120	100	80	1
2	80	90	110	1
3	110	140	120	1
Demand	1	1	1	3

The model formulation of this transportation problem is as follows:

Minimize $z = (120x_{11} + 100x_{12} + 80x_{13}) + (80x_{21} + 90x_{22} + 110x_{23}) + (110x_{31} + 140x_{32} + 120x_{33})$

subject to the constraints:

$$120x_{11} + 100x_{12} + 80x_{13} = 1,\ \ 80x_{21} + 90x_{22} + 110x_{23} = 1,\ \ 110x_{31} + 140x_{32} + 120x_{33} = 1$$

$$120x_{11} + 80x_{21} + 110x_{31} = 1,\ \ 100x_{12} + 90x_{22} + 140x_{32} = 1,\ \ 80x_{13} + 110x_{23} + 120x_{33} = 1$$

$$x_{ij} = 1 \text{ or } 0, \quad \text{for all } i = 1, 2, 3 \text{ and } j = 1, 2, 3.$$

Let u_i $(i = 1, 2, 3)$ and v_j $(j = 1, 2, 3)$ are the dual variables corresponding to the constraints of the above transportation problem. Then, the dual problem is:

Maximize $z^* = (u_1 + u_2 + u_3) + (v_1 + v_2 + v_3)$ subject to the constraints:

$$u_1 + v_1 \le 120,\quad u_1 + v_2 \le 100,\quad u_1 + v_3 \le 80$$

$$u_2 + v_1 \le 80,\quad u_2 + v_2 \le 90,\quad u_2 + v_3 \le 110$$

$$u_3 + v_1 \le 110,\quad u_3 + v_2 \le 140,\quad u_3 + v_3 \le 120.$$

$$u_i \ge 0\ (i = 1, 2, 3) \text{ and } v_j \ge 0\ (j = 1, 2, 3)$$

For the optimum values of dual variables, an optimum solution to the given problem is to be obtained by transportation method.

Therefore, using VAM for initial basic feasible solution and MODI method for optimality, the optimum solution to the given assignment problem is obtained, as displayed in the transportation (assignment) *Table 8.20*:

				u_i
	120	100	[1] 80	0
	[ε] 80	[1] 90	110	10
	[1] 110	140	[ε] 120	40
v_j	70	80	80	

Table 8.20

We observe that the optimum values of dual variables are: $u_1 = 0, u_2 = 10, u_3 = 40, v_1 = 70, v_2 = 80$ and $v_3 = 80$. Sum of these values is 280.

Also, Minimum $z = 1 \times 80 + \varepsilon \times 80 + 1 \times 90 + 1 \times 110 + \varepsilon \times 120$

$$= 280 + 200\,\varepsilon = 280 \text{ as } \varepsilon \to 0.$$

This shows that minimum z = maximum z^* = 280.

Further, it may be observed that optimum values of dual variables satisfy all the constraints and are not restricted to the nature of sign.

8:7. BRANCH AND BOUND METHOD

Branch and bound method is a systematic approach that determine an optimum solution for all feasible solutions of an assignment problem in a finite number of steps. The set of all feasible solutions are continuously branched into smaller sub-sets and lower bound (minimum cost of the cost matrix) is found within each sub-set. Next branching is made either from the branch with lowest bound or from the branch of the latest (higher order) active bound.

Computational Procedure of Branch and Bound Method

Given below are the steps for the solution of an assignment problem by this method are:

Step 1. Ignore the assumption of one job to one person and find the minimum time of each job. Then calculate the total time taken for the completion of all jobs. This number represents a lower bound on all the feasible solutions. It is not possible to have an assignment having lesser than the lower bound.

Step 2. Branch the feasible solution into 'n' sub-sets. Assign job J_1 to person P_1 and remaining jobs are assigned to remaining persons with minimum time. Calculate the minimum time.

Step 3. Next, assign job J_1 to P_2 and the remaining jobs to remaining persons with minimum time. Repeat until job J_1 is assigned to all the n persons. Calculate the total time at each step.

Step 4. Draw the tree diagram. Denote the total time below the corresponding node and mention the assignment over the arrow (branch). The next branching is made from the terminal node which have the least total time. If there is a tie, then the terminal node at the higher level is to be branched for unique solution.

Step 5. Go to *Step 2* and repeat the procedure for job J_2, job J_3, ..., job J_n in place of job J_1.

Step 6. If the minimum total time happens to be at any of the terminal node at the $(n-1)^{th}$ level, the optimality is reached. The optimum assignment is given by the path starting from the initial node to the terminal node together with the missing job and the assignment made.

Example 8.9. *Solve the assignment problem of assigning four jobs to four machines in such a way that only one machine is assigned to one job. The processing time (in hours) of all the machines for various jobs are as follows:*

Job	*Machine time (in hours)*			
	M_1	M_2	M_3	M_4
J_1	22	19	21	9
J_2	20	9	19	17
J_3	14	21	19	18
J_4	13	12	16	20

Solution. The least time for each job and the corresponding machine is:

Job:	J_1	J_2	J_3	J_4
Least time:	9	9	14	12
Corresponding machine:	M_4	M_2	M_1	M_2

Total time: 44 hours

We observe that the minimum total time is 44 hours. But this is not feasible, since, M_2 can not do more than one job. We now assign J_1 to M_1 and the remaining jobs to the remaining machines having least time. Thus, we have

Job:	J_1	J_2	J_3	J_4
Least time:	22	9	18	12
Corresponding machine:	M_1	M_2	M_4	M_2

Total time: 61 hours

Next, assign J_1 to M_2 and remaining jobs to remaining machines. Thus, we have

Job:	J_1	J_2	J_3	J_4
Least time:	19	17	14	13
Corresponding machine:	M_2	M_4	M_1	M_1

Total time: 63 hours

Further, assign j_1 to M_3 and remaining jobs to remaining machines. So, we have

Job:	J_1	J_2	J_3	J_4
Least time:	21	9	14	12
Corresponding machineL	M_3	M_2	M_1	M_2

Total time: 56 hours

Furthermore assign j_1 to M_4 and remaining jobs to remaining machines. So, we have

Job:	J_1	J_2	J_3	J_4
Least time:	9	9	14	12
Corresponding machine:	M_4	M_2	M_1	M_2

Total time: 44 hours

Making use of above information, the first branch of the solution tree is given in *Fig. 8.2:*

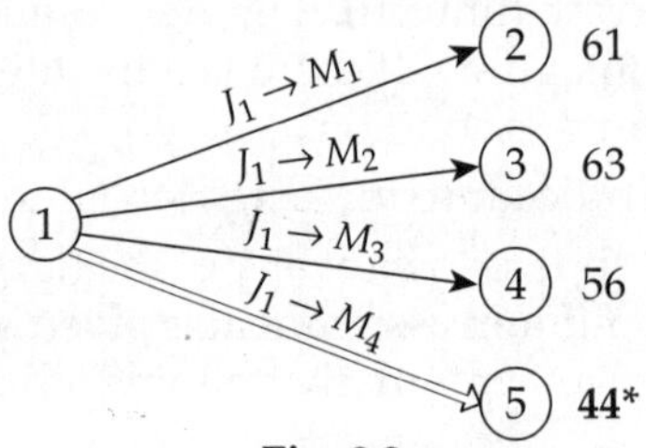

Fig. 8.2

Since, $J_1 \rightarrow M_4$ is the least fix $J_1 \rightarrow M_4$.

Reduced problem is now of three jobs J_2, J_3, J_4 and three machines M_1, M_2 and M_3.

Now, repeat the procedure when J_2 is assigned to M_1, M_2 and M_3 respectively. Thus, we have when

J_2 is assigned to M_1.

Job:	J_1	J_2	J_3	J_4
Least time:	9	20	19	16
Corresponding machine:	M_4	M_1	M_2	M_3

Total time: 64 hours

J_2 is assigned to M_2.

Job:	J_1	J_2	J_3	J_4
Least time:	9	9	14	13
Corresponding machine:	M_4	M_2	M_1	M_1

Total time: 45 hours

J_2 is assigned to M_3.

Job:	J_1	J_2	J_3	J_4
Least time:	9	19	14	12
Corresponding machine:	M_4	M_3	M_1	M_2

Total time: 54 hours

Using the above information, second branch of solution tree is given in *Fig. 8.3*.

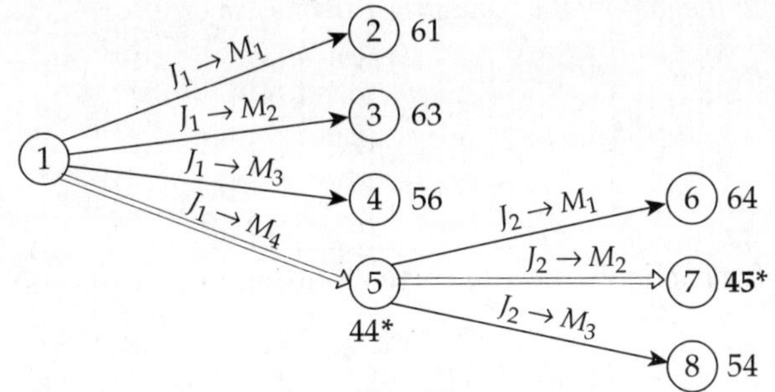

Fig. 8.3

Here, 45 is the least time. The nodes 2, 3, 4, 6 and 8 are having total time greater than 45, they are not considered for further branching and J_2 is assigned to M_2. When J_3 is assigned to M_1 and M_3, the least times are given below:

Job:	J_1	J_2	J_3	J_4
Time:	9	9	14	16
Corresponding machine:	M_4	M_2	M_1	M_3

Total time: 48 hours

and

Job:	J_1	J_2	J_3	J_4
Time:	9	9	19	13
Corresponding machine:	M_4	M_2	M_3	M_1

Total time: 50 hours

Third branch of solution tree is given in *Fig. 8.4:*

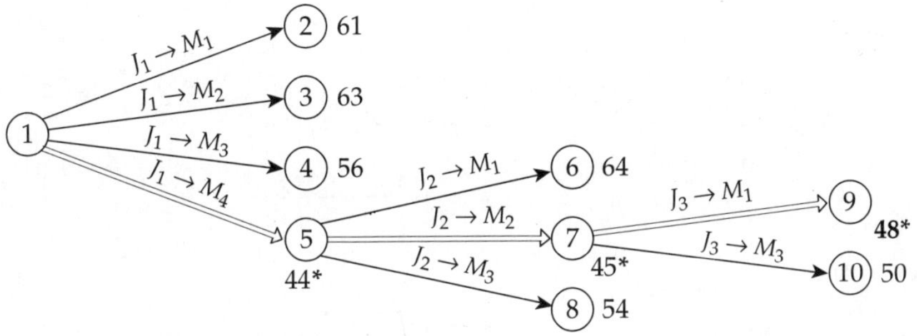

Fig. 8.4

Final branch of solution tree is given in *Fig. 8.5*.

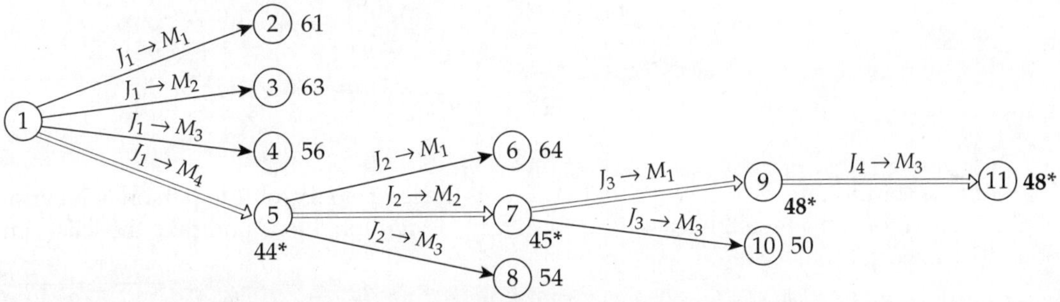

Fig. 8.5

The optimum assignment is:

$J_1 \to M_4$, $J_2 \to M_2$, $J_3 \to M_1$ and $J_4 \to M_3$ with minimum total time as 48 hours.

Example 8.10. *Find the optimum assignment of jobs and persons from the following problem, if the time taken (in days) for completing the jobs by the respective persons are given.*

		Persons A	B	C
	1	18	14	13
Job	2	19	16	15
	3	14	16	12

Solution. The least time for each job and the corresponding persons is:

Job:	1	2	3
Least time:	13	15	12
Corresponding persons:	C	C	C

Total time: 40 days

This implies that no assignment is possible which have less than 40 days. In other words lower bound is 40 days. We now assign Job *1* to person *A*, and Job 2 and 3 to the remaining persons. So, we have

Job:	1	2	3
Least time:	18	15	12
Corresponding person:	A	C	C

Total time: 45 days

When Job *1* is assigned to person *B*, we have

Job:	1	2	3
Least time:	14	15	12
Corresponding person:	B	C	C

Total time: 41 days

When Job *1* is assigned to person *C*, we have

Job:	1	2	3
Least time:	13	16	14
Corresponding time:	C	B	A

Total time 43 days

Using the above information, first branch of the solution tree is given in *Fig. 8.6:*

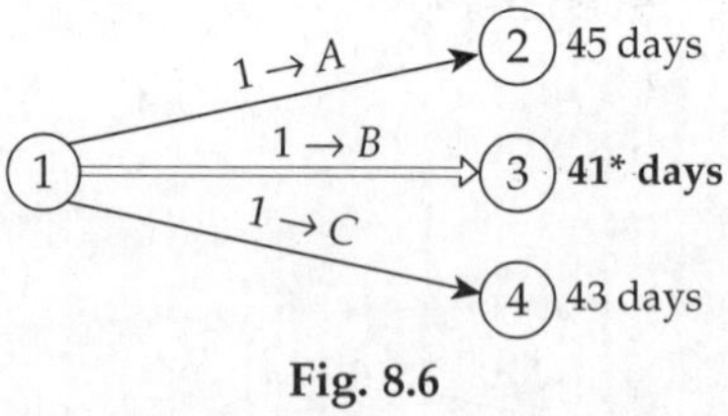

Fig. 8.6

From the solution tree, we observe that 41 days is the least, so fix Job *1* to person *B*. Next, assign Job 2 to *A* and for Job 3 take the least time. Similarly, assign Job 3 to *C* and take the least time for Job *A*. Thus, we get

Job:	1	2	3
Least time:	14	19	12
Corresponding person:	B	A	C

Total time: 45 days

and

Job:	1	2	3
Least time:	14	15	14
Corresponding time:	B	C	A

Total time 43 days

Second branch of the solution tree, therefore, is given in Fig. 8.7:

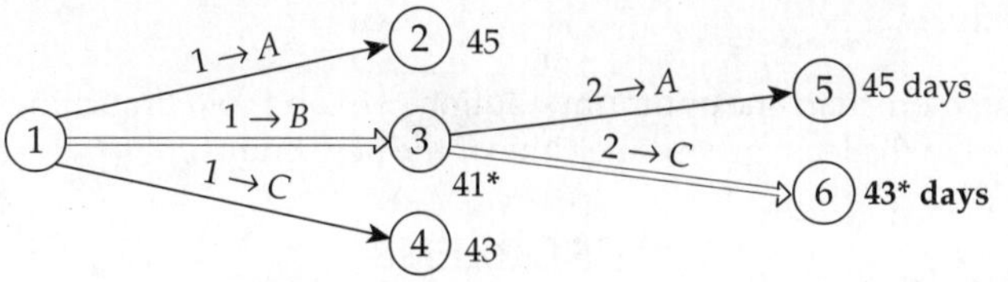

Fig. 8.7

Here, 43 is the least, the nodes 2 and 5 are not considered for further branching. Therefore, we fix Job 2 to person *C*. But node 4 also have 43 days, therefore, it can also be branched further.

Considering now, Job 3 to person *A*, the possible assignments are:

Job:	*1*	*2*	*3*
Time:	13	19	16
Corresponding person:	*C*	*A*	*B*
	Total time: 48 days		

and

Job:	*1*	*2*	*3*
Time:	13	16	14
Corresponding person:	*C*	*B*	*A*
	Total time: 43 days		

Thus, branching at node 4 also, solution tree is obtained and is given in *Fig. 8.8:*

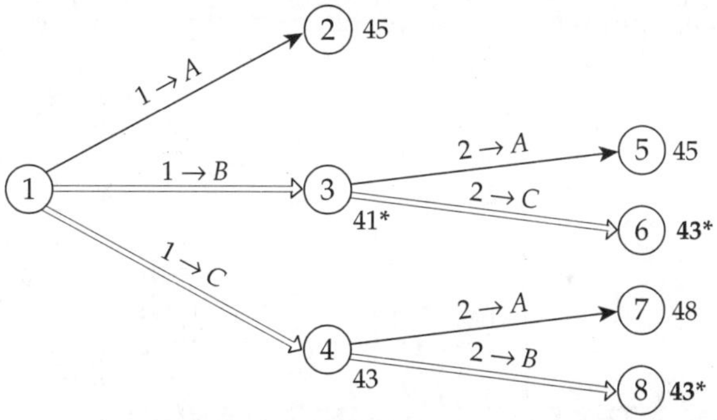

Fig. 8.8

Next branching can be started from the current active node 6 or from the least time node 8. Therefore, final branch solution tree is:

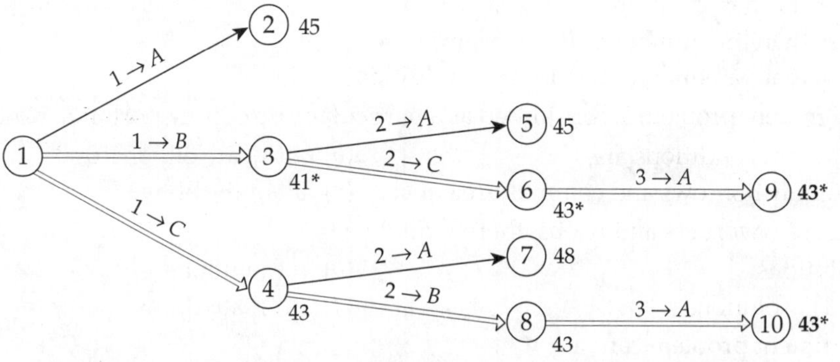

Fig. 8.9

In this solution tree, the nodes 2, 5 and 7 have timing greater than 43 days and hence cannot be branched further. At the same time node 9 and 10 have same timings. So, the possible two assignments are:

$$1 \to B,\ 2 \to C\ \ 3 \to A; \text{ and } 1 \to C, 2 \to B, 3 \to A;$$

with least time of 43 days.

Remark. To avoid having more than one optimum solution, we can avoid branching the terminal nodes at the lower level which is equal to the least processing time of the new branch.

PROBLEMS

8.39. A department head has four subordinates and four tasks for completion. The subordinates differ in their efficiency of performing the assigned work. His estimates of time (in hours) for each subordinate for different tasks is given below:

		Subordinates			
		A	*B*	*C*	*D*
Task	1	15	23	24	18
	2	26	25	21	23
	3	18	16	15	12
	4	26	23	21	17

How should the tasks be assigned to subordinates, so that the time of completion is minimum.

8.40. Find the optimum assignment of jobs whose processing time over the respective machines are given below:

		Machines			
		M_1	M_2	M_3	M_4
Jobs	J_1	25	24	23	22
	J_2	18	19	20	21
	J_3	17	16	16	17
	J_4	19	18	20	21

TEST YOUR UNDERSTANDING

1. The assignment problem is a special case of transportation problem in which number of origins
(*a*) equals the number of destinations
(*b*) is greater than the number of destinations
(*c*) is less than the number of destinations
(*d*) is less than or equal to the number of destinations.

2. An assignment problem is considered as a particular case of transportation problem, because
(*a*) all the rim conditions are 1 (*b*) all x_{ij} are either 1 or 0
(*c*) the number of rows and columns are equal (*d*) all of the above.

3. If there are *n* workers and *n* jobs, there would be
(*a*) *n* solutions (*b*) $n\,!$ solutions
(*c*) $(n-1)\,!$ solutions (*d*) $(n\,!)^n$ solutions.

4. An assignment problem can be
(*a*) designed and solved as a transportation problem
(*b*) of maximization type
(*c*) solved only if number of rows equals the number of columns
(*d*) all of the above.

5. The method used for solving an assignment problem is called
   (*a*) MODI method (*b*) reduced matrix method
   (*c*) Hungarian method (*d*) none of the above.
6. The minimum number of lines covering all zeros in a reduced cost matrix of order n can be
   (*a*) at the most n (*b*) at the least n
   (*c*) $n - 1$ (*d*) $n + 1$
7. The purpose of a dummy row or column in an assignment problem is to
   (*a*) obtain balance among all the activities and all the resources
   (*b*) prevent a solution from becoming degenerate
   (*c*) provide an unbalanced problem
   (*d*) obtain multiple optimum solutions.
8. An assignment problem can be solved by
   (*a*) simplex method (*b*) transportation method
   (*c*) branch and bound method (*d*) all of the above.
9. Every basic feasible solution of a general assignment problem, having a square pay-off matrix of order n should have assignments equal to
   (*a*) $n + n + 1$ (*b*) $n + n - 1$
   (*c*) $m + n - 1$ (*d*) $m + n$
10. For solving an assignment problem by MODI method, the number of dummy allocations need to be added are
   (*a*) n (*b*) $2n$
   (*c*) $n - 1$ (*d*) $2n - 1$
11. Which of the following statement is not correct?
   (*a*) For every prohibited assignment, the given cost element is replaced by M, which is a very large value,
   (*b*) Multiple optimum assignment schedules may have different objective function values,
   (*c*) The dual variables for an assignment problem are given by the u_i and v_j values of the optimum solution to the transformed transportation problem,
   (*d*) The variables of the dual problem of a given assignment problem may be unrestricted in sign.
12. Maximization assignment problem is transformed into a minimization problem by
   (*a*) subtracting each element of the profit matrix from the highest element of the matrix,
   (*b*) subtracting all the elements in a column from the highest element of that column,
   (*c*) subtracting all the elements in a row from the highest element of that row,
   (*d*) any of the above.

ANSWERS

1. (*a*)	2. (*d*)	3. (*b*)	4. (*d*)	5. (*c*)	6. (*a*)
7. (*a*)	8. (*d*)	9. (*b*)	10. (*c*)	11. (*b*)	12. (*a*)

REVIEW EXERCISE

1. What is an assignment problem? Explain.
[MBA, Panjab Tech. Univ. 2011; M.Com. (Nov.), Madurai Univ. 2003]
2. What is an assignment problem? Give two applications in health care administration.
[MBA (HCA) PT, Delhi Univ. 2000]
3. Give the mathematical formulation of an assignment problem.
[BCA (Nov.), Madras Univ. 2010]
4. "An assignment problem is a special case of a transportation problem". Explain.
[B.Tech. (Mech.), Panjab Tech. Univ. 2007]
5. Explain the difference between a transportation problem and an assignment problem.
[MBA (Dec.), Panjab Tech. Univ. 2012; PG Dip. in Glob. Bus. Oper., Delhi Univ. 2010]
6. Write short note on the assignment problem and its applications.
[MBA, Ranchi Univ. 2011; MBA, Hyderabad Univ. 2012]
7. Explain the Hungarian method in the context of assignment problem.
[MBA, Anantpur Univ. 2010; MBA, Annamalai Univ. 2009]
8. What is the optimality criterion in the assignment problem?
[B.Tech. (Mech.), HP Univ. 2012; B.Tech. (Mech.), Karnataka Univ. 2012]
9. What is the unbalanced assignment problem? How it is solved by Hungarian method of assignment? *[MBA, Osmania Univ. 2010; MBA, Panjab Tech. Univ. 2008]*
10. How will you solve an assignment problem, where a particular assignment is prohibited in case of minimization? Name the method used for solving a maximization assignment problem.
[MIB, Delhi Univ. 2011]
11. Describe a method of drawing minimum number of lines in the context of assignment problem. Name the method.
12. Explain the method of solving a maximization assignment model.
[MBA, Panjab Univ. 2009; MBA, Madras Univ. 2000]
13. What is meant by branch and bound method for solving an assignment problem?
14. State some application areas of branch and bound technique in the context of an assignment problem.

CHAPTER 9

Other Methods of Solving Linear Programming Problems

"We must learn to ask questions, challenge old systems and concepts, be innovative and cope with change"

This chapter introduces efficient computational procedures including the decomposition method, upper and lower bounding and parametric programming. The chapter then presents the problem of maximizing the ratio of two linear functions subject to a set of linear inequalities and non-negative constraints on the variables. Followed by this, a totally different Karmarkar interior point algorithm is given. The new algorithm is claimed to be more efficient than the simplex method in handling large linear programming problems. At the end is discussed travelling salesman problem, *i.e.*, a shortest cyclic route model.

9:1. BOUNDED VARIABLES LPP

Bounded variables linear programming problem may have, in addition to the given constraints, some (or all) variables with lower and upper limits. In such cases the standard form of LPP will look like

$$\text{Maximize} \quad z = \mathbf{cx} \quad \text{subject to the constraints:} \quad \mathbf{Ax} = \mathbf{b}, \quad \mathbf{l} \le \mathbf{x} \le \mathbf{u},$$

where $\mathbf{x} = [x_1, x_2, ..., x_n]$, $\mathbf{b}^T = [b_1, b_2, ..., b_m]$, $\mathbf{c}^T = (c_1, c_2, ..., c_n)$, $\mathbf{l} = [l_1, l_2, ..., l_n]$, $\mathbf{u} = [u_1, u_2, ..., u_n]$, and $\mathbf{A} = (a_{ij})$; $i = 1, 2, ..., m, j = 1, 2, ..., n$.

Here, $\mathbf{A}$ is an $m \times n$ real matrix, while $\mathbf{l}$ and $\mathbf{u}$ denote the lower and upper bounds of $\mathbf{x}$ respectively. In cases of unbounded variable, these limits are 0 and ∞ respectively.

The inequality constraints $\mathbf{l} \le \mathbf{x} \le \mathbf{u}$ can be converted into equality constraints by introducing slack and/or surplus variables $\mathbf{s}'$ and $\mathbf{s}''$ as shown below:

$$\mathbf{x} + \mathbf{s}' = \mathbf{u} \quad \text{and} \quad \mathbf{x} - \mathbf{s}'' = \mathbf{l},$$

where $\quad \mathbf{x} \ge \mathbf{0}, \ \mathbf{s}' \ge \mathbf{0} \ \text{and} \ \mathbf{s}'' \ge \mathbf{0}.$

The lower bound constraint can be written as $\mathbf{x} = \mathbf{l} + \mathbf{s}''$, and thus $\mathbf{x}$ can be eliminated from all the remaining constraints.

The upper bound constraint can be written as $\mathbf{x} = \mathbf{u} - \mathbf{s}'$. This, however, does not ensure the non-negative value of $\mathbf{x}$.

It is in this context that a special technique known as ***bounded variable simplex method*** was developed in order to overcome this difficulty.

In bounded variable simplex method the optimality condition for a solution is same as in the simplex method, discussed earlier. But, the inclusion of constraints $\mathbf{x} + \mathbf{s}' = \mathbf{u}$ in simplex table requires modification in the feasibility condition of the simplex method due to following reasons:

(*i*) A basic variable should become non-basic variable at its upper bound (in usual simplex method all non-basic variables are at zero level).

(*ii*) When a non-basic variable becomes basic variable, the value should not exceed its upper bound and also should not disturb the non-negativity and upper bound conditions of all existing basic variables.

The Computational Procedure

The iterative procedure for determining an optimum basic feasible solution to an LPP with bounded variables may be summarized as follows:

Step 1. Check whether objective function is of maximization or minimization. If it is of minimization, then convert it into that of maximization by writing Minimize $z = -$ Maximize z^*, where $z^* = -z$.

Step 2. If any of the b_i's (RHS of constraints) is negative, multiply the corresponding constraint by '–1' to make it positive.

Step 3. Convert the inequations of the constraints into equations and obtain an initial basic feasible solution.

Step 4. If any of the variables is at a positive lower bound, substitute it out at its lower bound.

Step 5. Compute the net evaluations $(z_j - c_j)$ and examine their sign. If all $(z_j - c_j) \geq 0$, then an optimum basic feasible solution has been attained. Otherwise, go to next step.

Step 6. Choose the most negative of $(z_j - c_j)$. Let it be $(z_r - c_r)$ for some $j = r$.

Step 7. Compute the quantities

$$\theta_1 = \underset{i}{\text{Min.}} \left\{ \frac{x_{Bi}}{a_{ir}},\ a_{ir} > 0 \right\},\quad \theta_2 = \underset{i}{\text{Min.}} \left\{ \frac{u_r - x_{Bi}}{-a_{ir}},\ a_{ir} < 0 \right\} \quad \text{and} \quad \theta = \text{Min.}\ \{\theta_1, \theta_2, u_r\}$$

where x_{Bi}'s correspond to the current basic feasible solution and u_r is the upper bound for a variable x_r. It is to be noted that if all $a_{ir} > 0$, then $\theta_2 = \infty$.

(*i*) If $\theta = \theta_1$ which corresponds to x_{Bk}, then x_r enters the basis and x_k leaves the basis, using regular row operations of the simplex method.

(*ii*) If $\theta = \theta_2$ which corresponds to x_{Bk}, then x_r enters the basis and x_k leaves the basis, x_{Bk} being non-basic variable at its upper bound, must be substituted by using

$$(x_{Bk})_r = (x_{Bk})'_r - x_{kr}\, u_r \qquad \text{where } 0 \leq (x_{Bk})'_r \leq u_r$$

and $(x_{Bk})_r$ denotes that the variable x_r was x_{Bk}. The non-basic variable x_r at upper bound is put at zero level by using

$$x_r = u_r - x'_r, \qquad 0 \leq x'_r \leq u_r$$

(*iii*) If $\theta = u_r$, x_r is substituted at its upper bound and remains non-basic, *i.e.*, it can be put at zero level by using

$$x_r = u_r - x'_r, \qquad 0 \leq x'_r \leq u_r.$$

Step 8. Go to *Step 5* and repeat the procedure, until we get an optimum basic feasible solution or there is an indication of an unbounded solution.

Example 9.1. *Using the bounded variable technique, solve the following LPP:*

Maximize $z = x_2 + 3x_3$ *subject to the constraints:*

$$x_1 + x_2 + x_3 \leq 10,\ \ x_1 - 2x_3 \geq 0,\ \ 2x_2 - x_3 \leq 10,\ \ 0 \leq x_1 \leq 8,\ \ 0 \leq x_2 \leq 4,\ \ x_3 \geq 0.$$

Solution.

Step 1. By introducing slack variables $s_1 \geq 0$, $s_2 \geq 0$ and $s_3 \geq 0$, in the set of inequations representing the constraints of the given problem, the reformulated LPP will be

Maximize $z = \mathbf{cx}$ subject to the constraints:

$$\mathbf{Ax} = \mathbf{b} \text{ and } 0 \leq x_1 \leq 8,\ 0 \leq x_2 \leq 4 \text{ and } x_3, s_1, s_2, s_3 \geq 0.$$

where

$$\mathbf{A} = \begin{pmatrix} 1 & 1 & 1 & 1 & 0 & 0 \\ -1 & 0 & 2 & 0 & 1 & 0 \\ 0 & 2 & -1 & 0 & 0 & 1 \end{pmatrix},\quad \mathbf{b} = \begin{pmatrix} 10 \\ 0 \\ 10 \end{pmatrix} \text{ and } \mathbf{c} = (0, 1, 3, 0, 0, 0).$$

Step 3. An obvious initial basic feasible solution is:

$$s_1 = 10,\ s_2 = 0 \ \text{and}\ s_3 = 10 \ \text{with}\ \mathbf{B} = \mathbf{I}_3 \ \text{as the initial basis.}$$

Step 4. Since, there are no upper bounds for basic variables s_1, s_2, s_3, and for non-basic variable x_3, we assume that all these have an upper bound at ∞, *i.e.*, $u_4 = u_5 = u_6 = u_3 = \infty$.

Step 5. Considering the objective function as one of the constraint, we write

$$0 \cdot x_1 - x_2 - 3x_3 + 0 \cdot s_1 + 0 \cdot s_2 + 0 \cdot s_3 + z = 0$$

Using the above information, initial simplex tableau is obtained as under:

Initial Iteration. Enter non-basic variable x_3 and drop basic variable s_2.

Coeff. of basic variable	*Basic variable*	x_1	x_2	x_3	s_1	s_2	s_3	z	*Solution*	$u_i - x_{Bi}$
0	s_1	1	1	1	1	0	0	0	10	$\infty - 10 = \infty$
0	$\leftarrow s_2$	–1	0	(2)	0	1	0	0	0	$\infty - 0 = \infty$
0	s_3	0	2	–1	0	0	1	0	10	$\infty - 10 = \infty$
	$(z_j - c_j)$	0	–1	–3 ↑	0	0	0	1	$z = 0$	
	u_j	8	4	∞	∞	∞	∞			

Step 6. Since $(z_3 - c_3)$ is the most negative, x_3 enters the basis.

Step 7. Now,

$$\theta_1 = \min.\left\{\frac{x_{Bi}}{a_{i3}},\ a_{i3} > 0\right\} = \min.\{10, 0\} = 0 \qquad \text{(corresponds to } s_2\text{)}$$

$$\theta_2 = \min.\left\{\frac{u_i - x_{Bi}}{-a_{i2}},\ a_{i2} < 0\right\} = \infty \qquad \text{(corresponds to } s_3\text{)}$$

and $\quad u_3 = \infty,$

$$\therefore \quad \theta = \text{Min.}\ (\theta_1, \theta_2, u_3) = \text{Min.}\ \{0, \infty, \infty\} = 0.$$

This shows that $\theta = \theta_1$ and hence, s_2 leaves the basis.

Thus, the element a_{23} becomes the leading element. Converting the leading element into unity and all other elements of the entering column to zero, we get:

First Iteration. Introduce x_1 and drop s_1.

Coeff. of basic variable	*Basic variable*	x_1	x_2	x_3	s_1	s_2	s_3	z	*Solution*	$u_i - x_{Bi}$
0	$\leftarrow s_1$	(3/2)	1	0	1	–1/2	0	0	10	$\infty - 10 = \infty$
3	x_3	–1/2	0	1	0	1/2	0	0	0	$\infty - 0 = \infty$
0	s_3	–1/2	2	0	0	1/2	1	0	10	$\infty - 10 = \infty$
	$(z_j - c_j)$	–3/2 ↑	–1	0	0	3/2	0	1	$z = 0$	
	u_j	8	4	4	∞	∞	∞			

Since, $(z_1 - c_1)$ is the most negative, x_1 enters the basis. Now,

$$\theta_1 = \underset{i}{\text{Min}}.\left\{\frac{x_{Bi}}{a_{i1}},\ a_{i1} > 0\right\} = \text{Min}.\left(\frac{10}{3/2}\right) = \frac{20}{3} \qquad \text{(corresponds to } s_1\text{)}$$

$$\theta_2 = \infty \quad \text{and} \quad u_1 = 8.$$

$$\therefore \quad \theta = \text{Min.}\ \{\theta_1, \theta_2, u_1\} = \text{Min.}\left\{\frac{20}{3}, \infty, 8\right\} = \frac{20}{3}.$$

This shows that $\theta = \theta_1$ and hence, s_1 leaves the basis.

Final Iteration. Optimum solution.

Coeff. of basic variable	Basic variable	x_1	x_2	x_3	s_1	s_2	s_3	z	Solution
0	x_1	1	2/3	0	2/3	−1/3	0	0	20/3
3	x_3	0	1/3	1	1/3	2/3	0	0	10/3
0	s_3	0	7/3	0	1/3	2/3	1	0	40/3
	$(z_j - c_j)$	0	0	0	1	2	0	1	$z = 10$

This tableau shows that an optimal solution has been attained. Hence, an optimum basic feasible solution is

$$x_1 = \frac{20}{3},\ x_2 = 0,\ x_3 = \frac{10}{3} \text{ and maximum } z = 10.$$

Example 9.2. *Using the bounded variable technique, solve the following LPP:*

Maximize $z = 3x_1 + 5x_2 + 2\,x_3$ *subject to the constraints:*

$$x_1 + 2x_2 + 2x_3 \le 14,\ \ 2x_1 + 4x_2 + 3x_3 \le 23,\ \ 0 \le x_1 \le 4,\ \ 2 \le x_2 \le 5,\ \ 0 \le x_3 \le 3.$$

Solution. Since x_2 has a positive lower bound, we let $x_2' = x_2 - 2$, *i.e.*, $x_2 = x_2' + 2$. Then, we shall have $0 \le x_2' \le 3$.

Modified LPP will be:

Maximize $z = 3x_1 + 5(x_2' + 2) + 2x_3 = 3x_1 + 5x_2' + 2x_3 + 10$ subject to the constraints:

$$x_1 + 2(x_2' + 2) + 2x_3 \le 14 \text{ or } x_1 + 2x_2' + 2x_3 \le 10$$

$$2x_1 + 4(x_2' + 2) + 3x_3 \le 23 \text{ or } 2x_1 + 4x_2' + 3x_3 \le 15$$

$$0 \le x_1 \le 4,\ 0 \le x_2' \le 3,\ 0 \le x_3 \le 3.$$

Now, by introducing slack variables $s_1 \ge 0$ and $s_2 \ge 0$, the set of inequations representing the constraints of the given problem are converted into a set of equations.

An obvious initial basic feasible solution is

$$s_1 = 10 \text{ and } s_2 = 15$$

with $\mathbf{B} = \mathbf{I}_2$ as the initial basis.

Since, there are no upper bounds specified for these basic variables, we arbitrarily assume that both have an upper bound at ∞, *i.e.*, $u_4 = u_5 = \infty$.

The starting simplex tableau is:

Coeff. of basic variable	Basic variable	x_1	x_2'	x_3	s_1	s_2	z	Solution	$u_i - x_{Bi}$
0	s_1	1	2	2	1	0	0	10	$\infty - 10 = \infty$
0	s_2	2	4	3	0	1	0	15	$\infty - 15 = \infty$
	$(z_j - c_j)$	−3	−5	−2	0	0	1	$z = 10$	
	u_j	4	3	3	∞	∞			

This tableau shows that x_2 enters the basis.

Now, $\theta_1 = \text{Min.}\left(\frac{10}{2}, \frac{15}{4}\right) = \frac{15}{4}$ (corresponding to s_2)

$$\theta_2 = \infty\ [\because a_{i2} > 0\ (i = 1, 2)] \text{ and } u_2 = 3$$

$$\therefore \quad \theta = \text{Min.}\ \{\theta_1, \theta_2, u_2\} = \text{Min.}\ (15/4, \infty, 3) = 3$$

This implies that $\theta = u_2$ and, therefore, x_2' is substituted at its upper bound and remains non-basic. Thus,

$$x_2' = u_2 - x_2'' = 3 - x_2'', \quad \text{where } 0 \le x_2'' \le 3.$$

Basic variables are, therefore, updated by using

$$x_{B1} = x'_{B1} - a_{12}\, u_2 = 10 - 2 \times 3 = 4$$

and

$$x_{B2} = x'_{B2} - a_{22}\, u_2 = 15 - 4 \times 3 = 3.$$

Using updated values of basic and non-basic variables, the initial simplex tableau becomes as given below:

Initial Iteration. Introduce x_1 and drop s_2.

Coeff. of basic variable	*Basic variable*	x_1	x_2''	x_3	s_1	s_2	z	*Solution*	$u_i - x_{Bi}$
0	s_1	1	–2	2	1	0	0	4	$\infty - 4 = \infty$
0	$\leftarrow s_2$	(2*)	–4	3	0	1	0	3	$\infty - 3 = \infty$
	$(z_j - c_j)$	–3 ↑	5	–2	0	0	1	$z = 15 + 10$ $= 25$	
	u_j	4	3	3	∞	∞			

This shows that x_1 enters the basis.

Now, $\theta_1 = \text{Min.}\left(\dfrac{4}{1}, \dfrac{3}{2}\right) = \dfrac{3}{2}$ (corresponding to s_2)

$$\theta_2 = \infty \quad \text{and} \quad u_1 = 4$$

$\therefore$ $\theta = \text{Min.}\left(\dfrac{3}{2}, \infty, 4\right) = \dfrac{3}{2}$

This shows that $\theta = \theta_1$ and hence s_2 leaves the basis.

First Iteration. Introduce x_2'' and drop x_1.

Coeff. of basic variable	*Basic variable*	x_1	x_2''	x_3	s_1	s_2	z	*Solution*	$u_i - x_{Bi}$
0	s_1	0	0	1/2	1	–1/2	0	5/2	$\infty - 5/2 = \infty$
3	$\leftarrow x_1$	1	(–2*)	3/2	0	1/2	0	3/2	$4 - 3/2 = 5/2$
	$(z_j - c_j)$	0	–1↑	5/2	0	3/2	1	$z = 25 + 9/2$ $= 59/2$	
	u_j	4	3	3	∞	∞			

This shows that x_2'' enters the basis.

Since, all the elements of column under x_2'' are ≤ 0, we have

$$\theta_1 = \infty, \ \theta_2 = \text{Min.}\left\{\infty, \frac{5/2}{-(-2)}\right\} = \frac{5}{4},$$

since, $u_2 - x_{B2} = 4 - \dfrac{3}{2}$ and $(-a_{22}) = -(-2)$.

$$u_2 = 3.$$

$\therefore$ $\theta = \text{Min.}\left\{\infty, \dfrac{5}{4}, 3\right\} = \dfrac{5}{4}.$

This shows that $\theta = \theta_2$ and hence x_1 leaves the basis.

Second Iteration. Update the values of basic variables.

Coeff. of basic variable	*Basic variable*	x_1	x_2''	x_3	s_1	s_2	z	*Solution*
0	s_1	0	0	1/2	1	–1/2	0	5/2
5	x_2''	–1/2	1	–3/4	0	–1/4	0	–3/4
	$(z_j - c_j)$	–1/2	0	7/4	0	5/4	1	$z = 115/4$
	u_j	4	3	3	∞	∞		

Since, upper bound for x_1 is 4, we update the basic variables using

$$x_{B1} = x'_{B1} - a_{11}\,u_1 = \frac{5}{2} - 0 \times 4 = \frac{5}{2}$$

$$x_{B2} = x'_{B2} - a_{21}\,u_1 = -\frac{3}{4} - \left\{-\frac{1}{2}\right\} \times 4 = \frac{5}{4}$$

Also, non-basic variable x_1 at upper bound can be put at zero level by substituting $x_1 = 4 - x_1'$ $(0 \le x_1' \le 4)$. Thus, using updated values of basic variables, the iterative simplex tableau is:

Final Iteration. Optimum solution.

Coeff. of basic variable	*Basic variable*	x_1'	x_2''	x_3	s_1	s_2	*Solution*
0	s_1	0	0	1/2	1	–1/2	5/2
5	x_2''	1/2	1	–3/4	0	–1/4	5/4
	$(z_j - c_j)$	1/2	0	7/4	0	5/4	$z = 123/4$
	u_j	4	3	3	∞	∞	

This table shows that an optimum solution has been attained. The solution is given by

$$x_1' = 0 \text{ and } x_2'' = \frac{5}{4}$$

Now, since, $\quad x_1' = 4 - x_1 \quad \text{and} \quad x_2' = 3 - x_2''$

$\therefore \quad x_1' = 0 \Rightarrow 4 - x_1 = 0 \text{ or } x_1 = 4$

and $\quad x_2'' = \frac{5}{4} \Rightarrow x_2' = 3 - \frac{5}{4} = \frac{7}{4}$

Also, since, $x_2 = x_2' + 2$, therefore, $x_2 = \frac{7}{4} + 2 = \frac{15}{4}$

Hence, an optimum basic feasible solution is

$$x_1 = 4,\quad x_2 = \frac{15}{4},\quad x_3 = 0 \text{ and maximum } z = \frac{123}{4}$$

PROBLEMS

Solve the following linear programming problems using bounded variable simplex method:

9.01. Maximize $z = 2x_1 + x_2$ subject to the constraints:

$$x_1 + 2x_2 \le 10,\; x_1 + x_2 \le 6,\; x_1 - x_2 \le 2,\; x_1 - 2x_2 \le 1,$$
$$0 \le x_1 \le 3 \text{ and } 0 \le x_2 \le 2.$$

9.02. Maximize $z = 3x_1 + 2x_2$ subject to the constraints:
$$x_1 - 3x_2 \le 3,\ x_1 - 2x_2 \le 4,\ 2x_1 + x_2 \le 20,$$
$$x_1 + 3x_2 \le 30,\ -x_1 + x_2 \le 6,\ 0 \le x_1 \le 8 \text{ and } 0 \le x_2 \le 6.$$

9.03. Maximize $z = x_2 + 3x_3$ subject to the constraints:
$$x_1 + x_2 + x_3 \le 10,\ x_1 - 2x_3 \ge 0,\ 2x_2 - x_3 \le 10,$$
$$0 \le x_1 \le 8,\ 0 \le x_2 \le 4, \text{ and } x_3 \ge 0.$$

9.04. Minimize $z = 6x_1 - 2x_2 - 3x_3$ subject to the constraints:
$$2x_1 + 4x_2 + 2x_3 \le 8,\ x_1 - 2x_2 + 3x_3 \le 7,$$
$$0 \le x_1 \le 2,\ 0 \le x_2 \le 2 \text{ and } 0\, x_3 \le 1.$$

9.05. Minimize $z = 3x_1 + 5x_2 + 2x_3$ subject to the constraints:
$$x_1 + 2x_2 + 2x_3 \le 10,\ 2x_1 + 4x_2 + 3x_3 \le 15,$$
$$0 \le x_1 \le 4,\ 0 \le x_2 \le 3,\ 0 \le x_3 \le 3.$$

9.06. Maximize $z = 4x_1 + 10x_2 + 9x_3 + 11x_4$ subject to the constraints:
$$2x_1 + 2x_2 + 2x_3 + 2x_4 \le 5,\ 48x_1 + 80x_2 + 160x_3 + 240x_4 \le 257$$
$$0 \le x_j \le 1 \text{ for } j = 1, 2, 3, 4.$$

9.07. Maximize $z = 3x_1 + x_2 + x_3 + 7x_4$ subject to the constraints:
$$2x_1 + 3x_2 - x_3 + 4x_4 \le 40,\ -2x_1 + 2x_2 + 5x_3 - x_4 \le 35,$$
$$x_1 + x_2 - 2x_3 + 3x_4 \le 100,$$
$$x_1 \ge 2,\ x_2 \ge 1,\ x_3 \ge 3,\ x_4 \ge 4.$$

9.08. Maximize $z = 4x_1 + 2x_2 + 6x_3$ subject to the constraints:
$$-x_1 + x_2 + 2x_3 \le 8,\ 4x_1 - x_2 \le 9,\ -3x_1 + x_2 + 4x_3 \le 12,$$
$$1 \le x_1 \le 3,\ 0 \le x_2 \le 5 \text{ and } 0 \le x_3 \le 2.$$

9:2. PARAMETRIC LINEAR PROGRAMMING

After formulation of linear programming model and solving it by simplex method, we get an optimum solution of an LPP. Sometimes the decision-maker desires to know how the solution will change if parameters of the problem, such as cost (or profit), availability of resources are changed. In order to consider the impact of such changes on the model after solving the problem, we have already discussed the need to perform a post-optimality (sensitivity) analysis in *chapter 6*. When various parameters of the linear programming model changes simultaneously, sensitivity analysis is carried out systematically. The parametric linear programming investigates the effect of predetermined simultaneous changes (continuous variations) in the parameters of the given problem on the optimum solution.

In this section, we shall present the parametric analysis of an LPP systematically by gradually changing the c_j parameters and then b_i parameters, *i.e.*, on cost or profit and on the availability of resources.

The purpose of this analysis is to reduce the computational time required to take care of changes on the optimum solution due to variation in linear programming model parameters over a range of variation. We start the analysis on optimum solution obtained at $\lambda = 0$ where λ is any scalar (constant). Then, using the optimality and feasibility conditions of the simplex method, we determine the range of λ for which the optimum solution at $\lambda = 0$ remains unchanged. Let λ lie between o and λ_1. It means $0 \le \lambda \le \lambda_1$ is the range of λ beyond which the current solution will become infeasible and/or non-optimum. In this case, λ_1 is referred to as a ***critical value***. Thus, at $\lambda = \lambda_1$ a new solution is determined which remains optimum and feasible in other interval, say $\lambda_1 \le \lambda \le \lambda_2$. Again, a new solution at $\lambda = \lambda_2$ is obtained. The process of determining the range of λ is repeated till a stage is reached beyond which the solution either does not change or exist.

Parametric Changes in the Objective Function Coefficients, *i.e.*, c

Let the cost vector **c** of LPP: Maximize, $\mathbf{z} = \mathbf{cx}$ subject to the constraints: $\mathbf{Ax} = \mathbf{b}$ and $\mathbf{x} \ge \mathbf{0}$ be replaced by $\mathbf{c}+\lambda\mathbf{c}'$; *i.e.*, $\mathbf{c}(\lambda) = \mathbf{c}+\lambda\mathbf{c}'$, where $\mathbf{c}'$ represents predetermined variation in **c** and $\lambda \ge 0$ is a scalar parameter. The aim here is to find the range (or critical) value of λ, so that the current optimum solution remains unchanged with a change in the coefficients c_j. The given LP problem is initially solved by using simplex method at $\lambda = 0$. Since, changes in cost coefficient c_j only affect the optimality of the current solution as λ changes, only $(z_j - c_j)$ values are affected. Thus, for the perturbed LP problem, let us calculate the shadow prices of non-basic variables which are computed as follows:

$$\begin{aligned} z_j(\lambda) - c_j(\lambda) &= \mathbf{c}_B(\lambda)\mathbf{y}_j - c_j(\lambda), \quad \text{where } \mathbf{y}_j = \mathbf{B}^{-1}\mathbf{a}_j \\ &= (\mathbf{c}_B + \lambda\mathbf{c}'_B)\mathbf{y}_j - (c_j + \lambda c'_j) \\ &= (z_j - c_j) + \lambda(\mathbf{c}'_B\mathbf{y}_j - c'_j); \quad \text{for all } j \end{aligned}$$

For a solution to be optimum for all values of λ, we must have

$$z_j(\lambda) - c_j(\lambda) \ge 0, \; i.e., \; (z_j - c_j) + \lambda(\mathbf{c}'_B\mathbf{y}_j - c'_j) \ge 0, \quad \text{where } \mathbf{y}_j = \mathbf{B}^{-1}\mathbf{a}_j$$

or

$$\lambda \ge -(z_j - c_j)/(\mathbf{c}'_B\mathbf{y}_j - c'_j); \quad \text{if } (\mathbf{c}'_B\mathbf{y}_j - c'_j) > 0$$

and

$$\lambda \le -(z_j - c_j)/(\mathbf{c}'_B\mathbf{y}_j - c'_j); \quad \text{if } (\mathbf{c}'_B\mathbf{y}_j - c'_j) < 0$$

These inequalities, for a given solution, can be used for determining the range $\lambda_1 \le \lambda \le \lambda_2$, within which the current solution remains optimum. Thus, we have

$$\max_j \left\{ \frac{-(z_j - c_j)}{(\mathbf{c}'_B\mathbf{y}_j - c'_j) > 0} \right\} \le \lambda \le \min_j \left\{ \frac{-(z_j - c_j)}{(\mathbf{c}'_B\mathbf{y}_j - c'_j) < 0} \right\}$$

Notes

(*i*) If $\lambda_1 \to \infty$, then the problem has an optimum solution for all $\lambda > \lambda_1$.

(*ii*) If λ_2 is finite, then we improve the range of λ beyond λ_2 and obtain

$$\lambda^* = \min_j \left\{ \frac{-(z_j - c_j)}{(\mathbf{c}'_B\mathbf{y}_j - c'_j) < 0} \right\}$$

When $\lambda = \lambda^*$, we get an alternative solution and when $\lambda > \lambda^*$, we improve the solution by using simplex method.

Remark. In simplex tableau, column under variable x_j stands for $\mathbf{B}^{-1}\mathbf{a}_j$ and is denoted by $\mathbf{y}_j$.

Parametric Changes in the availability of Resources, *i.e.*, b

Let the requirement vector **b** of the LPP be replaced by $\mathbf{b}+\lambda\mathbf{b}'$, *i.e.*, $\mathbf{b}(\lambda) = \mathbf{b}+\lambda\mathbf{b}'$, where $\mathbf{b}'$ represents predetermined variation in **b** and $\lambda \ge 0$ is a scalar parameter. It may be observed that the parametrized right-hand side **b** (λ) can only affect the feasibility of the problem. The critical values of λ are thus determined based on the following information:

$$\mathbf{x}_B(\lambda) \ge 0; \; i.e., \; \mathbf{B}^{-1}(\mathbf{b}+\lambda\mathbf{b}') \ge 0 \text{ or } \mathbf{x}_B + \lambda\mathbf{B}^{-1}\mathbf{b}' \ge 0$$

or

$$\lambda \ge \frac{-x_{Bi}}{\mathbf{B}^{-1}b'_i}, \quad \text{if } \mathbf{B}^{-1}b'_i > 0, \text{ for all } i$$

and

$$\lambda \le \frac{-x_{Bi}}{\mathbf{B}^{-1}b'_i}, \quad \text{if } \mathbf{B}^{-1}b'_i < 0, \text{ for all } i$$

These two inequalities for a given solution, can be used for determining the range $\lambda_1 \le \lambda \le \lambda_2$ within which the current optimum solution remains feasible. Thus, we have

$$\max_i \left\{ \frac{-x_{Bi}}{\mathbf{B}^{-1} b_i' > 0} \right\} \le \lambda \le \min_i \left\{ \frac{-x_{Bi}}{\mathbf{B}^{-1} b_i' < 0} \right\}$$

Notes

(*i*) If $\mathbf{B}^{-1} b_i' < 0$ for at least one i ($i = 1, 2, ..., m$), then there exists a critical value λ^* given by

$$\lambda^* \le \min_i \left\{ \frac{-x_{Bi}}{\mathbf{B}^{-1} b_i' < 0} \right\}$$

beyond which $\mathbf{x}_B(\lambda)$ becomes infeasible.

(*ii*) When $\lambda = \lambda^*$, an alternative solution is obtained and when $\lambda > \lambda^*$ an improved solution of the problem is obtained.

Remark. If $\lambda = \lambda_1$, the current solution remains optimum and at this solution the value of the objective function is given by $z = \mathbf{c}_B \mathbf{x}_B + \lambda \mathbf{c}_B \mathbf{x}_B'$. At λ_1, the current basis $\mathbf{x}_B$ (right-hand side) is replaced by $\mathbf{x}_B(\lambda) = \mathbf{B}^{-1}(\mathbf{b} + \lambda \mathbf{b}')$ and x_{Br} is removed from the basis by the usual simplex method. The process of finding the new range $[\lambda_1, \lambda_2]$ of values of λ is repeated over which the new basis is optimum. When $x_{Bi}' \left(= \mathbf{B}^{-1} b_i' \ge \mathbf{0}\right)$ for all i, the process is terminated. Hence, the current basis is optimum for all values of λ greater than or equal to the last value of λ.

Example 9.3. *For the following LPP:*

Maximize $z = (3 - 6\lambda)\, x_1 + (2 - 2\lambda)\, x_2 + (5 + 5\lambda)\, x_3$ *subject to the constraints:*

$$x_1 + 2x_2 + x_3 \le 430, \quad 3x_1 + 2x_3 \le 460, \quad x_1 + 4x_2 \le 420, \quad x_1, x_2, x_3 \ge 0;$$

find the range of λ *over which the solution remains basic feasible and optimum.*

[M.Sc. (Math.), CCS Univ. 2005]

Solution. Let $\lambda = 0$. Introducing slack variables $s_1 \ge 0$, $s_2 \ge 0$ and $s_3 \ge 0$, the given LPP is solved by simplex method. An optimum basic feasible solution is displayed in the following simplex tableau:

Coeff. of basic variable	*Basic variable*	x_1	x_2	x_3	s_1	s_2	s_3	z	*Solution*
2	x_2	–1/4	1	0	1/2	–1/4	0	0	100
5	x_3	3/2	0	1	0	1/2	0	0	230
0	s_3	2	0	0	–2	1	1	0	20
	$(z_j - c_j)$	4	0	0	1	2	0	1	$z = 1350$

When $\lambda \ne 0$, we write

$$c^* = \mathbf{c} + \lambda \mathbf{c}' = (3,\ 2,\ 5,\ 0,\ 0,\ 0) + \lambda(-6,\ -2,\ 5,\ 0,\ 0,\ 0)$$

$\therefore$ $\mathbf{c}_B' = (-2,\ 5,\ 0)$ and hence in order to find the first critical value (or range) of λ over which the solution in simplex table given above is optimum, we compute $(z_j' - c_j')$, *i.e.*, $\mathbf{c}_B' \mathbf{B}^{-1} \mathbf{a}_j - c_j$. Since, columns under x_2, x_3 and s_3 forms the basis $\mathbf{B}$, therefore columns under x_1, s_1 and s_2 does not form basis.

Now, $$z_j' = \mathbf{c}_B' \mathbf{B}^{-1} \mathbf{a}_j = (-2\ \ 5\ \ 0) \begin{bmatrix} -1/4 & 1/2 & -1/4 \\ 3/2 & 0 & 1/2 \\ 2 & -2 & 1 \end{bmatrix} = (8\ \ -1\ \ 3)$$

$$\therefore \quad z_j' - c_j' = (8,\ 0,\ 0,\ -1,\ 3,\ 0) - (-6,\ 0,\ 0,\ 0,\ 0,\ 0) = (14,\ 0,\ 0,\ -1,\ 3,\ 0)$$

Thus, the solution will remain optimum for those value of λ, for which

$$\text{Max.} \left\{ \frac{-(z_j - c_j)}{(\mathbf{c}_B' \mathbf{y}_j - c_j') > 0} \right\} \le \lambda \le \text{Min.} \left\{ \frac{-(z_j - c_j)}{(\mathbf{c}_B' \mathbf{y}_j - c_j') < 0} \right\}, \text{ where } \mathbf{y}_j = \mathbf{B}^{-1} \mathbf{a}_j$$

i.e., $$\text{Max.}\left\{\frac{-4}{14}, \frac{-2}{3}\right\} \le \lambda \le \text{Min.}\left\{\frac{-1}{-1}\right\} \text{ or } \frac{-2}{7} \le \lambda \le 1$$

This means that for $\lambda \in [0, 1]$, the solution given in the above simplex tableau remains optimum. The objective function value in this interval is given by

$$z(\lambda) = \mathbf{c}_B\mathbf{x}_B + \lambda\,\mathbf{c}'_B\mathbf{x}_B = 1350 + 950\lambda$$

The shadow prices of non-basic variables are given by

$$z_1(\lambda) - c_1(\lambda) = (z_1 - c_1) + \lambda(z'_1 - c'_1) = 4 + \lambda(8+6) = 4 + 14\lambda \ge 0 \;\Rightarrow\; \lambda \ge -2/7$$
$$z_4(\lambda) - c_4(\lambda) = (z_4 - c_4) + \lambda(z'_4 - c'_4) = 1 + \lambda(-1+0) = 1 - \lambda \ge 0 \;\Rightarrow\; \lambda \le 1$$
$$z_5(\lambda) - c_5(\lambda) = (z_5 - c_5) + \lambda(z'_5 - c'_5) = 2 + \lambda(3-0) = 2 + 3\lambda \ge 0 \;\Rightarrow\; \lambda \ge -2/3$$

At $\lambda = 1$, we observe that $z_4(\lambda) - c_4(\lambda) = 0$. But for $\lambda > 1$, $z_4(\lambda) < 0$ which indicates that the solution obtained above no longer remains optimum. So, we introduce s_1 into basis and thereby x_2 leaves the basis. Thus, when $\lambda = 1$, the improved simplex tableau is:

Coeff. of basic variable	*Basic variable*	x_1	x_2	x_3	s_1	s_2	s_3	*Solution*
0	s_1	–1/2	2	0	1	–1/2	0	200
5	x_3	3/2	0	1	0	1/2	0	230
0	s_3	1	4	0	0	0	1	420
	$(z_j - c_j)$	9/2	–2	0	0	5/2	0	$z = 1150$
	$(z'_j - c'_j) \rightarrow$	27/2	2	0	0	5/2	0	
	$z_j(\lambda) - c_j(\lambda) \rightarrow$	18	0	0	0	5	0	

Above simplex tableau indicate that when λ is increased above $\lambda^* = 1$, *i.e.,* for $\lambda > 1$, the new solution $x_1 = 0$, $x_2 = 0$ and $x_3 = 230$ becomes the unique optimum.

From the above simplex tableau, the shadow prices of the non-basic variables are given by

$$(z_1 - c_1) + \lambda(z'_1 - c'_1) = 9/2 + (27/2)\lambda \ge 0 \;\Rightarrow\; \lambda \ge -1/3$$
$$(z_2 - c_2) + \lambda(z'_2 - c'_2) = -2 + 2\lambda \ge 0 \;\Rightarrow\; \lambda \ge 1$$
$$(z_5 - c_5) + \lambda(z'_5 - c'_5) = (5/2) + (5/2)\lambda \ge 0 \;\Rightarrow\; \lambda \ge -1$$

Therefore, for $\lambda \ge 1$, the shadow prices of non-basic variables are non-negative, and hence the solution obtained above is optimum.

The optimum solution for the entire range of λ is as under:

λ	x_1	x_2	x_3	z
$0 \le \lambda \le 1$	0	100	230	$1350 + 950\lambda$
$\lambda \ge 1$	0	0	230	$1150 + 1150\lambda$

Example 9.4. *Consider the parametric linear programming problem:*

Maximize $z = 3x_1 + 2x_2 + 5x_3$ *subject to the constraints:*

$$x_1 + 2x_2 + x_3 \le 430 + \mu,$$
$$3x_1 + 2x_3 \le 460 - 4\mu,$$
$$x_1 + 4x_2 \le 420 - 4\mu \;\text{ and }\; x_1, x_2, x_3 \ge 0.$$

Determine the critical values (range) of μ for which the solution remains optimum basic feasible.

[M.Sc. (Math.), CCS Univ. 1995]

Solution. Let $\mu = 0$. Introducing slack variables $s_1 \ge 0$, $s_2 \ge 0$ and $s_3 \ge 0$; the given LPP is solved by simplex method. An optimum basic feasible solution is displayed in *Example 9.3.*

From the optimum simplex tableau, we write

$$\mathbf{B}^{-1} = \begin{pmatrix} 1/2 & -1/4 & 0 \\ 0 & 1/2 & 0 \\ -2 & 1 & 1 \end{pmatrix} \text{ and } \mathbf{x}_B = \begin{bmatrix} x_2 \\ x_3 \\ s_3 \end{bmatrix} = \begin{bmatrix} 100 \\ 230 \\ 20 \end{bmatrix}$$

In order to find the range over which the solution remains optimum, we first compute $\mathbf{B}^{-1}\mathbf{b}'$:

$$\therefore \qquad \mathbf{B}^{-1}\mathbf{b} = \begin{pmatrix} 1/2 & -1/4 & 0 \\ 0 & 1/2 & 0 \\ -2 & 1 & 1 \end{pmatrix}\begin{pmatrix} 1 \\ -4 \\ -4 \end{pmatrix} = \begin{pmatrix} 3/2 \\ -2 \\ -10 \end{pmatrix}$$

The first critical value of μ is given by $\mu^* = \text{Min.}\left\{\frac{-230}{-2}, \frac{-20}{-10}\right\} = 2$

This corresponds to x_{B3} (= s_3) and therefore the solution will remain optimum over the interval [0, 2]. In particular for any $\mu \in [0, 2]$, the objective function value and right-hand side values are given below:

$$z(\mu) = \mathbf{c}_B\,\mathbf{x}_B + \mu\,\mathbf{c}_B\,\mathbf{x}'_B = (2,\ 5,\ 0)\begin{bmatrix} 100 \\ 230 \\ 20 \end{bmatrix} + \mu(2,\ 5,\ 0)\begin{bmatrix} 3/2 \\ -2 \\ -10 \end{bmatrix} = 1350 - 7\mu$$

and

$$\mathbf{x}_B(\mu) = \mathbf{x}_B + \mu\,\mathbf{x}'_B = [100,\ 230,\ 20] + \mu[3/2,\ -2,\ -10]$$
$$= [100 + 3\mu/2,\ 230 - 2\mu,\ 20 - 10\mu]$$

At μ = 2, the optimum simplex tableau will be:

Coeff. of basic variable	*Basic variable*	x_1	x_2	x_3	s_1	s_2	s_3	z	*Solution*
2	x_2	–1/4	1	0	1/2	–1/4	0	0	103
5	x_3	3/2	0	1	0	1/2	0	0	226
0	s_3	2	0	0	–2	1	1	0	0
	$(z_j - c_j)$	4	0	0	1	2	0	1	z = 1336

When μ > 2, basic variable s_3 becomes negative and therefore by dual simplex method s_3 leaves and s_1 enters the basis. The improved simplex tableau (optimum basic feasible) becomes:

Coeff. of basic variable	*Basic variable*	x_1	x_2	x_3	s_1	s_2	s_3	z	*Solution*
2	x_2	1/4	1	0	0	0	1/4	0	103
5	x_3	3/2	0	1	0	1/2	0	0	226
0	s_1	0	0	0	1	–1/2	–1/2	0	0
	$(z_j - c_j)$	5	0	0	0	5/2	1/2	1	z = 1336

To determine the second critical value of μ, we compute

$$\mathbf{B}^{-1}\mathbf{b} = \begin{pmatrix} 0 & 0 & 1/4 \\ 0 & 1/2 & 0 \\ 1 & -1/2 & -1/2 \end{pmatrix}\begin{pmatrix} 1 \\ -4 \\ -4 \end{pmatrix} = \begin{pmatrix} -1 \\ -2 \\ 5 \end{pmatrix}$$

$$\therefore \qquad \mu^* = \text{Min.}\left\{\frac{-103}{-1}, \frac{-226}{-2}\right\} = 103$$

This corresponds to x_{B1} (= x_2), and therefore the solution will remain optimum over the interval [2, 103]. In particular for any $\mu \in [2, 103]$, the objective function value and right-hand side value are given as:

$$z(\mu) = (2, 5, 0)\begin{bmatrix} 103 \\ 226 \\ 0 \end{bmatrix} + \mu(2, 5, 0)\begin{bmatrix} -1 \\ -2 \\ 5 \end{bmatrix} = 1336 - 12\mu$$

and $\quad \mathbf{x}_B(\mu) = [103, 226, 0] + \mu[-1, -2, 5] = [103 - \mu, 226 - 2\mu, 0 + 5\mu]$

Evidently, for $\mu > 103$, the basic variable x_2 becomes negative and hence becomes infeasible. But, all entries in the row corresponding to this basic variable are positive, and, therefore, dual simplex method is not applicable. At $\mu = 103$, the optimum solution is:

$$x_1 = 0, \quad x_2 = 0, \quad x_3 = 20 \quad \text{and} \quad \text{maximum } z = 100.$$

Hence, the required range of μ for which the solution remains optimum is:

$\mathbf{x}^{(1)} = [0, 100, 230]$, if $0 \le \mu \le 2$

$\mathbf{x}^{(2)} = [0, 103, 226]$, if $2 \le \mu \le 103$

$\mathbf{x}^{(3)} =$ No feasible solution, if $\mu > 103$

PROBLEMS

9.09. Consider the parametric linear programming problem:

Maximize $z = (\lambda - 1)\, x_1 + x_2$ subject to the constraints:

$$x_1 + 2x_2 \le 10, \quad 2x_1 + x_2 \le 11, \quad x_1 - 2x_2 \le 3, \quad x_1, x_2 \ge 0.$$

Perform complete parametric programming analysis. Identify all critical values of the parameter λ and all optimum basic solutions.

9.10. Perform a complete parametric programming analysis and identify all the critical values of the parameter λ for the following LPP:

Maximize $z = (6 - \lambda)\, x_1 + (12 - \lambda)\, x_2 + (4 - \lambda)\, x_3$ subject to the constraints:

$$3x_1 + 4x_2 + x_3 \le 2, \quad x_1 + 3x_2 + 2x_3 \le 1, \quad x_1, x_2, x_3 \ge 0.$$

9.11. For the following LPP

Minimize $z = \lambda\, x_1 - \lambda\, x_2 - x_3 + x_4$ subject to the constraints:

$$3x_1 - 3x_2 - x_3 + x_4 \ge 5, \quad 2x_1 - 2x_2 + x_3 - x_4 \le 3; \quad x_1, x_2, x_3, x_4 \ge 0.$$

Find the range of λ over which the solution remains basic feasible and optimal.

[M.Sc. (Math.), CCS Univ. 2006]

9.12. Consider the LPP of maximizing

$z = (3 - 6\lambda)\, x_1 + (2 - 2\lambda)\, x_2 + (5 + 5\lambda)\, x_3$ subject to the constraints:

$$x_1 + 2x_2 + x_3 \le 40, \quad 3x_1 + 2x_3 \le 60, \quad x_1 + 4x_2 \le 30; \quad x_1, x_2, x_3 \ge 0.$$

Perform the complete parametric programming analysis and identify all the critical values of the parameter λ.

9.13. Consider the parametric linear programming problem:

Maximize $z = 3x_1 + 4x_2 + x_3 + 7x_4$ subject to the constraints:

$$8x_1 + 3x_2 + 4x_3 + x_4 \le 7 - \theta, \quad 2x_1 + 6x_2 + x_3 + 5x_4 \le 3 + 3\theta$$

$$x_1 + 4x_2 + 5x_3 + 2x_4 \le 8 + \theta; \quad x_1, x_2, x_3, x_4 \ge 0.$$

Perform a complete parametric programming analysis.

9.14. Given the LPP:

$$\text{Maximize } z = 7x_1 + 4x_2 + 6x_3 + 5x_4 \text{ subject to the constraints:}$$
$$2x_1 + x_2 + 2x_3 + x_4 \le 10 + \theta,\ \ 2x_1 - x_2 + 4x_3 + 4x_4 \le 26 - \theta$$
$$3x_1 + x_2 - 2x_3 + 3x_4 \le 45 - \theta;\ \ x_1, x_2, x_3, x_4 \ge 0.$$

Perform a complete parametric programming analysis and identify all the critical values of the parameter θ.

9.15. Consider the LPP:

$$\text{Maximize } z = (4 - 10\lambda)\, x_1 + (8 - 4\lambda)\, x_2 \text{ subject to the constraints:}$$
$$x_1 + x_2 \le 4,\ \ 2x_1 + x_2 \le 3 - \lambda;\ \ x_1 \ge 0 \text{ and } x_2 \ge 0.$$

Study the variations in the optimum solution with the parameter λ, where $-\infty < \lambda < \infty$.

9.16. Consider the parametric linear programming problem:

$$\text{Maximize } z = (3 - 6\lambda)\, x_1 + (2 - 2\lambda)\, x_2 + (5 + 5\lambda)\, x_3 \text{ subject to the constraints:}$$
$$x_1 + 2x_2 + x_3 \le 40 - \lambda,\ \ 3x_1 + 2x_3 \le 60 + 2\lambda,$$
$$x_1 + 4x_2 \le 30 - 7\lambda;\ \ x_1, x_2, x_3 \ge 0.$$

Determine the critical values (range) of λ for which the solution remains optimum basic feasible.

9:3. DECOMPOSITION METHOD

While solving linear programming problems, if there are only two decision variables it is easy to solve the LPP by graphical method. When decision variables are more than two, we solve the LPP by simplex method. But, when the decision variables exceeds four or five it becomes too complex to solve and require considerable computation time. So, we solve such problems by decomposition method due to Dantzig and Wolf. By this method, we split the original problem into partial problems and find the solution of each of the partial problem. After this, the solution to the given problem is obtained according to certain rules.

Let an enterprise has its central office in the capital of a country that controls the activities of its two branches located at metro cities. Certain part of the resources is directly controlled by the central office, while the other parts of resources are the responsibilities of its branches. Let the components of $\mathbf{b}$ denote the capacities of resources of the central office, and components $\mathbf{b}_1$ and $\mathbf{b}_2$ represent the capacities of resources of the branches. Further, the coefficient matrix $\mathbf{A}$ is partitioned into $\mathbf{A}_1$ and $\mathbf{A}_2$ representing the technological coefficients of central resources, while $\mathbf{B}_1$ and $\mathbf{B}_2$ denote technological coefficients of the resources of its branches. Then, the system of inequalities: $\mathbf{A\,x} \le \mathbf{b}$, and $\mathbf{x} \ge \mathbf{0}$ becomes

$$\mathbf{A}_1\,\mathbf{x}_1 + \mathbf{A}_2\,\mathbf{x}_2 \le \mathbf{b}_0;\ \ \mathbf{x}_1 \ge 0 \text{ and } \mathbf{x}_2 \ge 0 \qquad \text{(constraints of central system)}$$
$$\mathbf{B}_1\,\mathbf{x}_1 \le \mathbf{b}_1,\ \mathbf{x}_1 \ge \mathbf{0} \text{ and } \mathbf{B}_2\,\mathbf{x}_2 \le \mathbf{b}_2,\ \mathbf{x}_2 \ge \mathbf{0} \qquad \text{(constraints of the branches)}$$

Let the inner product $\mathbf{c}_1\mathbf{x}_1$ and $\mathbf{c}_2\mathbf{x}_2$ describe the economic results of the two branches. Then, $\mathbf{c}_1\,\mathbf{x}_2 + \mathbf{c}_2\,\mathbf{x}_2$ will represent the total result of the enterprise.

The complete linear programming model for this problem, therefore, can be expressed as:

$$\text{Maximize } z = \mathbf{c}_1\,\mathbf{x}_1 + \mathbf{c}_2\,\mathbf{x}_2 \text{ subject to the constraints:}$$
$$\mathbf{A}_1\,\mathbf{x}_1 + \mathbf{A}_2\,\mathbf{x}_2 \le \mathbf{b}_0,\ \ \mathbf{B}_1\,\mathbf{x}_1 \le \mathbf{b}_1,\ \mathbf{B}_2\,\mathbf{x}_2 \le \mathbf{b}_2;$$
$$\mathbf{x}_1 \ge \mathbf{0} \text{ and } \mathbf{x}_2 \ge \mathbf{0}.$$

It may be noted that for the overall benefit of the enterprise, the decomposition method reflects the aim of coordinating the activities of the branches. The model given above consists of two independent sets of inequalities. The first set, consisting of a compiling constraint, involves all the decision variables $\mathbf{x}_1$ and $\mathbf{x}_2$, while the second set consists of two different groups of constraints, wherein each group involving the decision variables $\mathbf{x}_1$ (or $\mathbf{x}_2$) of that group only.

The general form of the above linear programming model is:

$$\text{Maximize } z = c_1 x_1 + c_2 x_2 + \dots + c_n x_n \text{ subject to the constraints:}$$

$$\mathbf{A}_1 \mathbf{x}_1 + \mathbf{A}_2 \mathbf{x}_2 + \dots + \mathbf{A}_n \mathbf{x}_n \le \mathbf{b}_0$$

$$\mathbf{B}_1 \mathbf{x}_1 \le \mathbf{b}_1, \quad \mathbf{B}_2 \mathbf{x}_2 \le \mathbf{b}_2, \dots, \mathbf{B}_n \mathbf{x}_n \le \mathbf{b}_n,$$

$$\mathbf{x}_1 \ge \mathbf{0}, \quad \mathbf{x}_2 \ge \mathbf{0}, \quad \dots, \quad \mathbf{x}_n \ge \mathbf{0}.$$

In matrix notations, the LPP can be written as:

$$\text{Maximize } z = \mathbf{cx} \text{ subject to the constraints:}$$

$$\mathbf{Ax} \le \mathbf{b} \text{ and } \mathbf{x} \ge \mathbf{0},$$

where $\quad \mathbf{c} = (\mathbf{c}_1 \ \mathbf{c}_2 \ \dots \ \mathbf{c}_n), \ \mathbf{x} = [\mathbf{x}_1 \ \mathbf{x}_2 \ \dots \ \mathbf{x}_n], \ \mathbf{b} = [\mathbf{b}_0 \ \mathbf{b}_1 \ \mathbf{b}_2 \ \dots \ \mathbf{b}_n]$

and

$$\mathbf{A} = \begin{bmatrix} \mathbf{A}_1 & \mathbf{A}_2 & \cdots & \mathbf{A}_n \\ \mathbf{B}_1 & \mathbf{0} & \cdots & \mathbf{0} \\ \mathbf{0} & \mathbf{B}_2 & \cdots & \mathbf{0} \\ \vdots & \vdots & \vdots & \vdots \\ \mathbf{0} & \mathbf{0} & \cdots & \mathbf{B}_n \end{bmatrix}$$

A detailed discussion of the general decomposition principle is beyond the scope of this book, as it requires special computing hardware keeping in view of large number of decision variables involved in the problem. The interested readers may refer to the following:

"A Decomposition Principle for Linear Programming", J. of Operations Research, Vol. 8, No. 1 (pp. 101–111), 1960.

9:4. AN INTERIOR–POINT ALGORITHM

For the solution of linear programming problems, simplex method is a useful technique in determining an optimum solution. In this method we move along the edges of the solution space from one extreme point to the next. But when there are n variables in an LPP having 2^n extreme points of the solution space, it becomes difficult and time-consuming to derive the optimum solution by simplex method.

In order to derive the solution of extremely large linear programming problems, Narendra Karmarkar in 1984 developed a polynomial-time algorithm that cuts across the interior of the solution space.

Subsequently, many researchers are engaged in modifying *Karmakar's algorithm* to fully tap this potential. Much progress has been made and a number of powerful algorithms using the interior-point approach have been developed. Today, the more powerful software packages that are designed for solving huge linear programming problems (such as CPLEX) include at least one algorithm using the interior-point approach alongwith simplex method and its variants.

The basic idea behind Karmakar's algorithm and its subsequent variants that use the interior-point approach are now discussed herewith.

The Key Solution Concept. Although Karmakar's algorithm is radically different from simplex method, yet it have the same characteristics. The algorithm starts by identifying a feasible ***trial solution***. Then, it moves from the current trial solution to a better trial solution in the feasible region. The process of moving from current trial solution to next better trial solution continues until it reaches a trial solution that is optimal (optimum).

The difference between Karmakar's algorithm and simplex algorithm lies in the nature of these trail solutions. For simplex method, trial solutions are CPF (corner point feasible) solutions, so all the movements is along edges of the boundary of feasible region. For Karmakar's algorithm, the trial solutions are ***interior points***, *i.e.*, points inside the boundary of the feasible region.

Illustration. Consider the linear programming problem of maximizing $z = x_1$ subject to the constraints: $x_1 \le 4$ and $x_1 \ge 0$. Introducing the slack variable $x_2 \ge 0$, the standard form of the LPP is

$$\text{Maximize } z = x_1 + 0.x_2 \text{ subject to the constraints:}$$
$$x_1 + x_2 = 4 \text{ and } x_1 \ge 0, x_2 \ge 0.$$

The solution space (feasible region) of the problem is given by the linear segment *AB*. The direction of increase in *z* is in the positive direction of x_1.

We start with any arbitrary interior (non-extreme) point C in the feasible region. The gradient of the objective function at C is in the direction of fastest increase in *z*. If we locate an arbitrary point along the gradient and then project it perpendicularly on the feasible region, we obtain the new point *D* with a better objective value of *z*. Such improvement is obtained by moving in the direction of projected gradient *CD*. If we repeat the procedure at *D*, we will determine a new point *E* that is closer to the optimum solution at *A*. On the other hand, if the objective is of minimization (instead of maximization), the procedure of the projected gradient will move us away from point *A* towards the minimum at *B*, *i.e.*, $x_1 = 0$.

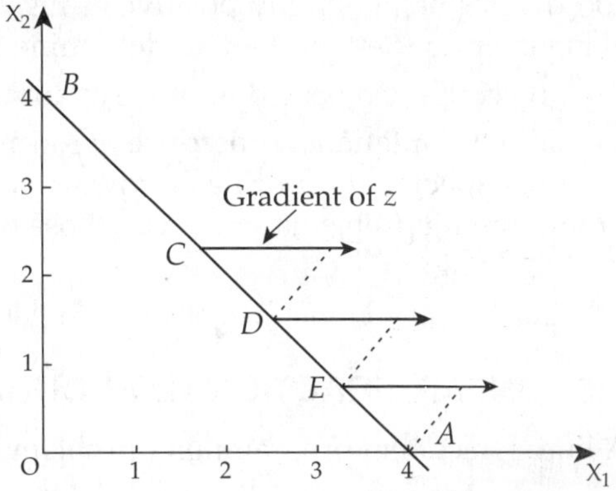

Fig. 9.1

The ideas used to describe the Karmarkar algorithm in the above illustration can be summarized as follows:

(*a*) Shoot through the interior of the feasible region toward an optimum solution.
(*b*) Move in a direction that improves the objective function value at the fastest possible rate.
(*c*) Transform the feasible region to place the current trial solution near its centre, thereby enabling a larger improvement when (*b*) is implemented.

Interior-Point Algorithm

One of the several variants of **Karmarkar's algorithm** is the ***Interior-Point Algorithm.*** It assumes that the LPP is given as:

$$\text{Minimize } z = \mathbf{cx} \text{ subject to the constraints: } \mathbf{Ax} = \mathbf{0},\ \mathbf{1x} = \mathbf{1} \text{ and } \mathbf{x} \ge \mathbf{0}.$$

Further, the constraints are homogeneous equations except for the constraint $\mathbf{1x} = \sum_{j=1}^{n} x_j = 1$, which defines an n-dimensional simplex. The validity of ***Karmarkar's algorithm*** rests on satisfying the following conditions:

(*i*) $\mathbf{x} = \left(\frac{1}{n}, \frac{1}{n}, \ldots, \frac{1}{n}\right)$ satisfies $\mathbf{Ax} = \mathbf{0}$ and (*ii*) Minimum $z = 0$.

Karmarkar provides algebraic transformations for converting the general LPP to this form. It also shows how the transformation results in $\mathbf{x} = \left(\frac{1}{n}, \frac{1}{n}, \ldots, \frac{1}{n}\right)$ being a feasible solution of $\mathbf{Ax} = \mathbf{0}$.

Karmarkar's principle idea is to start from an interior point represented by the centre of the simplex and, then advance in the direction of the projected gradient to determine a new solution point. The validity of the algorithm rests on the condition that the new solution point must be strictly an interior point.

For the new solution point to be strictly interior, it must not lie on the boundaries of the simplex. To guarantee this, a sphere with its centre coinciding with that of the simplex is inscribed tightly inside the simplex. In the n-dimensional case, the radius r of this sphere equals $1/\sqrt{n(n-1)}$. Now, a smaller sphere with radius ar $(0 < a < 1)$ will be a subset of the sphere. Any point in the intersection of the smaller sphere with the homogeneous system, $\mathbf{Ax} = \mathbf{0}$, will be an interior point,with all its coordinates being strictly positive. Thus, we can move as far as possible in this restricted space along the projected gradient to determine the new (necessarily improved) solution point.

Procedure is repeated until we attain the optimum extreme point (optimum solution).

The computational procedure of Karmarkar's algorithm is beyond the scope of this book, as it requires special computing hardware and thus is not a 'popular' method for solving linear programming problems. However, those who are interested in the details of algorithm may refer to the following:

"Karmarkar's Linear Programming Algorithm", *Interfaces*, Vol. 16, No. 4 (pp. 75 – 90), 1986.

9:5. LINEAR FRACTIONAL PROGRAMMING

A linear fractional programming problem is defined as follows:

Maximize or Minimize $R(\mathbf{x}) = (\mathbf{cx} + \alpha) / (\mathbf{dx} + \beta)$ subject to the constraints:

$$\mathbf{Ax} = \mathbf{b}, \quad \mathbf{x} \geq \mathbf{0}$$

where, (*i*) $\mathbf{x}$, $\mathbf{c}$ and $\mathbf{d}$ are $n \times 1$ row matrices,

(*ii*) $\mathbf{A}$ is an $m \times n$ matrix, where $\mathbf{A} = (a_{ij})$ $(i = 1, ..., m; j = 1, ... n)$

(*iii*) $\mathbf{b}$ is an $m \times 1$ column matrix,

and (*iv*) α, β are scalars.

Such type of problems were solved earlier by Charnes and Cooper by resolving it into two linear programming problems. Later, Kanti Swarup gave an algorithm for the solution of programming problems with linear fractional functionals without reducing it to linear programming problems.

Fractional Programming Algorithm

Consider the linear fractional programming problem:

Maximize $z = (\mathbf{cx} + \alpha) / (\mathbf{dx} + \beta)$ subject to the constraints: $\mathbf{Ax} = \mathbf{b}$ and $\mathbf{x} \geq \mathbf{0}$

with the additional assumption that ***denominator is positive for all feasible solutions.***

Let $\mathbf{x}_B$ the initial basic feasible solution is such that

$$\mathbf{B}\mathbf{x}_B = \mathbf{b} \quad \text{or} \quad \mathbf{x}_B = \mathbf{B}^{-1}\mathbf{b} \quad \text{with} \quad \mathbf{x}_B \geq 0,$$

where, $\mathbf{B} = (\mathbf{b}_1, \mathbf{b}_2, ..., \mathbf{b}_m)$. Further, let

$$z^1 = \mathbf{c}_B\mathbf{x}_B + \alpha \quad \text{and} \quad z^2 = \mathbf{d}_B\mathbf{x}_B + \beta$$

where, $\mathbf{c}_B$ and $\mathbf{d}_B$ have their components as the coefficients associated with the basic variables in the numerator and denominator of the objective function, respectively. In addition, we assume that for this basic feasible solution

$$\mathbf{y}_j = \mathbf{B}^{-1}\mathbf{a}_j, \quad z_j^1 = \mathbf{c}_B\mathbf{y}_j \quad \text{and} \quad z_j^2 = \mathbf{d}_B\,\mathbf{a}_j$$

are known for every column $\mathbf{a}_j$ of $\mathbf{A}$ not in $\mathbf{B}$.

We now wish to examine the possibility of finding another basic feasible solution with improved value of $z = z^1/z^2$. We shall confine our attention to those basic feasible solutions in which only one column of **B** is changed. If the new basic feasible solution is denoted by $\hat{\mathbf{x}}_B$, then $\hat{\mathbf{x}}_B = \hat{\mathbf{B}}^{-1}\mathbf{b}$, where $\hat{\mathbf{B}} = (\hat{\mathbf{b}}_1, \hat{\mathbf{b}}_2, ..., \hat{\mathbf{b}}_m)$, *i.e.*, a new non-singular matrix obtained from **B** by removing one of the columns of **B** and replacing it by $\mathbf{a}_j$. The columns of the new basis matrix $\hat{\mathbf{B}}$ are given by

$$\hat{\mathbf{b}}_i = \mathbf{b}_i \, (i \neq r) \quad \text{and} \quad \hat{\mathbf{b}}_r = \mathbf{a}_j$$

We, now, obtain the value of new basic variables in terms of the original ones and $\mathbf{y}_{ij}$, *i.e.*,

$$\hat{x}_{Bi} = x_{Bi} - x_{Br}\,(y_{ij} / y_{rj}), \quad \text{for } i \neq r$$

$$\hat{x}_{Br} = x_{Br} / y_{rj} = \theta \text{ (say)}, \quad \text{for } i = r$$

where, $\mathbf{a}_j = y_{1j}\mathbf{b}_1 + y_{2j}\mathbf{b}_2 + ... + y_{mj}\,\mathbf{b}_m$ and $\mathbf{B}^{-1}\mathbf{a}_j = [y_{1j} \;\; y_{2j} \;\; \cdots \;\; y_{mj}]$

Our main interest is in finding a new basic feasible solution with an improved value of the objective function. Having found a new basic feasible solution, we must determine whether, value of 'z' is improved. The value of the objective function for the original basic solution is $z = z^1 / z^2$. Let the new value of the objective function be

$$\bar{z} = \bar{z}^1 / \bar{z}^2$$

Now, we have

$$\bar{z}^1 = z^2 + \theta\left(c_j - z_j^1\right), \quad \text{and} \quad \bar{z}^2 = z^2 + \theta\left(d_j - z_j^2\right);$$

where, z_j^1 and z_j^2 refer to the original values for basic feasible solution.

The value of the objective function will improve, if $\bar{z} > z$.

This gives, $$\left[z^1 + \theta\left(c_j - z_j^1\right)\right] \Big/ \left[z^2 + \theta\left(d_j - z_j^2\right)\right] > z^1 / z^2,$$

or $$\left[z^1 + \theta\left(c_j - z_j^1\right)\right] \Big/ \left[z^2 + \theta\left(d_j - z_j^2\right)\right] - z^1 / z^2 > 0,$$

or $$z^2\left[z^1 + \theta\left(c_j - z_j^1\right)\right] - z^1\left[z^2 + \theta\left(d_j - z_j^2\right)\right] > 0$$

(z^2 and $\bar{z}^2$ are positive, since the denominator of the objective function is positive for all feasible solutions.)

or $$z^2\left(c_j - z_j^1\right) - z^1\left(d_j - z_j^2\right) > 0,$$

where, θ being positive in the non-degenerate case. If $\theta = 0$, then $\bar{z} = z$

Let $\Delta_j = -z^2\left(z_j^1 - c_j\right) + z^1\left(z_j^2 - d_j\right)$. Value of Δ_j is less than zero, if

(*a*) $$z_j^2 - d_j > 0 \quad \text{and} \quad \left(z_j^1 - c_j\right) \Big/ \left(z_j^2 - d_j\right) > z^1 / z^2$$

(*b*) $$z_j^2 - d_j < 0 \quad \text{and} \quad \left(z_j^1 - c_j\right) \Big/ \left(z_j^2 - d_j\right) < z^1 / z^2$$

(*c*) $$z_j^2 - d_j = 0 \quad \text{and} \quad z_j^1 - c_j > 0$$

We deduce that given a basic feasible solution $\mathbf{x}_B = \mathbf{B}^{-1}\mathbf{b}$, if for any column $\mathbf{a}_j$ in **A** but not in **B**, $\Delta_j < 0$ holds, and if at least one $y_{ij} > 0$ ($i = 1, 2, ..., m$), then it is possible to obtain a new basic feasible solution by replacing one of the columns in **B** by $\mathbf{a}_j$ and the new value of the objective function satisfies $\bar{z} \geq z$.

We, now, show that for any $\mathbf{a}_j \in \mathbf{A}$ not in **B** at least one $y_{ij} > 0$. If possible let all $y_{ij} \leq 0$ ($i = 1, 2, ..., m$).

We have the basic feasible solution: $\sum_{i=1}^{m} x_{Bi}\, b_i = \mathbf{b}$. Suppose that we add and subtract $\lambda\, \mathbf{a}_j$ (λ is any scalar) to this solution. Then, we obtain

$$\sum_{i=1}^{m} x_{Bi}\, b_i - \lambda\, \mathbf{a}_j + \lambda\, \mathbf{a}_j = \mathbf{b}.$$

But
$$-\lambda\, \mathbf{a}_j = -\lambda \sum_{i=1}^{m} y_{ij}\, \mathbf{b}_i$$

$\therefore$
$$\sum_{i=1}^{m} (x_{Bi} - \lambda\, y_{ij})\mathbf{b}_i + \lambda\, \mathbf{a}_j = \mathbf{b};$$

when $\lambda > 0$, we have $x_{Bi} - \lambda\, y_{ij} \geq 0$.

Since, by assumption $y_{ij} \leq 0$, (i = 1, 2, ..., m), therefore, $(x_{B1} - \lambda\, y_{1j}, \ldots, x_{Bm} - \lambda\, y_{mj})$ and λ is a feasible solution for all $\lambda > 0$. Thus, the set S is unbounded contrary to our hypothesis of regularity.

In the algorithm we have shown that if we start with a basic feasible solution and if there is a column vector $\mathbf{a}_j$ not in the basis having $\Delta_j < 0$, then there exists another basic feasible solution, such that $\bar{z} \geq z$.

When, degeneracy is not present, $\bar{z} > z$. Thus, we can move from one basis to another, changing one vector at a time so long as there is some $\mathbf{a}_j$ not in the basis with the condition of $\Delta_j < 0$ and at each step, z is increased.

This process cannot continue indefinitely, because there are only a finite number of basis and in the absence of degeneracy no basis can ever be repeated. Since, z is increased at every step and the same basis cannot yield two different values of z, while at the same time the maximum is to occur at one of the basic feasible solution.

This process will terminate only in one way, *i.e.*, when all $\Delta_j \geq 0$ for the columns of **A** not in the basis.

Now, for those columns of **A** that are in the basis

$$z_j^1 = \mathbf{c}_B\, \mathbf{y}_j = \mathbf{c}_B\, \mathbf{B}^{-1}\mathbf{a}_j = \mathbf{c}_B\, \mathbf{B}^{-1}\mathbf{b}_i = c_j$$

and
$$z_j^2 = \mathbf{d}_B\, \mathbf{y}_j = \mathbf{d}_B\, \mathbf{B}^{-1}\mathbf{a}_j = \mathbf{d}_B\, \mathbf{B}^{-1}\mathbf{b}_i = d_j$$

Therefore,
$$\Delta_j = z^1\left(z_j^2 - d_j\right) - z^2\left(z_j^1 - c_j\right) = 0$$

Thus, we summarize the result as follows:

Given a basic feasible solution $\mathbf{x}_B = \mathbf{B}^{-1}\mathbf{b}$ with $z_0 = (\mathbf{c}_B\, \mathbf{x}_B + \alpha)/(\mathbf{d}_B\, \mathbf{x}_B + \beta)$ to the LPP:

Maximize $z = (\mathbf{cx} + \alpha) / (\mathbf{dx} + \beta)$ subject to the constraints: $\mathbf{Ax} = \mathbf{b}$ and $\mathbf{x} \geq \mathbf{0}$

such that $\Delta_j \geq 0$ for every column $\mathbf{a}_j$ in **A**. Then, z_0 is the maximum value of z and the basic feasible solution is an optimum solution.

Example 9.5. *Solve the following LPP:*

Maximize $z = (5x_1 + 3x_2)/(5x_1 + 2x_2 + 1)$ *subject to the constraints:*

$3x_1 + 5x_2 \leq 15,\ \ 5x_1 + 2x_2 \leq 10,\ \ x_1 \geq 0$ *and* $x_2 \geq 0$.

Solution. After introducing slack variables, the problem in the standard form is:

Maximize $z = (5x_1 + 3x_2)/(5x_1 + 2x_2 + 1)$ subject to the constraints:

$3x_1 + 5x_2 + s_1 = 15,\ \ 5x_1 + 2x_2 + s_2 = 10$

$x_1 \geq 0, x_2 \geq 0, s_1 \geq 0$ and $s_2 \geq 0$.

After computing $z_j^1 - c_j$, $z_j^2 - d_j$, Δ_j and $\mathbf{B}^{-1}\,\mathbf{a}_j$, the initial basic feasible solution is given in the following tableau:

Initial Iteration. Introduce x_1 and drop s_2.

		c_j:	0	0	5	3
		d_j:	0	0	5	2

$\mathbf{d}_B$	$\mathbf{c}_B$	$\mathbf{x}_B$	s_1	s_2	x_1	x_2	*Solution*	*Ratio*
0	0	s_1	1	0	3	5	15	$15/3 = 5$
0	0	s_2	0	1	(5)	2	10	$10/5 = 2$
$z^2 = 1$	$z^1 = 0$	$z = z^1/z^2$					$z = 0$	
		$z_j^1 - c_j$	0	0	–5	–3		
		$z_j^2 - d_j$	0	0	–5	–2		
		Δ_j	—	—	–5	–3		

First Iteration. Introduce x_2 and drop s_1.

$\mathbf{d}_B$	$\mathbf{c}_B$	$\mathbf{x}_B$	s_1	s_2	x_1	x_2	*Solution*	*Ratio*
0	0	s_1	1	–3/5	0	(19/5)	9	45/19
5	5	x_1	0	1/5	1	2/5	2	5
$z^2 = 11$	$z = z^1 = 10$	z^1/z_2					$z = 10/11$	
		$z_j^1 - c_j$	0	1	0	–1		
		$z_j^2 - d_j$	0	1	0	0		
		Δ_j	—	1	—	–11		

Second Iteration. Introduce s_2 and drop x_1.

$\mathbf{d}_B$	$\mathbf{c}_B$	$\mathbf{x}_B$	s_1	s_2	x_1	x_2	*Solution*
2	3	x_2	5/19	–3/19	0	1	45/19
5	5	x_1	–2/19	(5/19)	1	0	20/19
$z^2 = 209/19$	$z^1 = 235/19$	$z = z^1/z^2$					$z = 235/209$
		$z_j^1 - c_j$	5/19	16/19	0	0	
		$z_j^2 - d_j$	0	1	0	0	
		Δ_j	1045/361	–11/21361	—	—	

Final Iteration. Optimum solution.

$\mathbf{d}_B$	$\mathbf{c}_B$	$\mathbf{x}_B$	s_1	s_2	x_1	x_2	*Solution*
2	3	x_2	1/5	0	3/5	1	3
0	0	s_2	–2/5	1	19/5	0	4
$z^2 = 7$	$z^1 = 9$	$z = z^1/z^2$					$z = 9/7$
		$z_j^1 - c_j$	3/5	0	–16/5	0	
		$z_j^2 - d_j$	2/5	0	–19/5	0	
		Δ_j	3/5	—	59/5	—	

Now, let us consider the above four tables. In the starting tableau, we find

$$\Delta_1 = -5,\ x_1 = 0 \text{ and } \Delta_2 = -3,\ x_2 = 0$$

We choose max. Δ_j (*i.e.*, Δ_1 here). Thus, z can be increased by taking x_1 into the basis. The method to determine departing variables and also the new values of y_{ij}, x_{Bi}, z_j^1, z_j^2 corresponding to new basic feasible solutions will be the same as for a linear programming problem. Thus, here x_4 is a departing variable.

In the optimum tableau all $\Delta_j \geq 0$. Thus, we have reached maximum $z = 9/7$ and the optimum solution is:

$$x_1 = 0, \text{ and } x_2 = 3$$

Example 9.6. **(*Animal Husbandry Problem*).** *ABC Agricultural Cooperative raises cows and sheep, and has a housing capacity of 50 stalls for cows and 200 stalls for sheep. In addition, the cooperative has 72 acres of pasture, whereas one cow needs one acre and a sheep needs only 0.2 acre. To care for the cattle the cooperative can provide only up to 10,000 labour hours per day. A sheep requires attention for 25 hours annually and a cow for 150 hours annually. The capital invested is ₹2,000 per cow and ₹200 per sheep. The annual fixed costs are ₹40,000. If the annual resultant profit is ₹450 per sheep and ₹2,500 per cow, determine the number of cows and sheep, the cooperative should decide to keep in order to maximize profitability (ratio of net profit to capital invested). Formulate the above problem as linear integer fractional programming problem.*

Model Formulation

Let x_1, x_2 denote the required number of cows and sheep, respectively. Then, the annual profit is: ₹ $(2{,}500x_1 + 450x_2)$ and the total capital invested is ₹ $(40{,}000 + 2{,}000x_1 + 200x_2)$. Thus, it is desired to maximize the profitability objective function

$$z = \frac{2{,}500x_1 + 450x_2}{40{,}000 + 2{,}000x_1 + 200x_2} \quad \text{subject to the constraints:}$$

$$x_1 \leq 50,\ x_2 \leq 200,\ x_1 + 0.2x_2 \leq 72,$$

$$150x_1 + 25x_2 \leq 10{,}000,\ x_1, x_2 \geq 0 \text{ and are integers.}$$

This is a linear fractional programming problem.

In fact, it is a linear integer fractional programme, since x_1 and x_2 are required to be non-negative integers.

PROBLEMS

Solve the following linear fractonal programming problems:

9.17. Maximize $z = \dfrac{5x_1 + 2x_2}{x_1 + 8x_2 + 1}$ subject to the constraints:

$$3x_1 + x_2 \geq 1,\ x_1, x_2 \geq 0.$$

9.18. Minimize $z = \dfrac{-6x_1 - 5x_2}{2x_1 + 7}$ subject to the constraints:

$$x_1 + 2x_2 \leq 3,\ 3x_1 + 2x_2 \leq 6,\ x_1, x_2 \geq 0.$$

9.19. Maximize $z = \dfrac{-3x_1 - x_2}{x_1 + 2x_2 + 5}$ subject to the constraints:

$$x_1 + x_2 \geq 1,\ 2x_1 + 3x_2 \geq 2,\ x_1, x_2 \geq 0.$$

9.20. Maximize $z = \dfrac{2x_1 + 3x_2}{x_1 + x_2 + 7}$ subject to the constraints:

$$3x_1 + 5x_2 \leq 15,\ 4x_1 + 3x_2 \leq 12,\ x_1, x_2 \geq 0.$$

9:6. SHORTEST CYCLIC ROUTE MODEL (TRAVELING SALESMAN PROBLEM)

Suppose a salesman has to visit a number of cities. The distance (or cost or time) between every pair of cities is known. He wishes to start from a particular city, visit each city once, and then return to his starting point. The problem is to find the route which is shortest in distance (or cost or time).

If the distance (or cost/time) between every pair of cities is independent of the direction of travel, the problem is said to be ***symmetrical***. When for one or more pair of cities the distance (or cost/time) varies with the direction, the problem is called ***asymmetrical***. For example, it takes more time to go up a hill between two points (stations) than come down the hill between these two points. Similarly, a flight may take more time against the wind direction as compared to that in the direction of wind.

There have been a number of applications of traveling salesman problem that have nothing to do with salesmen. For example, when a truck leaves a distribution centre to deliver goods to a number of locations, the problem of determining the shortest route for doing this is a traveling salesman problem. Another example involves the manufacture of printed circuit boards for wiring chips and other components. When many holes need to be drilled into a printed circuit board, the problem of finding the most efficient drilling sequence is a traveling salesman problem.

Let, the salesman is to visit 3 cities, say A, B and C starting from city A, the possible routes are: $A \to B \to C \to A$, $A \to C \to B \to A$, *i.e.*, there are two possible routes or $(3-1)!$ different routes. Similarly, for four cities there are 6 possible routes or $(4-1)!$ different routes. In general, when there are n cities, total number of different routes will be $(n-1)!$. It may be noted that since the salesman has to visit all the n cities, the shortest route will be independent of the selection of the starting city (or point). Our objective is to find out the shortest route without trying each of the route.

Unfortunately, there is no analytical method which can be used satisfactorily. However, a few computational techniques for solving such problems have been suggested, out of which the method of solving assignment problem is more suitable for solving these problems with an additional restriction of returning to starting assignment (point).

Formulation of Traveling Salesman Problem as an Assignment Problem

The traveling salesman problem is very similar to the assignment problem except that in the former, there is an additional restriction. A similar problem arises when n items say $A_i, i = 1, 2, ..., n$ are to be processed on a machine in a scheduled time. The problem then becomes of choosing the sequence of these items. Let the set-up cost of the machine when item A_i is followed by A_j be c_{ij}. Also, let $x_{ij} = 1$ if item A_i is followed by A_j directly and 0 otherwise. It is to be noticed that $c_{ij} = \infty$ when $i = j$, *i.e.*, item A_i is not processed again after A_i. It is important to note that only one $x_{ij} = 1$ for each value of i and for each value of j.

The traveling salesman problem, therefore, can be put in the form of an assignment problem as given below:

		To (item)				
		A_1	A_2	A_3	...	A_n
	A_1	∞	c_{12}	c_{13}	...	c_{2n}
From	A_2	c_{21}	∞	c_{23}	...	c_{2n}
(item)	A_3	c_{31}	c_{32}	∞	...	c_{3n}
	$\vdots$	$\vdots$	$\vdots$	$\vdots$	$\vdots$	$\vdots$
	A_n	c_{n1}	c_{n2}	c_{n3}	...	∞

The problem is to find the n items A_i $(i = 1, 2, ..., n)$ one in each row and in each column, so as to minimize the total of the associated costs.

In view of the above, the assignment problem can be solved and one may hope that the solution satisfies the additional restriction also.

If the solution to the assignment problem does not satisfy the additional restriction, then after solving the problem by assignment technique, we use the method of enumeration.

The procedure is best ***illustrated*** with the help of following examples:

Example 9.7. *A machine operator processes five types of items on his machine each week, and must choose a sequence for them. The set-up cost per change depends on the item presently on the machine and the set-up cost to be made according to the following table:*

From item	*To item*				
	A	*B*	*C*	*D*	*E*
A	∞	4	7	3	4
B	4	∞	6	3	4
C	7	6	∞	7	5
D	3	3	7	∞	7
E	4	4	5	7	∞

If he processes each type of item once and only once each week, how should he sequence the items on the machine in order to minimize the total set-up cost?

[B.E. (Dec.), Karnataka Univ. 2011; B.Tech. (Nov.), Hyderabad Univ. 2010]

Solution. Reduce the cost matrix and make assignments in rows and columns having single zero.

Initial Iteration. Draw minimum number of lines to cover all the zeros. See *Table 9.1*.

First Iteration. Modify the *Table 9.1* by subtracting the lowest element from all the elements not covered by lines and adding the same at the intersection of two lines. See *Table 9.2*. We observe that this table gives us an optimum assignment schedule.

∞	1	3	[0]	1	✓
1	∞	2	~~0~~	1	✓
2	1	∞	2	[0]	
~~0~~	[0]	3	∞	4	
[0]	~~0~~	~~0~~	3	∞	
			✓		

Table 9.1

∞	~~0~~	2	[0]	~~0~~
[0]	∞	1	~~0~~	~~0~~
2	1	∞	3	[0]
~~0~~	[0]	3	∞	4
~~0~~	~~0~~	[0]	4	∞

Table 9.2

The optimum assignment is $A \to D, B \to A, C \to E, D \to B$ and $E \to C$ with a minimum cost of 20.

This assignment schedule does not provide us the solution of travelling salesman problem as it gives $A \to D, D \to B, B \to A$, while B is not allowed to follow A unless C and E are processed.

Second Iteration. We try to find the next best solution which satisfies this extra restriction. The next minimum (non-zero) element in the matrix is 1. So we try to bring 1 into the solution. But, the element '1' occurs at two places. We shall consider all the cases separately until the acceptable solution is reached.

We start with making an assignment at (2, 3) instead of zero assignment at (2, 1). The resulting feasible solution, then will be

$$A \to D,\ D \to B,\ B \to C,\ C \to E,\ E \to A$$

When an assignment is made at (3, 2) instead of zero assignment at (3, 5), the resulting feasible solution will be

$$A \to E,\ E \to C,\ C \to B,\ B \to D,\ D \to A$$

The total set-up cost in both the programmes comes out to be 21.

Example 9.8. *A salesman is planning to tour cities B, C, D and E starting from his home city A. The inter-city distances are shown in the following table:*

City:	*A*	*B*	*C*	*D*	*E*
A	∞	*103*	*188*	*136*	*38*
B	*103*	∞	*262*	*176*	*52*
C	*188*	*262*	∞	*85*	*275*
D	*136*	*176*	*85*	∞	*162*
E	*38*	*52*	*275*	*162*	∞

How should he plan his tour, so that (i) he visits each of the cities only once, and (ii) travels the minimum distance. ***[MBA (Nov.), Delhi Univ. 2003]***

Solution. Since, the salesman has to visit all the cities before returning back to his home city, the assignments from the city to itself is prohibited, *i.e.*, we consider the huge penalty of assignment *A* to *A*, *B* to *B*, *C* to *C*, *D* to *D* and *E* to *E*.

Now, reduce the cost (distance) matrix and make assignments in rows and columns having single zeros.

Initial Iteration. Draw the minimum number of lines to cover all the zeros of the reduced matrix. See *Table 9.3.*

First Iteration. Modify *Table 9.3* by subtracting the lowest element from all the elements not covered by lines and assign the same at the intersection of two lines. See *Table 9.4.*

	A	*B*	*C*	*D*	*E*	
A	—	51	150	98	[0]	✓
B	51	—	210	124	~~0~~	✓
C	103	163	—	[0]	190	
D	51	77	[0]	—	77	
E	~~0~~	[0]	237	124	—	
					✓	

Table 9.3

	A	*B*	*C*	*D*	*E*
A	—	~~0~~	99	47	[0]
B	[0]	—	159	73	~~0~~
C	103	163	—	[0]	241
D	51	77	[0]	—	128
E	~~0~~	[0]	237	124	—

Table 9.4

The optimum assignment is $A \to E \to B \to A$ and $C \to D \to C$ or $A \to B \to E \to A$ and $C \to D \to C$ with a minimum distance of 363 in both the cases.

This assignment schedule does not provide us the solution of travelling salesman problem as in both the cases the salesman returns to his starting city *A* without visiting cities *C* and *D*.

Second Iteration. Now, we try to find the next best solution that satisfies the extra restriction. For this, we choose the next lowest element (in place of zero) in the second row, *i.e.*, for city *B*. Thus, we make the assignment at (2, 4) being the lowest element after 0 (not covered by lines). The resulting feasible solution, then will be: $A \to E,\ E \to B,\ B \to D,\ D \to C,\ C \to A$.

The total distance in this case comes out to be 539, *i.e.*, an increase of 176 over the optimum assignment.

PROBLEMS

9.21. A machine operator processes five types of items on his machine each week and must choose a sequence for the items. The set-up cost per change depends on the items presently on the machine and the set-up cost to be made, according to the following table:

	A	*B*	*C*	*D*
A	∞	4	7	3
B	4	∞	6	3
C	7	6	∞	7
D	3	3	7	∞

If he processes each type of item once and only once each week, how should he sequence the items on his machine? ***[B.Tech. (Nov.), Hyderabad Univ. 2010]***

9.22. Given the following matrix of set-up costs, show how to sequence production so as to minimize set-up cost per cycle:

From	*To*				
	A	*B*	*C*	*D*	*E*
A	∞	2	5	7	1
B	6	∞	3	8	2
C	8	7	∞	4	7
D	12	4	6	∞	5
E	1	3	2	8	∞

[B.E. (Mech.), Nagpur Univ. 2011; MBA (Nov.), Delhi Univ. 2004; MBA (PT), Delhi Univ. 2008]

9.23. Solve the following traveling salesman problem so as to minimize the cost per cycle:

From	*To*				
	A	*B*	*C*	*D*	*E*
A	—	3	6	2	3
B	3	—	5	2	3
C	6	5	—	6	4
D	2	2	6	—	6
E	3	3	4	6	—

[B.Tech., HP Univ. 2015; MBA, Panjab Univ. 2007]

9.24. Solve the traveling salesman problem given by the following data:

$$c_{12} = 20,\ c_{13} = 4,\ c_{14} = 10,\ c_{23} = 5,\ c_{34} = 6$$
$$c_{25} = 10,\ c_{35} = 6,\ c_{45} = 20,\ \text{where } c_{ij} = c_{ji}.$$

and there is no route between cities i and j if a value for c_{ij} is not shown.

[B.E. (Mech.), Panjab Univ. 2012]

9.25. The *ABC* Ice Cream Company has a distribution depot in Greater Kailash Part I for distributing ice-cream in South Delhi. There are four vendors located in different parts of South Delhi (call them *A*, *B*, *C* and *D*) who have to be supplied ice-cream every day. The following matrix displays the distances (in kilometres) between the depot and the four vendors:

From	*To*				
	Depot	*Vendor A*	*Vendor B*	*Vendor C*	*Vendor D*
Depot	—	3.5	3	4	2
Vendor *A*	3.5	—	4	2.5	3
Vendor *B*	3	4	—	4.5	3.5
Vendor *C*	4	2.5	4.5	—	4
Vendor *D*	2	3	3.5	4	—

What route should the company van follow so that the total distance traveled is minimized?

[MBA (Nov.), Delhi Univ. 2009]

9.26. A salesman must travel from one city to another city to maintain his accounts. This week he has to leave his home base and visit each other city and return home. The table given below shows the distance (in kilometres) between various cities. The home city is city *A*. Use the assignment method to determine the tour that will minimize the total distances of visiting all cities and returning home.

From city	*To city*				
	A	*B*	*C*	*D*	*E*
A	—	375	600	150	190
B	375	—	300	350	175
C	600	300	—	350	500
D	160	350	350	—	300
E	190	175	500	300	—

[MBA (PT), Delhi Univ. 2009]

TEST YOUR UNDERSTANDING

1. The bounded variable algorithm
   (*a*) is not efficient computationally, because it accounts for the bounds implicitly.
   (*b*) is used to solve an LPP when all other methods fail.
   (*c*) have the same optimality condition which is applicable for simplex method.
   (*d*) none of the above.

2. In bounded variable algorithm
   (*a*) lower bound of a decision variable can never be converted into non-negative decision variable.
   (*b*) upper bound of a decision variable can always be converted into non-negative decision variable.
   (*c*) lower and upper bounds of a decision variable are 0 and ∞ respectively in the case of unbounded variables.
   (*d*) the lower bound constraints $l_j \leq x_j$ is converted as $x'_j \ (= x_j - l_j) \leq 0$, where x'_j are slack variables.

3. In bounded variable algorithm
   (*a*) when a non-basic variable enters the basis, its value should exceed its upper bound.
   (*b*) if any of the variable is at a positive lower bound, substitute it out at its upper bound.
   (*c*) if the value of decision variable equals upper bound, the basic variable will change.
   (*d*) optimum solution to an LPP is unchanged, when continuous changes are carried out on the RHS of the constraints.

4. Parametric LPP is a technique to
   (*a*) investigate the effect of simultaneous changes of all components of RHS of the constraints and coefficient of variables in the objective function.
   (*b*) investigate the feasibility and optimality of an LPP.
   (*c*) minimize cost of operations.
   (*d*) determine an optimum solution to an LPP.
5. Parametric linear programming
   (*a*) maximizes the additional computational effort required to obtain the indicated results.
   (*b*) assumes that there can never be an unbounded solution.
   (*c*) helps in determining the feasible as well as optimum solution to an LPP.
   (*d*) does not investigate the behaviour of the optimum solution as a result of pre-determined linear variations in the parameters of the problem.
6. Linear fractional programming problem (LFPP)
   (*a*) may have linear constraints.
   (*b*) can be solved directly provided denominator is positive for all feasible solutions.
   (*c*) can be resolved into two linear programming problems.
   (*d*) all of the above.
7. A linear fractional programming problem can be solved using
   (*a*) graphic method (*b*) simplex method
   (*c*) modified simplex method (*d*) revised simplex method.
8. A linear fractional programming problem when solved
   (*a*) may have degeneracy in the solution.
   (*b*) will always have its optimum solution in integers.
   (*c*) can never have unbounded solution.
   (*d*) gives infeasible but optimum solution.
9. Karmarkar's algorithm
   (*a*) cuts across the interior of the solution space.
   (*b*) reduces the number of iterations.
   (*c*) is suitable for extremely large linear programming problems.
   (*d*) all of the above.
10. Which of the following is not correct?
   (*a*) Karmarkar's algorithm admits solution of an LPP at non-extreme point of the feasible region.
   (*b*) In Karmarkar's algorithm, constraints are assumed to be non-homogeneous equations except of the constraint $x_1 + x_2 + x_3 + ... + x_n = 1$.
   (*c*) Following are the two basic conditions on which the validity of Karmarkar's algorithm rests:

   $$(i)\ x = \left(\frac{1}{n}, \frac{1}{n}, \ldots, \frac{1}{n}\right) \text{ satisfying } \mathbf{Ax} = \mathbf{0}, \text{ and } (ii) \text{ minimum } z = 0.$$

   (*d*) Karmarkar's algorithm is also called the Interior-Point algorithm.
11. Decomposition method of solving an LPP is suitable when there are
   (*a*) only two decision variables (*b*) less than two variables
   (*c*) more than five decision variables (*d*) none of the above.

12. For a salesman, who has to visit *n* cities before returning to his home city, following are the number of ways of his tour plan:

(*a*) n, (*b*) $n\,!$

(*c*) $(n + 1)\,!$ (*d*) $(n - 1)\,!$

ANSWERS

1. (*c*)	2. (*c*)	3. (*d*)	4. (*a*)	5. (*b*)	6. (*d*)
7. (*c*)	8. (*a*)	9. (*d*)	10. (*b*)	11. (*c*)	12. (*d*)

REVIEW EXERCISE

1. State the computational procedure for resolving a linear programming problem with upper bound conditions.
2. Explain the basic idea behind the decomposition method for solving linear programming problems.
3. Write a short note on parametric linear programming.
4. Explain the basic difference between sensitivity analysis and parametric programming.
5. Explain parameterizaton of cost vector in an LPP.
6. Explain what you mean by linear fractional programming.
7. Derive an algorithm for solving a linear fractional programming problem.
8. Explain the basic idea of the interior-point used in Karmarkar Algorithm.
9. In simplex method, trial solutions are at corner points of the feasible region, whereas Karmarkar algorithm admits interior points of the feasible region. Discuss.
10. What is traveling salesman problem? *[MBA (Nov.), Annamalai Univ. 2010; MBA (Dec.), Gujarat Tech. Univ. 2010; B.Tech. (Nov.), HP Univ. 2015]*
11. How does the solution of traveling salesman problem differ from the solution of an assignment problem? *[BCA (Nov.), Madras Univ. 2010; B.Tech. (Mech.), HP Univ. 2012]*
12. Discuss the concept of parametric changes in the objective function of a linear programming problem.

CHAPTER 10

Operations Scheduling

"An order is to be observed in all things"

An important function of a business firm is ***scheduling*** which assigns load to each work centre and determines the time schedule for each operation. The utilization of the firm's facilities depends on the efficiency of its scheduling function. To keep the facilities busy, waiting time due to non-availability of work should be kept at minimum. Providing sufficient work is the responsibility of firm's sales department, while the responsibility of smooth flow of work through different work centres or departments of the firm rests with the production control department. It must ensure that facilities do no remain idle for want of work from the preceding work centres.

10:1. PROBLEM OF SEQUENCING

Sequencing of the jobs, *i.e.*, the order for loading of jobs to the firm's facilities, must be carefully planned so that facilities at work centres are fruitfully utilized and the jobs are turned out in the least possible time. Thus, when there are many work centres, scheduling involves establishing the amount of work to be done at each work centre and the time when each element of work will start at the work centre.

The ***technique of sequencing*** helps the management to determine the most economical job sequence (order), the one that will consume minimum ***elapsed time***. The effectiveness of the technique increases, if the firm has backlog of orders. The limitations of the technique, however, is that all jobs must have a fixed operation sequence (order).

10:2. BASIC TERMINOLOGY AND ASSUMPTIONS

Terminology

1. Number of machines means the activity — stages/service facilities/operations, through which a job must pass through before it is completed. For example, a book to be printed has to be processed through composing, printing, binding, finishing, etc. In this case the book constitutes the job and various activities like composing, printing, etc. constitute the number of machines.
2. Processing time means the time that each job requires at each machine/work centre.
3. Processing order means the order in which machines/work centres are required for the completion of job.
4. Total elapsed time is the time between starting of the first job and completion of the last one.
5. ***Idle time*** of a machine/work centre is the time that a machine/work centre remains idle during the total elapsed time.
6. Job sequence refers to the order for completion of jobs.

Assumptions

Basic assumptions to solve scheduling problems are the following:

1. Only one operation is carried out on a machine/service facility at a time.
2. Processing times on various machines are independent of the order in which different jobs are processed on them.
3. Time consumed by different jobs in going from one machine to another is negligible.
4. A job once started on a machine would be performed to the point of completion uninterrupted.
5. No passing rule is followed strictly, *i.e.*, same order of jobs is maintained over each machine/service facility.
6. A machine cannot process more than one job at a given point of time.
7. All machines are of different types.
8. All jobs are completely known and are ready for processing.
9. A job would start on a machine as soon as the job and the machine on which it is to be processed are both available, but in the specified order.

10:3. GANTT CHART

We shall illustrate the use of Gantt Charts with the help of following examples:

Illustration 1. Suppose there are two jobs J_1 and J_2, each requiring work on two machines A and B (in this order) with the required processing times as follows:

		Processing time (in hours)	
Job	:	J_1	J_2
Machine A	:	2	6
Machine B	:	8	5

What order of performance of the job will involve the least time?

For processing of two jobs, two orders are possible, *viz.*, $J_1 \rightarrow J_2$ and $J_2 \rightarrow J_1$. Both alternatives are evaluated in the figures (Gantt chart) given below:

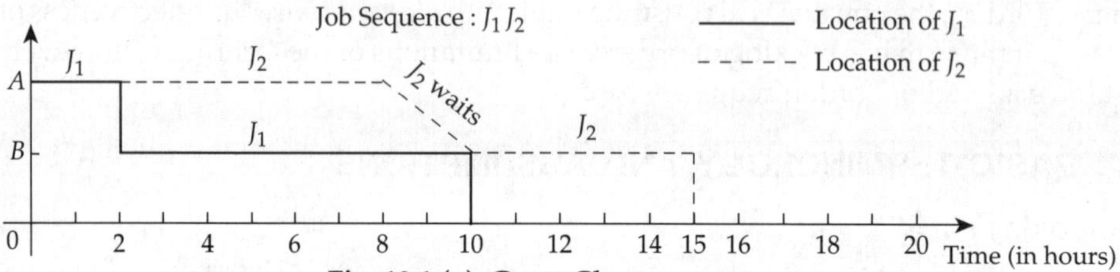

Fig. 10.1 (*a*) Gantt Chart

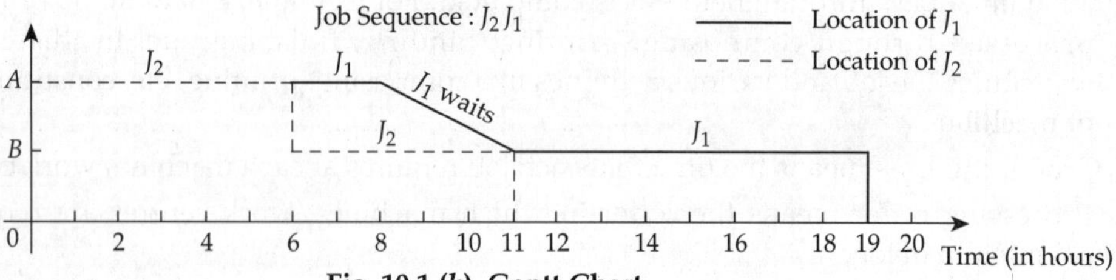

Fig. 10.1 (*b*) Gantt Chart

Fig. 10.1(a) depicts the job sequence $J_1 \rightarrow J_2$. According to this sequence, two jobs can be completed in 15 hours. The other sequence $J_2 \rightarrow J_1$ is shown in *Fig. 10.1(b)*, wherein it is clear that two jobs would take 19 hours to finish.

Evidently, job sequence $J_1 \rightarrow J_2$ is optimum, since total time T is least of the two timings.

Illustration 2. Three jobs *1*, *2*, and *3* are to be processed on two machines *A* and *B* in the order *AB* for which the processing times are given below:

		Processing time (in hours)		
Job	:	*1*	2	3
Machine *A*	:	2	5	4
Machine *B*	:	4	3	6

In what sequence should jobs be performed so that the total elapsed time, T, is the least?

With three jobs to be processed on two machines, we have six possible orders. These are: 123, 132, 213, 231, 312 and 321. Each of these sequences are evaluated in the chart shown in *Fig. 10.2*.

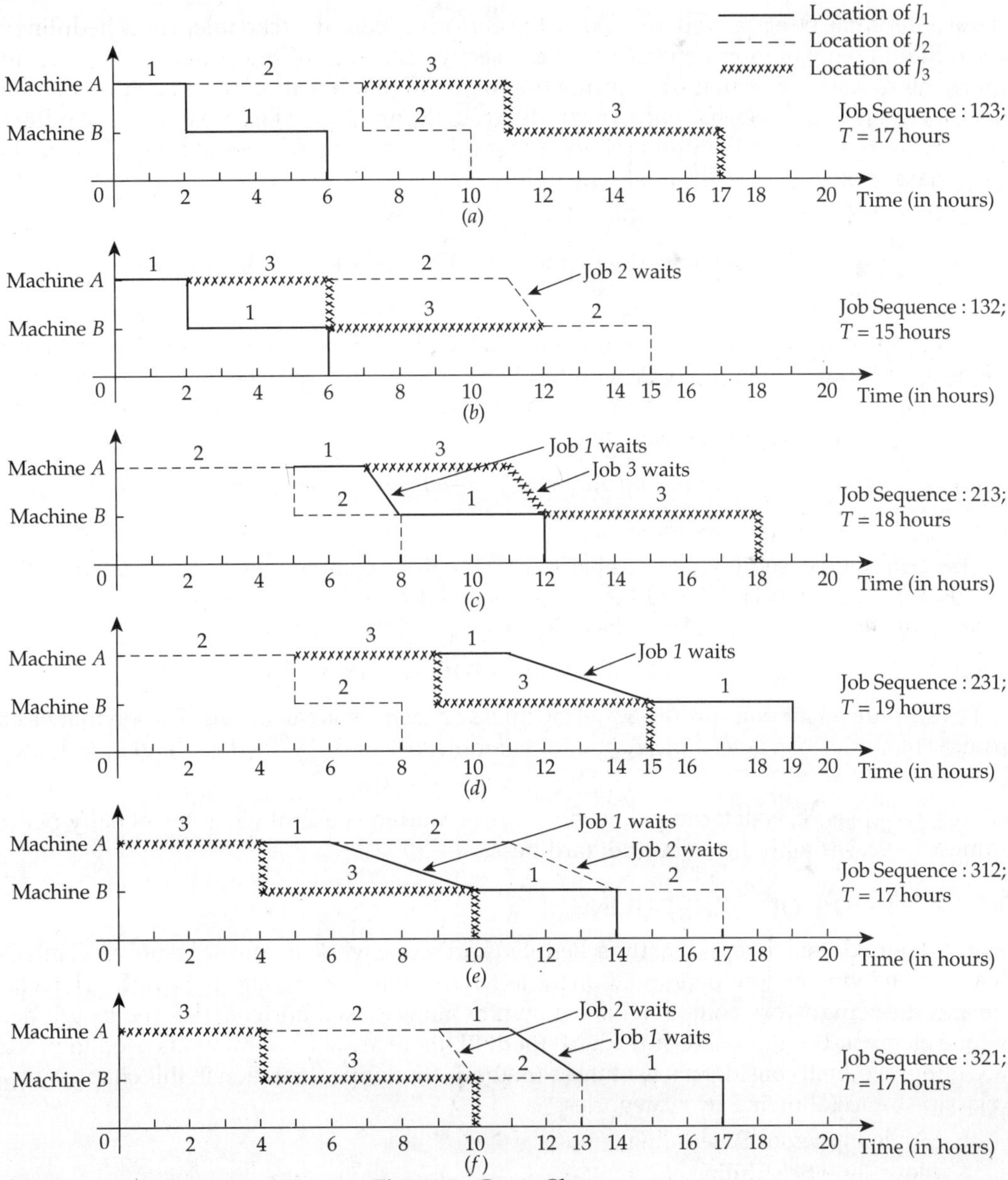

Fig. 10.2. Gantt Chart

Clearly, the sequence 132 takes the minimum processing time in aggregate, that is, 15 hours, and hence, is optimal.

The use of Gantt Chart becomes disproportionately difficult as the number of jobs increases. In fact, when there are n jobs, each of which is to be processed over two machines in the same order, there are n ! possible sequences. When there are 6 jobs to be processed, we shall have to evaluate 6 ! (= 720) possible orders in an attempt to find the optimal one and for 8 such jobs, the number of sequences would be 8 ! (= 40, 320). Hence, we need a more efficient method by which the sequence desired can be determined.

10:4. CRITERIA AND OBJECTIVE FOR SCHEDULING

Several criteria can be employed to evaluate the performance of the schedule. The scheduling criteria chosen in a given situation depends on the objective function of the manager. For example, the underlying objective function or cost function of a company/firm may be such that a penalty is associated with a tardy job, but once a job is delayed, the amount of tardiness does not influence the cost. In this situation, a scheduling criterion that minimizes the number of tardy jobs will be most appropriate for selecting an optimal schedule.

Let there be, n, jobs to be scheduled, then

(*a*) $$\text{Mean flow time} \equiv F = \frac{1}{n}(F_1 + F_2 + \ldots + F_n),$$

where, F_j denotes the flow time of job j.

This criterion measures the average amount of time that a job spends in the system. Minimization of, F, is appropriate when rapid turnaround is required and when the objective is to keep a low in-process inventory. Rapid turnaround may provide a comparative advantage in the firm/company, when customers are sensitive to fast deliveries.

(*b*) $$\text{Mean tardiness} \equiv T = \frac{1}{n}(T_1 + T_2 + \ldots + T_n),$$

where, T_j denotes the tardiness of job j.

This criterion is useful when the objective of the firm/company includes a penalty per unit of time, if job completion is delayed beyond a specified due date. For example, a penalty may be imposed for each job that is delayed beyond its specified due date.

(*c*) $$\text{Maximum tardiness} \equiv T_{\text{max.}} = \max_i . \{T_i\}$$

To compute ***maximum tardiness***, the tardiness of each job is calculated. The job that has largest tardiness of all the jobs determine $T_{\text{max.}}$. For example, if $T_1 = 3$, $T_2 = 5$, $T_3 = 1$ and $T_4 = 4$, then

$$T_{\text{max.}} = \text{max.} \{3, 5, 1, 4\} = 5$$

This, we observe, is determined by job 2. This criterion is useful when the penalty per day for tardiness increases with the amount of tardiness.

10:5. METHODS OF SCHEDULING

A simple method used to represent the scheduled and actual work, involves the use of 'Gantt Charts'. It is a chart showing various operations on the left in rows and, time scale on the right. The scheduled time and the actual work completed are shown by lines drawn horizontally, the length denoting the time element. Besides the 'Gantt Chart' method, there are several methods used in production scheduling. We shall consider a few of these methods for our presentation in this chapter which can be classified into following two categories:

1. Single Processor Scheduling, and
2. Flow Shop Scheduling.

10:6. SINGLE PROCESSOR SCHEDULING

When there are a number of jobs which are to be processed on a single machine, an optimal schedule is achieved according to processing time or due date of completion. The processing time is taken as the sum of waiting time and operation time both.

We shall now consider a static job shop in which a number of jobs with known processing times require processing on a single machine. The job shop is static in the sense that new job arrivals do not disturb the processing of current jobs. It may be assumed that the arrival of new job wait for being considered in the next batch of jobs (after the processing of current jobs).

1. Shortest Processing Time (SPT) Procedure

The shortest average processing time per job is determined and the job having shortest operation time is processed first and the job having next higher operation time is processed second and so on; *i.e.*, a sequence of jobs in non-decreasing order of processing times is obtained. It is referred to as shortest processing time (SPT) schedule and the schedule estimates mean flow time '*F*'. In addition, the SPT rule also minimizes mean lateness and mean waiting time.

When the importance of jobs (or tasks) is different, we make use of another method known as weighted-schedule-processing time (WSPT) procedure. In this procedure, weight W_i is assigned to each job '*i*', where a larger value indicates that the job has greater importance. The weighted mean flow time is obtained by using the following formula:

$$WMFT = \left(\sum_{i=1}^{n} W_i F_i \right) \Big/ \sum_{i=1}^{n} W_i$$

The 'WSPT' procedure for minimizing weighted mean flow time, sequence the '*n*' jobs, such that

$$\frac{t_{[1]}}{W_{[1]}} \leq \frac{t_{[2]}}{W_{[2]}} \leq \frac{t_{[3]}}{W_{[3]}} \leq \ldots \leq \frac{t_{[n]}}{W_{[n]}};$$

where the number in square brackets defines the position of the job in the optimal schedule.

Example 10.1. *A computer centre has four jobs to be performed during a shift. The list of jobs to be processed along with time (in minutes) for each job is given below:*

Job	:	1	2	3	4
Processing time (minutes)	:	4	7	2	5

Determine an optimal schedule that minimizes the total time to be taken for all jobs.

Solution.

Step 1. The smallest processing time is 2, which is for processing job *3*. So, process job *3* in the beginning.

Step 2. After job *3* has been set for processing, we are left with 3 jobs and their processing times as given below:

Job	:	1	2	4
Processing time	:	4	7	5

Step 3. Repeating *steps 1* and *2*, job *1* will be processed next to job *3*, since the minimum processing time is for job *1*. After job *1* has been set for processing, we are left with only job *2* and job *4*. Among these two, since the minimum processing time is for job *4*, it will be processed next to job *1* and then finally job *2*.

The optimal sequence, therefore, is as shown below:

3	1	4	2

Step 4. The calculations for obtaining the flow time for each job are given below:

Job	*Processing time* (t_i)	*Completion time* *In*	*Out*	*Flow time* (F_i)
3	2	0	2	2
1	4	2	6	6
4	5	6	11	11
2	7	11	18	18

Step 5. We observe that job *3* starts at time 0 and is over by time 2 minutes. Job *1* then starts after job *3* and is finished by taking 4 minutes. Thus, the completion time of job *1* is 6 minutes. The completion times for the remaining jobs are similarly computed. Since, all the jobs are available at time zero, the flow time for each job is identical to its completion time.

The mean flow time is computed by simply adding the flow time for each job and then dividing it by the number of jobs, *i.e.*, 4. Thus, we get

$$F = \frac{F_1 + F_2 + F_3 + F_4}{4} = \frac{2+6+11+18}{4} = \frac{37}{4} = 9 \text{ (approx.)}.$$

Flow time of jobs and number of jobs waiting in-process inventory is shown graphically in *Fig. 10.3*.

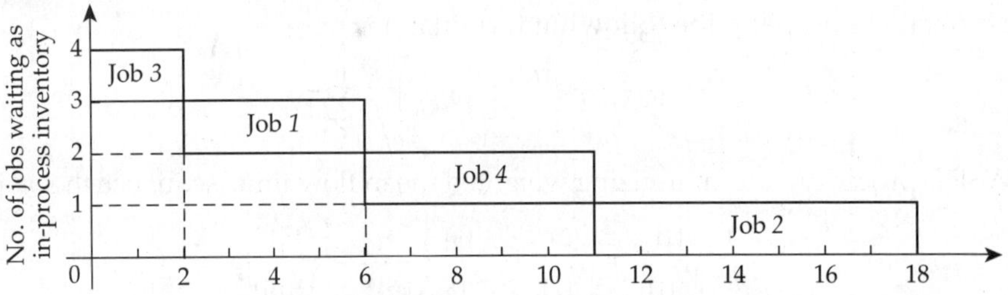

Fig. 10.3. Flow Time of Jobs

From the sequence as shown in *Fig. 10.3*, we observe that the number of jobs waiting as in-process inventory is seen to be '4 jobs' during time 0—2, '3 jobs' during time 2—6, '2 jobs' during time 6—11, and 'one job' during time 11—18.

Thus, the average in-process inventory

$$= \frac{4\times2+3\times4+2\times5+1\times7}{2+4+5+7} = \frac{37}{18} = 2 \text{ jobs (approx.)}$$

Example 10.2. *Given below is 8 jobs with processing times, due dates and importance weights. Use SPT procedure for finding the mean-flow time and average in-process inventory of jobs.*

Job	:	1	2	3	4	5	6	7	8
Processing time (t_i in hours)	:	5	8	6	3	10	14	7	3
Due date (d_i)	:	15	10	15	25	20	40	45	50
Importance weight (w_i)	:	1	2	3	1	2	3	2	1
Ratio (t_i/w_i)	:	5	4	2	3	5	4.7	3.5	3

Solution.

Step 1. The smallest processing time is 3, which is for processing of job *4* and job *8*. So, process either job *4* first and then job *8*, or job *8* first and then job *4*.

Step 2. After jobs *4* and *8* has been set for processing, we are left with six jobs and their processing times as given below:

Job	:	*1*	*2*	*3*	*5*	*6*	*7*
Processing time	:	5	8	6	10	14	7

Step 3. Repeating *steps 1* and *2*, job *1* will be processed next to jobs *4* and *8*; since the minimum processing time is for job *1*. After job *1* has been set for processing, we are left with five jobs. Repeating *Steps 1* and *2*, we get the following schedule (sequence or order) of processing all the eight jobs:

4	8	1	3	7	2	5	6

Step 4. The calculations for obtaining the flow time for each job are shown below:

Job	*Processing time* (t_i)	*Completion time* In	*Out*	*Flow time* (F_i)
4	3	0	3	3
8	3	3	6	6
1	5	6	11	11
3	6	11	17	17
7	7	17	24	24
2	8	24	32	32
5	10	32	42	42
6	14	42	56	56

The mean flow time is computed by simply adding the flow time for each job and then dividing it by total number of jobs, *i.e.*, 8. Thus, we get

$$F = \frac{3+6+11+17+24+32+42+56}{8} = \frac{191}{8} = 23.87 \text{ hours}$$

Flow time of jobs and number of jobs waiting in-process inventory is shown graphically in *Fig. 10.4*.

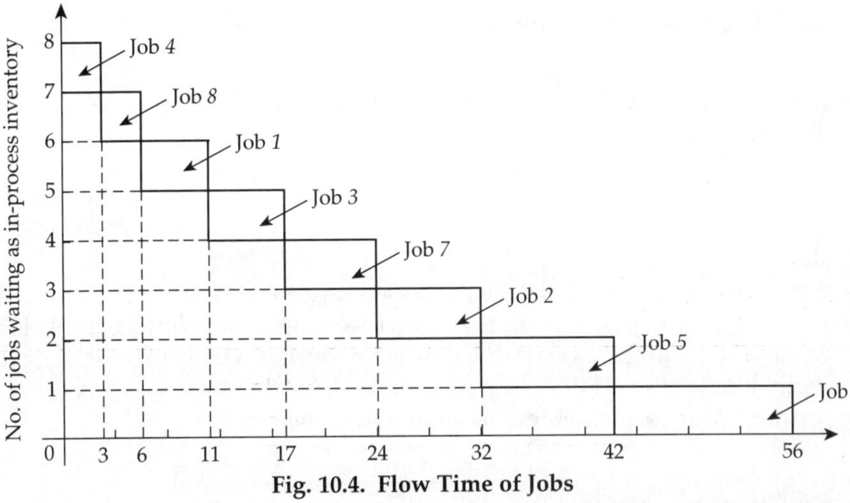

Fig. 10.4. Flow Time of Jobs

From the sequence shown in *Fig. 10.4*, we observe that the number of jobs waiting as in-process inventory is seen to be eight during time 0—3, seven during time 3—6, six during time 6—11, five during time 11—17, four during time 17—24, three during time 24—32, two during time 32—42 and one during time 42—56. Thus, the average in-process inventory is:

$$\frac{(8\times3)+(7\times3)+(6\times5)+(5\times6)+(4\times7)+(3\times8)+(2\times10)+(1\times14)}{3+3+5+6+7+8+10+14} = \frac{191}{56} = 3.41 \text{ jobs.}$$

Further, if the importance weights 'w_i' were to be considered, the WSPT could be used to minimize the weighted mean flow time (WMFT) to yield the following schedule:

3	4	8	7	2	6	1	5

This results by first choosing the job with minimum ratio (t_i/w_i), then choosing next from the minimum ratio (t_i/w_i) of the remaining seven jobs, and so on. The respective flow times of jobs in this sequence are: 6, 9, 12, 19, 27, 41, 46 and 56. The mean flow time equals

$$\frac{6+9+12+19+27+41+46+56}{8} = \frac{216}{8} \text{ or } 27 \text{ hours.}$$

The weighted mean flow time is equal to

$$\frac{(3\times6)+(1\times9)+(1\times12)+(2\times19)+(2\times27)+(3\times41)+(1\times46)+(2\times56)}{3+1+1+2+2+3+1+2} = \frac{412}{16} \text{ or } 27.47 \text{ hours.}$$

Remarks

(i) *It can be checked that no other sequence can produce a better mean flow time than the sequence obtained by the SPT rule.*

(ii) *In business situations, by finishing the shorter jobs first, both the turnaround time and the work-in-process inventory are reduced.*

2. Due Date (DD) Procedure

In the due date procedure, jobs are sequenced in the order of non-decreasing due dates. The due date procedure minimizes the maximum tardiness.

Example 10.3. *In a factory, there are four jobs to perform. The production time (in days) for the jobs and their due dates (in days) are given below:*

Job	:	1	2	3	4
Production time (days)	:	4	7	2	8
Due date (days)	:	6	9	19	17

Determine an optimal schedule that minimizes the maximum tardiness.

Solution.

Step 1. The smallest among the due dates is '6', which is for the due date of job 1. So, process job 1 in the beginning.

Step 2. After job 1 has been set for processing, we are left with 3 jobs and their due dates as given below:

Job :	2	3	4
Due date :	9	19	17

Step 3. Repeating *steps 1* and 2, job 2 will be processed next to job 1, since the least among the due dates is for job 2. After job 2 has been set for processing, we are left with only job 3 and job 4. Among these two, since, the least due date is for job 4, it will be processed next to job 3 and then finally job 3.

The optimal sequence, therefore, is as shown below:

1	2	4	3

Step 4. The calculations for obtaining the tardiness for each job are given below:

Job	*Production time (days)*	*Due date (days)*	*Completion time*		*Tardiness* (T_i)
			In	*Out*	
1	4	6	0	4	0
2	7	9	4	11	2
4	8	17	11	19	2
3	2	19	19	21	2

The due date procedure minimizes the maximum tardiness. In table, given above, the components for individual job tardiness, T_i, for the due date sequence are shown. We observe that the maximum tardiness is 2 days and, hence, the minimum of the maximum tardiness is 2 days. No other schedule can produce a tardiness of less than 2 days.

3. Moore Procedure

The Moore procedure minimizes the total number of tardy jobs. This procedure is best explained by considering the *Example 10.3* given above in the following steps:

Step 1. Arrange the jobs in non-decreasing order of their due dates. If this sequence yields one or zero tardy jobs, then the DD-schedule is optimum and the procedure stops. In *Example 10.3*, three jobs are tardy, so we proceed to *Step 2*.

Step 2. Identify the first tardy job in the DD-schedule. In this example, the first tardy job in the DD-schedule is job 2, which is marked by (*) as shown below:

DD-Schedule	:	*1*	2*	4	3
Completion time	:	4	11	19	21
Due date	:	6	9	17	19

Step 3. Identify the longest job from among the jobs including and on the left of the job marked (*) in the schedule of *Step 2*. That is, we pick the longest job among the jobs that are completed not later than the completion time of the first tardy job in *Step 2*. In our example, jobs *1* and 2 are candidates, and, since job 2 has the longest processing time of the two, it is identified.

The identified job is removed and the completion times for the remaining jobs are revised. Thus, we get the reduced scheduling problem:

DD-schedule	:	*1*	4	3
Completion time	:	4	12 (= 4 + 8)	14 (=12 + 2)
Due date	:	6	17	19

Step 4. We now repeat *step* 2. In this example, all jobs are now on time. So, we terminate the procedure. Thus, in this example, the Moore schedule is

1	4	3	2

which has been obtained by simple putting the jobs removed in *Step 3* at the end of the schedule.

PROBLEMS

10.01. There are six jobs which must go through machine *A*. Processing time (in days) and due date (in days) are given below:

Job	:	1	2	3	4	5	6
Machine *A*	:	8	10	9	16	12	21
Due date	:	7	12	10	14	13	9

Using SPT procedure, determine the optimal schedule and total mean flow time.

10.02. Seven jobs *X, Y, Z, P, Q, R* and *S* have arrived at one time to be processed on one machine. Using SPT procedure, find out the optimal job schedule when their operation time is given below:

Job	:	X	Y	Z	P	Q	R	S
Operation time (in minutes)	:	12	10	9	7	4	2	1

If jobs are processed in the order of their arrival, *i.e.*, FCFS; then determine the average processing time for processing all the jobs.

10.03. There are eight jobs to be processed through a single machine, the operation time for each of these jobs is given below:

Job	*A*	*B*	*C*	*D*	*E*	*F*	*G*	*H*
Operation time (in minutes) :	12	24	16	8	20	22	17	9

Find out the optimal schedule, total processing time and average processing time (or mean flow time).

10.04. Compare the jobs shown below in average completion time and average job lateness on application of each of the following rules: FCFS and SPT procedure:

Job	*A*	*B*	*C*	*D*	*E*
Processing time (days) :	7	4	5	2	6
Due date (days) :	8	3	7	9	6

10.05. A publisher has manuscripts of a number of different books. The time required to publish the books (in days) and the desired date of their publication (in days) are given below:

Book	*1*	*2*	*3*	*4*
Processing time (days) :	30	50	20	90
Due date (days) :	40	30	60	50

Using DD procedure, determine the optimal schedule and the maximum tardiness.

10.06. A ready-made garments manufacturer has to process 5 items. The time taken by each of these items (in days) and their due dates (in days) are given below:

Item	*1*	*2*	*3*	*4*	*5*
Processing time (in days) :	5	3	7	6	12
Due date (in days) :	12	17	15	19	18

(*a*) Find an order in which these items are to be processed so as to minimize the total processing time.

(*b*) Use Moore procedure to find the optimal schedule.

10:7. FLOW SHOP SCHEDULING

In many situations, there is more than one processor and a job consists of several operations that are to be performed in a specific order. Moving from single processor job shop to multiple processor job shop poses a formidable challenge. We first consider a special job shop case in which the flow of work is uni-directional.

A flow shop is a special type of job shop in which m machines are numbered as 1, 2, ..., m and a job may require a maximum of m operations — one operation on each machine. Further, for every job, if operation j precedes operation k, then the machine required for operation j has a lower number than the machine required for operation k.

An example of a flow shop is as under:

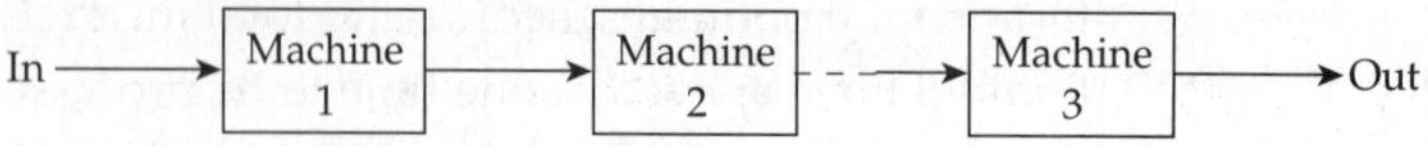

Fig. 10.5

Here, all the jobs require one operation on each processor — such a shop is called a "pure" flow shop. When the flow of work is uni-directional, a job may require fewer than m operations and the operations need not be performed on adjacent processors. A particular example of such a flow shop is as under:

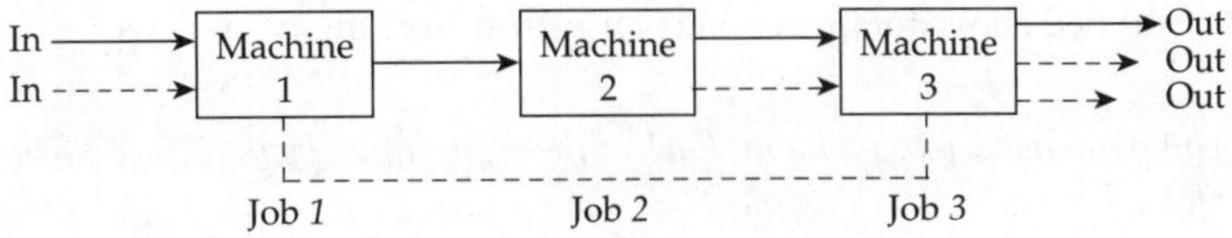

Fig. 10.6

Here, job *1* is processed on all the three processors, job 2 is processed on processors 2 and 3, and job *3* is processed on processors 1 and 3.

An example of a flow shop is an assembly line, where work progresses from one stage to the next in the same direction. In several manufacturing situations (*e.g.*, the manufacturing of printed circuit boards), the same sequence of operations is required for a large number of orders.

Two-machine Flow Shop

For processing of a number of jobs through 2 machines, Johnson's procedure minimizes the make span. This procedure works as follows:

Let there be n jobs to be processed on 2 machines, A and B in the order AB. That is, each job will go to machine A first and then to machine B or in other words passing off is not allowed. All the jobs are to be processed on machine A without any idle time. On the other hand, machine B is subject to its idle time at various stages. The processing time for each job on each machine is known and is given by

$$A_i = \text{Processing time for } i^{\text{th}} \text{ job on machine } A,$$

and

$$B_i = \text{Processing time for } i^{\text{th}} \text{ job on machine } B.$$

The iterative procedure for determining the optimal sequence for n jobs through 2 machines is as follows:

Step 1. Examine the A_i's and B_i's for $i = 1, 2, 3, ..., n$ and select the minimum of these, *i.e.*, Min. $\{A_i, B_i\}$. If there are two or more minimum processing times, then select any one of these arbitrarily.

Step 2. (*i*) If the minimum processing time is for machine A, process (do) that job first and place it at the beginning of the sequence.

(*ii*) If the minimum processing time is for machine B, process (do) that job last and place it at the end of the sequence.

Step 3. Cross off the jobs already assigned and repeat *Steps 1* and *2*, placing the remaining jobs next to first or next to last, until all the jobs have been assigned. The sequence, thus, obtained is known as optimal sequence.

Step 4. After finding the optimal sequence as stated above, we can find the overall or total elapsed time and also the idle time on machines A and B as under:

(*i*) Total elapsed time

= time between starting the first job of the optimal sequence on machine A and completing the last job of the optimal sequence on machine B.

(*ii*) Idle time for machine A

= time when the nth job in a sequence finishes on machine B – the time when the nth job in optimal sequence finishes on machine A.

(*iii*) Idle time for machine B

= time at which the first job in a sequence finishes on machine A + $\sum_{i=2}^{n}$ {time when the ith job in optimal sequence starts on machine B – time when the $(i-1)^{\text{th}}$ job in optimal sequence finishes on machine B}.

Note. The procedure outlined above for processing of *n* jobs on 2-machines gives us the minimum total elapsed time.

Example 10.4. *A company has 3 jobs on hand. Each of these jobs must be processed through two departments, the sequential order for which is*

Department A: Press Shop, and Department B: Finishing.

The table below lists the number of days required by each job in each department:

	Job I	*Job II*	*Job III*
Department A	8	6	5
Department B	8	3	4

Find the sequence in which three jobs should be processed so as to take minimum time to finish all the 3 jobs.

Solution.

Step 1. The smallest processing time in the two departments is 3.

Step 2. Since, the smallest processing time corresponds to Job *II* in department *B*, Job *II* will be processed in the last as shown below:

		II

Step 3. After Job *II* has been set for processing in the last, we are left with 2 jobs and their processing times as given below:

	Job I	*Job III*
Department *A*	8	5
Department *B*	8	4

Steps 1 **and** ***2.*** The minimum processing time in the reduced problem is 4 which corresponds to Job *III* in department *B*. Thus, Job *III* will be processed next to last and finally Job *I* will be processed in the beginning.

The optimal sequence, therefore, is as shown below:

I	*III*	*II*

Step 4. The minimum elapsed time for the two departments is, now, calculated as follows:

Job	*Department A*		*Department B*	
	In	*Out*	*In*	*Out*
I	0	8	8	16
III	8	13	16	20
II	13	19	20	23

Step 5. We observe that Job *I* starts at time 0 in department *A* and is over by time 8 days, when it passes to department *B* to be worked on till 16 days, Job *III* then starts in department *A* at time 8 days as this department is free at that time. It is completed at time 13 days and has to wait for 3 days before it is processed in department *B*, starting at time 16 days when this department is free and complete its finishing by time 20 days. Lastly, Job *II* starts in department *A* at time 13 days and is completed at time 19 days. Then it has to wait for one day before passing on to department *B*. It enters department *B* at time 20 days and works on till 23 days. Thus, the total minimum elapsed time comes out to be 23 days.

Step 6. Idle time in the two departments is given below:

Department *A*: 4 days (23 – 19) Department *B*: 8 days (8 – 0)

Total minimum elapsed time can also be shown by using *Gantt Chart:*

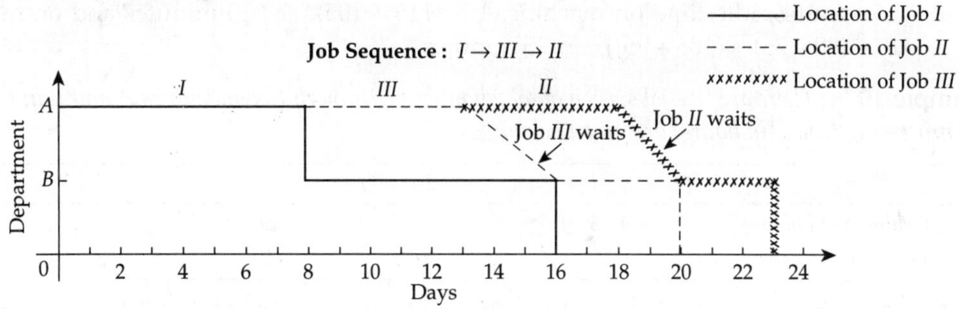

Fig. 10.7. Gantt Chart

Example 10.5. *Draw the schedule of following jobs on machine A and machine B (routing A to B) by using bar chart and work out the total elapsed time. Use Johnson's rule for sequence the jobs:*

Job	*Processing time (in minutes)*	
	Machine A	*Machine B*
P	15	20
Q	25	10
R	35	15
S	10	15
T	20	30

[BBA, GGSIP Univ. 2004]

Solution.

Step 1. The smallest processing time on the two machines is 10 on machine *A* as well as on machine *B*.

Step 2. Since, the smallest processing time corresponds to job *S* on machine *A*, and corresponds to job *Q* on machine *B*; job *S* will be processed in the beginning and job *Q* will be processed in the last as shown below:

S				*Q*

Step 3. The smallest processing time in the reduced problem is 15 on machine *A* as well as on machine *B*. This corresponds to job *P* and job *R* respectively. Therefore, job *P* will be processed next to job *S* and job *R* will be processed next to last as shown below:

S	*P*		*R*	*Q*

Now, the only job remaining is *T*. It shall be assigned to the only space left in the sequence. The optimal sequence, therefore, is:

$$S \to P \to T \to R \to Q.$$

Step 4. Using bar chart, now, the scheduling problem is depicted as follows:

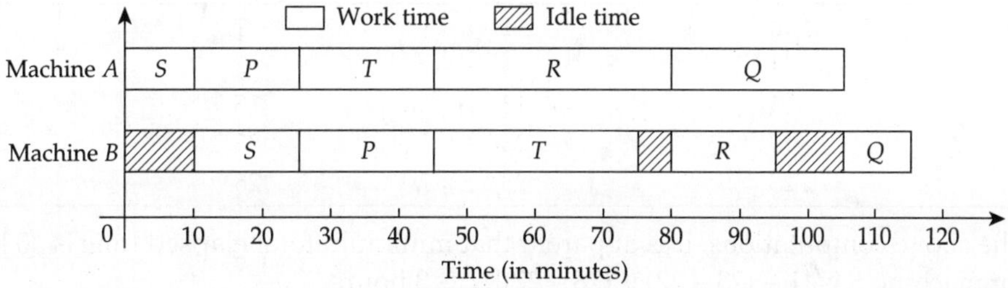

Fig. 10.8. Bar Chart

According to above Bar Chart, the total elapsed time to complete the schedule is 115 minutes. Idle time on machine *A* is (115 – 105), *i.e.*, 10 minutes and on machine *B* is [10 + (80 – 75) + (105 – 95)], *i.e.*, 25 minutes.

Example 10.6. *There are five jobs, each of which must go through two machines A and B in the order AB. The processing times (in hours) are given below:*

Job	*1*	*2*	*3*	*4*	*5*
Machine A (in hours)	*5*	*1*	*9*	*3*	*10*
Machine B (in hours)	*2*	*6*	*7*	*8*	*4*

Determine the sequence for five jobs that will minimize the total elapsed time T. Find T and also idle time on machine B. ***[MCA, Panjab Univ. 2007]***

Solution. The smallest processing time on the two machines is 1, that corresponds to machine *A*. Therefore, job 2 will be processed in the beginning. This is shown below:

2				

After job 2 has been set for processing in the beginning, we are left with four jobs and their processing times are given below:

Job	*1*	3	4	5
Machine *A*	5	9	3	10
Machine *B*	2	7	8	4

The smallest processing time in this reduced problem is '2' which corresponds to machine *B*, and therefore, job *1* will be processed in the last. After deleting job *1*, the smallest processing time in the reduced problem is '3' which corresponds to machine *A*. Therefore, job *4* will be processed after job 2 as shown below:

2	4			*1*

After deleting job *4*, now, we observe that the smallest processing time is '4' in the reduced problem. Since, this processing time corresponds to machine *B*, job *5* will be processed next to last. The only job remaining, now, is job *3*; so it shall be assigned in the only space left, *i.e.*, after job *4*. The optimal sequence, therefore, is:

$$2 \rightarrow 4 \rightarrow 3 \rightarrow 5 \rightarrow 1$$

The total minimum elapsed time for machines *A* and *B* is computed in the table below:

Job	*Machine A*		*Machine B*	
	Time in	*Time out*	*Time in*	*Time out*
2	0	1	1	7
4	1	4	7	15
3	4	13	15	22
5	13	23	23	27
1	23	28	28	30

From the above computations, it is apparent that minimum total elapsed time is 30 hours and idle time for machine *B* is {1 + (23 – 22) + (28 – 27)}, *i.e.*, 3 hours.

Example 10.7. *A manufacturing company processes 6 different jobs on two machines A and B. Number of units of each job and its processing times on machine A as well as on machine B are given below. Find the optimal sequence, total minimum elapsed time and idle time for each machine.*

Job	*1*	*2*	*3*	*4*	*5*	*6*
Number of units of each job	3	4	2	5	2	3
Machine A (in minutes)	5	16	6	3	9	6
Machine B (in minutes)	8	7	11	5	7.5	14

[B.Com., Panjab Univ. 2004]

Solution. Smallest processing time on machine *A* and machine *B* is of '3' minutes, which corresponds to job *4* on machine *A*. Therefore, job *4* is processed in the beginning. After assigning this job, smallest processing time in the reduced problem is '5' which corresponds to job *1*. Therefore, job *1* is processed after job *4*. Repeating the procedure, we get the following assignments:

4					

,

4	1				

,

4	1	3	6		

,

4	1	3	6	5	2

.

The optimal sequence, therefore, is:

$$4 \rightarrow 1 \rightarrow 3 \rightarrow 6 \rightarrow 5 \rightarrow 2.$$

Total minimum elapsed time on both the machines *A* and *B* is computed in the following table:

Job	*Number of units of the job*	*Machine A*		*Machine B*	
		Time in (minutes)	*Time out (minutes)*	*Time in (minutes)*	*Time out (minutes)*
4	1st unit	0	3	3	8
	2nd unit	3	6	8	13
	3rd unit	6	9	13	18
	4th unit	9	12	18	23
	5th unit	12	15	23	28
1	1st unit	15	20	28	36
	2nd unit	20	25	36	44
	3rd unit	25	30	44	52
3	1st unit	30	36	52	63
	2nd unit	36	42	63	74
6	1st unit	42	48	74	88
	2nd unit	48	54	88	102
	3rd unit	54	60	102	116
5	1st unit	60	69	116	123.5
	2nd unit	69	78	123.5	131
2	1st unit	78	94	131	138
	2nd unit	94	110	138	145
	3rd unit	110	126	145	152
	4th unit	126	142	152	159

From the above table, we observe that the minimum total elapsed time is of 159 minutes. Idle time on machine *A* is of (159 – 142), *i.e.*, 17 minutes and idle time on machine *B* is only 3 minutes.

PROBLEMS

10.07. Use Gantt Chart technique to obtain the sequence in which three jobs J_1, J_2 and J_3 should be performed on two machines *A* and *B* for completion in the minimum time. The processing on machines is to be done in the order *AB* and the timings (in hours) are given below:

Job	*Processing time (hours)*	
	Machine A	*Machine B*
J_1	5	7
J_2	6	3
J_3	4	6

10.08. In a factory, there are six jobs to perform, each of which should go through two machines *A* and *B*, in the order *AB*. The processing timings (in hours) for the jobs that would minimize the total elapsed time is *T*. What is the value of *T* ?

Job	J_1	J_2	J_3	J_4	J_5	J_6
Machine *A*	1	3	8	5	6	3
Machine *B*	5	6	3	2	2	10

[MBA, Madras Univ. 2004]

10.09. A company has six jobs on hand coded '*A*' to '*F*'. All jobs have to go through two machines '*M I*' and '*M II*'. The time required for each job on each machine, in hours, is given below:

Job	*A*	*B*	*C*	*D*	*E*	*F*
M I	3	12	18	9	15	6
M II	9	18	24	24	3	15

Draw a sequence table of scheduling the six jobs on two machines in order *M I* → *M II*.

[MBA, Panjab Univ. 2005]

10.10. In a factory, there are six jobs to process, each of which should go to machines *A* and *B* in the order *AB*. The processing timings (in minutes) are given below. Determine the optimal schedule and total elapsed time.

Job	J_1	J_2	J_3	J_4	J_5	J_6
Machine *A*	5	10	8	9	6	12
Machine *B*	7	8	13	7	11	10

[MBA (Nov.), Madras Univ. 2006]

10.11. Seven jobs are to be processed on two machines *A* and *B* in the order *A* → *B*. Each machine can process only one job at a time. The processing time (in hours) are given below:

Job	1	2	3	4	5	6	7
Machine *A*	10	12	13	7	14	5	16
Machine *B*	15	11	8	9	6	7	16

Suggest optimum sequence of processing the jobs and total elapsed time.

[M.Com., Delhi Univ. 2006]

10.12. A ready-made garment company has to process 7 items through two stages of production, *viz.*, cutting and sewing. The time taken by each of these items at different stages are given as follows:

Item	1	2	3	4	5	6	7
Cutting	4	8	3	5	5	12	7
Sewing	3	5	6	4	8	5	8

Find an order in which these seven items are to be processed so as to minimize the total processing time. ***[M.Com., Delhi Univ. 2010]***

10.13. Seven jobs are to be performed. Each of which has to go through machine *A* and machine *B* in the order *AB*. Processing time (in hours) is given as under:

Job	1	2	3	4	5	6	7
Machine *A*	3	12	15	6	10	11	9
Machine *B*	8	10	10	6	12	1	3

Determine the sequence (Schedule) of the jobs that will minimize the total elapsed time.

[B.Tech. (Comp. Sc.) (Dec.), Hyderabad Univ. 2011]

10.14. Find the sequence that minimizes the total elapsed time required to complete the following tasks on two machines, when the tasks are done first on Machine *I* and then on Machine *II*:

Task	T_1	T_2	T_3	T_4	T_5	T_6	T_7	T_8	T_9
Machine *I*	2	5	4	9	6	8	7	5	4
Machine *II*	6	8	7	4	3	9	3	8	11

[MBA, Hyderabad Univ. 2012; BBA, GGSIP Univ. 2010]

10.15. A book binder has one printing press, one binding machine, and the manuscripts of a number of different books. The time required to perform the printing and binding operations for each book are shown below. Determine the order in which books should be processed, in order to minimize the total time required to turn out all the books.

Book	1	2	3	4	5	6
Printing time (hours)	30	120	50	20	90	110
Binding time (hours)	80	100	90	60	30	10

[MBA, UP Tech. Univ. 2008]

10.16. A company plans to fill six positions. Since, the positions are known to vary considerably with respect to skill and responsibility, different types of aptitude tests and interviews are required for each. While the aptitude tests are conducted by people from the clerical positions, the job interviews are held by the personnel from the management cadre. The job interviews immediately follow the aptitude test. The time required (in minutes) by each of the positions are given as follows:

Position	P_1	P_2	P_3	P_4	P_5	P_6
Aptitude test	140	180	150	200	170	100
Job interview	70	120	110	80	100	90

In order to minimize the waiting time of the management personnel, what should be the order of filling the positions? ***[MBA, UP Tech. Univ. 2004]***

10.17. A company has categorized its investment proposals into seven types. The financial analysis are needed to analyze the risk and return characteristics of these types of investment proposals. The proposals are examined by a committee for approval. The time that the analysis and the committee takes is based on the size of the project. If it is required to minimize the time taken in evaluation of the seven proposals, how should they be scheduled? The requirements of time are given below:

Investment	*A*	*B*	*C*	*D*	*E*	*F*	*G*
Analysis (by analysts)	8	5	10	8	14	10	7
Evaluation (by committee)	5	3	7	4	8	6	5

What is the time needed for evaluation?

10.18. The maintenance crew of a factory is divided in two parts C_1 and C_2 which cares for the maintenance of seven machines $M_1, M_2, ..., M_7$ in the factory. Crew C_1 is responsible for replacement of the parts which are worn out while crew C_2 oils and resets the machines back for operation. The maintenance time (in hours) required by crews C_1 and C_2 on different machines is given below:

Machines	M_1	M_2	M_3	M_4	M_5	M_6	M_7
Crew C_1	24	14	22	20	30	28	16
Crew C_2	10	20	8	18	10	6	4

In what order the machines should be handled by crew C_1 and C_2 so that the total time taken is minimized. Also, determine what shall this time be?

10.19. In the machine Shop, 8 different products are being manufactured each requiring time (in minutes) on two machines *A* and *B* as given below:

Product	*I*	*II*	*III*	*IV*	*V*	*VI*	*VII*	*VIII*
Machine *A*	30	45	15	20	80	120	65	10
Machine *B*	20	30	50	35	36	40	50	20

Decide the optimum sequence of processing of different products in order to minimize the total manufacturing time for all the products. Name and discuss the scheduling model used.

[PG Dip. in Management, AIMA, June, 1996]

Three-Machine Flow Shop

General solution method is not available (except enumeration, of course) for processing of *n* jobs through three machines. However, we do have a method applicable under the condition that no passing of jobs is permissible (that is to say, the same order over each machine is maintained), and if ***either*** or ***both*** of the conditions stipulated below is/are satisfied.

Let there be *n* jobs, each of which is to be processed through three machines *A*, *B* and *C* in the order *ABC*. An optimum solution to this problem can be obtained, if either or both of the following two conditions hold:

(*i*) minimum $A_i \geq$ maximum B_i, for $i = 1, 2, ..., n$ and/or

(*ii*) minimum $C_i \geq$ Maximum B_i, for $i = 1, 2, ..., n$.

The steps of the method can be summarized as below:

Step 1. Replace the given problem with an equivalent problem involving *n* jobs with 2 fictitious machines *G* and *H*, and define the corresponding processing times G_i and H_i as follows:

(*a*) $G_i = A_i + B_i \ (i = 1, 2, ..., n)$, and (*b*) $H_i = B_i + C_i \ (i = 1, 2, ..., n)$

That is, the processing times on machine *G* is the sum of the processing times on machines *A* and *B*; while the processing times on machine *H* is the sum of the processing times on machines *B* and *C*.

Step 2. To the problem obtained in *Step 1*, apply the method discussed earlier for the processing of n jobs through 2 machines. The optimum sequence resulting from this shall also be optimum for the given (original) problem.

Remark. If the processing times of the fictitious machines are equal, *i.e.*, $G_i = H_i$ $(i = 1, 2, ..., n)$, then there will be n ! optimum sequences, each of which will yield same minimum total elapsed time.

K-Machine Flow Shop

Since, no general solution method is available for processing of n jobs through K-machines, the method of solving n jobs through three machines is further extended to K-machines.

Let there be n jobs, each of which is to be processed through machines $M_1, M_2, ..., M_K$ in the order $M_1 M_2 M_3 ... M_K$. The processing time for each job on each machine is known and is given by M_{ij}, such that

$$M_{ij} = \text{Processing time for } i^{\text{th}} \text{ job on machine } j,$$

where, $i = 1, 2, ..., n$ and $j = 1, 2, ..., K$.

An optimum solution to the problem can be obtained, if either or both of the following two conditions is satisfied:

(*i*) minimum $M_{i1} \geq$ maximum M_{ij} and (*ii*) minimum $M_{iK} \geq$ maximum M_{ij},

where, $j = 2, 3, ..., K - 1$ and $i = 1, 2, ..., n$.

In this way, we convert the K-machine problem into a 2-machine problem by introducing fictitious machines G and H, such that

$$G_i = M_{i1} + M_{i2} + ... + M_{iK-1}, \text{ and } H_i = M_{i2} + M_{i3} + ... + M_{iK}, \text{ for } i = 1, 2, ..., n.$$

Thus, the processing times on machine G is the sum of the processing times on machines M_1, $M_2, M_3, ..., M_{K-1}$; while the processing times on machine H is the sum of the processing times on machines $M_2, M_3, ..., M_K$.

Next Step. Apply the method of n job and 2-machine in determining the optimum sequence. The optimum sequence (or order) resulting from this shall also be the optimum for the original K-machine and 'n' job problem.

Remarks

1. If neither of the two conditions given above is satisfied, the method of finding optimum sequence (schedule) fails.
2. In addition to the two conditions given above, if $M_{i2} + M_{i3} + ... + M_{iK-1} =$ constant for $i = 1, 2, ..., n$, then determine the optimum sequence for n jobs and two machines M_1 and M_K in the order $M_1 M_K$.
3. In addition to the two conditions given above, if $M_{i1} = M_{iK}$ and $G_i = H_i$ for $i = 1, 2, ..., n$; then there will be n ! optimum sequences, each of which will yield minimum total elapsed time.
4. The above-mentioned procedure of solving the sequencing problems is applicable only to those sequencing problems in which the minimum cost (or time) of processing the jobs through first and/or last machine is greater than or equal to the cost (time) of processing the jobs through mediocre machines. There are many industrial operations, in which the machines are set in the order which does not obey this rule.

Example 10.8. *A book binder has one printing press, one binding machine and the manuscripts of a number of different books. The times required to perform the printing and binding operations for each book are known. Determine the order in which the books should be processed so as to minimize the total time required to process all the books. Find also the total time required.*

Book	*Processing time (in minutes)*				
	1	2	3	4	5
Printing time	40	90	80	60	50
Binding time	50	60	20	30	40

Suppose that an additional operation is added to the process described above, viz., finishing. The time required for this additional operation are given below:

Book	*Finishing time (in minutes)*				
	1	*2*	*3*	*4*	*5*
Finishing time	*80*	*100*	*60*	*70*	*110*

What is the order in which the books should be processed? Find also the minimum total elapsed time.

[BE, Karnataka Univ. 2012; MBA (June), Visvesvaraya Tech. Univ. 2011]

Solution. Since, the books first go for printing and then to the binder, therefore these will be processed in the order P_iB_i, where P_i denotes the processing time (in minutes) on printing press and B_i the processing time (in minutes) on the binding machine.

Using the method of processing 5 jobs through 2 machines, the optimal sequence is:

1	2	5	4	3

For minimum total elapsed time, we have

Book	*Printing machine*		*Binding machine*	
	Time in	*Time out*	*Time in*	*Time out*
1	0	40	40	90
2	40	130	130	190
5	130	180	190	230
4	180	240	240	270
3	240	320	320	340

Above table indicates that the minimum total elapsed time is 340 minutes (or 5 hours and 40 minutes), while idle time on printing machine is (340 – 320), *i.e.*, 20 minutes and on binding machine is {40 + (130 – 90) + (240 – 230) + (320 – 270)}, *i.e.*, (40 + 40 + 10 + 50) or 140 minutes.

Now, in addition to printing and binding machines, one more machine named 'finishing machine is introduced. Thus, the problem becomes that of 5 jobs on three machines with the processing times (in minutes) as given below:

Book	*1*	*2*	*3*	*4*	*5*
P_i	40	90	80	60	50
B_i	50	60	20	30	40
F_i	80	100	60	70	110

Here, minimum P_i = 40, minimum F_i = 60 and maximum B_i = 60. Since, the condition minimum $F_i \geq$ maximum B_i is satisfied, the problem can be converted into that of 5 jobs through 2 machines. Thus, if G and H are two fictitious machines, such that

$$G_i = P_i + B_i \text{ and } H_i = B_i + F_i; \quad (i = 1, 2, ..., 5),$$

the given problem can be rewritten as the problem of two machines and 5 books with their processing times as shown below:

Book	*1*	*2*	*3*	*4*	*5*
G_i	40 + 50 = 90	90 + 60 = 150	80 + 20 = 100	60 + 30 = 90	50 + 40 = 90
H_i	50 + 80 = 130	60 + 100 = 160	20 + 60 = 80	30 + 70 = 100	40 + 110 = 150

Using, now, the method of processing 5 jobs through 2 machines, the following optimal sequence (or schedule) can easily be obtained:

4	1	5	2	3

For minimum total elapsed time, we compute the following:

Book		4	1	5	2	3
Printing machine	In	0	60	100	150	240
	Out	60	100	150	240	320
Binding machine	In	60	100	150	240	320
	Out	90	150	190	300	340
Finishing machine	In	90	160	240	350	450
	Out	160	240	350	450	510

This table indicates that the minimum total elapsed time to complete all books is 510 minutes with idle time of (510 – 320), *i.e.*, 190 minutes on printing machine; {60 + (100 – 90) + (240 – 190) + (320 – 300) + (510 – 340)}, *i.e.*, (60 + 10 + 50 + 20 + 170) or 310 minutes on binding machine; and 90 minutes on finishing machine.

Example 10.9. *Six jobs have to be processed on three machines A, B and C in the order ACB. The time taken by each job on each machine (in hours) is given below. Each machine can process only one job at a time.*

Job	J_1	J_2	J_3	J_4	J_5	J_6
Machine A	12	8	7	11	10	5
Machine B	7	10	9	6	10	5
Machine C	3	4	2	5	1.5	4

Determine the sequence so as to minimize the processing time. **[B.Com., Guru Nanak Dev Univ. 2010]**

Solution. Here, minimum $A_i = 5$, minimum $B_i = 5$, and maximum $C_i = 5$.

Since, minimum A_i = maximum C_i and also minimum B_i = maximum C_i; both the conditions (*i*) and (*ii*) of the method are satisfied. So, we can convert the given problem into 'six' job and 'two' machine problem by considering *G* and *H* as two fictitious machines, where $G_i = A_i + C_i$ and $H_i = B_i + C_i$.

The, new, modified problem is:

Job	J_1	J_2	J_3	J_4	J_5	J_6
Machine *G*	12 + 3 = 15	8 + 4 = 12	7 + 2 = 9	11 + 5 = 16	10 + 1.5 = 11.5	5 + 4 = 9
Machine *H*	7 + 3 = 10	10 + 4 = 14	9 + 2 = 11	6 + 5 = 11	10 + 1.5 = 11.5	5 + 4 = 9

Using, now, the method of obtaining optimal sequence of *n* jobs through 2 machines, we get

$$J_3 \to J_6 \to J_5 \to J_2 \to J_4 \to J_1$$

Using the optimal sequence, the total elapsed time is computed as given below:

Job	*Machine A*		*Machine C*		*Machine B*	
	In	*Out*	*In*	*Out*	*In*	*Out*
J_3	0	7	7	9	9	18
J_6	7	12	12	16	18	23
J_5	12	22	22	23.5	23.5	33.5
J_2	22	30	30	34	34	44
J_4	30	41	41	46	46	52
J_1	41	53	53	56	56	63

From above computations, we observe that the minimum total processing time is 63 hours. Idle time on Machine $A = (63 - 53) = 10$ hours,

Machine $B = 9 + (23.5 - 23) + (34 - 33.5) + (46 - 44) + (56 - 52) = 16$ hours,

and Machine $C = 7 + (12 - 9) + (22 - 16) + (30 - 23.5) + (41 - 34) + (53 - 46) + (63 - 56) = 43.5$ hours.

Example 10.10. *We have 4 jobs each of which has to go through the machines M_j ($j = 1, 2, ..., 6$) in the order $M_1, M_2, ..., M_6$. Processing time (in hours) is given below:*

	Machines					
	M_1	M_2	M_3	M_4	M_5	M_6
Job A	18	8	7	2	10	25
Job B	17	6	9	6	8	19
Job C	11	5	8	5	7	15
Job D	20	4	3	4	8	12

Determine a sequence of these four jobs that minimizes the total elapsed time.

[M.Com., CCS Univ. 1998]

Solution. Here, minimum $M_{i1} = 11$ and minimum $M_{i6} = 12$; maximum of M_{i2}, M_{i3}, M_{i4} and M_{i5} are 8, 9, 6 and 10 respectively.

Since, the conditions minimum $M_{i1} \geq$ maximum M_{ij} and minimum $M_{i6} \geq$ maximum M_{ij}, for ($i = 1, 2, 3, 4$) are satisfied, the given problem can be written as under:

Job	A	B	C	D
Machine *G*	45	46	36	39
Machine *H*	52	48	40	31

Here, $G_i = t_{i1} + t_{i2} + t_{i3} + t_{i4} + t_{i5}$ and $H_i = t_{i2} + t_{i3} + t_{i4} + t_{i5} + t_{i6}$

where, $i = 1, 2, 3, 4$.

Using the method for optimum sequencing the following optimal sequence can easily be obtained:

C	A	B	D

The total elapsed time is computed in the following table:

Job	*Machines*					
	M_1	M_2	M_3	M_4	M_5	M_6
C	0-11	11-16	16-24	24-29	29-36	36-51
A	11-29	29-37	37-44	44-46	46-56	56-81
B	29-46	46-52	52-61	61-67	67-75	81-100
D	46-66	66-70	70-73	73-77	77-85	100-112

This table shows that the minimum total elapsed time is 112 hours.

Example 10.11. *Solve the following sequencing problem in the order ABCD, when passing is not allowed:*

Item	*Machine (Processing time in hours)*			
	A	B	C	D
I	15	5	4	15
II	12	2	10	12
III	16	3	5	16
IV	17	3	4	17

[BE (Comp. Sc.), Panjab Univ. 2008]

Solution. Here, minimum $A_i = 12$, maximum $B_i = 5$, maximum $C_i = 10$ and minimum $D_i = 12$.

Since, minimum $A_i \geq$ maximum B_i and maximum C_i, the given problem can be rewritten as:

Item	*I*	*II*	*III*	*IV*
Machine *G*	24	24	24	24
Machine *H*	24	24	24	24

Let $G_i = A_i + B_i + C_i$ and $H_i = B_i + C_i + D_i$.

Since, $G_i = H_i$ and $A_i = D_i$ are satisfied, therefore using the method for optimal sequencing, we get 4 ! (= 24) sequences, each giving us an optimal sequence:

I II III IV, *I II IV III*, *I III II IV*, *I III IV II*, *I IV II III*, etc.

Each of the above optimal sequences will yield us the same total elapsed time. For the elapsed time, we have

Item		*I*	*II*	*III*	*IV*
Machine *A*	In	0	15	27	43
	Out	15	27	43	60
Machine *B*	In	15	27	43	60
	Out	20	29	46	63
Machine *C*	In	20	29	46	63
	Out	24	39	51	67
Machine *D*	In	24	39	51	67
	Out	39	51	67	84

Above table shows that the minimum total elapsed time is 84 hours.

PROBLEMS

10.20. Consider the following 3 machines and 5 jobs flow shop problem. Check whether Johnson's rule can be extended to this problem. If so, what is the optimal schedule and the corresponding total time when passing is not allowed.

Job	*Machine 1*	*Machine 2*	*Machine 3*
1	11	10	12
2	13	8	20
3	15	6	15
4	12	7	19
5	20	9	7

[MBA, Madras Univ. 2006]

10.21. Determine the optimal sequence of jobs that minimizes the total elapsed time in hours based on the following information. Processing time in hours on machines are given in hours and passing is not allowed, *i.e.*, in the order *ABC*:

Job	1	2	3	4	5
Machine *A*	3	8	7	5	2
Machine *B*	3	4	2	1	5
Machine *C*	5	8	10	7	6

[M.Com., Delhi Univ. 1996]

10.22. Six jobs have to be processed on three machines *A*, *B* and *C* in the order *ABC*, the time taken by each job on each machine is indicated below. Each machine can process only one job at a time.

Job	*Processing time (in hours)*					
	J_1	J_2	J_3	J_4	J_5	J_6
Machine *A*	12	8	7	11	10	5
Machine *B*	7	10	9	6	10	5
Machine *C*	3	4	2	5	1.5	4

Determine the sequence so as to minimize the processing time.

[MCA, Guru Nanak Dev Univ. 2010]

10.23. Find the sequence that minimizes the total time required in performing the following jobs on three machines in the order *ABC*:

Job	*Processing time (in hours)*					
	1	2	3	4	5	6
Machine *A*	8	3	7	2	5	1
Machine *B*	3	4	5	2	1	6
Machine *C*	8	7	6	9	10	9

10.24. We have seven jobs, each of which must go through machines *A*, *B* and *C* in the order *ABC*. Processing times (in hours) are given in the following table:

Job	1	2	3	4	5	6	7
Machine *A* (A_i)	7	8	11	14	21	17	8
Machine *B* (B_i)	6	3	1	2	5	4	1
Machine *C* (C_i)	10	9	15	13	18	11	9

Determine the optimal sequence and total elapsed time associated with it.

[MBA (PT.), Delhi Univ. 2009]

10.25. A ready-made garments manufacturer has to process 7 items through two stages of production, *viz.*, cutting and sewing. The time taken for each of these items at the different stages are given below in appropriate units:

Item		1	2	3	4	5	6	7
Processing time	Cutting	5	7	3	4	6	7	12
	Sewing	2	6	7	5	9	5	8

(*a*) Find an order in which these items are to be processed through these stages so as to minimize the total processing time.

(*b*) Suppose a third stage of production is added, *viz.*, pressing and packing, with processing time for these items as follows:

Item	1	2	3	4	5	6	7
Processing time (Pressing and packing)	10	12	11	13	12	10	11

Find an order in which these seven items are to be processed so as to minimize the time taken to process all the items through all the three stages.

10.26. Determine the optimal sequence of jobs that minimizes the total elapsed time based on the following information:

Job	*1*	*2*	*3*	*4*	*5*	*6*	*7*
Machine *A*	10	8	12	6	9	11	9
Machine *B*	6	4	6	5	3	4	2
Machine *C*	8	7	5	9	10	6	5

Processing time on machines is given in hours and passing is not allowed.

[M.Com., Delhi Univ. 2008]

10.27. Determine the optimal sequence of jobs in the order $M_1M_2M_3$ that minimizes the total elapsed time based on the following information. Processing time on machines is given in hours and passing is not allowed:

Job	*A*	*B*	*C*	*D*	*E*	*F*	*G*
Machine M_1	3	8	7	4	9	8	7
Machine M_2	4	3	2	5	1	4	3
Machine M_3	6	7	5	11	5	6	12

[MBA (Dec.), Visvesvaraya Tech. Univ. 2011]

10.28. A firm works 40 hours a week and has a capacity of overtime work to the extent of 20 hours in a week. it has received seven orders to be processed on three machines *A*, *B* and *C*, in the order *ABC* to be delivered in a week's time from now. The process times (in hours) are recorded in the following table:

Job	*1*	*2*	*3*	*4*	*5*	*6*	*7*
Machine *A*	7	8	6	6	7	8	5
Machine *B*	2	2	1	3	3	2	4
Machine *C*	6	5	4	4	2	1	5

The manager, who, insists on performing the jobs in the sequence in which they are received, is refusing to accept an eighth order, which requires 7, 2 and 5 hours on machines *A*, *B* and *C*, respectively, because according to him the eight jobs would require a total of 61 hours for processing, that exceeds the firm's capacity. What is your advice to him.

10.29. Find the sequence, for the following eight jobs, that will minimize the total elapsed time for the completion of all the jobs. Each job is processed in the order *CAB*. Entries give the time in hours on the machines *A*, *B* and *C*.

Job	*I*	*II*	*III*	*IV*	*V*	*VI*	*VII*	*VIII*
Machine *A*	4	6	3	4	5	3	6	2
Machine *B*	8	10	7	8	11	8	9	13
Machine C	5	6	2	3	4	9	15	11

[BE (Dec.), Karnataka Univ. 2010; MBA, Bangalore Univ. 2017]

10.30. Solve the following sequencing problem, giving an optimal solution when passing is not allowed:

Job	*Machine (Processing time in hours)*			
	M_1	M_2	M_3	M_4
1	20	3	3	25
2	12	5	1	11
3	18	4	2	10
4	17	2	4	28

[MBA, Madras Univ. 1996]

10.31. When passing is not allowed, solve the sequencing problem giving an optimal solution:

Job	Machine (Processing time in hours)			
	M_1	M_2	M_3	M_4
A	20	10	9	20
B	17	7	15	17
C	21	8	10	21
D	25	5	9	25

[BE (Comm. & Elect.), Panjab Univ. 2008]

10.32. Solve the following sequencing problem when passing is not allowed:

Item	Machine				
	A	B	C	D	E
I	9	7	4	5	11
II	8	8	6	7	12
III	7	6	7	8	10
IV	10	5	5	4	8

[MBA, Madras Univ. 2010; MBA, Bangalore Univ. 2018]

10.33. Six jobs are to be processed on four machines *A, B, C* and *D* in that order. Using the processing time (in hours) given below, determine the optimal sequence(s) of the job performance:

Job	1	2	3	4	5	6
Machine *A*	12	8	13	9	10	12
Machine *B*	6	3	4	2	6	7
Machine *C*	6	6	6	4	7	2
Machine *D*	8	12	9	14	12	10

10.34. Solve the following sequencing problem, giving an optimal solution when passing is not allowed. Processing time on machine M_1, M_2, M_3 and M_4 is given in hours.

Machine	Job				
	A	B	C	D	E
Machine M_1	10	12	8	15	16
Machine M_2	3	2	4	1	5
Machine M_3	5	6	4	7	3
Machine M_4	14	7	12	8	10

10:8. PROCESSING OF TWO JOBS THROUGH '*m*' MACHINES

Let there be any two jobs, say *1* and *2* to be processed through '*m*' machines, say *A, B, C, ..., K*. The technological order of each of the two jobs through '*m*' machines is known. Also, this ordering may not be the same for both jobs. The actual processing times are known to be: $A_1, A_2; B_1, B_2; ..., K_1, K_2$ and each machine can work only one job at a time. The objective of the problem is to minimize the total elapsed time *T*, *i.e.*, to minimize the time from the start of first job to the completion of last job.

For solving such a problem, we present a graphical procedure as follows:

Step 1. Construct a two-dimensional graph, where the processing times corresponding to the first job are shown on the horizontal axis (*i.e.*, *X*-axis) and the processing times corresponding to second job are shown on vertical axis (*i.e.*, *Y*-axis). On each axis, mark off the operation times in the order in which the operations must be performed for that job.

Step 2. Shade the area corresponding to each machine at the intersection of the intervals marked for that machine on the two axis.

Step 3. Determine a path that starts at the lower left corner (*i.e.*, origin) and stops at the upper right corner, while avoiding the shaded area in the graph. In other words, the path is not allowed to pass through shaded areas which correspond to operating both jobs concurrently on the same machine.

Step 4. For determining a path mentioned in *Step* 3, we observe that

(*i*) The processing of both jobs are represented by a continuous path consisting of horizontal, vertical and 45 degree (*i.e.*, 45°) diagonal segment.

(*ii*) The horizontal movement indicates the progress of job *1*, while job 2 remains idle.

(*iii*) The vertical movement indicates the progress of job 2, while job *1* remains idle.

Step 5. The path (obtained in *Step* 3) with minimum vertical distance indicates the optimal solution. Note that this will be the same as the path with minimum horizontal distance.

Step 6. Determine the total elapsed time. This is being done as follows:

Elapsed time for job *1* = Processing time for job *1* + Idle time for job *1*.

Elapsed time for job 2 = Processing time for job 2 + Idle time for job 2.

Example 10.12. *Use graphical method to determine the minimum time needed to process two jobs on four machines A, B, C and D. The technological order for these jobs on machines is as follows:*

Job 1 :	A	B	C	D
Job 2 :	D	B	A	C

Processing time (in hours) are given as follows:

	A	B	C	D
Job 1 :	4	6	7	3
Job 2 :	4	7	5	8

Solution.

Step 1. Draw two perpendicular lines, horizontal one representing the processing time of job *1* on different machines, while job 2 remains idle; and vertical one representing the processing time of job 2, while job *1* remains idle.

Step 2. Mark the processing times for job *1* and job 2 on the horizontal and vertical lines respectively according to the given order of machines as shown in *Fig. 10.9*.

Step 3. Construct various blocks starting from origin by pairing the same machines until a point marked 'finish' is reached.

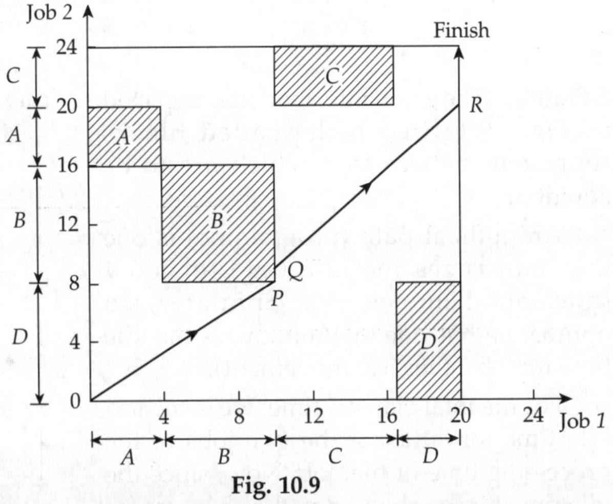

Fig. 10.9

Step 4. Draw a line starting from the origin to the point marked 'finish' by moving horizontally, vertically and diagonally along a line which makes an angle of 45° with the horizontal line. The horizontal segment of this line indicates that job *1* is under process while job 2 is idle. Similarly, the vertical segment of the line indicates that job 2 is under process while job *1* is idle. The diagonal segment of the line shows that both the jobs are under process simultaneously.

Since, simultaneous processing of both the jobs on a machine is not possible, diagonal movement is not allowed to pass through block *B* (shaded area).

Step 5. An optimal path is one that minimizes the idle time for both the jobs. Thus, we must choose that path on which diagonal movement is a maximum.

Step 6. The total elapsed time is obtained by adding the idle time for either job to the processing time for that job. In the present problem, the idle time for the chosen path is found to be 4 hours for job *1*, while job 2 remains busy till the completion of work. The total elapsed time is obtained as follows:

Elapsed time for job *1* = Processing time of job *1* + Idle time for job *1* = 20 + (2 + 2) = 24 hours.

Elapsed time for job 2 = Processing time of job 2 + Idle time for job 2 = 24 + 0 = 24 hours.

Using Gantt Chart, the scheduling problem is solved as under:

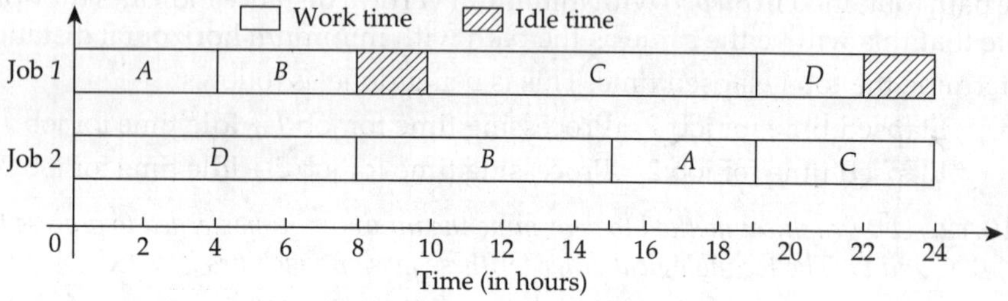

Fig. 10.10. Gantt Chart

According to above Gantt Chart, the total time required to complete the schedule is 24 hours, the minimum.

Example 10.13. *Use graphical method to minimize the time needed to process the following jobs on the machines shown, i.e., for each machine find the job which should be done first. Also, calculate the total elapsed time to complete both the jobs.*

Job 1	*Sequence*	:	*A*	*B*	*C*	*D*	*E*
	Time	:	3	4	2	6	2
Job 2	*Sequence*	:	*B*	*C*	*A*	*D*	*E*
	Time	:	5	4	3	2	6

[MBA, Panjab Univ. 2010]

Solution. The given information is depicted in *Fig. 10.11* in which shaded blocks represent the overlaps which are to be avoided.

An optimal path (programme) is one that minimizes the idle time for job *1* (horizontal movement). Similarly, an optimal path is one that minimizes the idle time for job 2 (vertical movement).

For the total elapsed time, we shall add idle time for either of the two jobs to the processing time of that job. Now, since the idle time for the choosen path is 5 hours for job *1* and 2 hours for job *2*, the total elapsed time is obtained as follows:

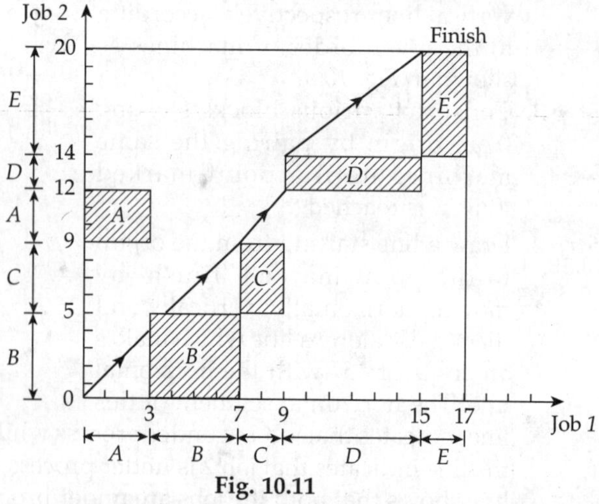

Fig. 10.11

Processing time of job *1* + idle time for job *1* = 17 + (2 + 3) = 22 hours.

Processing time of job 2 + idle time for job 2 = 20 + 2 = 22 hours.

Using Gantt Chart, the scheduling problem is solved as under:

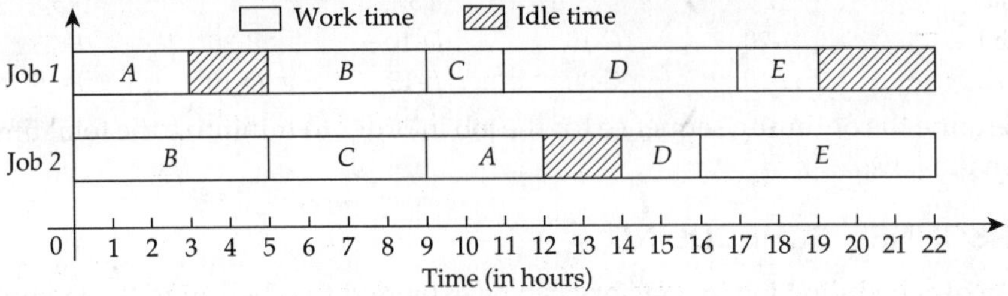

Fig. 10.12. Gantt Chart

From the above Chart, we observe that the total time required to complete the schedule is 22 hours, the minimum.

PROBLEMS

10.35. Using graphical method, calculate the minimum time needed to process the following jobs on the machines shown below, *i.e.*, for each machine find the job which should be processed first. Also, calculate the total time needed to complete both the jobs.

Job *1*	Sequence :	*B*	*C*	*D*	*E*
	Time (in hours) :	4	6	8	12
Job 2	Sequence :	*A*	*D*	*E*	*B*
	Time (in hours) :	8	10	6	4

10.36. Solve the following problem graphically:

Job *1*	Sequence :	*A*	*B*	*C*	*D*	*E*
	Time (in hours) :	1	2	3	5	1
Job 2	Sequence :	*C*	*A*	*D*	*E*	*B*
	Time (in hours) :	3	4	2	1	5

[MBA (Dec.), Panjab Tech. Univ. 2014]

10.37. A machine shop has four machines *A*, *B*, *C* and *D*. Two jobs must be processed through each of these four machines. The time (in hours) taken on each of the machines and the necessary sequence of jobs through the shop are given below:

Job *1*	Sequence :	*C*	*A*	*E*	*F*	*D*	*B*
	Time (in hours) :	2	3	4	5	6	1
Job 2	Sequence :	*B*	*A*	*E*	*F*	*C*	*D*
	Time (in hours) :	3	2	5	3	2	3

Use graphic method to obtain the total minimum elapsed time. ***[M.Com., Panjab Univ. 2004]***

10.38. Use graphic method to find the total minimum elapsed time sequence of 2 jobs through 5 machines, when we are given the following information:

				Machines		
Job *1*	Sequence :	*A*	*B*	*C*	*D*	*E*
	Time (in hours) :	6	8	4	12	4
Job 2	Sequence :	*B*	*C*	*A*	*D*	*E*
	Time (in hours) :	10	8	6	4	12

[BE, Karnataka Univ. 2011; BE (Nov.), Visvesvaraya Tech. Univ. 2020]

10.39. A machine shop has six machines *A, B, C, D, E* and *F*. Two jobs must be processed through each of the machines. Times on machines and the necessary sequence of the jobs through the shop are given below:

Order :	1	2	3	4	5	6
Job *I* :	*A*-20	*C*-10	*D*-10	*B*-30	*E*-25	*F*-16
Job *II* :	*A*-10	*C*-30	*B*-15	*D*-10	*F*-15	*E*-20

Determine the optimum sequence for the job in order to minimize the total time necessary to finish the jobs.

10:9. PERSONNEL SCHEDULING

The problem of scheduling job in a factory can be extended to scheduling the factory maintenance crews in such a way so as to minimize their idle time. Therefore, we have ***maintenance-crew scheduling*** problems.

Let a company have a set of different machines in its plant that need preventive maintenance. The crew team is divided into two groups *A* and *B*. First, crew *A* takes the machine and replaces the parts according to needs. Then, crew *B* resets the machine and puts it back into operation. If the service time of both the groups on the different machines are known, we can determine the optimal sequence in which the maintenance of jobs be done so that the idle time of crews is the least.

Like for sequencing problems given earlier, the rules when the service times are given in the rows form may be summarized as under:

(*i*) Choose the smallest of all values that appear in two rows.

(*ii*) If the value is in the first row, then the machine corresponding to the value shall be serviced by crew *A* in the beginning.

(*iii*) If it is in the second row, then that machine will be serviced by crew *A* in the end.

(*iv*) If there is a tie in the same row, either of the machines involved is selected and assigned the first, or the last, place accordingly as the tie is in the first or the second row.

(*v*) After making the assignment, cross out the machine involved.

(*vi*) Repeat steps (*i*) to (*v*) until all the assignments are made.

The resulting sequence shall ensure minimization of the idle time of maintenance crew.

Example 10.14. *The maintenance crew of a company is divided in two groups, C_1 and C_2, which cares for maintenance of the machines. Crew C_1 is responsible for replacement of parts which are worn out, while Crew C_2 oils, and resets the machine back for operation. The time (in hours) required by crews C_1 and C_2 on different machines which need working on them are as follows:*

Machine	M_1	M_2	M_3	M_4	M_5	M_6	M_7
Crew C_1	8	6	10	11	10	14	4
Crew C_2	5	3	7	12	8	6	7

In which order should the machines be handled by crew C_1 and C_2, so that the total time taken is minimum?

Solution. Treating the machines as seven items and crews as the service facilities, the optimal sequence can easily be obtained as

M_7	M_4	M_5	M_3	M_6	M_1	M_2

For the total elapsed time, we have

		M_7	M_4	M_5	M_3	M_6	M_1	M_2
Crew C_1	In	0	4	15	25	35	49	57
	Out	4	15	25	35	49	57	63
Crew C_2	In	4	15	27	35	49	57	63
	Out	11	27	35	42	55	62	66

Total time required by the crews to handle seven machines is 66 hours. The working and the idle time (in hours) of the crews C_1 and C_2 are shown in *Fig. 10.13.*

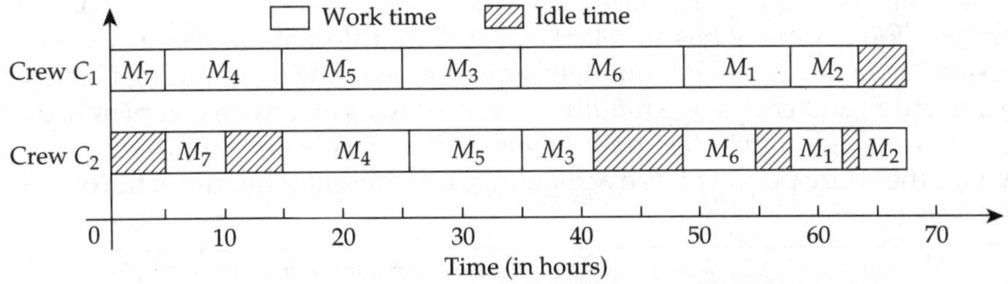

Fig. 10.13. Gantt Chart

Example 10.15. *A company has eight large machines which receive preventive maintenance. The maintenance team is divided into two crews A and B. Crew A takes the machine 'power' and replaces parts according to a given maintenance schedule. The second crew resets the machine and puts it back into operation. At all times no passing rule is considered to be in effect. The servicing times for each machine are given below:*

Machine	*a*	*b*	*c*	*d*	*e*	*f*	*g*	*h*
Crew A	5	4	22	16	15	11	9	4
Crew B	6	10	12	8	20	7	2	21

Determine the optimal sequence of scheduling the factory maintenance crews to minimize their idle time and represent it on a Gantt Chart.

Solution. Using the method of processing 8 jobs through 2 machines, an optimal sequence can easily be obtained as

b	*h*	*a*	*e*	*c*	*d*	*f*	*g*

For the total elapsed time, we have

Machine		*b*	*h*	*a*	*e*	*c*	*d*	*f*	*g*
Crew *A*	In	0	4	8	13	28	50	66	77
	Out	4	8	13	28	50	66	77	86
Crew *B*	In	4	14	35	41	61	73	81	88
	Out	14	35	41	61	73	81	88	90

Total time required by crews will be 90 hours. Idle time for crew *A* will be 4 hours and for *B* also idle time will be 4 hours. *Fig. 10.14* shows the idle time of crews *A* and *B*.

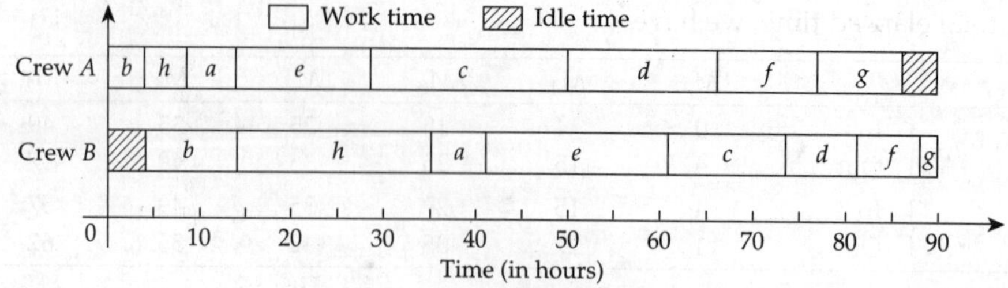

Fig. 10.14. Gantt Chart

PROBLEM

10.40. A refrigeration company has six plants located in different parts of a city. Every year, it is necessary for each plant to be completely overhauled. The overhauling is carried out in two stages *A* and *B*, and each stage requires a crew of workmen with completely different skills. The work on stage *B* can start only when stage *A* has been completed. A plant has to be closed for the entire period of its overhauling. The present schedule of the overhaul of the six plants is as follows:

Plant	*Time required by crews (days)*					
	P_1	P_2	P_3	P_4	P_5	P_6
Crew *A*	6	6	4	6	5	8
Crew *B*	4	2	10	5	3	6

(*a*) Determine the optimal sequence.

(*b*) If downtime of any of the six plants costs ₹ 800 per day for a plant, idle time for crew *A* costs ₹ 250 per day, and idle time for crew *B* costs ₹ 430 per day. Which of the two schedules, the present one and the one determined in (*a*), will be more economical? What are their respective costs?

10:10. ROUTING PROBLEMS AND SEQUENCING

Routing problems are the problems related to sequencing. Such problems usually occur in the areas of communication and transportation, where networks are used to solve this type of problems. A network involves the determination of a route to reach from one point (origin) to another point (destination) for which there exist a number of alternative paths at various stages to reach the final destination. The cost of journey (movement), which may be function of distance/cost, or time may be different for different routes (paths). The problem, then, is to find that route which gives the minimum cost. The procedure involves the determination of the cost of all the possible routes and selecting the one with least cost. But, it takes long time to decide about the best route as there are large number of alternative routes (paths). The necessity of selecting the best path (optimum route), therefore, arises.

There are many problems associated with networks, out of which the following two are most important:

1. Traveling salesman problem (already discussed in *Section 9:6* of *Chapter 9*).
2. Minimal path problem (to be discussed in detail in *Chapter 16*).

10:11. PROBLEMS OF COMPLEX SEQUENCING

We have discussed in this chapter only simple sequencing problems. But, in real life more complicated sequencing problems may exist which involve the following constraints:

Overlap. When a job consists of making a number of similar items, overlap happens. Here, a few first items coming out of one operation may go in for second operation before remaining items in the lot could go in for the first operation.

Rework. If one of the operations in sequence is an inspection operation, rework is said to have happened. The defective items are to be sent back to previous operation for rework causing either a delay or splitting the jobs into two lots.

Transportation Time. It may take a considerable amount of time to transport the items from one facility to another. The facilities may even be in different plants.

Variable Processing Time. The time required to perform a certain operation may vary from shift to shift as it frequently happens in multishift operations.

Material Shortages. The material required for performing a particular operation may run out.

Expediting. Because of urgent demand, a particular job may have to be moved out of sequence and speeded up.

Machine Breakdown. A machine may break down and operator may be absent or injured at work.

Further, alternative routes for the jobs on different machines may be permitted, where more than one machine of a given type exists. Moreover, machine times and/or costs may be of probabilistic nature. At present each of these complex situations have to be treated individually and a tailor-made solution is to be determined.

To handle these complex situations, there are two approaches. One approach is to split the problem into simpler sub-problems each of which can, then, be handled by a particular technique. This approach, however, is open to all the shortcomings to treating a system in paths, *i.e.*, of sub-optimization. This may lead to lower value of overall system effectiveness. As a rule whenever sub-optimization is unavoidable, because of inadequacy of available techniques, the operations researcher must keep his eye on the total system.

Another approach for complex sequencing problem is the use of ***Monte Carlo Technique***. This technique makes use of the probability distribution of the characteristics of the system such as processing time on each machine, availability of the machines, more than one machine of the same type, variations in working pace of operators in different shifts, alternative routes for the jobs, etc.

TEST YOUR UNDERSTANDING

1. In sequencing problems
   (*a*) the selection of an appropriate order for a series of jobs is to be done on a finite service facilities.
   (*b*) all the jobs must be processed on a first-come-first service basis.
   (*c*) a service facility can process more than one job at a time.
   (*d*) all the service facilities are not of different type.
2. In sequencing problems
   (*a*) all jobs are completely known and are ready for processing.
   (*b*) jobs are processed sequentially, *i.e.*, first on the first machine and then on the second machine, and so on.
   (*c*) total elapsed time is determined by the point of time at which the first of the *n* jobs goes to machine *A* until the time when the last job comes out of machine *B*.
   (*d*) all of the above.

3. A sequencing problem involving six jobs and three machines requires evaluation of:
   (*a*) (6 ! + 6! + 6!) sequences
   (*b*) $(6\,!)^3$ sequences
   (*c*) (6 × 6 × 6) sequences
   (*d*) (6 + 6 + 6) sequences
4. The general assumption which is not correct in solving a sequencing problem is that:
   (*a*) the time taken by different jobs in moving from one machine to another is negligible.
   (*b*) the processing times on various machines are independent of the order in which different jobs are processed on them.
   (*c*) a job once started on a machine would be performed to the point of completion uninterrupted.
   (*d*) a machine can process more than one job at a given point of time.
5. In sequencing problems
   (*a*) Gantt Charts can be used to determine optimum sequence for the relatively small-sized problems.
   (*b*) Gantt Charts provides a very effective technique for solving large sized sequencing problems.
   (*c*) a unique optimum solution exists.
   (*d*) first-come-first-served basis of job performance is the ideal one, since it involves a sense of fair play.
6. In '*n*' jobs and two machines (say *A* and *B*) sequencing problems in which the order of processing is *AB*
   (*a*) job having minimum time on machine *B* is processed first.
   (*b*) job having minimum time on machine *A* is processed in the last.
   (*c*) job having minimum time on machine *B* is processed in the last.
   (*d*) job having minimum time on machine *A* is processed in the last.
7. If A_i, B_i and C_i denote the processing times of *i*th job on three machines *A*, *B* and *C* respectively, then a *n*-job three machine problem can be reduced to an *n*-job and two machine problem, provided that:
   (*a*) Min. $A_i \geq$ Max. B_i and/or Min. $C_i \leq$ Max. B_i
   (*b*) Min. $A_i \geq$ Max. B_i and/or Min. $C_i \geq$ Max. B_i
   (*c*) Min. $A_i \leq$ Max. B_i and/or Min. $C_i \geq$ Max. B_i
   (*d*) Min. $A_i \leq$ Max. B_i and/or Min. $C_i \leq$ Max. B_i
8. For a number of jobs to process on a single machine, an optimal schedule is obtained by
   (*a*) SPT procedure
   (*b*) Moore procedure
   (*c*) Due date procedure
   (*d*) all of the above.
9. Sequencing problems involving processing of two jobs on '*n*' machines
   (*a*) can be solved graphically
   (*b*) cannot be solved graphically
   (*c*) have a condition that the processing of two jobs must be in the same order
   (*d*) none of the above.
10. Which of the following is not correct?
   (*a*) There is no general solution procedure available for solving sequencing problems involving processing of *n* jobs through three machines.
   (*b*) In case of two jobs needing processing on some machines in varying order, the problem may be solved graphically.

(*c*) Solution to a sequencing problem is found in such a manner that none of the machines is idle at any time.

(*d*) Maintenance crew scheduling problems can also be considered as sequencing problems.

ANSWERS

1. (*a*)	**2.** (*d*)	**3.** (*b*)	**4.** (*d*)	**5.** (*a*)
6. (*c*)	**7.** (*b*)	**8.** (*d*)	**9.** (*a*)	**10.** (*c*)

REVIEW EXERCISE

1. Briefly describe each of the following priority rules:
(*i*) SPT, (*ii*) FCFS, (*iii*) EDD. ***[M.Tech. (Dec.), Dayal Bagh Edu. Institute 2012]***

2. What is meant by sequencing? Explain in detail. ***[MBA, Panjab Tech. Univ. 2015]***

3. What is "no passing" rule in a sequencing algorithm?

4. Explain the principal assumptions made while dealing with sequencing problems. ***[BE, Karnataka Univ. 2011]***

5. Distinguish between "single processor scheduling" and "flow shop scheduling".

6. What is meant by mean flow time and mean tardiness?

7. Explain the various procedures of single processor scheduling. ***[BBA, GGSIP Univ. 2003]***

8. Write the assumptions for Johnson's rule. ***[MBA (Dec.), Panjab Tech. Univ. 2014]***

9. Give Johnson's procedure for n jobs on 2 machines. Give justification of the rule used in the procedure. ***[B.Tech. (Comp. Sc.), Hyderabad Univ. 2011; MCA (Feb.), Hyderabad Univ. 2012]***

10. When can we apply Johnson's algorithm for finding optimum ordering of n jobs through 3 machines sequencing problem. ***[B.Tech. (Comp. Sc.) (Dec.), Hyderabad Univ. 2011]***

11. Discuss the situations leading to multiple optimum solutions in sequencing problems. ***[B.Tech. (Comp. Sc.) (Dec.), Hyderabad Univ. 2011]***

12. How would you use Gantt Chart for solving scheduling problems? Why is it not employed for solving large problems?

13. Explain how you will solve the sequencing problems, when there are 2 jobs and m machines. ***[MBA (Dec.), Panjab Tech. Univ. 2014]***

14. How does the sequencing technique help the manager?

15. Describe various approaches for scheduling in personnel services. ***[BBA, GGSIP Univ. 2005]***

16. Write short note on operations scheduling. ***[BBA, GGSIP Univ. 2006]***

17. Define: Idle time, Optimum sequence and SPT rule. ***[BE (Nov.), Visvesvaraya Tech. Univ. 2020]***

18. Network routing problems are related to sequencing. Justify your answer.

CHAPTER 11

Integer Programming

"God made the integers, rest is the work of man"

The basic assumption of a linear programming problem is that the decision variables are allowed to take any value within the feasible region, *i.e.*, the variables may have fractional and/or integer values. However, there are situations when the fractional value of decision variables has no significance. For instance, it is meaningless to recruit 2.6 men in a project or to use 2.4 boilers in a thermal power station. A linear programming problem in which some or all of the variables in the optimum solution are restricted to assume non-negative integer values is called an ***integer linear programming problem*** abbreviated as *IPP*.

11:1. INTEGER LINEAR PROGRAMMING PROBLEM

A linear programming problem in which all the variables in the optimum solution are restricted to assume non-negative integer values is known as ***an all-integer linear programming problem*** (all-IPP) or ***pure-integer linear programming problem***. On the other hand, if only some of the variables in the optimum solution are restricted to assume non-negative integer values, while the remaining variables are free to take any non-negative value, then it is called a ***mixed-integer linear programming problem mixed***. Further, if all the variables in a linear programming problem are allowed to take values either 1 or 0, *i.e.*, $x_j = 1$ (if decision is to be taken) and $x_j = 0$ (if decision is not to be taken); then the linear programming problem is called ***zero-one programming problem*** or ***Binary integer programming problem*** (B-IPP). The variables, here, are termed as ***binary variables***.

It is worth mentioning that great mathematician Professor ***Egon Balas*** was pioneer in integer programming research retrospective. In the year 1960, he had developed an algorithm for the optimal allocation of transporting the firewood in Romania by introducing of 1 or 0 integer variables representing yes or no choice. Such type of problems are termed as ***zero-one linear programming problems***.

11:2. FORMULATION OF INTEGER LINEAR PROGRAMMING PROBLEM

Formulation of integer programming problem is carried out in the similar fashion as the formulation of linear programming model from the given information with an additional restriction of some or all of the non-negative variables to assume integer values. To illustrate the formulation of linear programming problem, wherein variables are assumed to have non-negative integer values, consider the following problem:

Example 11.1. *A company manufactures three products P_1, P_2 and P_3 which yield per unit profit of ₹ 200, ₹ 400 and ₹ 300 respectively. Each of these products is processed on three different machines. The time required* *on each machine per unit of the product is given below:*

Product	*Time required (in hours per unit)*		
	Machine 1	*Machine 2*	*Machine 3*
P_1	30	20	10
P_2	40	10	30
P_3	20	20	20
Time available (hours)	600	400	800

How many products of each type should be produced to maximize profit? Fractional units of product are not permissible. Formulate it as a linear IPP.

Integer Programming Formulation. Let x_1, x_2 and x_3 be the number of units of products P_1, P_2 and P_3 respectively. Then, considering the profitability of products P_1, P_2 and P_3 and the availability of time (in hours) on machine M_1, M_2 and M_3; the appropriate linear programming formulation of the given problem is:

Maximize $z = 200x_1 + 400x_2 + 300x_3$ subject to the constraints:

$$30x_1 + 40x_2 + 20x_3 \le 600 \quad \textit{(Time constraint on } M_1\textit{)}$$
$$20x_1 + 10x_2 + 20x_3 \le 400 \quad \textit{(Time constraint on } M_2\textit{)}$$
$$10x_1 + 30x_2 + 20x_3 \le 800 \quad \textit{(Time constraint on } M_3\textit{)}$$
$$x_1 \ge 0,\ x_2 \ge 0,\ x_3 \ge 0 \text{ and are integers.}$$

Since, non-negative variables are assumed to be in integers, the formulation of given problem in the format of LPP is known as ***linear integer programming problem***.

We shall now discuss the integer programming formulation of situations in which variables are inherently discrete in nature and revolve around the binary (*i.e.*, 0–1) variables. Although, there is a large number of situations giving rise to Binary (or 0–1) integer programming problems, a few of these are given below:

Example 11.2 (*Assignment Problem*). *Consider a situation, where 'n' activities are to be performed on 'n' facility centres with the condition that only one activity is to be performed at one facility centre. The cost of processing is represented by c_{ij} (i = 1, 2, ..., n and j = 1, 2, ..., n). Then, the optimal assignment policy which will turn out all the activities at minimum total cost can be formulated as an integer programming problem, viz.,*

Minimize $z = \sum_{i=1}^{n} (c_{i1}x_{i1} + c_{i2}x_{i2} + \ldots + c_{in}x_{in})$ subject to the constraints:

$$x_{i1} + x_{i2} + \ldots + x_{in} = 1; \quad \text{for } i = 1, 2, \ldots, n,$$
$$x_{1j} + x_{2j} + \ldots + x_{nj} = 1; \quad \text{for } j = 1, 2, \ldots, n,$$

$$x_{ij} = \begin{cases} 0, & \text{if } i^{\text{th}} \text{ facility is not assigned to } j^{\text{th}} \text{ activity.} \\ 1, & \text{if } i^{\text{th}} \text{ facility is assigned to } j^{\text{th}} \text{ activity.} \end{cases}$$

To illustrate the formulation of ***assignment problem*** as an integer programming problem, consider the following problem:

Suppose, we have an assignment problem with three workers and an equal number of jobs, and cost matrix is given below:

	J_1	J_2	J_3
W_1	5	7	3
W_2	4	5	8
W_3	6	6	9

Find the minimum total cost of assigning each worker at the most one job. Formulate this as an integer programming problem.

Integer Programming Formulation. Let x_{ij} represent the variables where i^{th} worker is assigned the jth job. Then, the appropriate integer programming problem of the above assignment problem is as follow:

$$\text{Minimize } z = (5x_{11} + 7x_{12} + 3x_{13}) + (4x_{21} + 5x_{22} + 8x_{23}) + (6x_{31} + 6x_{32} + 9x_{33})$$

subject to the constraints:

$$x_{11} + x_{12} + x_{13} = 1; \quad x_{11} + x_{21} + x_{31} = 1,$$
$$x_{21} + x_{22} + x_{23} = 1; \quad x_{12} + x_{22} + x_{32} = 1,$$
$$x_{31} + x_{32} + x_{33} = 1; \quad x_{13} + x_{23} + x_{33} = 1,$$

$$x_{ij} = \begin{cases} 0, & \text{if assignment is not carried out.} \\ 1, & \text{if assignment is carried out.} \end{cases}$$

Example 11.3 (*Knapsack Problem*). *An integer programming problem that have only one constraint and binary decision variables is known as* ***knapsack problem***.

To illustrate the formulation of ***Knapsack problem*** as an integer programming problem, consider the following problem:

Four items are considered for loading on an aeroplane, which has a capacity to load up to 25 tonnes. The weights and values of the items are as follows:

Item	A	B	C	D
Weight (tonnes)	2	7	5	3
Value (per unit)	10	36	25	14

Which items, and what quantities should be loaded on the plane, so as to maximize the value of the cargo transported? Formulate this as an IPP.

Integer Programming Formulation. Let x_1, x_2, x_3 and x_4 be the number of items A, B, C and D respectively, that are loaded on the plane. Accordingly, the mathematical formulation of the problem is:

$$\text{Maximize } z = 10x_1 + 36x_2 + 25x_3 + 14x_4 \text{ subject to the constraints:}$$

$$2x_1 + 7x_2 + 5x_3 + 3x_4 \leq 25$$

$$x_j = 0 \text{ or } 1 \qquad (j = 1, 2, 3, 4)$$

Here, decision variables are: 0 (do not load) and 1 (to load).

Example 11.4 (*Capital Budgeting Problem*). *Usually, the companies face the situation of selecting one or more research and development projects or other investment opportunities among several competing projects. The ones accepted would get the value '1' and those not accepted would be assigned '0', i.e.*

$$x_j = \begin{cases} 1, & \text{if accepted.} \\ 0, & \text{if not accepted.} \end{cases}$$

To illustrate the formulation of ***capital budgeting problem*** as an integer programming problem, consider the following problem:

A corporation is considering the four possible investment opportunities. The following table presents information about the investment (in ₹ thousand) profits:

Project	*Present value of expected return*	*Capital required (year-wise) by projects*		
		Year 1	*Year 2*	*Year 3*
1	650	700	550	400
2	700	850	550	350
3	225	300	150	100
4	250	350	200	—
Capital available for investment		1,200	700	400

In addition, projects 1 and 2 are mutually exclusive and project 4 is contingent on the prior acceptance of project 3. Formulate an integer programming problem to determine which projects should be accepted and which should be rejected to maximize the present value from accepted project. ***[MBA, Delhi Univ. 2008]***

Integer Programming Formulation. Let x_j ($j = 1, 2, 3, 4$) denote the project j ($j = 1, 2, 3, 4$). Then, since there are only two options, *i.e.*, the project is either accepted or not, the given situation will be treated as binary integer programming problem. The appropriate integer programming formulation is as follows:

$$\text{Decision Variables: } x_j = \begin{cases} 1, & \text{if project } j \text{ is accepted.} \\ 0, & \text{if project } j \text{ is rejected.} \end{cases}$$

Constraints on Expenditure

$$700x_1 + 850x_2 + 300x_3 + 350x_4 \le 1{,}200 \qquad \textit{(Capital required in year 1)}$$
$$550x_1 + 550x_2 + 150x_3 + 200x_4 \le 700 \qquad \textit{(Capital required in year 2)}$$
$$400x_1 + 350x_2 + 100x_3 \le 400 \qquad \textit{(Capital required in year 3)}$$
$$x_1 + x_2 \le 1, \quad x_3 - x_4 \ge 0 \quad \text{and} \quad x_j = 1, 0.$$

Objective Function

$$\text{Maximize (Total Profit)} = 650x_1 + 700x_2 + 225x_3 + 250x_4.$$

Example 11.5 (*Fixed Charge Problem*). *Consider a production planning problem wherein it is required to have n items. Let the production of jth item involve two types of set-up cost, namely the fixed cost k_j ($k_j \ge 0$) independent of the quantity produced and a variable cost c_j per unit. If x_j is the number of units of the jth item produced, the production cost function for the jth item may be written as*

$$c_j(x_j) = \begin{cases} K_j + c_j x_j, & x_j > 0 \\ 0\ , & x_j = 0 \end{cases}$$

Integer Programming Formulation. The appropriate integer programming problem can be formulated as under:

$$\text{Minimize } z = \sum_{j=1}^{n} (K_j + c_j x_j) \text{ subject to the constraints:}$$

$$\sum_{j=1}^{n} x_j \ge n \text{ and } x_j \ge 0 \text{ (integer).}$$

It may be noted that because of fixed cost K_j, the objective function is non-linear. In order to overcome this difficulty, we reformulate the given problem as integer programming problem with the introduction of auxiliary binary variables y_j, given by

$$y_j = 1 \text{ or } 0 \text{ according as } x_j > 0 \text{ or } x_j = 0, \text{ respectively.}$$

These conditions can be expressed as a single linear constraint:

$$x_j \le My_j,$$

where the symbol M denotes a number that is so large that whenever x_j is feasible, $x_j \le M$.

The general fixed cost problem, therefore, can be stated as follows:

$$\text{Minimize } z = \sum_{j=1}^{n} (c_j x_j + K_j y_j) \text{ subject to the constraints:}$$

$$\sum_{j=1}^{n} x_j \ge n; \; x_j - My_j \le 0; \qquad (j = 1, 2, ..., n)$$

$$x_j \ge 0 \text{ and integer, } y_j = 0 \text{ or } 1, \text{ for all } j.$$

Note that although the original fixed charge problem has nothing to do with integer programming problem, the 'transformed' problem is a zero — one (or Binary) mixed integer programming problem.

To illustrate the formulation of ***fixed charge problem*** as an integer programming problem, consider the following two situations:

1. *Consider the following production data:*

Product	*Profit per unit (₹)*	*Direct labour requirement (hours)*
1	6	12
2	8	11
3	5	14

Fixed cost (₹)	*Direct labour requirement*
8,000	up to 10,000 hours
16,000	10,000 to 20,000 hours
22,000	20,000 to 30,000 hours

Formulate an integer programming problem to determine the production schedule so as to maximize the total net profit. ***[MBA (Nov.), Delhi Univ. 2005]***

Integer Programming Formulation. Let x_1, x_2, x_3 be number of units of products 1, 2 and 3 respectively to be produced, and let y_j ($j = 1, 2, 3$) be fixed cost in rupees; where y_j are the binary integer variables corresponding to the fixed cost of ₹ 8,000; ₹ 16,000 and ₹ 22,000 respectively, such that $y_j = 0$ or 1. Then, the integer programming model can be formulated as follows:

Maximize $z = 6x_1 + 8x_2 + 5x_3 - 8{,}000y_1 - 16{,}000y_2 - 22{,}000y_3$ subject to the constraints:

$$12x_1 + 11x_2 + 14x_3 \leq 10{,}000y_1 + 20{,}000y_2 + 30{,}000y_3,$$

$$x_1 + x_2 + x_3 = 1, \quad \text{where } x_j \geq 0 \qquad (j = 1, 2, 3)$$

$$y_j = 0 \text{ or } 1 \quad (j = 1, 2, 3).$$

2. *Three machines are available to make 5,000 units which are required for the internal use of a company as components to another product. If the production is to be made by anyone of these machines, a set-up cost is incurred apart from the cost of making each unit on different machines. The cost data are given below:*

Machine	*Set-up cost (₹)*	*Cost per unit (₹)*	*Maximum production*
1	8,000	5	4,000
2	5,000	4	3,000
3	4,000	8	1,000

What is the best production strategy? Formulate only. ***[MBA, Delhi Univ. 1994]***

Integer Programming Formulation. Let x_1, x_2 and x_3 represent the quantities to be produced on machines 1, 2 and 3, respectively. Further, let y_1, y_2 and y_3 indicate whether a machine is to be used, *i.e.*, '1' or not used, *i.e.*, '0'. Accordingly, the fixed cost would be $8{,}000y_1 + 5{,}000y_2 + 4{,}000y_3$; while the variable cost would be $5x_1 + 4x_2 + 8x_3$. The integer programming formulation, therefore, will be as follows:

Minimize $z = (8{,}000y_1 + 5{,}000y_2 + 4{,}000y_3) + (5x_1 + 4x_2 + 8x_3)$ subject to the constraints:

$$x_1 + x_2 + x_3 \geq 5{,}000,$$

$$x_1 \leq 4{,}000y_1, \; x_2 \leq 3{,}000y_2, \text{ and } x_3 \leq 1{,}000y_3;$$

$$x_j \geq 0 \;\; (j = 1, 2, 3) \text{ and } y_j = 0 \text{ or } 1. \qquad (j = 1, 2, 3)$$

Example 11.6 (*Caterer Problem*). *A caterer is to organize garden parties for a week. He needs a total of 160, 120, 60, 80, 90, 100 and 105 fresh napkins during the seven days of the week. Each new napkin costs ₹ 2. He can also use soiled napkins after getting them washed from a laundry. Ordinarily, washing charges are ₹ 0.35 per napkin and they are returned after 4 days. However, the laundry also provides express service at a cost of ₹ 0.60 per napkin, in which case they are returned after 2 days. Determine the planning schedule the caterer should adopt to buy or send napkins to the laundry so as to minimize the total cost, if he has no napkins at the beginning of the week. Formulate the problem as an integer linear programming model.*

Integer Programming Formulation. Let u denote the number of purchased napkins and x_j, y_j denote the number of washed napkins obtained after ordinary washing and express washing respectively on any day. Then, the total supply S_j on any day will be:

$$S_j = u + \sum_{j=1}^{7} (x_j + y_j).$$

Therefore, the requirement for each day must be met can be expressed as

$$S_j \geq D_j; \qquad (j = 1, 2, 3, 4, 5, 6, 7)$$

where, D_j denotes the cumulative demand of napkins.

Thus, various constraints are:

$$S_1 \geq D_1, \quad S_2 \geq D_2, \ldots, S_7 \geq D_7,$$

i.e.,

$$u + x_1 + y_1 \geq 160, \quad u + (x_1 + x_2) + (y_1 + y_2) \geq 280,$$
$$u + (x_1 + x_2 + x_3) + (y_1 + y_2 + y_3) \geq 340,$$
$$u + \sum_{j=1}^{4} (x_j + y_j) \geq 420, \quad u + \sum_{j=1}^{5} (x_j + y_j) \geq 510,$$
$$u + \sum_{j=1}^{6} (x_j + y_j) \geq 610, \quad u + \sum_{j=1}^{7} (x_j + y_j) \geq 715.$$

Since, there is no supply of napkins given for ordinary washing for the first four days,

$$x_1 = x_2 = x_3 = x_4 = 0.$$

Likewise, since there is no supply of napkins given for express washing for first two days,

$$y_1 = y_2 = 0.$$

Accordingly, the above constraints reduce to

$$u \geq 160, \quad u \geq 280, \quad u + y_3 \geq 340, \quad u + y_3 + y_4 \geq 420,$$
$$u + x_5 + y_3 + y_4 + y_5 \geq 510, \quad u + x_5 + x_6 + y_3 + y_4 + y_5 + y_6 \geq 610,$$
$$u + x_5 + x_6 + x_7 + y_3 + y_4 + y_5 + y_6 + y_7 \geq 715.$$

Further, out of these first constraint becomes redundant and can be deleted.

The objective is to minimize the cost, *viz.*

$$\text{Minimize (total cost) } z = 2u + 0.35 \sum_{j=1}^{7} x_j + 0.60 \sum_{j=1}^{7} y_j$$
$$= 2u + 0.35(x_5 + x_6 + x_7) + 0.60(y_3 + y_4 + y_5 + y_6 + y_7)$$

Thus, the problem is to find non-negative integer values of u, x_j and y_j that minimize the above objective function, while satisfying the six constraints given above.

Remark. All transportation, assignment and travelling salesman problems are integer programming problems (IPP), since decision variables are either zero or one, *i.e.*, $x_{ij} = 0$ or 1.

All sequencing and routing decisions are IPP as it requires the integer values of the decision variables.

PROBLEMS

11.01. Consider the problem of assigning three jobs to three men. Each man is capable of doing all the jobs; however, the time taken by different men on each job is different and can be assumed to be known. The assignment has to be done, so that each job is assigned only once. Each man gets only one job and the total time taken by all the jobs is minimized. Formulate it as an integer programming problem with decision variables defined as $x_{ij} = 1$ (if i^{th} man is assigned job j) and $x_{ij} = 0$ (otherwise). ***[MCA, IGNOU 2001]***

11.02. A jeweller and her apprentice make silver pins and necklaces by hand. Each week they have 80 hours of labour and 360 grams of silver available. It requires 8 hours of labour and 20 grams of silver to make a pin; 10 hours of labour and 60 grams of silver to make a necklace. Each pin also contains a small gem of some kind. The demand for pins is not more than six per week. A pin earns the jeweller ₹ 4,000 as profit and a necklace earns ₹ 1,000. The jeweller wants to know how many of each item to make each week in order to maximize profit.

Formulate an integer linear programming model for this problem. ***[MBA, Delhi Univ. 2008]***

11.03. The Brilliant Investment Company is considering four investments. Investment *1* will yield a net present value (NPV) of ₹ 16,000; Investment 2, an NPV of ₹ 22,000; Investment 3, an NPV of ₹ 12,000; and Investment 4, an NPV of ₹ 8,000. Each investment requires a certain cash outflow at the present time: Investment *1*, ₹ 5,000; Investment 2, ₹ 7,000; Investment 3, ₹ 4,000; and Investment 4, ₹ 3,000. At present ₹ 14,000 are available for investment. Formulate this situation as an integer linear programming model.

11.04. ABC Construction Company is faced with the problem of determining which projects it should undertake over the next four years. The following table gives information about each project:

Project	*Estimated present value*	*Capital requirements (in thousand ₹)*			
		Year 1	*Year 2*	*Year 3*	*Year 4*
A	18,00,000	3,00,000	4,00,000	4,00,000	3,00,000
B	2,00,000	1,20,000	80,000	—	40,000
C	7,20,000	3,00,000	2,00,000	2,00,000	2,00,000
D	8,00,000	2,00,000	4,00,000	4,00,000	1,00,000
	Funds available	6,50,000	8,00,000	8,00,000	5,00,000

Formulate a zero-one programming model to maximize the estimated present value.

[MBA (Nov.), Delhi Univ. 2002]

11.05. A chemical manufacturing company has 4 independent investment projects and must allocate a fixed capital budget to one or more of them so that the company's total assets are maximized. The estimated investments and the anticipated cash outflows associated with these projects are given in the table as under:

Project	*Investment (₹ lakh)*	*Cash outflows (₹ lakh)*	
		1st year	*2nd year*
A	105	60	160
B	140	108	40
C	80	200	150
D	100	90	70

The company has earmarked ₹ 600 lakh for investment in first year and ₹ 700 lakh in the second year. If projects *A* and *C* are mutually exclusive, how should the investment be made so that the company's total assets are maximized?

Formulate the problem as an integer linear programming model.

11.06. An air-conditioning and refrigeration company has been awarded a contract for the air-conditioning of a new computer installation. The company has to make a choice between two alternatives: (*a*) hire one or more refrigeration technicians for 6 hours a day, or (*b*) hire one or more part time refrigeration apprentice technicians for 4 hours a day.

The rate of wages of the refrigeration technicians is ₹ 20 per hour, while the corresponding rate for apprentice technicians is ₹ 8 per hour. The company wants to engage the technicians on work for not more than 25 man-hours per day and also limit the charges to technicians to ₹ 240. The company estimates that the productivity of a refrigeration technician is 8 units and that of a part-time apprentice technician is 3 units. Formulate the integer programming problem to enable the company to select the optimum number of technicians and apprentices.

[MBA, IGNOU 2002]

11.07. ABC Associates is considering four investment projects. The net present value and resource requirement for capital and labour for each project are as follows:

Project	*Capital (₹ lakh)*	*Labour power (₹ lakh)*	*Net present value (₹ lakh)*
1	15	5	50
2	3	4	5
3	10	15	15
4	4	2	8

The resources available for capital and labour power are ₹ 100 million and ₹ 150 million respectively. For administrative reasons, project *1* and *4* are mutually exclusive and the project 2 can be selected only, if project 3 is selected. The Associates want to find a mix of projects having the greatest total value. Formulate this problem as an integer linear programming problem.

[MBA, Delhi Univ. 2009]

11.08. XYZ Construction Company has an opportunity to build five shopping malls during the next year. The expected net profit and expected cost for each of the shopping malls are shown in the following table:

Expected Net Profit and Expected Net Cost

Shopping mall	*Expected net profit (₹ '000)*	*Expected cost (₹ '000)*
1	200	150
2	150	90
3	140	50
4	125	80
5	180	100

The company has budgeted ₹ 3,00,000 for the construction of shopping malls during the next year. Also, due to various legal restrictions and marketing considerations, the following relationships among the projects must be met:

(*i*) Exactly one of the shopping malls 1, 2 and 5 must be constructed.

(*ii*) At most only one of the two shopping malls 1 and 4 may be constructed.

(*iii*) If shopping mall 4 is constructed, then shopping mall 5 must also be constructed.

Formulate the problem as an integer linear programming problem.

[MBA (Nov.), Delhi Univ. 2009]

11:3. METHODS OF SOLVING AN INTEGER PROGRAMMING PROBLEM

For solving an integer programming problem, we may think to solve it by graphic method or simplex method (ignoring the restriction of integrality) and then the integer valued solution, thus obtained will satisfy all the constraints. For example, consider the following integer programming problem:

$$\text{Maximize } z = 20x_1 + 8x_2 \text{ subject to the constraints:}$$
$$6x_1 + 8x_2 \leq 16; \text{ and } x_1, x_2 \geq 0 \text{ and integers.}$$

Using graphical method or simplex method, the optimum non-integer valued solution (ignoring the integrality condition) is obtained as: $x_1 = 2.67$ and $x_2 = 0$.

By rounding the solution to $x_1 = 3$ and $x_2 = 0$, we observe that it does not satisfy the constraint $6x_1 + 8x_2 \leq 16$. So, the rounded solution may not be feasible. Hence, there is a need to develop a systematic procedure for obtaining optimal solution to integer programming problems.

Given below are the most acceptable methods for solving an integer programming problem:

1. Search Method, and 2. Cutting Plane Method.

1. Search Method. It is an enumeration method in which all feasible integer points are enumerated. The widely used search method is the ***branch and bound method***.

2. Cutting Plane Method. A systematic procedure for solving an integer programming problem was first developed by R.E. Gomory. His method consists in first solving the integer programming problem by simplex method, ignoring the condition of integrality. Then, introduce additional constraints one after the another to cut (eliminate) certain part of the solution space until an integral solution is obtained.

11:4. BRANCH AND BOUND METHOD

The concept behind this method is to divide the entire feasible solution space of LP problem into smaller parts called sub-problems and then search each of them for an optimal solution. This approach is useful in those cases where there is large number of feasible solutions and enumeration of those becomes economically impractical or impossible.

The branch and bound method starts by imposing feasible and infeasible upper and/or lower bounds for the decision variables in each sub-problem. This helps in reducing the number of iterations to arrive at the optimum solution, because each sub-problem worse than the current feasible bound is discarded and only the remaining sub-problems are examined. At a point where no more sub-problems can be created, we will find an optimum solution.

The iterative procedure is as follows:

Step 1. Obtain an optimum solution of the given integer programming problem ignoring the restriction of integrality.

Step 2. Test the integrality of the optimum solution obtained in *Step 1*. There are two cases:

(*i*) If the solution is in integers, the current solution is optimum to the given integer programming problem.

(*ii*) If the solution is not in integers, go to next step.

Step 3. Let the optimum value x_j of the decision variable x_j be not an integer, then choose the largest fractional value x_j^* such that $x_j^* < x_j < x_j^* + 1$. Here, x_j^* and $x_j^* + 1$ are consecutive non-negative integers.

Any feasible integer value of x_j must satisfy one of the following two conditions:

$$x_j \leq [x_j^*] \text{ or } x_j \geq [x_j^*] + 1,$$

where, $[x_j^*]$ is the largest integer contained in x_j.

Step 4. Branch (or Partition) the original problem into two sub-problems having the additional constraints. These are as follows:

Sub-Problem 1. Given IPP with an additional constraint $x_j \le [x_j^*]$.

Sub-Problem 2. Given IPP with an additional constraint $x_j \ge [x_j^*]+1$.

It may be noted that if $x_2 = 7.65$, then we shall have $x_2 \le 7$ in sub-problem 1 and $x_2 \ge 7+1$ (= 8) in sub-problem 2.

Step 5. Solve the two sub-problems obtained in *Step 4*. There may arise three cases:

(*i*) If the optimum solution of two sub-problems are integral valued, then the required solution is one that gives larger value of z.

(*ii*) If the optimum solution of one sub-problem is integral and the other sub-problem has no feasible optimum solution, the required solution is same as that of the sub-problem having integer valued solution.

(*iii*) If the optimum solution of one sub-problem is integral, while that of the other is not integer valued, then record the integer valued solution and repeat *Steps* 3 and 4 for the non-integer valued sub-problem.

Step 6. Repeat *Steps* 3 to 5, until all integer valued solutions are recorded.

Step 7. Choose the solution amongst the recorded integer valued solutions that yields an optimum value of the objective function.

Example 11.7. *Use Branch and Bound Method to solve the following integer programming problem:*

$$\text{Maximize } z = 3x_1 + 4x_2 \text{ subject to the constraints:}$$
$$7x_1 + 16x_2 \le 52,\; 3x_1 - 2x_2 \le 18,$$
$$x_1 \ge 0,\; x_2 \ge 0 \text{ and integers.}$$

Solution.

Step 1. Ignoring the integer restriction, the optimum solution to the given linear programming problem can easily be obtained by using graphical (or simplex) method as under:

$$x_1 = 196/31 \text{ and } x_2 = 15/31; \text{ with maximum } z = 648/31.$$

Step 2. Since, the solution of given problem is not in integers, let us choose $x_2^* = 15/31$ being the largest fractional value.

Step 3. Since, 15/31 can be written as 0.48; therefore $x_2^* < x_2 < x_2^* + 1 \Rightarrow 0 < x_2 < 1$. This gives $x_2 \ge 0$ or $x_2 \ge 0 + 1$ (= 1).

Step 4. By inserting the additional constraints, the two sub-problems are:

Sub-Problem 1. Maximize $z = 3x_1 + 4x_2$ subject to the constraints:

$$7x_1 + 16x_2 \le 52,\; 3x_1 - 2x_2 \le 18,$$
$$x_2 \le 0 \text{ and } x_1, x_2 \ge 0.$$

Sub-Problem 2. Maximize $z = 3x_1 + 4x_2$ subject to the constraints:

$$7x_1 + 16x_2 \le 52,\; 3x_1 - 2x_2 \le 18,$$
$$x_2 \ge 1 \text{ and } x_1, x_2 \ge 0.$$

Step 5. Optimum solution to the sub-problems 1 and 2 are obtained as follows:

Sub-Problem 1. $x_1 = 6$ and $x_2 = 0$, with maximum $z = 18$.

Sub-Problem 2. $x_1 = 5.14$ and $x_2 = 1$, with maximum $z = 19.43$.

The solution to the sub-problem 1 is in integers and lower limit of the objective function is 18. Now, since solution to sub-problem 2 is not in integers, we sub-divide it into following two sub-problems:

Sub-Problem 3. Maximize $z = 3x_1 + 4x_2$ subject to the constraints:

$$7x_1 + 16x_2 \le 52, \quad 3x_1 - 2x_2 \le 18,$$
$$x_2 \ge 1, x_1 \le 5 \text{ and } x_1, x_2 \ge 0.$$

Sub-Problem 4. Maximize $z = 3x_1 + 4x_2$ subject to the constraints:

$$7x_1 + 16x_2 \le 52, \quad 3x_1 - 2x_2 \le 18,$$
$$x_2 \ge 1, x_1 \ge 5 + 1 \ (= 6), \text{ and } x_1, x_2 \ge 0.$$

Step 6. Optimum solution to the sub-problem 3 and 4 are obtained as follows:

Sub-Problem 3. $x_1 = 5$ and $x_2 = 1.063$, with maximum $z = 19.25$.

Sub-Problem 4. No feasible solution.

Again, since solution to sub-problem 3 is not in integers, we sub-divide it into following two sub-problems:

Sub-Problem 5. Maximize $z = 3x_1 + 4x_2$ subject to the constraints:

$$7x_1 + 16x_2 \le 52, \quad 3x_1 - 2x_2 \le 18,$$
$$x_2 \ge 1, x_1 \le 5 \text{ and } x_2 \le 1; \ x_1, x_2 \le 0.$$

Sub-Problem 6. Maximize $z = 3x_1 + 4x_2$ subject to the constraints:

$$7x_1 + 16x_2 \le 52, \quad 3x_1 - 2x_2 \le 18,$$
$$x_2 \ge 1, x_1 \le 5 \text{ and } x_2 \ge 1 + 1 \ (= 2); \ x_1, x_2 \ge 0.$$

Step 7. Optimum solution to the sub-problem 5 and 6 are obtained as follows:

Sub-Problem 5. $x_1 = 5$ and $x_2 = 1$; with maximum $z = 19$.

Sub-Problem 6. $x_1 = 2.85$ and $x_2 = 2$; with maximum $z = 16.57$.

The solution to sub-problem 5 is in integers with new lower bound of the objective function as 19.

In sub-problem 6, since the value of objective function is 16.57 which is less than the new lower bound of 19; therefore it can not be partitioned further.

Hence, the optimum solution to the given integer programming problem is:

$$x_1 = 5 \text{ and } x_2 = 1, \quad \text{with maximum } z = 19.$$

The entire Branch and Bound procedure for the given IPP is shown below:

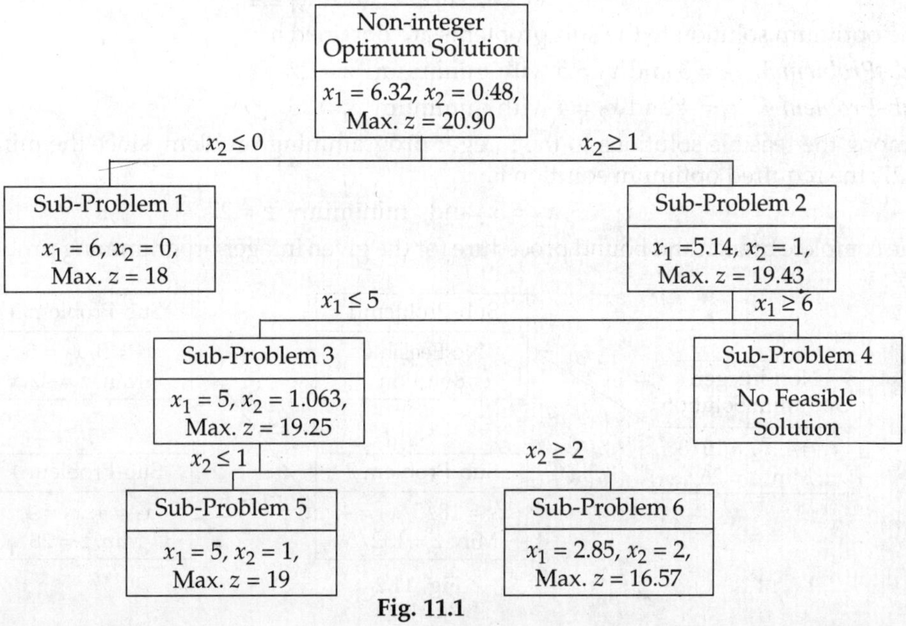

Fig. 11.1

Example 11.8. *Use branch and bound method to solve the following integer programming problem:*

$$\text{Minimize } z = 4x_1 + 3x_2 \text{ subject to the constraints:}$$
$$5x_1 + 3x_2 \geq 30, \quad x_1 \leq 4,$$
$$x_2 \leq 6, \quad x_1, x_2 \geq 0 \text{ and integers.}$$

[B.Sc. Comp. Sc., Pondicherry Univ. 2001]

Solution. Ignoring the integer restriction, the optimum solution to the linear programming problem can easily be obtained as:

$$x_1 = 4, \; x_2 = \frac{10}{3} \text{ and minimum } z = 26.$$

Since, the value of x_2 is not in integers, we branch on this variable. The two branches are $x_2 \leq 3$ and $x_2 \geq 4$, since $[x_2] = [10/3] = 3$. Thus, we have

Sub-Problem 1. Minimize $z = 4x_1 + 3x_2$ subject to the constraints:

$$5x_1 + 3x_2 \geq 30, \quad x_1 \leq 4,$$
$$x_2 \leq 6 \text{ and } x_2 \leq 3; \; x_1, x_2 \geq 0.$$

Sub-Problem 2. Minimize $z = 4x_1 + 3x_2$ subject to the constraints:

$$5x_1 + 3x_2 \geq 30, \quad x_1 \leq 4,$$
$$x_2 \leq 6 \text{ and } x_2 \geq 4; \; x_1, x_2 \geq 0.$$

The optimum solution to the sub-problems can easily be obtained as:

Sub-Problem 1. No feasible solution.

Sub-Problem 2. $x_1 = 18/5$ and $x_2 = 4$ with minimum $z = 132/5$.

Since, the value of x_1 in sub-problem 2 is not in integers, we branch on this variable. The two branches are $x_1 \leq 3$ and $x_1 \geq 4$, since $[x_1] = [18/5] = 3$. Thus, we have

Sub-Problem 3. Minimize $z = 4x_1 + 3x_2$ subject to the constraints:

$$5x_1 + 3x_2 \geq 30, \quad 0 \leq x_1 \leq 4,$$
$$0 \leq x_2 \leq 6, \; x_2 \geq 4 \text{ and } x_1 \leq 3.$$

Sub-Problem 4. Minimize $z = 4x_1 + 3x_2$ subject to the constraints:

$$5x_1 + 3x_2 \geq 30, \quad 0 \leq x_1 \leq 4,$$
$$0 \leq x_2 \leq 6, \; x_2 \geq 4 \text{ and } x_1 \geq 4$$

The optimum solution to the sub-problems are obtained as:

Sub-Problem 3. $x_1 = 3$ and $x_2 = 5$ with minimum $z = 27$.

Sub-Problem 4. $x_1 = 4$ and $x_2 = 4$ with minimum $z = 28$.

Among the feasible solutions to the i..teger programming problem, since the minimum value of z is 27; the required optimum solution is

$$x_1 = 3, x_2 = 5 \text{ and minimum } z = 27.$$

The complete branch and bound procedure for the given integer programming problem is shown below:

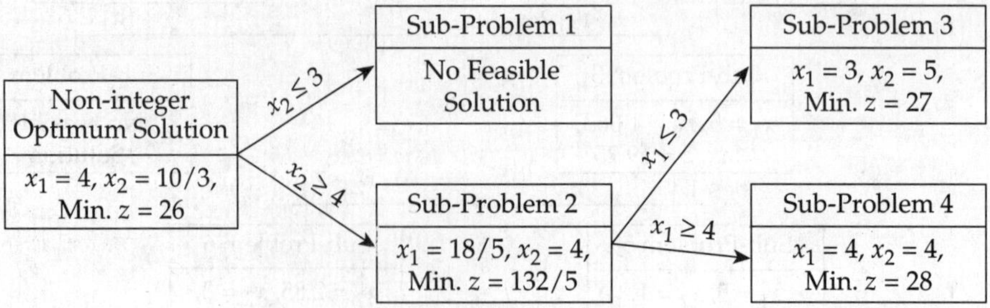

Fig. 11.2

Example 11.9. *Use branch and bound method to solve the following integer programming problem:*

$$\text{Maximize } z = 7x_1 + 6x_2 \text{ subject to the constraints:}$$
$$2x_1 + 3x_2 \le 12, \quad 6x_1 + 5x_2 \le 30,$$
$$x_1, x_2 \ge 0, \text{ and integers.}$$

Solution. Ignoring the integer restriction, the optimum solution to the linear programming problem can easily be obtained as:

$$x_1 = 3.75, x_2 = 1.5 \text{ and Maximum } z = 35.25.$$

Since, the optimum solution is not in integers, consider the value of z (= 35.25) as initial upper bound. The lower bound is obtained by rounding off the values of x_1 and x_2 to the nearest integers, *i.e.*, $x_1 = 3$ and $x_2 = 1$. The lower bound then will be $z = 27$.

Since, the largest fraction value corresponds to x_1, consider $x_1^* = 3.75$. Thus, $[x_1^*] = [3.75]$. The two sub-problems, therefore, are:

Sub-Problem 1. Maximize $z = 7x_1 + 6x_2$ subject to the constraints:

$$2x_1 + 3x_2 \le 12, \quad 6x_1 + 5x_2 \le 30,$$
$$x_1 \le 3 \text{ and } x_1 \ge 0, x_2 \ge 0.$$

Sub-Problem 2. Maximize $z = 7x_1 + 6x_2$ subject to the constraints:

$$2x_1 + 3x_2 \le 12, \quad 6x_1 + 5x_2 \le 30,$$
$$x_1 \ge 4 \text{ and } x_1 \ge 0, x_2 \ge 0.$$

The optimum solution to the sub-problems can easily be obtained as:

Sub-Problem 1. $x_1 = 3, x_2 = 2$; and maximum $z = 33$.

Sub-Problem 2. $x_1 = 4, x_2 = 1.2$; and maximum $z = 35.20$.

Since, the solution to sub-problem 2 is not in integers and $x_2^* = [1.2]$, we sub-divide it into following two sub-problems:

Sub-Problem 3. Maximize $z = 7x_1 + 6x_2$ subject to the constraints:

$$2x_1 + 3x_2 \le 12, \quad 6x_1 + 5x_2 \le 30,$$
$$x_1 \ge 4, x_2 \le 1; \; x_1 \ge 0, x_2 \ge 0.$$

Sub-Problem 4. Maximize $z = 7x_1 + 6x_2$ subject to the constraints:

$$2x_1 + 3x_2 \le 12, \quad 6x_1 + 5x_2 \le 30,$$
$$x_1 \ge 4, x_2 \ge 2; \; x_1 \ge 0, x_2 \ge 0.$$

The optimum solution to the two sub-problems can easily be obtained as:

Sub-Problem 3. $x_1 = 25/6, x_2 = 1$, and maximum $z = 35.16$.

Sub-Problem 4. No feasible solution.

In sub-problem 3, since the value of x_1 is not an integer, we branch on this variable. The two branches are: $x_1 \le 4$ and $x_1 \ge 5$, since $[x_1^*] = [4.16]$. Thus, we have

Sub-Problem 5. Maximize $z = 7x_1 + 6x_2$ subject to the constraints:

$$2x_1 + 3x_2 \le 12, \quad 6x_1 + 5x_2 \le 30,$$
$$x_1 \le 4, x_1 \ge 4, x_2 \le 1 \text{ and } x_1 \ge 0, x_2 \ge 0.$$

Sub-Problem 6. Maximize $z = 7x_1 + 6x_2$ subject to the constraints:

$$2x_1 + 3x_2 \le 12, \quad 6x_1 + 5x_2 \le 30,$$
$$x_1 \ge 5, x_1 \ge 4, x_2 \le 1 \text{ and } x_1 \ge 0, x_2 \ge 0.$$

The optimum solution to the sub-problems can easily be obtained as:

Sub-Problem 5. $x_1 = 4, x_2 = 1$; and maximum $z = 34$.

Sub-Problem 6. $x_1 = 5, x_2 = 0$; and maximum $z = 35$.

Among the feasible solutions to the integer programming problem, since the maximum value of z is 35, the required optimum solution is

$$x_1 = 5, x_2 = 0, \quad \text{and} \quad \text{maximum} \quad z = 35.$$

The complete branch and bound procedure for the given integer programming problem is shown below:

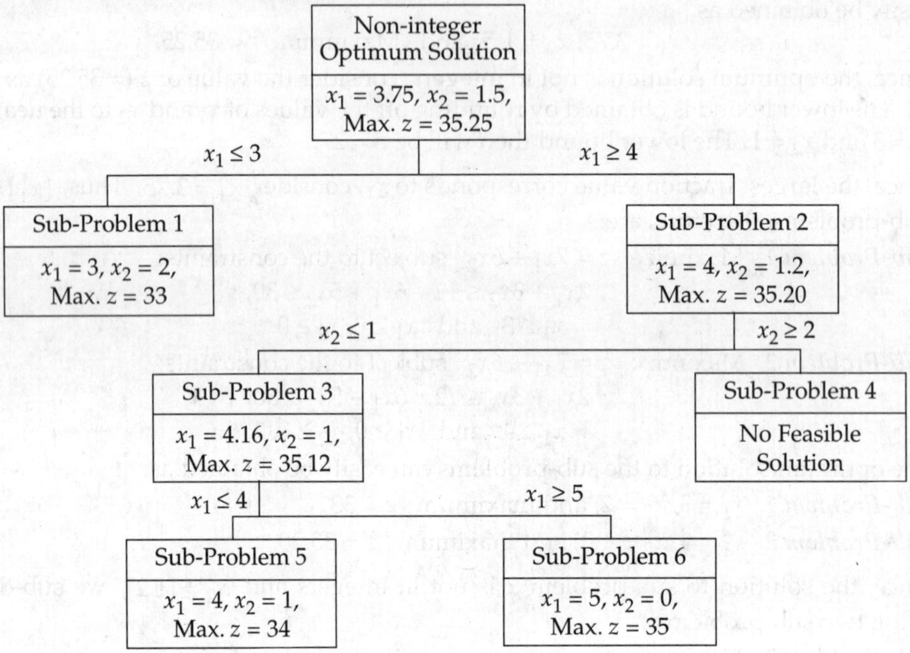

Fig. 11.3

PROBLEMS

Use Branch and Bound Method to solve the following integer linear programming problems:

11.09. (*a*) Maximize $z = x_1 + x_2$ subject to the constraints:

$$3x_1 + 2x_2 \leq 12, x_2 \leq 2;$$
$$x_1, x_2 \geq 0 \text{ and integers.}$$ ***[BE, Karnataka Univ. 2012]***

(*b*) Maximize $z = 2x_1 + 3x_2$ subject to the constraints:

$$6x_1 + 5x_2 \leq 25, \; x_1 + 3x_2 \leq 10;$$
$$x_1 \geq 0, x_2 \geq 0 \text{ and integers.}$$ ***[BE (Mech.), Jammu Univ. 2006]***

11.10. Maximize $z = 7x_1 + 10x_2$ subject to the constraints:

$$-x_1 + 3x_2 \leq 6, \; 7x_1 + x_2 \leq 35;$$
$$x_1, x_2 \geq 0 \text{ and integers.}$$ ***[BE (Dec.), Karnataka Univ. 2010]***

11.11. (*a*) Minimize $z = 9x_1 + 10x_2$ subject to the constraints:

$$x_1 \leq 9, \; x_2 \leq 8, \; 4x_1 + 3x_2 \geq 40;$$
$$x_1, x_2 \geq 0 \text{ and integers.}$$

(*b*) Maximize $z = x_1 + x_2$ subject to the constraints:

$$2x_1 + 5x_2 \leq 16, \; 6x_1 + 5x_2 \leq 30, \; x_2 \geq 0;$$
$$x_1 \geq 0 \text{ and integer.}$$ ***[BE (TIT) (Dec.), Panjab Univ. 2006]***

11.12. An exporter of ready-made garments makes two types of shirts: X and Y. He makes a profit of ₹ 10 and ₹ 40 per shirt on X and Y shirts, respectively. He has two tailors, say A and B, at his disposal to stitch these shirts. Tailors A and B can devote at the most 7 hours and 15 hours per day, respectively. Both these shirts are to be stitched by both the tailors. Tailor A and tailor B spend two hours and five hours respectively in stitching a X shirt, four hours and three hours respectively in stitching a Y shirt. How many shirts of both the types should be stitched in order to maximize daily profits? (A non-integer solution for this problem will not be accepted). *[MBA (PT) (Nov.), Delhi Univ. 2008]*

11.13. A manufacturer of toys makes two types of toys, say A and B. Processing of these toys is done on two machines X and Y. Toy A requires two hours on machine X and six hours on machine Y. Toy B requires four hours on machine X and five hours on machine Y. There are 16 hours of time per day available on machine X and 30 hours on machine Y. The profit obtained on both the toys is same, that is ₹ 5 per toy. What should be the daily production of each type of the two toys? (A non-integer solution for this problem will not be accepted). *[MBA (Nov.), Delhi Univ. 2008]*

11.14. The dietician at the local hospital is planning the breakfast menu for the maternity ward patients. She is primarily concerned with Vitamin E and iron requirements in planning the breakfast. According to the State Medical Association (SMA), new mothers must get at least 12 milligrams of Vitamin E and 24 milligrams of iron from breakfast. The SMA handbook reports that a scoop of scrambled egg contains 2 milligrams of Vitamin E and 8 milligrams of iron. The SMA handbook recommends that new mothers should eat at least two scoops of cottage cheese for their breakfast. The dietician considers this as one of the model constraints. The hospital accounting department estimate that a scoop of cottage cheese costs ₹ 2 and scoop of scrambled egg also costs ₹ 2. The dietician is attempting to determine the optimum breakfast menu that satisfies all the requirements and minimizes total cost. The cook insists that he can serve foods by only full scoop, thus necessitating an integer solution. Determine the optimum integer valued solution to the problem. *[MBA (HCA), Delhi Univ. 2000]*

11:5. CUTTING PLANE METHOD

For solving an all-IPP by cutting plane method or Gomory's ***fractional cut method***, the iterative procedure is summarized in the following steps:

Step 1. Write the given LPP into its standard form and determine an optimum basic feasible solution by using simplex method ignoring the integer value restrictions.

Step 2. Test the integrality of the optimum solution obtained in *Step 1*.

(*a*) If the optimum solution admits all integer values, *i.e.*, all $x_{Bi} \geq 0$ and are integers, an optimum solution obtained.

(*b*) If the optimum solution does not include all integer values, *i.e.*, at least one of the x_{Bi} is not an integer, while all other $x_{Bi} \geq 0$; go to next step.

Step 3. Rewrite each of the non-integer x_{Bi} as $[x_{Bi}] + f_i$, where $[x_{Bi}]$ is the integral part of x_{Bi} and f_i is the positive fractional part of x_{Bi}, $0 \leq f_i < 1$. Choose the largest fraction of x_{Bi}, *i.e.*, $\max_i . \{f_i\}$.

In case of tie, select arbitrarily. Let $\max_i . \{f_i\} = f_k$ corresponding to basic variable x_{Bk}. The kth row corresponding to f_k is called ***source row*** in the optimum simplex table.

Step 4. Express each of the negative fraction, if any, in the kth row (source row) as the sum of a negative integer and a non-negative fraction (*e.g.*, –5/8 can be written as –1 + 3/8).

Step 5. Generate the ***Gomorian constraint*** (fractional cut). From the kth row (source row), consider:

$$a_{k1}x_1 + a_{k2}x_2 + ... + a_{kn}x_n = x_{Bk}$$

This can be written as:

$$([a_{k1}] + f_{k1})\, x_1 + ([a_{k2}] + f_{k2})\, x_2 + ... + ([a_{kn}] + f_{kn})\, x_n = [x_{Bk}] + f_{k0}$$

From this constraint, the fractional cut can be written as:

$$f_{k1}x_1 + f_{k2}x_2 + ... + f_{kn}x_n \geq f_{k0}$$

or
$$-(f_{k1}x_1 + f_{k2}x_2 + ... + f_{kn}x_n) \leq -k_{k0}$$

or
$$-(f_{k1}x_1 + f_{k2}x_2 + ... + f_{kn}x_n) + G_1 = -f_{k0},$$

where, G_1 is the Gomorian slack and $0 \leq f_{kj} < 1$ $(j = 0, 1, 2, ..., n)$.

Step 6. Add the fractional cut (Gomorian constraint) generated in *Step 5* at the bottom of the optimum simplex table. Use the dual simplex method to find an improved optimum solution.

Step 7. Go to *Step 2* and repeat the procedure until an optimum integer solution is attained.

Example 11.10. *Find the optimum solution to the following integer programming problem using cutting-plane method:*

$$\textit{Maximize} \quad z = x_1 - 2x_2 \quad \textit{subject to the constraints:}$$

$$4x_1 + 2x_2 \leq 15; \quad x_1, x_2 \geq 0 \quad \textit{and integers.}$$

Solution. Introducing slack variable $s_1 \geq 0$ in the constraint of given integer programming problem and ignoring the condition of integrality, an initial basic feasible solution is $s_1 = 15$.

Now, using simplex method, we solve the given linear integer programming problem. The iterative simplex tableaus are as follows:

Initial Simplex Tableau. x_1 enters and s_1 leaves.

Basic variable	x_1	x_2	s_1	*Solution*	*Ratio*
$\leftarrow s_1$	(4)	2	1	15	15/4
Net-evaluation row	–1	2	0	0	

Since, the negative index number corresponds to first column, non-basic variable x_1 enters the basis and thereby the current basic variable s_1 leaves the basis. Thus, we get the next simplex tableau as under:

Final Simplex Tableau. Optimum solution.

Basic variable	x_1	x_2	s_1	*Solution*
x_1	1	1/2	1/4	15/4
Net-evaluation row	0	5/2	1/4	

In the optimum solution, since $x_1 = \frac{15}{4}\left(= 3 + \frac{3}{4}\right)$, we consider the fraction 3/4 to construct the fractional cut. Thus, we write

$$(1+0)\,x_1 + \frac{1}{2}x_2 + \frac{1}{4}s_1 = 3 + \frac{3}{4}$$

This implies
$$\frac{1}{2}x_2 + \frac{1}{4}s_1 \geq \frac{3}{4} \quad \text{or} \quad -\frac{1}{2}x_2 - \frac{1}{4}s_1 \leq -\frac{3}{4}$$

Therefore,
$$-\frac{1}{2}x_2 - \frac{1}{4}s_1 + F_1 = -\frac{3}{4} \qquad \text{(where } F_1 \geq 0\text{)}$$

is the fractional cut to be introduced in the final simplex method. Thus, we have

Modified Simplex Tableau. Drop F_1 and enters s_1.

Basic variable	x_1	x_2	s_1	F_1	*Solution*
x_1	1	1/2	1/4	0	15/4
$\leftarrow F_1$	0	–1/2	(–1/4)	1	–3/4
Net-evaluation row	0	5/2	1/4	0	15/4

As $x_{B_2}\left(=-\frac{3}{4}\right)$ is negative, using dual simplex method, the basic variable F_1 leaves the basis. Further, since $\max.\left\{\frac{5/2}{-1/2}, \frac{1/4}{-1/4}\right\}=-1$, the non-basic variable s_1 enters the basis. Thus, we get

Basic variable	x_1	x_2	s_1	F_1	*Solution*
x_1	1	0	0	1	3
s_1	0	2	1	–4	3
Net-evaluation row	0	2	0	1	3

This table shows that the optimum basic feasible solution has been attained in integers. Hence, the optimum solution to the given integer programming problem is

$$x_1 = 3, x_2 = 0; \text{ and maximum } z = 3.$$

Example 11.11. *Use cutting plane method to solve the following integer programming problem:*

$$\text{Maximize } z = x_1 + 4x_2 \text{ subject to the constraints:}$$
$$2x_1 + 4x_2 \le 7,\quad 5x_1 + 3x_2 \le 15,$$
$$x_1, x_2 \ge 0 \text{ and integer.}$$

Solution. After introducing slack variables $s_1 \ge 0$ and $s_2 \ge 0$, the initial basic feasible solution is $s_1 = 7$ and $s_2 = 15$.

Now, using simplex method, we solve the linear intger programming problem. The iterative simplex tableaus are given below:

Initial Simplex Tableau. x_2 enters and s_1 leaves.

Basic variable	x_1	x_2	s_1	s_2	*Solution*	*Ratio*
$\leftarrow s_1$	2	(4)	1	0	7	7/4
s_2	5	3	0	1	15	15/3
Net-evaluation row	–1	–4 ↑	0	0	0	

Final Simplex Tableau. Optimum solution.

Basic variable	x_1	x_2	s_1	s_2	*Solution*
x_2	1/2	1	1/4	0	7/4
s_2	7/2	0	–3/4	1	39/4
Net-evaluation row	1	0	1	0	7

To apply the cutting plane method, we begin by choosing the constraint in the final simplex tableau containing the optimum solution in which a basic variable has its value in fraction. In the final simplex tableau,

$$x_{B_1} = \frac{7}{4}\left(=1+\frac{3}{4}\right) \text{ and } x_{B_2} = \frac{39}{4}\left(=9+\frac{3}{4}\right).$$

Now, maximum $\{f_1, f_2\}$ = maximum $\{3/4, 3/4\} = 3/4$, *i.e.*, both f_1, f_2 are equal. So, we arbitrarily select any one of these.

Let us choose f_2. Therefore, we write,

$$\left(3+\frac{1}{2}\right)x_1 + 0\cdot x_2 + \left(-1+\frac{1}{4}\right)s_1 + 0\cdot s_2 = \left(9+\frac{3}{4}\right)$$

This implies $\quad \frac{1}{2}x_1 + \frac{1}{4}s_1 \geq \frac{3}{4}$ or $-\frac{1}{2}x_1 - \frac{1}{4}s_1 \leq -\frac{3}{4}$

Introducing slack variable F_1 (≥ 0) in this inequality, the fractional cut is obtained as

$$-\frac{1}{2}x_1 - \frac{1}{4}s_1 + F_1 = -\frac{3}{4}$$

Adding this additional constraint in the final simplex table obtained above, we get

Modified Simplex Tableau-I. Drop F_1 and enter x_1.

Basic variable	x_1	x_2	s_1	s_2	F_1	*Solution*
x_2	1/2	1	1/4	0	0	7/4
s_2	7/2	0	–3/4	1	0	39/4
$\leftarrow F_1$	(–1/2)	0	–1/4	0	1	–3/4
Net-evaluation row	1↑	0	1	0	0	7

As $x_{B_3}\left(=-\frac{3}{4}\right)$ is negative, using dual simplex method the basic variable F_1 leaves the basis.

Further, since max. $\left\{\frac{1}{-1/2}, \frac{1}{-1/4}\right\} = -2$, the non-basic variable x_1 enters the basis, *i.e.*, x_1 becomes basic variable in place of F_1. Thus, we get

Basic variable	x_1	x_2	s_1	s_2	F_1	*Solution*
x_2	0	1	0	0	1	1
s_2	0	0	–5/2	1	7	9/2
x_1	1	0	1/2	0	–2	3/2
Net-evaluation row	0	0	1/2	0	2	11/2

Here, since $x_{B_2} = \frac{9}{2}\left(=4+\frac{1}{2}\right)$ and $x_{B_3} = \frac{3}{2}\left(=1+\frac{1}{2}\right)$, for the second fractional cut, we write

$$0\cdot x_1 + 0\cdot x_2 + \left(-3+\frac{1}{2}\right)s_1 = \left(4+\frac{1}{2}\right)$$

Second fractional cut, therefore, is: $\frac{1}{2}s_1 \geq \frac{1}{2}$. This gives

$$-\frac{1}{2}s_1 \leq -\frac{1}{2} \quad \text{or} \quad -\frac{1}{2}s_1 + F_2 = -\frac{1}{2}$$

Adding this additional constraint in the new simplex Tableau-I, we get

Modified Simplex Tableau-II. Drop F_2 and enter s_1.

Basic variable	x_1	x_2	s_1	s_2	F_1	F_2	*Solution*
x_2	0	1	0	0	1	0	1
s_2	0	0	–5/2	1	7	0	9/2
x_1	1	0	1/2	0	–2	0	3/2
$\leftarrow F_2$	0	0	(–1/2)	0	0	1	–1/2
Net-evaluation row	0	0	1/2↑	0	2	0	11/2

As $x_{B_4}\left(=-\frac{1}{2}\right)$ is negative, using dual simplex method the basic variable F_2 leaves the basis and non-basic variable s_1 enters the basis, *i.e.*, s_1 becomes basic variable in place of F_2. Thus, we get

Basic variable	x_1	x_2	s_1	s_2	F_1	F_2	*Solution*
x_2	0	1	0	0	1	0	1
s_2	0	0	0	1	7	–5	7
x_1	1	0	0	0	–2	1	1
s_1	0	0	1	0	0	–2	1
Net-evaluation row	0	0	0	0	2	1	5

This simplex tableau shows that an optimum basic feasible integer solution has been attained. Hence, the optimum solution in integers is

$$x_1 = 1,\ x_2 = 1, \quad \text{and} \quad \text{maximum } z = 5.$$

Example 11.12. *Use cutting plane method to solve the following integer programming problem:*

$$\textit{Maximize} \quad z = 200x_1 + 300x_2 \quad \textit{subject to the constraints:}$$

$$2x_1 + 4x_2 \le 17, \quad 3x_1 + 3x_2 \le 15,$$

$$x_1, x_2 \ge 0 \quad \textit{and} \quad \textit{integer.}$$

Solution. Introducing slack variables $s_1 \ge 0$ and $s_2 \ge 0$ in the constraints of given IPP, an initial basic feasible solution is $s_1 = 17$ and $s_2 = 15$. Using simplex method, we now solve the linear integer programming problem. The iterative simplex tableaus are given below:

Initial Iteration. Optimum non-integer solution.

Basic variable	x_1	x_2	s_1	s_2	*Solution*
x_2	0	1	1/2	–1/3	7/2
x_1	1	0	–1/2	2/3	3/2
Net-evaluation row	0	0	50	100/3	1350

Here, $x_{B_1}(=x_2)=\frac{7}{2}\left(=3+\frac{1}{2}\right)$ and $x_{B_2}(=x_1)=\frac{3}{2}\left(=1+\frac{1}{2}\right)$. This shows that $f_1 = f_2 = 1/2$. So, max. $\{f_1, f_2\} = 1/2$ implies that we can select any one of f_1 and f_2. Let us choose f_1. Therefore, we write

$$x_2 + \frac{1}{2}s_1 + \left(-1+\frac{2}{3}\right)s_2 = \left(3+\frac{1}{2}\right).$$

This implies $\quad \frac{1}{2}s_1 + \frac{2}{3}s_2 \ge \frac{1}{2} \quad \text{or} \quad -\frac{1}{2}s_1 - \frac{2}{3}s_2 \le -\frac{1}{2}.$

Introducing a slack variable F_1 (> 0) in this inequality, the fractional cut is

$$-\frac{1}{2}s_1 - \frac{2}{3}s_2 + F_1 = -\frac{1}{2}.$$

By inserting this additional constraint, the modified optimum simplex tableau becomes:

First Iteration. Drop F_1 and enter s_2.

Basic variable	x_1	x_2	s_1	s_2	F_1	*Solution*
x_2	0	1	1/2	–1/3	0	7/2
x_1	1	0	–1/2	2/3	0	3/2
← F_1	0	0	–1/2	(–2/3)	1	–1/2
Net-evaluation row	0	0	50	100/3↑	0	1350

Making use of dual simplex method, we find from the above tableau that the basic variable F_1 leaves the basis and non-basic variable s_2 enters the basis. The new simplex tableau will be as under:

Basic variable	x_1	x_2	s_1	s_2	F_1	*Solution*
x_2	0	1	3/4	0	–1/2	15/4
x_1	1	0	–1	0	1	1
s_2	0	0	3/4	1	–3/2	3/4
Net-evaluation row	0	0	25	0	50	1325

Here, since $x_{B_1} = \frac{15}{4}\left(= 3 + \frac{3}{4}\right)$ and $x_{B_3} = \frac{3}{4}$, we write

$$x_2 + \frac{3}{4} s_1 + \left(-1 + \frac{1}{2}\right) F_1 = \left(3 + \frac{3}{4}\right)$$

Second fractional cut F_2, therefore, is

$$-\frac{3}{4} s_1 - \frac{1}{2} F_1 + F_2 = -\frac{3}{4}, \text{ since } \frac{3}{4} s_1 + \frac{1}{2} F_1 \geq \frac{3}{4} \text{ and } F_2\ (\geq 0) \text{ is slack variable.}$$

By inserting this additional constraint, the modified simplex tableau becomes:

Second Iteration. Drop F_2 and enter s_1.

Basic variable	x_1	x_2	s_1	s_2	F_1	F_2	*Solution*
x_2	0	1	3/4	0	–1/2	0	15/4
x_1	1	0	–1	0	1	0	1
s_2	0	0	3/4	1	–3/2	0	3/4
$\leftarrow F_2$	0	0	**(–3/4)**	0	–1/2	1	–3/4
Net-evaluation row	0	0	25 ↑	0	50	0	1325

Using dual simplex method, we find from the above simplex tableau that the basic variable F_2 leaves the basis and non-basic variable s_1 enters the basis. The new simplex tableau will be as follows:

Final Iteration. Optimum solution in integer.

Basic variable	x_1	x_2	s_1	s_2	F_1	F_2	*Solution*
x_2	0	1	0	0	–1	1	3
x_1	1	0	0	0	5/3	–4/3	2
s_2	0	0	0	1	–2	1	0
s_1	0	0	1	0	2/3	–4/3	1
Net-evaluation row	0	0	0	0	100/3	100/3	1300

From the above simplex tableau, we observe that an optimum basic feasible solution in integer has been obtained. Hence,

$$x_1 = 2, x_2 = 3, \text{ and maximum } z = 1300.$$

Example 11.13. *Solve the following integer programming problem using cutting plane method:*

$$\textit{Maximize } z = 3x_1 + x_2 + 3x_3 \textit{ subject to the constraints:}$$
$$-x_1 + 2x_2 + x_3 \leq 4,\quad 4x_2 - 3x_3 \leq 2,$$
$$x_1 - 3x_2 + 2x_3 \leq 3;\quad x_1, x_2, x_3 \geq 0 \textit{ and integer.}$$

[BE (Mech.), Jammu Univ. 2004]

Solution. Introducing slack variables $s_1 \geq 0, s_2 \geq 0$ and $s_3 \geq 0$ in the respective constraints, an initial basic feasible solution is $s_1 = 4, s_2 = 2$ and $s_3 = 3$. Using, now, simplex method to linear programming problem, an optimum non-integer solution is obtained as given in the following simplex tableau:

Initial Iteration. Non-integer optimum solution.

Basic variable	x_1	x_2	x_3	s_1	s_2	s_3	*Solution*
x_3	0	0	1	4/9	1/9	4/9	10/3
x_2	0	1	0	1/3	1/3	1/3	3
x_1	1	0	0	1/9	7/9	10/9	16/3
Net-evaluation row	0	0	0	2	3	5	87/3

Since, $x_{B_1}\ (= x_3) = \frac{10}{3}\left(= 3 + \frac{1}{3}\right)$ and $x_{B_3}\ (= x_1) = \frac{16}{3}\left(= 5 + \frac{1}{3}\right)$, we choose max. $\{f_1, f_3\} = 1/3$; *i.e.*, both f_1 and f_3 are equal. So, let us choose $f_1 = 1/3$.

Then, we write

$$x_3 + \frac{4}{9}s_1 + \frac{1}{9}s_2 + \frac{4}{9}s_3 = 3 + \frac{1}{3}$$

This implies that $\frac{4}{9}s_1 + \frac{1}{9}s_2 + \frac{4}{9}s_3 \geq \frac{1}{3}$ or $-\frac{4}{9}s_1 - \frac{1}{9}s_2 - \frac{4}{9}s_3 \leq -\frac{1}{3}$

Introducing now, Gomorian slack F_1 (≥ 0) in this inequality, the fractional cut is

$$-\frac{4}{9}s_1 - \frac{1}{9}s_2 - \frac{4}{9}s_3 + F_1 = -\frac{1}{3}$$

By inserting this additional constraint, the modified optimum simplex tableau becomes:

First Iteration. Drop F_1 and enter s_1.

Basic variable	x_1	x_2	x_3	s_1	s_2	s_3	F_1	*Solution*
x_3	0	0	1	4/9	1/9	4/9	0	10/3
x_2	0	1	0	1/3	1/3	1/3	0	3
x_1	1	0	0	1/9	7/9	10/9	0	16/3
$\leftarrow F_1$	0	0	0	(**–4/9**)	–1/9	–4/9	1	–1/3
Net-evaluation row	0	0	0	2↑	3	5	0	87/3

Since, $x_{B_4}\left(= -\frac{1}{3}\right)$ is only negative, F_1 leaves the basic. Using, now, dual simplex method, the new simplex tableau is as follows:

Basic variable	x_1	x_2	x_3	s_1	s_2	s_3	F_1	*Solution*
x_3	0	0	1	0	0	0	1	3
x_2	0	1	0	0	1/4	0	3/4	11/4
x_1	1	0	0	0	3/4	1	1/4	21/4
s_1	0	0	0	1	1/4	1	–9/4	3/4
Net-evaluation row	0	0	0	0	5/2	3	9/2	55/2

Here, $x_{B_2}\ (= x_2) = \frac{11}{4}\left(= 2 + \frac{3}{4}\right)$, $x_{B_3}\ (= x_1) = \frac{21}{4}\left(= 5 + \frac{1}{4}\right)$ and $x_{B_4}\ (= s_1) = \frac{3}{4}$

So, max. $\{f_2,\ f_3,\ f_4\} = \text{max.}\left\{\frac{3}{4}, \frac{1}{4}, \frac{3}{4}\right\} = \frac{3}{4}$, *i.e.*, both f_1 and f_4 are equal.

Let us choose $f_2 = 3/4$. This suggests that the second fractional cut will be

$$-\frac{1}{4}s_2 - \frac{3}{4}F_1 + F_2 = -\frac{3}{4},$$ where F_2 (≥ 0) is the second Gomorian slack.

By inserting this additional constraint in the modified simplex tableau, we get

Second Iteration. Drop F_2 and enter F_1.

Basic variable	x_1	x_2	x_3	s_1	s_2	s_3	F_1	F_2	*Solution*
x_3	0	0	1	0	0	0	1	0	3
x_2	0	1	0	0	1/4	0	3/4	0	11/4
x_1	1	0	0	0	3/4	1	1/4	0	21/4
s_1	0	0	0	1	1/4	0	–9/4	0	3/4
$\leftarrow F_2$	0	0	0	0	–1/4	0	(–3/4)	1	–3/4
Net-evaluation row	0	0	0	0	5/2	3	9/2↑	0	55/2

Using dual simplex method, we find from the above simplex tableau that the basic variable F_2 leaves the basis and non-basic variable F_1 enters the basis. The new simplex tableau will be as follows:

Final Iteration. Optimum solution in integer.

Basic variable	x_1	x_2	x_3	s_1	s_2	s_3	F_1	F_2	*Solution*
x_3	0	0	1	0	–1/3	0	0	4/3	2
x_2	0	1	0	0	0	0	0	1	2
x_1	1	0	0	0	2/3	1/9	0	1/3	5
s_1	0	0	0	1	1	1	0	–3	3
F_1	0	0	0	0	1/3	0	1	–4/3	1
Net-evaluation row	0	0	0	0	1	1/3	0	6	23

This simplex tableau shows that an optimum basic feasible integer solution has been reached. Hence, optimum integer valued solution is

$$x_1 = 5, x_2 = 2 \text{ and } x_3 = 2, \quad \text{with} \quad \text{maximum} \quad z = 23.$$

PROBLEMS

11.15. Solve the following all-integer linear programming problem:

$$\text{Maximize} \quad z = 2x_1 + x_2 \quad \text{subject to the constraints:}$$
$$x_1 \le 3/2, \quad x_2 \le 3/2;$$
$$x_1, x_2 \ge 0 \quad \text{and integer.}$$

11.16. Solve the following integer programming problem:

$$\text{Maximize} \quad z = 5x_1 + 3x_2 \quad \text{subject to the constraints:}$$
$$4x_1 + 2x_2 \le 25, \quad x_1 \le 5, x_2 \le 8;$$
$$x_1, x_2 \ge 0 \quad \text{and integers.}$$

[BE (Mech.), Kurukshetra Univ. 2012; BE (Mech.), Panjab Univ. 2012]

11.17. Find the optimum integer solution to the all-integer linear programming problem:

$$\text{Maximize} \quad z = x_1 + x_2 \quad \text{subject to the constraints:}$$
$$3x_1 + 2x_2 \le 5, \quad x_2 \le 2;$$
$$x_1 \ge 0, x_2 \ge 0 \quad \text{and integers.}$$

[BE, Karnataka Univ. 2010; B.Tech. (Comp.Sc.), Panjab Tech. Univ. 2009]

11.18. Solve the following all-integer linear programming problem:

$$\text{Maximize} \quad z = 2x_1 + 3x_2 \quad \text{subject to the constraints:}$$
$$-3x_1 + 7x_2 \le 14, \quad 7x_1 - 3x_2 \le 14;$$
$$x_1, x_2 \ge 0 \quad \text{and integers.}$$

11.19. Solve the following linear integer programming problem:

$$\text{Maximize } z = 3x_2 \text{ subject to the constraints:}$$
$$3x_1 + 2x_2 \le 7,\ -x_1 + x_2 \le 2;$$
$$x_1, x_2 \ge 0 \text{ and integers.}$$

[BE, Karnataka Univ. 2011; MBA, Annamalai Univ. 2009]

11.20. The ABC Electric Appliances Company produces two products: refrigerators and ranges. Production takes place in two separate departments. Refrigerators are produced in department *I* and ranges in department *II*. The company's two products are produced and sold on a weekly basis. The weekly production cannot exceed 25 refrigerators in department *I* and 35 ranges in department *II*, because of the limited available facilities in these two departments. The company regularly employs a total of 50 workers in the departments. A refrigerator requires two man-weeks of labour, while a range requires one man-week of labour. A refrigerator contributes a profit of ₹ 300 and a range gives profit of ₹ 200. How many units of refrigerators and ranges should the company produce to realize a maximum profit? (A non-integer solution for this problem will not be accepted.)

[MBA (PT) (Nov.), Delhi Univ. 2003]

11.21. A company produces two products, each of which requires stamping, assembly and painting operations. Total productive capacity by operation, if it were solely devoted to one product or the other is shown below:

Operation	*Productive capacity (units/week)*	
	Product A	*Product B*
Stamping	50	75
Assembly	40	80
Painting	90	45

Pro-rata allocation of productive capacity is permissible so also combination of production of the two products. Demand for the two products is unlimited. The profit on product *A* and product *B* is ₹ 150 and ₹ 120 respectively. Determine the optimal product-mix. (A non-integer solution will not be accepted).

[MBA (Nov.), Delhi Univ. 2009]

11.22. Use cutting plane method to solve the following integer programming problem:

$$\text{Maximize } z = 2x_1 + 20x_2 - 10x_3 \text{ subject to the constraints:}$$
$$2x_1 + 20x_2 + 4x_3 \le 15,\ 6x_1 + 20x_2 + 4x_3 = 20;$$
$$x_1, x_2, x_3 \ge 0 \text{ and integer.}$$

11.23. Use Gomory's method to solve the following integer programming problem:

$$\text{Maximize } z = 4x_1 + 5x_2 + x_3 \text{ subject to the constraints:}$$
$$3x_1 + 2x_2 \le 10,\ x_1 + 4x_2 \le 11,$$
$$3x_1 + 3x_2 + x_3 \le 13;\ x_1, x_2, x_3 \ge 0 \text{ and integer.}$$

11.24. Solve the following mixed integer programming problem:

$$\text{Maximize } z = 4x_1 + 6x_2 + 2x_3 \text{ subject to the constraints:}$$
$$4x_1 - 4x_2 \le 5,\ -x_1 + 6x_2 \le 5, -x_1 + x_2 + x_3 \le 5;$$
$$x_1, x_2, x_3 \ge 0;\ \text{where } x_1 \text{ and } x_3 \text{ are integers.}$$

[BE (TIT) (Nov.), Panjab Univ. 2006]

TEST YOUR UNDERSTANDING

1. The use of cutting plane method
   (*a*) require the use of standard linear programming approach between each cutting plane application.
   (*b*) yields better value of the objective function.
   (*c*) reduces the number of constraints in the given problem.
   (*d*) both (*b*) and (*c*).
2. In cutting plane algorithm, each cut which is made involves the introduction of
   (*a*) a less than or equal to constraint (*b*) a greater than or equal to constraint
   (*c*) an equality constraint (*d*) an artificial variable.
3. In a mixed integer programming problem
   (*a*) different objective functions are mixed together.
   (*b*) all the decision variables require integer solutions.
   (*c*) only few of the decision variables require integer solutions.
   (*d*) none of the above.
4. While solving an IPP, any non-integer variable in the solution is picked up to
   (*a*) enter the solution (*b*) leave the solution
   (*c*) obtain the cut constraint (*d*) all of the above.
5. Branch and Bound method divides the feasible solution space into smaller parts by
   (*a*) enumerating (*b*) branching
   (*c*) bounding (*d*) all of the above.
6. The 0 – 1 integer programming problem
   (*a*) requires the decision variables to have values either 0 or 1.
   (*b*) requires that decision variables have coefficients between zero and one.
   (*c*) requires that all constraints have coefficients between zero and one.
   (*d*) all of the above.

ANSWERS

1. (*a*) **2.** (*a*) **3.** (*c*) **4.** (*c*) **5.** (*b*) **6.** (*a*)

REVIEW EXERCISE

1. Write a short note on integer programming model. ***[PG Dip., Indian Inst. of Material Management 2019]***
2. What is integer linear programming? Explain the merits and demerits of "rounding off" a continuous optimum solution to an LPP to obtain an integer solution. ***[MBA, Osmania Univ. 2001]***
3. What is the need for integer programming? ***[MBA (Nov.), Annamalai Univ. 2002]***
4. Explain the methods used in solving integer programming problems. ***[B.E., Karnataka Univ. 2012]***
5. What is meant by mixed integer programming problem? ***[MBA, Annamalai Univ. 2010]***
6. Describe the Branch and Bound method for the solution of an integer programming problem. ***[MBA (June), IGNOU 2007; MBA, Annamalai Univ. 2010]***
7. Distinguish between linear programming problem and integer programming problem. ***[MBA, Annamalai Univ. 2009]***
8. Explain cutting plane method for pure and mixed integer programming problems. ***[MBA, Annamalai Univ. 2010]***
9. What is meant by zero-one programming problem? Explain with an appropriate example. ***[MBA, Delhi Univ. 2009]***
10. Explain some of the practical applications of integer programming problems.

CHAPTER 12

Goal Programming

"Success could be achieved easily when one has a definite goal"

Goal Programming is a management science technique that can be treated as an extension of linear programming and allows an organization to consider multiple objectives rather than a single objective as in the case of linear programming problems. There are situations, in today's business environment, where an organization may have multiple (possibly conflicting) goals such as maximization of profit, minimizing overtime or cost, employment stability, high product quality, minimizing noise level in the neighbourhood, industrial and labour relations, and so on.

12:1. WHY GOAL PROGRAMMING?

Goal programming allows handling multi-objective situations. It uses the concept of penalties in the format of linear programming. To apply goal programming, management is made to set some estimated targets for each of their goals (or objective) and rank them in the order of their priorities (or importance). Now, since all the goals are unlikely to be satisfied simultaneously, a penalty is assigned for each unit of deviation from the target value in each direction. Thus, a modified linear programming problem, using 'deviational' variables is formulated whose optimum solution comes ***as close as possible for achieving the stated goals*** in the sense that it minimizes the sum of penalties incurred for under-achieving or over-achieving the various goals. In this technique, we start with the most important goal and continue until the achievement of a less important goal.

Remark. In linear programming problem, there is an objective function (considered as goal) subject to certain constraints with non-negative restriction on decision variables. Any solution to this LPP must satisfy all the constraints prior to optimization of the objective function. Contrary to this, in goal programming, each goal enters the problem formulation as an equality constraint which contains slack variables, indicating either under-achievement or over-achievement of goals. The objective function then contains these deviational variables, and a solution will attempt to minimize them in order of priority.

Illustration. *Consider a firm which is engaged in producing two products, X and Y. Product X sells for a net profit of ₹ 60 per unit, while product Y sells for a net profit of ₹ 40 per unit. Goal of the firm is to earn ₹ 1,900 in the next week. Also, the management wants to achieve sales volume for the two products close to 17 units of product X and 15 units of product Y respectively.*

Express the problem in terms of goal programming model.

Goal Programming Model

Let x_1 and x_2 denote the number of units of product X and Y respectively. Then, the linear programming formulation of the problem is:

Maximize $z = 60x_1 + 40x_2$ subject to the constraints:

$x_1 \leq 17$ and $x_2 \leq 15$ (*Sales target goal*)

with $x_1 \geq 0$ and $x_2 \geq 0$. (*Non-negative restriction*)

Suppose, the manager of the firm has set the goal to achieve a profit of ₹ 1,900. But, from the linear programming problem given above, we observe that the maximum profit obtainable is: $60 \times 17 + 40 \times 15$, *i.e.*, ₹ 1,620. Thus, the problem is not feasible in terms of the requirement of ₹ 1,900 as profit. However, goal programming recognizes that any goal may be under-achieved, met exactly or over-achieved. For example, the goal of a profit of ₹ 1,900, in the above situation would be under-achieved, while if the goal stated in the above example is to attain a profit of ₹ 1,620, it would be met exactly. On the other hand, if the manager sets to achieve a profit of ₹ 1,000; it would be over-achieved. Thus, the requirement of achieving a profit of ₹ 1,900 can be expressed as: $60x_1 + 40x_2 = 1{,}900$ as a goal. If the manager's goal was instead to have a profit of at least ₹ 1,900; then the goal would have been expressed as: $60x_1 + 40x_2 \geq 1{,}900$.

Clearly, the goal constraint will be under-achieved. So, if we let d^- as ***deviational variable*** of under-achievement, we write: $60x_1 + 40x_2 + d^- = 1{,}900$. However, in a given problem, we would not know whether a goal will be under-achieved or over-achieved or exactly met, therefore, we introduce two deviational variables: one of under-achievement and the other of over-achievement. Thus, if we denote the deviational variable of over-achievement by d^+, then the goal constraint is:

$$60x_1 + 40x_2 + d^- - d^+ = 1{,}900;$$

where, (*i*) deviational variables d^- and d^+ are either positive or zero, (*ii*) either d^- or d^+ is zero, (*iii*) both d^- and d^+ are zero. This relationship can always be met. For example, if $x_1 = 0$ and $x_2 = 0$, then $d^- = 1{,}900$ and $d^+ = 0$ provides the solution.

Since, goal of the manager of the firm is to achieve a profit of ₹ 1,900 which is under-achieved; therefore, we want to be as close to the profit of ₹ 1,900 as possible and, hence, minimize d^-. Accordingly, the goal programming problem may be expressed as follows:

$$\text{Minimize } z = d^- \text{ subject to the constraints:}$$

$$x_1 \leq 17,\quad x_2 \leq 15,\quad 60x_1 + 40x_2 + d^- - d^+ = 1{,}900$$

$$x_1, x_2, d^-, d^+ \geq 0.$$

Remark. The goal programming problem given above is said to have a ***single goal***.

12:2. CATEGORIZATION OF GOAL PROGRAMMING PROBLEM

Goal programming problem may be categorized in terms of how the goals compare between ***non-pre-emptive*** and ***pre-emptive***. When all goals are of roughly comparable importance, goal programming is known as non-pre-emptive. Illustration given in the previous section is an example of non-pre-emptive goal programming problem. In case of a ***pre-emptive goal programming***, the goals are assigned priority levels. The goals are ranked from the most important, ***goal 1***, to the least important, ***goal n***, and the objective function coefficient assigned for the deviational variables representing goal *j* as P_j. It is assumed that $P_1 >>> P_2 >>> P_3 \ldots >>> P_n$, where the symbol >>> means 'more important than'. In fact, *P*'s are not assigned any actual values. Rather they are a convenient way of indicating that one goal is more important than the other. These coefficients indicate that weight of goal 1 is much larger than weight 2, which in turn is much larger than that of goal 3 and so on.

A lower priority goal is never sought to be achieved at the expense of higher priority goal. Thus, a highest priority goal is considered first and an attempt is made to achieve it as closely as possible. The next priority goal is considered only when an optimum solution with respect to the first priority goal is achieved. Similarly, the next-in-line goals are considered only when the optimality with respect to the higher priority goals are obtained. We continue in this manner until we can come closer to satisfying a goal, is to decrease the deviation from a higher priority goal.

12:3. FORMULATION OF LINEAR GOAL PROGRAMMING PROBLEM

Following are the major steps in the formulation of a linear goal programming problem:

Step 1. Identify the decision variables and right hand side constants of the constraints which may be either available resources or specified goal levels. It requires a careful analysis of the problem in order to identify all relevant variables that have some effect on the set of goals.

Step 2. Formulate all the objectives (or goals) of the problem. These are generally determined by (*i*) the desire of the decision maker, (*ii*) limited resources, (*iii*) any other restriction(s) either explicitly or implicitly placed on the desire of the decision maker.

Step 3. Reduce the number of goals by eliminating a few negligibly important (or redundant) goals.

Step 4. Express each goal in the form of constraint equation by introducing a negative and positive deviation variable (denoted by d_i^- and d_i^+ respectively), *i.e.*,

$$G_i : f_i(x_1, x_2, ..., x_n) - d_i^+ + d_i^- = b_i; \quad i = 1, 2, ..., m$$

where, d_i^- = negative deviation from i^{th} goal (under-achievement), and

d_i^+ = positive deviation from i^{th} goal (over-achievement).

Step 5. Assign the goals to priority levels. All absolute goals (*i.e.*, either $d_i^- = 0$ or $d_i^+ = 0$), if any exist, are assigned to top priority.

Step 6. Establish the achievement function. The priorities associated with each objective, along with the deviational variables are used to form what we call the achievement function.

Remarks

1. Since, both the under-achievement and over-achievement goals cannot be achieved simultaneously, one or both of the deviational variables d_i^- and d_i^+ must be zero in the solution. In other words, at optimality, if one assumes a positive value in the solution, the other must be zero and *vice-versa*.
2. The deviational variables d_i^+ (called ***surplus variables in linear programming problems***) are removed from the objective function of goal programming problem when over-achievement is acceptable. Similarly, if under-achievement is acceptable, d_i^- (called ***slack variables in linear programming problems***) are removed from the objective function of goal programming problem. But, if exact attainment of the goal is desired, then both d_i^- and d_i^+ are included in the objective function and ranked according to their priority factor.

The goal programming problem, in general, can be expressed as follows:

Minimize $z = P_1(d_1^+ + d_1^-) + P_2(d_2^+ + d_2^-) + ... + P_m(d_m^+ + d_m^-)$ subject to the constraints :

$$a_{i1}x_1 + a_{i2}x_2 + ... + a_{in}x_n + d_i^- - d_i^+ = b_i, \quad i = 1, 2, ..., m$$

$$x_j \geq 0, \quad d_i^- \geq 0 \text{ and } d_i^+ \geq 0; \quad (i = 1, 2, ..., m \text{ and } j = 1, 2, ..., n).$$

where, x_j is the decision variable j, P_i is the priority attached to goal i, d_i^- is the degree of under-achievement of goal i, and d_i^+ is the degree of over-achievement of goal i.

Example 12.1 (*Equally Ranked Multiple Goals*). *A mobile company producing two types of mobiles, namely Smart and Popular models, respectively. Both the models must be processed through two processes, say I and II. Total time available in process I is 96 hours and in process II is 78 hours. One unit of Smart mobile requires 8 hours in process I and 5 hours in process II. One unit of popular mobile requires 5 hours in process I and 8 hours in process II. The profit is ₹150 per unit of Smart mobile and ₹110 per unit for Popular mobile. The firm sets two equally ranked goals, i.e., to reach a profit goal of ₹3,000 and also to meet the production of 15 units of Smart mobiles. Formulate the given situation as a goal programming problem.*

Formulation of Goal Programming Problem

Let the company produce x_1 units of Smart mobiles and x_2 units of Popular mobiles. Then, total profit to the company will be $z\ (= 150x_1 + 110x_2)$ and the constraints on the availability of time will be: $8x_1 + 5x_2 \leq 96$ and $5x_1 + 8x_2 \leq 78$.

Since, the goals to the company are equally important or ranked, ₹ 1 deviation from the profit target is just as important as a deviation of one unit of Smart mobile in the Goal Programming model.

Let d_1^- and d_1^+ denote the under-achievement and over-achievement deviations, respectively of profit goal; d_2^- and d_2^+ denote the under-achievement and over-achievement deviations, respectively of Smart mobile goal. Then, the appropriate goal programming problem is formulated as follows:

$$\text{Minimize } z = d_1^- + d_1^+ \text{ subject to the constraints:}$$
$$8x_1 + 5x_2 \le 96,\ 5x_1 + 8x_2 \le 78,$$
$$150x_1 + 110x_2 + d_1^- - d_1^+ = 3{,}000,$$
$$x_1 + d_2^- - d_2^+ = 15,$$
$$x_1, x_2 \ge 0 \text{ and } d_1^-, d_2^-, d_1^+, d_2^+ \ge 0.$$

Example 12.2. *A firm produces two products A and B which yield a contribution margin of ₹ 400 and ₹ 600 per unit, respectively. The firm has a limited capacity in two departments where these products need processing. The availability and requirement of time in the production of two products per day are given below:*

Department	*Processing time (in hours)*		*Available time (in hours)*
	Product A	*Product B*	
I	7	4	60
II	8	7	85

The management of the firm has specified the following goals:

1. *Produce a product-mix to make a daily profit of at least 6,500 units: Priority P_1.*
2. *Achieve a daily sales of at least 30 units of product B: Priority P_2.*
3. *Achieve a daily sales of at least 25 units of product A: Priority P_3.*

Formulate the above situation as a goal programming problem.

Formulation of Goal Programming Problem

Let x_1 and x_2 be the daily output of products A and B, respectively. Further, define the following deviational variables:

d_1^- : under-achievement of the daily profit target, and
d_1^+ : over-achievement of the daily profit target;
d_2^- : under-achievement of the target daily sales of product B, and
d_2^+ : over-achievement of the target daily sales of product B;
d_3^- : under-achievement of the target daily sales of product A, and
d_3^+ : over-achievement of the target daily sales of product A.

Making use of the above mentioned variables, we find that the total profit to the firm can be written as a linear expression z (= $400x_1 + 600x_2$). The constraints on time are:

$$7x_1 + 4x_2 \le 60 \text{ and } 8x_1 + 7x_2 \le 85.$$

The constraints due to three goals are:

$$400x_1 + 600x_2 + d_1^- - d_1^+ = 3{,}000,$$
$$x_2 + d_2^- - d_2^+ = 30, \text{ and } x_1 + d_3^- - d_3^+ = 25$$

Hence, the complete goal programming problem is as follows:

$$\text{Minimize } z = P_1 d_1^- + P_2 d_2^- + P_3 d_3^- \text{ subject to the constraints:}$$

$$400x_1 + 600x_2 + d_1^- - d_1^+ = 3{,}000,$$

$$x_2 + d_2^- - d_2^+ = 30, \quad x_1 + d_3^- - d_3^+ = 25,$$

$$7x_1 + 4x_2 \leq 60, \quad 8x_1 + 7x_2 \leq 85;$$

$$x_1, x_2 \geq 0 \text{ and } d_i^- \geq 0, d_i^+ \geq 0 \qquad (i = 1, 2 \text{ and } 3)$$

Note. It may be noted that above examples fall under the category of non-pre-emptive goal programming. In these examples no weight-age or priority is attached to the over-achievement of any goal, and equal importance is given to the under-achievement of each of the goal, *e.g.*, if there are two goals — one is of profit and another is of work force, then both are weighted the same.

We, shall now discuss the next example on pre-emptive goal programming.

Example 12.3. *The manager of the record shop in a town has a decision problem that involves multiple goals. The record shop employs five full-time and four part-time salesman and the total available time for a full-time and a part-time salesman are 180 hours and 90 hours respectively. According to the performance records of salesmen, the average sales has been five records per hour for full-time salesman and two records per hour for part-time salesman. The average hourly wage rate ₹18 for full-time salesman and ₹12 for part-time salesman. Average profit from the sales of a record is ₹35.*

In view of past sales record and increased enrolment at the local college, the manager feels that the sales goal for the next month should be 7,200 records. Since, the shop is open six days a week, overtime is often required of salesman (not necessarily overtime but extra hours for the part-time salesman). The manger believes that a good employer-employee relationship is an essential factor of business success.

Therefore, he feels that a stable employment level with occasional overtime requirement is a better practice than an unstable employment level with no overtime. However, he feels that overtime of more than 100 hours among the full-time salesman should be avoided because of the declining sales effectiveness caused by fatigue.

The manager has set the following goals:

(*i*) *The first goal to achieve a sales goal of 7,200 records for the next month.*
(*ii*) *The second goal is to limit the overtime of full-time salesmen to 100 hours.*
(*iii*) *The third goal is to provide job security to salesman. The manager feels that full utilization of employees regular working hours (no layoffs) is an important factor for a good employer–employee relationship. However, he is twice as concerned with the full utilization of full-time salesman as with the full utilization of part-time salesman.*
(*iv*) *The last goal is to minimize the sum of overtime for both full-time and part-time salesman. The manager desires to assign differential weights to the minimization of over time according to the net marginal profit ratio between the full-time and part-time salesman.*

Formulate the problem as GP model.

Formulation of Goal Programming Problem

Let the decision variables for the manager of record shop are:

x_1 = full-time salesman hours per month, and x_2 = part-time salesman hours per month. As there are 5 full-time salesman and 4 part-time salesman, and the average sales has been 5 records per hour for full-time and 2 records per hour for part-time salesman, total number of sales per hour will be $5x_1 + 2x_2$.

It is required to achieve 7,200 sales hours in a month. Therefore, the constraint on sales hours will be:

$$5x_1 + 2x_2 + d_1^- - d_1^+ = 7{,}200$$

where, d_1^- and d_1^+ denote the under-achievement and over-achievement of the sales hours goal.

Time available for full-time salesman is 180 hours per month and that of part-time salesman is 90 hours. Therefore, constraints on regular working hours of the two types of salesman per hour will be:

$$x_1 + d_2^- - d_2^+ = 5 \times 180, \textit{ i.e.}, 900 \quad \text{and} \quad x_2 + d_3^- - d_3^+ = 4 \times 90, \textit{ i.e.}, 360$$

where, d_2^- and d_2^+ are the under-utilization time and over-utilization time, respectively for full-time salesman per month, d_3^- and d_3^+ are the under-utilization time and over-utilization time respectively for part-time salesman per month.

Manager's second goal is to limit the overtime of full-time salesman to 100 hours. In order to achieve this goal, since we do not have a deviational variable to minimize, we must introduce a new constraint. It may be noted that the minimization of overtime for full-time salesman is a part of the last goal. So, to limit the over-time of full-time salesmen to 100 hours, we introduce the following constraint:

$$d_2^+ + d_{21}^- - d_{21}^+ = 100$$

where, d_2^+ = actual overtime of full-time salesman.

d_{21}^- = difference between the actual overtime of full-time salesmen and desired 100 hours of overtime.

d_{21}^+ = overtime in excess of desired 100 hours.

It may be noted, here, that both the negative and positive deviation from the allowed 100 hours have been introduced because the actual overtime can be less than, equal to, or even more than 100 hours. Thus, we have to minimize the deviational variable d_{21}^+ to achieve the second goal.

The goal programming model for the given problem, therefore, is formulated as follows:

Minimize $z = P_1 d_1^- + P_2 d_{21}^+ + P_3(2d_2^- + d_3^-) + P_4(3d_3^+ + d_2^+)$ subject to the constraints:

$$5x_1 + 2x_2 + d_1^- - d_1^+ = 7{,}200,$$

$$x_1 + d_2^- - d_2^+ = 900, \quad x_2 + d_3^- - d_3^+ = 360,$$

$$d_2^+ + d_{21}^- - d_{21}^+ = 100;$$

$$x_1, x_2, d_1^-, d_1^+, d_2^-, d_2^+, d_{21}^-, d_{21}^+, d_3^-, d_3^+ \geq 0.$$

Example 12.4. *Bright Electronics manufactures two types of TV sets. One TV set, the Deluxe, requires 2 hours in assembly, while the other, the Supreme, requires 4 hours assembly time. The normal assembly operation is limited to 80 hours per week. Marketing surveys indicate that no more than 60 Deluxe and 30 Supreme TV sets should be produced each week. The net profit from the Deluxe model is ₹100 each and is ₹150 each from the Supreme model.*

The company president has stated the following objectives (goals) in order of priority:

1. *Maximize total profit.*
2. *Minimize overtime operation of the assembly line.*
3. *Sell as many TV sets as possible (this is not necessarily the same as maximizing profit). Since, the net profit from the Supreme model is 2 times that from the Deluxe model, the president has 2 times as much desire to maximize the sales of the Supreme model as he does for the Deluxe model.*

Formulate this as a Linear Goal Programming Problem.

Formulation of Goal Programming Problem

Let the decision variables are defined as:

x_1 = number of Deluxe TV sets built each week, and

x_2 = number of Supreme TV sets built each week.

The highest priority of the president is to maximize profit. Profit, in turn, can be written as a function of x_1 and x_2 as follows:

$$\text{Profit per week} = 100x_1 + 150x_2.$$

Setting profit to an arbitrarily high level of ₹ 5,000 per week, the goal is written as:

$$100x_1 + 150x_2 + d_1^- - d_1^+ = 5{,}000.$$

Note that d_1^- is the negative deviation measure, *i.e.*, the amount by which we ***underachieve*** our goal. On the other hand, d_1^+ is the amount by which we ***overachieve*** our target.

The second priority objective is the minimization of assembly time overtime operation. Since, assembly line time is ***normally*** 80 hours per week, our goal is:

$$2x_1 + 4x_2 + d_2^- - d_2^+ = 80.$$

Here, d_2^- is the amount of 'slack' time on the assembly line while d_2^+ represents the amount of 'overtime'. As our objective is measured by the minimization of overtime, we achieve this goal by minimizing d_2^+.

The third, and final priority is associated with a sales objective. Since, marketing has indicated that the demands for Deluxe and Supreme TV sets are 60 and 30 units, respectively per week, we naturally strive to satisfy these demands and our goal is:

$$x_1 + d_3^- - d_3^+ = 60, \quad \text{and} \quad x_2 + d_4^- - d_4^+ = 30.$$

Our third goal is achieved by minimizing d_3^- and fourth goal is achieved by minimizing d_4^-. Further, since the profit on Supreme TV is 2 times the profit on Deluxe TV, we shall have 2 times more desire to minimize d_4^- as to minimize d_3^-.

The achievement function for the problem, then is to find x_1 and x_2 so as to minimize

$$z = \{d_3^- + d_4^-, d_1^-, d_2^+, d_3^- + 2d_4^-\}$$

The final decision model is a linear goal programming problem given below:

Find x_1 and x_2 so as to:

Minimize $z = \{d_1^-, d_2^+, d_3^- + d_4^-, d_3^- + 2d_4^-\}$ subject to the constraints:

$$100x_1 + 150x_2 + d_1^- - d_1^+ = 5{,}000,$$
$$2x_1 + 4x_2 + d_2^- - d_2^+ = 80,$$
$$x_1 + d_3^- - d_3^+ = 60,$$
$$x_2 + d_4^- - d_4^+ = 30;$$
$$x_1, x_2, d_1^-, d_1^+, d_2^-, d_2^+, d_3^-, d_3^+, d_4^-, d_4^+ \geq 0.$$

PROBLEMS

12.01. A manufacturing firm produces two types of products — *A* and *B*. The unit profit of a product *A* is ₹ 70 and that of product *B* is ₹ 30. The goal of the firm is to earn a total profit of exactly ₹ 800 in the next week.

Formulate this problem as a Goal Programming model.

12.02. An office equipment manufacturing company produces two types of products: tables and chairs. The company realizes a profit of ₹ 250 on the table and ₹ 150 on the chair. Because of the strong demand for these two, during the coming days, the company believes it can sell, at the prevailing prices, all these two types of products it can produce. The company's productive capability is, however, limited, since both products have common production facilities which are considered to be scarce resources. These scarce resources pertain to the assembly department and the finishing department. The unit processing times and capacities for each department are shown in the following table:

Type of product	Required to process each product (in hours)		Unit profit contribution
	In assembly deptt.	In finishing deptt.	
Chair	1	1	150
Table	3	1	250
Hours available in each deptt./day	60	40	

However, the company during this period is facing a major reorganization change and feels that maximizing profit is not realistic objective. However, it would like to achieve some satisfactory level of profit during this difficult period. Management feels that a daily profit of ₹ 6,000 would be satisfactory and wishes to determine, given the production time constraints, the product mix that would yield this rate of profit contribution.

Formulate this problem as a Goal Programming model.

12.03. A rural clinic hires its staff from nearby cities and towns on a part-time basis. The clinic attempts to have a general practitioner (GP), a nurse and an internist on duty during at least a portion of each week. The clinic has a weekly budget of ₹ 1,200. A GP charges the clinic ₹ 40 per hour, a nurse charges ₹ 20 per hour, and an internist charges ₹ 150 per hour. The clinic has establish the following goals in order of priority:

(*i*) A nurse should be available for at least 30 hours per week.

(*ii*) The weekly budget of ₹ 1,200 should not be exceeded.

(*iii*) A GP and internist should be available at least 20 hours per week.

(*iv*) An internist should be available for at least 6 hours per week.

Formulate a goal programming model to determine the number of hours to hire each staff member in order to satisfy various goals. ***[MBA, Delhi Univ. 2004]***

12.04. Two products are manufactured on two sequential machines. The following table gives the machining times in minutes per unit for the two products:

Machine	*Product 1*	*Product 2*
X	7	5
Y	8	4

The daily production quota for the two products are 70 and 50 units, respectively. Each machine runs 8 hours a day. Overtime, though not desirable, may be used if necessary to meet the production quota. Formulate the problem as a goal programming model.

12.05. You have been retained as a consultant to formulate a model for job evaluation in an organization. The data collected for 5 representative benchmark jobs across the hierarchy of all blue collar jobs evaluated on a 5-point rating scale for four factors, *viz.*, skill, effort, working condition, and responsibility.

Job grade	*Skill*	*Effort*	*Working condition*	*Responsibility*
A	4	5	5	5
G	4	4	4	4
M	3	4	3	3
S	2	3	2	2
Z	1	2	2	1

Find weights to be assigned to the four factors satisfying the following broad guidelines:

(*i*) *Grade A* : Total points should be as near as possible to 600 but not in excess.

(*ii*) *Grade M* : Total points should be as near as possible to 360 but may be higher or lower.

(*iii*) *Grade Z* : Total points should be at least as near as possible to 120 but not lower.

(*iv*) The difference between two benchmark grades should be at least 120.

Formulate a goal programming model. *[MMS, Mumbai Univ. 2000]*

12.06. A farmer has 1,000 acres of land on which he intends to plant corn, wheat and soybeans. Each acre of corn costs ₹ 10,000 for preparation, requires 7 worker days of labour, and yields a profit of ₹ 3,000. An acre of wheat costs ₹ 12,000 for preparation, requires 10 worker days of labour, and yields a profit of ₹ 4,000. An acre of soybeans costs ₹ 7,000 for preparation, requires 8 worker days of labour, and yields a profit of ₹ 2,000. The farmer has taken a loan of ₹ 80,00,000 for crop preparation and has contracted with a union for 6,000 worker days of labour. A granary has agreed to purchase 200 acres of corn, 500 acres of wheat and 300 acres of soybeans. The farmer has established the following goals, in order of their importance:

(*i*) In order to maintain good relations with the union, the labour contract must be honoured; that is, the full 6,000 worker days must be used.

(*ii*) Preparation costs should not exceed the loan amount, so that additional loans will not have to be secured.

(*iii*) The farmer desires a profit of at least ₹ 1,05,00,000 in order to remain in good financial condition.

(*iv*) Contracting for excess labour should be avoided.

(*v*) The farmer would like to use as much of the available acreage as possible.

(*vi*) The farmer would like to meet the sales with the granary. However, the goal should be weighted according to the profit returned by each crop.

Formulate a goal programming model for determining the number of acres of each crop the farmer should plant in order to satisfy the goals in the best possible way.

[MBA, Delhi Univ. 2003]

12.07. Construct a goal programming model using the following data. Show the deviational variables in the constraints:

Resource type	*Product 1*	*Product 2*	*Available resource*
Labour hours	2	4	600
Material 1 (kg.)	4	5	1,000
Material 2 (kg.)	5	4	1,200
Profit/unit	₹ 20	₹ 32	

It is known that production of one unit of product 1 would require 0.3 person and production of one unit of product 2 would require 0.75 person. The manager has set-up two goals:

(*a*) Profit of ₹ 5,400, and

(*b*) Total staff of 108 persons. *[MBA (Dec.), Gujarat Tech. Univ. 2010]*

12:4. GRAPHICAL GOAL ATTAINMENT METHOD

The graphical goal attainment method is based on a well-defined set of logical steps. Following this systematic procedure, the given linear goal programming problem can be solved with a minimum amount of computational effort. The procedure may be summarized as follows:

Step 1. Formulate the appropriate linear programming problem.

Step 2. Construct the graph of all the objectives (goals) in terms of decision variables as follows: Treat each goal as if it were an equation with deviation (positive as well as negative) variables all set equal to zero, and for each goal equation arbitrarily select two sets of points. Plot each set of points and connect them with a straight line. Indicate the positive and negative deviations range by arrows → or ←, for each goal line.

Step 3. Identify the goal line corresponding to the goal having top (highest) priority. Locate the feasible (solution) region with respect to this goal at the first priority level.

Step 4. Move to goal(s) having the next highest priority and determine the 'best' solution(s) space with respect to the goal(s) at this next higher priority level provided that 'best' solution(s) do not degrade the solution(s) already achieved for higher priority goals.

Step 5. Repeat *Step 4* until all priority levels have been investigated.

Step 6. Identify the optimal solution which corresponds to the most acceptable 'best' value, located in *Step 5.*

Example 12.5. *Asian paints company manufactures two types of outdoor paints, one a latex and the other enamel paint. Each 200 gallons of the latex paint requires 20 man-hour of labour, while each 200 gallons of enamel paint requires 30 man hour labour. It is assumed that only 80 man-hours of labour are available each week and that the owner of the company definitely will not consider hiring additional help or utilizing overtime. Both the enamel and the latex paint produce a profit of ₹ 2 per gallon. The Managing Director of the company has conveyed the desire of the Board of Directors to achieve profit of ₹ 2,000 per week from these two paints. Finally the management has also agreed to supply, if possible, 1,400 gallons of enamel paint per week to a sister concern. Formulate this as an LGPP and solve it.*

Formulation of Goal Programming Problem

The objective and their priorities of the management are as follows:

P_1 : Avoid the utilization of overtime. This is given the top priority because it is stated in the problem as an absolute objective. *(First goal)*

P_2 : Achieve a weekly profit of at least ₹ 2,000. *(Second goal)*

P_3 : Supply the sister concern at least 1,400 gallons of enamel paint every week. *(Third goal)*

Let x_1 and x_2 designate the decision variables corresponding to the number of latex and enamel paints (200 gallons taken as a unit) required to be manufactured per week satisfying all the goals. If d_i^- denotes the negative deviation from the ith goal attainment and d_i^+ denotes the positive deviation from the ith goal attainment, then the various goals of the problem (in order of priority) are:

$$G_1 : \quad 20x_1 + 30x_2 + d_1^- - d_1^+ = 80$$
$$G_2 : \quad 200x_1 + 200x_2 + d_2^- - d_2^+ = 2{,}000$$
$$G_3 : \quad 200x_2 + d_3^- - d_3^+ = 1{,}400$$

The goal G_1 shall be completely attained, if overtime is minimized, that is, the positive deviation (d_1^+) is minimized (*i.e.*, over-utilization is minimized). The second goal shall be attained if profit is maximized which will happen if the negative deviation (d_2^-) from the goal is minimized (*i.e.*, under-achievement is minimized). The third goal shall be attained when the under-achievement, *i.e.*, d_3^- is minimized. The achievement function of the problem is thus

$$z = \{d_1^+, d_2^-, d_3^-\}$$

and the formulated linear goal programming problem is as follows:

Determine x_1 and x_2 so as to

$$\text{Minimize } z = \{d_1^+, d_2^-, d_3^-\} = P_1 d_1^+ + P_2 d_2^- + P_3 d_3^- \text{ and satisfy the goals:}$$

$$20x_1 + 30x_2 + d_1^- - d_1^+ = 80$$

$$200x_1 + 200x_2 + d_2^- - d_2^+ = 2{,}000$$

$$200x_2 + d_3^- - d_3^+ = 1{,}400$$

$$x_1, x_2, d_1^-, d_2^-, d_3^-, d_1^+, d_2^+, d_3^+ \geq 0.$$

Solution of GPP by Graphical Method

Since, there are only two decision variables x_1 and x_2, the graph of various goals will have two axes, horizontal one the X_1-axis and vertical one the X_2-axis. Each point in any of the four quadrants gives a value of any pair (x_1, x_2). But since the variables are not permitted to assume negative values, we need to confine to the first quadrant only. Since, the goal equations G_1, G_2 and G_3 are to be graphed as if there were no deviations from goal attainment, the three constraints are plotted merely as straight lines after treating each $d_i^- = d_i^+ = 0$. The three goals plotted as straight lines are shown in *Fig. 12.1*. Note that only the ***decision*** variables (x_1 and x_2) are used in the graph. However, the region of an increase in any positive or negative deviation variable ($d_1^-, d_2^-, d_3^-, d_1^+, d_2^+$ and d_3^+) is reflected by the arrows (→ and ←) across each goal line. The particular deviation variables to be minimized (*i.e.*, those in the achievement function) have been shown encircled.

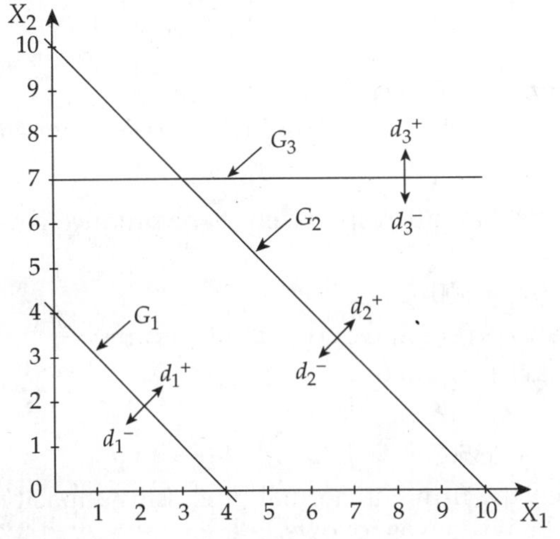

Fig. 12.1

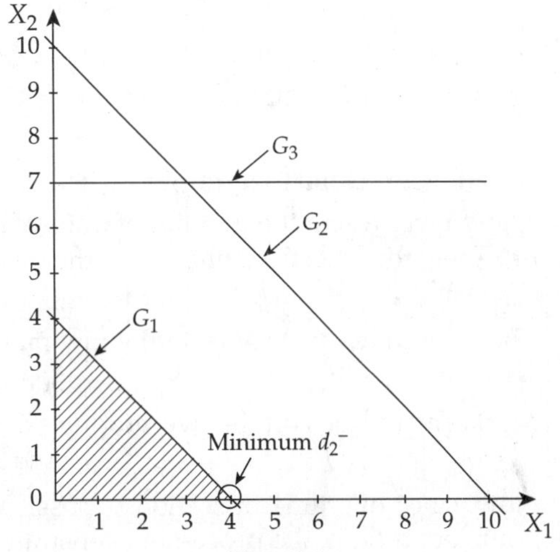

Fig. 12.2

Goal Attainment at Priority Level One

The goal with top priority 'G_1' is considered first. This goal shall be attained when d_1^+ is minimized. Since, this is the first goal considered, we may set $d_1^+ = 0$ as its minimum value. The region satisfying $d_1^+ = 0$ and $x_1, x_2 \geq 0$ is shown shaded in *Fig. 12.2*. Any pair of values (x_1, x_2) in this region shall satisfy G_1 and at the same time minimize d_1^+ at the priority level one.

Goal Attainment at Next Higher Priority Levels

We now move to the next higher priority, P_2, and consider the goals at that priority. This is simply 'G_2' and to satisfy G_2, we must minimize d_3^-. The reader can easil; note that d_2^- cannot be set to zero (as its minimum value) since this would degrade our previous solution, *viz.*, a higher priority goal. The minimum value of d_2^-, without affecting the previously attained higher priority goals is at $x_1 = 4$, $x_2 = 0$ which is the point indicated in *Fig. 12.2*.

Moving now to the last priority level we consider the goal G_3. To satisfy, G_3, we must minimize d_3^-. But we see that any minimization of d_2^- would require a movement from the solution previously determined and thus would degrade at least one higher priority goal already attained. Consequently the optimum solution to the problem is

$$x_1^0 = 4,\ x_2^0 = 0;\ z^0 = (0,\ 1200,\ 1400).$$

The optimum values of the achievement function z_0 simply indicate that the highest priority goal has been completely attained, while the second and third priority goals have not.

Example 12.6. *A firm is engaged in producing two products A and B, which yield a unit profit of ₹ 400 and ₹ 300, respectively. Both of these require a certain raw material whose supply is limited to 300 kg per month. The raw material requirement is 1 kg per unit of each product. Further, it is known that 1 unit of product A requires 2 hours of processing time, while each unit of product B needs one hour. The firm has a normal processing capacity of 400 hours per month. The marketing department has stated that according to the prevailing conditions, and likely to continue in the foreseeable future, the maximum number of units of products A and B which can be sold each month is 150 and 350, respectively. The manager of the firm has set the following goals in order of their importance and the priorities being pre-emptive.*

Goal 1. *Avoid under-utilization of normal productive capacity.*

Goal 2. *Sell the maximum number of units of products A and B. However, since the two products yield profit in the ratio 400 : 300, the manager would give the weightage to the sale of products in the same ratio.*

Goal 3. *Minimize overtime of processing capacity.*

You are required to formulate this as a goal programming problem and solve it graphically.

[MBA (Nov.), Delhi Univ. 2003]

Formulation of Goal Programming Problem

Let x_1 and x_2 represent the number of units of products A and B, respectively, to be produced every month. Then, the objective function of the firm is:

$$\text{Maximize}\quad z = 400x_1 + 300x_2.$$

The respective constraints on the raw material and processing of two products are:

$$x_1 + x_2 \le 300 \quad \text{and} \quad 2x_1 + x_2 \le 400.$$

Market conditions on the two products are:

$$x_1 \le 150 \quad \text{and} \quad x_2 \le 350.$$

Now, since overtime operation is possible when the firm has normal processing capacity of 400 hours per month, the processing constraint can be written as follows:

$$2x_1 + x_2 + d_1^- - d_1^+ = 400,$$

where, d_1^- is the number of hours unutilized capacity and d_1^+ is the number of hours overtime operation of the normal processing capacity.

Further, since the sales goals are given in terms of the maximum possible sales volume, they can be expressed as follows:

$$x_1 + d_2^- = 150, \quad \text{and} \quad x_2 + d_3^- = 350.$$

It may be noted, here, that since the given sales levels are the maximum ones, deviational variables are provided only for their under-achievement and not for over-achievement.

Thus, the complete goal programming problem is formulated as follows:

$$\text{Minimize}\quad z = P_1 d_1^- + P_2(4d_2^- + 3d_3^-) + P_3 d_1^+ \quad \text{subject to the constraints:}$$

$$2x_1 + x_2 + d_1^- - d_1^+ = 400,\quad x_1 + d_2^- = 150,$$

$$x_2 + d_3^- = 350,\quad x_1 + x_2 \le 300,$$

$$x_1, x_2, d_1^-, d_1^+, d_2^- \text{ and } d_3^- \ge 0.$$

Solution of GPP by Graphical Method

Since, there are only two decision variables x_1 and x_2, the graph of various goals will have only horizontal axis, say x_1 and vertical axis, say x_2. Further, since all the variables are non-negative, we confine to the solution only in the first quadrant of cartesian axis.

First of all, we plot the material constraint, as shown in *Fig. 12.3*, given by the region below in the line GH in the first quadrant. Then, the constraints involving goals are plotted (by setting deviational variables equal to zero). Since, the production capacity constraint may be under-achieved or over-achieved, the feasible region lies on both sides of the line CD, as indicated by d_1^- and d_1^+. Thus, it does not alter the feasible region. The first goal would, thus, be met and $d_1^- = 0$. Now, we move on to second goal.

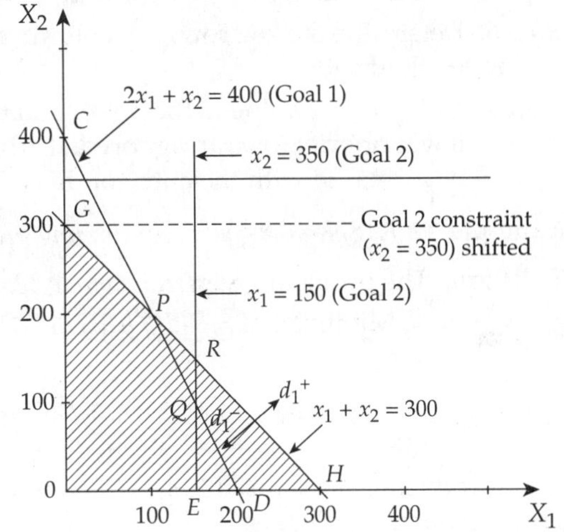

Fig. 12.3

From the lines representing the second constraint, we observe that the region HRE shows the undesirable deviation and is, therefore, not to be considered. On the other hand, the maximum output of the product *B* ($x_2 = 350$) lies beyond the feasible region. Accordingly, the goal constraint line is shown shifted until it touches the boundary of the feasible region at point *G*. As the product *A* yields a greater profit of ₹ 400 over the profit of product *B*, which yields profit of ₹ 300; a higher penalty is stipulated for deviating from the output target of product *A*. Between points *G*, *P* and *R*, we may obtain the optimum solution by computing the values of d_2^- and d_3^-, and the value of *z* there from. Thus, we have,

	x_1	x_2	d_2^-	d_3^-	$z\ (= 4d_2^- + 3d_3^-)$
G	0	300	150	50	750
P	100	200	50	150	650
R	150	150	0	200	600 Minimum

From the above computation, we observe that, the minimum value of *z* lies at *R*, *i.e.*, at (150, 150). Hence, the optimum solution is: $x_1 = 150$ and $x_2 = 150$, with minimum $z = 600$. Now, substituting these values of x_1 and x_2 in the Goal 1 constraint, we have $2 \times 150 + 150 - d_1^+ = 400$, as $d_1^- = 0$. This gives $d_1^+ = 50$ which implies that 50 hours of processing capacity shall be worked overtime. Further, since the second goal has not been met fully, there has been no consideration of the next priority goal.

12:5. SIMPLEX METHOD FOR SOLVING GOAL PROGRAMMING PROBLEM

The major steps for solving the linear goal programming problem by simplex method are given as follows:

Step 1. Identify the decision variables of the key decision and formulate the given situation as linear goal programming problem.

Step 2. Determine the initial basic feasible solution and set up initial simplex table. Compute $(z_j - c_j)$ values (or net evaluations or index numbers) separately for each of the ranked goals: $P_1, P_2, \ldots$ and enter at the bottom of the simplex table. These are shown from bottom to top, *i.e.*, first priority goal 'P_1' is shown at the bottom anu least priority goal at the top.

Step 3. Examine $(z_j - c_j)$ values (or net-evaluations) in the P_1-row first. If all $\{(z_j - c_j) \leq 0\}$ at the highest priority levels, then the optimum solution has been obtained. If at least one $\{(z_j - c_j) > 0\}$ at a certain priority level and there is no negative entry at the higher unachieved priority levels, in the same column, then the current solution is not optimum.

Step 4. If the target values of each goal in the solution column is zero, the current solution is optimum.

Step 5. Examine the positive values of net-evaluations of the highest priority, P_1, and choose the largest of these. The column corresponding to this value becomes the key column. Otherwise, move to the next higher priority, P_2, and select the largest positive value of net-evaluation for determining the key column.

Step 6. Determine the key row, key column and leading element in the same way as in the simplex method.

Step 7. Any positive value in the net-evaluation row which has negative net-evaluation under any lower priority rows are ignored. It is because deviations from the highest priority goal would be increased with the entry of this non-basic variable in the basis.

Example 12.7. *Solve the goal programming problem given in Example 12.6 using Simplex Method.*

Solution. The problem given in *Example 12.6* is reproduced below:

$$\text{Minimize } z = P_1 d_1^- + P_2(4d_2^- + 3d_3^-) + P_3 d_1^+ \text{ subject to the constraints:}$$

$$2x_1 + x_2 + d_1^- - d_1^+ = 400, \quad x_1 + x_2 \leq 300,$$

$$x_1 + d_2^- = 150, \quad x_2 + d_3^- = 350;$$

$$x_1, x_2, s_1, d_1^-, d_1^+, d_2^-, d_3^- \geq 0.$$

Introducing slack variable $s_1 \geq 0$ in the inequality: $x_1 + x_2 \leq 300$, the standard form of the goal programming problem is:

$$\text{Minimize } z = 0.x_1 + 0.x_2 + 0.s_1 + P_1 d_1^- + 4P_2 d_2^- + 3P_2 d_3^- + P_3 d_1^+$$

subject to the constraints:

$$2x_1 + x_2 + d_1^- - d_1^+ = 400, \quad x_1 + x_2 + s_1 = 300,$$

$$x_1 + d_2^- = 150, \quad x_2 + d_3^- = 350;$$

and

$$x_1, x_2, s_1, d_1^-, d_1^+, d_2^-, d_3^- \geq 0.$$

From the above constraints, an initial basic feasible solution is:

$$d_1^- = 400, \quad s_1 = 300, \quad d_2^- = 150 \text{ and } d_3^- = 350.$$

Using, now, Simplex Method, we have the following simplex tableau:

Initial Simplex Tableau. x_1 enters and d_2^- leaves.

Priority goal	c_j / *Basic variable*	0 x_1	0 x_2	P_1 d_1^-	P_3 d_1^+	0 s_1	$4P_2$ d_2^-	$3P_2$ d_3^-	*Solution* (b_i)	*Ratio* (b_i/a_{ij})
P_1	d_1^-	2	1	1	−1	0	0	0	400	400/2
0	s_1	1	1	0	0	1	0	0	300	300/1
← $4P_2$	d_2^-	(1)	0	0	0	0	1	0	150	150/1
$3P_2$	d_3^-	0	1	0	0	0	0	1	350	
Net-evaluations i.e., $(z_j - c_j)$	P_3	0	0	0	−1	0	0	0	0	
	P_2	4	3	0	0	0	0	0	1,650	
	P_1	2↑	1	0	−1	0	0	0	400	

For computing z_j values, consider the weights (coefficients) mentioned with P_1, P_2 and P_3 and then multiply these weights with the corresponding entries of the various columns of the simplex tableau. For example, considering P_1-row, $z_1 = 1 \times 2$, *i.e.*, 2. This gives $(z_1 - c_1) = 2 - 0 = 2$. Now, considering P_2-row, $z_1 = 4 \times 1 + 3 \times 0$, *i.e.*, 4. This gives $(z_1 - c_1) = 4 - 0 = 4$. There is no P_3-row, *i.e.*, P_3 is not in basis. Therefore, $(z_1 - c_1)$ corresponding to this row gives $(z_1 - c_1) = 0$. Similarly, $(z_4 - c_4)$ corresponding to P_3, we get $(z_4 - c_4) = 0 - 1 = -1$, because P_3 is not basic variable, *i.e.*, value of $P_3 = 0$.

To test the optimality, we first look the $(z_j - c_j)$ values in the bottom row, headed P_1. In this row, since $(z_1 - c_1)$ and $(z_2 - c_2)$ are positive, the solution is not optimum.

For the revised simplex tableau, we consider the most positive values of $(z_j - c_j)$ in the P_1-row. It being '2' for the column headed by x_1, therefore, x_1 will be basic variable in the revised simplex tableau. Using, now replacement ratio 'b_i/a_{i1}' and then selecting the least non-negative one, we find that d_2^- leaves the basis thereby 'a_{31}' becomes the leading element. Converting now the leading element into unity and all other elements of the entering column to zero, the revised simplex tableau is given below:

Priority goal	c_j Basic variable	0 x_1	0 x_2	P_1 d_1^-	P_3 d_1^+	0 s_1	$4P_2$ d_2^-	$3P_2$ d_3^-	Solution	Ratio
← P_1	d_1^-	0	(1)	1	–1	0	–2	0	100	100/1
0	s_1	0	1	0	0	1	–1	0	150	150/1
0	x_1	1	0	0	0	0	1	0	150	—
$3P_2$	d_3^-	0	1	0	0	0	0	1	350	350/1
Net-evaluations, *i.e.*, $(z_j - c_j)$	P_3	0	0	0	–1	0	0	0	0	
	P_2	0	3	0	0	0	–4	0	1,050	
	P_1	0	1↑	0	–1	0	–2	0	100	

In this tableau, since $(z_2 - c_2)$ corresponding to P_1-row is only positive, x_2 enters the basis. Further, the minimum of the replacement ratio corresponds to P_1-row, basic variable d_1^- becomes non-basic in the revised tableau. Thus, a_{12} (= 1) becomes the leading element. Converting this leading element to unity and all other elements of entering column to zero, second revised simplex tableau is as follows:

Second Simplex Tableau. d_1^+ enters and s_1 leaves.

Priority goal	c_j Basic variable	0 x_1	0 x_2	P_1 d_1^-	P_3 d_1^+	0 s_1	$4P_2$ d_2^-	$3P_2$ d_3^-	Solution	Ratio
0	x_2	0	1	1	–1	0	–2	0	100	
← 0	s_1	0	0	–1	(1)	1	1	0	50	50/1
0	x_1	1	0	0	0	0	1	0	150	
$3P_2$	d_3^-	0	0	–1	1	0	2	1	250	250/1
Net-evaluations, *i.e.*, $(z_j - c_j)$	P_3	0	0	0	–1	0	0	0		
	P_2	0	0	–3	3 ↑	0	2	0		
	P_1	0	0	–1	0	0	0	0		

From the above tableau, we observe that there are no positive $(z_j - c_j)$ value in the P_1-row and hence the optimum solution is achieved in relation to the top-priority goal P_1. So, we consider now the $(z_j - c_j)$ values in relation to P_2-row. Here $(z_4 - c_4)$ gives the most positive value which implies that d^+ will become basic variable in the revised simplex tableau. Further, s_1-row gives the least positive replacement ratio. Thus, s_1 leaves the basis and thereby a_{24} (= 1) becomes the leading element. Converting the leading element to zero and all other elements of entering column to zero, the next revised simplex tableau is as follows:

Final Simplex Tableau. Optimum solution.

Priority goal	Basic variable	0 x_1	0 x_2	P_1 d_1^-	P_3 d_1^+	0 s_1	$4P_2$ d_2^-	$3P_2$ d_3^-	Solution
0	x_2	0	1	0	0	1	–1	0	150
P_3	d_1^+	0	0	–1	1	1	1	0	50
0	x_1	1	0	0	0	0	1	0	150
$3P_2$	d_3^-	0	0	0	0	–1	1	1	200
Net-evaluations, i.e., $(z_j - c_j)$	P_3	0	0	0	0	1	1	0	
	P_2	0	0	0	0	–3	–1	0	
	P_1	0	0	–1	0	0	0	0	

The above simplex tableau gives the optimum solution in the sense that it enables the manager to attain the set goals as closely as possible within the given decision scenario and the priority structure. It may be seen that none of the $(z_j - c_j)$'s in the P_1 and P_2 rows is positive. However, there is a positive $(z_j - c_j)$ under the variable s_1 and d_2^- in the P_3-row. This indicates that non-basic variable s_1 or d_2^- enters the basis. But, we observe that this will be done only at the expenses of achieving the higher priority goal, since in the P_2-row, $(z_5 - c_5)$ is '–3'. The rule, then, is that if there is a positive $(z_j - c_j)$ at a lower priority level, then the variable in that column can not be selected as the incoming variable, if there is a negative $(z_j - c_j)$ value at a higher priority goal.

Hence, the optimum solution is: $x_1 = 150$, $x_2 = 150$, $d_1^+ = 50$ and $d_3^- = 200$. This implies that 150 units each of the products *A* and *B* be produced, so that the target of selling 350 units of product *B* shall be under-achieved by 200 units and an overtime work of 50 hours shall be involved.

Example 12.8. *A textile company produces two types of materials, A and B. The material A is produced according to direct orders from furniture manufacturers. The material B is distributed to retail fabric stores. The average production rates for the materials A and B are identical, viz., 1,000 metres per hour. By running two shifts, the operational capacity of the plant is 80 hours per week.*

The marketing department reports that the maximum estimated sales for the following week is 70,000 metres of material A and 45,000 metres of material B. According to the accounting department the profit from a metre of material A is ₹5 and from a metre of material B is ₹3.

The management of the company decides that a stable employment level is a primary goal for the firm. Therefore, whenever there is demand exceeding normal production capacity, management simply expands production capacity by providing overtime. However, management feels that overtime operation of the plant of more than 10 hours per week should be avoided because of the accelerating costs. The management has the following goals in order of their importance.

P_1 : *Avoid any under-utilization of production capacity.*

P_2 : *Limit the overtime allowed for plant operation to 10 hours per week.*

P_3 : *Achieve the sales target of 70,000 metres of material A and 45,000 metres of material B.*

P_4 : *Minimize overtime operation of the plant as much as possible.*

Formulate this problem as a goal programming problem and derive its solution by simplex method.

Formulation of Goal Programming Problem

Production Hours Constraint. Regular production hours available to the firm are limited to 80 hours. The management may allow some extra hours of work (overtime), if necessary. Thus, the production hours constraint could be setup as follows:

$$x_1 + x_2 + d_1^- - d_1^+ = 80,$$

where, x_1 = number of hours used for producing the material A.

x_2 = number of hours used for producing the material B.

d_1^- = amount of under-utilization of available normal production hours.

d_1^+ = amount of extra hours (overtime) beyond normal production hours allowed by the management.

Sales Constraint. Since, the maximum sales for the following week is 70,000 metres of material A and 45,000 metres of B, the sales constraint is

$$x_1 + d_2^- = 70 \quad \text{and} \quad x_2 + d_3^- = 45$$

where, d_2^- and d_3^- are the under-achievement of sales target of materials A and B, respectively.

Overtime Constraint. The overtime constraint is

$$d_1^+ + d_4^- - d_4^+ = 10,$$

where, d_4^- = amount of under-utilized hours between the actual overtime allowed and 10 hours of overtime, and d_4^+ = amount of overtime in excess of the allowed 10 hours. Both negative and positive deviations are included from the allowed 10 hours of overtime, because the actual overtime can be less than, equal to, or even more than 10 hours.

Substituting the value of d_1^+ in the production hours constraint, we have

$$x_1 + x_2 + d_4^- - d_4^+ = 90$$

In this constraint, under-utilization (d^-) is meaningless and hence has been ignored.

Achievement (Objective) Function. Since, sales goal for materials are equally important, the profit contribution ratio 5 : 3 between these two will be considered as different weights. The achievement function, therefore, is:

$$\text{Minimize } z = P_1 d_1^- + P_2 d_4^+ + 5P_3 d_2^- + 3P_3 d_3^- + P_4 d_1^+.$$

The complete goal programming problem, therefore, is as under:

$$\text{Minimize } z = P_1 d_1^- + 5P_3 d_2^- + 3P_3 d_3^- + P_2 d_4^+ + P_4 d_1^+ \quad \text{subject to the constraints:}$$

$$x_1 + x_2 + d_1^- - d_1^+ = 80,$$

$$x_1 + x_2 + d_4^- - d_4^+ = 90,$$

$$x_1 + d_2^- = 70, \quad \text{and} \quad x_2 + d_3^- = 45;$$

$$x_1, x_2, d_1^-, d_1^+, d_2^-, d_3^-, d_4^-, d_4^+ \geq 0.$$

Solution by Simplex Method. Considering the objective function (goal) as one of the constraint, the initial simplex tableau is:

Priority goal	c_j / Basic variable	0 x_1	0 x_2	P_1 d_1^-	$5P_3$ d_2^-	$3P_3$ d_3^-	0 d_4^-	P_4 d_1^+	P_2 d_4^+	*Solution*	*Ratio*
P_1	d_1^-	1	1	1	0	0	0	−1	0	80	80/1
0	d_4^-	1	1	0	0	0	1	0	−1	90	90/1
← $5P_3$	d_2^-	(1)	0	0	1	0	0	0	0	70	70/1
$3P_3$	d_3^-	0	1	0	0	1	0	0	0	45	—
Net-evaluations *i.e.*, $(z_j - c_j)$	P_4	0	0	0	0	0	0	−1	0	0	
	P_3	5	3	0	0	0	0	0	0	485	
	P_2	0	0	0	0	0	0	0	−1	0	
	P_1	1↑	1	0	0	0	0	−1	0	80	

Using simplex method for the calculation of z-value, we have

$$z = P_1 \times 80 + 0 \times 90 + 5P_3 \times 70 + 3P_3 \times 45 = 80P_1 + 485P_3$$

This implies that $P_1 = 80$, $P_2 = 0$, $P_3 = 485$ and $P_4 = 0$ represent the unachieved portion of each goal.

First Iteration. Since, $(z_1 - c_1)$ and $(z_2 - c_2)$ in the P_1-row are positive, the solution is not optimum and we choose the largest of these. But, at the P_1-row both $(z_1 - c_1)$ and $(z_2 - c_2)$ are same, so we check the next priority levels in x_1 and x_2 columns. We observe that at priority P_3, the largest net evaluation is 5 under x_1-column. Thus, the column under x_1 becomes the key column.

The key row is the row with the least replacement ratio value, which is obtained by dividing the solution column by the corresponding positive coefficients in the key-column. Thus, the third row (*i.e.*, d_2^--row) becomes the key row. The number '1' which is in the third row and first column becomes the leading element.

Using simplex method procedure, revised simplex tableau is obtained as under:

		0	0	P_1	$5P_3$	$3P_3$	0	P_4	P_2		
Priority goal	*Basic variable*	x_1	x_2	d_1^-	d_2^-	d_3^-	d_4^-	d_1^+	d_4^+	*Solution*	*Ratio*
$\leftarrow P_1$	d_1^-	0	(1)	1	−1	0	0	−1	0	10	10/1
0	d_4^-	0	1	0	−1	0	1	0	−1	20	20/1
0	x_1	1	0	0	1	0	0	0	0	70	—
$3P_3$	d_3^-	0	1	0	0	1	0	0	0	45	45/1
Net-evaluations *i.e.*, $(z_j - c_j)$	P_4	0	0	0	0	0	0	−1	0	0	
	P_3	0	3	0	−5	0	0	0	0	135	
	P_2	0	0	0	0	0	0	0	−1	0	
	P_1	0	1↑	0	−1	0	0	−1	0	10	

Second Iteration. Since, $(z_2 - c_2) > 0$ at P_1-level, column under x_2 becomes key column. Proceeding in the usual manner, first row becomes the key row, thereby the number '1' in the first row and second column becomes the leading element. The revised simplex tableau is obtained as follows:

		0	0	P_1	$5P_3$	$3P_3$	0	P_4	P_2		
Priority goal	*Basic variable*	x_1	x_2	d_1^-	d_2^-	d_3^-	d_4^-	d_1^+	d_4^+	*Solution*	*Ratio*
0	x_2	0	1	1	−1	0	0	−1	0	10	—
¬0	d_4^-	0	0	−1	0	0	1	(1)	−1	10	10/1
0	x_1	1	0	0	1	0	0	0	0	70	—
$3P_3$	d_3^-	0	0	−1	1	1	0	1	0	35	35/1
Net-evaluations, *i.e.*, $(z_j - c_j)$	P_4	0	0	0	0	0	0	−1	0	0	
	P_3	0	0	−3	−2	0	0	3↑	0	105	
	P_2	0	0	0	0	0	0	0	−1	0	
	P_1	0	0	−1	0	0	0	0	0	0	

From the above tableau, we observe that production of 70,000 metres of material *A* and 10,000 metres of material *B* is sufficient to achieve the first, second and fourth goals. Also, the value of $d_3^- = 35$ suggests that 35,000 metres of material is not achieved. Thus, first two priorities (P_1 and P_2) have been achieved. Therefore, to improve the current solution, the selection of key column is done at P_3-level. At P_3-level, since $(z_7 - c_7) > 0$, column under d_1^+ becomes the key column.

Now proceeding as usual, second row becomes the key row, the number '1' in the second row and under d_1^+-column becomes the leading element.

The revised simplex tableau is obtained as under:

		0	0	P_1	$5P_1$	$3P_3$	0	P_4	P_2	
Priority goal	*Basic variable*	x_1	x_2	d_1^-	d_2^-	d_3^-	d_4^-	d_1^+	d_4^+	*Solution*
0	x_2	0	1	0	–1	0	1	0	–1	20
P_4	d_1^+	0	0	–1	0	0	1	1	–1	10
0	x_1	1	0	0	1	0	0	0	0	70
$3P_3$	d_3^-	0	0	0	1	1	–1	0	1	25
Net-evaluations, *i.e.*, $(z_j - c_j)$	P_4	0	0	–1	0	0	1	0	–1	10
	P_3	0	0	0	–2	0	–3	0	3	75
	P_2	0	0	0	0	0	0	0	–1	0
	P_1	0	0	–1	0	0	0	0	0	0

From the above tableau, we observe that there is positive number at P_3 level under d_4^+ column. But, since there is a negative number '–1' at a higher priority level 'P_2' in the same column, this cannot become the key column. Also, there is one positive number '1' under d_4^- column at 'P_4' level and therefore, fourth goal cannot be achieved at the expenses of achieving the third goal. Hence, the optimum solution is:

$$x_1 = 70, \; x_2 = 20, \; d_1^+ = 10, \; d_3^- = 25, \text{ and } d_1^- = d_2^- = d_4^+ = 0.$$

PROBLEMS

12.08. A small furnishing company manufacture tables and chairs. Each chair requires 4 man-hours of labour and each table requires 5 man-hours of labour. If only 40 man-hours are available each week and owner of the company would neither hire additional labour nor utilize overtime. Formulate the linear goal programming problem and solve it graphically. Both the table and the chair fetch a profit of ₹ 100 each. The owner has a target to earn a profit of ₹ 2,000 per week. Also, he would like to supply 10 chairs, if possible per week to a sister concern.

[**Hint:** The goals of the company and their assigned priorities are:

P_1 : Avoid hiring extra labour or utilize overtime.

P_2 : Reach a profit goal of ₹ 2,000 a week.

P_3 : Supply 10 chairs a week to sister concern.]

12.09. A manufacturing firm produces two types of products *A* and *B*. According to past experience, production of either Product *A* or Product *B* requires an average of one hour in the plant. The plant has a normal production capacity of 400 hours a month. The marketing department of the firm reports that because of limited market, the maximum number of Product *A* and Product *B* that can be sold in a month are 240 and 300, respectively. The net profit from the sale of Product *A* and Product *B* are ₹ 800 and ₹ 400, respectively. The manager of the firm has set the following goals arranged in the order of importance (pre-emptive priority factors).

P_1 : He wants to avoid any under-utilization of normal production capacity.

P_2 : He wants to sell maximum possible units of Product *A* and Product *B*. Since, the net profit from the sale of Product *A* is twice the amount of Product *B*, therefore, the manager has twice as much desire to achieve sales for Product *A* than Product *B*.

P_3 : He wants to minimize the overtime operation of the plant as much as possible.

Formulate and solve the given problem by graphical method of goal programming.

12.10. A company produces motorcycle seats. The company has two production lines. The production rate for line 1 is 50 seats per hour, and for line 2 it is 60 seats per hour. The company has entered into a contract to supply 1,200 seats daily to another company. Currently, the normal operation period for each line is 8 hours. The production manager of the company is trying to determine the best daily operation hours for the two lines in order to achieve the following goals:

P_1 : Produce and deliver 1,200 seats daily.

P_2 : Limit the daily overtime operation hours of line 2 to 3 hours.

P_3 : Minimize under-utilization of the regular daily operation hours of each line. Assign differential weights based on the relative productivity rate.

P_4 : Minimize the daily overtime operation hours of each line as much as possible. Assign differential weights based on the relative cost of overtime. it is assumed that the cost of operation is identical for the two production lines.

Formulate a goal programming model for this problem and then solve it by using the graphical method.

12.11. Solve the following linear goal programming problem graphically:

$$\text{Minimize } z = G_1(d_3^+ + d_4^+) + G_2 d_1^+ + G_3 d_2^- + G_4\left(d_3^- + \frac{3}{2}d_4^-\right)$$

and satisfy the goals:

$$\begin{aligned} G_1 &: x_1 + x_2 + d_1^- - d_1^+ = 40 \\ G_2 &: x_1 + x_2 + d_2^- - d_2^+ = 100 \\ G_3 &: x_1 + d_3^- - d_3^+ = 30 \\ G_4 &: x_2 + d_4^- - d_4^+ = 15 \end{aligned}$$

$$d_i^-, d_i^+ \geq \text{ for all } i = 1, 2, 3, 4 \text{ and } x_1 \geq 0, x_2 \geq 0.$$

The goals have been listed in order of priority.

12.12. A production manager faces the problem of job allocation between two assembly lines. The production rate of assembly line 1 is 10 units per hour, and the production rate of assembly line 2 is 12 units per hour. The normal working period for both lines is eight hours per day. The production manager has set the following goals for the next day, listed in order of their importance:

(*i*) Avoid any under-achievement of the production level, which is set at 200 units of product.

(*ii*) Avoid any overtime operation of line 2 beyond four hours.

(*iii*) Avoid any under-utilization of regular working hours (assign differential weights according to the relative productivity of the two lines).

(*iv*) Minimize overtime in both assembly lines.

Formulate a goal programming model and solve it. ***[MBA, Delhi Univ. 2010]***

12.13. The Soft-Ad Company is planning for the next month an advertising campaign for one of its clients: Amco. The Company is considering advertising during sports programmes and during soap operas on TV. Amco has specified the following goals:

Goal 1 : Reach at least 20 lakh high-income males (HIM).

Goal 2 : Reach at least 30 lakh medium-income families (MIF).

Goal 3 : Reach at least 15 lakh high-income families (HIF).

The information about cost and viewers for each one-minute ad slots is given below:

Ad slot	Number of viewers per ad (in lakh)			Cost per slot (₹)
	HIM	MIF	HIF	
Sports	4	5	2	2,00,000
Soap opera	3	8	4	3,00,000

Further, the company has budgeted ₹ 12 lakh for these advertisements, which should not be exceeded. It has also estimated that each lakh of viewers by which it falls, short of goal 1 costs it ₹ 2 lakh on account of lost sales. Similarly, this penalty for goals 2 and 3 are ₹ 1.0 lakh and ₹ 0.8 lakh, respectively. How should Soft-Ad plan for Amco? Formulate and solve this as a linear goal programming problem.

12.14. The production manager of a company wants to schedule a week's production run for two products *A* and *B*, each of which requires the labour and materials as shown below:

Product	Labour (in hours)	Material M_1 (in kgs)	Material M_2 (in kg)
A	2	4	5
B	4	5	4
Available (per week)	600	1,000	1,200

The unit profit for products *A* and *B* are ₹ 20 and ₹ 32, respectively.

The manager would like to maximize profit, but he is equally concerned with maintaining workforce of the division at nearly constant level in the interest of employee morale. The workforce, which consists of people engaged in production, sales, distribution and other general staff is consisted of 108 persons in all. Also, it is known that the production of one unit of product *A* would maintain 0.3 persons in the workforce and one unit of product *B* would maintain 0.75 persons.

Had the production manager been considering only maximizing of profit, without regard to maintaining the workforce, he would do so by producing 167.67 units of product *A* and 66.67 units of product *B*. On the basis of available capacity, this would yield a profit of ₹ 5,486.67. However, this would maintain 100.3 people in the workforce. The manager feels that probably he could increase the workforce requirement to the desired level by accepting somewhat lower profit. So, the following two goals are to be achieved: (*a*) profit of ₹ 5,400 per week, and (*b*) workforce of 108 persons.

Formulate and solve the given problem as linear goal programming problem using simplex method. ***[MBA, Gujarat Technological Univ. 2010]***

12.15. Use simplex method to solve the following goal programming problems:

(*a*) Minimize $z = P_1 d_1^- + P_2 d_2^- + 2P_2 d_2^- + P_3 d_1^+$ subject to the constraints:

$$10x_1 + 10x_2 + d_1^- - d_1^+ = 400,$$
$$x_1 + d_2^- = 40, \quad x_2 + d_3^- = 30;$$
$$x_1, x_2, d_1^+, d_1^-, d_2^-, d_3^- \geq 0.$$

(*b*) Minimize $z = P_1 d_1^- + P_2(2d_2^- + d_3^-) + P_3 d_1^+$ subject to the constraints:

$$x_1 + x_2 + d_1^- - d_1^+ = 400, \quad x_1 + d_2^- = 240,$$
$$x_2 + d_3^- = 300; \; x_1, x_2, d_1^-, d_1^+, d_2^-, d_3^- \geq 0.$$

12.16. 'Mateel Ltd.' produces toy carriage, whose final assembly must include four wheels and two seats. The factory producing the parts operates three shifts a day. The following table provides the amounts produced of each part in three shifts *viz.*, are as follows:

Shift	*Units produced per run*	
	Wheels	*Seats*
1	500	300
2	600	280
3	640	360

Ideally, the number of produced wheel is exactly twice that of the number of seats (as, each carriage has four wheels and two seats). However, because the production rate vary from shift to shift, exact balance in production may not be possible. *Mateel* is interested in determining the number of production runs in each shift that minimizes the imbalance in production of the parts. The capacity limitations restrict the number of runs below 5 for shift 1, below 20 for shift 2, and below 5 for shift 3.

(*i*) Formulate the problem as a goal programming model.
(*ii*) Find the solution to the above model using any method known to you.

[MBA (PT), Delhi Univ. 2008]

12.17. Conceptual Products is a computer company that produces CP400 and CP500 computers. The computers use different mother-boards which are in abundant supply, but uses same cases and disc drives. CP400 models use two floppy disc drives and no zip drive, whereas CP500 models use one floppy disc drive and one zip drive.

The disc drives and cases are bought from vendors. There are 1,000 floppy disc drives, 500 zip drives and 600 cases available to Conceptual Products on a weekly basis. It takes one hour to manufacture a CP400 and its profit is ₹ 2,000, whereas it takes one and one-half hours to manufacture a CP500 and its profit is ₹ 5,000.

The company has three goals, which are given below:

Priority 1: Meet a contract of 200 models of CP400 per week (Goal 1).
Priority 2: Make at least 500 computers in total every week (Goal 2).
Priority 3: Make at least ₹ 25,00,000 in profit every week (Goal 3).

Formulate this as a goal programming problem and solve it. ***[MBA, Delhi Univ. 2009]***

12.18. The Electric Co., located in old town area, produces two products popular with home renovators: old fashioned chandeliers and ceiling fans. Both chandeliers and fans require a two-step production process involving wiring and assembly. It takes 2 hours to wire each chandelier and 3 hours to wire a ceiling fan. Final assembly of chandeliers and fans requires 6 and 5 hours, respectively. The production capability is such that only 12 hours of wiring time and 30 hours of assembly time are available. Each chandelier produced brings the company a profit of ₹ 70 and each fan a profit of ₹ 60. Harrison's management wants to achieve several goals, arranged in order of their importance to the company.

Goal 1: To produce profit of ₹ 300, if possible, during the production period.
Goal 2: To fully utilize the available wiring department hours.
Goal 3: To avoid overtime in the assembly department.
Goal 4: To meet a contract requirement of producing at least 7 ceiling fans.

Formulate and solve the problem as goal programming problem.

[MBA (Jan.), Gujarat Tech. Univ. 2011]

12.19. Mr. *X* has inherited ₹ 10,00,000 and seeks your advice concerning investing of his money. You have determined that 10 per cent can be earned on a bank account and 14 per cent by investing in certificates of deposit. In real estate, you estimate an annual return of 15 per cent, while on the stock market you estimate that 20 per cent can be earned annually. Mr. *X* has established the following goals in order of preference:

(*i*) Minimize risk; therefore, he wants to invest not more than 35 per cent in any one type of investment.

(*ii*) Must have ₹ 1,00,000 in bank account to meet any emergency.

(*iii*) Maximize annual cash return.

Formulate and solve (upto three iterations only) this problem as a goal programming problem that determines the amount of money which X should invest in each investment option.

[MBA, Delhi Univ. 2007]

12.20. A production manager faces the problem of job allocation among three of his teams. The processing rates of the three teams are 5, 6 and 8 units per hour, respectively. The normal working hours for each team is 8 hours per day. The production manager has the following goals for the next day in order of priority:

(*i*) The manager wants to avoid any under-achievement of production level, which is set at 180 units of product.

(*ii*) Any overtime operation of team 2 beyond two hours and team 3 beyond three hours should be avoided.

(*iii*) Minimize the sum of overtime.

Formulate and solve this problem as a goal programming problem.

[MBA (Nov.), Delhi Univ. 2009]

12:6. LIMITATIONS OF GOAL PROGRAMMING PROBLEMS

Major limitations in dealing with goal programming problems are as follows:

1. The model formulation stage is highly complex due to presence of large number of goals all of which must be properly accounted for.
2. Since, there are several goals which are to be evaluated, the task of evaluation is very tedious.
3. In the fast moving technology and science, the process of development is continuous, and cannot assume to be coming to an end. In such a situation, decision-making is a highly uncertain and risky management responsibility. So, the management must continuously evolve new strategies and drop the redundant ones to reflect the changing needs.
4. The effective use of advanced programming techniques, *i.e.*, GPP, presupposes the existence of a fully developed organizational structure, which is not true in most of the developing economies. Hence, GPP has very limited scope in a country like India.

TEST YOUR UNDERSTANDING

1. The technique of goal programming:

(*a*) allows to have multiple goals, with or without priorities.

(*b*) is an approach to achieve goal of a solution to all-integer programming problems.

(*c*) is an attempt to minimize deviations from targets.

(*d*) is used when goals are not satisfied in an ordinal sequence.

2. In goal programming problem, goals are assigned priorities, such that

(*a*) a lower priority goal will never be achieved at the expense of higher priority goal.

(*b*) the priority factors have the relationship of $P_1 >>> P_2 >>> \dots P_k >>> P_{k+1} \dots$, where the symbol >>> means 'more important than'.

(*c*) the coefficients, P_i's are not assigned any actual values.

(*d*) all of the above (*a*), (*b*) and (*c*) are true.

3. If d^+ and d^- are the two deviational variables with respect to over-achievement and under-achievement, respectively, then which of the following would not hold:
   (*a*) each of them can be positive or zero. (*b*) both are non-zero.
   (*c*) both are equal to zero. (*d*) either d^+ or d^- is zero.
4. Deviational variables in goal programming problem must satisfy following conditions:
   (*a*) $d_i^+ + d_i^- = 0$ (*b*) $d_i^+ - d_i^- = 0$
   (*c*) $d_i^+ \times d_i^- = 0$ (*d*) $d_i^+ / d_i^- = 0$.
5. In goal programming problem, a constraint having under-achieved variable is expressed as
   (*a*) a "less than or equal to" type constraint. (*b*) a "greater than or equal to" type constraint.
   (*c*) an equality constraint. (*d*) all of the above.
6. In goal programming problem
   (*a*) a multiple criteria, instead of single objective are used in decision-making.
   (*b*) each goal is either under-achieved or over-achieved.
   (*c*) goals are given an ordinal ranking in terms of their contributions or importance to the organization.
   (*d*) all of the above.

ANSWERS

1. (*c*)	2. (*d*)	3. (*b*)	4. (*c*)	5. (*b*)	6. (*d*)

REVIEW EXERCISE

1. What is goal programming? Clearly state its assumptions. ***[MBA (Nov.), Delhi Univ. 2004]***
2. Identify major differences between linear programming and goal programming. ***[MBA (Jan.), Gujarat Tech. Univ. 2011; MBA, Delhi Univ. 2009]***
3. Explain graphical method for solving linear goal programming problem. ***[B.Tech. (Comp. Sc.), Panjab Tech. Univ. 2009]***
4. "The basic method of solving a goal programming problem is to convert it into a linear programming problem." Discuss. ***[PG Dip. Indian Inst. of Mat. Management 2017]***
5. When goal programming is applicable? ***[MBA, Osmania Univ. 2010; MBA, Pondicherry Univ. 2007]***
6. "Goal programming appears to be the most appropriate, flexible and powerful technique for decision-making problems involving multiple conflicting objectives." Discuss. ***[MBA (Nov.), Delhi Univ. 2009]***
7. "In goal programming, we attempt to 'satisfy' or come as close as possible to satisfying, the various goals." Discuss. ***[MBA (Nov.), Delhi Univ. 2003]***
8. Explain (*a*) Deviational variables, (*b*) Pre-empltive priority factors, and (*c*) Cardinal value.
9. What does it mean to rank goals in Goal Programming? How does this affect the solution of the problem?
10. Explain the salient features of goal programming. ***[MBA (Dec.), Gujarat Tech. Univ. 2010]***
11. Give any two business applications of goal programming. ***[MBA (Nov.), Annamalai Univ. 2002]***
12. Write short note on linear goal programming.

CHAPTER 13

Scheduling Techniques

"Every modern corporation has channels of communication along which lines, from foot to crown, reports flow up and vetoes down"

A ***network*** is a graphic representation of a project's operations and is composed of activities and events that must be completed to reach the end objective of a project. Network analysis is being used for planning and scheduling large projects in the fields of construction, maintenance, fabrication, computer system installation, research and development design, etc.

13:1. WHY NETWORKS?

Given below are some of the ***advantages*** of using network:

1. Networks are most valuable and powerful for planning, scheduling and control of operations in large and complex projects.
2. Networks are useful to evaluate the level of performance by comparing actual performance against the planned targets.
3. These helps to determine the interdependence of various activities for proper integration and coordination of various operations.
4. These help us to evaluate the time-cost trade off and determine the optimum schedule.
5. These are simple and can be easily oriented towards computers.
6. The networks clearly designate the responsibilities of different supervisors. Supervisor of an activity knows his time schedule precisely, and also the supervisors of other activities with whom he has to coordinate.

Limitations in Using Network

Besides the above mentioned advantages, there are certain ***limitations*** in the construction of networks which are given below:

1. Construction of networks for complex projects is complicated and time consuming due to trial and error approach.
2. Difficulty in securing the realistic time estimates. In the case of new and non-repetitive type projects, the time estimates are often mere ficticious.
3. Estimation of reliable and accurate duration of various activities is a difficult exercise.
4. The natural tendency to oppose changes results in the difficulty of persuading the management to accept these techniques.
5. The planning and implementation of networks require personnel trained in the network methodology. Managements are reluctant to spare the existing staff to learn these techniques or to recruit trained personnel.
6. Developing a clear logical network is also troublesome. This depends upon the data input and thus the plan can be no better than the personnel who provides the data.

13:2. BASIC COMPONENTS OF NETWORK

The basic components of network analysis are given as under:

Activity. An *activity* is a task, or item, of work to be done that consumes time, effort, money or other resources. It lies between two events, called the 'preceding' and 'succeeding' ones. An activity is represented by an arrow with its head indicating the sequence in which the events are to occur.

Event. An *event* represent the start (beginning) or completion (end) of some activity and as such it consumes no time. It has no time duration and does not consume any of the resources. An event is nothing but a ***node*** and is generally represented on the network by a circle, rectangle, hexagon or some other geometric shape. An event is not complete until all the activities flowing into it are completed.

Activities are identified by the numbers of their starting (tail or initial) event and ending (head, or terminal) event. An arrow (i, j) extended between two events; the tail event i represents the start of the activity and the head event j represents the completion of the activity as shown below:

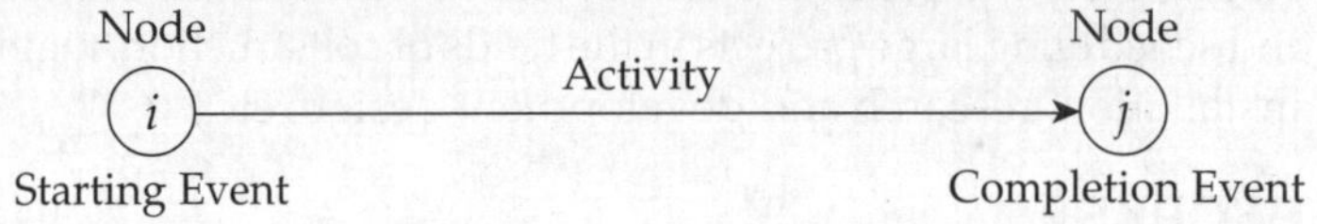

An event representing the joint completion of more than one activity is called a ***merge event***. If an event represents the joint initiation of more than one activity, it is called a ***burst event***.

The activities can be further classified into following three categories:

1. ***Predecessor Activity.*** An activity which must be completed before one or more other activities start is known as ***predecessor activity***.
2. ***Successor Activity.*** An activity which started immediately after one or more of other activities are completed is known as ***successor activity***.
3. ***Dummy Activity.*** An activity which does not consume any resource and time is known as ***dummy activity***. A dummy activity is depicted by dotted line in the network diagram.

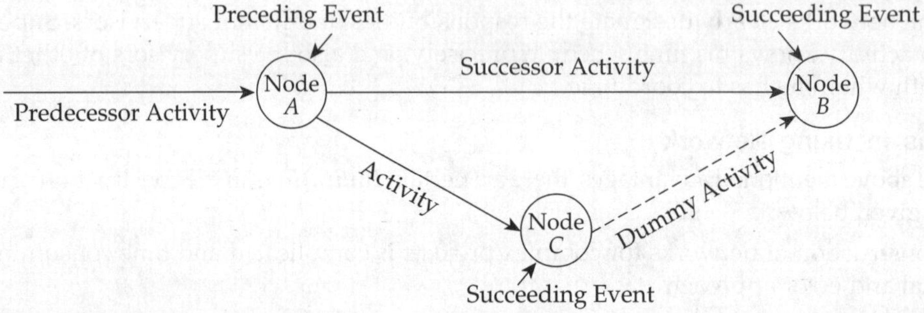

Fig. 13.1

Remarks

1. An activity must be uniquely identified by its starting and completion event which implies that
   (*a*) an event number should not get repeated or duplicated.
   (*b*) two activities should not be identified by the same completion event.
   (*c*) activities must be represented either by their symbols or by the corresponding ordered pair of starting–completion events.
2. A dummy activity in the network is introduced only to represent the given precedence relationships among activities of the project and is needed when (*a*) two or more parallel activities in a project have same head and tail events, or (*b*) two or more activities have some (but not all) of their immediate predecessor activities in common.

Concurrent Activity. Those activities which can be accomplished concurrently, *i.e.*, they may be done in part allowing the subsequent activities to commence before the preceding activity is fully completed are known as ***concurrent activities***.

Merge and Burst Events. An event that represents the completion of more than one activity is called ***merge event***, while an event which portrays the initiation of more than one activity is called the ***burst event***. For example, in *Fig. 13.2*(*a*) activities *B*, *C* and *D* must be completed before undertaking activity *F*. It may be noted that it only implies that the three activities should be completed before activity *F* begins, and not that they must be completed simultaneously. Similarly, in *Fig. 13.2*(*b*), activity *A* should be completed before activities *B*, *C* and *D* can be started.

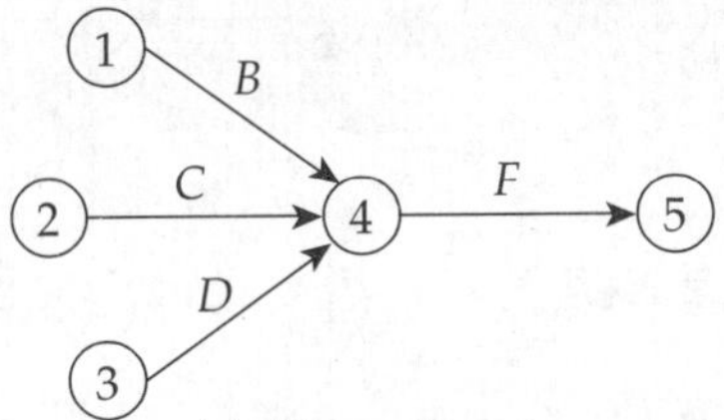

(*a*) Merge Event

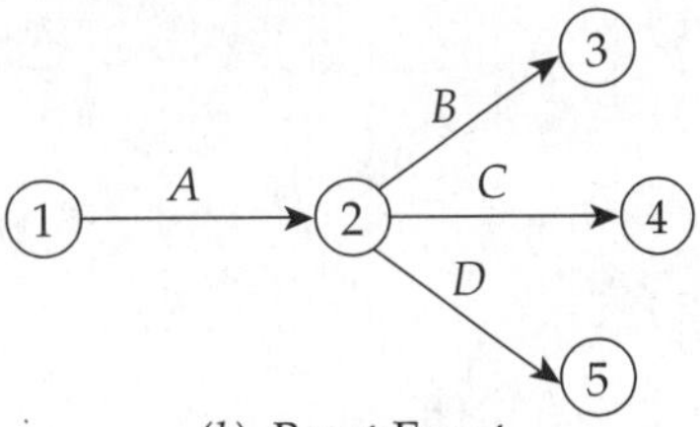

(*b*) Burst Event

Fig. 13.2

13:3. LOGICAL SEQUENCING

All the projects consist of certain activities that can begin only after certain others are completed. In fact, the entire project may be considered as a series of activities which may begin only after another activity or activities are completed. In a network schedule, such type of relationships are called ***constraints*** and are represented by inequalities. For, $A < B$, will indicate that activity *A* must be completed before the start of activity *B*. As an example, a project of laying a pipeline consists of the activities: trenching, laying pipe, and welding of pipe.

In logical sequencing, following three types of errors are most common, while drawing a network diagram:

1. **Looping.** If an activity is represented as going back in time, a ***closed*** loop would occur as shown in *Fig. 13.3*.

   A closed loop would produce an endless cycle in computer programmes without a built-in routine for detection or identification of the cycle. This situation can be avoided by checking the precedence relationship of the activities and also by numbering them in a logical order. Thus, one property of a correctly constructed network diagram is that it is "non-cyclic".

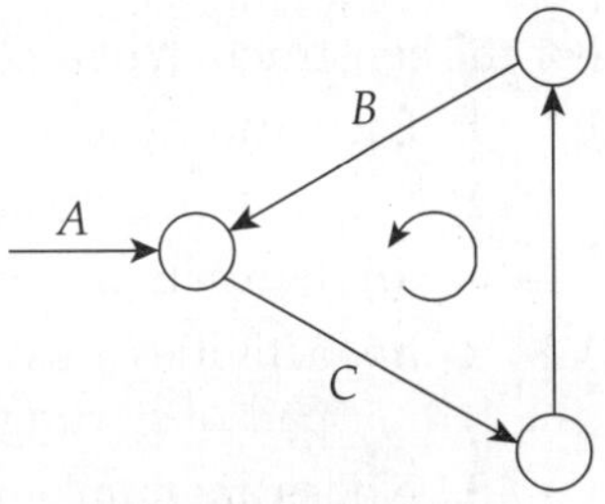

Fig. 13.3. Closed loop

2. **Dangling.** No activity should end without being joined to the end event. If it is not so, a dummy activity is introduced in order to maintain the continuity of the system. Such end-events other than the end of the project as a whole are called ***dangling*** events.

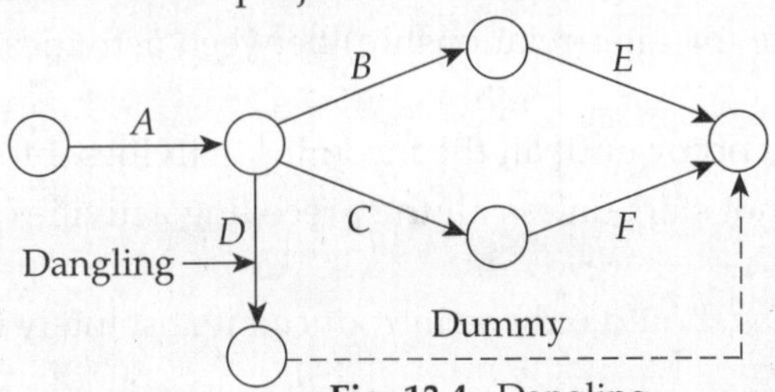

Fig. 13.4. Dangling

In the above network, activity *D* leads to dangling. A dummy activity is, therefore, introduced to avoid this dangling.

3. **Dependency Relationships.** When two chains of activities have a common event, wholly or partly independent of each other, a dummy activity is used to establish proper logical relationships. Given below are the two cases of dependency relationships:
   (*i*) Activity *C* is dependent upon both the activities *A* and *B*.
   (*ii*) Activity *D* is dependent upon activity *A* alone.

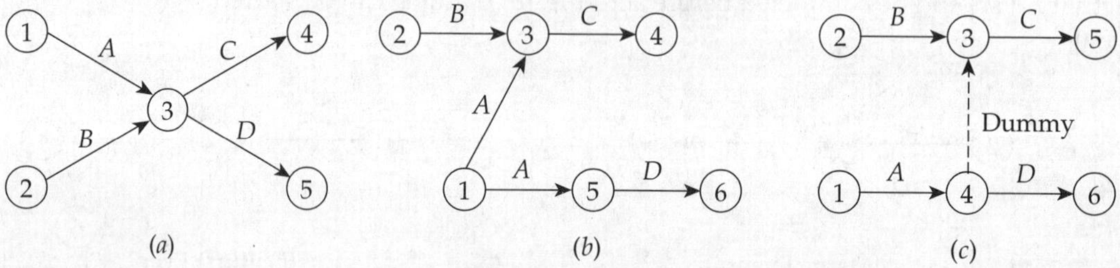

Fig. 13.5

In the context of above mentioned dependency relationships (*i*) and (*ii*), we observe that *Fig. 13.5(a)* is clearly wrong since it shows *D* as dependent upon not only *A* but also on *B* which is not desired. *Fig. 13.5(b)* is also wrong since *A* has been shown twice and thus contravenes the fundamental axioms of network that there must be one arrow for each activity. The way out to this dilemma is the representation by means of a dummy activity. *Fig. 13.5(c)* shows that *C* is dependent upon both *A* and *B* (*via* dummy) whereas *D* is dependent upon just *A*.

13:4. RULES OF NETWORK CONSTRUCTION

For the construction of a network, generally, the following rules are followed:

1. Each activity is represented by one and only one arrow.
2. Each activity must be identified by its ***starting*** and ***end*** node which implies that
   (*i*) two activities should not be identified by the same completion events, and
   (*ii*) activities must be represented either by their symbols or by the corresponding ordered pair of starting-completion events.
3. Nodes are numbered to identify an activity ***uniquely***. Tail node (starting point) should be lower than the head node (end point) of an activity.
4. Between any pair of nodes, there should be one and only one activity; however more than one activity may emanate from, and, terminate to a node.
5. Arrows should be kept straight and no curved or bent.
6. The logical sequence (or inter-relationship) between activities must follow the following rules:
   (*i*) An event cannot occur until all the incoming activities into it have been completed.
   (*ii*) An activity cannot start unless all the preceding activities on which it depends, have been completed.
   (*iii*) Dummy activities should only be introduced if absolutely necessary.

Numbering the Events

After the network is drawn in a logical sequence, every event is assigned a number. The number sequence must be such as to reflect the flow of the network. In event numbering, the following rules should be observed:

(*a*) Event numbers should be unique.
(*b*) Event numbering should be carried out on a sequential basis from left to right.
(*c*) The initial event which has all outgoing arrows with no incoming arrow is numbered 0 or 1.
(*d*) The head of an arrow should always bear a number higher than the one assigned at the tail of the arrow.
(*e*) Gaps should be left in the sequence of event numbering to accommodate subsequent inclusion of activities, if necessary.

Remark. The above procedure of assigning numbers to various events of a network is known as *Fulkerson's Rule*.

Illustration. *A television is manufactured in six steps, labelled A through F. Because of its size and complexity, the television is produced one at a time. The production control manager thinks that network scheduling techniques might be useful in planning future production. He recorded the following information:*

A is the first step and precedes B and C,
C precedes D and E,
F follows E, and D is successor of F,
B follows D, and precedes E.

(*a*) *Draw an activity-on-node diagram for the production control manager.*
(*b*) *On checking with the records, the production control manager corrects his last note to read, "D is a predecessor of F". Draw a new diagram for the revised network incorporating this new change.*
(*c*) *After pondering over the network and rechecking from the records, it was found that B was really a predecessor of D rather than vice versa. Draw a revised network incorporating this new change.*
(*d*) *Draw an arrow-diagram representation of the network in part (c). How many dummy activities did you use in your network?*

Solution.

(*a*) (*i*) Activity *A* is the first step and precedes B and C is shown in *Fig. 13.6*.
(*ii*) Activity C precedes D and E is shown in *Fig. 13.7*.

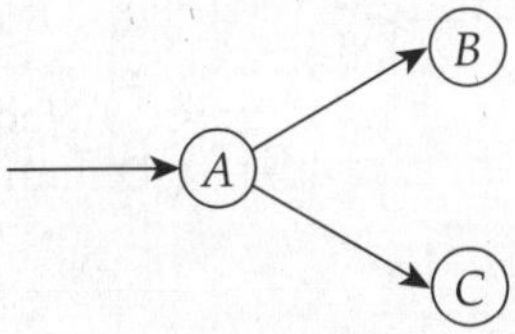

Fig. 13.6

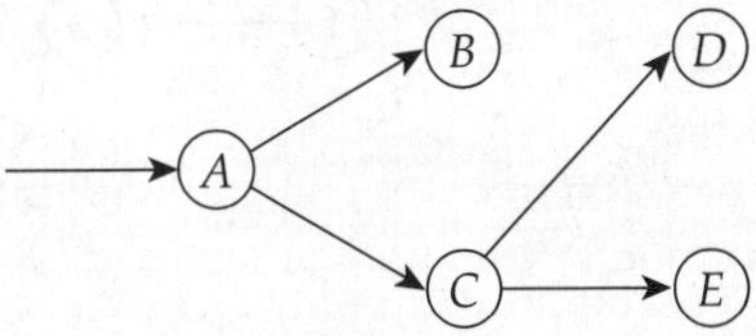

Fig. 13.7

(*iii*) Activity *F* follows *E*, and *D* is the successor of *F* is shown in *Fig. 13.8*.

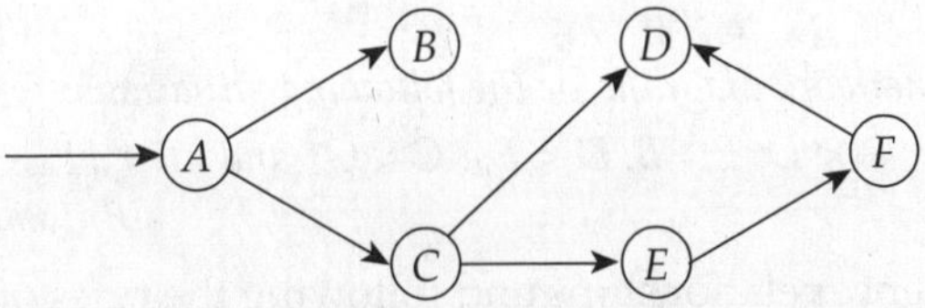

Fig. 13.8

(*iv*) Now, since activity *B* follows *D* and precedes *E*, the complete network drawing is shown in *Fig. 13.9*.

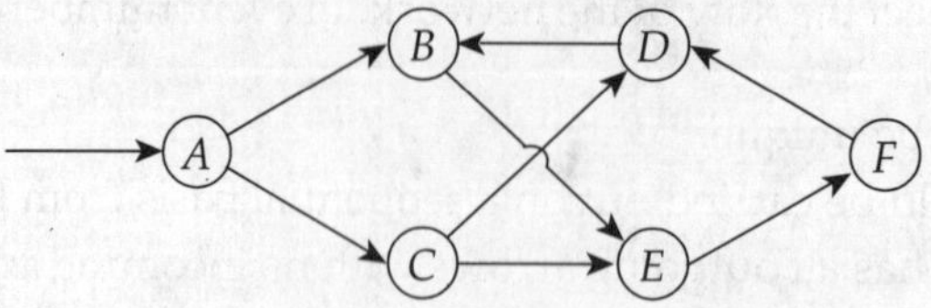

Fig. 13.9

Evidently, this network contains a cycle as shown in *Fig. 13.10* below:

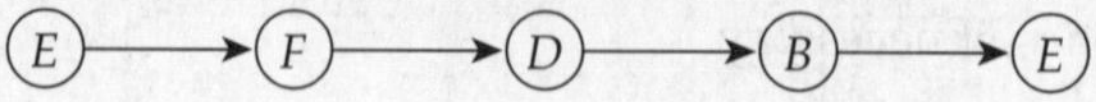

Fig. 13.10

(*b*) Revised network when *D* is a predecessor of *F* is shown in *Fig. 13.11*.

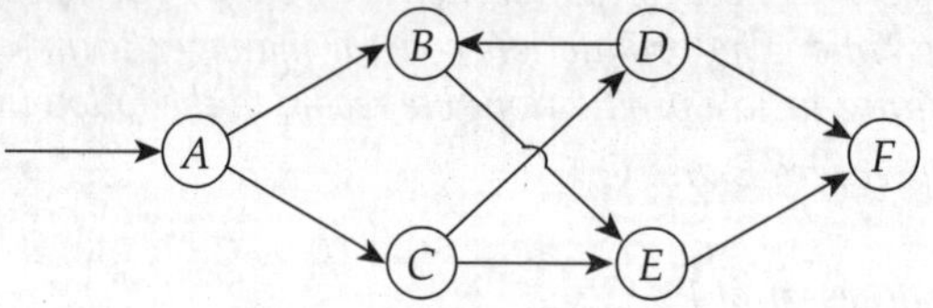

Fig. 13.11

(*c*) Incorporating the new change, when *B* is a predecessor of *D*, instead of *D* as a predecessor of *B*, revised network is shown in *Fig. 13.12*.

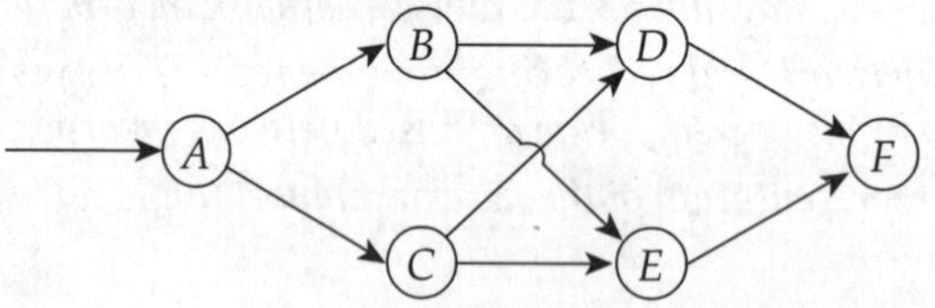

Fig. 13.12

(*d*) The arrow-diagram of the network presented in part (*c*) can be represented as:

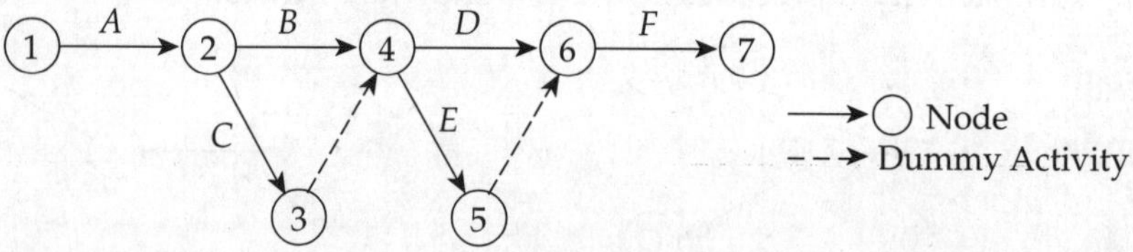

Fig. 13.13

The events of the diagram are numbered such that their ascending order indicates the direction of progress in the project.

Example 13.1. *Construct a network diagram for the following situations:*

$$A < D, E; \quad B, D < F; \quad C < G \text{ and } B < H$$

[B.Com. (April), Madras Univ. 2006]

Solution. Using the precedence relationships and following the rules of network construction, the required network is shown in *Fig. 13.14*.

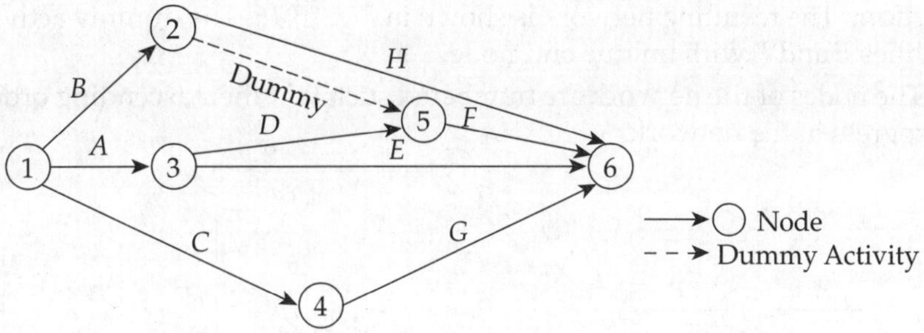

Fig. 13.14

It may be noted that a dummy activity is used to identify the activities *B* and *F* with unique end-nodes.

Example 13.2. *Draw a network diagram for the project whose activities and their predecessor relationships are given below:*

Activity	:	*A*	*B*	*C*	*D*	*E*	*F*	*G*	*H*	*I*	*J*	*K*
Predecessor Activity	:	—	—	—	*A*	*B*	*B*	*C*	*D*	*F*	*H, I*	*F, G*

[B.Com. (Nov.), Madras Univ. 2005]

Solution. Using the precedence relationships and following the rules of network construction, the required network diagram is shown in *Fig. 13.15*.

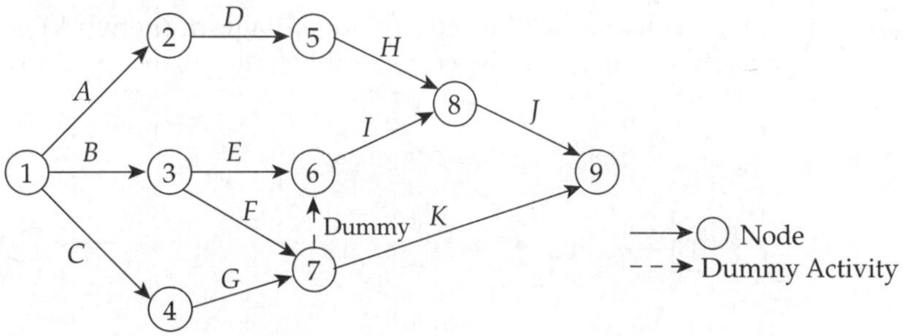

Fig. 13.15

It may be noted that a dummy activity is used to identify the activities *F* and *I* with unique end-nodes. The nodes of the project are numbered such that their ascending order indicates the direction of progress in the project.

Example 13.3. *Draw a network for the following project and number the events according to Fulkerson's Rule:*

1. *Activity A is the start event and activity K is the end event.*
2. *Activity J is the successor event to activity F.*
3. *Activity C and activity D are successor events to activity B.*
4. *Activity D is the preceding event to activity G.*
5. *Activity E and activity F occur after activity C.*
6. *Activity E precedes activity J.*
7. *Activity C restraints the occurrence of activity G and activity G precedes activity H.*
8. *Activity H precedes activity J.*
9. *Activity F restrains the occurrence of activity H.*
10. *Activity K succeeds event activity J.*

Solution. The resulting network is shown in *Fig. 13.16*. The dummy activity is used to identify the activities *E* and *F* with unique end node.

The nodes of the network are numbered such that their ascending order indicates the direction of progress in the network.

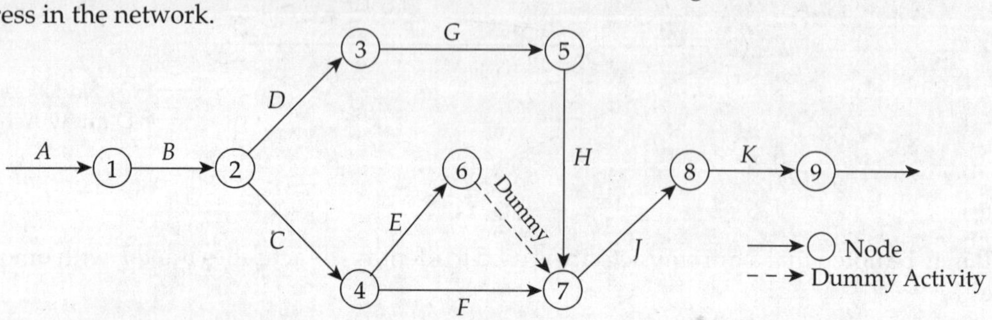

Fig. 13.16

Example 13.4. *A television is manufactured in six steps, labelled A through F. The television is produced one at a time. The manager thinks that network scheduling will improve the position. It is noted that A is the first step and A < B, A < C, C < D, C < E, D < B, B < E, E < F, D is a successor of F.*

(*a*) *Draw an activity node diagram and identify the cycle, if any.*

(*b*) *If D precedes F, show that the cycle gets eliminated.*

Solution.

(*a*) Using given constraints, the activity node diagram (network) is shown in *Fig. 13.17*. The dummy activities are introduced to establish the correct precedence relationships:

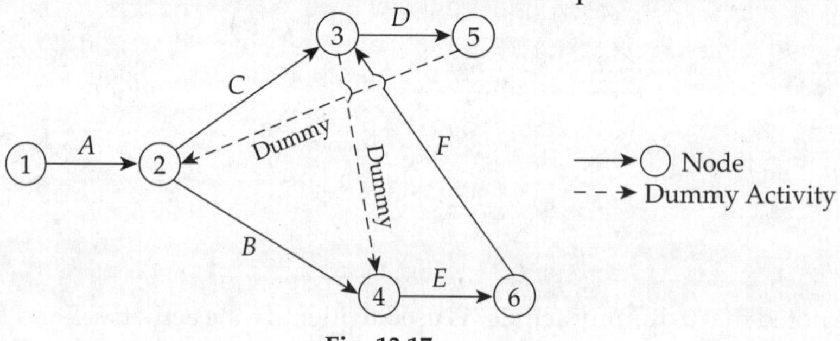

Fig. 13.17

The nodes are numbered in such a way that their ascending order indicates the direction of progress in the manufacturing process. Among the various activities involved, obviously we have the following cycle: $B \to E \to F \to D \to B$.

(*b*) If *D* precedes *F*, the revised network is as shown in *Fig. 13.18*.

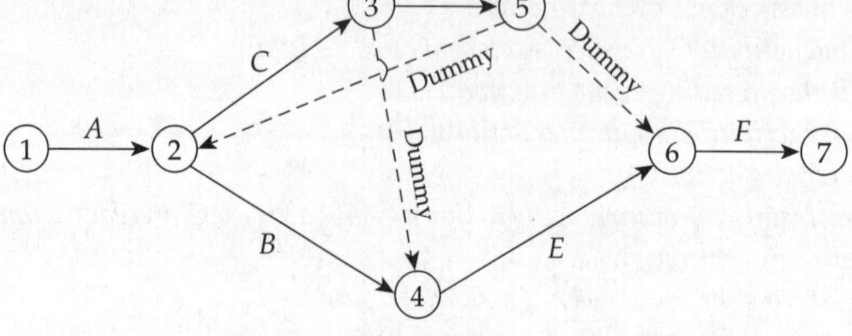

Fig. 13.18

PROBLEMS

13.01. Draw the network diagram to the following constraints:

$A < B, C;\ B, C < D;\ B < E, F;\ D < G.$ ***[B.Com. (Nov.), Madras Univ. 2007]***

13.02. Draw the network diagram for the following data:

Task :	*A*	*B*	*C*	*D*	*E*	*F*	*G*
Immediate Predecessor :	None	None	*B*	*B*	*B*	*E*	*A, C, D*

13.03. Draw a network diagram for the following description:

Activity :	*A*	*B*	*C*	*D*	*E*	*F*	*G*	*H*
Predecessor Activity :	—	*A*	*A*	*B*	*B, C*	*E*	*D, F*	*G*

[B.Com. (Nov.), Madras Univ. 2005]

13.04. A project consists of a series of tasks *A* to *I* with the following relationships ($W < U, V$) means *U* and *V* cannot start until *W* is completed ($U, V < W$) means *W* cannot start unless *U* and *V* are completed. Construct the network diagram if:

$$A < D, E;\ B, D < F;\ C < G;\ B < H;\ F, G < I.$$

13.05. Construct the network diagram comprising of activities *B, C, ..., Q*, such that the following constraints are satisfied:

$$B < E, F;\ C < G, L;\ E, G < H; L, H < I;\ L < M;\ H, M < N;\ H < J;\ I, J < P;\ P < Q.$$

The notation $X < Y$ means that the activity *X* must be finished before *Y* can begin.

[BE (Mech.), Jammu Univ. 2004]

13.06. A company has decided to redesign its electronic components. The project involves several activities which are listed in the following table. *First*, the electronic staff must finish the design of the component. *Second*, the marketing programme for promoting the component must be developed. *Third*, a new manufacturing process must be designed. *Fourth*, advertising media must be selected. *Fifth*, an initial production run must be successfully completed. *Sixth*, the component must be released to the market.

Activity	*Description of activity*	*Predecessor activity*
A	Finish component development	—
B	Design marketing programme	*A*
C	Design production system	*A*
D	Select advertising media	*B*
E	Initial production run	*C*
F	Release component to the market	*D, E*

[BBA (April), Guru Nanak Dev Univ. 2008]

13.07. Construct an arrow diagram comprising of activities *A, B, C, ...*; such that the following relationships are satisfied:

(*a*) Clear the site.
(*b*) Survey and layout.
(*c*) Rough grade.
(*d*) Excavate for sewer.
(*e*) Excavate for electrical manholes.
(*f*) Install sewer and backfill.
(*g*) Install electrical manholes.
(*h*) Construct the boundary wall.

[M.Com., Calicut Univ. 1995]

13.08. Construct the arrow diagram comprising activities *A, B, C* and *L*; such that the following relationships are satisfied:

(a) A, B and C, the first activities of the project, can start simultaneously.
(b) A and B precede D.
(c) B precedes E, F and H.
(d) F and C precede G.
(e) E and H precede I and J.
(f) C, D, F and J precede K.
(g) K precedes L.
(h) I, G and L are the terminal activities of the project.

13.09. Listed in the table are the activities and sequencing necessary for a maintenance of job on the heat exchangers in a refinery:

Activity	*Description*	*Predecessor activity*
A	Dismantle pipe connections	—
B	Dismantle header, closure, and floating front	A
C	Remove tube bundle	B
D	Clean bolts	B
E	Clean header and floating head front	B
F	Clean tube bundle	C
G	Clean shell	C
H	Replace tube bundle	F, G
I	Prepare shell pressure test	D, E, H
J	Prepare tube pressure test and reassemble	I

Draw a network diagram for the project. ***[MBA (PT.), Delhi Univ. 2008]***

13.10. An assembly is to be made from two parts X and Y. Both parts must be turned on a lathe, and Y must be polished, whereas X need not to be polished. The sequence of activities together with their predecessors is given below:

Activity	*Description*	*Predecessor activity*
A	Open work order	—
B	Get material for X	A
C	Get material for Y	A
D	Turn X on lathe	B
E	Turn Y on lathe	B, C
F	Polish Y	E
G	Assembly X and Y	D, F
H	Pack	G

Draw a network diagram for the project. ***[BMS, Lucknow Univ. 2008]***

13:5. NETWORK SCHEDULING

Programme Evaluation and Review Technique (PERT) and Critical Path Method (CPM) are the two basic network techniques which are widely used in planning and control of a project. Prior to the development of these techniques, the most popular technique for project scheduling was the Gantt Chart. This chart shows a graphic representation of work or a time scale. The main limitation of this technique is its inability to show the inter-relationship and inter-dependence among many activities. To overcome such limitations, PERT and CPM came into existence in late 1950's. Now a days, several variations of these techniques have been developed and out of which one such important variation being the Review Analysis of Multiple Projects (RAMP). This technique is useful for guiding the 'activities' of several projects at one time.

13:6. CRITICAL PATH ANALYSIS

The ***Critical Path Analysis (CPA)*** network which includes both ***CPM*** (***Critical Path Method***) and

***PERT* (*Programme Evaluation and Review Technique*)** are network techniques. The network approach help project managers in planning, scheduling and controlling.

The methodology of ***Critical Path Analysis (CPA)*** consist of the following five steps:

1. Analyse and breakdown the project in terms of specific activities and events.
2. Determine the interdependence and sequence of specific activities and prepare a network.
3. Assign estimates of time, cost or both to all the activities of network.
4. Identify the longest or critical path through the network.
5. Monitor, evaluate and control the progress of the project by replanning, rescheduling and reassignment of resources.

The central task in the control aspect of these techniques is to identify the longest path through the network. The longest path is the critical path because it equals the minimum time required to complete the project. All other paths other than the critical path (*i.e.*, non-critical or slack paths) offer flexibility in scheduling and transferring resources.

To accomplish the above-mentioned objectives, the following factors should be known in advance to prepare project scheduling:

1. Time schedule for each activity, *i.e.*, the time by which an activity must begin and the time before which it is completed.
2. Earlier and latest start time as well as earlier finish and latest finish time of each activity.
3. Float for each activity, *i.e.*, the spare time associated with each activity.
4. Critical activities and critical path for the network.

Remark. An activity in a network diagram is said to be ***critical***, if the delay in its start will further delay the project completion time. A ***non-critical*** activity allows some scheduling slack, so that the start time of the activity may be advanced or delayed within limits without affecting the completion date of the entire project.

Note. It may be noted that critical path analysis network helps the top management to concentrate their attention to the critical activities and their completion in time. Also,

(*a*) provides the knowledge of critical and non-critical activities which helps the management to divert the resources from non-critical to critical activities;

(*b*) provides a best way of planning and scheduling a large construction project;

(*c*) gives complete information about the importance, duration, size and performance of an activity;

(*d*) helps the management to get single time estimate and costs with higher accuracy.

Example 13.5. *The following data gives the characteristics of a project:*

Job	:	A	B	C	D	E	F	G	H
Predecessors	:	—	—	B	A, C	A, C	D	E	F, G
Duration (days)	:	10	5	3	4	6	6	5	5

Draw the network diagram and find the critical path.

Solution. Using the precedence relationship among various jobs, the resulting network is shown in *Fig. 13.19*.

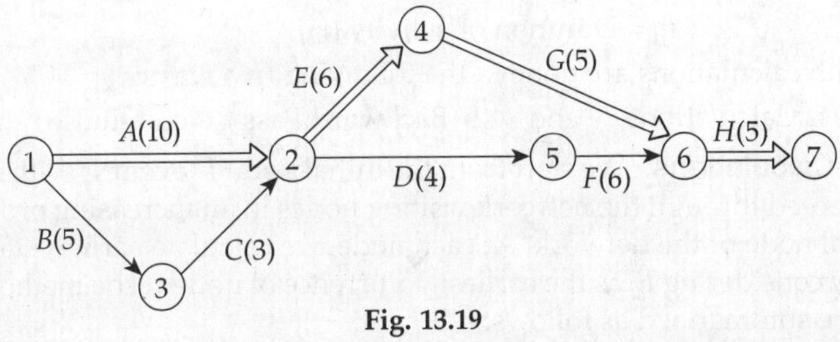

Fig. 13.19

Various paths of the project are:

1. $B \to C \to D \to F \to H$ with 23 days.
2. $B \to C \to E \to G \to H$ with 24 days.
3. $A \to D \to F \to H$ with 25 days.
4. $A \to E \to G \to H$ with 26 days.

Since, the longest path is the last one, the critical path will be $A \to E \to G \to H$. The project duration will be of 26 days.

Example 13.6. *A project consists of 9 jobs A to I with the following precedence relationships and estimates of time. Draw a project network and find the critical path with project duration.*

Job	:	A	B	C	D	E	F	G	H	I
Predecessor	:	—	—	A, B	A, B	B	D, E	C, F	D, E	G, H
Duration (days)	:	15	10	10	10	5	5	20	10	15

[BBA (May), Madras Univ. 1997]

Solution. Using the precedence relation among various jobs, the resulting network is shown in *Fig. 13.20*.

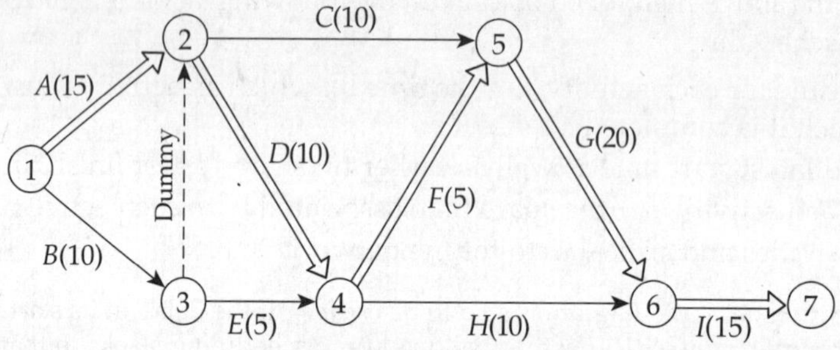

Fig. 13.20

Using the above network, various paths of the project and their duration are given below:

1. $A \to C \to G \to I$ with 60 days.
2. $A \to D \to F \to G \to I$ with 65 days.
3. $A \to D \to H \to I$ with 50 days.
4. $B \to C \to G \to I$ with 55 days.
5. $B \to D \to F \to G \to I$ with 60 days.
6. $B \to D \to H \to I$ with 45 days.
7. $B \to E \to F \to G \to I$ with 55 days.
8. $B \to E \to H \to I$ with 40 days.

Here, we observe that the longest path is of 65 days duration. Hence, the critical path is: $A \to D \to F \to G \to I$. The projection duration is of 65 days.

13:7. CRITICAL PATH CALCULATIONS

For the purpose of calculating various times of events and activities, the following symbols shall be used in critical path calculations:

E_i = Earliest occurrence time of event i

L_j = Latest occurrence time of event j

t_{ij} = Duration of activity (i, j)

The critical path calculations are done in the following two ways:

1. Forward Pass Calculations, and 2. Backward Pass Calculations.

1. Forward Pass Calculations. We start from the initial node *1* (event 1) with starting time of the project as zero. Proceed through the network visiting nodes in an increasing order of node number and end at the final node of the network. At each node, we calculate earliest start and finish times for each activity by considering E_i as the earliest occurrence of node *i*. The method for forward pass calculations may be summarized as follows:

Step 1. Set $E_1 = 0$; (initial node).

Step 2. Set the earliest start time for each activity that begins at node i as

$$ES_{ij} = E_i; \quad \text{for all activities } (i, j) \text{ that start at node } i.$$

Step 3. Compute the earliest finish time of each activity that beings at node i by adding the earliest start time of the activity to the duration of the activity. Thus,

$$EF_{ij} = ES_{ij} + t_{ij} = E_i + t_{ij}$$

Step 4. Move on to next node, say node j ($j > i$) and compute the earliest occurrence for node j, using

$$E_j = \max_i . \{EF_{ij}\} = \max_i . \{E_i + t_{ij}\}$$

for all immediate predecessor activities.

Step 5. If $j = n$ (final node), then the earliest finish time for the project is given by

$$E_n = \max. \{EF_{ij}\} = \max. \{E_{n-1} + t_{ij}\}, \text{ for } j = n$$

2. Backward Pass Calculations. We start from the final (last) node n of the network, proceed through the network visiting nodes in the decreasing order of node numbers and end at the initial node *1*. At each node, we calculate the least finish and start times for each activity by considering L_j as the latest occurrence of node j. The method may be summarized below:

Step 1. $L_n = E_n$; for $j = n$

Step 2. Set the latest finish time of each activity that ends at node j as

$$LF_{ij} = L_j$$

Step 3. Compute the latest occurrence times of all activities ending at j by subtracting the duration of each activity from the latest finish time of the activity. Thus,

$$LS_{ij} = LF_{ij} - t_{ij} = L_j - t_{ij}$$

Step 4. Proceed backward to the node in the sequence that decrease j by 1. Also, compute the latest occurrence time of node i ($i < j$) using

$$L_i = \min_j . \{LS_{ij}\} = \min_j . \{L_j - t_{ij}\},$$

for all immediate successor activities.

Step 5. If $j = 1$ (initial node), then

$$L_1 = \min. \{LS_{ij}\} = \min. \{L_2 - t_{ij}\}, \text{ for } i = 1$$

Based on the above calculations, an activity (i, j) will be ***critical***, if it satisfies the following conditions:

(*i*) $E_i = L_i$ and $E_j = L_j$

(*ii*) $E_j - E_i = L_j - L_i = t_{ij}$

An activity that does not satisfy the above conditions is termed as ***non-critical***.

Critical Path

The critical activities of a network that constitute an uninterrupted path which spans the entire network from start to finish is known as ***critical path***.

Notes

(*a*) It may be noted that critical path is based on the assumption of known time for each of the activity in the project which may not be true in real life situations.

(*b*) It does not incorporate statistical analysis in determining time estimates.

(*c*) It is difficult to use CPM as a controlling device for the simple reason that the entire evaluation of the project is to be repeated and a new critical path is to be determined each time when changes are made (introduced) in the network.

Float (or Slack) of an Activity and Event

The float of an activity is the amount of time by which it is possible to delay its completion time without affecting the total project completion time.

1. **Event Float.** The ***float*** (also called ***slack***) of an event is the difference between its latest time (L_i) and its earliest time (E_i). That is

$$\text{Event float} = L_i - E_i$$

   It is a measure of how much later than expected a particular event could occur without delaying the completion of the entire project.

2. **Activity Float.** As mentioned earlier, it is the float (or slack) in the activity time estimates. There are mainly three types of activity floats as discussed below:

   (*a*) ***Total Float.*** The ***total float*** of an activity represents the amount of time by which an activity can be delayed without delay in the project completion date. In other words, it refers to the amount of free time associated with an activity which can be used before, during or after the performance of this activity. Total float is the positive difference between the earliest finish time and the latest finish time, or the positive difference between the earliest start time and the latest start time of an activity depending upon which way it is defined.

   Thus, for each activity (i, j), the total float values are computed as follows:

$$\text{Total float} = L_j - (E_i + t_{ij}) = (L_j - t_{ij}) - E_i$$

   (*b*) ***Free Float.*** The ***free float*** is that portion of the total float within which an activity can be manipulated without affecting the float of subsequent activities. It is computed for an activity by subtracting the head event slack from its total float. The free float indicates the value by which an activity in question can be delayed without causing any delay in its immediate successor activities.

   Free float values for each activity (i, j) are computed as follows:

$$\text{Free float} = (E_j - E_i) - t_{ij} = E_j - (E_i + t_{ij})$$

   (*c*) ***Independent Float.*** The ***independent float*** is that portion of the total float within which an activity can be delayed for start without affecting floats of the preceding activities. It is computed by subtracting the tail event slack from the free float of the activity. If the result is negative, it is considered as zero.

   Independent float values for each activity (i, j) are computed as follows:

$$\text{Independent float} = (E_j - L_i) - t_{ij} = E_j - (L_i + t_{ij})$$

   Illustration. Consider the following activity of certain network:

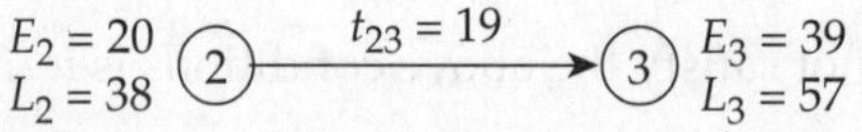

   Here,

$$\text{Total Float} = L_3 - (E_2 + t_{23}) = 57 - (20 + 19) = 18$$
$$\text{Free Float} = E_3 - (E_2 + t_{23}) = 39 - (20 + 19) = 0$$
$$\text{Independent Float} = E_3 - (L_2 + t_{23}) = 39 - (38 + 19) = -18$$

   which can be considered as zero, since it is negative.

   (*d*) ***Interfering Float.*** Utilization of the float of an activity can affect the float of subsequent activities in the network. Thus, interfering float can be defined as that part of the total float which causes reduction in the float of the successor activities. In other words, it can be defined as the difference between the latest finish time of the activity under consideration and the earliest start time of the following activity, or zero, whichever is larger. Thus, interfering float refers to that portion of the activity float which cannot be consumed without affecting adversely the float of the subsequent activity or activities.

Remarks

1. The basic difference between slack and float times is that slack which is used for events only, whereas float is applied for activities.
2. Latest occurrence time of an event is always greater than or equal to its earliest occurrence time, *i.e.*, $L_i \geq E_i$. This implies that

   Independent float ≤ Free float ≤ Total float.
3. An activity is critical, if its total float is zero, otherwise it is non-critical.
4. Once the float of an activity is disturbed, float of all other activities of the project is changed and should be recalculated.
5. The calculation of various floats can help the decision-maker in identifying the under-utilized resources, flexibility in the total schedule and possibilities of redeployment of resources.

13:8. PROCEDURE FOR DETERMINING CRITICAL PATH (CPM)

The procedure of determining the critical path involves the following eight steps:

Step 1. Break down the project into various activities systematically. Label all activities and arrange all the activities in a logical sequence.

Step 2. Number all the events and activities. Considering each activity to be deterministic, find their time. Indicate the activity times on the arrow diagram.

Step 3. Calculate the earliest start time, earliest finish time, latest start time and latest finish time. Tabulate the activity normal times, earliest times and latest times.

Step 4. Determine the total float for each activity by taking the difference between the earliest time and latest time, for each node (event).

Step 5. Identify the critical activities (the activities with zero float) and connect them with the starting node and the ending node in the network diagram by double arrow.

Step 6. Calculate the total project duration.

Step 7. If it is intended to reduce the total project duration, crash the critical activities of the network.

Step 8. Update the network and smooth the network resources.

Example 13.7. *Draw a network diagram and ascertain the critical path:*

Activities	:	*A*	*B*	*C*	*D*	*E*	*F*	*G*
Predecessor	:	—	—	—	*A*	*B*	*C*	*D, E, F*
Duration (days)	:	7	10	8	5	6	4	5

Solution. Using the precedence relationship among the activities, the resulting network is shown in *Fig. 13.21*. The nodes are numbered in such a way that their ascending order indicates the direction of progress in certain project.

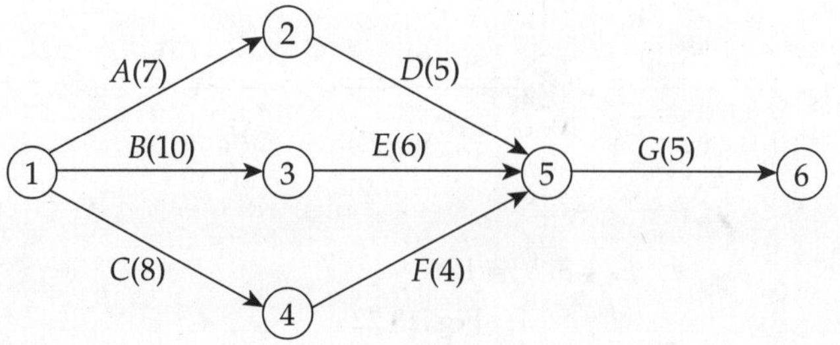

Fig. 13.21

To determine the critical path, we compute E_i and E_j for each of the activities. The critical path calculations as applied to *Fig. 13.21* are as follows:

Forward Calculations

Node 1. Set $E_1 = 0$

Node 2. $E_2 = E_1 + t_{12} = 0 + 7 = 7$

Node 3. $E_3 = E_1 + t_{13} = 0 + 10 = 10$

Node 4. $E_4 = E_1 + t_{14} = 0 + 8 = 8$

Node 5. $E_5 = \max\limits_{i=2,\,3,\,4}.\ \{E_i + t_{i5}\} = \max.\ \{E_2 + t_{25},\ E_3 + t_{35},\ E_4 + t_{45}\}$

$= \max.\ \{7 + 5, 10 + 6, 8 + 4\} = 16$

Node 6. $E_6 = E_5 + t_{56} = 16 + 5 = 21$

Backward Calculations

Node 6. Set $L_6 = E_6 = 21$

Node 5. $L_5 = L_6 - t_{56} = 21 - 5 = 16$

Node 4. $L_4 = L_5 - t_{45} = 16 - 4 = 12$

Node 3. $L_3 = L_5 - t_{35} = 16 - 6 = 10$

Node 2. $L_2 = L_5 - t_{25} = 16 - 5 = 11$

Node 1. $L_1 = \min\limits_{j=2,\,3,\,4}.\ \{L_j - t_{1j}\} = \min.\ \{L_2 - t_{12},\ L_3 - t_{13},\ L_4 - t_{14}\}$

$= \min.\ \{11 - 7, 10 - 10, 12 - 8\} = \min.\ \{4, 0, 4\} = 0$

To evaluate the critical nodes, all these calculations are put in the following table:

Activity	*Days*	*Earliest time*		*Latest time*		*Slack (Total float)*
		Start	*Finish*	*Start*	*Finish*	
A	7	0	7	4	11	4
B	10	0	10	0	10	0
C	8	0	8	4	12	4
D	5	7	12	11	16	4
E	6	10	16	10	16	0
F	4	8	12	12	16	4
G	5	16	21	16	21	0

Evidently critical activities are *B*, *E* and *G*. These are shown by double arrow in the network as shown in *Fig. 13.22*.

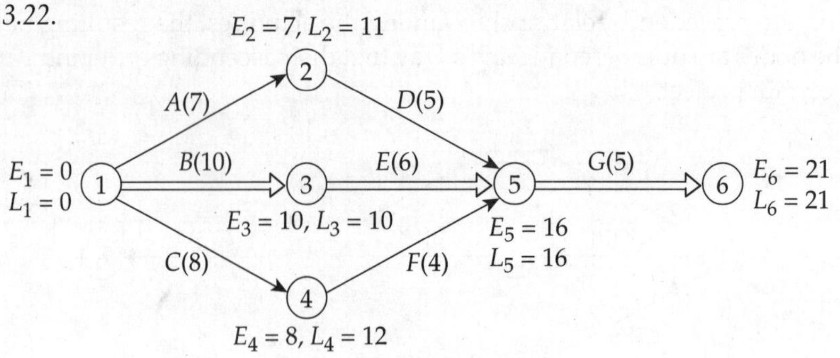

Fig. 13.22

Obviously, critical path is: $B \rightarrow E \rightarrow G$, with project duration of 21 days.

Example 13.8. *A project consists of a series of tasks labelled A, B, ..., H, I with the following relationships (W < X, Y) means X and Y cannot start until W is completed; (X, Y < W) means W cannot start until both X and Y are completed. With this notation construct the network diagram having the following constraints:*

$$A < D, E;\quad B, D < F;\quad C < G;\quad B, G < H;\quad F, G < I.$$

Find also the minimum time of completion of the project, when the time (in days) of completion of each task is as follows:

Task :	*A*	*B*	*C*	*D*	*E*	*F*	*G*	*H*	*I*
Time :	23	8	20	16	24	18	19	4	10

[M.Com., Delhi Univ. 2007; MBA, Madras Univ. 2010]

Solution. Using the given constraints, resulting network is shown in *Fig. 13.23*. The dummy activities D_1 and D_2 are introduced to establish the correct precedence relationships. The nodes of the projects are numbered in such a way that their ascending order indicates the direction of progress in the project:

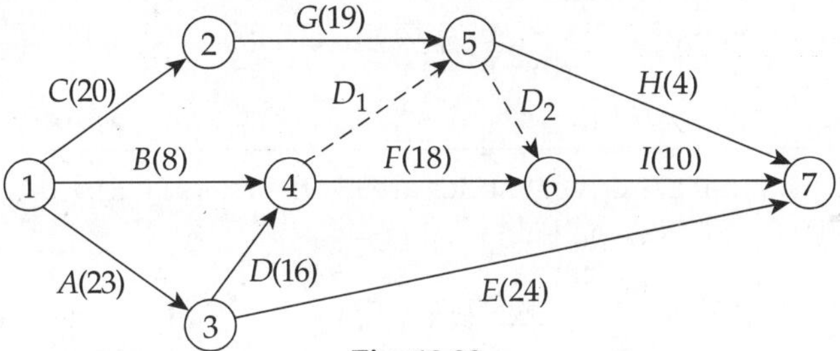

Fig. 13.23

To determine the minimum time of completion of the project (critical path), we compute E_j and E_i for each of the task (i, j) of the project. The critical calculations as applied to *Fig. 13.23* are as follows:

Forward Calculations

Node 1. Set $E_1 = 0$

Node 2. $E_2 = E_1 + t_{12} = 0 + 20 = 20$

Node 3. $E_3 = E_1 + t_{13} = 0 + 23 = 23$

Node 4. $E_4 = \max_{i=1,3}. \{E_i + t_{i4}\} = \max. \{0+8,\ 23+16\} = 39$

Node 5. $E_5 = \max_{i=2,4}. \{E_i + t_{i5}\} = \max. \{20+19,\ 39+0\} = 39$

Node 6. $E_6 = \max_{i=4,5}. \{E_i + t_{i6}\} = \max. \{39+18,\ 39+0\} = 57$

Node 7. $E_7 = \max_{i=3,5,6}. \{E_i + t_{i6}\} = \max. \{23+24,\ 39+4,\ 57+10\} = 67$

Backward Calculations

Node 7. Set $L_7 = E_7 = 67$

Node 6. $L_6 = L_7 - t_{67} = 67 - 10 = 57$

Node 5. $L_5 = \min_{j=6,7}. \{L_j - t_{5j}\} = \min. \{57-0,\ 67-4\} = 57$

Node 4. $L_4 = \min_{j=5,6}. \{L_j - t_{4j}\} = \min. \{57-0,\ 57-18\} = 39$

Node 3. $L_3 = \min_{j=4,7}. \{L_j - t_{3j}\} = \min. \{39-16,\ 67-24\} = 23$

Node 2. $L_2 = L_5 - t_{25} = 57 - 19 = 38$

Node 1. $L_1 = \min_{j=2,3,4}. \{L_j - t_{1j}\} = \min. \{38-20,\ 23-23,\ 39-8\} = 0$

To evaluate the critical nodes, all these calculations are displayed in the following table:

Task (i, j)	Normal time (days)	Earliest time Start (E_i)	Earliest time Finish (E_j)	Latest time State (L_i)	Latest time Finish (L_j)	Float Total	Float Free	Float Independent
(1, 2)	20	0	20	18	38	18	0	0
(1, 3)	23	0	23	0	23	0	0	0
(1, 4)	8	0	8	31	39	31	31	31
(2, 5)	19	20	39	38	57	18	0	–18
(3, 4)	16	23	39	23	39	0	0	0
(3, 7)	24	23	47	43	67	20	20	20
(4, 5)	0	39	39	57	57	18	0	0
(4, 6)	18	39	57	39	57	0	0	0
(5, 6)	0	39	39	57	57	18	18	0
(5, 7)	4	39	43	63	67	24	24	6
(6, 7)	10	57	67	57	67	0	0	0

The above table shows that the critical nodes are for the tasks (1, 3), (3, 4), (4, 6) and (6, 7). Thus, we have

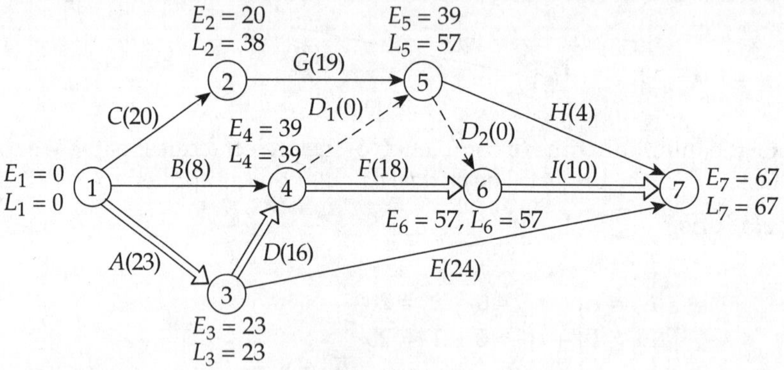

Fig. 13.24

From the above diagram, we observe that *CPM* is: $1 \to 3 \to 4 \to 6 \to 7$, and Project completion time is 67 days.

Example 13.9. *A small project consists of seven activities for which the relevant data is given below:*

Activity	*Preceding activities*	*Activity duration (days)*
A	—	4
B	—	7
C	—	6
D	A, B	5
E	A, B	7
F	C, D, E	6
G	C, D, E	5

(i) Draw the network and find the project completion time.
(ii) Calculate total float for each of the activities and highlight the critical path.
(iii) Draw the time scaled diagram.

[MBA, Panjab Univ. (UIAMS) 2015; B.Tech. (Mech.) (Nov.), HP Univ. 2007]

Solution.

(*i*) Using the precedence relationship among the activities, resulting network is shown in *Fig. 13.25*.

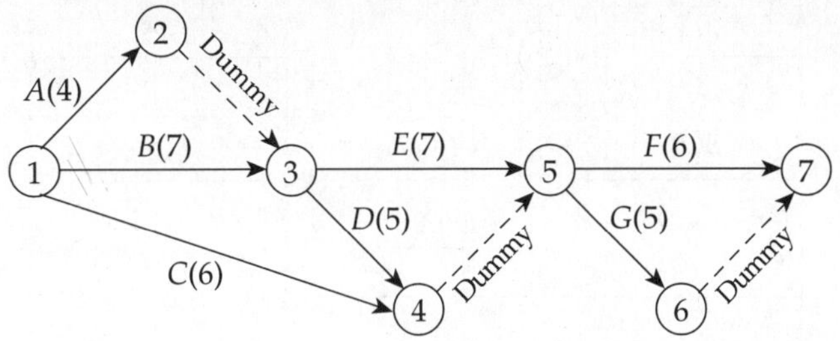

Fig. 13.25

(*ii*) To determine the project completion time, we compute the earliest time 'E_i' and latest finish time 'L_j' for each activity of the project. For it we proceed as follows:

Forward Calculations

Set $E_1 = 0$

$E_2 = E_1 + t_{12} = 0 + 4 = 4$

$E_3 = \max.\ (E_1 + t_{13}, E_2 + t_{23})$
$= \max.\ (0 + 7, 4 + 0) = 7$

$E_4 = \max.\ (E_1 + t_{14}, E_3 + t_{34})$
$= \max.\ (0 + 6, 7 + 5) = 12$

$E_5 = \max.\ (E_3 + t_{35}, E_4 + t_{45})$
$= \max.\ (7 + 7, 12 + 0) = 14$

$E_6 = E_5 + t_{56} = 14 + 5 = 19$

$E_7 = \max.\ (E_5 + t_{57}, E_6 + t_{67})$
$= \max.\ (14 + 6, 19 + 0) = 20$

Backward Calculations

Set $L_7 = E_7 = 20$

$L_6 = L_7 - t_{67} = 20 - 0 = 20$

$L_5 = \min.\ (L_6 - t_{56}, L_7 - t_{57})$
$= \min.\ (20 - 5, 20 - 6) = 14$

$L_4 = L_5 - t_{45} = 14 - 0 = 14$

$L_3 = \min.\ (L_4 - t_{34}, L_5 - t_{35})$
$= \min.\ (14 - 5, 14 - 7) = 7$

$L_2 = L_3 - t_{23} = 7 - 0 = 7$

$L_1 = \min.\ (L_2 - t_{12}, L_3 - t_{13}, L_4 - t_{14})$
$= \min.\ (7 - 4, 7 - 7, 14 - 6) = 0$

To evaluate the critical nodes and total float, the above calculations are displayed in the following table:

Activity (i, j)	*Normal time (days)*	*Earliest time*		*Latest time*		*Total float*
		E_i	$E_i + t_{ij}$	$L_j - t_{ij}$	L_j	
1–2	4	0	4	3	7	3
1–3	7	0	7	0	7	0
1–4	6	0	6	8	14	8
3–4	5	7	12	9	14	2
3–5	7	7	14	7	14	0
5–7	6	14	20	14	20	0
5–6	5	14	19	15	20	1

From the above table, we observe that the critical nodes (events) are: (1, 3), (3, 5) and (5, 7). Thus, we have

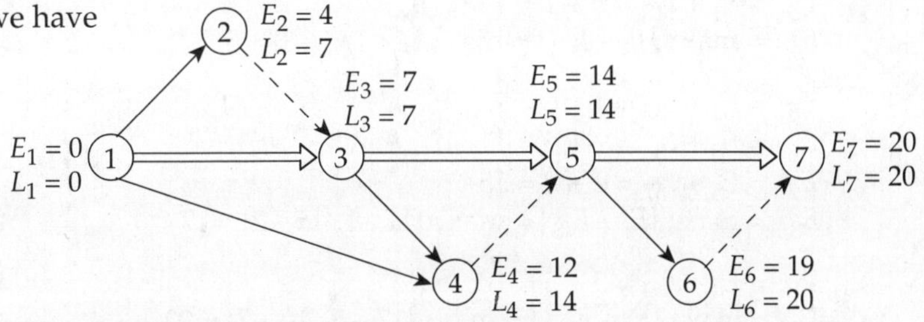

Fig. 13.26

From the given *Fig. 13.26*, *CPM* is $1 \to 3 \to 5 \to 7$ and Project completion time is 20 days.

(*iii*) Time scale diagram of the above network diagram is as under:

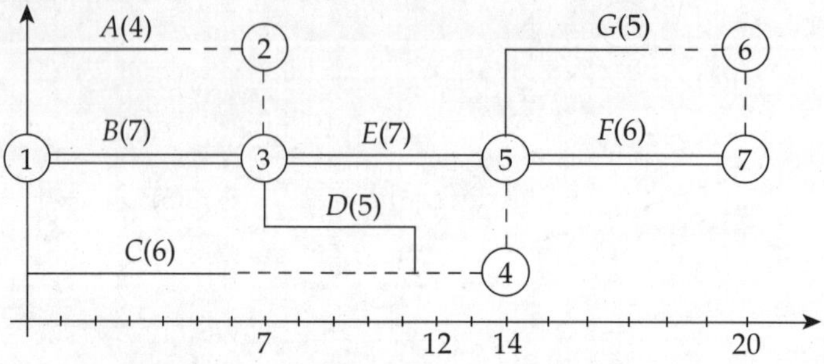

Fig. 13.27

Example 13.10. *A project schedule has the following characteristics:*

Activity	*Time*	*Activity*	*Time*
1–2	*4*	*5–6*	*4*
1–3	*1*	*5–7*	*8*
2–4	*1*	*6–8*	*1*
3–4	*1*	*7–8*	*2*
3–5	*6*	*8–10*	*5*
4–9	*5*	*9–10*	*7*

(*i*) *Construct PERT network,* (*ii*) *Compute* T_E, T_L, *for each event,* (*iii*) *Find the critical path.*

[PG Dipl. in Glob. Bus. Oper., Delhi Univ. 2011]

Solution. Using given information, the resulting network is shown in *Fig. 13.28*.

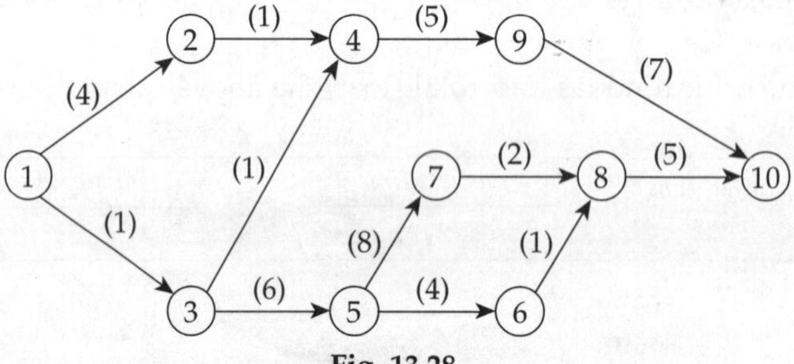

Fig. 13.28

To determine the minimum time for completion of the project (critical path), we compute earliest time E_i and for each of the activity of the project. The critical path calculations are:

Set $E_1 = 0$

$E_2 = E_1 + t_{12} = 0 + 4 = 4$

$E_3 = E_1 + t_{13} = 0 + 1 = 1$

$E_4 = \max_{i=2,3}. \{E_i + E_{i4}\} = \max. \{4+1,\ 1+1\} = 5$

$E_5 = E_3 + t_{35} = 1 + 6 = 7$

$E_6 = E_5 + t_{36} = 7 + 4 = 11$

$E_7 = E_5 + t_{57} = 7 + 8 = 15$

$E_8 = \max_{i=6,7}. \{E_i + E_{i8}\} = \max. \{11+1,\ 15+2\} = 17$

$E_9 = E_4 + t_{49} = 5 + 5 = 10$

$E_{10} = \max_{i=8,9}. \{E_i + E_{i10}\} = \max. \{17+5,\ 10+7\} = 22$

The values of L_j are now obtained:

Set $L_{10} = LS_{10} = 22$

$$L_9 = L_{10} - t_{9,\,10} = 22 - 7 = 15$$
$$L_8 = L_{10} - t_{8,\,10} = 22 - 5 = 17$$
$$L_7 = L_8 - t_{78} = 17 - 2 = 15$$
$$L_6 = L_8 - t_{98} = 17 - 1 = 16$$
$$L_5 = \underset{j=6,7}{\text{min.}} \{L_j - t_{5j}\} = \text{min.} \{16-4,\ 15-8\} = 7$$
$$L_4 = L_9 - t_{49} = 15 - 5 = 10$$
$$L_3 = \underset{j=4,5}{\text{min.}} \{L_j - t_{3j}\} = \text{min.} \{10-1,\ 7-6\} = 1$$
$$L_2 = L_4 - t_{24} = 10 - 1 = 9$$
$$L_1 = \underset{j=2,3}{\text{min.}} \{L_j - t_{1j}\} = \text{min.} \{9-4,\ 1-1\} = 0$$

For determining the critical nodes, all these calculations are displayed in the following table:

Activity	*Normal time*	*Earliest time*		*Latest time*		*Total float*
		Start	*Finish*	*Start*	*Finish*	
(1, 2)	4	0	4	5	9	5
(1, 3)	1	0	1	0	1	0
(2, 4)	1	4	5	9	10	5
(3, 4)	1	1	2	9	10	8
(3, 5)	6	1	7	1	7	0
(4, 9)	5	5	10	10	15	5
(5, 6)	4	7	11	12	16	5
(5, 7)	8	7	15	7	15	0
(6, 8)	1	11	12	16	17	5
(7, 8)	2	15	17	15	17	0
(8, 10)	5	17	22	17	22	0
(9, 10)	7	10	17	15	22	5

The above table shows that the critical nodes are for the activities (1, 3), (3, 5), (5, 7), (7, 8) and (8, 10). The critical path comprises these activities, *viz.*, $1 \to 3 \to 5 \to 7 \to 8 \to 10$:

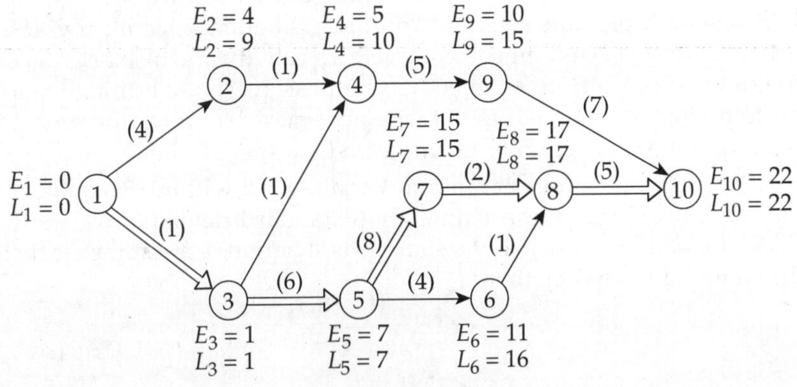

Fig. 13.29

13:9. PROGRAMME EVALUATION AND REVIEW TECHNIQUE (PERT)

PERT may be termed as the approach of multiple time estimates. It deals with the problem of uncertain activity time by the application of statistical analysis to the determination of estimated time for each activity concerning the project. Under *CPM*, we use the precise known time that an activity will take, but such a precise known time approach may not be reliable in real life situation. In order to arrive at a most reliable estimate of time, multiple time estimate approach may be adopted and *PERT* performs this task. Under *PERT*, each activity of the project is based upon three different time estimates, namely (*a*) the optimistic time, (*b*) the pessimistic time, and (*c*) the most likely time.

All the three time estimates are shown in the figure given below:

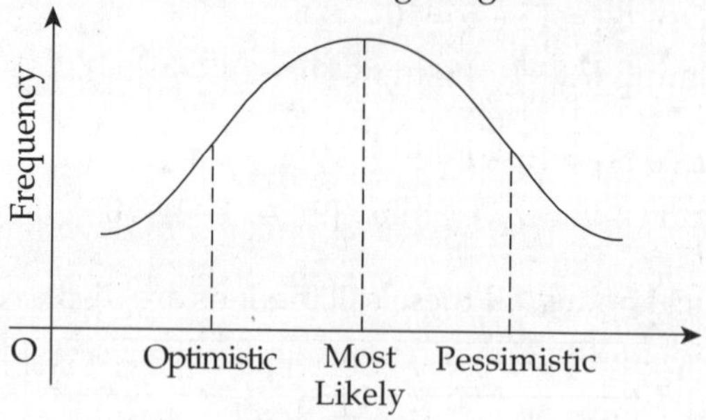

Fig. 13.30

The ***optimistic time*** (t_0) is the shortest possible time to complete the activity. If everything goes perfectly with no complications. The ***pessimistic time*** (t_p) is the longest time that an activity could take, if every thing goes wrong. the ***most likely time*** (t_m) is the estimate of the normal time an activity would take. If only one time were available, this would be it. Otherwise, it is the mode of probability distribution.

From the three time estimates concerning an activity, the expected time (t_e), which is usually adopted under *PERT*, is computed by the following formula:

$$t_e = \frac{(t_0 + 4t_m + t_p)}{6}$$

In fact, $(t_0 + t_p)/2$ is assumed to weight half as much as the most likely time 't_m'. Thus, the expected value of the activity duration can be approximated as the arithmetic mean of $(t_0 + t_p)/2$ and $2t_m$. Thus, we have

$$t_e = \frac{[2t_m + (t_0 + t_p)/2]}{3} = \frac{(t_0 + 4t_m + t_p)}{6}$$

The expected time 't_e' provides the time in which there is a fifty-fifty chance of the activity being completed. The multiple time estimate approach is better, since it provides quite often the improved estimates of the expected time to complete an activity and also allows us to consider the variability of the time for completion of an activity. A basic reason to estimate more than one time for an activity is to provide data by which management may determine the probabilities that each activity as well as the entire project will be completed by a specific data.

Since, almost (*i.e.*, 99%) all values of random variables fall within ±3 standard deviation from the mean, or fall within the range approximately 6 standard deviation in length, therefore, the interval (t_0, t_p) is assumed to enclose about 6 standard deviation of a symmetric distribution. Thus, if σ denotes the standard deviation, then

$$6\sigma \cong t_p - t_0 \quad \text{or} \quad \sigma = \frac{(t_p - t_0)}{6}$$

The, variance, therefore, is: $\text{Var.} \; (= \sigma^2) = \left[\frac{(t_p - t_0)}{6}\right]^2$

13:10. PROCEDURE FOR THE DEVELOPMENT OF *PERT*

Various steps involved for the development of *PERT* network to analyse any project are as follows:

Step 1. Draw a rough *PERT* network on the basis of following three questions for each activity:

(*a*) Which activity or activities precede the particular activity?

(*b*) Which activity or activities follow this activity?

(*c*) Which activities are concurrent with this activity?

Step 2. The network is then suitably sketched to conform to the rules and conventions. Events are numbered in ascending order from left to right.

Step 3. Find out the time estimates for each activity, *viz.*, the most likely time, t_m, pessimistic time, t_p; and optimistic time t_0.

Step 4. Compute the expected time, t_e; for each activity using the formula:

$$t_e = (t_0 + 4t_m + t_p)/6$$

Step 5. Using the expected activity time estimates, determine the earliest start time, earliest finish time, latest start time and latest finish time for each activity of the project.

Step 6. Compute the float associated with each activity. As the critical path activities are the activities with zero float, determine the critical path.

Step 7. Calculate the variance of each activity's time estimate using the formula:

$$\left[\frac{(t_p - t_0)}{6}\right]^2$$

Step 8. Use probability in the activity times to estimate the probability of the project completion date and then, using this estimate, compute the probability of meeting a specified completion date by using the standard normal equation:

$$Z = \frac{\text{Due date} - \text{Expected date of completion}}{\sqrt{\text{Project variance}}}$$

where, Z = standard normal deviate with mean zero, $\sqrt{\text{Project variance}}$ = number of standard deviations, the due date or target date (scheduled time) lies from the mean (or expected) date. Using the cumulative normal distribution tables, the corresponding value of the standard normal variate is read off. This will give the required probability of completing the project on scheduled time.

Step 9. If the critical path duration is not acceptable to the management, then crash the project, *i.e.*, compress the project duration by adding extra resources to an activity.

Example 13.11. *In a PERT network, the critical path comprises of 6 activities, whose estimated duration in days are given below:*

Activity	*Estimated duration (days)*		
	Optimistic (t_0)	*Most likely* (t_m)	*Pessimistic* (t_p)
1	3	6	12
2	8	9	10
3	6	8	10
4	2	5	8
5	3	9	12
6	3	4	5

If the project is scheduled for completion within 42 days, what is the probability of achieving the schedule?

Solution. The expected estimated duration (time) and variance of each critical activity is computed as follows:

Activity	Expected estimated time $t_e = \left[\frac{1}{6}(t_0 + 4t_m + t_p)\right]$	Standard deviation $\sigma = \left[\frac{(t_p - t_0)}{6}\right]$	Variance σ^2
1	6.50	1.50	2.25
2	9.00	0.33	0.11
3	8.00	0.67	0.45
4	5.00	1.00	1.00
5	8.50	1.50	2.25
6	4.00	0.33	0.11

Expected time for completion of the project, therefore, is given by

$$T_e = (6.5 + 9 + 8 + 5 + 8.5 + 4) = 41 \text{ days.}$$

Variance of the project time is given by

$$\sigma^2 = (2.25 + 0.11 + 0.45 + 1 + 2.25 + 0.11) = 6.17$$

The scheduled time of completing the project is 42 days. Therefore, for the probability of completing the project within the scheduled time is determined by using the standard normal deviate,

$$Z = \frac{\text{Due date} - \text{Expected date of completion}}{\sqrt{\text{Variance of project length}}}$$

$$= \frac{\text{Due date} - T_e}{\sqrt{\sigma^2}} = \frac{42 - 41}{\sqrt{6.17}} = \frac{1}{2.48} = 0.40$$

∴ The probability of meeting the due date is given by

$$P\,(Z \le 0.40) = 0.50 + \Phi\,(0.40) = 0.50 + 0.1554 = 0.6554$$

(from the tables under standard normal curve).

Hence, the probability of completing the project by schedule duration is 65.54%.

Example 13.12. *A project consists of eight activities with the following relevant information:*

Activity	*Immediate Predecessor*	*Estimated duration (days)*		
		Optimistic	*Most likely*	*Pessimistic*
A	—	1	1	7
B	—	1	4	7
C	—	2	2	8
D	A	1	1	1
E	B	2	5	14
F	C	2	5	8
G	D, E	3	6	15
H	F, G	1	2	3

(i) *Draw the PERT network and determine critical path. Find out the expected project completion time.*

(ii) *What duration will have 95% confidence for project completion?*

(iii) *If the average duration for activity F increases to 14 days, what will be its effect on the expected project completion time which will have 95% confidence?*

(For standard normal Z = 1.645, area under the standard normal curve from 0 to Z is 0.45).

[MBA, Panjab Univ. 2007; MBA, Bangalore Univ. 2016]

Solution. The expected time and variance of each activity is computed in table below:

Activity	t_0	t_m	t_p	$t_e = (t_0 + 4t_m + t_p)/6$	$[(t_p - t_0)/6]^2$
A	1	1	7	2	1
B	1	4	7	4	1
C	2	2	8	3	1
D	1	1	1	1	0
E	2	5	14	6	4
F	2	5	8	5	1
G	3	6	15	7	4
H	1	2	3	2	$(2/6)^2$

(*i*) Using the precedence relationship among the activities and the expected time t_e, the resulting network is shown in *Fig. 13.31*.

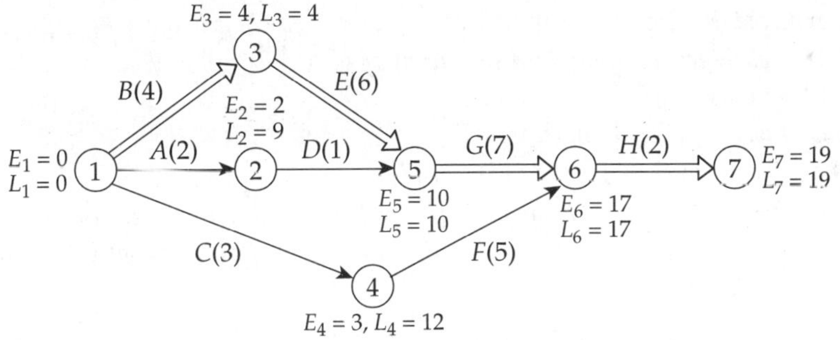

Fig. 13.31

From the above network diagram, we observe:

$$CPM:\ 1 \rightarrow 3 \rightarrow 5 \rightarrow 6 \rightarrow 7,\ \ i.e.,\ \ B \rightarrow E \rightarrow G \rightarrow H.$$

Expected duration of the project is 19 days.

Variance of the project length = 1 + 4 + 4 + 0.108 = 9.108

$\therefore$ The standard deviation of the project $= \sqrt{\text{Variance of the project}} = \sqrt{9.108} = 3.02$

(*ii*) We are given that the standard normal deviate Z = 1.645. Therefore, the probability of meeting the due date is given by

$$P\ (Z \leq 1.645) = 0.50 + 0.45 = 0.95 \text{ or } 95\%,$$

since area under standard normal curve from 0 to Z is 0.45

Now, since $$Z = \frac{\text{Due date} - \text{Expected date of completion}}{\sqrt{\text{Variance}}}$$

Therefore, Due date = 1.645 × 3.02 + 19 = 4.9679 + 19 = 24 days.

Hence, 24 days of project completion time will have 95% confidence of completion in the scheduled time.

(*iii*) When the average duration of activity *F* increases of 14, the path $C \rightarrow F \rightarrow H$ also becomes critical. The standard deviation of new critical path $= \sqrt{9.108 + 2} = 3.36$

$\therefore$ $P\ (Z < 1.645) = 0.95$ gives us

$$\text{Due date} = 19 + 3.36 \times 1.645 = 24.52 \text{ days.}$$

Hence, the project completion duration of 24.52 days will have 95% confidence.

Example 13.13. *A small project is composed of seven activities whose time estimates are listed in the table given below:*

Activity *i – j*	*Estimated duration (weeks)*		
	Optimistic	*Most likely*	*Pessimistic*
1–2	1	1	7
1–3	1	4	7
1–4	2	2	8
2–5	1	1	1
3–5	2	5	14
4–6	2	5	8
5–6	3	6	15

(*a*) *Draw the project network.*

(*b*) *Find the expected duration and variance of each activity. What is the expected project length?*

(*c*) *Calculate the variance and standard deviation of project length. What is the probability that the project will be completed:*

(*i*) *at least 4 weeks earlier than expected?*

(*ii*) *no more than 4 weeks later than expected?*

(*d*) *If the project due date is 19 weeks, what is the probability of not meeting the due date?*

[MBA, Panjab Tech. Univ. 2008; MBA, Madras Univ. 2010]

Solution.

(*a*) The problem is same as is given in *Example 13.12*, except that activity 6–7 has been deleted. As such the network diagram remains the same as given in *Fig. 13.31* after deleting activity 6–7, we have

(*b*) *CPM:* $1 \to 3 \to 5 \to 6$, Duration of project = 17 days.

Variance of project length = 1 + 4 + 4 = 9

and, therefore, standard deviation of the project $= \sqrt{9} = 3$

(*c*) The standard normal deviate is given by

$$z = \frac{\text{Due date} - \text{Expected date of completion}}{\sqrt{\text{Variance}}}$$

∴ We compute

(*i*) $$z = \frac{13-17}{3} = -\frac{4}{3} = -1.33$$

∴ $P(z \le -1.33) = 0.50 - \Phi(1.33) = 0.50 - 0.4082$ [*from normal tables*]

$= 0.0918$ or 9.18%

the probability of meeting the due date (4 weeks earlier than expected) is: 0.0918 or 9.18%.

(*ii*) $$z = \frac{21-17}{3} = \frac{4}{3} = 1.33$$

∴ The probability of meeting the due date (4 weeks later than expected) is given by

$P(z \le 1.33) = 0.50 + \Phi(1.33) = 0.50 + 0.4082$

$= 0.9082$ or 90.82%

(*d*) When the due date is 19 weeks,

$$z = \frac{19-17}{3} = \frac{2}{3} = 0.67$$

∴ Probability of meeting the due date is

$$P\ (z \le 0.67) = 0.5 + \Phi\ (0.67) = 0.7486 = 74.86\%$$

Hence, the probability of not meeting the due date is 1 – 0.7486, *i.e.*, 0.2514 or 25.14%.

PROBLEMS

13.11. The following are the details of estimated times of activities of certain project:

Activity	*Immediate predecessor*	*Normal time (days)*
A	—	16
B	—	20
C	A	8
D	A	10
E	B, C	6
F	D, E	12

(*a*) Find the critical path and the expected time of the project.

(*b*) Find the total and free-float for each activity. ***[C.A. Final, Nov. 2007]***

13.12. An architect has been awarded a contract to prepare plans for an hospital renovation. The job consists of the following activities and their estimated times:

Activity	*Description*	*Predecessors*	*Time (days)*
A	Prepare sketches	—	2
B	Outline specifications	—	1
C	Prepare drawings	A	3
D	Write specifications	A, B	2
E	Run off prints	C, D	1
F	Have specification	B, D	3
G	Assemble bid packages	E, F	1

(*a*) Draw the network diagram of the project, and (*b*) Indicate the critical path. What is total duration of the project? ***[MBA (HCA) (PT.), Delhi Univ. 2008]***

13.13. An established company has decided to add a new product to its line. It will buy the product from a manufacturing concern, package it, and then sell it to a number of distributors selected on a geographical basis. Market research has indicated the volume expected and the size of sales force required. The steps shown in the following table are to be planned:

Planning Activities for New Product Introduction

Activity	*Description*	*Time (weeks)*
A	Organize sales office	6
B	Hire salesmen	4
C	Train salesmen	7
D	Select advertising agency	2
E	Plan advertising agency	4

(*Contd.*)

F	Conduct advertising campaign	10
G	Design package	2
H	Set up packaging facilities	10
I	Package initial stocks	6
J	Order-stock from manufacturer	13
K	Select distributors	9
L	Sell to distributors	3
M	Ship-stocks	5

(*i*) Draw an arrow diagram for this project.

(*ii*) Indicate the critical path.

(*iii*) For each non-critical activity, find the total and free float. ***[MBA, Delhi Univ. 2008]***

13.14. Following table gives the list of various activities and their immediate predecessor involved in installation of CAT scanner in a hospital:

Activity	*Description*	*Immediate predecessors*	*Expected duration (days)*
A	Finalization of the layout plans	—	2
B	Demolition of the structures	A	6
C	Walls erection	B	12
D	Flooring	B	8
E	Electrical wiring	C	6
F	Air conditioning ducting	C	4
G	Fire alarm installation	C	3
H	False ceiling and light fittings	E, F, G	10
I	Wall plastering and painting	H	9
J	Equipment installation	D, I	6
K	Calibration and testing	J	3
L	Final finishing	K	2
M	Handling over	L	1

(*a*) Prepare a network diagram for this project.

(*b*) Find the total time required to install CAT scanner.

(*c*) Calculate the total float for various non-critical activities.

[MBA (HCA) (Nov.), Delhi Univ. 2003]

13.15. For the network:

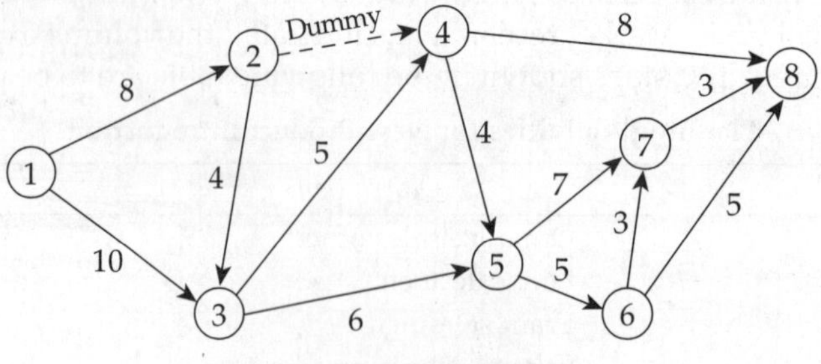

Fig. 13.32

Find the floats (all three) for the activities 1–3, 3–5 and 5–7. ***[B.Com., Panjab Univ. 2007]***

13.16. Draw the network for the following project:

Activity	*Preceded by activity*	*Duration (weeks)*	*Activity*	*Preceded by activity*	*Duration (weeks)*
A	—	10	G	C	8
B	A	9	H	F	8
C	A	7	I	D	4
D	B	6	J	G, H	11
E	B	12	K	E	5
F	C	6	L	I	7

Number the events by Fulkerson's rule and find the critical path. Also, find the time for completing the project. ***[M.Com., Bangalore Univ. 2017]***

13.17. Tasks $A, B, C, ..., H, I$ constitute a project. The notation $X < Y$ means that the task X must be finished before Y can begin. With this notation:

$$A < D,\ \ A < E,\ \ B < F,\ \ D < F,\ \ C < G,\ \ C < H,\ \ F < I,\ \ G < I.$$

Draw a graph to represent the sequence of tasks and find the minimum time of completion of the project, when the time (in days) of completion of each task is as follows:

Task :	A	B	C	D	E	F	G	H	I
Time :	8	10	8	10	16	17	18	14	9

[B.Tech., Panjab Tech. Univ. 2012; MBA (Nov.), Annamalai Univ. 2002; MBA, Madras Univ. 2000]

13.18. A small project consists of the following jobs, where precedence relationship is given below:

Job :	1–2	1–3	2–3	2–5	3–4	3–6	4–5	4–6	5–6	6–7
Duration (days) :	15	15	3	5	8	12	1	14	3	14

(*a*) Draw an arrow diagram representing the project.

(*b*) Find the critical path and the total project duration.

[BE, Karnataka Univ. 2010; MBA, UP Tech. Univ. 2006]

13.19. Draw the network diagram for the following activities and find the critical path and duration of the project:

Activities :	A	B	C	D	E	F	G	H	I	J	K	L
Immediate predecessor activity :	—	—	—	B, C	A	C	E	E	D, E, H	E	I, J	G
Duration (days) :	9	4	7	8	7	9	10	8	6	9	10	7

[MBA (Dec.), Panjab Tech. Univ. 2009]

13.20. The following data gives the list of activity with their duration in days:

Activity :	1–2	1–6	2–3	2–4	3–4	3–5	4–5	4–6	5–8	6–7	7–8
Duration (days) :	4	5	3	5	2	4	7	5	4	2	3

(*a*) Draw a project network and calculate early start, late start, early finish, late finish, total slack and free slack.

(*b*) Mark the critical path. What is the project duration?

[MBA (June), Visvesvaraya Univ. 2011]

13.21. Consider the data of the project given below. Find its critical path and project duration:

Activity :	A	B	C	D	E	F	G	H	I
Predecessor :	—	—	A	B	C, D	B	E	E	F, G
Duration (days) :	4	7	2	9	6	5	2	10	4

[MBA, Osmania Univ. 2001]

13.22. Draw the network for the data given below and compute:

(*i*) Critical path,

(*ii*) Early start and late start times for each activity, and

(*iii*) Total slack for each activity:

Activity :	A	B	C	D	E	F	G	H	I
Predecessor :	—	—	—	A	B	C	D, E	B	H, F
Estimated time (weeks) :	3	5	4	2	3	9	8	7	9

[PG Dip. in Glob. Bus. Oper., Delhi Univ. 2010]

13.23. A small maintenance project consists of the following twelve jobs whose precedence relations are identified with their node numbers:

Job (i, j) :	(1, 2)	(1, 3)	(1, 4)	(2, 3)	(2, 5)	(2, 6)
Duration (in days) :	10	4	6	5	12	9
Job (i, j) :	(3, 7)	(4, 5)	(5, 6)	(6, 7)	(6, 8)	(7, 8)
Duration (in days) :	12	15	6	5	4	7

(*i*) Draw an arrow diagram representing the project.

(*ii*) Calculate earliest start, earliest finish, latest start and latest finish time for all the jobs.

(*iii*) Find the critical path and project duration.

(*iv*) Tabulate total float, free float and independent float for each activity.

[MBA, Pune Univ. 2007]

13.24. A project has the following time schedule:

Activity	*Time in weeks*	*Activity*	*Time in weeks*
1–2	2	4–6	3
1–3	2	5–8	1
1–4	1	6–9	5
2–5	4	7–8	4
3–6	8	8–9	3
3–7	5		

Construct *PERT* network and compute: (*i*) Total float for each activity; and (*ii*) Critical path and its duration.
[M.Com., Karnataka Univ. 2013]

13.25. A project schedule has the following characteristics:

Activity	*Time in weeks*	*Activity*	*Time in weeks*
1–2	4	5–6	4
1–3	1	5–7	8
2–4	1	6–8	1
3–4	1	7–8	2
3–5	6	8–10	5
4–9	5	9–10	7

(*i*) Construct *PERT* network; (*ii*) Compute T_E, T_L for each event; (*iii*) Find the critical path, and its duration.
[PG Dip. in Glob. Bus. Oper., Delhi Univ. 2011; BE (Mech.), Kurukshetra Univ. 2012]

13.26. A small project consists of seven activities, the details of which are given below:

Activity	*Duration (in days)* *Most likely*	*Optimistic*	*Pessimistic*	*Immediate Predecessor*
A	3	1	7	—
B	6	2	14	*A*
C	3	3	3	*A*
D	10	4	22	*B, C*
E	7	3	15	*B*
F	5	2	14	*D, E*
G	4	4	4	*D*

(*i*) Draw the network, number the nodes, find the critical path, the expected project completion time and the next most critical path.
(*ii*) What project duration will have 95% confidence of completion?

[M.Com., Kerala Univ. 2008]

13.27. The manager of swimming club is planning club's swimming team programme. The first team practice is scheduled for August 1, 2010. The activities, their immediate predecessor(s) and the activity time estimates (in weeks) are given below:

Activity	*Immediate predecessor*	*Optimistic time*	*Most likely time*	*Pessimistic time*
A	—	1	1	2
B	*A*	4	6	8
C	*A*	2	4	6
D	*B, C*	1	2	3
E	*B*	2	3	4
F	*A*	1	2	3
G	*D*	1	2	3
H	*G*	1	2	3
I	*E, H, F*	1	1	1

(*i*) Draw the project network and develop activity schedule with earliest and latest time.
(*ii*) What are the critical activities, and what is the expected project completion time?
(*iii*) If the club manager plans to start the project on May 1, 2010, what is the probability that the swimming pool will be ready by the scheduled August 1, 2010 (within 13 weeks)?

[M.Com., Delhi Univ. 2010]

13.28. The following optimistic (*O*), pessimistic (*P*) and most likely (*M*) time estimates for each task have been given for a project:

Task	*Predecessors*	*Activity times (days)* *O*	*M*	*P*
A	—	10	12	15
B	—	6	10	16
C	*A, B*	3	5	10
D	*C*	8	12	17
E	*C*	4	7	12
F	*C*	3	4	6
G	*D, E*	5	8	13
H	*F, G*	5	7	10

(*i*) Construct the network diagram for this project and compute its expected completion time assuming all tasks take their expected completion times.
(*ii*) Compute the probability that the project will take more than 50 days to complete.

[MBA (PT), Delhi Univ. 2015]

13.29. A project consists of activities $A, B, C, D, ..., H, I$. The notation $X < Y$ means that the activity X must be completed before Y can start, and $X, Y < W$ means that W will start only after completion of the activities X and Y. With this notation construct the network diagram for the following constraints:

$$A < D;\ A < E;\ B < F;\ C < G;\ D < H;\ E, F < I.$$

The project has the following time schedules for the above activities:

Task :	*A*	*B*	*C*	*D*	*E*	*F*	*G*	*H*	*I*
Optimistic time :	5	18	26	16	15	6	7	7	3
Pessimistic time :	10	22	40	20	25	12	12	9	5
Most likely time :	8	20	33	18	20	9	10	8	4

Determine the following:
(*a*) Expected task times and their variance.
(*b*) Earliest expected times to reach each node.
(*c*) Draw the project network and identify the critical path.
(*d*) What is the probability that the jobs on the critical path will be completed in 41.5 days?
(*e*) The duration of the project that will have 95% chance of being completed.

[MBA, Madras Univ. 2006]

13.30. The critical activities of a project, their expected duration and standard deviation (in weeks) are given below:

Activity :	*B*	*E*	*G*	*H*	*K*	*L*
Expected duration :	12	3	8	7	5	6
Standard deviation :	2/3	1/3	2	5/3	4/3	0

(*i*) What is the probability that the project will be completed in (*a*) 41 weeks, (*b*) 45 weeks?
(*ii*) If the manager wants the probability to be 98 per cent of completing the project in time, how many weeks prior to the deadline should be needed to start the project?

[M.Com., Delhi Univ. 2009]

13.31. The following table list the jobs of a network along with their time estimates:

Job	*Duration (days)*		
(i–j)	*Optimistic*	*Most likely*	*Pessimistic*
1–2	3	6	15
1–6	2	5	14
2–3	6	12	30
2–4	2	5	8
3–5	5	11	17
4–5	3	6	15
6–7	3	9	27
5–8	1	4	7
7–8	4	19	28

(*a*) Draw the project network diagram.
(*b*) Calculate the length and variance of the critical path after estimating the earliest and latest event times for all nodes.

(*c*) Find the probability of completing the project before 31 days.
(*d*) What is the chance of project duration exceeding 46 days?
(*e*) What will be effect on the current critical path, if the most likely time of activity 3–5 gets revised to 14 instead of 11 days given above?
[MBA (Nov.), Delhi Univ. 2009; MBA (Nov.) Visvesvaraya Tech. Univ. 2020; MBA, Lucknow Univ. 2008; MBA, Bangalore Univ. 2018]

13.32. A project is composed of eight activities, the time estimates for which are given below:

Activity	*Immediate preceding activity*	*Time estimates (weeks)*		
		Optimistic	*Most likely*	*Pessimistic*
A	—	2	4	12
B	—	10	12	26
C	A	8	9	10
D	A	10	15	20
E	A	7	7.5	11
F	B, C	9	9	9
G	D	3	3.5	7
H	E, F, G	5	5	5

(*a*) Draw *PERT* network.
(*b*) Find out the critical path as well as expected project completion time.
(*c*) If a 30 weeks deadline is imposed, what is the probability that the project will be finished within time limit? *[B.Tech. (Nov.), HP Univ. 2015; MIB, Delhi Univ. 2005]*
(*d*) If the project manager wants to be 99% sure that the project is completed on the scheduled date, how many weeks before that date should he start the project work.
[Area between z = 0 and z = 0.41 is 0.1591 and the value of z that covers 49% of area is 2.33.]

13.33 A publisher has just signed a contract for the publication of a book. What is the earliest time by which the book can be ready for distribution? The tasks given in the table below are involved, with time estimates (in weeks).

Task	*Description*	*Precedence*	*Likely likely*	*Optimistic*	*Pessimistic*
A	Appraisal of book by reviewers	—	8	4	10
B	Initial pricing of book	—	2	2	2
C	Assessment of marketability	A, B	2	1	3
D	Revisions by author	A	6	4	12
E	Editing of final draft	C, D	4	3	5
F	Typesetting of text	E	3	3	3
G	Plates for artwork	E	4	3	5
H	Designing and printing of jacket	C, D	6	4	9
I	Printing and binding of book	F, G	8	6	16
J	Inspection and final assembly	I, H	1	1	1

(*i*) For this *PERT* network, find the expected task durations and the variance of task durations.
(*ii*) Draw a network and find the critical path. What is the expected length of the critical path, and what is its variance?
(*iii*) What is the probability that the length of the critical path does not exceed (*a*) 32 weeks? (*b*) 36 weeks? *[MBA, Bangalore Univ. 2017; MBA, Delhi Univ. 2010]*

13.34. An event management company must perform the tasks shown in the following table, before an international conference can be held (all durations are in days):

Activity	*Description*	*Predecessor*	*Optimistic estimate*	*Pessimistic estimate*	*Most likely estimate*
A	Find venue	—	3	5	4
B	Find sponsors	*A*	1	3	2
C	Inauguration guests	*A*	2	8	5
D	Set advertisements	*A*	1	3	2
E	Set up registration facility	*D*	1	3	2
F	Prepare electronics	*B*	2	4	3
G	Print advertising	*E*	1	1	1
H	Set up transportation	*C*	1	2	1
I	Rehearsals	*F, H*	1	3	2
J	Last-minute details	*I*	1	1	1

(*i*) Draw the project network.
(*ii*) Determine the critical path.
(*iii*) If the event manager wants to have a 99% probability of completing all preparations for the conference by February 20th of 2013, when should the work begin?

[MBA, Delhi Univ. 2012]

13.35 A sociologist plans a questionnaire survey consisting of the following tasks:

Activity	*Task*	*Precedence*	*Duration (in days)*		
			Likely	*Minimum*	*Maximum*
A	Design to questionnarie	—	5	4	6
B	Sampling design	—	12	8	16
C	Testing of questionnaire and refinements	*A*	5	4	12
D	Recruiting for interviewers	*B*	3	1	5
E	Training of interviewers	*D, A*	2	2	2
F	Allocation of areas to interviewers	*B*	5	4	6
G	Conducting interviews	*C, E, F*	14	10	18
H	Evaluation of results	*G*	20	18	34

(*i*) Draw the *PERT* network diagram.
(*ii*) Identify the critical path.
(*iii*) Determine the expected duration of the project and its variance.
(*iv*) Find the probability that the project is completed in 56 weeks.

[PG Dip. in Glob. Bus. Oper., Delhi Univ. 2011]

13.36. Given the following information:

Activity	*Nodes*	*Times estimates (weeks)*		
		Optimistic	*Pessimistic*	*Most likely*
A	1–2	3	15	8
B	1–3	4	8	6
C	1–4	6	20	12
D	2–4	6	6	6

(Contd.)

E	2–6	5	9	7
F	3–4	6	15	9
G	3–5	4	12	6
H	4–7	10	12	11
I	5–7	4	8	6
J	5–8	10	10	10
K	6–7	10	18	14
L	7–8	2	6	3

(*i*) Draw the project network. Obtain the standard deviation of completion time of each activity.
(*ii*) Calculate expected duration of the project and its standard deviation.
(*iii*) Find the probability that the project will be completed in 38 weeks.
(*iv*) If the project manager wants to be 90 per cent sure that the project should be completed in time, how many weeks prior to the deadline should he start it.

[M.Com., Delhi Univ. 2004]

13.37. Consider a project having the following activities and their time estimates:

Activity	*Preceding activity*	*Time estimates (weeks)*		
		Optimistic	*Most likely*	*Pessimistic*
A	—	3	4	5
B	—	4	8	10
C	B	5	6	8
D	A, C	9	10	15
E	B	4	6	8
F	D, E	3	4	5
G	D, E	5	6	8
H	D, E	1	3	4
I	G	2	4	5
J	F, I	7	8	10
K	G	4	5	6
L	H	8	9	13
M	J, K, L	6	7	8

(*i*) Draw the *PERT* network and determine the expected project completion time.
(*ii*) Identify the critical activities involved in the project.
(*iii*) If labour costs ₹ 1,500 per week, how much should the company budget an amount for labour costs that will be sufficient with 95% level of confidence?

[MIB, Delhi Univ. 2011]

13.38. The data for a *PERT* network is displayed in the table on the next page. Determine the critical path and the expected duration for completion of the entire project. Give answers to the following:

(*i*) What is the probability that the project duration will exceed 60 days?
(*ii*) What is the chance of completing the project between 45 days and 54 days?
(*iii*) It becomes known later that the three time estimates for activity 4–6 have to be revised to 14–20–30. What impact does this have on project duration? What will be the probability that the project can now be completed before 46 days?

Activity nodes	Time duration (days)		
	a	*b*	*c*
1–2	2	4	6
1–3	6	6	6
1–4	6	12	24
2–3	2	5	8
2–5	11	14	23
3–4	15	24	45
3–6	3	6	9
4–6	9	15	27
5–6	4	10	16

where, *a* is optimistic time; *b* is most likely time; and *c* is pessimistic time.

Given: z	0.00	0.25	0.50	0.75	1.00	1.50	1.67
$P(Z \leq z)$	0.500	0.599	0.692	0.778	0.841	0.933	0.953

[B.Tech. (Mech.) (Nov.), HP Univ. 2014; MBA, Osmania Univ. 2010; MBA (June), Visvesvaraya Univ. 2011]

13.39. In the metropolitan city of Bengaluru, the Bengaluru Development Authority is undertaking several construction projects. Many of these projects involve construction of huge flyovers and the organizations involved in the work do not have adequate prior experience of constructing similar flyovers. Moreover, the process is riddled with considerable uncertainty due to political and public interference. Consequently although estimation of activities is possible, estimation of the duration of the project is difficult. The following table gives the list of activities and three possible time estimates for the same. Use this information to analyse the problem of uncertainty by answering the questions given at the end of the table.

Activity	*Predecessor(s)*	*Optimistic (months)*	*Most likely (months)*	*Pessimistic (months)*
A	—	2	4	9
B	—	2	3	8
C	*B*	4	7	14
D	*A*	4	5	16
E	*B*	5	7	12
F	*C, D*	2	4	8
G	*C, D*	6	8	14
H	*E, F*	6	7	14
I	*G, H*	2	3	9

(*i*) Draw the network diagram of the project and identify alternative paths in the network.
(*ii*) For each path, compute the expected duration and hence identify the critical path and the near critical path of the problem.
(*iii*) What is the probability of completion of the project by 36 months?
(*iv*) If the organization wants to ensure 80% probability of completion, what date should it set for project completion?
(*v*) Compute the probability of completion of the project for the near critical path. Do you observe anything interesting in your results? Comment on the analysis.

[MBA, Delhi Univ. 2011]

13.40. A nationalized bank wishes to plan and schedule the development and installation of a new computerized cheque processing system. The changeover in cheque processing procedure requires employment of additional personnel to operate the new system, development of new systems (computer software), and modification of existing cheque sorting equipment. The activities required to complete the project along with three time estimates and the precedence relationship among the activities have been determined by bank management and are given in the following table:

Activity	*Description*	*Predecessor*	*Time estimates (days)*		
			Optimistic	*Most likely*	*Pessimistic*
A	Position recruiting	—	5	8	17
B	System development	—	3	12	15
C	System training	A	4	7	10
D	Equipment training	A	5	8	23
E	Manual system test	B, C	1	1	1
F	Preliminary system changeover	B, C	1	4	13
G	Computer-personnel interface	D, E	3	6	9
H	Equipment modification	D, E	1	2.5	7
I	Equipment testing	H	1	1	1
J	System debugging and installation	F, G	2	2	2
K	Equipment changeover	G, I	5	8	11

(*a*) Draw a network diagram for this project and find the critical path and its length.

(*b*) Calculate the total and free floats for non-critical activities.

(*c*) What is the probability that the length of the critical path does not exceed 40 days?

[MBA (Oct.), BHU, 2009; MBA (Nov.), Delhi Univ. 2005]

13.41. A publisher is preparing to produce the second edition of a textbook. The activities required and their estimated times are given in the following table:

Activity	*Description*	*Immediate Predecessor*	*Activity times (in months)*		
			Optimistic	*Pessimistic*	*Most likely*
A	Assess market	—	1	3	2
B	Get reviews from users	A	1	2	1.5
C	Revamp old material and add new material	B	3	9	5
D	Obtain review and prepare final draft	C	4	12	6
E	Revise and expand problems	B	2	7	4
F	Copy/edit final draft	D	1	2.5	1.5
G	Copy/edit problems	E	0.5	1.5	1
H	Set type, proof and print book	F, G	5	9	6
I	Prepare instructor's manual	E	2	4	3
J	Produce instructor's manual	I	1	2	1.5
K	Complete book and instructor's manual	H, J	0	0	0

(*a*) Draw an arrow diagram and identify the critical path.

(*b*) What is the probability that project will be completed within 21 months? 24 months? 27 months?

[MBA, Delhi Univ. 2009]

13.42. A civil engineering firm has to bid for the construction of a dam. The activities and time estimates are given below:

Activity		*Duration (in days)*	
	Optimistic	*Most likely*	*Pessimistic*
1–2	14	17	25
2–3	14	18	21
2–4	13	15	18
2–8	16	19	28
3–4 (dummy)			
3–5	15	18	27
4–6	13	17	21
5–7 (dummy)			
5–9	14	18	20
6–7 (dummy)			
6–8 (dummy)			
7–9	16	20	41
8–9	14	16	22

The policy of the firm with respect to submitting bids is to bid the minimum amount that will provide a 95% of probability of at best break-even. The fixed cost for the project is ₹ 8,00,000 and the variable cost is ₹ 9,000 spent every day for working on the project. The duration is in days and the costs are in terms of rupees.

What amount should the firm bid under this policy? (You may perform the calculations on duration, etc. upto two decimal places.)

[MBA, Osmania Univ. 2009; B.Tech. (Mech.), HP Univ. 2010]

13:11. TIME-COST RELATIONSHIP IN *PERT*

In many projects, the estimated project duration may exceed the target duration. It becomes necessary, therefore, to reduce (crash or compress) the critical path so as to complete the project by the target date. This process involve extra resources in terms of direct and indirect costs. The ***direct costs*** are associated with the individual activities such as manpower, loading, equipment utilized, etc. in respect of various activities and the ***indirect costs*** are associated with the overhead costs such as the administrative or supervision costs, loss of revenue, etc. Further, the direct cost increases, if the job duration is to be reduced and the indirect cost increases, if the job duration is to be increased. Therefore, the cost slope represents the rate of increase in the cost of performing the activity per unit decrease in time and is called a ***time-cost trade-off***.

The ***total project cost*** can be found out by adding the total of the project direct and indirect costs. Figure given below shows the time-cost relationship in a project.

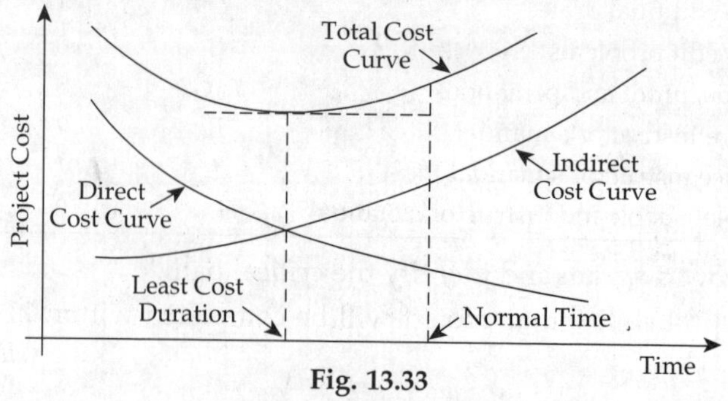

Fig. 13.33

The length of the project is to be shortended to the point where the savings in indirect project costs are equal to the cost of additional resources.

It may be noted that the process of shortening a project is known as ***project crashing*** that requires the determination of optimum time-cost relationship which involves the following major steps:

Step 1. Schedule a project with all its activities at their normal duration.

Step 2. Identify the critical path and critical activities.

Step 3. Compute the cost slope for each activity on the critical path by using the relationship:

$$\frac{\text{Crash cost} - \text{Normal cost}}{\text{Normal time} - \text{Crash time}} = \text{Cost slope}$$

The values of cost slope for critical activities indicate the direct extra cost required to incur on an activity per unit time.

Step 4. Rank the activities in the ascending order of cost slope.

Step 5. Crash the activities on critical path as per the ranking, *i.e.*, activity having lower cost slope would be crashed first to the maximum extent possible.

Step 6. If due to crashing of an activity time in *Step 5*, other paths in the network may become critical. Then identify and crash the activities on parallel critical paths.

Step 7. Crashing as per *Steps* 5 and 6, one reaches a point when further crashing is either not possible or does not result in the reduction of crashing of project duration.

Step 8. Compute the total project cost (*i.e.*, indirect cost plus direct cost) corresponding to different project durations. This is being done by adding the crashing cost commulatively to the normal cost.

Example 13.14. *The following table gives data on normal time-cost and crash time-cost for a project:*

Activity	*Normal*		*Crash*	
	Time (days)	*Cost (₹)*	*Time (days)*	*Cost (₹)*
1–2	6	600	4	1,000
1–3	4	600	2	2,000
2–4	5	500	3	1,500
2–5	3	450	1	650
3–4	6	900	4	2,000
4–6	8	800	4	3,000
5–6	4	400	2	1,000
6–7	3	450	2	800

The indirect cost per day is ₹100.

(*i*) *Draw the network and identify critical path.*

(*ii*) *What are the normal project duration and associated cost?*

(*iii*) *Crash the relevant activities systematically and determine the optimum project completion time and its cost.* **[M.Com., Bangalore Univ. 2016; B.Tech. (Mech.), HP Univ. 2014; M.Com., Delhi Univ. 2009]**

Solution. Using normal time duration:

(*i*) Resulting network with earliest and latest occurrence times at various nodes are as shown in *Fig. 13.34*.

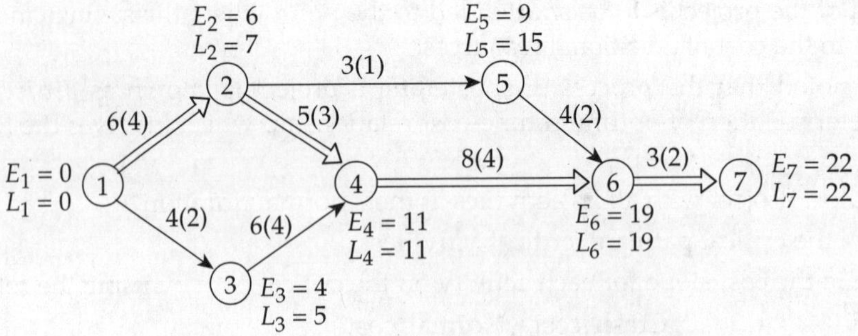

Fig. 13.34

CPM: $1 \to 2 \to 4 \to 6 \to 7.$

(*ii*) Normal project duration is 22 days. The associated total cost is computed below:

Direct cost = ₹ (600 + 600 + 500 + 450 + 900 + 800 + 400 + 450) = ₹ 4,700

Indirect cost = 100 × 22 = ₹ 2,200

Total cost = ₹ 6,900

(*iii*) For the optimum cost schedule, we first compute the cost slope (incremental cost of crashing) as follows:

Critical activity	*Cost slope*
1–2	(1,000 – 600)/(6 – 4) = 200
2–4	(1,500 – 500)/(5 – 3) = 500
4–6	(3,000 – 800)/(8 – 4) = 550
6–7	(800 – 450)/(3 – 2) = 350

Crashing of Activities. Since the activities lying on the critical path control the project duration, we crash the activities lying on the critical path.

Step 1. Activity (1, 2) on the critical path gives the lowest cost-slope. Therefore, we shall crash this activity by one day at an additional cost of ₹ 200 × 1 = ₹ 200:

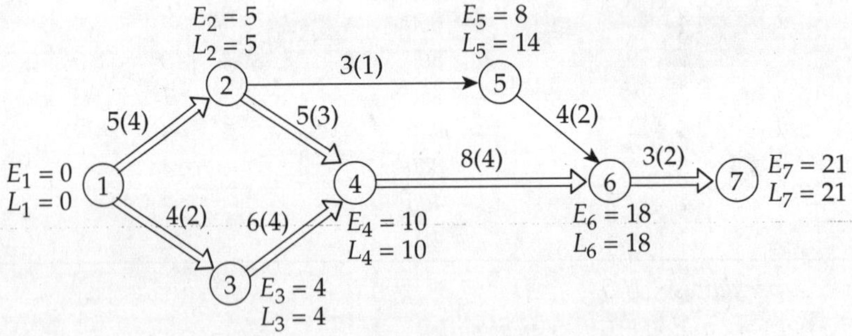

Fig. 13.35

With crashing of activity 1–2, the total project cost corresponding to 21 days

= ₹ 4,700 + ₹ 200 × 1 + ₹ 100 × 21 = ₹ 7,000

Step 2. In the network given in *Fig. 13.36*, there are two parallel critical paths, namely

$$1 \to 2 \to 4 \to 6 \to 7 \quad \text{and} \quad 1 \to 3 \to 4 \to 6 \to 7$$

Now, the least cost slope is for activity 6–7, therefore, we shall crash this activity by one day then crash activity 4–6 by four days. The additional cost will be:

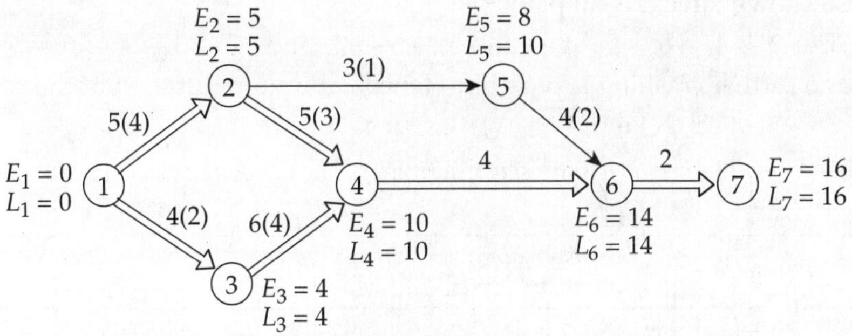

Fig. 13.36

With crashing of activities 4–6 and 6–7, the total project cost corresponding to 16 days

= ₹ 7,000 + ₹ 2,550 – ₹ 100 × 5 = ₹ 9,050

Step 3. We, now, crash activities 1–2 and 3–4 by one day each at additional cost of ₹ 550 × 1 + ₹ 200 × 1, *i.e.*, ₹ 750. Next, we crash activities 2–4 and 3–4 by one day each at additional cost of ₹ 500 × 1 + ₹ 550 × 1, *i.e.*, ₹ 1,050.

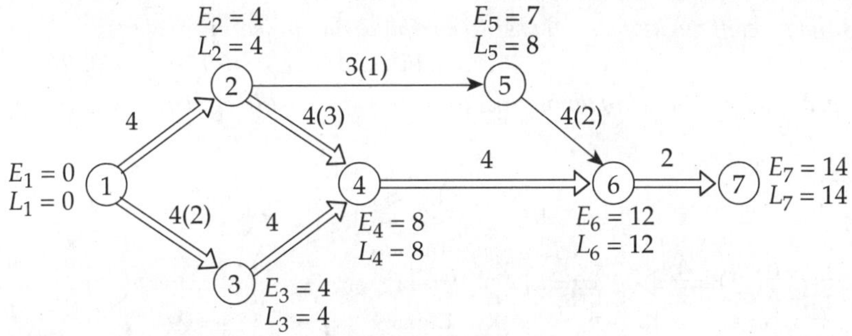

Fig. 13.37

We observe in the network of *Fig. 13.37*, that after the crashing of activities 1–2 by one day, 2–4 by one day and 3–4 by two days, the total project cost corresponding to 14 days is:

₹ 9,050 + ₹ 750 + ₹ 1,050 – ₹ 100 × 2, *i.e.*, ₹ 10,650

Step 4. Next, we crash activities 1–3 and 2–4 by one day each at additional cost of ₹ 700 × 1 + ₹ 500 × 1, *i.e.*, ₹ 1,200, since cost slope for activity 1–3 is ₹ 700.

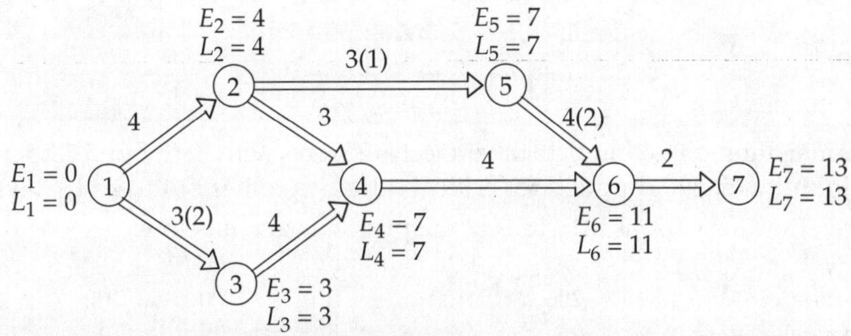

Fig. 13.38

After crashing the activities 1–3 and 2–4, the total project cost corresponding to 13 days is:

₹ 10,650 + ₹ 1,200 – ₹ 100 × 1, *i.e.*, ₹ 11,750

Fig. 13.38 shows three critical paths:

$$1 \to 2 \to 4 \to 6 \to 7, \quad 1 \to 2 \to 5 \to 6 \to 7 \text{ and } 1 \to 3 \to 4 \to 6 \to 7$$

Also, no further crashing is possible. Hence, minimum duration of the project is 13 days with associated cost of ₹ 11,750.

Example 13.15. *A small project is having seven activities. The relevant data about these activities is given below:*

Activity	*Dependence*	*Normal duration (days)*	*Crash duration (days)*	*Normal cost (₹)*	*Crash cost (₹)*
A	—	7	5	500	900
B	A	4	2	400	600
C	A	5	5	500	500
D	A	6	4	800	1,000
E	B, C	7	4	700	1,000
F	C, D	5	2	800	1,400
G	E, F	6	4	800	1,600

(*i*) *Find out the normal duration and the minimum duration.*

(*ii*) *What is the percentage increase in cost to complete the project in 21 days?*

[B.Tech. (Mech.) (Dec.), HP Univ. 2009; BBS, Delhi Univ. 2008]

Using the precedence relationship among the activities, the resulting network is given in Fig. 13.39.

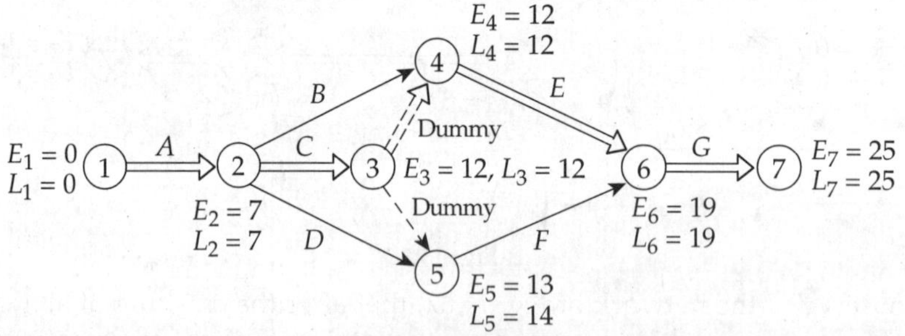

Fig. 13.39

Solution. The *E*-values and *L*-values based on normal duration of each activity are given at the nodes of the above network. The various paths through the network are:

$1 \to 2 \to 4 \to 6 \to 7$	with 24 days.
$1 \to 2 \to 3 \to 4 \to 6 \to 7$	with 25 days.
$1 \to 2 \to 3 \to 5 \to 6 \to 7$	with 23 days.
$1 \to 2 \to 5 \to 6 \to 7$	with 24 days.

(*i*) The normal duration to complete the project is 25 days with *A*, *C*, *E* and *G* as critical activities. The minimum (crash) duration for critical activities comes out to be 18 days.

The cost slope for various activities of the project is given below:

Activity :	A	B	C	D	E	F	G
Cost slope :	200	100	0	100	100	200	400
Minimum duration :	5	2	5	4	4	2	4

(*ii*) *First Crashing.* Among the critical activities, the least cost slope is for activity *E*. So, we reduce the activity (4, 6) by 2 days at an additional cost of ₹ 100 per day. The project duration, thus, reduces to 23 days and project total cost will be ₹ 4,500 + ₹ 100 × 2, *i.e.*, ₹ 4,700.

(*iii*) *Second Crashing.* Now, since there are two parallel critical paths, we reduce the activity (1, 2) by two days at an additional cost of ₹ 200 per day.

The project duration, thus, reduces to 21 days and project total cost will be:

₹ 4,700 + ₹ 2 × 200, *i.e.*, ₹ 5,100

Thus, the percentage increase in cost to complete the project in 21 days is:

(600 × 100)/4,500, *i.e.*, 13.3%.

Example 13.16. *A small marketing project consists of the jobs in the table given below. With each job is listed its normal time and a minimum or crash time (in days). The cost (in ₹ per day) of crashing each job is also given:*

Job (i–j)	*Normal duration (days)*	*Minimum (crash) duration (days)*	*Cost of crashing (₹ per day)*
1–2	9	6	20
1–3	8	5	25
1–4	15	10	30
2–4	5	3	10
3–4	10	6	15
4–5	2	1	40

(*a*) *What is the normal project length and the minimum project length?*

(*b*) *Determine the minimum crashing cost of schedules ranging from normal length down to, and including, the minimum length schedule, i.e., if L is the length of the normal schedule, find the cost of schedules which are L, L – 1, L – 2 and so on days long.*

Overhead costs total ₹ 60 per day. What is the optimal length schedule duration on each job for optimal solution? **[*C.A. Final, May, 2002*]**

Solution.

(*a*) Considering the normal duration of the project, the resulting network is given in *Fig. 13.40.*

The critical path comprises the activities 1–3, 3–4 and 4–5 with the normal duration of time as 20 days and the minimum project length is 12 days.

(*b*) Since the present schedule involves more time, we shorten the duration by crashing some of the activities. As the activities lying on the critical path control the project duration, we shorten the duration of some activities lying on the critical path.

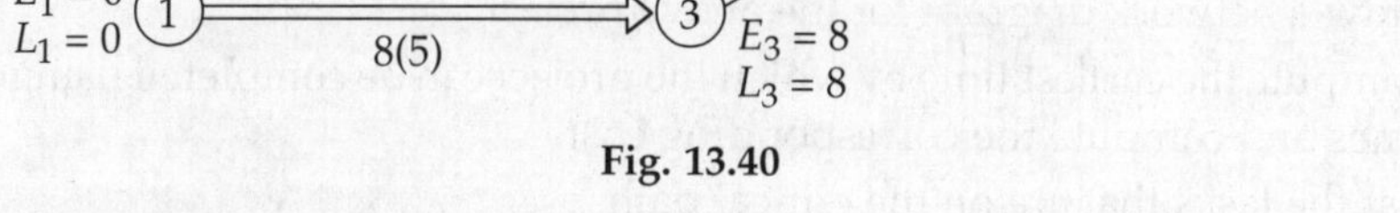

Fig. 13.40

Following table gives the project duration with crashing cost and total cost:

Activities crashed	*Project duration*	*Crashing cost (₹)*	*Overhead cost ₹60 per day*	*Total cost (₹)*
—	20	Nil	20 × 60	1,200
3–4	19	15	19 × 60	1,155
3–4	18	15 + 15 = 30	18 × 60	1,110
3–4	17	30 + 15 = 45	17 × 60	1,065
1–4, 3–4	16	45 + 30 + 15 = 90	16 × 60	1,050
4–5	15	90 + 40 = 130	15 × 60	1,030
1–3, 1–4, 2–4	14	130 + 65 = 195	14 × 60	1,035
1–3, 1–4, 2–4	13	195 + 65 = 260	13 × 60	1,040
1–3, 1–4, 1–2	12	260 + 75 = 335	12 × 60	1,055

Since, the minimum total cost occurs for 15 days of duration, the optimum length of schedule is 15 days.

PROBLEMS

13.43. For the data given in the table below, draw the network. Crash systematically the activities and determine the optimal project duration and cost.

Activity	*Normal*		*Crash*	
	Time (days)	*Cost (₹)*	*Time (days)*	*Cost (₹)*
1–2	8	100	6	200
1–3	4	150	2	350
2–4	2	50	1	90
2–5	10	100	5	400
3–4	5	100	1	200
4–5	3	80	1	100

Indirect cost = ₹ 70 per day. ***[MBA, Annamalai Univ. 2003]***

13.44. The following tasks, precedence relationships, normal task times, crash costs and minimum possible task times are given for a software development project:

Task	*Predecessors*	*Normal time*	*Crash time*	*Normal cost*	*Crash cost*
A	—	10	8	31	34
B	*A*	4	3	12	17
C	*A*	8	5	10	18
D	*B*	12	11	18	21
E	*C*	7	6	20	22
F	*D, E*	15	12	40	47
G	*E*	7	7	14	14
H	*F, G*	4	3	9	12

Times are in days and costs are in thousands of rupees.

(*i*) Draw a network diagram for the above project.

(*ii*) Compute the earliest time by which the project can be completed using the normal task times and compute the corresponding cost.

(*iii*) List the tasks that are on the critical path.

(*iv*) Suppose the project had to be completed in five days' less time than it is computed in (*ii*). Determine which tasks should be crashed and how much their task times should be reduced and compute the resulting project cost.

[MBA (PT.), Delhi Univ. 2015]

13.45. Noha industries is installing a new computer system. The activity, the activity times and the project network are as follows:

Activity :	A	B	C	D	E	F	G	H
Time (in weeks) :	3	6	2	5	4	3	9	3

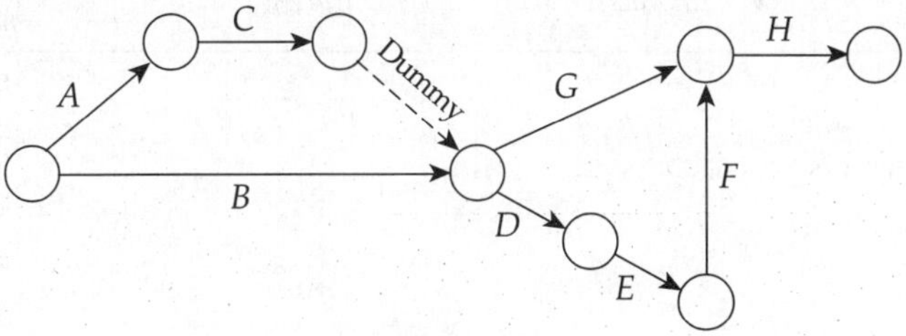

Fig. 13.41

The critical path calculations shows $B \rightarrow D \rightarrow E \rightarrow F \rightarrow H$ is the critical path with expected completion time for project as 21 weeks. However, management wishes to complete the project in 16 weeks with the following information:

Activity	*Crash time*	*Normal cost (₹)*	*Crash cost (₹)*
A	1	900	1,700
B	3	2,000	4,000
C	1	500	1,000
D	3	1,800	2,400
E	3	1,500	1,850
F	1	3,000	3,900
G	4	8,000	9,800
H	2	1,000	2,000

(*i*) Find the total floats associated with each activity.

(*ii*) Find out optimum crashing policy and corresponding crashing cost for desired completion time of the project. **[M.Com., Delhi Univ. 2010]**

13.46. The following table gives the activities in a construction project, and other relevant information:

Activity	*Immediate predecessor*	*Time (days)*		*Direct cost (₹)*	
		Normal	*Crash*	*Normal*	*Crash*
A	—	4	3	60	90
B	—	6	4	150	250
C	—	2	1	38	60
D	A	5	3	150	250
E	C	2	2	100	100
F	A	7	5	115	175
G	D, B, E	4	2	100	240

Indirect costs vary as follows:

Days :	15	14	13	12	11	10	9	8	7	6
Cost (₹) :	600	500	400	250	175	100	75	50	35	25

(*i*) Draw an arrow diagram for the project.

(*ii*) Determine the project duration which will result in minimum total project cost.

[MIB, Delhi Univ. 2011]

13.47. The following table gives the list of activities and other details for a project:

Activity	*Preceding activity*	*Duration (days)*	*Crashed duration and cost per day of crashing*
A	—	9	6 days; ₹ 10,000
B	*A*	3	2 days; ₹ 12,000
C	*A*	6	4 days; ₹ 9,000
D	*A*	11	7 days; ₹ 15,000
E	*B*	8	6 days; ₹ 11,000
F	*C*	12	8 days; ₹ 10,000
G	*C*	6	4 days; ₹ 12,000
H	*D*	7	4 days; ₹ 18,000
I	*B, E, F*	9	8 days; ₹ 22,000
j	*C, G, H*	10	6 days; ₹ 27,000
K	*F, G, I, J*	6	4 days; ₹ 19,000

The indirect cost is ₹ 45,000 per day.

(*i*) Draw a *PERT* network of the project after checking/removing redundancy.

(*ii*) Calculate the time-cost optimal trade-off project duration and corresponding cost.

[MBA, Delhi Univ. 2012]

13.48. Table below show jobs, their normal time and cost, crash time and cost for a project:

Job	*Normal time (days)*	*Normal cost (₹)*	*Crash time (days)*	*Crash cost (₹)*
1–2	6	1,400	4	1,900
1–3	8	2,000	5	2,800
2–3	4	1,100	2	1,500
2–4	3	800	2	1,400
3–4	Dummy	—	—	—
3–5	6	900	3	1,600
4–6	10	2,500	6	3,500
5–6	3	500	2	800

Indirect cost for the project is ₹ 300 per day.

(*i*) Draw the network of the project.

(*ii*) What is the normal duration and corresponding cost of the project?

(*iii*) If all activities are crashed, what will be the project duration and corresponding cost?

(*iv*) Find the optimum duration and minimum project cost.

[M.Com., Bangalore Univ. 2015]

13.49. *XYZ* construction firm has taken a contract to complete a project. To avoid the risk of time runover, the client has imposed some penalty structure. The necessary information about the project is given on next page:

Activity	*Precedence*	*Normal time*	*Crash time*	*Cost slope (₹)*
A	—	4	3	200
B	—	5	3	1,400
C	A	6	4	600
D	A	5	3	800
E	B, C	15	10	1,000
F	D	18	15	1,200
G	D	10	10	—
H	E, F	14	14	—

If project is completed within 36 days, there will be no penalty, but however, if it is delayed beyond 36 days, following structure of penalties will be applicable:

Days :	37	38	39	40	41	42
Penalty (₹) :	1,000	2,500	4,000	6,000	8,500	11,000

Suggest an optimum schedule for the project. ***[PG Dip. in Glob. Bus. Oper., Delhi Univ. 2009]***

13.50. Given the following information about a project:

Activity	*Node*	*Time (days)*		*Cost of reduction per day*
		Normal	*Crash*	
A	1–2	8	6	200
B	1–3	5	4	600
C	1–4	9	6	100
D	2–3	3	3	—
E	2–5	6	4	250
F	3–7	9	7	800
G	4–6	12	8	400
H	5–7	5	2	700
I	6–8	3	3	—
J	7–8	2	1	300

It is further given that the normal cost of doing all activities of the project in normal time is ₹ 18,400 and the overhead cost is ₹ 500 per day. Further, a penalty of ₹ 2,000 per day in excess of 21 days would be levied. You are required to calculate:

(*a*) The normal duration of the project and the associated cost.

(*b*) The minimum duration of the project and the associated cost.

(*c*) The minimum cost at which the project can be done and the time corresponding to it.

[M.Com., Delhi Univ. 2008]

13.51. The following table gives the list of various activities involved in the production of a wireless communication equipment, their immediate predecessor, their normal time and cost estimates at their crash time and cost estimates:

Activity identification	*Activity description*	*Time (months)*		*Cost (₹)*		*Immediate predecessor*
		Normal	*Crash*	*Normal*	*Crash*	
a	System calculations	3	1	45,000	63,000	—
b	Release of drawings	2	1	15,000	23,000	—
c	Procurement of PSU	12	10	85,000	105,000	*a, b*
d	Procurement of raw material	8	6	4,50,000	500,000	*a, b*
e	Production documentation	4	3	40,000	49,500	*a, b*

(*Contd.*)

f	Time study/shop order	3	1	25,000	41,000	*e*
g	PCB manufacture	6	5	50,000	59,000	*d, f*
h	Mechanical parts manufacture	7	6	2,00,000	2,20,000	*d, f*
i	Electronic assembly	2	2	43,000	43,000	*c, g, h*
j	Testing	4	2	50,500	75,000	*i*

Indirect cost per month is ₹ 20,000. Determine the optimal time *versus* cost schedule.

[MBA, Delhi Univ. 2010]

13.52. A company has recently won a contract for the installation of a die casting machine and its associated building construction work at a local factory of a large national firm of electronic engineers. The following table gives the various activities involved in this job, their normal time and cost estimates and their crash time and cost estimates:

			Normal		*Crash*	
Activity	*Description*	*Predecessor*	*Time (days)*	*Cost (₹)*	*Time (days)*	*Cost (₹)*
A	Prepare foundation and underground services, and errect building frame structure	—	30	90,000	25	1,05,000
B	Fabricate part and assemble steel frames to support the machine	—	25	1,80,000	20	1,90,000
C	Collect die costing machine and its associated gear from the manufacturers	—	10	50,000	8	54,000
D	Assemble and check control gear	C	10	7,500	7	9,000
E	Fit control gear on steel frames and install	B, D	10	4,200	10	4,200
F	Fit aluminium sheet wall claddings	A, E	20	20,000	16	30,000
G	Erect assembled plant on to prepared foundation and frame-work, and connect services	A, E	35	28,000	30	35,500
H	Erect mechanical handling plant	B, D	20	12,000	18	15,000
I	Fit ventilation and fire protection system	F	20	14,000	15	24,000

If the variable overhead costs are ₹ 5,000 per day, determine the optimal project duration.

[MBA, Delhi Univ. 2008]

13:12. DISTINCTION BETWEEN *PERT* AND *CPM*

Both ***PERT*** and ***CPM*** are managerial techniques for planning and control of large complex projects. Both are techniques to network analysis, wherein a network is prepared to analyse inter-relationships between different activities of a project. However, there are several differences between the two techniques:

1. ***CPM*** is used for ***repetitive jobs*** like planning the construction of a house. On the other hand, ***PERT*** is used for ***non-repetitive jobs*** like planning and assembly of the space platform.
2. ***PERT*** is a ***probabilistic model*** with uncertainty in activity duration. Multiple time estimates are made to calculate the probability of completing the project within scheduled time. On the contrary, ***CPM*** is a ***deterministic model*** with well-known activity (single) times based upon past experience. It, therefore, does not deal with uncertainty in project duration.

3. ***PERT*** is said to be ***event-oriented*** as the results of analysis are expressed in terms of events or distinct points in time indicative of progress. ***CPM*** is, on the other hand, ***activity-oriented*** as the results of calculations are considered in terms of activities or operations of the project.
4. ***PERT*** is applied mainly for planning and scheduling research programmes. On the other hand, ***CPM*** is employed in construction and business problems.
5. ***PERT*** incorporates ***statistical analysis*** and thereby enables the determination of probabilities concerning the time by which each activity and the entire project would be completed. On the other hand, ***CPM*** does not incorporate statistical analysis in determining time estimates because time is precise and known.
6. ***PERT*** serves a useful ***control device*** as it assists the management in controlling a project by calling attention through constant review to such delays in activities which might lead to a delay in the project completion date. But, it is difficult to use ***CPM*** as a controlling device for the simple reason that one must repeat the entire evaluation of the project each time the changes are incorporated into the network.

13:13. LINEAR PROGRAMMING FORMULATION AND NETWORK

Usually a project may consist of a large number of activities resulting in a number of constraints. Therefore, it is sometimes desirable to formulate the problem in linear programming format. Problems may consist of any one of the following situations:

(*a*) There are n activities associated with normal duration, crash duration and cost-slope of certain project. It is required to determine the duration of each activity so that in a fixed project duration time, total cost is a minimum.

(*b*) There are n activities associated with normal and crash times, cost slope and indirect cost of the project. It is required to determine the optimum project duration so as to make the total cost a minimum.

In general, let $x_1, x_2, ..., x_n$ represent n activities and $c_1, c_2, ..., c_n$ be the associated costs. Then a ***linear programming formulation*** of the problem is:

$$\text{Minimize} \quad z = c_1x_1 + c_2x_2 + ... + c_nx_n \quad \text{subject to the constraints:}$$

$$x_1 + x_2 + ..., x_n \leq \text{normal time}$$

$$x_1 + x_2 + ... + x_n \geq \text{crash time}$$

$$x_1, x_2, ..., x_n \geq 0.$$

TEST YOUR UNDERSTANDING

1. In the context of network, which of the following is not correct?
   (*a*) A network is a graphic representation of activities and nodes.
   (*b*) A project network cannot have multiple initial and final nodes.
   (*c*) An arrow diagram is essentially a closed network.
   (*d*) An arrow representing an activity may not have a length and shape.
2. The objective of network analysis is to
   (*a*) minimize total project cost
   (*b*) minimize total project duration
   (*c*) minimize production delays, interruption and conflicts
   (*d*) all of the above

3. Network problems have advantage in terms of project
   (*a*) scheduling (*b*) planning
   (*c*) controlling (*d*) all of the above
4. In critical path analysis, the word *CPM* mean
   (*a*) Critical Path Method (*b*) Crash Project Management
   (*c*) Critical Project Management (*d*) Critical Path Management
5. In critical path analysis, *CPM* is
   (*a*) event oriented (*b*) probabilistic in nature
   (*c*) deterministic in nature (*d*) dynamic in nature
6. The slack for an activity in network, is equal to
   (*a*) $LS - ES$ (*b*) $LF - LS$
   (*c*) $EF - ES$ (*d*) $EF - LS$
7. The term commonly used for activity slack time is
   (*a*) free float (*b*) independent float
   (*c*) total float (*d*) all of the above
8. An activity in the network
   (*a*) represents a task which has a definite beginning and a definite end.
   (*b*) cannot start unless all its immediate predecessors are completed.
   (*c*) have float zero if it is critical, otherwise total float as positive in case of non-critical activity.
   (*d*) all of the above.
9. If an activity has zero slack, it implies that
   (*a*) it is a dummy activity. (*b*) it lies on the critical path.
   (*c*) there are more than one critical paths. (*d*) the project is progressing well.
10. The activity which can be delayed without affecting the execution of the immediate succeeding activity is determined by
    (*a*) total float (*b*) free float
    (*c*) independent float (*d*) interfering float
11. A dummy activity is used in network diagram, when
    (*a*) two parallel activities have the same tail and head events.
    (*b*) the chain of activities may have a common event, yet be independent by themselves.
    (*c*) both (*a*) and (*b*). (*d*) none of the above.
12. In time-cost trade-off analysis
    (*a*) cost decreases linearly as time increases. (*b*) cost increases linearly as time increases.
    (*c*) cost at normal time is zero. (*d*) none of the above.
13. In crashing of activities
    (*a*) we should always concentrate on critical path and choose between critical activities only.
    (*b*) the minimum time in which a project can be completed would be given by the length of the critical path.
    (*c*) both (*a*) and (*b*). (*d*) none of the above.
14. In the context of crashing, which of the following is correct?
    (*a*) The process of shortening a project is called crashing.
    (*b*) When there are multiple critical paths, any of the them may be reduced in duration by crashing of any activity.

(c) An activity may or may not be critical, the one with the minimum crashing cost should be crashed first.

(d) The sum of crash costs of all the activities comprising the project is called project direct cost.

15. In the context of crashing of activities, which of the following is not correct?

(a) One should always focus on critical path and choose between critical activities only.

(b) Crashing cost per day is calculated as:

(Crash Cost – Normal Cost) / (Normal Time – Crash Time)

(c) Among all the activities of the project, we should choose the activity with least crashing cost.

(d) When there are multiple critical paths, one must crash them all.

ANSWERS

1. (d)	**2.** (b)	**3.** (d)	**4.** (a)	**5.** (c)
6. (a)	**7.** (d)	**8.** (d)	**9.** (b)	**10.** (b)
11. (c)	**12.** (a)	**13.** (c)	**14.** (a)	**15.** (c)

REVIEW EXERCISE

1. Explain the rules of network construction.
[M.Com., Kerala Univ. 2012; B.Tech., Panjab Tech. Univ. 2010]

2. "A network is a logical and chronological set of activities and events." Discuss.
[PG Dip., Indian Inst. of Mat. Management 2017]

3. What is meant by Burst event? *[M.Com., Bangalore Univ. 2018]*

4. Explain the rules devised by Fulkerson. *[B.Tech. (Mech.) (Sept.), HP Univ. 2013]*

5. What do you mean by looping and dangling errors in network? Show the errors using simple sketch. *[B.Tech. (Mech.), Panjab Tech. Univ. 2011; MBA (Dec.), Visvesvaraya Univ. 2011]*

6. When do you need to introduce a dummy activity into your *PERT* network? What are the time estimates for a dummy activity? *[MIB, Delhi Univ. 2011]*

7. State the objectives, advantages and disadvantages of network analysis.
[M.Com., Madurai Univ. 2003]

8. What is a critical path? Why is it so important in scheduling and controlling large projects? Can a critical path change during the course of a project? Why?
[M.Com., Bangalore Univ. 2016]

9. Explain reasons for incorporating dummy activities in a network diagram. In what way do these differ from the normal activities? *[MBA (HCA) (Nov.), Delhi Univ. 2003]*

10. Define EST, LST, EFT and LFT in network scheduling. *[MBA (Dec.), Panjab Tech. Univ. 2011]*

11. What do you mean by slack? Define critical path in the light of the definition of slack.
[MBA, Osmania Univ. 2010; MBA, Mumbai Univ. 2010; B.Tech. (Mech.) (Sept.), HP Univ. 2013]

12. Discuss all the four types of floats. Discuss their importance.
[BE (Dec.), Gujarat Tech. Univ. 2012; B.Tech. (Mech.), Panjab Tech. Univ. 2011]

13. Define 'critical path', 'slack time', 'total float', 'free float', 'independent float', 'activity variance', 'project variance' in the context of network models.
[PG Dip. in Glob. Bus Oper., Delhi Univ. 2010]

14. "The free float is a part of total float in an activity of a project." Discuss.
[MBA (Nov.), Visvesvaraya Tech. Univ. 2020; PG Dip., Indian Inst. of Mat. Management 2017]
15. Explain the following terms used in *PERT:*
(*i*) Pessimistic time, (*ii*) Optimistic time, and (*iii*) Most likely time.
[MBA (Dec.), Panjab Tech. Univ. 2010]
16. Distinguish between *PERT* and *CPM*. What is a critical path?
[MBA (Nov.), Visvesvaraya Tech. Univ. 2020; PG Dip. in Glob. Bus. Oper., Delhi Univ. 2011; M.Com., Delhi Univ. 2008]
17. Name the time estimates of an activity that follow normal distribution.
[PG Dip., Indian Inst. of Mat. Management 2019]
18. What are the major characteristics of the *PERT* model and the *CPM* model? What are their limitations, if any? Discuss. *[MBA, Delhi Univ. 2009; MBA (Nov.), Annamalai Univ. 2002]*
19. How will you find the probability of completing a *PERT* project by a particular date?
[MBA, Hyderabad Univ. 2012]
20. State and differentiate between *PERT* and *CPM*. What are the objectives of *PERT* and *CPM*? What are the assumptions on which *PERT/CPM* analysis is done for the projects? Which one of these (*PERT* and *CPM*) do you advocate for execution of ultra mega projects in India?
[MBA (Dec.), Panjab Tech. Univ. 2010]
21. Explain as to why cost slope is starting point of crashing activities in a network.
[PG Dip., Indian Inst. of Mat. Management 2018]
22. Explain crashing of project networks.
[MBA, Panjab Tech. Univ. 2015; B.Tech. (Mech.), Karnataka Univ. 2012]
23. What is time-cost trade-off in *CPM*? *[MBA, Osmania Univ. 2001]*
24. How does crashing of activities impact fixed costs and variable costs of a project?
[MBA (June), Panjab Tech. Univ. 2012]
25. What do you understand by optimal crashing in case of critical path? Using an example, explain the method. *[MBA (June), Annamalai Univ. 2012]*

CHAPTER 14

Resource Analysis in Project Management

"Optimal utilization of resources is service to humanity"

For each project, the project management have the following two objectives: (*a*) the project must be completed without any delay, and (*b*) it should use available man power and other resources as small as possible. In general, the project management involves the following three phases:

(*i*) Project Planning, (*ii*) Project Scheduling, and (*iii*) Project Controlling.

Out of these three phases, the first two are accomplished before the start of the actual project. The third phase comes into operation during the execution of the project. During all the three phases of the project management, the network techniques of ***PERT*** and ***CPM*** (discussed in previous chapter) are valuable in completing the project.

14:1. PROJECT PLANNING AND PROJECT SCHEDULING

Project planning has two important aspects, *viz.*, identification of the activities (jobs and tasks) to be performed and estimation of the required resources. Resources may be in terms of manpower, equipment (or machinery), material, financial, time and space. Since, resources are the starting point of any project, their proper planning and estimation helps to estimate the time and cost of various activities and hence of the entire project.

After the project network plan is completed and activity times are known, we consider the question as to how long the project would take to complete and resources required at different stages of time. It also includes identifying the tasks that are critical and resources that are limited so that the entire project proceeds in a logical, orderly and systematic manner.

14:2. UPDATING OF THE PROJECT

After the planning and scheduling of a project, when the actual execution starts, it may be observed that some of the activities take less time than anticipated, while others may take more time than estimated. There may be a number of reasons for this, such as the non-availability of the resources, breakdown of machinery, labour strikes, availability of additional labour during the progress of work. In addition, the planner may feel it desirable to re-estimate duration times of operations not yet undertaken. The process of incorporating the changes and re-scheduling or re-planning the project is called ***updating***.

Updating can be done in following two ways:

1. Use the revised time estimates of incomplete activities and calculate from the initial event, the earliest and latest completion time of each event in the usual manner to know the project completion time.

2. Change the complete work estimate to zero duration and represent all the activities already finished by an arrow called the ***elapsed time arrow***. Events in the revised network diagram are renumbered. The completion times of remaining activities are taken the revised times.

Example 14.1. *Consider the network diagram with its earliest 'E' and latest 'L' values as shown in Fig. 14.1:*

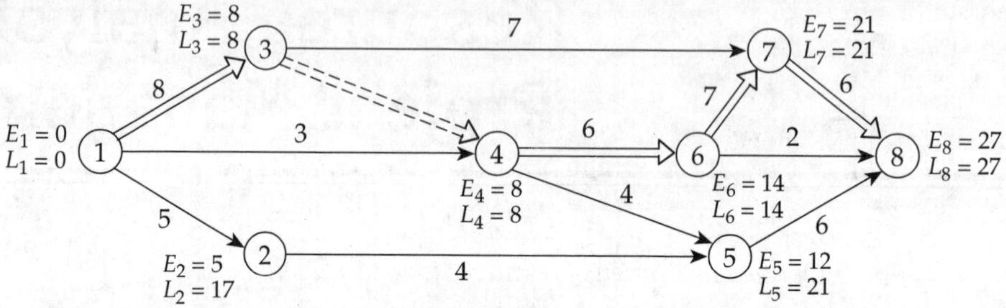

Fig. 14.1

CPM: 1 → 3 → 4 → 6 → 7 → 8 and project Duration: 27 days.

Suppose the progress is checked at the end of the 12th day and it is observed that:

(*i*) *Activity 1–4 was completed.*

(*ii*) *Activity 1–3 proceeded more rapidly and was completed in 5 days.*

(*iii*) *Activity 1–2, the supply of raw material to be used in activity 2–5, has been delayed drastically and it was estimated that the actual delivery will require an additional six days more, i.e., total 18 days.*

(*iv*) *Activity 4–5 started following the completion of activities 1–3 and 1–4, and was completed at the end of 9th day.*

(*v*) *Activity 4–6 started following the completion of the activities 1–3 and 1–4 at the end of 5th day, and will require 2 more days to complete, i.e., activity 4–6 will take 8 days instead of 6 days.*

(*vi*) *Activity 3–7 started following the completion of activity 1–3 and was completed at the end of 12th day.*

(*vii*) *The time to complete the activity 5–8 has been revised to 8 days.*

Draw the revised network after updating and determine the new critical path and the duration of project completion.

Solution. The information given above can be conveniently put in a tabular form as below:

Activity (days)	*Completed (additional days)*	*In progress (days)*	*Not started (days)*
1–2	—	6	—
1–3	5	—	—
1–4	3	—	—
2–5	—	—	4
3–4	0	—	—
3–7	12	—	—
4–5	9	—	—
4–6	—	2	—
5–8	—	—	8
6–7	—	—	7
6–8	—	—	2
7–8	—	—	6

Updating can now be done in two ways. In the first, we may use the revised time estimates and compute from the initial starting event. The second way, which is more convenient, is to initialize/update the completed work to zero duration. The events in the revised network are numbered in a different fashion. The time durations assigned to the activities become the revised times. Observe that in the revised network shown in *Fig. 14.2*, the earliest start time for the starting event has been entered as 12 days:

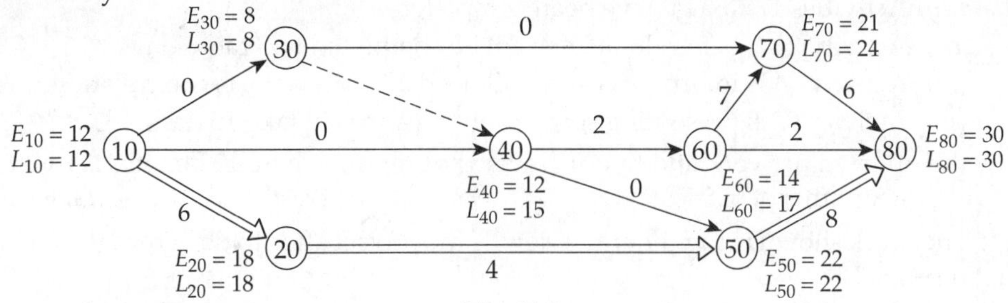

Fig. 14.2

The activities *10–30, 10–40, 30–40, 30–70* and *40–50* have been assigned zero completion time each. Activities *10–20* and *40–60* which are now in progress are assigned the times they need for their completion. Remaining activities which have not yet started are assigned their new time estimates.

After computing the earliest start times and latest allowable times, we find that the critical path has changed to *10–20–50–80*. The total duration has been increased by 3 days. In other words, the updated network indicates that the latest estimate projection completion time is 30 days measured from the original start of the project.

Remark. Decision as to when the network should be updated must be based on the following factors:

(*a*) Project of smaller time duration must be updated more often than long projects.

(*b*) For long projects, the frequency of updating should increase as the project progresses.

(*c*) If the activity time increases, corrective measures may be necessary and if it decreases, this may allow changes in the project not previously possible.

PROBLEMS

14.01. For the network shown in *Fig. 14.3*, a review of the project after 10 days reveals the following:

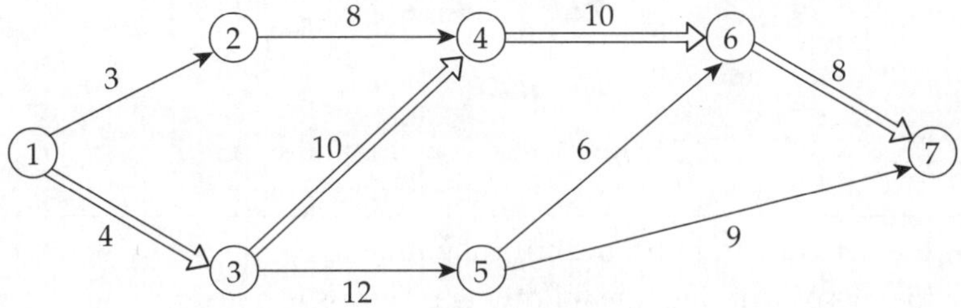

Fig. 14.3

(*a*) Activity *1–2, 1–3* and *2–4* are completed.

(*b*) Activity *3–4* is in progress and will take 6 more days.

(*c*) Activity *3–5* is in progress and will take 5 days more.

(*d*) It is estimated that due to arrival of new machine, activity *4–6* will take only 6 days.

Formulate a new network after updating and determine the new critical path.

14.02. A project has the following activities and their duration:

Activity :	12	13	14	24	25	34	36	47	57	67	68	78
Duration (days) :	13	15	9	10	27	7	18	30	12	10	10	9

(*a*) Draw the network of the project and find its duration.

(*b*) At the end of 25 days, it is observed that

(*i*) Activities 12, 13, 14 have been completed.

(*ii*) Activity 24 is being done and will be completed in 5 more days.

(*iii*) Activity 36 is in progress and will need 20 more days for completion.

(*iv*) Activity 67 is presenting some problem and will take 15 days.

Draw the updated network and find its revised duration, where 12 denotes activity *1–2* and same is true for other activities. ***[B.Tech. (Mech.), Karnataka Univ. 2012]***

14.03. For the network shown in *Fig. 14.4*, a review of the project after 15 days reveals the following situation:

(*a*) Activities *1–2*, *1–3*, *2–3*, *2–4* and *3–4* are completed.

(*b*) Activities *3–5* and *4–6* are in progress and need 2 and 4 days more respectively.

(*c*) The revised estimate shows that activity *8–9* will take only 8 days, but activity *7–9* will need 10 days.

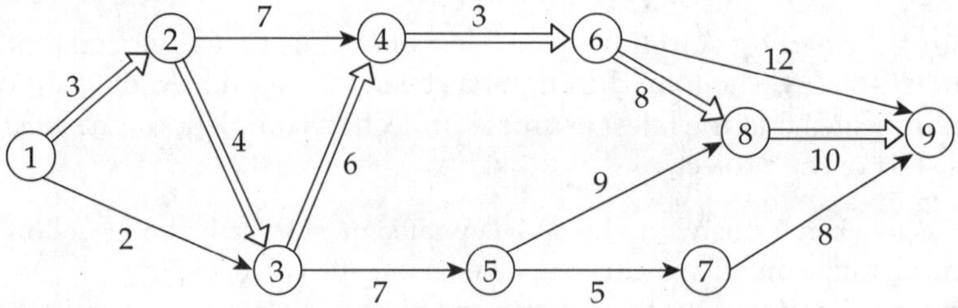

Fig. 14.4

Formulate a new network after updating and determine the new critical path.

14.04 A company had planned its operations as follows:

Activity	*Duration (days)*	*Activity*	*Duration (days)*
1–2	7	*3–4*	6
1–3	8	*3–6*	24
1–4	6	*4–7*	19
2–4	8	*5–7*	9
2–5	16	*6–8*	7
		7–8	8

(*i*) Draw the network and find the critical path.

(*ii*) After 15 days of working, the following progress is noted:

(*a*) Activities *1–2*, *1–3* and *1–4* completed as per original schedule.

(*b*) Activity *2–4* is in progress and will be completed in 4 more days.

(*c*) Activity *3–6* is in progress and will need 17 more days to complete.

(*d*) The staff at activity *3–6* are specialized. They are directed to complete *3–6* and undertake an activity *6–7*, which will require 7 days. This arrangement arose due to a modification in a specialization.

(*e*) Activity *6–8* will be completed in 4 days instead of originally planned 7 days.

(*f*) There is no change in other activities. Update the network diagram after 15 days of start of work based on the assumptions given above.

Indicate the revised critical paths along with their duration. *[C.A. Final (May), 2007]*

14:3. RESOURCE ALLOCATION AND SCHEDULING

A resource is defined by the physical variables such as crew-size (labour), finance (cost), equipment (shop facilities), etc. available to the management to achieve the objective of a project.

The first step in the ***Resource Allocation*** procedure is to determine the ***earliest start time*** (ES), and the ***latest start time*** (LS) for each of the activities of the project.

The resources are allocated by stepping through time, scheduling the various activities as soon as their respective predecessors are scheduled and the resources required for them are available. Two sets of activities are defined, namely the first comprises activities eligible for assignment as their predecessors have been scheduled (this may be termed as the ***eligible activity set*** 'EAS'). From the *EAS*, activities are selected, which can start on the particular time when resources are to be allocated, and are ordered in accordance with our usual criterion. These activities constitute the ***ordered activities set*** 'OAS'.

In general, there are two criteria by which scheduling can be finalized *vis-a-vis* resource deployment:

Resource Levelling. In the process of ***Resource Levelling***, the activities are critically sequenced, subject to the constraint of availability of resources; and also the minimum period of the project is re-determined accordingly.

Under this technique, the project manager may be able to lower the maximum resource requirements by shifting a non-critical activity between its earliest start time and latest finish time.

The two general rules normally used in scheduling non-critical activities are as under:

1. If the total float of a non-critical activity is equal to its free float, then it can be scheduled anywhere between its earliest start and latest completion times.
2. If the total float of a non-critical activity is more than its free float, then its starting time can be delayed relative to its earliest start time by no more than the amount of its free float without affecting the scheduling of its immediately succeeding activities.

Resource Smoothing. ***Resource Smoothing*** means maintaining the total project duration at the minimum level, by shifting resources from the activities having floats so that a uniform demand on resources is achieved. In other words, resource smoothing involves rescheduling of the activities of a project, in such a way as to utilize the resources in a fairly uniform manner, and achieving a minimum project duration time.

The procedure of carrying out resource smoothing can be summarized in the following steps:

1. Calculate the 'earliest start' and 'latest finish' times of each activity and then draw a time-scaled version (or squared) of the network. Critical path in the network is drawn along a straight line and non-critical activities on both sides of this line. Resource requirement of each activity is given along the arrows.
2. Draw the resource histogram by taking earliest start times or latest start times of activities on the *x*-axis and cumulative resources required on *y*-axis.
3. Shift 'start times' of non-critical activities first having largest float in order to smoothen the resources.

Example 14.2. *Given below is the crew-size (manpower) requirements for each activity in a project:*

Activity	*Normal time (days)*	*Crew-size (manpower required)*
0–1	2	4
1–2	3	3
1–3	4	3
2–4	2	5
3–5	4	3
3–6	3	4
4–7	6	3
5–7	6	6
6–8	5	2
7–9	4	2
8–9	4	9

(a) *Draw the network of the project activities.*

(b) *Re-arrange the activities suitably for reducing the existing total crew-size.*

(c) *If only 9 men are available for the execution of the project, then re-arrange the activities suitably for levelling the manpower resource.*

Solution.

(*a*) Using the precedence relationships among the activities, the network is given below:

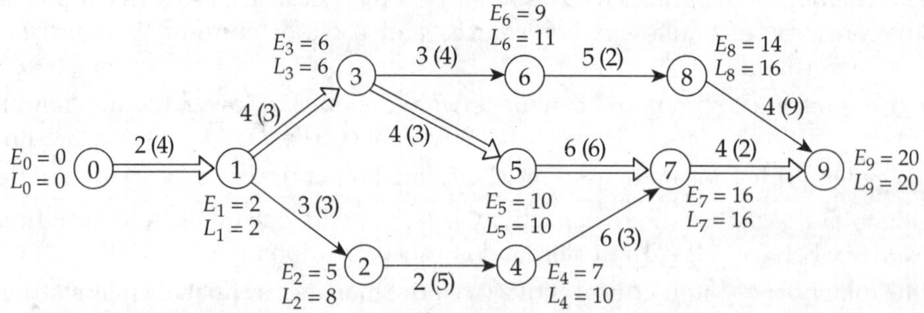

Fig. 14.5

Numbers in bracket with each activity arrow in the network represents manpower requirement of the activity.

The earliest and latest completion times of each activity are indicated along the nodes in *Fig. 14.5.* The critical path is: $0 \rightarrow 1 \rightarrow 3 \rightarrow 5 \rightarrow 7 \rightarrow 9$ and the project duration is of 20 days.

In the time-scaled version of the network, the critical path will be drawn along the straight line and non-critical activities are added with it as shown in *Fig. 14.6.*

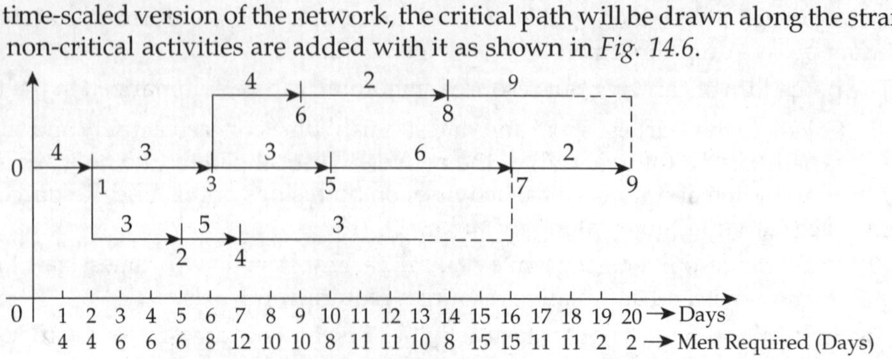

Fig. 14.6

(*b*) Number above the arrows indicate the crew size (the resource requirements), the numbers below the arrows indicate the days and the non-critical paths are added as shown in *Fig. 14.6.* The resource histogram can now be drawn by summing up the manpower requirements for each day. This is shown in Fig. *14.7.*

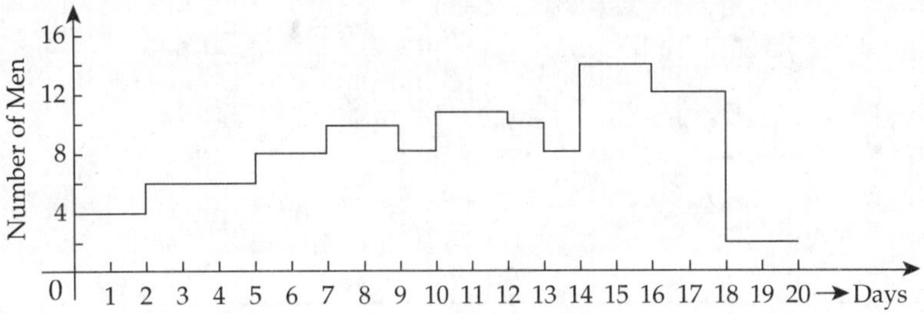

14.7

From *Fig. 14.7,* we observe that the maximum demand of 15 men occurs on 15th and 16th day. To somoothen the load, the activities will have to be shifted depending upon the floats. Now, since the path *3 → 6 → 8 → 9* has a float of two days, the activities *6–8* and *8–9* are shifted to the right. Thus, the start of each of these is delayed by 2 days. Similarly, the activity *4–7* can be shifted to the right so that it starts on 10th day instead of starting on 7th day. After, making the necessary shifting, the network is drawn as shown in *Fig. 14.8.*

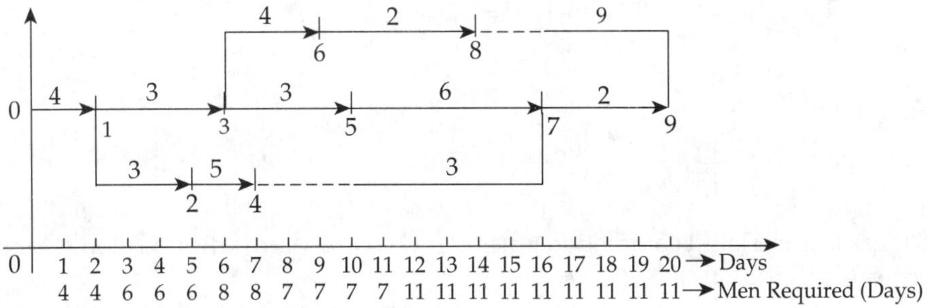

Fig. 14.8

Resource histogram for this squared network is given in *Fig. 14.9.* This indicates that the maximum manpower required is of 11 men.

(*c*) Since, only 9 men are available instead of 11, therefore, activities have to be rearranged and expanded in term of their completion time. This would increase the total project duration from 20 days to 25 days.

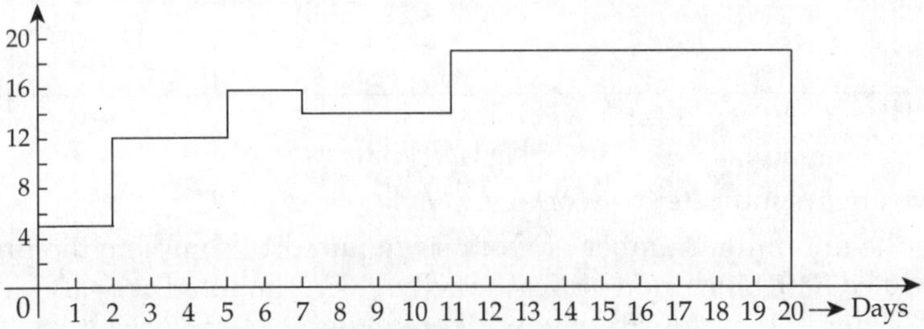

Fig. 14.9

Example 14.3. *The Madras Construction Company is bidding on a contract to install a line of microwave towers. It has identified the following activities, along with their expected times, predecessor restrictions, and work requirements:*

Activity	*Duration (weeks)*	*Predecessors*	*Crew size (workers)*
A	4	None	4
B	7	None	2
C	3	A	2
D	3	A	4
E	2	B	6
F	2	B	3
G	2	D, E	3
H	3	F, G	4

The contract specifies that the project must be completed in 14 weeks. This company will assign a fixed number of workers to the project for its entire duration and so it would like to ensure that the minimum number of workers is assigned and that the project will be completed in 14 weeks. Find the schedule which will do this. **[C.A. Final (May), 1995]**

Solution. Using the precedence relationship, the network diagram is given as under:

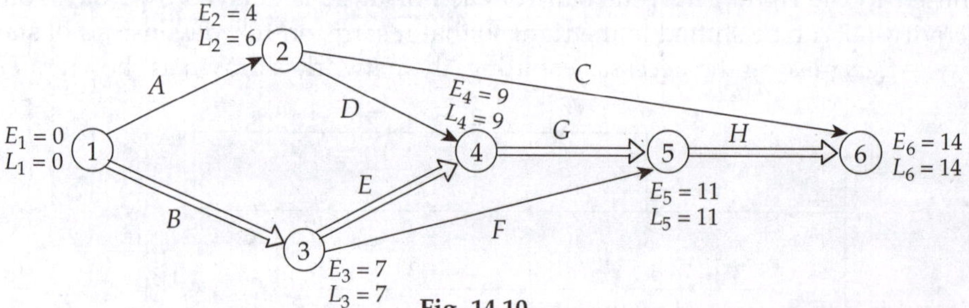

Fig. 14.10

The earliest and latest completion times of each activity are indicated along the nodes in *Fig. 14.10*. The critical path is:

$$B \rightarrow E \rightarrow G \rightarrow H, \text{ and the project duration is 14 weeks.}$$

The time-scaled variation of the network, showing the critical path along the straight line, is shown in *Fig. 14.11*.

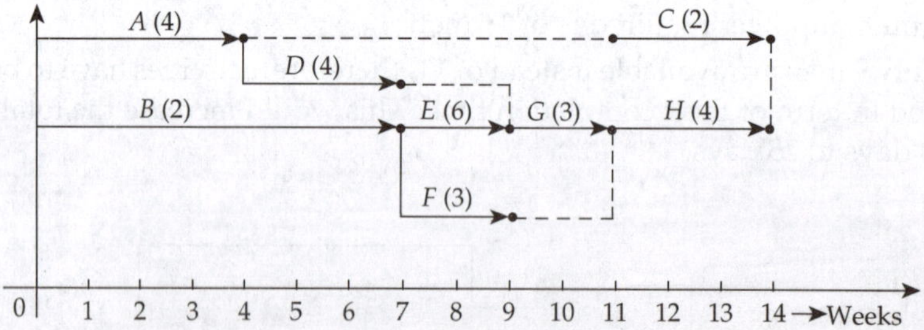

Fig. 14.11

Number above arrows indicate the workers.

To determine the minimum number of workers required to complete the project, the critical activities are scheduled first. Since, non-critical activity *'A'* is scheduled at $t = 0$ followed by the non-critical activities *'C'* and *'D'* on fourth day, workers required would rise to 8. The total float of activity *A* is more than its free float and also the earliest start time as well as free float is zero. Therefore, it has to be scheduled, otherwise it would result in delay of the project.

Activity '*C*' has a float of 7 weeks and activity '*F*' has a float of 2 weeks. Requirement of workers in 5th, 6th and 7th week is of 8 workers, and during 8th and 9th week is of 9 workers. This demand has to be smoothened. For this, we shift activity '*C*' to 11th week and activity '*F*' to 9th week. Thus, the total weekly requirement of workers after re-arranging the activities comes out to 6 workers during the entire project duration of 14 weeks as shown in *Fig. 14.12*.

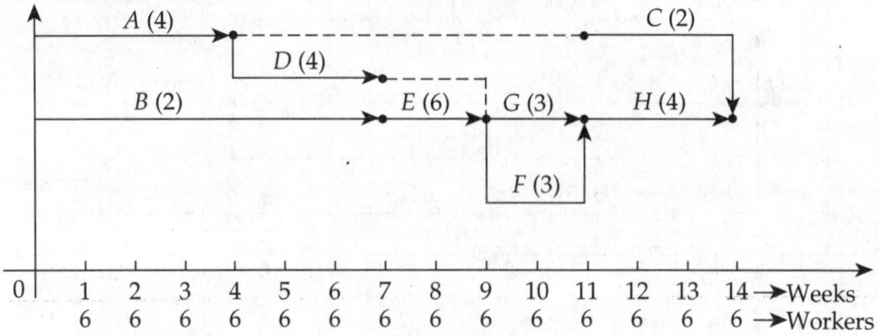

Fig. 14.12

Example 14.4. *Consider a project consisting of thirteen activities having duration and resource requirements shown below:*

Activity	*Preceding activity*	*Duration (days)*	*Resource required*	
			Labourer (L)	*Carpenter (C)*
A	—	2	2	—
B	A	4	2	2
C	A	4	—	4
D	A	3	2	—
E	A, B	4	—	6
F	A, C	7	2	—
G	A, C	2	—	4
H	A, D	5	2	4
J	F	2	—	2
K	H	5	—	2
L	J	2	2	—
M	E, G, L	3	4	—
N	M, K	2	4	—

(*i*) *Draw the network, identify the critical path and duration by which project is completed.*

(*ii*) *Analyse the project from the point of view of resources to bring out the necessary steps involved in the analysis and smoothing of resources.*

Solution.

(*i*) Using the precedence relationship, the required network is given below:

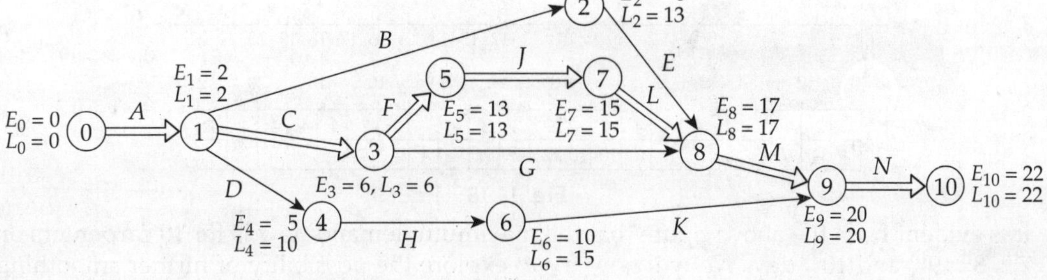

Fig. 14.13

The critical path for the above network is:

$0 \rightarrow 1 \rightarrow 3 \rightarrow 5 \rightarrow 7 \rightarrow 8 \rightarrow 9 \rightarrow 10$. The project duration is 22 days.

(*ii*) The time-scaled variation of the network, showing the critical path along the straight line is as shown in *Fig. 14.14*.

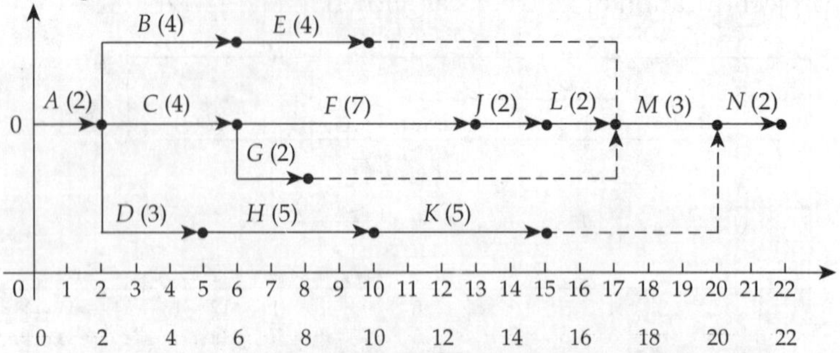

L	2	2	4	4	4	4	4	4	4	4	2	2	2	–	–	2	2	4	4	4	4	4
C	–	–	6	6	6	10	14	14	10	10	2	2	2	4	4	–	–	–	–	–	–	–

Fig. 14.14

The last two rows of *Fig. 14.14* give the number of labourers and carpenters required each day.

From the above figure, we observe that the demand on the resources is not uniform. On the 7th and 8th day, the demand for carpenters is as high as 14 carpenters, whereas on the 11th, 12th and 13th days it is only two carpenters. If the carpenters and labourers are to be hired for the entire project duration of 22 days, then during most of the days they will be idle and the company will have to hire at least 14 carpenters and 4 labourers. So, we re-schedule our activities in such a way as to utilize the resources in a fairly uniform manner.

As mentioned above, the maximum demand on the resources occurs on the 6th, 7th, 8th, 9th and 10th day. The activities on these days will have to be shifted depending upon their floats, such that the demand comes down. As can be seen from the above time-scaled version, activity *3–8* has maximum float, therefore we will try to shift activity *3–8* so that it starts on the 16th day instead of 7th day. This reduces the demand on carpenters from 14 to 10 on 7th and 8th day. The maximum demand for the carpenters on any day is now 10 and not 14. The modified resource accumulation table and the time-scaled version of the project is shown in *Fig. 14.15*.

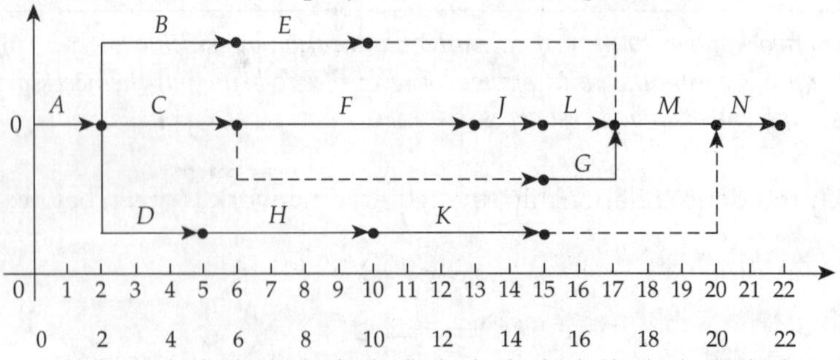

L	2	2	4	4	4	4	4	4	4	4	2	2	2	–	–	2	2	4	4	4	4	4
C	–	–	6	6	6	10	10	10	10	10	2	2	2	4	4	4	4	–	–	–	–	–

Fig. 14.15

It is evident from the above figure that the maximum demand now is for 10 carpenters on the 6th, 7th, 8th, 9th and 10th day. We will now try to explore the possibility of further smoothing the resources which is possible because activities *1–4*, *4–6* and *6–9* have a total float of 5 days. Thus, shifting the start of the activity *4–6* (or *H*) from 10th day instead of 5th day, the resource accumulation table is:

L	2	2	4	4	4	2	2	2	2	2	4	4	4	2	2	2	2	4	4	4	4	4
C	–	–	6	6	6	6	6	6	6	6	4	4	4	6	6	6	6	2	2	2	–	–

1 2 3 4 5 6 7 8 9 10 11 12 13 14 15 16 17 18 19 20 21 22

From the above table, we observe that the requirement for labourers is 4 and the requirement for carpenters reduces to 6 as against 14 carpenters originally estimated. Hence, by judiciously utilizing the float, we can smooth the demand on the resources.

PROBLEMS

14.05 The following data is given for a small project:

Activity	*Preceding activity*	*Duration (days)*	*Labour requirement (men)*
A	—	1	2
B	—	2	1
C	*A*	1	1
D	—	5	1
E	*B*	1	1
F	*C*	1	1

(*a*) Draw the ***PERT*** Chart.
(*b*) Indicate the critical path without taking account of resource constraint.
(*c*) Give the bar diagram on time-scaled network.
(*d*) From the time-scaled chart given in (*c*), prepare a resource aggregation profile.
(*e*) Now, prepare resource allocation profile with 2 men constraint and another diagram for 5 days constraint.

14.06. Consider the following data pertaining to a ***CPM*** network. Draw the network, and give the resource requirement diagram for early start schedule.

Activity :	*1–2*	*1–3*	*1–4*	*2–3*	*2–4*	*3–5*	*4–5*
Duration (days) :	6	10	6	10	4	6	6
Manpower required :	8	4	9	7	6	17	6

[MBA (Nov.), Annamalai Univ. 2002]

14.07. A project consists of 9 activities. Table below shows the duration for each activity and the corresponding labour requirements. Draw a network and establish critical path.

Also, draw a square network assuming that all the activities begin at the earliest start time. Adjust the project work, such that there will be smooth demand for this resource:

Activity	*Duration*	*No. of labourers*
1–2	3	5
2–4	2	3
2–3	3	7
3–4	0	0
3–5	3	2
4–5	7	2
3–6	2	1
5–6	6	6
4–6	5	5

14.08. Following are the manpower requirements for each activity in a project:

Activity	*Normal time (days)*	*Manpower required per day*
1–2	10	2
1–3	11	3
2–4	13	4
2–6	14	3
3–4	10	1
4–5	7	3
4–6	17	3
5–7	13	5
6–7	9	8
7–8	1	11

(*a*) Draw the network and find out total float and free float for each activity.

(*b*) The contractor stipulates that during the first 26 days, only 4 to 5 men and during remaining days 8 to 11 men only can be made available. Rearrange the activities suitably for levelling the manpower resources, satisfying the above condition.

14.09. The duration and requirement of work force for each activity in a project is given below:

Activity	*Duration (days)*	*Manpower required (per day)*
1–2	8	4
1–3	7	10
1–5	12	5
2–3	4	6
2–4	10	7
3–4	3	5
3–5	5	8
3–6	10	7
4–6	7	11
5–6	4	3

Draw the network diagram of the project and find its duration under the assumption of availability of necessary manpower. Find the maximum and minimum workforce requirement each day without changing the project duration.

14.10. The following personnel requirements are specified for the various activities of a project:

Activity	*Duration (days)*	*Number of workers*	*Activity*	*Duration (days)*	*Number of workers*
(0, 1)	2	0	(3, 5)	3	2
(0, 2)	3	5	(3, 6)	2	1
(1, 3)	2	0	(4, 5)	7	2
(2, 3)	3	7	(4, 6)	5	5
(2, 4)	2	3	(5, 6)	6	6

Find the minimum number of workers (as function of the project time) required during the scheduling of the project. Using resource levelling, estimate the maximum number of workers required.

14.11. For a project consisting of several activities, the duration and required resources for carrying out each of the activity and their availability are given below:

Activity (i–j)	*Resources required*		
	Equipment	*Operators*	*Duration (days)*
1–2	X	30	4
1–3	Y	20	3
1–4	Z	20	6
2–4	X	30	4
2–5	Z	20	8
3–4	Y	20	4
3–5	Y	20	4
4–5	X	30	6

Resource availability:

Equipment X = 1, Equipment Y = 1, Equipment Z = 1, Number of persons = 50.

(*a*) Draw the network, identify critical path and compute the total float for each of the activities.

(*b*) Find the project completion time under the given resource constraints.

14.12. A project consists of 10 activities, each of which require either, or both, of the two type of resources R_1 and R_2 for its performance. The duration of the activities and their resource requirements are as follows:

Activity	*Duration (days)*	*Resource requirements*	
		R_1	R_2
1–2	3	3	2
1–3	2	6	—
1–4	6	4	—
2–6	4	—	4
3–5	2	2	2
4–5	1	4	—
4–8	4	4	—
5–7	3	3	2
6–7	2	1	3
7–8	4	4	5

Resource availability: 7 units of R_1 and 5 units of R_2.

Determine the duration of the project under the given resources constraint. If the resources were not a problem, how long would the project take to complete in the normal course?

14.13. The following table gives for each activity of a project, its duration and corresponding resource requirements as well as total availability of each type of resources:

Activity	*Duration (days)*	*Resources required*	
		Machines	*Men*
1–2	7	2	20
1–3	7	2	20
2–3	8	3	30
2–4	6	4	30
3–6	9	2	20
4–5	3	2	20
5–6	5	2	20
Minimum available resources		4	40

(*i*) Draw the Network, compute earliest occurrence time and latest occurrence time for each event, the total float for each activity and identify the critical path assuming that there are no resources constraint.

(*ii*) Under the given resources constraint, find out the minimum duration to complete the project and compare the utilization of the resources for that duration.

14.14. The following table gives the activities in a small project and other relevant information:

Activity	*Duration (days)*	*Immediate predecessor*	*Resources required*	
			Operators	*Machines*
A	3	None	2	—
B	2	*A*	2	2
C	4	*A*	4	4
D	6	*A*	5	5
E	3	*B*	2	2
F	2	*E*	2	—
G	6	*C*	—	2
H	4	*D*	2	2
I	4	*G*	4	2
J	2	*D*	2	—
K	2	*J*	2	2
L	4	*F, H, I*	4	4

(*a*) Draw the Network, compute Earliest Start Time (*E*) and Latest Finish Time (*F*) for each of the activities and find out the project completion time and identify the Critical Path.

(*b*) Draw the Time-scaled diagram along with resource accumulation table. Comment on demand for the operators and machines for the entire project duration and suggest the method for smoothing the resources.

14:4. MANAGEMENT OPERATION SYSTEM TECHNIQUE (MOST)

Management Operation System Technique (MOST) is a graphical presentation of ***PERT*** to a time-scale. The target date is taken as a zero date of completion of the project. The expected date of start and completion of other activities based on zero date of completion are shifted to the left side of the graph indicating inter-dependencies of the activities by a line. The completion of various activities on a date of review are drawn on the same graph with solid horizontal lines, so that target and completion, even in parts, may be shown on the same graph.

This gives a direct impression to the management to assess the present status of the project and the expected date of completion of various activities as on the date of review. Floats are also indicated by dotted lines. Critical path may be shown by a shaded line. This technique can also be applied for large projects.

Example 14.5. *Consider the following network diagram involving seven activities with expected duration of each activity as indicated above the arrow. Discuss MOST for this network diagram.*

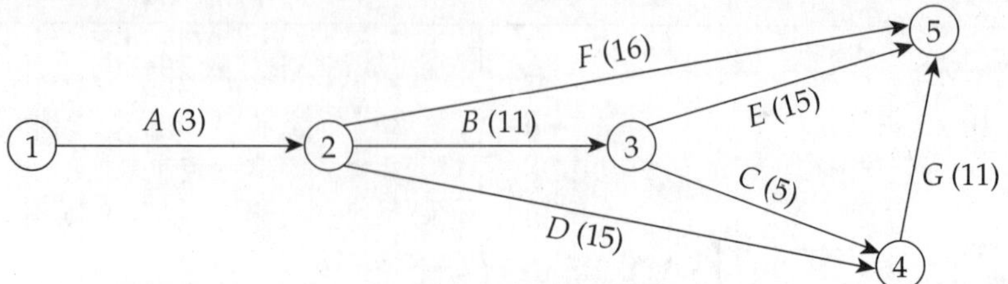

Fig. 14.16

Solution. The time-scale diagram of the given *PERT* network is shown in *Fig. 14.17.*

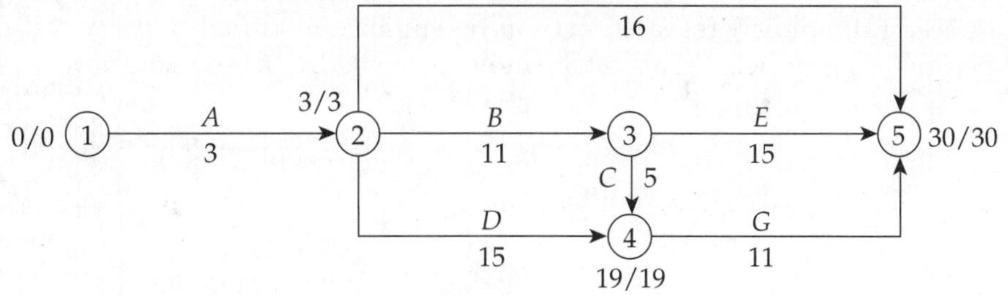

Fig. 14.17

Using *MOST* on *Fig. 14.17*, we get the required graphical representation of *Fig. 14.16*. This is shown in *Fig. 14.18*.

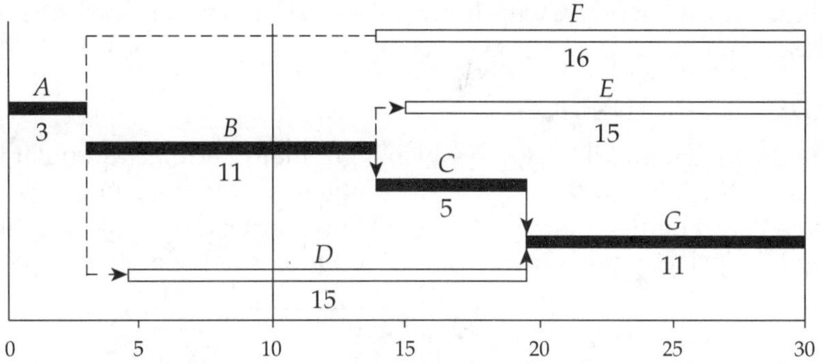

Fig. 14.18

Critical path is *A, B, C* and *G*. On 10th day, the progress is reviewed. *A* is completed, *B* is ahead of the schedule and *D* is behind.

14:5. GRAPHICAL EVALUATION AND REVIEW TECHNIQUE (GERT)

Graphical Evaluation & Review Technique (GERT) deals with probabilistic situations. Although ***PERT*** technique also deals with uncertainties in timing of the projects, its calculations are based on the assumption that each activity contemplated will materialize without any omission and there is no allowance for a probabilistic choice of a path between nodes. ***PERT*** cannot deal with a situation, where the activities already dropped at a subsequent stage to occur again with another alternative, *e.g.*, Research & Development projects where the outcome cannot be predetermined in a particular way. The ***GERT*** accommodates this kind of contingency by means of looping as shown below:

(*a*) Activities *A, B* to be completed and Activities *C, D* are to be started thereafter.

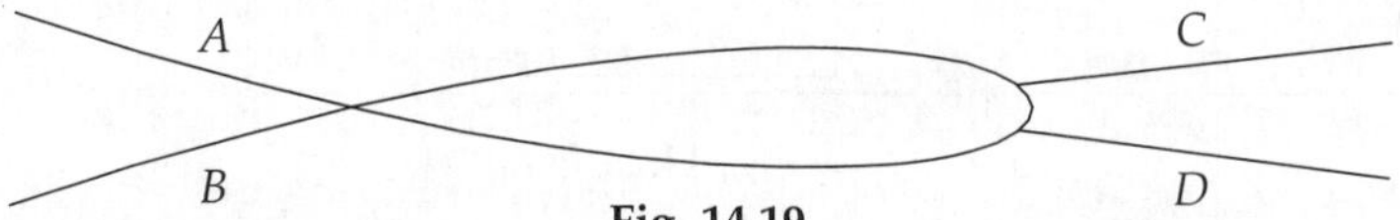

Fig. 14.19

(*b*) Activities *A, B, C* or *A, B, D* to be started but not *C, D* together.

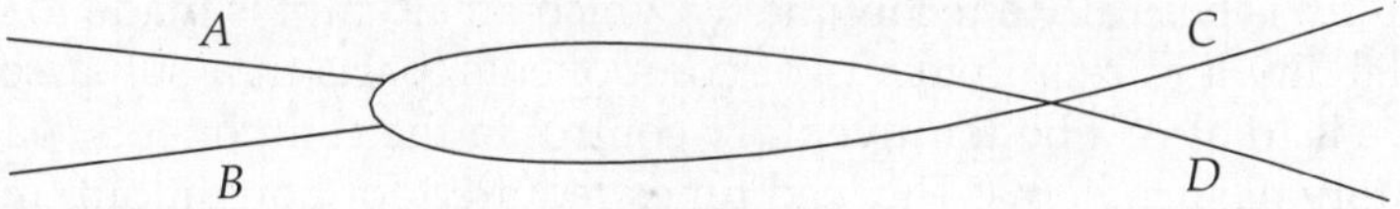

Fig. 14.20

(*c*) An applied research activity is conducted and table model experiment is executed. If the experiment is successful, a profit plan is designed and erected. If the experiment is unsuccessful, applied research activity is suitably modified and revised table model experiment is conducted. This can be indicated by ***GERT*** as shown below:

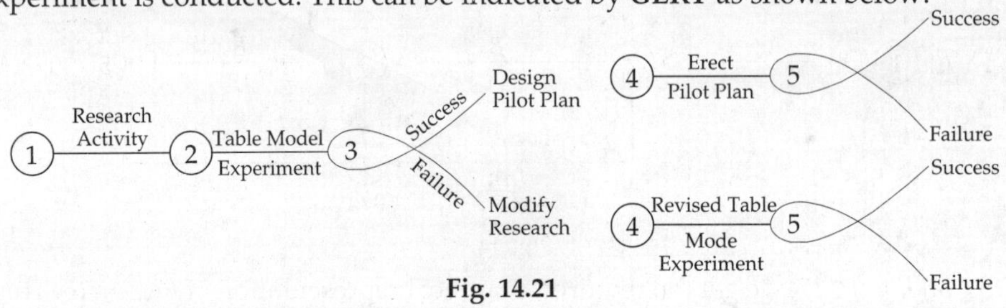

Fig. 14.21

Looping is done at Nodes 3 & 5, ***GERT*** is ***Prima facie*** a branching network, where at a decision node different branches are evolved with probabilistic estimates. It is, therefore, rightly said that "***GERT*** is in when ***PERT*** is out."

14:6. PRECEDENCE PLANNING

A new method of drawing network diagram which is gradually becoming popular is the ***Precedence planning/diagram method***. It is also known as the ***Method of Potentials*** or the ***Ray Method***.

Under this method activities are represented by boxes instead of arrows. Logical relationships are then represented by lines joining these boxes. A typical example is shown in *Fig. 14.22*.

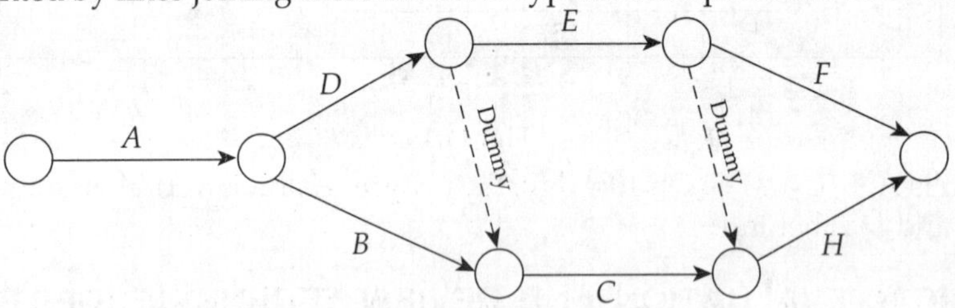

Fig. 14.22

Fig. 14.23, represents the precedence diagram. Such a diagram is very simple and can be easily understood. It may be noted that it does not involve dummies. Precedence planning diagrams are very useful in ***line balancing problems***. A further advantage of it is the inclusion of delay or lag time.

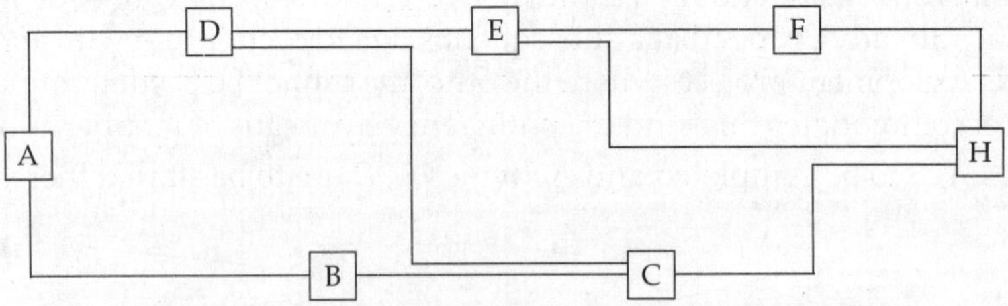

Fig. 14.23

14:7. LINE OF BALANCE (LOB)

Line of Balance (LOB) is basically a technique by which an attempt is made to maintain a balance between the availability and requirements of components/parts in a sub-assembly or assembly shop. The objective is to have a better inventory control in the components/parts and reduce the total cost of inventory of such items. The lead times required for bought out/manufactured items are ascertained. A time scale ***PERT*** diagram is drawn showing the various activities. The time is shown as if the product is required on zero date. The network diagram will indicate the starting time for each activity. From the diagram the cumulative requirements of components/parts at periodical intervals of time are indicated in a bar chart. The physical position of inventory is also depicted in the same bar chart so that management can identify as to what items are over/under stocked and take remedial action.

The procedure for line balancing may be summarized in the following six steps:

Step 1. Divide the total work (Project) into sub-tasks called work elements (or element tasks).

Step 2. Establish precedence requirements between the elemental tasks and construct a precedence diagram.

Step 3. Select a feasible cycle time and determine the minimum number of work stations needed.

Step 4. Apply assignment heuristic and assignment elemental tasks to individual work stations.

Step 5. Check whether further improvements can be made.

Step 6. Calculate effectiveness and efficiency of the proposed line.

PROBLEM

14.15. A toy manufacturer intends to produce 10,000 pieces of a particular toy per year. He has identified 10 work elements with the following precedence restrictions and durations:

Elements	:	*1*	*2*	*3*	*4*	*5*	*6*	*7*	*8*	*9*	*10*
Immediate predecessors	:	—	1	1	2, 3	4	4	6	5	7, 8	9
Duration (minutes)	:	5	8	3	2	5	5	8	2	3	6

(*a*) Draw a precedence diagram for the assembly of toys.

(*b*) Design the assembly line for a cycle time of 8 minutes.

(*c*) Calculate line efficiency, balance delay and smoothness index for assembly line under (*b*).

TEST YOUR UNDERSTANDING

1. In resource analysis

(*a*) Project costs can be classified into project direct costs and project indirect costs.

(*b*) The lowest possible direct cost required to complete an activity within its crashed time is known as crash cost.

(*c*) It is possible to reduce the normal duration of a project by employing additional resources.

(*d*) None of the above.

2. Which of the following is not correct?

(*a*) Information about the float times for various activities is essential in effecting resource levelling.

(*b*) It is possible to reduce an activity's duration below its crash time by allocating more resources and funds to it.

(*c*) A project scheduling problem can be formulated as an LPP.

(*d*) The method of drawing network diagram is known as the 'Method of Potentials' or the 'Ray Method'.

3. Which of the following is not correct?

(*a*) Resource allocation deals with the allocation of limited resources in the best possible manner, so that the project delay is kept at the minimum.

(*b*) The criteria for resource allocation are float, activity duration and quantum of resources needed, in that order.

(*c*) The commonly used approaches to resource planning are: (*i*) Resource allocation, and (*ii*) Resource levelling.

(*d*) Irrespective of the method of resource levelling used to satisfy resource limitations, the precedence relationships in the network is not necessary.

4. Which of the following is not correct?
 (*a*) While allocating resources, critical activities must be given preference over others.
 (*b*) The resource allocation programmes aim to allocate the available resource supplies to activities of a project with a view to complete the project in the maximum time.
 (*c*) Resource levelling aims at smoothening of the resource usage rate without changing the project duration.
 (*d*) Resource smoothing involves re-scheduling of the activities of a project in such a way that minimum project duration is achieved.

5. A project scheduling problem
 (*a*) cannot be formulated as a linear programming problem.
 (*b*) can be formulated as linear programming problem.
 (*c*) gives the least-duration of time in which it is completed.
 (*d*) updating of the project is not possible.

6. Which of the following is not correct?
 (*a*) After the planning and scheduling of a project, the process of incorporating the changes and re-scheduling the project is called updating.
 (*b*) ***GERT*** deals with probabilistic situations in project scheduling.
 (*c*) ***PERT*** is a branching network, where at a decision node different branches are evolved with probabilistic estimates.
 (*d*) ***MOST*** is a graphical representation of ***PERT*** to a time-scale.

7. Which of the following is correct?
 (*a*) ***GERT*** is a branching network, where at a decision node different branches are evolved with probabilistic estimates.
 (*b*) ***LOB*** technique balances between availability and requirement of components in a assembly shop.
 (*c*) Precedence planning diagrams are very useful in line balancing problems.
 (*d*) All of the above.

ANSWERS

1. (*a*) **2.** (*b*) **3.** (*d*) **4.** (*b*) **5.** (*b*) **6.** (*c*) **7.** (*d*)

REVIEW EXERCISE

1. Explain updating of the project networks. ***[MCA, Panjab Tech. Univ. 2010]***

2. What is resource levelling?

3. Explain how network analysis can be used for resource planning and levelling in project management.

4. Explain clearly the term Resource smoothing and Resource levelling.
[B.Tech., Nagpur Univ. 2005; MBA, Panjab Tech. Univ. 2005]

5. Give a procedure for resource smoothing using ***PERT/CPM***.

6. Differentiate between resource levelling and resource allocation in project scheduling.
[MBA, Panjab Tech. Univ. 2011]

7. Distinguish between 'Precedence Diagram' and 'Network Diagram'.

8. Write short note on the following:
 (*a*) MOST, (*b*) LOB, (*c*) GERT, and (*d*) Updating of network schedule.

CHAPTER 15

Dynamic Programming

"Repetition is the only form of permanence that nature can achieve"

Dynamic programming is an approach (technique) that decomposes a large decision-making problem into smaller problems for obtaining an optimum solution. Here, the problem is divided into sub-problems (known as ***stages***), and then solve each of these sub-problems. The decision of any sub-problem will depend upon the outcome of the preceding sub-problem (stage) and ultimately the overall objective function is optimized. Thus, dynamic programming is a multi-stage decision-making process and is largely based upon the ***principle of optimality*** due to Bellman, *viz.*,

"An optimal policy has the property that whatever the initial state and initial decisions are, the remaining decisions must constitute an optimal policy with regard to the state resulting from the first decision."

15:1. CHARACTERISTICS OF DYNAMIC PROGRAMMING PROBLEMS

The basic features that characterize the dynamic programming problems are as follows:

1. The problem can be sub-divided into stages with a policy decision required at each stage. Optimal solution to the given problem, then, can be obtained from the optimal solutions to the sub-problems.
2. Every stage consists of a number of states associated with the beginning of that stage. The states are the various possible conditions in which the system may find itself at that stage of the problem.
3. The policy decision at each stage *converts the* ***current state*** to a stage associated with the beginning of the ***next stage***. For example, to decide about the production policy, cost of production of material and labour may be taken at different stages.
4. The state of the system at a stage is generally described by a set of variables, called ***state variables***.
5. When the current state is known, an ***optimal policy*** *for the* ***remaining stages*** is independent of the policy decisions adopted in ***previous stages***. Thus, the optimal decision depends only on the current state and not on how you got there. This is the ***principle of optimality*** for dynamic programming.
6. The solution procedure begins by finding the ***optimal policy for the last stage***.

The optimal policy for the last stage prescribes the optimal policy decision for each of the possible states at that stage. The solution of this one-stage problem is usually trivial.

15:2. SHORTEST ROUTE PROBLEM

The dynamic programming approach to the shortest route problem essentially involves treating each node as, if it were on the optimal route making calculations accordingly. In doing so, we will work backward by starting at the terminal (final) node and calculating the shortest route from each node to terminal node until we reach the starting (origin) node. At this point, we will have solved the original problem of finding the shortest route from starting node to terminal node. To illustrate, consider the following situation:

Illustration. Suppose a traveler wants to go from city A to city B. Further, let the two cities are at a long distance from each other and also that there is no direct link between them. The traveler is, in fact, required to break his journey at least thrice during the course of his travel. At each break, there are options available, as shown in *Fig. 15.1*. Each circle in the figure represents a node (state) at particular stage. At each state, the traveler has to leave the old means of transport and shift to a new one. Determine the shortest route of travel.

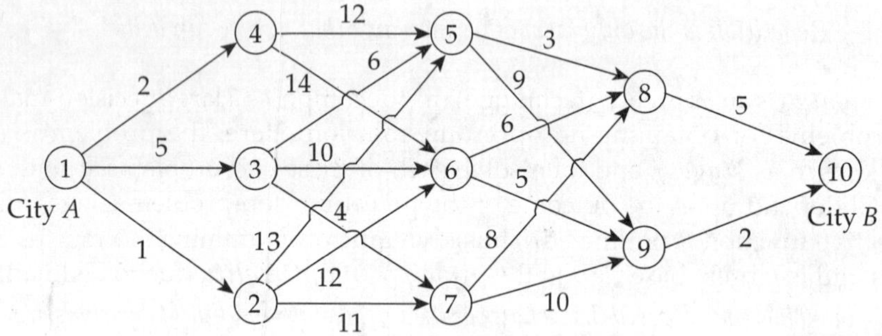

Fig. 15.1

Solution. Sub-divide the given problem of determining the route from city A to city B (node 1 to node 10) into four Stage problem as shown below:

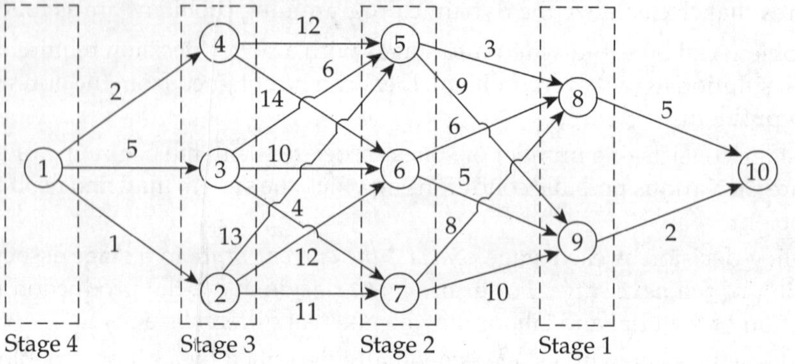

Fig. 15.2

In *Fig. 15.2*, first stage begins with nodes that are exactly one path away from the destination, and ends at the destination node. In dynamic programming technology, nodes 8 and 9 are considered to be the input nodes for Stage 1, and node 10 is considered to be the output node for Stage 1.

The second stage begins with all nodes that are exactly two paths away from the destination and ends with all nodes that are exactly one path away. Hence, nodes 5, 6 and 7 are the input nodes for Stage 2, and nodes 8 and 9 are the output nodes for Stage 2. Clearly, the output nodes for Stage 2 are the input nodes for Stage 1.

The input nodes for the third stage of the problem are all nodes that are exactly three paths away from the destination — that is, nodes 2, 3 and 4. The output nodes for Stage 3, all of which are one path closer to the destination, are nodes 5, 6 and 7. Finally, the input node for Stage 4 is node 1 and the output nodes are 2, 3 and 4.

Arbitrarily, we begin the stage 1 calculations with node 9 and move on to node 10. This route requires us to travel 2 kms. Similarly, only one path goes from node 8 to node 10. The shortest route from node 8 to node 10 is thus the length of that route, *i.e.*, of 5 kms. The stage 1 decision problem is, thus, solved. For each input node, we have identified an optimal decision — that is, the best path to travel over to reach the output node. The Stage 1 results are summarized as under:

Stage 1

Input node	*Path (decision)*	*Shortest distance (to node 10 in kms)*
8	8–10	5
9	9–10	2*

To begin the solution for the Stage 2 problem, we move to node 7. Two paths leave node 7 and are connected to input nodes for Stage 1; path 7–8, which has a length of 8 kms; and path 7–9, which have a length of 10 kms. If we select path 7–8, we will have a distance from node 7 to node 10 as 13 kms, that is, the length of path 7–8, plus the shortest distance to node 10 from node 8. Thus, the decision to select path 7–8 has a total associated distance of 8 + 5 (= 13) kms. With a distance of 10 kms for path 7–9 and Stage 1 results showing a distance of 2 kms from node 9 to node 10, the decision to select path 7–9 has an associated distance of 10 + 2 (= 12) kms. Thus, if we are at node 7, we should select path 7–9 because it is on the path that will reach node 10 in the shortest distance 12 kms. By performing similar calculations for nodes 5 and 6, we have the following Stage 2 results:

Stage 2

Input node	*Path (decision)*	*Associated path*	*Shortest distance to node 10 (in kms)*
5	5–8	5–8–10	3 + 5 = 8
	5–9	5–9–10	9 + 2 = 11
6	6–8	6–8–10	6 + 5 = 11
	6–9	6–9–10	5 + 2 = 7*
7	7–8	7–8–10	8 + 5 = 13
	7–9	7–9–10	10 + 2 = 12

To begin the third Stage, let us start with node 2. Note that three paths connect node 2 to the Stage 2 input nodes. Thus, to find the shortest route from node 2 to node 10, we must make three calculations. We select path 2–7 and then follow the shortest route to the end, we will have a distance of 11 + 12 (= 23) kms. Similarly, selecting path 2–6 requires 12 + 7 (= 19) kms, and selecting path 2–5 requires 13 + 8 (= 21) kms. Thus, the shortest route from node 2 to node 10 is 19 kms, which indicates that path 2–6 is the best decision, given that we are at node 2. Similarly, we find that the shortest route from node 3 to node 10 is given by Min. {4 + 12, 10 + 7, 6 + 8} = 14; the shortest route from node 4 to node 10 is given by Min. {14 + 7, 12 + 8} = 20. We compute the Stage 3 calculations with the following results:

Stage 3

Input node	*Path (decision)*	*Associated path*	*Shortest distance to node 10 (destination)*
2	2–5	2–5–8–10	13 + 8 = 21
	2–6	2–6–9–10	12 + 7 = 19
	2–7	2–7–9–10	11 + 12 = 23
3	3–5	3–5–8–10	6 + 8 = 14*
	3–6	3–6–9–10	10 + 7 = 17
	3–7	3–7–9–10	4 + 12 = 16
4	4–5	4–5–8–10	12 + 8 = 20
	4–6	4–6–9–10	14 + 7 = 21

In solving the Stage 4 sub-problem, we find that the shortest route from node 1 to node 10 is given by Min. {1 + 19, 5 + 14, 2 + 20} = 19. Thus, the optimal decision at State 4 is the selection of path 1–3. By moving through the network from Stage 4 to Stage 3 to Stage 2 to Stage 1, we can identify the best decision at each Stage and, therefore, the shortest route from node 1 to node 10. Thus, we have

Stage :	4	3	2	1
Path (decision) :	1–3	3–5	5–8	8–10

Hence, the shortest route is:

1–3–5–8–10 with a distance of 5 + 6 + 3 + 5 = 19 kms.

Remark. It may be noted that the calculations at each successive stage make use of the calculations at prior stages. This characteristic is an important part of dynamic programming. Further, evaluating several paths at each stage does not require us to enumerate all possible paths from node 1 to node 10.

15:3. FORMULATION OF DYNAMIC PROGRAMMING PROBLEM

Formulation of a dynamic programming problem requires the creation of recursive relationship. It needs us to define an optimum payoff function for every stage and state. The recursive relationship establishes the relationship between the optimum payoff functions at successive stages.

In dynamic programming, it is possible to write a recursive relationship for every problem. Once it is written, it becomes much easier to solve the given problem. In other words, the recursive relationship expresses the notion that optimum payoff at any given stage, for any given state, is given as the value of the best alternative, when each alternative includes the total of the immediate payoff and the optimum payoff obtained in the previous stage. Accordingly, a recursive relationship identifies the optimal policy for stage n, given the optimal policy for stage $(n + 1)$ available.

Using dynamic programming notation, we define the following:

N = number of stages,

n = index for the current stage ($n = 1, 2, ..., N$),

$(n + 1)$ = previous stage (it may be noted that successive stages are numbered in descending order),

s_n = state of the system in the current stage for which the recursive relation holds,

s_{n+1} = state of the previous stage,

x_n = decision variable at stage n,

x_n^* = optimum value of x_n (given s_n),

$f_n(s_n)$ = total payoff for each alternative, starting from state s_n, in stage n, to the end of process,

$f_n(s_n, x_n)$ = contribution of stages $n, n + 1, n + 2, ..., N$ to objective function, if system starts in state s_n at stage n, immediate decision is x_n, and optimal decision is made thereafter, the function is known as the stage transformation function.

$f_n^*(s_n)$ = optimal total payoff from state s_n in state n,

$f_{n+1}^*(s_{n+1})$ = optimal total payoff obtained in the previous stage.

The basic structure of the dynamic programming can be described diagrammatically is shown as under:

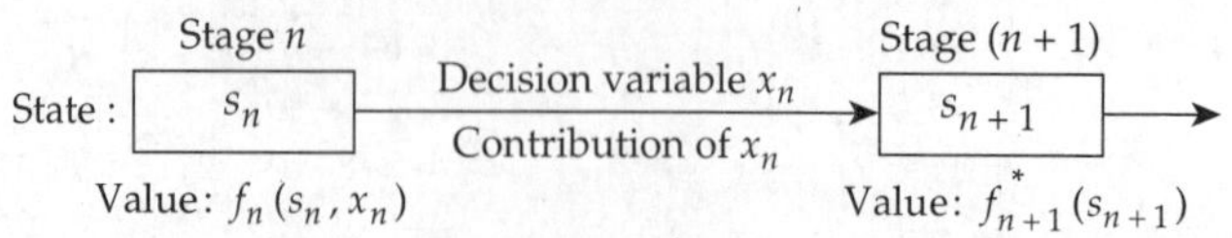

Making use of above notations, the ***recursive relationship*** (for minimization) for state s_n at Stage n may be stated as under:

$$f_n^*(s_n) = \min_{x_n} .\{p_n(x_n) + f_{n+1}^*(s_{n+1})\},$$

where, $p_n(x_n)$ is the immediate payoff in stage n when decision x_n is made for a specific value of s_n.

For dynamic programming problems in which the payoff is sought to be maximized, the ***recursive relationship*** would be:

$$f_n^*(s_n) = \max_{x_n} .\{p_n(x_n) + f_{n+1}^*(s_{n+1})\}$$

Remark. When we use recursive relationship in solving dynamic programming problems, the solution procedure starts at the end and moves ***backward*** stage by stage (***each time finding the optimal policy for that stage***) until it finds the optimal policy starting at the initial stage. This optimal policy yields an optimum solution for the given problem, namely, x_1^* for the initial Stage s_1, then x_2^* for the resulting Stage s_2, then x_3^* for the resulting Stage s_3 and so on to x_N^* for the resulting Stage s_N.

This backward movement was demonstrated by the shortest route problem, where the optimal policy was found successively beginning in each state at Stages 1, 2, 3 and 4, respectively. In fact for this problem, the solution procedure can move either backward or forward. However, for many dynamic programming problems (especially when the stages correspond to time periods), the solution procedure must move ***backward***.

15:4. SOLVING DYNAMIC PROGRAMMING PROBLEMS

Major steps in solving dynamic programming problems are as follows:

Step 1. Identify the problem — its decision variables, objective function to be optimized and constraints of the problem.

Step 2. Divide the given problem into number of sub-problems (called ***stages***).

Step 3. Identify the state variables at each stage whose value constitute the state at that stage.

Step 4. Specify the relationship that expresses the state at one stage as a function of the state and decision at the next stage.

Step 5. Write the recursive relationship for computing the optimal policy.

Step 6. Decide whether forward or backward approach is to be followed to solve the problem at different stages.

Step 7. The required results of appropriate stages are to be listed separately.

Step 8. Determine the overall optimal policy or optimum decision and its value at each stage. It may be noted that there may be more than one such optimal policy.

Example 15.1 (*Investment Problem*). *A company has ₹ 6,00,000 to invest, and a total of three investments are available to it. If x_j rupees (in lakh) are invested in investment j, then net present value (in lakh of rupees) of $v_j(x_j)$ is obtained. The expressions for $v_j(x_j)$'s are given below:*

$$v_1(x_1) = 9x_1 + 4,\ v_2(x_2) = 5x_2 + 9 \ \text{ and } \ v_3(x_3) = 2x_3 + 3;$$

where, $$x_1 \geq 0,\ x_2 \geq 0,\ x_3 \geq 0.$$

The amount placed in each investment has to be in multiples of ₹ 1 lakh only. How should the company invest ₹ 6,00,000 in order to maximize the net present value obtained from the investments?

Solution. To solve the problem using dynamic programming, we first define stages, states, and decision variables.

Stages. Consider each investment as a stage. Thus, the three investments can be considered as stage 1, stage 2, and stage 3 respectively.

States. Since, total amount to be invested is ₹6,00,000, for each of the stage 2 and 3, the states are 0, 1, 2, 3, 4, 5 and 6 (amount in lakh of rupees). For stage 1, there is only one state 6, the total amount of funds available.

Decision. The amount of funds to be invested in each of the three investment options.

If $f_n(s_n, x_n)$ denote the payoff at stage n, then the recursive relationship is

$$f_n^*(s_n) = \max_{x_n} . \{v_n x_n + f_{n+1}^*(s_n - x_n)\},$$

where,
s_n = amount of funds available for investment at stage s_n,
$v_n x_n$ = immediate payoff, *i.e.*, the payoff from current investment,
$f_n^*(s_n)$ = optimal payoff at stage n for stage s_n,
$f_{n+1}^*(s_n - x_n)$ = optimal payoff at the previous stage.

Using backward recurrence relationship, the solution starts with $f_3^*(s_3)$ and is completed when $f_1^*(s_1)$ is obtained.

Stage 3 (For n = 3). At this stage, the solution is trivial. Hence, the payoff functions for various states is defined in the table given below; wherein if no investment is made then the payoff is zero. For other investment amounts, the payoff are obtained by substituting the amount x_3 in the equation $2x_3 + 3$.

State s_3 *(Funds available)*	$f_3(s_3, x_3) = v_3x_3 = 2x_3 + 3$								
	$x_3 = 0$	$x_3 = 1$	$x_3 = 2$	$x_3 = 3$	$x_3 = 4$	$x_3 = 5$	$x_3 = 6$	$f_3^*(s_3)$	x_3^*
0	0							0	0
1	0	(2 + 3)						5	1
2	0	2 + 3	(4 + 3)					7	2
3	0	2 + 3	4 + 3	(6 + 3)				9	3
4	0	2 + 3	4 + 3	6 + 3	(8 + 3)			11	4
5	0	2 + 3	4 + 3	6 + 3	8 + 3	(10 + 3)		13	5
6	0	2 + 3	4 + 3	6 + 3	8 + 3	10 + 3	(12 + 3)	15	6

To illustrate, for $s_2 = 4$ and $x_3 = 5$, we have

$$f_3(4) = 2 \times 5 + 3 = 13. \quad \text{Similarly, } f_3(5) = 2 \times 5 + 3 = 13.$$

Stage 2 (For n = 2). At this stage we obtain x_2^*. This is being done by calculating and comparing $f_2(s_2, x_2)$ for the alternative values of x_2, namely $x_2 = 0, 1, 2, 3, 4, 5$. The computations for two stage problem, therefore, is as under:

s_2	$f_2(s_2, x_2) = (5x_2 + 9) + f_3^*(s_2 - x_2)$							$f_2^*(s_2)$	x_2^*
	$x_2 = 0$	$x_2 = 1$	$x_2 = 2$	$x_2 = 3$	$x_2 = 4$	$x_2 = 5$	$x_2 = 6$		
0	0 + 0							0	0
1	0 + 5	(14 + 0)						14	1
2	0 + 7	(14 + 5)	(19 + 0)					19	1, 2
3	0 + 9	14 + 7	(19 + 5)	(24 + 0)				24	2, 3
4	0 + 11	14 + 9	19 + 7	(24 + 5)	(29 + 0)			29	3, 4
5	0 + 13	14 + 11	19 + 9	24 + 7	(29 + 5)	(34 + 0)		34	4, 5
6	0 + 15	14 + 13	19 + 11	24 + 9	29 + 7	(34 + 5)	(39 + 0)	39	5, 6

***Stage 1* (For n = 1).** At this stage, we obtain x_1^*. This is being done by calculating and comparing $f_1(s_1, x_1)$ for the alternative values of x_1, namely $x_1 = 0, 1, 2, 3, 4, 5$. The computations for three stage problem are given below:

s_1	$f_1(s_1, x_1) = (9x_1 + 4) + f_2^*(s_1 - x_1)$							$f_1^*(s_1)$	x_1^*
	$x_1 = 0$	$x_1 = 1$	$x_1 = 2$	$x_1 = 3$	$x_1 = 4$	$x_1 = 5$	$x_1 = 6$		
6	0 + 39	13 + 34	22 + 29	31 + 24	40 + 19	(49 + 14)	58 + 0	63	5

The optimal solution is $x_1^* = 5$, which makes $s_2 = 6 - 5 = 1$. This gives $x_2^* = 1$ which implies $s_3 = 1 - 1 = 0$.

Hence, the optimal policy is to invest ₹ 5 lakh in first proposal, 1 lakh in second proposal and no investment in third proposal. The maximum *NPV* obtainable is ₹ 63 lakh.

Example 15.2 (*Allocation Problem*). *The owner of a chain of four grocery stores has purchased six loads of mangoes which he wants to send to four stores. Since, stores are located in different parts of the city, profitability differs in the four stores. The expected profit obtainable from selling different number of loads in the four stores (in thousands of rupees) is as given below:*

Number of loads	*Stores*			
	1	*2*	*3*	*4*
0	*0*	*0*	*0*	*0*
1	*4*	*2*	*6*	*2*
2	*6*	*4*	*8*	*3*
3	*7*	*6*	*8*	*4*
4	*7*	*8*	*8*	*4*
5	*7*	*9*	*8*	*4*
6	*7*	*10*	*8*	*4*

Use dynamic programming to determine how the loads should be divided between the four stores so as to maximize the profit.

Solution. To solve the problem using dynamic programming, we first define stages, states and decision variables.

Let the four stores be considered as four stages and the number of loads as states. For each of the stages 2, 3 and 4, the states are 0, 1, 2, 3, 4 and 5; whereas there is only one state 5 for stage 1. The decision variables x_j ($j = 1, 2, 3, 4$) denote the number of loads to be allocated to the jth stage.

If $f_n(s_n, x_n)$ be the profit associated with the optimum solution $f_n^*(s_n)$, then

$$f_n^*(s_n) = \max.\{P_n(x_n) + f_{n+1}(s_n - x_n)\}, \quad n = 1, 2, 3, 4$$

and
$$f_1^*(s_1) = \max_{x_1}.\{P_1(x_1)\}.$$

Here, s_n = availability of loads at stage n,

$P_n(x_n)$ = expected profit from allocation of x_n loads to store n, and

$f_n^*(s_n - x_n)$ = optimum profit at the previous stage.

Using backward recurrence relation, the solution starts with $f_4^*(s_4)$ and is completed when $f_1^*(s_1)$ is obtained.

***Stage 4* (For n = 4).** At this stage, the solution is trivial, because allocation of loads to store 4 only is to be analysed. For various number of loads available for allocation to store 4, the payoff function is defined in the following table:

s_4	$f_4(s_4, x_4) = P_4(x_4)$							$f_4^*(s_4)$	x_4^*
	$x_4 = 0$	$x_4 = 1$	$x_4 = 2$	$x_4 = 3$	$x_4 = 4$	$x_4 = 5$	$x_4 = 6$		
0	0							0	0
1	0	2						2	1
2	0	2	3					3	2
3	0	2	3	4				4	3
4	0	2	3	4	4			4	3, 4
5	0	2	3	4	4	4		4	3, 4, 5
6	0	2	3	4	4	4	4	4	3, 4, 5, 6

Stage 3 (For n = 3). At this stage, obtaining x_3^* requires calculating and comparing $f_3(S_3, x_3)$ for alternative values of x_3, namely $x_3 = 0, 1, 2, 3, 4, 5$. Thus, for a two stage problem, the computations are as under:

s_3	$f_3(s_3, x_3) = P_3(x_3) + f_4^*(s_3 - x_3)$							$f_3^*(s_3)$	x_3^*
	$x_3 = 0$	$x_3 = 1$	$x_3 = 2$	$x_3 = 3$	$x_3 = 4$	$x_3 = 5$	$x_4 = 6$		
0	0 + 0							0	0
1	0 + 2	(6 + 0)						6	1
2	0 + 3	(6 + 2)	(8 + 0)					8	1, 2
3	0 + 4	6 + 3	(8 + 2)	8 + 0				10	2
4	0 + 4	6 + 4	(8 + 3)	8 + 2	8 + 0			11	2
5	0 + 4	6 + 4	(8 + 4)	8 + 3	8 + 2	8 + 0		12	2
6	0 + 4	6 + 4	(8 + 4)	(8 + 4)	8 + 3	8 + 2	8 + 0	12	2, 3

Stage 2 (For n = 2). At this stage, we compute x_2^* and $f_2^*(s_2)$ as shown in the following table:

s_2	$f_2(s_2, x_2) = P_2(x_2) + f_3^*(s_2 - x_2)$							$f_2^*(s_2)$	x_2^*
	$x_2 = 0$	$x_2 = 1$	$x_2 = 2$	$x_2 = 3$	$x_2 = 4$	$x_2 = 5$	$x_2 = 6$		
0	0 + 0							0	0
1	(0 + 6)	2 + 0						6	0
2	(0 + 8)	(2 + 6)	4 + 0					8	0, 1
3	(0 + 10)	(2 + 8)	(4 + 6)	6 + 0				10	0, 1, 2
4	0 + 11	(2 + 10)	(4 + 8)	(6 + 6)	8 + 0			12	1, 2, 3
5	0 + 12	2 + 11	(4 + 10)	(6 + 8)	(8 + 6)	9 + 0		14	2, 3, 4
6	0 + 12	2 + 12	4 + 11	(6 + 10)	(8 + 8)	9 + 6	10 + 0	16	3, 4

Stage 1 (For n = 1). At this stage, we compute x_1^* and $f_1^*(s_1)$ as shown in the following table:

s_1	$f_1(s_1, x_1) = P_1(x_1) + f_1^*(s_1 - x_1)$							$f_1^*(s_1)$	x_1^*
	$x_1 = 0$	$x_1 = 1$	$x_1 = 2$	$x_1 = 3$	$x_1 = 4$	$x_1 = 5$	$x_1 = 6$		
6	0 + 16	(4 + 14)	(6 + 12)	7 + 10	7 + 8	7 + 6	7 + 0	18	1, 2

The optimal solution is $x_1^* = 1$ or 2. This gives us:

When $x_1^* = 1, s_2 = 6 - 1 = 5$. This implies $x_2^* = 2, 3, 4$,

$x_1^* = 2, s_2 = 6 - 2 = 4$. This implies $x_2^* = 1, 2, 3$.

When $x_2^* = 2, s_3 = 5 - 2 = 3$. This implies $x_3^* = 2$,
$x_2^* = 3, s_3 = 5 - 3 = 2$. This implies $x_3^* = 1, 2$,
$x_2^* = 4, s_3 = 5 - 4 = 1$. This implies $x_3^* = 1$.
When $x_2^* = 1, s_3 = 4 - 1 = 3$. This implies $x_3^* = 2$,
$x_2^* = 2, s_3 = 4 - 2 = 2$. This implies $x_3^* = 1, 2$,
$x_2^* = 3, s_3 = 4 - 3 = 1$. This implies $x_3^* = 1$.
When $x_3^* = 1, s_4 = 1 - 1 = 0$. This implies $x_4^* = 0$,
$x_3^* = 2, s_4 = 3 - 2 = 1$. This implies $x_4^* = 1$.

Above mentioned alternative optimum solution are re-arranged as follows:

	Decision	x_1^*	x_2^*	x_3^*	x_4^*
	1	1	2	2	1
	2	1	3	1	1
Possible	3	1	3	2	0
optimal	4	1	4	1	0
allocations	5	2	1	2	1
	6	2	2	1	1
	7	2	2	2	0
	8	2	3	1	0

Alternative Solution. Let x_1, x_2, x_3 and x_4 be the number of loads allocated to four stores and $f_1(x_1)$, $f_2(x_2)$, $f_3(x_3)$, $f_4(x_4)$ be the respective profits. Then, the given problem is to maximize $z = f_1(x_1) + f_2(x_2) + f_3(x_3) + f_4(x_4)$ subject to the constraint $x_1 + x_2 + x_3 + x_4 \leq 6$, where x_1, x_2, x_3 and x_4 are non-negative integers.

Stage 1. The estimated profits corresponding to different number of loads allocated to store 1 are as follows:

Number of loads (x_1) :	0	1	2	3	4	5	6
Profit $f_1(x_1)$:	0	4	6	7	7	7	7

Stage 2. Now, we consider stores 1 and 2 combined. Six loads can be divided among two stores in 7 different ways: as 6 in store 1 and 0 in store 2, 5 in store 1 and 1 in store 2, and so on. Each combination will have associated with its certain profits. The profits for all the total number of loads, *i.e.*, $f_1(x_1) + f_2(x_2)$ are shown in the following table:

	x_1	0	1	2	3	4	5	6
	$f_1(x_1)$	0	4	6	7	7	7	7
x_2	$f_2(x_2)$							
0	0	0 + 0 = 0*	0 + 4 = 4*	0 + 6 = 6*	0 + 7 = 7	0 + 7 = 7	0 + 7 = 7	0 + 7 = 7
1	2	2 + 0 = 2	2 + 4 = 6*	2 + 6 = 8*	2 + 7 = 9	2 + 7 = 9	2 + 7 = 9	
2	4	4 + 0 = 4	4 + 4 = 8*	4 + 6 = 10*	4 + 7 = 11	4 + 7 = 11		
3	6	6 + 0 = 6	6 + 4 = 10*	6 + 6 = 12*	6 + 7 = 13			
4	8	8 + 0 = 8	8 + 4 = 12*	8 + 6 = 14*				
5	9	9 + 0 = 9	9 + 4 = 13					
6	10	10 + 0 = 10						

For a particular number of loads, the profits for all possible combinations can be read along the diagonal. Maximum profits are marked by *. Thus, the optimum profits and corresponding allocations of loads to the two stores are as follows:

Number of loads	*0*	*1*	*2*	*3*	*4*	*5*	*6*
$f_2(x_2) + f_1(x_1)$	0	4	6	8	10	12	14
$x_2 + x_1$	0 + 0	0 + 1	1 + 1	2 + 1	3 + 1	4 + 1	4 + 2
			0 + 2	1 + 2	2 + 2	3 + 2	

Stage 3. Now, consider the distribution of six loads to three stores 1, 2 and 3. The decision at this stage will result in allocating certain number of loads to store 3 and to stores 2 and 1 combined. Then, following the backward process, they will be distributed to stores 2 and 1. The profits for all the total number of loads, such as 6, 5, 4, ..., 0 are shown in the following table:

Number of loads $(x_2 + x_1)$		*0*	*1*	*2*	*3*	*4*	*5*	*6*
$f_2(x_2) + f_1(x_1)$		*0*	*4*	*6*	*8*	*10*	*12*	*14*
x_3	$f_3(x_3)$							
0	0	0*	4	6	8	10	12	14
1	6	6*	10*	12*	14*	16*	18*	
2	8	8	12*	14*	16*	18*		
3	8	8	12	14	16			
4	8	8	12	14				
5	8	8	12					
6	8	8						

For any particular number of loads, the profits for all possible combinations can be read along the diagonal. Maximum profits are marked by *. Thus, the optimum profits and corresponding allocation of loads to the three stores are given below:

Number of loads	*0*	*1*	*2*	*3*	*4*	*5*	*6*
$f_3(x_3) + f_2(x_2) + f_1(x_1)$	0	6	10	12	14	16	18
$x_3 + (x_2 + x_1)$	0 + 0	1 + 0	1 + 1	2 + 1	2 + 2	2 + 3	2 + 4
				1 + 2	1 + 3	1 + 4	1 + 5

Stage 4. Next, we consider the distribution of 6 loads to four stores. The corresponding profits for all possible combinations are as follows:

$(x_3 + x_2 + x_1)$	*0*	*1*	*2*	*3*	*4*	*5*	*6*
$f_3(x_3) + f_2(x_2) + f_1(x_1)$	0	6	10	12	14	16	18
x_4	6	5	4	3	2	1	0
$f_4(x_4)$	4	4	4	4	3	2	0
$f_4(x_4) + f_3(x_3) + f_2(x_2) + f_1(x_1)$	4	10	14	16	17	18*	18*

From the above table, we observe that the maximum possible profit is 18 for $x_4 = 1$ or 0. Going back, we get 8 optimum allocations, yielding the same profit of 18.

Example 15.3 (*Employment Smoothening Problem*). *A company has trained five sales persons in the past three months, whom it wants to assign the three sales regions: North, West, and South. It has been estimated that increase in sales (in lakh of rupees) over the next year for the five salesmen in each region, were allocated to different number of salesmen, as indicated below:*

Sales persons	*Region*		
	North	*West*	*South*
0	0	0	0
1	61	36	66
2	86	61	86
3	106	91	96
4	121	126	116
5	136	166	146

Use dynamic programming to determine which allocation would maximize the increase in sales.

Solution. To solve the above problem using dynamic programming, we first define stages, states and decision variables. Thus, we treat the problem as a three stage problem taking each region as a stage. The three regions may be taken as three stages, n (n = 1, 2, 3). It is required to determine the number of sales persons to be allocated at a given stage, x_n. Further, at each stage, the 'states' represents the number of sales persons still available for allocation to the regions remaining. To illustrate, let at stage 1, $s_1 = 5$; at stage 2, $s_2 = 5 - x_1$, and at stage 3, $s_3 = 5 - x_2$.

If $f_n(s_n, x_n)$ denotes the number of salesmen allocated with the optimum solution $f_n^*(s_n)$, $n = 1, 2, 3$; then the recursive relationship is:

$$f_n^*(s_n) = \max_{x_n}.\{p_n(x_n) + f_{n+1}^*(s_n - x_n)\},$$

where, s_n = number of sales persons available for allocation at stage n,

$P_n(x_n)$ = the increment in sales resulting from current allocation of sales persons — the immediate payoff,

$f_n^*(s_n)$ = the optimal payoff at stage n, and

$f^*_{n+1}(s_n - x_n)$ = optimal payoff at previous stage.

Using backward recurrence relation, the solution start with $f_3^*(s_3)$ and is completed when $f_1^*(s_1)$ is obtained.

Stage 1 (For n = 1). At this stage, the solution is trivial, because allocation of sales persons to the South region only is to be analysed. For various number of sales persons available for allocation of South, the payoff function is defined in the following table:

s_3	$f_3(s_3, x_3) = f_3^*(s_3)$						$f_3^*(s_3)$	x_3^*
	$x_3 = 0$	$x_3 = 1$	$x_3 = 2$	$x_3 = 3$	$x_3 = 4$	$x_3 = 5$		
0	0						0	0
1	0	66					66	1
2	0	66	86				86	2
3	0	66	86	96			96	2
4	0	66	86	96	116		116	4
5	0	66	86	96	116	146	146	5

***Stage 2* (*For n* = 2).** At this stage obtain $x_2{}^*$. This is being done by calculating and comparing $f_2(s_2, x_2)$ for the alternative values of x_2, namely $x_2 = 0, 1, 2, ..., 5$. The computations for two stage problem, therefore, are as under:

s_2	$f_2(s_2, x_2) = p_2(x_2) + f_3{}^*(s_2 - x_2)$						$f_2{}^*(s_2)$	$x_2{}^*$
	$x_2 = 0$	$x_2 = 1$	$x_2 = 2$	$x_2 = 3$	$x_2 = 4$	$x_2 = 5$		
0	0 + 0						0	0
1	0 + 66	36 + 0					66	0
2	0 + 86	36 + 66	61 + 0				102	1
3	0 + 96	36 + 86	61 + 66	91 + 0			127	2
4	0 + 116	36 + 96	61 + 86	91 + 66	126 = 0		157	3
5	0 + 146	36 + 116	61 + 96	91 + 86	126 + 66	166 + 0	192	4

***Stage 3* (*For n* = 3).** At this stage obtain $x_1{}^*$. This is being done by calculating and comparing $f_1(s_1, x_1)$ for the alternative values of x_1, namely $x_1 = 0, 1, 2, ..., 5$. The computations for three stage problem, therefore, are as under:

s_1	$f_1(s_1, x_1) = p_1(x_1) = f_2(s_1 - x_1)$						$f_1{}^*(s_1)$	$x_1{}^*$
	$x_1 = 0$	$x_1 = 1$	$x_1 = 2$	$x_1 = 3$	$x_1 = 4$	$x_1 = 5$		
5	0 + 192	61 + 157	86 + 127	106 + 102	121 + 66	136 + 0	218	1

The optimal solution is $x_1{}^* = 1$, which makes $s_2 = 5 - 1 = 4$. This gives $x_2{}^* = 3$, which implies $s_3 = 4 - 3 = 1$. Thus, $x_2{}^* = 1$. This implies that the optimal solution calls for employing one salesman in North, three salesmen in West and one salesman in South. Further, since $f_1{}^*(5) = 218$, the allocation of 5 salesmen to the three regions will lead to a total of 218 lakh of rupees.

Alternative Solution. Let $f_n(x_n)$ be the expected increase in sales from allocation of x_n ($n = 0, 1, 2, ..., 5$) salesmen to region n (n = North, West, South).

We start with region: 'North'. The returns corresponding to different number of salesmen allocated to area 'North' only are given below:

Number of Salesman (x_1) :	0	1	2	3	4	5
Increase in Sales $f_1(x_1)$:	0	61	86	106	121	136

Now, we consider regions — North and West combined. Five salesmen can be divided into two regions in six different ways as shown below:

North (x_1)	0	1	2	3	4	5
West (x_2)	5	4	3	2	1	0
Sales $f_1(x_1)$	0	61	86	106	121	136
Sales $f_2(x_2)$	166	126	91	61	36	0

If 'p' denotes the total sales from region 'North' and 'West' combined, various alternatives are:

$$f_1(0) + f_2(5), f_1(1) + f_2(4), \text{etc.}$$

Thus, in general the problem is to maximize $f_1(x_n) + f_2(s_n - x_n)$, where s_n stands for total number of salesmen.

The above equation can be applied to determine the optimum distribution of any number of salesmen. This can be done in tabular form as follows:

x_1 / $f_1(x_1)$ → x_2, $f_2(x_2)$ ↓		0 0	1 61	2 86	3 106	4 121	5 136
0	0	0*	61*	86	106	121	136
1	36	36	97*	122*	142	157	
2	61	61	122*	147	167		
3	91	91	152*	177			
4	126	126	187*				
5	166	166					

The number marked * are the maximum along each diagonal.

Expected sales are computed for all the possible combinations. For a particular number of salesmen, the sales for different combinations can be read along the diagonal. For instance, if two salesmen are to be distributed between two regions, the sales for the combination 2 + 0, 1 + 1, 0 + 2 can be read along the diagonal 2 – 2. The maximum sales of 97 for two salesmen results from the combination of 'one' salesman in North and 'two' salesmen in West. The optimum returns from various combinations are summarized below:

s	0	1	2	3	4	5
$f_1(x_1) + f_2(x_2)$	0	61	97	122	152	187
$x_1 + x_2$	0 + 0	1 + 0	1 + 1	2 + 1 or 1 + 2	1 + 3	1 + 4

Now, we move to third stage and consider the distribution of five salesmen into three regions, namely North, West and South. So, we maximize

$$\{f_1(x_1) + f_2(x_2) + f_3(x_3)\}.$$

This is given in the following table:

$x_1 + x_2$ / $f(x)$ → x_3, $f_3(x_3)$ ↓		0 0	1 61	2 97	3 122	4 152	5 187
0	0	0*	61	97	122	152	187
1	66	66*	127*	163*	188*	218*	
2	86	86	147	183	208		
3	96	96	157	193			
4	116	116	177				
5	146	146					

The numbers marked * are the maximum along each diagonal.

From the above table, we observe that along the diagonal 5 – 5 the optimum (maximum) sales is of 218 lakh rupees. The optimum combination is of 4 salesmen to regions 'North' and 'West' combined and *one* salesman to South region.

Now, proceeding backward, four salesmen in region 'North' and 'West' implies *one* salesman in region 'North' and *three* salesmen in 'West'. Hence, we have:

Region	:	North	West	South
Number of salesmen to be allocated	:	1	3	1

Example 15.4. *A company has 6 salesmen and 3 market areas A, B, C. It is desired to determine the number of salesmen to allocate to each market area to maximize profit. The following table gives the profit from each market area as a function of the number of salesmen allocated:*

Area	*Salesmen*						
	0	1	2	3	4	5	6
A	38	41	48	58	66	72	83
B	40	42	50	60	66	75	82
C	60	64	68	78	90	102	101

Use dynamic programming technique to solve the above problem. ***[MBA (Nov.), Madurai Univ. 2008]***

Solution. Let the three areas be considered as three-stage problem. The decision variables x_j denote the number of salesmen and let $f_j(x_j)$ be the expected profit from the allocation of x_j salesmen to area j (j = A, B, C).

We start with area A. The returns corresponding to different number of salesmen allocated to area A only are:

No. of salesmen (x_1) :	0	1	2	3	4	5	6
Profit $f_1(x_1)$:	38	41	48	58	66	72	83

Now, we consider areas A and B combined. Six salesmen can be divided into two areas in seven different ways as shown below:

Area A :	x_1	0	1	2	3	4	5	6
Area B :	x_2	6	5	4	3	2	1	0
Profits	$f_1(x_1)$	38	41	48	58	66	72	83
	$f_2(s_2)$	82	75	66	60	50	42	40

If P denotes the total profits from area A and B considered, various alternatives are:

$$P = f_1(0) + f_2(6), \quad P = f_1(1) + f_2(5), \text{ etc.}$$

Thus, in general the problem is to

$$\text{Maximize } P = f_1(x_j) + f_2(s - x_j)$$

where, s stands for total number of salesmen.

The above equation can be applied to determine the optimum distribution of any number of salesmen. This can be best done in the tabular form as given below:

	x_1	0	1	2	3	4	5	6
	$f_1(x_1)$	38	41	48	58	66	72	83
x_2	$f_2(x_2)$							
0	40	78*	81*	88*	98*	106*	112	123*
1	42	80	83	90	100	108	114	
2	50	88*	91	98	108	116		
3	60	98*	101	108	118			
4	66	104	107	114				
5	75	113*	116					
6	82	120						

The number marked * are the maximum along each diagonal.

Expected profits are computed for all the possible combinations. For a particular number of salesmen, the profit for different combinations can be read along the diagonal. For instance, if two salesmen are to be distributed between two areas, the value of sales for the combination 2 + 0, 1 + 1, 0 + 2 can be read along the diagonal 2 – 2. The maximum profit of 88 for two salesmen results from the combination of 2 salesmen in area *A* and none in area *B* or 2 salesmen in area *B* and none in area *A*. The optimum returns from various combinations are summarized below:

s	0	1	2	3	4	5	6
$F(x) = f_1(x_1) + f_2(x_2)$	78	81	88	98	106	113	123
$x = x_1 + x_2$	0 + 0	1 + 0	2 + 0 0 + 2	3 + 0 0 + 3	4 + 0	0 + 5	6 + 0

Now, we move to the third stage and consider the distribution of 6 salesmen into three areas, namely *A*, *B* and *C*. So, we maximize

$$P = F(x_j) + f_3(s - x_j);\quad 0 \le x_j \le s.$$

	x	0	1	2	3	4	5	6
	$F(x)$	78	81	88	98	106	113	123
x_3	$f_1(x_3)$							
0	60	138*	141	148*	158*	166	173	183*
1	64	142*	145	152	162	170	177	
2	68	146	149	156	166	174		
3	78	156	159	166	176			
4	90	168*	171	178				
5	102	180*	183*					
6	101	109						

The number marked * are the maximum along each diagonal.

As we see along the diagonal 6 – 6, the optimum combination is 6 salesmen to area *A* and *B* combined and none in area *C*; or 1 salesman in area *A* and *B* combined and 5 salesmen in area *C*. This gives a maximum profit of 183.

Now, proceeding backwards from the previous table, we find that the optimum division is:

(*i*) 6 salesmen in *A* and *B* combined implies 6 salesmen in *A* and none in *B*.

(*ii*) 1 salesman in *A* and *B* combined implies 1 salesman in *A* and none in *B*.

Thus, the best possible allocation among all the three areas is as follows:

Area	:	A	B	C
No. of salesmen	:	6	0	0
To be allocated	:	1	0	5

The maximum allocation on the basis of this allocation is 183.

Example 15.5 (*Cargo Load Problem*). *A vessel is to be loaded with stocks of 3 items. Each unit of item 'n' has a weight w_n and value v_n. The maximum cargo weight, the vessel can take is 5 and the details of the three items are as follows:*

Item (n)	*Weight (w_n) in tons*	*Value (v_n)*
1	1	30
2	3	80
3	2	65

Develop the recursive equation for the above case and find the most valuable cargo load without exceeding the maximum cargo weight by using dynamic programming. **[B.Tech. (Nov.), Hyderabad Univ. 2010]**

Solution. We have to determine how many units of three items are to be loaded. So, it is a three stage problem. Define:

n = the stage (item) under consideration,
x_n = the number of units of item of type n (n = 1, 2, 3),
W = maximum capacity available,
s_n = state at stage n, indicating the weight available.

Let $f_n(x_n)$ denote the value of optimal allocation for the three types of items. If $f_n(s_n, x_n)$ be the value associated with optimum solution $f_n^*(s_n)$, n = 1, 2, 3; then

$$f_1^*(s_1) = \max_{x_1}.\ f_1(s_1, x_1)$$

and $$f_n^*(s_n) = \max_{x_n}.\{p_n(x_n) + f_{n-1}^*(s_n - x_n)\}, \qquad n = 1, 2, 3;$$

where, $p_n(x_n)$ denote the expected value obtained from allocation of x_n units of weight to item 'n'.

For one-stage problem (*i.e.*, for one item cargo loading), we write

$$f_1^*(s_1) = \max_{x_1}.\ \{30x_1\},$$

where the largest value of x_1 is $[W/w_1] = [5/1] = 5$.

We have the following tabular computations:

s_1	$f_1(s_1, x_1) = \{30x_1\}$						Optimum solution	
	$x_1 = 0$	$x_1 = 1$	$x_1 = 2$	$x_1 = 3$	$x_1 = 4$	$x_1 = 5$	$f_1^*(s_1)$	x_1^*
0	0						0	0
1	0	30					30	1
2	0	30	60				60	2
3	0	30	60	90			90	3
4	0	30	60	90	120		120	4
5	0	30	60	90	120	150	150	5

For 2-stage problem, the largest value of x_2 is $[5/3] = 1$, and

$$f_2^*(s_2) = \max_{x_2}.\ \{80x_2 + f_1^*(s_2 - 3x_2)\}$$

s_2	$f_2(s_2, x_2) = \{80x_2 + f_1^*(s - 3x_2)\}$		Optimum solution	
	$x_2 = 0$	$x_2 = 1$	$f_2^*(s_2)$	x_2^*
0	0 + 0 = 0		0	0
1	0 + 30 = 30		30	0
2	0 + 60 = 60		60	0
3	0 + 90 = 90	80 + 0 = 80	90	0
4	0 + 120 = 120	80 + 30 = 110	120	0
5	0 + 150 = 150	80 + 60 = 140	150	0

For 3-stage problem, the largest value of x_3 is $[5/2] = 2$, and

$$f_3^*(s) = \max_{x_3}.\ \{65x_3 + f_2^*(s - 2x_3)\}$$

s_3	$f_3(s_3, x_3) = \{65x_3 + f_2^*(s - 2x_3)\}$			Optimum solution	
	$x_3 = 0$	$x_3 = 1$	$x_3 = 2$	$f_3^*(s_3)$	x_3^*
0	0 + 0			0	0
1	0 + 30			30	0
2	0 + 60	65 + 0 = 65		65	1
3	0 + 90	65 + 30 = 95		95	1
4	0 + 120	65 + 60 = 125	130 + 0 = 130	130	2
5	0 + 150	65 + 90 = 155	130 + 30 = 160	160	2

Given $W = 5$, the optimum solution, therefore, is $x_3^* = 2$. This gives us the allocation of 2 units of item 3 (weighing 2 tons). The remaining $5 - 2 \times 2$, *i.e.*, one tone is allocated in an optimal manner according to stage 2.

For this quantity, we have $x_2^* = 0$, so that no unit of item 2 are included. Proceeding to stage 3, the optimal allocation is $x_1^* = 1$ (measuring 1 ton).

Hence, the optimum solution is to load 2 units of item 3, 1 unit of item 1 and no unit of item 2. The maximum value is $2 \times 65 + 1 \times 30$, *i.e.*, 160 units.

Example 15.6. *Four types of items are under consideration for loading on a plane. The weights and values of various type of items are given below:*

Item	*Weight (tons)*	*Value*
A	3	22
B	6	44
C	5	35
D	4	24

Determine as to which items should be loaded on the plane, and in what quantities, if the maximum capacity of the plane is 13 tons and the objective is to maximize the value of the shipment. Give a mathematical representation of the problem and solve it by dynamic programming.

Solution. Here, the stage may be represented by the items under consideration. Thus, if

n = the stage (item) under consideration,
x_n = the number of units of item of type n to load,
w_n = the weight of 1 unit of item of type n,
v_n = the value of 1 unit of item of type n,
K = maximum capacity available,

and s_n = state at stage n, indicating the weight available.

The mathematical formulation of the problem, therefore, is:

Maximize $z = 22x_1 + 44x_2 + 35x_3 + 24x_4$ subject to the constraints:

$$3x_1 + 6x_2 + 5x_3 + 4x_4 \leq 13$$

$x_1 \geq 0, x_2 \geq 0, x_3 \geq 0, x_4 \geq 0$ and are integers.

Let $f_n^*(s_n)$ is the optimal total payoff at stage n for state n, where s_n is the amount of weight available for allocation. Then, the recursive relationship is

$$f_n^*(s_n) = \max_{x_n} . \{v_n x_n) + f_{n+1}^*(s_n - w_n x_n)\}.$$

Using backward recursion, the solution of the problem is as follows:

Stage 1. We first consider item *D*. With a maximum capacity to load 13 tons and a load of 4 tons per unit, the maximum number of units of *D* which can be loaded is 3. Accordingly, x_4 can take values 0, 1, 2, 3. The payoff function is given by $v_4x_4 = 24x_4$, where x_4 designate the number of units of *D*.

Computations at this stage are as follows:

State	$f_4(s_4, x_4) = v_4x_4 = 24x_4$				Optimum solution	
s_4	$x_4 = 0$	$x_4 = 1$	$x_4 = 2$	$x_4 = 3$	$f_4^*(s_4)$	x_4^*
0	0				0	0
1	0				0	0
2	0				0	0
3	0				0	0
4	0	24			24	1
5	0	24			24	1
6	0	24			24	1
7	0	24			24	1
8	0	24	48		48	2
9	0	24	48		48	2
10	0	24	48		48	2
11	0	24	48		48	2
12	0	24	48	72	72	3
13	0	24	48	72	72	3

***Stage* 2.** At this stage, an allocation is made to item *C*, with the remaining weight utilized according to optimal policy recommended in *Stage 4*. The payoff corresponding to various states are shown in table given below. With 5 tons per unit, a maximum of 2 units of *C* can be loaded. Here, the total payoff function is given by:

$$f_3(s_n) = 35x_3 + f_4^*(s_3 - 5x_3)$$

Note that if x_3 units of *C* are loaded, $(s_3 - 5x_3)$ tons would be left.

The computations at this stage are:

State	$f_3(s_3, x_3) = 35x_3 + f_4^*(s_3 - 5x_3)$			Optimum solution	
s_3	$x_3 = 0$	$x_3 = 1$	$x_3 = 2$	$f_3^*(s)$	x_3^*
0	0			0	0
1	0			0	0
2	0			0	0
3	0			0	0
4	24			24	0
5	24	35 + 0 = 35		35	1
6	24	35 + 0 = 35		35	1
7	24	35 + 0 = 35		35	1
8	48	35 + 0 = 35		48	0
9	48	35 + 24 = 59		59	1
10	48	35 + 24 = 59	70 + 0 = 70	70	2
11	48	35 + 24 = 59	70 + 0 = 70	70	2
12	72	35 + 24 = 59	70 + 0 = 70	72	0
13	72	35 + 48 = 83	70 + 0 = 70	83	1

***Stage* 3.** At this stage a unit of product *B* involves 6 tons, where a maximum of 2 units of it can be loaded. The total payoff function, therefore, is $f_2(s_2) = 44x_2 + f_3^*(s_2 - 6x_2)$. The calculation for this stage are given below:

State	$f_2(s_2, x_2) = 44x_2 + f_3^*(s_2 - 6x_2)$			Optimum solution	
s_2	$x_2 = 0$	$x_2 = 1$	$x_2 = 2$	$f_2^*(s_2)$	x_2^*
0	0			0	0
1	0			0	0
2	0			0	0
3	0			0	0
4	24			24	0
5	35			35	0
6	35	44 + 0		44	1
7	35	44 + 0		44	1
8	48	44 + 0		48	0
9	59	44 + 0		59	0
10	70	44 + 24		70	0
11	70	44 + 35		79	1
12	72	44 + 35	88 + 0	88	2
13	83	44 + 35	88 + 0	88	2

Stage 4. In the final (first) stage, 13 tons are available to be loaded. If x_1 be the number of units of item *A*, then x_1 can take values 0, 1, 2, 3 or 4; since a maximum of 4 units of *A*, each of 3 tons, can be loaded. The payoff function at this stage is

$$f_1(s_1, x_1) = 22x_1 + f_2^*(s_1 - 3x_1)$$

The computations at this stage are:

State	$f_1(s_1, x_1) = 22x_1 + f_2^*(s_1 - 3x_1)$					Optimum solution	
s_1	$x_1 = 0$	$x_1 = 1$	$x_1 = 2$	$x_1 = 3$	$x_1 = 4$	$f_1^*(s_1)$	x_1^*
13	0 + 88 = 88	22 + 70 = 92	44 + 44 = 88	66 + 24 = 90	88 + 0 = 88	92	1

The optimal solution is $x_1^* = 1$, *i.e.*, one unit of item *A* (weighing 3 tons). The remaining $(13 - 3) = 10$ tons are allocated in an optimal manner according to stage 2. For this quantity, we have $x_2^* = 0$, so that no units of item *B* are included. Proceeding to stage 3, the optimal allocation is $x_3^* = 2$. This implies that 2 units of item *C* weighing $2 \times 5 = 10$ tons are included. With no capacity left, $x_4 = 0$. The maximum value, thus, is $(1 \times 22 + 2 \times 35)$, *i.e.*, 92 units.

PROBLEMS

15.01. A student has to take an examination in three courses *x*, *y* and *z*. He has three days available for study. He feels that it would be better to devote a whole day to study the same course, so that he may study a course for one day, two days or three days or not at all. His estimates of grades he may get according to days of study he puts-in are as follows:

Study (days)	Course		
	x	*y*	*z*
0	1	2	1
1	2	2	2
2	2	4	4
3	4	5	4

How should he distribute the available days to different courses, so that he maximizes his grade points.

15.02. A chairman of a certain political party is making plans for his election to Parliament. He has engaged the services of six volunteer workers and wishes to assign them three districts in such a way that their effectiveness is maximized. He feels that it would be inefficient to assign a worker more than one district, but he is also willing to assign no worker to any one of the districts judging by what the workers can accomplish in other districts.

The following table gives the estimated increase in the number of votes in favour of party's candidate, if it were allocated various number of workers:

Number of workers	*Districts*		
	1	2	3
0	0	0	0
1	25	20	33
2	42	38	43
3	55	54	47
4	63	65	50
5	69	73	52
6	74	80	53

How many of the workers should be assigned to each of the three districts in order to maximize the total number of votes in its favour? ***[MFC, Delhi Univ. 2001]***

15.03 A firm has divided its marketing area into three zones. The amount of sales depends upon the number of salesmen in each zone. The firm has been collecting the data regarding sales and salesmen in each area over a number of past years.

The information is summarized in the following table. For the next year, firm has only 9 salesmen and the problem is to allocate these salesmen to three different zones so that the total sales are maximum.

Profits in thousands of rupees

No. of salesmen	0	1	2	3	4	5	6	7	8	9
Zone 1	30	45	60	70	79	90	98	105	100	90
Zone 2	35	45	52	64	72	82	93	98	100	100
Zone 3	42	54	60	70	82	95	102	110	110	110

[MBA, Panjab Tech. Univ. 2002]

15.04. A cosmetics manufacturing company is interested in selecting the advertising media for its product and also the frequency of advertising in each media. The data collected over the past two years regarding the frequency of advertising in three media of newspaper, radio, and television. The related sales of the product give the following results:

Expected sales in thousands of rupees

Frequency/week	1	2	3	4
Television	220	275	325	350
Radio	150	250	300	320
Newspaper	100	175	225	250

The cost of advertising in newspaper is ₹ 500 per appearance, while in radio and in television it is ₹ 1,000 and ₹ 2,000 respectively. The budget provides ₹ 4,500 per week for advertisement. The problem is of determining the optimal combination of advertising media and advertising frequency. ***[M.Tech. (Nov.), Panjab Tech. Univ. 2012]***

15.05. An oil company has 8 units of money available for exploration of three sites. If oil is present at a site, the probability of finding it depends upon the amount allocated for exploring the site as follows:

Site	*Units of money allocated*								
	0	1	2	3	4	5	6	7	8
1	0	0.0	0.1	0.2	0.3	0.5	0.7	0.9	1.0
2	0	0.1	0.2	0.3	0.4	0.6	0.7	0.8	1.0
3	0	0.1	0.1	0.2	0.3	0.5	0.8	0.9	1.0

The probability that the oil exists at sites 1, 2 and 3 is 0.4, 0.3 and 0.2 respectively. Find the optimal allocation of money.

15.06. A company has decided to introduce a product in three phases. Phase-1 will feature making a special offer at a greatly reduced rate to attract the first-time buyers. Phase-2 will involve intensive advertising to persuade the buyers to continue purchasing at regular price. Phase-3 will involve a follow up advertising and promotion campaign.

A total of ₹ 5 million has been budgeted for this marketing campaign. If m is the market share captured in phase-1, fraction f_2 of m is retained in Phase-2, and fraction f_3 of market share in Phase-2 is retained in Phase-3. The expected values of m, f_2 and f_3 at different levels of money expected are given below. How should the money be allocated to the three phases to maximize the final share?

Money spent (million of ₹)	*m per cent*	*Effect on market share*	
		f_2	f_3
0	0	0.30	0.50
1	10	0.50	0.70
2	15	0.70	0.85
3	22	0.80	0.90
4	27	0.85	0.93
5	30	0.90	0.95

[*Hint:* The problem can be treated as a three-stage problem taking each phase as a stage and amount of money spent as the state of the system. Let

s_n = millions of rupees spent on marketing campaign at stage n.

x_n = amount of money allocated to Phase-n (stage n): $n = 1, 2, 3$.

$p_n(x_n)$ = market share captured in Phase-n; $n = 1\ 2, 3$.

If $f_n(x_n)$ is the value of optimum allocation for Phase-1 through Phase-3 (both inclusive), and $f_n(s_n, x_n)$ is the market share associated with the optimum solution $f_n{}^*(s_n)$, $n = 1, 2, 3$; then $f_n{}^*(s_n) = \max_{x_n}.\{p_n(x_n) \times f_{n-1}(s_n - x_n)\}$ is the recursive relation. Use the forward recursive relation to find the optimal solution starting with $f_1{}^*(s_1)$.]

15.07. The manufacturing process for a perishable product is such that the cost of changing the level of production from one month to the next month is thrice the square of the difference in production levels. Any production not sold by the end of the month is disposed of at a cost of ₹ 30 per unit. If the sales forecast given below must be met, determine a production schedule to minimize costs. Assume that the production of July was of 250 units:

Month	:	August	September	October
Sales Forecast	:	260	225	200

[*Hint:* Recursive equation is:

$$f_n(p) = \text{min.} \{3(x_j - p)^2 + 30(x_j - s) + f_{n-1}(x_j)\},$$

where, s = sales forecast for the current month,

x_j = production in the month j, j = month October, September, August and July, respectively.]

15.08. A ship is to be loaded with stock of 3 items. Each unit of item 'n' has a weight w_n and value v_n. The maximum cargo weight the ship can take is 5, and the details of the three items are as follows:

Item (n)	*Weight (w_n)*	*Value (v_n)*
1	2	7
2	3	10
3	1	3

Find the most valuable cargo load without exceeding the maximum cargo weight by using dynamic programming. ***[BE (Comp. Sc.) (Dec.), Panjab Univ. 2004]***

15.09. In a cargo loading problem, there are four items of different weights per unit and different value per unit, as given below:

Item	*Weight (kg./unit)*	*Value (rupees per unit)*
1	1	1
2	3	5
3	4	7
4	6	11

The maximum cargo load is restricted to 17 units. How many units of each item be loaded to maximize the value? ***[BE (Nov.), Panjab Univ. 2005]***

15.10. A truck can carry 20 tonnes of load. Four type of products are to be loaded. Their per unit weights in tonnes and value in thousands of rupees are given below:

Type	*Weight/unit*	*Value/unit*
A	1	20
B	2	30
C	3	50
D	4	85

At least one unit of each type of load is to be shipped. Determine the loading that will maximize the value of load.

15.11. A man is engaged in buying and selling identical items. He operates from a warehouse that can hold 500 items. Each month he can sell any quantity that he chooses to stock at the beginning of the month. Each month, he can buy as much as he wishes for delivery at the end so long as his stock does not exceed 500 items. For the next four months he has the following error-free forecasts of cost and sale prices:

Month	*1*	*2*	*3*	*4*
Cost (in ₹)	27	24	26	28
Sales price (in ₹)	28	25	25	27

If he currently has a stock of 200 units, what quantities should be sell and buy in the next four months? Find the solution using dynamic programming. ***[M.Sc. (Math), CCS Univ. 2004]***

[*Hint:* Analyse the problem by treating four months as four stages. Let W be the warehouse capacity (which is 500), and let

x_i = amount to be sold during the month i,
y_i = amount to be ordered during the month i,
b_i = stock level in the beginning of month i,
p_i = sales price in the month i,
c_i = purchase price in the ith month.

Solve the problem starting with 4th month and then proceed backward. If $f_n(b_n)$ is the maximum return when there are n more months to follow and the initial stock at the beginning of month 'n' is b_n, then 'n' varies from 1 to 4 as the months vary from 4th to 1st.

n :	4	3	2	1
Month :	*I*	*II*	*III*	*IV*

Recurrence equation for this problem is given by

$$f_1(b_n) = \max_{x_n,\, y_n}. [p_1 x_n - c_1 y_n], \text{ where } b_n \geq x_n \text{ and } b_n - x_n + y_n \leq H$$

and
$$f_n(b_n) = \max_{x_n,\, y_n}. [p_n x_n - c_n y_n + f_{n-1}(b_n - x_n + y_n).]$$

15:5. OPTIMAL SUB-DIVISION PROBLEM

When a given quantity is divided into a number of parts, then the maximization of the product of these parts or minimization of the sum of these parts subject to given constraints (if any) is called optimal sub-division problem.

Example 15.7 (*Optimal Sub-division Problem*). *Divide a positive quantity 'c' into 'n' parts in such a way that their products is maximum.* **[*M.Sc. (Math.), CCS Univ. 2010; B.Tech. (Nov.), Hyderabad Univ. 2010*]**

Or

$$\text{Maximize } z = y_1 \cdot y_2 \cdot \ldots \cdot y_n \text{ subject to the constraints:}$$
$$y_1 + y_2 + \ldots + y_n = c \text{ and } y_j \geq 0; \quad j = 1, 2, \ldots, n$$

Solution. Let y_j be the jth part of the positive quantity c $(j = 1, 2, \ldots, n)$, then each j corresponding to part y_j may be regarded as a stage. Now, since y_j may assume any non-negative value satisfying the constraint

$$y_1 + y_2 + \ldots + y_n = c,$$

the alternatives at each stage are infinite. This means that y_j may be considered to be continuous.

Let $f_n(c)$ denote the maximum attainable product when the quantity c is divided into n parts.

If we regard c as a fixed quantity and n as the number of stages, which varies over positive integers, then a recursive equation connecting $f_n(c)$ and $f_{n-1}(c)$ is

$$f_n(c) = \max_{0 < x \leq c}. \{x \cdot f_{n-1}(c - x)\}$$

For $n = 1$ (*i.e.*, one-stage problem), we write

$$f_1(c) = \max_{y_1 = c}. \{y_1\} = c \text{ (initially true)}$$

For $n = 2$ (*i.e.*, two-stage problem), the quantity c is divided into two parts, say $y_1 = x$ and $y_2 = c - x$. Then

$$f_2(c) = \max. \{y_1 \cdot y_2\} = \max_{0 < x \leq c}. \{x \cdot (c - x)\}$$
$$= \max_{0 < x \leq c}. \{x \cdot f_1(c - x)\}, \text{ since } f_1(c - x) = c - x$$

Similarly, for $n = 3$, the quantity c is divided into three parts, given the initial choice of x which leaves $c - x$ to be further divided into two parts. Denote the maximum possible product for $(c - x)$ into two parts by $f_2(c - x)$. Then, using the principle of optimality, we have

$$f_3(c) = \max_{0 < x \le c} . \{x \cdot f_2(c - x)\}$$

Continuing in a similar manner, the recursive equation for general value is given by

$$f_n(c) = \max_{0 < x \le c} . \{x \cdot f_{n-1}(c - x)\}$$

We now solve the recurrence equation formulated above.

For $n = 2$, the function $x \cdot (c - x)$ attains its maximum value at $x = c/2$ satisfying the condition $0 < x \le c$. Thus

$$f_2(c) = \frac{c}{2}\left(c - \frac{c}{2}\right) = \left(\frac{c}{2}\right)^2$$

∴ The optimal policy is $(c/2, c/2)$ and $f_2(c) = (c/2)^2$.

For $n = 3$, $$f_3(c) = \max_{0 < x \le c} . \left\{x \cdot \left(\frac{c - x}{2}\right)^2\right\}, \text{ since } f_2(c - x) = \left(\frac{c - x}{2}\right)^2$$

Now, since the maximum value of $x\left(\frac{c - x}{2}\right)^2$ is attained for $x = c/3$ satisfying the condition $0 < x \le c$; therefore

$$f_3(c) = \left\{\frac{c}{3} \cdot \frac{1}{4}\left(c - \frac{c}{3}\right)^2\right\} = \left(\frac{c}{3}\right)^3$$

Thus, for $n = 3$, we have

Optimal policy: $\left(\frac{c}{3}, \frac{c}{3}, \frac{c}{3}\right)$ and $f_3(c) = \left(\frac{c}{3}\right)^3$

In general, for n-stage problem, we assume that

Optimal policy: $\left(\frac{c}{n}, \frac{c}{n}, \ldots, \frac{c}{n}\right)$ and $f_n(c) = \left(\frac{c}{3}\right)^n$ for $n = 1, 2, \ldots, m$.

Now, for $n = m + 1$, the recursive equation is

$$f_{m+1}(c) = \max_{0 < x \le c} . \{x \cdot f_m(c - x) = \max_{0 < x \le c} . \left\{x \cdot \left(\frac{c - x}{m}\right)^m\right\}$$

$$= \left(\frac{c}{m + 1}\right)^{m+1}$$

as the maximum value of $x\left(\frac{c - x}{m}\right)^m$ is attained at $x = \frac{c}{m + 1}$, *i.e.*, the result is also true for $n = m + 1$.

Hence, by mathematical induction, the optimal policy is

$$\left(\frac{c}{n}, \frac{c}{n}, \ldots, \frac{c}{n}\right) \text{ and } f_n^0(c) = \left(\frac{c}{n}\right)^n$$

Example 15.8. *Use dynamic programming to show that*

$$z = p_1 \log p_1 + p_2 \log p_2 + \ldots + p_n \log p_n \quad \textit{subject to the constraints:}$$

$$p_1 + p_2 + \ldots + p_n = 1 \text{ and } p_j \geq 0 \;(j = 1, 2, \ldots, n)$$

is a minimum when $p_1 = p_2 = \ldots = p_n = 1/n$. **[M.Sc. (Math.), CCS Univ. 2000]**

Solution. The problem here is to divide unity into n parts so as to minimize the quantity $\sum_i p_i \log p_i$.

Let $f_n(1)$ denote the minimum attainable sum of $p_i \log p_i$, $(i = 1, 2, \ldots, n)$.

For $n = 1$ (stage 1), we have

$$f_1(1) = \min_{0 < x \leq 1} \{p_1 \log p_1\} = 1 \log 1$$

as unit is divided only into $p_1 = 1$ part.

For $n = 2$, the unity is divided into two parts p_1 and p_2, such that $p_1 + p_2 = 1$

If $p_1 = x$ and $p_2 = 1 - x$, then

$$f_2(1) = \min_{0 < x \leq 1} \{p_1 \log p_1 + p_2 \log p_2\}$$

$$= \min_{0 < x \leq 1} \{x \log x + (1 - x) \log (1 - x)\}$$

$$= \min_{0 < x \leq 1} \{x \log x + f_1(1 - x)\}$$

In general, for an n-stage problem, the recursive equation is

$$f_n(1) = \min_{0 < x \leq 1} \{p_1 \log p_1 + p_2 \log p_2 + \ldots + p_n \log p_n\}$$

$$= \min_{0 < x \leq 1} \{x \log x + f_{n-1}(1 - x)\}$$

We now solve this recursive equation.

For $n = 2$ (stage 2), the function $x \log x + (1 - x) \log (1 - x)$ attains its minimum value at $x = 1/2$ satisfying the condition $0 < x \leq 1$. Thus,

$$f_2(1) = \frac{1}{2} \log \frac{1}{2} + \left(1 - \frac{1}{2}\right) \log \left(1 - \frac{1}{2}\right) = 2\left(\frac{1}{2} \log \frac{1}{2}\right)$$

Similarly, for stage 3, the minimum value of the recursive equation is obtained as

$$f_3(1) = \min_{0 < x \leq 1} \{x \log x + f_2(1 - x)\}$$

$$= \min_{0 < x \leq 1} \left\{x \log x + 2\left(\frac{1-x}{2}\right) \log \left(\frac{1-x}{2}\right)\right\}$$

Now, since the minimum value of

$$x \log x + 2\left(\frac{1-x}{2}\right) \log \left(\frac{1-x}{2}\right)$$

is attained at $x = 1/3$ satisfying $0 < x \leq 1$. Therefore, we have

$$f_3(1) = \frac{1}{3} \log \frac{1}{3} + 2\left(\frac{1}{3} \log \frac{1}{3}\right) = 3\left(\frac{1}{3} \log \frac{1}{3}\right)$$

$\therefore$ Optimal policy is: $\quad p_1 = p_2 = p_3 = \dfrac{1}{3}$

In general, for n-stage problem we assume that

Optimal policy: $\quad p_1 = p_2 = \ldots = p_n = \dfrac{1}{n}$ and $f_n^0(1) = n\left\{\dfrac{1}{n} \log \dfrac{1}{n}\right\}$

This can be shown easily using mathematical induction.

For $n = m + 1$, the recursive equation is

$$f_{m+1}(1) = \min_{0<x\le 1}. \{x \log x + f_m(1-x)\}$$

$$= \min_{0<x\le 1}. \left[x \log x + m\left\{\frac{1-x}{m}\log\left(\frac{1-x}{m}\right)\right\}\right]$$

$$= \frac{1}{m+1}\log\frac{1}{m+1} + m\left\{\frac{1}{m+1}\log\frac{1}{m+1}\right\}$$

$$= (m+1)\left\{\frac{1}{m+1}\log\frac{1}{m+1}\right\},$$

since minimum of $x \log x + \frac{1-x}{m}\log\frac{1-x}{m}$ is attained at $x = \frac{1}{m+1}$

Hence, the required optimal policy is

$$\left(\frac{1}{n}, \frac{1}{n}, \ldots, \frac{1}{n}\right) \text{ with } f_n^0(1) = n\left(\frac{1}{n}\log\frac{1}{n}\right)$$

Example 15.9. *Use dynamic programming to solve the following problem:*

$$\text{Minimize } z = y_1^2 + y_2^2 + y_3^2 \text{ subject to the constraints:}$$

$$y_1 + y_2 + y_3 \ge 15 \text{ and } y_1, y_2, y_3 \ge 0.$$

[B.Tech., Hyderabad Univ. 2011; MBA, Panjab Tech. Univ. 2002]

Solution. Since, the decision variables are y_1, y_2 and y_3, the given problem is a three-stage problem defined as follows:

$$s_3 = y_1 + y_2 + y_3 \ge 15, \quad s_2 = y_1 + y_2 = s_3 - y_3 \text{ and } s_1 = y_1 = s_2 - y_2$$

Therefore, the function (recurrence) relation is

$$f_1(s_1) = \min_{0<y_1\le s_1}. \; y_1^2 = (s_2 - y_2)^2$$

$$f_2(s_2) = \min_{0<y_2\le s_2}. \left\{y_1^2 + y_2^2\right\} = \min_{0<y_2\le s_2}. \left\{y_2^2 + f_1(s_1)\right\}$$

and

$$f_3(s_3) = \min_{0<y_3\le s_3}. \left\{y_1^2 + y_2^2 + y_3^2\right\} = \min_{0<y_3\le s_3}. \left\{y_3^2 + f_2(s_2)\right\}$$

$\therefore$

$$f_2(s_2) = \min_{0<y_2\le s_2}. \left\{y_2^2 + (s_2 - y_2)^2\right\}$$

$$= \left(\frac{1}{2}s_2\right)^2 + \left(s_2 - \frac{1}{2}s_2\right)^2 = \frac{1}{2}s_2^2$$

since the function $y_2^2 + (s_2 - y_2)^2$ attains its minimum value at $y_2 = \frac{1}{2}s_2$

Again

$$f_3(s_3) = \min_{0<y_3\le s_3}. \left\{y_3^2 + f_2(s_2)\right\} = \min_{0<y_3\le s_3}. \left\{y_3^2 + \frac{1}{2}(s_3 - y_3)^2\right\}$$

or

$$f_3(15) = \min_{0<y_3\le 15}. \left\{y_3^2 + \frac{1}{2}(15 - y_3)^2\right\}, \text{ since } s_3 = (y_1 + y_2 + y_3) \ge 15$$

Since the minimum value of the function $y_3^2 + \frac{1}{2}(15 - y_3)^2$ occurs at $y_3 = 5$; we have

$$f_3(15) = \left\{ 5^2 + \frac{1}{2}(15-5)^2 \right\} = 75$$

Thus,
$$s_3 = 15 \Rightarrow y_3^o = 5$$
$$s_2 = s_3 - y_3 = 15 - 5 = 10 \Rightarrow y_2^o = \frac{1}{2} s_2 = 5$$
$$s_1 = s_2 - y_2 = 10 - 5 = 5 \Rightarrow y_1^o = s_1 = 5$$

Hence, the optimal policy is

$$(5, 5, 5) \text{ with } f_3^o(15) = 75$$

PROBLEMS

Use dynamic programming to find the value of

15.12. Maximize $z = y_1 \cdot y_2 \cdot y_3$ subject to the constraints:
$$y_1 + y_2 + y_3 = 5, \; y_1, y_2, y_3 \geq 0.$$
[BE (Mech.), Nagpur Univ. 2011; B.Tech., Hyderabad Univ. 2011]

15.13. Minimize $z = y_1 + y_2 + y_3 + \ldots + y_n$ subject to the constraints:
$$y_1 \cdot y_2 \ldots y_n = d; \; y_j \geq 0 \text{ for } j = 1, 2, \ldots, n.$$
[B.Sc., Bharthiar Univ. 2011]

15.14. Use the principle of optimality to find the maximum value of
$$z = b_1x_1 + b_2x_2 + b_3x_3 + \ldots + b_nx_n,$$
when $x_1 + x_2 + x_3 + \ldots + x_n = c$

and $x_1, x_2, x_3, \ldots, x_n \geq 0.$ *[M.Sc. (Math.), CCS Univ. 2009]*

15.15. Find the minimum value of
$$z = y_1^2 + y_2^2 + \ldots + y_n^2 \text{ subject to the constraints:}$$
$$y_1 \cdot y_2 \cdot \ldots \cdot y_n = c,$$
where $y_j \geq 0$ for $j = 1, 2, \ldots, n.$ *[BE (TIT), Panjab Univ. 2006]*

15.16. Use dynamic programming to solve:

(*a*) Minimize $z = x_1^2 + x_2^2 + \ldots + x_{10}^2$ subject to the constraints:
$$x_1 \cdot x_2 \cdot \ldots \cdot x_{10} = 8$$
$$x_1, x_2, \ldots, x_{10} \geq 0$$

(*b*) Minimize $z = x_1^3 + x_2^3 + x_3^3$ subject to the constraints:
$$x_1 \cdot x_2 \cdot x_3 = 27 \text{ and } x_i \geq 0, i = 1, 2, 3.$$
[ME (Structural), Madras Univ. 2000]

15.17. Use dynamic programming to solve:

Minimize $z = y_1^2 + y_2^2 + y_3^2$ subject to the constraints:
$$y_1 + y_2 + y_3 = 10 \text{ and}$$
(*a*) y_1, y_2 and y_3 are non-negative,

(*b*) y_1, y_2 and y_3 are non-negative integers.

15.18. Find the minimum value of

$$z = x_1^2 + 2x_2^2 + 4x_3 \quad \text{subject to the constraints:}$$
$$x_1 + 2x_2 + x_3 \geq 8; \quad x_1, x_2, x_3 \geq 0.$$

15.19. Find the maximum value of

$$z = x_1^2 + 2x_2^2 + 4x_3 \quad \text{subject to the constraints:}$$
$$x_1 + 2x_2 + x_3 \leq 8; \quad x_1, x_2, x_3 \geq 0.$$

[M.Sc. (Math.), CCS Univ. 1999]

15.20. Find the maximum value of

$$z = -x_1^2 - 2x_2^2 + 3x_2 + x_3 \quad \text{subject to the constraints:}$$
$$x_1 + x_2 + x_3 \leq 1; \quad x_1, x_2, x_3 \geq 0.$$

15:6. SOLVING LLP BY DYNAMIC PROGRAMMING

Dynamic programming can be used for solving linear programming problems as well, although only relatively small-sized problems can be handled this way. Even non-linear programming problems can also be handled by this method.

Consider the linear programming problem:

$$\text{Maximize} \quad z = c_1 x_1 + c_2 x_2 \quad \text{subject to the constraints:}$$
$$a_{11} x_1 + a_{12} x_2 \leq B_1,$$

and
$$a_{21} x_1 + a_{22} x_2 \leq B_2,$$

with
$$x_1 \geq 0 \text{ and } x_2 \geq 0.$$

In this problem, there are two inter-related decisions to be made. That is, the problem is to determine level of activity x_1 and activity x_2. In the context of dynamic programming, the two can be interpreted as two stages. Thus, we let j ($j = 1, 2$) as the activity. Accordingly, x_j represents the decision variable at stage j. The states, here, are the total amount of resources available for allocation to the activities remaining at any stage. Accordingly, $S_j = (B_1, B_2)$; where B_i $(i = 1, 2)$ is the amount of resource i remaining to be allocated. Therefore, $S_1 = (B_1, B_2)$ and $S_2 = (B_1 - a_1 x_1, B_2 - a_{21} x_1)$. However, when we begin to solve at stage 2, we do not know the value of x_1. Hence, we begin the solution by taking $S_2 = (B_1, B_2)$ at that stage.

We may define the payoff function as follows:

$$f_2 (B_1, B_2; x_2) = c_2 x_2,$$

and
$$f_1 (b_1, b_2; x_1) = c_1 x_1 + \max. (c_2 x_2)$$

with
$$a_{12} x_2 \leq B_1 - a_{11} x_1 \text{ and } a_{22} x_2 \leq B_2 - a_{21} x_1;$$

where,
$$x_1 \geq 0 \text{ and } x_2 \geq 0.$$

The optimal payoff at stage j ($j = 1, 2$) is:

$$f_j^* (B_1, B_2) = \max_{x_j}. f_j (B_1, B_2, x_j)$$

Therefore,
$$f_2^* (B_1, B_2) = \max_{x_2 \geq 0}. \{c_2 x_2\},$$

and
$$f_1^* (B_1, B_2) = \max_{x_1 \geq 0}. \{c_1 x_1 + f_2^* (B_1 - a_{11} x_1, B_2 - a_{21} x_1)\}$$

Remark. For the general linear programming problem, let $f_n (B_1, B_2, ..., B_m)$ be the maximum value for stages $x_1, x_2, ..., x_m$ for states $B_1, B_2, ..., B_m$. Then,

$$f_j(B_1, B_2, \ldots, B_m) = \max_{x_j \geq 0}. \{c_j x_j + f_{j-1}^*(B_1 - a_{1j} x_j, B_2 - a_{2j} x_j, \ldots, B_m - a_{mj} x_j)\}$$

Example 15.10. *Use dynamic programming to solve the following LPP:*

$$\text{Maximize } z = 3x_1 + 5x_2 \text{ subject to the constraints:}$$
$$x_1 \le 4,\ x_2 \le 6,\ 3x_1 + 2x_2 \le 18$$

and $x_1, x_2 \ge 0$. **[B.Tech. (Nov.), Hyderabad Univ. 2010]**

Solution. The problem consists of three resources and two decision variables. The states of the equivalent dynamic programming problem, therefore, are (B_1, B_2, B_3). Thus,

$$f_2(B_1, B_2, B_3) = \max.\ \{5x_2\},$$

where maximum is taken over $0 < x_2 \le 6$ and $0 < 2x_2 \le 18$.

i.e.,
$$f_2(B_1, B_2, B_3) = 5 \times \max.\ \{x_2\}$$
$$= 5 \times \min.\left\{6, \frac{18}{2}\right\} = 5 \times 6 = 30,$$

since maximum of x_2 satisfying the restrictions $x_2 \le 6$ and $x_2 \le \frac{18}{2}$ is the minimum of 6 and $\frac{18}{20}$.

$$\therefore \quad x_2^* = \min.\left\{6, \frac{18}{2}\right\} = 6$$

Now, $$f_1(B_1, B_2, B_3) = \max.\ \{3x_1 + f_2(B_1 - x_1,\ B_2 - 0,\ B_3 - 3x_1)\},$$

where maximum is taken over $0 < x_1 \le 4$ and $0 < 3x_1 \le 18$ or simply $0 < x_1 \le 4$, since then obviously $x_1 \le 6$, *i.e.*, $f_1(4, 6, 18) = \max.\ \{3x_1 + f_2(4 - x_1, 18 - 3x_1)\}$

$$= \max.\left\{3x_1 + 5 \times \min.\left(6, \frac{18 - 3x_1}{2}\right)\right\}$$

Now, $$\min.\left(6, \frac{18 - 3x_1}{2}\right) = \begin{cases} 6, & \text{if } 0 < x_1 \le 2 \\ \dfrac{18 - 3x_1}{2}, & \text{if } 2 < x_1 \le 4 \end{cases}$$

$$\therefore \quad 3x_1 + 5 \times \min.\left(6, \frac{18 - 3x_1}{2}\right) = \begin{cases} 3x_1 + 30, & \text{if } 0 < x_1 \le 2 \\ 3x_1 + 5 \times \left(9 - \frac{3}{2}x_1\right), & \text{if } 2 < x_1 \le 4 \end{cases}$$

Since, the maximum of both $3x_1 + 30$ and $3x_1 + 5\left(9 - \frac{3x_1}{2}\right)$ occurs at $x_1 = 2$, we have

$$f_1(4, 6, 18) = 3 \times 2 + 5 \times \min.\left(6, \frac{18 - 3x_1}{2}\right) = 6 + 30 = 36$$

$$\therefore \quad x_2^* = \min.\left\{6, \frac{18 - 3x_1^*}{2}\right\} = \min.(6, 6) = 6$$

Hence, the optimum solution is

$$x_1^* = 2,\ x_2^* = 6 \text{ and maximum } z = 36$$

Example 15.11. *Solve the following LPP by dynamic programming approach:*

$$\text{Maximize } z = 2x_1 + 5x_2 \text{ subject to the constraints:}$$
$$2x_1 + x_2 \le 43, \text{ and } 2x_2 \le 46,$$
$$x_1 \ge 0 \text{ and } x_2 \ge 0.$$ **[B.Tech. (Nov.), Hyderabad Univ. 2010]**

Solution. The problem consists of two resources and two decision variables. The states of the equivalent dynamic programming, therefore, are B_1 and B_2. Thus,

$$f_2(B_1, B_2) = \max. \{5x_2\},$$

where maximum is taken over $0 < x_2 \le 43$ and $0 < x_2 \le \frac{46}{2}$,

i.e., $$f_2(B_1, B_2) = 5 \times \max. (x_2\}$$
$$= 5 \times \min. \{43, 46/2\},$$

since $2x_2 \le 46 \Rightarrow x_2 \le 23$ and the maximum of x_2 satisfying the restrictions $x_2 \le 43$ and $x_2 \le 23$ is the minimum of 43 and 23.

$\therefore$ $$x_2^* = \min. \{43, 23\} = 23$$

Now, $$f_1(B_1, B_2) = \max. \{2x_1 + f_2(B_1 - 2x_1, B_1 - 0)\};$$

where maximum is taken over $0 < x_1 \le 43/2$.

As it is the last stage, we substitute the values of B_1 and B_2 as 43 and 46 respectively.

$\therefore$ $$f_1(43, 46) = \max. \{2x_1 + 5 \times \min. (43 - 2x_1, 46/2)\}$$

Now, $$\min.\left(43 - 2x_1, \frac{46}{2}\right) = \begin{cases} 23, & \text{if } 0 < x_1 \le 10 \\ 43 - 2x_1, & \text{if } 10 < x_1 \le 43/2 \end{cases}$$

$\therefore$ $$2x_1 + 5 \times \min.\left(43 - 2x_1, \frac{46}{2}\right) = \begin{cases} 2x_1 + 115, & \text{if } 0 < x_1 \le 10 \\ 215 - 8x_1, & \text{if } 10 < x_1 \le 43/2 \end{cases}$$

Since, the maximum of both $2x_1 + 115$ and $215 - 8x_1$ occurs at $x_1 = 10$, we have

$$f_1(43, 46) = 2x_1 + 5 \times \min. \{43 - 2x_1, 46/2\}$$
$$= 2 \times 10 + 5 \times \min. (43 - 2 \times 10, 46/2) = 135 \text{ and } x_1^* = 10.$$

$\therefore$ $$x_2^* = \min. \{43 - 2x_1^*, 46/2\} = \min. \{23, 23\} = 23$$

Hence, the optimum solution is

$$x_1^* = 10, x_2^* = 23 \quad \text{and} \quad \text{maximum} \quad z = 135$$

PROBLEMS

Use dynamic programming procedure to solve the following linear programming problems:

15.21. Maximize $z = 3x_1 + 7x_2$ subject to the constraints:

$x_1 + 4x_2 \le 8, x_2 \le 2,$ and $x_1 \ge 0, x_2 \ge 0$ ***[M.Sc. (Math.), Kerala Univ. 2001]***

15.22. Maximize $z = x_1 + 9x_2$ subject to the constraints:

$2x_1 + x_2 \le 25, x_2 \le 11,$ and $x_1 \ge 0, x_2 \ge 0$

15.23. Maximize $z = 3x_1 + 4x_2$ subject to the constraints:

$2x_1 + x_2 \le 40, 2x_1 + 5x_2 \le 180,$ and $x_1 \ge 0, x_2 \ge 0$

[BE (Prod.), Panjab Univ. 2001]

15.24. Maximize $z = 2x_1 + 4x_2$ subject to the constraints:

$2x_1 + 3x_2 \le 48, x_1 + 3x_2 \le 42, x_1 + x_2 \le 21$

and $x_1 \ge 0, x_2 \ge 0$ ***[B.Tech., Hyderabad Univ. 2011]***

15.25. Maximize $z = 8x_1 + 16x_2$ subject to the constraints:

$x_1 + x_2 \le 200, x_2 \le 125, 3x_1 + 6x_2 \le 900$

and $x_1 \ge 0, x_2 \ge 0$

15.26. Maximize $z = 10x_1 + 30x_2$ subject to the constraints:

$3x_1 + 6x_2 \le 168, 12x_2 \le 240; x_1 \ge 0$ and $x_2 \ge 0$

[BE (Mech.) (Sept.), Nagpur Univ. 2010; B.Tech., Hyderabad Univ. 2009]

TEST YOUR UNDERSTANDING

1. Dynamic programming deals with the
   (*a*) multi-stage decision making problems.
   (*b*) single-stage decision making problem.
   (*c*) time-independent decision making problem.
   (*d*) problems which fix the levels of different variables so as to maximize profit or minimize cost.
2. Dynamic programming approach
   (*a*) helps in reducing the computational effort in sequential decision making.
   (*b*) is based on the principle of optimality due to Bellman.
   (*c*) divides the given problem into a sequence of smaller sub-problems called the stages.
   (*d*) all of the above.
3. Dynamic programming problem
   (*a*) cannot be dealt with non-linear constraints.
   (*b*) can be solved by simplex method.
   (*c*) is solved starting from the initial stage to the next till the final stage is reached.
   (*d*) can be solved by using Monte-Carlo method.
4. Dynamic programming problem
   (*a*) can be solved only by recursive equation approach.
   (*b*) cannot be solved using recursive equation approach.
   (*c*) does not suggest an optimum solution.
   (*d*) under certainty involves those problems whose conditions at each stage are known with certainty.
5. The solution of recursive equation in dynamic programming does not involve the
   (*a*) backward computational procedure or forward computational procedure.
   (*b*) two types of computations, according as the system is continuous or discrete.
   (*c*) number of stages that provides an optimum solution or there is an indication of an unbounded solution.
   (*d*) classical methods of optimization or tabular computational scheme to achieve the optimum solution.
6. The payoff function in dynamic programming depends on
   (*a*) stages (*b*) states
   (*c*) alternatives (*d*) all of the above.
7. The payoff function to optimal sub-divide a positive quantity c into n parts is written as
   (*a*) max. $\{x \cdot f_n(c + x)\}$ (*b*) max. $\{x \cdot f_{n-1}(c - x)\}$
   (*c*) max. $\{x \cdot f_n(c - x)\}$ (*d*) max. $\{x \cdot f_{n-1}(c + x)\}$.
8. A stage in a dynamic programming problem represents
   (*a*) status of the system at a particular state. (*b*) number of decision alternatives.
   (*c*) different time periods in the planning period. (*d*) all of the above.

ANSWERS

1. (*a*)	2. (*d*)	3. (*c*)	4. (*d*)	5. (*c*)
6. (*b*)	7. (*b*)	8. (*c*).		

REVIEW EXERCISE

1. Differentiate between linear programming and dynamic programming.
[B.Tech. (Mech.), Hyderabad Univ. 2012; MBA (Dec.), Panjab Tech. Univ. 2010]
2. What is the need of dynamic programming and how is it different from linear programming problem? *[MBA, Panjab Tech. Univ. 2009]*
3. State and explain Bellman's principle of optimality. *[MBA, Delhi Univ. 2009]*
4. Discuss very briefly some of the business problems that are typically solved by dynamic programming. *[MBA (PT.), Delhi Univ. 2008]*
5. Explain dynamic programming problem. *[MBA (Nov.), Delhi Univ. 2007]*
6. State the "Principle of Optimality" in dynamic programming and give a mathematical formulation of a dynamic programming problem. *[MBA (Nov.), Delhi Univ. 2004]*
7. Explain the following in the context of dynamic programming:
(*a*) Principle of optimality, (*b*) State, (*c*) Stage. *[BE (Dec.), Gujarat Tech. Univ. 2012]*
8. What is recursive relationship in dynamic programming? Describe the general process of backward procedure in dynamic programming? *[MBA, Panjab Tech. Univ. 2003]*
9. Write short note on characteristics of dynamic programming.
[MBA (Dec.), Panjab Tech. Univ. 2012]
10. Write step by step procedure to solve a general problem by dynamic programming approach.
[B.Tech. (Mech.), Panjab Technical Univ. 2012]
11. What do you understand by forward and backward algorithms?
[B.Tech., Hyderabad Univ. 2011]
12. Explain the general algorithm for solving a dynamic programming problem.
[B.Sc., Bharthiar Univ. 2011]
13. What are the management applications of dynamic programming?
[MBA (Dec.), Panjab Tech. Univ. 2008]
14. Dynamic model is representation of the behaviour of real problem.
[PG Dip., Indian Inst. of Mat. Management 2018]

CHAPTER 16

Network Routing Problems

"Without logistics network we could not have present day goods services at affordable prices"

One of the earliest problem human being faced was of the proper route to reach from one place to another and hence network routing problems. This chapter highlights the origin and growth of network flow problems.

16:1. ORIGIN OF THE NETWORK FLOW PROBLEMS

Around 1650 AD the following picture of the city of Königsberg, Prussia (now Kaliningrad, Russia) appeared in a book by M. Zeiller[1]. The city is on the banks of River Pregolya, and included two large islands.

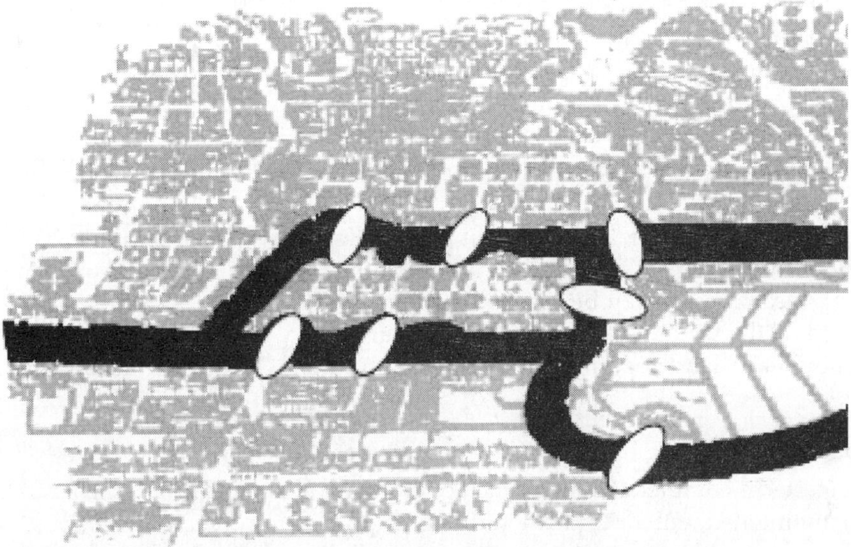

City of Königsberg

The picture revealed that the river that surrounds the islands was divided into two branches (mainlands). These branches were crossed (connected) by ***seven bridges*** as shown in the sketch (*Fig. 16.1*) on the next page. The city had a local sport to ***find a route*** that starts any place in the city, crosses each bridge once, ***and only once***, and finally reach back to the starting point. No one was able to find even one such route. In 1736, the famous mathematician Euler proved that such a route does not exist. His paper was the first that defined a ***network flow problem***. He formulated the general problem as follows (in his own words):

1. Reprinted in Biggs. N.L., et al., *Graph Theory 1736-1936*, Oxford University Press, 1976.

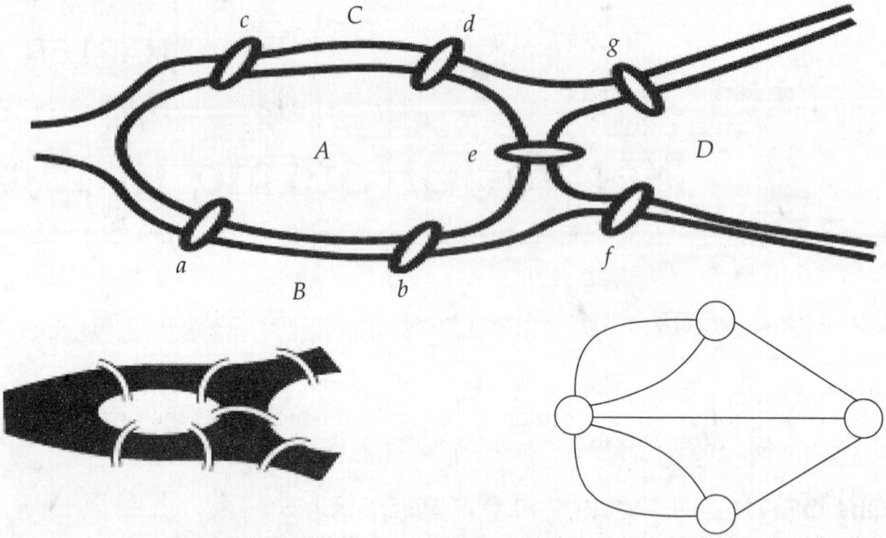

Fig. 16.1. The seven bridges of Königsberg

"whatever be the arrangement and division of the river into branches, however bridges there be, can one find out whether or not it is possible to cross each bridge exactly once?"

He denoted each region of land *A*, *B*, *C*, and *D* by a "point" (called a ***node*** in network flow problem), and each bridge *a*, *b*, *c*, *d*, *e*, *f*, and *g* connecting these regions by a "line", referred to as an ***edge***. With these notations the ***network*** for the problem of Königsberg bridges is shown in *Fig. 16.2*.

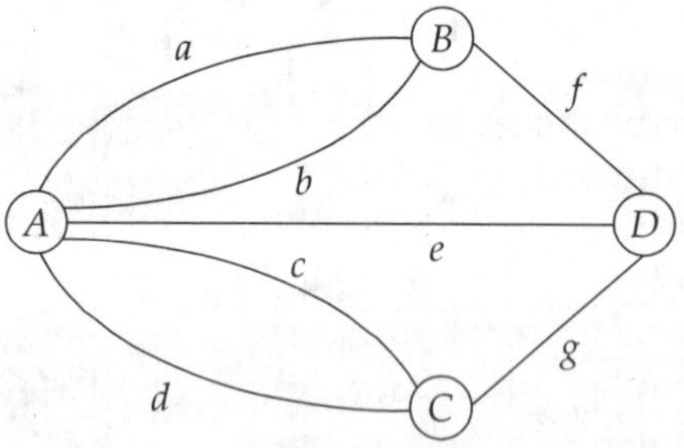

Fig. 16.2. Königsberg bridges network

Euler's proof concludes that for such a route (now called an ***Euler tour***) to exist, the network should be "connected" (*i.e.*, each region can be reached from every other by means of the bridges) and that the number of bridges touching each land region is seven. Certainly, this was not the case for the bridges of ***Königsberg***.

Chinese Postman Problem

Euler's work started a completely new branch of mathematics called ***graph theory***. Application of graph theory augmented with optimization theory to practical problems is the domain of network flow problems. A network flow problem related with ***Euler tours*** is the famous ***Chinese Postman Problem:*** "The distances of the bridges are known and we have to cross each bridge at least once. The problem is to find a route that comes back to the starting point with a minimum total distance traveled on the bridges".

Clearly, the Euler tour (if it exists) would be the optimal solution to the ***Chinese Postman Problem***. In other words the problem is to find a shortest path that visits every edge of a (connected) undirected network. When the graph of the given network does not satisfy the Euler's conditions, ***Chinese Postman Problem*** attempts to find a tour with minimal repetition of the edges. Many problems in trash collection, road sweeping, snow-plows, transmission line inspections, school bus routing, etc. can be modeled as ***Chinese Postman Problem***. It is so named as it was originally studied by the Chinese mathematician Mei Ko Kuan in 1962.

16:2. NETWORK FLOW PROBLEMS

The ***Chinese Postman Problem*** is only one among many network flow problems. For example, networks for information transmission, flow of water through pipes, and the distribution of goods and energy are part of our everyday life. Just think how our mobility is made possible by the highway, rail, and airline networks. Without effective distribution and logistics network, we could not have all the goods and services that are available now at affordable prices.

Features of Network Flow Models

The following three features are mainly responsible for the widespread use of network flow models in practice:

Visual Content. Network models allow us to depict a problem graphically. The pictorial appeals of these network diagrams make the problems easily understood.

Model Flexibility. There is inherent flexibility in network models. For example, models having similar structures can be used to analyze communication phenomena in sociology, to determine ancestral orderings in archaeology, to allocate library budgets to determine what type of publications to acquire, etc. in addition to the application areas mentioned earlier.

Solvability. Several computationally very efficient algorithms exist for solving most network flow problems. The special purpose network-based algorithms attain these efficiencies by exploiting the special structure inherent in the network flow problems.

There are a few "generic" network flow problems and a large number of problem types with variations that cater to the practical applications. In fact there is a multitude of OR situations that can be modeled as network flow problems. Some recent surveys report that as much as 75% of the real-world optimization problems can be represented by network flow models. We shall briefly discuss a few important general problem types, namely, the spanning tree problems, the shortest path problems, the maximum flow problems, and the minimum cost flow problems.

Network Notations and Definitions

Nodes, Links and Flows

A network consists of a set of ***points*** and a set of ***lines*** connecting different pairs of points. The points are called **nodes** and the lines are called **links** (or arcs or branches). We describe a network by the notation (N, L) where N is the set of nodes and L is the set of links. The links are labeled by naming the nodes at either end. To illustrate consider the following sets:

$$N = \{1, 2, 3, 4, 5\} \quad \text{and} \quad L = \{1–2, 1–3, 1–4, 2–3, 4–3, 3–5, 4–5, 5–4\}$$

The network (N, L) described by them is given in *Fig. 16.3*.

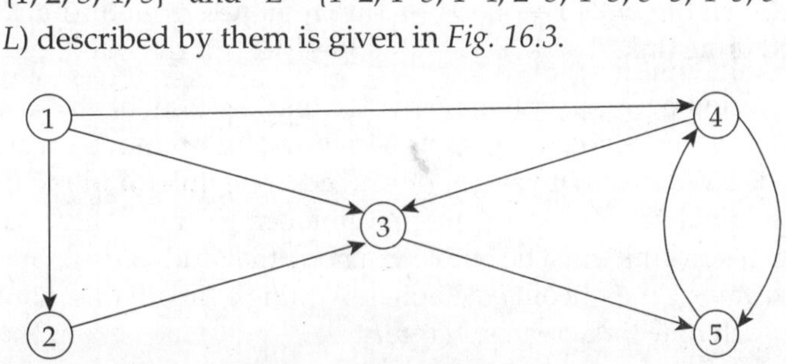

Fig. 16.3. Network (N, L)

The links in a network may have a flow of some type passing through them (*e.g.*, flow of traffic on highways). A link is said to be **directed**, if it allows positive flows in one direction and a ***zero*** flow in the opposite direction. For example, the link 1–2 in the network of *Fig. 16.3* has a flow in the direction indicated by 1–2 (*from* node 1 *to* node 2). On the other hand link 4–5 has flows in both the directions 4 →5 and 5 → 4. Clearly, the link 4–5 may be said to be ***undirected*** while all other links in the network are directed.

A network is to be a **directed network** if its entire links are directed. A mixture of directed and undirected links gives an undirected network as is the case in *Fig. 16.3.* It may be interesting to observe the following components of a typical network in real life:

Nodes	*Links*	*Flows*
Pumping Stations	Pipes	Oil, Gas or Water
Servers	Computer Networks	Data
Electric Joints	Wires	Electricity
Airports	Airlines	Airplanes
Office	Departments	Human Resource
Telephone Numbers	Calls made	Voice Data
Intersections	Roads	Traffic
Metro Stations	Rail Lines	Trains
Housing Pockets	TV Cable	TV Transmission
Factories	Distributors	Goods

Paths, Cycles and Trees

Path. A ***path*** is a sequence of ***distinct links*** that join two nodes through some other nodes. For example, there are ***four*** paths that connect node 1 to node 5 in the network of *Fig. 16.3,* namely: $1 \to 3 \to 5$, $1 \to 2 \to 3 \to 5$, $1 \to 4 \to 5$ and $1 \to 4 \to 3 \to 5$.

A **directed path** from node i to node j is a sequence of connecting links having flows towards node j. An ***undirected path*** may allow flows in either direction (towards node j or away from node j). The four path considered above are all directed paths connected node *1* and node *5*. On the other hand, the paths $2 \to 3 \to 1 \to 4$ is not a directed path. Nevertheless it happens to be an undirected path connecting node *2* to node *4*.

A path is said to form a **cycle** if it connects a node to itself through other nodes. For example the paths $4 \to 5 \to 4$ and $3 \to 5 \to 4 \to 3$ both from cycles. In fact both are ***directed cycles*** as they contain direct paths.

A **Connected network** is a network where every two distinct nodes are connected by at least one (directed) path. Note that the connecting path need not be directed even, if the network is directed. The network of *Fig. 16.3* is a connected network. It may be interesting to observe that it still remains connected even if the links $1 \to 4$ or $4 \to 3$ are removed from the network. However, it will cease to be connected if the links $1 \to 4$, $3 \to 4$ and $3 \to 5$ are all removed.

Tree. A ***tree*** is a connected network that may involve only a subset of all the nodes of the network, without having any cycle in between. To grow a tree we first remove (or hide) all the links and consider only the nodes. We start with at a node and add one link (or branch) to it by connecting it to a node from the original set of nodes. The first link can go anywhere to connect some pair of nodes. Thereafter, each new link must be between a node that is already connected (to other nodes) and a new node from among the unconnected ones. Adding a link in this manner avoids formation of a cycle. Note that each new link creates a larger tree.

A tree that links all the nodes of the network with no cycles allowed is called a **spanning tree**.

Link Capacity. The maximum amount of flow that can pass through a directed link is called a ***link capacity***. Both the spanning tree and the link capacity play an important role in network analysis.

Example 16.1. *Consider the network given in Fig. 16.4:*

Determine (a) two paths, (b) a cycle, (c) two directed cycles, (d) a tree, and (e) a spanning tree.

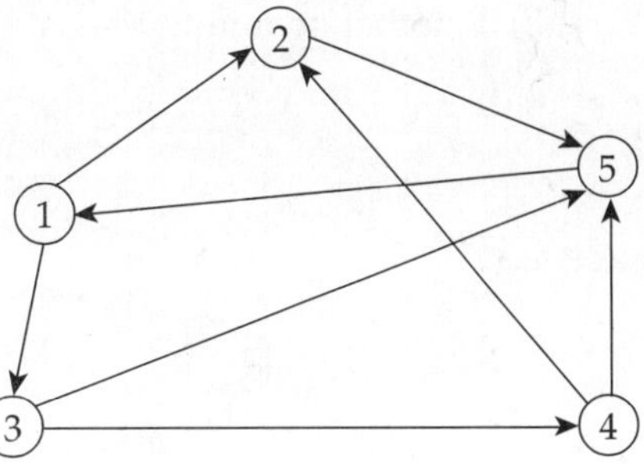

Fig. 16.4

Solution.

(*a*) Two paths are: (*i*) $1 \to 3 \to 4 \to 2$ and (*ii*) $1 \to 3 \to 4 \to 5$.

(*b*) The following (undirected) path forms a cycle: $1 \to 3 \to 4 \to 2 \to 1$.

(*c*) The following directed paths give directed cycles: $1 \to 3 \to 4 \to 5 \to 1$ and $1 \to 3 \to 5 \to 1$.

Note that $1 \to 2 \to 5 \to 1$ is another directed cycle.

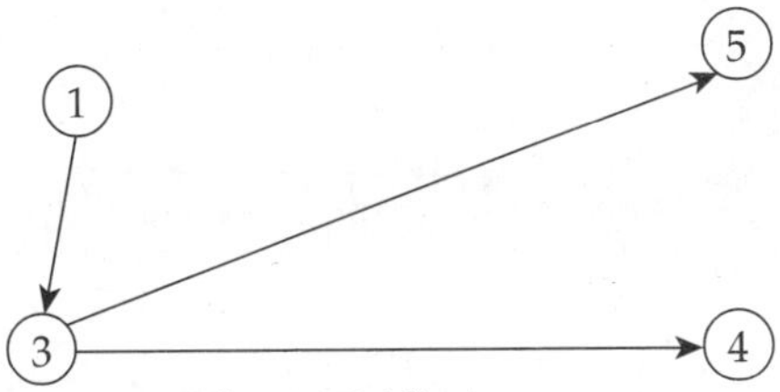

Fig. 16.4 (*d*) A tree

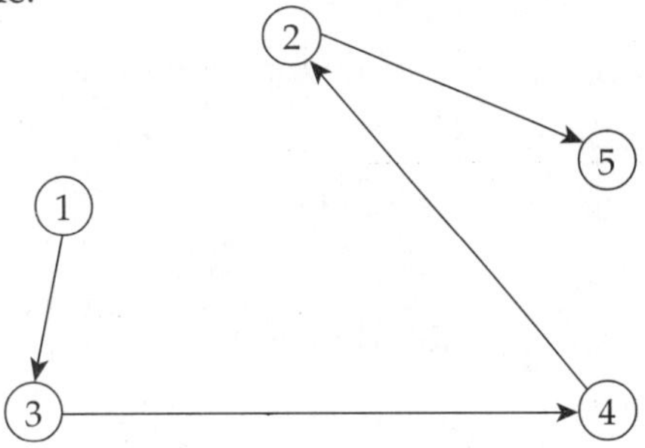

Fig. 16.4 (*e*) A Spanning Tree

Example 16.2. *For the directed network of* Fig. 16.5

(*a*) *Determine a directed path from node 1 to node 6. Identify three undirected paths connecting node 1 and node 6.*

(*b*) *Find three directed cycles.*

(*c*) *Grow a tree one link at a time until a spanning tree has been obtained.*

(*d*) *Repeat the process of (c) to obtain another spanning tree.*

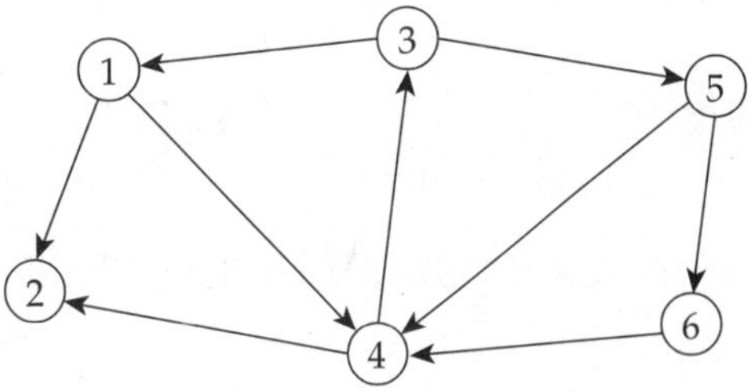

Fig. 16.5

Solution.

(*a*) A directed path from node *1* to node *6* is $1 \to 4 \to 3 \to 5 \to 6$. There are several undirected paths connecting node *1* and node *6*. Three of them are:

$1 \to 2 \to 4 \to 6$ $1 \to 4 \to 6$ and $1 \to 3 \to 5 \to 6$

(*b*) The following (directed) paths form directed cycles:

$1 \to 4 \to 3 \to 1$ $4 \to 3 \to 5 \to 4$ and $5 \to 4 \to 3 \to 5$

(*c*) Starting with node *1* we grow a tree one link at a time as follows:

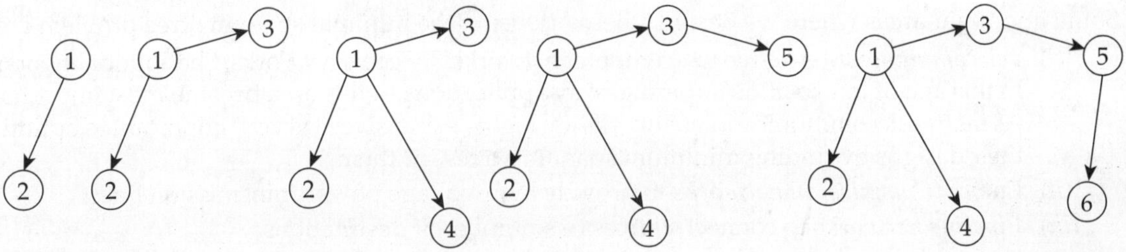

Fig. 16.6. Growing a Tree

(*d*) Another spanning tree is:

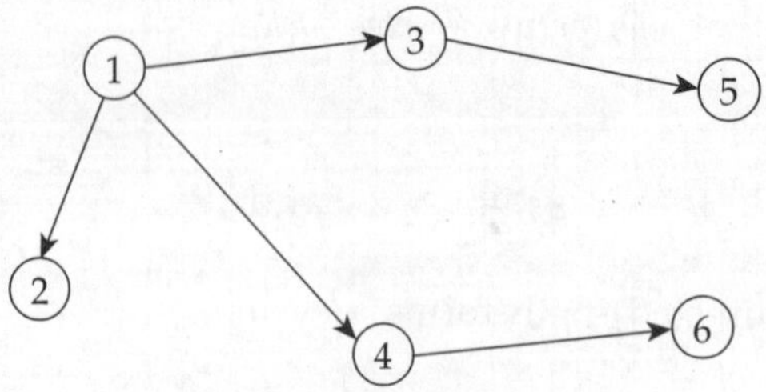

Fig. 16.7. A Spanning Tree

PROBLEMS

16.01. Consider the network as shown in *Fig. 16.8*:

Determine

(*a*) the sets *N* and *L*

(*b*) two paths

(*c*) an undirected cycle, a directed cycle

(*d*) a tree, and

(*e*) a spanning tree.

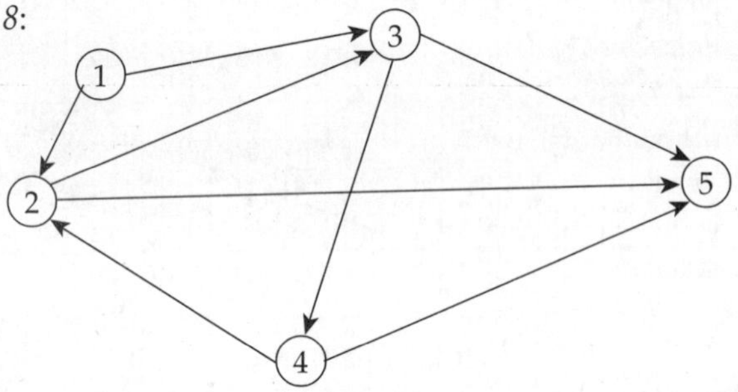

Fig. 16.8

16.02. Draw the network (*N*, *L*), where *N* and *L* are given by

$$N = \{1, 2, 3, 4, 5, 6\}$$

$$L = \{1\text{–}2, 1\text{–}5, 2\text{–}3, 2\text{–}4, 3\text{–}5, 3\text{–}4, 4\text{–}3, 4\text{–}6, 5\text{–}2, 5\text{–}6\}$$

Find a spanning tree for the network.

16:3. MINIMAL SPANNING TREE PROBLEM

A minimal spanning tree problem has the following characteristics:

- We are given the ***nodes*** of a network but not the links. Instead we are given the potential links between node pairs and the lengths (representing distance, time, cost, etc.) for each potential link.
- It is desired to chose enough potential links in such a way that there is a path connecting (directly or indirectly) all the nodes of the network. That is, no node is left out unconnected.
- The chosen links must provide a path between each pair of nodes in such a way that the total length of these links is a minimum.

Some Applications of the Minimal Spanning Tree Problem

Some important areas where we design a network using the minimal spanning tree problem are:

(*i*) ***Telecommunication networks*** (computer networks, TV cable networks, fiber-optic networks). In this age of information superhighway applications in this area become most important. As many telecommunication networks are highly expensive, it is very important to optimize their designs by finding minimum spanning trees for them.

(*ii*) ***Power transmission networks*** to provide high-voltage power transmission lines.

(*iii*) ***Pipeline networks*** to connect sources to a number of destinations.

(*iv*) ***Transportation networks*** to provide the links (roads, railway lines, etc.) at a minimum cost.

Minimal Spanning Tree Algorithm

Step 1. Start with any given node. Connect it (*i.e.*, insert a link) to the nearest distinct node. Tick (✓) all nodes that have been connected (starting node is always a connected node).

Step 2. Identify the unconnected (unticked) node that is closest to a connected (ticked) node and then connect these two nodes.

Step 3. Repeat *Step* 2 until all the nodes are connected (ticked).

The resulting network is guaranteed to be a minimal spanning tree.

There may be ties for the nearest distinct node (*Step* 1) or the closest unconnected node (*Step* 2). In such cases break the ties arbitrarily — the algorithm will still yield an optimal solution.

Example 16.3. *The Midwest TV Cable Company is in the process of providing cable services to five new housing development areas. The Fig. 16.9 depicts the potential TV linkages among the five areas. The cable miles are shown on each branch.*

Determine the most economical cable network for the Midwest Company.

[B.Sc. (Stat.), Delhi Univ. 2004]

Also determine the minimal spanning tree for the given network when

(*a*) *Nodes 2 and 5 cannot be linked.*

(*b*) *Nodes 5 and 6 are linked by a 2-mile cable.*

(*c*) *Nodes 6 and 2 are linked by a 4-mile cable.*

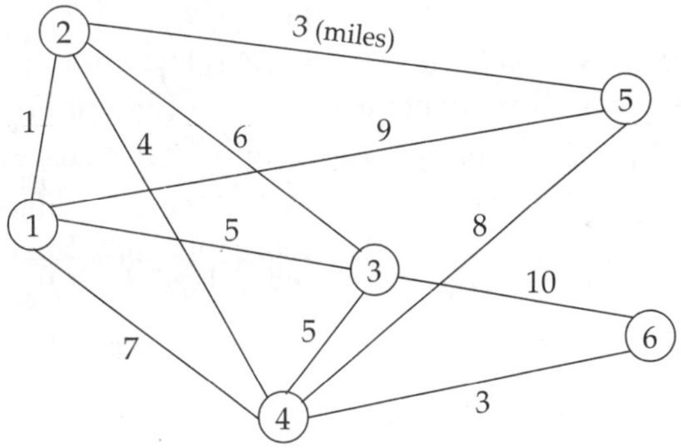

Fig. 16.9

Solution.

Step 1. We start with node *1*. Connect it (*i.e.*, insert a link) to the nearest distinct node (node *2*). Tick (✓) nodes *1* and *2* as both are now connected nodes (*Fig. 16.10*).

Step 2. Identify the unconnected (unticked) nodes that is closest to a connected (ticked) node. The unconnected node closest to either node *1* or node *2* is node *5* (closest to *2*). So we connect node *2* to node *5* and tick (✓) it to indicate that it is a connected node (*Fig. 16.11*).

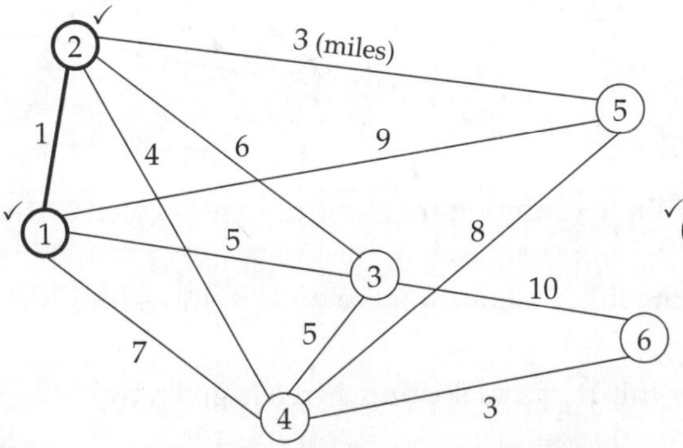

Fig. 16.10

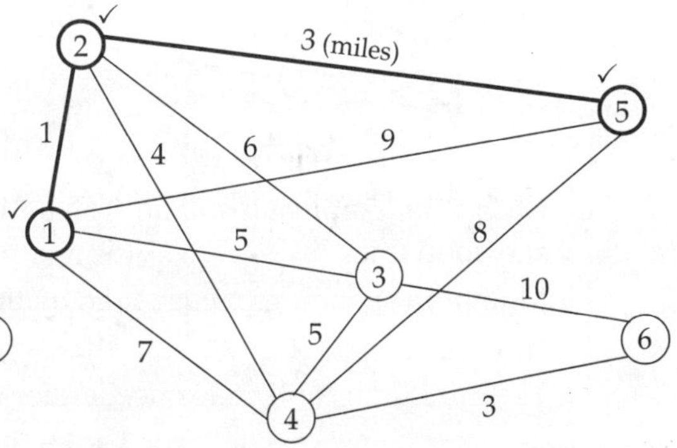

Fig. 16.11

Step 3. The unconnected node closest to any ticked node (node *1*, *2* or *5*) is node *4* (closest to node *2*). So we connect node *4* to node *2* (*Fig. 16.12*).

Step 4. The unconnected node closest to a ticked node is node *6* (closest to node *4*). So we connect node *6* to node *4* (*Fig. 16.13*).

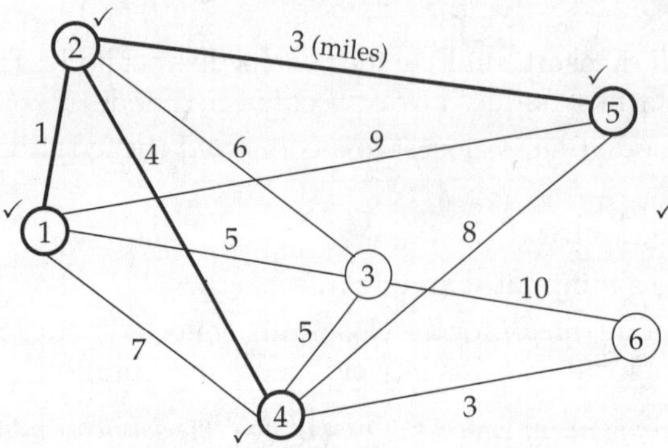

Fig. 16.12

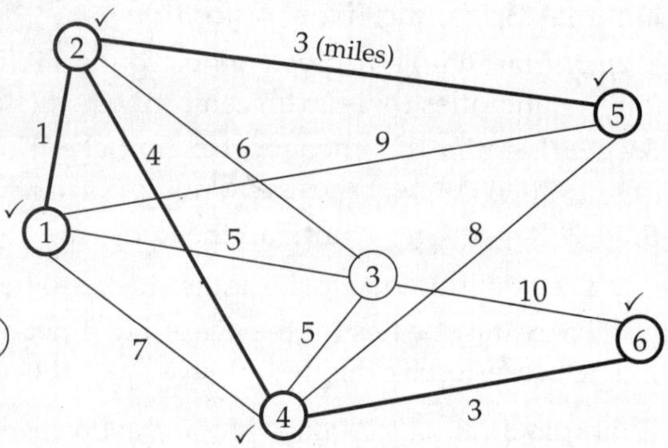

Fig. 16.13

Step 5. The unconnected node closest to a ticked node is node *3* (closest to node *1* or node *4*). So we connect node *3* to node *1* (*Fig. 16.14*).

Step 6. All nodes are now connected. Thus, the minimal spanning tree provides an optimal solution. The most economical cable network, *i.e., the minimal spanning tree* for the Midwest Company is given by the links connecting the node pairs as indicated:

1–2–5; 2–4–6; 1–3

The minimum cable miles needed for this spanning tree is given by

$$z = 1 + 3 + 4 + 3 + 5 = 16 \text{ miles.}$$

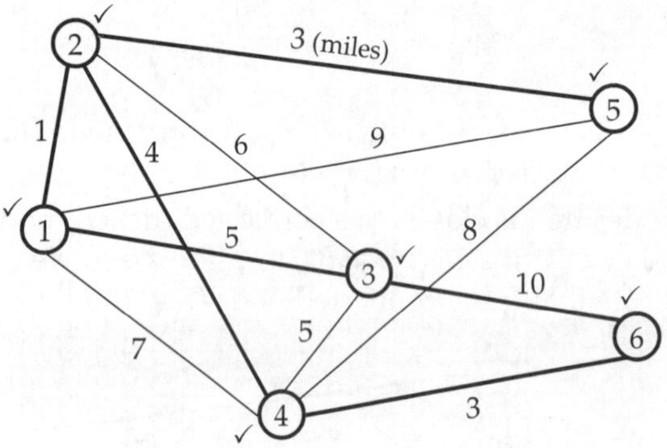

Fig. 16.14

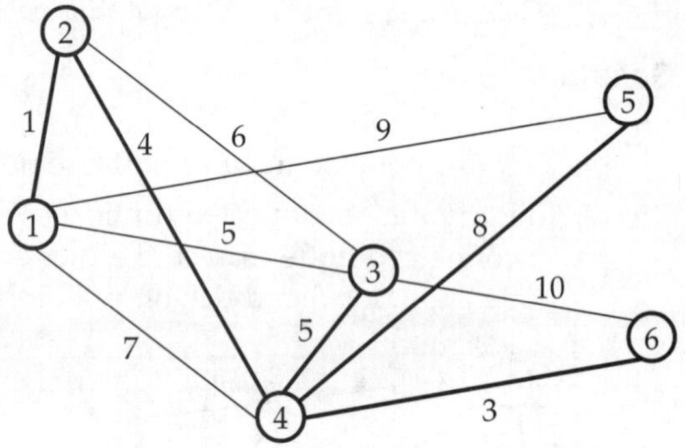

Fig. 16.15

(*a*) *Nodes 2 and 5 cannot be linked.* The minimal spanning tree for this case is given by the links shown in thick lines (*Fig. 16.15*).

Minimum spanning tree is given by the links connecting the node pairs as indicated:

1–2–4–6; 4–3; 4–5

The minimum cable miles needed for this minimal spanning tree is given by

$$z = 1 + 4 + 3 + 5 + 8 = 21 \text{ miles.}$$

(*b*) *Nodes 5 and 6 are linked by a 2-mile cable.* The minimal spanning tree is given by the links shown in thick lines (*Fig. 16.16*).

Links connecting node pairs as per order of iterations are: 1–2–5–6–4–3.

The minimum cable miles needed for this minimal spanning tree is given by

$$z = 1 + 3 + 2 + 3 + 5 = 14 \text{ miles.}$$

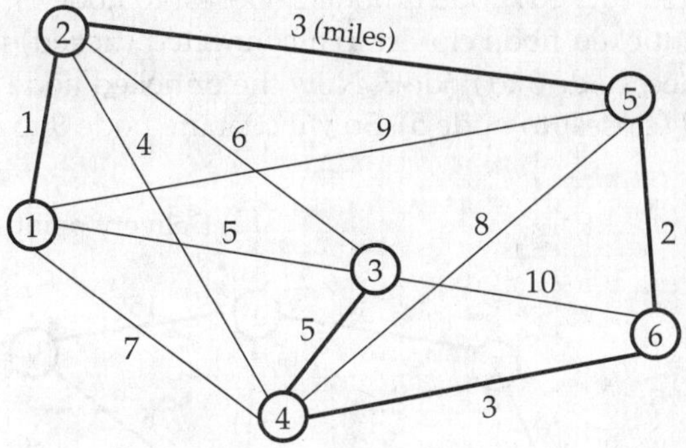

Fig. 16.16

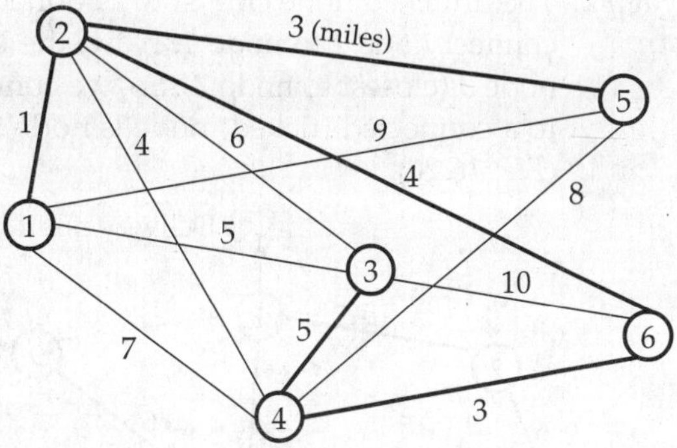

Fig. 16.17

(c) *Nodes 6 and 2 are linked by a 4-mile cable.* The minimal spanning tree is given by the links shown in thick lines (*Fig. 16.17*).

Links connecting node pairs as per order of iterations are: 1–2–5; 2–6–4–3.

The minimum cable miles needed for this minimal spanning tree is given by

$$z = 1 + 3 + 4 + 3 + 5 = 16 \text{ miles.}$$

Example 16.4 (*Minimal Pipeline Spanning*). *The Reliance Company is in the process of providing pipeline network that connects nine offshore natural gas wellheads with an inshore delivery point. The location of wellhead 1 is the closest to the shore, so it is equipped with adequate pumping and inventory capacity to pump the output of the remaining eight wellheads to the delivery point. The mileages of the feasible links (nodes 1 to 9) connecting to the inshore delivery point are shown in the network given in Fig. 16.18:*

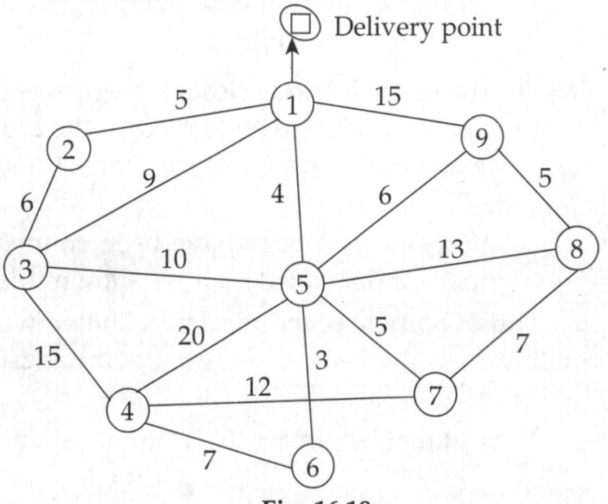

Fig. 16.18

Determine the most economical pipeline network for the Reliance Company that links the offshore wellheads to the inshore delivery point.

Solution.

Step 1. We start at node *1*. Connect node 5 to it being the nearest distinct node, the link having the minimum length of *4*. The unconnected node closest to a connected (ticked) node is node *6* (closest to node *5*). So, we connect nodes *6* to node *5*. Now, the unconnected node closest to a connected (ticked) node is node *7* (closest to node *5*). So, we connect node *7* to node *5* (*Fig. 16.19*).

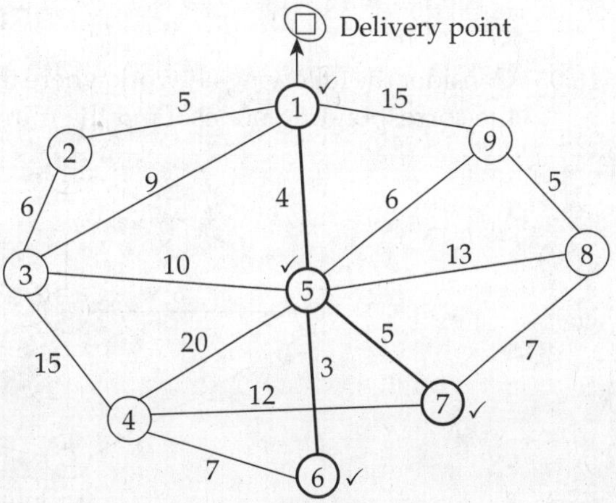

Fig. 16.19

***Step* 2.** The unticked node closest to a connected (ticked) node is node *2* (closest to node *1*). So, we connect node 2 to node *1*. Now, the unticked node closest to a connected (ticked) node is node *3* (closest to node 2). So, we connect node *3* to node 2. Now the unticked node closest to a connected (ticked) node is node *9* (closest to node *5*). So we connate node *9* to node *5* (*Fig. 16.20*).

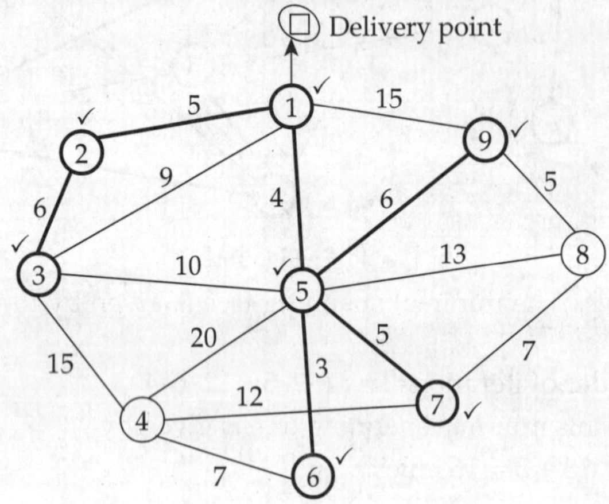

Fig. 16.20

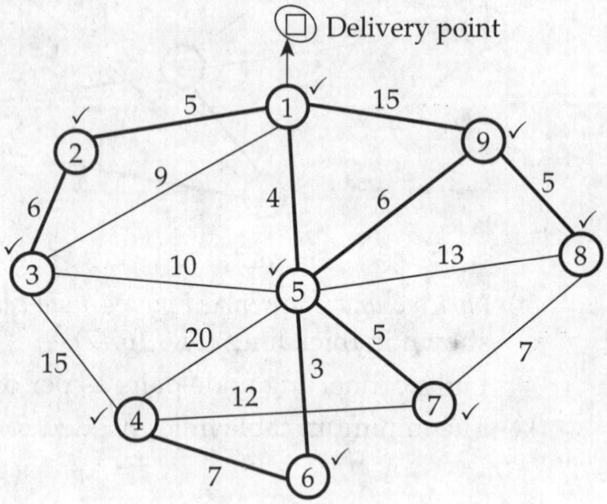

Fig. 16.21

***Step* 3.** The unticked node closest to a connected (ticked) node is node *8* (closest to node *9*). So, we connect node *8* to node *9*. Now, the only unticked node is node *4* with the closest connected node being node *6*. So, we connect node *4* to node *6* (*Fig. 16.21*).

Final Step.

Now, all the nodes have been connected. The minimal spanning tree connecting all the nodes is displayed by links shown thick in the network above.

Thus, the most economical pipeline network for the Reliance Company that links the offshore wellheads to the inshore delivery point is given by the nodes connected in the following order: 1–5–6; 5–7; 1–2–3; 5–9–8; 4–6.

The optimal length for this minimal spanning tree is given by

$$z = 4 + 3 + 5 + 5 + 6 + 6 + 5 + 7 = 41 \text{ miles.}$$

PROBLEMS

16.03 Consider the following network where the numbers on links represent actual distance between the corresponding nodes. Find the minimal spanning tree:

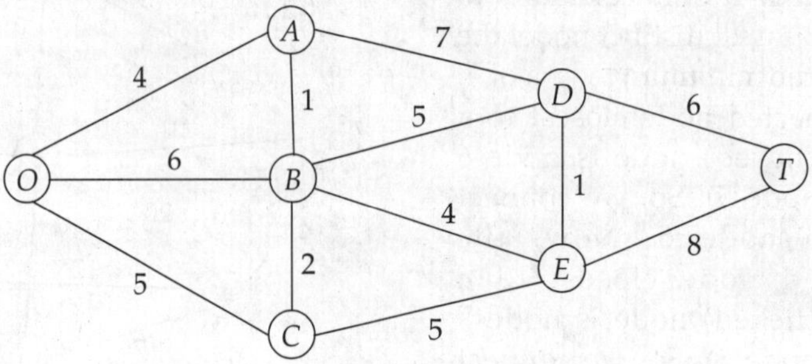

Fig. 16.22

16.04 The management of City Information Park desires to determine which road telephone cables should be installed to connect all 7 work stations with a minimum total length of cable. Nodes and distances of potential links are given in the network below. Use the minimal spanning tree algorithm to find the most economical cable tree:

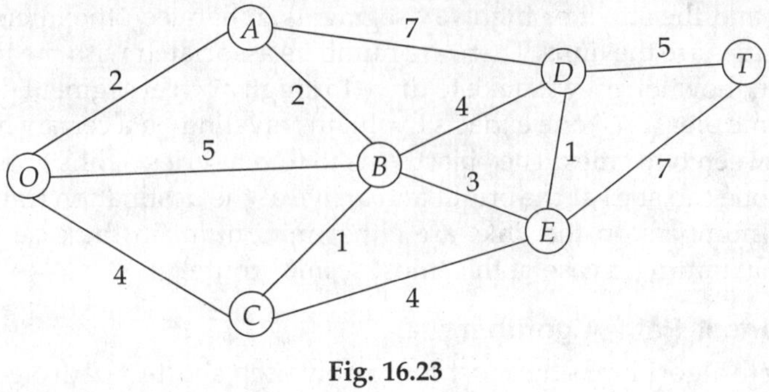

Fig. 16.23

16.05 For the Midwest TV Cable Company of *Example 16.3* determine the minimal spanning tree for the given network, when

(*a*) Nodes 3 and 5 are linked by a 2-mile cable.

(*b*) Node 2 cannot be linked directly to nodes 3 and 5.

(*c*) Nodes 1 and 2 are linked by a 8-mile cable.

16.06. For Reliance Company of *Example 16.4*, suppose the nine offshore wellheads can be considered to form two groups: Group *A* — a high pressure group consisting of wells 2, 3, 4, 6 and Group *B* — a low pressure group consisting of wells 5, 7, 8, 9. In view of pressure difference wellheads of groups *A* and *B* cannot be linked. Further, both groups *A* and *B* have to be connected to the delivery point through wellhead 1.

Determine the most economical pipeline network for the Reliance Company.

16.07. The Corporation Bank is in the process of installing computer terminals at each branch using optical telephone cables. However, each branch need not be connected directly to the head office to which it could be connected indirectly via another branch that is connected (directly or indirectly) to the head office. The cost of optical telephone cable is the number of miles involved multiplied by rupees in thousand. The distance in miles between various branches is given below:

H.O. / Branch	*Distance between branches*					
	H.O.	*Branch 1*	*Branch 2*	*Branch 3*	*Branch 4*	*Branch 5*
H.O.	—	190	70	115	270	160
Branch 1	190	—	100	110	215	50
Branch 2	70	100	—	140	120	220
Branch 3	115	110	140	—	175	80
Branch 4	270	215	120	175	—	310
Branch 5	160	50	220	80	310	—

The Bank desires to know which pairs of branches should be directly connected by optical telephone lines in order to connect every branch (directly or indirectly) to the H.O. at a minimum total cost.

Formulate this problem as a network flow problem and solve using the minimal spanning tree algorithm.

16:4. SHORTEST ROUTE PROBLEM

The shortest network flow problem is one that we all use in our daily lives: what is the fastest route to take between two locations in the city during the rush hour, what is the most scenic route to drive, or the cheapest route to fly, between two cities in our vacation. The street intersections or the cities are the nodes, and the street or highway segments in between the intersections or non-stop flights between the cities are the links. There are numbers associated with the links. These numbers can be the estimates of how long it will take to drive through a street segment during the rush hour, or a measure of "scenic pleasure" one expects to obtain traveling on a certain highway segment, or the cost of flying between two cities. The objective is to find a series of links connected with correct directions such that one can start at the origin and arrive at the destination traversing the links and that the sum of the numbers on the links are either minimum (in the case of the fastest or the cheapest route) or maximum (in case of the "most scenic" route).

The Dijkstra's Shortest Path Algorithm

Dijkstra's shortest path algorithm is the most commonly used shortest path algorithm. One can find the shortest paths from a given source to all nodes in a network at the same time; hence this problem is sometimes also called the ***single-source shortest path problem***. It happens to be relatively fast and efficient compared to other algorithms for solving the same problem.

Dijkstra's algorithm is a ***labeling algorithm***. It finds the shortest route from the starting node to any other node in the network. At each iteration a new node is assigned a ***final label***. This label gives the shortest distance from the start node to this node. Also, the ***working labels*** of all the nodes are improved at each stage (the working label gives the shortest distance found so far).

We shall use the following notation to represent a label assigned to a node *i*: (u_i, j), where u_i is the shortest distance of node *i* from the starting node *via* node *j*, the immediately preceding node. Note that $u_i - u_j$ = length of the link $i - j$.

The algorithm works as follows:

Step 1. Label the initial node as (0, —) as it represents the origin. The '—' indicates that this node has no preceding node. Now, as no further improvement is possible in this label, it becomes ***final node***. So, tick (✓) this node to indicate that this node has been solved.

Step 2. Update the working labels of all unticked (or unsolved) nodes that can be reached directly from the last solved (or ticked) node.

Step 3. Choose the unticked node with the smallest working label, *i.e.*, having smallest u_i, and record its working label as its final label. Tick (✓) this node to indicate that this node has been solved.

Step 4. Repeat *Steps* 2 and 3 until the destination node is ticked.

The working label for the destination node is always its best label as this node does not lead to any further node. The shortest route can now be found by tracking back.

Example 16.5. *Use Dijkstra's algorithm to determine a Shortest Path from A to C for the following network:*

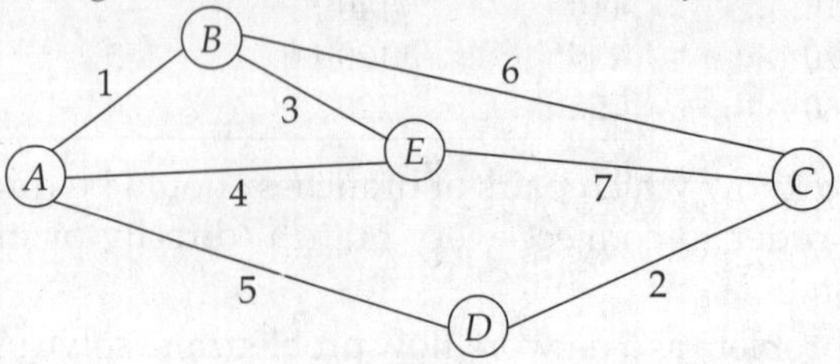

Fig. 16.24

Solution.

Step 1. Label the starting node A as (0, –) as it is the origin. As no further improvement is feasible for this node, we mark this label as ***final***. Tick (✓) this node to indicate that this node has been solved (*Fig. 16.25*).

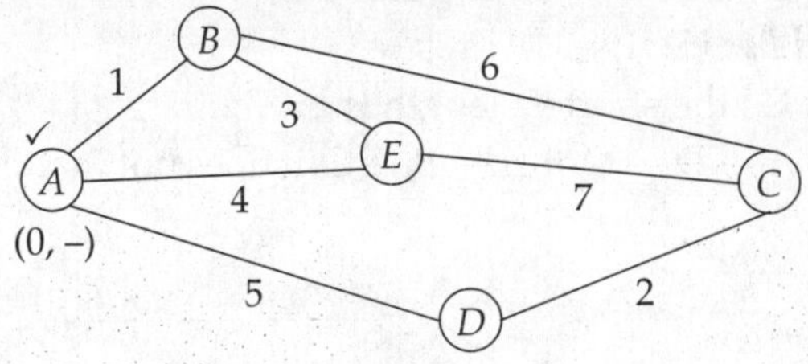

Fig. 16.25

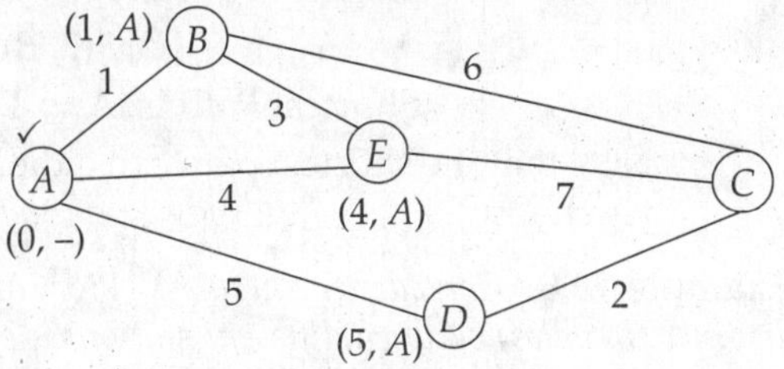

Fig. 16.26

Step 2. Nodes directly connected to the last ticked node (node A), are B, E and D (*Fig. 16.26*). We, compute the working labels for each one of them:

$$B = (1, A), \quad E = (4, A) \quad \text{and} \quad D = (5, A)$$

Step 3. Choose the unticked node with the smallest working label, *i.e.*, having smallest u_i, and record its working label as its final label. So, we tick (✓) node B to indicate that this node has been solved (*Fig. 16.27*).

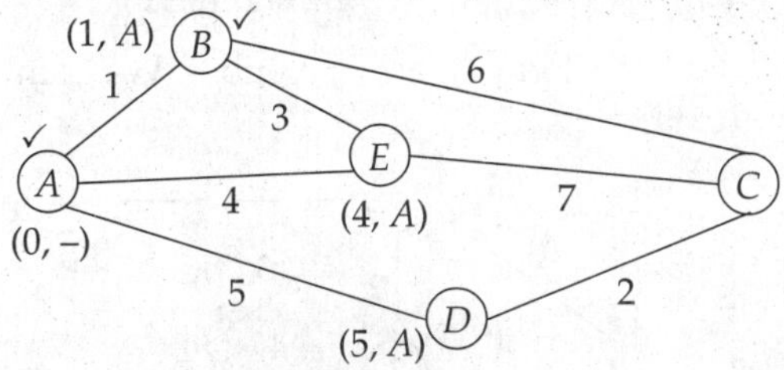

Fig. 16.27

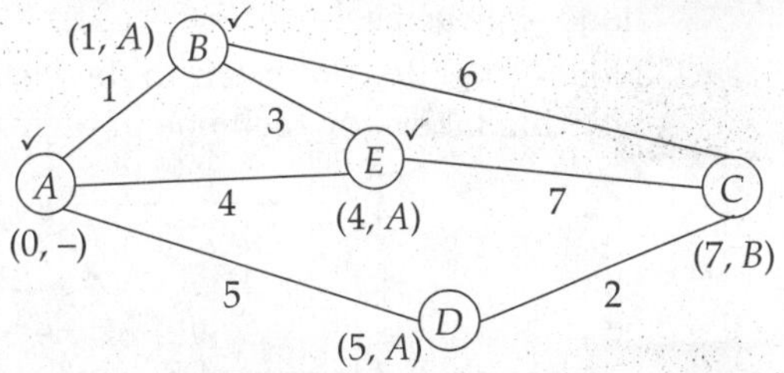

Fig. 16.28

Step 4. Unticked nodes directly connected to the last ticked node (node B) are nodes E and C. Compute/update the working labels for nodes E and C (*Fig. 16.28*).

Step 5. Choose the unticked node with smallest working label (node E) and mark its label as final. Tick (✓) node E to indicate that this node has been solved (*Fig. 16.29*).

Step 6. Update working labels of all unticked nodes (no change). Choose the unticked node with smallest working label (node D) and mark its label as final. So, we tick (✓) node D to indicate that this node has been solved.

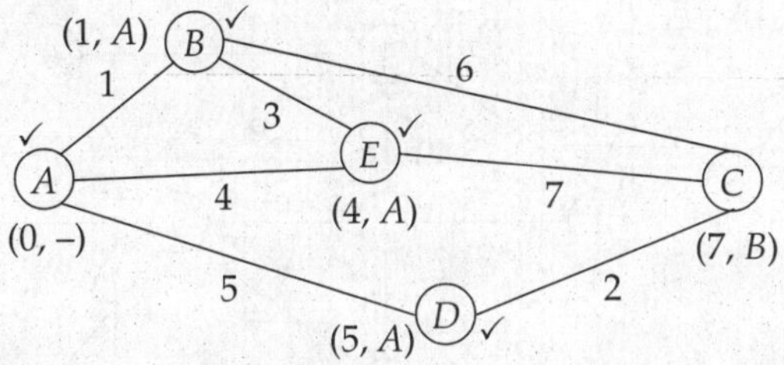

Fig. 16.29

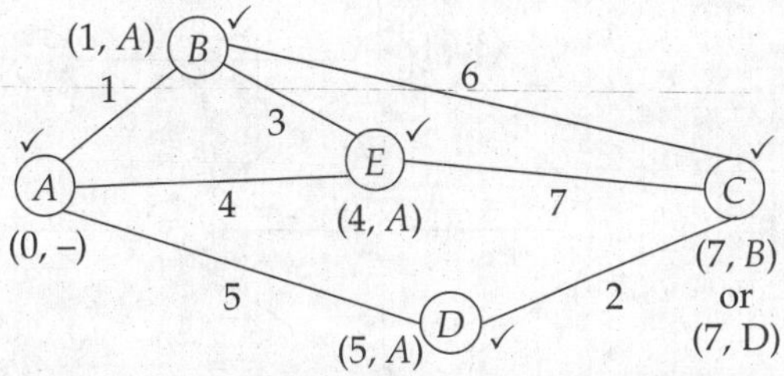

Fig. 16.30

Step 7. The unticked node directly connected to the last ticked node D is node C. Thus, node C gets another working label (7, D). However, the new label gives no improvement and is therefore as good as the previous label. Moreover, as node C happens to be the destination node, it does not least to any further node. So, we make its both the working labels final and tick node C to indicate this node having been solved (*Fig. 16.30*).

Step 8. All nodes from origin to destination have been ticked. So, the algorithm stops.

Step 9. The shortest path from node A to node C is found by back tracking as follows:

$$[C] : (7, B) \rightarrow [B] : (1, A) \rightarrow [A]$$

Shortest Path is: $\mathbf{A \Rightarrow B \Rightarrow C}$; the shortest length is $z = 7$

or

$$[C] : (7, B) \rightarrow [D] : (5, A) \rightarrow [A]$$

Shortest Path is: $\mathbf{A \Rightarrow D \Rightarrow C}$; the shortest length is $z = 7$

Note that the shortest path can either be $A \rightarrow D \rightarrow C$ or $A \rightarrow B \rightarrow C$ (it does not matter which we choose).

Example 16.6. *Consider the network (Fig. 16.31). The distance (in miles) between different stations is shown on each link. Determine the shortest route from station 1 to station 8.* ***[B.Sc. (Stat), Delhi Univ. 2003]***

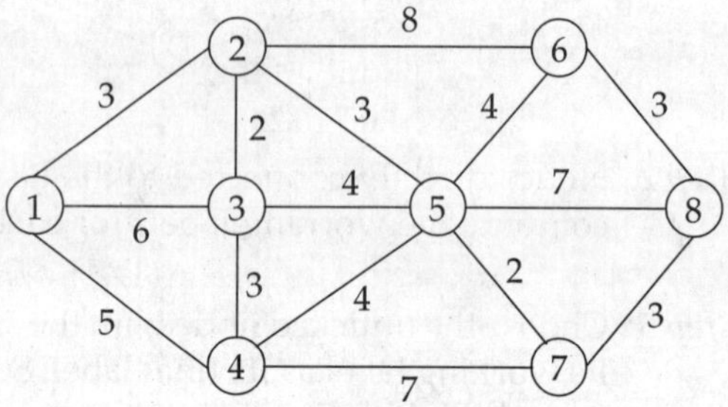

Fig. 16.31

Solution.

Step 1. Label the starting station node *1* as (0, –) as it is the origin. As no further improvement is feasible for this node, we mark this label as ***final***. Tick (✓) this node to indicate that this node has been solved (*Fig. 16.32*).

Step 2. Nodes directly connected to the last ticked node (node *1*), are *2*, *3* and *4*. We compute the working labels for each one of them (*Fig. 16.33*).

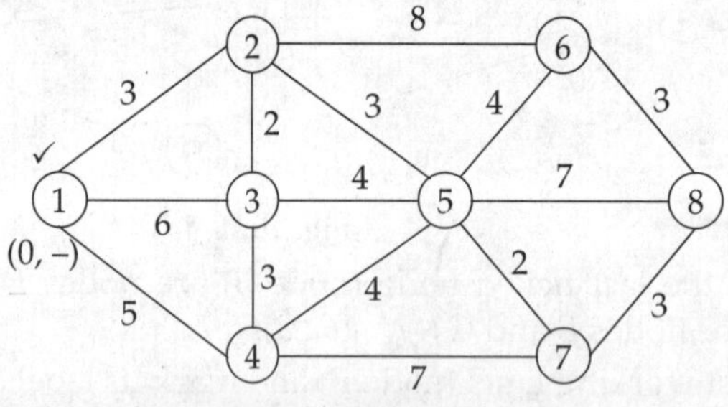

Fig. 16.32

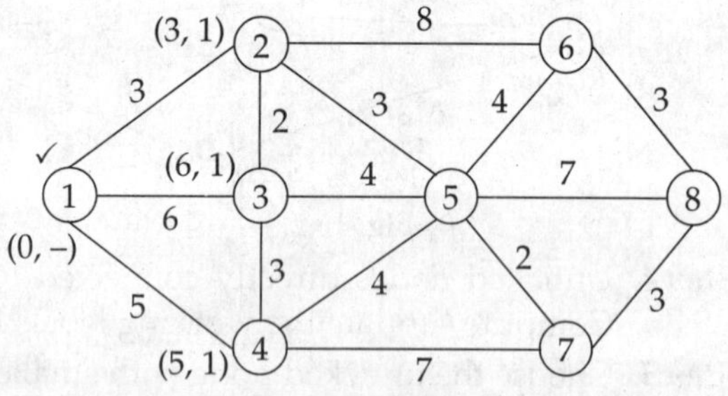

Fig. 16.33

Step 3. Choose the unticked node with the smallest working label, *i.e.*, having smallest u_i (node *2*). Its working label becomes its final label and therefore, we tick (✓) node *2* to indicate that this node has been solved (*Fig. 16.34*).

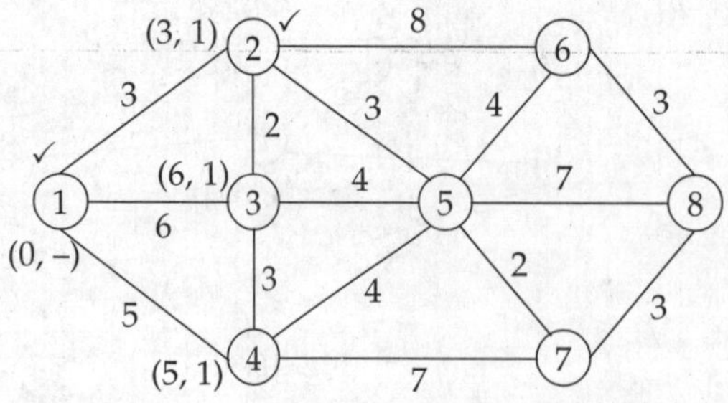

Fig. 16.34

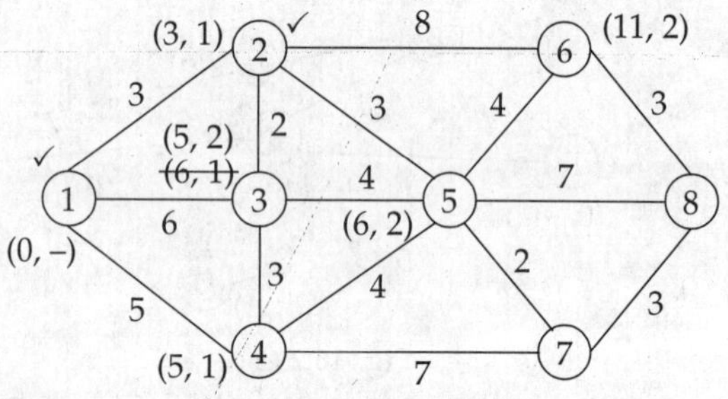

Fig. 16.35

Step 4. Unticked nodes directly connected to the last ticked node (node 2) are nodes *3*, *5* and *6*. So, we compute/update the working labels for nodes *3*, *5* and *6*. Note that the label of node *3* has been improved from (6, 1) to (5, 2) (*Fig. 16.35*).

Step 5. Choose the unticked node with smallest working label (node *3*) and mark its label as final. Tick (✓) node *3* to indicate that this node has been solved. Note that node *4* could have been selected in place of node *3* as it also has the smallest working label (*Fig. 16.36*).

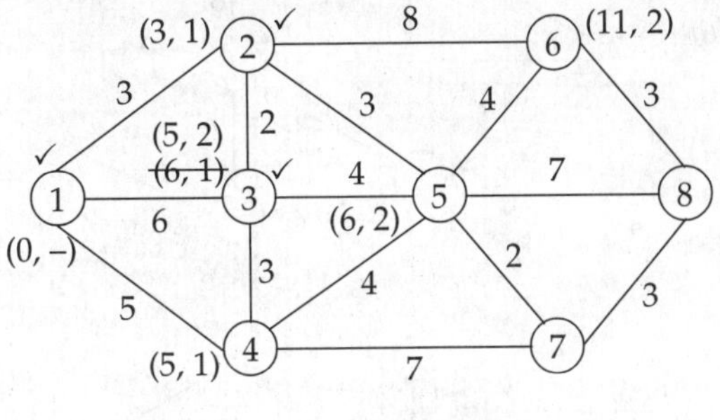

Fig. 16.36

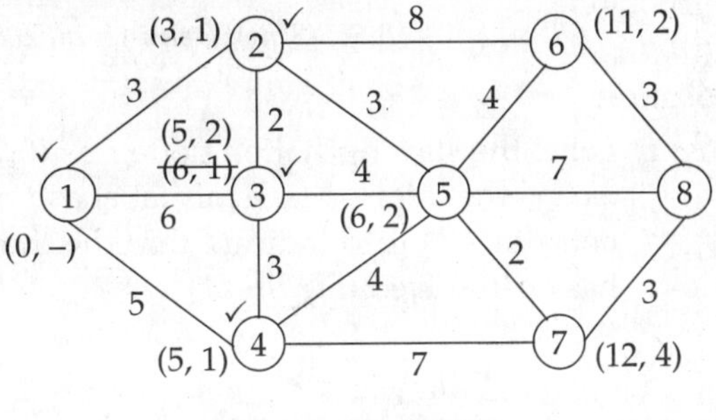

Fig. 16.37

Step 6. Update working labels of all unticked nodes (no change). Choose the unticked node with smallest working label (node *4*) and mark its label as final. So we tick (✓) node *4* to indicate that this node has been solved (*Fig. 16.37*).

Step 7. The unticked nodes directly connected to the last ticked node (node *4*) are nodes *5* and *7*. Compute/update the working labels of these nodes and choose the unticked node with smallest working label (node *5*) and mark its label as final. Tick (✓) node *5* to indicate that this node has been solved (*Fig. 16.38*).

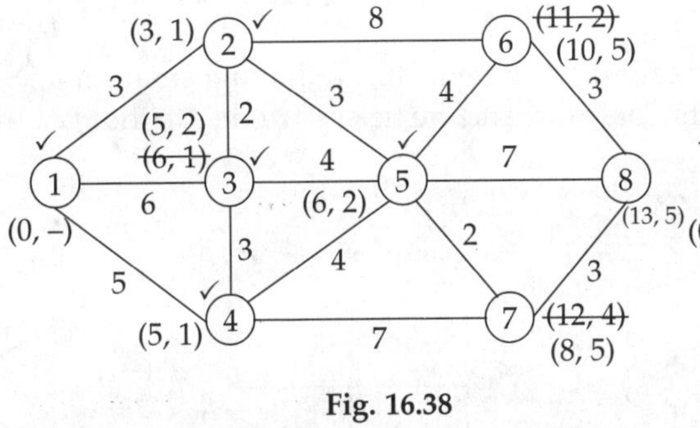

Fig. 16.38

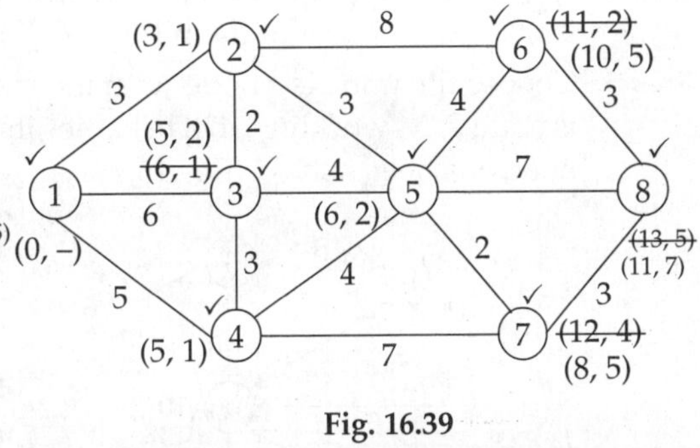

Fig. 16.39

Step 8. The unticked nodes directly connected to the last ticked node (node *5*) are node *6*, *7* and *8*. Compute/update the working labels of these nodes and choose the unticked node with smallest working label (node *7*) and mark its label as final. Tick (✓) node *7* to indicate that this node has been solved (*Fig. 16.39*).

Step 9. Compute/update the working labels of the remaining unsolved nodes (nodes *6* and *8*), choose the unticked node with smallest working label (node *6*) and mark its label as final. Tick (✓) node *6* to indicate that this node has been solved. Finally, we mark the working label of destination node *8* as final and tick (✓) it to end the algorithm.

Step 10. The shortest path from node *1* to node *8* is found by back tracking as follows:

$$[8] : (11, 7) \rightarrow [7] : (8, 5) \rightarrow [5] : (6, 2) \rightarrow [2] : (3, 1) \rightarrow [1]$$

Shortest Path is: **1 ⇒ 2 ⇒ 5 ⇒ 7 ⇒ 8**;

the shortest length is $z = 11$

Example 16.7. *The network given in Fig. 16.40 gives the permissibe routes and their lengths in miles between stations of city 1 (node 1) and six other cities (nodes 2–7).*

Determine the shortest route and hence the shortest distance from city 1 to city 7.

[B.Sc. (Stat), Delhi Univ. 2006]

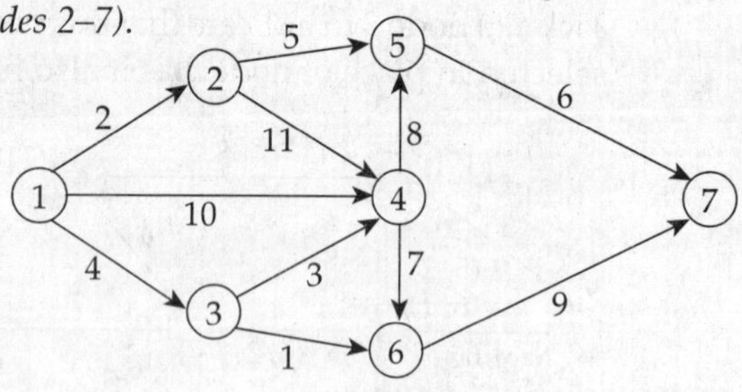

Fig. 16.40

Solution.

Step 1. Label the starting station node *1* as (0, –) as it is the origin. Mark this label as ***final*** and tick (✓) it to indicate that this node has been solved (*Fig. 16.41*).

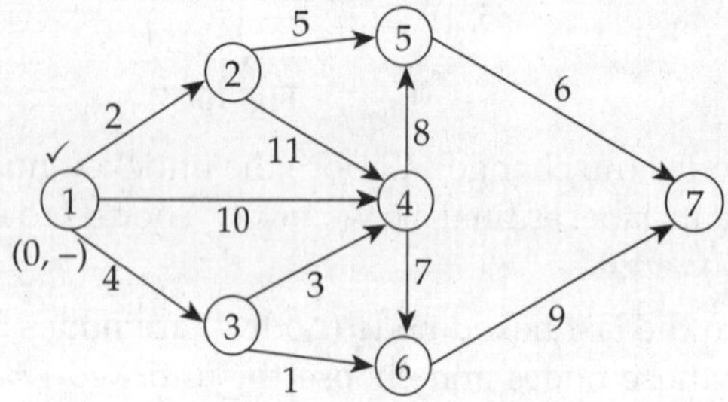

Fig. 16.41

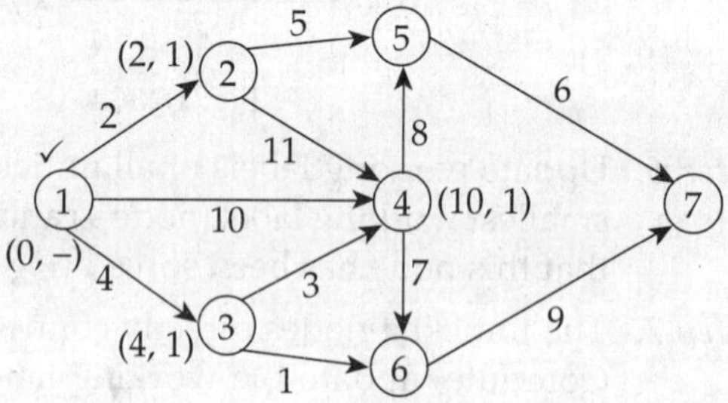

Fig. 16.42

Step 2. Nodes directly connected to the ticked node (node *1*), are *2*, *3* and *4*. We compute the working labels for each one of them (*Fig. 16.42*).

Step 3. Choose the unticked node with the smallest working label, *i.e.*, having smallest u_i. Since, it is node *2*, its working label becomes its final label and we tick (✓) node *2* to indicate that this node has been solved (*Fig. 16.43*).

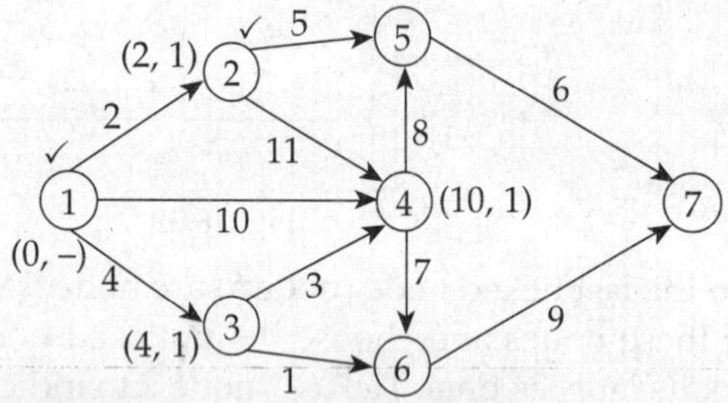

Fig. 16.43

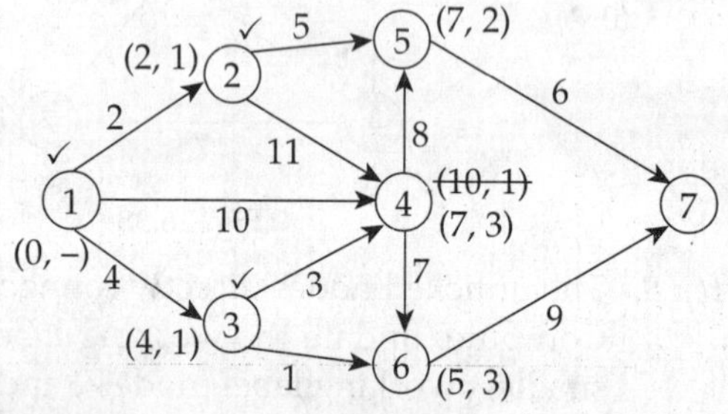

Fig. 16.44

Step 4. Unticked nodes directly connected to the last ticked node (node *2*) are nodes *4* and *5*. So, we compute/update the working labels for nodes *4* and *5*. Choose the unticked node with smallest working label (node *3*) and mark its label as final. Tick (✓) node *3* to indicate that this node has been solved (*Fig. 16.44*).

Step 5. The unticked nodes directly connected to the last ticked node (node *3*) are nodes *6* and *4*. Compute/update the working labels of these nodes and choose the unticked node with smallest working label (node *6*) and mark its label as final. Tick (✓) node *6* to indicate that this node has been solved (note that the working label of node *4* has improved) (*Fig. 16.45*).

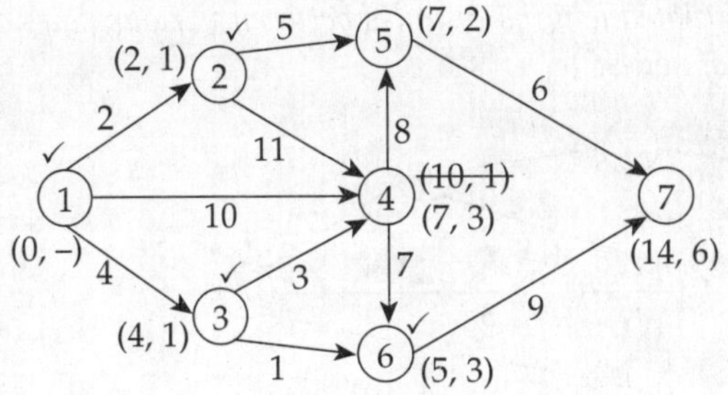

Fig. 16.45

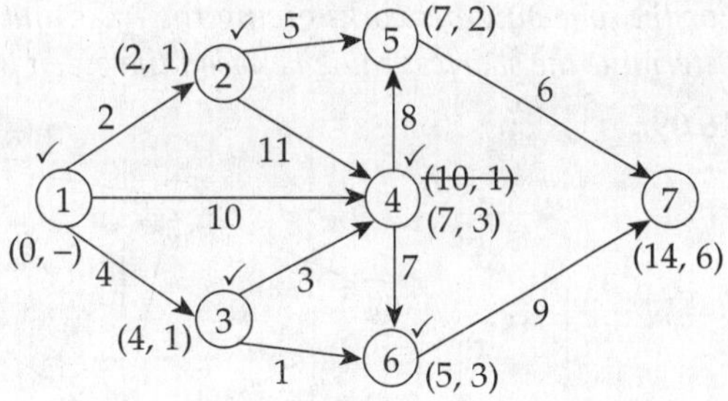

Fig. 16.46

Step 6. Unticked node directly connected to the last ticked node (node *6*) is node *7*, the destination. Its working label is (14, 6). Now choose the unticked node having smallest working label (node *4*) and mark its label as final. Tick (✓) node *4* to indicate that this node has been solved (*Fig. 16.46*).

Step 7. Update working labels of the remaining two unsolved nodes. The working label of node *5* being smaller, we mark its label as final. So, we tick (✓) node *5* to indicate that this node has been solved (*Fig. 16.47*).

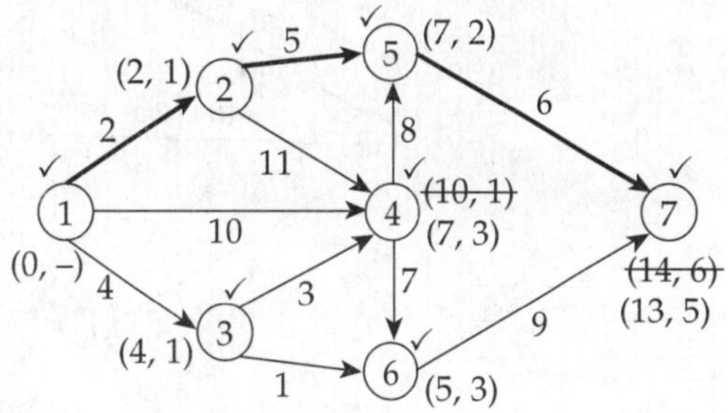

Fig. 16.47

Step 8. The working label of node *7* now gets improved. As no further movement is feasible, we mark the working label of destination node *7* as final and tick (✓) it to end the algorithm.

Step 9. The shortest path from node *1* to node *7*, found by back tracking is as follows:

$$[7] : (13, 5) \rightarrow [5] : (7, 2) \rightarrow [2] : (2, 1) \rightarrow [1]$$

Shortest Path is: $\mathbf{1 \Rightarrow 2 \Rightarrow 5 \Rightarrow 7}$; the shortest length is $z = 13$.

Remark. Another shortest path algorithm is: Floyd's Algorithm. In this algorithm process looks for the best path to find a final path that is better than all other paths. Depending on the size of the network this algorithm may take more time than Dijkstra's algorithm.

PROBLEMS

16.08. The following network gives the links and associated lengths in miles between city *1* (node *1*) and four other cities (nodes 2 to 5). Use Dijkstra's labeling procedure to determine the shortest route between city *1* and city *2:*

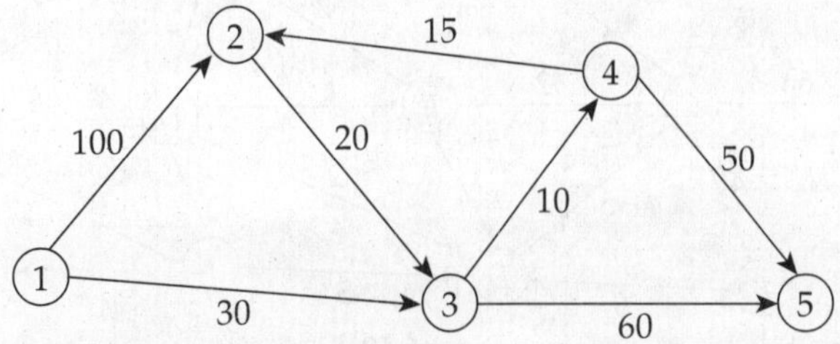

Fig. 16.48

For the following networks giving the links and associated lengths in miles between various pairs of nodes, determine the shortest route between the given pair of nodes:

16.09.

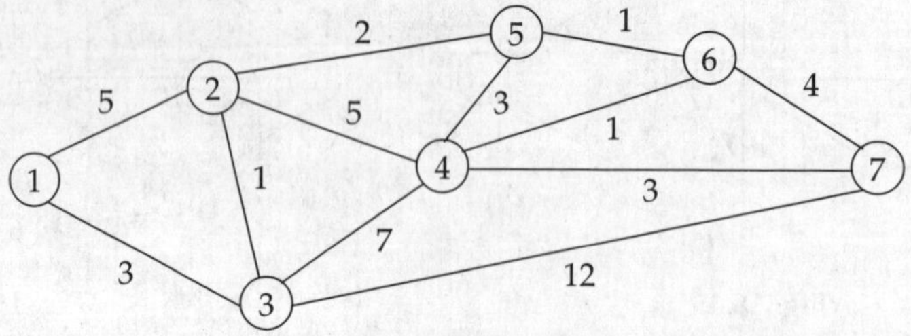

Fig. 16.49

(*a*) Node *1* to node *7*, (*b*) Node *7* to node *1*, (*c*) Node *6* to node *7*

16.10.

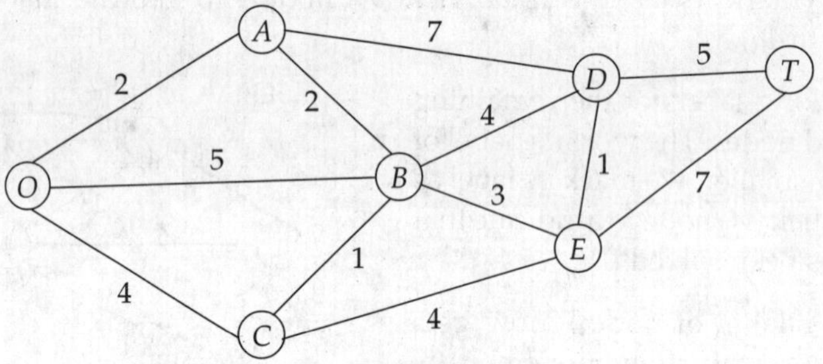

Fig. 16.50

Node *O* to node *T*

16.11.

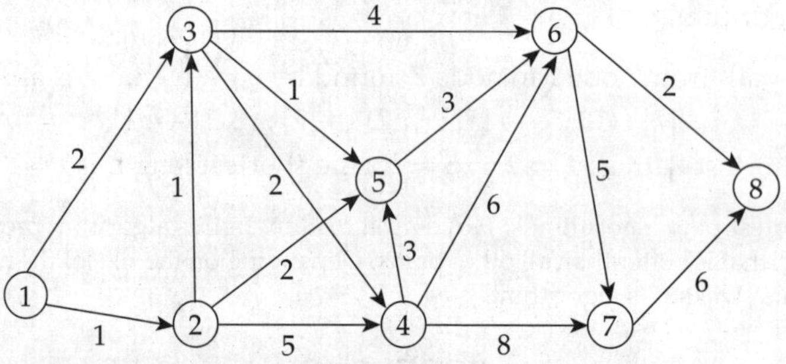

Fig. 16.51

(*a*) Node *1* to node *8*, (*b*) Node *1* to node *6*, (*c*) Node *4* to node *8*

16.12.

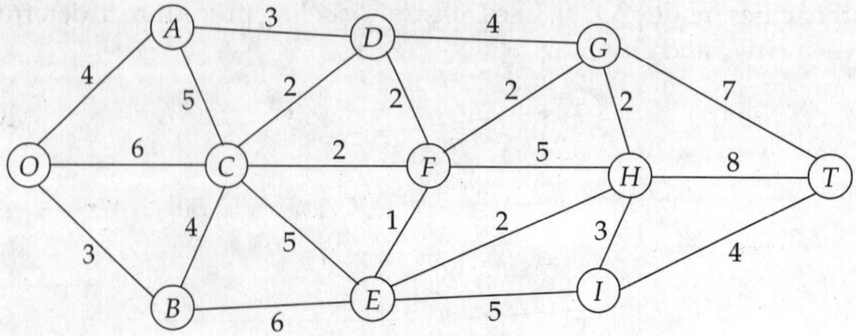

Fig. 16.52

Node *O* to node *T*

16.13. The following network gives the links and associated lengths in miles between city *1* and six other cities *2* to *7*. Use Dijkstra's labeling procedure to determine the shortest route between city *1* and every other node:

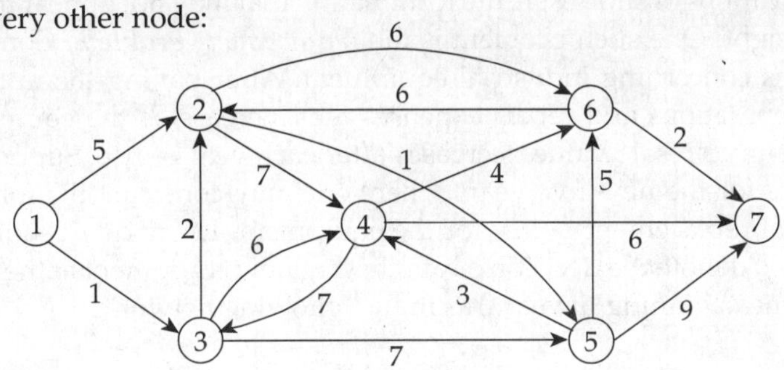

Fig. 16.53

16.14. India Mobile has license to service six regions of the country. The satellite distance (in miles) connecting the six region pairs are given in the network below. The nodes represent the regions. Determine the most efficient message route that India Mobile should establish between (*a*) region *1* and region *6*; and (*b*) region *1* and region *5:*

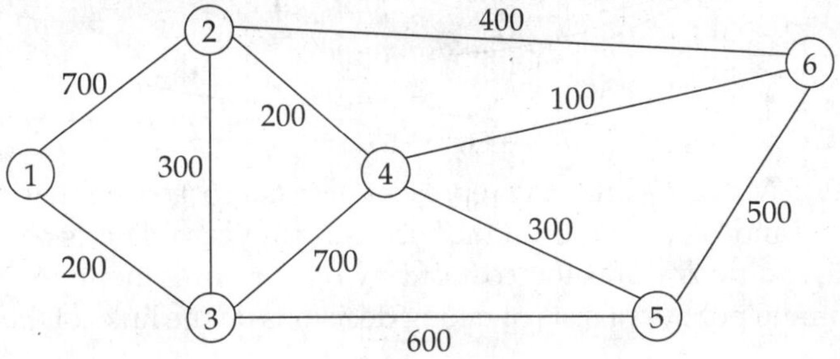

Fig. 16.54 *[MBA, Delhi Univ. 2008]*

16.15. Prachi is planning a trip by car to another city which she has never visited earlier. So, she is studying the road map to locate the shortest route to the destination. She finds that there are five other cities *A*, *B*, *C*, *D* and *E* that will come on the way. The map shows the road distances (in miles) directly connecting any two cities. This information is summarized below. A dash indicates that there is no road connecting the respective city:

City	*Distance (in miles) between adjacent cities*					
	A	*B*	*C*	*D*	*E*	*Destination*
Origin	40	60	50	—	—	—
A		10	—	70	—	—
B			20	55	40	—
C				—	50	—
D					10	60
E						80

(*a*) Formulate the problem as a shortest route problem, where nodes may represent cities, links represent roads and numbers represent the length of each link in miles. Determine the shortest route for Prachi.

(*b*) If the numbers in the table represent time (in minutes) rather than link distances, what will be the most efficient route for Prachi?

(*c*) What if Prachi wants to follow the most scenic route to enjoy the trip?

16:5. APPLICATIONS OF SHORTEST ROUTE PROBLEM

There are many applications of shortest route formulation that may not appear to be a network flow problem at the first sight. One such problem is the ***replacement problem***. Consider, for example, replacement decisions concerning industrial equipment. After paying the initial purchase price, each year there are operations and repair expenses that increase as the equipment ages. On the other hand, the trade-in or resale value decreases after each year of use. Suppose, it is possible to estimate those figures for the next five years. Then, one may construct the following network to determine the optimal replacement policy for the equipment. Let node i denote the beginning of year i, and the link (i, j) denote the decision of purchasing new equipment at the beginning of year i and keeping it until the beginning of year j, as in the following network:

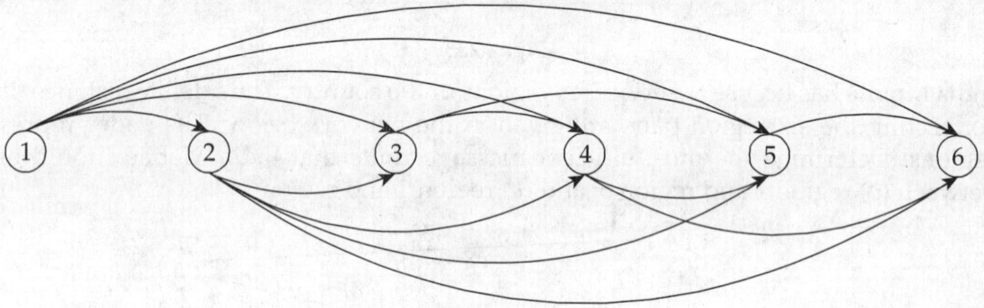

Fig. 16.55

The labels associated with each link may represent the total cost of purchasing a new equipment at the beginning if year i and keeping it until the beginning of year j. This is obtained by adding the yearly operating and repair costs and the cost of buying the equipment new minus its resale or trade-in value. The optimal policy consists of taking decisions on the links of the shortest route from node *1* and node *6*.

Example 16.8 (*Car Replacement Problem*). *The following table gives replacement costs as a function of the year a car was purchased and the number of years it has been in use:*

Purchased at start of	*Replacement cost (in ₹'000) after using till end of year* 1	2	3
2019	4.0	5.4	9.8
2020	4.3	6.2	8.7
2021	4.8	7.1	—
2022	4.9	—	—

At the start of each year a decision is made whether a car should be replaced. However, a car must be in service for at least one year and a maximum of three year.

Formulate the above as a network and determine the least cost replacement policy.

Solution. Let node i denote the beginning of year i, and the link (i, j) denote the decision of purchasing new equipment at the beginning of year i and keeping it until the beginning of year j, as shown in the network (*Fig. 16.56*).

The nodes *1* to *5* represent the years 2019 to 2023, respectively. The length of each link represents the replacement cost. That is, the total cost of purchasing a new car at the beginning of year i and keeping it until the beginning of year j, which is obtained by adding the yearly operating and repair costs and the cost of buying the equipment as new minus its resale or trade-in value.

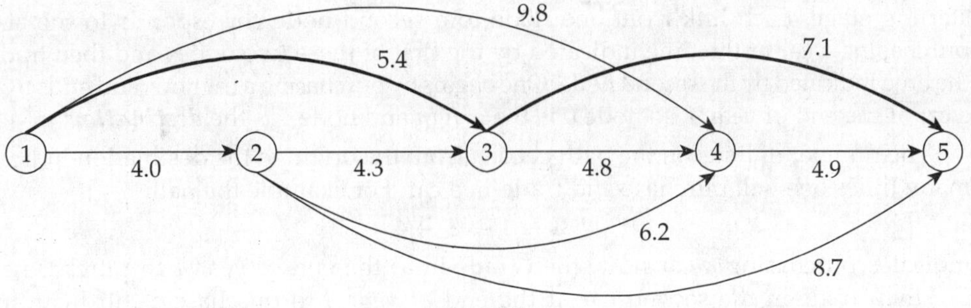

Fig. 16.56

The optimal replacement policy for this problem is equivalent to the shortest route between nodes *1* and *5*. Using the shortest route algorithm the optimal solution to the problem comes out to be the path **1 → 3 → 5**. Thus the optimal replacement policy is "A car purchased at the beginning of 2019 (node *1*) must be replaced after 2 years at the start of 2021 (node *3*). This replaced car should then be used until the end of 2022, *i.e.*, the start of 2023 (node *5*)". The total cost for this replacement would be 5.4 + 7.1 = 12.5 units (₹ '000).

PROBLEMS

16.16. *ABC* Limited has earmarked ₹ 210,000 to purchase and maintain a certain three-year-old used car. Since, operating and maintenance costs go up rapidly as the car ages. They are willing to trade-in the car on another three-year-old car one or more times during the next three summers if that would minimize the total net cost. The company will definitely purchase a brand new car after and as such they should definitely plan to trade-in the car then.

The following table gives the relevant data for ***each*** time the company purchase a three-year-old car.

Data for Purchasing a Three-Year-Old Car

Price (₹ '000)	*Operating and maintenance costs for ownership year (₹ '000)*				*Trade-in value at the end of ownership year (₹ '000)*			
	1	*2*	*3*	*4*	*1*	*2*	*3*	*4*
120	20	30	45	65	85	65	45	30

When should *ABC* Ltd. traded in the car (if at all) during the next three summers to minimize the total net cost of purchasing, operating, and maintaining the car over the next four years?

Formulate this problem as a shortest-path problem.

[***Hint:*** The network formulation of this problem as a shortest path problem is shown in *Fig. 16.57*. Nodes *1*, *2*, *3*, and *4* represent end of year *1*, *2*, *3*, and *4* respectively. Node *0* is the

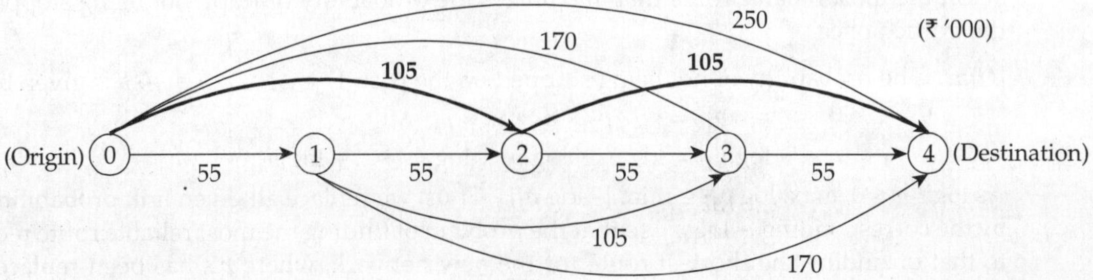

Fig. 16.57

starting stage. Each link from one node to a second node corresponds to the activity of purchasing a car at the time indicated by the first of these two nodes and then trade-in it at the time indicated by the second node. One begins by purchasing a car now, and ends by trade-in a car at the end of year *4*, so node *0* is the ***origin*** and node *4* is the ***destination.***

The number of links on the path chosen from the origin to the destination indicates how many times one will purchase and trade-in a car. For example the path

$$\mathbf{0 \to 1 \to 3 \to 4}$$

indicates purchasing a car now, then trade-in at the end of year *1* to purchase a second car, then trade-in the second car at the end of year *3* to purchase a third car, and then trade-in this third car at the end of year *4*. As the company wants to minimize the total net cost from now (node *0*) to the end of year *4* (node *4*), each link length needs to measure the net cost of that link's cycle of purchasing, maintaining, and trade-in a car. Thus, we must have

Link length = purchase price + operating and maintenance cost – trade-in value.

We obtain the following shortest path solution: $\mathbf{0 \to 2 \to 4}$

That is, trade-in the first car at the end of year *2*; and trade-in the second car at the end of year *4*.

The length of this path is z = ₹ 105,000 + ₹ 105,000 = ₹ 210,000.]

16.17. **(*Most Reliable Route*).** Kapil drives daily to work and tries to adopt a known shortest route. However, this route involves a number of intersections patrolled by traffic police. Kapil feels in view of all the fines paid for over speeding and other minor offences, the shortest route may not be economical and hence not reliable. He is thus looking for the next best route that will maximize his chance of not being stopped by the traffic police.

The network in *Fig. 16.58* shows the potential links between his home and office. The numbers on the links represent the probabilities of not being stopped on that link. For sequential links on the route, the associated probabilities for the preceding links just get multiplied. For instance, the probability of not being stopped by police on the route 1 → 3 → 5 → 7 is 0.9 × 0.3 × 0.25 = 0.0675:

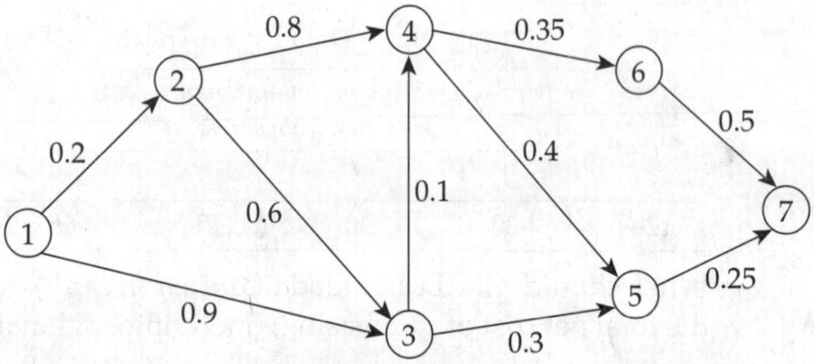

Fig. 16.58

Select the most reliable route that maximizes the probability of Kapil not being stopped by the traffic police.

[***Hint:*** The probability of not being stopped on the path 1 → 2 → 3 → ... → k is given by

$$p_{1k} = p_1 \times p_2 \times p_3 \times ... \times p_k$$

$$\Rightarrow \qquad \log p_{1k} = \log p_1 + \log p_2 + \log p_3 + ... + \log p_k$$

As $\log p_{1k} \le 0$, max. $\log p_{1k} \cong$ min. $\{-\log p_{1k}\}$. Thus, we replace all given link probabilities p_{ij} by the corresponding '$-\log p_{ij}$' so that the problem of finding the most reliable route reduces to that of finding the shortest route for the new network, where p's has been replaced by ($-\log p$'s) for each link.]

16.18. Excel Tele Services has laid the communication network between two metro stations: station *1* and station *7*. The probability that a link in the network will operate without fault is shown in the network given in *Fig. 16.59*.

If messages are sent from metro station *1* to metro station *7*, Excel Tele Services desires to determine the route that will maximize the probability of a successful transmission. Formulate it as a shortest route problem. Hence, solve it.

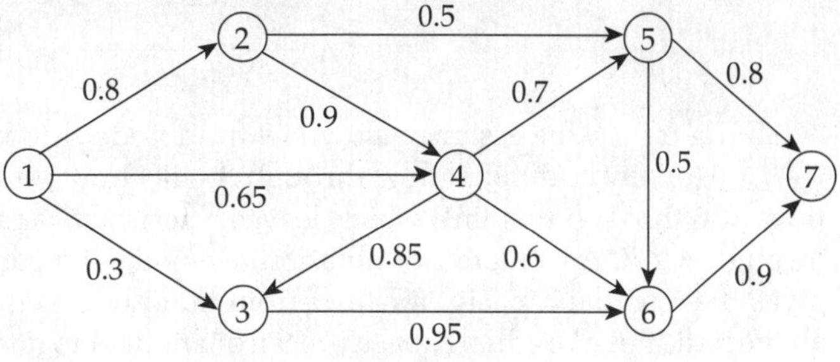

Fig. 16.59

16:6. MAXIMAL-FLOW PROBLEM

Here, the numbers on the links represent the "capacity" of the link. That is, these numbers are the maximum flow possible between the node from which the link is incident out of and the node it is incident into. The network can be an oil pipeline in which nodes are the pumping stations and the objective is to determine the maximum amount that can be pumped between an origin node and a destination node. Likewise, the network can represent a city's traffic flow network where the capacities on the links represent the maximum number of vehicles that can travel (per unit time) on a segment between two intersections. During heavy rush, we can increase some of the link capacities by making certain streets one-way, or changing the lane allocation in multi-lane highways or streets. The solution of maximal flow problem on these networks enables us to estimate how much we can ease the traffic congestion during the rush hours. Maximal flow problems also play an important role in the design and operation of telecommunication networks and computer networks like Internet and the company intranets. Some other typical applications aim at maximizing the flow of (*a*) water through a system of aqueducts, (*b*) data through a communication network, and (*c*) a company's product through its distribution network from its factories to its customers, etc.

A general ***maximal-flow problem*** may be described as follows:

(*i*) Let there be a directed and connected network. All flows through this network originate at one node, called the **Source** and terminate at another node called the **Sink**. All intermediate nodes are ***transshipment*** nodes.

(*ii*) The direction of flow through a link is indicated by pointing of arrowhead and the maximum amount of flow is given by the capacity of the link.

(*iii*) At the source all links point away from the node, while at the sink all links point into the node.

(*iv*) It is desired to maximize the total flow from the source to the sink. This maximal flow happens to be the amount of flow leaving the source or the amount of flow entering the sink.

As will be discussed later, a maximal flow problem can be formulated as an LPP and hence could be solved by usual simplex method. However, there are more efficient special algorithms available for solving this problem.

The Augmenting Path Method

The augmenting path method is based on two intuitive concepts.

Residual Network. After some flows have been assigned to the links, the ***residual network*** gives the remaining, (*i.e.*, residual) link capacities for assigning any additional flows. For example, consider a bi-directional link *A–B* having a link capacity of 9.

(*a*)

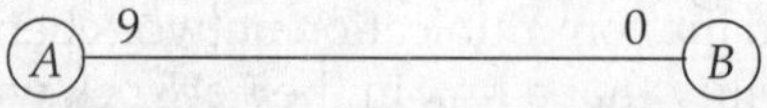

(*b*)

In figure (*a*), the assigned flow is 9 from node A to node B. This leaves a residual capacity of $9 - 9 = 0$ for any additional flow through $A - B$. However, in figure (*b*), the assigned flows include a flow of 6 through this link which leaves a residual capacity of $9 - 6 = 3$ for any additional flow through $A - B$. As a matter of notation, we make it a convention that the number nearing a node gives the residual capacity for flow from that node to the other node in the link. Thus, figure (*a*) above indicates a residual capacity of 9 from node A to node B, *i.e.*, $A \rightarrow B$. In figure (*b*), the number 6 on the left indicates a residual capacity of 6 for flows from A to B, whereas the number 3 on the right shows a residual capacity for flows from B to A. This flow in the opposite direction ($B \rightarrow A$) actually cancels some previously assigned flows in the positive direction ($A \rightarrow B$).

Augmenting Path. It is a directed path from source to the sink in a residual network, such that every link on this path has a strictly positive residual capacity in the positive direction of the link. The minimum of these residual capacities on the path is called the residual capacity of the augmenting path. This least residual capacity gives the amount of flow that can be added (or augmented) to the entire path.

Augmenting Path Algorithm

Every augmenting path gives an opportunity to further augment the flow by adding a flow equal to its residual capacity to the path. The algorithm sequentially selects some augmenting path (arbitrarily) and adds a flow equal to its residual capacity to the path. The process continues till there is no more augmenting path available. At that stage the algorithm stops and maximal flows have been attained.

The ***augmenting path algorithm*** may be carried out in the following manner:

Step 1. Change every link in the given network from a directed to an undirected one. Obtain the initial residual network for the problem. All link capacities in the original direction remain unchanged, whereas link capacities in the opposite direction are all zero.

Step 2. (*a*) Out of many possible paths, identify an augmenting path from the source to the link such that every link on this path has a strictly positive residual capacity. Go to *Step 3.*

(*b*) If no such augmenting path is available, the current net flows give an optimal flow pattern. Go to *Step 5.*

Step 3. The residual capacity of the augmenting path is given by

k^* = min. [positive residual capacities of the links (in +ve direction) on the path]

Increase the flow in this path by k^*. That is, add k^* to the net flow at the sink.

Step 4. Update all residual capacities by subtracting k^* from the positive residual capacity of each link (positive direction) on this augmenting path. Note that no resulting residual capacity becomes negative. Also, add k^* to the residual capacity of each link in the opposite direction on this augmenting path. Return to *Step 2.*

Step 5. The optimal flow pattern is obtained by comparing the final residual capacities with the original link capacities. There is flow along a link only if the residual capacity of the link is less than the original link capacity. The amount of this flow equals the difference between these capacities.

The algorithm is best illustrated by solving a problem.

Example 16.9. *For the network shown in Fig. 16.60, use the augmenting path algorithm to find the flow pattern giving the maximum flow from the source to the sink, given that the link capacity from node i to node j is the number nearest node i along the link between these nodes. Node O represents the source and node S is the Sink.*

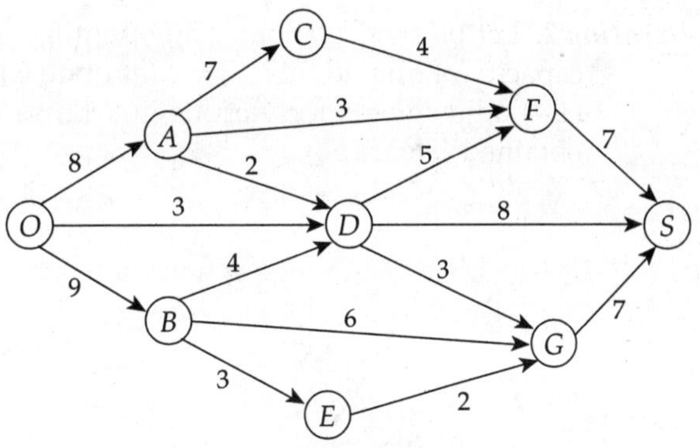

Fig. 16.60

Solution. Before assigning any flows, we change every directed link to an undirected one. However, the link capacity in the original direction remains unchanged and the link capacity in the opposite direction is zero.

Subsequently, whenever some flow is assigned to a link, that much flow is subtracted from the residual capacity in the same direction and added to the residual capacity in the opposite direction.

Iteration 0. The initial residual network is given in *Fig. 16.61:*

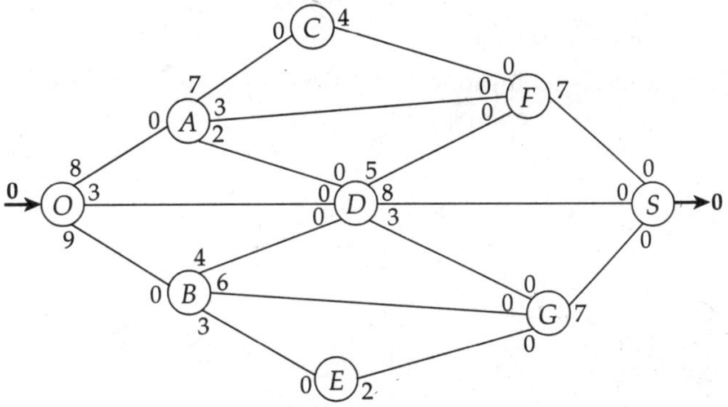

Fig. 16.61

Iteration 1. *Identify an augmenting path by finding some directed path from the source to the sink such that every link on this path has a strictly positive residual capacity.* One of the several augmenting paths is: $O \to B \to G \to S$, which has a residual capacity of min. $\{9, 6, 7\} = 6$. So, we assign a flow of 6 to this path and make (± 6) adjustments for various link flows on this path. The following residual network is obtained (*Fig. 16.62*):

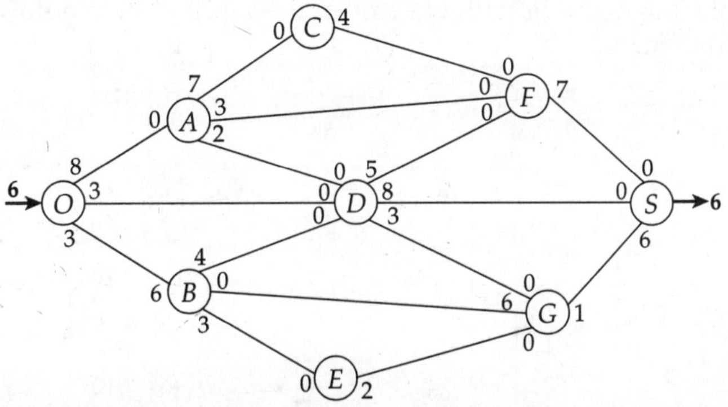

Fig. 16.62

Note that the source has supplied a flow of 6 units and the sink has received a flow of 6 units so far.

Iteration 2. Let us choose another augmenting path $O \to A \to C \to F \to S$, which has a residual capacity of min. {8, 7, 4, 7} = 4 (at node *C*). So, we assign a flow of 4 to this path and make (± 4) adjustments fc r various link flows on this path. The following residual network is obtained (*Fig. 16.63*):

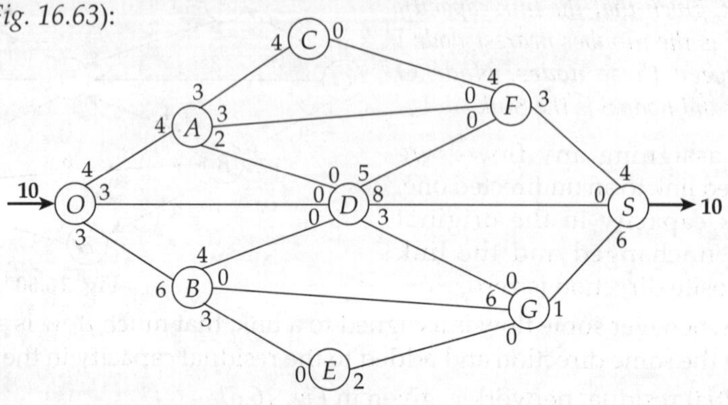

Fig. 16.63

Observe that flow to sink has increased to 10.

Iteration 3. Now, let us choose the augmenting path $O \to B \to D \to F \to S$. As minimum residual flow on this path is 3 (say at node *F*), we assign a flow of 3 to this path. Making (± 3) adjustments for various link flows on this path, the residual network is obtained (*Fig. 16.64*):

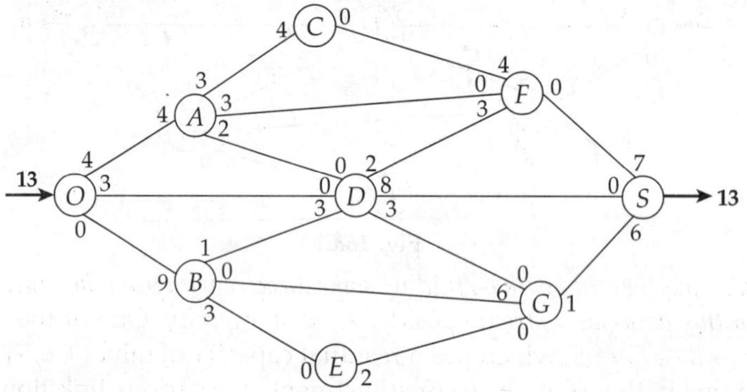

Fig. 16.64

Iteration 4. Choose the augmenting path $O \to A \to D \to S$. Assigning a flow of 2 to this path and making ± 2 adjustments for various residuals on the path, the residual network is obtained (*Fig. 16.65*):

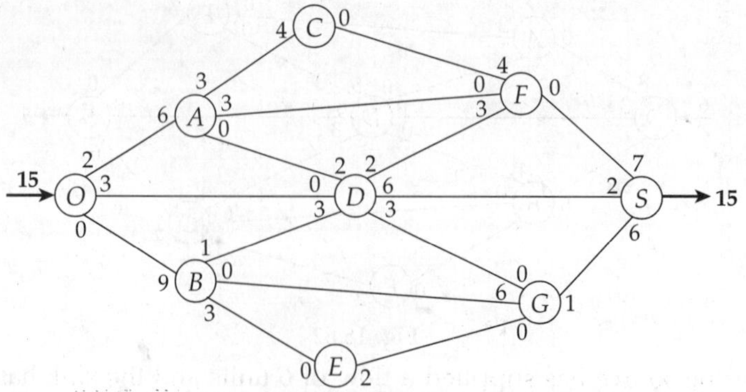

Fig. 16.65

Iteration 5. Now, we choose the augmenting path $O \rightarrow D \rightarrow S$. Assign a flow of 3 to this augmenting path. The residual network is obtained (*Fig. 16.66*):

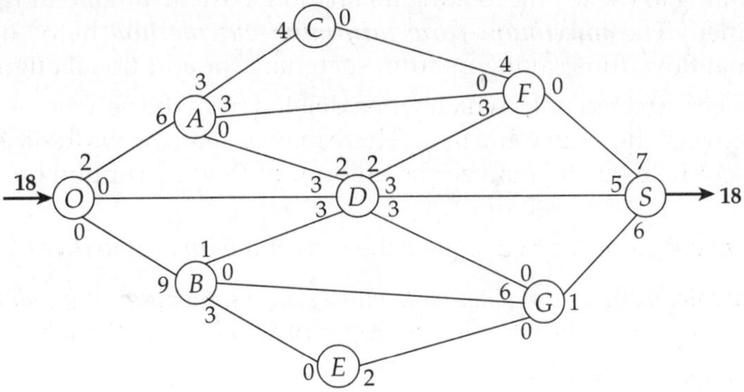

Fig. 16.66

Iteration 6. Choose the augmenting path $O \rightarrow A \rightarrow F \rightarrow D \rightarrow S$. Assigning a flow of 2 to this path, the residual network is obtained (*Fig. 16.67*):

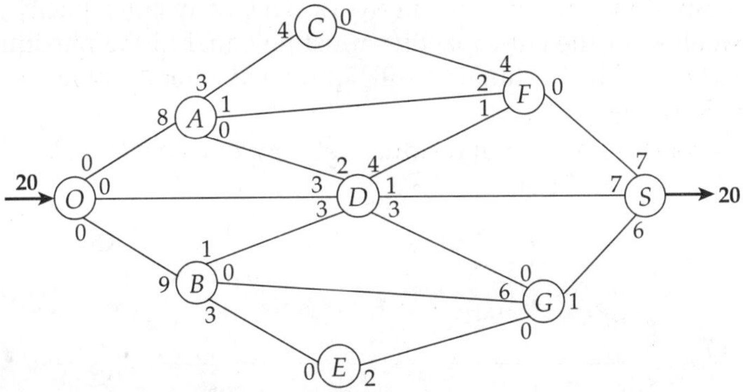

Fig. 16.67

Optimal Solution. There are no more augmenting paths available, so the current flow pattern is optimal. The maximal flow is 20 units. The maximal flow pattern across various links is identified by comparing the final residual capacities with the original link capacities. There is a flow along a link only if the final residual capacity is less than the corresponding original capacity. The amount of this flow is given by the difference in these (original versus residual) capacities. So, comparing the final residual network (last iteration) with the original network, we obtain the maximal flow pattern as shown in the network below:

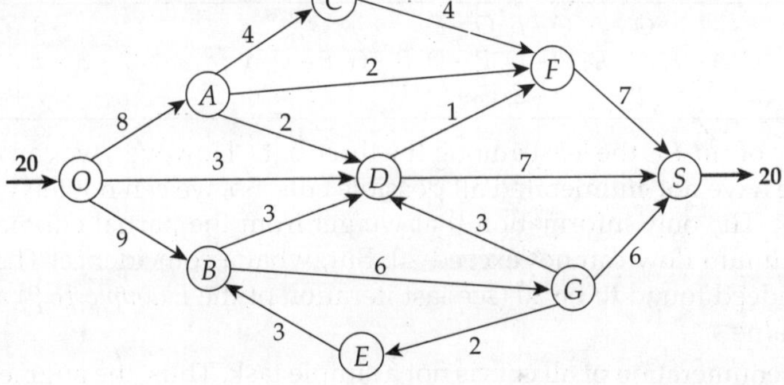

Fig. 16.68

16:7. MAXIMUM - FLOW MINIMUM - CUT METHOD

It is sometimes frustrating to waste time looking for an (non-existent) augmenting path when optimal flow has been attained. The ***maximum-flow minimum-cut method*** helps us to recognize the attainment of maximal flow, thus saving us from searching for additional augmenting paths.

A **cut** defines a set of directed links which when deleted from the network will cause a complete disruption of flow between the source and sink. There may be many ways to slice through a network to generate a cut. A cut helps us to analyze the amount of flow disruption by cutting all supplies midway.

The **cut capacity** or ***cut value*** for a cut equals the sum of the ***link capacities*** of the associated links.

Theorem (*max-flow min-cut*). *For any network with a single source and single sink, the maximum flow possible from source to sink is equal to minimum cut capacity for all cuts of the network.*

Proof. Beyond the scope of this book.

Clearly, the min-cut corresponds to a cut that disconnects the source and the sink and has the smallest possible cut capacity among all such cuts.

The above theorem bears an interesting implication. Let f denote the amount of flow from the source to the sink for any flow pattern in the network. Then, any cut-capacity provides an upper-bound to f and the smallest of the cut-capacities must be equal to the maximum value of f. As a cut-capacity can never be negative, optimality will be attained whenever there is a cut in the residual network with a cut value equal to zero.

To illustrate let us consider the initial residual network of *Example 16.9:*

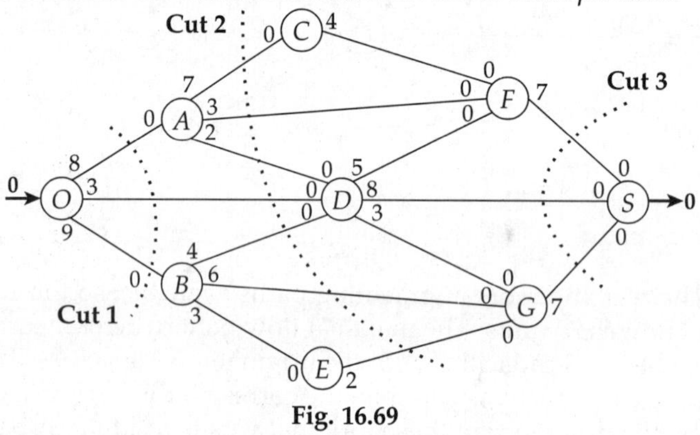

Fig. 16.69

The ***cut capacities*** for the three cuts indicated in the network are computed below:

Cut	*Associated links*	*Cut-capacity*
1	$O-A, O-B, O-D$	$8+3+9=20$
2	$A-C, A-F, A-D, O-D, B-D, B-G, E-G$	$7+3+2+3+4+6+2=27$
3	$F-S, D-S, G-S$	$7+8+7=22$

The cut-capacity of ***cut 1*** is the least among the three cuts. However, we can not say that it is the min-cut value as we have not enumerated all possible cuts. So, we can not tell what is the maximal flow in the network. The only information that we get from the partial enumeration of the three cuts is that the maximum flow can not exceed 20. But, what a coincidence! The maximal flow for this network was indeed found to be 20 (see last iteration of the *Example 16.9*) and the ***cut 1*** really gives the min-cut value.

The exhaustive enumeration of all cuts is not a simple task. Thus, the augmenting path method is considered more efficient.

PROBLEMS

16.19. In the network (*Fig. 16.70*), the bi-directional capacities are shown on the links using the normal convention. For example for the link 3–4 the flow limit is 10 units from node *3* to node *4* and it is 5 units from node *4* to node *3:*

(*a*) Illustrate three cuts on the network flows and obtain their capacities.

(*b*) Determine two additional cuts and find their capacities.

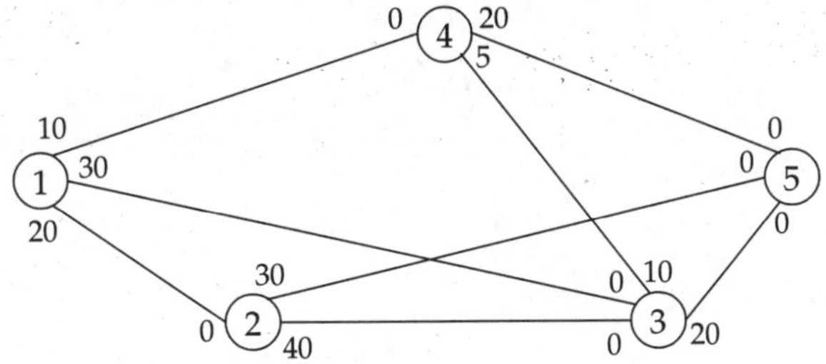

Fig. 16.70

16.20. Consider the maximal flow problem for the network given in *Fig. 16.71*. The source is node *1*, the sink is node *6* and the link capacities are shown on the directed links:

(*a*) Use the augmenting path algorithm to solve it.

(*b*) Determine any two cuts and find their capacities.

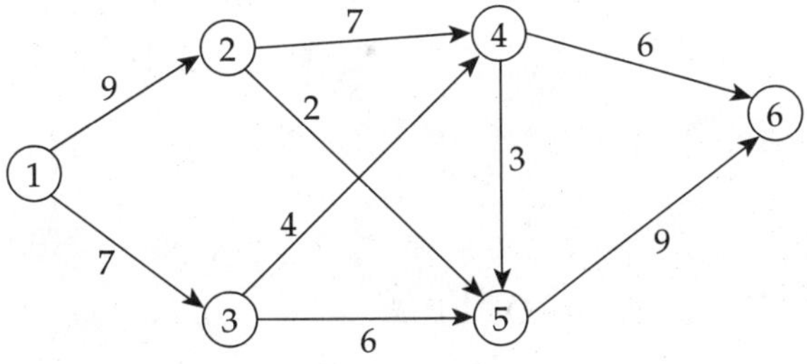

Fig. 16.71

16.21. Consider the residual network (*Fig. 16.72*) before any flows have been assigned. The source is node *A*, the sink is node *T* and the link capacities (in the original direction) are shown on the links by positive numbers. The 0′s on the links represent link capacities in the opposite direction. Use the augmenting path algorithm to solve it.

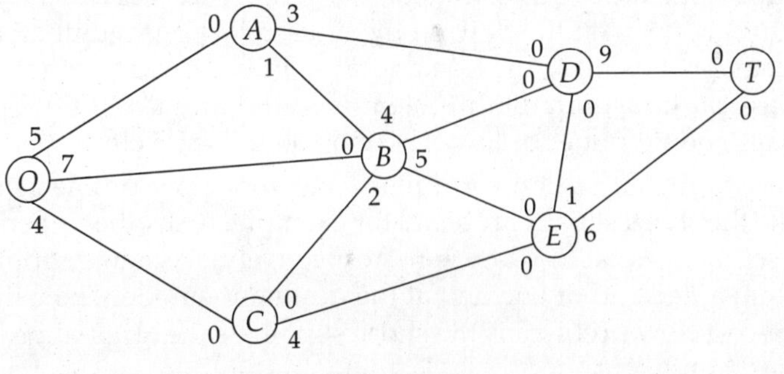

Fig. 16.72

16.22. Determine the maximum flow in the network of *Problem 16.19*. Also, determine:

(*a*) the surplus capacities for all the links.

(*b*) the amount of flow through node 2, 3 and 4.

(*c*) Can the network flow be increased by increasing the capacities in the directions 3 → 5, and 4 → 5?

16.23. For the following network, determine the maximal flow and the optimal flow in each link:

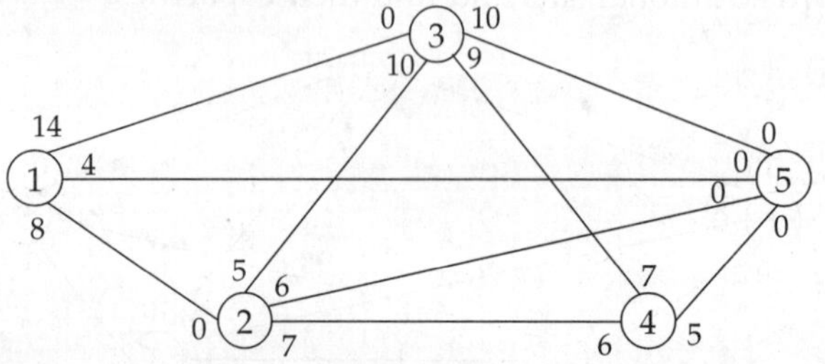

Fig. 16.73

16.24. Consider the network (*Fig. 16.74*). Find the flow pattern giving the maximal flow from source to sink, where the link capacities are given on the link nearing each node. The source is node *O* and the sink is node *T*:

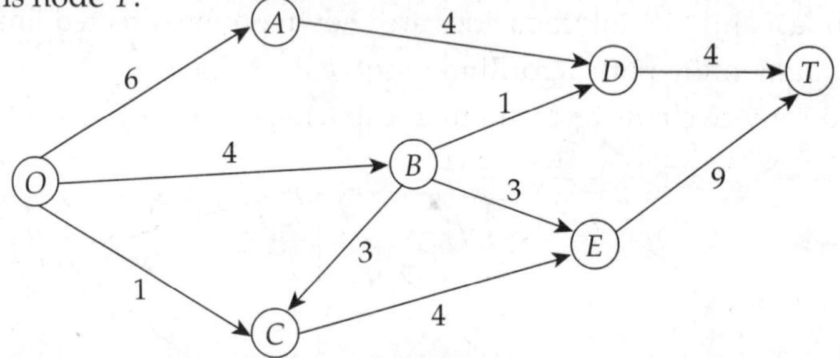

Fig. 16.74

16:8. MINIMUM-COST FLOW PROBLEM

The basis of maximal flow problems is that there exists a capacitated network and costs are constant regardless of how much flow is flowing in the links. But, if there is a unit cost associated with flow on each link, we have a minimum-cost flow problem. The scenario of a generic minimum-cost flow problem is the following:

At some nodes (supply nodes) there are an excess of some commodity, while some other nodes (demand nodes) require certain amounts of this commodity. The remaining nodes (trans-shipment nodes) neither require nor supply this commodity. There are lower (which can be zero) and upper (which can be infinity) bounds on the flow on the links, as well as a unit cost of transportation on the links. The objective is to satisfy the demand at a ***minimal cost*** without violating the bound constraints. The classical transportation problem is a minimum cost flow problem without any trans-shipment nodes and no upper or lower bounds on link capacities.

Interestingly, it is possible to transform many network flow problems into a minimum cost flow problems. Take the shortest path problem, for example. Relax the upper and lower bounds on the link flows (*i.e.*, set them equal to ∞ and zero, respectively). Assign a supply amount of one unit at the origin node and a demand of one unit at the destination node. Then, the original flow in the resulting minimum-cost flow problem is the shortest path of the original problem.

A maximal flow problem can be converted into a minimum-cost flow problem. It requires a little bit more effort. Assign a zero unit cost to all the links. Lower bounds are zero and the upper bounds are the link capacities. Define a "dummy" link from the origin to the destination node with a flow cost of one per unit and no upper or lower bounds. We also need an upper bound on the maximal flow quantity. Any "large" number will do, but this number is very easy to obtain. For example, the sum of the capacities of all links that are incident out of the origin node. Assign this quantity as the "supply" available at the origin, and let the same quantity be "demanded" at the destination node. In trying to find the minimum-cost flow, the solution procedure will push the maximum possible flow through the original network, since it costs nothing on these links. Only the overflow over and above the maximal flow possible in the original network is diverted through the "dummy" link from the origin to the destination node. The optimal objective function value of the minimum-cost flow problem is equal to the optimal flow amount on the "dummy" link. If this amount flows through the "dummy" link, the rest of the "supply" available at the origin must have been transported to the destination node over the original network. Thus, the difference between the estimated upper bound (*i.e.*, the supply amount at the origin) and the optimal objective function value is the maximum flow possible from the origin node to the destination node in the network.

16:9. TRANSPORTATION PROBLEM AS A MINIMUM-COST FLOW PROBLEM

Consider a transportation problem where three factories *A*, *B* and *C* can supply any of six customers *1, 2, 3, 4, 5* and *6* with a particular product. Let the demand for this product from each of the customers be 40, 35, 25, 20, 60 and 30 tons respectively. The maximum production at the three factories is 60, 70 and 80 tons respectively. The variable production cost per ton is: 11.3, 11.0 and 10.8 (₹ '000) at factories *A*, *B* and *C*, respectively. The transportation costs per ton from a factory to a customer are given below:

Source	*Transportation cost (₹ '000) per ton to customer*						*Supply*
	1	2	3	4	5	6	
A	1.5	1.8	2.1	4.2	2.5	3.0	60
B	2.2	4.6	3.5	2.4	1.8	4.0	70
C	3.6	4.8	1.6	4.4	2.8	2.0	80
Demand	40	35	25	20	60	30	

It is desired to determine the quantity of product to be supplied from each factory to each customer so as to minimize the total cost involved.

To treat this problem as a minimum cost network flow problem, we need first to find the cost for each factory/customer combination of producing and transporting one ton from a factory to customer. These costs are obtained by adding the variable production costs to the given transportation costs. Thus, we obtain the following table:

Source	*Production + Transportation cost (₹ '000) per tone to customer*						*Supply*
	1	2	3	4	5	6	
A	12.8	13.1	13.4	15.5	13.8	14.3	60
B	13.2	15.6	14.5	13.4	12.8	15.0	70
C	14.4	15.6	12.4	15.2	13.6	12.8	80
Demand	40	35	25	20	60	30	

The problem can be viewed as a minimum cost network flow problem as shown in *Fig. 16.75:*

(*a*) Each ***source*** (factory) can supply as much as the factory can produce.
(*b*) Each ***sink*** (customer) has a demand equal to the associated customer demand.
(*c*) Each link has a capacity equal to the demand of the customer it goes to and a cost equal to the combined production as well as transportation cost.

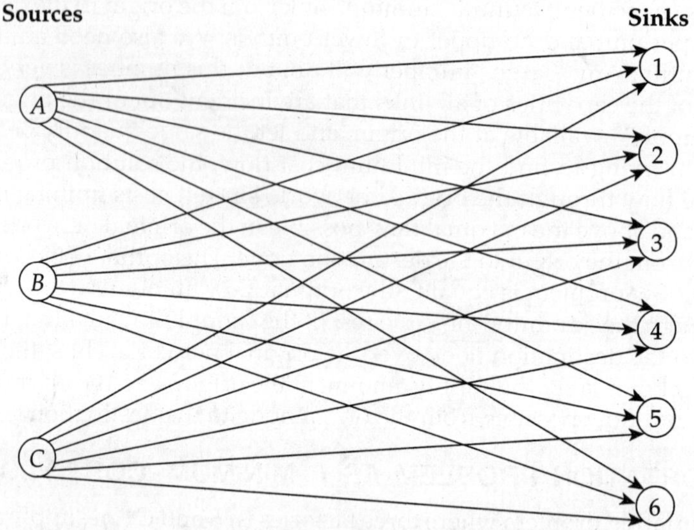

Fig. 16.75

It is easy to observe that:

(*i*) imposing a limit on the amount supplied from a particular factory to a particular customer is merely a matter of setting a link capacity appropriately.

(*ii*) links can be deleted to ensure that a particular factory do not supply to a particular customer.

Example 16.10. *Consider the following transportation problem:*

Source	*Destination* D_1	D_2	D_3	*Supply*
S_1	6	7	4	40
S_2	5	8	6	60
Demand	30	40	30	

Formulate the network representation of this problem as a minimum cost flow problem.

Solution. The network formulation of this problem is shown in *Fig. 16.76*, where the number next to each node is the net flow generated there and the number next to each link is the cost per unit flow through that link:

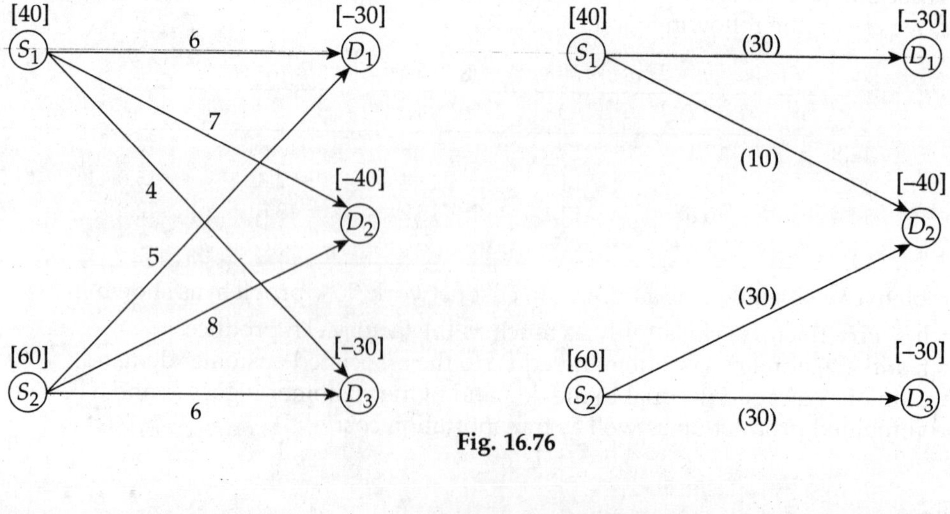

Fig. 16.76

16:10. MORE NETWORK FLOW PROBLEMS

1. Transhipment Problem

Usually goods are not transported directly from factories to customers but shipped (transhipped) *via* intermediate locations (depots or warehouses). For example, let us consider the transportation problem discussed in *Section 16:09* and add a new depot D (or source), where:

(*a*) *D* has a storage cost of 0.7 (₹ '000 per ton).

(*b*) the cost of shipping from factories *A*, *B* and *C* to *D* is 0.1, 0.3 and 0.7 (₹ '000 per ton) respectively.

(*c*) the cost of shipping from *D* to customers 1, 2, 3, 4, 5 and 6 is 0.7, 0.9, 1.1, 0.8, 0.6 and 0.9 (₹ '000 per ton) respectively.

We incorporate the new depot in the network flow representation as shown in *Fig. 16.77*:

It may be seen that we have added the following to the earlier network of *Fig. 16.75:*

(*i*) two new nodes — D_1 representing goods in to the depot (the "front door") and D_2 representing goods out of the depot (the "back door").

(*ii*) a link between D_1 and D_2 with capacity equal to the total factory capacity, representing the maximum amount that can reach the depot from the factories; and a cost 0.7 assigned to it.

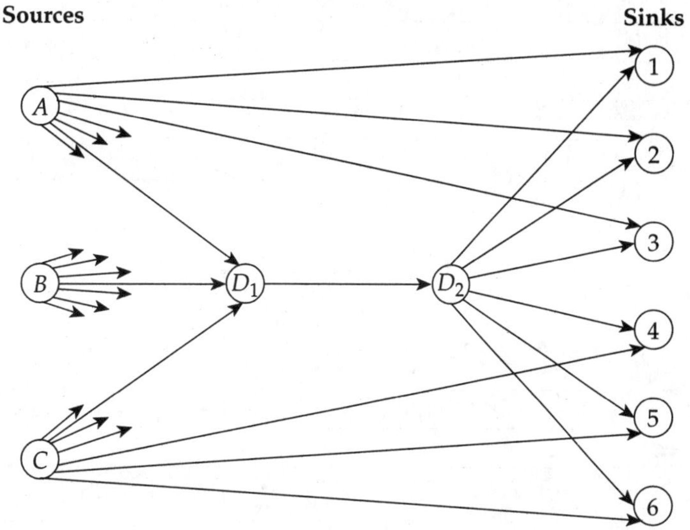

Fig. 16.77

(*iii*) links from the sources (factories) to D_1 with a capacity equal to the factory production capacity and a cost equal to the combined production as well as transportation cost (factory to depot), *i.e.*

Link	*Capacity*	*Cost (₹ '000 per ton)*
$A \to D_1$	60	11.3 + 0.1 = 11.4
$B \to D_1$	70	11.0 + 0.3 = 11.3
$C \to D_1$	80	10.8 + 0.7 = 11.5

(*iv*) links from D_2 to the customers with a capacity equal to the customer demand and a cost equal to the cost of shipping from the depot to the customer, *i.e.*

Link	*Capacity*	*Cost (₹ '000 per ton)*
$D_2 \to 1$	40	0.7
$D_2 \to 2$	35	0.9
$D_2 \to 3$	25	1.1
$D_2 \to 4$	20	0.8
$D_2 \to 5$	60	0.6
$D_2 \to 6$	30	0.9

These type of problems are called ***transhipment*** problems as they involve the transhipment of goods *via* intermediate locations rather than directly from factories to customers.

We solve this problem using a software package, where the problem is entered as a network flow problem with the input shown below:

From/To	*A*	*B*	*C*	D_1	D_2	*1*	*2*	*3*	*4*	*5*	*6*	*Supply*
A				11.4		12.8	13.1	14.4	15.5	13.8	14.3	60
B				11.3		13.2	15.6	14.5	13.4	12.8	15.0	70
C				11.5		14.4	15.6	12.4	15.2	13.6	12.8	80
D_1					0.7							0
D_2						0.7	0.9	1.1	0.8	0.6	0.9	0
1												0
2												0
3												0
4												0
5												0
6												0
Demand	0	0	0	0	0	40	35	25	20	60	30	

The following output is obtained:

From	*To*	*Flow*	*Unit cost*	*Total cost*
A	D_1	60	11.4	684
B	D_1	70	11.3	791
C	3	25	12.4	310
C	6	30	12.8	384
C	D_1	25	11.5	287.5
D_1	D_2	155	0.7	108.5
D_2	1	40	0.7	28
D_2	2	35	0.9	31.5
D_2	4	20	0.8	16
D_2	5	60	0.6	36
Total		Objective function value z =		2676.5

One may see that 155 units flow through the depot (from D_1 to D_2). If we restrict to say 100 units (because the depot cannot cope with more than that), we merely need to edit the Flows bounds in the software package appropriately.

2. Management of Foreign Exchange

Another interesting area where network flow is widely used is the management of foreign exchange dealings for banks, corporate, etc. To illustrate, let us consider a problem involving 4 currencies (Euro, £, USD and INR) as shown in *Fig. 16.78* below:

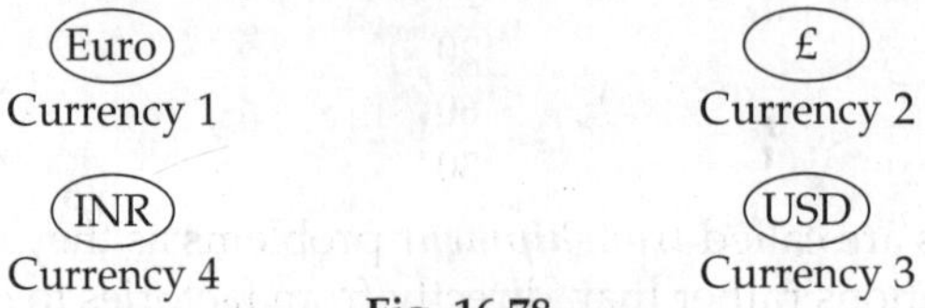

Fig. 16.78

The international money market rates are constantly available for converting any pair of these (commonly traded) currencies and these rates can be represented by the matrix (r_{ij}) where r_{ij} = the amount of currency j obtained in exchange for one unit of currency i. For example, the amount of INR (currency 4) obtained in exchange for one £ (currency 2) is r_{24}.

Let us consider a simple deal in which we exchange one £ to INR and then convert back. We end up with an amount £ $[r_{24}(r_{42})]$. We observe that $r_{24}(r_{42})$ does not equal 1, almost always. This is so, because rates are quoted only to a certain decimal places and truncation takes place. Although we would expect $r_{24}(r_{42}) < 1$, *i.e.*, we lose some amount on the deal, it is not certain that a complex series of transactions will really be unprofitable. For example, will converting one £ through the chain £ → USD → INR → Eruo → £ yielding £ $[r_{23}\, r_{34}\, r_{41}\, r_{12}]$ be necessarily unprofitable?

Foreign exchange transactions can be modeled as a network flow gains problem and the available network algrorithms enable us to identify profitable deals within seconds (necessary when rates are constantly changing).

We considered above a deal in the spot market (converting currency immediately). The same approach can be extended to include a time dimensions, involving both the forward market (converting currency at a given point in the future) and deposits (depositing currency for a certain time to gain interest). The inclusion of the time dimension may be illustrated by simply repeating the same picture as shown in *Fig. 16.78* for each time period. The links between currency nodes in different time periods naturally represent forward transactions and deposits. This is shown in *Fig. 16.79* below:

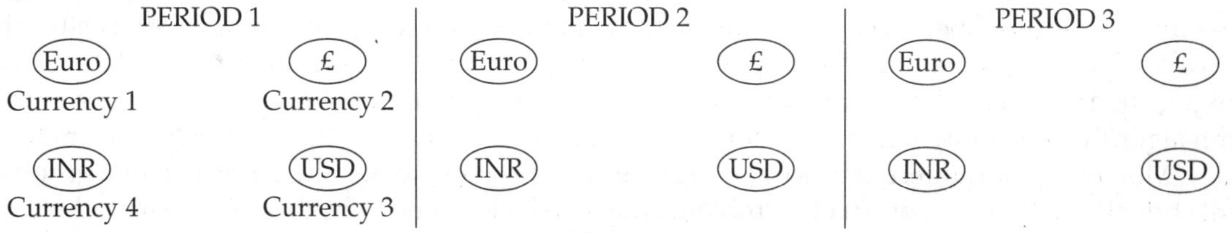

Fig. 16.79

Note that there can be two types of links in the above problem:

(*i*) those which are effectively **risk-free**; and (*ii*) those which are not.

For example, a risk-free link might be investing Euro in the current time period (period *1*) for a fixed time in a fixed interest instrument (issued by a reputable issuer). A risky link might be changing money on the spot market in period 2 (as we do not yet know what the spot market rate in period 2 will be). Apart form pure "speculation" to make money, such models can be used, for example, to balance the currency holdings of a company so as to meet its future requirements for foreign currency at a minimum cost.

3. Network Flow: State of the Art

Commercial computer codes for the minimum cost network flow problem are readily available and able to solve large problems. In solving such problems it is the number of links which (essentially) determine the solution time.

Computational results presented in the literature indicate that problems involving thousands of links can be solved within seconds (on a powerful computer).

4. Sensor Networks: Exposure Problem

A sensor network is a computer network of many, spatially distributed devices using sensors to monitor conditions at different locations, such as temperature, sound, vibration, pressure, motion, or pollutants. A fundamental property of sensor networks is the coverage: how well a sensor network can cover a certain terrain.

The exposure of an object in a sensor network depends on how well the sensor network can observe the object moving on an arbitrary path over a period of time. It is directly related to a sensor network's coverage. The analysis of exposure thus helps us to study a sensor network's coverage, giving us important information about the sensor network itself. Knowing the ***minimal exposure*** path is of great use. It is on this path that we should add new sensors, should we want to improve the general coverage efficiently.

Sensor networks are applied in a wide variety of areas, such as video surveillance, traffic monitoring, air traffic control, robotics, cars, home monitoring, manufacturing, and industrial automation.

5. Route Resolving Tools in Urban Analysis

Route resolving is used in many problem-solving and design situations. Such situations include route planning, market analysis, traffic planning and location planning. Urban areas set certain requirements to route resolving tools. In traffic planning for example, route resolving must take into account one-way streets and turning restrictions. To be able to solve traveling salesman problem correctly in urban areas, route resolving has to put points alongside the same street in correct order. This means that algorithm must be able to break street into pieces and use the exact point at the right side of street as a starting point for route resolving. This ability is usually called ***dynamic segmentation***.

Basic algorithm behind route resolving is Dijkstra's algorithm. One must describe street network as a graph with nodes and links. Traditionally, crossings are nodes and streets are the links. If every two-way street is described as two separate links, then one-way streets are handled very easily. The key for solving the problem then is to realize that we can turn the algorithm around. Rather than thinking streets as being links between crossing can be thought as being links between streets. All graph algorithms work in this new model with minor changes and handling of turning restrictions is easy. Moreover, the model enables traveling salesman problem to be solved not just for a group of points but also for a group of streets. Problems that can be handled with this feature include snow cleaning, post delivery or garbage collection.

6. Seeds, Chicken and Dog Puzzle

You want to cross the river with your belongings which are a chicken, a dog and some valuable seeds. While you're with them, you can control their behaviour but when they are left unsupervised they follow their natural instincts. The river may be crossed on a little boat in which there is room for only you and one of your belongings.

Some of your options are unwise, for example, if you cross the river with the dog leaving the chicken and the seeds behind, then the chicken will eat the seeds, and if you cross with the seeds then the dog will eat the chicken.

The goal is to get yourself and your belongings intact to the opposite shore in the minimum number of crossings.

This puzzle is modeled by the network displayed in *Fig. 16.80*. The notation is straightforward: 'S' stands for 'seeds', 'C' for 'chicken', 'D' for 'dog', '•' for yourself and '⊥' represents the river. The nodes represent the possible states of the puzzle and there is a link between

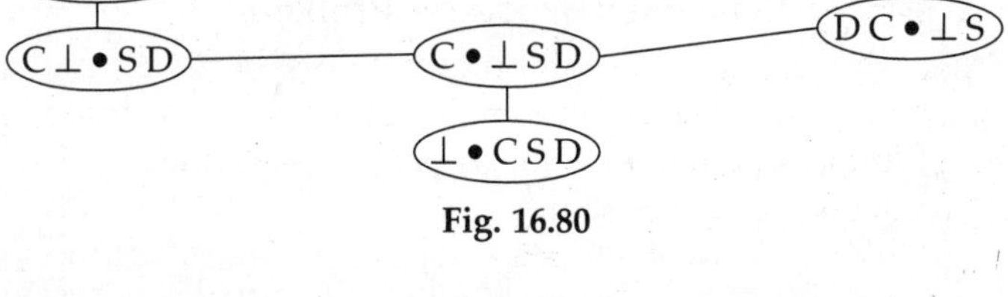

Fig. 16.80

two states if one can be reached from the other by a sensible river crossing. A solution is a path from the start node at the top, to the end node at the bottom. As you may have noticed, there are two possible solutions to this puzzle, and one path is shorter than the other.

7. The Three Glass Puzzle

This famous puzzle can also be modeled with networks in more than one way, and it goes as follows: You have three glasses '*A*', '*B*' and '*C*' with capacities 8, 5 and 3 gallons, respectively. Glass *A* is filled full with water while the other two are empty. We want to divide the water in two equal parts by pouring from one glass to the other, without using any measuring device. What is the smallest number of pourings needed to achieve this result?

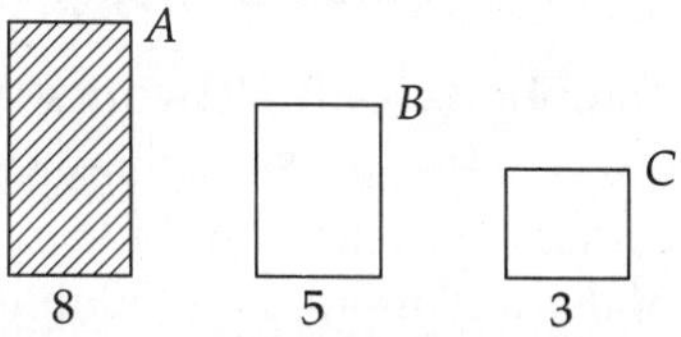

Fig. 16.81

The problem can be represented as shortest path problem. We don't really need three numbers to describe any possible distribution of water. Since, the sum of the amounts of water in the three glasses should equal 8 units, each distribution can be characterized by the quantities in *B* and *C* only. Initially *B* and *C* are empty so that start state (source node) is (0, 0) and the target is (4, 0), *i.e.*, the sink node:

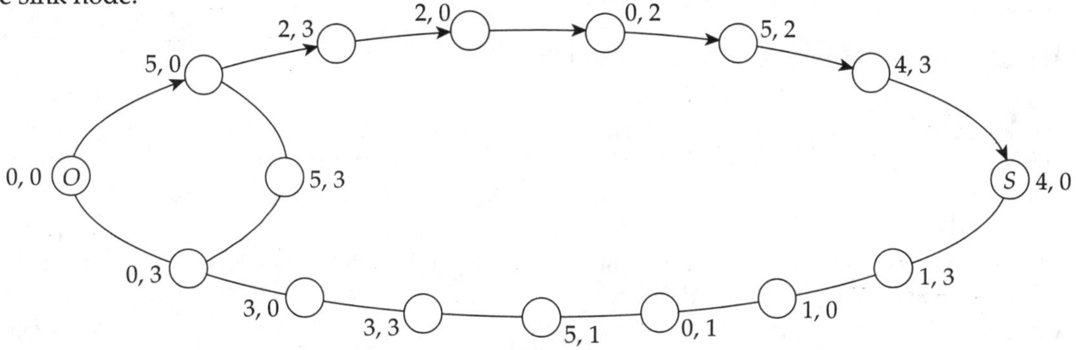

Fig. 16.82

In the network of *Fig. 16.82*, every node represents a possible water distribution and there is a link between two nodes, if the corresponding states can be reached from each other by pouring the contents of one glass on another. The above network shows different routes that lead from start node (0, 0) to end node (4, 0). Each link represents a single pouring and can be assumed to have a length of 1 unit. The problem, thus, reduces to finding the shortest path between start and end nodes.

Observe that there are two paths that go from the start to the end node, and that one is longer than the other. The optimal solution given by the highlighted path requires 7 pourings. It is also evident that both, the start and the target node are in the same cycle, so the solution may be reached by making sure we stay on the larger path. Many puzzles shares this feature.

16:11. INSIGHTS INTO BIG NETWORKS

Techniques of network flows have proved useful in many fields. Physicists analyze electric circuits in terms of networks, with wires as links and junction points as nodes. Chemists find a natural correspondence between networks and the structural diagrams of molecules: an atom is a node, and a link is a bond between atoms. Some studies describe even the neural networks of the brain. An economy is also a network: companies or industries are nodes, and links represent the transactions.

During the past few decades the network theory has become more statistical and algorithmic. A rich source of ideas has been the study of "random networks", which are typically formed by starting with isolated nodes and adding links to it one at a time. The idea was propagated by famous mathematician ***Paul Erdös*** who suggested the existence of a "giant component", a connected piece of the network spanning most of the nodes.

The recent work on the World Wide Web and other very large networks is also statistical and algorithmic in nature, and has close ties to the Erdös theory of random networks.

Telephone Call Networks

Can you imagine how large a real life network can be? A good example of a really big network comes from telephone billing records. A recently studied "call network" found as many as 53,767,087 nodes and more than 170 million links.

The call network is actually a directed multi-network; directed because the two ends of a call can be distinguished as originator and receiver, and a multi-network because a pair of telephones can exchange more than one call in a day. It is not a connected network but has 3.7 million separate components, most of them tiny; three-fourths of the components are pairs of telephones that called only each other. Yet the call network also has one giant connected component, with 44,989,297 nodes, or more than 80 per cent of the total. The emergence of a giant component is characteristics of Erdös random networks.

Social Networks: People Who Know People

Among the most interesting large networks are those in which we are the nodes. These "social networks" are associated with the idea that the acquaintanceship network connecting the entire human population has a diameter of six or less. We attribute this notion to Guglielmo Marconi, who had said decades ago: ". . . the wireless tele-networks would so contract the world that any two people could be connected by a chain of 5.83 intermediaries".

The theory of social networks is based on the idea of random networks. It is shown that any bias in the random placement of links tends to reduce the overall connectivity of the network and increases its diameter. Thus, social structures that bring people together in clusters have the side effect of pushing the clusters farther apart.

Several sub-networks of the acquaintanceship network have also been explored. The prototype is the "collaboration network" centered on Paul Endös. In this network distance from Erdös's node is termed the Erdös number. Erdös himself has an Erdös number 0. All those who co-authored a paper with him have Erdös number 1. Those who did not write a joint paper with Erdös but who are co-authors of a co-author have Erdös number 2, and so on. The network built up in this way by adding concentric layers of co-authors, can be viewed as a component of a larger network with a node for every contributor to the literature of science.

It may be interesting to consider another collaboration network that has move actors instead of scientists at the nodes. As feature films are a smaller universe than scientific publications, the structure of this "Hollywood network" can be determined with more clarity. According to a recent Internet movie database, the Hollywood network has 355,848 nodes, representing actors who have appeared in 170,479 films.

The World Wide Web Network

The World Wide Web has the advantage that it comes already encoded for computer analysis. The nodes and links do not have to be catalogued; any computer attached to the Internet can navigate through the network just by following links from node (a page) to node.

To estimate the diameter of the Web, Barabási and his colleagues at Notre Dame did not visit every node and traverse every link; they simply studied a small corner of the Web and extrapolated the findings to the rest of the network. The Barabási group used a software called "robot" to follow all the links on a starting page, then all the links on each page reached from that page, and so on. The same technique is employed by search engines to index the Web. The diameter of the network is an important statistic when you are trying to find something on the Web. A random search would typically have to examine half the 800 million documents before stumbling on the right one. But, the Notre Dame result suggests that from any reasonable starting point, there should be a path to the target page crossing only about 19 links.

A study group draws attention to two special kinds of nodes in the Web: "Hubs" that are nodes of high out-degree (pages that point to many other pages) and "Authorities" that have high in-degree (that are pointed to by many other pages, and especially by hubs). Their algorithm defines hubs and authorities by an iterative feedback process. An initial scan of the Web identifies pages of high out-degree and high in-degree, which form the initial sets of hubs and authorities. Then, these sets are refined by a recursive procedure that discards a hub candidate unless many of its outward links point to pages that are members of the authority set; likewise authorities are weeded out unless they are pointed to by many of the hubs. Repeated application of this algorithm narrows the focus to those hubs and authorities that are most densely connected to one another.

Similar principles are at work in the popular Web search engine called "Google". It employs a conventional text-based scan to create an index of the Web's content, but the pages recommended in response to a query are ranked according to information from the link analysis. A page (node) is rated highly if many pages point to it, and if many other pages point to those pages (nodes), and so on.

16:12. CONCLUDING REMARKS

Network flow problems can be formulated as linear programs. Further, as shown earlier, many network flow problems can be formulated as minimum cost flow problems. Like all mathematical programming problems, an algorithm that is specifically designed to solve a particular class of problem is more efficient on that class than a general-purpose algorithm. For example, shortest path problem can be solved as a minimum cost flow problem, which could be solved as a linear program using the usual Simplex Method. But it is more efficient to use the ***network simplex algorithm*** for this network flow problem. Finally, there are special purpose algorithms to solve just shortest path problems, which are the most efficient way of solving these specific problems.

It is the beyond the scope of this book to get into the theory behind algorithms for network flow problems. One may refer to excellent book by *Ahuja, Magnanti, & Orlin*, that discuss the solution algorithms as well as numerous real world applications of network flow problems. One may also refer to the original paper of *Dijkstra*. In addition, there are a number of informative resources on the Internet.

TEST YOUR UNDERSTANDING

1. Which of the following OR problems can not be expressed as a network flow problem?
(*a*) an assignment problem (*b*) a transportation problem
(*c*) a replacement problem (*d*) a queueing problem.

2. Which of the following situations can not be modeled as a network flow problem?
(*a*) Determining the shortest route between two cities
(*b*) Finding the maximum capacity of water flowing in pipelines
(*c*) Determining the optimal stock to keep for cattle
(*d*) Finding the most scenic driving route for the weekend.

3. Which of the following statements is not correct?
   (*a*) A directed network allows flows in one direction only
   (*b*) The network algorithms are more efficient than the simplex method
   (*c*) A connected network always has at least one path
   (*d*) A cycle in a network may be directed or undirected.
4. A minimal spanning tree involves
   (*a*) all the nodes with cycles allowed
   (*b*) all the nodes with cycles not allowed
   (*c*) the shortest path between starting and ending nodes
   (*d*) a connected network with all directed potential links.
5. A typical application of minimal spanning tree involves
   (*a*) construction of paved roads that link several towns
   (*b*) determining the least cost path between two cities
   (*c*) finding the shortest route communication network
   (*d*) determining a least cost flow pattern from refineries to various outlets.
6. Dijkstra's algorithm is designed to determine the shortest path between
   (*a*) any two nodes of the network
   (*b*) the source node and any other node
   (*c*) any node and the destination node
   (*d*) any pair of nodes of a unidirectional network.
7. Among all possible "cuts" of flow in a network, the minimal cut capacity corresponds to
   (*a*) the minimal network flow (*b*) the maximal network flow
   (*c*) no network flow (*d*) none of the above.
8. An augmenting path is an algorithm to find
   (*a*) a minimum cost flow pattern (*b*) a maximal cost flow pattern
   (*c*) a minimal spanning tree (*d*) a shortest path from source to sink.
9. A transportation problem can be represented as a network flow problem, where
   (*a*) origins represent sinks and distinctions the sources
   (*b*) origins represent sources and destinations the sinks
   (*c*) objectives is to maximize the network flow
   (*d*) per unit transportation costs become irrelevant.
10. Which of the following statements is not correct?
   (*a*) A physicist considers electric circuits as networks with wires as links and junctions as nodes
   (*b*) A chemist view structural diagrams of molecules as networks with an "atom" as a node and a "bond" between atoms as a link
   (*c*) A chess tournament is a network with players being nodes and matches the links
   (*d*) In a telephone call-network, the nodes are the calls made and the links are the telephone numbers.

ANSWERS

1. (*d*)	2. (*c*)	3. (*a*)	4. (*b*)	5. (*a*)
6. (*b*)	7. (*b*)	8. (*b*)	9. (*b*)	10. (*d*)

REVIEW EXERCISE

1. What do you mean by the following terms?
   (*a*) a node (*b*) a link
   (*c*) flow (*d*) a potential link
2. Explain the following:
   (*i*) a tree (*ii*) a spanning tree
   (*iii*) minimal spanning tree (*iv*) a cycle.
3. What is a network flow problem? Illustrate with an example.
4. Explain the difference between a directed and an undirected flow through a link.
5. What do you mean by a directed or an undirected network? What is a "connected" network?
6. Describe a procedure to determine a minimal spanning tree for a given set of nodes and potential links.
7. Briefly explain an algorithm to solve Shortest Path Problem.
8. "Dijkstra's algorithm is a labeling algorithm". Explain.
9. What is a maximal flow problem in networks? Give some examples to illustrate.
10. In the context of a maximal flow problem explain the following terms:
    (*i*) a source (*ii*) a sink
    (*iii*) trans-shipment nodes (*iv*) link capacity.
11. Explain what the objective in a maximal flow problem is.
12. A maximal flow problem is a single-source, single-sink problem". Explain.
13. What is a residual network? Explain its role in determining an augmenting path.
14. Briefly describe the augmenting path algorithm to solve a maximal flow problem.
15. Explain the term '**cut**' in a network flow problem. What is its role in studying the amount of flow passing through the effected links?
16. State the max-flow min-cut theorem in the context of a maximal flow problem. Explain its role in determining an optimal flow pattern.
17. "A transportation problem is a minimum cost flow problem". Explain.
18. Briefly describe the components of a minimum cost flow problem.
19. "The minimum cost flow problem holds a central position among network optimization models, as almost all network flow problems are its special cases". Explain.
20. "Networks provide a natural way of graphically displaying the flow of activities in a project". Explain.
21. Write a short note on the following:
    (*i*) Seven Bridges of Königsberg (*ii*) Chinese Postman Problem.
22. "An Euler tour lies at the foundation of network flow problems". Explain.
23. List some real-life application of network flow problem, clearly pointing out the nodes, links and flows in each case.
24. "A network provides a powerful visual and conceptual aid for showing relationships between components of systems in virtually every field of scientific, social and economic endeavor". Discuss.

REVIEW EXERCISES

CHAPTER 17

Probability

"Let men suspect your tale untrue, keep probability in view"

Operations Research techniques discussed in the preceding chapters assume that all the data are known with certainty. This assumption is not true always and may be far away from reality. For example, the demand for electric power during the summer months can vary randomly from year to year, depending on weather conditions.

In such cases we use observed or historical data to describe the demand by a probability distribution.

17:1. UNCERTAINTY AND PROBABILITY

If it were possible to predict the future with ***complete certainty***, the structure of the business world would radically be different from what it is these days. Had there been nothing like ***uncertainty***, there would be no excess production, no special clearance sales, no speculations on the stock market, etc. Unfortunately, we do not live in a world where failure could be predicted with complete certainty. Our desire to deal with uncertainty (*e.g.*, while taking decision under uncertainty) leads us to what is known as ***'Probability Theory'***. We talk of this theory only for those phenomena or experiments (the act of performing something) that involve uncertainty. Any phenomenon whose outcomes* cannot be predicted with complete certainty is usually referred to as a ***random phenomenon*** or a ***random experiment***. The following are some examples of a random experiment:

Experiment	*Outcome of interest*
1. A coin is tossed in the air.	Coin lands with head up.
2. A parachuter jumps from a plane.	His parachute will open in time.
3. A ceiling fan is installed.	It will fall down.

Could you predict with complete certainty about the occurrence of the outcome of interest in any of these experiments? Obviously no. There is some uncertainty involved in the happening of the outcomes of interest. However, we can make the following observations. We learn from past experience that it is very unlikely that a fan will fall down, therefore, there is a very small ***uncertainty*** in the case of experiment 3. Also, the past experience says that the parachute in experiment 2 did open successfully most of the time and, therefore, the uncertainty in this case is also small. But, you may observe that the uncertainty involved in this case is obviously higher as compared to what it is in the case of experiment 3. There is some uncertainty involved in experiment *1* also. Obviously, the amount of uncertainty in this case is much higher as compared to what it is in the other cases.

The above observation lead us to a hint that 'uncertainty' can be measured, because if it could not then how could we talk of 'higher' or 'lesser' uncertainty. A widely used measure of uncertainty is known as ***probability***, which we shall now define.

* An 'outcome' and the 'occurrence of an outcome' of a random experiment are considered as undefined terms in probability theory.

17:2. SAMPLE SPACE AND PROBABILITY

Definition 1 (*Sample Space*). *The set of all possible outcomes of a random experiment is called the* ***sample space*** *associated with the experiment. The possible outcomes (elements of the sample space) are called* ***sample points.***

Illustration 1. Consider the random experiment of tossing a coin once. There are three thinkable outcomes for the coin, *viz.*, to land on:

(*i*) Head, (*ii*) Tail, and (*iii*) Edge

To simplify our theory, we make it the convention that the coin will never land on edge. Thus, if we denote by H, the occurrence of head; by T, the occurrence of tail; and by S, the associated sample space, then

$$\mathbf{S} = \{H, T\}$$

2. If the random experiment consists of tossing a coin twice, the associated sample space is

$$\mathbf{S} = \{HH, HT, TH, TT\},$$

where, HT denotes the occurrence of H on the first coin and occurrence of T on the second, etc.

Definition 2 [*Discrete (Continuous) Sample Space*]. *A sample space that consists of a finite (or an infinite but countable) number of sample points is called a* ***discrete sample space.***

A sample space which is not discrete is called a ***continuous sample space.***

Illustration 1. Consider the number of customers arriving at a service counter within one hour. The sample space associated with this random experiment is

$$\mathbf{S} = \{0, 1, 2, ...\}$$

which is discrete.

2. The sample space associated with the inter-failure time 't' of electronic component is given by

$$\mathbf{S} = \{0 \le t < \infty\}$$

which is discrete.

Definition 3 (*Event*). *An* ***event*** *is a subset of the sample space.*

An event consisting of only a single sample point is called an ***elementary event.***

Illustration 1. Consider the random experiment of rolling a dice once. The associated sample space is

$$\mathbf{S} = \{1, 2, 3, 4, 5, 6\}$$

Then, the subsets

$$\mathbf{A} = \{1, 2\}, \quad \mathbf{B} = \{3, 4, 5, 6\} \quad \text{and} \quad \mathbf{C} = \{4\} \quad \text{etc.,}$$

all represent events of **S**. Clearly, **C** is an elementary event.

2. Consider once again the case of electronic component failure sample space

$$\mathbf{S} = \{0 \le t < \infty\}$$

The events $\mathbf{A} = \{0 \le t < 5\}$, $\mathbf{B} = \{t \le T\}$ of **S** represent the events that failure time is 5, and less than or equal to T, respectively.

Notes

1. The sample space itself is an event.
2. We shall use capital letters to denote events and lower letters for sample points.

Definition 4 (*Equally Likely Outcomes*). *Outcomes of a random experiment are said to be* ***equally likely,*** *if taking into consideration all the relevant evidences, there is no reason to expect one in preference to other.*

Illustration 1. As the result of drawing a card from a well-shuffled pack, any card may appear in a draw. Thus, the 52 possible outcomes are equally likely.

2. The falling and not falling down of a ceiling fan are not equally likely outcomes.

Definition 5 (*Favourable Outcomes*). *The outcomes of a random experiment which entail the occurrence of an event* **A** *are said to be* ***outcomes favourable*** *to* **A**.

Illustration. In tossing a dice, the number of outcomes favourable to the event

$$\mathbf{E} = \{\text{multiple of 3 appears}\}$$

is two, namely 3 and 6.

Note. In the case of a finite discrete sample space, the total number of sample points is said to be the *number of exhaustive outcomes*.

Definition 6 (*Mutually Exclusive Outcomes*). *The outcomes of a random experiment are said to be* ***mutually exclusive****, if they cannot occur simultaneously.*

Illustration. In the case of tossing a coin, H and T are mutually exclusive.

Classical Definition of Probability

Definition 7 (*Probability*). *If there are n exhaustive, pairwise mutually exclusive and equally likely outcomes of a random experiment and if m out of these are favourable to an event* **A**, *then the* ***probability*** *of the happening of* **A**, *denoted by P(***A***), is defined by*

$$P(\mathbf{A}) = \frac{m}{n} \qquad 0 \le m \le n,\ n > 0$$

Illustration 1. A dice is tossed once and the dots on the face turned up observed. The sample space associated with the experiment is

$$\mathbf{S} = \{1, 2, 3, 4, 5, 6\}$$

If **A** = the event that an even number occurs, then clearly $n = 6$ and $m = 3$. Thus, we have

$$P(\text{an even number occurs}) = \frac{3}{6} = \frac{1}{2}$$

2. A bag contains 4 red and 5 white balls and a ball is drawn at random. Let **A** be the event that a red ball is drawn, then clearly, $n = {}^9C_1$, $m = {}^4C_1$ so that

$$P(\mathbf{A}) = \frac{4}{9}$$

Notes

1. This definition is based on a ***priori*** knowledge of equally likely outcomes.
2. This definition fails when the sample space is infinite.

17:3. ALGEBRA OF EVENTS

Definition 1 (*Equality of Events*). *Two events* **A** *and* **B** *are said to be* ***equal****, if and only if, they consist of exactly the same sample points. We denote it by writing* **A** = **B**.

Illustration. Consider the random experiment of throwing a dice once. Then, the two events **A** and **B** defined by

$$\mathbf{A} = \text{An even number comes face up}, \quad \text{and} \quad \mathbf{B} = \{2, 4, 6\}$$

are equal.

Definition 2 (*Union of Two Events*). *Let* **A** *and* **B** *be two events of some sample space. Then, the set of all sample points that belong to either event* **A** *or event* **B** *(or both), is called the* **union** *of* **A** *and* **B** *and is denoted by* $\mathbf{A} \cup \mathbf{B}$.

In other words, the occurrence of the event $\mathbf{A} \cup \mathbf{B}$ *means that either event* **A** *occurs, or event* **B** *occurs (or both event* **A** *and event* **B** *occur).*

Illustration. Consider the events,

$$\mathbf{A} = \{1, 2\} \text{ and } \mathbf{B} = \{2, 3, 4\} \text{ of } \mathbf{S} = \{1, 2, 3, 4, 5, 6\}$$

Then,

$$\mathbf{A} \cup \mathbf{B} = \{1, 2, 3, 4\}$$

Definition 3 (*Intersection of Two Events*). *Let* **A** *and* **B** *be two events of some sample space. Then, the set of all sample points that belong to both the events* **A** *and* **B**, *is called the* **intersection** *of* **A** *and* **B** *and is denoted by* $\mathbf{A} \cap \mathbf{B}$.

In other words, the occurrence of the event $\mathbf{A} \cap \mathbf{B}$ *means that both the events.* **A** *and* **B** *occur simultaneously. This is also called the* **joint occurrence** *of* **A** *and* **B**.

Illustration. In the preceding example, $\mathbf{A} \cap \mathbf{B} = \{2\}$

Definition 4 (*Impossible Event*). *An event* **A** *that contains no sample point is called an* ***impossible event*** *and is usually denoted by*

$$\mathbf{A} = \phi \quad or \quad \mathbf{A} = \{\ \}$$

In other words, an impossible event cannot occur.

Illustration. A ball is drawn at random from a bag that contains 10 red and 5 black balls. Then, the event

$$\mathbf{A} = A \text{ green ball is drawn}$$

is an impossible event.

Definition 5 (*Complementary Event*). *The event consisting of all sample points not contained in the event* **A**, *is called the* ***complementary event*** *of* **A** *and is denoted by* **A′**.

In other words, **A′** *represents the event that* **A** *does not occur.*

Illustration. For any sample space

$$\mathbf{S}' = \phi$$

Definition 6 (*Mutually Exclusive Events*). *Two events* **A** *and* **B** *are said to be* ***mutually exclusive***, *if* $\mathbf{A} \cap \mathbf{B} = \phi$.

In other words, two mutually exclusive events cannot occur simultaneously.

Illustration. For the random experiment of tossing a coin, the two events defined by

$$\mathbf{A} = \text{Head appears;} \quad \text{and} \quad \mathbf{B} = \text{Tail appears,}$$

are mutually exclusive.

The events being the subsets of the sample space, satisfy certain laws of algebra of sets. These include

Commutative laws: $\mathbf{A} \cup \mathbf{B} = \mathbf{B} \cup \mathbf{A};\ \mathbf{A} \cap \mathbf{B} = \mathbf{B} \cap \mathbf{A}$

and *Associative laws:* $(\mathbf{A} \cup \mathbf{B}) \cup \mathbf{C} = \mathbf{A} \cup (\mathbf{B} \cup \mathbf{C}).$

$$(\mathbf{A} \cap \mathbf{B}) \cap \mathbf{C} = \mathbf{A} \cap (\mathbf{B} \cap \mathbf{C}).$$

These laws allow us to extend the definitions of union and intersection of events to cover more than two events.

Laws of Probability

The following laws of probability are easily established from the classical definition of probability:

1. $P(\phi) = 0$ *(The probability of an impossible event is zero)*
2. $P(\mathbf{A}') = 1 - P(\mathbf{A})$ *(The sum of the probabilities of an event and its complement is unity)*
3. $P(\mathbf{A} \cup \mathbf{B}) = P(\mathbf{A}) + P(\mathbf{B}) - P(\mathbf{A} \cap \mathbf{B})$ *(The Addition Theorem of Probability)*

An event **A** such that $P(\mathbf{A}) = 1$, is called ***a certain event***. Thus, it follows that sample space is a certain event.

7:4. CONDITIONAL PROBABILITY

Definition (*Conditional Probability*). *If $P(\mathbf{B}) > 0$, then the **conditional probability** of the occurrence of an event **A** given that the event **B** has already occurred, is defined by*

$$P(\mathbf{A} | \mathbf{B}) = \frac{P(\mathbf{A} \cap \mathbf{B})}{P(\mathbf{B})}$$

Illustration. An examination in Queuing was given to a management class of 50 students of which only 20 had studied probability earlier. The following result was obtained:

	Passed	*Failed*	*Total*
Students who studied probability earlier	17	3	20
Student who did not study probability	19	11	30
Total	36	14	50

A student was then selected at random from the class. If

A = the student had studied probability earlier, and

B = the student passed the examination.

then **A** ∩ **B** = the student had studied probability earlier and passed the examination.

Clearly, $P(\mathbf{B}) = 36/50$, $P(\mathbf{A} \cap \mathbf{B}) = 17/50$

Thus, $P(\mathbf{A} \mid \mathbf{B}) = P(\mathbf{A} \cap \mathbf{B})/P(\mathbf{B}) = 17/36$, is the probability that a student selected at random had studied probability given that he has passed the examination in Queuing.

Laws of Conditional Probability

The following laws of conditional probability are easily established for any events **A**, **B**, **C**, with $P(\mathbf{B}) > 0$:

1. $P(\phi \mid \mathbf{B}) = 0$
2. $P(\mathbf{A}' \mid \mathbf{B}) = 1 - P(\mathbf{A} \mid \mathbf{B})$
3. $P(\mathbf{A} \cup \mathbf{C} \mid \mathbf{B}) = P(\mathbf{A} \mid \mathbf{B}) + P(\mathbf{C} \mid \mathbf{B}) - P(\mathbf{A} \cap \mathbf{C} \mid \mathbf{B})$
4. $P(\mathbf{A} \cap \mathbf{B}) = P(\mathbf{A}) \cdot P(\mathbf{B} \mid \mathbf{A})$ *(Theorem of Compound Probability)*

Independence of Events

Definition (*Independence of Two Events*). *Two events **A** and **B** of the same sample space are said to be **independent**, if the occurrence (or non-occurrence) of one does not influence the probability of occurrence or non-occurrence of the other.*

Thus, if **A** and **B** are two independent events, we must have

$$P(\mathbf{A} \mid \mathbf{B}) = P(\mathbf{A}), \quad P(\mathbf{B}) > 0$$

and

$$P(\mathbf{B} \mid \mathbf{A}) = P(\mathbf{B}), \quad P(\mathbf{A}) > 0$$

This leads us to the following:

Criterion of Independence. *Two events* **A** *and* **B** *will be independent, if*

$$P(\mathbf{AB}) = P(\mathbf{A}) \cdot P(\mathbf{B}).$$

Illustration. Let two coins be tossed in the air. Then,

$$\mathbf{S} = \{HH, HT, TH, TT\}$$

If we define **A** = the event that a tail appears on the first coin; and **B** = the event that a head appears on the second coin, then

$$\mathbf{A} = \{TH, TT\}, \quad \text{and} \quad \mathbf{B} = \{TH, HH\}$$

Clearly, $P(\mathbf{A}) = 2/4 = P(\mathbf{B})$. Also, $\mathbf{AB} = \{TH\}$; $P(\mathbf{AB}) = 1/4$

Since, $P(\mathbf{AB}) = P(A) \cdot P(\mathbf{B})$, therefore, **A** and **B** are independent.

17:5. RANDOM VARIABLES

Discrete Random Variable

Definition 1 (*Random Variable*). *A* ***random variable*** *X is defined over the sample space* **S** *that assigns to every $e_i \in \mathbf{S}$, a real number $X(e_i)$.*

The set of all values $X(e_i)$ thus obtained is called the ***range space*** *of X and is denoted by R_X.*

Illustration. The sample space associated with the tossing of two coins is

$$\mathbf{S} = \{HH, HT, TH, TT\}$$

If we define the random variable X by

X = the number of heads obtained, then we have

$$X(HH) = 2, \quad X(HT) = 1, \quad X(TH) = 1, \quad \text{and} \quad X(TT) = 0$$

Thus, the set of all possible values of X, the range space, is given by

$$R_X = \{0, 1, 2\}$$

Note. It is quite possible that the outcomes of an experiment are themselves numerical. For example, in the case of die rolling, the corresponding random variable is represented by the set of outcomes {1, 2, 3, 4, 5, 6}.

Definition 2 (*Discrete Random Variable*). *Let X be a random variable with range space R_X. If R_X is finite or countably infinite, then X is called a* ***discrete random variable****. That is, possible values of X may be listed as $x_1, x_2, ..., x_n$. The list terminates in the finite case, and in the countably infinite case, the list continues indefinitely.*

Illustration. Consider the random experiment of tossing a coin until the head appears. The sample space is

$$\mathbf{S} = \{H, TH, TTH, TTTH, ...\}$$

If X = number of throws required to obtain a head, then the range space of X is given by $R_X = \{1, 2, 3, 4, ...\}$.

Thus, X is a discrete random variable (with countably infinite R_X).

Definition 3 (*Probability Distribution*). *Let X be a random variable with $R_X = \{x_1, x_2, ...\}$. The function $p_x(\cdot)$ defined over R_X by*

$$p_x(x_j) = P(X = x_j),$$

is called the ***point probability function*** *of X.*

The collection of pairs $\{x_i, p_x(x_i)\}$, i = 1, 2, ... is called the ***probability distribution*** *of X.*

The probability that a discrete random variable will be within certain interval can be computed, if its probability distribution is known.

Illustration. Let X be a random variable with its probability distribution given by

X_j	-1	0	1/4	1/2	2	5/2	4
$p_x(x_j)$	0.1	0.05	0.2	0.15	0.2	0.2	0.1

Then,

(*i*) $P(X \geq 2) = p_X(2) + p_X(5/2) + p_X(4) = 0.5$;

(*ii*) $P(0 \leq X \leq 2) = p_X(0) + p_X(1/4) + p_X(1/2) + p_X(2) = 0.6$; and

(*iii*) $P(0 < X < 1) = p_X(1/4) + p_X(1/2) = 0.35$.

Note. We must have

$$p_X(x_j) \geq 0 \text{ for all } x_j \in R_X \quad \text{and} \quad \sum_{x_j \in R_X} p_X(x_j) = 1.$$

Continuous Random Variables, Distribution Functions

To take into account the 'continuous type' variables, "the life of an electric bulb", "the height of an individual", etc., we require the following extension of the Definition 1 above.

Definition 4 (*Random Variable*). *A* ***random variable*** *X is a function whose domain is the sample space* **S** *and whose range R_X is a non-empty set of real numbers, such that for every real number x, the set of sample points e for which $X(e) \leq x$ is an event of* **S**.

This definition is a more general one and covers our earlier definition of a random variable.

Definition 5 (*Cumulative Distribution Function*). *Let X be a random variable. The function $F_X(\cdot)$, whose value of each real number x given by*

$$F_X(x) = P(X \leq x)$$

is called the ***cumulative distribution function*** *of the random variable X and is denoted by* ***cdf***.

Thus, if X is a discrete random variable, then

$$F_X(x) = \Sigma\, p(x_j) \quad \text{(Summation is taken over all indices j such that } x_j \leq x).$$

The real number x defines an event of R_X, namely the set

$$\mathbf{A} = \{y : y \in R_X,\ y \leq x\}$$

By extending R_X to all real numbers, if necessary, the set **A** becomes the interval $(-\infty, x]$, and the event **A** has a probability $F_X(x)$ attached to it. Consequently, the domain of F is the set of all real numbers, while the real numbers between 0 and 1 (both inclusive) constitute its range space.

Definition 6 (*Continuous Random Variable*). *A random variable X is said to* ***continuous***, *if its cdf $F_X(x)$ is a continuous function of x.*

Sometimes for the sake of simplicity, a continuous random variable is considered to be one whose range space is an interval or a collection of intervals on a real line. This certainly is not equivalent to the above definition, but works equally well. For example, if

$$X = \text{Life of an electric bulb before its failure,}$$

then $R_X = [0, \infty]$, X is a continuous random variable that can assume any value between $0 \leq X < \infty$. It can be shown that X satisfies the condition of ***definition 6***.

Remarks

1. For a continuous random variable X, the point probabilities $P(X = x_i)$, x_i being a specified value, vanish. Thus, we have

$$P(a \leq X \leq b) = P(a < X \leq b) = P(a \leq X < b) = P(a < X < b), \text{ for real } a < b.$$

2. The connection between the probability of the event $(a < x \leq b)$ and the ***cdf*** is given by

$$P(a < X \leq b) = F_X(b) - F_X(a)$$

Definition 7 (*Probability Density Function*). *Let X be a continuous random variable. A non-negative function f defined for all real numbers x such that*

$$F_X(x) = \int_{-\infty}^{x} f(t)\,dt$$

is called a ***probability density function****, denoted by* ***pdf****, for the random variable X whose cdf is* $F_X(\cdot)$.

The function *f* must have enough properties for the existence of the improper integral. In fact, it can be proved that any continuous function *f* satisfying

$$(a)\ f(x) \geq 0 \quad \text{and} \quad (b) \int_{R_X} f(x)\,dx = 1$$

for all x in R_X, qualifies as a *pdf* for X.

Illustration. Let *f* be a function defined by

$$f(x) = \begin{cases} 2x, & 0 < x < 1 \\ 0, & \text{otherwise} \end{cases}$$

Clearly, we observe that

$$R_X = (-\infty < X < \infty),$$

$f(x)$ is defined and continuous for all $x \in R_X$; and

(*a*) $$f(x) \geq 0 \quad \text{for all} \quad x \in R_X$$

(*b*) $$\int_{R_X} f(x)\,dx = \int_{-\infty}^{\infty} f(x)\,dx$$

$$= \int_{-\infty}^{0} f(x)\,dx + \int_{0}^{1} f(x)\,dx + \int_{1}^{\infty} f(x)\,dx = 0 + \int_{0}^{1} 2x\,dx + 0 = 1$$

Thus, $f(x)$ represents a *pdf*.

Remark. Since, $F_X(b) = \int_{-\infty}^{b} f(t)\,dt$ and $F_X(a) = \int_{-\infty}^{a} f(t)\,dt$, the probability $P(a < x \leq b) = F_X(b) - F_X(a)$ can be written in terms of the integrals as follows:

$$P(a < X \leq b) = \int_{-\infty}^{b} f(t)\,dt - \int_{-\infty}^{a} f(t)\,dt = \int_{-\infty}^{b} f(t)\,dt$$

Since, $f \geq 0$, we can also say that $P(a < X \leq b)$ is the area under the curve $y = f(x)$ bounded by the ordinates at $x = a$ and $x = b$, as shown in *Fig. 17.1*:

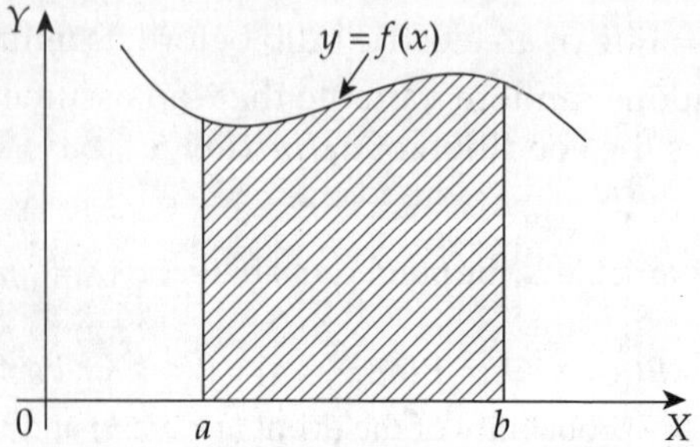

Fig. 17.1. Area under the curve $y = f(x)$

Theorem (*Obtaining pdf and Probability Distribution from cdf*).

(*a*) *Let X be a continuous random variable with cdf F and pdf f, then*

$$f(x) = \frac{dF(x)}{dx}$$

for all x at which derivative exists.

(*b*) *Let X be a discrete random variable with cdf F and*

$$R_X = \{x_1, x_2, \ldots \mid x_1 < x_2 < \ldots\}.$$

Then, $$p_X(x_j) = F(x_j) - F(x_{j-1}) \quad \text{for } j = 1, 2, \ldots$$

Proof. Beyond the scope of this book.

Illustration. The waiting time X for customer arrival at some service counter has been found to be a continuous random variable with *cdf*

$$F_X(x) = \begin{cases} 0 \quad , & \text{if } x < 0 \\ 1 - e^{-\alpha}, & \text{otherwise} \end{cases}$$

The *pdf* of X is obtained by

$$f(x) = \frac{dF_X(x)}{dx} \qquad \text{(assuming that the derivative exists for all } x\text{)}$$

Thus, $$f(x) = \begin{cases} 0 \quad , & \text{if } x < 0 \\ e^{-\alpha}, & \text{if } x > 0 \end{cases}$$

17:6. EXPECTATION OF A RANDOM VARIABLE

Definition (*Expectation of a Random Variable*). *Let X be a random variable with range space R_X. The **expectation** of X, denoted by E (X), is defined by*

$$E(X) = \begin{cases} \sum\limits_{i=1}^{n} x_i\, p(x_i)\,, & \text{if } X \text{ is discrete with } p(x_i) = P(X = x_i) \\ \int\limits_{-\infty}^{\infty} x\, f(x)\, dx, & \text{if } X \text{ is continuous with } R_X = (-\infty, \infty) \text{ and } pdf\ f(x) \end{cases}$$

E (X) is said to exist, if the above sum (integral) converges.

Illustration. Let a discrete random variable X have the probability distribution

x_i	0	1	2	3
$p(x_i)$	0.2	0.1	0.4	0.3

The Expectation of X is given by

$$E(X) = \sum_{i=1}^{4} x_i\, p(x_i) = 0(0.2) + 1(0.1) + 2(0.4) + 3(0.3) = 1.8$$

Note. E (X) is generally called the ***'expected value'*** or the ***'mean value'*** of X.

Remarks

1. The concept of expectation of a random variable X can be generalized to include functions of X also. In particular, if X is a random variable and *g* is a single valued function of X, then *g* (X) is also a random variable, and

$$E[g(X)] = \begin{cases} \sum_{i=1}^{n} g(x_i)\,p(x_i), & \text{if } X \text{ is discrete} \\ \int_{-\infty}^{\infty} g(x)\,f(x)\,dx, & \text{if } X \text{ is continuous} \end{cases}$$

For example, if $g(X) = \log X$, then $E[\log X] = \sum_i (\log x_i)\, P(X = x_i)$.

2. If X and Y are random variables defined over the sample space, then the combinations of X and Y such as

$$X \pm Y, \quad XY, \quad bX, \quad a \pm bX$$

where, a, b are constants, are also random variables.

17:7. CENTRAL TENDENCY AND DISPERSION

In summarizing the data of a single random variable X, measures of the central tendency and dispersion are commonly used. There are several measures of central tendency, such as a property for values of X to tend (cluster) towards the centre. The three measures of central tendency are most important. These are ***mean, mode*** and ***median***.

Definition 1 (*Mean*). *Let X be a random variable. Then,* ***mean*** *of X is given by its expectation, namely E (X), and is denoted by* $\bar{X}$.

Definition 2 (*Mode*). *The* ***mode*** *of a random variable X is that value of the variable X that corresponds to the greatest probability (density) and is denoted by* $\hat{X}$.

In the case of discrete X, the mode $\hat{x}$ is determined by the double inequality

$$P(X = x_i) \le P(X = \hat{x}) \quad \text{whenever } x_i \le \hat{x}$$

and

$$P(X = x_j) \le P(X = \hat{x}) \quad \text{whenever } x_j > \hat{x}$$

In the case of continuous distributions, it is determined by solving

$$\frac{df(x)}{dx} = 0 \quad \text{and} \quad \frac{d^2 f(x)}{dx^2} < 0 \text{ for all } x$$

Note. The mode of a distribution need not be unique.

A distribution with two modes is called a ***bimodal*** distribution.

Definition 3 (*Median*). *For the distribution of a random variable, discrete or continuous, the variate value x satisfying the pair of inequalities*

$$P(X \le x) = \frac{1}{2} \quad \text{and} \quad P(X \ge x) = \frac{1}{2}$$

is called the ***median***.

Remark. The mode can be obtained by the approximate empirical result

Mean – Mode = 3 (Mean – Median),

whenever mean and median are known.

The most important measure of dispersion around the central value is the "***variance***" of the distribution.

Definition 4 (*Variance*). *Let X be a random variable with E (X) = μ. The* ***variance*** *of X, denoted by V (X), is defined by*

$$V(X) = E\{(X - \mu)^2\}$$

Note. 1. $V(X)$ is usually represented by the symbol σ^2.

2. $V(X)$ is always non-negative.

Remark. We have

$$V(X) = E\{(X-\mu)^2\} = E\{X^2 + \mu^2 - 2X\mu\}$$
$$= E(X^2) + \mu^2 - 2\mu E(X) = E(X^2) + \mu^2 - 2\mu \cdot \mu$$
$$= E(X^2) - \{E(X)\}^2.$$

The positive square root of $V(X)$ is more popular in use.

Definition 5 (*Standard Deviation*). *The* ***standard deviation*** *of a random variable X is defined by*

$$S.D. = +\sqrt{V(X)}$$

and is generally denoted by σ.

Thus, for a given frequency distribution (x_i, f_i) (a probability distribution with $p_i = f_i / N$, for all $i = 1, 2, \ldots, n$, where $N = \sum f_i$), the variance and S.D. can be computed by the formulae,

$$\sigma^2 = \frac{1}{N}\sum_{i=1}^{n} f_i (x_i - \overline{x})^2, \quad \text{where } \overline{x} = \frac{1}{n}\sum_{i=1}^{n} f_i x_i$$

$$= \frac{1}{N}\sum_{i=1}^{n} f_i x_i^2 - \overline{x}^2 \quad \text{and} \quad \sigma = +\sqrt{\frac{1}{N}\sum_{i=1}^{n} f_i (x_i - \overline{x})^2}.$$

17:8. SOME PROBABILITY DISTRIBUTIONS

We are now in a position to consider some important probability laws that shall be used in the sequel.

The probability distributions generated by the probability laws shall be designated by some commonly known names.

Poisson Distribution

Definition 1 (*Poisson Distribution*). *The probability distribution of a discrete random variable X, with range space* $R_X = \{0, 1, 2, \ldots\}$, *that obey the Poisson probability law*

$$P(X = x) = \frac{e^{-\lambda}\lambda^x}{x!} \quad \text{for all } x \in R_X;\ \lambda > 0,$$

is called a ***Poisson distribution*** *with parameter* λ.

An important property of Poisson distribution is that its ***mean*** and ***variance coincide***. In fact, for a Poisson variable with parameter λ, the mean and variance both are equal to λ.

Note. The 'parameter λ' is generally referred to as 'mean λ'.

A probability chart for a Poisson Distribution has been shown in *Fig. 17.2* for $\lambda = 3$.

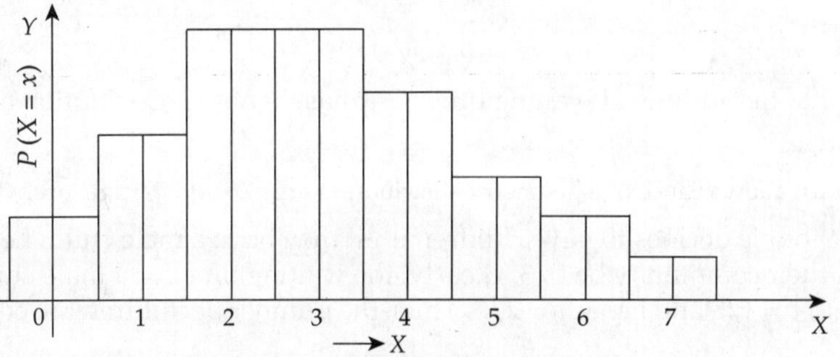

Fig. 17.2. Poisson Distribution for $\lambda = 3$

The Poisson distribution can be applied to situations, where the chance of an event's happening is very very small and a very large population is exposed to it. For example, the whole population in a certain city is exposed to committing suicide, but the chance is very small that a particular individual will commit suicide.

A typical application of the Poisson distribution occurs in analyzing Queuing problems, where customers arrive randomly at a service counter. In certain situations, the number of arrivals and departures to and from the service counter can be described by the Poisson distribution.

Another important property of Poisson distribution is ***its additivity***. Namely, if X and Y are two independent Poisson random variables with means λ and μ, respectively, then $X + Y$ is also a Poisson random variable with mean $\lambda + \mu$.

Illustration. The number of customers arriving at a facility for service between 10 A.M. and 11 A.M. is a random variable, say X_1, with Poisson distribution with mean 2. Similarly, the number of customers arriving between 11 A.M. and 12 noon, say X_2, has a Poisson distribution with mean 6. If X_1 and X_2 are independent, we are interested in knowing the probability that more than 5 customers will come between 10 A.M. and 12 noon.

Let $X = X_1 + X_2$. Then, by additivity property of Poisson distribution, X has a Poisson distribution with mean 2 + 6(= 8).

Thus, the probability that there will be x customers between 10 A.M. and 12 noon is given by

$$P(X = x) = \frac{e^{-\lambda}\lambda^x}{x!} = \frac{e^{-8}8^x}{x!}; \quad x = 0, 1, 2, \ldots$$

Probability that more than 5 customers will arrive between 10 A.M. and 12 noon is given by

$$P(X > 5) = 1 - P(X \le 5) = 1 - \sum_{x=0}^{5} \frac{e^{-8}8^x}{x!} = 1 - 0.1912 = 0.8088$$

Geometric Distribution

Definition 2 (*Geometric Distribution*). *A discrete random variable X is said to have a* ***geometric distribution****, if it assumes only the non-negative values and obeys the probability law given by*

$$P(X = x) = \begin{cases} q^x p, & x = 0, 1, \ldots \\ 0, & \text{otherwise} \end{cases}$$

where, $0 < p \le 1$ *and* $q = 1 - p$.

The geometric distribution is the waiting time distribution and can be regarded, in a sense, as describing the 'time' until a certain number of failures (with probability q) occur preceding the first success (whose probability of occurrence is p).

The most important property of geometric distribution is its ***Lack of Memory***. That is, if an event can occur at one of the times $t = 0, 1, 2, \ldots$ and the occurrence (waiting) time X has a geometric distribution, then

$$P(X = k + t) \mid X \ge k) = P(X = t),$$

which implies that the additional waiting time $X - k$ has the same distribution as the initial time to wait.

Remark. The mean and variance of a geometric distribution are q/p and q/p^2 respectively.

Illustration. A couple decides to have children until they have a male child. Let the probability of a male child in their community be $1/3$. Clearly, the waiting time for a male child has a geometric distribution with $p = 1/3$ and hence $q = 2/3$. Thus, the number of children the couple is expected to have before first male child is born is $\frac{q}{p} = 2$

Normal Distribution

Perhaps the most important and useful continuous *pdf* is the so-called ***normal density function.***

Not only does it arise naturally in many applications, it also services as a highly accurate approximate to other distributions. Almost all the probability distributions tend to normal distribution under certain limiting conditions.

Definition 3 (*Normal Distribution*). *A continuous random variable X having the pdf*

$$f(x) = \frac{1}{\sigma\sqrt{2\pi}} e^{-\frac{1}{2}\left(\frac{x-\mu}{\sigma}\right)^2} \qquad -\infty < x < \infty;$$

where, μ and σ are real constants is said to have a ***normal distribution*** *with parameters (constants) μ and σ.*

It is expressed by writing $X - N(\mu, \sigma^2)$. The constants π and *e* are irrational numbers, whose values are approximately 2.7187 and 3.1416, respectively.

Note. In fact, the parameters μ and σ happen to be the ***mean*** and ***S.D.*** of the distribution, respectively.

The graph of the normal *pdf* is the familiar bell-shaped curve, popularly known as the normal curve, shown in *Fig. 17.3:*

Observe that the normal curve has its peak at $x = \mu$, then it decreases on either side of $x = \mu$ and is asymptotic to the X-axis on both sides of $x = \mu$.

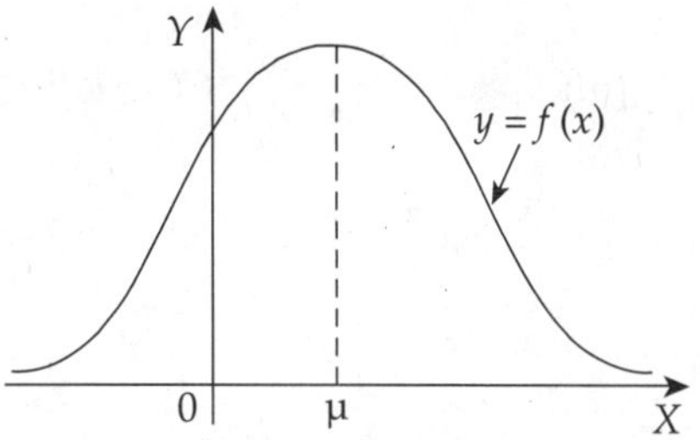

Fig. 17.3. Normal Curve

The most important properties of the normal distribution are:

(*i*) The normal curve is perfectly symmertical about the ordinate $x = \mu$. That is, if the curve is folded along the line $x = \mu$, the two halves will coincide.

(*ii*) The mean, mode and median all coincide at $x = \mu$.

(*iii*) The most probable limits for a normal variate are $\mu \pm 3\sigma$, namely,

$$P(\mu - 3\sigma \le X \le \mu + 3\sigma) = 0.9973$$

That is, P (X lies outside $\mu \pm 3\sigma$) = 0.0027, although the range for X is $(-\infty, \infty)$.

(*iv*) The sum or difference of independent normal variates is also a normal variable.

Normal Area Property

The mean μ and standard deviation σ uniquely identify a normal distribution. Thus, for different combinations of μ and σ, there will be different normal curves. We shall, therefore, require different tables for areas under the different normal curves. The complication is resolved by converting the normal variable 'X' into a Standard Normal Variable 'Z' by the simple transformation $Z = \frac{X - \mu}{\sigma}$.

Note that $E(Z) = 0$, and $V(Z) = 1$.

The standard normal variate X requires only a single table for areas under the normal curve, since values of mean and variance are now fixed. The area under the standard normal curve bounded by ordinates at Z = 0 and Z = z (shown shaded) is given in *Fig. 17.4.*

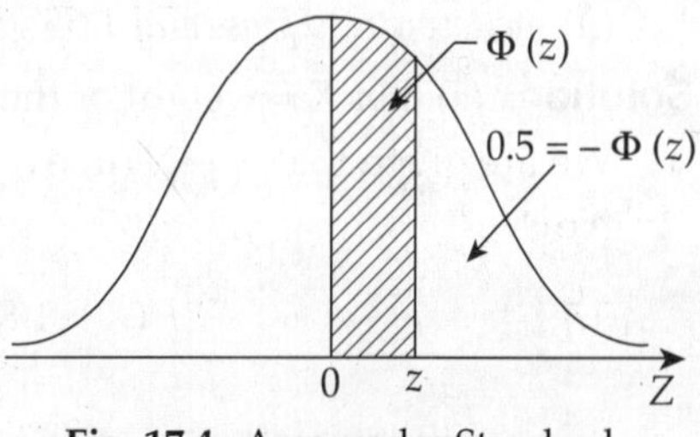

Fig. 17.4. Area under Standard Normal Curve

$$P(0 < X \le z) = \Phi(z) = \int_0^z f(t)\,dt = \frac{1}{\sqrt{2\pi}} \int_0^z e^{-\frac{1}{2}t^2}\,dt$$

The values of $\Phi(z)$ for different values of z are available in normal tables at the end of the book. These tables have been tabulated for areas to the right of $Z = 0$ only. This is so because in view of the symmetry of the normal curve, any area to the left of $Z = 0$ is exactly equal to the corresponding area to the right of $Z = 0$. Any normal probability for the standard normal variate can be computed using the tables for $\Phi(z)$. The following properties of $\Phi(z)$ are useful in using these tables:

(*i*) $$\Phi(-z) = \int_{-z}^{0} f(t)\,dt = \int_{0}^{z} f(t)\,dt = \Phi(z)$$

(*ii*) $P(a \le Z \le b) = \Phi(b) - \Phi(a);\; P(-a \le Z \le b) = \Phi(b) + \Phi(a);\; P(-a \le Z \le -b) = \Phi(a) - \Phi(b)$

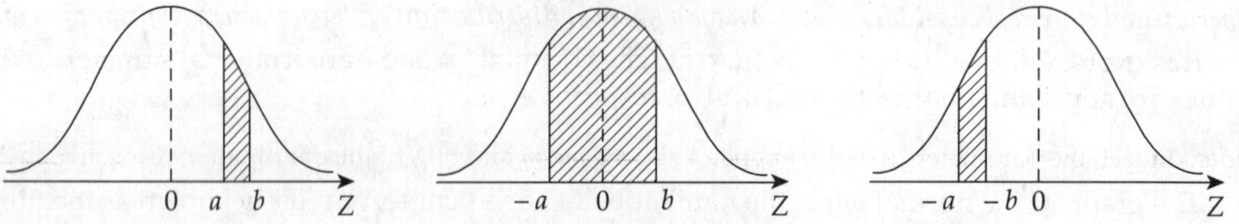

Fig. 17.5. Normal Probabilities

(*iii*) $$P(-a \le Z \le a) = \Phi(a) + \Phi(-a) = 2\Phi(a)$$

(*iv*) $$P(Z > a) = P(Z < -a) = 0.5 - \Phi(a)$$

(*v*) $$P(Z < b) = 0.5 + \Phi(b)$$

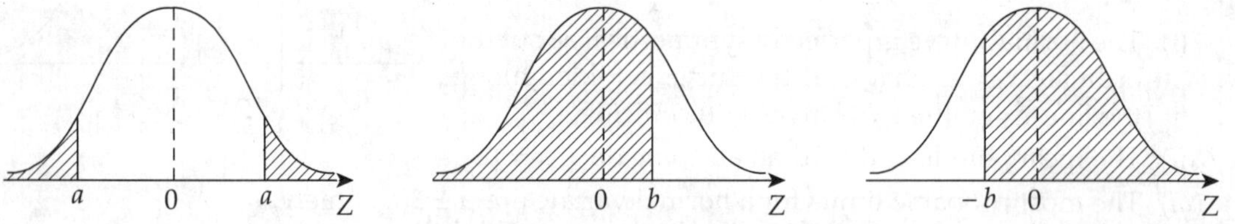

Fig. 17.6. Normal Probabilities

We note that $P(-3 < Z < 3) = 2\Phi(3) = 2 \times 0.4986 = 0.9972$ so that $P(|Z| > 3) = 0.0028$.

This indicates that although theoretically the range of Z is $(-\infty, \infty)$, but for all practical purposes Z lies within ± 3. It is for this reason, that the tail of the normal curve seems to approach Z-axis right from ± 3.

Example 17.1. *A manufacturer produces airmail envelopes whose weights have mean 1.50 gms. and standard deviation 0.15 gm. The envelopes are sold in lots of 500. Assuming that weights are normally distributed, find how many envelopes in a lot will weigh*

(*i*) *heavier than 1.85 gms.*

(*ii*) *between 1.20 gms. and 1.55 gms.*

Solution. (*a*) Let X = weight of the envelope (in gms.)

We are given that $X \sim N(\mu, \sigma)$, where $\mu = 1.5$ and $\sigma = 0.15$

Now,

(*i*) $$P(X > 1.85) = P\left(\frac{X - 1.5}{0.15} > \frac{1.85 - 1.5}{1.5}\right)$$

$$= P(Z > 2.33), \quad Z \text{ is an } S.N.V.$$

$$= 0.5 - \Phi(2.33) = 0.5 - 0.4901 = 0.009$$

Thus, the number of envelopes which weight more than 1.85 gms. is

$$500 \times P(X > 1.85) = 500 \times 0.0099 = 4.95 = 5$$

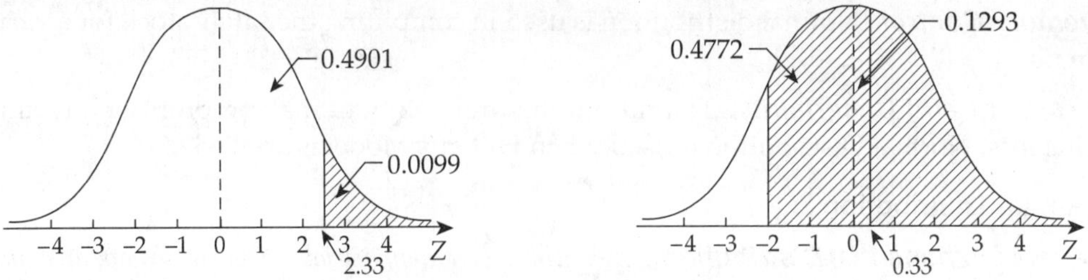

Fig. 17.7. Normal Probabilities

(*ii*)
$$P(1.20 < X < 1.55) = P\left(\frac{1.20 - 1.50}{0.15} < Z < \frac{1.55 - 1.055}{1.5}\right)$$
$$= P(-1 < Z < 0.33) = \Phi(2) + \Phi(0.33)$$
$$= 0.4772 + 0.1293 = 0.6065$$

Thus, the number of envelopes with weight between 1.20 gms and 1.55 gms is

$$500\, P(1.20 < X < 1.55) = 500 \times 0.6065 = 303.25 \approx 303$$

Example 17.2. *A banker claims that the life of a regular savings bank account opened with his bank averages 15 months with a standard deviation of 6.45 months. Assuming that life of a regular savings bank account is normally distributed, find what is the probability that*

(*a*) *there is balance in the savings bank account of a depositor at the end of 19 months?*

(*b*) *the savings account of a depositor will have been closed before 21 months?*

Solution. Let X = Life of a regular savings bank account. Then, we are given that $X \sim N(\mu, \sigma^2)$, where $\mu = 15$ and $\sigma = 6.45$.

(*a*)
$$P(X > 19) = P\left(\frac{X - \mu}{\sigma} > \frac{19 - 15}{6.45}\right)$$
$$= P(Z > 0.62), \quad Z \text{ is an } S.N.V.$$
$$= 0.5 - \Phi(0.62)$$
$$= 0.5 - 0.2324 \quad \text{(from tables)}$$
$$= 0.2676$$

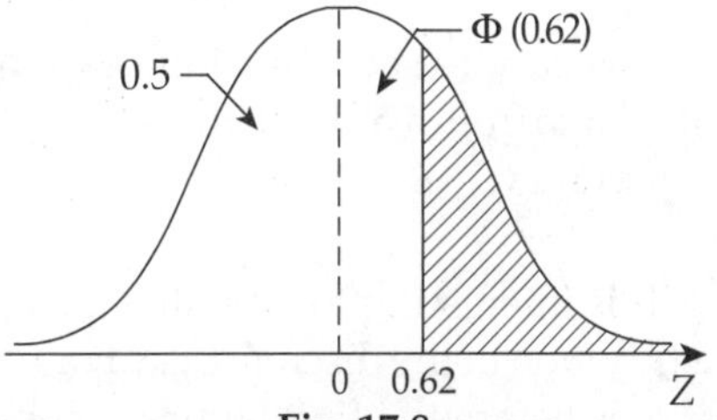

Fig. 17.8

(*b*)
$$P(X \le 21) = P\left(\frac{X - \mu}{\sigma} \le \frac{21 - 15}{6.45}\right)$$
$$= P(Z \le 0.93)$$
$$= 0.5 + \Phi(0.93)$$
$$= 0.5 + 0.3238 \quad \text{(from tables)}$$
$$= 0.8238$$

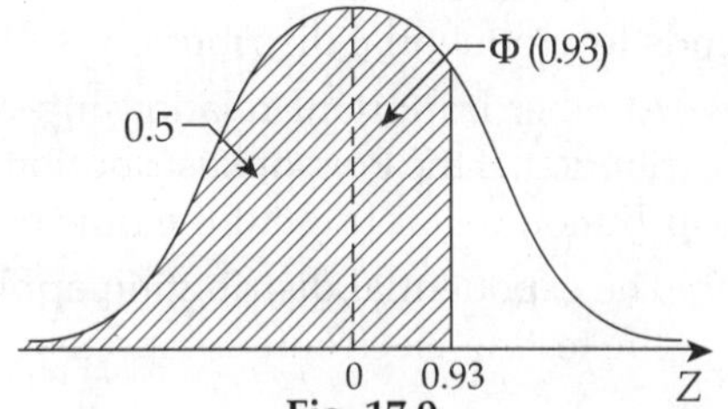

Fig. 17.9

Normal distribution is very useful for the experimental and physical sciences. It possesses many mathematical properties which makes its application possible in a wide variety of situations. It is very useful in Statistical Quality Control and in problems of Hypothesis Testing. It finds extensive applications in various OR problems, such as given below, to mention a few:

1. ***PERT.*** The average project length follows a normal distribution. The distribution is used to find out the probability of completing the project within a specified duration.

2. ***Inventory Control.*** Normal distribution is used in computing the safety stock for a variable for a given risk.

3. ***Investment and Uncertainty.*** To find out the probability of making profit of any amount or incurring loss, Normal Distribution is used when Hillier's Model is applied.

Exponential Distribution

Definition 4 (*Exponential Distribution*). *A continuous random variable X is said to have an **exponential distribution** with parameter* α*, if its **pdf** is given by*

$$f(x) = \begin{cases} \alpha e^{-\alpha x}, & \text{for } 0 < x < \infty \\ 0, & \text{otherwise} \end{cases}$$

where, α *is a positive constant and e = 2.7187.*

The graph of the *pdf* of the exponential distribution is given in *Fig. 17.10:*

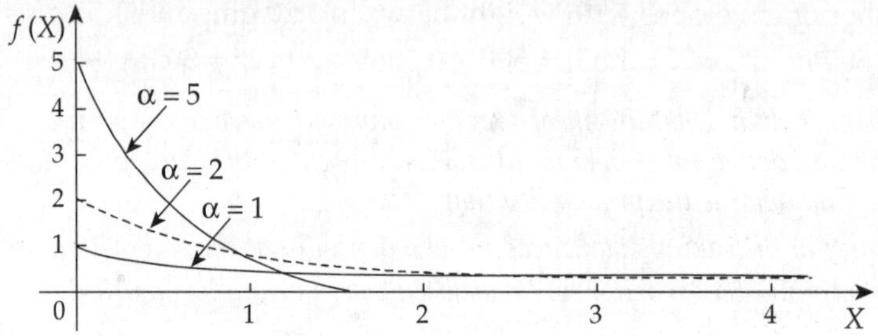

Fig. 17.10. Exponential Distribution

It is easy to see that (*i*) the exponential density function is an exponential decreasing function with a maximum at $x = 0$, the maximum value being α; (*ii*) the mean and variance of the distribution are given by

$$E(X) = 1/\alpha \quad \text{and} \quad V(X) = 1/\alpha^2$$

It is interesting to note that exponential distribution is the continuous ***analog*** to the geometric distribution in the discrete case. For example, if the geometric random variable represents the ***number*** of trials before the first failure occurs, its equivalence in the exponential distribution would be the waiting time for failure. In fact as $P \to 0$ and the inter-trial time $\to 0$, the geometric distribution tends to exponential distribution.

Another important relationship exists between the Poisson distribution and the exponential distribution. If the Poisson distribution describes the number of failures per unit time, the exponential distribution will represent the time between two successive failures.

The exponential distribution applies to events that are subject to a constant chance of failure, *e.g.*, life testing electronic components.

Exponential Distribution in Queuing Problems

Exponential distribution is often encountered in waiting line (queuing) problems as a probability model for ***service time***. Its mean α represents 'the rate of service', *i.e.*, the average number of customers served per unit of time. The random variable X represents the arrival time and α is the parameter whose value rate depends on the rate of performance of the service ($0 < x < \infty$).

Let T be the arrival time that follows an exponential distribution with parameter α. The distribution generates the probability distribution of ***inter-arrival*** times (time span between two successive arrivals). The probability of next arrival between times t_1 and t_2 is given by

$$P(t_1 < T < t_2) = F(t_2) - F(t_1) = e^{-\alpha t_1} - e^{-\alpha t_2},$$

since, $F(t) = P(T \le t) = \int_0^t e^{-\alpha x}\, dx = 1 - e^{-\alpha t}$.

Like geometric distribution, exponential distribution also lacks in memory. That is,

$$P(t \le T \le t+k \mid T \ge t) = P(0 \le T \le k).$$

It means that the probability of completing a service during time interval $(t, t+k)$ given that the service has been in progress for time t is the same as the ***unconditional*** probability of completing the service in time interval $(0, k)$.

Illustration 1. *At Standard, one of the largest restaurants in New Delhi, it normally takes 10 minutes to serve the order after receiving it. If the service time is exponentially distributed, what is the probability that the customer waiting time is (i) more than 8 minutes, (ii) 12 minutes or less, and (iii) between 5 and 10 minutes.*

Solution. Let T be the customer waiting time for the service. Then, T is exponential with average serving rate $\alpha = \frac{1}{10} = 0.1$ per minute. The required probabilities are

(*i*) $P(T > 8) = e^{-\alpha t} = e^{-0.1 \times 8} = e^{-0.8} = 0.449$

(*ii*) $P(T \le 12) = 1 - e^{-12\alpha} = 1 - e^{-1.2} = 1 - 0.301 = 0.699$

(*iii*) $P(5 \le T \le 10) = e^{-5\alpha} - e^{-10\alpha} = e^{-0.5} - e^{-1} = 0.607 - 0.368 = 0.239$

Illustration 2. *Customers arrive at a booking office window, being manned by a single individual at a rate of 30 per hour. The inter-arrival time of customers is exponentially distributed. For a time span of 10 minutes, calculata various probabilities of next arrival in a given time interval. Also, determine the mean inter-arrival time.*

Solution. We are given the average number of customers arriving at the booking office window, per minute,

$$\alpha = 30/60 = 0.5$$

The probability of next arrival between a time interval, $t_1 - t_2$, is given by

$$P(t_1 < T < t_2) = e^{-\alpha t_1} - e^{-\alpha t_2}$$

Considering time intervals of 1 minute duration each, the probability distribution of inter-arrival times and the mean inter-arrival time are calculated in the following table:

Inter-arrival time $t_1 - t_2$ *(minutes)*	*Mid-value of inter-arrival time (ii)*	*Inter-arrival probability* $e^{-\alpha t_1} - e^{-\alpha t_2}$ *(iii)*	*(ii)* × *(iii)*
0–1	0.5	0.393	0.1965
1–2	1.5	0.239	0.3585
2–3	2.5	0.145	0.3625
3–4	3.5	0.088	0.3080
4–5	4.5	0.053	0.2385
5–6	5.5	0.032	0.1760
6–7	6.5	0.200	0.1300
7–8	7.5	0.012	0.0900
8–9	8.5	0.007	0.0595
9–10	9.5	0.004	0.0380
		Mean inter-arrival time = 1.9575 minutes	

Rectangular Distribution

Definition 5 (*Rectangular or Uniform or Homogenous Distribution*). *A continuous random variable X is said to have rectangular or uniform distribution, if it has the **pdf** given by*

$$f(x) = \begin{cases} \dfrac{1}{b-a}, & \text{if } a < x < b \\ 0, & \text{otherwise} \end{cases}$$

where, a > 0 and b > 0 are some constants.

The distribution is called uniform, because its ***pdf*** is constant in the interval $a < x < b$. The distribution function $F(x)$ is given by

$$F(x) = \int_{-\infty}^{x} f(x)\,dx = \begin{cases} \dfrac{x-a}{b-a}, & \text{if } a < x < b \\ 0, & \text{if } x < a \text{ and is 1, if } x \geq b \end{cases}$$

The function $F(x)$ varies linearly in the interval $a < x < b$. *Fig. 17.11* shows the density function $f(x)$ in the interval $a < x < b$.

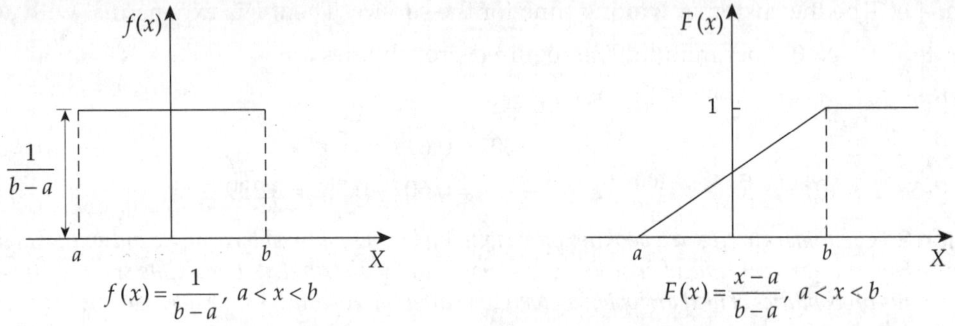

Fig. 17.11. Rectangular Distribution

It is easy to see that for rectangular distribution, mean $= \frac{1}{2}(a+b)$ and variance $= \frac{1}{2\sqrt{3}}(b-a)$.

Rectangular distribution finds its applications in statistical and inventory control problems.

Gamma Distribution

Definition 6 (*Gamma or Erlang Distribution*). *A continuous random variable X is said to have a **gamma** or **Erlang distribution**, if it has the pdf given by*

$$f(x) = \begin{cases} \dfrac{e^{-\lambda x} x^{n-1} \lambda^n}{\Gamma(n)}, & \text{if } 0 < x < \infty \\ 0, & \text{otherwise} \end{cases}$$

where, λ and n are positive constants (given parameters).

Note that by putting $n = 1$, the above ***pdf*** reduces to the exponential ***pdf***.

Given n independent and identically distributed exponential random variables, the distribution of the sum of these random variables gives rise to the Gamma distribution. Like Poisson distribution, the mean and variance of a Gamma (Erlang) distribution are also equal. Gamma distribution has the additivity property and is analogous to the negative binomial (sum of geometric random variables) distribution. For $n = 1$ and $\lambda = 0.5$, the graph for pdf of gamma distribution is shown in *Fig. 17.12.*

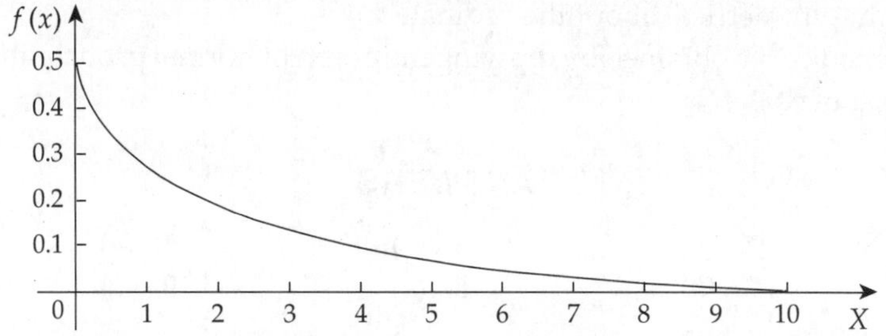

Fig. 17.12. Gamma Distribution

The Erlang distribution is widely used in studying the waiting time process.

TEST YOUR UNDERSTANDING

1. If an event cannot take place, its probability will be
   (*a*) 1 (*b*) –1
   (*c*) 0 (*d*) none of these
2. If **A** and **B** are independent events, then $P(\mathbf{A} \cap \mathbf{B})$ equals
   (*a*) $P(\mathbf{A}) + P(\mathbf{B})$ (*b*) $P(\mathbf{A}) \cdot P(\mathbf{B})$
   (*c*) $P(\mathbf{A} \mid \mathbf{B})$ (*d*) $P(\mathbf{B} \mid \mathbf{A})$
3. If two events **A** and **B** are dependent, the conditional probability of **B** given **A**, *i.e.*, $P(\mathbf{B} \mid \mathbf{A})$ is calculated as
   (*a*) $P(\mathbf{AB})/P(\mathbf{B})$ (*b*) $P(\mathbf{A})/P(\mathbf{B})$
   (*c*) $P(\mathbf{AB})/P(\mathbf{A})$ (*d*) $P(\mathbf{A})/P(\mathbf{AB})$
4. If the events **A** and **B** be mutually exclusive, then $P(\mathbf{A} + \mathbf{B})$ will be equal to
   (*a*) $P(\mathbf{A}) + P(\mathbf{B})$ (*b*) $P(\mathbf{A}) - P(\mathbf{B})$
   (*c*) $P(\mathbf{A}) . P(\mathbf{B})$ (*d*) $P(\mathbf{A})/P(\mathbf{B})$
5. If **A** and **B** are any two arbitrary events, then
   (*a*) $P(\mathbf{A} \cap \mathbf{B}) \geq P(\mathbf{A}) + P(\mathbf{B})$ (*b*) $P(\mathbf{A} \cap \mathbf{B}) \leq P(\mathbf{A}) + P(\mathbf{B})$
   (*c*) $P(\mathbf{A} \cap \mathbf{B}) = P(\mathbf{A}) + P(\mathbf{B})$ (*d*) none of these
6. If X be a random variable, then variance of X is
   (*a*) $E[X + E(X)]^2$ (*b*) $E(X^2) - [E(X)]^2$
   (*c*) $E[X - E(X)]^2$ (*d*) none of these
7. If X and Y be any two independent random variables, then $E(X + Y)$ equals
   (*a*) $E(X) \cdot E(Y)$ (*b*) $E(X) + E(Y)$
   (*c*) $E(X) + E(Y) - E(XY)$ (*d*) none of these
8. For a normal distribution
   (*a*) mean = median = mode (*b*) mean ≠ mode = median
   (*c*) mean ≠ median = mode (*d*) mean ≠ mode ≠ median
9. Area under standard normal curve within range, mean ±2 standard deviation, is about
   (*a*) 99.5% (*b*) 99.7%
   (*c*) 95.5% (*d*) 68%

10. The normal curve
   (*a*) always has its peak at $x = \mu$
   (*b*) is perfectly symmetrical about the ordinate $x = \mu$
   (*c*) is bell shaped curve obtained by drawing the graph of normal probability density function.
   (*d*) all of the above.

ANSWERS

1. (*c*)	**2.** (*b*)	**3.** (*c*)	**4.** (*a*)	**5.** (*b*)
6. (*c*)	**7.** (*b*)	**8.** (*a*)	**9.** (*c*)	**10.** (*d*)

REVIEW EXERCISE

1. What do you mean by the term probability? Discuss its importance in business decision-making. ***[M.Com., Dayalbagh Edu. Institute 2007]***
2. Define conditional probability and outline its properties. ***[MCA, Panjab Tech. Univ. 2010]***
3. Write short note on sample space and probability. ***[MCA, Panjab Tech. Univ. 2010]***
4. State Baye's theorem and explain its role in management decision-making.
5. Define multiplication theorem of probability.
6. What do you understand by the following terms:
   Random experiment, Event, Mutually exclusive events, exhaustive events, complementary events, equally likely events, and independent events? ***[MFC, Panjab Univ. 2002]***
7. Define Binomial distribution and indicate its chief characteristics. Under what conditions does it tend to Poisson distribution. ***[MFC, Panjab Univ. 2002]***
8. Define Poisson probability model and explain how Poisson distribution applies to a variable taking discrete values. ***[PG Dipl., Indian Inst. of Mat. Management 2019]***
9. Illustrate some situations where Poisson distribution may be successfully employed. ***[M.Com., Bharathidasan Univ. 2007]***
10. Write down the density function of a normal distribution and discuss its important statistics?
11. What is a normal distribution? Highlights its chief uses in Operations Research.
12. Illustrate that two tails of normal distribution will never touch X-axis. ***[PG Dipl., Indian Inst. of Mat. Management 2018]***
13. Explain why the normal distribution is considered the most important distribution in probability and statistics.
14. Define Gamma distribution for a continuous random variable. Discuss its use in Operations Research. ***[PG Dipl., Indian Inst. of Mat. Management 2019]***

CHAPTER 18

Markov Analysis

"Future is bright, whereas the present is gold"

Markov Analysis is a way of analysing the current movement of some variable in an effort to forecast its ***future movement***. As a management tool, Markov Analysis has been used during the last several years, mainly as a marketing aid for examining and predicting the behaviour of customers from the standpoint of loyalty to one brand and their switching pattern to other brands. Though this is a leading use of the technique, there are however other areas in which application of Markov Analysis has produced significant contributions. In the field of ***accounting***, it can be applied to the behaviour of accounts receivable.

In production, it is helpful in evaluating alternative maintenance policies, queuing systems and work assignments. The technique is also useful to the Personnel department in determining future manpower requirements of an organization.

18:1. MARKOV PROCESS

Markov process (chain or sequence of events) is a stochastic process which is used to analyse decision problems in which the occurrence of a specific event depends on the occurrence of the event immediately prior to the current event. Basically, Markov process helps us to identify

(*i*) a specific state of the system being studied, and

(*ii*) the state-transition relationship.

For example, in the case of market share of a particular product and its competitive brands, the states may be expressed in terms of the brand that a customer is using presently. In the example of production, machines used to manufacture a product may be described as a system, where the machine can be in one of the two states: working or not working at any point of time.

18:2. STATE AND TRANSITION PROBABILITIES

The occurrence of an event at a specified point in time (say, period n) puts the system in a given state, say E_n. If after the passage of one time unit, another event occurs (during time period $n + 1$), the system has moved from state E_n to state E_{n+1}.

For example, E_t may represent the number of persons in the queue at a Mother Dairy booth at time t. As time changes to $t + 1$, the state of the queue also changes to E_{t+1}. The probability of moving from one state to another or remain in the same state in a single time period is called ***transition probability***. Also, since the probability of moving from one state to another depends on the probability of the preceding state, transition probability is a ***conditional probability***.

From the point of view of marketing research, it is assumed that the number of states (brands) are finite and the decision of change of brand is taken periodically, so that such changes will occur over a period of time. In general, let:

E_i = finite number of possible outcomes ($i = 1, 2, ..., m$) of each of the sequence of events (or experiments).

p_{ij} = conditional probability of outcome E_j for any particular event (experiment), given that outcome E_i occurred for the immediately preceding extent (experiment).

The outcomes $E_1, E_2, ..., E_m$ are called states and numbers p_{ij} are called transition probabilities.

Whenever a new result or outcome occurs, process is said to have ***incremented*** one step. Each step represents a time period or any other condition which would result in another state. The symbol n will be used to indicate the number of steps or increments. For example, if $n = 0$, then it represents the ***initial state***.

18:3. CHARACTERISTICS OF A MARKOV PROCESS

The Markov chain analysis is based on the following assumptions:

(*i*) The given system has a finite number of states, none of which "absorbing" in nature (a state is said to be absorbing if a customer would never switch to a particular brand).

(*ii*) The states are both collectively exhaustive and mutually exclusive.

(*iii*) The probability of moving from one state to another depends only on the immediately preceding state.

(*iv*) Transition probabilities are stationary, *i.e.*, they are constant.

(*v*) The process has a set of initial probabilities, which may be given or determined.

(*vi*) The transition probabilities of moving to alternative states in the next time period, given a state in the current time period must sum to unity.

18:4. STATE TRANSITION MATRIX

A state-transition matrix is a rectangular array which summarizes the transition probabilities for a given Markov process. In such a matrix rows identify the current state of the system being studied and columns identify the alternative states to which the system can move.

Let E_i = state i of a stochastic process; ($i = 1, 2, ..., m$) and p_{ij} = transition probability of moving from state E_i to state E_j in one step.

Then, a one stage state-transition matrix **P** can be described as given below:

$$\mathbf{P} = \begin{array}{c} \\ E_1 \\ E_2 \\ \vdots \\ E_m \end{array} \begin{array}{c} \begin{array}{cccc} E_1 & E_2 & \cdots & E_m \end{array} \\ \begin{pmatrix} p_{11} & p_{12} & \cdots & p_{1m} \\ p_{21} & p_{22} & \cdots & p_{2m} \\ \vdots & \vdots & \vdots & \vdots \\ p_{m1} & p_{m2} & \cdots & p_{mm} \end{pmatrix} \end{array}$$

In the transition matrix of the Markov chain, $p_{ij} = 0$ when no transition occurs from state i to state j; and $p_{ij} = 1$ when the system is in state i, it can move only to state j at the next transition.

Each row of the transition matrix represents a one-step transition probability distribution over all states. This means

$$p_{i1} + p_{i2} + \cdots + p_{im} = 1 \quad \text{for all } i \quad \text{and} \quad 0 \le p_{ij} \le 1.$$

18:5. TRANSITION DIAGRAM

A ***transition diagram*** shows the transition probabilities that can occur in any situation. Such a diagram is given in *Fig. 18.1.*

The arrows from each state indicate the possible states to which a process can move from the given state. The following transition matrix corresponds to the *Fig. 18.1.*

$$\mathbf{P} = \begin{matrix} & E_1 & E_2 & E_3 \\ E_1 & 0 & p_{12} & 0 \\ E_2 & 0 & p_{22} & p_{23} \\ E_3 & p_{31} & 0 & p_{33} \end{matrix}$$

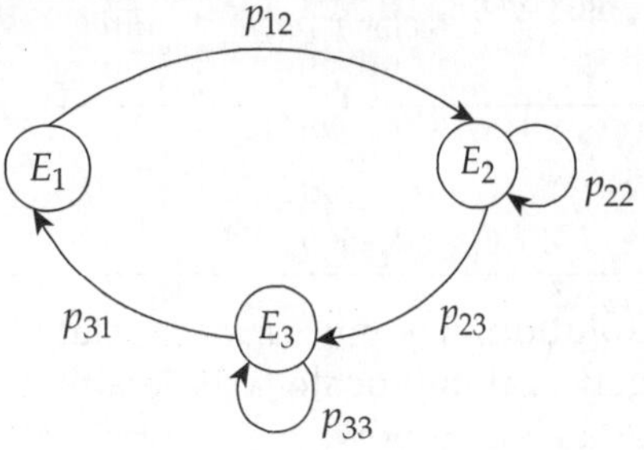

Fig. 18.1

A zero element in the matrix indicates that the transition is impossible.

Remark. Transition probabilities can also be represented by a probability tree diagram which shows the probabilities and movement from one step to another step, along with all possible branches or paths that may connect various states over a period of time.

18:6. CONSTRUCTION OF A STATE-TRANSITION MATRIX

Major steps of constructing a matrix of transition probabilities may be summarized as given below:

Step 1. Calculate retention probabilities. ***Retention*** probabilities are calculated by dividing the number of customers retained by the number of customers originally served.

Number retained = Original number served – Number lost.

Step 2. Calculate the probabilities associated with customer gains or losses. For customer gain probabilities, divide the number of customers each entity has gained by the number of customers served by the source of the gain. For customer loss probabilities, divide the number of customers each entity has lost by the original number of customers it served.

Step 3. Using the results of *Steps 1* and 2, construct the state-transition matrix. The retention probabilities are listed along the main diagonal (upper left to lower right). The remaining entries are determined as follows:

(*a*) If loss probabilities are used, insert them in the appropriate row cells of the state-transition matrix.

(*b*) If gain probabilities are used, insert them in the appropriate column cells of the state-transition matrix.

In either case, the resulting state-transition matrix will be the same.

Example 18.1. *In a small town with three advocates X, Y and Z each advocate knows that some clients switch back and forth, depending on which advocate is available at the time the client needs one. There are no new clients in the current legal market; however, none of the old clients are leaving the area. During a slack period, the three advocates collected data which identified the number of clients each advocate had seen during the preceding year. Tables given below summarize the results of this study, and the manner in which clients were gained or lost respectively. Construct the state-transition matrix that describes the problem at hand.*

Data Summary — Client Activity

Advocate	*Clients as on January 1, 2020*	*Change during year*		*Clients as on January 1, 2021*
		Gain	*Loss*	
X	400	75	50	425
Y	500	50	150	400
Z	500	100	25	575

Gain–Loss Summary

Advocate	*Clients as on Jan. 1, 2020*	*Gains*			*Losses*			*Clients as on Jan. 1, 2021*
		From X	*From Y*	*From Z*	*To X*	*To Y*	*To Z*	
X	400	0	50	25	0	50	0	425
Y	500	50	0	0	50	0	100	400
Z	500	0	100	0	25	0	0	575

Solution. The entries for the state-transition matrix that describes this problem are the probabilities that each advocate will retain, gain and lose clients in the coming year. We compute these probabilities now.

Step 1. Calculate retention probabilities. Retention probabilities are calculated by dividing the number of customers that were retained by the number of clients that were originally served. (The number of clients retained equals the original number of clients minus the number of customers that were lost.) The result is presented in the table given below:

Retention Probabilities

Advocate	*Clients as on Jan. 1, 2020*	*Number lost*	*Number retained*	*Probability of retention*
X	400	50	350	350/400 = 0.875
Y	500	150	350	350/500 = 0.700
Z	500	25	475	475/300 = 0.950

Step 2. Calculate the probabilities associated with customer gains or losses. The probabilities associated with client gains are calculated by dividing the number of clients gained (a column value) by the number of clients served at the source of the gain. The probabilities associated with client losses are calculated by dividing the number of clients served. The results are summarized in the following tables:

Gain Probabilities

Original number served	*Advocate*	*Probability of gain*		
		From X	*From Y*	*From Z*
400	X	0/400 = 0	50/500 = 0.100	25/500 = 0.050
500	Y	50/400 = 0.125	0/500 = 0	0/500 = 0
500	Z	0/400 = 0	100/500 = 0.200	0/500 = 0

Loss Probabilities

Original number served	*Advocate*	*Probability of loss*		
		To X	*To Y*	*To Z*
400	X	0/400 = 0	50/400 = 0.125	0/400 = 0
500	Y	50/500 = 0.10	0/500 = 0	100/500 = 0.2
500	Z	25/500 = 0.50	0/500 = 0	0/500 = 0

Step 3. Using the results of Steps 1 and 2, construct the state-transition matrix. List the retention probabilities along the main diagonal (under left to lower right). If loss probabilities are used, insert them in the appropriate row cells. If gain probabilities are used, insert them in the appropriate column cells. The results in the transition table should be the same. Also,

the sum of the probabilities in each row must be 1.00. On summarizing the state-transition probabilities the following state-transition matrix for the given problem is obtained:

Transition Matrix

From	*To* X	Y	Z
X	0.875	0.125	—
Y	0.100	0.700	0.200
Z	0.050	—	0.950

PROBLEMS

18.01. Two manufacturers *A* and *B* are competing with each other in a restricted market. Over the year, *A*'s customers have exhibited a high degree of loyalty as measured by the fact that customers using *A*'s product: 80 per cent of the time. Also, former customers purchasing the product from *B* have switched back to *A*'s 60 per cent of the time.

(*a*) Construct and interpret the state transition matrix in terms of (*i*) retention and loss, and (*ii*) retention and gain.

(*b*) Calculate the probability of a customer purchasing *A*'s and *B*'s product at the end of the second period.

18.02. There are three dairies *A*, *B* and *C* in a small town which supply all the milk consumed in the town. Assume that the initial consumer sample is composed of 1,000 respondents distributed over three dairies *A*, *B* and *C*. It is known by all the dairies that consumers switch from one dairy to another due to advertising, price and dissatisfaction. All these dairies maintain records of the number of their customers and the dairy from which they obtained each new customer. The following table illustrate the flow of customers over an observation period of one month. Assume that the matrix of transition probabilities remain fairly stable and at the beginning of period one, market shares are A = 25%, B = 45% and C = 30%. Construct the state transition probability matrix to analyse the problem.

Dairy	*Period one: number of customer*	*Changes during period* Gain	Loss	*Period two: number of customers*
A	250	62	50	262
B	450	53	60	443
C	300	50	55	295
Total	1,000	165	165	1,000

18:7. *n*-STEP TRANSITION PROBABILITIES

We are mainly interested in finding out the probabilistic behaviour of a system over a period of time. Let the system occupy state E_i at time 0. Then, the probability that the system moves to state E_j at time n (*i.e.*, after n steps) is an n-steps transition probability that may be denoted by $p_{ij}^{(n)}$. Note that $p_{ij}^{(1)} = p_{ij}$ for all i and j.

If $P_i(n)$ denote the probability that the system is in state E_i at time n, then the probability that the system will be in state E_j at time $n + 1$ can be obtained from the basic equation:

$$p_j^{(n+1)} = \sum_{i=1}^{m} p_i^{(n)} \, p_{ij}, \qquad n = 0, 1, 2, \ldots$$

where, p_{ij} is the transition probability of transition from E_i to E_j in one step.

For $j = 1, 2, ..., m$, this gives a system of m equations that can be written in matrix form as

$$\begin{bmatrix} p_1^{(n+1)} \\ p_2^{(n+1)} \\ \vdots \\ p_m^{(n+1)} \end{bmatrix} = \begin{pmatrix} p_{11} & p_{12} & \cdots & p_{1m} \\ p_{21} & p_{22} & \cdots & p_{2m} \\ \vdots & \vdots & \vdots & \vdots \\ p_{m1} & p_{m2} & \cdots & p_{mm} \end{pmatrix} \begin{bmatrix} p_1^{(n)} \\ p_2^{(n)} \\ \vdots \\ p_m^{(n)} \end{bmatrix}$$

or
$$\mathbf{p}^{(n+1)} = \mathbf{P}\,\mathbf{p}^{(n)},$$

where, $\mathbf{p}^{(n+1)}$, $\mathbf{p}^{(n)}$ are the state probability vectors at time $n + 1$ and n respectively, and $\mathbf{P}$ is one-stage state transition matrix.

Thus, if we know the initial state probabilities ($n = 0$), we can compute them at any time successively as below:

$$\mathbf{p}^{(1)} = \mathbf{P}\,\mathbf{p}^{(0)}$$
$$\mathbf{p}^{(2)} = \mathbf{P}\,\mathbf{p}^{(1)} = \mathbf{P}^2\,\mathbf{p}^{(0)}$$
$$\vdots$$
$$\mathbf{p}^{(n)} = \mathbf{P}\,\mathbf{p}^{(n-1)} = \mathbf{P}^n\,\mathbf{p}^{(0)}$$

Thus, it is possible to find the probability that the system occupies each of its states after n steps by pre-multiplying the initial state probability vector by the nth power of the transition matrix.

18:8. STEADY STATE (EQUILIBRIUM) CONDITIONS

One may observe that if no attempt is made to alter the transition probabilities, the system would eventually reach a point of ***equilibrium***, *i.e.*, for all practical purposes, no further changes would occur in the state probabilities. After a Markov process has been in operation for a long time (many steps), a given state will tend to occur a fixed per cent of time. There is a limiting probability that the system in a finite (but large) number of transitions will reach steady-state equilibrium. That is, when n becomes very large each p_{ij} tends to fixed limits and each state probability vector $\mathbf{p}^{(n)}$ approaches a constant value, *i.e.*,

$$\mathbf{p}^{(n+1)} = \mathbf{p}^{(n)} = \mathbf{p}, \text{ independent of } n.$$

Thus, in the limiting case

$$\lim \mathbf{p}^{(n+1)} = \lim \mathbf{p}\,\mathbf{p}^{(n)} \quad \text{becomes} \quad \mathbf{p} = \mathbf{P}\mathbf{p}$$

Therefore, in the limit as n approaches to infinity, $\mathbf{p}^{(n)}$ becomes constant (*i.e.*, independent of time) and the system is said to have reached a steady state equilibrium.

18:9. MARKOV ANALYSIS ALGORITHM

Markov analysis is used to provide management with answer to the following questions:

(*a*) What is the rate of grain and loss for the current market?

(*b*) For a given future period, what is the projected market share of each in the system?

(*c*) Assuming it could exist, what is the equilibrium state?

The following algorithm provides a procedure for answering these questions regardless of the area of application.

Step 1. Using the data of the problem, determine the probabilities associated with retention, gain, and losses.

Step 2. Using the data of *Step 1*, develop the appropriate state-transition matrix.

Step 3. Using the state-transition matrix of *Step 2*, determine the expected future market shares for the next period. The procedure is as follows:

$$\begin{pmatrix}\text{market shares at beginning}\\ \text{of period } 1\end{pmatrix}\begin{bmatrix}\text{state-transition}\\ \text{matrix}\end{bmatrix}=\begin{pmatrix}\text{expected market shares}\\ \text{at beginning of period } 2\end{pmatrix}$$

Step 4. If more than one expected market share calculation is required, apply

$$\begin{pmatrix}\text{expected market shares}\\ \text{at beginning of period } k\end{pmatrix}\begin{bmatrix}\text{state-transition}\\ \text{matrix}\end{bmatrix}=\begin{pmatrix}\text{expected market shares at}\\ \text{beginning of period } k+1\end{pmatrix};$$

where, k equals the number of time periods under consideration.

Step 5. Using the state-transition matrix of *Step 2*, determine the equilibrium conditions for the current problem.

Example 18.2. *Suppose there are two market products of brands A and B respectively. Let each of these two brands have exactly 50% of the total market in same period and let the market be of a fixed size. The transition matrix is given below:*

$$\begin{array}{cc} & \textit{To} \\ & \begin{array}{cc} A & B \end{array} \\ \textit{From} \begin{array}{c} A \\ B \end{array} & \begin{pmatrix} 0.9 & 0.1 \\ 0.5 & 0.5 \end{pmatrix} \end{array}$$

If the initial market share breakdown is 50% for each brand, then determine their market shares in the steady-state.

Solution.

Step 1. We are given the state-transition matrix

$$\begin{pmatrix} 0.9 & 0.1 \\ 0.5 & 0.5 \end{pmatrix}$$

Step 2. Using the state-transition matrix, we determine the expected market share for the next period. We are given that the initial market share for brands *A* and *B* are 50% each, *i.e.*, the market share at the beginning of period *1* are [0.5, 0.5]. After the promotional efforts applied to brands *A* and *B*, the transition matrix indicates that in the second period, brand *A* will retain 90% of its customers and take away 50% of *B*'s, thus market share for brand *A* in the second period will be:

$$0.5 \times 0.9 + 0.5 \times 0.5 = 0.7$$

and the corresponding market share for brand *B* in the second period will be:

$$0.5 \times 0.1 + 0.5 \times 0.5 = 0.3$$

Using the procedure stated in the algorithm, this could be obtained by

$$\begin{pmatrix}\text{market shares at}\\ \text{beginning of period } 1\end{pmatrix}\begin{bmatrix}\text{state-transition}\\ \text{matrix}\end{bmatrix}=\begin{pmatrix}\text{expected market shares at}\\ \text{beginning of period } 2\end{pmatrix}$$

The given data, therefore, gives:

Market share at beginning of period 1		*State-transition matrix*		*Expected market share at beginning of period 2*
(0.5 0.5)	×	$\begin{bmatrix} 0.9 & 0.1 \\ 0.5 & 0.5 \end{bmatrix}$	=	(0.7 0.3)
[50% 50%]				[70% 30%]

Step 3. If the same transition matrix holds from one period to another, then market shares of two brands for different periods are obtained as follows:

Market share at beginning of period 2		State-transition matrix		Expected market share at beginning of period 3
$(0.7 \quad 0.3)$	×	$\begin{bmatrix} 0.9 & 0.1 \\ 0.5 & 0.5 \end{bmatrix}$	=	$(0.78 \quad 0.22)$
[70% 30%]				[78% 22%]

Market share at beginning of period 3		State-transition matrix		Expected market share at beginning of period 4
$(0.78 \quad 0.22)$	×	$\begin{bmatrix} 0.9 & 0.1 \\ 0.5 & 0.5 \end{bmatrix}$	=	$(0.812 \quad 0.188)$
[78% 22%]				[81.2% 18.8%]

Proceeding in this way one obtains the following table:

Period	*Brand A (Market share)* %	*Brand B (Market share)* %
0	50	50
1	70	30
2	78	22
3	81.2	18.8
4	82.48	17.52
5	82.992	17.008
6	83	17

This table shows that starting with period 0, the market shares of Brand *A* and Brand *B* are 50% each. After 6 time periods the market shares tend to stabilize at about 83% and 17%, respectively.

Thus, the steady-state market shares of brand *A* and brand *B* will be 5/6 and 1/6, respectively of the total market share.

Remark. Equilibrium (or steady-state) position can also be determined by determining x and y which satisfy:

$$(x \quad y)\begin{pmatrix} 0.9 & 0.1 \\ 0.5 & 0.5 \end{pmatrix} = (x, \quad y) \quad \text{and} \quad x + y = 1.$$

This gives: $0.9x + 0.5y = x$, $0.1x + 0.5y = y$; and $x + y = 1$.

Solving these, we get $x = 5/6$ and $y = 1/6$.

Example 18.3. *The School of International Studies for Population found out by its survey that the mobility of the population of a state to the village, town and city is in the following percentages:*

From	*To*		
	Village	*Town*	*City*
Village	50%	30%	20%
Town	10%	70%	20%
City	10%	40%	50%

What will be the proportion of population in village, town and city after two years, given that the present population has proportions of 0.7, 0.2 and 0.1 in the village, town and city respectively? What will be the respective proportions in the long run?

Solution. We are given the state-transition matrix:

$$\begin{pmatrix} 0.50 & 0.30 & 0.20 \\ 0.10 & 0.70 & 0.20 \\ 0.10 & 0.40 & 0.50 \end{pmatrix}$$

Using the state-transition matrix, we determine the expected proportion of population in a village, town and city for the first year is obtained by multiplying the matrix of transition probabilities by the matrix of the present population proportion of the village, town and city. Thus, we get

$$\underset{\textit{Present proportion}}{[0.70 \quad 0.20 \quad 0.10]} \times \underset{\textit{Transition matrix}}{\begin{pmatrix} 0.50 & 0.30 & 0.20 \\ 0.10 & 0.70 & 0.20 \\ 0.10 & 0.40 & 0.50 \end{pmatrix}} = \underset{\textit{Proportion after one year}}{[0.38 \quad 0.39 \quad 0.23]}$$

This implies that the population after one year is:

Village = 38%, Town = 39% and City = 23%.

The expected proportion of population for the second year can be determined in the same manner considering the proportion of population after first year. Thus, we have

$$\underset{\textit{Proportion after first year}}{[0.38 \quad 0.39 \quad 0.23]} \times \begin{pmatrix} 0.50 & 0.30 & 0.20 \\ 0.10 & 0.70 & 0.20 \\ 0.10 & 0.40 & 0.50 \end{pmatrix} = \underset{\textit{Proportion after second year}}{[0.252 \quad 0.479 \quad 0.269]}$$

∴ Proportion of population after second year is obtained as:

Village = 25.2%, Town = 47.9% and City = 26.9%

In the long run, the proportion after 'n' years can be determined by using the relation given below:

$$\underset{\textit{Initial proportion}}{[0.70 \quad 0.20 \quad 0.10]} \times \underset{\textit{Transition matrix}}{\begin{pmatrix} 0.50 & 0.30 & 0.20 \\ 0.10 & 0.70 & 0.20 \\ 0.10 & 0.40 & 0.50 \end{pmatrix}^{n-1}} = \begin{bmatrix} \text{Proportion after} \\ n \text{ years} \end{bmatrix}$$

Example 18.4. *The price of an equity share of a company may increase, decrease or remain constant on any given day. It is assumed that the change in price on any day affects the change on the following day as described by the following transition matrix:*

		Change Tomorrow		
		Increase	*Decrease*	*Unchanged*
Change Today	*Increase*	0.5	0.2	0.3
	Decrease	0.7	0.1	0.2
	Unchanged	0.4	0.5	0.1

(*i*) *If the price of share increased today, what are chances that it will increase, decrease or remain unchanged tomorrow?*

(*ii*) *If the prices of the share decreased today, what are the chances that it will increase tomorrow?*

(*iii*) *If the price of the share remained unchanged today, what are the chances that it will increase, decrease, or remain unchanged day after tomorrow?*

Solution.

(*i*) Given that the price of share has increased today, the probability of its going up (today) is 1 and probability of each of events, decreasing or remaining unchanged is equal to zero.

Thus, $\mathbf{p}^{(0)} = [1 \quad 0 \quad 0]$ and $\mathbf{p}^{(1)} = \mathbf{p}^{(0)}\,\mathbf{P}$

$$\therefore \quad \underset{\textit{Today's proportion}}{[1 \quad 0 \quad 0]} \times \underset{\textit{Transition matrix}}{\begin{bmatrix} 0.5 & 0.2 & 0.3 \\ 0.7 & 0.1 & 0.2 \\ 0.4 & 0.5 & 0.1 \end{bmatrix}} = \underset{\textit{Tomorrow's proportion}}{[0.5 \quad 0.2 \quad 0.3]}$$

This implies that the price will increase, decrease or remain unchanged tomorrow are 50%, 20% and 30% respectively (given that it has increased today).

(*ii*) Here, let the probability of decrease is 1 and the probability of the other two events be zero. Then, $\mathbf{p}^{(0)} = [0 \quad 1 \quad 0]$. Using this, we get

$$[0 \quad 1 \quad 0] \times \begin{bmatrix} 0.5 & 0.2 & 0.3 \\ 0.7 & 0.1 & 0.2 \\ 0.4 & 0.5 & 0.1 \end{bmatrix} = [0.7 \quad 0.1 \quad 0.2]$$

This shows that the chances of price increase tomorrow are 70%, 10% and 20%, respectively.

(*iii*) Here, let the probability of prices remain unchanged today as 1 and the probability of the other two events as zero. Then, $\mathbf{p}^{(0)} = [0 \quad 0 \quad 1]$. Using this, we get

$$\underset{\textit{Today's proportion}}{[0 \quad 0 \quad 1]} \times \begin{bmatrix} 0.5 & 0.2 & 0.3 \\ 0.7 & 0.1 & 0.2 \\ 0.4 & 0.5 & 0.1 \end{bmatrix} = \underset{\textit{Tomorrow's proportion}}{[0.4 \quad 0.5 \quad 0.1]}$$

and

$$\underset{\textit{Tomorrow's proportion}}{[0.4 \quad 0.5 \quad 0.1]} \times \begin{bmatrix} 0.5 & 0.2 & 0.3 \\ 0.7 & 0.1 & 0.2 \\ 0.4 & 0.5 & 0.1 \end{bmatrix} = \underset{\textit{Day-after tomorrow's proportion}}{[0.59 \quad 0.18 \quad 0.23]}$$

This implies that the chances that the price will increase, decrease or remain unchanged day-after tomorrow are 59%, 18% and 23%, respectively.

Example 18.5. *A housewife buys three kinds of cereals: A, B and C. She never buys the same cereal on successive weeks. If she buys cereal A, then the next week she buys cereal B. However, if she buys either B or C, then in the next week she is three times as likely to buy A as the other brand. Obtain the transition probability matrix and determine how often she would buy each of the cereals in the long run.*

[MBA (PT.) Delhi Univ. 2015]

Solution. The given information is presented in the following table in the form of a transition probability matrix.

Cereals	*Cereals*		
	A	*B*	*C*
A	0	1	0
B	0.75	0	0.25
C	0.75	0.25	0

If x, y and z denote the steady-state probabilities of states A, B and C respectively, we have

$$[x \quad y \quad z] \times \begin{bmatrix} 0 & 1 & 0 \\ 0.75 & 0 & 0.25 \\ 0.75 & 0.25 & 0 \end{bmatrix} = [x \quad y \quad z]$$

where, $\quad x + y + z = 1 \quad$ and $\quad x \geq 0, \; y \geq 0, \; z \geq 0$

This gives $\quad 0.75y + 0.75z = x, \; x + 0.25z = y$

and $\quad 0.25y = z \;$ with $\; x + y + z = 1$

The above equations are re-written as follows:

$$x - 0.75y - 0.75z = 0$$
$$x - y + 0.25z = 0$$
$$0.25y - z = 0$$

and
$$x + y + z = 1$$

Solving these four equations in three variables, we get

$$x = 0.428, \quad y = 0.457 \quad \text{and} \quad z = 0.114$$

Thus, in the long run, the house wife would buy three cereals A, B and C with a frequency of 42.8%, 45.7% and 11.4%, respectively.

Example 18.6. *On January 1 (this year), Bakery A had 40% of its local market share, while the other two bakeries B and C had 40% and 20% respectively of the market share. Based upon a study by a marketing research firm, the following facts were compiled. Bakery A retains 90% of its customers, while gaining 5% of competitor B's customers and 10% of C's customers. Bakery B retains 85% of its customers, while gaining 5% of A's customers and 7% of C's customers. Bakery C retains 83% of its customers and gains 5% of A's customers and 10% of B's customers. What will each firm's share be on January 1, next year, and what will each firm's market share be at equilibrium?* ***[MBA (Nov.), Delhi Univ. 2001]***

Solution. Using the information given in the problem, we formulate the state-transition matrix as follows:

$$\begin{array}{cc} & \textit{Bakery} \\ & \begin{array}{ccc} A & B & C \end{array} \\ \textit{Bakery} \begin{array}{c} A \\ B \\ C \end{array} & \begin{pmatrix} 0.90 & 0.05 & 0.05 \\ 0.05 & 0.85 & 0.10 \\ 0.10 & 0.07 & 0.83 \end{pmatrix} \downarrow \begin{array}{c} \textit{Retention} \\ \textit{and} \\ \textit{gain} \end{array} \\ & \longrightarrow \textit{Retention and loss} \end{array}$$

On January 1, the market shares of the three Bakeries are 40, 40 and 20 per cent respectively. The management of three Bakeries is interested in knowing their market shares on January 1 next year.

The expected market shares for Bakeries A, B and C on January 1 next year are computed as below:

$$\begin{pmatrix} \text{Market share} \\ \text{on January 1} \\ \text{this year} \end{pmatrix} \times \begin{pmatrix} \text{State-} \\ \text{transition} \\ \text{matrix} \end{pmatrix} = \begin{pmatrix} \text{Expected market share} \\ \text{on January 1} \\ \text{next year} \end{pmatrix}$$

$$(0.40 \quad 0.40 \quad 0.20) \begin{pmatrix} 0.90 & 0.05 & 0.05 \\ 0.05 & 0.85 & 0.10 \\ 0.10 & 0.07 & 0.83 \end{pmatrix} = (0.400 \quad 0.374 \quad 0.226)$$

Thus, the market shares of Bakeries A, B and C on January 1 next year will be 40%, 37.4% and 22.6%, respectively.

Equilibrium market shares x, y and z for the three Bakeries must satisfy

$$(x \quad y \quad z) \begin{pmatrix} 0.90 & 0.05 & 0.05 \\ 0.05 & 0.85 & 0.10 \\ 0.10 & 0.07 & 0.83 \end{pmatrix} = (x \quad y \quad z)$$

where, $x + y + z = 1$ and $x, y, z \geq 0$.

This gives

$$x = 0.90x + 0.05y + 0.10z$$
$$y = 0.05x + 0.85y + 0.07z$$
$$z = 0.05x + 0.10y + 0.83z$$

The equations can be rewritten as

$$-0.10x + 0.05y + 0.10z = 0$$
$$0.05x - 0.15y + 0.07z = 0$$
$$0.05x + 0.10y - 0.17z = 0$$
$$x + y + z = 1$$

Solving these equations, we get

$$x = 0.43, \quad y = 0.28, \quad z = 0.29$$

Hence, the three firm's market shares at equilibrium are:

Bakery A : 43% of the total market,
Bakery B : 28% of the total market,
and Bakery C : 29% of the total market.

Example 18.7. *A certain piece of equipment is inspected at the end of each day and classified as just overhauled, good, fair or inoperative. Let us denote the four classifications as states 1, 2, 3 and 4 respectively. If item is inoperative it is overhauled, a procedure that takes one day. Assume that the working condition of the equipment follows a Markov process with the following state-transition matrix:*

$$\text{Today}\;\begin{matrix} \\ 1 \\ 2 \\ 3 \\ 4 \end{matrix}\overset{\textit{Tomorrow}}{\begin{matrix} 1 & 2 & 3 & 4 \\ \end{matrix}}\begin{pmatrix} 0 & 3/4 & 1/4 & 0 \\ 0 & 1/2 & 1/2 & 0 \\ 0 & 0 & 1/2 & 1/2 \\ 1 & 0 & 0 & 0 \end{pmatrix}$$

It costs ₹125 to overhaul a machine on an average and ₹75 is lost in production, if a machine is found inoperative. Use steady state probabilities to compute the expected per day cost of maintenance.

[MBA (Oct.), Delhi Univ. 2007]

Solution. The state-transition matrix indicates that a just overhauled machine is in good condition after a day's use 3/4 of the times; and 1/4 of the time it is in fair condition after a day's use. Also, it indicates that machine that is in good condition has an equal chance of remaining in good condition or fair condition after a day's use, while a machine in fair condition also has an equal chance of being in fair or inoperative condition after a day's use. Moreover, an inoperative machine will be overhauled the next day, so that it will have been just overhauled at the end of the day.

Let, the steady-state probabilities that the machine will be in states *1, 2, 3* and *4* be p_1, p_2, p_3 and p_4, respectively. Then, the steady-state equation $\mathbf{p} = \mathbf{P}\,\mathbf{p}$ gives

$$\begin{pmatrix} p_1 \\ p_2 \\ p_3 \\ p_4 \end{pmatrix} = \begin{pmatrix} 0 & 3/4 & 1/4 & 0 \\ 0 & 1/2 & 1/2 & 0 \\ 0 & 0 & 1/2 & 1/2 \\ 1 & 0 & 0 & 0 \end{pmatrix}\begin{pmatrix} p_1 \\ p_2 \\ p_3 \\ p_4 \end{pmatrix}, \quad \sum_{i=1}^{4} p_i = 1$$

This gives the system of equations:

$$p_1 = p_4$$
$$p_2 = (3/4)\,p_1 + (1/2)\,p_2$$
$$p_3 = (1/4)\,p_1 + (1/2)\,p_2 + (1/2)\,p_3$$
$$p_4 = (1/2)\,p_3 \quad \text{and} \quad p_1 + p_2 + p_3 + p_4 = 1$$

Solving these equations for p_1, p_2, p_3 and p_4, we find that the required steady state probabilities are:

$$p_1 = 2/11, \quad p_2 = 3/11, \quad p_3 = 4/11 \quad \text{and} \quad p_4 = 2/11$$

Thus, on average, two out of every 11 days the machine will be overhauled and two out of every 11 days it will be found inoperative at the end of the day.

Hence, the average cost per day of maintenance is

$$(2/11) \times ₹\,125 + (2/11) \times ₹\,75 = ₹\,36.36.$$

PROBLEMS

18.03. A market survey is made on two brands of breakfast food *A* and food *B*. Every time a customer purchases, he may buy the same brand or switch to another brand. The transition matrix is given below:

$$\text{From}\ \begin{matrix} & \textit{To} & \\ & A & B \\ A & \\ B & \end{matrix}\begin{pmatrix} 0.8 & 0.2 \\ 0.6 & 0.4 \end{pmatrix}$$

At present, it is estimated that 60 per cent of the people buy brand *A* and 40 per cent buy brand *B*. Determine the market share of brand *A* and brand *B* in the steady state.

18.04. (*a*) The purchase pattern of two brands of toothpaste can be expressed as a Markov process with the following transition probabilities:

	Formula X	*Formula Y*
Formula *X*	0.90	0.10
Formula *Y*	0.05	0.95

(*i*) Which brand appears to have most loyal customers? Explain.

(*ii*) What are the projected market share for the two brands?

[MBA (Nov.), Delhi Univ. 2004]

(*b*) Suppose that in problem (*a*) a new toothpaste brand enters the market such that the following transition probabilities exist:

	Formula X	*Formula Y*	*Formula Z*
Formula *X*	0.80	0.10	0.10
Formula *Y*	0.05	0.75	0.20
Formula *Z*	0.40	0.30	0.30

(*i*) What are the new long-run market shares?

(*ii*) Which brand will suffer most from the introduction of the new brand of toothpaste?

[MBA (Nov.), Delhi Univ. 2007]

18.05. A salesman's territory consists of three cities, *A*, *B* and *C*. He never sells in the same city on successive days. If he sells in city *A*, then the next day he sells in city *B*. However, if he sells in either *B* or *C*, then the next day he is twice as likely to sell in city *A* as in the other city. In the long run, how often does he sell in each of the cities?

18.06. A researcher is analyzing brand switching between different airlines, operating on the Delhi-Mumbai route by frequent fliers. On the basis of the data collected by her, the researcher has developed the following transition probability matrix:

$$\begin{array}{ccc} & & \textit{To Airline} & \\ & & AA \quad & BB \quad & CC \\ \text{From Airline} & \begin{array}{c} AA \\ BB \\ CC \end{array} & \left[\begin{array}{ccc} 0.90 & 0.03 & 0.07 \\ 0.15 & 0.80 & 0.05 \\ 0.20 & 0.30 & 0.50 \end{array}\right] \end{array}$$

It is found that currently the airlines *AA*, *BB* and *CC* have 20%, 50% and 30% of the market share, respectively.

(*i*) Obtain the market share for each airline in two months' time, and

(*ii*) Calculate the long-run market share for each airline. ***[MBA, Delhi Univ. 2011]***

18.07. There are three firms, *ABC*, *PQR* and *XYZ* having a market share of 40 per cent, 40 per cent and 20 per cent, respectively, on January 1, 2002. Over a year, the following developments take place:

ABC retains 80 per cent of its customers, loses 16 per cent to *PQR* and 4 per cent to *XYZ*.
PQR retains 84 per cent of its customers, loses 12 per cent to *ABC* and 4 per cent to *XYZ*.
XYZ retains 76 per cent of its customers, loses 18 per cent to *ABC* and 6 per cent to *PQR*.

Assuming that the market does not change:

(*a*) What share of the market shall be held by each firm on January 1, 2004?

(*b*) What would the long-run shares of the firms at equilibrium be, if the buying habits do not change? ***[MBA (PT.) (Nov.), Delhi Univ. 2008]***

18.08. A grocer stocks his store with three types of detergents *A*, *B* and *C*. When brand *A* is sold out, the probability is 0.7 that he stocks up with brand *A* again. When he sells out brand *B*, the probability is 0.8 that he will stock up with brand *B* again. Finally, when he sells out brand *C*, the probability is 0.6 that he will stock up with brand *C* again. When he switches to another detergent, he does so with equal probability for the remaining two brands. Find the transition matrix. In the long run how does he stock up with detergents? ***[MBA (PT.), Delhi Univ. 2009]***

18.09. Suppose the new razor blades were introduced in the market by three companies at the same time. When they were introduced, each company had an equal share of the market, but during the first year the following changes took place:

(*i*) Company *A* retained 90% of its customers, lost 3% to *B* and 7% to *C*.

(*ii*) Company *B* retained 70% of its customers, lost 10% to *A* and 20% to *C*.

(*iii*) Company *C* retained 80% of its customers, lost 10% to *A* and 10% to *B*.

Assuming that no changes in buying habits of the customer occur:

(*i*) What are the market shares of the three companies at the end of the first year and second year?

(*ii*) What are the long-run market shares of the three companies?

[MBA (Nov.), Delhi Univ. 2009; MBA (PT.), Delhi Univ. 2010]

18.10. *ABC* Company had 45% of the local market for its cosmetics, while the two other manufacturers, *XYZ* Company and *PQR* Company have 30% and 25% shares, respectively, in the local market, as on 1st January of this year. A study by a market research firm has disclosed the following:

Every year, *ABC* retains 80% of its customers, while it gains 3% and 6% of the customers from its two competitors, *XYZ* and *PQR* respectively. *XYZ* retains 90% of its customers, and gains 14% and 9% of customers respectively from *ABC* and *PQR*. *PQR* retains 85% of its customers and gains 7% and 6% of customers from *XYZ* and *ABC*, respectively.

What will be the share of each firm on January 1 next year? What will be the market share of each firm at equilibrium? ***[MBA (Nov.), Delhi Univ. 2003]***

18.11. (*a*) Given the following transition matrix, determine the equilibrium market share for each company:

$$\textit{Company}\ \begin{array}{c} \\ X \\ Y \\ Z \end{array} \overset{\textit{Company}}{\begin{array}{ccc} X & Y & Z \\ \end{array}} \begin{pmatrix} 0.25 & 0.25 & 0.50 \\ 0.05 & 0.85 & 0.10 \\ 0.10 & 0.20 & 0.70 \end{pmatrix}$$

(*b*) In an effort to improve its market share, company *X* consider two marketing strategies. The first is to provide a coupon with its products that customers can apply to his or her next purchase. Data indicate that this strategy affects the loyalty of company *X*'s own customers as reflected in the following transition matrix:

$$\begin{array}{cc} & \textit{Company} \\ & \begin{array}{ccc} X & Y & Z \end{array} \\ \begin{array}{r} X \\ \textit{Company}\ Y \\ Z \end{array} & \begin{pmatrix} 0.50 & 0.10 & 0.40 \\ 0.05 & 0.85 & 0.10 \\ 0.10 & 0.20 & 0.70 \end{pmatrix} \end{array}$$

Instead of improving/retaining of probabilities, the second strategy is designed to gain customers from company *Y* and company *Z*. The transition matrix resulting from the second market strategy is as follows:

$$\begin{array}{cc} & \textit{Company} \\ & \begin{array}{ccc} X & Y & Z \end{array} \\ \begin{array}{r} X \\ \textit{Company}\ Y \\ Z \end{array} & \begin{pmatrix} 0.25 & 0.25 & 0.50 \\ 0.10 & 0.85 & 0.05 \\ 0.20 & 0.10 & 0.70 \end{pmatrix} \end{array}$$

Suppose that the costs for both market strategies are the same. If you were the market analyst for company *X*, which market strategy would you recommend and Why?

18.12. Dharamshala is a small town located in Himachal Pradesh. The town has three grocery stores. Because of its location, there are no new customers; however the old ones are not leaving the area. During a slack period, the three grocery operators got together and conducted a study on the number of customers that each customer visited the grocery store in the past. The data is summarized below:

Grocer	*Customers as on January 1, 2020*	*Change during year*		*Customers as on January 1, 2021*
		Gain	*Loss*	
1	500	100	100	500
2	400	75	100	375
3	500	50	25	525

The gain-loss summary for the three grocers is as follows:

Customers as on January 1, 2020	*Gains from* Store 1	Store 2	Store 3	*Losses to* Store 1	Store 2	Store 3	*Customers as on January 1, 2021*
500	0	100	0	0	50	50	500
400	50	0	25	100	0	0	375
500	50	0	0	0	25	0	525

Construct the state-transition matrix that defines the market structure for the three gorcers of Dharamshala.

18.13. There are three dairies in a town, say *A*, *B* and *C*. They supply the milk consumed in the town. It is known by all the dairies that consumers switch from dairy to dairy over time, because of advertising, dissatisfaction with service and other reasons. All these dairies maintain records of the number of their customers and the dairy from which they obtained each new customer. Following table illustrates the flow of customers over an observation period of one month, say June:

Dairy	*June 1 customers*	*Gains from* A	B	C	*Losses to* A	B	C	*July 1 customers*
A	200	0	35	25	0	20	20	220
B	500	20	0	20	35	0	15	490
C	300	20	15	0	25	20	0	290

We assume that the matrix of transition probabilities remains fairly stable and that the July 1 market shares are $A = 22\%$, $B = 49\%$ and $C = 29\%$.

Managers of these dairies are willing to know:

(*i*) Market shares of their dairies on 1st August and 1st September, and

(*ii*) Their market shares in steady state.

18.14. A market research team has conducted a survey of consumer buying habits with respect to three brands of talcum powder in an area. It estimates that at present, 20% of the customers buy brand *A*, 50% of the customers buy brand *B* and 30% of them buy brand *C*. In addition, the team has analysed its survey and has determined the following brand switching matrix:

		Brand next bought A	B	C
	A	0.6	0.3	0.1
Brand just bought	*B*	0.4	0.5	0.1
	C	0.2	0.1	0.7

What will be the expected distribution of consumers two time periods later and also at equilibrium? ***[MBA, Delhi Univ. 2002]***

18.15. The number of units of an item that are withdrawn from inventory on a day-to-day basis is a Markov chain process in which requirements for tomorrow depend on today's requirements. A one-day transition probability matrix is given below:

Number of units withdrawn from inventory

		Tomorrow		
		5	10	12
	5	0.6	0.4	0.0
Today	10	0.3	0.3	0.4
	12	0.1	0.3	0.6

(*i*) Construct a tree diagram showing inventory requirements on two consecutive days.
(*ii*) Develop a two-day transition probability matrix.
(*iii*) Comment on how a two-day transition probability matrix might be helpful to a manager who is responsible for inventory management. *[MBA, Delhi Univ. 2004]*

18:10. SPECIAL CASES IN MARKOV ANALYSIS

Until now, we have analysed the situations where none of the states was absorbing in nature. In our real life, the accounts receivable situation of a firm may be regarded as an ***absorbing Markov-chain*** problem. In absorbing state, transition probability in the retention diagonal from upper left to lower right is equal to 1.

Illustration. *Consider the accounts receivable problem of a firm. Suppose that for this firm, an accounts receivable turns into bad debt, if the account is more than three months overdue. Now, at the beginning of each month, each of the accounts may be classified into one of the following states:*

Stage 1 : New Account
State 2 : Account is one month overdue for payment
State 3 : Account is two months overdue for payment
State 4 : Account is three months overdue for payment
State 5 : Account has been paid
State 6 : Account is written off as bad debt.

Suppose further that on the basis of the past data, the following matrix indicating how the status of an account changes from one month to the next, has been derived:

State	*State*					
	1	*2*	*3*	*4*	*5*	*6*
1	*0*	*0.7*	*0*	*0*	*0.3*	*0*
2	*0*	*0*	*0.6*	*0*	*0.4*	*0*
3	*0*	*0*	*0*	*0.4*	*0.6*	*0*
4	*0*	*0*	*0*	*0*	*0.8*	*0.2*
5	*0*	*0*	*0*	*0*	*1*	*0*
6	*0*	*0*	*0*	*0*	*0*	*1*

The various entries in the matrix can be interpreted in the usual way. For example, the probability of a new account to be paid by the beginning of the next month is 30%, while the probability of it being not paid and, therefore, become a one-month overdue is 70%; if an account is two months overdue at the beginning of a month, there is a 40% chance that at the beginning of the next month the account shall not be paid up and hence be three months overdue, and a 60% chance that the account would be paid up. It is assumed for simplicity that after the three-months period, an account is either paid up or is written off as bad debt. Thus, if and when an account is paid up or written off as bad debt, the account gets closed and no more transitions can occur. States 5 and 6, account paid up and account written off respectively, are therefore the absorbing states — every new account shall eventually be absorbed as either a paid up debt or a bad debt. The other states, 1 through 4, are transient states.

Answer. For the analysis of Markov chains having absorbing states, we first re-write the transition probability matrix **P** with the states listed as absorbing states and transient states, in that order. Thus,

$$\mathbf{P} = \begin{bmatrix} \mathbf{I} & \mathbf{O} \\ \mathbf{R} & \mathbf{Q} \end{bmatrix}$$

where, **I** = unit matrix containing absorbing states,
O = null matrix or zero matrix,
R = matrix of relationship between the transition and absorbing states,
Q = matrix of relationship among transient states.

After this, we calculate the fundamental matrix of the Markov chains, *viz.*, $(\mathbf{I} - \mathbf{Q})^{-1}$ and then $(\mathbf{I} - \mathbf{Q})^{-1} \times \mathbf{R}$. The elements of this matrix gives the expected number of periods that will be spent in transient state before reaching an absorbing state.

Thus, $(i, j)^{th}$ element of $(\mathbf{I} - \mathbf{Q})^{-1} \times \mathbf{R}$ gives the expected number of periods that will be spent in transient state t_i before reaching the state. Further, when the chain begins in a given state t_i, the probability that it shall be eventually absorbed in the absorbing state a_j would be given by the $(i, j)^{th}$ element of $(\mathbf{I} - \mathbf{Q})^{-1} \times \mathbf{R}$.

Transient State. A state is said to be ***transient***, if it is not possible to move to that state from any other state except itself. Otherwise, the state is termed as ***persistent***.

For example in the accounts receivable problem, all non-absorbing states are transient states, because all movement will be towards absorption and away from these states.

Periodic Process. A *periodic* (or cycling) Markov chain process is one in which transition matrix '**P**' contains all zero elements in the retention cells (diagonal elements of **P**), and all other elements are either '1' or '0'. For example.

$$\mathbf{P}^{(2)} = \begin{matrix} & X & Y \\ X & 0 & 1 \\ Y & 1 & 0 \end{matrix}\ ; \qquad \mathbf{P}^{(3)} = \begin{matrix} & X & Y & Z \\ X & 0 & 1 & 0 \\ Y & 0 & 0 & 1 \\ Z & 1 & 0 & 0 \end{matrix}$$

Here, the matrix $\mathbf{P}^{(3)}$ shows that system will move sequentially from state *X* to *Y*, *Y* to *Z* and *Z* to *X*, and so on. Thus, there can be no steady-state conditions for such Markov chains.

Example 18.8. *Consider the information of the illustration given above and answer the following questions:*

(*i*) *What is the probability that a new account shall eventually be collected?*
(*ii*) *What is the probability that new account shall be written off as bad debt?*
(*iii*) *What is the probability that an account which is one month overdue shall eventually be a bad debt?*
(*iv*) *What is the probability that a two-month overdue account shall finally be collected?*
(*v*) *If the firm makes a credit sales of ₹ 400,000 per month on the average, what amount can it expect to lose per annum by way of bad debts?*

Solution. Above questions are different versions of a general question, namely, if the chain begins in a given transient state, what is the probability of ending up in each absorbing state? Besides, we may also like to know that if a chain begins in a particular transient state, and before reaching a particular state, what is the expected number of times that each state will be entered, and how many period do we expect to spend in a transient state before absorption takes place.

Now, let the transient states be represented by t_i's and absorbing states by a_i's. Then, the transition probability matrix for the problem can be written as

$$\begin{bmatrix} \mathbf{I} & \mathbf{O} \\ \mathbf{R} & \mathbf{Q} \end{bmatrix} = \begin{array}{c} \\ a_1 \\ a_2 \\ t_1 \\ t_2 \\ t_3 \\ t_4 \end{array} \begin{array}{c} \begin{array}{cccccc} a_1 & a_2 & t_1 & t_2 & t_3 & t_4 \end{array} \\ \begin{bmatrix} 1 & 0 & 0 & 0 & 0 & 0 \\ 0 & 1 & 0 & 0 & 0 & 0 \\ 0.3 & 0 & 0 & 0.7 & 0 & 0 \\ 0.4 & 0 & 0 & 0 & 0.6 & 0 \\ 0.6 & 0 & 0 & 0 & 0 & 0.4 \\ 0.8 & 0.2 & 0 & 0 & 0 & 0 \end{bmatrix} \end{array}$$

Here, $$\mathbf{Q} = \begin{bmatrix} 0 & 0.7 & 0 & 0 \\ 0 & 0 & 0.6 & 0 \\ 0 & 0 & 0 & 0.4 \\ 0 & 0 & 0 & 0 \end{bmatrix} \quad \text{and} \quad \mathbf{R} = \begin{bmatrix} 0.3 & 0 \\ 0.4 & 0 \\ 0.6 & 0 \\ 0.8 & 0.2 \end{bmatrix}$$

$\therefore$ $$\mathbf{I} - \mathbf{Q} = \begin{bmatrix} 1 & -0.7 & 0 & 0 \\ 0 & 1 & -0.6 & 0 \\ 0 & 0 & 1 & -0.4 \\ 0 & 0 & 0 & 1 \end{bmatrix}$$

Now, $$(\mathbf{I}-\mathbf{Q})^{-1} = \frac{1}{[\mathbf{I}-\mathbf{Q}]} \text{ adj. } (\mathbf{I}-\mathbf{Q}) = \text{adj. } (I-Q), \quad \text{since } |\mathbf{I}-\mathbf{Q}| = 1$$

$\therefore$ $$\mathbf{I} - \mathbf{Q} = \begin{array}{c} \\ t_1 \\ t_2 \\ t_3 \\ t_4 \end{array} \begin{array}{c} \begin{array}{cccc} t_1 & t_2 & t_3 & t_4 \end{array} \\ \begin{bmatrix} 1 & 0.7 & 0.42 & 0.168 \\ 0 & 1 & 0.6 & 0.24 \\ 0 & 0 & 1 & 0.4 \\ 0 & 0 & 0 & 1 \end{bmatrix} \end{array} \quad \text{and, therefore}$$

$$(\mathbf{I}-\mathbf{Q})^{-1} \times \mathbf{R} = \begin{bmatrix} 1 & 0.7 & 0.42 & 0.168 \\ 0 & 1 & 0.6 & 0.24 \\ 0 & 0 & 1 & 0.4 \\ 0 & 0 & 0 & 1 \end{bmatrix} \begin{bmatrix} 0.3 & 0 \\ 0.4 & 0 \\ 0.6 & 0 \\ 0.8 & 0.2 \end{bmatrix}$$

$$= \begin{array}{c} \\ t_1 \\ t_2 \\ t_3 \\ t_4 \end{array} \begin{array}{c} \begin{array}{cc} a_1 & a_2 \end{array} \\ \begin{bmatrix} 0.9664 & 0.0336 \\ 0.9520 & 0.0480 \\ 0.9200 & 0.0800 \\ 0.8000 & 0.2000 \end{bmatrix} \end{array}$$

Hence,

(*i*) Probability that a new account (t_1) shall eventually be collected (a_1) = 0.9664.

(*ii*) Probability that a new account (t_1) shall be written off as bad debt (a_2) = 0.0336.

(*iii*) Probability that an element which is one-month overdue (t_2) shall eventually be written off as bad debt (a_2) = 0.048.

(*iv*) Probability that a two-month overdue account (t_3) shall finally be collected (a_1) = 0.92.

(*v*) For a credit sales of ₹ 4,00,000 per month, on the average, the annual sales shall be ₹ 48,00,000. The amount the firm can expect to lose by way of bad debts will be ₹ 48,00,000 × 0.0336 or ₹ 161,280.

Example 18.9. *Consider a firm that has developed the following probability transition matrix with states defined as:*

State :	a_1	t_1	t_2	t_3	a_2
—	*Account received*	*0–30 days due*	*31–60 days due*	*61–90 days due*	*Bad debt*

$$\begin{array}{llccccc} & & t_1 & t_2 & t_3 & a_1 & a_2 \\ & t_1 & 0 & 0.8 & 0 & 0.2 & 0 \\ \textit{Transient state} & t_2 & 0 & 0 & 0.5 & 0.5 & 0 \\ & t_3 & 0 & 0 & 0 & 0.7 & 0.3 \\ \textit{Absorbing state} & a_1 & 0 & 0 & 0 & 1 & 0 \\ & a_2 & 0 & 0 & 0 & 0 & 1 \end{array}$$

The accounts receivable (in ₹) after a prescribed time limits are:

$$\mathbf{F} = [₹ 1,00,000 \quad ₹ 2,00,000 \quad ₹ 50,000]$$

Determine the expected fund paid and debt.

Also, find the steady-state inventory if the firm requires ₹ 1,00,000 in each transaction period of 30 days.

Solution. Here, we are given

$$\mathbf{Q} = \begin{bmatrix} 0 & 0.8 & 0 \\ 0 & 0 & 0.5 \\ 0 & 0 & 0 \end{bmatrix} \text{ and } \mathbf{R} = \begin{bmatrix} 0.2 & 0 \\ 0.5 & 0 \\ 0.7 & 0.3 \end{bmatrix}$$

$$\therefore \quad (\mathbf{I}-\mathbf{Q})^{-1} = \begin{bmatrix} 1 & -0.8 & 0 \\ 0 & 1 & -0.5 \\ 0 & 0 & 0 \end{bmatrix}^{-1} = \begin{bmatrix} 1.0 & 0.8 & 0.4 \\ 0 & 1 & 0.5 \\ 0 & 0 & 1 \end{bmatrix}$$

The probability of being absorbed, given any starting state is determined as under:

$$(\mathbf{I}-\mathbf{Q})^{-1} \times \mathbf{R} = \begin{bmatrix} 1 & 0.8 & 0.4 \\ 0 & 1 & 0.5 \\ 0 & 0 & 1 \end{bmatrix} \begin{bmatrix} 0.2 & 0 \\ 0.5 & 0 \\ 0.7 & 0.3 \end{bmatrix} = \begin{bmatrix} 0.88 & 0.12 \\ 0.85 & 0.15 \\ 0.70 & 0.30 \end{bmatrix}$$

The first row of the new matrix implies that if the account is presently in the first month, there is a 0.88 probability that the amount is collected (paid) and 0.12 of being in a bad debt (uncollected).

The expected fund paid and debt is calculated as $\mathbf{F} \times [(\mathbf{I}-\mathbf{Q})^{-1} \times \mathbf{R}]$; *i.e.*,

$$[1,00,000 \quad 2,00,000 \quad 50,000] \begin{bmatrix} 0.88 & 0.12 \\ 0.85 & 0.15 \\ 0.70 & 0.30 \end{bmatrix} = \begin{matrix} a_1 & a_2 \\ [2,93,000 & 57,000] \end{matrix}$$

Further, the steady-state inventory is given by

$$[1,00,000 \quad 0 \quad 0] \begin{bmatrix} 1 & 0.8 & 0.4 \\ 0 & 1 & 0.5 \\ 0 & 0 & 1 \end{bmatrix} = [1,00,000 \quad 80,000 \quad 40,000]$$

PROBLEMS

18.16. A manufacturing company has two production departments, designated below as P_1 and P_2, and three service departments, designated below as S_1, S_2 and S_3, and wishes to allocate all factory overhead to production departments. A primary distribution of overhead to all departments has been made and is indicated below. The company makes the secondary distribution of overhead from service departments to production departments on a reciprocal basis, recognizing the fact that services of one service department are utilized by another. Data regarding costs and allocation percentages are as follows:

Department	*Direct cost* (₹)	*Percentage allocation of total cost of the department* S_1	S_2	S_3
S_1	25,000	0	30	10
S_2	1,24,000	20	0	10
S_3	77,000	10	10	0
P_1	1,20,000	40	30	30
P_2	2,90,000	30	30	50

For example, S_1 spends 20% of its time and efforts on S_2, 10% on S_3, 40% on P_1 and 30% on P_2. Similarly, the values in respect of S_2 and S_3 may be interpreted. Once the expenses of S_1 are apportioned, part of them would go to S_2 and S_3, while it would also receive a part of the expenses of each of them. The process of apportionment needs to be repeated over and over again until all the expenses get absorbed by the production departments.

Draft this problem as absorbing chains to figure out how to allocate the costs of service departments to production department and calculate the total cost for each of the two departments.

[*Hint:* The transition probability matrix is

$$\begin{array}{c} \\ S_1 \\ S_2 \\ S_3 \\ P_1 \\ P_2 \end{array} \begin{array}{c} \begin{array}{ccccc} S_1 & S_2 & S_3 & P_1 & P_2 \end{array} \\ \begin{bmatrix} 0 & 0.2 & 0.1 & 0.4 & 0.3 \\ 0.3 & 0 & 0.1 & 0.3 & 0.3 \\ 0.1 & 0.1 & 0 & 0.3 & 0.5 \\ 0 & 0 & 0 & 1 & 0 \\ 0 & 0 & 0 & 0 & 1 \end{bmatrix} \end{array} = \begin{bmatrix} \mathbf{Q} & \mathbf{R} \\ \mathbf{O} & \mathbf{I} \end{bmatrix}$$

18.17. A credit card company is attempting to determine a more effective set of credit control policies. It has traditionally classified all its accounts receivables into one of the five categories listed below:

Accounts receivable category	:	1	2	3	4	5
Status of accounts receivable	:	0-30 days late	31-90 days late	91+ days late	paid in full	defaulted

Based on the historical data, the company has derived the following transition matrix describing the behaviour of various categories on a weekly basis:

From category	To category y				
	1	2	3	4	5
1	0.4	0.2	0.1	0.2	0.1
2	0.3	0.4	0.1	0.1	0.1
3	0.2	0.4	0.1	0.1	0.2
4	0.0	0.0	0.0	1.0	0.0
5	0.0	0.0	0.0	0.0	1.0

Currently, the company has the following accounts receivable in various categories:

Category :	1	2	3
Amount :	₹ 2 million	₹ 4 million	₹ 3 million

You are required to calculate the expected amounts which may eventually be (*i*) collected, and (*ii*) defaulted.

18.18. Let $\{X_n, n \geq 0\}$ be a Markov chain having state space $S = \{1, 2, 3, 4\}$ and transition matrix

$$\mathbf{P} = \begin{bmatrix} 1/3 & 2/3 & 0 & 0 \\ 1 & 0 & 0 & 0 \\ 1/2 & 0 & 1/2 & 0 \\ 0 & 0 & 1/2 & 1/2 \end{bmatrix}$$

Find whether all the states are persistent or transient. Also, find the mean recurrence time for state 2 and whether state 2 is periodic or aperiodic.

[ME (Software), Annamalai Univ. 2002]

TEST YOUR UNDERSTANDING

1. Markov analysis is:
   (*a*) helpful in evaluating alternative maintenance policies in production department.
   (*b*) used to determine the optimal course of action in a given problem.
   (*c*) used in the situations involving multiple time periods.
   (*d*) none of the above.
2. Markov analysis is useful in
   (*a*) calculating transition probabilities at some future time.
   (*b*) predicting the state of the system at some future time.
   (*c*) both of the above.
   (*d*) none of the above.
3. In Markov analysis, state probabilities must
   (*a*) sum to one (*b*) be less than one
   (*c*) be greater than one (*d*) all of the above.
4. In a matrix of transition probability, the probability values should add up to unity in each
   (*a*) column (*b*) row
   (*c*) diagonal (*d*) all of the above.
5. While calculating equilibrium probabilities for a Markov process, it is assumed that
   (*a*) transition probabilities do not change. (*b*) there is a single non-absorbing state.
   (*c*) there is a single absorbing state. (*d*) none of the above.

6. In a matrix of transition probability
   (*a*) the element p_{ij} indicates the probability of the system to move from i^{th} state to j^{th} state, or *vice-versa*.
   (*b*) the sum of probabilities of each row and each column is equal to 1.
   (*c*) the element zero indicates that the transition is not possible.
   (*d*) all of the above.
7. If a matrix of transition probability is of the order $n \times n$, then the number of equilibrium equations would be
   (*a*) n (*b*) $n + 1$
   (*c*) $n - 1$ (*d*) n^2
8. In the long run, the state probabilities become 0 and 1 in
   (*a*) no case (*b*) some cases
   (*c*) all cases (*d*) cannot say.
9. In a matrix of transition probability, the elements p_{ij} (where $i = j$) is a
   (*a*) loss (*b*) gain
   (*c*) retention (*d*) none of the above.
10. The elements of n-step transition matrix are obtained by using the formula:
   (*a*) $\mathbf{p}^{(n)} = \mathbf{p}^{(n-1)} \times p^{(0)}$ (*b*) $\mathbf{p}^{(n)} = \mathbf{p}^{(n-1)} \mathbf{p}^{(1)}$
   (*c*) $\mathbf{p}^{(n)} = \mathbf{p}^{(n-2)} \times \mathbf{p}^{(2)}$ (*d*) $\mathbf{p}^{(n)} = \mathbf{p}^{(n-3)} \mathbf{p}^{(3)}$
11. Which of the following is not correct?
   (*a*) Markov analysis assumes that the given system has a finite number of states.
   (*b*) A unity element (*i.e.*, 1) in the transition matrix indicates that the transition is not possible.
   (*c*) The equilibrium probabilities does not depend only on the initial condition in which the system is.
   (*d*) The probabilities of switching from one state to another are called transition probabilities.
12. It is possible to reach S_5 from S_3, but not from S_5 to S_3. Then
   (*a*) S_3 is a transient state (*b*) S_5 is a transient state
   (*c*) Both S_3 and S_5 are transient states (*d*) None of them is a transient state.

ANSWERS

1. (*b*)	2. (*c*)	3. (*a*)	4. (*b*)	5. (*a*)	6. (*c*)
7. (*b*)	8. (*c*)	9. (*c*)	10. (*b*)	11. (*b*)	12. (*a*)

REVIEW EXERCISE

1. Define Markov process and give an example. *[ME (Software Engg.) Annamalai Univ. 2002]*
2. Explain the following terms:
   (*a*) Transition probabilities (*b*) Matrix of transition probabilities
   (*c*) Equilibrium of steady-state (*d*) Absorbing state.
3. What do you understand by transition probabilities? Is the assumption of stationary transition probabilities realistic? Why or why not? *[MBA (Nov.), Delhi Univ. 2003]*
4. What do you understand by Markov chains? In What areas of management can it be applied successfully? *[MBA (PT.) (Nov.), Delhi Univ. 2003]*

5. Describe how Markov-chain process is used in deciding the market share of a commodity of particular brand. *[PG Dipl., Indian Inst. of Mat. Management 2018]*
6. Explain how decision tree helps to understand the problem of Markov process. *[MBA, Delhi Univ. 2002]*
7. "The Markov chain method analyses the current behaviour of a process and relates the existing characters to the future." Elucidate this statement by taking an example from functional area of marketing. *[MBA, Delhi Univ. 2005; AMIE, 2005]*
8. Define Markov chain. Explain, with suitable example, the classification of states in a Markov chain. *[AMIE, 2004]*
9. Write short note on Markov-chain process. *[PG Dipl., Indian Inst. of Mat. Management 2017]*
10. Describe briefly the methods that are available for solving Markov process problems.
11. "A Markov analysis is a procedure that can be used to describe the behaviour of a system in a dynamic situation. Specially, it describes and predicts the movement of a system, among different system states, as time passes." Discuss.
12. Briefly discuss the general idea of Markovian decision analysis. *[MBA, Delhi Univ. 2004]*

CHAPTER 19

Decision Theory

"It is exciting to focus on developing capacity to make better informed decision"

Every day we make ***decisions*** and occasionally an important decision that can have immediate and long-term effect. Some of these decisions are difficult to make, because of the complexity of a decision situation. It is precisely this complexity that has led many decision-makers to analyse the process of decision-making. It is in this context that we need a decision theory which may be defined as a tool that can be used to assist management to select wisely, the best course of action from amongst several alternatives. In other words, the objective of decision theory is to provide a systematic way of approaching a decision to get the desired outcome, where the possibility of occurrence of different outcomes are evaluated in advance.

19:1. ELEMENTS OF DECISION-MAKING PROBLEM

Irrespective of the type of decision-making problem, the essential elements which are common to all such problems are:

1. **Decision-maker.** The decision-maker refers to an individual or a group of individuals responsible for making a choice of an appropriate course of action from the available courses of action.
2. **Decision Alternatives.** The decision-alternatives or courses of action are called ***actions, acts*** or ***strategies***. These are under control and known to the decision-maker.
3. **States-of-Nature.** States-of-Nature are sometimes called ***outcomes*** or ***events***. The decision-maker must develop an exhaustive list of all possible states-of-nature. However, these are dependent upon some factors beyond the control of the decision-maker.
4. **Uncertainty.** Sometimes there is indefiniteness regarding as to which event or outcome will occur. This uncertainty is indicated in terms of ***probability*** assigned to events.
5. **Pay-off.** It is the effectiveness (net benefit to the decision-maker) associated with a particular combination of a course of action and state-of-nature.

19:2. PAY-OFF TABLE AND OPPORTUNITY LOSS TABLE

Pay-off Table. Suppose the problem under consideration has n possible events (states-of-nature) denoted by $E_1, E_2, ..., E_n$ and m alternative acts (strategies) denoted by $A_1, A_2, ..., A_m$. Then, the pay-off corresponding to strategy A_j of the decision-maker under the event (state-of-nature) E_i is denoted by p_{ij} $(i = 1, 2, ..., n; j = 1, 2, ..., m)$.

Various pay-off elements may be summarized in the form of a pay-off table given below:

States-of-nature (Events)	*Decision alternatives (Course of action)*				
	A_1	A_2	A_3	...	A_m
E_1	p_{11}	p_{12}	p_{13}	...	p_{1m}
E_2	p_{21}	p_{22}	p_{23}	...	p_{2m}
E_3	p_{31}	p_{32}	p_{33}	...	p_{3m}
$\vdots$	$\vdots$	$\vdots$	$\vdots$		$\vdots$
E_n	p_{n1}	p_{n2}	p_{n3}	...	p_{nm}

Pay-offs may be calculated in terms of pay-off, cost, or opportunity loss. When such pay-offs are calculated in terms of profit, they are known as ***pay-offs***. A pay-off table shows the relation between all possible outcomes (states-of-nature), all possible courses of action (strategies) and the value associated with these.

In the pay-off table, columns designate the various courses of action out of which the decision-maker may choose, while the rows show the admissible states-of-nature under which the decision-maker has to take decision. The cell value p_{ij} shows pay-off resulting by taking course of action A_j, when the state-of-nature E_i for all $i = 1, 2, ..., n$ and $j = 1, 2, ..., m$. A pay-off table represents the economics of a problem, *i.e.*, a problem of revenue and costs. A pay-off may be thought of as a conditional value or conditional profit (or loss). It is conditional value in the sense that associated with each course of action, there is certain profit (or loss), given that a specific state-of-nature has occurred. Thus, a pay-off table contains all conditional values of all possible combinations of courses of action and states-of-nature.

Opportunity Loss or Regret Table. The opportunity loss has been defined to be the difference between the highest possible pay-off for a state-of-nature and the actual pay-off obtained for the particular course of action taken, *i.e.*, an opportunity loss occurs due to failure of not adopting the best available course of action (or strategy). The opportunity loss values are calculated separately for each state-of-nature (or outcome) by first locating the most favourable course of action for that state-of-nature and then determining the departure of the pay-off value for that course of action and pay-off value for the best possible course of action that could have been selected.

Consider a fixed state-of-nature E_i $(i = 1, 2, ..., n)$ for which the pay-off corresponding to the n courses of action are given by $p_{i1}, p_{i2}, ..., p_{im}$. Let M_i be the pay-off for the least possible course of action. Then, the opportunity loss table will be as follows:

States-of-nature (Events)	*Conditional opportunity loss (Decision alternatives)*				
	A_1	A_2	A_3	...	A_m
E_1	$M_1 - p_{11}$	$M_1 - p_{12}$	$M_1 - p_{13}$	...	$M_1 - p_{1m}$
E_2	$M_2 - p_{21}$	$M_2 - p_{22}$	$M_2 - p_{23}$	...	$M_2 - p_{2m}$
E_3	$M_3 - p_{31}$	$M_3 - p_{32}$	$M_3 - p_{33}$	...	$M_3 - p_{3m}$
$\vdots$	$\vdots$	$\vdots$	$\vdots$		$\vdots$
E_n	$M_n - p_{n1}$	$M_n - p_{n2}$	$M_n - p_{n3}$	...	$M_n - p_{nm}$

Illustration 1 (*Construction of Pay-off Table*). Consider a vendor who sells a weekly magazine for ₹ 10 and purchases the same for ₹ 8. At the end of week, the unsold copies of the magazine are disposed off for ₹ 3 each. According to past experience, the weekly demand for this magazine is between 70 and 75 copies.

Assuming that the order for this particular magazine can be placed only once during the week, the problem before the vendor is to decide as to how many copies of the magazine should be purchased for the next week.

Here, since the weekly demand varies between 70 and 75, the possible events are:

E_1 : Demand of 70 copies, E_2 : Demand of 71 copies, E_3 : Demand of 72 copies

E_4 : Demand of 73 copies, E_5 : Demand of 74 copies, E_6 : Demand of 75 copies.

Also, there are six possible courses of action, *viz.*,

A_1 : Purchase 70 copies, A_2 : Purchase 71 copies, A_3 : purchase 72 copies

A_4 : Purchase 73 copies, A_5 : Purchase 74 copies, A_6 : Purchase 75 copies.

To consider how the pay-off table can be constructed, let Q be the quantity (in units) for the magazine and D denote the demand (in units) of the magazine. Then, the profit P = (Total revenue from the sale of all copies that were purchased) – (Total cost of purchasing the copies of the magazine).

Therefore, if $D \geq Q$, $P = 10Q - 8Q$, *i.e.*, $P = 2Q$; and

if $D < Q$, $P = [10D + 3(Q - D)] - 8Q$, *i.e.*, $P = 7D - 5Q$

The calculation for pay-off for each pair of alternative courses of action (demand) and the states-of-nature (quantity purchased) are shown below:

$(E_1, A_1) = 2 \times 70 = 140$, $(E_1, A_2) = 7 \times 70 - 5 \times 71 = 135$

$(E_1, A_3) = 7 \times 70 - 5 \times 72 = 130$, $(E_1, A_4) = 7 \times 70 - 5 \times 73 = 125$

$(E_1, A_5) = 7 \times 70 - 5 \times 74 = 120$, $(E_1, A_6) = 7 \times 70 - 5 \times 75 = 115$

$(E_2, A_1) = 2 \times 70 = 140$, $(E_2, A_2) = 2 \times 71 = 142$

$(E_2, A_3) = 7 \times 71 - 5 \times 72 = 137$, $(E_2, A_4) = 7 \times 71 - 5 \times 73 = 132$

$(E_2, A_5) = 7 \times 71 - 5 \times 74 = 127$, $(E_2, A_6) = 7 \times 71 - 5 \times 75 = 122$

and so on. The pay-offs for all combinations of A_j's and E_i's are given in the following table (known as ***pay-off table***).

States-of-nature	*Courses of action*					
	$A_1 : 70$	$A_2 : 71$	$A_3 : 72$	$A_4 : 73$	$A_5 : 74$	$A_6 : 75$
$E_1 : 70$	140	135	130	125	120	115
$E_2 : 71$	140	142	137	132	127	122
$E_3 : 72$	140	142	144	139	134	129
$E_4 : 73$	140	142	144	146	141	136
$E_5 : 74$	140	142	144	146	148	143
$E_6 : 75$	140	142	144	146	148	150

Illustration 2 (*Opportunity Loss or Regret Table*). Consider the pay-off table obtained in *illustration 1*. We observe that, if demand is of 70 copies of the magazine, then the optimal act is to buy 70 copies to get a profit of ₹ 140. In this first row, the highest profit is ₹ 140 corresponding to act A_1. If any other strategy is adopted, the profit earned would be less and, the greater the departure from the optimal strategy the lesser the profit earned, and consequently, the greater the opportunity loss (or regret). With the strategy of buying 74 copies, the profit of ₹ 120 which is ₹ 20 less than the profit that could be earned at this level of demand.

Similarly, if the demand turns out for 72 copies, the optimal course would be to order for 72 copies, *i.e.*, the profit being ₹ 144. Any order size other than this causes lesser profit than this.

The opportunity loss (or regret) table is obtained, therefore, by subtracting all the values of a row from the highest profit value of that row, we get the opportunity loss table given below:

States-of-nature	*Courses of action*					
	$A_1 : 70$	$A_2 : 71$	$A_3 : 72$	$A_4 : 73$	$A_5 : 74$	$A_6 : 75$
$E_1 : 70$	140–140	140–135	140–130	140–125	140–120	140–115
$E_2 : 71$	142–140	142–142	142–137	142–132	142–127	142–122
$E_3 : 72$	144–140	144–142	144–144	144–139	144–134	144–129
$E_4 : 73$	146–140	146–142	146–144	146–146	146–141	146–136
$E_5 : 74$	148–140	148–142	148–144	148–146	148–148	148–143
$E_6 : 75$	150–140	150–142	150–144	150–146	150–148	150–150

19:3. DECISION-MAKING PROCESS

The decision-making process involves the following major steps:

Step 1. Identify and define the problem under consideration.

Step 2. Determine the various alternative courses of action from which the final decision is to be made.

Step 3. Identify the possible outcomes, called the ***states-of-nature*** or ***events*** for the decision-making problem. The events are beyond the control of the decision-maker.

Step 4. Determine the ***pay-off*** function which describes the consequences resulting from the different combination of the acts and events. The pay-offs may be designated as p_{ij}'s, the pay-off resulting from the combination of the event and j^{th} action.

Step 5. Apply an appropriate decision theory model to select best course of action from the given list on the basis of some criterion (measure of effectiveness) that result in the optimum (desired) pay-off.

19:4. DECISION-MAKING UNDER CERTAINTY

When all pertinent information relating to decision-making problem is known in advance with complete certainty, it is known as ***decision-making under certainty***. In decision-making under certainty, for each strategy there is one unique pay-off. In other words, there is only one outcome (state-of-nature) and one pay-off corresponding to it.

19:5. DECISION-MAKING UNDER UNCERTAINTY

Those decision situations where more than one outcome can result from any single decision fall under the category of ***decision-making under uncertainty***. Here, only pay-offs are known and nothing is known about the likelihood of each state-of-nature. So, the decision-maker must base his decision solely on the actual conditional pay-off values together with his attitude or anticipation. There are various criteria (approaches) to solve such type of decision-making problems under uncertainty. Some of these are as under:

1. Maximax or Minimin (or Optimistic) Criterion. This criterion is also known as ***optimistic decision criterion*** and is based upon bold, aggressive and optimistic outlook for the future event. In this criterion, the decision-maker ensures that he should not miss the opportunity to achieve the largest possible profit (maximax) or lowest possible cost (minimin). Thus, he selects the alternative (course of action) that represents the maximum among the maximum pay-offs (or minimum among the minimum pay-offs).

2. Maximin or Minimax (or Wald) Criterion. This criterion is also known ***pessimistic decision criterion*** and is based upon the conservative (or pessimistic) approach to assume that the worst possible is gong to happen. The decision-maker considers each strategy and locate the minimum pay-off (profit) for each; and then select that course of action which maximizes the minimum pay-off (profit).

When dealing with costs, the maximum cost associated with each alternative (strategy) is considered and choose that strategy which minimizes this maximum cost.

3. Regret (or Savage) Criterion. This criterion is also known as ***opportunity loss decision criterion,*** because the decision-maker feels regret after choosing a wrong decision resulting in an opportunity loss of pay-off. It is alternatively known as the criterion of minimax regret. Basic steps of this criterion are as under:

Step 1. Determine the amounts of regret for pay-off of each alternative for a particular event. The regret amount for the ith alternative when even j occurs, is given below:

$$\begin{matrix} i^{\text{th}} \\ \text{regret} \end{matrix} = \begin{cases} (\text{maximum pay-off} - i^{\text{th}} \text{ pay-off}) \text{ for the } j^{\text{th}} \text{ event, if the pay-offs represent profits} \\ (i^{\text{th}} \text{ pay-off} - \text{minimum pay-off}) \text{ for the } j^{\text{th}} \text{ event, if the pay-offs represent costs.} \end{cases}$$

Step 2. Determine the maximum regret amount for each alternative.

Step 3. Choose that alternative which corresponds to the minimum regrets.

Remark. The reader may observe that while the other decision rules do not take into account the cost of opportunity loss by making the wrong decision, the minimax regret criterion does so.

4. Laplace Criterion (or Equally Likely Decision Criterion). This criterion is based on simple philosophy that if we are uncertain about the various events, then we treat them as equally probable. Under this assumption, since the probabilities of future outcome (states-of-nature) are not known, each state-of-nature should be assigned the same probability. Basic steps of this criterion are given below:

Step 1. Determine the expected pay-off for each alternative.

Step 2. Select that alternative which corresponds to the maximum of the expected pay-offs.

In case, pay-offs are in terms of costs, select the alternative which corresponds to the minimum of the *expected pay-offs*.

5. Hurwicz (or Realism) Criterion. This criterion is a compromise between the maximax and maximin criteria. In reality, decision-makers are neither completely optimistic nor pessimistic. In fact, decision-makers are usually optimistic to a certain degree at the same time. The scale used for the coefficient of optimism, denoted by α, is 0 to 1, *i.e.*, $0 \le \alpha \le 1$. If $\alpha = 0$, the decision-maker is completely pessimistic (zero optimism) and if $\alpha = 1$, he is completely optimistic (zero pessimistic). Basic steps of Hurwicz criterion are:

Step 1. Choose an appropriate degree of optimism (or pessimism) of the decision-maker. Let α be his degree of optimism, then $(1 - \alpha)$ will be his degree of pessimism; $0 \le \alpha \le 1$.

Step 2. Determine the maximum as well as minimum pay-off for each alternative and obtain the quantities:

$$h = \alpha \times \text{maximum pay-off} + (1 - \alpha) \times \text{minimum pay-off for each alternative.}$$

If pay-offs represent profit, the alternative for which h is maximum is selected. When the pay-offs represent cost, the alternative for which h is least is selected.

Example 19.1. *Consider the following pay-off in terms of yearly net profits for each decision alternative:*

$$\text{Decision alternative (Event)} \overset{\textit{Net profit (Act)}}{\begin{bmatrix} 220 & 180 & 100 \\ 160 & 190 & 180 \\ 140 & 170 & 200 \end{bmatrix}}$$

Calculate (i) Maximax criterion, (ii) Minimax criterion, (iii) Maximin criterion, (iv) Laplace criterion, and (v) Regret criterion. **[BBA (Nov./Dec.), Periyar Univ. 2015]**

Solution.

(*i*) The maximum pay-offs associated with the different alternatives (strategies) are:

$$A_1 \text{ (First)} : 220, \quad A_2 \text{ (Second)} : 190 \quad \text{and} \quad A_3 \text{ (Third)} : 200$$

The highest pay-off being 220, therefore, the first alternative is selected corresponding to ***maximax criterion***.

(*ii*) Here, since the least pay-off among the maximum pay-offs associated with different alternatives is 190; therefore, according to ***minimax criterion***, second alternative is selected.

(*iii*) The minimum assured pay-offs associated with the different alternative are:

$$A_1 \text{ (First)} : 100, \quad A_2 \text{ (Second)} : 160, \quad \text{and} \quad A_3 \text{ (Third)} : 140$$

The highest pay-off being 160, therefore, the second alternative is selected corresponding to ***maximin criterion***.

(*iv*) The Laplace criterion assumes that all the three alternatives are equally likely. So, the associated probabilities are given by $P(E_i) = \frac{1}{3};\ i = 1, 2, 3.$

∴ Expected pay-off by selecting different alternatives are given by

$$E(A_1) = \frac{(220+160+140)}{3}, \quad E(A_2) = \frac{(180+190+170)}{3} \quad \text{and} \quad E(A_3) = \frac{(100+180+200)}{3},$$

i.e., $E(A_1)$ 140, $E(A_2) = 180$ and $E(A_3) = 160$.

Since, we are dealing with pay-offs as profits, the highest of the expected profit is 180. Hence, according to ***Laplace Criterion***, decision alternative A_2 is selected.

(*v*) Using regret criterion, we use opportunity loss pay-off table by computing the difference between the maximum pay-off and other pay-offs under a given outcome (state-of-nature) as shown below:

$$\begin{matrix} A_1 \\ A_2 \\ A_3 \end{matrix} \begin{bmatrix} 220-220 & 190-180 & 200-100 \\ 220-160 & 190-190 & 200-180 \\ 220-140 & 190-170 & 200-200 \end{bmatrix} \quad \text{or} \quad \begin{bmatrix} 0 & 10 & 100 \\ 60 & 0 & 20 \\ 80 & 20 & 0 \end{bmatrix}$$

The maximum opportunity loss pay-offs associated with the different strategies are:

$$A_1 : 100, \quad A_2 : 60 \quad \text{and} \quad A_3 : 80$$

Since, regret criterion is also known as minimax criterion, we choose the minimum (least) opportunity loss pay-off among the maximum opportunity loss pay-offs. As 60 is the least among the three, alternative A_2 is selected according to ***regret criterion***.

Example 19.2. *A Super Bazar must decide on the level of supplies it must stock to meet the needs of its customers during Diwali days. The exact number of customers is not known, but it is expected to be in one of the four categories; 300, 350, 400 or 450 customers. Four levels of supplies are thus suggested with level j being ideal (from the viewpoint of incurred costs), if the number of customers falls in category j. Deviations* *from the ideal levels results in additional costs either because extra supplies are stocked needlessly or because demand cannot be satisfied. The table given below provides these costs in thousands of rupees.*

Customer category	*Supplies level*			
	A_1	A_2	A_3	A_4
E_1	7	12	20	27
E_2	10	9	10	25
E_3	23	20	14	23
E_4	32	24	21	17

Determine the best level of inventory using (i) Laplace criterion, and (ii) Regret criterion.

Solution.

(*i*) The Laplace criterion assumes that E_1, E_2, E_3 and E_4 are equally likely. Thus, the associated probabilities are given $P(E_j) = 1/4$ ($j = 1, 2, 3, 4$), and the expected costs due to deviations from the best level, for different categories of customers are:

$$E(A_1) = \frac{(7+10+23+32)}{4} = 18.00 \qquad E(A_2) = \frac{(12+9+20+24)}{4} = 16.25$$

$$E(A_3) = \frac{(20+10+14+21)}{4} = 16.25 \qquad E(A_4) = \frac{(27+25+23+17)}{4} = 23.00$$

Since, we are dealing with costs incurred in having inventory, the best level of inventory is to have the supply level of A_2 or A_3.

(*ii*) Here, we determine the regrets as shown below, where the regret amounts when event E_j occurs, are computed by using the relation:

Regret pay-off = Pay-off – Minimum pay-off from E_j, $(j = 1, 2, 3, 4)$

Alternative	*Pay-off amount*				*Regret pay-off amount*				*Maximum regret*
	E_1	E_2	E_3	E_4	E_1	E_2	E_3	E_4	
A_1	7	10	23	32	0	1	9	15	15
A_2	12	9	20	24	5	0	6	7	7*
A_3	20	10	14	21	13	1	0	4	13
A_4	27	25	23	17	20	16	9	0	20
Minimum pay-off	7	9	14	17					

Since, the alternative A_2 corresponds to the minimum of the maximum possible regrets, the decision-maker would choose A_2.

Example 19.3. *A manufacturer makes a product of which the principal ingredient is a chemical X. At the moment, the manufacturer spends ₹1,000 per year on supply of X, but there is a possibility that the price may soon increase to four times its present figure, because of a worldwide shortage of the chemical. There is another chemical Y, which the manufacturer could use in conjunction with a third chemical Z, in order to give the same effect as chemical X. Chemicals Y and Z would together cost the manufacturer ₹3,000 per year; but their prices are unlikely to rise. What action should the manufacturer take? Apply the maximin and minimax regret criteria for decision-making and give two sets of solutions.*

If the coefficient of optimism is 0.4, find the course of action that minimizes the cost.

Solution. The information given in the problem may be summarized in the following pay-off table (negative figures in the table indicates profit):

State-of-nature	Courses of action	
	A_1 *(use of Y and Z)*	A_2 *(use of X)*
E_1 : Price of X increases	₹ 3,000	₹ 4,000
E_2 : Price of X does not increase	₹ 3,000	₹ 1,000

(i) *Maximin Criterion:* For the given pay-off matrix, the maximum cost for each alternative are:

$$A_1 : ₹\,3{,}000 \quad \text{and} \quad A_2 : ₹\,4{,}000$$

Since, the minimum of these maximum pay-offs is ₹ 3,000, the alternative A_1 is selected.

(ii) *Minimax Regret Criterion:* Since, the pay-off table deals with costs, an opportunity loss table is obtained as under:

State-of-nature	Course of action		
	A_1	A_2	*Maximum regret*
E_1	₹ 3,000 – ₹ 3,000 = 0	₹ 4,000 – ₹ 3,000 = ₹ 1,000	₹ 1,000
E_2	₹ 3,000 – ₹ 1,000 = ₹ 2,000	₹ 1,000 – ₹ 1,000 = 0	₹ 2,000

Since, minimum of the maximum regrets is ₹ 1,000, alternative A_2 is chosen.

(iii) Since, the coefficient of optimism is 0.4, the coefficient of ***pessimism*** will be: 1 – 0.4 (= 0.6). Then, according to ***Hurwicz criterion***, select the course of action that optimizes (maximum for profit and minimum for cost) the pay-off value:

$$H = \alpha \text{ (Best pay-off)} + (1 - \alpha) \text{ (worst pay-off)}$$
$$= \alpha \text{ (maximum in column of pay-off matrix)}$$
$$+ (1 - \alpha) \text{ (minimum in column of pay-off matrix)}$$

Thus, we get

Course of action	*Best pay-off*	*Worst pay-off*	*H*
A_1	3,000	3,000	0.4 × (3,000) + 0.6 × (3,000) = 3,000
A_2	1,000	4,000	0.4 × (1,000) + 0.6 × (4,000) = 2,800

Since, A_2 has the least cost of ₹ 2,800, *i.e.*, $E(A_2) = 2{,}800$, the manufacturer should adopt this course of action.

PROBLEMS

19.01. A toy company is bringing out a new type of toy. The company is attempting to decide whether to bring out a full, partial or minimal product line. The company has three levels of product acceptance. The management will make its decision on the basis of expected profit from the first year of production. The relevant data are shown in the following table:

Anticipated First Year Profit *(in '000 ₹)*

Product acceptance	*Product line*		
	Full	*Partial*	*Minimal*
Good	80	70	50
Fair	50	45	40
Poor	–25	–10	0

Take optimal decision under each of the following decision rules:

(*a*) Maximax, (*b*) Maximin, (*c*) Laplace, (*d*) Minimax regret.

[B.Com. (Hons.), Delhi Univ. 2010; MBA, GGSIP Univ. 2002]

19.02. For the following cost matrix, suggest the best decision according to (*a*) Maximin criterion, (*b*) Hurwicz criterion when $\alpha = 0.2$.

Decision	*State-of-nature (Act)*	
	A_1	A_2
D_1	30	35
D_2	–20	10

19.03. Using Hurwicz decision rule coefficient of optimism $\lambda = 0.6$, choose the best product from the following data:

Product	*Profit (in rupees), if the market is*		
	Good	*Fair*	*Poor*
A	10,000	8,000	2,000
B	12,500	9,000	– 1,000

19.04. A farmer wants to decide which of the three crops he should plant on his 100-acre farm. The profit from each is dependent on the rainfall during the growing season. The farmer has categorized the amount of rainfall as high, medium and low. His estimated profit for each crop is shown in the table given below:

Rainfall	*Estimated conditional profit (₹)*		
	Crop A	*Crop B*	*Crop C*
High	8,000	3,500	5,000
Medium	4,500	4,500	5,000
Low	2,000	5,000	4,000

If the farmer wishes to plant only one crop, decide which should be his 'best crop' using:

(*a*) Maximax criterion
(*b*) Maximin criterion
(*c*) Laplace criterion
(*d*) Minimax regret criterion
(*e*) Hurwicz criterion (farmer's degree of optimism being 0.6). **[*MBA, Pune Univ. 2010*]**

19.05. The estimated sales of proposed types of perfume are as under:

Types of perfume	*Estimated level of sales (units)*		
	₹20,000	*₹10,000*	*₹20,000*
A	25	15	10
B	40	20	5
C	60	25	3

(*a*) For each of the following decisions, state the optimal action and specify the value leading to its selection:

(*i*) Maximin criterion
(*ii*) Maximax criterion
(*iii*) Laplace criterion
(*iv*) Minimax regret criterion.

(*b*) What will be the optimal act, if the pay-off entries represent the costs instead of sales.

19.06. The research department of Hindustan Unilever Ltd. has recommended to the marketing department to launch a shampoo of three different types. The marketing manager has to decide one of the types of shampoo to be launched under the following estimated pay-offs for various level of sales:

Types of shampoo	Estimated level of sales (units)		
	15,000	10,000	5,000
Egg Shampoo	30	10	10
Clinic Shampoo	40	15	5
Deluxe Shampoo	55	20	3

What will be the marketing manager's decision if (*i*) Maximin, (*ii*) Minimax, (*iii*) Maximax, (*iv*) Laplace, and (*v*) Regret criteria are applied?

19.07. The ABC company is faced with four decision alternatives relating to investments in a capital expansion programme. Since, these investments are made in future, the company foresees different market conditions as expressed in the form of state-of-nature. The following table summarizes the decision alternatives, the various states-of-nature and the rate of return associated with each state-of-nature.

Decision	States-of-nature		
	θ_1	θ_2	θ_3
D_1	17%	15%	8%
D_2	18%	16%	9%
D_3	21%	14%	9%
D_4	19%	12%	10%

If the company has no information regarding probability of the occurrence of three states-of-nature, give the recommended decision for the decision criteria listed below:

(*i*) Maximax criterion. (*ii*) Maximin criterion.
(*iii*) Minimax regret criterion. (*iv*) Laplace rational criterion.
(*v*) Hurwicz alpha criterion ($\alpha = 0.75$). ***[MBA, BHU 2007]***

19.08. Given the following pay-off table:

States-of-nature	Strategies			
	a_1	a_2	a_3	a_4
E_1	50	400	–50	0
E_2	300	0	200	300
E_3	–150	100	0	300
E_4	50	0	100	0

Indicate the strategy to be selected under the following conditions:

(*i*) Maximax criterion, (*ii*) Maximin criterion,
(*iii*) Minimax regret criterion, (*iv*) Laplace criterion. ***[MBA, Rajasthan Univ. 2004]***

19.09. A bicycle repairman has an opportunity to purchase a stock of discontinued bicycles. They were originally supposed to be sold for ₹ 400 each. The repairman is offered all five bicycles for ₹ 500, which makes his cost for each bicycle ₹ 100. If he sells them, he believes that he can get ₹ 250 for each bicycle, thereby making a profit of ₹ 150. He has two options: either to buy all the discontinued bicycles or not to buy at all. There are six states-of-nature; these being the demand for 0, 1, 2, 3, 4 and 5 bicycles.

(*i*) Prepare the pay-off as well as regret tables for the problem.
(*ii*) If the repairman has the option of buying any number of bicycles 0 to 5, find the average expected pay-off and average expected regret for each stock action.

[MCA (Oct.), Pune Univ. 2007]

19.10. A steel manufacturing company is concerned with the possibility of a strike. It will cost an extra ₹ 20,000 to acquire an adequate stock-pile. If there is a strike and company has not stock-piled, management estimates an additional expenses of ₹ 60,000 on account of lost sales. Should the company stock-pile or not, if it is to use (*i*) Optimistic criterion, (*ii*) Wald criterion, (*iii*) Savage criterion, (*iv*) Hurwicz criterion for $\alpha = 0.4$, (*v*) Laplace criterion?

[B.Tech. (Mech.) (Nov.), Panjab Tech. Univ. 2012]

19:6. DECISION-MAKING UNDER RISK

Decision-making problems, where there can be more than one outcome and it is possible to assign a probability of occurrence to each one are known as the ***decision-making problems under risk***. In these problems, the decision-maker knows the probability of occurrence of each state and chooses that course of action which has the largest expected pay-off.

The most widely used criterion for evaluating the alternative courses of action is the expected monetary value (*EMV*), or expected opportunity loss (*EOL*)/expected utility.

1. Expected Monetary Value (EMV) for a specified course of action is the sum of the product of pay-off values with probabilities associated with each state-of-nature.

If a_{ij} is the pay-off associated with the course of action A_i ($i = 1, 2, ..., m$) and states-of-nature S_j ($j = 1, 2, ..., n$) with the corresponding probabilities of S_j as p_j, then

$$EMV = p_1 a_{i1} + p_2 a_{i2} + ... + p_n a_{in}, \quad i = 1, 2, ..., m$$

The *EMV* criterion may be summarized in the following two steps:

***Step* 1.** Construct pay-off table by listing all possible courses of action and states-of-nature. Write the conditional pay-off values associated with each possible course of action and states-of-nature along with probabilities of occurence of each state-of-nature.

***Step* 2.** Calculate *EMV* for each course of action by multiplying the conditional pay-offs by the associated probabilities and adding these weighted values for each course of action.

Following points are relevant in respect of decision-making under risk:

1. We must be able to construct the conditional pay-off table in order to compare the *EMV* of different courses of action. In fact the conditional pay-offs are obtained for each action by considering various act-event combinations.
2. The choice of strategy is purely on the basis of *EMV*. Sometimes, the opportunity loss or utility may be more appropriate than *EMV*.
3. The criterion of *EMV* does not take into account the quality of risk. Generally, the decision-maker goes by the trade-off between risk and return.

Example 19.4. *The pay-offs (in ₹) of three acts A_1, A_2 and A_3, are the possible states-of-nature S_1, S_2 and S_3 are given below:*

Acts	*States-of-nature*		
	S_1	S_2	S_3
A_1	–200	2,000	4,000
A_2	–500	–1,000	6,000
A_3	2,000	–500	3,000

The probabilities of the states-of-nature are: 0.3, 0.4 and 0.3 respectively.

Determine the optimal act using the principle of expectation, i.e., according to EMV criterion.

Solution. Using given information, the calculations of *EMV* are obtained as under:

$EMV\ (A_1) = -200 \times 0.3 + (2{,}000) \times 0.4 + 4{,}000 \times 0.3 = -60 + 800 + 1{,}200 = ₹\ 1{,}940$

$EMV\ (A_2) = -500 \times 0.3 + (-1{,}000) \times 0.4 + 6{,}000 \times 0.3 = -150 + (-400) + 1{,}800 = ₹\ 1{,}250$

$EMV\ (A_3) = 2{,}000 \times 0.3 + (-500) \times 0.4 + 3{,}000 \times 0.3 = 600 - 200 + 900 = ₹\ 1{,}300$

Since, maximum of *EMV* is ₹ 1,940, according to *EMV* criterion, the optimal act is A_1.

2. Expected Opportunity Loss (*EOL*). An alternative approach to maximizing expected monetary value (*EMV*) is to minimize ***expected opportunity loss (EOL)***. Expected opportunity loss or expected value of regrets are calculated in the same manner as the expected pay-offs in the *EMV* criterion. Major steps in the *EOL* criterion may be summarized as below:

Step 1. List the conditional pay-off (profit) table for each act-event combination, along with corresponding event probabilities.

Step 2. For each event, determine the ***conditional opportunity loss (COL)*** value by first locating the most favourable act (maximum pay-off) for that event; and then taking the difference between this conditional profit value and each conditional profit value for that event.

Step 3. For each act, determine the expected *COL* values and sum these values to get the expected opportunity los (*EOL*) for that act.

Step 4. Choose that act which corresponds to the ***minimum COL*** value.

Example 19.5. *Calculate the expected opportunity loss (EOL) from the following pay-off table:*

Events	*Probabilities*	*Act*		
		A_1	A_2	A_3
E_1	0.3	–40	–20	120
E_2	0.4	350	450	300
E_3	0.3	500	640	700

Solution. By subtracting the maximum pay-off of an event from pay-offs of that event, the conditional opportunity loss (or regret) values are given in the table given below:

A_1	A_2	A_3		A_1	A_2	A_3
120 – (–40)	120 – (–20)	120 – 120		160	140	0
450 – 350	450 – 450	450 – 300	or	100	0	150
700 – 500	700 – 640	700 – 700		200	60	0

Using this able, expected opportunity loss (*EOL*) values of acts A_1, A_2 and A_3 are as under:

$$EOL\ (A_1) = 0.3 \times 160 + 0.4 \times 100 + 0.3 \times 200 = 148$$

$$EOL\ (A_2) = 0.3 \times 140 + 0.4 \times 0 + 0.3 \times 60 = 60$$

and $$EOL\ (A_3) = 0.3 \times 0 + 0.4 \times 150 + 0.3 \times 0 = 60$$

Since, minimum of *EOL* is 60, according to *EOL* criterion, the decision should be A_2 or A_3.

3. Expected Value of Perfect Information (*EVPI*). The ***expected value of perfect information (EVPI)*** is the difference between expected profit of the optimal decision without perfect information with perfect information. Expected profit with perfect information (*EPPI*) may be defined as the maximum attainable expected monetary value (*EMV*) based on perfect information about the states-of-nature that will occur. Now, if *EMV** represents the maximum obtainable *EMV* without perfect information, then perfect information would increase expected profit from *EMV** up to the value of *EPPI*.

Thus, we have $$EVPI = EPPI - EMV^*$$

It may be noted that *EVPI* is always equal to *EOL* of selecting the optimal action under certainty. The identity *EPPI* = *EMV** + *EOL* follows from the result *EVPI* = *EOL* and *EVPI* = *EPPI* – *EMV**.

Remark. If *EVPI* is relatively larger than the cost involved in gathering additional information, it is advisable to gather the additional information regarding the states-of-nature.

Example 19.6. *Pay-offs of three acts: A, B and C, states-of-nature: X, Y and Z are given below:*

States-of-nature	*Acts*		
	A	*B*	*C*
X	*–20*	*–50*	*200*
Y	*200*	*–100*	*–50*
Z	*400*	*600*	*300*

The respective probabilities of the states-of-nature are: 0.3, 0.4 and 0.3. Calculate the EMV and select the best act. Also, find the expected value of perfect information, i.e., EVPI.

Solution. Using given information, *EMV* for various acts are calculated as under:

$$EMV\ (A) = -20 \times 0.3 + 200 \times 0.4 + 400 \times 0.3 = 194$$
$$EMV\ (B) = -50 \times 0.3 - 100 \times 0.4 + 600 \times 0.3 = 125$$
$$EMV\ (C) = 200 \times 0.3 - 50 \times 0.4 + 300 \times 0.3 = 130$$

Clearly, the maximum of *EMV* is 194, *i.e.*, best act is *A*. Thus, *EMV** = 194.

Now, to find *EVPI*, we first compute *EPPI* which is obtained form the following table:

States-of-nature	*Prob.*	*Acts*			*Max. of states-of-nature*	*Expected pay-off*
		A	*B*	*C*		
X	0.3	–20	–50	200	200	200 × 0.3 = 60
Y	0.4	200	–100	–50	200	200 × 0.4 = 80
Z	0.3	400	600	300	600	600 × 0.3 = 180

Using above information,

$$EPPI = 60 + 80 + 180 = 320.$$
$$\therefore \quad EVPI = EPPI - EMV^* = 320 - 194 = 126$$

Example 19.7. *A departmental store purchases biscuits which can be ordered in lots of 10 packets. Each lot of packet of biscuits costs ₹ 75 and sells at ₹ 90 each. Unsold lot of packets of biscuit, however, have no salvage value. The probability distribution obtained from past sales data is as under:*

Demand :	*10*	*20*	*30*	*40*	*50*
Probability :	*0.20*	*0.35*	*0.25*	*0.15*	*0.05*

(i) Construct the pay-off table as well as conditional profit table.
(ii) How much quantity should the departmental store buy to maximize its profit.
(iii) Compute EPPI and EVPI.

Solution. Since, the number of lot of packets (of biscuit) is under control of decision-maker, purchase of biscuits are considered as 'courses of action' or 'decision alternatives'. On the other hand the demand of biscuits is uncertain and is known with probability, therefore, it is considered as 'states-of-nature' or 'event'.

Let the decision alternatives A_1, A_2, A_3, A_4 and A_5 represents the purchase of 10, 20, 30, 40 and 50 packets respectively. The states-of-nature E_1, E_2, E_3, E_4 and E_5 represent the demand of 10, 20, 30, 40 and 50 packets, respectively.

Since, there are 5 decision alternatives and 5 states-of-nature, there should be 5 × 5 = 25 consequences.

Let Q = the number of biscuit packets purchased, and D = the number of biscuit packets demanded.

Then, the conditional profit P is obtained as under:

P = Marginal profit × Number of packets sold – Marginal loss × Number of packets not sold.

$$= \begin{cases} (₹\,90 - ₹\,75) \times Q, & \text{if } D \geq Q \\ (₹\,90 - ₹\,75) \times D - (Q - D) \times ₹\,75, & \text{if } D < Q \end{cases}$$

(*i*) The calculation of pay-off table is given below:

States-of-nature	*Conditional pay-off (₹)* A_1 : 10	A_2 : 20	A_3 : 30	A_4 : 40	A_5 : 50	*Maximum pay-off*
E_1 : 10	150	–600	–1,350	–1,100	–2,850	150
E_2 : 20	150	300	–450	–1,200	–1,950	300
E_3 : 30	150	300	450	–300	–1,050	450
E_4 : 40	150	300	450	600	–150	600
E_5 : 50	150	300	450	600	750	750

Here, for the combination of E_4 and A_3, *i.e.*, when Q = 30 and D = 40; we have

Pay-off = (₹ 90 – ₹ 75) × 30, since $D > Q$

= ₹ 15 × 30 = ₹ 450

For the combination of E_3 and A_4, *i.e.*, when D = 30 and Q = 40;

Pay-off = (₹ 90 – ₹ 75) × 30 – (40 – 30) × 75, when $D < Q$

= ₹ 15 × 30 – 10 × ₹ 75 = – ₹ 300

Now, using the pay-off table, the expected conditional pay-off (profit) table is computed as follows:

Event E_i	*Probability*	*Expected conditional pay-off (₹)* A_1	A_2	A_3	A_4	A_5
E_1	0.20	150 × 0.20	–600 × 0.20	–1,350 × 0.20	–1,100 × 0.20	–2,850 × 0.20
E_2	0.35	150 × 0.35	300 × 0.35	–450 × 0.35	–1,200 × 0.35	–1,950 × 0.35
E_3	0.25	150 × 0.25	300 × 0.25	450 × 0.25	–300 × 0.25	–1,050 × 0.25
E_4	0.15	150 × 0.15	300 × 0.15	450 × 0.15	600 × 0.15	–150 × 0.15
E_5	0.05	150 × 0.05	300 × 0.05	450 × 0.05	600 × 0.05	750 × 0.05
	EMV	150	120	–225	–595	–1,500

Expected monetary value (profit) called ***EMV*** can be computed by multiplying pay-off's to corresponding probabilities as given below for quantity of 10 packets as A_1.

So, $EMV(A_1) = 150 \times 0.2 + 150 \times 0.35 + 150 \times 0.25 + 150 \times 0.15 + 150 \times 0.05 = 150$

Similarly, we compute $EMV(A_2)$, and so on as shown above in the last row of the conditional profit table.

(*ii*) As the highest *EMV* belongs to A_1, the optimum number of packets of biscuit to be purchased is 10.

(*iii*) The expected profits with perfect information (*EPPI*) can be calculated by multiplying the maximum pay-offs for each states-of-nature by its probability. Now, we have the following information:

States-of-nature	E_1	E_2	E_3	E_4	E_5
Maximum pay-off	150	300	450	600	750
Probability	0.20	0.35	0.25	0.15	0.05

Using this information, we get

$$EPPI = 150 \times 0.20 + 300 \times 0.35 + 450 \times 0.25 + 600 \times 0.15 + 750 \times 0.05 = ₹\,375.$$

$$\therefore \quad EVPI = EPPI - EMV^* = ₹\,375 - ₹\,150 = ₹\,225$$

Example 19.8. *A grocery store with a bakery department is faced with the problem of how many cakes to buy in order to meet the day's demand. The grocer prefers not to sell day-old goods in competition with fresh products: leftover cakes are, therefore, a complete loss. On the other hand, if a customer desires a cake and all of them have been sold, the disappointed customer will buy from elsewhere and the sales will be lost. The grocer has, therefore, collected information of the past sales on a selected 100 days period as shown in the table below:*

Sales per day :	25	26	27	28
No. of days :	10	30	50	10
Probability :	0.10	0.30	0.50	0.10

A cake costs ₹ 80 and sells for ₹ 100. Construct the pay-off table and the opportunity loss table. What is the optimal number of cakes that should be bought each day? Also, find the interpret EVPI.

[MBA (Oct.), Delhi Univ. 2010]

Solution. Given $SP = ₹\,100$ per cake, $CP = ₹\,80$ per cake, Profit = ₹ 20 per cake,
Quantity unsold = Loss of ₹ 80 per cake

Let Q = number of cakes purchased and D = number of cakes demanded.

If demand is more than or equal to the quantity stocked, *i.e.*, $D \geq Q$, then it can sell all the quantity stocked. Therefore,

$$\text{Total profit} = \text{Total revenue} - \text{Total cost}$$
$$= 100Q - 80Q = 20Q$$

In case demand is less than quantity stocked, *i.e.*, $D < Q$, then

$$\text{Total profit} = [20D - 80(Q - D)] = 100D - 80Q$$

Using the above conditions, we obtain the following conditional profit (pay-off) table:

Event (Demand)	*Probability*	*Conditional pay-off (₹)*			
		$A_1 : 25$	$A_2 : 26$	$A_3 : 27$	$A_4 : 28$
$E_1 : 25$	0.10	500	420	340	260
$E_2 : 26$	0.30	500	520	440	360
$E_3 : 27$	0.50	500	520	540	460
$E_4 : 28$	0.10	500	520	540	560
	EMV	500	510	490	420

In the above table, *EMV* are computed by multiplying the pay-offs with the corresponding probabilities. For example, if stock is of 27 cakes, then

$$EMV\,(A_3) = 340 \times 0.10 + 440 \times 0.30 + 540 \times 0.50 + 540 \times 0.10 = 490$$

Now, since the highest *EMV* is seen in strategy A_2, according to *EMV* decision criterion, the store should stock 26 cakes.

The opportunity loss table is now computed from the given pay-off table as follows:

Event	*Probability*	*Act (Conditional opportunity loss)*			
		$A_1 : 25$	$A_2 : 26$	$A_3 : 27$	$A_4 : 28$
$E_1 : 25$	0.10	500 – 500 = 0	500 – 420 = 80	500 – 340 = 160	500 – 260 = 240
$E_2 : 26$	0.30	520 – 500 = 20	520 – 520 = 0	520 – 440 = 80	520 – 360 = 160
$E_3 : 27$	0.50	540 – 500 = 40	540 – 520 = 20	540 – 540 = 0	540 – 460 = 80
$E_4 : 28$	0.10	560 – 500 = 60	560 – 520 = 40	560 – 540 = 20	560 – 560 = 0
	EOL	32	22	42	112

In the above table, *EOL* are computed by multiplying the conditional opportunity loss entries with the corresponding probabilities. For example,

$$EOL\ (A_4) = 240 \times 0.10 + 160 \times 0.30 + 80 \times 0.50 + 0 \times 0.10 = 112$$

Here, the least *EOL* is seen in strategy A_2. Therefore, according to *EOL* decision criterion, the store should stock 26 cakes. Further,

$$EVPI = \text{minimum } EOL = 22$$

The value of *EVPI* is ₹ 22 and represents the worth of information or the amount we can afford to pay for obtaining such information.

Example 19.9. *An enterprising person plans to build a hotel. Usually he builds hotel having 25 or 50 or 100 beds, depending on whether anticipated demand is low, medium or high. He has been able to find out net profits which are expressed in the table given below with the corresponding probabilities:*

States-of-nature	*Probability*	*Courses of action*		
		A_1 : 25 beds hotel	*A_2 : 50 beds hotel*	*A_3 : 100 beds hotel*
E_1 : Low demand	0.2	20,000	–10,000	–30,000
E_2 : Medium demand	0.3	25,000	25,000	–5,000
E_3 : High demand	0.5	30,000	50,000	60,000

(*i*) *Compute expected profit (maximum EMV), EPPI and EVPI.*

(*ii*) *A research firm agrees to conduct a survey for ₹ 8,000, and provide him the information regarding states-of-nature. Should the survey be conducted?*

Solution.

(*i*) The expected profit or *EMV* are computed as under:

$$EMV\ (A_1) = ₹\ [0.2 \times 20{,}000 + 0.3 \times 25{,}000 + 0.5 \times 30{,}000] = ₹\ 26{,}000$$

$$EMV\ (A_2) = ₹\ [0.2 \times (-10{,}000) + 0.3 \times 25{,}000 + 0.5 \times 50{,}000] = ₹\ 30{,}500$$

$$EMV\ (A_3) = ₹\ [0.2 \times (-30{,}000) + 0.3 \times (-5{,}000) + 0.5 \times 60{,}000] = ₹\ 22{,}500$$

Clearly, the highest *EMV* is associated with the action A_2. Hence, the highest expected profit under certainty is ₹ 30,500.

To compute *EPPI*, we shall find out the highest pay-off for each action under certainty. Therefore, since the highest pay-off in each states-of-nature (*i.e.*, E_1, E_2 and E_3) corresponds to A_1, A_2 and A_3 respectively, we have

$$EPPI = ₹\ [0.2 \times 20{,}000 + 0.3 \times 25{,}000 + 0.5 \times 60{,}000] = ₹\ 41{,}500$$

$$\therefore \quad EVPI = EPPI - EMV^* = ₹\ 41{,}500 - ₹\ 30{,}500 = ₹\ 11{,}000$$

(*ii*) *EVPI* is relatively larger than the expenditure incurred in conducting a survey in order to collect further information regarding the states-of-nature. Hence, it is advisable to conduct the survey.

PROBLEMS

19.11. A physician purchases a particular vaccine on Monday of each week. The vaccine must be used within the week following, otherwise it becomes worthless. The vaccine costs ₹ 2 per dose and the physician charges ₹ 4 per dose. In the past 50 weeks the physician has administered the vaccine in the following quantities:

Dose per week :	20	25	40	60
Number of weeks :	5	15	25	5

On the basis of *EMV*, find how many doses the physician must purchase each week to maximize his profits.

19.12. An ice-cream retailer buys ice cream at a cost of ₹ 5 per cup and sells it for ₹ 8 per cup; any remaining unsold at the end of the day can be disposed off at a salvage price of ₹ 2 per cup. Past sales have ranged between 15 and 18 cups per day; there is no reason to believe that sales volume will take on any other magnitude in future. Find the *EMV*, if the sale history has the following probabilities:

Market size :	15	16	17	18
Probability :	0.10	0.20	0.40	0.30

[BCA, Panjab Tech. Univ. 2010]

19.13. Under an employment promotion programme, it is proposed to allow sale of newspapers on buses during off-peak hour. The vendor can purchase the papers at a special concessional rate of ₹ 2.5 and sell it for ₹ 4.00 (a piece). Any unsold copy is a dead loss. A vendor has estimated the following probabilities for the number of copies demanded:

No. of copies :	15	16	17	18	19	20
Probability :	0.04	0.19	0.33	0.26	0.11	0.07

Prepare a pay-off table and find out how many copies should be ordered, so that his expected profits will be maximum.

[B.Com. (Hons.), Delhi Univ. 2010; M.Tech., Panjab Tech. Univ. 2012]

19.14. The manager of a flower shop promises delivery within four hours on all flower orders. The flowers are purchased on the previous day and delivered to the manager by 8.00 AM the next morning. Manager's daily demand for roses is as follows:

Roses (in dozens) :	7	8	9	10
Probability :	0.1	0.2	0.4	0.3

The manager purchases roses for ₹ 10 per dozen and sells them for ₹ 30 per dozen. All unsold roses are donated to a local hospital. How many dozens of roses should the manager order each evening to maximize his profit? What is the optimum expected profit?

[MBA, Delhi Univ. 2002]

19.15. A small industry finds from the past data that the cost of making an item is ₹ 25, the selling price is ₹ 30, if it is sold within a week, and it could be disposed off at ₹ 20 per item at the end of the week:

Weekly sales :	< 3	4	5	6	7	≥ 8
No. of weeks :	0	10	20	40	30	0

Find the optimum number of items per week the industry should produce.

[MBA, Pune Univ. 2010; MCM, Lucknow Univ., 2008]

19.16. Following are the records of demand of an item for the past 300 days:

Demand of item :	10,000	11,000	12,000	13,000	14,000
No. of days :	18	90	120	60	12
Probability :	0.06	0.30	0.40	0.20	0.04

(*a*) What is the expected demand?

(*b*) It costs ₹ 15 to make an item which sells for ₹ 20 normally but at the end of the day any surplus has to be disposed off at ₹ 10 per item. What is the optimum output?

[B.Tech. (Mech.) (Dec.), Panjab Tech. Univ. 2011]

19.17. An investor is given the following investment alternatives and percentage rates of return:

Investment alternatives	*States-of-nature (Market conditions)*		
	Low	*Medium*	*High*
Regular shares	7%	10%	15%
Risky shares	–10%	12%	25%
Property	–12%	18%	30%

Over the past 300 days, 150 days have been medium market conditions and 60 days have had high market conditions. On the basis of these data, state the optimum investment strategy for the investment. ***[MBA, Nagpur Univ. 1999]***

19.18. A milk producing cooperative union desires to determine how many kilograms of butter it should produce on daily basis to meet the demand. Past record have shown the following pattern of demand:

Quantity demanded (kg)	:	*15*	*20*	*25*	*30*	*35*	*40*	*45*
Days when given demand occurred	:	4	16	20	80	40	30	10

Assume that the stock levels are restricted to the range 15–45 kg (a multiple of 5 kg), and that butter left unsold at the end of the day must be disposed off due to inadequate storing facilities. Butter costs ₹ 14 per kg and is sold at ₹ 20 per kg. (*a*) Construct a conditional profit table, (*b*) Determine the action alternative associated with the maximization of expected profit, and (*c*) Determine *EVPI*. ***[MBA, Gujarat Univ. 2008]***

19.19. The demand for a seasonal product is given below:

Demand	:	40	45	50	55	60	65
Probability	:	0.10	0.20	0.30	0.25	0.10	0.05

The product costs ₹ 60 per unit and sells at ₹ 80 per unit. If the units are not sold within the season, they will have no market value.

(*i*) Prepare a pay-off table and a regret table.

(*ii*) Determine the optimum number of units to be produced.

(*iii*) Calculate *EVPI* and interpret it. ***[MBA, Allahabad Univ. 2006]***

19.20. A cell phone company located in Bengaluru wants to determine the number of units of cell phone to be stocked on monthly basis. The past pattern of demand for cell phone units is as under:

Units sold per month	:	100	101	102	103	104	105
No. of months	:	20	40	5	10	20	5

Assume that the stock level of cell phones are restricted to the range 101–105, and that cell phones left unsold at the end of the month must be sent back to the Head Office at Mumbai. The cell phone costs ₹ 3,000 to the company, which it sells for ₹ 4,000.

From the above:

(*i*) Construct a conditional pay-off table.

(*ii*) Determine the action alternatives with the expected profit.

(*iii*) Determine *EVPI*. ***[MBA (Oct.), Bangalore Univ. 2009]***

19.21. A TV dealer finds that the cost of holding a television in stock for a week is ₹ 50. Customers who cannot obtain new television immediately tend to go to other dealers and he estimates that for every customer who cannot get immediate delivery he loses an average of ₹ 200. For one particular model of television, the probabilities of a demand of 0, 1, 2, 3, 4 and 5 TVs in a week are 0.05, 0.10, 0.20, 0.30, 0.20 and 0.15, respectively.

(*i*) How many televisions per week should the dealer order? Assume that there is no time lag between ordering and delivery.

(*ii*) Compute *EVPI.*

(*iii*) The dealer is thinking of spending on a small market survey to obtain additional information regarding the demand levels. How much should he be willing to spend on such a survey? ***[M.Tech. (Dec.), Panjab Tech. Univ. 2011; MBA, Delhi Univ. 2004]***

[***Hint:*** If D denotes the demand and Q the number of televisions in stock, then the conditional cost function is:

$$50Q + 200(D - Q), \text{ if } D \geq Q \text{ and } 50D + 50(Q - D), \text{ if } D < Q$$

∴ The conditional cost in ₹ are as given in the following table:

Event (Demand)	*Probability*	*Conditional cost in ₹/Act (Stock)*						*Minimum conditional cost (₹)*
		0	*1*	*2*	*3*	*4*	*5*	
0	0.05	0	50	100	150	200	250	0
1	0.10	200	50	100	150	200	250	50
2	0.20	400	250	100	150	200	250	100
3	0.30	600	450	300	150	200	250	150
4	0.20	800	650	500	350	200	250	200
5	0.15	1000	850	700	550	400	250	250
	Expected cost	590	450	330	250	230	250	

ECPI (expected cost under perfect information) =

$0 \times 0.05 + 50 \times 0.10 + 100 \times 0.20 + 150 \times 0.30 + 200 \times 0.20 + 250 \times 0.15 =$ ₹ 147.50]

19.22. A company manufacture goods for a market in which the technology of the products is changing rapidly. The research and development department produced a new product which appears to have potential for commercial exploitation. A further ₹ 60,000 is required for development testing. The company has 100 customers and each customer might purchase at the most one unit of the product. Market research suggests that a selling price of ₹ 6,000 for each unit with total variable costs of manufacturing and selling estimate are ₹ 2,000 for each unit. As a result of previous experience of this type of market, it has been possible to derive a probability distribution relating to the proportions of customers who will buy the product, as follows:

Proportion of customers	:	0.04	0.08	0.12	0.16	0.20
Probability	:	0.1	0.1	0.2	0.4	0.2

Determine the expected opportunity losses, given no other information than that stated above, and state whether or not the company should develop the product.

[***Hint:*** If the proportion of customers who purchase the new product by 'p' the conditional profit is equal to (₹ 6,000 – ₹ 2,000) × 100p – ₹ 60,000, *i.e.*, ₹ (400p – 60) × 1,000.

The conditional profit table is:

States-of-nature (Proportion of customers)	Probability	Alternative strategies	
		A_1 : Develop product	A_2 : Do not develop
E_1 : 0.04	0.1	–44,000	0
E_2 : 0.08	0.1	–28,000	0
E_3 : 0.12	0.2	–12,000	0
E_4 : 0.16	0.4	4,000	0
E_5 : 0.20	0.2	20,000	0

Compute *EOL* for both strategies and choose the least of these.]

19.7: DECISION TREE

A decision tree is a schematic and graphic representation of decision problem. It displays various decision alternative and sequence of events as of they were branches of a tree. A decision tree consists of the following four instructive parts:

1. **Decision Node.** It is usually denoted by symbol □ and represents the expected value with the highest expected value, *i.e.*, maximum of *EMV*.
2. **Chance Node.** It is usually denoted by symbol ○ and represents all possible outcomes that can be obtained form a chosen course of action/event.
3. **Alternative Branches (or Decision Alternatives).** These branches are represented by straight lines that state strategies/alternatives of the decision problem.
4. **Probabilistic Branches.** These branches are also represented by straight lines that state the states-of-nature of a decision problem.

A ***decision node*** represents the choice between different alternatives/strategies. From it come ***alternative branches*** that lead to chance nodes which represent factors in the decision situation over which the decision-maker has no control. From these chance nodes come probabilistic branches that state the actual or estimated numerical likelihood of various chances. Since, the states-of-nature (or events) are mutually exclusive and exhaustive, the probabilities associated with each event (states-of-nature) must add to one.

In order to evaluate a decision tree, it is necessary to evaluate each chance node usually denoted by alphabets *A*, *B*, *C*, ... and each decision node usually denoted by numerals. These two types of nodes are evaluated differently. Chance nodes are evaluated using the expected values, and the value of decision node is the expected value of the most desirable alternative action.

A decision-tree is highly useful to a decision-maker in multistage decision situations which involve a series of decisions each dependent on the preceding one. In decision tree, we compute *EMV* for each main branch of the decision tree. These values then become the conditional expected pay-offs for their respective branches; the process is repeated until the *EMV* for a given path has been determined. The optimal path (strategy) is one that corresponds to the maximum *EMV*. This technique is called ***Rolling Back Technique***. According to this technique, we take that decision which yields optimal *EMV* and then ***roll back*** to the preceding decision point. The ***rolling back process*** continues till the initial decision point (node) is reached.

Example 19.10. *A manager has a choice between (i) A risky contract promising ₹7 lakh with probability 0.6 and ₹4 lakh with probability 0.4, and (ii) A diversified portfolio consisting of two contracts with independent outcomes each promising ₹3.5 lakh with probability 0.6 and ₹2 lakh with probability 0.4.*

Construct a decision tree for using EMV criteria. Can you arrive at the decision using EMV criteria?

[MBA (Nov.), Annamalai Univ. 2010]

Solution. The conditional pay-off table for the problem may be constructed as follows:

(₹ *in lakh*)

Event E_i	*Probability* (*i*)	*Conditional pay-off (Decision)* Risky contract (*ii*)	Portfolio (*iii*)	*Expected pay-off (Decision)* Risky contract (*i*) × (*ii*)	Portfolio (*i*) × (*iii*)
E_1	0.6	7	3.5	4.2	2.1
E_2	0.4	4	2	1.6	0.8
			EMV	5.8	2.9

Using the *EMV* criterion, the manager must go in for the risky contract which will yield him a higher expected monetary value of ₹5.8 lakh. The decision tree for the problem is given in *Fig. 19.1.*

Since, the net *EMV* is higher for alternative 'risky contract', the optimal decision is to choose "risky contract".

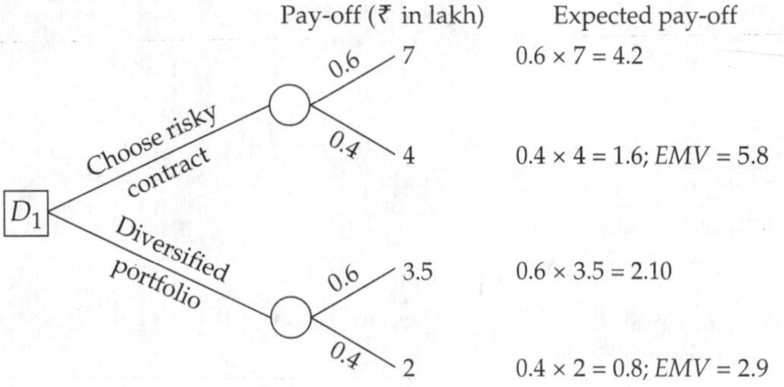

Fig. 19.1. Decision tree for the manager

Example 19.11. *A company is currently working with a process which after paying for materials, labour, etc., brings a profit of ₹12,000. The following alternatives are made available to the company:*

(*i*) *The company can conduct research (R_1) which is expected to cost ₹10,000 having 90% chances of success. If it proves a success, the company gets a gross income of ₹25,000.*

(*ii*) *The company can conduct research (R_2) which is expected to cost ₹8,000 having a probability of 60% success, the gross income will be ₹25,000.*

(*iii*) *The company can pay ₹6,000 as royalty for a new process which will bring a gross income of ₹20,000.*

(*iv*) *The company continues the current process.*

Because of limited resources, it is assumed that only one of the two types of research can be carried out at a time. Use decision-tree analysis to locate the optimal strategy for the company.

[*MBA, Panjab Univ. 2015; BFIA, Delhi Univ. 2004; MBA, Annamalai Univ. 2009*]

Solution. The various act-event combinations and resulting pay-offs of the problem are introduced in the decision-tree diagram (*Fig. 19.2*). The net *EMV* corresponding to various event/decision points are indicated in bold type in *Fig. 19.2.*

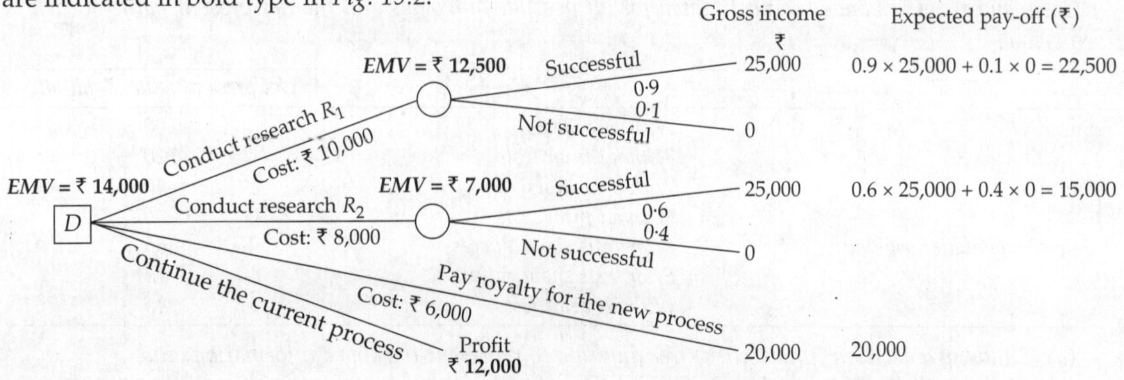

Fig. 19.2. Company's decision-tree

Decision Analysis at Point D

Decision	*Event*	*Probability*	*Gross income*	*Expected income*
1. Conduct research R_1	Successful	0.9	₹25,000	₹ 22,500
	Not successful	0.1	0	0
				Total expected income = ₹ 22,500
				Less: Cost ₹ 10,000
				Net *EMV* ₹ 12,500
2. Conduct research R_2	Successful	0.6	₹25,000	15,000
	Not successful	0.4	0	0
				Total expected income = ₹ 15,000
				Less: Cost ₹ 8,000
				Net *EMV* ₹ 7,000
3. Pay royalty for the new process	Certain	1	₹20,000	₹ 20,000
				Less: Cost ₹ 6,000
				Net *EMV* ₹ 14,000
4. Continue the current process	Certain	1	₹12,000	*EMV* ₹ 12,000

As the net *EMV* is highest for the alternative 'pay royalty for the new process', the optimal decision would be to procure new process on royalty basis.

Example 19.12. *An oil company has recently acquired rights in a certain area to conduct surveys and test drilling to lead to lifting oil, if it is found in commercially exploitable quantities.*

The area is considered to have good potential for finding oil in commercial quantities. At the outset, the company has the choice to conduct further geological tests or to carry out a drilling programme immediately. On the known conditions, the company estimates that there is a 70 : 30 chance of further tests showing a 'success'.

Whether the tests show the possibility of ultimate success or not or even if no tests are undertaken at all, the company could still pursue its drilling programme or alternatively consider selling its rights to drill in the area. Thereafter, however, it carries out the drilling programme, the likelihood of final success or failure is considered dependent on the foregoing stages. Thus,

(i) if 'successful' tests have been carried out, the expectation of success in drilling is given as 80 : 20.
(ii) if the tests indicate 'failure', then the expectation of success in drilling is given as 20 : 80.
(iii) if no tests have been carried out at all, the expectation of success in drilling is given as 55 : 45.

Costs and revenues have been estimated for all possible outcomes and the net present value of each is given below:

Outcome		*Net present value (₹ million)*
Success :	*With prior tests*	100
	Without prior tests	120
Failure :	*With prior tests*	–50
	Without prior tests	–40
Sale of exploitation rights :	*Prior tests show 'success'*	65
	Prior tests show 'failure'	15
	Without prior tests	45

(a) Draw up a decision (probability) tree diagram to represent the above information; and
(b) Evaluate the tree in order to advise the management of the company on its best course of action.

[MBA (Nov.), Delhi Univ. 2003]

Solution.

(*a*) The various act-event combination and resulting pay-offs of the problem are introduced in the following decision tree diagram. See *Fig. 19.3.*

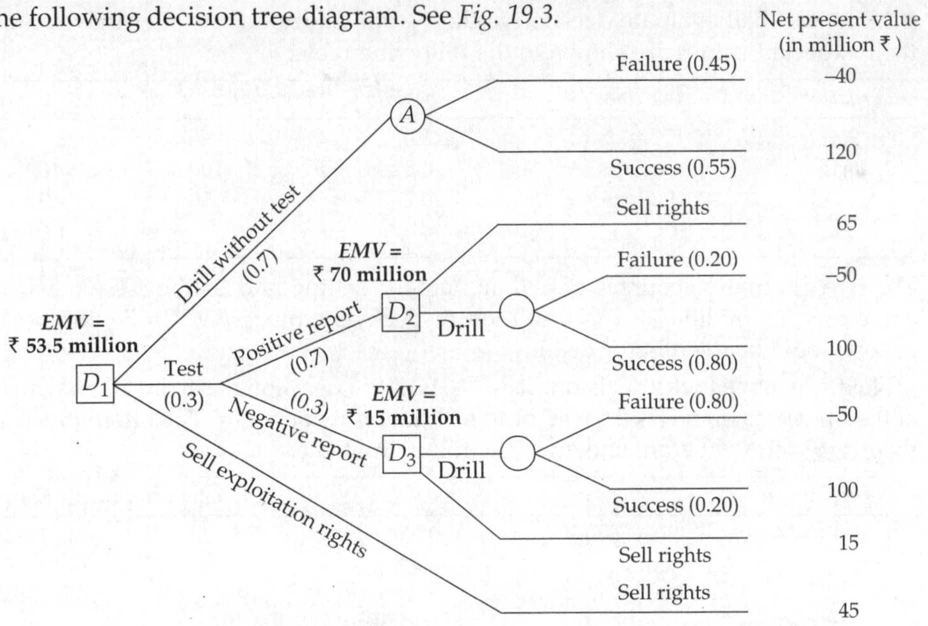

Fig. 19.3. Decision tree for oil company

It may be observed that there are several branches of the decision-tree which originate from squares or from circles. A square represents a decision node at which decision maker has to take a decision, while a circle represents a chance node at which events are branched out. At decision node D_1 there are three branches that represent drilling, testing for oil, and selling of rights — the decision maker has to select one.

If the company decides to drill without testing, there are two possible events: it may get oil or not, which are shown as branches emanating from *A*, whose probabilities are 0.55 and 0.45 respectively. A profit of ₹ 120 million would result in case oil is obtained and a loss of ₹ 40 million if it is not.

The second alternative at node D_1 is to go for a test, which may give positive or negative results with probabilities 0.7 and 0.3 respectively. In case of positive result, a choice has to be made as to whether to sell the rights for ₹ 65 million or to drill, which is likely to succeed and fail with probabilities 0.80 and 0.20 respectively. Therefore, a decision node D_2 is shown by a square. Similarly, for a negative result, there are two options — to sell for ₹ 15 million or to drill, which has a 20 per cent chance of success. Thus, decision node D_3 is shown here.

The third alternative is to sell the exploitation rights for ₹ 45 million.

(*b*) To evaluate the tree, we start with the evaluation of decision node D_3. Clearly, this point is reached after a 'negative' is indicated by the test. The expected values of each of the two alternatives, to drill and to sell, are obtained in the following table:

Alternative	*Outcome*	*Probability*	*Conditional value*	*Expected value*
1. Drill	Success	0.2	100	100 × 0.20 = 20
	Failure	0.8	–50	–50 × 0.80 = –40
				EMV = –20
2. Sell	—	1.0	15	15

The expected value of drilling option is a loss of ₹ 20 million, while if the site is sold, we can get ₹ 15 million. On the basis of *EMV*, our decision would be to sell the site.

Now, we shall evaluate decision node D_2. For this node, the alternatives together with their expected values are shown in the following table:

Alternative	*Outcome*	*Probability*	*Conditional value*	*Expected value*
1. Sell	—	1.0	65	65
2. Drill	Success	0.8	100	100 × 0.8 = 80
	Failure	0.2	–50	– 50 × 0.2 = –10
				EMV = 70

The two alternative courses of action, namely, selling and drilling, have expected value equal to ₹ 65 million and ₹ 70 million respectively. Since *EMV* = 70, the best course of action would be to go in for drilling.

Next, we move to the decision node D_1, where a decision has to be taken whether to drill at the outset, to undertake a test, or to sell the rights outright. The alternatives along-with their expected values are shown in the following table:

Alternative	*Outcome*	*Probability*	*Conditional value*	*Expected value*
1. Drill	Success	0.55	120	120 × 0.55 = 66
	Failure	0.45	–40	–40 × 0.45 = –18
				EMV = 48
2. Test	Positive	0.7	70	70 × 0.7 = 49.0
	Negative	0.3	15	15 × 0.3 = 4.5
				EMV = 53.5
3. Sell		1.0	45	45

From the above table, we observe that the expected value of the alternative for carrying out a test is ₹ 53.5 million, which is the highest among the three alternatives. On the basis of *EMV*, it is better to test before drilling. Hence, our decision is as follows:

"The test should be carried out. If it proves negative, the rights should be sold to give a return of ₹ 15 million. To proceed with drilling, if that happens, would expectedly lead to a loss. However, if the test proves positive, the drilling should be undertaken".

PROBLEMS

19.23. There is 40% chance that a patient admitted to the hospital, is suffering from cancer. A doctor has to decide, whether a serious operation should be performed or not. If the patient is suffering from cancer and the serious operation is performed, the chance that he will recover is 70%, otherwise it is 35%. If the patient is not suffering from cancer and the serious operation is performed; the chance that he will recover is 20%, otherwise it is 100%. Assume that recovery and death are the only possible results. Construct an appropriate decision tree. What decision should the doctor take?

19.24. Growfast Company is evaluating four alternative single-period investment opportunities whose returns are based on the state of the economy. The possible states of the economy and the associated probability distribution is as follows:

State :	*Fair*	*Good*	*Great*
Probability :	0.2	0.5	0.3

The returns for each investment opportunity and each state of economy are as follows:

Alternative	*State of economy*		
	Fair	*Good*	*Great*
W	₹ 1,000	₹ 3,000	₹ 6,000
X	₹ 500	₹ 4,500	₹ 6,800
Y	₹ 0	₹ 5,000	₹ 8,000
Z	₹ 4,000	₹ 6,000	₹ 8,500

Using the decision-tree approach, determine the expected return for each alternative. Which alternative investment proposal would you recommend, if the expected monetary value criterion is to be employed?

19.25. The *XYZ* manufacturing company must decide whether to manufacture a component part at its Pune plant or purchase the component part from a supplier. The resulting profit is depending upon the demand for the product. The following pay-off table shows the project profit in millions of rupees:

Decision alternative	*States-of-nature*		
	Low demand	*Medium demand*	*High demand*
Manufacture	–20	40	100
Purchase	10	45	70

The current estimates for probabilities of states-of-nature are:

Pr. (low demand) = 0.35, Pr. (medium demand) = 0.35, and Pr. (high demand) = 0.30

(*i*) Use decision tree to recommend a decision.

(*ii*) Use *EVPI* to determine whether *XYZ* company should attempt to obtain a better estimate of demand.

(*iii*) A test market study of the potential demand for the product is expected to report either a favourable (*F*) or unfavourable (*U*) condition. The relevant conditional probabilities are as follows:

Pr. (*F* | Low demand) = 0.10, Pr. (*F* | Medium demand) = 0.40

and Pr. (*F* | High demand) = 0.50

What is the probability that the market research report would be favourable.

[MBA, Delhi Univ. 2012]

19.26. A company is launching a new product in the market. Three alternative decisions are available for the management, which are given below:

D_1 : Publicity need to create the demand and its cost is ₹ 200 per day.

D_2 : Appoint marketing people, whose cost is ₹ 1,250 per day.

D_3 : Conducting exhibition, whose cost is ₹ 600 per day.

The cost of units associated with the respective decision alternatives are: ₹ 8, ₹ 6 and ₹ 7 respectively. The price per unit of product is fixed at ₹ 20. The expected demand for the item are as follows:

Demand :	100	200	500
Probability :	0.5	0.3	0.2

The company has to decide whether the new item can be manufactured or not? If yes, which decision alternative should be chosen?

[***Hint:*** Since, Profit = (Price of unit – Unit cost) × Demand, the pay-off table will be:

	A_1	A_2	A_3
D_1	1,200	2,400	6,000
D_2	1,400	2,800	7,000
D_3	1,300	2,600	6,500

19.27. A businessman has two independent investments *A* and *B* available to him, but he lacks the capital to undertake both the them simultaneously. He can choose to take *A* first and then stop, or if *A* is successful then take *B*, or *vice-versa*. The probability of success on *A* is 0.7, while for *B* it is 0.4. Both investments require an initial capital outlay of ₹ 2,000 and both return nothing, if the venture is unsuccessful. Successful completion of *A* will return ₹ 3,000 (over cost), successful completion of *B* will return ₹ 5,000 (over cost).

Draw a decision tree and determine the best strategy.

[MBA (Dec.), Panjab Tech. Univ. 2008]

19.28. A Finance Manager of certain company is considering drilling a well. In the past, only 70% of wells drilled were successful at 20 metres depth in that area. Moreover, on finding no water at 20 metres, some persons in that area drilled it further up to 25 metres, but only 20% struck water at that level. The prevailing cost of drilling is ₹ 500 per metre. The Finance Manager estimated that in case he does not get water in his own well, he will have to pay ₹ 15,000 to buy water from outside for the same period of getting water from the well. The following decision are considered:

(*i*) Do not drill any well, (*ii*) Drill up to 20 metres, and
(*iii*) If no water is found at 20 metres, drill further up to 25 metres.

Draw an appropriate decision tree and determine the Finance Manager's optimal strategy.

19.29. The Oil India Corporation is considering whether to go for an off-shore oil drilling contract to be awarded in Mumbai High. If they bid, value would be ₹ 600 million with a 65% chance of gaining the contract. They may set up a new drilling operation or move already existing operation, which has proved successful, to new site. The probability of success and expected returns are as follows:

Outcome	*New drilling operation*		*Existing operation*	
	Probability	*Expected revenue (₹ million)*	*Probability*	*Expected revenue (₹ million)*
Success	0.75	800	0.85	700
Failure	0.25	200	0.15	350

If the Corporation do not bid or lose the contract, they can use ₹ 600 million to modernize their operation. This would result in a return of either 5% or 8% on the sum invested with probabilities 0.45, and 0.55. (Assume that all costs and revenues have been discounted to present value.)

(*i*) Construct a decision tree for the problem showing clearly the courses of action.
(*ii*) By applying an appropriate decision criterion, recommend whether or not the Oil India Corporation should bid the contract. ***[MBA (PT), Delhi Univ. 2010]***

19.30. The investment company is considering four investment proposals for a client: shares, bonds, real estate and saving certificates. These investments will be held for one year. The past data regarding the four proposals are as follows:

Shares. There is 25 per cent chance that shares will decline by 10 per cent, a 30 per cent chance that they will remain stable and a 45 per cent chance that they will increase in value by 15 per cent. Also, the shares under consideration do not pay any dividends.

Bonds. These bonds stand a 40 per cent chance of increase in value by 5 per cent and 60 per cent chance of remaining stable and they yield 12 per cent.

Real Estate. This proposal has a 20 per cent chance of increasing in 30 per cent value, a 25 per cent chance of increasing in 20 per cent value, a 40 per cent chance of increasing in 10 per cent value, a 10 per cent chance of remaining stable and a 5 per cent chance of losing 5 per cent of its value.

Saving Certificates. These certificates yield 8.5 per cent with certainty.

Use a decision tree to structure the alternatives available to the investment staff, and using the expected criterion, choose the alternative with the highest expected value.

[MBA (Dec.), Panjab Tech. Univ. 2013; MBA, Delhi Univ. 2006]

19.31. A company has developed a new product in its R&D laboratory. The company has the option of setting up production facility to market this product straight away. If the production is successful, then over the three years expected product life, the returns will be ₹ 120 lakh with a probability of 0.70 . If the market does not respond favourable, then the returns will be only ₹ 15 lakh with probability of 0.30.

The company is considering whether it should test-market this product building a small pilot plant. The chance that the test-market will yield favourable response is 0.80. If the test-market gives favourable response, then the chance of a successful total market improves to 0.85.

If the test-market gives poor response, then the chance of success in the total market is only 0.30.

As before, the returns from a successful market will be ₹ 120 lakh and from an unsuccessful market only ₹ 15 lakh. The installation cost to produce for the total market is ₹ 40 lakh and the cost of the test-marketing pilot plant is ₹ 5 lakh. Using decision-tree analysis, draw a decision tree diagram and carry out necessary analysis to determine the optimal decision.

[MBA (PT) (Nov.), Delhi Univ. 2008]

19.32. A company is planning to launch a new product, which can be introduced initially in Western India or in the entire country. If the product is introduced only in Western India, the investment outlay will be ₹ 12 million. After two years, the company can evaluate the project to determine whether it should cover the entire country. For such expansion it will have to incur an additional investment of ₹ 10 million. To introduce the product in the entire country right in the beginning would involve an outlay of ₹ 20 million. The product, in any case, will have a life of 5 years, after which the plant will have zero net value.

If the product is introduced only in Western India, demand would be high or low with the probabilities of 0.8 and 0.2 respectively and annual cash inflows of ₹ 4 million and ₹ 2.5 million, respectively.

If the product is introduced in the entire country right in the beginning, the demand would be high or low with probabilities of 0.6 and 0.4 respectively and annual cash inflows of ₹ 8 million and ₹ 5 million, respectively.

Based on the observed demand in Western India, if the product is introduced in the entire country the following probabilities would exist for high and low demand on all All-India basis:

Western India	*Whole country*	
	High demand	*Low demand*
High demand	0.90	0.10
Low demand	0.40	0.60

The hurdle rate applicable to this project is 12%.

(*a*) Set up a decision tree for the investment situation.
(*b*) Advise the company on the investment policy it should follow. Support your advice with appropriate reasoning. ***[MBA (Nov.), Delhi Univ. 2005]***

19.33. A large steel manufacturing company has three options with regard to production: (*i*) produce commercially, (*ii*) build pilot plant, and (*iii*) stop producing steel. The management has estimated that their pilot plant, if built has 0.8 chance of high yield and 0.2 chance of low yield. If the pilot plant does show a high yield, management assigns a probability of 0.75 that the commercial plant will also have a high yield. If the pilot plant shows a low yield, there is only a 0.1 chance that the commercial plant will show a high yield. Finally, the management's best assessment of the yield on a commercial-size plant without building a pilot plant first has a 0.6 chance of high yield. A pilot plant will cost ₹ 3,00,000. The profits earned under high and low yield conditions are ₹ 1,20,00,000 and (–) ₹ 12,00,000 respectively. Find the optimum decision for the company. ***[B.Tech., Panjab Tech. Univ. 2006]***

19.34. The ore mining company is attempting to decide whether or not a certain piece of land should be purchased. The land cost is ₹ 3,00,000. If there are commercial ore deposits on the land, the estimated value of property is ₹ 5,00,000. If no ore deposits exist, however, the property value is estimated at ₹ 2,00,000. Before purchasing the land, the property can be cored at ₹ 20,000. The coring will indicate if conditions are favourable or unfavourable for ore mining. If coring report is favourable, the probability of recoverable ore deposits on the land is 0.8, while if the coring report is unfavourable, the probability is only 0.2. Prior to obtaining any coring information, management estimates that the odds are 50 : 50 that ore is present on the land. Management has also received coring report on pieces of land similar to the ore in question and found that 60% of the coring report were favourable.

Construct a decision tree and determine whether the company should purchase the land, decline to purchase it or take a coring test before making its decision. Specify the optimal course of action and *EMV*. ***[MBA, Delhi Univ. 2011]***

19:8. DECISION-MAKING WITH UTILITIES

So far problems were analyzed with probabilities assigned to various states-of-nature where selection of an optimal course of action was based on the criterion of expected profit (or loss) expressed in monetary terms. However, in many situations, such criteria involving expected monetary pay-off may not be appropriate. This is because of the fact that different individuals attach different utility to money under different conditions. The term utility is the measure of preference that individuals have for various alternatives available to them. The utility of a given alternative is unique to individual decision-makers and, unlike a simple monetary amount, can incorporate intangible factors or subjective standards from their own value systems.

Illustration. Mr. Ram has won ₹ 3,00,000 in the recently popular quiz programme (*KBC*). He is asked either to compete or quit. Now, he has two alternatives: (*a*) quit and take his winnings, (*b*) take a further chance in which he has 50-50 chance of winning ₹ 12,00,000 or nothing; then the question is: what would he do?

Answer. On *EMV* basis he has

$$EMV = 0.50 \times ₹\ 12{,}00{,}000 + 0.50 \times 0 = ₹\ 6{,}00{,}000$$

This amount is twice what he has already won. But, would he really give up a sure ₹ 3,00,000 for a 50-50 chance for ₹ 12,00,000 or nothing? Many individuals would not, because they would think of all the things they could do with ₹ 3,00,000, and how they would regret it if they end up with nothing. Hence, a new pay-off measure of utility reflecting the decision-maker's attitude and preference has to be introduced.

The basic axioms of utility may be stated as follows:

1. If outcome A is preferred to outcome B, then the utility $U(A)$ of outcome A is greater than the utility $U(B)$ of outcome B, and *vice-versa.* If both are equally preferred, then
$$U(A) = U(B)$$
2. If the decision-maker is indifferent between the two alternatives and outcome A is received with probability p and outcome C with probability $(1 - p)$, then
$$U(B) = p\,U(A) + (1 - p)\,U(C)$$

Under this alternative criterion, it is assumed that a rational decision-maker will choose that alternative which optimizes the ***expected utility*** rather than expected monetary one. Once we know the individual's utility function along with the probability assigned to outcomes in a particular situation, then total expected utility for each course of action can be obtained by multiplying utility values with their probabilities. The strategy which corresponds to optimum utility function is called the equal strategy.

When a utility function for money is incorporated into a decision analysis approach to a problem, this utility function must be constructed to fit the preferences and values of the decision-maker involved. The utilities are measured into utils and we have to find the expected utility value (*EUV*) rather than expected monetary value (*EMV*).

It may be noted that the decision-maker's utility function for money has the property that the decision-maker is indifferent between two alternative courses of action, if the two alternatives have same expected utility.

Example 19.13. *A manager must choose between two investments A and B that are calculated to yield net profits of ₹ 3,000 and ₹ 4,500 respectively with probabilities subjectively estimated at 0.85 and 0.50. Assume that the manager's utility function reveals that utilities for ₹ 3,000 and ₹ 4,500 are 35 and 40 utils, respectively. What is the best choice on the basis of the expected utility value (EUV)?*

Solution. Let U_i is the utility value of states-of-nature i, and p_i is the probability value of states-of-nature i.

Then, $$EUV(A) = 0.85 \times 35 = 29.75 \text{ utils}$$
and $$EUV(B) = 0.50 \times 40 = 20 \text{ utils}$$

Since, $EUV(A) > EUV(B)$, the best choice is to invest A.

19:9. DECISION-MAKING UNDER CONFLICT (COMPETITIVE GAME)

The situations where two or more rational opponents are involved under conditions of competition and conflicting interests fall under the category of decision-making under conflict. In the competitive situation, the action of one competitor depends upon the action taken by the opponent. For example, candidates for an election, advertising and marketing campaigns by competing business firms, countries involved in military battles, etc. have their conflicting interests. In a ***competitive situation*** the courses of action (alternatives) for such competitor may be either finite or infinite. A competitive situation will be called a 'Game', if it has the following properties:

(*i*) There are finite number of competitors (participants) called ***players***.
(*ii*) Each player has finite number of strategies (alternatives) available to him.
(*iii*) A play of the game takes place when each player employs his strategy.
(*iv*) Every game results in an outcome, *e.g.*, loss or gain or a draw, usually called ***pay-off***, to some player.

The main objective in the competitive situation (game) is to determine the rules of rational behaviour in which the outcomes are dependent on the actions of the inter-dependent players. The players in the game strive for ***optimal strategies***. An optimal strategy is such as to provide the best situation in the game in the sense that it involves maximal pay-off to the players.

19:10. TWO-PERSON ZERO-SUM GAME

When there are two competitors playing a game, it is called a '***two-person game***'. In case the number of competitors exceeds two, say n, then the game is termed as a '***n-person game***'.

A game of two persons (players), in which the gains of one player are the losses of the other is called a ***Two-person Zero-sum Game***, *i.e.*, in a two person game, the algebraic sum of the gains to both the players after a play is bound to be zero. In other words, the loss (gain) of one player is exactly equal to the gain (loss) of the other. If the sum of gains or losses is not equal to zero, then the game is of non-zero sum character or simply ***non-zero sum game***.

Illustration. Let there be two companies producing juice, with the branded name of 'Real Juice' and 'Frooti'. Both the companies are struggling for a larger share in the market. Any share of the market gained by Real Juice must be lost share of Frooti, and therefore, the sum of the gains and losses equals zero. In other words, if gain of Real Juice is 40%, the lost share of Frooti will be 40%. The sum of gain and loss is zero, *i.e.*, (+40%) + (–40%) = 0.

Some Basic Terms

1. **Player.** The competitors in the game are known as players. A player may be individual or group of individuals, or an organization.
2. **Strategy.** A strategy for a player is defined as a set of rules or alternative courses of action available to him in advance, by which player decides the course of action that he should adopt. A strategy may be of two types:
   (*a*) ***Pure Strategy.*** If players select the same strategy each time, then it is referred to as pure-strategy. In this case each player knows exactly what the other player is going to do, the objective of the player is to maximize his gains or to minimize his losses.
   (*b*) ***Mixed Strategy.*** When the players use a combination of strategies and each player always kept guessing as to which course of action is to be selected by the other player at a particular occasion, then the strategy is known as mixed strategy. Thus, there is a probabilistic situation and objective of the player is to maximize expected gains or to minimize expected losses.
3. **Optimum Strategy.** A course of action or play which puts the player in the most preferred position, irrespective of the strategy of his competitors, is called an ***optimum strategy***.
4. **Value of the Game.** It is the expected pay-off of play when all players of the game follow their optimum strategies. The game is called ***fair***, if value of the game is zero and unfair, if it is non-zero.
5. **Pay-off Matrix.** When players select their particular strategies, the pay-offs (gains or losses) can be represented in the form of a matrix called the ***pay-off matrix***. Since, the game is zero-sum,

therefore, gain of one player is equal to the loss of other and *vice-versa*. In other words, one player's pay-off table would contain the same amounts in pay-off table of other player with the sign changed. Thus, it is sufficient to construct pay-off only for one of the two players.

Let player A have m strategies $A_1, A_2, ..., A_m$ and player B have n strategies $B_1, B_2, ..., B_n$. Assume that each player has his choices from amongst the pure strategies. Also, it is assumed that player A is always gainer and player B is always the loser. That is, all pay-offs are assumed in terms of player A. Let a_{ij} be pay-off which player A gains from player B and if player A chooses strategy A_i and player B chooses strategy B_j, then the ***pay-off matrix*** to player A is:

$$\begin{array}{cc} & \textit{Player } B \\ & \begin{array}{cccc} B_1 & B_2 & \cdots & B_n \end{array} \\ \textit{Player } A \begin{array}{c} A_1 \\ A_2 \\ \vdots \\ A_m \end{array} & \begin{bmatrix} a_{11} & a_{12} & \cdots & a_{1n} \\ a_{21} & a_{22} & \cdots & a_{2n} \\ \vdots & \vdots & & \vdots \\ a_{m1} & a_{m2} & \cdots & a_{mn} \end{bmatrix} \end{array}$$

The pay-off matrix to player B is $(-a_{ij})$.

Illustration. Consider a two-person coin tossing game. Each player tosses an unbiased coin simultaneously. Player B pays ₹ 7 to A, if $\{H, H\}$ occurs and ₹ 4, if $\{T, T\}$ occurs; otherwise player A pays ₹ 3 to B. This two-person game is a zero-sum game, since the winning of one player is the loss for the other. Each player has his choices from amongst two ***pure*** strategies H and T. If we agree conventionally to express the outcome of the game in terms of the pay-offs to one player only, say A, then the above information yields the following pay-off matrix in terms of the pay-offs to the player A. Clearly, the entries in B's pay-off matrix will be just the negative of the corresponding entries in A's pay-off matrix, so that the sum of pay-off matrices for player A and player B is ultimately a null matrix. We generally display the pay-off matrix of that player who is indicated on the left side of the matrix. For example, A's pay-off matrix may be displayed as:

		Player B	
		H	*T*
Player A	*H*	7	–3
	T	–3	4

19:11. SOLUTION OF TWO-PERSON ZERO-SUM GAME

As stated earlier, a two-person zero-sum game is the one that involve two players (competitors) and where the gain made by one player equals the loss incurred by the other player. The objective of one player is to maximize his minimum expected gains, while the other player minimizes his maximum expected losses. To explain this ***maximin-minimax principle***, let us consider the following illustration.

Illustration. Suppose that two companies (competitors or players) are competing for market share of a particular product, wherein each one is considering as to what promotional strategy to be employed for having the maximum market share. Let the following pay-off matrix describes the increase in market share for ***Company A*** and at the same time decrease in the market share for ***Company B***:

		Player (Company) B		
		B_1	B_2	B_3
Player (Company) A	A_1	65%	62%	56%
	A_2	70%	72%	75%

From the above pay-off matrix, observe that if player A selects strategy A_1 and player B selects strategy B_3, then player A wins 56% of market share and player B loses 56% of market share. Similarly, if player A selects strategy A_2 and player B selects strategy B_1, then player A wins 70% of the market share, while player B loses 70% of the market share.

In this game, objective of player A is to select a strategy that enables him to gain as much as possible. In contrast, objective of player B is to select a strategy that enables him to lose as little as possible. In other words, player A will choose that row which ***maximizes*** his ***minimum expected pay-off*** (gain) among all rows, *i.e.*, we determine the ***maximin value***. On the other hand, player B will choose that column which ***minimizes*** his ***maximum expected pay-off*** (loss) among all the columns, *i.e.*, we determine the ***minimax value***.

If the maximin value equals the minimax value, then the two-person game is said to have ***pure strategies*** and the game is said to have a ***saddle point*** (equilibrium point). The corresponding strategies are called ***optimum strategies***. The amount of pay-off at saddle point (equilibrium point) is known as ***value of the game***, generally denoted by v. Further, the saddle point need not be unique. We shall denote the maximin value of the game by $\underline{v}$ and the minimax value of the game by $\overline{v}$.

Remarks.

1. A game is said to be fair, if $\underline{v} = 0 = \overline{v}$.
2. A game is said to be strictly determinable, if $\underline{v} = v = \overline{v}$.

Rule for Determining a Saddle Point

We may now summarize the procedure of locating the saddle point of a pay-off matrix as follows:

***Step* 1.** Select the minimum element of each row of the pay-off matrix and mark them [*].

***Step* 2.** Select the greatest element of each column of the pay-off matrix and mark them [†].

***Step* 3.** If there appears to be an element in the pay-off matrix marked [*] and [†] both, the position of that element is called saddle point of the pay-off matrix.

Example 19.14 (*A Game with Pure Strategy*). *Solve the game whose pay-off matrix is given as:*

		Player A		
		A_1	A_2	A_3
	B_1	1	3	1
Player B	B_2	0	−4	−3
	B_3	1	5	−1

Solution. From the pay-off matrix, we observe the following:

	Minimum of first row	= min. {1, 3, 1} = 1
	Minimum of second row	= min. {0, −4, −3} = −4
	Minimum of third row	= min. {1, 5, −1} = −1
	Maximin value	= maximum of {1, −4, −1} = 1
Further,	Maximum of first column	= max. {1, 0, 1} = 1
	Maximum of second column	= max. {3, −4, 5} = 5
	Maximum of third column	= max. {1, −3, −1} = 1
	Minimax value	= minimum of {1, 5, 1} = 1

$$\text{Maximin value } (\underline{v}) = 1 = \text{Minimax value } (\overline{v}), \text{ i.e., } \overline{v} = \underline{v}$$

As $\overline{v} = \underline{v}$, there exists saddle point and value of game is 1. But, the element 1 appears at positions (1, 1) and (1, 3), therefore, optimum strategies (solution) is not unique. Hence, the solution of the game is:

(*i*) Optimum strategy for player *A* are A_1 and A_3,
(*ii*) Optimum strategies for player *B* is B_1,
(*iii*) Value of game is 1 for player *A* and (–1) for player *B*.

Note. Since, $v \neq 0$, the game is not fair, although strictly determinable.

Example 19.15 (*A Game with Mixed Strategy*). *Consider a game with pay-off matrix presented in the following table:*

		Player B		
		B_1	B_2	B_3
Player A	A_1	50	35	55
	A_2	40	45	60

Solution. It is evident from the pay-off matrix, that

Minimum of first row = min. {50, 35, 55} = 35, and
Minimum of second row = min. { 40, 45, 60} = 40
∴ Maximin value = maximum of {35, 40} = 40, *i.e.*, $\underline{v} = 40$
Further, Maximum of first column = max. {50, 40} = 50
Maximum of second column = max. {35, 45} = 45, and
Maximum of third column = max. {55, 60} = 60
∴ Minimax value = Minimum of {50, 45, 60} = 45, *i.e.*, $\overline{v} = 45$

Thus, $\underline{v} = 40$ and $\overline{v} = 45$, *i.e.*, $\underline{v} \neq \overline{v}$. This shows that there is ***no saddle point*** and, therefore, we can not have pure strategies. In such cases, each player would like to mix up his strategies in a random selection. The random selection plan involves selecting each strategy a certain per-cent of the time, such that the player's expected gains (or losses) are equal, regardless of the opponent's selection of strategies. Selection of a strategy, a given per-cent of the time is analogous to the selection of a strategy with given probability. There are various methods to solve such type of games, but only the following three methods will be discussed here:

1. Equal Gain Method for 2 × 2 Games.
2. Calculus Method for 2 × 2 Games.
3. Graphical Method for *m* × 2 and 2 × *n* Games.

1. Equal Gain Method for 2 × 2 Games. Consider a 2 × 2 two-person zero-sum game without any saddle point having the following pay-off matrix:

$$\begin{array}{ccc} & & \textit{Player B} \\ & & \begin{array}{cc} B_1 & B_2 \end{array} \\ \textit{Player A} & \begin{array}{c} A_1 \\ A_2 \end{array} & \begin{bmatrix} a_{11} & a_{12} \\ a_{21} & a_{22} \end{bmatrix} \end{array}$$

Let the player A play his mixed strategy $S_A = \begin{bmatrix} A_1 & A_2 \\ p_1 & p_2 \end{bmatrix}$, where $p_1 + p_2 = 1$

If player B chooses strategy B_1, the net expected gain of player A will be

$$E_1(p) = a_{11}p_1 + a_{21}p_2$$
$$= a_{11}p_1 + a_{21}(1 - p_1)$$

If player B chooses strategy B_2, the net expected gain of player A will be

$$E_2(p) = a_{12}p_1 + a_{22}p_2$$
$$= a_{12}p_1 + a_{22}(1 - p_1)$$

By equating the two expected gains, we get

$$a_{11}p_1 + a_{21}(1 - p_1) = a_{12}p_1 + a_{22}(1 - p_1)$$

or $$[a_{11} + a_{22} - (a_{12} + a_{21})]\,p_1 = a_{22} - a_{21}$$

∴ $$p_1 = \frac{a_{22} - a_{21}}{a_{11} + a_{22} - (a_{12} + a_{21})}$$

and $$p_2 = 1 - p_1 = \frac{a_{11} - a_{12}}{a_{11} + a_{22} - (a_{12} + a_{21})}$$

Similarly, let player B plays his mixed strategy $S_B = \begin{bmatrix} B_1 & B_2 \\ q_1 & q_2 \end{bmatrix}$, where $q_1 + q_2 = 1$

Then, the net expected loss of player B will be:

$$E_1(q) = a_{11}q_1 + a_{12}q_2 = a_{11}q_1 + a_{12}(1 - q_1)$$

and $$E_2(q) = a_{21}q_1 + a_{22}q_2 = a_{21}q_1 + a_{22}(1 - q_1)$$

By equating the two net expected losses, we get

$$(a_{11} - a_{12})\,q_1 + a_{12} = (a_{21} - a_{22})\,q_1 + a_{22}$$

or $$q_1 = \frac{a_{22} - a_{12}}{a_{11} + a_{22} - (a_{12} + a_{21})}$$

and $$q_2 = 1 - q_1 = \frac{a_{11} - a_{21}}{a_{11} + a_{22} - (a_{12} - a_{21})}$$

Value of game is given by

$$v = E(p, q) = q_1(a_{11}p_1 + a_{21}p_2) + q_2(a_{12}p_1 + a_{22}p_2)$$
$$= p_1(a_{11}q_1 + a_{12}q_2) + p_2(a_{21}q_1 + a_{22}q_2)$$

Substituting the values of p_1, p_2, q_1 and q_2, we get

$$v = \frac{a_{11}a_{22} - a_{12}a_{21}}{(a_{11} + a_{22}) - (a_{12} + a_{21})}$$

Example 19.16. *For the game with following pay-off matrix, determine the optimum strategies and value of the game:*

Player B

		B_1	B_2
Player A	A_1	3	–5
	A_2	–1	1

[MBA, Lucknow Univ., 2008]

Solution. Clearly, the pay-off matrix of the game does not possess any saddle point, since maximin value ≠ minimax value, *i.e.*, –1 ≠ 1.

So, let the mixed strategies are:

$$S_A = \begin{bmatrix} A_1 & A_2 \\ p_1 & p_2 \end{bmatrix}, \text{ and } S_B = \begin{bmatrix} B_1 & B_2 \\ q_1 & q_2 \end{bmatrix},$$

where, $p_1 + p_2 = 1$ and $q_1 + q_2 = 1$.

Using the formula for expected gains of 2 × 2 game, we have

$$p_1 = \frac{a_{22} - a_{21}}{a_{11} + a_{22} - (a_{12} + a_{21})} = \frac{1-(-1)}{3+1-(-1-5)} = \frac{1}{5}$$

and

$$q_1 = \frac{a_{22} - a_{12}}{a_{11} + a_{22} - (a_{12} + a_{21})} = \frac{1-(-5)}{3+1-(-1-5)} = \frac{3}{5}$$

$$p_2 = 1 - p_1 = 1 - \frac{1}{5} = \frac{4}{5}, \text{ and } q_2 = 1 - q_1 = 1 - \frac{3}{5} = \frac{2}{5}$$

Value of game is given by

$$v = \frac{a_{11} \times a_{22} - a_{12} \times a_{21}}{a_{11} + a_{22} - (a_{12} + a_{21})} = \frac{3 \times 1 - (-5) \times (-1)}{3+1-(-1-5)} = -\frac{1}{5}$$

2. Calculus Method for 2 × 2 Games. This method is similar to the previous method except that instead of equating the two expected gains, the expected value for a player is maximized. To illustrate this method, let us consider the following 2 × 2 game:

		Player B	
		B_1	B_2
Player A	A_1	*a*	*b*
	A_2	*c*	*d*

Let x and y be the probabilities of adopting strategies A_1 and B_1 by players A and B respectively. Then, obviously player A will select strategy A_2 with probability $(1 - x)$ and player B will select strategy B_2 with probability $(1 - y)$. The expectation, therefore, is written as under:

$$E(x, y) = axy + bx(1 - y) + cy(1 - x) + d(1 - x)(1 - y).$$

If the above expectation is to be maximized, then

$$\frac{\partial}{\partial x} E(x, y) = 0 \text{ and } \frac{\partial}{\partial y} E(x, y) = 0$$

$$\therefore \quad \frac{\partial E}{\partial x} = 0 \Rightarrow ay + b(1-y) - cy - d(1-y) = 0,$$

i.e.,

$$(a-b-c+d)y = (d-b) \text{ or } y = \frac{(d-b)}{[(a-b)-(c-d)]}$$

and

$$\frac{\partial E}{\partial y} = 0 \Rightarrow ax - bx + c(1-x) - d(1-x) = 0,$$

i.e.,

$$(a-b-c+d)x = (d-c) \text{ or } x = \frac{(d-c)}{[(a-b)-(c-d)]}$$

Substituting the values of x, $(1 - x)$, y and $(1 - y)$ in the expression of expected value, the value of game is found to be

$$v = \frac{(ad - bc)}{[(a-b)-(c-d)]}$$

Example 19.17. *For the game with the following pay-off matrix, determine the optimum strategies and the value of game:*

		Player P_2	
		1	2
Player P_1	1	5	1
	2	3	4

[M.Com. (Nov.), Madurai Univ. 2002]

Solution. Suppose player P_1 selects his strategies 1 and 2 with probabilities p and $(1-p)$ respectively. Further, suppose player P_2 selects his strategies 1 and 2 with probabilities q and $(1-q)$, respectively. Then, player P_1's mathematical expectation is given by

$$E(p, q) = 5pq + p(1-q) + 3(1-p)q + 4(1-p)(1-q)$$
$$= 5pq - 3p - q + 4$$

If the above expectation is to be maximized, then

$$\frac{\partial}{\partial p} E(p, q) = 0 \quad \text{and} \quad \frac{\partial}{\partial q} E(p, q) = 0$$

$$\therefore \quad \frac{\partial}{\partial p} E(p, q) = 0 \Rightarrow 5q - 3 = 0 \text{ or } q = \frac{3}{5} \text{ and } 1 - q = 1 - \frac{3}{5} = \frac{2}{5}$$

$$\frac{\partial}{\partial q} E(p, q) = 0 \Rightarrow 5p - 1 = 0 \text{ or } p = \frac{1}{5} \text{ and } 1 - p = 1 - \frac{1}{5} = \frac{4}{5}$$

Substituting the values of p and q in the expression of mathematical expectation, the value of game is given by

$$v = 5 \times \frac{1}{5} \times \frac{3}{5} - 3 \times \frac{1}{5} - \frac{3}{5} + 4 = \frac{17}{5}$$

Alternative Method of Solution. For player P_1, the mathematical expectation is:

$$E(p, q) = 5pq - 3p - q + 4.$$
$$= 5\left(p - \frac{1}{5}\right)\left(q - \frac{3}{5}\right) + \frac{17}{5}$$

If player P_1 chooses $p = \frac{1}{5}$, he ensures that his expectation is at least $\frac{17}{5}$. He can not be sure of more than $\frac{17}{5}$, because by choosing $q = \frac{3}{5}$, player P_2 can keep $E(p, q)$ down to $\frac{17}{5}$. Therefore, player P_1 might as well settle for $\frac{17}{5}$ and player P_2 reconcile to $\frac{17}{5}$. Hence, the optimum strategies for players P_1 and P_2 are:

$$S_{P_1} = \begin{bmatrix} 1 & 2 \\ 1/5 & 4/5 \end{bmatrix}, \quad S_{P_2} = \begin{bmatrix} 1 & 2 \\ 3/5 & 2/5 \end{bmatrix}, \text{ and the value of game is } v = \frac{17}{5}$$

3. Graphical Method for *m* × 2 and 2 × *n* Games. The graphical method can be used to solve games where one of the players has only two alternatives (strategies) and the other had two or more alternatives, *i.e.*, pay-off matrix happens to be ***rectangular*** one, say $m \times 2$ or $2 \times n$.

Consider the following $2 \times n$ game:

$$\text{Player } A \quad \begin{array}{c} \\ A_1 \\ A_2 \end{array} \begin{array}{c} \text{Player } B \\ \begin{array}{cccc} B_1 & B_2 & \cdots & B_n \end{array} \\ \begin{pmatrix} a_{11} & a_{12} & \cdots & a_{1n} \\ a_{21} & a_{22} & \cdots & a_{2n} \end{pmatrix} \end{array}$$

It is assumed that the game does not have a saddle point. Let the optimum mixed strategy for player A be given by $S_A = \begin{bmatrix} A_1 & A_2 \\ p & 1-p \end{bmatrix}$. The average (expected) pay-off for player A, when he plays strategy A_1 against B's pure moves (strategies) $B_1, B_2, \ldots B_n$ are:

B's pure move	*A's expected pay-off E (p)*
B_1	$E_1(p) = a_{11}p + a_{21}(1-p) = (a_{11} - a_{21})p + a_{21}$
B_2	$E_2(p) = a_{12}p + a_{22}(1-p) = (a_{12} - a_{22})p + a_{22}$
$\vdots$	$\vdots$
B_n	$E_n(p) = a_{1n}p + a_{2n}(1-p) = (a_{1n} - a_{2n})p + a_{2n}$

According to maximin criterion for game with mixed strategy, player A should select the values of p and $(1 - p)$ so as to maximize his minimum expected pay-offs (gains). This may be done by plotting the following expected pay-off lines as function of p:

$$E_j(p) = (a_{1j} - a_{2j})p + a_{2j}, \qquad j = 1, 2, \ldots, n.$$

To plot the expected pay-off lines, we draw two parallel vertical lines one unit apart and mark a scale on each. These vertical lines represents the two strategies available to player A. Then, we draw lines to represent each of player B's strategies. To represent player B's first strategy, *i.e.*, B_1, we join a_{11} on scale I to a_{21} on scale II. This straight line will represent the expected pay-off of the line $E_1(p) = (a_{11} - a_{21})p + a_{21}$ with $p\,(= 0)$ on axis I and $p\,(= 1)$ on axis II. Similarly, other pay-off lines can be drawn. Common region of the pay-off lines represents the entire feasible area of expected pay-offs when p varies between 0 and 1. Player A will try to look for a pay-off on the ***upper boundary*** of this feasible area and player B will try to establish the pay-off at the lower boundary of the feasible area.

The highest point on the ***lower boundary***, no matter what player B does, the pay-off to player A is the maximum. However, it is also equal to the minimum pay-off, *i.e.*, the common point of the player B's strategies is the lowest on the upper boundary and the highest on the lower boundary. The two strategies of player B which passes through the ***maximin point*** identify the two ***critical moves*** of player B and yield the 2×2 pay-off matrix. This reduced pay-off matrix can be used to determine the optimum strategies of two players for the original game using the method of solving 2×2 games.

Example 19.18. *Use graphical method in solving the following game:*

		Player A			
		A_1	A_2	A_3	A_4
Player B	B_1	2	2	3	–2
	B_2	4	3	2	6

[BE (Dec.), Karnataka Univ. 2011; MBA (Dec.), Panjab Tech. Univ. 2010]

Solution. Since, maximin value ≠ minimax value in the above 2×4 game, there is no saddle point. Assume that player A plays strategy A_1 all the time. Then, the value for player B depends on what strategy he chooses. If player B plays strategy B_1 all the time, the value of the game will be 2 and if player B plays strategy B_2 all the time, the value of game will be 4. In case the player B mixes his strategies with a probability p and $(1 - p)$, then the expected value of the game will be:

$$E_1(p) = 2p + 4(1 - p) = -2p + 4$$

Similarly, if player A plays strategies A_2, A_3 and A_4 respectively all the time, the expected value of the game to player B will be:

$$E_2(p) = 2p + 3(1-p) = -p + 3$$
$$E_3(p) = 3p + 2(1-p) = p + 2$$
$$E_4(p) = -2p + 6(1-p) = -8p + 6$$

These expected pay-off values can be represented graphically by straight lines as shown in *Fig. 19.4*.

Two vertical lines one unit apart represent the strategies available to player B. Player B determines the expected pay-off for each alternative strategy available to him. If player A selects strategy A_1, player B will gain 2 units by selecting strategy B_1 and 4 units by selecting B_2. The value 4 is plotted along the axis I and value 2 along axis II. A straight line joining the two points is then drawn to represent E_1. Similarly straight lines E_2, E_3 and E_4 are drawn.

Since, player B wishes to maximize his minimum expected pay-off, we consider the highest point of intersection 'E' on the ***lower envelope*** of player B's expected pay-off equations. This point 'E' represents the maximin expected value of the game to player B. The straight lines E_3 and E_4 passing through E, define the two relevant moves A_3 and A_4 needs to play. Thus, the solution to the given 2 × 4 problem reduces to the solution of the similar game with 2 × 2 pay-off matrix:

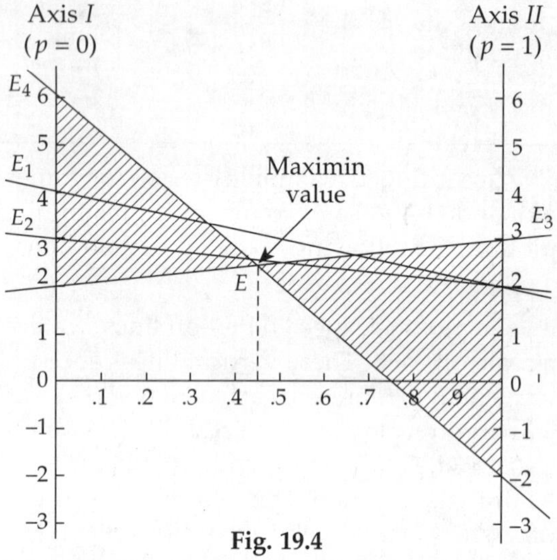

Fig. 19.4

	A_3	A_4
B_1	3	–2
B_2	2	6

The expected gain of player B is obtained by equating $E_3(p)$ and $E_4(p)$. Thus,

$$p + 2 = -8p + 6 \;\Rightarrow\; 9p = 4 \;\text{ or }\; p = \frac{4}{9} \;\text{ and }\; 1 - p = 1 - \frac{4}{9} = \frac{5}{9}$$

Further, expected loss to player A for selecting strategies A_3 and A_4 with probabilities q and $(1-q)$ are given by the following equations:

$$E_1(q) = 3q + (-2)(1-q) = 5q - 2$$

and
$$E_2(q) = 2q + 6(1-q) = 6 - 4q$$

By equating the two equations, we get,

$$5q - 2 = 6 - 4q \;\Rightarrow\; 9q = 8 \;\text{ or }\; q = \frac{8}{9} \;\text{ and }\; 1 - q = 1 - \frac{8}{9} = \frac{1}{9}$$

The value of game is given by

$$E(p, q) = 3pq - 2p(1-q) + 2(1-p)q + 6(1-p)(1-q)$$
$$= 9pq - 8p - 4q + 6 = 9\left(p - \frac{4}{9}\right)\left(q - \frac{8}{9}\right) + \frac{22}{9}$$

If player B chooses $p = \frac{4}{9}$, he ensures that his expectation is at least $\frac{22}{9}$. He, can not be sure of more than $\frac{22}{9}$, because by choosing $q = \frac{8}{9}$, player A can keep the expectation $E(p, q)$ down to $\frac{22}{9}$.

Hence, the optimum strategies of players are:

$$S_B = \begin{bmatrix} B_1 & B_2 \\ 4/9 & 5/9 \end{bmatrix} \text{ and } S_A = \begin{bmatrix} A_1 & A_2 & A_3 \\ 0 & 8/9 & 1/9 \end{bmatrix},$$

and value of game is given by $v = \frac{22}{9}$.

Dominance Property

Sometimes, it is observed that one of the pure strategies of either player is always ***inferior*** to at least one of the remaining ones. The superior strategies are said to ***dominate*** the inferior ones. Clearly, a player would have no incentive to use inferior strategies which are dominated by the superior ones. In such cases of dominance, we can reduce the size of the pay-off matrix by deleting those strategies which are dominated by the others. Thus, if each element in one row, say k^{th} of the pay-off matrix (a_{ij}) is less than or equal to the corresponding elements in some other row, say r^{th}, then player A will never choose k^{th} strategy.

In such a case the kth strategy is said to dominated by the rth one.

General rules for dominance are:

(*a*) If all the elements of a row, say k^{th}, are less than or equal to the corresponding elements of any other row, say r^{th}, then k^{th} row is ***dominated*** by r^{th} row.

(*b*) If all the elements of a column, say k^{th} are greater than or equal to the corresponding elements of any other column, say r^{th}, then k^{th} column is dominated by the rth column.

(*c*) Dominated rows and/or columns may be deleted to reduce the size of pay-off matrix, as the optimal strategies will remain unaffected.

The ***Modified Dominance Property.*** The dominance property is not always based on the superiority of pure strategies only. A given strategy can also be said to be dominated if it is inferior to an average of two or more other pure strategies. More generally, if some convex linear combination of some rows dominates the ith row, then i^{th} row will be deleted. Similar arguments follow for columns.

Example 19.19. *In a small town, there are two discount stores ABC and XYZ. They are the only stores that handle sundry goods. The total number of customers is equally divided between the two, because price and quality of goods sold are equal. Both stores have good reputation in the community, and they render equally good customer services. Assume that a gain of customers by ABC is a loss of XYZ, and vice-versa.*

Both stores plan to run annual pre-Diwali sales during the first week of October. Sales are advertised through the local newspaper, radio and television media. With the aid of an advertising firm, ABC store constructed the game matrix given below (Figures in the matrix represent a gain or loss of customer). Find the optimum strategies for both stores and value of the game:

Store ABC	*Store XYZ*		
	Newspaper	*Radio*	*Television*
Newspaper	30	40	–80
Radio	0	15	–20
Television	90	20	50

[MBA (Nov.), Delhi Univ. 2010]

Solution. From the pay-off matrix of the problem, we have

$$\underline{v} = \text{max. \{row minima\}} = \text{max. } \{-80, -20, 20\} = 20$$

and $$\overline{v} = \text{min. \{column maxima\}} = \text{min. } \{90, 40, 50\} = 40$$

Since, $\underline{v} \neq \overline{v}$, there is no saddle point and therefore, strategies of both the stores are mixed one.

If store *XYZ* chooses the media of advertisement as Television, he may gain 80, while if he chooses Newspaper instead of Television, then he is sure of not gaining anything. Thus, store *XYZ* will never choose Newspaper as its strategy. In other words first column is dominated by third column or third column dominates first column yielding the reduced size of the pay-off matrix. (It may be noted that all the elements of first column are greater than the corresponding elements of third column). After deleting the first column, we get the following 3× 2 pay-off matrix:

		Store XYZ	
		Radio	Television
Store ABC	Newspaper	40	–80
	Radio	15	–20
	Television	20	50

In the reduced pay-off matrix, we observe that store *ABC* will never choose Radio as its strategy, because by doing so he may lose 20, while by choosing strategy of Television, he is sure to gain at least 20. Thus, store *ABC* will never choose Radio (second row). In other words second row is dominated by third row. (Here, all the elements of second row are less than the corresponding elements of third row). After, deleting the second row, the reduced 2 × 2 pay-off matrix is:

		Store XYZ	
		Radio	Television
Store ABC	Newspaper	40	–80
	Television	20	50

Now, let the store *ABC* chooses strategy of newspaper with probability p and that of Television with probability $(1 - p)$, while store *XYZ* chooses strategy of Radio with probability q and that of Television with probability $(1 - q)$. Thus, the expected value of game is given by

$$E(p, q) = 40pq + 20(1-p)q + (-80)p(1-q) + 50(1-p)(1-q)$$

$$= 150\left(p - \frac{1}{5}\right)\left(q - \frac{13}{15}\right) + 24$$

If store *ABC* chooses $p = 1/5$, he ensures that his expectation is at least 24. He can not be sure of more than 24, because by choosing $q = 13/15$, store *XYZ* can keep the expectation $E(p, q)$ down to 24.

Hence, the optimum strategies of the stores are:

$$S_{ABC} = \begin{bmatrix} N.P & Radio & TV \\ 1/5 & 0 & 4/5 \end{bmatrix}, \quad S_{XYZ} = \begin{bmatrix} N.P. & Radio & TV \\ 0 & 13/15 & 2/5 \end{bmatrix}$$

and the value of game is given by $v = 24$.

PROBLEMS

19.35. Determine which of the following two-person zero-sum games are strictly determinable and fair. Give optimum strategies for each player in the case of strictly determinable games:

(*a*)

	Player B	
Player A	5	0
	0	2

(*b*)

	Player B	
Player A	1	1
	4	–3

(c)

		Player B	
		B_1	B_2
Player A	A_1	2	6
	A_2	–2	λ

[BBA, Madras Univ. 2012]

(d)

		Player B	
		B_1	B_2
Player A	A_1	0	2
	A_2	–1	4

19.36. Find the range of values of p and q which will make the entry (2, 2), a saddle point for the game:

	Player B		
	2	4	5
Player A	10	7	q
	4	p	8

[***Hint:*** Maximin value $(\underline{v}) = 7$ and minimax value $(\overline{v}) = 7$, when p and q are not considered. Therefore, $\underline{v} = \overline{v} = 7$, if $p \leq 7$ and $q \geq 7$.]

19.37. For the game with pay-off matrix:

$$\text{Player } A$$
$$\text{Player } B \begin{bmatrix} -1 & 2 & -2 \\ 6 & 4 & -6 \end{bmatrix},$$

determine the best strategies for players A and B, and also the values of the game for them. Is this game (*i*) fair? (*ii*) strictly determinable?

19.38. Solve the following 2-person zero-sum games:

(*a*)

	Player B		
	–3	1	2
Player A	1	2	1
	1	0	–2

[MBA, Panjab Tech. Univ. 2012]

(*b*)

	Player B		
	–2	12	–4
Player A	1	4	8
	–5	2	3

[MBA, Delhi Univ. 2011]

(c)

		Player B		
		B_1	B_2	B_3
Player A	A_1	8	4	1
	A_2	10	6	2
	A_3	12	8	3
	A_4	4	4	4

[MBA, Gujarat Univ. 2009]

(*d*)

		Player B			
		C_1	C_2	C_3	C_4
Player A	V_1	20	15	12	35
	V_2	25	14	5	10
	V_3	40	2	8	5
	V_4	–5	4	0	0

[PG Dip. in Glob. Bus. Oper., Delhi Univ. 2011]

19.39. Two competitive manufacturers are producing a new toy under licence from a patent holder. In order to meet the demand, they have the option of running the plant for 8, 16 or 24 hours a day. As the length of production increases, so does the cost. One of the manufacturers say *A*, has set up the matrix given below, in which he estimates the percentage of the market that he could capture and maintain the different production schedule:

Manufacturer A	*Manufacturer B*		
	C_1 : *8 hrs.*	C_2 : *16 hrs.*	C_3 : *24 hrs.*
S_1 : 8 hrs.	60%	56%	34%
S_2 : 16 hrs.	63%	60%	55%
S_3 : 24 hrs.	83%	72%	60%

(*i*) At which level should each manufacturer produce?
(*ii*) What percentage of the market will manufacturer *B* have?

[PG Dip. in Glob. Bus. Oper., Delhi Univ. 2011]

19.40. For the game with following pay-off matrix, determine the optimum strategies and value of the game:

(*a*)

	B_1	B_1
A_1	2	–1
A_2	–1	0

(*b*)

	B_1	B_1
A_1	5	1
A_2	3	4

[M.Com. (Nov.), Madurai Univ. 2002]

19.41. Consider a "modified" form of "matching biased coins" game problem. The matching player is paid ₹ 8.00, if two coins turn both heads and ₹ 1.00 if the coins turn both tails. The non-matching player is paid ₹ 3.00, when the two coins do not match. Given the choice of being the matching or non-matching player, which one would you choose and what would be your strategy?

[MBA, Delhi Univ. 2007]

[***Hint:*** The pay-off matrix for the matching player is: $\begin{array}{c} \\ H \\ T \end{array}\begin{array}{c} \begin{matrix} H & \;\; T \end{matrix} \\ \begin{bmatrix} 8 & -3 \\ -3 & 1 \end{bmatrix} \end{array}$.

19.42. In a game of matching coins with two players, suppose *A* wins one unit of value, when there are two heads, wins nothing when there are two tails and loses 1/2 unit of value when there are one head and one tail. Determine the pay-off matrix, the best strategies for each player and the value of game to *A*.

19.43. *A* and *B* play game in which each has three coins, a 5 p., 10 p. and a 20 p. Each selects a coin without the knowledge of the other's choice. If the sum of coins is an odd amount, *A* wins *B*'s coin. If the sum is even, *B* wins *A*'s coin. Find the best strategy for each player and the value of game.

[PGDM, Mumbai Univ. 2012; B.Tech. (Comp. Sc.) (Dec.), Hyderabad Univ. 2011]

[***Hint:*** *A*'s pay-off matrix is: $\begin{array}{c} \\ A_1 \\ A_2 \\ A_3 \end{array}\begin{array}{c} \begin{matrix} B_1 & \;\; B_2 & \;\; B_3 \end{matrix} \\ \begin{bmatrix} -5 & 10 & 20 \\ 5 & -10 & -10 \\ 5 & -20 & -20 \end{bmatrix} \end{array}$.

19.44. Two players *A* and *B* match coins. If coins match, then player *A* wins two units of value, and if the coins do not match, then player *B* wins 2 units of value. Determine the optimum strategies for the players and value of the game.

[M.Com., Delhi Univ. 2008; MBA (Nov.), Madras Univ. 2006]

19.45. A total of 90,000 customers frequent the ***Huge-Bazaar*** and ***Alliance-Mart Supermarkets***. To induce customers to enter, each store gives away a free item. Each week, the give away item is announced in the Monday newspaper. Of course, neither store knows which item the other store will choose to give away this week. ***Huge-Bazaar*** is considering giving away a 2-litre bottle of soft drink or a 100 gm packet of chocolate. ***Alliance-Mart*** is considering giving away 500 gm pack of butter or 1 litre orange juice. For each possible choice of items, the number of customers who will stop at ***Huge-Bazaar*** store during the current week is shown in the table given below. Each store wants to maximize its expected number of customers during the current week. Determine an optimal strategy for each store and value of the game.

Huge-Bazaar	*Alliance-Mart*	
	Butter	*Juice*
Soft drink	40,000	50,000
Chocolate	60,000	30,000

[MBA, Delhi Univ. 2012]

19.46. Use graphical method in solving the following games:

(*a*)

	B_1	B_2	B_3
A_1	6	–3	7
A_2	–3	0	4

[MBA, Jammu Univ. 2008]

(*b*)

	A_1	A_2	A_3	A_4
B_1	2	2	3	–2
B_2	4	3	2	6

[MBA (Dec.), Panjab Tech. Univ. 2010]

(*c*)

	B_1	B_2	B_3
A_1	1	3	11
A_2	8	5	2

[BE (Dec.), Karnataka Univ. 2010; B.Tech. Hyderabad Univ. 2011]

(*d*)

	B_1	B_2	B_3	B_4	B_5
A_1	0	4	–8	5	1
A_2	1	5	8	–4	0

[B.Tech., Panjab Tech Univ. 2012]

19.47. A soft drink company calculated the market share of two products against its major competitor having three products and found out the impact of additional advertisement in any one of its products against the other:

$$\begin{array}{cc} & \textit{Competitor B} \\ & \begin{array}{ccc} B_1 & B_2 & B_3 \end{array} \\ \textit{Company A} \begin{array}{c} A_1 \\ A_2 \end{array} & \begin{bmatrix} 6 & 7 & 15 \\ 20 & 12 & 10 \end{bmatrix} \end{array}$$

What is the best strategy for the company as well as the competitor? What is the pay-off obtained by the company and the competitor in the long run? Use graphical method to obtain the solution.

19.48. Use the notion of dominance to simplify the rectangular game with the following pay-off:

$$\begin{array}{cc} & \textit{Player A} \\ & \begin{array}{cccc} I & II & III & IV \end{array} \\ \textit{Player B} \begin{array}{c} 1 \\ 2 \\ 3 \\ 4 \end{array} & \begin{bmatrix} 18 & 4 & 6 & 4 \\ 6 & 2 & 13 & 7 \\ 11 & 5 & 17 & 3 \\ 7 & 6 & 12 & 2 \end{bmatrix} \end{array}$$

Find its solution using graphical method. ***[MBA, Madras Univ. 2000; MIB, Delhi Univ. 2011]***

19.49. Two firms A and B make LED and LCD television sets. Firm A can make either 150 LED sets in a week or an equal number of LCD sets, and make a profit of ₹ 400 per LED set and ₹ 300 per LCD set. Firm B, on the other hand, make either 300 LED sets, or 150 LED and 150 LCD sets, or 300 LCD sets per week. It also has the same profit margin on the two sets as of firm A. Each week there is a market of 150 LED sets and 300 LCD sets and the manufacturers would share market in the proportion in which they manufacture a particular type of set.

Write the pay-off matrix of firm A per week. Obtain graphically A's and B's optimum strategies and value of the game. *[MBA (Nov.), Delhi Univ. 2009]*

[***Hint:*** Let the strategies open to firms A and B are as follows:

A_1 : 150 LED sets, A_2 : 150 LCD sets;

B_1 : 300 LED sets, B_2 : 150 LED sets and 150 LCD sets, B_3 : 300 LCD sets.

Then, for A_1B_1 profit of firm $A = \left(\frac{150}{150+300}\right) \times 150 \times 400 = 20{,}000$ approx; wherein 150/(150 + 300) is the market share for firm A, 150 is the total market for LED sets and ₹ 400 is the profit per set.

Similarly,

$$A_1B_2 : \left(\frac{150}{150+150}\right) \times 150 \times 400 = 30{,}000, \quad A_1B_3 : \left(\frac{150}{150+0}\right) \times 150 \times 400 = 60{,}000,$$

$$A_2B_1 : \left(\frac{150}{150+0}\right) \times 300 \times 300 = 90{,}000, \quad A_2B_2 : \left(\frac{150}{150+150}\right) \times 300 \times 300 = 45{,}000,$$

and $$A_2B_3 : \left(\frac{150}{150+300}\right) \times 300 \times 300 = 30{,}000]$$

19.50. Two firms are competing for business under the similar condition, so that one firm's gain is another firm's loss. Firm A's pay-off matrix is given below:

		Firm B		
		No ad	*Medium ad*	*Heavy ad*
	No advertising	10	5	−2
Firm A	Medium advertising	13	12	15
	Heavy advertising	16	14	10

Suggest the optimum strategies for the two firms and net outcome thereof.

[MIB, Delhi Univ. 2011; MBA, Delhi Univ. 2009]

19.51. Find optimal strategies for two companies competing with each other on the basis of the following pay-off matrix showing gain or loss of customers:

Company A	*Company B*		
	Quantity discounts	*CSR initiatives*	*Event sponsorship*
Quantity discounts	20	60	30
CSR initiative	−10	30	25
Event sponsorship	40	50	−30

(*i*) State the required assumptions for finding optimal strategies of two companies,

(*ii*) Find out value of the game. Use dominance rule. *[M.Com., Delhi Univ. 2010]*

19.52. Use dominance property in solving the following 3 × 3 game:

Player B

		B_1	B_2	B_3
	A_1	3	–2	4
Player A	A_2	–1	4	2
	A_3	2	–2	6

[MBA, Madurai Univ. 2010]

19.53. In an election campaign, the strategies adopted by the ruling and opposition party alongwith pay-offs (ruling party's % share in votes polled) are given below:

Ruling party's strategies	*Opposition party's strategies*		
	Campaign one day in each city	*Campaign two days in large towns*	*Spend two days in large rural sectors*
Campaign one day in each city	55	40	35
Campaign two days in large towns	70	70	55
Spend two days in large rural sectors	75	55	65

Assume a zero-sum game. Find optimum strategies for both parties and expected pay-off to ruling party. *[MBA, Kurukshetra Univ. 2007]*

19.54. Two breakfast food manufacturers, *ABC* and *XYZ* are competing for an increased market share. The pay-off matrix, shown in the following table, describes the increase in market share of *ABC* and decrease in market share of *XYZ*:

ABC	*XYZ*			
	Give coupons	*Decrease price*	*Maintain present strategy*	*Increase advertising*
Give coupons	2	–2	4	1
Decrease price	6	1	12	3
Maintain present strategy	–3	2	0	6
Increase advertising	2	–3	7	7

Determine the optimum strategies for both the manufacturers and value of the game.

[MBA (Oct.), Delhi Univ. 2010; PG Dip. in Glob. Bus. Oper., Delhi Univ. 2010]

19.55. Is the following two person zero-sum game stable? (The pay-off is for player *A*). Solve the following game:

Player B

$$\textit{Player A}\quad\begin{bmatrix} 5 & -10 & 9 & 0 \\ 6 & 7 & 8 & 1 \\ 8 & 7 & 15 & 2 \\ 3 & 4 & -1 & 4 \end{bmatrix}$$

[MBA (Nov.), Annamalai Univ. 2010]

19.56. The following pay-off matrix describes the increase in sales of Firm *A* from each of its strategies A_1, A_2, A_3 and A_4 against each of the strategies B_1, B_2, B_3 and B_4 of the rival Firm *B*.

Firm B

		B_1	B_2	B_3	B_4
	A_1	43	73	33	13
Firm A	A_2	38	28	23	8
	A_3	48	58	8	18
	A_4	63	68	18	23

Assuming a zero-sum game, show that no optimal strategy-mix exists in pure strategies irrespective of whether we use the principle of dominance or that of maximin and minimax. The game, however, has a solution in mixed strategies but not without the aid of the principle of dominance. Solve the game and find its value. *[M.Com., Delhi Univ. 2007]*

19.57. A steel company is negotiating with its union for revision of wages to its employees. The management with the help of a mediator has prepared a pay-off matrix shown below. Plus sign represents wage increase, while negative sign stands for wage decrease. Union has also constructed a table which is comparable to that developed by management. The management does not have the specific knowledge of game theory to select the best strategy or strategies for the firm. You have to assist the management on the problem. What game value and strategies are available to the opposing group?

Steel company strategies	*Additional costs to steel company (₹)* *Union strategies*			
	U_1	U_2	U_3	U_4
C_1	0.25	0.27	0.35	−0.02
C_2	0.20	0.16	0.08	0.08
C_3	0.14	0.12	0.15	0.13
C_4	0.30	0.14	0.19	0.00

[MBA (PT.) (Dec.), Delhi Univ. 2015]

19.58. (*a*) Find the value of game having following pay-off matrix using the principle of dominance:

		Player B			
		I	*II*	*III*	*IV*
Player A	*I*	3	5	4	2
	II	5	6	2	4
	III	2	1	4	0
	IV	3	3	5	2

[PG Dipl., Indian Inst. of Mat. Man. 2019]

(*b*) Following is the pay-off matrix for player *A*:

		Player B				
		I	*II*	*III*	*IV*	*V*
Player A	*I*	2	4	3	8	4
	II	5	6	3	7	8
	III	6	7	9	8	7
	IV	4	2	8	4	3

Using dominance property, obtain optimal strategies for both the players and determine value of game. *[PG Dip. in Glob. Bus. Oper., Delhi Univ. 2011]*

19.59. (*a*) Solve the following game using dominance principle:

		Firm B			
		B_1	B_2	B_3	B_4
	A_1	35	65	25	5
Firm A	A_2	30	20	15	0
	A_3	40	50	0	10
	A_4	55	60	10	15

[MBA, Bangalore Univ. 2017]

(*b*) Given below is the pay-off matrix for player *A*; using dominance solve the two-person zero-sum game:

		Player B			
		I	*II*	*III*	*IV*
	I	3	2	4	0
Player A	*II*	3	4	2	4
	III	4	2	4	0
	IV	0	4	0	8

[MCA (Feb.), Hyderabad Univ. 2012; BE, Karnataka Univ. 2012; B.Tech. (Mech.), Panjab Tech. Univ. 2010]

[***Hint:*** Clearly, first row is dominated by third row, and then first column is dominated by third column. In the reduced 3 × 3 pay-off matrix, none of the pure strategies of player *B* is inferior to any of his other strategies. However, an average of strategies *III* and *IV* of player *B* dominates his strategy *II*, thereby reduces the 3 × 2 pay-off matrix as $\begin{bmatrix} 2 & 4 \\ 4 & 0 \\ 0 & 8 \end{bmatrix}$.
Again, since none of the pure strategies of player *A* is inferior to any of his other strategies, average of his strategies *III* and *IV* dominates strategy *II* of player *A*. Thus, the reduced pay-off matrix is a 2 × 2 pay-off matrix:

	III	*IV*
III	4	0
IV	0	8

.]

19.60. Solve the following two-person zero-sum game:

	B_1	B_2	B_3	B_4	B_5
A_1	8	12	14	10	11
A_2	9	11	15	10	13
A_3	7	8	6	11	12
A_4	10	9	7	9	9
A_5	12	13	10	12	10

[MBA, Madras Univ. 2010]

[***Hint:*** A_4 is dominated by A_5, then B_2 by B_1, B_4 by B_1, A_3 by A_2, B_3 by B_5 and A_1 by A_2.]

19.61. Two breakfast-food manufacturing firms, *A* and *B*, are competing for an increased market share. To improve its market share, both the firms decides to launch the following strategies:

A_1, B_1 : Give coupons; A_2, B_2 : Decrease price;
A_3, B_3 : Maintain present strategy; A_4, B_4 : Increase advertising.

The pay-off matrix, shown in the following table, describes the increase in market share for firm *A*, and decrease in the market share for firm *B*:

Firm A	*Firm B*			
	B_1	B_2	B_3	B_4
A_1	35	65	25	5
A_2	30	20	15	0
A_3	40	50	0	10
A_4	55	60	10	15

Determine the optimal strategies for each firm and the value of game. ***[MBA, Delhi Univ. 2010]***

19.62. Two firms *A* and *B* have for years been selling a competitive product which forms a part of both firms total sales. The marketing executive of firm '*A*' raised the question. "What should be the firm's strategies in terms of advertising for the product in question?" The market research team of firm *A* developed the following data for varying degrees of advertising:

(*i*) No advertising, medium advertising, and large advertising for both firms will result in equal market shares.

(*ii*) Firm *A* with no advertising: 40% of the market share with medium advertising by firm *B* and 28% of the market share with large advertising by firm *B*.

(*iii*) Firm *A* using medium advertising: 70% of market share with no advertising by firm *B* and 45% of the market share with large advertising by firm *B*.

(*iv*) Firm *A* using large advertising: 75% of the market share with no advertising by firm *B* and 47.5% of the market share with medium advertising by firm *B*.

(*a*) Based upon the foregoing information, answer the marketing executive's questions.

(*b*) What advertising policy should firm *A* pursue, when consideration is given to the above factors: selling price of product is ₹ 4.00 per unit; variable cost of product is ₹ 2.50 per unit; annual volume of 30,000 units for firm *A*; cost of annual medium advertising ₹ 5,000 and cost of annual large advertising ₹ 15,000? What contribution, before other fixed costs, is available to the firm? ***[MBA, Delhi Univ. 2000]***

[***Hint:*** The pay-off matrix of the game between firms *A* and *B* is:

Firm A	*Firm B*		
	No advertising	*Medium advertising*	*Large advertising*
No advertising	50	40	28
Medium advertising	70	50	45
Large advertising	75	47.5	50

19:12. ARITHMETIC METHOD FOR *n* × *n* GAME

When a game matrix cannot be reduced in size by graphical method or by using the principal of dominance, then for $n \times n$ games an easy method for finding the optimum strategies of the two players consists of the following steps:

Step 1. Let $\mathbf{A} = (a_{ij})$ be an $n \times n$ pay-off matrix. Obtain a new matrix **C**, whose first column is obtained from **A** by subtracting its 2nd column from 1st; second column is obtained by subtracting **A**'s 3rd column from 2nd and so on till the last column of **A** has been taken care of. Thus, **C** is an $n \times (n-1)$ matrix.

Step 2. Obtain a new matrix **R**, from **A**, by subtracting its successive rows from the preceding ones, in exactly the same manner as was done for columns in *Step 1*. Thus, **R** is an $(n-1) \times n$ matrix.

Step 3. Determine the magnitude of ***oddments*** corresponding to each row and each column of **A**. The oddment corresponding to i^{th} row of **A** is defined as the determinant $|\mathbf{C}_i|$, where $\mathbf{C}_i$ is obtained from **C** by deleting its ith row.

Similarly, oddment (j^{th} column of **A**) = $|\mathbf{R}_j|$, where $\mathbf{R}_j$ is obtained from **R** by deleting its jth column.

Step 4. Write the magnitude of oddments (after ignoring negative signs, if any, against their respective rows and columns).

Step 5. Check whether the sum of row oddments is equal to the sum of column oddments. If so, the oddments expressed as fractions of the grand total yield the optimum strategies. If not, the method fails.

Step 6. Calculate the expected value of game corresponding to the optimum mixed strategy determined above for the Row Player (against any move of the column player).

Example 19.20. *Consider the following 2 × 2 game:*

	C_1	C_2
R_1	4	7
R_2	6	5

(*a*) *Does it have a saddle point?*

(*b*) *Is it correct to state that the value of game G, will satisfy $5 < G < 6$?*

(*c*) *Determine the frequency of optimum strategies by matrix oddment method and find the value of game.*

Solution.

(*a*) From the given pay-off matrix, we have maximin value = 5 and minimax value = 6. Since maximin value ≠ minimax value, there does not exist saddle point.

(*b*) Since, maximin value ≤ G ≤ minimax value, we must have $5 \le G \le 6$.

(*c*) Subtracting the second column from first, we get

$$\mathbf{C} = \begin{bmatrix} -3 \\ 1 \end{bmatrix}$$

This gives $|\mathbf{C}_1| = 1$ and $|\mathbf{C}_2| = 3$

Now, subtracting the second row from first, we get

$$\mathbf{R} = (-2,\ 2)$$

Thus $|\mathbf{R}_1| = 2$ and $|\mathbf{R}_2| = 2$

∴ The augmented pay-off matrix is:

			|Row oddment|
	4	7	1
	6	5	3
|Column oddment|	2	2	4

Since, the sums of row oddments and column oddments each total 4, the optimum strategies of two players are:

$$\text{Row player} : \left(\frac{1}{4},\ \frac{3}{4}\right) \text{ and Column player} : \left(\frac{1}{2},\ \frac{1}{2}\right)$$

The value of the game is:

$$G = \frac{1}{4} \times 4 + \frac{3}{4} \times 6 = \frac{11}{2} \text{ or } 5.5$$

Example 19.21. *Use matrix oddment method to solve the following 3 × 3 game:*

$$\begin{array}{|ccc} 0 & 1 & 2 \\ 2 & 0 & 1 \\ 1 & 2 & 0 \end{array}$$

Solution. We compute the matrices **C** and **R** from the given pay-off matrix; and then obtain the row and column oddments. Thus, we have

$$\mathbf{C} = \begin{bmatrix} -1 & -1 \\ 2 & -1 \\ -1 & 2 \end{bmatrix} \text{ and } \mathbf{R} = \begin{bmatrix} -2 & 1 & 1 \\ 1 & -2 & 1 \end{bmatrix}$$

The oddments are:

$$\mathbf{C}_1 = \begin{vmatrix} 2 & -1 \\ -1 & 2 \end{vmatrix} = 3, \quad \mathbf{C}_2 = \begin{vmatrix} -1 & -1 \\ -1 & 2 \end{vmatrix} = -3, \quad \mathbf{C}_3 = \begin{vmatrix} -1 & -1 \\ 2 & -1 \end{vmatrix} = 3$$

and
$$\mathbf{R}_1 = \begin{vmatrix} 1 & 1 \\ -2 & 1 \end{vmatrix} = 3, \quad \mathbf{R}_2 = \begin{vmatrix} -2 & 1 \\ 1 & 1 \end{vmatrix} = -3, \quad \mathbf{R}_3 = \begin{vmatrix} -2 & 1 \\ 1 & -2 \end{vmatrix} = 3$$

∴ The augmented pay-off matrix is:

				|Row oddments|
	0	1	2	3
	2	0	1	3
	1	2	0	3
|Column oddments|	3	3	3	9

Since, the sum of row and column oddments are both equal to 9, the optimum strategies are:

$$\text{Row player}: \left(\frac{1}{3}, \frac{1}{3}, \frac{1}{3}\right) \text{ and Column player}: \left(\frac{1}{3}, \frac{1}{3}, \frac{1}{3}\right)$$

The value of the game $= \frac{1}{3} \times 0 + \frac{1}{3} \times 2 + \frac{1}{3} \times 1 = 1$

19:13. ITERATIVE METHOD OF APPROXIMATE SOLUTION OF GAMES

Iterative method can be applied to solve 3 × 3 or higher games which cannot be solved by Arithmetic method and are extremely tedious to be solved by method of linear programming. This method gives an approximate solution for the value of the game and the true value can be determined to any degree of accuracy. Optimum strategies can also be determined but not so satisfactorily. The method assumes that each player acts under the assumption that past is the best guide for the future and will play in such a manner, so as to maximize the expected gain or to minimize the expected loss. The method is best explained through an example given below:

Example 19.22. *Solve the following game approximately:*

$$\begin{array}{cc} & \textit{Player B} \\ \textit{Player A} & \begin{pmatrix} 1 & -1 & -1 \\ -1 & -1 & 3 \\ -1 & 2 & -1 \end{pmatrix} \end{array}$$

[M.Sc. (Math.), CCS Univ. 2003]

Solution. Let the player A select the second row, being the superior one of his other strategies and place it under the matrix. Player B examines this row and chooses a column corresponding to the ***smallest*** number of this row. This is first column. First column is then placed to the right of the matrix. Player A examines this column and chooses a row corresponding to the ***largest*** number in this column. This is the first row. First row is then added to the row last chosen. Player B then chooses the column corresponding to the smallest number in the new row and adds this column to the column last chosen. In case of a tie the player will select that row or column which is different from his last choice. The procedure is repeated for a finite number of iterations. Here, ten iterations are presented with the smallest elements in each succeeding row with the largest elements in each succeeding column being encircled.

Player B

	1	–1	–1	(1)	0	(1)	0	–1	(0)	–1	(0)	–1	0	4/10
Player A	–1	–1	3	–1	–2	–3	–4	–1	–2	(1)	0	(3)	(2)	3/10
	–1	2	–1	–1	(1)	0	(2)	(1)	0	–1	–2	–3	–4	3/10
	(–1)	–1	3											
	0	(–2)	2											
	(–1)	0	1											
	0	(–1)	0											
	–1	1	(–1)											
	–2	3	(–2)											
	–1	2	(–3)											
	(–2)	1	0											
	–1	0	(–1)											
	(–2)	–1	2											
	$\frac{4}{10}$	$\frac{2}{10}$	$\frac{4}{10}$											

The approximate strategies after 10 iterations are found by dividing the number of encircled elements in each row or column by total number of iterations. Thus, we shall have

$$S_A = \begin{bmatrix} A_1 & A_2 & A_3 \\ 4/10 & 3/10 & 3/10 \end{bmatrix} \text{ and } S_B = \begin{bmatrix} B_1 & B_2 & B_3 \\ 4/10 & 2/10 & 4/10 \end{bmatrix}$$

The upper bound for the value of the game can be determined by dividing the largest element '2', in the last column by the total number of iterations. Likewise, the lower bound can be determined by dividing the smallest element '–2', in the last row by the total number of iterations.

Thus, the approximate value of game is

$$-\frac{2}{10} \le v \le \frac{2}{10}, \quad i.e., \quad (-1/5) \le v \le (1/5)$$

PROBLEMS

19.63. Use matrix odd'ment method to solve the following games:

(*a*)

$$A \overset{B}{\begin{pmatrix} 0 & 2 \\ 1 & 1 \end{pmatrix}}$$

(*b*)

$$A \overset{B}{\begin{pmatrix} 2 & 0 & 0 \\ 0 & 0 & 4 \\ 0 & 3 & 0 \end{pmatrix}}$$

[B.Tech., Hyderabad Univ. 2011]

19.64. Solve the following 3 × 3 games:

(*a*)

$$\text{Player } A \overset{\text{Player } B}{\begin{bmatrix} 7 & 1 & 7 \\ 9 & -1 & 1 \\ 5 & 7 & 6 \end{bmatrix}}$$

(*b*)

$$\text{Player } A \overset{\text{Player } B}{\begin{pmatrix} 2 & -1 & -2 \\ 1 & 0 & 1 \\ -2 & -1 & 2 \end{pmatrix}}$$

[M.Sc. (Math.), CCS Univ. 2000]

19.65. Firm *X* is fighting for its life against the determination of firm *Y* to drive it out of the industry. Firm *X* has the choice of increasing price, leaving it unchanged, or lowering it. Firm *Y* has the same three options. Firm *X*'s gross sales in the event of each of the pairs of choices are shown below:

Firm X's pricing strategies	*Firm Y's Pricing strategies* *Increase price*	*Do not change*	*Reduce price*
Increase price	90	80	110
Do not change	110	100	90
Reduce price	120	70	80

Assuming firm *X* as the maximizing one, solve the problem.

[M.Com., Delhi Univ. 2009; MBA, Osmania Univ. 2002]

19.66. Find approximate value of the game and strategies by iterative method whose pay-off matrix is given below. (Try 10 iterations):

$$\begin{array}{cc} & \text{Player } B \\ & \begin{array}{ccc} 1 & 2 & 1 \end{array} \\ \text{Player } A \begin{array}{c} 1 \\ 2 \\ 3 \end{array} & \begin{bmatrix} 2 & 0 & 1 \\ 0 & 2 & 1 \\ 3 & 0 & 0 \end{bmatrix} \end{array}$$

[MBA, Madurai Univ. 2009]

19.67. Solve the following matrix game approximately:

$$\text{Player } A \overset{\text{Player } B}{\begin{pmatrix} 2 & 3 & -1 & 0 \\ 5 & 4 & 2 & -2 \\ 1 & 3 & 8 & 2 \end{pmatrix}}$$

[B.Tech. (Comp.Sc.), Panjab Tech. Univ. 2010]

19.68. In a well-known children's game, each player plays 'stone' or 'scissors' or 'papers'. If one plays 'store' and the other 'scissors' then the former wins a rupee. Similarly, 'scissors' beats 'paper' and 'paper' beats 'stone', *i.e.*, the player calling the former word wins a rupee. If the two players name the same item, then there is a tie, *i.e.*, there is no pay-off. Write down the pay-off matrix. Find the value of the game and hence write down the optimal strategies of both players. ***[AIMA Dip. in Management (Dec.), 1995; M.Sc. (Math.), CCS Univ. 2009]***

[***Hint:*** Pay-off matrix is:

$$\begin{array}{c} \\ \textit{Player } A \\ \\ \end{array} \overset{\textit{Player } B}{\begin{pmatrix} 0 & 1 & -1 \\ -1 & 0 & 1 \\ 1 & -1 & 0 \end{pmatrix}} \begin{array}{l} \text{Stone} \\ \text{Paper} \\ \text{Scissors} \end{array}$$

Stone Paper Sciessors]

19:14. COMPETITORS AGAINST PASSIVITY

In many a situation, the opponent may be somewhat ***passive*** in character. For example, the game theory can also be applied against nature which, by itself, cannot take any decision. Only the opponent would guess what nature can thrust upon the system. Let us take a simple example:

A particular day was fixed to launch a new rocket. The probability of rain on that day was estimated to be 0.1. It was also found that being caught in rain without firing the rocket was 5 times more inconvenient as having to fire the rocket on such a bad weather.

We can easily describe the given situation in terms of the following pay-off matrix:

$$\textit{Man} \begin{array}{l} \text{Fire Rocket} \\ \text{Fire No Rocket} \end{array} \overset{\begin{array}{cc} & \textit{Nature} \\ \textit{Rain} & \textit{No Rain} \end{array}}{\begin{pmatrix} 0 & -1 \\ -5 & 0 \end{pmatrix}}$$

We see that the man's expected gain if rocket is fired is –9/10 and the expected gain if rocket is not fired is –5/10. Therefore, firing the rocket won't be a good proposition. But this proposition itself would be flexible so long as the probability of rain changes.

TEST YOUR UNDERSTANDING

1. Essential characteristics of a decision-making model are:
   (*a*) pay-off (*b*) decision alternatives
   (*c*) states-of-nature (*d*) all of the above.
2. The decision-making criterion that should be used to achieve maximum long term pay-off is
   (*a*) *EMV* (*b*) *EOL*
   (*c*) Hurwicz (*d*) minimax
3. The decision-maker's knowledge and experience may influence the decision-making process, when using the criterion of
   (*a*) realism (*b*) maximin
   (*c*) maximax (*d*) minimax regret
4. A type of decision-making environment is
   (*a*) certainty (*b*) uncertainty
   (*c*) risk (*d*) all of the above

5. The value of the coefficient of optimism 'α' is needed, while using the criteria of
(*a*) equally likely (*b*) realism
(*c*) maximin (*d*) minimax

6. The minimum expected opportunity loss (*EOL*) is equal to
(*a*) *EMV* (*b*) *EVPI*
(*c*) minimum regret (*d*) both (*b*) and (*c*).

7. The expected value of perfect information (*EVPI*) is
(*a*) equal to expected regret of optimal decision under risk.
(*b*) maximum opportunity loss.
(*c*) the utility of additional information.
(*d*) *EPPI* + maximum of *EMV*.

8. In a decision-making situation
(*a*) total number of courses of action cannot be more than the number of events.
(*b*) Laplace principle is based on the premise of equally-likely occurrence of possible events.
(*c*) minimax is an optimist's choice, while minimum is a pessimist's criterion.
(*d*) for any utility function, the units of measuring utility should always range between zero and 100.

9. In the study of decision-making under uncertainty
(*a*) the decision-maker has no idea about the possible states-of-nature and their probabilities.
(*b*) events are also called states-of-nature.
(*c*) opportunity loss in the pay-off matrix is also called conditional pay-off.
(*d*) all of the above.

10. The concept of utility is used to
(*a*) measure the utility of money (*b*) take into account aversion of risk
(*c*) find the value of *EPPI* (*d*) both (*a*) and (*b*).

11. When sum of gains of one player is equal to sum of losses to another player, the situation is known as
(*a*) zero-sum game (*b*) fair game
(*c*) conflicting game (*d*) negotiable game

12. A game is said to be fair, if
(*a*) upper value is more than lower value of the game.
(*b*) upper and lower values of the game are not equal.
(*c*) upper and lower values of the game are same and zero.
(*d*) none of the above.

13. When maximin and minimax values of the game are same, then
(*a*) there is a saddle point. (*b*) solution does not exit.
(*c*) strategies are mixed. (*d*) none of the above.

14. The pay-off value for which each player in a game always selects the same strategy is called
(*a*) equilibrium point, (*b*) saddle point,
(*c*) both (*a*) and (*b*), (*d*) none of the above.

15. Games which involve more than two players are called
(*a*) biased games, (*b*) negotiable games,
(*c*) conflicting games, (*d*) *n*-person games.

16. In a pure strategy game
(*a*) any strategy may be selected arbitrarily.
(*b*) both players select their optimal strategy.
(*c*) a particular strategy is selected by each player.
(*d*) none of the above.

17. A mixed strategy game can be solved by
(*a*) matrix method, (*b*) algebraic method,
(*c*) graphical method, (*d*) all of the above.

18. The size of the pay-off matrix of a game can be reduced by using the principle of
(*a*) dominance, (*b*) rotation reduction,
(*c*) game inversion, (*d*) game transpose.

ANSWERS

1. (*d*)	**2.** (*a*)	**3.** (*a*)	**4.** (*d*)	**5.** (*b*)	**6.** (*d*)
7. (*a*)	**8.** (*b*)	**9.** (*d*)	**10.** (*d*)	**11.** (*a*)	**12.** (*c*)
13. (*a*)	**14.** (*c*)	**15.** (*d*)	**16.** (*b*)	**17.** (*d*)	**18.** (*a*)

REVIEW EXERCISE

1. Explain clearly the term decision in a decision-making problem. What are the basic steps of a decision-making process? ***[MBA (Dec.), Panjab Tech. Univ. 2011]***

2. Explain clearly the following terms:
(*i*) Course of action, (*ii*) States-of-nature, (*iii*) Pay-off table, (*iv*) Opportunity loss.
[MBA, HP Univ. 2010]

3. (*a*) What are the different environments in which decisions are made?
[B.Tech. (Mech.), HP Univ. 2012]
(*b*) List the different steps in decision-making.
[MBA, Madras Univ. 2008; MBA, Panjab Tech. Univ. 2013]

4. (*i*) Explain decision-making under uncertainty.
[M.Com., Kerala Univ. 2012; MBA, Annamalai Univ. 2010]
(*ii*) Explain decision-making under risk. ***[MCA, Panjab Tech. Univ. 2010]***

5. Discuss various methods useful for decision-making under uncertainty.
[M.Com., Banglore Univ. 2017; MBA (Dec.), Panjab Tech. Univ. 2011]

6. With suitable illustrative examples, explain the maximin, the minimax and the regret criterion in decision-making. ***[MBA, Mumbai Univ. 2012]***

7. State the nature of Laplace decision criterion. ***[M.Com., Bangalore Univ. 2018]***

8. (*a*) Write a short note on Hurwicz criterion. Define α the 'index of optimism' connected with it. For what values of α is the criterion too optimistic and too pessimistic?
(*b*) Define 'maximin criterion'. Illustrate with an example of 'Savage minimax regret criterion'.

9. What is *EMV*? How is it computed to be used as a criterion of decision-making and when?
[MBA, MD Univ. 2010]

10. Indicate the difference between decision-making under risk and uncertainty.
[MBA (UIAMS), Panjab Univ. 2015]

11. Explain the expected value of perfect information. ***[MBA (Dec.), Panjab Tech. Univ. 2010]***

12. Explain the difference between expected opportunity loss and expected value of perfect information. *[MBA, Panjab Tech. Univ. 2009]*
13. Construct maximax and maximin decision strategies. Under what circumstances is each appropriate? *[MBA (June), Panjab Tech. Univ. 2012]*
14. Explain the difference between *EMV*, *EVPI* and *EOL*. *[MBA, Delhi Univ. 2011]*
15. What is decision tree? Explain the steps involved in drawing a decision tree. *[M.Com., Bangalore Univ. 2018]*
16. What do you understand by decision tree? Explain how it will be useful in decision-making? *[M.Com., Bharthidasan Univ. 2007; MBA, Madras Univ. 2010]*
17. What is Decision Theory? Describe the four different states of decision environment in managerial applications. *[MBA, IGNOU 2012]*
18. Explain the assumptions made in game theory. How it is helpful for business? *[MBA, Bangalore Univ. 2018]*
19. What is game theory? Discuss its importance in business decisions. Explain its advantages and disadvantages. *[MBA, Panjab Tech. Univ. 2008]*
20. What are the characteristics of a game? *[MBA (Dec.), Panjab Tech. Univ. 2011]*
21. Explain two-person zero-sum game by suitable example. *[MBA (Dec.), Panjab Tech. Univ. 2012]*
22. (*a*) Explain the following terms in game theory:
    Pure strategy, Mixed strategy, Saddle point, Pay-off matrix, and Rectangular games. *[MBA, Panjab Tech. Univ. 2010]*
    (*b*) Define mixed strategy in a game. *[MBA, Annamalai Univ. 2010]*
23. Explain the following:
    (*i*) Minimax and maximin principle,
    (*ii*) Pure and Mixed strategies. *[B.Tech. (Mech.), Hyderabad Univ. 2012; MBA, Mumbai Univ. 2010]*
24. What do you mean by the term saddle point? How do you determine the saddle point in case of pure strategy of a game? *[MIB, Delhi Univ. 2011]*
25. Discuss dominance with reference to game theory. *[MBA, Annamalai Univ. 2009]*
26. Explain the graphical method of solving $2 \times n$ and $m \times 2$ games.
27. What are the major limitations and applications of game theory? *[MBA, Hyderabad Univ. 2012]*
28. State the assumptions underlying game theory. Discuss its importance in business decisions. *[M.Com., Kerala Univ. 2012; MBA, Anantpur Univ. 2010]*
29. What do you understand by zero-sum and non-zero-sum games? What do you mean by strategy and dominance? *[MBA, Mumbai Univ. 2010; MBA, Osmania Univ. 2010]*
30. State four properties which a competitive situation should have, if it is to be called a competitive game. *[MBA (Dec.), Panjab Tech. Univ. 2014; B.Tech., Hyderabad Univ. 2011]*

CHAPTER 20

Queuing Theory

"Delay is the enemy of efficiency" and "waiting is the enemy of utilization"

Queuing Theory is a resource management technique that attempts to explore, understand and compare various queuing situations for determining optimum number of service facilities approximately. The word ***queue*** (or ***waiting line***) stands for a number of customers waiting to be serviced and by the term 'customer', we mean the arriving unit that require some service to be performed. The queue does not include the customer being serviced. Queues may be of persons waiting at a doctor's clinic or at railway booking office, these may be of machines waiting to be repaired or of ships in the harbour waiting to be unloaded. A queuing problem arises in the absence of a perfect balance between the service facilities and customers, waiting is required either of service facilities or for the customer's arrival. Providing too much service capacity is costly (owing to idle employees or equipment) and providing too little service capacity is costly (owing to waiting members of the queue). The purpose of queuing analysis is to provide information which a manager can use to determine an acceptable level of service.

20:1. QUEUING SYSTEM

A queuing system can be completely described by the input (or arrival pattern), the service mechanism (or service pattern), the queue discipline, and customer's behaviour. Here, customers arrive at a service facility and are attended by one or more of the servers (service facilities). As soon as a customer is served (serviced) , it departs. Thus, a queuing system can be described as composed of customers arriving for service, waiting for service if it is not immediate, leaving the system after being serviced.

The general framework of a queuing system is shown in *Fig. 20.1.*

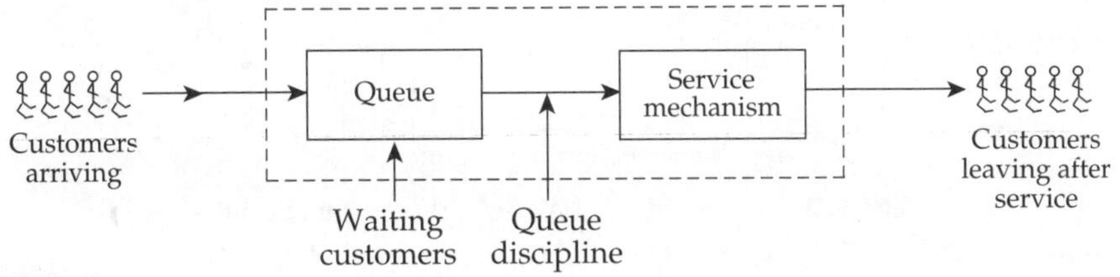

Fig. 20.1. Queuing System

1. Input (*or Arrival*) Pattern. The input describes the way in which customers arrive and join the queuing system. If the total number of customers requiring service are only few, then size of the input source is known as ***finite***. On the other hand, if customers requiring service are sufficiently large in number, then the input source is considered to be ***infinite***. In case, when more than one arrival is allowed to enter the system simultaneously, the input is said to occur in ***bulk*** or ***batches***. Families boarding a train/bus, ships discharging cargo at a dock are examples of bulk, or batch arrivals.

Customers may arrive in the system at known (regular or otherwise) times, or they may arrive in a random way. If the time between successive arrivals (inter-arrival times) is uncertain, the arrival pattern is measured by either ***mean arrival rate*** or ***inter-arrival time***. These are characterized by the probability distribution associated with this random process. Management scientists have found that the ***Poisson probability distribution*** provides a good description of the arrival pattern.

The Poisson probability function provides the probability of, x, arrivals in a specific time period. The probability function is defined as:

$$P(x) = \frac{\lambda^x e^{-\lambda}}{x!} \quad \text{for } x = 0, 1, 2, \ldots$$

where, x = the number of arrivals in the time period, and
λ = the mean number of arrivals per time period, e = 2.718213

2. Service Mechanism. The service mechanism is concerned with service time and service facilities. Service time is the time interval from the commencement of service to the completion of service of a customer. If there are infinite number of servers, then all customers are served instantaneously on arrival and there will be no queue. There are two aspects of service mechanism, namely (*i*) the structure of the service system, and (*ii*) the speed of service.

(*i*) ***Structure of the Service System.*** If the number of servers (service facilities) are finite, then customers are served according to specific order. Further, the customers may be served in batches of fixed size or of variable size rather than individually by the same server, such as a computer with parallel processing or people boarding a bus. The service system in this case is termed as ***bulk service system***.

Service facilities can be of following types:

(*a*) ***Single Queue-one Service Facility***, *i.e.*, one queue-one ***service channel***, wherein the customer waits till the service point is ready to take him in for service.

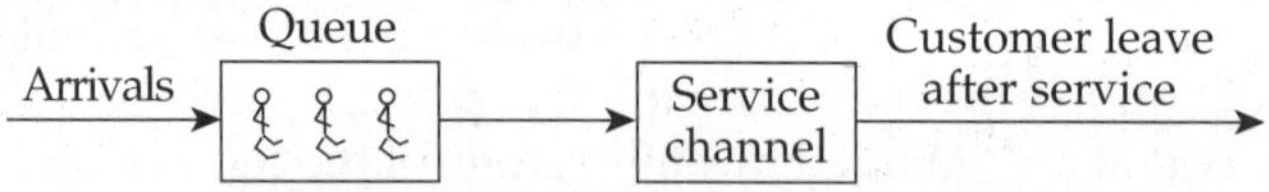

Fig. 20.2. (*a*) Single Queue-Single Server

(*b*) ***Single Queue and Multiple Parallel Service Facilities (Servers)***; wherein the customers wait in a single queue until one of the service channels is ready to take the customer in for service.

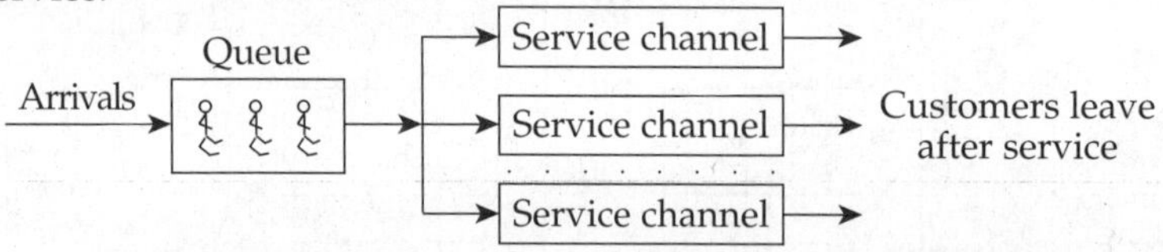

Fig. 20.2. (*b*) Single Queue, Multiple Parallel Servers

(*c*) ***Multiple Queues and Multiple Parallel Service Facilities (Servers)***; wherein there are several queues and the customers may join any one of these, but there is only one service channel for each queue.

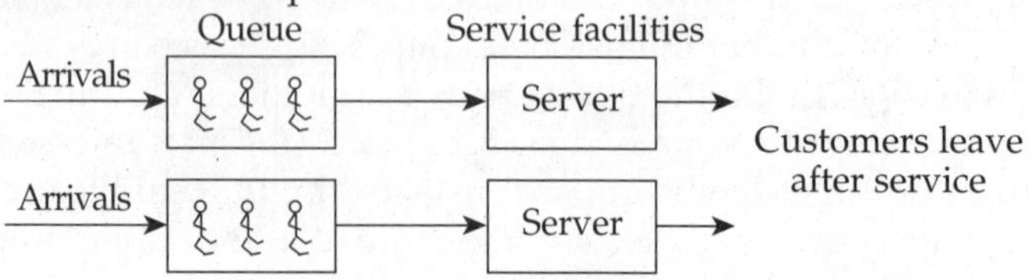

Fig. 20.2. (*c*) Multiple Queue, Multiple Parallel Servers

By ***parallel service channels***, we mean a number of service channels providing identical service facilities. Further, customers may wait in a single queue until one of the service channels is ready to serve, as in a barber shop, where many chairs are considered as different service channels; or customers may form separate queues in front of each service channel as in the case of booking counters at railway station.

(*ii*) ***Speed of Service.*** In a queuing system, the speed with which service is provided can be expressed in either of two ways — service rate or service time. The service rate describes the number of customers serviced during a particular time period. The service time indicates the amount of time needed to service a customer. Service rates and service times are reciprocals of each other and either of them is sufficient to indicate the capacity of service facility. Thus, if a server can attend, on the average, 10 customers in an hour; the service time would be expressed as 10 customers per hour and service time would be equal to 6 minutes per customer. Generally, however, we consider the service time only.

In case the service times are known exactly, the problem can be handled easily. But, when the service times are different and not known with certainty, the probability distribution for the service time can be assumed to follow an ***Exponential probability distribution***.

Using an exponential probability distribution, the probability that the service time will be less than or equal to a time of length *t* is given by

$$P(\text{service time} \le t) = 1 - e^{-\mu t},$$

where, μ = mean number of units that can be served per unit time period.

3. Queue Discipline. The queue discipline is a rule according to which customers are serviced when a queue has been formed. The most common queue discipline is "first come, first served" (FCFS) or the "first in, first out" (FIFO) according to which customers are serviced in the strict order of their arrival. Some of the other queue disciplines are: (*a*) "last come, first served" (LCFS). This queue discipline is practiced in most cargo handling situations, where the last item loaded is removed first. (*b*) "service in random order" (SIRO). According to this queue discipline, customer's service is done in preference over some other customers. For example, Speed Post service of Post Office, Express or Mail services of Railway, and so on.

4. Customer's Behaviour. For a customer, it is necessary to know his reaction upon entering the queuing system. A customer may decide to wait no matter how long the queue becomes (patient customer), or if the queue is too long to suit him, he may decide not to enter it or leave the system after joining the queue (impatient customer). Impatient customers generally behave in the following ways:

(*i*) ***Balking.*** A customer may decide not to enter the queue because of its length.

(*ii*) ***Reneging.*** A customer enters the queue, but after some time loses patience and decide to leave the queue.

(*iii*) ***Jockeying.*** A customer may move from one queue to another for his personal economic gains.

5. Capacity of the System. The source from which customers are generated may be finite or infinite. A ***finite source*** limits the customers arriving for service, *i.e.*, there is a finite limit to the maximum queue size. The queue can also be viewed as one with forced balking where a customer is forced to balk if he arrives at a time when queue size is at its limit. Alternatively, an ***infinite source*** is forever "abundant" as in the case of telephone calls arriving at a telephone exchange.

20:2. DETERMINISTIC QUEUING SYSTEM

A queuing system, wherein the customers arrive at regular intervals and the service time for each customer is known and constant is known as a ***deterministic queuing system***.

Let the customers come at the teller counter of a bank for withdrawal every 3 minutes. Thus, the interval between the arrival of any two successive customers, that is the inter-arrival time is exactly 3 minutes. Further, suppose that the incharge of that particular teller takes exactly 3 minute to serve a customer. This implies that the arrival and service rates are both equal to 20 customers per hour. In this situation there shall never be a queue and the incharge of the teller shall always be busy with servicing customers.

Now, suppose instead, that the incharge of the teller can serve 30 customers per hour, *i.e.*, he takes 2 minutes to serve a customer and then has to wait for one minute for the next customer to come for service. Here also, there would be no queue, but the teller is not always busy.

Further, suppose that the incharge of the teller can serve only 15 customers per hour, *i.e.*, he takes 4 minutes to serve a customer. Clearly, in this situation he would always be busy and queue length will increase continuously without limit with the passage of time. This implies that when the service rate is less than the arrival rate, the service facility cannot cope with all the arrivals and eventually the system leads to an ***explosive*** situation. In such situations, the problem can be resolved by providing additional service facilities, like opening parallel service counters. We can summarize the above situations as follows:

Let the arrival rate be λ customers per unit time and service rate be μ customers per unit time. Then,

1. if $\lambda > \mu$, the waiting line (queue) shall be formed and will increase indefinitely; the service facility would always be busy and the service system will eventually fail.
2. if $\lambda \leq \mu$, there shall be no queue and hence no waiting time; the proportion of time the service facility would be idle is $1 - \lambda/\mu$.

However, it is easy to visualize that the condition of uniform arrival and uniform service rate has a very limited practicability. Generally, the arrival and servicing time are both variable and uncertain. Thus, variable arrival rate and servicing time are more realistic assumptions. The probabilistic queuing models are based on these assumptions.

20:3. OPERATING CHARACTERISTICS OF QUEUING SYSTEM

Some of the operational characteristics of a queuing system, that are of general interest for the evaluation of the performance of an existing queuing system and to design a new queuing system are as follows:

1. ***Expected number of customers in the queuing system*** denoted by L_s or $E(n)$ is the average number of customers in the system, both waiting and in service. Here, n stands for the number of customers in the queueing system.
2. ***Expected number of customers in the queue*** denoted by L_q or $E(m)$ is the average number of customers waiting in the queue. Here, $m = n - 1$, *i.e.*, excluding the customers being served.
3. ***Expected waiting time in the queuing system*** denoted by W_s or $E(v)$ is the average total time spent by a customer in the queuing system. It is generally taken to be the waiting time plus servicing time.
4. ***Expected waiting time in queue*** denoted by W_q or $E(w)$ is the average time spent by a customer in the queue before the commencement of his service.

5. ***The server utilization factor (or busy period)*** denoted by $\rho\ (= \lambda/\mu)$ is the proportion of time that a server actually spends with customers. Here, λ stands for the average number of customers arriving per unit of time and μ stands for the average number of customers completing service per unit of time.

The server utilization factor is also known as ***traffic intensity*** or the clearing ratio.

20:4. PROBABILITY DISTRIBUTIONS IN QUEUING SYSTEM

It is assumed that customers joining the queuing system arrive in a random manner and follow a ***Poisson distribution*** or equivalently the inter-arrival time obey ***exponential distribution***. In most of the cases, ***service time*** is also assumed to be exponentially distributed. This implies that the probability of service completion in any short time period is constant and independent of the length of time that the service has been in progress.

Since, mean arrival rate λ is constant over a time, it follows that the probability of an arrival between time t and $t + dt$ is $\lambda \cdot dt$. This shows that the probability of an arrival in time dt is $\lambda\, dt$. When λ is the average number of arrivals per unit time, the expected number of arrivals in a time interval 0 to t will be λt. Poisson probability distribution function, therefore, is given by

$$P(x = n) = \frac{(\lambda t)^n e^{-\lambda t}}{n!} \quad \text{and} \quad P(x = 0) = \frac{(\lambda t)^0 e^{-\lambda t}}{0!} = e^{-\lambda t}$$

Let the random variable T be the time between successive arrivals (inter-arrival), then since an arrival can occur at any time, the random variable T must be continuous. The probability of no arrival in the time interval 0 to t will be equal to the probability that T exceeds t. Thus,

$$P(T > t) = P(\text{no arrival in time } t)$$

$$= P_0(t) = e^{-\lambda t}$$

The cumulative distribution function of T denoted by $F(t)$ is given by

$$F(t) = P(T \le t) = 1 - P(T > t) = 1 - e^{-\lambda t}, \quad t > 0$$

The density function $f(t)$ for inter-arrival times, therefore, is given by

$$f(t) = \frac{d}{dt} F(t) = \lambda e^{-\lambda t}, \quad t > 0$$

The ***expected*** (or mean) inter-arrival time is given by

$$E(t) = \int_0^{\infty} t \cdot f(t)\, dt = \int_0^{\infty} \lambda t e^{-\lambda t}\, dt = 1/\lambda,$$

where, λ is the ***mean arrival rate***.

Thus, T has the ***exponential distribution*** with mean $1/\lambda$. We would intuitively expect that, if the mean arrival rate is 'λ', then the mean time between arrivals is $1/\lambda$. Conversely, we can also show that, if the inter-arrival times are independent and have the same exponential distribution, then the arrival rate follows the Poisson distribution.

Similarly, ***mean service rate*** 'μ' is also assumed to be constant over time and independent of the number of units served. The probability that a service is completed between time t and $t + dt$, therefore, is $\mu \cdot dt$. It can be easily shown that the time 't' to complete the service of an arrival (or unit) follows the exponential distribution:

$$s(t) = \mu e^{-\mu t} \text{ (for } t > 0) \quad \text{and} \quad s(t) = 0 \text{ (for } t < 0),$$

where μ is the mean service rate for a particular service facility (service channel). It follows that service time follows the exponential distribution with mean $1/\mu$. Under the condition of continuous service, the probability of n complete services in time 't' is given by

$$P_n(t) = \frac{(\mu t)^n \cdot e^{-\mu t}}{n!}$$

Consequently, we can also show that

P (no service in dt) = $1 - \mu\, dt$ + other higher terms of dt,

and P (one service in dt) = $\mu\, dt$ + other higher terms of dt.

Since, Poisson distribution assumes that the time period dt is very small, other higher terms of dt tends to zero and can be ignored. Thus,

$$P_0(dt) + P_1(dt) = 1, \quad \text{or} \quad P_0(\Delta t) + P_1(\Delta t) = 1, \quad \text{and} \quad P_1(dt) = \mu \cdot dt$$

So, $$P_0(dt) = 1 - P_1(dt) = 1 - \mu \cdot dt.$$

20:5. TRANSIENT-STATE AND STEADY-STATE OF THE SYSTEM

A queuing system is said to be in ***transient state*** when its operating characteristics (behaviour of the system) vary with time. Usually a system is in ***transient period*** in the beginning (or start-up) period. The transient period ends when the system reaches the ***steady-state operation***, *i.e.*, its behaviour becomes independent of its initial conditions and time.

An essential condition for reaching the steady-state is that the total time since beginning of the operation must be sufficiently large (theoretically, it should be infinity). Queuing models describe the steady-state operating characteristics of a queue (waiting line).

So, if $P_n(t)$ denote the probability that there are n customers in the system at time t, then in the steady-state case, we shall have

$$\lim_{t \to \infty} P_n(t) = P_n \text{ (independent of } t\text{)}$$

20:6. CLASSIFICATION OF QUEUING MODELS

Generally ***queuing model*** may be completely specified in the following symbolic form:

$$(a/b/c) : (d/e)$$

The first and second symbols denote the type of distribution of inter-arrival times and of inter-service times, respectively. Third symbol specifies the number of servers, whereas fourth symbol stands for the capacity of the system and the last symbol denotes the queue discipline.

If we specify the following letters as:

$M \equiv$ Poisson arrival or departure distribution,

$E_k \equiv$ Erlangian or Gamma inter-arrival for service time distribution,

$GI \equiv$ General input distribution,

$G \equiv$ General service time distribution,

then $(M/E_k/1) : (\infty/FCFS)$ defines a queuing system in which arrivals follow Poisson distribution, service times are Erlangian, single server, infinite capacity and "first come, first served" queue discipline.

20:7. PROBABILISTIC QUEUING MODELS

Of the numerous queuing models available, we shall consider the following:

1. Poisson-exponential single server model with infinite population, *i.e.*, $(M/M/1) : (\infty/FCFS)$ model.

2. Poisson-exponential single server model with finite population, *i.e.*, $(M/M/1):(N/FCFS)$ model.
3. Poisson-exponential multiple server model with infinite population, *i.e.*, $(M/M/S):(\infty/FCFS)$ model.
4. Poisson-exponential multiple server model with maximum limit on arrivals, *i.e.*, $(M/M/S):(N/FCFS)$ model.
5. Poisson-Erlang service time single server model with *k*-phases, *i.e.*, $(M/E_k/1):(\infty/FCFS)$ model.

20:8. POISSON-EXPONENTIAL SINGLE SERVER AND INFINITE POPULATION, *i.e.*, $(M/M/1):(\infty/FCFS)$ MODEL

This model is based on the following assumptions:

(*a*) Arrivals follow Poisson distribution with mean arrival rate λ.

(*b*) Service time has exponential distribution with average (mean) service rate μ.

(*c*) Arrivals are from infinite population (or source).

(*d*) Customers are served on *FCFS* basis.

(*e*) There is only one service station (facility).

To evaluate this model, we must have $\lambda < \mu$, because for $\lambda = \mu$, there is no queue and for $\lambda > \mu$, the waiting line shall be formed and will increase indefinitely. The rate λ/μ $(= \rho)$ is defined as the traffic intensity and indicates that the proportion of time (or probability), the service station is busy. When there are no customer in the system, the probability of idle system is denoted by P_0 and is given by $(1 - \lambda/\mu)$ or $(1 - \rho)$.

The probability of having exactly one customer in the system is denoted by P_1, probability of having exactly two customers in the system is denoted by P_2, and so on. In general, the probability of having exactly n customers in the system is denoted by P_n.

Thus, $$P_1 = \left(\frac{\lambda}{\mu}\right)P_0 = \frac{\lambda}{\mu}\left(1-\frac{\lambda}{\mu}\right) = \rho(1-\rho),\quad P_2 = \left(\frac{\lambda}{\mu}\right)P_1 = \rho P_1 = \rho[\rho(1-\rho)] = \rho^2(1-\rho),$$

and so on. In general $P_n = \rho^n(1-\rho)$.

The model can explicitly be explained by using balance equations. For this, we consider the generalized queuing system, wherein there are n customers in the system, and let λ_n as the arrival rate of customers, μ_n as the departure rate of customers and P_n is the steady state probability of n customers.

Clearly, state n can change only to two possible states: state $(n - 1)$, when a departure occurs at the rate μ_n and state $(n + 1)$, when an arrival occurs at rate λ_n. State 0 can only change to state *l* when an arrival occurs at the rate λ_0. Since, no departure is possible when the system is empty, μ_0 is undefined.

Under steady-state conditions, for $n > 0$, the rates of flow into and out of state 'n' must be equal. This is illustrated in *Fig. 20.3*.

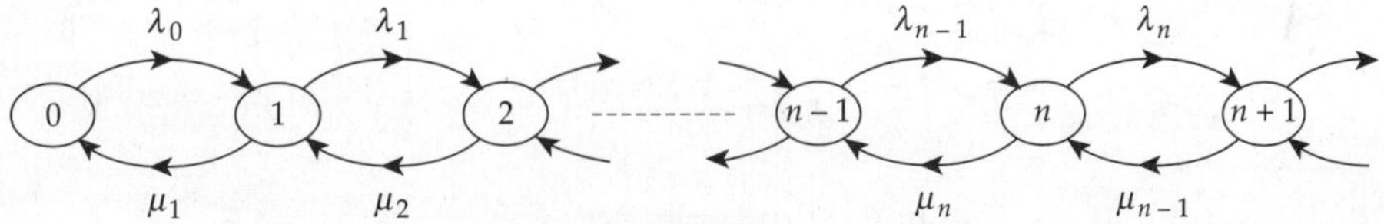

Fig. 20.3

The balance equation is:

Expected rate of flow into state n = Expected rate of flow out of state n

i.e., $$\lambda_{n-1} P_{n-1} + \mu_{n+1} P_{n+1} = \lambda_n P_n + \mu_n P_n, \qquad n \geq 1$$

and $$\mu_1 P_1 = \lambda_0 P_0 \qquad n = 0$$

Using the iterative procedure, we have

$$P_1 = \frac{\lambda_0}{\mu_1} P_0, \quad P_2 = \frac{\lambda_1 + \mu_1}{\mu_2} P_1 - \frac{\lambda_0}{\mu_2} P_0 = \frac{\lambda_1 \lambda_0}{\mu_2 \mu_1} P_0,$$

$$P_3 = \frac{\lambda_2 + \mu_2}{\mu_3} P_2 - \frac{\lambda_1}{\mu_3} P_1 = \frac{\lambda_2 \lambda_1 \lambda_0}{\mu_3 \mu_2 \mu_1} P_0$$

In general, we can write

$$P_n = \frac{\lambda_{n-1} \lambda_{n-2} \dots \lambda_0}{\mu_n \mu_{n-1} \dots \mu_1} P_0, \text{ for } n \geq 1$$

Now, since $\lambda_n = \lambda$ for $n \geq 0$ and $\mu_n = \mu$ for $n > 1$, therefore,

$$P_n = \rho^n P_0, \text{ for } n \geq 1; \text{ where } \rho = \lambda/\mu.$$

To obtain the value of P_0. we use the boundary condition

$$\sum_{n=0}^{\infty} P_n = 1 \quad \text{or} \quad P_0 + \sum_{n=1}^{\infty} P_n = 1$$

$$\therefore \quad \left[P_0 + \sum_{n=1}^{\infty} \rho^n P_0 \right] = 1 \quad \text{or} \quad P_0 = \left[1 + \sum_{n=1}^{\infty} \rho^{n-1} \right]^{-1} = 1 - \rho$$

Thus, $$P_n = \rho^n (1 - \rho), \text{ for } n \geq 0$$

Operating Characteristics

1. Probability that the number of customers in the system are exactly n is given by

$$P_n = \rho^n P_0 = \left(\frac{\lambda}{\mu}\right)^n \left(1 - \frac{\lambda}{\mu}\right); \quad \lambda \leq \mu$$

2. Probability that the number of customers in the system is greater than or equal to k is given by

$$P_{n \geq k} = \sum_{n=k}^{\infty} P_n = \sum_{n=k}^{\infty} \rho^n (1-\rho) = \rho^k (1-\rho) \sum_{n=k}^{\infty} \rho^{n-k}$$

$$= \rho^k (1-\rho) \sum_{x=0}^{\infty} \rho^x = \rho^k = \left(\frac{\lambda}{\mu}\right)^k, \quad \text{since } \sum_{x=0}^{\infty} \rho^x = \frac{1}{1-\rho}$$

3. Average number of customers in the system (customers in the queue as well as the one being served) is given by

$$L_s = \sum_{n=0}^{\infty} n P_n = \sum_{n=0}^{\infty} n \rho^n (1-\rho) = \rho(1-\rho) \sum_{n=0}^{\infty} n \rho^{n-1}$$

$$= \rho(1-\rho) \frac{d}{d\rho} \sum_{n=0}^{\infty} \rho^n = \rho(1-\rho) \times \frac{d}{d\rho} \left[\frac{1}{1-\rho}\right]$$

$$= \rho(1-\rho) \times \left[\frac{1}{(1-\rho)^2}\right] = \frac{\rho}{1-\rho} = \frac{\lambda}{\mu - \lambda}$$

4. Average queue length (customers waiting in the queue for service) is given by

$$L_q = \sum_{m=0}^{\infty} m P_n = \sum_{n-1=0}^{\infty} (n-1) P_n, \quad \text{where } m = n-1$$

$$= \sum_{n=1}^{\infty} n P_n - \sum_{n=1}^{\infty} P_n = \sum_{n=0}^{\infty} n P_n - \frac{\lambda}{\mu}, \quad \text{since } \sum_{n=0}^{\infty} P_n = 1$$

$$= \frac{\lambda}{\mu-\lambda} - \frac{\lambda}{\mu} = \frac{\lambda^2}{\mu(\mu-\lambda)}$$

5. Average length of non-empty queue

$$= \frac{\text{Average length of queue}}{\text{Probability of non-empty queue}}$$

$$= \frac{\left[\dfrac{\lambda^2}{\mu(\mu-\lambda)}\right]}{\left(\dfrac{\lambda}{\mu}\right)^2} = \frac{\lambda^2}{\mu(\mu-\lambda)} \times \left(\frac{\mu}{\lambda}\right)^2 = \frac{\mu}{\mu-\lambda}$$

6. Average time an arrival spends in the system (or the average waiting time of an arrival who waits) is given by

$$W_s = \frac{1}{\lambda} L_s = \frac{1}{\lambda} \times \frac{\lambda}{\mu-\lambda} = \frac{1}{\mu-\lambda}$$

7. Average waiting time of an arrival in the queue is given by

$$W_q = \frac{1}{\lambda} L_q = \frac{1}{\lambda} \times \frac{\lambda^2}{\mu(\mu-\lambda)} = \frac{\lambda}{\mu(\mu-\lambda)}$$

8. Average length of a non-empty system

= Average length of non-empty queue + Customer currently in service

$$= \frac{\mu}{\mu-\lambda} + \frac{\lambda}{\mu}$$

9. Average (expected) waiting time of an arrival who has to wait

$$= \frac{\text{Average waiting time of an arrival in the queue}}{\text{Probability of having a queue}}$$

$$= \frac{\left[\dfrac{\lambda}{\mu(\mu-\lambda)}\right]}{\left(\dfrac{\lambda}{\mu}\right)} = \frac{\lambda}{\mu(\mu-\lambda)} \times \frac{\mu}{\lambda} = \frac{1}{\mu-\lambda},$$

since probability of having a queue $= 1 - P_0 = 1 - (1-\rho) = \rho = \dfrac{\lambda}{\mu}$.

10. Probability that the time in the system is greater than or equal to $t = (\mu-\lambda)\, e^{-(\mu-\lambda)t}$.

11. Probability that a customer spends more than t units of time in the system is given by

$$P(\text{waiting time in the system} \geq t) = e^{-(\mu-\lambda)t}$$

12. Probability that a customer spends more than t units of time in the queue is given by

$$P(\text{waiting time in the queue} \geq t) = \frac{\lambda}{\mu} e^{-(\mu-\lambda)t}$$

Relationship among Operating Characteristics

The general relationship among various operating characteristics (performance measures) is as follows:

1. Expected number of customers in the system is equal to the expected number of customers in the queue plus a customer currently in service, *i.e.*,

$$L_s = L_q + \frac{\lambda}{\mu}$$

2. Expected waiting time of a customer in the system is equal to the expected waiting time in the queue plus the expected service time of a customer in service, *i.e.*,

$$W_s = W_q + \frac{1}{\mu}$$

3. Expected waiting time of a customer in the queue is:

$$W_q = \frac{L_q}{\lambda}$$

4. Expected number of customers in the system is equal to the average number of arrivals per unit of time multiplied by the average time spent by the customer in the system, *i.e.*,

$$L_s = \lambda W_s$$

5. Expected number of customers in the queue is equal to the average number of arrivals per unit of time multiplied by the average time spent by a customer in the queue, *i.e.*,

$$L_q = \lambda W_q$$

Note. Relations between Average queue length and Average waiting time are known as ***Little's Formulae.***

Example 20.1. *XYZ tailoring house has one tailor specialized in men's shirts. The number of customers requiring stitching of shirts appear to follow Poisson distribution with mean arrival rate of 12 per hour. Customers are attended by the tailor on a first come-first service basis, and they are willing to wait for service, if there be queue. The time the tailor takes to attend a customer is exponentially distributed with a mean of 4 minutes.*

Required:

(*a*) *The traffic intensity (i.e., utilization factor).*
(*b*) *The probability that the queuing system is idle.*
(*c*) *The average time the tailor is free on 8-hour working day.*
(*d*) *What is the probability that there shall be 5 customers in the shop (at a point of time)?*
(*e*) *What is the expected number of customers in shop?*
(*f*) *What is the expected number of customers waiting for tailor's services?*
(*g*) *What is the expected length of the non-empty queue?*
(*h*) *How much time a customer should expect to spend in the queue?*
(*i*) *How much time a customer should expect to spend in the shop?*
(*j*) *What is the probability that a customer shall spend more than 10 minutes for the tailor's service?*

Solution. Here, we are given

$$\lambda = 12 \text{ customers per hour, and } \mu = \frac{1}{4} \times 60 = 15 \text{ custermers per hour.}$$

(*a*) The traffic intensity: $\rho = \frac{\lambda}{\mu} = \frac{12}{15} = 0.8$

(*b*) The probability that the queuing system is idle, *i.e.*, no customer in the system:

$$P_0 = 1 - \frac{\lambda}{\mu} = 1 - 0.8 = 0.2$$

(*c*) The average time the tailor is free on 8-hour working day

$$= P_0 \times \text{number of hours} = 0.2 \times 8 = 1.6 \text{ hours.}$$

(*d*) The probability that there will be 5 customers in the shop:

$$P_5 = \rho^5 \times P_0 = \rho^5 \times (1-\rho) = (0.8)^5 \times (1-0.8) = 0.066$$

(*e*) Expected number of customers in the shop:

$$L_s = \frac{\lambda}{\mu - \lambda} = \frac{12}{15-12} = \frac{12}{3} = 4 \text{ customers.}$$

(*f*) Expected number of customers waiting for tailor's services:

$$L_q = \frac{\lambda^2}{\mu(\mu-\lambda)} = \left(\frac{\lambda}{\mu}\right) \times \left(\frac{\lambda}{\mu - \mu}\right)$$

$$= 0.8 \times 4 = 3.2 \text{ customers or 3 customers approx.}$$

(*g*) Expected length of non-empty queue

$$= \frac{\mu}{\mu - \lambda} = \frac{15}{15-12} = 5 \text{ customers.}$$

(*h*) Expected waiting time in the queue

$$= \frac{\lambda}{\mu(\mu-\lambda)} = \frac{12}{15(15-12)} = \frac{4}{15} \text{ hours or 16 minutes.}$$

(*i*) Expected time a customer spends in the system

$$= \frac{1}{\mu - \lambda} = \frac{1}{15-12} = \frac{1}{3} \text{ hours or 20 minutes.}$$

(*j*) Probability that a customer spends more than 10 minutes for service

$$= \frac{\lambda}{\mu} e^{-(\mu-\lambda)t} = \frac{12}{15} e^{-(15-12) \times \frac{10}{60}} = 0.8 \times e^{-1/2}$$

$$= 0.8 / e^{1/2} = 0.49, \text{ where } e^{1/2} = 1.6487$$

Example 20.2. *Customers arrive at a sales counter manned by a single person according to Poisson process with mean rate of 20 per hour. The time required to serve a customer has an exponential distribution with mean of 100 seconds. Find the average waiting time of a customer.*

[MBA, UP Tech. Univ. 2018; MBA, BHU 2008]

Solution. Here, we have

$$\lambda = 20 \text{ customers per hour, and } \mu = 36 \text{ customers per hour,}$$

since a customer takes 100 seconds to be serviced.

Average waiting time of a customer in the queue:

$$W_q = \frac{\lambda}{\mu(\mu-\lambda)} = \frac{20}{36(36-20)} = \frac{5}{36 \times 4} \text{ hours}$$

$$= \frac{5 \times 3600}{36 \times 4} \text{ Seconds} = 125 \text{ seconds}$$

Average waiting time of a customer in the system:

$$W_s = \frac{1}{\mu - \lambda} = \frac{1}{36 - 20} = \frac{1}{16} \text{ hours or } 225 \text{ seconds.}$$

Example 20.3. *A maintenance service facility has Poisson arrival rates, negative exponential service times, and operates on a first-come first-served queue discipline. Breakdowns occur on an average of three per day with a range of zero to eight. The maintenance crew can service on an average six machines per day with a range from zero to seven. Find the following:*

(i) Utilization factor of the service facility.
(ii) Mean time in the system.
(iii) Mean number in the system in breakdown or repair.
(iv) Mean waiting time in the queue.
(v) Probability of finding 2 machines in the system.
(vi) Expected number of machines in the queue. ***[MBA, Mahatma Gandhi Univ. 2010]***

Solution. Here, we have

$$\lambda = 3 \text{ breakdown per day, and } \mu = 6 \text{ machines per day.}$$

(*i*) Utilization factor of the service facility:

$$\rho = \frac{\lambda}{\mu} = \frac{3}{6} = 0.5 \text{ or } 50\%$$

(*ii*) Mean time in the system:

$$W_s = \frac{1}{\mu - \lambda} = \frac{1}{6 - 3} = \frac{1}{3} \text{ day.}$$

(*iii*) Mean number in the system in breakdown or repair:

$$L_s = \frac{\lambda}{\mu - \lambda} = \frac{3}{6 - 3} = 1 \text{ machine.}$$

(*iv*) Mean waiting time in the queue:

$$W_q = \frac{\lambda}{\mu} \times \frac{1}{\mu - \lambda} = \frac{3}{6} \times \frac{1}{6 - 3} = \frac{1}{6} \text{ day.}$$

(*v*) Probability of finding two machines in the system:

$$P_2 = \left(\frac{\lambda}{\mu}\right)^2 \left(1 - \frac{\lambda}{\mu}\right) = (0.5)^2 (0.5) = 0.125.$$

(*vi*) Mean number of machines in the queue:

$$L_q = \frac{\lambda}{\mu} \times \frac{\lambda}{\mu - \lambda} = \frac{3}{6} \times \frac{3}{6 - 3} = \frac{1}{2} \text{ machine.}$$

Example 20.4. *A telephone repairman finds that the time spent on his jobs has an exponential distribution with mean 30 minutes. If he repairs sets in the order in which they come in, and if the arrival of sets is approximately Poisson with an average rate of 10 per 8-hour day, what is repairman's expected idle time each day? How many jobs are ahead of the average set just brought in?*

[MBA, Madurai Univ. 2009; MBA, Bangalore Univ. 2018]

Solution. We are given,

$$\lambda = 10 \text{ sets per day, and } \mu = 16 \text{ sets per day}$$

$$\therefore \quad \rho = \frac{\lambda}{\mu} = \frac{10}{16} = 0.625$$

The probability for the repairman to be idle is:

$$P_0 = 1 - \rho = 1 - 0.625 = 0.375$$

(*i*) Expected idle time per day = 8 × 0.375 = 3 hours.

(*ii*) Expected (or average) number of TV sets in the system is:

$$L_s = \frac{\lambda}{\mu - \lambda} = \frac{10}{16-10} = \frac{5}{3} \text{ or 2 TV set approx.}$$

Example 20.5. *At Dr. Prachi's clinic patients arrive at an average of 6 patients per hour. The clinic is attended by Dr. Prachi herself. Some patients require only the required prescription. Some come for minor check-up, while some others require thorough inspection for the diagnosis. This takes the doctor six minutes per patient on an average. It can be assumed that arrivals follow a Poisson distribution and the Doctor's inspection time follows as Exponential distribution. Determine:*

(*i*) *The percentage of time that a patient can walk to the doctor without having to wait.*

(*ii*) *The average number of patients in the clinic.*

(*iii*) *The average number of patients in the queue.*

(*iv*) *The average waiting time of a patient in the clinic.*

Solution. Here, we have

$$\lambda = 6 \text{ patients per hour, and } \mu = \frac{1}{6} \times 60 = 10 \text{ patients per hour.}$$

(*i*) The probability that a patient can walk to a doctor without waiting or the probability of an idle system:

$$P_0 = 1 - \frac{\lambda}{\mu} = 1 - \frac{6}{10} = \frac{4}{10} (= 0.4) \text{ or } 40\%$$

(*ii*) Average number of patients in the system:

$$L_s = \frac{\lambda}{\mu - \lambda} = \frac{6}{10-6} = \frac{3}{2} \text{ patients or 2 patients approx.}$$

(*iii*) Average number of patients in the queue:

$$L_q = \frac{\lambda}{\mu} \times \frac{\lambda}{\mu - \lambda} = \frac{6}{10} \times \frac{3}{2} = \frac{9}{10} \text{ patients or 1 patient approx.}$$

(*iv*) Average waiting time of a patient in the system:

$$W_s = \frac{1}{\mu - \lambda} = \frac{1}{10-6} = \frac{1}{4} \text{ hour or 15 minutes.}$$

Example 20.6. *Workers come to a store-room to receive components for a particular project assigned to them. The average time between two arrivals is 60 seconds and arrivals are assumed to be in Poisson distribution. The average service time by the storekeeper is 40 seconds. Determine the following:*

(*i*) *Average queue length.*

(*ii*) *Average length of non-empty queue.*

(*iii*) *Average number of workers in the system including the worker being attended to.*

(*iv*) *Average waiting time of an arrival in the queue.*

(*v*) *Average waiting time of an arrival (worker) who waits.* **[B.Tech., Hyderabad Univ. 2011]**

Solution. We are given

λ = one worker per minute, since average time between two arrivals is 60 seconds, *i.e.*, $\lambda = 1$.

μ = 1.5 workers per minute, since the average service time of worker is 40 seconds, *i.e.*, 60/40 minutes.

(*i*) Average queue length:

$$L_q = \frac{\lambda}{\mu} \times \frac{\lambda}{\mu-\lambda} = \frac{1}{1.5} \times \frac{1}{(1.5-1)} = \frac{100}{75} = \frac{4}{3} \text{ or one worker approx.}$$

(*ii*) Average length of non-empty queue:

$$L_s + 1 = \frac{\lambda}{\mu-\lambda} + 1 = \frac{\mu}{\mu-\lambda} = \frac{1.5}{1.5-1} = 3 \text{ workers.}$$

(*iii*) Average number of workers in the system:

$$L_s = \frac{\lambda}{\mu-\lambda} = \frac{1}{1.5-1} = \frac{1}{0.5} = 2 \text{ workers.}$$

(*iv*) Average waiting time of an arrival in the queue:

$$W_q = \frac{\lambda}{\mu} \times \frac{1}{\mu-\lambda} = \frac{1}{1.5} \times 2 = \frac{4}{3} \text{ minutes.}$$

(*v*) Average waiting time of worker who waits:

$$W_s = \frac{1}{\mu-\lambda} = \frac{1}{1.5-1} = 2 \text{ minutes.}$$

Example 20.7. *A self-service store employs one cashier at its counter. Nine customers arrive on an average in every five minutes, while the cashier can serve 10 customers in 5 minutes. Assuming Poisson distribution for arrival rate and exponential distribution for service rate, find:*

(*i*) *Average number of customers in the system.*
(*ii*) *Average queue length.*
(*iii*) *Average time a customer spends in the system.*
(*iv*) *Average time a customer waits before being served.* **[B.Tech. (Mech.), Hyderabad Univ. 2012]**

Solution. Here, we have

$$\lambda = \frac{9}{5} \text{ or 1.8 customers per minute, and } \mu = \frac{10}{5} \text{ or 2 customers per minute.}$$

(*i*) Average number of customers in the system

$$L_s = \frac{\lambda}{\mu-\lambda} = \frac{1.8}{2-1.8} = 9 \text{ customers.}$$

(*ii*) Average number of customers in the queue:

$$L_q = \frac{\lambda^2}{\mu(\mu-\lambda)} = \frac{1.8}{2} \times \frac{1.8}{2-1.8} = 8.1 \text{ customers.}$$

(*iii*) Average time a customer spends in the system:

$$W_s = \frac{1}{\mu-\lambda} = \frac{1}{2-1.8} = 5 \text{ minutes.}$$

(*iv*) Average time a customer spends in the queue:

$$W_q = \frac{\lambda}{\mu} \times \frac{1}{\mu-\lambda} = \frac{1.8}{2} \times \left(\frac{1}{2-1.8}\right) = 4.5 \text{ customers.}$$

Example 20.8. *Customers arrive at the first class ticket counter of a theater at a rate of 12 per hour. There is one clerk to service the customers at a rate of 30 per hour.*

(*i*) *What is the probability that there is no customer at counter (Idle system)?*
(*ii*) *What is the probability that there are more than 2 customers in the counter?*

(*iii*) *What is the probability that there is no customer waiting to be served?*
(*iv*) *What is the probability that a customer is being served and no one is waiting?*

[M.Tech. (Dec.), Panjab Tech. Univ. 2011]

Solution. Here, we have

$$\lambda = 12 \text{ customers per hour, and } \mu = 30 \text{ customers per hour.}$$

$$\therefore \quad \rho = \text{Utilization rate (factor)} = \frac{\lambda}{\mu} = \frac{12}{30} = \frac{2}{5} = 0.4$$

(*i*) Probability of no customer at the counter:

$$P_0 = 1 - \rho = 1 - 0.4 = 0.6$$

(*ii*) Probability of more than two customers at the counter:

$$P(\geq 3) = \left(\frac{\lambda}{\mu}\right)^3 = (0.4)^3 = 0.064$$

(*iii*) Probability that no customer is waiting for service = Probability that at most one customer at counter:

$$P_0 + P_1 = P_0 + \left(\frac{\lambda}{\mu}\right)P_0 = \left(1 + \frac{\lambda}{\mu}\right)P_0$$

$$= (1 + 0.4) \times 0.6 = 1.4 \times 0.6 = 0.84$$

(*iv*) Probability that a customer being served and no one is waiting:

$$P_1 = \left(\frac{\lambda}{\mu}\right)P_0 = 0.4 \times 0.6 = 0.24$$

Example 20.9. *A child care shop dealing with children's requirements has one cashier who handles all customer's payments. The cashier takes on an average 4 minutes per customer. Customers come to cashier's counter in a random manner but on an average of 10 people per hour. The management received a large number of customer's complaints and decided to investigate the following questions:*

(*i*) *What is the average length of waiting line to be expected under the existing conditions?*
(*ii*) *What portion of his time is the cashier expected to be idle?*
(*iii*) *What is the average length of time that a customer would be expected to wait to pay for his purchase?*
(*iv*) *If it was decided that a customer would not tolerate a wait of more than 12 minutes, what is the probability that a customer would have to wait at least that length of time?*

Solution. From the given information, we have

Mean rate of arrival, $\lambda = 10$ customers per hour, and
Mean rate of service, $\mu = 15$ customers per hour.

(*i*) Average number of customers in the queue:

$$L_q = \frac{\lambda^2}{\mu(\mu - \lambda)} = \frac{10 \times 10}{15(15 - 10)} = \frac{4}{3} \text{ customers or 1 customer (approx.)}$$

(*ii*) Probability that the cashier will be idle:

$$P_0 = 1 - \frac{\lambda}{\mu} = 1 - \frac{10}{15} = \frac{1}{3} = 0.333$$

i.e., the cashier will be expected to be idle 33.3% of his time.

(*iii*) Average length of time that a customer is expected to wait in the queue:

$$W_q = \frac{\lambda}{\mu(\mu-\lambda)} = \frac{10}{15(15-10)} \text{ hour or 8 minutes.}$$

(*iv*) Probability that a customer has to wait at least 12 minutes before he can make payment is given by:

$$P(\text{waiting time} \geq \frac{1}{5} \text{ hour}) = \frac{\lambda}{\mu}e^{-(\mu-\lambda)t} = \frac{2}{3}e^{-(15-10)1/5} = \frac{2}{3}e^{-1} = \frac{2}{3}(0.368) = 0.245$$

Example 20.10. *A milk plant distributes its products by trucks, loaded at the loading dock. It has its own fleet plus the trucks of a transport company. This company has complained that sometimes trucks have to wait in the queue and thus the company loses money. The company has asked to management either to go in for a second loading dock or discount prices equivalent to waiting time. The data available is as follows:*

Average arrival rate = 3 trucks per hour, and average service rate = 4 trucks per hour. The transport company has provided 40% of the total number of trucks. Determine:

(*a*) *The probability that a truck has to wait.*
(*b*) *Expected waiting time of a truck.*
(*c*) *Expected waiting time for transport company's truck per day.*

[*MBA, Mahatma Gandhi Univ. 2010; MBA, Annamalai Univ. 2009*]

Solution. According to data of the problem, we have

$$\lambda = 3 \text{ trucks per hour, and } \mu = 4 \text{ trucks per hour.}$$

(*a*) Probability that a truck has to wait for service:

$$P(w > 0) = 1 - P_0 = 1 - \left(1 - \frac{\lambda}{\mu}\right) = \frac{\lambda}{\mu} = \frac{3}{4} \text{ or } 0.75$$

(*b*) Expected waiting time of a truck:

$$W_s = \frac{1}{\mu-\lambda} = \frac{1}{4-3} \text{ or 1 hour.}$$

(*c*) Expected waiting time of company's truck per day:

$$= \begin{pmatrix}\text{Number of} \\ \text{trucks per day}\end{pmatrix} \times \begin{pmatrix}\text{\% contractors} \\ \text{trucks}\end{pmatrix} \times \begin{pmatrix}\text{Expected waiting} \\ \text{time per truck}\end{pmatrix}$$

$$= (3 \times 24) \times \frac{40}{100} \times \frac{3}{4(4-3)} = 21.6 \text{ hours per day.}$$

Example 20.11. *Customers arrive at a one-window drive-in bank according to Poisson distribution with mean 10 per hour. Service time per customer is exponential with 5 minutes. The space in front of the window, including the customer in service can accommodate a maximum of three customers. Other customers can wait outside this space.*

(*i*) *What is the probability that an arriving customer can drive directly to the space in front of the window?*
(*ii*) *What is the probability that an arriving customer will have to wait outside the indicated space?*
(*iii*) *How long is an arriving customer expected to wait before being served?*

[*B.Tech.(Mech.), HP Univ. 2010; MBA, IGNOU 2002; B.Tech. (Dec.), Panjab Tech. Univ. 2011*]

Solution. Here, we have

$$\lambda = 10 \text{ customers per hour, and } \mu = \frac{60}{5} = 12 \text{ customers per hour.}$$

(*i*) Probability that an arriving customer can drive directly to the space in front of the window is given by:

$$P_0 + P_1 + P_2 = \left(1-\frac{\lambda}{\mu}\right) + \left(\frac{\lambda}{\mu}\right)\times\left(1-\frac{\lambda}{\mu}\right) + \left(\frac{\lambda}{\mu}\right)^2\times\left(1-\frac{\lambda}{\mu}\right)$$

$$= \left(1-\frac{\lambda}{\mu}\right)\left\{1+\frac{\lambda}{\mu}+\frac{\lambda^2}{\mu^2}\right\} = \left(1-\frac{10}{12}\right)\left\{1+\frac{10}{12}+\frac{100}{144}\right\} = \left(\frac{1}{6}\right)\times\left(\frac{364}{144}\right) = 0.42$$

(*ii*) Probability that an arriving customer will have to wait outside the indicated space:

$$= 1 - 0.42 = 0.58$$

(*iii*) Average waiting time of a customer in the queue:

$$W_q = \frac{\lambda}{\mu}\times\left(\frac{1}{\mu-\lambda}\right) = \frac{10}{12}\times\left(\frac{1}{12-10}\right) = \frac{5}{12} = 0.417 \text{ hour.}$$

Example 20.12. *In a railway marshalling yard, goods trains arrive at a rate of 30 trains per day. Assume that the inter-arrival time follows an exponential distribution and the service time distribution is also exponential with an average of 36 minutes. Calculate the following:*

(*i*) *the mean queue size (line length), and*

(*ii*) *the probability that the queue size exceeds 10.*

If the input of trains increases to an average 33 trains per day, what will be the change in (i) and (ii)?

[MBA, Panjab Tech. Univ. 2014; BE (Dec.), Karnataka Univ. 2011]

Solution. Here, we have

$$\lambda = 30 \text{ trains per day} = \frac{30}{24} \text{ trains per hour} = \frac{30}{24\times 60} = \frac{1}{48} \text{ trains per minute, and}$$

$$\mu = \frac{1}{36} \text{ trains per minute, since a train takes 36 minutes for its service.}$$

(*i*) Mean queue size (line length):

$$L_s = \frac{\lambda}{(\mu-\lambda)} = \left(\frac{1}{48}\right)\times\frac{36\times 48}{(48-36)} = 3 \text{ trains.}$$

(*ii*) Probability that the queue size exceeds 10:

$$P(n \geq 10) = \left(\frac{\lambda}{\mu}\right)^{10} = \left(\frac{36}{48}\right)^{10} = (0.75)^{10} = 0.06$$

Now, if the input increases to 33 trains per day, then we have

$$\lambda = \frac{33}{24\times 60} = \frac{11}{480} \text{ trains per minute, and } \mu = \frac{1}{36} \text{ trains per minute.}$$

So, the traffic intensity, $\rho = \frac{\lambda}{\mu} = \frac{11}{480}\times 36 = 0.83$

Then, recalculating the values of (*i*) and (*ii*), we get

$$L_s = \frac{\lambda}{\mu-\lambda} = \frac{\rho}{1-\rho} = \frac{0.83}{1-0.83} = 4.9 \text{ or } 5 \text{ trains (approx.)}$$

and $$P(n \geq 10) = \left(\frac{\lambda}{\mu}\right)^{10} = (0.83)^{10} = 0.2 \text{ (approx.)}$$

Example 20.13. *Arrivals at a telephone booth are considered to be Poisson, with an average time of 10 minutes between one arrival and the next. The duration of the phone call is assumed to be distributed exponentially with mean 3 minutes.*

(*a*) *What is the probability that a person arriving at the booth will have to wait?*

(*b*) *The telephone department will install a second booth when convinced that an arrival would expect waiting for at least 3 minutes for phone. By how much should the flow of arrivals increase in order to justify a second booth?*

(*c*) *Find the average number of arrivals in the system.*

(*d*) *Estimate the fraction of a day that the phone will be in use.*

(*e*) *What is the probability that it will take a person more than 10 minutes altogether to wait for phone and complete his call?* **[MBA (June), Visvesvaraya Tech. Univ. 2011; MBA, Kerala Univ. 2010; BE, Karnataka Univ. 2012; PG Dip. in Glob. Bus. Oper., Delhi Univ. 2010]**

Solution. Here, we have

$$\lambda = \text{one person in 10 minutes} = \frac{1}{10} (= 0.10) \text{ person per minute, and}$$

$$\mu = \text{one person in 3 minutes} = \frac{1}{3} (= 0.33) \text{ person per minute.}$$

(*a*) Probability that the person arriving at the booth will have to wait:

$$P(w > 0) = 1 - P_0 = \frac{\lambda}{\mu} = \frac{0.10}{0.33} = 0.3$$

(*b*) The installation of second booth will be justified, if the arrival rate is greater than the waiting time. Here, we are given the waiting time of customer in the queue as 3 minutes.

Let λ' be the increased arrival rate. Then, expected waiting time in the queue is:

$$W_q = \frac{\lambda'}{\mu(\mu - \lambda')}$$

$$\therefore \quad 3 = \frac{\lambda'}{0.33(0.33 - \lambda')} \quad \text{or} \quad 3 = \frac{\lambda'}{0.1089 - 0.33\lambda'}$$

or $\quad \lambda' = 0.16$ person per minute.

Hence, the arrival rate should become 0.16 per minute to justify the second booth, *i.e.*, it should increase by 0.16 per person per minute.

(*c*) Average number of arrivals (persons) in the system:

$$L_s = \frac{\lambda}{\mu - \lambda} = \frac{0.10}{0.33 - 0.10} = 0.43 \text{ person or 1 person (approx.)}$$

(*d*) Fraction of the day the phone will be busy, *i.e.*,

$$\text{Traffic intensity} = \frac{\lambda}{\mu} = \frac{0.10}{0.33} = 0.3$$

(*e*) The probability that the waiting time in a system is greater than and equal to *t*:

$$P(w > t) = e^{-(\mu - \lambda)t} = e^{-0.23 \times 10} = e^{-2.3}, \text{ since } t = 10$$

$$= e^{-2} \times e^{-0.3} = 0.135 \times 0.740 = 0.10$$

Example 20.14. *Arrivals of mechanists at a tool crib are considered to be Poisson distributed at an average rate of 6 per hour. The length of time the mechanists must remain at the tool crib is exponentially distributed with the average time being 0.05 hours.*

(*a*) *What is the probability that a mechanist arriving at the tool crib will have to wait?*

(*b*) *What is the average number of mechanists at the tool crib?*

(*c*) *The company will install a second tool crib, when convinced that a mechanist would have to spend 6 minutes of waiting and being served at the tool crib. By how much time the flow of mechanists to the tool crib should increase to justify the addition of a second tool crib?*

[MCA, Rajasthan Univ. 2008]

Solution. Here, we have

$$\lambda = 6 \text{ mechanists per hour, and } \mu = \frac{1}{0.50} = 20 \text{ mechanists per hour.}$$

(*a*) The probability that a mechanists arriving at the tool crib will have to wait, *i.e.*, service channel is busy:

$$P(w > 0) = 1 - P_0 = \frac{\lambda}{\mu} = \frac{6}{20} = 0.3$$

(*b*) Average number of a mechanists at the tool crib or average (expected) number of customers in the system:

$$L_s = \frac{\lambda}{\mu - \lambda} = \frac{6}{20 - 6} = 0.43 \text{ or } 1 \text{ mechanist (approx.)}$$

(*c*) Average time an arrival spends in the system is:

$$W_s = \frac{1}{\mu - \lambda'}; \text{ where } \lambda' \text{ is the increased arrival rate.}$$

or
$$\frac{6}{60} = \frac{1}{20 - \lambda'} \text{ or } \lambda' = 10 \text{ per hour.}$$

Hence, arrival rate should increase from 6 per hour to 10 per hour to justify an addition of a second tool crib.

Example 20.15. *On an average 96 patients per 24-hour day require the service of an emergency clinic. Also, on an average, a patient requires 10 minutes of active attention. Assume that the facility can handle only one emergency at a time. Suppose that it costs the clinic ₹100 per patient treated to obtain an average servicing time of 10 minutes, and that each minute of decrease in this average time would cost ₹10 per patient treated. How much would have to be budgeted by the clinic to decrease the average size of the queue from one and one-third patients to half a patient.* ***[MBA, Delhi Univ. 2008; MBA (Dec.), Dayal Bagh Edu. Inst. 2011]***

Solution. According to the given information:

$$\lambda = 4 \text{ patients per hour, and } \mu = 6 \text{ patients per hour.}$$

$\therefore$ Traffic intensity, $\rho = \frac{\lambda}{\mu} = \frac{4}{6} = \frac{2}{3}$

Average number of patients in the queue:

$$L_q = \frac{\lambda^2}{\mu(\mu - \lambda)} = \frac{4 \times 4}{6(6-4)} = \frac{4}{3} = 1\frac{1}{3} \text{ patients.}$$

Fraction of the time for which there are no patients:

$$P_0 = 1 - \frac{\lambda}{\mu} = 1 - \frac{2}{3} = \frac{1}{3}$$

When the average queue size is decreased from $\frac{4}{3}$ patients to $\frac{1}{2}$ patient, we are to determine the value of μ^*. So, we have

$$L_q = \frac{\lambda^2}{\mu^*(\mu^* - \lambda)} \Rightarrow \frac{1}{2} = \frac{4^2}{\mu^*(\mu^* - 4)}$$

or $$(\mu^*)^2 - 4\mu^* = 32 \text{ or } (\mu^*)^2 - 8\mu^* + 4\mu^* - 32 = 0$$

or $$(\mu^* - 8)(\mu^* + 4) = 0. \text{ This gives } \mu^* = 8 \text{ or } \mu^* = -4$$

Since, negative value of μ^* is not possible, we hall have $\mu^* = 8$ customers per hour.

The service rate of 8 customers per hour implies a service time equal to $\frac{60}{8}(= 7.5)$ minutes per customer.

Thus, Reduction in service time = 10 – 7.5 = 2.5 minutes.

Budget per patient = ₹ (100 + 2.5 × 10) = ₹ 125

Hence, in order to get the required size of the queue, the budget should be increased from ₹ 100 per patient to ₹ 125 per patient.

∴ Total budget = 96 × ₹ 125 = ₹ 12,000 per day.

Example 20.16. *An overhead crane move jobs from one machine to another and must be used every time when a machine requires loading or unloading. The demand for service is random. Data taken by recording the elapsed time between service calls followed an exponential distribution having a mean call every 30 minutes. In a similar manner, the actual service time of loading or unloading took an average of 10 minutes, if the machine time is valued at ₹ 8.50 per hour. How much does the downtime cost per day?*

Solution. According to given data, we have

$$\lambda = \frac{60}{30}(= 2) \text{ calls per hour, and } \mu = \frac{60}{10}(= 6) \text{ calls per hour.}$$

The downtime per machine is the average system time. Therefore,

$$W_s = \frac{1}{\mu - \lambda} = \frac{1}{6 - 2} = \frac{1}{4} = 0.25 \text{ per hour.}$$

The daily demand for service, assuming an 8-hour day, is 8 times the demand per hour.

∴ Daily demand = 8λ = 8 × 2 = 16 calls for service per day.

Since, each call requires an average downtime of 0.25 hour, total cost per day is

₹ (8.50) × (0.25 hour) × 16 calls, *i.e.*, ₹ 34.

Example 20.17. *Customers arrive at the rate of twenty per hour and the present serving arrangements can cope with thirty per hour for an eight-hour day. Calculate and state*

(*a*) *the average time in the queue,*

(*b*) *the implied value of customer's time, if the service has considered but rejected a faster service arrangement which would cost an extra ₹ 20 for an 8-hour day and would raise the service rate to 40 customers per hour.* ***[MBA, Lucknow Univ. 2009]***

Solution. From the given information, we have

λ = 20 customers per hour, and μ = 30 customers per hour.

(*a*) Average waiting time of a customer in the queue:

$$W_q = \frac{\lambda}{\mu(\mu - \lambda)} = \frac{20}{30(30 - 20)} = \frac{1}{15} \text{ hours or 4 minutes.}$$

(*b*) Average waiting time of a customer in the system:

$$W_s = \frac{1}{\mu - \lambda} = \frac{1}{30-20} = \frac{1}{10} \text{ hours or 6 minutes.}$$

When the service rate μ = 40 customers per hour, we have

$$W_s = \frac{1}{40-20} = \frac{1}{20} \text{ hours or 3 minutes.}$$

If the break-even value of customers waiting time is ₹ k per hour, then

$$8 \text{ hours} \times \left(\frac{6}{60} - \frac{3}{60}\right) \text{hours} \times \lambda \times k < ₹\, 20, \text{ where } \lambda = 20$$

i.e.,
$$8 \times \frac{3}{60} \times 20 \times k < 20 \text{ or } k < \frac{5}{2} (= 2.50)$$

Hence, the break-even value of customers waiting time is ₹ 2.50.

Example 20.18. *A repairman is to be hired to repair machines which breakdown at an average rate of 3 per hour. Breakdowns are distributed in time in a manner that may be regarded as Poisson. Non-productive time on any one machine is considered to cost the company ₹ 10 per hour. The company has narrowed the choice to two repairman: one slow but cheap, the other fast but expensive. The slow but cheap repairman demands ₹ 5 per hour and in return he will service breakdown machines exponentially at an average rate of 4 machines per hour. The second, fast but expensive repairman demands ₹ 8 per hour and will repair machines exponentially at an average rate of 6 machines per hour. Which repairman should be hired?*

[MBA, Madras Univ. 2000; MBA, Osmania Univ. 2010; B.Tech. Panjab Tech. Univ. 2012]

Solution. In this problem, we compare the total expected daily cost for both the repairman. This would equal the total wages paid plus the cost due to machine breakdown, *i.e.*, for the non-productive machine hours. We are given

Mean arrival rate, λ = 3 machines per hour, Cost of idle time of machines = ₹ 10 per hour, and Number of hours (assuming 8-hour day) = 8

Case 1. Slow and Cheap Repairman:

Mean service rate, μ = 4 machines per hour.

Wages of repairman = ₹ 5 per hour.

Average number of machines in the system:

$$L_s = \frac{\lambda}{\mu - \lambda} = \frac{3}{4-3} = 3 \text{ machines per hour}$$

Machine hours lost in a 8-hour shift = 3 × 8 = 24

Total cost = Hire charges of repairman + Cost of idle machines

= ₹ 5 × 8 + 24 × ₹ 10 = ₹ 280

Case 2. Fast amd Expensive Repairman:

Mean service rate, μ = 6 machines per hour

Wages of repairman = ₹ 8 per hour

Machine hours lost in a 8-hour shift = 1 × 8 = 8

Total cost = Hire charges + Cost of idle machines

= ₹ 8 × 8 + ₹ 10 × 8 = ₹ 144

From the above two cases, we conclude that the fast expensive repairman should be employed.

Example 20.19. *In the production shop of a company the breakdown of the machines is found to be Poisson with an average rate of 3 machines per hour. Breakdown time at one machine cost ₹40 per hour to the company. There are two choices before the company for hiring the repairmen. One of the repairman is slow but cheap, the other fast but expensive. The slow and cheap repairman demands ₹20 per hour and will repair the broken down machines exponentially at the rate of 4 per hour. The fast and expensive repairman demands ₹30 per hour and will repair machines exponentially at an average rate of 6 per hour. Which repairman should be hired?* ***[M.Com., Delhi Univ. 2006]***

Solution. In this problem, we compare the total expected daily cost for both the repairmen. This would equal the total wages paid plus the downtime cost.

Case 1. Slow and Cheap Repairman:

$$\lambda = 3 \text{ machines per hour, and } \mu = 4 \text{ machines per hour.}$$

$$\text{Average downtime of a machine } = \frac{1}{\mu - \lambda} = \frac{1}{4-3} = 1 \text{ hour.}$$

∴ The downtime of 3 machines that arrive in an hour = 1 × 3 = 3 hours.

Downtime cost = ₹ 40 × 3 = ₹ 120, Charges paid to the repairman = ₹ 20 × 3 = ₹ 60

Total cost = ₹ 120 + ₹ 60 = ₹ 180

Case 2. Fast and Expensive Repairman:

$$\lambda = 3 \text{ machines per hour and } \mu = 6 \text{ machines per hour.}$$

$$\therefore \quad \text{Average downtime of machine } = \frac{1}{\mu - \lambda} = \frac{1}{6-3} = \frac{1}{3} \text{ hour.}$$

$$\therefore \quad \text{The downtime of 3 machines that arrive in an hour } = \frac{1}{3} \times 3 = 1 \text{ hour.}$$

Downtime cost = ₹ 40 × 1 = ₹ 40, Charges paid to the repairman = ₹ 30 × 1 = ₹ 30

Total cost = ₹ 40 + ₹ 30 = ₹ 70

From the above two cases, the decision of the company should be to engage the fast expensive repairman.

Example 20.20. *Unique Tyre Company decided to hire a new mechanic to handle all tyre changes for customers ordering a new set of tyres. Two mechanics applied for the job. One mechanic has limited experience, can be hired for ₹14 per hour, and can service an average of three customers per hour. The other mechanic has several years of experience, and can service an average of four customers per hour, but needs to be paid ₹20 per hour. Assume that customers arrive at the company garage at the rate of two customers per hour, and the company works eight hours a day.*

Calculate the following characteristics of the queuing system using each mechanic assuming Poisson arrivals and exponential service times:

(i) Expected length of queue, (ii) Expected length of system, (iii) Expected waiting time in queue, and (iv) Expected waiting time in the system.

If the company assigns a customer waiting time cost of ₹30 per hour, which mechanic provides the lower operating cost. ***[M.Com., Delhi Univ. 2009]***

Solution. *Mechanic A:* In usual notations, it is given:

Arrival rate, $\lambda = 2$ customers per hour, Service rate, $\mu = 3$ customers per hour.

$$\therefore \quad \text{Expected length of queue,} \quad L_q = \frac{\lambda^2}{\mu(\mu - \lambda)} = \frac{4}{3(3-2)} = \frac{4}{3} \text{ customers,}$$

Expected length of system, $L_s = \dfrac{\lambda}{\mu - \lambda} = \dfrac{2}{3-2} = 2$ customers,

Expected waiting time in queue, $W_q = \dfrac{\lambda}{\mu(\mu - \lambda)} = \dfrac{2}{2(3-2)} = 1$ hour,

and Expected waiting time in system, $W_s = \dfrac{1}{\mu - \lambda} = \dfrac{1}{3-2} = 1$ hour.

Mechanic B: In usual notations, it is given:

$$\lambda = 2 \text{ customers per hour, and } \mu = 4 \text{ customers per hour.}$$

∴ Expected length of queue, $L_q = \dfrac{\lambda^2}{\mu(\mu - \lambda)} = \dfrac{4}{4(4-2)} = \dfrac{1}{2}$ customer,

Expected length of system, $L_s = \dfrac{\lambda}{\mu - \lambda} = \dfrac{2}{4-2} = 1$ customer,

Expected waiting time in queue, $W_q = \dfrac{\lambda}{\mu(\mu - \lambda)} = \dfrac{2}{4(4-2)} = \dfrac{1}{4}$ hour,

and Expected waiting time in system, $W_s = \dfrac{1}{\mu - \lambda} = \dfrac{1}{4-2} = \dfrac{1}{2}$ hour.

Total (daily) cost = wages of the mechanic + Cost of waiting time.

∴ For mechanic *A*, total cost = ₹ 14 × 8 + ₹ 30 × 1 = ₹ 142

and For mechanic *B*, total cost = ₹ 20 × 8 + ₹ 30 × 1/2 = ₹ 175

Evidently, mechanic *A* has lower operating cost.

Example 20.21. *A road transport company has one reservation clerk on duty at a time. He handles information of bus schedules and makes reservations. Customers arrive at a rate of 8 per hour and the clerk can service 12 customers on an average per hour. After stating your assumptions, answer the following:*

(i) What is the average number of customers waiting for service of the clerk?

(ii) What is the average time a customer has to wait before getting service?

(iii) The management is contemplating to install a computer system to handle the information and reservations. This is expected to reduce the service time from 5 to 3 minutes. The additional cost of having the new system works out to ₹ 50 per day. If the cost of goodwill of having to wait is estimated to be 12 paise per minute spent on waiting before being served, should the company install the computer system? Assume 8 hours working day. ***[MBA, Madurai Univ. 2010]***

Solution. We are given

$$\lambda = 8 \text{ customers per hour, and } \mu = 12 \text{ customers per hour.}$$

(*i*) Average number of customers waiting for service of the clerk (in the system):

$$L_s = \frac{\lambda}{\mu - \lambda} = \frac{8}{12-8} = 2 \text{ customers.}$$

Average number of customers waiting for service of the clerk (in the queue):

$$L_q = \frac{\lambda^2}{\mu(\mu - \lambda)} = \frac{8 \times 8}{12(12-8)} \text{ or } 1.33 \text{ customers.}$$

(*ii*) Average waiting time of a customer (in the system) before getting service:

$$W_s = \frac{1}{\mu - \lambda} = \frac{1}{12-8} \text{ hour or 15 minutes.}$$

Average waiting time of a customer (in the queue) before getting service:

$$W_q = \frac{\lambda}{\mu(\mu-\lambda)} = \frac{8}{12(12-8)} = \frac{1}{6} \text{ hour or 10 minutes.}$$

(iii) We, now, calculate the difference between the goodwill cost of customers with one system and the goodwill cost of customers with an additional computer system. This difference will be compared with the additional cost of ₹ 50 per day of installing another computer system.

Let an arrival waits for W_q hours before being served and there are λ arrivals per hour. Then, expected waiting time for all customers in an 8-hour day with one system is equal to

$$8\lambda \times W_q, \textit{ i.e., } 8 \times 8 \times \frac{1}{6} \text{ hours or } \frac{64}{6} \times 60 \text{ minutes, } \textit{i.e.,} \text{ 640 minutes.}$$

The goodwill cost per day with one system = 640 × ₹ 0.12 = ₹ 76.80

The expected waiting time of a customer before getting service (when there is an additional computer system) is:

$$W_q^* = \frac{8}{20(20-8)} = \frac{8}{20 \times 12} \text{ or } \frac{1}{30} \text{ hour or 2 minutes.}$$

Thus, expected waiting time of customers in an 8-hour day with an additional computer system is:

$$8\lambda \times W_q^* = 8 \times 8 \times 2 \text{ minutes} = 128 \text{ minutes.}$$

∴ The total goodwill cost with an additional computer system

= 128 × ₹ 0.12 = ₹ 15.36

Hence, reduction in goodwill cost with the installation of a computer system

= ₹ 76.80 – ₹ 15.36 = ₹ 61.44

Whereas the additional cost of a computer system is ₹ 50 per day, ₹ 61.44 is the reduction in goodwill cost when additional computer system is installed, hence there will be net saving of ₹ 11.44 per day. It is, therefore, worthwhile to install a computer.

Example 20.22. *A warehouse has only one loading dock manned by three persons crew. Trucks arrive at the loading dock at an average rate of 4 trucks per hour and the arrival rate is Poisson distributed. The loading of a truck takes 10 minutes on an average and can be assumed to be exponentially distributed. The operating cost of a truck is ₹ 20 per hour and the members of the loading crew are paid @ ₹ 6 each per hour. Would you advise the truck owner to add another crew of three persons (loaders)?* **[MIB, Delhi Univ. 2011; MBA (Dec.), Visvesvaraya Tech. Univ. 2011; MBA (Oct.), Pune Univ. 2007]**

Solution. From the data of the problem, we have

$$\lambda = 4 \text{ trucks per hour, and } \mu = 6 \text{ trucks per hour.}$$

For Existing Crew:

Loading crew cost = Number of loaders × Hourly wage rate

= 3 × ₹ 6 = ₹ 18 per hour.

$$\text{Cost of waiting time} = \begin{pmatrix}\text{Expected waiting} \\ \text{time per truck}\end{pmatrix} \times \begin{pmatrix}\text{Expected arrivals} \\ \text{per hour}\end{pmatrix} \times \begin{pmatrix}\text{Hourly waiting} \\ \text{cost}\end{pmatrix}$$

$$= \frac{1}{(6-4)} \times 4 \times ₹\ 20 = ₹\ 40 \text{ per hour.}$$

∴ Total cost = (₹ 18 + ₹ 40) or ₹ 58 per hour.

With Proposed Crew Addition:

Loading crew cost = 6 × ₹ 6 = ₹ 36 per hour.

$$\text{Cost of waiting time} = \frac{4}{12-4} \times ₹\,20 = ₹\,10 \text{ per hour.}$$

∴ Total cost = (₹ 36 + ₹ 10) = ₹ 46 per hour.

Since, total hourly cost after the additional of another crew of three persons is less than the existing cost, the truck owner must add a crew of another three loaders.

Example 20.23. *In a large maintenance department, fitters draw parts from the parts store which is at present staffed by one store's man. The maintenance foreman is concerned about the time spent by fitters getting parts and wants to know if the employment of a store's labourer would be worthwhile. On investigation, it is found that (a) a simple queue situation exist; (b) fitters cost ₹ 2.50 per hour; (c) the store's man costs ₹ 2 per hour and can deal, on average, with 10 fitters per hour; (d) a labourer could be employed at ₹ 1.75 per hour and would increase the capacity of the store to 12 per hour; (e) on an average 8 fitters visit the store each hour.*

[MBA, Pondicherry Univ. 2006]

Solution. According to given information, we have

Mean arrival rate, $\lambda = 8$ fitters per hour.

$$\text{Mean service rate, } \mu = \begin{cases} 10 \text{ fitters per hour (without labourer)} \\ 12 \text{ fitters per hour (with labourer)} \end{cases}$$

We calculate the average number of customers in the system before and after the labourer is employed and compare the reduction in the resulting queuing cost with the increase in service cost.

Without Labourer: Average number of fitters in the system:

$$L_s = \frac{\lambda}{\mu - \lambda} = \frac{8}{10-8} = 4 \text{ fitters.}$$

∴ Cost per hour = 4 × ₹ 2.50 = ₹ 10

With Labourer:

$$L_s = \frac{\lambda}{\mu - \lambda} = \frac{8}{12-8} = 2 \text{ fitters.}$$

∴ Cost per hour = 2 × ₹ 2.50 + ₹ 1.75 = ₹ 6.75

Since, the cost with labourer is less than the cost without labourer, it is worth while to engage labourer.

PROBLEMS

20.01. The computer lab at State University has a help desk to assist students working on computer spreadsheet assignments. The students patiently form a single line in front of the desk to wait for help. Students are served based on a first-come, first-served priority rule. On average, 15 students per hour arrive at the help desk. Student arrivals are best described using a Poisson distribution. The help desk server can help an average of 20 students per hour, with the service rate being described by an exponential distribution. Calculate the following operating characteristics of the service system:

(*a*) The average utilization of the help desk server,
(*b*) The average number of students in the system,
(*c*) The average number of students waiting in line,
(*d*) The average time a student spends in the system,
(*e*) The average time a student spends waiting in line.

[PG Dipl., Indian Inst. of Mat. Management 2019]

20.02. Customers arrive at a booking office window being manned by a single individual at a rate of 25 per hour. The time required to serve a customer has exponential distribution with a mean of 120 seconds. Find the average waiting time of a customer. *[MBA, Lucknow Univ. 2009]*

20.03. Customers arriving at bank wait in a single queue for the next available teller. Customer arrivals can be modeled by a Poisson distribution that has a mean of 70 per hour during the midmorning hours. A teller can process on an average of 100 customers per hour, which can be modeled by an exponential distribution. If there is one teller on duty, determine:

(*i*) the average time a customer waits in line for the next teller,
(*ii*) the average number of customers waiting in the queue,
(*iii*) average service time,
(*iv*) system utilization, and
(*v*) the probability of no customer waiting in queue. *[MBA (Oct.), Delhi Univ. 2010]*

20.04. A mechanic finds that time spent on his job has an exponential distribution with a mean 40 minutes. If he repairs the machine on first-come-first-served basis, and if the arrival of machines is with on an average rate of 8 per 8-hour day, then answer the following:

(*i*) What is his expected idle time each day?
(*ii*) What is the average number of machines in the system?
(*iii*) What is the probability of having more than 8 machines? *[M.Com., Delhi Univ. 2010]*

20.05. Trucks arrive at a bridge according to Poisson pattern. The time required to weight the truck is exponential. The mean arrival rate is 12 trucks per day and average service rate is 18 trucks per day. Find probability of no trucks. Utilization factor and probability that a truck waits.
[PG Dipl., Indian Inst. of Mat. Management 2016]

20.06. A cashier at a bank takes on an average of 3 minutes to serve a person, the service time being exponentially distributed. Customers arrive at his counter in Poisson fashion, the average arrival rate being one customer every 4 minutes. Find:

(*i*) the probability that there are more than two customers in the queue,
(*ii*) the probability that there is at least one customer in the queue,
(*iii*) the probability that there is no one at the counter, and
(*iv*) the average time a customer spends in the bank.

20.07. A departmental store has a single cashier. During rush hours, customers arrive at a rate of 20 customers per hour. The average number of customers that can be processed by the cashier is 24 per hour. Assume that the conditions for use of single-channel queuing model apply. Required:

(*i*) What is the probability that the cashier is idle?
(*ii*) What is the average number of customers in the queuing system?
(*iii*) What is the average time a customer spends in the system?
(*iv*) What is the average number of customers in the queue?
(*v*) What is the average time a customer spends in the queue waiting for service?
[MBA, Gujarat Univ. 2008]

20.08. At a public telephone booth in a Post Office arrivals are considered to be Poisson with an average inter-arrival time of 12 minutes. The length of a phone call may be assumed to be distributed exponentially with an average of 4 minutes.

(*a*) What is the probability that a fresh arrival will not have to wait for the phone?
(*b*) What is the probability that an arrival will have to wait more than 10 minutes before the phone is free?
(*c*) What is the average length of queue that form from time to time?

20.09. At a certain petrol pump, customers arrive in a Poisson process with an average time of 5 minutes between arrivals. The time interval between servers at the petrol pump follows an exponential distribution and the mean time taken to service an arrival is 2 minutes. Find the following:

(*a*) Expected average queue length.
(*b*) Average number of customers in the system.
(*c*) Average time a customer has to wait in the queue.
(*d*) Average time a customer has to spend in the system.
(*e*) By how much should the flow of customers be increased to justify the opening of another service point, where the customer has to wait for 5 minutes for the service?

[BFIA, Delhi Univ. 2005; M.Com., Madurai Univ. 2003]

20.10. Consider a self-service store with one cashier. Assume Poisson arrivals and exponential service times. Suppose that nine customers arrive on an average every 5 minutes, and the cashier can serve 10 customers in 5 minutes. Find:

(*a*) Average number of customers queuing for service,
(*b*) Probability of having more than 10 customers in the system, and
(*c*) Probability that a customer has to queue for more than 2 minutes.

[MBA (Jan.), Pune Univ. 2000]

20.11. A repair shop has a single mechanic to attend customers who bring small appliances for repairs. The mechanic inspects the appliances for defects and fixes them right away or renders a diagnosis. All this takes him 6 minutes on an average. If 4 customers visit the shop per hour on an average, and if arrivals and service times are exponential, determine:

(*i*) proportion of time during which the shop is empty,
(*ii*) probability of finding at least one customer in the shop,
(*iii*) the average number of customers in the shop, and
(*iv*) the average time spent by a customer in the shop. *[M.Com., Delhi Univ. 2008]*

20.12. Telephone users arrive at a booth following a Poisson distribution with an average time of 5 minutes between one arrival and the next. The time taken for a telephone call is on an average 3 minutes and it follows an exponential distribution. What is the probability that the booth is busy. When the customers would have to wait for at least seven minutes for their turn to make a call, how much should be the flow of arrivals in order to justify second booth?

[B.Tech. (Mech.), HP Univ. 2015; BE (Mech.), Karnataka Univ. 2010]

20.13. A fertilizer company distributes its products by trucks loaded at its only loading station. Both, company trucks and contractor's trucks are used for this purpose. It was found out that on an average every 5 minutes one truck arrived and the average loading time was 3 minutes, 50% of the trucks belong to the contractors. Making suitable assumptions, determine:

(*i*) the probability that a truck has to wait,
(*ii*) the waiting time of a truck that waits, and
(*iii*) the expected waiting time of contractor's trucks per day, assuming a 24 hours shift.

[MBA (Oct.), IGNOU 2008; M.Com., Delhi Univ. 2006 (C.C.)]

20.14. A foreign bank is considering opening a drive-in window for customer service. Management estimates that customers will arrive for service at the rate of 12 customers per hour. The teller whom it is considering to staff the window can serve customers at the rate of one every three minutes. Assuming Poisson arrivals and Exponential service, find:

(*i*) Utilization of teller,
(*ii*) Average number of customers in the system,

(*iii*) Average waiting time of a customer in the queue, and
(*iv*) Average waiting time of a customer in the system. ***[MBA (PT), Delhi Univ. 2010]***

20.15. A maintenance service facility has Poisson arrival rates, negative exponential service times, and operates on a first-come-first-served queue discipline. Breakdowns occur on an average of three per day with a range of zero to eight. The maintenance crew can service on an average six machines per day with a range from zero to seven. Find the following:
(*i*) Utilization factor of the service facility,
(*ii*) Mean number of machines in the system — breakdown or repair,
(*iii*) Mean waiting time of a machine in the queue,
(*iv*) Probability of finding two machines in the system, and
(*v*) Expected number of machines in the queue. ***[MBA, Mahatma Gandhi Univ. 2010]***

20.16. In the central railway station, 15 computerized reservation counters are available. A customer can book his/her ticket in any train on any day from any one of these computerized reservation counters. The average time spent per customer by each clerk is 5 minutes. Average arrivals per hour during three types of activity periods have been calculated and customers have been surveyed to determine how long they are willing to wait during each type of period.

Type of period	*Arrivals/hr*	*Customer's acceptable waiting time*
Peak	110	15 minutes
Normal	60	10 minutes
Low	30	5 minutes

Making suitable assumptions on this queuing process, determine how many counters should be kept open during each type of period. ***[PGDM, Mumbai Univ. 2012]***

20.17. City Hospital has one doctor to attend patients with diabetes. Arrival of patients is considered to follow Poisson distribution with an average time of 12 minutes between one arrival and the next. The length of time the patient spends with the doctor is 4 minutes and is considered to be distributed exponentially.
(*i*) What is the probability that a patient arriving at the hospital will have to wait?
(*ii*) What portion of the time doctor is likely to be idle?
(*iii*) How many patients would be expected to be in the hospital's service area?
(*iv*) What is the average length of the queue that form from time to time?
(*v*) The hospital will appoint a second doctor when convinced that patients would wait for at least 5 minutes. By how much the flow of patients must increase in order to justify the second doctor? ***[PG Dip. in Glob. Bus. Oper., Delhi Univ. 2008]***

20.18. Delhi Medical Store has only one cashier who handles all the payments. On an average, the cashier can serve 5 customers per 15 minutes, who arrive at his counter randomly on an average rate of 18 per hour. The management of the store noticed that the cashier was idle many times and some of the customers complained against long waits number of times. It was, therefore, decided (by making suitable assumptions) to investigate the following:
(*i*) What portion of the time the cashier is likely to be idle?
(*ii*) What is the average length of the waiting line to be expected under existing servicing conditions?
(*iii*) How many customers would be expected to be in the service area?
(*iv*) What is the average length of time that a customer would be expected to wait to pay for his purchases?

(*v*) Suppose that the management of the store will employ another cashier, if they are convinced that a customer has to wait for at least five minute for making payments. What arrival rate would justify the employment of second cashier?

[PG Dip. in Glob. Bus. Oper, Delhi Univ. 2011]

20.19. The XYZ Company's quality control department is managed by a single clerk, who takes on an average 5 minutes in checking parts of each of the machine coming for inspection. The machines arrive once in every 8 minutes on an average. One hour of the machine is valued at ₹ 15 and a clerk's time is valued at ₹ 4 per hour. What are the average hourly queuing system costs associated with the quality control department. *[B.Tech. (Mech.), HP Univ. 2014]*

20.20. A mechanic is to be recruited to repair scooters that breakdown at an average rate of 6 per hour. The breakdowns follow Poisson distribution. Non-productive time of a scooter costs ₹ 40 per hour. Mechanics *A* and *B* have been interviewed for the purpose. Mechanic *A* charges ₹ 20 per hour, and services scooters at the rate of 8 per hour. Mechanic *B* charges ₹ 28 per hour, and services 12 scooters per hour, on an average. Which one of the two should be recruited and why? Assume an 8 hours of work each day. *[M.Com., Delhi Univ. 2007]*

20.21. Goods trucks arrive randomly at a stockyard with a mean of 8 trucks/hour. A crew of four operatives can unload a truck in 6 minutes. Trucks waiting in queue to be unloaded are paid a waiting charge at the rate of ₹ 60 per hour. Operatives are paid a wage rate of ₹ 20 per hour. It is possible to augment the crew strength to 2 or 3 (of four operatives per crew) when the unloading time will be 4 minutes or 3 minutes respectively per truck. Find the optimal crew size. *[M.Tech., Panjab Tech. Univ. 2012; BE, Karnataka Univ. 2011]*

22.22. The average time between successive arrivals to a repair shop, which works 8 hours a day is 30 minutes. The shop has one mechanic who can repair the incoming vehicles at an average rate of 3 per hour. The mechanic is paid at ₹ 14 per hour, while the cost of waiting time (in terms of customer dissatisfaction and goodwill) is ₹ 20 per hour of the time spent waiting in the queue. The owner is contemplating to replace the mechanic by another one who demands ₹ 18 per hour and who can repair 4 vehicles on the average. Under conditions of single serve model, calculate the total cost per day at present, and the total cost per day if the present mechanic is replaced. Is it advisable to replace the existing mechanic?

[M.Com., Delhi Univ. 2005]

20:9. POISSON-EXPONENTIAL SINGLE SERVER AND FINITE POPULATION, *i.e.*, (*M*/*M*/1) : (*N*/*FCFS*) MODEL

This queuing model is based on the similar assumptions as discussed in *Section 20:8*, except that the input population is finite. For this model, we have a total of *N* customers, wherein a customer is either in the system or outside the system, in a sense of arriving. The system is consisting of a queue and a single service station (facility). When a customer is in the arriving condition, the time it takes to arrive is a random variable having exponential distribution with mean $1/\lambda$. When there are n customers in the system, then there are $(N - n)$ customers in the arriving state and, therefore, total arrival rate of customers in the system is $\lambda\,(N - n)$.

In previous *Section 20:8*, we had obtained the probability of n customers as $P_n = \rho^n P_0$, wherein population was considered as infinite, *i.e.*, $N \to \infty$. But, this model admits the condition that $n \le N$, which in turn will give different value of P_0. For obtaining the value of P_0, we make use of the boundary condition, $\sum_{n=0}^{N} P_n = 1$. Therefore,

$$1 = P_0 \sum_{n=0}^{N} \rho^n = P_0 \begin{cases} \dfrac{1-\rho^{N+1}}{1-\rho}, & \rho \neq 1 \\ N+1, & \rho = 1 \end{cases} \quad \text{or} \quad P_0 = \begin{cases} \dfrac{1-\rho}{1-\rho^{N+1}}, & \rho \neq 1 \\ \dfrac{1}{N+1}, & \rho = 1 \end{cases}$$

Hence, $$P_n = \begin{cases} \dfrac{(1-\rho)\rho^n}{1-\rho^{N+1}}, & \rho \neq 1 \\ \dfrac{1}{N+1}, & \rho = 1 \end{cases}$$

Remark. If $N \to \infty$, then $P_n = (1-\rho)\,\rho^n$, same as obtained in *Section 20:8*.

Operating Characteristics

When, N is the number of customers in the source population, $1/\lambda$ as the average inter-arrival time between successive arrivals, and μ as the service rate; the following formulas can be used to compute the steady-state operating characteristics for single channel queuing model:

1. The probability that no customer is in the system:

$$P_0 = \frac{1-\rho}{1-\rho^{N+1}}; \quad \text{for } \rho \neq 1 \text{ and } \rho < 1,$$

where, $\rho = \dfrac{\lambda}{\mu}$ is the utilization factor.

2. The probability that there shall be n customers in the system:

$$P_n = \left(\frac{1-\rho}{1-\rho^{N+1}}\right)\rho^n; \quad \rho \neq 1 \text{ and } n = 0, 1, 2, \ldots, N.$$

3. Average number of customers in the system:

$$L_s = \sum_{n=0}^{N} n P_n = \sum_{n=0}^{N} \frac{n(1-\rho)\rho^n}{(1-\rho^{N+1})}; \quad \rho \neq 1$$

$$= \left(\frac{1-\rho}{1-\rho^{N+1}}\right) \sum_{n=0}^{N} n\rho^n = \left(\frac{1-\rho}{1-\rho^{N+1}}\right) \rho \sum_{n=0}^{N} \frac{d}{d\rho}\rho^n$$

Now, $$\rho \sum_{n=0}^{N} \frac{d}{d\rho}\rho^n = \rho \frac{d}{d\rho}\left(\sum_{n=0}^{N} \rho^n\right) = \rho \frac{d}{d\rho}(1+\rho+\rho^2+\ldots+\rho^N)$$

$$= \rho + 2\rho^2 + 3\rho^3 + \ldots + N\rho^N$$

$\therefore$ $$L_s = \left(\frac{1-\rho}{1-\rho^{N+1}}\right)(\rho + 2\rho^2 + 3\rho^3 + \ldots + N\rho^N)$$

$$= \left[\frac{\rho}{(1-\rho)} - \frac{(N+1)\rho^{N+1}}{(1-\rho^{N+1})}\right]$$

$$= \frac{\rho\left[1-(N+1)\rho^N + N\rho^{N+1}\right]}{(1-\rho)(1-\rho^{N+1})}$$

4. Average number of customers in the queue:

$$L_q = \sum_{n=1}^{N} (n-1)P_n = \sum_{n=1}^{N} nP_n - \sum_{n=1}^{N} P_n$$

$$= L_s - (1-P_0) = L_s - \frac{\rho(1-\rho^N)}{1-\rho^{N+1}}, \text{ since } P_0 = \frac{1-\rho}{1-\rho^{N+1}}, \; \rho \neq 1$$

$$= \frac{\rho^2\left[1 - N\rho^{N-1} + (N-1)\rho^N\right]}{(1-\rho)(1-\rho^{N+1})}.$$

5. Average waiting time a customer spends in the system: $W_s = L_s / \lambda'$, where $\lambda' = \lambda(1-P_N)$ is the mean rate of customers entering the system.

6. Average waiting time of a customer in the queue:

$$W_q = W_s - 1/\mu \text{ or } L_q/\lambda\,(1-P_N)$$

7. The probability that an arriving customer has to wait for services is:

$$P(w > 0) = 1 - P_0$$

Example 20.24. *Consider a single server queuing system with Poisson arrival and exponential service time. Suppose, the mean arrival rate is 3 calling units per hour, the expected service time is 0.25 hours and the maximum permissible number of calling units in the system is 2. Derive the steady-state probability distribution of the number of calling units in the system. Also, calculate the expected number of calling units in the system.*

Solution. We are given:

$$\lambda = 3 \text{ units per hour, and } 1/\mu = 0.25 \text{ or } \mu = 4 \text{ units per hour.}$$

The utilization factor is:

$$\therefore \quad \rho = \frac{\lambda}{\mu} = \frac{3}{4} = 0.75$$

Also, we are given $N = 2$

$\therefore$ The steady-state probability distribution of the number of n customers (calling units) in the system is:

$$P_n = \frac{(1-\rho)\rho^n}{1-\rho^{N+1}} = \frac{(1-0.75)(0.75)^n}{1-(0.75)^{2+1}} = (0.43)(0.75)^n$$

and

$$P_0 = \frac{(1-\rho)}{1-\rho^{N+1}} = \frac{1-0.75}{1-(0.75)^{2+1}} = \frac{0.25}{1-(0.75)^3} = 0.43$$

The expected number of calling units in the system:

$$L_s = \frac{0.75}{1-0.75} - \frac{3\times(0.75)^3}{1-(0.75)^3} = \frac{0.75}{0.25} - \frac{3\times0.42}{1-0.42}$$

$$= 3 - \frac{3\times0.42}{0.58} = 3\left(1 - \frac{0.42}{0.58}\right) = \frac{3\times0.16}{0.58} = \frac{0.48}{0.48}$$

$$= 0.83 \text{ or (one approx.).}$$

Example 20.25. *At a railway station, only one train is handled at a time. The railway yard is sufficient only for two trains to wait, while other is given signal to leave the station. Trains arrive at the station at an average rate of 6 per hour and the railway station can handle them on an average of 12 per hour. Assuming Poisson arrivals and exponential service distribution, find the steady-state probabilities for the various number of trains in the system. Also, find the average waiting time of a new train coming into the yard.*

[B.Sc. (Math.) (Nov.), Madras Univ. 2012]

Solution. We are given:

$$\lambda = 6 \text{ trains per hour, and } \mu = 12 \text{ trains per hour.}$$

$\therefore$ The traffic intensity $\rho = \dfrac{\lambda}{\mu} = \dfrac{6}{12} = \dfrac{1}{2}$ or 0.5

The maximum queue length is of 2 trains, *i.e.*, the maximum number of trains in the system is 3 (= N). The probability that there is no train in the system (both waiting and in service) is given by

$$P_0 = \frac{1-\rho}{1-\rho^{N+1}} = \frac{1-0.5}{1-(0.5)^{3+1}} = \frac{0.5}{0.937} = 0.53$$

Now, since $P_n = \rho^n P_0$, we get

$$P_1 = \rho P_0 = (0.5) \times (0.53) = 0.27,$$
$$P_2 = \rho^2 P_0 = (0.5)^2 \times (0.53) = (0.5) \times 0.27 = 0.135,$$
$$P_3 = \rho^3 P_0 = (0.5)^3 \times (0.53) = (0.5) \times 0.27 \times 0.135 = 0.67,$$

The expected number of trains in the system:

$$L_s = \frac{0.5}{1-0.5} - \frac{4\times(0.5)^4}{1-(0.5)^4} = \frac{0.5}{0.5} - \frac{4\times 0.0625}{1-0.0625}$$
$$= 1 - \frac{0.250}{0.937} = \frac{0.687}{0.937} = 0.74$$

Thus, the average number of trains in the system is 0.74 and each train takes on an average $\dfrac{1}{12}$ (= 0.08) hours for getting service. As the arrival of new train expects to find an average of 0.74 trains in the system before it, the average waiting time of a new train coming in the yard is:

$$W_q = (0.74)(0.08) \text{ hours} = 0.0592 \text{ hours or } 3.5 \text{ minutes.}$$

Example 20.26. *Assume that goods trains are coming in a yard at the rate of 30 trains per day and suppose that the inter-arrival times follow an exponential distribution. The service time for each train is assumed to be exponential with an average of 36 minutes. If the yard can admit 9 trains at a time (there being 10 lines, one of which is reserved for shunting purposes), calculate the probability that the yard is empty and find the average queue length.*

Solution. We have

$$\lambda = \frac{30}{60\times 24} = \frac{1}{48} \text{ trains per minute, and } \mu = \frac{1}{36} \text{ trains per minute.}$$

$\therefore$ Traffic intensity $\rho = \dfrac{\lambda}{\mu} = \dfrac{36}{48} = 0.75$

The probability that the yard is empty:

$$P_0 = \frac{1-\rho}{1-\rho^{N+1}} = \frac{1-0.75}{1-(0.75)^{10}}, \text{ since } N = 9$$
$$= \frac{0.25}{0.94} = 0.27$$

Average queue length is given by

$$L_q = L_s - \frac{\lambda}{\mu} = \frac{\rho}{1-\rho} - \frac{(N+1)\rho^{N+1}}{1-\rho^{N+1}} - \rho$$

$$\therefore \quad L_q = \frac{0.75}{1-0.75} - \frac{(9+1)(0.75)^{9+1}}{1-(0.75)^{9+1}} - 0.75 = 3 - \frac{10\times(0.75)^{10}}{1-(0.75)^{10}} - 0.75$$

$$= 2.25 - \frac{0.563}{0.943} = 2.25 - 0.60 = 1.65$$

$$= 1.65 \text{ trains or } 2 \text{ trains (approx.)}.$$

Example 20.27. *If for a period of 2 hours in the day (8 to 10 a.m.) trains arrive at the yard every 20 minutes, but the service time continues to remain 36 minutes, then calculate for this period:*

(*a*) *The probability that the yard is empty.*

(*b*) *Average number of trains in the system, on the assumption that the line capacity of the yard is limited to 4 trains only.* **[B.Tech. (Mech.) (Nov.), HP Univ. 2010]**

Solution. Here, we have

$$\lambda = \frac{1}{20} \text{ trains per minute, and } \mu = \frac{1}{36} \text{ trains per minute.}$$

The traffic intensity, $\rho = \frac{\lambda}{\mu} = \frac{36}{20} = 1.8$

(*a*) The probability that the yard is empty:

$$P_0 = \frac{1-\rho}{1-\rho^{N+1}} = \frac{1-1.8}{1-(1.8)^{4+1}}, \quad \text{since } N = 4$$

$$= \frac{-0.8}{-17.9} = 0.04$$

(*b*) Average number of trains in the system:

$$L_s = \frac{\rho}{1-\rho} - \frac{(N+1)\rho^{N+1}}{1-\rho^{N+1}}$$

$$= \frac{1.8}{1-1.8} - \frac{(4+1)(1.8)^{4+1}}{1-(1.8)^{4+1}} = \frac{1.8}{-0.8} - \frac{5\times 18.9}{-17.9}$$

$$= \frac{94.5}{17.9} - \frac{1.8}{0.8} = 5.28 - 2.25 = 3.03 \text{ or } 3 \text{ trains.}$$

PROBLEMS

20.23. Let there be an automobile inspection situation with 3 inspection stalls. The station can accommodate 4 cars waiting (7 in station) at one time. The arrival pattern is Poisson with a mean of one car every minute. The service time is exponential with mean of 6 minutes per car. Find the (*a*) expected number of cars in the queue; (*b*) expected number of cars in the system; (*c*) expected waiting time in the system; and (*d*) expected number of cars per hour that can not enter the system.

20.24. A Parlor has space to accommodate only 10 customers. He can service only one person at a time. If a customer comes to his shop and finds it full, he goes to another parlor. Customers randomly, arrive at an average rate of 10 per hour and the service time of attendants at parlor is negative exponential with an average of 5 minutes per customer. Find P_0 and P_n.

[B.Tech. Hyderabad Univ. 2011]

20.25. Patients arrive at a clinic according to a Poisson distribution at a rate of 30 patients per hour. The waiting room does not accommodate more than 14 patients. Examination time per patient is exponential with mean rate of 20 per hour. Find:

(*i*) The effective arrival rate at the clinic.

(*ii*) What is the probability that an arriving patient will not wait?

(*iii*) What is the expected waiting time until a patient is discharged from the clinic?

[B.Tech. (Nov.), Hyderabad Univ. 2010; MBA (Nov.), Madras Univ. 2009]

20.26. There are 45 clients on a LAN with a single database server. Each client issues a request every 10 seconds on an average to the database server. The server processes the request in approximately 0.5 seconds. clients' requests are assumed to follow the Poisson distribution and server's processing time follows exponential distribution:

(*i*) What is the system utilization?

(*ii*) How long does an average request take on the server (waiting if any and processing)?

(*iii*) What is the average number of requests waiting to be processed?

[MBA, Delhi Univ. 2012]

20.27. In a car-wash service facility, cars arrive for service according to a Poisson distribution with mean 5 per hour. The time for washing and cleaning each car varies but is found to follow an exponential distribution with mean 10 minutes per car. The facility cannot handle more than one car at a time and has a total of 5 parking spaces.

(*i*) Find the effective arrival rate.

(*ii*) What is the probability that an arriving car will get service immediately upon arrival?

(*iii*) Find the expected number of parking spaces occupied.

20.28. A petrol station has a single pump and space for not more than 3 cars (2 waiting, I being served). A car arriving when the space is filled to capacity goes elsewhere for petrol. Cars arrive according to a Poisson distribution at a mean rate of one every 8 minutes. Their service time has an exponential distribution with a mean of 4 minutes.

The proprietor has the opportunity of renting an adjacent piece of land, which would provide space for an additional car to wait (He cannot build another pump.). The rent would be ₹ 10 per week. The expected net profit from each customer is ₹ 0.50 and the station is open 10 hours every day. Would it be protiable to rent the additional space?

20:10. POISSON-EXPONENTIAL MULTIPLE SERVER AND INFINITE POPULATION, *i.e.*, (*M*/*M*/*S*) : (∞/*FCFS*) MODEL

When multiple servers (service stations) are involved, there may be separate independent queue for each server or the customers form a single queue from which they are picked up for service. If there are on as many queues as the service stations, equal to, say '*S*', the system can be considered as of *S* different single server queuing systems. For our purpose, here, we assume that the customers form a single queue and as soon as a service station becomes free, the customer is picked up from the queue for service.

The model is based on the following assumptions:

1. The arrivals follow a Poisson probability distribution, the average arrival rate being λ.
2. The service time for each service station follows an exponential probability distribution.
3. There are *k* service stations, each of which provides identical service.
4. The mean service rate μ is same for each service station.

5. The arrivals wait in a single queue and then move to first open service station for service on a FCFS basis.
6. The input population is infinite.
7. The arrival rate is smaller than the combined service rate of all 'S' service facilities.

Operating Characteristics

The following formulas can be used to compute the steady-state operating characteristics for multiple service facilities and single queue model, where

λ = mean arrival rate for the system,

μ = mean service rate for each service station,

S = number of servers (service stations),

$S\mu$ = mean service rate for the multiple service facility system.

$\rho = \lambda/S\mu$, the utilization factor of the entire system.

1. Probability that no customer is in the system:

$$P_0 = \left[\sum_{n=0}^{S-1} \frac{(\lambda/\mu)^n}{n!} + \frac{(\lambda/\mu)^S}{S!}\left(\frac{S\mu}{S\mu-\lambda}\right)\right]^{-1}, \quad \text{provided } \frac{\lambda}{S\mu} < 1$$

2. Average number of customers in the queue:

$$L_q = \frac{(\lambda/\mu)^S \lambda\mu}{(S-1)!(S\mu-\lambda)^2} P_0$$

3. Average number of customers in the system: $L_s = L_q + \dfrac{\lambda}{\mu}$

4. Average time a customer spends in the queue: $W_q = \dfrac{L_q}{\lambda}$

5. Average time a customer spends in the system:

$$W_s = W_q + \frac{1}{\mu} \quad \text{or} \quad W_s = \frac{L_s}{\lambda}$$

6. Probability that an arriving customer has to wait for service:

$$P(N \geq S) = P_w = \frac{1}{S!}\left(\frac{\lambda}{\mu}\right)^S \left(\frac{S\mu}{S\mu-\lambda}\right) P_0$$

7. Probability that a customer enters the service without wait: $1 - P(n \geq S)$
8. Probability of n customers in the system:

$$P_n = \frac{1}{n!}\left(\frac{\lambda}{\mu}\right)^n P_0, \quad 1 \leq n \leq S, \quad \text{and} \quad P_n = \frac{1}{S!\,S^{n-S}}\left(\frac{\lambda}{\mu}\right)^n P_0, \quad n > S$$

Remark. The result obtained above is valid only if $\dfrac{\lambda}{S\mu} < 1$; that is, the mean arrival rate must be less than the mean maximum potential service rate of the system.

Example 20.28. *A telephone exchange has two long-distance operators. The telephone company finds that during the peak load, long-distance calls arrive in a Poisson fashion at an average rate of 15 per hour. The length of service on these calls in approximately distributed with mean length of 5 minutes.*

(a) *What is the probability that a subscriber will have to wait for his long distance call during the peak hours of the day?*

(b) *If the subscribers wait and are serviced in turn, what is the expected waiting time?*

Establish the formulae used.

Solution. We are given

$$S=2,\ \lambda=\frac{15}{60} \text{ or } \frac{1}{4} \text{ arrivals, and } \mu=\frac{1}{5} \text{ service per minute}$$

$$\therefore \quad P_0=\left[\sum_{n=0}^{2-1}\frac{1}{n!}\left(\frac{5}{4}\right)^n+\frac{1}{2!}\left(\frac{2\times 1/5}{2/5-1/4}\right)\right]^{-1}$$

$$=\left[1+\frac{5}{4}+\frac{1}{2}\times\left(\frac{5}{4}\right)^2\left(\frac{8}{8-5}\right)\right]^{-1}=\frac{3}{13}$$

(a)
$$P_w=\left[\frac{1}{S!}\left(\frac{\lambda}{\mu}\right)^S\frac{S\mu}{S\mu-\lambda}\right]P_0$$

$$=\left[\frac{1}{2!}\left(\frac{5}{4}\right)^2\times\frac{8}{3}\right]\times\frac{3}{13}=\frac{25}{52} \text{ or } 0.48$$

(b)
$$W_q=\frac{1}{\lambda}L_q=\frac{1}{\lambda}\times\frac{(\lambda/\mu)^S\,\lambda\mu}{(S-1)!(S\mu-\lambda)^2}P_0$$

$$=\left[4\times\left(\frac{5}{4}\right)^2\times\frac{1}{20}\times\left(\frac{20}{3}\right)^2\right]\times\frac{3}{13}=\frac{125}{39} \text{ or } 3.2 \text{ minutes.}$$

Example 20.29. *Given an average arrival rate of 20 customers per hour, is it better for a customer to get service at single service channel with mean service rate of 22 customers or at one of the two service channels in parallel, with mean service rate of 11 customers for each of the two service channels? Assume that both queues are of M/M/S type.*

Solution. For a single service channel, $\lambda = 20$ customers per hour, and $\mu = 22$ customers per hour.

$$\therefore \quad P_0=1-\rho=1-\frac{20}{22}=\frac{1}{11} \text{ or } 0.09$$

Average queue length:

$$L_q=\frac{\lambda^2}{\mu(\mu-\lambda)}=\frac{20\times 20}{22(22-20)}=\frac{100}{11} \text{ or } 9.09$$

Average waiting time of a customer in the queue:

$$W_q=\frac{L_q}{\lambda}=\frac{9.09}{20}=0.45$$

Now, when there are two parallel service channels, we have $\lambda = 20$, $\mu = 11$ and $S = 2$.

$$\therefore \quad P_0=\left[\sum_{n=0}^{2-1}\frac{1}{n!}\left(\frac{20}{11}\right)^n+\frac{1}{2!}\left(\frac{20}{11}\right)^2\left(\frac{2\times 11}{2\times 11-20}\right)\right]^{-1}$$

$$=\left[\frac{31}{11}+\frac{200}{11}\right]^{-1}=\frac{11}{231} \text{ or } 0.047$$

Average queue length:

$$L_q = \left[\frac{(20/11)^2 \times 20 \times 11}{(2-1)!(2 \times 11 - 20)^2}\right] \times 0.047 = \frac{2000}{11} \times 0.047 = 8.54.$$

Average waiting time of a customer:

$$W_q = \frac{L_q}{\lambda} = \frac{8.54}{20} = 0.43$$

On comparison, we observe that it is better for a customer to wait for two service channels, since waiting time of a customer in two service channels is less than the waiting time in case of a single service channel.

Example 20.30. *A super-market have two sales girls serving at the counters. The customers arrive in a Poisson fashion at the rate of 12 customers per hour. The service time for each customer is exponential with mean 6 minutes. Find:*

(*i*) *the probability that an arriving customer has to wait for service,*

(*ii*) *the average number of customers in the system, and*

(*iii*) *the average time spent by a customer in the super-market.*

[*MBA, Madras Univ. 2010; MBA, Delhi Univ. 2009*]

Solution. We are given $\lambda = 12$ customers per hour, $\mu = 10$ customers per hour, and $S = 2$ girls.

$$\therefore \qquad P_0 = \left[\sum_{n=0}^{2-1} \frac{1}{n!}\left(\frac{12}{10}\right)^n + \frac{1}{2!}\left(\frac{12}{10}\right)^2 \frac{2 \times 10}{20-12}\right]^{-1} = \frac{1}{4} \text{ (or 0.25)}$$

(*i*) Probability of having to wait for service:

$$P(w > 0) = \frac{1}{S!}\left(\frac{\lambda}{\mu}\right)^S \frac{S\mu}{(S\mu - \lambda)} P_0 = \frac{1}{2}\left(\frac{12}{10}\right)^2 \frac{20}{20-12} \times \frac{1}{4} = 0.45$$

(*ii*) Average queue length is:

$$L_q = \frac{\lambda\mu\left(\frac{\lambda}{\mu}\right)^S P_0}{(S-1)!(S\mu-\lambda)^2} = \frac{12 \times 10 \times (1.2)^2 \times 0.25}{(2-1)!(20-12)^2} = \frac{27}{40}$$

Average number of customers in the system:

$$L_s = L_q + \frac{\lambda}{\mu} = \frac{27}{40} + \frac{12}{10} = 1.87 \text{ (or 2 customers) approx.}$$

(*iii*) Average time spent by a customer in supermarket:

$$W_s = \frac{L_s}{\lambda} = \frac{1.87}{12} = 0.156 \text{ hours or 9.3 minutes.}$$

Example 20.31. *A bank has two tellers working on savings accounts. The first teller handles withdrawals only. The second teller handles deposits only. It has been found that the service time distribution for both deposits and withdrawals is exponential with mean service time of 3 minutes per customer. Depositors are found to arrive in Poisson fashion throughout the day with mean arrival rate of 16 per hour. Withdrawers also arrive in Poisson fashion with mean arrival rate of 14 per hour. What would be the effect on the average waiting time for depositors and withdrawers, if each teller could handle both withdrawals and deposits? What could be the effect, if this could be accomplished by increasing the mean service time to 3.5 minutes?*

[*Dip. in Management (Dec.), AIMA 1998*]

Solution. Initially we have two independent queuing systems for withdrawers and depositors with input as Poisson distribution and service as exponential distribution.

For Withdrawers:

$\lambda = 14$ customers per hour, and $\mu = 3$ customers per minute or 20 customers per hour.

Average waiting time in the queue:

$$W_q = \frac{14}{20(20-14)} = \frac{14}{20 \times 6} = \frac{7}{60} \text{ hour or 7 minutes.}$$

For Depositors:

$\lambda = 16$ customers per hour, and $\mu = 3$ customers per minute or 20 customers per hour.

Average waiting time in the queue:

$$W_q = \frac{16}{20(20-16)} = \frac{16}{20 \times 4} = \frac{1}{5} \text{ hour or 12 minutes.}$$

If each teller could handle both withdrawals and deposits, we have a common queue with two servers. The queuing system is, thus, with 2 service channels and $\lambda = 14 + 16 = 30$ customers per hour and $\mu = 20$ customers per hour.

$$\therefore \quad P_0 = \left[\sum_{n=0}^{2-1} \frac{1}{n!}\left(\frac{30}{20}\right)^n + \frac{1}{2!}\left(\frac{30}{20}\right)^2 \frac{2 \times 20}{(2 \times 20 - 30)} \right]^{-1} = \frac{1}{7} = 0.14$$

Average waiting time of arrivals in the queue:

$$W_q = \left(\frac{30}{20}\right)^2 \times \frac{20}{(40-30)^2} \times \frac{1}{7} = \frac{9}{140} \text{ hours or 3.86 minutes.}$$

When the service time is increased to 3.5 minutes,

$\lambda = 30$ customers per hour, and $\mu = 120/7$ customers per hour or 17.14 customers per hour.

$$\therefore \quad P_0 = \left[\sum_{n=0}^{2-1} \frac{1}{n!}\left(\frac{21}{12}\right)^n + \frac{1}{2!}\left(\frac{21}{12}\right)^2 \frac{2 \times 17.14}{(2 \times 17.14 - 30)} \right]^{-1} = \frac{1}{15}$$

Average waiting time of arrivals in the queue:

$$W_q = \left(\frac{21}{12}\right)^2 \times \frac{17.14}{(34.28-30)^2} \times \frac{1}{15} = \frac{343}{30} \text{ hours or 11.43 minutes.}$$

Example 20.32. *A petrol service station has two petrol pumps. The service time follows the exponential distribution with a mean of 4 minutes and the vehicles arrive for service as per Poisson distribution at the rate of 10 vehicles per hour. Find the probability that a customer has to wait for service. What is the expected time of waiting of customer at the petrol pump?* ***[MBA (Nov.), Annamalai Univ. 2010]***

Solution. Here, we have

$\lambda = 10$ vehicles per hour, and $\mu = 1/4$ vehicles per minute or 15 vehicles per hour.

Number of service channels = 2.

Probability of no vehicle at the petrol service station:

$$P_0 = \left[\sum_{n=0}^{2-1} \frac{1}{n!}\left(\frac{\lambda}{\mu}\right)^n + \frac{1}{2!}\left(\frac{10}{15}\right)^2 \left(\frac{2 \times 15}{30-10}\right) \right]^{-1}$$

$$= \left[1 + \frac{2}{3} + \frac{1}{3}\right]^{-1} = 2^{-1} \text{ or } \frac{1}{2}$$

Probability that a vehicle arriving for service has to wait:

$$P_w = \frac{1}{S!}\left(\frac{\lambda}{\mu}\right)^S \frac{S\mu}{(S\mu-\lambda)} P_0 = \left(\frac{10}{15}\right)^2 \left(\frac{30}{30-10}\right) \times \frac{1}{2}$$

$$= \frac{1}{2} \times \frac{4}{9} \times \frac{3}{2} \times \frac{1}{2} = \frac{1}{6}$$

Expected waiting time of vehicle at petrol pump:

$$W_q = \frac{\mu(\lambda/\mu)^S}{(S-1)!\,(S\mu-\lambda)^2} P_0 = \frac{15(2/3)^2}{(2-1)!\,(30-10)^2} P_0$$

$$= \frac{15 \times 4/9}{400} \times \frac{1}{2} = \frac{1}{120} \text{ hours or 30 seconds.}$$

Example 20.33. *A company currently has two tool cribs, each having a single clerk, in its manufacturing area. One tool crib handles only the tools for the heavy machinery, while the second one handles all other tools. It is observed that for each tool crib the arrivals follow a Poisson distribution with a mean of 20 mechanics per hour and the service time distribution is negative exponential distribution with a mean of 2 minutes.*

The tool manager feels that, if tool cribs are combined in such a way that either clerk can handle any kind of tool as demand arises, would be more efficient and the waiting problem could be reduced to some extent. It is believed that the mean arrival rate at the two tool cribs will be 40 mechanics per hour; while the service time will remain unchanged.

Compare the status of queue, and the proposal with respect to the total expected number of mechanics at the tool crib(s), the expected waiting time including service time for each mechanic and probability that he has to wait for more than five minutes. ***[MBA, Delhi Univ. 2012]***

Solution. For each tool crib, we have $\lambda = 20$ mechanics per hour, and $\mu = 30$ mechanics per hour. Therefore, expected number of mechanics in the system and in queue are:

$$L_s = \frac{\lambda}{\mu-\lambda} = \frac{20}{30-20} = 2 \text{ mechanics,}$$

and

$$L_q = \frac{\lambda^2}{\mu(\mu-\lambda)} = \frac{(20)^2}{30 \times (30-20)} = \frac{4}{3} \text{ or 1.33 mechanics.}$$

Expected waiting time in the system is given by

$$W_s = \frac{1}{\mu-\lambda} = \frac{1}{30-20} = \frac{1}{10} \text{ hour or 6 minutes.}$$

For the proposed system (when both the tool cribs are combined), we have $\lambda = 40$ mechanics per hour, and $\mu = 30$ mechanics per hour, $S = 2$.

$$\therefore \quad P_0 = \left[\sum_{n=0}^{2-1} \frac{1}{n!}\left(\frac{\lambda}{\mu}\right)^n + \frac{1}{2!}\left(\frac{\lambda}{\mu}\right)^2 \left(\frac{2\mu}{2\mu-\lambda}\right)\right]^{-1}$$

$$= \left[1 + \frac{40}{30} + \frac{1}{2!}\left(\frac{40}{30}\right)^2 \times \frac{2 \times 30}{2 \times 30 - 40}\right]^{-1} = \frac{1}{5} \text{ or } 0.2$$

$$L_s = \frac{\lambda\mu(\lambda/\mu)^2 P_0}{(2-1)!\,(2\mu-\lambda)^2} + \frac{\lambda}{\mu} = 2.4 \text{ mechanics.}$$

$$W_s = \frac{L_s}{\lambda} = \frac{2.4}{40} = 0.06 \text{ hour or 3.6 minutes.}$$

We observe that expected waiting time comes to be less, when tool cribs are combined. Therefore, by combining the tool cribs and asking both servers to attend the customers leads to a more efficient handling of the demand. This is not a surprising result. By separating the functions, certain advantages occur but by combining the two, the net waiting time is reduced because in the former system it is possible that one tool crib is idle, whereas the second tool crib may have a long queue.

PROBLEMS

20.29. A two-counter service station waiting line with Poisson arrivals has a mean rate of 50 per hour and exponential service with mean rate of 75 per hour for each service station. Find

(*i*) Probability of an empty system.
(*ii*) Probability that a arrival has to wait for getting service.

20.30. A post-office has two counters, the first one handles money-orders and registration of letters, while the second handles all other business. It has been found that service time distribution for both counters are exponential with a mean service time of 4 minutes per customer. The customers are found to arrive at the first counter in a Poisson fashion with mean arrival rate of 10 per hour, while the customers arrive at the second counter in a Poisson fashion with mean arrival rate of 12 per hour. What would be the effect on the average waiting time for the two type of customers, if each counter can handle all type of business?

[MBA (PT.) (Nov.), Delhi Univ. 2003]

20.31. A telephone company is planning to install telephone booths in a new airport. It has established the policy that a person should not have to wait more than 10 per cent of the times he tries to use a phone. The demand for use is estimated to be Poisson with an average of 30 persons per hour. The average phone call has an exponential distribution with a mean time of 5 minutes. How many phone booths should be installed?

20.32. An insurance company has three claim adjusters at its branch office. Policy holders are found to make claims against the company in a Poisson fashion at an average rate of 20 per 8 hours day. The time claim adjuster spends with a claimant is found to follow exponential distribution with mean service time of 40 minutes per claim. Claims are processed in the order of their submission. How many hours a claim adjuster is expected to spend with the claimants per week?

20.33. A tax consulting firm has four service stations (counters) in its office to receive people who have problems and complaints about their income, wealth and sales taxes. Arrivals average 80 persons in a 8-hour service day. Each tax adviser spends an irregular amount of time servicing the arrivals which have been found to have an exponential distribution. The average service time is 20 minutes. Calculate:

(*i*) the average number of customers in the system,
(*ii*) average number of customers waiting to be serviced,
(*iii*) average time a customer spends in the system,
(*iv*) average waiting time for a customer, and
(*v*) the probability that customer has to wait before he gets service.

[MBA (Dec.), Panjab Tech. Univ. 2010]

20.34. There are three inspectors to investigate accidents. One inspector takes 2 days on an average for completing a case. The number of accidents that occur in a year is 300. The accidents occur in Poisson distribution and the inspectors' performance is exponentially distributed. Find:

(*i*) Average number of accidents in the system, (*ii*) Average queue length,
(*iii*) Average waiting time, and (*iv*) Average waiting time in the system.

20.35. A company has two manufacturing shops and two tool cribs, one for each shop. Both the tool cribs handle almost identical jobs, gauges, and measuring instruments. Analysis of service time shows a negative exponential distribution with a mean of 2.5 minutes per workman. Arrivals of workmen follows Poisson distribution with mean of 18 per hour.

The production manager feels that if tool cribs are combined for both shops, efficiency will improve and waiting time in the queue will reduce. Do you agree with his opinion?

[MBA, Delhi Univ. 2010]

20.36. A bank has two counters for withdrawals. Counter *I* handles withdrawals of value less than ₹ 3,000 and counter *II* for ₹ 3,000 and above. Analysis of service time shows a negative exponential distribution with mean service time of 6 minutes per customer for both the counters. Arrival of customers follows Poisson distribution with mean 8 per hour for Counter *I* and 5 per hour for Counter *II*.

(*i*) What are the average waiting times per customer of each counter?

(*ii*) If each counter could handle all withdrawals irrespective of their value, how would the average waiting time change? *[MBA (Nov.), Delhi Univ. 2003]*

20.37. Trucks are required to pass through a weighing station so that they can be checked for weight violations. Trucks arrive at the station at the rate of 40 an hour between 7 p.m. and 9 p.m. Currently two inspectors are on duty during those two hours, each of whom can inspect 25 trucks an hour.

(*i*) How many trucks would you expect to see at the weighing station including those being inspected?

(*ii*) If a truck was just arriving at the station, about how many minutes could the driver expect to be at the station?

(*iii*) What is the probability that both inspectors would be busy at the same time?

(*iv*) How many minutes, on an average, would a truck that is not immediately inspected have to wait? *[MBA, Delhi Univ. 2011]*

20:11. POISSON-EXPONENTIAL MULTIPLE SERVER AND FINITE POPULATION, *i.e.*, (*M*/*M*/*S*) : (*N*/*FCFS*) MODEL

This queuing situation is essentially the same as we had discussed in *Section 20:10*, except that the maximum number of arrivals in the system is limited to N (say), where $N \geq S$ (the number of servers). For example, the parking area once full of its capacity, turns away the arriving vehicles.

In this case, we let

Mean arrival rate of the system as λ, and mean service rate as $n\mu$, provided $0 \leq n \leq S$.

When $n > N$, mean arrival rate $\lambda = 0$ and mean service rate is $S\mu$, for $S \leq n \leq N$.

Utilizing the steady-state probabilities of previous section, we get

$$P_n = \begin{cases} \dfrac{1}{n!}\left(\dfrac{\lambda}{\mu}\right)^n P_0 & ; \quad 0 \leq n < S \\[2ex] \dfrac{1}{S!\,S^{n-S}}\left(\dfrac{\lambda}{\mu}\right)^n P_0\,; & S \leq n \leq N \end{cases}$$

where, $$P_0 = \left\{\sum_{n=0}^{S-1} \frac{1}{n!}\left(\frac{\lambda}{\mu}\right)^n + \sum_{n=S}^{N} \frac{1}{S!S^{n-S}}\left(\frac{\lambda}{\mu}\right)^n\right\}^{-1}$$

$$= \begin{cases} \left[\sum_{n=0}^{S-1} \frac{1}{n!}\left(\frac{\lambda}{\mu}\right)^n + \frac{1}{S!}\left(\frac{\lambda}{\mu}\right)^S \left\{1-\left(\frac{\lambda}{S\mu}\right)^{N-S+1}\right\}\frac{S\mu}{S\mu-\lambda}\right]^{-1} ; & \frac{\lambda}{S\mu} \neq 1 \\ \left[\sum_{n=0}^{S-1} \frac{1}{n!}\left(\frac{\lambda}{\mu}\right)^n + \frac{1}{S!}\left(\frac{\lambda}{\mu}\right)^S (N-S+1)\right]^{-1} ; & \frac{\lambda}{S\mu} = 1 \end{cases}$$

Remark. If we take $N \to \infty$ and consider $\lambda/S\mu < 1$, then the reduced result corresponds to that of *Section 20:10*. Also, if we take $S = 1$ then the reduced result corresponds to that of *Section 20:9*.

Operating Characteristics

Given below are the formulas to be used to compute the steady-state operating characteristics for the queuing model with multiple servers and finite population:

1. The effective arrival rate, $\lambda_e = \lambda(1-P_N)$, and the effective traffic intensity, $\rho_e = \lambda_e/\mu$.
2. Expected number of customers in the queue is given by:

$$L_q = \sum_{n=S}^{N} (n-S)P_n = \sum_{n=S}^{N} (n-S)\frac{(\lambda/\mu)^n}{S!S^{n-S}}P_0$$

$$= \frac{1}{S!}\left(\frac{\lambda}{\mu}\right)^S P_0 \cdot \rho \sum_{x=0}^{N-S} x\cdot\rho^{x-1}, \quad \text{for } \rho = \frac{\lambda}{S\mu}$$

$$= \frac{(S\rho)^S \rho P_0}{S!} \sum_{x=0}^{N-S} \frac{d}{d\rho}\rho^x = \frac{(S\rho)^S \rho P_0}{S!} \cdot \frac{d}{d\rho}\left[\frac{1-\rho^{N-S+1}}{1-\rho}\right]$$

$$= \frac{P_0 (S\rho)^S \rho}{S!(1-\rho)^2}\left[1-\rho^{N-S+1}-(1-\rho)(N-S+1)\rho^{N-S}\right]$$

3. Now since,

$$L_q = \sum_{n=S}^{N}(n-S)P_n = \sum_{n=S}^{N} nP_n - S\sum_{n=S}^{N} P_n,$$

therefore, the average number of customers in the system is given by

$$L_s = \sum_{n=0}^{N} nP_n = \sum_{n=0}^{S-1} nP_n + \sum_{n=S}^{N} nP_n = \sum_{n=0}^{S-1} nP_n + L_q + S\sum_{n=S}^{N} P_n$$

$$= L_q + S\left[\sum_{n=0}^{N} P_n - \sum_{n=0}^{S-1} P_n\right] + \sum_{n=0}^{S-1} nP_n$$

$$= L_q + S - P_0 \sum_{n=0}^{S-0} \frac{(S-n)(\rho S)^n}{n!}$$

4. Average waiting time in the system can be obtained by using Little's formula, that is,

$$W_s = [E(n)]/\lambda', \quad \text{where } \lambda' = \lambda(1-P_N) \text{ is the effective arrival rate.}$$

Average waiting time in the queue can be obtained by using

$$W_q = W_s - 1/\mu \quad \text{or} \quad W_q = L_q/\lambda'$$

Example 20.34. *A car servicing station has 3 stalls where service can be offered simultaneously. The cars wait in such a way that when a stall becomes vacant, the car at the head of the line pulls up to it. The station can accommodate at most four cars waiting (seven in the station) at one time. The arrival pattern is Poisson with a mean of one car per minute during the peak hours. The service time is exponential with mean 6 minutes. Find the average number of cars in the service station during peak hours, the average waiting time and the average number of cars per hour that cannot enter the station because of full capacity.*

Solution. Here $\lambda = 1$ car per minute, $\mu = 1/6$ car per minute, $S = 3, N = 7, \rho = \lambda/\mu = 6$ and

$$P_0 = \left[\sum_{n=0}^{3-1} \frac{1}{n!}(6)^n + \sum_{n=3}^{7} \frac{1\times 6^n}{3!3^{n-3}}\right]^{-1} = \frac{1}{1141}$$

(*i*) Expected number of cars in the queue is:

$$L_q = \frac{(3\times 6)^3 \times 6}{3!(-5)^2} \cdot \frac{1}{1141}\left[1-(6)^5-(-5)(5)(6)^4\right] = 3.09 \text{ (or 3) cars.}$$

(*ii*) Expected number of cars in the service station

$$L_s = 3.09 + 3 - P_0 \sum_{n=0}^{2} \frac{(3-n)}{n!}(6)^n = 6.06 \text{ cars.}$$

(*iii*) Expected waiting time a car spends in the system

$$W_s = \frac{6.06}{1(1-P_7)} = \frac{0.06}{1-\dfrac{(6)^7}{3!3^4}\times\dfrac{1}{1141}} = 12.3 \text{ minutes}$$

since, $P_n = \dfrac{1}{S!S^{n-S}}\left(\dfrac{\lambda}{\mu}\right)^n P_0$ for $S \le n \le N$

(*iv*) Expected number of cars per hour that cannot enter the station is given by

$$60\lambda P_N = 60\times 1\times P_7 = 60\times\frac{(6)^7}{3!3^4}\times\frac{1}{1141} = 30.3 \text{ cars per hour.}$$

PROBLEMS

20.38. A barber shop has two barbers and three chairs for waiting customers. Assume that customers arrive in a Poisson fashion at a rate of 5 per hour and that each barber services customers according to an exponential distribution with mean of 15 minutes. Further, if a customer arrives and there are no empty chairs in the shop, he will leave. Find the steady-state probabilities. What is the probability that the shop is empty? What is the expected number of customers in the shop?

20.39. A car servicing station has two bays, where service can be offered simultaneously. Due to space limitation, only four cars are accepted for servicing. The arrival pattern is Poisson with 12 cars per day. The service time in both the bays is exponentially distributed with $\mu = 8$ cars per day per bay. Find the average number of cars in the service station, the average number of cars waiting to be serviced and the average time a car spends in the system.

20:12. POISSON-EXPONENTIAL MULTIPLE SERVER WITH MAXIMUM LIMIT ON ARRIVALS, *i.e.*, $(M/M/R) : (K/GD, K > R)$ MODEL

This queuing situation describes the system, wherein there are R service channels and the maximum limit imposed on incoming customers is fixed, say K, such that, $K > R$. A practical situation corresponding to this model is that of machine servicing where the calling population is that of machines and an arrival corresponds to a machine breakdown. The repairmen, of course, are the servers. Whenever a machine breaks down, it will result in loss in production, until it is repaired. Consequently, a broken machine cannot generate new calls, while in service. The model is generally known as "***Machine-Repairman***" model.

Let there be R servers and the service times be identical exponential random variables with mean $1/\mu$ and λ, the arrival rate. Then,

$$\text{Mean arrival rate} = (K-n)\,\lambda, \text{ if } 0 \le n \le K, \text{ and is: } 0, \text{ if } n > K$$

and
$$\text{Mean service rate} = \begin{cases} n\mu, & 0 \le n < R \\ R\mu, & R \le n \le K \end{cases}$$

Making use of steady-state probabilities, we get

$$P_n = \begin{cases} \dfrac{K\lambda\cdot(K-1)\lambda\cdot(K-2)\lambda\ldots(K-n+1)\lambda}{\mu\cdot 2\mu\ldots n\lambda} P_0, & 0 \le n < R \\ \dfrac{K\lambda\cdot(K-1)\lambda\cdot(K-2)\lambda\ldots(K-n+1)\lambda}{\underbrace{(R\mu)(R\mu)\ldots(R\mu)}_{n-R \text{ times}}(R\mu)(R-1)\mu(R-2)\mu\ldots(1)} P_0, & R \le n \le K \end{cases}$$

$$= \begin{cases} \dfrac{K!/(K-n)!}{n!}(\lambda/\mu)^n P_0, & \text{if } 0 \le n < R \\ \dfrac{K!/(K-n)!}{R!\,R^{n-R}}(\lambda/\mu)^n P_0, & \text{if } R \le n \le K \end{cases}$$

$$= \begin{cases} \dbinom{K}{n}(\lambda/\mu)^n P_0, & \text{if } 0 \le n < R \\ \dbinom{K}{n}\dfrac{n!}{R!\,R^{n-R}}(\lambda/\mu)^n P_0, & \text{if } R \le n \le K \end{cases}$$

To obtain the value of P_0, we use the boundary condition $\sum_{n=0}^{K} P_n = 1$

Therefore,

$$P_0 = \left[\sum_{n=0}^{R-1}\binom{K}{n}\left(\frac{\lambda}{\mu}\right)^n + \sum_{n=R}^{K}\binom{K}{n}\frac{n!}{R!\,R^{n-R}}\left(\frac{\lambda}{\mu}\right)^n\right]^{-1}, \text{ where } \binom{K}{n} = \frac{K!}{n!\,(K-n)!}$$

Operating Characteristics

1. $$L_s = P_0\left[\sum_{n=0}^{R-1} n\binom{K}{n}\left(\frac{\lambda}{\mu}\right)^n + \frac{1}{R!}\sum_{n=R}^{K} n\binom{K}{n}\frac{n!}{R^{n-R}}\left(\frac{\lambda}{\mu}\right)^n\right]^{-1}$$

2. $$L_q = L_s - R + \sum_{n=0}^{R-1}(R-n)\binom{K}{n}\left(\frac{\lambda}{\mu}\right)^n$$

3. $$W_s = \frac{L_s}{\lambda'} \text{ and } W_q = \frac{L_q}{\lambda'} \text{ where } \lambda' = \lambda[K - L_s]$$

Example 20.35. *We have 5 machines, each of which when running, suffers breakdown at an average rate of 2 per hour. There are 2 servicemen and only one man can work on a machine at a time. If n machines are out of order when n > 2, then (n – 2) of them wait until serviceman is free. Once a serviceman starts work on a machine, the time to complete the repair has an exponential distribution with mean 5 minutes. Find the distribution of the number of machines out of action at a given time. Find also the average time on out-of-action machine has to spend waiting for the repairs to start.* *[AMIE, 2005]*

Solution. We are given

$$K = \text{total number of machines in the system} = 5$$
$$R = \text{number of servicemen} = 2,\ \lambda = 2,\ \text{and}\ \mu = 60/5 = 12$$

Let n = number of machines out of order.

Then,
$$P_n = \begin{cases} \binom{5}{n}\left(\dfrac{2}{60/5}\right)^n P_0 & , \quad 0 \le n < 2 \\ \binom{5}{n}\dfrac{n!}{2!\,2^{n-2}}\left(\dfrac{2}{60/5}\right)^n P_0, & 2 \le n \le 5 \end{cases}$$

and
$$P_0 = \left[\sum_{n=0}^{2-1}\binom{5}{n}\left(\frac{2}{12}\right)^n + \sum_{n=2}^{5}\binom{5}{n}\frac{n!}{2!\,2^{n-2}}\left(\frac{2}{12}\right)^n\right]^{-1} = \frac{468}{1493} = 0.3135$$

Average number of machines out of action is given by

$$L_q = \sum_{n=2+1}^{5}(n-2)P_n = \sum_{n=3}^{5}(n-2)P_n = P_3 + 2P_4 + 3P_5$$
$$= \frac{165}{1493} = 0.110 \text{ hours.}$$

Average time an out-of-action machine has to spend waiting for the repairs start is:

$$W_q = L_q/\lambda'$$

But
$$\lambda' = \sum_{n=0}^{5}(5-n)P_n = \lambda[5P_0 + 4P_1 + 3P_2 + 2P_3 + P_4]$$
$$= \lambda \times 0.434 \times 9.189 = 8 \text{ machines,}$$

$\therefore$
$$W_q = \frac{0.110}{8} \text{ hours or } 0.013 \text{ hours.}$$

Example 20.36. *A mechanic services four machines. For each machine the mean time between service requirements is 5 hours and is assumed to form an exponential distribution. The repair time tends to follow the same distribution and has a mean time of 1 hour. When a machine is down for repairs, the time lost has a value of ₹ 25 per hour. The mechanic costs ₹ 55 per day. Find the following:*

(*a*) *What is the expected number of machines in operation?*
(*b*) *What is the expected downtime cost per day?*
(*c*) *Would it be desirable to provide two mechanics, each to service only two machines?*

[MBA (Nov.), Delhi Univ. 2011]

Solution.

(*a*) $\lambda = 1/5 = 0.20$ arrivals per hour and $\mu = 1$ hour per machine.
Also, it is given that there are four machines and one mechanic, *i.e.*, $K = 4$ and $R = 1$.

Then, $$P_0 = \left[\sum_{n=0}^{4} \binom{4}{n} \frac{n!}{1!(1)^{n-1}} \left(\frac{\lambda}{\mu}\right)^n\right]^{-1} = \left[\sum_{n=0}^{4} \frac{n!}{(4-n)!}\left(\frac{1}{5}\right)^n\right]$$

$$= \left[1 + 4(0.2) + 4(3)(0.2)^2 + 4(3)(2)(0.2)^3 + 4(3)(2)(1)(0.2)^4\right]^{-1} = 0.4$$

The expected number of machines in the system, therefore, is

$$L_s = \sum_{n=0}^{4} n P_n = 4 - (1 - P_0)\cdot\frac{\mu}{\lambda} = 4 - 5(1 - 0.4) = 1,$$

the number not running. The expected number of machines running, therefore, is 4 – 1, *i.e.*, 3.

(*b*) The expected cost of downtime per day is found as follows:

If an 8-hour day is assumed, the total downtime is:

8 × expected number of down time = 8 × 1 = 8 hours per day

Cost = (₹ 25 per hour)(8 hours per day) = ₹ 200 per day

(*c*) For comparison with two mechanics servicing two machines each, we compute.

If $K = 2$, then, $$P_0 = \left[\sum_{n=0}^{2} \frac{2!}{(2-n)!}\left(\frac{1}{5}\right)^n\right]^{-1} = [1 + 2(0.20 + 2(1)(0.2)^2 = (1.48) = 0.68$$

This assumes that each mechanic and his two machines constitute a separate system with no interplay. The expected number of machines in the system (per mechanic) is:

$$L_s = 2 - 5(1 - 0.68) = 0.4$$

The expected downtime per mechanic is (8)(0.4) = 3.2 hours per day.

Total downtime per day = 2(3.2) = 6.4 hours per day.

Total cost = 2(₹ 55 per mechanic) + (6.4 hours per day)(₹ 25 per hour) = ₹ 110 + ₹ 160 = ₹ 270 per day.

Thus, since ₹ 270 > ₹ 200 + ₹ 55 the use of two mechanics is not justified.

PROBLEMS

20.40. A repairman is to be hired to repair machines that breakdown at an average rate of 3 machines per hour. Breakdowns occur randomly (Poisson) over time. Non-productive time on any machine is considered to cost the company ₹ 10 per hour. Management has narrowed the choice to two repairmen; one slow but cheap, the other fast but expensive. The slow but cheap repairman has a salary of ₹ 5 per hour; in return he will service broken-down machines at an average rate of 4 machines per hour. The fast but expensive repairman has a salary of ₹ 7 per hour and will repair broken-down machines at an average rate of 6 machines per hour. Which repairman should the company hire? Assume exponential repair times for both repairmen. *[MBA, Madras Univ. 2000]*

20.41 A group of engineers has two terminals available to aid in their calculations. The average computing job requires 20 minutes of terminal time, and each engineer requires some computation about once every 0.5 hours, *i.e.*, the mean time between a call for service is 0.5 hours. Assume these are distributed according to an exponential distribution. If there are six engineers in the group, find:

(*a*) The expected number of engineers waiting to use one of the terminals.

(*b*) The total lost time per day.

20.42. In the production shop of a company, the breakdown of the machines is found to be Poisson distributed with an average rate of 3 machines per hour. Breakdown time at one machine costs ₹ 40 per hour to the company. There are two choices before the company for hiring the repairmen. One of the repairmen is slow but cheap, the other fast but expensive. The slow-cheap repairman demands ₹ 20 per hour and will repair the broken down machines exponentially at the rate of 4 per hour. The fast-expensive repairman demands ₹ 30 per hour and will repair machines exponentially at an average rate of 6 per hour, which repairman should be hired? *[MBA, Gujarat Univ. 1997; M.Com., Delhi Univ. 2006]*

20.43. A mechanic services four machines. For each machine the mean time between service requirements is 10 hours and is assumed to form an exponential distribution. The repair time tends to follow the same distribution and has a mean time of 2 hours. When a machine is down for repair, the time lost has a value of ₹ 20 per hour. The mechanic costs ₹ 50 per day. Find the following:

(*a*) What is the expected number of machines in operation?

(*b*) What is the expected cost of downtime per day?

(*c*) Would it be desirable to provide two mechanics, each to service only two machines?

[MBA (Nov.), Delhi Univ. 2009, 2011]

20.44. PQR Company has been experiencing machine breakdown for sometime and now wants to hire a mechanic on a permanent basis as against the present practice of engaging a mechanic as and when a breakdown occurs. The company has a large number of machines and, on an average, two machines breakdown per day. The machine downtime is estimated to cost the company ₹ 800 per day.

Two mechanics *A* and *B*, are available. Mechanic *A* will cost the company ₹ 40 per day and can on average repair three machines per day, whereas mechanic *B* will cost ₹ 50 per day and his average service rate is four machines per day. Describing any assumptions you might take, determine which repairman should be hired, so that total expected cost per day is minimized? *[MBA (Nov.), Delhi Univ. 2003]*

20:13. NON-POISSON QUEUING SYSTEMS

Queues in which arrivals and/or departures may not follow the assumptions of Poisson distribution are called ***non-Poisson queues***. The development of these queuing systems is more complicated, mainly because the Poisson assumptions no longer hold good. Following are the techniques which are usually employed in studying the non-Poisson queues:

(*a*) Phase technique, (*b*) The technique of Imbedded Markov Chain, and

(*c*) The Supplementary Variable technique.

The ***Phase technique*** is used when an arrival demands phases of service, say *k* in number. The technique by which non-Markovian queues are reduced to Markovian is termed as ***Imbedded Markov Chain technique***. When one or more random variables are added to convert a non-Markovian process into a Markovian one, the technique involved in this conversion is called the ***Supplementary Variable technique***. The detailed study of non-Poisson queues is beyond the scope of this book. However, we shall introduce the steady-state characteristics of Single server models with multi phase service time as well as of arbitrary service time.

Multi-Phase Service Queuing System

When a unit has to pass through *k* stages for its service, then these *k* stages of servicing are called *k*-phases. The distribution of servicing time in each of these phases will be an independent variable and the distribution of total servicing time of a unit in the system will be combined distribution of time in all these phases.

Until now, we have assumed that the servicing time of a unit follows the exponential distribution given by $\mu e^{-\mu t}$ with $1/\mu$ as the average servicing time. Since, the exponential distribution involves only one parameter (the parameter μ), it is known as one parameter distribution. A two-parameter generalization of the exponential distribution is called the ***Erlangian (Gamma) service time distribution***. The probability density function of this Erlangian service time distribution is defined by

$$g(t\,;\mu\,;k) = C_k\, t^{k-1} e^{-k\mu t}, \quad k = 1,\ 2,\ \ldots$$

where, $0 \le t < \infty$ and $C_k = (k\mu)^k/(k-1)!$,

where, μ = expected number of customers completing service per unit of time, and

k = a positive constant.

The expected value of the total service time and variance of Erlang distribution are: $k(1/k\mu)$, *i.e.*, $1/\mu$ and $k(1/k\mu)^2$, *i.e.*, $1/k\mu^2$ respectively.

The mode of service time, t, for the kth Erlang is $(k-1)/k\mu$.

Fig. 20.4 shows how the shape of the Erlang distribution changes for various values of k. When k = 1, it reduces to the exponential distribution, and for $1 < k < \infty$, it reduces to a constant distribution for customer inter-arrival times.

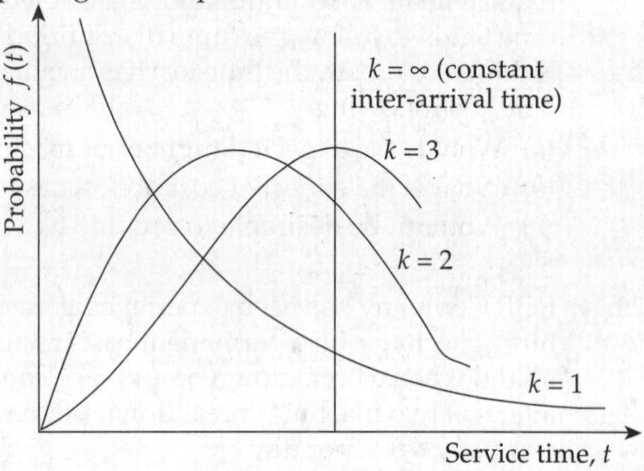

Fig. 20.4. Erlang Distribution

20:14. POISSON-ERLANGIAN SERVICE TIME, SINGLE SERVER AND INFINITE POPULATION, *i.e.*, ($M/E_k/1$) : ($\infty/FCFS$) MODEL

This model consist of a single service channel queuing system in which there are n phases in the system (waiting or in service). It has been assumed that a new arrival creates k-phases of service and departure of one customer reduces k-phases of service. Let

n = number of customers in the system,

$\lambda_n = \lambda$, constant arrival rate per unit time, and

$\mu_n = k\mu$, k-phases of service per unit time.

Operative Characteristics of the Queuing System

1. Average number of phases (not customers) in the system:

$$L_q(k) = \frac{1}{\mu} L_s(k) = \frac{k+1}{2} \cdot \frac{\lambda}{\mu(\mu-\lambda)}$$

$$L_s(k) = \frac{(k+1)}{2} \cdot \frac{\lambda}{\mu-\lambda}$$

2. Average number of customers (not phases) in the queue:

$$L_q = \frac{1}{k}[L_s(k) - \text{Average number of phases in service}]$$

$$= \frac{1}{k}\left[\frac{k+1}{2} \cdot \frac{\lambda}{\mu-\lambda} - \frac{k+1}{2} \cdot \frac{\lambda}{\mu}\right] = \frac{(k+1)}{2k} \cdot \left[\frac{\lambda^2}{\mu(\mu-\lambda)}\right]$$

Since, $1/\mu$ is the average service time per customer, and $(k+1)/2$ is the average number of phases of one customer in service, therefore, the time taken for serving a customer will be $(k+1)/2$. Thus, the average number of phases that arrive in this time would be $\lambda(k+1)/2\mu$.

3. Average waiting time of an arrival in the queue is:

$$W_q = \frac{k+1}{2k} \frac{\lambda}{\mu(\mu-\lambda)}$$

4. Average time an arrival spends in the system is:

$$W_s = W_q + \frac{1}{\mu} = \frac{k+1}{2k} \frac{\lambda}{\mu(\mu-\lambda)} + \frac{1}{\mu}$$

5. Average number of units in the system is given by

$$L_s = \lambda W_s = \frac{k+1}{2k} \frac{\lambda^2}{\mu(\mu-\lambda)} + \frac{\lambda}{\mu} = L_q + \frac{\lambda}{\mu}$$

Example 20.37. *A hospital clinic has a doctor examining every patient brought in for a general check-up. The doctor spends 4 minutes on each phase of the check-up, although the distribution of time spent on each phase is approximately exponential. If each patient goes through four phases in the check-up and if the arrivals of the patients to the doctor's clinic are approximately Poisson at the average rate of three per hour, what is the average time spent by a patient waiting in the doctor's clinic? What is the average time spent in the check-up? What is the most probable time spent in the check-up?*

[MBA (Nov.), Delhi Univ. 2002; M.Sc. (Math.), CCS Univ. 2004]

Solution. We are given

$k = 4$ and Mean arrival rate = 3 patients per hour, *i.e.*, $\lambda = 3$ per hour.

Service time per phase = $(1/4\mu) = 4$ minutes

$$\therefore \quad \mu = \frac{1}{4 \times 4} = \frac{1}{16} \text{ patients per minute or } \frac{60}{16}\left(=\frac{15}{4}\right) \text{hours}$$

and

$$W_s = \frac{4+1}{2\times 4}\left[\frac{3}{\frac{15}{4}\left(\frac{15}{4}-3\right)}\right] = \frac{2}{3} \text{ hours, } i.e., \text{ 40 minutes.}$$

Average time spent in the examination: $1/\mu = 16$ minutes.

Most probable time spent in the examination

$$= \frac{k-1}{k\mu} = \frac{4-1}{4\times\frac{1}{16}} = \frac{3}{1/4} = 12 \text{ minutes.}$$

Example 20.38. *An airline maintenance base has facilities for overhauling only one airplane engines at a time. Hence, to return the airplanes into use at the earliest, the policy is to stagger the overhauling of the 4 engines of each airplane. In other words only one engine is overhauled each time in airplane comes into the base. Under this policy, airplanes have arrivals according to Poisson process at a mean rate of one per day. The time required for an engine overhaul has an exponential distribution with a mean of 1/2 day.*

A proposal has been made to change the policy so as to overhaul all four engines consecutively each time an aeroplane comes into the shop. It is pointed out that although this will quadruple the expected service time, each plane would need to come into the shop only one-fourth time as often. Compare the two alternatives on a meaningful basis.

Solution. Two alternatives will be compared on the basis of the waiting time cost of airplanes requiring overhauling.

First alternative: Poisson arrival, exponential service and single service station.

Given that $\lambda = 1$ airplane per day; $\mu = 2$ airplanes per day.

Therefore, average number of airplanes in the system are

$$L_s = \frac{\lambda}{\mu - \lambda} = \frac{1}{2-1} = 1$$

Second alternative: Poisson arrival, multi-phase service and single service station.

Given that $\lambda = 1/4$ airplane per day; and $k = 4$

Since, service time per airplane is: $4 \times (1/2) = 2$ days, therefore mean service rate is:

$$\mu = 1/2 \text{ airplane per day.}$$

Average number of airplanes in the system are

$$L_s = \frac{k+1}{2k} \cdot \frac{\lambda^2}{\mu(\mu-\lambda)} + \frac{\lambda}{\mu} = \frac{4+1}{2 \times 4} \cdot \frac{(1/4)^2}{\frac{1}{2}\left(\frac{1}{2} - \frac{1}{4}\right)} + \frac{1/4}{\frac{1}{2}} = \frac{13}{16} \text{ or } 0.81$$

Since, L_s (= 0.81) in the second alternative is less than its value of the first alternative, therefore, the waiting cost for requiring overhauling in the second alternative will be less.

Hence, the proposal should be accepted.

Exampie 20.39. *Repairing a certain type of machine which breaks down in a given factory consists of 5 basic steps that must be performed sequentially. The time taken to perform each of the 5 steps is found to have an exponential distribution with mean 5 minutes and is independent of other steps. If these machines breakdown in a Poisson fashion at an average rate of two per hour and if there is only one repairman, what is the average idle time for each machine that has broken down?* ***[B.Tech. Hyderabad Univ. 2011]***

Solution. Here, we are given:

k (number of phases) = 5, and service time per phase = 5 minutes.

$\therefore$ service time per unit = 5 × 5 = 25 minutes.

$\therefore$ $\mu = \frac{1}{25}$ units per minute or $\frac{12}{5}$ units per hour and $\lambda = 2$ units per hour.

Average time of the machine = Average time spent by machine in the system

$$\therefore \quad W_s = \frac{k+1}{2k} \cdot \frac{\lambda}{\mu(\mu-\lambda)} + \frac{1}{\mu} = \frac{5+1}{2 \times 5} \cdot \frac{2}{\frac{12}{5}\left(\frac{12}{5} - 2\right)} + \frac{5}{12}$$

$$= \frac{1}{2} \times \frac{5}{2} + \frac{5}{12} = \frac{20}{12} = \frac{5}{3} \text{ hours or 100 minutes.}$$

PROBLEMS

20.45. A barber with a one-man takes exactly 25 minutes to complete one haircut. If customers arrive in Poisson fashion at an average rate of one every 40 minutes, how long on the average must a customer wait for service? Also, find the average time a customer spends in the barber shop. ***[B.Tech. (Nov.), Hyderabad Univ. 2010]***

20.46. A tailoring shop with one man takes exactly one day to stitch a suit. Customer's arrival follow Poisson pattern with mean rate of arrival of one customer in every two day. How long, on an average, a customer is expected to wait in such a situation?

20.47. The repair of a Lathe requires four steps to be completed one after another in a certain order. The time taken to perform each step follows exponential distribution with a mean of 10 minutes and is independent of other steps. Machine breakdown follows Poisson process with mean rate of 3 breakdowns per hour. Answer the following:

(*a*) What is the expected idle time of the machine, assuming there is only one repairman available in the workshop?

(*b*) What is the expected number of broken down machines in the queue?

20.48. A warehouse in a small State receive orders for a certain item and sends them by a truck as soon as possible to the customer. The orders arrive in Poisson fashion at a mean rate of 0.9 per day. Only one item at a time can be shipped by truck from the warehouse, which is located in the central part of the State. Because the customers are located in various places in the state, the distribution of service time in days has a distribution with probability density $4t\, e^{-2t}$. What is the expected delay between the arrival of an order and the arrival of the item to the customer? Service time here implies the time the truck takes to load, get to the customer, unload and return to the warehouse. Loading and unloading times are small compared with travel time.

20.49. A colliery working one shift per day uses a large number of locomotives which breakdown at random intervals; on average one fails per 8 hour shift. The fitter carries out a standard maintenance schedule on each faulty loco. Each of the five main parts of this schedule take on an average 1/2 hour but the time varies widely. How much time will the fitter have for the other tasks and what is the average time a loco is out of service?

20.50. In a certain bank, let the customers arrive according to a Poisson distribution with a mean of 4 customers per hour. From observations on the teller's performance, the mean service time is estimated to be 10 minutes, with a variance of 25 minutes. It is felt that the Erlang would be a reasonable assumption for the distribution of the teller's service time. Also, it is assumed that there is no limit on the number of customers. Bank officials wish to know, on the average, how long a customer must wait until he gets the service, and how many customers are waiting for service.

[***Hint:*** Use $\mu = 1/10$ and $\sigma^2 = 1/k\mu^2 = 25$ to obtain the value of k.]

20:15. POISSON ARRIVALS, ARBITRARY SERVICE TIMES, AND SINGLE SERVER, *i.e.*, (*M/G/*1) : (∞/*GD*) MODEL

This model consists of single service facility with Poisson arrivals. But service times is not an exponential distribution, rather it follows a general or unspecified probability distribution. Using Kendall notation, this model can be written as *M/G/1* model.

Operating Characteristics

The notations used to describe the operating characteristics for the $M/G/1$ model are:

λ = mean arrival rate, μ = mean service rate, and

σ = standard deviation of the service rate.

1. The probability that no customer is in the system: $P_0 = 1 - \lambda/\mu$
2. The average number of customers in the queue: $L_q = \dfrac{\lambda^2\sigma^2 + (\lambda/\mu)^2}{2(1-\lambda/\mu)}$
3. The average number of customers in the system: $L_s = L_q + \lambda/\mu$
4. The average time a customer spends in the queue: $W_q = L_q/\lambda$
5. The average time a customer spends in the system: $W_s = W_q + 1/\mu$
6. The probability that a customer has to wait for service: $P_w = \lambda/\mu$

Remark. The formulae for the average queue length and average number of customers in the system are known as ***Pollaczek-Khintechine*** (or *P-K*) formulae.

Example 20.40. *At certain milk booth customer arrivals are random, and the average arrival rate is 20 customers per hour. A study of the service process shows that the average or mean service time is 2 minutes per customer, with a standard deviation of 1.2 minutes. Find the operating characteristics of the queuing system.*

Solution. Given: $\lambda = 20$ customers per hour or $20/60$ (= 0.35) customers per minute.

$\mu = 1/2$ (= 0.50) customers per minute, and $\sigma = 1.2$ minutes.

1. The probability that no customer is in the system:

$$P_0 = 1 - \frac{\lambda}{\mu} = 1 - \frac{0.35}{0.50} = 0.30$$

2. The average number of customers in the queue:

$$L_q = \frac{(0.35)^2\,(1.2)^2 + (0.35/0.50)^2}{2(1-0.35/0.50)} = \frac{(0.12)\times(1.44)+0.49}{2\times 0.30} = \frac{0.66}{0.60}$$

= 1.1 customers or one customer (approx.).

3. The average number of customers in the system:

$$L_s = L_q + \frac{\lambda}{\mu} = 1.1 + 0.7 = 1.8 \text{ or } 2 \text{ customers (approx.).}$$

4. The average time a customer spends in queue:

$$W_q = \frac{L_q}{\lambda} = \frac{1.1}{0.35} = 3.14$$

5. The average time a customer spends in the system:

$$W_s = W_q + \frac{1}{\mu} = 3.14 + \frac{1}{0.50} = 5 \text{ minutes (approx.).}$$

6. The probability that a customer has to wait for service:

$$P_w = \frac{\lambda}{\mu} = \frac{0.35}{0.50} = 0.70$$

Example 20.41. *At certain health care centre, patients arrive at a mean rate of 4 patients per hour and they are checked by doctor at a mean rate of 5 patients per hour. The centre feels that service times have some unspecified positive skewed unimodal two-tailed distribution with a standard deviation of 0.05 hour (3 minutes).*

(*i*) *Determine the queuing characteristics for the health care centre.*

(*ii*) *How much the assumption of exponential service times would distort these values? Discuss.*

Solution. We have $\lambda = 4$ patients per hour, and $\mu = 5$ patients per hour, and $\sigma = 0.05$

$\therefore$ $\rho = 4/5 = 0.8$ hour.

For the single server model with exponential arrival and general service times, the values of various queuing characteristics will be as follows:

(*i*) Average queue length is given by

$$L_q = \frac{\lambda^2\sigma^2 + \rho^2}{2(1-\rho)}, \text{ where } \rho = \frac{\lambda}{\mu}$$

$$= \frac{4^2 \times (0.05)^2 + (0.8)^2}{2(1-0.8)} = 1.7 \text{ patients}$$

Average waiting time in queue:

$$W_q = \frac{1}{\lambda} L_q \pm \frac{1.7}{4} = 0.425 \text{ hours.}$$

Average number of patients in the system:

$$L_s = L_q + \rho = 1.7 + 0.8 = 2.5 \text{ patients.}$$

Average waiting time in the system:

$$W_s = \frac{1}{\lambda}L_q + \frac{1}{\mu} = 0.425 + 0.2 = 0.625 \text{ hour or } 37.5 \text{ minutes.}$$

(*ii*) If service time follows exponential distribution, then the results obtained will be:

$$L_q = 3.2 \text{ patients, } L_s = 4 \text{ patients, } W_s = 1 \text{ hour, and } W_q = 0.8 \text{ hour or 48 minutes.}$$

On comparison, we observe that patients waiting time and the number of patients are more in case (*ii*) as compared to case (*i*).

20:16. COST MODELS IN QUEUING

The ***service level*** in a queuing facility is a function of service rate μ, that balances the following two conflicting costs:

1. Cost of offering the service, and
2. Cost incurred due to delay in offering the service (customer waiting time).

Two type of costs conflict because an increase in the existing service facility would reduce the customer's waiting time (*i.e.*, cost of waiting). On the other hand, a decrease in the level of service would increase the cost of waiting. *Fig. 20.5* illustrates both types of costs as a function of level of service. The optimum service level is one that minimizes the sum of two costs.

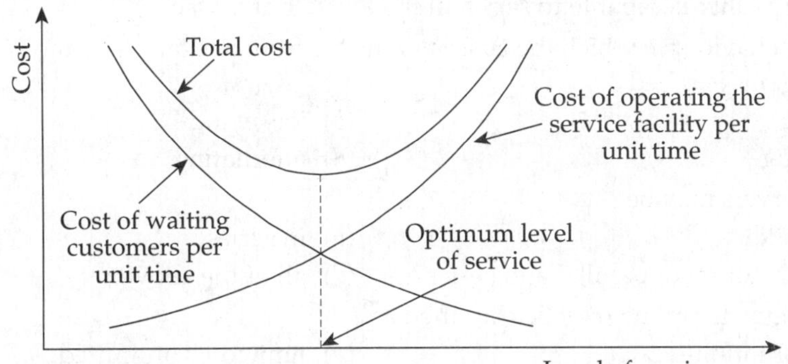

Fig. 20.5

The cost model can be expressed as

$$T = K\mu + C\,L_s,$$

where, K = cost of service rate per unit time,

C = cost of waiting customers per unit time, and

L_s = average number of customers in the system.

Optimum Service Rate. The procedure for determining the optimum service rate, μ, is based on model of *Section 20.8*.

Now, in the case of an $(M/M/1):(\infty/FCFS)$ model,

$$E(n) = \frac{\lambda}{\mu - \lambda}; \text{ and therefore } T = K\mu + \frac{C\lambda}{\mu - \lambda}$$

Since, the service rate μ is continuous, its optimum value can be obtained using the technique of maxima and minima. Thus, we get

$$\frac{d}{d\mu}(T) = K - \frac{C\lambda}{(\mu-\lambda)^2} \quad \text{and} \quad \frac{d^2}{d\mu^2}(T) = \frac{2\lambda C}{(\mu-\lambda)^3} > 0$$

Hence, the optimum value of μ is $\mu^0 = \lambda + \sqrt{\lambda C / K}$.

This result shows that the optimum value of μ is not only dependent on K and C, but on the arrival rate λ also.

Remark. In the case of Poisson arrival, Exponential Service and single Server model with limited capacity, the above cost model may be modified to reflect that the larger the value of N, the smaller will be the number of lost customer. The modified cost function is:

$$T = K\mu + C L_s + D_1 N + \lambda P_N D_2,$$

where, D_1 = cost of servicing each additional customer per unit time,
D_2 = cost per customer, and
λP_N = number of lost customers per unit time.

TEST YOUR UNDERSTANDING

1. Queue can be formed only when
   (*a*) arrivals exceed service capacity.
   (*b*) arrivals equals service capacity.
   (*c*) there are more than one service facilities.
   (*d*) service facility is capable to serve all the arrivals at a time.
2. Customer behaviour in which the customer moves from one queue to another for his personal economic gains is called
   (*a*) balking (*b*) jockeying
   (*c*) reneging (*d*) alternating
3. Multiple servers may be
   (*a*) in parallel (*b*) in series
   (*c*) in combination of parallel and series (*d*) all of the above.
4. Priority queue discipline may be classified as
   (*a*) finite or infinite. (*b*) limited or unlimited.
   (*c*) pre-emptive or non-pre-emptive. (*d*) all of the above.
5. For a "Poisson exponential, single server and infinite population" queuing model,
   (*a*) the system has a single service facility
   (*b*) the arrivals occur in a Poisson fashion
   (*c*) the service rate is according to exponential distribution
   (*d*) all of the above.
6. For a "Poisson-exponential, single server and infinite population" queuing model, which of the following is not correct?
   (*a*) $L_s = L_q - \lambda/\mu$ (*b*) $L_q = \lambda W_q$
   (*c*) $L_s = \lambda W_s$ (*d*) $W_s = W_q + 1/\mu$
7. The calling population is assumed to be infinite, when
   (*a*) arrivals are independent of each other. (*b*) arrivals are dependent upon each other.
   (*c*) capacity of the system is infinite. (*d*) service rate is faster than arrival rate.

8. In deterministic queuing model,
   (*a*) arrival rate is known and the service time is also certain.
   (*b*) arrival rate must not exceed the service rate.
   (*c*) the service rate and service time are reciprocals of each other.
   (*d*) if arrivals occur according to Poisson distribution, the inter-arrival times would be exponentially distributed.
9. Which of the following is not a key operating characteristic for a queuing system:
   (*a*) average time a customer spend waiting in the system and queue
   (*b*) utilization factor
   (*c*) per cent idle time
   (*d*) none of the above.
10. Service mechanism in a queuing system is characterized by
   (*a*) customer's behaviour
   (*b*) server's behaviour
   (*c*) customers in the system
   (*d*) all of the above.
11. Expected number of customers in the queue is given by
   (*a*) $L_q = \lambda/(\mu - \lambda)$
   (*b*) $L_q = [\lambda/(\mu - \lambda)] + (1/\mu)$
   (*c*) $L_q = \mu/(\mu - \lambda)$
   (*d*) $L_q = S\mu/(S\mu - \lambda)$.
12. Expected waiting time of a customer in the queuing system is:
   (*a*) $W_s = W_q + 1/\mu$
   (*b*) $W_q = L_q/\lambda$
   (*c*) $W_s = L_s/\lambda$
   (*d*) $W_q = \lambda L_q$.
13. Which of the cost estimates and performance measures are not used for economic analysis of a queuing system:
   (*a*) cost per server per unit of time.
   (*b*) cost per unit of time for a customer waiting in the system.
   (*c*) average waiting time of a customer in the system.
   (*d*) average number of customers in the system.
14. The cost of providing service in a queuing system decreases with
   (*a*) decreased arrival rate
   (*b*) increased arrival rate
   (*c*) decreased average waiting time in the queue
   (*d*) none of the above.

ANSWERS

1. (*a*)	**2.** (*b*)	**3.** (*d*)	**4.** (*c*)	**5.** (*d*)
6. (*a*)	**7.** (*a*)	**8.** (*b*)	**9.** (*d*)	**10.** (*b*)
11. (*c*)	**12.** (*a*)	**13.** (*c*)	**14.** (*d*)	

REVIEW EXERCISE

1. What do you understand by a queue? Explain various queue disciplines. ***[B.Tech., Hyderabad Univ. 2011]***
2. List the components of queuing system. ***[MBA, Annamalai Univ. 2010]***
3. Explain the essential features of queuing system. ***[MBA, Bangalore Univ. 2016]***
4. What is queuing theory? In what type of problem situations, it can be applied successfully? Discuss giving examples. ***[MBA, Delhi Univ. 2009]***

5. Explain the queuing process. *[MBA (Dec.), Panjab Tech. Univ. 2014]*
6. In what kind of situations queuing analysis is most appropriate? *[MBA (June), Panjab Tech. Univ. 2011]*
7. In respect to the queuing system, explain the following:
   (*i*) Input process, (*ii*) Queue discipline,
   (*iii*) Reneging, (*iv*) Balking, and
   (*v*) Jockeying. *[MBA (Dec.), Panjab Tech. Univ. 2010; B.Tech. (Mech.), Panjab Tech. Univ. 2011]*
8. Give an example from real life for each of the following queuing models:
   (*i*) First-come-first-served, (*ii*) Last-come-first-served,
   (*iii*) Random service.
9. Explain characteristics and classification of queuing models. *[MBA (Dec.), Panjab Tech. Univ. 2013; BE, Karnataka Univ. 2012]*
10. What are the various assumptions made in single channel queuing system? *[MBA, Panjab Tech. Univ. 2015; B.Tech., Nagpur Univ. 2005]*
11. What do you understand by (*i*) queue length, (*ii*) traffic intensity, and (*iii*) service channels. *[MBA, Annamalai Univ. 2009]*
12. Explain why do the arrivals and services follow Poisson and Exponential distribution, respectively. *[MBA, UP Tech. Univ. 2007]*
13. What are the operating characteristics of a single service channel waiting line with Poisson arrivals and exponential service times. *[MBA (Dec.), Panjab Tech. Univ. 2011]*
14. Queuing theory can be used effectively in determining optimal service level. Elucidate this statement with the help of an example. *[MBA, Delhi Univ. 2012]*
15. Show that for a single service station with Poisson arrival and exponential service time, the probability that exactly n calling units are in the queue system is

$$P_n = (1-\rho)\rho^n, \quad n \geq 0$$

    where ρ is the traffic intensity. Also, find the expected number of units in the system. *[M.Sc. (Math.), CCS Univ. 2000]*
16. Give some applications of queuing theory. *[MBA, Madras Univ. 2012; MBA, Panjab Tech. Univ. 2015]*
17. "Basic problem in queuing theory is to determine an optimal service level." Elucidate this statement. *[MBA (PT) HCA, Delhi Univ. 2002]*
18. Discuss the costs associated with queuing system. Also, explain the concept of optimum servicing rate and optimum cost. *[MBA (June), IGNOU 2007]*
19. Why do waiting lines form, even though a service system is underloaded? In a multi-channel system, what is the rationale for having customers wait in a single line, as is now being done in many banks and post offices, rather than multiple waiting lines? *[MBA, Delhi Univ. 2011]*
20. What are the components of total relevant cost in case of single channel queuing model? *[MIB, Delhi Univ. 2011]*

CHAPTER 21

Replacement Decisions

"Conduct is wise or foolish only in reference to its results"

The ***problem of replacement*** arises when the efficiency of an item/equipment diminishes with the ***passage of time*** or fail/breakdown ***all of a sudden***. For example, the old equipment might fail and work no more, or is worn out and needs substantial expenditure on its maintenance. In all such situations, it becomes essential to restore it by some remedial action, *i.e.*, determine the best policy to be adopted with respect to replacement of the item/equipment.

21:1. SITUATIONS WHICH MAKE THE REPLACEMENT OF ITEMS NECESSARY

The situations when the replacement of certain item/equipment is to be done are:

1. In respect of items/equipments have lost their utility because they have outlived their effective life, *i.e.*, deteriorate in performance over a period of time. Though they might be good technically yet on economic considerations, due to increasing maintenance costs, it is not worthwhile to continue with them and hence replacement may be called for. Examples are: machinery, equipment, building, etc.
2. In respect of units that perform adequately until sudden complete failure. The length of their service until failure varies randomly over some predictable range. Examples are: light bulbs, computer parts and some other electronic components.
3. In respect of the change in technology or process thereby making the old machinery obsolete.
4. In respect of staff of an organization which gradually diminishes due to death, retirement, and other reasons.

21:2. FAILURE MECHANISM OF ITEMS

The failure mechanism of items under replacement decisions are dealt with two type of failures, namely Gradual Failure and Sudden Failure.

1. Gradual Failure. The failure mechanism under gradual failure is progressive in nature. That is, as the life of an item increases, its efficiency decreases resulting in the decrease of productivity, decrease in the resale value or salvage value, increased operating cost which include the cost of new spares used, cost of repairs, preventive maintenance cost, cost of shortages due to breakdown and cost incurred on safety measures for life of human beings.

2. Sudden Failure. The items that fail after some period of giving desired service rather than deterioration, while in service, falls under the category of ***sudden failure***. The period of giving desired service is not constant, but follows some frequency distribution which may be ***progressive***, ***retrogressive*** or at ***random*** in nature.

(*i*) ***Progressive Failure.*** Under the mechanism of progressive failure, the probability of failure increases with increase in the life span of an item. For example, light bulbs, automobile parts, tyres of vehicles.

(*ii*) ***Retrogressive Failure.*** There are certain items that have more probability of failure in the beginning, but as time passes, the chances of its failure become less. In other words, if an item survives in the beginning, its expected life increases, *e.g.*, aircraft engines.

(*iii*) ***Random Failure.*** Under the mechanism of random failure, the constant probability of failure is associated with items that fail from random causes, such as physical shocks not related to age factor. For example, vacuum tubes in air-borne equipment have been found to fail at a rate independent of the age of the tube.

21:3. ASSUMPTIONS AND OBJECTIVES OF REPLACEMENT DECISIONS

Various assumptions of replacement decisions are:

1. If the scrap value of the equipment is zero, then replace the equipment when the maintenance cost becomes greater than the current average cost.
2. If we are given the resale value or the depreciation cost, the maintenance cost and the cost of equipment, then the optimum replacement period is determined by the minimum value of the average cost to date.
3. There is no change in technology of the asset/equipment under consideration.
4. The two categories of costs required to determine the optimum period for replacement of equipment having a gradual failure mechanism are: depreciation cost and operating cost.

The objective of replacement decisions are:

1. Time of replacement, *i.e.*, replace now or at a later stage.
2. How much to replace, *i.e.*, selective replacement of and equipment or total replacement.
3. How to replace, *i.e.*, if there is a choice of more than one equipment, then which equipment should be replaced.

21:4. TYPES OF REPLACEMENT PROBLEMS

1. Aging of Equipment / Machine. It includes the items that deteriorate with time, *i.e.*, maintenance cost increases with time and value of money (*a*) remains constant during the period, (*b*) changes with constant rate during the period.

2. Equipment Breaks-Down / Fails Suddenly. It includes the items that do not deteriorate gradually, but fail all of a sudden, like as light bulbs. Here, we have to determine the number of those items which fail in a particular time period and find the cost of replacement under (*a*) individual replacement policy, *i.e.*, cost of replacing the items as and when they fail, (*b*) group replacement policy, *i.e.*, the cost of replacing all the items at a certain fixed time period.

3. Staff Replacement. When people leave an organization for several reasons, such as retirement, resignation, death, etc., the problem of replacement arises. Here, we consider the life distribution for the service of staff members in an organization to adopt the replacement policy.

We shall now discuss the above-mentioned replacement situations.

21:5. REPLACEMENT POLICY FOR EQUIPMENT THAT DETERIORATE GRADUALLY

Generally, the maintenance cost of certain items always increases gradually with time and a stage comes when the maintenance cost becomes so large that it is better and economical to replace the item with a new one. For determining the optimum replacement period, we proceed in the following manner:

Case 1. When Time Value of Money is not Considered. Let C = capital cost of equipment, S = scrap value of equipment, $f(t)$ = maintenance cost of the equipment in year t, and n = number of years that equipment would be in use.

Total cost of the equipment (purchasing and maintaining) is given by

TC = Capital cost – Resale value of equipment after n years + Maintenance cost for n years.

$$= C - S(n) + \sum_{t=1}^{n} f(t), \text{ where } S(n) \text{ is the resale value at the end of } n^{\text{th}} \text{ year.}$$

Let $d(n)$ be the change in the resale value in the nth year, and $f(n)$ be the maintenance cost during the nth year. Then

$$d(1) = S(0) - S(1),\quad d(2) = S(1) - S(2),\quad d(3) = S(2) - S(3),\quad d(4) = S(3) - S(4), \text{ and so on,}$$

Thus,
$$d(n) = S(n-1) - S(n)$$

$$\therefore \quad \sum_{t=1}^{n} d(t) = d(1) + d(2) + d(3) + \ldots + d(n)$$

$$= S(0) - S(n) = C - S(n), \text{ where } S(0) = C$$

Therefore, total cost during n years is:

$$TC = C - S(n) + \sum_{t=1}^{n} f(t)$$

$$= \sum_{t=1}^{n} d(t) + \sum_{t=1}^{n} f(t) = \sum_{t=1}^{n} [d(t) + f(t)]$$

Let $e(t) = d(t) + f(t)$ be the effective maintenance cost in the nth year. Then,

$$TC = \sum_{t=1}^{n} e(t), \text{ and the average cost during the } n^{\text{th}} \text{ year is:}$$

$$A(n) = \frac{TC}{n} = \frac{1}{n} \sum_{t=1}^{n} e(t)$$

Now, $A(n)$ will be minimum for that value of n, for which

$$A(n+1) \geq A(n) \text{ and } A(n-1) \geq A(n)$$

$$\therefore \quad A(n+1) - A(n) = \frac{nA(n)}{n+1} + \frac{e(n+1)}{n+1} - A(n) = \frac{e(n+1) - A(n)}{n+1}$$

So,
$$A(n+1) - A(n) \geq 0 \;\Rightarrow\; e(n+1) \geq A(n)$$

Similarly,
$$A(n) - A(n-1) \leq 0 \;\Rightarrow\; e(n) \leq A(n-1)$$

This suggests the optimal replacement policy:

"Replace the equipment at the end of n years, if the effective maintenance cost in the $(n+1)^{th}$ year is more than or equal to the average total cost in the nth year and the nth year's effective maintenance cost is less than or equal to the previous year's average total cost".

Case 2. When Time Value of Money is Considered. When time value of money is considered, we have to discount the future payments on maintenance costs, so as to express them all in terms of present values.

If the rate of interest be 'r', then the present value of a rupee paid at the end of one year would be $1/(1 + r)$, *i.e.*, $(1 + r)^{-1}$, paid at the end of 2 years would be $(1 + r)^{-2}$, and so on. The present value of one rupee paid at the end of 1, 2, 3, ..., n years is called the ***present worth factor*** (pwf) in respect of 1, 2, ..., n years. The expression $(1 + r)^n$ is known as the ***payment compound amount factor*** (caf) of one rupee spent in n years time.

For the problem of replacement when time value of money is considered, we assume that

(*a*) the equipment has no salvage value, and

(*b*) the maintenance cost are incurred in the beginning of different time periods.

Working under these two assumptions, we first determine the ***annualized cost*** (weighted average of costs), for different years of the life of equipment. This is done in the following steps:

Step 1. Obtain the present value of maintenance cost for each of the years by multiplying the cost value by an appropriate present value factor which is based on the given rate of discount and time.

Step 2. Compute the cumulative present value for all the years and add the cost of equipment to each of these values.

Step 3. Find the cumulative present value factors up to each of the years 1, 2, 3, ...

Step 4. Find the annualized costs for various years by dividing the cost plus cumulative maintenance cost for each year, obtained in *Step 2*, by the corresponding cumulative present value factors in *Step 3*.

Step 5. Follow the following rules to decide on the replacement of an equipment:

(*a*) Do not replace, if the next period's cost is less than the weighted average of the previous costs, and

(*b*) Replace, if the next period's cost is greater than the weighted average of the previous costs.

Example 21.1. *A firm is considering replacement of a machine, whose cost price is ₹12,200 and the scrap value is ₹200. The running (maintenance and operating) cost in rupees are found from experience to be as follows:*

Year :	*1*	*2*	*3*	*4*	*5*	*6*	*7*	*8*
Running cost :	*200*	*500*	*800*	*1,200*	*1,800*	*2,500*	*3,200*	*4,000*

When should the machine be replaced?

[MBA, Madras Univ. 2012; B.Tech. (Comp. Sc.) (Dec.), Hyderabad Univ. 2011]

Solution. We are given the running cost, $f(n)$, the scrap value S = ₹200 and the cost of machine, C = ₹12,200. In order to determine the optimal time 'n' when the machine should be replaced, we calculate an average total cost per year during the life of the machine as shown in the table given below:

Year of service (n) (1)	*Running cost* (₹) $f(n)$ (2)	*Resale value* (₹) $S(n)$ (3)	*Change in resale value* (₹) $d(n)$ (4)	*Effective running cost* (₹) $f(n) + d(n) = e(n)$ (5)	*Total cost* (₹) $\Sigma e(n)$ (6)	*Average cost* (₹) $A(n)$ (6)/(1) = (7)
1	200	200	12,200 – 200 = 12,000	12,200	12,200	12,200
2	500	200	200 – 200 = 0	500	12,700	6,350
3	800	200	200 – 200 = 0	800	13,500	4,500
4	1,200	200	200 – 200 = 0	1,200	14,700	3,675
5	1,800	200	200 – 200 = 0	1,800	16,500	3,300
6	2,500	200	200 – 200 = 0	2,500	19,000	3,167
7	3,200	200	200 – 200 = 0	3,200	22,200	3,171
8	4,000	200	200 – 200 = 0	4,000	26,200	3,275

From the table, we observe that the average total cost $A(n)$ is minimum in the 6th year (₹ 3,167). Also, the effective maintenance cost in the 7th year, *i.e.*, $e(7)$ is greater than $A(6)$ and sixth year's effective maintenance cost $e(6)$ is less than the 5th year's average total cost $A(5)$, *i.e.*, $e(7) > A(6)$ and $e(6) < A(5)$. Hence, the machine should be replaced after every six year.

Example 21.2. *Following table gives the running costs per year and resale price of a certain equipment whose purchase price is ₹5,000:*

Year :	*1*	*2*	*3*	*4*	*5*	*6*	*7*	*8*
Running Cost (₹) :	*1,500*	*1,600*	*1,800*	*2,100*	*2,500*	*2,900*	*3,400*	*4,000*
Resale Value (₹) :	*3,500*	*2,500*	*1,700*	*1,200*	*800*	*500*	*500*	*500*

At what year is the replacement due? ***[MBA, Panjab Tech. Univ. 2006]***

Solution. We are given:

C = ₹ 5,000, $f(n)$ is the running cost, $S(n)$ is the resale value and $d(n)$ is the change in resale value in the nth year. Then, using the above data, the optimum period of replacement is computed in the following table:

Year (n)	*f(n)*	*S (n)*	*d (n)*	*f(n) + d (n) = e (n)*	*TC = Σ e (n)*	*A (n)*
1	1,500	3,500	5,000 – 3,500	1,500 + 1,500	3,000	3,000
2	1,600	2,500	3,500 – 2,500	1,600 + 1,000	5,600	2,800
3	1,800	1,700	2,500 – 1,700	1,800 + 800	8,200	2,733
4	2,100	1,200	1,700 – 1,200	2,100 + 500	10,800	**2,700**
5	2,500	800	1,200 – 800	2,500 + 400	13,700	2,740
6	2,900	500	800 – 500	2,900 + 300	16,900	2,817
7	3,400	500	500 – 500	3,400 + 0	20,300	2,900
8	4,000	500	500 – 500	4,000 + 0	24,300	3,037

We observe from the above table that minimum average cost occurs at the end of 4th year. Also, $e(5) > A(4)$ and $e(4) < A(3)$. Hence, the machine should be replaced after every four years.

Example 21.3. *Machine A costs ₹45,000 and the operating costs are estimated at ₹1,000 for the first year, increasing by ₹10,000 per year in the second and subsequent years. Machine B costs ₹50,000 and operating costs are ₹2,000 for the first year, increasing by ₹4,000 in the second and subsequent years. If we now have a machine of type A which is one year old, should we replace it with machine B? If so, when? Assume that both machines have no resale value and future costs are not discounted.*

[BE (Mech.), Nagpur Univ. 2011; B.Tech. Hyderabad Univ. 2011; BMS, Lucknow Univ. 2008]

Solution. Given: $S(n) = 0$, *i.e.*, no resale value for machines A and B both.

Machine A. Average total cost is computed below:

Year (n)	*f(n)*	*S (n)*	*d (n)*	*e (n)*	*TC = Σ e (n)*	*A(n)*
1	1,000	0	45,000 – 0	46,000	46,000	46,000
2	11,000	0	0	11,000	57,000	28,500
3	21,000	0	0	21,000	78,000	**26,000**
4	31,000	0	0	31,000	1,09,000	27,250
5	41,000	0	0	41,000	1,50,000	30,000

From the above table, we observe that $e(4) > A(3)$ and $e(3) < A(2)$. Hence, machine A should be replaced at the end of 3rd year.

Machine B. Average total cost for machine B is now computed below:

Year (n)	f (n)	S (n)	d (n)	e (n)	TC = Σ e (n)	A(n)
1	2,000	0	50,000 – 0	52,000	52,000	52,000
2	6,000	0	0	6,000	58,000	29,000
3	10,000	0	0	10,000	68,000	22,667
4	14,000	0	0	14,000	82,000	20,500
5	18,000	0	0	18,000	1,00,000	**20,000**
6	22,000	0	0	22,000	1,22,000	20,333

From the above table, we observe that $e(6) > A(5)$ and $e(5) < A(4)$. Hence, machine B should be replaced at the end of 5 years.

Further, since the minimum average cost for machine B is lower than that of machine A, machine B should be replaced by machine A.

Again, machine A is replaced by machine B at the time when its operating cost of the next year exceeds the minimum average yearly cost ₹ 20,000 of machine B.

On the one year old machine A, ₹ 11,000 would be required to be spent in the next year; while ₹ 21,000 would be needed in the following year. Since ₹ 21,000 > ₹ 20,000, we should keep the machine A for one year and replace it thereafter.

Example 21.4. *Let the value of money be assumed to be 10% per year and suppose that machine A is replaced after every 3 years, whereas machine B is replaced after every six years. The yearly cost of both the machines are given below:*

Year :	*1*	*2*	*3*	*4*	*5*	*6*
Machine A :	*1,000*	*200*	*400*	*1,000*	*200*	*400*
Machine B :	*1,700*	*100*	*200*	*300*	*400*	*500*

Determine which machine should be purchased. **[MBA (Dec.), Panjab Tech. Univ. 2012]**

Solution. Since, money carries the rate of interest, the present worth of the money to be spent over in a period of one year is:

$$v = \frac{100}{100+10} = \frac{10}{11} = 0.9091$$

∴ The total discounted cost (present worth) of machine A for 3 years is:

$$1{,}000 + 200 \times (0.9091) + 400 \times (0.9091)^2 = ₹\ 1{,}512 \text{ approx.}$$

Again, the total discounted cost of machine B for six years is:

$$1{,}700 + 100 \times (0.9091) + 200 \times (0.9091)^2 + 300 \times (0.9091)^3 + 400 \times (0.9091)^4 + 500 \times (0.9091)^5$$
$$= ₹\ 2{,}765.73.$$

Therefore,

$$\text{Average yearly cost of machine } A = ₹\ 1{,}512/3 = ₹\ 504$$
$$\text{Average yearly cost of machine } B = ₹\ 2{,}765/6 = ₹\ 461$$

This shows that the apparent advantage is with machine B. But, the comparison is unfair, since the periods for which the costs are considered are different. So, if we consider 6 years period for machine A also, then the total discounted cost of machine A will be

$$1{,}000 + 200 \times (0.9091) + 400 \times (0{,}9091)^2 + 1{,}000 \times (0.9091)^3 + 200 \times (0.9091)^4 + 400 \times (0.9091)^5,$$

i.e., 2,648.73. This is ₹ 117 less costlier than machine B. Hence, machine A should be replaced.

Example 21.5. *A person is considering to purchase a machine for his own factory. Relevant data about alternative machines are as follows:*

	Machine A	Machine B	Machine C
Present investment (₹) :	50,000	60,000	75,000
Total annual cost (₹) :	10,000	7,500	6,000
Life span :	10 years	10 years	10 years
Salvage value (₹) :	2,500	5,000	6,000

As an adviser to the buyer, you have been asked to select the best machine, considering 12% normal rate of return. You are given that:

(a) Single payment present worth factor (Pwf) at 12% interest for 10 years (= 0.322)

(b) Annual series present worth factor (Pwf) at 12% interest for 10 years (= 5.650)

Solution. The present value of total cost of each of the three machines for a period of ten years is computed in the following table:

Machine (1)	Present investment (2)	Present value of total annual cost (3)	Present value of salvage value (4)	Net cost (₹) (5) = (2) + (3) – (4)
A	50,000	10,000 × 5.65 = 56,500	2,500 × 0.322 = 805	1,05,695
B	60,000	7,500 × 5.65 = 42,375	5,000 × 0.322 = 1,610	**1,00,765**
C	75,000	6,000 × 5.65 = 33,900	6,000 × 0.322 = 1,932	1,06,968

From the information in the table, we observe that the present value of total cost for machine *B* is the least. Hence, machine *B* should be purchased.

Example 21.6. *A manufacturer is offered two machines A and B. Machine A is priced at ₹5,000, and running costs are estimated at ₹800 for each of the first five years, increasing by ₹200 per year in the sixth and subsequent years. Machine B, which has the same capacity as machine A, costs ₹2,500 but will have running cost of ₹1,200 per year for six years, increasing by ₹200 per year thereafter.*

If money is worth 10% per year, which machine should be purchased? (Assume that the machine will eventually be sold for scrap at a negligible price.) **[MBA, Madras Univ. 2010; MBA, Annamalai Univ. 2009]**

Solution. Since, the money is worth 10% per year, the discount rate for both the machines is given by

$$pwf \text{ or } v = \frac{1}{1+0.10} = 0.9091$$

For the solution of this problem, we compute the following tables for machines *A* and *B* separately, by using *Pwf* table:

For Machine A

Year (n)	f(t)	Pwf v^{n-1}	Present value $f(t) \times v^{n-1}$	Total cost	Cum. Pwf	Weighted average cost
1	800	1,000	800	5,800	1.000	5,800
2	800	0.909	727	6,527	1.909	3,418
3	800	0.826	661	7,188	2.735	2,627
4	800	0.751	601	7,789	3.486	2,234
5	800	0.683	546	8,335	4.169	1,998
6	1,000	0.621	621	8,956	4.791	1,869
7	1,200	0.564	677	9,633	5.355	1,799
8	1,400	0.513	718	10,351	5.868	1,764
9	1,600	0.466	746	11,097	6.335	**1,751**
10	1,800	0.424	763	11,860	6.759	1,754

For Machine B

Year (n)	f(t)	Pwf v^{n-1}	Present value $f(t) \times v^{n-1}$	Total cost	Cum. Pwf	Weighted average cost
1	1,200	1,000	1,200.00	3,700.00	1.000	3,700
2	1,200	0.909	1,090.91	4,790.91	1.909	2,509
3	1,200	0.826	991.98	5,782.59	2.735	2,114
4	1,200	0.751	901.56	6,684.15	3.487	1,917
5	1,200	0.683	819.60	7,503.75	4.169	1,799
6	1,200	0.621	745.08	8,248.83	4.791	1,728
7	1,400	0.564	790.30	9,039.13	5.355	1,688
8	1,600	0.513	821.12	9,860.25	5.868	(1,680)
9	1,800	0.466	839.70	10,699.95	6.335	1,689
10	2,000	0.424	848.20	11,548.15	6.759	1,708

From the above tables, we observe that for machine *A*, 1,600 < 1,751 < 1,800.

Since, the running cost of 9th year is ₹ 1,600 and that of 10th year is ₹ 1,800, and since 1,800 > 1,751, it is better to replace the machine *A* after 9th year.

Similarly, for machine *B* since 1,800 > 1,680, it is better to replace the machine *B* after 8th year.

Further, since the weighted average cost in 9 years of machine *A* is ₹ 1,751 and the weighted average cost in 8 years of machine *B* is ₹ 1,680, it is advisable to purchase machine *B*.

PROBLEMS

20.01. The cost of a machine is ₹ 6,100 and its scrap value is ₹ 100. The maintenance costs found from experience are as follows:

Year :	1	2	3	4	5	6	7	8
Maintenance cost (₹) :	100	250	400	600	900	1,200	1,600	2,000

When should the machine be replaced? ***[MBA, Bangalore Univ. 2016; M.Com., Kerala Univ. 2010; B.Tech. (Comp. Sc.) (Dec.), Hyderabad Univ. 2011]***

21.02 *XYZ* manufacturing company is using a machine whose purchase price is ₹ 65,000. The installation charges amount to ₹ 18,000 and the machine has a scrap value of only ₹ 8,000, because the firm has a monopoly of this type of work. The maintenance cost in various years is given in the following table:

Year :	1	2	3	4	5	6	7	8	9
Maintenance cost (₹) :	1,250	3,750	5,000	7,500	10,500	14,500	20,000	24,000	30,000

Determine after how many years should the machine be replaced, assuming that the machine replacement can be done only at the year end.

[MBA, UP Tech. Univ. 2008; M.Com., Delhi Univ. 2006]

21.03. The purchase price of a machine is ₹ 52,000. The installation charges amount to ₹ 14,400, and its scrap value is only ₹ 6,400. The maintenance cost in various years is given below:

Year :	1	2	3	4	5	6	7	8
Maintenance cost (₹) :	1,000	3,000	4,000	6,000	8,4000	11,600	16,000	19,200

After how many years should the machine be replaced? Assume that the machine replacement can be done only at the year ends. ***[B.Tech. (Mech.), Panjab Tech. Univ. 2011]***

21.04. The data on the operating costs per year and resale price of equipment *A*, whose purchase price is ₹ 10,000 are given below:

Year :	1	2	3	4	5	6	7
Operating cost (₹) :	1,500	1,900	2,300	2,900	3,600	4,500	5,500
Resale value (₹) :	5,000	2,500	1,250	600	400	400	400

(*i*) What is the optimum period for replacement?

(*ii*) When equipment *A* is 2 years old, equipment *B*, which is a new model for the same usage is available. The optimum period for replacement is 4 years with an average cost of ₹ 3,600. Should we change equipment *A* with that of *B*? If so when?

[B.Tech. (Mech.) (Nov.), Panjab Tech. Univ. 2012]

21.05. The data collected in running a machine, the cost of which is ₹ 60,000, are given below:

Year :	1	2	3	4	5
Resale value (₹) :	42,000	30,000	20,400	14,400	9,650
Cost of spares (₹) :	4,000	4,270	4,880	5,700	6,800
Cost of labour (₹) :	14,000	16,000	18,000	21,000	25,000

Determine the optimum period for replacement of the machine.

[MBA, Panjab Tech. Univ. 2008; BMS, Lucknow Univ. 2008]

21.06. A fleet owner finds from his past records that the costs per year of running a vehicle, whose purchase price is ₹ 50,000 are as under:

Year :	1	2	3	4	5	6	7
Running cost (₹) :	5,000	6,000	7,000	9,000	11,500	16,000	18,000
Resale value (₹) :	30,000	15,000	7,500	5,750	2,000	2,000	2,000

Thereafter, running cost increases by ₹ 2,000 per year but the resale value remains constant at ₹ 2,000. At what age is the replacement due? ***[B.Tech. (Mech.), Panjab Tech. Univ. 2012; B.Tech. (Comp. Sc.) (Dec.), Hyderabad Univ. 2011]***

21.07. (*a*) Machine *A* costs ₹ 9,000. Annual operating costs are ₹ 200 for the first year, and then increase by ₹ 2,000 every year. Determine the best age at which to replace the machine. If the optimum replacement policy is followed, what will be the average yearly cost of owning and operating the machine? ***[BE (Comp. Sc.), Madras Univ. 2002]***

(*b*) Machine *B* costs ₹ 10,000. Annual operating costs are ₹ 400 for the first year, and then increase by ₹ 800 every year. You now have a machine of type *A* which is one year old, should you replace it with machine *B*; if so when? ***[B.Tech., Hyderabad Univ. 2009]***

21.08. Machine *A* costs ₹ 3,600. Annual operating costs are ₹ 40 for the first year and then increase by ₹ 360 every year. Assuming that machine *A* has no resale value, determine the best replacement age.

Another machine *B*, which is similar to machine *A*, costs ₹ 4,000. Annual running costs are ₹ 200 for the first year and then increase by ₹ 200 every year. It has resale value of ₹ 1,500; ₹ 1,000 and ₹ 500, if replaced at the end of first, second and third years respectively. It has no resale value during fourth year and onwards.

Which machine would you prefer to purchase? Future costs are not to be discounted.

[MCA, Guwahati Univ. 1992]

21.09 (*a*) A transport manager finds from his past records that the cost per year of running a truck whose purchase price is ₹ 6,000 are given below:

Year :	1	2	3	4	5	6	7	8
Running cost (₹) :	1,000	1,200	1,400	1,800	2,300	2,800	3,400	4,000
Resale value (₹) :	3,000	1,500	750	375	200	200	200	200

Determine at what age is replacement due?

[B.Tech., Hyderabad Univ. 2009; PG Dipl. Indian Inst. of Mat. Man. 2018]

(*b*) Let the owner of a fleet have three trucks, two of which are two years old and the third truck is one year old. The cost price, running cost and resale value of these trucks are same as given in (*a*). Now, he is considering a new type of truck with 50% more capacity than one of the old ones at a unit price of ₹ 8,000. He estimates that the running costs and resale price for the truck will be as follows:

Year	1	2	3	4	5	6	7	8
Running cost (₹) :	1,200	1,500	1,800	2,400	3,100	4,000	5,000	6,100
Resale value (₹) :	4,000	2,000	1,000	500	300	300	300	300

Assuming that the loss of flexibility due to fewer trucks is of no importance, and that he will continue to have sufficient work for three of the old trucks, what should be his policy? ***[B.Tech., Hyderabad Univ. 2009]***

21.10. A taxi owner estimates from his past records that the costs per year for operating a taxi, whose purchase price when new is ₹ 60,000 are given below:

Age :	1	2	3	4	5
Operating cost (₹) :	10,000	12,000	15,000	18,000	20,000

After 5 years, the operating cost is ₹ 6,000 k, where $k = 6, 7, 8, 9, 10$ (k denoting age in years). If the resale value decreases by 10% of purchase price each year, what is the best replacement policy? Cost of money is zero. ***[MBA, Panjab Tech. Univ. 2014]***

21.11. For a machine the following data are available:

Year :	0	1	2	3	4	5	6
Cost of spares (₹) :	—	200	400	700	1,000	1,400	1,600
Salary of maintenance staff (₹) :	—	1,200	1,200	1,400	1,600	2,000	2,600
Losses due to breakdown (₹) :	—	600	800	700	1,000	1,200	1,600
Resale value (₹) :	1,200	6,000	3,000	1,500	800	400	400

Determine the optimum period for replacement of the above machine.

[MBA, Bangalore Univ. 2008]

[***Hint:*** Operating cost = Spares cost + Maintenance cost + Breakdown depreciation.]

21.12. A machine costs ₹ 500. Operation and maintenance costs are zero for the first year and increase by ₹ 100 every year. If money is worth 5% every year, determine the best age at which the machine should be replaced. The resale value of the machine is negligibly small. What is the weighted average cost of owning and operating the machine?

[PG Dipl. (Dec.), Indian Inst. of Mat. Man. 2017; BE (Dec.), Gujarat Tech. Univ. 2012]

21.13. A truck is priced at ₹ 60,000 and running costs are estimated at ₹ 6,000 for each of the first four years, increasing by ₹ 2,000 per year in the fifth and subsequent years. If money is worth 10% per year, when should the truck be replaced? Assume that the truck will eventually be sold for scrap at a negligible price. ***[M.Com., Kerala Univ. 2012]***

21.14. A new tempo costs ₹ 80,000 and may be sold at the end of any year at the following prices:

Year (End) :	1	2	3	4	5	6
Selling price (₹) *(present value)* :	50,000	33,000	20,000	11,000	6,000	1,000

The corresponding annual operating costs are:

Year (End) :	1	2	3	4	5	6
Cost/year (in ₹) *(at present value)* :	10,000	12,000	15,000	20,000	30,000	50,000

It is not only possible to sell the tempo after use, but also to buy a second hand tempo. It may be cheaper to do so than to replace by a new tempo.

Age of tempo	:	0	1	2	3	4	5
Purchase price (in ₹) *(at present value)*	:	80,000	58,000	40,000	26,000	16,000	10,000

What is the age to buy and to sell tempo so as to minimize average annual cost?

21.15. Let $v = 0.9$ and initial price is ₹ 5,000. Running cost varies as follows:

Year	:	1	2	3	4	5	6	7
Running cost (in ₹)	:	400	500	700	1,000	1,300	1,700	2,100

What would be the optimum replacement interval?

21.16. A manual stamper currently valued at ₹ 1,000 is expected to last 2 years and costs ₹ 4,000 per year to operate. An automatic stamper which can be purchased for ₹ 3,000 will last 4 years and can be operated at an annual cost of ₹ 3,000. If money carries the rate of interest 10% per annum, determine which stamper should be purchased.

21.17. The cost of a new machine is ₹ 5,000. The maintenance cost of nth year is given by $C_n = 500\ (n - 1);\ n = 1, 2, \ldots$. Suppose that the discount rate per year is 0.5. After how many years it will be economical to replace the machine by a new one?

[B.Tech. (Nov.), Hyderabad Univ. 2010; B.Tech. (Dec.), Nagpur Univ. 2004]

[***Hint:*** *Pwf* is: $v = (1 + 0.05)^{-1} = 0.9523$]

21.18. A pipeline is due for repairs. It will cost ₹ 10,000 and last for 3 years. Alternatively, a new pipeline can be laid at a cost of ₹ 30,000 and lasts for 10 years. Assuming cost of capital to be 10%, and ignoring salvage value, which alternative should be chosen?

[***Hint:*** *Pwf* is: $v = (1 + 0.1)^{-1} = 0.909$. Let k_n denote the discounted value of all future costs and replace the pipeline after n years. Then,

$$k_n = C + Cv^n + Cv^{2n} + Cv^{3n} + \ldots \infty = C/(1 - v^n), \text{ where } C \text{ is the initial outlay.}$$

$\therefore$ $k_3 = 10{,}000/[1 - (0.9091)^3]$, *i.e.*, ₹ 4,021 for the existing pipeline,

and $k_{10} = 30{,}000(1 - (0.9091)^{10}]$, *i.e.*, ₹ 48,820 for the new pipeline.

This shows that $k_3 < k_{10}$, *i.e.*, existing pipeline is better.]

21.19. The yearly cost of 2 machines A and B, when the money value is neglected is as follows:

Year	:	1	2	3
Machine A	:	1,800	1,200	1,400
Machine B	:	2,800	200	1,400

Find their cost patterns, if money value is 10% per year and hence find which machine is most economical.

[B.Tech., Nagpur Univ. 2006]

21.20. An individual is planning to purchase a car. A new car will cost ₹ 1,20,000. The resale value of the car at the end of the year is 85% of the previous year value. Maintenance and operating costs during the first year are ₹ 20,000 and they increase by 15% every year. The minimum resale value of car can be ₹ 40,000.

(*i*) When should the car be replaced to minimize average annual cost (ignore interest)?

(*ii*) If interest of 12% is assumed, when should the car be replaced?

21:6. REPLACEMENT POLICY FOR EQUIPMENT THAT BREAKS-DOWN/FAIL SUDDENLY

Generally a system contains a large number of low cost items (units) that are increasingly liable to fail with age (*e.g.*, television, computer, etc.). Very often the failure of an item may cause a total breakdown of the system. The costs of failure, in such a case, will be quite higher than the cost of item. For example, a condenser in an aircraft costs little, but its failure may result in total collapse of the aircraft.

Obviously, the replacement of individual units upon their failure, costs relatively high, whereas if the entire group is replaced, the unit cost works out to be comparatively a much lower value. Thus, we might be inclined to replace the whole lot at certain intervals together with the individual replacements as and when needed. If the group replacement is done at short intervals, the cost per period shall be high. As the time interval between group replacements increases, the cost of group replacement per period declines. But, the cost of individual replacements increases because of larger number of replacements needed. Thus, our problem is to determine the optimum interval of time to have group replacement. So, we consider the following two type of replacement policies:

Individual Replacement Policy. Under this policy, an item is replaced immediately after its failure.

Group Replacement Policy. Under this policy, we take decisions as to when all the items must be replaced, irrespective of the fact that items have failed or have not failed, with a provision that if any item fails before the optimum interval of time, it may be individually replaced.

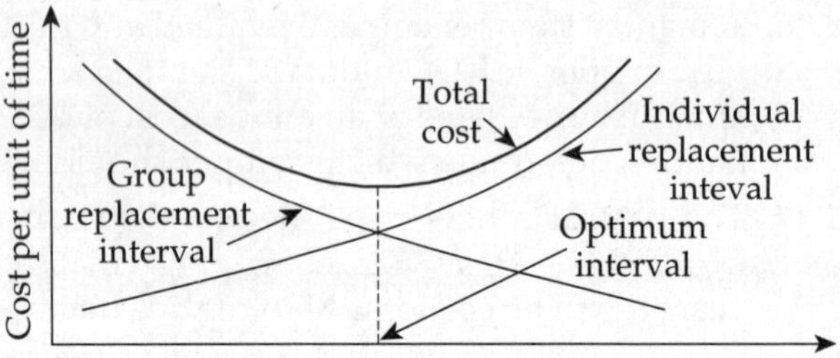

Fig. 21.1

Note. Given below are the main assumptions in group replacement:

(*a*) The items/units failing in every period (day/week/month) are replaced just before the end of that period.

(*b*) Sum of the cost of group replacement per unit of time and the cost of individual replacement of items which fail during the replacement interval (also per unit of time) represents the total cost of group replacement.

Mortality Tables. These are used to derive the probability distribution of the life span of an equipment. Let

$$M(t) = \text{number of survivors at any time } t,$$
$$M(t-1) = \text{number of survivors at any time } t-1, \text{ and}$$
$$N = \text{initial number of items/units.}$$

Then, the probability of failure during time period t is given by

$$p(t) = [M(t-1) - M(t)]/N$$

The probability that an equipment will survive till age 't' is given by: $p_s(t) = M(t)/N$

Example 21.7. *There are 1,000 bulbs in the system. Survival rate is given below:*

Week :	*0*	*1*	*2*	*3*	*4*
Bulbs in operation at the end of week :	*1,000*	*850*	*500*	*200*	*0*

The group replacement of 1,000 bulbs costs ₹ 2,000 and individual replacement is ₹ 5 per bulb. Suggest suitable replacement policy.

Solution. From the survival rate, we can find out the probability p_i, that a new bulb installed at a time zero fails during week i. Thus, we get

Week :	1	2	3	4
Survival probability :	0.85	0.50	0.2	0
Probability of failure :	(1 – 0.85) = 0.15	(0.85 – 0.50) = 0.35	(0.50 – 0.2) = 0.3	(0.2 – 0) = 0.2

The expected life of each bulb is

$$\Sigma\, i \times p_i = 1 \times 0.15 + 2 \times 0.35 + 3 \times 0.3 + 4 \times 0.2 = 2.55 \text{ weeks}$$

$\therefore$ Average number of failures per week = 1,000/2.55 = 392.16

Let N_i denote the number of replacements at the end of ith week. Then, we have

N_1 = number of bulbs being replaced by the end of first week

= N_0 × Probability of failure during first week

= 1,000 × p_1 = 1,000 × 0.15 = 150

N_2 = Number of bulbs being replaced by the end of 2nd week

= (Number of bulbs in the beginning) × (Probability of failure in 2nd week) + (Number of bulbs replaced in the 1st week) × (Probability of these failing in 1st week)

= $N_0 p_2 + N_1 p_1$ = 1,000 × 0.35 + 150 × 0.15 = 373

N_3 = Number of bulbs being replaced by the end of 3rd week

= (Number of bulbs initially replaced) × (Probability of failure in 3rd week) + (Number of bulbs replaced in 1st week) × (Probability of failure in 2nd week) + (Number of bulbs replaced in 2nd week) × (Probability of failure in 1st week)

= $N_0 p_3 + N_1 p_2 + N_2 p_1$

= 1,000 × 0.3 + 150 × 0.35 + 373 × 0.15 = 409

Similarly,

$N_4 = N_0 p_4 + N_1 p_3 + N_2 p_2 + N_3 p_1$

= 1,000 × 0.2 + 150 × 0.3 + 373 × 0.35 + 409 × 0.15 = 437

Individual replacement of bulbs would cost on the basis of expected (mean) life of 2.55 weeks, *i.e.*, (average number of failures per week) × ₹ 5. This comes out to be (392.16) × ₹ 5 or ₹ 1,960.80.

Now, since the replacement of all the bulbs simultaneously cost ₹ 2 per bulb and the replacement of individual bulb on failure cost ₹ 5, the average cost for different group replacement policies are obtained in the following table:

End of week	*Individual replacements*	*Total cost (₹) = Individual + Group*	*Average cost (₹)*
1	150	750 + 2,000 = 2,750	2,750.00
2	523	2,615 + 2,000 = 4,615	2,307.50
3	932	4,660 + 2,000 = 6,660	2,220.00
4	1,369	6,845 + 2,000 = 8,845	2,211.45

Since, the average cost is lowest against the end of week 4, the optimum interval of group replacements is by the end of 4th week. Further, since the average cost is more than ₹ 1,960.80 for individual replacement, the policy of individual replacement is better.

Example 21.8. *The following failure rates have been observed for a certain type of light bulbs:*

Week :	1	2	3	4	5
Percentage of failure by the end of the week :	10	25	50	80	100

There are 1,000 bulbs in use. It costs ₹ 2 to replace an individual bulb which has burnt out. If all bulbs are replaced simultaneously it costs 50 paise per bulb. It is proposed to replace all bulbs at fixed intervals, whether or not they have burnt out and to continue replacing burnt out bulbs as and when they fail. At what interval should all the bulbs be replaced? At what group replacement price per bulb would be, so that strictly individual replacement policy become preferable? **[BE (IT), Nagpur Univ. 2009; M.Com., Delhi Univ. 2010]**

Solution. Let p_i be the probability that a light bulb, which was new when placed in position for use, fails during the ith week of its life. Thus, we have

$p_1 = 0.10,\ \ p_2 = 0.25 - 0.10 = 0.15,\ \ p_3 = 0.50 - 0.25 = 0.25$

$p_4 = 0.80 - 0.50 = 0.30,\ \ \text{and}\ \ p_5 = 1.00 - 0.80 = 0.20$

If N_i denotes the number of replacements made at the end of the ith week, we have

$$N_0 = 1{,}000$$
$$N_1 = N_0 p_1 = 1{,}000 \times 0.10 = 100$$
$$N_2 = N_0 p_2 + N_1 p_1 = 1{,}000 \times 0.15 + 100 \times 0.10 = 160$$
$$N_3 = N_0 p_3 + N_1 p_2 + N_2 p_1 = 1{,}000 \times 0.25 + 100 \times 0.15 + 160 \times 0.10 = 281$$
$$N_4 = N_0 p_4 + N_1 p_3 + N_2 p_2 + N_3 p_1$$
$$= 1{,}000 \times 0.30 + 100 \times 0.25 + 160 \times 0.15 + 281 \times 0.10 = 377$$
$$N_5 = N_0 p_5 + N_1 p_4 + N_2 p_3 + N_3 p_2 + N_4 p_1$$
$$= 1{,}000 \times 0.2 + 100 \times 0.3 + 160 \times 0.25 + 281 \times 0.15 + 377 \times 0.10 = 350$$

The expected life of each light bulb is:

$$1 \times 0.10 + 2 \times 0.15 + 3 \times 0.25 + 4 \times 0.30 + 5 \times 0.20 = 3.35 \text{ weeks}$$

$\therefore$ Average number of failures per week $= \dfrac{1{,}000}{3.35} = 299$ bulbs per week.

Now, since the cost of individual replacement of a bulb is ₹ 2, the average cost of individual replacement = ₹ 299 × 2 = ₹ 598.

As the cost of all the 1,000 bulbs simultaneously costs 50 paise per bulb and the replacement of individual bulb on failure costs ₹ 2.00, the average cost for different group replacement policies is computed as follows:

End of week	*Individual replacement*	*Total cost (₹) = Individual + Group*	*Average cost (₹)*
1	100	200 + 500 = 700	700
2	260	520 + 500 = 1,020	510
3	541	1,082 + 500 = 1,582	527
4	918	1,836 + 500 = 2,336	584
5	1,268	2,536 + 500 = 3,036	607

Since, the average cost is lowest in the 2nd week, it is cheapest to have group replacement after every second week. Further, since the average cost is less than ₹ 598 (of individual replacement), the policy of group replacement is advisable.

if the price of a bulb is increased to 70 paise in group replacement, then the lowest average cost becomes ₹ 610, which is more than the individual replacement cost. Hence, in this case, individual replacement policy should be adopted.

Example 21.9. *An electronic equipment has 5,000 IC's. If any one IC fails, it has to be replaced. Cost of replacing a single IC is only ₹ 10. If all IC's are replaced simultaneously at a time, it would cost ₹ 3.50 per IC. The data of failures are as given below:*

Month	:	0	1	2	3	4	5	6
% of surviving at the end of each month	:	100%	97%	90%	70%	30%	15%	0%

Obtain the best replacement policy, i.e., replace as and when IC fails or replace all IC's at fixed intervals and replace the individual IC's which fail during the fixed interval. **[MBA, Bangalore Univ. 2006]**

Solution. Let $S(t)$ denote the percentage of survival and $p(t)$ is the failure of IC by the end of month. Then, we have

Month :	1	2	3	4	5	6
$p(t)$:	(100 – 97)/100 = 0.03	(97 – 90)/100 = 0.07	(90 – 70)/100 = 0.20	(70 – 30)/100 = 0.40	(30 – 15)/100 = 0.15	(15 – 0)/100 = 0.15

The expected, weighted average, life of each IC

$$= 1 \times 0.03 + 2 \times 0.07 + 3 \times 0.20 + 4 \times 0.40 + 5 \times 0.15 + 6 \times 0.15 = 4.02 \text{ months}$$

$$\therefore \quad \text{Cost of individual replacement} = \frac{5{,}000 \times ₹\,10}{4.02} = ₹\,12{,}437.81$$

If we consider the group replacement policy, we consider the concept of replacement interval when all the IC's are replaced in that lot. So, let N_i be the number of IC's replaced at the end of ith month. Then, different values of N_i are calculated as follows:

$$N_0 = 5{,}000 = 5{,}000$$
$$N_1 = N_0 p_1 = 5{,}000 \times 0.03 = 150$$
$$N_2 = N_0 p_2 + N_1 p_1 = 5{,}000 \times 0.07 + 150 \times 0.03 = 354.5 \quad \text{or} \quad 354 \text{ approx.}$$
$$N_3 = N_0 p_3 + N_1 p_2 + N_2 p_1 = 5{,}000 \times 0.20 + 150 \times 0.07 + 354.5 \times 0.03 = 1{,}021$$
$$N_4 = N_0 p_4 + N_1 p_3 + N_2 p_2 + N_3 p_1 = 2{,}085.45 \quad \text{or} \quad 2{,}085 \text{ approx.}$$
$$N_5 = N_0 p_5 + N_1 p_4 + N_2 p_3 + N_3 p_2 + N_4 p_1 = 1{,}014.94 \quad \text{or} \quad 1{,}015 \text{ approx.}$$
$$N_6 = N_0 p_6 + N_1 p_5 + N_2 p_4 + N_3 p_3 + N_4 p_2 + N_5 p_1 = 1{,}294.96 \quad \text{or} \quad 1{,}295 \text{ approx.}$$

From the above calculations, it is observed that the expected number of IC's failing each month increases till the 4th month and then start decreasing.

Under the group replacement policy along with individual replacement, the average cost of replacement is computed in the following table:

Month	*Number of IC's failed*	*Cumulative failures*	*Cost*		*Total cost*	*Average cost per month*
			Individual (₹)	*Group (₹)*	*(₹)*	*(₹)*
1	150	150	1,500	17,500	19,000	19,000
2	354	504	5,040	17,500	22,540	11,270
3	1,021	1,525	15,250	17,500	32,750	**10,917**
4	2,085	3,610	36,100	17,500	53,600	13,400
5	1,015	4,625	46,250	17,500	63,750	12,750
6	1,295	5,920	59,200	17,500	76,700	12,783

From the above table, we observe that total average cost is minimum in the 3rd month. Therefore, it is better to have a group replacement along with individual replacement by the end of third month. Further, since the average cost of group replacement, ₹ 10,917, is less than the average cost of individual replacement, ₹ 12,437.81, the policy of group replacement is advisable.

Example 21.10. *A computer has a large number of electronic tubes. They are subject to mortality as given below:*

Period	*Age of failure (hours)*	*Probability of failure*
1	0–200	0.10
2	201–400	0.26
3	401–600	0.35
4	601–800	0.22
5	801–1,000	0.07

If the electronic tubes are group replaced, the cost of replacement is ₹ 15 per tube. Group replacement can be done at fixed intervals in the night shift, when the computer is not normally used. Replacement of individual tubes which fail in service costs ₹ 60 per tube. How frequently should the tubes be replaced?

[B.Tech. (Mech.), Nagpur Univ. 2007; MBA, Bangalore Univ. 2018]

Solution. Consider each block of 200 hours as one period and assume that there are 1,000 tubes initially.

Let N_i be the number of replacements made at the end of the ith period, if all the 1,000 tubes are new initially. Then, the expected number of failures at different weeks can be calculated as shown below:

$$N_1 = N_0 p_1 = 1{,}000 \times 0.10 = 100$$
$$N_2 = N_0 p_2 + N_1 p_1 = 1{,}000 \times 0.26 + 100 \times 0.10 = 270$$
$$N_3 = N_0 p_3 + N_1 p_2 + N_2 p_1 = 1{,}000 \times 0.35 + 100 \times 0.26 + 270 \times 0.10 = 403$$
$$N_4 = N_0 p_4 + N_1 p_3 + N_2 p_2 + N_3 p_1 = 1{,}000 \times 0.22 + 100 \times 0.35 + 270 \times 0.26 + 403 \times 0.10 = 365$$
$$N_5 = N_0 p_5 + N_1 p_4 + N_2 p_3 + N_3 p_2 + N_4 p_1$$
$$= 1{,}000 \times 0.07 + 100 \times 0.22 + 270 \times 0.35 + 403 \times 0.26 + 365 \times 0.10 = 328$$

From the above calculations, we observe that the number of tubes failing in each period increases till the third period, and then starts decreasing. Thus, the value of N_i will oscillate till the system settles down to a steady state. In the steady state, the proportion of tubes failing during each period is the reciprocal of their average life.

Expected life of a tube = 1 × 0.10 + 2 × 0.26 + 3 × 0.35 + 4 × 0.22 + 5 × 0.07 = 2.90 periods.

$$\text{Expected number of failures per period} = \frac{1{,}000}{2.90} = 345$$

∴ Cost of individual replacements per period = 345 × 60 = ₹ 20,700

Now, since the replacement of all the 1,000 tubes simultaneously costs ₹ 15 per tube, and the replacement of an individual tube on failure costs ₹ 60; the average cost for different group replacement policies is given as follows:

End of period (hours)	*Individual replacement*	*Total cost (₹)* *Individual + Group*	*Average cost (₹)*
0–200	100	100 × 60 + 1,000 × 15 = 21,000	21,000
201–400	370	370 × 60 + 1,000 × 15 = 37,200	**18,600**
401–600	773	774 × 60 + 1,000 × 15 = 61,380	20,460
601–800	1,138	1,138 × 60 + 1,000 × 15 = 83,280	20,820

Since, the average cost is lowest against end of period 2, the optimum interval between group replacements is two periods, *i.e.*, after 400 hours. Further, since the average cost of group replacement (which is ₹ 18,600) is less ₹ 20,700 (cost of individual replacement), the policy of group replacement is better.

Example 21.11. *At time zero, all items in a system are new. Each item has a probability 'p' of failing immediately before the end of the first month of life, and a probability 'q = (1 – p)' of failing immediately before the end of the second month (i.e., all items fail by the end of the second month). If all items are replaced as they fail, show that the expected number of failures f(x) at the end of month x is given by*

$$f(x) = \frac{N}{1+q}\left[1-(-q)^{x+1}\right],$$

where, N is the number of items in the system.

If the cost per item of individual replacement is C_1, and the cost per item of group replacement is C_2, find the condition under which

(*a*) *A group replacement policy at the end of each month is the most profitable.*
(*b*) *No group replacement policy is better than a policy of pure individual replacement.*

[M.Sc. (Math.), CCS Univ. 2006]

Solution. Let

$$N = \text{number of items in the system in the beginning}$$
$$N_1 = \text{number of items expected to fail at the end of } 1^{st} \text{ month}$$
$$= N_0 p = N(1-q), \text{ since } p = 1-q$$
$$N_2 = \text{number of items expected to fail at the end of } 2^{nd} \text{ month}$$
$$= N_0 q + N_1 p = Nq + N(1-q)^2 = N(1-q+q^2)$$
$$N_3 = \text{number of items expected to fail at the end of } 2^{rd} \text{ month}$$
$$= N_1 q + N_2 p = N(1-q)\,q + N(1-q+q^2)(1-q) = N(1-q+q^2-q^3),$$

and so on. In general,

$$N_k = N\left[1-q+q^2-q^3+\ldots+(-q)^k\right],$$

$$\therefore \quad N_{k+1} = N_{k-1}q + N_k p$$
$$= N\left[1-q+q^2+\ldots+(-q)^{k-1}\right]q + N\left[1-q+q^2+\ldots+(-q)^k\right](1-q)$$
$$= N\left[1-q+q^2+\ldots+(-q)^{k+1}\right]$$

Hence, by mathematical induction, the expected number of failures at the end of month x will be given by

$$f(x) = N\left[1-q+q^2+\ldots+(-q)^x\right] = N\left[1-(-q)^{x+1}\right]/(1+q)$$

The value of $f(x)$ at the end of month x will vary for different values of $(-q)^{x+1}$ and it will reach the steady state as $x \to \infty$.

Hence, in the steady state case, the expected number of failures will be

$$\lim_{x\to\infty} f(x) = \lim_{x\to\infty} N\left[1-(-q)^{x+1}\right]/(1+q)$$
$$= N/(1+q); \text{ since } q<1 \text{ and } (-q)^{x+1} \to 0 \text{ as } x\to\infty$$
$$= \text{Total number of items in the system/mean age.}$$

Now, since C_1 is the cost of replacement per item individually and C_2 is the cost of an item in group, therefore

(*i*) If we have a group replacement at the end of each month, then the cost of replacement is NC_2.

(*ii*) If we have a group replacement policy at the end of every other month, then the cost is $NC_2 + NpC_1$.

The average cost per month, therefore, is $(NC_2 + NpC_1)/2$.

Average life of an item $= 1 \times p + 2 \times q = 1 \times (1-q) + 2q = 1 + q$.

Therefore, the average number of failures is $N/(1+q)$, and hence the cost of individual replacement is $NC_1/(1+q)$.

(*a*) A group replacement at the end of first month will be better than individual replacement, if total cost of group replacement is less than the average monthly cost of individual replacement. That is,

$$N(1-q)C_1 + NC_2 < NC_1/(1+q), \; i.e., \; C_2 < C_1 q^2/(1+q)$$

For a group replacement at the end of every second month, the total cost of replacement will be

$$(N_1+N_2)C_1 + NC_2 = N(2-2q+q^2)C_1 + NC_2$$

∴ Average monthly cost of group replacement at the end of second month is

$$\left[N(2-2q+q^2)C_1+NC_2\right]/2$$

In this case, the group replacement policy will be better than the individual replacement policy, if

$$\begin{bmatrix}\text{Average monthly cost of}\\ \text{group replacement}\end{bmatrix} < \begin{bmatrix}\text{Average monthly cost of}\\ \text{individual replacement}\end{bmatrix}$$

or $$\frac{1}{2}[N(2-2q+q^2)C_1+NC_2] < NC_1/(1+q),$$

or $$C_2 < [q^2(1-q)C_1]/(1+q)$$

(*b*) For the individual replacement policy to be better than any of the group replacement policies discussed above, we must have

$$C_2 > C_1 q^2/(q+1) \quad \text{and} \quad C_2 > C_1 q^2(1-q)/(q+1)$$

or $$C_1 < C_2(1+q)/q^2 \quad \text{and} \quad C_1 < C_2(1+q)/[q^2(1-q)]$$

But, $q < 1$, therefore $(1+q)/q^2 < (1+q)/q^2(1-q)$

Hence, $$C_1 < (1+q)C_2/q^2$$

PROBLEMS

21.21. An electronic equipment contains 500 resistors. When any resistor fails, it is replaced. The cost of replacing a resistor individually is ₹ 20. When all the resistors are replaced at the same time, the cost per resistor is ₹ 5. The percentage of surviving, $S(t)$ of resistor, at the end of month t is given below:

Month :	0	1	2	3	4	5
$S(t)$:	100	90	75	55	30	0

What is the optimum replacement plan? ***[MBA, Annamalai Univ. 2006]***

21.22. A computer contains 10,000 resistors. When any resistor fails, it is replaced. The cost of replacing a resistor individually is ₹ 1. If all the resistors are replaced at the same time, the cost per resistor would be reduced to 35 paise. The percent of surviving say $S(t)$ at the end of month t and $P(t)$ the probability of failure during the month t, are:

t :	0	1	2	3	4	5	6
$S(t)$:	100	97	90	70	30	15	0
$P(t)$:	—	0.03	0.07	0.20	0.40	0.15	0.15

What is the optimum replacement plan?

[M.Com., Kerala Univ. 2012; BE (Dec.), Karnataka Univ. 2011]

21.23. An electronic equipment has 5,000 IC's. If any one IC fails it has to be replaced. Cost of replacing a single IC is only ₹ 10. The data of failures is given below:

Month :	1	2	3	4	5	6
% Surviving at the end of each month :	0.03	0.10	0.30	0.40	0.15	0.02

Obtain the best replacement policy, *i.e.*, replace the individual IC's which fail during fixed interval. ***[MBA (Dec.), Panjab Tech. Univ. 2010]***

21.24. The following mortality rates have been observed for a certain type of fuses:

Week :	1	2	3	4	5
% failing by the end of week :	5	15	35	75	100

There are 1,000 fuses in use and it costs ₹ 5 to replace on individual fuse. If all fuses were replaced simultaneously, it would cost ₹ 1.25 per fuse. At what interval the groups replacement should be done? Which policy is better? *[MBA, Bangalore Univ. 2017]*

21.25. The following failure rates have been observed for a certain type of light bulbs:

End of week :	1	2	3	4	5	6	7	8
Probability of failure to date :	0.05	0.13	0.25	0.43	0.68	0.88	0.96	1.00

The cost of replacing an individual failed bulb is ₹ 1.25. The decision is made to replace all bulbs simultaneously at fixed intervals, and also to replace individual bulbs as they fail in service. If the cost of group replacement is 30 paise per bulb, what is the best interval between group replacements?

At what group replacement price per bulb would a policy of strictly individual replacement become preferable to the adopted policy.

[B.Tech., Panjab Tech. Univ. 2012; B.Tech., Hyderabad Univ. 2010]

21.26. India Electric Company operates three large baking ovens used to manufacture electric motors and generators. Each oven contains 2,400 heating resistors. It costs ₹ 80 to replace an individual resistor when it burns out. If all the resistors are replaced on Sunday night it costs ₹ 72,000. When a resistor fails, it must be replaced immediately. When group replacement is followed, all resistors are replaced at one time regardless of age. Statistical data provided by the research department of India Electric Company provide the following probability distribution for resistor failure:

Week :	1	2	3	4	5	6
Probability of failure :	0.02	0.08	0.15	0.30	0.25	0.20

Is individual or group replacement policy more desirable? *[MBA, Delhi Univ. 2007]*

21.27. A large computer system has 100 identical components, which are subject to random failure over time. The failure pattern is given below:

Month end :	1	2	3	4	5	6
Failure to date (Percentage) :	12	28	50	75	90	100

Failed components have to be replaced immediately for the effective functioning of computer. If they are replaced individually, as and when they fail, then the cost per replacement is estimated as ₹ 4. Alternatively, if all the components are replaced in one lot at periodical intervals, then the cost per unit is estimated as ₹ 1.50.

(*a*) Determine the more economical of the two policies to complete individual replacement or periodic group replacement.

(*b*) Would the above decision remain valid, if the unit cost in group replacement would be ₹ 2? Justify your answer.

21.28. A group of process plants in an oil refinery are fitted with valves. Over a period of time, the failure pattern of the 400 valves has been observed as follows:

Month :	1	2	3	4	5	6	7	8	*Total*
Number of valves failing :	8	20	48	104	120	56	32	12	400

It costs ₹ 100 to replace each valve individually. If all the valves are replaced at a time, it costs ₹ 50 per valve only.

The maintenance department is considering following replacement policies:

(*a*) Replace all valves simultaneously at fixed intervals, in addition to replacing valves as and when they fail.

(*b*) Replace valves as and when they fail.

Suggest the optimal replacement policy. *[MBA (June), Bangalore Univ. 1998]*

21.29. It has been suggested by a data processing firm that they adopt a policy of priodically replacing all the tubes in a certain piece of equipment. A given type of tube is known to have the mortality distribution shown in the table given below:

Tube failures per week :	1	2	3	4	5
Probability of failure :	0.3	0.1	0.1	0.2	0.3

There are approximately 1,000 tubes of this type in all the combined equipment. The cost of replacing the tubes on an individual basis is estimated to be ₹ 1.00 per tube and the cost of a group replacement policy average ₹ 0.30 per tube. Compare the cost of preventive replacement with that of remedial replacement. *[MBA (Nov.), Annamalai Univ. 2010]*

21.30. On its production line, Modern Food Company uses a battery of 400 small sprinklers for cleaning fresh vegetables, fruits, etc. before they are prepared for canning. Failure of sprinkler, which costs ₹ 50 per unit, does not disrupt production flow but may create quality control problem from insufficient cleaning. Such failures are reported by line foreman to the maintenance department, which must dis-assemble the arc-shaped battery frame from the production line before replacing a sprinkler at the end of the shift or on weekends. Due to the high setup cost of this task, management is considering a policy of group replacement. What replacement policy should the management adopt when the information regarding sprinkler breakdowns and replacement cost are given below:

Run time (Months) :	1	2	3	4	5	6
Probability of failure :	0.05	0.05	0.10	0.10	0.30	0.40

Replacement Cost

	Purchase	*Installation*	*Per unit*
Individual			
Group	₹ 50	₹ 175	₹ 225
	₹ 50	₹ 25	₹ 75

[BE (Mech.) (Sept.), Nagpur Univ. 2010]

21:7. REPLACEMENT POLICY OF STAFFING/MAN-POWER

The staff of an organization calls for replacement because people leave the organization for several reasons. For planning a suitable recruitment policy, historical data are collected to estimate the likely stay of individuals with the organization through time. Following examples illustrate the concept of staffing problems.

Example 21.12. *A company plans to have a strength of 50 sales persons over time. The people employed by the company leave their job, the distribution of which is assumed to be as follows:*

Year :	1	2	3	4	5	6	7	8	9	10
Percentage expected to leave by the end of period :	5	35	56	65	70	76	80	86	95	100

Using the above information, answer the following questions:

(a) What recruitment level per annum would be necessary to maintain the strength of 50 sales persons?

(b) If seven of the sales persons are to be promoted to the post of supervisor, for which length of service in the organization is the primary criterion, at what average length of service can a new entrant expect to be promoted?

Solution. Total number of sales persons is not given, so let the intake of sales persons every year is 100. Then, the distribution of the length of service for the sales persons, when equilibrium has been attained, shall be as follows:

Year :	0	1	2	3	4	5	6	7	8	9	10	Total
Number of persons continuing :	100	95	65	44	35	30	24	20	14	5	0	432

(*a*) If 100 sales persons are recruited every year, total number of sales persons 10 years hence will be 432. Therefore, to maintain a strength of 50 sales persons, we need to recruit 50 × (100/432), *i.e.*, 11.6 or 12 persons (approx.) per year.

In the above calculation, it is assumed that those persons who completed x years of service leave immediately before completing $(x + 1)$ years of service. If we assume that the persons leave as soon as they complete x years of service, the total number of sales persons would be (432 – 100), *i.e.*, 332. According to this, the recruitment necessary to have a strength of 50 sales persons will be 50 × (100/332), *i.e.*, 15 persons per year.

Further, since the sales persons leave at any time in the year, the number of recruitments should be (11.6 + 15)/2, *i.e.*, 13.3 or 13 persons approximately.

(*b*) With 50 persons in the sales team, one must recruit (50/432) × 100, *i.e.*,11.57 or 12 persons every year. Therefore, the distribution of the completed length of service would be as follows:

Year :	1	2	3	4	5
Number of persons :	12 × 0.95 = 11	12 × 0.65 = 8	12 × 0.44 = 5	12 × 0.35 = 4	12 × 0.30 = 4
Year :	6	7	8	9	10
Number of persons :	12 × 0.24 = 3	12 × 0.20 = 2	12 × 0.14 = 2	12 × 0.05 = 1	0

There are 7 posts of sales supervisors for which length of service is the main criterion. Now, from the above table, clearly the number of persons having longest service are: 1 in ninth year, 2 each in seventh and eighth year, and 3 in sixth year. Thus, the total number of persons are (1 + 2 + 2 + 3), *i.e.*, 8. Hence, sales persons can expect to be promoted after 5 years of service.

Example 21.13. *A public sector bank requires 400 clerks, 250 officers, and 50 managers for a certain state. The employees are recruited at the age of 21 years as clerks. The promotions take place only from within on the basis of the length of service. You are given the following life table:*

Age (Years) :	21	22	23	24	25	26	27	28	29	30
No. of employees in service :	1,000	640	560	450	400	355	300	205	190	182
Age (Years) :	31	32	33	34	35	36	37	38	39	40
No. of employees in service :	175	170	162	155	150	148	145	142	140	132
Age (Years) :	41	42	43	44	45	46	47	48	49	50
No. of employees in service :	125	118	112	105	101	100	95	90	88	85
Age (Years) :	51	52	53	54	55	56	57	58	59	60
No. of employees in service :	80	72	70	63	55	40	32	25	20	0

Determine:

(*a*) *How many employees should be recruited each year?*

(*b*) *At what age should the promotions take place?*

Solution. total number of employees recruited = (400 + 250 + 50) = 700. Total number of employees serving up to age of 59 years starting at the age of 21 years = 1,000 + 640 + 560 + ... + 25 + 20 = 7,277.

(*a*) The number of employees to be recruited each year in order to maintain the strength of

$$700 \text{ employees} = \frac{1{,}000}{7{,}277} \times 700 = 96$$

(*b*) Let the clerks be promoted at the age of x years. Then up to age of $(x - 1)$ years, number of clerks required are 400.

Since out of 700 employees there are 400 clerks, therefore, out of 1,000 employees the number of clerks comes out to be (400/700) × 1,000, *i.e.*, 571.4 or 571 approx.

From the life table, we observe that this number of 571.4 clerks is available up to the age of 22 years. Hence, the promotion of clerks will be due in 23rd year. Further, out of 700 employees, we need 250 officers, therefore, the number of officers will be (250/700) × 1,000, *i.e.*, 357. So, the total number of clerks and officers = (571 + 357) = 928. Thus, out of 1,000 employees, rest (1,000 – 928), *i.e.*, 72 managers are required. This number is available only up to the age of 52 years.

Hence, promotion to manager is due in 53rd year.

PROBLEMS

21.31. Calculate the probability of a staff resignation in each year from the following survival table:

Year	0	1	2	3	4	5	6	7	8	9	10
Staff at year end	1,000	940	820	580	400	280	190	130	70	30	0

21.32. A research team is planned to raise to a strength of 50 chemists and then to remain at that level. The number of recruits depends on their length of service which is as follows:

Year	1	2	3	4	5	6	7	8	9	10
Percentage left at year end	5	36	56	63	68	73	79	87	97	100

What is the recruitment per year necessary to maintain the required strength? There are 8 senior posts for which the length of service is the main criterion. What is the average length of service after which new entrant expects his promotion to one of these posts?

21:8. EQUIPMENT RENEWAL POLICY

Knowledge of probability distribution of failure time of certain equipments enables us to solve many practical problems of replacement. In statistical terminology, the term renewal mean either to insert a new equipment in place of an old one or to repair the old equipment, so that the *pdf* of its future life-time is that of a new equipment. Thus, the problem of replacement can be dealt with what is called the ***Renewal Policy*** where the life span of an equipment is considered to be a random variable.

The probability that a renewal occurs during the small time interval $(t, t + dt)$ is called the ***renewal rate*** at time t, where t is measured from the instant, the first equipment was started. It is generally denoted by $h(t)\,dt$ and $h(t)$ is called the ***renewal density function***.

Theorem (*Without Proof*). The renewal rate of a machine is asymptotically reciprocal of the mean life of the machine.

Example 21.14. *A piece of equipment either can fail completely, so that it has to be scrapped (no salvage value), or may suffer a minor defect which can be repaired. The probability that it will not have to be scrapped before age t is f(t). The conditional probability that it will need a repair in the instant t to t + dt, knowing that it was in running order at age t, is r(t)dt. The probability of a repair or complete failure is dependent only on the age of the equipment, and not on the previous repair history.*

Each repair costs ₹C, and complete replacement costs ₹K. For some considerable time, the policy has been to replace only on failure.

(*a*) *Derive a formula for the expected cost per unit time of the present policy of replacing only on failure.*

(*b*) *It has been suggested that it might be cheaper to scrap equipment at some fixed age T, thus avoiding the higher risk of repairs with advancing age. Show that the expected cost per unit time of such policy is*

$$\left[C\int_0^T f(u)\,r(u)\,du + K\right] \bigg/ \left[\int_0^T f(u)\,du\right]$$

[M.Sc. (Math.), CCS Univ. 1997]

Solution. Since, the probability that the equipment is not scrapped before age t is $f(t)$, therefore, evidently the equipment will fail at some time and hence

$$\int_0^\infty f(t)\,dt = 1$$

Also, when it is known that the equipment was running at age t, the probability that the equipment requires repair in time interval $(t, t + dt)$ is given by $r(t)\,dt$.

(*a*) Now, the probability that equipment requires repairs between the age u and $u + du$ is $f(u)\,r(u)\,du$.

∴ Expected cost of repair

$$= C \int_0^\infty f(u)\,r(u)\,du,$$

and the total expected cost

$$= \left[K + C \int_0^\infty f(u)\,r(u)\,du\right] \Big/ \left[\int_0^\infty f(u)\,du\right]$$

$$= K + C \int_0^\infty f(u) r(u)\,du; \quad \text{since} \quad \int_0^\infty f(u)\,du = 1$$

(*b*) If we have a policy of scrapping at the age T, the total expected cost up to age T

$$= K + C \int_0^T f(u) r(u)\,du$$

Hence, the expected cost per unit of time will be:

$$E(T) = \left[K + C \int_0^T f(u) r(u)\,du\right] \Big/ \left[\int_0^T f(u)\,du\right]$$

PROBLEM

21.33. A certain piece of equipment is extremely difficult to adjust. During a period, when no adjustment is made, the running cost increases linearly with time, at a rate b rupees each hour. The running cost immediately after an adjustment is not known precisely until the adjustment has been made. Before the adjustment, the resulting running cost x is deemed to be a random variable x with density function $f(x)$. If each adjustment costs k rupees, when should the replacement be made? *[M.Sc. (Math.), CCS Univ. 2004]*

21:9. RELIABILITY AND SYSTEM FAILURE RATES

Almost all physical and a lot of abstract conceptual systems may breakdown after sometime. Thus, the question of ***system reliability*** has become extremely important.

Reliability may be defined as the probability of a device (say p) performing its purpose adequately for a period of time under its operating conditions.

Some of the important aspects of reliability are given below:

1. Reliability is a function of time. We would not expect an almost worn out light bulb to be as reliable as one recently installed into service.

2. Reliability is a function of conditions of use. In very severe environments, we expect to encounter frequent system breakdown than in normal environment.
3. Reliability is expressed as a probability which helps us to quantify it and think of optimizing system reliability.

Failure Rates

The failure rate λ is defined as the number of failures in a given time interval, *i.e.*,

$$\lambda = \frac{\text{Number of failures}}{\text{Total units of operating hours}}$$

If λ remains constant, *i.e.*, the number of failures in equal time intervals do not change, probability of non-occurrence of failure, over a time period of t years will be $(1 - \lambda)^t$ which is equal to $e^{-\lambda t}$. Thus,

$$(1-\lambda)^t = e^{-\lambda t}.$$

Taking logarithm, this gives

$$t \log (1-\lambda) = -\lambda t \log e$$

Differentiating this equation with respect to t, we get

$$\frac{1}{\lambda} = -\frac{\log e}{\log (1-\lambda)} = m, \text{ a constant.}$$

$\therefore$ Reliability at any time t is given by

$$R(t) = e^{-\lambda t} = e^{-t/m},$$

where, $m = \dfrac{\text{Total number of survival hours}}{\text{Total number of failure}}.$

Bath-tub-shaped Failure Rate

The failure rate for most systems follow a typical "bath-tub" pattern. It starts with a high failure rate during the initial period of operation called the ***infant mortality*** on the debugging period as shown in *Fig. 21.2*. The operating period following this period has essentially a constant failure rate because failures during this period are typically due to chance. Upon reaching the workout period, the failures tend to increase quite rapidly. Failures, in this interval, are the result of gradual degradation of some property of the system essential to proper operation. For the purpose of reliability analysis, we can assume the failure rate to be constant which characterizes an experimental reliability function:

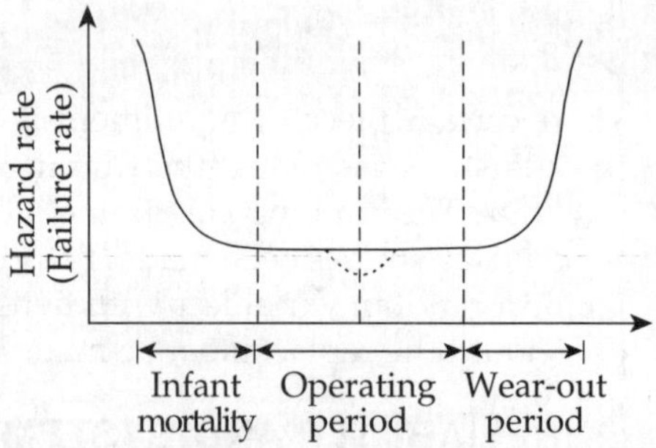

Fig. 21.2. Bath-tub-shaped Hazard Function

Instantaneous Failure Rate

Hazard rate or ***instantaneous failure rate*** is given by

$$H(t) = \lim_{h \to 0} \frac{R(t) - R(t+h)}{h R(t)} = \frac{1}{R(t)} \frac{d}{dt} R(t)$$

where, $R(t)$ and $R(t + h)$ are the reliabilities at time t and $t + h$, respectively.

Again, $R(t) = 1 - F(t)$, where $F(t)$ denotes failure probability distribution.

$$= 1 - \int_{-\infty}^{t} f(t)\, dt = \int_{t}^{\infty} f(t)\, dt,$$

where, $f(t)$ denotes the probability density function such that $\int_{-\infty}^{\infty} f(t)\, dt = 1$

Also, $$\frac{d}{dt} R(t) = -\frac{d}{dt}\left[\int_{t}^{\infty} f(t)\, dt\right] = -f(t),$$

where, $f(t)$ is referred to as ***failure distribution***.

Now, $$H(t) = \frac{d}{dt}\{(\log R(t)\} = -\frac{d}{dt}\{\log f(t)\}$$

$\therefore$ $$\frac{d}{dt}\{\log R(t)\} = -\lambda \quad \text{or} \quad \log R(t) = -\lambda t + C$$

When $t = 0,\ R(t) = R(0) = 1 \quad \therefore\ C = 0$

Hence, $$\log R(t) = -\lambda t \quad \text{or} \quad R(t) = e^{-\lambda t}$$

Mean Time Between Failure (*MTBF*)

Suppose that an equipment or a system is in operation. Assume that when the equipment fails it is repaired, the repair time being negligible. Then, the average failure time $E(t)$ is identical with the average time between successive failures. This is referred to as ***mean time between failure (MTBF)***.

The mean time between failures is related with reliability through the simple equation as under:

$$MTBF = E(t) = \int_{0}^{\infty} [1 - F(t)\, dt = \int_{0}^{\infty} R(t)\, dt$$

When failure distribution is exponential, $MTBF = 1/\lambda$. More is the *MTBF*, more reliable is the device.

Estimation of Reliability

The reliability of a component is estimated by a procedure known as ***life testing***. In this procedure a number of components are subjected to identical operational conditions with respect to various physical characteristics. The time of failure of each component is recorded till the last one fails. The reliability of the component at t is estimated as $\hat{R}(t)$, where

$$\hat{R}(t) = \frac{\text{Number of components surviving up to } t}{\text{Total number of components}}$$

The estimated failure rate $\hat{h}(t)$ is given by

$$\hat{h}(t) = \frac{\text{Number of failures in interval } (t,\ t+s)}{(\text{Number surviving at interval beginning}) \times (\text{Interval length})}$$

Reliability Improvement

System reliability can be improved either by improving the design or by providing redundancy in the system. The system with redundancy has a number of standby units which take over, if the other component of the system fails. Thus, a reserve stock of standby improves system's reliability but at a higher cost. The compromise between increased cost due to standby arrangement and increased given reliability will help to decide the optimal redundancy in the system.

Example 21.15. *An electronic circuit consists of 15 valves, 20 resistors and 10 capacitors all connected in a series. The components in each category are identical and their failure times are found to follow exponential distribution with the following mean failure times:*

	Valves	*Resistors*	*Capacitors*
Mean failure time (hrs.) :	*10,000*	*20,000*	*20,000*

What is the mean time between failure of the system? What is its reliability for 100 hours?

Solution. Let the failure rate of valves, resistors and capacitors be denoted by λ_1, λ_2 and λ_3, respectively, which are the parameters of exponential distribution. So, we have

$$\lambda_1 = 1/10{,}000, \quad \lambda_2 = 1/20{,}000, \quad \text{and} \quad \lambda_3 = 1/20{,}000$$

$$\therefore \qquad MTBF = 1/[15\lambda_1 + 20\lambda_2 + 10\lambda_3]$$

$$= \left[\frac{15}{10{,}000} + \frac{20}{20{,}000} + \frac{10}{20{,}000}\right]^{-1}$$

$$= \frac{20{,}000}{(30+20+10)} = 333.33 \text{ hours}$$

The reliability $R(t)$ of the system for 100 hours is

$$R(t) = e^{-100/333.33} = e^{-0.33} = 0.74082$$

PROBLEMS

21.34. The failure time of a component is exponentially distributed with mean life equal to 350 hours. The design is modified, so that the mean life is increased to 400 hours. What is the increase in reliability, if in both cases the component is tested for 800 hours?
[*Hint:* $R_1(800) = e^{-800/350}$ and $R_2(800) = e^{-800/400}$]

21.35. The failure time of the component of an equipment has an exponential distribution with common mean failure rate equal to 100 hours. What is the minimum number of components to be included in the design in parallel, so that the mean time between failure of the equipment is at least 200 hours?

[*Hint:* *MTBF* of equipment $= 100 \sum_{x=1}^{n} (-1)^{x-1}\, {}^nC_x / x$. Compute *MTBF* for $n = 2, 3$ and 4 and choose the minimum.]

TEST YOUR UNDERSTANDING

1. The problem of replacement is not concerned about the
   (*a*) items that deteriorate gradually.
   (*b*) items that fail suddenly.
   (*c*) determination of optimum replacement interval.
   (*d*) maintenance of an item to work out profitability.
2. For the items that deteriorate gradually,
   (*a*) operating and maintenance costs steadily increase with passage of time, whereas depreciation per year decreases with time.

(*b*) optimum replacement interval is the minimum time elapsing between the successive replacements.
(*c*) the annual maintenance costs and annual depreciation tend to decrease.
(*d*) all the the above.

3. Sudden failure of an item is seen as
(*a*) progressive (*b*) random
(*c*) retrogressive (*d*) all of the above.

4. Average annual cost will be minimized by replacing an item/equipment, when average cost to date is
(*a*) equal to the current maintenance cost. (*b*) less than the current maintenance cost.
(*c*) greater than the current maintenance cost. (*d*) all of the above.

5. When time value of money is considered,
(*a*) costs need to be discounted
(*b*) timing of the occurrence of cost is important
(*c*) present value factor serves as the weights
(*d*) all of the above.

6. If *r* is a the rate of interest, then the present value of one rupee spent in *n* years is given by
(*a*) $(1 + r)^n$ (*b*) $(1 - r)^n$
(*c*) $(1 + r)^{-n}$ (*d*) $(1 - r)^{-n}$.

7. The group replacement is suitable for identical low cost items which are likely to
(*a*) fail suddenly (*b*) fail over a period of time
(*c*) fail completely and all of sudden (*d*) none of the above.

8. Under group replacement policy
(*a*) group as well as individual replacements are done.
(*b*) all the items are replaced, irrespective of the fact that items have failed or have not failed.
(*c*) the optimal group replacement interval is determined at the point, where the sum of group replacement per unit of time and the cost of individual replacement is maximum.
(*d*) all of the above.

9. Staff replacement policy
(*a*) arises due to resignation, retirement, or death of a staff member from time to time.
(*b*) is likely replacement policy for items whose values deteriorate gradually.
(*c*) can be easily formulated, because people retire at known times.
(*d*) does not yield the optimum replacement interval.

10. Mortality problems
(*a*) are special type of problems, where failure is treated as birth and the replacement of an item on the failure is treated as death.
(*b*) uses mortality tables to derive the probability distribution of the life span of an equipment.
(*c*) are like replacement policies for items whose value does not deteriorate gradually.
(*d*) none of the above.

ANSWERS

1. (*d*)	**2.** (*b*)	**3.** (*d*)	**4.** (*a*)	**5.** (*d*)
6. (*c*)	**7.** (*b*)	**8.** (*b*)	**9.** (*a*)	**10.** (*b*)

REVIEW EXERCISE

1. (*a*) What is replacement problem? When does it arise? *[M.Sc. (Math.), CCS Univ. 2003]*
   (*b*) Discuss the assumptions of Replacement Theory. *[MBA, Bangalore Univ. 2018]*
2. Describe some important replacement situations and policies.
   [MBA, Panjab Tech. Univ. 2015; MBA, UP Tech. Univ. 2007]
3. What are the situations that make the replacement of items/equipment necessary?
   [M.Sc. (Math.), CCS Univ. 2010; BE (Dec.), Karnataka Univ. 2011; MBA, Mumbai Univ. 2010]
4. What are the reasons for replacement? List and explain. *[BE (Dec.), Karnataka Univ. 2011]*
5. Explain how the theory of replacement is used in replacement of items whose maintenance cost varies with time.
   [MBA (Nov.), Madras Univ. 2012; B.Tech. (Mech.), Panjab Tech. Univ. 2012]
6. Briefly explain costs which are relevant to decision for replacement of depreciable costs.
   [MBA, UP Tech. Univ. 2010]
7. What do you mean by 'Money value is not considered' in replacement analysis?
   [B.Tech., Hyderabad Univ. 2009]
8. What is group replacement? Give an example. *[MBA, Bangalore Univ. 2015; MBA, Annamalai Univ. 2010; B.Tech. (Mech.), Hyderabad Univ. 2012]*
9. Distinguish between individual replacement and group replacement policies.
   [B.Tech. (Comp. Sc.) (Dec.), Hyderabad Univ. 2011]
10. Explain how the theory of replacement is used in replacement of items that fail suddenly.
    [B.Tech. (Mech.), Panjab Tech. Univ. 2012; MBA (Nov.), Madras Univ. 2012]
11. Write short note on Mortality and Staffing problems in replacement decisions.
12. A large population is subject to a given mortality curve for a long period of time. All deaths are immediately replaced by births and there are no other entries or exits. Show that the age distribution becomes stable and that the number of deaths per unit time becomes constant.
    [M.Sc. (Math.), CCS Univ. 1998]

CHAPTER 22

Inventory Management — Deterministic

"None but the productive can be strong and none but the strong can survive"

Inventory consists of usable resources that have economic value and is maintained to fulfil the present and future needs of an organization. When the resource involved is of material or goods, it is referred as stock or simply as '***Inventory***'. The term is generally used to indicate raw materials in process, finished product, packaging, spares and others — stocked in order to meet an expected demand in the future. Though inventory of materials is an idle resource at present, it is almost essential to maintain the smooth functioning of an organization. Traditionally, inventory is viewed as a necessary evil — too little of it causes costly interruptions and too much of it results in idle capital. Inventory control aims at maintaining the balance between these two extremes.

22:1. TYPES OF INVENTORY

Inventories may be held for a variety of purposes, but, in general, there are following five types of inventory that an enterprise can use for serving these purposes:

Movement Inventories. Movement inventories, also called ***transit*** or ***pipeline inventories,*** arise due to shipment of inventory items to distribution centres and from various production centres. The amount of movement inventories depend on the time consumed in transportation and the nature of demand. For example, when coal is transported from coal fields to an industrial unit by trains, then the coal, while in transit, cannot provide any service to the customers for power generation or for burning.

Buffer Inventories. Buffer inventories are maintained to meet uncertainties of demand and supply. Such 'buffer' inventories which are in excess of those necessary to just meet the average demand during the ***lead time*** (the time elapsing between placing an order and having the goods in stock ready for use), held for protecting against the fluctuations in demand and are also termed as ***safety stocks***.

Anticipation (Seasonal) Inventories. Anticipation inventories, also known as ***seasonal inventories***, are held because a future demand for the product is anticipated. Production of specialized items like crackers well before Diwali, electric fans and air-conditioners when summers are approaching, are some examples of ***anticipation inventories***. The basic idea underlying is to smoothen the production process producing for a long duration on a continuous scale rather than operating with excessive overtime in one period and then let the system to be idle or close down for the reason of inadequate or no demand for another period.

Decoupling Inventories. If various production stages operate successively, then in the event of breakdown of one or any disturbance at some stage can affect the entire system. This kind of interdependence is not only costly but also disruptive for the entire system. The inventories used to reduce the inter-dependence of various stages of production system are known as ***decoupling inventories***.

Decoupling inventories are not only in the form of in-process inventories to 'decouple' successive production stages, but also in the form of raw material which are used to decouple the producers from suppliers, and finished goods to decouple the consumer from producer. Inventories may also be carried to take advantage of quantity discounts or raw material price in the event of an expected rise in price.

Lot-size Inventories. These are held for the reason that purchases are usually made in lots rather than for the exact quantity which may be needed at a point of time. Lot-size inventories are also called *cycle inventories*. Such inventories depend upon the production lot size, economical shipment quantities, storage space limitations, price-quantity discount schedules, inventory carrying cost, and so on.

22:2. REASONS FOR CARRYING INVENTORIES

Inventories, in general, are built up to provide a cushion between supply and demand. They are meant to take care of requirements of demand till next arrival, probable delay in delivery and sudden increase in demand. A manufacturing firm carries inventories because of following major reasons:

1. To ensure of an adequate supply of items to customers and avoid the shortages as far as possible at the minimum cost.
2. To reduce the risk of loss due to changes in prices of items stocked at the time of making the stock.
3. To take advantages of quantity discounts on bulk purchases.
4. To carry buffer stocks (reserve stocks) in case of delayed deliveries by the suppliers.
5. To help in minimizing the loss due to deterioration, obsolescence, damages or pilferage of goods, etc.
6. To make use of available capital and/or storage space in a most effective way and avoids an unnecessary expenditure on high inventories.
7. To utilize the benefits of price fluctuations. Thus, in conclusion, with a good inventory, a firm is able to make purchases in economic lots, maintain continuity of operations, avoid small-time consuming orders and guarantee for the prompt delivery of finished goods.
8. To satisfy other business constraints, such as:
   (*i*) ***Supplier's Condition of Minimum Quantity.*** Many times, a supplier insists on certain minimum quantity. This can happen under one of the following conditions:
   — The manufacturer has the monopoly and hence the order quantities are dictated to by the seller.
   — The set-up cost is very high.
   (*ii*) ***Government Regulations.*** Too much capital gets locked in inventories of items that are subjected to government regulations (*e.g.*, imported items).
   (*iii*) ***Seasonal Availability.*** There are certain items which are available in plenty and hence cheaply only during certain months (*e.g.*, ground nuts, coco-nuts, soyabeans, etc.) or are not available in certain months. The firm is thus forced to buy in large quantity and hence create inventories.

22:3. THE INVENTORY DECISIONS

Inventory control is the process of deciding what and how much of various items are to be kept in stock. It also determines the time and quantity of various items to be procured. The basic objective of inventory control is to reduce investment in inventories and ensuring that production process does not suffer at the same time. To attain various objectives in an inventory control situation, there are two basic questions to be answered. They are:

(*i*) How much to order? That is what is the optimum quantity of an item that should be ordered.

(*ii*) When should the order be placed?

The answer to the first question determines the ***economic order quantity(EOQ)*** by minimizing the total inventory cost, which is given by

Purchase cost + Set-up cost (Ordering cost) + Carrying cost + Shortage cost

When price discounts are not offered, the purchase cost remains constant and is independent of the quantity purchased. Total variable inventory cost is, then, given by

Ordering cost + Carrying cost + Shortage cost

The answer to the second question (when to order?) depends on the type of inventory system with which we are dealing. If the system requires ***periodic review*** (*e.g.*, every month or year), the time for receiving a new order coincides with the start of each period. Alternatively, if the system is based on continuous review, whenever the inventory level (inventory on hand plus on order) drops to a given level, say 's' or below, an order is placed for a fixed quantity Q. This policy is also known as a ***fixed-order quantity policy*** or ***re-order point policy***, generally denoted by (s, Q) policy. In case the order is placed for a sufficient quantity to bring the inventory level upto a pre-determined maximum level of S; then it is known as (s, S) policy.

Remark. In (s, S) policy, the inventory level (just after the replenishment of stock) is fixed, and the order quantity is variable.

22:4. OBJECTIVES OF SCIENTIFIC INVENTORY CONTROL

A fundamental objective of a good inventory control system is to determine '***what to order***', '***how much to order***', '***when to order***', and '***how much to carry in stock***' so as to gain economy in purchasing, storing, manufacturing and selling. These fundamental objectives may be amplified into the following objectives:

Service to the Customers. Sufficient stock of finished product is to be maintained to meet the requirements of customers.

Effective Use of Capital. The system should enable the management to make an effective use of its capital, *i.e.*, lock-up of capital should be barest minimum.

Continuity of Productive Operations. The system should be made to ensure the continuity of productive operations by ensuring uniform flow of materials and eliminating the possibility of stock outs.

Reduction of Administrative Work Load. The administrative work load on the purchasing, receiving, inspection, stores, accounts and other related departments should be barest minimum.

Economy in Purchasing. The system should enable the management to gain economy in bulk purchasing and take advantage of price discount.

Minimization of Risk Obsolescence and Deterioration. The possibility of the risk of loss on account of obsolescence and deterioration should be minimized. In-built checks in the system should enable the management to weed out obsolete and non-moving items periodically and automatically.

Up-to-date Accurate Records. In order to enable the company to prepare periodical financial statements, minimizing of discrepancies between physical stock and book balance, an up-to-date stock records must be maintained.

22:5. COSTS ASSOCIATED WITH INVENTORIES

Various costs associated with inventory control are often classified as follows:

Set-up Cost. The *set-up cost* associated with the setting up of machinery before starting production is generally assumed to be independent of the quantity order for or produced.

Ordering Cost. This is a cost associated with ordering of raw material for production purposes. Advertisements, consumption of stationery and postage, telephone charges, telegrams, rent for space used by the purchasing department, travelling expenditures incurred, etc. constitute the ***ordering cost***.

Purchase (or Production Cost). The cost of purchasing (or producing) a unit of an item is known as *purchase (or production) cost*. The purchase price will become important when quantity discounts are allowed for purchases above a certain quantity, or when economies of scale suggest that the per unit production cost can be reduced by a larger production run.

Carrying (or Holding) Cost. The *carrying cost* is associated with carrying (or holding) inventory. This cost generally includes the costs such as rent for space used for storage, interest on the money locked-up, insurance of stored equipment, production, taxes, depreciation of equipment and furniture used, etc.

Shortage (or Stock Out) Cost. The penalty for running out of stock (*i.e.*, when an item cannot be supplied on the customer's demand) is known as *shortage cost*. This cost includes the loss of potential profit through sales of items and loss of goodwill in terms of permanent loss of customers and its associated lost of profit in future sales.

Salvage Cost (or Selling Price). When the demand for certain commodity is affected by the quantity stocked, decision problem is based on a profit maximization criterion that includes the revenue from selling. *Salvage cost* may be combined with the cost of storage and hence is generally neglected.

Revenue Cost. When it is assumed that both the price and demand of the product are not under control of the organization, the revenue from the sales is independent of the company's inventory policy and may be neglected except for the situation, when the organization cannot meet the demand and hence sale is lost. Therefore, the ***revenue cost*** may or may not be included in the study of inventory policy.

22:6. FACTORS AFFECTING INVENTORY CONTROL

Besides the costs that determine the profitability, other factors which play an important role in the study of inventory control are the following:

Demand. The number of units required per period is called ***demand***. The demand pattern of a commodity may be either deterministic or probabilistic.

Lead Time. The time gap between placing of an order and its actual arrival in the inventory is known as ***lead time***.

Lead time has two components, namely the ***administrative lead time*** — from initiation of procurement action until the placing of an order, and the ***delivery lead time*** — from placing of an order until the delivery of the ordered material.

Order Cycle. The time period between placement of two successive orders is referred to as an *order cycle*. The order may be placed on the basis of following two types of inventory review systems:

(*a*) ***Continuous Review.*** The record of the inventory level is checked continuously until a specified point (called reorder point) is reached, where a new order is placed. This is often referred to as the ***two-bin system***. This divides the inventory into two parts and places it physically, or on paper, in two-bins. Items are drawn from only one bin and when it is empty, a new order is placed. Demand is then satisfied from the second bin until the order is received. Upon receipt of the order, enough items are placed in the second bin to make up the earlier total. The remaining items are placed in the first bin. This procedure is then repeated.

(*b*) ***Periodic Review.*** In this system the inventory levels are viewed at equal time intervals and orders are placed at such intervals. The quantity ordered each time depends on the available inventory level at the time of review.

Time Horizon. The time period over which the inventory level will be controlled is called the ***time horizon***. Time horizon may be finite or infinite depending upon the nature of the demand for the commodity.

Re-order Level. The level between maximum and minimum stock, at which purchasing (or manufacturing) activities must start for replenishment, is known as ***re-order level***.

Stock Replenishment. Although an inventory problem may operate with lead time, the actual replacement of stock may occur instantaneously or uniformly. Instantaneous replenishment occurs in case the stock is purchased from outside sources, whereas the uniform replenishment may occur when the product is manufactured by the company.

Inventory Turnover. Inventory turnover is defined as the ratio of the value of materials consumed to the average investment in inventories for the same period.

$$\text{Inventory turnover} = \frac{\text{Value of the material consumed}}{\text{Value of average inventory}}$$

Inventory turnover is a very useful index of inventory control. The higher the index, the lower is the inventory levels and lower the cost of maintaining the inventories. The top management can keep itself informed of the broad situation through the use of such an index.

Standardization. In inventory control, standardization is the determination of fixed sizes, shapes, quantity and dimensions of material. As it is desirable to encourage the use of standards prescribed for the type of industry in which the company operates, by keeping the material to standards, a great deal of reductions are possible in inventory space, obsolescence, handling costs, and inventory as well as improved quality.

22:7. AN INVENTORY CONTROL PROBLEM

The major steps to build-up a suitable inventory problem, and then to derive decision rules are as follows:

Step 1. Collect the data regarding the pattern of demand, the replenishment policy, time horizon, relevant inventory costs, etc.

Step 2. Build up an appropriate relationship using above information which can be representative of the behaviour (or objective) of the inventory system. Relationships so developed may have an objective function of minimizing the total inventory costs subject to changes in inventory reorder policy and constraints of limited resources (such as floor space for storage, capital investment, etc.). Thus, the problem would either be an unconstrained optimization problem or a mathematical programming problem depending upon whether constraints are imposed or not.

Step 3. Derive an optimal (optimum) inventory policy (*i.e.*, economic order quantity) by using an appropriate solution procedure, so as to balance among inventory costs.

Remark. When the future demand is known exactly, the inventory problem is said to be ***under certainty***. If the probability distribution of future demand (from past records) is known, the inventory problem is said to be ***under risk***. In case the levels of future demand are totally unknown, the inventory problem is termed to be ***under uncertainty***.

22:8. THE CONCEPT OF *EOQ*

The inventory problems in which demand is assumed to be fixed and completely pre-determined are usually referred to as the ***economic order quantity (EOQ)*** or ***lot size problems/models***.

By the 'order quantity', we mean the quantity produced or procured during one production cycle. When the size of order increases, the ordering costs (cost of purchasing, inspection, etc.) will decrease whereas the inventory carrying costs (cost of storage, insurance, etc.) will increase. Thus, in the production process there are two opposite costs, one encourages the increase in the order size and the other discourages. ***Economic Order Quantity (EOQ)*** is that size of order which minimizes total annual costs of carrying inventory and cost of ordering.

The two opposite costs can be shown graphically by plotting them against the order size as shown in *Fig. 22.1*.

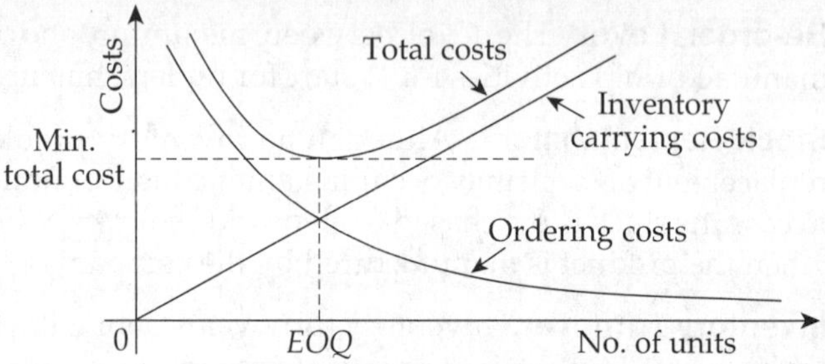

Fig. 22.1. Graph of *EOQ*

It is evident from above that the minimum total cost occurs at the point where the ordering costs and inventory carrying costs are equal.

22:9. ECONOMIC ORDER QUANTITY (*EOQ*) MODEL

The ***EOQ model*** is applicable when the demand for an item has a constant rate and also the entire quantity ordered arrives in inventory at one point in time. The term ***constant demand rate*** implies that the same number of units are taken from inventory each period of time such as 25 units every day or 175 units every week or 750 units every month.

The inventory pattern is shown in the *Fig. 22.2* below:

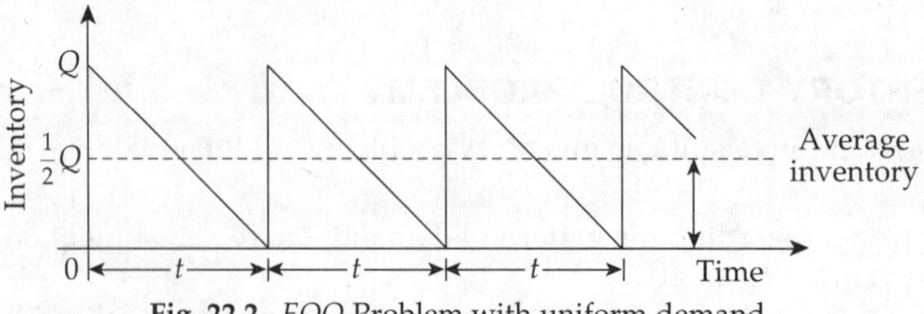

Fig. 22.2. *EOQ* Problem with uniform demand

If the average inventory during each cycle is $\frac{1}{2}Q$, the average inventory over the number of cycles is also $\frac{1}{2}Q$.

The ***holding cost*** can be calculated using the average inventory. We calculate the ***holding cost*** by multiplying the average inventory and the cost of carrying one unit in inventory for the stated period. The period selected for the *EOQ* model may be one week, one month, one year, or more. However, because the holding cost for most of the organization is expressed as ***annual***, the inventory models are developed on an ***annual cost basis***.

Let D = annual demand, *i.e.*, total number of units purchased/produced or supplied,
Q = Order quantity in each production run,
C_o = Cost of placing one order,
C_h = Annual cost of holding one unit in inventory, *i.e.*, C_h or IC, where I is the annual holding cost rate and C is the unit cost of item in inventory.

The general equation for the annual holding cost for the average inventory of $\frac{1}{2}Q$ units is:

Annual holding cost = Average inventory × Annual holding cost per unit

i.e.,
$$f(Q) = \frac{1}{2}Q \times C_h \quad \text{or} \quad \frac{1}{2}Q \times IC.$$

The general equation for the annual ordering cost is:

Annual ordering cost = Number of orders per year × Cost of one order

i.e.,
$$g(Q) = n \times C_o, \quad \text{where } D = nQ.$$

Since, the minimum total cost occurs at a point where ordering cost and total inventory cost are equal, we must have

$$\frac{1}{2}QC_h = nC_o = \frac{D}{Q}C_o, \quad \text{since } n = \frac{D}{Q}.$$

Hence, the optimum value of Q is given by

$$Q^o = \sqrt{\frac{2DC_o}{C_h}} \quad \text{or} \quad EOQ = \sqrt{\frac{2DC_o}{C_h}}.$$

This result is known as the ***economic (optimum)*** lot size formula due to ***R.H. Wilson***.

Total annual cost denoted by TC is:

$$TC = \text{Annual holding cost} + \text{Annual ordering cost}$$
$$= \frac{1}{2}Q^oC_h + \frac{D}{Q^o}C_o = \sqrt{2DC_hC_o}.$$

This represents the total annual cost associated with holding inventory when the optimum lot size, Q^o, is produced in each run. The optimum (minimum) total cost is demonstrated graphically in *Fig.* 22.3.

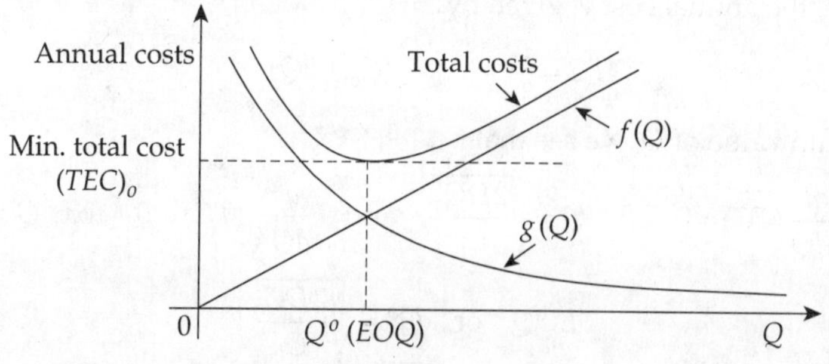

Fig. 22.3

Characteristics of EOQ Model

1. Optimum number of orders placed per year

$$n^o = \frac{D}{Q^o} = \sqrt{\frac{DC_h}{2C_o}}.$$

2. Optimum length of time between orders

$$t^o = \frac{T}{n^o} = T\sqrt{\frac{2C_o}{DC_h}} \quad \text{or} \quad \sqrt{\frac{2C_o}{DC_h}},$$

where, T (total time horizon) is one year.

The optimum length of time between orders is also called the '*economic review period*' (*ERP*).

Remarks

1. Since the holding cost $C_h > 0$, annual inventory costs are linear functions of Q with a positive slope; the smaller the average inventory, the lower are these costs. In contrast to these costs, ordering cost increases as Q increases.
2. If the carrying cost is given as a percentage of average value of inventory held, then total annual carrying cost may be expressed as $C_h = C \times I$ or $P \times I$. The total annual inventory cost then becomes

$$TC = \frac{1}{2}Q \times C \times I + DC_o / Q.$$

The optimum order quantity Q^0 will, then, be

$$Q^o = \sqrt{\frac{2DC_o}{CI}} \quad \text{or} \quad \sqrt{\frac{2AD}{CI}},$$

where, A is the cost of placing one order.

3. Generally, the annual demand for inventory item(s) is expressed in rupee value rather than in units. In such cases, the demand may be expressed in units as follows (provided unit cost of the item is known):

$$\text{Demand in units} = \frac{\text{Rupee value of demand}}{\text{Unit cost of the item}}$$

When the unit cost of the item is not known, then determine EOQ in rupee terms as follows:

$$Q^o = \sqrt{\frac{2 \times \text{Annual demand in rupees} \times \text{Ordering cost}}{\text{Inventory carrying cost}}}.$$

Corollary 1. *In the EOQ model discussed above, if the set-up (ordering) cost is $C_o + bQ$ instead of being fixed (where b is set-up cost per unit item produced), then there is no change in the optimum order quantity produced due to change in the set-up cost.*

Proof. In this case, the annual cost is given by

$$TC = \frac{1}{2}QC_h + \frac{D}{Q}(C_o + bQ).$$

For the optimum value of Q, we see that

$$\frac{d}{dQ}(TC) = 0 \;\Rightarrow\; Q = \sqrt{\frac{2DC_o}{C_h}} \quad \text{and} \quad \frac{d^2}{dQ^2}(TC) > 0 \quad \text{for} \quad Q > 0.$$

Hence, $$EOQ \;\text{ or }\; Q^o = \sqrt{\frac{2DC_o}{C_h}}.$$

This shows that there is no change in Q^o in spite of change in the set-up cost.

Corollary 2. *In the above EOQ model lead time has been assumed to be zero. But in most of the business situations, there exists a positive lead time, say L, from the time the order is placed until it is actually supplied.*

Proof. If the inventory consumption rate is 'K' units per day and L is the lead time in days, the total inventory requirements during the lead time will be 'LK'. Therefore, as soon as the inventory level becomes 'LK' an order Q is placed. This is called the re-order point $p = LK$. This is equivalent to continuously observing the level of inventory, until the re-order point is obtained. Because of this reasons, *EOQ* problem is sometimes called the ***continuous review problem***. *Fig. 22.4* shows the re-order points:

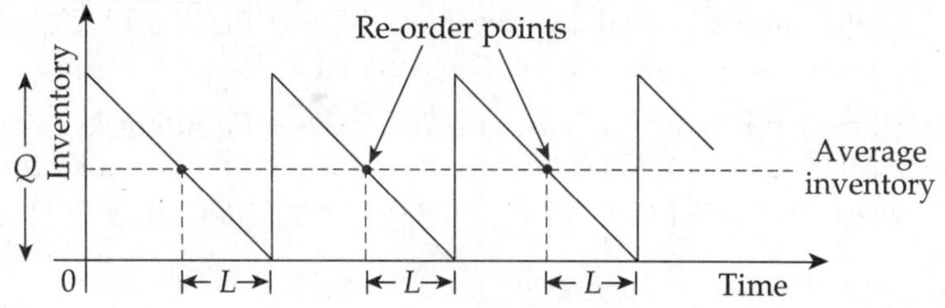

Fig. 22.4

Figure given above assumes that the lead time is less than the cycle length, which is not necessarily the case in general. To account for this situation, we define the effective lead time as

$$L_e = L - mt^o,$$

where, m is the largest integer not exceeding L/t^o. This result is justified because after m cycles of t^o each, the inventory situation acts as if the interval between placing an order and receiving another is L_e.

EOQ Model with Several Production Runs of Unequal Length

In this problem all the assumptions are same as in the *EOQ* model discussed above, except that the demand is uniform and the production runs differ in units.

Let $t_1, t_2, t_3, ..., t_n$ denote the times of successive production runs, such that

$$t_1 + t_2 + ... + t_n = 1 \text{ year}.$$

Thus, the fundamental situation can be represented graphically as shown in *Fig. 22.5:*

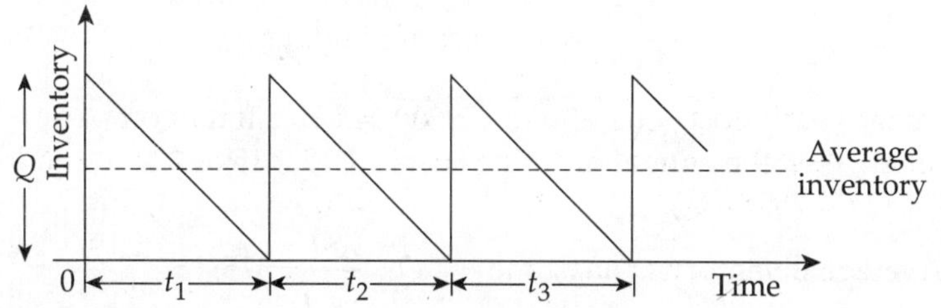

Fig. 22.5

Obviously, the annual inventory holding cost is given by

$$f(Q) = \left(\frac{1}{2}Qt_1\right)C_h + \left(\frac{1}{2}Qt_2\right)C_h + \ldots + \left(\frac{1}{2}Qt_n\right)C_h$$

$$= \frac{1}{2}Q(t_1 + t_2 + \ldots + t_n)C_h = \frac{1}{2}QC_h$$

and the set-up costs associated with runs of size Q are given by

$$g(Q) = \frac{D}{Q}C_o, \quad \text{since} \quad nQ = D.$$

∴ Total annual cost is:

$$TC = f(Q) + g(Q) = \tfrac{1}{2} QC_h + \frac{D}{Q} C_o.$$

This cost is the same as was obtained in *EOQ* model with production runs of equal length and hence the optimum quantities are

$$Q^o = \sqrt{2DC_o / C_h} \quad \text{and} \quad TC^o = \sqrt{2DC_h C_o}.$$

Remark. If the total time period is T instead of one year, then the optimum order quantity becomes: $Q^o = \sqrt{2C_o D / C_h T}$ and the minimum total cost becomes: $TC^o = \sqrt{2C_h C_o D / T}$. Thus, the uniform rate of demand is replaced by average rate of demand, *i.e.*, D is replaced by D/T.

Example 22.1. *An oil engine manufacturer purchases lubricants at the rate of ₹ 42 per piece from a vendor. The requirement of these lubricants is 1,800 per year. What should be the order quantity per order, if the cost per placement of an order is ₹ 16, and inventory carrying charge per rupee per year is only 20 paise.*

[B.Sc. (Stat.), Delhi Univ. 2000]

Solution. We are given

D = 1,800 lubricants per year, C_o = ₹ 16 per order, and C_h = ₹ 42 × ₹ 0.20 = ₹ 8.40

$$\therefore \quad Q^o = \sqrt{\frac{2 \times 1{,}800 \times 16}{8.40}} = 82.8 \text{ or } 83 \text{ lubricants.}$$

Example 22.2. *A manufacturing company purchases 9,000 parts of a machine for its annual requirements, ordering one month usage at a time. Each part costs ₹ 20. The ordering cost per order is ₹ 15 and the carrying charges are 15% of the average inventory per year. You have been assigned to suggest a more economical purchasing policy for the company. What advice would you offer and how much would it save the company per year?* **[B.Com. (Prof.), Guru Nanak Dev Univ. 2006; MBA, Dayal Bagh Ed. Inst. 2013]**

Solution. We are given,

D = 9,000 parts per year, C_o = ₹ 15 per order, and

C_h = 15% of the average inventory per year = ₹ 20 × 15/100 = ₹ 3 for each part per year

$$\therefore \quad Q^o = \sqrt{\frac{2 \times 15 \times 9{,}000}{3}} = 300 \text{ units}$$

and

$$t^o = \frac{300}{9{,}000} = \frac{1}{30} \text{ year or 12 days}$$

Minimum average yearly cost $= \sqrt{2 \times 3 \times 15 \times 9{,}000}$ = ₹ 900. If the company follows the policy of ordering every month, then the annual ordering cost is ₹ 12 × 15 or ₹ 180, and lot size of inventory each month = 9,000/12 = 750 (= Q).

$$\text{Average inventory at any time} = \tfrac{1}{2} Q = \frac{750}{2} = 375.$$

∴ Storage cost at any time = 375 × C_h = 375 × 3 = ₹ 1,125.

Total annual cost = ₹ 1,125 + ₹ 180 = ₹ 1,305.

Hence, the company should purchase 300 parts at time intervals of 1/30 year (or 12 days) instead of ordering 750 parts each month. The net saving of the company will be

₹ 1,305 – ₹ 900 = ₹ 405 per year.

Example 22.3. *A company operating 50 weeks in a year is concerned about its stock of copper cable. This costs ₹ 240 a metre and there is a demand for 8,000 metres a week. Each replenishment costs ₹ 1,050 for administration and ₹ 1,650 for delivery, while holding costs are estimated at 25 per cent of value held a year. Assuming no shortages are allowed, what is the optimal inventory policy for the company?*

How would this analysis differ if the company wanted to maximize profit rather than minimize cost? What is the gross profit if the company sell cable for ₹ 360 a metre.

Solution. From the data of the problem, we have

Demand rate (D) = 8,000 × 50 = 4,00,000 metres a year

Purchase cost (C) = ₹ 240 a metre; Ordering cost (C_o) = ₹ 1,050 + ₹ 1,650 = ₹ 2,700

Holding cost (C_h) = 0.25 × 240 = ₹ 60 a metre a year

$$\therefore \quad \text{Optimal order quantity, } Q^o = \sqrt{\frac{2DC_o}{C_h}} = \sqrt{\frac{2 \times 4,00,000 \times 2,700}{60}} = 6,000 \text{ metres,}$$

Total inventory cost (TC) = Purchase cost + Total expected cost

$$= D \times C + \left(\frac{1}{2} Q^o C_h + \frac{D}{Q^o} C_o \right)$$

$$= 4,00,000 \times ₹\,240 + \left(\frac{1}{2} \times 6,000 \times ₹\,60 + \frac{4,00,000}{6,000} \times ₹\,2,700 \right)$$

$$= ₹\,9,60,00,000 + ₹\,3,60,000 = ₹\,9,63,60,000.$$

With a turnover in excess of 96 million rupees a year, inventory costs are only ₹ 3,60,000 or 0.36 per cent. This figure is usually low but any well run organization should try to make all the waivings it can.

If the company wanted to maximize profit rather than minimizing of cost, the analysis used would remain exactly the same. This can be demonstrated by defining selling price (SP) per unit, so that gross profit per unit time becomes as under:

$$\text{Profit} = \text{Revenue} - \text{Cost} = D \times SP - \left[D \times C + \frac{D}{Q^o} C_o + \frac{1}{2} Q^o C_h \right]$$

$$= D \times SP - ₹\,9,63,60,000.$$

If the company sells the cable for ₹ 360 a metre, its revenue is

$$₹\,360 \times 4,00,000 = ₹\,14,40,00,000 \text{ per year.}$$

$$\therefore \quad \text{Gross profit} = ₹\,14,40,00,00 - ₹\,9,63,60,000 = ₹\,4,76,40,000 \text{ a year.}$$

Example 22.4. *A company plans to consume 760 pieces of a particular component. Past records indicate that purchasing department had used ₹ 12,000 for placing 15,000 orders. The average inventory was valued at ₹ 45,000 and the total storage cost was ₹ 7,650 which included wages, rent, taxes, insurance, etc. related to store department. The company borrows capital at the rate of 10% a year.*

If the price of a component is ₹ 12 and the order size is of 10 components, determine: purchase cost, purchase expenses, storage expenses, capital cost and total cost per year.

Solution. Using the given information, we get

$$\text{Ordering cost } (C_o) = \frac{₹\,12,000}{15,000} = ₹\,0.80$$

$$\text{Inventory carry cost } (I) = \frac{₹\,7,650}{₹\,45,000} \times 100 = 17\%$$

$$\therefore \quad \text{Purchase cost} = ₹\,12 \times 760 = ₹\,9,120$$

$$\text{Purchase expenses} = \frac{₹\,0.80 \times 760}{10} = ₹\,60.80$$

$$\text{Storage expenses} = \frac{1}{2} Q \times I \times C = \frac{1}{2} \times 10 \times 0.17 \times ₹\,12 = ₹\,10.20$$

$$\text{Capital cost} = \frac{1}{2}Q \times 0.10 \times 12 = \frac{1}{2} \times 10 \times 0.10 \times 12 = ₹\,6$$

$$\text{Total cost} = ₹\,(9{,}120 + 60.80 + 10.20 + 6.00) = ₹\,9{,}197.$$

Example 22.5. *Neon lights in an industrial park are replaced at the rate of 100 units per day. The physical plant orders the neon lights periodically. It costs ₹ 100 to initiate a purchase order. A neon light kept in storage is estimated to cost about ₹ 0.02 per day. The lead time between placing and receiving an order is 12 days. Determine the optimum inventory policy for ordering the neon lights.*

[M.Sc. (Math.), CCS Univ. 2008]

Solution. We are given

D = 100 units per day, C_o = ₹ 100 per order, C_h = ₹ 0.02 per unit per day, L = 12 days.

$$\therefore \quad Q^o = \sqrt{\frac{2 \times 100 \times 100}{0.02}} = 1{,}000 \text{ neon lights,}$$

$$\text{and} \quad t^o = \frac{Q^o}{D} = 10 \text{ days}$$

Now, since the lead time, L (= 12 days) exceeds the cycle length (= 10 days), we compute L_e.

$$\text{Thus,} \quad L_e = L - m t^o = 12 - 1 \times 10 = 2 \text{ days,}$$

where, m = (largest integer $\leq L/t^o$) = (largest integer $\leq 12/10$), *i.e.*, 1.

The re-order point, therefore, occurs when the inventory level drops to

$$L_e \times D = 2 \times 100 = 200 \text{ neon lights.}$$

Hence, the inventory policy for ordering neon lights is: Order 100 units whenever the inventory level drops to 200 units.

Example 22.6. *A chemical company holds its inventory of raw materials in special containers, with each container occupying 10 square feet of floor space. There are only 5,000 square feet of storage space available. Each year, this company uses 9,000 special containers of raw materials, paying ₹ 8 per container of raw material. If ordering cost is ₹ 40 per order and annual holding costs are 20 per cent of the average inventory value, how much is it worth for this company to increase its container of raw material storage area? For how many days (maximum) supply of inventory can be stored with the 5,000 square feet storage limitations, assuming that this company works a 300-day year?* **[MBA, Kurukshetra Univ. 2008]**

Solution. We are given

D = 9,000 containers, C_o = ₹ 40 per order, C_h = ₹ 8.00 × 0.20 = ₹ 1.60

$$\therefore \quad Q^o = \sqrt{\frac{2 \times 9{,}000 \times 40}{1.60}} = 670.8 \text{ or } 671 \text{ containers approx.}$$

Hence, the optimum order size is 671 containers. Since each container occupies 10 square feet of floor space, the company will require 6,710 square feet of floor space to store the optimum order quantity. The total storage space available with the company currently is only 5,000 square feet, therefore, it can order only 500 containers at a time. Hence, in order to calculate how much is worth for this company to increase the storage area, we will calculate the total variable cost that the company will have to incur for two alternatives, *viz.*, 671 containers per order and 500 containers per order.

Alternatives	*Annual ordering cost*	*Annual carrying cost*	*Total cost*
671 per order	$\frac{9{,}000}{671} \times ₹\,40 = ₹\,537$	$\frac{671}{2} \times ₹\,8 \times 0.2 = ₹\,537$	₹ 1,074
500 per order	$\frac{9{,}000}{500} \times ₹\,40 = ₹\,720$	$\frac{500}{2} \times ₹\,8 \times 0.2 = ₹\,400$	₹ 1,120

Thus, the company will save ₹ (1,120 – 1,074) = ₹ 46, if it follows economic order policy and it is ₹ 46 worth for the company to increase the storage space.

With 5,000 sq.ft. limitation only, 500 units can be ordered and each day usage is 30 (= 9,000/300) units and so nearly 500/30 or 17 days supply of inventory will be there in each order.

PROBLEMS

22.01. Given the annual consumption of material is 3,600 units, ordering costs are ₹ 400 per order. Cost per unit of material is ₹ 64 and storage cost or annual carrying cost is 50% of inventory value. Find the Economic Order Quantity, *EOQ* in (*i*) units, and (*ii*) rupees.
[MBA, Jammu Univ. 2008]

22.02 A manufacturer has to supply his customer with 600 units of his product per year. Shortages are not allowed and the storage cost amount to ₹ 0.60 per unit per year. The set-up cost per run is ₹ 80.00. Find the optimum run-size and the minimum average yearly cost.
[MBA (Nov.), Annamalai Univ. 2002]

22.03 A stockist purchases an item at the rate of ₹ 40 per piece from a manufacturer. 2,000 units of the item are required per year. What should be the order quantity per order, if the cost per order is ₹ 15 and the inventory charges per year is 20 paise?
[B.Tech. (Nov.), Hyderabad Univ. 2010]

22.04. A computer company sells a particular type of laptop. It costs the retail outlet ₹ 25,000 each time it places an order with the manufacturer. The annual carrying cost is ₹ 9,000. The sales manager estimates the annual demand of laptops to be of 1,200 units. Determine the optimal order quantity and the total minimum cost. *[MBA, Osmania Univ. 2010]*

22.05. A stockist has to supply 400 units of a product every Monday to his customers. He gets the product at ₹ 50 per unit from the manufacturer. The cost of ordering and transportation from the manufacturer is ₹ 50 per order. The cost of carrying inventory is 75% per year of the cost of the product. Find (*i*) economic lot size, and (*ii*) total optimal cost (including capital cost).
[B.Tech. (Mech.), Panjab Tech. Univ. 2010]

22.06. A manufacturer uses ₹ 10,00,000 worth of an item during the year. He has estimated the ordering cost as ₹ 2,500 per order and carrying cost as 12.5% of average inventory value. Find the optimal order size, number of orders per year, time period per order and total cost.
[PG Dipl. Indian Inst. of Mat. Man. 2017]

22.07. A certain item costs ₹ 235 per tonne. The monthly requirement is 5 tonnes and each time the stock is replenished, there is a set-up cost of ₹ 1,000. The cost of carrying inventory has been estimated at 10% of the value of the stock per year. What is the Optimal Order Quantity?
[B.Tech. (Comp. Sc.) (Dec.), Hyderabad Univ. 2011]

22.08. A company manufacturers a product having a monthly demand of 2,000 units. For one unit of product 2 kg of a particular raw material item is needed. The purchase price of the material is ₹ 20 per kg. The ordering cost is ₹ 120 per order and the holding cost is 10 per cent per annum. Calculate:

(*i*) Economic Order Quantity (*EOQ*), and

(*ii*) Annual cost of purchasing and storage of the raw material for the optimum quantity.
[MBA, Marathwada Univ. 2009]

22.09. An aircraft uses rivets at an approximately constant rate of 5,000 kg per year. The rivets cost ₹ 20 per kg and the company personnel estimate that it costs ₹ 200 to place an order, and the carrying cost of inventory is 10% per year.

(*i*) How frequently should orders for rivets be placed, and what quantities should be ordered for?

(*ii*) If the actual cost is ₹ 500 to place an order and 15% for carrying cost, the optimum policy would change. How much is the company losing per year because of imperfect cost information? ***[BFIA, Delhi Univ. 2007]***

22.10. A particular item has a demand of 9,000 units/year. The cost of one procurement is ₹ 100 and the holding cost per unit is ₹ 2.40 per year. The replacement is instantaneous and no shortages are allowed. Determine

(*i*) the economic lot size,

(*ii*) the number of orders per year,

(*iii*) the time between orders,

(*iv*) the total cost per year, if the cost of one unit is ₹ 1.

[B.Tech. (Mech.), Panjab Tech. Univ. 2013]

22.11. Find *EOQ* for the following data:

Annual usage = 1,000 pieces, Cost per piece = ₹ 250

Ordering cost = ₹ 6 per order, Expediting cost = ₹ 4 per order

Inventory holding cost = 20% of average inventory

Material holding cost = ₹ 1 per piece. ***[MBA, Shivaji Univ. 2009]***

22.12. The Deena Paints Ltd. would like to improve its inventory management policies for its supply of paint used for automobiles. Annual demand for such paint is 50,000 litres, and the paint which costs ₹ 20 per litre, is used at a constant rate. Annual carrying cost is estimated at 15 per cent of the value of paint held. Each order costs ₹ 80. Determine:

(*i*) How much paint should be ordered each time?

(*ii*) How often should paint be ordered?

(*iii*) Time between two consecutive orders.

(*iv*) What is the total annual cost associated with this policy? ***[MBA (Nov.), BHU 2009]***

22.13. (*a*) Compute *EOQ* and total variable cost for the following:

Annual demand :	25 units	Unit price :	₹ 2.50
Order cost :	₹ 4.00	Storage rate :	1% per year
Interest rate :	12% per year	Obsolescence rate :	7% per year

(*b*) Compute the order quantity and total variable cost that would result, if an incorrect price of ₹ 1.60 were used for the item. ***[M.Sc. (Math.), Rohilkhand Univ. 2011]***

[***Hint:*** (*a*) C_0 = Storage cost + Interest rate + Obsolescence rate = ₹ 0.50 per unit per year.

(*b*) Total variable cost = Ordering cost + Stock holding cost = ₹ 4 + ₹ 6.25.]

22.14. A factory follows an economic order quantity system for maintaining stocks of one of its component requirements. The annual demand is for 24,000 units, the cost of placing an order is ₹ 300, and the component cost is ₹ 60 per unit. The factory has imputed 24 per cent as the inventory carrying rate.

(*i*) Find the optimal interval for placing orders, assuming a year is equivalent to 360 days.

(*ii*) If it is decided to place only one order per month, how much extra cost does the factory incur per year as a consequence of this decision? ***[M.Com. (OC), Delhi Univ. 2010]***

22.15. A manufacturer of food processors has annual sales of 20,000 units. The manufacturer operates 250 days a year and produces 200 units per day. The set-up cost is ₹ 600 for a production run. The food processes cost ₹ 400 to produce. The inventory holding cost fraction is 0.025 per year.

(*a*) What is the most economical batch size?

(*b*) What is the minimum annual total of inventory holding cost and set-up cost?

(*c*) What is the number of production runs per year?

(*d*) How many days are required for the production of each batch?

(*e*) What is the maximum inventory on hand for the *EOQ*? ***[M.Com., Delhi Univ. 2008]***

22.16. Two items have the following characteristics:

Item	*Annual demand*	*Unit price*	*Order cost*	*Holding cost*
X	9 units	₹ 0.25	₹ 8.00	25%
Y	27 units	₹ 3.00	₹ 18.00	25%

(*a*) For each item, compute the annual ordering cost, the annual holding cost, and the annual total cost for order quantities of 24, 36, 48 and 60 units.

(*b*) Compute *EOQ* for each item, and verify the computation by computing total costs for one unit more than *EOQ* and one unit less than *EOQ*.

[***Hint:*** Total cost = Ordering cost + Holding cost $= \frac{D}{Q} \times C_o + \frac{1}{2} Q \times C \times I.$]

22.17. A purchase manager places order each time for a lot of 500 units of a particular item. From the available data the following results are obtained:

Inventory carrying cost = 40% Ordering cost per order = ₹ 600

Cost per unit = ₹ 40 Annual demand = 6 orders each of 200 units.

Find out the loss to the organization due to his ordering policy. ***[MBA, Osmania Univ. 1998]***

22.18. A manufacturing company of microwave ovens uses ₹ 75,000 worth of LED readout circuits annually in its production process. Cost per order is ₹ 45 and the carrying charges assessed against this classification of inventory are 25% of the average balance per year. This company follows an *EOQ* purchasing system and to date has not been offered any discounts on these circuits. Now, the supplier has indicated that if the company would buy its circuits four times a year in equal quantities, a discount of 1.5% off the list price would be given in return. Would you advise this company to accept this offer? In order to maintain the present total cost, what should be the minimum discount acceptable to the company if four orders of equal size are placed in a year? ***[MBA (May), Dayal Bagh Ed. Institute 2013]***

[***Hint:*** To earn discount, $Q = \frac{1}{4} \times 75{,}000 = 18{,}750$. Total cost when discount of 1.5% is available

$$= \frac{1}{2} \times 18{,}750 \times 0.985 \times 0.25 + 4 \times 45 + 75{,}000 \times 0.985 = ₹\, 76{,}363.60$$

Total cost (without discount) $= \sqrt{2 \times 75{,}000 \times 45 \times 0.25} + 75{,}000 = ₹\, 76{,}299.04$]

22:10. *EOQ* AND QUANTITY DISCOUNT

Sometimes discount is offered for bulk purchases or if the size of a single order is large due to discount in price per unit, and also the total price paid for the lot is reduced.

The discount in price affects inventory cost in following three ways:

1. When the price per unit purchased is reduced, total cost of the commodity will decrease to that extent.

2. To benefit from the discount in price when the number of units per order is increased, the number of orders is reduced and as a result total ordering cost will decrease.
3. With the increase in quantity per order, average inventory held goes up and due to this, storage cost will increase.

In order to decide about the ordered quantity to be raised, it is required to determine *EOQ*. Then, the total cost of inventory without quantity discount and with quantity discount (raising the level of ordered quantity to minimum required level) is determined. In case, the total inventory cost is less due to quantity discount, the discount offer is accepted, otherwise rejected. Therefore, the decision rule is:

Accept the discount offer, if *TC* (with discount) ≤ *TC*(*EOQ*),

and Reject the discount offer, if *TC* (with discount) > *TC*(*EOQ*).

Example 22.7. *An item sells for ₹ 4 per unit, but 10% discount is offered for lots of 150 units or more. A manufacturing unit that consumes this item at the rate of 20 items per day wants to decide whether or not to take the advantage of discount offer. The set-up cost for ordering a lot is ₹ 50 and holding cost per unit per day is ₹ 0.30. Will it be economical for the manufacturing unit to take advantage of the discount. (Assume that the factory works for all the 365 days of the year).*

Solution. We are given

$D = 20 \times 365 = 7{,}300$ items/year, $C_o = ₹\ 50$ per lot, and $C_h = ₹\ 0.30 \times 365 = ₹\ 109.50$.

Case 1. When there is no discount

$$Q^o \equiv EOQ = \sqrt{\frac{2 \times 7{,}300 \times 50}{109.50}} = 82 \text{ approx.},$$

Cost of purchasing the items in a year = 20 × 365 × ₹ 4 = ₹ 29,200.

$$\therefore \quad \text{Inventory holding cost} = \frac{1}{2} QC_h = \frac{1}{2} \times 82 \times 109.50 = ₹\ 4{,}489.50,$$

$$\text{Ordering cost} = C_o \times \frac{D}{Q} = \frac{50 \times 7{,}300}{82} = ₹\ 4{,}450 \text{ approx.}$$

TC (*EOQ*) = Cost of items + Inventory holding cost + Ordering cost

= ₹ 29,200 + ₹ 4,489.50 + ₹ 4,450 = ₹ 38,139.50.

Case 2. When there is discount

Cost of purchasing items for yearly requirement = 20 × 365 × ₹ 3.60 = ₹ 26,280,

since cost per unit of item = ₹ 4 less 10% of ₹ 4 = ₹ 4.00 – ₹ 0.40 = ₹ 3.60.

$$\text{Inventory holding cost} = \frac{1}{2} \times 150 \times ₹\ 109.50 = ₹\ 8{,}212.50, \text{ and}$$

$$\text{Ordering cost} = \frac{₹\ 50 \times 7{,}300}{150} = ₹\ 2{,}433 \text{ approx.}$$

∴ *TC* (10% discount) = ₹ 26,280 + ₹ 8,212.50 + ₹ 2,433 = ₹ 36,925.50.

On comparison of *case 1* and *case 2*, we observe that

TC (with discount) < *TC* (without discount at *EOQ*)

Hence, discount offer is accepted with a net saving of ₹ 1,214.

Example 22.8. *A factory requires 1,500 units of an item per month, each costing ₹ 27. The cost per order is ₹ 150 and the inventory carrying charges working out to 20% of the average inventory. Find the Economic Order Quantity and the number of orders per year.*

Would you accept a 2% price discount on a minimum supply quantity of 1,200 units? Compare the total costs in both the cases. **[PG Dip., Indian Institute of Mat. Man. 2016; M.Com., Bangalore Univ. 2015]**

Solution. Given: $D = 1{,}500 \times 12 = 18{,}000$ units per year, $C_o =$ ₹ 150 per order,
and $C_h = 20\%$ of ₹ 27 = ₹ 5.40.

Case 1. When there is no discount

$$Q^o \equiv EOQ = \sqrt{\frac{2 \times 18{,}000 \times 150}{5.40}} = 1{,}000 \text{ units.}$$

Cost of items in a year = 18,000 × ₹ 27 = ₹ 4,86,000.

$$\text{Number of orders per year} = \frac{D}{Q^o} = \frac{18{,}000}{1{,}000} = 18.$$

$$\text{Inventory holding cost} = \frac{1}{2} QC_h = \frac{1}{2} \times 1{,}000 \times ₹\ 5.40 = ₹\ 2{,}700.$$

$$\text{Cost of ordering} = C_o \times \frac{D}{Q} = \frac{₹\ 150 \times 18{,}000}{1{,}000} = ₹\ 2{,}700.$$

∴ $TC(EOQ) =$ ₹ 4,86,000 + ₹ 2,700 + ₹ 2,700 = ₹ 4,91,400.

Case 2. When there is a discount.

$$\text{Ordering cost} = \frac{₹\ 150 \times 18{,}000}{1{,}200} = ₹\ 2{,}250.$$

Cost of items in a year = 18,000 × ₹ 26.46 = ₹ 4,76,280,

where cost per unit of item = ₹ 27 less 20% of ₹ 27 = ₹ 27 – ₹ 0.54 = ₹ 26.46.

$$\text{Inventory holding cost} = \frac{1}{2} \times 1{,}200 \times 20\% \text{ of } ₹\ 26.46$$

= 600 × ₹ 5.29 = ₹ 3,175.20

∴ TC (2% discount) = ₹ 4,76,280 + ₹ 3,175.20 + ₹ 2,250 = ₹ 4,81,795.20.

On comparison of *Case 1* and *Case 2*, we observe that

TC (with discount) < TC (without discount at EOQ)

Hence, discount offer should be accepted with a net saving of ₹ 9,604.80.

Example 22.9. *A shopkeeper has a uniform demand of an item at the rate of 50 items per month. He buys from a supplier at a cost of ₹ 6 per item and the cost of ordering is ₹ 10 each time. If the stock holding cost is 20 per cent of stock value, how frequently should he replenish his stock?*

Suppose the supplier offers a 5% discount on orders between 200 and 999 items and 10% discount on orders exceeding or equal to 1,000, can shopkeeper reduce his cost by taking advantage of either of these discounts? **[MBA, Panjab Tech. Univ. 2011; MBA, Nagpur Univ. 2008; MIB, Delhi Univ. 2011]**

Solution. Given: $D = 50 \times 12 = 600$ units per year, $C_o =$ ₹ 10 per order.

Since, cost per order is ₹ 6 per unit, therefore, $C_h = 20\%$ of ₹ 6 = ₹ 1.20.

$$Q^o \equiv EOQ = \sqrt{\frac{2 \times 600 \times 10}{1.20}} = 100 \text{ items}$$

$$\text{Replenish cylcle} = \frac{EOQ}{\text{Monthly requirement}} = \frac{100}{50} = 2 \text{ months.}$$

$$TC(EOQ) = 600 \times ₹\ 6 + \frac{1}{2} \times 100 \times ₹\ 1.20 + \frac{600}{100} \times ₹\ 10$$

= ₹ 3,600 + ₹ 60 + ₹ 60 = ₹ 3,720.

For 5% Discount

Suppose that order is placed for a minimum required quantity to avail discount is of 200 units and as such 3 orders are placed per annum.

Cost of an item in this case = ₹ 6 – 5% of ₹ 6 = ₹ 5.70 per unit.

So, we get

$$\text{Total item cost} = 600 \times ₹\,5.70 = ₹\,3{,}420 \text{ per annum,}$$

$$\text{Ordering cost} = 3 \times ₹\,10 = ₹\,30.$$

$$\text{Holding cost} = \frac{1}{2} \times 200 \times ₹\,5.70 \times 20\% \text{ of stock value}$$

$$= ₹\,570 \times 0.20 = ₹\,114.$$

$$\therefore \quad TC\ (5\%\ \text{discount}) = ₹\,3{,}420 + ₹\,114 + ₹\,30 = ₹\,3{,}564 \text{ per year.}$$

On comparison, *TC* (with 5% discount) with *TC* (without discount), we observe that

TC (with discount) < *TC* (without discount at *EOQ*)

This shows that net saving by accepting 5% discount offer is of ₹ 156.

For 10% Discount

Here, suppose that order is placed for the minimum required quantity to avail discount is of 1,000 units and as such 600/1,000, *i.e.*, 0.6 orders are placed per year.

Cost of an item in this case = ₹ 6 – 10% of ₹ 6 = ₹ 5.40 per unit

So, we get

$$\text{Total item cost} = 600 \times ₹\,5.40 = ₹\,3{,}240 \text{ per annum}$$

$$\text{Ordering cost} = 0.6 \times ₹\,10 = ₹\,6, \text{ and}$$

$$\text{Holding cost} = \frac{1}{2} \times 1{,}000 \times ₹\,5.40 \times 20\% \text{ of stock value}$$

$$= ₹\,2{,}700 \times 0.2 \times 0.6 = ₹\,324$$

$$\therefore \quad TC = ₹\,3{,}240 + ₹\,324 + ₹\,6 = ₹\,3{,}570.$$

This shows that net saving by accepting 10% discount offer is of ₹ 150.

Since, net saving by accepting 5% discount offer is more than that of accepting 10% discount offer, discount offer of 5% should be accepted.

Example 22.10. *A company uses 8,000 units of a product as raw material, costing ₹10 per unit. The administrative cost per purchase is ₹40. The holding cost is 28% of the average inventory. The company is following an optimal purchase policy and places orders according to EOQ. It has been offered a quantity discount of one per cent, if it purchases its entire requirement only four times a year.*

Should the company accept the offer of quantity discount of one per cent? If not, what minimum discount should the company demand?

Solution. We are given

$$D = 8{,}000 \text{ units}, \quad C_o = ₹\,40 \text{ per order}, \quad C_h = 28\% \text{ of } ₹\,10 = ₹\,2.80 \text{ per unit per year}$$

$$\therefore \quad Q^o \equiv EOQ = \sqrt{\frac{2 \times 40 \times 8{,}000}{2.80}} = 478 \text{ units (approx.)}$$

Total cost associated with Q^o is obtained as:

$$TC = D \times C + \frac{DC_o}{Q^o} + \frac{1}{2} Q^o \times C_h$$

$$= 8{,}000 \times 10 + \frac{8{,}000}{478} \times 40 + \frac{1}{2} \times 478 \times 0.28 \times 10.$$

$$= 80{,}000 + 1{,}338.66 = ₹\,81{,}338.66.$$

When 1% discount is offered, the unit price will be 99% of ₹ 10, *i.e.*, ₹ 9.90. In this case,

$$TC = 8,000 \times 9.90 + \frac{8,000}{2,000} \times 40 + \frac{2,000}{2} \times 0.28 \times 9.90,$$

since the order quantity = 2,000 units, and C_h = ₹ 0.28 × 9.90 = 2.772.

Thus, TC = 79,200 + 2,932 = ₹ 82,132.

Since, the total cost comes out to be higher when one per cent discount is offered, the company should not accept the offer.

Now, let x be the percentage minimum discount acceptable to the company. Then, it can be determined by setting the total cost equal to the total cost associated with the policy of ordering *EOQ*. Accordingly,

$$8,000 \times \left(\frac{100 - x}{100}\right) \times 10 + \frac{8,000}{2,000} \times 40 + \frac{1}{2} \times 2,000 \times 0.28 \times \left(\frac{100 - x}{100}\right) = 81,338.66$$

or

$$\left(\frac{100 - x}{100}\right) \times 82,800 = 81,338.66 - 160 = 81,178.66$$

or

$$100 - x = 100 \times \frac{81,178.66}{82,800} = 98.04.$$

This gives $x = 100 - 98.04 = 1.97$ or 2 (approx.)

Hence, the minimum discount acceptable to the company is 2%.

PROBLEMS

22.19. The annual demand for an item of inventory whose price is ₹ 10 per unit is 2,400 units, ordering cost per order is ₹ 350 and inventory holding costs are 2% per month. The supplier offers a quantity discount of 7.5%, if the quantity ordered is 400 units or more. The rate of discount will increase to 12.5%, if the order is for 3,000 units or more. Find out the economic order quantity. ***[M.Com., Delhi Univ. 2006]***

22.20. A retailer gets discount for large orders. The discount is 4 per cent if the quantity ordered is 500 or more, and an additional 1 per cent discount is received if the quantity ordered is 1,000 or more. The product costs ₹ 50 and sells uniformly at the rate of 25,000 units a year. The fixed order cost is ₹ 50, while the inventory carrying cost is 20% of the value of the average inventory held. (*i*) What is the best order quantity? (*ii*) What is the minimum total annual cost, including the purchasing cost? ***[M.Com., Delhi Univ. 2004]***

22.21. The annual demand for a product is 64,000 units (or 1,280 units per week). The buying cost per order is ₹ 10 and the estimated cost of carrying one unit in stock for a year is 20%. The normal price of the product is ₹ 10 per unit. However, the supplier offers a quantity discount of 2% on an order of at least 1,000 units at a time, and a discount of 5% if the order is for at least 5,000 units. Suggest the most economic purchase quantity per order. ***[MBA, Pune Univ. 2007]***

22.22. A Purchase Manager has decided to place order for a minimum quantity of 500 numbers of a particular item in order to get a discount of 10%. From the past records, it was found out that in the last year, 8 orders each of size 200 units were placed. Given the ordering cost of ₹ 500 per order, inventory carrying cost of 40% of the inventory value and the price of the item of ₹ 400 per unit. Is the Purchase Manager justified in his decision? What is the effect of his decision to the company? ***[M.Com., Delhi Univ. 2006]***

22.23. The annual demand of a product is 1,000 units. Each unit costs ₹ 20 if orders are placed in quantities of 250 and inventory carrying cost is 40%. The following three strategies are available for the procurement:

(*i*) Place four orders of equal size every year.
(*ii*) Place the order for 500 units at a time and avail a discount of 10% on the cost of items.
(*iii*) Follow *EOQ* policy. ***[PG Dip. in Glob. Bus. Oper., Delhi Univ. 2010]***

22.24. A telecommunication company uses 1,200 units per month of a component each costing ₹ 2. Placing each order costs ₹ 50 and carrying cost is 6% per year of the average inventory.

(*a*) Find *EOQ*. If the company gets 5% discount on placing single order, should they accept the discount offer?
(*b*) Find break-even discount percentage which matches *EOQ* ordering for a single order.

[B.Com., Guru Nanak Dev Univ. 2008]

22:11. *EOQ* AND PRICE BREAKS

In *EOQ* problem, we have discussed so far, the production or purchased cost per unit was assumed to be constant. But in the real world, it is not always true that the unit cost of an item is independent of the quantity procured. So, we shall now consider a class of those inventory problems which depends upon the quantity manufactured or purchased. This generally happens when ***discounts*** are offered for the purchase of large quantities. These discounts take the form of price breaks. For example, the price break may be as follows:

₹ 1 per item for the purchase of item upto 99,
₹ 0.95 per item for the purchase of items from 100 upto 199,
₹ 0.0 per item for the purchase of items from 200 and above.

These can be expressed in the following manner:

$$C(Q) = \begin{cases} 1 \cdot Q & , \quad 0 \le Q \le 199 \\ 1 \times 200 + 0.95(Q - 200) & , \quad 200 \le Q \le 499 \\ 1 \times 200 + 0.95(500 - 200) + 0.90(Q - 500), & Q \ge 500. \end{cases}$$

In the inventory problem, such a variable cost must be considered. Since, this variable cost is more appropriate for purchased parts (because of quantity discounts), we shall refer only to purchased parts. The problem, then, is to determine

(*a*) how often the parts be purchased, and
(*b*) how many units should be purchased at any one time.

Case 1. Problem of EOQ with One Price Break

When there is only one price break (one quantity discount), the situation may be illustrated as follows:

Range of quantity	*Purchase cost per unit*
$0 \le Q_1 < b$	K_{11}
$b \le Q_2$	K_{12}

where, b is that quantity at and beyond which the quantity discount applies and $K_{12} < K_{11}$.

The procedure for obtaining *EOQ* may be summarized in the following steps:

Step 1. Compute Q_2^o, *i.e.*, optimum order quantity for the lowest price (highest discount), *i.e.*, K_{12} and compare it with the quantity b.

Step 2. If $Q_2^o \ge b$, then optimum order quantity will be Q^o, *i.e.*, $Q^o = Q_2^o$.

***Step* 3.** If $Q_2^o < b$, we cannot place order at the reduced price K_{12}. Therefore, in order to obtain the optimum order quantity, we need only to compare the total inventory cost for $Q^o = Q_1^o$ (at price K_{11}) with $Q = b$.

The values of $TC(Q_1^o)$ and $TC(b)$ may be determined as follows:

$$TC(Q_1^o) = DK_{11} + \frac{D}{Q_1^o}C_o + \frac{1}{2}Q_1^o \times K_{11} \times I$$

$$TC(b) = DK_{12} + \frac{D}{b}C_o + \frac{1}{2}b \times K_{12} \times I$$

If $TC(Q_1^o) > TC(b)$, then $Q^o = b$, otherwise $Q^o = Q_1^o$.

Example 22.11. *Find the optimum order quantity for a product, for which the price breaks are as follows:*

Quantity	*Unit cost*
$0 \le Q_1 < 500$	₹ 10.00
$500 \le Q_2$	₹ 9.25

The monthly demand for the product is 200 units, the cost of storage is 2% of the unit cost and cost of ordering is ₹ 350. ***[M.Com., Bharathidasan Univ. 2007]***

Solution. We are given

$C_o = ₹\,350$, $D = 200$ units per month, $I = ₹\,0.2$, $K_{11} = ₹\,10$ and $K_{12} = ₹\,9.25$.

***Step* 1.** The highest discount available is ₹ 9.25. So, we compute Q_2^o by taking into consideration K_{12} as ₹ 9.25.

$$\therefore \qquad Q_2^o = \sqrt{\frac{2 \times 350 \times 200}{9.25 \times 0.02}} = 870 \text{ units.}$$

Now, $Q_2^o = 870$ and $b = 500$ indicate that $Q_2^o > b$, *i.e.*, Q_2^o is within the range of $Q_2^o \ge 500$. Therefore, the optimum purchase quantity is given by

$$Q^o \equiv Q_2^o = 870 \text{ units.}$$

Example 22.12. *Find the optimum order quantity for a product, for which the price breaks are as follows:*

Quantity	*Unit cost*
$0 \le Q_1 < 800$	₹ 1.00
$800 \le Q_2$	₹ 0.98

The yearly demand for the product is 1,600 units per year, cost of placing an order is ₹ 5, the cost of storage is 10% per year.

Solution. We are given

$D = 1{,}600$ units per year, $C_o = ₹\,5$, $I = ₹\,0.10$, $K_{11} = ₹\,1.00$ and $K_{12} = ₹\,0.98$.

***Step* 1.** The highest discount available is K_{12} (= ₹ 0.98). So we compute Q_2^o at K_{12}.

$$\therefore \qquad Q_2^o = \sqrt{\frac{2 \times 5 \times 1{,}600}{(0.10) \times (0.98)}} = 404 \text{ units.}$$

Now, $Q_2^o = 404$ and $b = 800$ indicate that $Q_2^o < b$.

***Step* 2.** Considering now $K_{11} = ₹\,1.00$; the optimum order quantity Q_1^o is obtained as follows:

$$Q_1^o = \sqrt{\frac{2 \times 5 \times 1{,}600}{(0.10) \times (1.00)}} = 400 \text{ units.}$$

Since, $Q_1^o = 400$ and $b = 800$ indicate that $Q_1^o < b$, we compare the optimum cost of procuring the least quantity which will entitle us a price break (in this case, $b = 800$).

Now, $$TC(Q_1{}^o) = 1{,}600 \times 1 + \frac{1{,}600}{400} \times 5 + \frac{1}{2} \times 400 \times 1 \times 0.10 = ₹\,1{,}640.$$

$$TC(b) = 1{,}600 \times 0.98 + \frac{1{,}600}{800} \times 5 + \frac{1}{2} \times 800 \times 0.98 \times 0.10 = ₹\,1{,}617.20.$$

Since, $TC(Q_1{}^o) > TC(b)$; the optimum purchase quantity is given by

$$Q^o = b = 800 \text{ units.}$$

Case 2. Problem of EOQ with more than One Price Break

When there are n price breaks, the situation may be illustrated as follows:

Range of quantity	*Purchase cost per unit*
$0 \le Q_1 < b_1$	K_{11}
$b_1 \le Q_2 < b_2$	K_{12}
$\vdots$	$\vdots$
$b_{n-1} \le Q_n$	K_{1n}

where, $b_1, b_2, ..., b_{n-1}$ are those quantities which determine the price breaks.

The computational procedure for obtaining the optimum order quantity in this case will be as follows:

Step 1. Compute $Q_n{}^o$. If $Q_n{}^o \ge b_{n-1}$, the optimum order quantity is reached, *i.e.*, $Q_n{}^o$.

Step 2. If $Q_n{}^o < b_{n-1}$, compute $Q_{n-1}{}^o$. If $Q_{n-1}{}^o \ge b_{n-2}$, proceed as in the case of one price break; that is, the optimum order quantity is determined by comparing $TC(Q_{n-1}{}^o)$ with $TC(b_{n-1})$.

Step 3. If $Q_{n-1}{}^o < b_{n-2}$, compute Q_{n-2}. If $Q_{n-2}{}^o \ge b_{n-3}$, proceed as in the case of two price breaks; that is, the optimum order quantity is determined by comparing $TC(Q_{n-2}{}^o)$ with $TC(b_{n-2})$ and $TC(b_{n-2})$ and $TC(b_{n-1})$.

Step 4. If $Q_{n-2}{}^o < b_{n-3}$, compute $Q_{n-3}{}^o$. If $Q_{n-3}{}^o \ge b_{n-4}$, then compute $TC(Q_{n-3}{}^0)$ and proceed as in the case of three price-breaks, *i.e.*, the optimum order quantity is determined by comparing with $TC(b_{n-3})$, $TC(b_{n-2})$ and $TC(b_{n-1})$.

Step 5. Compute in this way, until $Q_{n-j}{}^0 \ge b_{n-(j+1)}$; $0 \le j \le n-1$; and then compare $TC(Q_{n-j}{}^o)$ with $TC(b_{n-j-2}), ..., TC(b_{n-1})$.

Example 22.13. *Find the optimal order quantity for a product for which the price breaks are as follows:*

Quantity	*Unit cost*
$0 \le Q_1 < 500$	*₹ 10.00*
$500 \le Q_2 < 750$	*₹ 9.25*
$750 \le Q_3$	*₹ 8.75*

The monthly demand for the product is 200 units, the cost of storage is 2% of the unit cost and the cost of ordering is ₹ 350. ***[B.Sc. (Math.), Madras Univ. 2012; MBA, Panjab Tech. Univ. 2002; M.Com., Bharthidasan Univ. 2007]***

Solution. We are given:

$$C_o = ₹\,350, \quad I = ₹\,0.02, \quad D = 200 \text{ units per month}$$

$$\therefore \qquad Q_3{}^o = \sqrt{\frac{2 \times 350 \times 200}{(8.75)(0.02)}} = 894 \text{ units.}$$

This shows that $Q_3{}^o > b_2$, since $b_2 = 750$.

Hence, the optimum order quantity is given by

$$Q^o \equiv Q_3{}^o = 894 \text{ units.}$$

Example 22.14. *Find the optimum order quantity of a product for which the price breaks are as follows:*

Quantity	*Purchasing cost per unit*
$0 \le Q_1 < 100$	₹ 20
$100 \le Q_2 < 200$	₹ 18
$200 \le Q_3$	₹ 16

The monthly demand for the product is 400 units. The storage cost is 20% of the unit cost of the product and the cost of ordering is ₹ 25.00 per month. **[M.Com., Calicut Univ. 2009]**

Solution. Here, $D = 400$ units, $I = 0.20$ and $C_o = ₹\ 25$.

$$\therefore \quad Q_3^{\,o} = \sqrt{\frac{2 \times 25 \times 400}{16 \times 0.20}} = 79 \text{ units.}$$

Since, $Q_3^{\,o} < b_2$ (= 200), we next compute $Q_2^{\,o}$. We have

$$Q_2^{\,o} = \sqrt{\frac{2 \times 25 \times 400}{18 \times 0.20}} = 75 \text{ units.}$$

Again, since $Q_2^{\,o} < b_1$ (= 100), we next compute $Q_1^{\,o}$. We have

$$Q_1^{\,o} = \sqrt{\frac{2 \times 25 \times 400}{20 \times 0.20}} = 70 \text{ units (approx.)}$$

Now,
$$TC(Q_1^{\,o}) = 25 \times \frac{400}{70} + 400 \times 20 + 20 \times (0.20) \times \frac{70}{2} = ₹\ 8{,}283.00.$$

$$TC(b_1) = 25 \times \frac{400}{100} + 400 \times 18 + 18 \times (0.20) \times \frac{100}{2} = ₹\ 7{,}480.00.$$

and
$$TC(b_2) = 25 \times \frac{400}{200} + 400 \times 16 + 16 \times (0.20) \times \frac{200}{2} = ₹\ 6{,}770.00.$$

Since, $TC(b_2) < TC(b_1) < TC(Q_1^{\,o})$, the optimum order quantity is: $Q^o \equiv b_1 = 200$ units.

Example 22.15. *A manufacturing concern requires 2,000 units of a material per year. The ordering costs are ₹ 10 per order. While carrying costs are ₹ 0.16 per year per unit of average inventory, the purchase price is ₹ 1 per unit. Find the economic order quantity, and the total inventory cost. If a discount of 5 per cent is available for orders of 1,000 units or more but less than 2,000 units, should the manufacturer accept this offer? Also, if he purchases a single lot of 2,000 units or more he has to pay ₹ 0.93 per unit. What purchase quantity would you recommend?* **[MBA (Nov.), Annamalai Univ. 2002]**

Solution. We are given

$D = 2{,}000$ units per year, $C_o = ₹\ 10$ per order, $C_h = ₹\ 1 \times 0.16 = ₹\ 0.16$.

When no discount is offered, the optimum order quantity is given by

$$Q^o = \sqrt{\frac{2 \times 2{,}000 \times 10}{0.16}} = 500 \text{ units.}$$

Total inventory cost is given by

$$TC = 2{,}000 \times 1 + \frac{2{,}000}{500} \times 10 + \tfrac{1}{2} \times 500 \times 0.16 = ₹\ 2{,}440.$$

When quantity discounts are offered, the following information is available:

Quantity	*Price per unit*
$0 < Q_1 < 1{,}000$	₹ 1.00
$1{,}000 \le Q_2 < 2{,}000$	₹ 0.95
$2{,}000 \le Q_3$	₹ 0.93

The optimum order quantity, Q_3^o, based on ₹ 0.93 per unit is given by

$$Q_3^o = \sqrt{\frac{2 \times 2,000 \times 10}{0.93 \times 0.16}} = 518 \text{ units (approx.)}$$

Since, $Q_3^o < b_2$ (= 2,000), we next compute Q_2^o. We have

$$Q_2^o = \sqrt{\frac{2 \times 2,000 \times 10}{0.95 \times 0.16}} = 513 \text{ units (axpprox.)}$$

Again, since $Q_2^o < b_1$ (= 1,000), we next compute Q_1^o

$$Q_1^o = \sqrt{\frac{2 \times 2,000 \times 10}{1.00 \times 0.16}} = 500 \text{ units.}$$

Now,
$$TC(Q_1^o) = 2,000 \times 10 + \frac{2,000}{500} \times 10 + \frac{1}{2} \times 500 \times 0.16 = ₹\,2,440$$

$$TC(b_1) = 2,000 \times 0.95 + \frac{2,000}{1,000} \times 10 + \frac{1}{2} \times 1,000 \times 0.95 \times 0.16 = ₹\,1,996.00$$

$$TC(b_2) = 2,000 \times 0.93 + \frac{2,000}{2,000} \times 10 + \frac{1}{2} \times 2,000 \times 0.93 \times 0.16 = ₹\,2,018.80.$$

Since, $TC(b_1) < TC(b_2) < TC(Q_1^o)$, the optimum order quantity is: $Q^o = b_1 = 500$ units.

Hence, the manufacturer should accept the offer of 5 per cent discount only, because in this case his net saving per year would be ₹ (2,440 – 1,996), *i.e.*, ₹ 444.

PROBLEMS

22.25. A soft drinks manufacturing company buys a large number of pallets every year which it uses in the warehousing of its bottled products. A local vendor has offered the following discount schedule for pallets:

Order quantity	*Unit price*
Up to 699	₹ 10.00
700 and above	₹ 9.25

The average yearly replacement is 2,400 pallets. The carrying cost is 12% of the average inventory and ordering cost per order is ₹ 100. Find out the optimum order quantity.

22.26. Find the optimal order quantity of a product for which the price breaks are as follows:

Quantity	*Unit cost*
$0 \le Q_1 < 50$	₹ 10.00
$50 \le Q_2 < 100$	₹ 9.00
$100 \le Q_3$	₹ 8.00

The monthly demand for the product is 200 units, the cost of storage is 25% of the unit cost and ordering cost is ₹ 20.00 per order. ***[M.Sc. (Math.), CCS Univ. 2003]***

22.27. A manufacturer of engines is required to purchase 4,800 castings per year. The requirement is assumed to be known as fixed. These castings are subject to quantity discounts. The price schedule is as follows:

Quantity	*Cost per unit*
Less than 500 units	₹ 150.00
500 or more but less than 750 units	₹ 138.75
750 or more units	₹ 131.25

Monthly holding cost expressed as a decimal fraction of the value of the unit is 0.02. Set-up cost associated with the procurement of purchased items is ₹ 750.00 per procurement. Find the optimum purchase quantity per procurement. **[B.Com., Guru Nanak Dev Univ. 2007]**

22.28. Determine the economic purchase quantity for the following situation:

Annual demand = 10,000 units, ordering cost = ₹ 28.80, carrying cost per unit per year = 20% of the unit price. The quantity *versus* unit price schedule is:

Quantity	*Unit price*
0–9,999	₹ 2.00
10,000–19,999	₹ 1.60
20,000 and above	₹ 1.40

[MMS, Mumbai Univ. 1994]

22.29. The estimated annual demand for a product is 10,000 units, the annual holding costs amount to 20% of the average inventory value, and the ordering costs are ₹ 100 per order. No shortage is to be allowed. The normal price charged by the supplier is ₹ 10. However, a discount of 2% is allowed on an order for at least 1,000 units. The discount rate will be raised to 5%, if the order is for 2,500 units or more.

Determine the most economical order quantity. If the discount slab of 1,000 units is raised to 1,500 units, how will you compute the most economical order?

[MBA, Bangalore Univ. 2009; MIB, Delhi Univ. 2011]

2.30. Find the optimal order quantity for a product, where the annual demand for the product is 500 units, the cost of storage per unit per year is 10% of the unit cost and ordering cost per order is ₹ 180.00. The unit costs are given below:

Quantity	*Unit cost*
$0 \leq Q_1 < 500$	₹ 25.00
$500 \leq Q_2 < 1{,}500$	₹ 24.80
$1{,}500 \leq Q_3 < 3{,}000$	₹ 24.60
$3{,}000 \leq Q_4$	₹ 24.40

[B.Tech. Hyderabad Univ. 2011; M.Sc. (Maths.), CCS Univ. 2010]

22.31. Assume that the following quantity discount schedule for a particular bearing is available to a retail store:

Order size (units)	*Discount*
0–49	0%
50–99	5%
100–199	10%
200 and above	12%

The cost of a single bearing with no discount is: ₹ 30. The annual demand is of 250 units, ordering cost is: ₹ 20 per order and annual inventory carrying cost is: ₹ 4 per unit. Determine the optional order quantity and the associated minimal total cost of inventory and purchasing costs, if shortages are not allowed. **[B.Tech. (Mech.) (Nov.), Panjab Tech. Univ. 2012]**

22.32. A diesel engine manufacturing company has planned its production schedule for the next year based on the forecasted demand, back orders and plant capacity. Instead of manufacturing the piston, that goes in final product, the company has decided to buy the piston from 'India Pistons Limited'. The number of pistons required are at the rate of 60 per day. Ordering costs have been estimated at ₹ 50 per order, and the carrying cost fraction is 0.15. All the assumptions of the basic *EOQ* model are applicable. The company, however, can take advantage of one of the several quantity discounts. The pricing schedule of India Pistons Limited is listed as follows:

Quantity ordered	*Unit price*
0–1,999	₹ 65
2,000–4,999	₹ 60
5,000–10,000	₹ 55
over 10,000	₹ 50

(*a*) What is the optimal order quantity?
(*b*) What is the minimum inventory cost?

22:12. *EOQ* MODEL WITH REPLENISHMENT

In the *EOQ* model discussed earlier, it has been assumed that the entire quantity of an item ordered for is received in a single lot. Here, we shall relax this assumption and consider situations in which the items are received for inventory at a constant rate over a time period and they are also being consumed at a constant rate. When production begins, a constant number of units of an item are supposed to be added to the inventory each time unit (say day or week or month) till the production run is completed. Simultaneously, the units of the same item would be demanded at a constant rate, as stipulated earlier. Obviously, the rate at which they are produced has to be higher than the rate of their depletion, for only then we can think of the inventory build up. Therefore, we assume that each production run of length t consists of two parts, say t_1 and t_2, such that

(*i*) the inventory is building up at a constant rate of $(k-r)$ units, per unit of time during t_1, $k > r$;

(*ii*) there is no replenishment (or production) during time t_2 and the inventory is decreasing at the rate or r per unit of time.

The graphical representation of the situation is shown in *Fig. 22.6*.

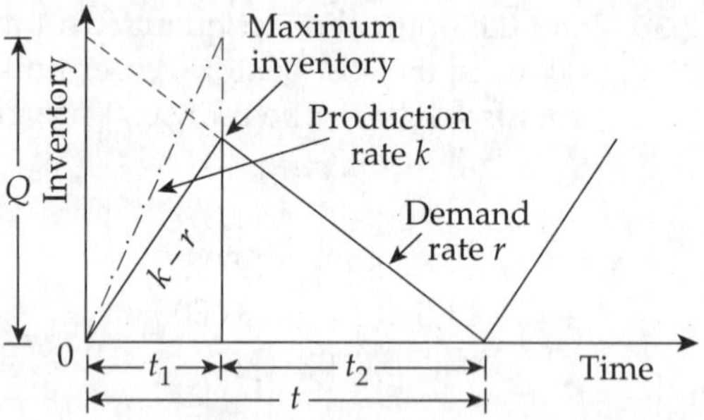

Fig. 22.6

Here, the total (order) quantity Q is produced over a period, t_1, which is defined by the production rate k. Since, inventory does not pile up in one slot but rather continuously over a time period and is also consumed simultaneously, the average inventory level would be determined not only by the lot size Q, but also be affected by the production rate k and depletion (demand) rate r.

To determine the average inventory, we proceed as follows:

Since, t_1 is the time required to produce Q at a rate k, we shall have

$$Q = kt_1 \quad \text{or} \quad t_1 = Q/k.$$

During production period t_1, inventory is increasing at the rate of k and simultaneously decreasing at the rate of r. Thus, inventory accumulates at the rate of $(k - r)$ units. Therefore, the maximum inventory level shall be equal to $t_1\,(k - r)$.

$$\therefore \qquad \text{Average Inventory} = \frac{1}{2}\, t_1 (k-r) = \frac{1}{2} Q(1-r/k), \qquad \text{since } t_1 = Q/k$$

Now, with C_h as the holding cost per unit per year, the total annual holding cost is given by

$$f(Q) = \text{Average inventory} \times C_h = \frac{1}{2} QC_h\left(1-\frac{r}{k}\right)$$

Annual ordering cost is given by

$$g(Q) = C_o \times D/Q,$$

since D is the total demand in a year.

Since, the minimum total cost occurs at the point where annual ordering cost and annual holding cost are equal, we must have

$$f(Q) = g(Q)$$

This implies that

$$\frac{1}{2}QC_h\left(1-\frac{r}{k}\right) = C_o\frac{D}{Q}$$

Hence, the optimum value of Q is:

$$Q^o = \sqrt{\frac{2DC_o}{C_h(1-r/k)}} = \sqrt{\frac{2DC_o}{C_h}\left(\frac{k}{k-r}\right)}.$$

Characteristics of Model with Replinishment

1. Optimum number of production runs per year

$$n^o = D/Q^o = \sqrt{\frac{DC_h}{2C_o}(1-r/k)}.$$

2. Optimum length of each lot size production run

$$t_1{}^o = Q^o/k = \sqrt{\frac{2DC_o}{C_h\,k(k-r)}}.$$

3. Total minimum production inventory cost

$$TC^o = \frac{C_o D}{Q^o} + \frac{1}{2}Q^o\left(1-\frac{r}{k}\right)C_h = \sqrt{2DC_o C_h(1-r/k)}.$$

Notes

1. When $k \to r$, then $C_h \to 0$. This shows that there will be no holding cost and no set-up cost.
2. When $k \to \infty$, *i.e.*, when the production rate becomes infinite, the above problem reduces to the one considered in *Section 22:9*. The inventory holding cost per unit of time is reduced from the cost discussed in *Section 22:9* in the ratio $(1 - r/k)$: 1 for minimum cost, although the set-up cost remains the same.

Example 22.16. *A contractor has to supply 10,000 bearings per day to an automobile manufacturer. He finds that, when he starts a production run, he can produce 25,000 bearings per day. The cost of holding a bearing in stock for one year is ₹ 2 and the set-up cost of a production run is ₹ 1,800. How frequently should production run be made?* **[B.Sc. (Stat.), Delhi Univ. 2006; MBA, Madras Univ. 2006]**

Solution. We are given

$D = 10{,}000 \times 300 = 30{,}00{,}000$ bearings (assuming 300 working days in a year),

C_h = ₹ 2 per bearing per year, C_0 = ₹ 1,800 per production run,

$r = 10{,}000$ bearings per day, $k = 25{,}000$ bearing per day.

$$\therefore \quad Q^o = \sqrt{\frac{2\times 30{,}00{,}000\times 1{,}800}{2}}\sqrt{\frac{25{,}000}{25{,}000-10{,}000}} = 94{,}868 \text{ bearings.}$$

$$t^o = \frac{94{,}868}{10{,}000} = 9.49 \text{ days}$$

$$\text{Length of production cycle} = \frac{94{,}868}{25{,}000} = 4 \text{ days (approx.)}$$

Thus, the production cycle starts at an interval of 9.49 days and production continues for 4 days, so that in each cycle a batch of 94,868 bearings are produced.

Example 22.17. *A manufacturing company needs 2,500 units of a particular component every year. The company buys it at the rate of ₹ 30 per unit. The order processing cost for this part is estimated at ₹ 15 and the cost of carrying a part in stock comes to about ₹ 4 per year.*

The company can manufacture this part internally. In that case, it saves 20% of the price of the product. However, it estimates a set-up cost of ₹ 250 per production run. The annual production rate would be of 4,800 units. However, the inventory holding costs remain unchanged.

(*i*) *Determine the EOQ and the optimal number of orders placed in a year.*
(*ii*) *Determine the optimum production lot size and the average duration of the production run.*
(*iii*) *Should the company manufacture the component internally or continue to purchase it from the supplier?* **[M.Com., Delhi Univ. 1993]**

Solution.

(*i*) Here, we have

$$D = 2{,}500 \text{ units}, \quad C_o = ₹\,15 \text{ per order}, \quad \text{and} \quad C_h = ₹\,4 \text{ per unit per year.}$$

$$\therefore \qquad Q^o = \sqrt{\frac{2DC_o}{C_h}} = \sqrt{\frac{2 \times 2{,}500 \times 15}{4}} = 137 \text{ units.}$$

$$n^o = \text{Optimal number of orders} = \frac{D}{Q^o} = \frac{2{,}500}{137} = 18.$$

(*ii*) We are given:

$$D = 2{,}500 \text{ units}, \quad C_o = ₹\,250 \text{ per production run}, \quad C_h = ₹\,4 \text{ per unit per year},$$
$$r = 2{,}500 \text{ units}, \quad \text{and} \quad k = 4{,}800 \text{ units.}$$

$$\therefore \qquad \text{Economic lot size, } Q_1{}^o = \sqrt{\frac{2DC_o}{C_h}} \times \sqrt{\frac{k}{k-r}}$$

$$= \sqrt{\frac{2 \times 2{,}500 \times 250}{4}} \times \sqrt{\frac{4{,}800}{4{,}800 - 2{,}500}} = 808 \text{ units.}$$

$$t^o = \text{Average duration of the production run} = \frac{808}{2{,}500} = 0.32 \text{ year.}$$

(*iii*) When item is purchased from outside:

$$\text{Total cost} = D \times C + \left(\frac{D}{Q^o} \times C_o + \frac{1}{2} Q^o C_h \right), \text{ where } C = ₹\,30 \text{ per unit.}$$

$$= 2{,}500 \times 30 + \left(\frac{2{,}500}{137} \times 15 + \frac{1}{2} \times 137 \times 4 \right)$$

$$= 75{,}000 + 547.72$$

$$= ₹\,75{,}547.72 \quad \text{or} \quad ₹\,75{,}548 \text{ approx.}$$

When item is produced internally:

$$\text{Cost per unit, } C = 80\% \text{ of } ₹\,30 = ₹\,24, \quad \text{and} \quad C_h = ₹\,250 \text{ per production run,}$$

$$\therefore \qquad \text{Total cost} = D \times C + \frac{D}{Q_1{}^o} \times C_o + \frac{1}{2} Q_1{}^o \times \left(\frac{k-r}{k} \right) \times C_h$$

$$= 2{,}500 \times 24 + \frac{2{,}500}{808} \times 250 + \frac{1}{2} \times 808 \times \left(\frac{4{,}800 - 2{,}500}{4{,}800} \right) \times 4$$

$$= 60{,}000 + 1{,}547.84 = ₹\,61{,}547.84.$$

From the above, we observe that the total cost of the item when produced internally is less than that of the item purchased from outside and hence, the company should manufacture the product internally.

Example 22.18.

(a) *At present a company is purchasing an item 'X' from outside supplier. The assumption is that the demand of the item is of 10,000 units/year. The cost of the item is ₹5 per unit and the ordering cost is estimated to be ₹100 per order. The cost of carrying inventory is 25%. If the consumption rate is uniform, determine the economic purchasing quantity.*

(b) *In the above problem, assume that company is going to manufacture the item with the equipment that is estimated to produce 100 units per day. The cost of the unit thus produced is ₹3.50 per unit. The set-up cost is ₹150 per set-up and the inventory carrying charge is 25 per cent. How has your answer changed?*

Solution. We are given

(a) $D = 10{,}000$ units per year, $C_o =$ ₹ 100 per order, and $C_h =$ ₹ 5 × 0.25 = ₹ 1.25.

$$\therefore \qquad EOQ \equiv Q^o = \sqrt{\frac{2DC_o}{C_h}} = \sqrt{\frac{2 \times 10{,}000 \times 100}{1.25}} = 1{,}265 \text{ units.}$$

(b) Here, $C_o =$ ₹ 150 per set-up, $C_h =$ ₹ 3.50 × 0.25 = ₹ 0.875, $k = 100$ units per day, and

$$r = \frac{10{,}000}{250} \text{ (= 40) units per day (assuming 250 working days in a year).}$$

$$\therefore \qquad EOQ \equiv Q^o = \sqrt{\frac{2 \times 10{,}000 \times 150}{0.875}} \sqrt{\frac{100}{100-40}} = 2{,}391 \text{ units.}$$

The increase in *EOQ* in case of (*b*) may be due to following reasons:

(*i*) Increased ordering cost from ₹ 100 to ₹ 150 per set-up.

(*ii*) Consumption of inventory simultaneously with production, reduces average inventory from $\frac{Q}{2}$ to $\left(1-\frac{r}{k}\right)\frac{Q}{2}$.

PROBLEMS

22.33. An item is produced at the rate of 50 items per day. The demand occurs at the rate of 25 items per day. If the set-up cost is ₹ 100.00 and holding cost is ₹ 0.01 per unit of item per day, find the economic lot size for one run, assuming that the shortages are not permitted. Also, find the time of cycle and minimum total cost for one run. *[B.Sc. (Math.), Madras Univ. 1995]*

22.34. A company has demand rate of 25 items per day and the supply rate is 40 items per day. Ordering cost per order is ₹ 60 and carrying cost is ₹ 73 in one year. Find *EOQ* and the total additional cost. It is given that an item costs ₹ 20. *[M.Com., Bangalore Univ. 2018]*

22.35. A contractor has to supply 20,000 units per day. He can produce 30,000 units per day. The cost of holding a unit in stock is ₹ 3 per year and the set-up cost per production run is ₹ 50. How frequently, and of what size, the production runs be made?

22.36. Amit manufactures 50,000 bottles of tomato ketchup in an year. The factory cost per bottle is ₹ 5, the set-up cost per production run is estimated to be ₹ 90, and the carrying costs on finished goods inventory amount to 20% of the cost per annum. The production rate is of 600 bottles per day, and sales amount to be of 150 bottles per day. What is the optimum production lot size and the number of production runs?

If the factory costs increase to ₹ 7.50 per bottle, what will be the optimum production lot size?

22.37. The details of a part to be machined are as follows:

Annual requirement	= 2,400 pieces
Machine rate	= 10 pieces/shift
No. of working days in a year	= 320 shifts
Cost of maching a component	= ₹ 100 per piece
Inventory carrying cost per annum	= 12% of value
Set-up cost per production run	= ₹ 400.

Find the optimum run size for machining. *[MBA (Nov.), Madras Univ. 2008]*

22.38. A manufacturing company uses an *EOQ* (Economic order quantity) approach in planning its production of gears. The following information is available:

Each gear costs ₹ 250 per unit, annual demand is 60,000 gears, set-up cost is ₹ 4,000 per set-up and the inventory carrying cost per month is established at 2 per cent of the average inventory value. When in production, these gears can be produced at the rate of 400 units per day and this company works only for 300 days in a year. Determine the economic lot size, the number of production runs per year and the total inventory cost.

[MBA, Gujarat Univ. 2009]

22.39. A product is to be manufactured on a machine. The cost, production and demand, etc. are as follows:

Fixed costs per lot	= ₹ 30
Variable costs per unit	= ₹ 0.10
Percentage charges for interest, taxes, insurance and storage	= 50%
Production rate	= 1,00,000 units per year
Demand rate	= 10,000 units per year.

Determine the economic manufacturing quantity.

22.40. Company *A* wants to know what production cost its major competitor, company *B* has assigned to product item p_7. After a bit of investigation, company *A* has collected the following data about company *B*'s production of item p_7:

Production lot size	= 2,600 units
Set-up cost	= ₹ 135
Annual demand	= 30,000 units
Daily demand	= 100 units
Production rate	= 200 units per day
Inventory holding costs	= 28% of the average value per year.

Company *A* has further learnt that company *B* produces according to 'economic lot size' model.

What is the company *B*'s cost of producing product item p_7?

[PG Dip. in Glob. Bus. Oper., Delhi Univ. 2011]

22:13. *EOQ* MODEL AND PLANNED SHORTAGES

There are inventory situations having shortages (or stock out). These shortages are likely to have the loss of customer goodwill, or reduction in future orders or unfavourable change in the market share. While in some situations, customers shift to other sources for their requirement and hence may be lost forever; in some other cases customers may wait until the next shipment arrives. This second situation of inventory is called the ***back-ordering situation***. In this section, we shall discuss only the ***back-order*** or ***planned shortages inventory model***.

When the shortages in any business concern occur, then these can be classified into following two categories:

(*a*) Shortages are lost sales.

(*b*) Shortages are to be carried out in the next production run.

In first category a customer moves to some other firm to fullfil his requirements. In second category, demand of the customer is met in the beginning of next production run. That is, we deal with those problems of shortages where back-orders are entertained. Three cases arise:

1. Instantaneous production with variable order cycle time.
2. Instantaneous production with fixed order cycle time.
3. Replenishment of inventory is finite.

Case 1. Instantaneous Production and Variable Order Cycle

This model is based on the assumptions of EOQ model discussed in *section 22.9*, with the difference that shortages are now permitted. Let C_b be the shortage cost per unit of time per unit quantity. The inventory situation can be illustrated graphically as shown in *Fig. 22.7:*

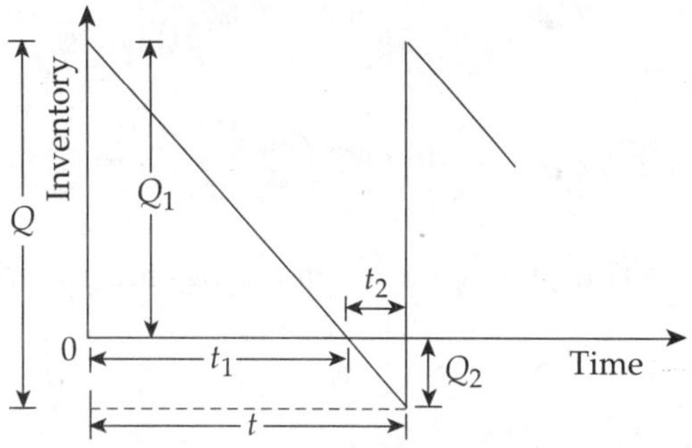

Fig. 22.7

Consider the total time period as one year and divide it into equal parts, say of interval t. Further, this time interval t is divided into two parts t_1 and t_2, such that $t = t_1 + t_2$.

During the interval t_1, the items are drawn from the inventory as needed and during t_2, orders for the item are being accumulated but not filled. Then, at the end of the interval t an amount Q is produced (or delivered). The amount Q has been divided into Q_1 and Q_2 such that $Q = Q_1 + Q_2$, where Q_1 denotes the amount which goes into inventory, and Q_2 denotes the amount which is immediately taken to satisfy past orders or unfilled demand.

Now, Total inventory over the time period t $= \frac{1}{2} Q_1 t_1$,

Average inventory at any time $= \frac{1}{t}\left(\frac{1}{2} Q_1 t_1\right)$,

Annual inventory holding cost $= \frac{1}{t} \times C_h \left(\frac{1}{2} Q_1 t_1\right)$,

Similarly,

Total amount of shortage over time period t $= \frac{1}{2} Q_2 t_2$,

Annual shortage costs $= \frac{1}{t} \times C_b \left(\frac{1}{2} Q_2 t_2\right)$,

Annual costs associated with runs of size $Q = nC_o = \frac{D}{Q} C_o$,

since, D/Q runs are produced each year.

$\therefore$ Total annual cost is = Holding cost + Shortage cost + Ordering cost

i.e.,
$$TC = \frac{1}{t} \times \left[C_h \left(\frac{1}{2} Q_1 t_1\right) + C_b \left(\frac{1}{2} Q_2 t_2\right)\right] + \frac{D}{Q} C_o.$$

Now, using the relationship for similar triangles, we have

$$\frac{t_1}{t} = \frac{Q_1}{Q} \quad \text{and} \quad \frac{t_2}{t} = \frac{Q_2}{Q}$$

i.e.,

$$t_1 = \frac{Q_1}{Q} t \quad \text{and} \quad t_2 = \frac{Q_2}{Q} t.$$

Making use of these values, we get

$$TC = \frac{1}{2} C_h \left(\frac{Q_1^2}{Q} \right) + \frac{1}{2} C_b \left[\frac{(Q - Q_1)^2}{Q} \right] + C_o \left(\frac{D}{Q} \right).$$

since, $Q_2 = Q - Q_1$.

For determining the optimum values of Q_1 and Q so as to optimize (minimize) TC, we have

$$\frac{\partial (TC)}{\partial Q_1} = 0 \Rightarrow Q_1 = \frac{C_2 Q}{(C_h + C_b)},$$

and

$$\frac{\partial (TC)}{\partial Q} = 0 \Rightarrow Q = \sqrt{\frac{2C_o D + C_h Q_1^2}{C_b} + Q_1^2}.$$

Also, $\dfrac{\partial^2 (TC)}{\partial Q_1^2} > 0$, and $\dfrac{\partial^2 (TC)}{\partial Q^2} > 0$ for these values of Q_1 and Q.

Thus, the optimum quantities are given by (on simplification):

$$Q^o = \sqrt{\frac{2C_o D}{C_h}} \sqrt{\frac{(C_h + C_b)}{C_b}}$$

and

$$Q_1^2 = \left(\frac{C_b}{C_h + C_b} \right) Q^o = \sqrt{\frac{C_b}{C_h + C_b}} \sqrt{\frac{2C_o D}{C_h}} \quad \textit{(Optimum Stock Level)}$$

Characteristics of Case I

1. Time between receipt of orders (when to order):

$$t^o = Q^o / D = \sqrt{\frac{2C_o}{DC_h}} \sqrt{\frac{C_h + C_b}{C_b}}$$

2. Total optimum inventory cost:

$$TC^o = \sqrt{2DC_o C_h} \sqrt{\frac{C_b}{C_h + C_b}}$$

3. Maximum inventory level:

$$Q^o - Q_1^o = Q^o \left(1 - \frac{C_b}{C_h + C_b} \right) = \sqrt{\frac{2C_o D}{C_h}} \sqrt{\frac{C_b}{C_h + C_b}}.$$

Remarks

1. If $C_h > 0$ and $C_b \to \infty$, ***shortages are prohibited.***

   In this case $Q_1^o = Q^o = \sqrt{2C_o D / C_h}$ and each batch of Q^o is used entirely for inventory.
2. If $C_h \to \infty$ and $C_b > 0$, ***inventories are prohibited.***

   In this case $Q_1^o = 0$, $Q^o = \sqrt{2C_o D / C_b}$ and each batch is used only to fill back orders.
3. If ***shortage costs are negligible***, then $C_h > 0$ and $C_b \to 0$. In this case $Q_1^o \to 0$ and $Q^o \to \infty$.

4. If ***inventory costs are negligible***, then $C_h \to 0$ and $C_b > 0$. In this case $Q^o \to \infty$ and $Q_1{}^o \to \infty$, *i.e.*, $Q_1{}^o \to Q^o$. Thus as inventory costs become very small, increasingly large batches should be produced and used entirely as inventory for future demands.
5. When ***inventories and shortages are equally costly***, we have

$$C_h = C_b \cdot \frac{C_b}{C_h + C_b} = \frac{1}{2}.$$

Thus, in this case $$Q^o = \sqrt{2}\sqrt{\frac{2C_o D}{C_h}} = (1.414)\sqrt{\frac{2C_o D}{C_h}}.$$

This shows that the lot size is 1.414 times as large as earlier, when no shortages were allowed.

Case 2. Instantaneous Production and Fixed Order Cycle

Let t be fixed, *i.e.*, inventory is to be replenished after every time period t. Also, let us assume that items are being supplied or produced at the rate of r units per unit of time during this fixed time period. Then,

$$\text{Total inventory over the time period } t_1 = \frac{1}{2} Q_1 t_1$$

and $$\text{Total amount of shortages over time period } t_2 = \frac{1}{2} Q_2 t_2$$

∴ Total production cost is given by

$$TC = \frac{1}{2} Q_1 t_1 C_h + \frac{1}{2} Q_2 t_2 C_b$$

Using, now, the relationship of similar triangles, *viz.*,

$$\frac{t_1}{t} = \frac{Q_1}{Q} \quad \text{and} \quad \frac{t_2}{t} = \frac{Q_2}{Q},$$

the cost equation reduces to

$$TC = \frac{1}{2Q} \cdot C_h Q_1^2\, t + \frac{1}{2Q} \cdot C_b Q_2^2\, t + \frac{1}{2r} C_h Q_1^2 + \frac{1}{2r} C_b (rt - Q_1)^2,$$

since $Q_2 = Q - Q_1$ and $Q = rt$.

The optimum value $Q_1{}^o$ is obtained as follows:

$$\frac{\partial (TC)}{\partial Q_1} > 0 \;\Rightarrow\; C_h Q_1 + C_b (Q_1 - rt) = 0 \quad \text{or} \quad Q_1 = \frac{rtC_b}{(C_h + C_b)}.$$

Now, $$\frac{\partial (TC)}{\partial Q_1^2} > 0 \text{ for all values of } Q_1, \text{ therefore,}$$

$$Q_1{}^o = rt\left(\frac{C_b}{C_h + C_b}\right). \qquad \textit{(Optimum Inventory Level)}$$

Note. In this problem, set-up cost is not considered, because of it being fixed as the time of one production run is fixed. Substituting the value of $Q_1{}^o$ in the total production cost equation, the optimum inventory cost is

$$TC^o = \left(\frac{C_h C_b}{C_h + C_b}\right) \cdot rt$$

Case 3. Replenishment (Production) is Finite

In this problem all the assumptions are same as in *Case 1* except that the rate of replenishment of inventory is finite, say k units per unit of time.

Assume that each production run of length t consists of two parts t_1 and t_2 which are further sub-divided into two parts say t_{11} and t_{12}, t_{21} and t_{22}, where:

(*i*) inventory is building up at a constant rate of $(k - r)$ units per unit of time during time t_{11},
(*ii*) no replenishment during time t_{12} and inventory is decreasing at the rate r per unit of time,
(*iii*) shortage is building up at a constant rate of r per unit of time during time t_{21},
(*iv*) shortage are being filled immediately at the rate of $(k - r)$ units per unit of time during time t_{22}.

The graphical representation of the situation is as follows:

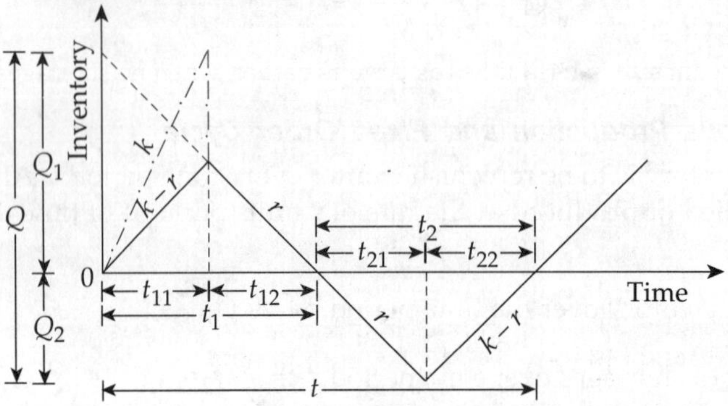

Fig. 22.8

From *Fig. 22.8*, we see that at the end of t_{11}, the level of inventory is Q_1 and at the end of period t_{12} inventory becomes nil. Now shortages start and suppose that the shortages build-up of quantity Q_2 up to time t_{21} and let then these shortages be filled up during time t_{22}.

Thus obviously,

$$Q_1 = t_{11}(k - r), \quad Q_1 = t_{12}r,$$
$$Q_2 = t_{21}r \text{ and } Q_2 = t_{22}(k - r).$$

Now, if Q is the lot size, then

$$Q_1 = Q - Q_2 - rt_{11} - rt_{22}.$$

Eliminating t_{11} and t_{22} from this, we have

$$Q_1 = Q - Q_2 - r\left(\frac{Q_1}{k-r} + \frac{Q_2}{k-r}\right) \quad \text{or} \quad Q_1 + Q_2 = (k-r)Q/k$$

Production cycle is

$$t = t_{11} + t_{12} + t_{21} + t_{22} = \frac{Q_1}{k-r} + \frac{Q_1}{r} + \frac{Q_2}{r} + \frac{Q_2}{k-r}$$
$$= k(Q_1 + Q_2)/r(k-r)$$

Substituting the value of $Q_1 + Q_2$, we get $t = Q/r$.

The average inventory and amount of shortage during production cycle time t are:

$$\text{Average inventory} = \tfrac{1}{2}Q_1(t_{11} + t_{12})/t, \quad \text{and} \quad \text{Average shortage} = \tfrac{1}{2}Q_2(t_{21} + t_{22})/t.$$

∴ The total inventory cost is:

$$TC = \tfrac{1}{2}Q_1C_h\frac{(t_{11} + t_{12})}{t} + \tfrac{1}{2}Q_2C_b\frac{(t_{21} + t_{22})}{t} + rC_o/Q$$
$$= \frac{1}{2Q} \times \frac{k}{k-r}\left[C_h\left(\frac{k-r}{k}Q - Q_2\right)^2 + C_bQ_2^2\right] + \frac{r}{Q}C_o,$$

since $Q_1 + Q_2 = \dfrac{r(k-r)}{k}t = \dfrac{r(k-r)}{k} \times \dfrac{Q}{r} = \dfrac{k-r}{k}Q.$

Now, $$\frac{\partial TC}{\partial Q_2} = 0 \Rightarrow Q_2 = \frac{C_h Q}{C_h + C_b}(1 - r/k)$$

and $$\frac{\partial TC}{\partial Q} = 0 \Rightarrow Q = \sqrt{\frac{2C_o(C_h + C_b)}{C_h C_b}}\sqrt{\frac{kr}{k-r}}$$

Since, $\frac{\partial^2 TC}{\partial Q^2} > 0$, and $\frac{\partial^2 (TC)}{\partial {Q_2}^2} > 0$ for all values of Q and Q_2, the optimum values of Q and Q_2 so as to minimize TC are given by

$$Q^o = \sqrt{\frac{2C_o(C_h + C_b)}{C_h C_b}}\sqrt{\frac{kr}{k-r}} \qquad \textit{(Optimum production lot size)}$$

and $$Q_2{}^o = Q^o\left(1 - \frac{r}{k}\right)\left(\frac{C_h}{C_h + C_b}\right)$$

$$= \sqrt{\frac{2rC_o C_h}{(C_h + C_b)C_b}}\sqrt{\frac{k-r}{k}}. \qquad \textit{(Optimum level of shortage)}$$

Characteristics of Case 3

1. Production cycle time is

$$t^o = \frac{Q^o}{r} = \sqrt{\frac{2C_o(C_h + C_b)}{C_h C_b}}\sqrt{\frac{k}{r(k-r)}}.$$

2. Maximum inventory level is

$$Q_1{}^o = \left(\frac{k-r}{k}\right)Q^o - Q_2{}^o = \sqrt{\frac{2C_b C_o}{C_h(C_h + C_b)}}\sqrt{r\left(1 - \frac{r}{k}\right)}.$$

3. Total minimum production inventory cost is

$$TC^o = \frac{1}{2Q^o} \times \left(\frac{k}{k-r}\right)(C_h Q_1^{o2} + C_b Q_2^{o2}) + \frac{r}{Q^o}C_o$$

$$= \sqrt{\frac{2C_h C_b C_o}{C_h + C_b}}\sqrt{r\left(1 - \frac{r}{k}\right)}.$$

Remarks

1. If $k \to \infty$, the problem is in complete agreement with the problem discussed in *case 1*.
2. If $C_b \to \infty$, the problem reduces to that of *Section 22:12*.
3. If $k \to \infty$ and $C_b \to \infty$, the problem reduces to that of *Section 22:9*.

Example 22.19. *The demand for a certain item is 16 units per period. Unsatisfied demand causes a shortage cost of ₹0.75 per unit per short period. The cost of initiating purchasing action is ₹15.00 per purchase and the holding cost is 15% of average inventory valuation per period. Item cost is ₹8.00 per unit. (Assume that shortages are being back ordered at the above mentioned cost). Find the minimum cost purchase quantity.*

Solution. We are given:

$$D = 16 \text{ units}, \quad C_b = ₹\,0.75 \text{ per unit}, \quad C_h = ₹\,8 \times 0.15 = ₹\,1.20, \quad C_o = ₹\,15.$$

$$\therefore \quad Q^o = \sqrt{\frac{2C_o D}{C_h}} \times \sqrt{\frac{C_h + C_b}{C_b}} = \sqrt{\frac{2 \times 15 \times 16}{1.20}}\sqrt{\frac{1.25 + 0.75}{0.75}} = 32 \text{ units (approx.)}$$

Minimum inventory cost is

$$TC = \sqrt{2DC_oC_h}\sqrt{\frac{C_b}{C_h + C_b}} = \sqrt{2\times16\times15\times1.20}\sqrt{\frac{0.75}{1.25+0.75}}$$

$$= \sqrt{576\times0.375} = \sqrt{216} = ₹\,15 \text{ approx.}$$

Example 22.20. *A dealer supplies you the following information with regard to a product dealt in by him:*

Annual demand: 10,000 units; Ordering cost: ₹10 per order; Price: ₹20 per unit.

Inventory carrying cost: 20% of the value of inventory per year.

The dealer is considering the possibility of allowing some back-order (stock-out) to occur. He has estimated that the annual cost of back-ordering will be 25% of the value of inventory.

(*i*) *What should be the optimum number of units of the product he should buy in one lot?*

(*ii*) *What quantity of the product should be allowed to be back-ordered, if any?*

(*iii*) *What would be the maximum quantity of inventory at any time of the year?*

(*iv*) *Would you recommend to allow back-ordering? If so, what would be the annual cost saving by adopting the policy of back-ordering.* **[M.Com., Delhi Univ. 1994]**

Solution. In the usual notations, we are given:

D = 10,000 units, C_o = ₹ 10 per order, C_h = 20% of ₹ 20 = ₹ 4 per unit per year, and

C_b = 25% of ₹ 20 = ₹ 5 per unit per year.

(*i*) (*a*) When stock-outs are not permitted:

$$Q^o = \sqrt{\frac{2DC_o}{C_h}} = \sqrt{\frac{2\times10{,}000\times10}{4}} = 223.6 \text{ units.}$$

(*b*) When back-ordering is permitted:

$$Q^o = \sqrt{\frac{2DC_o}{C_h}\times\frac{C_h+C_b}{C_b}} = \sqrt{\frac{2\times10{,}000\times10}{4}\times\frac{4+5}{5}} = 300 \text{ units.}$$

(*ii*) Optimum quantity of the product to be back-ordered is given by

$$Q_2{}^o = 300\times\frac{4}{4+5} = 133 \text{ units (approx.).}$$

(*iii*) Maximum inventory level = 300 – 133 = 167 units.

(*iv*) Minimum total variable inventory cost in cases (*a*) and (*b*) of (*i*) are:

$$TC\,(223.6) = \sqrt{2\times10{,}000\times10\times4} = ₹\,894.43$$

$$TC\,(300) = \sqrt{2\times10{,}000\times10\times4\times\frac{5}{4+5}} = ₹\,666.67.$$

Since, TC (223.6) > TC (666.67), the dealer should accept the proposal for back-ordering as this will result in a saving of (894.43 – 666.67), *i.e.*, ₹ 227.76 per year.

Example 22.21. *A contractor undertakes to supply diesel engines to a truck manufacturer at the rate of 25 per day. There is a clause in the contract penalizing him ₹10 per engine per day late for missing the scheduled delivery date. He finds that the cost of holding a complete engine in stock is ₹16 per month. His production process is such that each month, he starts a batch of engines through the shops, and all these engines are available for delivery any time after the end of the month. What should his inventory level be at the beginning of each month?* **[M.Sc. (Math.), CCS Univ. 2000]**

Solution. We are given,

C_h = ₹ 16 per engine per month, C_b = ₹ 10 per engine per day, r = 25 engines per day,

and t = one month (= 30 days)

The optimal value of Q_1 is

$$Q_1{}^o = rt\frac{C_b}{C_h + C_b} = 25 \times 30\frac{10}{10 + 16/30} = \frac{10 \times 25 \times 30 \times 30}{316} = 712 \text{ engines.}$$

Example 22.22. *The demand for an item in a company is 18,000 units per year, and the company can produce the items at a rate of 3,000 per month. The cost of one set-up is ₹ 500.00, and the holding cost of one unit per month is 15 paise. The shortage cost of one unit is ₹ 20.00 per month. Determine:*

(*i*) *Optimum production batch quantity and the number of shortages,*
(*ii*) *Optimum cycle time and production time,*
(*iii*) *Maximum inventory level in the cycle, and*
(*iv*) *Total associated cost per year, if the cost of the item is ₹ 20 per unit.*

[BE (Ind.), Karnataka Univ. 1994]

Solution. Here,

C_h = ₹ 0.15 per month, C_b = ₹ 20, C_o = ₹ 500, k = 3,000 units per month,

r = 18,000 units per year or 1,500 units per month.

(*i*) ∴ Optimum production batch quantity is given by

$$\therefore \qquad Q^o = \sqrt{\frac{2C_o(C_h + C_b)}{C_h C_b}} \cdot \sqrt{\frac{kr}{k - r}}$$

$$= \sqrt{\frac{2 \times 500(0.15 + 20)}{0.15 \times 20}} \cdot \sqrt{\frac{1{,}500 \times 3{,}000}{3{,}000 - 1{,}500}}$$

= 4,489 units approx.

Number of shortages is given by

$$Q_2{}^o = \frac{C_h}{C_h + C_b} Q^o \left(1 - \frac{r}{k}\right)$$

$$= \frac{0.15}{0.15 + 20} \times 4{,}489\left(1 - \frac{1{,}500}{3{,}00}\right) = 18 \text{ units (approx.)}$$

(*ii*) Optimum production time $= \dfrac{Q^o}{k} = \dfrac{4{,}489}{3{,}000} = 1.5$ months.

Optimum cycle time between set-ups

$$= \frac{Q^o}{\text{r}} = \frac{4{,}489}{1{,}500} = 3 \text{ months.}$$

(*iii*) Maximum inventory level is

$$Q_1{}^o = \frac{k - r}{k} Q^o - Q_2{}^o$$

$$= \frac{1{,}500}{3{,}000} \times 4{,}489 - 18 = 2{,}227 \text{ units (approx.).}$$

(*iv*) Total associated cost is ₹ 20 × 2,227 = ₹ 44,540.

PROBLEMS

22.41. The demand for a purchased item is 1,000 units/month, and shortages are allowed. If the unit cost is ₹ 1.50 per unit, the cost of making one purchase is ₹ 600, the holding cost for one unit is ₹ 2 per year, and the cost of one shortage is ₹ 10 per year, determine:

(*i*) The optimum purchase quantity, (*ii*) The number of orders per year,

(*iii*) The optimum total yearly cost.

22.42. The demand for product is 25 units per month and the items are withdrawn uniformly. The set-up cost each time a production run starts is ₹ 15. The inventory holding cost is ₹ 0.30 per item per month and production cost is ₹ 1 per item. Determine how often to make production run, if shortage cost is ₹ 1.50 per item per month and of what size it should be?

22.43. A manufacturer has to supply his customer with 24,000 units of his product every year. This demand is fixed and known. Since, the unit is used by the customer in an assembly operation and the customer has no storage space for units the manufacturer must supply a day's requirement each day. If the manufacturer fails to supply the required units, the shortage cost is ₹ 2 per unit per month. The inventory carrying cost is ₹ 1 per unit per month, and the set-up cost per run is ₹ 3,500. Determine the optimum run size (Q), the optimum level of inventory (S) at the beginning of any period, the optimum scheduling period, and the minimum total expected relevant yearly cost (TC). ***[MBA, Pune Univ. 2008]***

22.44. A dealer supplies you the following information with regard to an item of inventory:

Annual demand	:	5,000 units
Ordering cost	:	₹ 250 per order
Inventory-holding cost	:	30% of the value of inventory per year
Inventory stock-out cost	:	₹ 10 per unit per year
Price	:	₹ 100 per unit

Find out: (*i*) Economic order quantity, (*ii*) What quantity should be allowed to be back-ordered? (*iii*) What will be the maximum inventory at any particular time of the year?, and (*iv*) Cost of savings, if any, through back ordering. ***[M.Com., Delhi Univ. 2000]***

22.45. Suppose that Philips India has a product for which the assumptions of the inventory model with planned shortages are valid. The following data is given for the company:

Annual demand (D) = 2,000 units per year, Cost of unit (C) = ₹ 50 per unit,
Ordering cost (C_o) = ₹ 25 per order, Holding cost (C_h) = ₹ 10 per unit per year,
Back order cost (C_b) = ₹ 30 per unit per year.

Calculate: (*i*) Minimum cost order quantity, (*ii*) maximum number of back order units, (*iii*) maximum inventory level, (*iv*) time between orders, (*v*) total annual costs.

[MBA, Jammu Univ. 2009]

22.46. Higley Radio Components Company has a product for which the assumptions of the inventory model with backorders are valid. Information obtained by the company is as follows:

Annual demand = 20,000 units, Inventory carrying charge = 20 per cent per annum,
Unit cost = ₹ 50, Ordering cost = ₹ 250 per order,
Backorder cost = ₹ 30 per unit per year.

Calculate: (*i*) Economic order quantity, (*ii*) Maximum stock level, (*iii*) Maximum number of units backordered, and (*iv*) Total relevant cost per annum.

If the company decides to prohibit backorders and adopts the regular *EOQ* model, what effect will it have on order quantity, maximum stock level and total relevant cost.

[M.Com., Delhi Univ. 2009]

22.47. The demand of an item is uniform at a rate of 20 units per month. The fixed cost is ₹ 10 each time a production run is made. The production cost is ₹ 1 per item, and the inventory carrying cost is ₹ 0.25 per item per month. If the shortage cost is ₹ 1.25 per item per month, determine how often to make a production run and of what size it should be.

[MBA, Madras Univ. 1999]

22.48. A company has a demand of 12,000 units/year for an item, and it can produce 2,000 such items per month. The cost of one set-up is ₹ 400 and the holding cost per unit per month is ₹ 0.15. The shortage cost of one unit is ₹ 20 per year. Find the optimum lot size and the total cost per year, assuming the cost of 1 unit as ₹ 4. Also, find the maximum inventory, manufacturing time and total time. *[MBA, Mumbai Univ. 2010; B.Tech. (Mech.), Panjab Tech. Univ. 2012]*

22:14. MULTI-ITEM *EOQ* MODEL

When the inventories consist of several items under some ***limitations*** such as the availability of warehouse space, the total investment in inventories, or the total number of orders to be placed per year for all items, number of deliveries which can be accepted, size of delivery which can be handled, etc., then it is not possible to consider each item separately, since there exists a relation among the items. Then, the technique of Lagrange's multipliers is used to handle the simple cases.

In all such problems, first of all, we solve the problem ignoring the limitations and then consider the ***effect of limitations***.

Consider the problem of having n items in inventory, with instantaneous production and no lead time. Assume that the demand is known and is uniform at a rate of D_i per unit time for the ith item ($i = 1, 2, ..., n$). Let C_{hi} be the carrying cost or inventory holding cost per unit of the quantity of the ith item and C_{si} be the set-up cost per run for the ith item. If shortages are not allowed, then the total cost of ith item per unit of time is given by

$$\frac{1}{2} C_{hi}\, Q_i + \frac{D_i}{Q_i} C_{si},$$

where, Q_i is the quantity of ith item in inventory at the beginning of the period.

Thus, the total cost per unit of time is given by

$$TC = \sum_{i=1}^{n} \left\{ \frac{1}{2} C_{hi}\, Q_i + \frac{D_i}{Q_i} C_{si} \right\}$$

Case 1. Constraint on Warehouse Capacity (or Floor Space)

Let the storage space required per unit of time i is f_i, then total floor area (or volume) required by all inventory items must be less than or equal to total floor area (or volume) of the warehouse. Thus, the constraint is

$$\sum_{i=1}^{n} f_i\, Q_i \le W \quad \text{and} \quad Q_i \ge 0 \qquad \text{for } i = 1, 2, ..., n$$

This constraint indicates that even if all items reach their maximum inventory levels simultaneously, the warehouse will house them all. Here, it is assumed that all the items are received together. Thus, the problem becomes to minimize the total variable inventory cost for each item together under warehouse capacity constraint. Thus, we have to

$$\text{Minimize } (TC) = \sum_{i=1}^{n} \left\{ \frac{1}{2} C_{hi}\, Q_i + \frac{D_i}{Q_i} C_{si} \right\} \text{ subject to the constraint:}$$

$$\sum_{i=1}^{n} f_i\, Q_i \le W \quad \text{and} \quad Q_i \ge 0. \qquad (i = 1, 2, ..., n)$$

In order to find the optimum value for $Q_i (i = 1, 2, ..., n)$, so as to minimize (TC), we use the technique of Lagrange multiplier.

We write the Lagrangian function,

$$L(Q_i, \lambda) = \sum_{i=1}^{n} \left\{ \frac{1}{2} C_{hi} Q_i + \frac{D_i}{Q_i} C_{si} \right\} + \lambda \left\{ \sum_{i=1}^{n} f_i Q_i - W \right\},$$

where, λ is any non-negative Lagrange multilier.

The optimum values of λ and Q_i $(i = 1, 2, ..., n)$ are obtained by setting

$$\frac{\partial L}{\partial \lambda} = 0 \quad \text{and} \quad \frac{\partial L}{\partial Q_i} = 0.$$

Thus, $$\sum_{i=1}^{n} f_i Q_i - W = 0 \quad \text{or} \quad \sum_{i=1}^{n} f_i Q_i = W, \quad \text{since } \lambda \geq 0;$$

and $$\frac{1}{2} C_{hi} - \frac{D_i}{Q_i^2} C_{si} + \lambda f_i = 0 \quad \text{or} \quad Q_i^o = \sqrt{\frac{2DC_{si}}{C_{hi} + 2\lambda f_i}}.$$

Here, λ indicates an additional cost related to the storage area used by each unit of the item.

Remark. The procedure to find suitable value of λ so that it may satisfy the optimum values is given in the following steps:

Step 1. For $\lambda = 1$, compute EOQ for each item separately.

Step 2. If Q_i^o $(i = 1, 2, ..., n)$ satisfy the storage space constraint, then stop. Otherwise, go to the next step.

Step 3. For different values of $\lambda > 0$, compute EOQ for each item by systematic trial-and-error method until the additional constraint on storage space in satisfied.

Example 22.23. *A machine shop produces three products 1, 2 and 3 in lots. The shop has a warehouse, whose total floor area is 4,000 sq. metres. The relevant data for the three items is given below:*

Item	1	2	3
Annual demand (units/years)	500	400	600
Cost per unit (₹)	30	20	70
Set-up cost per lot (₹)	800	600	1,000
Floor area required (sq. metres)	5	4	10

The inventory carrying charges for the shop are 20% of the average inventory valuation per annum for each item. If no stock-outs are allowed and at no time can the warehouse capacity be exceeded, determine the optimum lot size for each item. **[BE (Ind.), Karnataka Univ. 1994]**

Solution. First of all, ignoring the restriction of total warehouse capacity, we get

$$Q_1^o = \sqrt{\frac{2 \times 500 \times 800}{0.20 \times 30}} = 365 \text{ units (approx.)}$$

$$Q_2^o = \sqrt{\frac{2 \times 400 \times 600}{0.20 \times 20}} = 346 \text{ units (approx.)}$$

$$Q_3^o = \sqrt{\frac{2 \times 600 \times 1{,}000}{0.20 \times 70}} = 292 \text{ units (approx.)}$$

Next, using these values we calculate the floor area required:

$$\sum_{j=1}^{n} Q_j^o f_j = 365 \times 5 + 346 \times 4 + 292 \times 10 = 1{,}825 + 1{,}384 + 2{,}920 = 6{,}129 \text{ sq. metres.}$$

As $\sum_{j=1}^{n} Q_j{}^o f_j > 4,000$ sq. metres, the floor area constraint is operative and Lagrange's multiplier technique has to be applied.

∴ For λ = 1, we get

$$Q_1{}^o = \sqrt{\frac{2 \times 500 \times 800}{0.20 \times 30 + 2 \times 5}} = 223 \text{ units}$$

$$Q_2{}^o = \sqrt{\frac{2 \times 400 \times 600}{0.20 \times 20 + 2 \times 4}} = 200 \text{ units}$$

$$Q_3{}^o = \sqrt{\frac{2 \times 600 \times 1,000}{0.20 \times 70 + 2 \times 10}} = 187 \text{ units}$$

and $$\sum_{j=1}^{3} Q_j{}^o f_j = 223 \times 5 + 200 \times 4 + 187 \times 10 = 3,785 \text{ sq. metres.}$$

This shows that the required space is less than the available space. Hence, we try for another value of λ. Trying for λ = 0.8, we get

$$Q_1{}^o = \sqrt{\frac{2 \times 500 \times 800}{0.20 \times 30 + 2 \times 0.8 \times 5}} = 239 \text{ units and } Q_1{}^o f_1 = 1,195 \text{ sq. metres.}$$

$$Q_2{}^o = \sqrt{\frac{2 \times 400 \times 600}{0.20 \times 20 + 2 \times 0.8 \times 4}} = 274 \text{ units and } Q_2{}^o f_2 = 856 \text{ sq. metres.}$$

$$Q_3{}^o = \sqrt{\frac{2 \times 600 \times 1,000}{0.20 \times 70 + 2 \times 0.8 \times 10}} = 200 \text{ units and } Q_3{}^o f_3 = 2,000 \text{ sq. metres.}$$

∴ $\sum Q_j{}^o f_j = 4,051$ sq. metres; which is slightly more than 4,000 sq. metres. Hence, the best value of λ should lie between 0.8 and 1.0. Trying for other values of λ in this range, we find that the most suitable value for λ is λ = 0.835, because at this value, we get $Q_1{}^o$ = 236 units, $Q_2{}^o$ = 211 units, and $Q_3{}^o$ = 197 units; and $\sum Q_j{}^o f_j = 3,994$ sq. metres which is very close to but less than 4,000 sq. metres.

It may be noted that for λ = 0.83, we get a value of $\sum Q_j{}^o f_j$ as 4,008 sq. metres, which is also very close to the given floor area of 4,000 sq. metres. However, as this value exceeds the available floor area, even though by a very small amount, this is not acceptable. Only those values of λ are acceptable which do not violate the constraint given by condition $\sum Q_j{}^o f_j \le f$.

Case 2. Constraint on Inventory Level

Since, the average number of units in the inventory of an item i is $Q_i/2$, and it is required that the total average number of units of all items held in the inventory should not exceed to number K. Therefore, we must have

$$\frac{1}{2} \sum_{i=1}^{n} Q_i \le K.$$

The problem of minimizing the total variable inventory cost under the above inventory constraint, then, is to

$$\text{Minimize } (TC) = \sum_{i=1}^{n} \left\{ \frac{1}{2} C_{hi} Q_i + \frac{D_i}{Q_i} C_{si} \right\} \text{ subject to the constraint:}$$

$$\frac{1}{2} \sum_{i=1}^{n} Q_i \le K \text{ and } Q_i \ge 0. \qquad (i = 1, 2, ..., n)$$

In order to find the optimum values for Q_i (i = 1, 2, ..., n) so as to minimize (TC), we use the technique of Lagrange multiplier.

The Lagrangian function is

$$L(Q_i, \lambda) = \sum_{i=1}^{n} \left\{ \frac{1}{2} C_{hi} Q_i + \frac{D_i}{Q_i} C_{si} \right\} + \lambda \left\{ \frac{1}{2} \sum_{i=1}^{n} Q_i - K \right\},$$

where, λ is any non-negative Lagrange multiplier.

The optimum values of λ and Q_i (i = 1, 2, ..., n) are obtained by setting

$$\frac{\partial L}{\partial \lambda} = 0 \text{ and } \frac{\partial L}{\partial Q_i} = 0.$$

Thus, we get $$Q_i^o = \sqrt{\frac{2 D_i C_{si}}{C_{hi} + \lambda}} \text{ and } \frac{1}{2} \sum_{i=1}^{n} Q_i = K.$$

Remark. The procedure to find a suitable value of 'λ' so that it may satisfy the total average inventory level is given in the following steps:

Step 1. For λ = 1, compute Q_i^o (i = 1, 2, ..., n).

Step 2. If Q_i^o (i = 1, 2, ..., n) satisfy the constraint $\sum_{i=1}^{n} Q_i^o \le K$, then stop. Otherwise, go to next step.

Step 3. Increase the value of λ, if value of LHS of constraint is more than available inventories, otherwise decrease the value of λ. Compute by systematic trial-and-error method Q_i (i = 1, 2, ..., n) until the constraint is satisfied.

Example 22.24. *A small shop produces three machine parts I, II and III in lots. The shop has limited storage space sufficient only for 500 units of all type of items. The relevant data for the three items is given below:*

Item	:	*I*	*II*	*III*
Demand rate (units/month)	:	600	1,200	1,500
Cost per unit (₹)	:	5	10	15
Set-up cost per lot (₹)	:	100	50	200

The inventory carrying charges for the shop are 20% of the average inventory valuation per month for each item. If stock-outs are not allowed, determine the optimum lot size for each item.

Solution. First of all, ignoring the restriction on inventory level, we get

$$Q_1^o = \sqrt{\frac{2 \times 600 \times 100}{0.20 \times 5}} = 346 \text{ units (approx.)}$$

$$Q_2^o = \sqrt{\frac{2 \times 1,200 \times 50}{0.20 \times 10}} = 250 \text{ units (approx.)}$$

$$Q_3^o = \sqrt{\frac{2 \times 1,500 \times 200}{0.20 \times 15}} = 447 \text{ units (approx.)}$$

$\therefore$ The average inventory level will be

$$\frac{1}{2} \sum_{i=1}^{3} Q_i^o = \frac{1}{2}(346 + 250 + 447) = 521.5 \text{ units.}$$

As $\sum_{i=1}^{3} Q_i^o > 250$ units, Lagrange's multiplier technique has to be applied.

For $\lambda = 1$, we get

$$Q_1^o = \sqrt{\frac{2 \times 600 \times 100}{0.20 \times 5 + 1}} = 245 \text{ units}$$

$$Q_2^o = \sqrt{\frac{2 \times 1,200 \times 50}{0.20 \times 10 + 1}} = 200 \text{ units}$$

$$Q_3^o = \sqrt{\frac{2 \times 1,500 \times 200}{0.20 \times 15 + 1}} = 387 \text{ units}$$

∴ Average inventory level, therefore, will be

$$\frac{1}{2} \sum_{i=1}^{3} Q_i = \frac{1}{2}(245 + 200 + 387) = 416 \text{ units.}$$

This shows that the average inventory level is less than the available storage space. So, we try for another value of λ. Trying for $\lambda = 0.1$, we get

$$Q_1^o = \sqrt{\frac{2 \times 600 \times 100}{0.20 \times 5 + 0.1}} = 316 \text{ units (approx.)}$$

$$Q_2^o = \sqrt{\frac{2 \times 1,200 \times 50}{0.20 \times 10 + 0.1}} = 230 \text{ units (approx.)}$$

$$Q_3^o = \sqrt{\frac{2 \times 1,500 \times 200}{0.20 \times 15 + 0.1}} = 435 \text{ units (approx.)}$$

∴ $$\text{Average inventory level} = \frac{1}{2}(316 + 230 + 435) = 491 \text{ units (approx.)}$$

This inventory level is very close to the available storage space. Hence, the optimum lot size of the three items are: 316 units, 230 units and 435 units respectively.

Case 3. Constraint on Investment

Inventory represents a substantial investment capital for many firms. Thus, the decision-maker places a ***limit on the amount of inventory*** to be carried. The inventory control policy must then be adjusted to meet this objective, if total investment exceeds the limit. If a monetary limit is placed on all items carried, then the constraint on investment can be expressed as

$$\sum_{i=1}^{n} k_i Q_i \le F,$$

where, k_i is the cost per unit of item i ($i = 1, 2, ..., n$), and F be total funds available to be invested on inventory.

The problem, then, becomes to minimize the total incremental cost under this budget constraints. Thus, we have to

$$\text{Minimize } (TC) = \sum_{i=1}^{n} \left\{ \frac{1}{2} C_{hi} Q_i + \frac{D_i}{Q_i} C_{si} \right\} \text{ subject to the constraint:}$$

$$\frac{1}{2} \sum_{i=1}^{n} k_i Q_i \le F, \quad \text{and} \quad Q_i \ge 0. \qquad (i = 1, 2, ..., n)$$

To find the optimum values for Q_i ($i = 1, 2, ..., n$), so as to minimize (TC), we use the technique of Lagrange multiplier.

The Lagrangian function, here, is

$$L(Q_i\,,\,\lambda) = \sum_{i=1}^{n} \left\{\frac{1}{2}C_{hi}\,Q_i + \frac{D_i}{Q_i}C_{si}\right\} + \lambda\left\{\sum_{i=1}^{n} k_i\,Q_i - F\right\},$$

where, λ is non-negative Lagrange multiplier.

The optimum values of λ and Q_i $(i = 1, 2, ..., n)$ are obtained by setting

$$\frac{\partial L}{\partial \lambda} = 0 \quad \text{and} \quad \frac{\partial L}{\partial Q_i} = 0.$$

Thus, $$\sum_{i=1}^{n} k_i\,Q_i = F \quad \text{and} \quad Q_i^{\,o} = \sqrt{\frac{2D_i\,C_{si}}{C_{hi} + 2\lambda k_i}}. \qquad (i = 1, 2, ..., n)$$

The procedure to find a suitable value of λ is same as discussed above.

Example 22.25. *Consider a shop which produces three items. The items are produced in lots. The demand rate for each item is constant and can be assumed to be deterministic. No back orders are to be allowed. The pertinent data for the items is given in the following table:*

Item	:	1	2	3
Holding cost (₹)	:	20	20	20
Set-up cost (₹)	:	50	40	60
Cost per unit (₹)	:	6	7	5
Yearly demand rate	:	10,000	12,000	7,500

Determine approximately the Economic Order Quantities when the total value of average inventory levels of these items does not exceed ₹1,000. **[M.Sc. (Math.), Rohilkhand Univ. 2001]**

Solution. By ignoring the restriction of total value for inventory level, we have

$$Q_1^{\,o} = \sqrt{\frac{2\times 10{,}000 \times 50}{20}} = 224 \text{ units (approx.)}$$

$$Q_2^{\,o} = \sqrt{\frac{2\times 12{,}000 \times 40}{20}} = 219 \text{ units (approx.)}$$

$$Q_3^{\,o} = \sqrt{\frac{2\times 7{,}500 \times 60}{20}} = 212 \text{ units (approx.)}$$

Corresponding value of average inventory at any time is

$$\frac{224}{2}\times 6 + \frac{219}{2}\times 7 + \frac{212}{2}\times 5 = \frac{3{,}937}{2} = ₹\,1{,}968.50.$$

This requirement of ₹ 1,968.50 is more than maximum fund available, *viz.*, ₹ 1,000. Therefore, the technique of Lagrange multiplier has to be applied. For $\lambda = 1$, we get

$$Q_1^{\,o} = \sqrt{\frac{2\times 10{,}000 \times 50}{20 + 2\times 1\times 6}} = 177 \text{ units (approx.)}$$

$$Q_2^{\,o} = \sqrt{\frac{2\times 12{,}000 \times 40}{20 + 2\times 1\times 7}} = 168 \text{ units (approx.)}$$

$$Q_3^{\,o} = \sqrt{\frac{2\times 7{,}500 \times 60}{20 + 2\times 1\times 5}} = 173 \text{ units (approx.)}$$

Based on these values, the investment over the average inventory at any time is given by

$$\frac{1}{2}\sum_{i=1}^{n} k_i\,Q_i = \frac{1}{2}[6\times 177 + 7\times 168 + 5\times 173] = ₹\,1{,}551.50.$$

Again, since ₹ 1,551.50 > ₹ 1,000; we try another value of $\lambda > 0$. Now, for $\lambda = 5$, we get

$$Q_1^o = \sqrt{\frac{2 \times 10{,}000 \times 50}{20 + 2 \times 5 \times 6}} = 112 \text{ units (approx.)}$$

$$Q_2^o = \sqrt{\frac{2 \times 12{,}000 \times 40}{20 + 2 \times 5 \times 7}} = 103 \text{ units (approx.)}$$

$$Q_3^o = \sqrt{\frac{2 \times 7{,}500 \times 60}{20 + 2 \times 5 \times 5}} = 113 \text{ units (approx.)}$$

Based on these values, the corresponding value of average inventory at any time is

$$\frac{1}{2} \sum_{i=1}^{n} k_i Q_i = \frac{1}{2}[6 \times 112 + 7 \times 103 + 5 \times 113] = ₹\ 979.$$

This shows that the average amount is less than the available amount. So, we try for another value of $\lambda < 5$.

For $\lambda = 4$, we shall get

$$Q_1^o = 121, \quad Q_2^o = 112, \quad Q_3^o = 123$$

and the cost of average inventory is ₹ 1,112.50.

This indicates that the average inventory cost exceeds the given amount.

Hence, the best value of λ lies between $\lambda = 4$ and $\lambda = 5$. The most suitable value of λ is 4.7, because at this value, we have

$$Q_1^o = 114, \quad Q_2^o = 105, \quad Q_3^o = 116$$

and the value of average inventory is ₹ 999.50. This is sufficiently close to the given amount of ₹ 1,000.

Case 4. Constraint on Number of Orders

There is a general approach which is not optimal, but considered 'reasonable' to find *EOQ* in case of multi-item inventory problem with a constraint on the number of orders per year. This approach can also be used, where ordering cost per order and carrying cost per unit per time period are not known. However, it works with the following assumptions:

(*a*) Ordering cost and carrying cost are same for all items, (*b*) Orders are received in lots,
(*c*) Demand is constant, and (*d*) Stockouts are not permitted.

Under these assumptions, the total number of orders per year (D/Q) for all items can be determined as under:

$$\text{Number of orders per year} = \frac{\sqrt{D \times C}}{\Sigma \sqrt{D \times C}} \times \frac{D}{Q},$$

where, D = demand in units, C = cost of units in (₹), and Q = lot size of units.

Example 22.26. *A company has to purchase four items A, B, C and D for the next year. The projected demand and unit price (in ₹) are as follows:*

Item	*Demand (units)*	*Unit price (₹)*
A	60,000	3
B	40,000	2
C	1,200	24
D	5,000	4

If the company wants to restrict the total number of orders to 40 for all the four items, how many orders should be placed for each item? ***[MBA, Delhi Univ. 2002]***

Solution. We are given the total number of orders of all the four items as 40. So, the computation of the allocation of the number of orders per year among the four items is given in the following table:

Item	*Demand 'D' (units)*	*Unit price 'C' (₹)*	$\sqrt{D \times C}$	$\frac{\sqrt{D \times C}}{\Sigma \sqrt{D \times C}}$	*Number of orders per year*
A	60,000	3	424.26	0.416	0.416 × 40 = 17
B	40,000	2	282.84	0.277	0.277 × 40 = 11
C	12,000	24	169.70	0.166	0.166 × 40 = 7
D	5,000	4	141.42	0.138	0.138 × 40 = 5
			Total: 1,018.22	—	Total: 40

PROBLEMS

22.49. Consider a shop which produces and stocks three items. The management desires never to have an investment in inventory of more than ₹ 15,000. The items are produced in lots. The demand rate for each item is constant and can be assumed to be deterministic. No back-orders are to be allowed. The pertinent data for the items are given in the following table. Determine the optimal lot size for each item:

Item	*Data for problem* 1	2	3
Demand rate (unit per year)	1,000	500	2,000
Cost per unit (₹)	20	100	50
Set-up cost per lot (₹)	50	75	100
Holding cost (₹)	20	20	20

22.50. A company producing three items has limited storage space of average 750 items of all types. Determine the optimal production quantities for each item separately, when the following information is given:

Product	1	2	3
Holding cost (₹)	0.05	0.02	0.04
Set-up cost (₹)	50	40	60
Demand rate (per unit)	100	120	75

22.51. A small shop produces three machine parts *I*, *II* and *III* in lots. The shop has only 650 sq.ft. of storage space. The appropriate data for the three items are represented in the following table:

Item	*I*	*II*	*III*
Demand rate (units/year)	5,000	2,000	10,000
Procurement cost (₹/order)	100	200	75
Cost per unit (₹)	10	15	5
Floor space required (sq.ft./unit)	0.70	0.80	0.40

The carrying charges on each item is 20 per cent of average inventory valuation per year. If no stockouts are allowed, determine the optimal lot size for each item.

22.52. A manufacturer produces and stocks three items which are produced in lots. The demand rate of each item may be considered constant and deterministic. Back orders are not permitted. The inventory holding cost is 30% and the policy is not to have an average investment in inventory above ₹ 25,000. Given the following additional data, find the optimal lot size for each item:

Item	1	2	3
Demand rate (units per year)	10,000	5,000	15,000
Cost per unit (₹)	50	100	75
Set-up cost per lot (₹)	100	150	125

22.53. Four different items are kept in a store. The demand rates are constant for the four items. Production rate is instantaneous and no shortages are permitted. The data for the problem is given in the following table:

Item	*Set-up cost (₹)*	*Demand rate (units)*	*Holding cost per unit time*	*Annual demand (units)*
1	50	20	0.2	5,000
2	100	10	0.1	10,000
3	20	10	0.1	5,000
4	90	5	0.2	7,500

Find the economic lot sizes, if the total number of orders per year for the four items is limited to 200 only.

22.54. The following data describe four inventory items. The company wishes to determine the *EOQ* for each of the four items, such that the total number of orders per year (365 days) is at most 150.

Item	*Set-up cost (₹)*	*Demand rate (units)*	*Holding cost (per unit time)*
1	100	10	0.1
2	50	20	0.2
3	90	5	0.2
4	20	10	0.1

Assuming that the only restriction is a limit of ₹ 10,000 on the amount of capital that can be invested in inventory. The purchase cost per unit of item 1, 2, 3 and 4 are ₹ 10, ₹ 5, ₹ 10 and ₹ 10 respectively.

22:15. DYNAMIC ORDER QUANTITY

Until now we have dealt with deterministic inventory problems where demand has been assumed to be uniform. But in real situations, we come across non-uniform demands such as rising (or falling) trend and/or depicting seasonal influences. When the demand is known, but have irregular lot sizes, the procedure of determining dynamic order quantity is given below:

For each month (or period) n, the condition $n^2 R_{n+1} < (C_o / C_h)$ is considered, where R_{n+1} is the requirement for the next month, C_o is the set-up (ordering) cost per production run and C_h is holding cost of a unit per unit time.

If the above condition is satisfied, n is increased by 1 to take the next month (period) into consideration. But, if the condition is not satisfied, there is an end of the particular grouping, and the following month (period) is taken as a new month 1 to proceed on in the like manner.

To ***illustrate*** the procedure, consider the following example.

Example 22.27. *The monthly requirement schedule for a product is given below:*

Month	:	1	2	3	4	5	6	7	8	9	10
Requirement (units)	:	200	300	20	140	180	360	196	200	100	270

Set-up cost is ₹360, cost of a unit is ₹15 and inventory carrying charges are 2.5% of the annual average inventory value. Determine an optimal plan of batch size.

Solution. We are given C_o = ₹ 360 and C_h = ₹ 0.025 × 15

$$\therefore \quad \frac{C_o}{C_h} = \frac{360}{0.025 \times 15} = 960.$$

Using given information, we get the following:

Month	*Requirement (units)*	*n*	$n^2 R_{n+1}$	*Is* $(n^2 R_{n+1} < 960)$?
1	200	1	300	Yes
2	300	2	80	Yes
3	20	3	1,260	No, consider n = 1 for month 4.
4	140	1	180	Yes
5	180	2	1,440	No, consider n = 1 for month 6.
6	360	1	196	Yes
7	196	2	800	Yes
8	200	3	900	No, consider n = 1 for month 9.
9	100	1	270	Yes
10	270	—	—	—

From the above table, we observe that there are four set-ups with respective sizes of 520, 320, 756 and 370 respectively. The last set-up could possibly take up some of the next year's requirements.

22:16. *EOQ* MODEL WITH UNCERTAIN DEMAND

In many practical situations, it is observed that neither the consumption rate of material (commodity) is constant throughout the year nor is the lead time. To face these uncertainties in consumption rate and lead time, an extra stock is maintained to absorb the uncertainty in both the demand as well as lead time. This extra stock is termed as ***'safety stock'***, ***'reserve stock'*** or ***'buffer stock'***. The reserve stock is maintained to take care of variation in demand during re-order period, while safety stock is maintained to take care of variation in lead time. Both reserve stock and safety stock are added to create a cushion to guard against risk of stock-outs and to provide better customer service.

The exact level of additional stock is based on the extent and nature of information available about the demand, lead time and stock-out cost. If demand remains constant and lead time is invariable, then there would be no fear of shortages and hence no need for additional stock. A fixed quantity would be reordered at fixed intervals called the ***reorder level*** (*ROL*). Here, the fluctuations in demand have no importance, but level of demand during lead time is significant.

Lead Time. The time interval between placement of an order and receipt of goods/commodity against the order is known as ***lead time***. It may be constant or not. Also, lead time may vary from stage to stage for the same item and from the same supplier. Because of this complexity, lead time is generally assume to be ***constant***.

Reorder Level. The stock level at which fresh orders should be placed with the supplier/producer for procuring additional inventory equal to the economic order quantity is known as ***reorder level (ROL)***. It is so fixed that the customers can be reasonably served from this stock until the replenishment of size 'Q' (the economic order quantity) arrives against the order placed (*Fig. 22.8*). Therefore, we have to fix the ROL with the operating policy in such a way that as soon as the stock crosses this level of order, a fresh order is placed with the vendor for additional stock sufficient to meet the demand during lead time. Although, some sophisticated reorder level formulae are available, but these are beyond the scope of this book. A simple formula to calculate the reorder level is based on the following assumptions:

(*a*) Constant daily usage of inventory, and (*b*) fixed lead time.

In other words, the formula assumes the condition of certainty. Thus,

$$ROL = \text{Average daily usage of inventory} \times \text{Lead time, } i.e., \text{ lead time demand.}$$

When safety stock is provided, then

$$ROL = DDLT + SS,$$

where, $DDLT$ denotes the demand during lead time and SS stands for safety stock.

An ideal inventory system, where the actual demand is same as the expected demand may be as depicted in *Fig. 22.9*.

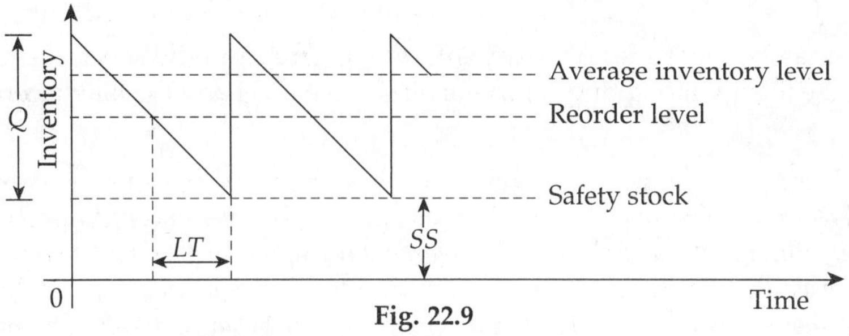

Fig. 22.9

In an ideal inventory system, we have

$$\text{Maximum inventory} = Q + SS,$$
$$\text{Minimum inventory} = SS,$$

and
$$\text{Average inventory} = \frac{1}{2}[(Q+SS)+SS] = \frac{1}{2}Q+SS.$$

Here, fresh supplies in inventory are received as soon as the stock-level reaches the safety stock level. Average stock held would be exactly equal to $\frac{1}{2}Q+SS$. However, a realistic inventory situation would look like the one shown in *Fig. 22.10*, wherein demand is not continuous and uniform. Instead, it is discrete and irregular. Displayed in the figure are three inventory cycles. In the first cycle, demand during lead time ($DDLT$) is just equal to the expected demand and as soon as the stock level reaches safety stock, fresh supplies are received. In the second cycle, $DDLT$ is more than the expected demand and, therefore, safety stock is used. Lastly, the third cycle illustrates a situation, wherein $DDLT$ is less than the expected demand.

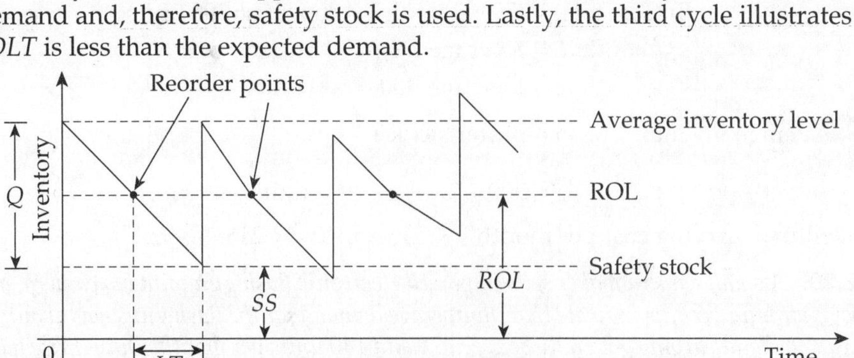

Fig. 22.10

Safety Stock. It may be noted that the *DDLT* would be sometimes more, and sometimes less than the expected demand. Therefore, an amount equal to the safety stock would always be maintained. The idea of keeping the safety stock is clearly to prevent stockouts and it is the amount of stock that an enterprise/organization would always like to preserve for meeting extraordinary situations.

Determination of Safety Stock

1. When the stockout costs are known and lead time is constant, one of the simple method for determining safety stock is to estimate the maximum lead time and the normal lead time for the item in question. Then,

   Safety stock = (Maximum lead time – Normal lead time) × Average demand

2. When stockout costs are not known, the amount of safety stock is to be kept in accordance with the level of service, the management aims to achieve. Further, the service level can be measured most commonly as follows:

$$\text{Service level } (S) = \frac{\text{Number of units supplied without delay}}{\text{Number of units demanded}}$$

The quantity may be measured over any interval of time, and the service level, *S*, would vary between 0 and 1, the former indicating complete disservice level and the later a cent percent service level.

Example 22.28. *The monthly demand for an electronic machine is approximately 600 units. Every time an order is placed, a fixed cost of ₹ 800 is incurred. The daily holding cost per unit inventory is ₹ 0.40. If the lead time is 10 days, determine EOQ, and the number of orders per month. In the past two months, the demand rate has gone as high as 50 units per day. For a reordering system based on inventory level, what should be the buffer stock? What should be the reorder level at this buffer stock? What should be the carrying cost?*

Solution. We are given,

$D = 600$ units per month, $C_o = ₹\ 80$ per order, $C_h = ₹\ 0.40$ per unit per day, and $LT = 10$ days.

$$\therefore \quad EOQ\,(= Q^o) = \sqrt{\frac{2DC_o}{C_h}} = \sqrt{\frac{2 \times 600 \times 80}{0.40}} = 490 \text{ units}$$

$$\text{Number of orders} = \frac{D}{Q^o} = \frac{600}{490} = \text{one order per month.}$$

$$\text{Average consumption} = \frac{600}{30} = 20 \text{ units per day.}$$

Since, maximum consumption is of 50 units per day,

$$\text{Buffer stock} = (50 - 20) \times LT = 30 \times 10 = 300 \text{ units}$$

$$ROL = \text{Average } DDLT + \text{Buffer stock}$$
$$= 20 \times 10 + 300 = 500 \text{ units}$$

$$\text{Average inventory level} = \text{Buffer stock} + \frac{1}{2}Q^o$$
$$= 300 + \frac{1}{2} \times 490 = 545 \text{ units}$$

Thus, inventory carrying cost per month = ₹ 545 × 0.40 = ₹ 218.

Example 22.29. *A small camera maker sells imported electronic flash gun with his camera, as an optional accessory. Last 3 month's records indicate that the average demand for the flash guns was about 100 units per month, the actual demand varying generally between 70 and 140 units per month. Only thrice had the demand exceeded 140 and was 150, 160 and 180 per month. The camera maker, by an agreement with a reliable* *overseas suppliers, receives 100 guns each month. Calculate the most economic buffer stock the supplier should hold. Assume inventory carrying charges of 20% and the landed cost of gun as ₹ 200 per unit. In case of excess demand, camera maker purchases extra units from other importers at a premium of ₹ 50 per unit.*

Solution. We are given,

D = 100 units per month, C_h = ₹ 200 × 0.20, and C_b (cost of shortage) = ₹ 50 per unit.

Let the camera-man decide to keep a buffer stock of 75 units. Then,

$$\text{Average inventory level} = \text{Buffer stock} + \frac{1}{2}Q^o$$
$$= 75 + \frac{1}{2} \times 100 = 125 \text{ units.}$$

Total average cost for 3 months = ₹ (125 + 200 × 0.20) × 3 = ₹ 15,000.

It is, now, obvious that with a starting stock of 175 units, only once in 3 months would there be a shortage (stock-out) of 5 units (when consumption rate is 180 units per month) and the same is purchased at premium of ₹ 50 each.

∴ Stock-out cost = 5 × ₹ 50 = ₹ 250.

Total inventory cost = Carrying cost + Stock-out cost
= ₹ 15,000 + ₹ 250 = ₹ 15,250.

The total inventory cost for 3 months for different levels of buffer stock is computed below:

Buffer stock (units)	*Maximum stock (BS + Q)*	*Inventory carrying cost*	*Stock-out cost*	*Total inventory cost*
75	175	₹ 15,000	₹ 250	₹ 15,250
65	165	₹ 13,800	₹ 750	₹ 14,550
60	160	₹ 13,200	₹ 1,000	₹ 14,200
50	150	₹ 12,000	₹ 1,500	**₹ 13,500**

From the above calculations, we observe that the minimum total inventory cost for 3 months is ₹ 13,500 and the corresponding most economic (optimum) buffer stock is around 50 units.

Remark. When the buffer stock maintained is very low, the inventory holding cost would be low but the shortages will occur very frequently and the cost of shortages would be very high. As against this, if the buffer stock maintained is rather large, shortages would be rather rare, resulting into low shortage costs but the inventory costs would be high. Hence, it becomes necessary to strike balance between the cost of shortages and cost of inventory holding to arrive at an optimum buffer stock termed as ***reserve stock***.

PROBLEMS

22.55. A health centre requires 2,000 units of a particular drug per month. Each unit of the drug costs ₹ 3 and the average procurement cost per order is estimated to be at ₹ 150. If the inventory carrying cost is 30% of the average inventory valuation per annum, what quantity of the drug should be ordered? If the procurement lead time is 6 days, what should reorder level be? ***[MBA (Nov.), Delhi Univ. 2006]***

22.56. The daily demand for an electronic machine is approximately 25 items. Every time an order is placed, a fixed cost of ₹ 25 is incurred. The daily holding cost per unit inventory is ₹ 0.40. If the lead time is 16 days, determine the economic lot size and the reorder point.

22.57. In an inventory model, suppose that the shortages are not allowed with production rate as infinite and

R = 600 units per years, I = 0.20, C_o = ₹ 80, P = ₹ 3, and Lead time is 1 years.

(*i*) Find the optimal order quantity, (*ii*) Reorder point, (*iii*) Minimum average yearly cost.

22.58. A company uses 10,000 units per year of an item. The purchase price is ₹ 1 per item. Ordering cost is ₹ 25 per order. Carrying cost per year is 12% of the inventory value.

(*i*) Find *EOQ*.

(*ii*) Find the number of orders/year.

(*iii*) If lead time is 4 weeks and assuming 50 working weeks per year, find the reorder point.

[BE, Annamalai Univ. 1993]

22.59. Following information in an inventory problem is available:

Annual demand	= 2,400 units	Unit price	= ₹ 2.40
Ordering cost	= ₹ 4.00	Storage cost	= 2% per year
Interest rate	= 10% per annum	Lead time	$= \frac{1}{2}$ month

Calculate: *EOQ*, reorder level, and total annual inventory cost.

How much does the total annual cost vary, if the unit price is changed to ₹ 5?

22.60. Obtain (*i*) Economic order quantity, (*ii*) Number of orders, (*iii*) Reorder level, and (*iv*) Safety stock for the following inventory problem:

Annual demand	= 36,000 units	Cost per unit	= 1
Ordering cost	= ₹ 25	Cost of capital	= 15%
Store charges	= 5%	Lead time	$= \frac{1}{2}$ month
Safety stock	= one month's consumption.		

22.61. The following relations to inventory costs have been established for *ABC* Ltd.:

(*i*) Orders must be placed in multiples of 100 units.

(*ii*) Requirement for the year are 3,00,000 units.

(*iii*) The purchase price per unit is ₹ 3.

(*iv*) Carrying cost is 25 per cent of the purchase price of goods.

(*v*) Cost per order placed is ₹ 20.

(*vi*) Desired safety stock is 10,000 units. This amount is on hand initially.

(*vii*) Three days are required for delivery.

Calculate:

(*i*) *EOQ*,

(*ii*) How many orders should the company place each year? and

(*iii*) At what inventory level should an order be placed?

22.62. Given the following data relating to one of the *A* class items, what inventory model do you suggest? What would be *EOQ*, *ROL* and the average inventory under the suggested model? Annual demand = 1,000 units. cost per item = ₹ 25, ordering cost per order = ₹ 20, and holding cost = 40%. Past lead times (days) are 10, 8, 12, 13 and 7.

22:17. SELECTIVE INVENTORY CONTROL TECHNIQUES

The inventory of an enterprise/firm generally comprises of large number of items. All items are not of equal importance to the firm in such terms as sales, profits, availability, etc. The firm, therefore, should pay more attention and care to those items whose usage value is high and less attention to those whose usage and consumption value is low. One way of exercising proper degree of control over all and various types of items held in inventory is to classify them into groups (or classes) on the basis of the degree of control.

By selectively applying inventory control policies to different groups, inventory objectives can be achieved with lower inventory levels than with a single policy applied to all the items, here, we shall consider the following techniques of selective control:

ABC, VED, SDE, HML, FNSD and *XYZ*

ABC (*Always, Better, Control*) Analysis

in this analysis, the inventory items in an organization are classified on the basis of their usage in monetary terms. The items are divided into three categories as follows:

A — high consumption value items; *B* — moderate consumption value items; and

C — low consumption value items.

The division of the items into various categories is accomplished by plotting the usage value of the items to obtain the ***ABC distribution curve*** given in *Fig. 22.11*.

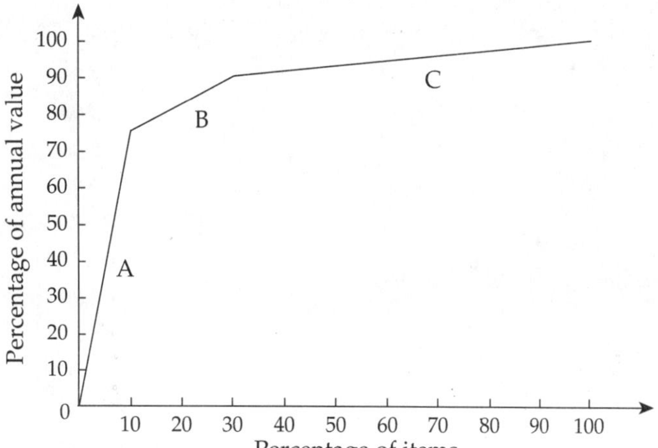

Fig. 22.11

Although, the break points between three categories, namely *A*, *B* and *C* vary according to individual business conditions, a common breakdown might be as follows:

(*i*) *A* type of items, only about 10% in number, account for 75% of the annual inventory usage value.

(*ii*) *B* type of items are the next 20% in number and account for the next 15% of the annual inventory usage value.

(*iii*) *C* type items are the next 70% in number and account only 10% of the annual usage value.

This type of classification is also known as the ***principle of law of Vital Few and Trival Many***.

The application of '*ABC*' analysis extends overall aspects of materials management like purchasing, inventory control, value analysis, etc. The items of '*A*' category require special managerial attention and, therefore, fixed-interval inventory control system might be used for these items. '*C*' category items can be managed in a more casual manner. For these items, a fixed-order quantity system might be used. The order quantities can be relatively large without incurring excessive costs. A large reserve stock can also be maintained, '*B*' category items are not so costly as to require special managerial attention, but these are also not so cheap that to ignore overstocking.

The procedure of *ABC* analysis may be summarized in the following steps:

***Step* 1.** Determine the number of units sold or consumed (used).

***Step* 2.** Determine the unit cost for each inventory item.

***Step* 3.** Compute the annual usage value of inventory items as follows:

Annual usage value = Annual usage rate × Unit cost.

***Step* 4.** Arrange the items in the descending order of usage value computed in *Step 3*.

***Step* 5.** Determine the cumulative annual usage value and the total number of items.

***Step* 6.** Express the annual usage value of each item as percentage of the total value of all items. Also compute the cumulative percentage of annual usage values.

Step 7. Obtain the percentage value for each of the items. That is, if there are 20 items involved in classification, then each item would represent 100/20 (= 5%) of the total items. Also, cummulate these percentage values.

Step 8. Draw a graph between percentage of items on x-axis and annual usage percentage of value on y-axis.

Step 9. Identify the cut-off points, where the curve sharply changes its shape. This provides three segments A, B and C.

Example 22.30. *What is selective inventory control? From the following details, draw a plan of ABC selective control:*

Item	*Units*	*Unit cost (in ₹)*
1	7,000	5.00
2	24,000	3.00
3	1,500	10.00
4	600	22.00
5	38,000	1.50
6	40,000	0.50
7	60,000	0.20
8	3,000	3.50
9	300	8.00
10	29,000	0.40
11	11,500	7.10
12	4,100	6.20

Solution. We first rank the items according to their usage value:

Item	*Units*	*Unit cost (₹)*	*Total cost (₹)*	*% of total cost*	*Ranking*
1	7,000	5.00	35,000	9.8	4
2	24,000	3.00	72,000	20.2	2
3	1,500	10.00	15,000	4.2	7
4	600	22.00	13,200	3.7	8
5	38,000	1.50	57,000	16.0	3
6	40,000	0.50	20,000	5.6	6
7	60,000	0.20	12,000	3.4	9
8	3,000	3.50	10,500	3.0	11
9	300	8.00	2,400	0.7	12
10	29,000	0.40	11,600	3.3	10
11	11,500	7.10	81,650	23.0	1
12	4,100	6.20	25,420	7.1	5
		Total	₹ 3,55,770	100.0	

We note that items ranked (1), (2) and (3) have ***large*** usage value, those ranked (4), (5), (6) and (7) a ***moderate*** usage value and those ranked (8) to (12) a ***small*** usage value. It will, therefore, be appropriate to classify these as "A", "B" and "C" items respectively . We can now compute the percentages of items and total usage value in each class as shown in the table below:

Original item no.	*Order of ranking*	*Number of items*	*Percentage of items*	*Value (₹)*	*Cumulative value (₹)*	*Cumulative %*	*% of total value*	*Category*
11	(1)			81,650	81,650	23.0		
2	(2)	3	25.0%	72,000	1,53,650	43.2	59.2%	*A*
5	(3)			57,000	2,10,650	59.2		
1	(4)			35,000	2,45,650	69.0		
12	(5)	4	33.3%	25,420	2,71,070	76.2	26.8%	*B*
6	(6)			20,000	2,91,070	81.8		
3	(7)			15,000	3,06,070	86.0		
4	(8)			13,200	3,19,270	89.7		
7	(9)			12,000	3,31,270	93.1		
10	(10)	5	41.7%	11,600	3,42,870	96.4	14.0%	C
8	(11)			10,500	3,53,370	99.3		
9	(12)			2,400	3,55,770	100.0		
Total		12		₹ 3,55,770			100%	

The cut-off points have been determined from the following graph:

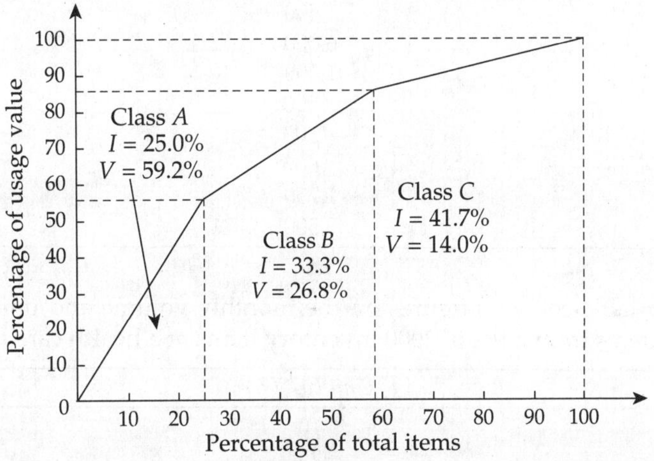

Fig. 22.12

PROBLEMS

22.63. Ten items kept in inventory by the School of Management Studies at State University are listed below. Which items should be classified as *'A'* items, *'B'* items and *'C'* items? What percentage of items is in each class? What percentage of total annual value is in each class?

Item	*Annual usage*	*Value per unit (₹)*	*Item*	*Annual usage*	*Value per unit (₹)*
1	200	40.00	6	1,200	0.80
2	100	360.00	7	120	100.0
3	2,000	0.20	8	2,000	0.70
4	400	20.00	9	1,000	1.00
5	6,000	0.04	10	80	400.00

[BFIA, Delhi Univ. 2004]

22.64. The following information is known about a group of items. Classify the materials in *A, B, C* classification:

Model number	*Annual consumption in pieces*	*Unit price (in paise)*
501	30,000	10
502	2,80,000	15
503	3,000	10
504	1,10,000	5
505	4,000	5
506	2,20,000	10
507	15,000	5
508	80,000	5
509	60,000	15
510	8,000	10

22.65. A corporation has the following 10 items in inventory. Based on the following data, classify these items under *A, B* or *C* categories. Represent the findings graphically.

Item	*Annual consumption*	*Unit cost (₹)*
A2	3,000	50
B8	4,000	12
C7	1,500	45
D1	6,000	10
E9	1,000	20
F3	500	500
G2	300	1,500
H2	600	20
I5	1,750	10
J8	2,500	5

[MBA, Delhi Univ. 2012]

22.66. The following table contains figures on the monthly volume and unit costs for a random sample of 16 items from a list of 2000 inventory items at a health care facility:

Item	*Unit cost (₹)*	*Usage*
P1	10	200
P2	25	600
P3	36	150
P4	16	25
P5	20	80
P6	80	200
P7	20	300
P8	30	800
P9	20	60
P10	10	550
P11	12	90
P12	15	110
P13	40	120
P14	30	40
P15	16	500
P16	10	30

(*i*) Develop an *ABC* classification for these items.

(*ii*) After reviewing your classification scheme, suppose the manager decides to place item *P16* into *A* category. What are some possible explanations for this decision?

[MBA (HCA) (Nov.), Delhi Univ. 2009]

VED Analysis

In this analysis, the items are classified on the basis of their criticality, *i.e.*, according to the cost of incurring a stock out. The '*V*' class items are ***vital*** without which the production process would come to standstill. '*E*' class items are ***essential*** and should be given second priority, as their non-availability would adversely affect the efficiency of the production system. The '*D*' class of items are ***desirable*** without which the production process is unaffected, but it is good, if they are available for the state of efficiency and less fatigue.

The *VED* analysis is done mainly in respect of spare parts. It can also be used in case of such raw material items whose availability is rare.

ABC analysis and *VED* analysis can also be combined to control the stocking of spare parts based on the desired customer service level as shown in the following matrix:

Item:	*V*	*E*	*D*
A	50%	Low	Very low
B	75%	50%	Low
C	95%	75%	50%

The above matrix shows the percentage service level to aim for in making arrangement for provisioning of items.

SDE Analysis

This type of analysis is useful in the study of those items which are ***scarce*** in availability. The "*S*" class items are ***scarce*** items, *e.g.*, imported items which are generally in short supply. The "*D*" class stands for ***difficult*** items, which are available in the market but not always traceable or immediately supplied, and "*E*" class items are ***easily*** available in the market.

HML Analysis

This type of analysis is similar to "*ABC*" analysis, except that ***cost*** per item (per unit) is taken. maximum or ***highest*** (*H*) costs are given top priority, then "*M*" (the ***medium*** cost items) and finally "*L*"(the ***low*** cost items). This usually helps to control over consumption at departmental level.

FNSD Analysis

In this type of analysis the items are classified in the descending order of their usage (movement) value, namely "*F*" class standing for ***fast*** moving items, "*N*" class for ***normal*** moving items, "*S*" class for ***slow*** moving items and "*D*" class for ***dead*** items, which are transferred to ***disposal*** cell. This method is used to combat ***obsolescence*** in all types of inventory. The analysis also helps in arranging stocks in the stores and deciding the distribution of inventory items.

XYZ Analysis

It is based on the closing inventory value of different items. Those items, whose inventory values are high, are classified as *X* items, while those with low investment in them are termed as *Z* items. Other items are the *Y* items whose inventory value is neither too high nor too low.

It can be easily visualized that the various types of analysis discussed are not mutually exclusive. They can be, and often are, used jointly to ensure better control over materials. For example, *ABC* and *XYZ* analysis may be combined to classify and control depending on whether the items are *AX*, *BY*, *CZ*, *AY*, ..., etc. Similarly, *XYZ-FNSD* combine classification exercise will help in timely prevention of obsolescence.

Summary of Selective Control Techniques

Different selective control techniques are suitable in different situations. It depends upon the nature of the inventories carried by an organization to decide which selective control technique must be chosen. For example, a spares consuming organization, like Indian Oil, would rely more on the *VED* classification. Moreover, as discussed earlier, sometimes it is more advantageous to combine two selective control methods. For instance, the *XYZ-FNSD* combine reduces obsolescence and the *ABC-VED* combine is useful in the case of spare parts, etc.

In the following table, we summarize the four selective control techniques, their basis of classification and their chief uses:

Selective control Techniques	*Basis of classification*	*Chief uses*
ABC	Consumption value	Controlling raw material components and work-in-progress inventories
VED	Criticality of the item	Determining the inventory levels of spare parts
XYZ	Value of items in storage	Reviewing the inventories and other uses
FNSD	Consumption rate (movement) of the item	Controlling obsolescence
SOS (Season and off season)	Nature of suppliers	Procurement and holding strategies for seasonal items like agricultural products.

TEST YOUR UNDERSTANDING

1. Inventories in general are build up to
   (*a*) satisfy demand during period of replenishment.
   (*b*) carry reserve stocks to avoid shortages.
   (*c*) keep pace with changing market conditions.
   (*d*) all of the above.
2. Economic order quantity (*EOQ*) results in
   (*a*) equalization of carrying cost and procurement cost,
   (*b*) minimization of set-up cost,
   (*c*) favourable procurement price,
   (*d*) reduced chances of stock outs.
3. If orders are placed with size determined by *EOQ*, then the re-order costs component is
   (*a*) greater than the holding cost component,
   (*b*) equal to the holding cost component,
   (*c*) less then the holding cost component,
   (*d*) either greater than or less than the holding cost component.
4. If small orders are placed frequently (rather than placing large orders infrequently), then total inventory cost is
   (*a*) reduced, (*b*) increased,
   (*c*) either reduced or increased, (*d*) minimized.

5. If *EOQ* is calculated, but an order is placed which is smaller than *EOQ*, then total inventory cost
   (*a*) decrease, (*b*) increase,
   (*c*) either decrease or increase, (*d*) no change.
6. Which costs can vary with order quantity?
   (*a*) unit cost only, (*b*) holding cost only,
   (*c*) re-order cost only, (*d*) all of these.
7. If the procurement cost used in the formula to compute *EOQ* is half of the actual procurement cost, the *EOQ* so obtained will be
   (*a*) half of *EOQ*, (*b*) 0.707 time *EOQ*,
   (*c*) 0.606 time *EOQ*, (*d*) one forth of *EOQ*.
8. If *EOQ* is calculated, but is found to be inappropriate size, would the total cost per unit
   (*a*) rise slowly around *EOQ*, (*b*) rise quickly around *EOQ*,
   (*c*) fall slowly around *EOQ*, (*d*) fall quickly around *EOQ*.
9. If the unit cost rises, will optimum order quantity
   (*a*) increases, (*b*) decreases,
   (*c*) either increase or decrease, (*d*) none of the above.
10. If the total investment in stock is limited, will the best order quantity for each item be
    (*a*) greater than *EOQ*, (*b*) equal to *EOQ*,
    (*c*) less than *EOQ*, (*d*) either greater or less than *EOQ*.
11. If we find a minimum on a total cost curve with discounted unit cost, then the optimum order size is
    (*a*) valid at this minimum, (*b*) at (or to) the left of this minimum,
    (*c*) at (or to) the right of this minimum, (*d*) anywhere.
12. When compared to instantaneous replenishment, does a finite replenishment rate lead to
    (*a*) the same size of batches, (*b*) smaller batches,
    (*c*) larger batches, (*d*) none of the above.
13. Operating decisions in an inventory system are concerned with
    (*a*) order quantity, (*b*) re-order level,
    (*c*) customer service level, (*d*) all of the above.
14. One of the important reason for carrying inventory is to
    (*a*) improve customer service, (*b*) get quantity discounts,
    (*c*) maintain operational capability, (*d*) all of the above.
15. Which of the following is correct?
    (*a*) In *ABC* analysis, the inventory items are classified into three categories on the basis of their usage values.
    (*b*) *HML* classifies items according as they are of high usage, medium usage or low usage.
    (*c*) *VED* analysis deals with classification of items on the basis of their availability.
    (*d*) *XYZ* analysis is based on the classification of items according to their unit cost.

ANSWERS

1. (*d*)	2. (*a*)	3. (*b*)	4. (*c*)	5. (*d*)
6. (*d*)	7. (*b*)	8. (*a*)	9. (*b*)	10. (*c*)
11. (*c*)	12. (*b*)	13. (*d*)	14. (*a*)	15. (*a*)

REVIEW EXERCISE

1. What is 'Inventory control'? List the relevant costs with brief explanations used in the development of inventory models. *[PG Dip. Glob. Bus. Oper., Delhi Univ. 2010]*
2. Describe the basic characteristics of an inventory system. *[MBA (Dec.), Panjab Tech. Univ. 2010]*
3. Describe briefly the *EOQ* concept. What are its limitations? Discuss. *[M.Com., Bangalore Univ. 2018]*
4. Give any four assumptions of classical *EOQ* model of Inventory Control. *[M.Com., Bangalore Univ. 2018; MIB, Delhi Univ. 2011]*
5. What are the costs associated with inventory? Distinguish between deterministic and stochastic models in inventory theory. *[BE (Mech.), Panjab Tech. Univ. 2011]*
6. What are the components of inventory cost? *[MBA, Madras Univ. 2012]*
7. Define the terms — *set-up cost, holding cost* and *shortage* or *penalty cost* as applied to an inventory problem. *[B.Tech. (Mech.) & BCA, Panjab Tech. Univ. 2011; M.Tech., Dayal Bagh Educ. Inst. 2012]*
8. Derive the *EOQ* formula for the manufacturing model without shortages. *[MBA, Panjab Tech. Univ. 2012; B.Tech. (Comp. Sc.), Hyderabad Univ. 2011]*
9. Prove that in the inventory problem of Economic Lot Size with uniform demand and unequal times of production run, the optimal lot size for each production run is given by $Q^o = \sqrt{2DC_o / C_h}$, and the optimal total cost is given by $\sqrt{2DC_h C_o}$, where D denotes the total number of units produced per unit time, C_o is the set-up cost per production run and C_h is the holding cost per unit of inventory per unit time. Production is assumed to be instantaneous and shortage cost is infinite. *[B.Tech. (Mech.) (Dec.), Panjab Tech. Univ. 2011]*
10. Derive *EOQ* formula for made in items, when lead time is zero and shortages are not allowed. *[MCA (Feb.), Hyderabad Univ. 2012]*
11. How does the basic inventory model change, when usage of inventories during their build-up is allowed, when stockouts are permitted and also when price breakups are available? *[PG Dip. in Glob. Bus. Oper., Delhi Univ. 2011]*
12. What do you mean by total relevant cost? What are the components of total revenue cost in case of stock out model of inventory control? *[MIB, Delhi Univ. 2011]*
13. "Although the classical inventory decision model, known as the *EOQ* model, is too oversimplified to represent many of the real-world situations, it is an excellent starting point to develop more realistic and complex inventory decision models." In the light of this statement, explain the major limitations of the *EOQ* model. Also, explain the major techniques developed by management scientists to overcome these. *[M.Com., Delhi Univ. 2009]*
14. Discuss *EOQ* model with price breaks. How do you determine optimum order quantity when only one price-break is available? Illustrate your result with the help of a graph. *[B.Sc. (Stat.), Delhi Univ. 2003; M.Sc. (Math.), Kerala Univ. 2011]*
15. Derive an inventory model with one price break and obtain the decision rule for finding optimal order quantity. Also, extend the decision rule for two price breaks. *[B.Sc. (Stat.), Delhi Univ. 1998]*
16. Describe the single item static model with any number of price breaks. *[M.Sc. (Stat.), Calicut Univ. 1995]*
17. Derive *EOQ* model for deterministic demand when replenishment rate is infinite and shortages are permitted. *[B.Tech., Hyderabad Univ. 2011]*
18. Write short note on inventory control techniques. *[B.Tech. (Comp. Sc.) (Dec.), Hyderabad Univ. 2011]*

19. What are the assumptions underlying the *EOQ* model, when stockouts are permitted? Show the inventory profile under this model. *[PG Dip. in Glob. Bus. Oper., Delhi Univ. 2011]*

20. (*a*) List down the important characteristics of *EOQ*. *[MBA (Dec.), Panjab Tech. Univ. 2011]*
    (*b*) What are the major weaknesses of *EOQ* model? *[MBA (Dec.), Panjab Tech. Univ. 2012]*

21. Explain the following terms in inventory control models:
    (*i*) Lead time, (*ii*) Reorder point, (*iii*) Safety stock. *[B.Tech., Hyderabad Univ. 2011]*

22. Explain the significance of lead time and safety stock in inventory control.
    [B.Sc. (Stat.), Delhi Univ. 2006; B.Sc. (Math.) (Nov.), Madras Univ. 2012]

23. What is the purpose of safety stock in inventory decisions? *[MBA, Panjab Tech. Univ. 2011]*

24. Explain buffer stock. *[BE (Mech.), Madras Univ. 1999]*

25. "*ABC* analysis is very useful approach for selective inventory control, but has some major limitations." Do you agree with this statement? Explain how these limitations, if any, can be removed. *[MBA, Delhi Univ. 2009]*

26. Explain *ABC* analysis. What are its advantages and limitations, if any?
    [BE (Dec.), Gujarat Univ. 2012; PG Dip., Indian Inst. of Mat. Man. 2019]

27. What are the limitations of *ABC* analysis? Explain how these limitations may be overcome by incorporating other selective inventory control techniques. *[MBA (Nov.), Delhi Univ. 2004]*

28. Explain *ABC* and *VED* inventory classification systems and indicate their advantages. How can these two systems be combined? *[MBA (PT.), Delhi Univ. 2008]*

29. With the help of an example, explain *ABC* analysis. How is it different from *VED* analysis and *FSN* analysis? Briefly discuss *MUSIC-3D* with reference to selective inventory control.
    [MBA (Nov.) Delhi Univ. 2009]

30. Compare among *ABC, VED* and *FSN* analysis of selective inventory control.
    [MBA, Delhi Univ. 2011]

CHAPTER 23

Inventory Management — Probabilistic

"Look for certain stores amid uncertain demands"

In the previous chapter 22, we discussed the deterministic inventory problems based on the assumption of constant and known demand for an item, as well as known lead time. When the demand or the lead time or both are not known with certainty and they need not be constant, then their random behaviour needs to be described with a known probability distribution (discrete or continuous).

Probabilistic inventory control problems can be classified into ***Static Demand Problems*** and ***Dynamic Demand Problems***. In this chapter we shall be dealing with the static demand probability problems only which fall in to following three categories:

1. Instantaneous demand without set-up cost.
2. Instantaneous demand with set-up cost.
3. Continuous demand without set-up cost.

The replenishment lead time and the cost of shortage during sales period are assumed to be known, while demand is represented by a discrete probability distribution.

23:1. SYSTEMS OF INVENTORY CONTROL

Basically, there are two types of inventory control systems, although both have numerous variations: One is known as "***Fixed order quantity system*** (Q-system) and the other "***Fixed order interval system***" (or Periodic review system or P-system).

1. Fixed Order Quantity System (Q-System)

This is also known as "***Two-bin System*** or ***Maximum-minimum System***". In this system, a reorder is placed when the stocks fall to a 'reorder level' which is equal to the lead time requirements plus buffer stocks. In the two-bin system, the review period is always fixed. Further, the ***bin*** of an item is of two types, namely the ***main-bin*** and ***reserve-bin***. From main-bin, we meet the demand that occurs before lead time and from the reserve-bin we meet the demand during lead time. When the main-bin is empty, a fresh order is placed which should arrive well before the reserve-bin is exhausted. The cycle of procurement and consumption is shown in *Fig. 23.1:*

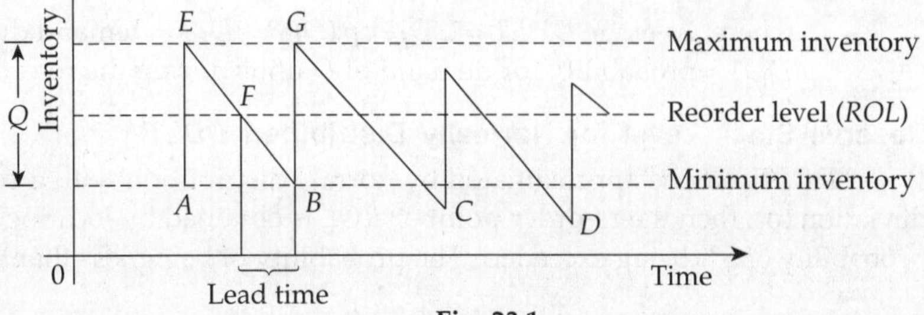

Fig. 23.1

In *Fig. 23.1*, it is shown that a supply equal to *EOQ* is received at point *A* and the quantity in stock reaches a point *E*. The materials (units of quantity) are then issued and at time *F*, when the stock reaches the *ROL*, an order is placed for quantity *Q* (= *EOQ*), and the issues continued. At point *B*, the supplies of order placed at point *F* are received and the stock (inventory) reaches point G. At point C there is a delay in receiving the supplies and we cut into the ***buffer stock***. Similarly, at point *D* due to heavy consumption we cut again into the buffer stock. Thus, for this system of ordering, we fix up the size of the order, *i.e.*, every time same quantity '*Q*' is ordered, but the time of placing the order is allowed to vary depending upon the actual usage or demand.

The probability of this variation in demand is controlled by raising and lowering the re-order level (*ROL*).

The procedure for determining an optimum solution is as under:

Step 1. Compute Q^0 based on the assumption of the fundamental problem of *EOQ*, *i.e.*, $Q^0 = \sqrt{2DC_o / C_h}$.

Step 2. Determine *ROL* so as to trade-off between stock-out (shortage) cost and carrying cost for the reserve stock.

The reserve stock is the number of units by which *ROL* is increased above the average demand during lead time (*DDLT*) to balance the shortage cost. However, if *DDLT* as well as lead time are normally distributed, then

$$ROL = \text{Average } DDLT + \text{Reserve stock} + \text{Safety stock},$$

where, Reserve stock = Standard deviation of *DDLT* × Service level factor,

and Safety stock = Average demand during maximum delay in lead time × Probability of delay.

The fixed order quantity system is also known by other names such as ***'perpetual inventory system'*** or ***'re-order point inventory system'*** or ***'(Q, R) system'***.

Determination of Reserve Stock

When the stock-out costs are known as well as unknown for a given customer service level and the lead time is constant, a discrete probability distribution is used to describe the demand during re-order lead time. The procedure may be summarized in the following steps:

Step 1. Determine the relative frequencies of the demand level during re-order lead time. If demand follows a continuous probability distribution, then convert it into discrete one by taking mean value of each class of the distribution. Otherwise, develop the discrete probability distribution of demand during the re-order lead time.

Step 2. Determine optimum re-order level in terms of least expected inventory cost as: Carrying cost + Shortage cost.

Step 3. Determine optimum reserve stock level as: Re-order level – Average *DDLT*,

where, Average $DDLT = \Sigma D_{LT} \cdot p(D_{LT})$, D_{LT} = demand during re-order lead time, and $p(D_{LT})$ = probability for demand of *D* units during the re-order lead time.

Reserve Stock Level for Normally Distributed *DDLT*

When *DDLT* may be approximated by normal distribution with a certain mean (μ) and standard deviation (σ), then the re-order point or *ROL* is obtained by locating the value of *DDLT* that has a probability of 5% being exceeded. The probability of being less than this *DDLT* level is 95%, when the desired service level is 95%. The reserve stock level would then be obtained by subtracting the average *DDLT* from this value. The concept is illustrated in *Fig. 23.2*:

$$\text{Reserve Stock } (RS) = ROL - \bar{D}_{LT} = X - \mu$$

$$\therefore \quad Z = (X - \mu)/\sigma = RS/\sigma \quad \text{or} \quad RS = Z\sigma,$$

where, X = Mean demand + Reserve stock.

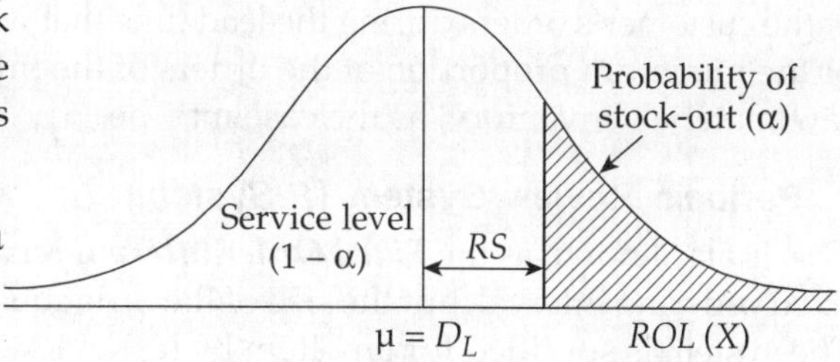

Fig. 23.2

Example 23.1. *The demand for an item during the re-order period is known to be normally distributed with mean = 600 units and standard deviation = 50 units. What service level can the firm expect to offer (a) when it has 80 units of reserve stock, and (b) when it provides for average demand only? Also, determine the reserve stock if it is sought to provide a 99% service level to the customers.* **[MBA (Nov.), Delhi Univ. 2003]**

Solution. We are given:

$\mu = 600$ units, $\sigma = 50$ units, and $X = 680$.

This information is depicted in *Fig. 23.3*.

(*a*) To determine the required probability, we shall find the area to the left of X = 680. We have

$$Z = \frac{X - \mu}{\sigma} = \frac{680 - 600}{50} = 1.6$$

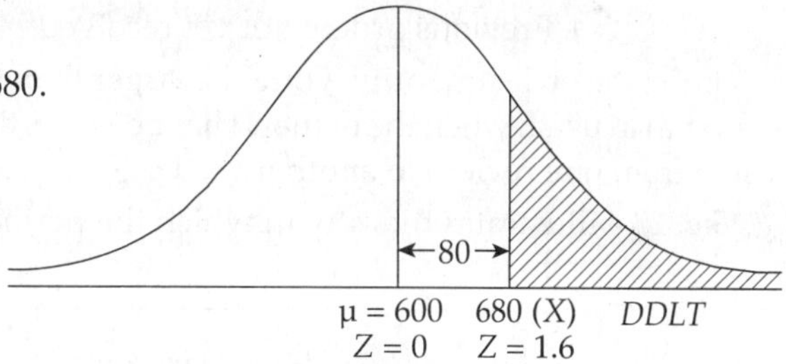

Fig. 23.3

From the normal area tables, the area between 0 and Z (= 1.6) is $\Phi(1.6) = 0.4452$.

∴ Area to the left of X (= 680) is equal to 0.5 + 0.4452, *i.e.*, 0.9452.

Thus, service level at 80 units of *RS* = 94.52%.

(*b*) When provision is made for average demand only, the service level is naturally 50%.

Now, since Z-value corresponding to 99% service level is 2.33, we have

$$RS = 2.33 \times 50 = 116.5 \text{ or } 117 \text{ units (approx.)} \quad \text{(because reserve stock} = Z\sigma)$$

Reserve Stock Level when Stock-out Costs are not Known

When the stock-out costs are not known, the amount of safety stock is to be kept in accordance with the level of service the management aims to achieve.

The service level can be measured in several ways. One of the methods of measuring service level is:

$$\text{Service level} = 1 - \text{Probability of a stock-out}$$

The service level is measured from 0 (no service) to 1 (hundred per cent service).

In order to determine the reserve stock level, it is necessary to know the probability of demand during lead time and desired service level. Setting a particular service level would mean the firm will not be out of stock of a particular item for that per cent of the time. For example, setting 85% service level means that the firm has committed to take a risk of stocking out only (100 – 85) = 15% of the time.

The service level can also be measured by using,

$$\text{Service level} = \frac{\text{Number of units supplied without delay}}{\text{Number of units demanded}}$$

A minimum service level in respect of *DDLT* may be established in advance on the basis of which the safety stock level would be fixed. This service level would mean the minimum proportion of the customer's orders during the lead time that would be filled out of the finished goods inventory, or the minimum proportion of the orders of the production department which shall be filled out of raw materials inventory, as the case may be.

2. Periodic Review System (*P-System*)

This is also known as the ***Fixed Order Interval System*** or ***Cyclic Review System***. This system has a fixed ordering interval, but the size of the order quantity may vary with the fluctuations in demand. The system is specified for any item by (*i*) Review period, and (*ii*) Replenishment level (R).

The operating procedure consists of reviewing the inventory position regularly, once in every t units of time. At each review period, an order is placed for an amount equal to the difference between a fixed replenishment level and the actual inventory level. Thus, the order quantity at the end of lead time is:

"Maximum inventory (replenishment) level – Inventory level on hand
+ Previous orders not yet received"

Since, the order quantity would be larger than usual, when the demand has been greater; and is smaller than usual when the demand has been less than the expectation, the order quantity is ***variable*** in size from one review to another.

Fig. 23.4 illustrates the way in which the periodic review system operates:

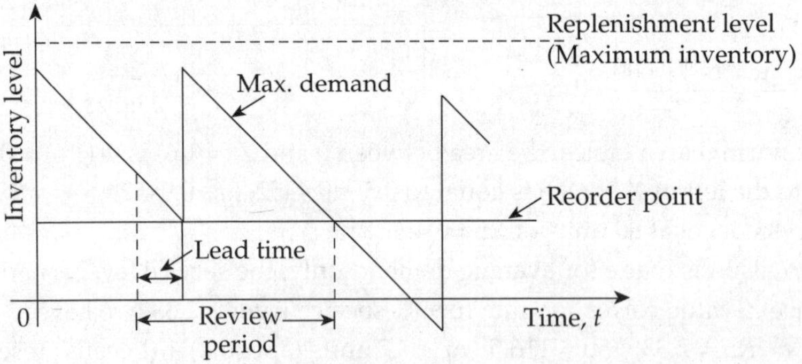

Fig. 23.4

The procedure for determining an optimum solution is as under:

Step 1. Compute economic review period between successive orders based on the assumptions of the fundamental problem of *EOQ*, *i.e.*, $t^0 = \sqrt{2C_0 / DC_h}$.

Step 2. Calculate the maximum inventory level so as to trade-off between stock-out (shortage) cost and carrying cost for the reserve stock.

In this case the maximum inventory level must satisfy the *DDLT* plus the review period '*t*'. A stockout will occur when average demand during ($t + LT$) is more than maximum inventory level. Reserve stock, therefore, will be the number of units by which maximum inventory level increased above the average demand during ($t + LT$).

The periodic review system is also known by other names, such as '***replenish system***' or '***periodic inventory system***', or '***P-system***'.

Comparison between *Q*-system and *P*-system

Following are the salient features in comparing the two systems:

1. In *Q*-system, there is continuous surveillance of the stock level. The moment the stock level on hand touches *ROL*, order equal to *EOQ* is placed. In *P*-system, the stock level is checked

at fixed regular interval and the order is placed for a quantity to replenish to the maximum pre-determined level.

2. In *Q*-system, the probability of stock-out is low, whereas in *P*-system the probability of stock-out could be higher.
3. The average inventory is less as compared to *P*-system, whereas it is expressed to be more comparable to *Q*-system.
4. *Q*-system is best suited, where continuous monitoring is possible. Also, it is well suited for bought out items. On the other hand, *P*-system is suited for non-critical items and easily available.
5. *Q*-system is used for each item individually and, therefore, the number of orders is expected to be high. On the other hand, *P*-system is ideally suited for joint replenishment orders.
6. In *Q*-system, the suppliers will not know exactly when the orders will be placed. Therefore, the supplier is forced to hold more inventory. In *P*-system, the suppliers will know when the orders would be placed. Accordingly the production planning is made. Thus, the inventory is expected to be low.

Example 23.2. *For a fixed order quantity system, find out the various parameters of an item with the following data:*

Annual consumption (D) = 10,000 units; Cost of one unit = ₹1.00;
C_o = ₹12 per production run; C_h = ₹0.24 per unit;
Past lead times: 15 days, 25 days, 13 days, 14 days, 30 days, 17 days.

Solution.

(*i*) $$Q^o = \sqrt{\frac{2 \times 10{,}000 \times 12}{0.24}} = ₹\,1{,}000 \text{ units.}$$

(*ii*) Maximum lead time = 30 days,
Average lead time = (15 + 25 + 13 + 14 + 30 + 17)/6 = 19 days
Demand rate = 10,000/360 = 27.8 units per day. (assuming 360 working days in a year)
Optimum buffer stock = (30 – 19) × 27.8 = 306 units (approx.)
Reserve stock = 19 × 27.8 = 528 units (approx.)
∴ *ROL* = 306 + 528 = 834 units.

(*iii*) Average inventory $= 306 + \frac{1}{2} \times 1{,}000 = 806$ units,
Maximum inventory = 306 + 1,000 = 1,306 units,
Minimum inventory = 306 + 0 = 306 units.

Example 23.3. *A company uses annually 50,000 units of an item each costing ₹1.20. Each order costs ₹45, and inventory carrying costs are 15% of the annual average inventory value.*

(*i*) *Find EOQ.*
(*ii*) *If the company operates 250 days a year, the procurement time is 10 days and safety stock is 500 units, find reorder level, maximum, minimum and average inventory.*

Solution. We are given
D = 50,000 units, C_o = ₹ 45 per order, C_h = ₹ 1.20 per unit, and I = ₹ 0.15.

(*i*) $$Q^o = \sqrt{\frac{2 \times 45 \times 50{,}000}{1.20 \times 0.15}} = 5{,}000 \text{ units.}$$

(*ii*) The company operates 250 days a year

$$\therefore \quad \text{Requirement per day} = \frac{50,000}{250} = 200 \text{ units,}$$

$$\text{Reserve stock} = 10 \times 200 = 2,000 \text{ units,}$$

$$\text{Safety stock} = 500 \text{ units.}$$

$$\therefore \quad ROL = 2,000 + 500 = 2,500 \text{ units.}$$

$$\text{Maximum inventory} = 5,000 + 500 = 5,5000 \text{ units.}$$

$$\text{Minimum inventory} = 5,000 \text{ units.}$$

$$\text{Average inventory} = \frac{1}{2} \times 5,000 + 500 = 3,000 \text{ units.}$$

Example 23.4. *Consider an item for which the following data are available:*

Annual average demand = 10,000 units; Standard deviation of demand per week = 50 units; Unit cost = ₹6; Ordering cost = ₹10 per order; Inventory carrying cost = 30 per cent; Average lead time = 4 weeks; Maximum delay = 3 weeks; Probability of delay = 0.20; Service level = 95 per cent.

Determine the buffer stock, reserve stock, safety stock and desirable maximum inventory level for this item.

Solution. Using the given information,

$$Q^o = \sqrt{\frac{2 \times 10,000 \times 10}{6 \times 0.30}} = 333 \text{ units (approx.)}$$

Buffer stock = (10,000/52) × 4 = 769 units (approx.) (one year = 52 weeks)

Safety stock = (10,000/52) × 3× 0.30 = 173 units (approx.)

Reserve stock $= 1.65 \times 50 \times \sqrt{4} = 165$ units, since, $P(z \leq 1.65) = 0.95$.

$\therefore$ ROL = 769 + (173 + 165) = 1,107 units.

Further, review period = (333/10,000) × 52 = 1.73 weeks or 2 weeks approx.

Since, the review period is of 2 weeks, we compute

Buffer stock = (10,000/52) × (2 + 4) = 1,153.8 units

Safety stock = (10,000/52) × 3× 0.30 = 173 units

Reserve stock $= 1.65 \times 50 \times \sqrt{6} = 202$ units

$\therefore$ Maximum inventory level = (10,000/52)(2 + 4) + (173 + 202)

= 1,528.8 or 1,529 units (approx.)

Example 23.5. *A manufacturing company requires a components at the annual average rate of 1,000 units. Placing an order costs ₹480 and has a 5-day lead time. Inventory holding cost is estimated at ₹15 per unit per year. The plant operates 250 days/year. It is assumed that daily demand is normally distributed with an average of 4 units and a standard deviation of 1.2 units.*

Suggest an inventory policy to control the inventory of the item based on a 95 per cent service level.

[MBA, Delhi Univ. 2002]

Solution. We are given

D = 1,000 units per year or 4 units per day, C_o = ₹ 480 per order, C_h = ₹ 15 per unit per year, $\sigma_D = 1.2$ units per day, $LT = 5$ days, and $\alpha = 5\%$ (95% service level).

Now, $\quad \sigma_{D_{LT}}{}^2 = \sigma_D{}^2 \times LT \quad$ or $\quad \sigma_{D_{LT}} = \sigma_D \times \sqrt{LT} = 1.2 \times \sqrt{5} = 2.676$ units.

$$\therefore \quad ROL = (\text{Demand per day}) \times LT + z \times \sigma_{D_{LT}}$$

$$= 4 \times 5 + 1.65 \times 2.676 = 25 \text{ units approx.,} \quad \text{since } P(z \leq 1.65) = 0.95.$$

$$\text{Reserve stock} = ROL - (\text{Daily demand}) \times LT$$
$$= 25 - 4 \times 5 = 5 \text{ units.}$$

Alternatively. The 0.95 service level corresponds to $z = 1.65$.

$$\therefore \qquad z = \frac{ROL - \mu_{LT}}{\sigma_{DLT}}$$

$$\Rightarrow \qquad 1.65 = \frac{ROL - 4 \times 5}{1.2 \times \sqrt{5}}$$

or $$ROL = 20 + 1.65 \times 1.2 \times \sqrt{5} = 25 \text{ units}$$

$$\text{Reserve stock} = ROL - \mu_{LT} = 25 - 20 = 5 \text{ units.}$$

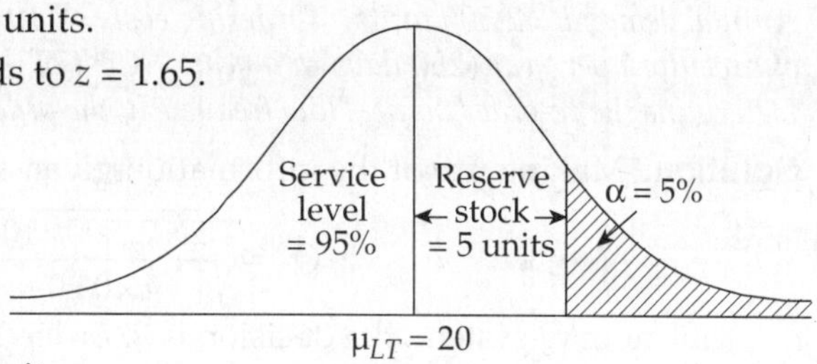

Fig. 23.5

Example 23.6. *A firm has experienced the probability distribution of inventory demand during the reorder period as recorded in the following table:*

Number of units :	30	40	50	60	70
Probability :	0.2	0.2	0.3	0.2	0.1

The firm is paying carrying cost per unit per year at the rate of ₹ 60, while stockout cost in the form of lost profits, etc. is estimated to be ₹ 400 per unit.

Find out the reserve stock level that would minimize the total annual expected cost.

[MBA, Delhi Univ. 2001]

Solution. The calculations of carrying cost and stock-out cost to determine the optimum reorder level are given in the following table:

DDLT	*Probability*	*Reorder level* 30	40	50	60	70
30	0.2	0	120	240	360	480
40	0.2	800	0	120	240	360
50	0.3	2,400	1,200	0	180	360
60	0.2	2,400	1,600	800	0	120
70	0.1	1,600	1,200	800	400	0
Expected annual cost:		7,200	4,120	1,960	1,180	1,320

Here, the cost matrix has been derived by considering each reorder level strategy, *DDLT* and the probability, *i.e.*,

Carrying cost = Extra unit carried × Probability × Cost per unit carried,

and Stock-out cost = Number of units short × Probability × Cost per unit short.

For example, when the re-order level is 60 units and *DDLT* is 40 units, the carrying cost will be (60 – 40) × 0.2 × ₹ 60, *i.e.*, ₹ 240. Similarly, when the reorder level is 40 units and *DDLT* is 50 units, the stock-out cost will be (50 – 40) × 0.3 × ₹ 400, *i.e.*, ₹ 1,200. In this way, all the elements of the above table are derived.

Next, for each of the strategies, we add up the carrying costs and stock-out costs to get the annual expected cost. The expected cost with reorder strategy of 60 units being the least and hence this is the optimum point. With this reorder level, reserve stock is:

Reserve stock = *ROL* – Average *DDLT* = 60 – 48 = 12 units, since

Average *DDLT* = 30 × 0.2 + 40 × 0.2 + 50 × 0.3 + 60 × 0.2 + 70 × 0.1 = 48 units.

Example 23.7. *Given the following information with regard to a product:*
Annual demand = 2,500 units, Ordering cost = ₹ 10 per order, Inventory carrying cost = 20% of the value of inventory per year (250 days in a year), Cost of the product = ₹ 4 per unit, and Lead time = 10 days. Determine the re-order point. How frequently the order is placed?

Solution. Making use of the information given above, we have

$$Q^o = \sqrt{\frac{2 \times 2{,}500 \times 10}{4 \times 0.20}} = 250 \text{ units}, \quad n^o = \frac{2{,}500}{250} = 10.$$

∴ Under Q-system, the decision is to order 250 units.

Now, $$ROL = \left(\frac{2{,}500}{250}\right) \times 10 = 100 \text{ units}.$$

and $$t^o = \sqrt{\frac{2 \times 10}{(2{,}500 \times 4 \times 0.20)}} = 0.1 \text{ years or } 25 \text{ days}.$$

∴ Under P-system, the decision is to order upto (25 + 10) × 10, *i.e.*, 350 units after every 25 days. It is due to the reason that the maximum inventory level is determined by the demand during review period.

In other words, 350 units of a product must be sufficient to meet the demand until the next order will arrive, *i.e.*, $t + LT = 35$ days after the current order is placed.

Example 23.8. *In addition to the information given in Example 23.7, suppose that demand is normally distributed with an average of 10 units per day and standard deviation is of 3 units. The acceptable service level is 95%. Find out the ROL and maximum inventory level.*

Solution. *Under Q-system:*

$$ROL = \text{Average } DDLT + \text{Reserve stock}$$
$$= 10 \times 10 + 1.65\sqrt{3^2 \times 10} = 100 + 15.65 = 115.65 \text{ units}.$$

Under P-system:

$$\text{Maximum inventory level} = \text{Average demand during } (t + LT) + \text{Reserve stock}$$
$$= 10 \times (25 + 10) + 1.65 \times 3\sqrt{35} = 379.28 \text{ units}$$

It may be observed that the amount of reserve stock in P-system is more than that in Q-system. It will cost ₹ (29.28 – 15.65) × 4 × 0.20; *i.e.*, ₹ 10.90 more per year.

PROBLEMS

23.01. The following data related to a particular item in stock:

Normal usage	= 110 units per day	Minimum usage	= 50 units per day
Maximum usage	= 140 units per day	Lead time	= 25-30 days
EOQ	= 5,000 units (calculated earlier)		

Using the above data, calculate the Reorder, Minimum and Maximum levels.

23.02. A company uses annually 24,000 units of a raw material which costs ₹ 1.25 per unit. Placing each order costs ₹ 22.5, and the carrying cost is 5.4% per year of the average inventory. Find the economic lot size and the total inventory cost (including the cost of material).

Should the company accept the offer made by the supplier of a discount of 5% on the cost price on a single order of 24,000 units?

Suppose the company works for 300 days a year. If the procurement time is 12 days and safety stock is 400 units, find the re-order point, the minimum, maximum and average inventory. ***[BE, Madras Univ. 1999]***

23.03. The daily demand for a commodity is approximately 100 units. Every time an order is placed, fixed cost of ₹ 100 is incurred. The daily holding cost per unit inventory is 2 paise. If the lead time is 12 days, determine the economic lot size and the reorder point.

Suppose that the demand follows normal distribution with mean 100 and standard deviation 10. Determine the size of the buffer stock, such that the probability of running out of stock during lead time is at most 0.05. *[BFIA, Delhi Univ. 2006]*

23.04. A pharmaceutical factory consumes annually 6,000 kg of a chemical costing ₹ 5 per kg. Placing each order costs ₹ 25 and the carrying cost is 6% per year per kg of average inventory.

(*i*) Find the Economic Order Quantity and the Total Inventory Cost (including the cost of chemicals.).

(*ii*) The factory works for 300 days a year. If the procurement time is 15 days and safety stock is 200 kg, find the reorder point, the maximum and average inventories.

(*iii*) If the supplier offers a discount of 5% on the cost price for a single order of annual requirement, should the factory accept it?

23.05. You are given the following information in regard to an item:

Annual usage	= 20,000 units	Ordering cost	= ₹ 160 per order
Carrying cost	= 20% of the average inventory	Unit cost	= ₹ 2
Lead time	= 10 working days	Total working days	= 250 per annum.

It is observed from the past data that the demand during lead time has been upto a maximum level of 150 units per day. Keeping this level in mind, what safety stock would you recommend? Also determine:

(*a*) Reorder level when the safety stock level suggested by you is kept in stock.

(*b*) Average level of inventory stock held,

(*c*) Ordering and carrying costs associated with this fixed order inventory policy.

[MBA, Bharthidasan Univ. 1996]

23.06. Calculate safety stock, reorder level, maximum, minimum and average inventory of a particular *A*-category item from the following data:

Annual demand	= 20,000 units	Inventory carrying cost	= 20% per year
Price per unit	= ₹ 10	Ordering cost per order	= ₹ 100
Standard deviation of demand per week	= 50 units	Service level	= 90%
		Average lead time	= 4 weeks.

[CA Final (Nov.), 1995]

23.07. Consider an item for which

Annual demand	= 1,000 units	Standard deviation of demand per week	= 10 units
Cost per unit of the item	= ₹ 5	Ordering cost/order	= ₹ 150
Inventory carrying cost	= 30 per cent	Average lead time	= 4 week
Maximum delay	= 3 weeks	Probability of delay	= 0.30
Service level	= 95 per cent		

Determine the buffer stock, reserve stock, safety stock and desirable maximum inventory level for this item. *[M.Com., Bangalore Univ. 2016; MBA (PT.) (Nov.), Delhi Univ. 2003]*

23.08. Daily demand of a product is normally distributed with mean = 60 units, and standard deviation of 6 units. The lead time is constant at 9 days (working). The cost of placing an order is ₹ 20 and the annual holding costs are 20% of the unit price of ₹ 10. A 95% service level is desired for the customers, who place orders during the reorder period. You are required to determine the order quantity and reorder level for the item in question, assuming that there are 300 working days during a year. *[B.Tech. (Mech.), Panjab Tech. Univ. 2008]*

23.09. The demand per month for a product is distributed normally with a mean of 100 and standard deviation 25. The lead time distribution is given below. What service level will be afforded by a reorder level of 500 units?

Lead time (months) :	1	2	3	4	5
Probability :	0.10	0.20	0.40	0.20	0.10

[***Hint:*** Normal value of z for the given distribution of lead time is given by

$$z = \frac{(ROL - \mu_{LT})}{\sigma_{D_{LT}}}, \text{ where}$$

μ_{LT} = average $DDLT \times LT = 100 \times LT$, and $\sigma_{D_{LT}} = \sigma_D \times \sqrt{LT} = 25 \times \sqrt{LT}$.]

23.10. A dealer has observed the distribution of demand during lead time of a certain brand of electric fans as given below:

Number of units :	4	5	6	7	8	9
Probability :	0.08	0.19	0.27	0.25	0.18	0.03

The dealer is paying carrying cost per unit per year at the rate of ₹ 50, while stock-out cost in the form of lost profits, etc. is estimated to be ₹ 200 per unit. Determine the reserve stock level that would minimize the total annual expected cost.

23.11. A company purchases a component from another factory. Based on the past records, the *EOQ* for the component is 3,600 units, with an average usage of 50 units a day. The normal lead time is six days and the carrying cost per unit is ₹ 10 per year. The cost of being out-of-stock for this component is estimated to be ₹ 50 per unit and the company places five orders a year.

An analysis of the past inventory records for the component has revealed the following usage pattern:

Usage during the reorder period (units) :	150	200	250	300	350	400	450
No. of times this quantity was used :	3	4	6	68	9	7	3

Determine the optimal safety stock and the optimal reorder level for the component.

[M.Com., Delhi Univ. 1998]

23.12. A company uses a component for which the forecast of usage is known to be normally distributed with mean absolute deviation (*MAD*) = 40 units. The company management adopts a policy which limits the stock-outs to be not more than one order cycle in a year. Required: (*a*) How much safety stock should be kept, if the order quantity is generally a two weeks' supply? (*b*) How much safety stock should be kept, if the order quantity is of four weeks' supply?

[***Hint:*** For normal distribution, $MAD = \sqrt{(2/\pi)}\,\sigma$. Therefore, $\sigma = 40 \times 1.25 = 50$.

(*a*) Number of order cycles = $52/2 = 26$. So, Prob. of being out of stock = $1/26$ and service level = $1 - 1/26 = 25/26$ or 0.96.

(*b*) Number of order cycles = $52/4 = 13$ and service level = $1 - 1/13 = 0.92$.

Using $P(z \le 1.75) = 0.96$ and $P(z \le 1.43) = 0.92$, find the value of X, where $z = (X - \mu)/\sigma$ with $\mu = 0$.]

23:2. ONE PERIOD PROBLEM

The inventory control systems that have been dealt so far are all for controlling over time. Now, we shall consider the situations, where the ordering quantity is to be decided for single periods only under the conditions of uncertain demand. Here, the demand of an item is given by probability distribution, and also the cost of holding as well as cost of stock-outs (shortages) are given. Further, since only one order is to be placed, the order size will have no bearing on the ordering cost.

Under these conditions, one of the most popular approach is the '***marginal analysis***', or '***incremental analysis***' and is employed to attain the desired result.

Marginal Analysis Approach. In this approach, let

MP = Marginal profit from each unit sold,

ML = Marginal loss from each unit that could not be sold,

$p(r \geq Q)$ = Probability that the demand is for 'Q' units or more,

$1 - p(r \geq Q)$ = Probability that the demand is for less than 'Q units.

For an optimum order quantity, obviously, we should have

Expected marginal profit ≥ Expected marginal loss,

i.e., $$p(r \geq Q) \times MP \geq [1 - p(r \geq Q)] \times ML$$

or $$p(r \geq Q) \geq \frac{ML}{MP + ML}.$$

The decision rule will be:

"Buy the maximum quantity 'Q' such that the probability of setting demand 'r' exceeding 'Q' is greater than or equal to the ratio ML/(MP + ML)".

Remarks

1. When cost price and selling price as well as salvage values are given, then we calculate MP and ML first. Thus, if C = cost per unit, P = selling price per unit, and S salvage value per unit, then $ML = C - S$ and $MP = P - C$.

$$\therefore \quad p(r \geq Q) \geq \frac{ML}{(MP + ML)} \Rightarrow p(r \geq Q) \geq \frac{C - S}{(P - C) + (C - S)} \Rightarrow p(r \geq Q) \geq \frac{C - S}{P - S}.$$

2. Sometimes the cost of holding (C_h) of a unit and the cost of stock-out (C_s) of a unit are given. Then, the optimum order quantity can be determined by using the following three steps:

   Step 1. Determine the cumulative probabilities for the demand of the item, r, to be greater than or equal to Q. Thus, find $p(r \geq Q)$.

   Step 2. Calculate the ratio $C_h/(C_h + C_s)$, where C_h is the cost of holding per unit for a period and C_s is the cost of shortage per unit for the period. This ratio is denoted by p_c, the critical probability.

   Step 3. Compare the cumulative probabilities with p_c. The largest value of Q for which the cumulative probability is equal to, or greater than p_c, would be the required ordering quantity.

   In general terms, the optimum ordering quantity

   $$= \max.\ p(r \geq Q) \geq p_c, \quad \text{where} \quad p_c = C_h/(C_h + C_s).$$

3. When the probability distribution is continuous, we represent p_c in terms of area under normal curve as shown by the shaded portion in the *Fig. 23.6:*

   Using the normal tables, we find the number of standard deviations, Z covered up to p_c. This Z value is multiple of the standard deviation of sales to convert it into units. This quantity has to be added to the mean sales (if $p_c < 0.50$) or subtracted from the mean sales (if $p_c > 0.50$) to get the EOQ.

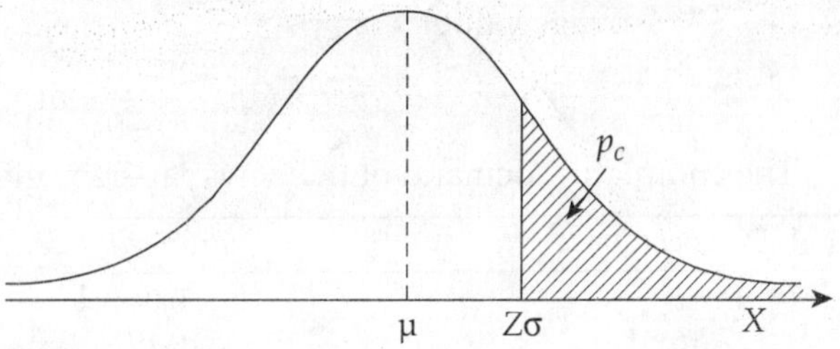

Fig. 23.6

   Thus, $$EOQ(Q^o) = \text{Mean sales } (\mu) \pm z \times \sigma,$$

   where, σ is the standard deviation of sales.

Example 23.9. *Agent Corner, an authorized dealer in domestic appliances, finds that the cost of holding a dish-washer in stock for a week is ₹ 50. Customers who can not get the new dish-washer immediately tend to go to other dealers for which his expected loss is ₹ 350 per customer.*

Probability distribution of demand is as follows:

No. of dish-washers :	0	1	2	3	4	5	6
Probability :	0.05	0.10	0.20	0.30	0.20	0.10	0.05

Assuming that there is no time lag between ordering and delivery, how many dish-washers should he order per week?

Solution. We are given

MP = ₹ 350 per customer, and ML = ₹ 50 per customer.

$$\therefore \quad p(r \geq Q) \geq \frac{50}{350+50}, \text{ i.e., } p(r \geq Q) \geq 0.125.$$

The cumulative probability distribution of the number of dish-washers is given as under:

Number of dish-washers (Q)	*Probability*	*Cumulative probability*		$p(r \geq Q)$
0	0.05	1.00	>	0.125
1	0.10	0.95	>	0.125
2	0.20	0.85	>	0.125
3	0.30	0.65	>	0.125
4	0.20	0.35	>	0.125
5	0.10	0.15	>	0.125
6	0.05	0.05	<	0.125

From the above table, we observe that at the level of 6 units, the cumulative probability (= 0.05) is less than the critical probability (= 0.125) and at the level of 5 units the cumulative probability (= 0.15) which is more than the critical probability. Hence, 5 dish-washers should be ordered per week.

Example 23.10. *Under an employment promotion programme, it is proposed to allow sale of newspapers on the buses during off-peak hours. The vendor can purchase the newspapers at a special concessional rate of 25 paise per copy against the selling price of 40 paise. Any unsold copies are, however, a dead loss. A vendor has estimated the following probability distribution for the number of copies demanded:*

Number of copies :	15	16	17	18	19	20
Probability :	0.04	0.19	0.33	0.26	0.11	0.07

How many copies should he order so that his expected profits will be a maximum?

[M.Com., Kerala Univ. 1996]

Solution. We are given:

P (= Selling price) = 40 paise per copy, C (= Purchase cost) = 25 paise per copy, and S (= Salvage value) = 0.

$$\therefore \quad p(r \geq Q) \geq \frac{25-0}{40-0} \text{ or } \frac{25}{40} \text{ or } 0.625$$

The cumulative demand of the newspapers is given as under:

Demand (Q)	*Probability*	*Cumulative probability*		$p(r \geq Q)$
15	0.04	1.00	>	0.625
16	0.19	0.96	>	0.625
17	0.33	0.77	>	0.625
18	0.26	0.44	<	0.625
19	0.11	0.18		
20	0.07	0.07		

From the above, we observe that 17 copies should be purchased (At the level of 18 copies, the cumulative probability is less than the critical probability).

Example 23.11. *A newspaper boy is thinking of selling a special one time edition of a sports magazine to his regular newspaper customers. Based on his knowledge about his customers, he believes that he can sell between 6 to 12 copies of the magazine with probabilities estimates as under:*

No. of copies :	6	7	8	9	10	11	12
Probability :	0.10	0.15	0.35	0.20	0.10	0.05	0.05

The magazines can be purchased at ₹ 80 each and can be sold for ₹ 120 each.

(*a*) *Magazines that are not sold can be returned to the publisher for a refund of 50%. Determine optimum quantity to be purchased?*

(*b*) *If the publisher does not take back the unsold magazines and the boy is forced to sell them as scrap at ₹ 5, what should be the order quantity?*

(*c*) *And if the boy gets magazines "on sale basis", what quantity should be ordered?*

Solution. We are given

P (= Selling price) = ₹ 120, C (= Purchase cost) = ₹ 80, S = salvage value @ 50% of purchase price = ₹ 40.

(*a*) $$p(r \geq Q) \geq \frac{80-40}{(120-80)+(80-40)} \left(= \frac{4}{8} \text{ or } 0.50\right)$$

The cumulative probability for the sports magazine is obtained as under:

Demand (Q)	*Probability*	*Cumulative probability*		$p(r \geq Q)$
6	0.10	1.00	>	0.50
7	0.15	0.90	>	0.50
8	0.35	0.75	>	0.50
9	0.20	0.40	<	0.50
10	0.10	0.20		
11	0.05	0.10		
12	0.05	0.05		

From the above table, we observe that the cumulative probability becomes less than the critical ratio, when $Q = 9$. Hence, the optimum order quantity is to buy 8 magazines.

(*b*) If the publisher does not take back unsold copies, S = ₹ 5.

$$\therefore \quad p(r \geq Q) \geq \frac{80-5}{120-5} \quad \text{or} \quad p(r \geq Q) \geq 0.652.$$

In this case also, since cumulative probability becomes less than critical ratio, when $Q = 9$; the optimal policy remains unchanged, *i.e.*, buy 8 magazines.

(*c*) In this case, unsold copies can be returned at full cost.

$$\therefore \quad p(r \geq Q) \geq \frac{80-80}{120-80}, \quad i.e., \; p(r \geq Q) \geq 0.$$

This shows that optimal policy is to buy 12 magazines.

Example 23.12. *A newspaper boy buys papers for 60 paise each and sell them for ₹ 1.40 each. He cannot return unsold newspapers. Daily demand has the following distribution:*

Number of customers :	23	24	25	26	27	28	29	30	31	32
Probability :	0.01	0.03	0.06	0.10	0.20	0.25	0.15	0.10	0.05	0.05

If each day's demand is independent of the previous day's, how many papers should be ordered each day?

[M.Sc. (Math.), CCS Univ. 2009]

Solution. Let Q be the number of newspapers ordered per day and let 'r' be the demand for it, *i.e.*, the number of newspapers actually sold per day.

Given: C_h = ₹ 0.60, and C_s = ₹ 1.40 – ₹ 0.60 (= ₹ 0.80).

$$\therefore \qquad p_c \text{ (Critical ratio)} = \frac{C_h}{C_h + C_s} = \frac{0.60}{0.60 + 0.80} = 0.428.$$

The optimum solution is obtained by developing the cumulative probability distribution of daily demand as follows:

r :	23	24	25	26	27	28	29	30	31	32
$p(r)$:	0.01	0.03	0.06	0.10	0.20	0.25	0.15	0.10	0.05	0.05
$p(r \geq Q)$:	0.01	0.04	0.10	0.20	0.40	0.65	0.80	0.90	0.95	1.00

Since, $p(r \geq Q)$ becomes less than p_c(= 0.428) when Q = 27, the optimum policy is Q^o = 28. Hence, the optimum number of newspapers to be ordered is 28.

Example 23.13. *A wholesaler in flowers estimates his expected demand in the long run at 14 baskets a day with a standard deviation of 6 baskets. Assuming that his sales are normally distributed, estimate the number of baskets he should stock if the purchase cost of a basket is ₹ 40 and the sale price is ₹ 90.*

Solution. Here, we have C_h = Purchase cost of a basket = ₹ 40,

C_s = Lost profit = ₹ 90 – ₹ 40 = ₹ 50,

μ = 14 baskets, σ = 6 baskets.

$$\therefore \qquad p_c = \frac{C_h}{C_h + C_s} = \frac{40}{50 + 40} = 0.44.$$

The Z value corresponding to 0.44 of the area under the normal curve can be read from tables as 0.15.

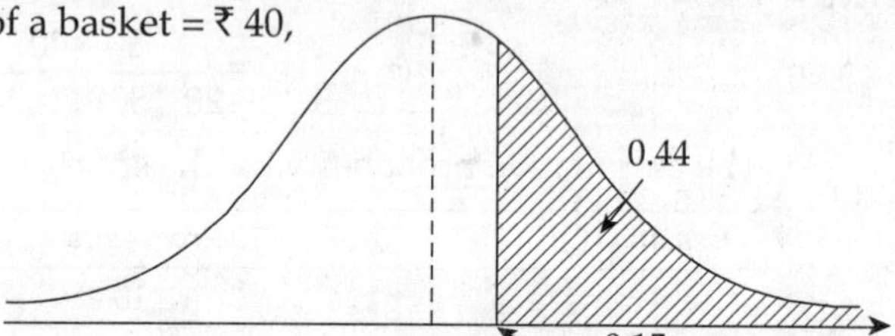

Fig. 23.7

$$\therefore \qquad EOQ\,(Q^o) = \mu + Z\sigma = 14 + 0.15 \times 6 = 15 \text{ baskets approximately.}$$

Example 23.14. *Demand for a certain product in 'a newspaper boy' model is normally distributed with mean 100 units and standard deviation 20. Lost profit is ₹ 8 per unit and salvage loss is ₹ 12. Find the optimal initial inventory.*

Solution. We are given: C_h = Salvage loss = ₹ 12, C_s = Loss profit = ₹ 8, Mean (μ) = 100 units, and Standard deviation (σ) = 20.

$$\therefore \qquad p_c = \frac{C_h}{C_s + C_h} = \frac{12}{8 + 12} = 0.60.$$

The Z value corresponding to 0.60 of the area under the normal curve can be read from tables as 0.25.

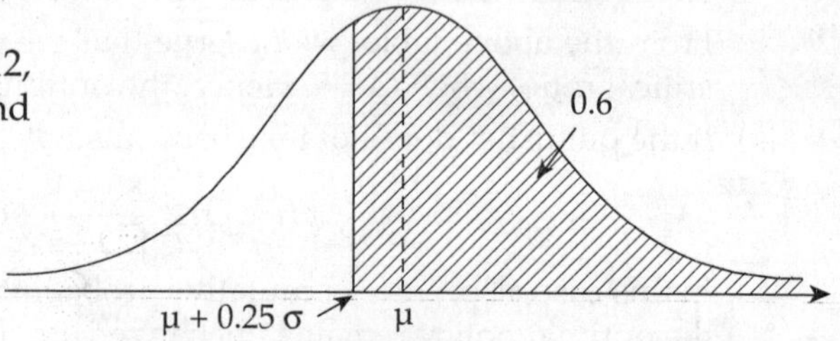

Fig. 23.8

$$\therefore \qquad EOQ\,(Q^o) = \mu - Z\sigma = 100 - 0.25 \times 20 = 95 \text{ units.}$$

PROBLEMS

23.13. A small industry finds from the past data that the cost of making an item is ₹ 25. The selling price of an item is ₹ 30, if it is sold within a week, and it could be dispose off at ₹ 20 per item at the end of the week.

Weekly sales :	≤ 3	4	5	6	7	≥ 8
No. of weeks :	0	10	20	40	30	0

Find the optimum number of items per week the industry should produce.

[*MBA (Nov.), Osmania Univ. 2002*]

23.14. A shop is about to order some air-conditioners of particular brand. The shop pays ₹ 10,000 for each AC and during the summer season it is sold for ₹ 20,000 each. Demand for the AC decline after summer season, and any unsold units are sold at ₹ 5,000. Past experience suggests the likely demand for AC's as follows:

Demand :	10	20	30	40	50
Probability :	0.20	0.30	0.30	0.10	0.10

How many AC's should the company buy?

23.15. A TV dealer finds that cost of holding a television in stock for a week is ₹ 20, customers who cannot obtain new television immediately tend to go to other dealers; and he estimates that for every customer who does not get immediate delivery he loses on an average ₹ 200. For one particular model of television the probabilities for a demand of 0, 1, 2, 3, 4 and 5 televisions in a week are 0.05, 0.10, 0.20, 0.30, 0.20 and 0.15, respectively. How many televisions per week should the dealer order? (Assume that there is no time lag between ordering and delivery.) ***[MBA (Nov.), Delhi Univ. 2009]***

23.16. Some of the spare parts of a ship cost ₹ 1,00,000 each. These spare parts can only be ordered together with the ship. If not ordered at the time when the ship is constructed, these parts cannot be available on need. Suppose that a loss of ₹ 10,00,000 is suffered for each spare part that is needed, when none is available in the stock. Further, suppose that the probabilities that the spare parts will be needed as replacement during the life-term of the class of the ship discussed are:

Spare parts repaired :	0	1	2	3	4	5 or more
Probability :	0.9488	0.0400	0.0100	0.0010	0.0002	0.0000

How many spare parts should be procured?

23.17. Under an employment promotion programme, it is proposed to allow sale of newspapers on the buses during off-peak hours. The vendor can purchase the newspapers at a special concessional rate of ₹ 1.25 per copy against the selling price of ₹ 2.00. Any unsold copies are, however, a dead loss. A vendor has estimated the following probability distribution for the number of copies demanded:

Number of copies :	15	16	17	18	19	20
Probability :	0.04	0.19	0.33	0.26	0.11	0.07

How many copies should he order so that his expected profits will be maximum? ***[M.Com., Kerala Univ. 1996]***

23.18. A company uses to order a new machine after a certain fixed time. It is observed that one of the parts of the machine is very expensive if it is ordered without machine. the cost of spare part when ordered with the machine is ₹ 500. The cost of down time of the machine and the cost of arranging the new part is ₹ 10,000. From the past record it is observed that spare part is required with the probabilities mentioned below:

Demand (r) :	0	1	2	3	4	5	6 or more
Probability P (r) :	0.90	0.05	0.02	0.01	0.01	0.01	0.00

Find the optimum number of spare parts that should be ordered with the order of the machine.

23.19. A grocery store estimates that he will sell 150 kg of an item in the next week. The distribution of demand is observed to be normally distributed with a standard deviation of 25 kg. The selling price of the item is ₹ 7 per kg. The store pays ₹ 3 per kg for the ingredients. Any amount of unsold item is given to someone at no cost. Find out the quantity of the item to be procured in order to maximize profit.

23:3. ONE PERIOD PROBLEM WITHOUT SET-UP COST

The marginal analysis approach discussed in the previous section is based on intuitive reasoning. We shall, now, use more formal approach to analyse the one-period (single period) inventory problems.

Let the demand of an item is of 'r' units per time period and the re-order time is fixed and known. The problem is to determine the optimum order level Q, where $r \le Q$ or $r > Q$ at the beginning of each period. Obviously, when $r \le Q$, no shortage occurs and hence we consider only holding cost C_h. But, when $r > Q$ both the holding cost C_h and stock-out cost C_s are considered.

While dealing with one-period inventory problems without set-up cost, following are the two types of inventory situations depending on the demand rate:

Case 1. Demand rate is uniform, and

Case 2. Demand rate is instantaneous.

In both cases, the units that build-up an inventory may consist of either discrete or continuous units.

Case 1. In this case, the problem is to determine the optimum order quantity 'Q' when the withdrawal of items from inventory is uniformly distributed over one-period, such that either $r \le Q$ or $r > Q$ at the beginning of each period. The time inventory graph of the two situations is given in *Fig. 23.9*.

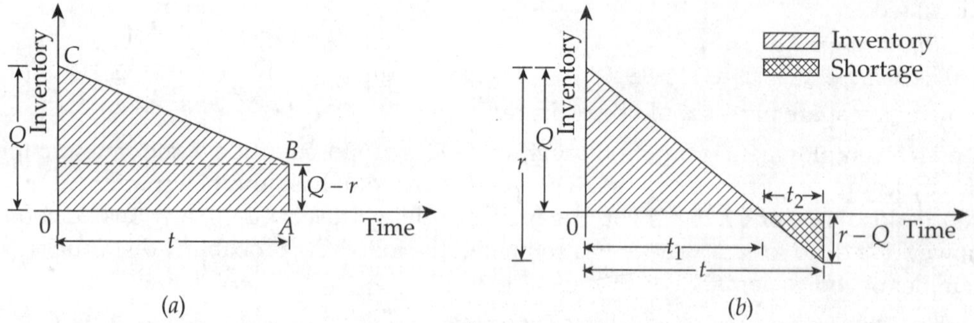

Fig. 23.9

Uniform Demand and Discrete Units

Let the demand for r units be estimated at discontinuous rate with probability $p(r)$, $r = 1, 2, ..., n$. That is, we may expect demand for 1 unit with probability $p(1)$; 2 units with probability $p(2)$; and so on. Since all the possibilities are to be taken care of, we must have

$$\sum_{r=0}^{\infty} p(r) = 1 \quad \text{and} \quad p(r) \ge 0.$$

When $r \le Q$, the total inventory is represented by the area of trapezium $OABC$, *i.e.*, $\frac{1}{2}(AB + OC) \times OA$. Hence, holding cost of carrying r units ($r \le Q$) will be $\frac{1}{2} C_h (Q - r + Q)t$.

$$\therefore \qquad \text{Expected carrying cost} = \frac{1}{2} C_h t \sum_{r=0}^{Q} (2Q - r) p(r)$$

When $r > Q$, both the costs are involved. The carrying cost is $\frac{1}{2} Q t_1 C_h$ and the shortage cost is $\frac{1}{2}(r - Q) C_s t_2$.

Further, from the similar triangles in *Fig. 23.9(b)*

$$\frac{t_1}{t}=\frac{Q}{r} \quad \text{and} \quad \frac{t_2}{t}=\frac{r-Q}{r}.$$

$\therefore$ Expected shortage cost $= \sum_{r=Q+1}^{\infty}\left[\frac{1}{2}QC_h\times\frac{Q}{r}t+\frac{1}{2}(r-Q)C_s\times\frac{r-Q}{r}t\right]p(r).$

Thus, the total expected cost (TC) associated with producing quantity Q is:

$$TC=C_h t\sum_{r=0}^{Q}\left(Q-\frac{r}{2}\right)p(r)+\sum_{r=Q+1}^{\infty}t\{C_hQ^2+C_s(r-Q)^2\}\frac{p(r)}{2r}.$$

$\therefore$ Average cost or total expected cost for one period is given by

$$TC(Q)=C_h\sum_{r=0}^{Q}\left(Q-\frac{r}{2}\right)p(r)+\sum_{r=Q+1}^{\infty}\{C_hQ^2+C_s(r-Q)^2\}\frac{p(r)}{2r}.$$

In order to find the optimum value of Q so as to minimize $TC(Q)$, the following conditions must hold:

$$TC(Q+1)>TC(Q) \quad \text{and} \quad TC(Q-1)>TC(Q)$$

i.e., $$TC(Q+1)-TC(Q)>0 \quad \text{and} \quad TC(Q)-TC(Q-1)<0$$

Now, substituting $Q+1$ for Q in the expression for $TC(Q)$, we have

$$TC(Q+1)=C_h\sum_{r=0}^{Q+1}\left(Q+1-\frac{r}{2}\right)p(r)+\sum_{r=Q+2}^{\infty}\{C_h(Q+1)^2+C_s(r-Q-1)^2\}\frac{p(r)}{2r}$$

$$=TC(Q)+(C_h+C_s)F(Q)-C_s,$$

where, $F(Q)=\sum_{r=0}^{Q}p(r)+\left(Q+\frac{1}{2}\right)\sum_{r=Q+1}^{Q}\frac{p(r)}{r}.$

Similarly, $$TC(Q-1)=TC(Q)-(C_h+C_s)F(Q-1)+C_s$$

$\therefore$ $$[TC(Q+1)-TC(Q)]>0 \Rightarrow (C_h+C_s)F(Q^o)-C_s>0$$

or $$F(Q^o)>\frac{C_s}{C_h+C_s}$$

and $$[TC(Q)-TC(Q-1)]<0 \Rightarrow (C_h+C_s)F(Q^o-1)-C_s<0$$

or $$F(Q^o-1)<\frac{C_s}{C_h+C_s}$$

Thus, $$F(Q^o-1)<\frac{C_s}{C_h+C_s}<F(Q^o).$$

Uniform Demand and Continuous Units

When certain demand is estimated as a continuous random variable, the cost equation of the inventory involves integrals instead of summation signs. The discrete point probabilities $p(r)$ are replaced by the probability differential $f(r)$ for small interval.

In this case, $\int_0^{\infty} f(r)\,dr=1$ and $f(r)\geq 0$. Proceeding exactly in the same manner, the expected sizes of over and under supply are as follows:

$$\int_0^Q \left(Q-\frac{r}{2}\right) f(r)\,dr \quad \text{and} \quad \int_Q^\infty \{C_h Q^2 + C_s (r-Q)^2\} \frac{f(r)}{2r}\,dr$$

∴ Average cost or total expected cost for one period is given as under:

$$TC(Q) = C_h \int_0^Q \left(Q-\frac{r}{2}\right) f(r)\,dr + C_h \int_Q^\infty Q^2 \frac{f(r)}{2r}\,dr + C_s \int_Q^\infty (r-Q)^2 \frac{f(r)}{2r}\,dr.$$

Now, $$\frac{d}{dQ}[TC(Q)] = C_h \int_0^Q f(r)\,dr + C_h \int_Q^\infty \frac{Q}{r} f(r)\,dr - C_s \int_Q^\infty \left(\frac{r-Q}{r}\right) f(r)\,dr$$

$$= C_h \int_0^Q f(r)\,dr + C_h \int_Q^\infty \frac{Q}{r} f(r)\,dr - C_s \left[\int_Q^\infty f(r)\,dr - \int_0^Q \frac{Q}{r} f(r)\,dr\right] + C_s \int_0^Q \frac{Q}{r} f(r)\,dr$$

$$= (C_h + C_s)\left[\int_0^Q f(r)\,dr + \int_Q^\infty \frac{Q}{r} f(r)\,dr\right] - C_s, \quad \text{since} \int_0^\infty f(r)\,dr = 1$$

∴ $$\frac{d}{dQ}[TC(Q)] = 0 \Rightarrow F(Q^o) = \frac{C_s}{C_h + C_s}, \quad \text{where } F(Q) = \int_0^Q f(r)\,dr + \int_Q^\infty \frac{Q}{r} f(r)\,dr.$$

Further, $$\left.\frac{d^2}{dQ^2} TC(Q)\right|_{Q=Q^o} = (C_1 + C_2)\int_Q^\infty \frac{f(r)}{r}\,dr > 0.$$

Hence, $F(Q^o)$ gives the condition for finding the optimum value of Q as Q^o.

Example 23.15. *The probability distribution of monthly sales of a certain item is as follows:*

Monthly sales :	0	1	2	3	4	5	6
Probability :	0.02	0.05	0.30	0.27	0.20	0.10	0.06

The cost of carrying inventory is ₹ 10 per unit per month. The current policy is to maintain a stock of four items at the beginning of each month. Assuming that the cost of shortage is proportional to both time and quantity short, obtain the inputed cost of a shortage of one item for one unit of time.

[M.Sc. (Math.), CCS Univ. 1993]

Solution. We are given

C_h = ₹ 10 per unit per month, Q (stock level) = 4 units.

Since, the demand is uniformly distributed over the month,

$$F(Q^o - 1) < \frac{C_s}{C_h + C_s} < F(Q^o)$$

Now, $$F(Q^o - 1) = F(4-1) = \sum_{r=0}^{4} p(r) + \left(4 - \frac{1}{2}\right) \sum_{r=4}^{6} \frac{p(r)}{4}$$

$$= 0.84 + \frac{7}{2}\left[\frac{0.20}{4} + \frac{0.10}{5} + \frac{0.06}{6}\right] = 0.92$$

Thus, the least value of C_s is given by

$$\frac{C_s}{C_h + C_s} = 0.92 \quad \text{or} \quad C_s = 0.92(10 + C_s)$$

This implies that C_s = ₹ 115.

Similarly, the greatest value of C_s is obtained by considering the right-hand side of $\dfrac{C_s}{C_h + C_s}$.

$$\therefore \quad \frac{C_s}{C_h + C_s} = \sum_{r=0}^{4} p(r) + \left(4 + \tfrac{1}{2}\right) \sum_{r=5}^{6} \frac{p(r)}{r} = 0.84 + \frac{9}{2}\left[\frac{0.10}{5} + \frac{0.06}{6}\right] = 0.975$$

$$\therefore \quad C_s = 0.975(10 + C_s), \quad \text{since } C_h = 10$$

or $$C_s = ₹\,390.$$

Hence, the imputed cost of shortage is given by

$$₹\,115 < C_s < ₹\,390.$$

Example 23.16. *A baking company sells cake by kg. It makes a profit of ₹5 per kg on every kg sold on the day it is baked. It disposes of all cakes not sold on the date it is baked: at a loss of ₹1.20 paise per kg. If demand is known to be rectangular between 2,000 and 3,000 kg determine the optimum daily amount baked.*

[M.Sc. (Math.), CCS Univ. 2008; B.Tech. (Mech.), Hyderabad Univ. 2010]

Solution. We are given: C_h = ₹ 1.20 (sale of leftovers); C_s = ₹ 5 (profit on sale); and demand 'r' is rectangular between 2,000 and 3,000 kgs.

Now, a distribution of *pdf*, $f(x)$ is said to be rectangular, if

$$f(x)\begin{cases} 1/Q, & 0 \le x \le a \\ 0\ , & \text{beyond } a \end{cases} \quad \text{and} \quad \int_0^{\infty} f(x)\,dx = \int_0^{a} \frac{1}{a}\,dx.$$

$$\therefore \quad \int_0^{Q} f(r)\,dr + \int_Q^{3,000} \frac{Q}{r} f(r)\,dr = \frac{C_s}{C_h + C_s},$$

where, $f(r) = 1/1,000;\ 2,000 \le r \le 3,000.$

Therefore, $$\int_{r=2,000}^{Q} \frac{1}{1,000}\,dr = \frac{1}{1,000}(Q - 2,000)$$

Thus, $$\frac{1}{1000}(Q - 2,000) = \frac{C_s}{C_h + C_s} = \frac{5.00}{1.20 + 5.00} = \frac{5.0}{6.2}$$

Hence, $$Q^o = \frac{5.0}{6.2} \times 1,000 + 2,000 = 2,806.45 \text{ kg}.$$

Case 2. In this case, the problem to be discussed is similar to the previous one, except that the withdrawal of items from inventory is not uniformly distributed over the period and the costs C_h, and C_s are independent of time. Here, also two situations arise, *viz.*, $r \le Q$ and $r > Q$.

The time inventory graph of these situations is given in *Fig. 23.10* below:

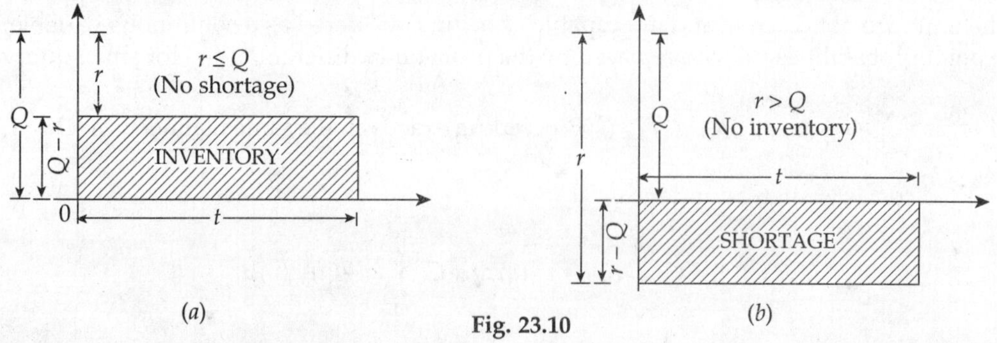

Fig. 23.10

Instantaneous Demand and Discrete Units

Let the demand for r units be estimated at a discontinuous rate with probability $p(r)$; $r = 1, 2, ..., n$.

When $r \le Q$, the expected carrying cost (if we have Q units on hand) at the beginning of the period will be $C_h \sum_{r=0}^{Q} (Q-r)p(r)$.

When $r > Q$, the expected shortage cost (if we have Q units on hand) at the beginning of the period will be $C_s \sum_{r=Q}^{\infty} (r-Q)p(r)$.

$\therefore$ Total expected cost is:

$$TC(Q) = C_h \sum_{r=0}^{Q} (Q-r)p(r) + C_s \sum_{r=Q+1}^{\infty} (r-Q)p(r)$$

The value Q^o of Q will be optimum, when

$$TC(Q+1) > TC(Q) \quad \text{and} \quad TC(Q) < TC(Q-1).$$

Now,

$$TC(Q+1) = C_h \sum_{r=0}^{Q+1} (Q+1-r)p(r) + C_s \sum_{r=Q+2}^{\infty} (r-Q-1)p(r)$$

$$= TC(Q) + (C_h + C_s) \sum_{r=0}^{Q} p(r) - C_s,$$

since

$$\sum_{r=Q+1}^{\infty} p(r) = \sum_{r=0}^{\infty} p(r) - \sum_{r=0}^{Q} p(r) = 1 - \sum_{r=0}^{Q} p(r).$$

Similarly,

$$TC(Q-1) = TC(Q) - (C_h + C_s) \sum_{r=0}^{Q-1} p(r) - C_s$$

$\therefore$ For the optimum value Q^o; we must have

$$(C_h + C_s) \sum_{r=0}^{Q} p(r) > C_s \quad \text{and} \quad (C_h + C_s) \sum_{r=0}^{Q-1} p(r) < C_s,$$

i.e.,

$$F(Q^o - 1) < \frac{C_s}{C_h + C_s} < F(Q^o), \quad \text{where} \quad F(Q^o) = \sum_{r=0}^{Q} p(r).$$

Instantaneous Demand and Continuous Units

When the units are not discrete and are capable of being considered as a continuous variable, the discrete point probabilities $p(r)$ are replaced by the probability differential $f(r)$ for small interval.

In this case, $\int_0^{\infty} f(r)\,dr = 1$ and $f(r) \ge 0$. Proceeding exactly in the same manner as in *Case 1*, the total expected cost is written as under:

$$TC(Q) = C_h \int_0^{Q} (Q-r) f(r)\,dr + C_s \int_Q^{\infty} (r-Q) f(r)\,dr$$

Now, $$\frac{d}{dQ}[TC(Q)] = C_h \int_0^Q f(r)\,dr - C_s \int_Q^\infty f(r)\,dr = (C_h + C_s)\int_0^Q f(r)\,dr - C_s$$

For the optimum value Q^o of Q, we have,

$$\frac{d}{dQ}TC(Q) = 0 \Rightarrow \int_0^Q f(r)\,dr = \frac{C_s}{C_h + C_s} \quad \text{and} \quad \left.\frac{d^2}{dQ^2}TC(Q)\right|_{Q=Q^o} = (C_h + C_s)F(Q^o) > 0.$$

Hence, $F(Q^o) = \dfrac{C_s}{C_h + C_s}$ gives the condition for finding the optimum value of Q as Q^o.

Remark. If $C_h = C_s$, the formula reduces to $\int_0^Q f(r)\,dr = \frac{1}{2}$, which shows that the optimum level of inventory in the beginning of the period is the median of the distribution of the demand.

Example 23.17. *A fish stall sells a variety of fish at the rate of ₹ 50 per kg on the day of the catch. If the stall fails to sell the catch on the same day, it pays for storage at the rate of ₹ 3 per kg and the price fetched is ₹ 45 per kg on the next day. Past record shows that there is an unlimited demand for one day old fish. The problem is to ascertain how much fish should be procured everyday so that the total expected cost is minimum. It has been found from the past record that daily demand forms a distribution with $f(x) = 0.06 - 0.0006x$; $20 \le x \le 100$.*

Solution. We are given

C_h (= Cost of carrying inventory) = ₹ 3 + ₹ (50 – 45) = ₹ 8.

C_s (= Cost of shortage) = ₹ 50, and $f(x) = 0.06 - 0.0006x$, where $20 \le x \le 100$.

$$\therefore \quad \int_0^Q f(x)\,dx = \frac{C_s}{C_h + C_s} \Rightarrow \int_0^Q (0.06 - 0.0006x)\,dx = \frac{50}{8 + 50} = 0.862$$

or $$\left[0.06x - 0.0006\frac{x^2}{2}\right]_0^Q = 0.862$$

or $$0.0003Q^2 - 0.06Q + 0.862 = 0$$

or $$Q^2 - 100Q + 1{,}436.67 = 0$$

On solving this quadratic equation,

$$Q = 17.4 \text{ kg} \quad \text{or} \quad 82.60 \text{ kg}$$

Hence, $Q^o = 82$ kg, since $20 \le x \le 100$.

Thus, 82.60 kg fish should be procured everyday.

PROBLEMS

23.20. A newspaper boy buys papers for ₹ 2.80 and sells them for ₹ 4.90 each. He cannot return unsold newspapers. Daily demand has the following distribution :

Customers :	25	26	27	28	29	30	31	32	33	34	35	36
Probability :	0.03	0.05	0.05	0.10	0.15	0.15	0.12	0.10	0.10	0.07	0.06	0.02

If each day's demand is independent of the previous day's demand, how many papers he should order each day?

23.21. Two products are stocked by a company. The company has limited space and cannot store more than 40 units. The demand distribution for the products are as follows:

For first product		*For second product*	
Demand (units)	*Probability of demand*	*Demand (units)*	*Probability of demand*
0	0.10	0	0.05
10	0.20	10	0.20
20	0.35	20	0.40
30	0.25	30	0.20
40	0.10	40	0.15

The inventory carrying costs are ₹ 5 and ₹ 10 per unit of the ending inventories for the first and second product, respectively. The shortage costs are ₹ 20 and ₹ 50 per unit of the ending shortages for the first and second product respectively. Find out the economic order quantities for both the product. *[MBA, Delhi Univ. 1997]*

23.22. The owner of a newspaper stall wants to determine the number of newspapers that must be stocked at the start of each day. It costs ₹ 3 to buy a copy, and the owner sells it for ₹ 7.50. The sale of the newspaper typically occurs between 7 A.M. and 8 A.M. Newspapers left at the end of the day are recycled for an income of 50 paise a copy. How many copies should the owner stock every morning assuming that the demand for the day can be described as:

(*a*) a normal distribution with mean 300 copies and standard deviation as of 20 copies.

(*b*) a discrete *pdf*, $f(r)$, defined as follows:

r :	*260*	*280*	*300*	*320*	*340*
$f(r)$:	0.10	0.20	0.40	0.20	0.10

23.23. Let the probability density of demand in respect of a certain item during a day is:

$$f(x) = \begin{cases} 0.1, & 0 \le x \le 10 \\ 0, & x > 10 \end{cases}$$

The demand is assumed to occur with a uniform pattern during the whole day. Let the unit carrying cost of the item in inventory be ₹ 0.50 per day and unit shortage cost be ₹ 4.50 per day. If ₹ 0.50 be purchasing cost per unit, determine the optimum order level of the inventory.

23.24. A baking company sells one of its types of cake by weight. It makes a profit of ₹ 9.50 on every kg of cake sold on the day it is baked. It disposes of all cakes not sold on the date these are baked at a loss of ₹ 1.50 per kg. If demand is known to be rectangular between 300 and 400 kg, determine the optimum amount baked. *[M.Sc. (Math.), CCS Univ. 1994]*

23.25. An ice-cream company sells one of its types of ice-cream by weight. If the product is not sold on the day it is prepared, it can be sold at a loss of 50 paise per pound. But there is an unlimited market for one-day old ice-cream. On the other hand, the company makes a profit of ₹ 3.20 on every pound of ice-cream sold on the day it is prepared. Past daily orders form as distribution with

$$f(x) = 0.02 - 0.0002\,x, \quad 0 \le x \le 100.$$

How many pounds of ice-cream should the company prepare every day?

[M.Sc. (Math.), CCS Univ. 1998]

23.26. A fish stall sells a variety of fish at the rate of ₹ 100 per kg on the day of the catch. If the stall fails to sell the catch fishes on the same day, it pays for storage at the rate of ₹ 6 per kg and the price fetched is ₹ 90 per kg on the next day. Past record shows that there is an unlimited demand for one-day old fish. The problem is to ascertain how much fish should be procured every day, so that the total expected cost is minimum. It has been found from the past records that daily demand follows an exponential distribution with

$$f(x) = 0.02 \exp.(-0.02)x; \quad 0 \le x \le \infty.$$

23.27. A shop owner places orders daily for goods which will be delivered 7 days later. On a certain day, the owner has 10 items in stock. Furthermore, on the six previous days, he has already placed orders for the delivery of 2, 4, 1, 10, 11, and 5 items in that order, over each of the next 6 days. The inventory cost per unit, per unit time is ₹ 0.15 and the shortage cost per unit is 95 paise. The distribution of requirement R over a 7-day period is: $f(R) = 0.02 - 0.002R$. Determine the number of items that should be ordered on the 7th day.

[***Hint:*** $Q^o = 10 + \sum_{i=1}^{6} Q_i + Q_7 = 10 + (2+4+1+10+11+5) + Q_7$ or $Q_7 = Q^o - 10$]

23:4. ONE PERIOD PROBLEM WITH SET-UP COST

The problem to be discussed here is similar to the previous one, *i.e.*, with instantaneous demand and continuous units. The only difference is that the fixed cost 'K' is associated with buying or producing items in a given time period. Let the initial inventory be y (before the starting of the period). This implies that an order of size $(Q - y)$ items will be placed to bring the on hand inventory of the item upto Q. Using the same notations, as earlier, the total expected cost per period will be:

$$TC^*(Q) = K + TC(Q) = K + C(Q-y) + C_h \int_0^Q (Q-r) f(r)\, dr + C_s \int_Q^\infty (r-Q) f(r)\, dr$$

As shown in the previous section, the optimum value Q^o must satisfy

$$F(Q \le Q^o) = \frac{C_s - C}{C_s + C_h},$$

where, $F(Q) = \int_0^Q f(r)\, dr$ and C is the cost per unit of an item.

Since, 'K' is constant, minimum value of $TC^*(Q)$ must also occur at Q^o.

We, now, introduce two new control variables S and s, where s denotes the re-order level and S represents the maximum stock level. Thus, $S = Q^o$, and the value of s $(< S)$ is determined from the equation

$$TC(s) = TC^*(S) = K + TC(S); \quad s < S$$

Since, the initial inventory before starting the period is y, for determining the order size to bring the on hand inventory of items upto Q^o, following three cases are analysed:

$$(a)\ y < s, \qquad (b)\ s \le y \le S, \qquad (c)\ y > S$$

Case 1. If we start the period with y units of inventory and do not buy or produce more, then the total expected cost is given by $TC(y)$. But, if we intend to buy additional $(Q - y)$ units, so as to bring inventory level upto Q^o, then $TC^*(Q^o)$ will include the set-up cost also.

Thus, for all $y < s$, the condition for ordering is:

$$\min_{Q > y}.\ \{TC^*(Q)\} = TC^*(S) < TC(y).$$

Thus, the optimum inventory policy in this case is to order $(S - y)$ units.

Case 2. In this case, the order size is determined by the condition

$$TC(y) = \min_{Q>y} TC^*(Q) = TC^*(S).$$

This implies that no ordering is less expensive than ordering. Hence $Q^0 = y$.

Case 3. If $Q > y$, then expected cost for an order upto Q will be more than total expected cost if no order is placed, *i.e.*, $TC(Q) \geq TC(y)$.

This condition indicates that it is more economical not to order in this case, *i.e.*, $Q^0 = y$. The optimal inventory policy, frequently referred to as ***Ss policy***, may be summarized as below:

(*i*) If $y < s$, order $(S - y)$ units, and (*ii*) if $y \geq s$, do not order.

Example 23.18. *If demand for a certain product has a rectangular distribution between 100 and 200, find the optimal order policy if carrying cost is ₹0.50 per unit, shortage cost is ₹5 per unit; number of units on hand before the beginning of the period are 10; set-up cost is ₹25 and unit price is ₹2.*

Solution. We are given:

$$f(r) = \frac{1}{200-100} = \frac{1}{100}; \; C_h = ₹\,0.50, \; C_s = ₹\,5, \; C = ₹\,2, \; K = ₹\,25, \text{ and } y = 10 \text{ units.}$$

$$\therefore \quad \int_0^Q f(r)\,dr = \frac{C_s - C}{C_h + C_s} \quad \Rightarrow \quad \int_0^Q \frac{1}{100}\,dr = \frac{5-2}{0.50+5}$$

i.e.,
$$\frac{Q}{100} = 0.54 \quad \text{or} \quad Q^0 = 54 \text{ units.}$$

The value of $TC(y)$ can be obtained by setting $Q \equiv y = 10$. That is,

$$TC(y) = 2(10-10) + 0.50 \int_0^{10} \frac{(10-y)}{100}\,dr + 5 \int_{10}^{100} \frac{(r-10)}{100}\,dr = ₹\,202.75$$

and
$$TC(Q^0) = 2(54-10) + 0.50 \int_0^{54} \frac{(54-r)}{100}\,dr + 5 \int_{54}^{100} \frac{(r-54)}{100}\,dr = ₹\,148.19$$

Thus, $\quad TC^*(Q^0) = K + TC(Q^0) = 25 + 148.19 = ₹\,173.19$

Since, $TC(y) > TC^*(Q^0)$, therefore, an order for $Q^0 - y = 54 - 10 = 44$ units must be placed.

23:5. INVENTORY CONTROL AND DYNAMIC PROGRAMMING

Most of the inventory problems are multi-stage optimization problems in the sense that costs are to be optimized over a number of periods and as such the technique of dynamic programming is quite useful here.

For any multi-stage problem, the solution by dynamic programming involves the following steps:

(*i*) Write the recurrence relation connecting optimum decision function for the $(n-1)$ stage process.

(*ii*) Write the relation giving the optimum decision function for a one-stage process and solve it.

(*iii*) Solve for the optimum decision function for the n-stage process.

(*iv*) If the number of stages is large, take a limiting process and solve the resulting functional equation.

At each stage, we have to find the values of one or more decisions like t, Q, etc.

As an illustration, consider a simple problem of stocking and supplying of only one commodity to meet an uncertain demand during a fixed interval of time. Suppose that the required units of the commodity are booked during intervals of equal length and are fulfilled immediately. Demand is unknown, but its probability distribution is known. This demand is satisfied as for as possible and the excess of demand incurred shortage cost. Assume that the following functions are known:

$f(x)$ = probability that demand will lie between x and $x + dx$.

$C(y)$ = cost of ordering (booking) y units initially to increase the inventory.

$E(y)$ = cost of ordering y units to meet an excess y units of demand over supply, *i.e.*, the shortage cost.

It is evident from these functions that they are independent of time. Furthermore, assume that the orders can be fulfilled immediately.

Now, if S denotes the inventory level at the initial stage, then problem is to determine how much quantity should be ordered so as to minimize the total expected cost.

Define,

$\phi_n(s)$ = Expected total cost of an n-stage process starting with an initial supply of s units and using an optimal policy of ordering.

To obtain a functional equation for this $\phi_n(s)$, we proceed as follows:

Considering the time interval as only one period, the minimum expected cost over that period will be:

$$\phi_n(s) = \min . \left\{ C(z-s) + \int_{x=z}^{\infty} E(x-z)\, f(x)\, dx \right\},$$

where the quantity ordered is $(z - s)$ units and the quantity $(x - z)$ units is to meet an excess demand of $(x - z)$ units over supply.

Now, as the optimum inventory level turns out to be a more basic quantity than the amount ordered, it is advisable to think of ordering up to a certain level of z.

For $n \geq 2$, the functional equation is:

$$\phi_n(s) = \min_{z \geq s} . \left\{ C(z-s) + \int_{x=z}^{\infty} E(x-z)\, f(x)\, dx + \phi_{n-1}(0) \int_{z}^{\infty} f(x)\, dx + \int_{0}^{\infty} \phi_{n-1}(z-x)\, f(x)\, dx \right\}.$$

This equation is the required recurrence relation which solves the given problems in a finite number of stages. Similarly, if lead time occurs, similar equations may be derived which solve the inventory problem in a finite number of stages.

23:6. INVENTORY CONTROL AND QUEUING THEORY

We now give an example suggesting a queuing theory approach to an inventory problem. In inventory techniques we have commodity input, its storage and finally the sales demand. We can imagine the queue to be consisting of demand orders waiting to be satisfied, but the real queue is often that of waiting for replenishment.

Consider a manufacturer of a very expensive item who uses the following inventory control procedure:

Suppose the manufacturer keep a buffer stock of Q units on hand. The customer demand is described by a Poisson distribution with mean λ. Whenever a demand for a unit comes (the customer arrives) the manufacturer places an order with the factory to manufacture another unit. Let the amount of time required to manufacture a unit be exponential with mean $1/\beta$. Further, let C_h be the carrying cost for inventory storage per unit per unit time. Whenever, the buffer stock depletes to zero, let C_s be the shortage cost per unit time. It is also assumed that in the case of shortage, the customer will wait until the stock is replenished by orders due. To look at it from another viewpoint, we may consider the shortage cost C_s as the discount allowed to the waiting customer for his not moving to some other firm.

The problem, now, is of determining the optimum value of Q so as to minimize the total expected cost.

Thus, $$TEC(Q) = C_h \sum_{r=1}^{Q} rp(r) + C_s \lambda \sum_{r=-\infty}^{0} p(r),$$

where, r is the on-hand inventory level. The positive range for r corresponds to the items in stock whereas the negative range for r corresponds to items in back order; and $p(r)$ is its point probability function. We observe that $\sum_{r=1}^{Q} p(r)$ is the average value of the safety stock and $\lambda \sum_{r=-\infty}^{0} p(r)$ is the expected number of breakdown per unit time.

This is so because $\sum_{r=-\infty}^{0} p(r)$ is the percentage of time during which there is on-stock safety stock and λ is the average demand rate.

If one is able to determine $p(r)$, $TEC(Q)$ could easily be optimized with respect to Q.

TEST YOUR UNDERSTANDING

1. Following type of inventories may be held for a variety of purposes:
   (*a*) Anticipation inventories, (*b*) Cycle inventories,
   (*c*) Buffer inventories, (*d*) all of the above.
2. Reorder level of an item is always
   (*a*) less than its minimum stock. (*b*) more than its minimum stock.
   (*c*) more than its maximum stock. (*d*) less than its maximum stock.
3. When the average *DDLT* is available, an additional stock is called
   (*a*) safety stock, (*b*) buffer stock,
   (*c*) reserve stock, (*d*) none of these.
4. If an item is ordered frequently, the risk of running out of stock is
   (*a*) the least, (*b*) very high,
   (*c*) either the least or high, (*d*) can't say.
5. Service level is defined as the probability of
   (*a*) stocking during each order cycle, (*b*) not stocking during each order cycle,
   (*c*) stocking during lead time, (*d*) not stocking during lead time.
6. Which of the following is not correct?
   (*a*) Lead time is the time elapsing between placing an order and having the goods ready in stock for use.
   (*b*) The amount of safety stock required depends on variations in the demand rate and lead time.

(*c*) Under the assumptions made in inventory problems, the stock is always replenished at the designated time and safety stock is needed.

(*d*) Inventory cycle is the time period occurring between successive procurement actions.

7. Which of the following is correct?

(*a*) Reorder quantity in a "fixed-order interval system" equals *EOQ*.

(*b*) Review period of the item is always kept higher than its lead time.

(*c*) The (*S*, *s*) system requires high safety stock volumes.

(*d*) In periodic review system of inventory control, the stock is usually replenished at unequal time intervals.

8. If demand during lead time (*DDLT*) is probabilistic and normally distributed, then *ROL* is given by

(*a*) average *DDLT* + reserve stock + safety stock.

(*b*) average *DDLT* + reserve stock – safety stock.

(*c*) average *DDLT* + buffer stock – safety stock.

(*d*) additional stock + reserve stock + safety stock.

9. Which of the following is not correct?

(*a*) When it is assumed that there are no variations in demand rate and lead time, the reorder level would be equal to the demand during lead time (*DDLT*), and no safety stock is needed to be kept.

(*b*) When *DDLT* is assumed to be normally distributed with a certain mean and standard deviation, the service level corresponding to a given safety stock shall be equal to a given safety stock and shall be equal to the area under the normal curve to the left of the reorder point.

(*c*) When demand is stochastic, *ROL* equals expected demand during lead time plus safety stock.

(*d*) Both mean and standard deviation are necessary to be known for determining the service level corresponding to a given amount of safety stock, when *DDLT* is given to be normally distributed.

10. Which of the following is not correct?

(*a*) The system of periodic reviews generally require larger provisions for safety stock than does the fixed order quantity system.

(*b*) Two-bin system is best suited for '*B*' class of items.

(*c*) Greater the variation in demand and/or lead time, greater the amount of safety stock needed.

(*d*) Most of the inventory problems are multi-stage optimization problems in the sense that costs are to be optimized over a number of periods and as such the technique of dynamic programming is quite useful.

ANSWERS

1. (*d*)	**2.** (*b*)	**3.** (*b*)	**4.** (*a*)	**5.** (*b*)
6. (*c*)	**7.** (*c*)	**8.** (*a*)	**9.** (*d*)	**10.** (*b*)

REVIEW EXERCISE

1. Explain the concept of Q-system, P-system and Two-Bin system for management of inventories by giving appropriate examples. *[MBA (PT.), Delhi Univ. 2008]*
2. (*a*) What are the characteristics of two-bin system? Write down the major advantages and disadvantages of this system.
   (*b*) Fixed period of review of inventory is two-bin cycle system. Discuss.
   [PG Dipl., Indian Inst. of Mat. Man. 2019]
3. Explain and compare fixed order quantity model and fixed order period model. Also, state their applications. *[B.Tech., Hyderabad Univ. 2011; MBA (Dec.), Dayal bagh Ed. Inst. 2012]*
4. Distinguish between the fixed order quantity system and the periodic review system. Also, explain the *Ss* system. *[MBA (Nov.), Delhi Univ. 2003; MBA (PT.), Delhi Univ. 2008]*
5. Discuss the problem of inventory control, when the stochastic demand is uniform, production of commodity is instantaneous and lead time is negligible (discrete case).
6. Show that for a probabilistic discrete inventory model with instantaneous demand and no set-up cost, the optimum stock level z can be obtained by
$$\sum_{d=0}^{z} p(d) \geq \frac{C_s}{C_h + C_s} \geq \sum_{d=0}^{z-1} p(d).$$
7. Explain and solve an inventory model with instantaneous stochastic demand with no set-up cost. *[M.Sc. (Math.), Garhwal Univ. 1995]*
8. Derive a single period (one period) probabilistic inventory model with instantaneous and continuous demand and no set-up cost.
9. Derive the rule that gives *EOQ* for a single period stochastic inventory system for which holding cost and shortage cost are proportional to time and quantity. Assume that the demand is discrete. *[M.Sc. (Stat.), Delhi Univ. 1995]*
10. Write a short note on inventory models under uncertainty. *[MBA (Nov.), Delhi Univ. 2007]*
11. Compare two bin-system, P-system and Q-system of inventory control.
    [MBA, Delhi Univ. 2009]
12. Explain the application of dynamic programming in inventory problems with an example.
    [MBA, Annamalai Univ. 2009]

CHAPTER 24

Simulation

"If man begins with certainties, he ends in doubts; but if he contents to begin with doubts, he shall end in certainties"

For various managerial problems discussed so far, we have been able to find a mathematical solution. However, in each of these situations the problem was simplified by certain assumptions so that the appropriate mathematical techniques could be employed. There are certain managerial situations which are so complex that these can not be converted into exact mathematical models and hence the mathematical solution is impossible. In such cases, simulation is a very useful technique.

24:1. WHAT IS SIMULATION?

Simulation is a quantitative procedure which describes a process by developing a ***model*** of that process and then conducting a series of organized ***trial and error*** experiments to predict the behaviour of the process over time. Observing the experiments is much like observing the process in operation. Simulation is thus a quantitative method or technique that utilizes a computerized symbolic model in order to represent actual decision-making under conditions of uncertainty of evaluating alternative courses of action based upon facts and assumptions.

A definition of simulation as given by Shannon is:

*"Simulation is the process of designing a **model** of a real system and conducting experiments with this model for the purpose of understanding the behaviour (within the limits imposed by a criterion or set of criteria) for the operation of the system."*

24:2. REASONS FOR USING SIMULATION

The major reasons for applying simulation technique to managerial problems are given below:

1. Simulation is an appropriate tool to use in solving a business problem, when ***experimenting on the real system*** (*a*) would be disruptive, (*b*) would be too expensive, (*c*) does not permit replication events, and (*d*) does not permit control over key variables.
2. Simulation is a desirable tool for solving a business problem, when a mathematical model (*a*) is too complex to solve, (*b*) is beyond the capacity of available personnel, and (*c*) is not detailed enough to provide information on all important decision variables.
3. Simulation may be the only method available, because it is difficult to observe the actual environment.
4. Without appropriate assumption, it is impossible to develop a mathematical solution.
5. Actual observation of a system may be too expensive.
6. There may not be sufficient time to allow the system to operate extensively. For example, one may not be able to wait for a number of years to study the long-run trends of sales of a firm.

24:3. STEPS IN SIMULATION PROCESS

All simulations vary in complexity from situation to situation. However, in general, the process of simulating a system consists of the following steps:

Step 1. **Define the Problem.** This involves understanding the problem to be studied, the objectives of the study, the specific issues that need to be addressed and the time constraints for performing the study.

Step 2. **Analyse Cost and Benefits.** Since, a simulation study can be very expensive, it is advisable to review the probable costs and possible benefits of such a study before proceeding further.

Step 3. **Abstract the real system into model**. Formulate an appropriate mathematical model of the given situation.

Step 4. Test the model, compare its behaviour with the behaviour of the actual problem environment.

Step 5. Identify and collect the data needed to test the model.

Step 6. Run the simulation model.

Step 7. Analyse the results of the simulation. If simulation process is complete, then select the best course of action (or alternative), otherwise make the desired changes in model decision variables, parameters or design.

Step 8. Go to *Step* 4, and repeat the procedure..

Step 9. Rerun the simulation to test the new solution.

Step 10. Validate the simulation, *i.e.*, increase the chances that any inferences you draw about the real situation from running the simulation will be valid.

It may be noted that the fundamental principle of simulation is to make use of some device that can represent the phenomenon of a real life system to enable us to understand the properties, behaviour and functional characteristics of the system. The device used can be any convenient means — a mathematical formula or a physical model.

24:4. GENERATION OF RANDOM NUMBERS

Random numbers are the numbers which have equal probability of being generated. For example, when we are interested in one digit numbers 0, 1, 2, ..., 9, there are in all ten numbers and each of the numbers should have $\frac{1}{10}$ probability of being generated. There are several manual methods of generating numbers, *e.g.*, the rolling of dice, tossing of coins, pinning of roulettee, etc. But, it is convenient to use a table of random numbers which in fact is prepared on the basis of some such physical phenomenon. The grouping of random numbers in tables has no significance and one should be concerned with individual digits only. The first random number could be picked at random from any point in the tables and the subsequent ones are to be selected proceeding sequentially either in a vertical or horizontal direction.

A convenient manual method for generation of random numbers may be summarized in the following steps:

Step 1. Collect the data related to the current problem.

Step 2. Construct a frequency distribution with these data.

Step 3. Construct the relative frequency distribution.

Step 4. Assign a coding system that relates the identified events to generated random numbers.

Step 5. Select a suitable method for obtaining the required random numbers.

Step 6. Match the random numbers to the assigned events and tabulate the results.

Step 7. Repeat *Step 6* until the desired number of simulation runs has been generated.

Remark. The numbers generated by the computer are always predictable and reproducible and hence cannot be treated as random. For this reason, they are sometimes given the name ***pseudo-random numbers***. However, they satisfactorily play the role of random numbers in simulation.

24:5. EVENT-TYPE SIMULATION

The concept of ***event-type simulation*** is best explained by the following illustration:

Consider a situation where customers arrive at a one-man barber shop for hair cutting. The problem is to analyse the system in order to evaluate the quality of service and the economic feasibility of offering the service. To measure the quality of service, one has to make the assessment of the average waiting time per customer and the percentage of time the barber remains idle.

For the construction of a model of this system one notices that changes pertinent to the analysis of the system can occur only, if a customer arrives for service or departs after completion of service. If a customer arrives at barber's shop, he will have to wait, if the server (barber) is busy. On the other hand, a departure of customer, after being served, indicates that the server is available to serve the waiting customers, if any. Thus, we conclude that there occur two events, namely an arrival and a departure. It indicates that as the simulator progresses on the time scale, one should pay attention to the system only when an event occurs.

Let '*a*' denote the arrival event; '*d*' the departure event, and T, the simulated period (time span). Then, the simulator starts at time $t = 0$ and progresses up to time $t = t_1$, then $t = t_2$, and so on until entire simulated period T is covered.

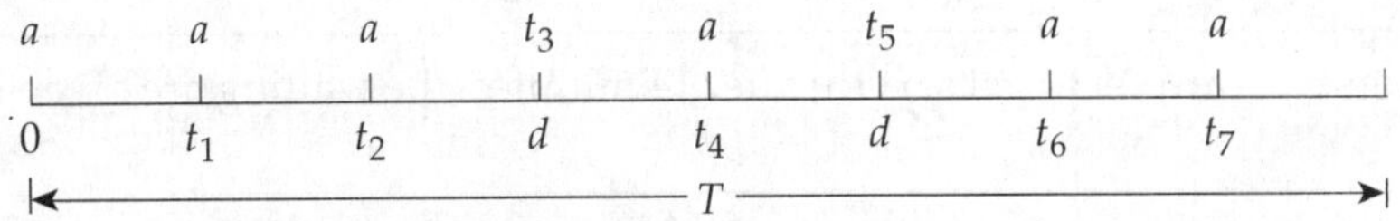

The above figure illustrates occurrences of '*a*' and '*d*' over T, where the simulation starts by generating '*a*' at t_1. Initially, as the facility is unoccupied, the customer starts service immediately. Then, the following two events must be generated:

(*a*) the next arrival may occur,

(*b*) the service of the customer may be completed.

The next arrival is determined from inter-arrival time. This determines '*a*' at t_2. Departure time of the customer in service is determined from service time and this generates '*d*' at t_3. Both '*a*' (at t_1) and '*d*' (at t_3) are now stored chronologically, so that the simulator recognizes that '*a*' occurs before '*d*'. The next event to be considered is '*a*' at t_2 and at this point '*a*' at t_1 is deleted from the stored list (because of past event).

The event '*a*' at t_2 generates '*a*' at t_4. Since, the facility is busy, the arriving customer '*a*' (at t_2) joins a waiting line. Now, '*a*' at t_4 is deleted from the list and '*d*' at t_3 is considered next. At this time, a customer is taken from the waiting line and departure event '*d*' at t_3 is generated. The process is repeated until the entire simulated period T is covered.

Example 24.1. *Customers arrive at a milk booth for the required service. Assume that inter-arrival and service times are constant and given by 1.5 and 4 time units, respectively. Simulate the system by hand computations for 14 time units. What is the average waiting time per customer? What is the percentage idle time of the facility? [Assume that the system starts at t = 0.]*

Solution. First customer starts service initially. Its departure time is $t = 0 + 4 = 4$ minutes. Next event (arrival) occurs at $t = 0 + 1.5 = 1.5$ minutes, which is listed before 'd' at $t = 4$. Now, since the facility is still busy, customer 2 is put in the queue and is first to be considered in this queue. A new arrival event (customer 3) occurs at $t = 1.5 + 1.5 = 3.0$ minutes which precedes 'd' at $t = 4$. Again, customer 3 is put in the queue and a new arrival event 'a' (customer 4) occurs at $t = 3.0 + 1.5 = 4.5$ minutes. This event succeeds 'd' at $t = 4$. At this point, first customer departs which leaves the service facility free. Customer 2, who was the first to join the queue, now gets service. The waiting time is computed as the time period from the instant he jointed the queue until his service is completed. This procedure is repeated until the simulated period is completed. The results of simulation are given in the following table:

Time (in minutes)	*Event arrival/departure*	*Customer No.*	*Waiting time*
0.0	*a*	1	
1.5	*a*	2	
3.0	*a*	3	
4.0	*d*	1 ⟶	4 – 1.5 = 2.5 (customer 2)
4.5	*a*	4	
6.0	*a*	5	
7.5	*a*	6	
8.0	*d*	2 ⟶	8 – 3.0 = 5.0 (customer 3)
9.0	*a*	7	
10.5	*a*	8	
12.0	*d*	3 ⟶	12 – 4.5 = 7.5 (customer 4)
13.0	*a*	9	
14.0	End	—	

For customers who are yet to get service after 14 minutes, the waiting are given in the following table:

Customer No. :	*5*	*6*	*7*	*8*	*9*	*10*
Waiting time (in minutes) :	14 – 6 = 8	14 – 7.5 = 6.5	14 – 9.0 = 5.0	14 – 10.5 = 3.5	14 – 12.0 = 2.0	14 – 13.5 = 0.5

From the simulation table given above, we observe that

(*i*) Average waiting time per customer is:

(2.5 + 5.0 + 7.5 + 8 + 6.5 + 5.0 + 3.5 + 2.0 + 0.5)/10, *i.e.*, 40.5/10 or 4.05 minutes.

(*ii*) Average waiting time per customer for those who must wait is:

40.5/9 or 4.5 minutes.

(*iii*) Percentage idle time of the facility is 0%, since the service facility is always busy.

PROBLEMS

24.01. At a telephone booth, suppose that the customers arrive with an average time of 1.2 time units between one arrival and the next. Service times are assumed to be 2.8 time units. Simulate the system for 12 time units by assuming that the system starts at $t = 0$. What is the average waiting time per customer? *[MCA, Purvanchal Univ. 1996]*

24.02. With the help of a single queuing model having inter-arrival and service time constantly 1.4 and 3 minutes respectively, explain discrete simulation technique taking 10 minutes as the simulation period. Find from this the average waiting time of a customer. (Assume that initially the system is empty and the first customer arrives at time $t = 0$.)

24.03. Suppose there are two type of customers. Type A customers have priority for service so that those of type B cannot be serviced until all waiting type A customers have completed their service. The service times for type A and type B customers are 2 and 1 time units, respectively. Further assume that the inter-arrival times for types A and B are 1.5 and 0.3 time units respectively. Simulate the system for 15 time units. Compute the average number of waiting customers for each type as well as the combined average. What is the average waiting time for each type of customer? [Assume that the system starts at $t = 0$.]

24:6. MONTE-CARLO SIMULATION

The *Monte-Carlo* technique is the most popular method of simulation. It uses random numbers and is employed to solve problems requiring decision-making under uncertainty and where mathematical solution is not possible. The general procedure of Monte-Carlo simulation comprises the following steps:

Step 1. **Define the Problem.** This involves defining clearly the objectives of the problem and identifying the variables and constraints which have maximum effect on the objectives of the decision-making problem.

Step 2. **Construct an Appropriate Model.**

(*a*) Specify the variables and parameters of the model.

(*b*) Formulate the appropriate decision rules, *i.e.*, state the conditions under which experiment is to be performed. Specify the number of runs of simulation to be made.

(*c*) Define the relationship between variables and parameters.

(*d*) Specify values of variables to be tested.

Step 3. **Establishing of Probability Distribution.** In the real world system, there are number of variables which are probabilistic in nature. These we shall like to simulate. The variables may specify: (*a*) Inventory demand on daily basis, (*b*) Lead time for inventory order to arrive, (*c*) Time between arrivals and users, (*d*) Service time, etc. Identify the probability distribution that will be used. The probability distribution can be Empirical, or Normal, or Poisson/Exponential.

Step 4. **Compute Cumulative Probability Distribution.** Determine the proper cumulative probability distribution of each variable selected in *Step 2*. Cumulative probability is the sum of numbers in the probability distribution column added to the previous cumulative probability.

Step 5. **Setting of Random Number Intervals.** From the cumulative probability distribution of each variable obtained in *Step 4*, generate a sequence of random numbers. Random numbers of a series of digits may be of one digit, two digits, and so on. For example, if we are interested in one digit number, then there are only ten such numbers and the chance of that number would be 1/10. These random numbers are called ***pseudo random numbers***.

Using cumulative probability distribution obtained in *Step 4*, set the intervals of random number for each variable/item.

Step 6. **Using Steps 1 through 5, Experiment the Model.**

(*a*) Select a random number generator and the random number to be used in ***simulation***. This random number can be selected from the random number table.

(*b*) Correlate the generated random number with the facts identified in *Steps 1 through 5*.

Step 7. Making use of *Step 6*, summarize and examine the results obtained in the form of an appropriate table.

Step 8. Evaluate the results of the simulation and select the best course of action.

Step 9. Formulate proposals for advice to management on the course of action to be adopted and modify the model, if necessary.

Example 24.2. *A tourist car operator finds that during the past few months, the use of car has varied so much that the cost of maintaining the car varied considerably. During the past 200 days, the demand for the car fluctuated as given below:*

Trips per week :	0	1	2	3	4	5
Frequency :	16	24	30	60	40	30

Using random numbers : 82, 95, 18, 96, 20, 84, 56, 11, 52, 03; simulate the demand for ten week period.

Solution. For various demand levels, random number interval is shown in the table given below:

Demand per week	*Frequency*	*Probability*	*Cumulative probability*	*Random number interval*
0	16	0.08	0.08	00–07
1	24	0.12	0.20	08–19
2	30	0.15	0.35	20–34
3	60	0.30	0.65	35–64
4	40	0.20	0.85	65–84
5	30	0.15	1.00	85–99

The simulated demand for the ten week period is as under:

Week :	*1*	*2*	*3*	*4*	*5*	*6*	*7*	*8*	*9*	*10*
Random no. :	82	95	18	96	20	84	56	11	52	03
Demand (cars) :	4	5	1	5	2	4	3	1	3	0

$$\text{Total demand} = 4 + 5 + 1 + 5 + 2 + 4 + 3 + 1 + 3 + 0 = 28$$

$$\therefore \quad \text{Average demand} = 28/10 = 2.8 \text{ car per week.}$$

Example 24.3. *The occurrence of rain in a city on a day is dependent upon whether or not it rained on the previous day. If it rained on the previous day, the rain distribution is:*

Event :	*No rain*	*1 cm rain*	*2 cm rain*	*3 cm rain*	*4 cm rain*	*5 cm rain*
Probability :	0.50	0.25	0.15	0.05	0.03	0.02

If it did not rain on the previous day, the rain distribution is:

Event :	*No rain*	*1 cm rain*	*2 cm rain*	*3 cm rain*
Probability :	0.75	0.15	0.06	0.04

Simulate the city's weather for 10 days and determine by simulation the total days without rain as well as total rainfall during the period. Use the following random numbers for simulation:

67, 63, 39, 55, 29 78, 70, 06, 78, and 76.

Assume that for the first day of the simulation it had not rained the day before.

[MBA, Rohilkhand Univ. 2011; MBA (June), Visvesvarayya Univ. 2011]

Solution. We simulate the city's weather with and without rainfall in the following steps:

Step 1. Previous day rain distribution:

Event	*Probability*	*Cumulative probability*	*Random number interval*
No rain	0.50	0.50	00–49
1 cm rain	0.25	0.75	50–74
2 cm rain	0.15	0.90	75–89
3 cm rain	0.05	0.95	90–94
4 cm rain	0.03	0.98	95–97
5 cm rain	0.02	1.00	98–99

Step 2. Previous day no rain distribution:

Event	*Probability*	*Cumulative probability*	*Random number interval*
No rain	0.75	0.75	00–74
1 cm rain	0.15	0.90	75–89
2 cm rain	0.06	0.96	90–95
3 cm rain	0.04	1.00	96–99

Step 3. Simulation of weather for 10 days using given random numbers:

Day	*Random number*	*Event*	*Cumulative rain*
1	67	No rain	
2	63	No rain	
3	39	No rain	
4	55	No rain	
5	29	No rain	
6	78	1 cm rain	1 cm
7	70	1 cm rain	2 cm
8	06	No rain	2 cm
9	78	1 cm rain	3 cm
10	76	2 cm rain	5 cm

During the simulated period it did not rain on 6 out of 10 days.

The total rainfall during the period is 5 cm.

Example 24.4. *An automobile company manufactures scooters. The daily production varies from 97 to 103 depending upon the availability of working conditions:*

Production (per day) :	97	98	99	100	101	102	103
Probability :	0.05	0.10	0.20	0.25	0.20	0.15	0.05

Produced scooters should be sent through a tempo which has space only for 100 scooters. If more than 100 scooters are produced, the extra scooters produced to be kept and clubbed with next day's production. Find the average empty space as well as average waiting of scooters for want of space.

Random nos. for 10 days : 1010, 0730, 7040, 6080, 9130, 4870, 5170, 2180, 8730, 9833.

Solution. For various quantity of production, random number interval is computed in the following table:

Production per day	*Probability*	*Cumulative probability*	*Random number interval*
97	0.05	0.05	00–04
98	0.10	0.15	05–14
99	0.20	0.35	15–34
100	0.25	0.60	35–59
101	0.20	0.80	60–79
102	0.15	0.95	80–94
103	0.05	1.00	95–99

Random numbers given to us are of 4 digits. Hence, we need to convert the previously generated random numbers into new ones, the formula being:

$$y_j = a + (b - a)\,\mu_j,$$

where, y_j = newly generated random number,

a = starting range of random number interval,

b = ending range of random number interval, and

μ_j = previously generated random number.

For 1010, $y_1 = 0 + (99 - 0) \times \frac{1010}{10,000} = 99 \times 0.1010 = 9.999$

For 0730, $y_2 = 0 + (99 - 0) \times \frac{0730}{10,000} = 99 \times 0.073 = 7.227$

Similarly, for the remaining 8 random numbers, we have

$y_3 = 99 \times 0.7040 = 69.69$, $y_4 = 99 \times 0.6080 = 60.19$, $y_5 = 99 \times 0.9130 = 90.38$,

$y_6 = 48.21$, $y_7 = 51.18$, $y_8 = 21.58$, $y_9 = 86.43$, $y_{10} = 97.35$.

Based on the ten new random numbers generated, the simulation of the production per day is as given below:

Day	*Random number*	*Production per day*	*Number of scooters waiting*	*Number of empty spaces*
1	9.99	98	Nil	2
2	7.23	98	Nil	2
3	69.69	101	1	—
4	60.19	101	1	—
5	90.39	102	2	—
6	48.21	100	Nil	—
7	51.18	100	Nil	—
8	21.58	99	–1	—
9	86.43	102	2 – 1 = 1	—
10	97.35	103	3	—
		Total :	8 scooters	4

Average empty space = 4/10 = 0.4 per day, and Average waiting = 8/10 = 0.8 per day.

Example 24.5. *A retailer deals in a perishable commodity. The daily demand and supply are variables. The data for the past 500 days, shows the following demand and supply:*

Supply		*Demand*	
Availability (kg)	*No. of days*	*Demand (kg)*	*No. of days*
10	40	10	50
20	50	20	110
30	190	30	200
40	150	40	100
50	70	50	40

The retailer buys the commodity at ₹20 per kg and sells it at ₹30 per kg. Any commodity remains at the end of the day, has no saleable value. Moreover, the loss (unearned profit) on any unsatisfied demand is ₹8 per kg. Given the following pair of random numbers, simulate 6 days sales, demand and profit.

(31, 18), (63, 84), (15, 79), (07, 32), (43, 75), (81, 27).

The first random number in the pair is for supply and the second random number is for demand, viz., in the first pair (31, 18), use 31 to simulate supply and 18 to simulate demand. **[CA Final (Nov.) 2000]**

Solution. The demand and supply patterns yield probability distributions. Random numbers are alloted for each level of supply and demand in the proportions indicated by the probabilities as shown below:

Supply				*Demand*			
Available (kg)	*Probability*	*Cumulative probability*	*Random number interval*	*Demand (kg)*	*Probability*	*Cumulative probability*	*Random number interval*
10	40/500 = 0.08	0.08	00–07	10	50/500 = 0.10	0.10	00–09
20	50/500 = 0.10	0.18	08–17	20	110/500 = 0.22	0.32	10–31
30	190/500 = 0.38	0.56	18–55	30	200/500 = 0.40	0.72	32–71
40	150/500 = 0.30	0.86	56–85	40	100/500 = 0.20	0.92	72–91
50	70/500 = 0.14	1.00	86–99	50	40/500 = 0.08	1.00	92–99

To determine the levels of supply and demand for each day, we now use the given random numbers in pairs alternatively. The first pair of random number is 31. Using this, the amount purchased on the first day is 30 kg. The second random number of this pair is 18. This corresponds to 20 kg of demand on first day. Continuing in this way, the simulation of supply and demand for six days is as under:

Day	*Random number*	*Supply (kg)*	*Random number*	*Demand (kg)*	*Cost ₹20 × supply*	*Revenue (₹)*	*Shortage (loss) or unsatisfied demand (₹)*	*Profit (₹)*
1	31	30	18	20	600	30 × 20 = 600	—	—
2	63	40	84	40	800	30 × 40 = 1,200	—	400
3	15	20	79	40	400	30 × 20 = 600	20 × 8 = 160	(200 – 160) = 40
4	07	10	32	30	200	30 × 10 = 300	20 × 8 = 160	100 – 160 = –60
5	43	30	75	40	600	30 × 30 = 900	10 × 8 = 80	300 – 80 = 220
6	83	40	27	20	800	30 × 20 = 600	—	400 – 600 = –200

During the simulated period of six days, the net profit of the retailer is:

₹ (400 + 40 – 60 + 220 – 200), *i.e.*, ₹ 400.

Example 24.6 (*Inventory Problem*). *A publishing house has bought out a new monthly magazine, which sells at ₹37.50 per copy. The cost of producing it is ₹30 per copy. A news-stand estimates the sales pattern of the magazine as follows:*

Demand (copies) :	0 < 300	300 < 600	600 < 900	900 < 1,200	1200 < 1,500	1500 < 1,800
Probability :	0.18	0.32	0.25	0.15	0.06	0.04

The news-stand has contract for 750 copies of the magazine per month from the publisher. The unsold copies are returnable to the publisher who will take them back at cost less ₹4 per copy for handling charges.

The news-stand manager wants to simulate the demand and profitability. The following random numbers may be used for simulation:

27, 15, 56, 17, 98, 71, 51, 32, 62, 83, 96, 69.

You are required to:

(*i*) *Allocate random numbers to the demand pattern forecast by the news-stand.*

(*ii*) *Simulate 12 months sales and calculate the monthly and annual profit/loss.*

(*iii*) *Calculate the loss on lost sales.* **[CA Final (Nov.), 2005]**

Solution.

(*i*) Allocation of random numbers:

Demand	*Probability*	*Cumulative probability*	*Random number interval*
0 < 300	0.18	0.18	00–17
300 < 600	0.32	0.50	18–49
600 < 900	0.25	0.75	50–74
900 < 1,200	0.15	0.90	75–89
1,200 < 1,500	0.06	0.96	90–95
1,500 < 1,800	0.04	1.00	96–99

(*ii*) Simulation of monthly pattern of demand and profitability:

Month	*R.No.*	*Demand (copies)*	*Sales (copies)*	*Return of copies 750-demand*	*Profit* on sale*	*Loss** on return*	*Net profit/ loss*	*Lost# sale*
1	27	450	450	300	3,375	1,200	2,175	—
2	15	150	150	600	1,125	2,400	(1,275)	—
3	56	750	750	—	5,625	—	5,625	—
4	17	150	150	600	1,125	2,400	(1,275)	—
5	98	1,650	750	—	5,625	—	5,625	900
6	71	750	750	—	5,625	—	5,625	—
7	51	750	750	—	5,625	—	5,625	—
8	32	450	450	300	3,375	1,200	2,175	—
9	62	750	750	—	5,625	—	5,625	—
10	83	1,050	750	—	5,625	—	5,625	300
11	96	1,650	750	—	5,625	—	5,625	900
12	69	750	750	—	5,625	—	5,625	—
				Total	54,000	7,200	46,800	2,100

* Profit per copy = ₹ 37.50 – ₹ 30 = ₹ 7.50 per copy.

** Loss per copy, if returned = ₹ 4 per copy.

Lost sale = Demand – Sale.

(*iii*) Loss due to lost sales = 2,100 × ₹ 7.50 = ₹ 15,750.

Example 24.7. *A car manufacturing company manufactures 40 cars per day. The sale of cars depends upon demand which has the following distribution:*

Sales of cars :	37	38	39	40	41	42
Probability :	*0.10*	*0.15*	*0.20*	*0.35*	*0.15*	*0.05*

The production cost and sale price of each car is ₹ 4 lakh and ₹ 5 lakh respectively. Any unsold car is to be disposed off at a loss of ₹ 2 lakh per car. There is a penalty of ₹ 1 lakh per car, if demand is not met. Using following random numbers, estimate total profit/loss for the company for the next ten days:

9, 98, 64, 98, 94, 01, 78, 10, 15, 19.

If the company decides to produce 39 cars per day, what will be its impact on profitability?

[MBA (Dec.), Visvesvarayya Univ. 2011]

Solution. Allocation of random numbers:

Sale of cars	*Probability*	*Cumulative probability*	*Random numbers*
37	0.10	0.10	00–09
38	0.15	0.25	10–24
39	0.20	0.45	25–44
40	0.35	0.80	45–79
41	0.15	0.95	80–94
42	0.05	1.00	95–99

Simulation of total profit/loss for the company:

Day	*Random numbers*	*Estimated sales*	*Profit* when production is 40 cars/day (₹ lakh)*	*Profit when production is 39 cars/day (₹ lakh)*
1	9	37	37 × 1 – 3 × 2 = 31	37 × 1 – 2 × 2 = 33
2	98	42	40 × 1 – 2 × 1 = 38	39 × 1 – 3 × 1 = 36
3	64	40	40 × 1 = 40	39 × 1 – 1 × 1 = 38
4	98	42	40 × 1 – 2 × 1 = 38	39 × 1 – 3 × 1 = 36
5	94	41	40 × 1 – 1 × 1 = 39	39 × 1 – 2 × 1 = 37
6	01	37	37 × 1 – 3 × 2 = 31	37 × 1 – 2 × 2 = 33
7	78	40	40 × 1 = 40	39 × 1 – 1 × 1 = 38
8	10	38	38 × 1 – 2 × 2 = 34	38 × 1 – 1 × 2 = 36
9	15	38	38 × 1 – 2 × 2 = 34	38 × 1 – 1 × 2 = 36
10	19	38	38 × 1 – 2 × 2 = 34	38 × 1 – 1 × 2 = 36
			Total : 359	Total : 359

* Profit = ₹ 5 lakh – ₹ 4 lakh = ₹ 1 lakh, Penalty = ₹ 1 lakh, loss due to unsold car = ₹ 2 lakh.

We observe that there is no additional profit or loss if the company decides to reduce production to 39 cars per day.

Example 24.8 (*Queuing Problem*). *With a view to improving the quality of customer services, a bank is interested in making an assessment of the waiting time of its customers coming to one of its branches located in a residential area. This branch has only one teller's counter. The arrival rate of the customers and service rate of the teller are given below:*

Time between two consecutive arrivals (in minutes)	:	3	4	5	6	7
Probability	:	0.17	0.25	0.25	0.20	0.13
Service time by the teller (in minutes)	:	3	4	5	6	7
Probability	:	0.10	0.30	0.40	0.15	0.05

You are required to simulate 10 arrivals of customers in the system starting from 11 A.M. and show the waiting time of customers and idle time of the teller.

Use following random numbers taking the first two random numbers in two digits each for first trial and so on:

11,	56,	23,	72,	94,	83,	83,	02,	97,	99,
83,	10,	93,	34,	33,	53,	49,	94,	37	97.

[CA Final (May) 2003]

Solution. Allocation of random numbers:

Time between two consecutive arrivals/service (in minutes)	*Arrivals*			*Service time*		
	Prob.	*Cum. Prob.*	*R.No.*	*Prob.*	*Cum. Prob.*	*R.No.*
3	0.17	0.17	00–16	0.10	0.10	00–09
4	0.25	0.42	17–41	0.30	0.40	10–39
5	0.25	0.67	42–66	0.40	0.80	40–79
6	0.20	0.87	67–86	0.15	0.95	80–94
7	0.13	1.00	87–99	0.05	1.00	95–99

Simulation of ten arrivals:

Sl.No.	Arrival			Service				Waiting time (in minutes	Idle time
	R.No.	Time (in minutes)	Time start	R.No.	Time (in minutes)	Time Start	Time End		
1	11	3	11.03	56	5	11.03	11.08	—	3
2	23	4	11.07	72	5	11.08	11.13	1	—
3	94	7	11.14	83	6	11.14	11.20	—	1
4	83	6	11.20	02	3	11.20	11.23	—	—
5	97	7	11.27	99	7	11.27	11.34	—	4
6	83	6	11.33	10	4	11.34	11.38	1	—
7	93	7	11.40	34	4	11.40	11.44	—	2
8	33	4	11.44	53	5	11.44	11.49	—	—
9	49	5	11.49	94	6	11.49	11.55	—	—
10	37	4	11.53	97	7	11.55	12.02	2	—
							Total	4	10

∴ Total waiting time of customers = 4 minutes, and total idle time of teller = 10 minutes.

Example 24.9. *Suppose that the demand for a particular item is normally distributed with a mean of 175 units and standard deviation of 25 units per day. Simulate the demand for the next 20 days.*

Random numbers : 98, 02, 80, 66, 96, 55, 50, 29, 58, 51, 04, 86, 24, 39, 47, 60, 65, 44, 93, 20.

Solution. Given that $\mu = 175$ and $\sigma = 25$. In normal distribution, 99.9% observations lie in $\mu \pm 3\sigma$ limits. Therefore, the demand can vary within $175 \pm 3 \times 25$, *i.e.*, from 100 to 250. Using this, random number interval is obtained in the table given below:

Demand	$z = \dfrac{x - \mu}{\sigma}$	Cumulative probability	Random number interval
100	–3	0.00	00
125	–2	0.02	00–01
150	–1	0.16	02–15
175	0	0.50	16–49
200	1	0.84	50–83
225	2	0.98	84–97
250	3	1.00	98–99

The simulated demand for the next 20 days is given in the table below:

Day	Random number	Demand	Day	Random number	Demand
1	98	250	11	04	150
2	02	150	12	86	225
3	80	200	13	24	175
4	66	200	14	39	175
5	96	225	15	47	175
6	55	200	16	60	200
7	50	200	17	65	200
8	29	175	18	44	175
9	58	200	19	93	225
10	51	200	20	20	175

Example 24.10 (*Inventory Problem*). *A company manufacture a component which requires a high degree of precision. Each unit of the component is, therefore, subjected to a strict quality control test to ascertain whether there is any defect in it. The defects are classified into three categories, viz., 'A', 'B' and 'C'.*

If defect 'A' occurs in the output, it is scrapped. If defect 'B' or 'C' occurs in the output, it is reworked to rectify the defect. The machine time required to rework defect 'B' component is 30 minutes and that for defect 'C' is 45 minutes. The probabilities of defect are as under:

		Defect A	Defect B	Defect C
Defect occurring	:	0.15	0.20	0.10
Defect not occurring	:	0.85	0.80	0.90

Using the following random numbers, simulate a study of 10 items of output and determine the number of items with no defects, number of items scrapped due to occurrence of defect A and the total machine time required for rework due to occurrence of defect 'B' or 'C':

Random number of defect A :	48,	55,	91,	40,	93,	01,	83,	63,	47,	52
Random number of defect B :	47,	36,	57,	04,	79,	55,	10,	13,	57,	09
Random number for defect C :	82,	95,	18,	96,	20,	84,	56,	11,	52,	03

[CA Final (May), 2004]

Solution. Since, the probabilities of occurrence of *A*, *B* and *C* defects are 0.15, 0.20 and 0.10 respectively, the random numbers 00 to 99 are allocated in proportion to the probabilities associated with each of the three defects:

Defect A		*Defect B*		*Defect C*	
Exists or not	*Random number interval*	*Exists or not*	*Random number interval*	*Exists or not*	*Random number interval*
Yes	00–14	Yes	00–19	Yes	00–09
No	15–99	No	20–99	No	10–99

Using given random numbers to determine the number of items without any defect, number of items scrapped and the total minutes of rework required; the simulation table is given below:

Item number	*Random numbers for*			*Whether any defect exists*	*Rework (in minutes)*	*Item scrapped*
	Defect A	*Defect B*	*Defect C*			
1	48	47	82	None	—	—
2	55	36	95	None	—	—
3	91	57	18	None	—	—
4	40	04	96	*B*	30	—
5	93	79	20	None	—	—
6	01	55	84	*A*	—	Scrap
7	83	10	56	*B*	30	—
8	63	13	11	*B*	30	—
9	47	57	52	None	—	—
10	52	09	03	*B*, *C*	30 + 45	—

From the above simulated table, we observe that no defect exists in 5 items, Defect *A* exists in one item, Defect *B* exists in 4 items and Defect *C* exists in one item. One item was scrapped and total work time is (30 + 30 + 30 + 30 + 45), *i.e.*, 165 minutes.

Example 24.11 (*Investment Problem*). *The investment required for introducing a new product is ₹10,000. The profits estimate for variable cost, selling price and annual sales volume are given. Simulate the situation to find out selling price, variable cost and annual sales volume by 5 trials using given random number and determine average profit.*

Variable cost		*Selling price*		*Annual volume*	
Value in ₹	*Probability*	*Value in ₹*	*Probability*	*Value in ₹*	*Probability*
1	0.20	3	0.20	4,000	0.20
2	0.50	4	0.60	6,000	0.50
3	0.30	5	0.20	8,000	0.30

Random numbers: (i) For variable cost: 9, 84, 41, 92, 65; (ii) For selling price: 24, 38, 73, 7, 4; (iii) For annual volume: 7, 48, 57, 64, 72.

Solution. Using the above information, random number interval for variable cost, selling price and annual volume is given in the table given below:

Variable cost			*Selling price*			*Annual volume*		
Value	*Cumulative probability*	*R.No. int.*	*Value*	*Cumulative probability*	*R.No. int.*	*Value*	*Cumulative probability*	*R.No. int.*
1	0.20	0–19	3	0.20	0–19	4,000	0.20	0–19
2	0.70	20–69	4	0.80	20–79	6,000	0.70	20–69
3	1.00	70–99	5	1.00	80–90	8,000	1.00	70–99

Simulation of 5 trials is given in the table below:

Trial	*Variable cost*		*Selling price*		*Volume*		*Profit (₹)*
	Random number	*Value*	*Random number*	*Value*	*Random number*	*Value*	
1	9	1	24	4	7	4,000	(4 – 1) × 4,000 = 12,000
2	84	3	38	4	48	6,000	(4 – 3) × 6,000 = 6,000
3	41	2	73	4	57	6,000	(4 – 2) × 6,000 = 12,000
4	92	3	7	3	64	6,000	(3 – 3) × 6,000 = 0
5	65	2	4	3	72	8,000	(3 – 2) × 8,000 = 8,000
							Total = ₹ 38,000

Average profit = ₹ 38,000/5 = 7,600.

Example 24.12 (*Investment and Budgeting Problem*). *A bookshop has studied the weekly receipts and payments over the past 200 days and has developed the following set of information:*

Daily receipts	*Probability*	*Weekly payments*	*Probability*
₹ 30,000	*0.20*	*₹ 40,000*	*0.30*
₹ 50,000	*0.30*	*₹ 60,000*	*0.40*
₹ 70,000	*0.40*	*₹ 80,000*	*0.20*
₹ 1,20,000	*0.10*	*₹ 1,00,000*	*0.10*

Using the following set of random numbers, simulate the weekly pattern of receipts and payments for the 15 days of the next fortnight, assuming further that the beginning bank balance is ₹ 80,000. What is the estimated balance at the end of 15 days period? What is the highest fortnightly balance during the fortnight? What is the average balance for the fortnight?

Random Numbers :

For receipts :	02,	93,	42,	58,	18,	49,	31,	45,	65,	71,	27,	20,	76,	07,	84
For payments :	64,	98,	32,	34,	07,	86,	48,	27,	83,	16,	74,	92,	04,	41,	81

Solution. According to given information, the random number interval is found to both the receipts and the payments as shown in table given below:

Receipts (₹)	*Probability*	*Cumulative Probability*	*Random number interval*	*Payments (₹)*	*Probability*	*Cumulative probability*	*Random number interval*
30,000	0.20	0.20	00–19	40,000	0.30	0.30	00–29
50,000	0.30	0.50	20–49	60,000	0.40	0.70	30–69
70,000	0.40	0.90	50–89	80,000	0.20	0.90	70–89
1,20,000	0.10	1.00	90–99	1,00,000	0.10	1.00	90–99

Based on random numbers, the receipts and payments as also the balance for various days can be simulated which is shown in table given below. From this table we observe that the balance at the end of 15 days period is a surplus of ₹ 10,000. The highest weekly balance during the fortnight is ₹ 70,000, while the average balance works out to be of ₹ 34,000.

Simulation Table

Day	*Receipts*		*Payments*		*Balance (₹)*
	R.No.	*Amount (₹)*	*R.No.*	*Amount (₹)*	
0	—	—	—	—	80,000
1	02	30,000	64	60,000	50,000
2	93	1,20,000	98	1,00,000	70,000
3	42	50,000	32	60,000	60,000
4	58	70,000	34	60,000	70,000
5	18	30,000	07	40,000	60,000
6	49	50,000	86	80,000	30,000
7	31	50,000	48	60,000	20,000
8	45	50,000	27	40,000	30,000
9	65	70,000	83	80,000	20,000
10	71	70,000	16	40,000	50,000
11	27	50,000	74	80,000	20,000
12	20	50,000	92	1,00,000	(–) 30,000
13	76	70,000	04	40,000	0
14	07	30,000	41	60,000	(–) 30,000
15	84	1,20,000	81	80,000	10,000

Example 24.13 (*Investment Problem*). *A washing powder manufacturing company wants to study an investment project based on three factors: market demand in units; contribution (= sales price – variable cost) per unit; and investment required. These factors are felt to be independent of each other. The company estimates the following probabilities:*

Annual demand		*Contribution per unit*		*Required*	
Units	*Probability*	*Profit (₹)*	*Probability*	*Investment (₹)*	*Probability*
20,000	0.05	3.00	0.10	17,50,000	0.25
25,000	0.10	5.00	0.20	20,00,000	0.50
30,000	0.20	7.00	0.40	25,00,000	0.25
35,000	0.30	9.00	0.20		
40,000	0.20	10.00	0.10		
45,000	0.10				
50,000	0.05				

Random Numbers:

For demand :	28	57	60	17	64	20	27	58	61	30
For profit :	19	07	90	02	57	28	29	83	58	41
For investment :	18	67	16	71	43	68	47	24	19	97

Using Monte-Carlo simulation for 10 runs, estimate the percentage of return on investment (ROI %) for each run as defined by

$$ROI\% = (Cash\ inflow \times 100)/Investment$$

Recommend an optimum investment strategy based on model value of ROI %. **[CA Final (Nov.) 1995]**

Solution. The yearly return can be determined by the formula:

Retrun on Investment (*ROI*) = (Contribution × Number of units demanded) / Investment)

To determine a cumulative probability corresponding to each of the three factors, we assign an appropriate set of random numbers representing each of the three factors as shown below:

Market Demand

Annual demand	*Probability*	*Cumulative probability*	*Random number*
20,000	0.05	0.05	00–04
25,000	0.10	0.15	05–14
30,000	0.20	0.35	15–34
35,000	0.30	0.65	35–64
40,000	0.20	0.85	65–85
45,000	0.10	0.95	86–94
50,000	0.05	1.00	95–99

Contribution

Contribution per unit (₹)	*Probability*	*Cumulative probability*	*Random number*
3.00	0.10	0.10	00–09
5.00	0.20	0.30	10–29
7.00	0.40	0.70	30–69
9.00	0.20	0.90	70–89
10.00	0.10	1.00	90–99

Investment

Investment required (₹)	*Probability*	*Cumulative probability*	*Random number*
17,50,000	0.25	0.25	00–24
20,00,000	0.50	0.75	25–74
25,00,000	0.25	1.00	75–99

We prepare the simulation worksheet for 10 trials. The simulated return (*ROI*) is also calculated by using the formula for *ROI*. The results of simulation are as under:

Simulation run	*RN for demand*	*Simulated demand ('000)*	*RN for profit*	*Simulated profit*	*RN for investment*	*Simulated investment ('000)*	*Simulated ROI* (%)*
1	28	30	19	5.00	18	1,750	8.57
2	57	35	07	3.00	67	2,000	5.25
3	60	35	90	10.00	16	1,750	20.00
4	17	30	02	3.00	71	2,000	4.50
5	64	35	57	7.00	43	2,000	12.25
6	20	30	28	5.00	68	2,000	7.50
7	27	30	29	5.00	47	2,000	7.50
8	58	35	83	9.00	24	1,750	18.00
9	61	35	58	7.00	19	1,750	14.00
10	30	30	41	7.00	97	2,500	8.40

* Simulated Return on Investment (*ROI*) (%) = $\dfrac{\text{Demand} \times \text{Profit per unit}}{\text{Investment}} \times 100$

The above table shows the highest likely *ROI*% of 20%, which is corresponding to the annual demand of 35,000 units resulting in a profit of ₹ 10 per unit. The required investment will be ₹ 17,50,000.

Example 24.14 (*Queuing Problem*). *A company has a single service station which has the following characteristics: The mean arrival rate of customers and the mean service time are 6.2 minutes and 5.5 minutes respectively. The time between an arrival and its services varies from one minute to seven minutes. The arrival and service time distributions are given below:*

Time (minutes) :	*1–2*	*2–3*	*3–4*	*4–5*	*5–6*	*6–7*
Arrival (probability) :	*0.05*	*0.20*	*0.35*	*0.25*	*0.10*	*0.05*
Service (probability) :	*0.10*	*0.20*	*0.40*	*0.20*	*0.10*	—

The queuing process starts at 11 A.M. and closes at 12 P.M. An arrival moves immediately into the service facility, if it is empty. On the other hand, if the service station is busy, the arrival will wait in the queue. Customers are served on the first come, first served basis.

If the clerk's wages are ₹6 per hour and the customer's waiting line costs ₹5 per hour, would it be economical for the manager to engage the second clerk? Use Monte-Carlo simulation technique.

[MCA, Madras Univ. 1999]

Solution. From the given frequency distribution of arrivals and service times, the probabilities and cumulative probabilities are first worked out as shown below. These, then, become the basis for generating arrival and service times in conjunction with a table of random numbers:

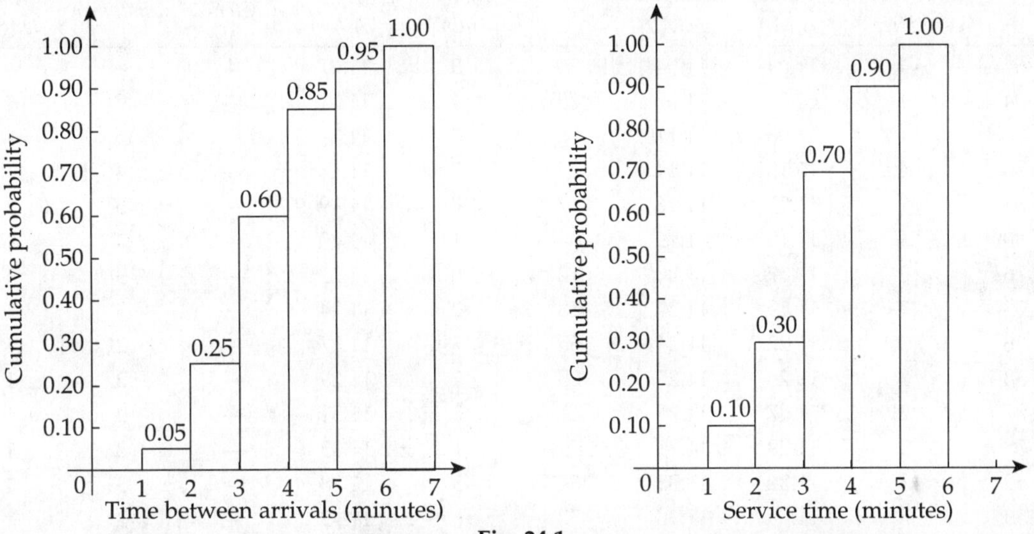

Fig. 24.1

Random Numbers Coding

Inter-arrival time (minutes)	*Cum. probability*	*RN allotted*	*Service time (minutes)*	*Cumulative Probability*	*RN allotted*
1–2	0.05	00–04	1–2	0.10	00–09
2–3	0.25	05–24	2–3	0.30	10–29
3–4	0.60	25–59	3–4	0.70	30–69
4–5	0.85	60–84	4–5	0.90	70–89
5–6	0.95	85–94	5–6	1.00	90–99
6–7	1.00	95–99			

A simulation worksheet is, now, developed in the following manner:

The random numbers developed above are related to the cumulative probability distributions of arrival and service times. The first random number of arrival time is 64. This number lies between

60 and 84 and indicates a simulated arrival time of 4 minutes. All simulated arrival and service times have been worked out in a similar fashion.

After generating the arrival and service times for a table of random numbers, the next step is to list the arrival time in the appropriate column. The first arrival comes in 4 minutes after the starting time. This means that the clerk waited for 4 minutes initially. It has been shown under the column — waiting time: clerk. The simulated service time for the first arrival is 3 minutes which results in the service begin completed by 11.07 A.M. The next arrival comes at 11.08 A.M. which indicates that no one has waited in the queue.

The second arrival comes at 11.05 A.M. But the services will begin only at 11.07 A.M., since the service of first arrival ends at 11.07 A.M. This means that the second arrival has waited for 2 minutes before the start of its service. One customer waiting in the queue is shown in the last column of the simulation table. The procedure is followed throughout the preparation of the simulation worksheet and is given as under:

Simulation Worksheet

RN	*Inter-arrival time*	*Arrival time (A.M.)*	*Service beings (A.M.)*	*RN*	*Service* *Time (min.)*	*Service* *Ends (A.M.)*	*Waiting time* *Clerk*	*Waiting time* *Customer*	*Line length*
64	4	11.04	11.04	30	3	11.07	4	—	—
04	1	11.05	11.07	75	4	11.11	—	2	1
02	1	11.06	11.11	38	3	11.14	—	5	1
70	4	11.10	11.14	24	2	11.16	—	4	1
03	1	11.11	11.16	57	3	11.19	—	5	1
60	4	11.15	11.19	09	1	11.20	—	4	1
16	2	11.17	11.20	12	2	11.22	—	3	1
08	2	11.19	11.22	18	2	11.24	—	3	1
36	3	11.22	11.24	65	3	11.27	—	2	1
38	3	11.25	11.27	25	2	11.29	—	2	1
07	2	11.27	11.29	11	2	11.31	—	2	1
08	2	11.29	11.31	79	4	11.35	—	2	1
59	3	11.32	11.35	61	3	11.38	—	3	1
53	3	11.35	11.38	77	4	11.42	—	3	1
03	1	11.36	11.42	10	2	11.44	—	6	1
62	4	11.40	11.44	16	2	11.45	—	4	1
36	3	11.43	11.46	55	3	11.49	—	3	1
27	3	11.46	11.49	52	3	11.52	—	3	1
97	6	11.52	11.52	59	3	11.55	—	—	—
86	5	11.57	11.57	63	3	12.00	2	—	—
Total	57				54		6	56	17

The following information can be obtained form the above simulation worksheet based on the period of one hour only:

(*a*) Average queue length

$$= \frac{\text{Number of customers in the waiting line}}{\text{Number}} = \frac{17}{20} = 0.85.$$

(*b*) Average waiting time of customer before service

$$= \frac{\text{Customer waiting time}}{\text{Number of arrivals}} = \frac{56}{20} = 2.80 \text{ minutes.}$$

(*c*) Average service time

$$= \frac{\text{Total service time}}{\text{Number of arrivals}} = \frac{54}{20} = 2.70 \text{ minutes.}$$

(*d*) Time a customer spends in the system

= Average service time + Average waiting time before service

= 2.70 + 2.80 = 5.50 minutes

Simulation worksheet developed in this problem also states that if one more clerk is added, there is no need for a customer to wait in the queue. But, before effecting any decision, the cost of having an additional clerk has to be compared with the cost due to customer waiting time. This can be worked out as follows:

One hour period		*Cost with one clerk*	*Cost with two clerks*
Customer waiting time (56 minutes × ₹ 5 per hour)	:	₹ 4.50	Nil
Clerk's cost	:	₹ 6.00	₹ 12
Total cost of one hour period	:	₹ 10.50	₹ 12

If the above analysis based on simulation for a period of one hour only is representative of the actual situation, then it may be concluded that the cost with one clerk is lower than what it is with two clerk. Hence, it would not be an economical proposition to engage an additional clerk.

Example 24.15 (*Maintenance Problem*). *A plant has a large number of similar machines. The machines break-down randomly and also break-downs are independent of each other. Once a machine breaks-down, it has to be taken out of production till the time it is repaired. On the basis of the past data, the following distributions have been constructed:*

No. of break-downs per hour	*Probability*	*Repair time in hrs. per breakdown*	*Probability*	*Random number interval*
0	0.900	1	0.100	000–099
1	0.085	2	0.240	100–399
2	0.012	3	0.450	340–789
3	0.003	4	0.165	790–954
		5	0.040	955–994
		6	0.005	995–999

Each hour that a machine remains idle due to being, or waiting, to be repaired, it costs the plant ₹ 80 per hour by way of lost production. If a repairman is paid at ₹ 8 per hour, how many repairmen should be hired by the company to service the machine breakdowns? For the purpose, simulate the system for a 50-hour period and use the following random numbers, reading row-wise starting with the NW corner:

For Breakdowns

100	375	084	990	128	660	310	852	635	737
985	118	834	886	995	654	801	743	699	098
914	803	441	125	636	611	154	945	424	235
044	005	359	598	460	321	692	195	451	948
980	331	809	797	186	740	541	116	483	690

For Repair Times

765	648	196	093	801	340	455	020	053	035
672	121	099	195	981	783	389	421	125	623

Solution. The random number coding for the hourly breakdowns and the repair times is shown in the table given below:

Breakdowns	*Probability*	*Random number interval*	*Repair time (hrs.)*	*Probability*	*Random number interval*
0	0.900	000–899	1	0.100	000–099
1	0.085	900–984	2	0.240	100–339
2	0.012	985–996	3	0.450	340–789
3	0.003	997–999	4	0.165	790–954
			5	0.040	955–994
			6	0.005	995–999

The simulation worksheet displays the needed data on hourly breakdowns and the repair times. Assuming that each breakdown occurs at the beginning of an hour, we observe that two machines would break at the beginning of the fourth hour, two at the beginning of the 15th hour and so on. It is estimated that in all 10 machines will breakdown during the period under review. The repair times simulated in respect of the breakdowns are shown in part (*b*) of the worksheet.

Waiting Time with One Repairman. When only one repairman is hired, the waiting time for the machines and the total cost involved would be obtained as follows. The first breakdown is estimated to occur in the beginning of the fourth hour. The breakdown M_1 would take 3 hours for repairs. The other breakdown, M_2 would have to wait for 3 hours before its repair starts (which would also take 3 hours to complete). Both these machines would be ready, however, before the machines M_3 and M_4 breakdown at the beginning of the 11th hour. Clearly, machine M_4 would have to wait for 2 hours during which M_3 would be repaired. The machines M_5 and M_6 are likely to fail at the beginning of the 15th hour and would require 4 and 3 hours respectively.

Of these, M_6 would wait for 4 hours before repair work can start on it. The machine M_7, failing at the start of the 21st hour, would have to wait for one hour before it is serviced. Note that none of the other machines will have to wait. Thus, the waiting time of the machines in the queue adds up to 10 hours:

Machine :	M_2	M_4	M_6	M_7
Waiting time :	3 hours	2 hours	4 hours	1 hour

Simulation Worksheet

(*a*) *Breakdowns*

Hour (1)	*Random No.* (2)	*Breakdowns* (3)	*Hour* (1)	*Random No.* (2)	*Breakdowns* (3)
1	100	0	9	635	0
2	375	0	10	737	0
3	084	0	11	985	2 M_3, M_4
4	990	2 M_1, M_2	12	118	0
5	128	0	13	834	0
6	660	0	14	886	0
7	310	0	15	995	2 M_5, M_6
8	852	0	16	654	0

(*Contd.*)

(1)	*(2)*	*(3)*	*(1)*	*(2)*	*(3)*
17	801	0	34	598	0
18	743	0	35	460	0
19	699	0	36	321	0
20	098	0	37	692	0
21	914	1 M_7	38	195	0
22	803	0	39	451	0
23	441	0	40	948	1 M_9
24	125	0	41	980	1 M_{10}
25	636	0	42	331	0
26	611	0	43	809	0
27	154	0	44	797	0
28	945	1 M_8	45	186	0
29	424	0	46	740	0
30	235	0	47	541	0
31	044	0	48	116	0
32	005	0	49	483	0
33	359	0	50	690	0

(b) *Repair Times*

Machine	*Random number*	*Repair time (hrs.)*
M_1	765	3
M_2	648	3
M_3	196	2
M_4	093	1
M_5	801	4
M_6	340	3
M_7	455	3
M_8	020	1
M_9	053	1
M_{10}	035	1

Total cost = Idle time cost + Repairman's wages

= Idle time × Hourly rate + Total hours × Hourly rate

Idle time = Repairing time + Waiting time.

In the present case, idle time = 22 + 10 = 32 hours.

∴ Total cost = 32 × 80 + 50 × 8 = 2,560 + 400 = ₹ 2,960.

For Two Repairmen. In this case, since not more than two machines have been found to breakdown simultaneously, none of the machines would have to wait. Also, each of the machines would be serviced before another breaks-down. Thus, the total idle time would be only the repairing time of 22 hours.

∴ Total cost = 22 × 80 + 50 × 8 × 2 = 1,760 + 800 = ₹ 2,560.

Thus, hiring more than 2 repairmen would only add to the total cost.

Hence, plant should hire one additional repairman.

Example 24.16 (*Operations Scheduling Problem*). *A job has to be processed over two machines, machine M_1 and machine M_2 in that order. The distribution of inter-arrival time of the jobs at the first machine is as follows:*

Time (minutes) :	1	2	3	4
Probability :	0.2	0.2	0.2	0.4

The processing times at the two machines are as follows:

Machine M_1		*Machine M_2*	
Time (minutes)	*Probability*	*Time (minutes)*	*Probability*
1	0.1	4	0.2
2	0.2	5	0.3
3	0.3	6	0.4
4	0.3	6	0.1
5	0.1		

On the basis of 10 simulation runs, find out the average queue length before machine M_1 and the average queue length before machine M_2.

Solution. The various steps for simulation are explained below:

1. Simulate the inter-arrival time of the ith job on machine M_1 with the help of a random number ($i = 1, 2, ...$).
2. Arrival time of ith job

   = Arrival time of $(i-1)^{th}$ job ($i = 2, 3, ...$) + Inter-arrival time of ith job.

   Arrival time of the first job

   = Inter-arrival time of the first job.
3. Simulate the processing time on machine M_1 by random numbers.
4. Departure time on machine M_1

   = Arrival time for processing on machine M_2

   = Max (Arrival time of ith job, process completion time of $(i-1)^{th}$ job on machine M_1)

   + Processing time of ith job on machine M_1.
5. If arrival time of ith job < process completion time of $(i-1)^{th}$ job, i^{th} job waits.

   If arrival time of ith job < process completion time of $(i-2)^{th}$ job, both i^{th} and $(i-1)^{th}$ jobs wait.
6. Simulate the processing time of machine M_2 with the help of random numbers.
7. Process completion by machine M_2 of ith job

   = Max. [Arrival time of ith job, process completion time of $(i-1)^{th}$ job]

   + Process time by machine M_2.
8. If arrival time of ith job < process completion time of $(i-1)^{th}$ job, then i^{th} job waits. If arrival time of ith job < process completion time of $(i-2)^{th}$ job, both i^{th} and $(i-1)^{th}$ jobs wait, and so on.

Using the procedure, the simulated processing data of 10 arrivals are displayed in the simulation worksheet:

Simulation Worksheet

Simulation run	*RN*	*Inter-arrival time*	*Arrival time on M_1*	*RN*	*Process time on M_1*	*Departure time on M_1*	*RN*	*Process time on M_2*	*Completion time on M_2*	*Queue size M_1*	*Queue size M_2*
1	8	4	4	0	1	5	3	5	10	—	—
2	7	4	8	6	4	12	0	4	16	—	—
3	8	4	12	9	5	17	0	4	21	—	—
4	9	4	16	3	3	20	0	4	25	1	1
5	4	3	19	7	4	24	0	4	29	1	1
6	4	3	22	0	1	25	2	5	34	1	1
7	0	1	23	0	1	26	4	5	39	2	2
8	6	4	27	5	3	30	5	6	45	—	2
9	8	4	31	6	4	35	1	4	49	—	2
10	7	4	35	4	3	38	2	5	54	—	3

On the basis of simulation study, the average queue length before machine M_1 is 0.5, and the average queue length before machine M_2 is 1.2.

Example 24.17 (*Network Scheduling Problem*). *A project schedule consists of five activities and the duration of these activities is non-deterministic with the following probability distribution:*

Activity	*Days*	*Probability*
	1	0.2
1–2	4	0.5
	8	0.3
	2	0.3
1–3	4	0.5
	7	0.2
	2	0.3
2–4	4	0.3
	6	0.4
	3	0.3
3–4	6	0.4
	8	0.3
	2	0.2
4–5	3	0.2
	4	0.6

Simulate the duration of the project 10 times and estimate the chances of various paths to be critical. What is the average duration of the project?

Solution. From the *PERT* network, we observe that there are two paths from the start to end of the project, namely $1 \to 2 \to 4 \to 5$ and $1 \to 3 \to 4 \to 5$. Using one-digit random number corresponding to the probability distribution of various activities, we obtain the simulated duration of each path and hence the simulated duration of the project. The table on the next page shows the results of the simulation:

Simulation Worksheet

Simulation run	*Activity* 1–2 *RN*	*D*	1–3 *RN*	*D*	2–4 *RN*	*D*	3–4 *RN*	*D*	4–5 *RN*	*D*	*Project duration*	*Critical paths*
1	8	8	9	2	8	6	3	6	6	4	18	$1\to2\to4\to5$
2	5	4	5	4	8	6	1	3	7	4	14	$1\to2\to4\to5$
3	5	4	6	4	7	6	4	6	7	4	14	$1\to2\to4\to5$, $1\to3\to4\to5$
4	7	8	5	4	1	2	9	8	5	4	16	$1\to3\to4\to5$
5	0	1	6	4	8	6	1	3	8	4	11	$1\to2\to4\to5$
6	8	8	3	4	0	2	4	6	3	3	13	$1\to2\to4\to5$, $1\to2\to5$, $1\to4\to5$
7	4	4	7	4	4	4	0	3	3	3	11	$1\to4\to5$
8	5	4	8	7	2	2	6	6	6	4	17	$1\to4\to5$
9	2	4	5	4	9	6	0	3	3	3	13	$1\to4\to5$
10	6	4	1	2	3	4	7	8	0	2	12	$1\to4\to5$

RN : Random number; *D* : Duration.

On the basis of above simulation we observe that there is a 70 per cent chance of the path $1\to4\to5$ being critical. The chance that the path $1\to4\to5$ will be critical is 60 per cent. Further, the average duration of the project is 13.9 days.

Example 24.18 (*Network Scheduling Problem*). *A project consists of 7 activities. The time for performance of each of the activities is a random variable with the respective probability distribution as given below:*

Activity	*Immediate predecessor*	*Time (in days) and its probability*				
A	—	3	4	5		
		0.20	0.60	0.20		
B	—	4	5	6	7	8
		0.10	0.30	0.30	0.20	0.10
C	A	1	3	5		
		0.15	0.75	0.10		
D	B, C	4	5			
		0.80	0.20			
E	D	3	4	5	6	
		0.10	0.30	0.30	0.30	
F	D	5	7			
		0.20	0.80			
G	E, F	2	3			
		0.50	0.50			

(*a*) *Draw the network diagram and identify the critical path using the expected activity times.*

(*b*) *Simulate the project using random numbers to determine the activity times. Find the critical path and the project duration.*

(*c*) *Repeat the simulation four more times. Is the same path critical in all the simulations?*

Solution.

(*a*) Using the given information, the resulting network is shown on the next page. The time for each of the activities is the expected time obtained by the summation of the products of the activity time and the corresponding probability.

For example, for activity A, the expected time = $3 \times 0.20 + 4 \times 0.60 + 5 \times 0.20 = 4$ days.

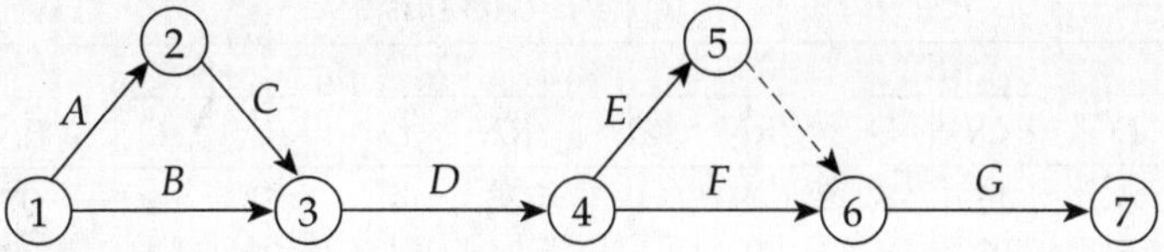

Observe that the critical path of the project is $1 \to 2 \to 3 \to 4 \to 6 \to 7$, with an expected duration of 20.2 days.

(*b*) and (*c*). To use simulation for obtaining the project length we first determine the random number intervals for each activity. Random number coding is shown in table given below. The five simulation runs, by taking random numbers, are then shown in simulation worksheet. For each run, the critical path and the project durations are obtained and the project time is obtained as follows:

Total time = Larger of times for activities A and C + Time for activity D
+ Larger of time for activities E and + Time for activity G

Random Number Coding

Activity	*Time*	*Probability*	*Cumulative probability*	*Random number interval*
	3	0.20	0.20	00–19
A	4	0.60	0.80	20–79
	5	0.20	1.00	80–99
	4	0.10	0.10	00–09
	5	0.30	0.40	10–39
B	6	0.30	0.70	40–69
	7	0.20	0.90	70–89
	8	0.10	1.00	90–99
	1	0.15	0.15	00–14
C	3	0.75	0.90	15–89
	5	0.10	1.00	90–99
D	4	0.80	0.80	00–79
	5	0.20	1.00	80–99
	3	0.10	0.10	00–09
E	4	0.30	0.40	10–39
	5	0.30	0.70	40–69
	6	0.30	1.00	70–99
F	5	0.20	0.20	00–19
	7	0.80	1.00	20–99
G	2	0.50	0.50	00–49
	3	0.50	1.00	50–99

Simulation Worksheet

Simulation run	A		B		C		D		E		F		G	
	RN	*D*	*RN*	*D*	*RN*	*D*	*RN*	*D*	*RN*	*D*	*RN*	*D*	*RN*	*D*
1	68	4	13	5	09	1	20	4	73	6	07	5	92	3
2	99	5	93	8	18	3	24	4	22	4	07	5	29	2
3	57	4	33	5	49	3	65	4	92	6	98	7	00	2
4	57	4	12	5	31	3	96	5	86	6	92	7	91	3
5	77	4	37	5	34	3	11	4	27	4	10	5	59	3

RN : Random Number, *D* : Duration.

We get the following information from the above table:

Simulation run	*Activity times*	*Project duration*	*Critical path(s)*
1	5 + 4 + 6 + 3	18 days	$1 \to 2 \to 3 \to 4 \to 5 \to 6 \to 7$ $1 \to 3 \to 4 \to 5 \to 6 \to 7$
2	8 + 4 + 5 + 2	19 days	$1 \to 2 \to 3 \to 4 \to 6 \to 7$ $1 \to 3 \to 4 \to 6 \to 7$
3	7 + 4 + 7 + 2	20 days	$1 \to 2 \to 3 \to 4 \to 6 \to 7$
4	7 + 5 + 7 + 3	22 days	$1 \to 2 \to 3 \to 4 \to 6 \to 7$
5	7 + 4 + 5 + 3	19 days	$1 \to 2 \to 3 \to 4 \to 6 \to 7$

It may be observed that the same path is not critical in each of the runs.

PROBLEMS

24.04. A retail store experiences the following distribution for sales of a product:

Sales (units) :	0	1	2	3	4	5	6
Probability :	0.08	0.12	0.28	0.24	0.14	0.10	0.04

Using the following random numbers, simulate demand for ten days and determine the average sales over the 10 days period:

87, 02, 37, 76, 56, 19, 99, 47, 78, 14.

[M.Com. (O.C.), Delhi Univ. 2010]

24.05. A confectioner sells confectionery items. Past data of demand per week (in hundred kilograms) with frequency is given below:

Demand/week :	0	5	10	15	20	25
Frequency :	2	11	8	21	5	3

Using the following sequence of random numbers, generate the demand for the next 10 weeks. Also, find the average demand per week:

35, 52, 90, 13, 23, 73, 34, 57, 35, 83, 94, 56, 67, 66, 60.

[B.Sc. (Stat.), Delhi Univ. 2003]

24.06 A car manufacturing company manufactures 40 cars per day. The sale of cars depend upon demand which has the following probabilities:

Sales of cars :	37	38	39	40	41	42
Probability :	0.10	0.15	0.20	0.35	0.15	0.05

The production cost and sale price of each car are ₹ 4 lakh and ₹ 5 lakh respectively. Any unsold car is to be disposed off at a loss of ₹ 2 lakh per car. There is a penalty of ₹ 1 lakh per car, if the demand is not met. Using the following random numbers, estimate total profit/loss for the company for the next ten days:

9, 98, 64, 98, 94, 01, 78, 10, 15, 19.

If the company decides to produce 39 cars per day, what will be its impact on profitability?

[MBA (Dec.), Visvesvarayya Univ. 2011]

24.07. A company manufactures 30 items per day. The sale of these items depends upon demand which has the following distribution:

Sales (units) :	*27*	*28*	*29*	*30*	*31*	*32*
Probability :	0.10	0.15	0.20	0.35	0.15	0.05

The production cost and sale price of each unit are ₹ 40 and ₹ 50 respectively. Any unsold product is to be dispose of at a loss of ₹ 15 per unit. There is a penalty of ₹ 5 per unit, if the demand is not met. Using the following random numbers estimate total profit/loss for the company for the next 10 days:

10, 99, 65, 99, 95, 01, 79, 11, 16, 20.

If the company decides to produce 29 items per day, what is the advantage or disadvantage to the company?

[MBA (Dec.), Rohilkhand Univ. 2011; B.Tech. (Prod.) (Nov.), Panjab Tech. Univ. 2012]

24.08. A book-store wishes to carry Systems Analysis and Design in stock. Demand is probabilistic and replenishment of stock take 2 days (*i.e.*, if an order is placed on March 1, it will be delivered at the end of day on March 3). The probabilities of demand are given below:

Demand (daily) :	*0*	*1*	*2*	*3*	*4*
Probability :	0.05	0.10	0.30	0.45	0.10

Each time an order is placed, the store incurs an ordering cost of ₹ 10 per order. The store also incurs a carrying cost of ₹ 0.50 per book per day. The inventory carrying cost is calculated on the basis of stock at the end of each day. The manager of the book-store wishes to compare two options for his inventory decision:

A. Order 5 books, when the inventory at the beginning of the day plus orders outstanding is less than 8 books.

B. Order 8 books, when the inventory at the beginning of the day plus orders outstanding is less than 8 books.

Currently (beginning of the 1*st day*) the store has stock of 8 books plus 6 books ordered 2 days ago and expected to arrive next day. Using Monte-Carlo simulation for 10 cycles, recommend which option the manager should choose?

The two-digit random numbers are given below:

89, 34, 78, 63, 61, 81, 39, 16, 13, 73.

[ICWA (June), 2006]

24.09. Bright Bakery keeps stock of a popular brand of cake. Previous experience indicates the daily demand as given below:

Daily demand :	*0*	*10*	*20*	*30*	*40*	*50*
Probability :	0.01	0.20	0.15	0.50	0.12	0.02

Consider the following sequence of random numbers:

48, 78, 19, 51, 56, 77, 15, 14, 68, 09.

Using this sequence, simulate the demand for the next 10 days. Find out the stock situation, if the owner of the bakery decides to make 30 cakes every day. Also, estimate the daily average demand for the cakes on the basis of simulated data. ***[MBA, Bangalore Univ. 2016; MBA, Hydrabad Univ. 2012; MBA, Madurai Univ. 2008]***

24.10. The automobile company manufactures around 150 scooters. The daily production varies from 146 to 154 depending upon the availability of raw materials and other working conditions:

Production (per day) :	146	147	148	149	150	151	152	153	154
Probability :	0.04	0.09	0.12	0.14	0.11	0.10	0.20	0.12	0.08

The finished scooters are transported in a specially arranged lorry accommodating 150 scooters. Using following random numbers:

80, 81, 76, 75, 64, 43, 18, 26, 10, 12, 65, 68, 69, 61, 57.

Simulate the process to find out:

(*i*) What will be the average number of scooters waiting in the factory?
(*ii*) What will be the average number of empty space on the lorry?

[MBA, Bangalore Univ. 2017]

24.11. A company manufactures around 200 mopeds depending upon the availability of raw materials and other conditions. The daily production has been varying from 196 mopeds to 204 mopeds, whose probabilities are as given below:

Production per day :	196	197	198	199	200	201	202	203	204
Probability :	0.05	0.09	0.12	0.14	0.20	0.15	0.11	0.08	0.06

The finished mopeds are transported in a specially designed three-storeyed lorry that can accommodate only 200 mopeds. Using following random numbers:

82, 89, 78, 24, 53, 61, 18, 45, 04, 23, 50, 77, 27, 54, 10;

simulate the process to find out:

(*i*) What will be the average number of mopeds waiting in the factory?
(*ii*) What will be the average number of empty spaces on the lorry?

[MBA, Rohilkhand Univ. 2011; ICWA Final (Dec.) 2006]

24.12. A dentist schedules all her patients for 30 minutes appointments. Some of the patients take more or less than 30 minutes depending on the type of dental work to be done. The following summary shows the various categories of work, their probabilities and the time needed to complete the work:

Category	*Time required (minutes)*	*Probability of category*
Filling	45	0.40
Crown	60	0.15
Cleaning	15	0.15
Extraction	45	0.10
Checkup	15	0.20

Simulate the dentist's clinic for four hours and determine the average waiting time for the patients as well as the idleness of the doctor. Assume that all the patients show up at the clinic at exactly their scheduled arrival times, starting at 8 A.M. Use the following random numbers for handling the above problem.

40, 82, 11, 34, 25, 66, 17 and 79.

[B.Tech., Panjab Tech. Univ. 2012; BE (Mech.), Nagpur Univ. 2011]

24.13. Two persons X and Y work on a two-station assembly line. The distributions of activity times at their stations are given below:

Time in seconds :	10	20	30	40	50	60	70	80
Time frequency for X :	4	7	10	15	35	18	8	3
Time frequency for Y :	2	3	6	8	12	9	7	3

(*a*) Simulate operation of the line for the eight items.

(*b*) Assuming *Y* must wait until X completes the first item before starting work, will he have to wait to process any of the other seven items? What is the average waiting time of items for *Y*? Use the following random numbers:

For X :	83,	70,	06,	12,	59,	46,	54	and	04.
For Y :	51,	99,	84,	81,	15,	36,	12	and	54.

(*c*) Determine the inventory of items between two stations.

(*d*) What is the average production rate? ***[BE (Mech.), Kurukshetra Univ. 2012]***

24.14. Haggins Plumbing and Heating maintains a stock of 30-gallon hot water heaters that it sells to home-owners and installs for them. The owner likes the idea of having large supply on hand, so as to meet over the past 50 weeks and note the following:

Hot water heater sales per week :	*4*	*5*	*6*	*7*	*8*	*9*	*10*
Number of weeks heaters were sold :	6	5	9	12	8	7	3

Using random numbers given below, simulate demand for 20 weeks and answer the following questions:

Random Numbers:

10,	24,	03,	32,	23,	59,	95,	34,	34,	51,
08,	48,	66,	97,	03,	96,	46,	74,	77	44.

(*i*) If Haggins maintains a constant supply of 8 hot water heaters in any given week, how many times will he be out of stock during the 20-week simulation period?

(*ii*) What is the average number of heaters demanded per week over the 20-week interval?

[M.Com. (OC), Delhi Univ. 2010]

24.15. Phillips India is engaged in manufacturing different type of equipments for various consumers. The company has two assembly lines to produce its product. The processing time for each of the assembly lines is regarded as a random variable and is described by the following distributions:

Processing time (min.) :	*40*	*42*	*44*	*46*	*48*
Assembly X :	0.10	0.15	0.40	0.10	0.25
Assembly Y :	0.20	0.40	0.20	0.15	0.05

Using the following random numbers, generate data on the processing times for 10 units of the product and compute the expected processing time for the product:

4236, 7573, 4943, 1283, 2014, 3604, 9344, 5316, 7606, 0089.

For the purpose, read the numbers horizontally, taking the first two digits for the processing time on assembly X and the last two digits for processing time on assembly *Y*.

[B.Tech. Panjab Tech. Univ. 2012]

[***Hint:*** Random numbers for *X* : 42, 75, 49, 12, 20, 36, 93, 53, 76, 00.
Random numbers for *Y* : 36, 73, 43, 83, 14, 04, 44, 16, 06, 89.]

24.16. The management of *ABC* Company is considering the question of marketing a new product. The fixed cost required in the project is ₹ 4,000. Three factors are uncertain, *viz.*, the selling price, variable cost and the annual sales volume. The product has a life of only one year. The management has the data on these factors as under:

Selling price (₹)	*Probability*	*Variable cost (₹)*	*Probability*	*Sales volume (units)*	*Probability*
3	0.2	1	0.3	2,000	0.3
4	0.5	2	0.6	3,000	0.3
5	0.3	3	0.1	5,000	0.4

Consider the following sequence of thirty random numbers:

81,	32,	60,	04,	46,	31,	67,	25,	24,	10,	40,	02,	39,	68,	08,
59,	66,	90,	12,	64,	79,	31,	86,	68,	82,	89,	25,	11,	98,	16.

Using the sequence (First 3 random numbers for the first trial, etc.), simulate the average profit for the above project on the basis of 10 trials. ***[B.Tech. (Prod.) Panjab Tech. Univ. 2012]***

24.17. A small scale factory has a machinist to process the jobs. The inter-arrival time for the job is not fixed. The processing time on the jobs also varies. These times are described by frequency distributions given below:

Inter-arrival time distribution		*Processing time distribution*	
Inter-arrival time (min.)	*Frequency*	*Processing time (min.)*	*Frequency*
4	10	3	5
5	20	4	30
6	40	5	30
7	20	6	30
8	10	7	5

Determine the idle time of the machinist and the waiting time of jobs in a period of 2 hours. If the machinist's wages are ₹ 20 per hour and the job's waiting time costs ₹ 30 per hour, would it be economical to engage a second attendant? ***[B.Tech. Panjab Tech. Univ. 2012]***

24.18. The following data reveals the details pertaining to the inter-arrival time of patients and service time by the clinic along with their respective probability distribution in respect of outpatient department (*OPD*) of a private hospital:

Inter-arrival time	*Probability*	*Service time*	*Probability*
10 minutes	0.5	10 minutes	0.6
15 minutes	0.4	20 minutes	0.1
20 minutes	0.1	30 minutes	0.3

Simulate the arrival and service times of 6 patients at the OPD by making use of the following random numbers:

Random numbers set for inter-arrival time	:	46,	63,	18,	08,	73,	59.
Random numbers set for service time	:	26,	18,	91,	76,	11,	18.

Estimate the average customer waiting time, average service time, average server idle time and the average time a patient spends in the OPD. ***[MBA, HCA (Nov.), Delhi Univ. 2009]***

24.19. After studying the weekly receipts and payments over the past 200 weeks, a retailer has developed the following information:

Weekly receipts (₹)	*Probability*	*Weekly payments (₹)*	*Probability*
3,000	0.20	4,000	0.30
5,000	0.30	6,000	0.40
7,000	0.40	8,000	0.20
12,000	0.10	10,000	0.10

Using the following set of random numbers, simulate the weekly pattern of receipts and payments for 12 weeks of the next quarter, assuming further that the beginning bank balance is ₹ 8,000. What is the estimated balance at the end of the 12 weekly period? What is the highest weekly balance during the quarter? What is the average weekly balance for the quarter?

Random Numbers

For receipts :	03,	91,	38,	55,	17,	46,	32,	43,	69,	72,	24,	22.
For payments :	61,	96,	30,	32,	03,	88,	48,	28,	88,	18,	71,	99.

[M.Com., Delhi Univ. 2010]

24.20. (*i*) A businessman is considering taking over a certain new business. Based on past information and his own knowledge of the business, he works out the probability distributions of the daily costs and sales revenue, as given below:

Cost (in ₹)	*Probability*	*Sales*	*Probability*
8,500	0.10	9,500	0.10
9,000	0.10	10,000	0.10
9,500	0.40	10,500	0.20
10,000	0.20	11,000	0.40
10,500	0.20	11,500	0.15
		12,000	0.05

Use the following sequences of random numbers to be used for estimating costs and revenues. Obtain the probability distribution of the daily net revenue.

Sequence 1 :	81,	83,	27,	81,	35,	91,	72,	90,	62,	28,
	26,	25,	91,	62,	82,	02,	12,	38,	10,	18.
Sequence 2 :	38,	71,	37,	28,	70,	82,	18,	71,	91,	58,
	48,	38,	71,	93,	02,	91,	73,	17,	09,	04.

(*ii*) Repeat the analysis in (*i*) by using the following random number streams:

Sequence 1 :	21,	64,	47,	17,	46,	42,	45,	67,	88,	27,
	79,	41,	30,	93,	22,	37,	58,	04,	29,	09.
Sequence 2 :	24,	58,	99,	85,	52,	30,	42,	12,	67,	31,
	42,	81,	63,	75,	65,	27,	42,	41,	98,	16.

[MBA (Nov.), Gujarat Univ. 1998]

24.21. An investment company wants to study the investment projects based on market demand, profit and the investment required, which are independent of each other. Following probability distributions are estimated for each of these three factors:

Annual demand (units in thousands) :	25	30	35	40	45	50	55
Probability :	0.05	0.10	0.20	0.30	0.20	0.10	0.05

Profit per unit in rupees :	3.00	5.00	7.00	9.00	10.00
Probability :	0.10	0.20	0.40	0.20	0.10

Investment required (in thousands of rupees) :	2,750	3,000	3,500
Probability :	0.25	0.50	0.25

Using simulation process, repeat the trial 10 times, compute the investment on each trial taking these factors into trial. What is the most likely return?

Use the following random numbers:

(30, 12, 16);	(59, 09, 69);	(63, 94, 26);	(27, 08, 74);	(64, 60, 61);
(28, 28, 72);	(31, 23, 57);	(54, 85, 20);	(64, 68, 18);	(32, 31, 87);

In the bracket above, the first random number is for annual demand, the second one is for profit and the last one is for the investment required. *[MBA (Nov.), Delhi Univ. 2009]*

24.22. A company trading in motor vehicle spares wishes to determine the level of stock it should carry for the item in its range. Demand is not certain and replenishment of stock takes 3 days. For one item X, the following information is obtained:

Demand (units per day) :	1	2	3	4	5
Probability :	0.1	0.2	0.3	0.3	0.1

Each time an order is placed, the company incurs an ordering cost of ₹ 20 per order. The company also incurs carrying cost of ₹ 2.50 per unit per day. The inventory carrying cost is calculated on the basis of average stock.

The manager of the company wishes to compare two options, for his inventory decisions.

(*i*) Order 12 units when the inventory at the beginning of the day plus order outstanding is less than 12 units.

(*ii*) Order 10 units when the inventory at the beginning of the day plus order outstanding is less than 10 units.

Currently (on first day) the company has a stock of 17 units.

The sequence of random numbers to be used is:

08, 91, 25, 18, 40, 27, 85, 75, 32, 52.

Using first number for day one, you are required to carry out a simulation run over a period of 10 days. Recommend which option the manager should choose.

[B.Tech., Panjab Tech. Univ. 2011]

24.23. The queue at a railway station ticket counter has the following characteristics:

Observed arrivals per minute		*Observed service time*	
Arrival	*Number*	*Minutes*	*Number*
0	80	1	40
1	10	2	30
2	8	3	20
3	2	4	10

Assume that all events occur at the end of a whole minute. Simulate the situation, by using random numbers, for 20 minutes and obtain the estimated probabilities of the waiting line consisting of 0, 1, 2, 3 persons.

[***Hint:*** Allot the random numbers between 00 and 99 in proportion to the probabilities of different number of arrivals per minute, and also for the observed service time. Then prepare a table to find the length of the line (queue) every minute.]

24.24. The observations of the study made for 100-day period by M/s Voltas (I) Ltd., an air-conditioner manufacturing and servicing unit, are tabulated below:

No. of air-conditioners requesting service	*Service requests*		*Servicing done*	
	Frequency	*Probability*	*Frequency*	*Probability*
0	5	0.05	8	0.08
1	10	0.10	2	0.02
2	15	0.15	30	0.30
3	20	0.20	20	0.20
4	25	0.25	15	0.15
5	15	0.15	20	0.20
6	10	0.10	5	0.05

Simulate a 10-day period arrival and service pattern. From this, find out the average backlog units per day.

We assume that M/s Voltas (I) Ltd. has only one service team and servicing is done on first come, first served basis for 24 hours.

24.25. A plant has a large number of similar machines. The machine break-downs or failures are random and independent.

The shift in-charge of the plant collected the data about the various time of machine breakdown and repair time required on hourly basis. The record for the past 100 observations was as follows:

Time between machine breakdown (hours)	*Probability*	*Repair time required (hours)*	*Probability*
0.5	0.05	1	0.28
1.0	0.06	2	0.52
1.5	0.16	3	0.20
2.0	0.33		
2.5	0.21		
3.0	0.19		

For each hour that one machine is down due to being or waiting to be repaired, the plant loses ₹ 70 by way of lost production. A repairman is paid at ₹ 20 per hour.

(*a*) Simulate this maintenance system for 15 breakdowns.

(*b*) How many repairmen should the plant hire for repair work?

24:7. ADVANTAGES OF SIMULATION

Simulation is the process of experimenting on the model rather than on the operation which the model represents. Following are some of the major advantages of simulation techniques:

1. It is flexible and straight forward technique and is easier to apply than analytical methods.
2. It facilitates to study the interactive system variables, and the effect of changes that take place in these variables, on the system performance in order to determine the desired results.
3. It may be the only method/technique, when it is difficult to observe the actual reality.
4. It can be used to analyse large and complex real world systems/models that cannot be solved by conventional quantitative methods.
5. By using simulation, we can investigate the consequences for a system of possible changes in parameters in terms of the model.
6. Once a model has been constructed, it may be used over and over again to analyse different possible situations.

24:8. LIMITATIONS OF SIMULATION

The limitations of the simulation are:

1. A good simulation model may be very expensive.
2. Simulation model does not produce answers by itself. It is 'run' rather than solved.
3. It does not generate optimum solutions to problems as do other quantitative methods.
4. Simulation is a time consuming exercise.
5. Not all situations can be evaluated using simulation. Only situations involving uncertainty are considered.

Conclusion

Simulation is indeed a versatile tool. It provides one statistical estimates rather than exact results and it only compares the alternatives rather than generating an optimal one. It is a slow and costly way to study a problem. Despite limitations, it is an invaluable tool in Operations Research. "When all else fails . . . Simulation."

TEST YOUR UNDERSTANDING

1. Simulation in the context of business problems
   (*a*) are comparatively more realistic than mathematical techniques.
   (*b*) does not produce optimum results.
   (*c*) comparatively costlier method of analysis.
   (*d*) all of the above.
2. The process of simulation
   (*a*) is a powerful mathematical technique.
   (*b*) is often referred to as "Monte-Carlo" simulation.
   (*c*) usually require use of computers to solve the problems.
   (*d*) involve the criterion wherein the output of a simulation model is independent of the simulation run.
3. While assigning random numbers in 'Monte Carlo' simulation, it is
   (*a*) not necessary to assign the exact range of random number interval as the probability.
   (*b*) necessary to develop a cumulative probability distribution,
   (*c*) necessary to assign particular appropriate random numbers.
   (*d*) all of the above.
4. Simulation should not be applied in all cases because, it
   (*a*) requires considerable talent for model building and extensive computer programming efforts.
   (*b*) provides at best approximate solution to the problem.
   (*c*) consumes much computer time. (*d*) all of the above.
5. As simulation is not an analytical model, therefore, result of simulation must be viewed as
   (*a*) approximate (*b*) exact
   (*c*) unrealistic (*d*) simplified.
6. Analytical results are taken into consideration before a simulation study, so as to
   (*a*) determine the optimum solution.
   (*b*) identify suitable values of the system parameters.
   (*c*) identify suitable values of decision variables for the specific choices of system parameters.
   (*d*) all of the above.
7. Large complicated simulation models are appreciated, because
   (*a*) their average costs are not well defined.
   (*b*) it is difficult to create the appropriate events.
   (*c*) they may be expensive to write and use as an experimental device.
   (*d*) all of the above.
8. Biased random sampling is made from among alternatives which have
   (*a*) equal probability (*b*) unequal probability
   (*c*) probability which do not sum to unity (*d*) none of the above.

9. In Monte-Carlo simulation
   (*a*) randomness is the key requirement.
   (*b*) the model is of deterministic nature.
   (*c*) random numbers can be used to generate the value of input variables only, if the sampled distribution is uniform.
   (*d*) None of the above.
10. Simulation is
   (*a*) descriptive in nature.
   (*b*) useful to analyse problems where analytical solution is difficult.
   (*c*) a statistical experiment as such its results are subject to statistical errors.
   (*d*) all of the above.
11. For solving a problem by simulation technique, important steps are:
   (*a*) test and validate the model
   (*b*) conduct the experiment
   (*c*) design the experiment
   (*d*) all of the above.
12. For making simulation more popular, we need to avoid
   (*a*) prolonged delays
   (*b*) large cost over runs
   (*c*) user dissatisfaction with simulation results
   (*d*) all of the above.

ANSWERS

1. (*d*)	2. (*b*)	3. (*b*)	4. (*d*)	5. (*a*)	6. (*c*)
7. (*b*)	8. (*b*)	9. (*a*)	10. (*d*)	11. (*d*)	12. (*d*)

REVIEW EXERCISE

1. Define simulation. What are the steps in simulation? *[MBA (Nov.), Annamalai Univ. 2010]*
2. Define discrete event simulation. *[MBA, Annamalai Univ. 2007]*
3. Write a brief note on simulation. *[PG Dipl., Indian Inst. of Mat. Man. 2019]*
4. Briefly explain step by step methodology of simulation process. Give four application areas, where this technique is used in practice.
   [BE (Dec.), Gujarat Tech. Univ. 2012; PG Dip. in Glob. Bus. Oper., Delhi Univ. 2010]
5. What is random variate? Discuss its role in simulation. *[MCA (Nov.), Madras Univ. 2002]*
6. Describe simulation process. What are the reasons for using simulation?
   [B.Tech. (Mech.), Panjab Tech. Univ. 2011; BE, Gujarat Tech. Univ. 2012; MBA, Rohilkhand Univ. 2011]
7. Explain Monte-Carlo method and give the situations, where these methods are useful.
   [B.Tech. (Mech.), Panjab Tech. Univ. 2012; B.Tech. & MBA, Hyderabad Univ. 2012]
8. Write a short note on Monte-Carlo simulation. *[PG Dipl., Indian Inst. of Mat. Man. 2017]*
9. What are the advantages and limitations of using simulation?
   [B.Tech. (Mech.), Hyderabad Univ. 2012; MBA, Gujarat Tech. Univ. 2011]
10. Mention any five reasons for using simulation for solving management problems. Also indicate five disadvantages of taking a simulation approach to solve a management problem.
   [PG Dip. in Glob. Bus. Oper., Delhi Univ. 2011]
11. Distinguish between analytical and simulation models. *[MBA, Bangalore Univ. 2016]*
12. Discuss in detail the application of simulation technique in corporate decision-making.
   [MBA, Annamalai Univ. 2010]

13. "Simulation is a powerful Operations Research technique for analysis and study of complex systems." In light of this statement, discuss how simulation can be useful for a Marketing Manager, while taking pricing decisions. *[MBA (Nov.), Delhi Univ. 2007]*
14. "Simulation is typically the process of carrying out sampling experiments on the models of the system rather than the system itself." Elucidate this statement by taking some examples. *[MBA (Nov.), Delhi Univ. 2009]*
15. Explain the application of simulation technique to inventory problems. *[B.Tech. (Mech.), Hyderabad Univ. 2012]*
16. Explain how do you simulate a *PERT*-network with the help of an example. *[B.Tech. (Comp. Sc.), Panjab Tech. Univ. 2010]*
17. "When it becomes difficult to use an optimization technique for solving a problem, one has to resort to simulation technique?" Discuss. *[AMIE 2005]*
18. Why is computer necessary in conducting a real world simulation?

CHAPTER 25

Non-linear Programming and Optimization

"Each venture is a new beginning to explore something hidden"

Like linear programming, ***Non-linear Programming*** is also a mathematical technique for determining the optimal (optimum) solution to many business problems. In ***non-linear programming problem***, either the objective function is non-linear, or one/more of the constraints have non-linear relationship or both.

25:1. FORMULATING A NON-LINEAR PROGRAMMING PROBLEM (*NLPP*)

Some of the real-life problems, that we shall formulate as non-linear programming problems (*NLPP*) are given below:

Example 25.1. *A company faces a responsive price-volume relationship for its products, the lower a product's price — the greater is the sales quantity, even in face of resultant price decreases by its competitors. If the sales-revenue does not vary proportionately with price, reflect this phenomenon in a non-linear objective function of the price.*

Formulation of the Problem

Let $x(p)$ represent the sales quantity as a function of price p, say in the product-mix problem. Clearly, the associated sales revenue is $px(p)$. Now, if the sales quantity function be given by the demand equation $x(p) = \alpha - \beta p$ for α and β as constants, over the range of p, then the sales revenue component in the objective function is quadratic, $z = px(p) = \alpha p - \beta p^2$. If each unit costs c to produce (where p and c are in the same units), then total profit P is given by

$$P = z - cx(p) = \alpha p - \beta p^2 - c\alpha + c\beta p = (\alpha + c\beta)p - c\alpha - \beta p^2.$$

Example 25.2 (*Product Allocation Problem*). *A manufacturing company produces two products: Mobiles and i-tablets. Sales-price relationships for these two products are given below:*

Product	*Quantity demanded*	*Unit price*
Mobiles	$1{,}500 - 5\,p_1$	p_1
i-tablets	$3{,}800 - 10\,p_2$	p_2

The total cost function for these two products are given by $200x_1 + 0.1x_1^2$ and $300x_2 + 0.1x_2^2$ respectively. The production takes place on two assembly lines. Mobiles are assembled on Assembly line I and i-tablets are assembled on Assembly line II. Because of the limitations of the assembly-line capacities, the daily production is limited to no more than 80 mobiles and 60 i-tablets. The production of both type of products require electronic components. The production of each of these sets require five units and six units of electronic equipment components respectively. The electronic components are supplied by another manufacturer, and the supply is limited to 600 units per day. The company has 160 employees, i.e., the labour supply amounts to 160 man-days.

The production of one mobile requires 1 man-day of labour, whereas 2 man-days of labour are required for one i-tablet. How many units of mobile and i-tablets should the company produce in order to maximize the total profit? Formulate the problem as a non-linear programming problem.

Formulation of the Problem as *NLPP*

Let us assume that whatever is produced is sold in the market. Let x_1 and x_2 stand for the quantities of mobiles and i-tablets respectively, manufactured by the firm. Then, we are given that

$$\begin{matrix} x_1 = 1{,}500 - 5p_1 \\ x_2 = 3{,}800 - 10p_2 \end{matrix} \quad \text{or} \quad \begin{cases} p_1 = 300 - 0.2x_1 \\ p_2 = 380 - 0.1x_2 \end{cases}$$

Further, if C_1, C_2 stand for the total cost of production of these units of mobiles and i-tablets respectively, then we are also given that

$$C_1 = 200x_1 + 0.1x_1^2 \quad \text{and} \quad C_2 = 300x_2 + 0.1x_2^2$$

Now, the revenue on mobiles is p_1x_1 and on i-tablets is p_2x_2. Thus, the total revenue R is measured by

$$R = p_1x_1 + p_2x_2$$

which can be written as

$$\begin{aligned} R &= (300 - 0.2x_1)x_1 + (380 - 0.1x_2)x_2 \\ &= 300x_1 - 0.2x_1^2 + 380x_2 - 0.1x_2^2. \end{aligned}$$

The total profit z is measured by the difference between the total revenue R and the total cost $C = C_1 + C_2$. Thus

$$z = R - C_1 - C_2 = 100x_1 - 0.3x_1^2 + 80x_2 - 0.2x_2^2.$$

The objective function thus obtained is a non-linear function.

In the present case, production is influenced by the available resources. Two assembly lines have limited capacity to produce mobiles and i-tablets. Since, no more than 80 mobiles can be assembled on assembly line I and 60 i-tablets on assembly line II per day, we have the restrictions: $x_1 \le 80$ and $x_2 \le 60$.

There is another side constraint in the daily requirement of the electronic components, so that $5x_1 + 6x_2 \le 600$. The number of available employees is limited to 160 man-days. Thus $x_1 + 2x_2 \le 160$. Also obviously, since the manufacturer cannot produce negative number of units, we must have $x_1 \ge 0$ and $x_2 \ge 0$.

Hence, the given problem can be put in the following mathematical format:

> Determine two real numbers, x_1 and x_2 so as to
> Maximize $z = 100x_1 - 0.3x_1^2 + 80x_2 - 0.2x_2^2$ subject to the constraints
> $x_1 \le 80,\; x_2 \le 60,\; 5x_1 + 6x_2 \le 600,$
> $x_1 + 2x_2 \le 160,\; x_1 \ge 0,\; x_2 \ge 0.$

This problem is a non-linear programming problem, since the objective function is non-linear in x_1 and x_2.

Remarks. In a non-linear programming problem, the objective function 'z' may be linear in x_1 and x_2, whereas the constraints are non-linear in x_1 and x_2, or both z and the constraints may be non-linear in x_1 and x_2. For example, the decision-making problem

$$\text{Maximize } f(x_1, x_2) = 3x_1 + 5x_2 \text{ subject to the constraints:}$$
$$x_1 + x_2 \le 3,\; x_1^2 + x_2^2 \le 10, \text{ and } x_1, x_2 \ge 0.$$

is a non-linear programming problem.

Example 25.3 (*Portfolio Selection Problem*). *An individual investor has an opportunity to invest a fixed amount of money in* **n** *different bonds and stocks. He wishes to maximize his anticipated returns, while considering variance of return as undesirable. Let x_j be the proportion of his assets invested in the jth security. Then, the vector* $\mathbf{x} = (x_1, x_2, ..., x_n)$ *is called a portfolio and the return R corresponding to a given portfolio* **x** *is a random variable. The investor is risk-averse and is, therefore, interested in determining a portfolio* **x** *that will minimize the variance of R subject to the restriction that his expected return is not less than some specified amount* **C** *(per unit invested). Formulate this portfolio selection problem as an NLPP.*

Formulation of the Problem as *NLPP*

Suppose the total funds available to investor is denoted by **B**. There are n channels of investment. The expected return per unit of investment from the ith channel is r_i, the variance of the ith investment is σ_i^2, while the covariance between i^{th} and j^{th} investment is σ_{ij}.

Thus, if an amount x_i $(i = 1, 2, ..., n)$ is invested in the ith type, then the expected return is $\sum_{i=1}^{n} r_i x_i$; while the variance of the investment is:

$$V = \sum_{i=1}^{n} \sum_{j=1}^{n} \sigma_{ij} x_i x_j.$$

As higher return and lower variance are desirable quantities from the investor's point of view, the objective function may be:

$$z = \sum_{i=1}^{n} r_i x_i - \sum_{i=1}^{n} \sum_{j=1}^{n} \sigma_{ij} x_i x_j.$$

Also, since the total amount the investor can spend is **B** and as his expected return must not be less than **C**; we have

$$\sum_{i=1}^{n} x_i = \mathbf{B} \quad \text{and} \quad \sum_{i=1}^{n} r_i \geq \mathbf{C}.$$

From the point of view of investor, z should be maximized subject to the constraints:

$$\sum_{i=1}^{n} x_i = \mathbf{B} \quad \text{and} \quad \sum_{i=1}^{n} r_i \geq \mathbf{C}, \quad \text{where } x_i \geq 0 \; (i = 1, 2, \ldots, n).$$

PROBLEMS

25.01. (***One-Potato, Two-Potato Problem***) A frozen-food company processes potatoes into packages of French fries, hash browns and flakes (for meshed potatoes). At the beginning of the manufacturing process, the raw potatoes are sorted by length and quality, and then allocated to the separate product lines.

The company can purchase its potatoes from two sources, which differ in their yields of various sizes and quality. Each source yields different fractions of the products, *viz.*, French fries, Hash Browns and Flakes. Suppose that it is possible, at different costs to alter these yields. Let f_1, f_2 and f_3 be the fractional yield per unit of weight of source *1* potatoes made into the three products. Similarly, let g_1, g_2 and g_3 be the yields for source 2. Suppose that each f_i and g_i can vary within ±10% of the yields shown below:

Product	*Source 1*	*Source 2*	*Purchase limitations*
French fries	0.2	0.3	1.8
Hash browns	0.2	0.1	1.2
Flakes	0.3	0.3	2.4
Relative profit	5	6	

Let $C_1(f_1, f_2, f_3)$ and $C_2(g_1, g_2, g_3)$ be the expense associated with obtaining these yields.

The problem is to determine how many potatoes should the company purchase from each source?

Formulate the problem as a non-linear programming problem.

25.02. A manufacturing concern operates its two available machines to polish its metal products. The two machines are equally efficient, although their maintenance costs are different. The daily maintenance and operation cost of the machines is given (in rupees) as the non-linear function:

$$f(x_1, x_2) = 100 - 1.2x_1 - 1.5x_2 + 0.3x_1^2 + 0.5x_2^2,$$

where, x_1 and x_2 are the number of hours of operation of machine *I* and machine *II*, respectively.

The past records of the firm indicate that the combined operating hours of two machines should be at least 35 hours a day in order to perform a satisfactory job. However, the production manager wishes to operate machine *I* at least 6 hours more than machine *II*, because of the higher repair cost of the latter. Find the optimal hours of operating the two machines and the minimum daily cost.

Formulate the problem as a non-liner programming problem.

25.03. A company manufacture two products *A* and *B*. It takes 30 minutes to process one unit of product *A* and 15 minutes for each unit of product *B*. The maximum machine time available is 35 hours per week. Products *A* and *B* require 2 kgs and 3 kgs of raw material per unit respectively. The available quantity of raw material is envisaged to be 180 kgs per week. The products *A* and *B*, which have unlimited market potential to sell these products for ₹ 200 nd ₹ 500 per unit respectively. If the manufacturing costs for products *A* and *B* are $2x^2$ and $3y^2$ respectively, find how much of each product should be produced per week, where

x = Quantity of Product *A* to be produced, and y = Quantity of Product *B* to be produced.

Formulate the problem as a non-linear programming problem.

25:2. GENERAL NON-LINEAR PROGRAMMING PROBLEM

Let z be a real valued function of n variables defined by

(*a*) $$z = f(x_1, x_2, \ldots, x_n),$$

and $\{b_1, b_2, \ldots, b_m\}$ be a set of constants, such that

(*b*) $$\begin{cases} g^1(x_1, x_2, \ldots, x_n) \ \{\leq, \geq \text{ or } =\} \ b_1 \\ g^2(x_1, x_2, \ldots, x_n) \ \{\leq, \geq \text{ or } =\} \ b_2 \\ \vdots \qquad\qquad\qquad \vdots \\ g^m(x_1, x_2, \ldots, x_n) \ \{\leq, \geq \text{ or } =\} \ b_m, \end{cases}$$

where, g^i's are real valued functions of n variables, $x_1, \ldots, x_n$ with

(*c*) $$x_j \geq 0, \quad j = 1, 2, \ldots, n.$$

If either $f(x_1, \ldots, x_n)$ or some $g^i(x_1, \ldots, x_n)$, $i = 1, 2, \ldots, m$; or both are non-linear, then the problem of determining the n-tupple $(x_1, x_2, \ldots, x_n)$ which makes z a maximum or minimum and satisfies (*b*) and (*c*), is called a ***general non-linear programming problem (GNLPP)***.

In matrix notations the *GNLPP* may be written as:

Determine $\mathbf{x} \in R^n$ so as to ***maximize or minimize*** the ***objective function*** $z = f(\mathbf{x})$, subject to the constraints:

$$g^i(\mathbf{x}) \ \{\leq, \geq \text{ or } =\} \ b_i, \quad \mathbf{x} \geq \mathbf{0} \qquad i = 1, 2, \ldots, m,$$

where, either $f(\mathbf{x})$ or some $g^i(\mathbf{x})$ or both are non-linear in $\mathbf{x}$.

Sometimes, it is convenient to write the constraints $g^i(x)$ {≤, ≥ or =} b_i as $h^i(\mathbf{x})$ {≤, ≥ or =} 0, for $h^i(x) = g^i(\mathbf{x}) - b_i$.

25:3. TYPES OF NON-LINEAR PROGRAMMING PROBLEMS

Non-linear programming problems are in many different shapes and forms. There is no simplex-like solution procedure for the solution of all these different types of such problems. However, numerous solution methods have been developed for various types of non-linear programming problems. The most important classes (special types) are introduced briefly in this and subsequent chapter. It may be noted that we shall assume throughout that the problems have been formulated in the form of general non-linear programming problem (*GNLPP*).

1. **Unconstrained Optimization.** ***Unconstrained optimization problems*** have no constraints, but only the objective function. So, the problem is to

$$\text{Maximize or Minimize} \quad z = f(\mathbf{x}),$$

for values of $\mathbf{x} = (x_1, x_2, \ldots, x_n)$. The ***necessary condition*** that a particular solution $\mathbf{x}(= \mathbf{x}^*)$ be optimum (optimal) is

$$\frac{\partial f}{\partial x_j} = 0 \text{ at } \mathbf{x} = \mathbf{x}^*, \qquad \text{for } j = 1, 2, \ldots, n$$

where, $f(\mathbf{x})$ is a differentiable function. This condition is also ***sufficient*** when $f(\mathbf{x})$ is a concave function for maximization problem and convex function for minimization problem. Thus, the solution can be obtained by solving n equations obtained by setting the n partial derivatives equal to zero. However, for a non-linear function these equations are often non-linear and it is not possible to solve these equations analytically. Some search procedures are then applied to solve such problems.

When a variable x_j is restricted to be non-negative, *i.e.*, $x_j \geq 0$, the preceding ***necessary and sufficient condition*** slightly changes to

$$\frac{\partial f}{\partial x_j} \leq 0 \text{ at } \mathbf{x} = \mathbf{x}^* (\text{if } x_j^* = 0), \text{ and } \frac{\partial f}{\partial x_j} = 0 \text{ at } \mathbf{x} = \mathbf{x}^* (\text{if } x_j^* > 0), \qquad \text{for all } j = 1, 2, \ldots, n$$

2. **Linearly Constrained Optimization.** In the ***constrained non-linear programming problems***, all the constraints $g(\mathbf{x})$ are linear, while the objective function $f(\mathbf{x})$ is non-linear. Such type of non-linear programming problems having the objective function as non-linear along with linear constraints having feasible solution are solved using some special algorithm.

3. **Quadratic Programming.** When the objective function $f(\mathbf{x})$ becomes ***quadratic*** and constraints are linear, then such type of non-linear programming problems fall in the category of ***quadratic programming problems***. Such problems arise very frequently in practical situations. Algorithms for solving quadratic programming problems are discussed in the next chapter.

4. **Convex Programming.** Special cases of various type of non-linear programming problems fall in the category of ***convex programming problems***, where the objective function $f(\mathbf{x})$ is a ***concave function*** to be maximized along with the constraints $g^i(\mathbf{x})$, which are convex functions. These assumptions ensure that a local maximum is also a ***global maximum***.

5. **Separable Programming.** A special case of convex programming is known as ***separable programming problems*** in which all the $f(\mathbf{x})$ and $g^i(\mathbf{x})$ are separable functions.

   A function is called separable, when it can be expressed as a sum of subfunctions, where each subfunction is a function of one variable only. For example, if $f(\mathbf{x})$ is a separable function, then it can be expressed as: $f(\mathbf{x}) = \sum_{j=1}^{n} f_j(x_j)$, where each $f_j(x_j)$ have the terms involving x_j ($j = 1, 2, ..., n$) only. Separable programming problems satisfy the assumptions of additivity, but violate the assumption of proportionality when any of the $f_j(x_j)$ are non-linear functions. Computational procedure for solving separable programming problems will be developed in the next chapter.

6. **Non-convex Programming.** A non-linear programming problem that does not satisfy the assumptions of convex programming, fall in the category of ***non-convex programming problems***. There is no algorithm that find an optimum (optimal) solution for all such problems. However, some specific type of non-convex programming problems can be solved by employing specific methods, such as ***Geometric Programming*** (to be discussed in subsequent chapter) and ***Fractional Programming*** (already discussed in chapter 9).

25:4. ONE-VARIABLE UNCONSTRAINED OPTIMIZATION

We now derive the solution procedure for solving un-constrained optimization with just single variable x, where the differentiable function $f(x)$ to be maximized is concave. So, the necessary and sufficient condition for a particular solution $x = x^*$ to be optimum (a global maximum) is:

$$\frac{df}{dx} = 0 \text{ at } x = x^*.$$

In case, the above equation can be solved directly for x^*, we get the global maximum or minimum. But, if $f(x)$ is not a simple function, then the derivative is not just linear or quadratic function, and hence we may not be able to solve the equation analytically. Here, we discuss only the case, when a ***global maximum*** or ***minimum*** occurs.

Computational Procedure

Given below are the steps involved in determining global maximum or minimum:

Step 1. Compute first derivative, $\frac{dy}{dx}$ and equate it to zero.

Step 2. Solve the equation $\frac{dy}{dx} = 0$ for $x = x^*$.

Step 3. Substitute the value $x = x^*$, $x = a$ and $x = b$ in the original equation and determine $f(x^*)$, $f(a)$ and $f(b)$, where x lies in the interval $a \le x \le b$.

Step 4. Compare these values to determine the global maximum and global minimum respectively.

Remarks

1. A local maximum of a concave function on a convex set is also a global maximum of that function.
2. A local minimum of a convex function on a convex set is also a global minimum of that function.
3. A local maximum (minimum) of a strictly concave function (convex function) on a convex set is also a unique global maximum (minimum) of that function.

Example 25.4. *A dealer receives x units of an item at the beginning of each month. The cost of carrying x units per month is given by*

$$C(x) = \frac{c_1 x^2}{2n} + \frac{c_2 (40n - x)^2}{2n}$$

where, c_1 = *cost per day of carrying a unit of item in stock (= ₹ 20)*

c_2 = *cost per day of shortage of a unit of item (= ₹ 300)*

n = *number of units of item to be supplied per day (= 50)*

Determine the order quantity x that minimizes the cost of inventory.

Solution. The necessary condition for a function to have either minimum or maximum value at a point is that its first derivative should be zero. Thus,

$$\frac{dC(x)}{dx} = \frac{c_1 x}{n} - \frac{c_2(40n - x)}{n} = 0$$

This gives
$$x = \frac{40nc_2}{c_1 + c_2} = \frac{40 \times 50 \times 300}{20 + 300} = 1{,}875.0$$

The nature of the extreme point given by x is determined by considering the second derivative.

$$\therefore \quad \frac{d^2C(x)}{dx^2} = \frac{c_1 + c_2}{n} = \frac{20 + 300}{50} = 6.4 > 0$$

Since, second derivative is positive, $x = 1{,}875.0$ is a local minimum point.

Using this value of x in the objective function, we get

$$C(1{,}875) = 7{,}50{,}000.$$

Hence, global minimum value for $C(x)$ occurs at $x = 1{,}875.0$.

Example 25.5. *The total profit y, in rupees, of a drug company from the manufacturing and sale of x drug bottles is given by*

$$y = -\left(\frac{x^2}{600}\right) + 3x - 60.$$

(a) *How many drug bottles must the company sell in order to achieve the maximum profit?*
(b) *What is the profit per bottle, when maximum is achieved?*

Solution. Given:
$$y = -\left(\frac{x^2}{600}\right) + 3x - 60.$$

$$\therefore \quad \frac{dy}{dx} = \left(\frac{-2x}{600}\right) + 3 = \left(\frac{-x}{300}\right) + 3$$

The first derivative condition for maximum value of y is

$$\frac{dy}{dx} = 0, \quad i.e., \quad \left(\frac{-2x}{600}\right) + 3 = 0 \quad \text{or} \quad x = 900.$$

Further, since $\frac{d^2y}{dx^2} = \frac{-1}{300} < 0$; the company must sell $x\,(= 900)$ drug bottles in order to achieve the maximum profit. The maximum profit, therefore, will be:

$$y = \frac{-(900)^2}{600} + 3 \times 900 - 60 = ₹\,1{,}290.$$

Example 25.6. *The cost of fuel for running a train is proportional to the cube of the speed generated in km per hour. When the speed is 12 km per hour, the cost of fuel is ₹ 32 per hour. If other charges are fixed, namely ₹ 2,000 per hour, find the most economical speed of the train for running a distance of 200 km.*

Solution. Let 'x' km per hour is speed of train. Then, the cost of fuel $= kx^3$, where k is constant of proportionality.

Given that it costs ₹ 32 per hour, when the speed is 12 km per hour. Therefore,

$$32 = k \times (12)^3 \quad \Rightarrow \quad k = 0.0185$$

$$\therefore \quad \text{Cost of fuel} = 0.0185x^3 \text{ rupees per hour.}$$

The fuel for running a distance of 200 km is given by:

$$0.0185x^3 \times \frac{200}{x}, \ \textit{i.e.,}\ 3.70x^2.$$

Also, fixed cost $= 2{,}000 \times \frac{200}{x}$. So, if C is the cost of running 200 km, then,

$$C = 3.7x^2 + \frac{4{,}00{,}000}{x}.$$

Now,
$$\frac{dC}{dx} = 7.4x - \frac{4{,}00{,}000}{x^2} \quad \text{and} \quad \frac{d^2C}{dx^2} = 7.4 + \frac{4{,}00{,}000}{x^3}$$

For maximum or minimum value of C,

$$\frac{dC}{dx} = 0 \ \Rightarrow \ 7.4x - \frac{4{,}00{,}000}{x^2} = 0 \ \Rightarrow \ x = 75 \text{ approx.}$$

For this value of x, $\frac{d^2C}{dx^2} > 0$, and hence C is minimum. Thus, the most economical speed of the train should be 75 km per hour.

25:5. MULTI-VARIABLE UNCONSTRAINED OPTIMIZATION

We now consider the optimization problems having objective function comprising of more than one variable and without any constraint. The multi-variable function should either be convex function or concave function in the specified region. If the multi-variable function $f(\mathbf{x})$ is concave, then the objective is to maximize $f(\mathbf{x})$, where $\mathbf{x} = (x_1, x_2, ..., x_n)$. In case, the multi-variable function $f(\mathbf{x})$ is convex, then the objective is to minimize $f(\mathbf{x})$, $\mathbf{x} = (x_1, x_2, ..., x_n)$.

For a single variable problem, the value of ordinary derivative was used to determine whether the value of the variable was to be increased or decreased for the next trial solution. The iterations were carried to reach a point where the value of the derivative was nearly zero. In case of multi-variable problem, there are number of directions in which the solution can move. Our objective is to reach a point, where all the partial derivatives are nearly zero. The value of each variable at this point is nearly optimum/optimal.

To optimize a multi-variable function $f(\mathbf{x})$, the concept of partial derivatives is used. This is because partial derivatives measure the change in the dependent variable due to unit change in one of the independent variables, while keeping constant the effect of other independent variables. So, the multi-variable function $f(\mathbf{x})$ posses a gradient $\nabla f(\mathbf{x})$, such that

$$\nabla f(\mathbf{x}^*) = \left(\frac{\partial f}{\partial x_1}, \frac{\partial f}{\partial x_2}, \ldots, \frac{\partial f}{\partial x_n}\right),$$

at any point $\mathbf{x} = \mathbf{x}^*$.

The necessary and sufficient conditions for local maximum (or minimum) of unconstrained multi-variable function are:

1. Necessary condition for the continuous function $f(\mathbf{x})$ to have an extreme point at $\mathbf{x} = \mathbf{x}^*$ is that $\nabla f(\mathbf{x}^*) = 0$; *i.e.,*

$$\frac{\partial f}{\partial x_1} = 0, \ \frac{\partial f}{\partial x_2} = 0, \ \ldots, \ \frac{\partial f}{\partial x_n} = 0 \ \text{ at } \mathbf{x} = \mathbf{x}^*.$$

2. A sufficient condition for a stationary point $\mathbf{x}^*$ to be an extreme point is that the ***Hessian matrix*** $H(\mathbf{x})$, evaluated at $\mathbf{x}^*$ is
   (*i*) positive definite, when $\mathbf{x}^*$ is a minimum point, and
   (*ii*) negative definite, when $\mathbf{x}^*$ is a maximum point.

Remark. The Hessian matrix $H(\mathbf{x})$ is defined as follows:

$$H(\mathbf{x}) = \begin{bmatrix} \dfrac{\partial^2 f}{\partial x_1^2} & \dfrac{\partial^2 f}{\partial x_1\, \partial x_2} & \cdots & \dfrac{\partial^2 f}{\partial x_1\, \partial x_n} \\ \dfrac{\partial^2 f}{\partial x_2\, \partial x_1} & \dfrac{\partial^2 f}{\partial x_2^2} & \cdots & \dfrac{\partial^2 f}{\partial x_2\, \partial x_n} \\ \vdots & \vdots & & \vdots \\ \dfrac{\partial^2 f}{\partial x_n\, \partial x_1} & \dfrac{\partial^2 f}{\partial x_n\, \partial x_2} & \cdots & \dfrac{\partial^2 f}{\partial x_n^2} \end{bmatrix}.$$

The leading principal minors of $H(x)$ are denoted by M_j, $(j = 1, 2, ...)$, such that

$$M_1 = \left| \frac{\partial^2 f}{\partial x_1^2} \right|, \; M_2 = \begin{vmatrix} \dfrac{\partial^2 f}{\partial x_1^2} & \dfrac{\partial^2 f}{\partial x_1\, \partial x_2} \\ \dfrac{\partial^2 f}{\partial x_2\, \partial x_1} & \dfrac{\partial^2 f}{\partial x_2^2} \end{vmatrix}, \text{ and so on.}$$

(*a*) When $M_1 > 0, M_2 > 0, ..., H(\mathbf{x})$ is positive definite.
(*b*) When $M_1 < 0, M_2 > 0, M_3 < 0, ..., H(x)$ is negative definite.
(*c*) When neither case (*a*) nor case (*b*) occur, $H(\mathbf{x})$ is indefinite.

To illustrate, let us consider

$$H(\mathbf{x}) = \begin{bmatrix} 4 & 3 & 0 \\ 3 & 2 & -1 \\ -2 & 0 & 5 \end{bmatrix}.$$

Here, the leading principal minors are:

$$M_1 = |\,4\,| = 4, \; M_2 = \begin{vmatrix} 4 & 3 \\ 3 & 2 \end{vmatrix} = -1, \; M_3 = \begin{vmatrix} 4 & 3 & 0 \\ 3 & 2 & -1 \\ -2 & - & 5 \end{vmatrix} = 9.$$

Since, $M_1 > 0, M_2 < 0$ and $M_3 > 0$; $H(\mathbf{x})$ is negative definite.

Example 25.7. *Consider:* $f(\mathbf{x}) = 2x_1x_2 + x_2 - x_1^2 - 2x_2^2$. *Determine the maximum or minimum point (if any) of the function* $f(\mathbf{x})$. **[M.Sc. (Math.), Rohilkhand Univ. 2011]**

Solution. For local optimum, the necessary condition is that gradiant $\nabla f(\mathbf{x}) = 0$, *i.e.*, $\dfrac{\partial f}{\partial x_1} = 0$ and $\dfrac{\partial f}{\partial x_2} = 0$. Therefore,

$$\frac{\partial f}{\partial x_1} = 0 \;\Rightarrow\; 2x_2 - 2x_1 = 0 \;\text{ or }\; x_2 = x_1; \text{ and}$$

$$\frac{\partial f}{\partial x_2} = 0 \;\Rightarrow\; 2x_1 + 1 - 4x_2 = 0 \;\text{ or }\; x_2 = \frac{(2x_1 + 1)}{4}.$$

Solving the two equation for x_1 and x_2, we get

$$x_1 = \frac{1}{2} \text{ and } x_2 = \frac{1}{2}, \; i.e., \; \mathbf{x}^* = \left(\frac{1}{2}, \frac{1}{2}\right).$$

The sufficient condition is verified by considering the Hessian matrix, $H(\mathbf{x})$. So, we write

$$H(\mathbf{x}) = \begin{bmatrix} \dfrac{\partial^2 f}{\partial x_1^2} & \dfrac{\partial^2 f}{\partial x_1 \partial x_2} \\ \dfrac{\partial^2 f}{\partial x_2 \partial x_1} & \dfrac{\partial^2 f}{\partial x_2^2} \end{bmatrix} = \begin{bmatrix} -2 & 2 \\ 2 & -4 \end{bmatrix}$$

Leading principal minors are:

$$|-2| \text{ and } \begin{vmatrix} -2 & 2 \\ 2 & -4 \end{vmatrix}, \textit{ viz.}, -2 \text{ and } 4 \text{ respectively.}$$

As the sign of leading principal minors are alternative starting with negative sign, $H(\mathbf{x})$ is negative definite. Hence, the point $\mathbf{x}^* = \left(\frac{1}{2}, \frac{1}{2}\right)$ is the local maximum of the function $f(x)$.

PROBLEMS

25.04. Examine the following functions of x for global optimum, *i.e.*, maximum or minimum:

(*a*) $f(x) = x^4 + 6x^2 + 12x$. (*b*) $f(x) = 75 - x^2$.

25.05. Examine the following functions for extreme points:

(*a*) $f(x_1, x_2) = x_1^2 + x_2^2 - 2x_1x_2$.

(*b*) $f(x_1, x_2) = x_1 + 2x_2 + x_1x_2 - x_1^2 - x_2^2$.

(*c*) $f(x_1, x_2) = 3x_1 + 2x_1^2 + 4x_2 + x_2^2 - 2x_1x_2$.

25.06. Determine the local maximum or minimum point (if any) of the following function:

$$f(x) = x_1^2 + 2x_2^2 + x_3^2 + x_1x_2 - 2x_3 - 7x_1 + 12.$$

25:6. CONSTRAINED OPTIMIZATION WITH EQUALITY CONSTRAINTS

If the non-linear programming problem is composed of some differentiable objective function and equality constraints, the optimization may be achieved by the use of ***Lagrange multipliers**** as illustrated below:

Consider the problem of maximizing or minimizing $z = f(x_1, x_2)$ subject to the constraints:

$$g(x_1, x_2) = c \text{ and } x_1, x_2 \geq 0,$$

where, c is a constant.

We assume that $f(x_1, x_2)$ and $g(x_1, x_2)$ are differentiable w.r.t. x_1 and x_2. Let us introduce a differentiable function $h(x_1, x_2)$, differentiable w.r.t. x_1 and x_2 and defined by $h(x_1, x_2) \equiv g(x_1, x_2) - c$. Then, the problem can be restated as

$$\text{Maximize } z = f(x_1, x_2) \text{ subject to the constraints:}$$

$$h(x_1, x_2) = 0 \text{ and } x_1, x_2 \geq 0.$$

To find the necessary conditions for a maximum (or minimum) value of z, a new function is formed by introducing a Lagrange multiplier λ, as under:

$$L(x_1, x_2, \lambda) = f(x_1, x_2) - \lambda h(x_1, x_2).$$

The number λ is an unknown constant, and the function $L(x_1, x_2, \lambda)$ is called the ***Lagrangian function with Lagrange multiplier*** $\boldsymbol{\lambda}$. The necessary conditions for a maximum or minimum (stationary value) of $f(x_1, x_2)$ subject to $h(x_1, x_2) = 0$ are thus given by

$$\frac{\partial L(x_1, x_2, \lambda)}{\partial x_1} = 0, \quad \frac{\partial L(x_1, x_2, \lambda)}{\partial x_2} = 0 \text{ and } \frac{\partial L(x_1, x_2, \lambda)}{\partial \lambda} = 0.$$

* The method of ***Lagrange multipliers*** is a systematic way of generating the necessary conditions for a stationary point.

Now, these partial derivatives are given by

$$\frac{\partial L}{\partial x_1} = \frac{\partial f}{\partial x_1} - \lambda \frac{\partial h}{\partial x_1}, \quad \frac{\partial L}{\partial x_2} = \frac{\partial f}{\partial x_2} - \lambda \frac{\partial h}{\partial x_2}, \quad \text{and} \quad \frac{\partial L}{\partial \lambda} = -h,$$

where, L, f and h stand for the functions $L(x_1, x_2, \lambda), f(x_1, x_2)$, and $h(x_1, x_2)$, respectively, or simply by

$$L_1 = f_1 - \lambda h_1, \quad L_2 = f_2 - \lambda h_2 \quad \text{and} \quad L_\lambda = -h.$$

The ***necessary conditions for maximum or minimum of*** $f(x_1, x_2)$ are thus given by

$$f_1 = \lambda h_1, \quad f_2 = \lambda h_2 \quad \text{and} \quad -h(x_1, x_2) = 0.$$

Note. These ***necessary conditions*** become sufficient conditions for a maximum (minimum), if the objective function is concave (convex) and the side constraints are in the form of equalities.

Example 25.8. *Obtain the set of necessary conditions for the non-linear programming problem:*

$$\text{Minimize } z = kx^{-1}y^{-2} \text{ subject to the constraints:}$$

$$x^2 + y^2 - a^2 = 0 \text{ with } x \ge 0, \; y \ge 0;$$

and hence find the minimum value of z.

Solution. The Lagrange function is

$$L(x, y, \lambda) = kx^{-1}y^{-2} + \lambda(x^2 + y^2 - a^2),$$

where, $f(x, y) = kx^{-1}y^{-2}$ and $h(x, y) = g(x, y) - C = (x^2 + y^2) - a^2$.

The necessary conditions for the minimum of $f(x, y)$ gives

$$\frac{\partial L}{\partial x} = 0 \Rightarrow -kx^{-2}y^{-2} + 2x\lambda = 0,$$

$$\frac{\partial L}{\partial y} = 0 \Rightarrow -2kx^{-1}y^{-3} + 2\lambda y = 0,$$

and

$$\frac{\partial L}{\partial \lambda} = 0 \Rightarrow x^2 + y^2 - a^2 = 0.$$

From the first two equations, we get

$$2\lambda = kx^{-3}y^{-2} = 2kx^{-1}y^{-4}. \text{ This yields } x = y/\sqrt{2}.$$

Using this value of x in the third equation,

$$y = a\sqrt{\frac{2}{3}} \text{ and therefore, } x = \frac{1}{\sqrt{2}} \times a\sqrt{\frac{2}{3}} = \frac{a}{\sqrt{3}}$$

$\therefore$

$$\text{Minimum } z = k \times \left(a/\sqrt{3}\right)^{-1}\left(a\sqrt{2/3}\right)^{-2} = 3\sqrt{3}k/2a^3.$$

Necessary Conditions for a General *NLPP*

Consider the general *NLPP:*

$$\text{Maximize (or Minimize) } z = f(x_1, x_2, ..., x_n) \text{ subject to the constraints:}$$

$$g^i(x_1, ..., x_n) = c_i \text{ and } x_i \ge 0, \qquad i = 1, 2, ..., m \; (< n)$$

The constraints can be reduced to

$$h^i(x_1, ..., x_n) = 0 \qquad \text{for } i = 1, 2, ..., m.$$

by the transformation $h^i(x_1, ..., x_n) = g^i(x_1, ..., x_n) - c_i$ for all $i = 1, 2, ..., m \; (< n)$.

The problem can be written in the matrix form as:

$$\text{Maximize (or Minimize) } z = f(\mathbf{x}), \; (\mathbf{x} \in \mathbf{R}^n) \text{ subject to the constraints:}$$

$$h^i(\mathbf{x}) = 0, \quad \mathbf{x} \ge \mathbf{0}.$$

To find the necessary conditions for a maximum or minimum of $f(\mathbf{x})$, the Lagrangian function $L(\mathbf{x}, \lambda)$, is formed by introducing m Lagrangian multipliers $\lambda = (\lambda_1, \lambda_2, ..., \lambda_m)$. This function is defined by

$$L(\mathbf{x}, \lambda) = f(\mathbf{x}) - \sum_{i=1}^{m} \lambda_i h^i(\mathbf{x}).$$

Assuming that L, f and h^i are all differentiable partially w.r.t. $x_1, x_2, ..., x_n$ and $\lambda_1, \lambda_2, ..., \lambda_m$, the necessary conditions for a maximum (minimum) of $f(\mathbf{x})$ are:

$$\frac{\partial L}{\partial x_j} \equiv \frac{\partial f}{\partial x_j} - \sum_{i=1}^{m} \lambda_i \frac{\partial h^i(\mathbf{x})}{\partial x_j} = 0, \qquad j = 1, 2, ..., n$$

and

$$\frac{\partial L}{\partial \lambda_i} \equiv -h^i(\mathbf{x}) = 0, \qquad i = 1, 2, ..., m$$

These, $(M + n)$ necessary conditions can be represented in the following abbreviated form:

$$L_j \equiv f_j - \sum_{i=1}^{m} \lambda_i h^i{}_j = 0 \quad \text{or} \quad f_j = \sum_{i=1}^{m} \lambda_i h^i{}_j; \qquad j = 1, 2, ..., n$$

and

$$-h^i = 0 \quad \text{or} \quad h^i = 0; \qquad i = 1, 2, ..., m$$

where, $f_j = \dfrac{\partial f(\mathbf{x})}{\partial x_j}$, $h^i = h^i(\mathbf{x})$ and $h^i{}_j = \dfrac{\partial h^i(\mathbf{x})}{\partial x_j}$.

Remark. These necessary conditions also become sufficient for a maximum (minimum) of the objective function, if the objective function is concave (convex) and the constraints are equality ones.

Example 25.9. *Obtain the set of necessary conditions for the non-linear programming problem:*

Maximize $z = x_1^2 + 3x_2^2 + 5x_3^2$ *subject to the constraints:*

$$x_1 + x_2 + 3x_3 = 2, \quad 5x_1 + 2x_2 + x_3 = 5$$

$$x_1, x_2, x_3 \geq 0.$$

[B.Sc. (Stat.), Delhi Univ. 2006]

Solution. Here, we have $\mathbf{x} = (x_1, x_2, x_3)$, $f(\mathbf{x}) = x_1^2 + 3x_2^2 + 5x_3^2$, $g^1(\mathbf{x}) = x_1 + x_2 + 3x_3$, $g^2(\mathbf{x}) = 5x_1 + 2x_2 + x_3$ and $c_1 = 2, c_2 = 5$. Defining $h^i(\mathbf{x}) = g^i(\mathbf{x}) - c_i$, $i = 1, 2$, we have the constraint $h^i(\mathbf{x}) = 0$ for $i = 1, 2$.

For necessary conditions of maximizing $f(\mathbf{x})$, we construct the Lagrangian function

$$L(\mathbf{x}, \lambda) = f(\mathbf{x}) - \lambda_1 h^1(\mathbf{x}) - \lambda_2 h^2(\mathbf{x}), \ \lambda = (\lambda_1, \lambda_2).$$

This yields the following necessary conditions:

$$\frac{\partial L}{\partial x_1} = 2x_1 - \lambda_1 - 5\lambda_2 = 0, \qquad \frac{\partial L}{\partial x_2} = 6x_2 - \lambda_1 - 2\lambda_2 = 0,$$

$$\frac{\partial L}{\partial x_3} = 10x_3 - 3\lambda_1 - \lambda_2 = 0, \qquad \frac{\partial L}{\partial \lambda_1} = -(x_1 + x_2 + 3x_3 - 2) = 0,$$

$$\frac{\partial L}{\partial \lambda_2} = -(5x_1 + 2x_2 + x_3 - 5) = 0.$$

Remark. In some cases it is not generally possible to solve the equations resulting from the set of necessary conditions, explicitly for the values of **x** and λ_i's. A convenient method in this case is to select successive numerical values of λ_i's and then solve the set of necessary conditions for **x**. This is repeated until for some values of λ_i, the resulting **x** satisfies all the constraints in equation form.

Sufficient Conditions for General *NLPP* with One Constraint

Let the Lagrangian function for a general *NLPP* involving n variables and one constraint be:

$$L(\mathbf{x}, \lambda) = f(\mathbf{x}) - \lambda h(\mathbf{x}).$$

The necessary conditions for a stationary point to be a maximum or minimum are

$$\frac{\partial L}{\partial x_j} \equiv \frac{\partial f}{\partial x_j} - \lambda \frac{\partial h}{\partial x_j} = 0, \text{ and } \frac{\partial L}{\partial \lambda} \equiv -h(x) = 0. \qquad (j = 1, 2, ..., n)$$

∴ The value of λ is obtained by

$$\lambda = \frac{\partial f / \partial x_j}{\partial h / \partial x_j} \qquad (\text{for } j = 1, 2, ..., n).$$

The sufficient conditions for a maximum or minimum require the evaluation at each stationary point, of $n - 1$ principal minors of the determinant given below :

$$D_{n+1} = \begin{vmatrix} 0 & \frac{\partial h}{\partial x_1} & \frac{\partial h}{\partial x_2} & \cdots & \frac{\partial h}{\partial x_n} \\ \frac{\partial h}{\partial x_1} & \frac{\partial^2 f}{\partial x_1^2} - \lambda \frac{\partial^2 h}{\partial x_1^2} & \frac{\partial^2 f}{\partial x_1 \partial x_2} - \lambda \frac{\partial^2 h}{\partial x_1 \partial x_2} & \cdots & \frac{\partial^2 f}{\partial x_1 \partial x_n} - \lambda \frac{\partial^2 h}{\partial x_1 \partial x_n} \\ \frac{\partial h}{\partial x_2} & \frac{\partial^2 f}{\partial x_2 \partial x_1} - \lambda \frac{\partial^2 h}{\partial x_2 \partial x_1} & \frac{\partial^2 f}{\partial x_2^2} - \lambda \frac{\partial^2 h}{\partial x_2^2} & \cdots & \frac{\partial^2 f}{\partial x_2 \partial x_n} - \lambda \frac{\partial^2 h}{\partial x_2 \partial x_n} \\ \vdots & \vdots & \vdots & & \vdots \\ \frac{\partial h}{\partial x_n} & \frac{\partial^2 f}{\partial x_n \partial x_1} - \lambda \frac{\partial^2 h}{\partial x_n \partial x_1} & \frac{\partial^2 f}{\partial x_n \partial x_2} - \lambda \frac{\partial^2 h}{\partial x_n \partial x_2} & \cdots & \frac{\partial^2 f}{\partial x_n^2} - \lambda \frac{\partial^2 h}{\partial x_n^2} \end{vmatrix}$$

If $D_3 > 0$, $D_4 < 0$, $D_5 > 0$, ..., the signs pattern being alternate, the stationary point is a local maximum.

If $D_3 < 0$, $D_4 < 0$, ..., $D_{n+1} < 0$, the sign being always negative, the stationary point is a local minimum.

Example 25.10. *Obtain the set of necessary and sufficient conditions for the following NLPP:*

Minimize $z = 2x_1^2 - 24x_1 + 2x_2^2 - 8x_2 + 2x_3^2 - 12x_3 + 200$ *subject to the constraints:*

$$x_1 + x_2 + x_3 = 11, \quad x_1, x_2, x_3 \geq 0.$$

Solution. We formulate the Lagrangian function as

$$L(x_1, x_2, \lambda) = 2x_1^2 - 24x_1 + 2x_2^2 - 8x_2 + 2x_3^2 - 12x_3 + 200 - \lambda(x_1 + x_2 + x_3 - 11)$$

The necessary conditions for the stationary point are

$$\frac{\partial L}{\partial x_1} = 4x_1 - 24 - \lambda = 0, \qquad \frac{\partial L}{\partial x_2} = 4x_2 - 8 - \lambda = 0,$$

$$\frac{\partial L}{\partial x_3} = 4x_3 - 12 - \lambda = 0, \qquad \frac{\partial L}{\partial \lambda} = -(x_1 + x_2 + x_3 - 11) = 0.$$

The solution of the simultaneous linear equations yields the stationary point

$$\mathbf{x}_0 = (x_1, x_2, x_3) = (6, 2, 3); \ \lambda = 0.$$

The sufficient condition for the stationary point to be minimum is that both the minors D_3 and D_4 should be negative. Now, we have

$$D_3 = \begin{vmatrix} 0 & 1 & 1 \\ 1 & 4 & 0 \\ 1 & 0 & 4 \end{vmatrix} = -8 \quad \text{and} \quad D_4 = \begin{vmatrix} 0 & 1 & 1 & 1 \\ 1 & 4 & 0 & 0 \\ 1 & 0 & 4 & 0 \\ 1 & 0 & 0 & 4 \end{vmatrix} = -48.$$

Since D_3 and D_4 both are negative, $\mathbf{x}_0 = (6, 2, 3)$ provides the solution to the *NLPP*.

Hence, the stationary point is a local minimum. Thus $x_0 = (6, 2, 3)$ provides the solution to the *NLPP*.

Exercise. Examine $z = 6x_1x_2$ for maxima and minima under the requirement $2x_1 + x_2 = 10$.

[***Hint:*** The stationary point is $x_0 = (2, 5, 5)$ which is a local maximum for z.]

Sufficient Conditions for a General *NLPP* with *m* (< *n*) Constraints

Introducing the m Lagrange multipliers $\lambda = (\lambda_1, \lambda_2, ..., \lambda_m)$, let the Lagrangian function for a general *NLPP* with more than one constaint be:

$$L(\mathbf{x}, \lambda) = f(\mathbf{x}) - \sum_{i=1}^{m} \lambda_i h^i(\mathbf{x}) \qquad (m < n)$$

The reader may verify that the equations

$$\frac{\partial L}{\partial x_i} = 0 \quad \text{and} \quad \frac{\partial L}{\partial \lambda_i} = 0 \qquad (i = 1, 2, ..., m; j = 1, 2, ..., n)$$

yield the necessary conditions for stationary points of $f(\mathbf{x})$. Thus, the optimization of $f(\mathbf{x})$ subject to $h(\mathbf{x}) = 0$ is equivalent to the optimization of $L(\mathbf{x}, \lambda)$.

We state here the sufficiency conditions for the Lagrange multiplier method of stationary point of $f(\mathbf{x})$ to be a maxima or minima (without proof). For this, we assume that the function $L(\mathbf{x}, \lambda)$, $f(\mathbf{x})$ and $h(\mathbf{x})$ all possess partial derivatives of order one and two w.r.t. the decision variables.

Let $$V = \left[\frac{\partial^2 L(x, \lambda)}{\partial x_i \, \partial x_j}\right]_{n \times n},$$

be the matrix of second order partial derivatives of $L(\mathbf{x}, \lambda)$ w.r.t. decision variables

$$U = \left[h^i{}_j(\mathbf{x})\right]_{m \times n},$$

where, $h^i{}_j(x) = \dfrac{\partial h^i(\mathbf{x})}{\partial \mathbf{x}_j}$, $i = 1, 2, \ldots, m;\; j = 1, 2, \ldots, n.$

Define the square matrix

$$\mathbf{H}^B = \begin{bmatrix} \mathbf{O} & \vdots & \mathbf{U} \\ \cdots & \cdots & \cdots \\ \mathbf{U}^T & \vdots & \mathbf{V} \end{bmatrix}_{(m+n) \times (m+n)},$$

where, $\mathbf{O}$ is an $m \times m$ null matrix.

The matrix $\mathbf{H}^B$ is called the ***bordered Hessian matrix***. The sufficient conditions for maximum and minimum stationary points are given below.

Consider, $(\mathbf{x}_0, \lambda_0)$ for the function $L(\mathbf{x}, \lambda)$ to be its stationary point. Let $\mathbf{H}^B{}_0$ be the corresponding bordered Hessian matrix computed at this stationary point. Then, $\mathbf{x}_0$ is a

(*a*) maximum point, if starting with principal minor of order $(2m + 1)$, the last $(n - m)$ principal minors of $\mathbf{H}^B{}_0$ from an alternating sign pattern starting with $(-1)^{m+n}$; and

(*b*) minimum point, if starting with principal minor of order $(2m + 1)$, the last $(n - m)$ principal minors of $\mathbf{H}^B{}_0$ have the sign of $(-1)^m$.

Remark. It may be observed that the above conditions are only sufficient for identifying an extreme point, but not necessary. That is, a stationary point may be an extreme point without satisfying the above conditions.

Example 25.11 (*Input-Allocation Problem*). *A manufacturing concern produces a product consisting of two raw materials, say A_1 and A_2. The production function is estimated as*

$$z = f(x_1, x_2) = 3.6x_1 - 0.4x_1^2 + 1.6x_2 - 0.2x_2^2,$$

where, z represents the quantity (in tons) of the product produced; x_1 and x_2 designate the input amounts of raw materials A_1 and A_2. The company has ₹50,000 to spend on these two raw materials. The unit price of A_1 is ₹10,000 and of A_2 is ₹5,000. Determine how much input amounts of A_1 and A_2 be decided so as to maximize the production output.

Solution. Since, the company must operate within the available funds, the budgetary constraint is

$$10{,}000x_1 + 5{,}000x_2 \le 50{,}000 \quad \text{or} \quad 2x_1 + x_2 \le 10.$$

We reduce this inequality constraint to an equality by imposing an additional assumption that the company has to spend every available single paisa on these raw materials. Then, the constraint is: $2x_1 + x_2 = 10$. Also, obviously $x_1 \ge 0$ and $x_2 \ge 0$. The problem of the company can, thus, be written as the following *NLPP:*

$$\text{Maximize} \quad z = f(x_1, x_2) = 3.6x_1 - 0.4x_1^2 + 1.6x_2 - 0.2x_2^2 \quad \text{subject to the constraints:}$$

$$2x_1 + x_2 = 10 \quad \text{and} \quad x_1, x_2 \ge 0$$

or

$$\text{Maximize} \quad z = f(x_1, x_2) \quad \text{subject to the constraints:}$$

$$h(x_1, x_2) = 0 \quad \text{and} \quad x_1, x_2 \ge 0$$

where, $h(x_1, x_2) = 2x_1 + x_2 - 10$. Observe that $f(x_1, x_2)$ and $h(x_1, x_2)$ are both differentiable w.r.t. x_1 and x_2.

Also, we observe that the objective function $z = f(x_1, x_2)$ is a concave function and the said constraint is an equality constraint. Therefore, the necessary and sufficient conditions for a maximum are

$$f_1 = \lambda h_1, \quad f_2 = \lambda h_2 \quad \text{and} \quad -h(x_1, x_2) = 0$$

That is,

$$3.6 - 0.8x_1 = 2\lambda, \quad 1.6 - 0.4x_2 = \lambda, \quad \text{and} \quad 2x_1 + x_2 = 10.$$

The first two of these yield

$$\lambda = 1.8 - 0.4x_1 = 1.6 - 0.4x_2$$

and, so the elimination of λ gives $0.4x_1 - 0.4x_2 - 0.2 = 0$.

Now, since $x_2 = 10 - 2x_1$, the last equation gives

$$0.4x_1 - 0.4(10 - 2x_1) - 0.2 = 0$$

or

$$1.2x_1 - 4.2 = 0, \quad \text{or} \quad x_1 = 3.5$$

Thus,

$$x_2 = 10 - 2x_1 = 10 - 2 \times 3.5 = 3.$$

The maximum value of the objective function is, thus, given by

$$z = f(3, 5, 3) = 3.6(3.5) - 0.4(3.5)^2 + 1.6(3) - 0.2(3)^2 = 10.7 \text{ (tonnes)}.$$

Thus, in order to have a maximum production of 10.7 tonnes, the company must input 3.5 units of raw material *A* and 3 units of raw material *B*.

Example 25.12. *Obtain the necessary and sufficient conditions for the optimum solution of the following NLPP:*

$$\text{Minimize } z = f(x_1, x_2) = 3e^{2x_1+1} + 2e^{x_2+5} \text{ subject to the constraints:}$$
$$x_1 + x_2 = 7, \quad x_1, x_2 \geq 0.$$ **[M.Sc. (Math.), Kerala Univ. 2001]**

Solution. Let us introduce a new differentiable Lagrangian function $L(x_1, x_2, \lambda)$ defined by

$$L(x_1, x_2, \lambda) = f(x_1, x_2) - \lambda(x_1 + x_2 - 7)$$
$$= 3e^{2x_1+1} + 2e^{x_2+5} - \lambda(x_1 + x_2 - 7),$$

where, λ is the Lagrangian multiplier.

Since, the objective function $z = f(x_1, x_2)$ is convex and the constraint an equality, the necessary and sufficient condition for the minimum of $f(x_1, x_2)$ are given by

$$\frac{\partial L}{\partial x_1} = 6e^{2x_1+1} - \lambda = 0 \quad \text{or} \quad \lambda = 6e^{2x_1+1}$$

$$\frac{\partial L}{\partial x_2} = 2e^{x_2+5} - \lambda = 0 \quad \text{or} \quad \lambda = 2e^{x_2+5}$$

$$\frac{\partial L}{\partial \lambda} = -(x_1 + x_2 - 7) = 0 \quad \text{or} \quad x_1 + x_2 = 7$$

using these three equations, we get

$$6e^{2x_1+1} = 2e^{x_2+5} \quad \text{or} \quad 3e^{2x_1+1} = e^{x_2+5} = e^{7-x_1+5}$$

or $$\log 3 + (2x_1 + 1) = 7 - x_1 + 5 \quad \text{or} \quad x_1 = \frac{1}{3}(11 - \log 3).$$

Thus, $$x_2 = 7 - \frac{1}{3}(11 - \log 3) = (10 + \log 3)/3.$$

Example 25.13. *Solve the non-linear programming problem:*

$$\text{Optimize } z = 4x_1^2 + 2x_2^2 + x_3^2 - 4x_1x_2 \text{ subject to the constraints:}$$
$$x_1 + x_2 + x_3 = 15, \quad 2x_1 - x_2 + 2x_3 = 20.$$ **[B.Sc. (Stat.), Delhi Univ. 2002]**

Solution. Here, we have

$$f(\mathbf{x}) = 4x_1^2 + 2x_2^2 + x_3^2 - 4x_1x_2, \quad h^1(\mathbf{x}) = x_1 + x_2 + x_3 - 15$$
$$h^2(\mathbf{x}) = 2x_1 - x_2 + 2x_3 - 20.$$

Construct the Lagrangian function

$$L(\mathbf{x}, \lambda) = f(\mathbf{x}) - \lambda_1 h^1(\mathbf{x}) - \lambda_2 h^2(\mathbf{x})$$
$$= (4x_1^2 + 2x_2^2 + x_3^2 - 4x_1x_2) - \lambda_1(x_1 + x_2 + x_3 - 15) - \lambda_2(2x_1 - x_2 + 2x_3 - 20)$$

The stationary point $(\mathbf{x}_0, \lambda_0)$ has, thus, gives the following necessary conditions:

$$\frac{\partial L}{\partial x_1} = 8x_1 - 4x_2 - \lambda_1 - 2\lambda_2 = 0, \qquad \frac{\partial L}{\partial x_2} = 4x_2 - 4x_1 - \lambda_1 + \lambda_2 = 0,$$

$$\frac{\partial L}{\partial x_3} = 2x_3 - \lambda_1 - 2\lambda_2 = 0, \qquad \frac{\partial L}{\partial \lambda_1} = -[x_1 + x_2 + x_3 - 15] = 0$$

$$\frac{\partial L}{\partial \lambda_2} = -[2x_1 - x_2 + 2x_3 - 20] = 0.$$

The solution to these simultaneous linear equations yields

$$\mathbf{x}_0 = (x_1, x_2, x_3) = (33/9, 10/3, 8) \text{ and } \lambda_0 = (\lambda_1, \lambda_2) = (40/9, 52/9).$$

The bordered Hessian matrix at this solution $(\mathbf{x}_0, \lambda_0)$ is given by

$$\mathbf{H}^B{}_0 = \left[\begin{array}{cc:ccc} 0 & 0 & 1 & 1 & 1 \\ 0 & 0 & 2 & -1 & 2 \\ \hdashline 1 & 2 & 8 & -4 & 0 \\ 1 & -1 & -4 & 4 & 0 \\ 1 & 2 & 0 & 0 & 2 \end{array}\right]$$

Here, since $n = 3$ and $m = 2$, therefore, $n - m = 1$, $(2m + 1 = 5)$. This means that one needs to check the determinant of $\mathbf{H}^B{}_0$ only and it must have the sign of $(-1)^2$.

Now, since det $\mathbf{H}^B{}_0 = 90 > 0$, $\mathbf{x}_0$ is a minimum point.

PROBLEMS

Solve the following non-linear programming problems, using method of Lagrangian multipliers:

25.07. Minimize $z = 6x_1^2 + 5x_2^2$ subject to the constraints:
$x_1 + 5x_2 = 3,\ x_1, x_2 \geq 0.$ *[M.Sc. (Math.), Kerala Univ. 2001]*

25.08. Maximize $f(x_1, x_2) = 3x_1^2 + x_2^2 + 2x_1x_2 + 6x_1 + 2x_2$ subject to the constraints:
$2x_1 - x_2 = 4,\ x, x_2 \geq 0.$

25.09. Maximize $z = 5x_1 + x_2 - (x_1 - x_2)^2$ subject to the constraints:
$x_1 + x_2 = 4,\ x_1, x_2 \geq 0.$

25.10. Maximize $z = 4x_1 + 6x_2 - 2x_1^2 - 2x_1x_2 - 2x_2^2$ subject to the constraints:
$x_1 + 2x_2 = 2$, and $x_1, x_2 \geq 0.$

25.11. Minimize $z = 2x_1^2 + x_2^2 + 3x_3^2 + 10x_1 + 8x_2 + 6x_3 - 100$ subject to the constraints:
$x_1 + x_2 + x_3 = 20,\ x_1, x_2, x_3 \geq 0.$

25.12. Minimize $z = x_1^2 + x_2^2 + x_3^2$ subject to the constraints:
$4x_1 + x_2^2 + 2x_3 = 14,\ x_1, x_2, x_3 \geq 0.$ *[B.Tech., Panjab Tech. Univ. 2011]*

25.13. Determine the optimum solution for the following *NLPP* and check whether it maximizes or minimizes the objective function:
$z = x_1^2 - 10x_1 + x_2^2 - 6x_2 + x_3^2 - 4x_3$ subject to the constraints:
$x_1 + x_2 + x_3 = 7$, and $x_1 \geq 0, x_2 \geq 0, x_3 \geq 0.$

25.14. Solve the following *NLPP:*
Optimize $z = 4x_1 + 9x_2 - x_1^2 - x_2^2$ subject to the constraints:
$4x_1 + 3x_2 = 15,\ 3x_1 + 5x_2 = 14;\ x_1 \geq 0, x_2 \geq 0.$

25.15. Minimize $z = x_1^2 + x_2^2 + x_3^2$ subject to the constraints:
$x_1 + x_2 + 3x_3 = 2,\ 5x_1 + 2x_2 + x_3 = 5,\ x_1, x_2, x_3 \geq 0.$
[M.E. (Nov.), Annamalai Univ. 2002; B.Sc. (Stat.), Delhi Univ. 2004]

25.16. Minimize $z = 6x_1 + 8x_3 - x_1^2 - x_2^2$ subject to the constraints:
$4x_1 + 3x_2 = 16,\ 3x_1 + 5x_2 = 15,\ x_1, x_2 \geq 0.$

25.17. A positive quantity b is to be divided into n parts in such a way that the product of n parts is to be maximum. Use Lagrangian multiplier method to obtain the optimal sub-division.
[***Hint:*** Let n parts be $x_1, x_2, ..., x_n$. Then, the Lagrangian function is:
$$L(x, \lambda) = x_1 \cdot x_2 \cdot x_3 \cdot ... \cdot x_n - \lambda(x_1 + x_2 + ... + x_n - b)]$$

25:7. CONSTRAINED OPTIMIZATION WITH INEQUALITY CONSTRAINTS

We shall derive the ***Kuhn-Tucker Conditions*** (necessary and sufficient) for the optimal solution of general *NLPP*. Consider the general *NLPP:*

$$\text{Optimize } z = f(x_1, x_2, ..., x_n) \text{ subject to the constraints:}$$
$$g(x_1, x_2, ..., x_n) \le C \text{ and } x_1, x_2, ..., x_n \ge 0,$$

where, C is a constant.

Introducing the function $h(x_1, x_2, ..., x_n) = g(x_1, x_2, ..., x_n) - C$, the constraint reduces to $h(x_1, x_2, ..., x_n) \le 0$. The problem, thus, can be written as

$$\text{Optimize } z = f(\mathbf{x}) \text{ subject to the constraints:}$$
$$h(\mathbf{x}) \le 0 \text{ and } \mathbf{x} \ge \mathbf{0}, \text{ where } \mathbf{x} \in \mathbf{R}^n.$$

We, now, slightly modify the problem by introducing new variable S, defined by $S^2 = -h(\mathbf{x})$, or $h(\mathbf{x}) + S^2 = 0$.

The new variable S is called a ***slack variable*** and appears as its square in the constraint equation so as to ensure its being non-negative. This avoids an additional constraint $S \ge 0$. Now, the problem can be restated as

$$\text{Optimize } z = f(\mathbf{x}), \ \mathbf{x} \in \mathbf{R}^n \text{ subject to the constraints:}$$
$$h(\mathbf{x}) + S^2 = 0 \text{ and } \mathbf{x} \ge \mathbf{0}.$$

This is a problem of constrained optimization in $n + 1$ variables with a single equality constraint and can thus be solved by the Lagrangian multiplier method.

To determine the stationary points, we consider the Lagrangian function defined by

$$L(\mathbf{x}, S, \lambda) = f(\mathbf{x}) - \lambda [h(\mathbf{x}) + S^2],$$

where, λ is the Lagrange multiplier. The necessary conditions for stationary points are

$$\frac{\partial L}{\partial x_j} \equiv \frac{\partial f}{\partial x_j} - \lambda \frac{\partial h}{\partial x_j} = 0 \quad (\text{for } j = 1, 2, ..., n)$$

$$\frac{\partial L}{\partial \lambda} \equiv -[h(\mathbf{x}) + S^2] = 0, \text{ and } \frac{\partial L}{\partial S} \equiv -2S\lambda = 0.$$

From the above, $\frac{\partial L}{\partial S} = 0$ requires either $\lambda = 0$ or $S = 0$. If $S = 0$, then it implies that $h(\mathbf{x}) = 0$. Thus, $\lambda h(\mathbf{x}) = 0$.

The variable S was introduced merely to convert the inequality constraint into an equality one, and, therefore, may be discarded. Moreover, since $S^2 \ge 0$, (2) gives $h(\mathbf{x}) \le 0$. Whenever $h(\mathbf{x}) < 0$, we get $\lambda = 0$ and whenever $\lambda > 0$, $h(\mathbf{x}) = 0$. However, λ is unrestricted in sign whenever $h(\mathbf{x}) = 0$.

The necessary conditions for the point $\mathbf{x}$ to be a point of maximum are thus restated (in the abbreviated form) as follows:

$$f - \lambda h_j = 0 \quad (j = 1, 2, ..., n)$$
$$\lambda h = 0, \ h \le 0 \text{ and } \lambda \ge 0^*$$
$$\text{Maximize } z = f \text{ subject to the constraints: } h \le 0.$$

The set of such necessary conditions is called ***Kuhn-Tucker Conditions***.

* More precisely, since $\lambda = \frac{f_j}{h_j}\left(= \frac{\partial f / \partial x_j}{\partial h / \partial x_j} = \frac{\partial f}{\partial h}\right)$ measures the rate of variation of f w.r.t. h, then as the right-hand side of $h(\mathbf{x}) \le 0$ increases about zero, the solution space becomes less constrained and hence $f(\mathbf{x})$ cannot decrease. This means that $\lambda \ge 0$.

A similar argument holds for the minimization of non-linear programming problem:

$$\text{Minimize } z = f(\mathbf{x}) \text{ subject to the constraints:}$$
$$g(\mathbf{x}) \geq C \text{ and } \mathbf{x} \geq \mathbf{0}.$$

Introduction of $h(\mathbf{x}) = g(\mathbf{x}) - C$ reduces the first constraint to $h(\mathbf{x}) \geq 0$. The new surplus variable S_0 can be introduced in $h(\mathbf{x}) \geq 0$, so that we may have the equality constraint $h(\mathbf{x}) - S_0^2 = 0$. The appropriate Lagrangian function is

$$L(\mathbf{x}, S_0, \lambda) = f(\mathbf{x}) - \lambda[h(\mathbf{x}) - S_0^2].$$

The following set of Kuhn-Tucker conditions is obtained:

$$f_j - \lambda h_j = 0 \quad (j = 1, 2, ..., n)$$
$$\lambda h = 0, \; h \geq 0 \text{ and } \lambda \geq 0$$
$$\text{Minimize } z = f \text{ subject to the constraints: } h \geq 0.$$

Theorem 25-1 (*Sufficiency of Kuhn-Tucker Conditions*). *The Kuhn-Tucker conditions for a maximization NLPP of maximizing $f(\mathbf{x})$ subject to the constraints: $h(\mathbf{x}) \leq 0$ and $\mathbf{x} \geq \mathbf{0}$, are sufficient conditions for a maximum of $f(\mathbf{x})$, if $f(\mathbf{x})$ is concave and $h(\mathbf{x})$ is convex.*

Proof. The result follows, if we are able to show that the Lagrangian function

$$L(\mathbf{x}, S, \lambda) = f(\mathbf{x}) - [h(\mathbf{x}) + S^2],$$

where, S is defined by $h(\mathbf{x}) + S^2 = 0$, is concave in $\mathbf{x}$ under the given conditions.

In that case the stationary point obtained from the Kuhn-Tucker conditions must be global maximum point.

Now, since $h(\mathbf{x}) + S^2 = 0$, it follows from the necessary conditions that $\lambda S^2 = 0$. Since, $h(\mathbf{x})$ is convex and $\lambda \geq 0$, it follows that $\lambda h(\mathbf{x})$ is also convex and $-\lambda h(\mathbf{x})$ is concave. Thus, we conclude that $f(\mathbf{x}) - \lambda h(\mathbf{x})$ and hence $f(\mathbf{x}) - \lambda[h(\mathbf{x}) + S^2] = L(\mathbf{x}, S, \lambda)$ is concave in $\mathbf{x}$.

Remark. By similar arguments, it can be shown that of the minimization *NLPP*, Kuhn-Tucker conditions are also the sufficient conditions for the minimum of the objective function, if the objective function $f(\mathbf{x})$ is convex and the function $h(\mathbf{x})$ is concave.

Example 25.14. *Maximize* $z = 3.6x_1 - 0.4x_1^2 + 1.6x_2 - 0.2x_2^2$ *subject to the constraints:*

$$2x_1 + x_2 \leq 10 \text{ and } x_1, x_2 \geq 0.$$

Solution. Here,

$$f(x) = 3.6x_1 - 0.4x_1^2 + 1.6x_2 - 0.2x_2^2$$
$$g(x) = 2x_1 + x_2, \text{ and } C = 10.$$
$$\therefore \quad h(x) = g(x) - C = 2x_1 + x_2 - 10.$$

The Kuhn-Tucker conditions are:

$$\frac{\partial f(\mathbf{x})}{\partial x_1} - \lambda\frac{\partial h(\mathbf{x})}{\partial x_1} = 0, \quad \frac{\partial f(\mathbf{x})}{\partial x_2} - \lambda\frac{\partial h(\mathbf{x})}{\partial x_2} = 0, \quad \lambda h(\mathbf{x}) = 0, \text{ and } h(\mathbf{x}) \leq 0, \; \lambda \geq 0,$$

where, λ is the Lagrangian multiplier.

That is,

$$3.6 - 0.8x_1 = 2\lambda \quad ...(1)$$
$$1.6 - 0.4x_2 = \lambda \quad ...(2)$$
$$\lambda[2x_1 + x_2 - 10] = 0 \quad ...(3)$$
$$2x_1 + x_2 - 10 = 0 \quad ...(4)$$
$$\lambda \geq 0 \quad ...(5)$$

From eq. (3) either $\lambda = 0$ or $2x_1 + x_2 - 10 = 0$.

Let $\lambda = 0$, then eqs. (2) and (1) yield $x_1 = 4.5$ and $x_3 = 4$. With these values of x_1 and x_2, however, eq. (4) cannot be satisfied. Thus, optimal solution cannot be obtained here for $\lambda = 0$. Let then $\lambda \neq 0$, which implies [from eq. (3)] that $2x_1 + x_2 - 10 = 0$. This together with eqs. (1) and (2) yields the stationary value

$$\mathbf{x}_0 = (x_1, x_2) = (3.5, 3)$$

Now, it is easy to observe that $h(\mathbf{x})$ is convex in $\mathbf{x}$, and $f(\mathbf{x})$ is concave in $\mathbf{x}$. Thus, Kuhn-Tucker conditions are the sufficient conditions for the maximum. Hence, $\mathbf{x}_0 = (3.5, 3)$ is the solution to the given *NLPP*. The maximum value of z (corresponding to $\mathbf{x}_0$) is given by $z_0 = 10.7$.

Kuhn-Tucker Conditions for General *NLPP* with *m* (< *n*) Constraints

Introducing $\mathbf{S} = (S_1, S_2, ..., S_m)$, let the Lagrangian function for a general *NLPP* with m ($< n$) constraints be

$$L(\mathbf{x}, \mathbf{S}, \lambda) = f(\mathbf{x}) - \sum_{i=1}^{m} \lambda_i \left[h^i(\mathbf{x}) + S_i^2 \right],$$

where, $\lambda = (\lambda_1, ..., \lambda_m)$ are the Lagrangian multipliers.

The necessary conditions for $f(\mathbf{x})$ to be a maximum are:

(1) $$\frac{\partial L}{\partial x_j} = \frac{\partial f}{\partial x_j} - \sum_{i=1}^{m} \lambda_i \frac{\partial h^i}{\partial x_j} = 0 \qquad \text{for } j = 1, 2, ..., n$$

(2) $$\frac{\partial L}{\partial \lambda_i} = h^i + S_i^2 = 0 \qquad \text{for } i = 1, 2, ..., m$$

(3) $$\frac{\partial L}{\partial S_i} = -2S_i \lambda_i = 0 \qquad \text{for } i = 1, 2, ..., m$$

where, $L = L(\mathbf{x}, \mathbf{S}, \lambda)$, $f = f(\mathbf{x})$ and $h^i = h^i(\mathbf{x})$.

Equation (3) states that either $\lambda_i = 0$ or $S_i = 0$. By an argument parallel to that considered in the case of single inequality constraint; the conditions (3) and (2) together are replaced by the conditions (4) given below.

(4) $$\lambda_i h^i = 0, \; h^i \leq 0, \text{ and } \lambda_i \geq 0 \qquad \text{for } i = 1, 2, ..., m$$

The Kuhn-Tucker conditions for a maximum may thus be restated as:

$$f_j = \sum_{i=1}^{m} \lambda_i h^i{}_j \qquad (j = 1, 2, ..., n)$$

$$\lambda_i h^i = 0, \; h^i \leq 0 \text{ and } \lambda_i \geq 0 \qquad (i = 1, 2, ..., m)$$

Maximize $z = f$ subject to the constraints: $h^i \leq 0$. $\quad (i = 1, 2, ..., m)$

Theorem 25-2 (*Sufficiency of Kuhn-Tucker Conditions*). *For the NLPP of maximizing $f(\mathbf{x})$, $\mathbf{x} \in \mathbf{R}^n$, subject to the inequality constraints: $h^i(\mathbf{x}) \leq 0$ ($i = 1, 2, ..., m$), the Kuhn-Tucker conditions are also the sufficient conditions for a maximum, if $f(\mathbf{x})$ is concave and all $h^i(\mathbf{x})$ are convex functions of $\mathbf{x}$.*

Proof. Exercise for the reader.

The Kuhn-Tucker conditions for a minimization non-linear programming problem may be obtained in a similar manner. These conditions in that case come out to be:

$$f_j = \sum_{i=1}^{m} \lambda_i h^i{}_j \qquad (j = 1, 2, ..., n)$$

$$\lambda_i h^i = 0, \; h^i \geq 0 \text{ and } \lambda_i \geq 0 \qquad (i = 1, 2, ..., m)$$

Minimize $z = f$ subject to the constraints: $h^i \geq 0$. $\quad (i = 1, 2, ..., m)$

It can be shown that for the minimization problem, Kuhn-Tucker conditions are also sufficient conditions for the minima, if $f(\mathbf{x})$ is convex and all $h^i(\mathbf{x})$ are concave in $\mathbf{x}$, that is, $-h^i(\mathbf{x})$ are also all convex.

Note. If $f(\mathbf{x})$ is strictly concave (convex), the Kuhn-Tucker conditions are sufficient conditions for an absolute maximum (minimum).

Remarks

1. We may consider $\mathbf{x} \geq \mathbf{0}$ or $-\mathbf{x} \leq \mathbf{0}$, to have been included in the inequality constraint $h^i(\mathbf{x}) \leq 0$.
2. In both the maximization and minimization *NLPP*, the Lagrange multiplier λ_i corresponding to the equality constraints $h^i(\mathbf{x}) = 0$ must be unrestricted in sign.
3. A general *NLPP* may contain the constraints of the '≥' or '=' or '≤' type. In the case of maximization *NLPP*, all constraints must be converted into those of '≤' type and in the case of minimization *NLPP*, into those of '≥' type by suitable multiplication by –1.

Example 25.15. *Determine* x_1, x_2 *and* x_3 *so as to*

Maximize $z = -x_1^2 - x_2^2 - x_3^2 + 4x_1 + 6x_2$ *subject to the constraints:*

$$x_1 + x_2 \leq 2, \quad 2x_1 + 3x_2 \leq 12, \quad x_1, x_2 \geq 0.$$

Solution. Here,

$$f(\mathbf{x}) = -x_1^2 - x_2^2 - x_3^2 + 4x_1 + 6x_2 \qquad \mathbf{x} \in \mathbf{R}^n$$

$$h^1(\mathbf{x}) = x_1 + x_2 - 2, \quad h^2(\mathbf{x}) = 2x_1 + 3x_2 - 12.$$

Clearly, $f(\mathbf{x})$ is concave* and $h^1(\mathbf{x})$, $h^2(\mathbf{x})$, are convex in $\mathbf{x}$. Thus, the Kuhn-Tucker conditions will be necessary and sufficient conditions for a maximum. These conditions are obtained by the partial differentiation of the Lagrangian function

$$L(\mathbf{x}, \mathbf{S}, \lambda) = f(\mathbf{x}) - \lambda_1\left[h^1(\mathbf{x}) + S_1^2\right] - \lambda_2\left[h^2(\mathbf{x}) + S_2^2\right]$$

where $\mathbf{S} = (S_1, S_2)$, $\lambda = (\lambda_1, \lambda_2)$, S_1, S_2 being slack variables and λ_1, λ_2 the Lagrange multipliers.

The Kuhn-Tucker conditions are given by

(1) $\quad f_j = \sum_{i=1}^{m} \lambda_i h^i{}_j \qquad (j = 1, 2, 3)$

(2) $\quad \lambda_i h^i = 0 \qquad (i = 1, 2)$

(3) $\quad h^i \leq 0 \qquad (i = 1, 2)$

(4) $\quad \lambda_i \geq 0 \qquad (i = 1, 2)$

Thus, in this problem, these are

(1) (*i*) $-2x_1 + 4 = \lambda_1 + 2\lambda_2$ (*ii*) $-2x_2 + 6 = \lambda_1 + 3\lambda_2$ (*iii*) $-2x_3 = 0$.

(2) (*i*) $\lambda_1(x_1 + x_2 - 2) = 0$ (*ii*) $\lambda_2(2x_1 + 3x_2 - 12) = 0$

(3) (*i*) $x_1 + x_2 - 2 \leq 0$ (*ii*) $2x_1 + 3x_2 - 12 \leq 0$

(4) $\lambda_1 \geq 0, \lambda_2 \geq 0$.

Now, there arise four cases:

Case 1. $\lambda_1 = 0$ *and* $\lambda_2 = 0$.

Equations (1) (*i*), (*ii*) and (*iii*) yield $x_1 = 2, x_2 = 3, x_3 = 0$.

However, this solution violates (3) [(*i*) and (*ii*) both], and it must, therefore, be discarded.

---

* The objective function is concave if the principal minors of bordered Hessian matrix, alternate in sign, beginning with the negative sign. If the principal minors are positive, the objective function is convex. In the present case

$$\mathbf{H}^B = \begin{bmatrix} -2 & 0 & 0 \\ 0 & -2 & 0 \\ 0 & 0 & -2 \end{bmatrix}, \; n = 3, \; m = 2, \; |\mathbf{H}^B| < 0. \text{ Thus, } f(\mathbf{x}) \text{ is concave.}$$

Case 2. *$\lambda_1 = 0$ and $\lambda_2 \neq 0$.*

Equations (2) yield $2x_1 + 3x_2 = 12$ and (1) (*i*) and (*ii*) yield $-2x_1 + 4 = 2\lambda_2$, and $-2x_2 + 6 = 3\lambda_2$. The solution of these simultaneous equations yields $x_1 = 2/13$, $x_2 = 3/13$, $\lambda_2 = 24/13 > 0$; also (1) (*iii*) gives $x_3 = 0$.

However, this solution violates (3) (*i*). This solution is also, thus, discarded.

Case 3. *$\lambda_1 \neq 0$ and $\lambda_2 \neq 0$.*

Equations (2) (*i*) and (*ii*) yield $x_1 + x_2 = 2$ and $2x_1 + 3x_2 = 12$. These together yield $x_1 = -6$ and $x_2 = 8$. Thus (1) (*i*), (*ii*) and (*iii*) gives $x_3 = 0$, $\lambda_1 = 68$, $\lambda_2 = -26$.

However, this solution is to be discarded since $\lambda_2 = -26$ violates (4).

Case 4. *$\lambda_1 \neq 0$ and $\lambda_2 = 0$.*

Equations (2) (*i*) yield $x_1 + x_2 = 0$. This together with (1) (*i*) and (*ii*) gives $x_1 = 1/2$ and $x_2 = 3/2$, $\lambda_1 = 3 > 0$. Further, from (1) (*iii*) $x_3 = 0$.

We observe that this solution does not violate any of the Kuhn-Tucker conditions.

Hence, the optimum (maximum) solution to the given *NLPP* is

$$x_1 = 1/2, \quad x_2 = 3/2, \text{ and } \quad x_3 = 0 \quad \text{with } \lambda_1 = 3, \text{ and } \lambda_2 = 0,$$

The maximum value of objective function being $z_0 = 17/2$.

Example 25.16. *Optimize* $z = 2x_1 + 3x_2 - (x_1^2 + x_2^2 + x_3^2)$ *subject to the constraints:*

$$x_1 + x_2 \leq 1, \quad 2x_1 + 3x_2 \leq 6 \quad \text{and} \quad x_1 \geq 0, \quad x_2 \geq 0.$$

Solution. Here, we have

$$f(x) = 2x_1 + 3x_2 - x_1^2 - x_2^2 - x_3^2, \quad h^1(x) = x_1 + x_2 - 1, \quad \text{and} \quad h^2(x) = 2x_1 + 3x_2 - 6.$$

Before applying the Kuhn-Tucker conditions, it is essential to determine whether the problem is of maximization or of minimization type. For that, we construct the bordered Hessian matrix:

$$\mathbf{H}^B = \begin{pmatrix} \mathbf{O} & \vdots & \mathbf{P} \\ \cdots & \cdots & \cdots \\ \mathbf{P}^T & \vdots & \mathbf{Q} \end{pmatrix}_{(m+n)\times(m+n)}$$

or
$$\mathbf{H}^B = \left(\begin{array}{cc|ccc} 0 & 0 & 1 & 1 & 0 \\ 0 & 0 & 2 & 3 & 0 \\ \hline 1 & 2 & -2 & 0 & 0 \\ 1 & 3 & 0 & -2 & 0 \\ 0 & 0 & 0 & 0 & -2 \end{array}\right) = -10. \qquad \begin{array}{r} n = 3, \\ m = 2, \\ n - m = 1, \\ 2m + 1 = 5. \end{array}$$

For maximization, the sign of $\mathbf{H}^B$ should be $(-1)^{m+n}$, *i.e.*, –*ve*, while for minimization, it should be $(-1)^m$, *i.e.*, +*ve*. Since, $\mathbf{H}^B < 0$, the solution point should maximize the objective function.

The Kuhn-Tucker conditions for this case are

$$2 - 2x_1 - \lambda_1 - 2\lambda_2 = 0, \quad \ldots (i) \qquad 3 - 2x_2 - \lambda_1 - 3\lambda_2 = 0, \quad \ldots (ii) \qquad -2x_3 = 0, \ \ldots (iii)$$

$$\left.\begin{array}{r} \lambda_1(2x_1 + 3x_2 - 6) = 0 \\ \lambda_2(x_1 + x_2 - 1) = 0 \end{array}\right\} \quad \ldots (iv) \qquad x_1 + x_2 - 1 \leq 0, \quad \ldots (v)$$

$$2x_1 + 3x_2 - 6 \leq 0, \quad \ldots (vi) \qquad \lambda_1, \lambda_2 \geq 0. \quad \ldots (vii)$$

Four solutions corresponding to the following values of λ_1 and λ_2 can be obtained:

Case 1. $\lambda_1 = 0, \lambda_2 = 0.$

Equations (*i*), (*ii*) and (*iii*) gives

$$x_1 = 1,\ x_2 = 3/2 \text{ and } x_3 = 0.$$

This solution does not satisfy equation (*vi*) and is, therefore, discarded.

Case 2. $\lambda_1 = 0, \lambda_2 \neq 0.$

The solution of equations (*i*), (*ii*), (*iii*) and (*v*) gives

$$x_1 = 12/13,\ x_2\ 18/13,\ x_3 = 0 \text{ and } \lambda_1 = 1/13.$$

This solution again does not satisfy equation (*vi*) and is, therefore, discarded.

Case 3. $\lambda_1 \neq 0, \lambda_2 = 0.$

The solution of equations (*i*), (*ii*), (*iii*) and (*iv*) yields

$$x_1 = 1/4,\ x_2 = 3/4,\ x_3 = 0 \text{ and } \lambda_1 = 3/2.$$

This solution satisfies all the conditions, and gives $z = 17/8$.

Case 4. $\lambda_1 \neq 0, \lambda_2 \neq 0.$

The solution of equations (*i*), (*ii*), (*iii*), (*iv*) and (*v*) gives

$$x_1 = -3,\ x_2 = 4,\ x_3 = 0,\ \lambda_1 = -34 \text{ and } \lambda_2 = 13.$$

This solution violates conditions (*vii*) and is, thus, infeasible and is, therefore, discarded.

Since, only one solution satisfies all the conditions, the same is the optimal (optimum) solution.

Hence, $x_1^0 = 1/4,\ x_2^0 = 3/4,\ x_3^0 = 0,$ and maximum $z = 17/8$.

PROBLEMS

Use Kuhn-Tucker conditions to solve the following non-linear programming problems:

25.18. Minimize $z = 2x_1^2 + 12x_1x_2 - 7x_2^2$ subject to the constraints:
$2x_1 + 5x_2 \leq 98,\ x_1, x_2 \geq 0.$

25.19. Maximize $z = 8x_1 + 10x_2 - x_1^2 - x_2^2$ subject to the constraints:
$3x_1 + 2x_2 \leq 6,\ x_1 \geq 0, x_2 \geq 0.$ *[IAS 1991]*

25.20. Minimize $z = x_1^2 + x_2^2 + x_3^2$ subject to the constraints:
$2x_1 + x_2 \leq 5,\ x_1 + x_2 \leq 2,\ x_1 \geq 1,\ x_2 \geq 2,\ x_3 \geq 0.$ *[IAS 1993]*

25.21. Minimize $z = 0.3x_1^2 - 2x_1 + 0.4x_2^2 - 2.4x_2 + 0.6x_1x_2 + 100$ subject to the constraints:
$2x_1 + x_2 \geq 4,\ x_1, x_2 \geq 0.$

25.22. Minimize $z = -\log x_1 - \log x_2$ subject to the constraints:
$x_1 + x_2 \leq 2$ and $x_1 \geq 0,\ x_2 \geq 0.$

25.23. Minimize $z = 2x_1 + 3x_2 - x_1^2 - 2x_2^2$ subject to the constraints:
$x_1 + 3x_2 \leq 6,\ 5x_1 + 2x_2 \leq 10,$ and $x_1 \geq 0,\ x_2 \geq 0.$
[M.Sc. (Math.), Rohilkhand Univ. 2011]

25.24. Maximize $z = 2x_1 - x_1^2 + x_2$ subject to the constraints:
$2x_1 + 3x_2 \leq 6,\ 2x_1 + x_2 \leq 4$ and $x_1, x_2 \geq 0.$
[BE (Comp. Sc.), Nagpur Univ. 2004]

25.25. Maximize $z = 7x_1^2 + 6x_1 + 5x_2^2$ subject to the constraints:

$$x_1 + 2x_2 \le 10, \; x_1 - 3x_2 \le 9, \; x_1 \ge 0 \text{ and } x_2 \ge 0.$$

[M.E. (Nov.), Annamalai Univ. 2002]

25.26. Maximize $z = 3x_1 + x_2$ subject to the constraints:

$$x_1^2 + x_2^2 \le 5, \; x_1 - x_2 \le 1 \text{ and } x_1 \ge 0, \; x_2 \ge 0.$$

[B.E. (Civil), Madras Univ. 1991]

25.27. Maximize $z = 8x_1^2 + 2x_2^2$ subject to the constraints:

$$x_1^2 + x_2^2 \le 9, \; x_1 \le 2 \text{ and } x_1, x_2 \ge 0.$$

25.28. Minimize $f(x_1, x_2) = (x_1 - 1)^2 + (x_2 - 5)^2$ subject to the constraints:

$$-x_1^2 + x_2 \le 4 \text{ and } -(x_1 - 2)^2 + x_2 \le 3.$$

25.29. A manufacturing firm produces a product *A*. The firm has the contract to supply 60 units at the end of the first, second and third months. The cost of producing *x* units of product *A* in any month is given by x^2. The firm can produce more units of product *A* in any month and carry them to subsequent month. However, carrying cost of ₹ 25 per unit is charged for carrying units of product *A* from one month to the next. Assuming that there is no initial inventory, determine the number of units of product *A* to be produced in each month so as to minimize the total cost.

[***Hint:*** Minimum (total cost) z = Production cost + Carrying cost

$$= x_1^2 + x_2^2 + x_3^2 + 60x_1 + 25\,(x_1 - 60) + 25\,(x_1 + x_2 - 120)$$

subject to the constraints:

$$x_1 + x_2 \ge 120, \; x_1 + x_2 + x_3 \ge 180, \; x_1 \ge 60; \; x_1, x_2, x_3 \ge 0$$

where, x_1, x_2 and x_3 = number of units of product *A* produced in first, second and third months, respectively.]

25:8. SADDLE POINT PROBLEMS

In Chapter 19, the ***saddle point*** of pay-off matrix was defined. Let (a_{ij}) be pay-off matrix for a two-person zero-sum game. If $\mathbf{a}_{i^*j}$ denote the pay-off minima at i^* over the rows and a_{ij^*} denote the pay-off minima at j^* over the columns, then the saddle point $a_{i^*j^*}$ is given by

$$a_{i^*j} \ge a_{i^*j^*} \ge a_{ij^*}.$$

We, now define the saddle point of functions. Let ϕ be a real valued function of several variables. Let $\mathbf{x} = (x_1, ..., x_n)$ and $\mathbf{u} = (u_1, ..., u_m)$, then for these variables we shall denote the function ϕ by $\phi\,(\mathbf{x}, \mathbf{u})$ $\mathbf{x} \in \mathbf{R}^n$, $\mathbf{u} \in \mathbf{R}^m$.

Definition 1 (*Saddle Point*). *Let $\phi\,(\mathbf{x}, \mathbf{u})$ be a function of $\mathbf{x} \in \mathbf{R}^n$ and $\mathbf{u} \in \mathbf{R}^m$. The function $\phi\,(\mathbf{x}, \mathbf{u})$ is said to have a* ***saddle point*** *at $(\mathbf{x}^0, \mathbf{u}^0)$, if and only if*

$$\phi(\mathbf{x}^0, \mathbf{u}) \ge \phi(\mathbf{x}^0, \mathbf{u}^0) \ge \phi(\mathbf{x}, \mathbf{u}^0).$$

Definition 2 (*Saddle Value Problem*). *Let $\mathbf{x} \in \mathbf{R}^n$, $\mathbf{u} \in \mathbf{R}^m$. The problem of determining saddle point value $\phi\,(\mathbf{x}^0, \mathbf{u}^0)$ under the constraints: $\mathbf{x} \ge \mathbf{0}$, and $\mathbf{u} \ge \mathbf{0}$, is called a* ***saddle value problem.***

We introduce the following notations:

$$\phi^0 = \phi\,(\mathbf{x}^0, \mathbf{u}^0)$$

Assume that $\phi\,(\mathbf{x}, \mathbf{u})$ is differentiable partially w.r.t. $\mathbf{x}$ and $\mathbf{u}$.

We define the following partial gradients of $\phi(\mathbf{x}, \mathbf{u})$:

$$\phi_{\mathbf{x}^0} = \left[\frac{\partial \phi^0}{\partial x_1}, \ldots, \frac{\partial \phi^0}{\partial x_n}\right] \quad \text{and} \quad \phi_{\mathbf{u}^0} = \left[\frac{\partial \phi^0}{\partial u_1}, \ldots, \frac{\partial \phi^0}{\partial u_m}\right]$$

as the column vectors, where the superscripts indicate the partial derivatives are evaluated at $(\mathbf{x}^0, \mathbf{u}^0)$.

Let $\mathbf{x} \in \mathbf{R}^n$, $\mathbf{u} \in \mathbf{R}^m$ and ϕ be a function of $\mathbf{x}$ and $\mathbf{u}$, respectively. Then, for a point $(\mathbf{x}^0, \mathbf{u}^0)$ to be a non-negative saddle point of $\phi(\mathbf{x}, \mathbf{u})$, it is necessary that

$$(a)\ \phi_{\mathbf{x}^0} \le \mathbf{0},\ \phi_{\mathbf{x}^0}{}^T \mathbf{x}^0 = 0 \ \text{ and } \ (b)\ \phi_{\mathbf{u}^0} \le \mathbf{0},\ \phi_{\mathbf{u}^0}{}^T \mathbf{u}^0 = 0, \ \text{ for } \mathbf{x}^0 \ge \mathbf{0} \text{ and } \mathbf{u}^0 \ge \mathbf{0}.$$

25:9. SADDLE POINTS AND *NLPP*

Let us consider the standard *NLPP:*

$$\text{Maximize } z = f(\mathbf{x}),\ \mathbf{x} \in \mathbf{R}^n \ \text{ subject to the constraints:}$$
$$h^i(\mathbf{x}) = 0,\ \mathbf{x} \ge \mathbf{0},\ i = 1, 2, \ldots, m$$

We formulate the associated Lagrangian function:

$$L(\mathbf{x}, \mathbf{u}) = f(\mathbf{x}) - \sum_{i=1}^{m} u_i h^i(\mathbf{x}),$$

where, $\mathbf{u} \in \mathbf{R}^m$ are Lagrange multipliers.

The problem of maximizing $f(\mathbf{x})$ is equivalent to that of maximizing $L(\mathbf{x}, \mathbf{u})$. We shall, now, investigate the relationship between maximization of $L(\mathbf{x}, \mathbf{u})$ and the saddle point problem for $\phi(\mathbf{x}, \mathbf{u})$, where ϕ s a real valued function of $\mathbf{x}$ and $\mathbf{u}$. We shall assume that $f(\mathbf{x})$ and all the constraint functions $h^i(\mathbf{x})$ possess partial differentials.

Theorem 25-3 (*Necessary Conditions for Saddle Point Correspondence*). *For $\mathbf{x}^0 \in \mathbf{R}^n$ to be a solution to NLPP, then it is necessary that $\mathbf{x}^0$ and some $\mathbf{u}^0$ satisfy the conditions for $\phi(\mathbf{x}, \mathbf{u}) = L(\mathbf{x}, \mathbf{u})$, where $\mathbf{u} \in \mathbf{R}^m$ are the Lagrange multipliers.*

Proof. Let $\mathbf{x}^0$ be a solution to *NLPP*. Then, it must satisfy the Kuhn-Tucker conditions:

(1) $\quad f_j = \sum_{i=1}^{m} u_i^0 h^i{}_j \qquad j = 1, 2, \ldots, n$

(2) $\quad u_i^0 h^i = 0 \qquad i = 1, 2, \ldots, m$

(3) $\quad h^i \le 0$ and $u_i^0 \ge 0, \qquad i = 1, 2, \ldots, m$

where, $f_j = \dfrac{\partial f(\mathbf{x})}{\partial x^0{}_j},\ h^i{}_j = \dfrac{\partial h^i(\mathbf{x})}{\partial x^0{}_j},\ h^i = h^i(\mathbf{x})$, etc.

Now, from (1), we may write

$$\left(f_j - \sum_{i=1}^{m} u_i^0 h^i{}_j\right) \le 0 \qquad \text{for } j = 1, 2, \ldots, n$$

and

$$\left(f_j - \sum_{i=1}^{m} u_i^0 h^i{}_j\right) x_j^0 = 0 \qquad \text{for } j = 1, 2, \ldots, n;\ i = 1, 2, \ldots, m.$$

or

$$\left[f_j - \sum_{i=1}^{m} u_i^0 h^i{}_j \,\middle|\, j = 1, \ldots, n\right] \le 0 \qquad \ldots (1)$$

and

$$\left[f_j - \sum_{i=1}^{m} u_i^0 h^i{}_j \,\middle|\, j = 1, \ldots, n\right]^T \mathbf{x} = 0.$$

Clearly $\mathbf{x}^0 \geq \mathbf{0}$, is a solution to the *NLPP*. From (3), we have

(*i*) $\qquad -h^i(\mathbf{x}) \geq 0 \qquad$ for $i = 1, 2, ..., m,$

and from (2), we have

(*ii*) $\qquad h^i(\mathbf{x})\, u_i^0 = 0 \qquad$ for $i = 1, 2, ..., m.$

Also, (3) allows us to write

(*iii*) $\qquad \mathbf{u}^0 \geq \mathbf{0}.$

Combining (*i*), (*ii*) and (*iii*), we may write

$$\left[-h^i(\mathbf{x}) \,\middle|\, i = 1, \ldots, m\right] \geq \mathbf{0} \text{ and } \left[-h^i(\mathbf{x}) \,\middle|\, i = 1, \ldots, m\right]^T \mathbf{u}^0 = \mathbf{0} \text{ for } \mathbf{x}^0 \geq \mathbf{0}. \qquad \ldots (2)$$

Now, if $\phi(\mathbf{x}^0, \mathbf{u}^0) \equiv L(\mathbf{x}^0, \mathbf{u}^0)$, then, we have

$$\frac{\partial \phi}{\partial {x^0}_j} = f_j - \sum_{i=1}^{m} u_i {h^i}_j \qquad \text{for } j = 1, 2, ..., n$$

and

$$\frac{\partial \phi}{\partial {u^0}_j} = -h^i(\mathbf{x}).$$

From these two equations, we rewrite as under:

$$\phi_{\mathbf{x}^0}{}^T \leq \mathbf{0} \text{ and } \phi_{\mathbf{x}^0}{}^T \mathbf{x}^0 = 0 \qquad \text{for } \mathbf{x}^0 \geq \mathbf{0}$$

$$\phi_{\mathbf{u}^0}{}^T \leq \mathbf{0} \text{ and } \phi_{\mathbf{u}^0}{}^T \mathbf{u}^0 = 0 \qquad \text{for } \mathbf{u}^0 \geq \mathbf{0}$$

These conditions are essentially the conditions of saddle point problems.

This completes the proof.

TEST YOUR UNDERSTANDING

1. In a non-linear programming problem,
   (*a*) the objective function is non-linear.
   (*b*) one or more of the constraints have non-linear relationship.
   (*c*) both (*a*) and (*b*).
   (*d*) none of the above.
2. In the context of constrained optimization with equality constraints,
   (*a*) a non-linear programming problem is solved using Lagrange multipliers.
   (*b*) the necessary conditions becomes sufficient for a maximum (or minimum) of the objective function, if the objective function is concave (convex) and the side constraints are equations.
   (*c*) the sufficient conditions can be verified by the bordered Hessian matrix.
   (*d*) all of the above.
3. In non-linear programming problem
   (*a*) a local minimum of a strictly convex function on a convex set is a unique global minimum of that function.
   (*b*) a local maximum of a strictly concave function on a convex set is not a global maximum of that function.
   (*c*) if the constraints of an *NLPP* are equations, optimum solution cannot be obtained by using Lagrange multipliers.
   (*d*) none of the above.

4. Unconstrained optimization problems have
   (*a*) only the differentiable objective function.
   (*b*) concave function for maximization and convex function for minimization.
   (*c*) the solution by setting the partial derivatives of objective function to zero.
   (*d*) all of the above.
5. The general *NLPP* with inequality constraints
   (*a*) are usually solved by simplex method.
   (*b*) can be solved by using Kuhn-Tucker conditions.
   (*c*) can be solved only by Lagrange's method.
   (*d*) can be solved only, if the constraints are of '≤' type.
6. In the context of general *NLPP*,
   (*a*) the constraints may be of '≤', '=' or '≥' type.
   (*b*) the concept of partial derivatives is used to optimize the multi-variable function.
   (*c*) if the objective function is strictly concave, the Kuhn-Tucker conditions are sufficient conditions for an absolute maximum.
   (*d*) all of the above.
7. When the function f consists of more than one variable, we form a Hessian matrix defined by

$$\mathbf{H}(\mathbf{x}_0) = \left(\frac{\partial^2 f}{\partial x_i\, \partial x_j}\right)_{\mathbf{x}=\mathbf{x}_0}$$ to decide about the nature of $f(\mathbf{x})$. Then,

   (*a*) $f(\mathbf{x})$ is relative minimum at $\mathbf{x} = \mathbf{x}_0$, if $\mathbf{H}(\mathbf{x}_0)$ is negative definite.
   (*b*) $f(\mathbf{x})$ is relative maximum at $\mathbf{x} = \mathbf{x}_0$, if $\mathbf{H}(\mathbf{x}_0)$ is positive definite.
   (*c*) there exists a point of inflexion at $\mathbf{x} = \mathbf{x}_0$, if $\mathbf{H}(\mathbf{x}_0)$ is indefinite.
   (*d*) there exists a saddle point, if $\mathbf{H}(\mathbf{x}_0)$ is indefinite.
8. Which of the following is not correct?
   (*a*) Like linear programming, non-linear programming is a mathematical technique for determining the optimum solutions to many business problems.
   (*b*) In case of maximization of *NLPP*, all constraints must be converted into '≤' type, and in the case of minimization *NLPP*, into '≥' type.
   (*c*) The objective function is concave, if the principal minors of bordered Hessian matrix, alternate in sign, beginning with the negative sign.
   (*d*) If the principal minors of bordered Hessian matrix are positive, the objective function is convex.

ANSWERS

1. (*c*)	2. (*d*)	3. (*a*)	4. (*d*)
5. (*b*)	6. (*d*)	7. (*c*)	8. (*a*)

REVIEW EXERCISE

1. What is a non-linear programming problem? Give two examples of *NLPP* stating clearly why do you so classify them?
2. Give the general form of *NLPP*. Also, explain unconstrained optimization.
   [M.Sc. (Math.), Kerala Univ. 2001]
3. What is the significance of Lagrange multipliers. *[B.Tech., Panjab Tech. Univ. 2010]*

4. Explain the Lagrange multiplier method of solving *NLPP* with equality constraints. ***[B.Sc., Bharthiar Univ. 2011]***

5. Explain the following:
   (*i*) Meaning of local optimum point,
   (*ii*) common characteristics of gradient based methods,
   (*iii*) significance of inflexion point. ***[B.Tech., Panjab Tech. Univ. 2010]***

6. State the Kuhn-Tucker sufficient conditions in non-linear programming problems. ***[ME, Annamalai Univ. 2004]***

7. Give the necessary conditions for a maximization non-linear programming problem in the form of Kuhn-Tucker conditions. ***[ME (Nov.), Annamalai Univ. 2002]***

8. Differentiate between constrained and unconstrained optimization problems with the help of an example. ***[B.Sc., Bharthiar Univ. 2011]***

9. Give the classification of optimization problems. ***[B.Tech., Panjab Tech. Univ. 2009]***

10. Write a short note on non-linear programming. ***[MBA (June), IGNOU 2007]***

CHAPTER 26

Solving Non-linear Programming Problems

"Let us recognize the fact that every complexity is explainable through echo of simplicity"

As discussed earlier a linear programming problem can easily be solved by simplex method or its variations. The optimum solution lies at one of the extreme points of the convex feasible region. But, in a non-linear programming problem (*NLPP*), the optimum solution can be found anywhere on the boundary of the feasible region and even at some interior point of it. In spite of the substantial advancement in the solution methods of *NLPP* in recent years, an efficient simplex like technique for a *GNLPP* is yet to be found. Some available techniques for solving some special cases of *GNLPP* shall be treated in this chapter.

26:1. GRAPHICAL SOLUTION OF NON-LINEAR PROGRAMMING PROBLEM

The ***graphical method*** for the solution of an *NLPP* involving only two variables is best illustrated by the following examples:

Example 26.1. *Minimize the distance of the origin from the convex region bounded by the constraints:*

$$x_1 + x_2 \geq 4,\ \ 2x_1 + x_2 \geq 5 \ \ \text{and} \ \ x_1 \geq 0,\ \ x_2 \geq 0.$$

Verify that the Kuhn-Tucker necessary conditions hold at the point of minimum distance.

Solution. The problem of minimizing the distance of the origin from the convex region is equivalent to minimizing the radius of a circle with origin as centre, say $r^2 = x_1^2 + x_2^2$ such that it touches (passes though) the convex region bounded by the given constraints. Thus, the problem is formulated as:

$$\text{Minimize} \ \ z\ (= r^2) = x_1^2 + x_2^2 \ \ \text{subject to the constraints:}$$
$$x_1 + x_2 \geq 4,\ \ 2x_1 + x_2 \geq 5,\ \ x_1, x_2 \geq 0.$$

Graphical Solution. Consider a set of rectangular cartesian axis OX_1X_2 in the plane. Each point has coordinates of the type (x_1, x_2) and conversely every ordered pair (x_1, x_2) of real numbers determine a point in the plane.

Clearly, any point which satisfies the conditions $x_1 \geq 0$ and $x_2 \geq 0$ lies in the first quadrant and conversely for any point (x_1, x_2) in the first quadrant, $x_1 \geq 0$, and $x_2 \geq 0$. Thus, our search for the number pair (x_1, x_2) is restricted to the points in the first quadrant only. Now, since $x_1 + x_2 \geq 4$ and $2x_1 + x_2 \geq 5$, the desired point must be somewhere in the unbounded convex region *ABC* (shown shaded) in *Fig. 26.1*. Since our search is for such a pair (x_1, x_2) which gives a minimum value of $x_1^2 + x_2^2$ and lies in the convex region,

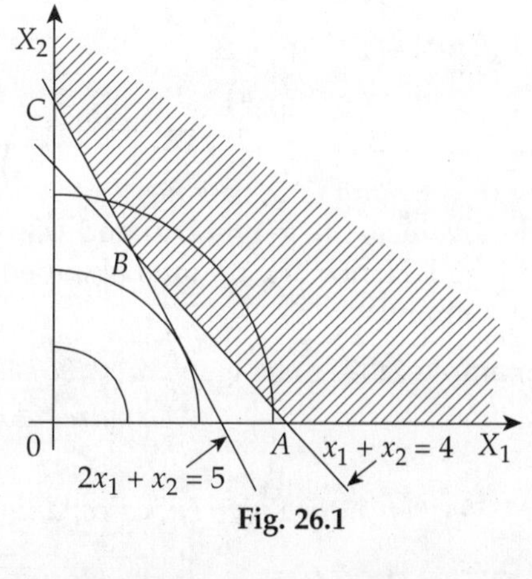

Fig. 26.1

the desired point will be that point of the region at which a side of the convex region is tangent to the circle. We proceed as follows:

Differentiating the equation of the circle, we have $2x_1\,dx_1 + 2x_2\,dx_2 = 0$, yielding

$$\frac{dx_2}{dx_1} = -\frac{x_1}{x_2} \qquad \text{... (1)}$$

Considering the equalities of constraints: $2x_1 + x_2 = 5$ and $x_1 + x_2 = 4$,

we have on differentiation

$$2dx_1 + dx_2 = 0 \quad \text{and} \quad dx_1 + dx_2 = 0.$$

These yield $\dfrac{dx_2}{dx_1} = -2$ and $\dfrac{dx_2}{dx_1} = -1$ respectively. ... (2)

Thus, from (1) and (2), we obtain

$$\frac{-x_1}{x_2} = -1 \ \text{ or } \ x_1 = x_2 \quad \text{and} \quad \frac{-x_1}{x_2} = -2 \ \text{ or } \ x_1 = 2x_2.$$

This shows that the circle $r^2 = x_1^2 + x_2^2$ has as a tangent to

(*i*) the line $x_1 + x_2 = 4$ at the point (2, 2) and (*ii*) the line $2x_1 + x_2 = 5$ at the point (2, 1).

The point (2, 1) does not lie in the convex region and hence is to be discarded. Thus, the minimum distance from the origin to the convex region bounded by the constraints is:

$$z_0 = 2^2 + 2^2 = 8 \quad \text{and is given by the point (2, 2).}$$

Verification of Kuhn-Tucker Conditions. We, now, verify that the above minima satisfy the Kuhn-Tucker conditions also. Here, we have

$$f(\mathbf{x}) = x_1^2 + x_2^2, \quad h^1(\mathbf{x}) = x_1 + x_2 - 4, \quad h^2(\mathbf{x}) = 2x_1 + x_2 - 5,$$

and the problem is that of minimizing $f(\mathbf{x})$ subject to the constraints: $h^1(\mathbf{x}) \ge 0$, $h^2(\mathbf{x}) \ge 0$ and $\mathbf{x} \ge \mathbf{0}$. The Kuhn-Tucker conditions for this minimization *NLPP* are:

$$f_j(\mathbf{x}) = \lambda_1 h_j^1(\mathbf{x}) + \lambda_2 h_j^2(\mathbf{x}), \qquad j = 1, 2$$

$$\lambda_i h^i(\mathbf{x}) = 0, \quad h^i(\mathbf{x}) \ge 0, \quad \lambda_i \ge 0 \qquad i = 1, 2$$

where, $f_j(\mathbf{x}) = \dfrac{\partial f(\mathbf{x})}{\partial x_j}$, $h_j^i(\mathbf{x}) = \dfrac{\partial h^i(\mathbf{x})}{\partial x_j}$, for ($j = 1, 2$), and λ_1, λ_2 are Lagrangian multipliers.

These conditions, thus, are as given below:

(*a*) $\begin{cases} 2x_1 = \lambda_1 + 2\lambda_2 \\ 2x_2 = \lambda_1 + \lambda_2 \end{cases}$ (*b*) $\begin{cases} \lambda_1[x_1 + x_2 - 4] = 0 \\ \lambda_2[2x_1 + x_2 - 5] = 0 \end{cases}$

(*c*) $\begin{cases} (x_1 + x_2 - 4) \ge 0 \\ (2x_1 + x_2 - 5) \ge 0 \end{cases}$ (*d*) $\lambda_1 \ge 0, \ \lambda_2 \ge 0.$

If the point (2, 2) satisfies these conditions, then we must have from (*a*), that $\lambda_1 = 4$, $\lambda_2 = 0$. With these values, $(x_1, x_2) = (2, 2)$ and $(\lambda_1, \lambda_2) = (4, 0)$, it is clear that the conditions (*b*), (*c*) and (*d*) are satisfied. Hence, the minima obtained by graphical method satisfies the Kuhn-Tucker conditions for a minima.

Example 26.2. *Solve graphically the following NLPP:*

Maximize $z = 2x_1 + 3x_2$ *subject to the constraints:*

$$x_1 x_2 \le 8, \quad x_1^2 + x_2^2 \le 20, \quad x_1, x_2 \ge 0.$$

Verify that the Kuhn-Tucker conditions hold for the maxima you obtain.

Solution. In this *NLPP*, the objective function is linear whereas the constraints are non-linear. Consider a set of rectangular cartesian axis OX_1X_2 in the first quadrant only.

Now, $x_1x_2 = 8$ represents a rectangular hyperbola with coordinate axis as its asymptotes; and $x_1^2 + x_2^2 = 20$ represents a circle of radius $\sqrt{20}$ with origin as its centre. Thus, since $x_1x_2 \le 8$ and $x_1^2 + x_2^2 \le 20$, the desired point may be somewhere in the convex feasible region *OABCD* (shown shaded) in *Fig. 26.2*. Since, our search is for such a pair (x_1, x_2) which gives a maximum value of $2x_1 + 3x_2$ and lies in the convex region, the desired point is obtained by moving parallel to $2x_1 + 3x_2 = k$, for some constant k, so long as $k = 2x_1 + 3x_2$ touches the extreme boundary point of feasible region. Thus, in our problem the boundary point $C = (2, 4)$ corresponds to maximum z. Hence, the optimal solution is $z_0 = 16, x_1 = 2, x_2 = 4$.

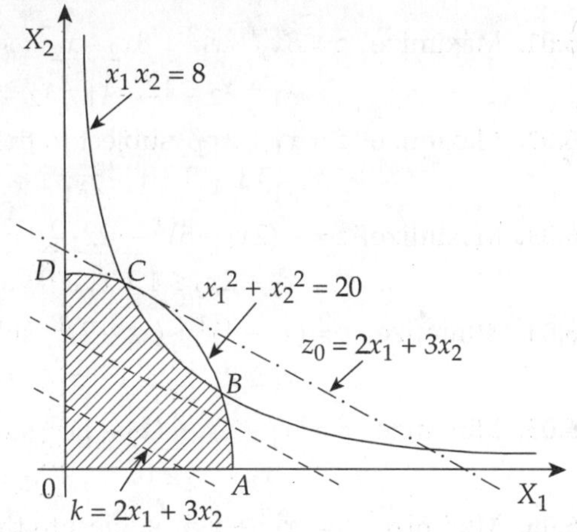

Fig. 26.2

Verification of Kuhn-Tucker Conditions. We, now verify that the above maximum solution satisfies the Kuhn-Tucker conditions also.

Here, we have

$$f(\mathbf{x}) = 2x_1 + 3x_2$$
$$h^1(\mathbf{x}) = x_1x_2 - 8$$
$$h^2(\mathbf{x}) = x_1^2 + x_2^2 - 20$$

and the problem is that of maximizing $f(\mathbf{x})$ subject to the constraints: $h^1(\mathbf{x}) \le 0$, $h^2(\mathbf{x}) \le 0$ and $\mathbf{x} \ge 0$.

The Kuhn-Tucker conditions for this maximizing *NLPP* are:

$$f_j(\mathbf{x}) = \lambda_1 h_j^1(\mathbf{x}) + \lambda_2 h_j^2(\mathbf{x}); \qquad j = 1, 2$$
$$\lambda_i h^i(\mathbf{x}) = 0 \qquad i = 1, 2$$
$$h^i(\mathbf{x}) \le 0 \qquad i = 1, 2$$
$$\lambda_i \ge 0 \qquad i = 1, 2$$

where, $f_j(\mathbf{x}) = \dfrac{\partial f(\mathbf{x})}{\partial x_j}$, $h_j^i(\mathbf{x}) = \dfrac{\partial h^i(\mathbf{x})}{\partial x_j}$ for $(j = 1, 2)$, and λ_1, λ_2 are Lagrangian multipliers.

These conditions are, thus, given as:

(*a*) $\begin{cases} 2 = \lambda_1 x_2 + 2\lambda_2 x_1 \\ 3 = \lambda_1 x_1 + 2\lambda_2 x_2 \end{cases}$ (*b*) $\begin{cases} \lambda_1 [x_1 x_2 - 8] = 0 \\ \lambda_2 [x_1^2 + x_2^2 - 20] = 0 \end{cases}$

(*c*) $\begin{cases} x_1 x_2 - 8 \le 0 \\ x_1^2 + x_2^2 - 20 \le 0 \end{cases}$ (*d*) $\lambda_1 \ge 0, \ \lambda_2 \ge 0.$

If the point (2, 4) satisfies these conditions, then we must have from (*a*) $\lambda_1 = \frac{1}{6}$ and $\lambda_2 = \frac{1}{3}$. With these values, namely $(x_1, x_2) = (2, 4)$ and $(\lambda_1, \lambda_2) = \left(\frac{1}{6}, \frac{1}{3}\right)$, it is obvious that the conditions (*b*), (*c*) and (*d*) are satisfied. Hence, the maxima obtained by graphical method satisfy Kuhn-Tucker conditions for a maxima.

PROBLEMS

Solve the following non-linear programming problems graphically:

26.01. Maximize $z = 8x_1 - x_1^2 + 8x_2 - x_2^2$ subject to the constraints:
$$x_1 + x_2 \leq 12, \quad x_1 - x_2 \geq 4 \text{ and } x_1, x_2 \geq 0.$$

26.02. Maximize $z = x_1 + 2x_2$ subject to the constraints:
$$x_1^2 + x_2^2 \leq 1, \quad 2x_1 + x_2 \leq 2, \quad x_1, x_2 \geq 0.$$

26.03. Maximize $z = -(2x_1 - 5)^2 - (2x_2 - 1)^2$ subject to the constraints:
$$x_1 + 2x_2 \leq 1, \quad x_1 \geq 0 \text{ and } x_2 \geq 0.$$

26.04. Minimize $z = (x_1 - 1)^2 + (x_2 - 2)^2$ subject to the constraints:
$$x_1 \leq 2, \quad x_2 \leq 1; \quad x_1 \geq 0 \text{ and } x_2 \geq 0.$$

26.05. Minimize $z = (x_1 - 2)^2 + (x_2 - 1)^2$ subject to the constraints:
$$-x_1 + x_2 \geq 0, \quad x_1 + x_2 \leq 0; \quad x_1 \geq 0 \text{ and } x_2 \geq 0.$$

26.06. Minimize $z = x_1^2 + x_2^2$ subject to the constraints:
$$x_1 + x_2 \geq 8, \quad x_1 + 2x_2 \geq 10, \quad 2x_1 + x_2 \geq 10; \quad x_1 \geq 0 \text{ and } x_2 \geq 0.$$

26.07. Maximize $z = x_1^2 + x_2^2$ subject to the constraints:
$$x_1 + x_2 \leq 8, \quad x_1 + 2x_2 \leq 10, \quad 2x_1 + x_2 \leq 10; \quad x_1 \geq 0 \text{ and } x_2 \geq 0.$$

26.08. Maximize $z = x_1$ subject to the constraints:
$$(1 - x_1)^3 - x_2 \geq 0, \quad x_1, x_2 \geq 0.$$

Also, show that the Kuhn-Tucker necessary conditions for the maxima do not hold. What do you conclude?

26.09. Maximize $z = x_1$ subject to the constraints:
$$(3 - x_1)^3 - (x_2 - 2) \geq 0, \quad (3 - x_1)^3 + (x_2 - 2) \geq 0, \quad x_1, x_2 \geq 0.$$

Also, show that the Kuhn-Tucker necessary conditions for the maxima do not hold in this case.

26.10. The manufacturing cost and raw material cost for making each of the two products *A* and *B* is proportional to the square of the quantity made. The products are made from a limited supply of a particular raw material and both are processed on the same machine.

It takes 40 minutes to process one unit of product *A* and 30 minutes for each unit of product *B*. The machine operates for a maximum of 40 hours a week. Product *A* needs 2 kg and product *B* needs 3 kg of raw material per unit which is in limited supply of 200 kg per week.

If net income from the products are ₹ 250 and ₹ 650 per unit and manufacturing costs are $2x^2$ and $3y^2$ respectively, find how much of each product should be produced, where x and y represent the number of units of product *A* and *B* respectively to be produced.

[MBA (Nov.), Delhi Univ. 2003]

26.11. The Operations Research team of the *ABC* Company has come up with the mathematical data needed for two products which the firm manufactures. It also has determined that this is a non-linear programming problem, having linear constraints and objective function is the sum of a linear and a quadratic form. The pertinent data, gathered by the OR team, is given on the next page:

Maximize (contribution) $z = 12x + 21y + 2xy - 2x^2 - 2y^2$ subject to the constraints:

$$8 - y \geq 0$$
$$10 - x - y \geq 0$$
$$x, y \geq 0$$

Find the maximum contribution and number of units that can be expected from these two products which are a part of the firm's total output, where x and y represent the number of units of the two products A and B respectively.

26:2. KUHN-TUCKER CONDITIONS WITH NON-NEGATIVE CONSTRAINTS

In the preceding chapter, we obtained the ***necessary*** conditions for a point $\mathbf{x}^0 \in \mathbf{R}^n$ to be relative maximum of $f(\mathbf{x})$ subject to the constraints: $h^i(\mathbf{x}) \leq 0, i = 1, 2, ..., m, \mathbf{x} \geq 0$. These conditions, called ***Kuhn-Tucker conditions***, were found by converting each inequality constraint to an equation through the addition of a squared slack variable, S_i^2, imposing the first-order conditions for maxima, on the first partial derivative of the Lagrangian function, and then simplifying the outcome. The following conditions resulted:

$$(a) \qquad f_j = \sum_{i=1}^{m} \lambda_i h_j^i \qquad j = 1, 2, ..., n$$

$$(b) \qquad -\lambda_i h^i(\mathbf{x}) = 0 \qquad i = 1, 2, ..., m$$

$$(c) \qquad h^i(\mathbf{x}) \leq 0 \qquad i = 1, 2, ..., m$$

$$(d) \qquad \lambda_i \geq 0 \qquad i = 1, 2, ..., m$$

The reader may have observed that in obtaining these conditions, the non-negativity constraints $\mathbf{x} \geq \mathbf{0}$ were completely ignored. However, we always had in mind to discard all such solutions of (*a*) to (*d*) that violate $\mathbf{x} \geq \mathbf{0}$.

Now, we shall consider the non-negativity constraint $\mathbf{x} \geq \mathbf{0}$ as one of the constraints, *viz.*, $h(\mathbf{x}) \geq \mathbf{0}$, where $h(\mathbf{x}) = \mathbf{x}$, and derive the Kuhn-Tucker conditions for the resulting problem.

We restate the problem as

$$\text{Maximize } z = f(\mathbf{x}) \quad \mathbf{x} \in \mathbf{R}^n \text{ subject to the constraints:}$$
$$h^i(\mathbf{x}) \leq 0, \; -\mathbf{x} \leq \mathbf{0}; \qquad i = 1, 2, ..., m$$

Clearly, there are $m + n$ inequality constraints, and thus we add the squares of $(m + n)$ slack variables $S_1, ..., S_m, S_{m+1}, ..., S_{m+n}$ in the inequalities so as to convert them into equations:

$$h^i(\mathbf{x}) + S_i^2 = 0 \qquad \text{for } i = 1, 2, ..., m$$

$$-x_j + S_{m+j}^2 = 0 \qquad \text{for } j = 1, 2, ..., n.$$

To find the necessary conditions for maximum of $f(\mathbf{x})$, we construct the associated Lagrangian function

$$L(\mathbf{x}, \mathbf{S}, \lambda) = f(\mathbf{x}) - \sum_{i=1}^{m} \lambda_i \left[h^i(\mathbf{x}) + S_i^2 \right] - \sum_{j=1}^{n} \lambda_{m+j} \left[-x_j + S_{m+j}^2 \right],$$

where, $\mathbf{S} = (S_1, S_2, ..., S_{m+n})$; and $\lambda = (\lambda_1, ..., \lambda_{m+n})$ are the Lagrangian multipliers.

The Kuhn-Tucker conditions are:

$$\frac{\partial L}{\partial x_j} = f_j - \sum_{i=1}^{m} \lambda_i h_j^i + \lambda_{m+j} = 0 \qquad \text{for } j = 1, 2, ..., n$$

$$\frac{\partial L}{\partial S_i} = -2\lambda_i S_i = 0 \qquad \text{for } i = 1, 2, ..., m$$

$$\frac{\partial L}{\partial S_{m+j}} = -2S_{m+j}\lambda_{m+j} = 0 \qquad \text{for } j = 1, 2, ..., n$$

$$\frac{\partial L}{\partial \lambda_i} = -\left(h^i(\mathbf{x}) + S_i^2\right) = 0 \qquad \text{for } i = 1, 2, ..., m$$

and

$$\frac{\partial L}{\partial \lambda_{m+j}} = -\left[-x_j + S_{m+j}^2\right] = 0 \qquad \text{for } j = 1, 2, ..., n.$$

The Kuhn-Tucker conditions, as obtained from these, upon simplification are:

(a)	$f_j = \sum_{i=1}^{m} \lambda_i h_j^i - \lambda_{m+j}$	$(j = 1, 2, ..., n)$	
(b)	$\lambda_i [h^i(\mathbf{x})] = 0$	$(i = 1, 2, ..., m)$	*Maximize f*
(c)	$-\lambda_{m+j} x_j = 0$	$(j = 1, 2, ..., n)$	*subject to:*
(d)	$h^i(\mathbf{x}) \le 0$	$(i = 1, 2, ..., m)$	$h^i(\mathbf{x}) \le 0$,
(e)	$\lambda_i, \lambda_{m+j}, x_j \ge 0$	$(i = 1, 2, ..., m; j = 1, 2, ..., n)$	$\mathbf{x} \ge \mathbf{0}$.

Remark. As before, these conditions are sufficient also if the objective function and the solution space satisfy certain conditions regarding convexity and concavity. These conditions are summarized in the following table:

Sense of optimization	*Required conditions*	
	Objective function	*Solution space*
Maximization	Concave	Convex set
Minimization	Convex	Convex set

It may be noted that it is simpler to verify whether a function is convex or concave than to see, if a solution space is convex.

26:3. QUADRATIC PROGRAMMING PROBLEM

Quadratic programming is concerned with the *NLPP* of maximizing (or minimizing) the quadratic objective function, subject to a set of linear inequality constraints.

Definition (*Quadratic Programming Problem*). *Let* $\mathbf{x}^T$ *and* $\mathbf{c} \in \mathbf{R}^n$ *and* $\mathbf{Q}$ *be a symmetric* $n \times n$ *real matrix. Then, the problem of maximizing*

$$f(\mathbf{x}) = \mathbf{cx} + \frac{1}{2}\mathbf{x}^T\mathbf{Q}\mathbf{x} \quad \textit{subject to the constraints:}$$

$$\mathbf{Ax} \le \mathbf{b} \quad \textit{and} \quad \mathbf{x} \ge \mathbf{0},$$

where, $\mathbf{b}^T \in \mathbf{R}^m$ *and* $\mathbf{A}$ *is an* $m \times n$ *real matrix, is called a* ***general quadratic programming problem****.*

Remark. $\mathbf{x}^T\mathbf{Q}\,\mathbf{x}$ represents a quadratic form. The reader may recall that a quadratic form $\mathbf{x}^T\mathbf{Q}\,\mathbf{x}$ is said to be *positive-definite (negative-definite)*, if $\mathbf{x}^T\mathbf{Q}\,\mathbf{x} > 0\ (< 0)$ for $\mathbf{x} \neq \mathbf{0}$ and *positive-semi-definite (negative-semi-definite)*, if $\mathbf{x}^T\mathbf{Q}\,\mathbf{x} \ge 0\ (\le 0)$ for all $\mathbf{x}$ such that there exist at least one $\mathbf{x} \neq \mathbf{0}$ satisfying $\mathbf{x}^T\mathbf{Q}\,\mathbf{x} = 0$.

The reader may easily show that

1. If $\mathbf{x}^T\mathbf{Q}\,\mathbf{x}$ is positive-semi-definite (negative-semi-definite), then it is convex (concave) in $\mathbf{x}$ over all of $\mathbf{R}^n$, and
2. If $\mathbf{x}^T\mathbf{Q}\,\mathbf{x}$ is positive-definite (negative-definite), then it is strictly convex (strictly concave) in $\mathbf{x}$ over all of $\mathbf{R}^n$.

These results help in determining whether the quadratic objective function $f(\mathbf{x})$ is concave (convex) and the implication of the same on the sufficiency of the Kuhn-Tucker conditions for constrained maxima (minima) of $f(\mathbf{x})$.

Now, let us consider a *QPP* written in the form

Maximize $z = f(x_1, x_2, \ldots, x_n) = \sum_{j=1}^{n} c_j x_j + \frac{1}{2} \sum_{j=1}^{n} \sum_{k=1}^{n} x_j d_{jk} x_k$ subject to the constraints:

$$\sum_{j=1}^{n} a_{ij} x_j \le b_i, \qquad i = 1, 2, \ldots, m.$$

and

$$x_j \ge 0, \qquad j = 1, 2, \ldots, n$$

where, $d_{jk} = d_{kj}$ for all j and k; and $b_i \ge 0$.

The Kuhn-Tucker conditions for the maximum are:

(*i*) $$f_j - \sum_{i=1}^{m} \lambda_i h_j^i + \lambda_{m+j} = 0 \qquad \text{for } j = 1, 2, \ldots, n$$

or $$\left\{c_j + \frac{1}{2}\left(2 \sum_{k=1}^{n} d_{jk} x_k\right)\right\} - \sum_{i=1}^{m} \lambda_i a_{ij} + \lambda_{m+j} = 0 \qquad (j = 1, 2, \ldots, n)$$

(*ii*) $$\lambda_i h^i(\mathbf{x}) = 0 \text{ or } \lambda_i \left(\sum_{j=1}^{n} a_{ij} x_j - b_i\right) = 0 \qquad (i = 1, 2, \ldots, m)$$

(*iii*) $$\lambda_{m+j}(-x_j) = 0 \text{ or } -\lambda_{m+j} x_j = 0 \qquad (j = 1, 2, \ldots, n)$$

(*iv*) $$h^i(\mathbf{x}) \le 0 \text{ or } \sum_{j=1}^{n} (a_{ij} x_j - b_i) \le 0 \qquad (i = 1, 2, \ldots, m)$$

(*v*) $$\lambda_i, \lambda_{m+j}, x_j \ge 0 \qquad (i = 1, 2, \ldots, m; j = 1, 2, \ldots, n)$$

Thus, the Kuhn-Tucker necessary conditions for an optimal solution to the *QPP* are:

(*a*) $$c_j + \sum_{k=1}^{n} d_{jk} x_k - \sum_{i=1}^{m} \lambda_i a_{ij} + \lambda_{m+j} = 0 \qquad (j = 1, 2, \ldots, n)$$

(*b*) $$\lambda_i \left(\sum_{j=1}^{n} a_{ij} x_j - b_i\right) = 0 \qquad (i = 1, 2, \ldots, m)$$

(*c*) $$(-x_j)\lambda_{m+j} = 0 \qquad (j = 1, 2, \ldots, n)$$

(*d*) $$\sum_{j=1}^{n} a_{ij} x_j \le b_i \qquad (i = 1, 2, \ldots, m)$$

(*e*) $$\lambda_i, \lambda_{m+j}, x_j \ge 0 \qquad (i = 1, 2, \ldots, m; j = 1, 2, \ldots, n).$$

These expressions can be simplified by introducing some additional notation. Let $S_i^2 \geq 0$ be slack variable introduced in the ith constraint (*d*), so that

$$(d)' \qquad \sum_{j=1}^{n} a_{ij}\,x_j + S_i^2 = b_i \qquad \text{for } i = 1, 2, \ldots, m$$

and let $u_j = \lambda_{m+j}$ for $j = 1, 2, \ldots, n$. Then (*b*) and (*c*) become, after simplification,

$$(b)' \qquad \lambda_i\,S_i^2 = 0 \qquad \text{for } i = 1, 2, \ldots, m$$

$$(c)' \qquad x_j\,u_j = 0 \qquad \text{for } j = 1, 2, \ldots, n.$$

Further, since $S_i^2, \lambda_i, x_j, u_j$ must be all non-negative for $i = 1, 2, \ldots, m; j = 1, 2, \ldots, n$. (*b*)′ and (*c*)′ can be expressed by a single constraint:

$$(b)'' \qquad \sum_{i=1}^{m} \lambda_i\,S_i^2 + \sum_{j=1}^{n} x_j\,u_j = 0. \qquad \textit{(Complementary slackness)}$$

Using the newly defined variables, u_j, we can rewrite (*a*) as

$$\sum_{k=1}^{n} d_{jk}\,x_k - \sum_{i=1}^{m} \lambda_i\,a_{ij} + u_j = -c_j, \qquad \text{for } j = 1, 2, \ldots, n.$$

If we assume that the quadratic form $\frac{1}{2}\sum\sum d_{jk}\,x_j\,x_k$ is negative-semi-definite, then $f(\mathbf{x})$ is concave in **x** and hence the conditions (*a*) to (*e*) above become necessary and sufficient conditions for an optimal solution to the *QPP*. Thus, under such assumption, if we are able to find non-negative λ_i, S_i^2, x_j and u_j such that (*d*)′, (*b*)″ and (*a*)′ are satisfied, then such x_j determines an optimal solution to the given *QPP*.

26:4. WOLFE'S MODIFIED SIMPLEX METHOD

The iterative procedure for the solution of a quadratic programming problem by Wolfe's modified simplex method can be summarized as follows:

Step 1. Convert the inequality constraints into equations by introducing slack variables S_i^2 in the ith constraint, $i = 1, 2, \ldots, m$, and the slack variables S_{m+j}^2 in the jth non-negativity constraint, for $j = 1, 2, \ldots, n$.

Step 2. Construct the Lagrangian function

$$L(\mathbf{x}, \mathbf{S}, \lambda) = f(\mathbf{x}) - \sum_{i=1}^{m} \lambda_i \left(\sum_{j=1}^{n} a_{ij}\,x_j - b_i + S_i^2 \right) - \sum_{j=1}^{n} \lambda_{m+j}\left(-x_j + S_{m+j}^2\right)$$

where, $\mathbf{x} = (x_1, x_2, \ldots, x_n)$, $\mathbf{S} = (S_1, \ldots, S_{m+n})$, $\lambda = (\lambda_1, \ldots, \lambda_{m+n})$.

Differentiate $\mathbf{L}\,(\mathbf{x}, \mathbf{S}, \boldsymbol{\lambda})$ partially with respect to the components of **x**, **S** and $\boldsymbol{\lambda}$ and equate the first order partial derivatives equal to zero. Derive the Kuhn-Tucker conditions from the resulting equations.

Step 3. Introduce the non-negative artificial variables $A_j, j = 1, 2, \ldots, n$ in the Kuhn-Tucker condition

$$c_j + \sum_{k=1}^{n} d_{jk}\,x_k - \sum_{i=1}^{m} \lambda_i\,a_{ij} + \lambda_{m+j} = 0$$

for $j = 1, 2, \ldots, n$ and construct an objective function

$$z = A_1 + A_2 + \ldots + A_n.$$

Step 4. Obtain an initial basic feasible solution to the *LPP*:

Minimize $z = A_1 + A_2 + \ldots + A_n$ subject to the constraints:

$$\sum_{k=1}^{n} d_{jk}\, x_k - \sum_{i=1}^{m} \lambda_i\, a_{ij} + \lambda_{m+j} + A_j = -c_j \qquad (j = 1, 2, \ldots, n)$$

$$\sum_{j=1}^{n} a_{ij}\, x_j + x_{n+i} = b_i \qquad (i = 1, 2, \ldots, m)$$

$$A_j,\ \lambda_i,\ \lambda_{m+j},\ x_j \geq 0 \qquad (i = 1, 2, \ldots, m;\ j = 1, 2, \ldots, n)$$

where, $x_{n+i} = S_i^2$, $i = 1, 2, \ldots, m$, and satisfying the ***complementary slackness conditions***

$$\sum_{j=1}^{n} \lambda_{m+j}\, x_j + \sum_{i=1}^{m} x_{n+i}\, \lambda_i = 0.$$

Step 5. Use two phase simplex method to obtain an optimum solution to the *LPP* of *Step 4*, the solution satisfying the complementary slackness condition.

Step 6. The optimum solution obtained in *Step 5* is an optimum solution to the given *QPP* also.

Note. If the given *QPP* is in the minimization form, convert it into that of maximization by appropriate adjustment in $f(x_1, x_2, \ldots, x_n)$.

Example 26.3. *Use Wolfe's method to solve the QPP:*

Maximize $z = 2x_1 + 3x_2 - 2x_1^2$ *subject to the constraints:*

$$x_1 + 4x_2 \leq 4, \quad x_1 + x_2 \leq 2$$

$$x_1, x_2 \geq 0.$$

[BE (Mech.), Kurukshetra Univ. 2012]

Solution. We convert the inequality constraints into equations by introducing slack variables S_1^2 and S_2^2, respectively. Considering $x_1 \geq 0$ and $x_2 \geq 0$ also as the inequality constraints, we convert them also into equations by introducing slack variables S_3^2 and S_4^2 in them. The problem thus becomes:

Maximize $z = 2x_1 + 3x_2 - 2x_1^2$ subject to the constraints:

$$x_1 + 4x_2 + S_1^2 = 4, \quad x_1 + x_2 + S_2^2 = 2,$$

$$-x_1 + S_3^2 = 0, \quad -x_2 + S_4^2 = 0.$$

Construct the Lagrangian function

$$\begin{aligned} L &= L(x_1, x_2, S_1, S_2, S_3, S_4, \lambda_1, \lambda_2, \lambda_3, \lambda_4) \\ &= (2x_1 + 3x_2 - 2x_1^2) - \lambda_1 (x_1 + 4x_2 + S_1^2 - 4) - \lambda_2 (x_1 + x_2 + S_2^2 - 2) \\ &\quad - \lambda_3 (-x_1 + S_3^2) - \lambda_4 (-x_2 + S_4^2). \end{aligned}$$

As $(-x_1^2)$ represents a negative semi-definite quadratic form; $z = 2x_1 + 3x_2 - 2x_1^2$ is concave in x_1, x_2. Thus, maxima of L will be maxima of $z = 2x_1 + 3x_2 - 2x_1^2$ and *vice versa*. To derive the necessary and sufficient conditions for maxima of L (and hence of z), we equate the first-order partial derivatives of L w.r.t. the variables x_1, x_2, S_i's and λ_i's. Thus, we have

$$\frac{\partial L}{\partial x_1} = 2 - 4x_1 - \lambda_1 - \lambda_2 + \lambda_3 = 0; \qquad \frac{\partial L}{\partial x_2} = 3 - 4\lambda_1 - \lambda_2 + \lambda_4 = 0;$$

$$\frac{\partial L}{\partial \lambda_1} = x_1 + 4x_2 + S_1^2 - 4 = 0; \qquad \frac{\partial L}{\partial \lambda_2} = x_1 + x_2 + S_2^2 - 2 = 0;$$

$$\frac{\partial L}{\partial \lambda_3} = -x_1 + S_3^2 = 0; \qquad \frac{\partial L}{\partial \lambda_4} = -x_2 + S_4^2 = 0;$$

$$\frac{\partial L}{\partial S_1} = -2\lambda_1 S_1 = 0; \qquad \frac{\partial L}{\partial S_2} = -2\lambda_2 S_2 = 0;$$

$$\frac{\partial L}{\partial S_3} = -2\lambda_3 S_3 = 0; \qquad \frac{\partial L}{\partial S_4} = -2\lambda_4 S_4 = 0.$$

Upon simplification and necessary manipulations these yield:

$$(1) \quad \begin{cases} 4x_1 + \lambda_1 + \lambda_2 - \lambda_3 = 2, & 4\lambda_1 + \lambda_2 - \lambda_4 = 3 \\ x_1 + 4x_2 + S_1^2 = 4, & x_1 + x_2 + S_2^2 = 2 \end{cases}$$

$$(2) \quad \lambda_1 S_1^2 + \lambda_2 S_2^2 + x_1 \lambda_3 + x_2 \lambda_4 = 0,\ x_1, x_2, S_1^2, S_2^2, \lambda_i \geq 0, \qquad i = 1, 2, 3, 4.$$

A solution x_j, $j = 1, 2$, to (1) above and satisfying (2) shall necessarily be an optimal one for maximizing L. To determine the solution to the above simultaneous equation (1), we introduce the artificial variables A_1 and A_2 (both non-negative) in the first two constraints of (1) and construct the dummy objective function $z = A_1 + A_2$.

Then, the problem becomes:

$$\text{Maximize } z = -A_1 - A_2 \text{ subject to the constraints:}$$

$$4x_1 + \lambda_1 + \lambda_2 - \lambda_3 + A_1 = 2$$
$$4\lambda_1 + \lambda_2 - \lambda_4 + A_2 = 3$$
$$x_1 + 4x_2 + x_3 = 4$$
$$x_1 + x_2 + x_4 = 2$$
$$x_1, x_2, x_3, x_4 \geq 0.$$
$$A_1, A_2, \lambda_i \geq 0, \qquad i = 1, ..., 4$$

satisfying the complementary slackness condition $\sum \lambda_i x_i = 0$, where we have replaced S_1^2 by x_3 and S_2^2 by x_4.

The optimum solution to the above *LPP* shall now be obtained by the two phase simplex method. An initial basic feasible solution to the *LPP* is clearly given by

$$\begin{bmatrix} 1 & 0 & 0 & 0 \\ 0 & 1 & 0 & 0 \\ 0 & 0 & 1 & 0 \\ 0 & 0 & 0 & 1 \end{bmatrix} \begin{bmatrix} A_1 \\ A_2 \\ x_3 \\ x_4 \end{bmatrix} = \begin{bmatrix} 2 \\ 3 \\ 4 \\ 2 \end{bmatrix}$$

The simplex iterations leading to an optimum solution are:

Initial Iteration. Enter x_1 and drop A_1.

Basic variable	x_1	x_2	x_3	x_4	λ_1	λ_2	λ_3	λ_4	A_1	A_2	*Solution* x_B	*Ratio*
$\leftarrow A_1$	(4)	0	0	0	1	1	–1	0	1	0	2	2/4
A_2	0	0	0	0	4	1	0	–1	0	1	3	—
x_3	1	4	1	0	0	0	0	0	0	0	4	4/1
x_4	1	1	0	1	0	0	0	0	0	0	2	2/1
$(z_j - c_j)$	–4 ↑	0	0	0	–5	–2	+1	+1	0	0	$z = -5$	

From the above table, we observe that x_1, λ_1 or λ_2 can enter the basis. But λ_1 and λ_2 will not enter the basis, because x_3 and x_4 are in the basis. This is in view of the complimentary slackness conditions $\lambda_1 x_3 = 0$ and $\lambda_2 x_4 = 0$. Hence, x_1 enter basis.

First Iteration. Enter x_2 and drop x_3.

Basic variable	x_1	x_2	x_3	x_4	λ_1	λ_2	λ_3	λ_4	A_2	*Solution* $\mathbf{x}_B$
x_1	1	0	0	0	1/4	1/4	–1/4	0	0	1/2
A_2	0	0	0	0	4	1	0	–1	1	3
$\leftarrow x_3$	0	(4)	1	0	–1/4	–1/4	1/4	0	0	7/2
x_4	0	1	0	1	–1/4	–1/4	1/4	0	0	3/2
$(z_j - c_j)$	0	0 ↑	0	0	–4	–1	0	0	0	$z = -3$

Here, we observe that either λ_1 or λ_2 can enter the basis. Since, x_3 and x_4 are still in the basis, therefore, these cannot enter the basis because of the complementary slackness conditions. However, since λ_4 is not in the basis, x_2 can enter the basis (because of the condition $\lambda_4 x_2 = 0$).

Second Iteration. Enter λ_1 and drop A_2.

Basic variable	x_1	x_2	x_3	x_4	λ_1	λ_2	λ_3	λ_4	A_2	*Solution* $\mathbf{x}_B$
x_1	1	0	0	0	1/4	1/4	–1/4	0	0	1/2
$\leftarrow A_2$	0	0	0	0	(4)	1	0	–1	1	3
x_2	0	1	1/4	0	–1/16	–1/16	1/16	0	0	7/8
x_4	0	0	–1/4	1	–3/16	–3/16	3/16	0	0	5/8
$(z_j - c_j)$	0	0	0	0	–4 ↑	–1	0	+1	0	$z = -3$

From this table, we see that λ_1 or λ_2 can enter the basis. But since x_4 is in the basis, λ_2 can't enter the basis and hence λ_1 enters the basis.

Final Iteration. Optimum solution.

Basic variable	x_1	x_2	x_3	x_4	λ_1	λ_2	λ_3	λ_4	*Solution* $\mathbf{x}_B$
x_1	1	0	0	0	0	3/16	–1/4	1/16	5/16
λ_1	0	0	0	0	1	1/4	0	–1/4	3/4
x_2	0	1	1/4	0	0	–3/64	1/16	–1/64	59/64
x_4	0	0	–1/4	1	0	–9/64	3/16	3/64	49/64
$(z_j - c_j)$	0	0	0	0	0	0	0	0	$z = 0$

The optimum solution is

$$x_1 = 5/16, \quad x_2 = 59/64 \quad \text{and} \quad \text{maximum } z = 3.19.$$

PROBLEMS

Use Wolfe's method to solve the following quadratic programming problems:

26.12. Maximize $z = 4x_1 + 6x_2 - 2x_1^2 - 2x_1x_2 - 2x_2^2$ subject to the constraints:
$x_1 + 2x_2 \le 2, \; x, x_2 \ge 0.$ *[MCA, Guwahati Univ. 2000]*

26.13. Maximize $z = 8x_1 + 10x_2 - x_1^2 - x_2^2$ subject to the constraints:
$3x_1 + 2x_2 \le 6, \; x_1, x_2 \ge 0.$

26.14. Maximize $z = 6x_1 + 3x_2 - 4x_1x_2 - 2x_1^2 - 3x_2^2$ subject to the constraints:
$x_1 + x_2 \le 1, \; 2x_1 + 3x_2 \le 4; \; x_1, x_2 \ge 0.$

26.15. Maximize $f(x_1, x_2) = 1.8x_1 + 3x_2 - 0.001x_1^2 - 0.005x_2^2 - 100$ subject to the constraints:
$$2x_1 + 3x_2 \le 2{,}500, \quad x_1 + 2x_2 \le 1{,}500, \quad x_1 \ge 0, x_2 \ge 0.$$

26.16. Maximize $z = 2x_1 + x_2 - x_1^2$ subject to the constraints:
$$2x_1 + 3x_2 \le 6, \quad 2x_1 + x_2 \le 4; \quad x_1, x_2 \ge 0.$$ ***[BE (Comp.), Nagpur Univ. 2004]***

26.17. Minimize $z = -x_1 - x_2 - x_3 + \frac{1}{2}(x_1^2 + x_2^2 + x_3^2)$ subject to the constraints:
$$x_1 + x_2 + x_3 \le 1, \quad 4x_1 + 2x_3 \le 7/3; \quad x_1, x_2, x_3 \ge 0.$$

26:5. BEALE'S METHOD

Beale has developed a technique of solving the quadratic programming problem that does not use the Kuhn-Tucker conditions in achieving the optimum solution. His technique involves partitioning of the variables into basic and non-basic ones and then using classical calculus results. At each iteration, the objective function is expressed in terms of only the non-basic variables.

Beale's Algorithm for *QPP*

Let the general quadratic programming problem be to maximize $f(\mathbf{x}) = \mathbf{cx} + \frac{1}{2}\mathbf{x}^T \mathbf{Qx}$ subject to the constraints: $\mathbf{Ax}$ {≤, ≥ or =} $\mathbf{b}$ and $\mathbf{x} \ge \mathbf{0}$; where $\mathbf{x} \in \mathbf{R}^n$, $\mathbf{A}$ is $m \times n$, $\mathbf{b}$ is $m \times 1$, $\mathbf{c}$ is $n \times 1$ and $\mathbf{Q}$ is a $n \times n$ symmetric matrix.

The Beale's iterative procedure of solving this problem can be summarized as follows:

Step 1. Convert the minimization $f(\mathbf{x})$ into that of maximization, if required. Introduce slack and/or surplus variables in the inequality constraints, if any, and put the *QPP* into standard form.

The quadratic programming problem in its standard form, therefore, will be:

$$\text{Maximize } f(\mathbf{x}) = \mathbf{cx} + \frac{1}{2}\mathbf{x}^T \mathbf{Qx} \text{ subject to the constraints:}$$
$$\mathbf{Ax} = \mathbf{b} \text{ and } \mathbf{x} \ge \mathbf{0}$$

Step 2. Choose arbitrarily any m variables as the basic variables, so that remaining $(n-m)$ variables become non-basic. Denote the basic and non-basic variables by

$$\mathbf{x}_B = (x_{B_1}, x_{B_2}, \ldots, x_{B_m}) \text{ and } \mathbf{x}_{NB} = (x_{NB_1}, x_{NB_2}, \ldots, x_{NB_{n-m}}) \text{ respectively.}$$

So, the constraints of *QPP* can be written as:

$$\mathbf{Bx}_B + \mathbf{Dx}_{NB} = \mathbf{b} \text{ or } \mathbf{x}_B = \mathbf{B}^{-1}\mathbf{b} - \mathbf{B}^{-1}\mathbf{Dx}_{NB},$$

where, $\mathbf{D}$ is an $m \times (n-m)$ matrix which contains the columns of $\mathbf{A}$ corresponding to non-basic variables $\mathbf{x}_{NB} \in \mathbf{R}^{n-m}$.

Thus, $$x_{B_i} = y_{i0} - \sum_{j=1}^{n-m} y_{ij}\, x_{NB_j}, \qquad i = 1, 2, \ldots, m$$

where, $$y_{i0} = (y_{10}\ y_{20}\ \cdots\ y_{m0})^T = \mathbf{B}^{-1}\mathbf{b} \text{ and } y_{ij} = \mathbf{B}^{-1}\mathbf{D}.$$

Step 3. Express each basic variable x_{B_i} entirely in terms of non-basic variables x_{NB_i}'s (and u_i's, if any) using given constraints. The non-basic variables u_i are defined by $\frac{1}{2}\frac{\partial f}{\partial x_{NB_i}}$.

Step 4. Express the objective function f, entirely in terms of non-basic x_{NB_i}'s (and u_i's, if any). Therefore, we get

$$f(\mathbf{x}) = f(\mathbf{x}_0) + \boldsymbol{\alpha}\mathbf{x}_{NB} + \mathbf{x}_{NB}^T \mathbf{G}\mathbf{x}_{NB},$$

where, $f(\mathbf{x}_0)$ = Value of objective function $f(x)$, when $\mathbf{x}_{NB} = \mathbf{0}$ and $x_{B_i} = y_{i0}$,

$\mathbf{G}$ = Symmetric matrix of order $(n - m) \times (n - m)$,

and $\boldsymbol{\alpha} = \alpha_1, \alpha_2, \ldots, \alpha_{n-m}$ (constants).

Step 5. Evaluate partial derivatives of $f(\mathbf{x})$ with respect to non-basic variables $\mathbf{x}_{NB}$. Thus, we get

$$\frac{\partial f(\mathbf{x})}{\partial x_{NB_k}} = \alpha_k + 2\sum_{j=1}^{n-m} g_{kj}\,\mathbf{x}_{NB_j}; \qquad k = 1, 2, \ldots, n-m.$$

Examine partial derivatives of $f(\mathbf{x})$ formulated above w.r.t. the non-basic variables at the point $\mathbf{x}_{NB} = \mathbf{0}$ (and $\mathbf{u} = \mathbf{0}$):

(*i*) If $\left(\dfrac{\partial f(\mathbf{x})}{\partial x_{NB_k}}\right)_{\substack{\mathbf{x}_{NB}=\mathbf{0}\\ \mathbf{u}=\mathbf{0}}} < 0$, for $k = 1, 2, \ldots, n-m$ and $\left(\dfrac{\partial f(\mathbf{x})}{\partial u_i}\right)_{\substack{\mathbf{x}_{NB}=\mathbf{0}\\ \mathbf{u}=\mathbf{0}}} = 0$,

then, the current basic solution is optimal. Go to *Step 8*.

(*ii*) If $\left(\dfrac{\partial f(\mathbf{x})}{\partial x_{NB_k}}\right)_{\substack{\mathbf{x}_{NB}=\mathbf{0}\\ \mathbf{u}=\mathbf{0}}} > 0$, for at least one k,

choose the most positive one. The corresponding ***non-basic variable*** will ***enter*** the basis.

(*iii*) If $\left(\dfrac{\partial f(\mathbf{x})}{\partial x_{NB_k}}\right)_{\substack{\mathbf{x}_{NB}=\mathbf{0}\\ \mathbf{u}=\mathbf{0}}} < 0$, for $k = 1, 2, \ldots, n-m$; but $\left(\dfrac{\partial f(\mathbf{x})}{\partial u_i}\right)_{\substack{\mathbf{x}_{NB}=\mathbf{0}\\ \mathbf{u}=\mathbf{0}}} \neq 0$,

for some $i = r$, introduce a new non-basic variable, u_i, defined by $u_i = \dfrac{1}{2}\dfrac{\partial f}{\partial u_r}$ and treat u_r as a basic variable (to be ignored later). Go to *Step 3*.

Step 6. Let $x_{NB_i} = x_k$ be entering variable identified in *Step 5 (ii)*. Compute the minimum of the ratios

$$\left\{\frac{\alpha_{h0}}{|\alpha_{hr}|}, \frac{\gamma_{k0}}{|\gamma_{kk}|}\right\}$$

for basic variable x_h, where α_{ho} is constant term and α_{hk} is the coefficient of x_k, in the expression of basic variable x_h when expressed in terms of the non-basic ones; and γ_{ko} is the constant term and γ_k is the coefficient of x_k in $\dfrac{\partial f}{\partial x_k}$.

(*i*) If minimum ratio occurs for some $\dfrac{\alpha_{h0}}{|\alpha_{hk}|}$, the corresponding basic variable, x_h, will leave the basis.

(*ii*) If minimum ratio occurs for some $\dfrac{\gamma_{k0}}{|\gamma_{kk}|}$, the exit criterion corresponds to a ***non-basic variable***. In this case, introduce an additional non-basic variable, called a ***free-variable*** defined by

$$u_i = \frac{1}{2}\frac{\partial f}{\partial x_k} \qquad (u_i \text{ is unrestricted}).$$

This relation becomes an additional constraint equation.

Step 7. Go to *Step 3* and repeat the procedure until an optimal basic solution is reached.

Step 8. Determine the optimal values of $\mathbf{x}_B$ and $f(\mathbf{x})$ by setting $\mathbf{x}_{NB} = \mathbf{0}$ in their expressions obtained in *Steps* 3 and 4.

Example 26.4. *Solve the following QPP by using Beale's method:*

$$\text{Maximize } f(x) = 2x_1 + 3x_2 - x_1^2 \text{ subject to the constraints:}$$

$$x_1 + 2x_2 \le 4, \quad x_1, x_2 \ge 0.$$

[MBA, Madurai Univ. 2009]

Solution. Introducing a slack variable x_3, the constraint becomes

$$x_1 + 2x_2 + x_3 = 4, \quad x_1, x_2 \ge 0.$$

We choose arbitrarily x_3 as a basic variable, so that we have

$$\mathbf{x}_B = (x_3), \quad \mathbf{x}_{NB} = (x_1, x_2) \text{ with } x_1 = x_2 = 0 \text{ and } x_3 = 4$$

Expressing the basic variable and $f(\mathbf{x})$ entirely in terms of non-basic $\mathbf{x}_{NB}$, we have

$$x_3 = 4 - x_1 - 2x_2 \text{ and } f = 2x_1 + 3x_2 - x_1^2$$

We evaluate the partial derivatives of f w.r.t. non-basic variables at $\mathbf{x}_{NB} = \mathbf{0}$.

$$\therefore \quad \left(\frac{\partial f}{\partial x_1}\right)_{\mathbf{x}_{NB}=\mathbf{0}} = (2 - 2x_1)_{\substack{x_1=0\\ x_2=0}} = 2, \quad \left(\frac{\partial f}{\partial x_2}\right)_{\mathbf{x}_{NB}=\mathbf{0}} = 3.$$

Since, both the partial derivates are positive, the current solution can be improved. As x_2 corresponds to the most positive partial derivative, it will ***enter*** the basis.

For determining the ***leaving*** basic variable, we compute the minimum of the ratios:

$$\min.\left\{\frac{\alpha_{h0}}{|\alpha_{hk}|}, \frac{\gamma_{k0}}{|\gamma_{kk}|}\right\} = \min.\left\{\frac{\alpha_{30}}{|\alpha_{32}|}, \frac{\gamma_{20}}{|\gamma_{22}|}\right\} = \left\{\frac{4}{|-2|}, \frac{3}{0}\right\} = 2$$

Since, the minimum occurs for $\dfrac{\alpha_{30}}{|\alpha_{32}|}$, x_3 will leave the basis.

Expressing the basic variable x_2 as well as the objective function f, in terms of the new non-basic variables (x_1, x_3), we have

$$x_2 = 2 - \frac{1}{2}x_1 - \frac{1}{2}x_3 \text{ and } f = 2x_1 + 3\left(2 - \frac{1}{2}x_1 - \frac{1}{2}x_3\right) - x_1^2$$

$$= 6 + \frac{1}{2}x_1 - \frac{3}{2}x_3 - x_1^2$$

The partial derivatives of this $f(\mathbf{x})$ w.r.t. the new $\mathbf{x}_{NB}$ are:

$$\left(\frac{\partial f}{\partial x_1}\right)_{\mathbf{x}_{NB}=\mathbf{0}} = \left(\frac{1}{2} - 2x_1\right)_{\substack{x_1=0\\ x_3=0}} = \frac{1}{2}, \text{ and } \left(\frac{\partial f}{\partial x_2}\right)_{\mathbf{x}_{NB}=\mathbf{0}} = -\frac{3}{2}$$

Clearly, the current basic solution is not optimal. The entry criterion qualifies x_1 to enter the basis. For the exit criterion, we compute the ratios

$$\min.\left\{\frac{\alpha_{20}}{|\alpha_{21}|}, \frac{\gamma_{10}}{|\gamma_{11}|}\right\} = \left\{\frac{2}{\left|-\frac{1}{2}\right|}, \frac{\frac{1}{2}}{|-2|}\right\} = \frac{1}{4}.$$

Since, the minimum of these ratios corresponds to $\dfrac{\gamma_{10}}{|\gamma_{11}|}$, no basic variable can be removed.

Thus, we introduce a ***free variable*** u_1 (unrestricted), as an additional non-basic variable, defined by

$$u_1 = \frac{1}{2}\frac{\partial f}{\partial x_1} = \frac{1}{2}\left(\frac{1}{2} - 2x_1\right) = \frac{1}{4} - x_1.$$

Note that currently we have $\mathbf{x}_{NB} = (x_3, u_1)$ and $\mathbf{x}_B = (x_1, x_2)$. Expressing the basic variable x_1 and $f(\mathbf{x})$ in terms of non-basic variables $\mathbf{x}_{NB}$:

$$x_1 = \frac{1}{4} - u_1, \quad x_2 = \frac{1}{2}(4 - x_1 - x_3) = \frac{15}{8} + \frac{1}{2}u_1 - \frac{1}{2}x_3,$$

and

$$f = 2\left(\frac{1}{4} - u_1\right) + 3\left(\frac{15}{8} + \frac{1}{3}u_1 - \frac{1}{2}x_3\right) - \left(\frac{1}{4} - u_1\right)^2 = \frac{97}{16} - \frac{3}{2}x_3 - u_1^2.$$

Now,

$$\left(\frac{\partial f}{\partial x_3}\right)_{\substack{\mathbf{x}_{NB} = 0 \\ \mathbf{u} = 0}} = -\frac{3}{2}, \quad \left(\frac{\partial f}{\partial u_1}\right)_{\substack{\mathbf{x}_{NB} = 0 \\ \mathbf{u} = 0}} = \left(-2u_1\right)_{\substack{x_3 = 0 \\ u_1 = 0}} = 0$$

Since, $\dfrac{\partial f}{\partial x_j} \leq 0$ for all x_j in $\mathbf{x}_{NB} = \mathbf{0}$, the current solution is optimum.

The optimum basic feasible solution to the given *NLPP* is thus:

$$x_1 = 1/4, \quad x_2 = 15/8, \quad \text{maximum} \quad z = 97/16.$$

Example 26.5. *Use Beale's method to solve the following NLPP:*

Minimize $z = 6 - 6x_1 - 2x_1^2 - 2x_1x_2 + 2x_2^2$ *subject to the constraints:*

$x_1 + x_2 \leq 2$ *and* $x_1, x_2 \geq 0$. **[M.Sc. (Math.), Rohilkhand Univ. 2011]**

Solution. Introducing a slack variable x_3, the constraint equation becomes $x_1 + x_2 + x_3 = 2$. Converting the minimization objective function into a maximization one, we have

$$f(\mathbf{x}) = -6 + 6x_1 - 2x_1^2 + 2x_1x_2 - 2x_2^2.$$

Choosing arbitrarily x_3 as the basic variable, we have

$$\mathbf{x}_B = x_3, \quad \mathbf{x}_{NB} = (x_1, x_2).$$

Expressing $\mathbf{x}_B$ and $f(\mathbf{x})$ in terms of $\mathbf{x}_{NB}$, we have

$$x_3 = 2 - x_1 - x_2$$
$$f = -6 + 6x_1 - 2x_1^2 + 2x_1x_2 - 2x_2^2.$$

The partial derivatives are:

$$\left(\frac{\partial f}{\partial x_1}\right)_{\mathbf{x}_{NB} = \mathbf{0}} = \left(6 - 4x_1 + 2x_2\right)_{\substack{x_1 = 0 \\ x_2 = 0}} = 6$$

$$\left(\frac{\partial f}{\partial x_2}\right)_{\mathbf{x}_{NB} = \mathbf{0}} = \left(2x_1 - 4x_2\right)_{\substack{x_1 = 0 \\ x_2 = 0}} = 0.$$

Clearly, x_1 will enter the basis.

Now,

$$\min.\left\{\frac{\alpha_{30}}{|\alpha_{31}|}, \frac{\gamma_{10}}{|\gamma_{11}|}\right\} = \min.\left\{\frac{2}{|-1|}, \frac{6}{|-4|}\right\} = \frac{6}{4}.$$

This corresponds to $\dfrac{\gamma_{10}}{|\gamma_{11}|}$. Thus, we cannot remove x_3 from the basis. We introduce a new non-basic free variable u_1, defined by

$$u_1 = \frac{1}{2}\frac{\partial f}{\partial x_1} = 3 - 2x_1 + x_2$$

The current basis is $\mathbf{x}_B = (x_3, x_1)$ and $\mathbf{x}_{NB} = (x_2, u_1)$. Expressing the current $\mathbf{x}_B$ and $f(\mathbf{x})$ in terms of $\mathbf{x}_{NB}$, we have

$$x_1 = \frac{3}{2} - \frac{1}{2}u_1 + \frac{1}{2}x_2, \quad x_3 = \frac{1}{2} + \frac{1}{2}u_1 - \frac{3}{2}x_2$$

and
$$f = -6 + \left(\frac{3}{2} - \frac{1}{2}u_1 + \frac{1}{2}x_2\right)(6 - 3 + u_1 - x_2 + 2x_2) - 2x_2^2$$
$$= -6 + \left(\frac{3}{2} - \frac{1}{2}u_1 + \frac{1}{2}x_2\right)(3 + u_1 + x_2) - 2x_2^2$$
$$= -\frac{3}{2} - \frac{1}{2}u_1^2 - \frac{3}{2}x_2^2 + 3x_2$$

Evaluating the partial derivatives of $f(\mathbf{x})$ w.r.t. $\mathbf{x}_{NB}$, we get

$$\left(\frac{\partial f}{\partial x_2}\right)_{\substack{\mathbf{x}_{NB} = \mathbf{0} \\ u_1 = 0}} = \left(3 - 3x_2\right)_{x_2 = 0} = 3,$$

and
$$\left(\frac{\partial f}{\partial u_1}\right)_{\substack{\mathbf{x}_{NB} = \mathbf{0} \\ u_1 = 0}} = \left(-\frac{u_1}{2}\right)_{u_1 = 0} = 0$$

Clearly, x_2 will enter the basis.

Again, we compute the ratios,

$$\text{min.}\left\{\frac{\alpha_{10}}{|\alpha_{12}|}, \frac{\alpha_{30}}{|\alpha_{32}|}, \frac{\gamma_{20}}{|\gamma_{22}|}\right\} = \text{min.}\left\{\frac{3/2}{1/2}, \frac{1/2}{|-3/2|}, \frac{3}{|-3|}\right\} = \frac{\alpha_{30}}{|\alpha_{32}|}$$

Thus, x_3 will leave the basis and the new variables are:

$$\mathbf{x}_B = (x_1, x_2) \text{ and } \mathbf{x}_{NB} = (u_1, x_3)$$

Expressing the new basis $\mathbf{x}_B$ and $f(\mathbf{x})$ in terms of $\mathbf{x}_{NB}$, we have

$$x_2 = \frac{1}{3} + \frac{1}{3}u_1 - \frac{2}{3}x_3, \quad x_1 = \frac{5}{3} - \frac{1}{3}u_1 - \frac{1}{3}x_3$$

and
$$f = -6 + \left(\frac{5}{3} - \frac{1}{3}u_1 - \frac{1}{3}x_3\right)\left(\frac{10}{3} + \frac{4}{3}u_1 - \frac{2}{3}x_3\right) - 2\left(\frac{1}{3} + \frac{1}{3}u_1 - \frac{2}{3}x_3\right)^2$$
$$= -\frac{2}{3} + \frac{2}{3}u_1 - \frac{4}{3}x_3 + \frac{2}{3}x_3 u_1 - \frac{2}{3}u_1^2 - \frac{2}{3}x_3^2.$$

The partial derivatives of $f(\mathbf{x})$ w.r.t. $\mathbf{x}_{NB}$ are given by:

$$\left(\frac{\partial f}{\partial x_3}\right)_{\substack{\mathbf{x}_{NB} = \mathbf{0} \\ u_1 = 0}} = \left(\frac{-4}{3} + \frac{2}{3}u_1 - \frac{4}{3}x_3\right)_{\substack{x_3 = 0 \\ u_1 = 0}} = -\frac{4}{3}$$

$$\left(\frac{\partial f}{\partial u_1}\right)_{\substack{\mathbf{x}_{NB} = \mathbf{0} \\ u_1 = 0}} = \left(\frac{2}{3} + \frac{2}{3}x_3 - \frac{4}{3}u_1\right)_{\substack{x_3 = 0 \\ u_1 = 0}} = \frac{2}{3}.$$

Since $\dfrac{\partial f}{\partial u_1} \neq 0$, current solution can be further improved. However, the entry criterion does not qualify x_3 to enter the basis. Thus, we introduce another non-basic free variable u_2, defined by

$$u_2 = \frac{1}{2}\frac{\partial f}{\partial u_1} = \frac{1}{3} + \frac{1}{3}x_3 - \frac{2}{3}u_1.$$

Treating u_1 as basic variable, and expressing the basic variables of $f(\mathbf{x})$ in terms of non-basic variables, we have

$$u_1 = \frac{1}{2} - \frac{3}{2}u_2 + \frac{1}{2}x_3, \quad x_2 = \frac{1}{2} - \frac{1}{2}u_2 - \frac{1}{2}x_3, \quad x_1 = \frac{3}{2} + \frac{1}{2}u_2 - \frac{1}{2}x_3$$

and
$$f = -\frac{2}{3} + \frac{2}{3}u_1(1 + x_3 - u_1) - \frac{4}{3}x_3 - \frac{2}{3}x_3^2$$
$$= -\frac{2}{3} + \frac{1}{6}(1 - 3u_2 + x_3)(1 + 3u_2 - x_3) - \frac{4}{3}x_3 - \frac{2}{3}x_3^2$$
$$= -\frac{2}{3} + \frac{1}{6}(1 - 9u_2^2 - 5x_3^2 + 6u_2 x_3 - 8x_1)$$
$$= -\frac{1}{2} + \frac{1}{6}(-9u_2^2 - 5x_3^2 + 6u_2 x_3 - 8x_3).$$

Now, we see that
$$\left(\frac{\partial f}{\partial x_3}\right)_{\substack{\mathbf{x}_{NB} = \mathbf{0} \\ u_2 = 0}} = \frac{1}{6}\left(-10x_3 + 6u_2 - 8\right)_{\substack{x_3 = 0 \\ u_2 = 0}} = -\frac{8}{6}$$
$$\left(\frac{\partial f}{\partial u_1}\right)_{\substack{\mathbf{x}_{NB} = \mathbf{0} \\ u_2 = 0}} = \frac{1}{6}\left(-18u_2 + 6x_3\right)_{\substack{x_3 = 0 \\ u_2 = 0}} = 0.$$

This shows that the current $\mathbf{x}_B$ provides an optimal solution.

Ignoring the free variable u_1, the optimal solution to the given *QPP* is:
$$x_1 = \frac{3}{2},\ x_2 = \frac{1}{2};\ \text{Minimum } z = -\left(-\frac{1}{2}\right) = \frac{1}{2}.$$

PROBLEMS

Solve the following quadratic programming problems by using Beale's method:

26.18. Maximize $f(\mathbf{x}) = 4x_1 + 6x_2 - x_1^2 - 3x_2^2$ subject to the constraints:
$x_1 + 2x_2 \le 4$ and $x_1, x_2 \ge 0$.

26.19. Maximize $f(\mathbf{x}) = \frac{1}{4}(2x_3 - x_1) - \frac{1}{2}(x_1^2 + x_2^2 + x_3^2)$ subject to the constraints:
$x_1 - x_2 + x_3 = 1$ and $x_1, x_2, x_3 \ge 0$.

26.20. Maximize $f(\mathbf{x}) = 2x_1 + 3x_2 - 2x_2^2$ subject to the constraints:
$x_1 + 4x_2 \le 4,\ x_1 + x_2 \le 2;\ x_1, x_2 \ge 0$. *[IAS, 1994]*

26.21. Minimize $z = -4x_1 + x_1^2 - 2x_1x_2 + 2x_2^2$ subject to the constraints:
$2x_1 + x_2 \ge 6,\ x_1 - 4x_2 \ge 0$ and $x_1, x_2 \ge 0$.

22.22. Maximize $z = 10x_1 + 25x_2 - 10x_1^2 - x_2^2 - 4x_1x_2$ subject to the constraints:
$x_1 + 2x_2 \le 10,\ x_1 + x_2 \le 9;\ x_1, x_2 \ge 0$.

26.23. Maximize $z = 6x_1 + 3x_2 - x_1^2 + 4x_1x_2 - 4x_2^2$ subject to the constraints:
$x_1 + x_2 \le 3,\ 4x_1 + x_2 \le 9;\ x_1, x_2 \ge 0$.

26.24. Minimize $z = 4t_0^2 + 4t_0t_1 + 6t_1^2 + 4t_0 - 56t_1 + 180$ subject to the constraints:
$t_0 \ge -3$ and $t_1 \le 6$.

26:6. SEPARABLE CONVEX PROGRAMMING

Separable Programming

Definition 1 (*Separable Programming Problem*). *An NLPP in which the objective function can be expressed as a linear combination of several different one-variable functions, of which some or all are non-linear, is called a* ***separable programming problem****.*

Definition 2 (*Convex Programming*). *Non-linear programming which has the problem of minimizing a convex objective function (or maximizing a concave objective function) in a convex set of points, is called* ***convex programming****.*

In both these definitions, nothing has been said about the constraints of the problem. So, in general we take them to be non-linear. A separable programming problem in which the separate functions are all convex can be called a convex programming problem with separable objective function. Thus, if $f(\mathbf{x})$ be the objective function, then for separable convex programming, $f(\mathbf{x})$ must be separable as $f(\mathbf{x}) = \sum_{i=1}^{n} f_i(x_i)$, where all $f_i(x_i)$ are convex ($i = 1, 2, ..., n$).

For example, if $f(\mathbf{x}) = 5x_1 + x_2^2 - 3x_1 + 5x_2$, then by letting

$f_1(x_1) = 5x_1^2 - 3x_1$ and $f_2(x_2) = x_2^2 + 5x_2$, we can write

$f(x) = f_1(x_1) + f_2(x_2)$, where both $f_1(x_1)$ and $f_2(x_2)$ are convex functions.

Piecewise Linear Approximations

In many decision-making situations, the non-linear profit or cost objective functions ar related by relatively smooth curves, rather than sharp curves obtaining breaking points. In such cases, the non-linear objective function, with a smooth curve for it, may be approximated by a series of ***piecewise linear segments***. This approximation will introduce some errors that, in many cases, may be negligibly small. In general, proper care is to be taken while approximating a non-linear function by linear segments, to reduce such errors. One way of reducing such errors is to increase the number of linear segments. But, excessive linear approximations will enlarge the size of the problem, thus increasing the computational time for an optimal solution.

Piecewise linear approximation can be made for both convex and concave functions. However, as we have seen earlier, if all the linear segments of the objective function are concave (convex), a valid optimal solution requires the objective function to be maximized (minimized). Thus, a *NLPP* becomes a *LPP** with piecewise linear approximations, and the simplex method can be applied to obtain an optimal solution.

We shall now study how piecewise linear approximations reduce a given separable convex (concave) *NLPP* to a *LPP* and then solve it by using the simplex method.

Consider the separable programming problem:

$$\text{Maximize } f(\mathbf{x}) = \sum_{j=1}^{n} f_j(x_j) \text{ subject to the constraints:}$$

$$\sum_{j=1}^{n} g_{ij}(x_j) \le b_i, \quad x_j \ge 0, \qquad (i = 1, 2, ..., m; j = 1, 2, ..., n)$$

where, some or all $g_{ij}(x_j)$, $f_j(x_j)$ are non-linear.

To solve this *NLPP* by piecewise linear approximation method, we shall reduce each non-linear $f_j(x_j)$ and $g_{ij}(x_j)$ to their piecewise linear approximations by fitting the linear segments to the original curve as shown in *Fig. 26.3* [For the function $f_j(x_j)$].

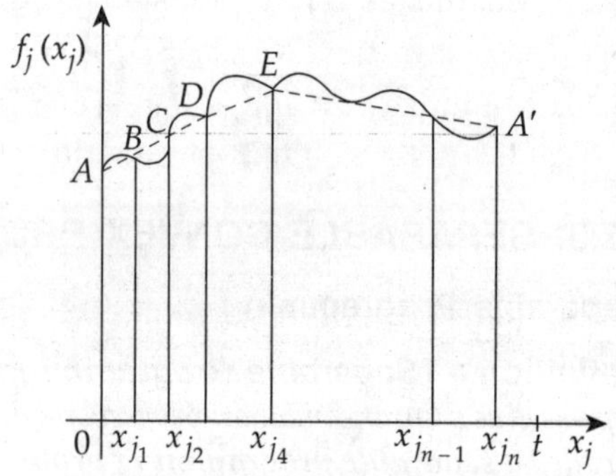

Fig. 26.3. Linear approximation of a function

This approximation can be carried out to any degree of accuracy by dividing the curve AA' into a suitable number of arcs, each are replaced by a linear segment. That is

$$\overset{\frown}{AB} = \overline{AB}, \quad \overset{\frown}{BC} = \overline{BC},$$

* This is so because the ***non-linear*** constraint functions can also be approximated by ***linear*** segments.

and so on. Let the point x_{jk} in the interval $0 \le x_j \le t$ be so determined that

$$x_{j0} = 0, \quad x_{j0} < x_{j1} < x_{j2} < \ldots < x_{jn} < t.$$

For each point x_{jk}, we calculate the function values $f_j(x_{jk})$. Connect the adjacent points

$$\left[x_{j1}, f_j(x_{j1})\right], \left[x_{j2}, f_j(x_{j2})\right], \ldots, \left[x_{jn}, f_j(x_{jn})\right] \ldots$$

by straight lines (shown dotted in *Fig. 26.3*) and denote this piecewise linear approximation to $f_j(x_j)$ by $f_j^*(x_j)$. Naturally $f_j^*(x_j)$ will be a better approximation to $f_j(x_j)$ if the distance between the adjacent x_{jk}'s is very small.

Assume that we have found an approximation of this kind for each $f_j(x_j)$ and $g_{ij}(x_j)$, $j = 1, 2, \ldots, n$. Let these approximated new functions be f_j^* and $g_{ij}^*(x_j)$ for $j = 1, 2, \ldots, n$, respectively. Thus, the given *NLPP* reduces to

$$\text{Maximize } f(\mathbf{x}) = \sum_{j=1}^{n} f_j^*(x_j) \text{ subject to the constraints:}$$

$$\sum_{j=1}^{n} g_{ij}^*(x_j) \le b_i, \quad x_j \ge 0, \qquad (i = 1, 2, \ldots, m; j = 1, 2, \ldots, n)$$

where, $f_j^*(x_j)$ and $g_{ij}^*(x_j)$ are now linear in x_j.

Now, evidently, we have

$$\frac{f_j(x_{j,k+1}) - f_j(x_{jk})}{x_{j,k+1} - x_{jk}} = \frac{f_j^*(x_j) - f_j(x_{jk})}{x_j - x_{jk}}$$

for any x_j in the interval $(x_{jk}, x_{j,k+1})$

This yields

$$f_j^*(x_j) = f_j(x_{jk}) + \frac{f_j(x_{j,k+1}) - f_j(x_{jk})}{x_{j,k+1} - x_{jk}} = (x_j - x_{jk})$$

Also, any x_j in the interval $x_{jk} \le x_j \le x_{j,k+1}$ can be expressed as a convex combination of the end points of the interval. Thus,

$$x_j = \alpha x_{j,k+1} + (1 - \alpha) x_{jk}.$$

Using it in $f_j^*(x_j)$ obtained above, we get

$$f_j^*(x_j) = \alpha f_j(x_{j,k+1}) + (1 - \alpha) f_j(x_{jk}).$$

The last result suggests the linear approximation $f_j^*(x_j)$ between the interval $x_{jk} \le x_j \le x_{j,k+1}$ as a linear combination of the values of the original function at the end points, *viz.*, $f_j(x_{jk})$ at x_{jk} and $f_j(x_{j,k+1})$ at $x_{j,k+1}$.

If we choose to write $\alpha = \alpha_{j,k+1}$ and $1 - \alpha = \alpha_{jk}$, then the above may be rewritten as

$$x_j = \alpha_{j,k+1} x_{j,k+1} + \alpha_{jk} x_{jk}, \qquad (\alpha_{j,k+1} + \alpha_{jk} = 1)$$

$$f_j^*(x_j) = \alpha_{j,k+1} f_j(x_{j,k+1}) + \alpha_{jk} f_j(x_{jk}).$$

These two relations with the additional restrictions

$$\alpha_{jk} + \alpha_{j,k+1} = 1, \ \alpha_{jk} \ge 0 \ \text{ and } \ \alpha_{j,k+1} \ge 0,$$

describe the approximating linear function $f_j^*(x_j)$ anywhere in the interval $x_{jk} \le x_j \le x_{j,k+1}$ in terms of the known values $f_j(x_{j,k+1})$ and $f_j(x_{jk})$.

Generalising this approximation concept for any x_j in the interval $0 \le x_j \le t$, we may write

$$x_j = \sum_{k=0}^{n} \alpha_{jk}\, x_{jk} \quad \text{and} \quad f_j^*(x_j) = \sum_{k=0}^{m} \alpha_{jk}\, f_j(x_{jk}),$$

where, $\sum_{k=0}^{m} \alpha_{jk} = 1$ and $x_{jk} \ge 0$ $(k = 0, 1, 2, \ldots, m)$ provided that

(*a*) no more than two α_{jk} are positive, and

(*b*) if two α_{hk} are positive they must correspond to adjacent points.

Now, assume that for each x_j in the given problem, a finite maximum is either given or can be determined, call it t_j. Then, the interval $0 \le x_j \le t_j$ is subdivided into mesh points x_{jk}, so that

$$x_{j0} = 0 \le x_{j1} < x_{j2} < \ldots < x_{jm} = t_j.$$

This shows the division of each specific variable x_j. Thus, if the same mesh points (the same intervals) are chosen for f and all constraints g_{ij}, for any particular x_j, we shall have

(*a*) $$x_j = \sum_{k=0}^{m_j} \alpha_{jk}\, x_{jk}$$

(*b*) $$f_j^*(x_j) = \sum_{k=0}^{m_j} \alpha_{jk}\, f_j(x_{jk}) \quad \text{and} \quad g_{ij}^*(x_j) = \sum_{k=0}^{m_j} \alpha_{jk}\, g_{ij}(x_{jk})$$

(*c*) $$\sum_{k=0}^{m_j} \alpha_{jk} = 1 \quad \text{and} \quad \alpha_{jk} \ge 0, \qquad (j = 1, 2, \ldots, n;\ k = 0, 1, \ldots, m)$$

provided that for each j, no more than two α_{jk} are positive and if two α_{jk} are positive, they must be adjacent.

Thus, the piecewise linear approximation of the given problem is:

$$\text{Maximize } f^*(\mathbf{x}) = \sum_{j=0}^{n} \sum_{k=0}^{m_j} f_j(x_{jk})\,\alpha_{jk} \text{ subject to the constraints:}$$

$$\sum_{j=0}^{n} \sum_{k=0}^{m_j} g_{ij}(x_{jk})\,\alpha_{jk} \le b_i \qquad (i = 1, 2, \ldots, m)$$

$$\sum_{k=0}^{m_j} \alpha_{jk} = 1,\ \alpha_{jk} \ge 0 \qquad (j = 1, 2, \ldots, n;\ k = 1, 2, \ldots, m)$$

with the additional restrictions:

(*i*) For each j, no more than two α_{jk} are positive, and

(*ii*) If two α_{jk} are positive, they must correspond to adjacent x_j's.

26:7. SEPARABLE PROGRAMMING ALGORITHM

Let the *NLPP* be:

$$\text{Maximize } f(\mathbf{x}) = \sum_{j=0}^{n} f_j(x_j) \text{ subject to the constraints:}$$

$$\sum_{j=0}^{n} g_{ij}(x_j) \le b_i,\ x_j \ge 0. \qquad (i = 1, 2, \ldots, m;\ j = 1, 2, \ldots, n)$$

The iterative procedure for solving such separable *NLPP* is given below in seven steps:

***Step* 1.** Convert the minimization objective function, into that of maximization, if it is in the minimization form.

***Step* 2.** Check whether the function $f_j(x_j)$ and $g_{ij}(x_j)$ satisfy the concavity (convexity) conditions required for the maximization *NLPP*. If the conditions are not satisfied the method fails, otherwise proceed on to the next step.

***Step* 3.** Divide the interval $0 \le x_j \le t_j$ $(j = 1, 2, ..., n)$ into a number of mesh-points x_{jk}, $(k = 0, 1, ..., m_j)$ such that, $x_{j0} = 0,\ x_{j0} < x_{j1} < \ldots < x_{jmj} = t_j$.

***Step* 4.** For each point x_{jk}, compute piecewise linear approximation for each $f_j(x_j)$ and $g_{ij}(x_j)$, $(j = 1, 2, ..., n; i = 1, 2, ..., m)$.

***Step* 5.** Using computations for *Step* 4, write down the piecewise linear approximation of the given *NLPP*.

Step* 6.** Solve the resulting *LPP* by two phase simplex method. In this method treat α_{i0} $(i = 1, 2, ..., m)$ as the artificial variables; and since the costs associated with them are not there, these are assumed to be zero, the ***phase I of the method is automatically completed. Thus, the initial simplex table of ***phase I*** becomes optimum and hence represents the starting simplex table for ***phase II***.

***Step* 7.** Obtain the optimum solution $\mathbf{x}^0$ of the given *NLPP* by using the relations

$$x_{j0} = \sum_{k=1}^{m_j} \alpha_{jk}\, x_{jk} \qquad (j = 1, 2, ..., n).$$

Example 26.6. *Use separable convex programming to solve the NLPP:*

$$\textit{Maximize } f(\mathbf{x}) = 3x_1 + 2x_2 \ \textit{ subject to the constraints:}$$
$$g(\mathbf{x}) = 4x_1^2 + x_2^2 \le 16; \ \textit{ and } \ x_1, x_2 \ge 0.$$

Solution. The constraints of the problem suggest that

$$x_1 \le 2 \ \text{ and } \ x_2 \le 4.$$

Let us suppose that $t_1 = 4$ and $t_2 = 4$ are the upper limits for the variables x_1 and x_2 respectively. Divide the closed interval [0, 4] into four equally sized sub-intervals for both x_1 and x_2. It is necessary that number of sub-intervals for x_1 and x_2 be same. Also, it is not essential that any set of them be of equal size.

Letting $f_1(x_1) = 3x_1$, $f_2(x_2) = 2x_2$, $g_{11}(x_1) = 4x_1^2$ and $g_{12}(x_2) = x_2^2$ (we observe that these functions satisfy the concavity-convexity conditions required for the maximization problem), the given problem reduces to

$$\text{Maximize } f(\mathbf{x}) = \sum f_j(x_j)\ (j = 1, 2) \ \text{ subject to the constraints:}$$

$$\sum_{j=1}^{2} g_{1j}(x_j) \le 16,\ x_1, x_2 \ge 0.$$

To obtain the approximating *LPP*, for this *NLPP*, let us divide the interval $0 \le x_j \le 4$ into a number of mesh-points x_{jk} such that $x_{j0} = 0, x_{j0} < x_{j1} < x_{j2} < x_{j3} < x_{j4} = 4$. For each point x_{jk}, compute the piecewise linear approximation for each of $f_j(x_j)$ and $g_{ij}(x_j)$.

Thus, for $j = 1, 2$, and $k = 0, 1, 2, 3, 4$, we have:

x_j	0	1	2	3	4
$f_1(x_1)$	0	3	6	9	12
$f_2(x_2)$	0	2	4	6	8
$g_{11}(x_2)$	0	4	16	36	64
$g_{12}(x_2)$	0	1	4	9	16
Mesh-points	x_{j0}	x_{j1}	x_{j2}	x_{j3}	x_{j4}

Using the information, the approximating *LPP* of the given *NLPP* may be written as

Maximize $f^*(\mathbf{x}) = (0 \cdot \alpha_{10} + 3\alpha_{11} + 6\alpha_{12} + 9\alpha_{13} + 12\alpha_{14}) + (0 \cdot \alpha_{20} + 2\alpha_{21} + 4\alpha_{22} + 6\alpha_{23} + 8\alpha_{24})$

subject to the constraints:

$$(0 \cdot \alpha_{10} + 4\alpha_{11} + 16\alpha_{12} + 36\alpha_{13} + 64\alpha_{14}) + (0 \cdot \alpha_{20} + \alpha_{21} + 4\alpha_{22} + 9\alpha_{23} + 16\alpha_{24}) \le 16$$

$$\alpha_{10} + \alpha_{11} + \alpha_{12} + \alpha_{13} + \alpha_{14} = 1$$

$$\alpha_{20} + \alpha_{21} + \alpha_{22} + \alpha_{23} + \alpha_{24} = 1$$

and $\alpha_{jk} \ge 0 \qquad (j = 1, 2; k = 0, 1, 2, 3, 4)$

with the additional restriction that

(*i*) for each j, no more than two α_{jk} are positive, and

(*ii*) if two α_{jk} are positive, they must correspond to adjacent points.

Introducing slack variable $S_1 \ge 0$, the inequation in the constraint is converted into an equation. The reduced *LPP* can now be solved by phase-II of the two phase simplex method, by treating α_{10} and α_{i0} as the artificial variables whose costs in the objective function are zero.

Thus, the initial simplex table for phase-II is as follows:

Phase-II

Initial Iteration.

	c_j :	0	3	6	9	12	0	2	4	6	8	0	
Coeff. of basic variable	*Basic variable*	α_{10}	α_{11}	α_{12}	α_{13}	α_{14}	α_{20}	α_{21}	α_{22}	α_{23}	α_{24}	s	*Solution*
0	s	0	4	16	36	64	0	1	4	9	16	1	16
0	α_{10}	1	1	1	1	1	0	0	0	0	0	0	1
0	α_{20}	0	0	0	0	0	1	1	1	1	(1)	0	1
		0	−3	−6	−9	−12	0	−2	−4	−6	−8	0	$z = 0$

This table shows that α_{14} must enter the basis and **s** leave the basis. But, this violates the additional restriction. So, we examine the next best variable to enter the basis. The table indicates that α_{13} can enter the basis. But it again creates the same difficulty. The next best entering is α_{24} and this indicates that either **s** or α_{20} can depart from the basis satisfying the additional restrictions. We choose that α_{20} leaves the basis.

First Iteration. Introduce α_{24} and drop α_{20}.

Coeff. of basic variable	*Basic variable*	α_{10}	α_{11}	α_{12}	α_{13}	α_{14}	α_{20}	α_{21}	α_{22}	α_{23}	α_{24}	s	*Solution*
0	s	0	(4)	16	36	64	−16	−15	−12	−7	0	1	0
0	α_{10}	1	1	1	1	1	0	0	0	0	0	0	1
8	α_{24}	0	0	0	0	0	1	1	1	1	1	0	1
		0	−3	−6	−9	−12	8	6	4	2	0	0	$z = 8$

This table indicates that any of the column corresponding to $(z_j - c_j)$, $j = 1, 2, 3, 4$ can enter the basis. But, in order to satisfy the additional restrictions, we find that α_{11} enters the basis, and **s** leaves basis.

Second Iteration. Introduce α_{11} and drop **s**.

	c_j :	0	3	6	9	12	0	2	4	6	8	0	
Coeff. of basic variable	*Basic variable*	α_{10}	α_{11}	α_{12}	α_{13}	α_{14}	α_{20}	α_{21}	α_{22}	α_{23}	α_{24}	s	*Solution*
3	α_{11}	0	1	4	9	16	−4	−15/4	−3	−7/4	0	1/4	0
0	α_{10}	1	0	−3	−8	−15	−4	15/4	3	(7/4)	0	−1/4	1
8	α_{24}	0	0	0	0	0	1	1	1	1	1	0	1
		0	0	6	18	36	−4	−21/4	−5	−13/4	0	3/4	$z = 8$

Third Iteration. Introduce α_{23} and drop α_{10}.

	c_j :	3	6	9	12	2	4	6	8	0	
Coeff. of basic variable	*Basic variable*	α_{11}	α_{12}	α_{13}	α_{14}	α_{21}	α_{22}	α_{23}	α_{24}	s	*Solution*
3	α_{11}	1	1	1	1	0	0	0	0	0	1
6	α_{23}	0	−12/7	−32/7	−60/7	15/7	12/7	1	0	−1/7	4/7
8	α_{24}	0	12/7	32/7	60/7	−8/7	−5/7	0	1	1/7	3/7
		0	3/7	22/7	67/7	12/7	4/7	0	0	9/14	$z = 69/7$

This shows that all $(z_j - c_j) \geq 0$. Therefore, an optimal solution has been achieved. The optimal solution is

$$\alpha_{11} = 1, \quad \alpha_{23} = \frac{4}{7} \quad \text{and} \quad \alpha_{24} = \frac{3}{7}.$$

Hence, the optimum solution to the given *NLPP* is given by

$$x_j^0 = \sum_{k=0}^{4} \alpha_{1k}\, x_{jk} \quad \text{for } j = 1, 2.$$

This gives

$$\begin{aligned} x_1^0 &= \alpha_{10}\, x_{10} + \alpha_{11}\, x_{11} + \alpha_{12}\, x_{12} + \alpha_{13}\, x_{13} + \alpha_{14}\, x_{14} \\ &= 0 \times 0 + 1 \times 1 + 0 \times 2 + 0 \times 3 + 0 \times 4 \\ &= 1. \end{aligned}$$

and

$$\begin{aligned} x_2^0 &= \alpha_{20}\, x_{20} + \alpha_{21}\, x_{21} + \alpha_{22}\, x_{22} + \alpha_{23}\, x_{23} + \alpha_{24}\, x_{24} \\ &= 0 \times 0 + 0 \times 1 + 0 \times 2 + \frac{4}{7} \times 3 + \frac{3}{7} \times 4 \\ &= \frac{24}{7}. \end{aligned}$$

Hence, the optimum solution to the given *NLPP* is

$$x_1^0 = 1, \; x_2^0 = \frac{24}{7} \text{ with max. } f(\mathbf{x}) = 3 + 2\left(\frac{24}{7}\right) = \frac{69}{7}.$$

PROBLEMS

Use the method of separable convex programming for solving the following non-linear programming problems:

26.25. Maximize $z = (x_1 - 2)^2 + (x_2 - 2)^2$ subject to the constraints:

$$x_1 + 2x_2 \le 4;\; x_1 \ge 0 \text{ and } x_2 \ge 0.$$

26.26. Maximize $f(\mathbf{x}) = x_1 + x_2^2$ subject to the constraints:

$$3x_1 + 2x_2^2 \le 9;\; x_1 \ge 0 \text{ and } x_2 \ge 0.$$

26.27. Maximize $f(\mathbf{x}) = 16 - 2(x_1 - 3)^2 - (x_2 - 7)^2$ subject to the constraints:

$$x_1^2 + x_2 \le 16,\; x_1, x_2 \ge 0.$$

[***Hint:*** Define, $f_1(x_1) = 8 - 2(x_1 - 3)^2$ and $f_2(x_2) = 8 - (x_2 - 7)^2$.]

26.28. Minimize $f(\mathbf{x}) = (x_1 - 2)^2 + (x_2 - 1)^2$ subject to the constraints:

(*a*) $-x_1^2 + x_2 \ge 0,\; -x_1 - x_2 + 2 \ge 0,\; x_1, x_2 \ge 0.$

(*b*) $x_1 - 2x_2 + 1 = 0,\; -\frac{1}{4}x_1^2 - x_2^2 + 1 \ge 0,\; x_1, x_2 \ge 0.$

26.29. Minimize $f(\mathbf{x}) = x_1^2 + x_2^2 + 5$ subject to the constraints:

$$3x_1^4 + x_2 \le 243,\; x_1 + 2x_2^2 \le 32;\; x_1 \ge 0 \text{ and } x_2 \ge 0.$$ *[AMIE, 2000]*

26.30. Given the non-linear programming problem,

Minimize $z = x_1 - 6x_2 + \dfrac{x_1^2 + x_2^2}{3}$ subject to the constraints:

$$2x_1 + 3x_2 \le 6,\; x_1 + 4x_2 \le 6;\; x_1 \ge 0, x_2 \ge 0.$$

Check whether the given function is convex. If so, find x_1 and x_2 and evaluate z.

[ME (Structural), Madras Univ. 2000]

26:8. STOCHASTIC PROGRAMMING

In linear programming model, all parameters are known with certainty. However, in real life value of model parameters may not be known. The basic reason for this is that often all or some of the parameters of the problem are influenced by random events in the decision environment.

So far no general solution like simplex method has been developed for *LP* problems with random parameters. There are two approaches for formulating *LP* problems under uncertainty so as to consider random effects on parameters explicitly in the solution of the model. The first is the ***stochastic programming*** approach in which attempt is made to solve the problem by making a few decisions by selecting model parameters at different points of time. The second approach is the ***chance-constrained programming*** in which an attempt is made to convert the probabilistic nature of the problem into an equivalent deterministic model and then it is solved by applying simplex method.

Given below are the mathematical model formulation of the two approaches for solving a stochastic programming problem:

1. **Sequential Stochastic Programming.** The general mathematical model of ***sequential stochastic programming problem*** is written as:

Maximize or Minimize $z = \sum_{j=1}^{K} E(c_j x_j) + \sum_{r=1}^{Q} P_r (c_{rj} x_{rj})$ subject to the constraints:

$$\sum_{j=1}^{K} a_{ij} x_j = b_i; \qquad i = 1, 2, ..., s$$

$$\sum_{j=1}^{K} a_{rij} x_j = \sum_{j=n+1}^{L} a_{rij} x_{rj} = b_{rj}$$

$$x_j \geq 0; \qquad i = s+1, s+2, ..., m$$

and $x_{rj} \geq 0;$ for all r and j

Here, x_j ($j = 1, 2, ..., K \leq n$) represents levels of **x** and for $j = K + 1, ..., n$, it represents levels of **x** after all random values are known. The constraint i contains only variables for deterministic a_{ij} and b_i.

Q = a finite number of possible sets of values for c_j ($j = K + 1, ..., n$), a_{ij} and b_i

P_r = probability of occurrence of an event.

2. **Chance-Constrained Programming.** The general mathematical model of a ***chance-constrained programming problem*** is written as:

Maximize or Minimize $z = c_1x_1 + c_2x_2 + ... + c_nx_n$ subject to the constraints:

$$P_r \,[(a_{i1}x_1 + a_{i2}x_2 + ... + a_{in}x_n) \geq b_i] \geq 1 - \alpha_i, \qquad i = 1, 2, ..., m$$

and $x_j \geq 0, \qquad i = 1, 2, ..., n$

where, α_i is the risk level for constraint i and $0 \leq \alpha_i \leq 1$ for each level of b_i. All parameters a_{ij}, b_i and c_j are normally distributed with known mean and variance.

The given chance-constrained programming is first transformed into an equivalent deterministic non-linear programming problem and then such problem is solved by separable programming.

TEST YOUR UNDERSTANDING

1. Which of the following is not correct?
   (*a*) An efficient simplex-like technique for general *NLPP* is yet to be found.
   (*b*) An optimum solution to a *NLPP* involving only two decision variables can conveniently be determined by graphical method.
   (*c*) Graphical method for the solution of *NLPP* is valid, only if the constraints have linear relationship.
   (*d*) In a *NLPP*, the optimum solution can be found anywhere on the boundary of the feasible region and even at some interior point of it.
2. The quadratic form $\mathbf{x}^T\mathbf{Q}\mathbf{x}$ is said to be negative-semi-definite, if
   (*a*) $\mathbf{x}^T\mathbf{Q}\mathbf{x} > 0$, (*b*) $\mathbf{x}^T\mathbf{Q}\mathbf{x} < 0$,
   (*c*) $\mathbf{x}^T\mathbf{Q}\mathbf{x} \geq 0$, (*d*) $\mathbf{x}^T\mathbf{Q}\mathbf{x} \leq 0$.
3. If $\mathbf{x}^T\mathbf{Q}\mathbf{x}$ is positive-semi-definite, then it is
   (*a*) strictly convex (*b*) strictly concave
   (*c*) convex (*d*) concave.

4. Quadratic programming is concerned with the *NLPP* of optimizing the quadratic objective function subject to
   (*a*) linear inequality constraints (*b*) non-linear inequality constraints
   (*c*) non-linear equality constraints (*d*) no constraint.
5. The general *QPP* (Quadratic Programming Problem) is defined as:

   Minimize $f(\mathbf{x}) = \mathbf{c}\mathbf{x} + \frac{1}{2}\mathbf{x}^T \mathbf{Q}\mathbf{x}$, subject to $\mathbf{A}\mathbf{x} \leq \mathbf{b}$ and $\mathbf{x} \geq 0$, where $\mathbf{b}^T \in \mathbf{R}^m$ and $\mathbf{A}$ is an $m \times n$ real matrix. This problem always has a unique optimum solution provided it has a feasible solution and the matrix $\mathbf{Q}$ is
   (*a*) negative definite (*b*) positive definite
   (*c*) negative semi-definite (*d*) positive semi-definite.
6. Which of the following methods of solving a quadratic programming problem is based on modified simplex method?
   (*a*) Wolfe's Method (*b*) Beale's Method
   (*c*) Frank-Wolfe Method (*d*) Fletcher's Method.

ANSWERS

1. (*c*)	**2.** (*d*)	**3.** (*c*)	**4.** (*a*)	**5.** (*b*)	**6.** (*a*)

REVIEW EXERCISE

1. What is meant by quadratic programming? How does quadratic programming problem differ from the linear programming problem? Discuss one method of solving it.
   [MBA (Nov.), Delhi Univ. 2003; B.Sc. (Stat.), Delhi Univ. 2005]
2. Is it correct to say that in the quadratic programming the objective function and the constraints both should be quadratic? If not, give your own comments.
3. Write short notes on:
   (*a*) Quadratic Programming, and (*b*) Separable Convex Programming.
4. Derive the Kuhn-Tucker conditions for an optimal solution to a quadratic programming problem. ***[ME (Mech. & Ind.), Andhra Pradesh Univ. 1996]***
5. Mention briefly the Wolfe's algorithm for solving a quadratic programming problem given in the usual notations:
   $$\text{Maximize } z = \mathbf{c}\mathbf{x} + \frac{1}{2}\mathbf{x}^T \mathbf{Q}\mathbf{x} \text{ such that } \mathbf{A}\mathbf{x} \leq \mathbf{b} \text{ and } \mathbf{x} \geq \mathbf{0}.$$
6. Describe briefly the Beale's Method for solving Quadratic Programming Problem.
7. What do you mean by separable and/or non-linear convex programming? How will you solve the separable non-linear programming problem:
   $$\text{Minimize } \sum f_{oj}(x_j) \text{ subject to the constraints:}$$
   $$\sum f_{oj}(x_j) \geq b_j \qquad (j = 1, 2, ..., m).$$
8. Show that if $f_{oj}(x_j)$ is strictly convex and $f_{ij}(x_j)$ is concave for $i = 1, 2, ..., m$; then we can discard the additional restriction in the approximated separable non-linear programming problem of previous question and solve the resulting *LPP* to find an approximate optimum solution to the given problem.

CHAPTER 27

Geometric Programming

"We have chosen a clear path of advance, but it needs a long and steep journey"

Geometric programming has been developed in connection with the problems of engineering design. Solution to engineering design problems requires a compromise among many alternatives. The engineer seeks an ideal compromise that optimizes some specific characteristics. The characteristics of an engineering design depend on the parameters both fixed and adjustable. The design engineer strikes to set the adjustable parameters for optimum design. A typical optimum design is that for which the system performs a given function *(having several parameters)* at minimum cost. An attempt to develop a rapid systematic method of designing at ***minimum*** cost has been the stimulus for the development of ***geometric programming***.

In geometric programming, posynomials are used to solve the engineering design problems having special type of functions, such as

$$f(\mathbf{x}) = c_j x_1^{a_{1j}} x_2^{a_{2j}} \dots x_m^{a_{mj}}, \text{ i.e., } c_j \prod_{i=1}^{m} (x_i)^{a_{ij}} \qquad (j = 1, 2, \dots, n)$$

where, $c_j > 0$ $(j = 1, 2, \dots, n)$ are constants, $x_i > 0$ $(i = 1, 2, \dots, m)$, and a_{ij} $(i = 1, 2, \dots, m; j = 1, 2, \dots, n)$ are unrestricted in sign. The polynomial is generally known as ***monomial*** in **x**.

Definition 1 (*Posynomial*). *A generalized polynomial (also called* ***signomial****) consists of a finite number of monomials such as*

$$f(\mathbf{x}) = \sum_{j=1}^{n} c_j \prod_{i=1}^{m} (x_i)^{a_{ij}}$$

is said to be ***posynomial****, provided all c_j $(j = 1, 2, \dots, n)$ are positive.*

Definition 2 (*General Form of Geometric Programming*). *General formulation of geometric programming is expressed as to*

$$\text{Minimize } f(\mathbf{x}) = c_1 u_1(x) + c_2 u_2(x) + \dots + c_n u_n(x) = \sum_{j=1}^{n} c_j u_j(x),$$

where, $c_j > 0$ *and* $u_j(\mathbf{x}) = \prod_{i=1}^{m} (x_i)^{a_{ij}}$, *with* a_{ij} *any real number.*

27:1. GEOMETRIC-ARITHMETIC MEAN INEQUALITY

To obtain the geometric-arithmetic mean inequality, we proceed as follows:

Consider the optimal sub-division problem:

$$\text{Maximize } f(\mathbf{x}) = x_1 x_2 \dots x_n \text{ subject to the constraints:}$$

$$x_1 + x_2 + \dots + x_n = C \qquad (C > 0).$$

Using method of recursive equations, we find that

$$\text{Maximum } f(\mathbf{x}) = \left(\frac{C}{n}\right)^n.$$

It then follows that

$$f(\mathbf{x}) \le \left(\frac{C}{n}\right)^n \quad \text{or} \quad \prod_{j=1}^{n} x_j \le \left(\sum_{j=1}^{n} x_j / n\right)^n$$

Taking the nth root of each side, the inequality becomes

$$\left(\prod_{j=1}^{n} x_j\right)^{\frac{1}{n}} \le \frac{1}{n} \sum_{j=1}^{n} x_j.$$

It is to be noted that the inequality becomes strict equation when $x_j = C/n$. The inequality is known as the ***geometric-arithmetic mean inequality*** and is fundamental for the development of geometric programming.

A more general form of this inequality can be written as

$$\left[\prod_{j=1}^{N} (V_j)^{\alpha_j}\right]^{\frac{1}{\lambda}} \le \frac{1}{\lambda} \sum_{j=1}^{N} \alpha_j V_j, \quad \text{... (1)}$$

where, $\lambda = \sum_{j=1}^{N} \alpha_j, V_j \ (j = 1, 2, \dots, N)$ are N positive numbers and α_j (j = 1, 2, ..., N) are N positive rational numbers. The numbers x_j are called the ***weight functions***.

Now, if we put $\alpha_j = \lambda\, \zeta_j$, the inequality becomes

$$\prod_{j=1}^{N} (V_j)^{\zeta_j} \le \sum_{j=1}^{N} \zeta_j V_j, \qquad \sum_{j=1}^{N} \zeta_j = 1. \quad \text{... (2)}$$

The numbers ζ_j (j = 1, 2, N) are called the ***normalized weight functions***.

Further, by putting $V_j\, \zeta_j = u_j$, we get

$$\prod_{j=1}^{N} \left(\frac{u_j}{\zeta_j}\right)^{\zeta_j} \le \sum_{j=1}^{N} u_j, \qquad \sum_{j=1}^{N} \zeta_j = 1. \quad \text{... (3)}$$

In terms of non-normalized weights α_j, the inequality (1) can be rewritten as

$$\prod_{j=1}^{N} \left(\frac{\lambda u_j}{\alpha_j}\right)^{\alpha_j/\lambda} \le \sum_{j=1}^{N} u_j$$

or

$$\prod_{j=1}^{N} \left(\frac{u_j}{\alpha_j}\right)^{\alpha_j} (\lambda)^{\lambda} \le \left(\sum_{j=1}^{N} u_j\right)^{\lambda}, \qquad \sum_{j=1}^{N} \alpha_j = \lambda. \quad \text{... (4)}$$

Note. It is important to note that (3) becomes an equation if all u_j/ζ_j are equal. Similarly (4) is an equation if all u_j/α_j are equal.

27:2. UNCONSTRAINED GEOMETRIC PROGRAMMING PROBLEM (GPP)

The unconstrained programming problem is defined as follows:

$$\text{Minimize } f(\mathbf{x}) = \sum_{j=1}^{n} c_j \prod_{i=1}^{m} x_i^{a_{ij}},$$

where, c_j (j = 1, 2, ..., n) > 0, x_i (i = 1, 2, ..., m) > 0 and a_{ij} (i = 1, 2, ..., m; j = 1, 2, ..., n) are unrestricted in sign.

Now, using the geometric-arithmetic mean inequality (3), the objective function defined above can be rewritten as

$$u_1 + u_2 + \ldots + u_n \geq \left(\frac{u_1}{\zeta_1}\right)^{\zeta_1} \left(\frac{u_2}{\zeta_2}\right)^{\zeta_2} \cdots \left(\frac{u_n}{\zeta_n}\right)^{\zeta_n} \qquad \ldots (5)$$

where, $u_j = u_j(\mathbf{x})$ for $j = 1, 2, \ldots, n$ and the weights ζ_j $(j = 1, 2, \ldots, n)$ satisfy the relation

$$\zeta_1 + \zeta_2 + \ldots + \zeta_n = 1.$$

The left-hand side of the inequality (5) is known as ***primal function,*** whereas the right-hand side is called the ***pre-dual function***.

Since, $u_j = c_j \prod_{i=1}^{m} x_i^{a_{ij}}$, the pre-dual function can be written as

$$\prod_{j=1}^{n} \left(\frac{u_j}{\zeta_j}\right)^{\zeta_j} = \prod_{j=1}^{n} \left[\frac{c_j \prod_{i=1}^{m} x_i^{a_{ij}}}{\zeta_j}\right]^{\zeta_j} = \prod_{j=1}^{n} \left(\frac{c_j}{\zeta_j}\right)^{\zeta_j} \prod_{j=1}^{n} \left(\prod_{i=1}^{m} x_i^{a_{ij}}\right)^{\zeta_j}$$

$$= \prod_{j=1}^{n} \left(\frac{c_j}{\zeta_j}\right)^{\zeta_j} \left\{\left(\prod_{i=1}^{m} x_i^{a_{i1}\zeta_1}\right)\left(\prod_{i=1}^{m} x_i^{a_{i2}\zeta_2}\right)\cdots\left(\prod_{i=1}^{m} x_i^{a_{in}\zeta_n}\right)\right\}$$

$$= \prod_{j=1}^{n} \left(\frac{c_j}{\zeta_j}\right)^{\zeta_j} \left\{\left(x_1^{\sum_{j=1}^{n} u_{1j}\zeta_j}\right)\left(x_2^{\sum_{j=1}^{n} u_{2j}\zeta_j}\right)\cdots\left(x_m^{\sum_{j=1}^{n} u_{mj}\zeta_j}\right)\right\}$$

$$= \prod_{j=1}^{n} \left(\frac{c_j}{\zeta_j}\right)^{\zeta_j} \prod_{i=1}^{m} x_i^{\sum_{j=1}^{n} a_{ij}\zeta_j}$$

Now, making use of the conditions $\sum_{j=1}^{n} a_{ij}\zeta_j = 0$, $(i = 1, 2, \ldots, m)$ together with the normality condition $\sum_{j=1}^{n} \zeta_j = 1$, the above expression reduces to

$$\prod_{j=1}^{n} \left(\frac{u_j}{\zeta_j}\right)^{\zeta_j} = \prod_{j=1}^{n} \left(\frac{c_j}{\zeta_j}\right)^{\zeta_j} \qquad \ldots (6)$$

Using (6), the inequality (5) becomes

$$u_1 + u_2 + \ldots + u_n \geq \left(\frac{c_1}{\zeta_1}\right)^{\zeta_1} \left(\frac{c_2}{\zeta_2}\right)^{\zeta_2} \cdots \left(\frac{c_n}{\zeta_n}\right)^{\zeta_n}$$

or

$$f(\mathbf{x}) \geq \phi(\zeta),$$

where, ζ is a vector with components $\zeta_1, \zeta_2, \ldots, \zeta_n$.

Since, equality is possible, we have from above

$$\text{Minimum } f(\mathbf{x}) = \text{Maximum } \phi(\zeta)$$

In this way, we have converted the problem of minimizing, $f(\mathbf{x})$ into a new problem of maximizing $\phi(\zeta)$. The original problem of minimization is referred to as ***primal*** one and the related problem of maximization as its ***dual***.

In the present discussion, $\phi(\zeta)$, is just constant, and therefore

$$\underset{\mathbf{x}}{\text{minimum}}\ f(\mathbf{x}) = \phi(\zeta)$$

This can be easily evaluated for given numerical values of c_j.

The method of solving unconstrained geometric programming problems is well illustrated through the following examples:

Example 27.1. *Solve the following GP problem:*

$$\text{Minimize } f(\mathbf{x}) = c_1 x_1 x_2 x_3 + c_2 x_1 x_2^{-1} + c_3 x_2 x_3^{-2} + c_4 x_1^{-3} x_3,$$

where, $c_j > 0$, $x_j > 0$; $j = 1, 2, 3, 4$ *and* $i = 1, 2, 3$.

Solution. Let $u_1 = c_1 x_1 x_2 x_3$, $u_2 = c_2 x_1 x_2^{-1}$, $u_3 = c_3 x_2 x_3^{-2}$ and $u_4 = c_4 x_1^{-3} x_3$.

Then, the objective function is: $u_1 + u_2 + u_3 + u_4$ or $\sum_{j=1}^{4} u_j$.

Using the geometric mean inequality, we can write

$$\prod_{j=1}^{4} \left(\frac{u_j}{\zeta_j}\right)^{\zeta_j} \le \sum_{j=1}^{4} u_j \text{ or } f(\mathbf{x}).$$

Substituting the value of u_j ($j = 1, 2, 3, 4$), this inequality gives

$$\prod_{j=1}^{4} \left(\frac{c_j}{\zeta_j}\right)^{\zeta_j} x_1^{(\zeta_1 + \zeta_2 - 3\zeta_4)}\, x_2^{(\zeta_1 - \zeta_2 + \zeta_3)}\, x_3^{(\zeta_1 - 2\zeta_3 + \zeta_4)} \le f(\mathbf{x}),$$

Here, the values of ζ_j are arbitrary and can be chosen according to convenience subject to

$$\zeta_j > 0 \quad \text{and} \quad \sum_{j=1}^{4} \zeta_j = 1.$$

Let us set the powers of x_i for each i equal to zero. Then, the right-hand side of the above inequality will become independent of x_i. For this we require

$$\zeta_1 + \zeta_2 - 3\zeta_4 = 0,\ \zeta_1 - \zeta_2 + \zeta_3 = 0 \text{ and } \zeta_1 - 2\zeta_3 + \zeta_4 = 0.$$

These are called the necessary conditions for orthogonality. The orthogonality conditions together with the normality condition gives $\sum_{j=1}^{4} \zeta_j = 1$ are sufficient to give the value of ζ_j.

The orthogonality-normality conditions yield

$$\zeta_1 = \frac{1}{5},\ \zeta_2 = \frac{2}{5},\ \zeta_3 = \frac{1}{5} \text{ and } \zeta_4 = \frac{1}{5}.$$

Using these values of ζ_j ($j = 1, 2, 3, 4$), the geometric-arithmetic mean inequality reduces to

$$\left(\frac{c_1}{1/5}\right)^{\frac{1}{5}} \left(\frac{c_2}{2/5}\right)^{\frac{2}{5}} \left(\frac{c_3}{1/5}\right)^{\frac{1}{5}} \left(\frac{c_4}{1/5}\right)^{\frac{1}{5}} \le f(\mathbf{x}).$$

Since, equality is possible, we have from above

$$\text{Minimum } f(\mathbf{x}) = \text{Maximum } \phi(\zeta)$$

$$= \left(\frac{c_1}{1/5}\right)^{\frac{1}{5}} \left(\frac{c_2}{2/5}\right)^{\frac{2}{5}} \left(\frac{c_3}{1/5}\right)^{\frac{1}{5}} \left(\frac{c_4}{1/5}\right)^{\frac{1}{5}}.$$

It is to be noted that the minimum value of the objective function has been determined without having determined the optimal value of the variables x_1, x_2 and x_3. To obtain the optimal value of x_i $(i = 1, 2, 3)$ we proceed as follows:

For equality in the geometric-arithmetic mean inequality all (u_j/ζ_j) must be equal. Therefore, if we let $(u_j/\zeta_j) = N$; $j = 1, 2, 3, 4$; we shall have

$$\sum_{j=1}^{4} u_j = N \sum_{j=1}^{4} \zeta_j = N.$$

Hence, $\quad$ Minimum $f(\mathbf{x}) = f(\mathbf{x}_0) \equiv N.$

Thus, $\quad u_j = \zeta_j f(\mathbf{x}_0).$

To evaluate x_i $(i = 1, 2, 3)$, we shall define new variables y_i by substituting

$$y_i = \log x_i, \qquad i = 1, 2, 3.$$

We have

$$y_1 + y_2 + y_3 = \log(u_1/c_1),$$
$$y_1 - y_2 = \log(u_2/c_2),$$
$$y_2 - 2y_3 = \log(u_3/c_3),$$
$$-3y_1 + y_3 = \log(u_4/c_4).$$

These are four equations in three unknowns and are not linearly independent. Since any three of them are linearly independent, the values of y_i and hence of x_i are uniquely determined. Thus, we get the optimal solution.

Now, if $c_1 = 16, c_2 = 4, c_3 = 2$ and $c_4 = 8$, the optimal solution of the above problem is obtained as follows:

$$\text{Minimum } f(\mathbf{x}) = f(\mathbf{x}_0) = \left(\frac{16}{1/5}\right)^{\frac{1}{5}} \left(\frac{4}{2/5}\right)^{\frac{2}{5}} \left(\frac{2}{1/5}\right)^{\frac{1}{5}} \left(\frac{8}{1/5}\right)^{\frac{1}{5}} = 20.$$

$$\therefore \quad 16x_1 x_2 x_3 = u_1 = \frac{1}{5}(20), \; 4x_1 x_2^{-1} = u_2 = \frac{2}{5}(20),$$

$$2x_2 x_3^{-2} = u_3 = \frac{1}{5}(20), \; 8x_1^{-3} x_3 = u_4 = \frac{1}{5}(20).$$

The solution of these equations gives us

$$x_1 = 1, \; x_2 = 0.5 \text{ and } x_3 = 0.5.$$

Example 27.2. *Solve the following GP problem:*

$$\textit{Minimize } f(\mathbf{x}) = c_1 x_1 x_2^{-1} + c_2 x_1^{-1} x_2 + c_3 x_1 + c_4 x_2^{-1},$$

where, $c_j > 0;\; x_i > 0;\; j = 1, 2, 3, 4$ *and* $i = 1, 2.$

Solution. Let $u_1 = c_1 x_1 x_2^{-1},\; u_2 = c_2 x_1^{-1} x_2,\; u_3 = c_3 x_1$ and $u_4 = c_4 x_2^{-1}$

Then, $\quad f(\mathbf{x}) = u_1 + u_2 + u_3 + u_4 = \sum_{j=1}^{4} u_j.$

Using the inequality $\quad \prod_{j=1}^{4} \left(\frac{c_j}{\zeta_j}\right)^{\zeta_j} \le \sum_{j=1}^{4} u_j,$ and $\sum_{j=1}^{4} \zeta_j = 1,$ we have

$$\prod_{j=1}^{4} \left(\frac{c_j}{\zeta_j}\right)^{\zeta_j} x_1^{\zeta_1 - \zeta_2 + \zeta_3} x_2^{-\zeta_1 + \zeta_2 - \zeta_4} \le f(\mathbf{x}),$$

with $\zeta_1 + \zeta_2 + \zeta_3 + \zeta_4 = 1.$

To make the right-hand side of the inequality independent of x_i $(i = 1, 2)$, we put

$$\zeta_1 - \zeta_2 + \zeta_3 = 0, \quad -\zeta_1 + \zeta_2 - \zeta_4 = 0 \quad \text{and} \quad \zeta_1 + \zeta_2 + \zeta_3 + \zeta_4 = 1.$$

Solving these three equations in four unknowns, we get

$$\zeta_1 = \frac{1}{2}(1 - 3\zeta_4), \quad \zeta_2 = \frac{1}{2}(1 - \zeta_4) \quad \text{and} \quad \zeta_3 = \zeta_4.$$

Using these values ζ_j $(j = 1, 2, 3, 4)$, the above-mentioned inequality reduces to

$$\prod_{j=1}^{4} \left(\frac{u_j}{\zeta_j}\right)^{\zeta_j} \le f(\mathbf{x}).$$

In this inequality, since equality is possible.

$$\text{Minimum } f(\mathbf{x}) = \text{Maximum } \phi(\zeta)$$

$$= \left(\frac{c_1}{\frac{1}{2}(1 - 3\zeta_4)}\right)^{\frac{1-3\zeta_4}{2}} \left(\frac{c_2}{\frac{1}{2}(1 - \zeta_4)}\right)^{\frac{1-\zeta_4}{2}} \left(\frac{c_3}{\zeta_4}\right)^{\zeta_4} \left(\frac{c_4}{\zeta_4}\right)^{\zeta_4}.$$

This is, now, the problem of maximization of one variable. So, the technique of differential calculus can easily be applied.

Taking logarithm on both sides, we get

$$\log \phi(\zeta) = \frac{1 - 3\zeta_4}{2} \log \frac{2c_1}{1 - 3\zeta_4} + \frac{1 - \zeta_4}{2} \log \frac{2c_2}{1 - \zeta_4} + \zeta_4 \log \frac{c_3}{\zeta_4} + \zeta_4 \log \frac{c_4}{\zeta_4}$$

Since, the given problem (primal) has a unique maximum, the value of ζ_4 maximizing $\log \phi(\zeta)$ must be unique. Hence, $\frac{\partial}{\partial \zeta_4}[\log \phi(\zeta)] = 0$ after simplification gives

$$-\log \frac{(2c_1)^{3/2}(2c_2)^{1/2}}{c_3 c_4} + \log \frac{(1 - 3\zeta_4)^{3/2}(1 - \zeta_4)^{1/2}}{\zeta_4^2} = 0$$

or

$$\frac{\sqrt{(1 - 3\zeta_4)^3 (1 - \zeta_4)}}{\zeta_4^2} = \frac{(2c_1)^{3/2}(2c_2)^{1/2}}{c_3 c_4}$$

This yields an approximate value of ζ_4, which is turn gives the unique values of ζ_3, ζ_2 and ζ_1. Using these values of ζ_1, ζ_2, ζ_3 and ζ_4, $\phi(\zeta_0)$ is attained and hence $f(\mathbf{x}_0)$ is obtained.

The optimal solution x_i $(i = 1, 2)$ is obtained by using the relation

$$u_j = \zeta_j f(\mathbf{x}_0), \qquad j = 1, 2, 3, 4.$$

Letting $c_1 = 5, c_2 = 2, c_3 = 5$ and $c_4 = 1$, the optimal solution of the above problem is thus obtained as follows:

$$f(\mathbf{x}_0) = \left(\frac{5}{\frac{1}{2}(1 - 3\zeta_4)}\right)^{\frac{1-3\zeta_4}{2}} \left(\frac{2}{\frac{1}{2}(1 - \zeta_4)}\right)^{\frac{1-\zeta_4}{2}} \left(\frac{5}{\zeta_4}\right)^{\zeta_4} \left(\frac{1}{\zeta_4}\right)^{\zeta_4}.$$

where, ζ_4 is obtained from the relation

$$\frac{\sqrt{(1 - 3\zeta_4)^3 (1 - \zeta_4)}}{\zeta_4^2} = \frac{10^{3/2} \times 2}{5} = 4\sqrt{10}.$$

After simplification, this yields $\zeta_4 \cong 0.16$ and therefore

$$\zeta_3 = 0.16, \quad \zeta_2 = 0.42 \quad \text{and} \quad \zeta_1 = 0.26.$$

Thus,

$$f(\mathbf{x}_0)=\left(\frac{5}{0.26}\right)^{0.26}\left(\frac{2}{0.42}\right)^{0.42}\left(\frac{5}{0.16}\right)^{0.16}\left(\frac{1}{0.16}\right)^{0.16}\cong 9.661$$

$$\therefore \qquad 5x_1=u_3=0.16(9.661)=1.546 \text{ and } x_2^{-1}=u_4=0.16(9.661)=1.546.$$

The solution of these equations yields

$$x_1=0.309 \text{ and } x_2=0.647.$$

PROBLEMS

Using geometric programming, solve the following problems:

27.01. Minimize $f(\mathbf{x})=c_1x_1^{-1}x_2^{-1}+c_2x_2x_3+c_3x_1x_3+c_4x_1x_2,$
where $c_j>0, x_i>0;\ i=1,2,3,4$ and $j=1,2,3.$

27.02. Minimize $z=2x_1^2x_2^{-3}+8x_1^{-3}x_2+3x_1x_2;\ x_1,x_2>0.$

27.03. Minimize $f(\mathbf{x})=5x_1x_2^{-1}x_3^2+x_1^{-2}x_3^{-1}+10x_2^3+2x_1^{-1}x_2x_3^{-3};\ x_1,x_2,x_3>0.$

27.04. Minimize $f(\mathbf{x})=7x_1x_2^{-1}+3x_2x_3^{-2}+5x_1^{-3}x_2x_3+x_1x_2x_3;\ x_1,x_2,x_3>0.$

27.05. Minimize $f(\mathbf{x})=2x_1+4x_2+\dfrac{10}{x_1x_2};\ x_1,x_2>0.$

27.06. Minimize $x_0=5x_1x_2^{-1}+2x_1^{-1}x_2+5x_1+x_2^{-1};\ x_1,x_2>0.$

27:3. CONSTRAINED GEOMETRIC PROGRAMMING PROBLEM

The constrained geometric programming problem, in general, is stated as follows:

$$\text{Minimize } f(\mathbf{x})=\sum_{j=1}^{n_0}c_{0j}\prod_{i=1}^{m}x_i^{a_{0ij}} \text{ subject to the constraints:} \qquad \ldots (7)$$

$$g_k(\mathbf{x})\equiv\sum_{j=1}^{n_k}c_{kj}\prod_{i=1}^{m}x_i^{a_{kij}}\ \{\le,\ =,\ \ge\}\ 1,$$

where, $x_i\ (i=1,2,\ldots,m)>0$, the coefficients $c_{0j}\ (j=1,2,\ldots,n_0)$ and $c_{kj}\ (k=1,2,\ldots,N; j=1,2,\ldots,n_k)$ are strictly positive, the exponents $a_{0ij}\ (i=1,2,\ldots,m; j=1,2,\ldots,n_0)$, and $a_{kij}\ (k=1,2,\ldots,N; i=1,2,\ldots,m; j=1,2,\ldots,n_k)$ are any real numbers.

The symbol N indicates the total number of constraints; n_0 represents the number of terms in the objective function and n_k denotes the number of terms in the kth constraint as follows:

For the solution of the above-mentioned constrained minimization problem, we rewrite the constraints:

$$f_k=\sigma_k[1-g_k(\mathbf{x})]\ge 0, \qquad k=1,2,\ldots,N \qquad \ldots (8)$$

where, σ_k is called the ***signum function*** and is introduced for the kth constraint, so that it takes on the value +1 or –1 depending on whether $g_k(\mathbf{x})$ is ≤ 1 or ≥ 1, respectively.

The problem now reduces to minimizing the objective function (7), subject to the constraints given in (8). This problem is called the primal problem and can be replaced by an equivalent problem (the dual problem) with linear constraints.

The method of solving constrained geometric programming problems is best illustrated through the following examples:

Example 27.3. *Solve the GP problem:*

$$\text{Minimize } f(\mathbf{x}) = \frac{4}{9} x_1^{-1} x_2 x_3^{-\frac{1}{2}} + x_1^{-1} x_3^{-1} \text{ subject to the constraint:}$$

$$\frac{2}{9} x_2^{\frac{1}{3}} x_3 + 5 x_1 x_2^{-\frac{2}{3}} = 1; \; x_i > 0, \; i = 1, 2, 3.$$

Solution. Let $u_1 = \frac{4}{9} x_1^{-1} x_2 x_3^{-\frac{1}{2}}$, $u_2 = x_1^{-1} x_3^{-1}$, $u_3 = \frac{2}{9} x_2^{\frac{1}{3}} x_3$ and $u_4 = 5 x_1 x_2^{-\frac{2}{3}}$.

Then,
$$f(\mathbf{x}) = \sum_{j=1}^{2} u_j \quad \text{and} \quad u_3 + u_4 = 1.$$

Using the inequalities

$$\prod_{j=1}^{2} \left(\frac{u_j}{\zeta_j}\right)^{\zeta_j} \le \sum_{j=1}^{2} u_j = f(\mathbf{x}), \quad \sum_{j=1}^{2} \zeta_j = 1,$$

and
$$\left(\frac{u_3}{\alpha_1}\right)^{\alpha_1} \left(\frac{u_4}{\alpha_2}\right)^{\alpha_2} \lambda^{\lambda} \le (u_3 + u_4)^{\lambda} = 1, \; \alpha_1 + \alpha_2 = \lambda,$$

We have

$$\left(\frac{u_1}{\zeta_1}\right)^{\zeta_1} \left(\frac{u_2}{\zeta_2}\right)^{\zeta_2} \le f(\mathbf{x}), \; \zeta_1 + \zeta_2 = 1,$$

and
$$\left(\frac{u_3}{\alpha_1}\right)^{\alpha_1} \left(\frac{u_4}{\alpha_2}\right)^{\alpha_2} (\alpha_1 + \alpha_2)^{(\alpha_1 + \alpha_2)} \le 1.$$

Multiplying the last inequality by $f(\mathbf{x})$ on both sides, we have

$$f(\mathbf{x}) \left(\frac{u_3}{\alpha_1}\right)^{\alpha_1} \left(\frac{u_4}{\alpha_2}\right)^{\alpha_2} (\alpha_1 + \alpha_2)^{(\alpha_1 + \alpha_2)} \le f(\mathbf{x})$$

or
$$\left(\frac{u_1}{\zeta_1}\right)^{\zeta_1} \left(\frac{u_2}{\zeta_2}\right)^{\zeta_2} \left(\frac{u_3}{\alpha_1}\right)^{\alpha_1} \left(\frac{u_4}{\alpha_2}\right)^{\alpha_2} (\alpha_1 + \alpha_2)^{(\alpha_1 + \alpha_2)} \le f(\mathbf{x}).$$

Substituting the values of u_j $(j = 1, 2, 3, 4)$, the inequality reduces to

$$\left(\frac{4}{9\zeta_1}\right)^{\zeta_1} \left(\frac{1}{\zeta_2}\right)^{\zeta_2} \left(\frac{2/9}{\alpha_1}\right)^{\alpha_1} \left(\frac{5}{\alpha_2}\right)^{\alpha_2} x_1^{d_1} x_2^{d_2} x_3^{d_3} \lambda^{\lambda} \le f(\mathbf{x}),$$

where, d_1, d_2, d_3 are functions of ζ_j $(j = 1, 2)$ and α_i $(i = 1, 2)$, are to be chosen in such a way that the left-hand side of the inequality becomes independent of x_i $(i = 1, 2, 3)$. Therefore, we have

$$d_1 \equiv -\zeta_1 - \zeta_2 + \alpha_2 = 0, \; d_2 \equiv \zeta_1 + \frac{1}{3}\alpha_1 - \frac{2}{3}\alpha_2 = 0, \; d_3 \equiv -\frac{1}{2}\zeta_1 - \zeta_2 + \alpha_1 = 0.$$

These three equations (orthogonality conditions) together with the normality condition $(\zeta_1 + \zeta_2 = 1)$, yield

$$\zeta_1 = 2, \; \zeta_2 = 3, \; \alpha_1 = 4 \text{ and } \alpha_2 = 5.$$

Substituting these values ζ_j $(j = 1, 2)$ and α_i $(i = 1, 2)$, the inequality further reduces to

$$\left(\frac{4}{18}\right)^2\left(\frac{1}{3}\right)^3\left(\frac{2}{9\times 4}\right)^4\left(\frac{5}{5}\right)^5 (4+5)^9 \le f(\mathbf{x})$$

or

$$\frac{27}{4} \le f(\mathbf{x}).$$

In this inequality, since equality is possible, we get

$$\text{Minimum } f(\mathbf{x}) = f(\mathbf{x}_0) = \frac{27}{4}.$$

The optimal values of the variables x_i $(i = 1, 2, 3)$ can easily be determined by using $u_j = \zeta_j f(\mathbf{x}_0); j = 1, 2$ and $u_3 = \frac{1}{\lambda}\alpha_1,\ u_4 = \frac{1}{\lambda}\alpha_2$.

Example 27.4. *Solve the problem:*

$$\text{Minimize } f(\mathbf{x}) = 40x_1^{-1} x_2^{-\frac{1}{2}} x_3^{-1} + 20x_1 x_3 + 20x_1 x_2 x_3 \text{ subject to the constraints:}$$

$$x_1^{-2} x_2^{-2} + 4x_2^{\frac{1}{2}} x_3^{-1} \le 3;\ x_1, x_2, x_3 > 0.$$

Solution. Let

$$u_1 = 40x_1^{-1} x_2^{-\frac{1}{2}} x_3^{-1},\ u_2 = 20x_1 x_3,\ u_3 = 20x_1 x_2 x_3,\ u_4 = \frac{1}{3}x_1^{-2} x_2^{-2} \text{ and } u_5 = \frac{4}{3}x_2^{\frac{1}{2}} x_3^{-1}.$$

$$f(\mathbf{x}) = u_1 + u_2 + u_3, \text{ and } g(\mathbf{x}) = u_4 + u_5$$

Using arithmetic-geometric inequality, we get

$$\left(\frac{u_1}{\zeta_1}\right)^{\zeta_1}\left(\frac{u_2}{\zeta_2}\right)^{\zeta_2}\left(\frac{u_3}{\zeta_3}\right)^{\zeta_3} \le u_1 + u_2 + u_3 = f(\mathbf{x})$$

and

$$\left(\frac{u_4}{\alpha_1}\right)^{\alpha_1}\left(\frac{u_5}{\alpha_2}\right)^{\alpha_2} \lambda^\lambda \le (u_4 + u_5)^\lambda = [g(\mathbf{x})]^\lambda,$$

where, $\zeta_1 + \zeta_2 + \zeta_3 = 1$ and $\alpha_1 + \alpha_2 = \lambda$.

Now, since the constraint is $g(\mathbf{x}) \le 1$, we write $1 \ge [g(\mathbf{x})]^\lambda$.

Multiplying it on both sides by $f(\mathbf{x})$, we get

$$f(\mathbf{x}) \ge f(\mathbf{x})[g(\mathbf{x})]^\lambda$$

or

$$f(\mathbf{x}) \ge \left(\frac{u_1}{\zeta_1}\right)^{\zeta_1}\left(\frac{u_2}{\zeta_2}\right)^{\zeta_2}\left(\frac{u_3}{\zeta_3}\right)^{\zeta_3}\left(\frac{u_4}{\alpha_1}\right)^{\alpha_1}\left(\frac{u_5}{\alpha_2}\right)^{\alpha_2} \cdot \lambda^\lambda$$

Substituting the values of u_j $(j = 1, 2, 3, 4)$, the inequality reduces to

$$\left(\frac{40}{\zeta_1}\right)^{\zeta_1}\left(\frac{20}{\zeta_2}\right)^{\zeta_2}\left(\frac{20}{\zeta_3}\right)^{\zeta_3}\left(\frac{1}{3\alpha_1}\right)^{\alpha_1}\left(\frac{4}{3\alpha_2}\right)^{\alpha_2} x_1^{d_1} x_2^{d_2} x_3^{d_3} \lambda^\lambda \le f(\mathbf{x}),$$

where, d_1, d_2, d_3 are functions of ζ_j $(j = 1, 2, 3)$ and α_i $(i = 1, 2)$ are to be chosen such that

$$d_1 \equiv -\zeta_1 - \zeta_2 + \zeta_3 - 2\alpha_1 = 0,\ d_2 \equiv -\frac{1}{2}\zeta_1 + \zeta_3 - 2\alpha_1 + \frac{1}{2}\alpha_2 = 0,\ d_3 \equiv -\zeta_1 + \zeta_2 + \zeta_3 - \alpha_1 = 0.$$

These three equations (orthogonality conditions) together with the normality condition $\zeta_1 + \zeta_2 + \zeta_3 = 1$, yield

$$\zeta_1 = 1 - (\alpha_2 + 1)/2,\ \zeta_2 = (1 + \alpha_2)/2,\ \zeta_3 = (1 + \alpha_2)/4 \text{ and } \alpha_1 = \alpha_2/2.$$

Substituting these values of ζ_j $(j = 1, 2, 3)$ and $\alpha_j = (1, 2)$, the above inequality further reduces to

$$\left(\frac{80}{1-\alpha_2}\right)^{\frac{1-\alpha_2}{2}} \left(\frac{40}{1+\alpha_2}\right)^{\frac{1+\alpha_2}{2}} \left(\frac{80}{1+\alpha_2}\right)^{\frac{1+\alpha_2}{4}} \left(\frac{2}{3\alpha_2}\right)^{\frac{\alpha_2}{2}} \left(\frac{4}{3\alpha_2}\right)^{\alpha_2} \lambda^{\lambda} \le f(\mathbf{x})$$

After simplification, we have

$$\phi(\alpha) \le f(\mathbf{x}).$$

Since equality is possible, we have minimum $f(\mathbf{x})$ = maximum $\phi(\alpha)$.

As $\phi(\alpha)$ becomes a function of a single variable α_2, the search for the maximum of $\phi(\alpha)$ will be a single variable search.

To determine the optimal values of the primal variables, we utilize the fact that $f(\mathbf{x})$ attains its minimum when $\phi(\alpha) = f(\mathbf{x})$. Also, it implies that $g(\mathbf{x}) = 1$. Thus, for the optimal solution we must have

$$u_1 + u_2 + u_3 = f(\mathbf{x}) = \phi(\alpha)$$

and
$$u_4 + u_5 = g(\mathbf{x}) = 1.$$

To determine the values of x_j, we take logarithms of u_i $(i = 1, 2, 3, 4, 5)$ and put $\log x_j = z_j$ $(j = 1, 2, 3)$, thereby getting five linear equations in z_j. Out of these only three are linearly independent which give us a unique solution.

PROBLEMS

Using geometric programming, solve the following problems:

27.07. Minimize $f(\mathbf{x}) = 5x_1^{-1} x_2^{-1} x_3^{-1} + 5x_2 x_3$ subject to the constraints:
$$2x_1x_3 + x_1x_2 = 4, \quad x_1, x_2, x_3 > 0.$$

27.08. Minimize $f(\mathbf{x}) = 2x_1 x_2^{-2} + 4x_1^{-1} x_2^{-2} + 10.7x_1 x_2$ subject to the constraints:
$$10x_1^{-1} x_2^{2} = 1; \; x_1, x_2 > 0.$$

27.09. Minimize $f(\mathbf{x}) = x_1^{-1} x_2^{-2} x_3^{-2}$ subject to the constraints:
$$x_1^2 + x_2^2 + x_3^2 \le 1, \;\; x_i > 0, \; i = 1, 2, 3.$$ ***[ME (Structural), Madras Univ. 2000]***

27.10. Minimize $f(\mathbf{x}) = 20x_1 x_3 + 40x_2 x_3 + 80x_1 x_2$ subject to the constraints:
$$8x_1^{-1} x_2^{-1} x_3^{-1} \le 1; \;\; x_1, \, x_2, \, x_3 > 0.$$

27.11. In certain reservoir pump installation, the first cost of the pipe is given by $(100D + 50D^2)$, where D is the diameter of the pipe in centimetres. The cost of the reservoir decreases with an increase in the quality of fluid handled and is given by $20/Q$, where Q is the rate at which the fluid is handled (cubic metres per second). The pumping cost is given by $(300Q^2/D^5)$. Find the optimal size of the pipe and the amount of fluid handled for minimum overall cost using Geometric Programming.

27:4. COMPLEMENTARY GEOMETRIC PROGRAMMING

The method of geometric programming has been extended by ***Avriel and Williams*** that include the terms of a rational function of posynomials and called the method as ***complementary geometric programming***. The case in which some of the terms may be negative will then become a special case of complementary geometric programming. While geometric programming problems have the remarkable property that every constrained local minimum is also a global minimum, no such claim can generally be made for complementary geometric programming. However, in many practical situations, it is sufficient to find a local minimum.

The algorithm that solves any complementary geometric programming problem (*CGPP*) consists of successive approximating rational functions of posynomial terms by posynomials. Thus, solving a complementory geometric problem by this algorithm, involve the solution of a sequence of ordinary geometric programming problems. The algorithm produces a sequence whose limit is a local minimum of the *CGPP* (except in a pathological case).

The complementary geometric programming problem, in general, is defined as follows:

Minimize $z = R_0(\mathbf{x})$ subject to the constraints:

$$R_k(\mathbf{x}) \le 1, \quad \mathbf{x} \ge \mathbf{0}, \qquad k = 1, 2, \ldots, n$$

where,
$$R_k(\mathbf{x}) = \frac{A_k(\mathbf{x}) - B_k(\mathbf{x})}{C_k(\mathbf{x}) - D_k(\mathbf{x})}, \qquad k = 1, 2, \ldots, n$$

where the terms $A_k(\mathbf{x})$, $B_k(\mathbf{x})$, $C_k(\mathbf{x})$ and $D_k(\mathbf{x})$ are posynomials in $\mathbf{x}$ and possibly some of them may be absent. However, we assume that $R_0(\mathbf{x}) > 0$, for all feasible $\mathbf{x}$. This assumption can always be satisfied by adding, if necessary, a sufficiently large constant of $R_0(\mathbf{x})$.

To solve the above-mentioned problem, we first introduce a new variable $x_0 > 0$, constrained to satisfy the relation $x_0 \ge R_0(\mathbf{x})$, *i.e.*, $\dfrac{R_0(\mathbf{x})}{x_0} \le 1$. The problem is, therefore, restated as follows:

Minimize x_0, subject to the constraints:

$$\frac{A_k(\mathbf{x}) - B_k(\mathbf{x})}{C_k(\mathbf{x}) - D_k(\mathbf{x})} \le 1, \qquad k = 1, 2, \ldots, n,$$

where, $A_0(\mathbf{x}) = R_0(\mathbf{x})$, $B_0(\mathbf{x}) = 0$, $C_0(\mathbf{x}) = x_0$ and $D_0(\mathbf{x}) = 0$.

It is to be noted that the constraints have meaning only, if $C_k(\mathbf{x}) - D_k(\mathbf{x})$ has a constant sign throughout the feasible region. Therefore, if $C_k(\mathbf{x}) - D_k(\mathbf{x})$ is positive for some feasible $\mathbf{x}$, it must be positive for all feasible $\mathbf{x}$. According as $C_k(\mathbf{x}) - D_k(\mathbf{x})$ is positive or negative, we rewrite the constraints of the problem as

$$\frac{A_k(\mathbf{x}) + D_k(\mathbf{x})}{B_k(\mathbf{x}) + C_k(\mathbf{x})} \le 1 \quad \text{or} \quad \frac{B_k(\mathbf{x}) + C_k(\mathbf{x})}{A_k(\mathbf{x}) + D_k(\mathbf{x})} \le 1.$$

Thus, any complementary geometric programming problem can be expressed in its standard form as follows:

Minimize $\mathbf{x}_0$, subject to the constraints:

$$\frac{P_k(\mathbf{x})}{Q_k(\mathbf{x})} \le 1, \qquad k = 1, 2, \ldots, n, \qquad \ldots (9)$$

and
$$\mathbf{x} = (x_0, x_1, \ldots, x_m) > \mathbf{0}.$$

where, $P_k(\mathbf{x})$ and $Q_k(\mathbf{x})$ are posynomials of the form

$$P_k(\mathbf{x}) = \sum_j p_{jk}(\mathbf{x}) = \sum_{j=1}^{N} c_{jk} \prod_{i=0}^{m} x_i^{a_{ijk}} \qquad \ldots (10)$$

and
$$Q_k(\mathbf{x}) = \sum_j q_{jk}(\mathbf{x}) = \sum_{j=1}^{N} d_{jk} \prod_{i=0}^{m} x_i^{b_{ijk}} \qquad \ldots (11)$$

27:5. COMPLEMENTARY GEOMETRIC PROGRAMMING ALGORITHM

The algorithm for solving complementary geometric programming problem is based on the assumption that with constraints of the type $[P_k(\mathbf{x})/Q_k(\mathbf{x})] \le 1$ $(k = 0, 2, \ldots, n)$. If we could approximate each of the $Q_k(\mathbf{x})$ by a ***posynomial term,*** we would then have constraints that require a posynomial to be less than or equal to 1. This follows because a posynomial divided by a posynomial term is again a posynomial. Thus, with this approximation, the problem reduces to an ordinary geometric programming problem.

The iterative procedure for solving any *CGP* problem is described in the following two steps:

Step 1. Approximate each of the posynomial $Q(\mathbf{x})$, omitting the subscript k, by a single term posynomial. Choose any $\overline{\mathbf{x}} > \mathbf{0}$ and let

$$u_j = q_j(\mathbf{x}),\ \zeta_j = \frac{q_j(\overline{\mathbf{x}})}{Q(\overline{\mathbf{x}})},$$

where, q_j denotes the jth term of the posynomial $Q(\mathbf{x})$.

Therefore, by using the arithmetic-geometric inequality, we write

$$Q(\mathbf{x}) \equiv \sum_{j=1}^{N} q_j(\mathbf{x}) \geq \prod_{j=1}^{N} \left[\frac{q_j(\mathbf{x})}{q_j(\overline{x})} Q(\overline{\mathbf{x}})\right]^{\zeta_j} \quad \text{... (12)}$$

Using (11) in (12), we get

$$Q(\mathbf{x}) \geq \overline{Q}(\mathbf{x}, \overline{\mathbf{x}}) \equiv Q(\overline{\mathbf{x}}) \prod_{i=0}^{m} (x_i / \overline{x}_i)^{\lambda},\ \text{ where } \lambda = \sum_{j=1}^{n} \{b_{ij}\, q_j(\overline{\mathbf{x}})/Q(\overline{\mathbf{x}})\},$$

where, the equality sign holds true if $x_i = \overline{x}_i$. The function $Q(\mathbf{x}, \overline{\mathbf{x}})$ is the approximation for $Q(\mathbf{x})$ at $\overline{\mathbf{x}}$.

Step 2. At any feasible point $\mathbf{x}^{(1)}$, replace $Q_k(\mathbf{x})$ in (9) by their approximation $\overline{Q}_k(\mathbf{x}, \mathbf{x}^{(1)})$, and solve the resulting ordinary geometric programming problem to obtain the next point $\mathbf{x}^{(2)}$. Continuing in this way, we generate a sequence $\{\mathbf{x}^{(\alpha)}\}$, where $\mathbf{x}^{(\alpha+1)}$ is an optimal solution for the ordinary geometric programming problem:

Minimize x_0, subject to the constraints:

$$\frac{P_k(\mathbf{x})}{\overline{Q}_k(\mathbf{x}, \mathbf{x}^{(\alpha)})} \leq 1, \qquad k \geq 0, 1, 2, \ldots, n \quad \text{... (13)}$$

and $\mathbf{x} = (x_0, x_1, \ldots, x_m) > \mathbf{0}$.

Avriel and Williams have shown that under certain mild restrictions, the sequence of points $\{X^{(\alpha)}\}$ converges to a local minimum of the complementary geometric programming problem.

The computational procedure of the complimentary geometric programming is illustrated through the following example:

Example 27.5. *Use the method of complementary geometric programming in solving the following problem:*

Minimize x_1, subject to the constraints:

$$-4x_1^2 + 4x_2 \leq 1,\ x_1 + x_2 \geq 1;\ x_1, x_2 > 0.$$

Solution. In terms of the complementary geometric programming, the problem is restated as follows :

Minimize x_1, subject to the constraints:

$$\left[4x_2/(1+4x_1^2)\right] \leq 1$$

$$\left[x_1^{-1}/(1+x_1^{-1}x_2)\right] \leq 1$$

and $x_1, x_2 > 0.$

Let $Q_1(\mathbf{x})$ and $Q_2(\mathbf{x})$ denote the denominators of the constraints, such that

$$Q_1(\mathbf{x}) = 1 + 4x_1^2 \quad \text{and} \quad Q_2(\mathbf{x}) = 1 + x_1^{-1} x_2.$$

Both these expressions can be approximated by a single term posynomial as

$$\bar{Q}_1(\mathbf{x}, \bar{\mathbf{x}}) = (1 + 4\bar{x}_1^2)(x_1/\bar{x}_1)^{x_1^2/(1+4x_1)}$$

and
$$\bar{Q}_2(\mathbf{x}, \bar{\mathbf{x}}) = (1 + \bar{x}_2/\bar{x}_1)(x_1/\bar{x}_1)^{-\alpha}(x_2/\bar{x}_2)^{\alpha} \qquad \alpha = \bar{x}_2/(\bar{x}_1, \bar{x}_2).$$

We now start the interative procedure from the point $\mathbf{x}^1 = (1, 1)$. CLearly, this point is feasible. By taking $\mathbf{x} = \mathbf{x}^1$, we have

$$\bar{Q}_1\left(\mathbf{x}, \bar{\mathbf{x}}^{(1)}\right) = 5x_1^{8/5} \quad \text{and} \quad \bar{Q}_2\left(\mathbf{x}_1, \bar{\mathbf{x}}^{(1)}\right) = x_1^{-1/2} x_2^{1/2}.$$

Using (13), we formulate the first ordinary geometric programming problem as:

Minimize x_1, subject to the constraints:

$$\frac{4}{5}x_1^{-8/5} x_2 \le 1, \quad \frac{1}{2}x_1^{-1/2} x_2^{-7/2} \le 1; \quad x_1, x_2 > 0.$$

The solution of this geometric programming problem can be obtained by solving a square system of linear equations namely,

$$\zeta_1 = 1, \quad \zeta_1 - \frac{8}{5}\zeta_2 - \frac{1}{2}\zeta_3 = 0, \quad \zeta_2 - \frac{7}{2}\zeta_3 = 0.$$

Solving these linear equations, we obtain

$$\zeta_1 = 1, \quad \zeta_2 = \frac{5}{13} \quad \text{and} \quad \zeta_3 = \frac{10}{13}.$$

Substituting these values of ζ_j $(j = 1, 2, 3)$ into dual objective function, we get

$$v(\zeta) = \left(\frac{4}{5}\right)^{5/13} \left(\frac{1}{2}\right)^{10/13} \cong 0.5385.$$

Making use of duality relations, therefore, we shall have

$$x_1^0 \cong 0.5385 \quad \text{and} \quad x_2^0 = \frac{5}{4}x_1^{8/15} \cong 9.4643.$$

Next, we choose $\mathbf{x}^{(2)}$ to be optimal solution of (ordinary geometric programming)$_2$, *i.e.*, $\mathbf{x}^{(2)} = (0.5385, 0.4643)$ and approximate Q_1 and Q_2 about this point, solve (ordinary geometric programming)$_2$, etc. The sequence of optimal solution of (ordinary geometric programming)$_\alpha$ as generated by the iterative procedure is shown below:

Iteration	x_1	x_2
0	1.0000	1.0000
1	0.5385	0.4643
2	0.5019	0.5007
3	0.5000	0.5000

The optimal values of the variables for the *CGP* are $x_1^* = 0.5$ and $x_2^* = 0.5$. It can be seen that in three iterations, the solution of the approximating geometric programming problems (OGP_α) is correct to four significant figures.

TEST YOUR UNDERSTANDING

1. Geometric programming
   (*a*) has been developed in connection with the problems of engineering design.
   (*b*) gives an overall picture of the relative importance of the various design parameters that emerge.

(*c*) put emphasis upon the relative magnitude of the term of objective function rather than the variables.
(*d*) all of the above.

2. The approach of geometric programming
(*a*) requires that all engineering characteristics be expressed quantitatively as generalized positive polynomials with adjustable parameters.
(*b*) often reduces a complicated optimization problem to one involving a set of simultaneous linear equations.
(*c*) both (*a*) and (*b*). (*d*) neither (*a*) nor (*b*).

3. Geometric programming technique
(*a*) is not suitable to digital computers.
(*b*) is used to minimize functions which are in the form of posynomials subject to constraints of the same type.
(*c*) is applicable to all non-linear programming problems.
(*d*) solves a special case of linear programming problem.

4. The geometric programming problem
(*a*) can be solved by simplex-like method. (*b*) without constraint cannot be solved.
(*c*) imposes engineering discipline on its user. (*d*) all of the above.

5. In the context of geometric programming, the objective function is usually referred to as a
(*a*) polynomial (*b*) posynomial
(*c*) linear (*d*) non-linear.

6. In the context of geometric programming
(*a*) if the objective function is of minimization, the decision variables are strictly positive.
(*b*) constraints are expressed in the form of geometric-arithmetic mean inequalities.
(*c*) the minimization problem is generally referred to as primal problem.
(*d*) all of the above.

ANSWERS

1. (*d*) **2.** (*c*) **3.** (*b*) **4.** (*c*) **5.** (*b*) **6.** (*d*)

REVIEW EXERCISE

1. Explain the terms 'Posynomial', 'Monomial', and 'Signomial' functions in Geometric Programming.
2. What is Geometric Programming?
3. Explain geometric programming and point out how it differs from other optimization techniques.
4. Define the degree of difficulty in Geometric Programming. ***[MCA, Madras Univ. 1999]***
5. What do you mean by a Complementary Geometric Programming problem?

CHAPTER **28**

Information Theory

"The most successful man in life is one who has the most reliable information"

We often encounter "information transmission" in our daily life. For example, information transmission takes place when we read newspapers, books, exchange letters, talk on telephone, work on internet, use of social media, attend and/or deliver a lecture, listen to radio, watch television, and so on. In all such cases, we notice that 'some' information is carried from one place/person to another.

However, we "seek" information only when we are in doubt. We go to enquiry (information) office to remove our doubts — we consult weather forecasts for information whether it will rain or not. On the other hand, if an event can occur in just one way, there is no uncertainty (or doubt) about it and no information is called for either. We get some information by the occurrence of an event only when there was some uncertainty before its occurrence. Naturally, the amount of ***information received by its occurrence*** must be equal to the amount of ***uncertainty prevailing before its occurrence***. Thus, uncertainty and information are two sides of the same coin.

An important aspect of information is the communication from source to receiver.

28:1. COMMUNICATION SYSTEM

The essential parts of a communication system are:

1. **Source (or Transmitter).** It is the source of *message* (either person or machine) which produces the information to be communicated or transmitted.
2. **Communication Channel.** It is the transmission network (or media) which carries the message from the source to receiver, *e.g.*, human voice, newspapers, books, etc. A communication channel can be with or without noise.
3. **Receiver.** It is the destination to which the message is conveyed from ***source*** (***or transmitter***) through a communication channel.

   Besides these, other parts of communication system are:

   (*a*) **Encoder.** It is an equipment which is used to improve the *efficiency* of the transmission channel through which a message is transmitted to the receiver.

   (*b*) **Noise.** It is the general term which creates ***interruptions or disturbances*** in the transmission of message (or information) from transmitter to receiver. For example, noise or disturbance in radio or television during the relay of a programme; error in newspaper printing, etc.

   (*c*) **Decoder.** It is used to transform encoded ***message*** into the original form at the receivers end.

The study of information theory is based upon the concept: ***"It is possible to transmit information through a noisy channel at any rate less than the channel capacity with an arbitrarily small probability of error."***

28:2. A MEASURE OF INFORMATION

Consider an event E with probability of occurrence p. Suppose, someone reliable comes and gives the ***message*** that E has occurred. The question is: What is the amount of information conveyed by this message? We try to answer this question intuitively. Observe that if p is close to 1 (say $p = 0.95$), then one may argue that the message has conveyed a very little information because it was virtually certain that E would occur. On the other hand, if p is close to 0 (say $p = 0.001$), then it is almost certain that E will not occur and consequently the message stating its occurrence is quite unexpected and hence contains a greater deal of information.

The above intuitive idea suggests that we should select a decreasing function of p, as a measure of the amount of information. The function proposed by Shannon is

$$h(p) = -\log p \qquad 0 < p \le 1$$

This function decreases from ∞ (infinite surprise and hence infinite information when probability prior to message is zero) to 0 (zero information when the event was sure to occur, *i.e.*, $p = 1$). The information function $h(p)$ measured in **bits**, is shown in *Fig. 28.1*.

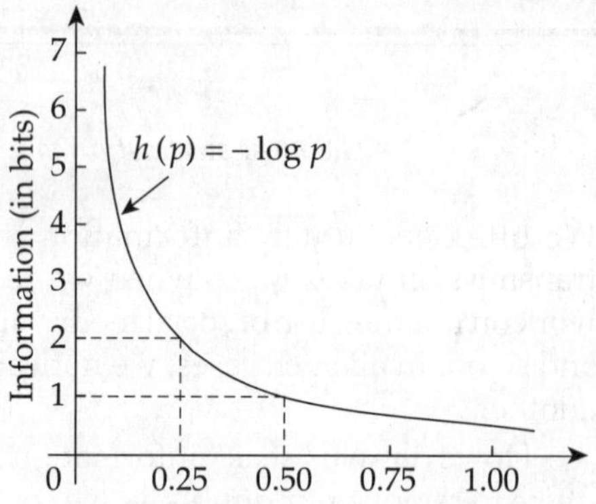

Fig. 28.1. Information $h(p)$ in Bits

Illustration. The probability that a seven is obtained, when a pair of dice is rolled, is $p = 6/36$, *i.e.*, 1/6. Thus, the amount of information conveyed by the message which states that the result is seven is given by

$$h(p) = -\log p = \log 6 \simeq 2.58 \text{ bits}.$$

Note. The unit of information is determined by the base of logarithm. Normally 2 is used as a base, which implies that $h\left(\frac{1}{2}\right) = \log_2 2 = 1$, and the information then is said to be measured in ***binary digits***.

Axiomatic Approach to Information

We shall demonstrate that the logarithmic form of $h(p)$ is the only possible definition when certain simple axioms are adopted, the most important one being that of additivity. We assume the following axioms concerning the amount of information conveyed by a message stating that an event E with probability p has occurred:

Axiom 1. The information depends only on p and hence can be written as $h(p)$, when h is some function of $p \in (0, 1)$.

Axiom 2. $h(p)$ is a monotonically decreasing continuous function of p in $(0, 1)$, *i.e.*,

$$h(p_1) > h(p_2), \quad \text{if } 0 < p_1 < p_2 \le 1. \qquad \text{... (1)}$$

Axiom 3. No information is conveyed by an event which is sure to occur, *i.e.*,

$$h(1) = 0. \qquad \text{... (2)}$$

Axiom 4. For two independent events E_1 and E_2, with probabilities of occurrences p_1 and p_2; the information content of the message which states that both have occurred is equal to the information of the message dealing with only E_1 plus that which deals with only E_2. That is

$$h(p_1, p_2) = h(p_1) + h(p_2) \qquad 0 < p_1, p_2 \le 1. \qquad \text{... (3)}$$

Remark. The function $h(p)$ satisfying the axioms 1 to 4 given above can only be of the form

$$h(p) = -c \log p, \; 0 < p \le 1,$$

where, $c > 0$ is an arbitrary constant and will be considered as equal to 1 (for the sake of simplicity) with logarithm to the base 2.

28:3. ENTROPY — THE EXPECTED INFORMATION

In the previous section, we considered the message which states that an event E has occurred. However, the information conveyed by this message is ***different*** from the one conveyed by its complementary message (stating that E has failed to occur). Actually, if p is the probability of occurrence of E, the information conveyed by the latter message is

$$h(1-p) = -\log(1-p).$$

The reader may observe that $h(p) \neq h(1-p)$, unless $p = 1/2$.

Thus, so far as the event E is concerned, the information to be received is either $h(p)$ or $h(1-p)$ and we do not know which one, so long as the message conveying occurrence or non-occurrence of E is not received. However, one can always compute ***expected information content*** of the message(s) prior to its arrival. The expected information content concerning the event E is given by

$$H(p, 1-p) = p\,h(p) + (1-p)\,h(1-p)$$
$$= -p\log p - (1-p)\log(1-p), \quad 0 < p < 1.$$

The function $H(p, 1-p)$ is known as ***entropy*** of the probability distribution $\{p, 1-p\}$. It may be observed that $H(p, 1-p)$ is symmetric in p and $1-p$; it is non-negative; and takes the value zero at $p = 0$ and $p = 1$ (if we define $x \log x \approx 0$ whenever $x = 0$). The function $H(p, 1-p)$ is shown in *Fig. 28.2*.

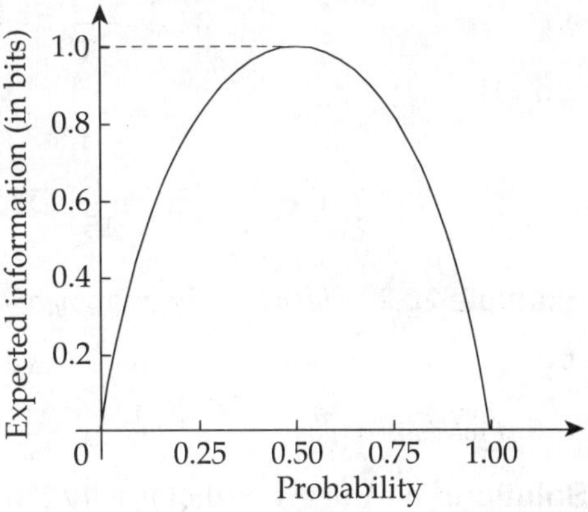

Fig. 28.2. The Entropy Function $H(p, 1-p)$

It may be seen from *Fig. 28.2* that $H(p, 1-p)$ is maximum at $p = 1/2$ and the value of this maximum is

$$H\left(\frac{1}{2}, \frac{1}{2}\right) = -\frac{1}{2}\log_2\frac{1}{2} - \frac{1}{2}\log_2\frac{1}{2} = \log_2 2 = 1 \text{ bit}.$$

We, now, consider more than two alternatives and extend the concept of entropy to several events.

Let there be n events $E_1, E_2, ..., E_n$ with probabilities $p_1, p_2, ..., p_n, p_i \geq 0, \sum p_i = 1$. The ***entropy*** of the probability distribution $\{p_1, p_2, ..., p_n\}$ is defined to be the mathematical expectation of information $h(p_i)$, concerning the occurrence of one of the events E_i. In other words, the entropy function is given by

$$H(p_1, p_2, \ldots, p_n) = p_1 h(p_1) + p_2 h(p_2) + \ldots + p_n h(p_n)$$
$$= -p_1 \log p_1 - p_2 \log p_2 - \ldots - p_n \log p_n\,;$$

where, $p_i \geq 0 \ (i = 1, 2, \ldots, n)$ and $p_1 + p_2 + \ldots + p_n = 1$.

This general form of entropy includes $H(p, 1-p)$ as a special case for $n = 2$; $p_1 = p$ and $p_2 = (1-p)$.

Remarks

1. The entropy function $H(p_1, p_2, ..., p_n)$ is non-negative, because $h(p_i) \geq 0$ and $p_i \geq 0$, where $p_i = 1$ for some i and $p_i = 0$ for $i \neq 1$.
2. The entropy function is symmetric when p_i $(i = 1, 2, ..., n)$ are interchanged with one another, *i.e.*,

$$H(p_i, 1-p_i) = H(1-p_i, p_i); \quad i = 1, 2, \ldots, n,$$

where $0 \leq p_i \leq 1$.

Example 28.1. *Evaluate the entropy associated with the events A, B and C, which are mutually exclusive with the probability distribution:*

Event :	*A*	*B*	*C*
Probability :	*1/5*	*4/15*	*8/15*

Solution. We are given: $p_1 = 1/5, p_2 = 4/15$ and $p_3 = 8/15$.

$$H(1/5, 4/15, 8/15) = -p_1 \log p_1 - p_2 \log p_2 - p_3 \log p_3$$

$$= -\frac{1}{5}\log\left(\frac{1}{5}\right) - \frac{4}{15}\log\left(\frac{4}{15}\right) - \frac{8}{15}\log\left(\frac{8}{15}\right)$$

$$= -\frac{1}{15}\left[3\log\frac{1}{5} + 4\log\left(\frac{4}{15}\right) + 8\log\left(\frac{8}{15}\right)\right]$$

$$= -\frac{1}{15}\left[-3\log 5 + 4(\log 4 - \log 15) + 8(\log 8 - \log 15)\right]$$

$$= -\frac{1}{15}\left[-3\log 5 - 4\log(3\times 5) - 8\log(3\times 5) + 4\log(2)^2 + 8\log(2)^3\right]$$

$$= \frac{1}{15}[15\log 5 + 12\log 3 - 32] = \log 5 + \frac{4}{5}\log 3 - \frac{32}{15}; \quad \log 2 = 1$$

Example 28.2. *Show that the entropy of the following probability distribution is* $2 - \frac{1}{2^{n-2}}$.

Events :	x_1	x_2	...	x_i	...	x_{n-1}	x_n
Probabilities (p_i) :	$\frac{1}{2}$	$\frac{1}{4}$	...	$\frac{1}{2^i}$	...	$\frac{1}{2^{n-1}}$	$\frac{1}{2^n}$

Solution. We are given that $p_i = 1/2^i$ for $i = 1, 2, ..., n-1$ and $p_n = 1/2^{n-1}$. Clearly, $\sum p_i = 1$. The entropy function of the probability distribution is given by

$$H = -\sum_{i=1}^{n} p_i \log p_i = -\sum_{i=1}^{n-1} p_i \log p_i - p_n \log p_n$$

$$= -\sum_{i=1}^{n-1} (1/2^i)\log(1/2^i) - (1/2^{n-1})\log(1/2^{n-1})$$

$$= \sum_{i=1}^{n-1}\left(\frac{1}{2^i}\right)\log 2^i + \left(\frac{1}{2^{n-1}}\right)\log 2^{n-1} \qquad \text{(taking log to base 2)}$$

$$= \sum_{i=1}^{n-1}(i/2^i) + (n-1)/2^{n-1}, \text{ since } \log_2 2 = 1$$

$$= \left\{\frac{1}{2} + \frac{2}{2^2} + \frac{3}{2^3} + \frac{4}{2^4} + \ldots + \frac{n-2}{2^{n-2}} + \frac{n-1}{2^{n-1}}\right\} + \frac{n-1}{2^{n-1}}$$

This gives

$$\frac{1}{2}H = \left\{\frac{1}{2^2} + \frac{2}{2^3} + \frac{3}{2^4} + \frac{4}{2^5} + \ldots + \frac{n-2}{2^{n-1}} + \frac{n-1}{2^n}\right\} + \frac{n-1}{2^n}$$

Thus, we get

$$H - \frac{1}{2}H = \frac{1}{2} + \frac{1}{2^2} + \frac{1}{2^3} + \ldots + \frac{1}{2^{n-1}} + \left[\frac{n-1}{2^{n-1}} - \frac{2(n-1)}{2^n}\right]$$

or

$$\frac{1}{2}H = \frac{1}{2} + \frac{1}{2^2} + \frac{1}{2^3} + \ldots + \frac{1}{2^{n-1}} = 1 - \left(\frac{1}{2}\right)^{n-1}$$

This implies

$$H = 2 - \frac{1}{2^{n-2}}.$$

PROBLEMS

28.01. Evaluate the entropy associated with the following probability distribution:

Event :	*A*	*B*	*C*	*D*
Probability :	1/2	1/4	1/8	1/8

28.02. An alphabet consists of four letters *A, B, C, D* with respective probabilities of transmission 1/3, 1/4, 1/4, 1/6. Find the average amount of information associated with the transmission of a letter.

28.03. For a random experiment, let there be *N* outcomes with probabilities

$$p_k = k^{-\alpha}\left(\sum_{k=1}^{N} k^{-\alpha}\right)^{-1}; \quad k = 1, 2, \ldots, N; \quad \sum_{k=1}^{N} p_k = 1; \quad 0 \le \alpha \le 1.$$

Show that the entropy of the probability distribution is given by $H = \log_2 N$ for $\alpha = 0$, and that for large *N* it is asymptotically given by

$$H = \begin{cases} \log_2 N & \text{for } 0 \le \alpha < 1 \\ \frac{1}{2}\log_2 N & \text{for } \alpha = 1. \end{cases}$$

28.04. Show that for large *N*, if $p_k = k^{-\alpha}\left(\sum_{k=1}^{N} k^{-\alpha}\right)^{-1}$, where $0 \le \alpha < 1$; then

$$H = \log_2 N - \frac{\alpha}{1-\alpha}\log_2 e.$$

28:4. ENTROPY AS A MEASURE OF UNCERTAINTY

Consider the probability distribution (0.98, 0.01, 0.01). There is ***little doubt*** about what will happen and hence little information is to be expected, ***on the average***. The word 'average' is important. The occurrence of the event with probability 0.01 gives an amount of information as high as – log (0.01) = 6.64 bits. However, this may happen only 2 out 100 times and in the other 98 cases, the information is as low as – log (0.98) = 0.0288 bits. The average (expected) information received is

$$H = 0.01\,(6.64) + 0.01\,(6.64) + 0.98\,(0.0288) = 0.161 \text{ bits},$$

which is small in accordance with the low degree of uncertainty.

Now, consider the probability distribution $\left(\frac{1}{3}, \frac{1}{3}, \frac{1}{3}\right)$, so that there is a ***great deal of doubt*** and hence much information is to be expected. When we have ten possibilities rather than three, each with probability 0.01, there is even more uncertainty and hence more information to be expected.

The above examples show that ***uncertainty*** and ***expected information (entropy)*** are two sides of the same coin. Uncertainty prevails prior to the arrival of the message and information is received when the message arrives. The more is the uncertainty prior to the message, the larger is the amount of information conveyed by it, at least on the average.

Thus, entropy introduced here as the expected information conveyed by the message that states which of the n possible outcomes is realized, may also be regarded as a measure of uncertainty associated with a probability distribution.

Notation

Let X be a discrete random variable with point probability distribution

X	x_1	x_2	...	x_n
P	p_1	p_2	...	p_n

; $p_i \ge 0, \quad p_1 + p_2 + \ldots + p_n = 1.$

The various values of X may be thought of as the observations of a random experiment. The following notations shall be used interchangeably for the terms "entropy of P/or X", "average uncertainty of P/or X", "average information of P/or X":

$$H(p_1, p_2, \ldots, p_n), \quad H(P), \ H(X);$$

each of which represents the quantity

$$-(p_1 \log p_1 + p_2 \log p_2 + \ldots + p_n \log p_n).$$

28:5. JOINT AND CONDITIONAL ENTROPIES

Relations between Joint and Marginal Entropies

Consider two sets of messages:

$$X = \{X_1, X_2, \ldots, X_M\} \quad \text{and} \quad Y = \{Y_1, Y_2, \ldots, Y_N\}.$$

where, X_i's are the messages sent (channel input) and Y_j's are the messages received (channel output).

Let $P_{ij} = P(X = X_i, Y = Y_j)$, where $i = 1, 2, \ldots, M$ and $j = 1, 2, \ldots, N$ denote the probability of the joint event that message X_i is sent and message Y_j is received. Our objective is to study relationship between ***joint, conditional*** and ***marginal (individual)*** information associated with the bivariate probability distribution (p_{ij}).

Let us define the marginal probability distribution of X and Y* by

$$p_{i0} = \sum_{j=1}^{N} p_{ij} \quad \text{and} \quad p_{0j} = \sum_{i=1}^{M} p_{ij} \qquad \text{for all } i \text{ and } j.$$

Then, naturally the ***marginal entropies*** of the two marginal distributions are given by

$$H(X) = -\sum_{i=1}^{M} p_{i0} \log p_{i0} \quad \text{and} \quad H(Y) = -\sum_{j=1}^{N} p_{0j} \log p_{0j}.$$

The entropy $H(X)$ measures the uncertainty of the message sent (irrespective of the message received) and $H(Y)$ performs the same role for the message received.

The ***joint entropy*** is the entropy of the joint distribution of the messages sent and received, and is therefore, given by

$$H(X, Y) = -\sum_{i=1}^{M} \sum_{j=1}^{N} p_{ij} \log p_{ij}.$$

$H(X, Y)$ measures the uncertainty of the messages sent and received simultaneously.

One may observe that

$$\text{Max. } H(X, Y) = \log MN = \log M + \log N = \text{Max. } H(X) + \text{Max. } H(Y)$$

Theorem 28-1. *$H(X, Y) \le H(X) + H(Y)$ with equality if and only, if X and Y are independent.*

Proof. We have

$$H(X) + H(Y) = -\sum_{i=1}^{M} p_{i0} \log p_{i0} - \sum_{j=1}^{N} p_{0j} \log p_{0j} = -\sum_{i=1}^{M}\left(\sum_{j=1}^{N} p_{ij}\right) \log p_{i0} - \sum_{j=1}^{N}\left(\sum_{i=1}^{M} p_{ij}\right) \log p_{0j}$$

$$= -\sum_{i=1}^{M}\left(\sum_{j=1}^{N} p_{ij}\right) \log (p_{i0}\, p_{0j}) = -\sum_{i=1}^{M}\left(\sum_{j=1}^{N} p_{ij}\right) \log q_{ij} \qquad q_{ij} = p_{i0}\, p_{0j} \qquad \ldots (1)$$

---

* We can very well treat X and Y at two discrete random variables with range spaces $R_X = \{X_1, X_2, \ldots, X_M\}$, $R_Y = \{Y_1, Y_2, \ldots, Y_N\}$ and joint probability distribution p_{ij} ($i = 1, 2, \ldots, M; j = 1, 2, \ldots, N$).

Also $$H(X, Y) = -\sum_{i=1}^{M} \sum_{j=1}^{N} p_{ij} \log q_{ij} \quad \text{... (2)}$$

We observe that

$$\sum_{i=1}^{M} \sum_{j=1}^{n} q_{ij} = \sum_{i=1}^{M} \sum_{j=1}^{N} p_{i0}\, p_{0j} = \left(\sum_{i=1}^{M} p_{i0}\right)\left(\sum_{j=1}^{N} p_{0j}\right) = 1 = \sum_{i=1}^{M} \sum_{j=1}^{N} p_{ij}$$

From equations (1) and (2) it follows that $H(X, Y) \le H(X) + H(Y)$ with equality, if and only if, $q_{ij} = p_{ij}$ for all i and j. The condition for equality obviously reduces to $p_{i0}\, p_{0j} = p_{ij}$, which means that X and Y are independent.

Conditional Entropies

Let $p_{j\,|\,i} = p_{ij}\, p_{i0}$ denote the conditional probability that Y_j is the message received when it is given that X_i is the message sent. If we vary Y_j over the set of messages received, for a fixed X_i, we get the conditional probability distribution $p_{1|i}, p_{2|i}, \ldots, p_{N|i}$. We, therefore, define the ***conditional entropy*** of Y given that $X = X_i$, as

$$H(Y \mid X = X_i) = -\sum_{j=1}^{N} p_{j|i} \log p_{j|i}.$$

Further, we define the ***average conditional entropy of Y*** given X as a weighted average of $H(Y \mid X = X_i)$, is

$$H(Y \mid X) = -\sum_{i=1}^{M} p_{i0}\, H(Y \mid X = X_i) = -\sum_{i=1}^{M} \sum_{j=1}^{N} p_{ij} \log p_{j|i}. \quad (\because p_{ij} = p_{i0}\, p_{j|i})$$

Theorem 28-2. $H(X, Y) = H(X) + H(Y \mid X) = H(Y) + H(X \mid Y)$.

Proof. We have

$$H(X) + H(Y \mid X) = -\sum_{i=1}^{M} p_{i0} \log p_{i0} - \sum_{i=1}^{M} \sum_{j=1}^{N} p_{ij} \log p_{j|i}$$

$$= -\sum_{i=1}^{M} \sum_{j=1}^{N} p_{ij} \log p_{i0} - \sum_{i=1}^{M} \sum_{j=1}^{N} p_{ij} \log p_{j|i}$$

$$= -\sum_{i=1}^{M} \sum_{j=1}^{N} p_{ij} \log (p_{i0}\, p_{j|i}) = -\sum_{i=1}^{M} \sum_{j=1}^{N} p_{ij} \log p_{ij}$$

$$= H(X, Y).$$

Similarly $$H(X, Y) = H(Y) + H(X \mid Y).$$

This theorem establishes the intuitively reasonable property that if X and Y are observed, but only the observations on X revealed, the remaining uncertainty about Y is $H(Y \mid X)$.

It is also intuitively reasonable to hope that revelation by the observations of X cannot increase the uncertainty about Y. This is established in the next theorem.

Theorem 28-3. *$H(Y \mid X) \le H(Y)$ with equality, if and only if, X and Y are independent.*

Proof. Follows from the *Theorems 28-1* and *28-2*.

Example 28.3. *A transmitter has an alphabet consisting of five letters $[x_1, x_2, x_3, x_4, x_5]$ and the receiver has an alphabet of four letters $[y_1, y_2, y_3, y_4]$. The joint probabilities for the communication are given below:*

$$\begin{array}{c} \\ x_1 \\ x_2 \\ x_3 \\ x_4 \\ x_5 \end{array} \begin{array}{c} \begin{array}{cccc} y_1 & y_2 & y_3 & y_4 \end{array} \\ \begin{bmatrix} 0.25 & 0.00 & 0.00 & 0.00 \\ 0.10 & 0.30 & 0.00 & 0.00 \\ 0.00 & 0.05 & 0.10 & 0.00 \\ 0.00 & 0.00 & 0.05 & 0.10 \\ 0.00 & 0.00 & 0.05 & 0.00 \end{bmatrix} \end{array}$$

Determine the different entropies for this channel (assume that 0 log 0 ≅ 0).

Solution. The channel is described here by the joint probabilities p_{ij}, $i = 1, 2, ..., 5$ and $j = 1, 2, ..., 4$. The various conditional and marginal probabilities $p_{i\,|\,j}$, p_{i0}, p_{0j}, etc. are easily computed from p_{ij}'s. These are:

$$p_{10} = 0.25 + 0.00 = 0.25,\quad p_{20} = 0.10 + 0.30 = 0.40,\quad p_{30} = 0.05 + 0.10 = 0.15,$$

$$p_{40} = 0.05 + 0.10 = 0.15,\quad p_{50} = 0.05 + 0.00 = 0.05.$$

Similarly,

$$p_{01} = 0.35,\quad p_{02} = 0.35,\quad p_{03} = 0.20,\quad p_{05} = 0.10.$$

The conditional probabilities are given in the following channel matrix (obtained by using $p_{j\,|\,i} = p_{ij}/p_{i0}$):

$$\begin{bmatrix} 1 & 0 & 0 & 0 \\ 1/4 & 3/4 & 0 & 0 \\ 0 & 1/3 & 2/3 & 0 \\ 0 & 0 & 1/3 & 2/3 \\ 0 & 0 & 1 & 0 \end{bmatrix}$$

The various entropies associated with channel are now computed, taking all logarithms to the base 2.

Marginal Entropies:

$$H(X) = -\sum_{i=1}^{5} p_{i0} \log p_{i0}$$

$$= -(0.25) \log (0.25) - (0.40) \log (0.40) - 2(0.15) \log (0.15) - (0.05) \log (0.05)$$

$$= \frac{1}{4} \log 4 + \frac{2}{5} \log \frac{5}{2} + \frac{3}{10} \log \frac{20}{3} + \frac{1}{20} \log 20$$

$$= 1.3260 \text{ bits.}$$

$$H(Y) = -\sum_{j=1}^{4} p_{0j} \log p_{0j}$$

$$= -2(0.35) \log (0.35) - (0.20) \log (0.20) - (0.10 \log (0.10)$$

$$= \frac{7}{10} \log (2.857) + \frac{1}{5} \log 5 + \frac{1}{10} \log 10$$

$$= 1.8556 \text{ bits.}$$

Conditional Entropies:

$$H(Y|X) = \sum_{i=1}^{5}\sum_{j=1}^{4} p_{ij} \log p_{j|i}$$

$$= (0.25)\log 1 + (0.10)\log 4 + (0.30)\log\frac{4}{3} + (0.5)\log 3 + (0.10)\log\frac{3}{2}$$

$$+(0.05)\log 3 + (0.10)\log\frac{3}{2} + (0.05)\log 1$$

$$= \frac{1}{10}\log 4 + \frac{3}{10}\log 4 - \frac{3}{10}\log 3 + \frac{2}{10}\log 3 + \frac{1}{10}\log 3 - \frac{1}{5}\log 2$$

$$= \frac{1}{10}[4\log 4 - 2\log 2] = \frac{6}{10} = 0.60 \text{ bits.}$$

$\therefore$ $$H(X|Y) = H(X) + H(Y|X) - H(Y) = 1.3260 + 0.6000 - 1.8556 = 0.0704.$$

Joint Entropy:

$$H(X, Y) = H(X) + H(Y|X) = 1.9260.$$

PROBLEMS

28.05. Consider the joint probabilities channel matrix $\begin{pmatrix} 0.25 & 0.25 \\ 0.50 & 0.00 \end{pmatrix}$ for two sets of messages $X = \{X_1, X_2\}$ and $Y = \{Y_1, Y_2\}$. Use it as a counter example to show that an individual conditional entropy may exceed the unconditional entropy. Explain the result in the light of *Theorem 28-3*.

28.06. If the face of the die is 1, 2, 3 or 4, an unbiased coin is tossed once. If the face of the die is 5 or 6, the coin is tossed twice. Find the conditional entropy of the probability distribution associated with the face of the die, given the number of heads obtained.
[***Hint:*** Let X_1 = (1, 2, 3 or 4), X_2 = (5 or 6); Y_1 = No head, Y_2 = 1 head; and Y_3 = 2 heads.]

28:6. MUTUAL INFORMATION

The Expected Mutual Information

Let us consider, the set of messages $X = \{X_1, X_2, ..., X_M\}$ sent and the set of messages $Y = \{Y_1, Y_2, ..., Y_N\}$ received. Then, the quantity

$$h(X_i, Y_j) = \log\frac{p_{ij}}{p_{i0}\, p_{0j}}, \qquad i = 1, 2, ..., M;\ j = 1, 2, ..., N$$

is known as the ***mutual information*** of the message sent X_i and the message received Y_j. The following observation on $h(X_i, Y_j)$ are obvious:

(*i*) $h(X_i, Y_j) = 0$, whenever X and Y are stochastically independent.

(*ii*) $h(X_i, Y_j) >$ or < 0 according as $p_{j\,|\,i} >$ or $< p_{0j}$ for a fixed X_i.

That is, according as Y_j is more or less frequently the message received, given that X_i is the message sent, than what the independent pattern implies.

Now, averaging all MN mutual information values with p_{ij}'s as weights, we obtain the ***Expected Mutual Information*** of X and Y, as

$$I(X, Y) = \sum_{i=1}^{M}\sum_{j=1}^{N} p_{ij}\log\frac{p_{ij}}{p_{i0}\, p_{0j}}.$$

Theorem 28-4. $I(X, Y) = H(X) - H(X \mid Y) = H(Y) - H(Y \mid X)$.

Proof. We have

$$H(X) - H(X \mid Y) = -\sum_{i=1}^{M}\sum_{j=1}^{N} p_{i0} \log p_{i0} + \sum_{i=1}^{M}\sum_{j=1}^{N} p_{ij} \log p_{i|j} = \sum_{i=1}^{M}\sum_{j=1}^{N} p_{ij} \log \frac{p_{i|j}}{p_{i0}}$$

$$= I(X, Y).$$

Similarly the other equality holds.

Remarks

1. The result of *Theorem 28-4* has an interesting interpretation, $H(X) - H(X \mid Y)$ may be considered as the reduction in uncertainty about X when Y is revealed. Thus, $I(X, Y)$ may also be called the *amount of information conveyed by Y about X*. Symmetry of $I(X, Y)$ then states that

   "The information conveyed about X by Y is the same as the information conveyed about Y by X."

   Also, $I(X, Y) = 0$ when X and Y are independent, that is, no information is conveyed by a variable about another which is stochastically independent of the former.
2. We can easily prove that $I(X, Y) = H(X) + H(Y) - H(X, Y)$.

Information Processed by a Channel and Channel Capacity

1. **Information Processed by a Channel.** Let a channel be described by the input alphabet $X = \{X_1, X_2, ..., X_M\}$, output alphabet $Y = \{Y_1, Y_2, ..., Y_N\}$ and the channel matrix $(p_{j \mid i})$, $i = 1, 2, ..., M; j = 1, 2, ..., N$. Then, the information processed by the channel is defined to be the ***mutual information***, $I(X, Y)$.

   Since, $I(X, Y) = H(X) - H(X \mid Y)$, the information processed by a channel depends upon the input probability distribution p_{i0}, one may vary the input distribution until $I(X, Y)$ reaches a maximum.
2. **Channel Capacity.** The ***channel capacity*** C of a channel is a number defined by

   $$C = \text{Max.}\, I(X, Y) = \text{Max.}\left\{(X) - H(X \mid Y)\right\}, \quad \text{for all } p_i$$

   But, for a noise tree channel, we have

   $$I(X, Y) = H(X) = H(Y) \quad \text{and} \quad I(X, Y) = H(X, Y)$$

   Therefore, channel capacity C in this case may be redefined as

   $$C = \text{Max.}\, I(X, Y) = \text{Max.}\ [H(X)]$$

   $$= \text{Max.}\left[-\sum_{i=1}^{m} p_i \log p_i\right] = -\log\left(\frac{1}{n}\right), \quad \text{where } p_i = P(x_i)$$

   $$= \log n \text{ bits per symbol (or second)}$$

   This is due to the fact that maximum of $H(X)$ occurs when $p_1 = p_2 = ... = p_n$.

Special Types of Channels

1. **Binary Symmetric Channel.** The channel capacity for the channel matrix

   $$\mathbf{P} = \begin{bmatrix} 1-p & p \\ p & 1-p \end{bmatrix} \text{ is given by } C = 1 - H(p, 1-p).$$
2. **Non-singular Channel Matrix.** For a square and non-singular channel **P**, the channel capacity is defined as

   $$C = \log_2 \sum_{j=1}^{n} \exp.\left\{-\sum_{i=1}^{m} a_{jk} H(Y \mid X = x_k)\right\},$$

   where, a_{jk} is the $(j, k)^{\text{th}}$ element of p^{-1}.

3. **Noiseless Channel.** A channel described by a channel matrix with one and only one non-zero element in each column is called a noiseless channel. For example,

$$\begin{pmatrix} 1/2 & 1 & 0 & 0 & 0 \\ 0 & 0 & 2/5 & 5/6 & 0 \\ 0 & 0 & 0 & 0 & 1 \end{pmatrix} \text{ is a noiseless channel.}$$

Example 28.4. *Find the capacity of the memoryless channel specified by the channel matrix,*

$$\mathbf{P} = \begin{bmatrix} 1/2 & 1/4 & 1/4 & 0 \\ 1/4 & 1/4 & 1/4 & 1/4 \\ 0 & 0 & 1 & 0 \\ 1/2 & 0 & 0 & 1/2 \end{bmatrix}.$$

Solution. The capacity of the memoryless channel is given by

$$C = \text{Max. } I(X, Y) = \text{Max. } \{H(X) + H(Y) - H(X, Y)\}$$

$$= -\sum_{i=1}^{4} p_{ij} \log p_{ij}; \quad j = 1, 2, 3, 4,$$

where, $p_{i1} = \left(\frac{1}{2}, \frac{1}{4}, \frac{1}{4}, 0\right)$, $p_{i2} = \left(\frac{1}{4}, \frac{1}{4}, \frac{1}{4}, \frac{1}{4}\right)$, $p_{i3} = (0, 0, 1, 0)$ and $p_{i4} = \left(\frac{1}{2}, 0, 0, \frac{1}{2}\right)$.

Given that, **P** is a square non-singular matrix,

$$\therefore \qquad \text{det. } \mathbf{P} = \frac{1}{16}, \quad \text{and} \quad \mathbf{P}^{-1} = \begin{pmatrix} 2 & -2 & 0 & 1 \\ 0 & 4 & -1 & -2 \\ 0 & 0 & 1 & 0 \\ -2 & 2 & 0 & 1 \end{pmatrix}$$

The channel capacity, C, is given by

$$C = \log_2 \sum_{j=1}^{4} \exp\left[\sum_{i=1}^{4} q_{ji} H(Y \mid X = X_i)\right],$$

where, q_{ji} is the $(j, i)^{\text{th}}$ element of $\mathbf{P}^{-1}$ and Y, X denote the channel output and channel input alphabets, respectively.

Since $\qquad H(Y \mid X = X_i) = -\sum_{j=1}^{4} p_{j|1} \log_{j|1}$, we have (taking logarithms to the base 2):

$$H(Y \mid X = X_1) = -\frac{1}{2}\log\frac{1}{2} - 2\left(\frac{1}{4}\log\frac{1}{4}\right) = \frac{1}{2} + 1 = \frac{3}{2},$$

$$H(Y \mid X = X_2) = -4\left(\frac{1}{4}\log\frac{1}{4}\right) = \log 2^2 = 2,$$

$$H(Y \mid X = X_3) = -\log 1 = 0,$$

and

$$H(Y \mid X = X_4) = -2\left(\frac{1}{2}\log\frac{1}{2}\right) = \log 2 = 1.$$

Thus, $\qquad H(Y \mid X = X_i) = \frac{3}{2} + 2 + 0 + 1 = \frac{9}{2}$ bits per symbol

Further, $\qquad -\sum_{j=1}^{4} a_{ji} H(Y \mid X = X_i) = -\frac{3}{2} q_{j1} - 2q_{j2} - q_{j4}$

This gives

$$\sum_{i=1}^{4} \exp\left[-\sum_{j=1}^{4} q_{ji}\, H(Y|X)\right] = 145/2^6.$$

Hence, $\quad C = \log_2 145/2^6 = 7.176 - 6.000 = 1.176$ bits per symbol.

28:7. ENCODING

Encoding may be defined as a transformation procedure of a message from source to receiver through a noise-less channel in some code language, *i.e.*, if $X = \{X_1, X_2, ..., X_m\}$ be the set of messages to be transmitted, then the ***codes*** may be defined as a relationship between all possible sequences of symbols of the set X with another set of messages, say $Y = \{Y_1, Y_2, ..., Y_n\}$ of ***code alphabet***.

Uses of Encoding

1. It is used to increase the efficiency of transmission.
2. It is used to minimize the expected code word length. If the code word associated with x_i is of length l_i, $i = 1, 2, ..., m$, then the expected length of messages is given by

$$L = l_1 p_1 + l_2 p_2 + \ldots + l_m p_m.$$

   The transmission for which L is minimum is considered to be efficient.
3. It is used to minimize the cost of transmission. If C_i $(i = 1, 2, ..., m)$ is the cost of transmission of some words W_i with probability of transmission W_i, then in a message of m words, the expected cost per message is given by

$$M = C_1 W_1 + C_2 W_2 + \ldots + C_m W_m.$$

   The transmission for which M is minimum is considered to be efficient.

Sub-Classes of Code

1. **Block Code.** A code which establishes a relationship with each of the symbols of the set **S** to a fixed sequence of symbols of the set **Q** is called a ***block code***. That is, each symbol s_i $(i = 1, 2, ..., m)$ is to be assigned a fixed sequence of symbols of **Q** called the ***code words***, associated with s_i. For example, s_1 may correspond to y_1, y_2 and s_2 may correspond to y_7, y_8, y_9, etc.
2. **Binary Code.** In particular, if the set $S = \{0, 1\}$, then a block code is said to be ***binary code***. An example of binary code is

$$s_1 \to 01,\quad s_2 \to 010,\quad s_3 \to 10,\quad s_4 \to 110.$$

3. **Non-singular Code.** A block code is said to be ***non-singular code*** if all words of the code are distinct. An example of non-singular code is

$$s_1 \to 00,\quad s_2 \to 01,\quad s_3 \to 10,\quad s_4 \to 11.$$

4. **Uniquely Decodable (Separable) Code.** A code is said to be ***uniquely decodable (separable) code***, if every finite sequence of symbols of the set **Q** is associated to at most one symbol of the set S. Examples of uniquely decodable (separable) code are given below:

   (*a*) $\quad s_1 \to 0,\quad s_2 \to 10,\quad s_3 \to 110,\quad s_4 \to 1110.$

   (*b*) $\quad s_1 \to 0,\quad s_2 \to 01,\quad s_3 \to 011,\quad s_4 \to 0111.$

   (*c*) $\quad s_1 \to 00,\quad s_2 \to 01,\quad s_3 \to 10,\quad s_4 \to 11.$

   It may be noted that the code (*c*) may be called a ***comma code***.

5. **Instantaneous Code.** A uniquely decodable code is said to be ***instantaneous***, if it is possible to decode each word in a sequence without reference to succeeding code symbols. In the preceding example; (*a*) and (*c*) are instantaneous codes whereas code (*b*) is not instantaneous. Various sub-classes of codes are indicated in *Fig. 28.3:*

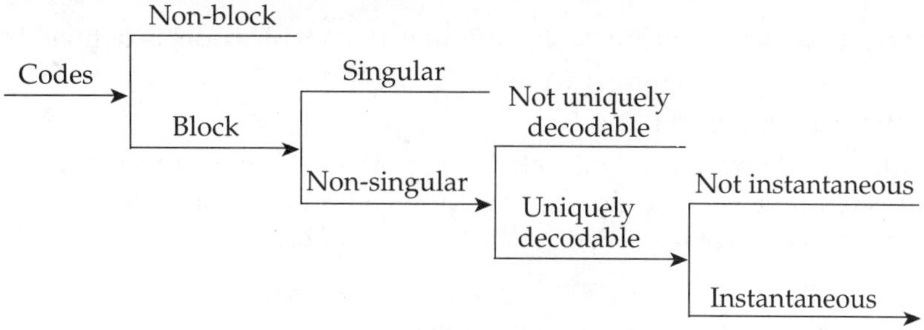

Fig. 28.3. Various Sub-classes of Codes

Prefix. *Let A and B be two finite (non-empty) sequences of code symbols. Then, we say that sequence A is the **prefix** of sequence B, if it is possible to write B as AC for some other sequence C of the code symbols.*

Example. *The code word 0111 has three prefixes — 011, 01, and 0.*

It may easily be observed that

"A necessary and sufficient condition for a code to be instantaneous is that no complete code word be prefix of some other."

Construction of Binary Instantaneous Codes

Consider a source $S = \{s_1, s_2, s_3, s_4, s_5\}$. Let us start by assigning 0 to symbol s_1, *i.e.*, $s_1 \to 0$. Then, all other code words must start with 1 (why?). We cannot let $s_2 \to 1$, because this would leave us with no symbols with which to start the remaining three code words. Thus, we might have $s_2 \to 10$. This in turn requires that remaining code words must start with 11. We avoid letting $s_2 \to 11$ (why?). Let $s_2 \to 110$. Now the only three-symbol prefix ***still unused*** is 111 and thus we might set $s_4 \to 1110$, $s_5 \to 1111$. Thus, we have constructed the instantaneous code

$$s_1 \to 0, \quad s_2 \to 10, \quad s_3 \to 110, \quad s_4 \to 1110, \quad s_5 \to 1111.$$

Note that in the above ***encoding procedure***, by selecting $s_1 \to 0$, we have restricted the flexibility of other codes. For example, if we would have taken $s_1 \to 100$, then we may set $s_2 \to 01$; and still have two prefixes of length 2 which are unused. In fact, we may construct the code

$$s_1 \to 00, \quad s_2 \to 01, \quad s_3 \to 10, \quad s_4 \to 110, \quad s_5 \to 111.$$

One may observe that in the first procedure (when we start with 0), the later codes are of larger length than that of these in the second procedure (when we start with 00).

Which of the two binary instantaneous codes constructed above is better? This question may not be resolved on the basis of the information available so far.

Average Length, Efficiency and Redundancy of a Code

Average Length. Consider a source alphabet $S = \{s_1, s_2, ..., s_M\}$ with input probabilities $p_1, p_2, ..., p_M$ $\left(p_i > 0, \sum p_i = 1\right)$. Let the symbol s_1 be assigned a code word of length l_i, $i = 1, 2, ..., M$. Then, obviously a sound definition for the ***average code-word length*** is

$$\overline{L} = p_1 l_1 + p_2 l_2 + \ldots + p_M l_M.$$

The basic problem of Noiseless Coding is to minimize average code-word length.

Efficiency of a Code. Let the average length of a uniquely decodable code for a source X be $\bar{L}$. Then, the ***efficiency***, η of the code is defined by

$$\eta = \frac{H(X)}{\bar{L}}.$$

Redundancy of a Code. *Redundancy*, β of a uniquely decodable code, is defined by

$$\beta = 1 - \eta,$$

where, η is the efficiency of the code.

Thus, the difference between the expected (or average) mutual information $I(X, Y)$ and its maximum value is defined as the redundancy of the communication system. The ratio of given redundancy to channel capacity is known as ***relative redundancy***, *i.e.*,

Redundancy for noise-free channel $\beta = C - I(X, Y) = \log n - H(X)$

$\therefore$ Relative redundancy for noise-free channel β

$$= \frac{\log n - H(X)}{\log n} = 1 - \frac{H(X)}{\log n} = 1 - \text{Efficiency of the system.}$$

28:8. SHANNON-FANO ENCODING PROCEDURE

In this method of encoding, binary numbers (0, 1) are used through a memoryless communication channel.

Let the ***ensemble*** $[S]$ of messages to be transmitted be given by

$$[S] = [m_1, m_2, ..., m_N]$$

and the corresponding probability distribution (P) of transmission, be given by

$$[P] = [p_1, p_2, \ldots, p_N], \qquad (p_i > 0,\ p_1 + p_2 + \ldots + p_N = 1)$$

We shall devise an encoding procedure of assigning efficient uniquely decodable binary codes to $[S]$. The following are the two necessary requirements:

(*i*) No complete code word can be the prefix of some other.

(*ii*) The binary digits in each code word appear independently with equal probability.

The ***Shannon-Fano encoding*** procedure is summarized in the following six steps:

Step 1. Arrange the messages in descending order of their probabilities. Without loss of generality, let

$$p_1 > p_2 > p_2 > \ldots > p_N.$$

Thus, we have

Message :	m_1	m_2	m_3	...	m_N
Probability :	p_1	p_2	p_3	...	p_N

Step 2. Partition the set of messages into most equiprobable groups, say S_1, S_2; *i.e.*,

Message	*Probability*
$S_1 : m_1, m_2$	$P(S_1) = p_1 + p_2,$
$S_2 : m_3, m_4, m_5, ..., m_N$	$P(S_2) = p_3 + p_4 + ... + p_N,$

such that $P(S_1) = P(S_2)$.

Step 3. Further, partition S_1 and S_2 into two most equiprobable sub-groups, say S_{11}, S_{12} and S_{21}, S_{22} respectively.

Step 4. Continue partitioning the resulting sub-groups into further two most equiprobable sub-groups till each sub-group contains exactly one message.

Step 5. Assign the binary digit 0 to the first position of the coded word for the messages of S_1 and assign the digit 1 to the first position of the code words for messages of S_2. The assignment of binary digits to other positions in the code words are similarly made in accordance with the further partitioning of S_1 and S_2.

Step 6. The partitioning and assigning binary digits 0 and 1 will continue till each sub-group contains only one message.

Example 28.5. *A source without memory has six characters with the following probabilities of transmission:*

A	B	C	D	E	F
1/3	1/4	1/8	1/8	1/12	1/12

Devise the Shannon-Fano encoding procedure to obtain uniquely decodable code to the above message ensemble. What is the average length, efficiency and redundancy of the code that you obtain?

Solution.

Step 1. We are already given the messages arranged in the descending order of their probabilities. We construct the table in which all the entries are to be made stepwise.

Characters	*Probability* p			*Equiprobable partitioning*	*Code-word*	*code-word length* l_i
A	1/3	S_1		S_{11}	00	2
B	1/4			S_{12}	01	2
C	1/8	S_2	S_{21}	S_{211}	100	3
D	1/8			S_{212}	101	3
E	1/12		S_{22}	S_{221}	110	3
F	1/12			S_{222}	111	3

Step 2. Partition the set of characters into two most equiprobable sets S_1, S_2. Here $S_1 = \{A, B\}$, $S_2 = \{C, D, E, F\}$ with $P(S_1) = 7/12$, and $P(S_2) = 5/12$.

Step 3. Further, partition S_1 and S_2 into two most equiprobable sub-groups, namely,

$$S_{11} = \{A\}, \quad S_{12} = \{B\}, \quad S_{21} = \{C, D\}, \quad \text{and} \quad S_{22} = \{E, F\}.$$

Step 4. Continue the partitioning till each sub-group has exactly one character. Namely,

$$S_{211} = \{C\}, \quad S_{212} = \{D\}, \quad S_{221} = \{E\}, \quad \text{and} \quad S_{222} = \{F\}.$$

Step 5. Assign binary digit 0 to the first position of all code words in S_1 and (hence) binary digit 1 to the first position of all code words in S_2. Assign symbol 0 to the second position of code word for S_{11} and symbol 1 to the second position of code word for S_{12}.

Assign digit 0 to the second position of all code words of S_{21} and digit 1 to those of S_{22}. The third position in the code words for S_{21} and S_{22} are assigned symbols 0 and 1 alternately. The reader may note that a very simple procedure of writing down the code words is apparent in the above. Consider any finally partioned (irreducible) sub-group S_{ijk}, $i, j, k \in (0, 1)$. Then, the subscripts i, j, k, not only determine the number and position of binary symbols in a code word, but also tells which position has which symbol. For example, i represents the first position, j — the second position, k — the third position, and so on; and subscript value 1 stands for symbol '0', 2 for symbol '1', etc. The code word in the table given above can then be written down directly by the subscripts of S_{ijk}'s just replacing $1 \to 0$, $2 \to 1$ everywhere.

Average Code Length. The average code length of the code constructed above is given by

$$\bar{L} = \sum_{i=1}^{5} p_i\, l_i = \frac{2}{3} + \frac{2}{4} + \frac{3}{8} + \frac{3}{8} + \frac{3}{12} + \frac{3}{12} = \frac{29}{12} \text{ bits/symbol.}$$

Entropy of Source

$$H(X) = -\frac{1}{3}\log\frac{1}{3} - \frac{1}{4}\log\frac{1}{4} - \frac{2}{8}\log\frac{1}{8} - \frac{2}{12}\log\frac{1}{12}$$

$$= \frac{1}{3}\log 3 + \frac{1}{4}\log 2^2 + \frac{1}{4}\log 2^3 + \frac{1}{6}\log(2^2 \times 3)$$

$$= \frac{19}{12} + \frac{1}{2}\log 3 = 2.3752 \text{ bits, since } \log_2 2 = 1.$$

Efficiency of the Code

$$\eta = \frac{H(X)}{\bar{L}} = \frac{12 \times 2.3752}{29} = 0.9828.$$

Redundancy of the Code

$$\beta = 1 - \eta = 0.0172.$$

Remark. The efficiency is slightly less than 100%, because of using approximate values, while partitioning into 'most' equiprobable groups.

Example 28.6. *Apply Shannon's encoding procedure to the following message ensemble:*

[X]	:	A	B	C	D
[P]	:	0.4	0.3	0.2	0.1

Solution. We are given the messages arranged in ascending order of their probabilities. We construct the table given below in which all the entries are to be made stepwise.

Characters	Probability p			Equiprobable partitioning	Code-word	Code-word length l_i
A	0.4	S_1		S_1	0	1
B	0.3	S_2	S_{21}	S_{21}	10	2
C	0.2		S_{22}	S_{221}	110	3
D	0.1			S_{222}	111	3

Partition the set of characters into two most equiprobable sets S_1, S_2; such that $S_1 = \{A\}$ and $S_2 = \{B, C, D\}$ with $P(S_1) = 0.4$ and $P(S_2) = 0.6$. Further, partition S_2 into two most equiprobable sub-groups, such as: $s_{21} = \{B\}$ and $S_{22} = \{C, D\}$. Continue partitioning the sub-groups, till each has exactly one character, such as: $S_{221} = \{C\}$ and $S_{222} = \{D\}$.

Assign binary digit 0 to the first position of all code words S_1 and binary digit 1 to the first position of the code words in S_2. Assign symbol 0 to the second position of code word for S_{21} and symbol 1 to the second position of code word for S_{22}. Again, assign symbol 0 to the third position of the code word for S_{221} and the symbol 1 to the third position of the code word for S_{222}.

Average code length is given by

$$\bar{L} = l_1 p_1 + l_2 p_2 + l_3 p_3 + l_4 p_4 = 1 \times 0.4 + 2 \times 0.3 + 3 \times 0.2 + 3 \times 0.1$$

$$= 0.4 + 0.6 + 0.6 + 0.3 = 1.9 \text{ bits per symbol.}$$

Entropy of source is given by

$$H(X) = -0.4\log 0.4 - 0.3\log 0.3 - 0.2\log 0.2 - 0.1\log 0.1$$

$$= \log 10 - 0.3\log 3 - \log 2 = \log 5 - 0.3\log 3.$$

PROBLEMS

28.07. Apply Shannon-Fano encoding procedure to the following message ensemble:

[X]	:	x_1	x_2	x_3	x_4	x_5	x_6	x_7	x_8	x_9
[P]	:	0.49	0.14	0.14	0.07	0.07	0.04	0.02	0.02	0.1

28.08. Apply a ternary partitioning technique (similar to the Shannon-Fano encoding procedure) for encoding the following messages into code words using a ternary alphabet [0, 1, 2]. Determine the average number of digits in the encoded messages and also code efficiency:

[m]	:	m_1	m_2	m_3	m_4	m_5	m_6
p [m]	:	3/8	1/6	1/8	1/8	1/8	1/12

[***Hint:*** Partition [*m*] into three most equiprobable sets S_1, S_2 and S_3. Further, partition each of these into three most equiprobable sub-groups. Let one of the final (irreducible) subsets be S_{212}. Then, assign the code word $S_{213} > 201$, etc.]

TEST YOUR UNDERSTANDING

1. Which of the following is not correct?

(*a*) Entropy is a measure of information.

(*b*) The entropy of the probability distribution $\{p_1, p_2, ..., p_n\}$ is given by $-\sum_{i=1}^{n} p_i \log p_i$.

(*c*) Information theory is confined to the studies in telecommunication engineering only.

(*d*) Shannon entropy find applications in diverse fields such as linguistics, accounting, psychology and economics, etc.

2. The expected value of information is given by

(*a*) $\sum_{i=1}^{m} p_i\, I(x_i)$ (*b*) $\sum_{j=1}^{n} q_j\, I(y_j)$

(*c*) $\sum_{i=1}^{m} p_i I(x_j);\ j=1,2,\dots,n$ (*d*) $\sum_{j=1}^{n} q_i I(x_j);\ i=1,2,\dots,m.$

3. An essential part of a communication system is

(*a*) source (*b*) communication channel

(*c*) receiver (*d*) all of the above.

4. Basic requirements of logarithmic entropy function is:

(*a*) additivity (*b*) grouping

(*c*) monotonical (*d*) all of the above.

5. In usual notations, if $H(X, Y)$ and $I(X, Y)$ represent measures of joint entropy and mutual information respectively, then we have:

(*a*) $H(X,Y) = \sum\sum p_{ij} \log p_{ij}$.

(*b*) $I(X, Y) = H(X) + H(Y) - H(X, Y)$.

(*c*) $H(X, Y) \geq H(X) + H(Y)$ with equality if *X* and *Y* are independent.

(*d*) $I(X, Y) \leq H(X) + H(Y) - H(X, Y)$ with equality if *X* and *Y* are independent.

6. Which of the following is not true?

Let $S = \{s_1, s_2, s_3, ...\}$ be the source alphabet, and $Q = \{a_1, a_2, ...\}$ the code alphabet.

(*a*) A code which maps symbols of *S* into a fixed sequence of symbols of *Q* is called a block code.

(*b*) The fixed sequences of symbols of Q are called code words.
(*c*) A code is said to be uniquely decodable, if every code word of Q corresponds to at most one source symbol S_i.
(*d*) Redundancy β of a uniquely decodable code is defined by $\beta = \eta - 1$, where η is the efficiency of the code.

ANSWERS

1. (*c*) 2. (*a*) 3. (*d*) 4. (*d*) 5. (*b*) 6. (*d*)

REVIEW EXERCISE

1. What do you mean by the value of information? *[MBA (Nov.), Annamalai Univ. 2002]*
2. Define entropy function and establish its formal requirements.
3. (*a*) Define the different entropies for a two part communication system and calculate them for a discrete channel with independent input-output.
(*b*) Give a measure for mutual information $I(X, Y)$, and show that
$$I(X, Y) = H(X) + H(Y) - H(X, Y).$$
4. Show that the entropy function is maximum when mutually exclusive events are equiprobable. Show, also, that the partitioning of events into sub-events cannot decrease the entropy of the system.
5. (*a*) Prove that
$$H(p_1, p_2, ..., p_n) \le \log_2 n.$$
where equality holds, if and only if $p_k = 1/n, k = 1, 2, ..., n$.
(*b*) Show that $H(p_1, p_2, ..., p_M)$ vanishes when the probability distribution is degenerate.
6. Give a brief account of memoryless schemes.
7. Show that all possible sets of binary codes with the prefix property for encoding the message ensemble (m_1, m_2, m_3) in words not more than three digits long.
8. (*a*) Define conditional entropy. In the usual notation, show that
$$H(Y) + H(X) \ge H(X, Y).$$
(*b*) What do you understand by channel capacity and redundancy, efficiency? Explain with an illustration.

CHAPTER 29

Investment Analysis and Break-Even Analysis

"Capital expenditure is extremely vital in financial decision-making"

Every individual or firm wishes to know how *'best'* to invest money to attain the maximum gain. To achieve this objective, a proper investment analysis is to be made. It involves the consideration of investment proposals, estimation of their cash flows, evaluation of cash flows, selection of the proposals based on some criterion and finally the continuous revaluation of these proposals (projects).

This chapter is focused on financial decision-making problems arising out of an investment decision, and provides detailed steps necessary for the analysis of such problems. The chapter begins with a consideration of the time value of money, followed by the ***investment analysis***, *i.e.*, the question of making capital budgeting decisions. Therefore, interrelations between cost, revenue, volume, and profit planning (*i.e.*, ***break-even analysis***) have been discussed.

29:1. TIME VALUE OF MONEY

One of the basic concepts of finance is the notion that money has ***time value***. This is because money today is more valuable than the same amount at some future date. We can always put available funds to some use and make them grow into higher sums, so that a larger sum would be available later on. A rational decision-maker would not value the opportunity to receive some amount of money now, equally, with the opportunity to have the same amount of money at some future date. This phenomenon is known as the decision maker's '***time preference of money***'.

The time preference for money is generally expressed by means of an interest rate. For example, if the time preference rate of an individual is 10%, he may forego the opportunity of receiving ₹ 100 now if he is offered ₹ 110 after one year. This amounts to pay him the current principal (₹ 100) plus the interest that would accrue on it after one year @10% p.a. Like individuals, firms also have time preference (or discount rate). They use this rate in evaluating the alternative decisions.

Compound Value

Consider the situation, when the investment involve more than one year. Let the interest, as it becomes due, is added to the principal, and the amount, thus, obtained forms the principal for the next time period. Then, the mode of interest computation is called ***Compound Interest***. The time period after which the interest is added each time to the principal to form the new principal is called ***Conversion Period*** and amount received after the last conversion period is called the ***Compound Value***. The difference of the compound value and the original sum borrowed is the amount of ***Compound Interest*** for given conversion periods.

Let P be the principal, r the rate of interest in percentage per conversion period, and n is the number of conversion periods. Then, the amount after 1st conversion period will be $P + P \times r$, *i.e.*,

$P(1 + r)$. Similarly, compound amount after 2nd conversion period will be $P(1 + r) + P(1 + r) \times r$, *i.e.*, $P(1 + r)^2$. In general, compound amount after n^{th} conversion period becomes $P(1 + r)^n$.

The factor $(1 + r)^n$ is also sometimes called the ***accumulation factor***.

Present Value

The ***present value*** (*PV*) is the amount of money that represents the sum of principal and interest, if *P* (*i.e.*, principal) is required to be invested now at a certain rate compounded over a number of time periods at a specific rate for each time period.

The present value of the amount $A(n)$ due at the end of *n* interest periods at the rate of *r* (in percentage per conversion period) may be obtained by solving for *P*, the following:

$$A(n) = P(1 + r)^n \quad \text{or} \quad P = A(n) \times (1 + r)^{-n}.$$

Continuous Compounding

In most of the investments, the interest is payable quarterly, half-yearly or at the end of some other fraction of the year. Let the annual year be composed of *f* such periods. Then, an *n*-year period shall be equivalent to $n \times f$ periods. If an amount *P* is invested at the interest rate of *r* % per $1/f$ year, the compound value at the end of *n* years shall be

$$A(n) = P\left(1 + \frac{r}{f}\right)^{n \times f}$$

By ***continuous compounding***, we mean compounding over an infinitely large number of periods (*i.e.*, $f \to \infty$). An example of it could be the rate of return that a firm earns on its initial investments, since it enables it to receive a continuous inflow of earnings throughout the year or over a specified time horizon.

Example 29.1. *A small industrialist borrows ₹50,000 as a loan from a bank. If the bank charges interest at the rate of 10% per annum, find the amount the industrialist has to return after 3 years, when the interest is compounded (i) annually, and (ii) quarterly.*

Solution.

(*i*) Here, $P = ₹\ 50{,}000$, $r = 10\%$ and $n = 3$.

$$\therefore \quad A = P(1 + r)^n \Rightarrow A = ₹\ 50{,}000 \times \left(1 + \frac{10}{100}\right)^3$$

$$\text{or} \quad A = ₹\ 50{,}000 \times (1.10)^3 = ₹\ 50{,}000 \times 1.331 = ₹\ 66{,}550.$$

(*ii*) Here, $P = ₹\ 50{,}000$, $r = 10\%$ annually or $\frac{10}{4}\%$ quarterly, *i.e.*, $\frac{0.10}{4}$ or 0.025 and $n = 3 \times 4 = 12$.

$$\therefore \quad A = P(1 + r)^n \Rightarrow A = ₹\ 50{,}000 \times (1 + 0.025)^{12}$$

$$= ₹\ 50{,}000 \times (1.025)^{12} = ₹\ 50{,}000 \times 1.3448 = ₹\ 67{,}240.$$

Example 29.2. *A person wishes to have ₹75,000 in a bank account for the college education of his son. How much amount the person should deposit now at 6.5% compounded annually, if he has to start college education in 8 years hence, from now?*

Solution. Let ₹ *P* be deposited in bank now. Then, the amount after 8 years (compounded annually at the rate *r* per cent per year) will be

$$A = P(1 + r)^n,$$

where, $P = ₹\ 75{,}000$, $r = 6.5\%$ p.a. or 0.065 and $n = 8$.

Therefore, $\quad 75{,}000 = P \times (1 + 0.065)^8$

Thus, $P = ₹ 75,000 \times (1.065)^{-8} = ₹ 75,000 \times \frac{1}{1.6550} = ₹ 45,317.$

Hence, the person should deposit ₹ 45,317 now to have ₹ 75,000 eight years hence.

29:2. ANNUITIES AND SINKING FUND

Instead of investing lump sum amount initially, one may invest a constant amount annually (or monthly) or at equal intervals for a designated period of time. The investment is generally assumed to be made at the end of each period. The constant periodic sums are called ***annuities***. A recurring deposit amount in Post Office/Bank, payment of insurance premium, etc., are the examples of annuity. It may be noted that:

(*a*) the size of each payment of an annuity is called the ***periodic payment***.

(*b*) an annuity which is payable for ever, *i.e.*, which never terminates is called ***perpetuity***.

(*c*) the total time from the beginning of the first payment period to the end of last payment period is known as the ***term of annuity***.

(*d*) the algebraic sum of payments and the accumulated interest is known as the ***amount of annuity***.

(*e*) the ***present value of an annuity*** is sum of the present values of its instalments.

Amount of an Annuity

The amount of annuity refers to the terminal value of the sequence of payments, including principal and compound interest, at the end of the term of annuity. To obtain the amount of annuity, let p = the periodic payment, r = the rate of interest, and n = the interval, or the term of annuity, or the number of payments.

The first payment, being made at the end of the first time period, shall carry accumulated interest for $(n-1)$ time periods, the second one would similarly carry interest for $(n-2)$ time periods, and so on. The last payment would not carry any interest, because it is made at the end of the term. Thus, the amount of an ordinary annuity of ₹ p per period for n periods at the rate of r per period is given by

$$A(n, r) = p(1+r)^{n-1} + p(1+r)^{n-2} + \ldots + p(1+r) + p$$

Multiplying this equation on both sides by $(1 + r)$, we get

$$(1+r)A(n,r) = p(1+r)^{n} + p(1+r)^{n-1} + \ldots + p(1+r)^{2} + p(1+r)$$

On subtraction, these two equations yields

$$[(1+r)-1]\,A(n,r) = p(1+r)^{n} - p$$

or
$$A(n,r) = \frac{p\left[(1+r)^{n}-1\right]}{r} = \frac{p(v^{n}-1)}{v-1},$$

where, $v = 1 + r$.

Present Value of an Annuity

The present value or capital value of an annuity 'A' is the sum of the present values of all payments. It represents the amount that must be invested now to purchase, the payments due in future.

It may be noted that the interest is compounded at the end of each payment period.

Consider an annuity of n payment of x (rupees) each, where the interest rate per period is 'r' and the first payment is due one period from now. Then, the present value of the annuity is given by

$$PV = x(1+r)^{-1} + x(1+r)^{-2} + \ldots + x(1+r)^{-n}$$

Multiplying the above equation on both sides by $(1 + r)^{-1}$ and then subtracting it from the same equation, we get

$$PV - PV(1+r)^{-1} = x(1+r)^{-1} - x(1+r)^{-n+1}$$

Hence, $$PV = x \times \frac{(1+r)^n - 1}{(1+r)^{n+1}} \times \frac{1+r}{r} = \frac{x}{r}\left[1-(1+r)^{-n}\right].$$

Present Value of a Perpetual Annuity

For a perpetuity, there is no end to the periodic payments. Under this condition, therefore, $n \to \infty$. This implies that $(1 + r)^{-n} \to 0$. So, the present value of a perpetuity of ₹ x per year @ r% per annum is given by

$$PV = \lim_{n \to \infty} \frac{x}{r}\left[1-(1+r)^{-n}\right] = \frac{x}{r}(1-0) = \frac{x}{r}.$$

Sinking Fund

A sinking fund is a fund that is created for a sequence of periodic payments over a time period at a specified interest rate. The amount in such a fund is usually used for repair of buildings, replacement of an equipment, expansion of an enterprise, and so on.

If A is the amount to be saved, and P, the periodic payment, then size of the sinking fund is computed by using the following formula

$$A = P \times A(n, r) = P \times \frac{(1+r)^n - 1}{r}.$$

Example 29.3. *Let the investor plans to invest ₹ 1,000 per year in a savings plan that earns 5% interest p.a. compounded annually. What is the sum of annuity payments at the end of 10 years?*

Solution. Given: $n = 10$ years, $r = 5\%$ p.a. $= 0.05$, and $p =$ ₹ 1,000.

∴ The required sum of annuity payments is given by

$$A(n, r) = \frac{p\left[(1+r)^n - 1\right]}{r}$$

or $$A(10,\ 0.05) = \frac{1{,}000 \times \left[(1.05)^{10} - 1\right]}{0.05} = \frac{1{,}000 \times (1.6288 - 1)}{0.05}$$

$$= 1{,}000 \times 12.576 = 12{,}576.$$

Hence, $S = A\ (10,\ 0.05) =$ ₹ 12,576.

Example 29.4. *A project offers an annual return of ₹ 2,50,000 for four years. If the cost of money is 14% p.a., what is the present value of the project?*

Solution. This is the case of an annuity of ₹ 2,50,000 for the next four years. We are given:

$x =$ ₹ 2,50,000 (annuity), $n = 4$ years and $r = 14\%$ p.a. $= 0.14$

∴ The present value of annuity is given by

$$PV = \frac{x}{r}\left[1-(1+r)^{-n}\right] = \frac{2{,}50{,}000}{0.14}\left[1-(1.14)^{-4}\right]$$

$$= 2{,}50{,}000 \times 2.914 = ₹\ 7{,}28{,}500.$$

Example 29.5. *A pipeline is due for repairs. It will cost ₹ 10,000 and lasts for 3 years. Alternatively, a new pipeline can be laid at a cost of ₹ 30,000 and lasts for 10 years. Assuming cost of capital to be 10% and ignoring salvage value, which alternative should be chosen?*

Solution. We compare the *PV* of the future costs associated with the two alternatives. The first alternative involve a replacement cost of ₹10,000 every 3 years and the second alternative involves a replacement cost of ₹30,000 every 10th year.

The present worth of the associated future cost is given by

$$PV = x + \frac{x}{(1+r)^n} + \frac{x}{(1+r)^{2n}} + \frac{x}{(1+r)^{3n}} + \ldots$$

$$= x\left[1+(1+r)^{-n}+(1+r)^{-2n}+(1+r)^{-3n}+\ldots\right]$$

$$= \frac{x}{1-(1+r)^{-n}}.$$

Here, $r = 0.10$ and $r = 3$ for alternative 1 and $n = 10$ for alternative 2.

Thus, for the alternative of existing pipeline,

$$PV = \frac{₹\,10{,}000}{1-(1+0.10)^{-3}} = ₹\left[\frac{10{,}000}{1-0.7513}\right] = ₹\,40{,}209.$$

For the alternative of the new pipeline,

$$PV = \frac{₹\,30{,}000}{1-(1+0.10)^{-10}} = ₹\left[\frac{30{,}000}{1-0.3855}\right] = ₹\,48{,}820.$$

Obviously, the existing pipeline should be preferred.

Example 29.6. *An equipment costs ₹46,000 and its effective life is estimated to be 10 years. A sinking fund is created for replacing the equipment at the end of its effective life when its scrap realizes a sum of ₹6,000 only. Calculate the amount which should be provided every year, for the sinking fund, if it accumulates at the compound interest rate of 7% per annum.*

Solution. Sinking fund created for replacing the equipment = Cost of the equipment,

Scrap value = ₹ 46,000 – ₹ 6,000 = ₹ 40,000.

Let the size of each annual deposit into the sinking fund be ₹ x. Then, using the formula for the amount of an ordinary annuity,

$$A(n,\,r) = \frac{x}{r}\left[(1+r)^n - 1\right],$$

where, $A\,(n, r)$ = ₹ 40,000, $n = 10$ and $r = 0.07$; we get

$$₹\,40{,}000 = \frac{x}{0.07}\left[(1+0.07)^{10} - 1\right] = x \times \frac{0.9671}{0.07}$$

or

$$x = ₹\,40{,}000 \times \frac{0.07}{0.9671} = ₹\,2{,}895.25$$

i.e., ₹ 2,895.25 should be deposited each year.

PROBLEMS

29.01. A trader borrowed ₹ 40,000 from two money-lenders. For one loan he paid interest at the rate of 12% per annum and for the other 14% per annum. After one year, he paid ₹ 5,100 as interest. How much did he borrow at these rates?

29.02. When a baby is born, ₹ 50,000 is placed to her credit in an account that pays interest at the rate of 6% (*i*) compounded annually, (*ii*) compounded quarterly, (*iii*) compounded monthly. If the amount is not disturbed, what amount will there be to her credit on her 20th birthday?

29.03. A person deposited ₹ 10,000 in a bank for 3 years offering interest at the rate of 6% compounded half yearly during first year, at the rate of 12% compounded quarterly during second year and at 10% compounded continuously during third year. Find his balance after three years. ***[B.Com., Delhi Univ. 2014]***

29.04. A sum of ₹ 60,000 is invested in a scheme which provides for interest at the rate of 6% per annum compounded half-yearly for the first three years and at 5% per annum compounded continuously for the next two years. Calculate the amount receivable on maturity. ***[B.Com., Delhi Univ. 2010]***

29.05. A person has borrowed ₹ 50,00,000 from a bank to buy a house at the rate of 9% p.a. What should be the uniform yearly instalment so as to repay the loan at the end of four years? ***[MBA, Mumbai Univ. 2010]***

29.06. Suppose a machine costing ₹ 90,000 is to be replaced at the end of 7 years, at which time it will have a salvage value of ₹ 8,000? In order to provide money at that time for a new machine costing the same amount, a sinking fund is set up. The amount in the fund at that time is to be the difference between the replacement cost and salvage value. If equal payments are placed in the fund at the end of each quarter and the fund earns 8% compounded quarterly, what should each payment be?

29.07. A company is considering proposal of purchasing a machine either by making full payment of ₹ 4,000 or by leasing it for 4 years at an annual rent of ₹ 1,250. Which course of action preferable, if the company can borrow money at the rate of 14% per annum?

29.08. A machine with useful life of 7 years costs ₹ 10,000, while another machine with useful life of 5 years costs ₹ 8,000. The first machine saves labour expenses of ₹ 1,900 annually and the second one saves labour expenses by ₹ 2,200 annually. Determine the preferred course of action. Assume cost of borrowing as 10% per annum.

29:3. METHODS OF INVESTMENT ANALYSIS

An efficient allocation of capital is one of the most important functions of a financial management. This function involves the firm's desire to commit its funds in long term assets and other profitable activities. Such investment decisions are of considerable importance, because they influence the wealth of the firm, determine its size, set up the pace and direction of its growth and also affect its business risk.

Investment decisions of a firm influence its wealth — it will increase if the investment proposals are profitable, otherwise it would decrease. Now, to evaluate an investment proposal, the study of the following components are necessary:

(*i*) The initial cash outflow.

(*ii*) The returns (or net cash-inflows calculated on yearly basis) over the economic life of the asset.

(*iii*) The economic life of the asset. Economic life is usually shorter than durable life due to technological advancements.

(*iv*) Minimum return desired on such investment, called "cut-off point".

The most widely used methods of evaluating an investment proposal can be grouped into following two major categories:

1. Traditional Methods. These include ***Payback Period method***, and ***Average Rate of Return method***.

(*i*) **Payback Period Method.** Payback period is the number of years required to recover the cash invested in a project. If annual cash inflows are same, then dividing the cash invested by the annual cash inflow; the payback period can be computed. If annual cash inflows are unequal, it can be calculated by adding up the cash inflows until the total is equal to the initial cost invested.

Illustration. *A project requires a cash outlay of ₹ 60,000 and yields an annual cash inflow of ₹ 12,000 for 7 years. What is its payback period?*

Solution. $$\text{Payback period } (p) = \frac{\text{Cash outlay } (P)}{\text{Average net cash inflow per year } (R)}$$

$$= \frac{60,000}{12,000} = 5 \text{ years.}$$

It may be noted that the proposal will be accepted, only if payback period is 5 years or less.

(*ii*) **Average Rate of Return Method.** The average rate of return is an accounting method. It is calculated as the ratio of the average annual book profits after taxes to the average investment in the project. For instance, if the average annual book profits (after taxes and depreciation) are estimated at ₹ 20,000 and net investment for a new equipment is ₹ 1,00,000; then the average rate of return would be:

₹ 20,000/₹ 1,00,000, *i.e.*, 0.20 or 20%.

After, computing the average rate of return, it is compared with a required rate of return to determine whether a particular project should be accepted or rejected.

Remark. Both the payback period method as well as average rate of return method, suffer from a serious limitation — they do not take into account the time value of money, while evaluating the investment worth of the projects.

2. Discounted Cash Flow (DCF) Methods. These include Net Present Value (*NPV*) method, Interest Rate of Return (IRR) method, and ***Discounted Payback*** Period method.

(*i*) **Net Present Value (NPV) Method.** This method takes into account the time value of money. It correctly postulates that cash flows arising at different time periods differ in value and are comparable only when their equivalent '***present values***' are found out.

The *NPV* of a proposal is computed by the following relation:

$$NPV = PV \text{ of cash inflows} - PV \text{ of cash outflows.}$$

The proposal is accepted, if $NPV > 0$, otherwise it is rejected. The *PV* of all cash outflows is determined using some pre-determined discounting rate, which generally equals the cost of capital (minimum rate of return or the interest rate).

Remarks

1. The *NPV* method can be used to select between mutually exclusive proposals by considering whether the incremental investment generates a positive *NPV*. Using this method, the proposal would be ranked in order of their *NPV* values.
2. The equation for *NPV*, assuming that all cash outflows are made in the initial year, can be written as

$$NPV = \frac{A_1}{1+k} + \frac{A_2}{(1+k)^2} + \ldots + \frac{A_n}{(1+k)^n} - C = \sum_{i=0}^{n} \frac{A_t}{(1+k)^t},$$

where, A_0 (= – C), A_1, A_2, ... represent cash inflows, k is firm's cost of capital, C is the cost of investment proposal and n is the expected life of the project (it is assumed that k is known).

(*ii*) **Internal Rate of Return (IRR) Method.** The internal rate of return is the rate at which an investment is repaid by proceeds from a project. In other words, it is the rate that equates the present value of the cash inflows with the present value of cash outflows of an investment. It is determined by using the following formula:

$$\sum_{i=1}^{n} \frac{A_i}{(1+r)^t} = C,$$

where, $A_1, A_2, ..., A_n$ represent cash inflows, C is the cost of investment project (or the *PV* of cash outflows), n is the expected life span of the project, and r is the internal rate of return (*IRR*).

The computed *IRR* is compared with the required rate of return. If the calculated rate exceeds the required rate, the proposal is accepted; otherwise it is rejected.

Note. The value of r in the above equation is determined by trial and error. We select any rate of interest to compute *PV* of cash inflows. If the calculated *PV* of cash inflows is lower than the *PV* of cash outflows, a lower rate of interest is to be tried. On the other hand, a higher rate should be tried, if the *PV* of cash inflows is higher than the *PV* of cash outflows. The process is to be repeated till the *NPV* becomes zero.

Remark. It may be observed that the *IRR* formula is same as used for the *NPV* method with the difference that in the *NPV* method, the required rate of return is assumed to be known and then the *NPV* is calculated, whereas in the *IRR* method, the value of '*r*' has to be determined such that the *NPV* is zero.

(*iii*) **Discounted Payback Period Method.** The payback period is defined as the number of years required to recover the original investment. In the case of discounted payback period, we consider the discounted present values of future cash inflows and find out the number of years required to recover the initial investment. If discounted payback period is less than the desired payback period, the project is accepted. To illustrate, let us consider the following investment proposal with an initial investment of ₹ 24,000.

Year	*Cash inflow* (₹)	*PV factor* @ 10%	*Discounted PV* (DPV) (₹)	*Cumulative* DPV (₹)
1	8,000	0.909	7,272	7,272
2	10,000	0.826	8,260	16,530
3	10,000	0.751	7,510	**24,040**
4	12,000	0.683	8,196	32,236
5	10,000	0.621	6,210	38,446

The number of years required to recover the initial amount is 3 years. Hence, the discounted pay-back period is 3 years.

Example 29.7. *Calculate the payback period for a project which requires a cash outlay of ₹ 30,000 and generates cash inflows of ₹ 11,000, ₹ 10,000, ₹ 7,000 and ₹ 6,000.*

Solution. When we add up the cash inflows, we find that ₹ 28,000 of the original outlay is recovered in the first three years. In the fourth year the balance ₹ 2,000 of the original outlay needs to be recovered and the cash inflow during the year is ₹ 6,000. Assuming that the cash inflows occur uniformly during the year, the time required to recover ₹ 2,000 will be $\frac{2,000}{6,000} \times 12 = 4$ months.

Thus, the payback period is 3 years and 4 months.

Example 29.8. *An investment proposal X initially costs ₹20,000 and generates year-end cash inflows of ₹5,000, ₹5,000, ₹10,000 and ₹10,000 during the next four years. If the required rate of return is assumed to be 10%, would you accept the proposal?*

Solution. Since, the required rate of return is 10%, we calculate present values of cash inflows using the discounting factor (or *PV* factor) @ 10% and then *NPV* of the proposal.

The calculation of *NPV* is given in the following table:

End of year	*PV factor @ 10%*	*Cash flow (₹)*	*Present value (₹)*
0	1.000	(–) 20,000	(–) 20,000
1	0.909	5,000	4,545
2	0.826	5,000	4,130
3	0.751	10,000	7,510
4	0.683	10,000	6,830
			NPV = 3,015

Since, *NPV* is positive, proposal is accepted.

Example 29.9. *The XYZ Company Ltd. is considering three proposed investments A, B and C. Each requires an investment of ₹4,900 and each has an economic life of three years, the total cash inflow over that period of investment is ₹6,000. The pattern of cash inflows for each proposal differs as below:*

Year	*Annual cash inflows*		
	A	*B*	*C*
1	*₹1,000*	*₹2,000*	*₹3,000*
2	*₹2,000*	*₹2,000*	*₹2,000*
3	*₹3,000*	*₹2,000*	*₹1,000*

(*i*) *Calculate the NPV of each proposal, if the required rate of return is 10%.*

(*ii*) *Calculate the NPV of each proposal, if the required rate of return is 6%.*

Explain why the result differs.

Solution. The *NPV* of three proposals at the given rates of return are given in the following tables:

(*i*) Rate of return is 10%.

Year	*PV factor @ 10%*	*Proposal A*		*Proposal B*		*Proposal C*	
		Cash flow (₹)	*PV (₹)*	*Cash flow (₹)*	*PV (₹)*	*Cash flow (₹)*	*PV (₹)*
0	1.000	(–) 4,900	(–) 4,900	(–) 4,900	(–) 4,900	(–) 4,900	(–) 4,900
1	0.909	1,000	909	2,000	1,818	3,000	2,727
2	0.826	2,000	1,652	2,000	1,652	2,000	1,652
3	0.751	3,000	2,253	2,000	1,502	1,000	751
			NPV: –86		*NPV:* 72		*NPV:* 230

(*ii*) Rate of return is 6%.

Year	*PV factor @ 6%*	*Proposal A*		*Proposal B*		*Proposal C*	
		Cash flow (₹)	*PV (₹)*	*Cash flow (₹)*	*PV (₹)*	*Cash flow (₹)*	*PV (₹)*
0	1.000	(–) 4,900	(–) 4,900	(–) 4,900	(–) 4,900	(–) 4,900	(–) 4,900
1	0.943	1,000	943	2,000	1,886	3,000	2,829
2	0.890	2,000	1,780	2,000	1,780	2,000	1,780
3	0.840	3,000	2,520	2,000	1,680	1,000	840
			NPV: 343		*NPV:* 446		*NPV:* 549

From the above tables, it is apparent that at 10% required rate of return 'proposal C' is the best, and also if the required rate of return is 6%, proposal C is again the best. The difference in the *NPV*'s is due to the fact that importance of early cash receipts diminishes as the rate of interest is lowered.

Example 29.10. *Consider the investment proposal of* **Example 29.8**. *Would you still accept the proposal, if the investor desires a minimum rate of return 12% p.a.? Use IRR method.*

Solution. Here, we are required to calculate *IRR* of the given series, so that

$$\frac{5,000}{(1+r)}+\frac{5,000}{(1+r)^2}+\frac{10,000}{(1+r)^3}+\frac{10,000}{(1+r)^4}=20,000.$$

The *IRR* (or *r*) value can be determined by trial and error method. A convenient way is to calculate *NPV* of the given information at various rate of returns and admit, only, that one which gives positive *NPV* and satisfies the condition of desired rate of return.

The calculation of *NPV* and *IRR* of the project proposal when rate of return is 20% are given below:

Year	*PV factor @ 20%*	*Cash flow (₹)*	*PV (₹)*
0	1.000	(–) 20,000	(–) 20,000
1	0.833	5,000	4,165
2	0.694	5,000	3,470
3	0.579	10,000	5,790
4	0.402	10,000	4,020
			NPV: –2,555

Since, *NPV* is negative, it indicates that r = 20% is a higher rate and, therefore, we should try lower rates. We try 18%, 16% and 14%. The *NPV* values at these rates of return are calculated below:

Year	*Cash flow (₹)*	*PV factor @ 18%*	*PV (₹)*	*PV factor @ 16%*	*PV (₹)*	*PV factor @ 14%*	*PV (₹)*
0	(–) 20,000	1.000	(–) 20,000	1.000	(–) 20,000	1.000	(–) 20,000
1	5,000	0.847	4,235	0.862	4,310	0.877	4,385
2	5,000	0.718	3,590	0.743	3,715	0.769	3,845
3	10,000	0.609	6,090	0.641	6,410	0.675	6,750
4	10,000	0.516	5,160	0.552	5,520	0.592	5,920
			NPV: –925		*NPV*: –45		*NPV*: 900

From the above *NPV* values, we observe that *IRR* lies very close to 16%. If we use 15%, *NPV* comes out to ₹ 450, hence, neglecting the small deviation, we take *IRR* (= *r*) to be 16%. Since, this rate is much higher than the minimum desired rate of 12%, the investment proposal must be accepted.

Example 29.11. *A company is considering which of the two mutually exclusive projects X and Y it should undertake. The Finance Director thinks that the project with the high NPV should be chosen, whereas the Managing Director thinks that the one with the higher IRR should be undertaken especially as both projects have the same initial outlay and length of life. The company anticipates a cost of capital of 10% and the, net after tax, cash flows of the projects are as follows:*

Year	:	0	1	2	3	4	5
Cash flow (₹ '000)							
Project X	:	(200)	35	80	90	75	20
Project Y	:	(200)	218	10	10	4	3

Required:
(a) Calculate the NPV and IRR of each project.
(b) State, with reasons, which project you would recommend.
(c) Explain the inconsistency in the ranking of the two projects.
The discount factors are as follows:

Year	:	0	1	2	3	4	5
Discount factors	:						
(10%)	:	1	0.91	0.83	0.75	0.68	0.62
(20%)	:	1	0.83	0.69	0.58	0.48	0.41

Solution.

(*a*) The calculation of *NPV* and *IRR* values of the two projects are given in the following tables:
For Project *X*:

Year	Cash flow (₹ '000)	PV factor @ 10%	PV (₹ '000)	PV factor @ 20%	PV (₹ '000)
0	(–) 200	1.00	(–) 200	1.00	(–) 200
1	35	0.91	31.85	0.83	29.05
2	80	0.83	66.40	0.69	55.20
3	90	0.75	67.50	0.58	52.20
4	75	0.68	51.00	0.48	36.00
5	20	0.62	12.40	0.41	8.20
			NPV: 29.15		*NPV:* –19.35

Above table clearly shows that a discount rate of 10% is too small to bring the *PV* of cash flows down to cash flows, while discounted rate 20% is too great. Therefore, the discount rate which will bring the discounted cash flows to just equal to *PV* of cash flows lies between 10% and 20% can be found out by using the following:

$$IRR(=r) = r_1 + \frac{|NPV|_1}{|NPV|_1 + |NPV|_2} \times (r_2 - r_1)$$

$$= 10 + \frac{29.15}{29.15 + 19.35} \times (20 - 10), \quad \text{since } |NPV|_2 = |-19.35| = 19.35$$

$$= 10 + \frac{29.15}{48.50} \times 10 = 16.01\%.$$

For Project *Y*:

Year	Cash flow (₹ '000)	PV factor @ 10%	PV (₹ '000)	PV factor @ 20%	PV (₹ '000)
0	(–) 200	1.00	(–) 200	1.00	(–) 200
1	218	0.91	198.38	0.83	180.94
2	10	0.83	8.30	0.69	0.90
3	10	0.75	7.50	0.58	5.80
4	4	0.68	2.72	0.48	1.92
5	3	0.62	1.86	0.41	1.23
			NPV: 18.76		*NPV:* –3.21

Here, to find out the discount rate (*IRR*) that will bring the discounted cash flows to just equal to *PV* of cash flows between 10% and 20%, we have

$$IRR(=r) = 10 + \frac{18.76}{18.76 + 3.21} \times (20 - 10) = 18.54\%$$

(*b*) Both the projects are acceptable because they generate the positive *NPV* at the company's cost of capital at 10%. However, the company will have to select Project X, because it has higher *NPV*. If the company follows *IRR* method, then Project *Y* should be selected, because of higher internal rate of return. But, when *NPV* and *IRR* give contradictory results, a project with high *NPV* is generally preferred because of higher return in absolute terms. Hence, Project X should be selected.

(*c*) The inconsistency in the ranking of the projects arise because of the difference in the pattern of cash flows. Project X's major cash flows occur mainly in the middle three years, whereas *Y* generates the major cash flows in the first year itself.

PROBLEMS

29.09. Calculate the *NPV* for a project which initially costs ₹ 5,000 and generates year-end cash-in flows of ₹ 1,800, ₹ 1,600, ₹ 1,400, ₹ 1,200 and ₹ 1,000 respectively in the five years of its life. Assume rate of return is 10%.

29.10. A project costs ₹ 32,400 and is expected to generate cash flows of ₹ 16,000, ₹ 14,000 and ₹ 12,000 over its life of three years. Calculate the internal rate of return of the project.

29.11. A machine is being considered for purchase at a cost of ₹ 30,000. The net cash flows resulting from its operation over its expected life have been tabulated as under:

Year :	*1*	*2*	*3*	*4*	*5*	*6*	*7*
Net cash flows at the year-end :	3,000	4,500	7,500	7,500	7,500	4,500	4,500

The net cash flows listed above do not include ₹ 5,000 which is the salvage value of the machine. What is the net present value of the investment proposal assuming 12% as the interest rate.

29.12. An enterprise is considering a capital investment of ₹ 10,00,000 in a project that is expected to yield the following cash flows:

Years :	*1*	*2*	*3*	*4*	*5*	*6*	*7*	*8*
Cash inflows (₹ '000) :	160	190	170	150	150	150	150	200

If the required rate of return is 14%, determine (*a*) *NPV* of the project, and (*b*) *IRR* of the project. Is the proposal an acceptable one?

29.13. Alfa Infotech Ltd. has two investment proposals with the following cash flows:

Year	*Proposal A* Cash flow (₹)	*Proposal B* Cash flow (₹)
0	– 4,00,000	– 7,20,000
1	1,12,000	2,60,000
2	1,40,000	2,60,000
3	1,64,000	2,60,000
4	1,80,000	2,60,000

The cost of capital is 10% p.a. Assuming that both the proposals are equally risky, which proposal should be accepted ? Justify your answer. ***[MBA, Gujarat Technological Univ. 2011]***

29.14. A company has two investment proposals, each involving an initial investment of ₹ 10,000. The net cash inflows expected from the projects are given as under:

	Net cash inflows (₹)						
Year:	1	2	3	4	5	6	7
Project *A* :	2,000	4,000	5,000	4,000	3,000	3,000	3,000
Project *B* :	4,000	6,000	8,000	—	—	—	—

If the company wants 20% return on investment, which of the project is selected?

29.15. A company whose cost of capital is 10%, is considering two projects A and B. The following data are available:

Year		:	*1*	*2*	*3*	*4*	*5*
Cash flow	Project A	:	10	20	30	50	55
(₹ '000)	Project B	:	50	40	20	10	10

If initial investment for both the projects is ₹ 70,000, select the most suitable project by computing net present value at 10%.

29.16. A company has ₹ 2,00,000 to invest. The following proposals are under consideration. The cost of capital for the company is estimated to be 15 per cent.

Project	*Initial outlay*	*Annual cash flow*	*Life of project*
A	₹ 1,00,000	₹ 25,000	10 years
B	₹ 70,000	₹ 20,000	8 years
C	₹ 30,000	₹ 6,000	20 years
D	₹ 50,000	₹ 15,000	10 years
E	₹ 50,000	₹ 12,000	20 years

Rank the above projects on the basis of:

(*i*) Payback method, (*ii*) NPV method, (*iii*) Profitability index method.

Present value of annuity of ₹ 1 received in steady stream discounted at the rate of 15% are:

8 years: 4.6586; 10 years: 5.1790; and 20 years: 6.3345.

[***Hint:*** (*i*) Payback period = (Initial Investment)/(Annual Cash flow).

(*ii*) $NPV = PV$ of Cash flow – Initial Investment.

(*iii*) Profitability Index = (PV of Cash Flow)/(Initial Investment)].

29.17. A company has two investment proposals requiring ₹ 12,000 and ₹ 30,000 of initial investment, respectively. The net cash inflows expected from the projects are detailed as under:

Project	*Capital outlay*	*Net cash inflows (₹)*						
		1	*2*	*3*	*4*	*5*	*6*	*7*
A	₹ 12,000	2,000	4,000	6,000	8,000	6,000	4,000	2,000
B	₹ 30,000	10,000	10,000	10,000	10,000	10,000	10,000	10,000

If the company wants 20% return on investment, which of the either project do you suggest?

29:4. INVESTMENT UNDER UNCERTAINTY (RISKY INVESTMENTS)

Proposed investment projects are not always risk-free. An investment is not risky, if we can specify certain sequence of cash flows. But, in most of the situations, we cannot anticipate with certainty the occurrence of the possible future events and consequently the resulting cash flows. For instance, we may decide to go in for an investment proposal on the basis of NPV or IRR methods, but after a year or so the economic situation or the statutory law changes, effect the demand of the new product produced. As a result, the earlier cash flow estimates become unusable and we find that under the new situation we would not have accepted the investment proposal. Because of the uncertainty of this event, the economic conditions, future cash flows become uncertain. There is always uncertainty involved in the cash inflows because of many events such as internal, external and political situation of the country, government's monetary policy, social conditions, etc.; new innovations, change in material, cost, etc.; change in the company's management, labour strike, natural disaster, etc. Investors should, therefore, take into account whether one project is more ***risky*** than another, while evaluating two different investment proposals.

Risk associated with an investment project may be defined as the project variability. The more variable the expected future returns are, the more riskier the investment is.

There are many ways in which risk can be taken into account, while making decision of investment. Basically, there are two approaches to risk adjustment. First, there is the ***certainty equivalent method*** which involves adjusting numerator of the equation of present values. Besides, there are three interrelated methods of analyzing the investment proposals involving risk. Second, there are: Statistical Distribution methods (also called the mean standard deviation approach), Decision-Tree method, and Simulation technique. Using these methods, an interval rather than a point estimate of the expected *NPV* is presented and, thus, they are more general and objective.

We shall, now, discuss various techniques of investment under uncertain conditions.

1. Certainty-equivalent Approach. In this method, risk is adjusted by applying some correction factor which reduces the estimated future cash flows to some conservative levels or to their certainty equivalents. According to this method,

$$NPV = \sum_{t=1}^{n} \frac{\alpha_t A_t}{(1+r)^t} + C,$$

where, α_t = (Certain Cash Flow)$_t$/(Risky Cash Flow)$_t$ is the risk-adjustment factor or certainty-equivalent factor; A_1, A_2, ... are the future cash inflows and C is the initial capital outlay. The α's value which lies between 0 and 1 are inversely related to the degree of risk involved.

The method has the merit that it recognizes the risk explicitly in each cash flow. However, following are the limitations of the method:

(*i*) Calculation and certainty-equivalent coefficients may be inconsistent from investment to investment, and

(*ii*) reducing the cash flows might result in inflation in anticipation.

Example 29.12. *The Beta Co. uses a certainty-equivalent approach in its evaluation of risk investments. Currently, the company is faced with two alternative proposals. The expected net cash flows for each project and risk analysis of the cash flow distribution are given below:*

Year	*Project X*		*Project Y*	
(t)	*Cash flow (₹)*	α_t	*Cash flow (₹)*	α_t
0	*(–) 30,000*	*1.00*	*(–) 40,000*	*1.00*
1	*15,000*	*0.95*	*25,000*	*0.90*
2	*15,000*	*0.85*	*20,000*	*0.80*
3	*10,000*	*0.70*	*15,000*	*0.70*
4	*10,000*	*0.65*	*10,000*	*0.60*

If the after tax risk free rate is 8% p.a., which of the two alternatives should be selected?

Solution. We compute the *NPV* for both the projects by using the certainty equivalent formula for *NPV*:

For Project X

Year (t)	*Cash flow (₹) (A_t)*	α_t	$\alpha_t A_t$	*PV factor @ 8%*	*Present value (₹)*
0	(–) 30,000	1.00	(–) 30,000	1.000	(–) 30,000.00
1	15,000	0.95	14,250	0.926	13,195.50
2	15,000	0.85	12,750	0.857	10,926.75
3	10,000	0.70	7,000	0.794	5,558.00
4	10,000	0.65	6,500	0.735	4,777.50
					NPV: 4,457.75

For Project *Y*

Year (*t*)	*Cash flow* (₹) (A_t)	α_t	$\alpha_t A_t$	*PV factor* @ 8%	*Present value* (₹)
0	(–) 40,000	1.00	(–) 40,000	1.000	(–) 40,000
1	25,000	0.90	22,500	0.926	20,835
2	20,000	0.80	16,000	0.857	13,712
3	15,000	0.70	10,500	0.794	8,337
4	10,000	0.60	6,000	0.735	4,410
				NPV:	7,294

Since, *NPV* for project *Y* is higher than project *X*, proposal *Y* should be preferred and accepted.

2. Statistical Distribution Method. Under this approach of considering risky investment proposals, the net cash flow from an investment in each period is viewed as a random variable which can assume any one of the possible values. The method requires that probability distribution of cash flows for each of the years be obtained and considered. Using the cash flow distributions, the expected value of the *NPV* distribution and its variance can be computed as discussed below:

(*a*) ***Expected Value of NPV.*** Consider an investment proposal with cash flows whose probability distributions are given for each of the *n* years of the project life. Let the cash flows have means equal to $\bar{x}_1, \bar{x}_2, \ldots, \bar{x}_n$ with standard deviations equal to $\sigma_1, \sigma_2, \ldots, \sigma_n$ respectively. Then, the expected value of the *NPV* distribution can be obtained by adding the discounted mean value of cash flows for each of the *n* time periods. Thus, we have

$$E(NPV) = \bar{x}_0 + \frac{\bar{x}_1}{(1+r)} + \frac{\bar{x}_2}{(1+r)^2} + \ldots + \frac{\bar{x}_n}{(1+r)^n}$$

or

$$\mu = \sum_{t=0}^{n} \frac{\bar{x}_t}{(1+r)^t},$$

where, *r* is the rate of interest and $\bar{x}_t$ is the expected monetary value (*EMV*) of the cash flows generated by the project during the tth period and is given by

$$\bar{x}_t = x_{t_1} p_{t_1} + x_{t_2} p_{t_2} + \ldots + x_{t_n} p_{t_n}.$$

Here, x_{t_j} $(j = 1, 2, \ldots, n)$ denote the possible cash flow values with corresponding probabilities p_{t_j} during the tth period.

Remark. If *NPV* value exceeds zero, the proposal becomes acceptable, while, if it is lower than zero, the proposal is rejected.

(*b*) ***Variance of NPV.*** For the variance of *NPV* distributions, we consider the following two cases:

(*i*) Cash flows between different periods are assumed to be independent of one another.

(*ii*) Cash flows between different periods are assumed to be perfectly correlated.

In *case* (*i*), since the cash flows of one period are not related to the cash flows of another period, it becomes difficult to calculate the variance of the distribution of cash flows. This is because independence substantially increases the number of possible outcomes. However, the complications in calculations can be avoided and instead the variance can be obtained directly using the following:

$$\text{Variance of } NPV\ (\sigma^2) = \sigma_0{}^2 + \frac{\sigma_1{}^2}{(1+r)^2} + \frac{\sigma_2{}^2}{(1+r)^4} + \ldots + \frac{\sigma_n{}^2}{(1+r)^{2n}}.$$

or

$$\text{Variance } (NPV) = \sum_{t=0}^{n} \frac{\sigma_t{}^2}{(1+r)^{2t}},$$

where, σ_t denotes the standard deviation of cash flows during the tth period and is given by

$$\sigma_t = \sqrt{\left[(x_{t1} - \bar{x}_t)^2 + (x_{t2} - \bar{x}_t)^2 + \ldots + (x_{tn} - \bar{x}_t)^2\right]}$$

Remark. Unless otherwise stated, the cash flows of a given project may be taken to be independent.

In *case* (*ii*), the assumption of perfect correlation between cash flows of the successive years implies that if random factors cause a cash flow x_j (for the jth year) to deviate from its mean value by a standard deviation, the same factors will cause the cash flow x_k (for the kth year) to deviate from its mean in the same direction by a standard deviation. In the situation, when the periodic net cash flows are given to be perfectly correlated, the variance of *NPV* distribution can be obtained using the following:

$$\text{Variance } NPV = \left[\sigma_0 + \frac{\sigma_1}{(1+r)} + \frac{\sigma_2}{(1+r)^2} + \ldots + \frac{\sigma_n}{(1+r)^n}\right]^2 = \left[\sum_{t=0}^{n} \frac{\sigma_t}{(1+r)^t}\right]^2.$$

Remark. The standard deviation value when the cash flows are perfectly correlated would be greater than when they are independent.

(*c*) ***Normally Distributed NPV.*** The normal distribution can be used to further analyse the risk in investment proposals. The basic assumption is that the future cash flows follow normal probability law.

In the situation, the *NPV* of the project would also be normally distributed, with a mean equal to the expected net present value and standard deviation equal to the square root of the variance of *NPV* distribution.

Using the table of area under the standard normal probability curve, all the probabilities of events like "*NPV* will be zero" or "*NPV* is negative" or the like can be easily computed.

(*d*) ***EVPI of a Project.*** The expression for *EVPI* can be written as follows:

$$EVPI = \sigma \times L_N(D),$$

where, σ = the standard deviation of the distribution and $L_N(D)$ = the normal loss integral or the unit normal loss function.

Example 29.13. *A company has determined the following probabilities for net cash flow generated by a project:*

Year 1		*Year 2*		*Year 3*	
Cash flow	*Probability*	*Cash flow*	*Probability*	*Cash flow*	*Probability*
₹2,000	0.10	₹2,000	0.20	₹2,000	0.30
₹4,000	0.20	₹4,000	0.30	₹4,000	0.40
₹6,000	0.30	₹6,000	0.40	₹6,000	0.20
₹8,000	0.40	₹8,000	0.10	₹8,000	0.10

Calculate the expected monetary values. Also, calculate the present value of the expected cash flow using 10% discount rate.

Solution. The following table shows the calculation of the expected monetary value:

Year 1			*Year 2*			*Year 3*		
Cash flow	*Prob.*	*Expected value*	*Cash flow*	*Prob.*	*Expected value*	*Cash flow*	*Prob.*	*Expected value*
₹ 2,000	0.10	₹ 200	₹ 2,000	0.20	₹ 400	₹ 2,000	0.30	₹ 600
₹ 4,000	0.20	₹ 800	₹ 4,000	0.30	₹ 1,200	₹ 4,000	0.40	₹ 1,600
₹ 6,000	0.30	₹ 1,800	₹ 6,000	0.40	₹ 2,400	₹ 6,000	0.20	₹ 1,200
₹ 8,000	0.40	₹ 3,200	₹ 8,000	0.10	₹ 800	₹ 8,000	0.10	₹ 800
	EMV:	₹ 6,000		*EMV:*	₹ 4,800		*EMV:*	₹ 4,200

Using *EMV* (or the expected cash flow), the expected *NPV* can be calculated as under:

Year	*Expected cash flow*	*PV factor @ 10%*	*Present value*
1	₹ 6,000	0.900	₹ 5,454.00
2	₹ 4,800	0.826	₹ 3,964.80
3	₹ 4,200	0.751	₹ 3,154.20
			NPV: ₹ 12,573.00

Example 29.14. *An investment proposal had the following discrete probability distribution of expected cash flows:*

State (j)	*Year 1*		*Year 2*		*Year 3*	
	Return	*Probability*	*Return*	*Probability*	*Return*	*Probability*
1	50	0.10	20	0.10	(–) 40	0.10
2	60	0.20	40	0.25	30	0.30
3	70	0.40	60	0.30	50	0.30
4	80	0.20	80	0.25	80	0.20
5	90	0.10	100	0.10	140	0.10

Give (a) the expected return each year and net present worth of the total expected return, (b) standard deviation of the expected returns in each year, and also standard deviation of present value of expected returns (discount rate = 10%).

Solution. The calculation of expected return and variance are shown in the following table:

Year 1				*Year 2*				*Year 3*			
X	p	pX	$p(X-\bar{X}_1)^2$	X	p	pX	$p(X-\bar{X}_2)^2$	X	p	pX	$p(X-\bar{X}_3)^2$
50	0.10	5	40	20	0.10	2	160	(–) 40	0.10	(–) 4	810
60	0.20	12	20	40	0.25	10	100	30	0.30	9	120
70	0.40	28	0	60	0.30	18	0	50	0.30	15	0
80	0.20	16	20	80	0.25	20	100	80	0.20	16	180
90	0.10	9	40	100	0.10	10	160	140	0.10	14	810
		$\bar{X}_1 = 70,\ \sigma_1^2 = 120$				$\bar{X}_2 = 60,\ \sigma_2^2 = 520$				$\bar{X}_3 = 50,\ \sigma_3^2 = 1{,}920$	

From the above table, clearly we have

$$\sigma_1 = \sqrt{120} = 10.95, \quad \sigma_2 = \sqrt{520} = 22.80 \text{ and } \sigma_3 = \sqrt{1{,}920} = 43.82.$$

The present values of the project and variance are computed in the following table:

Year (t)	*Expected value* $(\bar{X})$	*PV factor* @ 10%	*PV* $(1 + 0.10)^{-t}$	*Variance* σ_t^2	*PV factor* @ 10%	*PV* $(1 + 0.10)^{-2t}$
1	70	0.909	63.637	120	0.826	99.168
2	60	0.826	49.584	520	0.683	355.160
3	50	0.751	37.565	1,920	0.564	1,083.840
		EMV:	150.786		Variance:	1,538.168

From the above calculations, clearly the expected value of the project = 150.786 and standard deviation = $\sqrt{\text{Variance}} = \sqrt{1,538.168} = 39.22$.

It may be noted that cash flows have been assumed to be independent.

Example 29.15. *Two mutually exclusive investment proposals are being considered. The following information is available:*

	Project A	*Project B*
Cost	*₹20,000*	*₹20,000*
Life	*2 years*	*2 years*
Cash flow each year	*₹12,000*	*₹12,000*
Salvage value	0	0

Upon further analysis, it was found that probability of cash inflow each year for the project is as follows:

Project A		*Project B*	
Possible inflow	*Probability*	*Possible inflow*	*Probability*
₹10,000	0.20	*₹11,000*	0.20
₹12,000	0.60	*₹12,000*	0.60
₹14,000	0.20	*₹13,000*	0.20

Advise for selecting the proposal, assuming the cost of capital as 10%.

Solution. The *EMV* and the associated risk (standard deviation) are computed for each project as shown below:

Project	*Possible cash inflow* (X)	*Probability* (*p*)	*p*X	$p(X - \bar{X})^2$
A	10,000	0.20	2,000	8,00,000
	12,000	0.60	7,200	0
	14,000	0.20	2,800	8,00,000
			$\bar{X}_1$ = 12,000	σ_1^2 = 16,00,000
B	11,000	0.20	2,200	2,00,000
	12,000	0.60	7,200	0
	13,000	0.20	2,600	2,00,000
			$\bar{X}_2$ = 12,000	σ_2^2 = 4,00,000

Using the information given in the above table, the expected *NPV* and variance are shown in the following table:

Year	PV factor	Expected	PV	PV factor	Project A		Project B	
t	@ 10%	value $(\bar{X})$		$(1 + 0.10)^{-2t}$	Variance	PV	Variance	PV
0	1.000	(–) 20,000	(–) 20,000	1.000	0	0	0	0
1	0.909	12,000	10,908	0.826	16×10^5	13,21,600	40×10^4	3,30,400
2	0.826	12,000	9,912	0.683	16×10^5	10,92,800	40×10^4	2,73,200
			E (NPV) : 820		V (NPV) : 24,14,400		V (NPV) : 6,03,600	

$\therefore$ $\sigma_A (NPV) = \sqrt{2414400} = 1553.83$ and $\sigma_B (NPV) = \sqrt{603600} = 776.92.$

The relative risk degree of two projects can be determined by comparing their coefficient of variation. We know that

$$\text{Coefficient of Variation} = \frac{\text{Standard Deviation}}{\text{Expected Value}} \times 100$$

$$\therefore \qquad (CV)_A = \frac{1553.83}{820} \times 100 = 189.49\%$$

and

$$(CV)_B = \frac{776.92}{820} \times 100 = 94.75\%$$

Since CV of project B is smaller, project B must be preferred.

Example 29.16. *A company has developed the following discrete probability distribution of cash flows for a particular capital budgeting proposal:*

Year 1		Year 2		Year 3	
Cash flow	Probability	Cash flow	Probability	Cash flow	Probability
₹100	0.10	₹300	0.20	₹ 500	0.10
₹200	0.20	₹400	0.40	₹ 600	0.30
₹300	0.30	₹500	0.30	₹ 700	0.30
₹400	0.40	₹600	0.10	₹1,000	0.30

Assume that the probability distributions of the flows for different years are perfectly correlated over time; the initial cost of the project is ₹1,000; and the appropriate discount rate is 10%. Determine:

(*a*) *The mean and standard deviation of cash flows for each year.*
(*b*) *Expected NPV of the project.*
(*c*) *Probability that the project shall have an NPV greater than zero, assuming that the NPV distribution follows 'normal' pattern.*
(*d*) *Probability that the project would succeed, assuming that instead of being perfectly correlated, the successive flows are independent of each other.*
(*e*) *EVPI for the project under the assumption of (i) perfectly correlated, and (ii) independent cash flows.*

Solution.

(*a*) The calculations of expected values and variance of the project are shown in the following table:

Year 1				Year 2				Year 3			
Cash flow (X)	Prob. (p)	pX	$p(X-\bar{X}_1)^2$	Cash flow (X)	Prob. (p)	pX	$p(X-\bar{X}_2)^2$	Cash flow (X)	Prob. (p)	pX	$p(X-\bar{X}_3)^2$
100	0.10	10	4,000	300	0.20	60	3,380	500	0.10	50	5,760
200	0.20	40	2,000	400	0.40	160	360	600	0.30	180	5,880
300	0.30	90	0	500	0.30	150	1,470	700	0.30	210	480
400	0.40	160	4,000	600	0.10	60	2,890	1,000	0.30	300	20,280
	Total:	300	10,000			430	8,100			740	32,400

Thus, we have,

$$\bar{X}_1 = 300, \quad \sigma_1 = \sqrt{10,000} = 100; \quad \bar{X}_2 = 430, \quad \sigma_2 = \sqrt{8,100} = 90;$$

and $\bar{X}_3 = 740, \quad \sigma_3 = \sqrt{32,400} = 180.$

(*b*) The expected *NPV* is calculated in the following table:

Year t	*PV factor* $(1 + 0.10)^{-t}$	*Expected value*	*Present value*	σ_t	*Present value*
1	0.9091	300	₹ 272.730	100	₹ 90.910
2	0.8264	430	₹ 355.352	90	₹ 74.376
3	0.7513	740	₹ 555.962	180	₹ 135.234
		Total:	1,184.044		σ = 300.520
		less outflow:	1,000.000		
		Expected *NPV*:	184.044		

Thus, expected *NPV* of the project is 184.044.

(*c*) To determine the required probability, we shall find the area to the right of X = 0 under the normal curve with mean (expected *NPV*) = 184.044 and σ = 300.520 as shown in *Fig. 29.1:*

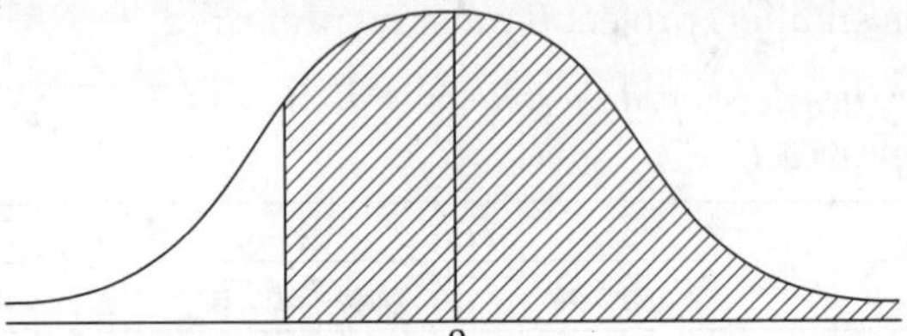

Fig. 29.1

$$z = \frac{0 - 184.044}{300.520} = (-)0.61.$$

From the table of area under standard normal curve, the area corresponding to $z = -0.61$ is $\phi(0.61) = 0.2291$.

Thus, $$p(X > 0) = P(z > -0.61) = 0.5000 + 0.2291 = 0.7291.$$

(*d*) If cash flows are assumed to be independent, then

$$\text{Variance } (NPV) = \sigma_1^2 + \sigma_2^2 + \sigma_3^2 = (90.910)^2 + (74.376)^2 + (135.234)^2 = 32,084.65$$

∴ Standard Deviation $= \sqrt{32,084.65} = 179.12$

For the area under standard normal curve to the right of X = 0, we have

$$z = \frac{0 - 184.044}{179.12} = (-)1.03.$$

From the table of area under standard normal curve, area corresponding to z = (–) 1.03 is 0.3485.

Thus, $$P(X > 0) = 0.5 + \phi(1.03) = 0.5000 + 0.3485 = 0.8485.$$

(*e*) (*i*) When cash flows are perfectly correlated, L_N (0.61) = 0.1659, using table of unit normal loss function.

∴ $$EVPI = 300.52 \times 0.1659 = ₹\ 49.86.$$

(*ii*) When cash flows are independent, L_N (1.03) = 0.07866, from table of unit normal loss function.

∴ $$EVPI = 179.12 \times 0.07866 = ₹\ 14.09.$$

PROBLEMS

29.18. A company uses the certainty-equivalent approach in its evaluation of risky investments. Currently, the company has two alternative approaches. The expected net cash flow for the two projects are as follows:

Year		:	0	1	2	3	4
Cash flow (in ₹)	Project *A*	:	(–) 15,000	7,500	7,500	5,000	5,000
	Project *B*	:	(–) 20,000	12,500	10,000	7,500	5,000

Risk analysis of the cash flow distribution has provided the following certainty-equivalents:

Year	:	0	1	2	3	4
Project *A*	:	1.00	0.95	0.85	0.70	0.65
Project *B*	:	1.00	0.90	0.80	0.70	0.60

If the after-tax risk-free rate is 5% p.a. which of the two projects should be selected?

29.19. A company uses certainty-equivalent approach in its evaluation of risky investments. The company has now to make a choice between two proposals X and Y. The necessary information is as follows:

Year	*Proposal X*		*Proposal Y*	
(*t*)	*Cash flow*	α_t	*Cash flow*	α_t
0	₹ (–) 40,000	1.00	₹ (–) 30,000	1.00
1	₹ 20,000	0.90	₹ 15,000	0.95
2	₹ 18,000	0.80	₹ 12,000	0.80
3	₹ 12,000	0.80	₹ 10,000	0.75
4	₹ 10,000	0.60	₹ 5,000	0.70

If the after tax risk-free rate is 8%, which of the two alternatives should be selected?

29.20. Omega Co. is presently considering an investment proposal. It was found that the probabilities for various net cash inflows generated by the proposal for the three years period are given below:

Event	*Year 1*		*Year 2*		*Year 3*	
	Cash flow	*Probability*	*Cash flow*	*Probability*	*Cash flow*	*Probability*
A	₹ 3,000	0.20	₹ 3,000	0.30	₹ 3,000	0.40
B	₹ 2,000	0.10	₹ 2,000	0.20	₹ 2,000	0.30
C	₹ 1,000	0.30	₹ 1,000	0.40	₹ 1,000	0.20
D	₹ 4,000	0.40	₹ 4,000	0.10	₹ 4,000	0.10

Compute the *EMV* for each year. Also, determine the expected present value of the investment proposal, using 10% discount rate.

29.21. The possible outcomes (in thousand rupees) for three alternative investment proposals and their probabilities of occurrence are given below:

Alternative A_1		*Alternative* A_2		*Alternative* A_3	
Outcome	*Probability*	*Outcome*	*Probability*	*Outcome*	*Probability*
125	0.20	225	0.30	200	0.40
200	0.40	400	0.50	500	0.50
300	0.40	500	0.20	1,000	0.10

Rank the three alternative in terms of (*i*) expected value of outcome, and (*ii*) risk.

29.22. *XYZ* Company is considering to make investment in a proposal which requires an outlay of ₹ 50,000. The project has a life of three years over which the following cash inflows are likely to be generated:

Year 1		*Year 2*		*Year 3*	
Cash flow	*Probability*	*Cash flow*	*Probability*	*Cash flow*	*Probability*
₹ 10,000	0.10	₹ 10,000	0.20	₹ 10,000	0.30
₹ 20,000	0.20	₹ 20,000	0.30	₹ 20,000	0.40
₹ 30,000	0.30	₹ 30,000	0.40	₹ 30,000	0.20
₹ 40,000	0.40	₹ 40,000	0.10	₹ 40,000	0.10

If the discount rate is 4% and the cash flows of the three years are independent, determine the expected *NPV* of the investment proposal and its standard deviation.

29.23. Obtain expected and the associated risk as measured by the standard deviation for the project *A*, if the cost of capital equals 10%, given probability distribution of cash flow is given below:

Year 1		*Year 2*		*Year 3*	
Cash flow	*Probability*	*Cash flow*	*Probability*	*Cash flow*	*Probability*
₹ 1,000	0.10	₹ 1,900	0.20	₹ 1,500	0.10
₹ 1,500	0.20	₹ 2,500	0.30	₹ 2,250	0.70
₹ 2,000	0.40	₹ 2,750	0.20	₹ 2,500	0.10
₹ 2,500	0.20	₹ 3,150	0.30	₹ 3,000	0.10
₹ 3,000	0.10				

29.24. Three mutually exclusive investment proposals are being considered. The following information is available:

	Project A	*Project B*	*Project C*
Cost	₹ 18,000	₹ 18,000	₹ 18,000
Life	2 years	2 years	2 years
Cash flow each year	₹ 12,000	₹ 12,000	₹ 12,000

Upon further analysis, it was found that the probability of cash inflow each year for project *A*, *B* and *C* is as follows:

Project A		*Project B*		*Project C*	
Possible inflow (₹)	*Probability*	*Possible inflow (₹)*	*Probability*	*Possible inflow (₹)*	*Probability*
6,000	0.3	8,000	0.3	10,000	0.3
12,000	0.4	12,000	0.4	12,000	0.4
18,000	0.3	17,000	0.3	14,000	0.3

Assuming cost of capital at 10%, advise the selection of project by giving the reasons for such action.

29.25. Assume a discount rate of 0.10 and an investment with the following projected cash flows:

Year	*Mean cash flow (in ₹)*	*SD of cash flow distribution (in ₹)*
0	(–) 1,600	700
1	2,000	1,000
2	1,000	1,500

The distribution is assumed to be normal and independent of each other.

(i) Compute the mean of the present value distribution. Using the mean value, would the investment be accepted?

(ii) Compute the standard deviation of the present value distribution.

(iii) Compute the probability of the investment having a net present value of zero or less.

29.26. An investment proposal involves an outlay of ₹ 2,20,000. The proposal is expected to have a life of 3 years only, and is likely to generate the following cash inflows at the year ends.

Year 1		*Year 2*		*Year 3*	
Cash inflow	*Probability*	*Cash inflow*	*Probability*	*Cash inflow*	*Probability*
₹ 50,000	0.2	₹ 70,000	0.2	₹ 1,40,000	0.3
₹ 80,000	0.7	₹ 1,00,000	0.5	₹ 1,50,000	0.3
₹ 1,20,000	0.1	₹ 1,30,000	0.3	₹ 1,62,500	0.4

(a) Determine the expected cash flow for each of the three years and its standard deviation.

(b) Using a 12% discount rate, calculate the expected net present value of the project.

(c) Determine the risk associated with the project as measured by the standard deviation, assuming that the cash flows of various years are independent of each other.

(d) Assuming the *NPV* is distributed normally, determine the probability that the project shall have a negative *NPV*.

(e) Obtain the *EVPI* of the project.

29:5. DECISION TREES AND INVESTMENT ANALYSIS

In some situations, cash flows are moderately correlated over time. That is, they are neither independent (or uncorrelated), nor perfectly correlated. Statistical methods discussed above fail in such cases so far as the determination of standard deviation of the project is concerned. Investment problems in such situations are tackled by using decision trees.

The decision tree approach to the evaluation of risk and uncertainty rests on the impact of all probabilistic estimates of potential outcomes. In other words, when using decision-tree analysis, every potential event is weighted in probabilistic terms and that is the basis of evaluation. We shall illustrate the procedure considering the following example:

Example 29.17. *A company is considering an investment proposal which costs ₹ 250 at time t = 0. The project is expected to generate the net cash flows as shown in the table given below:*

Year 1		*Year 2*			
Initial probability	*Net cash flow (in ₹)*	*Conditional probability*	*Net cash flow (in ₹)*	*Cash flow series*	*Joint probabilities*
0.25	(–) 100	0.40	(–) 400	1	0.10
		0.40	(–) 100	2	0.10
		0.20	200	3	0.05
0.50	200	0.20	(–) 100	4	0.10
		0.60	200	5	0.30
		0.20	500	6	0.10
0.25	500	0.20	200	7	0.05
		0.40	500	8	0.10
		0.40	800	9	0.10

Determine the expected value of NPV and its standard deviation, assuming a risk-free discount rate of 4%.

Solution. The last column in the given table shows the joint probability of a particular cash flow series. For instance, for series *4*, the joint probability of a ₹ 200 cash flow in period *1*, being followed by a cash flow of ₹ (–) 400 in period *2* is: 0.25 × 0.40 = 0.10. The above information can also be represented in the form of the following decision tree:

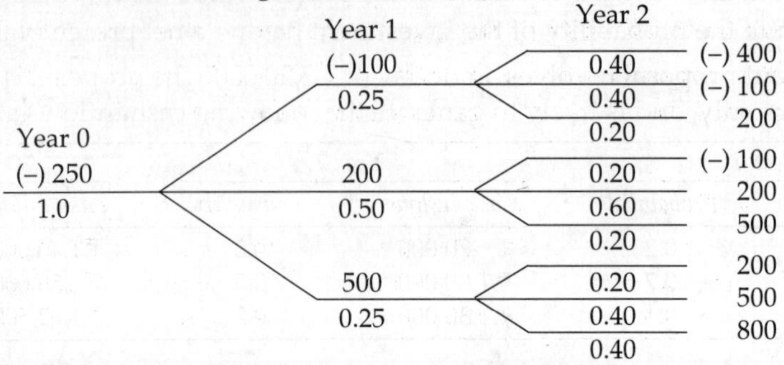

Fig. 29.2. Decision Tree for Cash Flows

The procedure to calculate the *NPV* for the project can be found out by discounting the cash flow of each series, multiplying the sum of the *PV* for each series by the joint probability to arrive at the expected value of the series and then subtracting the initial cost of the project from the sum of expected values of all the series. The calculations for the *NPV* of the project at a 4% risk-free rate are shown below:

Cash flow series *i*	*Discounted cash flow (PV)* Year 1	Year 2	*NPV of series (PV – ₹ 250) (in ₹)*	*Joint probability* p_i	*Expected NPV of series (in ₹)*
1	(–) 100 (0.962)	(–) 400 (0.925)	(–) 715.98	0.10	(–) 71.60
2	(–) 100 (0.962)	(–) 100 (0.925)	(–) 438.61	0.10	(–) 43.86
3	(–) 100 (0.962)	200 (0.925)	(–) 161.24	0.05	(–) 8.06
4	200 (0.962)	(–) 100 (0.925)	(–) 150.15	0.10	(–) 15.02
5	200 (0.962)	200 (0.925)	127.22	0.30	(–) 38.17
6	200 (0.962)	500 (0.925)	404.59	0.10	40.46
7	500 (0.962)	200 (0.925)	415.68	0.05	20.78
8	500 (0.962)	500 (0.925)	693.05	0.10	69.31
9	500 (0.962)	800 (0.925)	970.41	0.10	97.04
				Expected value of *NPV*:	₹ 127.22

The SD of the *NPV*'s is computed from the 9 *NPV*'s, their expected *NPV* and the joint probabilities by using the formula

$$\sigma = \sqrt{\sum_{i=1}^{9}\left(NPV_i - \overline{NPV}\right)^2 p_i},$$

where, NPV_i is the *NPV* of series *i* of net cash flow, covering all periods, $\overline{NPV}$ is the expected value of *NPV* of the project and p_i is the probability of occurrence of the ith series. Thus,

$$\sigma^2 = 0.10(-715.98 - 127.22)^2 + 0.10(-438.61 - 127.22)^2 + 0.05(-161.24 - 127.22)^2$$
$$+ 0.10(150.15 - 127.22)^2 + 0.30(127.22 - 127.22)^2 + 0.10(404.59 - 127.22)^2$$
$$+ 0.05(415.68 - 127.22)^2 + 0.10(693.05 - 127.22)^2 + 0.10(970.41 - 127.22)^2$$

This gives σ = ₹ 480.

PROBLEM

29.27. The relationship between a present decision, future events and their consequences in terms of cash flows and their probabilities for *ABC* company is shown below. Find the expected *NPV* and its SD.

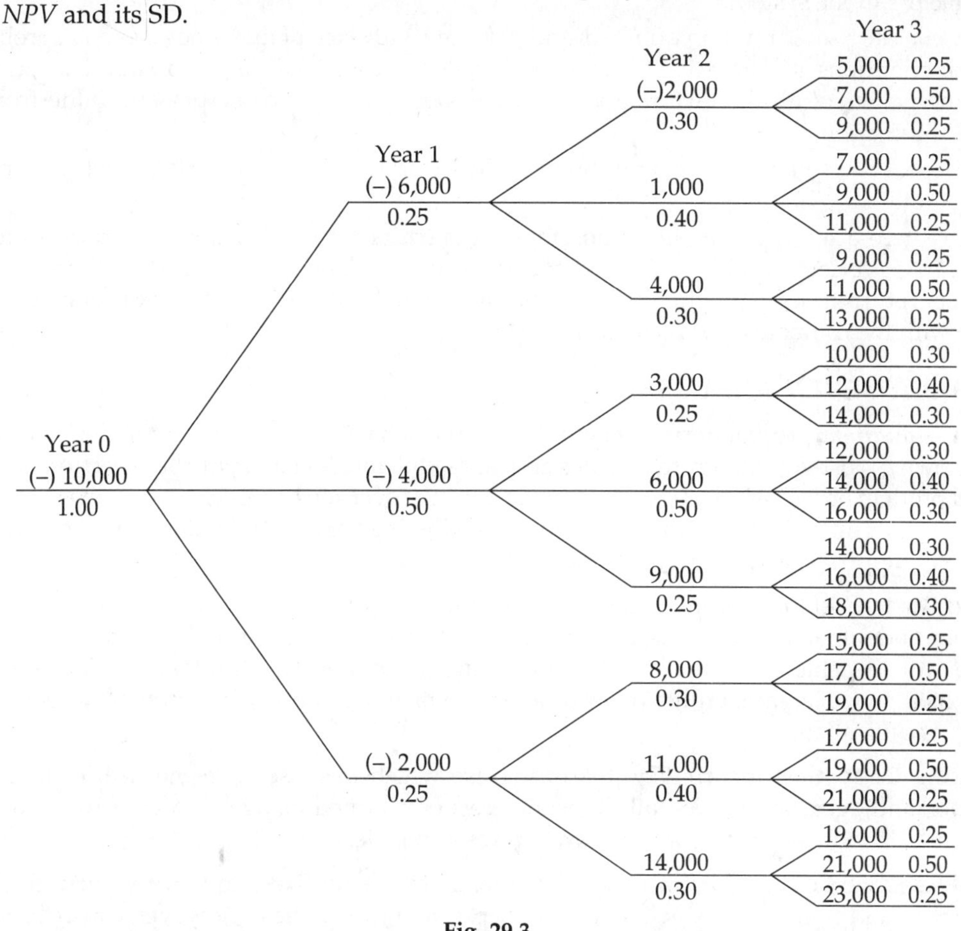

Fig. 29.3

29:6. SIMULATION APPROACH TO RISKY INVESTMENTS

While evaluating the risky investment proposals, simulation can be used to approximate the *NPV* or the expected return and its dispersion about the expected value. A simulation model proposed by ***Hertz**** considers the following factors, while evaluating an investment proposal:

1. Market analysis: (*a*) Market size, (*b*) Selling price, (*c*) Market growth rate, and (*d*) Share of market (which results in physical sales volume).
2. Investment Cost Analysis: (*a*) Investment required, (*b*) Residual value of investment.
3. Operating and Fixed Cost: (*a*) Operating costs, (*b*) Fixed costs, and (*c*) Useful life of facilities.

The probability distributions are assigned to each of the above nine factors, on the basis of the management's assessment of the probable outcomes. Once this is done, the net present value for the rate of return can be obtained that will result from a random combination of the above factors.

---

* ***David B. Hertz***, "Risk Analysis in Capital Investment", 42 (Jan.-Feb. 1964), pp. 95-106.

A simulation model consists of the following three steps:

Step 1. Estimate the range of values for each of the above nine factors; and within that range the likelihood of occurrence of each value, *i.e.*, develop the probability distribution of uncertain factors in the system.

Step 2. Generate a set of random numbers and associate with each of them the respective probability distributions, to determine the expected values for each factor in the model. Then, combine the values of all the factors and compute the rate of return or present value from that combination.

Step 3. Repeat *Steps 1* and *2* number of times to define and evaluate the odds of each possible rate of return.

Since, there are large number of possible combination of values, it is important to test the likelihood that various specific returns on the investment will occur. The result will be a listing of the rates of return that one might achieve, ranging from a loss to whatever maximum gain, that is possible with the estimates that have been made.

29:7. BREAK-EVEN ANALYSIS

Break-even analysis provides an understanding of the behaviour of profits in relation to output. It is also known as cost-volume-profit analysis and deals with inter-relationship between total revenue, cost, and profit in order to determine the potential effect of decisions necessary to achieve or maintain various levels of financial structure of a company. Following are the four factors involved in a study of cost-volume-profit relationship:

1. Fixed Cost. Cost incurred by the enterprise which do not vary with increase or decrease of its production and sale activity. The cost continue to be incurred irrespective of the volume of production and sale actually achieved by the enterprise (of course within defined limits). They include capital assets which, however, are a function of time with which they depreciate through irrespective of intensity of utilization.

2. Variable Cost. Cost incurred by the enterprise which increase with the rise in the level of production and decrease with its fall. It incudes cost of production, raw materials and bought out components, labour, electricity, water, power taxes and levies.

3. Semi-variable Costs. These are the costs which are partly fixed and partly variable such as maintenance cost of machines and equipment, depreciation, etc. These, however, can be broken up into variable and fixed portions.

4. Sales Price. The cost volume-profit relationship is given by

$$\frac{\text{Sale} - \text{Variable cost}}{\text{Sale}}.$$

It shows by how much profit will change with a given change in sales.

Another concept which is of significance in break-even analysis is that of ***contribution margin***, which is defined as the excess of selling price over variable cost. Thus,

$$\text{Contribution margin} = \text{Total sales} - \text{Total variable cost}.$$

Also, $\text{Contribution margin} = \text{Fixed cost} + \text{Profit}.$

When the contribution margin is related to the sales either on a unit basis or on the aggregate basis, we obtain the ***Profit/Volume ratio*** or ***the P/V ratio*** or ***the contribution margin ratio***. If S denotes the selling price per unit and V, the variable cost per unit, then

$$P/V \text{ ratio} = \frac{\text{Contribution margin per unit}}{\text{Selling price}} \times 100 = \frac{S-V}{S} \times 100.$$

Break-even Point

Break-even analysis represents the level of sales at which costs and revenues are in equilibrium; the ***equilibrium point*** is called the ***break-even point*** (*BEP*). It may be expressed in terms of number of units or sales in rupees. Thus, we have

$$BEP \text{ (in units)} = \frac{\text{Fixed costs}}{\text{Selling price per unit} - \text{Variable cost per unit}}$$

$$= \frac{F}{S-V} \text{ or } \frac{\text{Fixed costs}}{\text{Unit contribution margin}}$$

and
$$BEP \text{ (in rupees)} = \frac{F}{S-V} \times S = \frac{F}{1-(V/S)}$$

Margin of Safety

It is the excess of actual (or budgeted) sales over break-even sales. Also, margin of safety ratio can be calculated as the margin of safety expressed as a percentage of actual (or budgeted) sales. Thus,

$$\text{Margin of safety} = \text{Actual sales} - \text{Break-even sales}$$

Also,
$$\text{Margin of safety ratio} = \frac{\text{Actual sales} - \text{Break-even sales}}{\text{Actual sales}} \times 100$$

Here,
$$BEP \text{ (Sales)} = \frac{F}{P/V \text{ ratio}},$$

where,
$$P/V \text{ ratio} = \frac{\text{Total sales revenue} - \text{Total variable cost}}{\text{Total sales revenue}} \times 100$$

Sales to earn a given level of profit are obtained as follows:

$$\text{Sales (in rupees)} = \frac{\text{Fixed cost} + \text{Profit}}{P/V \text{ ratio}} = \frac{\text{Contribution}}{P/V \text{ ratio}},$$

and
$$\text{Sales (in units)} = \frac{\text{Fixed cost} + \text{Profit}}{\text{Contribution per unit}}.$$

Break-even Chart

The determination of the break-even point can be depicted graphically also. For this, the level of activity (in terms of the number of units, capacity level, sales volume in rupees, etc.) is shown on the *X*-axis, and the *Y*-axis is taken to show costs and revenue. The fixed cost is shown, first of all, by a straight line parallel to the horizontal axis at a height given by the amount of fixed cost. The variable cost component is then added to it, to get the total cost line. Then, the sales revenue function is plotted on the graph. The point where two lines intersect, gives the break-even point. A typical break-even chart is as shown in *Fig. 29.4*.

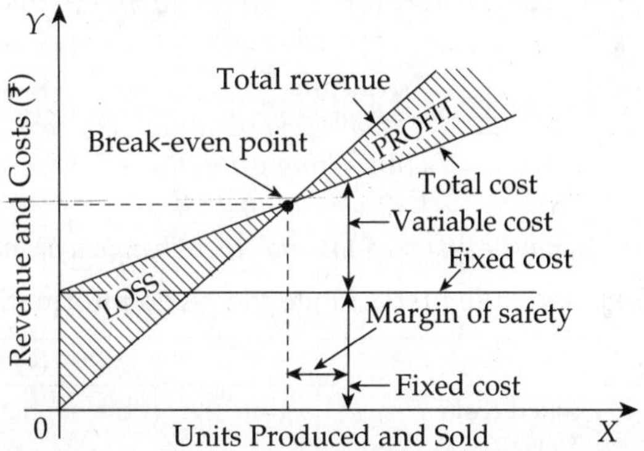

Fig. 29.4

It may be noted that the ***break-even chart*** or ***the cost-volume-profit graph*** gives more information than the break-even point and the margin of safety. These charts are often used for determining the following:

(*i*) Break-even production volume.
(*ii*) Profit appropriation for a given level of output.

(*iii*) Choice of optimum level of output, and
(*iv*) Impact of the rate of change in sales on costs and profits.

Example 29.18. *A manufacturing firm produces a single product whose selling price is ₹16 per unit and the variable cost per unit is ₹12. If the annual fixed costs of the firm are estimated as ₹1,20,000; find the break-even point in units, in rupees, and as a percentage of capacity, if the firm has an estimated capacity of 50,000 units of the product. What is the margin of safety?*

Solution. We are given: $S =$ ₹ 16, $V =$ ₹ 12 and $F =$ ₹ 1,20,000.

$$\therefore \quad \text{Break-even point (units)} = \frac{F}{S-V} = \frac{1,20,000}{16-12} = 30,000 \text{ units.}$$

$$\text{Break-even point (rupees)} = \frac{F}{1-V/S} = \frac{1,20,000}{1-12/16} = ₹\,4,80,000.$$

$$\text{Break-even point (as \% of capacity)} = \frac{\text{Break-even sales}}{\text{Capacity of sales}}$$

$$= \frac{30,000}{50,000} \times 100 = 60\%.$$

$$\text{Margin of safety} = 50,000 - 30,000 = 20,000 \text{ units.}$$

Example 29.19. *Break-even production of a firm is of 5,000 units, the fixed cost is ₹50,000, the variable cost per unit is ₹25; find out the price of the product. How much the firm should produce to earn profit of ₹25,000?*

Solution. Here, we are given: BEP (units) = 5,000, $F =$ ₹ 50,000 and $V =$ ₹ 25 per unit.

Let p is the price of a unit. Then,

$$BEP \text{ (units)} = \frac{F}{p-V}$$

$$\therefore \quad 5,000 = \frac{50,000}{p - ₹\,25} \quad \text{or} \quad p = \frac{1,75,000}{5,000} = ₹\,35.$$

To find out the sale of the product in units, so as to earn ₹ 25,000 as profit, we have

$$\text{Sales} = \frac{\text{Fixed cost + Profit}}{\text{Contribution margin per unit}} = \frac{50,000+25,000}{35-25} = 7,500 \text{ units.}$$

Example 29.20. *Standard Rolling Machines Corporation is considering the installation of a new machine to its present assembly unit. The installation of the new machine is expected to reduce variable cost by 15% per unit, but is likely to add ₹60,000 to the existing fixed costs. Given the following current data, determine whether or not the new machine should be purchased:*

Fixed costs	*₹2,00,000*
Variable cost per unit	*₹20*
Unit selling price	*₹30*
Expected annual sales	*30,000 units.*

It may be assumed that no other change will take place by the introduction of the new machine.

Solution. We first arrange the given information as below:

	Before the new machine	*After the new machine*
Fixed costs	₹ 2,00,000	₹ 2,60,000
Variable cost per unit	₹ 20	₹ 17
Unit selling price	₹ 30	₹ 30
Estimated annual sales	30,000 units	30,000 units

Now, break-even point is given as

$$BEP = \frac{\text{Fixed cost}}{\text{Unit selling price} - \text{Unit variable cost}}$$

(a) *BEP* (prior to the acquisition of new machine) $= \frac{2,00,000}{30-20} = 20,000.$

(b) *BEP* (after the acquisition of new machine) $= \frac{2,60,000}{30-17} = 20,000.$

Clearly, the break-even analysis suggests that the corporation should be indifferent with regard to the purchase of the new machine.

Remark. The above example shows that break-even analysis alone may not be sufficient to solve a decision problem. This necessitates to seek help of some other methods. A method which is sometimes considered as an alternative analysis method, is the ***cost-profit analysis method***.

Alternative Approach

The profit for each indicated assumption in the above example can be obtained as follows:

Profit = Total revenue – Total cost, where

Total revenue = Unit selling price × Units sold.

Total cost = Total fixed cost + Total variable cost

= Total fixed cost + (Unit variable cost) × (Units produced or sold)

Thus, in the above example, substituting for these costs gives:

(a) Before the acquisition of new machine

Total revenue = 30 × 30,000 = ₹ 9,00,000

Total cost = 2,00,000 + 20 × 30,000 = ₹ 8,00,000

∴ Profit = 9,00,000 – 8,00,000 = ₹ 1,00,000.

(b) After the acquisition of new machine

Total revenue = ₹ 9,00,000

Total cost = 2,60,000 + 17 × 30,000 = ₹ 7,70,000

∴ Profit = ₹ 9,00,000 – ₹ 7,70,000 = ₹ 1,30,000.

Thus, on the basis of cost profit analysis, the corporation is bound to benefit by a profit of ₹ 30,000, if the new machine is installed.

Example 29.21. *From the following data you are required to calculate break-even point and net sales value at this point:*

Direct material cost per unit	*₹ 8*
Direct labour cost per unit	*₹ 5*
Fixed overheads	*₹ 24,000*
Variable overheads	*@ 60% on direct labour*
Selling price per unit	*₹ 25*
Trade discount	*4%*

If sales are 15% and 20% above the break-even volume, determine the net profit.

Solution.

(i) *Break-even Point:*

Selling price per unit = ₹ 25

Trade discount @ 4% = ₹ 25 × 0.04 = ₹ 1

Net realization per unit sold = ₹ 25 – ₹ 1 = ₹ 24.

Cost varying with per unit output:

Direct material cost = ₹ 8, and Direct labour cost = ₹ 5

Variable overheads cost = ₹ 5 × 0.60 = ₹ 3

∴ Total cost = ₹ (8 + 5 + 3) = ₹ 16.

Contribution per unit = $S - V$ = ₹ 24 – ₹ 16 = ₹ 8.

$$\therefore \quad BEP \text{ sales (units)} = \frac{\text{Fixed cost}}{\text{Contribution per unit}} = \frac{₹\,24{,}000}{₹\,8} = 3{,}000 \text{ units.}$$

(*ii*) *Net sales values at BEP sales:*

Sales value = Unit sold × Sales price per unit

= 3,000 × ₹ 25 = ₹ 75,000

Less: Trade discount @ 4% = ₹ 75,000 × 0.04 = ₹ 3,000

∴ Net sales value = ₹ 75,000 – ₹ 3,000 = ₹ 72,000.

(*iii*) *Net profit, if sales are 15% above BEP volume:*

Units sold at *BEP* = 3,000 units

$$\text{Units sold at 15\% above } BEP = 3{,}000 \times \frac{115}{100} = 3{,}450 \text{ units}$$

Sales value of 3,450 units = ₹ 25 × 3,450 = ₹ 86,250.

$$P/V \text{ ratio} = \frac{\text{Contribution per unit}}{\text{Selling price per unit}} \times 100 = \frac{8}{25} \times 100 = 32\%$$

$$\text{Now,} \quad \text{Sales value} = \frac{\text{Fixed cost} + \text{Profit}}{P/V \text{ ratio}}$$

∴ Profit = Sales value × P/V ratio – Fixed cost

$$= \left(₹\,86{,}250 \times \frac{32}{100}\right) - ₹\,24{,}000 = ₹\,3{,}600.$$

(*iv*) *Net profit, if sales are 20% above BEP sales:*

$$\text{Units sold at 20\% above } BEP = 3{,}000 \times \frac{120}{100} = 3{,}600 \text{ units}$$

Sales value of 3,600 units = ₹ 25 × 3,600 = ₹ 90,000

∴ Profit = (Sales value × P/V ratio) – Fixed cost

= 90,000 × 0.32 – 24,000 = ₹ 4,800.

Example 29.22. *A manufacturer has three products with the following data:*

Product	*Price*	*Variable cost/unit*	*Per cent of sales*
Book case	*₹ 60*	*₹ 40*	30%
Table	*₹ 100*	*₹ 60*	20%
Chair	*₹ 200*	*₹ 120*	50%
	Total fixed costs: ₹ 75,000		*Total sales: ₹ 2,50,000*

The manufacturer also has the following alternative with the following data:

Product	*Price*	*Variable cost/unit*	*Per cent of sales*
Book case	*₹ 60*	*₹ 40*	44%
Bed	*₹ 216*	*₹ 180*	16%
Chair	*₹ 200*	*₹ 110*	40%
	Total fixed cost: ₹ 75,000		*Total sales: ₹ 2,70,000*

Which of the two alternatives do you recommend and why?

Solution. For each alternative, we compute the profit obtained. Various calculations are shown below:

Product	*Sales value (in ₹)*	*Quantity sold*	*Variable cost per unit (in ₹)*	*Total variable cost (in ₹)*
First Alternative				
Book cases	2,50,000 × 0.30 = 75,000	75,000/60 = 1,250	40	50,000
Tables	2,50,000 × 0.20 = 50,000	50,000/100 = 500	60	30,000
Chairs	2,50,000 × 0.50 = 1,25,000	1,25,000/200 = 625	120	75,000
			Total:	1,55,000
Second Alternative				
Book cases	2,70,000 × 0.44 = 1,18,800	1,18,800/60 = 1,980	40	79,200
Beds	2,70,000 × 0.16 = 43,200	43,200/216 = 200	180	36,000
Chairs	2,70,000 × 0.40 = 1,08,000	1,08,000/200 = 540	110	59,400
			Total:	1,74,600

Using the information given in the above table, we get

Profit (first alternative) = Total sales – Total variable cost – Fixed cost
= 2,50,000 – 1,55,000 – 75,000 = ₹ 20,000.

Profit (second alternative) = 2,70,000 – 1,74,600 – 75,000 = ₹ 20,400.

Hence, second alternative is recommended.

Example 29.23. *Construct a break-even chart on the basis of the following data:*

Budgeted plan	*80,000 units*
Fixed expenses	*₹ 4,00,000*
Variable expenses per unit	*₹ 10*
Selling price per unit	*₹ 20*

Solution. Let x be number of units sold. Then, the cost function is $f(x) = 10x + 4,00,000$, and the revenue function is: $f(x) = 20x$. Plotting them on the graph, the break-even chart is obtained as under:

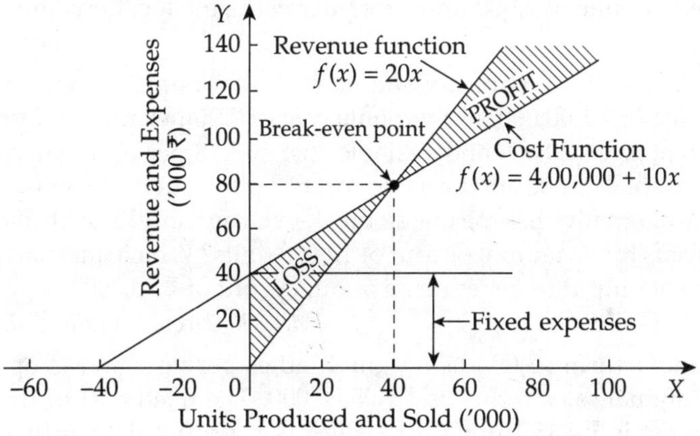

Fig. 29.5

PROBLEMS

29.28. Tuticorn Industries produces and sells toy transistors. Each unit has a variable cost of ₹ 5 and fixed costs total ₹ 2,000. If each toy sells for ₹ 7, what is the break-even point?

29.29. The fixed cost for a manufacturing process is ₹ 5,000 a year. Revenue per piece is ₹ 4. When 3,000 or fewer pieces are being made, the variable cost can be held to ₹ 2 each but when volume rises above this level, there is overtime and the cost rises to ₹ 2.40 each.
Find (*i*) The break-even point, and (*ii*) The profit at 4,000 pieces.

29.30. Aggarwal & Company sold in two successive years 7,000 and 9,000 units and incurred a loss of ₹ 10,000 and earned ₹ 10,000 as profit respectively. The selling price per unit is ₹ 100. Calculate:

(*a*) the amount of fixed cost,
(*b*) the number of units to break-even, and
(*c*) the number of units to earn a profit of ₹ 50,000.

29.31. A ball bearing manufacturing company is planning to install an additional plant which will require leasing new equipment for a monthly payment of ₹ 60,000. Variable cost would be ₹ 20 per item and each item would retail for ₹ 70.

(*i*) How many ball bearings must be sold in order to break-even?
(*ii*) What would be the profit (or loss), if 1,000 items are made and sold in a month?
(*iii*) How many items must be sold to realize a profit of ₹ 40,000?

[MBA (June), IGNOU 2004]

29.32. Atlas Co. produces three types of cycle pedals *A*, *B* and *C*. They were produced at 100% capacity last year, which represented a ₹ 20,00,000 sales volume. All the units produced were sold. Fixed cost for the firm were ₹ 2,50,000. It was observed that type *A* pedals cost ₹ 12 per unit and sold at ₹ 20 per unit contributed ₹ 10,00,000 to the total sales; type *B* pedals cost ₹ 9 per unit and sold at ₹ 12 per unit contributed ₹ 6,00,000 to the total sales, while type C pedals cost ₹ 4 per unit and sold for ₹ 8 per unit contributed ₹ 4,00,000 to total sales. Determine (*i*) the break-even point, (*ii*) the loss that would be expected if capacity were reduced by 10%.

29.33. Three locations are being considered for a new plant. Fixed costs per year for the three locations are ₹ 30,000, ₹ 60,000 and ₹ 1,10,000 respectively. Variable costs per unit are ₹ 75, ₹ 45 and ₹ 25 respectively. The selling price of the product is ₹ 120. Find the most economical location for an expected volume of 2,000 units per year using the locational break-even analysis.

[MBA (June), IGNOU 2000]

29.34. A firm is considering three capacity alternatives: *A*, *B* and C. Alternative *A* would have an annual fixed cost of ₹ 1,00,000 and variable cost of ₹ 20 per unit. Alternative *B* would have annual fixed cost of ₹ 1,20,000 and variable cost of ₹ 18 per unit. Alternative C would have annual fixed cost of ₹ 80,000 and variable cost of ₹ 30 per unit. Revenue is expected to be ₹ 50 per unit. Which alternative has the lowest break-even quantity? Which alternative will produce the highest profits for an annual output of 10,000 units? Which alternative would require the lowest volume of output to generate an annual profit of ₹ 50,000?

[M.Tech. (Dec.), Dayalbagh Educational Inst. 2012]

29.35. A plant is manufacturing 3,000 heavy duty lathes per year and is operating at 75% of its capacity. The annual sales return is ₹ 1,05,00,000. The fixed cost of the plant is ₹ 40,00,000 and variable cost is ₹ 4,150 per unit. There is a proposal to utilize space capacity by manufacturing precision lathes which would increase the fixed cost by ₹ 8,00,000, but reduce the variable cost by ₹ 750 per unit.

(*a*) Is the proposal economical? Give reasons for your answer.
(*b*) If a reduction in selling price by ₹ 500 per unit requires the plant to be run at 90% of its capacity to break-even, would this be better proposal than the earlier one?

29.36. A manager has the option of purchasing one, two or three machines. Fixed costs and potential volumes are as follows:

Number of machines	*Total annual fixed costs*	*Corresponding range of output (no. of units)*
1	₹ 96,000	0 to 300
2	₹ 1,50,000	301 to 600
3	₹ 2,00,000	601 to 900

Variable cost is ₹ 100 per unit and revenue is ₹ 400 per unit.

(*i*) Determine the break-even point for each range.

(*ii*) If projected annual demand is between 580 and 660 units, how many machines should the manager purchase? *[MBA (Dec.), IGNOU 2004]*

TEST YOUR UNDERSTANDING

1. Discounting of cash flows is a technique to
(*a*) compute present value of a (series of) future payment(s).
(*b*) compute compounded value of a stream of cash flows.
(*c*) find discount offered in a deal.
(*d*) none of the above.

2. Following is not a method of evaluating an investment proposal under uncertainty:
(*a*) Simulation.
(*b*) Discounted cash flow technique.
(*c*) Certainty equivalent approach.
(*d*) Statistical distribution approach.

3. Method of *IRR* is considered to be inferior to the *NPV* method, because
(*a*) it may give multiple rates of return.
(*b*) it is relatively more difficult to compute.
(*c*) the underlying assumption of varying re-investment rates is unrealistic.
(*d*) all of the above.

4. The *P/V* ratio is the ratio of
(*a*) contribution margin to sales.
(*b*) profit to the number of units sold.
(*c*) profit to the value of goods sold.
(*d*) sales to contribution margin.

5. Contribution margin refers to the excess of the selling price over
(*a*) unit fixed cost.
(*b*) unit total cost.
(*c*) unit variable cost.
(*d*) none of the above

6. Simulation approach to evaluate risky investments
(*a*) computes the period within which entire investment cost would be recovered.
(*b*) identifies various factors affecting cash flows and their probability distribution to compute estimated returns.
(*c*) tests the impact of changes in factors affecting investments on investment decisions.
(*d*) increases the rate of return to be used for discounting so as to compensate for risk involved.

7. Margin safety is the excess of
(*a*) total sales over total variable cost.
(*b*) actual sales over break-even sales.
(*c*) sales over total cost.
(*d*) selling price per unit over fixed cost per unit.

8. Break-even point is that level of production at which
   (*a*) there is neither profit nor loss.
   (*b*) total revenue equals total cost.
   (*c*) total contribution is just sufficient to recover the total fixed cost.
   (*d*) all of the above.

ANSWERS

1. (*a*)	**2.** (*b*)	**3.** (*d*)	**4.** (*a*)	**5.** (*c*)
6. (*b*)	**7.** (*b*)	**8.** (*d*).		

REVIEW EXERCISE

1. What is an annuity? Distinguish between various types of annuities on the basis of (*a*) their terms, and (*b*) the payment times.
2. What is investment analysis? Why is it of great significance to a firm?
3. Discuss the various methods of evaluating investment proposals.
4. Discuss in brief 'Discounted Cash Flow' method and its role in investment analysis.
5. What do you mean by *NPV* and *IRR*?
6. Critically examine the *NPV* and *IRR* methods used in appraising capital projects. Also, discuss a few simple approaches for incorporating risk factor in such decision-making.
7. What do you mean by risk investment proposals? How can it be measured?
8. How can the *EVPI* of an investment proposal be obtained?
9. What do you understand by decision-tree analysis? Give its role for investment decisions.
10. What is break-even analysis? How is it useful? Also, explain as to how the break-even point of a multi-product firm be calculated.

CHAPTER 30

Business Forecasting

"With the help of computers forecast the future is an impressive, ever-expanding array of quantitative capabilities"

As the economic and business conditions vary over time, an organization is closely related to how well management is able to anticipate the future and develop appropriate strategies. Good judgement, intuition, and an awareness of the state of the economy may give a manager a rough idea or "feeling" of what is likely to happen in the future. However, converting this feeling into a number that can be used as next quarter's sales volume or next year's raw material cost per unit (*i.e.*, forecasting) often is difficult. The purpose of this chapter is to introduce various forecasting methods.

Suppose that we have been asked to provide quarterly forecasts of the sales volume of a particular product during the coming year. Production schedules, raw material purchasing plans, inventory policies and sales quotas will be affected by the quarterly forecasts that we provide. Consequently, poor forecasts may result in increased costs for the firm.

30:1. FORECASTING METHODS

Forecasting methods can be classified into ***'qualitative'*** and ***'quantitative methods'***. The qualitative methods use personal judgement and involve qualities like intuition and experience as the basis of forecasts, and are subjective by their very nature. On the other hand, quantitative methods are objective in nature and they employ numerical information as the basis of making forecasts. In quantitative methods, a forecast can be developed using time series method or casual method.

The ***time series methods*** attempt to predict the future values using the historical data. The objective is to discover a pattern in the historical data and then extrapolate this pattern into the future; the forecast is based solely on past values of the variable that we are trying to forecast and/or on past forecast errors.

Casual forecasting methods are based on the assumption that the variable we are trying to forecast exhibits a cause-effect relationship with one or more other variables. For example, the demand for a product may be seen to be dependent upon the price charged, income of the people, amount spent on the advertisement and so on. The analysis is done in terms of what is called as regression analysis.

Qualitative methods generally involve the use of expert judgement to develop forecasts. For instance, a panel of experts might develop a consensus forecast of the prime rate for a year from now. An advantage of qualitative procedure is that they can be applied when the information on the variable being forecast cannot be quantified and when historical data either are not applicable or available.

Fig. 30.1 given below provides an overview of the types of forecasting methods:

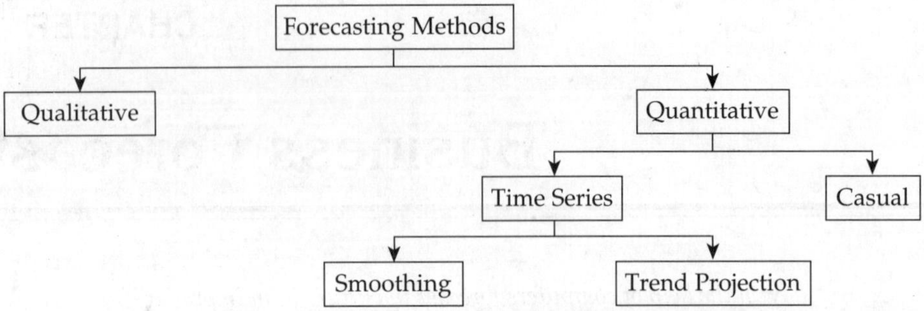

Fig. 30.1. An Overview of Forecasting Methods

30:2. COMPONENTS OF TIME SERIES

A ***time series*** is a set of measurements of a variable that are ordered through time. The measurements may be taken as an hour, day, week, month, or year, or at any other regular interval. Examples of time series are: 'the daily closing price of shares in the National Stock Exchange', 'monthly production of steel', 'weekly attendance of workers in a public limited company', 'sales of personal computers', and so on. The pattern or behaviour of the past data in a time series has several components. The usual assumptions is that four separate components — trend, cyclical, seasonal, and irregular or random combine to provide specific values for the time series.

1. Trend Component. The time series data generally exhibit random fluctuations, but inspite of this the time series may still show gradual shifts or movements to relatively higher or lower values over a longer period of time. The gradual shifting of the time series is referred to as the ***trend*** in the time series. For example, change of technology increase prices over a period of years, the sales of a commodity may decrease over a period of years, because of better (improved) products coming in the market. The increase or decrease in the movements of a time series is called a ***secular trend***.

2. Cyclical Component. ***Cyclic movements*** are recurrent upward or downward in time series, but the period of cycle is more than a year. The sales of a company, for example, may be high because the level of economic activity may be high. Similarly, the sales may be low due to overall subdued economic activity. A business cycle, in general, has to pass through the prosperity, recession, depression and recovery.

3. Seasonal Component. In time series the fluctuations that occur periodically — the movements recurring within a definite period, may be every week, or month or quarter — with reasonably high degree of predictability are called ***seasonal components***. For example, daily traffic volume data show within-the-day "seasonal" behaviour, with peak levels during rush hours, moderate flow during the rest of the day, and light flow from midnight to early morning. As another example, the sales of soft drink are likely to be relatively high every summer and low every winter.

4. Irregular Component. The fluctuations in time series that are short in duration, erratic in nature and follow no regularity in occurrence pattern are known as ***irregular variations***. These are also referred to as ***residual variations***, since by definition they represent what is left out in a time series after trend, cyclic and seasonal variations. Irregular or random fluctuations results due to the occurrence of unforeseen events like earthquakes, wars, floods, and so on.

30:3. SMOOTHING METHODS IN FORECASTING

In this section, we discuss three forecasting methods: moving averages, weighted moving averages, and exponential smoothing. The objective of each of these methods is to "smooth out" the random fluctuations caused by the irregular component of the time series. Therefore, they are referred to as ***smoothing methods***. Smoothing methods are easy to use and generally provide a high level of accuracy for short-range forecasts such as a forecast for the next time period.

Moving Averages

The ***moving averages*** method uses the average of the most recent n data values in the time series as the forecast for the next period.

$$\therefore \qquad \text{Moving average} = \left[\sum (\text{Most recent } n \text{ data values})\right]/n.$$

The term ***moving*** indicates that, as a new observation becomes available for the time series, it replaces the oldest observation in the above expression, and a new average is computed. As a result, the average will change, or move, as new observations become available.

Moving averages are usually calculated for 3 weeks/months/years, 5 weeks/months/years, and so on. In calculating 3-months moving average forecast of a time series data, the average of the first three months is taken to represent the forecast value for the fourth month. Leaving the first data value of time series, we take the average of the next three time series data to represent the forecast value for the fifth month. In this way, the moving average forecast values are calculated. In a similar way, we determine the 4-month moving average forecast, 5-month average forecast, and so on.

After calculating the moving average forecast, we compute forecast error. For this, we compute the difference between the observed value of the time series and the forecast value. This forecast error may be positive or negative, depending on whether the forecast is too low or too high.

Remark. The average of the sum of squared forecast errors is known as the ***mean squared error*** (*MSE*). The *MSE* is often used as measure of the accuracy of a forecasting method. In addition to *MSE*, there are two other measures for judging the accuracy of forecasts: ***mean absolute deviation*** (*MAD*), and ***mean absolute percentage error*** (*MAPE*).

Example 30.1. *Compute the 3-week moving average forecast for the following data, showing weekly sales (in '000) rupees of a company for three years:*

Week :	1	2	3	4	5	6	7	8	9	10
Sales ('000₹) :	8	9	8	10	9	12	13	14	12	15

Find the forecast errors and mean square error. What is the forecast for week 11?

Solution. Computation of the 3-week moving average is as follows:

Week	*Times series value*	*3-week moving total*	*Moving average forecast*	*Forecast error*	*Squared forecast error*
1	8	—	—		
2	9	—	—		
3	8	—	—		
4	10	25	8.33	1.67	2.788
5	9	27	9.00	0.00	0.000
6	12	27	9.00	3.00	9.000
7	13	31	10.33	2.67	7.128
8	14	34	11.33	2.67	7.128
9	12	39	13.00	–1.00	1.000
10	15	39	13.00	2.00	4.000
11	—	41	13.67	—	—
			Total	11.01	31.000

$MSE = 31/7 = 4.43$. Forecast for week 11 is ₹ 13,670.

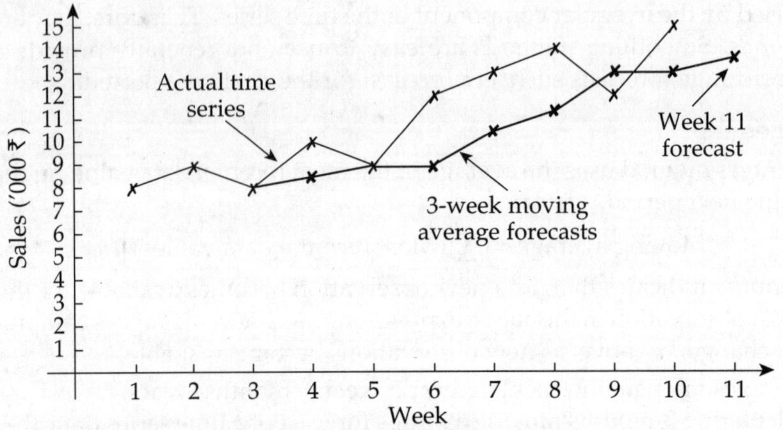

Fig. 30.2. Graph of Sales Time-series and 3-week Moving Average Forecast

Example 30.2. *From the following data, calculate the forecast values using four-yearly moving averages and plot them on a graph:*

Year :	*1*	*2*	*3*	*4*	*5*	*6*	*7*	*8*	*9*
Values (in '000 ₹) :	*506*	*620*	*1,036*	*673*	*588*	*696*	*1,116*	*738*	*663*

Further, compute forecast errors and find MAD.

Solution. Computation of the 4-yearly moving averages is as follows:

Year	Values (Y_i)	4-yearly moving average	Forecast error (F_i)	Mean absolute deviation
1	506			
2	620			
3	1,036			
4	673			
5	588	719.00	– 131.00	131.00
6	696	738.75	– 42.75	42.75
7	1,116	758.25	357.75	357.75
8	738	776.37	– 38.37	38.37
9	663	806.37	– 143.37	143.37
10	—	803.25	—	—
		Total	2.26	713.24

$$MAD = \frac{\sum |Y_i - F_i|}{5} = 142.6$$

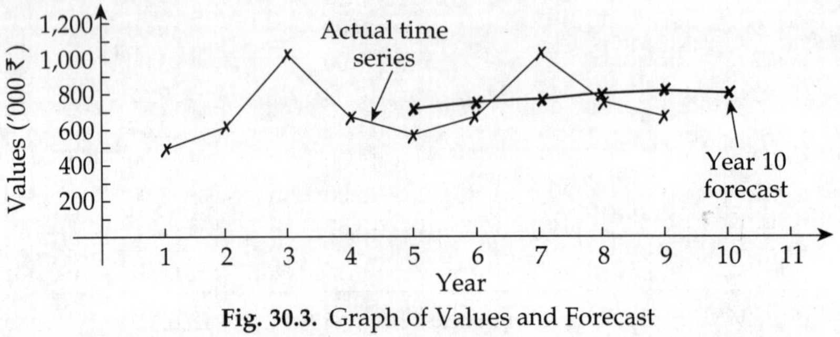

Fig. 30.3. Graph of Values and Forecast

Weighted Moving Averages

In the moving averages method, each observation in the calculation receives the same weight. One variation, known as ***weighted moving averages***, involve selecting different weights for each data value and then computing a weighted average of the most recent n data values as the forecast. In most of the cases, the most recent observation receives the most weight, and the weight decreases for older observations. For example, in *Example 30.1*, a three-weekly moving average may be calculated by assigning weightage of 3, 2 and 1 respectively for the three weeks' values.

∴ Weighted moving averages forecast for week 5 is 9.16 as shown below:

$$\text{Weighted moving average} = 10 \times \frac{3}{6} + 8 \times \frac{2}{6} + 9 \times \frac{1}{6} = 9.16.$$

It may be noted that the weights used for the three weeks are, respectively, 3/6, 2/6 and 1/6; such that the sum of weights is equal to 1. For a four-week average, the weights would be 4/10, 3/10, 2/10 and 1/10.

Example 30.3. *Calculate 5-year moving averages forecast for the following data, using weights 1, 2, 3, 1 and 1 respectively.*

Year :	1	2	3	4	5	6	7	8	9	10
Sales (₹ in lakh) :	40	33	72	81	76	68	91	87	98	97

What is the forecast for 11th year?

Solution. Calculation of weighted moving averages forecast are as under:

Year	*Sales (₹ in lakh)*	*5-year weighted moving average forecast*
1	40	
2	33	
3	72	
4	81	
5	76	
6	68	(76 × 1 + 81 × 2 + 72 × 3 + 33 × 1 + 40 × 1)/8 = 65.875
7	91	(68 × 1 + 76 × 2 + 81 × 3 + 72 × 1 + 33 × 1)/8 = 71.000
8	87	(91 × 1 + 68 × 2 + 76 × 3 + 81 × 1 + 72 × 1)/8 = 76.000
9	98	(87 × 1 + 91 × 2 + 68 × 3 + 76 × 1 + 81 × 1)/8 = 78.750
10	97	(98 × 1 + 87 × 2 + 91 × 3 + 68 × 1 + 76 × 1)/8 = 85.125
11		(97 × 1 + 98 × 2 + 87 × 3 + 91 × 1 + 68 × 1)/8 = 89.125

From the above computations, clearly the forecast for 11th year is ₹ 89,12,500.

Sum of weights = 1 + 2 + 3 + 1 + 1 = 8.

Exponential Smoothing

It is a special case of the weighted moving averages method, where the forecast for the next period is calculated as weighted average of all the previous observations. It is based on the premise that the most recent observation is the most important for predicting the future value.

The basic exponential smoothing model is:

$$F_{t+1} = \alpha Y_t + (1 - \alpha) F_t,$$

where, F_{t+1} = forecast of the time series for period $t + 1$,
Y_t = actual value of the time series in period t,
F_t = forecast of the time series for period t,
α = smoothing constant, $(0 \le \alpha \le 1)$.

The above equation is rewritten as:

$$F_{t+1} = F_t + \alpha(Y_t - F_t).$$

The difference $(Y_t - F_t)$ represents the error in the previous forecast.

To start the calculations, we let F_1 equal the actual value of the time series in period 1, *i.e.*, $F_1 = Y_1$. Hence, the forecast for period 2 is:

$$F_2 = \alpha Y_1 + (1-\alpha)F_1 = F_1 + \alpha(Y_1 - F_1) = Y_1 + \alpha(Y_1 - Y_1) = Y_1, \text{ since } F_1 = Y_1.$$

This shows that the exponential smoothing forecast for period 2 is equal to the actual value of the time series in period 1.

The forecast for period 3 is:

$$F_3 = \alpha Y_2 + (1-\alpha)F_2 = \alpha Y_2 + (1-\alpha)Y_1$$

Similarly,

$$F_4 = \alpha Y_3 + (1-\alpha)F_3 = \alpha Y_3 + (1-\alpha)[\alpha Y_2 + (1-\alpha)Y_1]$$

$$= \alpha Y_3 + \alpha(1-\alpha)Y_2 + (1-\alpha)^2 Y_1.$$

Hence, F_4 is a weighted average of the first three time series values. The sum of the coefficients, or weights, for Y_1, Y_2 and Y_3 equals 1. A similar argument can be made to show that in general, any forecast F_{t+1} is a weighted average of all previous time series values.

It may be noted that despite the fact that exponential smoothing provides a forecast, *i.e.*, a weighted average of all past observations, all the past data do not need to be saved in order to compute the forecast for the next period.

The smoothing constant, α, may be altered to change relative weightage of different values entering into the forecast. A high value of the constant would give a high weightage to the recent value, while its low value would give a relatively large weightage to the past data. For example, if $\alpha = 0.4$, then the two most recent values would get a weightage of 0.4 and 0.4 (1 – 0.4), *i.e.*, 0.24. The new forecast is based practically on the values of the last two periods.

Example 30.4. *The following are weekly sales data, in thousands of units, for micro-computer disks:*

113, 101, 98, 107, 120, 132, 110, 117, 112 and 125.

Use $\alpha = 0.2$, $\alpha = 0.5$ and $\alpha = 0.8$ to produce an exponential smoothing model for these data. Which value of α produces better forecasts, assuming the initial forecast as 108? Explain.

Solution. The forecasts using each of the above values are shown calculated in the table given below.

Week	*Sales (in '000 units)*	$\alpha = 0.2$		$\alpha = 0.5$		$\alpha = 0.8$	
t	Y_t	F_t	$\lvert Y_t - F_t \rvert$	F_t	$\lvert Y_t - F_t \rvert$	F_t	$\lvert Y_t - F_t \rvert$
1	113	108.0	5.0	108.0	5.0	108.0	5.0
2	101	109.0	8.0	110.5	9.5	112.0	11.0
3	098	107.4	9.4	105.8	7.8	103.2	5.2
4	107	105.5	1.5	101.9	5.1	199.0	8.0
5	120	105.8	14.2	104.4	15.6	105.4	14.6
6	132	108.7	23.3	112.2	19.8	117.1	14.9
7	110	113.3	3.3	122.1	12.1	129.0	19.0
8	117	112.7	4.3	116.1	0.9	113.8	3.2
9	112	113.5	1.5	116.5	4.5	116.4	4.4
10	125	113.2	11.8	114.3	10.7	112.9	12.1
11	—	115.6	—	119.6	—	122.6	—
		Total	82.3		91.0		97.4

To illustrate for $t = 1$, let $F_1 = 108$. Now, with $Y_1 = 113$ and $\alpha = 0.2$, the forecast for $t = 2$ is: $F_2 = F_1 + 0.2\,(Y_1 - F_1)$, *i.e.*, $F_2 = 108 + 0.2\,(113 - 108) = 108 + 0.2 \times 5 = 109$.

For each case, the absolute difference between actual and forecasted values are obtained and averaged to get *MAD*.

From the table given on previous page, (*i*) if $\alpha = 0.2$, then $MAD = 82.3/10 = 8.23$, (*ii*) if $\alpha = 0.5$, then $MAD = 91/10 = 9.1$, (*iii*) if $\alpha = 0.8$, then $MAD = 97.4/10 = 9.74$. Since, *MAD* is found to be minimum for $\alpha = 0.2$, therefore, this smoothing constant is most preferable of these.

Example 30.5. *Calculate mean squared error (MSE), and mean absolute percentage error (MAPE) for the forecasts obtained in Example 30.4 using* $\alpha = 0.8$.

Solution. The above measures of forecast error are calculated in the following table:

Week t	*Sales (in '000 units)* Y_t	*Forecast value* $F_t\ (\alpha = 0.8)$	$\lvert Y_t - F_t \rvert$	$(\lvert Y_t - F_t \rvert)^2$	$\frac{\lvert Y_t - F_t \rvert}{Y_t} \times 100$
1	113	108.0	5.0	25.00	4.42
2	101	112.0	11.0	121.00	10.89
3	098	103.2	5.2	27.04	5.30
4	107	199.0	8.0	64.00	7.47
5	120	105.4	14.6	213.16	12.16
6	132	117.1	14.9	222.01	11.28
7	110	129.0	19.0	361.0	17.27
8	117	113.8	3.2	10.24	2.73
9	112	116.4	4.4	19.36	3.93
10	125	112.9	12.1	146.41	9.68
			Total	1,209.22	85.13

Using the above table, we get

$$MSE = \frac{1{,}209.22}{10} = 120.9 \quad \text{and} \quad MAPE = \frac{85.13}{10} = 8.5.$$

PROBLEMS

30.01. Production of steel for ten years is recorded as under:

Year :	1	2	3	4	5	6	7	8	9	10
Production (in '000 tonnes) :	21	22	23	25	24	22	25	26	27	26

Using 3-year moving averages, forecast the next year's production of steel. Compute the forecast error and hence determine *MSE*.

30.02. The demand for an item is observed for 15 months and recorded below:

Month	*Demand (in units)*	*Month*	*Demand (in units)*
1	280	9	309
2	288	10	315
3	266	11	320
4	295	12	332
5	302	13	310
6	310	14	308
7	303	15	320
8	328		

Calculate (*i*) 3-monthly, and (*ii*) 4-monthly moving average forecast. What is the forecast for month 16 for each one?

30.03. Infrastructure Bond interest rates in percentage for 12 consecutive months are:

9.5, 9.3, 9.4, 9.6, 9.8, 9.7, 9.8, 10.5, 9.9, 9.7, 9.6, and 9.6.

(*i*) Develop a 3-month and 4-month moving averages for this time series. Which moving average provides the better forecasts? Explain.

(*ii*) What is the moving average forecast for the next month?

30.04. The sale of Gasoline for 12 weeks is observed and recorded as under:

Week :	1	2	3	4	5	6	7	8	9	10	11	12
Sale ('000 gallons) :	17	21	19	23	18	16	20	18	22	20	15	22

(*i*) Compute 4-week and 5-week moving averages for the time series.

(*ii*) Compute *MSE* for the 4-week and 5-week moving averages forecast.

(*iii*) What appears to be the best number of weeks of past data to use in the moving average computation?

30.05. For the following data, calculate 5-year weighted moving averages forecast using the weights: 1, 2, 2, 2, 1 respectively:

Month :	1	2	3	4	5	6	7	8	9	10	11
Sales (₹ in lakh) :	5	3	7	6	4	8	9	10	8	9	9

30.06. You are given the following information about demand of an item:

Month :	1	2	3	4	5	6	7	8	9	10	11
Demand (units) :	220	228	217	219	258	241	239	244	256	260	265

Calculate forecast values using (*i*) 5-monthly moving averages, and (*ii*) 4-monthly weighted moving averages with weights 4, 3, 2 and 1; the largest weight being for the most recent value.

30.07. Use the gasoline sales time series data from *Problem 30.04* given above to show the exponential smoothing forecasts using $\alpha = 0.1$. Using the mean squared error criteria, would you prefer a smoothing constant of $\alpha = 0.1$ or $\alpha = 0.2$?

30.08. For a manufacturing company, the monthly percentages of all shipments that were received on time over the past 12 months are:

80, 82, 84, 83, 83, 84, 85, 84, 82, 83, 84, and 83.

(*i*) Compare a 3-month moving averages forecast with an exponential smoothing forecast for $\alpha = 0.2$. Which provides the better forecasts?

(*ii*) What is the forecast for next month?

30:4. TREND PROJECTION IN FORECASTING

In the trend projection method of forecasting, the value of time series exhibits a long-term linear trend. This trend line may be linear (straight line) or curvilinear (non-linear) in nature. But, our focus in this chapter is on linear trend only which is determined by graphical means. For this, the given time series data are plotted on the graph that represent the time scale on X-axis. Then, a straight line is drawn through these points in such a way, that (*i*) the sum of derivations above this line is equal to the sum of deviations below this line, so that the sum of deviations is equal to zero, and (*ii*) the sum of the squares of these (vertical) deviations is the minimum. In other words, the trend line is drawn on the basis of the ***principle of least squares***. Such a line, like any other straight line, is represented by the equation

$$y_t = a + bx,$$

where, a = intercept at the y-axis, x = independent variable, *i.e.*, time, and
b = slope of the trend line, y_t = the trend value (which is to be computed).

The value of constant 'a' and 'b' of the trend line drawn on the principle of least squares are obtained by using the following pair of normal equations:

$$\sum y = Na + b\sum x, \quad \text{and} \quad \sum xy = a\sum x + b\sum x^2$$

Alternatively, $$b = \frac{\sum xy - N\bar{x}\,\bar{y}}{\sum x^2 - N(\bar{x})^2} \quad \text{and} \quad a = \bar{y} - b\bar{x}$$

Here, $\sum x$ = summation of the values of independent variables, *i.e.*, time,

$\sum y$ = summation of the values of dependent variables (the variable whose values are to be forecasted),

$\sum xy$ = summation of the products of x and corresponding y values,

$\sum x^2$ = summation of the squares of values of the independent variables,

$\bar{y}$ = the average value of the time series, and

N = number of periods.

The value of b in the equation indicates the change in the variable under consideration per unit time.

Note. If we take the middle time period or the mid-point of the two time periods as the starting point, x will be equal to zero and the pair of normal equations becomes:

$$\sum y = Na \quad \text{and} \quad \sum xy = b\sum x^2$$

or $$a = \frac{\sum y}{N} \quad \text{and} \quad b = \frac{\sum xy}{\sum x^2}.$$

Example 30.6. *Demand (in '000 metric tonnes) for sugar of a sugar factory is given below:*

Year :	*2013*	*2014*	*2015*	*2016*	*2017*	*2018*	*2019*
Demand :	*77*	*88*	*94*	*85*	*91*	*98*	*90*

(*i*) *Using the principle of least squares, fit a straight line trend. Show that the sum of deviations is equal to zero. Also, determine the sum of squares of deviations.*

(*ii*) *Calculate trend values and plot observed (actual) and trend values on a graph.*

(*iii*) *Obtain the forecast of demand for the years 2021 and 2024.*

Solution. Let the straight line trend equation is given by

$$y_t = a + bx,$$

where, a and b are constants to be determined.

The given data is reproduced in the table given below. Also, shown in the table is the calculation about the inputs for determining the values of a and b. Notice that the years have been replaced by time scale with values 0, 1, 2, 3, 4 and so on.

Year	x (*Time*)	y (*Demand*)	xy	x^2	y_t	$(y - y_t)$	$(y - y_t)^2$
2013	0	77	0	0	83	–6	36
2014	1	88	88	1	85	3	9
2015	2	94	188	4	87	7	49
2016	3	85	255	9	89	–4	16
2017	4	91	364	16	91	0	0
2018	5	98	490	25	93	5	25
2019	6	90	540	36	95	–5	25
Total	21	623	1,925	91		0	160

We have $\sum y = Na + b\sum x \Rightarrow 623 = 7 \times a + b \times 21$

and $\sum xy = a\sum x + b\sum x^2 \Rightarrow 1,925 = a \times 21 + b \times 91.$

Solving these two equations, we get

$$b = 2 \text{ and } a = 83.$$

(*i*) The trend equation, therefore, is: $y_t = 83 + 2x$. Using this equation, trend values for various years are obtained which are given in the 6th column of the table. The next column gives the deviations of y-values from the corresponding y_t-values. Their sum is found to be equal to zero.

(*ii*)

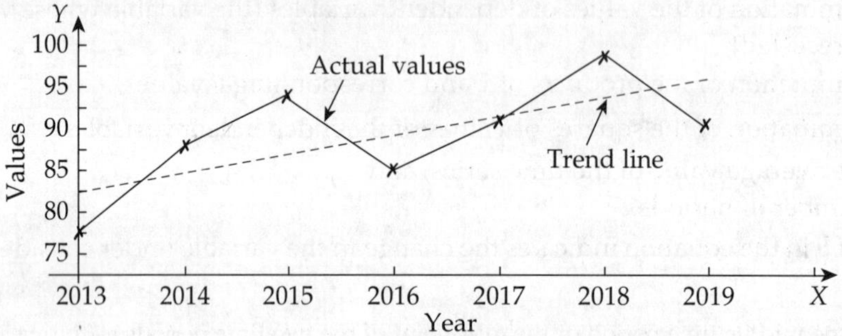

Fig. 30.4

(*iii*) *Forecasting values:*

$y_{2021} = 83 + 2 \times 8 = 99$ (in '000 metric tonnes) (since 2021 – 2013 = 8)

and $y_{2024} = 83 + 2 \times 11 = 105$ (in '000 metric tonnes).

Example 30.7. *The annual production of a commodity is given as follows:*

Year	2012	2013	2014	2015	2016	2017	2018	2019
Production ('000 quintals)	56	55	51	47	42	38	35	32

(*i*) *Using the principle of least squares, fit a straight line trend. Obtain the forecast for the year 2021. Plot the actual and trend values on graph.*

(*ii*) *Convert the annual trend equation into monthly trend equation.*

Solution. Let the straight line trend equation is given by

$$y_t = a + bx,$$

where, a and b are constants to be determined.

Calculation of trend equation values and deviations by using the given information is obtained in the following table:

Year	x	y (*Production*)	xy	x^2	y_t
2012	0	56	0	0	57.5
2013	1	55	55	1	53.8
2014	2	51	102	4	50.1
2015	3	47	141	9	46.4
2016	4	42	168	16	42.7
2017	5	38	190	25	39.0
2018	6	35	210	36	35.3
2019	7	32	224	49	31.6
Total	28	356	1,090	140	

We have $\sum y = Na + b\sum x \Rightarrow 356 = 8a + 28b$

and $\sum xy = a\sum x + b\sum x^2 \Rightarrow 1,090 = 28a + 140b$

Solving the two equations for a and b, we get

$$a = 57.5 \quad \text{and} \quad b = -3.7.$$

(*i*) The trend equation, therefore, is $y_t = 57.5 - 3.7x$. Using this equation, trend values for various years are obtained as: 57.5, 53.8, 50.1, 46.4, 42.7, 39.0, 35.3 and 31.6 respectively.

Forecast value for the year 2021 is obtained as:

$$y_{2021} = 57.5 - 3.7 \times 9, \qquad \text{(since 2021 - 2012 = 9)}$$
$$= 24.2, \text{ i.e., } 24,200 \text{ quintals.}$$

(*ii*)

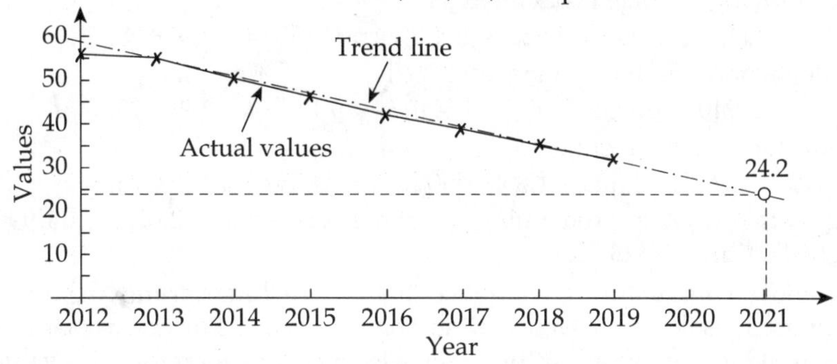

Fig. 30.5

(*iii*) In the trend equation obtained in (*i*), we had data on annual production. When this annual trend is to be converted into monthly, we have to divide the value of a by 12 and that of b by 144. Thus, the trend equation on monthly production is given by

$$y_t = \frac{1}{12} \times 57.5 - 3.7 \times \frac{1}{144} x$$
$$= 4.79 - 0.0256x.$$

30:5. SEASONAL COMPONENTS IN FORECASTING

In measuring seasonal variation of a time series, seasonal index is used. The seasonal index is expressed as percentage and is obtained by finding the percentage which the seasonal averages bear to the average of the seasonal averages. A consideration of monthly or quarterly time series data makes it easy to study seasonal patterns. Seasonal indices, which are used to study seasonal variations, can be obtained by using a number of methods. Here, we demonstrate the method of ***simple averages only***.

In fact seasonal indices give us a clear idea about the relative position of each month or quarter in time series data relating to such matters as production, sales, employment, and so on. Furthermore, seasonal indices may be used for short-term forecasting, which is quite necessary for planning the future course of action. For example, in studying the production figures of a company's overtime, the seasonal indices may be used to plan for the hiring of personnel for peak periods, to accumulate an inventory of raw materials, to allocate vacation time.

Example 30.8. *The quarterly sales data (number of copies sold) for a college text book over the past three years are as follows:*

Quarter	*Year I*	*Year 2*	*Year 3*
1	1,690	1,800	1,850
2	940	900	1,100
3	2,625	2,900	2,930
4	2,500	2,360	2,615

(*a*) *Calculate the 4-quarter moving average values for this time series. Plot both the original time series and moving averages on the same graph.*
(*b*) *Compute seasonal indices.*
(*c*) *When does the text book publisher experience the largest seasonal index? Does this result appear to be reasonable? Explain.*
(*d*) *Forecast the sales for all the four quarters of year 4.*

Solution.

(*a*) First quarter moving average is obtained as

(1,690 + 940 + 2,625 + 2,500)/4 = 7,755/4 = 1,938.75.

Similarly, the second quarter moving average is

(940 + 2,625 + 2,500 + 1,800)/4 = 7,865/4 = 1,966.25.

The third quarter moving average is

(2,625 + 2,500 + 1,800 + 900)/4 = 7,825/4 = 1,956.25.

Proceeding in this way, the remaining quarter averages are: 2,025.00, 1,990.00, 2,002.50, 2,052.50, 2,060.00 and 2,123.75.

From the above computations, we notice that the first quarter moving average represents an average quarterly sales value (across all seasons) for year 1. Further, as there are four quarters in a year, the moving average allow for no middle value and, therefore, this value is considered to correspond to the last half of quarter 2 and the first half of quarter 3. Similarly, if we go to the second moving average value of 1,966.25, the middle corresponds to the last half of quarter 3 and the first half of quarter 4. By computing the mid-points of successive moving averages we shall have ***centered moving averages***. For example, (1,938.75 + 1,966.25)/2 = 1,952.5 represents the first centered moving average. Other centered moving averages are computed as follows:

1,961.25, 1,990.62, 2,007.5, 1,996.25, 2,027.5, 2,056.25, 2,091.87.

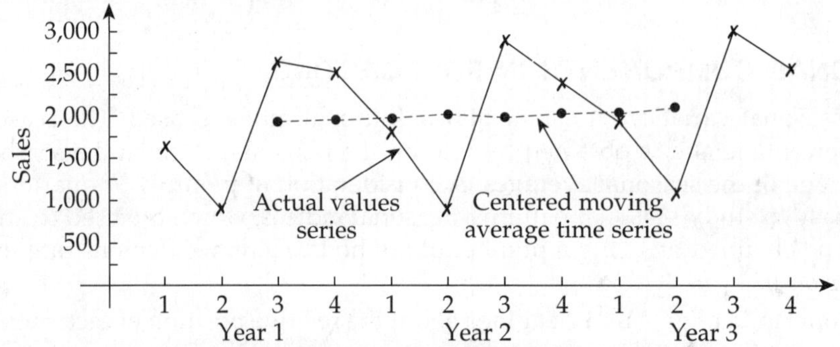

Fig. 30.6. Quarter/Year

(*b*) To obtain seasonal indices, we first calculate average sales for each of the four quarters. This is being done as follows:

Year	*Quarter*				*Total*
	I	*II*	*III*	*IV*	
1	1,690	940	2,625	2,500	7,755
2	1,800	900	2,900	2,360	7,960
3	1,850	1,100	2,930	2,615	8,495
Total	5,340	2,940	8,455	7,475	
Average	1,780	980	2,818.33	2,491.66	$\frac{8,069.99}{4} = 2,017.5$
Seasonal Index	88.2	48.6	139.7	123.5	

The overall mean of the quarterly averages is obtained as (1,780 + 980 + 2,818.33 + 2,491.66)/4, *i.e.*, 2,017.5. Each of the quarterly values has been expressed as a percentage of the overall mean. Thus, for quarter *I*, we have (1,780/2,017.5) × 100, *i.e.*, 88.2. Similarly, for other three quarters respective percentages are: 48.6, 139.7 and 123.5.

(*c*) The largest seasonal effect is in the third quarter, which corresponds to the demand of text book during July, August and September.

(*d*) ***Forecast*** for the sales in year 4 is: $y_4 = a + bx$, where x assumes the values 0, 1, 2 and 3 respectively. The computation is done as follows:

Year	x	y	xy	x^2
1	0	7,755	0	0
2	1	7,960	7,960	1
3	2	8,495	16,990	4
Total	3	24,210	24,950	5

Now, since $\sum y = Na + b\sum x$, and $\sum xy = a\sum x + b\sum x^2$,

we have 24,210 = 3*a* + 3*b* and 24,950 = 3*a* + 5*b*

Solving these two equations, we get

$$a = 7{,}700 \quad \text{and} \quad b = 370.$$

∴ The forecast of sales during year 4 is

$$7{,}700 + 370 \times 3 = 8{,}810, \qquad \text{(since } x = 3\text{)}$$

As the estimated sales is of 8,810 copies, the corresponding average sales will be 8,810/4, *i.e.*, 2,202.5 copies of text book. Based on the seasonal indices, the quarterly sales are estimated as follows:

Quarter	*Year 4*
I	2,202.5 × 0.882 = 1,942.6
II	2,202.5 × 0.486 = 1,070.4
III	2,202.5 × 1.397 = 3,077.0
IV	2,202.5 × 1.235 = 2,720.0

PROBLEMS

30.09. Apply the method of least squares to obtain the trend values from the following sales data of a commodity and forecast the sales for the year 2021 and 2022:

Year :	*2015*	*2016*	*2017*	*2018*	*2019*
Sales ('000 units) :	100	120	110	140	80

30.10. Fit a straight line trend by the method of least squares to the following data of production of paddy in a village of West Bengal and forecast the production there in the year 2021:

Year :	*2015*	*2016*	*2017*	*2018*	*2019*
Production ('000 tonnes) :	21	23	28	22	24

30.11. Fit a straight line trend to the following data on average monthly domestic demand (in million of barrels) for motor fuel:

Year :	1	2	3	4	5	6	7	8	9	10	11
Demand :	61	66	72	76	82	90	96	100	103	110	114

What is the forecast of demand for the 12th year?

30.12. The annual production of a commodity is given as follows:

Year :	*2013*	*2014*	*2015*	*2016*	*2017*	*2018*	*2019*
Production (in tonnes) :	70	80	90	95	102	110	115

(*a*) Fit a straight line trend to the above data.
(*b*) Convert the annual trend equation into a monthly trend equation.
(*c*) What is the forecast of production for the year 2021?

30.13. Given the following data about the demand of an item, use exponential smoothing to forecast demand taking $\alpha = 0.1$ and $\alpha = 0.3$. Also, plot the original data as well as the forecast values on graph.

Year	*Quarter*			
	I	*II*	*III*	*IV*
2016	70	160	110	200
2017	90	120	60	110
2018	100	190	150	300
2019	270	350	320	

Further, calculate *MAD* for each of the series and suggest which value of α is more appropriate. Does the method of exponential smoothing appear to be satisfactory one in this case?

30.14. Assuming that trend is absent, determine if there is any seasonality in the data given below:

Year	*Quarter I*	*Quarter II*	*Quarter III*	*Quarter IV*
1	3.7	4.1	3.3	3.5
2	3.7	3.9	3.6	3.6
3	4.0	4.3	3.3	3.1

What are the seasonal indices for various quarters?

30.15. The sales of a company, in millions of rupees, are given below:

Year	*Quarterly sales (in millions of rupees)*				*Total*
	I	*II*	*III*	*IV*	
1	18	27	23	14	82
2	17	30	20	13	80
3	21	26	23	20	90
4	21	29	26	16	92
5	19	28	22	14	83
6	22	30	26	16	94
7	24	32	24	19	99
8	21	29	29	13	92

Forecast the sales for the years 9 and 10 by using the method of least squares. Also, determine (*a*) seasonal indices, and (*b*) forecast the sales for the years 9 and 10 on a quarterly basis.

30.16. The seasonal indices of the sales of a popular brand of a coloured television by a company in Delhi are given below:

Year 2020 :	*Quarter I*	*Quarter II*	*Quarter III*	*Quarter IV*
Seasonal index :	130	90	75	105

If the total sales for the first quarter amounted to ₹ 5,85,000, forecast the worth of the televisions of this brand to be kept in store by the company to meet the demand in other quarters.

30:6. CASUAL FORECASTING METHOD

The casual forecasting method consider the situations, where the variable to be forecasted, called the ***dependent*** variable, is related to an ***independent*** variable. For example, production of goods in a manufacturing company depends on raw material, manpower, time and so on. Similarly, sales of a particular product depends on the price of product, competitors price, level of economic activity, and so on. The forecasting of dependent variable is done by developing a statistical relationship between it and the independent variable. This is done in terms of regression analysis.

Regression analysis involving one independent variable and one dependent variable for which the relationship between the variables is approximated by a straight line is called ***simple linear regression***. Regression analysis involving two or more independent variables is called ***multiple regression analysis***.

Forecasting Using Simple Regression Analysis

The simple regression analysis is employed when there is one independent variable. Here, an estimate of the dependent variable, say y is made corresponding to a given value of the independent variable, say x by placing the relationship between the two variables in the form of a ***regression line***. If we can develop the mathematical expression for this line, we may be able to use it to forecast the value of y corresponding to each possible value of x. The resulting equation of the line is called the ***estimated regression equation***.

Using the least square method of forecasting, we develop the estimated regression equation:

$$y_e = a + b\,x,$$

where, y_e = estimated (forecast) value of dependent variable,
a = y-intercept of the regression line,
b = slope of the regression line,
x = the independent variable.

As in the case of trend line, the values of a and b are obtained as follows:

$$b = \frac{\sum xy - n\,\bar{x}\,\bar{y}}{\sum x^2 - n(\bar{x})^2} \quad \text{and} \quad a = \bar{y} - b\,\bar{x},$$

where, $\bar{x}$ = mean value of the independent variable,
$\bar{y}$ = mean value of the dependent variable,
n = total number of observations.

The parameter b is called the ***regression coefficient***.

Example 30.9. *A manufacturer of optical lenses has the following data on the cost per unit (in ₹) of a certain company that makes lenses and the number of units in each order is given below:*

Order No.	:	1	2	3	4	5	6
Number of units	:	1	3	5	7	10	12
Cost per unit	:	58	52	46	40	37	22

Obtain the regression equation on the cost per unit. Use this equation to estimate the cost of 15 units of lenses.

Solution. Let the cost of an optical lense is denoted by y and the number of units of lenses by x. Then, we are to fit the equation of regression line y on x, *viz.*, $y_e = a + bx$.

The calculations are shown in table given below:

Order	x	y	xy	x^2
1	1	58	58	1
2	3	52	156	9
3	5	46	230	25
4	7	40	280	49
5	10	37	370	100
6	12	22	264	144
Total	38	255	1,358	328

We have, $\bar{x} = \frac{\sum x}{n} = \frac{38}{6} = 6.33,$ and $\bar{y} = \frac{\sum y}{n} = \frac{255}{6} = 42.5.$

$$b = \frac{\sum xy - n\bar{x}\bar{y}}{\sum x^2 - n(\bar{x})^2} = \frac{1{,}358 - 6 \times 6.33 \times 42.5}{328 - 6 \times (6.33)^2} = -\frac{256.15}{87.58} = -2.92.$$

$$a = \bar{y} - b\bar{x} = 42.5 - (-2.92) \times 6.33 = 42.5 + 18.48 = 60.98.$$

∴ The regression equation of y on x is given by

$$y_e = 60.98 - 2.92 \times x.$$

Forecast: When $x = 15$. $\quad y_e = 60.98 - (2.92) \times 15 = 17.18$

Thus, for 15 units of optical lenses, the cost of a lense is ₹ 17.18.

Forecasting Using Multiple Regression Analysis

The multiple regression is an extension of the simple regression analysis when there is more than one independent variable. We state here the case which has only two independent variables. A two-variable regression equation is:

$$y_e = a + b_1 x_1 + b_2 x_2,$$

The parameters (constants) a, b_1, and b_2 are obtained by solving the following three equations simultaneously:

$$\sum y = na + b_1 \sum x_1 + b_2 \sum x_2,$$

$$\sum x_1 y = a \sum x_1 + b_1 \sum x_1^2 + b_2 \sum x_1 x_2,$$

$$\sum x_2 y = a \sum x_2 + b_1 \sum x_1 x_2 + b_2 \sum x_2^2.$$

Example 30.10. *A company keeps track of its expenditure on advertising (x_1) and its spending on in-store displays (x_2) over a period of 10 weeks. The resulting sales for each week are recorded as the dependent variable y. Use the data given below to estimate the least squares equation relating these variables. If $x_1 = 17$ and $x_2 = 11$, what sales would you predict?*

Sales (y) in units	:	72	76	78	70	68	80	82	65	62	90
Advertising (x_1) in ₹	:	12	11	15	10	11	16	14	8	8	18
Expenditure (x_2) in ₹	:	5	8	6	5	3	9	12	4	3	10

[M.Com., Delhi Univ. 2008]

Solution. A two variable regression equation is $y_e = a + b_1x_1 + b_2x_2$, where a, b_1 and b_2 are to be determined. Calculations for regression equation are obtained in the following table:

y	x_1	x_2	x_1y	x_2y	x_1x_2	x_1^2	x_2^2
72	12	5	864	360	60	144	25
76	11	8	836	608	88	121	64
78	15	6	1,170	468	90	225	36
70	10	5	700	350	50	100	25
68	11	3	748	204	33	121	9
80	16	9	1,280	720	144	256	81
82	14	12	1,148	984	168	196	144
65	8	4	520	260	32	64	16
62	8	3	496	186	24	64	9
90	18	10	1,620	900	180	324	100
743	123	65	9,382	5,040	869	1,615	509

The constants a, b_1 and b_2 are obtained by solving the following simultaneous linear equations:

$$\sum y = na + b_1 \sum x_1 + b_2 \sum x_2,$$

$$\sum x_1 y = a \sum x_1 + b_1 \sum x_1^2 + b_2 \sum x_1 x_2,$$

$$\sum x_2 y = a \sum x_2 + b_1 \sum x_1 x_2 + b_2 \sum x_2^2.$$

Substituting the calculated values from the above table, we get

$$743 = 10a + 123b_1 + 65b_2$$

$$9,382 = 123a + 1,615b_1 + 869b_2$$

$$5,040 = 65a + 869b_1 + 509b_2$$

Solving these three equations simultaneously, we get

$$a = 47.165, \quad b_1 = 1.599 \quad \text{and} \quad b_2 = 1.149.$$

Hence, the required regression equation is:

$$y_e = 47.165 + 1.599x_1 + 1.149x_2$$

Forecast: When $x_1 = 17$ and $x_2 = 11$,

$$y_e = 47.165 + 1.599 \times 17 + 1.149 \times 11$$

$$= 47.165 + 27.183 + 12.639 = 86.987 \text{ (or 87)}$$

Hence, the estimated sales will be of ₹ 87, if expenditure on advertising is ₹ 17 and expenses on in-store is ₹ 11.

PROBLEMS

30.17. Height of father and son are given below. Find the height of son when the height of father is 70 inches.

Father's height (in inches) :	71	68	66	67	70	71	70	73	72	65	66
Son's height (in inches) :	69	64	65	63	65	62	65	64	66	59	62

30.18. A car manufacturer has recently held 3-day road side exhibits on the introduction of a new model of its deluxe cars. The number of sales personnel employed at each of a sample of

10 exhibitions and the number of cars booked at each one are given as follows:

No. of salesmen :	5	8	6	8	9	3	5	4	6	6
No. of cars booked :	132	160	148	156	168	102	142	98	152	142

Using above data, regress the number of cars booked on the number of salesmen, and obtain the regression equation. Estimate the number of cars booked, if 10 salesmen are employed on an exhibition.

30.19. Find the regression equation of Y on X, where Y and X are the marks obtained by 10 students which are given below:

Y :	20	60	55	45	75	35	25	90	10	50
X :	20	45	65	40	55	35	15	80	25	50

Use the regression equation to forecast the value of Y, when $X = 73$.

30.20. Obtain the least squares regression equation of Y on X from the following data:

X :	89	86	74	65	64	63	66	67	72	79
Y :	92	91	84	75	73	72	71	75	78	84

Use the regression equation to forecast values of Y when (*i*) $X = 70$, and (*ii*) $X = 85$.

30.21. Ashish, owner of a business unit is concerned about the sales behaviour of his product. He realizes that there are many factors that might help explain sales, but believes that advertising and prices are major determinants. He has collected the following data:

Sales (units sold) :	33	61	70	82	17	24
Advertising (no. of ads) :	3	6	10	13	9	6
Price (₹) :	125	115	140	130	145	140

(*i*) Calculate the regression equation to predict sales from advertising and price.

(*ii*) If advertising is given as 7 and price is ₹ 132, what sales would you predict?

30.22. Given the following data:

Performance Evaluation (Y) :	28	33	21	40	38	46
Aptitude Test Score (X_1) :	74	87	69	69	81	97
Prior Experience (X_2) :	5	11	4	9	7	10

(*i*) Develop the estimating equation best describing these data.

(*ii*) If an employee scored 83 in the aptitude test and had a prior experience of 7, what performance evaluation would be expected? ***[M.Com., Delhi Univ. 1999]***

30.23. A company recently embarked on an in-store promotional campaign, with displays of its antacid featured prominently in the supermarkets. The company also ran its usual radio and television commercials. Over a period of 10 weeks, the company kept track of its expenditure on radio and television advertising by variable X_1 and its expenditure on in-store displays, variable X_2. The resulting sales for each week in the area studied were recorded as the dependent variable Y. The company analyst hypothesized a linear regression linking sales volume with the two independent variables. Using the following data (all values in thousands of rupees), fit the desired regression equation:

Y :	72	76	78	70	68	80	82	65	62	90
X_1 :	12	11	15	10	11	16	14	8	8	18
X_2 :	5	8	6	5	3	9	12	4	3	10

Now, estimate the sales, if X = ₹ 13,000 and X_2 = ₹ 7,000. ***[M.Com., Delhi Univ. 2000]***

30:7. QUALITATIVE METHODS OF FORECASTING

Time series methods of forecasting discussed in the preceding sections require historical data. But, if historical data are not available, managers may use a qualitative technique to develop forecasts. Some of the qualitative methods of forecasting are discussed below:

1. Delphi Method. One of the most commonly used qualitative forecasting technique is the ***Delphi method***. This technique attempts to develop forecasts through "group consensus." In this method the experts are not known to each other and are asked to respond to a series of questionnarie. The responses from the first questionnaire are tabulated and used to prepare a second questionnaire that contains the information and opinion of the entire group. Each respondent is then asked to reconsider and possibly review his or her previous response in light of the group information provided. This process continues until the coordinator feels that some degree of consensus has been reached.

2. Sales Force Composite. In this method of forecasting, all the members comprising sales force of a company are asked to estimate the likely sales in their respective areas. The estimates are then reviewed to ensure that they are realistic. After this, the estimates are combined at the district, regional, and national level to obtain the overall forecast.

3. Consumer Panel Survey. Some marketing research methods employ ***consumer panels*** for making forecasts. Here, a consumer panel is maintained and consumers on such a panel are questioned about their purchase plans. The goal here is to forecast demand for products and services on the basis of the subjective judgements of the subjects involved, about possible purchases.

4. Scenario Writing. The qualitative procedure referred to as ***scenario writing*** consists of developing a conceptual scenario of the future based on a well-defined set of assumptions. Different sets of assumptions lead to different scenarios. The job of the decision-maker is to decide how likely each scenario is, and then to make decisions accordingly.

TEST YOUR UNDERSTANDING

1. Exponential smoothing method of forecasting is
   (*a*) qualitative method (*b*) quantitative method
   (*c*) Delphi method (*d*) regression-based method.
2. Components of time series include
   (*a*) trend component (*b*) cyclic component
   (*c*) seasonal component (*d*) all of the above.
3. The trend is linear, if
   (*a*) growth rate is constant (*b*) growth rate is not constant
   (*c*) growth is positive (*d*) growth is geometrical.
4. The trend can be found by the method of least squares, if the
   (*a*) nature of trend is known (*b*) trend is parabolic
   (*c*) trend is linear (*d*) trend is bi-quadratic.
5. If the trend is absent, seasonal indices are known by
   (*a*) simple average method (*b*) ratio-to trend method
   (*c*) ratio-to moving average method (*d*) none of these.
6. Cyclic variation is caused by
   (*a*) boom and recession (*b*) unpredictable events
   (*c*) seasonal events (*d*) none of these.

7. Seasonal variation is caused by
   (*a*) increase of population (*b*) strike and lockout
   (*c*) weather and social customs (*d*) none of these.
8. The accuracy of forecasting method is judged by
   (*a*) *MSE* (*b*) *MAPE*
   (*c*) *MAD* (*d*) all of these.
9. The most commonly used qualitative forecasting method is:
   (*a*) Delphi method (*b*) expert judgement
   (*c*) scenario writing (*d*) intuitive approaches.
10. Choose the statement which is correct:
    (*a*) Opinions and judgements play no role in quantitative forecasting.
    (*b*) *MAPE* represents the mean of absolute forecast errors expressed as a percentage of the forecasted values.
    (*c*) In regression-based forecast, the standard error of estimate gives the maximum error of forecast.
    (*d*) Multiple regression allows to have multiple dependent and independent variables.

ANSWERS

1. (*b*)	2. (*d*)	3. (*a*)	4. (*c*)	5. (*d*)
6. (*a*)	7. (*d*)	8. (*d*)	9. (*a*)	10. (*a*)

REVIEW EXERCISE

1. What is the difference between quantitative and qualitative techniques of forecasting?
2. What are the components of time series? Discuss the moving averages method of forecasting.
3. "Exponential smoothing method of forecasting is better than the moving average method." Discuss.
4. Discuss the method of obtaining trend-adjusted exponential smoothing forecast.
5. What is standard error of estimate? What does it indicate?

CHAPTER 31

Learning Curve Theory

"Learning is the process that gives efficiency in production"

In many production processes, production efficiency increases with experience. As cumulative production output increases, the average labour time required per unit declines. A graphical expression of this phenomenon is called a ***learning curve***. On this learning curve, when cumulative output doubles, the average labour time per unit declines at a constant percentage, which normally ranges from 10% to 40%. The percentages are commonly called cost reductions. The cost reduction percent designated on the curve is the complement of the learning rate. Therefore, if average labour time required per unit declines by 30% after production doubles, the curve is referred to as 70% learning curve. The period during which the output per hour increases is known as the ***learning stage***.

Learning curves have been used extensively in such industries as aircraft production, ship-building, and electronics to assist cost analysts in predicting labour costs. These cost predictions are then used in scheduling production, budgeting, setting product prices, and other managerial decisions.

31:1. GRAPHICAL AND TABULAR PRESENTATION OF LEARNING CURVE

The graphic presentation of learning curve is illustrated in *Fig. 31.1*.

Graphical presentation of learning curve shows that when the cumulative output doubles, the average labour time per unit declines by 20%. The total labour time and average labour time per unit for various levels of cumulative output is displayed in tabular form as given below:

Cumulative output (in units)	*Average labour time per unit (hours)*	*Total labour time (hours)*
5	100.00	500
10	80.00	800
20	64.00	1,280
40	51.20	2,048
80	40.96	3,276.8

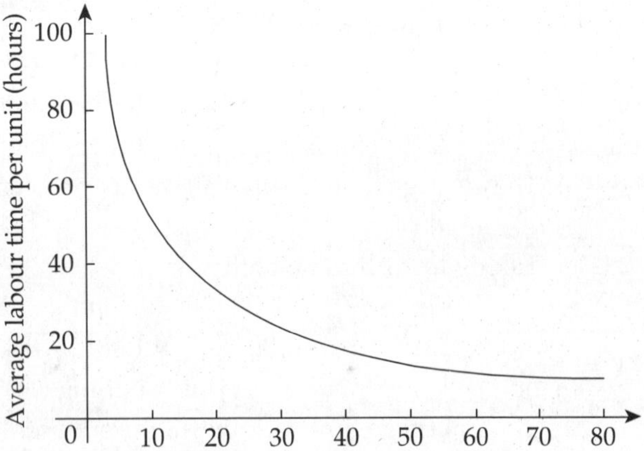

Fig. 31.1. Learning Curve

As cumulative output doubles from 5 to 10 units, the average labour time per unit declines by 20%, from 100 hours per unit to 80 hours per unit. As manufacturer gains experience with a product, estimates of the cost of direct labour should be adjusted downward to take this learning effect into account.

Note. If the concept of learning curve is applied to a broader set of costs than just labour costs, the phenomenon is referred to as an *experience curve*.

31:2. LEARNING CURVE EQUATION

The learning curve equation is usually expressed as:

$$y = ax^{-b},$$

where a = amount of time spent on the first unit produced,

x = total number of units produced,

b = coefficient, that describes the amount of learning that takes place,

y = average time to produce all the x units, *i.e.*, average number of direct labour hours per unit.

The learning curve equation is best explained with an example.

Example 31.1. *Prepare a depreciation schedule over the life of machine using a particular depreciation method in six stages (problems). It is presumed that it takes 48 minutes to solve one of these problems, when it is encountered for the first time. However, if a second problem is solved, the solution time is much less, say 24 minutes. Answer the following questions:*

(*i*) *What is the average time to solve a total of six problems?*

(*ii*) *How long would it take for a student to solve the last (sixth) problem?*

Solution. We are given

a = time to solve the first problem = 48 minutes,

x = total number of problems to be solved = 2,

y = average time to solve the two problems = (48 + 24)/2 = 36 minutes.

Substituting the value of a, x and y in the learning curve equation, we get

$$36 = 48 \times 2^{-b}$$

Taking logarithm on both sides,

$$\log 36 = \log 48 + (-b) \log 2$$

or $$b \log 2 = \log 48 - \log 36 = 1.6812 - 1.5563 = 0.1249$$

or $$b = 0.1249/\log 2 = 0.1249/0.3010 = 0.4150$$

(*i*) The average time to solve all the six problems is:

$$y = 48 \times 6^{-0.4150}$$

Taking logarithm on both sides,

$$\log y = \log 48 \times 6^{-0.4150} = \log 48 + (-0.4150) \log 6$$
$$= 1.6812 - 0.4150 \times 0.7782 = 1.3582$$

or $$y = \text{antilog}\ (1.3582) = 22.82 \text{ minutes}$$

Thus, average time to solve all the six problems = 6 × 22.82 = 136.9 minutes

(*ii*) Now, the average time to solve the first five problems is:

$$y = 48 \times 5^{-0.4150}$$

$\therefore$ $$\log y = \log 48 \times 5^{-0.4150} = \log 48 - (0.4150) \log 5$$
$$= 1.6812 - 0.4150 \times 0.6989 = 1.3911$$

This gives $$y = \text{antilog}\ (1.3911) = 24.61$$

Thus, average time to solve the first five problems = 5 × 24.61 = 123.05

Hence, time to solve the sixth problem = 13.8 minutes.

This shows that if learning continues at the same rate as occurred between problem 1 and problem 2, then the 6th problem will be solved in 13.8 minutes.

31:3. SPECIFIC LEARNING CURVES

Specific learning curves are derived from the effect that learning has an average production time, when production is doubled. Assume that a firm produces x units of a particular product and let y is the average time to produce all these x units. Then, as soon as production reaches $2x$, the value of y changes to y^* (say) and thereby the learning curve is named after the relationship y^*/y. Thus, if y^* is 90% of y, we say that we have a 90% learning curve. For example, if it takes 10 hours to produce the first unit of product and average of 9 hours each to produce 2 units, then it is said that we have a 90% learning curve. To make the concept clear, let us consider another example. Let 100 units of a product are produced by taking on an average 40 hours. Then, if we experience an 80% learning curve, the average time to build 200 units should be 32 hours.

It should be noted that the relationship between the 'name of the learning curve' and 'b' is:

$$b = \frac{-\log (\text{name})}{\log 2}$$

Thus, for an 80% learning curve, we can find the value of b as follows:

$$b = -\frac{\log (0.80)}{\log 2} = \frac{-1+0.9031}{0.3010} = \frac{-0.0969}{0.3010} = -0.3219.$$

When the value of coefficient is positive, it will be written in formula as negative and *vice-versa*.

∴ An 80% learning curve can be represented as follows:

$$y = a x^{-0.3219}$$

Formula for finding out Total Time

Total time to complete x units can be found out directly with the following formula:

$$t = a x^{1-b},$$

where, a, x and b are the same as are given in the learning curve equation.

Illustration. Let a = 48 units and b = 0.4150. Then, the total time to be taken for 6 units is given as:

$$t = a x^{1-b} = 48 \times (6)^{1-0.4150} = 48 \times (6)^{0.585}$$

Taking logarithm on both sides,

$$\log t = \log 48 + (0.585) \log 6$$

$$= 1.6812 + 0.4552 = 2.1364$$

$$\therefore \quad t = \text{antilog } (2.1364) = 136.9 \text{ minutes}$$

Formula for finding out Marginal Time to Complete Specific Unit

For finding out the time that a specific unit consume to complete, we compute the total time for x units and $x - 1$ units, and then take the difference of the two. This can also be done by using the following formula:

$$m = a(1-b)x^{-b}.$$

Illustration. Let $a = 48$ units, $b = 0.4150$ and $x = 6$. Then, the marginal time to complete the sixth unit is given by

$$m = 48 \times (1 - 0.4150) \times (6)^{-0.4150}$$
$$= 28.08 \times (6)^{-0.4150}$$

Taking logarithm on both sides,

$$\log m = \log 28.08 - (0.4150) \log 6$$
$$= 1.4484 - 0.4150 \times 0.7782$$
$$= 1.4484 - 0.3229 = 1.1254$$

$$\therefore \quad m = \text{antilog}\,(1.1254) = 13.3 \text{ minutes}$$

Note. This formula is quite inaccurate for small values of x because it is a continuous approximation of a discrete function. The formula should be used only for values of x larger than 300.

Learning Curve Ratio

In the initial stage of a new product/process, the learning effect pattern is so regular that the rate of decline established at the outset can be used to predict labour cost in advance.

$$\text{Learning curve ratio} = \frac{\text{Average labour cost of first } 2N \text{ units}}{\text{Average labour cost of first } N \text{ units}}.$$

For example, if the average labour cost for the first 500 units of a product is ₹ 25 and the average labour cost of first 1,000 units is ₹ 20, then

$$\text{Learning curve ratio} = \frac{₹\,20}{₹\,25} \times 100 \quad \text{or} \quad 80\%.$$

This learning curve ratio of 80% means that every time output doubles, the averages cost declines to 80% of the previous amount.

31:4. REGRESSION AND LEARNING CURVE

Regression analysis can be used to determine the learning curve rates as follows:

Learning curve rate is expressed by $y = a x^{-b}$.

Taking logarithm on both sides, we get

$$\log y = \log a + (-b) \log x$$

Now, let $\log y = Y$, $\log a = A$, $(-b) = B$ and $\log x = X$.

Then, $$Y = A + BX.$$

This is simple regression model. It can be proved using differential calculus that

$$B = \frac{n \sum XY - (\sum X)(\sum Y)}{n \sum X^2 - (\sum X)^2},$$

where n equals the sample size.

and $$A = \bar{Y} - B\bar{X}.$$

Example 31.2. *Given below are the following cumulative average costs. Determine the learning curve and hence the appropriate learning rate:*

Number of units (x) :	1	2	3	4	5	6	7	8
Cumulative average cost (y) :	49	33	29	25	20	18	17	17

Solution. Taking logarithm of x values and y values, we get

	1	2	3	4	5	6	7	8
$\log x$ (= X) :	0	0.3010	0.4771	0.6021	0.6990	0.7782	0.8451	0.9031
$\log y$ (= Y) :	1.690	1.5190	1.4620	1.3980	1.3010	1.2550	1.2300	1.2300

$\therefore \quad \Sigma X = 4.6056, \ \Sigma Y = 11.086, \ \Sigma XY = 6.0337, \ \Sigma X^2 = 3.3046$ and $\Sigma y^2 = 15.549$

Now, since $Y = A + BX$ is the equation of learning curve, we compute the following:

$$B = \frac{n\Sigma XY - (\Sigma X)(\Sigma Y)}{n\Sigma X^2 - (\Sigma X)^2} = \frac{8 \times 6.0337 - (4.6056)(11.086)}{8 \times 3.3046 - (4.6056)^2}$$

$$= \frac{48.2696 - 51.0577}{26.4368 - 21.2116} = -0.5335.$$

$$\therefore \quad A = \bar{Y} - B\bar{X} = \frac{11.086}{8} - (-0.5335) \times \frac{4.6056}{8}$$

$$= 1.3858 - (-0.5335) \times (0.5757)$$

$$= 1.3858 + 0.3071 = 1.6929 \text{ or } 1.693$$

Now, since $A = 1.693$ (= $\log a$) and $B = -0.5335 = (-b)$; we have

$$a = \text{antilog}\ (1.693) = ₹\ 49.32 \text{ and } b = 0.5335.$$

Now, $\quad b = -\log L/\log 2$ or $\log L = -b \log 2 = \log 2^{-b}$

$\therefore \quad L = 2^{-b} = 2^{-0.5335} = 0.6908$

Hence, learning curve rate is 69.1%.

31:5 LEARNING CURVE TABLE (When Production is not Exactly Doubled)

When production is not exactly doubled, a table of factors can be derived showing relationships for intermediate points on the learning curve. Given below is learning curve table for 80% learning curve:

x	y (%)	x	y (%)
1.0	100.0	2.1	78.9
1.1	96.9	2.2	77.8
1.2	93.3	2.3	76.8
1.3	91.7	2.4	76.0
1.4	89.5	2.5	74.9
1.5	87.6	2.6	74.0
1.6	86.1	2.7	73.2
1.7	84.4	2.8	72.3
1.8	83.0	2.9	71.5
1.9	81.5	3.0	70.7
2.0	80.0	3.1	70.0

When $x = 4$, y would be 80% of 80% (*i.e.*, 64%) and when $x = 8$, y would be 80% of 64% (*i.e.*, 51.2%). Similarly, when $x = 16$, y would be 80% of 51.2% (*i.e.*, 40.9%), and so on. The above table may be used to calculate the cost of each extra unit made, when we don't have a convenient 'doubling' of output.

If X represents total output and Y represents cumulative average time per unit, then Y can be derived by using the formula: $Y (\%) = X^b$.

Here, $\quad Y (\%) = X^{-0.3219} = 2^{-0.3219}$ (*i.e.*, 80% of the time for one unit).

By using Learning Curve Table, it is possible to produce a table of indices for all levels of output. Thus, if it is desired to estimate labour times for total output of 3 units, we find the factor for 3 units (*i.e.*, 70.7%) and multiply it by the time for the first unit (*i.e.*, 100 hours) to get the average time per unit, which will be 70.7 hours. Three units will take 3 × 70.7 hours, *i.e.*, 212.1 hours.

Example 31.3. *A firm produces 50 units of a product in 80 hours. The firm now wishes to estimate, what average time per unit will be, if the total output for the product is 125 units. An 80% learning curve applies.*

Solution. We are given:

Initial output = 50 units, and Total output = 125 units.

$$\therefore \qquad X = \frac{125}{50} \times 100 = 250\%,$$

i.e., 2.5 times the initial output of 50 units.

Now, if $X = 2.5$, then from the learning curve table, $Y = 74.9\%$.

So, total time required for 125 units = (2.5 × 80 hours) × 74.9% = 149.8 hours.

average time per unit = 149.8/125 = 1.2 hours (approx.)

Example 31.4. *A customer requires separate price quotations for each of the following possible order:*

Order :	*First*	*Second*	*Third*
Units :	*100*	*60*	*40*

It is expected that in first order '4' hours per unit will be consumed and an 80% learning curve applies. Find out the average time per unit

(a) when two orders are expected. (b) when three orders are expected.

Solution.

(*a*) Cumulative total of first order of 100 units and second order of 60 units is obtained as (100 + 60) units, *i.e.*, 160 units.

Since, in first run 100 units are expected to be produced, relationship of 160 units to 100 units is obtained as 160/100, *i.e.*, 1.6 or $x = 1.6$.

∴ For 160 units, the average time when first two orders are expected is given by

$$y = 4 \times (1.6)^{-0.3219} = 4 \times 0.861 = 3.44 \text{ hours.}$$

If we use learning curve table, then for $x = 1.6$, the relevant % factor is 86.1%. Therefore, the average time per unit when two orders are expected will be: 0.861 × 4 hours, *i.e.*, 3.44 hours.

(*b*) Cumulative total of the first order of 100 units, second order of 60 units and third order of 40 units is obtained as (100 + 60 + 40) units, *i.e.*, 200 units.

Since, in first run 100 units are expected to be produced, relationship of 200 units to 100 units is obtained as 200/100, *i.e.*, 2.00 or $x = 2$.

∴ For 200 units, the average time when first three orders are expected is given by

$$y = 4 \times (2)^{-0.3219} = 4 \times 0.8 = 3.2 \text{ hours.}$$

If we use learning curve table, then for $x = 2.0$, the relevant % factor is 80. Therefore, the average time per unit when three orders are expected = 4 hours × 0.8 = 3.2 hours.

Example 31.5. *In a particular industry, a 60% learning curve applies. The first item of a new product took 80 hours to make. How long should the* 6th *item take?*

Solution. Learning index is given by

$$b = \frac{-\log(0.60)}{\log 2} = -\frac{0.2218}{0.3010} = -0.7368$$

For 6 units, $y = 80\ (6)^{-0.7368}$

$\therefore$ $\log y = \log 80 \times (6)^{-0.7368} = \log 80 - (0.7368) \log 6$

$= 1.9031 - (0.7368) \times (0.7782)$

$= 1.9031 - 0.5734 = 1.3296$

or $y = \text{antilog}\ (1.3296) = 21.36$

Total time for 6 units = 21.36 × 6 = 128.16 hours

For 5 units, $y = 80 \times (5)^{-0.7368}$

$\therefore$ $\log y = \log 80 - (0.7368) \log 5$

$= 1.9031 - (0.7368) \times (0.6990) = 1.3881$

or $y = \text{antilog}\ (1.3881) = 24.44$

Total time for 5 units = 24.44 × 5 = 122.2 hours

Thus, time for 6th unit = (128.16 – 122.2) hours = 5.96 hours.

Example 31.6. *A company developing a new product makes a model for testing, and then a demonstration model. After demonstration model, the company goes for regular production. The time taken to make the model is 300 hours. From past experience of similar models, it is known that a 90% learning curve applies. The average time for each of the first two production models will be*

(1) 270 hours, (2) 243 hours, (3) 216 hours, and (4) 219 hours

Support the correct figure with calculation.

Solution. Average time per unit for the two production model are computed as follows:

Cumulative output	*Average time per unit (hours)*	*Total time (hours)*
1	300	300
2	300 × 0.90 = 270	270 × 2 = 540
4	270 × 0.90 = 243	243 × 4 = 972
8	243 × 0.90 = 219	219 × 8 = 1,752

Average time for third and fourth units produced is (972 – 540) hours, *i.e.*, 432 hours or an average of 216 hours. Therefore, third is correct answer, while 1st, 2nd and 4th answers are not correct.

Example 31.7. *Config Systems Ltd., observe that an 80% learning curve applies to production of item ABC. By the end of December 2019, 115 units of item ABC had been produced. Budgeted production for January 2020 was 35 units. The time taken to produce the very first unit of the item in August 2019 was 60 hours. Find the*

(i) budgeted total labour time for January 2020?

(ii) the average expected labour time per unit in January 2020.

Solution. Since, 80% learning curve is in operation, coefficient of learning (or learning index) is

$$b = \frac{-\log (0.80)}{\log 2} = \frac{-0.0969}{0.3010} = -0.3219$$

$\therefore$ For 115 units of item *ABC*,

$$y = 60 \times (115)^{-0.3219}$$

Taking logarithm on both sides,

$$\log y = \log 60 + (-0.3219) \log 115$$

$$= 1.7781 - 0.6633 = 1.1148$$

$\therefore$ $y = \text{antilog}\ (1.1148) = 13.02$

Cumulative time needed for 115 units = 115 × 13.02 = 1497 hours.

Since, 115 units of *ABC* have been produced to the end of December, 2019 and 35 units in January, 2020; the total production will be 150 units. Therefore, for 150 units, we get

$$y = 60 \times (150)^{-0.3219}$$

or $$\log y = \log 60 + (-0.3219) \log (150)$$

$$= 1.7781 - 0.7005 = 1.0776$$

$\therefore$ $$y = \text{antilog}\,(1.0776) = 11.96$$

Total time for producing 150 units = 150 × 11.96 = 1,794 hours

(*i*) Budgeted total labour time for January, 2020, *i.e.*, incremental time for 35 units is equal to (1794 – 1497) hours or 297 hours.

(*ii*) Average expected labour time per unit in January, 2020 = (297/35), *i.e.*, 8.48 hours per unit.

Example 31.8. *A company has observed that 80% learning curve ratio applies to all labour related costs relating to Model X, which has been recently introduced in the market. It is expected that first unit will take a time of 10 minutes. What is the expected time per unit (i) when cumulative production is 30 units, and (ii) when cumulative production is 40 units?*

Solution. Since, 80% learning curve ratio applies, coefficient of learning is given by

$$b = \frac{-\log (0.80)}{\log 2} = \frac{-0.0969}{0.3010} = -0.3219$$

(*i*) For x = 30 units, learning curve equation is

$$y = 10 \times (30)^{-0.3219}, \quad \text{since } a = 10$$

$\therefore$ $$\log y = \log 10 + (-0.3219) \log 30$$

$$= 1.0000 - (0.3219) \times 1.4771 = 0.5245$$

So, y = antilog (0.5245) = 3.346 minutes

(*ii*) For x = 40 units, learning curve equation is

$$y = 10 \times (40)^{-0.3219}, \quad \text{where } a = 10,$$

$\therefore$ $$\log y = \log 10 + (-0.3219) \log 40$$

$$= 1.0000 - (0.3219) \times (1.6020) = 0.4842$$

So, y = antilog (0.4842) = 3.05 minutes.

Example 31.9. *You have been asked about the application of the learning curve as a management accounting technique. You are required to illustrate the use of learning curve for calculating the expected average unit cost of making (i) 4 machines; and (ii) 8 machines, using the data given below:*

Direct labour needed to make the first machine is 1,000 hours. Learning curve is 80%, Direct labour cost is ₹3 per hour, and Direct materials cost is ₹1,800 per machine. Fixed cost for either size order = ₹8,000.

Solution. Since, 80% learning curve is in operation,

$$b = \frac{-\log (0.80)}{\log 2} = -0.3219$$

(*i*) For x = 4 units, $$y = (1{,}000) \times (4)^{-0.3219}$$

$\therefore$ $$\log y = \log 1{,}000 - (0.3219) \times \log 4$$

$$= 3.0000 - (0.3219) \times (0.6021) = 2.8062$$

So, y = antilog (2.8062) = 640 hours per unit of cumulative production at this level.

(*ii*) For $x = 8$ units, $y = (1{,}000) \times (8)^{-0.3219}$

$\therefore$ $\log y = \log 1{,}000 + (-0.3219) \times \log 8$

$= 3.0000 - (0.3219) \times (0.9031) = 2.7093$

So, y = antilog (2.7093) = 512 hours per unit of cumulative production at this level

Average unit cost for making 4 machines:		Average unit cost for making 8 machines:	
Direct materials cost	= ₹ 1.800	Direct materials cost	= ₹ 1,800
Direct labour (₹ 3 × 640 hours)	= ₹ 1,920	Direct labour (₹ 3 × 512 hours)	= ₹ 1,536
Fixed cost (₹ 8,000/4)	= ₹ 2,000	Fixed cost (₹ 8,000/8)	= ₹ 1,000
Total	₹ 5,720	Total	₹ 4,336

Example 31.10. *A company has 10 direct workers, who work for 25 days a month of 8 hours per day. The estimated down time is 25% of the total available time. The company received an order for a new product. The first unit of the new product requires 40 direct labour hours to manufacture the product. The company expects 80% (index is – 0.322) learning curve for this type of work. The company uses standard absorption costing and the cost data are as under:*

Direct materials	*₹ 60 per unit*
Direct labour	*₹ 6 per direct labour hour*
Variable overheads	*₹ 1 per direct labour hour*
Fixed overheads	*₹ 7,500 per month*

Required:

(*i*) *Calculate the cost per unit of the first order of 30 units.*

(*ii*) *If the company receives a repeat order for 20 units, what price will be quoted to yield a profit of 25% on selling price?* **[C.A. Final (Nov.) 2002]**

Solution. Since, the company expects 80% learning curve

$$y = ax^{-b} = ax^{-0.322},$$

where, x is the cumulative number of units, y is the cumulative average unit time of these x units, 'a' is the average time of the first unit and 'b' is the learning coefficient (or the index of learning).

$$\therefore \quad \log y = \log a + (-0.322) \log x$$

(*a*) When $x = 30$ units and $a = 40$ direct labour hours, then

$$\log y = \log 40 - (0.322) \log 30$$

$$= 1.6021 - (0.322) \times (1.4771) = 1.1264$$

$$y = \text{antilog } (1.1264) = 13.38 \text{ hours.}$$

$\therefore$ Total time required for 30 units = 13.38 hours × 30 units = 401.40 hours.

(*b*) When $x = 50$ units and $a = 40$ direct labour hours, then

$$\log y = \log 40 - (0.322) \log 50$$

$$= 1.6021 - (0.322) \times (1.6990) = 1.055$$

$$y = \text{antilog } (1.055) = 11.35 \text{ hours}$$

$\therefore$ Total time required for 50 units = 11.35 hours × 50 units = 567.5 hours

(*i*) Computation of cost per unit of the first order of 30 units:

Direct materials cost	(30 units × ₹ 60)	= ₹ 1,800.00
Direct labour cost	(401.40 hours × ₹ 6)	= ₹ 2,408.40
Variable overheads	(401.40 hours × ₹ 1)	= ₹ 401.40
Fixed overheads	(401.40 hours × ₹ 5)	= ₹ 2,007.00
	Total	₹ 6,616.80

$\therefore$ Cost per unit = ₹ 6,616.80/30 = ₹ 220.56.

(*ii*)	Direct materials cost	(20 units × ₹ 60)	= ₹ 1,200.00
	Direct labour cost	(567.5 – 401.4) hours × ₹ 6	= ₹ 996.60
	Variable overheads	(567.5 – 401.4) hours × ₹ 1	= ₹ 166.10
	Fixed overheads	(567.5 – 401.40) hours × ₹ 5	= ₹ 830.50
		Total	₹ 3,193.20

∴ Cost per unit = ₹ 3,193.20/20 = ₹ 159.66.

Thus, the price to be quoted to yield a profit of 25% on selling price is as under:

If selling price is ₹ 100, then profit is ₹ 25 and cost is ₹ 75

Thus, selling price per unit $= \frac{100}{75} \times 159.66 =$ ₹ 212.88.

Example 31.11. *A company has designed a prototype Electronic Starter for which the following information is available:*

Direct labour hours = 260 hours,
Direct materials cost = ₹ 30,000,
Direct labour rate = ₹ 20 per hour,
Variable overheads = 130% of direct labour,
Fixed overhead = 70% of direct labour.

Based on the demonstration of the prototype, the company has received order for 50 units during first six months and another 75 units during the following six months thereafter.

A learning curve effect of 80% is applicable. It is expected that in view of large volume, a 50% discount on material cost will be available in first six months and a 10% discount thereafter. The rates of overhead will remain unchanged and the same percentages would apply.

The company sets the selling price with a 40% mark-up on costs. Determine the selling price per unit for the first 50 units and the next 75 units.

(The index of learning rate at learning curve effect of 80% is 0.3219). **[ICWA Final (Dec.) 2002]**

Solution. Since, a learning curve effect of 80% is applicable,

$$y = ax^{-b},$$

where, $a = 260$ units and $b = 0.3219$.

(*i*) *When x = 51,* $y = 260 \times (51)^{-0.3219}$

∴ $\log y = \log 260 + (-0.3219) \log 51$

$= 2.4150 - (0.3219) \times (1.7076) = 1.8653$

or $y = \text{antilog } (1.8653) = 73.33$ hours

∴ Total time required for 51 units = 51 × 73.33 hours = 3,740 hours (approx.)

Incremental total hours = (3,740 – 260) hours = 3,480 hours

Incremental hours per unit = 3,480/50 = 69.6 hours.

(*ii*) *When x = 126,* $y = 260 \times (126)^{-0.3219}$

∴ $\log y = \log 260 + (-0.3219) \log 126$

$= 2.4150 - (0.3219) \times (2.1004) = 1.7389$

or $y = \text{antilog } (1.7389) = 54.81$ hours

∴ Total time required for 126 units = 126 × 54.81 hours = 6,906 hours

Incremental total hours = (6,906 – 3,740) hours = 3,166 hours

Incremental hours per unit = 3,166/75 = 42.21 hours.

Computation of Selling Price (₹ per unit)

	1st lot of 50 units			2nd lot of 75 units		
Direct material	₹ 30,000	× ₹ 0.95	= ₹ 28,500	₹ 30,000	× ₹ 0.90	= ₹ 27,000
Direct labour	69.6 hours	× ₹ 20	= ₹ 1,392	42.21 hours	× ₹ 20	= ₹ 844
Variable overhead	₹ 1,392	× 130%	= ₹ 1,810	₹ 844	× 130%	= ₹ 1,097
Fixed overhead	₹ 1,392	× 70%	= ₹ 974	₹ 844	× 70%	= ₹ 591
Total cost			₹ 32,676			₹ 29,532
40% make-up on total costs			₹ 13,070			₹ 11,813
Selling price per unit: (Total cost + Make up 40% cost)			₹ 45,746			₹ 41,345

Example 31.12. *A company has accepted an order for making 15 items of a specialized machine at a price of ₹ 4 lakh each. The delivery is to be completed within 4 months. The company works 23 days a month and the normal direct wages per day amount to ₹ 10,000. However, in case of need, the company can work overtime upto 8 days during the said period at double the normal rate of wages. Overheads amount to ₹ 12,000 per normal working day but no overheads are charged on overtime working days. The material cost is ₹ 2,40,000 per machine. The company has estimated that it will take 10 working days to manufacture the first machine. The company is expected to experience a learning effect of 90% (b = 0.152). The contract stipulates a penalty of ₹ 40,000 per machine delivered beyond the schedule of 4 months.*

You are required to calculate the costs and advise the company whether it is preferable to work only during normal working days and pay penalty for any delayed delivery of the machines or to work overtime to avoid paying penalty. ***[ICWA Final (June) 2000]***

Solution. Since, delivery is to be completed within 4 months, total number of working days is equal to 23 × 4, *i.e.*, 92 days.

Time for first machine = 10 days, *i.e.*, $a = 10$ and $b = 0.90$

(*i*) Average time for 15 days is given by

$$y = 15 \times 10 \times (15)^{-0.152}$$

Taking logarithm, we get

$$\log y = \log 15 + \log 10 - (0.152) \log 15$$
$$= 1.1761 + 1.000 - (0.152) \times (1.1761) = 1.9973$$

or $$y = \text{antilog}\,(1.9973) = 99.4 \text{ days}$$

(*ii*) Average time for 14 days is given by

$$y = 14 \times 10 \times (14)^{-0.152}$$

Taking logarithm, we get

$$\log y = \log 14 + \log 10 - (0.152) \times \log 14$$
$$= 1.1461 + 1.000 - (0.152) \times (1.1461) = 1.9719$$

or $$y = \text{antilog}\,(1.9719) = 93.7 \text{ days.}$$

(*iii*) Average time for 13 days is given by

$$y = 13 \times 10 \times (13)^{-0.152}$$

Taking logarithm, we get

$$\log y = \log 13 + \log 10 - (0.152) \log 13$$
$$= 1.1139 + 1.000 - (0.152) \times (1.1139) = 1.9446$$

or $$y = \text{antilog}\,(1.9446) = 88 \text{ days.}$$

Since, only 92 days are available, company cannot supply 15 machines or 14 machines. It can supply only 13 machines in time. For the remaining 2 machines, only alternative is to supply either with penalty or work overtime. Computations for these alternatives are given below:

	Overtime is not worked (in ₹)	*Overtime is worked (in ₹)*
Direct material	15 × 2,40,000 = 36,00,000	15 × 2,40,000 = 36,00,000
Direct labour	99.4 × 10,000 = 9,94,000	92 × 10,000 = 9,20,000
Overheads	92 × 12,000 = 11,04,000	92 × 12,000 = 11,04,000
Penalty	2 × 40,000 = 80,000	7.4* × 20,000 = 1,48,000
Total cost (15 machines):	57,78,000	57,72,000
Sales (15 machines × ₹ 40,000):	60,00,000	60,00,000
Profit (15 machines):	2,22,000	2,28,000

* Overtime for (99.4 – 92) days, *i.e.*, 7.4 days.

From the comparison of above computations, it is apparent that company should work overtime which will help the company to make additional profit of ₹ 6,000 (*i.e.*, ₹ 2,28,000 – ₹ 2,22,000).

Example 31.13. *ABC & Co. has given the following data:*

80% Average — Time Curve

Cumulative units (x)	*Average time (hours)*	*Total time (hours)*	*Marginal time (hours)*
1	100	100	100
2	80	160	60
3	?	?	?
4	64	256	?

It is required to fill in blanks.

Solution. We know that leaving curve is: $y = ax^{-b}$, where a = time for the first unit, y = average time per unit, x = cumulative number of units, and b = the learning index.

The coefficient of learning 'b' is determined as follows:

$$b = \frac{\log(1 - \%\text{ decrease})}{\log 2} = \frac{\log(1 - 0.20)}{\log 2}$$

$$= \frac{\log 0.80}{\log 2} = \frac{0.0969}{0.3010} = 0.3219$$

For $x = 3$, $\quad y = 100 \times (3)^{-0.3219} = 100 \times 0.7020 \text{ or } 70.2$

Now, it is possible to complete the above information, *viz.*,

Cumulative units (x)	*Average time (hours)*	*Total time (hours)*	*Marginal time (hours)*
1	100	100	(100 – 0) = 100
2	80	80 × 2 = 160	160 – 100 = 60
3	70.2	70.2 × 3 = 210.6	210.6 – 160 = 50.6
4	64	64 × 4 = 256	256 – 210.6 = 45.4

Example 31.14. *An electronics firm which has developed a new type of fire-alarm system has been asked to quote for a prospective contract. The customer requires separate price quotations for each of the following possible orders:*

Order :	*First*	*Second*	*Third*
Number of fire-alarm systems :	100	60	40

The firm estimates the following cost per unit for the first order:

Direct materials: ₹500, *Variable overheads: 20% of direct labour*

Direct labour: Deptt. A (highly automatic): 20 hours at ₹10 per hour

Deptt. B (Skilled labour): 40 hours at ₹15 per hour

Fixed overheads absorbed: Deptt. A: ₹8 per hour

Deptt. B: ₹5 per hour

Determine a price per unit for each of the three orders, assuming the firm uses a mark up of 25% on total costs and allows for an 80% learning curve. Extract from 80% learning curve table:

X :	1.0	1.3	1.4	1.5	1.6	1.7	1.8	1.9	2.0
Y (%) :	100.0	91.7	89.5	87.6	86.1	84.4	83.0	81.5	80.0

X represents the cumulative total volume produced to date expressed as a multiple of the initial order.

Y is the learning curve factor for a given X value, expressed as a percentage of the cost of the initial order.

[C.A. Final (May) 2005]

Solution.

(*i*) *Price per unit for first order of 100 units*

Direct materials		₹ 5,00.00
Direct labour	Deptt. *A*: 20 hours @ ₹ 10 = ₹ 200	
	Deptt. *B*: 40 hours @ ₹ 15 = ₹ 600	₹ 800.00
Variable overhead	20% of ₹ 800	₹ 160.00
Fixed overhead	Deptt. *A*: 20 hours @ ₹ 8 = ₹ 160	
	Deptt. *B*: 40 hours @ ₹ 5 = ₹ 200	₹ 360.00
Total cost		₹ 1,820.00
Profit @ 25% on total cost		₹ 455.00
Selling price per unit		₹ 2,275.00

(*ii*) *Price per unit for second order of 60 units*

Learning effect will be applicable only in department *B*.

Cumulative output becomes 100 units + 60 units (= 160 units), *i.e.*, 2 times for which learning is 86.1% from the learning curve table.

Therefore, total hours for 160 units = 160 units × 40 hours × 0.861 = 5,510.4 hours

Therefore, hours for 60 units = Hours for 160 units less hours for 100 units

= 5,510.4 hours – (40 × 100) hours = 1,510.4 hours.

$$\therefore \quad \text{Hours per unit} = \frac{1,510.4}{60} = 25.17$$

Calculation of Selling Price per Unit

Direct materials			₹ 500.00
Direct labour	Deptt. *A*: 20 hours @ ₹ 10	= ₹ 200.00	
	Deptt. *B*: 25.17 hours @ ₹ 15	= ₹ 377.55	₹ 577.55
Variable overhead	20% of ₹ 577.55		₹ 115.51
Fixed overhead	Deptt. *A*: 20 hours @ ₹ 8	= ₹ 160.00	
	Deptt. *B*: 25.17 hours @ ₹ 5	= ₹ 125.85	₹ 285.85
Total cost			₹ 1,478.91
Profit @ 25% on total cost			₹ 369.73
Selling price per unit			₹ 1,848.64

(*iii*) *Price per unit for third order of 40 units*

Cumulative output (100 + 60 + 40) units = 200 units, *i.e.*, 2 times for which learning is 80% from the table.

Total hours for 200 units = 200 units × 40 hours × 0.80 = 6,400 hours

∴ Hours for 40 units = Hours for 200 units less hours for 160 units
= 6,400 hours – 5,510.4 hours = 889.6 hours

Therefore, Hours per unit $= \dfrac{889.6}{40} = 22.24$

Calculation of Selling Price per Unit

Direct materials			₹ 500.00
Direct labour	Deptt. *A*: 20 hours @ ₹ 10	= ₹ 200.00	
	Deptt. *B*: 22.24 hours @ ₹ 15	= ₹ 333.60	₹ 533.60
Variable overhead	20% of ₹ 533.60		₹ 106.72
Fixed overhead	Deptt. *A*: 20 hours @ ₹ 8	= ₹ 160.00	
	Deptt. *B*: 22.24 hours @ ₹ 5	= ₹ 111.20	₹ 271.20
Total cost			₹ 1,411.52
Profit 20%			₹ 352.88
Selling price per unit			₹ 1,764.40

PROBLEMS

31.01. The usual learning curve equation is $y = ax^{-b}$, where y is the average time per unit for x units, a is the time for first unit and b is the learning coefficient. Given that $a = 10$ hours and learning rate is 80%, calculate the following:

(*i*) average time for 20 units, (*ii*) total time for 30 units, and (*iii*) time for 31 units to 40 units.

31.02. (*a*) Your Company has been approached by a customer to supply four units of a new product made to the customer's individual specification. The company experiences a 90% learning rate. The estimated labour time for the first unit of this product is 150 hours and the company's direct labour cost is ₹ 5 per hour. Estimate the labour cost for this order.

(*b*) After receiving the first order, if the customer places a repeat order, what will be the labour cost for the second order?

(*c*) If the customer had ordered all eight units at the same time, calculate the labour cost per unit for the combined order.

31.03. Manual Skill Limited is a company engaged in the provision of small quantities of specialized equipment to industrial customers. The work involved in the manufacture of the output calls for a high level of manual skill, and since no product has a long technological life, it is thought that a 70% learning curve applies to all output made by the company.

A customer has asked for the delivery of eight units of a product which has never before been made. An estimate of the direct cost of the first unit has been derived as follows:

Materials: ₹ 400, Direct labour (250 hours): ₹ 1,250

The company adds a contribution margin of 160% on direct costs, to cover fixed costs and provide a profit margin. *Required:* (*a*) State the formula for 70% learning curve, and (*b*) Use the formula to calculate the estimated direct costs and sales price of the ten units of the product.

31.04. A firm have been asked to give a quote for the production of 50 units of a new product. Estimates for the assembly of the first unit are:

Direct materials: ₹ 100, Direct labour: ₹ 50, Variable overhead: ₹ 30.

1. Labour costs are ₹ 5 an hour.
2. Variable overhead is based on labour hours (activity based) at a rate of ₹ 3 per labour hour.
3. The company quotes prices on a 20 per cent mark-up on the estimated costs to cover fixed overheads and profits.
4. The company operates an 80 per cent learning curve.

The company wishes to estimate the marginal cost of making 50 units and hence calculate the quoted price per unit.

31.05. A company manufacturing small tools has just completed an order for 100 units of a made-to-order special tool for a structural construction firm at a contract price of ₹ 50,000. This firm has made enquiries for a repeat order for another 200 units of the special tool.

The cost details in respect of the first order are as follows:

Labour:	₹ 300	per unit
Material:	₹ 50	per unit
Tools:	₹ 10	per unit
Overhead:	₹ 100	(20% fixed)
Cost per unit:	₹ 460	
Profit:	₹ 40	per unit
Price per unit:	₹ 500	

The company expects 80% learning rate in respect of labour and variable overhead, but there will be a 10% escalation in these and material costs. The amount of ₹ 1,000 spent on tools will not be required for the second order. Not satisfied with the present rate of profit, the company desires to raise it to 20% on cost.

On the basis of the data given above you are required to work out the contract price for the repeat order.

31.06 Dynamo, a manufacturer of aircraft parts has been asked to bid for 900 units of a particular type of component. Four months ago the company had produced a first lot of 400 units for another customer. The cost details of this first lot are given on next page:

Repeated assembly of this type of component experiences a learning effect of 85%. The cost benefit of this will be reflected in the bid price. Dynamo follows a policy of selling price to earn 30% gross margin. Tooling cost have been fully recovered from the first lot sold. Determine the selling price per unit for the second lot indented.

Direct materials	₹ 30,00,000	
Direct labour	₹ 20,00,000	8,000 hours @ ₹ 250
Tooling costs	₹ 4,80,000	
Variable overheads	₹ 9,00,000	Proportional to direct labour
General overheads	₹ 12,00,000	Proportional to direct labour
Total	₹ 75,80,000	*[ICWA Final (Dec.) 2001]*

31:6. USES OF LEARNING CURVE

Learning curve theory can be used to

(*i*) Prepare realistic budget.
(*ii*) Compare budgeted and actual costs and report cost variances.
(*iii*) Calculate the incremental (marginal) cost of making extra units of a product.
(*iv*) Increase market share of a company.
(*v*) Increase production of a particular company.
(*vi*) Quote selling prices for a contract, where prices are calculated at cost plus a percentage mark-up for profit.

31:7. COSTS AFFECTED BY LEARNING CURVE

(*i*) Direct costs and labour time are obviously affected by learning curve effect, output capacity increases as the workforce gains experience.
(*ii*) Material costs are usually unaffected by using learning curve, although it is conceivable that material handling might improve and so wastage costs might reduce.
(*iii*) Variable overhead costs will also be affected by the learning curve, but only to the extent that variable overheads vary with direct labour hours worked.
(*iv*) Fixed overhead expenditure should often remain unaffected by learning curve, although some problems might be caused in an organization, that uses absorption.

31:8. LEARNING CURVE THEORY IN MANUFACTURING ORGANIZATION

The areas in which the application of learning curve can help a manufacturing organization are:

(*i*) Learning curve provides better ***cost predictions to enable price quotations*** to be preferred for potential orders.
(*ii*) Learning curve enables to ***predict the inputs required*** more effectively and helps in the preparation of accurate delivery schedules.
(*iii*) Using learning curve theory the ***performance of workers improve,*** time taken per unit reduces and thus his productivity goes up.
(*iv*) If budgets and standards are set without considering learning curve, it is meaningless because variances will arise.
(*v*) To take full advantage of learning effect, idle production time should be avoided and work scheduling should pay regard to the expected learning effect.
(*vi*) To reduce unit variable costs, more units are to be produced.

31:9. ADVANTAGES OF LEARNING CURVE THEORY

Given below are the salient features of learning curve theory:

(*i*) Learning curve helps to analyse cost-volume-profit relationship during familiarization phase of product or process and thus it is very useful for cost estimates.

(*ii*) Learning curve is extremely useful in setting standards in learning phase.

(*iii*) It helps design engineers in making decisions based upon expected rates of improvement.

(*iv*) The use of cost data adjusted for learning effect helps in development of advantageous pricing policy.

(*v*) In the process of budgeting and profit planning, the incorporation of the learning curve introduces new complexity and sophistication.

(*vi*) It is helpful to Government in negotiations about the contracts.

31:10. LIMITATIONS OF LEARNING CURVE THEORY

Although there are certain distinct advantages of incorporating learning curve theory in practice, yet there are certain limitations, *viz.*,

(*i*) It is not always possible to get enough accurate data to decide what the learning curve is.

(*ii*) It assumes stable conditions at work, but in practice it is not always practicable specially due to labour turnover.

(*iii*) It provides that constant stage follows the learning stage. But, it is not always easy to ascertain as at what point of time, "constant stage" begins.

(*iv*) Time-break between repeating production should not be too long or the workers will forget and learning process will have to begin all over again.

(*v*) When the trade unions do not accept gradual reduction in production time per unit, management may try to establish a low standard time per unit from the onset. This will lead to adverse efficiency variances, until learning effect has taken place.

(*vi*) If there is productivity bonus, incorporation of learning effect may frustrate workers in learning stage, as it may appear as a threat to the size of bonus, that they may earn.

TEST YOUR UNDERSTANDING

1. If cumulative output in learning curve doubles, the average labour time per unit

(*a*) increases at a constant percentage (*b*) declines at a constant percentage

(*c*) remains unchanged (*d*) none of the above.

2. The percentages in learning curve are commonly called

(*a*) incremental costs (*b*) marginal costs

(*c*) cost reductions (*d*) none of the above.

3. The learning curve equation is expressed as

(*a*) $y = ax^{-b}$ (*b*) $y = ax^{b}$

(*c*) $y = -ax^{-b}$ (*d*) $y = -ax^{b}$.

4. Formula for finding out total time in learning curve is

(*a*) ax^{1+b} (*b*) ax^{1-b}

(*c*) x^{1+b}/a (*d*) x^{1-b}/a.

5. Formula for finding out total time in learning curve is

(*a*) $a(a-b)x^{-b}$ (*b*) $a(a-b)x^{b}$

(*c*) $a(a+b)x^{-b}$ (*d*) $a(a+b)x^{b}$.

6. Learning curve ratio is obtained by computing the average labour cost of

(*a*) $\dfrac{\text{first } N \text{ units}}{\text{first } 2N \text{ units}}$ (*b*) $\dfrac{\text{first } 2N \text{ units}}{\text{first } N \text{ units}}$

(*c*) $\dfrac{\text{first } (N-1) \text{ units}}{\text{first } (N+1) \text{ units}}$ (*d*) $\dfrac{\text{first } (N+1) \text{ units}}{\text{first } (N-1) \text{ units}}$.

7. By using learning curve,
   (*a*) Material costs are usually affected, while variable overhead costs will also be affected.
   (*b*) Material costs are usually unaffected, while variable overhead costs will also be unaffected.
   (*c*) Direct costs and labour time are unaffected.
   (*d*) Direct costs and labour time are affected.
8. Choose the statement which is not correct:
   (*a*) Learning curve is extremely useful in setting standards in learning phase.
   (*b*) Learning curve helps design engineers in making decisions based upon expected rates of improvement.
   (*c*) Learning curve theory does not increase production of a particular company.
   (*d*) Learning curve theory can be used to prepare realistic budget.

ANSWERS

1. (*b*)	**2.** (*c*)	**3.** (*a*)	**4.** (*b*)
5. (*a*)	**6.** (*b*)	**7.** (*d*)	**8.** (*c*)

REVIEW EXERCISE

1. Explain the concept of learning curve and discuss its relevance to pricing decisions. ***[C.A. Final (May) 2004]***
2. Briefly explain the learning curve ratio. ***[C.A. Final (Nov.) 2006]***
3. Discuss the application of the learning curve. ***[C.A. Final (May) 2007]***
4. Why does the management use the learning curve?
5. Explain the application of learning curve in business situations.
6. State the areas in which the application of learning curve theory can help a manufacturing organization. ***[C.A. Final (May) 2003]***
7. What are the distinctive features of learning curve theory in manufacturing environment? ***[C.A. Final (Nov.) 2007]***
8. Explain the learning curve and discuss its relevance in setting standards.
9. What do you understand by the concept of learning curve? Explain its relevance in planning for profits.
10. Explain the usefulness of the following cost measurements obtained in a learning curve analysis: (*a*) Cumulative average cost, (*b*) Total cost, (*c*) Incremental cost, and (*d*) Marginal cost.
11. What are the major advantages of learning curve theory?
12. Discuss the limitations and uses of learning curve.

CHAPTER 32

Statistical Quality Control

"Success of any firm/enterprise depends on quality of its products/services"

For any firm, it is essential to survive by maintaining the quality of its products because of tough competition by its opponents. This can be done by maintaining the quality of its products through inspection of its each item. But it is very costly and time consuming. Therefore, now a days, we make use of Statistics for quality control and the method is known as ***Statistical Quality Control*** (*SQC*).

32:1. MEANING OF STATISTICAL QUALITY CONTROL (*SQC*)

Quality control is the process of verification or corrections in the quality of the product when deviations in the quality are found to be more than expected. When the application of statistical techniques are used in all the areas of quality control in order to achieve the desired quality at a cost which the customer can afford to pay, the procedure is known as statistical quality control abbreviated by *SQC*.

32:2. USES OF QUALITY CONTROL

Given below are some of the uses of quality control:

1. Achieve better utilization of raw materials and equipments.
2. Take necessary corrective measures for maintaining the quality of products.
3. Ensure that sub-standard products do not reach to customers.
4. Locate and identify the process faults in order to control the defective, scrap and waste.
5. Most important one is to set-up standards of quality acceptable to the customers.

32:3. VARIATION IN QUALITY

Variation in ***quality*** can be due to ***chance causes*** or ***assignable causes:***

Random or Chance Variation. This is the variation which results from many minor causes that vary at random, independent of each other, and it is difficult as well as uneconomical to detect or eliminate their causes. For instance, slight variations in temperature, humidity, pressure and similar factors may interact randomly to produce slight variation in the quality of the product. This type of variation is inevitable in a manufacturing process and is beyond human control. There is no way to eliminate this variation and is, therefore, called ***permissible*** or ***allowable*** variations.

Assignable or Controllable Variation. This is the variation which results from some non-random causes that can be identified. For example, variations in the product dimensions may be observed due to several factors such as improper machine set-up, wear and tear of the equipment, worker's fatigue, change in the character of input raw materials, etc. This type of variation can be detected and eliminated, and is, therefore, called ***controllable*** or ***preventable*** variation.

A process is said to be out of statistical control, whenever it indicates the presence of some assignable variation. The objective of *SQC* is to devise statistical methods that would help us in isolating assignable variation from random variation thus enabling us to detect, identify and eliminate the assignable causes of variation.

Eliminating the assignable causes of variation is described as bringing a process under statistical control.

32:4. TECHNIQUES OF *SQC*

For controlling the quality of a product, in general, there are two ways, *viz.*,

1. Process control, and 2. Product control.

1. Process Control. This is concerned with controlling the quality of the product during the process of production. It ensures that a product of a requisite standard is to be produced. This is achieved largely through 'control chart' device.

2. Product Control. This is concerned with classification of raw materials or finished goods into acceptable or non-acceptable and also deciding whether another sample has to be tested. It is concerned with the inspection of goods already produced to ascertain whether they are fit to be dispatched.

Various techniques of *SQC*, in popular use, may be shown schematically as shown below:

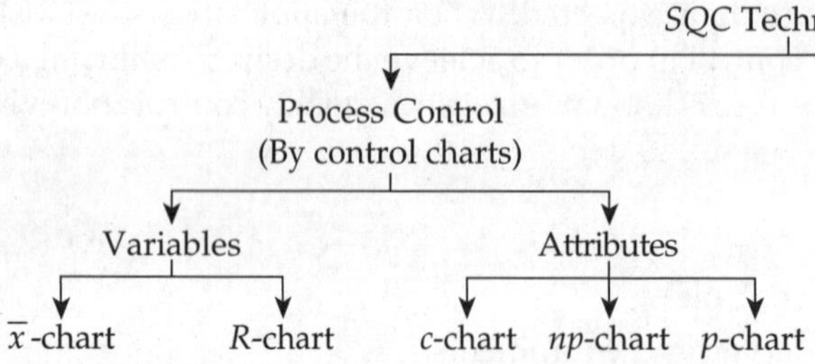

32:5. CONTROL CHARTS

A ***control chart*** is an important statistical tool used for the study and control of repetitive processes and is based on the fact that variability does exist in all repetitive processes.

A control chart consists of:

(*a*) ***a control*** or ***centre line*** (CL) that indicates the desired control level of the process.

(*b*) ***an upper control limit*** (UCL) that indicates the upper limit of tolerance.

(*c*) ***a lower control limit*** (LCL) that indicates the lower limit of tolerance.

The control chart has a horizontal scale that represents the consecutive sample number and a vertical scale representing the characteristic quality of each sample. This is shown in *Fig. 32.1.*

The data are collected on a series of samples taken during a production process at different time intervals and are plotted on the graph. As long as the points fall within the control limits, the process is under statistical control and we do not question the quality of the product. However, if a plotted point falls outside the control limits, this alerts the production manager to the possibility that the quality of the product is unacceptable and that the process is not under statistical control. So, he tries to determine whether variation is due to chance causes or some assignable causes.

If the control chart indicates that the observed variation is due to chance causes alone, the process is said to be 'in control'. If on the other hand, the control chart indicates that the observed variation is not likely to be due to chance, it can be said that the manufacturing process is 'out of control'. In that event, the process is halted and effort is made to seek and correct the possible cause.

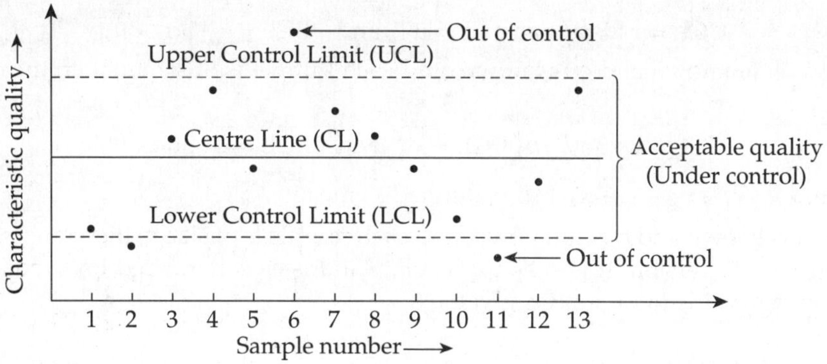

Fig. 32.1

The control charts can be divided under two headings:

1. Control Charts for Variables. They are used to achieve and maintain an acceptable quality level for a process whose product can be subjected to quantitative measurements such as diameter of a hole, thickness of a pipe, specific resistance of a wire, and strength of a yarn, etc.

2. Control Charts for Attributes. They are used to achieve and maintain an acceptable quality level for a process whose product can not be subjected to quantitative measurement, but can be classified as 'good' or 'bad', as 'acceptable' or 'not acceptable'. For instance surface finish of a table, colour or brightness of an article, etc. is either acceptable or not acceptable.

32:6. CONTROL CHARTS FOR VARIABLES

Control charts are used for measurable quality characteristics. They can be divided as under:

1. Control charts for sample means ($\overline{\mathbf{x}}$-charts).
2. Control charts for sample ranges (**R**-charts).

These charts are designed to achieve and maintain quality level for the process whose product is amenable to maintain quantitative measurement and can be constructed as follows:

Step 1. A random sample of size 'n' is taken during a manufacturing process and the quality measurements of variables $x_1, x_2, \ldots, x_n$ are noted.

Step 2. The sample mean $\overline{\mathbf{x}}$ and sample range **R** are calculated using

$$\overline{\mathbf{x}} = (x_1 + x_2 + \ldots + x_n)/n \quad \text{and} \quad \mathbf{R} = x_{\text{max.}} - x_{\text{min.}},$$

where $x_{\text{max.}}$ and $x_{\text{min.}}$ are the largest and smallest values respectively of measurements $x_1, x_2, \ldots, x_n$.

Step 3. If the process is found to be stable, k successive samples are selected and for each sample, mean $\overline{\mathbf{x}}$ and range **R** are calculated. Then $\overline{\overline{\mathbf{x}}}$ and $\overline{\mathbf{R}}$ are computed using

$$\overline{\overline{\mathbf{x}}} = (\overline{\mathbf{x}}_1 + \overline{\mathbf{x}}_2 + \ldots + \overline{\mathbf{x}}_k)/k \quad \text{and} \quad \overline{\mathbf{R}} = (\mathbf{R}_1 + \mathbf{R}_2 + \ldots + \mathbf{R}_k)/k.$$

$\overline{\mathbf{x}}$ - Chart

If the quality characteristic variable x be assumed to be $N(\mu, \sigma^2)$, then for the process to be in control, we must have

$$E(\overline{\mathbf{x}}) = \mu \quad \text{and} \quad V(\overline{x}) = \sigma/\sqrt{n},$$

where, $\overline{\mathbf{x}}$ is the arithmetic mean (AM) of the sample.

Case 1. When mean μ and standard deviation σ are known and each sample is of size n, then the control limits are:

$$\text{UCL} = \mu + 3\sigma/\sqrt{n}, \quad \text{CL} = \mu, \quad \text{and} \quad \text{LCL} = \mu - 3\sigma/\sqrt{n},$$

Case 2. When μ is unknown and σ is known, and each sample is of size n, then the control limits are:

$$\text{UCL} = \bar{\bar{x}} + 3\sigma/\sqrt{n}, \quad \text{CL} = \bar{\bar{x}}, \quad \text{and} \quad \text{LCL} = \bar{\bar{x}} - 3\sigma/\sqrt{n},$$

where, $\bar{\bar{x}} = (\bar{x}_1 + \bar{x}_2 + \ldots + \bar{x}_k)/k$ is defined as mean.

Case 3. When μ is known and σ is unknown, and each sample is of size n, then range **R** is used as a measure of dispersion. It is easy to calculate and is closely related to standard deviation. The average value of range **R** is given by

$$\bar{\mathbf{R}} = (\mathbf{R}_1 + \mathbf{R}_2 + \ldots + \mathbf{R}_k)/k.$$

The control limits in this case are:

$$\text{UCL} = \mu + A_2\bar{\mathbf{R}}, \quad \text{CL} = \mu, \quad \text{LCL} = \mu - A_2\bar{\mathbf{R}},$$

where, A_2 is found from the table of factors used in $\bar{x}$ and **R** quality control charts.

Case 4. When both μ and σ are unknown with sample of size 'n', then in such cases, mean is $\bar{\bar{x}}$ and standard deviation is represented by $\bar{\mathbf{R}}$. The control limits will be:

$$\text{UCL} = \bar{\bar{x}} + A_2\bar{\mathbf{R}}, \quad \text{CL} = \bar{\bar{x}}, \quad \text{and} \quad \text{LCL} = \bar{\bar{x}} - A_2\bar{\mathbf{R}}.$$

Case 5. When specified limits are given and sample is of size n. Then, if USL and LSL are upper and lower specified limits, mean is given by $\bar{\bar{x}} = (\text{USL} + \text{LSL})/2$. In this case the control limits are:

$$\text{UCL} = \bar{\bar{x}} + A_2\bar{\mathbf{R}}, \quad \text{CL} = \bar{\bar{x}}, \quad \text{and} \quad \text{LCL} = \bar{\bar{x}} - A_2\bar{\mathbf{R}}.$$

R-Chart

Here, centre line $= \bar{\mathbf{R}}$, the upper and lower control limits are set using

$$\text{UCL} = D_4\bar{\mathbf{R}}, \quad \text{and} \quad \text{LCL} = D_3\bar{\mathbf{R}},$$

where, D_3 and D_4 are also found from Table 32.1 depending on the sample size n.

Table 32.1. Factors used in the $\bar{x}$ and R Quality Control Charts

No. of units in a sample (n)	*Factor for $\bar{x}$-chart (A_2)*	*Factors for R-chart*		*Factor for estimating σ from $\bar{\mathbf{R}}$ (d_2)*
		Lower control limit (D_3)	*Upper control limit (D_4)*	
(1)	(2)	(3)	(4)	(5)
2	1.88	0	3.27	1.13
3	1.02	0	2.57	1.69
4	0.73	0	2.28	2.06
5	0.58	0	2.11	2.33
6	0.48	0	2.00	2.53
7	0.42	0.08	1.92	2.70
8	0.37	0.14	1.86	2.85
9	0.34	0.18	1.82	2.97
10	0.31	0.22	1.78	3.08

(contd.)

Table 32.1 (*Contd.*)

(1)	(2)	(3)	(4)	(5)
11	0.29	0.26	1.74	3.17
12	0.27	0.28	1.72	3.26
13	0.25	0.31	1.69	3.34
14	0.24	0.33	1.67	3.41
15	0.22	0.35	1.65	3.47
16	0.21	0.36	1.64	3.53
17	0.20	0.38	1.62	3.59
18	0.19	0.39	1.61	3.64
19	0.19	0.40	1.61	3.69
20	0.18	0.41	1.59	3.74

Remark. The upper and lower process tolerance limits (also called natural tolerance limits) for individual values of **x** are computed by using

$$\bar{\bar{\mathbf{x}}} + 3\bar{\mathbf{R}}/d_2 \quad \text{and} \quad \bar{\bar{\mathbf{x}}} - 3\bar{\mathbf{R}}_2, \text{ respectively.}$$

Here, d_2 can be found from the *Table 32.1* and depends on the sample size. If these natural tolerance limits fall within the customer's specifications, the process is said to be capable of meeting the customer specifications.

The process capacity is: $$6\sigma = 6\bar{\mathbf{R}}/2.$$

Example 32.1. *From a mass production of a product, consider a sample of size 50. Assume that the average sample mean was found to be 20 cm. Also, assume that from the past experience, the population standard deviation is 6.5. Determine the control limits.*

Solution. Given, $\text{Mean} \equiv \bar{\bar{\mathbf{x}}} = 20, \ \sigma = 6.5 \ \text{ and } \ n = 50.$

Then
$$\text{UCL} = \bar{\bar{\mathbf{x}}} + \sigma/\sqrt{n} = 20 + 6.5/\sqrt{50} = 20 + 6.5/7.07 = 20.92$$
$$\text{CL} = 20,$$
and
$$\text{LCL} = \bar{\bar{\mathbf{x}}} - \sigma/\sqrt{n} = 20 - 6.5/\sqrt{50} = 20 - 6.5/7.07 = 19.08$$

Example 32.2. *It is decided to use* $\bar{\mathbf{x}}$ *and* **R** *charts for a product. Following information is obtained after examining 30 samples each with sample size of 4:*

$$\sum \bar{\mathbf{x}} = 613.50 \quad \text{and} \quad \sum \mathbf{R} = 147.$$

Calculate 3σ *control limits for* $\bar{\mathbf{x}}$ *and* **R** *charts.*

Solution. Here, we have

$$\bar{\bar{\mathbf{x}}} = \sum \bar{\mathbf{x}}/30 = (613.50/30) = 20.45, \quad \text{and} \quad \bar{\mathbf{R}} = \sum \mathbf{R}/30 = (147/30) = 4.9$$

1. The 3σ control limits for the $\bar{\mathbf{x}}$ chart are as under:

$$\text{UCL} = \bar{\bar{\mathbf{x}}} + A_2\,\bar{\mathbf{R}} = 20.45 + 0.729 \times 4.9 = 24.02, \quad \text{since } A_2 = 0.729$$
$$\text{CL} = \bar{\bar{\mathbf{x}}} = 20.45,$$
and
$$\text{LCL} = \bar{\bar{\mathbf{x}}} - A_2\,\bar{\mathbf{R}} = 20.45 - 0.729 \times 4.9 = 16.82.$$

2. The 3σ control limits for the **R**-chart are:

$$\text{UCL} = D_4\,\bar{\mathbf{R}} = 2.282 \times 4.9 = 11.18, \qquad \text{since } D_4 = 2.282$$
$$\text{CL} = \bar{\mathbf{R}} = 4.9,$$
and
$$\text{LCL} = D_3\,\bar{\mathbf{R}} = 0, \qquad \text{since } D_3 = 0$$

Example 32.3. *A company manufactures screws to a nominal diameter 0.500 ± 0.030 cm. Five samples were taken randomly from the manufactured lot and 3 measurements were taken on each sample at different lengths. The reading are shown below:*

Sample number	*Measurements per sample (cm) x*		
	1	*2*	*3*
1	*0.488*	*0.489*	*0.505*
2	*0.494*	*0.495*	*0.499*
3	*0.498*	*0.515*	*0.487*
4	*0.492*	*0.509*	*0.514*
5	*0.490*	*0.508*	*0.499*

Calculate the control limits on $\overline{\mathbf{x}}$ *and* **R***-charts and draw the charts.*

[BBM, Dayalbagh Edu. Institute 2010]

Solution. We are given five samples. Their mean is calculated as follows:

$$\overline{\mathbf{x}}_1 = (0.488 + 0.489 + 0.505)/3 = 0.494,$$

$$\overline{\mathbf{x}}_2 = (0.494 + 0.495 + 0.499)/3 = 0.496.$$

Similarly $\overline{\mathbf{x}}_3 = 0.500,\ \overline{\mathbf{x}}_4 = 0.505,$ and $\overline{\mathbf{x}}_5 = 0.499.$

$$\therefore \quad \overline{\overline{\mathbf{x}}} = (\overline{\mathbf{x}}_1 + \overline{\mathbf{x}}_2 + \overline{\mathbf{x}}_3 + \overline{\mathbf{x}}_4 + \overline{\mathbf{x}}_5)/5 = 0.499$$

For the calculation of range, we have

$$\mathbf{R}_1 = (x_1)_{max.} - (x_1)_{min.} = 0.505 - 0.488 = 0.017, \quad \text{for sample 1,}$$

$$\mathbf{R}_2 = 0.499 - 0.494 = 0.005,\ \mathbf{R}_3 = 0.515 - 0.487 = 0.028,$$

$$\mathbf{R}_4 = 0.514 - 0.492 = 0.022, \text{ and } \mathbf{R}_5 = 0.508 - 0.490 = 0.018$$

$$\overline{\mathbf{R}} = (\mathbf{R}_1 + \mathbf{R}_2 + \mathbf{R}_3 + \mathbf{R}_4 + \mathbf{R}_5)/5 = 0.018$$

Control limits for $\overline{\mathbf{x}}$*-chart*

Let $A_2 = 1.02$ and $n = 3$. then from *Table 32.1*, we have

$$\text{CL} = \overline{\overline{\mathbf{x}}} = 0.499,$$

$$\text{UCL} = \overline{\overline{\mathbf{x}}} + A_2\,\overline{\mathbf{R}} = 0.499 + 1.02 \times 0.018 = 0.5174, \quad \text{for } n = 3$$

$$\text{LCL} = \overline{\overline{\mathbf{x}}} - A_2\,\overline{\mathbf{R}} = 0.499 - 1.02 \times 0.018 = 0.4807, \quad \text{for } n = 3$$

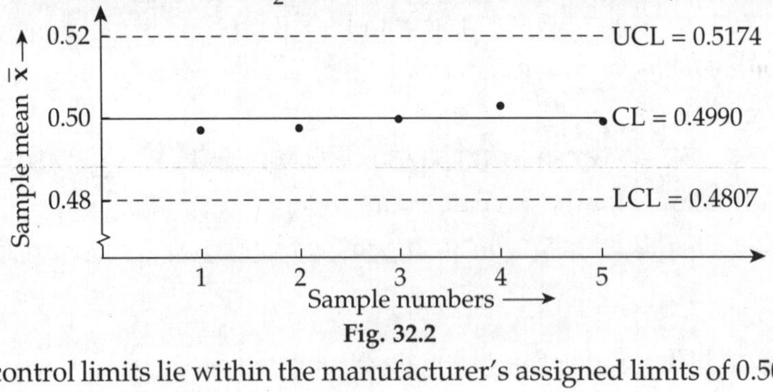

Fig. 32.2

Clearly, the control limits lie within the manufacturer's assigned limits of 0.50 + 0.03 = 0.53 and 0.50 – 0.03 = 0.47.

Control limits for **R***-chart*

$$\text{CL} \equiv \overline{\mathbf{R}} = 0.018,$$

$$\text{UCL} \equiv D_4\,\overline{\mathbf{R}} = 2.57 \times 0.018 + 0 = 0.0463, \quad \text{for } n = 3$$

$$\text{LCL} \equiv D_3\,\overline{\mathbf{R}} = 0 \times 0.18 - 0 = 0, \quad \text{for } n = 3$$

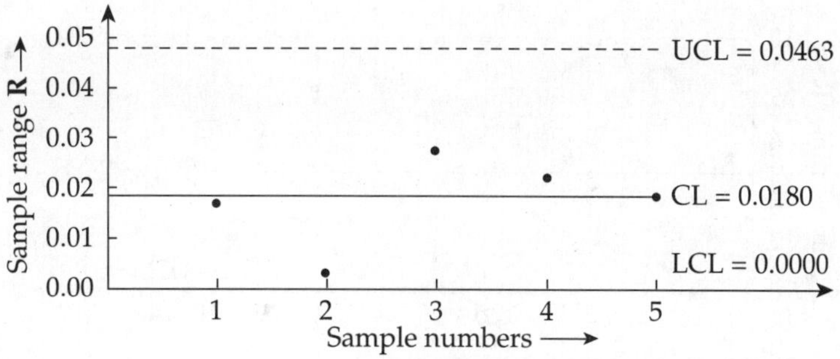

Fig. 32.3

We observe that no value of $\overline{\mathbf{x}}$ and **R** is out of control in *Fig. 32.2* and *32.3*. Therefore, the trial limits can be considered as actual control limits. The process is in statistical control, *i.e.*, it is operating free from assignable causes of variations and is only under the influence of chance causes of variations.

Example 32.4. *The following are the mean lengths and ranges of lengths of a finished product from 10 samples each of size 5. The specification limits for length are 200 ± 5 cm. Construct* $\overline{\mathbf{x}}$ *and* **R***-charts and examine whether the process is under control and state your recommendations.*

Sample no.	:	1	2	3	4	5	6	7	8	9	10
Mean $\overline{\mathbf{x}}$	:	201	198	202	200	203	204	199	196	199	201
Range **R**	:	5	0	7	3	4	7	2	8	5	6

Assume for $n = 5, A_2 = 0.577, D_3 = 0, D_4 = 2.115.$ ***[M.Com. (Dec.), Dayalbagh Edu. Institute 2008]***

Solution. The specification limits for length are given to be 200 ± 5 cm. Hence, mean is known whereas standard deviation is unknown.

Control limits for $\overline{\mathbf{x}}$*-chart* are as follows:

$$\text{CL} = \mu = 200,$$

$$\text{UCL} = \mu + A_2\,\overline{\mathbf{R}}, \text{ where } \overline{\mathbf{R}} = (\mathbf{R}_1 + \mathbf{R}_2 + \ldots + \mathbf{R}_{10})/10 = 47/10 = 4.7.$$

$$= 200 + 0.577 \times 4.7 = 202.71, \quad \text{since } A_2 = 0.577$$

and

$$\text{LCL} = \mu - A_2\,\overline{\mathbf{R}} = 200 - 0.577 \times 4.7 = 197.29.$$

Control limits for **R***-chart* are as follows:

$$\text{CL} = \overline{\mathbf{R}} = 4.7,$$

$$\text{UCL} = D_4\,\overline{\mathbf{R}} = 2.115 \times 4.7 = 9.941, \quad \text{since } D_4 = 2.115$$

$$\text{LCL} = D_3\,\overline{\mathbf{R}} = 0 \times 4.7 = 0, \quad \text{since } D_3 = 0$$

The $\overline{\mathbf{x}}$ and **R**-charts are given in *Fig. 32.4* and *Fig. 32.5* respectively.

$\overline{\mathbf{x}}$*-chart*

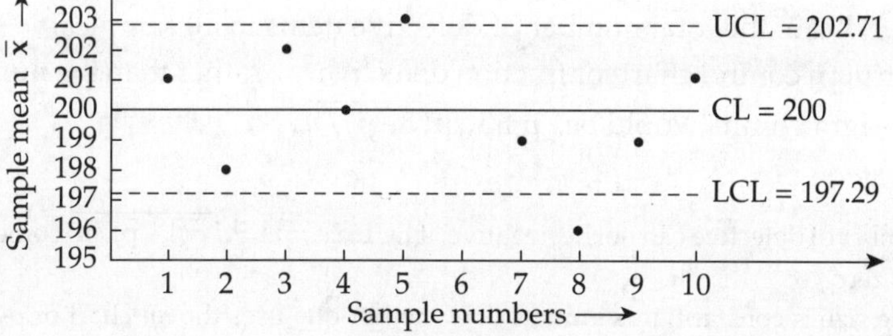

Fig. 32.4

R-*chart*

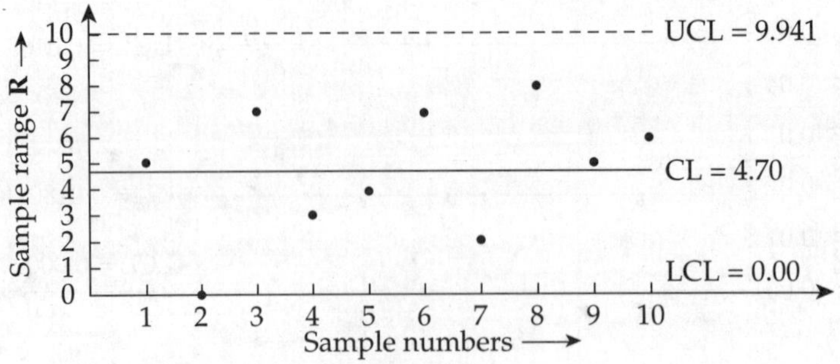

Fig. 32.5

It can be seen that all points lie within the control limits of **R**-chart. The process variability is, therefore, under control. However, three points corresponding to sample no. 5, 6 and 8 lie outside the control limits of $\bar{x}$-chart. The process is, therefore, not in statistical control and hence the process should be halted to check whether there are any assignable causes. If they are found, then the process should be readjusted to remove them, otherwise fluctuations are going to be there.

32:7. CONTROL CHARTS FOR ATTRIBUTES

Control charts for attributes are used to study process control when the quality characteristic happens to be an attribute (*e.g.*, brightness, colour, appearance, etc.). The quality of the product may be classified as either acceptable (non-defective) or not acceptable (defective) with respect to the quality attribute. Each sample item recorded as either defective or non-defective, it is classified as defective, if it does not posses the given attribute and non-defective otherwise.

Control Chart for Number of Defectives (*np* - Chart)

Suppose random samples of size n are taken from the product output at some time intervals. If d be the number of defectives in a sample, then the fraction defective in that sample is given by $p = d/n$ or $d = np$.

We may suppose that the number of defective $d = np$ is binomial with

$$E(np) = np \quad \text{and} \quad V(np) = \sqrt{np(1-p)}$$

where, p is the population fraction defective.

The process will be in statistical control, if p remains same for all the samples. This will be so, if the statistic np lies within the 3-sigma limits $np \pm 3\sqrt{np(1-p)}$.

Control Chart for Fraction Defectives (*p* - Chart)

If $\bar{p}$ denote the proportion of defectives produced by the entire process (also called the fraction defective in lot), then we write

$$\bar{p} = (p_1 + p_2 + \ldots + p_k)/k,$$

where, p_i $(i = 1, 2, \ldots, k)$ are the total number of defective items in all samples.

Here, we construct a control chart for fraction defective 'p', rather than the number of defectives 'np'. In this case, 3-sigma limits would be: $\bar{p} \pm 3\sqrt{\bar{p}(1-\bar{p})/n}$.

Remarks

1. Since the number of defective can not be negative, if LCL, *i.e.*, $\bar{p} - 3\sqrt{\bar{p}(1-\bar{p})/n}$ comes out to be negative, it is taken as zero.
2. When sample size is constant, it is immaterial whether one uses the *np*-chart or *p*-chart.
3. When the sample size varies, it is preferable to use the *p*-chart.

Control Chart for Number of Defects (*c* - chart)

It may be worthwhile to emphasize on the distinction between a defective and a defect: "***defective*** is an item that fails to satisfy some given specification(s), whereas a ***defect*** is an event of the item's lack of conformity to given specifications."

Thus, every defective item contains one or more defects. For example, these defects may be the defective rivets in an aircraft, exposed wires in a refrigerator, surface defects in a roll of paper, crumpled pages in a book, loose screws in a bicycle, etc.

In most of the cases, there are numerous opportunities for defects to occur, even though there is an almost negligible chance for defect to occur in a particular spot. Thus, the number of defects may be assumed to be a Poisson variate. Therefore, in all such cases, control limits are based on the assumption of Poisson distribution. Further, since the mean and variance of a Poisson distribution are equal, the standard deviation is given by $\sqrt{\bar{\mathbf{c}}}$, where $\bar{\mathbf{c}}$ denotes the average number of defects in a sample, *i.e.*, $\bar{\mathbf{c}} = (c_1 + c_2 + \ldots + c_n)/n$.

The control limits for **c**-chart are:

$$\text{UCL} = \bar{\mathbf{c}} + 3\sqrt{\bar{\mathbf{c}}}, \quad \text{CL} = \bar{\mathbf{c}}, \quad \text{and} \quad \text{LCL} = \bar{\mathbf{c}} - 3\sqrt{\bar{\mathbf{c}}}$$

Control Chart for Per Unit Defects (*u* - Chart)

The **c**-chart relates to samples of constant size. In the case of varying sample sizes, it is more convenient to study the control charts for the number of defects per unit, $u = \mathbf{c}/n$. The control limits for a **u**-chart are given by

$$\text{UCL} = \bar{\mathbf{u}} + 3\sqrt{\bar{\mathbf{u}}/n}, \quad \text{CL} - \bar{\mathbf{u}}, \quad \text{and} \quad \text{LCL} = \bar{\mathbf{u}} - 3\sqrt{\bar{\mathbf{u}}/n},$$

where, $\bar{\mathbf{u}}$ is the mean per unit defects.

Thus, $\bar{\mathbf{u}} = (u_1 + u_2 + \ldots + u_n)/(n_1 + n_2 + \ldots + n_n)$, u_i $(i = 1, 2, \ldots, n)$ being the per unit defects in the ith sample of size n.

Example 32.5. *20 samples each of size 10 were inspected. The number of defectives detected in each of them are given below:*

Sample no. :	1	2	3	4	5	6	7	8	9	10	11	12	13	14	15	16	17	18	19	20
No. of defectives :	0	1	0	3	9	2	0	7	0	1	1	0	0	3	1	0	0	2	0	0

Construct number defective chart and establish quality standard for the future.

[MBA (Dec.), Dayalbagh Edu. Institute 2006]

Solution. We are given

Sample size, $n = 10$, Number of samples $k = 20$,

and Total of defectives = 1 + 3 + 9 + 2 + 7 + 1 + 1 + 3 + 1 + 2 = 30.

∴ Average number of defectives per sample is given by

$$np = \frac{30}{20} = 1.5,$$

or $$p = 0.15,\ 1 - p = 0.85; \qquad \text{(since } n = 10\text{)}$$

Control limits for *np-chart* are:

$$\text{UCL} = np + 3\sqrt{np(1-p)} = 1.5 + 3\sqrt{1.5 \times 0.85}$$
$$= 1.5 + 3.39 = 4.89,$$
$$\text{CL} = 1.5,$$

and $$\text{LCL} = np - 3\sqrt{np(1-p)} = 1.5 - 3\sqrt{1.5 \times 0.85}$$
$$= 1.5 - 3.39 = -1.89 \simeq 0.$$

On plotting the given sample values against sample numbers, the following *np*-chart is obtained:

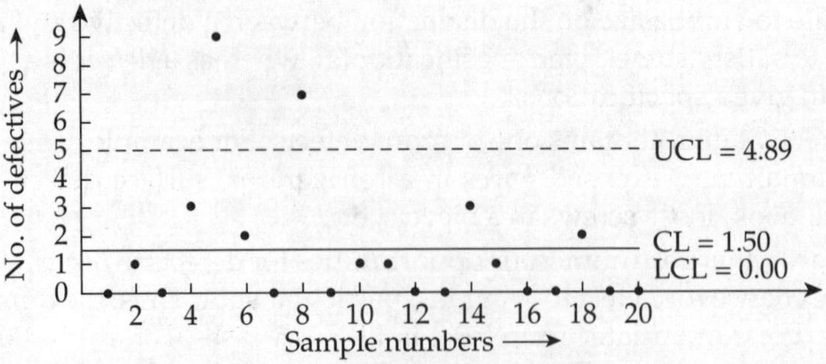

Fig. 32.6

Here, we observe that two points corresponding to sample nos. (5, 9) and (8, 7) lie outside the control limits, therefore, the process is out of control. To establish quality control ***for the future***, we homogenize the data. So, deleting sample number 5 and 8, we compute again new values of the control limits, *viz.*,

$$\therefore \quad p_1 = \frac{30-(9+7)}{200-(10+10)} = \frac{14}{180} = 0.078,$$

$$np_1 = 0.078 \times 10 = 0.78.$$

New control limits for *np*-chart are:

$$CL_1 = np_1 = 0.78,$$

$$UCL_1 = np_1 + 3\sqrt{np_1(1-p_1)} = 0.78 + 3\sqrt{0.78(1-0.078)}$$

$$= 0.78 + 3 \times 0.848 = 3.324$$

$$LCL_1 = 0.78 - 3 \times 0.848 = -1.764 \cong 0.$$

The *np*-chart is shown in *Fig. 32.7*.

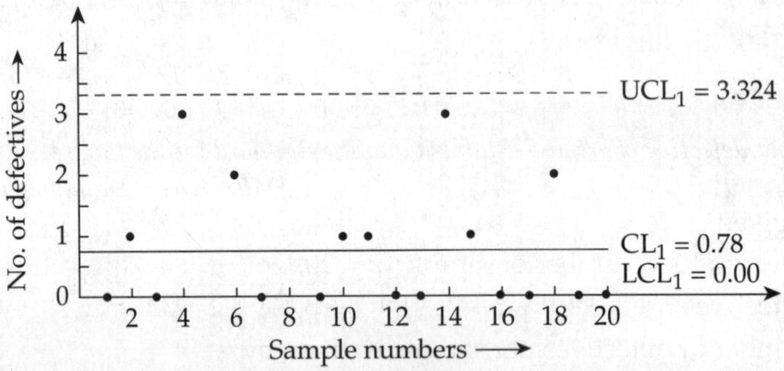

Fig. 32.7

Since, in *Fig. 32.7*, all the points (except the two which have been deleted) lie within the control limits, we take these new limits, along with the new centre line as standards for controlling production in the future.

Example 32.6. *The number of customer complaints received daily by an organization is given below:*

Day :	1	2	3	4	5	6	7	8	9	10	11	12	13	14	15
Complaints :	2	3	0	1	9	2	0	0	4	2	0	7	0	2	4

Does it mean that the number of complaints is under statistical control? Establish a control scheme for the future. ***[BBM (May), Dayalbagh Edu. Institute 2013]***

Solution. Given: $n = 15$, and c = Total number of complaints = 36.

$\therefore$ Average number of complaints per day, $\bar{c} = \frac{36}{15} = 2.4.$

The control limits of the **c**-chart are given by

$$CL = 2.4,$$
$$UCL = \bar{c} + 3\sqrt{\bar{c}} = 2.4 + 3\sqrt{2.4} = 7.05,$$
$$LCL = \bar{c} - 3\sqrt{\bar{c}} = 2.4 - 3\sqrt{2.4} = -2.25 \cong 0.$$

On inspection of the data, it is found that the process is not under control, since observation of day 5 lies outside the upper control limit.

In order to establish a control scheme for the future, it is necessary to homogenize the data by deleting the 5th observation. The new control values, therefore, are:

$$\bar{c} = \frac{36-9}{15-1} = \frac{27}{14} = 1.93.$$

$\therefore$
$$CL = 1.93,$$
$$UCL = 1.93 + 3\sqrt{1.93} = 6.10,$$
$$LCL = 1.93 - 3\sqrt{1.93} = -2.24 \cong 0.$$

We find that the process is again not under control, since observation no. 12 is still outside the new UCL. It is, therefore, necessary to again homogenize the data by deleting the 12th observation. So, by deleting the 12th observation also, new control values are:

$\therefore$
$$\bar{c} = \frac{27-7}{14-1} = \frac{20}{13} = 1.54.$$

$\therefore$
$$CL = 1.54,$$
$$UCL = 1.54 + 3\sqrt{1.54} = 5.26,$$
$$LCL = 1.54 - 3\sqrt{1.54} = -2.18 \cong 0.$$

The corresponding **c**-chart is drawn in *Fig. 32.8*. Now, all the thirteen observations lie within the control limits. Thus, the control scheme for the future is to operate this **c**-chart which is given in *Fig. 32.8*.

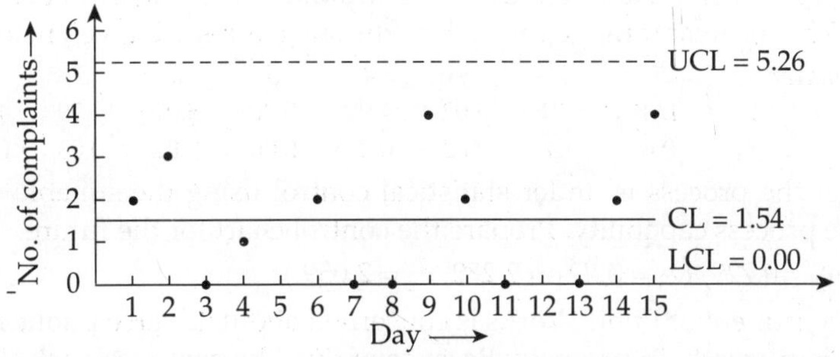

Fig. 32.8

PROBLEMS

32.01. A new shearing machine is set to cut off a piece of steel from a long bar. For various reasons the machine at times cuts off a piece that is too long or too short. These unacceptable pieces are automatically dropped in a box, and the operator of the shearing machine is to count these defective pieces after 100 (one hundred) pieces are sheared off. The record after the first day of operation is given below:

Number sheared off pieces :	100	100	100	100	100
Number of defective pieces :	5	6	7	4	8

Set out upper and lower control limits.

32.02. In a blade manufacturing factory 1,000 blades are examined daily. Following information shows the number of defective blades obtained thereto. Draw the *np*-chart and give you findings.

Day (i) :	1	2	3	4	5	6	7	8	9	10	11	12	13	14	15
No. of defective blades (p_i) :	9	10	12	8	7	15	10	12	10	8	7	13	14	15	16

[BBM (May), Dayalbagh Edu. Institute 2001]

32.03. A nationwide mail order house desires to verify the accuracy of its clerical work in completing invoices. Subgroups of 200 are taken each day for 30 consecutive days for inspection. A defective is defined as an invoice containing at last one of a number of possible errors. The number of defectives found in each of these 30 groups are as follows:

2	4	3	5	9	5	9	1	4	8
8	12	5	8	7	4	4	7	9	9
5	4	5	6	6	3	10	6	2	6

Construct a control chart of the data and comment on the state of control.

[BBM (May), Dayalbagh Edu. Institute 2008]

32.04. A machine is set to deliver packets of a given weight. Ten samples of size five each were examined and the following results were obtained:

Sample :	1	2	3	4	5	6	7	8	9	10
Mean :	15	17	15	18	17	14	18	15	17	5
Range :	7	7	4	9	8	7	12	4	11	5

Calculate the values for the central line and the control limits for the mean ($\overline{\mathbf{x}}$) chart and range (**R**) chart. Comment on the state of control.

[MBA (June), IGNOU 2000; BBM (May), Dayalbagh Edu. Institute 2014]

32.05. A task performer takes 10 rational subgroups, each of size 4 at regular interval of 30 minutes. He has to control the diameter of the component for which specification limits are 5.00 ± 0.05. The mean and range for each of the subgroups are given below:

Subgroup number :	1	2	3	4	5	6	7	8	9	10
Mean x_i :	4.98	5.01	5.02	4.99	5.00	5.01	4.99	4.99	5.02	5.00
Range R_i :	0.6	0.4	0.02	0.06	0.00	0.02	0.03	0.05	0.00	0.02

Test whether the process is under statistical control using the suitable control charts and compute the process capability. Prepare the control chart for the future.

(For $n = 4, A_2 = 0.729, D_3 = 0, D_4 = 2.282, d_2 = 2.059$.)

32.06. A town's department of Pubic Works is concerned about adverse public reaction to a sewer project that is currently in progress. Because of this, the commissioner of Public Works has authorized a weekly survey to be conducted of the residents. Each week, a sample of 100 residents is questioned on their feelings towards the project. The results to date are shown below. Analyse this data using an appropriate control chart with 5% risk of Type 1 error. Is the community sentiment stable?

Week :	1	2	3	4	5	6	7	8
Number opposed :	10	8	12	6	4	14	8	10

[MBA (May), Dayalbagh Edu. Institute 2010]

32.07. The number of mistakes made by an accounts clerk per week was recorded for 15 weeks and the data are given below:

Week number :	1	2	3	4	5	6	7	8	9	10	11	12	13	14	15
No. of mistakes :	2	3	0	1	9	2	0	0	4	2	0	7	0	2	4

Does it mean that the number of mistakes is under statistical control? Establish a control scheme for the future. **[M.Com. (Dec.), Dayalbagh Edu. Institute 2013]**

32.08. Ten samples of different sizes were taken and the number of defectives in each sample are given below:

Sample number :	1	2	3	4	5	6	7	8	9	10
Sample size :	100	300	200	150	200	250	300	150	100	250
No. of defectives :	7	12	6	5	18	10	18	6	3	20

Draw a control chart for fraction of defectives taking ± 3σ control limits.

[MBA (May), Dayalbagh Edu. Institute 2014]

32.09. A company manufacture bushings. After every one hour a sample of 125 finished bushings is drawn from the output. The data on 10 consecutive samples is given below:

Sample number :	1	2	3	4	5	6	7	8	9	10
Defectives :	15	13	16	11	13	14	20	25	30	45

Use a proper control chart to ascertain whether the system is under control. Find control limits for future. **[MBA (Dec.), Dayalbagh Edu. Institute 2007]**

32.10. There were 60 observations and 22 runs. What can the analyst conclude from this information? Construct the appropriate control chart for the sample observations listed below and determine, if the process is in control using 3σ limits.

Observation :	1	2	3	4	5	6	7	8	9	10	11
No. of defects/unit :	9	4	3	5	6	3	4	2	3	2	3

[MBA (May), Dayalbagh Edu. Institute 2011]

32.11. The number of defects per unit inspected in rational sub-groups of size five and for a sample of size 10 are as follows:

Sample	*Number of defects*					c_i	u_i
1	2	1	0	1	2	6	6/5
2	1	1	0	0	1	3	3/5
3	2	2	2	1	0	7	7/5
4	1	0	0	0	1	2	2/5
5	3	1	2	1	1	8	8/5
6	3	1	2	0	2	8	8/5
7	1	1	0	0	1	3	3/5
8	2	2	1	1	2	8	8/5
9	2	1	1	0	1	5	1
10	3	2	3	1	0	9	9/5

Set-up an appropriate control chart and give your comments.

32.12. Construct the appropriate control chart for the following data. 12 samples of five cookies each during two weeks were considered.

Sample number			*Chips per cookie*		
1	2	3	3	4	3
2	5	3	6	2	1
3	4	3	3	2	2
4	6	1	5	3	3
5	2	4	1	4	4
6	5	1	3	3	3
7	2	3	3	2	1
8	1	1	3	1	2
9	6	3	3	3	3
10	6	7	5	5	6
11	6	1	1	3	2
12	5	5	3	1	3

[*Hint:* Construct **u**-chart.] *[MBA, Osmania University 2010]*

32:8. INSPECTION

Inspection is the process of measuring the quality of a product or service in terms of established standards. It tries to ensure that items produced remain within the specified limits of variability. As a result of inspection, corrective actions may be initiated to reduce the final rejections and hence reduce the production cost.

The basic purpose of inspection is to identify the defects as soon as they occur and to ensure that they do not repeat in future operations. Given below are the types of inspection that may be carried out:

1. ***Purchaser or Vendor Place Inspection.*** Here, inspection is carried out mainly to ascertain the quality of raw materials.
2. ***In-process Inspection.*** This type of inspection is done to identify the defective items as quickly as possible so that timely remedial measures may be adopted to avoid future rejections.
3. ***Finished Producer's Inspection.*** In this case, finished products are inspected and tested to ascertain their quality standards.
4. ***Post-Sales Quality Evaluation.*** There are chances that sometimes approved items of quality standards may not provide reliable and satisfactory service to the customer. In such cases, the item must be thoroughly inspected, faults must be found and the necessary improvements must be incorporated.

32:9. ACCEPTANCE SAMPLING

Product control is concerned with classification of raw materials, semi-finished goods or finished goods into acceptable or rejectable items. It is concerned with the inspection of goods already produced to judge whether these are accepted or rejected.

One simple way to control the quality of product is to conduct 100% inspection, but it may be uneconomical or even impossible because of the following reasons:

1. Cost of inspection may be very high.
2. Inspection may be hazardous or even dangerous.
3. Handling of the product may induce defects or result in its deterioration.

Because of these reasons, some kind of acceptance sampling is often required. In acceptance sampling, the decision about the quality of batches or lots are made on the inspection of only a portion of the total number of items. If the sample of items conforms to the desired quality level, then the whole batch is accepted otherwise the whole batch is rejected or subjected to further inspection.

While developing an acceptance sampling plan, the following two terms are also used:

(a) ***The Acceptable Quality Level (AQL).*** This is the quality level which the consumer regards as acceptable. It represents a relatively small fraction defective, p, such that

$$P \text{ (rejecting a lot of quality } p) = \text{very small, say } 0.05, \text{ and}$$
$$P \text{ (accepting a lot of quality } p) = \text{very high, say } 0.95.$$

This is, a lot having quality *AQL* shall be accepted on the basis of sampling inspection.

(b) ***The Lot Tolerance Proportion Defectives (LTPD) or the Rejected Quality Level (RQL).*** This represents the maximum proportion defective (quality control) in the lot that the consumer may tolerate.

Thus, *AQL* refers to minimum percentage of defective items in a lot below which the lot should be accepted, whereas *LTPD* refers to a maximum percentage of defective items in a lot beyond which the lot should be rejected.

Besides the above two terms used in acceptance sampling plan, following two terms are also used.

(c) ***Average Outgoing Quality (AOQ).*** This is to protect the consumer from the inferior product by ensuring him a certain quality level after inspection (regardless of what quality level is being maintained by the producer). The fraction defective remaining in the lot after inspection is known as ***outgoing quality*** of the lot. The average fraction defective remaining in the lot after inspection is termed as ***average outgoing quality*** (*AOQ*).

Obviously, *AOQ* is a function of the ***incoming quality***, p, the fraction defective maintained by the producer. The ***average outgoing quality limit*** (*AOQL*) is defined by

$$AOQL = \max_{0 \le p \le 1} AOQ$$

The consumer's interest is taken care of by specifying the *AOQL*.

(d) ***Average Amount of Total Inspection (ATI).*** It is the average number of items inspected per lot to arrive at a decision in the sampling plan, calling for 100% inspection of the rejected lots. It is also a function of the incoming lot quality p. It is easy to observe that

$$ATI = ASN + \text{Average remaining size of the rejected lots.}$$

32:10. SINGLE SAMPLING PLAN

Single sampling plan consists of the following four steps:

1. From a batch of N, we draw n samples and inspect them ($n \le N$).
2. In case the number of defectives in the sample of size, n, are less than or equal to, c, accept the batch of N.
3. In case the number of defectives in the sample of size, n, are greater than, c, inspect all the remaining $(N - n)$ products.
4. Correct or replace all the defective pieces found.

For example, let $N = 800$, $n = 40$ and $c = 4$. Then, 40 number of samples would be chosen at random out of 800 and inspected. If number of defectives are 4 or less, the whole batch of 800 will be accepted. On the other hand, if the number of defectives are more than 4, then all the remaining (800 – 40), *i.e.*, 760 would be inspected. In this case, the defective are either rectified or replaced by good ones and only then the batch of 800 would be accepted.

32:11. OPERATING CHARACTERISTICS CURVE (*OC* - CURVE)

The ***Operating Characteristic Curve (OC-Curve)*** of an acceptance sampling plan shows the ability of the plan to distinguish between good and bad lots. An *OC*-curve is a graphic representation of the relationship between the probability of accepting a lot, P_a, and the incoming lot quality p (the fraction defective in the lot). Naturally, steeper the *OC*-curve, greater is the protection to the consumer. A good sampling plan is one, where *OC*-curve takes into account the protection desired by both the producer and the consumer. An ideal *OC*-curve would be one for which all good lots are accepted and all bad ones are rejected. But, unfortunately, such a plan can never be attained.

The steepness of the *OC*-curve depends upon the sample size. The larger the sample, the steeper the curve, and the smaller the zone between the qualities that are almost always accepted and the qualities that are almost always rejected.

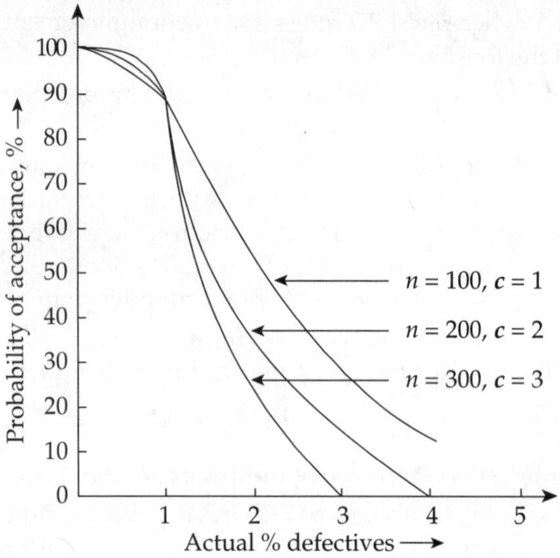

Fig. 32.9. Typical OC-Curves for Different Sample Sizes (n)

Construction of *OC*-curve. To construct an *OC*-curve, an agreement has to be reached between the producer and the consumer on the following four points:

1. ***Acceptable Quality Level (AQL).*** This is the maximum proportion of defectives that will make the lot acceptable.
2. ***Lot Tolerance Percentage Defective (LTPD).*** This is the maximum proportion of defectives that will make the lot definitely unacceptable.
3. ***Producer's is Risk (α).*** This is the risk, the producer is willing to take that lots of the quality level will be rejected, even though, they are acceptable. Usually, $\alpha = 5\%$.
4. ***Consumer's Risk (β).*** This is the risk, the consumer is willing to take that lots of quality level *LTPD* will be accepted, even though, they are actually unacceptable. Usually, $\beta = 10\%$.

With the above four points known, *OC*-curve for a sampling plan can be constructed.

32:12. DOUBLE SAMPLING PLAN

If the first sample is neither good enough, nor bad enough to accept or reject a lot (batch size), the accept-reject decision is taken on the basis of a second sample from the lot or batch size. Such a plan, based on two random samples, is called a ***double sampling plan***. The plan is completely specified by five notations, *viz.*, N, n_1, c_1, n_2 and c_2, where

N = number of products in a given batch (lot),
n_1 = number of products in the first random sample,
c_1 = acceptance number for the first random sample,
n_2 = number of products in the second random sample, and
c_2 = acceptance number for both the samples combined.

The plan may be described schematically as follows:

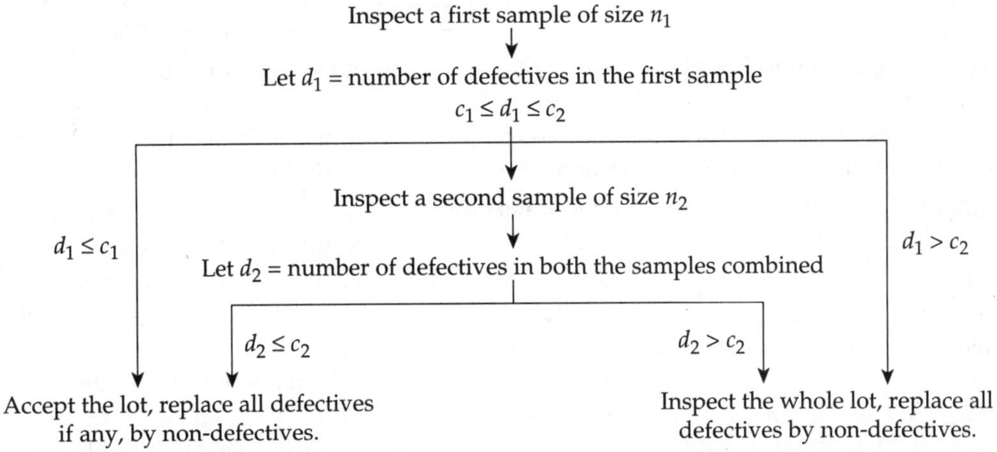

As in single sampling plan, here also there are two approaches for determining the values of n_1, n_2, c_1 and c_2, *viz.*, approach of lot quality protection and that of average quality protection.

Dodge and Roming obtained the most economical double sampling plans after providing adequate protection to producer and consumer such that:

1. Average total inspection is minimum, and
2. The probability of acceptance on the basis of first sample is same as the probability of acceptance on the basis of second sample.

Comparison between Single and Double Sampling Plans

The following points highlight the various merits and demerits of the two sampling plans:

1. Single sampling plans are simple, easy to design, and carry out. Further, since a sample can be plotted on a control chart, they provide maximum information concerning the lot.
2. The most important advantage of a double sampling plan is that it involves less amount of inspection than the single sampling plan, for the same quality assurance. Usually double sampling plan requires 25% to 30% less inspection, on the average than single sampling.

   However, this does not mean that the double sampling plan would be less costlier than the single sampling plan. This is so, because the double sampling plans are more complicated and the need of a second sample being unpredictable, the unit inspection cost may be higher.
3. There is a psychological advantage in favour of double sampling plans. For example, to a layman, it seems unfair to reject a lot just on the basis of one sample and appears more convincing to reject it on the basis of more than one sample.

REVIEW EXERCISE

1. Explain the term 'Statistical Quality Control'. How is the process control achieved with the help of control charts? What are the fundamentals underlying the construction of a quality control chart?
[B.Tech. (Mech.), Panjab Tech. Univ. 2013; M.Com. (Dec.), Dayalbagh Edu. Institute 2007]
2. Distinguish between process control and product control.
3. State the assumptions behind the control chart technique.
4. Why both $\bar{x}$-chart and **R**-chart are drawn for one job? Also, list the applications of **p**-chart and **c**-chart. *[B.Tech. (Mech.), Panjab Tech. Univ. 2013]*
5. What do you understand by sampling inspection plans? How are they used in controlling the quality of a manufactured product? State different types of acceptance sampling plans.
[BBM, Dayalbagh Edu. Institute 2005]
6. Discuss various sampling plans. Also, write note on acceptance sampling.
[MBA (June), IGNOU 2000]
7. What are the basic issues in inspection? What amount of inspection is optimal? Distinguish between random variations and assignable variation. *[BBM, Dayalbagh Edu. Institute 2007]*
8. Explain the following terms used in acceptance sampling:
(*i*) *AOQL*, (*ii*) *LTPD*, (*iii*) Consumer's risk, (*iv*) Single sampling, and (*v*) Double sampling.
9. Draw an *OC*-curve. Show its different regions and also discuss its characteristics.
[MBA (June), IGNOU 2001]
10. Mention the advantages of double sampling plan over single sampling plan.
[BBM, Dayalbagh Edu. Institute 2005]
11. Discuss the importance of statistical quality control in business with suitable examples.
[MBA, Osmania Univ. 2010]
12. Write a short note on Statistical Quality Control.

CHAPTER 33

Some Case Problems

"Practice unfolds many secrets of the problem"

A few illustrative case problems in Quantitative Methods/Operations Research/Management Science are presented in this chapter. These will help the reader to become familiar with the practical situations occurring in our day-to-day life.

33:1. CASE PROBLEMS IN LINEAR PROGRAMMING

1. Advertising Strategy

Anupam Motors company has hired marketing services firm to develop an advertising strategy for promoting Anupam's used-car sales. The marketing firm has recommended that Anupam company must use spot announcements on both television and radio as the advertising media for the proposed promotional campaign. Advertising strategy guidelines are expressed as follows:

(*i*) Use at least 36 announcements for combined television and radio coverage.

(*ii*) Do not use more than 28 radio announcements.

(*iii*) The number of radio announcements can not be less than the number of television announcements.

The television relay centre has quoted a cost of ₹ 2,500 per spot announcement, and the radio station has quoted a cost of ₹ 640 per spot announcement. Advertisement budget of the company has been set at ₹ 65,000. The marketing services firm has rated the various advertising media in terms of audience coverage and recall power of the advertisement. For company's media alternatives, the television announcement is rated at 900, and the radio announcement is rated at 500. The managing director of the company would like to know how many television and how many radio spot announcements should be used to maximize the overall rating of the advertising campaign.

The managing director of the company believes the television station will consider running the spot announcement of the motor company on its highly rated evening news programme (at the same cost), if the company will consider using additional television announcements.

Answer the following questions:

1. What should be the optimum number of television and radio spot announcements?
2. What are the relative merits of each advertising medium?
3. What should be the number of television spots that should be purchased, if the news programme is rated highly enough to make increasing the number of television spot?
4. What should be the best use of any possible increase in the advertising budget?

2. Production/Sales Budget

A small firm produces and sells three types of products *P*, *Q* and *R*. While capacity to sell the products is unlimited, the production is constrained by the limited supplies of key raw material and the time available on a special machine. The firm has not planned any expansion in machine capacity for the near future. It is also not possible to work overtime to enhance machine-time availability.

The firm is in the process of formulating its sales/production budget for the coming year. Data on operations/costs of current year are given below:

	Products				*Products*		
	P	*Q*	*R*		*P*	*Q*	*R*
Sales in units	25,000	10,000	6,000	Variable SDA	0.70	0.35	1.05
Selling price (₹ per unit)	14.00	7.00	21.00	Fixed FOH	2.00	1.00	3.20
Cost (₹ per unit):				Fixed SDA	1.40	0.70	2.10
Direct materials	5.00	2.60	11.60	Total costs	12.00	6.10	22.65
Direct labour	2.40	1.20	3.90	Operating profit	2.00	0.90	(1.65)
Variable FOH	0.50	0.25	0.80				

Note. There were no inventory balance; FOH: Factory Overheads; SDA: Selling-Distribution-Administration.

The supplies of the critical raw material and the machine time are expected to be at the same levels as those used in current year. The usage of these resource in the products are:

	P	*Q*	*R*
Material (kg per unit)	1/2	1/4	3/2
Machine time (hour per unit)	1/2	3/8	1/3

In preliminary discussion, the accountant suggests that since the product *R* results in a net loss, this product should be dropped from consideration for next year's plan and further, the resources available should be to produce the products *P* and *Q* in the ratio of 2 : 1 (units) similar to their operating profits.

Marketing manager strongly resists this approach contending that some major customers are joint consumers of all the three products. If *R* is not supplied, then the sales for *P* and *Q* may be adversely affected. On discussions, marketing indicates that the following minimum quantities of the three products should be made available for next year. Excess resources, if any, may then be allocated only between *P* and *Q* as per the accountant's suggestion.

Minimum requirements: P = 10,000 units; Q = 4,000 units; R = 2,400 units

Answer the following questions:

1. Determine the production/sales plan on each of the basis of the agreed approach between marketing and the accountant and using the method of linear programming problem. (To simplify the linear programming approach, it may be assumed that the minimum sales requirements will be met as a policy and only the excess resources will be allocated using linear programming method.)
2. Prepare a condensed statement of revenue-cost-profit for the current year, and budget for next year under each of the two plans.

   Information contained in the following simplex table may be used as needed:

c_j	*Basic variable*	5.40 *P*	2.60 *Q*	3.65 *R*	0 S_1	0 S_2	*RHS (Solution)*
3.65	*R*	0	3/28	1	6/7	−6/7	2,957
5.40	*P*	1	23/28	0	−4/7	18/7	19,930
	$(c_j - z_j)$	0	−1.445	0	−0.043	−10.757	

3. Investment Strategy

An investment advisory firm that manages more than ₹ 240 million in funds for its numerous clients. The firm uses an asset allocation model that recommends the portion of each client's portfolio to be invested in a growth stock fund, an income fund, and a money market fund.

To maintain diversity in each client's portfolio, the firm places limits on the percentage of each portfolio that may be invested in each of the three funds. General guidelines indicate that the amount invested in the growth fund must be between 25% to 45% of the total portfolio value. Similar percentage for the other two funds stipulate that between 25% and 50% of the total portfolio value must be in the income fund and at least 30% of the total portfolio value must be in the money market fund.

In addition, the firm attempts to assess the risk tolerance of each client and adjust the portfolio to meet the needs of the individual investor. For example, the investment company has just contracted with a new client who has ₹ 7,50,000 to invest. Based on an evaluation of the client's risk tolerance, the firm has assigned a maximum risk index of 0.05 for the client. The firm's risk indicators show the risk of the growth fund at 0.12, the income fund at 0.08, and the money market fund at 0.02. An overall portfolio risk index is computed as a weighted average of the risk rating for the three funds, where the weights are the fraction of the client's portfolio invested in each of the funds.

Additionally, the investment firm is currently forecasting annual yields of 18% for the growth fund, 12% for the income fund, and 8% for the money market fund. Based on the information provided, how should the new client be advised to allocate the ₹ 7,50,000 among the growth, income, and money market funds? Develop a linear programming model that will provide the maximum yield for the portfolio.

Answer the following questions:

1. How much of ₹ 7,50,000 should be invested in each of the three funds namely the growth fund, income fund, and money market fund?
2. What is the annual yield by adopting the investment recommendation?
3. If client's risk index could be increased to 0.055, how much would the yield recommendation change?
4. If annual yield for the growth fund were revised downward to 16% or even to 14%, would your investment recommendation change?

4. Technical Accounting*

A company manufactures two products a 10 cu.ft. and 6 cu.ft. refrigerators. The demand for the former has been estimated at 15,000 and for the later at 24,000.

The production process is broken down into three stages:

(*i*) shell production,
(*ii*) motor assembly, and
(*iii*) refrigerator assembly.

The production manager estimates that the shell production unit can through put up to 36,000 smaller refrigerators, but only half quantity of the 10 cu.ft. product. Similarly, the motor assembly unit can through put upto 30,000 smaller motors or 80% of this quantity of the other product. The assembly department has no restrictions as unskilled labour only is required. The accounting function has produced the standard cost breakdown, as follows:

* "Linear Programming and Traditional Accounting", by W.A. Layne, *The Singapore Accountant*, Vol. 14 (1979).

Standard Costs of Production (in rupees)

	10 cu.ft.	*6 cu.ft.*
Direct materials	30	24
Direct labour	16	10
Variable overheads	14	11
Marginal cost	60	45
Selling price	90	65
Net contribution	30	20

Fixed overhead charges are estimated at ₹ 4,00,000.

Traditional accounting would entail the manufacture of 15,000 of 10 cu.ft. refrigerators before the shell production unit capacity is exceeded. Only 6,000 smaller refrigerators would be manufactured. On the basis of direct labour, ₹ 3,20,000 of fixed overheads would be attributed to the 10 cu.ft. and ₹ 80,000 to the 6 cu.ft.

The net profit would be as follows:

	10 cu.ft.	*6 cu.ft.*	*Total (in '000 rupees)*
Contribution	450	120	570
Overheads allocation	320	80	400
Net profit *('000 ₹)*	130	40	170

But, this is an over simplification when a resource is limited. The scarce resource is shell production and contribution per unit is ₹ 20 for the 6 cu.ft. and ₹ 15 for the 10 cu.ft. This suggests that the 6 cu.ft. product is more profitable and should be produced to its demand of 24,000 refrigerators. For this, the net profit would stand at ₹ 2,24,000.

But, both are misleading because profitability is not directly related to contribution per unit product or per unit resource. An examination from the Linear Programming viewpoint illustrate this. The problem can be formulated mathematically as follows:

Maximize $z = 30x_1 + 20x_2$ subject to the constraints: *(Directive function)*

$2x_1 + x_2 \leq 36{,}000$ *(Shell production)*

$1.25x_1 + x_2 \leq 30{,}000$ *(Motor assembly)*

$x_1 \leq 15{,}000$ *(Demand)*

$x_2 \leq 20{,}000$ *(Demand)*

$x_1 \geq 0$ and $x_2 \geq 0$;

where, x_1 = Production of the 10 cu.ft. refrigerators and x_2 = Production of the 6 cu.ft. refrigerators.

Solving the linear programming problem by simplex method, the optimum simplex table is:

c_j	*Basic variable*	30 x_1	20 x_2	0 s_1	0 s_2	0 s_3	0 s_4	*Quantity*
20	x_2	0	1	1/3	8/3	0	0	20,000
0	s_3	0	0	–4/3	4/3	1	0	7,000
30	x_1	1	0	4/3	–4/3	0	0	8,000
0	s_4	0	0	–1/3	2/3	0	1	4,000
	$(z_j - c_j)$	0	0	140/3	40/3	0	0	6,40,000

It follows that the production plan is to produce 8,000 of the larger and 20,000 of the smaller refrigerators. The contribution is now ₹ 6,40,000, and the net profit would then stand at ₹ 2,40,000.

Conclusion. Decision costs are thus not a measure of the intrinsic worth of a product or resource. They are a measure of the implications involved in moving a company from one position to another.

Linear programming is designed to solve complex situations involving larger numbers of interacting variables where an action or policy undertaken at one point in the process or in relation to one product, has repercussions throughout the business complex. Traditional accounting techniques cannot produce this information as the measure of profitability and worth derived do not reflect the conflict in the use of resources. Planning should be approached by consideration of all constraints defining the firm's environment, where contribution is but one of the decision measurements.

33:2. CASE PROBLEMS IN TRANSPORTATION

1. South India Soaps Ltd.

The South India Soaps Ltd. (SISOL) operated three factories from which it shipped soaps to regional warehouses. In 2000, the demand for soaps was of 24,000 tonnes distributed as follows:

Region	*Demand in '000 tonnes*
Kochi	5
Nellore	4
Salem	4
Madurai	11

One-shift production capacity in each of the three factories was as follows:

Chennai	12
Coimbatore	7
Bengaluru	7

Estimated transport costs (in hundred rupees per thousand tonnes) are given below:

Factory	*Regional warehouses*			
	Kochi	*Nellore*	*Salem*	*Madurai*
Chennai	95	105	80	15
Coimbatore	115	180	40	30
Bengaluru	155	180	95	70

SISOL followed a policy of decentralization under which each of the four regional warehouses was under direct supervision of a regional sales manager and he was responsible for the profitability of operation under his control.

Over a period of time, this procedure led to increasing friction in the organization. There were questions whether this procedure achieved the objective of minimizing transport costs; also there was no coordination. For instance, in 1999, the sales manager of Madurai and Nellore placed their orders with the Chennai Factory which did not have the capacity to meet all demands. This led to inefficient and duplicate orders, friction, etc. The final pattern that emerged in 1999 was as follows:

	Kochi	*Nellore*	*Salem*	*Madurai*
Chennai	0	1	2	9
Coimbatore	3	0	0	2
Bengaluru	2	3	2	0

The general manager of SISOL called meeting of the executives at the central office. Some executives proposed that all orders should be routed through the central office which would determine the optimal programme. Others protested that this would seriously conflict with the firm's philosophy of decentralization.

You have been hired as a consultant by the general manager. Prepare a minimum cost distribution schedule for SISOL. Compare this schedule with the present schedule. Which is better?

2. Asian Games*

For maintaining law and order during the Asian Games, 1982, police force had been requisitioned from the various central police organizations like CRPF, BSF, ITBP and other states such as Haryana, Himachal Pradesh, Uttar Pradesh, Odisha and Gujarat. Besides this, there are 10 battalions of Delhi Police available for deployment. The outside and local force have been stationed at various places in the Union Territory of Delhi. The figures are in the number of sections (10 men) available for deployment.

Place	*Abbreviation*	*Force (Section)*
New Police Lines	NPL	400
Jaroda Kalan	JK	160
Pitam Pura	PP	160
Model Town	MT	90
Police Training School	PTS	90
Old Police Lines	OPL	120
New Kotwali	NK	20
Kamla Market	KMT	120
Parliament Street	PST	240
Rajpura Lines	RPL	120
Shakarpur	SP	120
Moti Nagar	MN	120
Mehram Nagar	MRN	240
	Total	2,000

The force is to be deployed at various stadia on the day of events. The requirement of the force has been calculated keeping in view the factors like capacity of the stadium, traffic problems, security of VIPs and participants, etc. The requirement of force is also mentioned in number of sections.

Stadium	*Abbreviation*	*Force (Section)*
Chhatrasal	CS	60
Delhi University	DU	40
Indraprastha	IP	400
Yamuna Velodrome	YV	40
Talkatora Indoor	TKI	80
Talkatora Swimming	TKS	80
National	NS	240
Hall of Sports	HS	60
Harbaksh	BH	60
Shivaji	SV	60
Jawaharlal Nehru	JLN	500
Hauz Khas	HK	20
Games Village	GV	60
Ambedkar	AK	200
Tughlakabad Range	TR	40
Golf Club	GC	20
Karnail Singh	KS	40
	Total	2,000

* Based on the data prepared by Mr. M.S. Updahya, IPS.

The cost of transportation of one section, *i.e.*, 10 men from every location to each stadium has been calculated as shown in the matrix below.

It is assumed that the availability and requirement of force is equal, *i.e.*, 2,000 sections. It will not be possible on any day during the games as all stadia are not having games on all the days. Hence, there will always be a surplus force available. Consequently dummy stadium will have to be planned everyday so as to absorb the unutilized force. The cost (in ₹ 100s for 10 persons — one section) matrix for the movement of the police force is given below:

	CS	*DU*	*IP*	*YV*	*TKI*	*TKS*	*NS*	*HS*	*HB*	*JLN*	*SV*	*HK*	*GV*	*AK*	*TR*	*GC*	*KS*	*Force availability*
NPL	2	1	13	12	15	15	16	14	25	20	14	22	24	11	35	16	10	400
JK	28	31	36	37	130	30	35	36	16	42	30	35	36	35	50	35	36	160
PP	15	14	26	25	28	28	28	17	38	33	17	35	37	24	48	29	23	160
MT	4	3	15	14	17	17	18	16	27	22	16	24	26	13	37	18	12	90
PTS	26	24	15	16	13	13	12	15	15	12	15	33	3	16	10	17	16	90
OPL	7	2	8	7	10	10	11	13	20	15	9	17	13	6	30	11	6	120
NK	11	6	4	3	6	6	7	9	16	11	5	13	15	2	26	7	4	20
KMT	13	8	3	4	4	4	5	5	21	16	2	25	25	1	24	10	2	120
PST	15	12	5	6	2	2	4	5	16	10	12	12	12	4	22	4	3	240
RPL	4	1	11	10	13	13	14	14	23	18	12	20	22	9	33	14	8	120
SP	19	16	3	4	7	7	6	5	21	12	8	18	18	5	30	8	8	120
MN	10	12	16	17	13	13	15	16	20	20	15	25	25	14	35	18	12	120
MRN	30	28	18	19	15	15	15	16	6	22	15	12	12	20	20	20	20	240
Force requirement	60	40	400	40	80	80	180	240	60	60	500	60	20	200	40	20	40	2,000

Suggest an optimal transportation schedule of the police force so as to minimize the total transportation cost.

33:3. CASE PROBLEM IN INTEGER PROGRAMMING

Text Book Publishing

Sultan Chand & Sons, a leading publisher of college text books, must make a decision regarding which books to publish next year. The books under consideration are listed in the following table, along with the projected 3-year sales expected from each book.

Book	*Type of the book*	*Projected sales (Number of books)*
Business English	Revision	20,000
Business Mathematics	New	30,000
Business Statistics	New	15,000
Information Technology	Revision	10,000
Macro Economics	New	25,000
Managerial Economics	Revision	18,000
Managerial Accounting	New	25,000
Cost Accounting	Revision	50,000
Financial Accounting	Revision	20,000
Quantitative Methods	New	30,000

The books that are listed as revisions are texts that the publisher already has under contract, these texts are being considered for publication as new editions. The books that are listed as new have been reviewed by the publisher, but contracts have not yet been signed.

The publisher has three individuals who can be assigned to these projects, all of whom have varying amounts of time available. Mr. Mehta has 60 days available, whereas Mr. Rao and Mr. Arora both have 40 days available. The days required by each person to complete each of the project, are shown in the following table. For instance, if Business Mathematics book is published, it will require 24 days of Mehta's time. 30 days of Rao's time, and 40 days of Arora's time. Symbol '×' indicates that the person will not be used on the project. Note that at least two staff members will be assigned to each project except the Managerial Accounting book.

The publisher will not publish more than two Economics books or more than one Accounting book in a single year. In addition, management has decided that one of the Economics books (Macro Economics or Managerial Economics) must be published, but not both.

Book	*Metha*	*Rao*	*Arora*
Business English	16 days	24 days	×
Business Mathematics	24 days	30 days	40 days
Business Statistics	30 days	40 days	×
Information Technology	20 days	×	24 days
Macro Economics	16 days	×	10 days
Managerial Economics	×	×	16 days
Managerial Accounting	18 days	26 days	×
Cost Accounting	×	34 days	18 days
Financial Accounting	40 days	36 days	30 days
Quantitative Methods	×	50 days	46 days

Prepare a report for Managing Director of Sultan Chand & Sons, that describes your findings and recommendations regarding the best publication strategy for next year. While carrying out analysis, assume that the fixed costs and the sales revenues per unit are approximately equal for all books. In fact management is interested primarily in maximizing the total sales volume.

In case of following possible changes, answer your suggestions:

1. If Mr. Rao can be moved to another project to allow to work 14 more days, would it be advantageous.
2. If one or more of the revisions could be postponed for another year, should they be? Clearly the publisher will risk losing market share by postponing revision.

33:4. CASE PROBLEMS IN GOAL PROGRAMMING

1. Production Allocation

A quality fabric producing company produces two types of line Products *A* and *B*. The Product *A* is produced according to issue orders from furniture manufacturers. The Product *B*, on the other hand, is distributed to retail fabric stores. The average production rates for Product *A* and for Product *B* are identical, *i.e.*, 1,000 metres per hour. By running two shifts, the operational capacity of the plant is 80 hours per week.

The marketing department reports that the maximum estimated sales for the following week is 70,000 metres of Product *A* and 45,000 metres of Product *B*. According to the accounting department, the approximate profit from a metre of Product *A* is ₹ 25, and from a metre of Product *B* is ₹ 15.

The President of the company believes that a good employer-employee relationship is an important factor for business success. Hence, he decides that a stable employment level is the primary goal for the firm. Therefore, whenever there is a demand exceeding normal production capacity, he simply expands production capacity by providing overtime. However, he also feels that overtime operation of the plant for more than 10 hours per week should be avoided because of the accelerating costs. The President of the company has the following four goals:

1. The first goal is to avoid any under-utilization of production capacity (*i.e.*, to maintain stable employment at normal capacity).
2. The second goal is to limit the overtime operation of the plant to 10 hours.
3. The third goal is to achieve the sales goals of 70,000 metres of Product *A* and 45,000 metres of Product *B*.
4. The last goal of the President is to minimize the overtime production of the plant as much as possible.

Since, the above problem involves multiple incompatible goals, the President of the company is very much keen to know the optimum production schedule. He also desires to know the effect on production schedule, if there would have been simply one goal, *i.e.*, profit maximization.

2. Kwality Office Supplies

Management of Kwality Office Supplies establishes monthly goals, or quotas, for the types of customers contacted. For the next 6 weeks, Kwality's customer contact strategy calls for the force, which consists of 5 salespeople, to make 400 contacts with customers who have previously purchased supplies from the firm. In addition, the strategy calls for 250 contacts of new customers. The purpose of this later goal is to ensure that the sales force is continuing to investigate new sources of sales.

Making allowances for travel waiting time, as well as demonstration and direct sales time, the firm has allocated 4 hours of sales force efforts to each contact of a previous customer. Customer contacts tend to take longer and require 2 hours per contact. Normally, each salesperson works 48 hours per week, or 288 hours over six week planning horizon; under a normal work schedule, the 5 salespeople will have 5 × 288 = 1,440 hours of sales force time available for customer contacts.

Management is willing to use some overtime, if needed, or to accept a solution that uses less than the scheduled 1,440 hours available. However, management desires a solution that will limit both overtime and under-utilization time of the work force to at most 48 hours over the 6 week period.

In addition to the customer contact goals the firm has established a goal regarding sales volume. Based on past experience, Kwality estimates that each previous customer contacted will generate ₹ 5,000 of sales and each new customer contacted will generate ₹ 3,000 of sales. Management would like to generate sales of at least ₹ 2,60,000 for next month.

After further consideration of these goals, management has concluded that the most important goals are the labour utilization goals. That is, given firm's small sales force and the short time frame involved, management has decided that the overtime goal and the labour utilization goal are both priority level-1 goals. Management also concluded that the ₹ 2,60,000 sales goal should be a priority level-2 goal and that the three customer contact goals should be priority level-3 goals.

Help the Kwality Office Supplies to decide how best to utilize the resources to meet the goals it has set.

33:5. CASE PROBLEMS IN NETWORK SCHEDULING THROUGH *PERT*/CPM

1. Warehouse Expansion

WH Abdi Inc., distributes a variety of food products that are sold through grocery stores and supermarket outlets. The company receive orders directly from the individual outlets, with a typical order requesting the delivery of anywhere from 30 to 60 different products. Under the company's current warehouse operation, warehouse clerks dispatch order-picking personnel to fill each order and have the goods moved to the warehouse shipping area. Because of the high labour costs and relatively low productivity of hand order-picking, management has decided to automate the warehouse operation by installing a computer-controlled order-picking system, along with conveyor system for moving goods from storage to the warehouse shipping area.

WH Abdi's director of material management has been named the project manager in charge of the automated warehouse system. After consulting with members of the engineering staff and warehouse management personnel, the director has compiled a list of activities associated with the project. The optimistic, most probable, and pessimistic times (in weeks) have also been provided for the each activity.

Activity	*Description*	*Immediate predecessor*
A	Determine equipment needs	—
B	Obtain vendor proposals	—
C	Select vendor	A, B
D	Order system	C
E	Design new warehouse layout	C
F	Design warehouse	E
G	Design computer interface	C
H	Interface computer	D, F, G
I	Install system	D, F
J	Train system operators	H
K	Test-system	I, J

Activity times (in weeks) are in the table given below:

Activity	*Optimistic*	*Most-probable*	*Pessimistic*
A	5	7	9
B	7	9	17
C	3	5	7
D	9	11	25
E	8	11	14
F	5	7	9
G	5	7	21
H	5	7	9
I	5	7	15
J	4	5	6
K	3	5	7

1. The management of the company have a target of 40-week completion time for the project. Can this completion time be achieved? Include probability information in your discussion. What recommendation do you have, if the 40-week completion time is required?
2. Suppose that management of the company requests that activity times be shortened to provide an 80% chance of meeting the 40-week completion time. If the variance in the project completion time is the same as above, how much should the expected project completion time be shortened to achieve the goal of an 80% chance of completion within 40-weeks?
3. Using the expected activity times as the normal times and the following crashing information, determine the activity crashing decisions and revised activity schedule for the warehouse expansion project.

Activity	*Crashed activity time (in week)*	*Normal cost (in ₹)*	*Crash cost (in ₹)*
A	4	1,000	1,900
B	7	1,000	1,800
C	2	1,500	2,700
D	8	2,000	3,200
E	7	5,000	8,000
F	4	3,000	4,100
G	5	8,000	10,250
H	4	5,000	6,450
I	4	10,000	12,400
J	3	4,000	4,400
K	3	5,000	5,500

2. Construction of New Hospital*

The equivalent of a new kindergarten class is born every day at Orlando's Arnold Palmer Hospital. With more than 13,000 births in the mid-2000s in a hospital that was designed 15 years earlier for a capacity of 6,500 births a year, the newborn intensive care unit was stretched to the limit. Moreover, with continuing strong population growth in central Florida, the hospital was often full. It was clear that new facilities were needed. After much analysis, forecasting and discussion, the management team decided to build a new 273-bed building across the street from the existing hospital. But, the facility had to be built in accordance with the hospital's Guiding Principles and its uniqueness as a health centre dedicated to the specialized needs of women and infants. The Guiding Principles are: Family-centered focus, a healing environment, where privacy and dignity are respected, sincere and dedicated staff providing the highest quality care and patient-centered flaw and function.

The Vice President of business development, Karl Hodges, wanted a hospital that was designed from the inside out by the people who understood the Guiding Principles who knew most about the current system and who were going to use the new system, namely, the doctors and nurses. Hodges and his staff spent 13 months discussing expansion needs with this group, as well as with patients and the community, before developing a proposal for the new facility. An administrative team created 35 user groups, which held over 1,000 planning meetings (lasting from 45 minutes to a whole day). They even created a "Supreme Courts" to deal with conflicting views on the multifaceted issues facing the new hospital.

Funding are regulatory issues added substantial complexity to this major expansion and Hodges was very concerned that the project stay on time and within budget. Tom Hyatt, director of facility development, was given the task of onsite manager of the $100 million project, in addition to overseeing ongoing renovations, expansions, and other projects. The activities in the multi-year project for the new building at Arnold Palmer are given below:

Expansion Planning and Arnold Palmer Hospital construction Activities and Times

Activity	*Scheduled time*	*Precedence activity*
1. Proposal and review	1 month	—
2. Establish master schedule	2 weeks	1
3. Architect selection process	5 weeks	1

(contd.)

* MBA, Panjab Technical University 2018.

Table (*contd.*)

	Activity	*Scheduled time*	*Precedence activity*
4.	Survey whole campus and its needs	1 months	1
5.	Conceptual architect's plans	6 weeks	3
6.	Cost estimating	2 months	2, 4, 5
7.	Deliver plans to board for consideration/decision	1 month	6
8.	Surveys/regulatory review	6 weeks	6
9.	Construction manager selection	9 weeks	6
10.	State review of need for more hospital beds ("Certificate of Need")	3.5 months	7, 8
11.	Design drawings	4 months	10
12.	Construction documents	5 months	9, 11
13.	Site preparations/demolish existing building	9 weeks	11
14.	Construction start/building pad	2 months	12, 13
15.	Relocate utilities	6 weeks	12
16.	Deep foundations	2 months	14
17.	Building structure in place	9 months	16
18.	Exterior skin/roofing	4 months	17
19.	Interior buildout	12 months	17
20.	Building inspections	5 weeks	15, 19
21.	Occupancy	1 month	20

Answer the following questions:

1. Develop the network for planning and construction of the new hospital at Arnold Palmer.
2. What is the critical path and how long is the project expected to take?
3. Why is the contraction of this 11-story building any more complex than construction of an equivalent office building?
4. What percent of the whole project duration was spent in planning that occurred prior to the proposal and reviews? "Prior to the actual building construction". Why?

3. Divekar Industries Pvt. Ltd.

Divekar Industries Pvt. Ltd. specializes in the manufacture of measuring instruments. The company has received an order for an instrument which is meant to determine critical thickness and density of certain material within tolerance.

This device utilizes radiations and is capable to measuring the thickness and density of sheets, plates, layers of powder and like materials without physical contact with the materials and so without their destruction.

A. Main Parts of the Device

(1) Head, (2) Arms, (3) Ionixation chamber components (purchased), (4) Meter (purchased), (5) Redioactive source: Radium or Stronium 90, or Cobalt 60.

B. Major Activities

Major activities required to be performed for each of the parts of the device are:

1. Design entire device.
2. Head: (*a*) Draft head (*b*) Make pattern for head (*c*) Cash head (*d*) Machine head

3. Arms:	(*a*) Draft arms	(*b*) Make pattern for arms
	(*c*) Cast arms	(*d*) Machine arms
4. Metre:	(*a*) Purchase metre	(*b*) Calibrate and test metre
5. Radioactive source:	(*a*) Purchase	
6. Ionization chamber parts:	(*a*) Purchases components	(*b*) Assemble components

C. Assembly Procedure

Head: After machining head, radium stronium 90 or cobalt 60 may be assembled to head.

Arms: After arms are machined, the assembled ionization chamber is assembled to them.

Metre: After sub-assembly of radium to head, metre is assembled to unit in final assembly along arms sub-assembly.

As soon as metre is calibrated and tested, test run plans may be drawn up.

After final assembly, test run of device is undertaken.

Time - Cost Data

Activity		*Normal*		*Crash*	
		Time	*Cost*	*Time*	*Cost*
A	Design	150	3	—	—
B	Buy radium	6	300	6	—
C	Buy ionization chamber	4	100	2	300
D	Assemble ionization chamber	3	60	1	300
E	Draft arms	2	80	1	120
F	Make pattern (armps)	3	60	1	260
G	Cast arms	2	40	1½	60
H	Machine arms	1	½	½	100
I	Assemble ionization chamber to arms	3	100	2	120
J	Draft head	2	80	1	120
K	Make pattern (Head)	3	60	1	260
L	Cast head	2	40	1½	60
M	Machine head	1	40	½	100
N	Sub-assembly of radium and head	2	40	1	120
O	Buy metre	2	150	1	200
P	Calibrate and test metre	2	20	2	—
Q	Final assembly of arms sub-assembly, head sub-assembly and metre	2	40	1½	130
R	Prepare for test run	2	70	1	150
S	Test run	2	20	1	10
			1,490		

Overhead cost for varying time requirements:

Normal = ₹ 1,900, 16 units = ₹ 1,410, 14 units = ₹ 1,200, 18 units = ₹ 1,675,
15½ units = ₹ 1,270, 13 units = ₹ 1,180, 17 units = ₹ 1,520, 15 units = ₹ 1,250.

Answer the following questions:

1. Prepare a network, identify critical path and compute project duration.
2. Use cost data to determine crashed time, crashed cost, optimum time and optimum cost.
3. Make a square network for the project.

4. Machine Erectors (P) Limited (MEPL)

MEPL undertake consultancy work on planning of projects related to construction and equipment erection work. They are recently approached by a party to work out a plan for replacement of some old machinery by a new Computer Numerical Controlled Machinery (CNCM), to be erected at a site different from the place occupied by the old machines.

The planning group (PG) of MEPL found that because of its versatility, the CNCM needs more space, better ventilation, etc. Hence, apart from a strong foundation, an extension of the existing building also was necessary. Estimate of conventional approach of erection required concrete insulation casting of 8 days (average) followed by 21 days of curing. The PG listed the activities and estimated their duration, wherever possible in deterministic form and elsewhere the three time estimates. *Table I* given below shows the activities and the time estimates. It also shows the precedence relationship worked out by them in the initial stage of replacement plant (Stage 1).

Table I: Precedence Relationship of Activities for Erection of CNCM

Activity	*Description*	*Duration (days)*	*Precedence relationship stages of deliberations* 1	2	3	4
A	Indent for CNCM	1	—	—	—	—
B	Transfer of CNCM from stores to plant	1-2-3	*A*	*A*	*A*	*A*
C	Remove old equipment	3-4-6	—	—	—	—
D	Transfer old equipment to stores	1-2-3	*C*	*C*	*C*	*C*
E	Extend existing building (casting)	5-8-12	*C*	*C*	*C*	n.r.
F	Wait for curing	21	*E*	*E*	*E*	n.r.
P	Prefabricated shed	8-10-12	n.r.	n.r.	n.r.	*C*
G	Install CNCM	3	*F, B*	*F, B*	*F, J*	*P, J*
H	Check alignments	1	*G*	*G, I*	*G, I*	*G, I*
I	Complete piping and electrification	8-8-14	*H*	*C*	*C*	*C*
J	Clean and lubricate CNCM	2	1	*H*	*B*	*B*
K	Test CNCM trial run	2-3-10	*J*	*J*	*H*	*H*
L	Check vibration, noise level and insulation	1-3-5	*K*	*K*	*K*	*K*
M	Full load test	2	*L*	*L*	*K*	*K*
N	Touch up painting CNCM	3	*M*	*M*	*L, M*	*L, M*

Note. n.r. — Not relevant.

The PG met to consider the possibilities of reducing the project duration by changing the precedence relationship of the activities without sacrificing the technological feasibility. The result of their deliberations is shown in Stage 2. Stage 3 shows the results after a further review. After the Stage 3 was listed, project leader suggested that a separate shed with precast modules be erected instead of extending the existing building and then waiting for parts of it to cure (a new technology). The project leader said that the erection of the prefabricated shed would take only 10 days (average) as compared to the site construction which would take 21 days for curing alone and an average of 8 days for erection casting. A comparison was immediately made between the additional expenses that would be incurred on the prefabricated shed and the saving of project duration. The project leader has estimated the daily project running administration cost at ₹ 5,000 per day, to cover mainly overheads. However, the prefabricated shed would involve an additional cost of ₹ 1,50,000.

Work out the cost-time trade-off between all the possibilities.

The PG decided to use the prefabricated shed instead of construction at site. The Stage 4 (final stage) activity relationships were the same as those for Stage 3, except that activities *E* and *F* for site construction were replaced (in Stage 4) by a single deterministic activity, *i.e.*, the erection of prefabricated shed, which was scheduled to take an average of 10 days. *Table II* shows the skilled and unskilled labour requirements needed for each activity as estimated by the PG.

Table II: Manpower Resource Levels for Activities for Erection in respect of 'CNCM'

Activity	*Labourer*	*Fitter*	*Welder*	*Painter*
A	—	—	—	—
B	—	2	—	—
C	8	2	—	—
D	8	—	—	—
E	10	—	—	—
F	2	—	—	—
P	10	2	—	—
G	6	4	—	—
H	2	4	—	—
I	4	6	2	—
J	4	4	—	—
K	2	2	—	—
L	2	2	—	—
M	2	2	—	—
N	2	1	—	2

Answer the following questions:

Analyse the following and other issues:

(*a*) Trade-off costs between Stages 3 and 4.

(*b*) Critical paths and durations of Stages 3 and 4.

(*c*) Expected time of complete erection of CNCM.

(*d*) The experience of PG had been that on 80% of occasions their estimates and promises were fulfilled. Keeping this in mind, the group wishes to know the total project duration to be quoted for the erection of CNCM.

(*e*) The client party wanted CNCM to be erected within 20 days from the commencement of the work. The PG wished to know their chances of achieving this.

(*f*) Since, skilled manpower resources available to the PG for work of this nature were limited, they wished to know daily requirements of skilled man-power resources, so that they could examine their implications to the project completion. If there is any constraint, what contingent plans you suggest?

(*g*) Draw the *PERT* network for Stages 3 and 4.

33:6. CASE PROBLEMS IN DECISION THEORY

1. Ken Group of Industries

Ken Group of Industries is faced with the problem of deciding the size of the proposed plant for the manufacture of handlooms. It has two options available at present:

- Build a large plant to accommodate any possible demand in future.
- Build a small plant to accommodate a low demand and consider the possibility of expanding when the demand increases.

The management has got the market research conducted which has categorized the demand as low, medium, or high with corresponding probabilities of 0.4, 0.3 and 0.3, respectively. The following data has been collected:

(*i*) If large plant is built at a cost of ₹ 10 lakh, it will accommodate any possible demand in future and there will be no need for expansion. The discounted operating returns for low, medium and high demand are estimated at ₹ 8 lakh, ₹ 18 lakh and ₹ 22 lakh, respectively.

(*ii*) If a small plant is built initially at a cost of ₹ 4 lakh, it will meet only low demand and will require to be expanded when demand becomes medium or high.

(*iii*) Depending upon the demand in future, small plant may require no expansion (*i.e.*, if the demand is low), or may require small expansion at a cost of ₹ 4 lakh (*i.e.*, if the demand is medium), or may require large expansion at a cost of ₹ 8 lakh (*i.e.*, if the demand is high).

(*iv*) Since, expansion of the plant will not be possible suddenly, some revenue is likely to be lost. The operating returns to be realized in case of small plant expansion, while large plant expansion are projected at ₹ 16 lakh and ₹ 19 lakh, respectively.

Answer the following questions:

(*a*) Represent all decisions and their economic consequences into a decision tree.

(*b*) Decide on the size of the plant and optimal plan of action.

2. Allocation of Facilities in Hospital

City Hospital is a 800-bed hospital, located in the rapidly growing region of city. It was built in the year 2010 as a 100-bed hospital and since that time has expanded to keep paces with the growing community.

In the past, expansion plans were routinely approved. Financing was never a problem. But, this has changed in few years as the public has become increasingly reluctant to accept the spiraling costs of medical care. Pressure build up, including insurance companies, politicians, and the area residents, have all voiced their concern over the problem. They insisted that the hospital hold the line on expenses, while providing at a minimum the same quality of service.

In response to this pressure, a new hospital administrator, Dr. PK Ghosh, was hired 3 months ago. In the short period of time he has been with the hospital, he has proved a tough administrator. Several employees have been terminated, while those who have remained have been working harder. Already costs have been reduced by 10 per cent. The hospital, however, is still operating at a loss.

Ever since Dr. Ghosh joined the hospital staff, there has been a moratorium on both hospital construction and investment in new equipment. Several proposals have been submitted, but no action has been taken.

Yesterday Dr. Ghosh received a call from Mr. K. Natrajan, Chairman of the board of trustees. Mr. Natrajan told Dr. Ghosh that several physicians had expressed their concern to him about the delay on three proposals that had been submitted during the past 3 months. Both Mr. Natrajan and Dr. Ghosh agreed that there were sufficient funds to accept one but not all three of these proposals. Mr. Natrajan closed the conversation by asking Dr. Ghosh to recommend one of the proposals to the board of trustees at its next meeting in 2 weeks.

After the phone call, Dr. Ghosh removed the proposals from his file cabinet and began to examine them carefully. The first was for the expansion of the coronary care facilities from 5 to 10 beds, the second was for the addition of a fourth operating room, and the third was for the addition of radiology facilities. The following paragraphs were taken from these proposals.

Coronary Care

The coronary care facilities in City Hospital are inadequate to meet the needs of the community, and as the community continues to grow, this problem will become even more acute. Current facilities operate at an occupancy rate of 90 per cent. This is higher than the occupancy rate for other units in the hospital. On many occasions patients must be accommodated in the corridor or sent to nearby hospitals. There is no question that improved facilities could save lives.

According to rough estimates, if aggregate demand for the facilities at hospital average 12,000 patient-days per month, the excess of the revenues over all costs for this project will be ₹ 12,00,000 per year. If demand is 10,000 patient-days per month, revenues will exceed costs by ₹ 60,000, but, if demand is 8,000 patient-days, then costs will exceed revenues by ₹ 3,00,000 per year.

Operating Room

Hospital currently maintains three rooms. Use of these facilities is high and on several occasions elective surgery has been postponed to accommodate surgery of an emergency nature. New facilities would not only alleviate this problem, but would end the reluctance of many surgeons to schedule elective surgery at Hospital. An important by-product of this project, one not considered in the economic data, would be the increase in the demand for hospital beds in the surgical wards. Since, these wards are currently operating at a net loss, this would be a welcomed addition to revenue.

If the aggregate demand for facilities at City Hospital averages 12,000 patient-days per month, revenues associated with this project will exceed costs by ₹ 19,00,000 per year. If demand averages 10,000 patient-days per month, revenues will exceed costs by ₹ 7,00,000 per year, and if demand averages 8,000 patient-days per month, costs will exceed revenues by ₹ 6,00,000 per month.

Radiology

The radiology department desires to purchase equipment used in cancer therapy. At the present time patients who require this therapy must be referred to a hospital 12 kilometres away.

If the aggregate demand for facilities at City Hospital averages 12,000 patient-days per month, revenues associated with this project will exceed costs by ₹ 14,00,000 per year. If demand averages 10,000 patient-days per month, revenues will exceed costs by ₹ 9,00,000 per year, and if demand averages 8,000 patient-days per month revenues will exceed costs by ₹ 6,00,000 per year.

The Recommendation

After reading these proposals, Dr. Ghosh realized that this decision depended to some extent on his ability to estimate the future demand for hospital facilities. But several factors made this impossible. First, he had been with the hospital just 3 months and as yet had not been able to detect any trend in demand. Second, the hospital had never made demand forecasts before and there was no one with this expertise on his staff. Third, there was not enough time to made an intelligent forecast.

The decision was, therefore, made on the basis of available information. Dr. Ghosh recommended that the board of trustees accept the operating room proposal. He felt that it had the advantage of contributing more than the other projects towards the elimination of the hospital's financial difficulties.

Answer the following questions:

1. What criterion did Dr. Ghosh use in comparing the alternative proposals?
2. Construct a conditional pay-off table.
3. Identify the proposal that would have been recommended, had the following methods been used: (*a*) Maximin, (*b*) Maximax, (*c*) Laplace, (*d*) Minimax regret.

4. Which method did Dr. Ghosh informally use to reach his conclusion? Was he justified in using this method?
5. Do you think that the choice of a criterion is an issue in this case? If so, why is it an issue? What criterion or criteria would you use?
6. Suppose that you had to make a recommendation to the board of trustees. Which proposal would you choose, why?
7. Suppose that the probabilities associated with each of the demand levels had been estimated and is given below:

Demand :	8,000	10,000	12,000
Probability :	0.30	0.50	0.20

Is this still a decision problem under uncertainty? Find the expected monetary value of each alternative. Which alternative would you recommend?

3. The Shipyard Company

The management of the Shipyard Company will soon be negotiating a new contract with the union that represents the company's employees. (The union's contract is renegotiated each year.) The one issue involved at this time is the wage rate. As has been the case for the past twenty years, both the company and the union have agreed to submit their dispute to an arbitration board.

The tripartite arbitration board consists of 3 members, namely, a company representative, a union representative, and an impartial member appointed by the Arbitration Association who acts as chairman.* In most cases the final decision is usually made by the chairman, since the union representative seldom feels compelled to vote with management and *vice versa*.

The selection of a management representative is of utmost importance, since his actions may well influence the final outcome of the dispute, even though the chairman is actually impartial. The chairman's decision may well be influenced by the choices made by both the union and management as to their respective representatives. For example, if management selects as its representative an individual who is a radical company man, and the union chooses an individual who is either unbiased or does not express his favouritism toward the union vocally, the chairman may have a negative reaction against the company and favour the union in his final decision.

Management is considering 5 different types of individuals to represent the company on the arbitration board. These individuals are indicated below by the symbols M_1, M_2, M_3, M_4 and M_5.

Management representative	*Characteristics*
M_1	An unbiased individual.
M_2	An individual whose voting record in arbitration cases indicates a slight bias in favour of management, but who on occasions has voted against management.
M_3	A company man who utilizes a 'soft-sell' approach.
M_4	An outspoken company man.
M_5	A radical company man.

The union has a similar set of individuals whom it is considering to represent it in the dispute. The individuals are indicated below by U_1, U_2, U_3, U_4 and U_5.

* The Arbitration Association maintains a national panel of persons, each of whom is an expert in his chosen field or profession and qualified to render impartial decision in arbitration cases.

Union representative	*Characteristics*
U_1	An unbiased individual.
U_2	An individual whose voting record in arbitration cases indicates a slight bias in favour of the union, but who on occasions has voted against the union.
U_3	A union man who utilizes a 'soft-sell' approach.
U_4	An outspoken union man.
U_5	A radical union man.

Past experience in arbitration cases has indicated that the average/monthly wage increase (in rupees) associated with the type of representative that management employs at the bargaining table is completely dependent on the type of representative that the union chooses. And likewise, the average/monthly wage increase (in rupees) associated with the type of representative that management chooses. (It is convenient to refer to the choices available to the union and management as strategies).

It should be emphasized that neither the union nor management knows the other's final strategy decision until the arbitration hearing actually begins. Neither party can alter its choice of a representative once the representative's name has been submitted to the chairman of the arbitration board.

Twenty years of arbitration experience indicates that the following table is an approximate indication of the expected average wage increases associated with the different possible combinations of union and management representatives chosen to serve on the arbitration board:

Different Combination of Union-Management

	M_1	M_2	M_3	M_4	M_5
U_1	0	5	10	10	30
U_2	10	5	15	15	35
U_3	25	15	10	20	40
U_4	15	5	5	15	20
U_5	–5	0	5	10	10

Although the average wage increases in this table are not an exact representation of the wage increase associated with the different union-management strategies utilized over the past twenty years, management does feel that it will prove to be a useful guide in selecting the type of individual to represent the company in future wage negotiations. Furthermore, both the union and management know the approximate wage increases associated with each union-management strategy combination.

Although management would like to select that strategy which will minimize the average monthly wage increase during the forthcoming contract period, it is more concerned with selecting a set of strategies that will enable it to minimize the increase in wage rates over an extended period of time. In other words, management would like to select the particular (pure) strategy that will minimize wage increases in long run. Management may find that its optimum strategy is one in which it does not choose the same particular (pure) strategy each year. Instead it may select strategies randomly according to some predetermined optimal probability distribution imposed over its pure strategies: M_1, M_2, M_3, M_4 and M_5. Such a strategy is called a mixed strategy.

The union is searching for an optimal pure strategy or mixed strategy that will maximize wage increases over an extended period of time.

Answer the following questions:

1. What assumptions are necessary to express this bargaining problem as a two-person zero-sum game? Are these assumptions realistic?
2. According to table given above, which three strategies will not be used at all by management? Why?
3. Which three strategies will not be chosen by union? Why?
4. Assuming that the table given above is an approximate representation of the union-management bargaining situation for the next 10 years, use linear programming to obtain an approximation to the optimum mixed strategy for management. What will be the expected wage increase during each contract period using this optimum mixed strategy for the next 10 years.
5. What is the union's optimum mixed strategy?
6. If the assumptions in (1) and (4) are not completely realistic, is this approach to union-management bargaining problems invalidated? Why?
7. How would management decide which five individuals to consider?
8. How would management know that the union was considering?
9. How would past experience in union-management negotiations be organized to generate a pay-off table such as given in the table given on the previous page?
10. Is it really appropriate for management to focus so heavily on minimizing long-run wage increases?

33:7. CASE PROBLEMS IN QUEUING THEORY

1. Construction of Facilities at Port

A Government undertaking, which operates its own fleet of ships to import foodgrains, is considering the construction of port facilities to support the import of foodgrains. The management wishes to determine the number of unloading berths and the type of installations so that total unloading costs are minimized.

A maximum of three berths can be built, and it is required that the same type of unloading installation be in each berth that is built. The choice of unloading installation lies among types *A*, *B* and C, for which the following information is available:

Installation type	*Fixed cost per day*	*Operating cost per day*	*Capacity average tonnage unloading per day*
A	₹ 600	₹ 840	3,600 tonnes
B	₹ 1,200	₹ 1,350	5,800 tonnes
C	₹ 1,500	₹ 1,600	6,400 tonnes

Fixed costs include such items as the amortization of the original cost of the installation over its expected life, general maintenance, etc.; they apply to all days, whether the equipment is used or not. Operating costs are incurred only during the time intervals that the uploading equipment is actually in use.

Ships to be unloaded each carry 8,000 tonnes of foodgrains, and are considered to arrive in a Poisson fashion throughout the year with a mean arrival rate of five ships a week. Service times for a given type of installation are considered to be exponential, with mean service rate corresponding to the average capacity column in the table given above.

If time spent in the unloading system (waiting time plus unloading time) is considered to cost the company ₹ 2,000 per ship per day, what type of unloading installation should be chosen and how many berths should be constructed?

2. Number of Crews at Warehouse

A single crew is provided for unloading and/or loading each truck that arrives at the loading deck of a warehouse. These trucks arrive according to Poisson input process at a mean rate of one per hour. The time required by a crew to unload and/or load a truck has an exponential distribution (regardless of the crew size). The expected time required by a one man crew would be two hours.

The cost of providing each additional member of the crew is ₹ 100 per hour. The cost that is attributable to having a truck not in use (*i.e.*, truck standing at the loading deck) is estimated to be ₹ 400 per hour. Assume that the mean service rate of the crew is proportional to its size. What should the size be in order to minimize the expected total cost per hour?

The general routine is that the order pickers assemble orders placed by stores and load them on one of the company trucks waiting at the single loads on a first-come, first-served basis. They, then proceed to the store for which the order is destined, unload and return for another order. Because of the many different routes and distances, and traffic problems at different times of the day, the time between arrivals of the trucks at the dock is random (assumed to follow an exponential distribution) averaging 30 minutes. The loading time also follows an exponential distribution and averages 15 minutes.

Truck drivers are paid ₹ 100 per hour and the crew of two loaders are each paid ₹ 40 per hour. Truckers have complained about the long waiting, so a sample was taken showing that truckers did indeed wait for an average of 32 minutes. The warehouse manager knows that a second truck dock would probably solve the problem, but the large capital expenditure plus the disruption of operations during construction are deterrents to this solution.

Tests are made with different crew patterns and it is found that a crew of three can be used to advantage, reducing the loading time to 10 minutes.

As management trainee working with the corporation, help it in finding suitable answers to the following questions:

1. Is the crew of three loaders more economical than the crew of two?
2. What is the probability that an arriving truck will find at least one truck already in the system?
3. How much of the time is the crew idle?
4. If another truck dock is available with the present operating characteristics, how will it affect the total expected cost per hour?

33:8. CASE PROBLEMS IN SIMULATION

1. Store Design and Staffing Plan

An organic milk supply company operates a chain of milk and milk products supply stores in South India. In each store customers enter at one end of the store and exit at the other end. Customers pick up milk, butter, cheese, ice-cream, etc. before leaving the store. When a new customer arrives at the store, the customer waits until the preceding customer's order is complete and then enter the store for service.

Typically, three employees operate each store during peak periods; two clerks take and fill orders, and a third clerk serves as cashier as well supervisor of the store. The organic milk company is considering a revised store design in which computerized order-taking and payment are integrated with specialized warehousing equipment. Management hopes that the new design will permit operating of each store with only one clerk. To determine whether the new design is beneficial, management has decided to build a new store using the revised design.

Organic Milk Company's new store will be located near a major shopping centre. Based on experience at other locations, management believes that during the peak late afternoon and evening hours, the time between arrivals follows an exponential probability distribution with a mean of 6 minutes. These peak hours are the most critical time period for the company; most of their profit is generated during these peak hours.

An extensive study of times required to fill orders with a single clerk has led to the following probability distribution of service times:

Service time (minutes) :	2	3	4	5	6	7	8	9
Probability :	0.24	0.20	0.15	0.14	0.12	0.08	0.05	0.02

In case customer's wait time prove too long with a single clerk, Organic Milk Company's management is considering two alternatives: add a second clerk to help with bagging, taking orders, and related tasks or appoint an helper to the clerk so that two customers can be served simultaneously, *i.e.*, a two channel system. With either of these options, two clerks will be needed. With the two-channel option, service times are expected to be the same for each channel. With the second clerk helping with a single channel, service times will be reduced. The following probability distribution describes service times given that option:

Service time (minutes) :	1	2	3	4	5
Probability :	0.20	0.35	0.30	0.10	0.05

Management of the milk supply company would like you to develop a spreadsheet simulation model of the new system and use it to compare the operation of the system using the following three designs:

Design A: One channel, one clerk,
Design B: One channel, two clerks,
Design C: Two channels, each with one clerk.

Management is especially concerned with how long customers have to wait for service. Research has shown that 30% of the customers will wait no longer than 6 minutes and that 90% will wait no longer than 10 minutes. As a guideline, management requires the average wait time to be less than 1.5 minutes.

Prepare a report that discusses the general development of the spreadsheet simulation model, and make any recommendation that you have regarding the best store design and staffing plan for the Organic Milk Company. One additional consideration is that the design allowing for a two-channel system will cost an additional ₹ 50,000 to build.

1. List the information the spreadsheet simulation model should generate so that a decision can be made on the store design and the desired number of clerks.
2. Run the simulation for 1,000 customers for each alternative considered.

2. Planning and Controlling of Quality Costs

The Managing Director of a manufacturing company based on past data from the cost accounting information system (adjusted for anticipated current and future activities), derived the probability distributions shown in *Exhibit-I* for each of 4 quality cost random variables. (Note: For the sake of simplicity and ease of computation, each of the 4 quantity cost random variables will be limited to 5 possible, arbitrarily chosen outcomes):

EXHIBIT - I

Probability Distribution for the 5 Quality Cost Random Variables

Prevention costs				*Appraisal costs*			
Cost ('000 ₹)	*Probability*	*Cumulative probability*	*RN interval*	*Cost ('000 ₹)*	*Probability*	*Cumulative probability*	*RN interval*
500	0.20	0.20	0–19	150	0.20	0.20	0–19
700	0.35	0.55	20–54	250	0.30	0.50	20–49
800	0.25	0.80	55–79	400	0.35	0.85	50–84
850	0.10	0.90	80–89	500	0.10	0.95	85–94
900	0.10	1.00	90–99	550	0.05	1.00	95–99
Internal failure costs				*External failure costs*			
Cost ('000 ₹)	*Probability*	*Cumulative probability*	*RN interval*	*Cost ('000 ₹)*	*Probability*	*Cumulative probability*	*RN interval*
90	0.05	0.05	0–4	150	0.10	0.10	0–9
220	0.25	0.30	5–29	190	0.20	0.30	10–29
240	0.35	0.65	30–64	260	0.25	0.55	30–54
370	0.15	0.80	65–79	410	0.30	0.85	55–84
400	0.20	1.00	80–99	650	0.15	1.00	85–99

The first step in the Monte-Carlo simulation is to determine the performance measure. The performance measure is 'Total quality costs'.

In the second step, the problem must be expressed mathematically. The mathematical expression for total quality cost is:

$$TQC = PC + AC + IFC + EFC,$$

where, PC = Prevention Costs, AC = Appraisal Costs,
IFC = Internal Failure Costs, and EFC = External Failure Costs.

In the third and fourth steps, the probability distribution and cumulative probability distributions for each of the 4 quality cost random variables must be specified. The probability distributions for prevent in costs, appraisal costs, internal failure costs, and external failure costs, are shown in the second column of *Exhibit-I*, respectively. In a similar manner the cumulative probability distributions are shown in the third column of *Exhibit-I*.

In the fifth step, two-digit representative numbers, ranging from 00–99 are assigned to each possible outcome of the probability distributions for the 4 quality cost random variables; they are shown in the fourth column of *Exhibit-I*.

In the sixth step, trial one, a random number for each of the four quality cost random variables is obtained from a random number table. The first random number selected is 23 for prevention costs; the second random number selected is 05 for appraisal costs; the third random number selected is 38 for internal failure costs; and the fourth random number selected is 14 for external failure costs. These random numbers are posted in *Exhibit-II* on line 1 to columns 2 through 5, respectively.

In the seventh step, trial one, each outcome or value of the quality cost random variable is determined by each random number selected in *Step 6*. Given the random number of 23 for prevention costs, 05 for appraisal costs, 38 for internal failure costs, and 14 for external failure costs the corresponding values are posted in *Exhibit-II* on line 1 to columns 6 through 9 and equals ₹ 7,00,000, ₹ 1,50,000, ₹ 2,40,000 and ₹ 1,90,000, respectively.

EXHIBIT - II

Simulation Worksheet for 20 Trials

(All costs in '000 ₹)

Trials	*Random number for*				*Values for*				*Performance measure (total quality cost)*
	PC	*AC*	*IFC*	*EFC*	*PC*	*AC*	*IFC*	*EPC*	*TQC*
(1)	(2)	(3)	(4)	(5)	(6)	(7)	(8)	(9)	(10)
1	23	05	38	14	700	150	240	190	1,280
2	97	11	93	43	900	150	400	260	1,710
3	49	36	43	07	700	250	240	150	1,340
4	61	31	09	57	800	250	220	410	1,680
5	97	93	61	72	900	500	240	410	2,050
6	97	89	81	25	900	500	400	190	1,990
7	11	64	71	10	500	400	370	190	1,460
8	60	37	38	47	800	250	240	260	1,550
9	73	32	74	97	800	250	370	650	2,070
10	56	49	16	43	800	250	220	260	1,530
11	48	96	77	38	700	550	370	260	1,880
12	17	94	47	47	500	500	240	260	1,500
13	57	08	43	44	800	150	240	260	1,450
14	34	95	50	13	700	550	240	190	1,680
15	36	71	80	27	700	400	400	190	1,690
16	20	47	19	93	700	250	220	650	1,820
17	54	50	69	87	700	400	370	650	2,120
18	67	78	61	00	800	400	240	150	1,590
19	83	45	47	45	850	250	240	260	1,600
20	28	49	90	66	700	250	400	410	1,760

In the eighth step, trial one, the value of the performance measures developed in *Step 2*, is computed using the corresponding value of each quality cost random variable found in *Step 7*. Total quality costs posted in *Exhibit-II* to line 1, column 10, for trial one, equal ₹ 12,80,000.

In step nine, *Steps 6-8* are repeated 1,000 of times. However, for illustrative purposes *Step 6-8* are only repeated for 19 more trials. The results of all 20 trials are shown in *Exhibit-II*.

In the tenth step, a probability distribution and a cumulative probability distribution and on the basis of the value for the performance measure for each.

Trial recorded in *Step 8*. On the basis of the results of 20 trials shown in *Exhibit-II*, a probability distribution for total quality costs based on an arbitrarily chosen range of ₹ 20,00,000 was constructed; it is shown in *Exhibit-III* given below:

EXHIBIT - III

Probability Distribution and Cumulative Probability Distribution for Simulation Results

Total quality cost range	*Number of trials**	*Probability#*	*Cumulative probability##*
1,28,000–1,48,000	4	0.20	0.20
1,48,000–1,68,000	7	0.35	0.55
1,68,000–1,88,000	5	0.25	0.80
1,88,800–2,08,000	4	0.20	1.00

* Number of trials in which the outcome was within the specified range.

Probability is equal to (Number of trials)/(Total number of trials).

Probability of attaining a total quality cost less than or equal to the designated total quality cost range.

In the eleventh and most important step, the cumulative probability distribution constructed in *Step 10* is analysed. The management now has information as to the possible amounts of total quality costs that are likely to be incurred. With this information, management can rationally plan and control its quality costs accordingly depending upon the degree of assurance that management requires in its decision-making.

Based on *Exhibit-III*, the management in its planning and controlling efforts know, for example, that the probability of total quality costs exceeding ₹ 16,80,000 is 45%. If ₹ 16,80,000 of total quality costs was critical amount relative to the total estimated profits of the company. Alternative courses of action could be explored and temporarily or even permanently adopted, if necessary to mitigate the highly probable adverse effect of ₹ 16,80,000 of total quality costs on company profits. Therefore, by carefully examining the cumulative probability distribution in *Exhibit-III*, management can assess the economic attractiveness of its originally planned quality cost programme and its risks.

However, the benefits of Monte-Carlo simulation for the planning and controlling of quality costs do not end here. A Monte-Carlo simulation model can serve management in much the same way as a laboratory serves the physical scientist. By making changes in the model's various parameters (*i.e.*, prevention costs, appraisal costs, internal failure costs, and external failure costs), the management can observe the results of the simulation and for different combinations of individual cost components, their estimated rupee amounts and related probabilities, for each of the 4 quality cost random variables would behave under circumstances. Management will not find an optimum solution through the use of this descriptive model to the problem of planning and controlling quality costs; instead, management will find a good solution to a quality cost problem, that is far too complex to be solved by any of the available optimization techniques despite what might be inferred from past and present articles in the management accounting literature.

It should be noted, of course, that Finance Directors are in no way precluded from, and in fact are urged to continue, what has so far proven to be a fruitless search for an optimal solution, continued research combined with a better understanding of quality costs may possibly lead to either a new or a known but not-as-yet used optimization technique whose informational output exceeds the informational output of the descriptive technique suggested.

33:9. CASE PROBLEM IN REPLACEMENT DECISIONS

Allied Industries (Pvt.) Ltd.

Allied Industries Pvt. Ltd. is considering the replacement of their existing 8 years old centre lathe with copying attachment by a full fledged copying lathe. The existing lathe was purchased for ₹ 1,20,000 and ₹ 15,000 were spent to install it. The proposed copying lathe is available for a sum of ₹ 2,90,000 and a sum of ₹ 45,000 shall have to be spent for its installation. The proposed machine is expected to have a useful life of 10 years and be worth ₹ 1,00,000 ten years hereafter.

The existing centre lathe can be sold for an estimated amount of ₹ 80,000. Its retention for one more year will reduce its resale price by ₹ 20,000.

Advantages from the proposed machine are as under:

1. Income from higher production: ₹ 80,000 per year are expected to be realized on account of faster rate of production.
2. Income from greater precision: ₹ 30,000 per year are expected to be earned due to greater precision.
3. Labour cost: The existing machine is run on two shifts per day, each shift of 8 hours duration. The proposed machine too shall be run on two shift basis. In all, considering 16 hours working and after excluding legitimate losses of time, 3,600 hours are available to operate existing as well as

the proposed machine. The labour cost on either machine is ₹ 40 per hour. Production wise, the proposed machine is 40% more productive than the existing machine.

4. Fringe benefits: Fringe benefits namely provident fund, gratuity bonus, ESIC, subsidized food, etc., being paid to the workmen have been computed at an average of 40% of the employee's basic wage.
5. Indirect wages: No change is expected in the requirement of indirect work men.
6. Power cost: The existing machine is driven by a motor having rated capacity of 10 Kws and the proposed machine by a motor of 18 Kws. The power cost, the function of Kw rating of the machine for the existing machine is ₹ 80,000 per annum.
7. Cost of maintenance and repairs: The annual cost of the maintenance and repairs of the old machine are ₹ 60,000 and those of the new machine are expected to ₹ 15,000.
8. Cost of supplies: The proposed machine has pneumatic controls. the cost of providing supplies is expected at ₹ 12,000 per year.
9. Taxes and insurance: The company plant and machinery is covered by the insurance. Average taxes and insurance are estimated at 3% of the resale value of ₹ 80,000 for the old lathe and of the installed cost of ₹ 3,35,000 for the new lathe.
10. Rejection data: The old lathe gives 3% rejection and the new lathe is expected to give 0.5% rejection. Materials worth ₹ 32,800 were rejected on the existing machine last year.
11. Work-in-process: The average work-in-process at the existing machine is evaluated at ₹ 98,000 which is expected to be reduced to ₹ 30,000 on account of faster rate of production. The inventory carrying charges are calculated at 20% of the average inventory investment.

Should the company have the replacement now considering annual interest rate of 12%?

33:10. CASE PROBLEM IN DYNAMIC PROGRAMMING

Usefulness of Kit in Defence

Anita & Company was engaged in making repair kits for defence since 1970. The defence department was happy with company management, as it was able to get its supply within time. The company was putting three items *A*, *B* and *C* in the repair kit. The cost and usefulness of each item differs. As these kits are needed in the field, the usefulness of each item in a given kit decreases as more are added to it. To defence, the item *A* costs ₹ 80, whereas item *B* costs ₹ 120 and item *C* costs ₹ 200. The usefulness of the repair kit is based upon multiplying the value in use by the probability of its use in the field. The usefulness of the items are given as under:

No. of units in repair kit	*Usefulness of item A*	*Usefulness of item B*	*Usefulness of item C*
1	60	140	200
2	120	200	300
3	180	250	380
4	230	290	440
5	275	320	480
6	310	345	510
7	345	365	540
8	375	380	560

There are mainly two issues faced by the company management (*i*) to determine the number of each type of item to pack in a field kit which will maximize the usefulness of the kit to Defence, subject to that each kit must contain at least one unit of item *A* as well as of *B*. (*ii*) Defence department, recently faced with financial problems, is pressing the company to provide them repair kit of maximum usefulness, but not exceeding the cost ₹ 600, of course maintaining the requirement of at least one unit each of item *A* and *B*.

33:11. CASE PROBLEMS IN INVENTORY CONTROL

1. Bright Star Company

Bright Star Company manufactures several products. One of its main products requires an electric motor. The Management of the Company uses the economic order quantity formula to determine the optimum number of motors to order. Management, now, wants to determine how much safety stock to keep.

It uses 30,000 electric motors annually (300 working days) and orders 3,000 motors at a time (*EOQ* norm). the lead time for an order is five days. The annual cost of carrying one motor in safety stock is ₹ 10.

Management has also estimated that the cost of being out of stock is ₹ 20 for each motor they are short. The company has analysed the usage during past reorder periods by examining the inventory records. This information is given below:

Usage during lead time	*No. of times quantity was used*
440	6
460	12
480	16
500	130
520	20
540	10
560	6

The company is searching for the safety stock level that will minimize the expected total of the costs of carrying additional inventory and costs associated with sufficient inventories (Stock out cost). From your analysis of the situation, recommend the optimum safety stock and the corresponding reorder level.

2. Spare Parts Inventory

Mr. Sohil Gupta is an assistant engineer in the Irrigation Department of certain State. He is responsible for the maintenance and upkeep of the various mechanical equipments needed during the construction work, *e.g.*, Bulldozers, loaders, dump trucks, and so on. For the maintenance of these equipments proper inventory control of spare parts is needed. In all, there are about 10,000 spare parts to be stocked. If a part is available in stock then the equipment is down for maintenance and it may hamper the productive activities. Overstocking of an item leads to blockage of capital and increases the inventory carrying costs.

After applying *A*, *B*, *C* analysis, Mr. Gupta has chosen a critical Part *I–100* from group *A* for further analysis. This requirement for this part is not uniform in every week. The part is needed whenever the operating equipment fails. Mr. Gupta has analysed the past records of the weekly requirements of this product. He has the data available for the last 150 weeks. The maximum demand during any one week was 150 and minimum was 20. The data is given as follows:

Weekly Demand for Part I-100 for 150 Weeks

100	—	130	140
80	90	110	90
—	80	80	150
70	30	—	120
40	—	80	70
100	100	60	—
60	—	90	70
80	70	40	50
50	100	70	110
100	50	110	60
60	80	50	70
70	—	80	80
30	60	120	40
40	90	30	70
90	20	100	120
—	70	70	60
100	100	30	—
50	40	40	110
140	60	70	60
70	30	50	70
90	100	80	140
60	70	120	80
90	110	50	—
40	50	80	—
—	80	130	—
70	90	70	50
110	40	60	50
60	70	80	80
70	100	60	70
100	60	100	100
70	90	—	80
60	80	80	90
100	—	70	80
70	80	120	110
100	70	70	60
80	100	110	90
90	60	80	70
130	70	90	120
70	110	140	—
—	80	70	130
80	90	99	90
130	70	—	—

Cost of the item *1–100* is ₹ 100 and inventory carrying cost is 20% of the average value of the item per year. Mr. Gupta has also estimated that in case the part is demanded and it is not available, the shortage cost amounts to ₹ 10 per part per week. The normal lead time for the procurement of this part is one week and ordering cost is ₹ 15. Mr. Gupta wants to have a proper inventory policy for this item. By using scientific approach, how will you like to advise him on a proper inventory policy.

33:12. CASE PROBLEM IN INVESTMENT ANALYSIS

Long Range Plan in Investment Analysis

Directors of a pubic limited company propose to install a central heating system in an office building. As part of the company's long range plan the office building will be sold on 31st March 2024 and Directors believe that whichever system is installed, will add some ₹ 1,00,000 to the resale value at that time.

Three systems, namely gas, oil and solid fuel are regarded as feasible. The Directors estimate that the costs of installing and running the 3 systems are as follows:

(i) *Installation Cost* (Payable on 1st April, 2022)

	Gas	*Oil*	*Solid fuel*
Cost :	₹ 1,70,000	₹ 1,50,000	₹ 1,40,000

(ii) *Annual Fuel Cost* (Payable at the end of each year). Annual fuel costs depend on the rate of increase of fuel prices. At the prices expected to exist during 2022, annual fuel cost would be as given below:

Prices	*Gas*	*Oil*	*Solid fuel*
Severe weather	₹ 40,000	₹ 53,000	₹ 45,000
Mild weather	₹ 24,000	₹ 37,000	₹ 36,000

Each year there is a 70% chance of severe weather and a 30% chance of mild weather. The chance of a particular weather in any one year is independent of the whether in other years. Fuel prices during 2023 and 2024 are expected to increase at either 15% per annum (Prob. 0.4) or 25% per annum (Prob. 0.6). At whichever rate price increases obtain in 2023 will be repeated in 2024.

(iii) *Maintenance Costs* (Payable at the end of year in which they occur)

	Gas (per annum)	*Oil (per annum)*	*Solid fuel (in 2023)*
Maintenance cost	₹ 2,500	₹ 2,000	₹ 10,000

All maintenance costs are fixed by contract when system installed.

The Directors feel that the systems are equivalent for heating purposes. The cost of capital is 20% per annum in money terms.

You are required to:

(a) Prepare calculations showing which central heating system would be installed assuming that the decision will be based on expected present value.

(b) Brief notes on additional factors that may affect the decision.

(c) Discuss the limitations of using expected value as criterion for making investment decisions under uncertainty.

		Severe weather		*Mild weather*		*Expected cost*
Gas	:	(₹ 40,000 × 0.70)	+	(₹ 24,000 × 0.30)	=	₹ 35,200
Oil	:	(₹ 53,000 × 0.70)	+	(₹ 37,000 × 0.30)	=	₹ 48,200
Solid fuel	:	(₹ 45,000 × 0.70)	+	(₹ 36,000 × 0.30)	=	₹ 42,300

Fuel cost expected to increase at the expected rate of

$$(15\% \times 0.4) + (25\% \times 0.60) = 21\%.$$

(Cash flow ₹ '000)

Year	*0*	*1*	*2*	*3*
Gas Installation:				
Fuel cost (increase @ 21%):	170	35.2	42.59	51.54
Moist:		2.5	2.5	2.5
Net cash flow:	170	37.7	45.09	54.04
Discount factor:		0.833	0.694	0.574
NPV = ₹ 2,63,986				
Oil:				
Fuel cost (increase @ 21%):	150	48.2	58.32	70.57
Moist:		2.0	2.0	2.0
Net cash flow:	50	50.2	60.32	72.57
NPV = ₹ 2,75,690				
Solid Fuel Installation:				
Fuel cost:	140	42.3	51.18	61.93
Moist:			10	
Net cash flow:	140	42.3	61.18	61.93
NPV = ₹ 2,53,552				

(*d*) PV of the building enhancement = ₹ 1,00,000 × 0.579 = ₹ 57,900

Net Cost of Solid Fuel = ₹ 2,53,552 – ₹ 57,900 = ₹ 1,95,652

Thus, installation of solid fuel system will only be worth while, if present value of the existing heating cost is greater than ₹ 1,95,652.

Other consideration relating to comfort, health and safety regulation are excluded.

33:13. A CASE PROBLEM IN MANAGEMENT SCIENCE APPLICATION

Games Ltd.

Background

Games Ltd. manufactures and sells a variety of board games and toys. The board games use dice, counters, cards, paper money, etc., and are based on holidays, football, careers, horse racing, murder mysteries, etc. The toys are classified as *soft* or *hard*. the soft toys are made from non-toxic, flame-resistant polyester fibre, *e.g.*, teddy bears, fur animals, etc. The hard toys are made from non-toxic plastic and wood. Games Ltd.'s main customers are large retailers and wholesalers. It employs over 100 people.

The company has four main board games — *W*, *X*, *Y* and *Z*. Their quarterly sales since 2018, given in hundreds of units, are shown below:

	2018				*2019*				*2020*				*2021*
	Q_1	Q_2	Q_3	Q_4	Q_1	Q_2	Q_3	Q_4	Q_1	Q_2	Q_3	Q_4	Q_1
W	52	47	50	51	48	49	52	51	46	51	52	51	50
X	75	80	110	175	92	96	122	210	111	116	164	259	135
Y	73	74	100	153	77	76	98	149	73	74	102	151	76
Z	150	143	181	256	123	116	148	213	100	95	120	175	80

Previous research has established that the sales of *W*, *X*, *Y* and *Z* are all subject to the same seasonal pattern as follows:

Quarter :	*1*	2	3	4
Seasonality :	–25%	–25%	0	+50%

All four games are currently sold to the trade for ₹ 200 each. Games Ltd.'s management accountant has estimated that the contribution to profit per unit of *W*, *X*, *Y* and *Z* is ₹ 30, ₹ 20, ₹ 40 and ₹ 40 respectively.

Production of Board Games

Two of the games, *X* and *Y*, are manufactured in the same way on the same equipment. There are three stages in the manufacture: boarding, printing, and packaging. the manufacturing times required for each game (in minutes), and the total time the equipment is available each week (in hours) are as follows:

	Boarding	*Printing*	*Packaging*
Game *X*	2 minutes	3 minutes	1 minutes
Game *Y*	2 minutes	6 minutes	2 minutes
Total time available each week	70 hour	120 hours	50 hours

The current stock and demand position is such that, for the next few weeks, at least twice as many of game *X* must be produced as game *Y*. Company policy is to maximize contribution to profit.

Stock Costs of Board Game W

Special boards for *W* are bought from a local supplier. The management accountant estimates that the order costs (administration and delivery, etc.) of placing any order are ₹ 5,000, regardless of its size. The holding costs (storage, etc.) of keeping one game in stock for a year are 10% of its ex-factory price to the trade. The supplier is willing to deliver the order on same-day, *i.e.*, with no lead time. At present, Games Ltd., adopts a policy of ordering a constant amount of boards at regular intervals, although the supplier could provide up to 160 boards a day, if necessary. Games Ltd., trades for 250 days a year. (The proportion of boards scrapped during production is negligible.)

Packaging of Toys

Many of the toys produced are of high-quality and need to be packed with care. Management is concerned about the amount of time which this takes. For several weeks, quality control staff have been engaged in "time and motion" studies, and analysing work practices with the operatives directly involved. The result of random sampling of one operation is given below:

No. of checks of one operation within packaging	*Time taken (seconds)*		*Packaging done correctly*	
	Mean	*Standard deviation*	*Yes*	*No*
400	95	40	324	76

Analysis of Customers for Toys

Games Ltd. has just started to analyse its customers for toys, with a view to reviewing its pricing policy for large orders and/or type of customer. A scrutiny of invoices for the last two quarters, which is thought by the Marketing Department to be representative, reveals the following table, in which customers are classified by order size:

Order size	*< ₹ 5,000*	*₹ 5,000 < ₹ 25,000*	*₹ 25,000 < ₹ 50,000*	*Over ₹ 50,000*
Retailers	10	50	90	150
Wholesalers	50	10	10	30

Modernization Plans

The Board is working on its next medium-term and long-term plans. To meet increased demand, the Board is considering the introduction of semi-automatic or fully automatic equipment for soft toys that are currently made by hand. At present annual cash outflows on toys are about ₹ 18,50,000 and annual cash inflows are about ₹ 22,50,000, a situation they cannot foresee changing. The management accountant and the marketing manager have jointly provided the following estimates of the future likely cash flows resulting from the use of new equipment, given the scenario the Board specified. All equipment would need replacing in 2028. The cost of capital is 10%.

Projected Cash Flows

	2021	*2022*	*2023*	*2024*	*2025*	*2026*
Semi-automatic	– 50	0	+60	+100	+100	+100
Fully automatic	–150	–50	+50	+150	+200	+200

Games Ltd., is currently repaying a fixed interest rate of 10% p.a., ₹ 20,00,000 mortgage by annual (year-end) amounts, which has just over four years to run. (This has already been taken into account in the estimates above.)

Answer the following questions :

1. (*a*) Calculate for the special boards for game *W*: (*i*) the economic order quantity (*EOQ*) and order frequency; (*ii*) the economic batch quantity (*EBQ*) and order frequency; (*iii*) the total annual stock costs for the *EOQ* and *EBQ*.
   (*b*) Explain the importance of your answers for management.
2. (*a*) Determine the optimum weekly contribution to profit from games *X* and *Y* by formulating, and graphically solving, the linear programming production model.
   (*b*) Explain your solution to (*a*) above for management, including recommendations for optimal product mix and profit.
3. (*a*) Evaluate the *three* options in Games Ltd.'s modernization plans: (*i*) no change, (*ii*) semi-automatic equipment, and (*iii*) fully automatic equipment, using net present value criteria, and recommend with reasons and comments, a course of action.
   (*b*) Calculate how much is outstanding now on Games Ltd.'s ₹ 20,00,000 mortgage.
   (*c*) Explain how network analysis could be used by Games Ltd., if it were to introduce semi-automatic production systems for the manufacture of soft toys.

APPENDEX A

Prerequisite — Linear Algebra

"I cherish both faces of Mathematics, the pure as a beautiful retreat from reality and the applied as an ardent hope for life"

A:1. MATRICES

Definition (*Matrix*). A system of mn numbers arranged in the form of an ordered set of m rows, each row consisting of an ordered set of n numbers and are enclosed by parentheses [ ] or ( ) or ‖ ‖ is called an $m \times n$ matrix (to be read as 'm' by 'n' matrix).

A matrix, in general, is denoted by a capital letter such as **A**, **B**, **C**, etc., while the numbers forming a matrix are called its elements and are denoted by small letters a, b, c, etc.

Thus, if a_{ij} stands for an element of a matrix which is in its ith row at j^{th} position, the matrix of order $m \times n$ can be written as

$$\mathbf{A} = \begin{bmatrix} a_{11} & a_{12} & \cdots & a_{1j} & \cdots & a_{1n} \\ a_{21} & a_{22} & \cdots & a_{2j} & \cdots & a_{2n} \\ \vdots & \vdots & \vdots & \vdots & \vdots & \vdots \\ a_{i1} & a_{i2} & \cdots & a_{ij} & \cdots & a_{in} \\ \vdots & \vdots & \vdots & \vdots & \vdots & \vdots \\ a_{m1} & a_{m2} & \cdots & a_{mj} & \cdots & a_{mn} \end{bmatrix} = (a_{ij}),$$

where, $i = 1, 2, ..., m$ *and* $j = 1, 2, ..., n$.

The numbers a_{ij} ($i = 1, 2, ..., m; j = 1, 2, ..., n$) in matrix **A** are called its ***elements (or entries)***.

Some Special Types of Matrices

(*i*) A matrix of order $n \times n$ is called a ***square matrix*** of order n.

(*ii*) An $1 \times n$ matrix is called a ***row vector***.

(*iii*) An $m \times 1$ matrix is called a ***column vector***.

(*iv*) A matrix whose every element is zero is called a ***null matrix*** and is denoted by **0**.

(*v*) A square matrix, in which all principle ***diagonal elements*** are unity, while all other elements are zeros, is called an ***identity matrix***. It is generally denoted by **I**. For example, a 3 × 3 identity matrix is given by

$$\mathbf{I}_3 = \begin{pmatrix} 1 & 0 & 0 \\ 0 & 1 & 0 \\ 0 & 0 & 1 \end{pmatrix}.$$

(*vi*) The ***transpose*** of a matrix $\mathbf{A} = (a_{ij})$ denoted by $\mathbf{A}^T$, is such that the $(i, j)^{\text{th}}$ element of $\mathbf{A}^T$ is the same as the $(j, i)^{\text{th}}$ element of **A**. For example, if

$$\mathbf{A} = \begin{bmatrix} 2 & 4 \\ 1 & 3 \\ 6 & 5 \end{bmatrix}, \quad \text{then} \quad \mathbf{A}^T = \begin{bmatrix} 2 & 1 & 6 \\ 4 & 3 & 5 \end{bmatrix}.$$

(*vii*) Two matrices $\mathbf{A} = (a_{ij})$ and $\mathbf{B} = (b_{ij})$ are said to be ***equal***, if and only if, they are of the same order and $a_{ij} = b_{ij}$ for all i and j.

A:2. PARTITIONED MATRICES

Matrix **A** can always be put in a more compact form, as a matrix whose elements are submatrices of **A**. This is achieved by drawing imaginary, horizontal and vertical lines to divide the elements of **A** into a number of blocks.

Thus,

$$\mathbf{A} = \left[\begin{array}{cc:c} a_{11} & a_{12} & a_{13} \\ a_{21} & a_{22} & a_{23} \\ \hdashline a_{31} & a_{32} & a_{33} \end{array}\right] = \begin{pmatrix} \mathbf{A}_{11} & \mathbf{A}_{12} \\ \mathbf{A}_{21} & \mathbf{A}_{22} \end{pmatrix}$$

where, $\mathbf{A}_{11} = \begin{pmatrix} a_{11} & a_{12} \\ a_{21} & a_{22} \end{pmatrix}$, $\mathbf{A}_{12} = \begin{pmatrix} a_{13} \\ a_{23} \end{pmatrix}$, $\mathbf{A}_{21} = (a_{31} \quad a_{32})$, $\mathbf{A}_{22} = (a_{33})$.

Matrix **A** so represented is called a ***partitioned matrix***.

Addition of Partitioned Matrices

Let $\mathbf{A} = (a_{ij})$ and $\mathbf{B} = (b_{ij})$ be two $m \times n$ matrices and suppose that they have been partitioned in the same way (***identically partitioned***), so that

$$\mathbf{A} = \begin{pmatrix} \mathbf{A}_{11} & \mathbf{A}_{12} \\ \mathbf{A}_{21} & \mathbf{A}_{22} \end{pmatrix} \quad \text{and} \quad \mathbf{B} = \begin{pmatrix} \mathbf{B}_{11} & \mathbf{B}_{12} \\ \mathbf{B}_{21} & \mathbf{B}_{22} \end{pmatrix}$$

where, $\mathbf{A}_{ij}$, $\mathbf{B}_{ij}$ are like matrices.

Then, it is easy to verify that

$$\mathbf{A} + \mathbf{B} = \begin{pmatrix} \mathbf{A}_{11} + \mathbf{B}_{11} & \mathbf{A}_{12} + \mathbf{B}_{12} \\ \mathbf{A}_{21} + \mathbf{B}_{21} & \mathbf{A}_{22} + \mathbf{B}_{22} \end{pmatrix}$$

Product of Partitioned Matrices

Let **A**, **B** be matrices of order $m \times n$ and $n \times p$ respectively, so that **AB** is defined. Suppose, **A**, **B** have been partitioned in the same way. Then,

$$\mathbf{A} = \begin{pmatrix} \mathbf{A}_{11} & \mathbf{A}_{12} \\ \mathbf{A}_{21} & \mathbf{A}_{22} \end{pmatrix} \quad \text{and} \quad \mathbf{B} = \begin{pmatrix} \mathbf{B}_{11} & \mathbf{B}_{12} \\ \mathbf{B}_{21} & \mathbf{B}_{22} \end{pmatrix},$$

where, $\mathbf{A}_{ij}$, and $\mathbf{B}_{ij}$ are like matrices. Then, it may be easily verified that

$$\mathbf{AB} = \begin{pmatrix} \mathbf{A}_{11}\mathbf{B}_{11} + \mathbf{A}_{12}\mathbf{B}_{21} & \mathbf{A}_{11}\mathbf{B}_{12} + \mathbf{A}_{12}\mathbf{B}_{22} \\ \mathbf{A}_{21}\mathbf{B}_{11} + \mathbf{A}_{22}\mathbf{B}_{21} & \mathbf{A}_{21}\mathbf{B}_{12} + \mathbf{A}_{22}\mathbf{B}_{22} \end{pmatrix}$$

A:3. DETERMINANT OF A SQUARE MATRIX

From the elementary course in Algebra, the reader is familiar with determinants of second and third order. For example:

$$\begin{vmatrix} a_{11} & a_{12} \\ a_{21} & a_{22} \end{vmatrix} = a_{11}a_{22} - a_{12}a_{21}.$$

$$\begin{vmatrix} a_{11} & a_{12} & a_{13} \\ a_{21} & a_{22} & a_{23} \\ a_{31} & a_{32} & a_{33} \end{vmatrix} = a_{11}\begin{vmatrix} a_{22} & a_{23} \\ a_{32} & a_{33} \end{vmatrix} - a_{12}\begin{vmatrix} a_{21} & a_{23} \\ a_{31} & a_{33} \end{vmatrix} + a_{13}\begin{vmatrix} a_{21} & a_{22} \\ a_{31} & a_{32} \end{vmatrix}$$

We proceed to define det **A** for any square matrix by induction on n, the order of **A**.

Definition (*Determinant*). *Let* **M** *be the set of all square matrices and matrix* **A** ∈ **M**. *Then* ***determinant*** *of* **A**, *to be denoted by* **det A**, *is the number found as follows:*

(*i*) *If* $\mathbf{A} = (a)$ *is* 1×1 *matrix, then* $\det \mathbf{A} = a$,

(*ii*) *For* $n > 1$, *if* $\mathbf{A} = (a_{ij})$ *is an* $n \times n$ *matrix and,* $\mathbf{M}_{ij}$ *denotes the* $(n-1)$ *rowed sub-matrix of* **A**, *obtained by striking off the* i^{th} *row and* j^{th} *column of* **A**, *then*

$$\det \mathbf{A} = \sum_{j=1}^{n} (-1)^{i+j} a_{ij} \det \mathbf{M}_{ij} \qquad (i \text{ fixed})$$

We shall also denote det **A** by the more convenient symbol $|\mathbf{A}|$.

For each pair, $(i, j)^{\text{th}}$ number

$$\mathbf{A}_{ij} = (-1)^{i+j} \det \mathbf{M}_{ij}$$

is called the ***co-factor*** of a_{ij} in $|\mathbf{A}|$. Thus, we may also write as

$$|\mathbf{A}| = \det \mathbf{A} = \sum_{j=1}^{n} a_{ij} \mathbf{A}_{ij} = \sum_{j=1}^{n} (-1)^{i+j} a_{ij} \mathbf{M}_{ij} \qquad (i \text{ fixed})$$

Properties of the Determinant

The following properties follow immediately from the definition:

(*i*) The value of the determinant is not changed, if its rows and columns are interchanged.

(*ii*) The value of the determinant is zero, if every element of a row or column is zero.

(*iii*) The value of the determinant is zero, if any two columns (or rows) of the determinant are identical.

(*iv*) If **B** is obtained from a square matrix **A** by an interchange of two columns (or rows), then $|\mathbf{B}| = -|\mathbf{A}|$.

(*v*) If every element of a column (or row) of a determinant is multiplied by a scalar λ, the value of the determinant is multiplied by λ.

(*vi*) The value of the determinant remains the same, if scalar λ times a row (or column) is added to another row (or column).

(*vii*) For any two square matrices **A**, **B** of the same order

$$|\mathbf{AB}| = |\mathbf{A}| \cdot |\mathbf{B}|.$$

A:4. INVERSE OF A SQUARE MATRIX

Definition (*Inverse*). *Let* **A** *be a square matrix, and if there exists a matrix* **B**, *such that*

$$\mathbf{AB} = \mathbf{I} = \mathbf{BA};$$

then **B** *is called an* ***inverse*** *of* **A** *and is denoted by* $\mathbf{A}^{-1}$.

It can be easily shown that the ***inverse of a square matrix, if it exists, is unique.***

From the relation $\mathbf{AA}^{-1} = \mathbf{I}$, it follows that

$$1 = |\mathbf{I}| = |\mathbf{AA}^{-1}| = |\mathbf{A}| . |\mathbf{A}^{-1}|$$

which shows that if **A** has an inverse, then necessarily $|\mathbf{A}| \neq 0$.

A square matrix **A** is called ***singular*** if $|\mathbf{A}| = 0$; and if $|\mathbf{A}| \neq 0$, it is called ***non-singular***. Thus, we shall talk of the inverse of a non-singular matrix only.

Elementary Transformations of a Matrix

An elementary row (column) transformation of a matrix is one of the following types:

Type 1. Interchange of any two rows (columns).

Type 2. Multiplying the elements of a row (column) by a non-zero scalar.

Type 3. Adding to any row (column), a scalar multiple of another row (column).

Computation of the Inverse of a Square Matrix by Elementary Transformations

Let **A** be a non-singular matrix, so that **A** has an inverse. We write

$$\mathbf{I\,A} = \mathbf{A}.$$

By successive elementary row transformations on the pre-factor of the product **I A** and corresponding transformations on the matrix **A** on the right, we reduce the above matrix equation to the form

$$\mathbf{B\,A} = \mathbf{I}$$

Then clearly, $\mathbf{B} = \mathbf{A}^{-1}$.

Similarly, if we write **A I** = **A** then using successive elementary column transformations on the post-factor of **A I**, and corresponding transformations on the matrix **A** on the right, we reduce the above matrix equation to the form **A C** = **I**. This implies that $\mathbf{C} = \mathbf{A}^{-1}$.

A:5. SOLUTION OF LINEAR EQUATIONS

Let there be a system of m linear equations in n unknowns, $(m < n)$:

$$\begin{aligned} a_{11}x_1 + a_{12}x_2 + \ldots + a_{1n}x_n &= b_1 \\ a_{21}x_1 + a_{22}x_2 + \ldots + a_{2n}x_n &= b_2 \\ \vdots \qquad\quad \vdots \qquad\quad & \vdots \quad\ \vdots \\ a_{m1}x_1 + a_{m2}x_2 + \ldots + a_{mn}x_n &= b_m \end{aligned}$$

The system can be put in a more convenient form, by setting

$$\mathbf{A} = \begin{bmatrix} a_{11} & a_{12} & \cdots & a_{1n} \\ a_{21} & a_{22} & \cdots & a_{2n} \\ \vdots & \vdots & \vdots & \vdots \\ a_{m1} & a_{m2} & \cdots & a_{mn} \end{bmatrix}, \quad \mathbf{X} = \begin{bmatrix} x_1 \\ x_2 \\ \vdots \\ x_n \end{bmatrix}, \quad \mathbf{b} = \begin{bmatrix} b_1 \\ b_2 \\ \vdots \\ b_m \end{bmatrix}$$

Then, the system can be represented by the single matrix equation

$$\mathbf{A\,X} = \mathbf{b},$$

where, **A** is an $m \times n$ matrix, **X** an $n \times 1$ matrix and b an $m \times 1$ matrix.

This system of linear equations is called ***homogeneous***, if **b** = **0** and ***non-homogeneous*** otherwise.

A set of values x_j $(j = 1, 2, ..., n)$ which satisfy the given set of equations is called a ***solution*** of the system of equations.

The coefficient matrix obtained for the system of non-homogeneous linear equations in the form (**A** : **b**) is called an ***augmented matrix***.

If $m = n$, we have a system of n equations in n unknowns. Here, if $|\mathbf{A}| \neq 0$, then we have a unique solution

$$\mathbf{X} = \mathbf{A}^{-1}\,\mathbf{b}.$$

In particular, if **b** = **0**, and $|\mathbf{A}| \neq 0$, then the only solution possible is the trivial one, that is

$$x_1 = x_2 = \ldots = x_n = 0.$$

A:6. VECTORS

Row or column matrices are often referred to as ***vectors***. A row matrix is called a ***row vector*** and a column matrix is called a ***column vector***. The vectors will generally be symbolized by lower case bold face letters such as **a, b, x, y**, etc.

Definition 1 (*Vector*). *An **n-vector** **a** is an ordered n-tuple of real numbers*

$$\mathbf{a} = (a_1, a_2, ..., a_n)$$

The members a_i ($i = 1, 2, ..., n$) are called the components of **a**.

Definition 2 (*Equality of two n-vectors*). *The two n-vectors $\mathbf{a} = (a_1, a_2, ..., a_n)$ and $\mathbf{b} = (b_1, b_2, ..., b_n)$ are said to be **equal**, if $a_i = b_i$, for all $i = 1, 2, ..., n$.*

Note that if **a** = **b** then **b** = **a**.

Definition 3 (*Sum of Two n-vectors*). *The **sum** of two n-vectors $\mathbf{a} = (a_1, a_2, ..., a_n)$ and $\mathbf{b} = (b_1, b_2, ... b_n)$, written as $\mathbf{a} + \mathbf{b}$, is defined to be the n-vector*

$$\mathbf{a} + \mathbf{b} = (a_1 + b_1, a_2 + b_2, ..., a_n + b_n).$$

The subtraction of **b** *from* **a** *denoted by* **a** – **b** *is given by*

$$\mathbf{a} - \mathbf{b} = (a_1 - b_1, a_2 - b_2, ..., a_n - b_n).$$

Definition 4 (*Multiplication of a Vector by a Scalar*). *The **product** of an n-vector $\mathbf{a} = (a_1, a_2, ..., a_n)$ by a scalar λ written as $\lambda\mathbf{a}$, is defined to be the n-vector*

$$\lambda\mathbf{a} = \lambda(a_1, a_2, ..., a_n)$$
$$= (\lambda a_1, \lambda a_2, ..., \lambda a_n).$$

Definition 5 (*Row and Column Vectors*). *An n-vector*

$$\mathbf{a} = (a_1, a_2, ..., a_n)$$

*is called a **row vector**. Corresponding to each row vector $\mathbf{a} = (a_1, a_2, ..., a_n)$, there corresponds another n-vector written as*

$$\mathbf{a}^T = \begin{pmatrix} a_1 \\ a_2 \\ \vdots \\ a_n \end{pmatrix} \quad \text{or} \quad \mathbf{a} = [a_1, a_2, ..., a_n],$$

*which we shall call a **column vector**.*

Remark. While row vectors and column vectors are both ***n*-vectors**, the two types are distinct. A row (column) vector can only be added to a row (column) vector.

Definition 6 (*Inner Product*). *The **Inner Product** of two n-vectors **a** and **b** is the scalar*

$$a_1b_1 + a_2b_2 + ... + a_nb_n = \sum_{i=1}^{n} a_i b_i$$

and is denoted by $\mathbf{a} \cdot \mathbf{b}^T$ or $\mathbf{b} \cdot \mathbf{a}^T$.

Definition 7 (*Length of a Vector*). *The **length** or **magnitude** of a vector **a**, written as $|\mathbf{a}|$ is defined by*

$$|\mathbf{a}| = \left[\mathbf{a} \cdot \mathbf{a}^T\right]^{\frac{1}{2}} = \left[\sum_{i=1}^{n} a_i^2\right]^{\frac{1}{2}} = \sqrt{a_1^2 + a_2^2 + ... + a_n^2}.$$

Note that for any scalar λ, $|\lambda\mathbf{a}| = |\lambda| \cdot |\mathbf{a}|$.

Definition 8 (*Null Vector*). *The **null vector** is a vector whose all components are zero. It is usually denoted by* **0**.

Definition 9 (*Unit Vector*). *The* ***unit vector*** $\mathbf{e}_i$ *is a vector with its* i^{th} *component unity and all other components zero.*

For n-component vectors, there are n unit vectors

$$\mathbf{e}_1 = (1, 0, 0, ..., 0), \quad \mathbf{e}_2 = (0, 1, 0, ..., 0), \quad ..., \quad \mathbf{e}_n = (0, 0, ..., 1).$$

Definition 10 (*Euclidean Space*). *An* ***n-dimensional euclidean space*** *is defined to be the collection of all n-vectors.*

The n-dimensional euclidean space will be denoted by $\mathbf{R}^n$. The vectors that will occur in our study, shall always belong to some euclidean space.

Note. The term 'point' and 'vector' in space $\mathbf{R}^n$, will be used without distinguishing between them.

A:7. VECTOR INEQUALITIES

Definition 1 ($\geq$ *Order*). *Let* $\mathbf{a} = (a_1, a_2, ..., a_n)$ *and* $\mathbf{b} = (b_1, b_2, ..., b_n)$, *then* $\mathbf{a}$ *is said to be* ***greater than or equal to*** $\mathbf{b}$, *if and only if,* $a_i \geq b_i$ *for* $i = 1, 2, ..., n$. *This is denoted by writing* $\mathbf{a} \geq \mathbf{b}$.

Definition 2 ($\leq$ *order*). *Vector* $\mathbf{a} = (a_1, a_2, ..., a_n)$ *is said to be* ***less than or equal to*** $\mathbf{b} = (b_1, b_2, ..., b_n)$, *if and only if,* $\mathbf{b} \geq \mathbf{a}$. *This is denoted by writing* $\mathbf{a} \leq \mathbf{b}$.

Remark. Two vectors **a** and **b** are ***order comparable***, if and only if, they have an equal number of components.

Illustration. Let $\mathbf{a} = (1, 2, 3)$ and $\mathbf{b} = (0, 1, -1)$. Clearly, $\mathbf{a} \geq \mathbf{b}$, or equivalently $\mathbf{b} \leq \mathbf{a}$.

The ***vector inequality*** $\mathbf{x} \geq \mathbf{0}$ simply indicates that **x** is a vector whose each component is non-negative. Such vector inequalities shall frequently occur in our study.

A:8. LINEAR COMBINATION OF VECTORS

Given two vectors **a** and **b**, we can always perform operations of addition and scalar multiplication to obtain vectors of the form $\mathbf{a} + \mathbf{b}$, $\mathbf{a} - \mathbf{b}$, $5\mathbf{a} + 3\mathbf{b}$, $2\mathbf{a} - 7\mathbf{b}$, etc. Such vectors are called ***linear combinations*** of **a** and **b**. In general, for any scalars λ and μ, the vector $\lambda\,\mathbf{a} + \mu\,\mathbf{b}$ is a linear combination of **a** and **b**.

Thus, if $\{\mathbf{a}_1, \mathbf{a}_2, ..., \mathbf{a}_k\}$ is a set of k vectors from $\mathbf{R}^n$ and $(\lambda_1, \lambda_2, ..., \lambda_k)$ is a set of k scalars, then the vector

$$\lambda_1\,\mathbf{a}_1 + \lambda_2\,\mathbf{a}_2 + ... + \lambda_k\,\mathbf{a}_k$$

is known as a linear combination of the given set of vectors $\mathbf{a}_1, \mathbf{a}_2, ..., \mathbf{a}_k$.

Definition 1 (*Linear Dependence and Independence of Vectors*). *A set of vectors* $\mathbf{a}_1, \mathbf{a}_2, ..., \mathbf{a}_k$ *of* $\mathbf{R}^n$, *is said to be* ***linearly dependent***, *if there exist scalars* $\lambda_1, \lambda_2, ..., \lambda_k$ *not all zero, such that* $\lambda_1\mathbf{a}_1 + \lambda_2\mathbf{a}_2 + ... + \lambda_k\mathbf{a}_k = \mathbf{0}$.

When, it is not possible to have the above equality without all λ_i's being zero, the given set of vectors $\mathbf{a}_1, \mathbf{a}_2, ..., \mathbf{a}_k$ is said to be ***linearly independent***.

Illustration 1. The vectors $\mathbf{a}_1 = (1, 2)$ and $\mathbf{a}_2 = (2, 4)$ are linearly dependent, since there exist $\lambda_1 = 2$ and $\lambda_2 = -1$ for which $\lambda_1\mathbf{a}_1 + \lambda_2\,\mathbf{a}_2 = 0$.

Illustration 2. The vectors $\mathbf{e}_1 = (1, 0, 0)$, $\mathbf{e}_2$ $(0, 1, 0)$ and $\mathbf{e}_3 = (0, 0, 1)$ are linearly independent since $\lambda_1\,\mathbf{e}_1 + \lambda_2\,\mathbf{e}_2 + \lambda_3\,\mathbf{e}_3 = \mathbf{0}$.

$$\Rightarrow \quad \lambda_1(1, 0, 0) + \lambda_2(0, 1, 0) + \lambda_3(0, 0, 1) = (0, 0, 0)$$

$$\Rightarrow \quad (\lambda_1, \lambda_2, \lambda_3) = (0, 0, 0)$$

$$\Rightarrow \quad \lambda_1 = 0, \ \lambda_2 = 0 \ \text{and} \ \lambda_3 = 0.$$

Note. If a set of vectors is linearly independent, then, any subset of it is also linearly independent. If a set of vectors is linearly dependent, then a superset of it is also linearly dependent.

Definition 2 (*Basis Set*). *A linearly independent set of vectors* $\mathbf{a}_1, \mathbf{a}_2, \ldots, \mathbf{a}_k$ *of* $\mathbf{R}^n$, *having the property that any vector of* $\mathbf{R}^n$ *can be expressed as linear combination of these vectors, is called a* ***basis set*** *for* $\mathbf{R}^n$.

Theorem A-I. *The set of n unit vector* $\mathbf{e}_i$ $(i = 1, 2, \ldots, n)$ *forms a basis for* $\mathbf{R}^n$.

Proof. We shall show that the unit vectors $\mathbf{e}_1, \mathbf{e}_2, \ldots, \mathbf{e}_n$ of $\mathbf{R}^n$ are linearly independent, and that any vector of $\mathbf{R}^n$ can be expressed as a linear combination of these unit vectors.

Consider,

$$\sum_{i=1}^{n} \lambda_i \mathbf{e}_i = \mathbf{0}, \quad \text{where } \lambda_i \ (i = 1, 2, \ldots, n) \text{ are scalars}$$

or
$$\lambda_1 (1, 0, 0, \ldots, 0) + \lambda_2 (0, 1, 0, \ldots, 0) + \ldots + \lambda_n (0, 0, \ldots, 1) = 0$$

or
$$(\lambda_1, \lambda_2, \lambda_3, \ldots, \lambda_n) = (0, 0, 0, \ldots, 0).$$

This implies that $\lambda_i = 0$ for all $i = 1, 2, \ldots, n$.

Thus, the n unit vectors $\mathbf{e}_i$; $i = 1, 2, \ldots, n$ are linearly independent.

Let $\mathbf{x}$ be any other vector of $\mathbf{R}^n$ different from the unit vectors $\mathbf{e}_i$ $(i = 1, 2, \ldots, n)$. This can be written as

$$\mathbf{x} = (x_1, x_2, \ldots, x_k, \ldots, x_n) = x_1 \mathbf{e}_1 + x_2 \mathbf{e}_2 + \ldots + x_n \mathbf{e}_n = \sum x_i \mathbf{e}_i,$$

since, $\mathbf{e}_i = (0, 0, \ldots, 0, 1, 0, \ldots, 0)$ where unity is at the ith place.

This happens to be a linear combination of the n unit vectors $\mathbf{e}_i$ $(i = 1, 2, \ldots, n)$. Since, $\mathbf{x}$ is any vector of $\mathbf{R}^n$ different from the n unit vectors, therefore, every vector of $\mathbf{R}^n$ can be expressed as a linear combination of these unit vectors.

Hence $\{\mathbf{e}_i : i = 1, 2, \ldots, n\}$ forms a basis for $\mathbf{R}^n$.

Definition 3 (*Standard Basis*). *The set of vectors* $\mathbf{e}_i$, $i = 1, 2, \ldots, n$ *is called the* ***standard basis*** *for* $\mathbf{R}^n$.

Remark. We may state without proof the important fact, that in any basis of $\mathbf{R}^n$, there are exactly n vectors.

Theorem A-2. *Let* $\{\mathbf{a}_1, \mathbf{a}_2, \ldots, \mathbf{a}_n\}$ *be any basis set for* $\mathbf{R}^n$ *and* $\mathbf{b} \neq \mathbf{0}$ *is any other vector in* $\mathbf{R}^n$. *If* $\mathbf{b}$ *is expressible as:*

$$\mathbf{b} = \sum_{i=1}^{n} \delta_i \mathbf{a}_i, \quad \text{where } \delta_i \neq 0 \text{ for some } i,$$

then the vector $\mathbf{a}_k$ *(for which* $\delta_k \neq 0$*) can be replaced by the vector* $\mathbf{b}$. *The new set of n vectors* $\mathbf{a}_1, \mathbf{a}_2, \ldots, \mathbf{a}_{k-1}, \mathbf{a}_{k+1}, \mathbf{a}_{k+2}, \ldots, \mathbf{a}_n, \mathbf{b}$ *is also a basis set for* $\mathbf{R}^n$.

Proof. Let $\delta_k \neq 0$. Consider the new set of vectors $\mathbf{a}_1, \mathbf{a}_2, \ldots, \mathbf{a}_{k-1}, \mathbf{a}_{k+1}, \ldots, \mathbf{a}_n, \mathbf{b}$.

To prove that this set is a basis for $\mathbf{R}^n$, we must show

(*i*) $\mathbf{a}_1, \mathbf{a}_2, \ldots, \mathbf{a}_{k-1}, \mathbf{a}_{k+1}, \mathbf{a}_n, \mathbf{b}$ are linearly independent, and

(*ii*) any vector of $\mathbf{R}^n$ can be expressed as a linear combination of the vectors $\mathbf{a}_1, \mathbf{a}_2, \ldots, \mathbf{a}_{k-1}, \mathbf{a}_{k+1}, \ldots, \mathbf{a}_n, \mathbf{b}$.

Let us suppose that the set $\{\mathbf{a}_1, \mathbf{a}_2, \ldots, \mathbf{a}_{k-1}, \mathbf{a}_{k+1}, \ldots, \mathbf{a}_n, \mathbf{b}\}$ is not a linearly independent set. Then, the equation

$$\sum_{i=1, i \neq k}^{n} \lambda_i \mathbf{a}_i + \lambda \mathbf{b} = \mathbf{0}$$

will hold even with some non-zero scalars λ_i $(i = 1, 2, \ldots, k-1, k+1, \ldots, n)$ and λ.

Now $\lambda = 0$ implies

$$\sum_{i=1, i \neq k}^{n} \lambda_i \mathbf{a}_i = \mathbf{0} \quad \text{with some } \lambda_i \neq 0.$$

This contradicts the fact that the set $\{\mathbf{a}_1, \mathbf{a}_2, \ldots, \mathbf{a}_{k-1}, \mathbf{a}_{k+1}, \ldots, \mathbf{a}_n\}$ is linearly independent (being a subset of a linearly independent set of vectors).

Thus, $\lambda \neq 0$, and we have

$$\sum_{i=1, i \neq k}^{n} \lambda_i \mathbf{a}_i + \lambda \sum_{i=1}^{n} \delta_i \mathbf{a}_i = \mathbf{0}, \quad \text{since} \quad \mathbf{b} = \sum_{i=1}^{n} \delta_i \mathbf{a}_i$$

or

$$\sum_{i=1, i \neq k}^{n} (\lambda_i + \lambda \delta_i) \mathbf{a}_i + \lambda \delta_k \mathbf{a}_k = \mathbf{0}.$$

Since, $\lambda \neq 0$ and $\delta_k \neq 0$, therefore the set of vectors $\mathbf{a}_1, \mathbf{a}_2, \ldots, \mathbf{a}_n$ is linearly dependent. This contradicts the hypothesis.

Thus, $\{\mathbf{a}_1, \mathbf{a}_2, \ldots \mathbf{a}_{k-1}, \mathbf{a}_{k+1}, \ldots, \mathbf{a}_n, \mathbf{b}\}$ is a linearly independent set.

Now, let $\mathbf{x}$ be any vector in $\mathbf{R}^n$, other than the given basis vectors $\mathbf{a}_1, \mathbf{a}_2, \ldots, \mathbf{a}_n$. Then, we can express $\mathbf{x}$ as

$$\mathbf{x} = \sum_{i=1}^{n} \lambda_i \mathbf{a}_i, \quad \text{where } \lambda_i \; (i = 1, 2, \ldots, n) \text{ are scalars}$$

Now, since

$$\mathbf{b} = \sum_{i=1}^{n} \delta_i \mathbf{a}_i = \sum_{i=1, i \neq k}^{n} \delta_i \mathbf{a}_i + \delta_k \mathbf{a}_k, \quad \delta_k \neq 0$$

we have

$$\mathbf{a}_k = \frac{1}{\delta_k}\left[\mathbf{b} - \sum_{i=1, i \neq k}^{n} \delta_i \mathbf{a}_i\right] = \frac{\mathbf{b}}{\delta_k} - \sum_{i=1, i \neq k}^{n} \frac{\delta_i}{\delta_k} \mathbf{a}_i$$

This gives,

$$\mathbf{x} = \sum_{i=1, i \neq k}^{n} \lambda_i \mathbf{a}_i + \lambda_k \left[\frac{\mathbf{b}}{\delta_k} - \sum_{i=1, i \neq k}^{n} \frac{\delta_i}{\delta_k} \mathbf{a}_i\right]$$

$$= \sum_{i=1, i \neq k}^{n} \left(\lambda_i - \frac{\lambda_k}{\delta_k} \delta_i\right) \mathbf{a}_i + \frac{\lambda_k}{\delta_k} \mathbf{b}.$$

Thus, $\mathbf{x}$ can be expressed as a linear combination of the vectors $\mathbf{a}_1, \mathbf{a}_2, \ldots, \mathbf{a}_{k-1}, \ldots, \mathbf{a}_n, \mathbf{b}$.

Hence, the set of vectors $\{\mathbf{a}_1, \mathbf{a}_2, \ldots, \mathbf{a}_{k-1}, \mathbf{a}_{k+1}, \ldots, \mathbf{a}_n, \mathbf{b}\}$ forms a basis for $\mathbf{R}^n$.

PROBLEMS

A.01. If vectors $\mathbf{a}_1, \mathbf{a}_2, \ldots, \mathbf{a}_k$ constitute a linearly independent set, and vectors $\mathbf{a}_1, \mathbf{a}_2, \ldots, \mathbf{a}_{k+1}$ constitute a linearly dependent set, then show that the vector $\mathbf{a}_{k+1}$, is expressible as a linear combination of the vectors $\mathbf{a}_1, \mathbf{a}_2, \ldots, \mathbf{a}_k$.

A.02. Given the basis vectors

$$\mathrm{e}_1 = (1, 0, 0), \quad \mathrm{e}_2 = (0, 1, 0) \quad \text{and} \quad \mathrm{e}_3 = (0, 0, 1) \quad \text{for } \mathbf{R}^3.$$

Find a vector that can be removed from the basis and be replaced by a vector $\mathbf{b} = (4, 3, 6)$ such that the new set is also a basis set for $\mathbf{R}^3$.

A.03. Do the following vectors form a basis for $\mathbf{R}^3$?

(*a*) $\mathbf{a}_1 = [3, 0, 2]$, $\quad \mathbf{a}_2 = [7, 0, 9]$, $\quad \mathbf{a}_3 = [4, 1, 2]$

(*b*) $\mathbf{a}_1 = [1, 1, 0]$, $\quad \mathbf{a}_2 = [3, 0, 1]$, $\quad \mathbf{a}_3 = [5, 2, 1]$

(*c*) $\mathbf{a}_1 = [1, 5, 7]$, $\quad \mathbf{a}_2 = [4, 0, 6]$, $\quad \mathbf{a}_3 = [1, 0, 0]$.

A:9. HYPERPLANES AND HYPERSPHERES

Definition 1 (*Line Segment in* $\mathbf{R}^n$). *The **line segment** joining two points* $\mathbf{x}_1 = \left(x_1^{(1)}, x_2^{(1)}, \ldots, x_n^{(1)}\right)$, *and* $\mathbf{x}_2 = \left(x_1^{(2)}, x_2^{(2)}, \ldots, x_n^{(2)}\right)$, *is the collection of all points* $\mathbf{x} = (x_1, x_2, \ldots, x_n)$, *where*

$$\mathbf{x} = \lambda \mathbf{x}_1 + (1-\lambda)\mathbf{x}_2, \qquad \text{for all values of } \lambda \text{ such that } 0 \le \lambda \le 1.$$

Definition 2 (*Hyperplane*). *In* $\mathbf{R}^n$, *the set of points* $\mathbf{x} = (x_1, x_2, \ldots, x_n)$ *satisfying*

$$c_1 x_1 + c_2 x_2 + \ldots + c_n x_n = z \qquad (\text{not all } c_i = 0)$$

*defines a **hyperplane** for given values of* c_i*'s and* z.

Remarks

1. If $z = 0$, then the hyperplane is said to pass through the origin and its equation is then written as
$$c_1 x_1 + c_2 x_2 + \ldots + c_n x_n = 0 \quad \text{or} \quad \mathbf{c}\mathbf{x} = 0.$$
This implies that $\mathbf{c}$ is orthogonal to every vector $\mathbf{x}$ in the hyperplane.
2. If $z \ne 0$ and $P(\mathbf{x}_1)$, $Q(\mathbf{x}_2)$ are any two points on the hyperplane, then
$$\mathbf{c}\mathbf{x}_1 = z \quad \text{and} \quad \mathbf{c}\mathbf{x}_2 = z.$$
This yields $\mathbf{c}(\mathbf{x}_1 - \mathbf{x}_2) = 0$, which implies that $\mathbf{c}$ is normal to the vector $\overline{\mathbf{QP}}$, *i.e.*, $(\mathbf{x}_2 - \mathbf{x}_1)$, and is thus in the direction of the normal to the hyperplane.

In general, $\mathbf{c}$ is called the ***normal*** to the hyperplane and $\pm \dfrac{\mathbf{c}}{|\mathbf{c}|}$ are called ***unit normals***.

Definition 3 (*Parallel Hyperplanes*). *Two hyperplanes are said to be **parallel**, if they have the same unit normals. The hyperplanes* $\mathbf{c}_1\mathbf{x} = z_1$ *and* $\mathbf{c}_2\mathbf{x} = z_2$ *are parallel, if* $\mathbf{c}_1 = \lambda\mathbf{c}_2$ *for some* $\lambda \ne 0$.

Definition 4 (*Closed and Open Half Space*). *In* $\mathbf{R}^n$, *every hyperplane* $\mathbf{c}\mathbf{x} = z$ *determines two open half spaces and two closed half spaces. The sets defined by*

$$\mathbf{H}_+^0 = \{\mathbf{x} : \mathbf{c}\mathbf{x} > z\} \text{ and } \mathbf{H}_-^0 = \{\mathbf{x} : \mathbf{c}\mathbf{x} < z\} \text{ are called \textbf{open half spaces}}$$

and

$$\mathbf{H}_+ = \{\mathbf{x} : \mathbf{c}\mathbf{x} \ge z\} \text{ and } \mathbf{H}_- = \{\mathbf{x} : \mathbf{c}\mathbf{x} \le z\} \text{ are called \textbf{closed half spaces}}.$$

Illustration. Let $\mathbf{c} = \{2, 3\}$ and $z = 7$. Then

$\mathbf{H}_+ = \{2x_1 + 3x_2 \ge 7\}$ gives a positive closed half space, and

$\mathbf{H}_- = \{2x_1 + 3x_2 \le 7\}$ gives a negative closed half space.

Definition 5 (*Polytope*). *The intersection of a finite number of closed half spaces is called a **polytope**. The hyperplanes producing the half spaces are called the generating hyperplanes of the polytope.*

Definition 6 (*Hypersphere*). *A **hypersphere** in* $\mathbf{R}^n$ *with centre at* $\mathbf{a}$ *and radius* $\varepsilon > 0$ *is defined to be the set of points*

$$\mathbf{S} = \{\mathbf{x} : |\mathbf{x} - \mathbf{a}| = \varepsilon\}.$$

Thus, the equation for hypersphere in $\mathbf{R}^n$ is

$$\sum_{i=1}^{n} (x_i - a_i)^2 = \varepsilon^2.$$

The inside of a hypersphere with centre at $\mathbf{a}$ and radius ε is the set of points

$$S_i = \{\mathbf{x} : |\mathbf{x} - \mathbf{a}| < \varepsilon\}.$$

A:10. CONVEX SETS AND THEIR PROPERTIES

Definition 1 (*Convex Set*). *A subset* $\mathbf{S} \subset \mathbf{R}^n$, *is said to be* ***convex***, *if for any two points* $\mathbf{x}_1, \mathbf{x}_2$ *in* **S** *the line segment joining the points* $\mathbf{x}_1$ *and* $\mathbf{x}_2$ *is also contained in S.*

In other words, a subset $\mathbf{S} \subset \mathbf{R}^n$ is convex, if and only if

$$\mathbf{x}_1, \mathbf{x}_2 \in \mathbf{S} \;\Rightarrow\; \lambda \mathbf{x}_1 + (1-\lambda)\mathbf{x}_2 \in \mathbf{S}; \quad 0 \le \lambda \le 1.$$

Some convex and non-convex sets in $\mathbf{R}^2$ are given below:

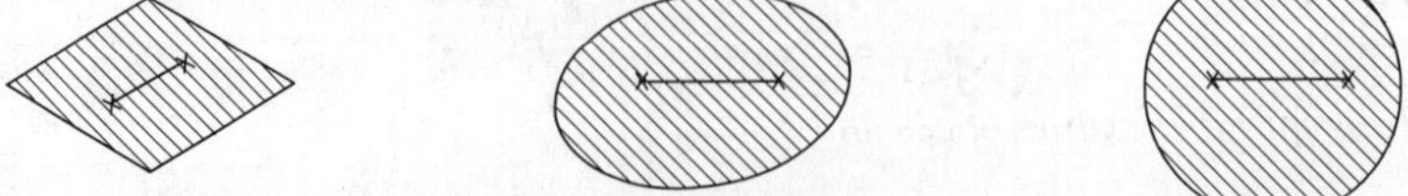

Fig. A.1. Convex Sets

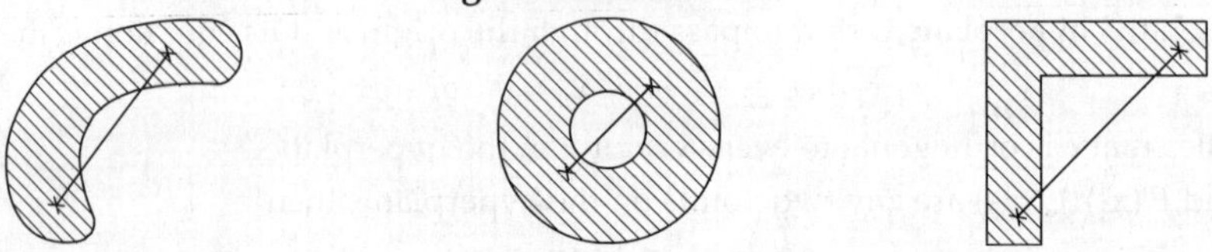

Fig. A.2. Non-convex Sets

Illustration. Show that the set

$$S = \left\{(\mathbf{x}_1, \mathbf{x}_2) : 3x_1^2 + 2x_2^2 \le 6\right\}$$

is convex.

Let $\mathbf{x}, \mathbf{y} \in \mathbf{S}$, where $\mathbf{x} = (x_1, x_2)$ and $\mathbf{y} = (y_1, y_2)$.

The line segment joining $\mathbf{x}$ and $\mathbf{y}$ is the set

$$\{\mathbf{u} : \mathbf{u} = \lambda \mathbf{x} + (1-\lambda)\mathbf{y},\; 0 \le \lambda \le 1\}.$$

For some λ, $0 \le \lambda \le 1$, let $\mathbf{u} = (u_1, u_2)$ be a point of this set, so that

$$u_1 = \lambda x_1 + (1-\lambda) y_1, \quad \text{and} \quad u_2 = \lambda x_2 + (1-\lambda) y_2.$$

Now,
$$\begin{aligned} 3u_1^2 + 2u_2^2 &= 3[\lambda x_1 + (1-\lambda) y_1]^2 + 2[\lambda x_2 + (1-\lambda) y_2]^2 \\ &= \lambda^2\left(3x_1^2 + 2x_2^2\right) + (1-\lambda)^2\left[3y_1^2 + 2y_2^2\right] + 2\lambda(1-\lambda)(3x_1 y_1 + 2x_2 y_2) \\ &\le 6\lambda^2 + 6(1-\lambda)^2 + 12\lambda(1-\lambda), \end{aligned}$$

since
$$(3x_1 y_1 + 2x_2 y_2) \le \sqrt{\left(x_1\sqrt{3}\right)^2 + \left(x_2\sqrt{2}\right)^2}\,\sqrt{\left(y_1\sqrt{3}\right)^2 + \left(y_2\sqrt{2}\right)^2},$$

Thus, $3u_1^2 + 2u_2^2 \le 6$; and hence $\mathbf{u} = (u, u_2)$ is also a point of S.

Hence, S is a convex set.

The following results easily follow from the definition of convex set:

(*i*) A hyperplane in $\mathbf{R}^n$ is a convex set.

(*ii*) The sets $\mathbf{H}_+$, $\mathbf{H}_-$, $\mathbf{H}_+{}^0$ and $\mathbf{H}_-{}^0$ are convex sets.

(*iii*) A closed ball in $\mathbf{R}^n$, namely, $\{\mathbf{x} : |\,\mathbf{x} - \mathbf{x}_0\,| \le r\}$, where $r > 0$ and $\mathbf{x}_0, \mathbf{x} \in \mathbf{R}^n$, is a convex set.

Definition 2 (*Extreme Point or Vertex of a Convex Set*). *An* ***extreme point (vertex)*** *of a convex set is a point of the set which does not lie on any segment joining to other points of the set.*

Thus, a point $\mathbf{x}$ of a convex set S is an extreme point of the set, if there does not exist any pair of points $\mathbf{x}_1, \mathbf{x}_2$, in S, such that

$$\mathbf{x} = \lambda \mathbf{x}_1 + (1-\lambda)\mathbf{x}_2, \quad 0 \le \lambda \le 1$$

Note. We should note the strict inequality imposed on λ. An extreme point is a boundary point of the set; however, not all boundary points of a convex set are necessarily extreme points.

Definition 3 (*Convex Combination of Vectors*). *Given a set of vectors* $\{\mathbf{x}_1, \mathbf{x}_2, ..., \mathbf{x}_k\}$, *a linear combination*

$$\mathbf{x} = \lambda_1 \mathbf{x}_1 + \lambda_2 \mathbf{x}_2 + \ldots + \lambda_k \mathbf{x}_k$$

is called a ***convex combination*** *of the given vectors, if*

$$\lambda_1, \lambda_2, \ldots, \lambda_k \geq 0 \quad \textit{and} \quad \sum_{i=1}^{k} \lambda_i = 1.$$

The reader may recall that we defined a line segment joining two points $\mathbf{x}_1$ and $\mathbf{x}_2$, as the set of points

$$\{\mathbf{x} \mid \mathbf{x} = \lambda \mathbf{x}_1 + (1-\lambda)\mathbf{x}_2, \quad 0 \leq \lambda \leq 1\}.$$

Thus, a line segment joining $\mathbf{x}_1$ and $\mathbf{x}_2$ is the set of all convex combination of $\mathbf{x}_1$ and $\mathbf{x}_2$. Also, clearly, a line segment is a set. You may wonder if we ask : given three vectors $\mathbf{x}_1$, $\mathbf{x}_2$ and $\mathbf{x}_3$, will the set of linear combination of these vectors be a convex set?

Theorem A-3. *The set of all convex combination of a finite number of points of* $\mathbf{S} \subset \mathbf{R}^n$ *is a convex set.*

Proof. Let

$$\mathbf{S} = \left\{ \mathbf{x} : \mathbf{x} = \sum_{i=1}^{m} \lambda_i \mathbf{x}_i \geq 0, \quad \sum_{i=1}^{m} \lambda_i = 1 \right\}, \quad \mathbf{x}_i \in \mathbf{S}$$

We have to show that **S** is a convex set.

Let $\mathbf{x}'$ and $\mathbf{x}''$ be in **S**, so that

$$\mathbf{x}' = \sum_{i=1}^{m} \lambda_i' \mathbf{x}_i, \quad \text{where } \lambda_i' \geq 0, \sum_{i=1}^{m} \lambda_i' = 1$$

and

$$\mathbf{x}'' = \sum_{i=1}^{m} \lambda_i'', \quad \text{where } \lambda_i'' \geq 0, \sum_{i=1}^{m} \lambda_i'' = 1.$$

Consider now the vector

$$\begin{aligned} \mathbf{x} &= \lambda \mathbf{x}' + (1-\lambda)\mathbf{x}'', \qquad 0 \leq \lambda \leq 1 \\ &= \lambda \sum_{i=1}^{m} \lambda_i' \mathbf{x}_i + (1-\lambda) \sum_{i=1}^{m} \lambda_i'' \mathbf{x}_i \\ &= \sum_{i=1}^{m} [\lambda \lambda_i' + (1-\lambda)\lambda_i''] \mathbf{x}_i = \sum_{i=1}^{m} \mu_i \mathbf{x}_i \end{aligned}$$

where, $\mu_i = \lambda \lambda_i' + (1-\lambda)\lambda_i''$, $i = 1, 2, \ldots, m$.

Since, $0 \leq \lambda \leq 1$, $\lambda_i' \geq 0$, and $\lambda_i'' \geq 0$, therefore, $\mu_i \geq 0$ for each i.

Also,
$$\begin{aligned} \sum_{i=1}^{m} \mu_i &= \sum_{i=1}^{m} [\lambda \lambda_i' + (1-\lambda)\lambda_i''] = \lambda \sum_{i=1}^{m} \lambda_i' + (1-\lambda) \sum_{i=1}^{m} \lambda_i'' \\ &= \lambda + (1-\lambda) = 1. \end{aligned}$$

Hence, **x** is a convex combination of the vectors $\mathbf{x}_1$, $\mathbf{x}_2$, ..., $\mathbf{x}_m$, or $\mathbf{x} \in \mathbf{S}$.

Thus, for each pair of points $\mathbf{x}'$, $\mathbf{x}''$ in **S**, the line segment joining them is contained in the set.

Hence, **S** is a convex set.

Definition 4 (*Convex Polyhedron*). *The set of all convex combination of a finite number of linearly independent vectors is called a* ***convex polyhedron****.*

The convex polyhedron generated by the finite set of linearly independent vectors $\mathbf{x}_1, \mathbf{x}_2, ..., \mathbf{x}_m$, is the set

$$\left\{ \mathbf{x} : \mathbf{x} = \sum_{i=1}^{m} \lambda_i \mathbf{x}_i, \ \lambda_i \geq 0, \text{ and } \sum_{i=1}^{m} \lambda_i = 1 \right\}.$$

Definition 5 (*Convex Cone*). *A non-empty subset* $\mathbf{C} \subset \mathbf{R}^n$, *is called a* ***cone****, if for each* $\mathbf{x}$ *in* $\mathbf{C}$, *and* $\lambda \geq 0$, *the vector* $\lambda \mathbf{x}$ *is also in* $\mathbf{C}$.

A cone is called a ***convex cone****, if it is a convex set.*

Illustration. If $\mathbf{A}$ be an $m \times n$ matrix, then the set of n-vectors $\mathbf{x}$ satisfying $\mathbf{Ax} \geq \mathbf{0}$ is a convex cone in $\mathbf{R}^n$. It is a cone, because if $\mathbf{A\,x} \geq \mathbf{0}$, then $\mathbf{A}\,\lambda\mathbf{x} \geq \mathbf{0}$ for non-negative λ. It is convex, because if $\mathbf{Ax}^{(1)} \geq \mathbf{0}$ and $\mathbf{Ax}^{(2)} \geq \mathbf{0}$, then $\mathbf{A}\,[\lambda\,\mathbf{x}^{(1)} + (1-\lambda)\,\mathbf{x}^{(2)}] \geq \mathbf{0}$.

Definition 6 (*Convex Hull of a Set*). *Given a subset* $\mathbf{Y} \subset \mathbf{R}^n$, *the smallest convex set containing* $\mathbf{Y}$ *is called the* ***convex hull*** *of* $\mathbf{Y}$, *and is denoted by* $< \mathbf{Y} >$.

It can easily be seen that the convex hull of a set is the intersection of all convex sets containing $\mathbf{Y}$.

Illustration. Let $\mathbf{A} = \{\mathbf{x}_1, \mathbf{x}_2\}$. Then the line segment joining $\mathbf{x}_1, \mathbf{x}_2$ is a convex set. Also, if $\mathbf{S}$ is any convex set containing $\mathbf{A}$, then it must contain the line segment joining $\mathbf{x}_1$ and $\mathbf{x}_2$.

$$\therefore \quad <\mathbf{A}> = \{\mathbf{x} : \mathbf{x} = \lambda \mathbf{x}_1 + (1-\lambda)\mathbf{x}_2, \ 0 \leq \lambda \leq 1\}.$$

Theorem A-4. *If* $\mathbf{A}$ *is any finite subset of vectors in* $\mathbf{R}^n$, *then the convex hull of* $\mathbf{A}$ *is the set of all convex combination of vectors in* $\mathbf{A}$.

Proof. Let $\mathbf{A}$ be a finite subset of vectors in $\mathbf{R}^n$ and $< \mathbf{A} >$ be its convex hull. Also, let $\mathbf{S}$ be the set of all convex combination of vectors in $\mathbf{A}$, then we know that $\mathbf{S}$ is a convex set containing $\mathbf{A}$, and hence $< \mathbf{A} > \subset \mathbf{S}$.

Again, let $< \mathbf{A} >$ contains $\mathbf{A}$. We shall show that this implies that $< \mathbf{A} > \supset \mathbf{S}$.

We use finite induction on k, the number of vectors in $\mathbf{A}$, to prove this result.

For $k = 2$, let $\mathbf{x}_1, \mathbf{x}_2$ be two vectors in $\mathbf{A}$. Then $< \mathbf{A} >$ contains $\mathbf{x}_1, \mathbf{x}_2$: and is a convex set

$$\therefore \quad \mathbf{x} = \lambda \mathbf{x}_1 + (1-\lambda)\mathbf{x}_2 \qquad 0 \leq \lambda \leq 1$$

is also in $< \mathbf{A} >$.

$\therefore$ All possible convex combination of vectors $\mathbf{x}_1$, and $\mathbf{x}_2$ are in $< \mathbf{A} >$.

Assume that, for any positive integer k, the result is true for a set of at most $k-1$ vectors.

Consider the set $\mathbf{A} = \{\mathbf{x}_1, \mathbf{x}_2, ..., \mathbf{x}_k\}$. Then, $< \mathbf{A} >$ is a convex set containing $\mathbf{A}$, and in particular $\{\mathbf{x}_1, \mathbf{x}_2, ..., \mathbf{x}_{k-1}\}$. Let

$$\mathbf{x}_1 = \left\{ \mathbf{x} \,|\, \mathbf{x} = \sum_{i=1}^{k-1} \mu_i \mathbf{x}_i ; \ \mu_i \geq 0, \ \sum_{i=1}^{k-1} \mu_i = 1 \right\}$$

Thus, by induction hypothesis $< \mathbf{A} > \supset \mathbf{x}_1$.

Now, $< \mathbf{A} >$ contains $\mathbf{x}_1$, as well as $\{\mathbf{x}_k\}$.

$\therefore$ $< \mathbf{A} >$ contains all segments joining $\mathbf{x}_k$ to a point of $\mathbf{x}_1$.

or
$$\mathbf{x} = \delta \mathbf{x}_k + (1-\delta) \sum_{i=1}^{k-1} \mu_i \mathbf{x}_i, \ \mu_i \geq 0, \ \sum_{i=1}^{k-1} \mu_i = 1,$$

is a point in $< \mathbf{A} >$.

Now, $$\mathbf{x} = \delta \mathbf{x}_k + (1-\delta) \sum_{i=1}^{k-1} \mu_i \mathbf{x}_i \Rightarrow \mathbf{x} = \delta \mathbf{x}_k + \sum_{i=1}^{k-1} \beta_i \mathbf{x}_i,$$

where, $\beta_i = (1-\delta)\mu_i$, for $i = 1, 2, ..., k-1$ and $\beta_k = \delta$.

Since, $\mu_i \geq 0$ for each i, and $0 \leq \delta \leq 1$, $\therefore$ $\beta_i \geq 0$ for $i = 1, 2, ..., k$.

Also, $$\sum_{i=1}^{n} \beta_i = \delta + (1-\delta) \sum_{i=1}^{k-1} \mu_i = \delta + (1-\delta), \text{ since } \sum_{i=1}^{k-1} \mu_i = 1.$$

$\therefore$ $\mathbf{x} = \sum_{i=1}^{n} \beta_i \mathbf{x}_i$ is a convex combination of the vectors $\mathbf{x}_1, \mathbf{x}_2, ..., \mathbf{x}_k$ and is in $< \mathbf{A} >$, *i.e.*, $< \mathbf{A} > \supset \mathbf{S}$.

We conclude that $$< \mathbf{A} > = \mathbf{S}.$$

In view of the above theorem, the convex polyhedron may equivalently be defined as the convex hull of a set point **S**, consisting of a finite number of points in $\mathbf{R}^n$.

For example, convex hull of eight vertices of a cube is a convex polyhedron.

Remark. If a convex set **S** is closed and bounded with a finite number of points, then any point in the set can be expressed as a convex combination of its extreme points. This means **S** represents a convex hull of its extreme points.

Definition 7 (*Simplex*). *A* ***simplex*** *in n-dimension is a convex polyhedron having exactly (n + 1) vertices.*

For example, a simplex in zero dimension is a point, in one dimension it is a line segment, in two dimensions it is a triangle, in three dimensions it is a tetrahedron, and so on.

A:11. SUPPORTING AND SEPARATING HYPERPLANES

Definition (*Supporting Hyperplane*). *Let* $\mathbf{S} \subset \mathbf{R}^n$ *be any closed convex set and* $\mathbf{w} \in \mathbf{S}$ *be a boundary point. Then a hyperplane* $\mathbf{cx} = z$ *is called a* ***supporting hyperplane*** *of* **S** *at* **w**, *if*

(*i*) $\mathbf{cw} = z$ *and* (*ii*) $\mathbf{S} \subset \mathbf{H}$, *or* $\mathbf{S} \subset \mathbf{H}_-$, *where* $\mathbf{H}_+ = \{\mathbf{x} : \mathbf{cx} \geq z\}$ *and* $\mathbf{H}_- = \{\mathbf{x} : \mathbf{cx} \leq z\}$.

Remarks

1. The supporting hyperplane need not be unique.
2. **S** may intersect the supporting hyperplane in more than one boundary points.

Theorem A-5. *Let* **S** *be a closed convex set which is bounded from below. Then,* **S** *has extreme points in every supporting hyperplane.*

Proof. Let **w** be a boundary point of a closed convex set **S**.

Also, let $\mathbf{cx} = z$ be supporting hyperplane at $\mathbf{w} \in \mathbf{S}$. Further, let $\mathbf{B} = \mathbf{S} \cap \{\mathbf{x} \mid \mathbf{cx} = z\}$.

Then, **B** is a closed convex set and $\mathbf{B} \neq \phi$, for $\mathbf{w} \in \mathbf{B}$.

We claim that every extreme point of **B** is also an extreme point of **S**.

Let us assume to the contrary that an extreme point **b** of **B**, is not an extreme point of **S**. Then, there exist $\mathbf{x}_1, \mathbf{x}_2 \in \mathbf{S}$, such that

$$\mathbf{b} = \lambda \mathbf{x}_1 + (1-\lambda)\mathbf{x}_2, \quad 0 \leq \lambda \leq 1$$

$$\therefore \quad \mathbf{cb} = \lambda \mathbf{cx}_1 + (1-\lambda)\mathbf{cx}_2. \quad ...(i)$$

Since, $\mathbf{cx} = z$ is a supporting hyperplane at **w** and $\mathbf{x}_1, \mathbf{x}_2 \in \mathbf{X}$,

$$\mathbf{cx}_1 \leq z \text{ and } \mathbf{cx}_2 \leq z \text{ or } \mathbf{cx}_1 \geq z \text{ and } \mathbf{cx}_2 \geq z. \quad ...(ii)$$

From (*i*) and (*ii*)

$$\mathbf{cb} \leq \lambda z + (1-\lambda)z \quad \text{or} \quad \mathbf{cb} \geq \lambda z + (1-\lambda)z$$

These implies $\mathbf{cb} \le z$ or $\mathbf{cb} \ge z$.

$\therefore$ **b** is not a point of **B**.

This is a contradiction.

Therefore, every extreme point of **B** is also an extreme point of **S**.

Theorem A-6 (*Separating Hyperplane*). *Let* $\mathbf{S} \subset \mathbf{R}^n$ *be a closed convex set. Then, for any point* **y** *not is* **S**, *there is a Hyperplane containing* **y** *such that* **S** *is contained in one of the open half spaces determined by the hyperplane.*

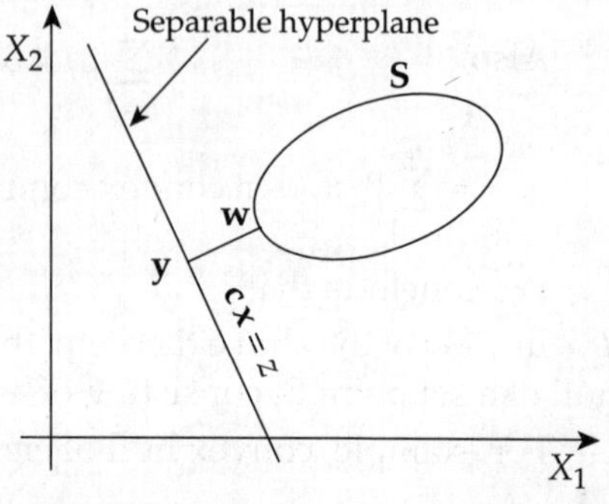

Fig. A.3. Separating Hyperplane $\mathbf{R}^2$

Proof. We are given that $\mathbf{y} \notin \mathbf{S}$.

Since, **S** is a closed set, there exist $\mathbf{w} \subset \mathbf{S}$, such that

$$|\mathbf{w}-\mathbf{y}| = \min_{\mathbf{x} \in S} |\mathbf{x}-\mathbf{y}| \quad \text{... } (i)$$

Let **u** be any point of **S**. Then, since **S** is a convex set

$$[\lambda \mathbf{u} + (1-\lambda)\mathbf{w}] \in \mathbf{S}, \ \text{ for } 0 \le \lambda \le 1 \quad \text{... } (ii)$$

From (*i*) and (*ii*)

$$|\lambda \mathbf{u} + (1-\lambda)\mathbf{w} - \mathbf{y}| \ge |\mathbf{w}-\mathbf{y}|$$

$$\Rightarrow \quad |(\mathbf{w}-\mathbf{y}) + \lambda(\mathbf{u}-\mathbf{w})|^2 \ge |\mathbf{w}-\mathbf{y}|^2$$

$$\Rightarrow \quad \lambda^2 |\mathbf{u}-\mathbf{w}|^2 + |\mathbf{w}-\mathbf{y}|^2 + 2\lambda(\mathbf{w}-\mathbf{y})(\mathbf{u}-\mathbf{w}) \ge |\mathbf{w}-\mathbf{y}|^2$$

$$\Rightarrow \quad \lambda^2 |\mathbf{u}-\mathbf{w}|^2 + 2\lambda(\mathbf{w}-\mathbf{y})(\mathbf{u}-\mathbf{w}) \ge 0.$$

Letting $\lambda \to 0$, and $\mathbf{c} = (\mathbf{w} - \mathbf{y})$, such that $|\mathbf{c}| \neq 0$; we get

$$(\mathbf{w}-\mathbf{y})(\mathbf{u}-\mathbf{w}) \ge 0 \quad \text{or} \quad \mathbf{c}(\mathbf{u}-\mathbf{w}) \ge 0$$

Further, $$\mathbf{c}(\mathbf{u}-\mathbf{y}) \ge \mathbf{c}(\mathbf{w}-\mathbf{y}) = |\mathbf{c}|^2$$

This implies that $\mathbf{c}(\mathbf{u}-\mathbf{y}) \ge 0$, since $|\mathbf{c}| \neq 0$.

Hence, $$\mathbf{cu} \ge \mathbf{cy} \quad \text{or} \quad \mathbf{cu} \ge z, \text{ for } z = \mathbf{cy}.$$

Thus, **y** lies on the hyperplane $\mathbf{cx} = z$ and for all $\mathbf{u} \in \mathbf{S}$, $\mathbf{cu} \ge z$.

This completes the proof.

PROBLEMS

A.04. Given the hyperplane $3x_1 + 2x_2 + 4x_3 + 6x_4 = 7$. Find in which half space is the point $\mathbf{x} = \{6, 1, 7, 2\}$.

A.05. Prove that the set $\mathbf{x} = \{(x_1, x_2) \mid x_1^2 + x_2^2 \le 4\}$ is a convex set.

A.06. Examine whether the following set is convex or not:

(*i*) $S = \{(x_1, x_2) : x_1x_2 \ge 1, x_1 \ge 0, x_2 \ge 0\}$.

(*ii*) $S = \{(x_1, x_2) : 5x_1 + 2x_2 \ge 10, 2x_1 + 5x_2 \ge 10\}$.

(*iii*) $S = \{(x_1, x_2) : x_1^2 + x_2^2 \le 1, x_1 + x_2 \ge 1\}$.

A.07. Show that $S = \{(x_1, x_2, x_3) : 2x_1 - x_2 + x_3 \le 4\} \subset \mathbf{R}^3$ is a convex set.

A.08. Show that in $\mathbf{R}^3$, the closed ball $x_1^2 + x_2^2 + x_3^2 \le 1$ is a convex set.

A.09. Let $S = \{(x_1, x_2) : x_1 - x_2 \le 4, x_1 + x_2 \ge -3, x_2 \le 8\}$. Find all the extreme points of S and represent $\mathbf{x} = (2, 1)$ as convex combination of the extreme points.

A.10. Determine the convex hull of the following sets:

(*i*) $A = \{(x_1, x_2) : \mid x_1^2 + x_2^2 = 1\}$, (*ii*) $A\{(x_1, x_2)\}$.

A.11. Graph the convex hull of the points

$$(0, 0),\ (0, 1),\ (1, 2),\ (1, 1),\ (4, 0).$$

Which of these points is an interior point of the convex hull? Express it as a convex combination of the extreme points.

A:12. CONVEX FUNCTIONS

Definition 1 (*Convex Function*). *Let* **S** *be a non-empty convex subset of* $\mathbf{R}^n$. *A function* $f(\mathbf{x})$ *on* **S** *is said to be* ***convex***, *if for any two vectors* $\mathbf{x}_1$ *and* $\mathbf{x}_2$ *in* **S**,

$$f[\lambda \mathbf{x}_1 + (1-\lambda)\mathbf{x}_2] \le \lambda f(\mathbf{x}_1) + (1-\lambda) f(\mathbf{x}_2), \quad 0 \le \lambda \le 1$$

Definition 2 (*Strictly Convex Function*). *Let* **S** *be a non-empty convex subset of* $\mathbf{R}^n$. *A function* $f(\mathbf{x})$ *on* **S** *is said to be* ***strictly convex***, *if for any two different vectors* $\mathbf{x}_1$ *and* $\mathbf{x}_2$ *in* **S**,

$$f[\lambda \mathbf{x}_1 + (1-\lambda)\mathbf{x}_2] < \lambda f(\mathbf{x}_1) + (1-\lambda) f(\mathbf{x}_2), \quad 0 < \lambda < 1$$

It follows from the above two definitions that every strictly convex function is also convex. The graph of a strictly convex function has been illustrated in *Fig. A.4.*

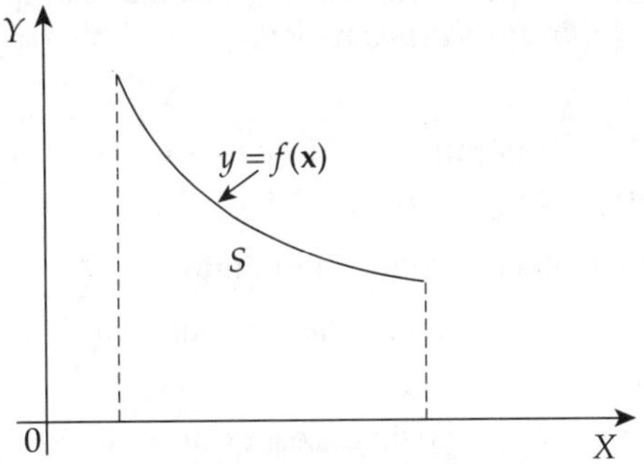

Fig. A.4. Strictly Convex Function

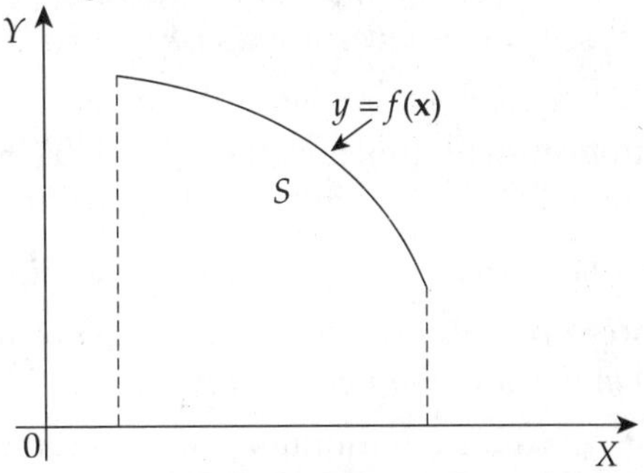

Fig. A.5. Strictly Concave Function

Definition 3 [*Concave (Strictly Concave) Function*]. *A function* $f(\mathbf{x})$ *on a non-empty subset* **S** *of* $\mathbf{R}^n$ *is said to be concave, if the inequality '≤' is reversed to '≥' and strictly concave if the inequality < is reversed to >.*

Clearly, every 'strictly concave function' is also concave. The graph of a strictly concave function has been illustrated in *Fig. A.5.*

It is possible for a function to be both convex and concave. For example $f(\mathbf{x}) = \mathbf{x}$ is such a function (*Fig. A.6*). The function in *Fig. A.7* is strictly convex for $\mathbf{x} > \mathbf{x}_0$ but not strictly convex for $\mathbf{x} < \mathbf{x}_0$.

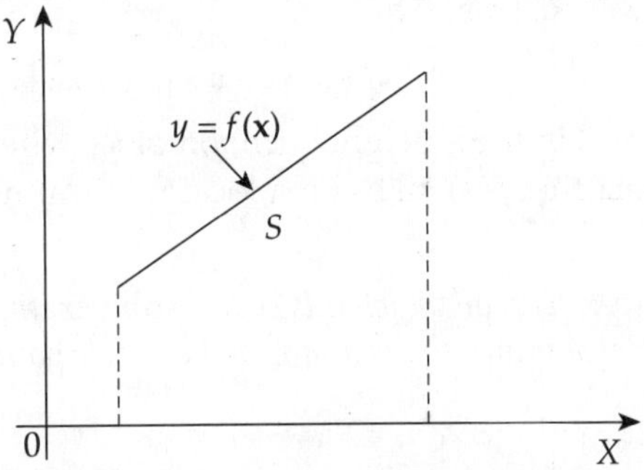

Fig. A.6. Both Convex and Concave Functions

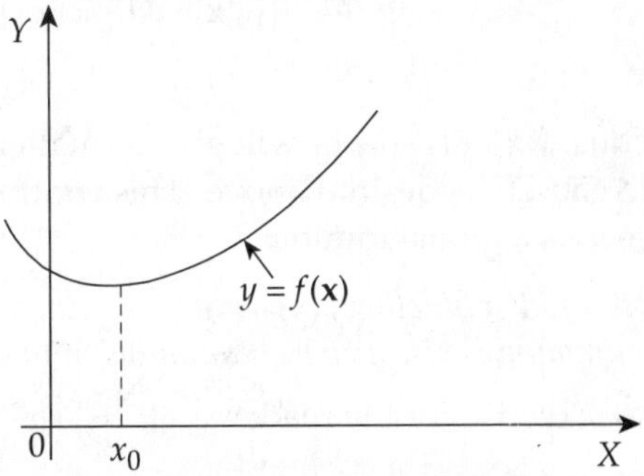

Fig. A.7. Strictly Convex for $\mathbf{x} > \mathbf{x}_0$ but not strictly convex for $\mathbf{x} < \mathbf{x}_0$

The following results are the immediate consequences of the above definitions:

(*i*) *A linear function* $z = \mathbf{c}\,\mathbf{x}$, $\mathbf{x} \in \mathbf{R}^n$ *is a convex (concave) function but not strictly convex (concave).*

(*ii*) *The sum of convex (concave) functions is convex (concave), and if at least one of the functions is strictly convex (concave) then so is their sum.*

Note. In what follows, we shall deal with convex functions only. However, all the results remain valid if we deal with concave functions instead.

A:13. LOCAL AND GLOBAL EXTREMA

In the problems of constrained optimization, one is interested in determining a vector **x** that minimizes the function $f(\mathbf{x})$ [or maximizes '$-f(\mathbf{x})$'] subject to the 'constraints' $g_i(\mathbf{x}) \le 0$ $(i = 1, 2, ..., m)$. The set of the vectors **x** satisfying these constraints is usually called the 'feasible region'. These terms shall be defined in a more formal way later.

Definition 1 (*Global Minima*). *A **global minimum** of the function $f(\mathbf{x})$ is said to be attained at $\mathbf{x}_0$, if $f(\mathbf{x}_0) \le f(\mathbf{x})$ for all **x** in the feasible region.*

Illustration. Function $f(x) = x_1^2$, subject to the constraint $x_1 \ge 0$, has a minimum at $x_1 = 0$.

Definition 2 (*Local Minima*). *A **local** or **relative minimum** $f(\mathbf{x}_0)$ of function $f(\mathbf{x})$ is said to be attained at $\mathbf{x}_0$, if there exists a positive ε such that $f(\mathbf{x}_0) \le f(\mathbf{x})$ for all **x** in the feasible region which also **satisfy** the condition $|\mathbf{x}_0 - \mathbf{x}| \le \varepsilon$.*

Illustration. The function $f(\mathbf{x}) = x_1^2 - x_1^3$ subject to the constraint $x_1 \le 0$, has a local or relative minimum at $x_1 = 0$ (with $\varepsilon = 2/3$). Note that $f(x)$ has no global minimum at all.

Note. The word extremum (plural extrema) is used to indicate either maximum or minimum.

Theorem A-7. *Let $f(\mathbf{x})$ be a convex function of a convex set S. If $f(\mathbf{x})$ has a local minimum on S, then this local minimum is also a global minimum on S.*

Proof. Let a local minimum $f(\mathbf{x}_0)$ be attained at $\mathbf{x}_0$. Then, there exists at least one $\mathbf{x}_1$ in S $(\mathbf{x}_1 \ne \mathbf{x}_0)$ such that $f(\mathbf{x}_1) < f(\mathbf{x}_0)$. Since, $f(\mathbf{x})$ is a convex function on S, we have

$$f[\lambda \mathbf{x}_1 + (1-\lambda)\mathbf{x}_0] \le \lambda f(\mathbf{x}_1) + (1-\lambda) f(\mathbf{x}_0)$$

Also,
$$\lambda f(\mathbf{x}_1) + (1-\lambda) f(\mathbf{x}_0) < \lambda f(\mathbf{x}_0) + (1-\lambda) f(\mathbf{x}_0)$$

or
$$\lambda f(\mathbf{x}_1) + (1-\lambda) f(\mathbf{x}_0) < f(\mathbf{x}_0)$$

Thus,
$$f[\lambda \mathbf{x}_1 + (1-\lambda)\mathbf{x}_0] \le f(\mathbf{x}_0).$$

Now, for any $\varepsilon > 0$, we observe that

$$|[\lambda \mathbf{x}_1 + (1-\lambda)\mathbf{x}_0] - \mathbf{x}_0| = \lambda |\mathbf{x}_1 - \mathbf{x}_0| < \varepsilon,$$

so long as $\lambda < \varepsilon < |\mathbf{x}_1 - \mathbf{x}_0|^{-1}$ and obviously such a $\lambda < 1$ does exist, whenever $\varepsilon < |\mathbf{x}_1 - \mathbf{x}_0|$.

Thus, $\lambda \mathbf{x}_1 + (1-\lambda)\mathbf{x}_0$ will give a smaller value for $f(\mathbf{x})$ in the ε-neighbourhood of $\mathbf{x}_0$, whenever λ is so chosen as desired above. This contradicts the fact that $f(\mathbf{x})$ takes on a local minimum at $\mathbf{x}_0$. Hence $\mathbf{x}_0$ is a global minima.

Corollary. *If a function $f(\mathbf{x})$ has a local minimum on a convex set S on which it is strictly convex, then this local minimum is also a global minimum on that set. This global minimum is attained at a single point.*

Proof. Exercise for the reader.

Theorem A-8. *Let $f(\mathbf{x})$ be a convex function on a convex set S. Then, the set of points in S at which $f(\mathbf{x})$ takes on its global minimum, is a convex set.*

Proof. The result is obvious, if the global-minimum is attained at just a single point. Let us assume that the global minimum is attained at two different points $\mathbf{x}_1$ and $\mathbf{x}_2$ of S. Then, $f(\mathbf{x}_1) = f(\mathbf{x}_2)$.

Now, since $f(\mathbf{x})$ is convex,

$$f[\lambda \mathbf{x}_2+(1-\lambda)\mathbf{x}_1]\le \lambda f(\mathbf{x}_2)+(1-\lambda)f(\mathbf{x}_1)=f(\mathbf{x}_2),\ 0\le\lambda\le 1$$

$\therefore$
$$f[\lambda \mathbf{x}_2+(1-\lambda)\mathbf{x}_1]\le f(\mathbf{x}_2)$$

or
$$f[\lambda \mathbf{x}_2+(1-\lambda)\mathbf{x}_1]\le f(\mathbf{x}_1),\ \text{since}\ f(\mathbf{x}_2)=f(\mathbf{x}_1).$$

Thus, every point $\mathbf{x} = \lambda \mathbf{x}_2 + (1 - \lambda)\mathbf{x}_1$ corresponds to a global minima. The set of all such points is, obviously, a convex set.

Corollary. *If the global minimum is attainable at two different points of S, then it is attainable at an infinite number of points of S.*

Proof. Exercise for the reader.

Theorem A-9. *Let $f(\mathbf{x})$ be differentiable in its domain. If $f(\mathbf{x})$ is defined in an open convex set S, then $f(\mathbf{x})$ is convex, if and only if*

$$f(\mathbf{x}_2)-f(\mathbf{x}_1)\ge(\mathbf{x}_2-\mathbf{x}_1)^T\ \nabla f(\mathbf{x}_1)$$

for all $\mathbf{x}_1, \mathbf{x}_2 \in S$.

Proof. (*Sufficiency*). Here, we shall prove that if

$$f(\mathbf{x}_2)-f(\mathbf{x}_1)\ge(\mathbf{x}_2-\mathbf{x}_1)^T\ \nabla f(\mathbf{x}_1),$$

then $f(\mathbf{x})$ is convex.

Since, $\mathbf{x}_1, \mathbf{x}_2 \in S,\ \mathbf{x}_0 = \lambda\mathbf{x}_2 + (1-\lambda)\mathbf{x}_1$ for $0\le\lambda\le1$ implies that $\mathbf{x}_0 \in S$.

Using the above condition for $\mathbf{x}_1$ and $\mathbf{x}_0$, we have

$$f(\mathbf{x}_1)-f(\mathbf{x}_0)\ge(\mathbf{x}_1-\mathbf{x}_0)^T\ \nabla f(\mathbf{x}_0) \quad \text{... }(i)$$

Similarly, for $\mathbf{x}_2$ and $\mathbf{x}_0$,

$$f(\mathbf{x}_2)-f(\mathbf{x}_0)\ge(\mathbf{x}_2-\mathbf{x}_0)^T\ \nabla f(\mathbf{x}_0) \quad \text{... }(ii)$$

Multiplying (*ii*) by λ and (*i*) by $(1 - \lambda)$ and then adding, we get

$$\lambda f(\mathbf{x}_2)+(1-\lambda)f(\mathbf{x}_1)\ge f(\mathbf{x}_0)+\left[\lambda\mathbf{x}_2^T+(1-\lambda)\mathbf{x}_1^T\right]\nabla f(\mathbf{x}_0)-\mathbf{x}_0^T\ \nabla f(\mathbf{x}_0)$$

Using the definition of $\mathbf{x}_0$, this yields

$$\lambda f(\mathbf{x}_2)+(1-\lambda)f(\mathbf{x}_1)\ge f[\lambda\mathbf{x}_2+(1-\lambda)\mathbf{x}_1]$$

which implies that $f(\mathbf{x})$ is convex.

(*Necessity*). We, now prove that if $f(\mathbf{x})$ is convex, we must have

$$f(\mathbf{x}_2)-f(\mathbf{x}_1)\ge(\mathbf{x}_2-\mathbf{x}_1)^T\ \nabla f(\mathbf{x}_1)$$

For $0\le\lambda\le1$, we have for $\mathbf{x}_1, \mathbf{x}_2 \in S$ that

$$\lambda f(\mathbf{x}_2)+(1-\lambda)f(\mathbf{x}_1)\ge f[\lambda\mathbf{x}_2+(1-\lambda)\mathbf{x}_1]$$

i.e.,
$$\lambda f(\mathbf{x}_2)-\lambda f(\mathbf{x}_1)\ge f[\lambda\mathbf{x}_2+(1-\lambda)\mathbf{x}_1]-f(\mathbf{x}_1)$$

or
$$f(\mathbf{x}_2)-f(\mathbf{x}_1)\ge[f\{\lambda\mathbf{x}_2+(1-\lambda)\mathbf{x}_1\}-f(\mathbf{x}_1)]/\lambda.$$

Taking the limit of the right hand side as $\lambda\to0$, we obtain

$$f(\mathbf{x}_2)-f(\mathbf{x}_1)\ge(\mathbf{x}_2-\mathbf{x}_1)^T\ \nabla f(\mathbf{x}_1).$$

This completes the proof.

Theorem A-10. *Let $f(\mathbf{x}), \mathbf{x}\in \mathbf{R}^n$ be a convex continuously differentiable function defined on an open convex set S. Let $\mathbf{x}_0$ be in S. Then, $f(\mathbf{x}_0)\le f(\mathbf{x})$ for all $\mathbf{x}$ in S, if and only if*

$$(\mathbf{x}-\mathbf{x}_0)^T\ \nabla f(\mathbf{x}_0)\ge 0.$$

Proof. (*Necessary Condition*). Let $f(\mathbf{x})$ be a convex function on a convex set S, and let $\mathbf{x}_0 \in S$, such that $f(\mathbf{x}_0) \leq f(\mathbf{x})$ for all $\mathbf{x} \in S$. Since, S is open, the point $\mathbf{x}_0$ is an interior point of S.

$$\therefore \qquad f(\mathbf{x}_0) \leq f[\lambda \mathbf{x} + (1-\lambda)\mathbf{x}_0]$$

for all $\mathbf{x}$ in S and $0 \leq \lambda \leq 1$.

For $\lambda > 0$,
$$\frac{f[\mathbf{x}_0 + \lambda(\mathbf{x}-\mathbf{x}_0)] - f(\mathbf{x}_0)}{\lambda} \geq 0.$$

Taking limit as λ approaches zero, we get
$$(\mathbf{x}-\mathbf{x}_0)^T \nabla f(\mathbf{x}_0) \geq 0.$$

(*Sufficient Condition*). Given that $(\mathbf{x}-\mathbf{x}_0)^T \nabla f(\mathbf{x}_0) \geq 0$ for all $\mathbf{x}$ in S, by convexity of $f(\mathbf{x})$, we have
$$f(\mathbf{x}) - f(\mathbf{x}_0) \geq (\mathbf{x}-\mathbf{x}_0)^T \nabla f(\mathbf{x}_0) \geq 0.$$

Hence, $f(\mathbf{x}) \geq f(\mathbf{x}_0)$ for all $\mathbf{x}$ and $\mathbf{x}_0$ minimizes $f(\mathbf{x})$ over S.

Remark. From a geometric point of view, the above theorem states that $\mathbf{x}_0$ is a minimum point, if the angle between the vectors $\nabla f(\mathbf{x}_0)$ and $(\mathbf{x}-\mathbf{x}_0)^T$ for all $\mathbf{x}$ in S is less than or equal to $\pi/2$.

A:14. QUADRATIC FORMS

Definition 1 (*Quadratic Form*). *Let $\mathbf{A} = (a_{ij})$ be an n-rowed square matrix. Then, given an $n \times 1$ matrix $\mathbf{x} = (x_1, x_2, ..., x_n)^T$, the function*

$$\mathbf{Q}(\mathbf{x}) = \mathbf{x}^T \mathbf{A}\mathbf{x} = \sum_{i=1}^{n} \sum_{j=1}^{n} a_{ij}\, x_i\, x_j$$

is called a quadratic form over the domain of all $n \times 1$ matrices $\mathbf{x}$.

Illustration. The quadratic form

$$\mathbf{Q}(\mathbf{x}) = x_1^2 + 2x_2^2 - 7x_3^2 - 4x_1x_2 + 6x_1x_3 - 5x_2x_3$$

can be written as

$$[x_1 \quad x_2 \quad x_3] \begin{bmatrix} 1 & -2 & 3 \\ -2 & 2 & -5/2 \\ 3 & -5/2 & -7 \end{bmatrix} \begin{bmatrix} x_1 \\ x_2 \\ x_3 \end{bmatrix}.$$

Note. The matrix $\mathbf{A}$ can always be assumed symmetric, since each element of every pair of coefficients a_{ij} and a_{ji} $(i \neq j)$ can be replaced by $[(a_{ij} + a_{ji})/2]$ without changing the value of $\mathbf{Q}(\mathbf{x})$.

Properties of a Quadratic Form

(*i*) ***Positive Definite.*** The quadratic form $\mathbf{Q}(\mathbf{x})$ is positive definite, if and only if, $\mathbf{Q}(\mathbf{x}) > 0$ (positive) for all $\mathbf{x} \neq \mathbf{0}$.
For example, $\mathbf{Q}(\mathbf{x}) = x_1^2 + 5x_2^2 + 7x_3^2$ is positive definite as it is positive for all values of x_1, x_2, x_3, except when $x_1 = x_2 = x_3 = 0$.

(*ii*) ***Positive Semi-definite.*** The quadratic form $\mathbf{Q}(\mathbf{x})$ is positive semi-definite, if and only if, $\mathbf{Q}(\mathbf{x}) \geq 0$ (non-negative) for all $\mathbf{x}$ and there exist at least one non-zero vector $\mathbf{x} \neq \mathbf{0}$ for which $\mathbf{Q}(\mathbf{x}) = 0$.
For example, $\mathbf{Q}(\mathbf{x}) = (x_1 - x_2)^2 + 2x_3^2$ is positive semi-definite, as it is zero for $x_1 = x_2$ and $x_3 = 0$ and also positive for all values of x_1, x_2, x_3, except when $x_1 = x_2 = x_3 = 0$.

(*iii*) ***Negative Definite.*** The quadratic form **Q** (**x**) is negative definite, if and only if, '– **Q** (**x**)' is positive definite. That is, **Q** (**x**) is negative definite when **Q** (**x**) is negative when **x** ≠ **0**.

(*iv*) ***Negative Semi-definite.*** The quadratic form **Q** (**x**) is negative semi-definite, if and only if, '– **Q** (**x**)' is positive semi-definite. That is, **Q** (**x**) is negative semi-definite when **Q** (**x**) ≤ 0 for all **x** and zero for at least one **x** ≠ **0**.

For example, if $\mathbf{Q}(\mathbf{x}) = -x_1^2 + 2x_1x_2 - x_2^2 = -\left(x_1^2 - 2x_1x_2 + x_2^2\right) = -(x_1 - x_2)^2$, then **Q** (**x**) ≤ 0 for all **x** but zero for $x_1 = x_2$.

(*v*) ***Indefinite.*** The quadratic form **Q** (**x**) is indefinite, if **Q** (**x**) is positive for some **x** and negative for some other **x**.

For example, $\mathbf{Q}(\mathbf{x}) = x_1^2 - 3x_2$ can be positive for $x_1 \neq 0$ and $x_2 = 0$ and negative for $x_1 = 0$ and $x_2 \neq 0$.

PROBLEMS

A.12. Show that the following functions are convex over $\mathbf{R}^1$:

(*i*) $f(\mathbf{x}) = 3\mathbf{x}^2$, (*ii*) $f(\mathbf{x}) = e^x$ and (*iii*) $f(\mathbf{x}) = |\,\mathbf{x}\,|$.

A.13. Show that the function $f(\mathbf{x}) = 2x_1^2 + x_2^2$ is a convex function over all of $\mathbf{R}^2$.

A.14. Write the following quadratic forms in the form $\mathbf{x}^T\mathbf{A}\mathbf{x}$:

(*i*) $x_1^2 + 8x_1x_2 + 16x_2^2 - 3x_3^2$ (*ii*) $2x_1^2 - 6x_1x_2 + 2x_1x_3 + 2x_2^2 + 6x_2x_3 - 5x_3^2$

A.15. Write down the quadratic forms whose associated matrices are:

(*i*) $\begin{bmatrix} 2 & -3 & 1 \\ -3 & 4 & 2 \\ 1 & 2 & -6 \end{bmatrix}$ (*ii*) $\begin{bmatrix} 1 & 2 & 4 \\ 2 & 6 & -2 \\ 4 & -2 & 14 \end{bmatrix}$

A.16. Determine whether the quadratic form

$$x_1^2 + 2x_2^2 + 2x_3^2 - 2x_1x_2 - 2x_2x_3$$

is positive definite.

TEST YOUR UNDERSTANDING

1. If **A** and **B** are invertible square matrices of the same order, then

(*a*) $(\mathbf{AB})^{-1} = \mathbf{B}^{-1}\mathbf{A}^{-1}$, (*b*) $(\mathbf{A}^T)^{-1} = (\mathbf{A}^{-1})^T$,

(*c*) $(\mathbf{A}^T\mathbf{B}^T)^{-1} = (\mathbf{B}^{-1})^T(\mathbf{A}^{-1})^T$, (*d*) all of the above.

2. The inverse of $\begin{pmatrix} 1 & 1 \\ 0 & 1 \end{pmatrix}$ is

(*a*) $\begin{pmatrix} 1 & 0 \\ -1 & 0 \end{pmatrix}$, (*b*) $\begin{pmatrix} -1 & 1 \\ 0 & 1 \end{pmatrix}$,

(*c*) $\begin{pmatrix} 1 & -1 \\ 0 & 1 \end{pmatrix}$, (*d*) $\begin{pmatrix} 1 & 0 \\ 0 & 1 \end{pmatrix}$.

3. Which of the following statements is not true?

(*a*) A row (column) matrix is called a row (column) vector.

(*b*) The terms a 'point' in $\mathbf{R}^n$ and a 'vector' in $\mathbf{R}^n$ are used interchangeably meaning the same thing.

(*c*) A row (column) vector cannot be added to any other now (column) vector.
(*d*) Two vectors are order comparable, if and only if, they have an equal number of components.

4. Select the wrong statement:
(*a*) If a set of vectors is linearly independent, then any subset of it is linearly dependent.
(*b*) The set of vectors $\mathbf{e}_i$, $i = 1, 2, ..., n$ is called the standard basis of $\mathbf{R}^n$.
(*c*) If $c_1x_1 + c_2x_2 + ... + c_nx_n = 0$, then the hyperplane is said to pass through the origin.
(*d*) The intersection of a finite number of closed half spaces is called a polytope.

5. The set $S = \{(x_1, x_2) : x_1x_2 \geq 1 \; ; x_1 \geq 0 \text{ and } x_2 \geq 0\}$ is
(*a*) convex (*b*) not convex
(*c*) concave (*c*) not conave.

6. Which of the following statements is not correct?
(*a*) A cone is called a convex cone, if it is a convex set.
(*b*) The smallest convex set containing **Y** is called the convex hull of **Y**.
(*c*) A simplex in *n*-dimension is a convex polyhedron having *n* vertices.
(*d*) The supporting hyperplane need not be unique.

7. Choose the correct statement:
(*a*) A supporting hyperplane must be unique.
(*b*) A simplex in two dimension is a triangle.
(*c*) A function cannot be both convex and concave.
(*d*) Every strictly convex function may be concave.

8. A function $f(\mathbf{x})$ at point $\mathbf{x} = \mathbf{x}_0$ is said to have achieved global minima, if
(*a*) $f(\mathbf{x}) \geq f(\mathbf{x}_0)$ (*b*) $f(\mathbf{x}) \leq f(\mathbf{x}_0)$
(*c*) $f(\mathbf{x}) = f(\mathbf{x}_0)$ (*d*) none of these

9. Identify the incorrect statement: A quadratic form $\mathbf{Q}(\mathbf{x})$ is:
(*a*) positive definite, if and only if, $\mathbf{Q}(\mathbf{x}) > 0$.
(*b*) negative definite, if and only if, $\mathbf{Q}(\mathbf{x}) < 0$.
(*c*) indefinite, if $\mathbf{Q}(\mathbf{x}) > 0$ for some $\mathbf{x}$ and $\mathbf{Q}(\mathbf{x}) < 0$ for some other $\mathbf{x}$.
(*d*) positive definite as well negative definite irrespective of sign of $\mathbf{Q}(\mathbf{x})$.

10. The quadratic form with the associated matrix $\begin{pmatrix} 1 & 2 & 4 \\ 2 & 6 & -2 \\ 4 & -1 & 14 \end{pmatrix}$ is:

(*a*) $x_1^2 + 6x_2^2 + 14x_3^2 + 4x_1x_2 + 8x_1x_3 - 4x_2x_3$,

(*b*) $x_1^2 + 6x_2^2 + 14x_3^2 + 4x_1x_3 + 8x_2x_3 - 4x_1x_2$,

(*c*) $x_1^2 + 6x_2^2 + 14x_3^2 + 4x_2x_3 + 8x_1x_3 - 4x_1x_2$,

(*d*) $x_1^2 + 6x_2^2 + 14x_3^2 + 8x_1x_2 + 4x_1x_3 - 4x_2x_3$.

ANSWERS

1. (*d*)	2. (*c*)	3. (*c*)	4. (*a*)	5. (*a*)
6. (*c*)	7. (*b*)	8. (*a*)	9. (*d*)	10. (*a*).

APPENDEX B

Answers to Problems

Chapter 2. Linear Programming

2.01. Maximize $z = 5x_1 + 7x_2$ subject to the constraints:

$$4x_1 + 3x_2 \le 960,\ 3x_1 + 2x_2 \le 1{,}200,\ x_2 \le 1{,}500;\ x_1 \ge 0 \text{ and } x_2 \ge 0,$$

where x_1, x_2 = number of units of product *A* and *B*, respectively.

2.02. Minimize $z = x_1 + x_2$ subject to the constraints:

$$2x_1 + x_2 \ge 12,\ 5x_1 + 8x_2 \ge 74,\ x_1 + 6x_2 \ge 24;\ x_1 \ge 0 \text{ and } x_2 \ge 0,$$

where x_1, x_2 = number of pills of size *A* and size *B*, respectively.

2.03. Maximize $z = 3(3x_1 + 5x_2) - 10(x_1 + x_2)$ subject to the constraints:

$$x_1 + x_2 \le 20,\ 5x_2 - x_1 \ge 60,\ 2x_1 + 5x_2 \le 80;\ x_1 \ge 0 \text{ and } x_2 \ge 0,$$

where x_1, x_2 = number of old and young hens, respectively.

2.04. Maximize $z = 4{,}00{,}000x_1 + 9{,}00{,}000x_2 + 5{,}00{,}000x_3 + 2{,}00{,}000x_4$ subject to the constraints:

$$40{,}000x_1 + 75{,}000x_2 + 30{,}000x_3 + 15{,}000x_4 \le 8{,}00{,}000,$$

$$3{,}00{,}000x_1 + 4{,}00{,}000x_2 + 2{,}00{,}000x_3 + 1{,}00{,}000x_4 \ge 20{,}00{,}000,$$

$$40{,}000x_1 + 75{,}000x_2 \le 5{,}00{,}000,$$

$$x_1 \ge 3,\ x_2 \ge 2,\ 5 \le x_3 \le 10,\ 5 \le x_4 \le 10;\ x_1, x_2, x_3, x_4 \ge 0,$$

where x_1, x_2 = number of advertising units on Prime time and Normal time of TV, respectively; x_3, x_4 = number of advertisement units on Radio and Magazine, respectively.

2.05. Maximize $z = 22.50x_1 + 40x_2 + 15x_3$ subject to the constraints:

$$0.01x_1 + 0.03x_2 + 0.02x_3 \le 120,\ 0.2x_1 + 0.5x_2 + 0.3x_3 \le 3{,}000,$$

$$x_1 + x_2 + x_3 \ge 5{,}000,\ x_1 \le 3{,}000,\ x_2 \le 2{,}000,\ x_3 \le 6{,}000;\ x_1, x_2, x_3 \ge 0,$$

where x_1, x_2, x_3 = number of issues of Daily Life, Agriculture Today and Surf's Up per week, respectively.

2.06. Maximize $z = 5{,}000x_1 + 10{,}000x_2 + 25{,}000x_3$ subject to the constraints:

$$0.2x_1 + 0.3x_2 + 0.6x_3 = 4{,}000,\ 0.4x_1 + 0.4x_2 + 0.1x_3 = 5{,}000,$$

$$0.1x_1 + 0.2x_2 + 0.3x_3 = 1{,}500,\ 0 \le x_1 \ge 2{,}500,\ 0 \le x_2 \ge 4{,}000;$$

$$0 \le x_3 \ge 1{,}500,\ x_3 \ge 0.25(x_1 + x_2 + x_3),$$

where x_1, x_2, x_3 = number of manual, semi-automatic and automatic machines, respectively.

2.07. Minimize $z = 0.2x_1 + 0.5x_2 + 0.18x_3$ subject to the constraints:

$$x_1 + x_2 + x_3 = 1,\ x_2 + x_3 \le 2x_1,\ x_3 \ge 0.5x_2;\ x_1, x_2, x_3 \ge 0,$$

where x_1, x_2, x_3 = proportion of equity capital, debentures, and term loan, respectively.

2.08. Maximize $z = 3x_1 + 8x_2 + 2x_3$ subject to the constraints:

$$3x_1 + 4x_2 \le 18,\ 3x_1 + 5x_2 \le 21,\ x_3 = 3x_2,\ x_1 \ge 0,\ x_2 \ge 0,\ 5 \ge x_3 \ge 0,$$

where x_1, x_2, x_3 are the number of units of products *A*, *B* and *C*, respectively.

2.09. Minimize $z = x_1 + x_2 + x_3 + x_4 + x_5 + x_6 + x_7 + x_8 + x_9 + x_{10} + x_{11} + x_{12}$ subject to the constraints:

$$x_1 + x_2 + x_3 + x_4 \geq 50,\ x_2 + x_3 + x_4 + x_5 \geq 60,\ x_3 + x_4 + x_5 + x_6 \geq 80,$$
$$x_4 + x_5 + x_6 + x_7 \geq 80,\ x_5 + x_6 + x_7 + x_8 \geq 70,\ x_6 + x_7 + x_8 + x_9 \geq 70,$$
$$x_7 + x_8 + x_9 + x_{10} \geq 60,\ x_8 + x_9 + x_{10} + x_{11} \geq 50,\ x_9 + x_{10} + x_{11} + x_{12} \geq 50,$$
$$x_{10} + x_{11} + x_{12} + x_1 \geq 30,\ x_{11} + x_{12} + x_1 + x_2 \geq 20,\ x_{12} + x_1 + x_2 + x_3 \geq 40;$$
$$x_j \geq 0\ (j = 1, 2, 3, \ldots, 12),$$

where x_j $(j = 1, 2, \ldots, 12)$ denote the number of nurses required during different time periods starting from 6:00 AM onwards.

2.10. Maximize $z = 1{,}00{,}000x_1 + 80{,}000x_2 + 50{,}000x_3$ subject to the constraints:

$$7{,}500x_1 + 6{,}250x_2 + 5{,}000x_3 \leq 2{,}50{,}000,$$
$$x_1 \leq 12,\ x_2 \geq 5,\ 6 \leq x_3 \leq 10;\ x_1, x_2, x_3 \geq 0.$$

2.11. Minimize $z = x_1 + x_2 + x_3 + x_4 + x_5 + x_6 + x_7$ subject to the constraints:

$$x_1 + x_4 + x_5 + x_6 + x_7 \geq 35,\ x_2 + x_5 + x_6 + x_7 + x_1 \geq 55,\ x_3 + x_6 + x_7 + x_1 + x_2 \geq 60,$$
$$x_4 + x_7 + x_1 + x_2 + x_3 \geq 50,\ x_5 + x_1 + x_2 + x_3 + x_4 \geq 60,\ x_6 + x_2 + x_3 + x_4 + x_5 \geq 50,$$
$$x_7 + x_3 + x_4 + x_5 + x_6 \geq 45;\ 0 \leq x_j \leq 40\ (j = 1, 2, \ldots, 7).$$

where x_j $(j = 1, 2, \ldots, 7)$ denote the number of doctors required on week days, *i.e.*, on Sunday, Monday, ..., to Saturday, respectively.

2.12. Minimize $z = x_1 + x_2 + x_3 + x_4 + x_5 + x_6$ subject to the constraints:

$$x_1 + x_2 \geq 77,\ x_2 + x_3 \geq 85,\ x_3 + x_4 \geq 68,\ x_4 + x_5 \geq 25,$$
$$x_5 + x_6 \geq 23,\ x_6 + x_1 \geq 72;\ x_j\ (j = 1, 2, \ldots, 6) \geq 0,$$

where x_j $(j = 1, 2, \ldots, 6)$ denote the number of doctors starting work on shift j $(j = 1, 2, 3, 4, 5, 6)$.

2.13 Minimize $z = x_1 + x_2 + x_3 + x_4 + x_5 + x_6$ subject to the constraints:

$$x_1 + x_2 \geq 20,\ x_2 + x_3 \geq 14,\ x_3 + x_4 \geq 20,\ x_4 + x_5 \geq 10,$$
$$x_5 + x_6 \geq 5,\ x_6 + x_1 \geq 7;\ x_j \geq 0\ (j = 1, 2, 3, 4, 5, 6),$$

where x_j $(j = 1, 2, 3, 4, 5)$ is the number of workers joining duty at 12 a.m., 8 a.m., 12 p.m., 4 p.m. and 8 p.m., respectively.

2.14. Maximize $z = 4{,}000x_1 + 3{,}500x_2 + (3{,}500 - 1{,}000)x_3 + 2{,}000x_4$ subject to the constraints:

$$0.30x_1 + 0.15x_2 + 0.15x_3 + 0.75x_4 \leq 600,\ \frac{1}{2}x_1 + \frac{1}{3}x_2 + \frac{1}{5}x_4 \leq 100,$$
$$\frac{1}{3}x_3 \leq 40,\ \frac{1}{12}x_1 + \frac{1}{12}(x_2 + x_3) + \frac{1}{12}x_4 \leq 80,\ x_1 \leq 100,$$
$$x_2 + x_3 \leq 400,\ 200 \leq x_4 \leq 600;\ x_1, x_2, x_3, x_4 \geq 0.$$

where x_1, x_2, x_4 = quantity (in kiloliters) produced of Venus, Diana and Aurora, respectively. x_3 denotes the quantity in kiloliters of Diana by hired facility.

2.15. Maximize $z = 200A + 300B - \{(30(x_{1A} + x_{1B}) + 40(x_{2A} + x_{2B}) + 50(x_{3A} + x_{3B})\}$ subject to the constraints:

$$\begin{cases} 0.2x_{1A} + 0.1x_{2A} + 0.5x_{3A} \leq 0.8A, \\ 0.1x_{1A} + 0.2x_{2A} + 0.05x_{3A} \leq 0.3A, \\ 0.3x_{1A} + 0.3x_{2A} + 0.7x_{3A} \geq 0.5A. \end{cases} \quad \text{(alloy specification for } A\text{)}$$

$$0.6x_{1A} + 0.6x_{2A} + 0.8x_{3A} = A. \quad \text{(material balance for } A\text{)}$$

$$\begin{cases} 0.1x_{1B} + 0.2x_{2B} + 0.05x_{3B} \geq 0.4B,\ 0.1x_{1B} + 0.2x_{2B} + 0.05x_{3B} \leq 0.6B, \\ 0.3x_{1B} + 0.3x_{2B} + 0.07x_{3B} \geq 0.3B,\ 0.3x_{1B} + 0.3x_{2B} + 0.2x_{3B} \leq 0.7B. \end{cases} \quad \text{(alloy specification for } B\text{)}$$

$$0.7x_{1B} + 0.8x_{2B} + 0.95x_{3B} = B. \quad \text{(material balance for } B\text{)}$$

$$x_{1A} + x_{1B} \leq 1{,}000,\ x_{2A} + x_{2B} \leq 2{,}000,\ x_{3A} + x_{3B} \leq 3{,}000,$$
$$A,\ B,\ x_{iA},\ x_{iB} \geq 0\ (i = 1,\ 2,\ 3)$$

2.16. Maximize $z = (74{,}400x_{11} + 49{,}600x_{12}) + (46{,}800x_{21} + 31{,}200x_{22}) + (48{,}300x_{31} + 32{,}200x_{32})$ subject to the constraints:

$$x_{11} + x_{12} \le 12,\ x_{21} + x_{22} \le 24,\ x_{31} + x_{32} \le 12,\ x_{11} + x_{12} \ge 5,\ x_{21} + x_{22} \ge 4,\ x_{31} + x_{32} \ge 5,$$
$$18x_{11} + 12x_{12} + 16x_{21} + 10x_{22} + 19x_{31} + 15x_{32} \le 500;\ x_{ij} \ge 0\ (i = 1, 2, 3 \text{ and } j = 1, 2),$$

where x_{11} = number of colour adv. in M_1 and x_{12} = number of black and white adv. in M_1. Similarly, x_{21} and x_{22} for M_2, x_{31} and x_{32} for M_3.

2.17. Maximize $z = 20x_1 + 40x_2 + 20x_3 + 15x_4 + 30x_5$ subject to the constraints:

$$5x_1 + 4x_2 + 3x_3 + 7x_4 + 8x_5 \le 25,\ x_1 + 7x_2 + 9x_3 + 4x_4 + 6x_5 \le 25,$$
$$8x_1 + 10x_2 + 2x_3 + x_4 + 10x_5 \le 25;\ x_j \ge 0\ (j = 1, 2, 3, 4, 5),$$

where x_j $(j = 1, 2, 3, 4, 5)$ designate the respective project.

2.18. Maximize $z = (90y_1 + 60y_2 + 75y_3) - (75x_1 + 75x_2 + 60x_3)$ subject to the constraints:

$$y_1 \le x_1 \le 150,\ y_2 \le x_1 + x_2 - y_1 \le 150,\ y_3 \le x_1 + x_2 + x_3 - y_1 - y_2 \le 150,$$
$$x_1 + x_2 + x_3 = y_1 + y_2 + y_3;\ x_j \ge 0 \text{ and } y_j \ge 0\ (j = 1, 2, 3),$$

where $y_j\,(j = 1, 2, 3)$ and $x_j\,(j = 1, 2, 3)$ designate the number of units to buy and sell, respectively.

2.19. Maximize $z = 9.5x_1 + 8.5x_2 + 12.0x_3 + 15.0x_4 + 32.5x_5 + 35.0x_6$ subject to the constraints:

$$x_1 + x_2 + x_3 + x_4 + x_5 + x_6 = 1,\ 6x_1 + 4x_2 + 3x_3 + 5x_4 + 3x_5 + 10x_6 \ge 5,$$
$$0.02x_1 + 0.01x_2 + 0.08x_3 + 0.25x_4 + 0.45x_5 + 0.40x_6 \le 0.20,\ x_5 + x_6 \le 0.25;$$

$x_j \ge 0\ (j = 1, 2, 3, 4, 5, 6)$, where x_j $(j = 1, 2, 3, 4, 5)$ designate the proportion of total funds invested in Bank deposit, Treasury notes, corporate deposit, Blue-chip stock, speculative stock, and Real estate, respectively.

2.20. Maximize $z = 0.14x_1 + 0.19x_2 + 0.23x_3 + 0.12x_4$ subject to the constraints:

$$x_1 + x_2 + x_3 + x_4 \le 20{,}00{,}000,\ x_4 \ge 2{,}00{,}000,$$
$$\frac{12x_1 + 24x_2 + 48x_3 + 6x_4}{x_1 + x_2 + x_3 + x_4} \le 42 \text{ or } -30x_1 - 18x_2 + 6x_3 - 36x_4 \le 0,$$
$$x_3 \le 0.2(x_1 + x_2 + x_3 + x_4) \text{ or } -0.2x_1 - 0.2x_2 + 0.8x_3 - 0.2x_4 \le 0;$$

$x_j\ (j = 1, 2, 3, 4) \ge 0$, where x_j $(j = 1, 2, 3, 4)$ designate the amount of funds to be invested in Government bonds, Blue-chip bonds, Speculative bounds and Short-term deposits.

2.21. Maximize $z = (x_1' - x_1'') - 3(x_2' - x_2'') + 0 \cdot s_1$ subject to the constraints:

$$-(x_1' - x_1'') + 2(x_2' - x_2'') + s_1 = 15,\ (x_1' - x_1'') + 3(x_2' - x_2'') = 10;$$
$$x_1' \ge 0,\ x_1'' \ge 0,\ x_2' \ge 0,\ x_2'' \ge 0 \text{ and } s_1 \ge 0.$$

2.22. Maximize $z = 3x_1 + 2x_2 + 5x_3 + 0 \cdot s_1 + 0 \cdot s_2 + 0 \cdot s_3$ subject to the constraints:

$$2x_1 - 3x_2 + s_1 = 3,\ x_1 + 2x_2 + 3x_3 - s_2 = 5,\ 3x_1 + 2x_3 + s_3 = 2;\ x_j \ge 0 \text{ and } s_j \ge 0\ (j = 1, 2, 3).$$

2.23. Maximize $z = 2x_1 + x_2 - 6x_3 - 4x_4 + 0 \cdot s_1 + 0 \cdot s_2 + 0 \cdot s_3 + 0 \cdot s_4$ subject to the constraints:

$$3x_1 + x_4 + s_1 = 25,\ x_1 + x_2 + x_3 + x_4 = 20,\ 4x_1 + 6x_2 - s_2 = 5,\ x_1 + 3x_3 + 2x_4 + s_3 = 3,$$
$$x_1 + 3x_3 + 2x_4 - s_4 = 2;\ x_j \ge 0 \text{ and } s_j \ge 0\ \ (j = 1, 2, 3, 4).$$

2.24. Minimize $z = -3x_1 + x_2 + x_3' - x_3'' + 0 \cdot s_1 + 0 \cdot s_2$ subject to the constraints:

$$x_1 - 2x_2 + x_3' - x_3'' + s_1 = 11,\ -4x_1 + x_2 + 2(x_3' - x_3'') - s_2 = 3,$$
$$2x_1 - (x_3' - x_3'') = -1;\ x_1 \ge 0,\ x_2 \ge 0,\ x_3' \ge 0,\ x_3'' \ge 0,\ s_1 \ge 0 \text{ and } s_2 \ge 0.$$

2.25. Maximize $z = 2x_1 + x_2 - 5x_3 + 3x_4$ subject to the constraints:

$$3x_1 + 2x_2 + s_1 = 15,\ 4x_1 + 5x_2 - s_2 = 20,\ x_1 + x_2 - x_3 + 2x_4 = 10,\ 2x_1 + 4x_2 - x_3 + s_3 = 30,$$
$$2x_1 + 4x_2 - x_3 - s_4 = 2;\ x_j \geq 0 \text{ and } s_j \geq 0\ (j = 1, 2, 3, 4),$$

where s_1 and s_3 are slack variables, while s_2 and s_4 are surplus variables.

Chapter 3. Solving Linear Programming Problems

3.01. $x_1 = 6$ and $x_2 = 2$ with maximum $z = 18$.

3.02. $x_1 = 4$ and $x_2 = 2$ with maximum $z = 26$.

3.03. $x_1 = 2$ and $x_2 = 0$ with minimum $z = -2$.

3.04. $x_1 = 6$ and $x_2 = 11$ with maximum $z = 225$.

3.05. $x_1 = 0$ and $x_2 = 30$ with maximum $z = 2{,}700$.

3.06. $x_1 = 4$ and $x_2 = 4/3$ with maximum $z = 56/3$.

3.07. $x_1 = 5$ and $x_2 = 2$ with maximum $z = 1$.

3.08. $x_1 = 0$ and $x_2 = 1$ or $x_1 = 1$ and $x_2 = 0$ with minimum $z = 1$.

3.09. (*a*) Infeasible solution, (*b*) Unbounded solution.

3.10. No feasible solution.

3.11. (*i*) Maximize $z = 0.60x_1 + 0.70x_2$ subject to the constraints:

$$2.4x_1 + 3x_2 \leq 1{,}200;\ 0 \leq x_1 \leq 300 \text{ and } 0 \leq x_2 \leq 240,$$

where x_1 and x_2 are the units of product *A* and *B*, respectively.

(*ii*) Produce 300 units of product *A* and 160 units of product *B*; maximum profit = ₹ 292.

3.12. Produce 9 pieces of product P_2 only; maximum profit = ₹ 450.

3.13. *Formulation.* Minimize $z = 5x_1 + 4x_2$ subject to the constraints:

$$40x_1 + 20x_2 \geq 100,\ 40x_1 + 20x_2 \geq 200,\ 20x_1 + 60x_2 \geq 200,\ x_1 + x_2 \leq 400;\ x_1 \geq 0, x_2 \geq 0;$$

where x_1, x_2 = number of valves purchased from supplier *A* and *B*, respectively.

Optimum solution. Purchase 4 valves from supplier *A* and 2 valves from supplier *B* at a minimum cost of ₹ 28.

3.14. *Formulation.* Minimize $z = 600x_1 + 400x_2$ subject to the constraints:

$$3x_1 + 3x_2 \geq 40,\ 3x_1 + x_2 \geq 40,\ 2x_1 + 5x_2 \geq 44;\ x_1 \geq 0 \text{ and } x_2 \geq 0,$$

where x_1, x_2 = number of days of work at plant G and *J*, respectively.

Optimum Solution. Plant G and *J* should run for 12 days and 4 days respectively to have a minimum production cost of 8,800 monetary units.

3.15. Produce 450 units of product *A* and no unit of product *B*, or produce 300 units of product *B* only to get the maximum profit of ₹ 2,700.

3.16. Produce 18 dresses of type *A* and 12 dresses of type *B* to get the maximum profit of ₹ 5,040.

3.17. Optimum product mix : 3 kgs of ingredient B_1 and 2 kgs of ingredient B_2 to have the minimum cost of ₹ 31.

3.18. Product 20 units each of product *A* and *B* to have the minimum total cost of ₹ 3,000.

3.19. *Formulation.* Maximize $z = 60(9x_1 + 5x_2) + 90(7x_1 + 9x_2)$ subject to the constraints:

$$9x_1 + 5x_2 \geq 500,\ 7x_1 + 9x_2 \geq 300,\ 5x_1 + 3x_2 \leq 1{,}500,\ 7x_1 + 9x_2 \leq 1{,}900,$$
$$2x_1 + 4x_2 \leq 1{,}000;\ x_1 \geq 0 \text{ and } x_2 \geq 0,$$

where x_1, x_2 = number of units of quality *R* and *S* petrol, respectively.

Optimum solution. Produce only 1,900/7 (= 271.5) units of Petrol of quality *R* only to get the maximum revenue of ₹ 3,17,571.40.

3.20. Product 400/13 units under process 1 and 150/13 units under process 2 to achieve the maximum profit of ₹ 1,80,000/13.

3.21. Purchase 2,500 kgs of scrap metal from supplier *A* and 500 kgs of scrap metal from supplier *B* to get the minimum cost of ₹ 4,500.

3.22. Produce 10 units of model X and 2 units to model *Y* to achieve the maximum profit of ₹ 560.

3.23. To achieve the minimum cost of ₹ 440, 100 kgs of type *I* and 80 kgs of type *II* fertilizer should be used.

3.24. (*i*) *Formulation.* Maximize $z = 40x_1 + 80x_2$ subject to the constraints:

$$2x_1 + 3x_2 \le 48;\ 0 \le x_1 \le 15 \text{ and } 0 \le x_2 \le 10,$$

where x_1, x_2 = number of units of AM radio and AM-FM radio.

(*ii*) *Optimum solution.* Produce 9 units of AM radio and 10 units of AM-FM radio to achieve the maximum profit of ₹ 1,160.

3.25. Firm should make 3 units of product X and 6 units of product *Y* to achieve maximum profit of ₹ 960.

3.26. For the aeroplane, 40 tickets of first class and 160 tickets of economy class must be sold to achieve the maximum profit of ₹ 64,000.

3.27. *Formulation.* Maximize $z = 90x_1 + 60x_2$ subject to the constraints:

$$5x_1 + 8x_2 \le 2{,}000,\ 7x_1 + 4x_2 \le 1{,}400;\ 0 \le x_1 \le 175 \text{ and } 0 \le x_2 \le 255,$$

where x_1, x_2 = number of outboard motors of type *A* and type *B*, respectively.

Optimum solution. Produce (800/9) units of type *A* and (1,750/9) units of type *B* motors to achieve the maximum profit of ₹ 19,667 approx.

3.28. Produce 30 units of product *X* and 50 units of product *Y* to achieve the maximum profit of ₹ 6,500.

3.29. (*a*) $x_1 = 3$ and $x_2 = 9$; maximum $z = 330$.

(*b*) $x_1 = 0$ and $x_2 = 20$ or $x_1 = 75/4$ and $x_2 = 25/2$; maximum $z = 200$.

3.30. (*a*) Infeasible solution.

(*b*) $x_1 = 15/2$ and $x_2 = 9/4$ or $x_1 = 2$ and $x_2 = 5$; maximum $z = 12$.

3.31. $x_1 = 0, x_2 = 8$ and $x_3 = 10$; maximum $z = 160$.

3.32. $x_1 = 6, x_2 = 0$ and $x_3 = 8$; maximum $z = 86$.

3.33. $x_1 = 15, x_2 = 5$ and $x_3 = 0$; maximum $z = 25$.

3.34. (*a*) $x_1 = 200/11, x_2 = 1{,}000/11$; maximum $z = 46{,}000/11$.

(*b*) $x_1 = 100/3, x_2 = 200/3, x_3 = 0$; maximum $z = 22{,}000/3$.

3.35. $x_1 = 100/3, x_2 = 50/3, x_3 = 50/3$; maximum $z = 1{,}650$.

3.36. $x_1 = 1, x_2 = 1, x_3 = 1/2$ and $x_4 = 0$; maximum $z = 13/2$.

3.37. Unbounded solution.

3.38. (*a*) $x_1 = 0$ and $x_2 = 3$; maximum $z = 60$.

(*b*) $x_1 = 18$ and $x_2 = 0$; minimum $z = 36$.

(*c*) $x_1 = 3/5$ and $x_2 = 6/5$; minimum $z = 18/5$.

3.39. $x_1 = x_2 = x_3 = 5/2$ and $x_4 = 0$; maximum $z = 15$.

3.40. $x_1 = 23/3, x_2 = 5$ and $x_3 = 0$; maximum $z = 85/3$.

3.41. Maximize $z = 30x_1 + 20x_2$ subject to the constraints:

$$x_1 + x_2 \le 8,\ 6x_1 + 4x_2 \ge 12,\ 5x_1 + 8x_2 = 20;\ x_1, x_2, x_3 \ge 0.$$

Optimal solution: $x_1 = 4$ and $x_2 = 0$; maximum $z = 120$.

3.42. (*a*) Minimize $z = 50x_1 + 60x_2$ subject to the constraints:

$$x_1 + x_2 = 1{,}000,\ x_1 \le 300,\ x_2 \ge 150 \text{ and } x_1 \ge 0,\ x_2 \ge 0,$$

where x_1, x_2 = no. of kgs of chemical *A* and *B*, respectively.

(*b*) Blend 300 kg of chemical *A*, and 700 kg of chemical *B* to achieve the least cost of ₹ 57,000.

3.43. Manufacture 30 trucks and 15 automobiles per week; maximum profit will be of ₹ 12,000.

3.44. (*i*) Maximize $z = 25x_1 + 40x_2$ subject to the constraints:

$$0.50x_1 + 0.25x_2 \le 2{,}200,\ 4x_1 + 7x_2 \le 28{,}000, \text{ and}$$
$$0.20x_1 + 0.20x_2 \le 1{,}400;\ x_1, x_2 \ge 0.$$

(*ii*) *Optimum solution:* $x_1 = 3{,}360$, $x_2 = 2{,}080$; maximum z = ₹ 1,67,200.

(*iii*) Used and unused resources:

Resources	*Used*	*Unused*	*Total*
Leather	2,200 kgs	0 kg	2,200 kgs
Wood	28,000 sq. metres	0 sq. metres	28,000 sq. metres
Glue	1,080 litres	312 litres	1,392 litres

(*iv*) *Shadow prices:* ₹ 6, ₹ 5.5 and ₹ 0.00 for leather, wood and glue, respectively.

3.45. Produce 89/41 units of belt *A*, 50/41 units of belt *B* and 62/41 units of belt *C*; maximum profit will be of ₹ 765/41.

3.46. (*i*) Maximize $z = 10x_1 + 25x_2 + 3x_3$ subject to the constraints:

$$3x_1 + 10x_2 + x_3 \le 30{,}000,\ 5x_1 + 8x_2 + x_3 \le 40{,}000,\ x_1 + 6x_2 + x_3 \le 1{,}200;\ x_1, x_2, x_3 \ge 0.$$

(*ii*) $x_1 = 1{,}200$, $x_2 = x_3 = 0$ with maximum $z = 12{,}000$.

(*iii*) The problem does not have multiple solutions.

3.47. (*i*) Minimize $z = 2x_1 + 3x_2$ subject to the constraints:

$$x_1 + x_2 \ge 350,\ x_1 \ge 125,\ 2x_1 + x_2 \le 600;\ x_1 \ge 0 \text{ and } x_2 \ge 0.$$

(*ii*) $x_1 = 250$ and $x_2 = 100$; minimum $z = 800$.

(*iii*) The LPP have unique optimum solution, since none of the non-basic variables have $(z_j - c_j) = 0$.

3.48. *Optimum solution:* Manufacture 100 units of product 2; 230 units of product 3; while no unit of product 1, maximum profit will be of ₹ 1,350.

3.49. (*i*) Yes, because each of the basic variables is non-negative.

(*ii*) Yes; $x_1 = 0$, $x_2 = 30$ and $x_3 = 10$; maximum $z = 1{,}500$.

(*iii*) No, because there exist an optimum solution.

(*iv*) No, since all basic variables are non-negative.

(*v*) No, because corresponding to non-basic variables, none of $(z_j - c_j)$ is zero.

(*vi*) Because its marginal contribution is (–8) rupees per unit.

(*vii*) In order to produce product X, price of the product should be increased by ₹ 8.

3.50. (*a*) Maximize $z = 250x_1 + 300x_2 + 400x_3$ subject to the constraints:

$$5x_1 + 3x_2 + 5x_3 \le 100,\ 3x_1 + 5x_2 + 5x_3 \le 80;\ x_1, x_2, x_3 \ge 0,$$

where x_1, x_2 and x_3 denote the Soft, Strong and hard alloy respectively.

(*b*) Produce 10 tons of Soft alloy as well as of Strong alloy only to get the maximum profit of ₹ 6,500.

3.51. Produce only 23.75 filing cabinets and no other items, maximum revenue will be of ₹ 9,500.

3.52. 15 gams of food *A* and 20 grams of food *B* should be consumed for a total minimum cost of ₹ 25.

3.53. Diet should contain 5 units of food *A* and 30 units of food *B*. Minimum cost will be of ₹ 110.

3.54. Produce 30 kgs of product *5-10-5* and 1,185 kgs of product *5-5-10*, while product *20-5-10* should not be produced, maximum profit will be of ₹ 20,625.

3.55. (*i*) The model formulation of the LPP is

Maximize $z\ 40{,}000x_1 + 55{,}000x_2$ subject to the constraints:

$$1{,}000x_1 + 1{,}500x_2 \le 20{,}000;\ 0 \le x_1 \le 12,\ 0 \le x_2 \ge 5,$$

where x_1, x_2 = annual number of insertions/messages for media *A* and media *B*, respectively.

(*ii*) Advertise 12 messages in media *A* and 5.3 messages in media *B*; maximum effective audience will be 77,333.

3.56. Allocate 250 acres to corn and 625 acres to wheat only, while soyabean should not be grown; maximum profit will be of ₹ 32,500.

3.57. Firm should give 6 advertisements in TV and 5 advertisements in magazine; maximum sales will be of 1,47,000.

3.58. Three programmes each of Radio as well as on TV to reach the maximum total effective audience of 24,00,000; Type *A* audience : 15,90,000 and Type *B* audience: 8,10,000.

3.59. $x_1 -1$ and $x_2 = 3$.

3.60. $x_1 = 5, x_2 = 7, x_3 = 2$ and $x_4 = 0$.

Chapter 4. Theory of Simplex Method

4.01. $x_1 = 0$ and $x_3 = 7$ with $x_2 = 0$ (non-basic),
$x_1 = 0$ and $x_2 = 7/2$ with $x_3 = 0$ (non-basic).

4.02. (*i*) $x_1 = 0$ and $x_2 = 1/2$ with $x_3 = x_4 = 0$ (non-basic),
$x_2 = 1/2$ and $x_3 = 0$ with $x_1 = x_4 = 0$ (non-basic),
$x_2 = 1/2$ and $x_4 = 0$ with $x_1 = x_3 = 0$ (non-basic).

(*ii*) $x_1 = 2$ and $x_2 = 2$ with $x_3 = 0$ (non-basic),
$x_1 = 3$ and $x_3 = 1$ with $x_2 = 0$ (non-basic).

4.03. (*i*) $x_1 = 1,\ x_2 = 2$ (basic) and $x_3 = x_4 = 0$ (non-basic),
(*ii*) $x_1 = 22/9,\ x_4 = 7/9$ (basic) and $x_2 = x_3 = 0$ (non-basic),
(*iii*) $x_2 = 45/16,\ x_3 = 7/16$ (basic) and $x_1 = x_4 = 0$ (non-basic),
(*iv*) $x_3 = 44/17,\ x_4 = 45/17$ (basic) and $x_1 = x_2 = 0$ (non-basic).

4.04. $x_1 = 0,\ x_2 = 3,\ x_3 = 0,\ x_4 = 10,\ x_5 = 0,\ x_6 = 1$; and
$x_1 = 4,\ x_2 = 5,\ x_3 = 0,\ x_4 = 0,\ x_5 = 0,\ x_6 = 11$.

4.05. (*i*) $x_1 = 2, x_2 = 2$ and $x_3 = 0$, (*ii*) $x_1 = 0, x_2 = 0$ and $x_3 = 2$, (*iii*) $x_1 = 0, x_2 = 0$ and $x_3 = 2$.

4.06. (*i*) $x_1 = 3,\ x_2 = 2$ and $x_3 = 0$, (*ii*) $x_1 = 0,\ x_2 = 1/2$ and $x_3 = 3/2$.

4.07. $x_1 = 1,\ x_2 = 2$; Maximum $z = 5$.

4.08. $x_1 = 8/5,\ x_2 = 3/5$; maximum $z = 11/5$.

4.09. $x_1 = 2/7,\ x_2 = 9/7$; maximum $z = 13/7$.

4.10. $x_1 = 0.43,\ x_2 = 0.14$; maximum $z = 3.86$.

4.11. $x_1 = 3.33,\ x_2 = 1.33;$ maximum $z = 12.67.$

4.12. $x_1 = 4,\ x_2 = 6,\ x_3 = 0;$ maximum $z = 12.$

4.13. Unbounded solution.

4.14. Unbounded solution.

4.15. $x_1 = 3/5,\ x_2 = 6/5;$ minimum $z = 12/5.$

4.16. $x_1 = 59/17,\ x_2 = 36/17,\ x_3 = 97/17;$ maximum $z = 6/17.$

Chapter 5. Duality in Linear Programming

5.01. (*a*) Minimize $z^* = 4y_1 - y_2 - 2y_3$ subject to the constraints:

$$y_1 - y_3 \ge 2,\ y_2 + 2y_3 \le -3;\ y_1 \ge 0 \text{ and } y_2 \ge 0.$$

(*b*) Minimize $z^* = 10y_1 + 2y_2 + 6y_3$ subject to the constraints:

$$y_1 + 2y_2 + 2y_3 \ge 1,\ y_1 - 2y_3 \ge -1,\ y_1 - y_2 + 3y_3 \ge 3;\ y_1, y_2, y_3 \ge 0.$$

5.02. Maximize $z^* = 40y_1 + 20y_2 + 30y_3$ subject to the constraints:

$$2y_1 + 20y_2 + 4y_3 \le 20,\ 20y_1 + 3y_2 + 15y_3 \le 40;\ y_1, y_2, y_3 \ge 0.$$

5.03. Maximize $z^* = -160y_1 + 30y_2 + 10y_3$ subject to the constraints:

$$-2y_1 + y_2 + y_3 \le 1,\ -4y_1 - y_2 \le 2;\ y_1, y_3 \ge 0 \text{ and } y_2 \text{ unrestricted.}$$

5.04. Maximize $z^* = -100y_1 + 50y_2 - 150y_3$ subject to the constraints:

$$-2y_1 + 8y_2 - 5y_3 \le 80,\ -5y_1 + 9y_2 - 6y_3 \le 90,\ 3y_1 - 10y_3 \le 100;\ y_1, y_2, y_3 \ge 0.$$

5.05. Minimize $z^* = 10y_1 - 15y_2$ subject to the constraints:

$$y_1 - 4y_2 \ge 3,\ y_1 + y_2 \ge 5,\ 3y_1 - 2y_2 = 7;\ y_1 \ge 0 \text{ and } y_2 \ge 0.$$

5.06. Maximize $z^* = -5y_1 + 9y_2 + 8y_3$ subject to the constraints:

$$-2y_1 + 4y_2 - 8y_3 \le 3,\ 3y_1 - 2y_2 + 4y_3 \le -2,\ -y_1 + 3y_3 = 1;$$
$$y_1 \ge 0,\ y_2 \ge 0 \text{ and } y_3 \text{ unrestricted.}$$

5.07. Maximize $z^* = 2y_1 + 3_2 - 5y_3$ subject to the constraints:

$$2y_1 + 3y_2 - y_3 \le 2,\ 3y_1 + y_2 - 4y_3 \le 3,\ 5y_1 + 7y_2 - 6y_3 \le 4;$$
$$y_1 \ge 0,\ y_3 \ge 0 \text{ and } y_2 \text{ unrestricted.}$$

5.08. Minimize $z^* = 50y_1 - 40y_2 + 75y_3$ subject to the constraints:

$$y_1 - 2y_2 + 3y_3 \ge 7,\ 2y_1 - 3y_2 + 4y_3 \ge 2,\ 3y_1 + 4y_2 + 5y_3 = -5;$$
$$y_1 \ge 0,\ y_2 \ge 0 \text{ and } y_3 \text{ unrestricted.}$$

5.09. Maximize $z^* = 15y_1 + 20y_2 - 10y_3$ subject to the constraints:

$$y_1 + 2y_2 - y_3 \le 1,\ 2y_1 + y_2 - 2y_3 \le 2,\ 3y_1 + 3y_2 - y_3 \le 3;$$
$$y_1 \ge 0,\ y_2 \ge 0,\ y_3 \text{ unrestricted.}$$

5.10. *Primal:* Maximize $z = 30x_1 + 20x_2$ subject to the constraints:

$$2x_1 + x_2 \le 800 \text{ and } x_1 + 2x_2 \le 1{,}000;\ x_1 \ge 0 \text{ and } x_2 \ge 0,$$

where x_1 and x_2 designate the units of product *A* and *B* respectively.

Dual: Minimize $z^* = 800y_1 + 1{,}000y_2$ subject to the constraints:

$$2y_1 + y_2 \ge 30 \text{ and } y_1 + 2y_2 \ge 20;\ y_1 \ge 0 \text{ and } y_2 \ge 0,$$

where y_1 and y_2 designate the number of units of chemical and compound respectively.

5.11. (*a*) Both primal as well as dual have no feasible solution.

(*b*) *Primal:* $x_1 = 5/2$ and $x_2\ 3/2$ with maximum $z = 9.$

Dual: $y_1 = 1,\ y_2 = 2$ and $y_3 = 0$ with minimum $z^* = 9.$

5.12. *Primal:* $x_1 = 4$ and $x_2 = 0$; maximum $z = 8$.
Dual: $y_1 = 0, y_2 = 0$ and $y_3 = 1$; minimum $z^* = 8$.

5.13. *Primal:* $x_1 = 7$ and $x_2 = 0$; maximum $z = 21$,
Dual: $y_1 = 0, y_2 = 3, y_3 = y_4 = 0$; minimum $z^* = 21$.

5.14. *Primal:* $x_1 = 0, x_2 = 11/12, x_3 = 5/4$; minimum $z = 67/12$,
Dual: $y_1 = 7/12$ and $y_2 = 2/3$; maximum $z^* = 67/12$.

5.15. *Primal:* $x_1 = 23/3, x_2 = 5, x_3 = 0$; maximum $z = 85/3$,
Dual: $y_1 = 0, y_2 = 5/3, y_3 = 14/3$; minimum $z^* = 85/3$.

5.16. *Primal:* $x_1 = 45, x_2 = 15, x_3 = 0$; minimum $z = 675$.
Dual: $y_1 = 5/2, y_2 = 0, y_3 = 15/2$; maximum $z^* = 675$.

5.17. *Dual:* Minimize $z^* = 120y_1 - 90y_2 + 100(y_3' - y_3'')$ subject to the constraints:

$$y_1 - 3y_2 + 2(y_3' - y_3'') \geq 5,\ y_1 + 2y_2 + 4(y_3' - y_3'') \geq 8,\ 2y_1 + y_2 + 2(y_3' - y_3'') \geq 10;$$

$$y_1, y_2, y_3', y_3'' \geq 0.$$

Introducing surplus variables; $s_1 \geq 0, s_2 \geq 0, s_3 \geq 0$ and artificial variables: $A_1 \geq 0, A_2 \geq 0$, and $A_3 \geq 0$; the initial basic feasible solution is $A_1 = 5$, $A_2 = 8$ and $A_3 = 8$. Initial basic feasible solution is displayed in simplex tableau as follows:

Zeroth Iteration. y_3' enters and A_2 leaves.

Basic variable	Coefficients of										Solution
	y_1	y_2	y_3'	y_3''	s_1	s_2	s_3	A_1	A_s	A_3	
A_1	1	–3	2	–2	–1	0	0	1	0	0	5
$\leftarrow A_2$	1	2	$\boxed{4}$	–4	0	–1	0	0	1	0	8
A_3	2	1	2	–2	0	0	–1	0	0	1	10
$(z_j - c_j)$	$4M - 120$	90	$8M\uparrow - 100$	$-8M + 100$	$-M$	$-M$	$-M$	0	0	0	$23M$

Since, it is minimization problem, all $(z_j - c_j)$ must be ≤ 0 for optimal solution.

First Iteration. y_1 enters and A_1 leaves.

Basic variable	Coefficients of											Solution
	y_1	y_2	y_3'	y_3''	s_1	s_2	s_3	A_1	A_s	A_3	z^*	
$\leftarrow A_1$	$\boxed{1/2}$	–4	0	0	–1	1/2	0	1	–1/2	0	0	1
y_3'	1/4	1/2	1	–1	0	–1/4	0	0	1/4	0	0	2
A_3	3/2	0	0	0	0	1/2	–1	0	–1/2	1	0	6
$(z_j - c_j)$	$2M\uparrow - 95$	$-4M - 140$	0	0	$-M$	$M - 25$	$-M$	0	$-M + 25$	0	1	$7M + 200$

5.18. *Dual Formulation:* Minimize $z^* = 100y_1 + 80y_2$ subject to the constraints:

$$2y_1 + y_2 \geq 100,\ 6y_1 + 5y_2 \geq 400,\ 3y_1 + 5y_2 \geq 250;\ y_j \geq 0\ (j = 1, 2, 3).$$

Primal Solution: $x_1 = 5, x_2 = 15, x_3 = 0$; maximum $z = 6{,}500$.
Dual Solution: $y_1 = 25$ and $y_2 = 50$; minimum $z^* = 6{,}500$.

5.19. *Primal:* $x_1 = 20, x_2 = 0, x_3 = 5$; maximum $z = 1{,}050$.
Dual: $y_1 = 0, y_2 = 10, y_3 = 10$; minimum $z^* = 1{,}050$.

5.20. *Primal Formulation:* Minimize $z = 10x_1 + 36x_2 + 24x_3$ subject to the constraints:

$$x_1 + 4x_2 + 2x_3 \geq 20,\ 2x_1 + 9x_2 + 5x_3 \geq 30;\ x_j \geq 0\ (j = 1, 2, 3).$$

Optimum Solution: $x_1 = 0, x_2 = 5, x_3 = 0$; minimum $z = 180$.

Dual Formulation: Maximize $z^* = 20y_1 + 30y_2$ subject to the constraints:

$$y_1 + 2y_2 \leq 10,\ 4y_1 + 9y_2 \leq 36,\ 2y_1 + 5y_2 \leq 24;\ y_j \geq 0\ (j = 1, 2, 3).$$

Optimum Solution: $y_1 = 9, y_2 = 0$; maximum $z^* = 180$.

5.21. *Primal:* Maximize $z = 800x_1 + 500x_2$ subject to the constraints:

$$2x_1 + 3x_2 \leq 42,\ 7x_1 + 7x_2 \leq 70,\ 7x_1 + 5x_2 \leq 70;\ x_j \geq 0\ (j = 1, 2).$$

Dual: Minimize $z^* = 42y_1 + 70y_2 + 70y_3$ subject to the constraints:

$$2y_1 + 7y_2 + 7y_3 \geq 800,\ 3y_1 + 7y_2 + 5y_3 \geq 500;\ y_j \geq 0\ (j = 1, 2, 3).$$

Optimum Output: Produce 10 units of model *A* and no unit of model *B*; maximum profit is of ₹ 8,000.

Dual variables denote the worth of the resources of the company. The variables y_1, y_2 and y_3 represent the marginal profitability of the capacity in department *I*, *II* and *III*, respectively. If the company were to rent out its capacity, the minimum rental for each department would be equal to the marginal profitability. Since, the department *I* and *III* have unused capacity, their marginal profitability is nil, while for department *II*, it is ₹ 800/7 per hour. Hence, the value of dual variables are: $y_1 = 0$, $y_2 = 800/7$ and $y_3 = 0$. Therefore, minimum of rental will be: $z^* = 42 \times 0 + 70 \times 800/7 + 70 \times 0 =$ ₹ 8,000. Thus, max. z = min. z^* = ₹ 8,000.

5.22. (*a*) Minimize $z = 3x_1 + 2x_2$ subject to the constraints:

$$14x_1 + 4x_2 \geq 60,\ 10x_1 + 8x_2 \geq 40,\ 4x_1 + 16x_2 \geq 32;\ \text{and}\ x_1 \geq 0,\ x_2 \geq 0.$$

(*b*) Maximize $z^* = 60y_1 + 40y_2 + 32y_3$ subject to the constraints:

$$14y_1 + 10y_2 + 4y_3 \leq 3,\ 4y_1 + 8y_2 + 16y_3 \leq 2;\ y_j \geq 0\ (j = 1, 2, 3).$$

Economic Interpretation: The optimal values of the dual variables y_1, y_2 and y_3 would indicate the imputed value of one unit each of vitamins *A*, *B* and *C*, respectively. Obviously, the total value imputed to 14 units of *A*, 10 units of *B*, and 4 units of *C* should not exceed ₹ 3, because each unit of food F_1 contains as much quantity as the three vitamins and costs ₹ 3. Similarly, a unit of food F_2 contains, respectively, 4, 8 and 16 units of vitamins *A*, *B* and *C*, and costs ₹ 2. Thus, the combined imputed value of these quantities of vitamins should not exceed ₹ 2. The total value would be equal to $60y_1 + 40y_2 + 32y_3$ and is to be maximized.

(*c*) *Optimum Solution:* $x_1 = 4, x_2 = 1$; minimum $z = 14$.

5.23. *Optimum mix:* $x_1 = 100/3$, $x_2 = 20/3$, $x_3 = 0$; maximum profit = ₹ 7,333.33; where x_1, x_2 and x_3 be the shipment (per 100 units) of transistors, resistors and electron, respectively.

Dual: Minimum cost $= 100y_1 + 600y_2 + 300y_3$ subject to the constraints:

$$y_1 + 10y_2 + 2y_3 \geq 100,\ y_1 + 4y_2 + 2y_3 \geq 60,$$
$$y_1 + 5y_2 + 6y_3 \geq 40;\ \text{and}\ y_1 \geq 0, y_2 \geq 0, y_3 \geq 0.$$

where y_1, y_2, and y_3 denote the engineering service, direct labour, and administrative service in hours, respectively.

Economic Interpretation: It may be observed that the maximum profitability of three resources is engineering time — ₹ 100/3 per hour, labour time — ₹ 20/3 per hour, and administrative time is nil. These are the minimum rentals that this firm would seek, if the capacity to be rented out. The minimum total rental would be of ₹ (100 × 100/3) + ₹ (600 × 20/3) + ₹ (300 × 0), *i.e.*, ₹ 7,333.33 which is the same as the maximum profit.

5.24. (*i*) Models A and B = none; Model C = 15 cars; maximum profit = ₹ 4,50,000.

(*ii*) Production process I : ₹ 150; Range: 0 to 10,000/3.

Production process II : nil; Range: 5,400 to ∞.

Production process III : nil; Range: 2,400 to ∞.

(*iii*) Minimum $z^* = 3{,}000y_1 + 6{,}000y_2 + 3{,}000y_3$ subject to the constraints:

$$60y_1 + 100y_2 + 80y_3 \geq 7{,}500,\ 100y_1 + 240y_2 + 100y_3 \geq 15{,}000,$$
$$200y_1 + 360y_2 + 160y_3 \geq 30{,}000;\ y_1 \geq 0,\ y_2 \geq 0 \text{ and } y_3 \geq 0,$$

where y_1, y_2 and y_3 are dual variables.

Optimum Solution: $y_1 = 150$, $y_2 = y_3 = 0$,

(*iv*) $x_1 = 10$, $x_2 = 10$ and $x_3 = 10$.

5.25. (*a*) Minimum $z^* = 550y_1 + 700y_2 + 200y_3$ subject to the constraints:

$$1.5y_1 + 4y_2 + 2y_3 \geq 4,\ 2y_1 + y_2 + 3y_3 \geq 6,\ 4y_1 + 2y_2 + y_3 \geq 3,$$
$$3y_1 + y_2 + 2y_3 \geq 1;\ y_j \geq 0\ (j = 1, 2, 3).$$

(*b*) $x_1 = 0$, $x_2 = 25$, $x_3 = 125$, and $x_4 = 0$; maximum $z = 525$.

(*c*) Loss will be of ₹ 7.20. (*d*) No effect on optimal product mix.

5.26. (*i*) Yes, since all $c_j - z_j \leq 0$.

(*ii*) The optimal product mix is: $x_1 = 0$, $x_2 = 8/3$, and $x_3 = 56/3$ units.

(*iii*) Department II should be given priority.

(*iv*) Shadow prices of three departments are 1/6, 11/3 and 0, respectively.

(*v*) Minimize $z^* = 96y_1 + 40y_2 + 60y_3$ subject to the constraints:

$$6y_1 + 2y_2 + 5y_3 \geq 2,\ 8y_1 + y_2 + 3y_3 \geq 5,\ 4y_1 + 2y_2 + 2y_3 \geq 8;\ y_1, y_2, y_3 \geq 0.$$

(*vi*) $y_1 = 1/6$, $y_2 = 11/3$ and $y_3 = 0$.

5.27. $x_1 = 21/13$, $x_2 = 10/13$; minimum $z = 31/13$.

5.28. $x_1 = 1$ and $x_2 = 0$; minimum $z = 1$.

5.29. $x_1 = 3/5$, $x_2 = 6/5$; maximum $z = -12/5$,

5.30. $x_1 = 1/4$, $x_2 = 5/4$, $x_3 = 0$; minimum $z = 10$.

5.31. $x_1 = 0$, $x_2 = 30/11$, $x_3 = 16/11$, $x_4 = 0$; minimum $z = 258/11$.

5.32. $x_1 = 65/23$, $x_2 = 0$, $x_3 = 20/23$; minimum $z = 215/23$.

5.33. $x_1 = 0$, $x_2 = 0$, $x_3 = 1/2$, $x_4 = 3/2$, $x_5 = 0$; minimum $z = 24$.

5.34. $x_1 = 0$, $x_2 = 4/3$, $x_3 = 0$; maximum $z = 4/3$.

5.35. $x_1 = 6$, $x_2 = 2$ and $x_3 = 0$; minimum $z = 10$.

5.36. $x_1 = x_2 = 0$ and $x_3 = 4$; minimum $z = 16$.

Chapter 6. Post-Optimal Analysis

6.01. (*i*) *Optimum Solution:* $x_1 = 27.1$, $x_2 = 13.3$; Maximum $z = 1{,}005$;

(*ii*) $2.8 \leq c_1 \leq 112.5$.

6.02. $12 \leq b_1 \leq 60$ and $6 \leq b_2 \leq 30$.

6.03. $1.25 \leq c_3 \leq 11.5$, $c_4 \leq 1.10$; $3.75 \leq b_2 \leq 17.42$.

6.04. (*a*) $x_1 = 185.71,\ x_2 = 28.57,\ x_3 = 0$; maximum $z = 485.70$.
(*b*) $x_1 = 0,\ x_2 = 17.65,\ x_3 = 56.47$; maximum $z = 240$.

6.05. (*a*) $x_1 = 14,\ x_2 = 0,\ x_3 = 10$; minimum $z = 52$.
(*b*) $x_1 = 7,\ x_2 = 0$ and $x_3 = 5$; minimum $z = 26$.

6.06. (*a*) Minimum $z^* = 7y_1 + 12y_2 + 10y_3$ subject to the constraints:
$$-3y_1 + 2y_2 + 3y_3 \le 1,\ -y_1 + 4y_2 + 3y_3 \ge 3,\ -2y_1 - 8y_3 \le 2;\ y_j \ge 0\ (j = 1, 2, 3).$$
(*b*) $y_1 = 1/5,\ y_2 = 4/5$ and $y_3 = 0$.
(*c*) $c_5 \le -2/5$. (*d*) $-3 \le b_1 \le \infty$.

6.07. (*a*) $x_1 = 0,\ x_2 = 8/3,\ x_3 = 8/3,\ x_4 = 0$; maximum $z = 40/3$.
(*b*) No feasible solution.
(*c*) $x_1 = 49/2,\ x_2 = 0,\ x_3 = 0,\ x_4 = 11/2$; maximum $z = 87/2$.
(*d*) $x_1 = 0,\ x_2 = 21/2,\ x_3 = 11/10,\ x_4 = 47/10$; maximum $z = 431/10$.

6.08. (*a*) $x_1 = 3/2$ and $x_2 = 2$; maximum $z = 8.5$.
(*b*) Resources *1* and *2* are scarce, while resource *3* is in abundance.
(*c*) $x_1 = 7/2$ and $x_2 = 7/8$; maximum $z = 12.25$.

6.09. (*a*) $(15/16) \le a_{12},\ (-5/6) \le a_{22}$ and $(7/8) \le a_{32}$.
(*b*) $-3 \le \Delta a_{11} \le 17/21$.

6.10. (*a*) $x_1 = 5/3,\ x_2 = 56/27,\ x_3 = 0,\ x_4 = 5/27$; maximum $z = 643/27$.
(*b*) $-1 \le \Delta a_{11} \le 56/25$.
(*c*) $(-41/15) \le a_{13} < \infty,\ (39/40) \le a_{23} < \infty,\ (2/21) \le a_{33} < \infty$.

6.11. $x_1 = 0,\ x_2 = 25/19,\ x_5 = 9/19$; maximum $z = 30$.

6.12. (*a*) $2.8 \le c_1 \le 56,\ 3.45 \le c_4 \le 7.50,\ (-1/9) \le c_7 \le 1,\ c_2 \le 8.45,\ c_3 \le 1.50$.
(*b*) $x_1 = 11/14,\ x_2 = 5/21,\ x_3 = 0$; maximum $z = 139/42$.
(*c*) No change in the optimum solution.
(*d*) $x_1 = 33/38,\ x_2 = 0,\ x_3 = 0,\ x_4 = 1/19$; maximum $z = 113/38$.

6.13. (*a*) *Optimum Solution:* 2 rings and 3 brackets; maximum profit is: ₹ 230.
(*i*) 1.6 rings and 3.2 brackets; maximum profit is: ₹ 224.
(*ii*) 1.8 rings and 3.6 brackets; maximum profit is: ₹ 252.
(*iii*) Solution remains unchanged, but maximum profit increases to ₹ 260.
(*b*) (*i*) *Optimum Solution:* 16/5 units of product *1* and 8/5 units of product *2*.
(*ii*) Dual prices for machines *A* and *B* are: ₹ 14 and ₹ 2, respectively.
(*iii*) No change in optimum solution. Required range for product 2 is: ₹ 15 to ₹ 90.

6.14. (*a*) Maximize $z = 4x_1 + 6x_2 + 3x_3 + x_4$ subject to the constraints:
$$1.5x_1 + 2x_2 + 4x_3 + 3x_4 \le 550,\ 4x_1 + x_2 + 2x_3 + x_4 \le 700,$$
$$2x_1 + 3x_2 + x_3 + 2x_4 \le 200;\ x_j \ge 0\ (j = 1, 2, 3, 4).$$
(*c*) $x_1 = 0,\ x_2 = 25,\ x_3 = 125$, and $x_4 = 0$; maximum profit = ₹ 525.
(*d*) First and third constraints. (*e*) Machine *Y* have excess capacity, 425 hours.
(*f*) Product-mix will not change. (*g*) Product-mix will not change.
(*h*) Machine *X*: ₹ 0.3, Machine *Y*: Nil, Machine *Z*: ₹ 0.18.

6.15. (*a*) Range for aluminium: 600 to 1,200. So, the change is within limits and, therefore, the additional profit is 150 × ₹ 4, *i.e.*, ₹ 600.
(*b*) Range for labour: 375 to 750. So, the change is within range and, therefore, reduction of machine hours by 150 will lead to reduction of profit to the extent of 150 × ₹ 12, *i.e.*, ₹ 1,800.
(*c*) ₹ 60.
(*d*) Maximum allowable decrease is ₹ 30, which falls within allowable limits and, therefore, basis will not change. The reduction in total profit will be: 90 × ₹ 15 = ₹ 1,350.
(*e*) Profit reduction on proposed model = ₹ 4 × 3 + ₹ 12 × 3, *i.e.*, ₹ 48. But, profit contribution of product is only ₹ 40. So, not advisable to introduce new model.

6.16. (*a*) Current basis does not change.
(*b*) No change.
(*c*) Current basis will change.
(*d*) Current basis will remain optimum and shadow prices will remain valid.
(*e*) Current basis may not remain optimum.
(*f*) No change in the current basis.

6.17. (*a*) $x_1 = 3.12,\ x_2 = 28.13$ and $x_3 = 0$; maximum $z =$ ₹ 162.5.
(*b*) No change in the optimum solution.
(*c*) When current capacity of 150 is increased to 160, the increase in profit is 10 × 0.25, *i.e.*, 2.5 (because shadow price of constraint 1 is 0.25). Therefore, total profit will be ₹ (162.5 + 2.5), *i.e.*, ₹ 165.
(*d*) No change in the value of x_3.
(*e*) Shadow prices are 0.25, 1.25 and 0 respectively.

6.18. (*a*) Minimize $z = 14{,}000x_1 + 30{,}000x_2 + 11{,}000x_3$ subject to the constraints:

$$2x_1 + 5x_2 + x_3 \geq 50 \quad \text{(Constraint on new product)}$$
$$3x_1 + 2x_2 + 9x_3 \leq 75 \quad \text{(Constraint on ingredient 1)}$$
$$x_1 + 7x_2 + 2x_3 \leq 60 \quad \text{(Constraint on ingredient 2)}$$
$$x_1, x_2, x_3 \geq 0,$$

where x_1, x_2 and x_3 denote the number of times for processes *1*, *2* and *3*, respectively.
Optimum Solution: $x_1 = 5.56, x_2 = 7.78$ and $x_3 = 0$; with minimum $z =$ ₹ 101.18 thousands.
(*b*) Total cost to produce 70 tons = ₹ 3,26,351.
(*c*) No effect on the optimum solution.
(*d*) No, since process 3 is not currently used.
(*e*) Maximize $z^* = 50y_1 + 75y_2 + 60y_3$ subject to the constraints:

$$2y_1 + 3y_2 + y_3 \leq 14{,}000,\ 5y_1 + 2y_2 + 7y_3 \leq 30{,}000,$$
$$y_1 + 9y_2 + 2y_3 \leq 11{,}000, \text{ with } y_1 \geq 0,\ y_2 \geq 0 \text{ and } y_3 \geq 0,$$

where y_j $(j = 1, 2, 3)$ are the dual variables.

Chapter 7. Transportation Problem

7.01. $x_{11} = 15,\ x_{12} = 4,\ x_{22} = 9,\ x_{23} = 4$ and $x_{33} = 7$; Total cost = 304.

7.02. $x_{13} = 8,\ x_{21} = 5,\ x_{24} = 8$ and $x_{32} = 11$; Total cost = 271.

7.03. $x_{11} = 1,\ x_{12} = 5,\ x_{24} = 1,\ x_{31} = 6,\ x_{33} = 3$ and $x_{34} = 1$; Total cost = 102.

7.04. $x_{11} = 40,\ x_{13} = 40,\ x_{14} = 120,\ x_{22} = 40,\ x_{23} = 140,$ and $x_{31} = 110$; Total cost = 13,220.

7.05. (*i*) $x_{12}=6,\ x_{13}=11,\ x_{14}=17,\ x_{21}=15,\ x_{31}=6,\ x_{33}=6$ and $x_{42}=19$; Total cost = 336.

(*ii*) $x_{11}=6,\ x_{12}=6,\ x_{13}=17,\ x_{14}=5,\ x_{21}=15,\ x_{34}=12$ and $x_{42}=19$; Total cost = 324.

7.06. Minimize $z=(450x_{11}+400x_{12}+350x_{13})+(350x_{21}+550x_{22}+300x_{23})$

$$+(200x_{31}+150x_{32}+600x_{33}) \text{ subject to the constraints:}$$

$$x_{11}+x_{12}+x_{13}=1{,}500,\ x_{21}+x_{22}+x_{23}=1{,}750,\ x_{31}+x_{32}+x_{33}=2{,}750,$$

$$x_{11}+x_{21}+x_{31}=2{,}000,\ x_{12}+x_{22}+x_{32}=3{,}000,\ x_{13}+x_{23}+x_{33}=1{,}000;$$

$$x_{ij}\geq 0 \text{ for } i=1,2,3 \text{ and } j=1,2,3.$$

Initial basic feasible solution by Vogel's Approximation Method:

$x_{11}=250,\ x_{12}=250,\ x_{13}=1{,}000,\ x_{21}=1{,}750$ and $x_{32}=2{,}750$; Total cost = 15,87,500.

7.07. (*a*) $x_{11}=20,\ x_{21}=10,\ x_{24}=30,\ x_{31}=6,\ x_{32}=18$ and $x_{33}=11$;
minimum total cost = ₹ 832.

(*b*) $x_{11}=3{,}500,\ x_{12}=1{,}500,\ x_{22}=2{,}500,\ x_{23}=2{,}000,\ x_{24}=1{,}500$ and $x_{31}=2{,}500$;
minimum total cost = ₹ 39,500.

7.08. $AW=15{,}000,\ AZ=30{,}000,\ BW=20{,}000,\ BY=30{,}000,\ CW=15{,}000,\ CX=40{,}000$;
minimum total cost = ₹ 26,80,000.

7.09. $x_{11}=15,\ x_{14}=5,\ x_{22}=19,\ x_{24}=9,\ x_{33}=13$ and $x_{34}=4$; minimum total cost = ₹ 209.

7.10. $x_{11}=5,\ x_{14}=2,\ x_{22}=3,\ x_{23}=7,\ x_{32}=5$ and $x_{34}=13$; minimum total cost is ₹ 785, *i.e.*, a saving of ₹ 215.

7.11. $x_{12}=10,\ x_{13}=20,\ x_{15}=10,\ x_{21}=20,\ x_{31}=5,\ x_{34}=30$ and $x_{35}=5$;
minimum total cost = ₹ 3,600.

7.12. $x_{13}=25, x_{14}=30, x_{21}=10, x_{22}=20, x_{23}=15, x_{31}=30, x_{43}=10$ and $x_{45}=40$; or $x_{12}=20$, $x_{13}=5, x_{14}=30, x_{21}=10, x_{23}=35, x_{31}=30, x_{43}=10$ and $x_{45}=40$;
minimum total cost = ₹ 695.

7.13. (*a*) No. (*b*) $x_{12}=12, x_{13}=2, x_{14}=8, x_{23}=15, x_{31}=7$ and $x_{34}=1$; minimum cost = ₹ 104.

(*c*) If clerk decides to despatch all the 8 units from C to *II* instead of 7 units to 1 and one unit to *IV*, then the transportation cost c_{32} must be reduced from present 6 to at least 4 in order to have the improved minimum total shipping cost.

7.14. (*i*) No, it is not degenerate since the number of occupied cells are 3 + 4 – 1, *i.e.*, 6.

(*ii*) Yes, it is optimal. It has unique solution, since none of the opportunity costs of unoccupied cells is zero.

(*iii*) ₹ 15, (*iv*) The cost would increase by ₹ 8, since the opportunity cost of unoccupied cell (2, 2) is 8.

7.15. (*a*) Allocation is not optimum, since at least one of the opportunity costs of the occupied cell is positive. Optimum allocation is obtained as follows :

A to *F*: 100, *B* to *F*: 70; *B* to *J*: 80; *C* to *A*: 90; *C* to *G*: 50; *C* to *J*: 40; *D* to *H*: 210, and *D* to *J*: 70; minimum total cost = ₹ 6,810.

(*b*) No change in the optimum schedule.

7.16. $(P_1, 1)$ = 400 tons, $(P_2, 3)$ = 300 tons, $(P_3, 2)$ = 500 tons and $(P_3, 3)$ = 500 tons;
minimum total cost = ₹ 4,98,000.

7.17. $(A, E) = 10$ units, $(B, D) = 60$ units, $(B, E) = 10$ units, $(B, F) = 10$ units and $(C, D) = 15$ units. Also, 40 units are supplied to F at the cost of ₹ 2 per unit, which the company is unable to supply. Therefore, there will be a penalty of ₹ 2 × 40, *i.e.*, of ₹ 80.

7.18. $(A, D) = 75$ units, $(A, E) = 175$ units, $(B, D) = 125$ units, $(B, G) = 175$ units, $(C, F) = 275$ units, and $(C, G) = 125$ units. Also, additional 50 units produced utilizing over time are supplied to warehouse E from factory A; minimum total cost = ₹ 12,525.

7.19. $x_{11} = 50,\ x_{12} = 100,\ x_{24} = 25,\ x_{25} = 125,\ x_{31} = 30,\ x_{33} = 75,$ and $x_{34} = 20$; maximum total profit will be ₹ 3,170.

7.20. $x_{12} = 4,\ x_{13} = 1,\ x_{14} = 5,\ x_{21} = 6,\ x_{23} = 3,\ x_{32} = 6$; minimum total cost = ₹ 472.

Alternative solution: $x_{12} = 9,\ x_{13} = 1,\ x_{21} = 6,\ x_{23} = 3,\ x_{32} = 1$ and $x_{34} = 5$.

7.21. $(P_1, W_4) = 200,\ (P_1, W_5) = 600,\ (P_2, W_3) = 200,\ (P_3, W_2) = 400,\ (P_3, W_4) = 200,$ $(P_4, W_1) = 100,$ and $(P_5, W_1) = 100$; minimum total cost = ₹ 11,800.

7.22. $x_{11} = 70,\ x_{12} = 100,\ x_{14} = 30$ and $x_{23} = 100$; or $x_{11} = 75,\ x_{12} = 95,\ x_{14} = 30$ and $x_{23} = 100$; maximum total profit will be ₹ 31,600.

7.23. $x_{12} = 200,\ x_{22} = 120,\ x_{23} = 100,\ x_{24} = 280,\ x_{31} = 180$ and $x_{34} = 120$; maximum total profit will be ₹ 15,400.

7.24. $(A, 1) = 100$ kg, $(B, 2) = 20$ kg, $(C, 3) = 60$ kg, $(D, 1) = 20$ kg and $(D, 2) = 60$ kg; maximum profit = ₹ 680.

7.25. $(W, B) = 25,\ (W, D)) = 50;\ (W, E) = 200;\ (X, A) = 150,\ (Y, B) = 75,\ (Y, C) = 75$ and $(Z, D) = 200$; maximum profit = ₹ 2,06,875.

7.26. From $F_1 \to 10$ units to S_2, from $F_2 \to 90$ units to S_2 and 60 units to S_3, from $F_3 \to 50$ units to S_3, from $F_4 \to 80$ units to S_1 and 20 units to S_2; maximum profit = ₹ 480.

7.27. $(1, P) = 40,\ (1, Q) = 30,\ (2, Q) = 20,\ (2, R) = 20,\ (3, R) = 40,\ (3, S) = 50$ and $(4, S) = 10$; maximum return = ₹ 1,30,00,000.

7.28. Minimize $z = (3x_{11} + 6x_{12} + 5x_{13} + 2x_{14}) + (4x_{21} + 5x_{23} + 4x_{24}) + (3x_{31} + 4x_{32} + 4x_{33} + 4x_{34})$ subject to the constraints:

$$x_{11} + x_{12} + x_{13} + x_{14} = 40,\ x_{21} + x_{23} + x_{24} = 50,\ x_{31} + x_{32} + x_{33} + x_{34} = 30,$$
$$x_{11} + x_{21} + x_{31} = 50,\ x_{12} + x_{32} = 30,\ x_{13} + x_{23} + x_{33} = 40,\ x_{14} + x_{24} + x_{34} = 20;$$
$$x_{ij} \geq 0\ (i = 1, 2, 3 \text{ and } j = 1, 2, 3, 4).$$

Initial basic feasible solution by VAM:

$$x_{13} = 20,\ x_{14} = 20,\ x_{21} = 30,\ x_{23} = 20 \text{ and } x_{32} = 30.$$

7.29. (*i*) $x_{13} = 90,\ x_{21} = 30,\ x_{23} = 20,\ x_{32} = 80,\ x_{41} = 40$ and $x_{42} = 20$; minimum total cost = 1,59,000.

(*ii*) $x_{13} = 90,\ x_{21} = 30,\ x_{23} = 20,\ x_{31} = 40,\ x_{32} = 40$ and $x_{42} = 60$.

7.30. $x_{14} = 200,\ x_{22} = 15,\ x_{25} = 160,\ x_{32} = 75,\ x_{34} = 30,\ x_{41} = 110$ and $x_{43} = 120$; minimum total cost = 7,330.

7.31. $x_{12} = 20,\ x_{16} = 40,\ x_{21} = 10,\ x_{23} = 10,\ x_{33} = 30,\ x_{34} = 20,\ x_{35} = 40$ and $x_{41} = 50$; minimum total cost = 1,120.

7.32. $x_{11} = 20,\ x_{12} = 20,\ x_{22} = 10,\ x_{23} = 20,\ x_{24} = 20,\ x_{33} = 30$ and $x_{44} = 20$; minimum total cost = 2,210; where 1, 2, 3 and 4 stand for the respective month.

7.33. $x_{11} = 200,\ x_{12} = 25,\ x_{22} = 225,\ x_{33} = 150,\ x_{44} = 80$ and $x_{55} = 120$; minimum total cost is: ₹ 22,880, where 1, 2, 3, 4 and 5 stand for the months of January, February, March, April and May, respectively.

7.34. (*i*) In different age groups exposure level through TV: 25 million in the age group 26–35, and 10 million in the age group 51 and older; Newspapers: 30 million in the age group 18–25; and Magazine: 15 million in the age group 36–50; minimum total cost will be ₹ 71 lakh.

(*ii*) The revised plan of exposure level is through TV: 4 million in the age group 18–25, 25 million in the age group 26–35 and 10 million in the age group 51 and older; Newspapers: 26 million in the age group 36–50; and Magazines: 15 million in the age group 36–50.

This exposure would involve a total cost of ₹ 71.8 lakh.

7.35. (*a*) $x_{12} = 12,\ x_{13} = 2,\ x_{14} = 8,\ x_{23} = 12,\ x_{31} = 7$ and $x_{34} = 1$; minimum total cost = 101.

(*b*) No, the optimal solution will not change, because the changes are confined to occupied cells only.

(*c*) Yes, the new optimal solution is to ship 12 units from *A* to *II*, 10 units from *A* to *IV*, 1 unit from *B* to *I*, 17 units from *B* to *III*, 6 units from *C* to *I*, and 2 units from *C* to *IV*.

7.36. *Optimum solution:* $x_{11} = 200,\ x_{24} = 150,\ x_{31} = 25,\ x_{32} = 75$ and $x_{33} = 150$; minimum total cost = 7,550.

(*i*) The optimum solution is degenerate, since the number of basic variables are 5, which are less than 3 + 4 – 1.

(*ii*) Yes, there are multiple optimum solutions, because one of the unoccupied cell has opportunity cost equal to zero.

(*iii*) No increase in the cost, because opportunity cost of the cell (3, 4) is zero.

(*iv*) Yes, because there will be no increase in cost as the opportunity cost of the cell (3, 4) is zero. On the other hand, due to discount of 20%, actual transportation cost per unit will decrease by ₹ 4 per unit.

(*v*) New optimum solution is: $x_{11} = 199,\ x_{24} = 150,\ x_{31} = 25,\ x_{32} = 75$ and $x_{33} = 150$; minimum total cost = ₹ 7,540, *i.e.*, a reduction of ₹ 10 in the cost.

7.37. (*i*) Yes, the solution is optimal and unique, since the number of basic cells are 6, and no opportunity cost is zero.

(*ii*) No, the solution is not degenerate, since neither of the basic variable gives its value as zero.

(*iii*) Total cost decreases by 8.

(*iv*) Not advisable, since total cost will increase,

(*v*) Yes, the cost will reduce by 10.

7.38. Transport 100 units from *X* to *A* and *B* each, 50 units from *Y* to *B* and 250 units from *Y* to *C*.

7.39. (*i*)

	A	*B*	*C*	*D*	*E*	*F*	*Available*
A	0	3	2	6	4	1	200
B	3	0	4	3	8	7	190
C	2	4	0	4	4	2	210
D	6	3	4	0	2	5	150
E	4	8	4	2	0	1	150
F	1	7	2	5	1	0	150
Required	150	150	150	170	245	185	

(*ii*) Transport 15 units from *A* to *E*, 35 units from *A* to *F*, 40 units from *B* to *D*, 60 units from *C* to *E*, and 20 units from *D* to *E*; minimum total cost = ₹ 495.

7.40. Transport from $P_1 \to 40$ units to W_2 and 200 units to W_3, from $P_2 \to 80$ units to W_1 and W_2 each; total cost is: 36,400.

7.41. Transport 4 units from S_1 to D_2, 5 units from S_2 to D_2, and then 3 units from D_2 to D_1; total minimum cost will be 21.

7.42.

	2	4	5	6	7	*Available ('000 gallons)*
1	20	0	0	0	3	110+50
3	0	30	0	0	9	110+60
5	0	2	0	4	10	110
6	8	0	4	0	0	110
7	40	0	10	0	0	110
Demand ('000 gallons)	110+90	110+20	110	110	110	

Chapter 8. Assignment Problem

8.01. (*i*) Minimize $z = [(10x_{11} + 16x_{12} + 32x_{13}) + (14x_{21} + 22x_{22} + 40x_{23}) + (22_{31} + 24x_{32} + 34x_{33}) + (14x_{41} + 18x_{42} + 36x_{43})]$ subject to the constraints:

$x_{i1} + x_{i2} + x_{i3} = 1,\ x_{1j} + x_{2j} + x_{3j} + x_{4j} = 1$ and $x_{ij} = 0$ or 1; where $i = 1, 2, 3, 4;\ j = 1, 2, 3.$

(*ii*) Gita $\to 1$, Sita $\to 3$, Ajay $\to 2$ and Vijay will not be assigned any client; minimum total time will be: 62 days.

8.02. $A \to CG$, $B \to SM$, $C \to PM$ and $D \to OR$; minimum total time for preparation will be: 71 hours.

8.03. $1 \to C,\ 2 \to D,\ 3 \to B,\ 4 \to A$ or $1 \to B,\ 2 \to C,\ 3 \to D,\ 4 \to A$; minimum total time will be: 17 hours.

8.04. (*a*) $I \to 1,\ II \to 2,\ III \to 3,\ IV \to 4$ or $I \to 2,\ II \to 3,\ III \to 4,\ IV \to 1$; minimum total cost will be: 20.

(*b*) $W_1 \to M_3,\ W_2 \to M_2,\ W_3 \to M_1,\ W_4 \to M_4$; minimum total cost will be: 23.

8.05. $1 \to A,\ 2 \to B,\ 3 \to C,\ 4 \to D$ or $1 \to A,\ 2 \to D,\ 3 \to B,\ 4 \to C$; minimum total time will be: 31 minutes.

8.06. $J_1 \to O_5,\ J_2 \to O_1,\ J_3 \to O_2,\ J_4 \to O_3,\ J_5 \to O_4$; minimum total cost will be: 71 hours.

8.07. $I \to B,\ II \to E,\ III \to C,\ IV \to A$ and $V \to D$; minimum total time will be: 55 hours.

8.08. $A \to III,\ B \to V,\ C \to IV,\ D \to I,\ E \to II$ or $A \to III,\ B \to V,\ C \to I,\ D \to IV,\ E \to II$; minimum total time will be: 13 hours.

8.09. $A \to 5,\ B \to 1,\ C \to 4,\ D \to 3,\ E \to 2$; minimum total cost will be: 9.

8.10. $A \to A',\ B \to C',\ C \to D',\ D \to B'$ and $E \to E'$; minimum total distance will be: 39 kilometres.

8.11. $M_1 \to E,\ M_2 \to D,\ M_3 \to B,\ M_4 \to A,\ M_5 \to C$; minimum total time is: 13 hours.

8.12. $J_1 \to M_3,\ J_2 \to M_2,\ J_3 \to M_1,\ J_4 \to M_4$ and $J_5 \to M_5$; solution is not unique. Alternative optimum solution is: $J_1 \to M_3,\ J_2 \to M_5,\ J_3 \to M_1,\ J_4 \to M_4$ and $J_5 \to M_2$; minimum total cost will be: 365.

8.13. $1 \to A,\ 2 \to B,\ 3 \to C$ or $1 \to A,\ 2 \to C,\ 3 \to B$; minimum total cost will be: 50.

8.14. $1 \to 11,\ 2 \to 8,\ 3 \to 7,\ 4 \to 9,\ 5 \to 10,\ 6 \to 12$; minimum total distance will be: 125 kms.

8.15. $A \to IV,\ B \to II,\ C \to III$ and $D \to I$; sales territory V remains unassigned.

8.16. $A \to 2,\ B \to 6,\ C \to 3,\ D \to 1$ and $E \to 4$; minimum total distance will be: 38 kilometres.

8.17. $A \to III,\ B \to II,\ C \to VI,\ D \to I$ and $E \to IV$; zone V should be managed by junior manager.

8.18. (*i*) $C_1 \to R_3,\ C_2 \to R_1, C_4 \to R_2$ and $C_5 \to R_4$; minimum total cost will be: ₹ 54 lakh.
(*ii*) Supplementary grant of ₹ 5 lakh will be needed.
(*iii*) Contractor C_3 will be unsuccessful in bid.

8.19. $P_1 \to J_2,\ P_2 \to J_3,\ P_3 \to J_1,\ P_4 \to J_4$, job J_5 will not be assigned to any person; minimum total cost will be: 75.

8.20. $A \to I,\ B \to IV,\ C \to III$; maximum sales will be: ₹ 247 lakh.

8.21. Dr. $A \to 03$, Dr. $B \to 04$, Dr. $C \to 01$, while operation of type *02* is not assigned; maximum utilization of operation theaters will be: 46 hours.

8.22. (*i*) $A \to IV,\ B \to I,\ C \to III,\ D \to II$ or $A \to IV,\ B \to III,\ C \to II,\ D \to I$; total maximum sales will be: ₹ 745 lakh.
(*ii*) Optimum assignment will not change and total sales will remain the same, *i.e.*, of ₹ 745 lakh.

8.23. $A \to II,\ B \to III,\ C \to V,\ D \to I$; while territory IV is not assigned to any salesman; maximum total sales will be: ₹ 68 lakh.

8.24. $1 \to B,\ 2 \to A,\ 3 \to E,\ 4 \to C,\ 5 \to D$ or $1 \to B,\ 2 \to E,\ 3 \to A,\ 4 \to C,\ 5 \to D$ or $1 \to B,\ 2 \to A,\ 3 \to E,\ 4 \to D,\ 5 \to C$ or $1 \to B,\ 2 \to E,\ 3 \to A,\ 4 \to D,\ 5 \to C$;
maximum expected profit will be: ₹ 191.

8.25. (*i*) $P \to V,\ Q \to I,\ R \to IV,\ S \to III,\ T \to II$; maximum runs will be: 232.
(*ii*) Batsman U will be included in the team at position II and he will replace batsman S, Total run will be: 263.

8.26. $2 \to A,\ 3 \to B,\ 4 \to D,\ 6 \to C$; Job *1* and *5* are not assigned; maximum profit will be: 28.

8.27. $P \to$ Zone A, $Q \to$ Zone B, $R \to$ Zone C and $S \to$ Zone D or
$P \to$ Zone A, $Q \to$ Zone C, $R \to$ Zone B and $S \to$ Zone D;
maximum sales will be: ₹ 99,000.

8.28. $A \to 2,\ B \to 4,\ C \to 1,\ D \to 3$; maximum profit will be: ₹ 22,000.

8.29. (*i*) $A \to 1,\ B \to 2,\ C \to 4,\ D \to 5,\ E \to 3$; maximum profit will be: ₹ 128.
(*ii*) $A \to 2,\ B \to 5,\ C \to 4,\ E \to 3$, while tailor D is absent; maximum profit will be: ₹ 106.

8.30. (*i*) $A \to 4,\ B \to 3,\ C \to 2$ and Job *1* is not assigned to any person; minimum total cost will be: ₹ 131 lakh.
(*ii*) No, the method is not applicable.

8.31. $A \to 1,\ B \to 2,\ C \to 4,\ D \to 5,\ E \to 3$; maximum total preference score will be: 40.

8.32. $M_1 \to A,\ M_2 \to B,\ M_3 \to E,\ M_4 \to D$; while place C will remain vacant; total assignment cost will be: ₹ 12.

8.33. (*i*) $O_1 \to J_3,\ O_2 \to J_2,\ C_3 \to J_4,\ O_4 \to J_1$; minimum total time will be: 34 hours.
(*ii*) $O_1 \to J_2,\ O_2 \to J_4,\ O_3 \to J_1,\ O_4 \to J_3$; minimum total time will be: 36 hours.

8.34. $A \to 2,\ B \to 3,\ C \to 4,\ D \to 1$; maximum revenue will be: ₹ 35,000.

8.35. Company 1 → Route 1, Company 2 → Route 2, Company 3 → Route 3, Company 4 → Route 4; minimum total cost will be: ₹ 15,000.

8.36. Room 301 → Manager 5, Room 302 → Manager 1, Room 303 → Manager 3, Room 304 → Manager 2, Room 305 → Manager 4; minimum total preference ranking is: 7.

8.37. *Optimum Solution:*

Crew	*Residence at*	*Route number*	*Waiting time (hrs)*
1	Chandigarh	$d-1$	4.5
2	Delhi	$e-2$	9.5
3	Delhi	$a-3$	9.0
4	Chandigarh	$b-4$	5.0
5	Delhi	$c-5$	5.5

Total minimum waiting time is: 33.5 hours.

8.38. *Optimum Solution:*

Flight pair	*Crew based at*	*Lay-over time*
103 – 1	Jaipur	17.75 hours
2 – 104	Delhi	8.50 hours
3 – 101	Delhi	17.50 hours
102 – 4	Jaipur	8.75 hours

Total minimum lay-over time is: 52.50 hours.

8.39. $1 \to A,\ 2 \to C,\ 3 \to B,\ 4 \to D$; minimum total time will be: 69 hours.

8.40. $J_1 \to M_4,\ J_2 \to M_1,\ J_3 \to M_3,\ J_4 \to M_2$; minimum total time will be: 74 hours.

Chapter 9. Other Methods of Linear Programming

9.01. $x_1 = 3$ and $x_2 = 2$; maximum $z = 8$.

9.02. $x_1 = 7$ and $x_2 = 6$; maximum $z = 33$.

9.03. $x_1 = 20/3,\ x_2 = 0$ and $x_3 = 10/3$; maximum $z = 10$.

9.04. $x_1 = 0,\ x_2 = 3/2$ and $x_3 = 1$; minimum $z = -6$.

9.05. $x_1 = 4,\ x_2 = 7/4$ and $x_3 = 0$; minimum $z = 83/4$.

9.06. $x_1 = 9/16,\ x_2 = 1,\ x_3 = 15/16$ and $x_4 = 0$; maximum $z = 331/16$.

9.07. $x_1 = 71/4,\ x_2 = 1,\ x_3 = 29/2$ and $x_4 = 4$; maximum $z = 387/4$.

9.08. $x_1 = 3,\ x_2 = 5$ and $x_3 = 2$; maximum $z = 34$.

9.09. $x^{(1)} = [0,\ 5]$, if $0 \le \lambda \le 3/2$; $x^{(2)} = [4,\ 3]$, if $3/2 \le \lambda \le 3$ and $x^{(3)} = [5,\ 1]$, if $\lambda \ge 3$.

9.10. $x^{(1)} = [2/5,\ 1/5]$, if $0 \le \lambda \le 3$; $x^{(2)} = [0,\ 1/3]$, if $3 \le \lambda \le 9$ and $x^{(3)} = [0,\ 0]$, if $9 \le \lambda$.

9.11. $-2 \le \lambda \le 3$.

9.12. $x^{(1)} = [0,\ 5,\ 30]$, if $0 \le \lambda \le 1$ and $x^{(2)} = [0,\ 0,\ 30]$, if $\lambda \ge 1$.

9.13. $x^{(1)} = [16/19,\ 0,\ 0,\ 5/19]$, if $0 \le \theta \le 4$; $x^{(2)} = [0,\ 0,\ 0,\ 3]$, if $4 \le \theta \le 7$ and $x^{(3)}$ = No feasible solution for $7 \le \theta$.

9.14. $x^{(1)} = [0,\ 0,\ 0,\ 10]$, if $0 \le \theta \le 2$; $x^{(2)} = [0,\ 0,\ 0,\ 12]$, if $2 \le \theta \le 11/2$; $x^{(3)} = [0,\ 7/2,\ 0,\ 12]$, if $11/2 \le \theta \le 35/2$; $x^{(4)} = [0,\ 55/2,\ 0,\ 0]$, if $35/2 \le \theta \le 59/2$ and $x^{(5)}$ = No feasible solution, if $59/2 \le \theta$.

9.15. $x^{(1)} = [4,\ 0]$, if $-\infty \le \lambda \le -5$; $x^{(2)} = [0,\ 5]$, if $-5 \le \lambda \le -1$; $x^{(3)} = [0,\ 3]$, if $-1 \le \lambda \le 2$; $x^{(4)} = [0,\ 0]$, if $2 \le \lambda \le 3$ and $x^{(5)}$ = No feasible solution, if $\lambda > 3$.

9.16. $x^{(1)} = [0,\ 5,\ 30]$, if $0 \le \lambda \le 1$; $x^{(2)} = [0,\ 0,\ 30]$, if $1 \le \lambda \le 30/7$ and $x^{(3)}$ = No feasible solution for $\lambda \ge 30/7$.

9.17. $x_1 = 0$ and $x_2 = 1$; maximum $z = 2/9$.

9.18. $x_1 = 3/2$ and $x_2 = 3/4$; minimum $z = -51/40$.

9.19. $x_1 = 0$ and $x_2 = 1$; maximum $z = -1/7$.

9.20. $x_1 = 0$ and $x_2 = 3$; maximum $z = 9/10$.

9.21. $A \to D \to B \to C \to A$ or $A \to C \to B \to D \to A$; minimum cost = 19.

9.22. $A \to B \to C \to D \to E \to A$; corresponding cost = 15.

9.23. $A \to D \to B \to C \to E \to A$ or $A \to E \to C \to B \to D \to A$; corresponding cost = 16.

9.24. $1 \to 4 \to 5 \to 2 \to 3 \to 1$; corresponding cost = 49.

9.25. Depot → Vendor *D* → Vendor *A* → Vendor *C* → Vendor *B* → Depot;
Total distance covered in this sequence = 15 kms.

9.26. $A \to E \to B \to C \to D \to A$; minimum distance = 1,165 km.

Chapter 10. Operations Scheduling

10.01. Optimal Schedule: $1 \to 3 \to 2 \to 5 \to 4 \to 6$; mean flow time is 37 days.

10.02. Optimal Schedule: $S \to R \to Q \to P \to Z \to Y \to X$; mean flow time is 18 minutes.
Another Schedule (using FCFS rule): $X \to Y \to Z \to P \to Q \to R \to S$; Average processing time is 33.4 minutes.

10.03. Optimal Schedule: $D \to H \to A \to C \to G \to E \to F \to B$; total processing time is 475 minutes, mean flow time is 59.4 minutes.

10.04. Optimal Schedule: $D \to B \to C \to E \to A$; average processing time is 12 days. Another schedule (using FCFS rule) is $A \to B \to C \to D \to E$ with average processing time of 15.2 days.

10.05. Optimal Schedule: $2 \to 1 \to 4 \to 3$; maximum tardiness is of 13 days.

10.06. (*a*) Optimal Order: $2 \to 1 \to 4 \to 3 \to 5$; minimum total processing time is 33 days.
(*b*) Optimal Schedule: $1 \to 3 \to 2 \to 4 \to 5$.

10.07. Optimal Schedule: $J_3 \to J_1 \to J_2$; minimum total time is 20 hours.

10.08. Optimal Schedule: $J_1 \to J_6 \to J_2 \to J_3 \to J_4 \to J_5$; minimum total elapsed time is 29 hours.

10.09. Optimal Schedule: $A \to F \to D \to B \to C \to E$; minimum total time is 96 hours.

10.10. Optimal Schedule: $J_1 \to J_5 \to J_3 \to J_6 \to J_2 \to J_4$; minimum total time is 61 hours.

10.11. Optimal Sequence: $6 \to 4 \to 1 \to 7 \to 2 \to 3 \to 5$; minimum total elapsed time is 83 hours.

10.12. Optimum Sequence: $3 \to 5 \to 7 \to 6 \to 2 \to 4 \to 1$; minimum total elapsed time is 47 hours.

10.13. Optimum Schedule: $1 \to 4 \to 5 \to 3 \to 2 \to 7 \to 6$; minimum total elapsed time is 67 hours.

10.14. Optimum Sequence: $T_1 \to T_3 \to T_9 \to T_8 \to T_2 \to T_6 \to T_4 \to T_7 \to T_5$ or
$T_1 \to T_3 \to T_9 \to T_2 \to T_8 \to T_6 \to T_4 \to T_7 \to T_5$; minimum total elapsed time is 61 hours.

10.15. Optimum Order: $4 \to 1 \to 3 \to 2 \to 5 \to 6$; minimum total elapsed time is 430 hours.

10.16. Optimum Order: $P_2 \to P_3 \to P_5 \to P_6 \to P_4 \to P_1$; minimum total time is 16 hours and 50 minutes. Waiting time for management personnel would be 260 minutes.

10.17. Optimum Schedule: $E \to C \to F \to G \to A \to D \to B$ or $E \to C \to F \to A \to G \to D \to B$; required time for evaluation is 51 hours and waiting time of committee is 13 hours.

10.18. Optimum Order: $M_2 \to M_4 \to M_1 \to M_5 \to M_3 \to M_6 \to M_7$ or $M_2 \to M_4 \to M_5 \to M_1 \to M_3 \to M_6 \to M_7$; minimum total time is 158 hours.

10.19. Optimum Sequence: $VIII \to III \to IV \to VII \to VI \to V \to II \to I$; minimum total time of manufacturing is 6 hours and 45 minutes.

10.20. Optimum Schedule: $4 \to 3 \to 1 \to 2 \to 5$; minimum total time is 92.

10.21. Optimum Sequence: $1 \to 4 \to 5 \to 3 \to 2$; minimum total time is 42 hours.

10.22. Optimum Sequence: $J_2 \to J_5 \to J_4 \to J_3 \to J_1 \to J_6$; minimum total time is 64 hours.

10.23. Optimum Sequence: $4 \to 5 \to 6 \to 2 \to 1 \to 3$; minimum total time is 53 hours.

10.24. Optimum Sequence: $7 \to 2 \to 1 \to 3 \to 5 \to 6 \to 4$; minimum total time is 102 hours.

10.25. (*a*) Optimum Order: $3 \to 4 \to 5 \to 7 \to 2 \to 6 \to 1$; minimum total elapsed time is 46 hours.

(*b*) Optimum Order: $1 \to 4 \to 3 \to 6 \to 2 \to 5 \to 7$; minimum total elapsed time is 86 hours.

10.26. Optimum Sequence: $4 \to 5 \to 1 \to 2 \to 3 \to 6 \to 7$ or $4 \to 5 \to 1 \to 3 \to 2 \to 6 \to 7$; minimum total time is 72 hours.

10.27. Optimum Sequence: $A \to D \to G \to B \to F \to C \to E$; minimum total time is 59 hours.

10.28. Optimum Sequence: $7 \to 1 \to 8 \to 4 \to 2 \to 5 \to 3 \to 6$; total elapsed time is 57 hours. Since total elapsed time is less than 61 hours, all the eight jobs can be processed within the given capacity.

10.29. Optimum Sequence: $III \to IV \to I \to V \to II \to VIII \to VII \to VI$; minimum total time is 79 hours.

10.30. Optimum Sequence: $4 \to 1 \to 2 \to 3$; minimum total elapsed time is 97 hours.

10.31. Optimum Sequence: $A \to B \to C \to D$ or $A \to B \to D \to C$, etc., 4! such sequences; minimum total time is 122 hours.

10.32. Optimum Sequence: $I \to III \to II \to IV$; minimum total elapsed time is 66 hours.

10.33. Optimum Sequence: $4 \to 2 \to 5 \to 1 \to 3 \to 6$ or $4 \to 2 \to 5 \to 1 \to 6 \to 3$.

10.34. Optimum Sequence: $C \to A \to E \to D \to B$; minimum total elapsed time is 76 hours.

10.35. Total minimum time is 40 hours with idle time of 6 hours for Job *1*.

10.36. Total minimum time is 15 hours with idle time of 3 hours for Job *1*.

10.37. Total minimum time is 23 hours with idle time of 5 hours for Job *1* and 5 hours for Job *2*.

10.38. Total minimum time is 44 hours with idle time of 10 hours for Job *1* and 4 hours for Job *2*.

10.39. Minimum total time is 150 hours.

10.40. (*a*) Sequence: $P_3 \to P_6 \to P_4 \to P_1 \to P_5 \to P_2$.

(*b*) Present schedule cost = ₹ 64,630; and optimum schedule cost = ₹ 60,310, Optimum schedule will be economical.

Chapter 11. Integer Programming

11.01. Minimize $z = t_{11}x_{11} + t_{22}x_{22} + t_{33}x_{33}$ subject to the constraints:

$$x_{i1} + x_{i2} + x_{i3} = 1, \quad x_{1j} + x_{2j} + x_{3j} = 1; \text{ where } i, j = 1, 2, 3.$$

$x_{ij} = 1$, if i^{th} job is assigned to j and $x_{ij} = 0$, otherwise.

11.02. Maximize $z = 4{,}000x_1 + 1{,}000x_2$ subject to the constraints:

$$8x_1 + 10x_2 \le 80,\ \ 20x_1 + 60x_2 \le 360,\ \ x_1 \le 6;$$

x_1, x_2 are 0 or 1; where x_1 = number of pins produced per week and x_2 = number of necklaces produced per week.

11.03. Maximize $z = 16{,}000x_1 + 22{,}000x_2 + 12{,}000x_3 + 8{,}000x_3$ subject to the constraints:

$$5{,}000x_1 + 7{,}000x_2 + 4{,}000x_3 + 3{,}000x_4 \le 14{,}000;$$

$$x_1, x_2, x_3 \text{ and } x_4 \text{ are 0 or 1.}$$

11.04. Maximize $z = 18{,}00{,}000x_1 + 2{,}00{,}000x_2 + 7{,}20{,}000x_3 + 8{,}00{,}000x_4$ subject to the constraints:

$$300x_1 + 120x_2 + 300x_3 + 200x_4 \le 650,\ \ 400x_1 + 80x_2 + 200x_3 + 400x_4 \le 800,$$

$$400x_1 + 200x_3 + 400x_4 \le 800,\ \ 300x_1 + 40x_2 + 200x_3 + 100x_4 \le 500;$$

where x_j $(j = 1, 2, 3, 4)$ are variables indicating projects *A*, *B*, *C* and *D*, respectively. Also $x_j = 1$ or 0.

11.05. Maximize $z = 105x_1 + 140x_2 + 80x_3 + 100x_4$ subject to the constraints:

$$60x_1 + 108x_2 + 200x_3 + 90x_4 \le 600,\ \ x_1 + x_3 = 1,\ \ 160x_1 + 40x_2 + 150x_3 + 70x_4 \le 700;$$

and $x_j = 1$ or 0, where x_j $(j = 1, 2, 3, 4)$ denote the project *A*, *B*, *C* and *D*, respectively.

11.06. Maximize $z = 8x_1 + 3x_2$ subject to the constraints:

$$6x_1 + 4x_2 \le 25,\ \ 120x_1 + 32x_2 \le 240;$$

$x_1, x_2 \ge 0$ and integers, where x_1 denotes the refrigeration technician and x_2 denote the part time apprentice technician.

11.07. Maximize $z = 50x_1 + 5x_2 + 15x_3 + 8x_4$ subject to the constraints:

$$15x_1 + 3x_2 + 10x_3 + 4x_4 \le 100,\ \ x_2 - x_3 \le 1,$$

$$5x_1 + 4x_2 + 15x_3 + 2x_4 \le 150;\ \ x_1 + x_4 \ge 1,$$

where $x_j = 1$, if investment j $(j = 1, 2, 3, 4)$ is made, and zero, if investment is not made.

11.08. Maximize $z = 200x_1 + 150x_2 + 140x_3 + 125x_4 + 180x_5$ subject to the constraints:

$$150x_1 + 90x_2 + 50x_3 + 80x_4 + 100x_5 \le 300,\ \ x_4 - x_5 \le 0;$$

$$x_1 + x_2 + x_5 = 1,\ \ x_1 - x_4 \ge 0,\ x_j\ (j = 1, 2, 3, 4, 5) \text{ are 1 or 0,}$$

where x_j $(j = 1, 2, 3, 4, 5)$ denote the shopping malls.

11.09. (*a*) $x_1 = 2$ and $x_2 = 2$; maximum $z = 4$.
(*b*) $x_1 = 1$ and $x_2 = 3$; maximum $z = 11$.

11.10. $x_1 = 4$ and $x_2 = 3$; maximum $z = 58$.

11.11. (*a*) $x_1 = 9$ and $x_2 = 2$; minimum $z = 101$.
(*b*) $x_1 = 4$ and $x_2 = 1.2$; maximum $z = 5.2$.

11.12. One shirt of type *X* as well as of *Y* each to get a maximum profit of ₹ 50.

11.13. Two toys of type *A* and three toys of type *B* to get a maximum profit of ₹ 25.

11.14. Two scoops of cottage cheese and three scoops of scrambled egg must be used to have minimum cost of ₹ 10.

11.15. $x_1 = 1$ and $x_2 = 1$; maximum $z = 3$.

11.16. $x_1 = 2$ and $x_2 = 8$; maximum $z = 34$.

11.17. $x_1 = 0$ and $x_2 = 2$; maximum $z = 2$.

11.18. $x_1 = 3$ and $x_2 = 3$; maximum $z = 15$.

11.19. $x_1 = 0$ and $x_2 = 2$; maximum $z = 6$.

11.20. Produce 13 refrigerators and 34 ranges to have maximum profit of ₹ 10,700.

11.21. Produce 24 units of product A and 32 units of product B to get maximum profit of ₹ 7,440.

11.22. $x_1 = 2,\ x_2 = 0$ and $x_3 = 2$; maximum $z = -16$.

11.23. $x_1 = 2,\ x_2 = 2$ and $x_3 = 1$; maximum $z = 19$.

11.24. $x_1 = 2,\ x_2 = 1$ and $x_3 = 6$; maximum $z = 26$.

Chapter 12. Goal Programming

12.01. Minimize $z = \{d_1^-,\ d_1^+\}$ subject to the constraints:

$$70x_1 + 30x_2 + d_1^- - d_1^+ = 800;\ x_1, x_2, d_1^-, d_1^+ \geq 0.$$

12.02. Minimize $z = \{d_1^-,\ d_1^+\}$ subject to the constraints:

$$x_1 + 3x_2 \leq 60,\ x_1 + x_2 \leq 40 \text{ (Resource constraints)},$$
$$150x_1 + 250x_2 + d_1^- - d_1^+ = 6{,}000;\ x_1, x_2, d_1^-,\ d_1^+ \geq 0.$$

12.03. Minimize $z = P_1 d_1^- + P_2 d_2^+ + P_3 d_3^- + P_4 d_4^-$ subject to the constraints:

$$x_2 + d_1^- - d_1^+ = 30,\ 40x_1 + 20x_2 + 150x_3 + d_2^- - d_2^+ = 1{,}200,\ x_1 + x_3 + d_3^- - d_3^+ = 20,$$
$$x_3 + d_4^- - d_4^+ = 6;\ x_j \geq 0 \text{ and } d_i^+, d_i^- \geq 0, \text{ for } i = 1, 2, 3, 4 \text{ and } j = 1, 2, 3.$$

12.04. Minimize $z = \{d_1^+,\ d_2^+,\ d_3^-,\ d_4^-\}$ subject to the constraints:

$$x_1 - d_1^+ + d_1^- = 70,\ x_2 - d_2^+ + d_2^- = 50,\ 7x_1 + 5x_2 - d_3^+ + d_3^- = 480,$$
$$8x_1 + 4x_2 - d_4^+ + d_4^- = 480;\ x_1 \geq 0, x_2 \geq 0, d_i^+ \geq 0, d_i^- \geq 0\ (i = 1, 2, 3, 4).$$

12.05. Minimize $z = d_1^+ + d_2^+ + d_2^- + d_3^- + d_4^- + d_5^- + d_6^- + d_7^-$ subject to the constraints:

$$4x_1 + 5x_2 + 5x_3 + 5x_4 + d_1^- - d_1^+ = 600,\ 3x_1 + 4x_2 + 3x_3 + 3x_4 + d_2^- - d_2^+ = 360,$$
$$x_1 + 2x_2 + 2x_3 + x_4 + d_3^- - d_3^+ = 120,\ x_2 + x_3 + x_4 + d_4^- - d_4^+ = 120,$$
$$x_1 + x_3 + x_4 + d_5^- - d_5^+ = 120,\ x_1 + x_2 + x_3 + d_6^- - d_6^+ = 120,$$
$$x_1 + x_2 + x_4 + d_7^- - d_7^+ = 120; \text{ all variables are non-negative.}$$

12.06. Minimize $z = P_1 d_1^- + P_2 d_2^+ + P_3 d_3^- + P_4 d_1^+ + P_5 d_4^- + 3P_6 d_5^- + 4P_6 d_6^- + 2P_6 d_7^-$
subject to the constraints:

$$7x_1 + 10x_2 + 8x_3 + d_1^- - d_1^+ = 6{,}000,\ 10{,}000x_1 + 12{,}000x_2 + 7{,}000x_3 + d_2^- - d_2^+ = 80{,}00{,}000,$$
$$3{,}000x_1 + 4{,}000x_2 + 2{,}000x_3 + d_3^- - d_3^+ = 1{,}05{,}00{,}000,\ x_1 + x_2 + x_3 + d_4^- - d_4^+ = 1{,}000,$$
$$x_1 + d_5^- - d_5^+ = 200,\ x_2 + d_6^- - d_6^+ = 500,\ x_3 + d_7^- - d_7^+ = 300; \text{ all variable are non-negative.}$$

12.07. Minimize $z = d_1^- +\ d_2^-$ subject to the constraints:

$$20x_1 + 32x_2 + d_1^- - d_1^+ = 5{,}400,\ 0.3x_1 + 0.75x_2 + d_2^- - d_2^+ = 108,\ 2x_1 + 4x_2 \leq 600,$$
$$4x_1 + 5x_2 \leq 1{,}000,\ 5x_1 + 4x_2 \leq 1{,}200;\ x_1, x_2, d_1^-, d_2^-, d_1^+, d_2^+ \geq 0.$$

12.08. Minimize $z = P_1 d_1^+ + P_2 d_2^- + P_3 d_3^-$ subject to the constraints:

$$4x_1 + 5x_2 + d_1^- - d_1^+ = 40, \ x_2 + d_3^- - d_3^+ = 10,$$
$$100x_1 + 100x_2 + d_2^- - d_2^+ = 2{,}000;$$
$$x_1, x_2, d_1^-, d_2^-, d_3^-, d_1^+, d_2^+, d_3^+ \geq 0.$$

where x_1 and x_2 are the number of chairs and tables, respectively.

Optimum Solution: $x_1 = 10, \ x_2 = 0, \ d_1^+ = 0, \ d_2^- = 1{,}000$ and $d_3^- = 10$.

12.09. Minimize $z = P_1 d_1^- + P_2 (2d_2^- + d_3^-) + P_3 d_1^+$ subject to the constraints:

$$x_1 + x_2 + d_1^- - d_1^+ = 400, \ x_1 + d_2^- - d_2^+ = 240,$$
$$x_2 + d_3^- - d_3^+ = 300; \text{ all variables are non-negative.}$$

Optimum Solution: $x_1 = 240, \ x_2 = 300, \ d_1^+ = 140, \ d_1^- = d_2^- = d_3^- = 0$.

12.10. Minimize $z = P_1 d_1^- + P_2 d_4^+ + P_3 (5d_2^- + 6d_3^-) + P_4 (6d_2^+ + 5d_3^+)$ subject to the constraints:

$$50x_1 + 60x_2 + d_1^- - d_1^+ = 1{,}200, \ x_1 + d_2^- - d_2^+ = 8, \ x_2 + d_3^- - d_3^+ = 8,$$
$$x_2 + d_4^- - d_4^+ = 11; \text{ all variables are non-negative.}$$

Optimum Solution: $x_1 = 10.8, \ x_2 = 11, \ d_2^+ = 3.8, \ d_3^+ = 3, \ d_1^- = d_4^+ = d_2^- = d_3^- = 0$.

12.11. Optimum Solution: $x_1 = 25, \ x_2 = 15, \ d_1^- = 0, \ d_2^- = 60$ and $d_3^- = 5$;

$$d_1^+ = d_2^+ = d_3^+ = d_4^- = d_4^+ = 0.$$

12.12. Minimize $z = P_1 d_1^- + P_2 d_2^+ + P_3 (10d_4^- + 12d_2^-) + P_4 (d_2^+ + d_4^+)$ subject to the constraints:

$$10x_1 + 12x_2 + d_1^- - d_1^+ = 200, \ x_2 + d_2^- + d_3^- - d_3^+ = 12,$$
$$x_1 + d_4^- - d_4^+ = 8; \text{ all variables are non-negative.}$$

Optimum Solution: $x_1 = 8$ hours, *i.e.*, line 1 for 8 hours, $x_2 = 12$ hours, *i.e.*, line 2 for 12 hours and $d_1^+ = 24$; all other variables equals zero; output = 224 units.

12.13. Minimize $z = 2d_1^- + d_2^+ - 0.8d_3^-$ subject to the constraints:

$$4x_1 + 3x_2 + d_1^- - d_1^+ = 20, \ 5x_1 + 8x_2 + d_2^- - d_2^+ = 30,$$
$$2x_1 + 4x_2 + d_3^- - d_3^+ = 15, \ 2x_1 + 3x_2 \leq 12; \text{ all variables being non-negative.}$$

Optimum Solution: $x_1 = 4, \ x_2 = 4/3, \ d_2^+ = 2/3, \ d_3^- = 5/3; \ d_1^+ = d_1^- = d_2^- = d_3^+ = 0$.

12.14. Minimize $z = d_1^- + d_2^- + 0 \cdot s_1 + 0 \cdot s_2 + 0 \cdot s_3 + 0 \cdot d_1^+ + 0 \cdot d_2^+$ subject to the constraints:

$$2x_1 + 4x_2 + s_1 = 600, \ 4x_1 + 5x_2 + s_2 = 1{,}000,$$
$$5x_1 + 4x_2 + s_3 = 1{,}200, \ 20x_1 + 32x_2 + d_1^- - d_1^+ = 5{,}400,$$
$$0.3x_1 + 0.75x_2 + d_2^- - d_2^+ = 108; \text{ all variables being non-negative.}$$

Optimum Solution: $x_1 = 150, \ x_2 = 75, \ d_2^- = 27/4$ or $6.75, \ s_2 = 25$ and $s_3 = 150$; minimum $z = 6.75$.

12.15. (*a*) $x_1 = 70, \ x_2 = 30, \ d_1^+ = 300; \ d_1^- = d_2^- = d_3^- = 0$.

(*b*) $x_1 = 240, \ x_2 = 300, \ d_1^+ = 140; \ d_1^- = d_2^- = d_3^- = 0$.

Chapter 13. Network Analysis

13.01.

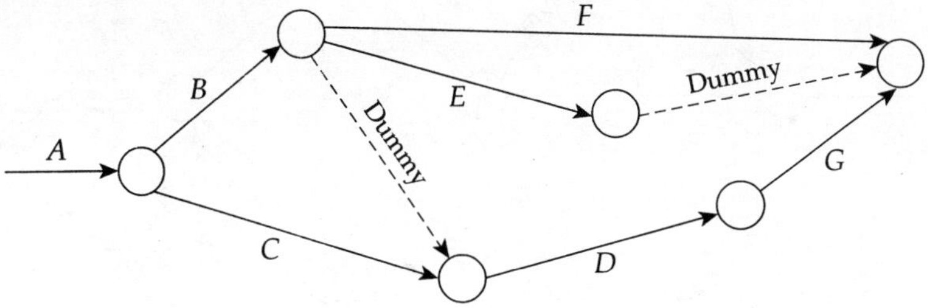

13.02.

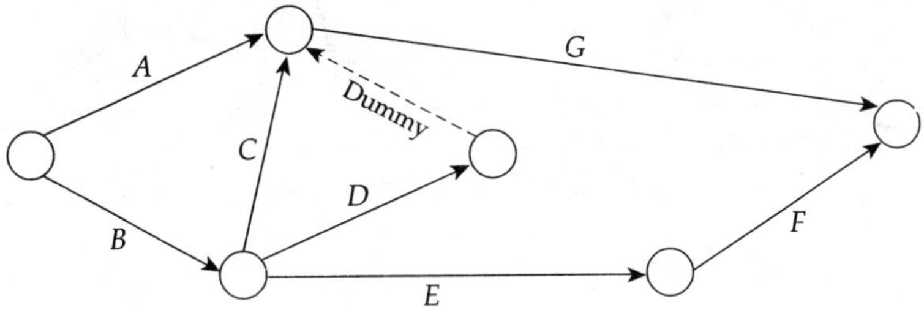

13.03.

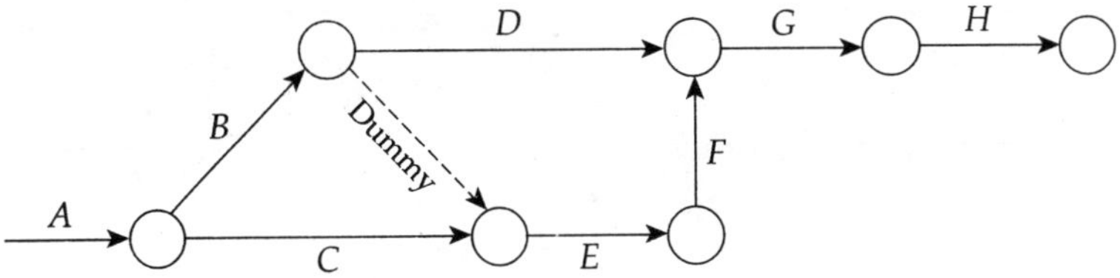

13.04.

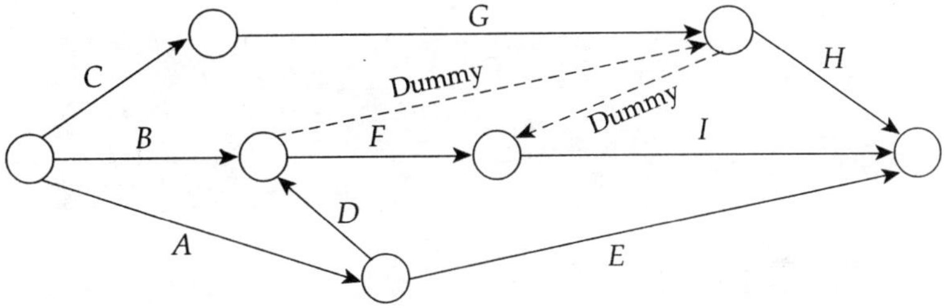

13.05.

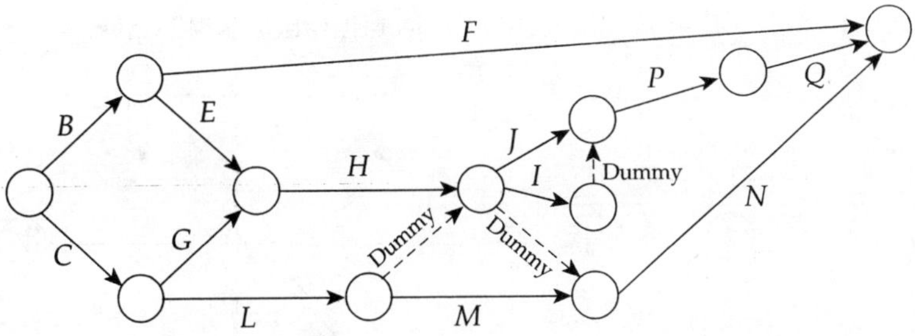

13.06.

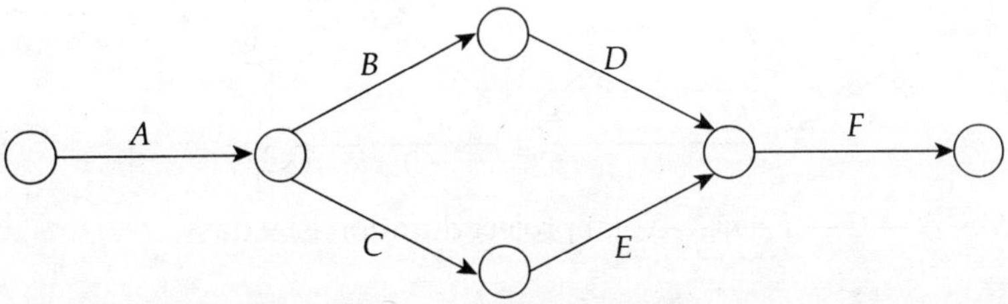

13.07.

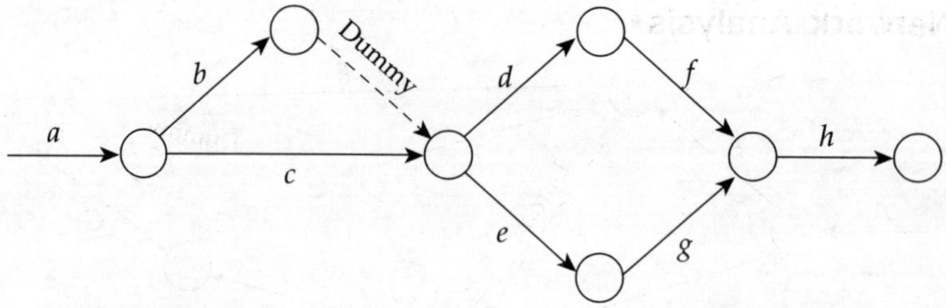

13.08.

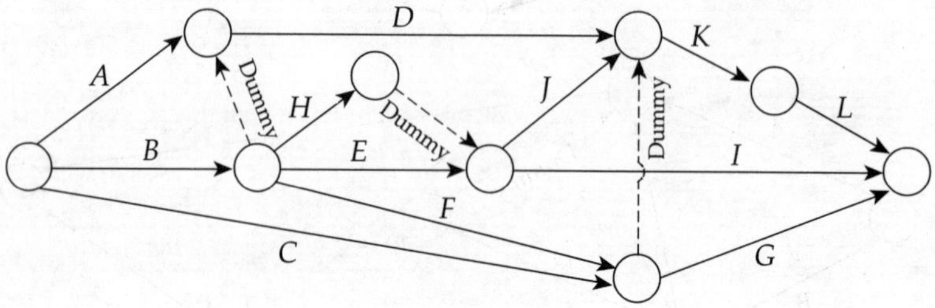

13.09.

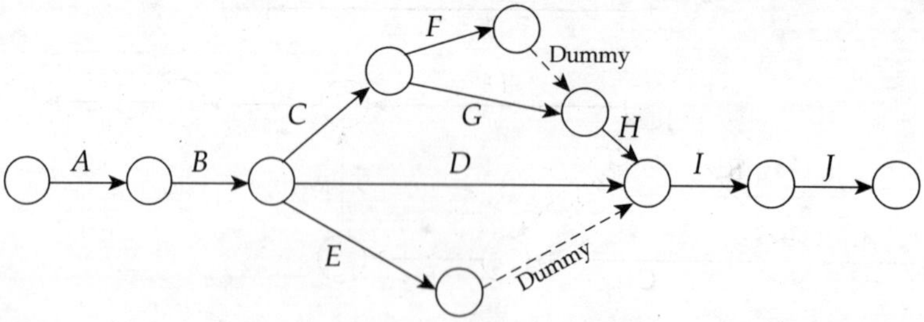

13.10.

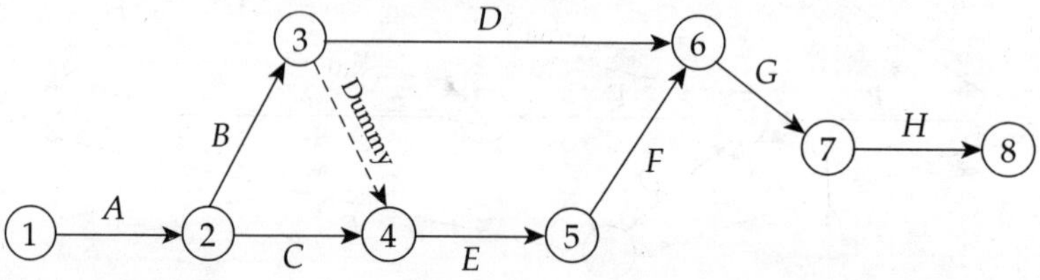

13.11. (*a*) *CPM:* $A \to C \to E \to F$. Expected project duration is: 42 days.

(*b*)

Activity	:	A	B	C	D	E	F
Total Float	:	0	4	0	4	0	0
Free Float	:	0	4	0	4	0	0

13.12. (*a*)

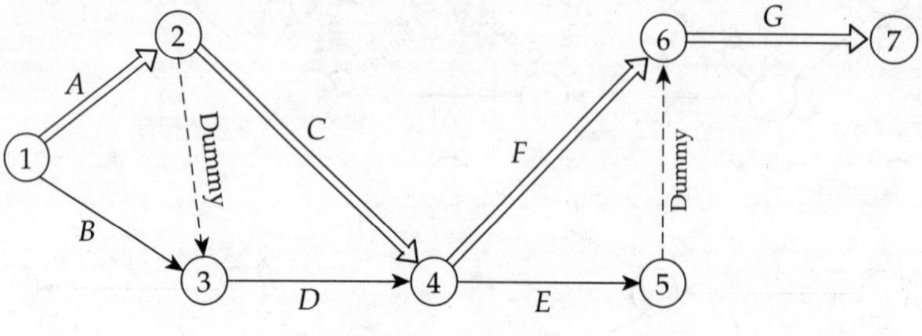

(*b*) *CPM:* $A \to C \to F \to G$. Total project duration is: 9 days.

13.13. (*i*)

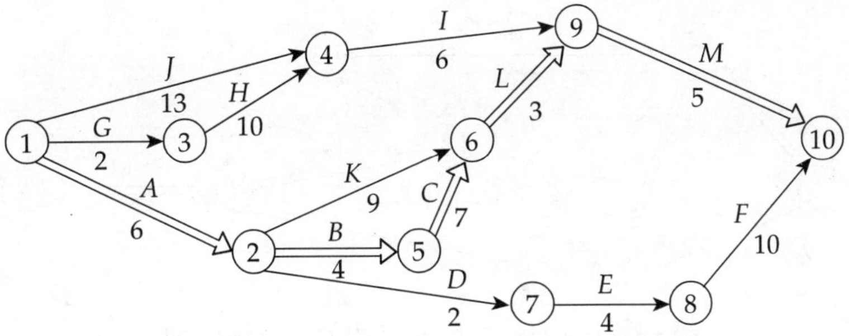

(*ii*) *CPM:* $A \to B \to C \to L \to M$. Project duration is of 25 weeks.

(*iii*)

Activity	:	*G*	*J*	*K*	*D*	*H*	*I*	*E*	*F*
Total Float	:	2	1	2	3	2	1	3	3
Free Float	:	0	0	2	0	1	1	0	3

13.14. (*a*)

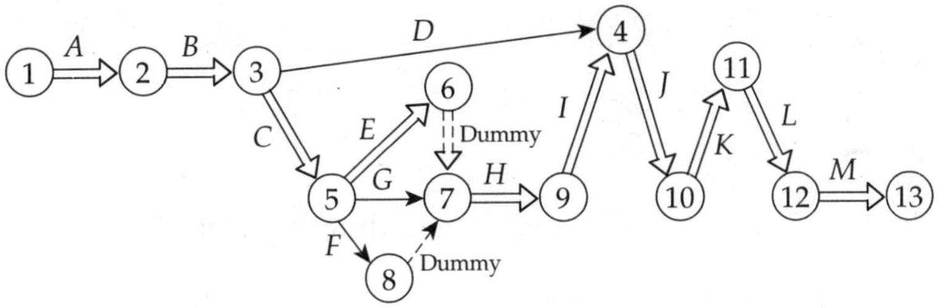

(*b*) *CPM:* $A \to B \to C \to E \to H \to I \to J \to K \to L \to M$. Total time for installation of scanner: 57 days.

(*c*) Total float for non-critical activities will be: *D* : 20 days, *G* : 3 days and *F* : 2 days.

13.15. *CPM:* $1 \to 2 \to 3 \to 4 \to 5 \to 6 \to 7 \to 8$. Total project duration is of 32.

Activity	:	*1–3*	*3–5*	*5–7*
Total Float	:	2	3	1
Free Float	:	2	3	1
Independent Float	:	2	3	1

13.16.

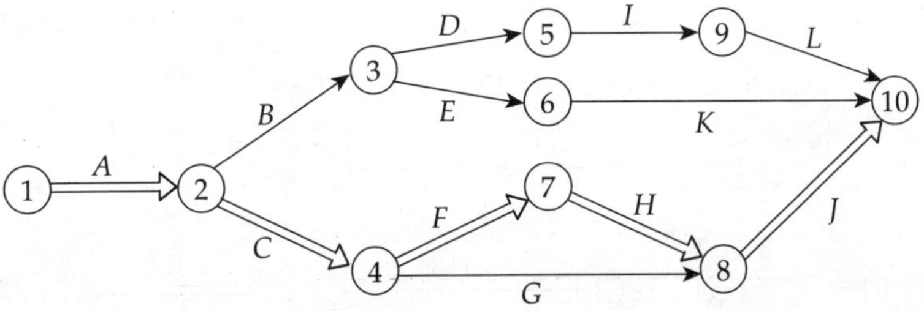

CPM: $A \to C \to F \to H \to J$. Total time for completion of project is: 42 weeks.

13.17.

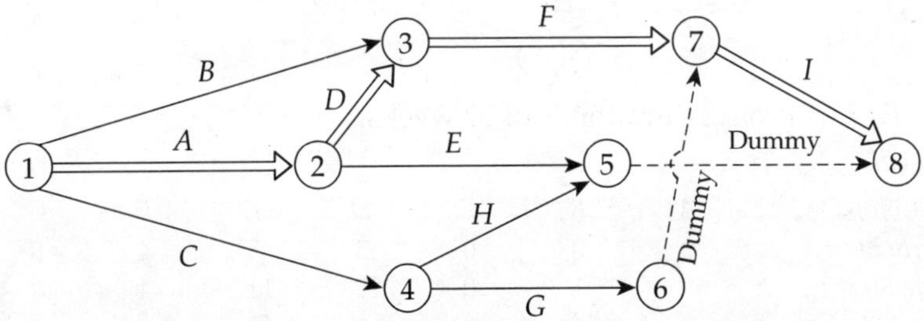

CPM: $1 \to 2 \to 3 \to 7 \to 8$. Minimum time to complete the project is of 44 days.

13.18. (*a*)

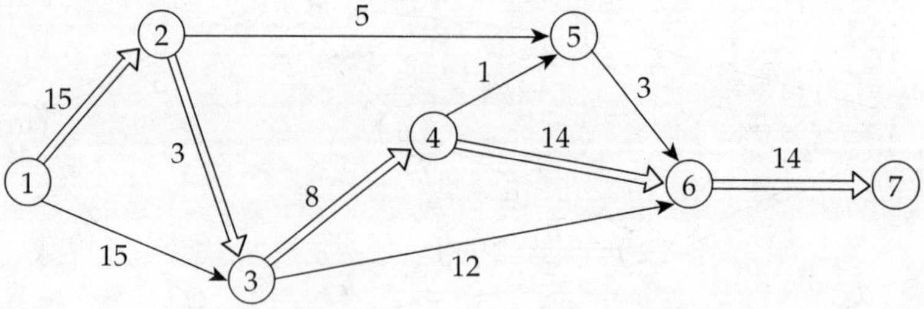

(*b*) *CPM:* $1 \to 2 \to 3 \to 4 \to 6 \to 7$. Project duration is of 54 days.

13.19. (*i*)

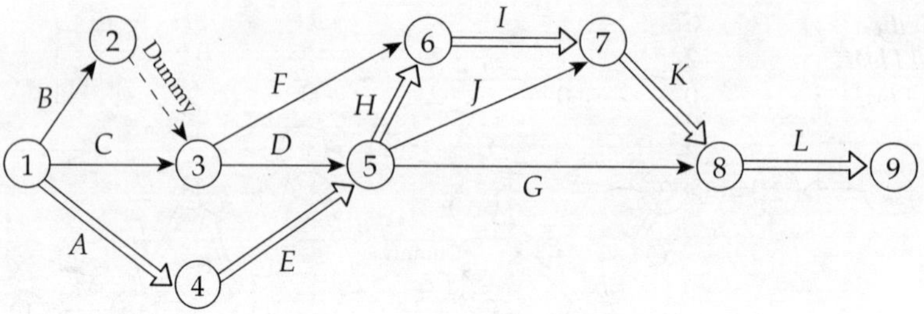

(*ii*) *CPM:* $A \to E \to H \to I \to K \to L$. Project duration is of 47 days.

13.20. (*a*)

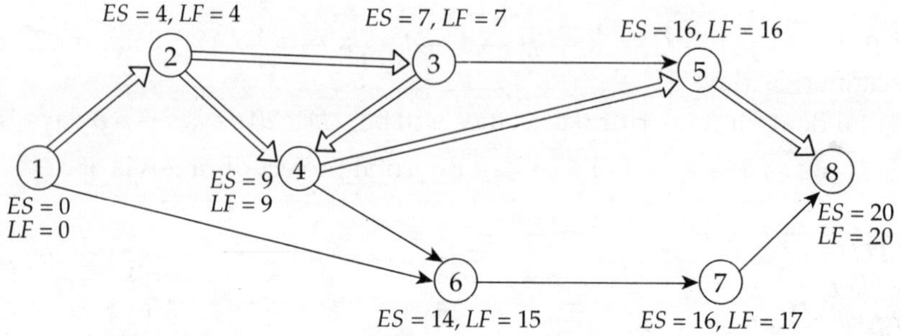

(*b*) *CPM:* $1 \to 2 \to 4 \to 5 \to 8$ or $1 \to 2 \to 3 \to 4 \to 5 \to 8$. Project duration is of 20 days.

13.21. *CPM:* $B \to D \to E \to H$. Project duration is of 32 days.

13.22.

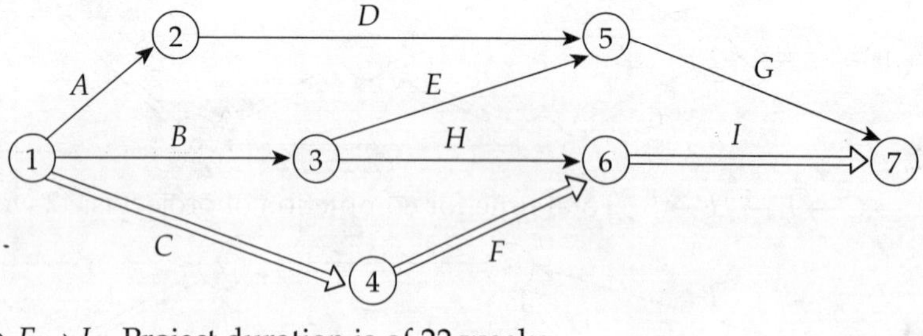

(*i*) $C \to F \to I$. Project duration is of 22 weeks.

(*ii*) and (*iii*).

Activity :	*A*	*B*	*C*	*D*	*E*	*F*	*G*	*H*	*I*
Early Start :	0	0	0	3	5	4	8	5	13
Late Start :	9	1	0	12	11	4	14	6	13
Total Slack :	9	1	0	9	6	0	6	1	0

13.23. (*i*)

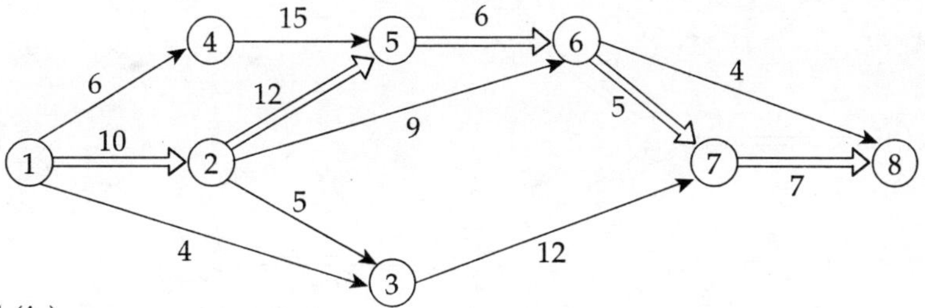

(*ii*) and (*iv*)

Activity	*Earliest*		*Latest*		*Float*		
	Start	*Finish*	*Start*	*Finish*	*Total*	*Free*	*Independent*
1–2	0	10	0	10	10	0	0
1–3	0	4	17	21	17	11	11
1–4	0	6	1	7	1	0	0
2–3	10	15	16	21	6	0	0
2–5	10	22	10	22	0	0	0
2–6	10	19	19	28	9	9	9
3–7	15	27	21	33	6	6	0
4–5	6	21	27	22	1	1	0
5–6	22	28	22	28	0	0	0
6–7	28	33	28	33	0	0	0
6–8	28	32	36	40	8	8	8
7–8	33	40	33	40	0	0	0

(*iii*) *CPM:* $1 \rightarrow 2 \rightarrow 5 \rightarrow 6 \rightarrow 7 \rightarrow 8$. Project duration: 40 days.

13.24.

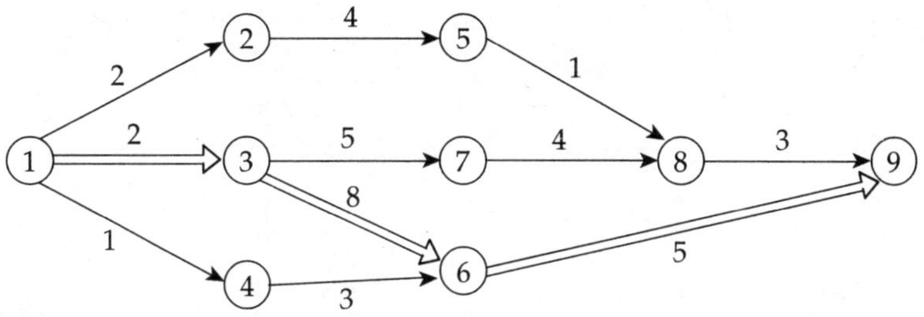

(*i*)

Activity :	1–2	1–3	1–4	2–5	3–6	3–7	4–6	5–8	6–9	7–8	8–9
Total Float :	5	0	6	5	0	1	6	5	0	1	1

(*ii*) *CPM:* $1 \rightarrow 3 \rightarrow 6 \rightarrow 9$. Duration of the project is of 15 weeks.

13.25. (*i*)

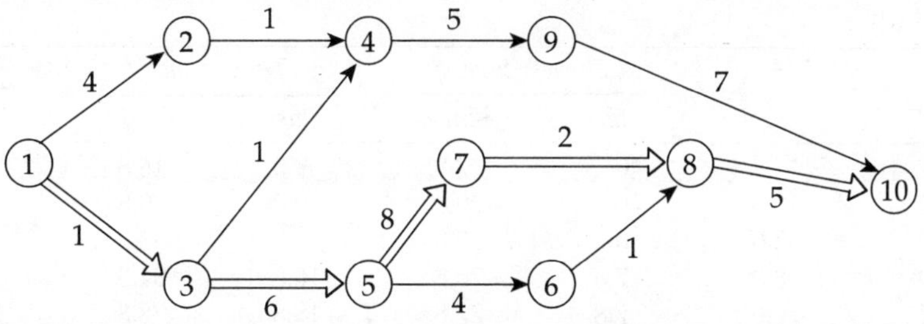

(*ii*)

Activity :	1–2	1–3	2–4	3–4	3–5	4–9	5–6	5–7	6–8	7–8	8–10	9–10
T_E :	4	1	5	2	7	10	11	15	12	17	22	17
T_L :	0	1	10	10	7	15	16	15	17	17	22	22

(*iii*) *CPM:* $1 \rightarrow 3 \rightarrow 5 \rightarrow 7 \rightarrow 8 \rightarrow 10$. Duration of the project is of 22 weeks.

13.26. (*i*)

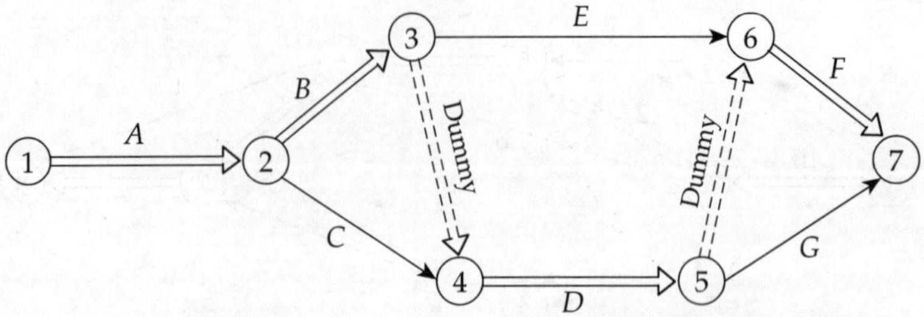

Critical Path: $A \to B \to D \to F$. Expected project duration is 27.7 days.

Next Critical Path: $A \to B \to D \to G$, with duration of 25 days.

(*ii*) Project duration which has 95% probability of completion is:

$$P(z)\,(=1.64) = \frac{T_S - 27}{\sqrt{18}} \quad \text{or} \quad T_S = 1.64 \times \sqrt{18} + 27 = 34 \text{ days.}$$

13.27. (*i*)

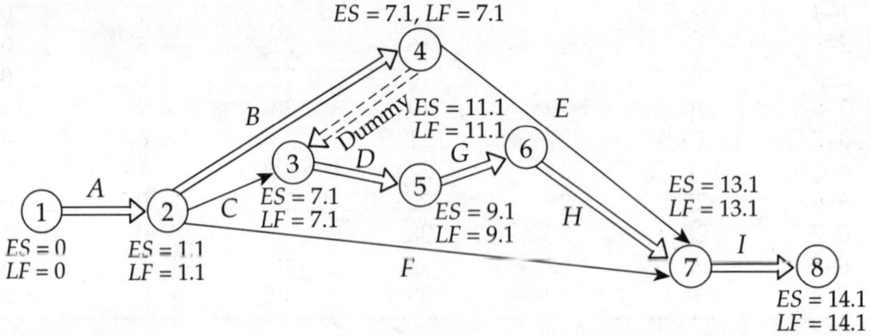

(*ii*) *Critical Activities: A, B, D, G, H* and *I*. Expected project completion time is 14.1 weeks.

(*iii*) *Required Probability, i.e.,* the completion of project within 13 weeks = 0.3015 or 30.15%.

13.28. (*i*)

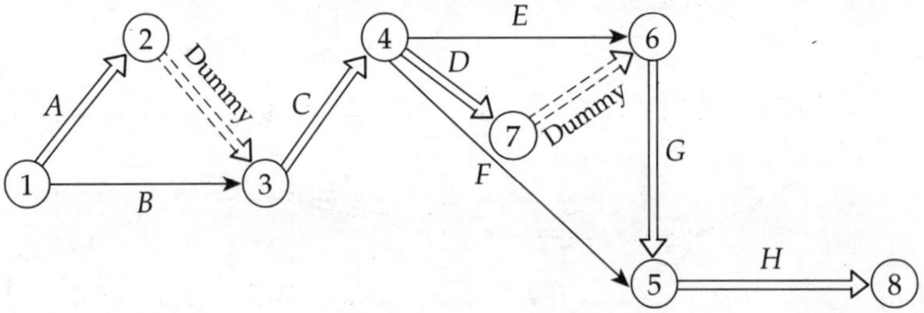

CPM: $A \to C \to D \to G \to H$ and expected completion time of the project is 45.3 days.

(*ii*) Probability of completing the project for more than 50 days is 0.9699 or 96.99% (*i.e.*, 97%).

13.29. (*a*) and (*b*)

Task	t_e	*Earliest time*		*Latest time*		*Total Float*	σ^2
		Start	*Finish*	*Start*	*Finish*		
A	7.8	0	7.8	9.0	16.8	9.0	0.69
B	20.0	0	20.0	9.0	29.8	9.8	0.44
C	33.0	0	33.0	0	33.0	0	5.34
D	18.0	7.8	25.8	16.8	34.8	9.0	0.44
E	20.0	7.8	27.8	18.8	38.8	11.0	2.78
F	9.0	20.0	29.0	29.8	38.8	9.8	1.00
G	9.8	33.0	42.8	33.0	42.8	0	0.69
H	8.0	25.8	33.8	34.8	42.8	9.0	0.11
I	4.0	29.0	33.8	38.8	42.8	9.8	0.11

(*c*)

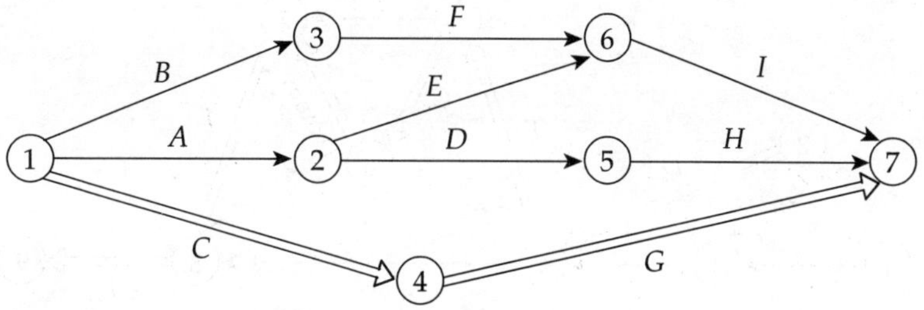

CPM: $C \to G$, *i.e.,* $1 \to 4 \to 7$.

(*d*) Probability of completing the project in 41.5 weeks is: 0.30.

(*e*) Duration of project having 95% chance of completion is: 46.85 weeks.

13.30. (*i*) (*a*) Probability of completing the project in 41 weeks is: 0.5 or 50%.

(*b*) Probability of completing the project in 45 days is: 0.7486 or 74.86%.

(*ii*) Project should start prior to deadline is (53.4 – 41.0) weeks or 12.4 weeks.

13.31. (*a*)

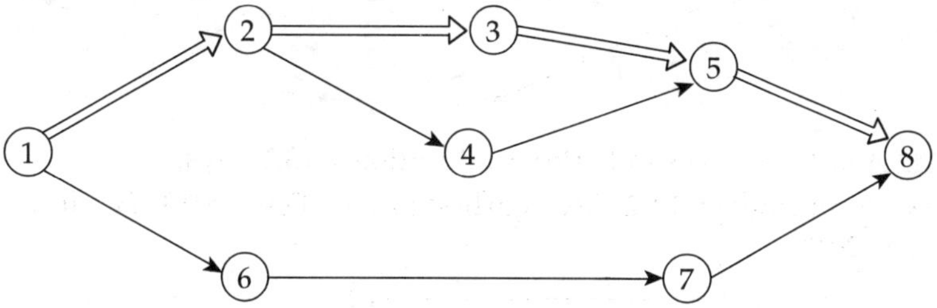

(*b*) *Critical Path:* $1 \to 2 \to 3 \to 5 \to 8$. Project duration is expected to be of 36 days; and variance of critical path is: 25 days.

(*c*) Probability of completion the project before 31 days is: 0.1587 or 15.87%.

(*d*) Probability of completion exceeding 46 days is: 0.9772 or 97.72%.

(*e*) Project will be delayed by 2 days.

13.32. (*a*)

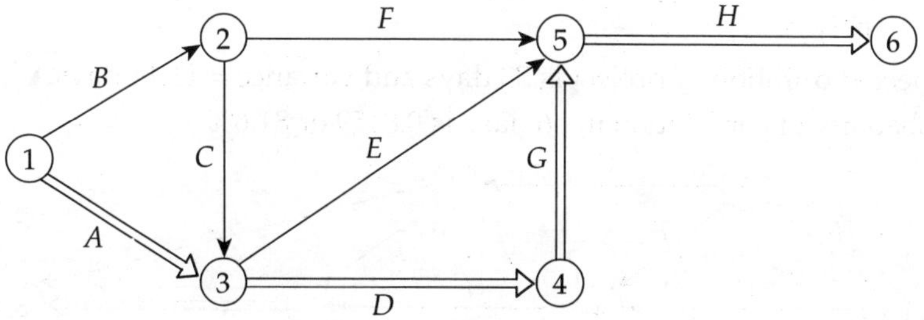

(*b*) *CPM:* $A \to D \to G \to H$. Expected project duration is: 30 weeks.

(*c*) Probability of completing the project within time is: 0.6591 or 65.91%.

(*d*) Project should be started 34.7 or 35 weeks before the scheduled date.

13.33. (*i*)

Task	:	*A*	*B*	*C*	*D*	*E*	*F*	*G*	*H*	*I*	*J*
T_e	:	7.7	2	2	6.7	4	3	4	6.2	9	1
Variance	:	1.0	0	0.11	1.78	0.11	0	0.11	0.69	2.78	0

(*ii*)

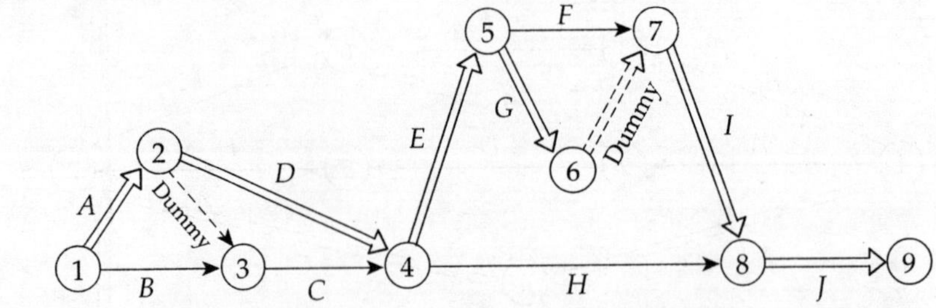

CPM: $A \to D \to E \to G \to I \to J$. Expected project length = 32.3 weeks.

Project variance = 5.78 weeks and σ = 2.40 weeks.

(*iii*) Probability of not exceeding 32 weeks is 0.4364 or 43.64% and probability of not exceeding 36 weeks is: 0.9302 or 93%.

13.34. (*i*)

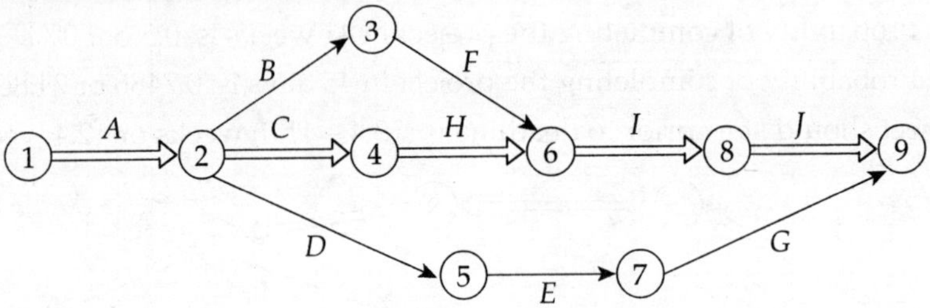

(*ii*) *CPM*: $A \to C \to H \to I \to J$. Project duration = 13.1 days.

(*iii*) Work should begin 14.2 days earlier to 20th Feb., 2013, *i.e.*, it should work since 7th Feb., 2013.

13.35. (*i*)

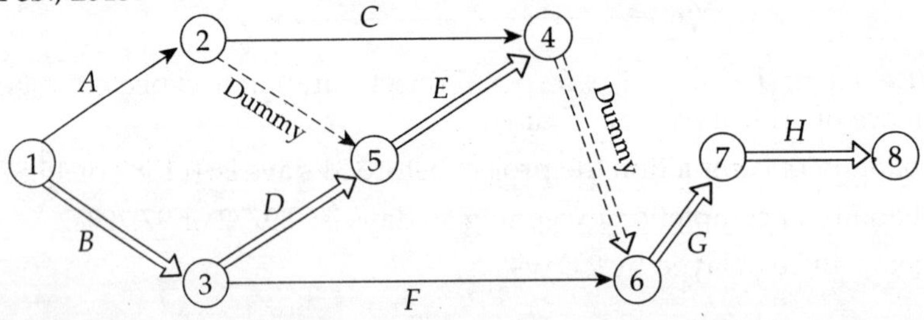

(*ii*) *CPM*: $B \to D \to E \to G \to H$.

(*iii*) Expected duration of project is 53 days and variance = 11.11 days.

(*iv*) Probability of completion in 56 days is 0.8159 or 81.6%.

13.36. (*i*)

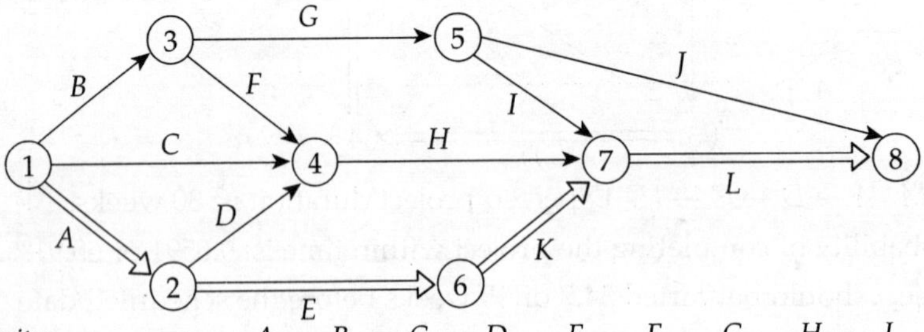

Activity	:	*A*	*B*	*C*	*D*	*E*	*F*	*G*	*H*	*I*	*J*	*K*	*L*
Standard deviation	:	2	0.8	1.5	0	0.8	1.2	1.1	0.6	0.8	0	1.1	0.8

(*ii*) *CPM*: $A \to E \to K \to L$. Project duration = 32.6 weeks, standard deviation = 4.7.

(*iii*) Probability of completing the project in 38 weeks = 0.8749 or 87.49%.

(*iv*) Project should be started 35.3 weeks prior to deadline.

13.37. (*i*)

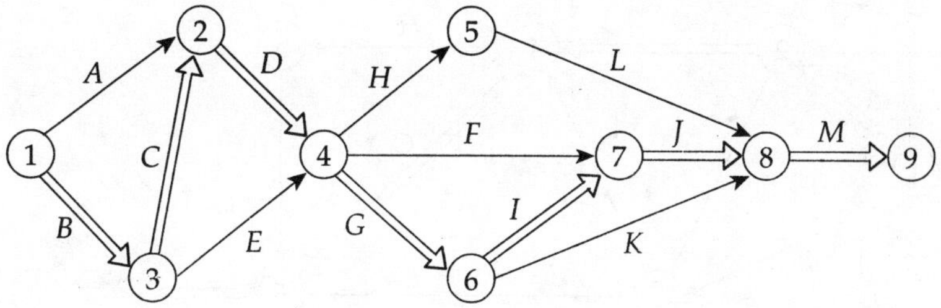

Expected project completion time is: 52.2 weeks.

(*ii*) *Critical Activities:* *B, C, D, G, I, J* and *M*.

(*iii*) To ensure 95% level of confidence, project duration should be of 54.6 weeks at labour cost of ₹ 81,913.

13.38.

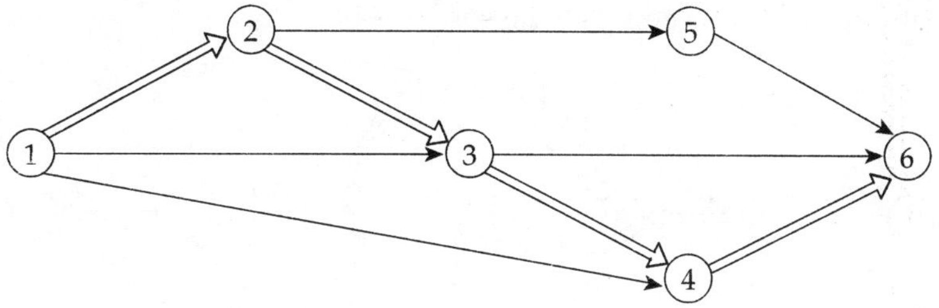

CPM: $1 \rightarrow 2 \rightarrow 3 \rightarrow 4 \rightarrow 6$. Expected project completion time is: 51 days;

Variance of critical path is 5.95 or 6 days approx.

(*i*) Probability for the project duration to exceed 60 days is: 0.9332 or 93.32%.

(*ii*) Probability for the project to complete between 45 and 54 days is: (0.6915 – 0.1587), *i.e.*, 0.5328 or 53.28%.

(*iii*) When the time estimates of activity 4-6 are revised, the project duration will be of 56 days and now probability for the project completion before 46 days is: 0.047.

13.39. (*i*)

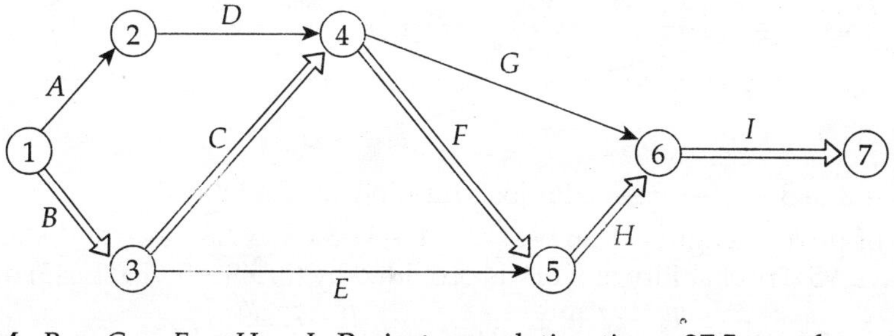

(*ii*) *CPM:* $B \rightarrow C \rightarrow F \rightarrow H \rightarrow I$. Project completion time: 27.5 months.

Alternative Paths: $B \rightarrow C \rightarrow G \rightarrow I$ with project completion duration: 23.8 months;

$B \rightarrow E \rightarrow H \rightarrow I$ with project completion duration: 23 months;

$A \rightarrow D \rightarrow F \rightarrow H \rightarrow I$ with project completion duration: 27.3 months;

$A \rightarrow D \rightarrow G \rightarrow I$ with project duration: 23.67 months.

(*iii*) Probability of completion by 36 months: 0.9987 or 99.87%.

(*iv*) To ensure 80% probability of completion, project duration will be 30 months.

(*v*) Probability of completion for the near critical path is: 0.9976 or 99.76.

13.40. (*a*)

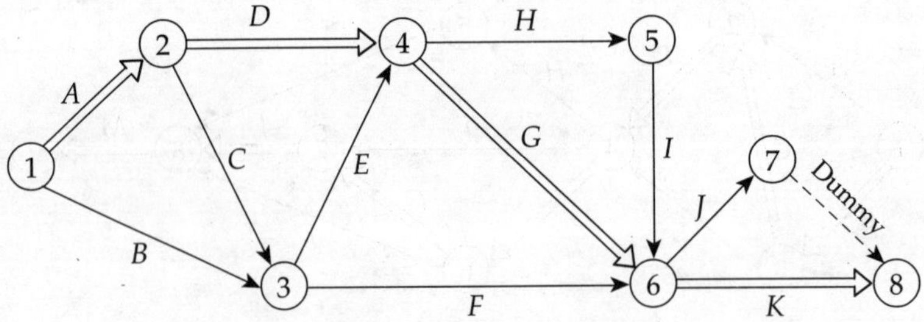

CPM: $A \to D \to G \to K$. Project duration: 33 days.

(*b*)

Non-critical activities :	B	C	E	F	H	I	J
Total Float :	7	2	2	4	2	2	6
Free Float :	5	0	2	4	0	2	0

(*c*) Probability of not exceeding 40 days is: 0.9649 or 96.5%.

13.41. (*a*)

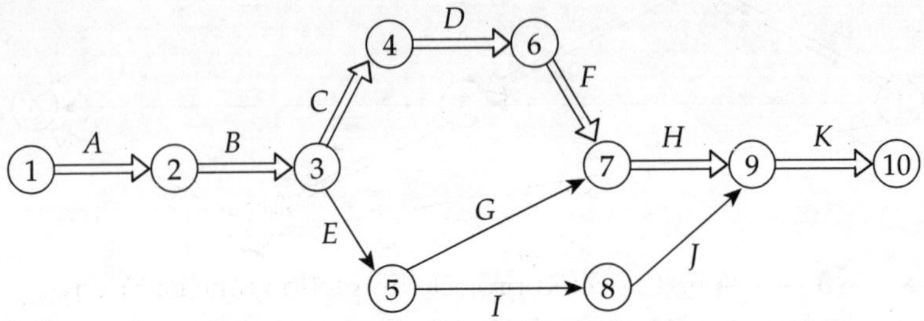

CPM: $A \to B \to C \to D \to F \to H \to K$. Project duration is: 23.4 months.

(*b*) Probability of completion in 21 months is : 0.0985 or 9.85%, in 24 months is: 0.6255 or 62.55% and in 27 months is: 0.9738 or 97.38%.

13.42.

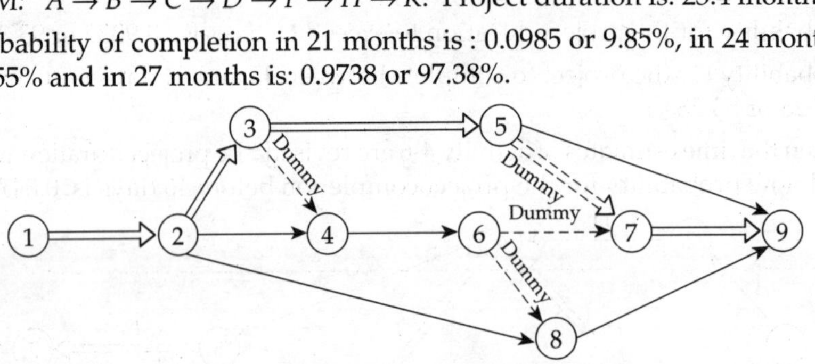

CPM: $1 \to 2 \to 3 \to 5 \to 7 \to 9$. Project duration is: 77.49 days.

Variance of project length = 3.36 + 1.36 + 4 + 17.36 = 26.08 days, and standard deviation is: 5.11 days. 95% probability of work is provided by (77.49 + 5.11 × 1.645) days, *i.e.*, 85.89 or 86 days, because $P\left(z \le \frac{T_s - 77.49}{5.11}\right) = 0.95 \;\Rightarrow\; \frac{T_s - 77.49}{5.11} = 1.645.$

Amount of bid = ₹ 8,00,000 + ₹ 9,000 × 86 = ₹ 15,74,000.

13.43.

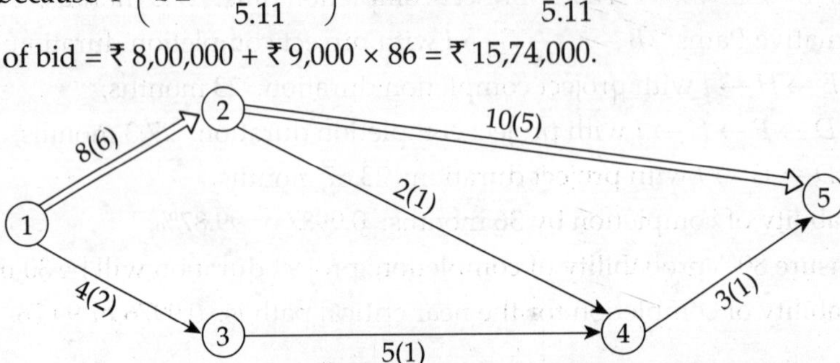

Critical Activities are: (1, 2) and (2, 5), and Normal project duration: 18 days.
Optimal project duration: 12 days and Associated project cost

= Normal cost + Crash cost + Indirect cost
= ₹ 580 + (2 × 50 + 4 × 60) + 12 × ₹ 70 = ₹ 1,760.

13.44. (*i*)

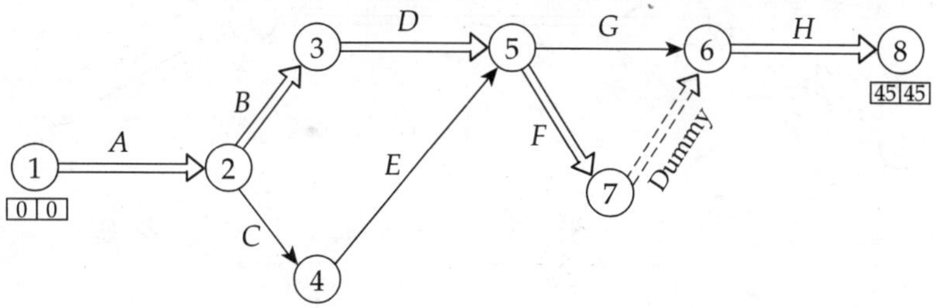

(*ii*) Earliest time to complete the project is: 45 days, and corresponding cost is: ₹ 110 thousand of rupees.

(*iii*) *A*, *B*, *D*, *F* and *H* are the critical tasks.

(*iv*) Task *A* should be crashed for 2 days and task *F* by 3 days. So, the resulting project cost is: ₹ 120 thousands of rupees.

13.45. (*i*)

Activity :	*A*	*B*	*C*	*D*	*E*	*F*	*G*	*H*
Total float :	1	0	4	0	0	0	9	0

(*ii*) Optimum project duration: 14 weeks and Associated project cost: ₹ 49,350.

13.46. (*i*)

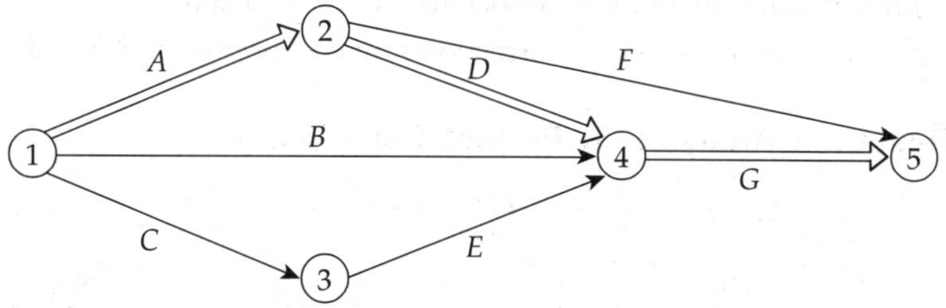

(*ii*) Critical Activities: *A*, *D* and *G* with Normal project duration: 13 days.
Optimum project duration: 10 days with Associated project cost: ₹ 943.

13.47. (*i*)

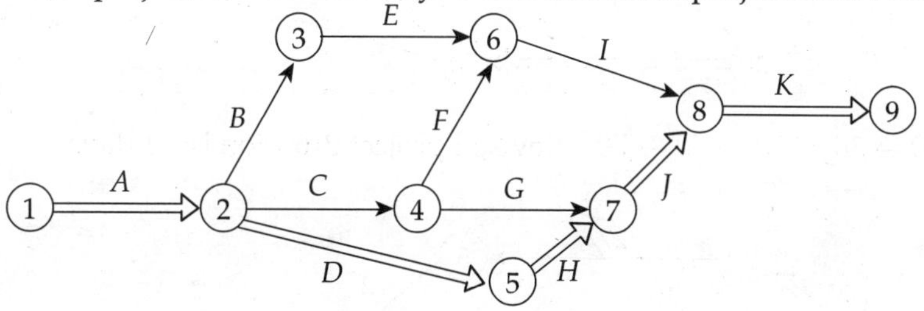

(*ii*) Normal project duration: 43 days; Optimal trade-off project duration: 33 days and Corresponding cost is: ₹ 16,87,000.

13.48. (*i*)

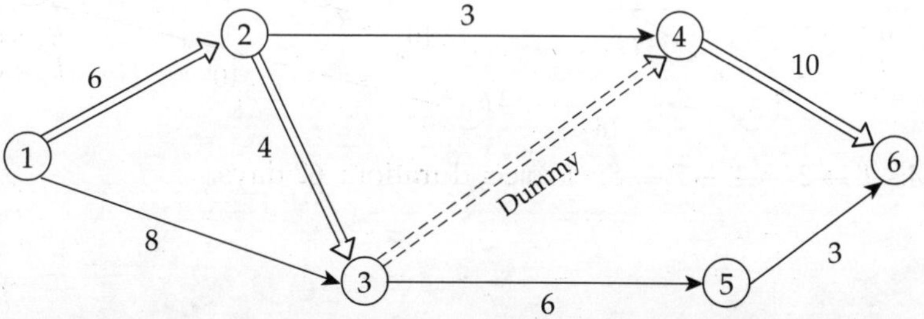

(*ii*) Normal duration of project: 20 days and Corresponding cost: ₹ 15,200.

(*iii*) Least project length: 12 days and Corresponding cost: ₹ 16,493.

(*iv*) Optimum duration of project: 17 days and Corresponding cost: ₹ 15,000.

13.49.

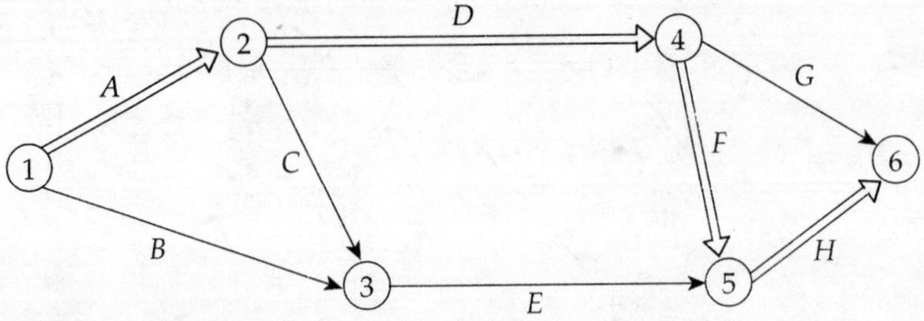

Normal project schedule: $A \to D \to F \to H$. Normal project duration: 41 days.
Optimum project schedule: $A \to D \to F \to H$ and $A \to C \to E \to H$. Minimum duration of project: 35 days.

13.50. (*a*) Normal duration of project: 24 days. Associated cost: ₹ 36,400.

(*b*) Minimum duration of project: 17 days. Associated cost: ₹ 31,250.

(*c*) Minimum cost of project: ₹ 29,400. Corresponding duration of project: 21 days.

13.51. Normal project duration: 24 months. Normal cost: ₹ 4,80,000.
Optimal time schedule: 20 months. Associated cost: ₹ 14,46,000.

13.52. Optimal project duration: 61 days. Corresponding cost of project: ₹ 3,41,466.

Chapter 14. Resource Analysis in Project Scheduling

14.01.

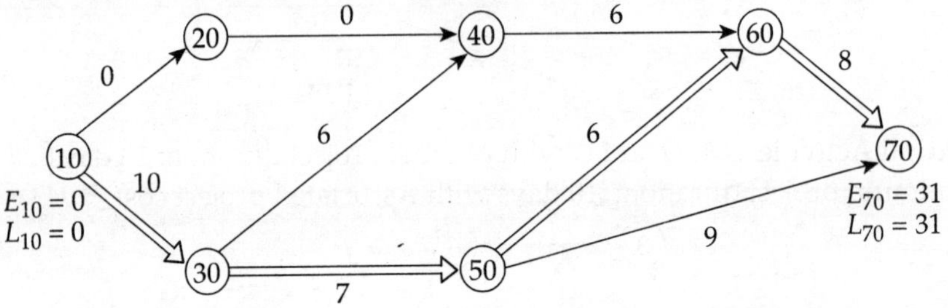

CPM: $10 \to 30 \to 50 \to 60 \to 70$; Revised project duration is: 31 days.

14.02. (*a*)

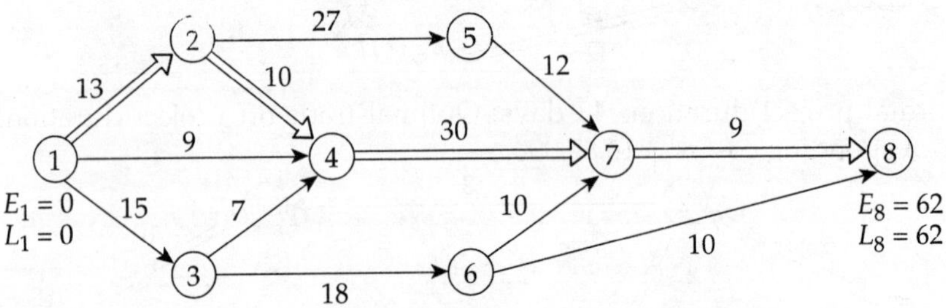

CPM: $1 \to 2 \to 4 \to 7 \to 8$; Project duration: 62 days.

(*b*) Revised updated networks:

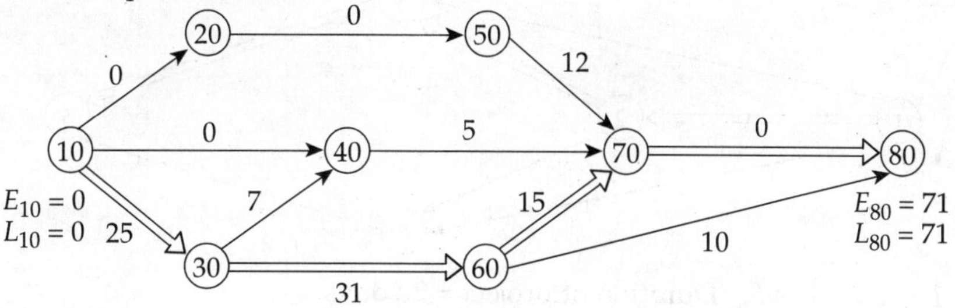

Revised Critical Path: *10 → 30 → 60 → 70 → 80*; Revised project duration: 71 days.

14.03. Updated network:

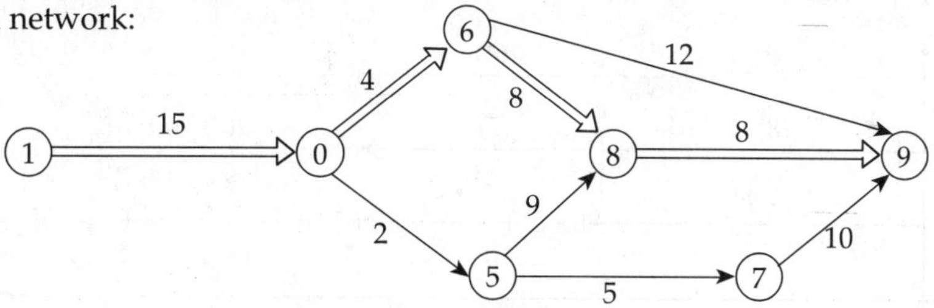

Revised critical path: *1 → 0 → 6 → 8 → 9*; Project duration = 35 days.

14.04. (*i*)

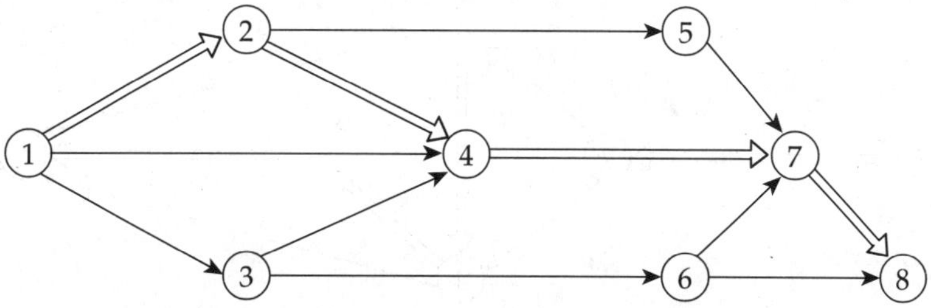

CPM: *1 → 2 → 4 → 7 → 8*; Project duration = 42 days.

(*ii*) Revised critical path: *1 → 3 → 6 → 7 → 8*; Project duration = 47 days.

14.05. (*a*)

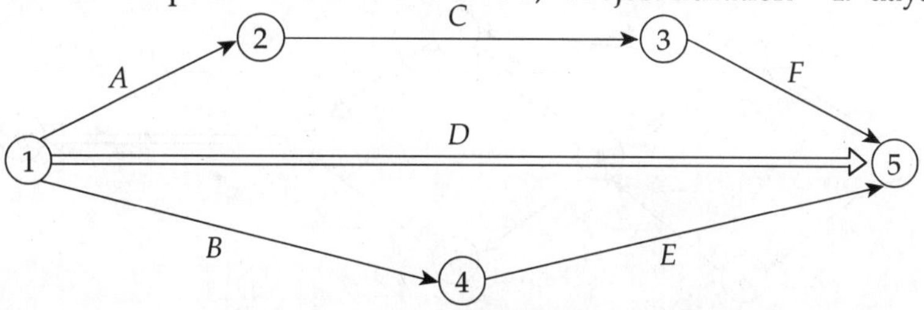

(*b*) *CPM*: Activity *D*, *i.e.*, *1 → 5*; Duration of project = 5 days.

(*c*) & (*d*) Time scale diagram (Labour required on arrows).

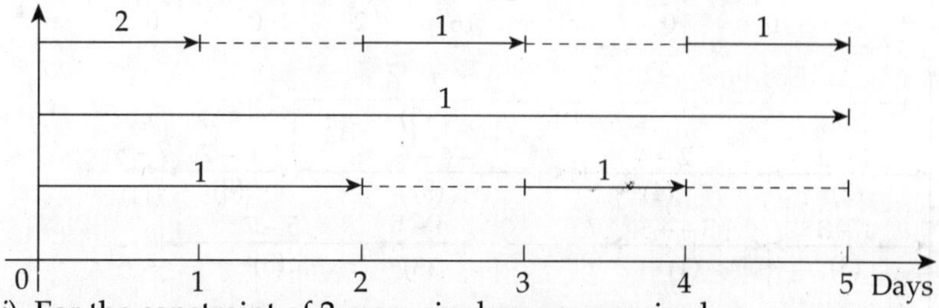

(*e*) (*i*) For the constraint of 2 men, six days are required.

(*ii*) For the constraint of 5 days, four men are required on day 1.

14.06.

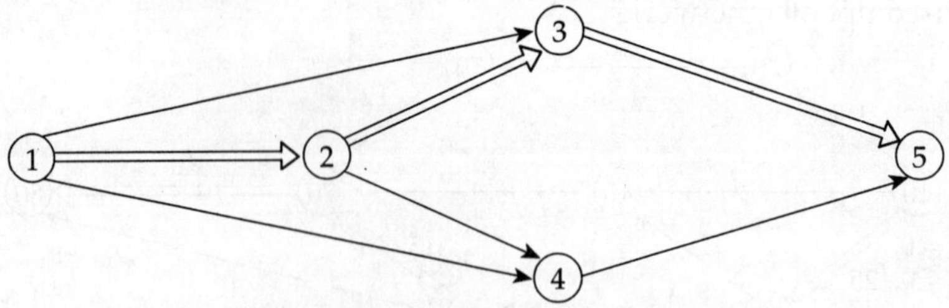

CPM: $1 \rightarrow 2 \rightarrow 3 \rightarrow 5$. Duration of project = 22 days.

Resource requirement (time scaled) diagram:

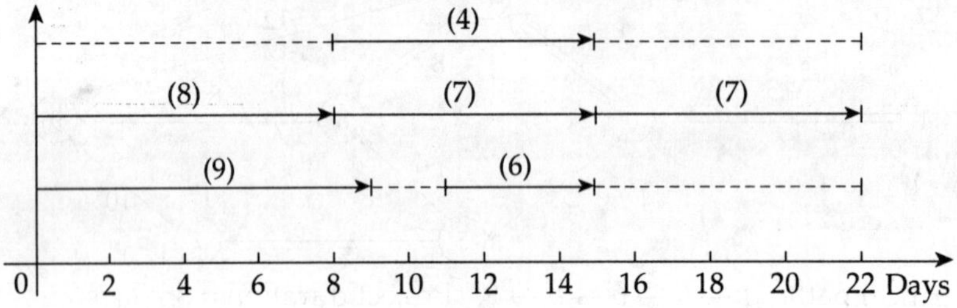

14.07.

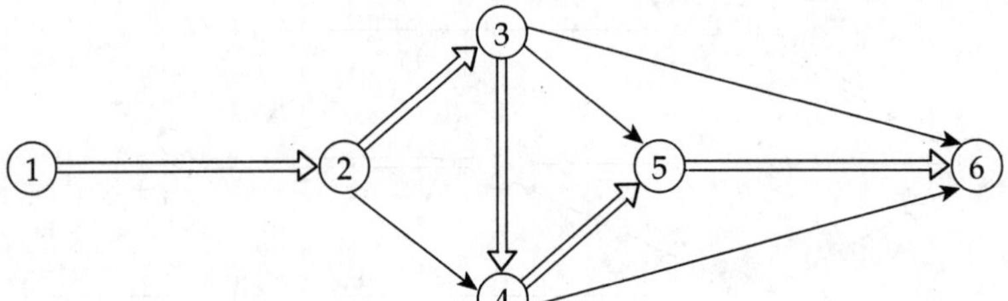

CPM: $1 \rightarrow 2 \rightarrow 3 \rightarrow 4 \rightarrow 5 \rightarrow 6$. Duration of project = 19 days.

14.08. (*a*)

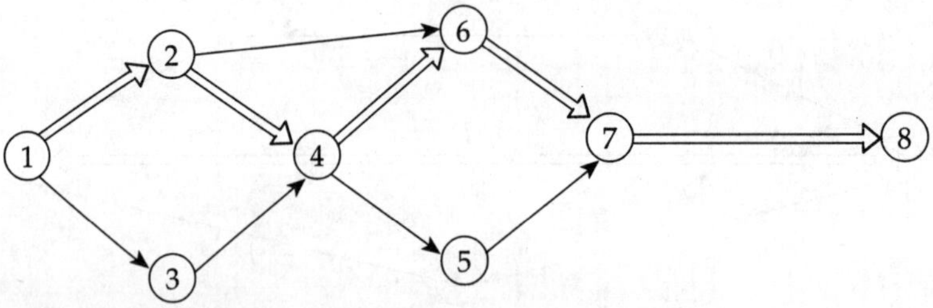

Activity :	1–2	1–3	2–4	2–6	3–4	4–5	4–6	5–7	6–7	7–8
Total float :	0	2	0	16	2	6	0	6	0	0
Free float :	0	0	0	16	2	0	0	6	0	0

(*b*)

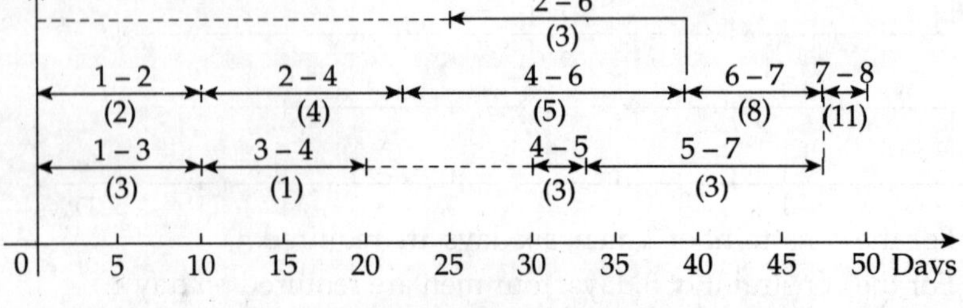

Manpower requirement:

t	*Days*	*Requirement (men)*	
0–10	1st to 10th	5	
10–11	11th	7	(overtime required)
11–21	12th to 21st	5	
21–23	22nd to 23rd	4	
23–26	24th to 26th	5	
26–46	27th to 46th	11	
46–49	47th to 49th	8	
49–50	50th	11	

14.09.

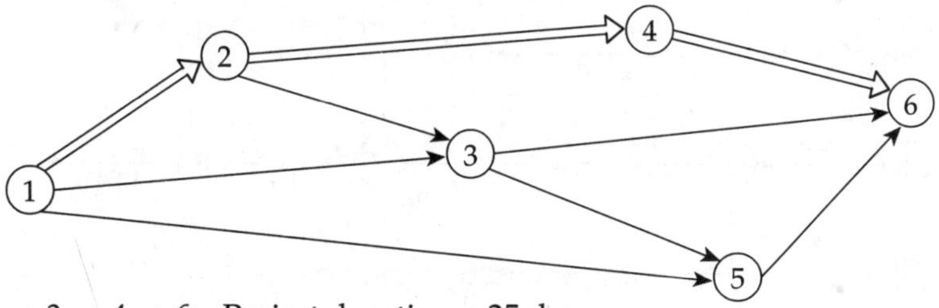

CPM: $1 \to 2 \to 4 \to 6$. Project duration = 25 days.

Maximum manpower required = 22 men, and Minimum manpower required = 9 men.

14.10. Minimum number of workers: 2 from 10th to 12th day.

Maximum number of workers: 6 from 12th to 18th day.

14.11. (*a*)

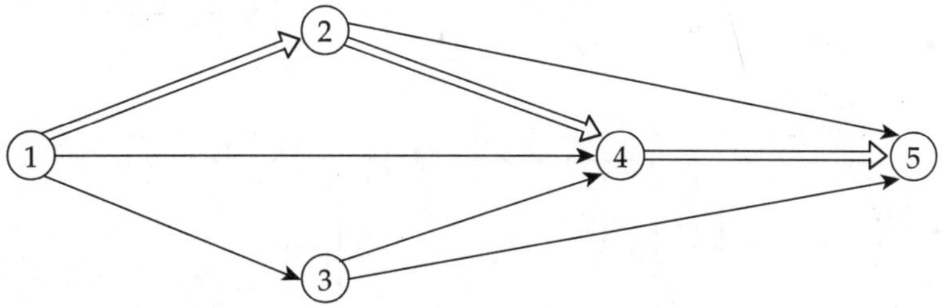

CPM: $1 \to 2 \to 4 \to 5$. Project duration = 14 days.

Activity :	1–2	1–3	1–4	2–4	2–5	3–4	3–5	4–5
Total float :	0	1	2	0	2	1	7	0

(*b*) *Time (days)* :	1 to 8	9 to 12	13 to 18	19 to 21
Operators :	50	40	50	20

14.12. Project duration is 18 days with given constraints and 14 days without constraints.

14.13. (*i*)

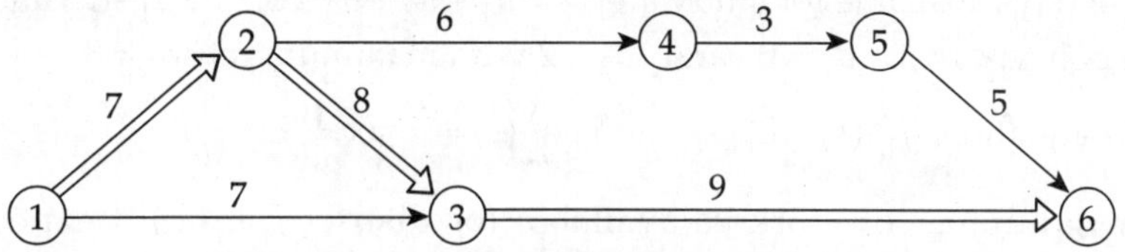

CPM: $1 \to 2 \to 3 \to 6$. Minimum project duration = 24 days.

Activity :	1–2	1–3	2–3	2–4	3–6	4–5	5–6
E_T :	7	15	15	13	24	16	24
L_T :	7	15	15	16	24	19	24
Total float :	0	0	0	3	0	3	0

14.14. (*a*)

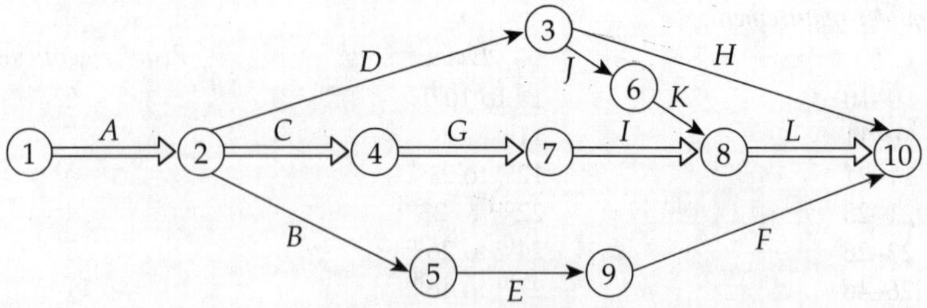

CPM: $A \to C \to G \to I \to L$. Project completion time = 21 days.

(*b*) $a \mid b \Rightarrow a$ for operators required, and b for mechanics required:

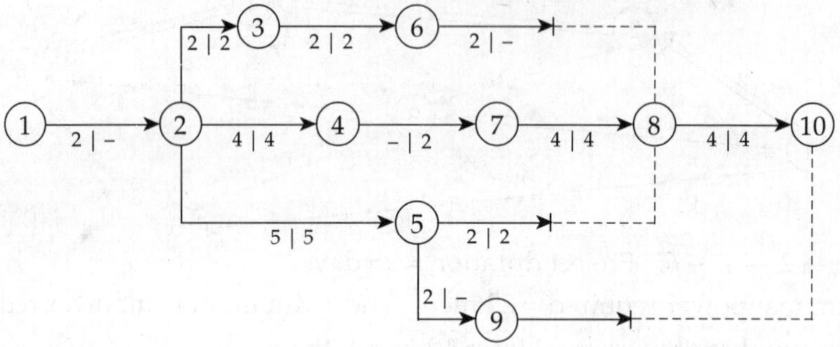

	1	2	3	4	5	6	7	8	9	10	11	12	13	14	15	16	17	18	19	20	21
O	2	2	2	11	11	11	11	7	7	6	4	4	4	4	4	4	4	4	4	4	4
M	—	—	—	11	11	11	11	9	7	4	4	6	6	2	2	2	2	4	4	4	4

14.15. Precedence diagram is:

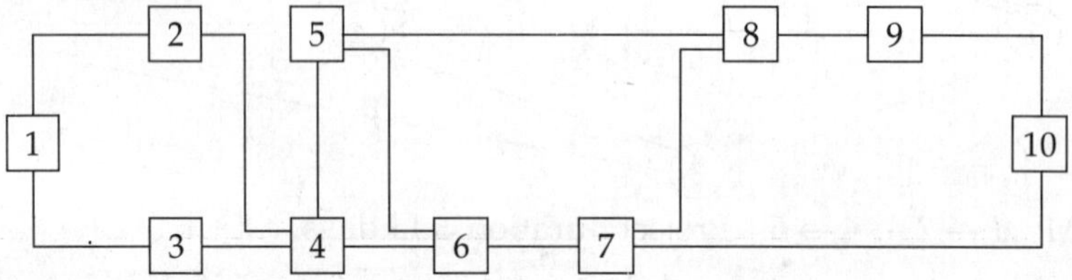

Chapter 15. Dynamic Programming

15.01. Recursive equation: $f_1^*(s_1) = f_1(d_1)$ and $f_j^*(s_j) = \max.\left\{f_j(d_j) + f_{j-1}^*(s_j - 1)\right\}$, $j = 2, 3$;
where $f_j(d_j)$ = grades earned by devoting d_j number of days ($j = 1, 2, 3$); and s_j = state variables; number of days available for study are: $s_3 = d_1 + d_2 + d_3 \le 3$, $s_2 = d_1 + d_2$ and $s_1 = d_1$.
Optimal policy: $d_1 = 1$, $d_2 = 0$ and $d_3 = 2$ with maximum return = 8.

15.02. Recursive equation: $f_1^*(s) = E_1(s_1)$ and $f_j^*(s) = \max\limits_{0 < x_j \le s}\left\{E_j(x_j) + f_{j+1}^*(s - x_j)\right\}$, $j = 1, 2, 3$;

where x_j = number of workers available for district j, $E_j(x_j)$ = expected number of votes, when x_j workers are assigned to district j, and s be the total number of workers, *i.e.*, $x_1 + x_2 + x_3 = 6$.
Optimal policy: $x_1 = 2$, $x_2 = 3$ and $x_3 = 1$; maximum increase in the number of votes are 129.

15.03. Optimal policy to get maximum profit for 9 salesman is: ₹ 2,10,000; when five salesman are allotted to Zone 3, one salesman allotted to Zone 2 and three salesman are allotted to Zone 1.

15.04. Optimal policy is to purchase two advertisements in newspaper, one advertisement in radio and television each to get the maximum expected sales of ₹ 5,95,000.

15.05. If x_1, x_2 and x_3 be the units of money allocated to sites 1, 2 and 3 respectively; $p_1(x_1)$, $p_2(x_2)$ and $p_3(x_3)$ be the corresponding probabilities of finding oil, if it exists. Then, actual probabilities of finding oil at the three sites are $(0.4) \times p_1(x_1)$, $(0.3) \times p_2(x_2)$ and $(0.2) \times p_3(x_3)$. The problem, therefore, is to

Maximize $z = [(0.4) \times p_1(x_1) + (0.3) \times p_2(x_2) + (0.2) \times p_3(x_3)]$ subject to the constraints:

$$x_1 + x_2 + x_3 \le 8,$$

where the decision variables x_1, x_2 and x_3 are non-negative integers.

Optimal policy: $x_3 = 0$, $x_2 = 0$ and $x_1 = 8$; maximum probability is 0.40, *i.e.*, 40%.

15.06. Optimal policy is to obtain maximum market share of 7.7 per cent. This is obtained by any one of the following:

(*a*) Spend 3, 2 and 0 million rupees in first, second and third phase (stage) respectively; or

(*b*) Spend 3, 1 and 1 million rupees in first, second and third phase (stage) respectively.

15.07. Optimal policy: 260 units in August, 250 units in September and 245 units in October.

15.08. Optimal policy: $x_3^* = 1,\ x_2^* = 1$ and $x_1^* = 0$ or $x_3^* = 2,\ x_2^* = 0$ and $x_1^* = 1$; maximum value of load = 17.

15.09. Optimal policy: $x_4^* = 1,\ x_3^* = 0,\ x_2^* = 1$ and $x_1^* = 2$; maximum value of cargo = 30.

15.10. Optimal policy: $x_4^* = 3,\ x_3^* = 1,\ x_2^* = 1$ and $x_1^* = 3$; maximum value of load = 3,95,000.

15.11. Optimal policy:

Month :	1	2	3	4
Number (n) :	4	3	2	1
Purchase (units) :	0	500	0	0
Sale (units) :	200	0	0	500

Maximum return = 7,100.

15.12. $y_1 = 5/3,\ y_2 = 5/3,\ y_3 = 5/3$; maximum $z = 125/27$.

15.13. $y_1 = d^{1/n},\ y_2 = d^{1/n},\ \ldots,\ y_n = d^{1/n}$; minimum $z = nd^{1/n}$.

15.14. $x_1 = 0,\ x_2 = 0,\ \ldots,\ x_{n-1} = 0,\ x_n = c$; maximum $z = b_n c$.

15.15. $y_1 = c^{1/n},\ y_2 = c^{1/n},\ \ldots,\ y_n = c^{1/n}$; minimum $z = nc^{2/n}$.

15.16. (*a*) $x_1 = 2^{3/10},\ x_2 = 2^{3/10},\ \ldots,\ x_{10} = 2^{3/10}$; minimum $z = 10 \times 2^{3/5}$.

(*b*) $x_1 = 3,\ x_2 = 3$ and $x_3 = 3$; minimum $z = 81$.

15.17. (*a*) $y_1 = 10/3,\ y_2 = 10/3$ and $y_3 = 10/3$; minimum $z = 100/3$,

(*b*) $y_1 = 3, y_2 = 4$ and $y_3 = 3$ or $y_1 = 3, y_2 = 3$ and $y_3 = 4$ or $y_1 = 4, y_2 = 3$ and $y_3 = 3$; minimum $z = 34$.

15.18. $x_1 = 2,\ x_2 = 2$ and $x_3 = 2$; minimum $z = 20$.

15.19. $x_1 = 8,\ x_2 = 0$ and $x_3 = 0$; maximum $z = 64$.

15.20. $x_1 = 0,\ x_2 = 1/2$ and $x_3 = 1/2$; maximum $z = 3/2$.

15.21. $x_1 = 8$ and $x_2 = 0$; maximum $z = 24$.

15.22. $x_1 = 7$ and $x_2 = 11$; maximum $z = 106$.

15.23. $x_1 = 2.5$ and $x_2 = 35$; maximum $z = 147.5$.

15.24. $x_1 = 6$ and $x_2 = 12$; maximum $z = 60$.

15.25. $x_1 = 50$ and $x_2 = 125$ or $x_1 = 100$ and $x_2 = 100$; maximum $z = 2,400$.

15.26. $x_1 = 16$ and $x_2 = 20$; maximum $z = 760$.

Chapter 16. Network Routing Problems

16.01. (*a*) $N = \{1, 2, 3, 4, 5\}$, $L = \{1–2, 1–3, 2–3, 2–5, 3–4, 3–5, 4–2, 4–5\}$

(*b*) $1 \to 2 \to 3 \to 5$, $1 \to 3 \to 4 \to 5$. (*c*) 2–3–5–2, $2 \to 3 \to 4 \to 2$.

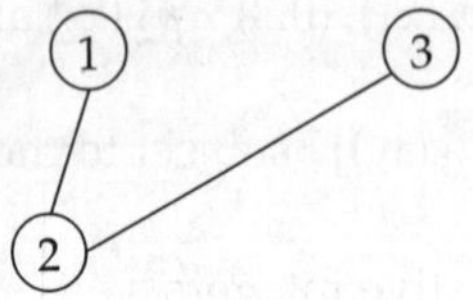

(*d*) Tree

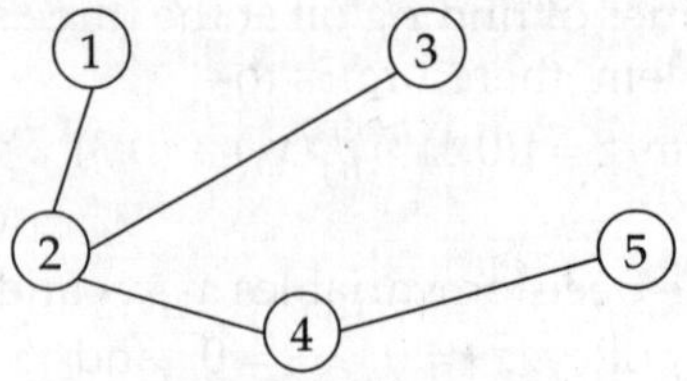

(*e*) Spanning tree

16.02.

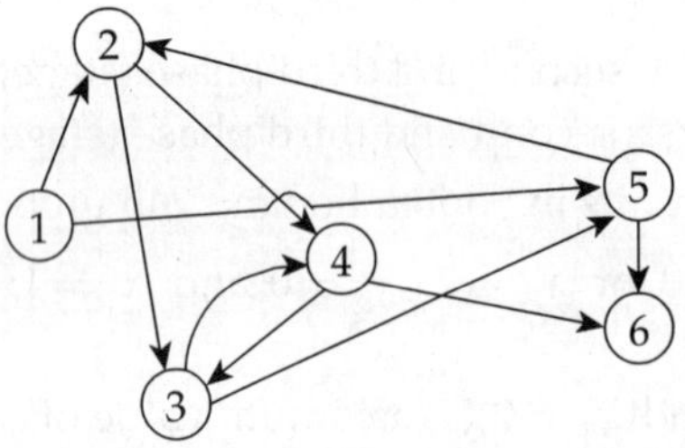

A spanning tree for the network is given below:

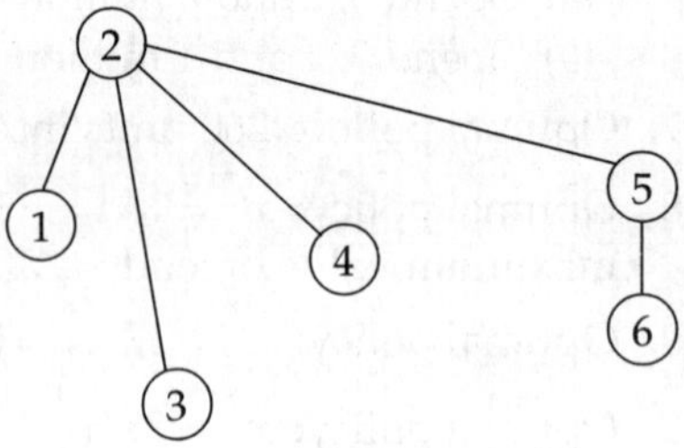

16.03. *O–A–B–C; B–E–D–T* with minimum length $z = 18$.

16.04. *O–A–B–C; B–E–D–T* with minimum length $z = 14$.

16.05. (*a*) *1–2–5–3; 2–4–6* with minimum length as $z = 13$.

(*b*) *1–2–4; 4–3; 4–5* with minimum length as $z = 21$.

(*c*) *1–3–4–6; 4–2–5* with minimum length as $z = 20$.

16.06. Group *A*: *1–2–3–4–6*; Group *B*: *1–5–7* and *5–9–8* with minimum length as $z = 53$.

16.08. *1–3–4–2*; $z = 55$ miles.

16.09. (*a*) *1–3–2–5–6–7*; $z = 11$, (*b*) *7–6–5–2–3–1*; $z = 11$, (*c*) *6–7*; $z = 4$.

16.10. *O–A–B–E–D–T* with shortest length of $z = 13$.

16.11. (*a*) *1–2–3–6–8*; $z = 8$ miles, (*b*) *1–2–3–6*; $z = 6$ miles, (*c*) *4–6–8*; $z = 8$ miles.

16.12. *O–C–F–G–T*; $z = 17$.

16.13. *1–3–2*; $z = 3$, *1–3*; $z = 1$, *1–3–4*; $z = 7$, *1–3–2–5*; $z = 4$, *1–3–2–6*; $z = 9$, *1–3–2–6–7*; $z = 11$.

16.14. (*a*) *1–3–2–4–6*; $z = 800$ miles, (*b*) *1–3–5*; $z = 800$ miles.

16.15. (*a*) *1–3–5–7*; $z = 165$ miles, (*b*) *1–3–5–7*; $z = 165$ minutes.

16.20. Maximum flow in the network is 15. The optimal flow pattern for various links are shown in the network below:

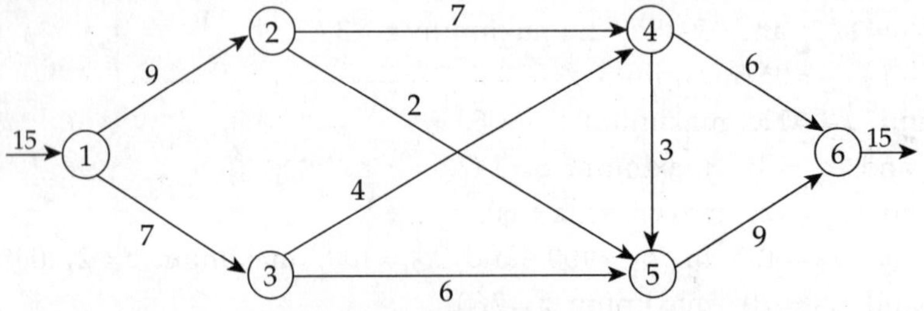

16.21. The optimal flow pattern is shown in the network below:

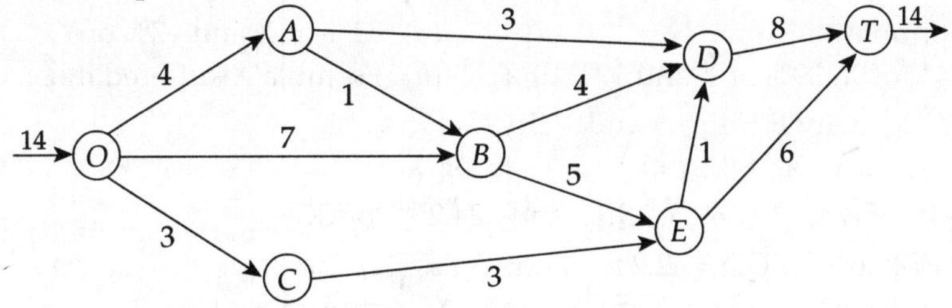

16.22. (*a*) Surplus capacities are: *2–3:* 40 units, *2–5:* 30 units and *4–3:* 5 units.

(*b*) A amount of flow through nodes are: *2:* 20 units, *3:* 30 units and *4:* 20 units.

(*c*) No, because there is no surplus capacity out of node *1*.

16.23. Maximum flow in the network is 25. The optimal flow patterns for various links are shown in the network below:

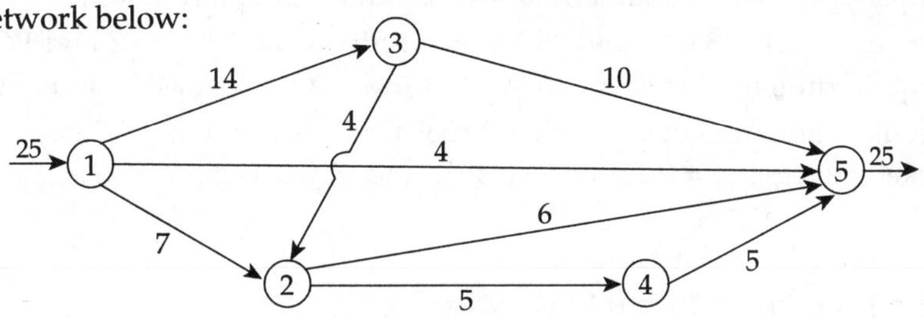

16.24. Maximum flow in the network is 9 units. The optimal flow patterns for various links are shown in the network below:

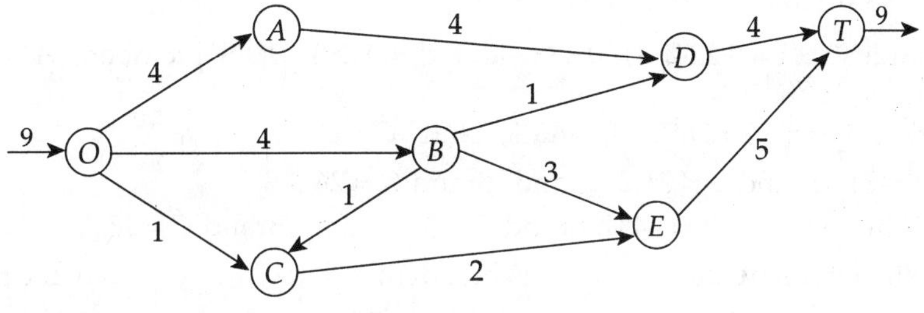

Chapter 18. Markov Analysis

18.01. (*a*)

$$\text{Present Purchase}\quad \begin{matrix} & \text{Next Purchase} \\ & \begin{matrix} A & \quad B \end{matrix} \\ \begin{matrix} A \\ B \end{matrix} & \begin{pmatrix} 0.80 & 0.20 \\ 0.60 & 0.40 \end{pmatrix} \end{matrix} \downarrow \text{Retention and Gain}$$

$\longrightarrow$ Retention and Loss

(*b*) 76% of *A* and 24% of *B*.

18.02.

$$\begin{matrix} & \begin{matrix} A & \quad B & \quad C \end{matrix} \\ \begin{matrix} A \\ B \\ C \end{matrix} & \begin{pmatrix} 0.80 & 0.100 & 0.100 \\ 0.08 & 0.866 & 0.055 \\ 0.09 & 0.093 & 0.810 \end{pmatrix} \end{matrix} \downarrow \text{Retention and Gain}$$

$\longrightarrow$ Retention and Loss

18.03. Brand *A:* 75% and Brand *B:* 25%.

18.04. (*a*) (*i*) Formula *Y*; (*ii*) 83% of *X* and 17% of *Y*.
(*b*) (*i*) 44% of *X*, 39% of *Y* and 17% of *Z*, (*ii*) Formula *X* suffered most.

18.05. City *A* = 40%, City *B* = 45%, and City *C* = 15%.

18.06. (*i*) 39.6% of *AA*, 46.3% of *BB*, and 14.1% of *CC*.
(*ii*) 64.86% of *AA*, 10.86% of *BB*, and 24.28% of *CC*.

18.07. (*a*) *ABC* = 40.6%, *PQR* = 42.2%, and *XYZ* = 17.2%.
(*b*) *ABC* = 40.18%, *PQR* = 45.54%, and *XYZ* = 14.28%.

18.08. *A* = 30.77%, *B* = 46.15%, and *C* = 23.08%.

18.09. (*i*) First year: Company *A* = 36.7%, Company *B* = 27.7%, and Company *C* = 35.6%;
Second year: Company *A* = 39.33%, Company *B* = 24.03%, and Company *C* = 36.64%.
(*ii*) Company *A* = 50%, Company *B* = 16%, and Company *C* = 34%.

18.10. (*i*) Next year: 38.4%, 35.4% and 26.2%, respectively, of *ABC*, *XYZ* and *PQR* companies.
(*ii*) At equilibrium market share: *ABC* = 34.16%, *XYZ* = 31.74%, and *PQR* = 34.09%.

18.11. (*a*) At equilibrium market share: *X* = 0.833, *Y* = 0.583, and *Z* = 0.333,
(*b*) Market strategy: *X* = 0.1584, *Y* = 0.495, and *Z* = 0.3466.

18.12.

$$\text{Grocer}\begin{matrix} 1 \\ 2 \\ 3 \end{matrix}\overset{\begin{matrix} & \text{Grocer} & \\ 1 & 2 & 3 \end{matrix}}{\begin{pmatrix} 0.80 & 0.10 & 0.10 \\ 0.25 & 0.75 & 0.00 \\ 0.00 & 0.05 & 0.95 \end{pmatrix}}$$

18.13. (*i*) 1st August: *A* = 0.234, *B* = 0.483, and *C* = 0.283. 1st September: *A* = 0.245, *B* = 0.477, and *C* = 0.278.
(*ii*) Market share: *A* = 28%, *B* = 45%, and *C* = 27%.

18.14. Brand *A* = 42%, Brand *B* = 31.2%, and Brand *C* = 26.8%.
At equilibrium: Brand *A* = 44%, Brand *B* = 31%, and Brand *C* = 25%.

18.15. (*i*) When demand at day 0: 5 units: When demand at day 0: 10 units: When demand at day 0: 12 units:

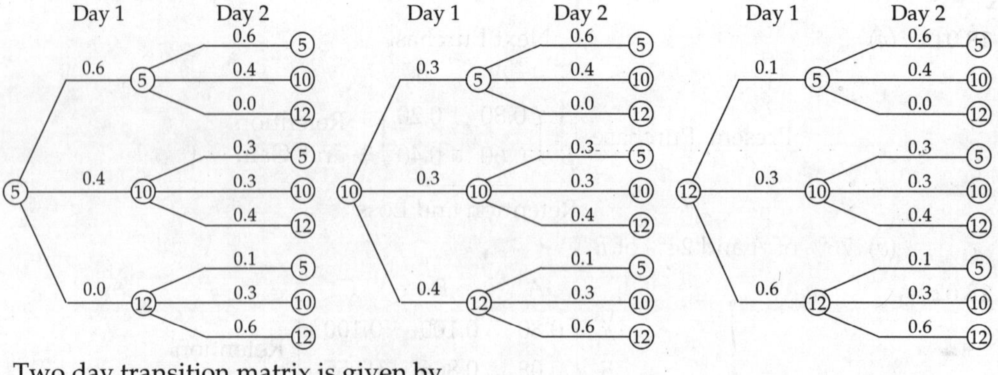

(*ii*) Two day transition matrix is given by

$$\begin{bmatrix} 0.6 & 0.4 & 0.0 \\ 0.3 & 0.3 & 0.4 \\ 0.1 & 0.3 & 0.6 \end{bmatrix}\begin{bmatrix} 0.6 & 0.4 & 0.0 \\ 0.3 & 0.3 & 0.4 \\ 0.1 & 0.3 & 0.6 \end{bmatrix} = \begin{bmatrix} 0.48 & 0.36 & 0.16 \\ 0.31 & 0.33 & 0.36 \\ 0.21 & 0.31 & 0.48 \end{bmatrix}.$$

(*iii*) Transition matrix may be useful to the inventory manager in planning inventory replenishments. If the order be placed every day on the basis of demand for the day, and lead time be two days, then this transition matrix may be used for guiding ordering decisions.

18.16. P_1 and P_2 would get ₹ 1,07,000 and ₹ 1,19,000, respectively, as the share of overhead charges of the service departments S_1, S_2 and S_3.

18.17. (*i*) ₹ 48,34,862 (*ii*) ₹ 41,65,138.

Chapter 19. Decision Theory

19.01. (*a*) Full product line of ₹ 80,000, (*b*) Minimum product line of ₹ 0, (*c*) Full or partial product line of ₹ 35,000, (*d*) Partial product line of ₹ 5.

19.02. (*a*) D_1 with maximum value of 30, (*b*) D_2 with least $E(D_i)$ as (–14).

19.03. Best product is *B* with $E(B)$ = ₹ 7,100.

19.04. (*a*) Crop *A*, (*b*) Crop *C*, (*c*) Crop *A*, (*d*) Crop *A* or Crop *C*, (*e*) Crop *A*.

19.05. (*a*) (*i*) Perfume *A*, since maximum of minimum sales is of 10 units,
(*ii*) Perfume C, since maximum of maximum sales is of 60 units,
(*iii*) Perfume C, since expected sales is of 29.33 units,
(*iv*) Perfume C, since minimum of maximum regret is of 7 units.

(*b*) Perfume *A* is selected by minimax cost criterion as well as by Laplace criterion and regret criterion, Perfume *C* is selected by minimin cost criterion.

19.06. (*i*) Egg shampoo, since maximin pay-off is of 10 units.
(*ii*) Egg shampoo, since minimax pay-off is of 30 units.
(*iii*) Deluxe shampoo, since maximax pay-off is of 55 units.
(*iv*) Deluxe shampoo, using 'Laplace criterion' is of 55 units.
(*v*) Deluxe shampoo, since minimax pay-off in opportunity loss pay-off table is of 7 units.

19.07. (*i*) D_3 is selected, since maximax return is 21%.
(*ii*) D_4 is selected, since maximin return is 10%.
(*iii*) D_3 is selected, since minimax return is 2% by using opportunity loss table.
(*iv*) D_3 is selected, since D_3 gives the maximum of the expected value among all the four decisions.
(*v*) D_3 is selected, since the maximum return according to Hurwicz criterion is 18%.

19.08. (*i*) Strategy a_2, (*ii*) Strategy a_3, (*iii*) Strategy a_2 or a_4, (*iv*) Strategy a_4.

19.09. (*i*)

Conditional Pay-off Table

Demand (Event) :	0	1	2	3	4	5
Buy none (₹) :	0	0	0	0	0	0
Buy all (₹) :	–500	–250	0	250	500	750

Regret Table

Demand (Event) :	0	1	2	3	4	5
Buy none (₹) :	0	0	0	250	500	750
Buy all (₹) :	500	250	0	0	0	0

(*ii*)

Demand (Event) :	0	1	2	3	4	5
Average pay-off :	0	108.3	175	200	183.3	125
Average regret :	375	266.6	200	175	191.6	250

Optimum stock level is: 3 discontinued cycles.

19.10. (*i*) Company should not stock pile and associated cost is Nil, according to minimin criterion.

(*ii*) Company should stock-pile, according to minimax criterion. Associated cost will be: ₹ 20,000.

(*iii*) Company should stock-pile, according to minimax regret of ₹ 20,000.

(*iv*) According to Hurwicz criterion with ($\alpha = 0.4$), company should stock-pile and associated cost will be: ₹ 20,000.

(*v*) According to Laplace criterion, company should stock pile and associated cost will be: ₹ 20,000.

19.11. Purchase 40 doses of vaccine, because maximum of *EMV* is: ₹ 54.

19.12. Maximum of *EMV* is ₹ 48.60 for stock size of 17 ice-cream cups each day.

19.13. For maximum expected profit, 17 copies of newspaper should be ordered.

19.14. Maximum of *EMV* is ₹ 168 for the order of 9 dozen of roses.

19.15. Maximum of *EMV* is ₹ 26 for the production of 6 items.

19.16. (*a*) Expected demand is of 11,860 items.

(*b*) Optimum demand is of 12,000 items, and *EMV* = ₹ 55,800.

19.17. Invest in property, since the maximum expected return (*EMV*) is: 18.6%.

19.18. (*a*)

Event (Demand)	*Prob.*	*Conditional pay-off (₹) Act (Stock)*							*Maximum conditional pay-off*	*Expected pay-off*
		15	20	25	30	35	40	45		
15	0.02	90	20	–50	–120	–190	–260	–330	90	1.8
20	0.08	90	120	50	–20	–90	–160	–230	120	9.6
25	0.10	90	120	150	80	10	–60	–130	150	15.0
30	0.40	90	120	150	180	110	40	–30	180	72.0
35	0.20	90	120	150	180	210	140	70	210	42.0
40	0.15	90	120	150	180	210	240	170	240	36.0
45	0.05	90	120	150	180	210	240	270	270	13.5
EMV		90	118	138	**148***	118	68	3		189.90

(*b*) 30 kilograms of butter, (*c*) *EVPI* = ₹ (189.9 – 148) = ₹ 41.90.

19.19. (*i*)

Event (Demand)	*Prob.*	*Act (Produce): Pay-off*							*Act (Produce): Regret*					
		40	45	50	55	60	65		40	45	50	55	60	65
40	0.10	800	500	200	–100	–400	–700		0	300	600	910	1200	1500
45	0.20	800	900	600	300	0	–300		100	0	300	600	0	1200
50	0.30	800	900	1000	700	400	100		200	100	0	300	600	900
55	0.25	800	900	1000	1100	800	500		300	200	100	0	300	600
60	0.10	800	900	1000	1100	1200	900		400	300	200	100	0	300
65	0.05	800	900	1000	1100	1200	1300		500	400	300	200	100	0
	EMV :	800	**860***	840	700	460	180	*EOL:*	220	**160***	180	221	380	840

(*ii*) Produce 45 units, since the maximum expected profit (*EMV*) is: ₹ 860.

(*iii*) *EVPI* = ₹ 860 – ₹ 700, *i.e.*, ₹ 160.

19.20. (*i*) Conditional Pay-off Table:

State-of-nature	*Probability*	*Courses of action*					
		100	*101*	*102*	*103*	*104*	*105*
S_1	0.20	100	97	94	91	88	85
S_2	0.40	100	101	98	95	92	89
S_3	0.05	100	101	102	99	96	93
S_4	0.10	100	101	102	103	100	97
S_5	0.20	100	101	102	103	104	101
S_6	0.05	100	101	102	103	104	105

(*ii*) Stock 101 units of the cell phone every month, expected profit will be: ₹ 1,00,200.

(*iii*) *EVPI* = *EPPI* – ₹ 1,00,200 = ₹ 1,650.

19.21. (*i*) Order 4 TV per week.

(*ii*) *EVPI* = *EMC* – *ECPI* = ₹ 230.00 – ₹ 147.50 = ₹ 82.50.

(*iii*) Dealer should not be willing to spend more than ₹ 82.50 for the market survey.

19.22. *EOL* (Develop the product)

$$= 0.1 \times 44{,}000 + 0.1 \times 28{,}000 + 0.2 \times 12{,}000 + 0.4 \times 0 + 0.2 \times 0 = ₹\ 9{,}600.$$

EOL (Do not develop) = $0.1 \times 0 + 0.1 \times 0 + 60.2 \times 0 + 0.4 \times 4{,}000 + 0.2 \times 20{,}000$ = ₹ 5,600.

Since, minimum of the *EOL* corresponds to ₹ 5,600, the company should not develop the product.

19.23.

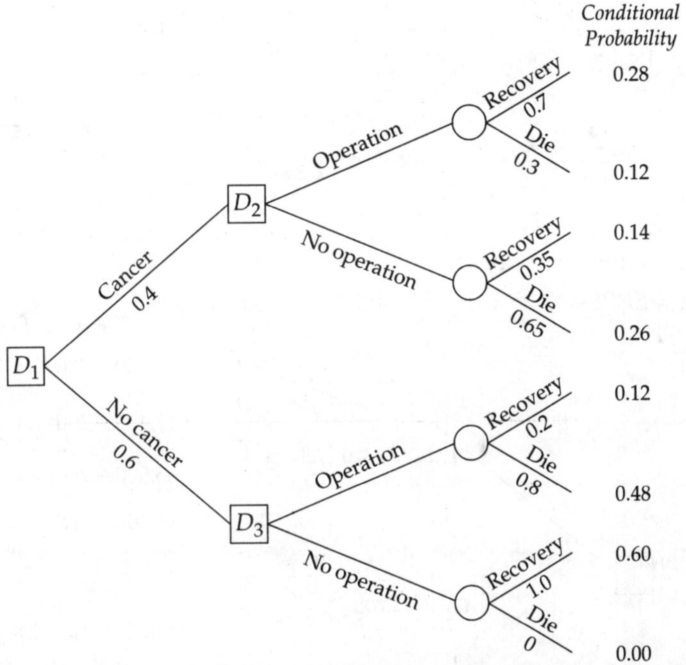

Irrespective of cancer or no cancer, the chance of recovery will be (0.14 + 0.60) when there is no operation, whereas the chance of recovery are (0.28 + 0.12) when there is operation.

Since, 0.14 + 0.60 > 0.28 + 0.12, do not take decision for operation.

19.24.

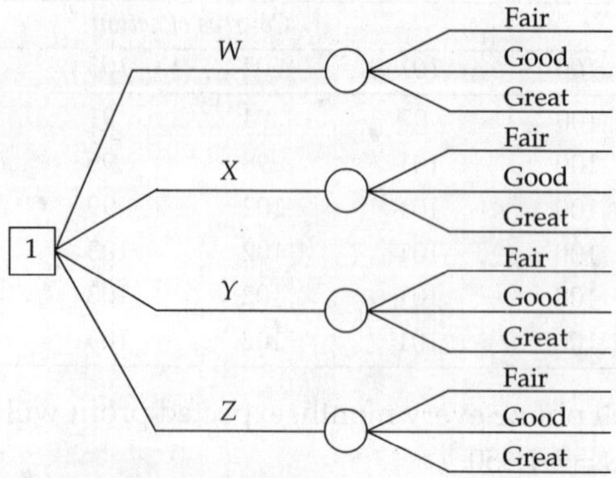

Monetary value (₹)	Probabilities	EMV
1,000	0.2	
3,000	0.5	
6,000	0.3	₹ 3,500
500	0.2	
4,500	0.5	
6,800	0.3	₹ 4,390
0	0.2	
5,000	0.5	
8,000	0.3	₹ 4,900
4,000	0.2	
6,000	0.5	
8,500	0.3	₹ 4,750

$EMV(W) =$ ₹ $1{,}000 \times 0.2 +$ ₹ $3{,}000 \times 0.5 +$ ₹ $6{,}000 \times 0.3 =$ ₹ $3{,}500$
$EMV(X) =$ ₹ $500 \times 0.2 +$ ₹ $4{,}500 \times 0.5 +$ ₹ $6{,}800 \times 0.3 =$ ₹ $4{,}390$
$EMV(Y) =$ ₹ $0 \times 0.2 +$ ₹ $5{,}000 \times 0.5 +$ ₹ $8{,}000 \times 0.3 =$ ₹ $4{,}900$
$EMV(Z) =$ ₹ $4{,}000 \times 0.2 +$ ₹ $6{,}000 \times 0.5 +$ ₹ $8{,}500 \times 0.3 =$ ₹ $4{,}750$

Alternative Y is the best with maximum return of ₹ 4,900.

19.25. (*i*)

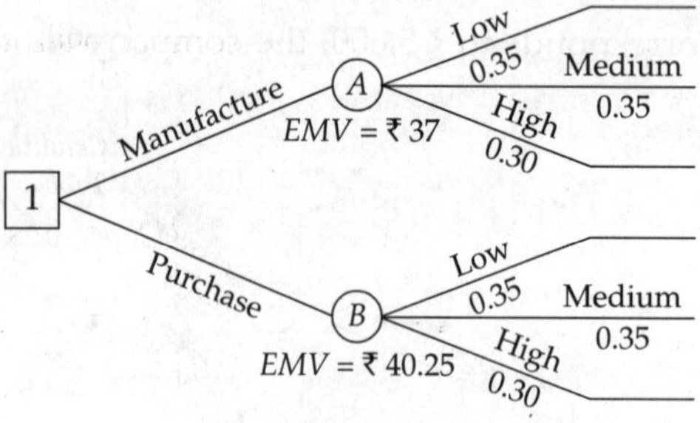

Demand	Expected pay-off	EMV
–20	–7.00	
40	14.00	
100	30.00	37.00
10	3.50	
45	15.75	
70	21.00	40.25

Since expected profit for purchase is higher, the company should take the decision to purchase the component.

(*ii*) $EVPI = EPPI - 40.25 = 49.25 - 40.25 =$ ₹ 9.00.

19.26.

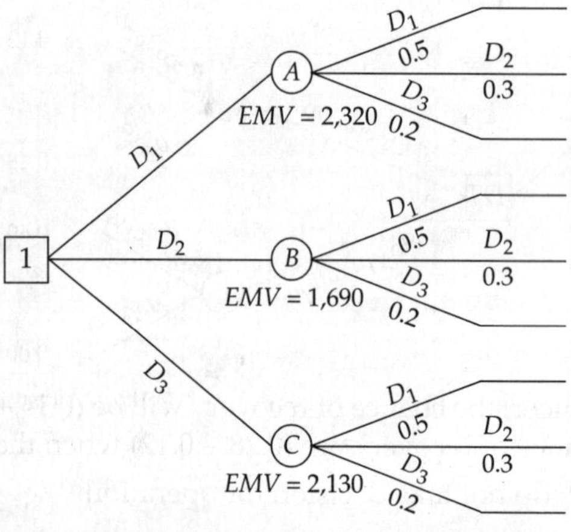

Expected pay-off

$(1{,}200 - 200) \times 0.5 = 500$
$(2{,}400 - 200) \times 0.3 = 660$
$(6{,}000 - 200) \times 0.2 = 1{,}160$
$(1{,}400 - 1{,}250) \times 0.5 = 75$
$(2{,}800 - 1{,}250) \times 0.3 = 465$
$(7{,}000 - 1{,}250) \times 0.2 = 1{,}150$
$(1{,}300 - 600) \times 0.5 = 350$
$(2{,}600 - 600) \times 0.3 = 600$
$(6{,}500 - 600) \times 0.2 = 1{,}180$

Since, highest *EMV* is at node *A*, the best decision is D_1 to manufacture the new product.

19.27.

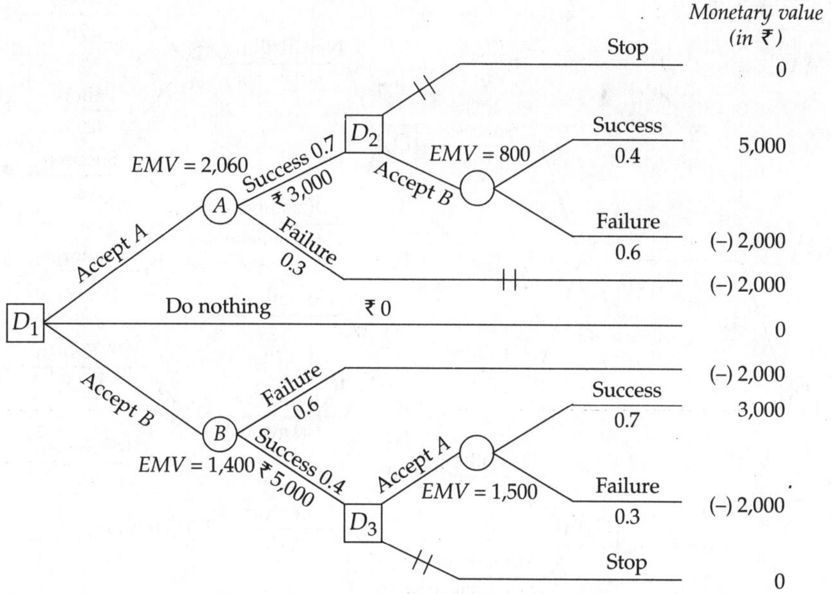

Decision node	*Options*	*EMV (in ₹)*	*Decision*
D_2	Accept *B*	0.4 × 5,000 + 0.6 × (–2,000) = 800	Accept *B*
	Stop	0	
D_3	Accept *A*	0.7 × 3,000 + 0.3 × (–2,000) = 1,500	Accept *A*
	Stop	0	
D_1	Accept *A*	0.7 × (3,000 + 800) – 0.3 × 2,000 = 2,060	Accept *A*
	Accept *B*	0.4 × (5,000 + 1,500) – 0.6 × 2,000 = 1,400	Accept *B*
	Do nothing	0	

Optimal decision is to accept investment *A*, and if successful, then accept *B*.

19.28.

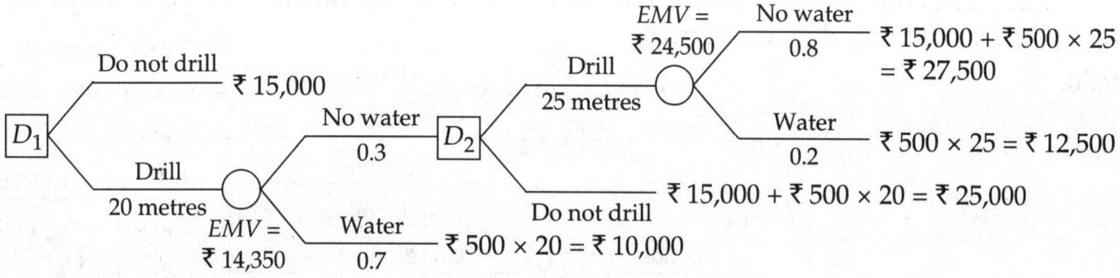

Decision node	*Options*	*EMV (in ₹)*	*Decision*
D_2	Drill up to 25 metres	0.8 × 27,500 + 0.2 × 12,500 = 24,500	Drill up to 25 metres
	Do not drill	25,000	
D_1	Drill up to 20 metres	0.3 × 24,500 + 0.7 × 10,000 = 14,350	Drill up to 20 metres
	Do not drill	15,000	

The optimal strategy is to drill up to 20 metres, and if water is not found, then drill further up to 25 metres.

19.29. (*a*)

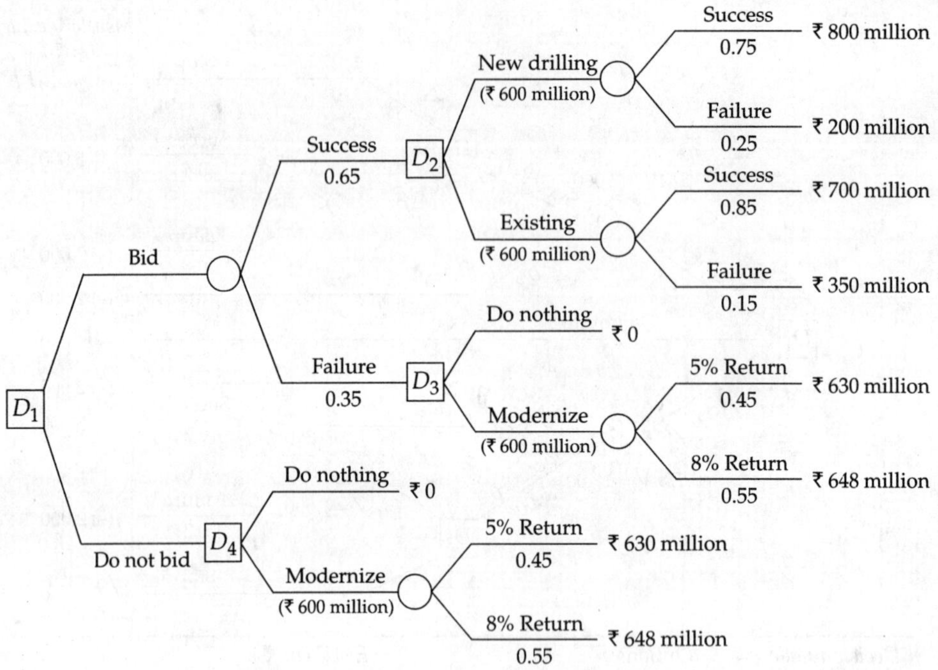

Decision node	*Option*	*EMV (million ₹)*	*Decision*
D_2	New drilling	(0.75 × 800 + 0.25 × 200) – 600 = 50	New drilling
	Existing operation	(0.85 × 700 + 0.15 × 350) – 600 = 47.5	
D_3	Do nothing	0	
	Modernize	(0.45 × 630 + 0.55 × 648) – 600 = 39.9	Modernize
D_4	Do nothing	0	
	Modernize	(0.45 × 630 + 0.55 × 648) – 600 = 39.9	Modernize
D_1	Bid the contract	(0.65 × 50 + 0.35 × 39.9) = 46.46	Bid the contract
	Do not bid		

(*b*) The corporation should bid the contract. If successful, then set-up new drilling operation and if not, then modernize.

19.30.

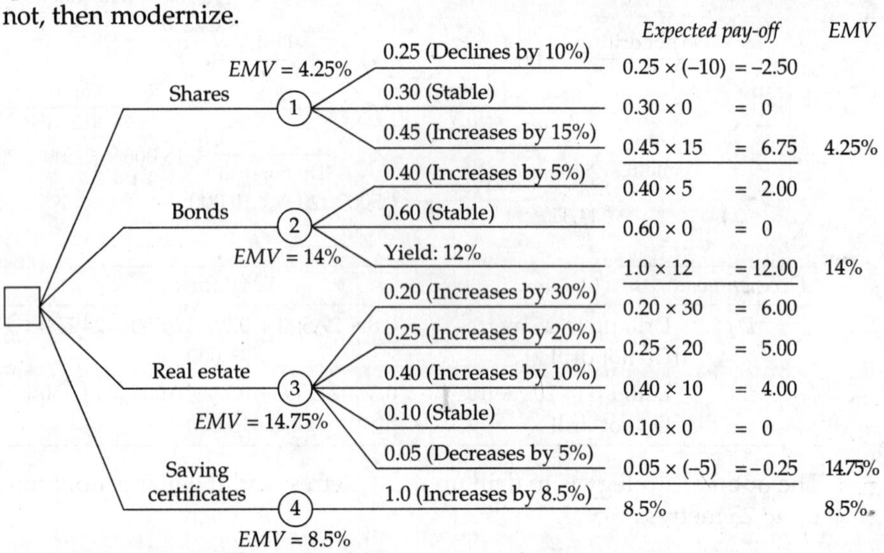

As maximum *EMV* is at node 3, one should invest in real estate.

19.31.

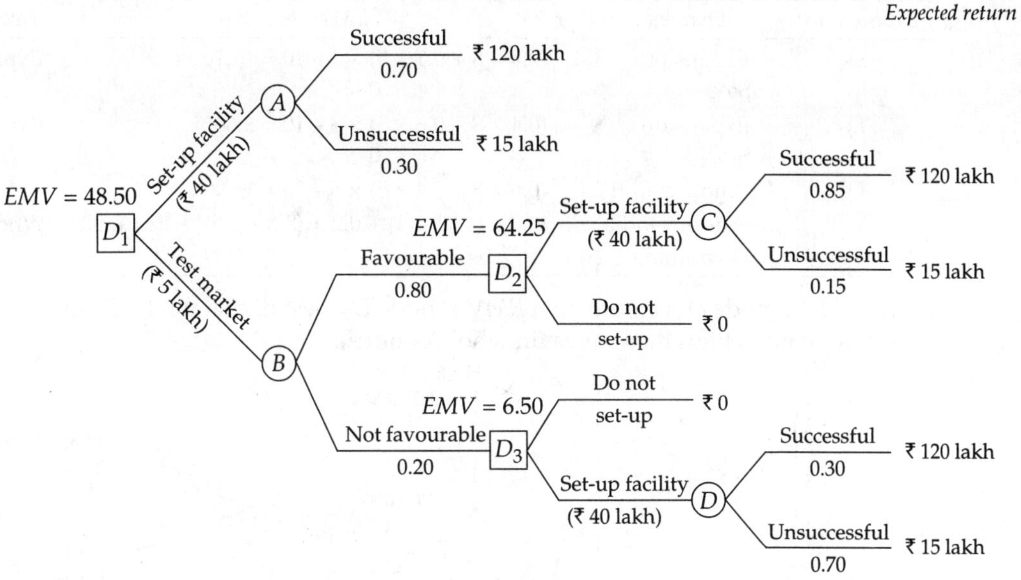

Decision node	*Options*	*EMV (in lakh of ₹)*	*Decision*
D_2	Set-up facility	0.85 × 120 + 0.15 × 15 – 40 = 64.25	Set-up facility
	Do not set-up	0	
D_3	Set-up facility	0.30 × 120 + 0.70 × 15 – 40 = 6.50	Set-up facility
	Do not set-up	0	
D_1	Set-up facility	0.70 × 120 + 0.30 × 15 – 40 = 48.5	Set-up facility
	Test market	0.80 × 64.25 + 0.2 × 6.50 – 5 = 47.7	

The company should set-up production facility straight way and not undertake test market.

19.32.

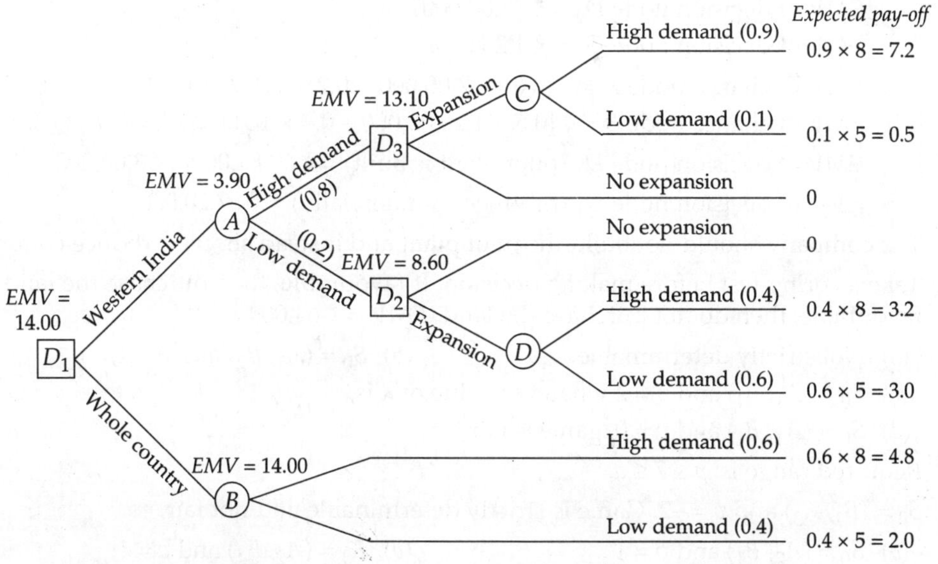

Decision node	*Options*	*EMV*	*Decision*
D_3	Expansion	$(0.9 \times 8 + 0.1 \times 5) \times 3 - 10 = 13.10$	Expansion
	Stop	0	
D_2	Expansion	$(0.4 \times 8 + 0.6 \times 5) \times 3 - 10 = 8.60$	Expansion
	Stop	0	
D_1	Whole country	$(0.6 \times 8 + 0.4 \times 5) \times 5 - 20 = 14.00$	
	Western India	$[0.8 \times (13.1 + 4) + 0.2 \times (8.6 + 2.5)] - 12 = 3.90$	Whole country
	Do nothing	0	

Since, *EMV* at node D_3 is more than *EMV* at node D_2, best decision is to decide for expansion and if demand is high then invest in whole country.

19.33.

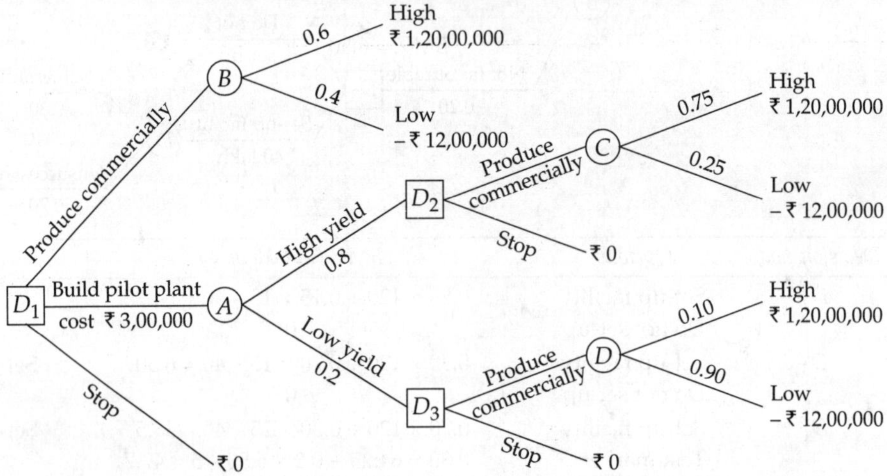

$\left\{ \begin{array}{l} EMV \text{ of chance node } C = ₹\,[0.75 \times 1{,}20{,}00{,}000 - 0.25 \times 12{,}00{,}000] = ₹\,87{,}00{,}000. \\ EMV \text{ of chance node } D = ₹\,[0.1 \times 1{,}20{,}00{,}000 - 0.9 \times 12{,}00{,}000 = ₹\,1{,}20{,}000. \end{array} \right.$

∴ *EMV* of decision node D_2 = ₹ 87,00,000.

EMV of decision node D_3 = ₹ 1,20,000.

$\left\{ \begin{array}{l} EMV \text{ of chance node } A = ₹\,[0.8 \times 87{,}00{,}000 + 0.2 \times 1{,}20{,}000] = ₹\,69{,}84{,}000. \\ EMV \text{ of chance node } B = ₹\,[0.6 \times 1{,}20{,}00{,}000 - 0.4 \times 12{,}00{,}000] = ₹\,67{,}20{,}000. \end{array} \right.$

EMV of decision node D_1 (pilot plant is built) = ₹ 69,84,000 – ₹ 3,00,000 = ₹ 66,84,000.

∴ *EMV* of decision node D_1 (produce commercially) = ₹ 67,20,000.

The company should not build the pilot plant and instead should produce commercially.

19.34. Take a coring test before making decision. If favourable, then purchase the land and if not favourable, then do not purchase the land. *EMV* = ₹ 64,000.

19.35. (*a*) Not strictly determinable, (*b*) $S_0 = (A_1, B_2)$ and $v = 1$,

(*c*) $S_0 = (A_1, B_1)$ and $v = 2$ whatever value of λ is,

(*d*) $S_0 = (A_1, B_1)$ and $v = 0$, game is fair.

19.36. Required range is: $p \le 7 \le q$.

19.37. $S_0 = (B_1, A_3)$ and $v = -2$. Game is strictly determinable and not fair.

19.38. (*a*) $S_0 = (A_2, B_1)$ and $v = 1$. (*b*) $S_0 = (A_2, B_1)$ and $v = 1$,

(*c*) $S_0 = (A_4, B_3)$ and $v = 4$, (*d*) $S_0 = (V_1, C_3)$ and $v = 12$.

19.39. (*i*) Both manufacturers *A* and *B* should produce at the level of 24 hours per day.
(*ii*) Manufacturer *B* will have 60% and manufacturer *A* will have 40% share of the market.

19.40. (*a*) $S_A = \begin{bmatrix} A_1 & A_2 \\ 1/4 & 3/4 \end{bmatrix}$, $S_B = \begin{bmatrix} B_1 & B_2 \\ 1/4 & 3/4 \end{bmatrix}$; and $v = -1/4$.

(*b*) $S_A = \begin{bmatrix} A_1 & A_2 \\ 1/5 & 4/5 \end{bmatrix}$, $S_B = \begin{bmatrix} B_1 & B_2 \\ 3/5 & 2/5 \end{bmatrix}$; and $v = 17/5$.

19.41. $S_{\text{Matching}} = \begin{bmatrix} H & T \\ 4/15 & 11/15 \end{bmatrix}$, $S_{\text{Non-matching}} = \begin{bmatrix} H & T \\ 4/15 & 11/15 \end{bmatrix}$; and $v = -1/15$.

19.42. $S_A = S_B = \begin{bmatrix} H & T \\ 1/4 & 3/4 \end{bmatrix}$; and $v = -1/8$.

19.43. $S_A = \begin{bmatrix} A_1 & A_2 & A_3 \\ 1/2 & 1/2 & 0 \end{bmatrix}$, $S_B = \begin{bmatrix} B_1 & B_2 & B_3 \\ 2/3 & 1/3 & 0 \end{bmatrix}$; and $v = 0$.

19.44. $S_A = S_B = \begin{bmatrix} H & T \\ 1/2 & 1/2 \end{bmatrix}$; and $v = 0$.

19.45. $S_{HB} = \begin{bmatrix} H_1 & H_2 \\ 3/4 & 1/4 \end{bmatrix}$, $S_{AM} = \begin{bmatrix} A_1 & A_2 \\ 1/2 & 1/2 \end{bmatrix}$; and $v = 45{,}000$ customers.

19.46. (*a*) $S_A = \begin{bmatrix} A_1 & A_2 \\ 1/4 & 3/4 \end{bmatrix}$, $S_B = \begin{bmatrix} B_1 & B_2 & B_3 \\ 1/4 & 3/4 & 0 \end{bmatrix}$; and $v = -3/4$.

(*b*) $S_B = \begin{bmatrix} B_1 & B_2 \\ 4/9 & 5/9 \end{bmatrix}$, $S_A = \begin{bmatrix} A_1 & A_2 & A_3 & A_4 \\ 0 & 0 & 8/9 & 1/9 \end{bmatrix}$; and $v = 22/9$.

(*c*) $S_A = \begin{bmatrix} A_1 & A_2 \\ 3/11 & 8/11 \end{bmatrix}$, $S_B = \begin{bmatrix} B_1 & B_2 & B_3 \\ 0 & 9/11 & 2/11 \end{bmatrix}$; and $v = 49/11$.

(*d*) $S_A = \begin{bmatrix} A_1 & A_2 \\ 4/5 & 1/5 \end{bmatrix}$, $S_B = \begin{bmatrix} B_1 & B_2 & B_3 & B_4 & B_5 \\ 0 & 0 & 3/5 & 2/5 & 0 \end{bmatrix}$; and $v = -24/5$.

19.47. $S_A = \begin{bmatrix} A_1 & A_2 \\ 1/2 & 1/2 \end{bmatrix}$, $S_B = \begin{bmatrix} B_1 & B_2 & B_3 \\ 0 & 1/2 & 1/2 \end{bmatrix}$; and $v = 11$.

19.48. $S_B = \begin{bmatrix} 1 & 2 & 3 & 4 \\ 0 & 4/9 & 0 & 5/9 \end{bmatrix}$, $S_A = \begin{bmatrix} I & II & III & IV \\ 0 & 5/9 & 0 & 4/9 \end{bmatrix}$; and $v = 38/9$.

19.49. Pay-off matrix of firm *A*: $\begin{bmatrix} 20{,}000 & 30{,}000 & 60{,}000 \\ 90{,}000 & 45{,}000 & 30{,}000 \end{bmatrix}$

$S_A = \begin{bmatrix} A_1 & A_2 \\ 1/3 & 2/3 \end{bmatrix}$, $S_B = \begin{bmatrix} B_1 & B_2 & B_3 \\ 0 & 2/3 & 1/3 \end{bmatrix}$; and $v = 40{,}000$.

19.50. $S_A = \begin{bmatrix} NA & MA & HA \\ 0 & 4/7 & 3/7 \end{bmatrix}$, $S_B = \begin{bmatrix} NA & MA & HA \\ 0 & 5/7 & 2/7 \end{bmatrix}$; and $v = 90/7$.

19.51. (*i*) $S_A = \begin{bmatrix} \text{Quant discount} & \text{CSR} & \text{Sponsor} \\ 7/8 & 0 & 1/8 \end{bmatrix}$, $S_B = \begin{bmatrix} \text{Quant discount} & \text{CSR} & \text{Sponsor} \\ 3/4 & 0 & 1/4 \end{bmatrix}$,

(*ii*) Value of game = 45/2.

19.52. $S_A = \begin{bmatrix} A_1 & A_2 & A_3 \\ 1/2 & 1/2 & 0 \end{bmatrix}$, $S_B = \begin{bmatrix} B_1 & B_2 & B_3 \\ 3/5 & 2/5 & 0 \end{bmatrix}$; and $v = 1$.

19.53. $S_R = \begin{bmatrix} R_1 & R_2 & R_3 \\ 0 & 2/5 & 3/5 \end{bmatrix}$, $S_0 = \begin{bmatrix} O_1 & O_2 & O_3 \\ 2/5 & 0 & 3/5 \end{bmatrix}$; and $v = 61$.

19.54. $S_{ABC} = \begin{bmatrix} \text{Coupons} & \text{Decrease} & \text{Maintain} & \text{Increase} \\ 0 & 1/2 & 1/2 & 0 \end{bmatrix}$,

$S_{XYZ} = \begin{bmatrix} \text{Coupons} & \text{Decrease} & \text{Maintain} & \text{Increase} \\ 1/10 & 9/10 & 0 & 0 \end{bmatrix}$; and $v = 3/2$.

19.55. $S_A = \begin{bmatrix} A_1 & A_2 & A_3 & A_4 \\ 0 & 0 & 5/18 & 13/18 \end{bmatrix}$, $S_B = \begin{bmatrix} B_1 & B_2 & B_3 & B_4 \\ 0 & 0 & 1/9 & 8/9 \end{bmatrix}$; and $v = 31/9$.

19.56. $S_A = \begin{bmatrix} A_1 & A_2 & A_3 & A_4 \\ 1/5 & 0 & 0 & 4/5 \end{bmatrix}$, $S_B = \begin{bmatrix} B_1 & B_2 & B_3 & B_4 \\ 0 & 0 & 2/5 & 3/5 \end{bmatrix}$; and $v = 21$.

19.57. $S_C = \begin{bmatrix} C_1 & C_2 & C_3 & C_4 \\ 0 & 7/13 & 6/13 & 0 \end{bmatrix}$, $S_U = \begin{bmatrix} U_1 & U_2 & U_3 & U_4 \\ 1/13 & 0 & 12/13 & 0 \end{bmatrix}$; and $v = 3/13$.

19.58. (*a*) $S_A = \begin{bmatrix} A_1 & A_2 & A_3 & A_4 \\ 0 & 3/5 & 0 & 2/5 \end{bmatrix}$, $S_B = \begin{bmatrix} B_1 & B_2 & B_3 & B_4 \\ 0 & 0 & 2/5 & 3/5 \end{bmatrix}$; and $v = 16/5$.

(*b*) Optimal strategies are: Player *A*: *III* and Player *B*: *I*; value of game is 6.

19.59. (*a*) $S_A = \begin{bmatrix} A_1 & A_2 & A_3 & A_4 \\ 1/5 & 0 & 0 & 4/5 \end{bmatrix}$, $S_B = \begin{bmatrix} B_1 & B_2 & B_3 & B_4 \\ 0 & 0 & 2/5 & 3/5 \end{bmatrix}$; and $v = 9$.

(*b*) $S_A = \begin{bmatrix} I & II & III & IV \\ 0 & 0 & 2/3 & 1/3 \end{bmatrix}$, $S_B = \begin{bmatrix} I & II & III & IV \\ 0 & 0 & 2/3 & 1/3 \end{bmatrix}$; and $v = 8/3$.

19.60. $S_A = \begin{bmatrix} A_1 & A_2 & A_3 & A_4 & A_5 \\ 0 & 1/3 & 0 & 0 & 2/3 \end{bmatrix}$, $S_B = \begin{bmatrix} B_1 & B_2 & B_3 & B_4 & B_5 \\ 1/2 & 0 & 0 & 0 & 1/2 \end{bmatrix}$; and $v = 11$.

19.61. $S_A = \begin{bmatrix} A_1 & A_2 & A_3 & A_4 \\ 1/5 & 0 & 0 & 4/5 \end{bmatrix}$, $S_B = \begin{bmatrix} B_1 & B_2 & B_3 & B_4 \\ 0 & 0 & 2/5 & 3/5 \end{bmatrix}$; and $v = 13$.

19.62. (*a*) $S_A = \begin{bmatrix} A_1 & A_2 & A_3 \\ 0 & 1/3 & 2/3 \end{bmatrix}$, $S_B = \begin{bmatrix} B_1 & B_2 & B_3 \\ 0 & 2/3 & 1/3 \end{bmatrix}$; and $v = 145/3$, *i.e.*, 48.3% of market share.

(*b*) Market share of firm *A*, when annual value is of 30,000 units is :

		Firm B		
		No Adv.	*Medium Adv.*	*Large Adv.*
Firm A	*No Adv.*	0.50 × 30,000	0.40 × 30,000	0.28 × 30,000
	Med. Adv.	0.70 × 30,000	0.50 × 30,000	0.45 × 30,000
	Large Adv.	0.75 × 30,000	0.475 × 30,000	0.50 × 30,000

i.e.,

$$\begin{bmatrix} 15,000 & 12,000 & 8,400 \\ 21,000 & 15,000 & 13,500 \\ 22,500 & 14,250 & 15,000 \end{bmatrix}.$$

Net profit = (Sales price – Cost price) × Sales volume – Adv. expenditure
= ₹ 1.50 × Sales volume – Adv. expenditure

Net profit to firm A is obtained as under:

$$\begin{array}{ll} & \textit{Firm B} \\ & \begin{array}{ccc} \textit{No Adv.} & \textit{Medium Adv.} & \textit{Large Adv.} \end{array} \\ \textit{Firm A} \begin{array}{r} \textit{No Adv.} \\ \textit{Med. Adv.} \\ \textit{Large Adv.} \end{array} & \begin{bmatrix} 1.5\times 15,000-0 & 1.5\times 12,000-0 & 1.5\times 8,400-0 \\ 1.5\times 21,000-5,000 & 1.5\times 15,000-5,000 & 1.5\times 13,500-5,000 \\ 1.5\times 22,500-15,000 & 1.5\times 14,250-15,000 & 1.5\times 15,000-15,000 \end{bmatrix}, \end{array}$$

$$i.e., \begin{bmatrix} 22,500 & 18,000 & 12,600 \\ 26,500 & 17,500 & 15,250 \\ 18,750 & 6,375 & 7,500 \end{bmatrix}.$$

Based on the above information, firm A must adopt the policy of "Medium advertising" to gain the maximum profit of ₹ 15,250 and spend ₹ 5,000 for advertising.

19.63. (*a*) $S_A = \begin{bmatrix} A_1 & A_2 \\ 0 & 1 \end{bmatrix}, \ S_B = \begin{bmatrix} B_1 & B_2 \\ 1/2 & 1/2 \end{bmatrix};$ and $v = 1$.

(*b*) $S_A = \begin{bmatrix} A_1 & A_2 & A_3 \\ 6/13 & 3/13 & 4/13 \end{bmatrix}, \ S_B = \begin{bmatrix} B_1 & B_2 & B_3 \\ 6/13 & 4/13 & 3/13 \end{bmatrix};$ and $v = 12/13$.

19.64. (*a*) $S_A = \begin{bmatrix} A_1 & A_2 & A_3 \\ 1/10 & 1/10 & 1/10 \end{bmatrix}, \ S_B = \begin{bmatrix} B_1 & B_2 & B_3 \\ 38/60 & 14/60 & 8/10 \end{bmatrix};$ and $v = 5.6$.

(*b*) $S_A = \begin{bmatrix} A_1 & A_2 & A_3 \\ 1/4 & 1/2 & 1/4 \end{bmatrix}, \ S_B = \begin{bmatrix} B_1 & B_2 & B_3 \\ 1/4 & 1/2 & 1/4 \end{bmatrix};$ and $v = 1/2$.

19.65. $S_X = \begin{bmatrix} \text{Increase} & \text{Not change} & \text{Reduce} \\ 1/4 & 3/4 & 0 \end{bmatrix}, \ S_Y = \begin{bmatrix} \text{Increase} & \text{Not change} & \text{Reduce} \\ 0 & 1/2 & 1/2 \end{bmatrix};$ and $v = 95$.

19.66. $S_A = \begin{bmatrix} A_1 & A_2 & A_3 \\ 4/10 & 6/10 & 0 \end{bmatrix}, \ S_B = \begin{bmatrix} B_1 & B_2 & B_3 \\ 1/10 & 2/10 & 7/10 \end{bmatrix};$ and $(9/10) \le v \le (11/10)$.

19.67. $S_A = \begin{bmatrix} A_1 & A_2 & A_3 \\ 0 & 1/10 & 9/10 \end{bmatrix}, \ S_B = \begin{bmatrix} B_1 & B_2 & B_3 & B_4 \\ 4/10 & 0 & 0 & 6/10 \end{bmatrix};$ and $(14/10) \le v \le (16/10)$.

19.68. $S_A = \begin{bmatrix} A_1 & A_2 & A_3 \\ 1/3 & 1/3 & 1/3 \end{bmatrix}, \ S_B = \begin{bmatrix} B_1 & B_2 & B_3 \\ 1/3 & 1/3 & 1/3 \end{bmatrix};$ and $v = 0$.

Chapter 20. Queuing Theory

20.01. (*a*) $\rho = 0.75$, (*b*) $L_s = 3$ students, (*c*) $L_q = 2$ students,
(*d*) $W_s = 12$ minutes, (*e*) $W_q = 9$ minutes.

20.02. $W_q = 10$ minutes and $W_s = 12$ minutes.

20.03. (*i*) $W_q = 1.38$ minutes, (*ii*) $L_q = 1.6$ or (approx. 2 customers),
(*iii*) $W_s = 1.98$ minutes, (*iv*) $\rho = 0.7$, (*v*) $P_0 = 0.3$.

20.04. (*i*) $P_0 = 5.33$ hours, (*iii*) $L_s = 2$ machines, (*iii*) $P(\ge 8) = 0.026$.

20.05. $P_0 = 0.25$, $\rho = 0.75$, and $P_1 = 0.45$.

20.06. (*i*) $P(> 2) = 0.56$, (*ii*) $P_1 = 0.75$, (*iii*) $P_0 = 0.25$, (*iv*) $W_s = 12$ minutes.

20.07. (*i*) $P_0 = 0.167$, (*ii*) $L_s = 5$ customers, (*iii*) $W_s = 15$ minutes,
(*iv*) $L_q = 4$ customers (approx.) (*v*) $W_q = 12.5$ minutes.

20.08. (*a*) $1 - P(w > 0) = 1 - \rho = 0.67$, (*b*) $P(w > 10) = \int_{10}^{\infty} \frac{\lambda}{\mu}(\mu - \lambda)e^{-(\mu-\lambda)t}\,dt = 0.063.$

(*c*) $[L_q/P(n > 1)] = 1.5$ or 2 arrivals (approx.).

20.09. (*a*) $L_q = 4/15$ customers, (*b*) $L_s = 2/3$ customers,
(*c*) $W_q = 4/3$ minutes or 80 seconds, (*d*) $W_s = 10/3$ minutes,
(*e*) $\lambda' = 150/7$ customers per hour.

20.10. (*a*) $L_s = 9$ customers, (*b*) $P(\geq 10) = (0.9)^{10}$, (*c*) $P(w > 2) = 0.67$.

20.11. (*i*) $P_0 = 0.6$, (*ii*) $P_1 = 0.4$,
(*iii*) $L_s = 2/3$ or one customer (approx.), (*iv*) $W_s = 10$ minutes.

20.12. $P(w > 0) = 0.6$, $W_q = 0.75$, and second booth is justified, when arrivals increase to 14 per hour.

20.13. (*i*) $P(w > 0) = 0.6$, (*ii*) $W_q = 7.5$ minutes,
(*iii*) Exp. waiting time = (no. of trucks) × (% of contractor trucks) × $W_q = 10.8$ hours per day.

20.14. (*i*) $\rho = 0.60$, (*ii*) $L_s = 3/2$ customers,
(*iii*) $W_q = 4.5$ minutes, (*iv*) $W_s = 7.5$ minutes.

20.15. (*i*) $\rho = 0.50$, (*ii*) L_s = one machine, (*iii*) $W_q = 1/6$ day or 4 hours,
(*iv*) $P_2 = 0.125$, (*v*) $L_q = 1/2$ machine.

20.16. Arrivals in peak hours, normal period and low period are: 9 per hour, 8 per hour and 6 per hour, respectively, using W_q. Number of counters in respective periods are: 110/9 (or 13 approx.), 60/8 (or 8 approx.) and 30/6 (*i.e.*, 5), respectively.

20.17. (*i*) $P(w > 0) = 0.33$, (*ii*) $P_0 = 0.67$, (*iii*) $L_s = 0.5$ patients, (*iv*) $L_q = 0.2$ patients,
(*v*) Second doctor is justified when average waiting becomes 14 minutes.

20.18. (*i*) $P_0 = 0.10$, (*ii*) $L_q = 0.45$ customers, (*iii*) $L_s = 9$ customers, (*iv*) $W_s = 30$ minutes,
(*v*) Justified, when arrival rate is more than 12 customers per hour.

20.19. Average cost per hour is ₹ 25 and average cost of clerk per hour is ₹ 4. Total cost to the deptt. is ₹ 29.

20.20. For mechanic *A*, total cost is ₹ 544 per day and for mechanic *B*, total cost is ₹ 1,120 per day. Since cost of mechanic *A* is less than that of mechanic *B*, mechanic *A* is to be recruited.

20.21. (*i*) $\mu = 1/6 \times 60$ or 10 trucks per hour; waiting time cost = $8 \times W_q \times 60$ = ₹ 192.
(*ii*) $\mu = 1/4 \times 60$ or 15 trucks per hour; waiting time cost = $8 \times W_q \times 60$ = ₹ 36.60.
(*iii*) $\mu = 1/3 \times 60$ or 20 trucks per hour; waiting time cost = $8 \times W_q \times 60$ = ₹ 16.
Since, total cost per hour = waiting time cost + Operative cost, in the three cases, we have: ₹ (192 + 80), ₹ (36.60 + 160), and ₹ (16 + 240). This shows the optimal crew size is of 2.

20.22. Total cost of present mechanic is: ₹ 325.33 and that of new mechanic is: ₹ 224. It is advisable to replace the existing mechanic.

20.23. (*a*) $L_q = 3$ cars, (*b*) $L_s = 6$ cars, (*c*) $W_s = 12.3$ minutes,
(*d*) 30 cars per hour (approx.).

20.24. $P_0 = 0.1926$ and $P_n = (0.1926) \times (5/6)^n$; $n = 0, 1, 2, ..., 10$.

20.25. (*i*) $\lambda_e = 19.98$ or 20 patients approx., (*ii*) $P_0 = 0.67$, (*iii*) $W_s = 39$ minutes.

20.26. (*i*) $\rho = 0.2$ seconds, (*ii*) $W_s = 2.5$ seconds, (*iii*) $L_q = 0.5$ clients.

20.27. (*i*) $\lambda_e = 4.5$ (or 5) cars per hour, (*ii*) $P_0 = 0.17$, (*iii*) L_s = one car approx.

20.28. $P_0 = 0.533$, $P_3 = 0.067$ (when $N = 3$), *i.e.*, proportion of lost customer (present system) is 0.067. $P_0 = 0.516$, $P_4 = 0.032$ (when $N = 4$), *i.e.*, proportion of lost customer (proposed system) is 0.032. Increase in cars served per hour = λ (0.067 – 0.032) = (1/8) × 60 × 0.035 per hour, *i.e.*, 0.0262 per hour. So, increase in profit per week = ₹ 0.50 × 0.262 × 10 × 7, *i.e.*, ₹ 9.17. So, not economical, since the rent is ₹ 10 per week.

20.29. (*i*) 0.14, (*ii*) 0.65.

20.30. Average waiting time will reduce to 4.65 minutes from 8 minutes on Ist counter and 16 minutes on IInd counter.

20.31. $P_0 = 0.08$, $P_5 = 0.13$ and $P_6 = 0.047$, *i.e.*, customer has to wait as $0.047 < 0.10$, six telephone booths be installed.

20.32. Expected weekly time an adjuster spends with claimant is 22.2 hours.

20.33. (*i*) $L_s = 6.6$ (or 7) customers, (*ii*) $L_q = 3.3$ (or 3) customers,
(*iii*) $W_s = 40$ minutes, (*iv*) $W_q = 20$ minutes, (*v*) $P(> 4) = 0.66$,
(*vi*) Expected number of idle tax advisers at any specified time = $4P_0 + 3P_1 + 2P_2 + 1.P_3$ = 2/3 or 0.67 advisers.

20.34. $P_0 = 0.178$, (*i*) $L_s = 0.35$ accidents, (*ii*) $L_q = 1.99$ (or 2) accidents,
(*iii*) $W_q = 0.43$ days, (*iv*) $W_s = 2.42.$ days.

20.35. Yes, since the waiting time reduces from 7.5 minutes to 3.75 minutes.

20.36. (*i*) 24 minutes at counter I, and 6 minutes at counter II. (*ii*) $W_q = 2.72$ minutes.

20.37. (*i*) $L_s = 4.8$ trucks, (*ii*) $W_s = 7.2$ minutes, (*iii*) $P_0 = 0.87$, (*iv*) $W_q = 4.8$ minutes.

20.38. $P_0 = 0.28$, and $L_s = 3$ customers approx.

20.39. $P_0 = 0.196$, $L_s = 6.28$ cars, $L_q = 5.26$ cars, and $W_s = 35.8$ minutes.

20.40. Cost of cheap repairman is: ₹ 35 and that of expensive repairman is: ₹ 17. This shows that fast repairman should be hired because of lower cost.

20.41. (*a*) $W_q = 1.40$ (or 2 engineers), (*b*) Total lost time = 11.2 hours or 672 minutes.

20.42. Since the cost of slow but cheap repairman is ₹ 140 and that of fast but expensive repairman is: ₹ 70; fast repairman should be employed.

20.43. (*a*) L_s: 3 machines, (*b*) Expected cost of down time = ₹ 200 per day,
(*c*) Provision of two mechanics is not economical.

20.44. Mechanic *B* should be hired. Minimum total cost will be ₹ 1,050 per day.

20.45. $W_q = 20.8$ minutes, and $W_s = 45.8$ minutes.

20.46. W_q = half day.

20.47. (*a*) Idle time = 16.25 minutes, (*b*) $L_q = 0.3$ machines.

20.48. $W_q = 6.75$ days, and Average time a locomotive is out of order = 7.75 days.

20.49. Time the filter will have for other task = 5.5 hours, Average time a locomotive is out of service = 3.18 hours.

20.50. $W_q = 12.5$ minutes, and $L_q = 5/6$ or one customer approx.

Chapter 21. Replacement Decisions

21.01. Replace the machine at the end of 6 years, since minimum average cost in 6th year is ₹ 1,575.

21.02. Replace the machine after every 6th year.

21.03. Replace the machine after 6th year, since minimum average cost is ₹ 15,667.

21.04. (*i*) Replace the equipment after 5 years.
(*ii*) Yes, change the equipment *A* by equipment *B* after two years.

21.05. Optimum period of replacement of the machine is after 4 years.

21.06. Vehicle is due for replacement after six years, since minimum average cost in sixth year is ₹ 17,030.30.

21.07. (*a*) Minimum average yearly cost of machine *A* is 3rd year, also the average yearly cost of owning and operating cost for this period is ₹ 5,200.
(*b*) Minimum average yearly cost of machine *B* is ₹ 4,000, and hence machine *A* should be replaced by machine *B*. Further, since machine *A* is one year old, replace it after one year.

21.08. Replace machine *A* at the end of 5th year and machine *B* at the end of 6th year; machine *B* should be purchased.

20.09. (*a*) Replacement of truck is due at the end of 5 years.
(*b*) All the three old trucks should be replaced after two years.

21.10. Best replacement policy is to replace the taxi after one year, since total average minimum cost is ₹ 16,000 in first year.

21.11. Optimum period of replacement is the end of 5th year.

21.12. Machine should be replaced at the end of third year; since weighted average cost in this year is ₹ 217.60.

21.13. Truck should be replaced after every 9 years.

21.14. New tempo as well as second-hand tempo should be replaced at the end of 4th year. Since, minimum total average cost of second-hand tempo (₹ 30,250) is less than the minimum total average cost of the new tempo (₹ 31,500), replace the existing tempo with the second-hand tempo. Net saving will be of ₹ 1,250.

21.15. Optimum replacement interval (period) is 6 years.

21.16. Purchase an automatic stamper, since present worth of future costs for the automatic stamper is less than that of manual stamper.

21.17. Replace the machine by new one by the end of 5th year.

21.18. Existing pipeline should be continued, since the discounted value of all future costs of the existing pipeline is ₹ 4,021 (for 3 years) and of the new pipeline is ₹ 48,820 (for 10 years).

21.19. Machine *A* should be purchased, since it is more economical.

21.20. (*i*) Car should be replaced after first year, since minimum average annual cost is minimum in first year.

21.21. Group replacement at the end of 3rd month, since the cost of group replacement (₹ 2,475) < cost of individual replacement (₹ 2,860).

21.22. Group replacement at the end of 3rd months, the cost of group replacement is ₹ 2,183.60.

21.23. Replace the IC's at the end of every month.

21.24. Replace all the fuses after two weeks, *i.e.*, group replacement is better.

21.25. Group replacement policy will be better. Replace all bulbs after third week. At ₹ 0.49 per bulb in group replacement, individual replacement policy becomes preferable.

21.26. Group replacement is better and replace all the resistors by the end of 6th week.

21.27. (*a*) Group replacement is superior and replace all the identical components at the end of 3rd month.

(*b*) Group replacement is still superior, but optimum replacement interval becomes six months.

21.28. Replace all the valves simultaneously, *i.e.*, group replacement policy is better.

21.29. Group replacement policy is preferable at the end of the fourth week, since cost of group replacement (₹ 321) is less than the cost of individual replacement (₹ 323).

21.30. Group replacement policy by the end of 4th week, since group replacement cost is less than the cost individual replacement.

21.31. $p_1 = 0.06$, $p_2 = 0.12$, $p_3 = 0.24$, $p_4 = 0.18$, $p_5 = 0.12$, $p_6 = 0.09$, $p_7 = 0.06$, $p_8 = 0.06$, $p_9 = 0.04$ and $p_{10} = 0.03$.

21.32. Twelve chemists every year; promotion of new entrants will start by the end of 5th year.

21.33. Replacement is due when $z > X$, where z is the running cost with $0 \le x \le X$, and X is the maximum value of x.

21.34. Increase in reliability is: 0.03407.

21.35. MTBF = 2,08,933 hours.

Chapter 22. Inventory Control Problems — Deterministic

22.01. (*i*) $EOQ = 300$ units, (*ii*) $EOQ =$ ₹ 19,200.

22.02. $EOQ = 400$ units, and $TC =$ ₹ 240.

22.03. $EOQ = 87$ units per order.

22.04. $EOQ = 81.6$ or 82 (approx.) units, $TC =$ ₹ 7,34,757.50.

22.05. (*i*) EOQ (Q^0) = 235.5 units, (*ii*) $TC =$ ₹ 2,169.84 per week.

22.06. (*i*) $EOQ =$ ₹ 20,000, $n^0 = 5$ and $t^0 = 73$ days, $TC =$ ₹ 10,02,500.

22.07. $EOQ = 71.5$ tons.

22.08. (*i*) $EOQ = 100$ units per month, (*ii*) $TC =$ ₹ 4,800.

22.09. (*i*) $t^0 = 73$ days and $n^0 = 5$ orders, with $EOQ = 1{,}000$ units.

(*ii*) Loss per year will be of ₹ 1,873.

22.10. (*i*) $EOQ = 866$ units, (*ii*) $n^0 = 10.4$ orders/year,

(*iii*) $t^0 = 1.15$ months between procurement, (*iv*) $TC =$ 11,080/year.

22.11. $EOQ = 20$ pieces.

22.12. (*i*) $EOQ = 1{,}633$ units, (*ii*) $n^0 = 30.6$ per year,

(*iii*) $t^0 = 0.033$ year, (*iv*) $TC =$ ₹ 10,04,899.

22.13. (*a*) $EOQ = 20$ units, and $TC =$ ₹ 10.

(*b*) Q (not optimum) = 25 units and $TVC =$ ₹ 10.25.

22.14. (*i*) $t^0 = 15$ days, (*ii*) Extra cost to the factory is ₹ 3,600.

22.15. (*a*) EOQ = 2,000 units, (*b*) TC = ` 1,200 units, (*c*) n^0 = 10 production runs,
(*d*) t^0 = 10 days, (*e*) 1,200 units to meet the requirement of EOQ.

22.16. (*a*)

Order quantity	*Ordering cost (₹)*		*Holding cost (₹)*		*Total cost (₹)*	
	X	Y	X	Y	X	Y
24	3.00	20.25	0.75	9.00	3.75	29.25
36	2.00	13.50	1.12	13.50	3.12	27.00
48	1.50	10.12	1.50	18.00	3.00	28.12
60	1.20	8.10	1.88	22.50	3.08	30.60

(*b*) EOQ (for X) = 48 units, and EOQ (for Y) = 36 units.

22.17. EOQ = 300 units and TC (EOQ) = ₹ 4,800.
Total cost under the existing system, TC (200) = ₹ 5,200.
Loss to the organization = ₹ (5,200 – 4,800) = ₹ 400.

22.18. TC (without discount) = ₹ 76,299.04, TC (when discount is availed) = ₹ 76,363.60, Company should not accept the 1.5% discount. Minimum per cent discount acceptable to the company will be 1.584% or 1.6%.

22.19. EOQ = 870 units.

22.20. (*i*) EOQ = 510 units, (*ii*) TC = ₹ 12,04,898.98.

22.21. EOQ = 5,000 units.

22.22. Planing an order of 500 units is justified.
TC (Q^0 = 100) = ₹ 6,56,000 and TC (500) = ₹ 6,13,600.
∴ Net saving of ₹ 42,400 by following Q = 500 units.

22.23. (*i*) TC = ₹ 22,600, (*ii*) TC = ₹ 20,600, (*iii*) When Q = 316 units, TC = ₹ 22,503.
Follow strategy (*ii*), because TC is least in this case.

22.24. (*a*) EOQ = 3,464 units, TC = ₹ 29,216. When discount offer is accepted, TC = ₹ 28,230.80. Accept discount offer, since TC cost is the least.
(*b*) Break-even discount percentage is 1.68%.

22.25. EOQ = 700 units.

22.26. EOQ = 100 units.

22.27. EOQ = 1,656 units approx.

22.28. EOQ = 20,000 items.

22.29. EOQ = 1,010 units. When discount slab is raised to 1,500 units, EOQ = 1,000 units.

22.30. EOQ = 268 units, TC = ₹ 13,170.82.

22.31. EOQ = 50 units (without discount), EOQ = 100 units (with discount of 10%), TC = ₹ 7,000.

22.32. (*a*) EOQ = 10,000 units, (*b*) TC = ₹ 9,37,590.

22.33. EOQ = 1,000 units, t^0 = 40 days and TC = ₹ 200.

22.34. EOQ = 200 items, t^0 = 8 working days, and TC (EOQ) = ₹ 120.

22.35. EOQ = 1,414 units, t^0 = 1.68 hours.

22.36. EOQ = 3,464 units.

22.37. EOQ = 800 pieces.

22.38. EOQ = 4,000 gears, n^0 = 15 production runs/year, TC (EOQ) = ₹ 1,20,000.

22.39. EOQ = 3,651 units.

22.40. EOQ = 7,615 units, TC = ₹ 1,119.60.

22.41. (*i*) EOQ = 3,600 units, (*ii*) n^0 = 50 orders per year, (*iii*) TC = ₹ 6,000.

22.42. $EOQ = 54.7$ units, and $t^0 = 2.16$ months.

22.43. (*i*) $EOQ = 4{,}583$ units, (*ii*) $S^0 = 3{,}055$ units,
(*iii*) $t^0 = 2.29$ months, (*iv*) $TC =$ ₹ 3,000.

22.44. (*i*) $EOQ = 577.35$ units, (*ii*) Maximum shortage level = 433 units,
(*iii*) Maximum inventory = 144 units, (*iv*) Cost of saving = ₹ 4,330.12.

22.45. (*i*) $EOQ = 115$ units, (*ii*) $Q_2^0 = 29$ units,
(*iii*) Maximum inventory = 86 units, (*iv*) $t^0 = 21$ days, (*v*) $TC =$ ₹ 866.

22.46. (*i*) $EOQ = 1{,}155$ units, (*ii*) Maximum inventory = 866 units,
(*iii*) $Q_2^0 = 289$ units, (*iv*) $TC =$ ₹ 15,000.
When back ordering is not permitted, Q^0 (or EOQ) = 1,000 units and $TC =$ ₹ 17,320.

22.47. $EOQ = 44$ items, $t^0 = 2.2$ months. **22.48.** $EOQ = 3{,}413$ units, $TC =$ ₹ 51,336.

22.49. $Q_1^0 = 70.71$ units, $Q_2^0 = 61.23$ units, and $Q_3^0 = 141.40$ units.

22.50. $Q_1^0 = 428$ units, $Q_2^0 = 628$ units, and $Q_3^0 = 444$ units.

22.51. $Q_1^0 = 324$ units, $Q_2^0 = 263$ units, and $Q_3^0 = 531$ units;
Total storage space required = 649.6 sq.ft.

22.52. $Q_1^0 = 127$ units, $Q_2^0 = 78$ units, and $Q_3^0 = 141$ units.
Corresponding value of average inventory = ₹ 24,725.

22.53. $Q_1^0 = 78.42$ items, $Q_2^0 = 20.49$ items, $Q_3^0 = 119.58$ items, and $Q_4^0 = 110.57$ items.

22.54. $Q_1^0 = 141.4$ units, $Q_2^0 = 100$ units, $Q_3^0 = 67.3$ units, and $Q_4^0 = 63.24$ units.

22.55. (*i*) $EOQ = 4{,}899$ items, (*ii*) Re-order point = 7,100 units.

22.56. $EOQ = 56$ items, $t^0 = 2$ days, Re-order point = 350 items.

22.57. (*i*) $EOQ = 400$ units, $t^0 = (2/3)$ years, (*ii*) Re-order point = 200 units, (*iii*) $TC =$ ₹ 240.

22.58. (*i*) $EOQ = 2{,}083$ units, (*ii*) $n^0 = 5$ orders, (*iii*) Re-order point = 800 units.

22.59. $EOQ = 258$ units, $ROL = 100$ units, $TC =$ ₹ 5,834.37, Increase in cost = ₹ 6,278.03.

22.60. (*i*) $EOQ = 300$ units, (*ii*) $n^0 = 12$ orders, (*iii*) $ROL = 4{,}500$ units,
(*iv*) Safety stock = 3,000 units.

22.61. (*i*) $EOQ = 4{,}000$ units, (*ii*) $n^0 = 75$ orders, (*iii*) $ROL = 12{,}500$ units.

22.62. $EOQ = 63$ units, Safety stock = 8.83 units, $ROL = 36$ units, and average inventory = 40 units.

22.63. 'A' class items: 2 and 10, 20% of items, Annual value = 68%.
'B' class items: 7, 1 and 4, 30% of items, Annual value = 28%.
'C' class items: 8, 9, 6, 3 and 5, 50% of items, Annual value = 4%.

22.64. 'A' class: Model No. 502 and 506, 20% of models.
'B' class: Model No. 509, 504 and 508, 30% of models.
'C' class: Model No. 501, 510, 507, 503 and 507, 50% of models.

22.65. 'A' class items: F_3 and G_2; *i.e.*, 64.37 items.
'B' class items: *A*2, *B*8, *C*7 and *D*1; *i.e.*, 29.92 items.
'C' class items: *E*9, *H*2, *I*5 and *J*8; *i.e.*, 5.71 items.

Chapter 23. Inventory Control Problems — Probabilistic

23.01. *ROL* = 4,200 units, Maximum level = 1,175 units, and Minimum level = 7,950 units.

23.02. *EOQ* = 4,000 units, *TC* = ₹ 30,270, *TC* = ₹ 29,292 (when discount is 5% on the cost price). Discount offer should be accepted, *ROL* = 1,360 units, Average inventory = 2,400 units, Maximum inventory = 4,400 units, and Minimum inventory = 400 units.

23.03. *EOQ* = 1,000 units, *ROL* = 200 units, Buffer stock = 223 units.

23.04. (*i*) *EOQ* = 1,000 kg, t^0 = 2 months, *TC* = ₹ 30,300.
(*ii*) *ROL* = 500 kg, Maximum inventory = 1,200 kg, Average inventory = 700 kg,
(*iii*) Discount offer should be accepted.

23.05. Maximum *DDLT* = 1,500 units, Average *DDLT* = 800 units, Safety stock = 700 units.
(*a*) *ROL* = 1,500 units, (*b*) Average inventory level = 2,700 units,
(*c*) Ordering cost = ₹ 800, Carrying cost = ₹ 1,080.

23.06. Safety stock =233 units, *ROL* = 1,773 units, Maximum inventory = 1,647 units, Minimum inventory = 233 units, Average inventory = 940 units.

23.07. *Under Q-system:* Buffer stock = 77 units, Reserve stock = 33 units, Safety stock = 17 units, *ROL* = 127 units.

Under P-system: Buffer stock = 519 units, Reserve stock = 86 units, Safety stock = 17 units, Maximum inventory level = 622 units.

23.08. *EOQ* = 600 units, *ROL* = 570 units, where *DDLT* = 540 units and safety stock = 30 units.

23.09.

Lead time (months)	*Value of z for ROL = 500 units*	*Probability of having stock for different values of z (%)*	*Probability of lead time (%)*	*Conditional probability of not running out of stock (%)*
1	$\frac{500-100\times 1}{25\sqrt{1}} = 16.0$	100.0	0.10	10.0
2	$\frac{500-100\times 2}{25\sqrt{2}} = 8.49$	100.0	0.20	20.0
3	$\frac{500-100\times 3}{25\sqrt{3}} = 4.49$	100.0	0.40	40.0
4	$\frac{500-100\times 4}{25\sqrt{4}} = 2.00$	97.7	0.20	19.5
5	$\frac{500-100\times 5}{25\sqrt{5}} = 0.00$	50.0	0.10	5.0
			Total	94.5

A reorder level of 500 units will achieve 94.5% service level.

23.10. Reserve stock = 2 fans, Minimum total cost = ₹ 90.

23.11. Optimum safety stock = 100 units, *ROL* = 400 units.

23.12. (*a*) 88 units, (*b*) 72 units.

23.13. Produce 6 items per week.

23.14. Purchase 30 A.C.'s.

23.15. Four TV sets per week.

23.16. Three spare parts should be procured.

23.17. Purchase 17 copies of newspaper.

23.18. Order 2 machines.

23.19. Procure 154.5 kg.

23.20. Order 30 newspapers.

23.21. Optimum order quantity is 30 units of first product as well as of second product.

23.22. (*a*) Stock 298 copies, (*b*) Stock 300 copies.

23.23. Optimum order level = 8 units.

23.24. Optimum quantity = 386.4 kg.

23.25. Prepare 63.3 pounds.

23.26. Procure 100 kg of fresh fish.

23.27. Order (63 – 43), *i.e.* 20 items.

Chapter 24. Simulation

24.01. Average time = 3.36 time units per customer.

24.02. Average time = 2.35 minutes per customer.

24.03. Average time = 1.07 time units per customer for type *A*, and 2.50 time units per customer for type *B*.

24.04. Average sales will be of 2.8 (or 3 approx.) units.

24.05. Demand level (in hundred kgs) : 10, 15, 20, 05, 05, 15, 10, 15, 10, 15, respectively.
Average demand: 1,200 kg per week.

24.06. No additional profit or loss, if 39 cars are produced by the company instead of 40 cars, *i.e.*, total profit remains at ₹ 359 lakh.

24.07. Profit will be of ₹ 2,695, if the company manufactures 30 items per day. No additional profit or loss, if the company produce 29 items per day.

24.08. Option *A:* Total cost = ₹ 59.50 and Option *B:* Total cost = ₹ 42.50. So, option *B* must be chosen.

24.09. Stock situation (units per day): 0, 0, 20, 20, 20, 20, 40, 60, 60 and 80, respectively.
Average demand: 22 units per day.

24.10. (*i*) Average number of scooters waiting = 1.4 per day.
(*ii*) Average number of empty spaces = 0.6 per day.

24.11. (*i*) 10/15 or 0.67 mopeds per day, (*ii*) 14/15 or 0.93 empty spaces on the lorry.

24.12. Average waiting time for the patients = 285/8 or 35.6 minutes, and average idleness of the dentist is nil.

24.13. (*b*) Average waiting time of item: 80 seconds.
(*c*) Between two stations, six items are waiting,
(*d*) Total time taken to process 8 items: 8 minutes, and average production rate is: 8/8, *i.e.*, one item per minute.

24.14. (*i*) Number of weeks to be out of stock are three,
(*ii*) Average demand of heaters per week will be 6.75 or 7 approximately.

24.15. Total time for 10 units are : 866 minutes; and Expected time for the product is 866/10, *i.e.*, 86.6 minutes.

24.16. Average profit for the project is: ₹ 2,100.

24.17. Total idle time for the machinist and total waiting time for the jobs in 2 hours period are 21 minutes and 9 minutes, respectively.

(*i*) Average idle time for the machinist per hour is 10.5 minutes, and idle time cost of the machinist per hour is ₹ 3.50.

(*ii*) Average waiting time for the job per hour is 4.5 minutes, and waiting time cost of the jobs per hour is ₹ 2.25. As idle-time cost of the machinist per hour > Waiting time cost of the jobs per hour, not economical to engage 2nd attendant.

24.18. Average waiting time is 5/6 minutes or 50 seconds, and average service time is 16.6 minutes. Average server idle time is of 5 minutes.

24.19. (*i*) Balance at the end of 12-weekly period is a deficit of ₹ 3,000.

(*ii*) Highest weekly balance during quarter = ₹ 7,000. (*iii*) Average balance = ₹ 3,750.

24.20.

Net revenue	(*i*) *Frequency*	(*i*) *Probability*	(*ii*) *Frequency*	(*ii*) *Probability*
–1,000	1	0.05	0	0.00
–500	0	0.00	0	0.00
000	3	0.15	1	0.05
500	6	0.30	2	0.10
1,000	3	0.15	6	0.30
1,500	5	0.25	6	0.30
2,000	1	0.05	1	0.05
2,500	0	0.00	4	0.20
3,000	1	0.05	0	0.00

24.21. The highest return is 13.33%, which is corresponding to the annual demand of 40,000 units and a profit of ₹ 10 per unit.

24.22. Option *II* is better, since total cost in this option is ₹ 228.75 as compared to ₹ 276.25 in option *I*.

24.23.

Length of queue during a minute :	*0*	*1*	*2*	*3*	*4*
Number of minutes :	4	5	6	4	1
Probability :	0.20	0.25	0.30	0.20	0.05

24.24. Arrival rate = 4.2 airconditioners, approximately, Service rate = 2.9 airconditions per day. Average backlog = 6.8 airconditions per day.

24.25. Total current maintenance cost = ₹ 4,777,
Total cost with two additional repairmen = ₹ 4,052.
Plant should hire one additional repairman.

Chapter 25. Non-Linear Programming and Optimization

25.01. Maximize $z = 5(f_1, f_2, f_3) + 6(g_1, g_2, g_3)$ subject to the constraints:

$$0.2f_1 + 0.3g_1 \le 1.8,\ \ 0.2f_2 + 0.1g_2 \le 1.2,\ \ 0.3f_3 + 0.3g_2 \le 2.4;\ \ f_1, f_2, f_3 \ge 0.$$

25.02. Minimize $z = 100 - 1.2x_1 - 1.5x_2 + 0.3x_1^2 + 0.5x_2^2$ subject to the constraints:

$$6x_1 + x_2 \ge 35;\ \ x_1 \ge 0 \text{ and } x_2 \ge 0.$$

25.03. Maximize $z = (200 - 2x^2) + (500 - 2y^2)$ subject to the constraints:

$$0.5x + 0.25y \le 35,\ \ 2x + 3y \le 180;\ \ x \ge 0 \text{ and } y \ge 0.$$

25.04. (*a*) Global minimum, (*b*) Global maximum.

25.05. (*a*) Local minimum at all points. (*b*) Local maximum, and $x^* = (4/3, 5/3)$.
(*c*) Local maximum.

25.06. Local minimum, and $x^* = (4, -1, 1)$.

25.07. $x_1 = 3/31,\ x_2 = 18/31$; minimum $z = 54/31$.

25.08. $x_1 = 1$ and $x_2 = -2$; maximum $f(x_1, x_2) = 5$.

25.09. $x_1 = 5/2,\ x_2 = 3/2$; maximum $z = 13$.

25.10. $x_1 = 1/3,\ x_2 = 5/6$; maximum $z = 4.16$.

25.11. $x_1 = 5,\ x_2 = 11$ and $x_3 = 4$; minimum $z = 281$.

25.12. $x_1 = 2,\ x_2 = 2$ and $x_3 = 1$; minimum $z = 9$.

25.13. $x_1 = 4,\ x_2 = 2,\ x_3 = 1$; minimum $z = -35$.

25.14. $x_1 = 3,\ x_2 = 1$; minimum $z = 11$.

25.15. $x_1 = 0.81,\ x_2 = 0.35,\ x_3 = 0.28$; minimum $z = 0.857$.

25.16. $x_1 = 35/11,\ x_2 = 12/11$; minimum $z = 281$.

25.17. $L(x, \lambda) = (x_1 \cdot x_2 \cdot \ldots \cdot x_n) - \lambda(x_1 + x_2 + \ldots + x_n - b)$;
$x_1 = x_2 = x_3 = \ldots = x_n = b/n$; $\lambda = n(x_1 x_2 x_3 \ldots x_n)/b$; maximum $z = (b/n)^n$.

25.18. $x_1 = 44,\ x_2 = 2$; minimum $z = 4{,}900$.

25.19. $x_1 = 4/13,\ x_2 = 33/13$; maximum $z = 21.3$.

25.20. $x_1 = 1,\ x_2 = 2,\ x_3 = 0$; minimum $z = 5$.

25.21. $x_1 = 1.3,\ x_2 = 2$; minimum $z = 96.3$.

25.22. $x_1 = 1,\ x_2 = 1$; minimum $z = 0$.

25.23. $x_1 = 1,\ x_2 = 3/4$; minimum $z = 17/8$.

25.24. $x_1 = 2/3,\ x_2 = 14/9$; maximum $z = 22/9$.

25.25. $x_1 = 48/5,\ x_2 = 1/5$; maximum $z = 703.72$.

25.26. $x_1 = 1.43,\ x_2 = 0.48$; maximum $z = 4.77$.

25.27. $x_1 = 2,\ x_2 = \sqrt{5}$; maximum $z = 42$.

25.28. $x_1 = 1,\ x_2 = 5$; minimum $f(x_1, x_2) = 0$.

Chapter 26. Solving Non-linear Programming Problems

26.01. $x_1 = 6,\ x_2 = 2$; maximum $z = 24$.

26.02. $x_1 = 0.6,\ x_2 = 0.8$; maximum $z = 2.20$.

26.03. $x_1 = 0,\ x_2 = 0.5$; maximum $z = -25$.

26.04. $x_1 = 0,\ x_2 = 1$; minimum $z = 2$.

26.05. $x_1 = 1,\ x_2 = 1$; minimum $z = 1$.

26.06. $x_1 = 4,\ x_2 = 4$; minimum $z = 32$.

26.07. $x_1 = 4,\ x_2 = 2$ or $x_1 = 2,\ x_2 = 4$; maximum $z = 20$.

26.08. $x_1 = 1,\ x_2 = 0$; maximum $z = 1$. Constraint qualification is not satisfied.

26.09. $x_1 = 3,\ x_2 = 2$; maximum $z = 3$. Constraint qualification is not satisfied.

26.11. For achieving the maximum profit of ₹ 88, company must produce 2 and 8 units of two products, respectively.

26.12. $x_1 = 1/3,\ x_2 = 5/6$; maximum $z = 25/6$.

26.13. $x_1 = 4/13,\ x_2 = 33/13$, maximum $z = 267/13$.

26.14. $x_1 = 1,\ x_2 = 0$; maximum $z = 4$.

26.15. $x_1 = 500,\ x_2 = 500$; maximum $z = 800$.

26.16. $x_1 = 2/3,\ x_2 = 14/9$; maximum $z = 22/9$.

26.17. $x_1 = x_2 = x_3 = 1/3$; minimum $z = -5/6$.

26.18. $x_1 = 2,\ x_2 = 1,\ x_3 = 0$; maximum $f = 7$.

26.19. $x_1 = 1/8,\ x_2 = 0,\ x_3 = 7/8$; maximum $f(x) = 1/64$.

26.20. $x_1 = 0,\ x_2 = 3/4$; maximum $z = 9/8$.

26.21. $x_1 = 38/13,\ x_2 = 2/13$; minimum $z = 116/13$.

26.22. $x_1 = 0,\ x_2 = 5$; maximum $z = 100$.

26.23. $x_1 = 2,\ x_2 = 1$; maximum $z = 15$.

26.24. $t_0 = -3,\ t_1 = 17/3$; minimum $z = 34/3$.

26.25. $x_1 = 1.6,\ x_2 = 1.2$; maximum $z = 0.8$.

26.26. $x_1 = 0,\ x_2 = 2.1$; maximum $f(x) = 22.5$.

26.27. $x_1 = 3,\ x_2 = 7$; maximum $f(x) = 16$.

26.28. (*a*) $x_1 = 1,\ x_2 = 1$; minimum $f(x) = 1$,

(*b*) $x_1 = 0.823,\ x_2 = 0.94$; minimum $f(x) = 1.393$.

Chapter 27. Geometric Programming

27.01. Minimum $z = (5c_1/2)^{2/5}(5c_2)^{1/5}(5c_3)^{1/5}(5c_4)^{1/5}$.

27.02. $x_1 = 1.39$ and $x_2 = 1.13$.

27.03. Minimum $f(x) = 10.28$; $x_1 = 1.26,\ x_2 = 0.41$, and $x_3 = 0.59$.

27.04. Minimum $f(x) = 15.22$; $x_1 = 1.315,\ x_2 = 1.21$, and $x_3 = 1.20$.

27.05. Minimum $f(x) = 6\sqrt[3]{10}$; $x_1 = \sqrt[3]{10}$, and $x_2 = \sqrt[3]{10/2}$.

27.06. Minimum $x_0 = 9.66$; $x_1 = 0.31$ and $x_2 = 0.65$.

27.07. Minimum $f(x) = 15\sqrt[3]{2}$.

27.08. Minimum $f(x) = 49.2$.

27.10. Minimum $f(x) = 480$; $x_1 = 2,\ x_2 = 1$, and $x_3 = 4$.

Chapter 28. Information Theory

28.01. $H = 1.75$ bits.

28.02. $H = 1/2 \log 3 + 7/6$.

28.07. $\bar{L} = 2.33,\ H(x) = (1.60)\log 2 + 2\log 5 - (1.40)\log 7$.

Chapter 29. Investment Analysis and Break-even Analysis

29.01. ₹ 25,000 @ 12% p.a., and ₹ 15,000 @ 14% p.a.

29.02. (*i*) ₹ 1,60,355, (*ii*) ₹ 1,64,533, (*iii*) ₹ 1,65,510.

29.03. ₹ 13,196.67 after 3 years.

29.04. ₹ 79,176.52.

29.05. ₹ 15,43,320.8.

29.06. ₹ 2,213.22.

29.07. Leasing is preferable, because *PV* of machine on lease is ₹ 3,642.66.

29.08. Second machine is preferable, because *PV* of annual savings for first machine is ₹ 9,250, and that of second machine is ₹ 8,335.40.

29.09. *NPV* = ₹ 450.

29.10. *IRR* = 15%.

29.11. *NPV* = (–) ₹ 2,797.10.

29.12. (*a*) ₹ 2,04,906, (*b*) *IRR* = 6.62%; Not acceptable.

29.13. Proposal should be selected, because *NPA* of proposal *B* is ₹ 1,03,940 and that of *A* is ₹ 63,552.

29.14. Project *B* should be selected, since *NPV* of project *B* is ₹ 2,129.20 and that of project *A* is ₹ 2,035.50.

29.15. Project *A* is preferred, since *NPV* of project *A* > *NPV* of project *B*; *i.e.*, ₹ 46,458 > ₹ 36,576.

29.16. Projects: *A, B, C, D, E*, respectively; Ranking: (*i*) *C, B, E, A, D*; (*ii*) *A, D, E, B, C*; (*iii*) *D, C, E, A, B*.

29.17. Project *A*, because profitability index of project *A* is 0.3; while that of project *B* is 0.21.

29.18. Project *B* should be selected, since *NPV* of project *B* > *NPV* of project *A*.

29.19. Proposal *Y* should be selected, since *NPV* of proposal *Y* is higher than that of proposal *X*.

29.20. $(EMV)_1$ = ₹ 2,700, $(EMV)_2$ = ₹ 2,300 and $(EMV)_3$ = ₹ 2,400; Expected (*NPV*) = ₹ 5,156.50.

29.21. (*i*) A_3, A_2, A_1 ; since σ_1 = ₹ 67.08, σ_2 = ₹ 153.72, and σ_3 = ₹ 236.85.
(*ii*) A_1, A_2, A_3 ; since (*CV*) for A_1, A_2 and A_3 are: 29.81%, 41.83%, and 55.08%, respectively.

29.22. Expected (*NPV*) = ₹ 19,729; S.D. of the project = ₹ 1,532.

29.23. Expected (*NPV*) = ₹ 5,695; S.D. of the project = ₹ 675.10.

29.24. Project C is selected, because variance of C is less than the variance of *A* and *B* both.

29.25. (*i*) Expected (*NPV*) = ₹ 1,004, proposal is accepted, since expected (*NPV*) > 0.
(*ii*) S.D. of (*NPV*) = ₹ 1,688, (*iii*) Probability {(*NPV*) < 0} = 0.2676.

29.26. (*a*) $\bar{X}_1$ = ₹ 78,000, $\bar{X}_2$ = ₹ 1,03,000, and $\bar{X}_3$ = ₹ 1,52,000; σ_1 = ₹ 18,330, σ_2 = ₹ 21,000, and σ_3 = ₹ 9,407,
(*b*) ₹ 39,951.40, (*c*) ₹ 24,350.50, (*d*) 0.0505, (*e*) ₹ 514.80.

29.28. *BEP* = ₹ 1,000.

29.29. (*i*) *BEP* = 2,500 units, (*ii*) Profit = ₹ 2,600.

29.30. (*a*) Fixed cost = ₹ 80,000, (*b*) *BEP* = 8,000 units, (*c*) 13,000 units.

29.31. (*i*) *BEP* = 1,200 ball bearings, (*ii*) Loss of ₹ 10,000, (*iii*) 2,000 items.

29.32. (*i*) *BEP* = 6,667units, (*ii*) Loss of ₹ 75,000.

29.33. Location 2, with fixed cost = ₹ 60,000, variable cost = ₹ 90,000, and profit = ₹ 90,000.

29.34. Alternative *A* with *BEP* = 3,333.33 units, profit for *A* as well as for *B* is: ₹ 2,00,000 each, and alternative *A* with 5,000 units will require lowest volume of output.

29.35. (*a*) Economical, since it reduces the loss by ₹ 15,50,000.
(*b*) Not acceptable, since it reduces into heavy losses.

29.36. (*i*) *BEP:* 320 units, 500 units, and 666.67 units, respectively.
(*ii*) Manager should purchase two machines.

Chapter 30. Business Forecasting

30.01.

Year	:	4	5	6	7	8	9	10
Forecast of production	:	22	23.33	24	23.67	23.67	24.33	26
Forecast error	:	0	–0.33	1	0.33	–1.67	0.67	0

Forecast for production of steel = 26,330 tonnes, and *MSE* = 0.636.

30.02. (*i*) 3-month moving average forecast:
278, 283, 287.7, 302.3, 305, 313.7, 313.3, 317.3, 314.7, 322.3, 320.7 and 316.7, respectively.
(*ii*) 4-month moving average forecast:
282.3, 287.8, 293.3, 302.5, 310.8, 312.5, 313.8, 318, 319.3, 319 and 317.5, respectively.

Demand for 16th month is: (*i*) 312.7 units, and (*ii*) 317.5 units.

30.03. (*i*) 3-month moving average forecast:
9.40, 9.43, 9.60, 9.70, 9.77, 10, 10.07, 10.03 and 9.73 respectively.
4-month moving average forecast:
9.45, 9.53, 9.63, 9.73, 9.95, 9.98, 9.97 and 9.92, respectively.
(*ii*) *MSE* (3-month) = 1.08/9 = 0.12 and *MSE* (4-month) = 1.09/8 = 0.14.

Since 0.12 is less than 0.14, use a 3-month moving average.

30.04. (*i*) 4-week moving average forecast: 20, 20.25, 19, 19.25, 18, 19, 20, and 18.75, respectively.
5-week moving average forecast: 19.60, 19.40, 19.20, 19, 18.80, 19.20, and 19, respectively.
(*ii*) *MSE* (4-week) = 77.18/8 = 9.65, and *MSE* (5-week) = 51.84/7 = 7.41.
(*iii*) For the limited data provided, the 5-week moving average provides the smallest *MSE*.

30.05. Moving weighted average forecast: 5.125, 5.625, 6.500, 7.250, and 8.875, respectively.
Weighted moving average forecast for 12th month is: ₹ 9,00,000.

30.06. (*i*) 5-monthly moving average forecast: 228.4, 232.6, 234.8, 240.2, 247.6, 248, and 252.8.
(*ii*) 4-monthly weighted moving average forecast:
157.2, 199.2, 220.3, 235.1, 239.3, 241.4, 243.3, and 247.5.

30.07. For $\alpha = 0.1$, forecast for 2nd week on wards is: 17, 17.40, 17.56, 18.10, 18.09, 17.88, 18.10, 18.09, 1 8.48, 18.63, and 18.27; *MSE* = 101.72/11 = 9.25.
But, $\alpha = 0.2$ provides a lower *MSE*. Therefore, $\alpha = 0.2$ is better than $\alpha = 0.1$.

30.08. (*i*) 3-month moving average forecast: 82, 83, 83.33, 83.33, 84, 84.33, 83.67, 83, and 83.
When $\alpha = 0.2$, forecast for 2nd month onwards is:
80, 80.40, 81.12, 81.50, 81.80, 82.24, 82.79, 83.03, 82.83, 82.86, and 83.09.
MSE (3-month moving average forecast) = 11.12/9 = 1.24, and
MSE (Forecast when $\alpha = 0.2$) = 39.06/11 = 3.55.
Since, 1.24 is less than 3.55, use a 3-month moving average forecast.
(*ii*) (83 + 84 + 83)/3 = 83.3.

30.09. $Y_t = 114 - 2X$ (Origin: year 2015 with X unit = 1 year, and Y unit = sales in 1,000 units)

$\Rightarrow Y_0 = 114,\ Y_1 = 112,\ Y_2 = 110,\ Y_3 = 108$ and $Y_4 = 106$.

Estimated sales for the years 2021 and 2022 are 1,02,000 units and 1,00,000 units, respectively.

30.10. $Y_t = 23.6 + 0.5X$ (Origin: year 2015 with X unit = 1 year and Y unit = Production in '000 tonnes). Estimated production in 2021 is of 25.6 thousand tonnes.

30.11. $Y_t = 1{,}058.18 + 65.02X$ (Origin: 6th year, X unit = 1 year, Y unit = Annual demand)

Estimated production in 12th year is of 1,513.32 million barrels.

30.12. (*a*) $Y_t = 94.57 + 7.39X$ (Origin: year 2013, X unit = 1 year, and Y unit = production in tonnes)

(*b*) $Y_t = \dfrac{94.57}{12} + \dfrac{7.39}{144}X = 7.88 + 0.05X$, where X unit = 1 month.

(*c*) Estimated production in the year 2021 is of 131.52 tonnes.

30.13. For $\alpha = 0.1$: forecast values are: 70, 79, 82.10, 93.89; 93.50, 96.15, 92.54, 94.28; 94.85, 104.37, 108.93, 128.04; 142.23, 163.01, 178.71.

For $\alpha = 0.3$: forecast value are: 70.97, 100.90, 130.63, 118.44; 118.91, 101.24, 103.87, 102.71; 128.89, 135.23, 184.66; 210.26, 252.18, 272.53.

For $\alpha = 0.1$, $MAD = 83.37$, and for $\alpha = 0.3$, $MAD = 62.99$.

30.14. Seasonal Indices: 103.40, 111.56, 92.52, and 92.52 respectively.

30.15. Estimated sales for the year 9 and 10 are: 98 million and ₹ 100 million, respectively.

(*a*) Seasonal indices: 91.6, 129.8, 108.4 and 70.2, respectively.

(*b*) Estimated sales of Quarterly sales (in million rupees):

Year 9 :	22.44,	31.80,	22.56,	and	17.20 respectively
Year 10 :	22.90,	32.45,	27.10,	and	17.55 respectively.

30.16. Estimated worth of televisions:

Quarter :	*II*	*III*	*IV*
Seasonal effect :	0.90	0.75	1.05
Estimated sales (₹) :	4,50,000 × 0.90 = 4,05,000	4,50,000 × 0.75 = 3,37,500	4,50,000 × 1.05 = 4,72,500

30.17. Height of son will be 65 inches.

30.18. Regression Equation: $Y_e = 72.5 + 11.25X$, 185 cars when $X = 10$.

30.19. $Y_e = 14.68 + 0.74X$, $Y = 68.7$ (when $X = 73$).

30.20. $Y_e = 20.94 + 0.81X$; $Y = 77.48$ for $X = 70$ and $Y = 89.60$ for $X = 85$.

30.21. (*i*) $Y_e = 219.23 + 6.381X_1 - 1.671X_2$, (*ii*) 43.25 units.

30.22. (*i*) $Y_e = 4.364 + 0.1978X_1 + 1.858X_2$, (*ii*) $Y_e = 33.79$.

Chapter 31. Learning Curve Theory

31.01. (*i*) 3.81 hours, (*ii*) 100.35 hours, (*iii*) 21.61 hours.

31.02. (*a*) ₹ 2,430, (*b*) ₹ 1,944, (*c*) ₹ 4,374.

31.03. (*a*) $y = a\,e^{-0.5146}$, (*b*) Direct cost = ₹ 7,821.50, Sales price = ₹ 20,335.90.

31.04. Marginal cost = ₹ 147.26 per item.

31.05. Contract price = ₹ 71.00.

31.06. Selling price = ₹ 20,250.

Chapter 32. Statistical Quality Control

32.01. $UCL = 13.125$, and $LCL = 0$ by *np*-chart.

32.02. $CL = 11$, $UCL = 20.89$, and $LCL = 1.105$ by *np*-chart, Process is under control.

32.03. $CL = 0.0293$, $UCL = 0.0651$, and $LCL = 0$, Process is completely under control using *p*-chart.

32.04. $CL = 16.2$, $UCL = 20.5$, and $LCL = 11.9$ by $\overline{\mathbf{x}}$-chart,
$CL = 7.4$, $UCL = 15.65$, and $LCL = 0$ by **R**-chart.
One sample mean lies outside the control limits.

32.05. $CL = 5.00$, $UCL = 5.02187$, and $LCL = 4.97813$, by $\overline{\mathbf{x}}$-chart;
$CL = 0.03$, $UCL = 0.06846$, and $LCL = 0$ by **R**-chart.

32.06. $CL = 0.09$, $UCL = 0.137$, and $LCL = 0.043$, community sentiment is not stable as week 5 and 6 lies outside the control limits by *p*-chart.

32.07. $CL = 1.54$, $UCL = 5.26$, and $LCL = 0$ by *c*-chart. Number of mistakes will be in control for future is to operate by the *c*-chart.

32.08. $CL = 0.525$, $UCL = 0.0998$, and $LCL = 0.0052$. Process is under control by *p*-chart.

32.09. $CL_1 = 20.2$, $UCL_1 = 32.545$, and $LCL_1 = 7.855$;
$CL_2 = 17.44$, $UCL_2 = 29.09$, and $LCL_2 = 5.79$.

32.10. $CL = 4$, $UCL = 8$, and $LCL = 0$. Process is not under statistical control.

32.11. $CL = 1.18$, $UCL = 2.64$, and $LCL = 0$. Process is under control by *u*-chart.

32.12. $CL = 3.17$, $UCL = 5.56$, and $LCL = 0.78$. Process is not under ontrol by *u*-chart.

Appendix A. Prerequisite — Linear Algebra

A.02. Any one of $\mathbf{e}_1$, $\mathbf{e}_2$ or $\mathbf{e}_3$.

A.03. (*a*) Yes, (*b*) Yes, (*c*) Yes.

A.04. Open half space.

A.06. (*i*) Convex, (*ii*) Convex, (*iii*) Convex.

A.09. Extreme points: $\mathbf{x}_1 = \left(\frac{1}{2}, \frac{-7}{2}\right)$, $\mathbf{x}_2 = (-11, 8)$, $\mathbf{x}_3 = (12, 8)$; $(2, 1) = \frac{14}{23}\mathbf{x}_1 + \frac{3}{23}\mathbf{x}_2 + \frac{6}{23}\mathbf{x}_3$.

A.10. (*i*) $<\mathbf{A}> = \left\{(x_1, x_2) \mid x_1^2 + x_2^2 \le 1\right\}$, (*ii*) $<\mathbf{A}> = \{\mathbf{x} \mid \mathbf{x} = \lambda\mathbf{x}_1 + (1-\lambda)\mathbf{x}_2 ;\ 0 \le \lambda \le 1\}$.

A.11. (1, 1) is the interior point of convex hull; $(1, 1) = (0, 0)\times\frac{1}{16} + (0, 1)\times\frac{1}{2} + (1, 2)\times\frac{1}{4} + (4, 0)\times\frac{3}{16}$.

A.14. (*i*) $\begin{bmatrix} x_1 & x_2 & x_3 \end{bmatrix}\begin{pmatrix} 1 & 4 & 0 \\ 4 & 16 & 0 \\ 0 & 0 & -3 \end{pmatrix}\begin{bmatrix} x_1 \\ x_2 \\ x_3 \end{bmatrix}$, (*ii*) $\begin{bmatrix} x_1 & x_2 & x_3 \end{bmatrix}\begin{pmatrix} 2 & -3 & 1 \\ -3 & 2 & 3 \\ 1 & 3 & -5 \end{pmatrix}\begin{bmatrix} x_1 \\ x_2 \\ x_3 \end{bmatrix}$.

A.15. (*i*) $2x_1^2 + 4x_2^2 - 6x_3^2 - 6x_1x_2 + 2x_1x_3 + 4x_2x_3$,
(*ii*) $x_1^2 + 6x_2^2 + 14x_3^2 + 4x_1x_2 + 8x_1x_3 - 4x_2x_3$.

A.16. Yes, it is positive definite.

APPENDEX C

Statistical Tables

TABLE 1. Values of e^x and e^{-x}

x	e^x	e^{-x}	x	e^x	e^{-x}
0.00	1.000	1.000	3.00	20.086	0.050
0.10	1.105	0.905	3.10	22.198	0.045
0.20	1.221	0.819	3.20	24.533	0.041
0.30	1.350	0.741	3.30	27.113	0.037
0.40	1.492	0.670	3.40	29.964	0.033
0.50	1.649	0.607	3.50	33.115	0.030
0.60	1.822	0.549	3.60	36.598	0.027
0.70	2.014	0.497	3.70	40.447	0.025
0.80	2.226	0.449	3.80	44.701	0.022
0.90	2.460	0.407	3.90	49.402	0.020
1.00	2.718	0.368	4.00	54.598	0.018
1.10	3.004	0.333	4.10	60.340	0.017
1.20	3.320	0.301	4.20	66.686	0.015
1.30	3.669	0.273	4.30	73.700	0.014
1.40	4.055	0.247	4.40	81.457	0.012
1.50	4.482	0.223	4.50	90.017	0.011
1.60	4.943	0.202	4.60	99.484	0.010
1.70	5.474	0.183	4.70	109.95	0.009
1.80	6.050	0.165	4.80	121.51	0.008
1.90	6.686	0.150	4.90	134.29	0.007
2.00	7.389	0.135	5.00	148.41	0.007
2.10	8.166	0.122	5.10	164.02	0.006
2.20	9.025	0.111	5.20	181.27	0.006
2.30	9.974	0.100	5.30	200.34	0.005
2.40	11.023	0.091	5.40	221.41	0.005
2.50	12.182	0.082	5.50	244.69	0.004
2.60	13.464	0.074	5.60	270.43	0.004
2.70	14.880	0.067	5.70	298.87	0.003
2.80	16.445	0.061	5.80	330.30	0.003
2.90	18.174	0.055	5.90	365.04	0.003
3.00	20.086	0.050	6.00	403.43	0.002

TABLE 2. Area Under Standard Normal Curve

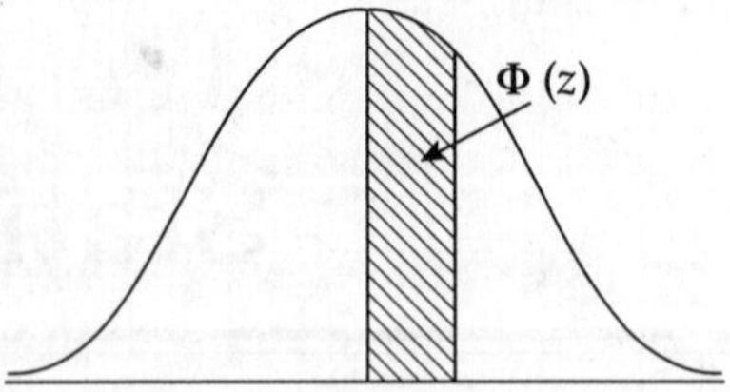

z	0.00	0.01	0.02	0.03	0.04	0.05	0.06	0.07	0.08	0.09
0.0	0.0000	0.0040	0.0080	0.0120	0.0160	0.0199	0.0239	0.0279	0.0319	0.0359
0.1	0.0398	0.0438	0.0478	0.0517	0.0557	0.0596	0.0636	0.0675	0.0714	0.0753
0.2	0.0793	0.0832	0.0871	0.0910	0.0948	0.0987	0.1026	0.1064	0.1103	0.1141
0.3	0.1179	0.1218	0.1255	0.1293	0.1331	0.1368	0.1406	0.1443	0.1480	0.1517
0.4	0.1554	0.1591	0.1628	0.1664	0.1700	0.1736	0.1772	0.1808	0.1844	0.1879
0.5	0.1915	0.1950	0.1985	0.2019	0.2054	0.2088	0.2123	0.2157	0.2190	0.2224
0.6	0.2257	0.2291	0.2324	0.2357	0.2389	0.2422	0.2454	0.2486	0.2517	0.2549
0.7	0.2580	0.2611	0.2642	0.2673	0.2703	0.2734	0.2764	0.2794	0.2823	0.2852
0.8	0.2881	0.2910	0.2939	0.2967	0.2995	0.3023	0.3051	0.3078	0.3106	0.3133
0.9	0.3159	0.3181	0.3212	0.3238	0.3264	0.3289	0.3315	0.3340	0.3365	0.3389
1.0	0.3413	0.3438	0.3461	0.3485	0.3508	0.3531	0.3554	0.3577	0.3599	0.3621
1.1	0.3643	0.3665	0.3686	0.3708	0.3729	0.3749	0.3770	0.3790	0.3810	0.3830
1.2	0.3849	0.3869	0.3888	0.3907	0.3925	0.3944	0.3962	0.3980	0.3997	0.4015
1.3	0.4032	0.4049	0.4066	0.4082	0.4099	0.4115	0.4131	0.4147	0.4162	0.4177
1.4	0.4192	0.4207	0.4222	0.4236	0.4251	0.4265	0.4279	0.4292	0.4306	0.4319
1.5	0.4332	0.4345	0.4357	0.4370	0.4302	0.4394	0.4406	0.4418	0.4429	0.4441
1.6	0.4452	0.4463	0.4474	0.4484	0.4405	0.4505	0.4515	0.4525	0.4535	0.4545
1.7	0.4554	0.4564	0.4573	0.4582	0.4591	0.4599	0.4608	0.4616	0.4625	0.4633
1.8	0.4641	0.4649	0.4656	0.4664	0.4671	0.4678	0.4686	0.4693	0.4699	0.4706
1.9	0.4713	0.4719	0.4726	0.4732	0.4738	0.4744	0.4750	0.4756	0.4761	0.4767
2.0	0.4772	0.4778	0.4783	0.4788	0.4793	0.4798	0.4803	0.4808	0.4812	0.4817
2.1	0.4821	0.4826	0.4830	0.4834	0.4838	0.4842	0.4846	0.4850	0.4852	0.4857
2.2	0.4861	0.4864	0.4868	0.4871	0.4875	0.4878	0.4881	0.4884	0.4887	0.4890
2.3	0.4893	0.4896	0.4898	0.4901	0.4904	0.4906	0.4909	0.4911	0.4913	0.4916
2.4	0.4918	0.4920	0.4922	0.4925	0.4927	0.4926	0.4931	0.4932	0.4934	0.4936
2.5	0.4938	0.4940	0.4941	0.4943	0.4945	0.4946	0.4948	0.4949	0.4951	0.4952
2.6	0.4953	0.4955	0.4956	0.4957	0.4959	0.4960	0.4961	0.4962	0.4963	0.4964
2.7	0.4965	0.4966	0.4967	0.4968	0.4969	0.4970	0.4971	0.4972	0.4973	0.4974
2.8	0.4974	0.4975	0.4976	0.4977	0.4977	0.4978	0.4979	0.4979	0.4980	0.4981
2.9	0.4981	0.4982	0.4982	0.4983	0.4984	0.4985	0.4985	0.4985	0.4986	0.4986
3.0	0.4987	0.4987	0.4987	0.4988	0.4988	0.4989	0.4989	0.4989	0.4990	0.4990

An entry in the table is the proportion under the entire curve which is between $z = 0$ and a positive value of z. Areas for negative values of z are obtained by symmetry :

$$P(Z \leq 1.35) = 0.5 + \Phi(1.35) = 0.5 + 0.4115 = 0.9115$$

$$P(Z \leq -1.35) = 0.5 - \Phi(1.35) = 0.5 - 0.4115 = 0.0885.$$

TABLE 3. Single-payment Compound Amount Factor (CAF)

Number of years (n)	Annual interest rates						
	3%	4%	5%	6%	7%	8%	10%
1	1.030	1.040	1.050	1.060	1.070	1.080	1.100
2	1.061	1.082	1.103	1.124	1.145	1.166	1.210
3	1.093	1.125	1.158	1.191	1.225	1.260	1.331
4	1.126	1.170	1.216	1.262	1.311	1.360	1.464
5	1.159	1.216	1.276	1.338	1.403	1.469	1.661
6	1.194	1.265	1.340	1.419	1.501	1.587	1.774
7	1.230	1.316	1.407	1.504	1.606	1.714	1.949
8	1.267	1.369	1.477	1.594	1.718	1.858	2.144
9	1.305	1.423	1.551	1.689	1.838	1.999	2.358
10	1.344	1.480	1.629	1.791	1.967	2.159	2.694
11	1.384	1.539	1.710	1.898	2.105	2.332	2.853
12	1.426	1.601	1.796	2.012	2.252	2.518	3.138
13	1.469	1.665	1.886	2.133	2.410	2.720	3.452
14	1.513	1.732	1.980	2.261	2.579	2.937	3.797
15	1.558	1.801	2.079	2.397	2.759	3.172	4.177
16	1.605	1.874	2.183	2.540	2.952	3.426	4.595
17	1.653	1.948	2.292	2.693	3.159	3.700	5.054
18	1.702	2.026	2.407	2.854	3.380	3.996	5.500
19	1.754	2.107	2.527	3.026	3.617	4.316	6.116
20	1.806	2.191	2.653	3.207	3.870	4.661	6.727
21	1.860	2.279	2.786	3.400	4.141	5.034	7.400
22	1.916	2.370	2.925	3.604	4.430	5.437	8.140
23	1.974	2.465	3.072	3.820	4.741	5.871	8.594
24	2.033	2.563	3.225	4.049	5.072	6.341	9.850
25	2.094	2.666	3.386	4.292	5.427	6.348	10.835

TABLE 4. Single-payment Present Worth Factor (PWF)

Number of years (n)	Annual interest rates						
	3%	4%	5%	6%	7%	8%	10%
1	0.9709	0.9615	0.9524	0.9434	0.9346	0.9259	0.9090
2	0.9420	0.9246	0.9070	0.8900	0.8734	0.8573	0.8264
3	0.9151	0.8890	0.8638	0.8396	0.8163	0.7938	0.7513
4	0.8885	0.8548	0.8227	0.7921	0.7629	0.7350	0.6830
5	0.8626	0.8219	0.7835	0.7473	0.7130	0.6806	0.6209
6	0.8375	0.7903	0.7462	0.7050	0.6663	0.6302	0.5645
7	0.8131	0.7599	0.7107	0.6651	0.6227	0.5834	0.5132
8	0.7894	0.7307	0.6768	0.6274	0.5820	0.5403	0.4665
9	0.7664	0.7026	0.6446	0.5919	0.5439	0.5002	0.4241
10	0.7441	0.6756	0.6139	0.5584	0.5083	0.4632	0.3855
11	0.7224	0.6496	0.5847	0.5268	0.4751	0.4289	0.3505
12	0.7016	0.6246	0.5568	0.4970	0.4440	0.3971	0.3186
13	0.6810	0.6006	0.5303	0.4688	0.4149	0.3677	0.2897
14	0.6611	0.5775	0.5051	0.4423	0.3878	0.3405	0.2633
15	0.6419	0.5553	0.4810	0.4173	0.3624	0.3152	0.2394
16	0.6232	0.5339	0.4581	0.3936	0.3387	0.2919	0.2176
17	0.6050	0.5133	0.4363	0.3714	0.3166	0.2703	0.1978
18	0.5874	0.4936	0.4155	0.3503	0.2959	0.2502	0.1799
19	0.5703	0.4746	0.3957	0.3305	0.2765	0.2317	0.1635
20	0.5537	0.4564	0.3769	0.3118	0.2584	0.2145	0.1486
21	0.5375	0.4388	0.3589	0.2942	0.2515	0.1987	0.1351
22	0.5219	0.4220	0.3418	0.2775	0.2257	0.1839	0.1228
23	0.5067	0.4057	0.3256	0.2618	0.2109	0.1703	0.1117
24	0.4919	0.3901	0.3101	0.2470	0.1971	0.1577	0.1015
25	0.4776	0.3751	0.2953	0.2330	0.1842	0.1460	0.0923

TABLE 5. Random Numbers

27767	43584	85301	88977	29490	69714	64015	64874	32444	48277
13025	14338	54066	15243	47724	66733	74108	88222	88570	74015
80217	36292	98525	24335	24432	24896	62880	87873	95160	59221
10875	62004	93391	61105	67411	06368	11784	12102	80580	41867
54127	57326	26629	19087	24472	88779	17944	05600	60478	03343
60311	42824	37301	42678	45990	43242	66067	42792	95043	52680
49739	71484	92003	98086	76668	73209	54244	91030	45547	70818
78626	51584	16453	94614	39014	97066	30945	57589	31732	57260
66692	13986	99837	05582	81232	44987	69170	37403	86995	90307
44071	28091	07362	97703	76447	42537	08345	88975	35840	85771
59820	96163	78851	16499	87064	13075	73035	41207	74699	09310
25704	91035	26313	77463	5538	72681	47431	43905	31048	56699
23304	90314	78438	66276	18396	73538	43277	58874	11466	16082
25852	58905	55018	56374	35824	71708	30540	27886	61732	75454
46780	56487	75211	10271	36633	68424	17374	52003	70707	70214
69849	96169	87195	46092	26787	60939	59202	11973	02902	33250
47670	07654	30342	40277	11049	72049	83012	09832	25571	77628
94304	71803	73465	0981	58869	35220	09504	96412	90193	79568
09105	59987	21437	36786	49226	77837	98524	97831	65704	09514
64281	61826	18555	64937	64654	25843	41145	42820	14924	39650
66847	70495	32350	02985	01755	14750	48968	38603	70312	05682
72461	33230	21529	53424	72877	17334	39283	04149	09850	64618
21032	91050	13058	16218	06554	07850	73950	79552	24781	89683
95362	67011	06651	16136	57216	39618	49856	99326	40902	05069
49712	97380	10404	55452	09971	59481	37006	22186	72682	07385
58275	61764	97586	54716	61459	21647	87417	17198	21443	41080
89514	11788	68224	23417	46376	25366	94746	49580	01176	28838
15472	50669	48139	36732	26825	05511	12459	91314	80482	71944
12120	86124	51247	44302	87112	21476	14713	71181	13177	55292
95294	00556	70481	06905	21785	41101	49386	5440	23604	23554
66986	34099	74474	20740	47458	64809	06312	88940	15995	69321
80620	51790	11436	38072	40405	68032	60942	00307	11897	92674
55411	85667	77535	9992	71209	92061	92329	98932	78284	46347
95083	06783	28102	57816	85561	29671	77936	6374	31384	51924
90726	57166	98884	08583	95889	57067	38101	77756	11657	13897
68984	83620	89747	98882	92613	89719	39641	69457	91339	22502
36421	16489	18059	51061	67667	60631	8404	40455	99396	63680
92638	40333	67054	16067	24700	71594	47468	03577	57649	63266
21036	82808	77501	97427	76479	68562	43321	31370	28977	23896
13173	33365	41468	85149	49554	17994	91178	10174	29420	90438
86716	38746	94559	37559	49678	53119	98189	81851	29651	84215
92581	02262	41615	70360	64114	58660	96717	54244	10701	41393
12470	56500	50273	93113	41794	86861	39448	93136	25722	08564
01016	00857	41396	80504	90670	08289	58137	17829	22751	36518
30030	60726	25807	24260	71529	78920	47648	13885	70669	93406
50259	46345	06170	97965	88302	98041	11947	56203	19324	20504
73659	76145	60808	5444	74412	8105	69181	96845	38525	11600
46874	37088	80940	44893	10408	36222	14004	2313	69249	05747
60883	52109	19516	90120	46759	71643	62342	07589	08899	05895

APPENDEX D

Select References

Ackoff, R.L. and M.W. Sasieni (1991), *Fundamentals of Operations Research*, New York: Wiley.

Ahuja, R.K., T. Magnati and J. Orlin (1993), *Network Flows: Theory, Algorithms and Applications*, Upper Saddle River, N.J. : Prentice-Hall.

Anderson, David R, Sweeney, Dennis J. and William, Thomas, A. (2005), *An Introduction to Management Science*, Thomson-South Western.

Asmussen, S. (2003), *Applied Probability and Queues, 2/e*, New York: Springer.

Asmussen, S. and P.N. Glynn (2007), *Stochastic Simulation*, New York: Springer.

Avriel, M. and A.C. Williams (1970), "Complementary Geometric Programming", *SIAM J. Appl. Math.*, pp. 125-141.

Axsäter, S. (2006), *Inventory Control, 2/e*, New York: Springer.

Balas, Egon (1960), *Optimal Transportation Plan for Firewood in Romania*, Pittsburg, USA.

Balas, Egon (1995), *Integer Programming and Combinatorial Optimization*, Springer.

Bazaraa, M., J.J. Arvis and H. Sherali (1990), *Linear Programming and Network Flows, 2/e*, New York: Wiley.

Bazaraa, M., H. Sherali and C. Shetty (1993), *Non-linear Programming: Theory and Algorithms, 2/e*, New York: Wiley.

Beale, E.M.L. (1959), "On Quadratic Programming", *Naval Research Logistics Quarterly*, Vol. 5, pp. 227-243.

Bertsekas, D. (1987), *Dynamic Programming: Deterministic and Stochastic Models*, Upper Saddle River, N.J.: Prentice-Hall.

Bradley, S., A. Hax and T. Magnati (1977), *Applied Mathematical Programming*, Reading Mass: Addison-Wesley.

Churchman, C.W., Ackoff R.L. and E.L. Arnoff (1957), *Introduction to Operations Research*, New York: Wiley.

Cooper, L. and D. Steinberg (1974), *Methods and Applications of Linear Programming*, Philadelphia W.B. Saunders.

Dano, S. (1975), *Non-Linear Programming and Dynamic Programming*, New York: Springer.

Dantzig, G.B. and M.N. Thapa (2003), *Linear Programming and Extensions*, New York: Springer.

Dantzig, G.B. and Wolfe, P. (1960), "A Decomposition Principle for Linear Programming", *Operations Research*, Vol. 8, No., 1 (pp. 101-111).

Denardo, E. (1982), *Dynamic Programming Theory and Applications*, Upper Saddle River, N.J.: Prentice-Hall.

Duffin, R.J., E.L. Peterson and C. Zener (1967), *Geometric Programming (Theory and Applications)*, New York: Wiley.

Evans, J. (1991), *Creative Thinking in the Decision and Management sciences*, Ohio : South-Western Publishing.

Evans, J.R. and E. Minieka (1992), *Optimization Algorithms for Networks and Graphs, 2/e*, New York: Marcel Dekker.

Gass, S.I. and A.A. Assad (2005), *An Annotated Timeline of Operations Research*, New York: Springer.

Gomory, R.E. (1963), "An Algorithm for Integer Solutions to Linear Programmes", *Recent Advances in Mathematical Programming* (ed. R. Graves and P. Wolfe), New York: Mc-Graw Hill.

Gupta, P.K. and Man Mohan (2003), *Linear Programming and Theory of Games, 9/e*, New Delhi: Sultan Chand & Sons.

Gupta, S.P. and P.K. Gupta (2022), *Business Statistics and Business Mathematics*, New Delhi: Sultan Chand & Sons.

Gupta Shiv K. and J.H. Cozzolino (1974), *Fundamentals of Operations Research for Management: An Introduction to Quantitative Methods*, San Francisco: Holden Day.

Hadley, G. (1962), *Linear Programming,* Massachusetts: Addison-Wesley.

Hadley, G. and T. Whitin (1968), *Analysis of Inventory Systems,* Englewood Cliffs, N.J.: Prentice-Hall.

Hertz, David B. (1964), "Risk Analysis in Capital Investment", 42 (Jan.-Feb.), pp. 95-106.

Hiller, F.S. and G.J. Lieberman (2012), *Introduction to Operations Research, 9/e,* New Delhi: Tata Mc-Graw Hill.

Hiller, F.S. and M.S. Hiller (2008), *Introduction to Management Science, 3/e,* Irwin: Mc-Graw Hill.

Hooker, J.N. (1986), "Karmakar's Linear Programming Algorithm", *Inter faces,* Vol. 16 (No. 4), pp. 75-90.

Hooker, J.N. (2007), *Integrated Methods of Optimization,* New York: Springer.

Hughe, A.J. and D.E. Grawoig (1972), *Linear Programming: An Emphasis on Decision Making,* New York: Collier-MacMillan.

Ignizio, J.P. and T.M. Cavalier (1994), *Linear Programming,* Upper Saddle River, N.J.: Prentice-Hall.

Kanti Swarup (1965), "Linear Fractional Functional Programming", *Operations Research,* Vol. 12, pp. 1029-1036.

Kunzi, H.P., W. Krelle and W. Oettli (1966), *Non-linear Programming,* Massachusetts : Blaisdell Publishing Co.

Law, A. and Keltron, W. (1991), *Simulation Modeling & Analysis, 2/e,* New York: Mc-Graw Hill.

Lee, S.M. (1972), *Goal Programming for Decision Analysis,* Philadelphia : Auerback Publishers Inc.

Lew, and H. Mauch (2007), *Dynamic Programming: A Computational Tool,* New York: Springer.

Luenberger, D. and Y. Ye (2008), *Linear and Non-linear Programming, 3/e,* New York: Springer.

Mendelson, E. (2005), *Introducing Game Theory and Applications,* CRC Press: Chapman and Hall.

Meyerson, R. (1991), *Game Theory: Analysis of Conflicts,* Cambridge, Mass: Harvard University Press.

Midgley, Ochoe-Avian and **Alejandro** (2004), *Community Operational Research,* Springer.

Miller, D.W. and M.K. Starr (1973), *Executive Decisions and Operations Research, 2/e,* N.J.: Prentice-Hall.

Mischael, P. Johnson (2011), *Community Based Operations Research,* Springer.

Moder, J. and C. Phillips (1970), *Project Management with CPM and PERT, 2/e,* New York: Van-Nostrand.

Morse, P.M. (1958), *Queues, Inventories and Maintenance,* New York: Wiley.

Nering, E. and A. Tucker (1992), *Linear Programming and Related Problems,* Boston: Academic Press.

Radrin, D. (1998), *Optimization in Operations Research,* N.J.: Prentice-Hall.

Prekopa, A., ed. (1975), *Progress in Operations Research,* Amsterdam: North-Holland.

Rao, S.S. (1978), *Optimization — Theory and Applications,* New Delhi: Wiley Eastern.

Ross, S. (1990), *A Course in Simulation,* New York: Macmillan.

Saaty, T.L. (1983), *Elements of Queuing Theory with Applications,* New York: Dover.

Saaty, T.L. (1994), *Fundamentals of Decision Making,* Pittsburgh: RWS Publications.

Salkin, H. and K. Mathur (1989), *Foundations of Integer Programming,* New York: North Holland.

Silver, E. and R. Peterson (1985), *Decision Systems for Inventory management and Production Planning, 2/e*: Wiley.

Smith, D. (1991), *Dynamic Programming — A Practical Introduction,* London: Ellis Horwood.

Sniedovich, M. (1991), *Dynamic Programming,* New York: Marcel Dekker.

Starr, M.K. and D.W. Miller (1974), *Inventory Control — Theory and Practice, 2/e,* N.J.: Prentice-Hall.

Taha, H.A. (2017), *Operations Research — An Introduction, 10/e,* Pearson Education, Prentice-Hall.

Takacs, L. (1972), *Introduction to the Theory of Queues,* Oxford: Oxford University Press.

Vajda, S. (1960), *An Introduction to Linear Programming and Theory of Games,* New York: Wiley.

Vanderbei, R.J. (2008), *Linear Programming: Foundations and Extensions, 3/e,* New York: Springer,

Vohra, N.D. (2010), *Quantitative Techniques in Management, 4/e,* New Delhi: Tata Mc-Graw Hill.

Wagner, H.M. (1974), *Principles of Operations Research,* New Delhi: Prentice-Hall of India.

Waters, C. (1992), *Inventory Control and Management,* New York: Wiley.

Wolfe, P. (1959), "The Simplex Method for Quadratic Programming", *Econometrica,* Vol. 27, pp. 382-398.

William, H. (1990), *Model Building in Mathematical Programming, 3/e,* New York: Wiley.

Index

A

ABC analysis, 767
Acceptance sampling, 1012
Accounting prices, 69
Ackoff, R.L., 3
Action phase, in OR, 9
Activity in *PERT/CPM*, 388
    concurrent, 389
    dummy, 388
    predecessor, 388
    successor, 388
Addition theorem of probability, 535
Advertising problem, 23
Allocation problem, 463
Agriculture problem, 22
All integer programming problem, 339
    cutting plane algorithm, 353
An interior point algorithm, 288
Analytical methods in OR, 6
Annuity, 929
    amount of, 929
    perpetual, 929
    present value of, 929
Arithmetic method, for games, 622
Arnoff, E.L., 3
Arrival pattern, 631
    rate, mean, 635
Artificial variables, 74
Assignable variation, 999
Assignment method, 245
Assignment problem, 243, 340
    assumptions in, 244
    a typical, 257
    branch & bound method, 267
    complete enumeration method, 244
    dual of, 265
    Hungarian method for, 245
    maximisation cases in, 255
    model formulation of, 243
    prohibited assignments, in, 257
    simplex method, 245
    special cases in, 255
    transportation method, 245
Associated cost vector, 46
Augmenting path, 512
    algorithm, 512
    method, 511
    optimal solution of, 515
    residual network of, 511
Average, length of a code, 921
    queue length, 639
    waiting time, 639

B

Backward pass calculations, 399
Balas, Egon, 339
Bar chart, 315
Basic feasible solution, 46, 110
    and extreme point, 111
    degenerate, 46, 110
    improved, 46
    optimality conditions for, 110, 119
    optimum, 46, 110
Basic solution, 45, 110
    reduction of a F.S. to, 110
Basic variables, 46
Basis matrix, 46
Basis set, 1055
    standard, 1055
Basis vectors, 46
    replacement of, 116
Bath-tub-shaped failure rate, 710
Beale's algorithm for QPP, 880
    free variable in, 881
Bellman's principle of optimality, 457
Big M method, 76
Big networks, 525
    insights of, 525
    social, 526
    telephone call of, 526
    world wide web of, 526
Binary code, 920
    construction of instantaneous, 921
    digit, 910
Binary symmetric channel (BSC), 918
Bits, 910
Blending problem, 21, 23
Block code, 920
Boarded Hessian matrix, 855
Bounded variables, 275
    simplex method, 275
Branch-and-Bound method, 267, 347
Brand switching model, 552
Break-even analysis, 927, 952
    chart, 953
    point, 953
    safety margin, 953
Buffer stock, 762
    optimum, 765, 778
Bussiness forecasting, 961
Busy period, in queues, 635

C

Canonical form of LPP, 42

Capacity of a system, 633
  finite source, 633
  infinite source, 633
Capital budgeting problem, 341
Cargo load problem, 29
Car replacement problem, 508
Casual forecasting, 975
Caterer problem, 344
Central tendency and dispersion, 540
Chance-constrained programming, 893
  variation, 999
Channel, 918
  binary symmetric, 918
  capacity of, 918
  information processed by, 918
  matrix non-singular, 918
  noiseless, 919
  parallel, 633
  service, 632
Chinese postman problem, 490
Churchman, C.W., 3
Closed half space, 1057
Code, 920
  alphabet, 920
  binary, 920
  block, 920
  comma, 920
  efficiency of, 921
  instantaneous, 921
  non-singular, 920
  prefix, 921
  redundancy of, 921
  uniquely decodable, 920
Column vector, 1049
Comma code, 920
Communication system, 909
Community, OR, 2
Competitive situation, 603
Competitors against passivity, 627
Complementary geometric algorithm, 905
  programming, 904
Complementary slackness, 876
  conditions, 877
  optimum solution property, 148
  solution property, 148
Components of time series, 962
Compound amount factor, 690
  interest, 927
  value, 927
Concave function, 1063
Conditional, probability, 535
  entropies, 914, 915
  opportunity loss, 576
Constant demand rate, 720
Constrained maxima & minima, 850
  all equality constraints, 850
  necessary conditions for, 851
  with inequality constraints, 858
Constraints, of LPP, 15, 41
Consumer panel survey, 979
Continuous, random variable, 537
  compounding, 928
Control charts, 1000
  attributes, 1001
  variables, 1001
Control, process, 1000
  product, 1000
Controllable variation, 999
Conversion period, 927
  accumulation factor, 928
Convex combination of vectors, 1059
Convex cone, 1060
Convex function, 1063
Convex, hull, 1060
  polyhedron, 1060
  programming, 885
Convex set, 1058
  extreme point of, 1058
Cost models, in queuing, 683
  optimum service rate, 683
Cost-slope, 425
Cost-time trade off, 424
Crash cost, 425
  time, 425
Crashing of activities, 426
Critical moves, 611
  activity, 397
  path, 396, 398
Critical path, analysis, 396
  calculations, 398
  procedure for, 401
  versus *PERT*, 396
Crop and livestock problem, 24
Cumulative distribution function, 537
Cut capacity in network, 516
Cutting plane method, 347, 353
Cyclical movements, 962

D

Dangling, 389
Dantzig, George, 1
Decision, 575
  alternatives, 575
  pay-off, 575
  under certainty, 578
  under conflict, 603
  under risk, 585
  under uncertainty, 575, 578
Decision-making, problem, 575
  cases in, 1031
  process, 578
  with utilities, 602
Decision tree, 594
  and investment analysis, 949
  chance node in, 594
  decision node in, 594
  rolling-back technique, 594
Decoder, 909
Decomposition method, 287

Degeneracy in TP, 194
Degenerate solution, 110
Delphi method, 979
Demand rate, constant, 720
Dependency relationship, 390
Destination in TP, 191
Determinant, 1050
  properties of, 1051
Deterministic models, 5
Diagonal elements, 1049
Dijkstra's shortest path algorithm, 500
Discrete random variable, 536
Distribution, Poisson, 541
  exponential, 546, 635
  functions, 537
  gamma, 548
  geometric, 542
  normal, 543
  of service times, 635
  rectangular, 548
Dominance property, 613
  modified, 613
Dominant strategy, 613
Double sampling plan, 1015
Dual problem, 133
  of assignment problem, 265
  properties of, 139
Dual simplex method, 155
Dual variables, 133
Duality, 133
  and simplex method, 139
  economic interpretation of, 148, 194
  existence property, 148
  fundamental theorem of, 148
  imputed price, 150
  in assignment problem, 265
  in LPP, 133
  in TP, 194
  shadow prices in, 150
  strong property, 148
Due date procedure, 310
Dummy, activity in *PERT/CPM*, 388
  constraints, 389
  destination in TP, 211
  source in TP, 211
Dynamic programming problem, 457
  a case in, 1042
  and inventory control, 800
  backward computational procedure, 461
  by linear programming, 484
  characteristics of, 457
  formulation of, 460
  forward computational procedure, 460
  recursive equation approach to, 461
  solving of, 461

E

Earliest time in *PERT/CPM*, 398
  finish (EF), 399
  starting (ES), 399

Economic, lot size, 720
Efficiently of a code, 922
Egon, Balas, 339
Elapsed time, 303
Elementary transformations, 1052
  computation of inverse by, 1052
Eligible activity set (EAS), 443
Employment smoothening problem, 467
EMV criterion, 585
Encoder, 909
Encoding, 920
  procedure, Shannon-Fano, 922
Entropy, 911
  as a measure of uncertainty, 913
  conditional, 915
  joint, 914
  marginal, 914
EOL criterion, 586
EOQ model, 720
  and quantity discount, 729
  with constant demand rate, 720
  with finite replenishment, 740, 747
  with fixed order cycle, 747
  with limitations, 753
  with no shortages, 720
  with price breaks, 734
  with planned shortage, 744
  with unequal production runs, 723
  with uniform demand, 720
  with variable order cycle time, 745
EPPI, 586
Equality of vectors, 1053
Equilibrium point of a pay-off matrix, 606
Equipment renewal policy, 708
Erlang distribution, 548
  characteristics of, 678
  of service time, 678
Estimated regression equation, 975
Estimation of reliability, 711
Euclidean space, 1054
Event(s), 532
  burst, 389
  complementary, 534
  elementary, 532
  equality of, 533
  equally likely, 532
  impossible, 534
  independence of, 535
  intersection of, 534
  merge, 389
  mutually exclusive, 534
  PERT/CPM, in, 388
  union of, 534
EVPI, 586
Existence property in duality, 148
  solution in TP of, 192
Expectation of a random variable, 539
Expected information, 911

Expected value, 539
monetary value, 585
of perfect information, 586
opportunity loss, 586
pay off, 575
Experience curve, 982
Exponential distribution, 546
in queuing problems, 635
Extreme point, of a convex set, 1058

Failure rates, 710
bath-tub shaped, 710
instantaneous, 710
Feasible solution, 45
basic (b.f.s.), 46
reduction to b.f.s., 110
Fixed-charge problem, 342
order quantity policy, 717
Float, in networks, 400
free, 400
independent, 400
interfering, 400
total, 400
Flow shop scheduling, 312
k-machines, 321
three machines, 320
two machines, 313
FNSD analysis, 771
Forecasting, 961
casual methods, 975
estimated regression equation, 975
moving average methods, 963
qualitative, 961
quantitative, 961
smoothing methods, 962
using multiple regression, 976
using simple regression, 975
weighted averages method, 965
Forward pass calculations, 398
Fractional cut, 353
method for all IPP, 353
Fractional programming, linear, 290
Free variable, 881
Fulkerson's rule, 391
Function, convex, 1063
concave, 1063
objective, 41
Fundamental theorem of duality, 148
Funding problem, 32

Game, 604
a case in, 1034
against passivity, 627
approximate solution of, 624
arithmetic method for, 622
calculus method, 609
equal gain method, 607
graphical solution of, 610
matrix oddment method for, 622
maximin value of, 606
minimax value of, 606
mixed strategies, in, 607
of strategy, 604
payoff matrix of, 604
player in, 604
pure strategies, in, 604
rectangular, 610
saddle point of, 606
two-person zero-sum, 605
value of, 604
without saddle point, 607
Gamma distribution, 548
Gantt chart, 304
General LPP, 41
constraints of, 41
feasible solution to, 45
non-negative restrictions of, 41
objective function of, 41
optimum solution to, 45
General NLPP, 844
Kuhn-Tucker conditions for, 858
necessary conditions for, 858
sufficient conditions for, 859
with m ($< n$) constraints, 860
with one constraint, 853
General structure of TP, 192
Generation of random numbers, 806
Geometric-arithmetic mean inequality, 895
Geometric distribution, 542
Geometric programming, 895
algorithm for complimentary, 905
complimentary, 904
constrained problem, 901
pre-dual function in, 897
primal-dual problems, 897
unconstrained problem, 896
weight functions in, 896
George, Dantzig, 1
GERT, 453
Global maximum, 845
minima, 1064
Goal programming, 363
cases in, 1024
categorization of, 364
formulation of, 365
graphical method, 371
limitations of, 385
simplex method for, 375
Gomorian constraints, 354
construction of, 354
Gomory's, all-IPP method, 353
Graphical solution, of LPP, 49
iso-profit approach, 51
of 2 × n and m × 2 games, 607, 610
no solution, 55
unbounded solution, 54

Group replacement policy, 698

Hazard rate, 710
Hertz, David B., 951
Hessian matrix, 848, 849
HML analysis, 771
Holding cost, 718
Hungarian assignment method, 245
Hurwicz criterion, 579
Hyperplane, 1057
 parallel, 1057
 separating, 1062
 supporting, 1061
 unit normals, 1057
Hyper sphere, 1057

Identity matrix, 1049
Idle time, 303
Improved b.f.s., 46, 116
Imputed price, 150
Independence of events, 535
Individual replacement policy, 698
Infeasible solution, 45
Information, 909
 a measure of, 910
 axiomatic approach to, 910
 expected, 911
 mutual, 917
 processed by a channel, 918
 source, 909
 theory, 909
Initial b.f.s. to a TP, 197
 least cost method, 198
 north-west-corner rule, 197
 Vogel's approximation method, 199
Inner product of vectors, 1053
Input allocation problem, 855
Input pattern, in queues, 631
 balking, 633
 bulk, 631
 customer behaviour, 633
 jockey, 633
 renege, 633
Inspection, 1012
Instantaneous, codes, 921
 construction of, 921
 failure rates, 710
Integer programming problem, 339
 all integer, 339
 assignment problem, 340
 branch-and-bound method of, 347
 capital budgeting problem, 341
 case in, 1023
 caterer problem, 344
 cutting plane, algorithm for all, 353
 fixed charge problem, 342
 for mixed, 339
 formulation of integer programming problems, 339
 Knapsack problem, 341
 solution methods, 347
Inter-arrival time, 632
 distribution of, 635
Interior point, concept of, 288
 Karmakar algorithm, 289
Inventory, 715
 anticipation, 715
 buffer or safety stock in, 715, 762
 control, 717, 718
 costs associated with, 718
 decisions, 717
 decoupling, 715, 716
 demand, 718
 holding cost in, 718
 lead time in, 718
 lot size, 716
 management, 715, 777
 movement, 715
 objectives, 717
 ordering cost in, 718
 pipeline, 715
 production cost, 718
 reorder level in, 719
 revenue cost in, 718
 salvage cost in, 718
 seasonal, 715
 set-up cost in, 718
 shortage cost in, 718
 stock replenishment, 719
 time horizon, 719
 transit, 715
Inventory control, systems, 777
 and dynamic programming, 800
 cases in, 1043
 continuous reveiw, 719
 factors affecting, 718
 P-system, 780
 periodic review, 719
 Q-system, 777
 reserve bin, 777
 two bin, 777
 with known demand, 720
 with uncertain demand, 762
Inventory control techniques, selective, 766
 ABC analysis, 767
 criticality (VED), 771
 FNSD analysis, 771
 HML analysis, 771
 SDE analysis, 771
 XYZ analysis, 771
Inventory decisions, 717
 deterministic, 715
 economic lot size (EOQ), 720
 marginal analysis, 787

multi-item-deterministic, 753
one period, 786
probabilistic, 783, 787
simulation of, 814
with limitations on floor space, 753
with limitations on investment, 757
with no shortage, 720
with planned shortage, 744
with price breaks, 734
with set-up cost, 799
with uncertain demand, 762
with unequal production runs, 723
without set-up cost, 792
Inverse, of a square matrix, 1051
Investment analysis, 927
a case in 1045
and decision trees, 949
methods of, 932
under risk, 939
under uncertainty, 939
Investment, and budgeting in simulation, 818, 819
with limitations on, 757
Irregular variations, 962
Iterative method in games, 624

Jockey in queues, 633
Johnson, Michael P., 2
Joint and conditional entropies, 914
relations between, 914
Judgement phase, in OR, 8

Karmarkar algorithm, 289
interior point algorithm, 289
Key column, 68
number, 69
row, 69
Knapsack problem, 341
Kuhn-Tucker conditions, 858
for general NLPP, 860
sufficiency of, 859
verification of, 870
with non-negative constraints, 873

Lagrange Multipliers, 850
Laplace criterion, 579
Largrangian function, 850
Latest time in *PERT/CPM*, 398
finishing (LF), 398
starting (LS), 398
Laws of probability, 535
of conditional probability, 535
Lead time, in inventory, 718, 762
administrative, 718
delivery, 718
Learning curve, 981
advantages, 996
and regression, 984
costs affected by, 996
equation, 982
limitations, 997
marginal time, 983
ratio, 984
specific, 983
stage, 981
table, 985
uses, 996
Length of a vector, 1053
Limitations of inventories, 753
of linear programming, 17
of operations research, 11
of the floor space, 753
on capital investment, 757
Line of balance, 454
Line segment in R^n, 1057
Linear algebra, 1049
Linear combination of vectors, 1054
Linear dependence, of vectors, 1054
independence of vectors, 1054
Linear equations, 1052
homogeneous, 1052
Linear fractional programming, 290
application of, 294
Kanti Swarup's algorithm for, 290
Linear programming problem, 15
additivity, 16
associated cost vector of, 46
assumptions, 16
basic feasible solution of (b.f.s.), 46
basic solution, 45
canonical form of, 42
cases in, 1017
certainty, 16
condition of optimability, 119
constraints of, 15
decision variables, 15
degenerate b.f.s., 46
divisibility, 16
dual problem, 133
feasible solution, 45
fundamental theorem of, 116
general, 41
graphical solution of, 49
limitations of, 17
improved b.f.s. of, 46
infeasible solution, 45
model formulation of, 17
objective function of, 15, 41
optimal solution of, 45
proportionality, 16
revised simplex method for, 120
simplex algorithm for, 68
simplex method for, 66
solution by dynamic programming, 484

standard form of, 43
three stages of, 16
two-phase simplex method for, 75
unbounded solution, 46
Link capacity in network, 516
Little's formula, 640
LOB, 454
Local extrema, 1064
minima, 1064
Logical sequencing, 389
Loops, in network, 389
Lower boundary, in games, 611

M

Machine-repairman model, 674
Mahalanobis, 2
Main bin, in inventory, 777
Maintenance, crew scheduling, 332
problems, simulation of, 823
Management science application, a case in, 1046
Manpower schedule problem, 28
Marginal entropies, 914
Markov, analysis, 551
algorithm, 556
imbedded chain, 677
process, 551
Matrices, 1049
addition of, 1050
augmented, 1052
determinant of, 1051
elementary transformation of, 1052
equality of, 1050
identity, of, 1049
inverse of, 1051
non-singular, 1051
-homogeneous, 1052
partitioned, 1050
product of, 1050
transpose of, 1049
Maxima-Minima of in NLPP, 844
necessary conditions for, 845, 848
objective function, 844
sufficient conditions for, 845, 848
Maximal flow problems, 511
Maximax criterion, 578, 580
Maximin criterion, 579, 580
value, 606
Maximin-minimax principle, 605
Maximum, tardiness, 306
Maximum-flow, Minimum-cut method, 516
McClosky, 1
Mean arrival rate, 635
flow time, 306
service rate, 635
tardiness, 306
Mean time between failure (MTBF), 711
Media selection problem, 23
Median, 540
Message, 909
Method of penalties, 76
Method of potentials, 454
Methodology of OR, 6
Methods of investment analysis, 932
average rate of return, 933
discounted cash flow, 933
discounted payback, 933
IRR, 934
NPV, 933
pay-back period, 933
traditional, 932
Midgley, 2
Minimal spanning tree, 494
alogorithm of, 495
application of, 494
pipeline of, 497
problem, 494
Minimax criterion, 579
Minimin criterion, 578
Minimum cost flow problem, 518
transhipment problems of, 521
transportation problems of, 519
Mixed integer programming, 339
cutting plane algorithm, 353
Mixed strategies, 604
Mode, 540
Models in OR, 4
advantages of, 5
analogue, 4
classification of, 4
descriptive, 4
deterministic, 5
dynamic, 5
general solution methods for, 5
heuristic, 5
iconic, 4
normative, 4
physical, 4
predictive, 4
probabilistic, 5
simulation, 5
specific, 5
static, 5
symbolic, 5
Modified dominance property, 613
MODI method, 210
Monte Carlo, simulation, 809
Moore procedure, 311
MOST, in *PERT*, 452
Most likely time in *PERT*, 408
Moving averages, 963
weighted, 965
Multi-item deterministic problems, 753
Multi-variable unconstrained optimization, 848
Mutual information, 917
expected, 917

Nature and characteristics of OR, 3
Net evaluations, in simplex tableau, 68
  in *uv* method, 211
Network, 387
  activity in, 388
  closed loop in, 389
  components of, 388
  connected, 492
  critical path in, 396
  dangling in, 389
  definitions of, 491
  directed, 492
  dummy activity in, 388
  event in, 388
  floats in, 400
  formulation as LPP, 435
  links of, 491
  looping in, 389
  node in, 388, 491
  notation, 491
  problem, 31
  simulation of, 827
Network construction, rules of, 390
  numbering the events in, 391
Network flow, 523
  chinese postman problem of, 490
  features of, 491
  Königsberg bridges, 489
  problems, 489, 491
  sensor of, 523
Network scheduling, 396
  cases in, 1025
  cost slope in, 425
  cost-time trade-off in, 424
  crash costs, time in, 425
  crashing in, 425
  normal cost, time in, 425
  precedence planning in, 454
  problem, 827
  project's costs in, 424
  resource analysis in, 439
  time-cost optimization algorithm, 425
Nodes, 388
  end, 390
  starting, 390
Noiseless channel, 919
Non-linear programming, 841
  formulation of, 841
  general NLPP, 844
  graphical solution of, 869
  necessary conditions for, 851
Non-singular, code, 920
  channel matrix, 918
  matrix, 1051
Normal distribution, 543
  area property of, 543
  probability density function, 543
North-west corner rule, 197
NPV method, 933
  expected value of, 941
  normally distributed of, 942
  statistical distribution, 941
  variance of, 941
Null matrix, 1049
Numerical methods in OR, 6

Objective, in OR, 3
  in inventory control, 717
OC-curve, 1014
Oddments, 623
One-potato, two-potato problem, 843
Open half spaces, 1057
Operations scheduling, 303
  problem, 826
Opportunity loss, 576
  table, 576
Optimal solution, of a LPP, 45
  of a DPP, 457, 461
  of a TP, 207
  of an assignment problem, 246
Optimal sub-division problem, 479
Optimistic time in *PERT*, 408
Optimum, lot size, 721
  simplex tableau, 69
  strategies, 604, 606
Operations research, 1
  and business/industry, 9
  and community, 2
  and decision making, 10
  and scientific quantification, 10
  development and growth in, 1
  methodology of, 6
  models in, 4
  nature and characteristic features of, 3
  opportunity costs, 211
Order cycle, in inventory, 719
Ordered activity set (OAS), 443
Ordering cost, 718
Outcomes, of a random experiment, 531
  equally likely, 532
  favourable, 533
  mutually exclusive, 533

P-system, 780
Parametric linear programming, 281
  changes in **b**, 282
  changes in **c**, 282
Partitioned matrices, 1050
  addition of, 1050
  product of, 1050
Pay off, 575
  expected, 579
  matrix, 605
  table, 576

Penalty method, 76
Personnel scheduling, 332
PERT, 408
  MOST in, 452
  most likely time in, 408
  optimistic time in, 408
  pessimistic time in, 408
  probability considerations in, 409
  procedure for, 409
  time-cost relationship in, 424
  versus *CPM*, 434
Phases of OR, 8
  action, 9
  judgement, 8
  research, 8
Phase technique, in queues, 677
Piecewise linear approximations, 886
  linear segments, 886
Pivot element, 69
Poisson distribution, 541, 632
  of arrivals, 635
Pollaczek-Khintchine (P-K) formula, 681
Polytope, 1057
Portfolio selection problem, 843
Post-optimal analysis, 163
  applications of, 181
Posynomial, 895
  term, 905
Precedence planning, 454
  line of balance, 454
Pre-dual function, 897
Prefix in codes, 921
Present value, 928
Present-worth factor, 690
Price-breaks, in inventory, 734
Primal, problem, 133
  variables, 135
primal-dual pair, formulation of, 134, 135
  standard form of, 135
  symmetric form of, 136
  unsymmetric form of, 136
Principle of optimality, 457
Probability, 531
  conditional, 535
  considerations in *PERT*, 409
  definition of, 533
  density function, 538
  laws of, 535
Probabilistic inventory problems, 777
  with instantaneous demand, 796
  with set-up cost, 799
  without set-up cost, 792
Probability distribution, 541
  exponential, 546
  gamma, 548
  geometric, 542
  normal, 543
  Poisson, 541
Problems of complex sequencing, 334
Process control, 1000
Processing of 2 jobs through *m* machines, 328
Product, allocation problem, 20
  mix problem, 18, 19
Production problem, with no shortages, 740
  with shortages, 745
Programming, 15
Project controlling, 439
  management, 439
  planning, 439
  scheduling of, 439
  updating of, 439
Pure strategies, 604

Q

Q-system, 777
Quadratic form, 1066
  indefinite, 1067
  negative definite, 1067
  positive definite, 1066
Quadratic programming problem, 874
  Beale's algorithm for, 880
  Kuhn-Tucker conditions for, 873
  Wolfe's modified simplex method, 876
Quality control, statistical of, 999
  uses of, 999
  variations in, 999
Queues, 631
  single, 631, 632
  multiple, 632
Queuing model, 636
  classification of, 636
  cost models, 683
  probabilistic queuing models, 636
  $\{(M/M/1) : (\infty/FIFS)\}$, 637
  $\{(M/M/1) : (N/FIFS)\}$, 659
  $\{(M/M/S) : (\infty/FIFS)\}$, 664
  $\{(M/M/S) : (N/FIFS)\}$, 671
  $\{(M/M/R) : (K/GD, K > R)\}$, 673
  $\{(M/E_k/1) : (\infty/FIFS)\}$, 678
  $\{(M/G/1) : (\infty/GD)\}$, 681
  non-Poisson queuing systems, 677
Queuing system, 631
  arrival pattern, 631
  balking of, 633
  capacity of, 633
  characteristics of, 634
  customer's behaviour, 633
  deterministic of, 634
  input pattern of, 631
  Jockeying of, 633
  multi-phase of, 678
  Poisson-exponential, 636
  probability distributions in, 635
  queue discipline of, 633
  reneging, 633
  service channels in, 632
  service mechanism of, 632
  simulation of, 821
  single server, 636
  steady state of, 636
  transient state of, 636

Queuing theory, 631
and inventory control, 801
cases in, 1036

RAMP, 396
Random experiment, 531
Random numbers, generation of, 806
pseudo, 807
Random variables, 536
cdf of, 537
continuous, 537
discrete, 536
expectation of, 539
probability density function of, 538
variation, 999
Receiver, 909
Rectangular distribution, 548
games of, 610
graphical solution method, 610
Redundancy, of a code, 922
Regret table, 576
Reliability, of a system, 709
estimation of, 711
improvement of, 711
Renewal, density function, 708
policy, 708
rate, 708
Re-order, level, 719, 762
point of, 717
Replacement, decisions, 687
of equipment/asset, 687
when money value changes, 689
when money value does not change, 689
which deteriorates gradually, 688
which fails suddenly, 697
Replacement policy, 697
group, 698
individual, 698
mortality tables of, 698
staffing/man-power of, 706
Replacement problem, 688
a case in, 1041
equipment renewal policy, 708
Research phase in OR, 8
Reserve stock, 762
Residual variations, 962
Resource, allocation, 443
analysis in management, 439
EAS set in, 443
levelling, 443
OAS set in, 443
smoothing, 443
Retention and lost, customers, 553
Revenue cost, 718
Revised simplex method, 120
computational procedure of, 121
Risky investments, 939
certainly equivalent approach, 940
simulation approach, 951
statistical distribution method, 941
Rolling back technique, 594
Routing problems and sequencing, 334
Row vector, 1053

Saddle point, of a payoff matrix, 606
games without, 607
problems, 864
rule for determining, 606
Saddle points, and NLPP, 865
necessary conditions for, 865
Safety stock, 762
Sales force composite, 979
Salvage cost, 718
Sample space, 532
continuous, 532
discrete, 532
points, 532
Savage criterion, 579
Scenario writing, 979
Scheduling, 303
assumptions, 304
basic terminology, 303
objectives for, 306
methods of, 306
techniques, 387
Scheduling of activities, 443
ES time of, 443
eligible activity set (EAS) of, 443
LS time of, 443
ordered activity set (OAS) of, 443
Scientific method, in OR, 3
SDE analysis, 771
Search method in IP, 347
Seasonal components, 962
Secular trend, 962
Sensitivity analysis, 163
changes in the activity coeff., 171
changes in the cost coeff., 164
changes in the RHS constants, 167
structural changes of, 177
Separable codes, 920
Separable programming, 885
algorithm of, 888
convex, 885
Separating hyper-plane, 1062
Sequencing problem, 303
complex, 334
elapsed time, 303
idle time of a machine in, 303, 313
processing order of jobs in, 303
processing time of a job in, 303
simulation of, 826
with n jobs and k machines, 321
with n jobs and two machines, 313
with 2 jobs and m machines, 328

Sequential programming of, 892
Service channels/servers, 632
 in parallel, 632, 633
 level of, 683
Service mechanism, 632
 bulk service system, 632
 rate, optimum, 683
Service time, 546
 distribution of, 678
Set-up cost, 718
Shadow prices, 150
Shannon-Fano encoding, 922
Shortest, labeling algorithm, 500
 applications of, 508
 cyclic route model, 295
 path algorithm, 500
 route problem, 457, 500, 508
Simplex, 1061
 algorithm, 68
 tableau, 67
Simplex method for LPP, 1, 66
 feasibility conditions in, 67
 foundations of, 109
 for goal programming problem, 375
 insights into, 66
 optimality conditions in, 67
 solution of linear equations, 102
 terminology of, 110
 theory of, 109
Simulation, 805
 cases in, 1037
 advantages of, 837
 event type, 807
 in investment and budgeting, 818
 limitations of, 837
 model, 805
 Monte Carlo, 809
 of a queuing problem, 815, 821
 of inventory problem, 813, 816
 of maintenance problem, 823
 of networks scheduling problem, 827, 828
 of operations scheduling problem, 826
 process of, 806
 reasons for using of, 805
Single processor scheduling, 307
 DD procedure of, 310
 Moore procedure of, 311
 SPT procedure of, 307
Single sampling plan, 1013
Sinking fund, 930
Slack variables, 42
Smoothing, 962
 exponential of, 965
Solution methods for OR models, 5
 feasible, 45
 of simultaneous linear equations, 102
 optimum, 45
Solving dynamic programming problems, 461
Source, 909
 finite, 631
 infinite, 631
 information, 909
 of message, 909
 shadow prices, 69
Specific learning curve, 983
Speed of service, 633
SPT procedure, in scheduling, 307
Square matrix, 1049
Standard basis, 1055
Standard deviation, 541
Standard form of LPP, 43
 primal-dual, 134, 135
Statistical quality control, 999
 techniques of, 1000
Steady state, queueing system, 636
 equilibrium conditions of, 556
Stepping stone method, 208
Stochastic programming, 892
Strategy, optimum, 606
Strong duality property, 148
Structural changes, in an LPP, 177
 addition of a constraint in, 178
 addition of activity in, 177
 deletion of a constraint in, 179
 deletion of a variable in, 179
Sum of vectors, 1053
Supplementary variable technique, 677
Supporting hyperplane, 1061
Surplus variable, 42
System, failure rate, 709
 capacity of, 633
 reliability of, 709
System of inventory control, 777
 continuous review, 719
 fixed order quantity (Q), 777
 periordic review, 719, 780
 two-bin, 777

Time-cost relationship procedure, 425
Time-cost trade-off, 424
Time series, 961
 components, 962
 cyclical component, 962
 irregular component, 962
 seasonal component, 962
 trend, 962
 trend projection, 968
Time-value of money, 927
 accumulation factor in, 928
 compound interest in, 927
 present value in, 929
 preference of money, 927
Total project cost, 424
 direct cost, 424
 indirect cost, 424
Traffic intensity, in queues, 635
Transient state, 636

Transition diagram, 553
Transition matrix, state, 552
construction of, 553
Transition probability, 551
Transmitted message, 909
Transpose of a matrix, 1049
Transportation algorithms/methods, 207
MODI method, 210
north-west corner method, 197
uv method, 210
stepping stone method, 208
Vogel's approx, method (VAM), 199
Transportation problem, 191
cases in, 1021
alternate optimal solution in, 211, 215
basic feasible solution of, 193, 197
degeneracy in, 194
duality in, 194
existence of fasible solution in, 192
general structure of, 192
initial b.f.s. of, 197
LP formulation of, 192
maximization, 211
post-optimal analysis in, 232
solution of, 196
time minimization in, 235
unbalanced, 192, 211
Transportation table, 195
cells in, 195
destinations in, 192
dummy destination, in, 211
dummy source, in, 211
origins in, 195
Transhipment problems, 237
Traveling salesman problem, 295
formulation as an AP, 295
Tree in network, 492
Tree problem-minimal spanning, 494
applications of, 494
Trefthen, 1
Trend component, 962
projection of, 968
Trim loss problem, 25
Two-bin system, 777
Two-person zero-sum game, 604
Two-phase simplex method, 75
Types of non-linear programming problems, 845
convex programming, 845
quadratic programming, 845
seprable programming, 845

U

Unbalanced, transportation problem, 211
Unbounded solution, 46, 55, 109, 118
Uncertainty, 531
and probability, 531
in decisions, 575
Unconstrained optimization, 845
one variable of, 846
multi-variable of, 848
Uniquely decodable codes, 920
comma, 920
instantaneous, 921
non-singlar, 920
prefix, 921
Unrestricted variable, 42
Upper envelope (boundary) in games, 611
Utilities, decision-making with, 602
uv method, in TP, 210

V

Value of the game, 606
maximin value of, 606
minimax value of, 606
Variables, artificial, 74
basic, 46
dual, 133
slack, 42
surplus, 42
unrestricted, 42
Variance, 540
Vector inequalities, 1054
Vectors, 1053
convex combination of, 1059
equality of, 1053
inner product of, 1053
length of, 1053
linear combination of, 1054
multiplication by a scalar, 1053
null, 1053
row, column, 1053
sum of, 1053
unit, 1054
VED analysis, 771
Vogel's approximation method, 199

W

Waiting line, 631
Weak duality property, 147
Weight functions in GP, 896
normalised, 896
Wholistic approach, 3
Wilson, R.H., 721
Wolfe's modified simplex method, 876
World wide web network, 526

X

XYZ analysis, 771

Z

Zero-one programming problem, 339
in capital budgeting problem, 341
Zero-sum game, 604

Principles and Practice of Management

L.M. Prasad

About the Book

The book in its 11th edition has been thoroughly restructured and revised. All chapters of the present edition have been rewritten not only to incorporate the latest developments in management but also to make presentation of subject matter more lucid and crisp. Some chapters of the previous edition have been merged with other related chapters to put the focus properly. This has resulted in reduced number of chapters inspite of adding new chapter dealing with management practices of prominent countries and business leaders. Thus, the present edition Is ideally suited with MBA/PGDM, M.Com. Students as well as management practitioners, particularly those who have not gone through formal management education.

Salient Features

- Subject matter presented in simple and lucid style with suitable Figures, Tables, and Exhibits.
- An Opening case study so that the concerned concepts can be related to practices easily.
- Multiple choice questions, short answer questions, and discussion/application questions at the end of each chapter to enable the reader to assess her/his behavioral pattern
- Case study at the end of each chapter to enable the reader to assess her/his behavioral pattern
- Case study at the end of each chapter for class room discussion to sharpen diagnostic and analytical skills of the readers
- Guidelines for learning through cases at the end of the book to enable the readers to make full use of learning through cases.
- Glossary at the end of the book for bird's eye view.

Statistical Methods

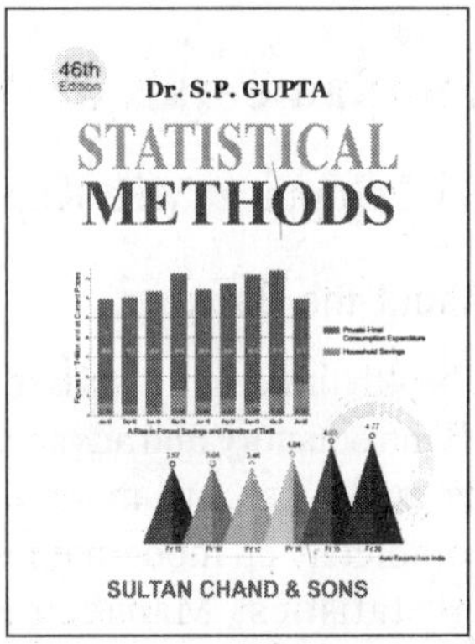

S.P. Gupta

About the Book

Statistical Methods are playing an ever-increasing role in framing suitable policies in a large number of diversified fields covering natural, physical and social sciences. Statistics today has become indispensable in all phases of human endeavour. This book is primarily meant for undergraduate students of Commerce and Economics. This book entitled "Statistical Methods" was first published in 1969.

Salient Features

- The book is written in a very simple and lucid style.
- For the first time the book is being published in the broader form on the initiative of the publisher, Sultan Chand & Sons.
- It caters to the requirements of under-graduate as well as post-graduate students of Commerce, Economics and Management.
- The Engineering, Medical and Research students can also refer to the book to meet their specific requirements.
- The book is non-mathematical in character.
- Besides good theory it contains more than 1,500 Illustrations and about more than 2,000 exercise questions for better understanding of the subject.
- A large number of questions have been picked up from the examination papers of leading Universities.
- Hints to ticklish problems have been given wherever necessary.

Edition: 46th, 2021; *Pages:* xxiv + 1304; *Size:* 180 × 240 mm; ISBN: 978-93-5161-176-9 (TC-506)

Ordinary and Partial Differential Equations

Chaitanya Kumar

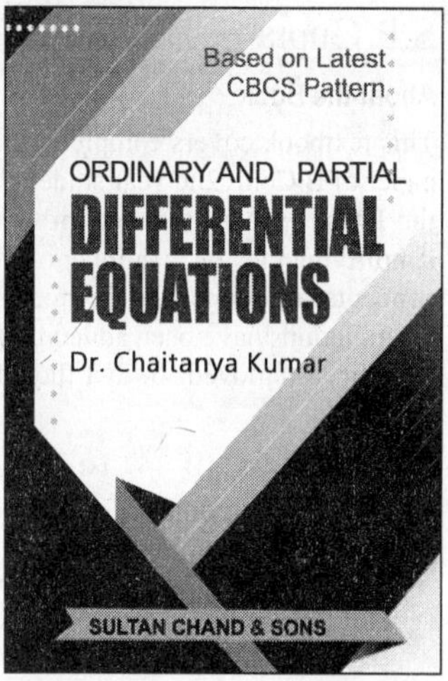

About the Book

This book has been designed for B.A. (Prog.) III Year, Semester-V, B.Sc. (Hons.) (Mathematics and Physics) and GE-3 (for all honours courses) students of all universities of India following Choice Based Credit System (CBCS).

It is also useful for B.Tech students of various universities and various competitive examinations.

The students of open and distance education courses will find the book most useful.

Salient Features

- Written in a lucid style and simple language.
- Written with a view to present a qualitative understanding of the subject.
- Comprehensive step by step explanations for easier understanding of the subject.
- A large number of solved examples and unsolved problems have been drawn from recent examination papers of universities.
- Answers to all the problems in each exercise are given immediately after the exercise.
- This book is based on latest syllabus under CBCS pattern.
- Delhi University latest syllabus, Question papers with their solutions are also provided in the book.

Edition: 1st, 2021; *Pages:* xvi + 320; *Size:* 135 × 215 mm; ISBN: 978-93-5161-197-4 (TC-1238)

Operations Research – Concepts, Problems & Solutions

V.K. Kapoor

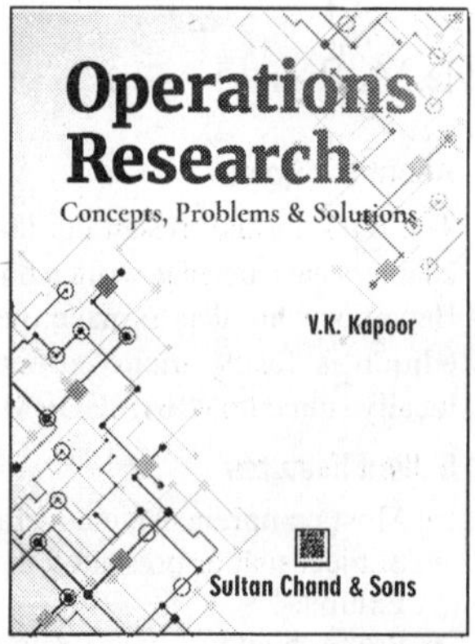

About the Book

The revised edition of this book offers a fine blend and conceptual finesse of practical examples of Operations Research techniques as well as extensive coverage of theoretical details. It is intended to serve as a comprehensive manual for undergraduate and postgraduate students of Commerce and management courses, i.e., MBA / PGDBM, M.Com., CA. It also suits the requirement of students of MCA, MIT, M.Sc., B.E., B.Tech. who need practical knowledge of Operations Research. The book shall also be useful to Research Scholars and students preparing for competitive examinations.

Salient Features

- The book elucidates the key concepts and methods of Operations Research.
- One of the noble features of the manual is that all the major steps in the computational procedure of important techniques are presented in a step by step algorithm format.
- Large number of numerical problems from the real business world have been included.
- The revised edition of the book included practical problems with complete workout solutions and self-assessment review problems with hints and answers for recapitulation and better grasp over the subject.
- To enhance the understanding of the subject by the students of various disciplines, a conceptual rather than mathematical approach has been adopted.

Edition: 5th; *Pages:* xvi + 640; *Size:* 185 × 240 mm; ISBN: 978-81-8054-854-3 (TC-532)

Business Statistics & Business Mathematics

S.P. Gupta • P.K. Gupta

About the Book

This textbook covers complete syllabus in Business Mathematics & Business Statistics paper of B.Com. 2nd year students. The Book has been designed strictly according with the latest updated syllabus prescribed by University of Delhi. Besides revising and simplifying the text, a number of illustrations and examples are added to explain various concepts introduced in the text. Also, new problems mostly from recent university examinations have been added in this edition. At the same time old stereotype problems have been removed. In fact the whole book has been rewritten and given a new look altogether.

Salient Features

- Book covers complete syllabus in Business Mathematics & Business Statistics paper of B.Com. examination.
- There are detailed self-contained chapters on all the syllabus elements.
- Part I of the book starts with the introduction to Statistics and discusses measures of central tendency and of variation, correlation & regression analysis, index numbers, time series and interpretation of data.
- Part II of the book begins with introducing the concept of matrices and determinants. Their applications to business and economic problems are discussed in the chapter.
- Concept of differentiation & integration along with their applications are given in chapters 3 & 4 of part II.
- The language used in text is simple and the subject matter has been presented in a lucid and straightforward style.
- Special care has been taken to develop the concepts in an easy to understand manner and are self explanatory.
- The book does not require any previous knowledge of the subject.

Edition: 5th, 2022; *Pages:* xxxvi + 740; *Size:* 185 × 240 mm; ISBN: 978-93-5161-143-1 (TC-1131)

Organizational Behaviour

L.M. Prasad

About the Book

The book in the present edition has been thoroughly revised and restructured. All the chapters have been re-written not only to incorporate new developments in Organizational Behaviour but also to make the subject-matter more lucid and crisp. Thus, the present edition is vastly improved both in terms of contents and presentation. The book is ideally suited to MBA/PGDBM and other relevant courses.

Salient Features

- Most comprehensive coverage of subject-matter with latest development.
- Subject-matter presented in simple and lucid style with suitable Figures, Tables and Exhibits.
- Proper blend of theory and practice.
- An opening case of organizational behaviour so that the concerned concepts can be related to practices easily.
- Multiple choice questions, short answer questions, discussion/application questions at the end of each chapter to test the assimilation of subject-matter concerned.
- Awareness Developer at the end of each chapter to enable the reader to assess her/his behavioural pattern.
- Case at the end of each chapter for class-room discussion to sharpen diagnostic and analytical skills of the readers.
- Guidelines for learning through cases at the end of the book to enable the readers to make full use of learning through cases.
- Glossary at the end of the book for bird's eye view.

Edition: 6th, 2019; *Pages:* xviii + 550; *Size:* 185 × 240 mm; ISBN: 978-93-5161-146-2 (TC-211)